The Physics of Energy

The Physics of Energy provides a comprehensive and systematic introduction to the scientific principles governing energy sources, uses, and systems.

This definitive textbook traces the flow of energy from sources such as solar power, nuclear power, wind power, water power, and fossil fuels through its transformation in devices such as heat engines and electrical generators, to its uses including transportation, heating, cooling, and other applications. The flow of energy through the Earth's atmosphere and oceans, and systems issues including storage, electric grids, and efficiency and conservation are presented in a scientific context along with topics such as radiation from nuclear power and climate change from the use of fossil fuels.

Students, scientists, engineers, energy industry professionals, and concerned citizens with some mathematical and scientific background who wish to understand energy systems and issues quantitatively will find this textbook of great interest.

Robert L. Jaffe holds the Morningstar Chair in the Department of Physics at MIT. He was formerly director of MIT's Center for Theoretical Physics and recently chaired the American Physical Society's Panel on Public Affairs. Jaffe is best known for his research on the quark substructure of the proton and other strongly interacting particles, on exotic states of matter, and on the quantum structure of the vacuum. He received his BA from Princeton and his PhD from Stanford. In recognition of his contributions to teaching and course development at MIT, Jaffe has received numerous awards including a prestigious MacVicar Fellowship. Jaffe is a member of the American Academy of Arts and Sciences.

Washington Taylor is a Professor of Physics at MIT, and is currently the Director of MIT's Center for Theoretical Physics. Taylor's research is focused on basic theoretical questions of particle physics and gravity. Taylor has made contributions to our understanding of fundamental aspects of string theory and its set of solutions, including connections to constraints on low-energy field theory and observable physics and to new results in mathematics. Taylor received his BA in mathematics from Stanford and his PhD in physics from UC Berkeley. Among other honors, Taylor has been an Alfred P. Sloan Research Fellow and a Department of Energy Outstanding Junior Investigator, and has received MIT's Buechner faculty teaching prize.

A long awaited book which comprehensively covers the fundamentals that engineers, scientists and others specializing in energy related fields need to master. Wonderfully written, it unlocks and presents the science behind energy systems in a pure yet accessible manner, while providing many real world examples to help visualize and frame this knowledge. This book would serve as an excellent text for a foundational course in energy engineering.

Khurram Afridi, Department of Electrical, Computer and Energy Engineering, University of Colorado Boulder

Finding the energy to power a civilization approaching 10 billion people without unacceptable consequences to the environment is the greatest challenge facing humanity this century. This book develops all of the fundamental concepts in physics underlying a quantitative understanding of energy sources, interconversion, and end usage, which are essential parts of meeting this challenge. It will serve as unique and authoritative textbook for the teaching of these topics. ... Overall it is a masterful exposition of the fundamental concepts of physics and their application to the energy-environment problem.

Michael J Aziz, Gene and Tracy Sykes Professor of Materials and Energy Technologies,
Harvard John A. Paulson School of Engineering and Applied Sciences

The textbook by Jaffe and Taylor is an invaluable resource, for instructors and students alike, discussing the physics of energy, a subject that is most important for humankind. ... The book has great potential as a teaching text for emerging courses on energy physics and promises to become a classic for years to come.

Katrin Becker and Melanie Becker, Texas A&M University

Jaffe and Taylor have produced in a single volume a comprehensive text on energy sources, energy conversion technologies, and energy uses from the unifying vantage of physics. Either in a course or in self-study *The Physics of Energy* can serve as the foundation for an understanding of conventional and renewable energy technologies.

Paul Debevec, Professor Emeritus, Department of Physics, University of Illinois

Jaffe and Taylor have compiled a comprehensive treatise that covers all aspects of energy: its fundamental role in physics, its sources and its uses. In addition to serving as the backbone for a variety of courses, this book should be an invaluable resource for anyone interested in the physics of energy in all of its forms.

David Gross, Chancellor's Chair Professor of Theoretical Physics, Kavli Institute for Theoretical
Physics, University of California, Santa Barbara, Joint Winner of the Nobel Prize for Physics, 2004

The book can be very useful as a mid-level textbook, as a survey for self-instruction for the serious-minded energy policy analyst, or as a desk reference covering the physics of the full range of energy topics – everything from the energy content of biofuels, to safe nuclear reactor design, to efficient design and placement of wind turbines, to geothermal energy flow, and dozens more topics ... This book very effectively fills a gap between the plentiful simplistic treatments of energy issues and books for full time professionals in the various energy areas.

Rush Holt, CEO of the American Association for the Advancement of Science, former Member of Congress

We live in an age of wonders, when a designer in almost any engineering field can find a dizzying assortment of tools, materials, components, and construction technologies for building. ... *The Physics of Energy* answers the question of where to begin. No engineer's library will be complete without a copy of this literary and intellectual masterpiece. A brilliant story of the foundations of everything.

Steven Leeb, Professor of Electrical Engineering and Computer Science, Massachusetts Institute of Technology

The book is the only comprehensive discussion of energy sources, flows, and uses that I know of. ... It is designed as a text for a college level course, or as a refresher for those who already have the background, and is successful in achieving its goal of introducing the student to the science of energy.

Burton Richter, Paul Pigott Professor in the Physical Sciences, Emeritus and Director Emeritus,
Stanford Linear Accelerator Center, Joint Winner of the Nobel Prize for Physics, 1976

This is a unique textbook: broad, deep, and crucially important for our society. ... [Students] are also inspired by new insights into nature and everyday life: no other energy book covers heat pumps, spark ignition engines, climate change, wave/particle duality and the Big Bang.

Joshua Winn, Princeton University

The Physics of Energy

ROBERT L. JAFFE

Massachusetts Institute of Technology

WASHINGTON TAYLOR

Massachusetts Institute of Technology

CAMBRIDGE UNIVERSITY PRESS

CAMBRIDGE
UNIVERSITY PRESS

University Printing House, Cambridge CB2 8BS, United Kingdom

One Liberty Plaza, 20th Floor, New York, NY 10006, USA

477 Williamstown Road, Port Melbourne, VIC 3207, Australia

314–321, 3rd Floor, Plot 3, Splendor Forum, Jasola District Centre, New Delhi – 110025, India

79 Anson Road, #06–04/06, Singapore 079906

Cambridge University Press is part of the University of Cambridge.

It furthers the University's mission by disseminating knowledge in the pursuit of education, learning, and research at the highest international levels of excellence.

www.cambridge.org
Information on this title: www.cambridge.org/9781107016651
DOI: 10.1017/9781139061292

First published 2018

Printed in the United Kingdom by Bell and Bain Ltd, Glasgow, January 2018

A catalogue record for this publication is available from the British Library.

ISBN 978-1-107-01665-1 Hardback

To our parents, our teachers,
our spouses and, most of all,
to our children

Contents

Contents

Preface

This book provides a comprehensive introduction to energy systems for individuals interested in understanding the fundamental scientific principles underlying energy processes from sources to end uses.

Origins and Uses

The Physics of Energy emerged from a one-semester course with the same title that we developed and have taught at MIT (Massachusetts Institute of Technology) since 2008. The course serves as an introduction to energy science in a new energy curriculum at MIT and is open to students who have completed freshman calculus, a year of physics with calculus, and a term of college chemistry. Although particularly suited to students majoring in hard sciences or engineering, many students who have taken the course were interested primarily in economics and policy. The MIT course also serves as an elective for physics majors, where the concepts developed in more formal courses are brought together into a broader context and applied to real-world problems.

Finding no existing book that matched the level of our course and the approach we wished to take, we began to generate lecture notes for "The Physics of Energy" in 2007. Those lecture notes slowly expanded in scope and depth into this book, which has developed a coherent structure in its own right. Because the students in the course have a wide range of backgrounds, we do not assume an extensive amount of physics background. Our goal in the course, and in the book, is to take students from a basic freshman physics background to a qualitative and quantitative understanding of both the physical principles underlying all energy systems and an appreciation of how these ideas are interconnected and relevant for practical energy systems.

This book can be used in many ways. It can serve as a textbook for a single-semester course for undergraduates in a broad range of science and engineering fields or as a "capstone course" for physics undergraduates; these are the audiences for the MIT course. A course using this book could also be tailored primarily to focus on various subsets of the material; depending upon the goals, one could focus for example on the basic physics underlying energy sources, on the aspects most relevant to climate and twenty-first century energy choices, on renewable energy sources, etc. Some specific one-semester paths through the material are suggested below. While we continue to teach the material from this book as a one-semester course at MIT, with roughly one lecture per chapter, this requires a rather fast pace and many topics are not treated in the full depth of the material in the book. With more careful and thorough treatment, the book can also serve as the text for a more advanced two-semester course in energy science, for example in conjunction with a graduate certificate in energy studies. The book is also designed to be self-contained and suitable for self-study. Anyone with a serious interest in energy and some background in basic mathematics and physics should hopefully find most of the book accessible.

We hope that *The Physics of Energy* also will serve as a "desk reference" for those who work in energy-related fields of science, engineering, economics, or policy, who need to be aware of the scientific foundations of technologies beyond their own field of specialization.

Scope and Focus

Providing energy for the world to use in a sustainable fashion is a major challenge for the twenty-first century. Economic considerations and policy decisions will be central to any global attempt to address this energy challenge. For individuals, organizations, and nations to make rational choices regarding energy policies, however, a clear understanding of the science of energy is essential. Decisions made in the absence of good scientific understanding have the potential to waste vast amounts of effort and money and to adversely affect countless lives and large ecosystems.

This book aims to provide an in-depth introduction to energy systems, from basic physical principles to practical consequences in real systems. The goals of the book are to provide a clear picture of what energy is and how it flows through Earth systems (Part II), how humans use energy (Part I), and how energy systems take energy from natural sources to end uses (Part III). In each of these areas our goal is to build a basic understanding of the underlying science, develop a global picture of how the parts fit together into a coherent whole, and to develop sufficient technical understanding to assess energy systems in a quantitative fashion.

Introductory textbooks on energy issues typically mix science, economics, and policy, with the aim of presenting a unified introduction to the subject. Throughout this book, however, we focus solely on the science of energy and energy systems and refrain from addressing issues of economics and politics. This is not because we believe economics and policy are unimportant, but instead because we believe that the science is best conveyed without constant reference to the economic and political context. Those aspects of energy studies are well presented in many other books and in other courses in a typical curriculum in energy studies. Our goal is to help to provide the scientific understanding of energy systems that is a prerequisite for any informed discussion of energy choices in the economic and political realm.

As the title indicates, our starting point for viewing energy sources and uses is primarily through the lens of physics. Indeed, a secondary goal of the book is to illustrate how the concept of energy unifies virtually all of modern physics into a coherent conceptual framework. It is not possible, however, to provide a comprehensive overview of energy science and systems without widening the discussion to include areas that are traditionally considered the domain of chemistry, biology, earth science, and many fields of engineering. Energy science is, in fact, an excellent example of a subject where traditional academic boundaries are blurred in nearly every application. Rather than limit our perspective and therefore the usefulness of this book, we have chosen to include a significant amount of material that might not ordinarily be construed as "physics." Overlap with other fields of science is particularly significant in chapters on energy in matter, fossil fuels, geothermal energy, energy and climate, and biological energy, while chapters on engines, heat extraction devices, nuclear reactors, wind turbines, and photovoltaic solar cells contain some material usually found in engineering courses. Some topics such as biofuels and fuel cells, at the edges of the conceptual framework of the book and at the limit of our own expertise, receive only limited discussion. We have, however, included a more extensive discussion of some aspects of earth science particularly as it relates to issues of present and future climate, since this topic is of central importance to the future trajectory of human energy use.

Throughout the book, we strive to emphasize the underlying physical principles that govern different energy systems. In particular, we develop the basic ideas of quantum mechanics, thermodynamics, fluid mechanics, the physics of oceans and of the atmosphere, and other basic physical frameworks in enough technical detail that their application to various energy systems can be understood quantitatively as well as qualitatively. We also strive, in so far as possible, to provide explanations of phenomena in relatively simple terms from first principles and to maintain a consistent introductory level throughout the book. A reader interested in any specific topic can gain some initial understanding from the treatment here of the underlying principles and learn how the topic connects with the rest of the energy landscape, but will have to go elsewhere for a more advanced and more detailed treatment of the subject; we have attempted to give a sampling of references that can lead the interested reader further into any particular topic.

Confronting such a wide range of topics, one possibility would have been to invite other authors, experts in those areas farthest from our own expertise, to write about those topics. Instead, we have chosen to try to educate ourselves sufficiently to cover the subject in its entirety. This approach has the advantages that the level and "feel" of the book is more coherent and uniform, that the pedagogical progression is orderly, and that the unity of energy science is more apparent. The disadvantage, of course, is that we have written extensively about areas in which neither of us is an expert. We have had the help of many colleagues, who have helped us grasp the subtleties of their fields and corrected many of our misconceptions. Along the way we have come to appreciate more deeply the wisdom captured by the basic paradigms in many fields of engineering and physical science and to recognize the role that fundamental physics plays in those paradigms.

Prerequisites

As mentioned at the outset, we assume that readers have, at some point in their lives, had a two-semester course in calculus at the college level or the equivalent. We also assume that students have encountered the elementary concepts of probability theory. Some of the more advanced areas of a first-year calculus course, such as elementary vector calculus, differential operators, line and surface integrals, and series expansions, are reviewed in Appendix B, as are the basic notions of complex numbers. We try to avoid all but the most elementary differential equations. Some simple aspects of linear vector spaces, Fourier series, and tensor analysis are used in isolated chapters and are summarized briefly in Appendix B.

We also assume that the reader has been exposed to mechanics and electrodynamics from a calculus-based perspective, and has had some introductory chemistry. Recognizing that some readers may have studied introductory physics long ago and that not all physics courses cover the same syllabus, we have organized the book to provide self-contained introductory reviews of the basics of mechanics, electromagnetism, heat, heat transfer, and chemical reactions early in the book as we describe their energy-related applications.

We *do not* assume that readers have any previous exposure to the concept of entropy, or to quantum mechanics or fluid dynamics.

Structure and Outline

This book is organized into three parts, addressing energy *uses*, *sources*, and *systems*. The first part plays a dual role of introducing many of the basic physics principles used throughout the text. Within this global framework, the book has a structure somewhat like a tree; the first ten chapters form the trunk of the tree on which the remaining material is supported. Many of the subsequent chapters branch off in different directions, and within chapters specialized topics and particular examples of energy technologies form the twigs and leaves of the tree. The dependence of each chapter on previous chapters is indicated in the *Reader's guide* at the beginning of each chapter. This enables a somewhat modular structure for the book, so that readers and instructors can follow a variety of paths and select a sequence of material based on their interest.

After an introduction (§1), *Part I: Basic energy physics and uses* describes the uses of mechanical (§2), electromagnetic (§3, §4), and thermal energy (§5, §6), reviewing and introducing concepts from these subjects as needed. To proceed further it is necessary to introduce basic notions of quantum mechanics (§7) and explain the concepts of entropy and thermodynamics in depth (§8). These chapters are intended to be self-contained introductions to these subjects for students with no previous exposure. Readers with some familiarity with these subjects may find the chapters useful reviews. With these concepts in hand we describe the flow of energy in chemical processes (§9),

and the interconversion of work and thermal energy in engines and heat extraction devices (§10–§13). Although fundamental concepts such as the second law of thermodynamics and the quantization of energy are used throughout the book, we recognize that these are conceptually difficult ideas. We have endeavored to structure the book so that a reader who reads §7 and §8 lightly and is willing to accept those concepts as given will find almost all of the material in the book accessible and useful.

Part II: Energy sources focuses on the sources of primary energy. The most extensive sections are devoted to nuclear energy (§16–§19), solar energy (§22–§25), and wind energy (§28–§30). Part II begins with an interlude (§14) in which the four forces that govern all known natural processes are described and with a closer look (§15) at some aspects of quantum mechanics such as tunneling that are necessary in order to understand aspects of both nuclear and solar energy. The study of nuclear power is followed by an introduction to the nature and effects of ionizing radiation (§20).

Before delving into solar power, we pause to consider the nature of energy and its role in our universe at a fundamental level (§21). Much of Part II is devoting to following the flow of solar energy as it enters other natural systems from which energy can be extracted. Solar energy (§22–§25) is the primary source of energy that is stored in biological systems (§26) and drives oceanic (§27) and atmospheric (§28) circulations. After introducing elementary concepts from fluid dynamics (§29), we describe the harvesting of wind power by wind turbines (§30) and the utilization of water power from rivers, waves, ocean currents, and the tides (§31). We then review Earth's internal structure and the origins of geothermal power (§32). Part II ends with an introduction (§33) to the nature, occurrence, characteristics, and uses of the primary fossil fuels: coal, oil, and natural gas.

In *Part III: Energy system issues and externalities* we turn to some of the complex issues associated with energy systems. First we describe the way the flow of energy on Earth's surface affects our climate (§34), the evidence for past climate change, and prospects for future climate change (§35). Next we turn to a quantitative discussion of energy efficiency, some case studies of conservation, and an overview of possible energy sources that may replace fossil fuels and how energy systems may evolve in the coming centuries (§36). We then turn to energy storage (§37) and in the final chapter (§38), we analyze the production and transmission of electrical energy and the structure of electric grids.

In addition to the mathematical appendix, other appendices include a list of symbols used in the book (Appendix

A), and tables of units, conversion factors, and fundamental constants (Appendix C). Various data including global and national energy and carbon dioxide information are summarized in Appendix D.

Each chapter (except the introduction) begins with a *Reader's guide* that previews the content of the chapter so that readers can decide where to concentrate their efforts. Throughout each chapter, *Key concept boxes* provide a concise summary of the major points as they arise. The key concept boxes not only identify essential points for the careful reader but also provide a gloss of the chapter for a reader seeking an overview of the subject. In each chapter we include sample calculations or analyses set off as *Examples* and we present some further explanations or specialized material in *Boxes*.

Questions and Problems

We have included questions for discussion/investigation and problems at the end of each chapter. Many of these questions and problems were developed for the course taught at MIT. The questions for discussion and investigation may be suitable for discussion groups or recitation sections associated with the course or they may form the topics for term papers. The problems serve a variety of purposes: some fill in details left unstated in the text, some are designed to deepen theoretical understanding, and many are useful as applications of the ideas developed in the text. The problems can be used for homework, evaluations, or self-study. We have used labels to indicate the character of some of the problems: [T] for more *theoretically* oriented problems; [H] for more challenging (*hard*) problems; and [C] for occasional problems that require extensive analytic or computer-based *computation*. Some problems require students to find and use data about specific systems that are readily available on the web but are not provided in the text. Solutions to the problems are available for teachers upon request.

Paths Through This Book

The flexible and modular structure of *The Physics of Energy* makes it possible to use this book in many ways. As an alternative to proceeding linearly through the book, the reader interested in a particular area of energy science can focus on a given chapter or set of chapters in the later part of the book, guided by the interdependencies outlined in the *Reader's guides*. Teachers interested in structuring a course can choose a variety of paths through the material based on the background and interest of the students. And the reader interested in attaining a global understanding of energy science and systems can proceed linearly

through the book, skimming any initial chapters that contain familiar material, delving as deeply or as lightly as they are interested into the technical aspects of the material, and focusing attention on later chapters of particular interest. For a worker in one field of energy research, policy, or business, the book can be used to gain an overview of the fundamental concepts of other energy related fields. Thus, for example, §32 (Geothermal energy) provides a relatively self-contained summary of the origins, harvesting, and scope of geothermal energy, and §38 (Electricity generation and transmission) gives an introduction to the basic principles of electric grids. Larger sections of the book form comprehensive, stand-alone treatments of major fields of energy science. Thus §22 (Solar energy) through §25 (Photovoltaics) and §14 (Forces of nature) through §20 (Radiation) provide introductions to the physics of solar and nuclear energy accessible to scientists and engineers with an elementary knowledge of quantum and thermal physics.

The book's modularity suggests ways that sections might be organized as a text for a course focusing on a particular speciality within energy studies. Teachers or readers primarily interested in electrical engineering, for example, would find that

§3 (Electromagnetism)

§4 (Waves and light)

§7 (Quantum mechanics)

§21 (Energy in the universe)

§22 (Solar energy)

§23 (Insolation)

§24 (Solar thermal energy)

§25 (Photovoltaics)

§36 (Systems)

§37 (Energy storage), and

§38 (Electricity generation and transmission)

form a relatively self-contained sequence focused on electromagnetic physics with energy applications.

Similarly, those interested primarily in earth sciences could use

§5 (Thermal energy)

§6 (Heat transfer)

§8 (Entropy and temperature)

§22 (Solar energy)

§23 (Insolation)

§27 (Ocean energy)

§28 (Wind energy)

§31 (Water energy)

§32 (Geothermal energy)

§33 (Fossil fuels)

§34 (Energy and climate), and

§35 (Climate change)

to provide an earth-science focused introduction to energy physics. Variants focused on nuclear, mechanical, chemical, and environmental engineering are also possible.

It is also possible to follow a path through the book that introduces new physics concepts in conjunction with related systems and uses. Such a course might begin with a review of mechanics (§2 and §4) followed by chapters on wind and water energy (§27, §28, §29, §30, and §31), which require only basic mechanics as a prerequisite. Next, after introducing basic thermal and quantum physics (§5–§8), the focus would shift to chemical energy (§9), engines and thermal energy transfer (§10–§13), geothermal and fossil fuel energy (§32, §33), climate (§34, §35), and efficiency and conservation (§36). Depending on the time available and the interests of the instructor and class, the focus could then turn to electromagnetic and solar energy (§3, §21–§25, and §38) or nuclear energy and radiation (§14–§20) or both.

A "capstone" course for physics majors could go over §1–§7 fairly quickly, emphasizing the energy applications, and then delve more deeply into the nuclear, solar, wind, and climate sections, adding in other chapters according to the interest of the instructor and students.

A course aimed at a broad range of students with a greater interest in implications for economics and policy could skip some of the more technical subjects of Part I, focus in Part II on material focused on giving the bottom line and overview of each energy source, such as §16, §23, §28 and the corresponding overview sections of the chapters on other resources. The course could then cover much of the material from Part III on climate and systems issues, referring back to the earlier part of the book for technical background as needed.

Units

Throughout this book we use SI units, which are introduced in §1. Each specialized area of energy studies has its own colloquial units, be they *billion barrels of oil equivalent*, or $ft^2 \, {}^\circ F \, hr/BTU$ (a unit of thermal resistance used in the US), or any others from a long list. These units have often evolved because they enhance communication within a specific community of energy practice. In a book that seeks a unified overview of energy science, there is great virtue, however, in having a single system of units. Familiarity with a single universal system of units also has the benefit of facilitating communication among specialists in different disciplines. In keeping with our interdisciplinary emphasis, with rare exceptions, we therefore keep to the SI system. Conversion tables from many colloquial units to SI units are given in Appendix C.

Further Reading

There exists a huge range of other books that cover various aspects of energy science and related issues. For each of the specific subjects covered in the individual chapters of this book, we have given a handful of references for readers interested in further study. Some more general references that cover a variety of energy science issues from a technical perspective compatible with this book, some of which we found very helpful in learning about these subjects ourselves, include [1, 2]. A good introduction to energy issues at a significantly less technical level is given by MacKay in [3]. For further reading on energy issues as connected to policy and economics, the MIT text [4] is a good starting point, as well as any of the many excellent books by Smil, such as [5]. The text by Vanek, Albright, and Angenent [6] provides an engineering perspective on energy issues, and the text by Andrews and Jelley [7] provides another perspective on energy science from a physicist's point of view. Finally, thanks to the efforts of innumerable unnamed individuals, Wikipedia is perhaps the best starting point for investigation of specific aspects of energy systems and technologies. In particular, for most topics in this book, Wikipedia provides excellent references for further reading.

Acknowledgments

Completing a project of this magnitude would not have been possible without the assistance and support of many people.

A few words here cannot suffice to adequately thank our spouses: Diana Bailey and Kara Swanson. We could never have undertaken to write this book without their unwavering support and endless patience over the better part of a decade. And we would like to thank our children, Clarissa, Logan, Rebecca, and Sam for their encouragement and understanding. Writing this book meant much time not spent with family and many other obligations neglected and opportunities foregone.

RLJ would like to thank the Rockefeller Foundation for a Residency at its Bellagio Center and WT would like to thank the Kavli Institute for Theoretical Physics and Stanford University, where each of us conceived an interest in learning more and teaching about the physics of energy. We would like to thank Ernie Moniz, former head of the MIT Energy Initiative (MITEi) for encouraging our early interest in this subject, and Ernie and Susan Hockfield, former President of MIT, for their early leadership of MIT's work in energy research and education that inspired our own efforts in this area.

We are grateful for financial support and encouragement from MIT's Department of Physics and MITEi during both the development of our course and the writing of this book. We thank Marc Kastner, Edmund Bertschinger, and Peter Fisher, heads of the physics department as we have worked on this project, for ongoing support. Thanks also to Bob Silbey, Marc Kastner, and Michael Sipser, deans of the School of Science for their support. Bob Armstrong, currently director of MITEi, has continued to support us through to the conclusion of our work. And we would like to extend our appreciation to Rafael Reif for his continued support for basic science as well as environment and energy studies at MIT. We also acknowledge financial support from the MIT Energy Initiative through a grant from the S.D. Bechtel, Jr. Foundation. WT would also like to thank the Aspen Center for Physics, where part of the work on this book was done.

Thanks also to Joyce Berggren, Scott Morley, and Charles Suggs of the Center for Theoretical Physics and to Amanda Graham of MITEi for supporting our work in countless ways.

We are grateful to many of our friends and colleagues, who offered their time and expertise to read and provide us feedback on sections of the book at various stages in its development: Patrick Brown, Ike Chuang, Ibrahim Cissé, Kerry Emanuel, John Heywood, Andrew Hodgdon, Chris Llewellyn Smith, Ian Taylor, and Andrew Turner.

We would like to acknowledge conversations and input on many aspects of the subject matter of this book with a host of friends, colleagues, and experts in other fields, including Allan Adams, Khurram Afridi, Richard Alley, Dimitri Antoniadis, Bob Armstrong, Angela Belcher, Vladimir Bulovic, Tonio Buonassisi, Wit Busza, Jacopo Buongiorno, Peter Dourmashkin, John Deutch, Ahmed Ghoniem, Stephen Jaffe, Joel Jean, Alejandro Jenkins, Andrew Kadak, Scott Kemp, Dan Kleppner, Steve Koonin, Steve Leeb, Don Lessard, Seth Lloyd, Daniel Nocera, Frank O'Sullivan, Dave Prichard, Dan Rothman, Leigh Royden, Ignacio Perez-Arriaga, Don Sadoway, Sara Seager, Susan Silbey, Francis Slakey, Larry Schwartz, Jeffrey Tester, Ethan Taylor, and Alexandra von Meier. We have had useful discussions about energy-related topics with many others not listed, to all of whom we are grateful. Much of what is valuable in this book we learned from someone else. The errors, however, are all our own.

We would also like to thank the graduate student teaching assistants who worked on our course or on the preparation of lecture notes during the early stages of what became this book: Alex Boxer, Aidan Crook, Eric Edlund,

Christopher Jones, Teppo Jouttenus, Misha Leybovich, Jeremy Lopez, Bonna Newman, and Stephen Samouhos.

Also thanks to all the *Physics of Energy* students whose enthusiasm for the subject and feedback on the material made this book far better than it otherwise would have been.

We are particularly grateful to Josh Winn, who read (and taught the *Physics of Energy* from) the entire book in nearly final form and provided us with innumerable excellent suggestions and corrections. Our editor, Simon Capelin, at Cambridge University Press has been a steadfast and patient supporter throughout a long process. Also Lindsay Barnes, Pippa Cole, Rachel Cox, Rosemary Crawley, Sarah Lambert, Caroline Mowatt, and Ali Woollatt at CUP helped generously at various stages in the preparation of the book. WT would also like to thank James and Janet Baker, Dragon Systems and Nuance for producing Dragon Dictate for Windows and Dragon NaturallySpeaking, which were used extensively in the production of this book.

Finally, we want to acknowledge our gratitude to our own teachers and mentors: Orlando Alvarez, Gerry Brown, Sarah Ann Cassiday, Paul Cohen, Sid Drell, Bill Gerace, David Gross, Francis Low, Silvia Moss, Kathy Rehfield, Isadore Singer, Kevin Sullivan, Viki Weisskopf, and James Wisdom, who shared so much of their own time, energy, and insight, and helped us both along the paths that have led to this book.

Part I

Basic Energy Physics and Uses

Introduction

Energy is a precious resource for humankind. Throughout history, technological ingenuity and heroic efforts have been devoted to the extraction of energy from moving water, coal, petroleum, uranium, sunlight, wind, and other **primary energy** sources. We hear constantly of the rate at which we "consume" energy – to power our cars, heat our homes, and wash our clothes. From a scientific point of view, however, energy is not lost when we use it. Indeed, the essence of energy is that it is *conserved*.

Energy is not easy to define – we return to this subject in §21 – but for now we observe that in any physical system free from outside influences, *energy* does not change with time. This fact is referred to as **conservation of energy**, and this is energy's most fundamental attribute. In fact, energy is important in science precisely because it is conserved. While it moves easily and often from one system to another and changes form in ways that are sometimes difficult to follow, energy can neither be created nor destroyed. It is possible to understand the behavior of most physical systems by following the flow of energy through them.

When we say casually that energy has been "consumed," what we mean more precisely is that it has been degraded into less useful forms – particularly into thermal energy that merely increases the ambient temperature. What exactly is meant by a "less useful form" of energy is a theme that plays an important role throughout this book, beginning with the introduction of the notion of *entropy* in §8, and continuing through the discussion of *exergy* in §36.

Energy is ubiquitous. We see it in the motion of the world around us – the flow of water, wind, and waves, the violence of volcanic eruptions and the intense activity of small animals – and we feel it in the warmth of the air and water in our environment. Most of the energy that powers organisms, ecosystems, and air and water circulation on Earth arrived here as solar radiation produced by *nuclear fusion* within the Sun. Just one hundredth of one percent

The Universe is Full of Energy

There is no shortage of energy in the world around us. Energy in solar radiation hits the Earth at a steady rate 10 000 times greater than the rate at which humanity uses energy. The challenge is to find practical and economical ways of channeling energy to human use from its natural state.

(1 part in 10 000) of the solar energy that hits the Earth would be sufficient to supply all current human energy needs, if this energy could be effectively harnessed. Even greater quantities of energy are contained in the physical objects that surround us, as matter itself is a form of energy. From Albert Einstein's famous relation $E = mc^2$, it is easy to compute that the mass of a pair of typical pickup trucks ($\sim$2750 kg each) contains enough energy to power human civilization for a year. This type of energy, however, is impossible to extract and put to use with any current or foreseeable technology. (One would need a matching pair of pickup trucks made of *antimatter*.)

The energy problem, then, is not a shortage of energy, but rather a shortage of *usable* energy. The challenge for humanity is to identify mechanisms for transforming solar and other large repositories of available energy into useful forms in an economically and technologically practical fashion.

The flow of energy through human activities, from sources to end uses and through all the conversions in between, forms a complicated system with many interdependencies (see Figure 1.2). Understanding human energy use requires understanding not only each individual part of the energy system but also how these parts are connected. Physical principles place limits not only on how much

Figure 1.1 On Earth energy is at work around us all the time. (Waterfall image by Stefan Krasowski, hummingbird image by Brady Smith).

energy is available from each possible resource, but also on the efficiency with which energy can be converted from one form to another. For example, only about 25% of the thermal energy released in a typical automobile engine is actually used to power the car; the remainder is lost to the environment as heat. Understanding the physical limitations on energy conversion efficiencies helps to guide efforts to improve the efficiency of existing systems.

Beyond economic and technological limitations, there are also broader impacts associated with the use of energy from some sources. The burning of fossil fuels leads to emission of *carbon dioxide* (CO_2). Atmospheric CO_2 absorbs outgoing infrared radiation, affecting the global radiation budget and Earth's climate. Use of nuclear power generates radioactive waste and can lead to accidents in which substantial quantities of radiation and/or radioactive material are released into the environment. Mining and burning coal can have serious effects on human health and the local environment. Most renewable resources, such as solar and wind power, are relatively diffuse, so that large-scale reliance on these resources will require substantial land areas, potentially conflicting with other human activities and native ecosystems. Sensible energy choices involve a careful weighing and balancing of these

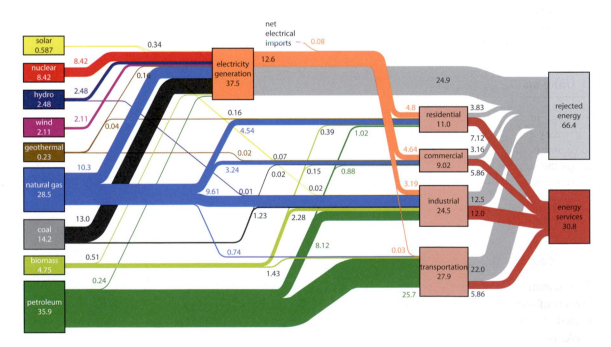

Figure 1.2 Energy flow from sources to end uses in the US for the year 2016. Many physical processes are involved in this complex flow. A substantial fraction of the energy extracted is lost. In some cases, physical principles limit the fraction of energy that can be transformed into useful form. The units in this figure are *quads*; see Table C.1 for translation of quads to other units. Redrawn from [11].

kinds of broader impacts and risk against the advantages of any given energy source.

The rate at which humanity uses energy has increased steadily and rapidly over the last century (see Figure 1.3). As population continues to grow, and per capita energy use increases, global demands on limited energy resources become more intense. Unless a dramatic, sudden, and unexpected change occurs in the human population or human energy use patterns, the quest for usable energy to power human society will be a dominant theme throughout the twenty-first century.

Many of the questions and issues related to energy choices are fundamentally economic and political in nature. To understand energy systems and to make rational economic and political choices regarding energy, however, requires a clear understanding of the science of energy. Without understanding how energy systems work, how they are connected, and the relative scope and limitations of different energy processes and resources, it is impossible to make informed and intelligent decisions regarding extraction, transformation, or utilization of energy resources on any scale large or small. This book is devoted to explaining the scientific principles underlying energy systems, with a focus on how these general principles apply in specific and practical energy-related contexts, and on the interconnections among the variety of terrestrial energy systems. Economic and political aspects of energy systems are generally avoided in this book.

1.1 Units and Energy Quantities

To engage in any meaningful discussion of energy, it is necessary to use a system of units for computation and communication. Reflecting its many forms, energy is perhaps the single quantity for which the widest variety of distinct units are currently used. For example, *calories*, *electron volts*, *British Thermal Units* (BTU), *kilowatt*

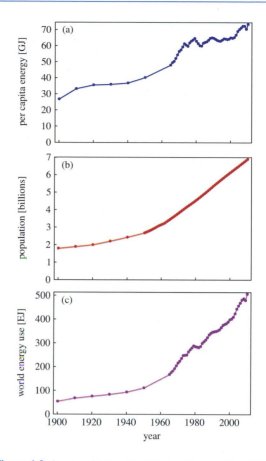

Figure 1.3 Graphs of (a) yearly global energy use per capita; (b) total human population; (c) yearly total global energy use, which is the product of the preceding two quantities. Data cover the last century, over which energy use per person has nearly tripled and population has quadrupled, giving a factor of twelve increase in total energy use. Data on population since 1950 from [8], energy use since 1965 from [9], estimates for earlier years from [5, 10].

hours, and *barrels of oil equivalent* are all standard energy units in widespread use. A summary of many different units used for energy can be found in Appendix C (see Table C.1), along with conversion factors, fundamental constants, and other useful data.

In this book we use the **SI** (**Système International**) unit system, known colloquially in the US as the **metric system**. This unit system is in general use throughout the world (except in Liberia, Myanmar, and the US) and is used globally for scientific work. It has the convenient feature that standard units for any physical quantity differ by powers of ten in a common fashion denoted by prefixes, so that even unfamiliar units are readily manipulated with a small amount of practice (see Table C.4).

Science, Economics, and Policy

The scientific, social, economic, and political aspects of human energy use are deeply interconnected. We put economics and policy aside in this book, not because these issues are unimportant, but because we believe that understanding the science of energy is a precondition for an intelligent discussion of policy and action.

Basic SI units for time, length, and mass are the *second*, *meter*, and *kilogram*. The **second** (s) is defined as the time required for a fixed number of oscillations of the electromagnetic wave emitted when a specific *quantum transition* (§7) occurs in a cesium atom. The **meter** (m) is *defined* so that the speed of light in a vacuum (§4) is precisely

$$c = 2.997\,924\,58 \times 10^8 \text{ m/s} . \qquad (1.1)$$

The **kilogram** (kg) is a mass equal to that of a specific sample of material kept by the International Bureau of Weights and Measures in France, though this may change in the near future.

Given units of length, time, and mass, we can define the fundamental SI unit of energy, the **joule** (J),

$$1 \text{ joule} = 1 \text{ kg m}^2/\text{s}^2 . \qquad (1.2)$$

One joule is roughly equivalent to the *kinetic energy* (§2) of a tennis ball (mass $\cong 0.057$ kg) after falling from a height of 2 m.

Dimensional analysis is a useful approach to understanding qualitative features of many physical systems and relationships between quantities based on their unit structure. We denote the units associated with a given quantity by putting the quantity in brackets. For example, the units of energy are

$$[\text{energy}] = [\text{mass}] \times [\text{distance}]^2/[\text{time}]^2 . \qquad (1.3)$$

A *force* has units known as **newtons** (N); in terms of the basic units of length, time, and mass, 1 newton $= 1$ kg m/s^2, and we write

$$[\text{force}] = [\text{mass}] \times [\text{distance}]/[\text{time}]^2 . \qquad (1.4)$$

Multiplying a force by a distance gives us a quantity with units of energy. This is one of the basic equations of elementary mechanics: *work = force × distance*. As we review in the next chapter, work represents a transfer of energy from one system to another.

Another important quantity in the discussion of energy is **power**. Power is the rate at which energy is used, or transformed from one form to another. It has units of energy per unit time,

$$[\text{power}] = [\text{energy}]/[\text{time}]$$
$$= [\text{mass}] \times [\text{distance}]^2/[\text{time}]^3 . \qquad (1.5)$$

The SI unit of power is the **watt** (W),

$$1 \text{ W} = 1 \text{ J/s} = 1 \text{ kg m}^2/\text{s}^3 . \qquad (1.6)$$

It is important to keep units of energy distinct from units of power. For example, a refrigerator that uses electrical power at an average of 300 W will use roughly 300 W × 24 h × 3600 s/h $= 25\,920\,000$ J $\cong 26$ MJ of energy

---

SI Units

The SI international unit system is used globally in scientific work. Basic SI units include the meter, second, and kilogram. The SI unit of energy is the *joule* (1 J $= 1$ kg m^2/s^2); the unit of power is the *watt* (1 W$= 1$ J/s $= 1$ kg m^2/s^3).

---

each day. A popular unit of energy is the **kilowatt-hour** (kWh),

$$1 \text{ kWh} = (1 \text{ kW}) \times (3600 \text{ s}) = 3.6 \text{ MJ} . \qquad (1.7)$$

One further quantity that arises frequently in energy systems, and that illustrates the utility of dimensional analysis, is **pressure**. Pressure is a force per unit area acting at right angles to a surface. The units of pressure are also the units of energy per unit volume,

$$[\text{pressure}] = [\text{force}]/[\text{area}] = \frac{[\text{mass}]}{[\text{distance}][\text{time}]^2}$$
$$= [\text{energy}]/[\text{volume}] . \qquad (1.8)$$

The connection between force per unit area and energy per unit volume, which is suggested by their units, figures in the dynamics of gases and liquids (see §5 and §29). The SI unit of pressure is the **pascal** (Pa),

$$1 \text{ pascal} = 1 \text{ N/m}^2 = 1 \text{ kg/m s}^2 . \qquad (1.9)$$

One **atmosphere** (atm) of pressure is defined as

$$1 \text{ atm} = 101\,325 \text{ Pa} . \qquad (1.10)$$

This is roughly the average pressure exerted by Earth's atmosphere at sea level.

The SI units for many other quantities, ranging from familiar ones such as temperature (degrees Centigrade or Kelvins) to specialized ones such as permeability (darcys), are introduced as the physics that requires them is encountered throughout the book.

1.2 Types of Energy

Energy is present in the world in many different forms. While chemical energy, thermal energy, mass energy, and potential energy may seem intuitively very different from the simple notion of kinetic energy of a falling tennis ball, they all represent a common physical currency. Each form of energy can be measured in joules, and, with less or more effort, each form of energy can be converted into every other form. Much of the first part of this book is devoted to a systematic development of the physical principles underlying the varied forms of energy and the application of these principles to understanding a variety of energy systems. We briefly summarize some of the principal forms of energy here and point forward to the chapters where each form is introduced and described in further detail.

Mechanical kinetic and potential energy Kinetic energy, mentioned above, is the energy that an object has by virtue of its motion. The kinetic energy of an object of mass m moving at a speed v is

$$\text{kinetic energy} = \tfrac{1}{2}mv^2. \qquad (1.11)$$

For example, the kinetic energy of a 3000 kg rhinoceros charging at 50 km/hour ($\cong$ 14 m/s $\cong$ 30 miles/hour) is roughly 300 kJ (see Figure 1.4).

Potential energy is energy stored in a configuration of objects that interact through a force such as the gravitational force. For example, when the tennis ball discussed above is held at a height of 2 m above the ground, it has potential energy. When the ball is released and falls under the influence of the gravitational force, this potential energy is converted to kinetic energy. Mechanical kinetic and potential energies are reviewed in §2.

Thermal energy Thermal energy is energy contained in the microscopic dynamics of a large number of molecules, atoms, or other constituents of a macroscopic material or fluid. Thus, for example, the thermal energy of the air in a room includes the kinetic energy of all the moving air molecules, as well as energy in the vibration and rotation of the individual molecules. Temperature provides a measure of thermal energy, with increasing temperature indicating

greater thermal energy content. As an example, the thermal energy of a kilogram of water just below the boiling point (100 °C) is greater than the thermal energy at room temperature (20 °C) by roughly 335 kJ. Thermal energy is introduced in §5, and temperature is defined more precisely in §8.

Electromagnetic energy Electromagnetism is one of the four fundamental forces in nature; the other three are gravity and the strong and weak nuclear forces. (We give a short introduction to the four forces in §14.) Electrically charged particles produce electric and magnetic fields that in turn exert forces on other charged particles. Electromagnetic energy can be stored in a configuration of charged particles such as electrons and protons in much the same way that gravitational potential energy is stored in a configuration of massive objects. Electromagnetic energy can be transmitted through electrical circuits, and provides a convenient way to distribute energy from power plants to homes, businesses, and industries over great distances. More fundamentally, electromagnetic energy is contained in electric and magnetic fields, and can propagate through space in the form of electromagnetic radiation such as visible light.

A light bulb provides a simple example of several aspects of electromagnetic energy. A 100 W incandescent bulb draws 100 J of energy per second from the electric grid. This energy is converted into thermal energy by the electrical resistance of the filament in the bulb; the heated filament then radiates energy as visible light at around 2.6 W, and the remainder of the energy is lost as heat. By comparison, a compact fluorescent light (CFL) can produce the same amount of energy in visible light while drawing 20 to 30 W from the grid, and a light emitting

Figure 1.4 A charging rhino carries a lot of kinetic energy. Pushing a boulder uphill stores potential energy.

Figure 1.5 Incandescent, LED, and fluorescent bulbs, all with roughly the same output of energy as visible light, draw 100 W, 16 W, and 27 W respectively from an electrical circuit.

diode (LED) emits roughly the same amount of visible light, but draws only 16 W (see Figure 1.5). We cover basic aspects of electromagnetism and electromagnetic energy in §3 and §4, including electromagnetic fields, charges, circuits, electrical resistance, and electromagnetic waves. Thermal radiation is described in later chapters.

Chemical energy Chemical energy is energy stored in chemical bonds within a material. The energy in these bonds originates in the electromagnetic interactions between atoms at the molecular level, which must be described in the framework of quantum mechanics. We introduce some basic notions of quantum mechanics in §7, and describe chemical energy in §9. A simple example of chemical energy is the energy contained in hydrocarbon bonds in food and fossil fuels. Most of the chemical energy in an apple or a liter of gasoline is contained in the bonds connecting carbon atoms within the material to other carbon atoms or to hydrogen atoms. When the apple is eaten or the gasoline is burned, this energy is released and can be used to power a person walking down the street or an automobile driving along a highway. The energy in a typical chemical bond is a few *electron volts*, where an **electron volt** (eV) is the energy needed to move a single electron across a one volt electric potential difference (electric potentials are reviewed in §3),

$$1\,\text{eV} = 1.602\,18 \times 10^{-19}\,\text{J}. \qquad (1.12)$$

In contrast, the standard unit of energy in food is the **kilocalorie** (kcal) or **Calorie** (Cal), with 1 kcal = 1 Cal = 4.1868 kJ. Thus, consuming one Calorie of food energy

corresponds to harvesting the energy in something like 10^{22} chemical bonds. One kilogram of apples contains roughly 500 Cal $\cong$ 2.1 MJ of energy, while one kilogram of gasoline contains roughly 44 MJ of energy. While the chemical bonds in these materials are similar, apples are about 85% water, which is why – among other reasons – we do not burn apples in our cars.

Nuclear binding energy Just as the atoms in a molecule are held together by electromagnetic forces, similarly the protons and neutrons in an atomic nucleus are held together by the strong nuclear force. Nuclear binding energies are roughly a million times greater than molecular bond energies, so typical nuclear processes emit and absorb millions of electron volts (10^6 eV = 1 **MeV**) of energy.

Small nuclei can fuse together, releasing energy in the process. *Nuclear fusion* in the core of the Sun combines four hydrogen nuclei (protons) into a helium nucleus, generating heat that in turn produces solar radiation. The part of this solar radiation that reaches Earth powers photosynthesis and drives biological processes and the dynamics of the atmosphere and oceans.

Larger nuclei, such as uranium nuclei, become unstable as the electromagnetic repulsion between charged protons opposes the strong nuclear binding force. Their decay into smaller parts – a process known as *nuclear fission* – provides a compact and carbon-free power source when harnessed in a nuclear reactor.

The ideas of quantum physics developed in §7 and §9 are elaborated further in §15, and used as the basis for understanding the physics of nuclear, solar, and geothermal power in later chapters.

Mass energy Mass itself is a form of energy. According to quantum physics, each particle is an excitation of a quantum field, just as a single photon of light is a quantum excitation of the electromagnetic field. It is difficult to convert mass energy into useful form. This can be done by bringing a particle in contact with an *antiparticle* of the same type. The particle and antiparticle then annihilate and liberate some of their mass energy as electromagnetic radiation and/or as kinetic energy of less massive particles that are products of the annihilation reaction. Antimatter is not found naturally in the solar system, however, so mass energy does not represent a practical energy source.

Einstein's formula gives the energy equivalent of a mass m,

$$E = mc^2, \qquad (1.13)$$

where c is the speed of light from eq. (1.1). Thus, for example, the energy released when a proton (mass $M_p \cong$

1.67×10^{-27} kg) and an antiproton (of the same mass) annihilate is

$$E_{p+\bar{p}} = 2M_p c^2 \cong 3 \times 10^{-10}\,\text{J} \cong 1877\,\text{MeV}. \quad (1.14)$$

While irrelevant for most day-to-day purposes, mass energy is important in understanding nuclear processes and nuclear power. Because the energies involved in nuclear binding are a noticeable fraction of the masses involved, it has become conventional to measure nuclear masses in terms of their energy equivalent. The systematics of mass energy and nuclear binding are developed in §15 and §17, and eq. (1.13) is explained in §21.

The zero of energy Although we often talk about energy in absolute terms, in practical situations we only measure or need to consider **energy differences**. When we talk about the potential energy of a tennis ball held 2 m above the ground, for example, we are referring to the *difference* between its energies in two places. When we talk about the binding energy of an atom or a nucleus, we refer to its energy compared to that of its isolated constituents. So the proper answer to a question like "What is the energy of a bucket of water?," is "It depends." We return to this question in §9 (see Question 1.5). Situations (such as astrophysics and cosmology) in which an absolute scale for energy is relevant are discussed in §21.

1.3 Scales of Energy

As we study different energy systems through this book, it will be helpful if the reader can develop an intuition for energy quantities at different scales. Some energy systems function at a scale relevant to an individual human. Other energy systems operate at a scale relevant for a country or the planet as a whole. Still other energy systems are microscopic, and are best understood on a molecular or atomic scale. We conclude this chapter with a brief survey of some energy quantities characteristic of these scales.

Energy at the human scale Energies that a person might encounter in day-to-day life are generally in the range from joules to gigajoules (1 GJ = 10^9 J). The falling tennis ball discussed above has kinetic energy of 1 joule, while the average daily energy use for a US citizen in 2010 was roughly 1 GJ. The global average per capita energy use in 2010 was roughly 200 MJ/day. A number that may give a qualitative feel for human energy scales is the food energy eaten by a single person in a day. A 2400 Calorie diet corresponds to just over 10 MJ/day. Much of this food energy is used for basic metabolism – like the automobile engine that we discuss in the next chapter, our bodies can only transform a fraction of food energy (roughly 25%) into

---

Scales of Energy

Some useful energy numbers to remember are (in round numbers, circa 2010):

10 MJ:	daily human food intake (2400 Cal)
200 MJ:	average daily human energy use
500 EJ:	yearly global energy use
15 TW:	average global power use

---

mechanical work. A manual laborer who works at a rate of 100 W for eight hours does just under 3 MJ of work per day (and probably needs more than 2400 Calories/day to comfortably sustain this level of output). One can thus think of modern technology as a way of harnessing the energy equivalent of 60 or 70 servants to work for each individual on the planet (200 MJ/3 MJ $\cong$ 66).

Energy at the global scale Energy quantities at the global scale are measured in large units like exajoules (1 EJ = 10^{18} J) or **quads** (1 quad = 10^{15} Btu = 1.055 EJ), and power is measured in terawatts (1 TW = 10^{12} W). Total world oil consumption in 2014, for example, was about 196 EJ. The total energy used by humanity in that year was close to 576 EJ. This represents a sustained power usage of around 17 TW. Energy flow through many natural systems at the global scale is conveniently measured in units of terawatts as well. For example, solar energy hits the Earth at a rate of roughly 173 000 TW (§22). The total rate at which wave energy hits all the world's shores is only a few (roughly 3) TW (§31.2).

Energy at the micro scale To understand many energy systems, such as photovoltaic cells, chemical fuels, and nuclear power plants, it is helpful to understand the physics of microscopic processes involving individual molecules, atoms, electrons, or photons of light. The electron volt (1.12) is the standard unit for the micro world. When an atom of carbon in coal combines with oxygen to form CO_2, for example, about 4 eV ($\cong 6.4 \times 10^{-19}$ J) of energy is liberated. Another example at the electron volt scale is a single photon of green light, which carries energy $\cong 2.5$ eV. Photovoltaic cells capture energy from individual photons of light, as we describe in detail in §25.

Discussion/Investigation Questions

1.1   Given that energy is everywhere, and cannot be destroyed, try to articulate some reasons why it is so hard to get useful energy from natural systems in a clean and affordable way. (This question may be worth revisiting occasionally as you make your way through the book.)

1.2 A residential photovoltaic installation is described as producing "5000 kilowatt hours per year." What is a kilowatt hour per year in SI units? What might be the purpose of using kWh/y rather than the equivalent SI unit?

1.3 Try to describe the flow of energy through various systems before and after you use it in a light bulb in your house. Which of the various forms of energy discussed in the chapter does the energy pass through?

1.4 Give examples of each of the types of energy described in §1.2.

1.5 Discuss some possible answers, depending on the context, to the question posed in the text, "What is the energy of a bucket of water?"

1.6 Compare the global average rate of energy use per person to typical human food energy consumption. What does this say about the viability of biologically produced energy as a principal energy solution for the future?

Problems

1.1 Confirm eq. (1.14) and the estimate for the rhinoceros's kinetic energy below eq. (1.11) by explicit calculation.

1.2 How much energy would a 100 W light bulb consume if left on for 10 hours?

1.3 In a typical mid-latitude location, incident solar energy averages around 200 W/m^2 over a 24-hour cycle. Compute the land area needed to supply the average person's energy use of 200 MJ/day if solar panels convert a net 5% of the energy incident over a large area of land to useful electrical energy. Multiply by world population to get an estimate of total land area needed to supply the world energy needs from solar power under these assumptions. Compare this land area to some relevant reference areas – the Sahara Desert, your native country, etc.

1.4 The US total energy consumption in 2014 was 98.0 quads. What is this quantity in joules? About 83% of US energy comes from fossil fuels. If the whole 83% came from oil, how many tons of oil equivalent (toe) would this have been? How many barrels of oil (bbl) is this equivalent to?

1.5 The gravitational potential energy of an object of mass m at a distance h above ground is given by $E = mgh$. Use dimensional analysis to compute the units of g. What does this result suggest about the behavior of objects near Earth's surface?

1.6 The energy emitted or absorbed in chemical processes is often quoted in kilojoules per *mole* (abbreviated mol) of reactants, where a mole contains $N_A \cong 6.022 \times 10^{23}$ (*Avogadro's number*) molecules (§5). Derive the conversion factor from eV/molecule to kJ/mol.

1.7 The energy available from one kilogram of ^{235}U is 82 TJ. Energy is released when each uranium nucleus splits, or *fissions*. (The fission process is described in §16, but you do not need to know anything about the process for this problem.) 235 grams of ^{235}U contain approximately Avagadro's number (see Problem 1.6) atoms of ^{235}U. How many millions of electron volts (MeV) of energy are released, on average, when a ^{235}U nucleus fissions? How many kilograms of gasoline have the same energy content as one kilogram of ^{235}U?

1.8 The US total electrical power consumption in 2010 was 3.9 TkWh. Utilities try to maintain a capacity that is twice the average power consumption to allow for high demand on hot summer days. What installed generating capacity does this imply?

Mechanical Energy

The systematic study of physical laws begins both logically and historically with the basic notions of classical mechanics. The laws of classical mechanics, as formulated by the English physicist and mathematician Isaac Newton, describe the motion of macroscopic physical objects and their interaction through forces (see Box 2.1). The mathematical framework of calculus provides the necessary tools with which to analyze classical mechanical systems. In this chapter, we review the fundamental principles of mechanics – including kinetic energy, forces, potential energy, and frictional energy loss – all in the context of the use of *energy in transport*.

In 2016, approximately 29% of the energy used in the US – roughly 29 EJ – was used for transportation [12], including personal and commercial, land, water, and air transport (Figure 2.1). This energy use led to CO_2 emissions of almost half a gigaton, or about one third of total US emissions [12]. Transport of people, as well as food, raw materials, and other goods, presents a particular challenge for clean and efficient energy systems. Because cars, airplanes, and trucks are all mobile and not (at least currently) directly connected to any kind of energy grid, they must carry their fuel with them. Historically this has favored the use of fossil fuels such as gasoline, which have high energy density and are easily combusted. In later chapters we examine other options for transportation energy sources. Here we focus on how energy is actually used in transport. Studying how energy is used to put a vehicle in motion (kinetic energy), take a vehicle up and down hills (potential energy), and keep a vehicle in motion in our atmosphere (air resistance), gives insight into how energy needs for transport might be reduced, independent of the fuel option used.

To introduce the principles of mechanics in the context of energy usage, we analyze a specific example throughout much of this chapter. Imagine that four friends plan to

Reader's Guide

This chapter contains a concise review of the basic elements of classical mechanics that are needed for the rest of the book. The core principle of energy conservation serves as a guide in developing and connecting the key concepts of mechanics. While this chapter is fairly self-contained, we assume that the reader has some previous exposure to classical mechanics and to elementary calculus.

The centerpiece of the chapter is a study of energy use in transport.

drive from MIT (Massachusetts Institute of Technology) in Cambridge, Massachusetts to New York City in a typical gasoline-powered automobile, such as a Toyota Camry. The distance from Cambridge to New York is approximately 210 miles (330 km).[1] The Toyota Camry gets about 30 miles per gallon (30 mpg $\cong$ 13 km/L) highway mileage, so the trip requires about 7 gallons (27 L) of gasoline. The energy content of gasoline is approximately 120 MJ/gallon (32 MJ/L) [13], so the energy needed for the trip amounts to (7 gallons) $\times$ (120 MJ/gallon) $\cong$ 840 MJ. This is a lot of energy, compared for example to the typical daily human food energy requirement of 10 MJ. Where does it all go? As we describe in more detail in later chapters, automobile engines are far from perfectly efficient. A typical auto engine only manages to convert about 25% of its gasoline fuel energy into mechanical energy when driving long distances on a highway. Thus, we can only expect about 210 MJ of delivered mechanical energy from the 7 gallons of gasoline used in the trip. But 210 MJ is still a substantial amount of energy.

---

[1] In this chapter and some other examples throughout the book we use colloquial US units, suited to the examples, as well as SI units.

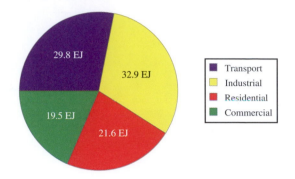

Figure 2.1 2016 US energy use by economic sector. Roughly 29% of US energy use is for transport, including air, land, and water transport of people and goods [12].

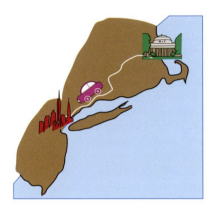

Figure 2.2 In this chapter we illustrate the basics of mechanics by analyzing the energy used by an automobile driving from Cambridge, Massachusetts to New York City.

The principles of mechanics developed in this chapter enable us to give a rough "back-of-the-envelope" estimation of how the 210 MJ of energy is used. In addition to the basic elements of kinetic energy, potential energy, forces, work, and power, we also need to address questions of friction and air resistance. The *drag coefficient* that arises in studying air resistance of a moving vehicle is an example of a **phenomenological parameter** that captures complicated details of a physical system in a single number. Such parameters can be estimated or measured even though they may be prohibitively difficult to compute from first principles. Parameters of this type are common in science and engineering and appear frequently throughout this book.

We assume that the reader has had prior exposure to the basics of classical mechanics and integral calculus. A brief review of some aspects of multi-variable calculus is given in Appendix B. Readers who are interested in a more comprehensive pedagogical introduction to elementary mechanics and/or calculus should consult an introductory textbook such as [14] or [15] for mechanics or [16] for calculus.

2.1 Kinetic Energy

The simplest manifestation of energy in physics is the kinetic energy associated with an object in motion. For an object of mass m, moving at a speed v, the motional or **kinetic energy** is given by[2]

$$E_{kin} = \frac{1}{2}mv^2. \tag{2.1}$$

In the context of the road trip from Cambridge to New York, the kinetic energy of the automobile carrying three passengers and the driver, at a total mass of 1800 kg ($\cong$ 4000 lbs), moving at a speed of 100 km/h, or 62 mph (miles per hour), is

$$E_{kin} = \frac{1}{2}mv^2 \cong \frac{1}{2}(1800\,\text{kg})(100\,\text{km/h})^2 \cong 700\,\text{kJ}. \tag{2.2}$$

If we assume that the road is roughly flat from Cambridge to New York, and if we neglect friction and air resistance as is often done in elementary mechanics courses, it seems that only 0.7 MJ of energy would be needed to make the trip. The driver needs only to get the car rolling at 100 km/h, which takes 0.7 MJ, and the vehicle could then coast all the way with no further energy expenditure. We should include the effect of driving on city streets to reach the freeway, where it is occasionally necessary to stop at a red light. After every stop the car must get back to full speed again. This is why fuel efficiency in the city is very different from highway driving. But even assuming a dozen stops with acceleration to 50 km/h between each stop, we only need about 2 MJ of additional energy (kinetic energy at 50 km/h is 1/4 of that at 100 km/h). This is far less than the 210 MJ of fuel energy we are using. So what do we need the other 208 MJ for?

> **Kinetic Energy**
>
> The kinetic energy of an object of mass m moving at a speed v is given by
>
> $$E_{kin} = \frac{1}{2}mv^2.$$

---

[2] Note that in many classical mechanics texts, the symbol T is used for kinetic energy. In this text we reserve T for temperature, and use E_{kin} for kinetic energy throughout.

Example 2.1   Kinetic Energy of a Boeing 777–300ER

Consider a Boeing 777–300ER aircraft, loaded with fuel and passengers to a mass of $350\,000\,\text{kg}$, at a cruising speed of $900\,\text{km/h}$. The plane has a kinetic energy of

$$E_{\text{kin}} = \frac{1}{2}mv^2 \cong \frac{1}{2}(350\,000\,\text{kg})(900\,\text{km/h})^2 \cong 11\,\text{GJ}\,.$$

This represents the daily food energy intake of a thousand people, or the full average daily energy consumption of over 50 people.

$V(z) = mgz$

Figure 2.3 As the car goes up hill, kinetic energy is changed into potential energy. Some of this energy is returned to kinetic form as the car goes back down the hill, but some is lost in braking to moderate the car's speed.

Of course, the road is not really completely flat between Cambridge and New York City. There is some moderately hilly terrain along the way that we must account for (see Figure 2.3). (For small hills, the car regains the energy used to go up when the car goes down again, but for large hills energy is lost to braking on the way down.) To include the effects of hills, we need to review the notion of mechanical potential energy.

2.2 Potential Energy

As mentioned in §1, energy can be transformed from one form to another by various physical processes, but never created or destroyed. As a ball that is thrown directly upward slows due to the force of gravity, it loses kinetic energy. The lost kinetic energy does not disappear; rather, it is stored in the form of *gravitational potential energy*. We can use the fundamental principle of conservation of energy to understand the relationship between kinetic and potential energy in a precise quantitative fashion.

Potential energy is energy stored in a configuration of objects that interact through forces. To understand potential energy we begin by reviewing the nature of forces in Newtonian physics. The exchange of energy between potential and kinetic leads us to the concepts of work and power. Describing forces and potential energy in

three-dimensional systems leads naturally to a review of vectors and to momentum. In the end we find a more significant contribution to the energy cost of the road trip.

Potential energy plays an important role in many aspects of practical energy systems. For example, mechanical energy, such as that produced by a windmill, can be stored as potential energy by using the mechanical energy to pump water uphill into a reservoir.

2.2.1 Forces and Potential Energy

Newton's second law of motion describes the action of a force F on an object of mass m. For the moment we assume that the object of mass m is moving in a single dimension along a trajectory $x(t)$, describing the position x of the object as a function of the time t. Newton's law in this context is

$$F = m\frac{d^2}{dt^2}x(t) = m\ddot{x}\,. \tag{2.3}$$

(We often use the shorthand notations $\dot{x} \equiv dx(t)/dt$ and $\ddot{x} \equiv d^2x(t)/dt^2$.)

According to eq. (2.3), an object that is acted on by any force will experience an *acceleration*. Recall that the **velocity** of an object is the rate of change of position (e.g. $\dot{x}(t)$), and the **acceleration** is the rate of change of velocity (e.g. $\ddot{x}(t)$). A **force** is an influence on an object from another object or field. In Newtonian physics, eq. (2.3) can be taken as the *definition* as well as a description of the effect of a force. The four fundamental forces of nature are described in §14, but for most of this book we are only concerned with gravitational and electromagnetic forces. As an example of Newton's second law, consider a ball thrown up in the air near Earth's surface. The ball experiences a constant downward acceleration of magnitude $g = 9.807\,\text{m/s}^2$ towards Earth, indicating the presence of a force of magnitude $F = mg$. If we denote the height of the ball at time t by $z(t)$, then

$$m\ddot{z} = F = -mg\,. \tag{2.4}$$

Example 2.2 Airplane at Altitude

$E_{\text{kin}} = \frac{1}{2}mv^2 \cong 11\,\text{GJ}$

$V = mgh \cong 41\,\text{GJ}$

Recall that a 777–300ER flying at 900 km/h has 11 GJ of kinetic energy. How much potential energy does it have when flying at an altitude of 12 000 meters (39 000 feet)?

Using (2.9), we have

$$V = mgh \cong (350\,000\,\text{kg})(9.8\,\text{m/s}^2)(12\,000\,\text{m}) \cong 41\,\text{GJ}.$$

Getting the plane up to cruising altitude takes about four times as much energy as getting it to cruising speed.

As the ball rises, the force of gravity reduces the speed, and therefore also the kinetic energy, of the ball. Here the principle of energy conservation comes into play: this energy is not lost. Rather, it is stored in the potential energy associated with the position of the ball relative to earth. After the ball reaches the top of its trajectory and begins to fall, this potential energy is converted back into kinetic energy as the force (2.4) accelerates the ball downward again.

With this example in mind, we can find the general expression for potential energy simply by applying conservation of energy to an object subject to a known force or set of forces. First, how does a force cause kinetic energy to change?

$$\frac{dE_{\text{kin}}}{dt} = \frac{d}{dt}\left(\frac{1}{2}m\dot{x}^2\right) = m\dot{x}\ddot{x} = F\frac{dx}{dt}. \qquad (2.5)$$

This change can represent a transfer of kinetic energy to or from potential energy. We denote potential energy by V. In the absence of other forms of energy, conservation of energy requires $V + E_{\text{kin}} = \text{constant}$. The change in potential energy in time dt is then given by

$$dV = -dE_{\text{kin}} = -Fdx. \qquad (2.6)$$

If the force acts in the same direction as the particle is moving, the kinetic energy increases and the potential energy must decrease by the same amount. If the force is acting in the opposite direction, the kinetic energy decreases and the potential energy increases by the same amount. Integrating eq. (2.6), the potential energy change when moving from an initial position x_1 to another position x_2 can be described by a function $V(x)$ that satisfies

$$V(x_2) = -\int_{x_1}^{x_2} dx\, F(x) + V(x_1). \qquad (2.7)$$

We have thus *defined* the potential energy of a system subject to known forces using the principle of conservation of energy. A potential energy function V can be defined in this way for any force that depends only on the coordinates describing the physical configuration of the system (e.g. in a one-dimensional system F depends only on x and not on $\dot{x}$ or t).

While we have derived the formula (2.7) for potential energy assuming knowledge of the relevant force(s), it is often convenient to turn the relation around, and write the force as the derivative of the potential

---

Box 2.1 Newton's Laws

Newton's three laws of motion are:

1. An object in motion remains in motion with no change in velocity unless acted on by an external force.
2. The acceleration $\boldsymbol{a}$ of an object of mass m under the influence of a force $\boldsymbol{F}$ is given by $\boldsymbol{F} = m\boldsymbol{a}$.
3. For every action there is an equal and opposite reaction.

We assume that the reader has previously encountered these laws in the context of introductory physics. In this chapter we describe how these laws can be understood in the context of energy. A more detailed discussion of the *inertial reference frames* in which Newton's laws hold and analysis of apparent forces in accelerating reference frames are given in §27 (Ocean Energy Flow).

$$F(x) = -\frac{dV(x)}{dx}.\qquad(2.8)$$

Note that the force resulting from this equation is unchanged if $V(x)$ is shifted by an additive constant. It is clear from this analysis that only potential energy *differences* matter for describing forces, consistent with the discussion at the end of §1.2.

Returning to the example of a ball in Earth's gravitational field near the surface, the force is a constant $F = -mg$, so the gravitational potential when the ball is at height z is

$$V(z) = \int_0^z dz' \, mg = mgz,\qquad(2.9)$$

where we have set $V(0) = 0$. Thus, gravitational potential energy grows linearly with height.

Another commonly encountered type of force grows linearly with distance, $F = -kx$ around a point of equilibrium $x = 0$. This is the force associated with a **harmonic oscillator**, such as a spring; the linear relation between the force and a spring's extension is known as **Hooke's law**. The potential energy associated with this force can be found from eq. (2.7), and is $V(x) = \frac{1}{2}kx^2$. Newton's law then becomes

$$m\ddot{x} = F = -\frac{dV(x)}{dx} = -kx.\qquad(2.10)$$

Although they do not figure directly in the physics of transport studied here, harmonic oscillators occur in many energy systems studied in later chapters and are described further in Box 2.2.

2.2.2 Work and Power

When energy is transferred to an object by the action of a force, the transferred energy is called **work**. If the object gains energy, we say that the force does work on the object; if the object loses energy, then we say that the force does negative work on the object, or equivalently, that the object does work on whatever generates the force. When you throw a tennis ball, you do work on it and its kinetic energy increases. If it rises, its potential energy increases as well. If your dog catches it, the ball does work on the dog.

From eqs. (2.5) and (2.6), we see that as an object moves a distance dx under the influence of a force F, the work done is

$$dW = F dx.\qquad(2.11)$$

If the force has the same sign as the direction of motion, then dW is positive and the force does work on the object. Work can increase potential as well as kinetic energy when the object is subject to additional forces. For example, if your hands exert an upward force on a book to lift it at

a constant speed, they do work on the book by increasing its gravitational potential energy although its kinetic energy remains unchanged. From conservation of energy, eq. (2.11) also describes the work done in this case.

The rate at which work is done is measured by **power**, as introduced in §1.1. From eq. (2.11) we see that the instantaneous power exerted by a force F on an object moving with velocity $v = dx/dt$ is

$$P = \frac{dW}{dt} = F\frac{dx}{dt} = Fv.\qquad(2.12)$$

The connection between power and force, mediated by velocity, is key in many energy applications. An agent may exert great force on an object, but it does no work unless the object is in motion.

2.2.3 Forces and Potential Energy in Three Dimensions

Up to now, for simplicity we have treated a one-dimensional physical system. It is straightforward to generalize the concepts of kinetic energy, forces, and potential energy to higher-dimensional systems. The position of a particle moving in three dimensions is described by a vector (see Figure 2.4)

$$\mathbf{x} = (x, y, z) = x\hat{\mathbf{x}} + y\hat{\mathbf{y}} + z\hat{\mathbf{z}},\qquad(2.13)$$

where depending upon context we may use bold font ($\mathbf{x}$), triplet notation (x, y, z), or an expansion in unit vectors $\hat{\mathbf{x}}, \hat{\mathbf{y}}, \hat{\mathbf{z}}$ to denote vectors in three dimensions.

Newton's second law in three dimensions reads

$$\mathbf{F} = m\mathbf{a} = m\frac{d^2\mathbf{x}}{dt^2} = m\ddot{\mathbf{x}}.\qquad(2.14)$$

As in the one-dimensional case, in the absence of other forms of energy, conservation of energy requires the sum of the kinetic and potential energy of the particle to be

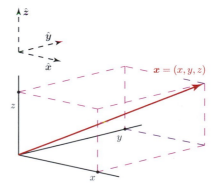

Figure 2.4 A vector $\mathbf{x}$ in three dimensions and the unit vectors along the x-, y-, and z-directions.

Box 2.2   The Harmonic Oscillator

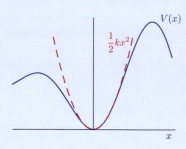

Suppose a one-dimensional system is described by a potential energy function $V(x)$, such as the one shown to the left, with a (local) minimum at $x = 0$. If $V(x)$ is smooth near $x = 0$, we can expand it in a Taylor series, $V(x) = V(0) + \frac{1}{2}kx^2 + \ldots$, where k must be nonnegative for $x = 0$ to be a minimum. For k positive, therefore, the potential looks like a harmonic oscillator with a restoring force $F = -kx$ that keeps x close to the location of the minimum. If k were exactly zero, then the behavior of the oscillator would be different, but this is rarely the case for real physical systems. For a mass m moving in such a potential, Newton's second law is $F = m\ddot{x} = -kx$, and $E_{kin} = m\dot{x}^2/2$. Setting $V(0) = 0$, since we are free to choose the zero of energy, conservation of energy tells us that

$$E = \frac{1}{2}m\dot{x}^2 + \frac{1}{2}kx^2\,.$$

What sort of motion satisfies this conservation law? Up to some multiplicative coefficients, the square of the function $x(t)$ plus the square of its derivative $\dot{x}(t)$ must add to a constant. This is a generic property of trigonometric functions: recall that $\cos^2 z + \sin^2 z = 1$ and $d(\cos z)/dz = -\sin z$. Substituting a function of the form $x(t) = A\cos\omega t$ into the equation for E above, we find a solution when $A = \sqrt{2E/k}$ and $\omega = \sqrt{k/m}$. (The same solution can be found directly from $m\ddot{x} = -kx$ by noting that $d^2(\cos z)/dz^2 = -\cos z$.) This is not the most general solution, however. In particular, for this solution x is maximized and $\dot{x} = 0$ at $t = 0$, but we could consider other initial conditions, such as $\dot{x}$ nonzero and $x = 0$ at $t = 0$. The most general solution is found by shifting the solution by an arbitrary time t_0, so that the maximum value of x occurs at that time,

$$x(t) = A\cos\omega(t - t_0) = A\cos(\omega t - \phi)\,.$$

This general solution describes a stable, harmonic (meaning sinusoidal) oscillation about the point of stability, $x = 0$. The solution repeats itself in a time $T = 2\pi/\omega$, the **period**, and $1/T = \nu$ is the **frequency** of the oscillation, measured in **Hertz** (Hz), or oscillations per second. ω is referred to as the **angular frequency** of the oscillation, and is measured in radians/s (or simply s^{-1} since radians are dimensionless). The solution is unchanged if the **phase** $\phi = \omega t_0$ is replaced by $\phi + 2\pi$, so ϕ can be thought of as an angle that is defined over the interval $-\pi < \phi \leq \pi$. Some aspects of harmonic motion are illustrated to the right. Part (a) shows the solution $x(t) = A\cos(\omega t - \phi)$, and part (b) illustrates the feature that the total energy E oscillates back and forth between kinetic energy $E_{kin} = E\sin^2(\omega t - \phi)$ and potential energy $V = E\cos^2(\omega t - \phi)$, each carrying half the energy on average.

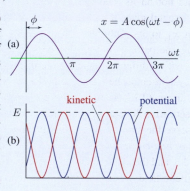

Harmonic oscillators arise as an approximate description of a vast range of physical systems, and appear repeatedly throughout this book. For example, vibrational excitations of many molecules can be approximated as harmonic oscillators (§9), and the associated quantum energy spectrum (§7) plays an important role in the emission of electromagnetic radiation (§22).

---

constant. The change in potential energy can be determined from the force $\mathbf{F}(x, y, z)$. The rate of change of kinetic energy E_{kin} for an object subject to the force $\mathbf{F}$ becomes

$$\frac{dE_{kin}}{dt} = \frac{d}{dt}\left(\frac{1}{2}m\dot{x}^2\right) = m\ddot{x}\cdot\dot{x} = \mathbf{F}\cdot\dot{x}\,, \qquad (2.15)$$

where "·" denotes the usual vector *dot product* (Appendix B.1.1). Demanding conservation of energy as in eq. (2.6),

we define the differential change in the potential energy V when the force $\mathbf{F}$ moves the object through the differential displacement $d\mathbf{x}$ to be

$$dV = -dE_{kin} = -\mathbf{F}\cdot d\mathbf{x}\,. \qquad (2.16)$$

Integrating a differential like this requires specifying the path P along which the integral is taken (see Figure 2.5) and defines a *line integral* (Appendix B.1.4). Here

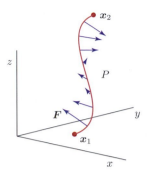

Figure 2.5 The integral of $F \cdot dx$ along a path P from x_1 to x_2 gives the work done by F as an object moves from x_1 to x_2. This also gives the potential energy difference between the two points $V(x_1) - V(x_2)$.

$dx = (dx, dy, dz)$ is the tangent to the path $x(\tau) = (x(\tau), y(\tau), z(\tau))$ at each point, where τ parameterizes the path. If we integrate along path P from point x_1 to x_2, we obtain

$$V(x_2) = -\int_P dx \cdot F + V(x_1). \qquad (2.17)$$

For a force F that depends only on an object's position x and not on its velocity $\dot{x}$ or explicitly on the time t, this integral gives the same answer for any path P taken from x_1 to x_2. Otherwise, by traveling from x_1 to x_2 along one path, then returning back to x_1 along another path with a different value for the integral, the energy of the system could be changed. Forces for which the integral in eq. (2.17) is independent of path are known as **conservative forces** in recognition of the fact that they conserve the sum of kinetic and potential energy. Examples of *non-conservative* forces include friction, discussed in §2.3, and the electromotive force induced by a changing magnetic field, discussed in §3.5. Note that the dot product $F \cdot dx$ arises because the only component of the force that changes the energy is the part parallel to the motion of the particle. Note also the sign: if the particle moves against (with) the force, $F \cdot dx$ is negative (positive) and the potential energy increases (decreases).

Inverting the formula (2.17) we obtain the force in terms of the derivatives of the potential energy

$$F = -\nabla V = -\left(\frac{\partial V}{\partial x}, \frac{\partial V}{\partial y}, \frac{\partial V}{\partial z} \right), \qquad (2.18)$$

where ∇V is the *gradient* (Appendix B.1.3) of V. The gradient is a vector pointing in the direction in which the function increases most steeply, with a magnitude proportional to the rate of increase of the function in that direction. Thus, the force points in the direction in which the potential energy decreases. Finally, the formula for the

instantaneous power (2.12) generalizes to

$$P(t) = F(t) \cdot v(t). \qquad (2.19)$$

As a simple example of a potential in three dimensions, consider the classical gravitational potential energy between two objects of masses M and m. This potential is given by

$$V(r) = -\frac{GMm}{r}, \qquad (2.20)$$

where $r = |r|$ is the distance between the centers of mass of the two objects and $G = 6.674 \times 10^{-11} \, \mathrm{m}^3/\mathrm{kg}\,\mathrm{s}^2$ is **Newton's constant**. The zero of this potential is defined to be at infinite distance $r \to \infty$; as two objects originally far apart are pulled together by the gravitational force they acquire (positive) kinetic energy and their potential energy becomes negative. The force on the object of mass m arising from this potential is (assuming that the object of mass M is at the origin, and the object of mass m is at position $r = (x, y, z)$)

$$F = -\nabla V = -\frac{GMm}{(x^2 + y^2 + z^2)^{3/2}}(x, y, z)$$
$$= -\frac{GMm}{r^3}r = -\frac{GMm}{r^2}\hat{r}. \qquad (2.21)$$

This gives a vector of magnitude GMm/r^2 pointing towards the object of mass M. There is an equal and opposite force on the object of mass M, giving an example of Newton's third law. In a situation where $m \ll M$, we can make the approximation that the object of mass M is fixed at the origin, since it accelerates much less than the smaller mass. The general potential (2.20) can be related to the gravitational potential on Earth's surface (2.9). For M we use Earth's mass $M_\oplus \cong 5.97 \times 10^{24}$ kg. Near Earth's surface we can write $r = R_\oplus + z$, where Earth's radius is $R_\oplus \cong 6370$ km. Expanding[3] eq. (2.20) to linear order in z then reproduces eq. (2.9) with $g = GM_\oplus/R_\oplus^2$ (Problem 2.5).

2.2.4 Momentum

Momentum is an important concept in mechanics that is related to kinetic energy. The momentum of a particle of mass m moving with velocity $\dot{x}$ is

$$p = m\dot{x}. \qquad (2.22)$$

Like energy, the total momentum of an isolated system is conserved.[4] Thus, for example, when two particles collide, the total momentum of the two particles must be the same after the collision as before, provided that the influence of

---

[3] See Appendix B.4.1 for a quick review of series expansions.

[4] See §21 for further explanation of momentum conservation.

Example 2.3   Gravitational Potential Energy of the Moon

Consider the potential energy of the Moon in its orbit around Earth. The mass of the Moon is $m_{\text{moon}} \cong 7.3 \times 10^{22}$ kg. The Moon orbits at a radius of $r_{\text{orbit}} \cong 3.8 \times 10^8$ m, so the gravitational potential energy is

$$V = -\frac{GM_\oplus m_{\text{moon}}}{r_{\text{orbit}}} \cong -7.7 \times 10^{28} \text{ J}.$$

The Moon also has kinetic energy as it revolves around the Earth. Approximating its orbit as circular, it is not hard to show that $E_{\text{kin}} = -\frac{1}{2}V \cong 3.8 \times 10^{28}$ J (Problem 2.20). So the total energy of the moon is $E(\text{moon}) = E_{\text{kin}} + V = \frac{1}{2}V \cong -3.8 \times 10^{28}$ J.

The zero of potential energy is defined so that V vanishes as $r \to \infty$, so work equal to $\frac{1}{2}|V|$ would have to be done on the Earth–Moon system to send the Moon out of orbit into interplanetary space. Were the Moon centered at Earth's surface, its potential energy would be

$$V = -\frac{GM_\oplus m_{\text{moon}}}{R_\oplus} \cong -4.6 \times 10^{30} \text{ J}.$$

The difference between these two values for the Moon's potential energy, roughly 4.5×10^{30} J, gives an enormous energy potential, roughly 10 billion times the annual worldwide energy use of humans.

---

other particles can be ignored. Newton's second law (2.14) can be expressed in terms of momentum as

$$F = m\ddot{x} = \frac{dp}{dt}. \tag{2.23}$$

Newton's third law follows from conservation of momentum. Suppose that two particles with momenta p_1 and p_2 exert forces F_{12} and F_{21} on one another. Since the total momentum is conserved ($\dot{p}_{\text{total}} = (\dot{p}_1 + \dot{p}_2) = 0$) it follows that $F_{12} = -F_{21}$.

It is sometimes convenient to express kinetic energy in terms of momentum

$$E_{\text{kin}} = \frac{1}{2}m\dot{x}^2 = \frac{p^2}{2m}. \tag{2.24}$$

This relation enters into the basic framework of quantum mechanics in §7.

Potential Energy

Potential energy is described by a function $V(x)$, where the coordinates x describe the physical configuration of a system. V gives rise to a force through

$$F = -\nabla V.$$

Potential energy can be found by integrating a (conservative) time-independent force

$$V(x_2) = V(x_1) - \int_{x_1}^{x_2} dx \cdot F,$$

where this integral is independent of the path from x_1 to x_2. Potential energy can be shifted by an overall constant, without affecting the resulting forces; this constant is usually chosen so that potential energy vanishes at a convenient location.

Two simple examples of potential energy are a mass m at height z above Earth's surface, and a mass connected to a spring with spring constant k,

$$V(z) = mgz \text{ in a uniform gravitational field,}$$

$$V(x) = \frac{1}{2}kx^2 \text{ for a mass on a spring.}$$

2.2.5 Back to the Trip: The Effect of Hills

Armed with the concept of potential energy, let us return to the subject of the car trip from Cambridge to New York. We found that accelerating the 1800 kg (4000 lb) vehicle to 100 km/h (62 mph) requires 0.7 MJ, a tiny fraction of the 210 MJ of mechanical energy delivered by the combustion of the 7 gallons (27 L) of gasoline that the trip consumes. What about the energy needed to go up and down hills encountered along the way? While there is no significant net elevation gain between Cambridge and New York (both places are close to sea level), the topography varies from flat to hilly along the route, a section of which is shown in Figure 2.6. After studying the terrain, we can make a very rough estimate that on average in every mile of horizontal distance covered, the car gains (and loses) roughly

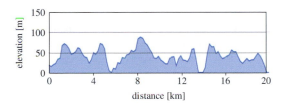

Figure 2.6 Elevation changes along a relatively hilly 20 kilometer section of the route from Cambridge to New York City.

15 m (50 ft) of elevation, corresponding to a grade (slope) of $15/1609 \cong 1\%$.

This estimate is certainly not an accurate characterization of the detailed topography of any highway route from Cambridge to New York, but simplifies the calculation while preserving the essential physics. This is an example of a simplifying assumption that enables us to make the kind of back-of-the-envelope estimate we proposed at the outset. The resulting model preserves the essential features relevant for the physics while eliding details that would make the problem too difficult for a quick calculation.

Proceeding with this estimate, we see that for each mile of travel, the additional energy needed to raise the vehicle up the 15 meters of elevation gain is

$$V = mgh = (1800\,\text{kg})(9.8\,\text{m/s})(15\,\text{m}) \cong 260\,\text{kJ}. \quad (2.25)$$

Of course, the mechanical energy used to move the car up a hill is not lost. It is stored as potential energy that can be converted back to kinetic energy as the car descends on the other side. If the hills were small, and if we could tolerate small fluctuations in vehicle speed, then the vehicle would simply speed up when going down a hill and slow down again when going up the next hill. In the absence of friction and air resistance, no further energy would be needed. In fact, however, hills are often steep enough and

long enough that a vehicle traveling at a reasonable speed at the top of the hill would achieve an illegal and/or dangerous speed before reaching the bottom of the hill unless brakes were used to slow it down. For our example of a car traveling at 62 mph at the top of a hill of height 15 m, full conversion of its potential energy to kinetic energy would result in a speed of 74 mph at the bottom (Problem 2.9), which is above the common US highway speed limit of 65 mph.

Keeping in mind that this is a back-of-the-envelope calculation, we estimate that roughly half of the excess kinetic energy from hills will be lost to braking, so roughly 130 kJ per mile might be used in surmounting hills. With this estimate, over 210 miles the total energy required would be 130 kJ/mi × 210 mi $\cong$ 27 MJ.

The 27 MJ of energy we have estimated for getting over hills is much more than the energy needed to get the car moving at full speed at the beginning of the trip, even including a dozen or so stops at traffic lights, but together they are still nearly an *order of magnitude*[5] less than the 210 MJ of energy produced by the engine. Clearly, we are still missing the biggest piece of the puzzle.

2.3 Air Resistance and Friction

2.3.1 Air Resistance
We next consider the effect of air resistance. As the car moves forward, it collides with air molecules, pushing them out of the way, and generating a complex flow of local air currents that sweep around the car (see Figure 2.7). Behind the car, a mass of air follows in the wake of the vehicle, containing air molecules that have been swept up in the flow around the car as it passes, and that are now moving at a velocity comparable to that of the car.

Figure 2.7 Cartoon of air flow around a car.

---

5 An **order of magnitude** corresponds roughly to a factor of 10. Thus, an *order-of-magnitude* estimate is one that has approximately the correct number of digits but can be off by a multiplicative factor of 10 or more. Similarly, a number "of order 1" could be anywhere in or near the range from 0.3 to 3.

Eventually, after the car has passed by, the energy in the swirling flow of air ends up as random motions of air molecules associated with a slight warming of the mass of air. The work done by the car in moving the air molecules aside has ended up as *thermal energy*. Thermal energy is a topic for a later chapter (§5). For now we note that resistance produces a different kind of force from those we have considered so far – a **dissipative force**, which converts organized kinetic energy into more-or-less random motion of molecules. In particular, we do not keep track of the positions of all the air molecules, so the **drag force** on the vehicle arising from air resistance cannot be described as the gradient of a potential function.

The precise details of how energy is lost to air resistance are highly complex and depend upon the geometry of the vehicle and many other variables. But up to a numerical coefficient that is often within an order of magnitude of one, it is possible to give a simple formula that approximately characterizes the rate at which the vehicle loses kinetic energy.

A Model of Air Resistance   When an automobile travels a distance dx, it passes through a volume of space (see Figure 2.8) $dV = A\,dx$, where A is the cross-sectional area of the car projected onto a plane perpendicular to the direction of the car's motion. The area A is roughly the height of the car times its width. If we assume that every molecule of air in the volume dV is accelerated by its interaction with the passing vehicle to the velocity v of the car, and that no other air molecules are affected, then the total mass of air affected would be $A\rho\,dx$, where ρ is the mass density of air (mass per unit volume). Thus, the kinetic energy transferred by the car to the air would be $\frac{1}{2}A\rho v^2\,dx$. In a real system, the flow of air is much more complicated. We can parameterize our ignorance of the details by introducing a dimensionless phenomenological parameter, called the **drag coefficient** c_d. The constant c_d combines into a single number all of the complex details of the automobile's shape and the pattern of the airflow. Although the value of the drag coefficient can vary with large changes in the car's velocity or the air's density, c_d can be regarded

as a constant in practical situations. Including this parameter as an overall proportionality constant, the amount of energy transferred to the air when the car moves a distance dx is given by

$$dE_{\text{air}} = \frac{1}{2}c_d A\rho v^2 dx. \tag{2.26}$$

Since the car travels a distance dx in time vdt, the rate at which the car loses kinetic energy to air resistance is

$$\frac{dE_{\text{air}}}{dt} = \frac{1}{2}c_d A\rho v^3. \tag{2.27}$$

This can be thought of as arising from a *drag force* $\mathbf{F} = -\frac{1}{2}c_d\rho A v^2\hat{\mathbf{v}}$ that acts in the direction opposite to the velocity of the car. Note that the rate (2.27) scales as v^3 while the total energy transferred in going a distance D (from integrating eq. (2.26)) scales as v^2

$$\Delta E = \frac{1}{2}c_d A D\rho v^2. \tag{2.28}$$

The two equations (2.27) and (2.28) are consistent since total energy transfer is given by the rate of transfer times the time taken for the vehicle to travel the distance D and the time taken $T = D/v$ is inversely proportional to the velocity.

Air Resistance

The rate of energy loss due to air resistance on a moving object is

$$\frac{dE_{\text{air}}}{dt} = \frac{1}{2}c_d A\rho v^3,$$

where v is the speed of the object, ρ is the mass density of the air, A is the cross-sectional area of the object, and c_d is the *drag coefficient*, a dimensionless quantity of order one.

Dimensional Analysis   Dimensional analysis gives some further perspective on the drag formula (2.27). From the qualitative physics of air resistance, we expect the rate of energy loss to depend on the cross-sectional area A of the car and the car's speed v. The density of the air ρ also matters – the denser the air, the more of the car's momentum goes into moving it. One variable that does not matter is the mass of the car, since a car made of plastic or one made of lead has to move the same amount of air out of the way. With these considerations it is natural to expect that the rate of energy loss to air resistance, dE/dt, has the form

$$\frac{dE_{\text{air}}}{dt} = \frac{1}{2}c_d\rho^\alpha v^\beta A^\gamma, \tag{2.29}$$

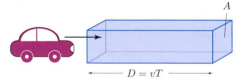

Figure 2.8 Air affected by a car's motion is roughly that in a volume given by the cross-sectional area A of the vehicle times the distance D traveled.

where α, β, and γ are constants, and c_d is a dimensionless constant. To find the values of α, β, and γ we balance dimensions on both sides of eq. (2.29),

$$\left[\frac{\text{kg m}^2}{\text{s}^3}\right] = \left[\frac{\text{kg}}{\text{m}^3}\right]^{\alpha} \left[\frac{\text{m}}{\text{s}}\right]^{\beta} \left[\text{m}^2\right]^{\gamma} . \tag{2.30}$$

The units only balance if $\alpha = 1$, $\beta = 3$, and $\gamma = 1$, which reproduces eq. (2.27), $dE_{\text{air}}/dt = c_d A \rho v^3/2$. This relation is a fundamental characteristic of the power transfer between solid surfaces and fluids, and plays a central role in wind power and other forms of energy production (§28).

2.3.2 Computing the Drag Coefficient c_d

The drag coefficient c_d depends upon the precise shape of the vehicle. A box-like vehicle with a shape that does not allow the air to flow smoothly around it will have a larger drag coefficient, while an aerodynamically designed vehicle that produces a smooth flow will have a smaller drag coefficient. Because the drag coefficient and cross-sectional area generally appear together, often it is convenient to treat the product $c_d A$ as an **effective area** parameterizing the drag of the vehicle,

$$A_{\text{eff}} = c_d A . \tag{2.31}$$

The drag coefficient c_d is difficult to compute from first principles. Although in principle the collision of the automobile with each air molecule can be described through classical mechanics, the number of molecules involved is so vast (with $\sim 10^{26}$ collisions/minute) that a detailed computation is beyond the reach of any computer imaginable. The methods of *fluid mechanics* (developed in later chapters beginning with §29) can be used to model airflow around a moving vehicle, but even this approach to estimating c_d is computationally highly intensive. In most situations it is more practical to estimate c_d experimentally, for example by measuring the rate of kinetic energy loss of the vehicle moving at a variety of speeds. To give some qualitative sense of the typical numbers for drag coefficients, a perfectly cylindrical block moving parallel to its axis in an idealized diffuse atmosphere can be shown to have a drag coefficient of $c_d = 4$ (Problem 2.16). For real systems, the drag coefficient is generally smaller than 4, but is of order one unless the object is very streamlined. A typical passenger auto has a drag coefficient of roughly 1/3; a more aerodynamic automobile like the Toyota Prius has a drag coefficient of about 1/4, and a typical passenger aircraft has a drag coefficient around 0.03.

2.3.3 Back to the Trip: The Effect of Air Resistance

We can now estimate air resistance losses on the road trip from Cambridge to New York. The Toyota Camry has a width of about 1.8 m and a height of about 1.5 m, so $A \cong 2.7\,\text{m}^2$. We estimate its drag coefficient at $c_d(\text{Camry}) \cong 1/3$. Air density at sea level is approximately $1.2\,\text{kg/m}^3$. Putting all these together, we find

$$\begin{aligned} \Delta E_{\text{air}} &= \frac{1}{2} c_d A D \rho v^2 \\ &\cong \frac{1}{2}(0.33)(2.7\,\text{m}^2)(330\,\text{km})(1.2\,\text{kg/m}^3)(28\,\text{m/s})^2 \\ &\cong 138\,\text{MJ} . \end{aligned} \tag{2.32}$$

This is almost two thirds of the 210 MJ of energy available from the engine. At last, we see where much of the energy is going.

2.3.4 Friction, Rolling Resistance, and a Final Accounting

We have assembled many pieces of the puzzle of how the car's energy is lost. Before doing a final accounting, we should consider frictional effects other than air resistance. Frictional effects within the engine and parts of the drive train are associated with thermal energy losses from the engine. Such effects are included in the $\sim$25% overall efficiency of the engine in converting chemical to mechanical energy that we took into account at the outset. There are further frictional losses characterized as **rolling resistance** that are associated with friction between the tires and the road, deformation of the rolling tires, and other losses that are independent of the drive system. To a first approximation, these losses are independent of velocity (but proportional to the vehicle's mass). They can be estimated roughly by placing the vehicle on an inclined plane and adjusting the slope until the vehicle rolls down the incline at a constant speed. Then, neglecting air resistance losses, which are small at low speeds, the rate of energy loss to rolling resistance corresponds to the rate of change of potential energy as the car rolls down the incline. For a typical passenger vehicle, rolling resistance corresponds to a grade of roughly 0.01 (vertical/horizontal distance traveled). Thus, the rate of energy loss due to rolling resistance is approximately equivalent to the extra energy needed to increase potential energy when driving the vehicle on an uphill grade of 1%.

In §2.2.5 we found an energy cost for potential energy of hills equivalent to 54 MJ for traversing an effective 1% grade throughout the trip from MIT to New York. Thus, we

Table 2.1 Contributions to energy consumption on a road trip from Cambridge to New York City.

Source	Energy consumption (MJ)
Kinetic energy (including 12 stoplights)	3
Potential energy from hills	27
Rolling resistance	54
Air resistance	138
Total	222

Energy Used by Automobiles

In highway driving of a typical car, over half of the mechanical energy delivered by the engine is used to accelerate the air through which the vehicle travels.

expect that roughly an additional 54 MJ of energy would also be lost to rolling resistance.

Finally we can assemble all the effects that consume the energy used to carry the 1800 kg automobile with four passengers from Boston to New York. The results are summarized in Table 2.1. Our final estimate for energy used is quite close to 1/4 of the 840 MJ fuel energy of the 7 gallons of gasoline used, matching the expected approximate 25% efficiency of the engine. Of course, all of the approximations made in this example have been rather rough, and some of these numbers are perhaps off by 20% or more, but we end up with a good qualitative sense of the relative energy cost of various aspects of highway driving. In city driving, velocities are lower so air resistance loss is less, and a higher fraction of energy is lost to rolling resistance. More energy is spent on developing kinetic energy through acceleration after stoplights (Problem 2.18) and lost in braking before them. In a modern passenger car, some energy is used by system electronics, perhaps amounting to several megajoules for the trip in the example discussed here. On a hot summer day, air conditioning can use a substantial amount of energy. We discuss heating and cooling energy costs in more detail in §5.

Despite their approximate nature, the calculations we have performed here illustrate the basic role of different aspects of elementary mechanics in transport energy, and give a sense of the relative amounts of energy lost to effects such as air resistance, acceleration, potential energy differences, and engine losses in a typical vehicle. A key question, which is explored further in the Questions in this

chapter and in §36, is the extent to which energy losses in transport can be avoided. The simple form of air resistance losses (2.28), for example, shows that the dominant losses in long-distance transport can be reduced substantially by moving at a lower velocity and streamlining vehicles to decrease the drag coefficient c_d.

2.4 Rotational Mechanics

As a final component of this chapter, we review the basic mechanics of rotating objects. The equations describing rotational motion are very similar in form to those of linear motion, with mass replaced by the *moment of inertia*, velocity replaced by *angular velocity*, and momentum replaced by *angular momentum*. These equations are needed for topics in many later chapters, including for example energy storage in flywheels, dynamics of ocean currents, and the large-scale circulation of the atmosphere.

For simplicity, consider first an object revolving in a circle of radius r around the origin in the xy-plane with constant **angular velocity** ω (see Figure 2.9). We assume that the object has mass m and is very small compared to the radius r, so that it can be considered as a point mass. The motion of the object can then be described by a time-dependent vector

$$\boldsymbol{r}(t) = (x(t), y(t)) = (r \cos \omega t, r \sin \omega t), \qquad (2.33)$$

where we choose the time coordinate so that the object is at position $\boldsymbol{r}(0) = (x(0), y(0)) = (r, 0)$ at time $t = 0$. In terms of this parameterization, the velocity vector at time t is given by

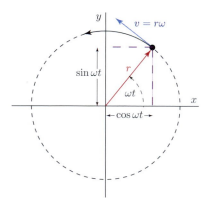

Figure 2.9 Angular motion in the plane. An object at distance r from the origin revolves with constant angular velocity ω about the origin with speed $v = r\omega$. Its (x, y) coordinates are $(r \cos \omega t, r \sin \omega t)$.

Example 2.4   Energy Storage in a Flywheel

Attempts have been made to develop commercially viable passenger vehicles that store braking energy in massive *flywheels* inside the vehicle. Consider a flywheel made from a uniform steel disk with mass $m = 5\,\text{kg}$ and radius $R = 0.3\,\text{m}$. *How much energy is stored in a flywheel of this type rotating at 40 000 rpm?*

The moment of inertia of a solid disk of density ρ, radius R, and height Z is

$$I = \int_0^R dr\, 2\pi r Z \rho r^2 = \frac{\pi}{2}\rho Z R^4,$$

while the mass is

$$m = \int_0^R dr\, 2\pi r Z \rho = \pi \rho Z R^2.$$

Thus, we have $I = \frac{1}{2}mR^2 = 0.225\,\text{kg m}^2$. At 40 000 rpm, we have

$$\omega = 40\,000 \times 2\pi/60 \cong 4200\,\text{s}^{-1},$$

so the stored energy is roughly

$$E_{\text{kin}} = \frac{1}{2}I\omega^2 \cong 2\,\text{MJ},$$

which is more than enough to store the braking energy of a passenger vehicle at highway speeds. We return to the topic of flywheel energy storage in §37.

ω
$R = 0.3$ m
$m = 5$ kg

$$v = \dot{r} = r\omega(-\sin\omega t, \cos\omega t). \qquad (2.34)$$

The acceleration vector

$$a = \dot{v} = -\omega^2 r \qquad (2.35)$$

has magnitude $r\omega^2$ and points towards the center of the circle; such acceleration is known as **centripetal acceleration**. The vector equation (2.35) is true in general for rotational motion about an axis perpendicular to r and is independent of any particular coordinate parameterization such as (2.33). According to Newton's second law the centripetal acceleration must be provided by a force that holds the particle in circular motion, $F = -m\omega^2 r$, without which the particle would move off in a straight line tangent to the circle.

It is useful to regard angular velocity as a *vector* pointing in the direction of a third dimension perpendicular to the plane of motion; in the case described above of motion in the xy-plane this is the z-direction, $\boldsymbol{\omega} = \omega\hat{z}$.

Then in general we have the vector relations,

$$v = \boldsymbol{\omega} \times r,$$

$$a = \boldsymbol{\omega} \times (\boldsymbol{\omega} \times r) = (\boldsymbol{\omega}\cdot r)r - \omega^2 r = -\omega^2 r, \quad (2.36)$$

where "$\times$" denotes the usual vector *cross product* (Appendix B.1.1), and the last line makes use of the vector identity $a \times (b \times c) = (a \cdot c)b - (a \cdot b)c$ and $r \cdot \boldsymbol{\omega} = 0$.

From eq. (2.34) we see that the object is always moving with speed $v = r\omega$ and has a momentum of magnitude

$$p = mv = mr\omega. \qquad (2.37)$$

The **angular momentum** of the object around the origin, which is given in three-dimensional vector form by $L = r \times p$, in this case has magnitude

$$L = rp = mr^2\omega. \qquad (2.38)$$

L points along the axis perpendicular to the plane of motion and can be written in terms of $\boldsymbol{\omega}$, $L = mr^2\boldsymbol{\omega}$. The rotational kinetic energy of the object is

$$E_{\text{rot}} = \frac{1}{2}mv^2 = \frac{1}{2}mr^2\omega^2. \qquad (2.39)$$

For a system composed of multiple parts with masses m_i rotating with uniform angular velocity ω at radii r_i about the z-axis, the angular momentum and kinetic energy can be written as

$$L = \sum_i m_i r_i^2 \omega = I\omega,$$

$$E_{\text{rot}} = \frac{1}{2}\sum_i m_i r_i^2 \omega^2 = \frac{1}{2}I\omega^2, \qquad (2.40)$$

where we have defined the **moment of inertia** of the system to be

$$I = \sum_i m_i r_i^2. \qquad (2.41)$$

For a continuous three-dimensional mass distribution $\rho(\boldsymbol{x})$ rotating around the z-axis, the same formulae hold for angular momentum and kinetic energy, except that we replace the sum with an integral in the equation for the moment of inertia

$$I = \int d^3\boldsymbol{x}\, \rho(\boldsymbol{x})r^2, \qquad (2.42)$$

where $r^2 = x^2 + y^2$ and the notation $d^3\boldsymbol{x}$ indicates an integral over the three components of the vector $\boldsymbol{x}$, $d^3\boldsymbol{x} = dx\,dy\,dz$.

The analog of Newton's second law for rotational motion relates the time rate of change of the angular momentum to the **torque**, $\boldsymbol{\tau} = \boldsymbol{r} \times \boldsymbol{F}$, exerted on the object through

$$\boldsymbol{\tau} = d\boldsymbol{L}/dt. \qquad (2.43)$$

If there is no torque acting on a system, then both the angular momentum $\boldsymbol{L}$ and the kinetic energy of rotation are conserved. The rate of power transferred through the torque is

$$P = \boldsymbol{\tau} \cdot \boldsymbol{\omega}, \qquad (2.44)$$

in analogy with eq. (2.12). These equations for rotational motion essentially follow by taking the cross product of $\boldsymbol{r}$ with the equations for linear motion.

Discussion/Investigation Questions

2.1   The analysis of a road trip from Cambridge, Massachusetts to New York suggests many ways that the energy cost of transport could be reduced including (a) slower speed; (b) a more efficient engine; (c) keeping tires properly inflated; (d) a more streamlined car; (e) carpooling. What are the relative advantages and disadvantages of these steps? Which are more effective for city or for highway driving?

2.2   Is there a minimum energy necessary for transport? Do the laws of physics dictate some minimum energy required to transport, say, 2000 kg of material from Cambridge, Massachusetts to New York? How would you minimize the energy required for transport?

2.3   Using the ideas developed in this chapter, consider the relative energy use of other modes of transport such as airplane or train travel compared to the automobile. Use the internet or other resources to estimate numbers for mass, cross-sectional area, and drag coefficient for a train or a plane, do a back-of-the-envelope calculation of relative energy costs per person for a trip using each of these modes of transport, and discuss the differences.

2.4   **Regenerative brakes** in many hybrid cars such as the Toyota Prius are designed to store the vehicle's kinetic energy in the battery for later use in acceleration. Such brakes currently can absorb roughly 50% of the kinetic energy lost through braking. Most cars get better gas mileage in highway driving than city driving, but the Prius gets better gas mileage in city driving (51 mpg city versus 48 mpg highway for the 2012 model). Can regenerative braking explain this? If so, how? Give a quantitative explanation for your answer.

Problems

2.1   Estimate the kinetic energy of the ocean liner *Queen Mary 2*, with mass (displacement) 76 000 tons, moving at a cruising speed of 26 knots. Compare with the change in potential energy of the *Queen Mary 2* when lifted by a three foot tide.

2.2   [T] A mass m moves under the influence of a force derived from the potential $V(x) = V_0 \cosh ax$, where the properties of $\cosh x$ and other *hyperbolic functions* are given in Appendix B.4.2. What is the force on the mass? What is the frequency of small oscillations about the origin? As the magnitude of the oscillations grows, does their frequency increase, decrease, or stay the same?

2.3   [T] An object of mass m is held near the origin by a spring, $\boldsymbol{F} = -k\boldsymbol{x}$. What is its potential energy? Show that $\boldsymbol{x}(t) = (x_0, y_0, z_0)\cos \omega t$ is a possible classical motion for the mass. What is the energy of this solution? Is this the most general motion? If not, give an example of another solution.

2.4   Estimate the potential energy of the international space station (mass 370 000 kg) relative to Earth's surface when in orbit at a height of 350 km. Compute the velocity of the space station in a circular orbit and compare the kinetic and potential energy.

2.5   [T] Relate eq. (2.20) to eq. (2.9) and compute g in terms of G, $M_\oplus$, $R_\oplus$ as described in the text.

2.6   Make a rough estimate of the maximum hydropower available from rainfall in the US state of Colorado. Look up the average yearly rainfall and average elevation of Colorado and estimate the potential energy (with respect to sea level) of all the water that falls on Colorado over a year. How does this compare with the US yearly total energy consumption?

2.7   Choose your favorite local mountain. Estimate how much energy it takes to hike to the top of the mountain (ignoring all the local ups and downs of a typical trail). How does your result compare to a typical day's food energy intake of 10 MJ, taking into account the fact that human muscles are not 100% efficient at converting energy into work?

2.8 Use any means at your disposal (including the internet) to justify estimates of the following (an order-of-magnitude estimate is sufficient): (a) The kinetic energy of the Moon in its motion around Earth. (b) The kinetic energy of a raindrop before it hits the ground. (c) The potential energy of the water in Lake Powell (relative to the level of the Colorado river directly beneath the dam). (d) Energy needed for you to power a bicycle 15 km at 15 km/h on flat terrain. (e) The energy lost to air resistance by an Olympic athlete running a 100 m race. (f) The gravitational potential energy of a climber at the top of Mount Everest.

2.9 Verify the claim that conversion of the potential energy at the top of a 15 m hill to kinetic energy would increase the speed of the Toyota Camry from 62 to 74 mph.

2.10 Consider a collision between two objects of different masses M, m moving in one dimension with initial speeds $V, -v$. In the *center-of-mass frame*, the total momentum is zero before and after the collision. In an *elastic* collision both energy and momentum are conserved. Compute the velocities of the two objects after an elastic collision in the center-of-mass frame. Show that in the limit where $m/M \rightarrow 0$, the more massive object remains at rest in the center-of-mass frame ($V = 0$ before and after the collision), and the less massive object simply bounces off the more massive object (like a tennis ball off a concrete wall).

2.11 As we explain in §34.2.1, the density of air decreases with altitude roughly as $\rho \sim \rho_0 e^{-z/H}$, where z is the height above the Earth's surface and $H \cong 8.5$ km near the surface. Compute the ratio of air resistance losses for an airplane traveling at 750 km/h at an altitude of 12 000 m compared to an altitude of 2000 m. What happens to this ratio as the speed of the airplane changes? How do automobile air resistance losses at 2000 m compare to losses at sea level?

2.12 Consider an airplane with mass 70 000 kg, cross-sectional area 12 m^2, and drag coefficient 0.03. Estimate the energy needed to get the plane moving at 800 km/h and lift the plane to 10 000 m, and estimate air-resistance losses for a flight of 2000 km using the formula in the previous problem. Do a rough comparison of the energy used per person to do a similar trip in an automobile, assuming that the plane carries 50 passengers and the automobile carries two people.

2.13 In the American game of baseball, a *pitcher* throws a *baseball*, which is a round sphere of diameter $b = 0.075$ m, a distance of 18.4 m (60.5 feet), to a *batter*, who tries to hit the ball as far as he can. A baseball has a mass close to 0.15 kg. A radar gun measures the speed of a baseball at the time it reaches the batter at 44.7 m/s (100 mph). The drag coefficient c_d of a baseball is about 0.3. Give a semi-quantitative estimate of the speed of the ball when it left the pitcher's hand by (a) assuming that the ball's speed is never too different from 100 mph to compute roughly how long it takes to go from the pitcher to the batter, (b) using (a) to estimate the energy lost to air resistance, and (c) using (b) to estimate the original kinetic energy and velocity.

2.14 Estimate the power output of an elite cyclist pedaling a bicycle on a flat road at 14 m/s. Assume all energy is lost to air resistance, the cross-sectional area of the cyclist and the bicycle is 0.4 m^2, and the drag coefficient is 0.75. Now estimate the power output of the same elite cyclist pedaling a bicycle up a hill with slope 8% at 5 m/s. Compute the air resistance assuming the drag coefficient times cross-sectional area is $c_d A = 0.45$ m^2 (rider is in a less aerodynamic position). Compute the ratio of the power output to potential energy gain. Assume that the mass of the rider plus bicycle is 90 kg.

2.15 Compare the rate of power lost to air resistance for the following two vehicles at 60 km/h and 120 km/h: (a) General Motors EV1 with $c_d A \cong 0.37$ m^2, (b) Hummer H2 with $c_d A \cong 2.45$ m^2.

2.16 [T] Consider an idealized cylinder of cross-sectional area A *moving along its axis* through an idealized diffuse gas of air molecules with vanishing initial velocity. Assume that the air molecules are pointlike and do not interact with one another. Compute the velocity that each air molecule acquires after a collision with the cylinder, assuming that the cylinder is much more massive than the air molecules. [Hint: assume that energy is conserved in the reference frame of the moving cylinder and use the result of Problem 2.10.] Use this result to show that the drag coefficient of the cylinder in this idealized approximation is $c_d = 4$.

2.17 One way to estimate the effective area (see eq. (2.31)) of an object is to measure its limiting velocity v_∞ falling in air. Explain how this works and find the expression for A_{eff} as a function of m (the mass of the object), v_∞, g, and the density of air. The drag coefficient of a soccer ball (radius 11 cm, mass 0.43 kg) is $c_d \approx 0.25$. What is its limiting velocity?

2.18 If the vehicle used as an example in this chapter accelerates to 50 km/h between each stoplight, find the maximum distance between stoplights for which the energy used to accelerate the vehicle exceeds the energy lost to air resistance. (You may assume that the time for acceleration is negligible in this calculation.) How does the result change if the vehicle travels at 100 km/h between lights? Why?

2.19 Estimate the rotational kinetic energy in a spinning yo-yo (a plastic toy that you can assume is a cylinder of diameter 5.7 cm and mass 52 g, which rotates at 100 Hz). Compare to the gravitational potential energy of the yo-yo at height of 0.75 m.

2.20 Verify the assertion (see Example 2.3) that $E_{kin} = -\frac{1}{2}V$ for the Moon in a circular orbit around

Earth. [Hint: the magnitude of the centripetal acceleration for circular motion (2.35) can be rewritten $a = v^2/r$.]

2.21 Estimate Earth's kinetic energy of rotation (the moment of inertia of a uniform sphere is $\frac{2}{5}MR^2$).

Electromagnetic Energy

The discovery of the laws of electromagnetism and the development of devices that use electromagnetism to transform, transmit, and employ energy represent arguably the greatest advance in the use of energy since the discovery of fire. This achievement was largely a product of nineteenth-century science and engineering. Devices that use electromagnetic energy are now compact, efficient, clean, and convenient. They have become pervasive components of modern civilization. Of the 103 EJ (97.2 quads) of US primary energy use in 2016, 40 EJ (37.5 quads) was associated with the electric power sector – 13 EJ (12.6 quads) in distributed electric power and 26 EJ (24.9 quads) in losses (see Figure 1.2). Other data on world and US electric energy production and consumption are summarized in Figure 3.1. As Figures 3.1(b, c) show, the production of electricity dominates over all other energy uses in the US, but most energy used for this purpose is "lost" in conversion from thermal energy to electricity. An explanation of these losses awaits our study of entropy and thermodynamics in §8.

Electromagnetic phenomena relate to energy in ways that extend well beyond electrical devices and circuitry devised by humans; indeed, many natural forms of energy are inherently electromagnetic in nature. Chemical transformations (such as combustion of fossil fuels) are governed by the electromagnetic interactions of atomic electrons. Furthermore, solar energy is transmitted to Earth in the form of electromagnetic waves. This electromagnetic energy source ultimately underlies almost all terrestrial energy sources, including fossil fuels, biomass, wind, hydro, and wave power.

In general, human use of electromagnetic energy involves four stages: *generation*, *transmission*, *storage*, and *utilization*.

Generation Almost all energy used by humanity is initially transformed into thermal or mechanical form before

Reader's Guide
This chapter reviews the laws of electromagnetism with an emphasis on electromagnetic energy, its uses, transformation, transmission, and storage. While this chapter is relatively self-contained, some previous acquaintance with the basics of the subject is assumed.

Some topics, such as Coulomb's law and simple electrical circuits, require only a high school background in physics. Others, such as magnetic induction and the transmission of electromagnetic energy, are typically covered in a first year college physics course. We introduce Maxwell's equations using the mathematics of vector calculus. In this language, many of the equations of electromagnetism take a simple form that is echoed in the basic equations of heat transport, fluid mechanics, and other subjects encountered in later chapters. A summary of the basic formulae of vector calculus can be found in Appendix B.

use. Fossil fuels (§33), nuclear energy (§19), and solar power (§24) can all be used to produce thermal energy. Thermal energy can be converted into mechanical energy with an efficiency that is limited by the laws of thermodynamics (§8, §10). Mechanical energy from such conversion, or directly from sources such as wind (§28) or hydropower (§31) can be converted to electromagnetic energy with relatively high efficiency. The absence of a thermodynamic limit on conversion from mechanical to electromagnetic energy and *vice versa* is one of the outstanding features of electromagnetic energy. The principles underlying the generators and motors that make this conversion possible are developed in §3.4. Solar power can also be directly converted to electromagnetic energy using photovoltaic cells, which – as we describe in §25 – have their own limits to efficiency.

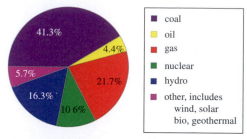

(a) world electricity generation by source (83.9 EJ total)

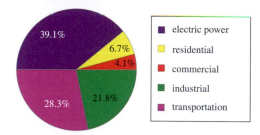

(b) US primary energy consumption by sector (102.9 EJ total)

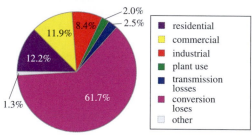

(c) US electricity end use by sector (40.1 EJ total)

Figure 3.1 (a) World net (not including losses) electricity generation by source (2013) [236]. Note the dominance of fossil fuels. (b) US primary energy consumption by sector, with electric power (including losses) separated from other sectors (2015) [12]. The electric power sector dominates all others in the US. (c) US electricity end use (including losses) by sector (2015) [12]. Note the dominance of conversion losses and the relatively small role played by losses in transmission.

Transmission Another outstanding feature of electromagnetic energy is the relative ease with which electric power can be transmitted with minimal losses across large distances. Transmission is necessary whenever electromagnetic energy is consumed away from the location of its generation. Before the development of modern electricity transmission, watermills and windmills converted the primary mechanical energy of fluid flow directly into useable work, but mills and factories were constrained to be on the river or at the windmill. Coal-fired steam engines powered factories and locomotives, at the cost of burning coal near workers' homes or along often densely populated railroad

lines. The development of electrical *transmission networks* allowed for the generation of energy near fuel sources or safely away from high population concentrations. The physics of *electric currents* and inefficiency in transmission due to *resistive losses* are developed in §3.2. More general aspects of large-scale electric grids are discussed later in the book, in §38.

Storage While in some cases electromagnetic energy can be used immediately, in other situations energy must be stored before use. Storage is required for portable use, such as for laptops and cell phones and in the transportation sector, particularly for electric vehicles. Storage is also essential for grid-scale use of intermittent renewable energy sources such as solar and wind energy. Large-scale storage of electromagnetic energy is a major challenge for the future. As described in §3.1, the direct storage of energy in electromagnetic form – in capacitors for example – is limited to a relatively low energy density[1] by the fundamental properties of electromagnetism. Thus, most approaches to storing electromagnetic energy involve conversion to other forms. The greatest efficiency can be realized with conversion to mechanical energy, which may subsequently be stored in flywheels (for small-scale storage) or by pumping water uphill (for large-scale storage). Conversion to chemical energy, as in battery storage or fuel cell technology, is slightly less efficient but is more suitable for mobile applications. Other approaches, such as compressed air energy storage (CAES) and conversion to thermal energy, are also possible, but in general have lower efficiency due to thermodynamic constraints. We return to the basic issues of storage and distribution of electromagnetic energy in §37 and §38.

Utilization The key feature of electromagnetic energy that has driven the worldwide development of large-scale power grids is the ease and efficiency with which electromagnetic energy can be put to use. *Motors* (§3.4), in particular, convert electromagnetic to mechanical energy at high efficiency. The *Premium energy efficiency program* of the (US) National Electrical Manufacturers Association [17], for example, requires efficiencies above 95% for motors rated above 250 hp. Electromagnetic energy can easily be converted directly to heat through resistive

---

[1] Energy density in this context is the **volumetric energy density**, meaning energy per unit volume. Elsewhere in this text, energy density generally refers to **gravimetric energy density** $\varepsilon = E_{\text{total}}/\text{mass}$, also known as **specific energy**, though occasionally we deal with volumetric energy density. If not otherwise specified, the two types of energy density can be distinguished by units.

heating (§3.2) at almost 100% conversion efficiency. Electric lighting is clean, safe, and increasingly efficient (§36), and played an important role in driving the development of the first large electric grids. The versatility of electromagnetic power in driving complex circuits, from sound systems to computer networks, has contributed to a steady growth in global electrical use. The compact nature of electrically driven systems and the absence of emissions at the point of use contribute to their convenience and appeal. The energy sector in which electromagnetic energy has had the least direct impact so far is transport. Further advances currently occurring in storage technologies may allow electromagnetic energy to replace fossil fuels as the principal energy source for many personal vehicles in the near future, though trains have used electric power for many years.

In this chapter we review the basic physics of electromagnetism in a unified fashion, focusing on the principles needed to understand the various aspects of generation, transmission, storage, and utilization summarized above. We begin in §3.1 with a look at electrostatics and capacitors as energy storage devices. In §3.2 we describe steady currents and resistors, focusing on resistive energy loss and the transformation of electromagnetic energy into thermal energy. In §3.3 we describe magnetic fields and forces, which make possible the motors and generators that we describe in §3.4. Induction and transformers are described in §3.5. Finally, in §3.6 we gather all the equations governing electromagnetism together into the unifying form of *Maxwell's equations*.

In the following chapter (§4), we explore the wave solutions that describe the propagation of light and electromagnetic energy across space. We develop the ideas of induction and transformers further in §38, where we describe some key aspects of modern electrical grids.

Readers desiring a more extensive introduction to electromagnetism can find it in [15], [18], or other introductory college-level textbooks.

3.1 Electrostatics, Capacitance, and Energy Storage

3.1.1 Electric Fields, Forces, and Potential

Electromagnetic interactions between charged particles are similar in many ways to the gravitational interactions of massive bodies. According to **Coulomb's law**, two charges exert forces on one another that are equal in magnitude but opposite in direction,

$$F_{12} = -F_{21} = \frac{1}{4\pi\epsilon_0}\frac{Q_1 Q_2}{r_{12}^2}\hat{r}_{12}. \tag{3.1}$$

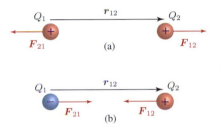

Figure 3.2 Like charges repel and unlike charges attract with forces that are equal and opposite.

Here Q_1 and Q_2 are quantities of **electric charge**, $r_{12} = x_2 - x_1$ is the vector separating the two charges, and F_{12} is the force exerted by Q_1 on Q_2. In SI units, charge is measured in **coulombs (C)**. The constant ϵ_0 is known as the **permittivity of the vacuum** with $1/4\pi\epsilon_0 = 8.988 \times 10^9$ Nm^2/C^2. There is a close parallel between eq. (3.1) and the gravitational force equation (2.21). Structurally, the equations are identical. The important difference is that while masses of physical objects are always positive, electromagnetic charges can have either sign. From eq. (3.1), it follows that same-sign charges repel and opposite-sign charges attract (Figure 3.2).

The forces between charges are very strong. Typical electric currents in household wiring transport multiple coulombs of charge per second, but the force between two coulombs of charge separated by one meter is $\sim 9 \times 10^9$ N – strong enough to lift three buildings, each with the mass of the Empire State Building! As a consequence of the strength of electromagnetic forces, only very small quantities of charge can be separated easily from one another by macroscopic distances, and matter, on average, is very nearly charge-neutral. Electromagnetic processes are capable of delivering vast amounts of energy in rather compact form. This is a significant advantage of electromagnetic

Figure 3.3 An aerial photograph of the Palo Verde Nuclear Power Plant in Arizona, US shows the sprawling power plant and the relatively diminutive electric transmission lines (upper right) that transmit almost 4000 MW of electric power. (Image: C. Uhlik)

energy transmission: relatively modest investments of materials, land, and effort can produce infrastructure capable of conveying huge amounts of energy. While a coal or nuclear power plant may cover hundreds of hectares (1 ha $= 10^4$ m^2) with concrete and metal, the wires that transmit power from the plant to end users are often no more than a few centimeters in diameter (see Figure 3.3). Transmitting energy at the same rate using steam, fossil fuel, or flowing water would require far more extravagant investments of material and capital.

Equation (3.1) describes the force produced on a charged particle by another charged particle in a fashion that appears as an instantaneous *action at a distance*. One of the conceptual advances in physics over the years has been the development of a set of **local physical laws**, which describe interactions between separated particles in terms of local disturbances that propagate continuously through the intervening space to effect the interaction. In the case of forces between electrically charged particles, the local mechanism that mediates the force is the *electric field* (for moving charges, *magnetic fields* are also involved, as we discuss later in this chapter). An **electric field** is described by a vector $E(x, t)$ at every point in space and time. For now, we focus on time-independent electric fields $E(x)$. A system of charged particles gives rise to an electric field E, and the field in turn exerts a force F on any other charge q,

$$F = qE. \tag{3.2}$$

The electric field produced at a point x in space by a single stationary charge Q at the point x_0 is given by

$$E = \frac{Q}{4\pi\epsilon_0 r^2} \hat{r}, \tag{3.3}$$

where $r = x - x_0$. It is easy to see that combining eq. (3.3) with eq. (3.2) gives Coulomb's law (3.1). The electric field

produced by a set of charges is the vector sum, or **superposition**, of the fields due to each charge individually. For charges Q_i at positions x_i, the electric field at position x is thus

$$E(x) = \frac{1}{4\pi\epsilon_0} \sum_i Q_i \frac{x - x_i}{|x - x_i|^3}. \tag{3.4}$$

The electric field can be represented graphically by **field lines** that are parallel to the direction of E. The field lines begin on positive charges and end on negative ones, and they cannot cross – since that would correspond to two different values of E at the same point. Some electric field lines associated with simple charge distributions are shown in Figure 3.4.

If work is done by moving a charged particle against the force exerted by an electric field, it is stored as **electrostatic potential energy**. This is a special case of the relation between forces and work that was introduced in §2. Suppose, as shown in Figure 3.5(a), that a charge q is carried from x_1 to x_2 along path P in the presence of an electric field generated by a static charge distribution. Then the work done is

$$W(P) = -\int_P dx \cdot F = -q \int_P dx \cdot E. \tag{3.5}$$

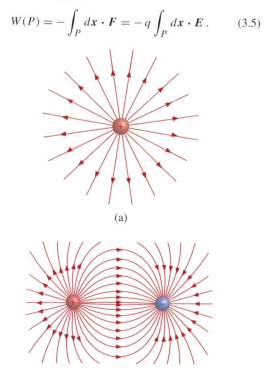

(a)

(b)

Figure 3.4 Electric field lines produced by (a) a point charge, and (b) a pair of opposite charges. The arrows show the direction of E. The more concentrated the lines, the stronger the field. ("Lines of Force for Two Point Charges" from the Wolfram Demonstration Project, contributed by S. M. Blinder)

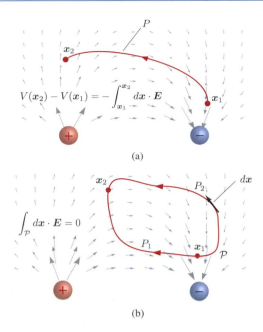

Figure 3.5 (a) When a charge q is carried along a path P from x_1 to x_2 in the electric field generated by fixed charges, the work done is $W_{x_1 \to x_2} = -q \int_P dx \cdot E = q V(x_2)$ $-q V(x_1)$ and is independent of the path taken. (b) The integral of the electric field around the closed path $\mathcal{P}$ vanishes. This can be viewed as the difference between the integrals from x_1 to x_2 on paths P_1 and P_2, which are equal.

For any electric field that is produced by a stationary charge distribution, as in eq. (3.4), the integral (3.5) depends only on the end points and not on the choice of path from x_1 to x_2, in accord with the principle of conservation of energy. This allows us to associate a potential energy with the position of a charge in an electric field. The difference in electromagnetic potential energy when the charged particle is moved from point x_1 to point x_2 is

$$\Delta E_{\text{EM}} = -q \int_{x_1}^{x_2} dx \cdot E . \qquad (3.6)$$

Electromagnetic potential energy is conventionally defined in units of energy per unit charge, or **voltage**. Dividing out the charge q from eq. (3.6), we obtain the difference in **electrostatic potential**[2] $V(x)$ (voltage) between the points x_1 and x_2

---

[2] Note the difference between mechanical potential energy – for example of a massive object in a gravitational field, which has units of energy – and electrostatic potential, which is measured in units of energy per unit charge. The symbol V is widely used for both kinds of potential; we follow this standard terminology in the book but the reader is cautioned to be careful to keep track of which units are relevant in any given case.

$$V(x_2) - V(x_1) = \Delta E_{\text{EM}}/q = -\int_{x_1}^{x_2} dx \cdot E . \qquad (3.7)$$

The electric field is therefore the negative gradient of $V(x)$,

$$E(x) = -\nabla V(x) , \qquad (3.8)$$

and the force on a charge is $F = qE = -q\nabla V(x)$. Thus, positive charges accelerate in the direction of decreasing potential V, while negative charges accelerate in the direction of increasing V. The SI units of voltage are **volts** (V), where 1 V = 1 J/C.

As for potential energy in mechanics, we can choose the zero of the electrostatic potential function $V(x)$ for convenience. The zero of the potential is generally chosen to be that of the surface of Earth at any particular location, and is called **ground voltage**.

3.1.2 Continuous Charge Distributions and Gauss's Law

In many physical systems, the number of charged particles is extremely large and it is mathematically convenient to approximate the charge distribution as a **charge density** $\rho(x)$, with units C/m^3, that is continuously distributed throughout space. In such a situation, the electric field produced by the charge distribution is computed by adding up the contributions of all the pieces in an integral, in a direct generalization of eq. (3.4),

$$E(x) = \frac{1}{4\pi\epsilon_0} \int d^3x' \rho(x') \frac{x - x'}{|x - x'|^3} . \qquad (3.9)$$

The complete set of laws of electromagnetism are described by *Maxwell's equations*. These laws of nature are most clearly described through local equations that relate the time-dependent electric and magnetic fields to the distribution of electric charges and currents. We develop the various pieces of Maxwell's equations throughout this chapter, and summarize the full set of equations in §3.6. The first of Maxwell's equations, known as **Gauss's law**, relates the electric field to the distribution of charges. Gauss's law can be formulated as a relationship between the charge Q_R in an arbitrary bounded region R and the integral of the electric field over the surface S that forms the boundary of R,

$$\oint_S dS \cdot E = \frac{1}{\epsilon_0} Q_R = \frac{1}{\epsilon_0} \int_R d^3x \, \rho(x) . \qquad (3.10)$$

In this equation, $\int dS \cdot E$ is a *surface integral* (Appendix B.1.4). dS is a vector-valued differential element of surface area, illustrated in Figure 3.6. Its magnitude is an infinitesimal bit of area $dS = |dS|$, and it points in the direction $\hat{n}$

Example 3.1   Potential of a Point Charge

If a charge q is moved radially inward from r_1 to $r_2 < r_1$ in the electric field of a point charge Q, the work done is

$$W(r_1, r_2) = \Delta E_{\text{EM}} = -\int_{r_1}^{r_2} dr \, \frac{qQ}{4\pi\epsilon_0 r^2} = \frac{qQ}{4\pi\epsilon_0}\left(\frac{1}{r_2} - \frac{1}{r_1}\right) = q(V(r_2) - V(r_1)),$$

so the electrostatic potential of a point charge is $V(r) = Q/4\pi\epsilon_0 r$. To avoid confusion regarding the signs in the relations of field, force, and voltage, remember that moving a positive charge *against* the direction that an electric field points requires doing work on the charge, and therefore raises its potential.

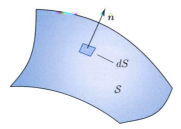

Figure 3.6 A surface area element $d\boldsymbol{S} = dS\,\hat{\boldsymbol{n}}$ on a piece of a surface $\mathcal{S}$.

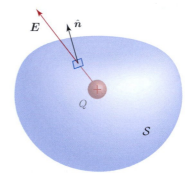

Figure 3.7 A charge Q and one of the electric field lines emanating from it. The charge is surrounded by a surface $\mathcal{S}$. The vector surface area element $d\boldsymbol{S}$ points normal to the surface, and the integral of $\boldsymbol{E} \cdot d\boldsymbol{S}$ over the surface is related by Gauss's law to the charge inside.

---

normal to the surface at each point, so $d\boldsymbol{S} = dS\,\hat{\boldsymbol{n}}$. Since $\mathcal{S}$ is a closed surface that surrounds the region R, we can distinguish outward and inward directions without ambiguity; by convention we always choose $\hat{\boldsymbol{n}}$ to point in the outward direction when computing surface integrals. A surface integral of the form $\int_{\mathcal{S}} d\boldsymbol{S} \cdot \boldsymbol{U}$ measures the **flux** of the vector field $\boldsymbol{U}$ through the surface $\mathcal{S}$. This is one of two uses of the term *flux* that figure importantly in electromagnetism (see Box 3.1).

According to Gauss's law, positive (negative) electric charges give rise to electric fields that diverge from (converge into) the regions of space where the charges lie. Although for static charge distributions Gauss's law follows from Coulomb's law (see Problem 3.2), Gauss's law is more general since it also holds for charge distributions that are changing in time. The *divergence theorem* of vector calculus (B.28) allows us to replace the surface integral of $d\boldsymbol{S} \cdot \boldsymbol{E}$ in eq. (3.10) by the volume integral of the *divergence* of $\boldsymbol{E}$,

$$\oint_{\mathcal{S}} d\boldsymbol{S} \cdot \boldsymbol{E} = \int_{R} d^3x \, \nabla \cdot \boldsymbol{E}(\boldsymbol{x}) = \frac{1}{\epsilon_0} \int_{R} d^3x \, \rho(\boldsymbol{x}), \quad (3.11)$$

where the *divergence operator* "$\nabla\cdot$" is defined in Appendix B.1.3. Since this relation holds in every region of space R, the integrands must be equal everywhere,

$$\nabla \cdot \boldsymbol{E}(\boldsymbol{x}) = \frac{\rho(\boldsymbol{x})}{\epsilon_0}. \quad (3.12)$$

Thus, Gauss's law appears as a *local* relationship between the divergence of $\boldsymbol{E}$ and the charge density ρ. Gauss's law alone is not sufficient to determine the electric field produced by a given charge distribution. For this, we need another differential relation on the electric field that forms part of another one of Maxwell's equations. Recall that the integral from $\boldsymbol{x}_1$ to $\boldsymbol{x}_2$ of the electric field produced by a static charge distribution does not depend on the path between the two points. From this it follows that the integral of the field around a closed loop $\mathcal{P}$ vanishes, as illustrated in Figure 3.5(b),

$$\oint_{\mathcal{P}} d\boldsymbol{x} \cdot \boldsymbol{E} = 0. \quad (3.13)$$

This implies that electric fields generated by stationary charges do not form closed loops. This is another relation that can be expressed locally at each point $\boldsymbol{x}$, in this case using *Stokes' theorem* (B.27), as

Electrostatics

The electric field produced by a static charge distribution $\rho(\boldsymbol{x})$ satisfies

$$\nabla \cdot \boldsymbol{E} = \frac{\rho}{\epsilon_0}, \qquad \nabla \times \boldsymbol{E} = 0.$$

The electrostatic voltage difference between two points $\boldsymbol{x}_1$ and $\boldsymbol{x}_2$ is

$$V(\boldsymbol{x}_2) - V(\boldsymbol{x}_1) = -\int_{\boldsymbol{x}_1}^{\boldsymbol{x}_2} d\boldsymbol{x} \cdot \boldsymbol{E}.$$

The force exerted by the electric field on a charge q is

$$\boldsymbol{F} = q\boldsymbol{E}.$$

$$\nabla \times \boldsymbol{E}(\boldsymbol{x}) = 0, \qquad (3.14)$$

where "$\nabla \times$" is the *curl operator* (Appendix B.1.3). Together, Gauss's law (3.12) and eq. (3.14) are sufficient to uniquely determine the electric field given by any static charge distribution as that described by Coulomb's law.[3]

3.1.3 Capacitance and Capacitive Energy Storage

Energy is stored in any system of separated charges. This energy is commonly characterized through the electrostatic potential of the charge configuration. Alternatively, this energy can be thought of as being contained in the electromagnetic field itself. In this section we describe a *capacitor* – a basic electrical circuit element that stores energy in the electrostatic field of separated charges.

A **conductor** is a material containing charges that are free to move. In a typical conductor such as a metal, electrons are loosely bound and can move in response to electric fields. The positively charged atomic nuclei in the metal are held in place by strong interatomic forces.[4] Consider a conductor of an arbitrary (connected) shape, that is given a net charge Q by adding or removing

some electrons. It follows from eq. (3.2) that in equilibrium, when all the charges have stopped moving, $\boldsymbol{E} = 0$ everywhere within the conductor. This in turn implies that all points of the conductor must be at the same potential – in equilibrium the conductor is an **equipotential** region. For example, if a spherical conductor of radius R carries a charge Q, it follows from symmetry and Gauss's law that the potential everywhere within the conductor and on its surface is $V = Q/4\pi\epsilon_0 R$. This is the same value that the potential would have at a distance R from a point charge Q. Notice that an **insulator**, in which charges are *not free to move about*, need not be an equipotential.

In general, if we set up two conductors fixed in space, and move a charge Q from one conductor to the other, the voltage difference between the conductors will be proportional to Q. This follows because the equilibrium charge distribution and the electric field produced by the charge Q are linearly proportional to Q, and hence so is the potential difference given by the integral of the electric field. The coefficient of proportionality between the charge and the voltage is called the **capacitance**, C,

$$Q = CV. \qquad (3.15)$$

A **capacitor** is an electrical circuit element containing two separated conductors. A capacitor can be charged by connecting the two conducting elements to a device such as a battery that creates a potential difference V. The charge on the capacitor is then given by eq. (3.15).

The classic and simplest example of a capacitor is the **parallel plate capacitor**, consisting of two conducting plates of area A separated by a gap of width d with $d \ll \sqrt{A}$ (Figure 3.8(a)). The gap may be filled with vacuum, or air, or with a **dielectric material** or **dielectric**, inside which the laws of electrostatics are modified by replacing the *permittivity of the vacuum* with a more general **permittivity** ϵ of the dielectric material,

$$\epsilon_0 \rightarrow \epsilon \equiv \kappa\epsilon_0. \qquad (3.16)$$

A selection of capacitors is shown in Figure 3.9(a), along with a look inside a *canister capacitor*, which is a parallel plate capacitor rolled up into a cylinder (Figure 3.9(b)).

Although the **dielectric constant** κ for air is very close to one ($\kappa_{\mathrm{air}} = 1.00059$ at 1 atm), κ can be very large for certain materials. The capacitance of a parallel plate capacitor is (see Problem 3.4)

$$C = \kappa\epsilon_0 A/d. \qquad (3.17)$$

Although this formula is specific to the parallel plate configuration, the dependence on the variables κ, A, and

---

[3] This can be shown under the assumptions that the region containing charges is bounded and the electric field approaches zero at large distances outside the charge region.

[4] **Plasmas**, which form at the extremely high temperatures found in the interiors of stars and in plasma fusion research devices here on Earth, are an exception where both positive ions and negative electrons can move about. Plasmas are discussed further in §19.

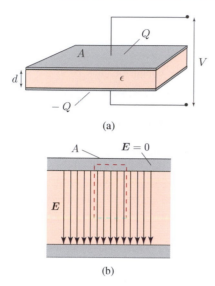

(a)

(b)

Figure 3.8 (a) A parallel plate capacitor with area A, separation d, and dielectric constant ϵ. When the potential difference between the plates is V, the plates carry charges $\pm Q = \pm \epsilon A V/d$. (b) A cross section through a parallel plate capacitor showing the electric field E and a Gaussian surface (dotted) that can be used to compute E (Problem 3.4).

Figure 3.9 (a) A selection of capacitors. (Credit: Eric Schrader, reproduced under CC-BY-SA 2.0 license via Wikimedia Commons) (b) A look inside a *canister capacitor*. (Redrawn from Tosaka reproduced under CC-BY-SA 3.0 license via Wikimedia Commons.)

d holds more broadly: to increase capacitance one can increase the effective area of the conductors, decrease the separation between them, or increase the dielectric constant (increasing ϵ effectively shields the charges from one another, allowing more charge to build up without increasing the voltage).

The SI unit of capacitance is the **farad** (F), with $1\,\mathrm{F} = 1\,\mathrm{C/V}$. The universal symbol for a capacitor in an electrical circuit is ⎯┤├⎯. The fact that electromagnetic forces are so strong (remember the magnitude of $1/4\pi\epsilon_0$), means

that even a small separation of charge requires a large voltage, so capacitances are typically very small numbers in SI units. If we measure A in cm^2 and d in mm, then for a parallel plate capacitor, the capacitance measured in farads is

$$C\ [\mathrm{F}] = \frac{\epsilon A}{d} = 0.885 \times 10^{-12} \times \frac{\kappa\,A\ [\mathrm{cm}^2]}{d\ [\mathrm{mm}]}. \quad (3.18)$$

The energy stored in a capacitor can be described either in terms of the stored charge Q or in terms of the voltage V. The capacitor can be charged by gradually moving charge from one conductor to the other. When charge q has been moved, the voltage difference is $V(q) = q/C$. When the next infinitesimal element of charge dq is moved, the work done is $dW = V\,dq = (q/C)dq$. Integrating this from the initially uncharged state, we find that the work done to charge the capacitor to total charge Q, now stored as energy in the capacitor, is

$$E_{\mathrm{EM}} = W = \frac{1}{C}\int_0^Q dq\,q = \frac{Q^2}{2C} = \frac{1}{2}CV^2. \quad (3.19)$$

The total energy stored in a capacitor can also be expressed in terms of the integral of the square of the electric field strength

$$E_{\mathrm{EM}} = \int d^3x\,\frac{\epsilon}{2}\,|E(x)|^2. \quad (3.20)$$

Although the general derivation of this result requires some further analysis, the fact that it holds for the specific case of a parallel plate capacitor is part of Problem 3.4. Note that eq. (3.19) and eq. (3.20) are two alternative descriptions of the same electrostatic energy quantity. In fact the integrand in eq. (3.20), $(\epsilon/2)|E|^2$, can be identified as the *local energy density* that is stored in an electric field E.

The *breakdown voltage* of air limits the (volumetric) energy density that can be stored in a capacitor using air as a dielectric to roughly $50\,\mathrm{J/m}^3$ (see Example 3.2). From eq. (3.20), it is clear that for a capacitor charged to a given electric field, the stored energy density is increased by using a dielectric material with large dielectric constant κ. Furthermore, for certain materials the breakdown voltage is significantly higher than for air, further increasing the possible energy storage density. The quest for electrostatic energy storage has stimulated the search for materials with large dielectric constant and high breakdown voltage. While capacitors are as yet relatively limited in energy density, they play an important role in electronic systems and are useful in many situations where high power density and durable components are required. We return to this topic in §37 (Energy storage).

Example 3.2   A Simple Capacitor

Consider a capacitor made from two square metal plates that are 1 cm on a side, separated by 1 mm. *Given that air suffers a **dielectric breakdown** and conducts electric charge in the presence of an electric field of magnitude exceeding $E_{max} \cong 3.3 \times 10^6$ V/m, determine the maximum energy that can be stored in the capacitor and the associated energy density.*

The capacitor has capacitance $C = \epsilon A/d \cong 0.885 \times 10^{-12}$ F. The maximum possible voltage to which the capacitor can be charged is $E_{max} \times 1$ mm $\cong 3.3 \times 10^3$ V, for a maximum total energy of $E_{EM} = CV^2/2 \cong 4.8 \times 10^{-6}$ J. The associated energy density is $\cong 50$ J/m^3, very small compared for example to the energy density of gasoline, which is around 32 GJ/m^3. Note that the energy density of an electric field is proportional to $|\boldsymbol{E}|^2$ (eq. (3.20)), so that this bound on capacitive energy density is independent of the capacitor geometry. As discussed in the text, the energy density of a capacitor can be increased by using a material with large dielectric constant κ and higher breakdown field; for typical materials κ is in the range 1–100.

Capacitance

The capacitance C of a pair of conductors is the ratio between stored charge and voltage, $Q = CV$. For the simplest *parallel plate* capacitor

$$C_{\text{parallel plate}} = \epsilon_0 \kappa A/d.$$

The energy stored in a capacitor can be expressed as

$$E_{EM} = \frac{1}{2}CV^2 = \frac{1}{2C}Q^2 = \int d^3x \frac{\epsilon}{2}|\boldsymbol{E}|^2.$$

3.2 Currents, Resistance, and Resistive Energy Loss

3.2.1 Current, Resistance, and Ohm's and Joule's Laws

Two primary features of electromagnetic energy are the ease with which it can be transmitted over large distances and the wide range of devices that can be driven by electric power. To understand these applications of electromagnetic energy we must study the motion of charges in **electric currents**. In an electric circuit, the electric current measures the net rate at which charge passes a given point in a wire, or more generally, the rate at which charge flows across a fixed surface,

$$I = dQ/dt. \tag{3.21}$$

Since electrons have negative charge, their flow corresponds to a negative current, or equivalently to a positive current in the opposite direction. The net charge density in a wire carrying a steady current is zero even though charge is moving, because the charge of the electrons is balanced by the opposite charge of the positive ions. Earlier we defined a conductor as a material in which charges are free to move and an insulator as a material in which they are not. Insulators are an idealization: even the best insulators (except for the vacuum) allow some charge to flow when they are placed in an electric field. For now, however, we divide materials into idealized conductors and insulators, and leave more exotic materials like semiconductors for later chapters.

In SI units, electric current is measured in **amperes**. In fact, the ampere – rather than the coulomb – is taken as the fundamental electrical unit in the SI system. One ampere is defined to be the current that produces a *magnetic force* per unit length of 2×10^{-7} N/m between two straight, parallel wires of negligible radius separated by 1 m. Magnetic forces between moving charges are discussed later in this chapter. One ampere works out to a flow of approximately 6.242×10^{18} electrons per second, so the coulomb, which is an ampere-second (1 C = 1 A s), corresponds to the magnitude of the charge of that number of electrons.

As described in §3.1.1, when a charge is moved across a voltage difference, work is done. From eq. (3.7) we see that moving a differential amount of charge dQ across a voltage difference V requires work $dW = VdQ$. The sign in this relation is such that moving a positive charge up to a higher voltage requires doing work on the charge. If charge is flowing continuously, then the power (work per unit time) is equal to the voltage times the current,

$$P(t) = \frac{dW}{dt} = V(t)\frac{dQ}{dt} = V(t)I(t). \tag{3.22}$$

This result is universal, and applies whatever the source of the voltage difference may be. If the charge is moving from a higher voltage to a lower voltage, then the charge does work as it moves.

When electrons flow through matter they collide with the molecules in the material and lose energy, so a current will not continue to flow unless driven by an electric field.[5] The rate of current flow I through any conductor is proportional to the voltage drop V across the conductor

$$V = IR. \tag{3.23}$$

This relation is known as **Ohm's law**, and the proportionality constant R is the **resistance**, which in SI units is measured in **ohms** (Ω), with $1\,\Omega = 1\,\text{V/A}$. **Resistors** are denoted in electrical circuits by the symbol ⎓⋀⎓ (see Figure 3.10). Equation (3.23) can be read in two different ways. First, if an *electromotive force* associated with a voltage difference V is established between the ends of a wire, then a current is driven in response with proportionality constant $1/R$. It does not matter what generates

the voltage. The wire could connect the plates of a capacitor, or could be connected to a battery, or – as we describe later – it could be moving through a magnetic field. Second, when a current I flows through a wire or other object with resistance R, the voltage across the resistance falls by an amount IR. When both I and V depend on time, the linear relationship (3.23) still holds, at least for time scales that occur in electric power applications,

$$V(t) = I(t)R. \tag{3.24}$$

When an amount of charge $I\Delta t$ is driven through a resistor by a voltage difference V, the work that is done, $IV\Delta t$, is dissipated in the collisions between the electrons and the molecules that make up the material. The energy that these collisions transfer eventually ends up in the random motion of molecules that we describe as *thermal energy* (see §5). From Ohm's law, we find that the power dissipated in the resistor is

$$P_{\text{resistance}}(t) = V(t)I(t) = I^2(t)R = \frac{V^2(t)}{R}. \tag{3.25}$$

This is known as **Joule's law** and the dissipation of power in a resistor is called **Joule heating**.

In many situations, such as power-line losses that we discuss later in this chapter, resistive heating represents lost energy. Resistive heating also, however, provides a way of transforming electrical energy to heat with almost perfect (100%) efficiency. Resistive heating is the principle underlying electric stoves, toasters, and other kitchen devices, and is also frequently used for household heating. While the high efficiency of electric to thermal conversion through resistive heating may make electric home heating

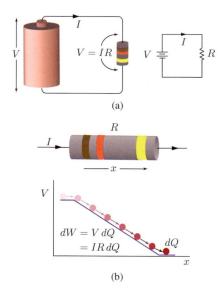

(a)

(b)

Figure 3.10 (a) Left: A battery sets up a voltage drop V across a resistor, leading to a current $I = V/R$; Right: The circuit diagram. (b) A charge dQ moves downhill through a resistor carrying current I. The battery does work $dW = V\,dQ = IR\,dQ$ on the charge. This energy is dissipated as thermal energy as the charge passes through the resistor.

---

[5] Unless the material is a **superconductor**, which has no electrical resistance and which can carry a current indefinitely without an external voltage to drive it. Superconductors are discussed further in §37 (Energy storage).

Power, Resistance, and Energy Dissipation

When a current $I(t)$ flows through a voltage difference $V(t)$, it consumes power

$$P(t) = I(t)V(t).$$

Resistance results from electron interactions with the fixed atoms in a conductor, and is governed by Ohm's law,

$$V = IR.$$

Electromagnetic energy is transformed into thermal energy in a resistor, at a rate

$$P(t) = I(t)V(t) = I^2(t)R = V^2(t)/R.$$

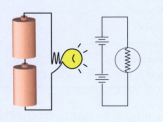

Example 3.3 A Flashlight

An old-fashioned flashlight takes two AA batteries, connected in series. Each battery has a voltage drop of $V = 1.5$ V between terminals and stores roughly 10 kJ of energy. The flashlight bulb has a resistance of roughly 5 Ω. *Determine the rate of power use and the approximate time that a pair of batteries will last.*

With the batteries connected in series, their voltages add to 3 V. The rate of power dissipation is $P = V^2/R \cong 1.8$ W. So the batteries should last for roughly $20\,000\,\text{J}/1.8\,\text{W} \cong 11\,200\,\text{s} \cong 3$ hours. The physics of batteries is described in more detail in §37.3.1.

seem attractive, most electricity is itself produced at a very low efficiency from thermal energy – for example from a coal-powered plant with roughly 30% thermal-to-electric conversion efficiency. In contrast, modern gas furnaces can have a 95% or higher efficiency in transforming chemical energy in the fuel into household heat. A more efficient way to use electrical energy for heating is with a *heat pump*, as discussed in §12 (Phase-change energy conversion) and §32 (Geothermal energy). The general question of comparative efficiency of heating methods is covered in more detail in §36 (Systems).

3.2.2 Current Density and Charge Conservation

Current, as described so far, refers to one-dimensional flow through a wire or to total charge flow through a fixed surface. If we want to describe continuous flow of charge through space we need a generalization that can be defined locally at each point in space, in much the same way that charge density generalizes the concept of point charge. For example, we may wish to understand how the flow of charge is distributed throughout the cross section of a wire, as shown in Figure 3.11.

The **current density** $j(x, t)$ is a vector quantity that is defined as the charge per unit time per unit area flowing past the point x at time t, with the vector direction of j

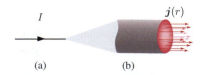

(a)   (b)

Figure 3.11 Current flow in a wire: (a) From far away where only the net current is seen. (b) Looked at more closely, the current density varies with distance from the wire's center. If the wire carries an alternating current (see §3.2.3), the current density decreases inward from the wire's surface – the darker the shading, the larger the current density.

being the direction in which net positive charge is flowing. To relate j to more familiar concepts, imagine charges in motion described by a charge density ρ and a local average velocity vector v that may vary both in space and time. Then consider, as shown in Figure 3.12, a tiny volume dV (still containing a vast number of charges) in which ρ and v can be approximated as constant. The volume dV has cross-sectional area dS perpendicular to the direction that the charges are moving, and its length $dl = |v|dt$ is the distance that the charges move in an infinitesimal time dt. All the charges in the volume dV cross the surface dS in the time dt, so the rate of charge flow across dS is

$$dQ = \rho\, dV = \rho\, dS |v| dt . \qquad (3.26)$$

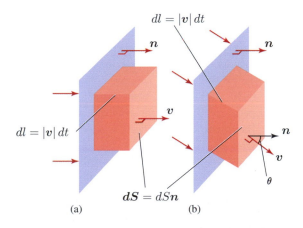

(a)   (b)

Figure 3.12 Flow of charge through a mathematical surface. (a) When the flow is perpendicular to the surface, all the charge in the shaded box flows through the surface in a time dt, $dQ = \rho dV = \rho |v| dt dS = |j||dS| dt$. (b) When the flow makes an angle θ with respect to the normal to the surface, less charge flows through, $dQ = \rho dV = \rho |v| dt dS \cos \theta = j \cdot dS\, dt$.

Table 3.1 Resistivities of some common conductors.

Substance	Resistivity ($\Omega\,$m)
Silver	1.59×10^{-8}
Copper	1.68×10^{-8}
Gold	2.44×10^{-8}
Aluminum	2.82×10^{-8}
Iron	1.0×10^{-7}

In this case the current density j is given by

$$j(x, t) = \rho(x, t)v(x, t), \tag{3.27}$$

and we can rewrite eq. (3.26) as

$$\frac{dQ}{dt} = j \cdot dS, \tag{3.28}$$

where $dS = dS\hat{n}$ is a differential surface area like the one used earlier in connection with Gauss's law (Figure 3.6). So conceptually, current = current density × area. Notice that if the surface is not perpendicular to the direction of current flow then the current through the surface is reduced by the cosine of the angle between $\hat{j}$ and $\hat{n}$, as shown in Figure 3.12(b).

What determines the direction of j, or equivalently of the velocity v of the charges? Since charges move in response to the electric field it is reasonable to expect that the direction of current flow in a conducting material is parallel to the applied electric field. Indeed, the current and electric field are related by a local version of *Ohm's law*,

$$j = E/\varrho = \sigma E, \tag{3.29}$$

where $\varrho(\sigma)$ is the **electric resistivity (conductivity)** of the material. Resistivity is measured in $\Omega\,$m in SI units. The resistivities of common substances range from $\varrho \sim 10^{-8}\,\Omega\,$m for excellent conductors like silver and copper, to $\sim 10^{18}\,\Omega\,$m for excellent insulators like quartz. The resistivities of some conductors are listed in Table 3.1.

We can use the local form (3.29) to derive Ohm's law (3.23) by considering current flowing uniformly through a volume with surface A perpendicular to j and length l (similar to the volume depicted in Figure 3.12(a), but with $dS \rightarrow A, dl \rightarrow l$). Using $I = jA$ and $V = El$, these two equations are equivalent when the resistance of the volume is

$$R = \varrho l/A. \tag{3.30}$$

Thus the resistance of a wire varies in direct proportion to its length and inversely with its area. We discuss the fundamental physics underlying eq. (3.29) in more detail,

Current Density and Charge Conservation

Moving charges are described locally by the *current density* $j(x, t)$, the charge per unit time per unit area moving in the direction $\hat{j}$.

The local form of Ohm's law is $j = \sigma E = E/\varrho$, where σ is the conductivity and $\varrho = 1/\sigma$ is the resistivity.

Charge conservation is expressed locally through

$$\frac{\partial \rho(x, t)}{\partial t} + \nabla \cdot j(x, t) = 0.$$

If the charge density at a point is decreasing, the current density must be diverging from that point.

and give a more careful proof of eq. (3.30) in §6.2.4, in the closely related context of heat conduction.

The basic principles underlying eq. (3.29) have close parallels in many other areas of physics. The sustained motion of charges at a constant velocity, and without net acceleration, in the presence of electric fields and resistance is reminiscent of the limiting velocity experienced by an object falling in the presence of Earth's gravitational field and air resistance. The same mathematical structure appears in the viscous flow of fluids through pipes, and in other contexts that we encounter in later chapters.

The concept of a local current density makes it possible to formulate **charge conservation** locally in a way that also has parallels in many other areas of energy physics. Experiments have shown (to a very high degree of precision) that net electric charge can neither be created nor destroyed – although opposite charges can cancel one another. Thus, the total charge within a given region R can only change with time if charge is flowing in or out of the region. The total flow of charge out of a region can be expressed as an integral of the normal component of the current density over the surface S bounding the region, so charge conservation may be written

$$\frac{dQ_R}{dt} = -\oint_S dS \cdot j, \tag{3.31}$$

where Q_R is the charge in the region R, and the minus sign accounts for the fact that when Q_R is decreasing, the integral on the right is positive. Using the divergence theorem (B.28), we can write the surface integral as a volume integral of $\nabla \cdot j$,

---

Box 3.1 Two Kinds of Flux

The term *flux* has two distinct though related meanings in physics, both of which appear in this chapter and recur throughout the book.

Flux I: The flow of a quantity per unit area per unit time Suppose Y is a physical attribute of matter (or radiation). Particle number, mass, energy, charge, momentum, and angular momentum are all examples. Even though they may ultimately be properties of individual atoms or molecules, we can define a local density $\rho_Y(x, t)$ of each of these attributes by averaging over the multitude of particles that occupy even a relatively small macroscopic volume. A cubic micron of water, for example, contains over 10^{10} molecules. Just as for electric charge, we can define a vector density $j_Y(x, t)$ that measures the amount of Y per unit time that flows across a surface perpendicular to $\hat{j}_Y$. This is known as the *flux* of Y. The concept of flux is ubiquitous in energy physics, from the flux of energy in wind or water to the flux of neutrons in a nuclear reactor. If Y happens to be conserved, like charge, then the conservation law takes the general form

$$\frac{\partial \rho_Y(x, t)}{\partial t} + \nabla \cdot j_Y(x, t) = 0.$$

If Y is not conserved, then the right-hand side is not zero, but instead gives the rate at which Y is created or destroyed at the point x.

Flux II: The surface integral of a vector field In formulating Gauss's law, we encountered an integral of the electric field over a surface, $\int_S dS \cdot E$, which measures the extent to which the electromagnetic field penetrates through the surface S. This is known as the *flux* of the electromagnetic field through the surface, sometimes abbreviated as the *electric flux*. The surface need not be closed;[a] $\int_S dS \cdot E$ defines the flux through any surface S. Notice that this flux is a **scalar**, it does not point in any direction, whereas the flux we defined earlier is a vector, carrying the direction of flow. We encounter magnetic flux later in this chapter.

Note that often a flux j of type I is integrated over a surface, giving a flux of type II. In this case we refer to $\int_S dS \cdot j$ as the *total flux* or *integrated flux* through the surface.

---

[a] A *closed surface*, like a closed path, is one without a boundary (Appendix B.1.4)

$$\frac{d}{dt} Q_R = \frac{d}{dt} \int_R d^3x \, \rho(x, t) = -\int_R d^3x \, \nabla \cdot j. \quad (3.32)$$

When the region R does not depend on time, we can move the time derivative inside the volume integral – being careful to change it to a partial derivative since it now acts on ρ, which depends both on x and on t. Since the resulting expression holds for any region R, it must hold for the integrands, so

$$\frac{\partial \rho(x, t)}{\partial t} + \nabla \cdot j(x, t) = 0. \quad (3.33)$$

This simple and local formulation of charge conservation states that the only way that the charge density at a point can decrease (increase) is if current density is diverging from (converging into) that point.

A vector field such as $j(x, t)$, which measures the quantity and direction of a substance flowing past a point per unit area per unit time, is called the **flux** of that substance. Fluxes of other physical quantities, such as particle number, mass, energy, and momentum, figure in many aspects of energy physics and appear in later chapters. Note

that this is a slightly different use of the term *flux* than we encountered earlier in connection with a surface integral of a vector field (see Box 3.1).

3.2.3 Resistive Energy Loss in Electric Power Transmission

Whenever electric current flows through a resistive element, energy is lost to Joule heating. There are energy losses to resistive heating as electric currents flow through wires, and inside electrical devices from motors to computers to cell phones. An important case is the resistive loss that occurs during transmission of electricity from power plants to customers over high-voltage power lines. Such power lines sometimes stretch for hundreds of kilometers, in which case resistive heating represents a significant energy loss. The technology of electric power transmission is complicated. We study electrical transmission grids in more detail in §38, where we introduce concepts like *impedance* and *reactance* that are needed for deeper understanding. Here, we make a start with a first look at resistive energy loss in transmission lines.

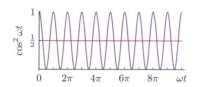

Figure 3.13 The average of $\cos^2 \omega t$ is $\langle \cos^2 \omega t \rangle = 1/2$. This follows from $\cos^2 \omega t = \frac{1}{2}(1 + \cos 2\omega t)$, or from $\sin^2 \omega t + \cos^2 \omega t = 1$ and $\langle \cos^2 \omega t \rangle = \langle \sin^2 \omega t \rangle$.

Though direct current is used for some long-distance lines, most electrical transmission lines use **alternating current** (**AC current**), arising from a sinusoidally varying voltage[6]

$$V(t) = V_0 \cos \omega t. \tag{3.34}$$

From Ohm's law (3.24), such an alternating voltage placed across a resistance R gives a sinusoidally varying current that oscillates with the same time dependence, $I(t) = (V_0/R) \cos \omega t$. If we average over time and use the fact that $\langle \cos^2 \omega t \rangle = 1/2$ (where "$\langle \ldots \rangle$" means time-averaged; see Figure 3.13), we find that the power lost to Joule heating is

$$\langle P \rangle_{\text{Joule}} = \langle I(t)V(t) \rangle = \frac{1}{2}\frac{V_0^2}{R}. \tag{3.35}$$

It is convenient to introduce the **root mean square** voltage, defined by $V_{\text{RMS}} = \sqrt{\langle V(t)^2 \rangle}$, so $V_{\text{RMS}} = V_0/\sqrt{2}$ for the sinusoidally varying voltage (3.34), and

$$\langle P \rangle_{\text{Joule}} = \frac{V_{\text{RMS}}^2}{R}. \tag{3.36}$$

The basic model for electric power transmission is shown in Figure 3.14. A power plant generates an AC voltage and powers a circuit that consists of the **load**, where useful work is done, and the **transmission lines**, which present a resistance R. We assume for now that the load is simply a resistance R_L. In general, the load consists of all kinds of electrical devices which provide not only resistance, but also capacitance and inductance (inductance is discussed in §3.5.2). When these effects are included, as is done in §38, we find that transmission losses are in fact minimized (for a fixed power to the load) when the load looks like a resistance R_L alone.

The transmission line and load resistances in Figure 3.14 are in series and therefore add (Problem 3.10), so the

---
[6] Other forms of oscillating voltages occur, in particular *three-phase AC* is quite common and is described further in §38. Here we limit ourselves to sinusoidally oscillating voltage in a single wire, and use the term AC voltage to refer to a voltage of the form eq. (3.34).

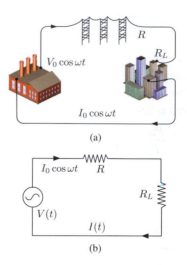

(a)

(b)

Figure 3.14 Power loss in electric transmission. (a) A schematic drawing of the circuit taking power through transmission lines with resistance R to a purely resistive load (in this case a city) with resistance R_L. (b) A standard circuit diagram of the same situation. (Credit: www.aha-soft.com)

current in the circuit is

$$I(t) = \frac{V(t)}{R + R_L} = \frac{V_0}{R + R_L} \cos \omega t = I_0 \cos \omega t, \tag{3.37}$$

where $I_0 \equiv V_0/(R + R_L)$. The voltage drops across the transmission line and the load are $V_R(t) = RI(t)$ and $V_L(t) = R_L I(t)$ respectively, and the time-averaged power lost in transmission is

$$P_{\text{lost}} = \langle V_R(t)I(t) \rangle = \frac{1}{2}I_0^2 R \tag{3.38}$$

The power through the load P_L takes a similar form with R replaced with R_L. When we make the further approximation that the resistance in the transmission line is much smaller than that in the load, $R \ll R_L$, we have $I_0 \cong V_0/R_L$, and

$$P_L = \frac{1}{2}I_0^2 R_L \cong \frac{V_0^2}{2R_L}. \tag{3.39}$$

We can use this relation to eliminate R_L and determine the ratio of the power lost in transmission to the power delivered to the load,

$$\frac{P_{\text{lost}}}{P_L} = \frac{R}{R_L} \cong 2\frac{RP_L}{V_0^2} = \frac{RP_L}{V_{\text{RMS}}^2}. \tag{3.40}$$

Expressing the transmission line resistance in terms of the length l, cross-sectional area A, and resistivity ϱ of the wires, we obtain

Resistive Losses

When AC power is transmitted to a resistive load R_L through wires with resistance R, the ratio of power lost in Joule heating to the power delivered to the load is approximately

$$\frac{P_{\text{lost}}}{P_L} \cong \frac{R P_L}{V_{\text{RMS}}^2} = \frac{P_L \varrho l}{A V_{\text{RMS}}^2}.$$

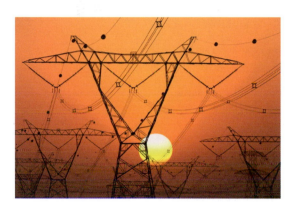

Figure 3.15 Electric power transmission lines.

$$\frac{P_{\text{lost}}}{P_L} \cong \frac{P_L \varrho l}{A V_{\text{RMS}}^2}. \tag{3.41}$$

The factor of $1/V_{\text{RMS}}^2$ in eq. (3.6) explains why electric power is transmitted at as high a voltage as possible. This factor can be understood as arising because Joule heating is proportional to the square of the current, and the current is inversely proportional to voltage for a fixed power output P_L. More complex considerations (see §38) limit power-line voltages to less than roughly 1 MV. In practice, long-distance electric power transmission uses (RMS) voltages ranging from 100 kV to 1200 kV, but voltages in excess of $\sim$500 kV are unusual. We return to the question of electric power transmission in more detail in §38.

3.3 Magnetism

Magnetism plays a role in a wide range of energy systems, and is inextricably intertwined with the dynamics of electric currents and charges. In particular, to describe how mechanical energy is transformed into electrical energy using generators, and back to mechanical energy using motors, as we do in the following section (§3.4), we need to understand not only electric fields and forces but also magnetic phenomena.

3.3.1 Magnetic Forces and Fields

Magnets were well known to the ancients, but the absence of elementary *magnetic charges* makes magnetic forces harder to describe than electric forces. We now know that moving charges produce **magnetic fields**, and that a magnetic field, in turn, exerts a force on moving charged particles. This is closely parallel to the way in which a static charge configuration produces an electric field that in turn exerts a force on electrically charged particles. In fact, electric and magnetic fields are unified in Einstein's *special theory of relativity*; for example, an electric field seen by one observer appears as a combination of electric and magnetic fields when viewed by another observer who is moving relative to the first. We describe this connection in more detail in §21, but in the rest of the book we treat magnetic fields and forces as distinct phenomena from electric fields and forces, as they were understood in the nineteenth century.

Magnetic Forces It is easier to describe magnetic forces than the origins of magnetic fields, so we begin with forces. Like the electric field, a magnetic field $\boldsymbol{B}(\boldsymbol{x}, t)$ is described by a vector at each point in space and time. The force acting on a charge q moving with velocity $\boldsymbol{v}$ in a magnetic field $\boldsymbol{B}$ is

$$\boldsymbol{F} = q\boldsymbol{v} \times \boldsymbol{B}, \tag{3.42}$$

as shown, for example, in Figure 3.16. If, instead of a single charge, we consider a small volume dV containing current density $\boldsymbol{j}(\boldsymbol{x})$, then the force per unit volume on the moving charges is $d\boldsymbol{F}/dV = \boldsymbol{j}(\boldsymbol{x}) \times \boldsymbol{B}(\boldsymbol{x})$. If this current is flowing in a wire, we can integrate over the wire's cross section to get the force per unit length on the wire,

$$d\boldsymbol{F}(\boldsymbol{x}) = I\,d\boldsymbol{x} \times \boldsymbol{B}(\boldsymbol{x}), \tag{3.43}$$

where $d\boldsymbol{x}$ is the tangent vector to the wire at the point $\boldsymbol{x}$.

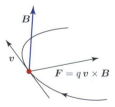

Figure 3.16 The magnetic force on a charged particle moving with velocity $\boldsymbol{v}$ is given by $q\boldsymbol{v} \times \boldsymbol{B}$; the direction of $\boldsymbol{F}$ is determined by the **right-hand rule** for vector cross products (Appendix B.1.1).

Example 3.4   Specifying a Transmission Cable

A (fairly typical) power transmission cable of radius 1.75 cm consists of 45 strands of aluminum (reinforced by 7 strands of structural steel that do not carry current). The effective area of the aluminum conductors is $8.0\,\text{cm}^2$. This cable is used to carry power at $V_{\text{RMS}} = 345\,\text{kV}$ over a distance of 200 km. *How much power can be transmitted by this cable if the maximum acceptable resistive loss in the cable is 2%?*

First compute the resistance of the cable:

$$R = \frac{\varrho l}{A} = 2.8 \times 10^{-8}\,\Omega\,\text{m} \times \frac{200 \times 10^3\,\text{m}}{8 \times 10^{-4}\,\text{m}^2} \cong 7.0\,\Omega.$$

Then, according to eq. (3.40), $0.02 = PR/V_{\text{RMS}}^2$. Solving for P,

$$P = 0.02 \times \frac{(3.45 \times 10^5\,\text{V})^2}{7\,\Omega} \cong 340\,\text{MW}.$$

So a single 1.75 cm radius cable can deliver 340 MW with a loss of 2% to Joule heating. Since power is usually carried in three cables (see §38), cables similar to these can carry the full output of a 1 GW power plant.

As described by the vector cross product, the magnetic force on a charge is perpendicular to the velocity. Thus, magnetic forces do not change the speed of a charged particle and do no work on the charges they accelerate. In fact (Problem 3.14), in a constant magnetic field a charge moves in a circle at a constant speed in the plane perpendicular to $\boldsymbol{B}$ while it drifts at a constant rate in the direction parallel to $\boldsymbol{B}$. The combined effects of electric and magnetic fields are summarized in the **Lorentz force law**,

$$\boldsymbol{F}(\boldsymbol{x}) = q\boldsymbol{E}(\boldsymbol{x}) + q\boldsymbol{v} \times \boldsymbol{B}(\boldsymbol{x}) \quad \text{(particle)}, \quad (3.44)$$

$$\frac{d\boldsymbol{F}}{dV} = \rho(\boldsymbol{x})\boldsymbol{E}(\boldsymbol{x}) + \boldsymbol{j}(\boldsymbol{x}) \times \boldsymbol{B}(\boldsymbol{x}) \quad \text{(distribution)}, \quad (3.45)$$

which completely determines the motion of charges under electric and magnetic forces.

Production of Magnetic Fields   Magnetic fields are produced by moving charges, and also by the quantum spins of particles in permanent magnets. There are no elementary **magnetic charges** – or **magnetic monopoles** as they are usually called – that source magnetic fields in the same way that electric charges source electric fields. This is an experimental statement – nothing we know about electromagnetism forbids the existence of objects with magnetic charge, but no one has ever found one in a reproducible experiment. The absence of magnetic charges allows us to draw some conclusions about the structure of magnetic fields. According to *Gauss's law*, the integral of a field over a closed surface counts the charge inside. With no magnetic charges, we have

$$\oint_S d\boldsymbol{S} \cdot \boldsymbol{B} = 0, \quad (3.46)$$

for any closed surface S. Because S is arbitrary, the divergence theorem of vector calculus (B.28) gives the local relation

$$\boldsymbol{\nabla} \cdot \boldsymbol{B}(\boldsymbol{x}) = 0. \quad (3.47)$$

This condition should be compared with eq. (3.12). With no magnetic charges, the lines of magnetic field have no place to begin or end, and must therefore be closed curves, unlike the electric field lines produced by stationary charges.

In 1820 the French physicist André-Marie Ampere formulated the law by which electric currents generate magnetic fields. Ampere's law, like Gauss's law, can be written either in integral or differential form. In the integral form, it describes the way that magnetic fields **circulate** around wires. Imagine, for example, a wire carrying a steady current I, and any closed curve C that encircles the wire, as shown in Figure 3.17. **Ampere's law** states

$$\oint_C d\boldsymbol{x} \cdot \boldsymbol{B} = \mu_0 I = \mu_0 \int_S d\boldsymbol{S} \cdot \boldsymbol{j}, \quad (3.48)$$

where S is any surface with boundary C. The proportionality constant $\mu_0 = 4\pi \times 10^{-7}\,\text{N/A}^2$, known as the **magnetic permeability of the vacuum**, relates the magnitude of the current to the strength of the magnetic field it produces. The direction of the normal $\hat{\boldsymbol{n}}$ in $d\boldsymbol{S} = dS\hat{\boldsymbol{n}}$ is correlated with the direction of the integral by the *right-hand rule* (see Appendix B.1.5 and Figure 3.17(a)). Using Stokes' theorem (B.27), Ampere's law can be written as a local relation between the curl of the magnetic field and the current density,

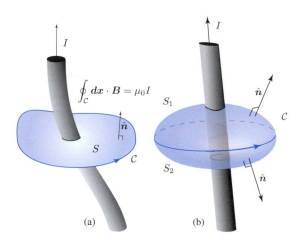

Figure 3.17 (a) The setup for Ampere's law: a curve C spanned by a surface S encircles a wire carrying current I. The direction of the normal $\hat{n}$ is determined from the orientation of the curve C by the *right-hand rule*. (b) The closed surface S consists of two surfaces S_1 and S_2, each of which spans the curve C. The current passing through S_1 must be the same as that passing through S_2 if no charge is building up inside S.

$$\nabla \times \boldsymbol{B}(\boldsymbol{x}) = \mu_0 \, \boldsymbol{j}(\boldsymbol{x}) \, . \qquad (3.49)$$

Ampere's law in the form of eq. (3.48) or (3.49) is only valid when current and charge densities do not change with time. One hint of this restriction can be found in an apparent ambiguity in eq. (3.48): what surface S should be used to evaluate the integral on the right hand side? Figure 3.17(b) shows two possible surfaces, S_1 and S_2, that span a closed curve C. If the integrals of the current density through S_1 and S_2 were different, Ampere's law would not be consistent. The difference between the two integrals $\int_{S_1} - \int_{S_2}$ is an integral of $d\boldsymbol{S} \cdot \boldsymbol{j}$ over a closed surface S, which in turn gives the time derivative of the charge enclosed within S (an application of charge conservation, eq. (3.31)), so the surface integral in eq. (3.48) could give different results if the charge density were changing with time. Thus, Ampere's law as stated in eqs. (3.48, 3.49) can be applied only to constant currents and charge densities. We extend Ampere's law to time-dependent situations in §3.6, where we complete Maxwell's equations.

Ampere's law determines the curl of $\boldsymbol{B}$, and the absence of magnetic charges determines its divergence. This is enough to uniquely determine the magnetic field associated with any time-independent current distribution.

Just as electric fields are affected by the permittivity ϵ of a medium, magnetic fields are similarly influenced by the **permeability** μ, which is significantly greater in

some materials than the vacuum permeability μ_0. As for dielectrics, in the presence of a material with uniform permeability, the equations governing magnetic fields are modified by taking $\mu_0 \to \mu$.

3.3.2 Examples of Magnetic Fields

Magnetic field around a wire One simple example of a magnetic field is the field around a long straight wire carrying constant current I,

$$\boldsymbol{B}_{\text{wire}} = \frac{\mu_0}{2\pi} I \frac{\hat{z} \times \hat{\rho}}{\rho} \, , \qquad (3.50)$$

where, as shown in Figure 3.18, $\hat{z}$ points along the wire and $\hat{\rho}$ points radially outward from the wire. The field (3.50) is easily determined from Ampere's law and the condition that the magnetic field is symmetric under rotation about the wire (Problem 3.15(a)). The resulting magnetic field circulates around the wire in the right-hand sense (if the thumb of the right hand points along the direction of current flow, the fingers wrap in the direction of the magnetic field). There is also a magnetic field inside the wire. Although its strength and direction depend on the way

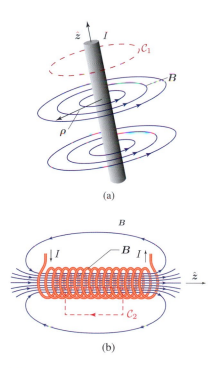

Figure 3.18 The magnetic field associated with (a) a long straight wire (eq. (3.50)) and (b) a long solenoid (eq. (3.51)). The contours C_1, C_2 shown in red are useful for computing $\boldsymbol{B}$ from Ampere's law (Problem 3.15).

Box 3.2   Electric and Magnetic Dipole Fields

For electric charges, the most important type of field is the radial electric field (3.3) produced by a point charge. Even for a complicated configuration of many positive and negative charges, if the net charge is nonzero, the dominant term in the electric field at large distances is of this form, where Q is the net charge. This is referred to as the **electric monopole field**.

For electric charge configurations with vanishing net charge, however, there is no monopole field. In this case, the dominant electric field at large distances is of the form of a **dipole field**

$$E = -\nabla \frac{d \cdot \hat{r}}{4\pi \epsilon_0 r^2} = -\frac{d - 3(d \cdot \hat{r})\hat{r}}{4\pi \epsilon_0 r^3} \, ,$$

where $d = \sum_i Q_i r_i$ is the **electric dipole moment** vector. The simplest charge arrangement giving rise to a dipole field is a configuration of two charges $\pm Q$ at positions $\pm \xi/2$, with dipole moment $d = Q\xi$ (Problem 3.8). When viewed at distances large compared to $|\xi|$, this configuration produces a dipole field.

Since there are no magnetic monopole charges, the dominant magnetic field at large distances from a configuration of moving charges is the **magnetic dipole field**

$$B = -\mu_0 \nabla \frac{m \cdot \hat{r}}{4\pi r^2} = -\mu_0 \frac{m - 3(m \cdot \hat{r})\hat{r}}{4\pi r^3} \, ,$$

where m is the **magnetic dipole moment** of the configuration of currents. A current I moving around a planar loop of area A has magnetic dipole moment $m = I A \hat{n}$ where $\hat{n}$ is the unit normal to the current loop with the usual right-hand orientation. Except for their overall magnitude, the long-range electric and magnetic dipole fields are identical. In contrast, the short-range magnetic field near a current loop and the electric field near a pair of opposite charges are dramatically different. The dipole magnetic field far away from a current loop is shown at right, where the dipole moment m is shown in orange.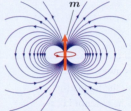

Magnetic dipoles appear in a variety of situations related to energy physics. They are useful in understanding the dynamics of motors and generators, and we encounter them later in this book in the context of quantum mechanics (§7) and Earth's magnetic dipole field, which plays a role in §32 (Geothermal energy) and §34 (Energy and climate). (Figure credit: Lines of Force for Two Point Charges. From the Wolfram Demonstrations Project. Contributed by S. M. Binder).

---

that current is distributed within the wire, it still circulates around the current in a right-hand sense.

Magnetic dipole fields   While there is no magnetic monopole charge in nature, a simple and important type of magnetic field is the **magnetic dipole** field produced in the ideal approximation of an infinitesimally small current loop. Magnetic dipoles play an important role in a number of aspects of energy physics discussed in later chapters, and are described in Box 3.2.

Magnetic dipoles describe the leading order behavior of magnetic fields at large distances from a system of moving charges. They also describe magnetic fields of elementary particles. For example, the electron carries negative charge, and when described quantum mechanically (§7) always has some angular momentum (*spin*). This gives the electron a small magnetic dipole moment oriented along the spin axis that cannot be described in terms of classical moving charges. In certain materials, dipole moments of

the electrons can align either spontaneously or in response to an external magnetic field. Such materials form **magnets** with a macroscopic magnetic field, although the source of the field is quantum in nature. **Permanent magnets** formed from such materials are used in a variety of electronic devices including motors and speaker systems.

Solenoid   Another important example of a simple magnetic field is the **solenoid**, shown in Figure 3.18(b). A solenoid is a long wire wound N times in a helix around a cylinder of cross-sectional area A and length l, which may be empty or filled with a material of permeability μ. Often, a magnetic material like iron is used, which has $\mu \gg \mu_0$. When the solenoid has many tightly packed coils, then away from the ends, the magnetic field is small immediately outside the solenoid, and constant both in magnitude and direction inside. The interior field points along the solenoid's axis and its strength is given by

Currents and Magnetic Fields

For time-independent current and charge densities, the magnetic field satisfies

$$\nabla \times \boldsymbol{B}(\boldsymbol{x}) = \mu_0 \boldsymbol{j}(\boldsymbol{x}), \quad \nabla \cdot \boldsymbol{B}(\boldsymbol{x}) = 0.$$

The Lorentz force on a charged particle or charge distribution is

$$\boldsymbol{F} = e\boldsymbol{E} + q\boldsymbol{v} \times \boldsymbol{B}, \quad \frac{d\boldsymbol{F}}{dV} = \rho\boldsymbol{E} + \boldsymbol{j} \times \boldsymbol{B}.$$

The magnetic fields of a long wire and a long solenoid are

$$\boldsymbol{B}_{\text{wire}} = \frac{\mu_0}{2\pi} I \frac{\hat{\boldsymbol{z}} \times \hat{\boldsymbol{\rho}}}{\rho}, \quad \boldsymbol{B}_{\text{solenoid}} = \mu n I \hat{\boldsymbol{z}}.$$

$$\boldsymbol{B}_{\text{solenoid}} = \mu n I \hat{\boldsymbol{z}}, \tag{3.51}$$

where $n = N/l$ is the number of turns of wire per unit length along the solenoid (Problem 3.15(b)). At large distances away from the solenoid (compared to its length and radius) the magnetic field is a dipole field with $\boldsymbol{m} = NIA\hat{\boldsymbol{z}}$, since the solenoid is essentially a combination of N simple current loops.

Magnetic forces between wires   Knowing how currents produce magnetic fields and how magnetic fields exert forces on moving charges, we can combine the two effects to find the force exerted by one wire carrying current I_1 upon another carrying current I_2. For the simple case where the wires are parallel, the force per unit length is perpendicular to the wires and has strength

$$\frac{dF}{dl} = -\frac{\mu_0}{2\pi d} I_1 I_2, \tag{3.52}$$

where d is the separation between the wires. The sign is such that if the currents are parallel, the wires attract, and if they are antiparallel, the wires repel (see Problem 3.17).

Magnetic forces are much weaker than electric forces. While the repulsion between two 1 C charges separated by one meter is 9×10^9 N, eq. (3.52) gives a magnetic attraction between two wires each carrying a current of 1 A of only 2×10^{-7} N.[7] On the other hand, the force exerted on the windings of a solenoid by the magnetic field inside can be quite substantial (Problem 3.16).

_____

[7] This, as noted above, provides the *definition* of the ampere in SI units.

3.4 Electric Motors and Generators

The basic principles of both the motor and the generator can be understood from the action of magnetic fields on moving charges (3.42). A motor is created by placing a loop carrying current in an external magnetic field. The magnetic field exerts forces on the charge flowing in the loop that produce a torque on the loop. The loop is positioned in such a way that the torque drives rotational motion in a uniform direction. A generator is the same device in reverse – moving the loop in the external magnetic field produces forces on the charges that drive an external current. In §3.4.1 and §3.4.2 we describe in more quantitative detail how these designs are realized.

3.4.1 An Electric Motor

Consider a wire carrying current I, formed into a rectangular loop of length l, width w, and area $A = lw$, that is free to rotate about an axis $\hat{\boldsymbol{y}}$ passing through the middle of the sides of length w in a constant magnetic field $\boldsymbol{B} = B\hat{\boldsymbol{z}}$. The arrangement is shown in Figure 3.19. According to the Lorentz force law (3.43), the two sides of the wire parallel to the y-axis experience a force per unit length $dF/dl = IB$ in a direction perpendicular to both the wire and the magnetic field – i.e. parallel to the x-axis. The magnetic forces on opposite sides of the loop give rise to torques $\boldsymbol{r} \times \boldsymbol{F}$ on the loop that sum to

$$\boldsymbol{\tau} = -IAB \sin\theta \, \hat{\boldsymbol{y}}, \tag{3.53}$$

where $A = lw$ is the area of the current loop. Note that this term can be written as $\boldsymbol{\tau} = \boldsymbol{m} \times \boldsymbol{B}$, where $\boldsymbol{m} = IA\hat{\boldsymbol{n}}$ is the magnetic dipole moment vector of the current loop (see Box 3.2), and θ is the angle from $\boldsymbol{m}$ to $\boldsymbol{B}$. The torque

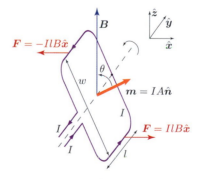

Figure 3.19 A rectangular current-carrying wire loop in a magnetic field experiences a Lorentz force IB per unit length along the sides parallel to $\hat{\boldsymbol{y}}$. (The forces on the other sides are irrelevant.) The wire loop in the diagram experiences a torque $IBA \sin\theta$, which causes the loop to rotate about the y-axis.

Box 3.3   Magnetic Units

In the SI system, magnetic fields are measured in **tesla** (T), named after the Serbian-American physicist and engineer Nikola Tesla. The tesla is not a fundamental unit, and is defined to be $1\,\mathrm{Vs/m^2}$. Equation (3.66), for example, shows that these are the correct units. An older unit for magnetic fields, which may be more familiar, is the **gauss** (G), with $1\,\mathrm{T} = 10\,000$ gauss. For reference, Earth's magnetic field near the surface ranges from ~ 0.25 to $\sim 0.65\,\mathrm{G}$ and a cheap alnico (aluminum-nickel-cobalt) magnet can produce a magnetic field as high as $1500\,\mathrm{G}$ between its poles. Magnetic flux has units of magnetic field × area or $\mathrm{T\,m^2}$. The SI unit of magnetic flux is the **weber** (Wb) with $1\,\mathrm{Wb} = 1\,\mathrm{T\,m^2}$. Sometimes magnetic fields are quoted in units of $\mathrm{Wb/m^2}$.

---

will cause the loop to rotate clockwise (as seen by a viewer looking along the y-axis as in the figure) if $\sin\theta < 0$ or counterclockwise if $\sin\theta > 0$. If, however, the direction of the current is reversed when $\sin\theta$ changes sign, then the torque will always act to accelerate the loop in the same direction. The reversal of the direction of current can be accomplished in various ways, for example by using an AC current or by having the ends of the wires make contact only by pressure with a housing that has a potential difference between the sides. The result is a basic **motor** – a machine that transforms electrical into mechanical energy (see Box 3.4 for more about motors).

The mechanical power output produced by the motor is equal to the electrical power input needed to keep the charge flowing. From eq. (2.44), the power exerted by the motor is

$$P = \boldsymbol{\tau}\cdot\boldsymbol{\omega} = IAB\dot{\theta}\,\sin\theta\,, \qquad (3.54)$$

since $\boldsymbol{\omega} = -\dot{\theta}\,\hat{\boldsymbol{y}}$. To maintain this power output, an external voltage must drive the current through the wire. To compute the electrical power needed, consider the Lorentz force exerted on the charges in the wire while the loop is in motion, as illustrated in Figure 3.20. A charge q in a section of the wire moving with velocity $\boldsymbol{v}$ experiences a force $\boldsymbol{F} = q\boldsymbol{v}\times\boldsymbol{B}$. On the two sides of the loop parallel to the $\hat{\boldsymbol{y}}$-axis this force acts opposite to the direction that the current is flowing. Using $\boldsymbol{v} = w\dot{\theta}\,\hat{\boldsymbol{\theta}}/2$, where $\hat{\boldsymbol{\theta}}$ is the unit vector in the direction of rotation, we find that the magnitude of the force is

$$F = qvB\sin\theta = \frac{w}{2}qB\dot{\theta}\,\sin\theta\,. \qquad (3.55)$$

To carry the charge around the loop requires work to be done against this force,

$$W = \oint_{\mathrm{loop}} d\boldsymbol{x}\cdot\boldsymbol{F} = 2Fl = qAB\dot{\theta}\,\sin\theta\,. \qquad (3.56)$$

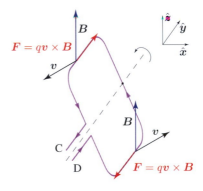

Figure 3.20 When the wire in Figure 3.19 rotates with angular velocity $\dot{\theta}$, the motion of the wire leads to a force $(w/2)qB\dot{\theta}\sin\theta$ on the charges in both sections of the wire parallel to the rotation axis. In a motor, the force opposes the direction of current flow. In a generator, this force produces an *EMF* that drives an external current.

Since $I = dq/dt$, the power required to keep the current moving around the loop is

$$P = \frac{dW}{dt} = IAB\dot{\theta}\,\sin\theta\,, \qquad (3.57)$$

which is precisely the rate (3.54) at which work is done by the motor. This work must be supplied by an external electrical power supply. Integrating the work done in rotating the loop around a half-cycle from $\theta = 0$ to $\theta = \pi$ gives

$$W = \int dt\,P = 2IAB. \qquad (3.58)$$

Since the current reverses when $\sin\theta$ changes sign, the average power output when the motor is rotating at frequency ν is then $\bar{P} = 4IAB\nu$. In most real motors the wire loop is wrapped many times, which multiplies the power by the number of windings. Note that there is no theoretical limit to the efficiency of an electric motor that converts electromagnetic to mechanical energy, though any

Box 3.4   Electric Motors

Electric motors are used in a wide range of devices, including cell phone vibrators, disk drives, fans, pumps, compressors, power tools, appliances, elevators, and for propulsion of ships and other vehicles – almost anything that moves and is driven by electricity contains a motor of some kind. Motors generally contain two magnets, the rotating **rotor** and the stationary **stator**. In the simplest motors the rotor

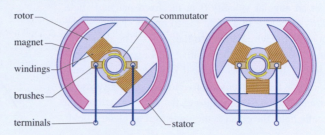

is an electromagnet powered by DC current with a *commutator* and *brushes* switching the direction of current flow depending on the angle of the rotor (see figures), while the stator uses a permanent magnet. The rotor often contains three, or sometimes five, sets of windings around an iron core. More complex designs use AC current to power the stator giving a varying magnetic field that drives the rotor (see §38 for further discussion).

Consider as a simple example a small motor, where the stator has a permanent magnet with field 0.02 T, and the rotor has an electromagnet formed by wrapping a wire 600 times around a single cylinder of radius 2 cm.

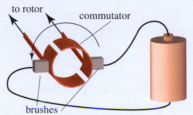

If the motor uses 300 mA of current and runs at 6000 rpm, what is the average power output of the motor (neglecting frictional losses)?

The average power output is $\bar{P} = 4NIABv \cong 4 \times 600 \times (300 \text{ mA}) \times \pi \times (0.02 \text{ m})^2 \times (0.02 \text{ T}) \times (100 \text{ s}^{-1}) \cong 1.8 \text{ W}$.

Box 3.5   Magnetic Dipoles in Magnetic Fields

In general, as in the special case of a current loop analyzed in the text, a magnetic dipole m in a homogeneous magnetic field B experiences a torque $\tau = m \times B$. This torque pushes the dipole into alignment with the field B. While analyzed in the text in terms of the Lorentz force, this force can be interpreted in terms of an interaction energy between the dipole and the field

$$E = -m \cdot B .$$

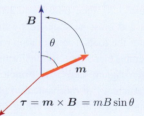

This interaction energy between magnetic dipoles and magnetic fields is relevant in later chapters, particularly in understanding simple quantum systems (§7) and the detailed dynamics of generators (§38).

real device will have losses, for example between the *commutator* and the *brushes* (see Box 3.4).

The power (3.57) required to keep the current flowing in the loop can be expressed as

$$P = \frac{dW}{dt} = -\frac{d}{dt} IAB \cos\theta = -\frac{d}{dt} m \cdot B . \quad (3.59)$$

Conservation of energy then leads us to the conclusion that the energy stored in the interaction of the current loop and the magnetic field is

$$E = -m \cdot B . \quad (3.60)$$

This result is, in fact, quite general (see Box 3.5), and corresponds to the everyday experience that a bar magnet reaches its minimum energy state when it aligns with an external magnetic field, such as when a compass aligns with Earth's magnetic field.

3.4.2 An Electric Generator

The motor described in the previous section can be reversed and used to transform mechanical into electrical energy. The device is then called a **generator**. Suppose that a mechanical agent forces the wire loop of Figure 3.20 to turn at an angular velocity $\dot{\theta}$ in the fixed magnetic field $\boldsymbol{B}$. This gives rise to a force (3.55) on the charges in the sections of the wire parallel to the y-axis. The negatively charged electrons and the positive ions experience opposite forces, but the ions are held in place by interatomic forces while the electrons are free to move and give rise to a current. Although the origin of the force is motion in a magnetic field, the effect is the same as if an electric field

$$|\boldsymbol{E}| = \frac{F}{q} = \frac{w}{2} B\dot{\theta} \, \sin\theta \qquad (3.61)$$

were present in the wire. The integral of this quantity around the loop is the work that is done on a unit of charge that is moved completely around the loop. This is known as the **electromotive force** or **EMF**, denoted $\mathcal{E}$,

$$\mathcal{E} = \oint d\boldsymbol{x} \cdot \frac{F}{q} = wlB\dot{\theta} \, \sin\theta \, . \qquad (3.62)$$

The electromotive force $\mathcal{E}$ generalizes the notion of voltage difference from electrostatics, and is applicable in time-dependent circuits as well as in static situations where a static voltage source like a battery provides an EMF. Note that the force acting on charges moving around the loop in the motor described in the last subsection can also be thought of as an electromotive force, known as the **back-EMF**.

   If a load is connected across the terminals C and D in Figure 3.20, then power equal to $\mathcal{E}I$ is delivered to the load and must be supplied by the mechanical agent that rotates the loop. The wire loop rotating in a magnetic field generates electricity; it transforms mechanical into electric power. Again, there is in principle no theoretical limit to the efficiency of mechanical to electric energy conversion though any real system using this mechanism will have some losses.

   Note that a generator of this type naturally produces an electromotive force that varies sinusoidally with time when the loop is rotated at a constant angular velocity. Just as the wire loop in Figure 3.21 rotates at an angular frequency $\omega = \dot{\theta}$, the EMF between terminals marked a and b is proportional to $AB\omega \sin\omega t$, where A is the area of the loop. As for the motor, winding a wire N times around the loop will multiply the EMF, and hence the output power at a given rate of rotation, by N.

3.5 Induction and Inductors

We have described how electric and magnetic fields are generated by and exert forces on charges and currents.

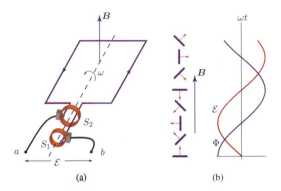

Figure 3.21 (a) A sketch of the physical arrangement of an AC generator. As the wire loop rotates, each end slides over a *slip ring* ($S_{1,2}$), maintaining electrical contact. The EMF between terminals a and b oscillates sinusoidally. (b) The orientation of the current loop is shown together with the EMF and the magnetic flux Φ through the loop as functions of time.

There are still missing pieces, however, in our description of how these fields are produced. In particular, *electric fields are produced when magnetic fields change* and *vice versa*. These relationships underlie the phenomenon of *light waves* – electromagnetic waves that propagate through the vacuum (§4). In this section we describe how electric fields must be produced when magnetic fields change; this is *Faraday's law of induction*, which is closely related to the mechanisms behind the motors and generators we have just studied. Like magnetic fields and forces in general, Faraday's law is most clearly explained as a simple application of Einstein's special theory of relativity. Here, however, we introduce induction empirically, as it was first discovered by Faraday; Box 3.6 provides a synopsis of the argument based on relativity.

3.5.1 Faraday's Law of Induction

By means of a series of ingenious experiments in the mid-nineteenth century, the English physicist Michael Faraday discovered that changing the magnetic flux through a wire loop induces an EMF that drives a current in the loop. Figure 3.22 shows two different ways in which this might occur. In case (a), the loop C moves through a stationary magnetic field whose strength varies with position. In case (b) the loop is stationary but the strength of the magnetic field increases with time as, for example, when the current in a nearby wire is increased. In both cases the flux through the loop,

$$\Phi_S = \int_S d\boldsymbol{S} \cdot \boldsymbol{B} \, , \qquad (3.63)$$

Box 3.6   Induction and Relativity

While Faraday's law was discovered empirically before Einstein's special theory of relativity, the clearest logical explanation for Faraday's law follows from
a simple application of relativity, which we describe here. A more detailed discussion of Einstein's theory is given in §21.1.4, and the *inertial reference frames*
to which this theory applies are described in §27.2.

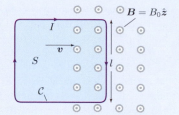

Consider a wire in the shape of a rectangular loop C in the xy-plane that is
translated horizontally as shown in the figure at right, moving from a region
without a magnetic field into a region with a field. (This could be done, for
example, by passing the loop between two of the coils of a large solenoid.)
While only one side of the rectangle is within the magnetic field $B = B_0\hat{z}$, the
electrons on that side of the rectangle will experience a Lorentz force, giving a net EMF $\mathcal{E} = lB_0v$ that drives a current
clockwise around the wire loop, where l is the length of the rectangle perpendicular to its motion, and v is the loop's
speed. This is analogous to the force experienced on the charges in the wire loop in the motor and generator described
in the previous section.

How does this force arise from the point of view of a physicist who is moving along with the loop (or alternatively,
as seen from a stationary loop when the device producing the magnetic field is moved across the loop at a constant
velocity v)? There cannot be any Lorentz force on the charges in a stationary loop since there is no velocity. But
Einstein's principle of special relativity requires the physics in this reference frame to be the same as in the frame in
which the loop is moving, so something must generate a force. The only electromagnetic force that acts on stationary
charges is that from an electric field. Thus, there must be an electric field along the part of the wire that is within the
magnetic field region. The integral of this electric field around the loop is

$$\oint_C d\boldsymbol{x} \cdot \boldsymbol{E} = -lB_0v\,,$$

where the sign arises since the integral is taken counterclockwise around the loop C by convention. The right-hand side
of this equation is precisely minus the rate of change of the flux $\Phi(S) = \int_S d\boldsymbol{S} \cdot \boldsymbol{B}$ of the magnetic field through the
planar surface S bounded by the wire loop,

$$\mathcal{E}(C) = \oint_C d\boldsymbol{x} \cdot \boldsymbol{E} = -\frac{d\Phi(S)}{dt} = -\frac{d}{dt}\int_S d\boldsymbol{S} \cdot \boldsymbol{B}\,.$$

This is precisely *Faraday's law* (3.64). While the argument just given relies on a wire loop of a particular shape, the
same argument generalizes to an arbitrary curve C, and the result does not depend upon the presence of an actual wire
loop around the curve C. Faraday's law is valid for any closed loop through which the magnetic flux changes, either in
magnitude or direction.

increases with time (here S is a surface bounded by the
loop C as in eq. (3.10)). Faraday observed that the EMF
induced in the loop was proportional to the time rate of
change of the flux,

$$\mathcal{E}(C) = \oint_C d\boldsymbol{x} \cdot \boldsymbol{E} = -\frac{d\Phi_S}{dt} = -\frac{d}{dt}\int_S d\boldsymbol{S} \cdot \boldsymbol{B}\,. \quad (3.64)$$

Equation (3.64), known as **Faraday's law**, describes the
induction of an electric field by a time-dependent magnetic flux. Since the integral of the electric field around a
closed path is not zero in this situation, it is no longer possible to define an *electrostatic* potential as we did in eq. (3.7).
As mentioned earlier, the EMF generalizes the concept of
voltage to include time-dependent phenomena of this type.

The minus sign in eq. (3.64) indicates that a current
produced by the EMF would act to reduce the magnetic
flux through the loop (Question 3.4). If it were the other
way, the EMF would produce a current that would further increase the EMF, leading to a runaway solution that
violates conservation of energy. This sign, and the general
fact that an induced current acts to reduce any imposed
magnetic field, is known as **Lenz's law**.

Faraday's law is closely related to the mechanism underlying motors and generators. In these devices, including
generators of the type described above, an EMF develops
in a loop rotating in a constant magnetic field. While the
flux through the loop changes with time, eq. (3.64) does
not apply in this situation since there is no appropriate

Example 3.5   AC Power

Generators provide a compact and efficient means of transforming mechanical to electrical power. Suppose you seek to generate an RMS voltage of 120 V at 60 Hz using a permanent magnet for a rotor, in a roughly constant magnetic field of 1000 gauss. *What area (or $A \times N$) must the current loop span?*

With $|\boldsymbol{B}| = 0.1$ T, $\omega = 2\pi \times 60\,\mathrm{s}^{-1}$, and $\mathcal{E} = 120\sqrt{2}$ V, we find from eq. (3.62), $AN = \mathcal{E}/B\omega = 10\sqrt{2}/\pi\,\mathrm{m}^2 \cong$ 4.5 m^2. A wire wound $N = 100$ times around a sturdy *armature* of radius 12 cm would satisfy these specs. This is a fairly compact and simple device: a magnet of modest strength and a mechanical system of modest dimensions. Note, however, that there is no free lunch: you must supply enough mechanical power to turn the armature. If, for example, you demand an RMS current of 10 A from this generator, you would have to supply an RMS power of $10 \times 120 =$ 1200 W, which is nearly two horsepower.

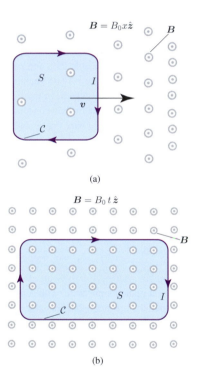

(a)

(b)

Figure 3.22 Two examples of EMF induced by changing magnetic flux. (a) A wire loop in the $z = 0$ plane moves into an increasing magnetic field. (b) A fixed wire and a magnetic field that is constant in space, but grows with time. In both cases, the direction of the EMF (shown by the arrows) is determined by Lenz's law.

reference frame in which the loop is stationary.[8] Nonetheless, generators that work on this principle are often spoken of as working *by induction*, in part because Faraday's law was understood before Einstein's theory of relativity. Other

---

[8] For a further discussion of reference frames and Einstein's special theory of relativity, see §21.

types of generators that *do* use induction as described by Faraday's law (3.64) are discussed in §38.

Faraday's law can be rewritten in a local form that emphasizes the direct relationship between electric and magnetic fields. First, note that the surface integral is the same no matter what surface is chosen to span the curve $\mathcal{C}$ (Problem 3.21), so the flux Φ can be associated with the curve $\mathcal{C}$ independent of the choice of S. Next, apply Stokes' theorem (B.27) to the left-hand side of eq. (3.64) giving

$$\int_S d\boldsymbol{S} \cdot (\nabla \times \boldsymbol{E}) = -\frac{d}{dt} \int_S d\boldsymbol{S} \cdot \boldsymbol{B} \qquad (3.65)$$

for any surface S spanning the current loop $\mathcal{C}$. Since S is arbitrary, the integrands must be equal, giving the local form

$$\nabla \times \boldsymbol{E} = -\frac{\partial \boldsymbol{B}}{\partial t}, \qquad (3.66)$$

which replaces eq. (3.14) if time-varying magnetic fields are present.

An important feature of Faraday's law is that a wire wound N times around the curve $\mathcal{C}$ captures N times the flux, enhancing the EMF by a factor of N. So, a succinct summary of Faraday's law for a wire with N windings is

$$\mathcal{E} = -N\frac{d\Phi}{dt}. \qquad (3.67)$$

3.5.2 Inductance and Energy in Magnetic Fields

In an electrical circuit, time-varying currents produce time-varying magnetic fields, $d\boldsymbol{B}/dt \propto dI/dt$. Time-varying magnetic fields in turn produce electromotive force, $\mathcal{E} \propto -d\boldsymbol{B}/dt$. Combining these two effects, we see that a time-varying current through a circuit or a part of a circuit leads to an EMF proportional to dI/dt. The coefficient of proportionality, denoted by L, is called the **inductance** of the circuit or circuit element, so that there is a contribution to

Faraday's Law of Induction

A time-varying magnetic field gives rise to a circulating electric field according to Faraday's law,

$$\nabla \times \boldsymbol{E} = -\frac{\partial \boldsymbol{B}}{\partial t} \, ,$$

or in integral form,

$$\oint_C d\boldsymbol{x} \cdot \boldsymbol{E} = -\frac{d}{dt} \int_S d\boldsymbol{S} \cdot \boldsymbol{B} = -\frac{d\Phi}{dt} \, .$$

Figure 3.23 A sample of simple inductors, basically loops of wire wound around an iron core with $\mu \gg \mu_0$ to enhance the inductance by a factor μ/μ_0. (Credit: Manutech, Haiti)

the voltage across a circuit component with inductance L given by

$$V = -L\frac{dI}{dt} \, . \tag{3.68}$$

The minus sign implements Lenz's law: with L positive, the voltage acts to oppose the increase in current. Inductance is measured in **henrys** (H), defined by 1 H = 1 V s/A, and named in honor of the American physicist Joseph Henry, the co-discoverer of induction. The electric circuit symbol for an inductor is —.

Any circuit will have some inductance by virtue of the magnetic fields created by the current flowing through its wires. An **inductor** is a circuit element designed principally to provide inductance. Typically, an inductor consists of a winding of insulated wire surrounding a core of magnetizable material like iron. Some examples of inductors are shown in Figure 3.23.

A classic example of an inductor is a long solenoid of length l with a wire winding N times around a core made of material with permeability μ. Ignoring end effects, the

magnetic field in the solenoid is given by eq. (3.51), so $d|\boldsymbol{B}|/dt = \mu n \, dI/dt$, and

$$\mathcal{E}_{\text{solenoid}} = -NA\mu n\frac{dI}{dt} \equiv -L_{\text{solenoid}}\frac{dI}{dt} \, , \tag{3.69}$$

with $n = N/l$, so the inductance of a solenoid is $L_{\text{solenoid}} = N^2 A\mu/l = n^2 \mathcal{V}\mu$, where $\mathcal{V} = lA$ is the volume.

Because they generate an EMF opposite to dI/dt, inductors act to slow the rate of change of the current in a circuit. They also store energy. The power expended to move a current I through a voltage V is $P = VI$. Since the voltage across an inductor is $V = -L \, dI/dt$, the power needed to increase the current through an inductor is $P = LI \, dI/dt$, where the sign reflects the fact that work must be done to increase the current. This work is stored in the magnetic fields in the inductor. Starting with zero current at $t = 0$ and increasing the current to I, the energy stored in the inductor is

$$E_{\text{EM}} = \int_0^t dt' \, P = L\int_0^t dt' I'\frac{dI'}{dt'}$$

$$= L\int_0^I dI' I' = \frac{1}{2}LI^2 \, . \tag{3.70}$$

E_{EM} can be expressed in terms of the integral of the magnetic field strength over all space, in a form analogous to eq. (3.20),

$$E_{\text{EM}} = \frac{1}{2\mu}\int d^3x \, |\boldsymbol{B}(\boldsymbol{x})|^2 \, . \tag{3.71}$$

This is a general result for the energy contained in a magnetic field. Although the derivation of this result in a completely general context would require further work, the fact that it holds for the specific case of a solenoid is easy to demonstrate (Problem 3.22). As indicated by eq. (3.71), if $\boldsymbol{B}$ is held fixed, less energy is stored in a magnetic material when the permeability μ is larger. It is more natural, however, to compare stored energy with currents held fixed, since they can be controlled externally. Then, as can be seen from eq. (3.70) and the fact that $L \propto \mu$, for the same current, more energy is stored in a system with larger permeability. In the following chapter (§4), we give a more detailed description of the storage and propagation of energy in electric and magnetic fields.

3.5.3 Mutual Inductance and Transformers

If two independent circuits or circuit components are in close spatial proximity, then when the current through one changes, the magnetic flux through the other changes proportionately. By Faraday's law this induces an EMF in the

> ### Inductance, Inductors, and Energy
>
> An inductor is a circuit element that produces a voltage proportional to the time rate of change of the current passing through it,
>
> $$V = -L\,dI/dt,$$
>
> where L is the inductance, measured in *henrys*, $1\,H = 1\,V\,s/A$.
>
> The inductance of a solenoid with n turns per unit length, volume $\mathcal{V}$, and core permeability μ is approximately
>
> $$L_{\text{solenoid}} \cong n^2 \mathcal{V} \mu.$$
>
> The energy stored in an inductor is $E_{\text{EM}} = LI^2/2$. This is a special case of the general expression for energy stored in magnetic fields,
>
> $$E_{\text{EM}} = \frac{1}{2\mu} \int dV\, |\boldsymbol{B}|^2.$$

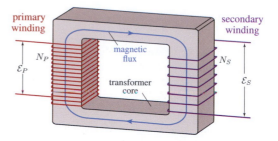

Figure 3.24 A simple transformer. Two coils are wound around a core with high magnetic permeability (typically iron), which acts as a *magnetic circuit*, channelling the magnetic flux through both coils. In the ideal limit where all magnetic flux goes through both coils, the ratio of secondary to primary voltages $\mathcal{E}_S/\mathcal{E}_P$ is given by the ratio of the number of turns, in this case roughly 1:2.

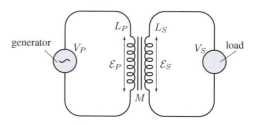

Figure 3.25 Two circuits, the *primary* (left) and the *secondary* (right), are coupled by mutual inductance. The resulting device, a *transformer*, transforms the AC voltage on the left V_P to $\sqrt{L_S/L_P}\,V_P$ on the right.

second circuit, proportional to the rate of change of the current in the first,

$$\mathcal{E}_2 = -M_{21}\frac{dI_1}{dt}. \qquad (3.72)$$

Similarly, a time-dependent current in circuit two leads to an EMF in circuit one: $\mathcal{E}_1 = -M_{12}dI_2/dt$. It is possible to show using conservation of energy (Problem 3.23) that this relationship is symmetric: $M_{12} = M_{21} \equiv M$, where M is known as the **mutual inductance** of circuits one and two. The sign of M depends on the relative orientation of the two circuits. In general $M^2 \le L_1 L_2$, and the ideal limit $M^2 = L_1 L_2$ is reached only when all the magnetic flux loops both circuits – as it would for two coincident ideal solenoids of the same area and length. A conceptual model of such a system is shown in Figure 3.24. The symbol for a mutual inductor is ⎽⎼⎼⎼⎼⎼, two inductors with lines denoting an iron core that channels the magnetic flux between them.

Mutual inductance enables the construction of one of the most common and useful electrical devices: the **transformer**. Consider the AC circuit shown in Figure 3.25. The left-hand circuit (the **primary**) is coupled by mutual inductance to the right-hand circuit (the **secondary**). In the ideal limit where $M^2 = L_P L_S$, all magnetic flux lines go through both inductors, so $\mathcal{E}_P = -N_P d\Phi/dt$, $\mathcal{E}_S = -N_S d\Phi/dt$, and,

$$\frac{\mathcal{E}_S}{\mathcal{E}_P} = \frac{V_S}{V_P} = \sqrt{\frac{L_S}{L_P}} = \frac{N_S}{N_P} \qquad (3.73)$$

(see Problem 3.25). Power is transmitted from the generator on the left to the load on the right.

Transformers play an essential role in the manipulation of AC power in electric power distribution networks. In the absence of imperfections such as resistance in the coils and incomplete containment of the magnetic flux, the efficiency of a transformer would be 100%. Those used in electric power distribution reach 98% or higher. We describe the role of transformers in electrical grids further in §38.

3.6 Maxwell's Equations

Up to this point, we have introduced four equations that describe the way that electric and magnetic fields interact with matter and with one another. Gauss's law (3.12) and Ampère's law (3.49) prescribe the way that electric and magnetic fields are produced by charges and currents. Two other equations – eq. (3.47), associated with the absence of magnetic charge, and Faraday's law (3.66) – constrain E and B. Although we stated these laws under special conditions – for example, Gauss's law was derived only for static charges – all but one of these equations continue to hold under all known conditions.

The one equation that needs further modification is Ampere's law, which we have described only in the conditions of steady currents and constant charge distribution. As discussed earlier, Ampere's law as stated in eq. (3.49) is not consistent with charge conservation. To see this, take the divergence of eq. (3.49). Since $\nabla \cdot (\nabla \times B) = 0$ (B.21), eq. (3.49) implies $\nabla \cdot j = 0$, which only holds if $\partial \rho / \partial t = 0$ (see eq. (3.33)). When current and charge densities change with time, this is not true. Around 1860, the English physicist James Clerk Maxwell recognized this problem, and solved it by subtracting a term proportional to $\partial E / \partial t$ from the left-hand side of Ampere's law. Maxwell's alteration of Ampere's law reads,

$$\nabla \times B - \mu_0 \epsilon_0 \frac{\partial E}{\partial t} = \mu_0 j. \qquad (3.74)$$

The demonstration that this equation is consistent with current conservation is left to Problem 3.26. Note that this modification of Ampere's law, which states that a time-varying electric field gives rise to a magnetic field, is closely parallel to Faraday's law of induction stating that a time-varying magnetic field gives rise to an electric field. Indeed, in the absence of electric charges and currents, the laws of electromagnetism are invariant under a symmetry exchanging B and $\sqrt{\mu_0 \epsilon_0} E$.

Maxwell's modification of Ampere's law completed the unification of electricity and magnetism that began with Faraday's discovery of induction. Together with the Lorentz force law (3.45), which describes the way that electric and magnetic fields act on charges and currents,

Maxwell's equations give a complete description of electromagnetic phenomena. These equations paved the way for Einstein's special theory of relativity, which further illuminated the unified structure of electromagnetism. Maxwell's achievement initiated great advances in understanding and application of electromagnetic theory, including the realization that light is an electromagnetic wave and that other forms of electromagnetic radiation could be produced from oscillating charges and currents.

Discussion/Investigation Questions

3.1 Think of some devices that generally are not run on electrical power. Discuss the reasons why other power sources are favored for these devices.

3.2 Can you identify the largest component of your personal electric power consumption?

3.3 Most countries provide AC electricity at $V_{\mathrm{RMS}} \approx 220$–$240$ V to residential customers. In the US and parts of South America $V_{\mathrm{RMS}} \approx 110$–$120$ V is provided. How do resistive losses in household wires compare between the two? Given the result, why do you think even higher voltages are not used?

3.4 Discuss the statement that Lenz's law follows from conservation of energy. Consider, for example, the scenario of Figure 3.22(b) if the sign of eq. (3.64) were reversed.

3.5 Currents produce magnetic fields, and magnetic fields exert forces on currents. When a current flows through a wire, does the magnetic field it produces act to make the wire expand or contract? What about the force on the wires in a solenoid? Does it act to make the solenoid expand or contract? These are significant effects for wires carrying very large currents.

Problems

3.1 The electric field outside a charged conducting sphere is the same as if the charge were centered at its origin. Use this fact to calculate the capacitance of a sphere of radius R, taking the second conductor to be located at infinity. What is the most charge you can store on an otherwise isolated spherical conductor of radius R without a breakdown of the surrounding air as discussed in Example 3.2? What is the maximum energy you can store on a conducting sphere of radius 1 mm?

3.2 [T] Prove Gauss's law from Coulomb's law for static charge distributions by showing that the electric field of a single charge satisfies the integral form of Gauss's law and then invoking linearity.

3.3 How much energy can you store on a parallel plate capacitor with $d = 1\,\mu\mathrm{m}$, $A = 10\,\mathrm{cm}^2$, and $\epsilon = 100\epsilon_0$,

---

Maxwell's Equations	
Gauss's law	$\nabla \cdot E = \dfrac{\rho}{\epsilon_0}$
No magnetic charge	$\nabla \cdot B = 0$
Faraday's law	$\nabla \times E + \dfrac{\partial B}{\partial t} = 0$
Ampere–Maxwell law	$\nabla \times B - \mu_0 \epsilon_0 \dfrac{\partial E}{\partial t} = \mu_0 j$

assuming that the breakdown field of the dielectric is the same as for air?

3.4 [T] Starting from Gauss's law and ignoring edge effects (i.e. assume that the plates are very large and the electric field is uniform and perpendicular to the plates), derive the formula for the capacitance of a parallel plate capacitor, $C = \epsilon_0 A/d$. Refer to Figure 3.8. You can assume that the electric field vanishes outside the plates defining the capacitor. Show that the energy stored in the capacitor, $\frac{1}{2}CV^2$, can be written as the integral of $\epsilon_0|\boldsymbol{E}|^2/2$ over the region within the capacitor (as asserted in eq. (3.20)).

3.5 Suppose that a capacitor with capacitance C is charged to some voltage V and then allowed to discharge through a resistance R. Write an equation governing the rate at which energy in the capacitor decreases with time due to resistive heating. Show that the solution of this equation is $E(t) = E(0)e^{-2t/RC}$. You can ignore the internal resistance of the capacitor. Show that the heat produced in the resistor equals the energy originally stored in the capacitor.

3.6 The dielectrics in capacitors allow some *leakage current* to pass from one plate to the other. The leakage can be parameterized in terms of a *leakage resistance* R_L. This limits the amount of time a capacitor can be used to store electromagnetic energy. The circuit diagram describing an isolated capacitor, slowly leaking charge, is therefore similar to the one analyzed in Problem 3.5. The Maxwell BCAP0310 ultracapacitor (see §37.4.2) is listed as having a capacitance of 310 F with a voltage up to 2.85 V, and a maximum leakage current of 0.45 mA when fully charged. Take this to be the current at $t = 0$. What is the leakage resistance of the BCAP0310? Estimate the time scale over which the charge on the capacitor falls to $1/e$ of its initial value.

3.7 A cloud-to-ground lightning bolt can be modeled as a parallel plate capacitor discharge, with Earth's surface and the bottom of the cloud forming the two plates (see Example 3.2). A particular bolt of lightning passes to the ground from a cloud bottom at a height of 300 m. The bolt transfers a total charge of 5 C and a total energy of 500 MJ to the ground, with an average current of 50 kA. How long did the lightning bolt last? What was the electric field strength in the *cloud–earth capacitor* just before it discharged? How does this electric field compare with the breakdown field of air (3 MV/m)? (It is now known that various effects cause lightning to begin at fields that are considerably smaller than the breakdown field.)

3.8 [T] Consider an electric dipole composed of two charges $\pm Q$ at positions $\pm \boldsymbol{\xi}/2$. Write the exact electric field from the two charges and show that the leading term in an expansion in $1/r$ matches the $\boldsymbol{E}$-field quoted in Box 3.2. [Hint: consider the binomial expansion, eq. (B.66).]

3.9 If each of the batteries used in the flashlight in Example 3.3 has an internal resistance of 0.5 Ω (in series with the circuit), what fraction of power is lost to Joule heating within the batteries?

3.10 [T] Consider two resistors placed in **series**, one after the other, in an electric circuit connected to a battery with voltage V. Show that the effective resistance of the pair is $R_1 + R_2$ by using the fact that the current through both resistors is the same while voltages add. Now connect the resistors in **parallel**, so that the voltage across both resistors is V. Compute the total current and show that the effective resistance satisfies $1/R = 1/R_1 + 1/R_2$.

3.11 An appliance that uses 1000 W of power is connected by 12 gauge (diameter 2.053 mm) copper wire to a 120 V (RMS) AC outlet. Estimate the power lost per meter, dP_{lost}/dL, (in W/m) as resistive heating in the wire. (Remember that the wire to the appliance has two separate current-carrying wires in a single sheath.)

3.12 Electrical power is often used to boil water for cooking. Here are the results of an experiment: a liter of water initially at 30 °C was boiled on an electric stove top burner. The burner is rated at 6.67 A (maximum instantaneous current) and 240 V_{RMS}. It took 7 minutes 40 seconds to reach the boiling point. The experiment was repeated using an "electric kettle." The kettle is rated at 15 A (maximum instantaneous current again) and uses ordinary (US) line voltage of 120 V_{RMS}. This time it took 4 minutes and 40 seconds to reach boiling. What are the power outputs and resistances of the burner and the kettle? Compare the efficiency for boiling water of the stove top burner and the kettle. To what do you attribute the difference?

3.13 [T] Use the magnetic force law (3.42) and the definition of work $W = -\int_a^b d\boldsymbol{x} \cdot \boldsymbol{F}$ to show that magnetic forces do no work. [Hint: Consider the vector identity (B.6).]

3.14 [T] Show that a charged particle moving in the xy-plane in the presence of a magnetic field $\boldsymbol{B} = B\hat{\boldsymbol{z}}$ will move in a circle. Compute the radius of the circle and frequency of rotation in terms of the speed v of the particle. Show that adding a constant velocity $u\hat{\boldsymbol{z}}$ in the z-direction still gives a solution with no further acceleration.

3.15 [T] Review how magnetic fields are calculated from Ampere's law by computing (a) the magnetic field due to a straight wire and (b) the magnetic field in the interior of a very long solenoid. The contours $\mathcal{C}_1, \mathcal{C}_2$ shown in red in Figure 3.18 will help. You may use the fact that the magnetic field vanishes just outside the outer boundary of the solenoid in (b).

3.16 [T] Show that the force per unit area on the windings of an air-core solenoid from the magnetic field of the solenoid itself is of order $F/A \sim B^2/\mu_0$. Check that the dimensions of this expression are correct and estimate F/A in pascals if $|B| = 1$ T.

3.17 [T] Derive eq. (3.52) from eq. (3.43) and eq. (3.50). Make sure you get the both the direction and magnitude.

3.18 An electric motor operates at 1000 rpm with an average torque of $\tau = 0.1$ Nm. What is its power output? If it is running on 1.2 A of current, estimate the back-EMF from the rotor.

3.19 Consider the motor described in Box 3.4. If the resistance in the wire wrapping the rotor is 1 Ω, compute the energy lost under the conditions described. What fraction of energy is lost to Joule heating in this situation? If the current is doubled but the rotation rate is kept fixed, how do the output power, Joule heating losses, and fraction of lost energy change?

3.20 [T] In §3.4.2 we derived the EMF on a wire loop rotating in a magnetic field using the Lorentz force law to compute the forces on the mobile charges. Although Faraday's law of induction (3.64) does not apply in a rotating reference frame, show that the same relation (3.62) follows from Faraday's law.

3.21 [T] Explain why the integral $\int_S dS \cdot B$ that appears in eq. (3.65) is independent of the choice of surface S. [Hint: make use of eq. (3.46).]

3.22 [T] Consider a long, hollow solenoid of volume $\mathcal{V}$. Show that, ignoring end effects, its inductance is $L = n^2 \mathcal{V} \mu_0$, and that the magnetic energy it stores $E_{EM} = LI^2/2$ can be written in the form of eq. (3.71).

3.23 [T] Prove that the mutual inductance is a symmetric relation, $M_{12} = M_{21}$, by computing the energy stored when current I_1 is established in loop 1 and then current I_2 is established in loop 2. Then set up the same currents in the opposite order.

3.24 Design a transmission system to carry power from wind farms in North Dakota to the state of Illinois (about 1200 km). The system should handle Illinois's summertime electricity demand of 42 GW. Land for transmission towers is at a premium, so the system uses very high voltage ($V_{RMS} = 765$ kV). Assume that the electricity is transmitted as ordinary alternating current, although in practice *three-phase power* (see §38.3.1) would be used. Assume that each tower can carry 36 aluminum cables (nine lines, each consisting of four conductors separated by non-conducting spacers). The conductors are 750 mm^2 in cross section. How many separate strings of transmission towers are needed if the transmission losses are to be kept below 5%? Assume a purely resistive load.

3.25 [T] Consider the transformer in Figure 3.25. Suppose the load is a resistor R and that the transformer is ideal, with $M^2 = L_S L_P$ and all magnetic flux lines passing through both inductors. Show that the voltage drop across the resistor is $V_S = \sqrt{L_S/L_P} V_P = N_S V_P/N_P$ and that the time-averaged power consumed in the resistor is $P = (L_S/L_P)(V_P^2/2R) = N_S^2 V_P^2/2N_P^2 R$.

3.26 [T] Take the divergence of both sides of eq. (3.74), use Coulomb's law on the left and current conservation on right to show that the equation is consistent.

Waves and Light

Waves have the capacity to propagate energy over great distances. Ocean waves can transmit energy halfway around the world, even though the molecules of water do not themselves experience any significant net motion. Sound waves in air and seismic waves propagating through Earth's interior likewise transmit energy without net motion of material. Indeed, many of the energy systems described in this book depend at a fundamental level upon energy propagation by waves. Most notably, all of the energy that is transmitted from the Sun to Earth comes in the form of electromagnetic waves, composed of propagating excitations of the electric and magnetic fields (§22). This incident electromagnetic solar energy is the source of almost all energy used by humanity. Other examples of energy systems based on waves include technologies that harvest the energy in ocean surface waves (§31.2), the use of seismic waves to probe the structure of Earth's interior in search of geothermal (§32) or fossil fuel (§33) resources, and the vibrational waves in crystals called *phonons* that facilitate the conversion of light into electricity in photovoltaic cells (§25). In the atomic and subatomic world, particles themselves are in many ways best characterized in terms of a quantum mechanical *wavefunction* that replaces the classical position of the particle with a diffuse probability distribution (§7).

In this chapter we introduce some of the basic properties of waves, focusing on wave solutions of Maxwell's equations that describe the propagation of light. The theory of waves is a complex and many-faceted subject. Rather than treating the subject comprehensively at the outset, we introduce the basics here and develop the theory of waves further as needed throughout the book. In §4.1 we describe the simplest type of waves and the wave equation they obey. In §4.2 we use the example of waves on a stretched string to introduce some of the fundamental features of wave propagation. We introduce electromagnetic

> **Reader's Guide**
> Waves propagate in many media, transporting energy without net motion of material. In this chapter we introduce the basic physics of waves, and characterize electromagnetic waves (light) and the transport of energy through EM waves. Along the way, we give a more complete characterization of energy in electromagnetic fields.
>
> Prerequisites: §2 (Mechanics), §3 (Electromagnetism).
>
> The introductory material on waves in this chapter serves as a basis for the development of wave-related phenomena in many subsequent chapters, including §7 (Quantum mechanics), §9 (Energy in matter), §22 (Solar energy), §31 (Water energy), §32 (Geothermal energy), §33 (Fossil fuels).

waves in §4.3, and discuss energy in electromagnetic fields and energy and momentum propagation in electromagnetic waves in §4.4. In §4.5 we conclude with a brief overview of some further wave-related phenomena that are encountered later in the book.

4.1 Waves and a Wave Equation

In the simplest kind of wave motion, a disturbance in some medium propagates through space at a constant velocity without changing form. Consider, for example, a long string, stretched taut along the x-axis. The vibrations of the string in a transverse direction y depend on both x and t, and can be described by a function $y = f(x, t)$. A wave of fixed form propagating along the x-axis with constant velocity v is described by

$$f(x, t) = f(x - vt). \tag{4.1}$$

At some fixed time, say $t = 0$, the wave has a profile $f(x)$. At a time $t = \Delta t$ later, the whole profile is shifted a

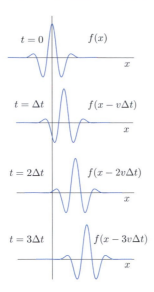

Figure 4.1 A right-moving waveform shown at time $t = 0$ and at equal time intervals Δt later.

distance $\Delta x = v\Delta t$ to the right (for $v > 0$), as illustrated in Figure 4.1.

4.1.1 The Basic Wave Equation

Waves occur in physical systems when excitations at a point are coupled locally to nearby points in such a way that disturbances propagate through space. Such systems are described by *partial differential equations* in which both space and time derivatives occur. Because the wave form (4.1) is only a function of $x - vt$, the derivatives of f with respect to x and t are proportional, $\partial f/\partial t = -v\partial f/\partial x$.

It follows that any function of the form $f(x, t) = f(x - vt)$ satisfies the **wave equation**

$$\frac{\partial^2 f}{\partial t^2} = v^2 \frac{\partial^2 f}{\partial x^2} . \tag{4.2}$$

This is the simplest of many partial differential equations that have wave-like solutions. Just as the differential equation $\ddot{x} = -kx$ for the harmonic oscillator describes oscillations in time of a wide class of physical systems, so eq. (4.2) describes the propagation of fluctuations in many systems that are distributed through a region of space. Physical effects such as damping and forcing, as well as nonlinearities, can complicate the behavior of propagating waves and require modifications of the simplest wave equation (4.2), just as similar effects can complicate the harmonic oscillator. The wave equation (4.2) can easily be generalized to three dimensions, where

$$\frac{\partial^2}{\partial t^2} f(\boldsymbol{x}, t) = v^2 \, \nabla^2 f(\boldsymbol{x}, t) , \tag{4.3}$$

where the *Laplacian operator* "∇^2" is defined in eq. (B.17). The wave equations (4.2) and (4.3) are *linear* in the function f. This has the important consequence that if f_1 and f_2 are separately solutions to eq. (4.2) or (4.3), and if a_1 and a_2 are arbitrary constants, then the **linear superposition**

$$f(x, t) = a_1 f_1(x, t) + a_2 f_2(x, t) , \tag{4.4}$$

is also a solution to the same equation. Linearity is a key characteristic of wave equations. Among its consequences are that waves propagating in a medium can pass through one another without interacting and that complicated wave forms can be built up out of simple components, such as *sine waves*.

4.1.2 Sinusoidal Wave Solutions

A particularly useful class of waveforms are **sine waves**, which vary sinusoidally both in space and in time,

$$f(x, t) = A \sin \left[2\pi \left(x/\lambda - vt \right) - \phi \right] . \tag{4.5}$$

This function has a spatial dependence that repeats periodically when $x \rightarrow x + \lambda$, so we refer to λ as the **wavelength** of the sine wave. Like the periodic solutions of the harmonic oscillator, eq. (4.5) is periodic in time, with a frequency of oscillation v and period $T = 1/v$; just as in the general harmonic oscillator solution (see Box 2.2), ϕ represents a phase shift that can be set to zero by an appropriate choice of time coordinate t. (Note that by changing the phase a sine function can be converted to a cosine or vice versa. For this reason, both cosine and sine functions are often generically referred to as *sine waves*.) It is often convenient to write a sine wave (4.5) in the alternative form

$$f(x, t) = A \sin(kx - \omega t - \phi) , \tag{4.6}$$

where $k \equiv 2\pi/\lambda$ is the **wave number** and $\omega = 2\pi v$ is the angular frequency.

The sine wave (4.6) solves the wave equation (4.2) provided that the speed of propagation v is related to λ and v or k and ω by

$$v = \lambda v = \frac{\omega}{k} . \tag{4.7}$$

In some media, the wave equation is modified so that waves with different wavelengths propagate at different velocities. In such situations, waves change shape as they propagate; this phenomenon is known as *dispersion*, and is discussed further in §4.5 below and in §31.2. In general, the velocity $v = \omega/k$ at which a wave of a given wavelength propagates is known as its **phase velocity**.

Table 4.1 Parameters describing sine waves. A sine wave $f(\boldsymbol{x},t) = A\sin(\boldsymbol{k}\cdot\boldsymbol{x}-\omega t-\phi)$ that satisfies a wave equation such as (4.3) can be characterized by these parameters.

Parameter	Symbol
Amplitude	A
Wave vector	$\boldsymbol{k}$
Wave number	$k = |\boldsymbol{k}|$
Direction of propagation	$\hat{\boldsymbol{k}} = \boldsymbol{k}/k$
Wavelength	$\lambda = 2\pi/k$
Angular frequency	ω
Frequency	$\nu = \omega/2\pi$
Speed of propagation	$v = \omega/k = \nu\lambda$
Phase shift	ϕ

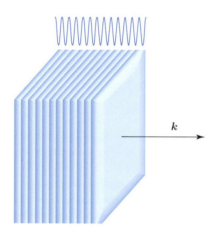

Figure 4.2 A section of a plane wave propagating in the $\boldsymbol{k}$-direction.

The simplest waveforms in three dimensions, known as **plane waves**, vary sinusoidally in the direction of propagation, but have constant values in the two perpendicular directions,

$$f(\boldsymbol{x},t) = A\sin(\boldsymbol{k}\cdot\boldsymbol{x} - \omega t - \phi) = A\sin(kx_{||} - \omega t - \phi).$$
$$(4.8)$$

Here $\boldsymbol{k}$ is the **wave vector**, with the unit vector $\hat{\boldsymbol{k}}$ specifying the direction of propagation and $k = |\boldsymbol{k}|$ the wave number as before. A is the **amplitude** of the wave, and $x_{||}$ is the coordinate along the direction of propagation. Figure 4.2 shows a sketch of a plane wave. The many parameters that can be used to describe a plane wave of definite frequency are summarized in Table 4.1.

A typical physical wave – for example, a water wave produced by throwing a rock into calm waters – is not a

---

Waves

A *wave* is an excitation of some medium that transports energy without transporting matter. The simplest form of wave is an oscillation that is periodic in both space and time. Such a *plane wave*

$$f(\boldsymbol{x},t) = A\sin(\boldsymbol{k}\cdot\boldsymbol{x} - \omega t - \phi)$$

can be characterized by the parameters listed in Table 4.1. The simplest wave equation in one dimension has the form

$$\frac{\partial^2 f}{\partial t^2} = v^2 \frac{\partial^2 f}{\partial x^2}.$$

An equation of this form governs propagation of electromagnetic waves as well as many other types of physical waves.

Any wave satisfying a *linear* wave equation can be written as a linear superposition of plane waves, though the relationship between ω and k depends on the specific wave equation.

The energy density in a wave varies quadratically with its amplitude A.

---

simple sinusoid. Even the most complicated waveforms, however, can be considered as *superpositions* of sine waves with a distribution of frequencies and phases. Thus, understanding the behavior of the periodic sine wave solutions is often all that is necessary to analyze the physics of any specific wave phenomenon. The most familiar example of this may be sound waves. Sound waves are compression waves in air. The human ear hears sine waves of different frequencies as different **pitches**, and a general sound wave such as that produced by a musical instrument can be described as a superposition of different pitches, each carrying a different amount of energy. The decomposition of a general waveform into sine waves is described further in the following section.

4.2 Waves on a String

A simple example of a physical system that exhibits wave behavior is a stretched string under tension. Imagine again an infinite string with mass density (per unit length) ρ that is in equilibrium when it is stretched straight along the x-axis. (We ignore effects of gravity.) If the displacement of the string in the y-direction is described by a function $y(x, t)$, then Newton's law of motion leads to the wave equation (4.2) for $y(x, t)$ (Problem 4.3)

$$\rho \frac{\partial^2}{\partial t^2} y(x, t) = \tau \frac{\partial^2}{\partial x^2} y(x, t) , \qquad (4.9)$$

where τ is the tension on the string. The solutions of this equation consist of waveforms $y = f(x \pm vt)$ propagating to the right ($-$) or left ($+$) with speed $v = \sqrt{\tau/\rho}$.

The moving string carries energy. The mass ρdx in an interval dx moves with velocity $\dot{y}(x, t) = \partial y/\partial t$, and therefore carries kinetic energy $dE_{\text{kin}} = \frac{1}{2}\rho\dot{y}^2 dx$. Similarly, the string stores potential energy $dV = \frac{1}{2}\tau y'^2 dx$ ($y' = \partial y/\partial x$) originating in the work performed against the tension when it deforms away from a straight line (Problem 4.3). These combine to give a total energy per unit length of

$$u(x, t) = \frac{d}{dx}(E_{\text{kin}} + V) = \frac{1}{2}\rho\dot{y}(x, t)^2 + \frac{1}{2}\tau y'(x, t)^2 . \qquad (4.10)$$

Note that the energy density of a wave on a string depends *quadratically* on the amplitude of the wave. This is a general feature of wave phenomena, from electromagnetic radiation emitted by hot objects to wind-driven waves on the ocean's surface.

4.2.1 Finite String: Mode Analysis

The potentially complicated motions of a vibrating string are simplified enormously when we focus on waveforms whose shape remains fixed as their amplitude changes. For such waveforms, the amplitude oscillates sinusoidally with a definite frequency. These motions are known as the **normal modes** of oscillation. Any solution to the wave equation (4.9) can be written as a linear superposition of normal modes. To see how this works explicitly, we consider a string of finite extent, such as a violin string, with the string held fixed at the endpoints, so that $y(x, t)$ is defined for x from $x = 0$ to $x = L$, and satisfies the boundary conditions that $y(0, t) = y(L, t) = 0$ for all times t.

For the finite string, the normal modes take the form

$$y_n(x, t) = y_n(t) \sin\left(\frac{n\pi}{L}x\right) , \qquad (4.11)$$

where n is any positive integer and the corresponding wave number $k_n = n\pi/L$ is fixed by the boundary condition that $y(L, t) = 0$. Substituting this expression into the wave equation (4.9), we obtain an *ordinary* differential equation for $y_n(t)$,

$$\ddot{y}_n(t) = -k_n^2 \frac{\tau}{\rho} y_n(t) . \qquad (4.12)$$

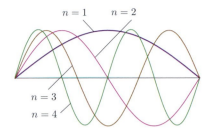

Figure 4.3 The first few modes ($n = 1, 2, 3, 4$) on a violin string of length L. The amplitude of the vertical displacement is exaggerated for clarity; the linear wave equation holds for oscillations that are very small compared to the length of the string.

This is the equation of a harmonic oscillator with frequency $\omega_n = vk_n = n(\pi/L)\sqrt{\tau/\rho}$. Thus, each normal mode corresponds to an *independent mode of oscillation* with its own frequency. As described in Box 2.2, the solution to eq. (4.12) is of the form $y_n(t) = c_n \cos(\omega_n t - \phi_n)$, where c_n is the amplitude and ϕ_n is the phase of the oscillator. The first four normal modes on a violin string are shown in Figure 4.3.

Any motion of the string can be written as a linear combination of the normal modes. For the finite string, this corresponds to the *Fourier series expansion* of the displacement $y(x, t)$ in terms of sine wave modes of the string (Appendix B.3.4), which allows us to write a general solution in the form

$$y(x, t) = \sum_{n=1}^{\infty} c_n \sqrt{\frac{2}{L}} \cos(\omega_n t - \phi_n) \sin(k_n x) . \qquad (4.13)$$

Note that the normalization factor $\sqrt{2/L}$ for the modes is chosen to simplify the expression for physical quantities such as the energy (4.16).

There are several ways to determine the amplitudes c_n and phases ϕ_n of the different modes given the physical configuration of the string. If, for example, the shape of the string and its initial velocity are given at some time, say $t = 0$, then eq. (4.13) can be solved for c_n and ϕ_n in terms of $y(x, 0)$ and $\dot{y}(x, 0)$ using eq. (B.55). An example of this procedure can be found in Example 4.1.

4.2.2 The Infinite String and Fourier Transforms

In the case of an infinite string, we can perform a similar mode analysis. While it is possible to do this in terms of sine and cosine modes, it is in many ways simpler and more transparent to use *complex numbers* (Appendix B.2). A complex mode e^{ikx} is a linear combination of cosine and

---

Example 4.1   Modes and Tones on a Violin String

Consider a violin string that is initially at rest but stretched into a symmetric triangular profile,

$$y(x, 0) = f(x) = \begin{cases} ax & \text{for } 0 \leq x \leq L/2, \\ a(L - x) & \text{for } L/2 \leq x \leq L, \end{cases}$$

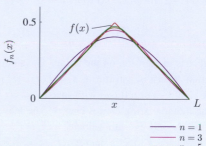

as shown in in red in the figure at right, where the amplitude of the waveform is enlarged for better viewing. According to the decomposition into normal modes (4.13), the displacement and velocity profiles at $t = 0$ are $y(x, 0) = \sum_{n=1}^{\infty} c_n \cos \phi_n \sqrt{2/L} \sin(n\pi x/L)$ and $\dot{y}(x, 0) = -\sum_{n=1}^{\infty} \omega_n c_n \sqrt{2/L} \sin \phi_n \sin(n\pi x/L)$. Since $\dot{y}(x, 0) = 0$, but $y(x, 0)$ is not everywhere vanishing, we deduce that $\phi_n = 0$ for all n. Then $y(x, 0) = \sum_{n=1}^{\infty} c_n \sqrt{2/L} \sin(n\pi x/L)$, so the problem of describing the string's motion reduces to finding the Fourier coefficients c_n that describe the string's initial, triangular shape. The c_n are determined by eq. (B.55),

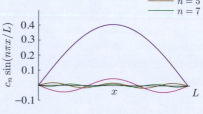

$$c_n = \sqrt{\frac{2}{L}} \int_0^L dx \, y(x, 0) \sin(n\pi x/L)$$

$$= a\sqrt{\frac{2}{L}} \left(\int_0^{L/2} dx \, x \sin(n\pi x/L) + \int_{L/2}^L dx \, (1 - x) \sin(n\pi x/L) \right)$$

$$= \frac{\sqrt{2L}a}{n^2\pi^2} \sin(n\pi/2) \, .$$

The c_n vanish for even n. The contributions from $n = 1$, 3, 5, and 7 are shown in the lower figure above and the successive approximations to $y(x, 0) = f(x)$ obtained by summing these terms is shown in the upper figure.

The vibrations of a violin string are perceived as musical tones. The lowest mode of the string, with wavelength $2L$ and frequency $\omega_1 = v\pi/L$ is known as the *fundamental tone*. The second mode, with frequency $\omega_2 = 2\omega_1$, known as the *first overtone*, is one *octave* higher than the fundamental, and the third mode or *second overtone* with frequency $\omega_3 = (3/2)\omega_2$ is the *perfect-fifth* above the second mode. The Fourier coefficients determine how much energy is fed into these tones. The energy in the nth normal mode is proportional to $n^2c_n^2$, so in this case the first overtone is not excited at all ($c_n = 0$ for even n), and the energy in the major-fifth ($n = 3$) is one-ninth the energy in the fundamental ($n = 1$).

---

sine modes. Using eq. (B.35), we can write

$$e^{ikx} = \cos(kx) + i\sin(kx) \, . \tag{4.14}$$

The wave equation is satisfied when the mode with wave number k oscillates with frequency $\omega(k) = v|k|$, so the general wave solution has a mode expansion

$$y(x, t) = \int_{-\infty}^{\infty} dk \left(Y_+(k)e^{ik(x-vt)} + Y_-(k)e^{ik(x+vt)} \right), \tag{4.15}$$

where $Y_\pm(k)$ is the *Fourier transform* (Appendix B.3.5) of the right-/left-moving parts of the wave form $y(x, t) = y_+(x - vt) + y_-(x + vt)$[1].

In order that $Y(x, t)$ be a real function, the Fourier transform must obey $\overline{Y}_\pm(k) = Y_\pm(-k)$, where $\overline{Y}$ denotes the *complex conjugate of Y* (Appendix B.2). Figure 4.4 shows the waveform of Figure 4.1 and its Fourier transform. *Fourier analysis*, which enables us to decompose a wave into a set of periodic modes with characteristic frequencies, plays a central role in analyzing many energy-related phenomena, from solar and wave energy to electricity transmission, geophysics, and Earth's climate.

Energy in the String   The energy of a finite oscillating string can be expressed in terms of the Fourier coefficients

---

[1] Note that the decomposition of $y(x, t)$ into right- and left-moving parts requires knowledge of both the wave form $y(x, t)$ and the time derivative $\dot{y}(x, 0)$ since the wave equation is a second-order differential equation.

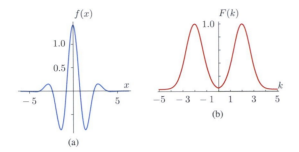

Figure 4.4 (a) A waveform $f(x)$, and (b) its Fourier transform $F(k)$ defined in eq. (B.60). Note that both $f(x)$ and $F(k)$ are real and satisfy $f(x) = f(-x)$, $F(k) = F(-k)$.

4.3 Electromagnetic Waves

Soon after formulating the complete set of equations for electromagnetism, Maxwell showed that light is a wave of undulating electric and magnetic fields – an electromagnetic wave. Once a light wave has been generated, it propagates independently of its source, carrying energy and momentum across the vacuum at a speed of $c = 299\,792\,458$ m/s. We often tend to think of electromagnetic waves (such as microwaves, TV, radio, and wifi signals) as carriers of information rather than energy. More fundamentally, EM waves – ranging from very low-energy radio waves, through infrared, visible, ultraviolet, and on to X-rays and ultra-high-energy gamma rays – carry energy over distances ranging from subatomic to astronomical. The role of electromagnetic waves as carriers of energy features prominently in many later parts of this book, from chemical energy and solar energy to Earth's climate.

c_n by substituting eq. (4.13) for $y(x, t)$ into eq. (4.10) for the energy density and integrating over x,

$$E = \int_0^L dx\, u(x, t) = \frac{\rho}{2} \sum_{n=1}^{\infty} c_n^2 \omega_n^2. \qquad (4.16)$$

In the case of an infinite string,

$$E = \int_{-\infty}^{\infty} dx\, u(x, t) = 2\pi\rho \int_0^{\infty} dk\, |Y(k)|^2 \omega^2 \qquad (4.17)$$

$$\equiv \int_0^{\infty} d\omega \frac{dE}{d\omega}, \text{ with } \frac{dE}{d\omega} = 2\pi \frac{\rho}{v} |Y(k)|^2 \omega^2. \qquad (4.18)$$

These results express the energy in a wave as a sum over contributions of different frequencies. For light waves, frequency corresponds to spectral color; thus, $dE/d\omega$ is called the **energy spectrum** of the wave. When we discuss energy transport by waves, for example in solar energy, the spectrum plays a central role. Illustrations of energy spectra can be found in §23 for the Sun's energy seen on Earth (Figure 23.9) and in §31 for the energy in ocean waves (Figure 31.13).

The normal modes on a string, $\cos(\omega t) \sin(kx)$, are examples of **standing wave** solutions to the wave equation. Unlike traveling waves, such as $f(x - vt)$, the shape of a standing wave stays unchanged, with only the overall amplitude changing in time. As explored in Problem 4.4, traveling waves transport energy, but standing waves do not. Using a trigonometric identity such standing waves can be written as a linear superposition of left-moving and right-moving waves

$$\cos(\omega t)\sin(kx) = \frac{1}{2}\left(\sin(k(x - vt)) + \sin(k(x + vt))\right). \qquad (4.19)$$

This kind of standing wave solution plays an important role in quantum physics (§7).

Electromagnetic waves are a bit more complicated than those described in the previous section because E and B are vectors, and must be related in a particular way in an electromagnetic wave. In empty space where the charge and current densities ρ and j vanish, *Maxwell's equations* (§3.6) give rise to wave equations for E and B. This can be seen for the electric field by computing the second time derivative

$$\frac{\partial^2 E}{\partial t^2} = \frac{1}{\mu_0 \epsilon_0} \nabla \times \frac{\partial B}{\partial t} = -\frac{1}{\mu_0 \epsilon_0} \nabla \times (\nabla \times E)$$

$$= \frac{1}{\mu_0 \epsilon_0} \nabla^2 E, \qquad (4.20)$$

where we have used the Ampere–Maxwell relation (3.74), Faraday's law (3.66), the vector derivative identity (B.22), and Gauss's law (3.12) in the successive steps of the computation. This yields a wave equation for E where the wave velocity is the **speed of light** $c \equiv 1/\sqrt{\mu_0 \epsilon_0}$. A similar equation for B can be derived starting with the time derivative of Faraday's law.

A plane wave solution to eq. (4.20) is of the form $E(x, t) = E_0 \sin(k \cdot x - \omega t - \phi)$, where $\omega = ck$. The condition $\nabla \cdot E = 0$ then requires that $k \cdot E = 0$, and eq. (3.66) gives $B(x, t) = \frac{1}{c}\hat{k} \times E(x, t)$, so that the magnetic field B, the electric field E, and the direction of propagation of the wave k are mutually perpendicular. The relations between the direction of propagation and the directions of E and B are illustrated in Figure 4.5. For any direction of propagation k, there are two independent choices for the directions of E and B. These correspond to two possible **polarizations** of light. For example, if we choose the direction of propagation $\hat{k}$ to be $\hat{x}$, then the two different polarizations

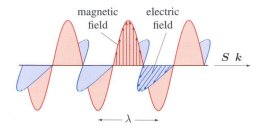

magnetic field   electric field

$S\ k$

λ

Figure 4.5 A snapshot of a propagating electromagnetic wave.

propagating along the x axis are described by (choosing the phase shift $\phi = 0$)

$$E^{(1)} = E_0 \sin(kx - \omega t)\hat{y}, \quad B^{(1)} = \frac{E_0}{c}\sin(kx - \omega t)\hat{z},$$

$$E^{(2)} = E_0 \sin(kx - \omega t)\hat{z}, \quad B^{(2)} = -\frac{E_0}{c}\sin(kx - \omega t)\hat{y}.$$
$$(4.21)$$

An electromagnetic wave is typically a superposition of components with different frequencies and polarizations. Using the language of Fourier transforms (as in the previous section and Appendix B.3.5) we can describe a general electromagnetic plane wave as an integral over wave number (or frequency) and a sum over the two independent polarizations. For a plane wave propagating in the $\hat{k}$-direction, the contribution of a specific frequency ω and polarization is determined by the appropriate Fourier component in the Fourier transform of the waveform. This type of **frequency decomposition** of an electromagnetic wave breaks down any given beam of electromagnetic radiation into a spectral distribution containing components such as **visible light** (wavelengths λ in the range 400–750 nm), **infrared radiation** ($\lambda > 750$ nm), and **ultraviolet radiation** ($\lambda < 400$ nm).

4.4 Energy and Momentum in Electric and Magnetic Fields

In the previous chapter, we expressed the energy stored in a capacitor as the integral of the square of the electric field (eq. (3.20)) and the energy stored in a solenoid as the integral of the magnetic field strength squared (eq. (3.71)). Although we derived these results under restricted circumstances – with stationary charges and constant currents – they in fact hold quite generally. The general expression for the energy density in an arbitrary configuration of electric and magnetic fields is

$$u(\mathbf{x}, t) = \frac{dU}{dV} = \frac{1}{2}\epsilon_0 |\mathbf{E}(\mathbf{x}, t)|^2 + \frac{1}{2\mu_0}|\mathbf{B}(\mathbf{x}, t)|^2.$$
$$(4.22)$$

This provides a universal and completely local characterization of how energy is stored in electromagnetic fields.

To understand the flow of energy in electromagnetic fields, we can derive a local equation expressing the conservation of energy analogous to the formulation of charge conservation in eq. (3.33). First, differentiate $u(\mathbf{x}, t)$ with respect to time and then use Maxwell's equations (Problem 4.11) to replace the time derivatives of $\mathbf{E}$ and $\mathbf{B}$, leading to the relation

$$\frac{\partial u(\mathbf{x}, t)}{\partial t} + \nabla \cdot \mathbf{S}(\mathbf{x}, t) = -\mathbf{j}(\mathbf{x}, t) \cdot \mathbf{E}(\mathbf{x}, t), \quad (4.23)$$

where $\mathbf{j}$ is the current density and $\mathbf{S}$ is the **Poynting vector**

$$\mathbf{S}(\mathbf{x}, t) = \frac{1}{\mu_0}\mathbf{E}(\mathbf{x}, t) \times \mathbf{B}(\mathbf{x}, t). \quad (4.24)$$

Equation (4.23) describes conservation of energy in electromagnetic fields. The Poynting vector $\mathbf{S}$ corresponds to the flux of energy in the electromagnetic fields, and is measured in energy per unit time per unit area. The term on the right-hand side of eq. (4.23) gives the rate at which electric fields do work on charges, describing a transfer of energy from the electromagnetic fields to mechanical energy through acceleration of charges, or to thermal energy through Joule heating. For example, if the current density $\mathbf{j}$ is flowing through a resistor then the power loss integrated over the volume of the resistor is $\int d^3x\, \mathbf{j} \cdot \mathbf{E} = IV$, in agreement with the formula (3.22) (Problem 4.15). Note again that magnetic fields, which exert forces at right angles to the velocity of charged particles, do no work.

The Poynting vector (4.24) describes transport of energy by electromagnetic waves. For an electromagnetic plane wave of the form (4.21), the contributions to the energy density (4.22) from the electric and magnetic fields are equal, and the corresponding Poynting vector is

$$\mathbf{S}(\mathbf{x}, t) = u(\mathbf{x}, t)c\hat{\mathbf{x}}, \quad (4.25)$$

in accord with the expectation that all energy in the plane wave is moving along the x-axis at the speed of light. Averaging over time, the power incident on a perpendicular surface is $\langle \hat{\mathbf{n}} \cdot \mathbf{S} \rangle = \epsilon_0 c E_0^2/2$.

Electromagnetic waves possess momentum as well as energy. The momentum density (momentum per unit volume) can be shown to be proportional to the Poynting vector,

$$\frac{d\mathbf{p}}{dV} = \frac{1}{c^2}\mathbf{S} = \epsilon_0 \mathbf{E} \times \mathbf{B}. \quad (4.26)$$

When light is absorbed by matter, not only energy, but also momentum is absorbed. Ordinary mechanics tells us that momentum absorbed per unit time is a force, so when light – or indeed any form of radiation – is absorbed, it exerts a force. This is usually described in terms of the force per unit area known as **radiation pressure**.

While the formula (4.26) can be derived from classical electromagnetism using more sophisticated methods for describing conserved quantities like momentum along the line of ideas developed in §21, perhaps the simplest way to understand the relationship between energy and momentum in electromagnetic waves is in terms of the quantum nature of light. For a fixed time interval and a fixed area, the absorbed momentum $|\boldsymbol{p}|$ (from eq. (4.26)) is equal to the absorbed energy E (from eq. (4.25)) divided by the speed of light, so $E = |\boldsymbol{p}|c$. If we compare this with the relation between energy and momentum from Einstein's special theory of relativity (§21), $E = \sqrt{|\boldsymbol{p}|^2 c^2 + m^2 c^4}$, then we see that the energy–momentum relation for light is what we would expect for particles with $m = 0$. Indeed, light behaves not only like a wave, but also in some ways like a stream of massless particles! This relation, which underlies the connection between eq. (4.22) and eq. (4.26), surfaces again in §7.

4.5 General Features of Waves and Wave Equations

The simple linear wave equation (4.3) is capable of generating a rich range of physical phenomena. When complexities such as dispersion, dissipation, and nonlinearities are

$$u(\boldsymbol{x}, t) = \frac{1}{2}\epsilon_0 |\boldsymbol{E}(\boldsymbol{x}, t)|^2 + \frac{1}{2\mu_0}|\boldsymbol{B}(\boldsymbol{x}, t)|^2$$

$$\boldsymbol{S}(\boldsymbol{x}, t) = \frac{1}{\mu_0}\boldsymbol{E}(\boldsymbol{x}, t) \times \boldsymbol{B}(\boldsymbol{x}, t)$$

respectively, and energy conservation states that

$$\frac{\partial u(\boldsymbol{x}, t)}{\partial t} + \boldsymbol{\nabla} \cdot \boldsymbol{S}(\boldsymbol{x}, t) = -\boldsymbol{j}(\boldsymbol{x}, t) \cdot \boldsymbol{E}(\boldsymbol{x}, t).$$

Electromagnetic fields and radiation also carry momentum

$$\frac{d\boldsymbol{p}}{dV} = \frac{1}{c^2}\boldsymbol{S}.$$

Figure 4.6 Complex surface waves on water, made visible by reflected light. (Credit: Alex Bruda)

Electromagnetic Waves

Maxwell's equations have traveling wave solutions that propagate at the speed $c = \sqrt{1/\mu_0\epsilon_0}$. There are two independent plane wave solutions for light traveling in a given direction, corresponding to different *polarizations*. For a wave propagating along the $\hat{\boldsymbol{x}}$-axis, one polarization is

$$\boldsymbol{E}^{(1)}(\boldsymbol{x}, t) = E_0 \sin(kx - \omega t)\hat{\boldsymbol{y}},$$

$$\boldsymbol{B}^{(1)}(\boldsymbol{x}, t) = B_0 \sin(kx - \omega t)\hat{\boldsymbol{z}},$$

with $B_0 = E_0/c$. The second polarization has $\boldsymbol{E} \sim \hat{\boldsymbol{z}}, \boldsymbol{B} \sim \hat{\boldsymbol{y}}$.

The density and flux of energy in electromagnetic fields are given by

---

included, the diversity of wave phenomena becomes even more extravagant. The fascinating complexity of wave behavior may be familiar to anyone who has watched small ripples in calm water propagate, reflect, and diffract around obstructions (Figure 4.6). We conclude this chapter with a brief description of a number of general features of waves and wave equations that are relevant in the context of various physical phenomena described later in the book. After summarizing some features of the simplest wave equation (4.3), we add in the effects of boundaries and media, and then turn to dispersion, dissipation, and nonlinearities, which can arise in other types of wave equations.

4.5.1 Superposition and Interference

Fourier's theorem (B.54), and its generalizations to two and three dimensions and to electromagnetic fields, indicate that any waveform, no matter how complex, can be

Example 4.2   A Solar-powered Satellite

The relevance of some of the ideas of this chapter to energy systems is illustrated by a solar-powered satellite in near-Earth orbit. Solar radiation in the region of Earth's orbit has an energy flux of roughly $|S| \cong 1366 \, \text{W/m}^2$ (§23). If the satellite has a solar panel of area $10 \, \text{m}^2$, then the incident power of solar radiation is roughly $13.7 \, \text{kW}$. Radiation pressure (eq. (4.26)) exerts a force on the satellite of $|S|A/c \cong 5 \times 10^{-5} \, \text{N}$. Though small, such forces must be considered when computing detailed satellite trajectories. As we discuss in more detail in §22, solar radiation has a frequency distribution that is roughly described by a *blackbody* radiation spectrum for an object radiating at 5780 K. This power spectrum, depicted in Figure 22.3, is peaked in the region of visible light. The materials used in the satellite's solar panel are chosen to optimize the fraction of incident solar energy that can be converted to electromagnetic power given the solar spectrum. The satellite shown in the figure is the Japanese GOSAT (Greenhouse Gases Observing Satellite), with high-efficiency triple junction photovoltaic solar panels (§25.9) that can generate $3.8 \, \text{kW}$ of power. (Image credit: GOSAT (Greenhouse Gases Observing Satellite) by JAXA).

Figure 4.7 Circular waves created by a drop falling on the water's surface. (Credit: David Arky/Getty Images)

regarded as a superposition of plane waves. This is often not, however, the most economical or convenient way of addressing a problem. For example, the waves produced by a localized excitation, as when a drop falls into still water (Figure 4.7), spread outward in a circular pattern. The waveforms that describe circular waves (in two dimensions) or spherical waves (in three dimensions) emanating from a point source can be simply expressed as superpositions of special functions, known as *Bessel functions* and *spherical harmonics*, that play a role analogous to that of sines and cosines in plane waves. The study of these functions and the analogous *Fourier–Bessel* expansions that make use of them go beyond the scope of this book. Expansions of this type, however, are used for a variety of energy-related applications, such as for global models of Earth's tides (§31.3) and climate (§34).

When multiple wave forms combine together, the superposition of their fluctuations produces a phenomenon known as **interference**. The behavior of the wave pattern at any point depends on the relative phases of the wave forms that are combined. If the waves are **in phase** – all reaching their maximum at the same time at the point of interest – then the amplitudes add (**constructive interference**), while if the waves are **out of phase** the waves will partially or completely cancel (**destructive interference**). Consider, for example, two waves with the same frequency and wave number moving along the x- and y-axes respectively. The total wave form is

$$f(x, y, t) = A \sin(kx - \omega t) + B \sin(ky - \omega t). \quad (4.27)$$

At the point $x = y = 0$, the waves constructively interfere

$$f(0, 0, t) = -(A + B) \sin(\omega t), \quad (4.28)$$

so the oscillations at that point have the combined amplitude of the two waves, while at the point $x = 0$, $y = \pi/k$ the waves destructively interfere

$$f(0, \pi/k, t) = A \sin(-\omega t) + B \sin(\pi - \omega t)$$
$$= (B - A) \sin(\omega t). \quad (4.29)$$

In the latter case, if the amplitudes are equal, $A = B$, then there is no oscillation whatever in f at the point $(0, \pi/k)$. Since, as discussed above, the energy density in a wave is typically proportional to its amplitude squared, the energy density at the origin in this example is four times as great (when $A = B$) as the energy density in either wave alone.

Figure 4.8 An interference pattern formed by two circular surface waves produced by local excitations of a medium. (Credit: David Hoult, Open Door).

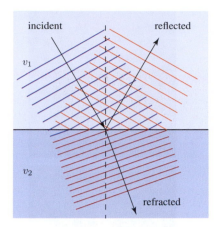

Figure 4.9 As waves pass through a boundary between media in which the waves propagate with different velocities, the direction of propagation and the wavelength of the waves change. At such an interface, some energy is propagated forward in *refracted* waves, while some energy is reflected from the boundary. In the figure, for clarity, the wavefronts of the incident, reflected, and refracted waves are shown in different colors.

Figure 4.8 illustrates a pattern of constructive and destructive interference. Superposition and interference play a particularly important role in quantum physics (§7).

4.5.2 Reflection, Refraction, and Diffraction

A variety of familiar wave phenomena occur when waves described by eq. (4.3) propagate through an inhomogeneous medium. Inhomogeneities may originate in barriers that obstruct propagation, boundaries between media with different wave propagation speeds, or media in which the wave speed changes smoothly as a function of position.

Reflection occurs when a wave encounters a barrier that prevents the wave from passing but does not absorb any energy from the wave. **Refraction** refers to a change in the direction of propagation of a wave front, which can occur when the speed of propagation changes either continuously or abruptly within a medium. For example, for ocean waves in shallow water the phase velocity depends upon the depth of the water, so that the direction of wave propagation changes depending upon undersea topography. This leads to refraction as waves from the deep ocean enter shallow waters (see §31.2).

Both reflection and refraction occur when a wave passes a boundary across which the speed of propagation changes discontinuously. This occurs for light, for example, when passing between media with different values of the permittivity ϵ and permeability μ. The speed of light in a general medium is $v = 1/\sqrt{\epsilon\mu}$. Since $\mu\epsilon$ need not equal $\mu_0\epsilon_0$, the speed of light in a medium can differ from c. For example, the speed of light in water is roughly $v \cong c/1.33$.

When light is incident at an angle on a boundary separating two media, such as a boundary between air and glass or water, the speed of propagation of the waves is different on the two sides of the boundary (see Figure 4.9). Since the electric field on the boundary oscillates with a fixed frequency, the frequency ω is the same on both sides of the boundary. From eq. (4.7), it follows that the wave number k changes as the inverse of the speed of propagation of the light wave as a wave passes through the boundary, leading to refraction (Problem 4.17).

For related reasons, the energy in a wave propagated across a sharp boundary between two media with different phase velocities will in general be reduced from the energy in the incident wave. To conserve energy, some of the incident energy in the wave must be reflected (or absorbed). Light incident from air onto water, for example, is reflected to an extent that depends upon the angle of incidence. Similar reflection occurs for other kinds of waves. In particular, the reflection of seismic waves when incident on an interface between materials, such as when passing from rock into a coal seam, plays an important role in exploration for underground geothermal and fossil fuel resources, as discussed further in §32 and §33.

When a propagating wave encounters obstructions or passes through one or more narrow openings, a complicated interference pattern can be produced when the scale of the obstructing object(s) and/or opening(s) is

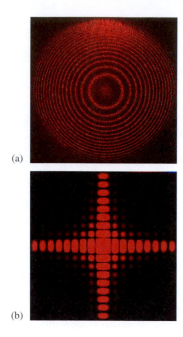

(a)

(b)

Figure 4.10 Diffraction of laser light. (a) From a circular (actual data). (Credit: yuriz/iStock/Getty Images) (b) A square (numerical simulation) aperture. (Source: V81/English Wikipedia)

commensurate with the wavelength of the propagating wave. This phenomenon is known as **diffraction**, and the resulting interference pattern is a **diffraction pattern**. A diffraction pattern encodes information about the shape and arrangement of the obstructing objects and can be used to reconstruct their properties. The diffraction patterns formed by light passing through a circular hole and a square hole are shown in Figure 4.10. One of the remarkable features of quantum physics (§7) is that measurements of individual particles can demonstrate a complex diffraction pattern, arising from the fact that in quantum physics the trajectories of particles are described by waves that can exhibit interference.

4.5.3 Dispersion and Dissipation

Dispersion and *dissipation* are effects that cause a wave form to change shape and lose energy in certain media. Both of these effects can occur in systems that obey linear wave equations, so that the effects can be analyzed for individual modes, which can then be combined through superposition.

In the simplest wave equations such as (4.2) and (4.20), the relation between angular frequency ω and wave number k is linear, $\omega = vk$, and waves of all wavelengths propagate at the same speed v. Some physical systems are

governed by wave equations for which ω has a more complicated functional dependence on k. This gives rise to **dispersion**, where waves of different wavelengths propagate at different velocities

$$v(k) = \frac{\omega(k)}{k}. \qquad (4.30)$$

The relation between k and ω characterized by $\omega(k)$ is known as a **dispersion relation**. The dispersion relation of deep water gravity waves, for example, is $\omega(k) = \sqrt{gk}$ (§31.2). Dispersion causes the shape of a wave form built from a superposition of sine wave modes to change over time as it propagates forward. Dispersion also leads to the phenomenon that the speed of propagation of a localized waveform as a whole, known as the *group velocity*, differs from the phase velocity of the component waves. For example, for ocean surface waves, depicted in Figure 31.12, the wave packet as a whole moves slower than the individual wave crests, making it seem as though wave crests move forward though the packet. Dispersion is described in more detail in §31.2, and the relation to group velocity is described in Box 31.2.

Dissipation refers to a gradual transfer of energy from the macroscopic motion of a wave to random molecular motion as the wave propagates. As the wave loses energy, its amplitude decreases. In general, the rate of energy loss is proportional to the energy in the wave. As a result, the usual sinusoidal oscillation of a wave of definite frequency ω is modulated by a falling exponential function e^{-at}. This can be described mathematically by including a negative imaginary component in the frequency

$$\omega \to \omega - ia \qquad (4.31)$$

when writing the general solution in the form (4.15). The exponential decay of a beam of light passing through an absorptive medium such as water or the atmosphere is a familiar example that is discussed in more detail in §23 and §34. An example of a waveform suffering both dispersion and dissipation as it propagates is shown in Figure 4.11.

4.5.4 Nonlinearity

The wave equations of primary interest in this book are linear differential equations. In some cases, however, nonlinearities play an important part in the physics of waves. Nonlinearities arise even in simple mechanical systems. The equation of motion for a pendulum of length l, for example, is $l\ddot{\theta} = -g\sin\theta$. If θ is small enough, only the first term in the power series expansion $\sin\theta = \theta - \theta^3/3! + \theta^5/5! + \cdots$ needs to be retained, and the pendulum reduces to a simple harmonic oscillator. If the second term cannot be ignored, however, then the pendulum equation

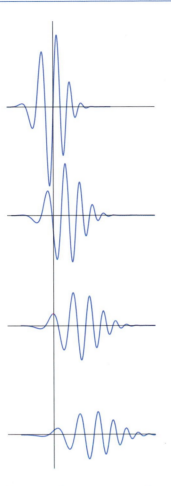

Figure 4.11 Snapshots of a wave packet at four equally spaced intervals in time. Dissipation causes the amplitude of the wave to decrease with time. Dispersion causes the shape of the wave packet to change with time.

$l\ddot{\theta} \approx -g\theta + g\theta^3/6$ is no longer linear in θ and is more difficult to solve. Similar effects enter wave equations, often when the displacement from equilibrium is large compared to the wavelength of the oscillation. In particular, for ocean surface waves nonlinearity plays an important role both in the development of waves and in the breaking of waves on the shore (§31.2). When the wave equation is not linear, plane waves of different frequencies do not propagate independently, and energy is transferred between modes.

While nonlinearities play some role in many wave systems, two important exceptions are electromagnetic waves and quantum mechanical systems, which are governed by *linear* wave equations for all purposes relevant for this book.

Discussion/Investigation Questions

4.1 Explain why the strings that play lower frequency notes on a guitar or violin are generally thicker and less tightly strung than the strings playing higher notes.

4.2 Suppose an electromagnetic plane wave propagates in a direction perpendicular to a long straight wire. Can you see why a current develops in the wire with the same frequency as the wave? Qualitatively, how does the current depend on the orientation of the wire and the polarization of the light? This is one of the basic ingredients in both transmission and reception of broadcast radio, television, etc.

4.3 Polarized sunglasses are designed to transmit only one polarization of light – the polarization with electric fields vertical – since glare arising from reflections on horizontal snow and water surfaces is preferentially polarized with horizontal electric fields. LCD displays such as the view screens on many digital cameras (and laptops, etc.) also emit polarized light. Examine the image on a digital camera screen as you rotate it through 90° while wearing polarizing sunglasses. Explain what you observe.

Problems

4.1 Sound waves travel in air at roughly 340 m/s. The human ear can hear frequencies ranging from 20 Hz to 20 000 Hz. Determine the wavelengths of the corresponding sine wave modes and compare to human-scale physical systems.

4.2 A violin A-string of length $L = 0.33$ m with total mass 0.23 g has a fundamental frequency (for the lowest mode) of 440 Hz. Compute the tension on the string. If the string vibrates at the fundamental frequency with maximum amplitude 2 mm, what is the energy of the vibrational motion?

4.3 [T] Derive the equation of motion for the string (4.9) from a microscopic model. Assume a simple model of a string as a set of masses Δm spaced evenly on the x axis at regular intervals of Δx, connected by springs of spring constant $k = \tau/\Delta x$. Compute the leading term in the force on each mass and take the limit $\Delta x \to 0$ to get the string wave equation, where $\rho = \Delta m/\Delta x$. In the same limit show that the energy density of the string is given by eq. (4.10).

4.4 [T] Show that the energy density on a string $u(x, t)$, defined in eq. (4.10), obeys the conservation law $\partial u/\partial t + \partial S/\partial x = 0$, where $S(x, t) = -\tau \dot{y} y'$ is the *energy flux*, the energy per unit time passing a point x. For the traveling wave $y(x, t) = f(x - vt)$, find $u(x, t)$ and $S(x, t)$ and show that energy flows to the right (for $v > 0$) as the wave passes a point x. Show that the total energy passing each point is equal to the total energy in the wave.

4.5 Compute the maximum energy flux possible for electromagnetic waves in air given the constraint that the electric field cannot exceed the breakdown field described in Example 3.2.

4.6 The strongest radio stations in the US broadcast at a power of 50 kW. Assuming that the power is broadcast uniformly over the hemisphere above Earth's surface, compute the strength of the electric field in these radio waves at a distance of 100 km.

4.7 [T] Derive the wave equation for $\mathbf{B}$ analogous to eq. (4.20).

4.8 Suppose an electromagnetic plane wave is absorbed on a surface oriented perpendicular to the direction of propagation of the wave. Show that the pressure exerted by the radiation on the surface is $p_{\mathrm{rad}} = W/c$, where W is the power absorbed per unit area. Solar radiation at the top of Earth's atmosphere has an energy flux $|S| = 1366\,\mathrm{W/m^2}$. What is the pressure of solar radiation when the Sun is overhead? What is the total force on Earth exerted by solar radiation?

4.9 It has been proposed that solar collectors could be deployed in space, and that the collected power could be beamed to Earth using microwaves. A potential limiting factor for this technology would be the possible hazard of human exposure to the microwave beam. One proposal involves a circular receiving array of diameter 10 km for a transmitted power of 750 MW. Compute the energy flux in this scenario and compare to the energy flux of solar radiation.

4.10 [T,H] Consider two electromagnetic plane waves (see eq. (4.21)) one with amplitude E_0^a and wave vector $\mathbf{k}^a$ and the other with amplitude E_0^b and wave vector $\mathbf{k}^b$. These waves are said to add **coherently** if the average energy density u in the resulting wave is proportional to $|E_0^a + E_0^b|^2$ or **incoherently** if the average energy density is proportional to $|E_0^a|^2 + |E_0^b|^2$. Show that two electromagnetic plane waves are coherent only if they are propagating in the same direction with the same frequency and the same polarization.

4.11 [T] Derive eq. (4.23) by taking the time derivative of eq. (4.22) and using Maxwell's equations. [Hint: see eq. (B.23).]

4.12 [T] A string of length L begins in the configuration $y(x) = A[\frac{1}{3}\sin k_1 x + \frac{2}{3}\sin k_2 x]$ with no initial velocity. Write the exact time-dependent solution of the string $y(x,t)$. Compute the contribution to the energy from each mode involved.

4.13 [TH] A string of length L is initially stretched into a "zigzag" profile, with linear segments of string connecting the $(x, f(x))$ points $(0,0)$, $(L/4, a)$, $(3L/4, -a)$, $(L, 0)$. Compute the Fourier series coefficients c_n and the time-evolution of the string $y(x, t)$. Compute the total energy in the tension of the initially stretched string. Compute the energy in each mode and show that the total energy agrees with the energy of the stretched string.

4.14 [T] What is the pressure exerted by a beam of light on a perfect mirror from which it reflects at normal (perpendicular) incidence? Generalize this to light incident at an angle θ to the normal on an imperfect mirror (which reflects a fraction $r(\theta)$ of the light incident at angle θ).

4.15 [T] Consider a cylindrical resistor of cross-sectional area A and length L. Assume that the electric field $\mathbf{E}$ and current density $\mathbf{j}$ are uniform within the resistor. Prove that the integrated power transferred from electromagnetic fields into the resistor $\int d^3 x\, \mathbf{j} \cdot \mathbf{E}$ is equal to IV. Compute the electric and magnetic fields on the surface of the resistor, and show that the power transfer is also given by the surface integral of the Poynting vector, so that all energy dissipated in the resistor is transferred in through electric and magnetic fields.

4.16 As stated in the text, the *dispersion relation* relating the wave number and angular frequency of ocean surface waves is $\omega = \sqrt{gk}$, where $g \cong 9.8\,\mathrm{m/s^2}$. Compute the wavelength and speed of propagation (phase velocity) for ocean surface waves with periods 6 s and 12 s.

4.17 [T] A wave satisfying eq. (4.2) passes from one medium in which the phase velocity for all wavelengths is v_1 to another medium in which the phase velocity is v_2. The incident wave gives rise to a reflected wave that returns to the original medium and a refracted wave that changes direction as it passes through the interface. Suppose that the interface is the plane $z = 0$ and the incoming wave is propagating in a direction at an angle θ_1 to the normal $\hat{z}$. Prove the **law of specular reflection**, which states that the reflected wave propagates at an angle $\pi - \theta_1$ with respect to the normal $\hat{z}$. Also prove **Snell's law**, which states that the wave in the second medium propagates at an angle θ_2 from the normal $\hat{z}$, where $\sin\theta_2 / \sin\theta_1 = v_2/v_1$. Use the fact that the wave must be continuous across the interface at $z = 0$.

4.18 A wave travels to the right on a string with constant tension τ and a mass density that slowly increases from ρ on the far left to ρ' on the far right. The mass density changes slowly enough that its only effect is to change the speed with which the wave propagates. The waveform on the far left is $A\cos(kx - \omega t)$ and on the far right is $A'\cos(k'x - \omega t)$. Find the relation between k' and k and then use conservation of energy to find the amplitude A' on the far right. You may find it helpful to use the result of Problem 4.4 ($S(x, t) = -\tau \dot{y} y'$).

Thermodynamics I: Heat and Thermal Energy

Thermal energy has played a central role in energy systems through all of human history. Aside from human and animal power, and limited use of wind and hydropower, most energy put to human use before 1800 AD came in the form of heat from wood and other biomass fuels that were burned for cooking and warmth in pre-industrial societies (and still are in many situations). The development of the steam engine in the late eighteenth century enabled the transformation of thermal energy into mechanical energy, vastly increasing the utility of thermal energy. Humankind has developed ever-increasing reliance on the combustion of coal and other fossil fuels as sources of thermal energy that can be used to power mechanical devices and to generate electricity. Today, over 90% of the world's energy relies either directly or in an intermediate stage on thermal energy, the major exception being hydropower.

Thermal energy, and its conversion into mechanical and electrical energy, is an important theme in this book. The scientific study of the transformation of heat into mechanical energy and *vice versa*, which began soon after the discovery of the steam engine, led to the discovery of energy conservation and many fundamental aspects of energy physics. Thermal energy belongs to a rich and surprisingly subtle subject – *thermodynamics* – that we develop systematically beginning in this chapter and continuing in §8.

In contrast to mechanical kinetic and potential energy, thermal energy involves *disordered* systems, and a new concept, *entropy*, is needed to provide a quantitative measure of disorder. The definition of entropy (as well as the precise meaning of temperature) can be best appreciated with the help of some basic knowledge of quantum physics. We introduce quantum mechanics in §7. In §8 we define and explore the concepts of entropy and *free energy*, leading to the fundamental result that there is a physical limit on the efficiency with which thermal energy can be converted to mechanical energy. We apply these ideas to

Reader's Guide
In this chapter, we take a first look at thermal energy and the associated concepts of heat and temperature. We assume minimal scientific background in this area beyond everyday experience. This chapter initiates a systematic treatment of these subjects, and introduces many related ideas including internal energy, thermodynamic equilibrium, thermal equilibrium, state functions, enthalpy, heat capacity, phase transitions, and the first law of thermodynamics. These concepts play a central role in engines, power plants, and energy storage. As always, we follow the role of energy and energy applications as we navigate a passage through a large and varied subject.

Prerequisites: §2 (Mechanics).

The material in this chapter forms a basis for later chapters on various aspects of thermodynamics, particularly §6 (Heat transfer), §8 (Entropy and temperature), §10 (Heat engines), and §12 (Phase-change energy conversion).

chemical reactions in §9, to engines in §10 and §11, and to power generation in §12 and §13. As illustrated in Figure 1.2, more than half of the thermal energy released from fossil fuel combustion and nuclear reactors is lost in the current US energy stream. A clear understanding of thermodynamics and entropy is needed to distinguish between the fraction of this energy loss that is unavoidable and the fraction that is due to the less-than-optimal performance of existing power systems.

While thermal energy is often harnessed as a means of powering mechanical devices or generating electricity, it is also an end in itself. Indeed, more than a third of US energy use currently goes to heating (and cooling) buildings, water, food, and other material goods. After developing some basic notions of heat, heat content, and heat capacity in this chapter, in §6 we develop tools for

Figure 5.1 At the Nesjavellir geothermal power station in Iceland, thermal energy originating deep within Earth's crust is used to produce 120 MW of electric power and also to heat homes and businesses in the capital, Reykjavík. (Image: G. Ívarsson)

studying heat transfer in materials and analyzing thermal energy requirements in situations such as the heating of a building in a cold climate.

We begin in §5.1 with a general introduction to some of the central concepts underlying thermodynamics. Precise definitions of some quantities are left to later chapters. Ideal gases are introduced as a simple paradigm for thermodynamic systems. In §5.2, we describe the transformation of thermal energy to mechanical energy. This leads to the formulation of the first law of thermodynamics in §5.3. The last part of the chapter describes the addition of heat to materials through heat capacity (§5.4), *enthalpy* (§5.5), and phase transitions (§5.6). For a more detailed introduction to thermal energy, heat, and other concepts introduced in this chapter, see [19] or [20]. A more sophisticated treatment of thermal physics is given in [21], incorporating entropy early in the discussion.

5.1 What is Heat?

5.1.1 Thermal Energy and Internal Energy

Heat is surely one of the most basic concepts in human experience. From the time our early ancestors learned to control fire, humankind has experimented with heat, and tried to understand the precise nature of the distinction between that which is hot and that which is cold. Now that we understand materials at a microscopic level, we know that the notions of heat and temperature refer to processes and properties that are most clearly described in terms of energy.

Thermal energy refers to the collective energy contained in the relative motions of the large number of microscopic particles comprising a macroscopic whole. A

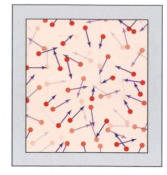

Figure 5.2 Thermal energy carried by the motion of N molecules in a volume V. For a typical macroscopic system, $N \sim 10^{23}$.

gas of N identical molecules, where N is a very large number, confined to a volume V (see Figure 5.2) provides a simple example of a system with thermal energy. The molecules move about randomly, colliding with one another and with the walls of the container. The kinetic energy associated with the motion of these molecules relative to the fixed walls of the container is thermal energy. Unless the gas is monatomic, like helium or argon, the molecules will also be rotating and vibrating, and these random motions also contribute to the thermal energy. In a solid the atoms are locked into place in a fixed structure, such as a crystal. Though they are not free to move about or to rotate, they can still vibrate. Their kinetic and potential energies of vibration constitute the thermal energy of the solid.

Qualitatively, **temperature** is a relative measure of the amount of thermal energy in a system (or part of a system). Increasing the temperature of a system – making it hotter – generally requires adding thermal energy. **Heat** refers to a *transfer of thermal energy* from one system to another. To proceed quantitatively, it is useful to define the notion of thermal energy, and the closely related concept of total *internal energy*, more precisely.

Internal energy The **internal energy** U of any physical system is the sum total of all contributions to the energy of the system considered as an isolated whole. Note that U does not include kinetic or potential energy associated with the motion of the system as a whole relative to external objects such as Earth. So, for example, the collective kinetic energy of wind or a mass of water due to its motion at a fixed velocity relative to Earth's surface *is not* included in its internal energy. The internal energy U does, however, include contributions from chemical and/or nuclear binding and from rest mass energies of the constituent particles.

Internal and Thermal Energy

The *internal energy U* of any physical system is the sum total of all contributions to the energy of the system considered as an isolated whole. It does not include the kinetic energy of bulk motion or the potential energy of the center of mass.

Thermal energy is that contribution to the internal energy of a system beyond the energy that the system would have if cooled to a temperature of absolute zero.

Heat

Heat is thermal energy that is transferred from one system (or part of a system) to another system (or part of a system).

Thermal energy Thermal energy, on the other hand, is that contribution to the internal energy of a system beyond the energy that the system would have if cooled to a temperature of **absolute zero**. Absolute zero is defined to be a state from which no further energy can be removed without changing the nature of the constituents or their chemical or nuclear binding. Thus, thermal energy *does not include* particle rest masses, nuclear binding energy, or chemical bond energies, but does include kinetic and potential energies associated with the relative motion of the constituent molecules, their rotations and vibrations, as well as the energy needed to separate molecules into a liquid or gaseous state, if applicable.

Note that the internal energy and thermal energy of a system only differ by an additive constant, namely the internal energy of the system at absolute zero. Often we are only interested in energy differences, so we can without confusion use the same symbol U to denote both quantities. Indeed, in some elementary physics texts internal and thermal energy are equated. In this chapter we do not consider chemical or nuclear reactions, so U can unambiguously refer to thermal energy. As soon as chemical and/or nuclear bonds in the system can change, however, as in §9 or §16, the notion of thermal energy becomes ambiguous, and one should always use internal energy, not thermal energy, for a consistent description of the system.

Now that we have a definition of thermal energy, we can give a useful definition of heat. *Heat is thermal energy that is transferred from one system (or part of a system) to another system (or part of a system).* Note that modern definitions of heat always invoke a transfer of energy from one system to another, so that there is no meaning to the "heat content" of a single system. Our common experience is that heat always flows from regions at higher temperature to regions at lower temperature, so temperature can be thought of as the capacity of a system to generate heat.

5.1.2 Thermal Equilibrium States and Heat

Before going further we need to specify more carefully the kind of systems to which these ideas apply. For the discussion that follows, we do not need a more precise definition of temperature beyond the naive statement that it is what is measured by a thermometer. In §5.1.3, we give an exact quantitative meaning to temperature for a particular class of simple systems, though a careful definition of temperature for general systems must wait until §8.

We are interested in **macroscopic systems**, composed of a very large number of particles. Atoms and molecules are very small, so even an amount of material that is tiny by human standards is made up of a huge number of individual particles. Each **mole** (mol) of a pure substance contains an **Avogadro's number**,[1] $N_A = 6.022 \times 10^{23}$, of molecules. Thus, while a cubic micron of water, for example, is far too small to see, this volume of water nevertheless contains about 3×10^{10} molecules. Concepts such as temperature are generally used to describe aggregate properties of systems containing a very large number of molecules. When we discuss $T(x)$, the temperature of a macroscopic system *at a point x*, we are actually referring to a local average over a small region that still contains an immense number of molecules.

Another important concept is that of *thermodynamic equilibrium* and the related notion of *thermal equilibrium*. Heuristically, a system reaches thermodynamic equilibrium when it has been left undisturbed long enough for it to "settle down" and cease to change qualitatively with time. More precisely, when a system is in **thermodynamic equilibrium** there is no net flow of energy or material at macroscopic scales. Thermodynamic equilibrium implies *thermal equilibrium*, *chemical equilibrium*, *mechanical equilibrium*, and *radiative equilibrium*, referring to the absence of net heat transfer, net chemical reaction changes

---

[1] Avogadro's number N_A is defined as the number of carbon atoms in 12 g of pure carbon-12, and has been measured experimentally to take approximately the value stated. The **molar mass** of a substance is the mass of one mole of the substance, generally quoted in units of g/mol. Molecular and molar masses, as well as the isotope carbon-12, are discussed further in §9 and §17.

in composition, net macroscopic motion, and net energy flow through radiation respectively at any macroscopic distance scale. In particular, a system in *thermal equilibrium* has constant temperature throughout. Other properties, such as density, may not be constant in thermodynamic or thermal equilibrium. For example, a jar containing oil and water can come to thermodynamic equilibrium with the oil and water forming separate layers. Although the temperature is uniform and there are no net chemical reactions occurring, the density (as well as many other properties) is not uniform. In many of the systems we study, such as a homogeneous gas in a confined volume, there are no chemical processes, radiation, or other non-thermal mechanisms for energy transfer, so that thermodynamic and thermal equilibrium are equivalent. Thus, for such systems we often simply refer to *thermal equilibrium* although the system is really in complete thermodynamic equilibrium.

In thermodynamics the word "state" is often used to refer to a configuration of a physical system in thermodynamic equilibrium – an **equilibrium state**.[2] In this context, the state of a system is characterized by various **state functions**, also known as **state variables**, that are macroscopic characteristics of the system. State functions do not depend upon how the system got to its current configuration. A system may require other variables beyond the temperature T to fully specify its state. The thermodynamic state of a bottle of pure oxygen, for example, is not uniquely specified by its temperature T. Additional information, such as the volume of the bottle V and the number of molecules N or moles n of oxygen, must also be specified. Density ρ and pressure p are other state functions. A mixture of oxygen and nitrogen requires another state variable – for example, the fractional abundance of oxygen – to fully specify the state of the system.

It is important to realize that thermodynamic equilibrium is a dynamic, not a static, situation. Even though the macroscopic (thermodynamic) state does not change, the molecules in a bottle of oxygen are constantly moving and the precise values of local functions, such as the density, are constantly fluctuating. Because the systems we study are so large, however, the fluctuations are usually too small to be noticed.

Like many important concepts in physics, the state of thermodynamic or thermal equilibrium is an idealization. The temperature of a real bottle of oxygen changes

---

States and State Functions

Thermodynamics applies to macroscopic systems containing huge numbers of particles. When a system settles down into thermodynamic equilibrium, its state is described by various *state functions* such as temperature, pressure, volume, and density.

---

slightly as drafts of cool or warm air circulate around it; atmospheric pressure changes slowly with the weather. Many situations *approximate* thermal equilibrium. For example, a pot of water heated *very slowly* from $20\,°C$ to $80\,°C$ can be viewed as passing through a smooth sequence of thermal equilibrium states.

5.1.3 Ideal Gases and the Equipartition Theorem

A simple example may be helpful to clarify the concepts just introduced. An **ideal gas** is assumed to be composed of molecules that move about *classically*.[3] The molecules collide elastically (preserving energy and momentum), and have no other intermolecular interactions. A **monatomic ideal gas** is an ideal gas in which the molecules are single atoms that have no internal dynamics such as rotations or vibrations. Like thermodynamic equilibrium, an ideal gas is an idealization that is only approximated by real gases. At room temperature and moderate densities, many gases are quite well approximated as ideal gases. The monatomic gases helium and argon, for example, behave as monatomic ideal gases at room temperature.

All of the thermal energy of an ideal monatomic gas comes from the kinetic energy of the molecules' motion in three dimensions. If we number the molecules $i = 1, \ldots, N$, then the energy of the ith-molecule is given by

$$E_i = \frac{1}{2} m v_i^2 , \qquad (5.1)$$

where v_i is the velocity of the ith-molecule, and all molecules have identical mass m. The thermal energy of the gas is then

---

[2] A thermodynamic state is a macroscopic characterization of a system consistent with many possible microscopic configurations of the molecules or other constituents (*microstates*), as discussed further in §8.1.

[3] The physics that preceded quantum mechanics is now called **classical physics**. So when we say that the molecules move about classically, we mean that their translational motion can be described by Newton's Laws; their internal dynamics may nonetheless be strongly affected by quantum physics at low temperatures.

$$U = \sum_{i=1}^{N} E_i = \sum_{i=1}^{N} \frac{1}{2}m\mathbf{v}_i^2 = \frac{m}{2}\sum_{i=1}^{N}\left(v_{ix}^2 + v_{iy}^2 + v_{iz}^2\right),$$

$$(5.2)$$

where v_{ix}, v_{iy}, v_{iz} are the Cartesian components of $\mathbf{v}_i$. Denoting averages over all the molecules by $\langle \ldots \rangle$, the total thermal energy is $U = N\langle E \rangle$, where the average molecular kinetic energy $\langle E \rangle$ is given by

$$\langle E \rangle = \frac{1}{N}\sum_i E_i = \langle \frac{1}{2}mv^2 \rangle$$

$$= \frac{1}{2}m\left(\langle v_x^2 \rangle + \langle v_y^2 \rangle + \langle v_z^2 \rangle\right) = \frac{3}{2}m\bar{v}^2, \quad (5.3)$$

with $\bar{v}^2 \equiv \langle v_x^2 \rangle = \langle v_y^2 \rangle = \langle v_z^2 \rangle = \frac{1}{3}v_{\text{RMS}}^2$, where v_{RMS} is the root mean square speed of the molecules, and we have used the fact that the motions of the molecules are random and have no preferred direction.

Temperature is a measure of thermal energy of a system. We expect that $\langle E \rangle$ generally increases with increasing temperature. For the monatomic ideal gas we can take eq. (5.3) as a working definition of temperature through

$$\langle E \rangle = \frac{3}{2}m\bar{v}^2 \equiv \frac{3}{2}k_{\text{B}}T. \quad (5.4)$$

Here

$$k_{\text{B}} = 1.381 \times 10^{-23} \, \text{J/K} \quad (5.5)$$

is the **Boltzmann constant**, which is essentially a conversion factor from units of temperature to units of energy. As a result, the thermal energy of N atoms or n moles of a monatomic ideal gas at temperature T is

$$U = \frac{3}{2}Nk_{\text{B}}T = \frac{3}{2}nRT, \quad (5.6)$$

where $R = N_A k_{\text{B}} = 8.314 \, \text{J/mol K}$ is the **gas constant**. The definition of the temperature of an ideal gas given by eq. (5.4), including the factor of 3/2, agrees with the general definition of temperature given in §8. In particular, for an ideal monatomic gas $\langle E \rangle$ varies linearly with temperature, as we prove in §8.7.4.

In general, physicists deal with individual molecules (N), while chemists deal with moles (n). Conversion from one to the other follows from

$$Nk_{\text{B}} = nR. \quad (5.7)$$

We use either notation, depending on application and context, throughout this book.

The components of motion in the x-, y-, and z-directions contribute equally to the internal energy, so eq. (5.6) assigns an energy $\frac{1}{2}k_{\text{B}}T$ (on average) to motion of each molecule in each direction. This is an example of a fundamental property of thermal energy known as the *equipartition theorem*. Before we can state this theorem precisely,

Degrees of Freedom and Equipartition of Energy

A degree of freedom is an independent contribution to the kinetic or potential energy of an object. In any system in thermal equilibrium at a temperature high enough that quantum effects are not important and low enough that atomic excitations are not relevant, each degree of freedom of each molecule contributes $\frac{1}{2}k_{\text{B}}T$ to the internal energy.

we need to introduce the concept of a **degree of freedom** of a mechanical system. Although the precise definition is more complicated, we can regard a degree of freedom as an independent contribution to the kinetic or potential energy of an object. A monatomic molecule is free to move (or *translate*) in three directions and has no potential energy associated with this motion, so it has three *translational* degrees of freedom associated with the contributions to kinetic energy from the three components of its velocity vector. A *diatomic molecule* can, in addition, rotate about two independent axes perpendicular to the line joining the atoms, adding two degrees of freedom. (Again, there is no potential energy associated with this motion.) A rotation *about* the axis joining the atoms does not contribute a degree of freedom in the approximation that the atoms themselves are pointlike and structureless.[4] Finally, the atoms in a diatomic molecule can vibrate like a spring along the axis that joins them, adding one kinetic and one potential degree of freedom. So a diatomic molecule has, in all, $3 + 2 + 2 = 7$ degrees of freedom (see Figure 5.3 and Table 5.1 for more information).

The **equipartition theorem** states that in any system in thermal equilibrium at a temperature high enough that quantum effects are not important, each degree of freedom of a molecule contributes $\frac{1}{2}k_{\text{B}}T$ to the internal energy. At low temperatures, quantum mechanical effects *freeze out* (i.e. render inactive) rotational or vibrational degrees of freedom of gas molecules. This effect is discussed further in §5.4.1 and §8.7.

Another simple ideal system is an ideal monatomic crystalline solid, in which the atoms are held in a regular array by strong forces. The atoms have no translational

---

[4] Rotations of an atom (and of a diatomic molecule around its axis) can occur at very high temperatures, along with other quantum excitations that eventually lead to *ionization* of the atom. These effects are described in §9 (Energy in matter).

Table 5.1 The number of degrees of freedom of some simple molecules. A molecule has at most three translational and three rotational degrees of freedom. A diatomic or linear triatomic molecule has only two rotational degrees of freedom. The number of vibrational degrees of freedom increases with the complexity of the molecule. According to the equipartition theorem, the heat capacity at high temperature is proportional to the total number of degrees of freedom (§5.4).

Degrees of freedom of simple molecules						
Molecule	Example	Translational	Rotational	Vibrational		Total
		kinetic	kinetic	kinetic	potential	
Monatomic	He	3	0	0	0	3
Diatomic	O_2, CO	3	2	1	1	7
Triatomic (linear)	CO_2	3	2	3	3	11
Triatomic (nonlinear)	H_2O	3	3	3	3	12
Polyatomic	CH_4, C_2H_5OH	3	3	many	many	> 12
Pure crystalline solid	Au	0	0	3	3	6

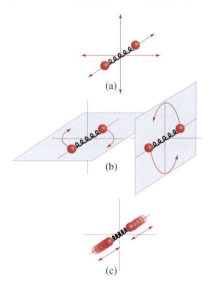

Figure 5.3 The seven degrees of freedom of a diatomic molecule: (a) translational kinetic energy of the center of mass in three independent directions denoted by the arrows; (b) rotational kinetic energy about the two axes perpendicular to the axis of the molecule; (c) vibrational kinetic and potential energy.

or rotational degrees of freedom, but each can vibrate with kinetic and potential energy in three directions. When quantum effects can be ignored, an ideal monatomic crystalline solid will thus have internal energy

$$U = 6 \times \tfrac{1}{2} N k_B T = 3nRT \, . \tag{5.8}$$

This result, known as the **law of Dulong and Petit**, was first discovered empirically by the French physical chemists Pierre Dulong and Alexis Petit. This law does a remarkably good job for pure elements like sulfur, copper, aluminum, gold, and iron at room temperature.

In §8.7.3 we derive eq. (5.8) and show how it arises as a high-temperature limit of a more accurate theory due to Einstein.

5.2 Pressure and Work

Mechanical energy is quite easily transformed into thermal energy. In §2, we discussed energy losses to air resistance. After being transferred to the kinetic energy of the many air molecules flowing around a vehicle, energy lost to air resistance is gradually transformed into random motion of all the air molecules in the surrounding atmosphere and ends up as thermal energy. Similarly, friction in the brakes generates thermal energy that spreads through the adjacent parts of the car. The opposite transformation, from thermal energy to usable mechanical energy, is harder to accomplish in an efficient fashion. As discussed in the introduction to this chapter, however, over 90% of current primary energy consumption relies on thermal energy at some stage. This thermal energy must be transformed into mechanical energy to be used either for transport or to power electrical generators. To better understand how this transformation can be carried out, it is useful for us to understand the microscopic origin of pressure and its relation to work.

5.2.1 Pressure and the Ideal Gas Law

Pressure, first introduced in §1, is force per unit area directed perpendicular, or *normal*, to a surface. Just as the moving molecules in a gas carry kinetic energy, which appears macroscopically as thermal energy, they also carry momentum that gives rise to pressure on the macroscopic scale. Molecules continually collide with any surface bounding a gas, transferring momentum. Newton's laws tell us that momentum transferred per unit time is a

---

Example 5.1   Thermal Energy of Argon Gas

Argon is a monatomic gas that accounts for approximately 1% of the Earth's atmosphere (§34.2).

Treating argon as an ideal gas, estimate the difference in thermal energy between one mole of argon gas at 300 K and at 600 K.

Approximating argon as an *ideal* monatomic gas, the thermal energy at temperature T is given by eq. (5.6), so at $T = 300$ K we have $U = 3RT/2 \cong 3.7$ kJ. The thermal energy at 600 K is twice this amount, so the thermal energy *difference* between 300 K and 600 K is approximately 3.7 kJ. (This agrees with the actual value to within 2% [22].)

Note that the thermal energy of real argon at a temperature T includes all the energy necessary to heat solid argon from 0 K to the temperature T, including the energy needed to melt and vaporize the argon as it passes through the melting and boiling points (83.8 K and 87.3 K at 1 atm). The heat capacity of solid and liquid argon is quite different from the heat capacity of the gas. Therefore a constant representing the other contributions should be added to the ideal gas formula when computing the total internal energy of real argon gas at any temperature above the boiling point. The thermal energy *difference* between 300 K and 600 K is not, however, affected by this overall additive constant.

At a temperature of 300 K and atmospheric pressure, one mole of argon fills a volume of approximately 25 L. Fixing this volume, the difference in thermal energy density between argon at 300 K and 600 K is roughly 150 kJ/m³. Note, for comparison, that this is far higher than the maximum electrostatic energy density of air-based capacitors (~40 J/m³, see Example 3.2), but significantly lower than the volumetric energy density of typical batteries (§37).

---

force, $dp/dt = F$, and macroscopically, the resulting force per unit area perpendicular to the bounding surface is the pressure exerted by the gas. Pressure can also be defined *locally* at any point within the gas, as the force per unit area on the surface of a small cavity or object inserted at that point. For example, the tremendous pressure deep in Earth's oceans that is generated by the overlying water (see §34.2.1) is readily measured on the surface of a deep-water submersible.

To describe the pressure of a gas quantitatively in terms of the microscopic motion of its constituents, we study an ideal gas enclosed in a cylindrical container of length l and cross-sectional area A, as shown in Figure 5.4. Consider the effects of the impacts of molecules of the gas on a boundary perpendicular to the x-axis, which is the face of a movable piston. Moving the piston increases or decreases the volume containing the gas. When a molecule with momentum mv_x in the x-direction hits the piston, which is much more massive and moving much more slowly than the molecule (see Problem 2.10), it bounces back with momentum $-mv_x$ in the x-direction, and transfers momentum

$$\Delta p_x = 2mv_x \qquad (5.9)$$

to the piston. If we ignore collisions between molecules, this molecule will next collide with the piston after traveling a distance $2l$ across the box and back, with speed v_x in the x-direction. The time between impacts is therefore

$$\Delta t = \frac{2l}{v_x}. \qquad (5.10)$$

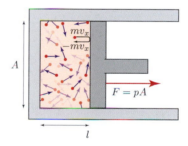

Figure 5.4 A cross section through a container of gas of length l and cross-sectional area A. The collisions of molecules of gas with the piston on the right transfer momentum $2mv_x$, which combine to create pressure on the piston at the macroscopic scale.

The rate at which this molecule transfers momentum to the piston is thus $\Delta p_x/\Delta t = mv_x^2/l$. Summing over all the molecules gives rise to a continuous rate of momentum transfer and therefore a force on the piston,

$$F = \frac{dp_x}{dt} = \sum_i \frac{mv_{ix}^2}{l} = Nm\frac{\langle v_x^2 \rangle}{l}, \qquad (5.11)$$

where, as in eq. (5.3), $\langle v_x^2 \rangle = \bar{v}^2$ denotes the average of v_x^2 over all the molecules. The pressure (force per unit area) is then given by[5]

---

[5] Note that we use the scalar quantity p here and elsewhere to denote pressure, while the vector quantity $\boldsymbol{p}$ or component p_x denotes momentum.

$$p = \frac{F}{A} = \frac{N}{Al}m\langle v_x^2\rangle = \frac{Nm}{V}\langle v_x^2\rangle = \rho\langle v_x^2\rangle, \quad (5.12)$$

where $V = Al$ is the volume containing the gas and $\rho = mN/V$ is the mass density. A more thorough analysis would take account of the collisions between molecules, but the result, eq. (5.12) for an ideal gas, does not change.

Note that the collisions of the molecules with the face of the piston do not transfer any momentum in the directions tangent (parallel) to the surface. In general, we can break up a force per unit area on a surface into a component normal to the surface, which is the pressure, and a component parallel to the surface, which is known as the *shear stress*. A gas in thermodynamic equilibrium exerts pressure, but no shear stress. Shear stress becomes relevant in the context of wind energy and is discussed in more detail in §29 (Fluids).

For an ideal gas, we can relate the pressure to the volume and temperature. Multiplying eq. (5.12) by V gives $pV = Nm\bar{v}^2$, and using eq. (5.4) we can express this in terms of N and T, leading to the famous **ideal gas law**

$$pV = Nk_BT = nRT. \quad (5.13)$$

Note that although the analysis leading to (5.12) assumed a particular box shape and size, the pressure depends only upon the local properties of the gas. So eq. (5.13) holds for an ideal gas in any volume. Equation (5.13) is a simple example of an **equation of state** that provides a dynamical relation between state variables. In general, knowing n, V, and T is enough to determine the pressure p for a pure (single species) gas. Only for an ideal gas, however, is the relation as simple as eq. (5.13). This derivation of the empirically known ideal gas law (5.13) confirms that the working definition of temperature used in eq. (5.4) matches that used in the ideal gas law.

5.2.2 Work from Pressure and Heat

Adding heat to a gas in a fixed volume raises the temperature and internal energy of the gas. This leads to a corresponding increase in pressure through eq. (5.13). The pressure of a hot gas can be used to do mechanical work. In the process, however, some energy is inevitably lost to the environment. We can use the relation between pressure and force to begin to understand these processes in a more quantitative fashion.

First, we consider the relation between pressure and work. Consider again an ideal monatomic gas in a container as shown in Figure 5.4. The pressure of the gas exerts a force pA on the piston. If the piston moves then this force does work on the piston

$$dW = pAdx = pdV, \quad (5.14)$$

Ideal Gases

An *ideal gas* is an idealized model of a gas of molecules that can scatter elastically, but exert no other forces on one another. An ideal gas obeys an *equation of state* known as the *ideal gas law*

$$pV = Nk_BT = nRT.$$

The thermal energy content of a *monatomic* ideal gas containing N molecules at a temperature T is

$$U = \frac{3}{2}Nk_BT.$$

Work and Pressure

When the volume of a system under pressure p expands by an infinitesimal amount dV, the system does work $dW = pdV$.

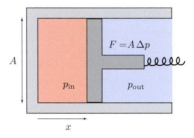

Figure 5.5 Gas in a closed container exerting force on a piston does some work on a spring attached to the piston, but also loses energy to gas at lower pressure outside the container.

---

where $dV = Adx$ is the volume swept out by the piston. The relationship (5.14) underlies the observation made in §1 that the units of energy and pressure × volume are the same.

Now let us consider in a more concrete situation how thermal energy added as heat to the gas in the cylinder shown in Figure 5.5 can be converted to mechanical work. Suppose we begin with the same gas at equal pressures and temperatures on the inside and outside, $p_{in}^{(initial)} = p_{out}$ and $T_{in}^{(initial)} = T_{out}$, so that the system is in thermal and mechanical equilibrium. We suppose that the region outside the piston remains at fixed pressure and temperature throughout. We add heat to the gas inside the container, thereby increasing its internal energy. This could be accomplished, for example, by placing the wall of the cylinder opposite the piston in contact with a material at a higher temperature. We assume that the piston is held fixed by an external agent while the heat is added.

The gas inside the container now has a higher pressure (and temperature) than the gas outside,

$$p_{\text{in}} = p_{\text{out}} + \Delta p, \tag{5.15}$$

leading to a net outward force on the piston,

$$F_{\text{net}} = A(p_{\text{in}} - p_{\text{out}}) = A\Delta p. \tag{5.16}$$

If we allow the piston to move a differential distance dx, then the gas can do useful work – for example by compressing the spring shown in Figure 5.5. The useful work is proportional to the net force,

$$dW_{\text{useful}} = F_{\text{net}}dx = \Delta p(Adx) = \Delta pdV. \tag{5.17}$$

The work done on the piston by the gas inside is proportional to p_{in}, not Δp, and is therefore larger than the useful work done

$$dW = p_{\text{in}}dV > dW_{\text{useful}}. \tag{5.18}$$

Where does the remaining energy go? As the piston moves to the right, it does indeed do some mechanical work. But in addition, as the moving piston impacts the molecules of the gas outside the cylinder, it increases their kinetic energies and thereby transfers energy to the gas outside the piston. The work done by the piston on the outside gas is just $p_{\text{out}}\Delta V$, as in eq. (5.14). Only the excess of the work done by the gas inside over the work done on the gas outside, expressed in eq. (5.17), is available to do useful mechanical work.

In this simple example energy has been lost to the outside environment in order to perform useful work. To quantify this, it is useful to define the *efficiency* η of such a conversion process as the *ratio of useful work over total thermal energy used*

$$\eta = \frac{dW_{\text{useful}}}{dW}. \tag{5.19}$$

For simplicity let us assume that the gas is ideal. Then the ideal gas law tells us that $p_{\text{in}} = Nk_BT_{\text{in}}/V$, and from the initial conditions $p_{\text{out}} = Nk_BT_{\text{out}}/V$. It follows that

$$\eta = \frac{p_{\text{in}} - p_{\text{out}}}{p_{\text{in}}} = \frac{T_{\text{in}} - T_{\text{out}}}{T_{\text{in}}}. \tag{5.20}$$

Remarkably, eq. (5.20) is actually the maximum efficiency with which mechanical work can be extracted from thermal energy by any process that can be repeated without relying on additional external inputs (e.g. a *heat engine* (§8.5)). A

full understanding and proof of this limitation requires the concepts of *entropy* and the *second law of thermodynamics*, which are developed in §8 (see Problem 10.10). The related concept of *exergy*, or *available work*, developed in §36, can also be used to show more directly from the 2nd Law that eq. (5.20) is the maximum conversion efficiency possible for processes such as the one described above (see Problem 36.11). In this discussion we have referred to the *useful work* done by a system in contrast to the work that cannot be put to use but is required to move the ambient atmosphere out of the way. While here we have primarily focused on mechanical work done, more generally the concept of useful work can be defined in terms of the increase in *exergy* of another system (§36.4.1).

Note that in the process we have considered here, the change dx was taken to be infinitesimally small, so that during the expansion the internal pressure and temperature did not change significantly. For a finite motion of the piston, the pressure and temperature will drop during the expansion; this is one of the issues that complicates the design of a practical engine realizing the maximal thermodynamic conversion efficiency (5.20).

5.3 First Law of Thermodynamics

The **first law of thermodynamics** is simply a restatement of the fundamental principle of conservation of energy in a form that explicitly includes thermal energy transferred as heat. The internal energy of a system increases if an infinitesimal quantity of heat dQ is added to it and it decreases if the system does work dW,

$$dU = dQ - dW. \tag{5.21}$$

In this book we are primarily concerned with the work done by liquids and gases through the action of pressure as a volume expands or contracts, in which case $dW = p\,dV$. So for most purposes we use the first law in the form

$$dU = dQ - p\,dV. \tag{5.22}$$

Since it appears so frequently, we often use the abbreviated form "1st **Law**" for the "first law of thermodynamics" throughout this book.

The pressure, volume, and internal energy that appear in eq. (5.22) are *state functions*. They are properties of an equilibrium state of the system. Heat, however, is different. It is not a state function. There is no meaning to the "heat content" of a system. It took many years to figure this out. When scientists first tried to understand heat, they thought it was a substance (known as "caloric") that flowed from one body to another when heat was transferred, and

The First Law of Thermodynamics

When a quantity of heat dQ is added to a system and mechanical work dW is done by the system, then, by conservation of energy, the change in the system's internal energy is $dU = dQ - dW$. If the work is done by the action of pressure as a volume expands or contracts, then the 1st Law reads

$$dU = dQ - pdV .$$

The internal energy U of a system is a state function, though heat and work are not.

that temperature was a measure of the amount of *caloric* in a body. This idea was abandoned as thermodynamics was developed in the early nineteenth century and it was realized that heat is a form of energy. A system can gain some heat Q_{in}, do some work W, expel some heat $Q_{out} < Q_{in}$, and end up back where it started – i.e. *in the same state*. The initial and final states of the system are identical, but more heat was added than was removed, so heat cannot be a state function.[6] Such cycles are discussed in depth in §8 (Entropy and temperature) and §10 (Heat engines).

5.4 Heat Capacity

In general, when heat is added to a system its temperature rises, and when heat is removed its temperature falls. As mentioned at the outset of this chapter, over a third of energy use in the US is directed toward raising or lowering the temperature of materials like water, air, steel, and concrete. The quantity of heat needed per unit of temperature change is a fundamental property of any material, known as the **heat capacity** C of the material,

$$dQ = CdT .$$ (5.23)

The value of C in this equation depends upon how other variables such as temperature and pressure are varied with the addition of the heat dT. In order to make eq. (5.23)

---

[6] In some texts, a special notation, $đQ$ and $đW$, is introduced for the differential of quantities for which there is no corresponding state function. We use the simple notation dQ, dW and ask the reader to remember the fact that neither the "heat content" nor the "work content" of a system are meaningful concepts. The quantities Q and W similarly always refer in this text to a quantity of energy ΔE transferred through heat or by work.

meaningful, we must specify enough other information to uniquely determine the amount of heat transferred. Two conventional ways to do this are to add the heat while the system is kept at *constant volume* or at *constant pressure*. Both are important, so we study each in turn.

5.4.1 Heat Capacity at Constant Volume

Suppose the volume of a system is kept fixed while it is heated (see Figure 5.6). Such a process is called **isometric** (same volume) heating. Since $dV = 0$, the 1st Law tells us that $dQ = dU$. Labeling the circumstances explicitly,

$$dQ|_V = dU|_V ,$$ (5.24)

where the $|_V$ indicates that the volume is fixed. With this constraint, the quantity $dQ|_V$ is well-defined. Heat added at constant volume thus goes entirely into increasing the internal energy. Comparing this with eq. (5.23) we can define the **heat capacity at constant volume** C_V through

$$C_V = \left.\frac{dQ}{dT}\right|_V = \left.\frac{\partial U}{\partial T}\right|_V ,$$ (5.25)

where in the last step we have used partial derivative notation because in general U will depend on both V and T. The case of an ideal gas, where U depends on T alone, is special. As a simple example, the monatomic ideal gas of N molecules described by (5.6) has $U = (3Nk_B/2)T$, giving a heat capacity

$$C_V (\text{monatomic ideal gas}) = \tfrac{3}{2}Nk_B = \tfrac{3}{2}nR .$$ (5.26)

Non-monatomic ideal gases, such as O_2, CO_2, and water vapor, have more complicated, temperature-dependent heat capacities that reflect the *thawing of*

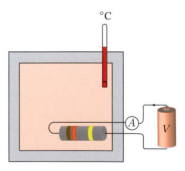

Figure 5.6 A scheme for measuring the heat capacity C_V of a gas. A quantity of gas is placed in an insulated container. A known amount of heat is added to the gas by running a known current (note the *ammeter*) through a resistor, and the temperature rise of the gas is recorded. To measure C_p, one wall of the container could be replaced by a piston maintaining constant pressure.

Example 5.2   Compression of an Ideal Gas

An ideal gas provides an excellent laboratory for thinking about the quantities that appear in the 1st Law. The internal energy of an ideal gas $U = \hat{c}_V N k_B T$ depends only on its temperature. Imagine starting with a liter of an ideal gas at **Standard Temperature and Pressure (STP)**,[a]

$$\text{STP:}\quad T = 0\,°\text{C} = 273.15\,\text{K}, \quad p = 10^5\,\text{Pa}.$$

Compress the gas rapidly down to half a liter. It will heat up significantly. Now, without changing its volume, let the gas lose heat gradually to its surroundings until it returns to room temperature. The internal energy of the gas is the same as at the starting point, but work has been added and heat has been extracted from it. Energy is conserved – the heat given off exactly equals the work done in compressing the gas – but there is no meaning to the "heat content" or "work content" of the gas, and therefore neither heat nor work can be thought of as state functions.

   For a real (non-ideal) gas, the internal energy also depends on the volume, due to the interactions between molecules and also to their finite size. Nevertheless, when a typical non-ideal gas has returned to the ambient temperature in a situation such as that of this example, the change in its internal energy is small compared to the amount of work done on it or the amount of heat given off.

---

[a] Another standard set of conditions is **Normal Temperature and Pressure** (NTP), where $T = 20\,°\text{C}$ and $p = 1\,\text{atm}$. Measurements are also often made at $25\,°\text{C}$ and 1 atm; we use each of these conditions at various places in the text, depending on context.

---

degrees of freedom as the temperature rises. As mentioned in §5.1.3, quantum mechanical effects suppress internal degrees of freedom of gas molecules at low temperatures. As the temperature increases, these degrees of freedom gradually become *active* and increase the heat capacity, which approaches $\frac{1}{2}k_B$ per degree of freedom per molecule at high temperature. The details of this mechanism are discussed in more detail in §8.7. Thus we write the heat capacity of an ideal gas in general as

$$C_V(T) = \hat{c}_V(T)N k_B = \hat{c}_V(T)nR, \qquad (5.27)$$

where $\hat{c}_V(T)$ is 1/2 the number of *active* degrees of freedom of each molecule in the ideal gas at temperature T. (Examples of degrees of freedom for common molecules are listed in Table 5.1.)[7]

   As an example, the heat capacity per molecule of H_2 in the gaseous state is shown as a function of temperature in Figure 5.7. At low temperatures, $C_V/N = \frac{3}{2}k_B$. At a temperature just over 100 K, the rotational degrees of freedom (of which there are two) increase the heat capacity per molecule from $\frac{3}{2}k_B$ to $\frac{5}{2}k_B$. Above 1000 K the two vibrational degrees of freedom (see Table 5.1) kick in and C_V/N rises to $\frac{7}{2}k_B$. At even higher temperatures C_V/N continues to rise as hydrogen molecules begin to dissociate and ionize (§9.3). Violations of the ideal gas law coming

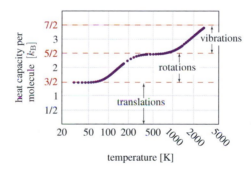

Figure 5.7 The constant volume heat capacity per molecule of H_2 as a function of temperature, showing the thawing of rotational and vibrational degrees of freedom. Data from [23] and [24].

from the finite sizes of the molecules and their interactions lead to further complications in practical calculations of heat capacities from first principles.

   The law of Dulong and Petit (5.8) states that the heat capacity of a pure metal should be $3k_B N = 3nR$. The heat capacities of some materials important for energy issues are given in Table 5.2. Although the heat capacities *per kilogram* vary over a wide range, the heat capacities per mole of many common substances cluster around 20–30 J/mol K. In particular, aluminum, copper, gold, and iron all lie quite close to $3R \cong 24.9$ J/mol K as predicted by the law of Dulong and Petit.

---

[7] In general, we use $\hat{c}_V \equiv C_V/N k_B$ to denote the heat capacity of a substance per molecule, in units of k_B.

Table 5.2 Specific heat capacities at constant pressure (kJ/kg K) of some interesting solids, liquids, and gases. For all materials, the heat capacity is given at constant pressure (1 atm). For solids and liquids, heat capacity is roughly the same at constant volume, while for gases the constant volume heat capacity is generally lower. For pure substances the molar heat capacity (kJ/mol K) is given as well. All heat capacities are at 25 °C unless otherwise noted.

Substance	Specific heat capacity c_p (kJ/kg K)	Molar heat capacity (J/mol K)
Ice (-10 °C)	2.05	38.1
Dry ice (CO_2) (-100 °C)	0.71	31.2
Aluminum	0.90	24.2
Copper	0.39	24.4
Gold	0.13	25.4
Iron	0.45	25.1
Steel	0.52	–
Window glass	≈ 0.84	–
Granite	≈ 0.80	–
Wood	1.2–2.3	–
Dry soil	≈ 0.8	–
Water (15–60 °C)	4.18	75
Ethanol (l)	2.44	112
Sodium (l)	1.39	32
Steam (100 °C)	2.08	37.5
Dry air (STP)	1.0034	29.2
Air (23 °C, 41% rel. humidity)	1.012	29.2
Hydrogen	14.3	28.8
Helium	5.19	20.8
CO_2	0.84	36.9

5.4.2 Heat Capacity at Constant Pressure

While gases are often heated and cooled at fixed volume, there are many situations in which a system or material is heated in an environment at fixed pressure. For example, air heated in a house can usually escape through cracks around the doors and windows, so that the gas that remains in the house maintains a constant pressure as the temperature increases. Processes that take place at fixed pressure are known as **isobaric**.

The **constant pressure heat capacity** C_p of a system is defined to be the amount of heat that must be added per unit temperature change at constant pressure,

$$C_p = \left. \frac{dQ}{dT} \right|_p . \tag{5.28}$$

According to the 1st Law, when heat is added at constant pressure, both the internal energy and the volume of the system change,

$$dQ|_p = (dU + p\,dV)|_p , \tag{5.29}$$

thus

$$C_p = \left. \frac{\partial U}{\partial T} \right|_p + p \left. \frac{\partial V}{\partial T} \right|_p . \tag{5.30}$$

Specifying constant pressure conditions makes dQ well-defined. Note, however, that the heat added does not go to an increase in internal energy alone: some of the thermal energy added at constant pressure does work on the environment by increasing the volume of the system. In general, therefore, C_p is greater than C_V. (There are systems that contract when heated, though they are relatively unusual. Water, which contracts from its melting point up to ~ 4°C, is a relevant example. For such substances, $C_p < C_V$.)

We know from common experience that solids and liquids do not change their volumes significantly with temperature, so we expect that $C_p \cong C_V$ for solids and liquids. Gases generally expand substantially when heated, however, so we expect C_p to be larger than C_V for a gas. For an ideal gas we can easily relate C_p to C_V. Since, for an ideal gas, U is a function of T alone (see §5.1.3), $dU/dT|_p = dU/dT|_V = C_V(T)$. Substituting this into eq. (5.30) and using $dV/dT|_p = Nk_B/p$, which follows from the ideal gas law (5.13), we find

$$C_p(T) = C_V(T) + Nk_B = (\hat{c}_V + 1)Nk_B . \tag{5.31}$$

In particular, for a monatomic ideal gas

$$C_p = C_V + Nk_B = \tfrac{3}{2}Nk_B + Nk_B = \tfrac{5}{2}Nk_B = \tfrac{5}{2}nR . \tag{5.32}$$

The *noble gases* helium, neon, argon, krypton, and xenon, for example, have constant pressure molar heat capacities within 1% of $5R/2 = 20.79$ J/mol K at STP. For non-ideal gases, the relationship between C_p and C_V can be derived in a similar fashion to eq. (5.31) when the equation of state is known. The ratio $\gamma \equiv C_p/C_V$ is quite important in the study of engines, and is known as the **adiabatic index**, also sometimes called the **isentropic index**. For an ideal monatomic gas, $\gamma = 5/3$.

5.4.3 Specific Heat Capacity

Since the heat capacity of a substance scales proportionally to the quantity of material, it is traditional to quote heat capacities *per unit mass*. Such quantities are known as **specific heat capacities** (**specific heats** for short), and are denoted c_V or c_p. Specific heats have units of J/kg K.

Example 5.3 How Much Energy Does it Take to Heat the Water for a Shower?

According to the US uniform plumbing code, a shower should run at about 9.5 L/min. Suppose a shower takes 8 minutes and the water for the shower is heated from 10 °C to 40 °C (104 °F), so $\Delta T = 30$ K. The thermal energy needed for the shower water is then

$$Q = c_p m \Delta T \cong (4.18 \text{ kJ/kg K})(8 \times 9.5 \text{ L})(1 \text{ kg/L})(30 \text{ K}) \cong 9.5 \text{ MJ} .$$

So, running a typical shower takes about as much energy as the food energy in a typical 2400 Calorie daily diet, or about 1% of the average American's total daily energy use. Low-flow showerheads can easily reduce this energy use by a factor of 2, as can shorter (or colder!) showers (Problem 5.7).

Example 5.4 Heat Transport With Helium in a Nuclear Power Plant

An advanced nuclear power plant design uses helium to transport heat from the reactor core out to turbines that generate electricity. The efficiency for converting reactor heat to transmitted electricity is about 40%. The core temperature is about 1000 °C and the spent helium is at about 100 °C.

If the cycle time for helium in the reactor is about 5 minutes, how much helium is needed to transport heat in a 100 MWe nuclear power plant?[a] How does this compare to the world's annual helium production?

To produce 100 MW of electric power at an efficiency of 40%, the power plant must transport 250 MW of power from the hot reactor core to the electricity-generating turbines. The helium is heated by $\Delta T \cong 900$ K. With a specific heat capacity of approximately 5.2 kJ/kg K, each kilogram of He delivers $5.2 \times 900 \cong 4.7$ MJ of thermal energy. To transport 250 MW requires $(250 \text{ MJ/s}) \div (4.7 \text{ MJ/kg}) \cong 50$ kg/s of helium. The entire helium cycle through the reactor is estimated to take about 5 minutes. So the "charge" of helium in the reactor must be $300 \text{ s} \times 50 \text{ kg/s} \cong 15\,000$ kg of He.

In 2010, estimated world helium production was $150 \times 10^6 \text{ m}^3$ measured at 15 °C and 1 atm [25]. Each cubic meter is about $(4 \text{ g/mol}) \times (1000 \text{ L}) \div (24 \text{ L/mol}) \cong 0.17$ kg. So the reactor requires $1.5 \times 10^4 \text{ kg} \div 0.17 \text{ (kg/m}^3) \cong 88\,000 \text{ m}^3$ of He – a tiny fraction ($\lesssim 0.06\%$) of the world's annual production.

---

[a] MWe refers to electric power output as opposed to the thermal power generated by the reactor, which is often denoted MWth. See §13 for further discussion.

Other useful definitions include **molar heat capacities** (in units of J/mol K), denoted $\hat{C}_V$ and $\hat{C}_p$, heat capacities per molecule $\hat{c}_V k_B$ and $\hat{c}_p k_B$ as already introduced earlier, and **volumetric heat capacities** (in units of J/m^3 K). Specific and molar heat capacities for a few common substances are given in Table 5.2. Heat capacities change with variations in temperature, pressure, etc., particularly near phase transitions (§5.6) and when quantum effects are relevant (as in Figure 5.7). Specific heat capacities are therefore usually quoted under certain standard conditions, for example at STP or at NTP.

Note that water, hydrogen, and helium have high specific heats. Water plays an important role in many energy systems in storing and transporting thermal energy, for example in radiators used for household heating. Light elements such as helium and lithium have high specific heat

capacities because, while their molar heat capacities are typical, their molar masses are very small. Helium, for example, with a molar mass of 4.00, has a very large specific heat of 5.19 kJ/kg K. This, along with the fact that it is chemically neutral and cannot be made radioactive (see §16), makes it an excellent material for transporting heat in nuclear reactors (Example 5.4).

5.5 Enthalpy

When heat is added to a system at constant volume, the internal energy U changes by exactly the amount of heat that is added. It proves extremely useful to define a new state function called **enthalpy**,

$$H \equiv U + pV , \tag{5.33}$$

Example 5.5   Enthalpy Versus Internal Energy: Adding Heat at Constant Pressure or Constant Volume

A quantity of heat Q is added to a volume V of an ideal monatomic gas initially at temperature T and pressure p. *Compare the results if the heat is added (a) at constant volume or (b) at constant pressure.*

Constant Volume The heat increases the internal energy, $\Delta U = Q$, and the change in the temperature is $\Delta T = \Delta U/C_V = \frac{2}{3}Q/Nk_B$. The pressure increases by $\Delta p = Nk_B\Delta T/V = \frac{2}{3}Q/V$. Because the volume remains fixed, no work is done, and the increase in enthalpy is $\Delta H = \Delta U + V\Delta P = \frac{5}{3}Q$.

Constant Pressure The heat increases the enthalpy, $\Delta H = Q$, and the change in the temperature is $\Delta T = \Delta H/C_p = \frac{2}{5}Q/Nk_B$. The volume increases by $\Delta V = Nk_B\Delta T/p = \frac{2}{5}Q/p$. The work done is $p\Delta V = \frac{2}{5}Q$, and by the 1st Law, the increase in the internal energy of the gas is $\Delta U = \Delta H - p\Delta V = \frac{3}{5}Q$.

Thus, a given amount of heat results in a 67% larger temperature and internal energy increase if delivered at constant volume compared to constant pressure. The internal energy increases less at constant pressure because some of the added heat is converted to work as the gas expands.

Internal Energy, Enthalpy, and Heat Capacities

For heat added to a system at constant volume, no work is done, and

$$dQ = dU = C_V dT, \qquad C_V = (\partial U/\partial T)|_V.$$

For heat added to a system at constant pressure,

$$dQ = dH = C_p dT, \qquad C_p = (\partial H/\partial T)|_p,$$

where H is the *enthalpy*

$$H = U + pV.$$

For most systems, dU is smaller than dH because work $dW = pdV$ is done through expansion as heat is added, so $C_p > C_V$.

For an ideal gas, $C_p - C_V = Nk_B$.

which plays the same role for processes at constant pressure. *When heat is added to a system at constant pressure, the enthalpy changes by exactly the amount of heat that is added.* To derive this result we need only combine the differentials

$$d(pV) = pdV + Vdp, \qquad (5.34)$$

with the 1st Law applied at constant pressure (5.29),

$$dQ|_p = dU|_p + pdV|_p$$
$$= d(U + pV)|_p \equiv dH|_p. \qquad (5.35)$$

Enthalpy is clearly a state function because it is composed of U, V, and p, all of which are state functions. The relation $dQ|_p = dH|_p$ is exactly analogous to $dQ|_V = dU|_V$ (5.24), and in a fashion similar to eq. (5.25) it follows that

$$C_p = \left.\frac{dQ}{dT}\right|_p = \left.\frac{\partial H}{\partial T}\right|_p. \qquad (5.36)$$

Because enthalpy is most useful when pressure is the variable under external control, H is most naturally regarded as a function of T and p, where V can be determined from T and p by an equation of state such as eq. (5.13).

The concept of enthalpy plays a key role in analyzing many energy systems that involve thermal energy conversion, such as power plants (§12), geothermal (§32) and fossil (§33) energy sources, and energy storage (§37).

5.6 Phase Transitions

The definition of heat capacity (5.23) suggests that a gradual flow of heat into a system results in a gradual increase in its temperature. If C_V is finite, then $dU = C_V dT$ implies that any increase in internal energy, no matter how small, results in a correspondingly small but nonzero increase in temperature. This relationship breaks down when materials undergo **phase transitions**, such as melting or boiling. When a material is at a temperature associated with a phase transition, the addition or removal of heat has the effect of changing the relative fractions of the material in different phases, without changing the temperature of the material. Thus, for example, adding heat to a cup containing ice and water melts ice, but does not raise the temperature until all the ice is melted. Such phase transitions play an important role in many energy systems –

including power plants that use steam to drive turbines, as well as air conditioning and refrigeration systems that use more unusual materials with lower boiling points than water. We introduce the basics of phase transitions here and develop and use these ideas further in §9 and §12.

5.6.1 Melting and Boiling Transitions

Consider a block of ice at a pressure of 1 atm. At temperatures below $0\,°C$, the ice has a heat capacity of about $2\,kJ/kg\,K$ (Table 5.2). Adding thermal energy increases the temperature of the ice until it reaches $0\,°C$ (at 1 atm), and then *the temperature ceases to increase further*. Instead, as more thermal energy is added, an increasing fraction of the once-solid block of ice turns into liquid water while the temperature remains fixed at $0\,°C$. The ice *melts* – or *fuses*, to use the term preferred by chemists. (This type of fusion is not to be confused with *nuclear fusion* described in §16.)

One might have thought that as ice is heated, the H_2O molecules would slowly become more mobile, first starting to rotate in place, and then beginning to wander about. If this were true, the ice would slowly soften and become more plastic, eventually flowing freely in a liquid state. In reality, however, solids remain quite rigid as T increases, until they abruptly turn to liquid *at their melting point*. Phase transitions such as the change from solid ice to liquid water, where a finite amount of energy is added while the system stays at a fixed temperature, are known as **first-order** phase transitions. The theory and classification of phase transitions is a rich and fascinating branch of physics that goes beyond the scope of this text.

Once the entire block of ice has melted, the smooth relation between thermal energy and temperature resumes. As more thermal energy is added, the temperature, internal energy, and enthalpy of the water increase, though the heat capacity ($c_V \cong 4.2\,kJ/kg\,K$) is substantially higher for liquid water than for ice. When the temperature reaches $100\,°C$, water undergoes another first-order phase transition – it turns to **water vapor** or **steam** through the process of **vaporization**. Below $100\,°C$, the attraction between the molecules of H_2O is sufficient to keep them close enough that they move by slithering around one another at a roughly constant separation. Above $100\,°C$, the kinetic energy of thermal motion is enough to overcome the intermolecular attraction, and the molecules bounce around as a gas, filling as much space as is allotted to them. Figure 5.9 gives a schematic sense of the configuration of H_2O molecules in solid, liquid, and gas states.

The transitions from solid to liquid and liquid to gas are typical of many substances. The melting and boiling points of different materials depend upon the strength of

Figure 5.8 At atmospheric pressure ice undergoes a phase transition to liquid water at $0\,°C$, while *dry ice*, solid CO_2, *sublimes* directly into a vapor at $-78.5\,°C$.

Figure 5.9 H_2O molecules in solid (left), liquid (center), and gas (right) states of water. The regular, crystalline array in the solid actually has more intermolecular space than the liquid in which the molecules move around but remain in close proximity. Unlike the liquid and solid phases, in the gas phase the spacing of molecules varies strongly with the pressure and temperature.

their intermolecular forces and the molecular weights of the chemical compounds. As a rule of thumb, elements or molecules that are both chemically relatively inert and low in molecular weight have the lowest melting and boiling points. Compounds with strong intermolecular forces and high molecular weight melt and vaporize at relatively high temperature. Methane, which has roughly the same molecular weight (16) as water (18), melts and vaporizes at a lower temperature because water has strong intermolecular forces (hydrogen bonds) that methane does not have. Examples of these regularities are shown in Figure 5.10, where the boiling points of the **noble gases** (monatomic, almost non-reactive gases with closed outer

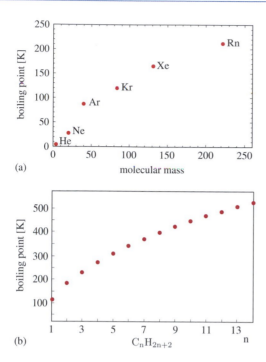

(a)

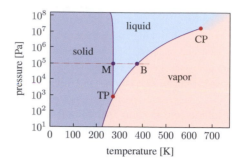

Figure 5.10 (a) Boiling points of the noble gases; (b) boiling points of the linear hydrocarbons C_nH_{2n+2} as a function of n, illustrating the increase of boiling point with molecular mass. (Based on data by Techstepp reproduced under CC-BY-SA 3.0 license via Wikimedia Commons)

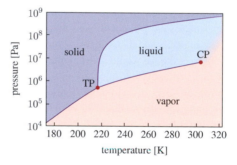

Figure 5.11 The phase diagram of water showing the first-order phase transitions where ice melts and water boils. At the *triple point* (TP), solid, liquid, and gaseous H_2O can all coexist in equilibrium. At pressures and temperatures above the *critical point* (CP), liquid and vapor are no longer distinguished. The normal melting (M) and boiling (B) points at 1 atm are also marked.

Figure 5.12 The phase diagram of carbon dioxide. Note that CO_2 has no liquid phase at atmospheric pressure (10^5 Pa), as anyone who has encountered dry ice can attest.

electron shells – in the right-most column of the Periodic Table D.1) and the simple linear carbon-hydrogen chain molecules (*paraffins*, see §11.2.1) with composition C_nH_{2n+2} (methane, ethane, butane, propane, etc.) are shown.

The temperatures at which melting (fusion) and boiling (vaporization) occur depend upon the pressure. The solid-to-liquid transition point is usually only weakly affected by pressure, but the liquid-to-gas transition is quite sensitive. At a pressure of half an atmosphere, such as one would encounter on a six kilometer high mountain, the boiling point of water is only 82 °C, while at a pressure of 10 atmospheres, water boils above 180 °C. Information about the phase of water at different pressures and temperatures can be displayed graphically in a **phase diagram** as shown in Figure 5.11. The domains where water is a solid, liquid, and gas are separated by curves that mark the solid-to-liquid and liquid-to-gas transitions. At very low pressure, a solid-to-gas transition known as **sublimation** can also occur. (Solid CO_2 sublimes at atmospheric pressure, see Figure 5.8.) Figure 5.11 is a simplified diagram: it displays the liquid and vapor phases correctly, but at very

low temperatures there are many distinct solid phases of ice differentiated by their crystal structure.

Phase diagrams contain much interesting and important information. From Figure 5.11 we see that solid, liquid, and vapor coexist at the **triple point**, where $T = 273.16$ K and $p = 612$ Pa. We also see that the melting point of water *decreases slightly* with pressure. This is unusual – it increases for most substances – and is associated with the fact that the density of water increases when it melts. The phase diagram of water also shows that the sharp distinction between water liquid and vapor ceases at the **critical point**, which occurs at 647 K and 22.1 MPa. Phase diagrams are known in varying detail for all common substances. For example, the phase diagram of CO_2 is shown in Figure 5.12. Some of the problems explore other features of phase diagrams.

Note that the phase diagrams in Figures 5.11 and 5.12 do not fully specify the state of a system on the curves that

---

Example 5.6   Internal Energy Change When Water Boils

Given the latent heat of vaporization, how does the internal energy change when water boils? The measured latent heat of vaporization of water is $\Delta h_{\text{fusion}} = 2257\,\text{kJ/kg}$ (at $100\,°C$ and 1 atm). Δu_{fusion} can be computed from $\Delta h = \Delta u + p\Delta v$. To compute the change in volume when water boils, we need the densities of water and steam at $100\,°C$, which are $957\,\text{kg/m}^3$ and $0.597\,\text{kg/m}^3$ respectively.

$$\Delta u_{\text{boiling}} = \Delta h_{\text{boiling}} - p\left(\frac{1}{\rho_{\text{steam}}} - \frac{1}{\rho_{\text{water}}}\right)$$
$$= 2257\,\text{kJ/kg} - 101\,325\,\text{Pa} \times \left(1.68\,\text{m}^3/\text{kg} - 0.001\,\text{m}^3/\text{kg}\right)$$
$$= 2257\,\text{kJ/kg} - 170\,\text{kJ/kg} = 2087\,\text{kJ/kg}$$

This confirms that the change in internal energy is less than the heat added because some of the heat does the work of expanding the newly created steam. The amount is small, only about 7.5%, but definitely not negligible.

---

mark a phase transition. A substance at its boiling point at some fixed temperature and pressure, for example, is represented by a single point on the diagram. Depending on how much energy is added to the substance, it may exist as a pure liquid, a pure vapor, or any mixture of the two. Likewise, at the melting point, the relative proportions of solid and liquid are not specified. When we study the use of fluids to convert between heat and mechanical energy in §12, we introduce other ways of looking at the change of phase that include this information as well.

5.6.2 Latent Heat and Enthalpies of Fusion and Vaporization

The amount of energy needed to turn a solid to a liquid or a liquid to a vapor is called the **latent heat** of fusion or vaporization. If the phase change takes place at constant pressure, then according to eq. (5.35) the added thermal energy equals the increase in *enthalpy* of the substance. Thus, the latent heats of fusion and vaporization are usually denoted ΔH_{fusion}, ΔH_{vap}, and are referred to as the **enthalpy of fusion** and the **enthalpy of vaporization**. Enthalpies of vaporization are characteristically much greater than enthalpies of fusion. The reasons for this are elaborated in §9, where the enthalpies of fusions and vaporization of some common and/or interesting substances can be found in Tables 9.2 and 9.3. Since p is constant and $H = U + pV$, ΔH_{vap} is related to the change in internal energy, ΔU_{vap}, by $\Delta H_{\text{vap}} = \Delta U_{\text{vap}} + p\Delta V_{\text{vap}}$. Usually we deal with enthalpies, internal energies, and volumes *per unit mass*, known as *specific* enthalpies, etc., of phase change, and denoted Δh, Δu, and Δv.

Water has a remarkably large enthalpy of vaporization. This makes boiling water a very effective way to store

Melting and Vaporization

Melting (fusion) and vaporization are first-order phase transitions. Although heat is added, the temperature stays fixed until all material has changed phase. Enthalpies of vaporization are typically greater than enthalpies of fusion. Water has an exceptionally large enthalpy of vaporization.

energy. The specific enthalpy density added by boiling water at one atmosphere, 2.257 MJ/kg, is greater than the density at which energy can be stored in most chemical batteries. For comparison, lithium-ion batteries store about 0.5 MJ/kg and TNT stores about 4.6 MJ/kg in chemical energy. On the other hand, water vapor is quite diffuse at ordinary pressures so the energy density per unit volume is much less than that of either a battery or TNT. Steam engines and turbines have used water and the liquid/vapor phase transition as a way of storing and transforming energy for centuries. These mechanisms dominate the production of electricity from thermal energy sources such as fossil fuels. We return to this topic in §12.

Discussion/Investigation Questions

5.1   A one liter box containing 0.1 L of liquid water at NTP is lifted from the floor and placed on a table one meter high. How much has the energy of the water changed? How much has its internal energy changed? The water is now heated until it is all converted to vapor. Has all this heat gone into the internal energy of the water?

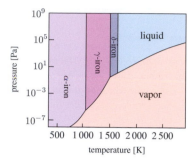

Figure 5.13 A phase diagram for pure iron. The α, γ, and δ phases are solids with different crystal structure. See Question 5.4.

(You can assume that the box itself has negligible heat capacity.)

5.2   A mixture of sodium and potassium *salts* is used for energy storage in certain solar energy plant designs. Such a mixture is used in a temperature range between 300 °C and 500 °C where it is a liquid and where its specific heat is approximately 1.5 kJ/kg K. Why would engineers choose a substance like this over steam?

5.3 CO_2 is a colorless gas. Why do you suppose the CO_2 vapor in Figure 5.8 streaming away from the dry ice appears white?

5.4   The phase diagram for pure iron is shown in Figure 5.13. Iron has three different solid phases, called α, γ, and δ-iron, with different crystal structures. Discuss the behavior of iron as it is heated at different pressures, at 1 atm or at 10^{-6} atm, for example. Under what conditions can several phases of iron coexist? Does iron have a triple point similar to water's?

Problems

5.1   A cylinder initially contains $V_0 = 1\,\text{L}$ of argon at temperature $T_0 = 0\,°C$ and pressure $p_0 = 1$ atm. Suppose that the argon is somehow made to expand to a final volume $V = 2\,\text{L}$ in such a way that the pressure rises proportionally to the volume, finally reaching $p = 2$ atm. How much work has the argon done? What is its final temperature T and how much thermal energy has been added to the gas?

5.2   Non-rigid airships known as *blimps* have occasionally been used for transportation. A blimp is essentially a balloon of volume V filled with helium. The blimp experiences a **buoyancy force** $F = (\rho_{\text{atm}} - \rho_{\text{He}})Vg$, where ρ_{atm} is the density of the surrounding air and $g \cong 9.8\,\text{m/s}^2$. The blimp maintains its shape because the pressure of the helium p_{He} is kept higher than the pressure of the surrounding atmosphere p_{atm}. A modern blimp with a volume $V = 5740\,\text{m}^3$ can lift

a total mass $m = 5825\,\text{kg}$ at temperature $0\,°C$ and $p_{\text{atm}} = 1$ atm. Assuming both air and helium behave as ideal gases, estimate the pressure of the helium gas inside the blimp. Assuming the blimp's volume is kept constant, how much mass can it lift at $20\,°C$?

5.3   When air is inhaled, its volume remains constant and its pressure increases as it is warmed to body temperature $T_{\text{body}} = 37\,°C$. Assuming that air behaves as an ideal gas and that it is initially at a pressure of 1 atm, what is the pressure of air in the lungs after inhalation if the air is initially (a) at room temperature, $T = 20\,°C$; (b) at the temperature of a cold winter day in Boston, $T = -15\,°C$; (c) coming from one person's mouth to another's during cardiopulmonary resuscitation (CPR). For CPR, assume the air is initially at body temperature and at $p = 1$ atm.

5.4   **[H]** Everyday experience indicates that it is much easier to compress gases than liquids. This property is measured by the **isothermal compressibility**, $\beta = -\frac{1}{V}\frac{\partial V}{\partial p}\big|_T$, the fractional change in a substance's volume with pressure at constant temperature. What is the isothermal compressibility of an ideal gas? The smaller β is, the less work must be done to pressurize a substance. Compare the work necessary at $20\,°C$ to raise the pressure of a kilogram of air and a kilogram of water ($\beta_{\text{water}} \cong 4.59 \times 10^{-10}\,\text{Pa}^{-1}$) from 1 atm to 10^4 atm.

5.5   **[HT]** A cylindrical tube oriented vertically on Earth's surface, closed at the bottom and open at the top (height 100 m, cross-sectional area $1\,\text{m}^2$) initially contains air at a pressure of one atmosphere and temperature 300 K. A disc of mass m plugs the cylinder, but is free to slide up and down without friction. In equilibrium, the air in the cylinder is compressed to $p = 2$ atm while the air outside remains at 1 atm. What is the mass of the disc? Next, the disc is displaced slightly in the vertical (z) direction and released. Assuming that the temperature remains fixed and that dissipation can be ignored, what is the frequency with which the disc oscillates?

5.6   How much energy does it take to heat 1 liter of soup from room temperature ($20\,°C$) to $65\,°C$? (You may assume that the heat capacity of the soup is the same as for water.)

5.7   A "low-flow" showerhead averages 4.8 L/min. Taking other data from Example 5.3, estimate the energy savings (in J/y) if all the people in your country switched from US code to low-flow shower heads. (You will need to estimate the average number of showers per person per year for the people of your country.)

5.8   A solar thermal power plant currently under construction will focus solar rays to heat a molten salt working fluid composed of sodium nitrate and potassium nitrate. The molten salt is stored at a temperature of $300\,°C$, and heated in a power tower to $550\,°C$. The salt has a specific heat capacity of roughly 1500 J/kg K in the temperature range of interest. The system can store 6000 metric tonnes of molten salt for power generation when no solar energy is available. How much energy is stored in this system?

5.9 A new solar thermal plant being constructed in Australia will collect solar energy and store it as thermal energy, which will then be converted to electrical energy. The plant will store some of the thermal energy in graphite blocks for nighttime power distribution. According to the company constructing the plant, the plant will have a capacity of 30 MkWh/year. Graphite has a specific heat capacity of about 700 J/kg K at room temperature (see [26] for data on the temperature dependence of the specific heat of graphite). The company claims a storage capacity of 1000 kWh/tonne at 1800 °C. Is this plausible? Explain. If they are planning enough storage so that they can keep up the same power output at night as in the day, roughly how much graphite do they need? Estimate the number of people whose electrical energy needs will be supplied by this power plant.

5.10 A start-up company is marketing steel "ice cubes" to be used in place of ordinary ice cubes. How much would a liter of water, initially at 20 °C, be cooled by the addition of 10 cubes of steel, each 2.5 cm on a side, initially at −10 °C? Compare your result with the effect of 10 ordinary ice cubes of the same volume at the same initial temperature. Would you invest in this start-up? Take the density of steel to be 8.0 gm/cm^3 and assume the heat capacities to be constants, independent of temperature.

5.11 Roughly 70% of the 5×10^{14} m^2 of Earth's surface is covered by oceans. How much energy would it take to melt enough of the ice in Greenland and Antarctica to raise sea levels 1 meter? Suppose that 1% of energy used by humans became waste heat that melts ice. How long would it take to melt this quantity of ice? If an increase in atmospheric CO_2 led to a net energy flux of 2000 TW of solar energy absorbed into the Earth system of which 1% melts ice, how long would it take for sea levels to rise one meter?

5.12 Carbon dioxide sublimes at pressures below roughly 5 atm (see Figure 5.12). At a pressure of 2 atm this phase transition occurs at about −69 °C with an enthalpy of sublimation of roughly 26 kJ/mol. Suppose a kilogram of solid CO_2 at a temperature of −69 °C is confined in a cylinder by a piston that exerts a constant pressure of 2 atm. How much heat must be added to completely convert it to gas? Assuming CO_2 to be an ideal gas, how much work was done by the CO_2 in the course of vaporizing? If the ambient pressure outside the cylinder is 1 atm, how much useful work was done? Finally, by how much did the internal energy of the CO_2 change when it vaporized?

Heat Transfer

The transfer of heat from one system to another, or within a single system, is an issue of central importance in energy science. Depending on the circumstances, we may want to impede heat flow, to maximize it, or simply to understand how it takes place in natural settings such as the oceans or Earth's interior. In practical situations, such as the transfer of heat from burning coal to steam flowing through pipes in a power plant boiler, the systems are far from equilibrium and consequently hard to analyze quantitatively. Although quantitative descriptions of heat transfer are difficult to derive from first principles, good phenomenological models have been developed since the days of Isaac Newton to handle many important applications. In this chapter we focus on those aspects of heat transfer that are related to energy uses, sources, and the environment. A thorough treatment of heat transfer including many applications can be found in [27].

Reader's Guide
This chapter describes the transfer of thermal energy by conduction, convection, and radiation. The principal focus is on conduction, but radiation and both free and forced convection are also characterized. The basic concepts are applied in the context of heat loss from buildings. The heat equation is introduced and applied to the problem of annual ground temperature variation.
    Prerequisites: §2 (Mechanics), §3 (Electromagnetism), §4 (Waves and light), §5 (Thermal energy).

6.1 Mechanisms of Heat Transfer

Heat is transferred from one location or material to another in three primary ways: *conduction* – transport of heat through material by energy transfer on a molecular level, *convection* – transport of heat by collective motion of material, and *radiation* – transport by the emission or absorption of electromagnetic radiation. Before delving into these mechanisms in detail, we summarize each briefly and explain where it enters into energy systems described in later chapters.

6.1.1 Conduction

Within a solid, liquid, or gas, or between adjacent materials that share a common boundary, heat can be transferred by interactions between nearby atoms or molecules. As described in §5, at higher temperatures particles have more energy on average. When a faster-moving atom from a

warmer region interacts with a slower-moving atom from a cooler region, the faster atom usually transfers some of its energy to the slower atom. This leads to a gradual propagation of thermal energy without any overall motion of the material. This is **heat conduction**.

Conduction is the principal mechanism for heat transfer through solids, and plays a role in many energy systems. For example, conduction determines the way that heat flows in the solid outer layers of Earth's crust (§32). It is also responsible for heat loss through the solid walls of buildings. Heat can be conducted in gases and liquids as well, but convection is – under many circumstances – more effective at moving heat in fluids. Heat transfer by conduction is the simplest of the three mechanisms to describe quantitatively. The fundamental theory was developed by Joseph Fourier in the early 1800s, and has quite a bit in common with the theory of electricity conduction that was described in §3.

6.1.2 Convection

Collective motion of molecules within a fluid can effectively transport both mass and thermal energy. This is the mechanism of **convection**. Examples of convection include currents or eddies in a river, flow in a pipe, and

(a)

(b)

Figure 6.1 (a) A metal spoon used to stir boiling water quickly conducts heat making it impossible to hold without a plastic handle. (b) Convective plumes of fog and steam rising from power plant cooling towers carry heat upward quickly. (Credit: Bloomberg/Bloomberg/Getty Images)

the rising currents of air in a thermal updraft over warm land. As the constituent molecules composing a fluid move through convection, they carry their thermal energy with them. In this way, heat can be transported rapidly over large distances. Convection generally transports heat more quickly in gases than does conduction.

Convection in energy applications can be divided into two general categories: **forced convection** and **free convection**.

Forced convection occurs when a fluid is forced to flow by an external agent (such as a fan). Forced convection is a very effective means of transporting thermal energy. Often heat transport through forced convection occurs in a situation where the fluid is forced to flow past a warmer or colder body, and the rate of heat transfer is driven by the temperature difference ΔT between the object and the fluid. Although the underlying fluid dynamics can be complex [27], forced convection is often described well by a simple empirical law first proposed by Newton and known as *Newton's Law of cooling*. Forced convection plays a significant role in heat extraction devices such as air-conditioners and in steam engines, where heat must be transported from one location to another as quickly and as efficiently as possible. We analyze forced convection in §6.3.1 below.

Free convection – also known as **natural convection** – occurs when a fluid rises or sinks because its temperature differs from that of its surroundings. Convective heat transfer occurs, for example, when air directly over a surface feature (such as a large blacktop parking lot) is heated from below. The air near the surface warms, expands, and begins to rise, initiating a convective upward flow of heat. Free convection is believed to be the principal method of heat transfer in Earth's liquid mantle and outer core (§32). Convective heat transport in the atmosphere plays an important role in global energy balance and temperature regulation (§34). Free convection usually arises spontaneously as a result of instability and is more difficult to analyze than forced convection. We touch briefly upon free convection here only to the extent that it enters a qualitative discussion of the effectiveness of insulation; we return to this topic later in the context of atmospheric physics.

6.1.3 Radiation

Finally, any material at a nonzero temperature radiates energy in the form of electromagnetic waves. These waves are generated by the microscopic motion of the charged particles (protons and electrons) composing the material. The higher the temperature, the more vigorous the motion of the particles in matter, and the more electromagnetic waves they radiate. As reviewed in §4, electromagnetic waves carry energy. Thus, the warmer an object, the more energy it radiates.

The fundamental law governing thermal radiation, the *Stefan–Boltzmann law*, is relatively simple, but its implementation in practice can be complex. The situation is simplest when a hot object is radiating into empty space. When matter is present, conduction and convection often dominate over radiation. Only radiation, however, can transfer energy across a vacuum. In particular, virtually all of the Sun's energy that reaches Earth comes via radiative heat transfer. To remain in approximate thermal equilibrium, Earth must radiate the same amount of energy back into space. These applications are described in more detail in later chapters. In fact, radiation is also the primary cooling mechanism for a human body at rest: a typical human radiates at about 1000 W, though in comfortable temperature ranges most (∼90%) of this energy is returned by radiation from the surroundings.

6.2 Heat Conduction

Many aspects of heat conduction can be understood using the physical concepts developed in §2, §3, and §5. In particular, an analogy to conduction of electricity in the

Figure 6.2 A traditional glass blower's kiln. The interior is heated to about 1100 °C, at which temperature the kiln is filled with radiation that appears yellow to our eyes.

Mechanisms of Heat Transfer

The principal methods of heat transfer are (1) conduction, where energy of molecular motion is transferred between particles through a material; (2) convection, where thermal energy is carried by the collective motion of a fluid; and (3) radiation, where the motion of hot molecules gives rise to radiation that carries away energy.

---

presence of resistance, developed in this section, helps visualize heat conduction and simplifies the analysis of practical applications.

6.2.1 Heat Flux

Heat conduction takes place when energy is transferred from more energetic particles to less energetic ones through intermolecular or interatomic interactions. When molecules with differing amounts of energy interact, energy will generally be transferred from the more energetic molecule to the less. This leads to a gradual flow of thermal energy away from regions of higher temperature towards regions of lower temperature. In liquids and gases at room temperature, thermal energy resides in the kinetic energy of translation and rotation and to a lesser extent in molecular vibration. Heat is transferred by intermolecular collisions and through the diffusive propagation of individual molecules. In a solid, the thermal energy resides principally in the vibrational energy of the atoms within the solid. The same interatomic forces that hold the atoms in place also transmit the vibrational energy between neighboring particles. A model for this system might be a web of interconnected springs. If springs on one side of the web are set in motion, their energy rapidly diffuses toward the other end.

The temperature in an object through which heat is flowing can vary as a function of position x and time t, $T(x, t)$. As discussed in §5, T at a point x refers to a local average over a small volume, a cubic micron for example, centered at x that contains a great number of individual molecules. The flow of heat through an object is described by a vector-valued function that specifies the amount of thermal energy flowing in a particular direction per unit time and per unit area. These are the characteristics of a *flux* as described in Box 3.1. We therefore define the **heat flux** $q(x, t)$ to be the thermal energy per unit time per unit area, or power per unit area (W/m^2 in SI units), flowing across a surface normal to the vector $\hat{q}$. Like T, q can depend on both x and t. It is also useful to define the rate that thermal energy flows across a surface $dQ/dt = \int_S dS \cdot q(x, t)$, which has the units of power.

6.2.2 Fourier's Law

Everyday experience suggests that the greater the temperature difference across an object, the greater the rate that heat flows through it. Fingers lose heat through insulating gloves, for example, much more rapidly when the ambient temperature is -40 °C than when it is 0 °C. At the molecular level, since temperature and internal energy are roughly proportional, a larger temperature gradient increases the rate at which molecular collisions transfer energy. If the temperature is locally constant, $\nabla T = 0$, then nearby molecules have the same average energy and molecular interactions on average transfer no energy. So it is reasonable to expect a linear relation of the form $q \propto \nabla T$, at least when ∇T is small. In the early nineteenth century, the French mathematician and physicist Joseph Fourier proposed just such a relation on the basis of empirical observations,

$$q(\mathbf{x}, t) = -k \, \nabla T(\mathbf{x}, t). \qquad (6.1)$$

This equation, known as **Fourier's law of heat conduction**, provides an excellent description of conductive heat flow in most practical situations. The constant of proportionality k, which is always positive, is called the **thermal conductivity** of the material. Note the *minus sign* in eq. (6.1). The gradient of T points in the direction in which heat *increases* most rapidly, and heat flows in the opposite direction, toward lower temperature. If the material is homogeneous and isotropic (rotationally invariant), k is independent of x and has no directional dependence. In general, k is temperature-dependent. In solids, the temperature dependence of k is fairly weak, and k can either increase or decrease with increased temperature, depending upon the material. The thermal conductivities of gases are more sensitive to temperature. For most purposes, a value at the average of the temperature range studied is sufficient.

The thermal conductivity of a material is the fundamental property that governs its ability to transport heat. Materials with large k are good conductors of heat; those with small k are thermal insulators. The units of k can be read off from eq. (6.1), $[k] = $ W/m K. The thermal conductivities of a variety of common materials are listed in Table 6.1. Thermal conductivities vary over many orders of magnitude, as illustrated in Figure 6.3. Excellent conductors of electricity, such as copper, silver, and gold, are also excellent heat conductors. This is no accident: the mobility of electrons that makes them good electricity conductors also allows them to transport heat very quickly.

6.2.3 The Analogy Between Conduction of Heat and Electricity

There is a useful analogy between the flow of steady currents in electric circuits and *steady-state* heat conduction. In the former case, charge – a conserved quantity – flows in the direction of the local electric field, which is the negative gradient of the time-independent potential $V(\boldsymbol{x})$. The mathematical description of heat flow is precisely parallel: the conserved quantity is thermal energy, which flows along the negative gradient of the temperature $T(\boldsymbol{x})$.

Table 6.1 Thermal conductivities k and thermal diffusivities a for various substances, including some common building materials, measured near room temperature. Unless otherwise noted, thermal conductivities are from [28]. Thermal diffusivity is calculated from its definition $a = k/\rho c_p$ (see §6.5) unless otherwise noted. Since the values of k and a vary significantly among sources, we quote only two significant figures, and in the case of materials with variable composition, only one.

Material	Thermal conductivity k (W/m K)	Thermal diffusivity a (10^{-7} m^2/s)
Air (NTP)	0.026*	205†
Argon (NTP)	0.018*	190
Krypton (NTP)	0.0094*	100
Xenon (NTP)	0.0057*	61
Water	0.57†	1.4†
Hardwood (white oak)	0.18	0.98
Softwood (white pine)	0.11	1.5
Sandy soil (dry/saturated)	0.3/2.2†	2.4/7.4†
Peat soil (dry/saturated)	0.06/0.50†	1.0/1.2†
Rock	≈ 3†	≈ 14†
Brickwork (common)	≈ 0.7	≈ 4
Window glass	1.0	5.5
Stone concrete	≈ 0.9	≈ 6
Carbon steel	45	120
Aluminum	240	900

* from [29], † from [30]

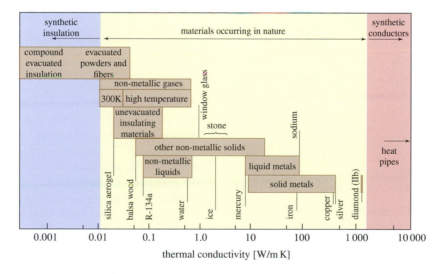

Figure 6.3 The approximate ranges of thermal conductivity of various substances. (All in the neighborhood of room temperature unless otherwise noted.) Data from [27].

Box 6.1   A Derivation of Fourier's Law?

Deriving the thermal conductivity of a material from first principles turns out to be surprisingly difficult – even more difficult in some cases than deriving the drag coefficient c_d for a complicated object like an automobile. For good conductors such as copper, silver, or aluminum, electrons are primarily responsible for both electrical (σ) and thermal (k) conductivity, and the methods of quantum mechanics and statistical physics provide a relationship between σ and k. This relation, $k = \sigma L T$, is known as the **Wiedemann–Franz–Lorenz law**. Here T is the temperature and L is a constant known as the **Lorenz number**, $L = (\pi k_B)^2/3e^2 = 2.44 \times 10^{-8}\,\text{W}\,\Omega/\text{K}^2$. The Wiedemann–Franz–Lorenz law explains the everyday observation that excellent conductors of electricity are also excellent conductors of heat.

For insulating solids such as quartz or salt, the analysis is more difficult, and in fact after 200 years of work there is still no convincing first-principles derivation of Fourier's law for (electrically) insulating solids. The German-born British theoretical physicist Rudolph Peierls once said, "It seems there is no problem in modern physics for which there are on record as many false starts, and as many theories which overlook some essential feature, as in the problem of the thermal conductivity of non-conducting crystals" [31]. Despite the absence of a satisfying derivation, empirical evidence confirms that all solids obey Fourier's law, and thermal conductivities can easily be measured experimentally.

Heat Flux and Fourier's Law

The flow of heat through material is described by a heat flux $q(x, t)$ with units W/m².

For most materials and over a wide range of temperatures, the flux of heat follows *Fourier's law of heat conduction*

$$q(\mathbf{x}, t) = -k\nabla T(\mathbf{x}, t).$$

The coefficient of proportionality k is the *thermal conductivity*, which ranges from very small ($\sim$0.02 W/m K for excellent insulators) to very large ($\sim$300 W/m K for the best metallic conductors).

The local flux of charge is described by the *electric current density* $j(x)$. Ohm's law (3.29) describes flow through a material with electric conductivity σ,

$$j(x) = \sigma E = -\sigma \nabla V(x). \tag{6.2}$$

If we replace j by q, $V(x)$ by $T(x)$, and σ by k, we obtain Fourier's law (6.1) for time-independent heat flow,

$$q(x) = -k\nabla T(x). \tag{6.3}$$

For heat conduction, temperature thus plays the role of a "potential." Its gradient drives the flow of energy that is described by the heat flux q. Thermal conductivity connects the heat flux to the temperature gradient in exactly the same way that electric conductivity relates the current density to the gradient of the potential. The ingredients in this analogy are summarized in Table 6.2.

Just as conservation of charge is described by the local equation (3.33), a similar equation can be used to give a local description of energy conservation in the context of heat flow. Assuming that the material through which heat is flowing is fixed in space so that no energy is transported through convection or material deformation, the same logic that led to eq. (3.31) gives

$$\frac{d}{dt}\int_V d^3x\, u(x, t) = -\oint_S dS \cdot q, \tag{6.4}$$

where u is the internal energy per unit volume. Using Gauss's law to convert this to a local form, we obtain a conservation law relating the density of internal energy to heat flux,

$$\frac{\partial u(x, t)}{\partial t} = -\nabla \cdot q(x, t). \tag{6.5}$$

For compressible materials – gases, in particular – held at constant pressure, the internal energy density u in eq. (6.5) must be replaced by the enthalpy density h; since in this case the material expands or contracts, and additional terms must be added for a complete description (see e.g. [27]).

Because the mathematical analogy between electric and heat flows is exact, we can carry over many results from the study of currents, resistances, and voltage differences to the study of conductive heat flow. In many cases this allows us to determine heat flow for cases of interest simply by borrowing the analogous result from elementary electric current theory.

6.2.4 A Simple Example: Steady-State Heat Conduction in One Dimension

The simplest and perhaps most frequently encountered application of Fourier's equation is in the context of heat

Table 6.2 The analogy between steady-state heat conduction and steady electric currents. There is a mathematical correspondence between the quantities and equations governing these two phenomena, which are precisely parallel except that in the electric case one is primarily interested in the *total* current (the flow of charge per unit time), while in the case of heat the quantity of interest is generally the heat flux (i.e. the flow of energy *per unit area* per unit time). Therefore the electric resistance $R_{electric}$ includes a factor of 1/(Area), while the thermal resistance $R_{thermal}$ does not. The places where the factor of area enters are highlighted in red.

Quantity	Electricity	Heat conduction
Conserved quantity	Q = charge	U = thermal energy
Flux	$j(x)$ = current density	$q(x)$ = heat flux
Motive force	$V(x)$ = electric potential	$T(x)$ = temperature
Conductivity	σ	k
Physical law	Ohm's law	Fourier's law
Differential form	$j = -\sigma \nabla V$	$q = -k\nabla T$
		$q = \dfrac{dU}{dA\,dt} = \Delta T / R_{thermal}$
Integral form	$I = \dfrac{dQ}{dt} = \Delta V / R_{electric}$	$\dfrac{dQ}{dt} = A\Delta T / R_{thermal}$
Block resistance	$R_{electric} = L/\sigma A$	$R_{thermal} = L/k$
Conservation Law	$\nabla \cdot j + \partial \rho / \partial t = 0$	$\nabla \cdot q + \partial u / \partial t = 0$

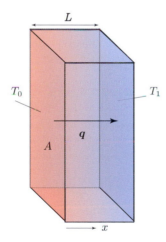

Figure 6.4 A homogeneous block of material of thickness L and cross-sectional area A that conducts heat from temperature T_0 on one end to T_1 on the other end. If the sides of the block are insulated, or if the block is very thin compared to its height and width, then the temperature and heat flux depend only on x.

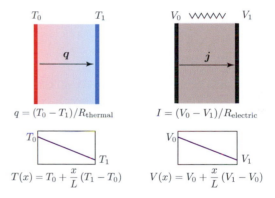

$$q = (T_0 - T_1)/R_{thermal} \qquad I = (V_0 - V_1)/R_{electric}$$

$$T(x) = T_0 + \frac{x}{L}(T_1 - T_0) \qquad V(x) = V_0 + \frac{x}{L}(V_1 - V_0)$$

Figure 6.5 The analogy between thermal resistance and electrical resistance. The only significant difference is one of notation: thermal resistance is defined such that ΔT divided by $R_{thermal}$ gives the heat flow per unit time *per unit area*, in contrast to electrical resistance, which is defined relative to total electric current.

flow across a homogeneous block of material with boundaries fixed at two different temperatures. Situations like this arise in computing heat loss through the walls of a house, for example. Consider a block of material of thickness L in the x-direction and cross-sectional area A in the directions perpendicular to x (see Figure 6.4). The material has thermal conductivity k. One side of the block

is maintained at temperature T_0 and the other side is held at temperature T_1, with $T_0 > T_1$. This setup is precisely parallel to the situation described in §3.2.2 of a homogeneous block of conducting material with a voltage difference ΔV between the two sides (Figure 6.5).

If the thickness of the block is much smaller than its other dimensions, then to a good approximation the edges are unimportant and the temperature and the heat flux will depend only on x, so this is essentially a one-dimensional problem with $q = q(x)\hat{x}$. We assume that there is an infinite reservoir for thermal energy on each side of the block, so that energy can be continuously removed or added on the boundaries through heat flux, without changing the boundary temperatures. Independent of the initial temperature profile, after some initial transient time-dependent changes in the heat flow, the system settles into a *steady-state flow*, in which the temperature at each point ceases to change with time, $T(x, t) \to T(x)$. In the steady-state flow, the heat flux q must be constant, independent of both x and t. If $\nabla \cdot q = \partial q / \partial x$ were nonzero at some point x, eq. (6.5) would require $\partial u / \partial t$ to be nonzero at x, and the temperature at x could not be constant. So eq. (6.1) reduces to

$$q = -k\frac{dT}{dx}. \tag{6.6}$$

Thus, in steady state $T(x)$ is a linear function of x with slope $-q/k$. There is a unique linear function $T(x)$ satisfying $T(0) = T_0$, $T(L) = T_1$, namely

$$T(x) = T_0 + \frac{x}{L}(T_1 - T_0). \tag{6.7}$$

Demanding that the slope be $-q/k$ fixes the heat flux q in terms of T_0, T_1, and k,

$$q = -k\frac{dT}{dx} = k\frac{T_0 - T_1}{L}. \tag{6.8}$$

The total rate at which heat passes through the block is then

$$\frac{dQ}{dt} = qA = (T_0 - T_1)\frac{kA}{L} = \Delta T\frac{kA}{L}. \tag{6.9}$$

This equation is the heat flow analog of Ohm's law (3.23) for a block of material with homogeneous resistance (3.30).

6.2.5 Thermal Resistance

While the analogy between electrical flow and heat flow described in §6.2.3 is mathematically precise, the notion of *resistance* is generally treated somewhat differently in these two contexts. In the case of heat conduction, we are often interested in heat conduction *per unit area*, while for electricity we are generally interested in net current. This has led to a slight, but potentially confusing, difference in notation on the two sides of the analogy. For a given configuration of materials, the **thermal resistance** R_{thermal} is conventionally defined after dividing out the factor of A in eq. (6.9),

$$R_{\text{thermal}} = \frac{L}{k}, \tag{6.10}$$

so that eq. (6.9) reads $q = \Delta T / R_{\text{thermal}}$. Thus, L/k is defined to be the thermal resistance for a block of material carrying a uniform heat flux, even though L/kA is the quantity that would be strictly analogous to $R_{\text{electrical}}(= L/\sigma A)$. This notational inconsistency in the heat/electricity analogy is highlighted in Table 6.2.

Analogy Between Conduction of Heat and Electricity

The steady conduction of heat through materials is analogous to the steady flow of electric current through resistive media, with $\Delta V \Leftrightarrow \Delta T$, $j \Leftrightarrow q$, $R_{\text{electric}} \Leftrightarrow R_{\text{thermal}}/A$. Thermal resistances in series and parallel combine into effective resistances in the same way as electrical resistors in series and parallel.

The units of thermal resistance are $[R] = [T]/[q] = $ m^2 K/W in SI units. In the US, thermal resistance is measured in ft^2 °F hr/BTU. The conversion factor

$$1\,\frac{\text{m}^2\,\text{K}}{\text{W}} = 5.678\,\frac{\text{ft}^2\,^\circ\text{F hr}}{\text{BTU}} \tag{6.11}$$

is very useful in practical calculations. Literature and specifications relating to insulation often refer to R_{thermal} as the **R-value** of an insulating layer. R-values are often quoted without specifying the units, which must then be determined from context.

The thermal resistance of geometries in which materials are combined in series or in parallel can be determined in analogy with the electric case.

Thermal resistance in series Suppose several layers with thermal resistances R_j, $j = 1, 2, \ldots$, all with the same area, are stacked in sequence as in Figure 6.6(a). This is analogous to electrical resistors in series (Problem 3.10 in §3). Thus, *thermal resistances in series add*,

$$R_{\text{eff}} = \sum_j R_j. \tag{6.12}$$

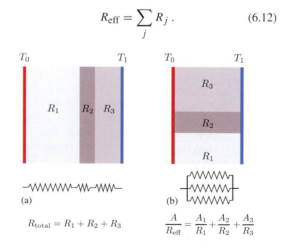

Figure 6.6 Thermal resistances combine in series and in parallel in analogy to electrical resistances.

The temperature drop across each layer is given by $\Delta T_j = q R_j$, where the same thermal current q flows through all the thermal resistances. In analogy with the case of electrical resistors, the total temperature difference is given by $\Delta T_{\text{total}} = \sum_j \Delta T_j = q R_{\text{eff}}$.

Thermal resistance in parallel Suppose thermal resistances R_j with areas A_j, $j = 1, 2, \ldots$ are placed alongside one another, separating materials with a common temperature difference ΔT, as in Figure 6.6(b). This is analogous to a system of electrical resistors in parallel, and the inverse resistances add,

$$\frac{A_{\text{tot}}}{R_{\text{eff}}} = \sum_j \frac{A_j}{R_j} \tag{6.13}$$

with a heat flux through each thermal resistance of $q_j = \Delta T / R_j$, and a total heat flow of $dQ/dt = \Delta T A_{\text{tot}} / R_{\text{eff}}$. The area factors in eq. (6.13) correct for the factors that were divided out in the definition of R_{thermal}.

More complex thermal "circuits" can be analyzed using further electric circuit analogies. Although the heat–electricity analogy is very useful, a few caveats are in order: wires are excellent electric conductors. In well-designed electric circuits, currents stay in the wires and do not wander off into surrounding materials. In contrast, one is often interested in heat transport through poor heat conductors – for example, in the evaluation of building insulation – which may lose heat to adjacent materials more effectively than they conduct it.

A more complex set of issues arise from the fact that the thermal conductivities of actual materials and building products measured under realistic conditions usually differ significantly from those determined from the theoretically ideal values of the thermal resistance $R_{\text{thermal}} = L/k$. There are several reasons for this. For one thing, in practice convection and radiation also contribute to heat transport in building materials. For small temperature differences, the rates of heat flow from both of these mechanisms are, as for conduction, proportional to ∇T (Problem 6.16). Thus, their contributions can be included in an **apparent thermal conductivity** k_a defined by $q \approx -k_a \nabla T$. Because heat transport by convection and radiation add to conduction, their effect is to lower the *apparent thermal resistance* compared to the thermal resistance computed from eq. (6.10). In general, for insulating materials in practical situations, empirically determined R-values are more accurate than values of R computed from $R = L/k$.

A further aspect of these issues is that the presence of **boundary layers** (*films*) of still air near surfaces can significantly increase the measured thermal resistance of a layer of conducting material separating two volumes of air at different fixed temperatures because of air's small thermal conductivity (see Table 6.1). This is particularly important for thin layers such as glass windows, which have little thermal resistance themselves, and mainly serve to create layers of still air (see Example 6.1). See §29 (Fluids) for more discussion of the behavior of fluids at boundaries.

Finally, many building systems, particularly windows and other fenestration products, are complex ensembles

---

Example 6.1   The R-value of a Pane of Glass

Glass has a relatively large thermal conductivity, $k \cong 1.0$ W/m K. Equation (6.10) predicts that the thermal resistance of a 1/8 inch thick pane of glass should be $R \cong (1/8\text{ in}) \times (0.025\text{ m/in}) \div (1.0\text{ W/m K}) \cong 0.0031\text{ m}^2$ K/W. The measured U-factor of a 1/8 single pane of glass listed in Table 6.3 is 5.9 W/m² K, corresponding to a nominal R-value of $R = 1/U \cong 0.17\text{ m}^2$ K/W. How can these facts be reconciled?

A moving fluid such as air comes to rest, forming a thin stationary film called a *boundary layer* at the surface of an object that obstructs the fluid's flow. This phenomenon is discussed further in §28 and §29. The layers of still air on either side of a window pane add additional thermal resistance in series with the glass itself. The layer on the interior (heated) side, where air movement is less, is typically thicker than the layer on the exterior side, where it is eroded by wind. Standard estimates are $R_{\text{air}} \approx 0.12\text{ m}^2$ K/W for the interior layer and 0.03 m² K/W for the exterior layer [28]. Together these account for almost the entire insulating value of the pane of glass, and are included in the quoted U-factor.

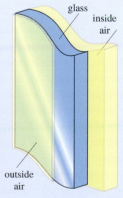

Note that the apparent thermal conductance of the layers of still air includes significant effects from radiative heat transfer (see §6.3, and Problem 6.9). Thus it would be a mistake to estimate the thickness of the layers of still air by $L_{\text{air}} \approx R_{\text{air}} \times k_{\text{air}}$, which ignores the effects of radiative heat transfer.

Table 6.3 Effective R-values for common building materials and U-factors for fenestration products [28]. The U-factors quoted here for windows are measured at the center of the glass surface and include the thermal resistance of air-films.

Material	Thickness	R-value (SI) $(m^2\,K/W)$	R-value (US) $(ft^2\,{}^\circ F\,hr/BTU)$
Gypsum drywall	1/2"	0.079	0.45
Plywood	3/4"	0.19	1.08
Hollow-backed vinyl siding		0.11	0.62
Wood shingles		0.15	0.87
Fiberglass batt	3.5"/8"	2.0/4.6	12/26
Blown cellulose		0.21–0.24/cm	2.9–3.4/in
Rigid (polyisocyanurate) foam		0.37–0.49/cm	5.3–7.1/in
Mineral fiber		≈ 0.2/cm	≈ 2.75/in
Common brick	4"	0.14	0.80
Concrete masonry unit (CMU)	4"/8"	0.14/0.19	0.80/1.11

Window type		U-factor (SI) $(W/m^2\,K)$	U-factor (US) $(BTU/ft^2\,{}^\circ F\,hr)$
Single pane glazing	1/8"	5.9	1.04
Double pane glazing with $\frac{1}{4}/\frac{1}{2}/\frac{3}{4}$" air space		3.1/2.7/2.3	0.55/0.48/0.40
Double glazing with $\frac{1}{2}$" air space and emissivity $\varepsilon = 0.2$ coating		2.0	0.35
Double glazing, $\frac{1}{2}$" space, argon filled, $\varepsilon = 0.2$ coating		1.7	0.30
Quadruple glazing, $\frac{1}{4}$" space, argon filled, $\varepsilon = 0.1$ coating		0.97	0.17
Quadruple glazing, $\frac{1}{4}$" space, krypton filled, $\varepsilon = 0.1$ coating		0.68	0.12

of materials (e.g. window, frame, spacers, and other hardware), each allowing conductive, convective, and/or radiative heat transfer. Such complex systems – including layers of still air as appropriate – are usually described by a **U-factor**, an empirically measured total rate of thermal energy transfer per unit area per unit temperature difference,

$$U = q/\Delta T. \tag{6.14}$$

The units of U are $[U] = W/m^2\,K$ or $BTU/ft^2\,{}^\circ F\,hr$.

U-factors are measured under standardized conditions. The U-factors of windows quoted in Table 6.3, for example, were measured at the center of the glass surface between an "indoor" environment containing still air at $25\,^\circ C$ and an "outdoor" environment at $0\,^\circ C$ with a wind velocity of 7 m/s; these U-factors include the thermal resistance of the films of still air that abut the glass surface [28]. When used under other conditions, the actual U-factor of a fenestration product will depart from these standard values.

For a pure material placed between two surfaces at fixed temperatures, and in the absence of convection or radiation, there is an inverse relationship between the U-factor and R-value, $U = 1/R$. In most realistic situations, however, an empirically measured value of R is used to characterize conductive heat flow for homogeneous materials such as insulation, while U incorporates the full set of factors involved in thermal energy transport through a given building component. In the US, U-factors are generally quoted for windows, while R-values are generally used for insulation, walls, etc. [32]. Table 6.3 gives measured R-values for some common building materials and the U-factors for some representative windows of specific designs.

6.3 Heat Transfer by Convection and Radiation

The ways that convection and radiation transfer heat were described qualitatively at the beginning of this chapter. Here we explore them at an elementary quantitative level. As mentioned earlier, both convective and radiative heat transfer can be quite complex, though we avoid most of the complexity in the situations treated in this book.

6.3.1 Forced Convection

Forced convection is a common way to transfer considerable heat quickly. We use it when we cool an overheated frying pan under a stream of water or when we blow on a

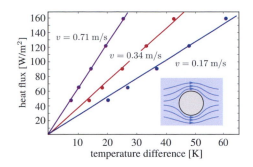

Figure 6.7 Measurements of heat flux ($dQ/dA\,dt$) from an isothermal sphere to a steady (*laminar*) flow of water around the sphere (see inset) as a function of $T - T_0$, for three values of the asymptotic flow velocity. The approximately linear slope illustrates Newton's law of cooling. Data from [33].

burned finger. A cold fluid forced to flow over a hot surface carries away heat efficiently. Isaac Newton is credited with the empirical observation, known as **Newton's law of cooling**, that the rate of heat transfer in situations involving forced convection is proportional to the temperature difference between the object (T) and the fluid (T_0),

$$\frac{dQ}{dt} = \bar{h}A(T - T_0), \tag{6.15}$$

where dQ/dt is the net rate of heat transfer from the object to the fluid (integrated over the surface of the object), A is the object's surface area, and $\bar{h}$ is a constant characterizing the object and the fluid flow. $\bar{h}$ is known as the **heat transfer coefficient**, and has units $[\bar{h}] = \mathrm{W/m^2\,K}$. $\bar{h}$ can be thought of as the average over the object's surface of a *local heat transfer coefficient h*.

Equation (6.15) can be understood as combining the effects of conduction and mass transport. When a fluid flows over an object, the fluid closest to the object actually adheres to its surface. The fluid only maintains its bulk flow rate and bulk temperature T_0 some distance from the surface. The object, therefore, is surrounded by a thin *boundary layer*, through which the temperature changes from T at the object's surface to T_0 in the bulk of the fluid. Heat is *conducted* from the object through the layer of fluid that is at rest at its surface. The thinner the boundary layer, the larger the temperature gradient at the object's surface and the faster the heat is conducted from the surface into the fluid. The role of convection is to transport the heated fluid away from the top of the boundary layer, thereby keeping the temperature gradient large. Since the conductive heat flux is proportional to the temperature gradient at the surface – which in turn is proportional to $T - T_0$ – the heat transferred from the surface to the fluid is proportional to $T - T_0$, as Newton suggested. Figure 6.7 shows data that support Newton's law of cooling. In reality, Newton's law is only approximately valid and works best when ΔT is small. When ΔT grows large, the heat flow itself affects

the dynamics of the boundary layer, introducing additional temperature dependence into eq. (6.15).

All of the complex physics of the fluid in the boundary layer, as well as heat and mass transport through it, enter into the heat transfer coefficient $\bar{h}$. Like the drag coefficient c_d and thermal conductivity k, $\bar{h}$ is a phenomenological parameter that is quite difficult to calculate from first principles and is instead usually taken from experiment. Any physical effect that influences the thickness of the boundary layer modifies the rate of heat transfer and thus changes $\bar{h}$. The value of $\bar{h}$ is much larger, for example, when the fluid flow is *turbulent* as opposed to *laminar*,[1] because the boundary layer is thinner for turbulent flow. Another case where $\bar{h}$ is large is when T is so high that it causes the fluid in contact with the object to boil. When the fluid begins to boil, the buoyancy of the vapor carries it rapidly away from the surface, enhancing heat transfer. This is one of several reasons why phase change is used in large-scale heat transfer devices such as refrigerators and steam boilers (see §12).

Forced convection can be a much more effective means of heat transfer than conduction. To quantify this, we can compare the local (convective) heat transfer coefficient h at a surface with the thermal conductivity of the fluid k. Since it only makes sense to compare quantities with the same dimensions (otherwise the comparison depends on the choice of units), it is customary to multiply h by a length scale characteristic of the surface over which heat is being transferred. Thus the ratio

---

[1] These terms are defined in §29. For now it suffices to think of laminar flow as smooth and steady and turbulent flow as chaotic and changing in time, without fixed pattern.

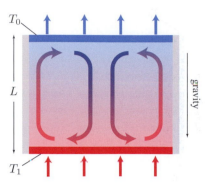

Figure 6.8 Two Bénard cells convecting heat between horizontal plates at temperatures T_1 and $T_0 < T_1$. The horizontal length scale for the convective cells is set by the separation L between the plates.

Forced Convection

Forced convection transfers heat effectively through motion of a fluid forced to flow past an object at a different temperature. Newton's law of cooling,

$$dQ/dt = \bar{h}A\Delta T \,,$$

parameterizes the rate of energy transfer in terms of a *heat transfer coefficient* $\bar{h}$, which depends on the properties of the flowing fluid and the object. Under a wide range of conditions, forced convection is a more efficient way to transfer heat than conduction.

$$\mathrm{Nu_L} = \frac{hL}{k} \,, \qquad (6.16)$$

known as the **Nusselt number**, is a measure of (forced) convective relative to conductive heat transport at a surface. The subscript L on Nu indicates which length scale has been used in the definition. The Nusselt number for water flowing at 2 m/s over a 60 mm long flat plate, for example, is found to be approximately 60 – an indication of the greater effectiveness of heat transfer by forced convection in this situation compared with conduction [27].

6.3.2 Free Convection

The most familiar form of free convection is driven by the force of gravity. When a packet of fluid is heated, it expands. The resulting reduction in gravitational force per unit volume (*buoyancy*) then pushes the expanded fluid upwards relative to the surrounding cooler and denser fluid. This drives a circulation pattern that transports heat away from the source. Such free convection would not take place in the absence of a gravitational field. Other density-dependent forces, real or apparent, however, can also drive free convection. For example, the *centrifugal force* felt by material whirling in a centrifuge, and the *Coriolis force* that figures in the large-scale circulation of the atmosphere and ocean currents (§27.2) can drive free convection patterns that transport heat in different environments.

The structure of convection patterns depends strongly on the geometry of the system. In this book we encounter two quite different idealized situations: one in which a fluid is bounded by horizontal planes, and the other in which the boundaries are vertical planes. We encounter the first case, for example, when studying convection in Earth's atmosphere. This situation can be modeled in the laboratory by placing a fluid between two horizontal plates held

at temperatures T_1 below and T_0 above. If T_0 is higher than T_1, no convection occurs since the warmer, less dense air is above the cooler denser air already in equilibrium. Even when T_0 is lower than T_1, convection does not start until a critical value of $\Delta T = T_1 - T_0$ is reached. This is described in more detail in the context of Earth's atmosphere in §34.2. Above $\Delta T_{\mathrm{critical}}$, convection cells known as **Bénard cells** appear. When ΔT is just above $\Delta T_{\mathrm{critical}}$, these cells are few in number and orderly in form, but as ΔT increases they become more numerous and the flow in the cells becomes turbulent. Figure 6.8 shows a cartoon of two Bénard cells between plates.

The second case, convection in a region bounded by one or more vertical surfaces, is relevant to convective heat transfer near and within walls and windows. A basic model is given by two vertical plates held at different temperatures T_0 and T_1 respectively with a fluid in between. In contrast to the horizontal case, there is always convection in this geometry, with warm fluid flowing upwards in the vicinity of the hotter surface and cool fluid flowing downward along the colder surface. The fluid velocity goes to 0 at the boundary, as discussed above in the context of forced convection. The velocity profile is shown in Figure 6.9. Like forced convection, this flow transports heat very efficiently from the hotter to the colder surface. This kind of convection is a serious source of heat loss in and near the walls and windows of buildings, where convection is driven at a single boundary fronting the large heat reservoir provided by interior or exterior spaces.

6.3.3 Radiative Heat Transfer

Heat transfer by radiation obeys quite different rules than conduction and convection. In particular, radiation is much less sensitive to the material characteristics of the emitting

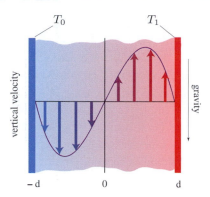

Figure 6.9 Velocity of vertical convection as a function of distance between vertical plates separated by $2d$. The scale of the vertical velocity depends upon the properties of the fluid, the temperature difference, and the distance between the plates [34].

object and its environment. In an ideal (but still useful) limit the radiant power dQ/dt emitted by a hot object depends only on its temperature T and surface area A. This is summarized by the *Stefan–Boltzmann law*,

$$\frac{dQ}{dt} = \sigma T^4 A, \tag{6.17}$$

where σ is the Stefan–Boltzmann constant $\sigma \cong 5.670 \times 10^{-8}\ \mathrm{W/m^2\,K^4}$. The Stefan–Boltzmann law is derived and described in more detail in §22 (Solar energy), after we have introduced the basic notions of quantum mechanics (§7) and entropy (§8), and given a more precise definition of thermal equilibrium (§8). The Stefan–Boltzmann law seems quite uncanny at first sight, since it requires that hot objects radiate electromagnetic energy simply by virtue of being hot, but independent of the density or other material characteristics of the radiating body. This result follows, however, from the quantum nature of thermodynamic equilibrium between any material containing charged matter (such as electrons) and the electromagnetic field. The frequency spectrum of thermal radiation has a characteristic shape, which is derived in §22. The Sun radiates most of its energy as visible light, but thermal radiation from objects at temperatures below 1000 K peaks in the infrared part of the electromagnetic spectrum where it is invisible to the unaided eye; only the high-frequency tail of the spectrum is visible as a red or orange glow for sufficiently hot objects.

Actually, eq. (6.17) holds only for **perfect emitters**, which absorb and re-emit all radiation that impinges on them. Materials with this property are also known as **black bodies**, and their radiation is known as **blackbody radiation**. For less ideal materials, the Stefan–Boltzmann law is

modified by a factor ε, the *emissivity*, which is less than or equal to one. The Sun, as an example, radiates approximately like a black body with temperature 5780 K and emissivity $\varepsilon_{\mathrm{sun}} \approx 1.0$ (Problem 6.4).

When electromagnetic radiation falls on a surface, it can be absorbed, reflected, or transmitted. As explained in detail in Box 22.1, conservation of energy requires that the fractions of the light absorbed (the *absorptivity*), reflected (*reflectivity*), and transmitted (*transmittance*) must sum to one at each wavelength of light. An **opaque object** is one that transmits no light, so for an opaque object the absorptivity and reflectivity sum to one. Furthermore, the laws of thermodynamics require that the absorptivity of an opaque object is equal to its emissivity for every wavelength of light (*Kirchhoff's law*). This explains the peculiar term "black body": in order to be a perfect emitter of radiation, a body must also be a perfect absorber, i.e. "black." Since the transmittance of a sample varies with its thickness, the absorptivity (= emissivity) usually quoted for a specific material is measured for a sample thick enough that negligible light is transmitted.

Emissivities in general depend on radiation frequency, but in many situations can be approximated as frequency-independent constants. Most non-conducting materials have emissivities of order one over the range of frequencies that dominate thermal radiation by objects at room temperature. A list of common materials with emissivity ε greater than 0.9 includes ice (0.97), brick (0.93), concrete (0.94), marble (0.95), and most exterior paints. In contrast, polished metals with high electrical conductivity are highly reflective and have low emissivity. They do not emit like black bodies. Examples include aluminum alloys ($\varepsilon \sim 0.1$), brass (0.03), gold (0.02), and silver (0.01).

The optical properties of materials can vary dramatically with the wavelength of light. A 5 mm thick sheet of ordinary glass, for example, transmits more than 90% of the visible light that falls on it at perpendicular incidence; most of the rest is reflected, and very little is absorbed. In contrast, the same sheet of glass is virtually opaque to the infrared radiation emitted by objects at room temperature, with an emissivity of about 0.84. Films exist that reflect, rather than absorb, most infrared radiation yet transmit visible light. Such films, as well as metal foils with high reflectivity, play an important role in the control of thermal radiation.

If two objects are at different temperatures, radiation transfers energy between them. When $\Delta T = 0$, the objects are in *thermal* and *radiative equilibrium* (described in more detail in §8, §22, §34). A typical (male human) adult with surface area 1.9 m^2 at a temperature of 37 °C radiates energy at a rate of $dQ_{\mathrm{human}}/dt = 5.67 \times 10^{-8} \times (310)^4 \times$

Radiative Heat Transfer

An object at temperature T radiates thermal energy according to the *Stefan–Boltzmann law*,

$$\frac{dQ}{dt} = \varepsilon \sigma T^4 A,$$

where $\sigma = 5.67 \times 10^{-8}\ \mathrm{W/m^2\,K^4}$ is the Stefan–Boltzmann constant and A is the object's surface area. ε is the object's *emissivity*, which equals its absorptivity, the fraction of incident light that it absorbs. $\varepsilon = 1$ defines a *black body*.

Most non-conducting materials have emissivity close to one. Highly reflective materials such as polished metal surfaces have $\varepsilon \approx 0$ and can be used to block thermal radiation.

Minimizing Heat Loss from Buildings

Conduction, convection, and radiation all contribute to heat loss from buildings. Conduction can be minimized by integrating *thermal insulation* into a building's exterior surfaces. Thermal radiation can be obstructed by a very thin layer of metal foil, often included as a facing for insulation. *Infiltration*, a form of convection, can be reduced by sealing gaps. Interior convection loops near walls and windows can be obstructed with drapes, wall hangings, and other furnishings.

1.9 $\sim$ 1000 W. If, however, he sits in an environment at 28 °C, energy radiates back to him from the environment at $dQ_{\mathrm{environment}}/dt \sim 900$ W, so the net radiative energy loss is $\sim$100 W.

6.4 Preventing Heat Loss from Buildings

Keeping buildings warm in a cold environment is a problem as old as human civilization. Enormous quantities of energy are used in the modern era for space heating. Roughly $\sim$30 EJ, or about 6% of human energy consumption, was used for residential space heating in 2009 [35]. The fraction goes up to roughly 1/6 in temperate climates. This is one area, however, where understanding the basic physics, making use of modern materials, and adopting ambitious standards have already led to considerable energy savings – and much more is possible. In 1978, 7.3 EJ were used to provide space heating for 76.6 million housing units in the US compared with 4.5 EJ for 111.1 million units in 2005 – an almost 60% reduction in the energy use per housing unit over less than three decades [36].[2]

We can identify the critical steps to reducing heat transfer from the warm interior of a building to the colder environment: (1) eliminate **infiltration**, the most dramatic form of convection in which colder air simply flows into the building as warmer air flows out; (2) use materials for

walls, windows, and roofs that have high thermal resistance to minimize heat conduction; (3) include a layer of highly reflective metal foil between the building's insulating layers and outer walls to obstruct outgoing radiation (see §6.4.2); and (4) obstruct convection loops that would otherwise form within or near walls and windows.

The first step in keeping a building at a higher temperature than its environment is to prevent air from escaping into the environment. Energy lost to in-/exfiltration accounts for roughly one third of thermal energy losses for US buildings. In many cases, older and/or poorly fitted windows and doors are principal culprits for excess infiltration. Some air exchange, however, is necessary for ventilation; the most ambitious ultra-efficient designs employ *heat exchangers* that use forced convection to transfer energy from exiting to entering air. Even with such devices, however, some thermal energy is lost as air leaves the building. In the following section (§6.4.1), we examine the problem of heat conduction through a building's walls, and in §6.4.2, we look briefly at ways to reduce radiative and convective heat loss.

6.4.1 Insulation

Traditional structural materials such as concrete, brick, steel, glass, and to a lesser extent wood, have relatively high thermal conductivities (see Table 6.1). With one of these materials on the outside and only drywall on the inside, the thermal losses through building walls would be enormous (see Example 6.2 for a quantitative estimate). Other materials should therefore be sandwiched between the interior and exterior walls to suppress heat conduction. A look at Table 6.1 suggests that air is a wonderful thermal insulator. (Noble gases such as argon are better, but much more expensive.) Filling a wall cavity with air, however, would result in unacceptable thermal loss through convection loops that would arise when the interior and exterior

---

[2] In addition to materials and design efficiency, more efficient heating equipment and migration to warmer regions have contributed to this reduction.

Example 6.2   Insulating a Simple House

To get a qualitative sense of the magnitude of heat transfer through walls and the importance of insulation, consider a series of progressively more ambitious insulation efforts. First, consider an uninsulated 20 cm thick concrete wall with the temperature on the inside at 20 °C and on the outside at 0 °C. Taking $k_{concrete} \approx 0.9$ W/m K from Table 6.1, we find $R = L/k \approx 0.2/0.9 \approx 0.2$ m^2 K/W, and the rate of heat flow is given by

$$q = \Delta T/R \approx 20/0.2 \approx 100 \, \text{W/m}^2 \, .$$

This is a lot of energy loss. Consider a relatively small single-story dwelling with a floor area of 120 m^2, horizontal dimensions 10 m × 12 m, and a height of 3 m. We can estimate the combined surface area of walls and ceiling (ignoring losses through the floor) to be 250 m^2. So the power needed to maintain the building at a temperature 20 °C warmer than the environment is 250 m^2 × 100 W/m^2 ≈ 25 kW. The rate of heat loss works out to about 2 GJ per day, about 10 times the world average daily per capita energy use, or roughly twice the per capita use of the average American. We must conclude that conductive energy losses from a concrete (or stone or brick) building are unacceptably large.

We consider a series of steps to improve the situation. First, we consider walls half as thick made of (hard) wood. Then, using $k \approx 0.18$ W/m K from Table 6.1, we find $R \approx 0.1/0.18 \approx 0.56$ m^2 K/W, and the rate of conductive energy loss drops by a factor of 2.8. Finally, if the hardwood were replaced by a sandwich of 3 cm of wood on the inside and outside with 20 cm of fiberglass insulation in between, the R-value increases to $R \approx 2(0.3/0.18) + 0.2(22.7) \approx$ 4.6 m^2 K/W (where 22.7 m^2 K/W is the thermal resistance of one meter of fiberglass, which can be read off of Table 6.3) and the conductive energy loss drops by another factor of about nine to about one kW.

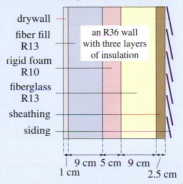

The moral of this example is, as might have been expected, that insulation is a great way to save energy. Insulation standards have been increasing steadily in recent years. The US Energy Efficiency and Renewable Energy Office recommends standards for home insulation. In 2010 those standards for the climate of Cambridge, Massachusetts were R38–R60 for attic space, R15.5–R21 for cavity walls plus insulated sheathing, and R25–R30 for floors (all in US units) [37]. The wall shown at right has been upgraded to R36 with several layers of insulation (Problem 6.7) (Image credit: Mark Bartosik, www.netzeroenergy.org)

surface temperatures differ significantly. In fact, the fundamental paradigm of insulation is to trap air (or another low-conductivity gas) in small cavities to avoid convection, using as low a density of material as possible to minimize heat conduction in the material itself. These goals are realized effectively by fiberglass, plastic foams, and shredded cellulose, the most common forms of insulation. Fiberglass, for example, is a fibrous material spun from glass. The thermal conductivity of glass is not too different from that of concrete (Table 6.1), but fiberglass is largely air, and its thermal conductivity is about 25 times lower than that of glass.

Applying effective insulation thus gives a simple way of dramatically reducing thermal energy losses from buildings. Insulation is one of many ways of saving energy that may help reduce human energy needs if applied broadly; we explore this and other approaches to *saving energy* in §36. Note that gaps in insulation are particularly wasteful.

According to the electric circuit analogy, the thermal resistance of *parallel* heat paths add inversely (see eq. (6.13)), so a path with zero resistance acts like a thermal short circuit, allowing a great deal of heat to escape. Problem 6.8 explores this phenomenon, finding that a gap of 1/60th the area of a wall reduces the effective thermal resistance of the wall by ∼22%.

6.4.2 Preventing Radiative and Convective Heat Transfer

A thin metal film provides a very effective barrier to radiative energy transport between regions or materials at different temperatures. With high reflectivity and low emissivity, such a film works in two ways. First, the radiation incident from the hotter region is reflected back, and second the film radiates only weakly into the colder region because of its low emissivity. Such films have many uses (Figure 6.10).

Figure 6.10 Marathoners avoid hypothermia by wrapping in radiation reflecting films after the 2005 Boston Marathon. (Credit: Stephen Mazurkiewicz, RunnersGoal.com)

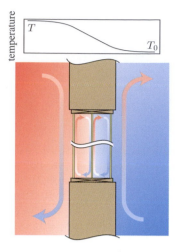

Figure 6.12 A sketch of a section of a triple-glazed window in an exterior wall and its immediate environs, showing the convection loops that can form near and within the window.

Figure 6.11 Foil-faced 120 mm fiberglass batts insulate the ceiling of this home in Italy where heat load on the roof is an important consideration. (Credit: Alec Shalinsky, Vetta Building Technologies)

Reflective films are used in building construction where radiative transfer (in either direction) would otherwise be large and other structural considerations, such as the need to allow the flow of water vapor, allow it. It is common, for example, to have reflective films installed below roof insulation in hot climates (Figure 6.11). Convection does not transmit heat downward from a hot roof, but radiation does. A metallic film reflects the heat back and does not radiate downward into the living space. In colder climates, a reflective film on the outside of insulating layers similarly helps to trap heat within the living space.

Windows cannot be obstructed by metal films. Instead, glass coatings have been developed that are transparent to visible light, but reflect the infrared radiation (see §4.3) that dominates the spectrum of thermal radiation at room temperature. Windows with $\varepsilon \approx 0.2$, known as *low-e* windows, are commonly available.

Finally, a few words about convection. Convective loops form wherever air is free to circulate along vertical surfaces

at a temperature different from that of the ambient air. Conventional bulk insulation blocks convection within (and infiltration and conduction through) walls. More problematic are the convection loops that can form both inside and outside a structure near windows and outside by the cool outside walls of buildings. In these situations, the motion of air increases the rate of heat transfer in a similar fashion to that discussed earlier for forced convection. Some of the paths for convection near and within a multi-pane window are shown in Figure 6.12. Convection within the window can be reduced by decreasing the thickness of the air space (which unfortunately also diminishes its thermal resistance), by evacuating the air from the space between the glass panes, or by using a gas such as *argon* that has higher viscosity than air.[3] The convection loops on the inside can be suppressed by window treatments such as shades or drapes that can also have insulating value.

In most real situations, convection and radiative heat transfer combine with conduction to give a total rate of heat transfer through given building components that is given by the U-factor described in §6.2.5. For example, the U-factors for different types of windows given in Table 6.3 can be understood as coming from a composition of all three types of heat transfer (Problems 6.10, 6.11).

6.5 The Heat Equation

In our analysis so far, we have assumed that the temperature in the materials we study does not change over time.

---

[3] *Viscosity*, discussed in §29, is a measure of the "thickness" of a fluid and it inhibits convection.

Often this is not the case. A hot object cools in a cold environment; heat introduced at one end of an object flows throughout its bulk. To study such systems we need to understand the dynamics of heat flow in time-dependent situations. We explore this phenomenon in the context of *conductive heat transfer* in a homogeneous medium that is fixed in place (as in an incompressible solid).

6.5.1 Derivation of the Heat Equation

The key ingredient in the dynamics of conductive heat transfer is the conservation law (6.5) that relates heat flow to thermal energy density,

$$\frac{\partial u(\boldsymbol{x}, t)}{\partial t} = -\boldsymbol{\nabla} \cdot \boldsymbol{q}(\boldsymbol{x}, t). \qquad (6.18)$$

When other parameters are held fixed, u is a function of the temperature T, with

$$\frac{\partial u}{\partial t} = \frac{\partial u}{\partial T} \frac{\partial T}{\partial t} = \rho c_V(T) \frac{\partial T}{\partial t}, \qquad (6.19)$$

where we have replaced the volumetric heat capacity $\partial u / \partial T$ by the product of the specific heat capacity c_V and the density ρ (§5.4.3). With this replacement, eq. (6.18) becomes

$$\rho c_V(T) \frac{\partial T}{\partial t} = -\boldsymbol{\nabla} \cdot \boldsymbol{q}. \qquad (6.20)$$

When the heat flux is due only to *conduction*, Fourier's law, $\boldsymbol{q}(\boldsymbol{x}, t) = -k \boldsymbol{\nabla} T(\boldsymbol{x}, t)$, allows us to eliminate $\boldsymbol{q}$ from this equation and obtain a single equation relating the space and time derivatives of the temperature,

$$\rho c_V \frac{\partial T(\boldsymbol{x}, t)}{\partial t} = \boldsymbol{\nabla} \cdot (k \boldsymbol{\nabla} T(\boldsymbol{x}, t)). \qquad (6.21)$$

In general ρ, c_V, and k could depend on the temperature and thus on $\boldsymbol{x}$ and t. If the temperature does not depart too much from some mean value T_0, however, we can evaluate ρ, c_V, and k at T_0 and treat them as constants. In that case, k may be moved out in front of the space derivatives, and combined with ρ and c_V into a new constant $a \equiv k/\rho c_V$, known as the **thermal diffusivity**. The equation then simplifies to

$$\frac{\partial T(\boldsymbol{x}, t)}{\partial t} = a \boldsymbol{\nabla} \cdot \boldsymbol{\nabla} T(\boldsymbol{x}, t) = a \nabla^2 T(\boldsymbol{x}, t), \qquad (6.22)$$

where ∇^2 is the Laplacian operator (B.17). This is the **heat equation**. Note that the heat equation is sometimes used in situations where a material is held at constant pressure, in which case $a = k/\rho c_p$; in this case a careful treatment reveals additional terms that must be added to eq. (6.22) to incorporate the expansion or compression of the material involved. In solids, where the heat equation is most frequently applied, $c_V \cong c_p$ and the distinction is irrelevant.

The Heat Equation

Time-dependent conductive heat flow in a rigid homogeneous medium is described by the heat equation

$$\frac{\partial T(\boldsymbol{x}, t)}{\partial t} = a \nabla^2 T(\boldsymbol{x}, t),$$

where $a = k/\rho c_p$ is the *thermal diffusivity* of the material.

6.5.2 Solving the Heat Equation

Equation (6.22) is one of the most famous and widely studied equations in all of physics. Although it is known in this context as the *heat equation*, the same equation arises in many other circumstances, for example in the study of diffusion of one material into another, where it is known as the *diffusion equation*. The information about the medium through which the heat is moving is summarized in a single parameter, the *thermal diffusivity* a. Values of a for some interesting materials are summarized in Table 6.1. Like the thermal conductivity, the values of a range over many orders of magnitude. Since the time rate of change of T is proportional to a, materials with large thermal diffusivities, such as metals, conduct heat rapidly.

Like the wave equation studied in §4, the heat equation is linear and homogeneous; i.e. each term is linear in T. Therefore, like waves, solutions to the heat equation can be superposed: if $T_1(\boldsymbol{x}, t)$ and $T_2(\boldsymbol{x}, t)$ are two solutions of eq. (6.22), then so is $T(\boldsymbol{x}, t) = b_1 T_1(\boldsymbol{x}, t) + b_2 T_2(\boldsymbol{x}, t)$ for any constants b_1 and b_2. Unlike waves, which propagate over long distances, local temperature variations die away in characteristic solutions to the heat equation. The magnitude of initial temperature inhomogeneities in a material with a fixed spatially averaged temperature decreases exponentially in time as the temperature through the material equilibrates. Similarly, if a material initially contains localized high-temperature regions, the thermal energy dissipates away through the material over time. From the perspective of energy conservation, thermal energy moves from regions of high temperature to regions of low temperature as the system approaches thermal equilibrium.

As an example, consider a one-dimensional world where at some initial time, $t = 0$, the temperature varies sinusoidally in space about a constant value T_0, so $T(x, 0) = T_0 + \Delta T \sin kx$. Then it is straightforward to check that the time-dependent function

$$T(x, t) = T_0 + \Delta T \sin kx \, e^{-ak^2 t} \qquad (6.23)$$

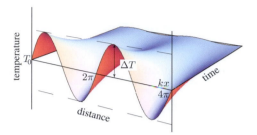

Figure 6.13 A sinusoidal temperature distribution at $t = 0$ dies away with t according to $T(x, t) = T_0 + \Delta T \sin kx \, e^{-ak^2 t}$.

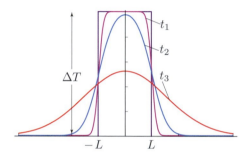

Figure 6.14 At time zero the temperature in the region between $-L$ and L is elevated by ΔT above the surroundings. The three curves, computed numerically, show the time-dependent temperature distribution that solves the heat equation (6.22) at times $t_1 = 0.01 L^2 / a$, $t_2 = 0.1 L^2 / a$, and $t_3 = L^2 / a$.

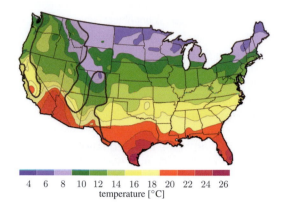

4   6   8   10 12   14 16   18 20   22 24   26
temperature [°C]

Figure 6.15 A map of annual mean ground surface temperatures in the contiguous US; the small isolated region in the western US indicates an area with high geothermal heat flux (§32). (Credit: Tester, J.W. et al., The Future of Geothermal Energy)

6.5.3 Seasonal Variation of Ground Temperature

An application of the heat equation with relevance to energy systems is the computation of ground temperature at various depths in response to daily and seasonal fluctuations on the surface. A contour map of annual mean ground surface temperature in the continental United States is shown in Figure 6.15. The temperature of the soil at depths greater than a few meters is roughly constant throughout the year, and matches very closely with the average annual surface temperature. The seasonal variation in soil temperature within the first few meters of the surface is an important consideration in building foundation design and in the specification of ground source heat pumps (see §32 (Geothermal energy)).

As a simple model of annual fluctuations in soil temperature, we assume that the temperature at the soil/air interface varies sinusoidally over the year with an average of T_0 and a maximum variation of $\pm \Delta T$,

$$T(0, t) = T_0 + \Delta T \sin \omega t , \tag{6.24}$$

where $\omega = 2\pi/\text{y}$. (In §36 (Systems) we fit empirical data to show that this is a good approximation to seasonal variations in surface temperature.) We neglect diurnal (daily) fluctuations for now, but return to this issue at the end of this analysis. We model the soil as a semi-infinite mass of material with thermal diffusivity a, beginning at the surface, where $z = 0$, and running to arbitrarily large (positive) depth z. We assume that there is no net heat flux into or out of the deep earth; the depth at which this assumption breaks down depends upon the local geothermal heat flux, which is discussed further in §32 (see also Problem 6.15).

agrees with the initial condition at $t = 0$ and satisfies eq. (6.22). The shape of the sinusoidal temperature variation thus remains fixed in place as its amplitude decays away to zero. Such a solution is sketched in Figure 6.13. The larger the thermal diffusivity, the faster the heat dissipates.

From Fourier's theorem (B.3) any initial condition for the heat equation can be written as a linear combination of sine wave modes. Each of these modes evolves in time according to eq. (6.23). Since the heat equation is linear, this means that the solution for any initial condition can be computed as a superposition of the solutions for the sine wave components. In fact, it was in this context that Fourier developed the methods that now bear his name.

Another example of a solution to eq. (6.22) that illustrates the dissipation of thermal energy is shown in Figure 6.14. In this example, the temperature is initially elevated above the surroundings by ΔT in the interval $-L \leq x \leq L$. As time goes on, the heat diffuses outward. This kind of problem can be solved using Fourier's methods.

Box 6.2 Solving the Heat Equation Numerically

Generally, we wish to solve for the temperature T in a medium of fixed extent. In this case, in addition to the initial temperature distribution, we need to know the behavior of the temperature or the heat flux at the boundaries of the medium – information known as *boundary conditions*. When boundary conditions are complicated, or when a is spatially dependent, it is often desirable to solve eq. (6.22) numerically. Tremendous amounts of computer time (and therefore energy) are consumed worldwide in solving such partial differential equations. The fact that the heat equation involves only one time derivative simplifies this task. Knowledge of $T(\mathbf{x}, t)$ at any instant in time is sufficient to determine the temperature function at all future times. In contrast, an equation that involves a second time derivative, such as Newton's Law, $\mathbf{F} = m\ddot{\mathbf{x}}$, requires knowledge of both $\mathbf{x}$ and $\dot{\mathbf{x}}$ at an initial time.

Suppose $T(\mathbf{x}, t)$ is known for all $\mathbf{x}$. Then the derivatives that appear in ∇^2 can be computed (in Cartesian coordinates, for example, $\nabla^2 = \partial_x^2 + \partial_y^2 + \partial_z^2$), and eq. (6.22) can be used to compute the temperature at a slightly later time, $t + \Delta t$,

$$T(\mathbf{x}, +\Delta t) = T(\mathbf{x}, 0) + a\Delta t \, \nabla^2 T(\mathbf{x}, 0) + \mathcal{O}(\Delta t)^2 \,.$$

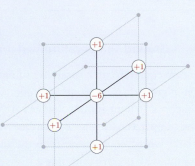

If Δt is sufficiently small, then the higher-order terms can be neglected in a numerical solution. In practice, $T(\mathbf{x}, t)$ can be computed approximately on a mesh of lattice points, $\mathbf{x} = (m\delta, n\delta, p\delta)$, where m, n, p are integers and δ is a small lattice spacing. In this case, a simple discrete update rule can be used to compute the leading approximation to eq. (6.22) in the limit of small Δt and δ. Writing $T(m\delta, n\delta, p\delta, t)$ as $T_{m,n,p}(t)$, we have

$$T_{m,n,p}(t + \Delta t) = T_{m,n,p}(t) + a\Delta t \, \delta^2 \left[T_{m+1,n,p}(t) + T_{m-1,n,p}(t) \right.$$
$$+ T_{m,n+1,p}(t) + T_{m,n-1,p}(t) + T_{m,n,p+1}(t)$$
$$\left. + T_{m,n,p-1}(t) - 6T_{m,n,p}(t) \right].$$

This algorithm can be improved and made more efficient in a variety of ways, but this simple iterative numerical procedure is a prototype for the vast range of computer simulations of partial differential equations currently in use. Numerical solutions of partial differential equations are important for many energy systems, in particular for simulating complex fluid dynamics systems (§29), and for modeling Earth's global climate (§34).

We also assume that the temperature distribution does not depend on the coordinates x and y that run parallel to Earth's surface. So we need to solve the one-dimensional heat equation

$$\frac{\partial}{\partial t} T(z, t) = a \frac{\partial^2}{\partial z^2} T(z, t), \qquad (6.25)$$

subject to the condition that $T(0, t)$ is given by eq. (6.24). The general theory of differential equations is beyond the level of this book; we proceed simply by trying a form for the solution that captures some features of the solutions discussed in the previous section: we look for a solution that oscillates in time with frequency ω and decays exponentially with z. Because we expect that the temperature at depth lags behind the surface temperature that is driving it, we anticipate that the phase of the time oscillation will vary with depth in the soil. All of this leads to a guess of the form,

$$T(z, t) = T_0 + \Delta T e^{-\alpha z} \sin(\omega t - cz), \qquad (6.26)$$

where α and c are parameters adjusted to solve eq. (6.25). Indeed, eq. (6.26) satisfies eq. (6.25) if

$$c = \alpha = \sqrt{\frac{\omega}{2a}}. \qquad (6.27)$$

All aspects of the temperature profile (for fixed ω) are determined by a single physical parameter, the thermal diffusivity a, which varies according to soil type and moisture content. As an example, for dry sandy soil $a \cong 2.4 \times 10^{-7}$ m^2/s. Substituting in this value, with $\omega = 2\pi/(3.15 \times 10^7)$ s^{-1}, we find $\alpha = 0.64$ m^{-1}. At a depth of 1 meter, then, the seasonal variation falls in amplitude by a factor of $e^{-0.64} = 0.53$. So a few meters below the surface the temperature is more or less constant all year round. The phase lag is substantial. At a depth of 1 meter, the phase of the sinusoidal oscillation lags behind the surface temperature by $\Delta\phi = -0.64$ radians $\cong -(12 \text{ months}) \times 0.64/2\pi \cong -1.2$ months. (This fact is observed by gardeners in many locations.) Some temperature profiles predicted by this simple model are shown in Figure 6.16. If we were to consider *diurnal* temperature

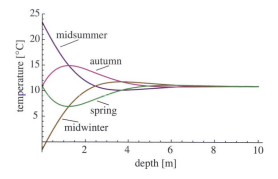

Figure 6.16 Annual temperature variation as a function of depth in the soil for Boston where $T_0 = 10.9\,°C$ and $\Delta T = 12.4\,°C$.

Ground Temperature Variation

The heat equation can be used to estimate the variation in ground temperature as a function of depth in response to temperature changes at the surface. If the temperature at the soil surface varies sinusoidally with frequency ω, then the temperature variation with depth z and time t is given by

$$T(z, t) = T_0 + \Delta T e^{-\alpha z} \sin(\omega t - \alpha z),$$

where $\alpha = \sqrt{\omega/2a}$ and a is the thermal diffusivity of the soil.

---

fluctuations, the frequency would be larger by a factor of 365, so $\alpha_{\mathrm{diurnal}} \cong 12\,\mathrm{m}^{-1}$, which means that the effects of diurnal variation are damped in 10 or 20 cm of soil. Again using the linearity of the heat equation, we can superpose the effects of diurnal fluctuations with annual fluctuations by simply adding the corresponding solutions of the heat equation.

Discussion/Investigation Questions

6.1 People usually discuss their experience of heat and cold in terms of temperature. Closer consideration, however, suggests that it is *heat conduction* rather than temperature that matters. What do you think? Have you ever taken a sauna? Or touched a metal surface at $-30\,°C$?

6.2 Look up the effective R-value of wall insulation (for example, *R-11 Mineral Fiber*) when the structural support for the wall is either 2×4 wood studs or 2×4 metal studs. Explain the difference.

6.3 Although it was not mentioned in the text, Newton's law of cooling requires that the thermal conductivity k of the object be "high enough," otherwise the surface of

the object will cool below its bulk temperature T. Then the convective heat transfer from the surface would be suppressed and the temperature of the object would not be uniform. Suppose L characterizes the distance from the center of the object, where the temperature is T, to its surface. What dimensionless constant must be small to avoid this problem?

6.4 A *water bed* replaces an ordinary mattress with a plastic bag filled with water. If the ambient temperature is below body temperature, water beds must either be heated or covered with a relatively thick layer of insulating foam. Why should this be necessary for a water bed, but not for an ordinary mattress?

6.5 Sketch the magnitude and direction of the heat flux under the ground driven by annual temperature variations as a function of depth for the four times of the year shown in Figure 6.16.

Problems

6.1 Suppose a small stoneware kiln with surface area $5\,\mathrm{m}^2$ sits in a room that is kept at $25\,°C$ by ventilation. The 15 cm thick walls of the kiln are made of special ceramic insulation, which has $k = 0.03\,\mathrm{W/m\,K}$. The kiln is kept at $1300\,°C$ for many hours to fire stoneware. The room ventilating system can supply $1.5\,\mathrm{kW}$ of cooling. Is this adequate? (You can assume that the walls of the kiln are thin compared to their area and ignore curvature, corners, etc.)

6.2 Two rigid boards of insulating material, each with area A, have thermal conductances U_1 and U_2. Suppose they are combined in series to make a single insulator of area A. What is the heat flux across this insulator as a function of ΔT? Now suppose they are placed side-by-side. What is the heat flux now?

6.3 The heat transfer coefficient $\bar{h}$ for air flowing at $30\,\mathrm{m/s}$ over a 1 m long flat plate is measured to be $80\,\mathrm{W/m^2\,K}$. Estimate the relative importance of heat transfer by convection and conduction for this situation.

6.4 Given the Sun's power output of $384\,\mathrm{YW}$ and radius $695\,500\,\mathrm{km}$, compute its surface temperature assuming it to be a black body with emissivity one.

6.5 Humans radiate energy at a net rate of roughly $100\,\mathrm{W}$; this is essentially waste heat from various chemical processes needed for bodily functioning. Consider four humans in a roughly square hut measuring $5\,\mathrm{m} \times 5\,\mathrm{m}$, with a flat roof at 3 m height. The exterior walls of the hut are maintained at $0\,°C$ by the external environment. For each of the following construction materials for the walls and ceiling, (i) compute the *R-value* ($\Delta T/q$) for the material, (ii) compute the equilibrium temperature in the hut when the four people are in it: (a) $0.3\,\mathrm{m}$ (1 foot) concrete walls/ceiling; (b) $10\,\mathrm{cm}$ (4") softwood walls/ceiling; and (c) $2.5\,\mathrm{cm}$ (1") softwood, with $0.09\,\mathrm{m}$ (3.5") fiberglass insulation on interior of walls and ceiling. Take thermal conductivities and/or R-values from Tables 6.1 and 6.3.

6.6 Consider a building with 3200 ft^2 of walls. Assume the ceiling is well-insulated and compute the energy loss through the walls based on the following materials, assuming an indoor temperature of 70 °F and an outdoor temperature of 30 °F: (a) walls are composed of 4″ thick hard wood; (b) walls composed of 3/4″ of plywood on the inside and outside, with 8″ of fiberglass insulation between; and (c) same as (b) but with 18 single-pane R-1 windows (US units), each with area 0.7 m^2 (remaining area is surrounded by walls as in (b)). Take thermal conductivities and/or R-values from Tables 6.1 and 6.3.

6.7 Estimate the R-value of the wall shown in the figure in Example 6.2.

6.8 Assume that a 60 m^2 wall of a house is insulated to an R-value of 5.4 (SI units), but suppose the insulation was omitted from a 1 m^2 gap where only 6 cm of wood with an R-value of 0.37 remains. Show that the effective R-value of the wall drops to $R = 4.4$ leading to a 23% increase in heat loss through the wall.

6.9 In Example 6.1, a film of still air was identified as the source of almost all of the thermal resistance of a single-pane glass window. Determine the thickness of the still air layers on both sides of the window if radiative heat transfer is ignored. It is estimated that about half the thermal conductance of the window is due to radiative heat transfer; estimate the thickness of the still air layers if this is the case.

6.10 According to [28], a double pane window with an emissivity $\varepsilon = 0.05$ coating and a 1/4″ air gap has a measured (center of glass) U-factor of 2.32 W/m^2 K. Assume that this coating is sufficient to stop all radiative heat transfer, so that the thermal resistance of this window comes from conduction in the glass, the air gap, and the still air layers on both sides of the window. Taking $R_{\text{still air}} \approx 0.17$ m^2 K/W from Example 6.1, compute the U-factor of this window ignoring convection in the air gap. How does your answer compare with the value quoted in [28]? Use the same method to compute the U-factor of a similar window with a 1/4″ *argon-filled* gap (measured value $U = 1.93$ W/m^2 K). Repeat your calculation for a 1/2″ air gap (measured U-factor 1.70 W/m^2 K). Can you explain why your answer agrees less well with the measured value in this case?

6.11 A building wall is constructed as follows: starting from the inside the materials used are (a) 1/2″ gypsum wallboard; (b) wall-cavity, 80% of which is occupied by a 3.5″ fiberglass batt and 20% of which is occupied by wood studs, headers, etc.; (c) rigid foam insulation/sheathing ($R = 0.7$ m^2 K/W); and (d) hollow-backed vinyl siding. The wall is illustrated in Figure 6.17 along with an equivalent circuit diagram.

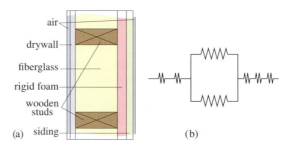

(a)

(b)

Figure 6.17 (a) A section through a wall in a wood-frame building. (b) A useful equivalent circuit diagram.

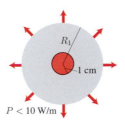

Figure 6.18 Cross section through an insulated copper pipe. The pipe carries fluid at 100 °C. The radius of the insulation must be 3.6 times the radius of the pipe in order to keep the power loss to less than 10 W/m.

Calculate the equivalent R-value of the wall. Do not forget the layers of still air on both inside and outside.

6.12 [H] An insulated pipe carries a hot fluid. The setup is shown in Figure 6.18. The copper pipe has radius $R_0 = 1$ cm and carries a liquid at $T_0 = 100$ °C. The pipe is encased in a cylindrical layer of insulation of outer radius R_1; the insulation has been chosen to be *closed-cell polyurethane spray foam* with an R-value of 1.00 m^2 K/W per inch of thickness. How large must R_1 be so that the heat loss to the surrounding room (at 20 °C) is less than 10 W/m? [Hint: First explain why the heat flux must be $q = c\hat{r}/r$. You may find it useful to make the analogy between heat propagation through the insulator and current flowing outward through a cylindrical conductor. Note also that $\nabla f(r) = \hat{r} f'(r)$.]

6.13 Ignoring convective heat transfer, estimate the change in U-factor by replacing argon by krypton in the quadruple glazed windows described in Table 6.3. You can ignore radiative heat transfer since these are low-emissivity windows.

6.14 In areas where the soil routinely freezes it is essential that building foundations, conduits, and the like be buried below the frost level. Central Minnesota is a cold part of the US with yearly average temperature of $T_0 \sim 5$ °C and variation $\Delta T \sim 20$ °C. Typical peaty soils in this region have thermal diffusivity

$a \sim 1.0 \times 10^{-7}$ m²/s. Building regulations require foundation footings to exceed a depth of 1.5 m. Do you think this is adequate?

6.15 Consider a region with average surface temperature $T_0 = 10\,°C$, annual fluctuations of $\Delta T = 30\,°C$, and surface soil with $a \cong 2.4 \times 10^{-7}$ m²/s, $k \cong 0.3$ W/m K. If the local upward heat flux from geothermal sources is 80 mW/m², compute the depth at which the heat flux from surface fluctuations is comparable. What is the answer if the area is geothermally active and has a geothermal heat flux of 160 mW/m²?

6.16 **[T]** When an object is in *radiative equilibrium* with its environment at temperature T, the rates at which it emits and absorbs radiant energy must be equal. Each is given by $dQ_0/dt = \varepsilon \sigma T^4 A$. If the object's temperature is raised to $T_1 > T$, show that to first order in $\Delta T = T_1 - T$, the object loses energy to its environment at a rate $dQ/dt = 4 \varepsilon \sigma \Delta T\, T^3 A$.

Introduction to Quantum Physics

Quantum physics is one of the least intuitive and most frequently misunderstood conceptual frameworks in physical science. Nonetheless, it is the universally accepted foundation of modern physics and is supported by an enormous range of experiments performed over the past hundred years. While quantum mechanics may seem bizarre, and applies primarily to systems much smaller than those we deal with on a day-to-day basis, it has great relevance to energy systems. Almost all of the energy that we use can be traced back to nuclear processes that originally took place in the Sun – or in some other star, in the case of geothermal or nuclear energy. Such nuclear processes can only be understood using quantum mechanics. A basic grasp of quantum mechanics is also necessary to understand the emission of light from the Sun and its absorption by carbon dioxide and water molecules in Earth's atmosphere (§22). The photovoltaic technology we use to recapture solar energy is based on the quantum mechanical behavior of semiconductors (§25). Nuclear fission reactors – currently used to generate substantial amounts of electricity (§18) – as well as nuclear fusion reactors – which may perhaps some day provide an almost unlimited source of clean energy – depend upon the peculiar ability of quantum particles to *tunnel* through regions where classical mechanics forbids them to go. The chips in your cell phone and in the computers that were used to write this book operate by quantum mechanical principles. Finally, quantum physics provides the basis for a rigorous understanding of thermodynamics. Concepts such as entropy and temperature can be defined in a precise way for quantum systems, so that a statistical treatment of systems with many degrees of freedom naturally leads to the laws of thermodynamics.

We begin our exploration of quantum mechanics in §7.1, with a famous pedagogical *thought experiment*, due to American physicist Richard Feynman, that illustrates the counterintuitive behavior of quantum particles and provides some suggestion of how the theory of

> **Reader's Guide**
> Quantum mechanics is a rich and complex subject that requires substantial study for a full understanding. Here we attempt to give the reader a self-contained introduction to the basic ideas and some important applications of this theory. By focusing on the role of energy, we can systematically and efficiently introduce many of the most important concepts of quantum mechanics in a way that connects directly with energy applications. We do not expect the reader to fully absorb the implications of these concepts in this chapter; as the results of quantum mechanics are applied in later chapters, however, we hope that the reader will become increasingly familiar with and comfortable with this counterintuitive physical framework.
>
> We begin this chapter (§7.1–§7.2) with an informal introduction to the main ideas of quantum mechanics and end (§7.7–7.8) with some applications. §7.3–§7.6 contain a somewhat more precise, but also more abstract introduction to the basics of the subject. Readers interested primarily in energy applications may wish to skim the middle sections on first reading and return when questions arise in later chapters. The quantum phenomenon of *tunneling* is of central importance in nuclear energy, and is introduced in a chapter (§15) closer to its application in that context.
>
> Prerequisites: §2 (Mechanics), §4 (Waves and light). We do not assume that the reader has had any previous exposure to quantum mechanics. Some familiarity with basic aspects of complex numbers is needed (Appendix B.2). Although not required, readers who have some familiarity with linear algebra (reviewed in Appendix B.3) will find it useful in the middle sections of this chapter.

quantum mechanics can be formulated. Feynman's thought experiment leads us in two complementary directions: first to *wave mechanics*, which enables us to understand the quantum mechanical propagation of particles through space, and second to a *space of states*, where the dynamics of quantum mechanics plays out in a remarkably simple way and where energy is the organizing principle. These two ways of thinking about quantum mechanics have been around since the mid-1920s when Austrian physicist Erwin Schrödinger proposed his *Schrödinger wave equation* and German physicist Werner Heisenberg introduced what he called *matrix mechanics*. At the time, the relation between the two formalisms was obscure, but a few years later the English physicist Paul Dirac showed that the two approaches are equivalent and that they give complementary insights into the working of the quantum world.

In §7.2 we follow the thread of wave mechanics. We introduce wavefunctions and Schrödinger's wave equation. We develop a physical interpretation for the wavefunction and describe and interpret some simple solutions to the Schrödinger equation. We then put wave mechanics aside and take a more abstract look at the results of Feynman's thought experiment. This leads us to a description of the foundations of quantum mechanics based on the concept of discrete *quantum states* labeled by their energy. We present this formulation in the next four sections, §7.3–§7.6. In these sections we introduce the basic notions needed to understand modern quantum mechanics, including the concept of the *state space* of a system and the *Hamiltonian*, which both describes the time-evolution of a quantum system and characterizes the energy of a state. Using these ideas, we can understand more deeply the structure of the Schrödinger equation, the solution of which gives us the spectrum of allowed energies of atomic and molecular systems. These *quantum spectra* play a fundamental role in many applications discussed later in this book.

In a sense, energy is the key to understanding the fundamental structure of quantum mechanics. In any physical quantum system, energy is *quantized* and only takes specific values. The set of quantum states with definite energy form a favored class of states, in terms of which all possible quantum states can be described through *quantum superposition*. The way in which quantum states with fixed energy evolve is extremely simple – these states are unchanged in time except for a periodic *phase oscillation* with a frequency that is related to their energy. Thus, by understanding the spectrum of energies and associated quantum states we can give a simple and complete characterization of any quantum system.

A concise set of axioms for quantum mechanics is developed in §7.3–§7.6. These axioms are obeyed by any quantum system and form the "rules of the game" for quantum physics. These rules provide a nearly complete definition of quantum mechanics; by focusing on energy we can simply characterize general quantum systems while

avoiding some complications in the story. Although sometimes classical physics is used to motivate parts of quantum mechanics, there is no sense in which quantum mechanics can be derived from classical physics. Quantum mechanics is simply a new set of rules, more fundamental than classical physics, which govern the dynamics of microscopic systems. With the basic axioms of the theory in hand, we return to wave mechanics in §7.7 and §7.8 where we explore the quantum dynamics of free particles and particles in potentials. For readers who would like to delve more deeply into quantum mechanics we recommend the texts by French and Taylor [38], at an introductory level, and Cohen-Tannoudji, Diu, and Laloë [39], for a somewhat more complete presentation.

7.1 Motivation: The Double Slit Experiment

7.1.1 Classical and Quantum Physics

The laws of classical physics have a structure that fits naturally with human intuition for how the world operates. In classical mechanics, objects move through space, residing at a specific point in space at every moment of time. Classical mechanics and classical electromagnetism are **deterministic**: two identical systems prepared in identical initial configurations will evolve identically in the absence of outside interference under the time evolution given by Newton's laws of mechanics or Maxwell's equations.

Underlying the apparent classical world, however, the natural world operates according to a much more alien set of principles. At short distances the quantum laws of nature violate our classical intuition. Quantum particles do not have well-defined positions or momenta. And at the quantum level, the laws of physics do not uniquely predict the outcome of an experiment, even when the initial conditions are precisely fixed – the laws of quantum physics are **non-deterministic** (see Box 7.3).

Another way in which classical and quantum physics differ is that many quantities, including energy, that can take arbitrary values in classical physics are restricted to discrete **quantized** values in quantum physics. One of the first places in which this became apparent historically is in the context of blackbody radiation (§6.3.3). From a classical point of view, one might expect that fluctuations of definite frequency ν in the electromagnetic field could hold an arbitrary, continuously variable amount of energy. Problems with the classical description of blackbody radiation (discussed further in §22) led the German physicist Max Planck to propose in 1900 that the amount of energy contained in light of a given frequency is *quantized* in units proportional to the frequency,

<hr>

Box 7.1   The World Needs Quantum Mechanics

Quantum physics may seem rather obscure and largely irrelevant to everyday life. But without quantum mechanics, atomic structure, chemical interactions between molecules, materials that conduct electricity, energy production in stars, and even the existence of structure in the universe would not be possible in anything like their current form.

- Without quantum mechanics, matter as we know it would not be stable. An electron circling a proton in a hydrogen atom would gradually radiate away energy until the atomic structure collapsed. Quantum mechanics allows the electron to remain in a lowest energy *ground state* from which no further energy can be removed.
- The rich variety of chemical structures and interactions depends critically on the atomic orbital structure of the atoms within each molecule (§9). Water, DNA, and neurotransmitters would neither exist nor function in a purely classical world.
- The production of energy in stars occurs through *nuclear fusion*, which is mediated by the strong nuclear force at short distances (§18). Without quantum physics, protons in the Sun would not overcome their electrostatic repulsion and come close enough to be affected by strong nuclear interactions until the temperature was nearly 1000 times greater (Box 7.2).
- The existence of structure in our universe – galaxies, nebulae, stars, and planets – is believed to have arisen from microscopic quantum fluctuations in the early universe that expanded over time and seeded the inhomogeneous distribution of matter that we see currently in the universe (§21).

Quantum mechanics also plays an important role in separating energy scales in a way that helps scientists to systematically understand physical systems. Quantum mechanics allows us to ignore levels of structure deep within everyday objects (§9.2). The fact that nuclei, for example, are complex systems with internal degrees of freedom is irrelevant when we warm a cup of coffee in the microwave, because microwaves do not have enough energy to make a quantum excitation of the nucleus. In classical systems, any energy, no matter how small, can excite a system at any scale. This quantum separation of energy scales not only makes life easier for scientists, but also avoids structural problems such as the *UV catastrophe* (§22.2.2) that would arise if electromagnetism were governed by classical physics at short distance scales.

(Galaxy image credit: NASA and STSCI)

<hr>

$$E = h\nu = \hbar\omega, \qquad (7.1)$$

where h is **Planck's constant**, which is often expressed as

$$\hbar = h/2\pi = 1.055 \times 10^{-34} \, \text{J s}. \qquad (7.2)$$

This relation between energy and frequency of oscillation plays a central role in quantum physics. In fact, almost all of the physics of quantum mechanics can be captured by the statement that every quantum system is a linear system in which states of fixed energy oscillate with a frequency given by eq. (7.1); much of the rest of this chapter involves unpacking this statement.

There is no inconsistency between classical and quantum physics. At a fundamental level, physical systems are governed by quantum mechanics; classical physics emerges as an approximation to the quantum world that is highly accurate under everyday conditions.

In the remainder of this section we consider a set of parallel scenarios in which the differences between classical and quantum processes are highlighted.

7.1.2 The Double Slit Experiment

Feynman described a pedagogical "thought experiment" in [40] to illustrate the weirdness of quantum mechanics and suggest ways to think about the quantum world. Feynman's "experiment" actually consists of the same type of experiment repeated three times: first with "bullets" (i.e. classical particles), second with water waves, and third with electrons. The experimental apparatus is a screen with two holes (or slits) through which the bullets, waves, or electrons can pass. On one side is an appropriate source: a gun, an ocean, or a cathode ray tube. On the other side is a screen that can register the arrival of the bullets, wave energy, or electrons.

Quantum Physics

Classical mechanics and electromagnetism are only approximate descriptions of a more fundamental physical reality described by quantum physics.

Quantum systems violate our classical intuition in many ways. Quantum particles do not have definite position or momentum, but rather are described by distributed *wavefunctions*.

Unlike the deterministic laws of classical physics, quantum processes are non-deterministic, so that even with perfect knowledge of initial conditions, the outcome of experiments involving quantum systems cannot be predicted with certainty.

Energy and other quantities that classically take arbitrary values in a continuous range are *quantized* and take only certain discrete values in quantum physics. For example, energy in an electromagnetic wave of angular frequency ω is quantized in units

$$E = \hbar\omega.$$

This relation between energy and frequency is fundamental to quantum physics.

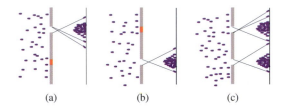

(a)   (b)   (c)

Figure 7.1 Double slit experiment with bullets. When either hole is open, the bullets pile up behind the gap. A few ricochet off the edges (some of those trajectories are shown) and they end up at the edges of the pile. When both gaps are open (c) the pattern is the sum of the single hole patterns (a) and (b). Some bullets in flight are shown, but the many bullets stopped by the shield are not.

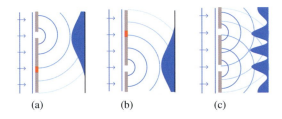

(a)   (b)   (c)

Figure 7.2 Double slit experiment with water waves. When either gap is open ((a) or (b)) a broad diffraction pattern is forms on the observing screen. When both gaps are open (c) the waves coming from the two gaps give rise to an interference pattern (see Figure 4.8). The incident water waves are shown, but the waves reflected from the barrier are not.

The scale of the apparatus is different for the three cases. For water waves, we imagine a calm harbor and a breakwater that is impenetrable to water except at two gaps separated by several times the gap length. Waves impinge on the breakwater with their crests parallel to the barrier; the wavelength between successive peaks is comparable to the size of the gaps. For bullets, we imagine an impenetrable shield with holes several times the size of a bullet. The bullets are sprayed out randomly by a poor marksman, covering the shield with a more or less uniform distribution of hits. For electrons, the screen is also impenetrable except for two slits; like the bullets, the electrons are sprayed over the screen more or less evenly. The size of the electron apparatus is discussed below.

First, consider the bullets. They move in straight lines and either go through a hole or are blocked. If they get through, they arrive on the observing screen in a location determined by simple geometry. Perhaps a few deflect slightly off the edges of the holes, but most end up behind the hole they passed through. The bullets arrive one by one and eventually build up patterns as shown in Figure 7.1.

If only one hole is open, the pattern of bullets that hit the screen is centered behind the open hole (a or b), while if both holes are open the distribution of bullets corresponds to the sum of the patterns from the two individual holes (c).

Next, consider water waves. First block one gap. As described in §4.5, the water waves *diffract* through the other gap and fan out in a circular pattern. The wave energy *arrives continuously* on the observing screen, forming a *diffraction pattern* similar to Figure 4.10(a). When the wavelength of the water waves is large compared to the size of the gap, the pattern is dominated by a single broad peak as in Figure 7.2(a). If we block the other gap instead, a similar pattern appears (Figure 7.2(b)). If, however, both gaps are open, quite a different pattern is observed on the screen: the circular waves coming from the two gaps *interfere*. When they meet crest-to-crest or trough-to-trough the intensity is four times as great as in the individual waves; when they meet crest-to-trough there is zero energy delivered to the screen. An *interference pattern*, shown in Figure 7.2(c), is formed on the observing screen. The pattern is most striking when the wavelength of the waves is comparable to the distance between the slits. This is what classical wave mechanics predicts.

To be more explicit, the amplitude of the wave pattern at the observing screen when the upper gap is open can be described by a time-dependent *waveform* $f_1(x, t)$, where x is the position on the observing screen. The energy density in a wave is proportional to the (time-averaged) square of the amplitude, $\langle f_1(x)^2 \rangle$, which (unlike the amplitude) is always positive. When the bottom gap is open, the wave pattern on the screen is described by a different waveform $f_2(x, t)$. And when both are open, the amplitudes of the waves add or *superpose*, $f(x, t) = f_1(x, t) + f_2(x, t)$. The energy density in the interference pattern, proportional to $\langle f(x)^2 \rangle = \langle f_1(x)^2 + f_2(x)^2 + 2f_1(x)f_2(x) \rangle$, is given by the sum of energies in the two separate patterns *plus an interference term* proportional to $\langle 2f_1(x)f_2(x) \rangle$. The interference term can be either positive or negative and gives rise to the spatial oscillations observed in Figure 7.2(c).

Finally, consider electrons (see Figure 7.3). Remarkably, if the slits are small enough, the electrons *diffract like waves, but arrive on the observing screen like particles*. First, suppose one slit is blocked. The electrons arrive on the screen one by one. Eventually they build up a pattern peaked roughly behind the open slit. This could be explained by electrons deflecting off the edges of the slit, or it could be a diffraction pattern. If the other slit is blocked and the first opened, a similar pattern builds up behind the open gap. The surprise occurs when both slits are open: Although the electrons arrive one by one like bullets, they build up an *interference pattern on the arrival screen* like the one formed by water waves. Even if the electrons are fired so slowly that only one is in the apparatus at a time, still the interference pattern is formed. A single electron in the double slit experiment thus behaves like a wave that has passed through both slits and interfered with itself. Here,

perhaps, is where quantum mechanics seems most alien to the classically trained mind. Our classical intuition suggests that a given electron either passes through one slit or through the other, but not both. Any attempt, however, to follow the electron trajectories in detail – by shining light on them, for example – disturbs the electrons' motion so much that the interference pattern is destroyed.

The appearance of an interference pattern for a single electron suggests that the motion of the electron should be described in some way in terms of the propagation of a wave. **Electron diffraction**, however, is only observable on very small distance scales. When American physicists Clinton Davisson and Lester Germer first demonstrated electron diffraction in 1927, they found – as proposed earlier by the French physicist Louis de Broglie – that the spacing between slits needed to see the interference effect is inversely proportional to the electron momentum p. Since interference in the diffraction pattern is most easily seen when the wavelength and slit separation are comparable, as for water waves, this suggests that electrons are characterized by waveforms with wavelength inversely proportional to momentum. Experiments show that the wavelength of a **matter wave**, known as the **de Broglie wavelength**, is given by

$$\lambda = h/p = 2\pi\hbar/p \,. \tag{7.3}$$

The small size of Planck's constant in SI units indicates that quantum effects are directly observable only at very small distances. For an electron traveling at 100 m/s, for example, the wavelength is $\lambda = 2\pi\hbar/p \cong 7.27 \times 10^{-6}$ m, so the size of slit needed to see interference for such an electron is quite small – but easily within experimental range. An image of an actual electron diffraction pattern is shown in Figure 7.4.

The **quantum wavefunction** description of particles is the only explanation that naturally fits the observed phenomenon of electron diffraction and other experimental data. Unlike classical physics, which in principle allows us

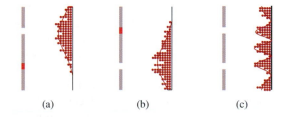

(a)　　　　　(b)　　　　　(c)

Figure 7.3 Double slit experiment with electrons. When either slit is open ((a) or (b)) the electrons pile up behind the slit in a broad peak. Though they arrive one by one, when both slits are open (c) the distribution of electrons displays the same kind of interference pattern as water waves. Neither the incident electrons nor those reflected from the screen are shown.

Figure 7.4 A diffraction pattern formed by electrons with energy 50 keV passing through two slits separated by 2.0 μm [41].

Matter Waves

Quantum particles propagate like waves but are detected as localized lumps of energy.

The wavelength of the matter wave is related to the particle's momentum by de Broglie's relation,

$$\lambda = h/p = 2\pi\hbar/p.$$

to compute the exact trajectory of a particle given its initial position and velocity, quantum physics provides only a probabilistic prediction for where a given electron will impact the screen.

In quantum physics, an electron is described by a wavefunction $\psi(x, t)$ that propagates according to a quantum wave equation. If we denote by $\psi_{1,2}(x)$ the form of the wavefunction at the screen at a fixed time, in a situation where only one slit or the other is open, then the wavefunction becomes the superposition $\psi(x) = \psi_1(x) + \psi_2(x)$ when both slits are open. *The experimental observations can then be explained by applying the rule that the probability of observing an electron in an interval dx is given by a **probability density** proportional to the square of $\psi(x)$.* In particular, the electron distribution on the screen displays interference when both slits are open, just as in the case of water waves.

We can describe the results of the electron diffraction experiment somewhat more abstractly by labeling the **state** of the electron by the slit it has passed through. We use $|1\rangle$ for an electron that has passed through the upper slit when only that slit is open, and $|2\rangle$ for the lower. The state of the electron at a fixed time t after passing through both open slits can then be expressed abstractly as

$$|\psi\rangle \sim |1\rangle + |2\rangle. \tag{7.4}$$

This notation for the state of a quantum system – "$| \ldots \rangle$" – was introduced by Dirac. It is a shorthand where all the information that defines the quantum state – usually a list of integers known as **quantum numbers** – is sandwiched between "$|$" and "$\rangle$". Notice that eq. (7.4) can be interpreted as expressing the electron's state as a superposition in a linear **quantum state space**. Dirac's notation, and the associated abstraction of quantum wavefunctions to general states in a linear space, underlies the framework of *matrix mechanics*, which gives a simple perspective on many quantum systems. In the following section we describe quantum wavefunctions in somewhat more detail, after which we return to the more abstract description of quantum physics in terms of the linear quantum state space.

7.2 Quantum Wavefunctions and the Schrödinger Wave Equation

As suggested in the previous section, the quantum description of a particle's evolution in space-time is given by a wavefunction $\psi(x, t)$. The wavefunction ψ is a *complex-valued* function, as discussed further below. A physical interpretation of $\psi(x, t)$ is that if the position of the particle is measured at time t, then the absolute square of the wavefunction, $|\psi(x, t)|^2$, gives the probability dP that the particle will be detected in a small volume dV centered at x,

$$\frac{dP}{dV} = |\psi(x, t)|^2. \tag{7.5}$$

Since the particle must be somewhere, the integral of this probability density over all space must be one,

$$\int d^3x \, |\psi(x, t)|^2 = 1. \tag{7.6}$$

This interpretation of ψ was first proposed by German physicist Max Born in 1926.

Like the electromagnetic waves described in §4, the matter waves of quantum mechanics obey a wave equation. This wave equation takes the place of Newton's laws; it dictates the space-time evolution of a quantum system. The wave equation for the quantum evolution of a single particle moving under the influence of a potential $V(x)$ was proposed by Erwin Schrödinger in 1926,

$$i\hbar\frac{\partial}{\partial t}\psi(\mathbf{x}, t) = -\frac{\hbar^2}{2m}\nabla^2\psi(\mathbf{x}, t) + V(\mathbf{x})\psi(\mathbf{x}, t). \tag{7.7}$$

Note that when $V = 0$ this (time-dependent) **Schrödinger equation** is very similar in form to the heat equation (6.22), but has a factor of the imaginary number i on the left-hand side. The factor of i changes the dissipative heat equation into a wave equation with oscillatory solutions (Problem 7.2). Since eq. (7.7) is a first-order differential equation in time (one time derivative), knowledge of $\psi(x)$ at one instant in time uniquely determines the wavefunction at all later times.

One striking feature of the wavefunction description of a particle is that the position of the particle, which we think of as fundamental in classical mechanics, is not well-defined in the quantum world. Any physical quantum wavefunction extends over a range of points in space, so the particle cannot be thought of as localized at a single point. The particle is instead spread out over the region where $\psi(x, t)$ is nonzero. Surprisingly, this fuzzy quantum mechanical picture seems to be a more fundamental and complete description of nature than the classical one,

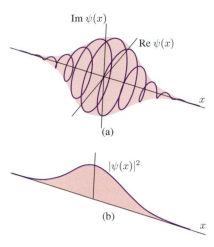

Figure 7.5 A quantum particle is described by a complex valued wavefunction $\psi(x)$. (a) The real and imaginary parts of $\psi(x)$ are plotted versus x. (b) The probability of finding the particle at x, given by $|\psi(x)|^2$, is nonzero over a range of values of x, suggesting that the classical notion of position has only approximate validity.

Quantum Wavefunction

The quantum behavior of a particle is described by its *wavefunction* $\psi(x, t)$, which obeys the *Schrödinger equation*,

$$i\hbar \frac{\partial}{\partial t}\psi(x, t) = -\frac{\hbar^2}{2m}\nabla^2 \psi(x, t) + V(x)\psi(x, t).$$

The absolute square of the wavefunction $\psi(x, t)$ gives the **probability density** (per unit volume) for observing the particle,

$$\frac{dP}{dV} = |\psi(x, t)|^2 .$$

This probability density is *normalized* to one: $\int d^3x\, |\psi(x, t)|^2 = 1$.

showing that our intuitive notions of position and velocity are really only approximations to the true underlying nature of reality.

7.2.1 Origins of the Schrödinger Equation

Schrödinger was motivated to write down his equation by the energy–momentum relation for a classical free particle, $E = p^2/2m$. To follow his logic, we consider a (complex) *plane wave*. For simplicity, we orient the wave so that it propagates along the x-axis,

Plane Wave Solutions to the Schrödinger Equation

The one-dimensional Schrödinger wave equation for a free particle has plane wave solutions of the form $e^{ikx-i\omega t}$ when

$$\hbar\omega = \frac{\hbar^2 k^2}{2m}.$$

This is the quantum version of the energy–momentum relation $E = p^2/2m$, with the identifications $E = \hbar\omega$, $p = \hbar k$.

$$\psi(x, t) = e^{ikx-i\omega t} = \cos(kx - \omega t) + i\sin(kx - \omega t). \tag{7.8}$$

If we substitute into eq. (7.7) (recalling that $\nabla^2 = \partial^2/\partial x^2$ in one dimension), we find that the plane wave satisfies the Schrödinger equation when

$$\hbar\omega = \frac{\hbar^2 k^2}{2m}. \tag{7.9}$$

Remembering de Broglie's proposal, $\lambda = 2\pi\hbar/p$, and the connection of wave number k and wavelength, $\lambda = 2\pi/k$, we identify $\hbar k$ with the momentum p. Using the Planck relation $\hbar\omega = E$ for a single unit of energy, the relation (7.9) for the free Schrödinger plane wave becomes $E = p^2/2m$, precisely the energy–momentum relation of classical mechanics.[1]

When $V(x)$ is added back in, Schrödinger's equation can thus be thought of as a transcription of the classical energy–momentum relation $E = p^2/2m + V(x)$ into the form of a wave equation, with the identification of wave number and frequency with momentum and energy respectively,

$$p = \hbar k, \qquad E = \hbar\omega. \tag{7.10}$$

Although this discussion has been somewhat heuristic, the relations (7.10) are central to the quantum physics of matter particles; we revisit this identification in the sections that follow. Finally, note that were it not for the factor of i in eq. (7.7), it would not be possible for the frequency of the wave to be proportional to the square of its wave number as is needed to match the energy–momentum relation (7.9) for a free particle.

---

[1] Note that the plane wave (7.8) does not satisfy the normalization condition (7.6); this issue is addressed at the end of §7.7.1.

7.2.2 Waves and Frequency

The Schrödinger equation differs in some ways from the wave equations studied in §4 that describe violin strings and light waves. In particular, the Schrödinger equation has only one time derivative and is explicitly complex due to the factor of i. Nonetheless, as we have seen above, the free ($V=0$) Schrödinger equation admits plane wave solutions (7.8). The Schrödinger equation also shares two of the other most significant features of general wave equations: *linearity* and the physical importance of *frequency*.

Linearity Like the other wave equations we have considered, the Schrödinger equation (7.7) is linear in ψ, so that if $\psi_1(\mathbf{x}, t)$ and $\psi_2(\mathbf{x}, t)$ are solutions, then any weighted sum of ψ_1 and ψ_2 with (in general, complex) constant coefficients,

$$\psi(\mathbf{x}, t) = a\,\psi_1(\mathbf{x}, t) + b\,\psi_2(\mathbf{x}, t)\,, \qquad (7.11)$$

is also a solution. Linearity is an essential ingredient in quantum mechanics, where it plays a fundamental role in our understanding of the structure of the theory. Linearity allows us to regard the wavefunction $\psi(\mathbf{x})$ at a point in time as living in an (infinite-dimensional) *vector space* of functions (Appendix B.3). In general, eq. (7.11) defines a **linear combination** of ψ_1 and ψ_2. In physical systems, a linear combination of solutions such as eq. (7.11) is referred to as a **superposition** of solutions, and in particular as a **quantum superposition** in the case of a quantum system.

When two or more solutions to a wave equation are added as in eq. (7.11), the resulting solution can exhibit constructive or destructive *interference* depending on the relative phases (or signs) of the solutions being combined, as described in §4.5.1. For example, in Figure 7.6 the sum of two time-dependent solutions gives another solution. At some times and in some regions of x the two solutions have the same sign and interfere constructively; in other regions the solutions have opposite sign and partially or completely cancel. This kind of constructive and destructive interference is what gives rise to the interference patterns seen in the double-slit experiment sketched in Figure 7.3(c) and Figure 7.4.

Frequency As described in §4.2, the physics of systems exhibiting wave-like behavior is often illuminated by decomposing the wave into *modes* whose behavior is periodic in time. In most physical systems, these oscillation modes and the associated frequencies have important physical significance. For light waves, for example, the frequency encodes the color of light, and for sound waves, the frequency of oscillation encodes the pitch. For solutions to the Schrödinger equation, the frequency is proportional

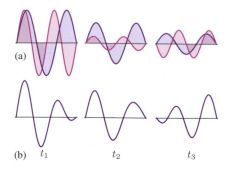

(a)

(b) t_1 t_2 t_3

Figure 7.6 Two (real) solutions (a) to Schrödinger's equation shown at three different times. Addition of the two solutions gives a third solution (b). At some times, the wavefunctions have the same sign in some regions and interference is constructive (e.g. on the left at t_1). At other times (e.g. on the left at t_3), the signs are opposite and the interference in the same region is destructive.

to the energy of the system, as we saw for the simplest plane wave solutions in eq. (7.10). Thus, in general, quantum states of fixed energy are associated with solutions of eq. (7.7) that are periodic in time, where the angular frequency of oscillation is related to the energy through Planck's relation $E = \hbar\omega$. This relationship between energy and frequency is fundamental to quantum physics and is present even for quantum systems with no wave interpretation, as we discuss later in this chapter. To understand these features of the Schrödinger equation more clearly, however, at this stage it is helpful to consider the simple example of a quantum particle in a box.

7.2.3 A Quantum Particle in a Box

The problem of finding the modes and energies of a quantum particle confined to a one-dimensional box is very closely related to the problem of finding the modes and frequencies of a violin string (§4.2.1). The reader may find it useful to review the analysis in that section briefly before proceeding.

Consider the Schrödinger equation (7.7) for a particle in a finite one-dimensional box $0 \leq x \leq L$. By definition, the particle has no probability of being outside the box, so we set $\psi(x, t) = 0$ for all $x \leq 0$ and $x \geq L$. Just as for the violin string, the spatial wavefunction can be decomposed into modes $\sin(n\pi x/L)$, where $n \geq 1$ is a positive integer. The single time derivative in the Schrödinger equation, however, forces us to choose a *complex exponential* for the time dependence,

$$\psi_n(x, t) = c_n\sqrt{\frac{2}{L}} \sin\left(\frac{n\pi x}{L}\right) e^{-i\omega_n t}\,, \qquad (7.12)$$

Quantum Particles

The wavefunction of a quantum particle can be written as a sum over modes, each with its own wavefunction $\psi_n(x)$ and frequency ω_n. The frequency is related to the energy of the mode by $E_n = \hbar\omega_n$, so that the general time-dependent solution is of the form

$$\psi(x,t) = \sum_n c_n \psi_n(x) e^{-iE_n t/\hbar}.$$

For the case of a particle of mass m moving in a one-dimensional box of length L,

$$\psi(x,t) = \sum_n c_n \sqrt{\frac{2}{L}} \sin\left(\frac{n\pi x}{L}\right) e^{-i\omega_n t},$$

where ω_n is related to the energy through

$$E_n = \hbar\omega_n = \frac{n^2\pi^2\hbar^2}{2mL^2}.$$

where c_n is a complex constant (which absorbs a possible phase in the exponential). If we substitute $\psi_n(x,t)$ into the Schrödinger equation (7.7) with $V(x) = 0$, a solution is obtained if

$$\hbar\omega_n = \frac{\hbar^2 n^2 \pi^2}{2mL^2}. \tag{7.13}$$

Although the mode analysis for the quantum particle in a one-dimensional box is very similar to that of the violin string, the interpretation of frequency is entirely different. While for the violin string the frequency determines the pitch of the resulting sound wave, as discussed earlier, for the quantum particle the frequency is associated with the energy of the state. In contrast, the energy of the vibrating violin string depends on its amplitude c_n (see eq. (4.16)) while the amplitude of the quantum wave is fixed by the normalization imposed by the probability interpretation (7.6) and does not affect the energy. Thus, in eq. (7.13) we associate an energy $\hbar\omega_n$ with the nth *quantum state* of a particle in a one-dimensional box. A few wavefunctions for a quantum particle in a box are shown in Figure 7.7, along with the energy spectrum and probability densities. Since in the Schrödinger equation there is only one $\partial/\partial t$ on the left-hand side, the frequency scales as n^2 instead of n. As in the case of the violin string, the general state of the quantum particle is given by an arbitrary wavefunction written as a linear combination of the modes (7.12). Such a

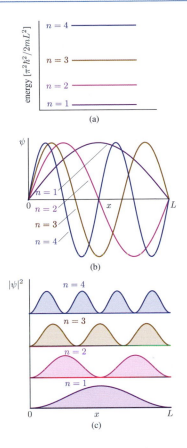

(a)

(b)

(c)

Figure 7.7 The first four energy modes of a quantum particle in a one-dimensional box of length L. (a) The energy in units of $\pi^2\hbar^2/2mL^2$; (b) the wavefunctions $\psi_n(x)$; (c) the probability distributions $|\psi_n(x)|^2$.

quantum superposition of states does not have a fixed value of energy.

The example of a particle in a box illustrates some key features of quantum mechanics that apply even for systems that are not described by wavefunctions in space. All quantum systems are linear, and every quantum state can be described as a linear combination of states that have fixed energies related to the frequency of oscillation through $E = \hbar\omega$. These universal features allow us to give a systematic description of a general quantum theory in the following sections.

One final comment may be helpful before leaving the discussion of wavefunctions. It may be unclear at this point how our familiar description of a particle as a pointlike object with position and momentum arises from the kind of linear combination of quantum states described in (7.11). In fact, given enough modes, we can find a combination in

---

Box 7.2   Quantum Tunneling and Solar Energy

Quantum mechanics allows processes to occur that are forbidden by the rules of classical mechanics. Such processes are critical ingredients in some energy systems. While classically a particle can be kept in or out of a particular region of space by a sufficiently high potential barrier, the quantum wavefunction of a particle cannot be stopped in this way. The quantum wavefunction for a particle with fixed energy will in general be nonzero, though small, even in regions where the particle is classically forbidden to go. This can allow the quantum particle to *tunnel* through a classical barrier.

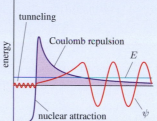

The core of the Sun contains protons (ionized hydrogen atoms) at high temperature and pressure. Coulomb repulsion between the protons acts to keep them apart. If the protons can get sufficiently close, however, nuclear forces take over and they can combine through the process of *nuclear fusion* into a more massive particle (a *deuteron*), giving off a great deal of energy. The potential barrier keeping the protons apart is so high that protons at the temperature of the solar core could not fuse without the benefit of quantum tunneling. The cartoon at right shows the *Coulomb barrier*, outside which the potential is repulsive, and, at shorter distances, a region of nuclear attraction. The kinetic energy E of the protons is not high enough to surmount the barrier classically, but the quantum wavefunction has a small amplitude inside the barrier. Thus quantum mechanics literally enables the Sun to shine as it does. The mechanism of tunneling is described in more detail in §15, and the application to solar fusion is described in §22.

---

which destructive interference between the modes causes the linear combination to cancel everywhere except in a very small spatial region. The classical "point particles" with which we are familiar correspond to linear combinations of many basis modes, resulting in a wavefunction that is non-negligible only in a region of space small compared with the precision of our measuring apparatus. We give a more detailed discussion of how particles can be localized in quantum mechanics in Box 7.6, and we describe the connection between quantum and classical physics for systems of this type in §21 (see in particular Box 21.1).

7.3 Energy and Quantum States

At this point we begin a systematic description of a general quantum system. In this section and the three that follow, we develop four axioms on which quantum mechanics rests. After stating each axiom, we explore the interpretation and give examples of the implications of the axiom. These axioms hold for any isolated quantum system governed by physical laws that are unchanging in time; the significance of these caveats is discussed further in §21. We start with the role of energy in quantum mechanics, generalizing the discussion of the previous section.

In a classical system (such as a stone thrown by a child) the *state* of the system is defined by giving the positions and velocities (or momenta) of all independent objects in the system (e.g. the position of the stone and its

velocity vector).[2] The energy of the system is then computed from the positions and velocities. For the example of a stone, with speed v and height h above sea level, $E = \frac{1}{2}mv^2 + mgh$. In classical systems, the parameters specifying the state (here v and h) are free to take arbitrary values in some continuous range, so the energy can take any value. The notion of states and energies in quantum mechanics is different from classical physics. Quantum mechanics is in some ways simpler, though less familiar.

Quantum Mechanics Axiom 1

Any physical quantum system has a discrete[3] set of energy basis states $|s_1\rangle$, $|s_2\rangle$, . . ., *where the energy of the state* $|s_i\rangle$ *is* E_i.

---

[2] Note that the "state" of a classical or quantum system refers to a unique configuration of the system, unlike a thermodynamic equilibrium "state" as discussed in §5.1.2, which refers to an ensemble of microscopic states with the same macroscopic properties; this distinction is discussed further in the following chapter (§8).

[3] The number of energy basis states is actually finite for any physical system. This fact is not generally emphasized, but is nonetheless true and simplifies the mathematics involved. For those who know some quantum mechanics and are surprised by this statement, further discussion can be found in Box 7.5. Note that the plane wave and particle-in-a box solutions discussed earlier, with infinite spectra, are mathematical idealizations that accurately approximate real physical systems with large but finite spectra at experimentally relevant energy scales.

The energy basis states $|s_i\rangle$ represent a set of possible quantum states of the system. These states are also often referred to as **energy eigenstates**. An example of a set of such states is given by the modes $\psi_n(x)$ of the particle in a one-dimensional box described in §7.2.3. Because the energy basis states $|s_i\rangle$ form a discrete set, these states should be thought of very differently from the states of a classical system, which generally involve specifying *continuous* variables such as position and velocity.

The set of possible energy values E_i associated with the energy basis states of a system is called the (energy) **spectrum** of the system. Sometimes there are multiple independent basis states with the same energy, $E_i = E_j$. In this case we say that the spectrum is **degenerate**. A central problem in quantum mechanics, which is relevant for many energy applications, is the determination of the spectrum of allowed energy values for a given physical system.

To illustrate the idea of energy basis states and the discrete quantum spectrum we give two simple examples.

7.3.1 The Qubit – The Simplest Quantum System

A familiar example of a system that has only a discrete set of *states* available is a register in a computer. Information in electronic devices is encoded in a series of **bits** that each represent a binary value 0 or 1. While bits in conventional computers are stored classically, there is a closely analogous kind of quantum system known as a **quantum bit**, or **qubit**, which has only two independent energy basis states. An example of such a system is provided by the *intrinsic spin of an electron*.

Before the advent of quantum mechanics, elementary particles were thought to be pointlike in nature, and it was presumed that they did not possess any intrinsic angular momentum. We now know that all the fundamental particles from which ordinary matter is made possess an intrinsic angular momentum (**spin**) of magnitude $\hbar/2$. (Note that Planck's constant has units of angular momentum.) If the intrinsic spin of a particle were a classical property, its component along any coordinate axis could vary continuously between $+\hbar/2$ and $-\hbar/2$. Quantum states, however, are discrete, and it turns out that the spin of an electron measured along any coordinate direction *can only take values $+\hbar/2$ and $-\hbar/2$* – no intermediate values are possible.

The quantum behavior of electron spin can be seen physically through the interaction of electrons with magnetic fields. Because they are spinning and charged, electrons – like macroscopic current loops – produce magnetic dipole fields. According to classical electromagnetism (see

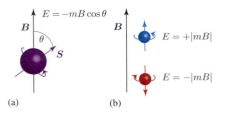

Figure 7.8 (a) A classical spinning (positively) charged sphere can be oriented at any angle in a magnetic field. Its energy depends on the angle, $E = -mB\cos\theta$. (b) An electron has only two spin states with respect to a magnetic field, with energies $\pm|mB|$. Note that the arrow on the electron denotes the spin direction, which is *opposite* to the direction of the magnetic dipole moment.

Box 3.2), the energy of a magnetic dipole m in a magnetic field B is

$$E = -\mathbf{B} \cdot \mathbf{m} = -m\,\hat{\mathbf{S}} \cdot \mathbf{B} = -mB\cos\theta, \qquad (7.14)$$

where for a charged particle m points along the axis of spin angular momentum, $\mathbf{m} = m\hat{\mathbf{S}}$. Note that for the electron, since the charge is negative the magnetic dipole moment points in the opposite direction to $\hat{\mathbf{S}}$, so m is negative. Thus, classical electromagnetism would suggest that an electron in a constant magnetic field should have an energy that varies continuously in proportion to the cosine of the angle θ between S and B (see Figure 7.8(a)).

The component of an electron's spin along a given axis can be measured using an apparatus developed in 1922 by German physicists Otto Stern and Walther Gerlach. In a **Stern–Gerlach experiment**, a particle is passed through a specially shaped magnetic field that has the effect of displacing particles vertically according to the direction of their spin. (Actually, Stern and Gerlach used neutral atoms in their original experiment but the same type of experiment can be done with electrons.) Electrons passing through the field are deflected in only two distinct directions, demonstrating that the component of the electron's spin along the axis defined by B can only take the possible values $\pm\hbar/2$. The situation is summarized in Figure 7.9: the electron has two possible spin states $|\pm\rangle$ along the vertical axis, known as "**spin-up**" or "**spin-down**," with energies $E_+ = \pm|m|B_z$ respectively. (Note that with the given conventions, for the states in the figure $E_+ < E_-$ since $B_z < 0$.)

While less familiar conceptually than the particle in a box, the electron spin is an ideal example of a quantum system because it has only two states; it is a *qubit*. Qubits are the building blocks of **quantum computers**, which are a subject of much current research. Although

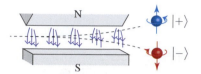

Figure 7.9 In a *Stern–Gerlach* type experiment, a beam of electrons is passed through an inhomogeneous magnetic field that has the effect of displacing the electrons by a distance proportional to the component of their spin along the axis defined by the magnetic field. Contrary to classical expectations, the electron is found to have only two possible values for spin, corresponding to angular momentum $\pm\hbar/2$.

they have yet to be realized at a useful scale in practice, quantum computers can in principle perform certain calculations, such as the factorization of large numbers, with exponentially fewer elementary operations than classical computers. Note that the spin of the electron is unrelated to its position and momentum, whose quantum behavior was surveyed in §7.2 and to which we return in §7.7.

7.3.2 The Hydrogen Atom

A discrete energy spectrum also arises in more complicated quantum systems. Consider, for example, a hydrogen atom, which consists of a negatively charged electron in "orbit" around a positively charged proton. Classically, the energy of this system is described by a potential proportional to $-1/r$, where r is the distance between the electron and the proton. A classical bound orbit in a potential of this type can have any negative value of the energy. In the quantum system, on the other hand, there is a discrete spectrum of possible energy values ε_0/n^2, where $\varepsilon_0 \cong -13.61\,\mathrm{eV} \cong -2.180 \times 10^{-18}\,\mathrm{J}$, and n is a positive integer. Since n appears to be able to take any positive value, it seems that this system has an infinite number of states. This is a convenient idealization. As n becomes large, however, the size of the region containing the electron grows and interactions with other matter cannot be ignored (see Box 7.5). The spectrum of energies for a hydrogen atom is shown in Figure 7.14, and the hydrogen atom is discussed in more detail in §7.8. Note that there are multiple states associated with each of the energy levels $-\varepsilon_0/n^2$. Even in the *ground state*, where $n = 1$ (the lowest-energy state), an electron can be in either of the two possible spin states. In chemistry, this is the *1s-shell*, common to all atoms, which can admit one or two electrons. Note also that there are corrections to this simple spectrum coming from detailed features of quantum physics. For example, interactions between the spins of the proton and electron cause a **hyperfine splitting**

of the ground state of hydrogen. (An analogous hyperfine splitting in the ground state of the cesium atom is used to define the second in SI units, as the time needed for 9 192 631 770 oscillations of the electromagnetic wave associated with a transition between the two lowest energy quantum states of the cesium atom.)

7.3.3 Photons and Transitions Between Energy States

Like the hydrogen atom, every atomic or molecular system has its own characteristic spectrum of allowed energy values. So, too, do nuclei. Under certain circumstances, a quantum system can make a **transition** from one energy state to another by emitting a **photon**, which is a quantized packet of electromagnetic radiation. Energy conservation requires that the energy of the photon is equal to the difference between the initial and final energy of the system that emitted it, $E_{\mathrm{photon}} = E_i - E_f$. Since the frequency of the photon is related to its energy by the Planck relation (7.1), $E = \hbar\omega$, the frequency of the radiation emitted in a transition encodes information about the energy spectrum of the system that emitted it,

$$\omega = \frac{1}{\hbar}\left(E_i - E_f\right) . \qquad (7.15)$$

This explains the selective absorption of certain frequencies of radiation by specific kinds of molecules, such as the absorption of infrared radiation from Earth by water vapor and CO_2 in the atmosphere, which plays a key role in the discussion of climate change in §34. Just as atoms and molecules have specific energy spectra, so do solid materials (see Example 7.2).

Electromagnetic radiation itself occurs over a wide range of energies and wavelengths, and takes many qualitatively different forms, ranging from low-energy, long-wavelength radio waves up to very-high-energy gamma rays (see Figure 9.2). The spectrum of electromagnetic radiation is described further in §9, in connection with the range of energy scales that characterize quantum transitions in matter. The quantum nature of radiation and thermodynamics are combined in §22 (Solar energy) to derive the detailed spectrum of blackbody radiation.

7.4 Quantum Superposition

The first axiom of quantum mechanics introduced a set of states of a system in which energy takes specific values. This is not the full story, however. An essential feature of quantum mechanics, which leads to some of its most surprising effects, is the fact that a system can exist in a combination of different energy basis states at one time.

Example 7.1   Spectra Provide Clues to Structure

When an atom containing multiple electrons is in its ground state, the electrons sit in the lowest energy states available. When the atom is excited, for example by heating the material containing it, electrons jump from the lowest energy levels to higher-energy states.

The electrons can then drop back down again from one energy level to another, emitting light at characteristic frequencies equal to the difference in the levels' energies divided by Planck's constant (eq. (7.15)). Observations of these **spectral lines** emitted by hot gases were a crucial tool in developing quantum mechanics and the theory of atomic structure. Emission spectra from a few elements are shown in the upper figure to the right.

Nuclei also emit electromagnetic radiation with characteristic frequencies. Because their energies are so much larger than atomic energies, their emissions are in the energy range known as gamma rays (see Figure 9.2). In the mid-twentieth century, studies of the gamma-ray spectra of nuclei helped physicists understand the principles of nuclear physics. The lower figure shows a portion of the spectrum of gamma rays emitted by a carbon nucleus.

(Image credit: (Top) NMSU, N. Vogt, reproduced under CC-BY-SA 4.0 license, (Bottom) NNDC)

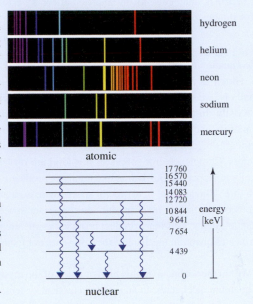

Example 7.2   Energy Spectra of Materials and Photovoltaics

Each physical system – whether an atom, a molecule, or a lump of coal – has a discrete *spectrum* of possible quantum energy levels. This spectrum determines the frequencies of electromagnetic radiation that can be absorbed or emitted by the system. Many materials, such as *crystalline silicon*, have a spectrum characterized by *bands* of many closely spaced energy levels for electrons separated by *gaps* where no energy states exist. This structure plays an important role, for example, in the conversion of solar radiation to electrical power using *photovoltaic* solar cells. Crystalline silicon has a band gap of about 1.1 eV separating the filled energy levels from the empty ones. Incoming photons with more than 1.1 eV of energy (wavelength below ~ 1130 nm, which includes visible light) can excite electrons across this band gap, allowing solar energy to be collected in the form of electrostatic potential energy. This key mechanism underlying photovoltaic solar power and the general structure of the energy spectrum for crystalline materials are developed in §22.

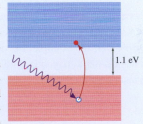

This extends to general quantum systems the concept of superposition that we identified in §7.2.2 for solutions of the Schrödinger wave equation.

Quantum Mechanics Axiom 2

The state of a quantum system at any time is a linear combination, or **quantum superposition**, *of energy basis states, and can be written in the form*

$$|s\rangle = z_1|s_1\rangle + z_2|s_2\rangle + \cdots + z_n|s_n\rangle, \qquad (7.16)$$

where $|s_i\rangle$ are the energy basis states, and z_i are complex numbers.

The physical properties of the state are independent of an overall scaling factor $|s\rangle \rightarrow \lambda|s\rangle$, so it is conventional to **normalize** the coefficients $\{z_j\}$ such that

$$|z_1|^2 + |z_2|^2 + \cdots + |z_n|^2 = 1. \qquad (7.17)$$

What is a linear superposition of quantum states like eq. (7.16) supposed to mean? For wavefunctions, this is just the linearity notion of eq. (7.11) in different notation. More abstractly, from a mathematical point of view eq. (7.16) is analogous to the expression for a vector in three dimensions in terms of unit basis vectors $\hat{x}, \hat{y}, \hat{z}$

$$\boldsymbol{v} = v_x\hat{\boldsymbol{x}} + v_y\hat{\boldsymbol{y}} + v_z\hat{\boldsymbol{z}}. \quad (7.18)$$

The equation (7.18) expresses the fact that any vector can be written as a linear combination of these basis vectors. Similarly, we can think of the states $|s_i\rangle$ as basis vectors in a multi-dimensional space and eq. (7.16) as an expression of the fact that any state of a quantum system can be written as a (complex) linear combination of those basis vectors. Complex vector spaces of this kind are reviewed briefly in Appendix B.3.1. Just like the inner product $\boldsymbol{v} \cdot \boldsymbol{w}$ in a three-dimensional (real) vector space, there is an analogous concept of an inner product on the vector space of quantum states. This inner product provides a structure that permeates quantum mechanics, though we do not focus on it here.

Within the second axiom is hidden much of the mystery of quantum mechanics. In an isolated classical system, the energy of the system at a given instant in time is a completely well-defined quantity. Consider, for example, a gas of classical pointlike atoms in a box. If we knew the positions and velocities of all the atoms, we could compute the total energy by summing the contributions of kinetic and potential energies. The second axiom states that this is not true for quantum mechanics. Even a very simple system, such as a qubit, can exist in a well-defined quantum state that is a superposition of basis states with different energies. Consider, for example, the qubit describing the spin of an electron in a magnetic field, which may be given at some instant by

$$|\psi\rangle = \frac{1}{\sqrt{2}}|+\rangle + \frac{1}{\sqrt{2}}|-\rangle. \quad (7.19)$$

This is an equal superposition of the state with spin-up and energy E_+ and the state with spin-down and energy E_-. This quantum state has no single well-defined energy. There is no classical analog for a state like this.

To make sense of this, we need to understand the notion of measurement on a quantum system, which leads naturally to a brief discussion of the concept of *quantum entanglement* between two systems.

7.5 Quantum Measurement

We have asserted that a quantum system can be in a superposition like eq. (7.16) of energy basis states. What happens if we measure the energy of such a state? Quantum mechanics postulates that in this situation the measurement yields the energy of one of the states in the superposition.

The Axioms of Quantum Mechanics

The fundamental principles of quantum mechanics can be summarized in four axioms:

Axiom 1: Any physical quantum system has a discrete set of *energy basis states* $|s_i\rangle$, $i = 1, \ldots, N$, with energies E_i.

Axiom 2: The state of the system at any point in time is a *quantum superposition* of energy basis states

$$|s\rangle = z_1|s_1\rangle + z_2|s_2\rangle + \cdots + z_n|s_n\rangle,$$

with $\sum_i |z_i|^2 = 1$.

Axiom 3: If the energy of the system is measured, with probability $|z_i|^2$ the energy will be found to be E_i. After such a measurement the state will change to $|s_i\rangle$ (or a combination of states if there are multiple independent states with energy E_i). An analogous statement holds for other observable features of the system.

Axiom 4: The state $|s_i\rangle$ evolves in time t through multiplication by a phase $e^{-iE_it/\hbar}$, and time evolution is linear in any superposition of states.

Which one of the possible energies is measured is determined only probabilistically, however. Furthermore, once the energy has been measured, the system changes its state. No longer a superposition, the state after the measurement simply becomes the energy basis state corresponding to the value that was measured. This is summarized in the third axiom.

Quantum Mechanics Axiom 3

For a system in a quantum state

$$|s\rangle = z_1|s_1\rangle + z_2|s_2\rangle + \cdots + z_n|s_n\rangle,$$

a measurement of the energy of the system gives the result E_i with probability $|z_i|^2$. Furthermore, if the energy is measured to be E_i, then at the time of the measurement the state becomes $|s\rangle = |s_i\rangle$.

This axiom is stated for a system with a nondegenerate spectrum. If there are multiple states $|s_j\rangle$ with the same energy E, the probability of measuring that energy is given by the sum of the terms $|z_j|^2$ for the relevant states, and the state after the measurement is given by the superposition $\sum_{j:E_j=E} z_j|s_j\rangle$ with the appropriate normalization.

Axiom 3 is even more wildly at odds with our classical intuition for how physical processes should work than the first two axioms. Nonetheless, by this time countless experiments have verified that this is indeed the correct description of physical processes in quantum mechanics.

Axiom 3 is, in fact, a slight simplification of what is needed for a full-fledged formulation of quantum mechanics. Energy is not the only property of a system that can be measured. Other *observables* include position, momentum, angular momentum, and other properties of particles. The rules for measurement of each of these quantities are essentially the same as for energy; the computation of probabilities, however, requires expressing the state as a linear combination of basis states with definite values of the observable of interest. These new basis states are, in turn, linear combinations of energy basis states. This is analogous to what happens when we take a vector described in terms of components along (two) Cartesian coordinate directions $(\hat{\boldsymbol{x}}, \hat{\boldsymbol{y}})$, and change to a new coordinate system $(\hat{\boldsymbol{x}}', \hat{\boldsymbol{y}}')$ rotated by an angle θ with respect to the first (see Figure 7.10). The vector has not changed, but its components along the coordinate unit vectors have. Similar linear transformations (Appendix B.3.2) are needed to define basis changes for quantum states and to formulate more general measurements in a precise fashion. An example of this kind of linear recombination appears in Box 7.4 below. In most of this book, we focus on the energy basis and do not need the more general formalism for arbitrary observables.

7.5.1 Measurement in Quantum Mechanics

At this point, it is natural to ask what precisely is meant by "measurement." Scientists, philosophers, and other confused individuals have argued at length on this question since the inception of quantum mechanics. The fact that something discontinuous seems to occur when a measurement is made raises at least two difficult questions:

first, when does a measurement occur? And second, by what process does the state of the system before the measurement abruptly change into a state of definite energy (or other observable) after the measurement? With respect to the first question, it suffices for our purposes to say that what we mean by a measurement is just what you would think: a measurement is performed when you do an experiment, detect something with instruments, and record the results. With respect to the second question, the so-called "collapse of the wavefunction," quantum mechanics neither gives nor requires any dynamical explanation for this apparently instantaneous change in the state of a quantum system at the moment a measurement is performed. Whether or not there is a more fundamental explanation of this, the fact that systems behave *as if* the wavefunction collapses is not in doubt. Fortunately, it is not necessary to go further than this in order to use quantum mechanics in the study of energy physics. Instead, we give a pair of examples of the way measurement works in quantum mechanics. The second example also illustrates the notions of quantum entanglement and energy conservation in quantum systems.

7.5.2 Measurement Example: A Quantum Superposition

Consider a Stern–Gerlach apparatus as shown in Figure 7.11, where the magnetic field points along the z-axis and the electron moves along the y-axis. Suppose that the state of the electron spin is given by

$$|s\rangle = \frac{1}{\sqrt{2}} (|+\rangle + |-\rangle) , \qquad (7.20)$$

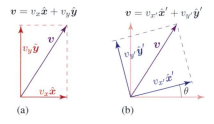

$$\boldsymbol{v} = v_x\hat{\boldsymbol{x}} + v_y\hat{\boldsymbol{y}} \qquad \boldsymbol{v} = v_{x'}\hat{\boldsymbol{x}}' + v_{y'}\hat{\boldsymbol{y}}'$$

(a)   (b)

Figure 7.10 A example of a change of basis in a two-dimensional vector space. A vector $\boldsymbol{v}$ can be written as a linear combination of components (a) along the x- and y-axes, or (b) along x'- and y'-axes that are rotated by an angle θ. Sometimes in quantum mechanics, an analogous basis change is used to work with different choices of quantum basis states.

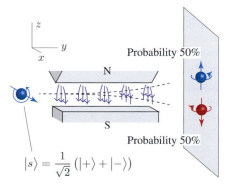

$$|s\rangle = \frac{1}{\sqrt{2}} (|+\rangle + |-\rangle)$$

Figure 7.11 A quantum state $|s\rangle$, prepared to have spin-up along the x-axis by a Stern–Gerlach apparatus (not shown) with its magnetic field along $\hat{\boldsymbol{x}}$, enters a Stern–Gerlach apparatus that measures the spin along the $\hat{\boldsymbol{z}}$-direction. The measurement yields spin-up and spin-down, each with 50% probability, as required by Axiom 3.

Box 7.3   Radioactive Decay and Non-determinism in Quantum Physics

According to quantum mechanics, even if two systems are  in exactly the same quantum state, the result of performing the same measurement on each of the two systems need not be the same. Suppose that a quantum system is prepared in a superposition of two states $|\psi\rangle = z_1|1\rangle + z_2|2\rangle$, and suppose that the measurement of an observable A – such as energy – on $|1\rangle$ would give a value a_1 and on $|2\rangle$ would give a_2. Then according to Axiom 3, measuring A on a set of similarly prepared systems would give a_1 or a_2 randomly with probabilities $|z_1|^2$ and $|z_2|^2$ respectively. No additional probing of the system would allow an observer to obtain more certainty about this measurement. This is a radical departure from classical physics, where any system, in principle, can be described in terms of *deterministic* equations.

Radioactive decay provides a simple example of a random process in quantum mechanics – one that can be observed with the aid of a simple radiation detector such as a *Geiger counter*, which measures the passage of energetic charged particles. To be specific, consider a particular nucleus known as plutonium-238 (denoted ^{238}Pu), which is used as an energy source on space exploration missions that extend far from the Sun (see Box 10.1). ^{238}Pu decays by emitting α particles. The α particles are stopped by other atoms in the plutonium, generating heat at a rate of $\sim$500 W/kg. The average lifetime of a ^{238}Pu nucleus is about 127 years. The inset figure shows a histogram, in one-second bins, of Geiger counter hits made by α particles from the decay of roughly 10^{10} nuclei of ^{238}Pu. It is a random sequence of integers with a mean of approximately two per second (see Box 15.1 for more about this distribution).

In quantum mechanics, the ^{238}Pu nucleus has a small, but non-vanishing probability of being in a state consisting of a ^{234}U nucleus plus an α particle. The probability controls the rate at which ^{238}Pu decays are recorded by the Geiger counter. We study nuclear decays in more detail in §17. Similar radioactive processes are central to the operation of nuclear reactors (§19), are the source of a significant fraction of geothermal energy (§32), and underlie mechanisms for the measurement of Earth's climate in past eras (§35).

One might imagine additional degrees of freedom within the nucleus whose detailed dynamics could be responsible for the occurrence and timing of nuclear decay in an entirely classical framework. These are examples of the *hidden variable theories* mentioned in §7.5.3. There is no experimental evidence for any such degrees of freedom, and they are not needed or present in the quantum description of the decay process. At this point our best explanation for the easily observed, but apparently random, phenomenon of radioactive decay lies in the non-deterministic world of quantum physics.

---

a quantum superposition of the up and down spin states. Since spin-up (down) electrons are deflected up (down) in the z-direction, and states of different spin have different energy in the presence of a magnetic field, the location where the electron lands on an observing screen amounts to a measurement of both its spin and its energy. For any individual electron, there is no way to predict which way it will be deflected. According to Axiom 3, however, there is a probability of 50% ($(1/\sqrt{2})^2 = 1/2$) that any given electron will deflect up, corresponding to spin-up, and a probability of 50% that it will deflect down, corresponding to spin-down. After the measurement, the electron will either be in the state $|+\rangle$ or the state $|-\rangle$, depending on the measurement.

7.5.3 Measurement Example: An Entangled State

How can energy conservation possibly work in quantum mechanics if the energy of a system takes a randomly determined value upon measurement? In this example, we

see that this apparent puzzle is resolved by considering a larger system in which the energy of the particle being measured is correlated with the energy of another particle elsewhere.

Consider a pair of electrons. Both electrons are placed in a magnetic field $\mathbf{B} = B_z \hat{z}$. We assume that the electrons are far apart from one another, so that they do not interact. Each electron has two energy basis states $|\pm\rangle$, with energies $E_\pm$. Including all combinations, the system composing both electrons then has four energy basis states, which we denote

$$|++\rangle, |+-\rangle, |-+\rangle, |--\rangle. \qquad (7.21)$$

For example, $|+-\rangle$ is the state where the first electron is in spin state $|+\rangle$ and the second is in spin state $|-\rangle$. The states in (7.21) have energies $2E_+$, $E_+ + E_-$, $E_+ + E_-$, and $2E_-$ respectively. Suppose that the full system is in the quantum state

Box 7.4   Preparing a Quantum Superposition

How can the state (7.20) be constructed physically?

Imagine a Stern–Gerlach apparatus (SG_x) with its magnetic field oriented along the x-axis. The apparatus separates an electron beam moving in the $+\hat{y}$-direction into two separate beams, with spin $\pm\hbar/2$ along the x-axis. Select an electron in the beam with spin $S = +\frac{\hbar}{2}\hat{x}$. We label this electron's state by $|+_x\rangle$; it is

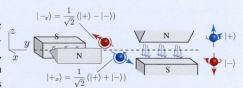

the unique state with spin $+\hbar/2$ along the x-axis. Now, let this electron pass into another Stern–Gerlach apparatus with its magnetic field oriented along the z-axis (SG_z). According to Axiom 2, the state of the electron $|+_x\rangle$ must be a linear combination $|+_x\rangle = \alpha|+\rangle + \beta|-\rangle$ of the energy basis states $|+\rangle$, $|-\rangle$ for the electron in the SG_z apparatus. To determine α and β, suppose that the state $|+_x\rangle$ is rotated by 180° about the x-axis. Rotation by a fixed angle about the x-axis does not change the component of angular momentum along that axis, and therefore $|+_x\rangle$, being uniquely defined by its angular momentum, can at most change by an overall phase. Rotation by 180° about the x-axis, however, reverses the state's orientation relative to the z-axis, and thus changes the sign of a state's spin along that axis. Thus, $|+\rangle \leftrightarrow |-\rangle$ under this rotation (again up to a phase). This implies that the components of $|+_x\rangle$ with respect to the basis states $|+\rangle$ and $|-\rangle$ must have the same magnitude, $|\alpha| = |\beta| = \frac{1}{\sqrt{2}}$ (since $|\alpha|^2 + |\beta|^2 = 1$). The phases of the states $|\pm\rangle$ can be chosen arbitrarily to define the basis; the standard convention is to choose these phases so that $\alpha = \beta = 1/\sqrt{2}$. Thus, the state $|+_x\rangle$ is precisely the state (7.20).

It may seem puzzling that the state $|+_x\rangle$ is a linear combination of the basis states $|+\rangle$ and $|-\rangle$, since $|+_x\rangle$ is itself an energy basis state in the presence of the magnetic field of the apparatus SG_x. There is, however, no contradiction here: the states $|\pm_x\rangle$ are the appropriate energy basis states for the physical setup of SG_x and $|\pm\rangle$ are the appropriate energy basis states for the setup of SG_z. These are just different bases for the same vector space of quantum states. Any state expressed in one basis is a linear combination of the states in the other basis (just as in Figure 7.10, where $\hat{x}$ or $\hat{y}$ can be expressed in terms of rotated unit vectors $\hat{x}'$ and $\hat{y}'$). When the electron leaves SG_x, the magnetic field goes to zero and the two spin states have the same energy, so any basis is good. Once the electron enters SG_z, however, only the basis states $|\pm\rangle$ have definite energies. The system, however, is still in the state (7.20) that was selected by the first apparatus, which is a linear combination of the energy basis states $|+\rangle$ and $|-\rangle$.

In a full treatment of quantum mechanics, such transformations between different bases play an important role in the theory – they are not needed, however, for most of the applications studied in this book.

$$|s\rangle = \frac{1}{\sqrt{2}}\left(|+-\rangle + |-+\rangle\right) . \qquad (7.22)$$

In this situation, the total energy is well-defined and takes the value $E_+ + E_-$. A state such as (7.22) is known as a *correlated* or **entangled** state because the spins of the two electrons are not independent; if one is up the other is down and vice versa. Imagine that we measure the spin/energy of the first electron in a Stern–Gerlach device. With probability 50%, we pick out the first term in eq. (7.22), so the energy of the first electron is E_+. In this case the state of the system becomes $|+-\rangle$, so the energy of the second electron is E_-. The other half of the time, the first electron has energy E_- and the second has energy E_+. So, in measuring the first electron it looks like the energy is randomly determined. But when we consider the whole system consisting of two electrons, the energy of the second electron compensates so that the energy of the whole system is conserved!

This story may seem strange. It appears that when the spin of the first electron is measured, the second electron goes from an indeterminate state to one of fixed spin. This apparent effect of a measurement on a distant object was referred to as a "spooky action at a distance" by Einstein. Over the years following the development of quantum mechanics, Einstein and other scientists troubled by this implication of quantum physics made various efforts to describe quantum mechanics in terms of additional structure (so-called **hidden variable theories**) in terms of which quantum physics might have a classical deterministic explanation. Such alternatives, however, give rise to predictions that differ from those of quantum physics unless classical information is allowed to propagate faster than the speed of light. Numerous experiments have confirmed the predictions of quantum theory over alternative deterministic models, so that the description given in this chapter represents our best understanding of the way the world works.

7.6 Time Dependence

While Axiom 2 provides a complete description of the space of allowed states of any quantum system at a fixed point in time, we have not yet discussed in general how a quantum state evolves in time. In §7.2.1 we saw that a solution to the Schrödinger equation with definite energy oscillates in time with a frequency proportional to the energy, $\omega = E/\hbar$. The same is true for all quantum states of definite energy.

Quantum Mechanics Axiom 4

If the state $|s(t_0)\rangle = |s_0\rangle$ of a system at time t_0 has definite energy E_0, then at time t the state will be

$$|s(t)\rangle = e^{-iE_0(t-t_0)/\hbar}|s(t_0)\rangle. \qquad (7.23)$$

Furthermore, time evolution is linear in the state.

This axiom describes the dynamics of any quantum system. Once we know a set of energy basis states, any state of the system takes the form of eq. (7.16). Axiom 4 implies that the time evolution of such a general state is given by

$$|s(t)\rangle = z_1 e^{-iE_1(t-t_0)/\hbar}|s_1\rangle + \cdots + z_n e^{-iE_n(t-t_0)/\hbar}|s_n\rangle. \qquad (7.24)$$

Note that Planck's constant $\hbar$ appears here for the first time in the set of quantum axioms, providing a scale in units of energy × time where quantum mechanics becomes relevant.

7.6.1 Energy and Time Dependence

The formulation of quantum mechanics in terms of a set of energy basis states provides a deeper explanation for the relationship between energy and the time dependence of quantum systems. Assuming, for simplicity, that all the energy basis states have distinct energies, conservation of energy implies that a system that is in a state $|s_i\rangle$ of fixed energy at one point in time must stay in that state as time advances. Only the phase can change with time, since this does not affect the energy. For a small change in time Δt, the change in phase can be written as $e^{i\Delta t\phi}$. Assuming that the laws of physics are the same at every point in time, the change in phase over time $n\Delta t$ must then be $e^{in\Delta t\phi}$. This is the only possible way in which a state of fixed energy can evolve given time-independent physical laws.

Thus, Axiom 4 can be seen as a consequence of the conservation of energy in quantum mechanics. We can *define* the energy of the state $|s_i\rangle$ in terms of the phase change under time development through $E_i = \hbar\Delta\phi_i/\Delta t$, and all of our other associations with the notion of energy can be derived from this principle. This connection between energy and time development in systems with time-independent physical laws is discussed further in §21.

7.6.2 Matrix Mechanics

The general solution (7.24) can be described in a more abstract (but extremely useful) form using the language of **matrix mechanics**. We can write a general time-dependent quantum state in the form

$$|s(t)\rangle = z_1(t)|s_1\rangle + \cdots + z_n(t)|s_n\rangle. \qquad (7.25)$$

Taking the time derivatives of eq. (7.24) and eq. (7.25) and equating the two resulting expressions yields

$$i\hbar\frac{\mathrm{d}}{\mathrm{d}t}|s(t)\rangle = i\hbar\left(\dot{z}_1(t)|s_1\rangle + \cdots + \dot{z}_n(t)|s_n\rangle\right)$$
$$= z_1(t)E_1|s_1\rangle + \cdots + z_n(t)E_n|s_n\rangle. \qquad (7.26)$$

Thus, the coefficients z_i in ψ each satisfy a first-order differential equation $i\hbar\dot{z}_i = E_i z_i$, whose solution (7.24) gives the time development of the system. This equation can be succinctly and suggestively summarized in matrix notation (Appendix B.3.2). To write the time-evolution rule (7.24) in matrix form, we write the state $|s(t)\rangle$ as a (column) vector

$$|s(t)\rangle = \begin{pmatrix} z_1(t) \\ z_2(t) \\ \vdots \\ z_n(t) \end{pmatrix}, \qquad (7.27)$$

and define an $n \times n$ matrix H known as the **Hamiltonian matrix**, in which the only nonzero entries are the energies E_j placed on the diagonal,

$$H = \begin{pmatrix} E_1 & 0 & \cdots & 0 & 0 \\ 0 & E_2 & & 0 & 0 \\ \vdots & & \ddots & & \\ 0 & 0 & & E_{n-1} & 0 \\ 0 & 0 & & 0 & E_n \end{pmatrix}. \qquad (7.28)$$

Using this notation, we can rewrite eq. (7.26) in vector form

$$i\hbar\frac{\mathrm{d}}{\mathrm{d}t}|s(t)\rangle = H|s(t)\rangle. \qquad (7.29)$$

This is the general form of the *time-dependent* Schrödinger equation, the first-order differential equation that governs the evolution of all quantum systems. The Schrödinger wave equation (7.7) that we discussed when studying wave mechanics in §7.2 is a particular case of this more general equation. We have presented this equation here using a basis of states where H is diagonal. Often, in the treatment of complicated quantum systems, it is natural to use another basis. Much of the work in a full course in quantum mechanics involves learning how to solve eq. (7.29) when working in a basis where the Hamiltonian H is not diagonal.

Example 7.3 Spin Precession in a Magnetic Field

In classical electromagnetism, a magnetic moment $\boldsymbol{m}$ placed in a constant magnetic field $\boldsymbol{B}$ experiences a torque (§3.4.1) $\boldsymbol{\tau} = \boldsymbol{m} \times \boldsymbol{B}$. A particle of mass M and charge q in general has a magnetic moment parallel to its spin, $\boldsymbol{m} = gq\boldsymbol{S}/2Mc \equiv m\hat{\boldsymbol{S}}$, where g is a constant known as the particle's *gyromagnetic ratio*. Newton's second law for rotational motion (2.43), $\boldsymbol{\tau} = d\boldsymbol{S}/dt$, leads to the result that the particle's spin **precesses** – the vector $\boldsymbol{S}$ moves in a circle – with angular frequency $\Omega = gq|\boldsymbol{B}|/2Mc$ around the magnetic field direction, as illustrated below.

The quantum description of this situation provides a simple example of time dependence in a quantum system. Consider an electron spin prepared in the state $|+_x\rangle$ at time $t = 0$ (see Box 7.4), $|\psi(0)\rangle = |+_x\rangle = \frac{1}{\sqrt{2}}(|+\rangle + |-\rangle)$. Placed in a magnetic field in the z-direction, the energies of the $\pm$ states are $E_\pm = \pm Bm = \pm\epsilon$ as described in §7.3.1. (Remember that the electron's charge is $-e$.) At time t, the state will have evolved to

$$|\psi(t)\rangle = \frac{1}{\sqrt{2}}\left(e^{-i\epsilon t/\hbar}|+\rangle + e^{i\epsilon t/\hbar}|-\rangle\right).$$

Thus, the relative phases of the $\pm$ components of the state change with time. Remember, as explained in Box 7.4, that the initial state with a relative $+$ sign between the two components corresponds to an electron with spin oriented along the $+\hat{\boldsymbol{x}}$-direction. Similarly, the linear combination with a relative minus sign corresponds to the state $|-_x\rangle$, so after time $t_{1/2} = \pi\hbar/2\epsilon$ has passed, the state has evolved into $|\psi(t_{1/2})\rangle = -i|-_x\rangle$. Thus, the spin of the electron has rotated from the $+\hat{\boldsymbol{x}}$-direction to the $-\hat{\boldsymbol{x}}$-direction. (The overall phase factor of

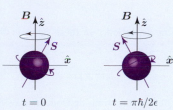

$t = 0$ $t = \pi\hbar/2\epsilon$

$-i$ does not affect the properties of the state.) When $t = \pi\hbar/\epsilon = T$, $|\psi(T)\rangle = -|+_x\rangle$, so the state has returned to its original form. Similar considerations at arbitrary t make it clear that the spin of the electron *precesses* about the direction of the magnetic field with angular frequency $\Omega = 2\epsilon/\hbar$, which is the *same frequency as found in classical electromagnetism* (Problem 7.9). Precession of quantum spins in magnetic fields is the fundamental mechanism behind magnetic resonance imaging (MRI) used for medical diagnostics.

Time-dependence of Quantum States

Axiom 4 can be reformulated in matrix form: every quantum system evolves according to the equation

$$i\hbar\frac{d}{dt}|\psi(t)\rangle = H|\psi(t)\rangle.$$

7.7 Quantum Mechanics of Free Particles

Now that we have a systematic formulation of the general rules of quantum mechanics, we can re-examine the quantum mechanical motion of a particle through space and time from a more general perspective. How do the wavefunctions described in §7.2 relate to the general rules for quantum systems introduced in §7.3–§7.6? And how do either of these descriptions correspond to our familiar notion of an object moving through space with a specific value of position and momentum? In this section we show how the Schrödinger equation for a particle can be deduced based on the axioms of quantum mechanics.

7.7.1 Free Particle

Consider first the simple case of a particle moving in a single dimension in the absence of any force or potential energy. We seek to construct the wavefunction $\psi(x)$ corresponding to an energy basis state for this system. The free particle has a *symmetry* that enables this to be done easily. Since there is no potential, shifting x to $x + \delta$ (for any number δ) does not change the system and therefore cannot change the energy of a state. We say that the system is *invariant under translation* by an arbitrary amount δ.[4] Thus, for any state $\psi_E(x)$ of definite energy E, the wavefunction $\psi_E(x + \delta)$ must have the same energy E, and therefore[5] must be proportional to $\psi_E(x)$. The constant of

---

[4] Note: this is not true when there is a potential because $V(x + \delta)$ cannot be equal to $V(x)$ for arbitrary δ unless V is constant.

[5] There is a slight subtlety here. There may be several wavefunctions corresponding to states with the same energy E, so that when one of these wavefunctions is translated it might transform into another wavefunction or combination of the other wavefunctions. But it is always possible to choose the energy basis states so that they do not mix with one another under a translation that is a symmetry of the system.

proportionality must be a pure phase – a complex number $e^{i\theta}$ of magnitude one. Thus, an energy basis state must satisfy the relation

$$\psi_E(x + \delta) = e^{i\theta(\delta)}\psi_E(x) \qquad (7.30)$$

for every possible δ, where we have noted that θ may depend on the parameter δ. The only functions that satisfy eq. (7.30) for all possible translations δ are the periodic functions e^{ikx} (Problem 7.8). It is convenient to introduce a factor of $\hbar$, through a change of variables $k = p/\hbar$, so that the wavefunctions with the desired translation property are

$$\psi_{E,p}(x) = e^{ipx/\hbar}. \qquad (7.31)$$

The time dependence of these states is dictated by Axiom 4 (eq. (7.23)),

$$\psi_{E,p}(x, t) = e^{ipx/\hbar}e^{-iEt/\hbar}. \qquad (7.32)$$

These are the time-dependent wavefunctions for the energy basis states of a free particle in one dimension. Each state is characterized by two real constants that are **conserved**, i.e. they do not change with time. The role of the energy E is by now familiar. p is another constant of the motion that appears only for free particles. From the experimentally derived de Broglie relation $\lambda = h/p$ (or $k = p/\hbar$), we see that the conserved quantity p is in fact the quantity that we interpret physically as the momentum of the particle. Starting from the axiomatic approach, however, has enabled us to derive the existence of this conserved quantity from first principles. Arguments along these lines occur frequently in fundamental physics: the existence of a symmetry (*translation by an arbitrary amount*) implies the existence of a conserved quantity (*momentum*). As explained further in §21, this connection between symmetries and conserved quantities plays a fundamental role in the structure of physical theories, both classical and quantum.

Although we have obtained the wavefunction corresponding to a basis state of definite energy E and momentum p, we have not derived any specific relation between E and p. Indeed, this relationship cannot be fixed without further information. For different systems, this relation takes different forms. For a particle moving at a small velocity $v \ll c$, the relation is the same as in classical mechanics, $E = p^2/2m$. At a more fundamental level this comes from the relation $E = \sqrt{m^2c^4 + p^2c^2}$ (Problem 7.14) which is true for particles with any velocity, and which follows from the even greater symmetry of *special relativity* discovered by Einstein (§21). As explained in §25, electrons propagating within solid crystals – that have a more limited translation symmetry – have an energy described by a more complicated function of momentum.

The relation $E = p^2/2m$ can be rewritten as a relationship between derivatives of eq. (7.32), reproducing the Schrödinger equation for the free particle as discussed in §7.2.1,

$$i\hbar\frac{\partial}{\partial t}\psi(x, t) = -\frac{\hbar^2}{2m}\frac{\partial^2}{\partial x^2}\psi(x, t). \qquad (7.33)$$

Rewriting this equation in the form of eq. (7.29), we define

$$H = -\frac{\hbar^2}{2m}\frac{\partial^2}{\partial x^2} \qquad (7.34)$$

to be the *Hamiltonian* for the free particle. Note that in this idealized situation – a free particle moving in infinite space – the Hamiltonian becomes a *differential operator* (Appendix B.3.6) acting on the space of all possible wavefunctions instead of a matrix acting on a finite-dimensional vector space.

It is straightforward to generalize this discussion to a particle in three dimensions. Analogous to (7.31), there are plane wave states

$$\psi_{\boldsymbol{p}}(\boldsymbol{x}) = e^{i\boldsymbol{p}\cdot\boldsymbol{x}/\hbar} = e^{i(p_xx+p_yy+p_zz)/\hbar}, \qquad (7.35)$$

that change under an arbitrary translation $\boldsymbol{x} \rightarrow \boldsymbol{x} + \boldsymbol{\delta}$ by only a phase. These are quantum particle states with momentum $\boldsymbol{p}$. They obey the Schrödinger equation (7.7) with $V(\boldsymbol{x}) = 0$. For the free particle in three dimensions, the Hamiltonian is the differential operator $H = -(\hbar^2/2m)\nabla^2$.

One subtlety that we have not addressed concerns the *normalization* of the plane wave states. In §7.2, Schrödinger wavefunctions were normalized through eq. (7.6), in keeping with the probability interpretation of $|\psi(x, t)|^2$ and in agreement with the more general normalization condition (7.17). This condition clearly cannot be imposed on the energy basis states (7.31) or (7.35), since $|\psi(x)|^2$ is constant everywhere in space, and has a divergent integral. These states correspond to a mathematical idealization of an infinite system; though formal mathematical machinery exists to extend the normalization condition to such wavefunctions, this is irrelevant for the purposes of this book. Any real physical system satisfies finite boundary conditions that lead to normalizable energy basis states for the quantum system.

7.7.2 Particle in a Box Revisited

We briefly reconsider the problem of a particle in a one-dimensional box in the context of the general axiomatic quantum framework, and then generalize to a three-dimensional cubical box. In §7.2.3 we found the wavefunctions and energies of the energy basis states for a particle of mass m confined to the interval $0 \leq x \leq L$. The properly normalized wavefunctions are (7.12)

$$\psi_n(x) = \sqrt{\frac{2}{L}} \sin\left(\frac{n\pi}{L}x\right), \qquad (7.36)$$

where $n = 1, 2, 3, \ldots$, and the constant in front is fixed by the condition that $\int_0^L |\psi_n(x)|^2 = 1$. The energies of these states are given by eq. (7.13),

$$E_n = \frac{\pi^2\hbar^2 n^2}{2mL^2}. \qquad (7.37)$$

The wavefunctions and energies of the first few energy levels are sketched in Figure 7.7. The **ground state** of the system is the state with the lowest energy, E_1. A general property of the ground state of a single-particle system is that the wavefunction has no **nodes** where the function vanishes (aside from $x = 0$ and $x = L$). The first excited state has one node, the second excited state has two nodes, etc. Note that since $\sin(ax) = -i(\exp(iax) - \exp(-iax))/2$, the standing wave modes in the box (7.36) can be thought of as linear combinations of left-moving and right-moving plane waves with $p_n = \pm\pi n\hbar/L$.

How are these results from wave mechanics to be interpreted in light of the axioms introduced in the preceding sections? The energy basis states $|n\rangle$ are represented by their Schrödinger wavefunctions $\psi_n(x)$. As dictated by Axiom 2, any state in the box can be represented as a sum over these energy basis states, $|\psi\rangle = z_1|1\rangle + z_2|2\rangle + \cdots$. In terms of wavefunctions, this is just the *Fourier series* expansion introduced in §4.2.1. The statement that the probability of finding the particle in any small region dx is given by $|\psi^2(x)| dx$ is a generalization of Axiom 3. Like energy, the position of a particle is a physically observable quantity, and obeys a rule for measurement similar to Axiom 3 for energy. Finally the time dependence of the energy basis states is reflected in the time dependence of the Schrödinger wavefunctions, $\psi_n(x, t) = e^{-iE_n t/\hbar}\psi_n(x)$.

It is straightforward to generalize the analysis of the particle in a box to three dimensions. The normalized energy basis state wavefunctions for a particle confined to a three-dimensional box $0 \leq x, y, z \leq L$ are simply products of the one-dimensional wavefunctions (7.31)

$$\psi_{n,m,l}(x, y, z) = \psi_n(x)\psi_m(y)\psi_l(z) \qquad (7.38)$$

$$= \left(\frac{2}{L}\right)^{3/2} \sin\left(\frac{p_n x}{\hbar}\right) \sin\left(\frac{p_m y}{\hbar}\right) \sin\left(\frac{p_l z}{\hbar}\right),$$

with $p_n = n\pi\hbar/L$. The energies of these states are

$$E_{n,m,l} = \frac{p_n^2 + p_m^2 + p_l^2}{2m} = \frac{\pi^2(n^2 + m^2 + l^2)\hbar^2}{2mL^2}, \qquad (7.39)$$

where $n, m, l = 1, 2, \ldots$. We have used three integers here to denote the basis states and corresponding energies, though this is a countable set of indices so we could enumerate them using a single index if desired. Unlike the one-dimensional box, the spectrum of the three-dimensional box has *degeneracies*. For example, there are three states with energy $6(\pi^2\hbar^2/2mL^2)$. The number of states in a given energy range grows quickly with E. This figures in the derivation of the spectrum of thermal radiation in §22 (Solar energy).

7.8 Particles in Potentials

Atoms in a molecule, electrons in atoms, and protons in a nucleus move under the influence of forces that can be described – to a good approximation – in terms of potentials. To study such systems it is necessary to solve the Schrödinger equation when a potential energy $V(x)$ is included in the Hamiltonian, as in eq. (7.7). Since we are looking for a state of definite energy E, the wavefunction $\psi(x, t)$ is of the form $\psi(x, t) = e^{-iEt/\hbar}\psi(x)$. Substituting this into eq. (7.7), we obtain the **time-independent Schrödinger equation** in one dimension

$$H\psi(x) = \left(-\frac{\hbar^2}{2m}\frac{d^2}{dx^2} + V(x)\right)\psi(x) = E\psi(x), \qquad (7.40)$$

or in three dimensions,

$$H\psi(x) = \left(-\frac{\hbar^2}{2m}\nabla^2 + V(x)\right)\psi(x) = E\psi(x). \qquad (7.41)$$

To conclude this introduction to quantum mechanics, we describe the spectrum of energies arising from the solution of these equations for two important physical systems: the simple harmonic oscillator and the (idealized) hydrogen atom.

7.8.1 Quantum Harmonic Oscillator

The harmonic oscillator potential in one dimension is

$$V(x) = \frac{1}{2}kx^2. \qquad (7.42)$$

This potential plays a central role in the analysis of a wide range of physical systems. As explained in Box 2.2, any smooth potential with a local minimum at which the second derivative $V''(x)$ is non-vanishing can be approximated by a harmonic oscillator for small fluctuations

---

Box 7.5   Finite Quantum Spectra

The solutions to free particle motion (7.32), the particle in a box (7.38), and the idealized hydrogen atom (7.52) lead to an infinite set of energy basis states. As stated in Axiom 1, however, any real quantum system has a *finite and discrete* spectrum of possible energies and associated energy basis states. In fact, the examples listed above, and all other cases where an infinite number of states appear, are idealizations that simplify calculations but obscure the fact that – both in practice and in principle – all systems can be described in terms of a finite number of energy basis states. This fact simplifies the conceptual framework needed for understanding physical systems; in particular, it guarantees that the *entropy* of any physical system, as defined in the following chapter (§8) using quantum physics, is finite.

At a practical level, the fact that only a finite set of states is possibly relevant in a real physical system relies on straightforward physics. Infinite families of states are generally associated with either energy values that are arbitrarily large (e.g. the box at large n), or wavefunctions that spread out over arbitrarily large distances (e.g. the plane waves that describe a free particle or the hydrogen atom at large n). These are idealizations – the energies and regions of space involved in any physical process are bounded. Although it may be convenient in practice to employ an infinite number of states, one could truncate the set of states to a finite number for any practical calculation at any useful level of accuracy.

As a matter of principle, the reasons for the finiteness of the spectrum are perhaps more surprising and probe physics at the boundaries of current understanding. Given sufficiently large energy, a system of any fixed size collapses into a black hole. Physicists do not fully understand the quantum mechanical behavior of black holes, but there are strong indications that black holes with any fixed size and energy (mass) have a finite number of possible quantum states. Thus, the physics of black holes puts an upper bound on the number of quantum states available to any system with bounded size!

The *size* of physical systems is also bounded. Recent observations (see §21) indicate that in the present phase of development of the observable universe, distances larger than roughly 14 billion light years are not experimentally accessible. This provides a maximum box size. Together, the nature of black holes and the large-scale structure of the universe conspire to ensure that, even in principle, any physical system subject to experimental analysis has a finite number of possible basis states in the framework of quantum mechanics.

---

around the minimum. This applies, for example, to the vibrational modes of molecules (see Figure 9.5), where the atoms may be coupled by a complicated inter-atomic potential.

The time-independent Schrödinger equation (7.40) for the harmonic oscillator potential (7.42) is

$$H\psi = \left(-\frac{\hbar^2}{2m}\frac{d^2}{dx^2} + \frac{1}{2}kx^2 \right)\psi(x) = E\psi(x). \quad (7.43)$$

Determining the solutions to equations like this is a standard problem treated in textbooks and courses on differential equations. The general solution methods involved are beyond the scope of this book; in this section we explain why equations like this have discrete spectra and then simply present the harmonic oscillator energy wavefunctions and spectrum, with some more detailed calculations relegated to problems. We return to this equation in §15.

We have already encountered second-order differential equations in the context of classical mechanics. Newton's law for a harmonic oscillator, $m\ddot{x} = -kx$, is a familiar example. Equations like this and eq. (7.43) have two independent solutions for every value of the parameters

$(m, k, \text{and } E)$. For the classical harmonic oscillator, for example, the two solutions to Newton's law are $\sin\omega t$ and $\cos\omega t$, where $\omega = \sqrt{k/m}$. Just as the solution to Newton's law is completely fixed by choosing the initial position $x(0)$ and velocity $\dot{x}(0)$, so too, the solution to eq. (7.43) is completely determined when we fix $\psi(0)$ and $\psi'(0)$.

The solution to eq. (7.43), however, has another important constraint: because $|\psi(x)|^2$ is a probability density, its integral over all of space must be one (eq. (7.6)). Figure 7.12 shows solutions of eq. (7.43) for several choices of energy close to each of the two values $E_0 = \hbar\omega/2$ and $E_1 = 3\hbar\omega/2$. The plots illustrate the fact that, except at the precise values E_0 and E_1, the integral of $|\psi(x)|^2$ over all x diverges. In fact, normalizable solutions of (7.43) are only possible for certain *discrete* value of the energy E. The normalization condition on the integral of $|\psi|^2$ plays the role of a *boundary condition* on the set of possible solutions, which leads to a discrete spectrum, just as the boundary condition $\psi = 0$ does for a particle in a box.

An elementary analytic argument (Problem 7.15) shows that a normalized solution to eq. (7.43) is given by

---

Box 7.6   Classical Objects and the Uncertainty Principle

The simple case of a particle in a box provides an opportunity to see how classical particle behavior can emerge from quantum mechanics. The low-lying energy basis states of a particle in a box are highly non-classical, as can be seen in Figure 7.7; these wavefunctions are distributed non-uniformly throughout the box and are not localized as one would expect of a classical point particle. From Axiom 2, we know that the states of a quantum system consist of all possible complex linear combinations of the energy basis states. By taking appropriate linear combinations of the basis wavefunctions $\psi_n(x)$ we can produce wavefunctions that are localized to a small region of space. Fourier's theorem (B.54) states that it is possible to expand any (square-integrable) function that vanishes at $x = 0$ and L as a sum of sine waves,

$$\psi(x) = \sum_{n=1}^{N} c_n \psi_n(x) = \sum_{n=1}^{N} c_n \sqrt{\frac{2}{L}} \sin\left(\pi n x / L\right) = \sum_{n=1}^{N} \frac{-ic_n}{\sqrt{2L}} \left(e^{ip_n x/\hbar} - e^{-ip_n x/\hbar}\right),$$

where we have used the decomposition of the sine wave into left- and right-moving momentum modes from (§7.7.2), and the approximation can be made arbitrarily good by taking sufficiently large N. Classical particle behavior emerges when many energy states are superposed to make a **wave packet** that is localized within the accuracy of a classical measuring apparatus. An example is shown in the figure. The two wavefunctions shown at the top of the figure are localized around $x = L/4$ and have zero average momentum. The coefficients c_n used to construct the wavefunctions are also shown. Note that the more one attempts to localize the particle, the greater the range of momentum $p_n = \pm \pi n \hbar / L$ one must include in the sum. Note that the range of momentum that appears in the sum can be seen visually by reflecting the set of c_n coefficients shown in the figure across the vertical axis at $n = 0$.

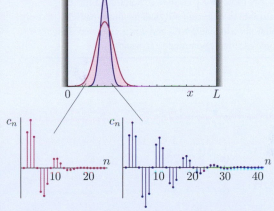

This is a general feature of quantum mechanics: the more we try to localize a particle's position, the more uncertainty is generated in its momentum. Thus, we cannot simultaneously specify both position and momentum to arbitrary precision with the quantum wavefunction. This result can be made precise and is known as the **Heisenberg uncertainty relation**,

$$\Delta x \Delta p \geq \hbar/2,$$

where Δx, Δp are the uncertainties in the measurement of position and momentum of the particle. The Heisenberg uncertainty relation follows directly from a standard relation in Fourier analysis, but we will not go into the details here. A related effect leads to an **energy/time uncertainty relation** so that over very short times the energy of a system cannot be measured precisely; this is relevant in absorption and emission processes by atoms (Example 23.2).

In another sense, the high-energy quantum states of the particle in the box are more like an *average* of a typical particle trajectory. A classical particle in a box with any energy greater than zero simply moves back and forth, bouncing off the walls. When averaged over a long time, the particle has *equal probability to be found anywhere in the box,* so $dP/dx = 1/L$. For the quantum states with large values of n, the quantum probability oscillates very rapidly about an average value of $1/L$, so an instrument that averages over the oscillations of the quantum probability distribution would obtain the classical result.

$$\psi_0(x) = \left(\frac{m\omega}{\pi\hbar}\right)^{1/4} e^{-m\omega x^2/2\hbar}. \qquad (7.44)$$

This is the *ground state* of the simple harmonic oscillator. Substituting into eq. (7.43) confirms that this wavefunction obeys the Schrödinger equation with energy $E = \frac{1}{2}\hbar\omega$. Just like the ground state of the particle in a one-dimensional box, this ground state has no *nodes* where

$\psi(x) = 0$. The excited states of the harmonic oscillator are given by (normalized) wavefunctions that are polynomials in x multiplying the Gaussian form $e^{-m\omega x^2/2\hbar}$ of the ground state. The first few excited state wavefunctions can be written compactly by introducing a characteristic length scale for the quantum oscillator, $\ell \equiv \sqrt{\hbar/m\omega}$, and defining $\xi \equiv x/\ell$ (Problem 7.17),

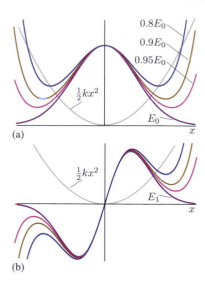

(a)

(b)

Figure 7.12 Solutions to the Schrödinger equation for the one-dimensional harmonic oscillator for various values of the energy. (a) Solutions with $\psi(0) = 1$, $\psi'(0) = 0$, and energy close to $E_0 = \hbar\omega/2$. (b) Solutions with $\psi(0) = 0$, $\psi'(0) = 1$, and energy close to $E_1 = 3\hbar\omega/2$. Solutions must be normalizable to represent physical quantum particles, so only the solutions with energy E_0 and E_1 are allowed. The harmonic oscillator potential is also shown.

$$\psi_1(x) = \frac{1}{(\pi\ell^2)^{1/4}}\sqrt{2}\,\xi e^{-\xi^2/2}, \tag{7.45}$$

$$\psi_2(x) = \frac{1}{(\pi\ell^2)^{1/4}}\frac{1}{\sqrt{2}}(2\xi^2 - 1)e^{-\xi^2/2}, \tag{7.46}$$

$$\psi_3(x) = \frac{1}{(\pi\ell^2)^{1/4}}\frac{1}{\sqrt{3}}(2\xi^3 - 3\xi)e^{-\xi^2/2}. \tag{7.47}$$

The energy of the nth excited state of the harmonic oscillator is

$$E_n = \left(n + \frac{1}{2}\right)\hbar\omega \text{ for } n = 0, 1, 2\ldots. \tag{7.48}$$

Note that the increment between successive energies is $\hbar\omega$, matching Planck's quantization condition (7.1). The spectrum and first few energy basis functions for the quantum simple harmonic oscillator are depicted in Figure 7.13.

7.8.2 Idealized Hydrogen Atom

Finally, we consider a simplified model of the hydrogen atom. The proton that forms the nucleus of the hydrogen atom is much more massive than the electron ($m_p/m_e \cong 1835$). Thus, we can treat the proton as stationary. Ignoring magnetic effects arising from the electron and proton spins,

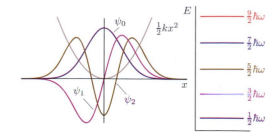

Figure 7.13 Spectrum $E_n = (n + 1/2)\hbar\omega$ and first few states of quantum simple harmonic oscillator.

Quantum Harmonic Oscillator Spectrum

The spectrum of allowed energies for a one-dimensional simple harmonic oscillator with frequency ω is

$$E_n = (n + 1/2)\hbar\omega, \ n = 0, 1, \ldots.$$

which can be shown to give corrections of less than 1% to the energies, we need only consider the electron moving in a potential

$$V(r) = -\frac{e^2}{4\pi\varepsilon_0 r}. \tag{7.49}$$

This leads to the time-independent Schrödinger equation

$$\left(-\frac{\hbar^2}{2m}\nabla^2 - \frac{e^2}{4\pi\varepsilon_0 r}\right)\psi(x, y, z) = E\psi(x, y, z). \tag{7.50}$$

Again, this is a standard type of partial differential equation. Like the harmonic oscillator, the solutions are normalizable only for specific, discrete values of the energy. Rather than explain the general solution of this equation here, we discuss some features of the solutions, and present the spectrum and the wavefunctions of some of the lowest energy states. A detailed solution of the Schrödinger equation for the hydrogen atom can be found in almost any standard quantum mechanics text.

The key to finding an analytic solution of eq. (7.50) is the rotational symmetry of the system. Since the potential depends only on $r = |\mathbf{r}|$ and the differential operator ∇^2 treats all directions equally, it can be shown that any solution to eq. (7.50) can be factored into the product of a function of r times a function of the dimensionless ratios x/r, y/r, and z/r. The simplest solutions, known as the *s*-waves or *s*-orbitals, are functions of r alone. The first few *s*-wave solutions $\psi_{ns}(r)$ are

$$\psi_{1s}(r) = \frac{1}{\sqrt{\pi}}\left(\frac{1}{a_0}\right)^{3/2} e^{-r/a_0},$$

$$\psi_{2s}(r) = \frac{1}{\sqrt{2\pi}}\left(\frac{1}{a_0}\right)^{3/2}\left(\frac{1}{2}-\frac{r}{4a_0}\right)e^{-r/2a_0},$$

$$\psi_{3s}(r) = \frac{1}{\sqrt{3\pi}}\left(\frac{1}{a_0}\right)^{3/2}\left(\frac{1}{3}-\frac{2r}{9a_0}+\frac{2}{81}\frac{r^2}{a_0^2}\right)e^{-r/3a_0},$$

$$(7.51)$$

where $a_0 = 4\pi\varepsilon_0\hbar^2/me^2 \cong 0.5292\times 10^{-10}$ m, known as the **Bohr radius**, sets the distance scale that characterizes the hydrogen atom. The energy of the nth s-state is

$$E_n = -\frac{me^4}{32\pi^2\varepsilon_0^2\hbar^2 n^2} \cong -\frac{13.61\text{ eV}}{n^2}. \qquad (7.52)$$

The next simplest set of solutions, known as *p-orbitals*, consists of functions of r multiplying either x/r, y/r, or z/r

$$\psi_{2p}(r) = \frac{1}{4\sqrt{2\pi}}\left(\frac{1}{a_0}\right)^{3/2} e^{-r/2a_0}\left\{\frac{x}{r},\frac{y}{r},\frac{z}{r}\right\},$$

$$\psi_{3p}(r) = \frac{1}{81}\sqrt{\frac{2}{\pi}}\left(\frac{1}{a_0}\right)^{3/2}\left(6-\frac{r}{a_0}\right)e^{-r/3a_0}\left\{\frac{x}{r},\frac{y}{r},\frac{z}{r}\right\},$$

$$(7.53)$$

where the notation $\{x/r, y/r, z/r\}$ means that any of these three choices is a solution. The three solutions in each set all have the same energy and together they do not select any special direction in space – as a set they span a space that is **rotationally invariant**, meaning that any rotation of any linear combination of the three functions in that set gives another linear combination of the same basis functions. The energies of the nth p-states are also given by eq. (7.52). Note, however, that the p-orbitals begin with $n = 2$. Next come states with quadratic dependence on x/r, y/r, and z/r, with the condition that no function proportional to $x^2/r^2 + y^2/r^2 + z^2/r^2 = 1$ is included. There are five independent *d-wave* states or *d*-**orbitals** for each $n \geq 3$. *f*-**orbitals** are cubic functions of x/z, y/z, and z/r. There are seven *f*-orbitals and their energies begin with $n = 4$. And so on. The spectrum of the idealized hydrogen atom is given in Figure 7.14, along with sketches of the electron probability densities for some of the orbitals.

This analysis gives a good approximation to the energy levels of the hydrogen atom. A zeroth order description of many-electron atoms can be obtained by ignoring the interactions between the electrons and simply filling hydrogen-like orbitals in an atom with nuclear charge Ze. The resulting atomic **shell model** is the basis of the periodic table of the elements. An essential ingredient in this description of atoms is the **Pauli exclusion principle**,

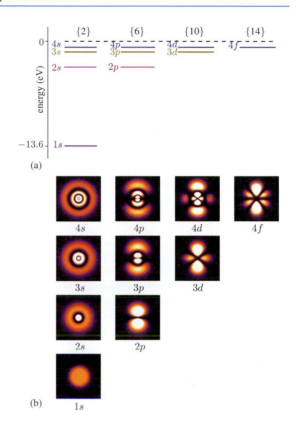

(a)

(b)

Figure 7.14 (a) Spectrum of the hydrogen atom. The number of electron states (including spin states) allowed in each orbital (s, p, d, f) is given at the top. (b) Plots of the electron probability density for the same orbitals. For p, d, and f orbitals, one combination of the many possible states is shown. (Source: PoorLeno/English Wikipedia)

which states that at most two electrons (one spin-up, one spin-down), can occupy a given orbital. Thus the s-orbitals can each hold two electrons, the p-orbitals six, and so forth. This provides a good starting point for analyzing the energy spectrum of atoms, though interactions between the electrons in larger atoms produce corrections that need to be considered for a more precise analysis.

In many places in this book, when we need to understand features of the quantum spectrum of atomic, molecular, and condensed matter (solid) systems, we use a Schrödinger equation (7.41) with an appropriate potential.

Discussion/Investigation Questions

7.1 We make a big deal about the linearity of quantum mechanics. Is classical mechanics linear? Consider, for example, two solutions to Newton's laws for motion in Earth's gravitational field, $x_1(t)$ and $x_2(t)$. Both obey $m\ddot{x}(t) = -GmM/|x|$. Show that a linear combination of x_1 and x_2 is not a solution. Can you think of a type of force for which solutions *do* obey superposition?

7.2 A suggestion rather than a question: a video recording was made in 1964 of Richard Feynman describing the double slit experiment. It can be found at www.cornell.edu/video/playlist/richard-feynman-messenger-lectures where it is the main part of Lecture 6: *The Quantum Mechanical View of Nature*. It is entertaining as well as enlightening.

7.3 Here is an energy basis state of *three* electrons in a magnetic field (in the notation of §7.5.3),

$$\psi = \frac{1}{\sqrt{3}} \left(|+ + -\rangle + |+ - +\rangle - |- + +\rangle \right). \quad (7.54)$$

Are all three of these electrons entangled? What is the probability that a measurement of the first electron's spin yields $+\frac{1}{2}\hbar$? If the first electron's spin is measured to be $+\frac{1}{2}\hbar$, what would a measurement of the second and third electrons spins yield? Does this result change if the first electron's spin is measured to be $-\frac{1}{2}\hbar$?

Problems

7.1 Planck's constant is very small on the scale of human affairs. Compare the angular momentum of a child's marble spinning at 60 rpm with Planck's constant. Compare the energy of the same marble attached to a spring with spring constant $k = 10\,\text{kg/s}^2$ and oscillating with an amplitude of 1 cm with the ground state energy $\frac{1}{2}\hbar\omega$ of this harmonic oscillator. Quantum effects are important in systems where $L \approx \hbar$ or $E \approx \hbar\omega$. (The moment of inertia of a uniform sphere of mass m and radius R is $I = \frac{2}{5}MR^2$.)

7.2 [T] In one dimension, compare the plane wave solutions to the heat equation $\partial T/\partial t = a\partial^2 T/\partial x^2$ with plane wave solutions to the free Schrödinger equation

$$i\partial\psi/\partial t = -(\hbar/2m)\partial^2\psi/\partial x^2 .$$

Show that the heat equation has real solutions that oscillate in space and decay in time, whereas the Schrödinger equation has complex solutions that oscillate in both space and time.

7.3 Nuclear forces are so strong that they keep protons and neutrons in a spherical region a few femtometers in radius (1 fm $= 10^{-15}$ m). To get a crude idea of the energy scales involved in nuclear physics, model the nucleus as a cube 5 fm on a side. What is the ground state energy of a proton (mass $m_p \cong 1.7 \times 10^{-27}$ kg) in this box? This is the energy scale of nuclear interactions.

7.4 Suppose an electron, constrained to move only in the vertical direction, sits on an impenetrable table in Earth's gravitational field. Write down the (one-dimensional) Schrödinger equation that determines its

wavefunction as a function of height z. Use dimensional analysis to estimate its ground state energy up to a multiplicative factor. Estimate its average height above the table top, also up to a multiplicative factor.

7.5 [T] A particle of mass m in a potential $V(x)$ has an energy basis state with wavefunction $\psi(x) = C/\cosh^2(\sqrt{mV_0}x/\hbar)$. The properties of $\cosh x$ and other *hyperbolic functions* are summarized in Appendix B.4.2. What is the potential and what is the energy of this state? What is the value of the constant C? [Hint: $\psi(x)$ must obey the Schrödinger equation.]

7.6 [T] A quantum particle is restricted to a one-dimensional box $0 \le x \le L$. It experiences no forces within the box, but cannot escape. At time $t = 0$, the particle is in the state $\sqrt{1/3}\psi_1(x) + \sqrt{2/3}\psi_2(x)$, where $\psi_i(x)$ are the (correctly normalized) energy basis states from eq. (7.36). (a) Compute the probability at $t = 0$ that the particle has energy $\epsilon = E_1 = \pi^2\hbar^2/2mL^2$. (b) Compute the probability at $t = 0$ that the particle is in the left half of the box ($x < L/2$). (c) Compute the state of the particle at time $t = \pi\hbar/\epsilon$. What is the probability at this time that the particle has energy $\epsilon = E_1$? What is the probability that the particle is in the left half of the box?

7.7 [T] In the text we introduced electron spin states with spin $\pm\frac{1}{2}$ in the $\hat{z}$-direction, $|\pm\rangle$, and states with spin $\pm\frac{1}{2}$ in the $\hat{x}$-direction, $|\pm_x\rangle = (|+\rangle \pm |-\rangle)/\sqrt{2}$. Measuring the z-component of spin in the state $|\pm_x\rangle$ gives $\pm\frac{1}{2}$, each with 50% probability. The analogous states with spin $\pm\frac{1}{2}$ in the $\hat{y}$-direction are $|\pm_y\rangle = (|+\rangle \pm i|-\rangle)/\sqrt{2}$. Show that a measurement of the spin in the $\hat{x}$- or $\hat{z}$-direction on these two states also yields $\pm\frac{1}{2}$ each with 50% probability.

7.8 [T] Show that the set of functions $\psi_k(x) = e^{ikx}$ (for real k) are the solutions to the condition $\psi(x + \delta) = e^{i\theta(\delta)}\psi(x)$ (for real θ). [Hint: use sequential translations by δ_1 and δ_2 to show that θ must depend linearly on δ, then expand this condition about $\delta = 0$.] Show that if k were complex, then the probability $dP/dx = |\psi_k(x)|^2$ would grow without bound as x approaches either $+\infty$ or $-\infty$, corresponding to unphysical boundary conditions.

7.9 [T] A particle with spin S has magnetic moment $m = qgS/2Mc$ (see Example 7.3). Solving for the classical motion of S in a constant magnetic field B, we found that S precesses around B with angular frequency $\Omega = qg|B|/2Mc$. We found that the quantum state of an electron $|\psi(0)\rangle = |+_x\rangle$ precesses with angular frequency $\omega = 2\epsilon/\hbar$, where ϵ is the energy of the electron in the field B. Show that $\omega = \Omega$.

7.10 The interatomic potential for diatomic molecules like O_2 and N_2 can be approximated near its minimum by a simple harmonic oscillator potential $\frac{1}{2}kx^2$. Consider the classical motion of a diatomic molecule composed of two atoms of equal mass m. Show that near the minimum of the interatomic potential this classical motion is that of a single mass $m/2$ on a spring with spring constant k. For the O=O double bond in O_2, the spring constant near the bottom of the potential is roughly $1140\,\text{N/m}$. Estimate the energy gap between the ground state of the molecule and the first vibrational quantum excitation of the molecule, corresponding to the first excited state of the associated harmonic oscillator potential.

7.11 **[T]** Consider a system of two independent harmonic oscillators, each with natural angular frequency ω. We denote states of this system by $|n\,m\rangle$ where the non-negative integers n, m denote the excitation levels of the two oscillators. The energy of this system is just the sum of the energies of the two oscillators. What is the ground state energy of this system? How many distinct states $|nm\rangle$ have total energy $10\,\hbar\omega$? If there are three independent oscillators, how many distinct states have total energy $\frac{9}{2}\hbar\omega$?

7.12 Consider a macroscopic oscillator, given by a mass of $m = 0.1\,\text{kg}$ on a spring with spring constant $k = 0.4\,\text{N/m}$. What is the natural frequency ω of this oscillator? Considering this as a quantum system, what is the spacing of the energy levels? Do you expect that we could observe the energy of the mass on the spring with sufficient precision to determine which quantum state the system is in? Classically, consider extending the spring by moving the mass a distance $d = 1\,\text{cm}$ from equilibrium, and letting go. Describe the classical motion of the mass. What is the energy of this system? Roughly what excitation number n characterizes quantum states with this energy?

7.13 **[T]** Show that the solution to Schrödinger's equation for a three-dimensional harmonic oscillator is

$$\psi_{lmn}(\boldsymbol{x}) = \psi_l(x)\psi_m(y)\psi_n(z),$$

where $\psi_n(x)$ is a solution to the one-dimensional oscillator with energy $(n + 1/2)\hbar\omega$. Show that the energies of the three-dimensional oscillator are $(N + 3/2)\hbar\omega$ for $N = 0, 1, 2, \ldots$ and find the degeneracies of the first four energy levels. Do you recognize the sequence?

7.14 **[T]** Show that the usual relation $E = p^2/2m$ arises as the leading momentum dependence of the energy in an expansion of the relativistic relation $E = \sqrt{m^2 c^4 + p^2 c^2}$ for small $p \ll mc$.

7.15 **[T]** Find the ground state of the quantum harmonic oscillator. Start by showing that the Hamiltonian from eq. (7.43) can be "factorized" similar to $(a^2 - b^2) = (a - b)(a + b)$ in the form

$$H = \frac{1}{2}\hbar\omega\left(\left(-\ell\frac{d}{dx} + \frac{x}{\ell}\right)\left(\ell\frac{d}{dx} + \frac{x}{\ell}\right) + 1\right),$$

where $\ell = \sqrt{\hbar/m\omega}$. Show that the Schrödinger equation is satisfied if $(\ell\,d/dx + x/\ell)\psi_0(x) = 0$. Show that $\psi_0(x) = Ne^{-ax^2}$ is such a solution for the right choice of a. Find N such that $\psi_0(x)$ is normalized. What is the energy of this state?

7.16 The energy required to remove an electron from an atom is called the *ionization energy* (§9.3.3). Predict the energy needed to ionize the last electron from an atom with *atomic number* Z. (The atomic number is the charge on the atom's nucleus (§17).)

7.17 **[T]** The polynomials appearing in eqs. (7.45)–(7.47) are known as *Hermite polynomials*. Compute the second, third, and fourth excited wavefunctions by assuming quadratic, cubic, and quartic polynomials and determining the coefficients by solving eq. (7.43). Compute the associated energy values. [Hint: Rewrite eq. (7.43) in terms of the dimensionless variable ξ and explain why the solutions must be either even or odd in $\xi \to -\xi$.]

Thermodynamics II: Entropy and Temperature

The complex flow of energy through natural and man-made systems is structured and constrained in ways that are greatly illuminated by the concept of *entropy*. Entropy is a quantitative measure of disorder or randomness in a physical system. The role of entropy is of central importance in thermodynamics and in a wide range of energy systems, from power plants to automobile engines and toe warmers. While it is possible to introduce entropy entirely at the macroscopic level without reference to the underlying quantum structure of matter, a more fundamental understanding of this subject can be attained by studying it from the ground up – by exploring the *statistical mechanics* of systems of many particles and their quantum states. We begin this chapter with a qualitative description of the basic principles underlying entropy and the *second law of thermodynamics*, and then give a more systematic and quantitative presentation of these ideas. A more detailed outline of the remainder of the chapter is given at the end of §8.1.

8.1 Introduction to Entropy and the Second Law

No system, no matter how cleverly designed, that extracts heat from a source at temperature T_+ and operates in an ambient environment at temperature T_- can convert thermal energy to mechanical energy in a sustained fashion with efficiency greater than

$$\eta = \frac{T_+ - T_-}{T_+} . \tag{8.1}$$

This **thermodynamic efficiency limit** for the conversion of thermal energy into useful work is one of the most important principles in energy physics. To understand the origin of this limitation, we need to explore the notion of *entropy*. Understanding entropy allows us to define the

> **Reader's Guide**
> Perhaps no other concept is as important in energy processes or as often misunderstood as *entropy*. Entropy and the second law of thermodynamics are the focal points of this chapter. These concepts are central to many later parts of the book. Even readers with some knowledge of thermodynamics are encouraged to review this chapter, as some of the ideas presented here may provide a new perspective on these concepts.
>
> Prerequisites: §5 (Thermal energy), particularly heat, enthalpy, and the first law of thermodynamics; §7 (Quantum mechanics), particularly quantization of energy.
>
> The central ideas developed in this chapter – entropy, the second law, temperature, thermal equilibrium, the Boltzmann distribution, and free energy – form part of the foundation for understanding the science of energy systems, and reappear frequently throughout the rest of the book.

concepts of temperature and thermal equilibrium in a rigorous way. Entropy also determines when a process may take place spontaneously and when it cannot.

Entropy can be thought of as a measure of ignorance regarding the precise state of a physical system. Consider, for example, a gas of many particles in a closed container at room temperature. We measure a few physical properties such as the temperature and pressure of the gas, but we are ignorant of the precise position and velocity of the individual particles in the gas. Indeed, even if we could measure the coordinates of each of the $\mathcal{O}(10^{23})$ molecules in the gas, no existing or realistically imaginable computer could store even a fraction of this information. That the seemingly abstract concept of ignorance should have profound practical consequences for the efficiency of energy systems is one of the marvels of the field of *statistical physics*.

To understand the concept of entropy in a more quantitative fashion, it is important to remember what is meant by a "state" in thermodynamics, in contrast to classical or quantum physics. In thermodynamics, as explained in §5, a *state* is characterized by a set of *state functions* such as temperature and pressure that reflect macroscopic properties of a system in *thermodynamic equilibrium* and give only a small amount of information about the precise configuration of the microscopic degrees of freedom of the system. By contrast, in classical or quantum physics a *state* of the system is a specific configuration of the system in which all microscopic physical features are precisely determined in a unique way. For clarity, in the context of thermodynamics we refer to a complete specification of all possible information describing a system at a specific moment in time as defining a **microstate** of the system. For example, a microstate of the gas mentioned above would be a particular quantum state with a fixed energy (a classical microstate would be defined by giving the positions and momenta of all the classical particles in the system). While we are often interested in studying the bulk (thermodynamic) properties of a gas, we are in general ignorant of its microstate. Entropy quantifies this ignorance. For example, as we show in §8.4, for a system in thermodynamic equilibrium at fixed energy, the entropy is proportional to the logarithm of the number of microstates available to the system that are compatible with our knowledge of its bulk properties. Notice that quantum mechanics is required to properly define entropy as a finite quantity. In quantum mechanics, the number of possible states of a finite system is finite. For even the simplest classical systems – such as a single particle moving on a line – there are an infinite and uncountable number of possible positions and velocities.

The most important feature of entropy, which follows directly from its interpretation as quantifying ignorance of the microstate of a system, is that entropy cannot decrease as any isolated system evolves over time. The laws that determine the time development of isolated classical or quantum physical systems are both *causal* and *reversible*. Recall that in quantum physics the time evolution of a state of definite energy is described by multiplying the state by a phase $\exp(-iEt/\hbar)$, and that any state can be written as a superposition of energy states. This evolution is **causal**: a given state in the present evolves into a unique state in the future; and it is **reversible**: a state is evolved into the past simply by replacing t with $-t$ in the time-evolution equation. These features of time evolution mean that if a system is isolated, our ignorance of the details of the system cannot be reduced over time without further measurement. For example, if at time t_1 there are 137 microstates of a system compatible with our knowledge

based on previous measurements, then simply allowing the system to evolve cannot lead to a situation at a later time t_2 where there are only 42 possible states of the system. If it could, then reversing the direction of time evolution, 42 states could evolve into 137, which is not compatible with the fact that time evolution is causal. Thus, the extent of our ignorance of an isolated system cannot decrease with time. This means that the entropy of an isolated system cannot decrease.

One might think that this argument would also require that entropy cannot increase with time, but this is not so for thermodynamic systems. When macroscopic systems evolve in time, we generally only maintain knowledge of macroscopic (thermodynamic) state information, and lose track of information relevant to the set of available microstates of the system. Consider, for example, a container partitioned into two equal volumes (Figure 8.1). Initially the left side holds a gas at room temperature and some pressure; the right side is a vacuum. The system has a vast number of microstates. When the partition between the two sides is removed, the gas will spontaneously expand to fill the whole volume. After some time,

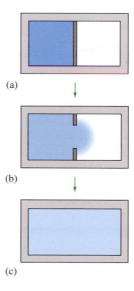

(a)

(b)

(c)

Figure 8.1 The second law at work. The gas, initially confined to the left half of a box (a), expands (b) to fill the whole box (c) when the partition is removed. Any microstate of (c) arising in this fashion would, if reversed in time, return to the left-hand side. *In practice*, the gas, once distributed throughout the box, is never observed to move entirely to one side of the container, since the set of microstates of (c) that do this are a minute fraction of all possible microstates for the gas when it fills the whole box.

Entropy and the Second Law of Thermodynamics

Entropy provides a quantitative measure of our ignorance of the microstate of a system, given our knowledge of its bulk properties.

For a system with fixed energy in thermodynamic equilibrium, the entropy is proportional to the logarithm of the number of microstates consistent with its bulk properties.

The second law of thermodynamics states that the entropy of an isolated system cannot decrease, but can increase with time.

it will settle down to a state described by a new pressure and temperature. Although each microstate that was once confined to the left side evolves into a unique state in which the gas fills the whole volume, these states are only a minute fraction of the full set of microstates available when the gas is in equilibrium in the whole volume. In practice, we cannot follow the evolution of the initial microstate in complete detail; we can only keep track of thermodynamic quantities such as the final volume, temperature, and pressure. Our inability to follow the detailed time evolution of an isolated system means that we lose information about the microstates of the system. This corresponds to an increase in entropy. The fact that the entropy of an isolated system cannot decrease, but can increase with time is the **second law of thermodynamics**.

At this stage it may not be obvious what these considerations have to do with heat, engines, and a fundamental limitation on efficiency for conversion of thermal energy into mechanical energy. Much of the rest of this chapter is devoted to making this connection precise. In §8.2 we give a quantitative definition of entropy in the context of discrete information systems. We then relate this definition to the entropy of physical systems in §8.3, using the fact that underlying every physical system is a quantum description with a discrete set of possible states of fixed energy. We show that when thermodynamic entropy is defined in this way it satisfies the second law of thermodynamics and gives us a very general way of defining the temperature of a system (§8.4). We then show (§8.5) that the second law of thermodynamics leads to the efficiency limit (8.1).

In the final parts of this chapter we develop some of the basic ideas of *statistical mechanics*. We use the microscopic description of the entropy and temperature of a system to compute the probability that a quantum-mechanical system is in any specific state when it is in thermal equilibrium at temperature T (§8.6). This probability distribution on the states of the system is known as the *Boltzmann distribution*, and plays a fundamental role in many branches of physics. It figures in many energy-related contexts, from understanding the rate of energy radiation from the Sun and Earth to analyzing nuclear reaction rates. In §8.7 we use the Boltzmann distribution to find the heat capacity and entropy of a metal and an ideal gas. Finally, in §8.8 we introduce the concept of *free energy*, which allows us to study the dynamics of systems in equilibrium with an environment at fixed temperature. An introduction to the origins of the 2nd Law focused on conceptual issues can be found in [42]. Much of the material in the later sections of this chapter is covered in detail, from a physicist's perspective in [21] and [43], and from an engineering perspective in [19].

8.2 Information Entropy

The notions of information, ignorance (lack of information), and entropy may seem rather imprecise and abstract. These days, however, we encounter these notions, at least indirectly, every time we check email, browse the web, make a telephone call, listen to recorded music, or interact with any of the other modern technological devices that rely on compression for efficient storage and communication of digital data. In fact, the concept of entropy has perhaps the clearest and simplest formulation in the realm of information and communication theory. In this section we follow this approach to give a mathematical definition of entropy that applies very generally to physical systems as well as to information systems.

The information-theoretic formulation of entropy we present here is not only a robust and precise way of formulating entropy quantitatively, but also provides a way of thinking about entropy that sheds light on many physical systems that we encounter later in the book. Often, in introductory books, the expression (8.10) is taken simply as a definition for thermodynamic entropy. Nevertheless, we believe that understanding the relationship of this formula to information entropy sheds substantial light on this often elusive concept.

The information content of simple images, quantified in terms of the number of *bits* (0 or 1) needed to digitally encode an image as described in Box 8.1, illustrates the relation between complexity and information content. The notion of *information entropy* was introduced by the American mathematician Claude Shannon in 1948 [44] in the context of his work on how information is transmitted over a (possibly noisy) communication channel. Entropy

Box 8.1   Information Content of an Image and `.gif` **Encoding**

As a simple example of how the concept of entropy is relevant to information systems, we consider briefly the encoding of a digital image. Consider the four images to the right. In each image a certain number of dots of a fixed size were placed randomly on a 300 × 300 pixel background of uniform color. The images were then stored as `.gif` files. The compression to gif format is done by an algorithm that uses the fact that certain pixel colors and combinations of adjacent colors are more frequent than others in these images to reduce the total storage space for the image. The utility of gif encoding arises from the fact that images produced by humans are not just random combinations of pixels. An image with little variation can be stored through gif encoding using less computer memory than a highly varied image. As shown in the graph below, the number of bytes (1 byte = 8 bits) required to store such images grows roughly linearly in the number of dots N, approximately following the curve $B = 1480 + 26N$. It requires roughly ~ 26 additional bytes of storage for each dot added to the image in this particular situation.

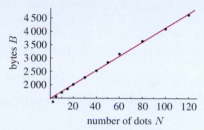

This example illustrates in a quantitative fashion the fact that for more complicated systems, more information is needed to precisely encode the details of the system. We can make a rough analogy between these images of dots and a physical system such as a gas in a box. The number of dots in the box, N, is analogous to the number of molecules in the gas. A "microstate" of the image "system" is specified by giving the precise location, shape, and size of every dot in the box. Specifying only the number of particles N does not completely determine the microstate; much more information is needed. The amount of information needed to describe the positions, shapes, and sizes of the dots is roughly measured by the number of bytes of the gif compressed image after subtracting a constant related to the number of bytes necessary to describe the empty box. The approximate linear fit suggests that for these images $\sim 26N$ bytes of information were required to specify the microstate using the gif encoding mechanism. This quantity of information essentially represents the entropy of the "system" described by the image containing N dots. The entropy is the amount of information needed to precisely specify the microstate of a system. This entropy has the character of a state function, in that it depends only on the macroscopic property N characterizing the number of dots in the box, and not on the specific microstate.

This example, while illustrative, is somewhat imprecise. The gif encoding algorithm depends upon the structure of the image being encoded and for small images does not realize the optimal possible information-theoretic compression. For example, to encode only images consisting of identically shaped dots on a uniform background, only the x- and y-coordinates of each dot need to be stored; the power of the gif encoding algorithm is that it works for *any* image, even though it is not optimal for small images fitting a precise pattern. Nonetheless, the basic principle underlying the analogy between entropy of images and entropy of a gas in a box is correct.

as used in physics can be regarded as an application of Shannon's theory.

Information entropy provides the answer to the question: What is the minimum number of bits required, on average, to **encode** a symbol s chosen from a fixed set of possibilities according to a particular underlying probability distribution so that the precise value of s can be reproduced from the given bits? For example, the symbol s might be "H" (heads) or "T" (tails) according to the flip of a fair coin, with a probability of 50% for each outcome. It is clearly possible to communicate the outcome of such

a coin flip with a single bit. We simply use 0 to represent H, and 1 to represent T. This is the simplest example of an *encoding*. Not surprisingly, there is no more efficient way to transmit information about flips of a fair coin – if Alice flips a fair coin 100 times, she cannot communicate the results of the flips to her colleague Bob with fewer than 100 bits of information. If, on the other hand, the coin is not fair, the situation is different. If the coin is weighted so that one result is more probable than the other, then *on average*, fewer than 100 bits of information are needed to transmit the results of a sequence of 100 coin flips.

To give a precise definition to information entropy, we need to clarify what is meant by the number of bits needed "on average." We begin by defining an **ensemble** of symbols to be a set $\mathcal{S} = \{s_1, \ldots, s_k\}$ of symbols, together with a probability p_i for each symbol s_i, normalized so that $\sum_i p_i = 1$. We now consider a sequence of N symbols chosen independently from the ensemble according to the given probability distribution. The **information entropy** per symbol, σ, is the minimum number of bits required to encode a symbol, averaged over the sequence of length N, in the limit as $N \rightarrow \infty$. Information entropy can be thought of as quantifying the *randomness* or *uncertainty* associated with a symbol chosen according to the given probability distribution.

As an example, consider the ensemble defined by the set of symbols $\mathcal{S} = \{A, B, C, D\}$ with the uniform probability distribution $p_i = 0.25$, $i = 1, 2, 3, 4$ that selects each symbol with probability 25%. If we encode each symbol by a pair of bits according to the mapping

$$A \rightarrow 00, \quad B \rightarrow 01, \quad C \rightarrow 10, \quad D \rightarrow 11, \quad (8.2)$$

then a typical sequence of symbols ABDBCAD... might be encoded as

$$
\begin{array}{ccccccccc}
& A & B & D & B & C & A & D & \ldots \\
\rightarrow & 00 & 01 & 11 & 01 & 10 & 00 & 11 & \ldots
\end{array}
$$
$$(8.3)$$

As in the case of a fair coin, this encoding of four uniformly distributed symbols is optimal. The entropy – the number of bits required per symbol – in this case is 2.

We can generalize this example to 2^k distinct symbols, each occurring with probability $p = 1/2^k$, which are optimally encoded in k bits. In this case the entropy σ is k, the number of bits required per symbol. Notice that in all these examples, the information entropy can be written in the form

$$\sigma = -\sum_{i=A,B,C,\ldots} p_i \log_2 p_i = \frac{k}{2^k} \sum_{i=A,B,C,\ldots} 1 = k,$$
$$(8.4)$$

where the sum is over the symbols and p_i is the probability of the ith symbol. For the example with four symbols, each occurring with probability 0.25, and each requiring $-\log_2(0.25) = 2$ bits to code, we have $\sigma = 4 \times 0.25 \times 2 = 2$.

The problem becomes more interesting if different symbols occur with different probabilities. For example, consider an ensemble of three symbols A, B, and C that occur with probabilities

$$p_A = 0.5, \quad p_B = 0.25, \quad p_C = 0.25. \quad (8.5)$$

For this probability distribution, the optimal encoding is obtained by assigning different numbers of bits to different symbols; specifically, fewer bits are used to encode symbols that occur more frequently. In this particular case, we can use two layers of the single bit code described above. First, we assign the bit 0 to A and 1 to *either* B or C, since each of these two alternatives occurs with probability 50%. Then, we assign a second bit to be 0 for B or 1 for C, so that the full encoding is

$$A \rightarrow 0, \quad B \rightarrow 10, \quad C \rightarrow 11. \quad (8.6)$$

A typical sequence of symbols ABABCAA... would then be encoded as

$$
\begin{array}{cccccccc}
A & B & A & B & C & A & A & \ldots \\
0 & 10 & 0 & 10 & 11 & 0 & 0 & \ldots
\end{array}
$$
$$(8.7)$$

This may seem ambiguous – when decoding the sequence, how do we know whether to decode one bit or two bits? In fact there is no ambiguity: if the first bit is 0, we decode it as A and consider the next bit as the initial bit of the next encoded symbol. If the first bit is 1, we decode the following bit to determine whether the symbol is B or C, and then move on to decode the next symbol.

We can easily compute the number of bits needed on average to communicate a symbol using the above encoding. For A, the probability is 0.5 and a single bit is required; for each of B and C the probability is 0.25 and two bits are required. Thus, the average number of bits required is

$$\sigma = 0.5(1) + 0.25(2) + 0.25(2) = 1.5$$
$$= -\sum_{i=A,B,C} p_i \log_2 p_i, \quad (8.8)$$

where the final expression uses the fact that $-\log_2 p_i$ bits were required to encode the symbol that occurred with probability p_i. This type of encoding system generalizes easily to any set of symbols with probabilities that are powers of $1/2$, giving a number of bits needed per symbol that always obeys the formula

$$\sigma = -\sum_i p_i \log_2 p_i. \quad (8.9)$$

Indeed, it is not hard to prove that this *information entropy* formula gives a number of bits that are sufficient, on average, to encode symbols chosen according to any probability distribution, even if the probabilities are not powers of $1/2$.[1] It is also possible to prove that no encoding

---

[1] The proof that eq. (8.9) gives a number of bits that is adequate to describe a symbol taken from a general ensemble proceeds by encoding large blocks of symbols with correspondingly small probabilities all at once, for example by considering

Information (Shannon) Entropy

Information entropy is a quantitative measure of randomness or uncertainty. For an ensemble consisting of a set of symbols $\mathcal{S} = \{s_i, i = 1, 2, \ldots, k\}$ and a probability distribution p_i, then the information entropy of a symbol chosen from this ensemble is given by

$$\sigma = -\sum_i p_i \log_2 p_i.$$

This is the number of bits (0/1) of information that are needed "on average" to encode or communicate a symbol chosen from the ensemble.

can be found that uses fewer bits per symbol (on average) than eq. (8.9). This result was first proved by Shannon and is known as his **source coding theorem**.

Shannon's information-theoretic definition of entropy is a useful way of characterizing ignorance or randomness in a variety of systems. The concept of information entropy provides a quantitative measurement of the uncertainty or variability associated with a probability distribution on a set of possible events. Related mathematical questions arise in a variety of other fields, particularly those related to information storage and communications, as well as in physics, chemistry, and statistics. While physical scientists first encountered the notion of entropy in the domain of thermodynamics, the lack of a microscopic/quantum description obscured the meaning of this quantity for some time. In fact, the definition (8.9) is precisely the measure of randomness that is needed to characterize the entropy of thermodynamic systems.

Before turning to the entropy of physical systems, we mention briefly how information entropy applies in more general contexts. One often wishes to communicate a sequence of symbols that are not drawn independently from a random distribution. For example, consider a work of literature in the English language. The individual characters are not chosen independently; the character "h," for example, is much more likely to be followed by "e" than by

---

every block of m sequential symbols as a single symbol in a larger ensemble. For each such block we use a string of bits of length corresponding to the lowest power of $1/2$ that is smaller than the probability of the block. The difference between the number of bits per symbol in the resulting compression and the optimum given in eq. (8.9) goes as $1/m$, becoming vanishingly small as we take m to be arbitrarily large.

"z." In this case, a correct measure of information entropy must incorporate such correlations in the probability distribution. This can be done by considering larger blocks of text, along the lines discussed in footnote 1. When correlations are considered, the entropy of the English language turns out to be roughly 1.1–1.2 bits/character. Similarly, for digital images the colors of adjacent pixels are not independent. The gif encoding mechanism discussed in Box 8.1 operates by finding an encoding for typical blocks of characters, making use of correlations between nearby pixels to compress frequently occurring blocks (such as a long sequence of pixels of the background color) into very short bit strings. For a more complete introduction to entropy in the context of information theory, see [45, 46, 47].

8.3 Thermodynamic Entropy

The notion of information entropy developed in the previous section can be applied directly to thermodynamics, by extending the concept of an *ensemble* to physical systems. As discussed in §5 and §8.1, a thermodynamic system is characterized by thermodynamic *state variables*, such as temperature, pressure, and volume, that characterize macroscopic properties of a system without specifying its *microstate*. A **thermodynamic ensemble** consists of a family of microstates $\mathcal{S} = \{s_\alpha, \alpha = 1, \ldots, \mathcal{N}\}$, all with the same macroscopic thermodynamic properties, distributed according to a characteristic probability distribution p_α. For example, we could consider the ensemble of states describing $N = 10^{24}$ oxygen molecules in a box of volume V with fixed internal energy U. The number $\mathcal{N}$ of microstates satisfying these conditions is finite, but enormously large. Each such microstate is physically different at a microscopic scale, but indistinguishable from the thermodynamic point of view. Note that each distinct microstate corresponds to a different "symbol" in information-theoretic terminology. For physical systems, the number of such microstates is generally far larger than the number of symbols in the information-theoretic examples described above.

The probability distribution on the set of states for a physical system at fixed energy in thermodynamic equilibrium is very simple. As we show in the following section, in such an ensemble there is an equal probability for each of the $\mathcal{N}$ possible states, so $p_\alpha = 1/\mathcal{N}$ for $\alpha = 1, \ldots, \mathcal{N}$. Later in this chapter we characterize the probability distributions for different kinds of thermodynamic systems, but for now we simply assume that there is some well-defined probability distribution p_α on the set of microstates in a given thermodynamic ensemble, keeping the oxygen gas at fixed energy in mind as an example.

Thermodynamic Entropy

The thermodynamic entropy of an ensemble of possible microstates $\{s_\alpha\}$ with fixed macroscopic thermodynamic properties and a distribution of microstate probabilities p_α is given by

$$S = -k_B \sum_\alpha p_\alpha \ln p_\alpha \,.$$

Given the notion of a thermodynamic ensemble, we can immediately conclude that the information entropy for the thermodynamic ensemble is $\sigma = -\sum_\alpha p_\alpha \log_2 p_\alpha$. The *thermodynamic entropy* differs from the information-theoretic entropy by a multiplicative constant: for a thermodynamic ensemble characterized by a probability distribution p_α on states s_α, the **thermodynamic entropy** is defined to be

$$S = k_B \ln 2 \times \sigma$$
$$= -k_B \sum_\alpha p_\alpha \ln p_\alpha \,. \qquad (8.10)$$

Here k_B is Boltzmann's constant (5.5), with units of J/K; note that $\ln 2 \times \log_2(x) = \ln(x)$. For the oxygen gas at fixed energy, for example, with equal probability $p_\alpha = 1/\mathcal{N}$ for the system to be in any of $\mathcal{N}$ microstates, the thermodynamic entropy is

$$S = k_B \ln \mathcal{N} \,. \qquad (8.11)$$

Had thermodynamic entropy been discovered after information entropy, it might more naturally have been defined as a dimensionless quantity, with $k_B = 1$ or $1/\ln 2$, and with temperature defined in units of energy. Entropy was, however, identified as an effective measure of disorder long before the discovery of the underlying quantum structure that gives meaning to the concept of discrete states of the system. In the context of its discovery, it was natural to measure entropy in units of energy/temperature, and Boltzmann's constant is the resulting experimentally determined proportionality constant.[2]

To understand the probability distributions p_α associated with thermodynamic ensembles, we need a more precise definition of temperature and of thermodynamic and thermal equilibrium. These concepts are developed in the next two sections.

---

[2] Actually the gas constant $R \equiv N_A k_B$ came first, and when the atomic structure of matter was understood, the factor of Avogadro's number N_A was divided out.

8.4 Thermal Equilibrium and Temperature

8.4.1 The Second Law and Thermodynamic Equilibrium for Isolated Systems

In §8.1, we argued that our ignorance (i.e. the entropy) of a closed system cannot decrease over time, due to the causal and reversible nature of the microscopic laws of physics. While that argument relied on a somewhat loose notion of entropy based on the number of accessible states of the system, the same logic applies for thermodynamic entropy as quantified more precisely in eq. (8.10). Just as the number of accessible states of a system cannot decrease through causal and reversible dynamics, the information entropy associated with our lack of knowledge about an isolated thermodynamic system cannot decrease over time. If the number of bits needed to specify the microstate of a system at time t is σ, then we cannot specify the microstate at time $t' > t$ with $\sigma' < \sigma$ bits; otherwise we would have a contradiction, since we could specify the state at time t' and use the reversible dynamics to run backwards and thus specify the state at time t using only σ' bits. This proves the second law of thermodynamics for thermodynamic entropy as defined through eq. (8.10).

Likewise, as discussed in §8.1, although entropy cannot decrease over time, it can increase, as in the case where the gas originally confined to one side of the box is allowed to expand into the full box. If entropy can only increase, it is natural to ask: how large can it get? What is the *maximum possible value of the entropy* for a given system? As a simple example, consider a system that can be in one of only two possible states, which we label $+$ and $-$. Such a system can describe, for example, the spin states of a particle in a magnetic field. An ensemble for this system is characterized by probabilities p_+ and $p_- = 1 - p_+$. The entropy of this system is then

$$S = -k_B (p_+ \ln p_+ + p_- \ln p_-)$$
$$= -k_B (p_+ \ln p_+ + (1 - p_+) \ln(1 - p_+)) \,, \quad (8.12)$$

which is graphed in Figure 8.2 as a function of p_+. The entropy is maximum when $p_+ = p_- = 1/2$ and $S = -k_B \ln 1/2 = k_B \ln 2$ (Problem 8.4).

Thus, for a simple two-state system, the maximum entropy is realized when the probability of each state is equal. This is true more generally for an isolated system with any number of available states. In fact, it is easy to see from the result for the two-state system that a probability distribution on any number of states in which each state is equally likely should give a maximum of S. Consider any probability distribution on $\mathcal{N}$ possible states. For any pair of these states a, b, if all other probabilities are kept fixed then the entropy depends on the probabilities

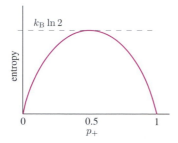

Figure 8.2 Entropy S of a single bit as a function of p_+.

p_a, p_b in a similar fashion to the two-state system graphed in Figure 8.2, and is maximized only when $p_a = p_b$. Since this is true for all pairs of states, the overall maximum of entropy only occurs when all states are equally probable.

Thus, *the entropy of a closed system is maximized when all microstates are equally likely*. Since the sum of probabilities over all $\mathcal{N}$ microstates must add to one, $p_\alpha = \bar{p} = 1/\mathcal{N}$. For a system of N particles with energy U, and $\mathcal{N}(U, N)$ accessible states,[3] the maximum entropy is then

$$S = k_{\mathrm{B}} \ln \mathcal{N}(U, N). \tag{8.13}$$

The fact that the maximum entropy occurs when all states are equally likely is intuitively sensible, as this is the situation in which we have the least information about the system.

As described in §8.1, our knowledge of a system left to itself can only decrease with time. For a very wide class of systems we can go further than this and assert that they evolve with time toward a state of maximum thermodynamic entropy, where as we have just shown, the probability of being in any specific microstate is the same. This tendency of complicated systems to randomize and lose order is familiar from everyday experience. Milk stirred into coffee rapidly mixes until the fluids are thoroughly intermingled, increasing entropy. An ice cube dropped into warm water rapidly melts as thermal energy from the water first warms and then melts the ice.

When a system reaches the state of maximum entropy, it no longer changes in any macroscopic way even though the constituent particles are changing their positions and velocities all the time. This is the state of *thermodynamic equilibrium* first introduced in §5. Systems in thermodynamic equilibrium have definite entropy, given for example by eq. (8.13) when the system has a fixed total energy.

Entropy and Thermodynamic Equilibrium

In *thermodynamic equilibrium* the entropy of a system is maximum. For an isolated system with energy U in thermodynamic equilibrium, all microstates are equally likely and the entropy is a *state function* given by

$$S = k_{\mathrm{B}} \ln \mathcal{N}(U),$$

where $\mathcal{N}$ is the number of microstates that are consistent with the macroscopic (thermodynamic) state of the system.

Therefore *entropy is a state function*, like pressure and volume; entropy characterizes the equilibrium state and does not depend on the system's history. Thermodynamic equilibrium is quite different from the equilibrium of a simple classical system like a mass acted upon by a force. In that case, *static* equilibrium is attained when the mass is at rest. The state of maximum entropy is one of *dynamic equilibrium*, where appearances do not change even though the system is sampling all possible microstates uniformly over time. In §8.4.3 we show how this characterization of thermodynamic equilibrium leads to a more rigorous definition of *temperature*.

8.4.2 The Approach to Thermodynamic Equilibrium

The approach to thermodynamic equilibrium can also be characterized in a fairly precise mathematical fashion. For most real macroscopic systems, the dynamics of the system has the property that, regardless of the initial configuration,[4] over a long time the system uniformly samples all possible microstates. A system with this property is called **ergodic**. For an ergodic system, even if an initial state is chosen that has some special structure or additional properties, when the system is observed at a random later time, the probability of finding the system in any given microstate is uniform (and thus given by the thermodynamic equilibrium distribution). The assumption that a given physical system is ergodic is known as the **ergodic hypothesis**. Many ergodic systems, including most macroscopic systems of many particles, have the stronger property of **mixing**, meaning that microstates with initial conditions that are very similar will evolve into

---

[3] We suppress, for the moment, any other state variables such as the volume or pressure, on which the number of microstates might depend.

[4] With some exceptions in idealized systems for very special initial configurations whose likelihood is vanishingly small in any real circumstances.

unrelated microstates over a sufficiently long time. For an ergodic system that exhibits mixing, there is generally a characteristic time scale (the **mixing time**) over which an initial class of states with some ordered structure (beyond that associated with the thermodynamic variables) will distribute uniformly over the space of all possible states of the system. Some systems, like a gas in a box, randomize and lose memory of an original ordered state quickly. Others take very long times to randomize. When an ergodic system with mixing begins in an ordered initial state, even small perturbations of the system modify the dynamics in small ways whose impact on the state of the system grows over time. After a time much larger than the mixing time, for all intents and purposes the system must be considered to be in a state of thermodynamic equilibrium, since the detailed dynamics of the time development and the effects of small perturbations of the system cannot be followed. We assume that all thermodynamic systems treated in this book are ergodic and mixing, so that over time they approach the state of thermodynamic equilibrium.

To illustrate the loss of order and the associated increase in entropy in a system with ergodic and mixing behavior consider once again the gas of N molecules (Figure 8.1), initially confined to the left half of a box by a partition. If the partition is suddenly removed, the gas molecules are free to move through the whole box. Collisions between the gas molecules, and between the gas molecules and the walls of the box, rapidly spread the gas throughout the box. In a short time the original order in the system is lost. In this process our ignorance about the state of the system effectively increases. Although in principle the final state of the system arose by means of causal and reversible dynamics from an initial state in which all particles began on the left-hand side of the box, this information is of no practical utility, since this state cannot be distinguished experimentally from any other typical state in which the gas molecules fill the box. If it were possible to follow the time evolution of the system exactly and if the motion of the system could be precisely reversed in time, the system would revert to the initial state of higher order. In such a situation we would not have lost information about the system's state and the entropy would not increase. This is only conceivable for very simple systems under highly controlled circumstances. For macroscopic systems with immense numbers of particles, it is generally not possible, and given only the thermodynamic information about the final state there is an increase in entropy from the initial conditions.

We can precisely quantify the increase in ignorance/entropy from the *free expansion* of the gas in the idealization of an ideal gas. In a simple classical picture,

The Second Law and the Approach to Thermal Equilibrium

The second law of thermodynamics states that the thermodynamic entropy of an isolated system cannot decrease over time, though it can increase. If a system is ergodic and exhibits mixing – properties of essentially all macroscopic physical systems – then it will evolve to a thermodynamic equilibrium state where its entropy is maximized.

the information needed to describe the position of each particle requires one additional bit (0/1 if the particle is on the left/right side of the box) once equilibrium is reached following the removal of the partition. Since the kinetic energies of the particles in an ideal gas do not change when the partition is removed, the velocity distribution and associated entropy contribution does not change as the gas spreads through the box. The only increase in entropy comes from the increased uncertainty in the particles' positions; adding one bit of entropy for each particle leads to an increase in thermodynamic entropy of $Nk_B \ln 2$. While this argument is somewhat heuristic and based on classical reasoning, a more careful quantum mechanical argument leads to the same conclusion. At the end of this chapter we compute the entropy of an ideal gas in a box using quantum mechanics, and verify that doubling the box size increases its entropy by $Nk_B \ln 2$.

8.4.3 Temperature

Next we turn to the question of how entropy behaves when systems are allowed to interact with one another. This leads to a general definition of *temperature*. Our discussion is guided by the fundamental principle that the changes that take place when systems interact cannot decrease our ignorance of the details of the microstates of the combined system.

First, we show that entropy is an **extensive variable**. This means that the entropy of a system consisting of two independent and non-interacting parts is just the sum of the entropy of the parts. Suppose that two systems A and B, shown in Figure 8.3(a) have independent probability distributions p_α^A and p_β^B over $\mathcal{N}^A$ and $\mathcal{N}^B$ possible microstates respectively. The entropies of the systems are $S_A = -k_B \sum_\alpha p_\alpha^A \ln p_\alpha^A$ for system A and similarly for system B. Now consider the systems A and B together as one system AB. The microstates of AB can be labeled by (α, β) and, according to the rules of probability, the likelihood of finding AB in the state (α, β) is $p_{\alpha, \beta}^{AB} = p_\alpha^A \times p_\beta^B$.

Example 8.1 Entropy of Mixing

In Figure 8.1, N molecules of an ideal gas initially confined in a volume $V/2$ are allowed to expand into a volume V. Intuitively, we expect that we know less about each molecule after the expansion than before, and therefore that the entropy of the gas must increase. As discussed in the text at the end of §8.4.2, exactly one additional bit of information per molecule is needed after the expansion (indicating whether each particle is in the left or right half of the box) to fully specify the microstate of the gas. The thermodynamic entropy of the system there-

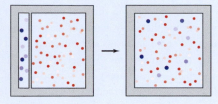

fore increases in the expansion by $\Delta S = Nk_B \ln 2$. More generally, if the gas begins in a volume cV, with $c \leq 1$, then an additional $-\log_2 c$ bits of information are required to specify each particle's location when the gas is allowed to expand to a volume V, so the entropy of the physical system increases by $\Delta S = -Nk_B \ln c$.

The same argument enables us to compute the entropy increase when two (ideal) gases, initially separated and at the same temperature and pressure, are allowed to mix. Suppose, as shown in the figure above, a container holds $N_1 = c_1 N$ molecules of an ideal gas in volume $V_1 = c_1 V$ and $N_2 = c_2 N$ molecules of a second ideal gas in volume $V_2 = c_2 V$, where $c_1 + c_2 = 1$. Since $N_1/V_1 = N_2/V_2 = N/V$, the pressures in each region (determined by the ideal gas law) are equal. When the partition is removed, each gas expands to fill the whole volume V, and the entropy of the system increases by the **entropy of mixing**,

$$\Delta S_{\text{mixing}} = k_B (N_1 \ln V/V_1 + N_2 \ln V/V_2)$$

$$= -Nk_B (c_1 \ln c_1 + c_2 \ln c_2).$$

The variables c_1 and c_2 are referred to as the **concentrations**, i.e. the fractional abundances by number, of the two types of gas in the final mixture.

Conversely, when a mixture is separated into its components, the entropy of the system decreases. As explored in Problem 8.5, this gives rise to a lower bound on the energy required to separate a mixture into its components. In §34 (Energy and climate), we use this idea to obtain a lower bound on the energy required to remove carbon dioxide from fossil fuel power plant emissions.

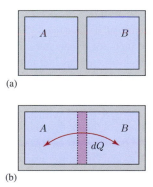

Figure 8.3 (a) Two systems are isolated and non-interacting. (b) The two systems are in *thermal contact* (denoted by the purple shaded interface), so that thermal energy can flow between them, allowing their internal energies to change, but their particle numbers are fixed.

With a little algebra, we can prove that the entropy of the system AB, defined in the usual way, is the sum of the entropies of system A and system B,

$$S_{AB} = -k_B \sum_{\alpha, \beta} p_{\alpha, \beta}^{AB} \ln p_{\alpha, \beta}^{AB}$$

$$= -k_B \sum_{\alpha, \beta} p_\alpha^A p_\beta^B \left(\ln p_\alpha^A + \ln p_\beta^B \right)$$

$$= -k_B \sum_\beta p_\beta^B \sum_\alpha p_\alpha^A \ln p_\alpha^A - k_B \sum_\alpha p_\alpha^A \sum_\beta p_\beta^B \ln p_\beta^B$$

$$= -k_B \sum_\alpha p_\alpha^A \ln p_\alpha^A - k_B \sum_\beta p_\beta^B \ln p_\beta^B = S_A + S_B.$$

$$(8.14)$$

In going to the last line we have used the relation $\sum_\alpha p_\alpha^A = \sum_\beta p_\beta^B = 1$. Note that the fact that the systems are independent was needed when we assumed that the probabilities of system A are independent of the state of B and *vice versa*. We turn next to the question of what happens when energy and entropy can flow between systems.

Consider two systems A and B that are put into contact with one another in such a way that thermal energy can flow between them, but particles cannot, so that the distribution of internal energy between A and B can change, but the number of particles in each system remains fixed. This situation is illustrated in Figure 8.3(b). We refer to the systems as being in **thermal contact**. For definiteness we assume that the volumes of the systems are fixed, although the same analysis can be carried out at constant pressure. The two systems together define a total system AB, to which we can apply the second law of thermodynamics: if left undisturbed, AB will evolve to a state of maximum entropy – a thermodynamic equilibrium. In such a situation, when the systems A and B are in a combined state of thermodynamic equilibrium and they are in thermal contact but no other transfer of energy or matter is possible between the systems, we say that A and B are in **thermal equilibrium** with one another. More generally, we often refer to a system as being in *thermal equilibrium* when the different parts of the system are in thermal equilibrium with one another and other mechanisms for energy transfer or matter transfer (such as mechanical, diffusive, or chemical processes) are not relevant or proceed on a much slower time scale than thermal energy transfer. Note, in particular, that since their volumes are fixed, the pressures of systems A and B can be different when they are in thermal equilibrium. If, on the other hand, the partition separating A and B was a perfect insulator but was free to move, then the systems would come into *mechanical equilibrium* without reaching thermal equilibrium. In neither case are A and B in *thermodynamic equilibrium*, which could be reached only if *all* forms of energy transfer between A and B were allowed.

When the total system AB is in thermodynamic equilibrium, implying that A and B are in thermal equilibrium, the total energy available is distributed between the two component systems; we denote the internal energy of system A by U_A and that of system B by U_B. While the precise distribution of energy between the systems fluctuates, when the number of degrees of freedom is very large these fluctuations become infinitesimal compared to the total amount of energy available.[5] This means that the entropy of the total system can be decomposed into contributions from the two subsystems

$$S_{A+B} = S_A(U_A) + S_B(U_B), \qquad (8.15)$$

just as if the systems were independent. Here $S_A(U_A)$ is the entropy of system A in thermodynamic equilibrium with energy U_A and similarly for system B. The corrections to eq. (8.15) from fluctuations of energy between the systems become vanishingly small as the number of degrees of freedom becomes large. Since S_{A+B} is maximized, this entropy cannot be increased by a different distribution of energy between the systems. The difference in the total entropy if the energy of system A were smaller by dQ and the energy of system B were larger by the same amount must therefore vanish,

$$dS_{A+B} = -\left.\frac{\partial S_A}{\partial U}\right|_V dQ + \left.\frac{\partial S_B}{\partial U}\right|_V dQ = 0. \qquad (8.16)$$

It follows that

$$\left.\frac{\partial S_A}{\partial U}\right|_{V_A} = \left.\frac{\partial S_B}{\partial U}\right|_{V_B} \qquad (8.17)$$

when A and B are in thermal equilibrium (where we have explicitly noted that the volume remains fixed).

Thus, when two systems come into thermal equilibrium, the quantity $\partial S/\partial U$ is the same for both. Two systems that are in thermal equilibrium are said to have the same **temperature**. Indeed, $\partial S/\partial U$ provides a universal definition of temperature,[6]

$$\frac{1}{T} \equiv \left.\frac{\partial S}{\partial U}\right|_V . \qquad (8.18)$$

It may be surprising that there is no constant of proportionality in this equation. Remember, however, that the constant of proportionality k_B (Boltzmann's constant) is incorporated in the definition (8.10) of entropy. k_B is chosen so that the units of T defined in eq. (8.18) coincide with the conventional units for temperature. Note that dimensional consistency now requires that the units of entropy are energy/temperature as asserted earlier. The ability to define temperature without any reference to a particular system such as an ideal gas or a tube of mercury shows the power of the laws of thermodynamics.

Equation (8.18) is not, however, a familiar definition of temperature. We should check, for example, that this definition agrees with the definition of temperature for ideal gases and simple solids that was used in §5. We do this in §8.7, after exploring some of the consequences of the definition (8.18).

---

[5] This statement can be made mathematically precise using the **law of large numbers**, which states that when N values are chosen from a given distribution, the average of the values chosen deviates from the average of the distribution by an increasingly small fraction of the average as $N \to \infty$; for example, when N coins are flipped, the number of heads will go as $(N/2)(1 + \mathcal{O}(1/\sqrt{N}))$ as $N \to \infty$.

[6] The analogous argument at fixed pressure leads to $\partial S/\partial H|_p = 1/T$ which can be shown to be equivalent to eq. (8.18) using the definition $H = U + pV$. The central concept here is that the change in entropy with thermal energy, however it is added to the system, defines $1/T$ (see Problem 8.14).

Thermodynamic Definition of Temperature

Temperature can be *defined* as the rate of change of entropy when a small quantity of energy dU is added to a system in thermodynamic equilibrium at fixed volume,

$$\frac{1}{T} \equiv \frac{\partial S}{\partial U}\bigg|_V .$$

This definition of temperature matches with other definitions for specific systems but is more general.

Notice that the definition (8.18) of temperature is *transitive*. If system A is in thermal equilibrium with systems B and C, then systems B and C are in thermal equilibrium with each other. This follows immediately from eq. (8.16). If $\partial S_A/\partial U = \partial S_B/\partial U$ and $\partial S_A/\partial U = \partial S_C/\partial U$ then certainly $\partial S_B/\partial U = \partial S_C/\partial U$. This property is often referred to as the **zeroth law of thermodynamics**.

8.4.4 Quasi-equilibrium and Reversible Processes

Equation (8.18) not only defines temperature, but relates the concepts of entropy and heat at a fundamental level. It enables us to compute the change in entropy of a system in certain circumstances, tying into the original thermodynamic definition of entropy. Suppose a small amount of thermal energy – i.e. heat – passes into a system in thermodynamic equilibrium. If the volume is fixed, then $dQ = dU$ and eq. (8.18) implies that the entropy of the system increases,

$$dS = \frac{\partial S}{\partial U}\bigg|_V dQ = \frac{dQ}{T} . \tag{8.19}$$

As heat is added, the temperature of the system may rise. We can integrate eq. (8.19), however, to compute the total change in entropy when a finite quantity of heat is added, as long as the change in state occurs in such a way that the system can be approximated as being in thermodynamic equilibrium at each intermediate point in the process. Such a process is known as a **quasi-equilibrium process**. Quasi-equilibrium heat transfer to a system can be effected by coupling it to a second system in equilibrium at a temperature $T + dT$ that is infinitesimally higher. The change in the total entropy of the two systems,

$$dS_{\text{total}} = \frac{dQ}{T} - \frac{dQ}{T + dT} \approx dQ\frac{dT}{T^2} , \tag{8.20}$$

becomes vanishingly small as $dT \to 0$, even when dQ is integrated to give a finite quantity of heat transfer.

Heat transfer under quasi-equilibrium conditions can be reversed without additional energy input or entropy production by changing the sign of dT, in the limit where $dT \to 0$. Heat transfer under these conditions is called **reversible** heat transfer, for which

$$dS = \frac{dQ_{\text{reversible}}}{T} . \tag{8.21}$$

When other sources of entropy production are present, quasi-equilibrium processes are not necessarily reversible, though the converse – reversible processes are necessarily quasi-equilibrium – still holds. Consider, for example, slow compression of a gas by a piston that experiences friction with the cylinder in which it moves. If the compression proceeds slowly enough, the system can be kept arbitrarily close to thermodynamic equilibrium throughout the process. The process is not, however, reversible, since frictional losses cannot be reversed. To be *reversible*, the entropy of the system plus its environment must not increase, since the reverse process must not violate the 2nd Law.

Like many other concepts in thermodynamics, *quasi-equilibrium* and *reversibility* are idealizations. In any real system, small deviations from thermal equilibrium, such as temperature gradients in the fluid, violate the quasi-equilibrium assumption and lead to generation of additional entropy. Thus, for a real system heat transfer is never really reversible in the sense defined above.

Equation (8.21) provides a convenient way to compute the entropy difference between two states of a thermodynamic system. If A and B are two equilibrium states and P_{AB} is a path from A to B in the space of thermodynamic states, along which the system remains close to equilibrium, then eq. (8.21) can be integrated along P_{AB} to obtain the entropy difference, $S(A) - S(B)$,

$$S_B - S_A = \int_{P_{AB}} \frac{dQ_{\text{reversible}}}{T} . \tag{8.22}$$

In §8.1 and in Figure 8.1 we examined the entropy increase when an ideal gas is allowed to expand freely from volume V to volume $2V$. A free expansion is *irreversible* and therefore eq. (8.22) cannot be used directly to compute the entropy change. As illustrated in Example 8.2, however, we can find a *reversible* path that leads to the same final state, and this enables us to compute the entropy change in a free expansion, and to verify the conclusion that entropy increases by $Nk_B \ln 2$.

In a reversible process, the total entropy of the system plus the environment does not change. As we find in the following section, maximizing efficiency of thermal energy conversion involves minimizing the production of additional entropy. Thus, reversible processes are often

Example 8.2   Entropy Change in the Free Expansion of an Ideal Gas, Revisited

Suppose, as illustrated in Figure 8.1, N molecules of an ideal gas, initially at temperature $T_A = T$ are held in a volume $V_A = V$. By the ideal gas law, the initial pressure is $P_A = Nk_BT/V$. This equilibrium state is marked A in the plot of the pressure–volume (pV) plane at right.

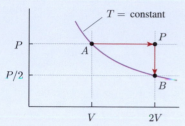

Suddenly the gas is allowed to expand irreversibly until its volume has doubled to $V_B = 2V$, and then the gas is allowed to settle once again into thermal equilibrium. No work was done and no heat was added to the gas, so its internal energy did not change. Since the internal energy of an ideal gas depends only on its temperature (eq. (5.6)), its temperature has not changed either: $T_B = T_A = T$. The pressure has decreased to $P_B = P/2$. This final equilibrium state is marked B in the figure. *By how much has its entropy changed?*

Note that the free expansion cannot be denoted by a path in the pV-plane because the system does not pass through a sequence of near-equilibrium states with well-defined temperature and pressure as it expands. Nevertheless, since entropy is a state function, the change in entropy of the gas does not depend upon the process used to get from A to B, so we may choose a *reversible* path from A to B and use eq. (8.22) to compute the entropy difference $S(B) - S(A)$. In particular, we can start by adding heat reversibly to the gas at constant pressure, allowing it to expand from V to $2V$ – the path marked AP in the figure. We can then remove heat from the gas at constant volume, allowing the pressure to drop from P to $P/2$ – the path marked PB in the figure. Along AP, $dQ_{\text{reversible}} = C_p(T)dT$, while along PB, $dQ_{\text{reversible}} = C_V(T)dT$, where $C_{p,V}(T)$ are the (possibly temperature-dependent) heat capacities of the ideal gas at constant pressure and volume. Thus,

$$S(B) - S(A) = \int_T^{T_P} \frac{dT'}{T'} C_p(T') - \int_T^{T_P} \frac{dT'}{T'} C_V(T') = \int_T^{T_P} \frac{dT'}{T'} (C_p(T') - C_V(T'))$$

$$= Nk_B \int_T^{T_P} \frac{dT'}{T'} = Nk_B \ln(T_P/T),$$

where we have used the fact that $C_p(T) - C_V(T) = Nk_B$ for an ideal gas (eq. (5.31)). Since the ideal gas law also fixes $T_P = 2T$, we find

$$S(B) - S(A) = Nk_B \ln 2,$$

in agreement with the information-theoretic argument of §8.4.2.

---

used as a starting point for constructing idealized thermodynamic cycles for converting thermal energy to mechanical energy. We return to this subject in §10.

8.5 Limit to Efficiency

We have alluded repeatedly to the thermodynamic limit on the efficiency with which heat can be converted into work. We now have the tools to derive this limit. The clearest and most general statement of the efficiency limit applies to a **cyclic process** in which an apparatus takes in heat energy Q_+ from a reservoir $\mathcal{R}_+$ that is in thermal equilibrium at a high temperature T_+, uses this heat energy to perform useful mechanical (or electromagnetic) work W, and then returns to its original state, ready to perform this task again. Such an apparatus is called a **heat engine**. The engine may also (indeed must) expel some heat Q_-

to the environment, which we describe as another reservoir $\mathcal{R}_-$ in thermal equilibrium at temperature T_-. The reservoirs are assumed to be so large that adding or removing heat does not change their temperatures appreciably. §10 is devoted to the study of some specific types of heat engines. For the moment we leave the details unspecified and consider the engine schematically, as depicted in Figure 8.4.

Since both energy and entropy are state functions, and the process is cyclic, both the energy and entropy of the apparatus must be *the same at the beginning and end of the cycle*. First let us trace energy conservation. We suppose that the engine absorbs heat energy Q_+ from $\mathcal{R}_+$, does an amount of work W, and expels heat energy Q_- to $\mathcal{R}_-$. Conservation of energy requires

$$Q_+ = W + Q_- . \tag{8.23}$$

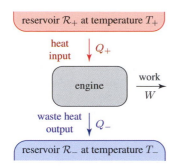

Figure 8.4 A schematic depiction of a heat engine. The entropy added to the engine is greater than or equal to Q_+/T_+ and the entropy dumped is less than Q_-/T_-.

Quasi-equilibrium and Reversible Processes

When a system is changed gradually, so that it remains infinitesimally close to thermodynamic equilibrium throughout the transformation, the process is called a *quasi-equilibrium* process. When heat dQ is transferred to a system in a quasi-equilibrium fashion and no other sources of entropy production are present, the process is *reversible*, and the entropy of the system increases by

$$dS = \frac{dQ_{\text{reversible}}}{T}.$$

Reversible processes do not increase the *total* entropy of the systems involved.

Additional entropy is generated when heat is transferred irreversibly.

Note the signs: $W > 0$ means the engine does work, $Q_+ > 0$ means the engine takes in heat from $\mathcal{R}_+$, and $Q_- > 0$ means the engine expels heat to $\mathcal{R}_-$.

Now consider the flow of entropy. If the transfer of heat from the high-temperature reservoir were reversible, then the entropy of the engine would increase by Q_+/T_+. In practice, the entropy of the engine increases by more than this, due for example to entropy creation in the friction of moving parts or free expansion of fluids. Thus, the entropy added to the engine when it absorbs heat from the reservoir at temperature T_+ can be written as

$$\Delta S_+ = \frac{Q_+}{T_+} + \Delta S_+', \tag{8.24}$$

where $\Delta S_+' \geq 0$ (as required by the 2nd Law).

Similarly, if heat were transferred reversibly into the low-temperature reservoir, then the entropy of the engine would decrease by Q_-/T_-, which is the amount by which the entropy of the low-temperature reservoir increases. In practice, the entropy of the engine does not decrease by this much because entropy is created by irreversibilities in the processes involved. Thus, the amount by which the engine's entropy drops when it expels heat at T_- can be written as

$$\Delta S_- = \frac{Q_-}{T_-} - \Delta S_-', \tag{8.25}$$

where $\Delta S_-' \geq 0$.

Since, by hypothesis, the engine is cyclic, the net change in its entropy must be zero, so $\Delta S_+ = \Delta S_-$. Combining eqs. (8.24) and (8.25) with this relation, and remembering that $\Delta S_\pm' \geq 0$, gives

$$\frac{Q_-}{T_-} \geq \frac{Q_+}{T_+} \quad \Rightarrow \quad \frac{Q_-}{Q_+} \geq \frac{T_-}{T_+}, \tag{8.26}$$

where the equality holds only in the limit that the engine operates reversibly throughout its cycle.

We define the **efficiency** of a heat engine to be the ratio of the work performed, W, to the input heat, Q_+,

$$\eta \equiv W/Q_+. \tag{8.27}$$

Combining eq. (8.23) (which follows from the 1st Law) and eq. (8.26) (which follows from the 2nd Law), we find

$$\eta = \frac{W}{Q_+} = \frac{Q_+ - Q_-}{Q_+} \leq \frac{T_+ - T_-}{T_+}. \tag{8.28}$$

This thermodynamic limit on the efficiency of a heat engine is known as the **Carnot limit**, after Sadi Carnot, the nineteenth-century French physicist who discovered it. This law is of essential importance in the practical study of energy systems. The fact that the relations eqs. (8.24) and (8.25) are inequalities rather than equalities expresses the important physical reality that a heat engine cannot operate perfectly reversibly. Indeed, it is extremely difficult to prevent additional entropy production during an engine's cycle.

Note that we do not associate any entropy change in the engine with the work W. Indeed, the essential difference between thermal energy and macroscopic mechanical or electromagnetic energy is the presence of entropy in the thermal case. A mechanical piston, for example, such as the one depicted in Figure 5.4, is idealized as a single macroscopic object with given position and velocity, and no entropy associated with ignorance of hidden degrees of freedom. In reality, some entropy will be created in the piston – through, for example, frictional heating. Energy lost in this fashion can be thought of as simply decreasing W,

The Carnot Limit

A heat engine that takes heat from a source at a high temperature T_+, does work, expels heat to a sink at low temperature T_-, and returns to its initial state, ready to begin the cycle again, has an efficiency η that cannot exceed the Carnot limit

$$\eta = \frac{W}{Q_+} \leq \eta_C = \frac{T_+ - T_-}{T_+}.$$

and hence the efficiency, further below the Carnot limit. Or, thinking of the whole process as part of the engine, this can be construed as an additional contribution to ΔS_+, with a corresponding increase in ΔS_- and Q_-.

The analysis leading to the Carnot limit has a number of other interesting implications. For example: no heat engine can transform heat into work without expelling heat to the environment; and no heat engine can remove heat from a low-temperature source, do work, and expel heat at a higher temperature. These constraints are so fundamental that in the early days of thermodynamics they served as statements of the 2nd Law. The proofs of these assertions are left to the problems. It is important to note that the Carnot limit applies only to heat engines. It does not, for example, limit the efficiency with which electrical energy can be converted into useful work. The Carnot limit also does not say anything directly regarding the efficiency with which a non-cyclic process can convert heat into work; in later chapters, however, we develop the notion of *exergy* (§37, §36), which places a 2nd law bound on the energy that can be extracted in a situation where a substance is hotter or colder than the environment without reference to cyclic processes.

The practical consequences of the efficiency limit (8.28) are significant. Most energy used by humans either begins or passes an intermediate stage as thermal energy, and must be converted into mechanical energy to be used for purposes such as transport or electrical power. Thus, in systems ranging from car engines to nuclear power plants, much effort is made to maximize the efficiency of energy conversion from thermal to mechanical form (see Box 8.2). Treatment of these conversion systems forms a substantial component of the material in the remainder of this book. In §10, we describe a range of heat engines and the thermodynamic processes involved in their operation more explicitly.

One clear implication of the Carnot limit is that the greatest thermodynamic efficiency can be achieved when thermal energy is extracted from a source at a temperature

T_+ that is as high as possible compared to the temperature T_- of the environment into which entropy must be dumped. Note, however, that for the distinct processes within a thermodynamic cycle, the maximization of efficiency for the cycle as a whole requires that thermal energy *transfer* from one part of the system to another must occur across temperature differentials that are as small as possible to avoid production of extra entropy.

8.6 The Boltzmann Distribution

Now that we have a rigorous definition of temperature, we can give a more precise description of a system in thermal equilibrium at a fixed temperature. While a system in thermodynamic equilibrium at fixed energy is equally likely to be in each of its available microstates, a system that is in thermodynamic equilibrium itself and in thermal equilibrium with an environment at fixed temperature T is characterized by a more complicated probability distribution known as the *Boltzmann distribution*. This probability distribution, often referred to as a *thermal distribution*, has a universal form $\sim e^{-E/k_B T}$ in which the probability of a given microstate depends only on the energy E of the state and the temperature T. In this section we derive this distribution. One application of this result, which we use in the following section, is to a system of many particles, such as in an ideal gas. For such a system, each particle can be thought of as being in thermal equilibrium with the rest of the system and has a probability distribution on states given by the Boltzmann distribution, even when the total energy of the gas itself is fixed.

8.6.1 Derivation of the Boltzmann Distribution

Once again, as in §8.4.3 where we gave a precise definition of temperature, we consider two systems that are allowed to come into thermal equilibrium with one another at temperature T. This time, however, as depicted in Figure 8.5, we imagine that one system, the **thermal reservoir** $\mathcal{R}$, is very large – effectively infinite – and that the other system $\mathcal{S}$ is much smaller (and could consist even of only a single particle, in which case Figure 8.5 exaggerates the size of $\mathcal{S}$ dramatically).

Together, $\mathcal{R}$ and $\mathcal{S}$ have a fixed total energy $U(\mathcal{RS}) = U(\mathcal{R}) + U(\mathcal{S})$. Energy can move back and forth between $\mathcal{S}$ and $\mathcal{R}$, so the energy apportioned to states of $\mathcal{S}$ is not fixed. The question we wish to address is then: What is the probability distribution on states of $\mathcal{S}$ when $\mathcal{S}$ is in thermal equilibrium with $\mathcal{R}$ at temperature T?

Since the combined system $\mathcal{RS}$ is in thermodynamic equilibrium and has fixed total energy, all states of the combined system with energy U are equally probable. Suppose

---

Box 8.2   Heat Engine Efficiency: Ideal vs. Actual

Many primary energy resources provide thermal energy, either directly or indirectly. For use by mechanical or electrical systems, thermal energy must be transformed into another form using a heat engine or equivalent system. The Carnot limit gives an absolute upper bound on the fraction of thermal energy that can be extracted. For most real systems, the actual efficiency is significantly below the Carnot limit. Some of the systems studied later in the book are tabulated here with (very rough) estimates of the temperature of the thermal source, the Carnot efficiency, and approximate values of the actual efficiency for typical systems in the US using standard technology. In most cases the environment can be roughly approximated as 300 K.

Energy source	T_+	T_-	η_C	$\eta_{typical}$
Ocean thermal energy	25 °C	5 °C	7%	2–3%
Steam turbine nuclear plant	600 K	300 K	50%	33%
Steam turbine coal plant	800 K	300 K	62%	33%
Combined cycle gas turbine	1500 K	300 K	83%	45%
Auto engine (Otto cycle)	2300 K	300 K	87%	25%
Solar energy (photovoltaic module)	6000 K	300 K	95%	15–20%

Throughout the rest of the book we consider many different types of thermal energy conversion, and explore the reasons why their actual efficiency falls short of the Carnot limit.

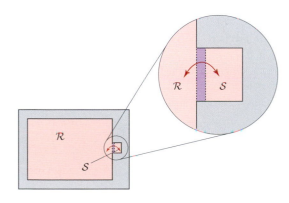

Figure 8.5 A large thermal reservoir, $\mathcal{R}$, and a small system $\mathcal{S}$ in thermal contact, but otherwise isolated from the environment.

$\mathcal{S}$ is in a state s with energy E that occurs with probability p. This leaves energy $U - E$ available to $\mathcal{R}$. According to eq. (8.11), the number of microstates of $\mathcal{R}$ with energy $U - E$ is

$$\mathcal{N}_\mathcal{R}(U - E) = e^{S_\mathcal{R}(U-E)/k_B} . \qquad (8.29)$$

Since each combined state of $\mathcal{RS}$ is equally probable, the probability of the given state s is proportional to the number of compatible microstates of $\mathcal{R}$

$$p(s) = p(E) = C e^{S_\mathcal{R}(U-E)/k_B} , \qquad (8.30)$$

where C is a constant. Note that this probability only depends on the energy E of the state s, so we can simply write it as $p(E)$.

To explicitly compute $p(E)$, we differentiate with respect to E

$$\frac{dp}{dE} = -\frac{1}{k_B} \frac{\partial S}{\partial U} \times \left(C e^{S_\mathcal{R}(U-E)/k_B} \right) = -\frac{p}{k_B T} . \qquad (8.31)$$

This simple differential equation for the probability $p(E)$ has the solution

$$p(E) = \frac{1}{Z} e^{-E/k_B T} , \qquad (8.32)$$

where Z is a constant. Since $p(E)$ is a probability, it must sum to one when all possible states of $\mathcal{S}$ are considered. This fixes Z to be

$$Z = \sum_j e^{-E_j/k_B T} , \qquad (8.33)$$

where j is an index over all possible states of the system $\mathcal{S}$. Z, which is a function of the temperature, is known as the **partition function** and is an important characteristic of systems in thermal equilibrium. The probability distribution (8.32) with this overall normalization is the **Boltzmann distribution** and the factor $e^{-E_j/k_B T}$, known as the **Boltzmann factor**, suppresses the probability of finding states with energy much greater than $k_B T$. The Boltzmann distribution gives the probability of finding the

---

The Boltzmann Distribution

If a system S is in equilibrium with a (much larger) thermal reservoir at temperature T, we say that S is *in thermal equilibrium at temperature T*. The probability of finding S in a state with energy E is then

$$p(E) = \frac{1}{Z} e^{-E/k_B T}, \text{ where } Z = \sum_j e^{-E_j/k_B T},$$

where Z is known as the *partition function*.

---

system S in a given state with energy E when S is *in thermal equilibrium at temperature T*. Note that $p(E)$ is *not necessarily* the probability of finding the system S with energy E, since there may be (and often are) many states with energy E.

8.6.2 Ensemble Averages

We have constructed several types of ensembles of states with specific probability distributions. Isolated systems of definite energy U that have reached thermodynamic equilibrium form the **microcanonical ensemble**. The system has $\mathcal{N}(U)$ microstates, each of which occurs with equal probability $1/\mathcal{N}(U)$. On the other hand, systems in thermal equilibrium with a heat reservoir at temperature T form a different ensemble known as the **canonical ensemble**. This ensemble has a probability distribution given by the Boltzmann distribution (8.32).

Often, when we observe a system, we would like to know its average properties. For example, although the precise value of the energy for a system in thermal equilibrium at temperature T changes over time as small fluctuations move energy between the system and the reservoir, we may want to know the average value of this energy. For a system in thermal equilibrium at temperature T, what we mean by the "average" of a quantity is the average over many measurements of that quantity when we pick elements at random from the *canonical ensemble*. For a system in thermal equilibrium at fixed energy, we take the average over the *microcanonical ensemble*. For an ergodic system, this is the same as the time-averaged value of the quantity.

Suppose that an ensemble is characterized by a probability distribution p_α of finding a system in (micro)state s_α, and suppose that Q is a physical quantity that takes value Q_α in state s_α. Then we define the **ensemble average** of the quantity Q by averaging over all states with the given probability distribution

$$\langle Q \rangle = \sum_\alpha p_\alpha Q_\alpha. \tag{8.34}$$

As a simple example, consider the average value rolled on a 6-sided die. Assuming that the die is fair, the probabilities are equal $p_1 = p_2 = \cdots = p_6 = 1/6$, and the values of the faces are $Q_1 = 1, Q_2 = 2, \ldots, Q_6 = 6$ so the ensemble average of the value rolled is

$$\langle Q \rangle = \frac{1}{6}(1 + 2 + 3 + 4 + 5 + 6) = 3.5. \tag{8.35}$$

Since §5, we have referred to the *internal energy* of a system at temperature T. We can now define this quantity precisely as the ensemble average of the energy in a thermodynamic system in equilibrium with a heat reservoir at temperature T,

$$\langle U \rangle = \sum_\alpha p_\alpha E_\alpha = \sum_\alpha \frac{e^{-E_\alpha/k_B T}}{Z} E_\alpha. \tag{8.36}$$

Since $Z = \sum_\alpha e^{-E_\alpha/k_B T}$, we can relate $\langle U \rangle$ to the temperature derivative of the partition function,

$$\langle U \rangle = \frac{1}{Z} k_B T^2 \frac{dZ}{dT} = k_B T^2 \frac{d \ln Z}{dT}. \tag{8.37}$$

Another example of great importance is the entropy of a thermodynamic ensemble. As in eq. (8.9), the entropy associated with an ensemble is $S = -k_B \sum_\alpha p_\alpha \ln p_\alpha$. This is precisely the ensemble average of the quantity $-k_B \ln p_\alpha$.

For a system in thermal equilibrium with a heat reservoir at temperature T the probabilities are given by the Boltzmann distribution, and we have

$$
\begin{aligned}
S &= -k_B \sum_\alpha \frac{e^{-E_\alpha/k_B T}}{Z} \ln\left(\frac{e^{-E_\alpha/k_B T}}{Z}\right) \\
&= -k_B \sum_\alpha \frac{e^{-E_\alpha/k_B T}}{Z}\left(-\frac{E_\alpha}{k_B T} - \ln Z\right) \\
&= k_B \ln Z + \frac{\langle U \rangle}{T}. \tag{8.38}
\end{aligned}
$$

The quantity $\langle U \rangle - TS$ arises often in thermodynamics and is known as the **free energy** or, more specifically, the **Helmholtz free energy**

$$F \equiv U - TS = -k_B T \ln Z. \tag{8.39}$$

As is customary, we have dropped the ensemble averaging notation, with the understanding that $U = \langle U \rangle$ and other state variables are ensemble averages. From eqs. (8.37)–(8.39), it is clear that the partition function is a very useful quantity. In the next section we see how to evaluate Z and find the thermodynamic properties of simple systems.

Energy, Entropy, and Free Energy in the Canonical Ensemble

A system in equilibrium with its environment at temperature T is described by the *canonical ensemble*. Its thermodynamic state variables fluctuate and must be interpreted as *ensemble averages*, $\langle Q \rangle = \sum_\alpha p_\alpha Q_\alpha$. The internal energy, entropy, and Helmholtz free energy of the system can all be described as ensemble averages and related to the partition function Z,

$$U = \langle U \rangle = k_B T^2 \frac{d \ln Z}{dT},$$

$$S = \langle -k_B \ln p \rangle = k_B \ln Z + \frac{U}{T},$$

$$F = \langle F \rangle = U - TS = -k_B T \ln Z.$$

8.7 The Partition Function and Simple Thermodynamic Systems

In this section we use the partition function to study the properties of some specific thermodynamic systems. We begin with a single quantum harmonic oscillator to illustrate the basic ideas involved, and then describe the thermodynamics of simple quantum solids (crystals) and gases (ideal monatomic). The harmonic oscillator and crystal are useful examples that illustrate the *freezing out* of degrees of freedom at low temperatures due to quantum effects that we have discussed in §5 in a qualitative fashion. The ideal monatomic gas gives us a quantitative understanding of the extensive nature of entropy for gases. Along the way, we also confirm the connection between the general definition of temperature given in this chapter and the more intuitive definition used earlier.

8.7.1 Thermodynamics of a Single Simple Harmonic Oscillator

The simple harmonic oscillator provides an elementary example of the Boltzmann distribution, and illustrates the way in which quantum degrees of freedom *freeze out* at low temperatures. From the spectrum of harmonic oscillator energies (7.48), we see that according to the Boltzmann distribution, the probability that an oscillator with frequency ω is in the nth excited state is

$$p_n = \frac{1}{Z} e^{-E_n/k_B T} = \frac{1}{Z} e^{-\hbar\omega(n+1/2)/k_B T}. \qquad (8.40)$$

Here the partition function, which normalizes the probability distribution, is given by

$$Z = \sum_{n=0}^{\infty} e^{-E_n/k_B T} = e^{-x/2} \sum_{n=0}^{\infty} e^{-nx} = \frac{e^{-x/2}}{1 - e^{-x}}, \qquad (8.41)$$

where $x = \hbar\omega/k_B T$, and we have used the summation formula $\sum_{n=0}^{\infty} y^n = 1/(1-y)$ when $y < 1$ (eq. (B.63)).

From eq. (8.40), we can see that the Boltzmann distribution behaves very differently for low temperatures $T \ll \hbar\omega/k_B$ ($x \gg 1$) and for high temperatures $T \gg \hbar\omega/k_B$ ($x \ll 1$). At low temperatures, the partition function is dominated by the leading term, so that the system is overwhelmingly likely to be in its lowest energy state. On the other hand, at high temperatures where $x \ll 1$, many of the states have non-negligible probability. This corresponds to the appearance of a classical degree of freedom as $k_B T \gg \hbar\omega$ (see Example 8.3).

Using the partition function (8.41) and eqs. (8.37) and (8.38), a little algebra yields expressions for the energy, entropy, and heat capacity of the oscillator,

$$U = \hbar\omega \left(\frac{1}{2} + \frac{1}{e^x - 1} \right), \qquad (8.42)$$

$$S = k_B \left(-\ln(1 - e^{-x}) + \frac{x}{e^x - 1} \right), \qquad (8.43)$$

$$C = \frac{dU}{dT} = k_B \frac{x^2 e^x}{(e^x - 1)^2}. \qquad (8.44)$$

At high temperature, where $x \to 0$ and therefore $e^x - 1 \sim x$, the leading terms in these expressions are

$$U \xrightarrow[\hbar\omega \ll k_B T]{} k_B T + \mathcal{O}(1/T), \qquad (8.45)$$

$$S \xrightarrow[\hbar\omega \ll k_B T]{} k_B \ln T + \text{constant}, \qquad (8.46)$$

$$C \xrightarrow[\hbar\omega \ll k_B T]{} k_B + \mathcal{O}(1/T^2). \qquad (8.47)$$

Thus, we see that, at high temperatures, the energy agrees with the equipartition theorem (§5.1.3) for a system with two degrees of freedom (kinetic and potential energies).

At low temperatures, where $x \to \infty$, the entropy (8.43) goes to zero, the energy (8.42) goes to the ground state energy $E_0 = \hbar\omega/2$, and the heat capacity (8.44) vanishes exponentially, $C \sim k_B x^2 e^{-x}$. These behaviors illustrate the **freezing out** of quantum degrees of freedom at low temperature. In particular, although excitations of the oscillator contribute k_B to the heat capacity at high temperature, their contribution to the heat capacity is exponentially suppressed at low temperature. This is a general characteristic of quantum systems with a gap Δ between the ground state and first excited state energies; the heat capacity is suppressed at low temperature by a factor of

Example 8.3   The Vibrational Mode of Oxygen Gas

A simple example of the freezing out of degrees of freedom at low temperature is given by the vibrational mode of the O_2 molecule, which can be approximated as a quantum harmonic oscillator with $\hbar\omega \cong 3.14 \times 10^{-20}$ J (see Figure 9.4). This gives $\hbar\omega/k_B \cong 2350$ K. In thermodynamic equilibrium at temperature T the probability p_n that an oxygen atom is in the nth excited vibrational state is given by eqs. (8.40) and (8.41),

$$p_n = y^n (1 - y),$$

where $y = e^{-\hbar\omega/k_B T} = e^{-(2350\,\text{K})/T}$. Thus, at room temperature $T \cong 300$ K, where $y = e^{-7.8} \cong 0.0004$, each molecule has a probability greater than 99.9% to be in the vibrational ground state. At 750 K, $y = e^{-3.1} \cong 0.044$ and the ground state probability is 96%. At $T = 2350$ K, $y = e^{-1} \cong 0.37$, and the ground state probability decreases to 63%, while $p_1 = y(1 - y) \cong 0.23$ is the probability to be in the first excited state.[a] These probabilities are depicted graphically in the figure below, with the area of each circle proportional to the probability that the atom is in a particular state.

A short calculation shows that the vibrational contribution to the heat capacity of a molecule of O_2 is only 2% of the equipartition value of k_B at 300 K, but by 750 K it has reached 47% of k_B, and by 2350 K it is within 8% of k_B. Note that by 750 K the heat capacity has reached nearly half its asymptotic value even though the probability of finding a molecule in an excited state is less than 5%; this corresponds to the fact that the excited states make an important contribution to the *rate of change of the energy* even at temperatures where their population is small. A similar phenomenon can be noted in the heat capacity of hydrogen gas as graphed in Figure 5.7, where $\hat{c}_V$ begins to increase above 5/2 at around 1000 K, well below the characteristic temperature of $\hbar\omega/k_B \cong 6000$ K for vibrational excitations of the hydrogen molecule.

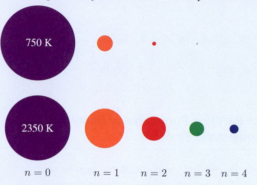

[a] For the purpose of this example we continue to model the vibrations of the O_2 molecule as a harmonic oscillator at 2350 K, even though dissociation of the molecule into oxygen atoms has begun to affect its heat capacity at that temperature.

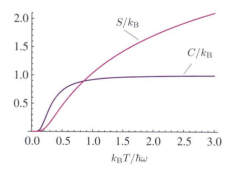

Figure 8.6 The entropy and heat capacity of a quantum harmonic oscillator as functions of $k_B T/\hbar\omega$. Both go to zero as $T \to 0$, a signature of the *freezing out* of the oscillator kinetic and potential energy degrees of freedom.

$e^{-\Delta/k_B T}$. The temperature dependence of the entropy and heat capacity of a single harmonic oscillator is shown in Figure 8.6.

The results derived in this section for a single simple harmonic oscillator are used to derive analogous thermodynamic formulae for a crystalline solid in §8.7.3 and for blackbody radiation in §22.3.

8.7.2 Many-particle Systems

Many of the thermal systems of interest to us are composed of a large number of essentially identical units, which we refer to in the following discussion as "particles." These particles might be protons in the core of the Sun or molecules of water in steam. More generally, systems can be mixtures of several types of particles, like air or salt water.

In many situations, the interactions between the particles do not contribute significantly to the energy of the total system, and the particles can be treated as independent, though the interactions often play a role in allowing the particles to come into thermal equilibrium. In such a situation, if the quantum states available to a single particle are $|a\rangle$ with energy E_a $(a = 1, 2, 3, \ldots)$ and the state of the ith particle is denoted by $|a_i\rangle$, then the total energy of a system of N particles can be written

$$E = \sum_{i=1}^{N} E_{a_i} . \qquad (8.48)$$

When such a system of independent particles is in thermal equilibrium, the probability of finding a single particle in the state $|a\rangle$ is given by the single-particle Boltzmann distribution

$$p_a = \frac{1}{z} e^{-E_a/k_B T},$$

where

$$z = \sum_a e^{-E_a/k_B T} . \qquad (8.49)$$

Here, z plays the role of the partition function for states of the single particle. This gives a simple form for the partition function of a total system consisting of N identical and independent particles with negligible interaction energies such as an ideal gas in thermodynamic equilibrium at temperature T,

$$Z = \sum_{a_1,\ldots,a_N} e^{-(E_{a_1}+\cdots+E_{a_N})/k_B T}$$

$$= \left(\sum_a e^{-E_a/k_B T} \right)^N = z^N . \qquad (8.50)$$

The expression (8.50) gives a correct description of the partition function for a system in which each of the particles is *distinguishable*, so that, for example, a configuration with the first particle in a specific state a and the second particle in state b is distinct from the configuration where the first particle is in state b and the second particle is in state a. In quantum physics, however, identical particles are *indistinguishable*. The state of an N-particle system can be labeled by the state of each particle, $|a_1, a_2, a_3, \ldots, a_N\rangle$. If particles 1 and 2 are identical, then $|a_1, a_2, \ldots\rangle$ and $|a_2, a_1, \ldots\rangle$ are equivalent quantum mechanical states. If all the particles are identical but are in *different single particle states*, then eq. (8.50) overcounts by a factor of the number of permutations of the N particles, $N! = N(N-1)(N-2)\ldots 1$, so this factor must be divided out of the partition function for the total system,

$$Z_{\text{indistinguishable}} = \frac{1}{N!} z^N = \frac{1}{N!} Z_{\text{distinguishable}} . \qquad (8.51)$$

This $N!$ correction is known as the **Gibbs factor**, and was introduced by the nineteenth-century American scientist J. Willard Gibbs. Gibbs was motivated to introduce this factor by the empirical observation that entropy is an extensive quantity, which does not hold without the $N!$ factor (see eq. (8.65) and Problem 8.15). At the time, there was some confusion as to why fundamental particles should be treated as indistinguishable, since this idea conflicts with classical notions of physical reality. In retrospect, from the information-theoretic point of view, Gibbs's result can be seen as one of the first harbingers of quantum physics.

The expression (8.51) is adequate at temperatures high compared to the spacing of quantum energy levels, where all particles are generally in distinct quantum states. If there is a non-negligible probability that some particles are in the same single-particle state, however, then the correction factor becomes more complicated. In that case, the system is called **degenerate**, and must be analyzed using different methods. At low temperatures, where particles crowd into the lowest available energy levels, two possibilities occur: for one kind of identical particle, known as *bosons*, an arbitrary number of particles can be in the same state. The other type of particle, known as *fermions*, obey the *Pauli exclusion principle*, which allows at most one particle to occupy each quantum state. Bosons, fermions, and the Pauli exclusion principle are described in more detail in §15.5. In the examples discussed in this section, either the particles are distinguishable or the temperature is high enough to ignore degeneracy. We return to the topic of *degenerate quantum gases* in later chapters.

8.7.3 Thermodynamics of a Crystal: Einstein's Model of Specific Heats

In §5.1.3 we discussed briefly the internal energy and heat capacity of an ideal three-dimensional crystal consisting of N atoms that are each free to oscillate in three directions. The *equipartition theorem* predicts that the heat capacity of such a system should be $C_V = 3Nk_B$. This is the *law of Dulong and Petit* (§5.1.3), which – as mentioned in §5 – gives a good approximation for metals like copper at room temperature. As the temperature decreases, however, measurements show that a metal's heat capacity starts to decrease rapidly and approaches zero at $T \to 0$. This gives empirical evidence that the equipartition theorem is only a high-temperature approximation. Here we re-examine the heat capacity of a metal from the point of view of statistical mechanics. We use a simple model first introduced by Einstein, who modeled a crystal as a regular array of atoms, each held in place by a spring force in each of the three dimensions with angular oscillation frequency ω_E. A one-dimensional cross section of Einstein's model is shown in Figure 8.7. The frequency ω_E defines a natural temperature scale, $T_E = \hbar\omega_E/k_B$, the **Einstein temperature**, below which quantum effects become increasingly important and deviations from the law of Dulong and Petit appear. The value of T_E for copper, for example, is about 270 K.

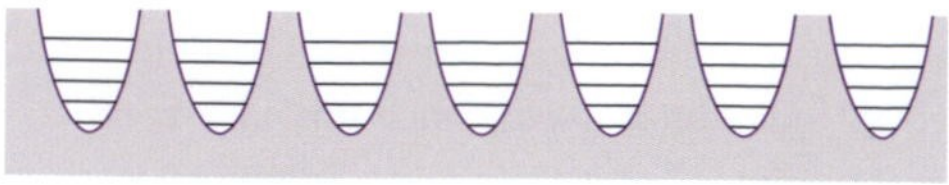

Figure 8.7 A schematic one-dimensional slice through an Einstein solid showing the harmonic oscillator potentials that hold each atom in place. After [48].

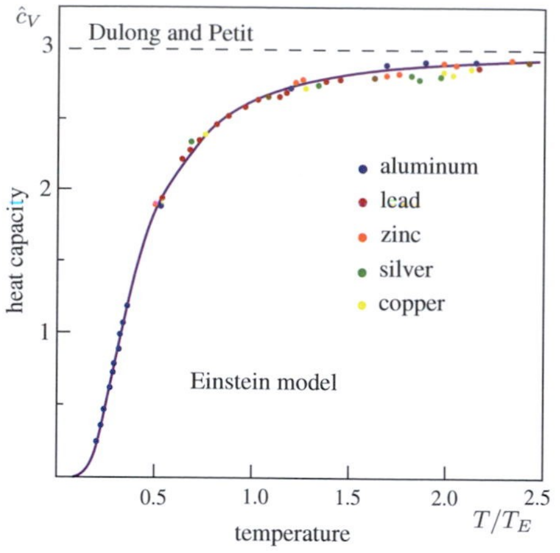

Figure 8.8 The heat capacity of some metals as a function of temperature. Temperature is given in units of the Einstein temperature, $T_E = \hbar\omega_E/k_B$. Note that the heat capacities per mole approach $3N_A k_B$ as $T \to \infty$, as predicted by the law of Dulong and Petit, but vanish as $T \to 0$, and that the Einstein model provides a good fit over the range shown. After [49].

We can use the single-particle partition function introduced in the previous section to find the thermodynamic properties of the Einstein model, because the energy of the system is simply the sum of the energies of the independent oscillators. Thus, many of the necessary formulae are simple generalizations of the results from §8.7.1 for the thermodynamics of a single quantum harmonic oscillator. In this case, we have three independent oscillators for each particle, one for each of the three dimensions, so the state is characterized by the excitation numbers l, m, n of the three oscillators, and the energy spectrum is

$$E(l, n, m) = \hbar\omega_E \left(l+m+n+3/2\right), \quad l, m, n = 0, 1, 2, \ldots.$$

$$(8.52)$$

The single-particle partition function z is therefore

$$z_{\text{3D-HO}} = z_{\text{HO}}^3 = \left(\frac{e^{-x/2}}{1 - e^{-x}} \right)^3, \qquad (8.53)$$

where $x = \hbar\omega_E/k_B T = T_E/T$. Writing $Z_E = z_{\text{HO}}^{3N}$,[7] the energy and entropy are simply those of $3N$ oscillators

$$U = 3N\hbar\omega_E \left(\frac{1}{2} + \frac{1}{e^x - 1} \right), \qquad (8.54)$$

$$S = 3Nk_B \left(-\ln(1 - e^{-x}) + \frac{x}{e^x - 1} \right), \qquad (8.55)$$

$$C = \frac{dU}{dT} = 3Nk_B \frac{x^2 e^x}{(e^x - 1)^2}. \qquad (8.56)$$

At high temperature, in parallel with the results for the single oscillator (8.47),

$$C \xrightarrow[\hbar\omega_E \ll k_B T]{} 3Nk_B, \qquad (8.57)$$

which confirms the law of Dulong and Petit at high temperature. One can furthermore verify that the entropy obeys the formula $\Delta S = \int dT\, C/T$ used in Example 8.2.

Einstein's model of the specific heat of solids is also valid at low temperatures, where quantum effects are

important.[8] This can be seen in Figure 8.8, where the heat capacities of various metals are compared with the prediction of the Einstein model. As the temperature goes well below T_E, we expect that all the oscillators should settle into their ground states, with energy $3\hbar\omega_E/2$ for each particle. This follows directly from the Boltzmann weights for the oscillator states entering eq. (8.53), and can be seen in eq. (8.54) by taking $T \to 0$.

As $T \to 0$, and all oscillators are increasingly likely to be in their ground state, we become certain of the microstate of the system. Thus, the entropy goes to zero as $T \to 0$, as can be seen from eq. (8.56). This result is often called the **third law of thermodynamics**: The entropy of a perfect crystal goes to zero as the temperature approaches absolute zero. This follows directly from our probabilistic definition of entropy. A corollary of this is that the heat capacity of a perfect crystal vanishes as $T \to 0$ (Problem 8.13).

This system also provides a further illustration of the freezing out of quantum degrees of freedom at low temperatures. The contribution of a degree of freedom to the heat capacity of a solid crystal reaches its classical value

---

[7] We do not have to worry about extra factors of $N!$ because each atom is distinguished from the others by its specific position in the crystal lattice.

[8] Actually, at *very* low temperatures Einstein's model fails and must be replaced by the slightly more complicated *Debye model* described in §9.2.1 (which also predicts that the entropy and heat capacity of a perfect crystal vanish as $T \to 0$).

Freezing Out Degrees of Freedom

When $k_B T$ is small compared to the typical spacing between energy levels of a quantum system, then the contribution of that system to the heat capacity and entropy is suppressed. The degrees of freedom are *frozen out*. When the temperature is large compared to the quantum energy scale, the degrees of freedom contribute to the heat capacity as dictated by the equipartition theorem; $\frac{1}{2} k_B$ for each degree of freedom.

of $\frac{1}{2} k_B$ only when the temperature of the system becomes large compared to the quantum level spacing.

8.7.4 Thermodynamics of an Ideal Monatomic Gas

As a final example, we return to the monatomic ideal gas, which was discussed previously in §5. In that chapter, we asserted that when the temperature is high enough that quantum discretization of energy is not relevant, the energy of each degree of freedom in the ideal gas is $\frac{1}{2} k_B T$. The internal energy of the system is thus $\frac{3}{2} N k_B T$, taking into account three directions of motion for each of N particles. Using the Boltzmann distribution and the partition function we can now derive this result from first principles.

In an ideal gas, the interparticle interactions are sufficient to enable the gas to come into equilibrium, but weak enough to give negligible contribution to the energy. We therefore model an ideal *monatomic* gas in thermal equilibrium as a collection of N non-interacting particles confined in an $L \times L \times L$ box. We quantized a particle in a box in §7.7.2, and found that the energy levels are labeled by three integers, l, m, n, with

$$E(l, m, n) = \frac{\hbar^2 \pi^2}{2m L^2} \left(l^2 + m^2 + n^2 \right). \tag{8.58}$$

Thus, the single-particle partition function is given by

$$z_{box} = \sum_{l,m,n=1}^{\infty} e^{-x(l^2 + m^2 + n^2)}, \tag{8.59}$$

where $x = \hbar^2 \pi^2 / (2m L^2 k_B T)$. Here, in contrast to the cases of the harmonic oscillator and the Einstein model of a solid, the spacing of energy levels is very small compared to $k_B T$ even at relatively low temperatures. For example, for helium atoms in a box one centimeter on a side, $\hbar^2 \pi^2 / (2m L^2 k_B) \sim 10^{-14}$ K, so even at a temperature of 1 K, $x \sim 10^{-14}$. Since the discrete contributions to the

sum in eq. (8.59) are so closely spaced we can approximate the sum by an integral with great accuracy,

$$\sum_{n=0}^{\infty} e^{-xn^2} \cong \int_0^{\infty} dn\, e^{-xn^2} = \sqrt{\frac{\pi}{4x}}. \tag{8.60}$$

The resulting partition function per particle is

$$z_{box} = \left(\frac{\pi}{4x} \right)^{3/2} = \left(\frac{m L^2 k_B T}{2\pi \hbar^2} \right)^{3/2}. \tag{8.61}$$

The probability that any two particles are in the same state is negligibly small (Problem 8.16), so the system is not degenerate and its partition function is given by eq. (8.51),

$$Z_{ideal} = \frac{1}{N!} z_{box}^N = \frac{1}{N!} \left(\frac{m L^2 k_B T}{2\pi \hbar^2} \right)^{3N/2}, \tag{8.62}$$

where, following the discussion after eq. (8.50), we have included the factor of $N!$ required when the gas particles are indistinguishable.

As in the previous examples, the internal energy and entropy follow from eqs. (8.37) and (8.38),

$$U_{ideal} = \frac{3}{2} N k_B T, \tag{8.63}$$

$$S_{ideal} = N k_B \left(\frac{3}{2} + \ln V + \frac{3}{2} \ln \frac{m k_B T}{2\pi \hbar^2} \right) - k_B \ln(N!), \tag{8.64}$$

where we have replaced L^3 by V. This result for the internal energy agrees with the equipartition theorem (§5.1.3), and confirms that the definition of temperature we have used throughout this chapter, $\partial S / \partial U = 1/T$, coincides with the phenomenological definition (5.6) of temperature used in §5.

Note that the Gibbs factor of $N!$ in Z_{ideal} does not affect the internal energy, since it appears as an additive constant in $\ln Z$ and is removed by the T derivative in the computation of U (see eq. (8.37)). The $N!$ does, however, contribute a term $-k_B \ln(N!)$ to the entropy. S_{ideal} can be simplified by using *Stirling's approximation* $\ln(N!) \approx N \ln N - N$ (B.73),

$$S_{ideal} = k_B N \left(\frac{5}{2} + \ln V + \frac{3}{2} \ln \frac{m k_B T}{2\pi \hbar^2} - \ln N \right)$$

$$= k_B N \left(\frac{5}{2} + \frac{3}{2} \ln \frac{m k_B T}{2\pi \hbar^2} - \ln(N/V) \right), \tag{8.65}$$

where N/V is the number density of particles. The importance of the extra N-dependence introduced by the factor of $N!$ is explored in Problem 8.15. This equation for S_{ideal} is known as the **Sackur–Tetrode equation**.

S_{ideal} displays the volume dependence of the entropy that we anticipated earlier in this chapter and derived in

Example 8.2: it increases by $Nk_B \ln 2$ if the volume of the system is doubled (while keeping N fixed). The temperature dependence of S_{ideal} reflects the relation to the heat capacity (see Example 8.2). One flaw in the expression for S_{ideal} is that it does not vanish as $T \to 0$; instead it seems to go to $-\infty$. This is because we ignored the quantization of the energy levels when we approximated the sum over n by an integral. This is not a physically relevant issue, however, since all gases condense to a liquid or solid at temperatures well above those where this approximation breaks down.

8.8 Spontaneous Processes and Free Energy

The laws of thermodynamics determine which physical (and chemical) processes can occur spontaneously and which cannot. In many cases the time scales for these processes are long enough that the initial and final states can be regarded as states of thermal equilibrium, characterized by state functions such as entropy. In these cases the laws of thermodynamics can be applied straightforwardly to determine which processes will occur spontaneously. This approach is particularly important in the study of chemical changes and changes of phase. Iron, for example, will rust – forming hydrated iron oxide $Fe_2O_3(H_2O)_n$ when exposed to moisture and oxygen. The time scale for this reaction is long enough that it is reasonable to consider iron, water, oxygen, and iron oxide all to be in thermal equilibrium in fixed proportions at any given time. With this assumption, it is possible to compute the entropy of these ingredients and the surrounding environment and to determine whether the 2nd Law allows this process to proceed or not.

The 1st Law requires that energy is conserved, though internal energy (including energy stored in chemical bonds – see §9) can be transformed into heat or *vice versa* in chemical and phase changes. Thus, for example, when iron rusts, heat is given off into the environment as chemical bonds between oxygen and iron are formed. Such a reaction is **exothermic**. In contrast, when water evaporates, heat from the environment is transformed into latent heat of vaporization and the surface from which the water evaporates is cooled. This is an **endothermic reaction**. Both exothermic and endothermic reactions can conserve energy, and both can occur spontaneously if they satisfy the 2nd Law.

Before turning to the conditions for spontaneity imposed by the laws of thermodynamics, we stress that these are necessary, but not sufficient conditions for a process to occur. **Activation barriers** may prevent a process from proceeding even though it is allowed by both the first and second laws of thermodynamics. Such barriers depend upon the details of the reaction and can be studied with the tools of chemistry. A good example is the combustion of carbon (e.g. coal), $C + O_2 \to CO_2$. The laws of thermodynamics allow carbon to undergo a combustion

---

Example 8.4   Internal Energy and Entropy of Helium

The Sackur–Tetrode equation (8.65) allows us to explore the way that the entropy of a simple monatomic gas such as helium changes with temperature, pressure, and volume.

First, consider heating a sample of helium from T_1 to T_2 at constant volume. We expect the entropy to increase because we are adding disorder to the system by increasing the average energy of the particles. According to eq. (8.65), $\Delta S = \frac{3}{2} Nk_B \ln(T_2/T_1)$, in agreement with the computation in Example 8.2.

Next, consider heating a sample of helium from T_1 to T_2 at constant pressure. Because the gas expands – according to the ideal gas law, $V_2/V_1 = T_2/T_1$ – the increase in information needed to specify the microstate increases even more than in the constant volume case. So we expect a greater increase in entropy in this case. From eq. (8.65) we find $\Delta S = Nk_B(\frac{3}{2} \ln(T_2/T_1) + \ln(V_2/V_1)) = \frac{5}{2} Nk_B \ln(T_2/T_1)$. We can get the same result by following the heat flow using $dS = dQ_{reversible}/T$ and conservation of energy, $dU = dQ - pdV$, or $dQ|_p = dH = C_p dT$, where H is the enthalpy (see §5). Then $\Delta S = \int_{T_1}^{T_2} C_p dT/T = C_p \ln(T_2/T_1)$, which agrees with eq. (8.65) because $C_p = C_V + Nk_B = \frac{5}{2} Nk_B$ for helium.

Next, consider doubling the pressure on a sample of helium at constant temperature. This halves its volume, reducing the set of accessible states for the gas molecules, and should decrease the total entropy. According to eq. (8.65), indeed, $\Delta S = -k_B N \ln 2$ when the volume is halved.

Finally, note that the entropy per particle S/N, which is a local property of an ideal gas, depends only on the density (N/V) and temperature, which are both also local properties, and not on V and N separately. The factor $N!$ in the partition function is needed for this form of the dependence, as seen in eq. (8.65) .

reaction in the presence of oxygen at room temperature. Nevertheless, coal does not *spontaneously* ignite because carbon atoms and oxygen molecules repel one another – an example of an *activation barrier*. Instead, for combustion to occur the coal–oxygen mixture must be heated so that the kinetic energy of the oxygen molecules overcomes the activation barrier, allowing them to get close enough to the carbon atoms to react.

Keeping the possible presence of activation barriers in mind, we investigate the implications of the laws of thermodynamics. The role of the first law is relatively simple: the process must conserve energy. We assume henceforth that this is the case, and turn to the more complex question of the effects of the 2nd Law.

8.8.1 Spontaneous Processes at Constant Volume or Constant Pressure

Assuming that energy is conserved, entropy alone determines in which direction an *isolated* system can evolve. Most processes of interest to us, however, are not isolated, but rather occur in the presence of thermal reservoirs, such as the ambient environment or the interiors of heated reactors. Under these conditions it is necessary to include not only the change in entropy of the system, but also the entropy that can be taken up from or dumped into the environment. In such circumstances, the concept of *free energy* is extremely useful. Free energy can be used to determine, among other things, whether a given process can occur spontaneously if a system is kept in contact with a heat reservoir at constant temperature. There are two versions of free energy; these apply to processes that take place at constant volume (*Helmholtz free energy*) or constant pressure (*Gibbs free energy*). We encountered Helmholtz free energy in eq. (8.39). Most processes of interest in energy physics take place at fixed pressure, so the Gibbs free energy is of greater interest to us. Therefore we focus here on the constant pressure case and quote the constant volume result at the end.

When a system undergoes a transformation such as a chemical reaction (e.g. silver tarnishing) or a change in phase (e.g. ice melting), while kept at constant T and p, the entropy of the system changes by an amount ΔS. In addition, an amount of heat $-\Delta H$ equal (but opposite in sign) to the change in enthalpy of the system is given off to the surroundings. This transfer of heat also adds entropy $\Delta S = -\Delta H/T$ to the environment. The reaction can only proceed if the total change in entropy is positive. Signs are important here: if the system transforms to a lower enthalpy state ($\Delta H < 0$), then heat is given off, so $-\Delta H$ is positive. The *total change in entropy*, including both the system and its surroundings, is

Gibbs and Helmholtz Free Energy

Gibbs free energy is a state function defined by $G = H - TS$. A reaction can occur spontaneously at constant temperature and pressure if the change in G is negative, $G_{products} - G_{reactants} < 0$. Helmholtz free energy, $F = U - TS$, plays the same role at constant temperature and volume.

The change in enthalpy and free energy have been tabulated for many physical and chemical reactions including chemical reactions, mixing of solids, liquids, and gases, and dissolving of solids in water and other liquids. This data makes it possible to analyze the implications of thermodynamics for these processes.

$$\Delta S_{tot}|_{T,p} = \Delta S_{system} + \Delta S_{surroundings}$$
$$= \Delta S - \frac{\Delta H}{T}, \tag{8.66}$$

where $|_{T,p}$ denotes the fact that this result holds at constant p and T. The condition that a reaction can occur spontaneously, $\Delta S_{tot} > 0$, is then

$$\Delta S - \frac{\Delta H}{T} > 0, \quad \text{or} \quad \Delta H - T\Delta S < 0. \tag{8.67}$$

To make use of eq. (8.67), we define a new *state function*, the **Gibbs free energy**

$$G \equiv H - TS. \tag{8.68}$$

At *constant p and T*, a transformation can occur spontaneously only if

$$\Delta G = \Delta H - T\Delta S < 0. \tag{8.69}$$

The implications of eq. (8.69) can be quite counterintuitive. Even an *endothermic* reaction – one in which the reactants absorb enthalpy from their surroundings ($\Delta H > 0$) – can occur if the entropy of the system increases enough. A commercially available *cold pack* is a familiar example: it consists of ammonium nitrate (NH_4NO_3) and water, initially isolated from one another. When the package is crushed, the NH_4NO_3 mixes with the water, dissolves and absorbs heat from the environment. The reaction proceeds because the change in entropy as the NH_4NO_3 dissolves is large and positive, so even though ΔH is positive, $\Delta G = \Delta H - T\Delta S < 0$.

If the process of interest is restricted to occur at constant volume, then everywhere that *enthalpy* appeared in the preceding discussion, *internal energy* should be substituted instead. The criterion for a spontaneous reaction at constant temperature and volume is

$$\Delta S_{tot}|_{T,V} = \Delta S_{system} + \Delta S_{surroundings}$$

$$= \Delta S - \frac{\Delta U}{T} > 0 . \qquad (8.70)$$

From the definition (8.39) of Helmholtz free energy, $F \equiv U - TS$, at *constant V and T* this criterion is

$$\Delta F = \Delta U - T\Delta S < 0 . \qquad (8.71)$$

In §8.6.2 we found that $F = -k_B T \ln Z$. Because of the simple relation between F and Z, if we wish to compute G, it is often easier to compute F from fundamental principles and then use $G = F + pV$ to obtain G.

We return to the subject of free energy and use it extensively when we discuss energy in matter in §9. Free energy plays a role in energy systems ranging from batteries (§37.3.1) to carbon sequestration (§35.4.2).

Discussion/Investigation Questions

8.1 In the game of *Scrabble*™, the letters of the English alphabet are inscribed on tiles and a prescribed number of tiles are provided for each letter. Consider an ensemble of *Scrabble*™ tiles with a probability distribution defined by the frequency of tiles in the box. Explain how to calculate the information entropy of this ensemble. It is not necessary to actually compute the entropy, but you can if you wish.

8.2 Explain why the statements "No cyclic device can transform heat into work without expelling heat to the environment" and "No device can move thermal energy from a low-temperature reservoir to a high-temperature reservoir without doing work" follow from the laws of thermodynamics.

8.3 Living systems are highly ordered and therefore relatively low entropy compared to their surroundings; for example, a kilogram of wood from a pine tree has less entropy then the corresponding masses of water and carbon dioxide placed in a container in equilibrium at room temperature. When living systems grow they expand these highly ordered (relatively low entropy) domains. How can this be reconciled with the second law of thermodynamics?

8.4 An inventor claims to have constructed a device that removes CO_2 from the air. It consists of a box with an opening through which the ambient air wafts. The CO_2 is separated out by a complex system of tubes and membranes, and ends up in a cylinder, ready for disposal. The device consumes no energy. Do you think that this is possible?

8.5 According to the Sackur–Tetrode formula (8.65), the entropy of an ideal monatomic gas grows with the mass of the atoms. Thus the entropy of a volume of argon is much greater than the same volume of helium (at the same T and p). Can you explain this in terms of information entropy?

Problems

8.1 What is the information entropy of the results of flipping a biased coin 1000 times, if the coin comes up tails with probability 5/6 and heads with probability 1/6? For this weighted coin, can you find an encoding for sequential pairs of coin flips (e.g. HH, HT, etc.) in terms of sequences of bits (possibly different numbers of bits for different possible pairs) that takes fewer than 2 bits on average for each pair of flips? How many bits on average does your encoding take, and how close is it to the information-theoretic limit?

8.2 What is the information entropy of the results of spinning a roulette wheel with 38 possible equally likely outcomes (36 numbers, 0 and 00) 10 times? How many fair coins would you need to flip to get this much information entropy?

8.3 Consider an amount of helium gas at atmospheric pressure and room temperature (you can use $T = 300\,\text{K}$) enclosed in a 1-liter partition within a cubic meter. The volume outside the partition containing the helium is evacuated. The helium gas is released suddenly by opening a door on the partition, and flows out to fill the volume. What is the increase in entropy of the gas?

8.4 **[T]** Prove analytically that the entropy of a two-state system (8.12) is maximized when $p_+ = p_- = 1/2$.

8.5 Consider air to be a mixture of 78% nitrogen, 21% oxygen, and 1% argon. Estimate the minimum amount of energy that it takes to separate a cubic meter of air into its constituents at STP, by computing the change in entropy after the separation and using the 2nd Law.

8.6 A heat engine operates between a temperature T_+ and the ambient environment at temperature 298 K. How large must T_+ be for the engine to have a possible efficiency of 90%? Can you find some materials that have melting points above this temperature, out of which you might build such an engine?

8.7 A nuclear power plant operates at a maximum temperature of 375 °C and produces 1 GW of electric power. If the waste heat is dumped into river water at a temperature of 20 °C and environmental standards limit the effluent water to 30 °C, what is the minimum amount of water (liters per second) that must feed through the plant to absorb the waste heat?

8.8 In a typical step in the CPU of a computer, a 32-bit register is overwritten with the result of an operation. The original values of the 32 bits are lost. Since the laws of physics are reversible, this information is added to the environment as entropy. For a computer running at 300 K that carries out 10^{11} operations per second, give a lower bound on the power requirement using the second law of thermodynamics.

8.9 **[T]** A simple system in which to study entropy and the Boltzmann distribution consists of N independent two-state subsystems coupled thermally to a reservoir at temperature T. The two states in each subsystem have energies 0 (by definition) and ϵ. Find the partition function, the internal energy, and the entropy of the total system as a function of temperature. Explain why the internal energy approaches $N\epsilon/2$ as $T \to \infty$. Is this what you expected? It may help if you explain why the entropy approaches $Nk_B \ln 2$ in the same limit.

8.10 Repeat the calculations of Example 8.3 for the H_2 molecule, with $\hbar\omega/k_B \sim 6000$ K. In particular, compute the probabilities that a given molecule is in the ground state or first excited state at temperatures $T = 500, 1000$, and 2000 K, compute the vibrational contribution to the heat capacity at each of these temperatures, and compare your results with Figure 5.7.

8.11 Consider a simple quantum system consisting of twenty independent simple harmonic oscillators each with frequency ω. The energy of this system is just the sum of the energies of the 20 oscillators. When the energy of this system is $U(n) = (n + 10)\hbar\omega$, show that its entropy is $S(n) = k_B \ln((n + 19)!/(19!\,n!))$. What is the entropy of the system in thermal equilibrium for $n = 1, 2$, and 3? Assume that $\hbar\omega \cong 0.196$ eV $\cong 3.14 \times 10^{-20}$ J, corresponding to the vibrational excitation mode of O_2. Beginning in the state with $U(1) = 11\hbar\omega$, by how much does the entropy increase if the energy is increased to $U(2) = 12\hbar\omega$? Use this result to estimate the temperature of the system by $1/T \equiv \Delta S/\Delta U$. Repeat for the shift from $U(2)$ to $U(3)$, and estimate the heat capacity by comparing the change in U to the change in estimated T. Compare the heat capacity to k_B to compute the effective number of degrees of freedom. Do your results agree qualitatively with the analysis of Example 8.3?

8.12 **[T]** Check the results quoted in eqs. (8.42)–(8.44) for a single harmonic oscillator. Show that as $T \to 0$, the entropy and the heat capacity both vanish. Show that $S(T_1) - S(T_0) = \int_{T_0}^{T_1} C(T)dT/T$.

8.13 **[T]** Show that in general the condition that $S(T_1) - S(T_0) = \int_{T_0}^{T_1} C_V(T)dT/T$ and the requirement that the entropy be finite at all temperatures (we cannot be infinitely ignorant about a finite system), require that $C(T) \to 0$ as $T \to 0$ for all systems.

8.14 **[T]** Use the Sackur–Tetrode equation (8.64) to show that $(\partial S/\partial U)|_V = 1/T$ for an ideal monatomic gas. Likewise show that $(\partial S/\partial H)|_p = 1/T$, verifying eq. (8.18) and the assertion of footnote 6.

8.15 **[T]** In §8.7.2 it was asserted that in order for entropy to be an extensive quantity, the classical partition function of a system of indistinguishable particles must be divided by $N!$. Consider a box evenly partitioned into two halves, each with volume V, and each containing N atoms of a monatomic ideal gas at a temperature T. Compute the entropy from eq. (8.64) omitting the last term. Now remove the partition, and compute the entropy of the resulting system of $2N$ particles in a volume $2V$ again from eq. (8.64) with the last term omitted. *We have not lost any information about the system*, but the entropy seems to have increased if we compute it without the Gibbs factor! Next, restore the $N!$ factor and repeat the analysis.

8.16 **[T]** Compute the *partition function per particle* (8.61) for helium gas at NTP confined in a cube of side 1 cm. What quantum state has the highest probability of being occupied? What is the probability that a given helium atom is in this state? Given the number of atoms in the box, what is the expected average occupation number of this state?

8.17 **[T]** Real crystals contain impurities, which lead to non-zero entropy at $T = 0$. Consider a crystal consisting of N atoms. The crystal is primarily composed of element A but contains $M \ll N$ atoms of an impurity B. An atom of B can substitute for an atom of A at any location in the crystal. Compute the entropy of this crystal at $T = 0$.

8.18 **[T]** Show that when a system comes into thermal equilibrium while in contact with a heat reservoir at temperature T, its free energy (Helmholtz, if V is held fixed; Gibbs if p is held fixed) is minimized.

8.19 **[T]** Show that the free energy of an ideal monatomic gas of N_3 particles in a volume V can be written as $F = -k_B N_3 T \left(1 - \ln(N_3\lambda^3/V)\right)$, where $\lambda = \sqrt{2\pi\hbar^2/mk_B T}$ is a *thermal length scale*. Show that a *two-dimensional* ideal gas of N_2 particles moving in a bounded area A has free energy $F_2 = -k_B N_2 T \left(1 - \ln(N_2\lambda^2/A)\right)$.

8.20 **[H]** The atoms of an ideal monatomic gas of N particles confined in a box of volume V can be *adsorbed* onto the surface of the box (surface area A), where they are bound to the surface with binding energy ϵ, but they can move around on the surface like an ideal two-dimensional gas. Use the results of Problem 8.19 to write the (Helmholtz) free energy of this system as a function of N_3 and N_2, the numbers of gas atoms in the bulk and the surface respectively. Minimize the free energy (Problem 8.18) to find the ratio N_2/N_3 when the system reaches thermal equilibrium. What happens in the limits where $T \to 0$, $T \to \infty$, and $\hbar \to 0$?

Energy in Matter

A hot water bottle, a chocolate bar, a liter of gasoline, a kilogram of uranium reactor fuel – all represent relatively concentrated sources of energy. In each case, the energy is in a form that is not as obvious as the kinetic energy of an automobile or the potential energy of water held behind a dam. Instead, it is conveniently stored in the internal structure of matter: in the motions of individual hot water molecules, the chemical bonds of hydrogen, carbon, and oxygen in gasoline and chocolate, or in the configuration of protons and neutrons in a uranium nucleus. With the right tools we can transform this internal energy into other forms and put it to use. The amount of energy that can be stored in or extracted from matter varies greatly depending upon the type of material structure involved. Only about 2.2 J is extracted by cooling a gram of gasoline by 1 °C, while burning the same gram would yield about 44 kJ, and annihilating a gram of gasoline with a gram of anti-gasoline (if you could do it) would yield 180 TJ, equivalent to over four days' output of a 1 GW power plant.

The interplay between the internal energy stored in matter and other forms of energy is ubiquitous, yet it is easy to lose sight of its central role in energy use and transformation. All the energy extracted from fossil fuels and biofuels comes from internal energy stored in chemical bonds. Such bonds in turn are formed through photosynthesis, which converts radiant energy from the Sun into the molecular bond energy of sugar and other molecules. Batteries store energy in the atomic binding energy of metals and the molecular binding energy of salts. *Heat engines* such as the Carnot and Stirling engines that we analyze in the next chapter (§10) transform energy back and forth between thermal energy in matter and mechanical energy. Nuclear energy is extracted from reactions that modify the binding energy of nuclei and liberate the excess as radiation and heat.

> **Reader's Guide**
> This chapter focuses on the storage and transformation of energy in materials. These topics are largely the subject of chemistry, so this chapter has considerable overlap with parts of an introductory college level chemistry course.
>
> The chapter consists of two parts. First is a tour of places and forms in which internal energy is stored in matter, from the thermal energy of molecular motion to nuclear binding. Second is a quick overview of how energy is transformed in chemical reactions, including such important processes as combustion.
>
> Prerequisites: §5 (Thermal energy), §7 (Quantum mechanics), §8 (Entropy and temperature). The background on thermodynamics presented in these earlier chapters is indispensable for this chapter, particularly the concepts of (thermodynamic) state functions, internal energy, enthalpy, entropy, free energy, and the connection between energy and frequency for photons – the quanta of electromagnetic radiation.

In the first part of this chapter (§9.1–§9.3) we explore the internal energy of matter by drilling down into its substructure to expose where energy is stored and how it can be released. It is useful to have a tangible system in mind to keep us on track – ours will be a *block of ice* like the one in Figure 9.1. We begin in §9.1 by characterizing the landscape of energy scales using three different measures: energy itself, temperature, and the frequency or wavelength of light. This gives us the opportunity to consider the spectrum of electromagnetic radiation in terms of the energies and frequencies of photons that are emitted or absorbed when the internal energy of matter changes. Next, in §9.2 and §9.3 we take a tour of the contributions to the internal energy of the block of ice. We start at 0 K and add energy, revealing successively deeper

Figure 9.1 What are the contributions to the internal energy of a block of ice? (Credit: Blue Lotus, reproduced under CC-BY-SA 2.0 license)

layers of structure and finding higher scales of energy as we go. We pause in some places along the tour to explain some of the details of molecular and atomic structure that we need for future topics.

The second part of this chapter (§9.4–§9.5) explores how energy, enthalpy, and free energy flow in chemical reactions. In §9.4 we introduce the concept of *enthalpy of formation* and show how to use it to compute the heat given off or absorbed by chemical reactions. We also describe the *free energy of reaction*, which tells us when a reaction can occur spontaneously. Finally in §9.5 we apply these concepts by studying some explicit examples of chemical reactions that are relevant in energy processes. For a more comprehensive introduction to these subjects, we recommend an introductory level college chemistry text such as [50].

9.1 Energy, Temperature, and the Spectrum of Electromagnetic Radiation

Quantum mechanics and thermodynamics introduce two natural ways to measure energy. Quantum mechanics connects energy and frequency via Planck's formula

$$E = \hbar\omega = h\nu. \qquad (9.1)$$

This relation is particularly relevant when electromagnetic energy is radiated as a system makes a transition from one quantum energy level to another. In this case, the wavelength and frequency of the emitted radiation are associated through the Planck relation with the energy scale E of the radiated photons

$$\lambda = \frac{c}{\nu} = \frac{hc}{E} \cong \frac{1.24 \times 10^{-6}}{E[\text{eV}]} \, \text{m}. \qquad (9.2)$$

For example, when an electron in a hydrogen atom shifts from the $n = 2$ level to the $n = 1$ level, it emits a photon carrying energy $E = 13.6 \times (1 - 1/4) \cong 10.2 \, \text{eV}$ (see eq. (7.52)). From eq. (9.2), this radiation has a wavelength of about 1.22×10^{-7} m, which is in the ultraviolet part of the electromagnetic spectrum (see Figure 9.2).

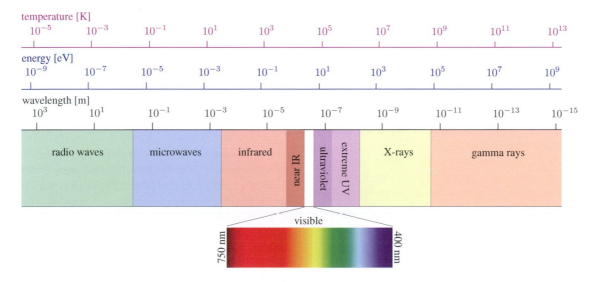

Figure 9.2 The spectrum of electromagnetic radiation. The relationships among temperature, wavelength, and energy, given by $E = hc/\lambda$ and $E = k_\text{B}T$ are shown. The exact boundaries between different forms of electromagnetic radiation – between infrared and microwave, for example – are imprecisely defined.

Thermodynamics, on the other hand, relates energy and temperature via Boltzmann's constant,

$$T = \frac{E}{k_B} \cong E[\text{eV}] \times 1.16 \times 10^4 \text{ K}. \qquad (9.3)$$

The significance of this relation lies in its role in the Boltzmann distribution. A system at temperature T has a characteristic energy scale of order $k_B T$. Excitations of states with energies much lower than this value will be common, while excitation of states with energies far higher will be substantially suppressed by the Boltzmann factor $e^{-E/k_B T}$. Thus, for example, in a gas at temperature T, collisions between molecules have enough energy to excite quantum states of the molecule with energies of order $k_B T$ but not much higher.

As we tour the possible internal excitations of matter in the following sections, we exploit the connections between radiation wavelength, temperature, and energy to determine the circumstances in which various forms of internal energy are most important. For example, excitations of electrons from one orbital to another within an atom typically involve energies of a few electron volts. We can immediately conclude that transitions like these emit and absorb *visible light* with wavelengths of a few hundred nanometers. Similarly, we can conclude that to significantly excite these transitions thermally requires temperatures of order 10^4 K.

If we combine the quantum and thermodynamic relations (9.1, 9.3), we can define a **thermal wavelength**

$$\lambda_{\text{th}} = \frac{hc}{k_B T} . \qquad (9.4)$$

In §5 we introduced the notion of *thermal radiation*, which is discussed in considerably greater detail in §22 (Solar energy). The physical significance of λ_{th} is that it sets the scale for the wavelengths of light emitted as thermal radiation from an object with temperature T. For example, the power emitted by an object at temperature T is peaked in the frequency range around $\omega_{\text{max}} \cong 2.82\,\omega_{\text{th}}$ (as we prove in §22), where $\omega_{\text{th}} = 2\pi c/\lambda_{\text{th}}$.

Figure 9.2 shows the relationship between wavelength, energy, and temperature graphically, and identifies the colloquial name for the type of electromagnetic radiation associated with given wavelengths.

9.2 A Tour of the Internal Energy of Matter I: From Ice to Vapor

What is the energy content of a block of ice? The answer depends on the context. A huge amount of energy would be given off in the assembly process if we were able to construct the block of ice out of a collection of initially separated protons, neutrons, and electrons – the fundamental constituents of matter. In this context we could say that the *internal energy* of the ice is very large and negative, because energy must be *added* to separate the ice back into its constituents. At the other extreme, if we are interested, for example, in the physics of ice in our oceans and atmosphere, it would be more appropriate to consider the energy content of the ice at a temperature T to be the (much smaller and positive) quantity of *thermal energy* that is needed to heat it from absolute zero to temperature T. Because we only measure *energy differences*, it is often convenient to ignore those components of the internal energy that are not involved in a specific process.

To give a full picture, we begin here by considering *all* of the contributions to the internal energy of a system. We can then focus attention on the components of interest for any particular process. We imagine starting with the block of ice in the state where it has the lowest possible internal energy while preserving the identities of its atomic constituents – at temperature absolute zero – and then slowly adding the energy needed to deconstruct it into its constituent parts. This leads us first to *warm*, then *melt*, then warm some more, and then to *vaporize* the ice into a gas of water molecules. Adding more energy *dissociates* the molecules into hydrogen and oxygen atoms. Still more energy ionizes the electrons from the atoms yielding eventually a fully ionized *plasma* of electrons, protons (the nuclei of hydrogen), and oxygen nuclei. Even more energy would eventually break up each oxygen nucleus into eight protons and eight neutrons. These steps and the associated energy scales are summarized in Table 9.1, which may serve as a reference as we work our way through the deconstruction of our ice cube. We have already discussed the initial stages of warming, melting, and vaporization in §5. Here we revisit these processes, focusing on a deeper understanding of what is happening at the molecular level.

Energy, Temperature, and Wavelength

Fundamental aspects of quantum mechanics and thermodynamics relate energy, temperature, and wavelength (or frequency),

$$E = \hbar\omega = \frac{hc}{\lambda} = k_B T .$$

These relationships connect the energy scale of electromagnetic radiation to quantum spectra, transitions, and thermal processes.

Table 9.1 Contributions to the internal energy of matter: Deconstructing a block of ice by adding energy.

Energy	Source	Physical process	Actual energy input per H_2O molecule
Thermal	Motions of molecules	Warm	0.39/0.79/0.35 meV/molecule K (water solid/liquid/vapor at 0 °C)
Latent heat	Intermolecular binding energy	Melt/ vaporize	0.062 eV/molecule 0.42 eV/molecule
Molecular binding	Binds atoms into molecules	Decompose	9.50 eV/molecule
Atomic binding	Binds electrons	Ionize	2.1 keV/molecule
Nuclear binding	Binds protons and neutrons into nuclei	Disassemble into p and n	~140 MeV/molecule
Rest mass	The mass itself!	Annihilate	17 GeV/molecule

In the next section we describe the processes of molecular dissociation, ionization, and nuclear dissociation that occur when additional energy is added. Note that mass energy, also contained in Table 9.1, is somewhat different from the other types of energy discussed here, in that mass energy cannot be changed simply by heating the block of ice, as discussed in §9.3.4.

The notion, introduced in §8.7.3, that degrees of freedom are *frozen out* by quantum mechanics at low temperatures simplifies the description of matter. In particular, it allows us to ignore levels of structure at energies higher than the scale that we wish to probe. Specifically, any given aspect of internal structure can be ignored as long as $k_B T$ is very small compared to the minimum energy needed to excite that structure. The fact that atoms themselves are complex systems that can be excited or torn apart, for example, is thus of no significance when we add heat to a block of ice at a very low temperature. Similarly, the internal structures of nuclei do not matter while we are unravelling the molecular and atomic structure of water.

9.2.1 Thermal Energy (Heating)

We start with a block of ice at absolute zero. The molecules of H_2O are held in place by intermolecular forces. In practice, a real block of ice will not be a perfectly formed crystal; imperfections in the crystal structure that decay only over very long time scales raise the energy slightly above the quantum ground state, but we can ignore this subtlety for the discussion here.[1] Also, note that in the

lowest energy state there is nominally a quantum zero-point energy (like the $\hbar\omega/2$ ground state energy of the quantum harmonic oscillator), but this just gives a constant shift to the system energy, so we can take the ground state energy to define the zero point of energy for the system.

As thermal energy is added to the ice, it goes initially into the excitations that cost the least energy, which in the case of a crystalline solid are vibrations of the lattice. Other excitations, such as those that cause a molecule to rotate or move through the lattice, require breaking the bonds that hold the water molecules in place in the solid and therefore involve more energy than vibrational modes. These higher-energy excitations are frozen out until the block of ice reaches a higher temperature. Excitations of the individual electrons in the atoms of hydrogen and oxygen require higher energy still (see §9.3.2).

We assume that the block of ice is kept at a constant pressure (say one atmosphere) as it is heated. The heat capacity at constant pressure, C_p, measures the rate of change of the *enthalpy* of the system. The molar heat capacity $\hat{C}_p$ for ice is shown in Figure 9.3 over the range 0 K to 270 K.

The heat capacity of a crystalline solid must vanish as $T \rightarrow 0$ (see Problem 8.13), so at first the temperature increases very quickly as energy is added. According to the *Debye model*,[2] which improves on the Einstein

---

[1] There is a further complication in this story from the fact that there are actually many different (16 known) solid phases of water ice, some of which can form at low temperatures at atmospheric pressure. We also ignore this subtlety here, which is irrelevant for the rest of the discussion in this chapter.

[2] Einstein's model, presented in the previous chapter, is based on the idea that all bonds are characterized by a single frequency ω_E. It predicts that heat capacities of solids vanish exponentially at small T. Instead, they are observed to vanish like T^3. A more sophisticated model with a range of frequencies is required to describe this behavior. Soon after Einstein, the Dutch physicist Peter Debye constructed such a model, which includes a range of vibrational frequencies. Like

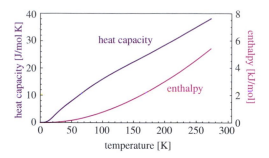

Figure 9.3 The (molar) heat capacity $\hat{C}_p$ and enthalpy of ice relative to absolute zero, as a function of temperature from 0 K to 270 K [51].

model for specific heats (§8.7.3), the heat capacity grows proportionally to T^3 at small T. Specifically, $\hat{c}_p \cong \hat{c}_V = (12\pi^4/5)(T/T_D)^3$ for $T \ll T_D$, where the *Debye temperature* T_D corresponds to the highest-energy lattice vibrations available. T_D for ice is approximately 222 K [51], corresponding to a energy of about 0.02 eV (from eq. (9.3)). The roughly cubic growth of the heat capacity as a function of T captured by the Debye model holds only up to around $0.1\,T_D \cong 20$ K, above which temperature the Einstein and Debye models roughly match (Problem 9.4), and the heat capacity becomes approximately linear in T.

By the time we reach 0 °C, the heat capacity of ice at atmospheric pressure is about 38 J/mol K, corresponding to $\hat{c}_p \cong 4.57$. The total increase in the enthalpy of ice as it is heated from 0 K to 0 °C is $\Delta H = \int_{0\,\mathrm{K}}^{273\,\mathrm{K}} dT\,\hat{C}_p(T) \cong$ 5.4 kJ/mol. Ice has a rather large, but not exceptional, heat capacity for a solid. As noted in §5 (and derived in §8), the heat capacities of the simplest solids – pure elements like copper, gold, iron, etc. – start at zero at $T = 0$, and increase to approximately $3R \cong 25$ J/mol K, at which point their vibrational motions contribute fully to the heat capacity. Because their degrees of freedom are limited, simple solids do not, in general, make good repositories for storing heat. This is reflected in their small heat capacities.

---

the Einstein model, the **Debye model** depends on a single parameter, the **Debye temperature** T_D. The Debye temperature corresponds to the temperature at which the most energetic (shortest wavelength) modes of the lattice are excited. At temperatures below the Debye temperature, the solid acts as a *phonon gas*, and is described in a very similar fashion to a *photon gas*, which is analyzed in §22.3. For solids where the Einstein model is good, T_D and T_E agree up to a constant factor of order one.

9.2.2 Melting and Vaporization

When the block of ice has been warmed to 0 °C, it begins to undergo a *phase transition* as described in §5. The temperature ceases to increase while the solid ice changes to liquid water as thermal energy is added. The energy that is put into a solid when it melts goes into loosening the grip that the molecules exert on one another. These intermolecular forces are weaker than the forces that bind atoms into molecules (see §9.3.2). Furthermore, when a material melts, the molecules remain relatively close to one another, so the density does not change substantially and the intermolecular bonds do not entirely break.

The energy needed per molecule for the solid-to-liquid transition can be obtained by expressing the *enthalpy of fusion* (§5.6.2) in units of eV/molecule.[3] For water, $\Delta H_{\mathrm{fusion}} \cong 6.01$ kJ/mol $\cong 0.062$ eV/molecule. This is greater than all of the internal energy added to ice in warming it from 0 K to its melting point ($\cong 5.4$ kJ/mol = 0.056 eV/molecule). The energy per molecule required to melt ice is several times greater than the energy associated with vibrations of the crystal lattice ($k_B T_D \approx 0.02$ eV), but roughly two orders of magnitude smaller than a typical chemical bond energy (see below). Like heating it, melting a solid is not a particularly efficient way to store energy. The latent heats of fusion of some other common or interesting materials are listed in Table 9.2.

Once the block of ice is completely melted, if we continue to heat the resulting liquid water at constant pressure, the temperature once again increases, and the enthalpy and internal energy of the water increase with temperature in proportion to the heat capacity of liquid water (§5.4). Note that the heat capacity of liquid water, given in various units as $c_V \cong 4.18$ kJ/kg K, $\hat{C}_V \cong 75$ kJ/mol K, $\hat{c}_V \cong \hat{c}_p \cong 9$, or $\hat{c}_V k_B \cong 0.78$ meV/molecule, is about twice that of the solid ice phase (at 0 °C). At 100 °C we encounter another first-order phase transition, as the liquid water begins to vaporize. As emphasized earlier, one of water's remarkable characteristics is its relatively large *latent heat (enthalpy) of vaporization*. ΔH_{vap} for water is compared with the enthalpy of vaporization for some other substances in Table 9.3. Hydrogen bonding between H_2O molecules provides unusually strong intermolecular forces that must be overcome in order to vaporize water. These bonds are also responsible for water's high heat capacity as a liquid, and for its rather high boiling point compared to other substances with comparable atomic weight, such as ammonia and methane (see Table 9.3). Converting the enthalpy of

---

[3] The conversion factor 1 eV/molecule = 96.49 kJ/mol is useful here.

Table 9.2 Latent heat (enthalpy) of fusion for some common or interesting substances in decreasing order (kJ/kg). All measurements have been made at $p = 1$ atm. Notice that enthalpies of fusion are generically less than enthalpies of vaporization, Table 9.3.

Substance	Chemical formula	Molecular weight (g/mol)	Melting point at $p = 1$ atm	Heat of fusion (kJ/kg)	Molar heat of fusion (kJ/mol)
Water	H_2O	18	273 K	334	6.01
Ammonia	NH_3	17	195 K	332	5.65
Sodium nitrate	$NaNO_3$	85	583 K	185	15.7
Methanol	CH_3OH	28	175 K	113	3.16
Methane	CH_4	16	91 K	59	0.94
Argon	Ar	40	84 K	30	1.2
Mercury	Hg	201	234 K	11	2.29

Table 9.3 Latent heat (enthalpy) of vaporization for some common or interesting liquids in decreasing order (kJ/kg). All measurements have been made at $p = 1$ atm. R-134a (tetrafluoroethane) is a commonly used, ozone-friendly refrigerant. Note the exceptionally high enthalpy of vaporization of water per unit mass.

Substance	Chemical formula	Molecular weight (g/mol)	Boiling point at $p = 1$ atm	Heat of vaporization (kJ/kg)	Molar heat of vaporization (kJ/mol)
Water	H_2O	18	373 K	2257	40.7
Ammonia	NH_3	17	240 K	1371	23.4
Methanol	CH_3OH	28	338 K	1104	35.3
Methane	CH_4	16	112 K	510	8.18
Hydrogen	H_2	2	20 K	452	0.46
Mercury	Hg	201	630 K	294	59.1
R-134a	$(CH_2F)(CF_3)$	120	247 K	217	26.0
Argon	Ar	40	87 K	161	6.43

vaporization of water into atomic units, 40.65 kJ/mol $\rightarrow$ 0.42 eV/molecule. This is roughly an order of magnitude greater than typical energies of fusion, although still an order of magnitude smaller than typical chemical bond energies.

Because water expands greatly when it vaporizes, there is a significant difference between the enthalpy of vaporization (the heat delivered to the water at constant pressure) and the increase in internal energy. We computed the difference, $\Delta H_{vap} - \Delta U_{vap} = p\Delta V_{vap}$, in Example 5.6, where we found $\Delta U \cong 2087$ kJ/kg, which, compared with $\Delta H_{vap} \cong 2257$ kJ/kg, gives a difference of about 7.5%.

> **Melting and Vaporization**
>
> Melting (fusion) and vaporization are first-order phase transitions. Although heat is added, the temperature stays fixed until all material has changed phase. Enthalpies of vaporization are typically greater than enthalpies of fusion. Typical enthalpies of fusion and vaporization are roughly one to two orders of magnitude smaller than the energy that binds atoms into molecules. Water has an exceptionally large enthalpy of vaporization.

9.3 A Tour of the Internal Energy of Matter II: Molecular Vibrations, Dissociation, and Binding Energies

Our block of ice is just a memory. It has melted and vaporized. The molecules of water are now free to move around and to rotate. These degrees of freedom contribute to the heat capacity of water vapor as described in §5. Further addition of energy excites vibrations of the water molecules that eventually lead to their dissociation into hydrogen and oxygen atoms. Still further addition

(a)   (b)

Figure 9.4 (a) A sketch of the O_2 molecule in which the interatomic separation and atomic size are represented in the correct proportion. (b) The vibrational mode of O_2.

of energy separates the electrons from the nuclei of the hydrogen and oxygen atoms.

9.3.1 Molecular Vibrations and Spectra

Atmospheric absorption of infrared radiation by rotational and vibrational modes of H_2O and CO_2 gives rise to the *greenhouse effect* (§24, §34) that keeps Earth 20 °C or more warmer than it would be otherwise (see Example 34.1). Thus, it is important for us to understand what is happening at the molecular level when these modes are excited, as well as when H_2O dissociates into hydrogen and oxygen. The rotational excitations of water vapor are described in Box 9.1. We begin an analysis of vibrational dynamics by considering the simple case of a diatomic molecule like O_2, which is shown schematically along with its vibration mode in Figure 9.4. The physics is quite similar for H_2O and CO_2.

Molecules are bound by *interatomic forces*, which are attractive at long distances but strongly repulsive when the atoms try to get too close. The force between the two atoms in a *diatomic molecule*, which is essentially quantum in nature, can be described approximately by a potential $V(r)$, where r is the separation of the nuclei. A very useful model is given by the **Morse potential**, first studied by the twentieth-century American theoretical physicist, Philip Morse

$$V_{\mathrm{Morse}}(r) = D_e\left[\left(1 - e^{-\alpha(r-r_e)}\right)^2 - 1\right]. \qquad (9.5)$$

The Morse potential is plotted in Figure 9.5, along with its quantum energy levels and those of a harmonic oscillator approximation around the minimum. Although it was not derived from a fundamental theory, the Morse potential makes a good approximation to actual interatomic potentials if its parameters are chosen appropriately. $V_{\mathrm{Morse}}(r)$ has the right qualitative behavior for small r, where it is strongly repulsive, and at large r, where it is attractive. It has its minimum at r_e, which is chosen to agree with the measured equilibrium separation about which the atoms oscillate. Near the bottom of the potential $V_{\mathrm{Morse}}(r)$ looks parabolic, like a harmonic oscillator. In low-lying energy states, the two

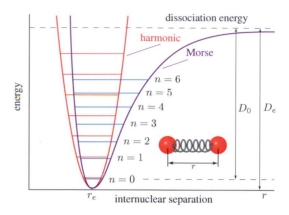

Figure 9.5 The Morse potential, eq. (9.5), for a diatomic molecule. The energies of the quantum states of vibrational motion are denoted by the blue lines marked with the quantum number n. Also shown is the harmonic oscillator approximation and its (evenly spaced) energy levels in red. Note that the lowest energy states of the Morse potential are quite evenly spaced, but that the energy levels get closer together higher in the potential.

atoms thus vibrate with a frequency ω_{vib} determined by the curvature of the potential near its minimum and the reduced mass of the atomic system (Problems 9.6, 9.7, and 9.9). If the potential were exactly harmonic, the vibrational energy levels of the molecule $E_{\mathrm{vib}}(n)$ would be evenly spaced with energies $E_{\mathrm{vib}}(n) = (n + \frac{1}{2})\hbar\omega_{\mathrm{vib}}$. Indeed, as shown in Figure 9.5, the first few levels in the Morse potential are nearly evenly spaced, as are the low-lying vibrational excitations of real molecules. More highly excited vibrational states oscillate with increasing amplitude about an average separation that increases with excitation energy, until finally, when the excitation energy exceeds $D_0 \cong D_e - \hbar\omega_{\mathrm{vib}}/2$, the atoms are no longer bound. The parameters D_e and α in the Morse potential are chosen to agree with the measured binding energy and the excitation energy of the first vibrational energy level (Problem 9.7).

According to quantum mechanics, the vibrating molecule can gain or lose energy by making a transition from a state with energy $E_{\mathrm{vib}}(n)$ to one with energy $E_{\mathrm{vib}}(n \pm 1)$, emitting or absorbing a photon with energy $|E_{\mathrm{vib}}(n) - E_{\mathrm{vib}}(n \pm 1)|$. For small n, this energy difference is approximately $\hbar\omega_{\mathrm{vib}}$, and is independent of n. In reality the fact that the energy levels are not exactly evenly spaced and the fact that the molecule changes its rotational energy state when it makes a *vibrational transition* through photon emission or absorption (due to angular momentum conservation), result in a **vibration–rotation band** of frequencies of light around the original value of $\hbar\omega_{\mathrm{vib}}$ that

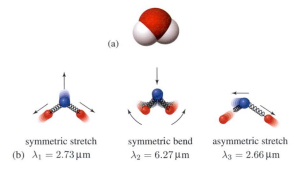

(a)

symmetric stretch   symmetric bend   asymmetric stretch
(b) $\lambda_1 = 2.73\,\mu m$ $\lambda_2 = 6.27\,\mu m$ $\lambda_3 = 2.66\,\mu m$

Figure 9.6 (a) A sketch of the water molecule showing the atomic sizes, separations, and bond angles at the proper relative scales; (b) the three characteristic vibrational modes of the water molecule, and the wavelengths of light that excite them.

can be absorbed or emitted by the oscillating molecule (Problem 9.9).[4] Also, at temperatures greater than or on the order of $\hbar\omega_{vib}/k_B$, collisions between molecules can transfer sufficient energy to excite or de-excite molecular vibrations.

With these general concepts in mind, we return to water. H_2O is a nonlinear *triatomic molecule*, as portrayed in Figure 9.6(a), so its modes of vibration are more complicated than those of O_2. As for O_2, however, each mode of oscillation of the H_2O molecule has an associated frequency at which it absorbs and/or emits radiation. The water molecule has three characteristic oscillations, shown in Figure 9.6(b). In the first mode the two hydrogen atoms oscillate in phase relative to the oxygen atom ("symmetric stretch"), while in the third they oscillate out of phase ("asymmetric stretch"). Energetically, these modes are quite similar, since the H–O bond is being stretched in a similar fashion in both cases. In the second mode, the molecule bends ("symmetric bend"), with the hydrogens rotating oppositely to one another while the oxygen moves to keep the center of mass fixed. The bonds are not as significantly stretched in this mode, with the result that the oscillation frequency is less than half that of the others. The frequencies of these vibrational modes of a water molecule in the gas phase are given in Table 9.4,

---

[4] Actually diatomic molecules made of the same atoms, like O_2 or N_2, are exceptional: because they are symmetric upon interchanging the atoms, they have no permanent electric dipole moment. This means that they cannot emit or absorb single photons and make a transition from one vibrational energy level to another. H_2O and CO_2, which do not have this symmetry, do not have this restriction. See §34 for further discussion.

along with the associated wavelengths and characteristic temperatures. The role of these modes, as well as related vibrational modes of CO_2, in atmospheric absorption of radiation is discussed more detail in §34.2.3 and §34.4.3.

9.3.2 Molecular Dissociation

Two quite distinct energy scales are associated with the potential that binds atoms into molecules. First, the spacing $\hbar\omega_{vib}$ between vibrational levels, and second, the much greater binding energy D_0. For O_2, for example, $D_0/\hbar\omega_{vib} \sim 25$. As the temperature of a sample is increased, the first manifestation of molecular structure occurs when the vibrational excitations start to contribute to the heat capacity (see Box 9.1). This occurs at temperatures somewhat below $\hbar\omega_{vib}/k_B$ (see Example 8.3).

As the temperature increases further, higher and higher vibrational energy levels are excited, and eventually the molecules begin to **dissociate** into atoms. Dissociation is a chemical reaction,

$$H_2O \rightarrow 2H + O, \qquad (9.6)$$

albeit a very simple one. Dissociation of water can be described in more detail as a sequence of two chemical reactions. First, one hydrogen is dissociated, $H_2O \rightarrow H + OH$, requiring 493 kJ/mol (or 5.11 eV/molecule); second, the OH molecule is dissociated, $OH \rightarrow O + H$, requiring 424 kJ/mol (or 4.40 eV/molecule).

Unlike melting or boiling, dissociation is a continuous process. The fraction of water molecules that have dissociated in a given sample grows gradually as the temperature increases. For example, by a temperature of 2500 K about 3% of water molecules are dissociated into a mixture of H, O, H_2, O_2, and OH, and above 3300 K more than half of the water molecules are dissociated. We rarely encounter temperatures this high in earth-bound energy systems. At even higher temperatures, H_2 and O_2 molecules also largely dissociate. Such temperatures occur inside the Sun, where few bound molecules are found at all, and in plasma confinement fusion devices, where temperatures comparable to the interiors of stars are produced.

Although we introduced the notion of molecular dissociation as a thermal process, it can also be instigated by absorption of a photon with energy greater than the dissociation energy. So, for example, photons (denoted by γ) with energy greater than 5.11 eV can dissociate water by the reaction

$$\gamma + H_2O \rightarrow OH + H. \qquad (9.7)$$

An energy of 5.11 eV corresponds to a wavelength of $\lambda = hc/E = 242$ nm, which is well into the ultraviolet (Figure 9.2). Such energetic photons are rare in sunlight, so

Table 9.4 Vibrational modes of the water molecule in a purely gaseous state: For each mode, the frequency of oscillation ν is given (in standard units of cm^{-1}), as well as the wavelength of associated radiation, excitation energy of a quantum (in electron volts), and the associated temperature. Note that in liquid water the frequencies of the vibrational modes are slightly different due to interactions with nearby hydrogen bonds.

Mode	Wavelength λ	Frequency $\nu/c = 1/\lambda$	Excitation energy $E = hc/\lambda$	Associated temperature $T = hc/\lambda k_B$
Symmetric stretch	$\lambda_1 = 2.73\,\mu m$	$3657\,cm^{-1}$	$0.454\,eV$	$T_1 = 5266\,K$
Symmetric bend	$\lambda_2 = 6.27\,\mu m$	$1595\,cm^{-1}$	$0.198\,eV$	$T_2 = 2297\,K$
Antisymmetric stretch	$\lambda_3 = 2.66\,\mu m$	$3756\,cm^{-1}$	$0.466\,eV$	$T_3 = 5409\,K$

Box 9.1   Rotations, Vibrations, and The Heat Capacity of Water Vapor

Once vaporized, water molecules are free to rotate without constraint. As mentioned in §5.1.3 and illustrated in Figure 5.3, a *diatomic molecule* such as O_2 or HO can rotate about either of the two axes perpendicular to the line joining the atoms. *Polyatomic molecules* such as water can rotate about three independent axes. To a good approximation, the rotational degrees of freedom can be separated from vibrations by making the approximation that the molecule is a *rigid rotor*. The rotational energy levels of polyatomic molecules are then quantized in units of $E_{rot} = \hbar^2/I$, where I is the moment of inertia of the molecule about the rotation axis. The three moments of inertia of water lie in the range from ~ 1–$3 \times 10^{-47}\,kg\,m^2$, so the scale of rotational excitations is $E_{rot} = \hbar^2/I = (1.05 \times 10^{-34}\,J\,s)^2/(1$–$3 \times 10^{-47}\,kg\,m^2) \approx 0.3$–$1.0 \times 10^{-21}\,J \approx 2$–$6 \times 10^{-3}\,eV$. This energy corresponds to a temperature $T_{rot} = E_{rot}/k_B \approx 25$–$70\,K$ and a wavelength $\lambda_{rot} = hc/E_{rot} \approx 0.2$–$0.6\,mm$, in the microwave portion of the electromagnetic spectrum (Figure 9.2). Thus rotational degrees of freedom are *thawed* at temperatures well below 373 K and contribute $3 \times \frac{1}{2}k_B$ per molecule to the heat capacity of water vapor. Water molecules can make transitions from one rotational level to another by emitting or absorbing microwave radiation, a feature exploited in microwave ovens.

In Earth's atmosphere, at temperatures of order 300 K very highly excited rotational energy levels of water vapor are populated. Pure *rotational transitions* among these energy levels give rise to absorption bands in Earth's atmosphere extending down to wavelengths around $12\,\mu m$ that play an important role in the greenhouse effect (§34).

Below the characteristic temperatures $T_{1,2,3}$ of the vibrational modes (see Table 9.4) of water, vibrational modes are frozen out and only translations (three degrees of freedom) and rotations (three degrees of freedom) contribute to the heat capacity. According to the equipartition theorem, we expect $\frac{1}{2}k_B$ per degree of freedom, leading to a heat capacity of $\hat{c}_p k_B = \hat{c}_V k_B + k_B = 6 \times \frac{1}{2}k_B + k_B = 4k_B$ per molecule. Then as the temperature rises, first one, then three vibrational modes "thaw," each contributing $2 \times \frac{1}{2}k_B$ to $\hat{c}_p k_B$. The heat capacity $\hat{c}_p k_B$ should thus slowly increase from $4k_B$ to $7k_B$ per molecule as T rises above the characteristic vibrational temperatures. The data on $\hat{c}_p$ for water vapor shown in Figure 9.7 confirms this.

---

that little water is dissociated in Earth's atmosphere. On the other hand, the bonds that hold water molecules together are broken in chemical processes going on all around us all the time. In a chemical reaction some bonds are broken, absorbing energy, but others are formed, releasing energy. The energetics of chemical reactions are discussed in more detail in the next section, but as a rule the order of magnitude of the energy released in a chemical reaction is ~ 1–$10\,eV$ per bond (again remembering the conversion factor, $1\,eV/molecule = 96.49\,kJ/mol$). Planck's relation

(9.2) determines the wavelength of the light associated with energy releases of this magnitude, $\lambda = hc/E$; so $\lambda \sim 120$–$1200\,nm$. Referring back to Figure 9.2, we see that these wavelengths *include the range of visible light* and extend into the nearby infrared and ultraviolet as well.

We explore energy and entropy flow in chemical reactions more thoroughly in §9.4, but first we continue our journey through the structure of water at ever higher energies and shorter distances.

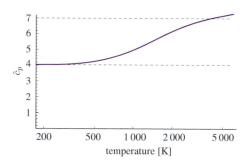

Figure 9.7 The heat capacity at constant pressure of water vapor, per molecule, in units of k_B, plotted versus temperature. Well below the characteristic temperatures for vibrations $\hat{c}_p \approx 4$, while at high temperature $\hat{c}_p \approx 7$. $\hat{c}_p$ continues to increase at very high temperatures as dissociation and electronic excitations start to contribute [24].

Energy in Chemical Bonds

The energy scale for breaking molecular bonds is $\sim$1–10 eV, which sets the scale for the energy of chemical reactions. Vibrations of molecules about their equilibrium positions are characterized by much lower energies. Vibrations prefigure dissociation and are responsible for a slow increase of the heat capacity with temperature and for absorption of infrared light by simple molecules like H_2O and CO_2. Water dissociates gradually as the temperature goes from $\sim$ 2000 K to $\sim$ 4000 K.

9.3.3 Atomic Binding

We next consider the dissociated gas of hydrogen and oxygen atoms that was once our block of ice at 0 K. Temperatures in excess of several thousand degrees Kelvin are needed to dissociate water molecules into hydrogen and oxygen atoms. What happens if we continue to add energy and raise the temperature still higher? The collisions between atoms, and the interactions of thermal radiation with the atoms, eventually have sufficient energy to excite electrons in the atomic orbitals. At even higher energies, electrons are knocked out completely and the hydrogen and oxygen atoms are **ionized**. The forces that bind electrons in atoms are in general stronger than intermolecular forces, so the associated **ionization energies** are usually larger than the energy required to break molecular bonds.

In §7 we studied the energy levels of the hydrogen atom, and learned that the electron in its ground state is bound by

an energy of $B_0 = me^4/(32\pi^2\epsilon_0^2\hbar^2) \cong 13.6\,\text{eV}$. In other words, 13.6 eV must be added to the electron in hydrogen to separate it entirely from the Coulomb attraction of the proton. The minimum energy needed to excite an electron, from the ground state to the $2s$ or $2p$ states, is not much smaller – 10.2 eV – so electronic excitations become relevant only at temperatures at which ionization already begins to occur. The temperature associated with the ionization energy of hydrogen is $T = B_0/k_B \cong 160\,000\,\text{K}$. Like dissociation, ionization is a gradual process. At a pressure of one atmosphere, a gas of monatomic hydrogen is already 50% ionized at a temperature of about 16 000 K due to collisions.[5] Temperatures like this do not occur on Earth's surface, so *thermal* ionization of hydrogen and other atoms is not significant in earth-bound energy generating processes.

While thermal ionization is not relevant for most terrestrial systems, photons – electromagnetic radiation – with energy above 13.6 eV can be found in nature or produced by artificial means. By Planck's relation, $E = h\nu = hc/\lambda$, 13.6 eV corresponds to a wavelength of 91.2 nm, which is well into the ultraviolet region. About 5 parts per billion of the energy of sunlight (as it arrives at the top of the atmosphere) is carried by photons with energies above 13.6 eV, so hydrogen and other atoms high in Earth's atmosphere can be ionized by sunlight. Also, medical X-rays and environmental radiation from cosmic rays and from radioactive nuclei often include energetic photons that can ionize atoms. In fact, for the purposes of discussions of radiation safety (§20), electromagnetic radiation is divided according to whether it is energetic enough to ionize atoms. *Ionizing radiation*, usually defined as radiation whose quanta have energy greater than roughly 20 eV, is known to be an environmental hazard, while lower-energy (longer wavelength) radiation is believed to be less of a concern.[6]

It takes 13.6 eV to remove hydrogen's one electron, but much more energy is required to completely ionize heavier elements like oxygen. The binding of a single electron to a nucleus with charge Z is $B_0(Z) = Z^2me^4/(32\pi^2\varepsilon_0^2\hbar^2)$, so if seven of the eight electrons in oxygen were removed,

---

[5] The fraction ionized x is determined by the **Saha–Langmuir equation**, which in this case gives $x = \frac{1}{2}(\sqrt{4z + z^2} - z)$, where $z = \frac{4\,780}{p}(k_BT)^{5/2}e^{-13.6/k_BT}$, where k_BT is in eV and p is in atmospheres. Note that the large fraction of ionized hydrogen atoms at roughly 10% of B_0/k_B arises for a similar reason to the breakdown of the Debye T^3 scaling of the specific heat at a temperature of roughly $0.1T_D$.

[6] There is a vigorous ongoing discussion of the possible health risks of non-ionizing radiation.

Table 9.5 Ionization energies for the eight electrons in the oxygen atom.

Electron	Ionization energy (kJ/mol)	eV/molecule
1st	13 149	13.62
2nd	3 388	35.12
3rd	5 301	54.93
4th	7 469	77.41
5th	10 990	113.9
6th	13 327	138.1
7th	71 330	739.4
8th	84 090	871.4

Ionization Energies

Removing the electrons from an atom takes energies in the range of 5–100 keV. The outermost electron is most easily removed, and the last electron is the hardest, with a binding energy of $13.6 \times Z^2$ eV. At atmospheric pressure a significant fraction of the hydrogen and oxygen atoms from water are ionized at temperatures exceeding 10 000 K.

the last electron would be bound by $64 \times 13.6 \cong 870$ eV, well into the domain of X-rays (Figure 9.2). The energies required to ionize each of the eight electrons in oxygen are listed in Table 9.5. The first (outermost) electron is the easiest to remove, with an ionization energy close to that of the single electron in hydrogen. The binding energy grows quickly with successive ionizations. If we add up the energy necessary to ionize the two hydrogen atoms and one oxygen atom in H_2O, we find[7]

$$H_2O \rightarrow 2\,^{1}\underline{H} + \,^{16}\underline{O} + 10e^{-} \qquad \Delta H_r = 199.8 \text{ MJ/mol.}$$
$$(9.8)$$

The electromagnetic radiation associated with ionization ranges from about 4 eV/photon, which can ionize the most weakly bound electron in the cesium atom and corresponds to a wavelength of 300 nm, up to the ionization energy of

Nuclear Binding

The energy necessary to remove a proton or neutron from a typical nucleus is roughly 8 MeV. This sets the scale for the energy available per nucleus in nuclear reactions. To thermally separate the protons and neutrons in the oxygen nuclei from water, a temperature on the order of 10^{11} K would be needed.

the innermost electron in a heavy element like uranium, which requires a jolt of over 100 keV, corresponding to a wavelength of 1/100 nm, at the upper limit of X-rays.

9.3.4 Nuclear Binding and Beyond

The binding of protons and neutrons into nuclei is a subject for a later chapter (§17), but for completeness we mention it here. Protons and neutrons are bound by about 8 MeV per particle in a typical nucleus with many protons and neutrons.[8] The binding energy per particle increases with the number of protons Z in the nucleus up to $Z \approx 26$, which is iron (^{56}Fe – 26 protons and 30 neutrons), and then decreases slowly thereafter. So it is energetically favorable for nuclei with Z significantly below 26 to combine (*fuse*) together and for nuclei with Z significantly above 26 to break apart (*fission*). These processes play important roles in nuclear power systems described in later chapters. The total binding energy of oxygen – the total energy given off if 8 protons and 8 neutrons could be brought together to form ^{16}O – is $\sim 3 \times 10^{12}$ J/mol. Although this is not a practical nuclear reaction, it gives some idea of the immense energies available when nuclear processes are involved. The temperature needed to thermally separate an oxygen nucleus into its component protons and neutrons is around $T \sim 8 \text{ MeV}/k_B \cong 10^{11}$ K. Such temperatures occurred in the very early moments of our observable universe, less than one second after the big bang.

At even higher temperatures, the neutrons and protons themselves dissociate into more fundamental objects called *quarks*. And at even higher temperatures, yet more fundamental constituents may be revealed that are at present unknown. Some of the concepts of subnuclear phenomena are introduced in §14 and §21. But for the purposes of terrestrial energy applications, all relevant aspects

---

[7] Here we anticipate the notation introduced in §17 and §18, where a neutral atom is labeled by the chemical species (which depends on the number of protons) and the total number of protons and neutrons A, in the form ^{A}El, and the fully ionized nucleus of the atom is labeled by $^{A}\underline{El}$.

[8] *Light nuclei* with fewer than $\sim$15–20 *nucleons* (neutrons + protons) are more variable.

of internal energy have been explored by the time the ice has dissociated into individual neutrons, protons, and electrons.

Finally, the ultimate energy content of matter is dictated by Einstein's famous relation, $E = mc^2$. If we could somehow liberate all its mass as energy, one gram of matter would emit $\sim 10^{-3}$ kg $\times (3 \times 10^8$ m/s$)^2 = 9 \times 10^{13}$ J of energy. We cannot liberate these energies, however, in any practical way, because all matter (at least in our part of the universe) is made using particles like neutrons and protons, and there is no observed naturally occurring (anti-)matter composed of antineutrons or antiprotons. Furthermore, the total number of *baryons* (§14), including neutrons, protons, and some other exotic types of matter, is conserved. So there is no way to produce antimatter without producing corresponding matter at the same time. Thus, as far as we know, protons and neutrons cannot be made to disappear into pure energy without using at least as much energy to produce antiprotons and antineutrons in the first place. In particular, no matter how much we heat the block of ice, at least at temperatures that can be understood using existing physical theories, there is no point at which the matter composing the ice will completely annihilate into pure energy.

9.4 Internal Energy, Enthalpy, and Free Energy in Chemical Reactions

It is time to return to earth, so to speak, and explore in further detail the flow of energy and entropy in matter in the course of chemical reactions. Our main aims here are to explain how to compute the energy released or absorbed in a chemical reaction and to establish the criteria that determine whether a chemical reaction can proceed spontaneously.

The concepts of *enthalpy* (§5.5) and *free energy* (§8.8) were originally developed to describe the flow of energy in chemical reactions. In most cases of interest to us, chemical reactions occur at fixed pressure, so the reactants can do pdV work during the reaction. This means that some of the heat that flows in or out of the reactants goes into pdV work and the rest goes into the internal energy of the system. Heat equates to *enthalpy* under these conditions. This explains some common terminology: chemists refer interchangeably to "**enthalpy of reaction**" and "**heat of reaction**." We use the former term throughout. The enthalpy of reaction ΔH_r is thus defined to be the amount of energy that is absorbed (if $\Delta H_r > 0$) or released (if $\Delta H_r < 0$) as heat in the course of a reaction that takes place at constant pressure. Since enthalpy is a state

function, ΔH_r also refers to the change in value of the enthalpy of the system as it goes through the reaction.

The entropy audit for a reaction that takes place at temperature T must include both the change in entropy of the reactants ΔS_r and the entropy delivered to the surroundings, $-\Delta H_r/T$. The *total* entropy change of the system and environment combined is related to the change in the *Gibbs free energy* ΔG_r (eq. (8.68))

$$\Delta S_{\text{tot}} = \Delta S_r - \frac{\Delta H_r}{T} = -\frac{\Delta G_r}{T}. \qquad (9.9)$$

The sign of ΔG_r determines whether the reaction can occur spontaneously or not. The *Gibbs condition* states that if ΔG_r is negative, then the entropy of the reactants plus the environment increases and the reaction may occur spontaneously. If ΔG_r is positive, it cannot.[9]

9.4.1 Enthalpy and Free Energy of Reaction: An Example

To make the discussion concrete, we first look at a specific example of a simple chemical reaction. In the next section (§9.4.2) we set up a general approach. The reaction we choose to follow is known as **calcination**.[10] In this reaction, *limestone* (calcium carbonate = $CaCO_3$) is heated to drive off carbon dioxide, leaving behind *quicklime* (calcium oxide = CaO). This reaction is one of the essential steps in making *cement*, which is in turn used in the production of concrete. The enormous amount of concrete produced worldwide is currently responsible for about 5% of all the CO_2 added to the atmosphere by human activity; roughly 60% of this CO_2 can be attributed to calcination in concrete production.

The first step in analyzing calcination is to write a correctly balanced chemical equation,

$$CaCO_3 \rightarrow CaO + CO_2 . \qquad (9.10)$$

Table 9.6 gives all of the relevant thermodynamic information about this reaction. Let us examine the implications of the numbers in this table. First, U, H, G, and S are all state functions. Thus, the table lists the *changes in the properties of the system*. For example, the enthalpy of the system *increases* by 178.3 kJ/mol when this reaction takes place, which makes eq. (9.10) an *endothermic* reaction: heat must be supplied to make it go.

---

[9] As discussed in §8.8, a negative reaction free energy is a necessary, but not sufficient, condition for a reaction to proceed. Activation barriers often inhibit reactions that are thermodynamically allowed.

[10] Although originally named for the reaction of eq. (9.10), *calcination* generally refers to any process in which volatile material is driven off from an ore by heating it.

Table 9.6 Thermodynamic information about the reaction $CaCO_3 \rightarrow CaO + CO_2$ (at $T = 25\,°C$ and $p = 1$ atm).

	Quantity	Value
ΔU_r	Reaction internal energy	+175.8 kJ/mol
ΔH_r	Reaction enthalpy	+178.3 kJ/mol
ΔS_r	Reaction entropy	+160.6 J/mol K
ΔG_r	Reaction free energy	+130.4 kJ/mol

Next, since ΔG_r is positive, the reaction *does not* occur spontaneously. In fact, the reverse reaction *does* occur spontaneously: quicklime, left in the presence of CO_2, will slowly absorb it, forming calcium carbonate.[11] Note that ΔS_r is *positive*. The entropy of the reactants increases in this reaction, so one might have thought it would go spontaneously. However so much heat is taken out of the environment that ΔS_{tot}, given by eq. (9.9), is *negative* and the reaction does not occur spontaneously at 25 °C.

Finally, note that the variables are related (Problem 9.10) by (1) $\Delta G_r = \Delta H_r - T\Delta S_r$ and (2) $\Delta H_r = \Delta U_r + p\Delta V_r = \Delta U_r + RT\Delta n_r$. The last equality uses the ideal gas law, $pV = nRT$, to replace $p\Delta V_r$ by $RT\Delta n_r$. Thus, when one mole of limestone undergoes calcination, $\Delta n_r = 1$ mole of CO_2 is emitted.

How can we use the numbers from Table 9.6? First, they tell us the energy required to drive this reaction. A heat input of $\Delta H_r = 178.3$ kJ/mol is needed to drive the CO_2 out of limestone. Conversely, if quicklime reacts with carbon dioxide to make calcium carbonate, 178.3 kJ/mol is given off.[12] Calculations such as this enable us to compute the energy consumption of manufacturing processes and compare the energy content of different fuels.

Another extremely useful application of the information in Table 9.6 is to estimate *the temperature to which limestone must be raised to allow this reaction to occur spontaneously*. Notice that the second term in $\Delta G_r = \Delta H_r - T\Delta S_r$ depends explicitly on the temperature, and ΔS_r is positive. Thus, as the temperature increases, ΔG_r *decreases*, and at a high enough temperature, ΔG_r is

---

[11] Note that some fraction of the CO_2 emitted through calcination in concrete production is reabsorbed in the lifetime of concrete. This fraction can be a half or more with proper recycling techniques; variations in manufacturing technique can also increase the reabsorption of CO_2 in the setting process.

[12] Notice that the numbers quoted in Table 9.6 are all referred to the standard temperature of 25 °C. To apply these ideas quantitatively in practice these values must be determined for different temperatures.

Reaction Enthalpy and Free Energy

Reaction enthalpy ΔH_r determines how much heat is given off ($\Delta H_r < 0$) or absorbed ($\Delta H_r > 0$) during a constant pressure chemical reaction.

Reaction (Gibbs) free energy ΔG_r is related to the total entropy change in a reaction (at constant pressure). When $\Delta G_r < 0$, the entropy of the reactants plus environment increases and a reaction can (but may not) proceed spontaneously; if $\Delta G_r > 0$, the reverse reaction may proceed spontaneously.

---

negative and the reaction can occur spontaneously. This is an essential feature of calcination reactions and other reactions of the form *solid* $\rightarrow$ *solid* + *gas*. When a gas is produced from a solid, the reaction entropy change is typically positive. This means that, even when the reaction is endothermic, raising the temperature enough may drive the free energy of reaction to be negative, so that the reaction can occur spontaneously.

The crossover temperature T_0 at which a reaction can proceed spontaneously occurs when $\Delta G_r(T_0) = 0$. To estimate T_0 for calcination we first assume that ΔH_r and ΔS_r do not change significantly with temperature. Then, setting $\Delta G_r = 0$, we find $T_0 = \Delta H_r/\Delta S_r \cong 178\,\mathrm{kJ}/(0.161\,\mathrm{kJ/K}) \cong 1100\,\mathrm{K} \cong 825\,°C$. When the T dependence of ΔS_r and ΔH_r is included, the precise value of T_0 is closer to 890 °C [52]. Thus, we can expect the calcination reaction to occur when limestone is heated above $T_0 \approx 890\,°C$.

Making quicklime from limestone generates CO_2 in several ways. Since the reaction is endothermic, $\Delta H_r \cong 177.5$ kJ/mol must be supplied to the limestone, usually by burning a fossil fuel. The reactants must furthermore be heated above 890 °C for the reaction to take place, again requiring heat from fossil fuels (though in principle this energy could be recaptured as the quicklime and CO_2 cool, and could be reused as heat energy as is done in *cogeneration* (§13.5.3)). Finally, the reaction itself releases one molecule of CO_2 for each molecule of CaO created. Given the huge amounts of concrete used in the modern world, it is no wonder that 5% of anthropogenic CO_2 comes from making it.

9.4.2 Hess's Law and Enthalpy/Free Energy of Formation

The example of making quicklime from limestone illustrates the importance of the enthalpy and free energy of reaction. Chemical reactions are ubiquitous in energy-related applications, such as the refining of ores to obtain metals, the manufacture of fertilizer, and the refining of crude oil into more useful products. *Combustion*, however, is particularly important, since it is the principal way in which we harvest energy from fossil fuels (and biofuels). In most familiar circumstances, the process of combustion is burning a substance in air. More generally, **combustion** pertains to a class of exothermic reactions in which a compound reacts completely with an oxidant.[13] We deal here with combustion primarily in the context of organic compounds reacting with oxygen. In an idealized complete combustion reaction of this sort, every carbon atom in the original compound ends up in CO_2, every hydrogen atom ends up in H_2O, and every nitrogen in N_2. Note that nitrogen is, in general, not oxidized, and the fate of other, less common elements must be specified if it is important. In practice, combustion is usually not complete. Some carbon and other elements may be left as ash, and some carbon may be partially oxidized to carbon monoxide (CO).

Chemists do not tabulate ΔH_r and ΔG_r for every possible chemical reaction. Instead, for all compounds of interest, they tabulate the **enthalpy of formation** and the **free energy of formation**, denoted ΔH^f and ΔG^f. These quantities are defined using the specific reaction that builds up a chemical compound out of its constituents. For any more general reaction, ΔH^f and ΔG^f can be used to compute ΔH_r and ΔG_r, using the fact that H and G are state functions and therefore depend only on the state of a substance and not on how it is made.

The **formation reaction** for a given compound begins with the constituent elements in the form that they are found at standard values of temperature and pressure, usually 25 °C and 1 atm. Elements that are gases under these conditions are assumed to be provided in the gaseous state. Only bromine and mercury are liquids at 25 °C and 1 atm. Note that gases that are found as *diatomic* molecules are taken in the molecular state. This applies to O_2, H_2, N_2, Cl_2, etc. Table 9.7 lists a few formation reactions in standard form.

Notice that the reaction equations in Table 9.7 are normalized to produce exactly one molecule of the desired compound. This may require, for example, reacting "half a molecule" of a diatomic gas. Note also that the state (solid, liquid, gas) of the resulting compound must be indicated. The enthalpy of formation for a compound in a gaseous state is greater (less negative) than a liquid because some

---

[13] For example, the term *combustion* is used to describe sulfur burning in the powerful oxidant fluorine, $S + 3F_2 \rightarrow SF_6$.

Table 9.7 Examples of formation reactions. These reactions illustrate the conventions of formation reactions: elements are assumed to be supplied in the phase (solid, liquid, or gas) that is stable under standard conditions and in the most stable molecular form. In particular, those that form diatomic molecules, O_2, H_2, N_2, F_2, Cl_2, Br_2, and I_2, are assumed to be supplied as such.

Compound	Formation reaction
Water	$H_2(g) + \frac{1}{2}O_2(g) \rightarrow H_2O\ (l)$
CO_2	$C + O_2(g) \rightarrow CO_2(g)$
Methane	$C + 2H_2(g) \rightarrow CH_4(g)$
Sodium bromide	$Na + \frac{1}{2}Br_2(l) \rightarrow NaBr(s)$
Glucose	$6C + 6H_2(g) + 3O_2(g) \rightarrow C_6H_{12}O_6(s)$

Table 9.8 Standard enthalpy and free energy of formation for some useful chemical compounds, all at 25 °C and 1 atm (from [50]).

Compound	Chemical formula	ΔH^f (kJ/mol)	ΔG^f (kJ/mol)
Diatomic gases	$H_2, O_2, Cl_2,\dots$	0*	0*
Methane	$CH_4(g)$	−75	−51
Water vapor	$H_2O(g)$	−242	−229
Water liquid	$H_2O(l)$	−286	−237
Octane (l)	$C_8H_{18}(l)$	−250	+6
Ethanol (g)	$C_2H_5OH(g)$	−235	−168
Ethanol (l)	$C_2H_5OH(l)$	−278	−175
Carbon dioxide (g)	$CO_2(g)$	−394	−394
Carbon monoxide (g)	CO	−111	−137
Calcium oxide	CaO	−635	−604
Iron ore (hematite)	Fe_2O_3	−824	−742
Calcium carbonate	$CaCO_3$	−1207	−1129
Glucose	$C_6H_{12}O_6$	−1268	−910
Ammonia	NH_3	−46	−16.4

* by definition

extra pdV work must be done when the gas is formed. Compare, for example, the enthalpies of formation of liquid and gaseous ethanol, or of water liquid and vapor, in Table 9.8, where a more extensive list of ΔH^f and ΔG^f for energy-related chemical compounds is given.

Given the enthalpies of formation for various compounds, the strategy for computing the enthalpies of reaction is simple. Because enthalpy is a state function, the enthalpy of reaction for a given reaction does not depend upon how the reaction actually occurs. We can choose to view the reaction as a two-step process in which the reactants are first taken apart into their constituent

Hess's Law

The reaction enthalpy for any reaction is equal to the sum of the enthalpies of formation of the products minus the enthalpy of formation of the reactants,

$$\Delta H_r = \sum_{\text{products}} \Delta H^f - \sum_{\text{reactants}} \Delta H^f.$$

elements, and then these elements are formed into the products. The reaction enthalpy, therefore, is the sum of the enthalpies of formation of the products minus the sum of the enthalpies of formation of the reactants,

$$\Delta H_r = \sum_{\text{products}} \Delta H^f - \sum_{\text{reactants}} \Delta H^f. \qquad (9.11)$$

This is the essence of **Hess's law**. Example 9.1 illustrates Hess's law for the calcination reaction considered above. The *free energy of reaction* can be computed in a precisely analogous fashion.

9.4.3 Higher and Lower Heats of Combustion

Combustion reactions must frequently be considered in comparisons of fossil fuels with one another and with renewable energy sources. There is a small, but sometimes confusing, complication regarding the enthalpy of reaction in combustion reactions: when water is produced in a combustion process, the enthalpy of combustion depends upon whether the water ends up as a vapor or a liquid. The difference may be significant because the latent heat of condensation of water is large and because less pdV work is done when water is produced as a liquid than as a vapor at fixed pressure. Both of these effects work in the same direction: the enthalpy of reaction (combustion) is larger – more heat is given off – when the produced water is a liquid rather than as a gas.

The **higher heating value** (HHV) or **gross calorific value** of a fuel is defined by the US Department of Energy [53] as "the amount of heat released by a specified quantity (initially at 25 °C) once it is combusted and the products have returned to a temperature of 25 °C." Thus defined, the HHV is the negative of the reaction enthalpy at 25 °C and 1 atm, and therefore includes the latent heat of condensation of the water, which is a liquid under those conditions.

The **lower heating value** (LHV) or **net calorific value** of a fuel is a measure of the heat released by combustion when the water vapor is *not* condensed. Unfortunately, there is more than one definition of LHV in common use. The one used by most sources, which we adopt as well, is the same as HHV except that the water is taken to be a

---

Example 9.1 Hess's Law Example: Calcination

Consider again the decomposition of limestone into CaO and CO_2:

$$CaCO_3 \rightarrow CaO + CO_2 \,.$$

We can view this as a sequence of two steps. In the first step, $CaCO_3$ is "unformed," with enthalpy of reaction equal to *minus* its enthalpy of formation (Table 9.8),

$$CaCO_3 \rightarrow Ca + C + \tfrac{3}{2}O_2, \qquad\qquad \Delta H_r = -\Delta H^f(CaCO_3) = 1207 \,\text{kJ/mol}\,.$$

In the second step, the products are combined to make CaO and CO_2,

$$Ca + \tfrac{1}{2}O_2 \rightarrow CaO, \qquad\qquad \Delta H_r = \Delta H^f(CaO) = -635 \,\text{kJ/mol},$$
$$C + O_2 \rightarrow CO_2, \qquad\qquad \Delta H_r = \Delta H^f(CO_2) = -394 \,\text{kJ/mol}.$$

Even though these are impractical reactions, we can imagine making CaO and CO_2 from $CaCO_3$ this way, and that is all that is necessary to compute the enthalpy of reaction. The overall heat of reaction is clearly the sum of the enthalpies of reaction for the three reactions above,

$$\Delta H_r(CaCO_3 \rightarrow CaO + CO_2) = (1207 - 635 - 394) \,\text{kJ/mol} = 178 \,\text{kJ/mol}\,,$$

which agrees with the data quoted in Table 9.6. This gives an explicit example of Hess's law (9.11). The same reasoning applies to the calculation of free energies of reaction.

Example 9.2   HHV and LHV of Methane

Compute the HHV and LHV for combustion of methane (CH_4), the dominant component of natural gas, and show how the two heating values are related.

HHV: The chemical reaction is $CH_4 + 2O_2 \rightarrow CO_2(g) + 2H_2O(l)$, where we have denoted that the water is liquid. The standard enthalpies of formation of CH_4, CO_2, and $H_2O(l)$ (at 1 atm and 25 °C) are −74.8 kJ/mol, −393.5 kJ/mol, and −285.8 kJ/mol, respectively. The HHV is obtained from Hess's Law: $\Delta H_{reaction} \cong 2(-285.8) - 393.5 + 74.8 \cong$ −890.3 kJ/mol. So HHV $\cong$ 890 kJ/mol. Note that enthalpies of combustion are generally negative, as energy is released.

LHV: Now assume that the water is created as a vapor. The enthalpy of formation of water vapor (at 25 °C) is −241.8 kJ/mol, so the enthalpy of reaction (still at 25 °C and 1 atm) is $2(-241.8) - 393.5 + 74.8 \cong -802.3$ kJ/mol.

The difference between the HHV and HLV should be the latent heat of condensation of two moles of water at 25 °C, which is measured to be $\Delta H_{vap}(25\,°C) = 44.00$ kJ/mol. This accounts exactly for the 88.0 kJ/mol difference between the HHV and LHV of methane, as it must.

Note that the difference between the LHV and HHV is a 10% effect, which is not insignificant.

---

vapor at 25 °C. The difference between LHV and HHV is, then, exactly the enthalpy of vaporization of the produced water at 25 °C. An explicit calculation of HHV and LHV is worked out in Example 9.2.[14] Which to use, HHV or LHV, in a given situation depends on the circumstances and on convention. In an internal combustion engine, for example, condensation of water generally takes place outside of the engine and cannot contribute to useful work, thus it seems logical to use the LHV; in principle, however, the additional enthalpy of condensation of water could be captured by a more sophisticated system. HHV is often used in any case for motor vehicle fuels, and we follow this convention here since this is really the theoretical maximum of energy that could be extracted from the combustion process. The HHV may well be appropriate for an external combustion engine.

9.5 Chemical Thermodynamics: Examples

The fundamentals of chemical thermodynamics introduced in this chapter shed light on many issues that arise when we consider sources of energy and energy systems in subsequent chapters. Here we illustrate the power of these tools with three practical examples, addressing the following questions: first, how does the energy content of ethanol compare to that of gasoline? Second, how does the commonly used energy unit a *tonne of TNT* get its value of $= 4.184$ GJ? And third, how are metals refined from oxide ores? Other applications can be found among the problems at the end of the chapter.

---

[14] Our convention differs, however, from one offered by the US Department of Energy [53] (Problem 9.14).

9.5.1 Ethanol Versus Gasoline

Ethanol (C_2H_5OH), or *grain alcohol*, is an alternative motor fuel that can be produced from biological crops. In §26 we analyze some of the general aspects of biofuels and examine different methods for producing ethanol and its role as a replacement for fossil fuels. Here, we focus on a specific question that can be addressed using the methods of this chapter: how does the energy density of ethanol compare to that of gasoline? This analysis also leads to a simple comparison of the carbon output from these two fuel sources.

We begin by computing the HHV for each fuel. We assume complete combustion of ethanol and gasoline (we take gasoline here for simplicity to be 100% octane, C_8H_{18}; a more detailed discussion of the composition and properties of actual gasoline mixtures is given in §11 and §33), so we have the chemical reactions

$$C_2H_5OH + 3O_2 \rightarrow 2CO_2(g) + 3H_2O(l),$$

$$C_8H_{18} + \tfrac{25}{2}O_2 \rightarrow 8CO_2(g) + 9H_2O(l). \tag{9.12}$$

Hess's law gives

$$\Delta H_c(C_2H_5OH)$$
$$= 2\Delta H^f(CO_2(g)) + 3\Delta H^f(H_2O(l)) - \Delta H^f(C_2H_5OH)$$
$$\cong 2(-394\,\text{kJ/mol}) + 3(-286\,\text{kJ/mol}) - (-278\,\text{kJ/mol})$$
$$= -1368\,\text{kJ/mol}, \tag{9.13}$$

$$\Delta H_c(C_8H_{18})$$
$$= 8\Delta H^f(CO_2(g)) + 9\Delta H^f(H_2O(l)) - \Delta H^f(C_8H_{18})$$
$$\cong 8(-394\,\text{kJ/mol}) + 9(-286\,\text{kJ/mol}) - (-250\,\text{kJ/mol})$$
$$= -5476\,\text{kJ/mol}. \tag{9.14}$$

Thus, octane produces far more energy per mole than ethanol. Octane, however, also has a greater molar mass: 114 g/mol versus 46 g/mol for ethanol. If we divide the enthalpy of combustion by the molar mass, we get the specific energy density E_c for the fuels (which is positive by convention). The results are

$$E_c(C_2H_5OH) \cong 29.7 \, \text{MJ/kg},$$

$$E_c(C_8H_{18}) \cong 47.5 \, \text{MJ/kg}.$$

So ethanol stores less energy than octane per kilogram. Burning ethanol produces only $29.7/47.5 = 63\%$ as much energy per unit mass as burning octane.

Now let us consider the CO_2 production from each of these fuels. Burning one mole of ethanol produces two moles of CO_2, whereas one mole of octane produces eight moles of CO_2. From the point of view of minimizing CO_2 output for a given amount of energy use, the relevant quantity to compare is the **CO_2 intensity**, defined as the amount of CO_2 per unit of energy. In this respect, ethanol and octane are almost identical: $1.46 \, \text{mol}(CO_2)/\text{MJ}$ for ethanol and $1.48 \, \text{mol}(CO_2)/\text{MJ}$ for octane. In fact, as we discuss further in §33, the common fuel with the lowest CO_2 intensity is methane (CH_4), which produces $1.12 \, \text{mol}(CO_2)/\text{MJ}$ when burned.

We see from their almost identical CO_2 intensities that the relative environmental impact of gasoline versus ethanol depends *entirely on how these two fuels are produced*. Gasoline, refined from petroleum, is a fossil fuel. Burning it introduces net CO_2 into the atmosphere. Bioethanol, as it grows, uses photosynthesis to capture CO_2 from the atmosphere. Therefore burning it would appear to be neutral with respect to the CO_2 content of the atmosphere. The complete story is more complex, however, since the energy inputs and consequent carbon footprint associated with ethanol production vary widely depending upon how the ethanol is produced. We explore these questions in more detail in §26.

9.5.2 A Tonne of TNT?

TNT, or trinitrotoluene, is a commonly used explosive with the chemical formula $C_6H_2(NO_2)_3CH_3$. TNT has given its name to one of the standard units of energy, the **tonne of TNT**, $1 \, \text{tTNT} = 4.184 \, \text{GJ}$, often used to characterize the destructive power of weapons of war.

When TNT explodes, it does not burn in the ambient atmosphere. Such a process would occur too slowly to generate an explosion. Instead, the pressure wave created by a detonator causes the TNT to decompose, liberating gases that expand violently. The "tonne of TNT" refers to the enthalpy liberated when TNT disintegrates, not to TNT's enthalpy of combustion.

Combustion of TNT is described by the reaction

$$C_6H_2(NO_2)_3CH_3 + \tfrac{21}{4} O_2 \rightarrow 7\,CO_2 + \tfrac{5}{2} H_2O(g) + \tfrac{3}{2} N_2. \tag{9.15}$$

Using $\Delta H_f(\text{TNT}) \cong -63 \, \text{kJ/mol}$ [22] and other enthalpies of formation from Table 9.8, Hess's law gives

$$\Delta H_c = 7(-394) + \tfrac{5}{2}(-242) - (-63) \cong -3300 \, \text{kJ/mol}.$$

The specific energy density of TNT (molecular mass 227 gm/mol) is then

$$E_c(\text{TNT}) \cong 3300 \, \text{kJ/mol} \div (0.227 \, \text{kg/mol})$$

$$\cong 14.5 \, \text{GJ/tonne} \cong 3.5 \, \text{tTNT/tonne}. \tag{9.16}$$

Clearly, a "tonne of TNT" is much less than the enthalpy of combustion of one tonne of TNT.

Instead, when TNT explodes, it disintegrates into water, carbon monoxide, nitrogen and left over carbon,

$$C_6H_2(NO_2)_3CH_3 \rightarrow \tfrac{5}{2}H_2O + \tfrac{7}{2}CO + \tfrac{7}{2}C + \tfrac{3}{2}N_2. \tag{9.17}$$

(Other pathways can occur, including $2(\text{TNT}) \rightarrow 5\,H_2 + 12\,CO + 2\,C + 3\,N_2$, but for an estimate, we consider only the first reaction.) A calculation similar to the first gives $\Delta H_r \approx -930 \, \text{kJ/mol}$ and a specific energy density of $E_{\text{reaction}} = 4.1 \, \text{GJ/t}$ (see Problem 9.15), which is within 2% of the defined value. Note that the *tonne of TNT* is a small fraction of the specific energy of oil (41.9 GJ/t) or coal (29.3 GJ/t). Unlike TNT, however, the *decomposition* of familiar fuels such as methane, ethanol, or gasoline is endothermic. They require oxygen to liberate energy (by combustion) and do not explode unless their vapors are mixed with air.

9.5.3 Energy and Metals

Mining and refining of metals represents a small but significant contribution to world primary energy consumption. In the US, for example, mining and refining of metals accounted for 582 PJ or about 2.6% of US industrial primary energy consumption in 2007 [12, 54]. The discoveries of metals label major divisions of human history such as the *Bronze Age* and the *Iron Age*, and metals continue to play an essential role in modern societies.

With this in mind, our last example of chemical thermodynamics comes from the field of *metallurgy*, the chemistry and material science of metals. Metals are often obtained from *ore* deposits, where they are found in relatively simple *mineral compounds*, mixed with commercially worthless rock referred to as *gangue*. Sulfide minerals, such as *iron pyrite* (FeS), *chalcopyrite* (FeCuS$_2$), and *galena* (PbS), are valuable sources of iron, copper, and lead. Oxides such as *cassiterite* (SnO$_2$), *hematite* (Fe$_2$O$_3$),

---

Example 9.3   Refining Tin From Cassiterite

Estimate the temperature at which oxygen can be spontaneously driven out of SnO_2, *(a)* $SnO_2 \rightarrow Sn + O_2$, *and compare this with the temperature at which the reaction (b)* $SnO_2 + 2\,CO \rightarrow Sn + 2CO_2$ *can spontaneously occur.*

The basic idea is similar to the calcination reaction described in §9.4.1: although the reactions do not proceed spontaneously at room temperature, the reaction entropy is positive, so the reaction free energy decreases with increasing temperature and eventually goes negative, allowing the reaction to proceed spontaneously at an elevated temperature. We estimate this temperature by assuming that the reaction free energy and reaction entropy are independent of temperature. The first step is to compute the entropy of formation of SnO_2 using $\Delta G_f = \Delta H_f - T\Delta S_f$, which yields $\Delta S_f(SnO_2) = -205$ J/mol K. Then, combining the data in Tables 9.8 and 9.9, we compute the reaction free energy and reaction entropy $\Delta S_r = (\Delta H_r - \Delta G_r)/T$ and construct the free energy as a function of T,

$$\text{Reaction (a):} \quad \Delta G_r(T) = \Delta H_r - T\Delta S_r = (581 - 0.205T)\ \text{kJ/mol}\,,$$

$$\text{Reaction (b):} \quad \Delta G_r(T) = \Delta H_r - T\Delta S_r = (15 - 0.030T)\ \text{kJ/mol}\,.$$

From these we estimate that reaction (a) can proceed spontaneously above $T_0(a) = \Delta H_r/\Delta S_r = 581/0.205 \sim 2800$ K, while reaction (b) can proceed spontaneously above $T_0(b) = 15/0.03 \sim 500$ K. Thus, it is not practical to obtain tin from cassiterite simply by driving off the oxygen. In contrast, since tin melts at 505 K, a stream of carbon monoxide, such as might be available from a smoldering campfire, would easily reduce cassiterite in surrounding rocks to a puddle of molten tin!

---

and *alumina* (Al_2O_3, produced, e.g., from *bauxite* ore) are primary sources of tin, iron, and aluminum. The first step in refining a sulfide ore is generally to convert it to an oxide. This can be done by *roasting* – heating in an oxidizing environment so that oxygen replaces the sulfur – a reaction with negative reaction free energy (Problem 9.21).

In order to recover the pure metal from an oxide, it is necessary to drive off the oxygen. At first sight, the situation seems analogous to the *calcination* reaction described above, in which heat is used to drive CO_2 out of limestone. This turns out to be impossible for most oxides, however, because oxygen binds too strongly to metals. Consider, for example, cassiterite. Tin was one of the first metals refined by man – (adding tin to copper initiated the *Bronze Age* in the fourth millennium BC), which suggests that it is relatively easy to refine. The free energy of formation of SnO_2 is negative at 25 °C (see Table 9.9), so the opposite reaction, $SnO_2 \rightarrow Sn + O_2$, will not occur spontaneously at that temperature. Recall that calcination is performed by raising the temperature high enough to change the sign of the free energy of reaction, so that the reaction proceeds spontaneously. It is natural then to ask, at what temperature does SnO_2 decompose spontaneously? As shown in Example 9.3, due primarily to the large enthalpy of formation of SnO_2, cassiterite does not spontaneously decompose until temperatures of order 2800 K are realized. This is high – even for modern technology.

Table 9.9 Free energies and enthalpies of formation (kJ/mol) and molar entropies (J/mol K) for Sn, Fe, and Al and their oxides at 25 °C (data from [50]).

Element or compound	ΔH^f (kJ/mol)	ΔG^f (kJ/mol)	Entropy of oxide (J/mol K)	Entropy of metal (J/mol K)
SnO_2	−581	−520	52	52
Fe_2O_3	−824	−742	87	27
Al_2O_3	−1676	−1582	51	28

How, then, did the ancients obtain tin from cassiterite? The trick was to use carbon, or more likely carbon monoxide (CO), to steal away the oxygen from the ore. As shown in Example 9.3, the reaction $SnO_2 + 2\,CO \rightarrow Sn + 2CO_2$ can proceed spontaneously at $T_0 \sim 200$ °C. So tin can be obtained from SnO_2 by reacting it with carbon monoxide at temperatures that are easily achievable with "campfire technology." The use of carbon monoxide as a *reducing agent* to remove oxygen from metal-oxide ores remains widespread today. It is, for example, the basic chemical process used to refine iron within a *blast furnace*. The method, however, does not work for aluminum (Problem 9.22), which is why aluminum was not isolated and used until modern times.

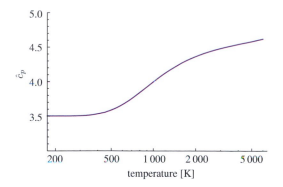

Figure 9.8 The heat capacity per molecule of CO at constant pressure as a function of temperature [24].

Discussion/Investigation Questions

9.1 The heat capacity per molecule of carbon monoxide at constant pressure is shown in Figure 9.8 from 175 K to over 5000 K. Explain why $\hat{c}_p \cong (7/2)k_B$ for small T and explain its subsequent increase with T.

9.2 H_2O has a *triple point* near $0\,°C$, where all three phases, solid (ice), liquid (water), and gas (water vapor), can coexist. So there are three possible phase transitions, solid $\to$ liquid (melting), water $\to$ gas (vaporization), and solid $\to$ vapor (*sublimation*). Each has an associated enthalpy of transition. Can you suggest a relation among ΔH_{fusion}, $\Delta H_{vaporization}$, and $\Delta H_{sublimation}$? Check your conjecture by looking up data. Be sure to get data at the same temperature.

9.3 The Iron Age began around 1200 BCE, more than 2000 years after the first appearance of bronze in human history. It may come as a surprise, therefore, that the reaction $Fe_2O_3 + 3CO \to 2Fe + 3CO_2$ is *exothermic* ($\Delta H_r = -25\,kJ/mol$) and has *negative* reaction free energy ($\Delta G_r = -29\,kJ/mol$) at NTP. So hematite can be quite easily reduced to iron by the CO produced in a hot, oxygen starved fire. Why, then, was iron not discovered earlier?

Problems

9.1 As described in §21, the universe is filled with electromagnetic radiation left over from the Big Bang. This radiation has a characteristic temperature of $\approx 2.7\,K$. At what wavelength does the power spectrum of this radiation peak? What is the energy of a photon with this wavelength?

9.2 A *blue supergiant* star may have a surface temperature as high as 50 000 K. Estimate the wavelength at which it emits maximum power. Why does the star appear blue?

9.3 When the spin of the electron in a hydrogen atom flips from parallel to antiparallel to the direction of the proton's spin, a photon of energy $E \cong 9.41 \times 10^{-25}\,J$ is

emitted. What is the photon's energy in electron volts? What is its frequency? Its wavelength? Above roughly what temperature would you expect this transition to be thermally excited? This emission line of neutral hydrogen plays a central role in modern astronomy.

9.4 [T] The cubic growth of the heat capacity of a solid as a function of temperature continues only up to roughly one tenth of the *Debye temperature* associated with the shortest wavelength excitations in the solid. This may seem surprising, since this implies that at this point the heat capacity becomes sensitive to the absence of modes whose probability of excitation is on the order of e^{-10}. A derivation of the Debye model parallel to that of the photon gas in §22.3 shows that the heat capacity in the Debye model takes the form

$$c_v/k_B = 9(T/T_D)^3 \int_0^{T_D/T} dx\, \frac{x^4 e^x}{(e^x - 1)^2},$$

where x is a rescaled energy parameter. At large x, $e^x(e^x - 1)^{-2} \to e^{-x}$ gives the Boltzmann suppression of states with increasing energy, while x^4 measures the increase in the number of states available with increasing energy. Compute (approximately) the minimum temperature T/T_D at which the factor $x^4 e^{-x}$ exceeds one for some x in the range of integration.

9.5 The power output of air conditioners is measured in "tons," an ancient nomenclature dating back to the days when air was cooled by blowing it over blocks of ice. A **ton of air conditioning** is defined to be the energy required to melt one ton (2000 pounds) of ice at $32\,°F$ distributed over a period of one day. What is the power equivalent (in kilowatts) of one ton of air conditioning?

9.6 Find the frequency of the lowest vibrational mode of a diatomic molecule in terms of the parameters of the Morse potential, eq. (9.5).

9.7 The Morse potential parameters for oxygen are given by $D_e = 5.211\,eV$, $\alpha = 2.78\,Å^{-1}$, and $r_e = 1.207\,Å$ [55]. Using the result from the previous problem, estimate the energy necessary to excite the oxygen molecule out of its ground state. Compare with your answer to Problem 7.10.

9.8 [T] Another commonly used model for the potential energy of a diatomic molecule is the **Lennard–Jones** (LJ) **potential** $V(r) = \epsilon((r_m/r)^{12} - 2(r_m/r)^6)$. Using the results of Problem 9.6, find the choice of the parameters r_m and ϵ that gives the same equilibrium separation between atoms and the same harmonic oscillator frequency as the Morse potential. Unlike the Morse potential, the depth of the LJ potential is not an independent parameter. What is the relation between the depth D_e, equilibrium separation r_e, and harmonic oscillator frequency ω of the LJ potential?

9.9 [T] Approximating the interatomic potential as a harmonic oscillator with angular frequency ω_{vib}, the vibrational-rotational energy levels of a diatomic molecule are given by $E(n, J) = \hbar\omega_{vib}(n + 1/2) +$

$\hbar^2 J(J+1)/(2\mu r_e^2)$. Here r_e is the equilibrium inter-atomic separation, μ is the reduced mass, and $J = 0, 1, 2, \ldots$ is the rotational quantum number. The CO bond in carbon monoxide has $r_e \cong 1.13 \times 10^{-10}$ m and $\hbar\omega_{vib} = 0.2657$ eV. Ignoring rotations, what is the wavelength λ_{vib} (in microns) of the radiation absorbed when the CO molecule makes a transition from vibrational level n to $n+1$? Vibrational transitions are always accompanied by a transition from rotational level J to $J \pm 1$. The result is a sequence of equally spaced absorption lines centered at λ_{vib}. What is the spacing between these vibrational-rotational absorption lines?

9.10 [T] Show that the data in Table 9.6 satisfy the definitions of Gibbs and Helmholtz free energy, (1) $\Delta G_r = \Delta H_r - T\Delta S_r$ and (2) $\Delta H_r = \Delta U_r + p\Delta V_r$.

9.11 [T] Suppose reaction data, ΔH_r° and ΔG_r° (and therefore ΔS_r°), on a chemical reaction are all known at a temperature T_0 and pressure p, but you want to know the reaction enthalpy and free energy at a different temperature T_1. Suppose the heat capacities of all the reactants (at pressure p) are known throughout the range $T_0 < T < T_1$. Find expressions for $\Delta H_r(T_1)$ and $\Delta G_r(T_1)$. Express your answers in terms of ΔC_p, the sum of the heat capacities of the products minus the heat capacities of the reactants, and the reaction data at T_0.

9.12 Estimate the amount of CO_2 produced per kilogram of $CaCO_3$ in the calcination reaction (9.10). In addition to the CO_2 released in the reaction, include the CO_2 emitted if coal is burned (at 100% efficiency) to supply the enthalpy of reaction and the energy necessary to heat the reactants to 890 °C. You can assume that the heat capacity of $CaCO_3$ is $\hat{C}_p \cong 0.09$ kJ/mol K, and assume that the coal is pure carbon so $\Delta H^f = 0$.

9.13 Look up the standard enthalpies of formation and the entropies of solid ice and liquid water and verify that ice may spontaneously melt at NTP (20 °C, 1 atm).

9.14 The convention we have used for LHV differs from the one offered by the US Department of Energy, which defines LHV as "the amount of heat released by combusting a specified quantity (initially at 25 °C) and *returning the temperature of the combustion products to 150 °C*, which assumes the latent heat of vaporization of water in the reaction products is not recovered" [53]. According to the DOE convention, the LHV of methane is 789 kJ/mol, compared with 802 kJ/mol from the convention we use in Example 9.2. Account for this difference.

9.15 Find the enthalpy of reaction for the two pathways of decomposition of TNT mentioned in §9.5.2.

9.16 Wood contains organic polymers such as long chains of cellulose $(C_6H_{10}O_5)_n$, and is commonly used as a biofuel.[15] Write a balanced equation for the complete combustion of one unit of the cellulose polymer to water and CO_2. Note that hydrogen and oxygen are already present in the ratio 2:1. Make a crude estimate of the enthalpy of combustion of cellulose by assuming that the conversion of the H and O to water does not contribute significantly, so that the energy content can be attributed entirely to the combustion of carbon. Compare your estimate to the true enthalpy of combustion of cellulose, 14.9 MJ/kg.

9.17 Estimate the lower heating value of a typical cord of wood by first estimating the mass of the wood and then assuming that this mass is completely composed of cellulose (see Problem 9.16). Compare your answer with the standard value of 26 GJ. Real "dried" firewood (density 670 kg/m³) contains a significant amount of water in addition to cellulose (and lignin and hemicellulose). When it burns, the water must be driven off as vapor. Compute the fraction y of water by mass in firewood that would account for the discrepancy you found in the first part of this problem.

9.18 Acetylene (used in welding torches) C_2H_2, sucrose (cane sugar) $C_{12}H_{22}O_{11}$, and caffeine $C_8H_{10}O_2N_4$, are all popular energy sources. Their heats of combustion are 310.6 kcal/mol, 1348.2 kcal/mol, and 1014.2 kcal/mol, respectively. Which substance has the highest specific energy density (kJ/kg)?

9.19 Thermite is a mixture of powdered metallic aluminum and iron oxide (usually Fe_2O_3). Although stable at room temperature, the reaction $2\,Al + Fe_2O_3 \rightarrow Al_2O_3 + 2\,Fe$ proceeds quickly when thermite is heated to its ignition temperature. The reaction itself generates much heat. Thermite is used as an intense heat source for welding and is particularly useful because it does not need an oxygen supply. What is the energy density in megajoules per kilogram of thermite?

9.20 An inexpensive hand warmer uses an exothermic chemical reaction to produce heat: iron reacts with oxygen to form *ferric oxide*, Fe_2O_3. Write a balanced

Table 9.10 Data for Problem 9.21.

Compound	ΔH^f (kJ/mol)	ΔG^f (kJ/mol)
PbO(s)	−219.4	−188.6
PbS(s)	−98.3	−96.7
SO_2(g)	−296.8	−300.2

---

[15] Wood also contains other organic polymers such as lignin (see §26 (Biofuels)), but the enthalpy of combustion of the other polymers is close enough to that of cellulose for the purpose of this problem.

chemical reaction for this oxidation process. Compute the energy liberated *per kilogram* of iron. Assuming that your hands each weigh about 0.4 kilogram and assuming that the specific heat of your hands is the same as water, what mass of iron would be needed to warm your hands from 10 °C to body temperature, ~38 °C?

9.21 Show that roasting lead ($2\,PbS + 3\,O_2 \rightarrow 2\,PbO + 2\,SO_2$) is an exothermic reaction and compute the free energy of this reaction under standard conditions. See Table 9.10 for data.

9.22 Show that the method of refining tin explained in Example 9.3 will not work for alumina.

Thermal Energy Conversion

Heat engines – devices that transform thermal energy into mechanical work – played a fundamental role in the development of industrial society. In the early eighteenth century, English ironmonger Thomas Newcomen developed the first practical steam engine. Newcomen's design, shown schematically in Figure 10.1, employed a *reciprocating piston*, i.e. a piston that is forced to move back and forth as heat energy is converted to mechanical energy. Later improvements by Scottish engineer James Watt led to the widespread use of the reciprocating-piston *steam engine* in a variety of applications, from pumping water out of mines and driving factory machinery to powering steamboats and locomotives. The ability to easily and relatively effectively convert heat from combustion of coal or other fuels into usable mechanical energy via steam engines provided the power needed for the Industrial Revolution that transformed society in the early nineteenth century. Many new types of engines followed in the late nineteenth and twentieth centuries. *Internal combustion engines* were developed through the latter half of the nineteenth century, and the modern *steam turbine* was invented and implemented in the 1880s.

In the mid twentieth century, *heat extraction devices* such as refrigerators, which use mechanical work to move thermal energy from a colder material to a warmer material, came into widespread use. The history and importance of refrigerators, air conditioners, and heat pumps are surveyed in §13, where these devices are discussed in detail. Heat extraction devices rely on the same principles as heat engines, and are essentially heat engines run in reverse.

Modern society depends critically upon the dual uses of heat engines in electric power generation and transport. As can be seen from Figure 3.1(a), over two-thirds of the world's electric power in 2013 came from thermal energy produced by combustion of coal, oil, and natural gas. And most vehicles used currently for transport, from

> **Reader's Guide**
> This chapter and the three that follow describe the physics of heat engines and heat extraction devices, which convert thermal energy into work or move thermal energy from low to high temperature. In this chapter we describe the basic thermodynamics behind these devices in the context of gas phase closed-cycle systems. We focus on the idealized *Carnot* and *Stirling* engines that have simple theoretical models. In the following chapters we turn to the more complex devices that dominate modern applications. Internal combustion engines are described in §11. Phase change and its utility in engines, refrigerators, air conditioners, and power plant turbines is described in §12, and the corresponding engine cycles are surveyed in §13.
>
> Prerequisites: §5 (Thermal energy), §8 (Entropy and temperature). In particular, this chapter relies heavily on the first and second laws of thermodynamics, the notion of heat capacity, and the Carnot limit on efficiency.
>
> Aside from their role as a foundation for §11–§13, the ideas developed in this chapter are relevant to the practical implementation of any thermal energy source, including §19 (Nuclear reactors), §24 (Solar thermal energy), §32 (Geothermal energy), and §33 (Fossil fuels).

automobiles to jet airplanes, rely on fossil fuel powered internal combustion engines. Even non-fossil fuel energy sources such as nuclear and solar thermal power plants depend on heat engines for conversion of thermal energy to mechanical and electrical energy.

In this chapter we focus on simple heat engines that employ a confined gas, known as the **working fluid**, to absorb and release heat in the process of doing work. In such **closed-cycle engines**, the working fluid is kept within the containment vessel and is not replaced between cycles. By contrast, in **open-cycle engines** (such as those in most automobile engines, described in §11) some or all of the working fluid is replaced during each cycle. Closed-cycle

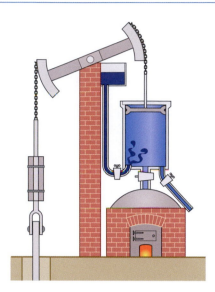

Figure 10.1 Thomas Newcomen's eighteenth-century steam engine was originally designed to pump water out of coal mines that were subject to frequent flooding. (Credit: Redrawn from Newton Henry Black, Harvey Nathaniel Davis, via Wikimedia Commons)

engines employ a sequence of processes such that after a single complete cycle, the working fluid and containment vessel return to their initial state. In an idealized implementation without friction or other losses, the only effects of the engine cycle are to transform some heat energy into mechanical energy, and to transfer some heat energy from a region of higher temperature to a region of lower temperature (generally the surrounding environment). In the engines we consider in this chapter, the working fluid is assumed to be a gas that does not change phase through the engine cycle. Engines and heat transfer devices in which the fluid changes phase, including steam engines and practical refrigerators and air conditioners, are described in §12 and §13.

Any heat engine cycle incorporates some basic component processes:

Heat input Heat is transferred into the working fluid from a substance at a higher temperature (T_+).

Expansion Thermal energy within the fluid is transformed into mechanical energy through $p\,dV$ work as the working fluid expands.

Heat output Entropy, associated with waste heat, must be dumped to a substance at a lower temperature (T_-) to satisfy the second law.

Contraction To close the cycle, the containment vessel contracts to its original size.

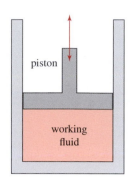

Figure 10.2 In a simple heat engine, a working fluid is contained within a cylindrical vessel. A reciprocating piston moves back and forth as the working fluid expands and contracts, converting heat energy into mechanical energy.

These processes need not be completely distinct. For example, the first two processes can be combined by allowing the working fluid to expand while adding heat to maintain a constant temperature. This process is known as *isothermal expansion* (§10.2.1). We can describe the sequence of processes involved in a heat engine by following the state of the working fluid using state variables such as the pressure p, volume V, and temperature T.

While in this chapter we focus on the conversion of thermal energy to mechanical energy through *heat engines*, which can in turn be used to power electrical generators, there are also mechanisms that can convert thermal energy directly into electrical energy. The process of *thermoelectric* energy conversion is described in Box 10.1. Such mechanisms can be useful in niche applications where size and reliability are primary considerations; due to the relatively low conversion efficiency of existing technologies, however, they are not suitable for large-scale use in electric power production.

We begin this chapter (§10.1) with a review of some of the basic thermodynamic relations and approximations that are useful in analyzing idealized heat engine cycles, and an introduction to the graphical interpretation of cycles in the pV- and ST-planes. We next analyze systematically (§10.2) the thermodynamics of a number of component processes that can be used in heat engines, and then combine these processes into complete cycles associated with two standard heat engines: the *Carnot cycle* (§10.3) and the *Stirling cycle* (§10.4). Both of these cycles realize the maximum Carnot efficiency (8.28). The Carnot cycle is generally used only as a textbook example, while the Stirling cycle has a number of real-world applications at present and may have more in the future. After discussing some of the limits on the efficiency of real engines (§10.5), we reverse the order of operation of the heat engines considered in this chapter to describe the related idealized heat extraction devices (§10.6) and introduce the appropriate measures of performance for these devices.

A basic introduction to thermal energy conversion can be found in most thermodynamics texts. Physics texts such as Kittel and Kroemer [21] can be supplemented by the more practical engineering approach found in, for example, Çengel and Boles [19] or Wark and Richards [56].

10.1 Thermodynamic Variables, Idealizations, and Representations

10.1.1 Thermodynamic Relations

To analyze an idealized heat engine cycle, we follow the state of the working fluid over time by tracking the *thermodynamic state variables*. For a fixed number of molecules N of a gas composed of a single molecular species in thermodynamic equilibrium, the state variables include pressure p, volume V, internal energy U, enthalpy H, entropy S, and of course temperature T. These state variables are not all independent. Generally, only two, e.g. pressure and volume, are independent, and the others can be computed as functions of these two.

When heat dQ is added or removed from the gas in our prototype engine or when the gas does work $dW = p\,dV$ or has work done on it, the changes in state variables are related by the differential form of the laws of thermodynamics. Specifically, the 1st Law (5.22) relates dQ and dW to the change in internal energy,

$$dU = dQ - p\,dV.\qquad(10.1)$$

In the analysis of the simplest engines and heat transfer devices we make some idealizations, all of which need to be replaced by more realistic assumptions in order to analyze and design practical, working systems. In this section we summarize the principal idealizations. Effects that go beyond these idealizations are described in §10.5.

Quasi-equilibrium/reversibility As described in §8.4.4, a process in which a system is infinitesimally close to thermodynamic equilibrium at every point in time is called a *quasi-equilibrium* process. In this chapter we assume that all processes involved in an engine cycle are quasi-equilibrium processes. Without the quasi-equilibrium assumption, the state variables of the system would not be well defined at each point in time, so this assumption is necessary for the kind of thermodynamic analysis used in this chapter.

From a practical point of view, it is helpful to design engines in which the processes are *reversible* (§8.4.4),

since this increases the efficiency of the engine. The Carnot limit (8.28) can only be achieved by using reversible processes since any additional entropy generated by non-reversible processes must be ejected to the environment as extra waste heat. The idealized engine cycles we describe in this chapter are all composed of reversible processes. In real engines, rapid expansion, contraction, and heat transfer processes deviate significantly from the reversible ideals, leading to production of extra entropy and an associated reduction in engine efficiency.

When heat dQ is *reversibly* added to or removed from a working fluid that is at (or infinitesimally close to) thermodynamic equilibrium at temperature T, the change in the entropy of the working fluid follows from the definition of temperature, as in eq. (8.21), $dQ|_{\text{reversible}} = T\,dS$, and the 1st Law can be written

$$dU = T\,dS - p\,dV.\qquad(10.2)$$

Ideal gas law We assume in this chapter that the engine's working fluid obeys the ideal gas equation of state $pV = Nk_{\text{B}}T$. For a real gas, in some regimes (for example just above the boiling point), corrections due to the finite size of molecules and interactions between molecules are important and modify the equation of state. When we consider phase-change cycles in §13 and later chapters, and need the equation of state of liquids at and below their boiling points, we must resort to tabulated data.

C_V *and* C_p In general, the heat capacity of the working fluid depends on the temperature and the volume or pressure of the fluid. As shown in §5.4, the heat capacity of an ideal gas is independent of volume and pressure, and furthermore the constant pressure and constant volume heat capacities are related by $C_p(T) = C_V(T) + Nk_{\text{B}}$. For simplicity, we assume that the heat capacity of the working fluid is constant, independent of temperature.

10.1.2 Graphical Representation of Thermodynamic Processes

When a gas expands or contracts in a quasi-equilibrium process, it can be characterized by thermodynamic state variables at each moment in time. We can choose any two state variables to describe the state of the system throughout the process. The system then follows a path – a curve – in the plane defined by these two variables. A cyclic device like a heat engine executes a closed curve. Although any pair of variables will do, we consider only (p, V) and (S, T) throughout this book. A hypothetical cycle is shown in both of these planes in Figure 10.3. Although this cycle is not realistic – implementing the steps in a quasi-equilibrium fashion would be difficult – it illustrates

Box 10.1   Thermoelectric Energy Conversion

An electric current passing through a resistor results in *Joule heating*, transforming low-entropy electromagnetic energy into higher-entropy thermal energy. This is only the simplest example, however, of a more complex set of *thermoelectric* relations between electric and thermal energy transfer. One of these, the **Seebeck effect**, which transforms thermal into electrical energy, forms the basis for exceptionally stable and long-lived *thermoelectric generators* that find application in circumstances where the engines described in §10–§13 are not sufficiently reliable.

The Seebeck effect, named after the nineteenth-century German physicist Thomas Seebeck, is the appearance of an electric field within a bar of metal when the ends of the bar are held at different temperatures as shown in the figure below. The Seebeck effect can be characterized by the equation

$$E = -S\nabla T,$$

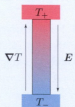

where the **Seebeck coefficient** S can be either positive or negative. This relation is somewhat of an over-simplification – the Seebeck coefficient depends strongly on temperature – but it captures the essence of the physical relationship between temperature gradient and current. For most common materials, the Seebeck coefficient at room temperature is very small, with $|S|$ of order a few μV/K, though it is higher for some substances, including, bismuth ($-72\,\mu$V/K), silicon ($440\,\mu$V/K), tellurium ($500\,\mu$V/K), and selenium ($900\,\mu$V/K), and for *doped semiconductors* (see §25).

The physics underlying the Seebeck effect relies on quantum aspects of the physics of solid materials, which we develop in §25. In brief, the idea is that when a material is at a higher temperature, the Boltzmann distribution (8.32) increases the number of electrons in excited higher-energy states, while leaving empty orbitals (*holes*) in lower-energy states. The rate at which excited electrons diffuse away from the higher-temperature region can differ from the rate at which electrons diffuse back in to fill the vacated low-energy orbitals, and the difference between these rates can lead to a net current characterized by the field $E \propto -\nabla T$.

A simple **thermoelectric generator** can be constructed by connecting two materials with different Seebeck coefficients (for example, $S_1 > 0$ and $S_2 < 0$) to heat reservoirs at temperatures T_+ and T_- as shown in the figure below. The difference in the electric fields in the two materials generates a potential difference between the terminals A and B that can drive a current through an external circuit. Ideally a thermoelectric generator should have a large difference of Seebeck coefficients, $S_1 - S_2$, good electrical conductivity (to minimize resistive losses), and poor thermal conductivity (to minimize thermal losses).

Since they have no working parts, thermoelectric generators are extremely robust and can function in challenging environments. Furthermore, they can be quite small – much smaller than any practical heat engine. These features make thermoelectric systems desirable for applications such as space exploration, where reliability and small size are at a premium. *Radioisotope thermoelectric generators* (RTGs), powered by thermal energy from radioactive sources (see §17 and §20), have been used in many satellites and interplanetary space probes since the 1960s, including the Voyager 1 and 2 probes, the Galileo and Cassini-Huygens spacecraft, and the Mars Curiosity rover. Thermoelectric generators are also used in a variety of other remote situations, such as for gas pipelines, on mountain tops, or underwater where solar photovoltaic energy is not a practical solution.

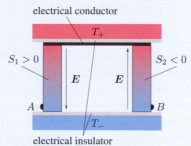

Thermoelectric generators are generally much less efficient than existing heat engines. NASA's *Multimission Thermoelectric Generator (MMTG)*, which powers the Mars Curiosity rover, is powered by the decay of plutonium-238 (see Table 20.7). It operates between $T_+ \cong 800$ K and $T_- \cong 470$ K and has a maximum efficiency of 6%, only 15% of the Carnot efficiency. While it seems unlikely that thermoelectric generators will ever be used for large-scale power production, one interesting potential application is in making use of some of the waste heat from systems such as automobile engines and photovoltaic solar cells to improve system efficiency.

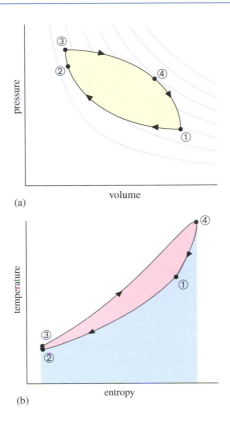

(b)

Figure 10.3 The path described by a hypothetical heat engine in the pV- and ST-planes, with the working fluid taken to be an ideal gas. Some notable corresponding points are labeled ① to ④. See the text for explanations of the shaded areas.

the methods that are applied to other more familiar cycles later in this chapter. Some contours of constant temperature, with $pV = $ constant, are shown in the pV-plane, and some corresponding points in the two graphs are labeled ① to ④.

Graphs like Figure 10.3 are useful in a number of ways:

The path Graphs in the pV- and ST-planes enable one to follow the history of the system in a simple visual and intuitive fashion. In step {12},[1] we can see that the gas is compressed as entropy (heat) is removed from it (dumped into the environment) and its temperature drops, while in step {23} its pressure rises rapidly as it begins to heat up while still being compressed without losing entropy (heat) to the environment. In {34} heat is added (from a source of thermal energy) as the gas expands and does work, and

---

[1] Steps in a cycle are labeled with the initial and final points in braces.

finally in {41} the gas continues to expand and does more work, although heat is no longer added and its temperature drops.

Visualizing work flow The pV graph gives an explicit visualization of the work performed by the system: because $dW = p\,dV$, the area under the portions of the curve $p(V)$ where V is increasing measures the *work done by the system*, and the area under portions where V is decreasing measures the *work done on the system*. Thus, the area contained within the system's path in the pV-plane equals *the net work W* performed by the system in a reversible cycle; this area, $\oint p\,dV$, is shaded in yellow in Figure 10.3(a).

Visualizing heat flow Similarly, the ST graph of a cycle composed of reversible processes gives us information about the flow of heat into or out of the system. Because $dQ|_{\text{reversible}} = T\,dS$, the area under the portions of the curve $T(S)$ where S increases (the sum of the [lower] blue and [upper] red shaded areas in Figure 10.3(b)) measures *heat added to the system*, while the area under portions where S decreases (shaded blue) measures *heat expelled by the system*. Combining these, $\oint S\,dT$, the area within the system's path in the ST-plane (shaded red) gives *the net thermal energy Q* consumed by the system in a reversible cycle. Since the internal energy of the system does not change over a cycle, $\Delta U = Q - W = 0$; the areas within the paths in the pV- and ST-planes both equal W.

Note that if an engine does not operate through quasi-equilibrium processes, it cannot be accurately represented by a point in the pV- or ST-planes at every stage of its cycle. For example, a gas that undergoes free expansion as described in §8.4 cannot be described as having a well-defined pressure and volume at intermediate stages of the process, though the state variables are well defined before and after this process. Graphically, irreversible steps can be denoted by dotted lines that connect points where the system is at or near thermodynamic equilibrium. These indicate that during the step in question the system does not follow a well-defined path through the space of thermodynamic states. It is important to remember that neither heat transferred nor work done can be computed by integrating $T\,dS$ or $p\,dV$ over *non-quasi-equilibrium* portions of the system's cycle.

10.2 Thermodynamic Processes in Gas Phase Engines

Using the idealizations and graphical tools outlined in the previous section, we describe several of the principal

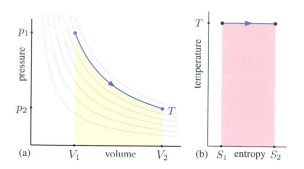

Figure 10.5 Graph of isothermal expansion in (a) the pV-plane, with some lines of constant temperature (*isotherms*) obeying $pV = $ constant shown; (b) the ST-plane. The areas under the paths give the work done $\int p\,dV$ and heat absorbed $\int T\,dS$, respectively (which are equal). In this figure, and throughout §10–§13, isotherms are denoted in blue.

processes used in gas phase engines: isothermal and adiabatic expansion (or contraction) and isometric heating.

10.2.1 Isothermal Expansion/Contraction

Isothermal expansion (i.e. constant temperature expansion) of an ideal gas is illustrated in Figure 10.4(a). The gas in a cylinder starts out in thermal equilibrium with a large

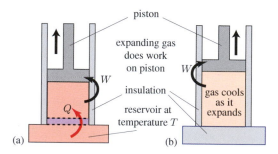

Figure 10.4 Isothermal (a) and adiabatic (b) expansion of gas in a cylinder. In isothermal expansion the gas expands, doing work on the piston. The gas is kept at a fixed temperature T by coupling to a thermal reservoir at that temperature. Thermal energy passes through the boundary, shaded in purple, between the thermal reservoir and the cylinder. In adiabatic expansion, the temperature of the gas drops as it does work on the piston because the gas is isolated from its environment by insulating boundaries denoted in gray.

heat reservoir at temperature T. Heat, but no matter, can flow back and forth between the cylinder and the reservoir. The initial volume of the gas is V_1 and its pressure is therefore $p_1 = Nk_B T/V_1$, where N is the number of molecules in the gas. We then allow a piston to move out, so that the gas expands and does work on the piston. Expansion continues until the volume reaches V_2, which defines the final state of the system.

Throughout the process, the gas stays in (or infinitesimally close to) thermal equilibrium with the reservoir at temperature T. As the gas expands and does work on the piston, the internal energy would decrease in the absence of any heat input ($dU = -p\,dV$ if $dQ = 0$). But the thermal coupling to the external heat reservoir maintains the temperature of the fluid, so $dU = C_V dT = 0$. It follows then from eq. (10.1) that the heat added is

$$dQ = p\,dV\,, \tag{10.3}$$

which is precisely the amount of work $dW = p\,dV$ done on the piston by the expanding fluid. Note that this work includes both mechanical work performed by the piston, such as compressing a spring or turning a crankshaft, and work done by the piston on the environment, such as displacing the air outside the piston, which is generally at a lower pressure and temperature than the working fluid inside the cylinder.

The isothermal expansion process is shown graphically in the pV- and ST-planes in Figure 10.5. The path in the pV-plane is an **isotherm** (constant temperature curve) taking the form of a hyperbola $pV = Nk_B T = $ constant. The area under the path in the pV-plane is the work done by the gas as it expands,

Isothermal Expansion

In the isothermal expansion of an ideal gas, T is constant and

$$dU = 0, \qquad dQ = dW,$$

$$pV = \text{constant}.$$

For an ideal gas expanding from V_1 to V_2,

$$W = Q = Nk_B T \ln(V_2/V_1) \text{ and}$$

$$\Delta S = Nk_B \ln(V_2/V_1).$$

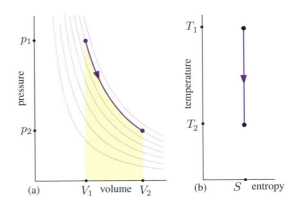

(a) V_1 volume V_2 (b) S entropy

Figure 10.6 Graph of adiabatic expansion in (a) the p-V plane. Some other lines of constant entropy $pV^\gamma = \text{constant}$ (*adiabats*) are shown; (b) the S-T plane. γ is the adiabatic index of the working fluid. In this figure, and throughout §10–§13, adiabats are denoted in purple.

$$W = \int_{V_1}^{V_2} p\, dV = Nk_B T \int_{V_1}^{V_2} \frac{dV}{V}$$

$$= Nk_B T \ln \frac{V_2}{V_1} = p_1 V_1 \ln \frac{V_2}{V_1}. \qquad (10.4)$$

The path in the ST-plane is extremely simple: since T is constant, the path is a straight line from S_1 to S_2, and the area under the line, $T\Delta S$, is the heat absorbed from the reservoir at temperature T. It is straightforward to compute the total change in entropy from $\Delta S = Q|_{\text{reversible}}/T$. From the 1st Law, $Q = W = Nk_B T \ln(V_2/V_1)$. Since T is constant throughout, and the process is assumed to be carried out reversibly,

$$\Delta S = \frac{Q}{T} = Nk_B \ln \frac{V_2}{V_1}. \qquad (10.5)$$

Isothermal compression can be idealized as the reverse of the process just described. Exerting an external mechanical force on the piston, we can compress the fluid. As the piston does work on the fluid, the temperature of the fluid rises slightly, and heat is then transferred from the fluid to the reservoir. As in isothermal expansion, $W = Q$, except that now W represents work done *on* the system and Q represents heat *expelled by* the system (both positive). The graph of the isothermal compression process in the pV- and ST-planes is identical to that shown in Figure 10.5, except that the directions of the arrows are reversed.

To approach the idealized processes of isothermal expansion and compression in a real system such as an engine cycle, it is necessary to have heat transfer occur on a rapid time scale comparable to the cycle time. This requires very efficient heat exchangers or a slow engine cycle.

10.2.2 Adiabatic Expansion

Next we consider (reversible) **adiabatic expansion**. In this process, the gas is allowed to expand reversibly – for example, by reducing the external pressure very slowly – but is completely thermally isolated, so that $dQ = 0$

throughout. This process is shown schematically in Figure 10.4(b) and in the pV- and ST-planes in Figure 10.6. As the gas expands, its volume increases while its pressure and temperature decrease. Since $dQ = 0$, it follows that $dS = dQ/T = 0$ so that the entropy is constant. Thus, reversible adiabatic expansion is often referred to as *isentropic* expansion.[2]

In the adiabatic expansion process, the fluid does work on the piston $W = \int p\, dV$, given by the area under the curve in Figure 10.6(a). In this process, the internal energy decreases by the amount of work done, given infinitesimally by

$$dU = -p\, dV. \qquad (10.6)$$

Therefore, as the gas expands its temperature drops, and pV is not constant as it is for an isothermal expansion. To relate pressure and volume in adiabatic expansion, we combine the 1st Law, $C_V dT = dU = -p\, dV$ (since $dQ = 0$) with the differential form of the ideal gas equation of state, $p\, dV + V\, dp = Nk_B dT$ to give

$$(C_V + Nk_B)p\, dV + C_V V\, dp = 0. \qquad (10.7)$$

Using the ideal gas relation $C_p = C_V + Nk_B$, eq. (10.7) can be integrated to give

$$pV^\gamma = \text{constant}, \qquad (10.8)$$

---

[2] Any process for which $\Delta S = 0$ is **isentropic**, while **adiabatic** in general refers to processes in which there is no heat transfer. Adiabatic *reversible* processes are isentropic. A free expansion (see §8.4) is adiabatic, but is irreversible and not isentropic.

Example 10.1　Isothermal Expansion of Air

Consider a closed cylinder of volume 0.5 L containing air. The cylinder is immersed in a container of boiling water and the air is brought to 100 °C and 1.3 atm of pressure. A piston is released, allowing the air to slowly expand while the temperature is kept at 100 °C by heat flowing into the cylinder from the boiling water. The expansion continues isothermally until the internal pressure reaches 1 atm.

What is the final volume of the air in the cylinder and what is the total heat input from the water?

During the isothermal expansion the air obeys the ideal gas law $pV = Nk_\mathrm{B}T$, with T held constant, so $p_1V_1 = p_2V_2$, and

$$V_2 = \frac{p_1}{p_2}V_1 = \left(\frac{1.3\,\mathrm{atm}}{1\,\mathrm{atm}}\right)0.5\,\mathrm{L} = 0.65\,\mathrm{L}.$$

From eq. (10.5) the heat input is $p_1V_1\ln(V_2/V_1) = 0.65\,\mathrm{atm\,L} \times 0.262 = 17.2\,\mathrm{J}$. Note that this is the same as the work done by the piston, from eq. (10.3).

---

where the *adiabatic index* $\gamma \equiv C_p/C_V$ is always greater than one (see §5.4.2), so p falls more steeply with V in adiabatic expansion than in isothermal expansion. This is illustrated in Figure 10.8, where both isotherms and **adiabats** (curves of constant entropy) are plotted. For air, which is dominantly composed of diatomic molecules ($> 99\%$ O_2 and N_2) and can be approximated as an ideal gas with $C_V \cong \frac{5}{2}Nk_\mathrm{B}$ at 300 K,

$$\gamma_\mathrm{air} \cong 1.4. \tag{10.9}$$

To compute the work done in adiabatic expansion we integrate $p\,dV$ with $pV^\gamma = $ constant from V_1 to V_2 (see Problem 10.2),

$$W = \Delta U = \int_{V_1}^{V_2} p\,dV = \frac{1}{\gamma - 1}(p_1V_1 - p_2V_2)$$

$$= \frac{1}{\gamma - 1}Nk_\mathrm{B}(T_1 - T_2). \tag{10.10}$$

Adiabatic compression is just adiabatic expansion run backwards. The graphs of adiabatic compression are identical to Figure 10.6, but with the arrows reversed.

In contrast to isothermal expansion and compression, the idealized processes of adiabatic expansion and compression require that heat transfer occur on a *much slower* time scale than the expansion or compression process. Thus, for implementations of these processes in real systems, either shorter time scales or strongly insulating materials are used.

10.2.3 Isometric Heating

The two processes considered in §10.2.1 and §10.2.2 are characterized by constant temperature and constant entropy respectively. There are many other thermodynamic processes we might consider that are relevant for various

---

Adiabatic Expansion

In reversible adiabatic (isentropic) expansion of an ideal gas, S is constant and

$$dQ = 0 \qquad\qquad dU = -p\,dV$$
$$pV^\gamma = \text{constant}, \qquad \gamma = C_p/C_v.$$

For an ideal gas expanding from (p_1, V_1) to (p_2, V_2),

$$W = -\Delta U = \frac{p_1V_1 - p_2V_2}{\gamma - 1} = \frac{1}{\gamma - 1}Nk_\mathrm{B}(T_1 - T_2).$$

---

kinds of engine systems. For the engines we consider in this chapter, however, the only other relevant process is heating at constant volume: **isometric heating**.

In an isometric heating process, as described in §5.4.1, the working fluid is gradually heated from an initial temperature T_1 to a final temperature T_2 while the volume V is held constant. This process is graphed in Figure 10.7. The pV curve is a straight line with constant V, and for an ideal gas $\Delta p = Nk_\mathrm{B}\Delta T/V$. The S–T relation for this process can be determined from the specific heat relation for the working fluid. Since the volume is kept fixed, the fluid does no work and $dU = dQ$ throughout the process. Thus,

$$dS = \frac{dQ}{T} = \frac{dU}{T} = C_V\frac{dT}{T}. \tag{10.11}$$

Integrating, we see that

$$S_2 = S_1 + C_V\ln\frac{T_2}{T_1}. \tag{10.12}$$

Example 10.2 Adiabatic Expansion of Air

Consider again the cylinder of heated air from the previous example, which begins at $100\,°C$ and $1.3\,$atm of pressure, volume $0.5\,$L. Now, let the air expand adiabatically (no heat added) until the pressure is $1\,$atm.
What is the final volume of the air? How much work is done?

During the adiabatic process the quantity pV^γ stays constant, with $\gamma \cong 1.4$ so

$$(V_2)^\gamma = \frac{p_1}{p_2}(V_1)^\gamma \quad \Rightarrow \quad V_2 \cong V_1 (1.3)^{1/1.4} \cong (0.5\,\text{L})(1.20) \cong 0.6\,\text{L}.$$

Note that the final volume is less than in the isothermal expansion (Example 10.1) and therefore less work is done on the piston,

$$W = \frac{p_1 V_1 - p_2 V_2}{\gamma - 1} \cong \frac{1}{(1.4 - 1)}\left((1.3\,\text{atm}) \times (0.5\,\text{L}) - (1\,\text{atm}) \times (0.6\,\text{L})\right) \cong 12\,\text{J},$$

compared with $17.2\,$J in the corresponding isothermal expansion. A concrete energy application where the difference between adiabatic expansion and isothermal expansion is relevant is compressed air energy storage (CAES), see §37.2.2.

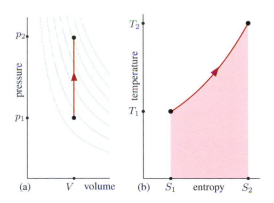

Figure 10.7 Graph of isometric heating in (a) the pV-plane, with some isotherms shown; (b) the ST-plane, where the area under the path gives the heat added during the expansion. The ST curve is of the form $S = a + C_V \ln T$. In this figure, and throughout §10–§13, constant volume paths are denoted in red.

For example, for a monatomic ideal gas with $C_V = (3/2)Nk_B$, we can move the $(3/2)$ inside the logarithm, where it becomes an exponent, so $S = Nk_B \ln T^{3/2} +$ constant, where the constant is independent of T.

Performing isometric heating in a reversible fashion is not easy. If we simply place a gas at temperature T in contact with a heat reservoir at temperature $T_+ > T$, then the two are not near equilibrium and transfer of heat from the reservoir to the gas is an irreversible process. If thermal energy dQ is transferred to the gas then its entropy increases by $dS_g = dQ/T$, which is more than the magnitude of the entropy lost from the reservoir, $|dS_r| = dQ/T_+$. The total change in entropy is then positive, $dS_{\text{tot}} = dS_g + dS_r > 0$, and the process is

irreversible. Following the logic of §8.4.4 (see eq. (8.20)), a process such as isometric heating, in which the temperature changes in the course of the process, can only be accomplished in a quasi-equilibrium and reversible fashion by ensuring that the heat transferred into the gas at each temperature T comes from a source at a temperature only slightly higher than T. In §10.4 we show how this can be done with the help of a device known as a *regenerator*. The regenerator is the key mechanism behind the *Stirling engine*, a reversible heat engine that can, in principle, reach Carnot efficiency and holds promise as a useful practical heat engine for some applications (§10.4).

Isobaric heating is heating at constant pressure, a process similar in some ways to isometric heating, except that $p\,dV$ work is performed. We leave the details to Problem 10.3, but summarize the results in the accompanying Concept Box along with the isometric case.

10.3 Carnot Engine

The thermodynamic processes described in the previous section can be combined in various ways to describe idealized heat engines. We begin with the **Carnot engine**, which is based on a thermodynamic cycle first described by Carnot in 1824. The **Carnot cycle** achieves the maximum possible efficiency (8.28) for any heat engine operating between fixed maximum and minimum temperatures T_+ and T_-. The Carnot engine has long been the standard textbook example of an efficient heat engine, although for reasons discussed below, it is not particularly useful for practical applications. Nonetheless, it provides a clear and simple example of how the working cycle of a heat engine can be treated thermodynamically.

Isometric and Isobaric Heating

For reversible *isometric* (constant volume) heating,

$$dU = dQ,$$
$$\Delta S = C_V \ln(1 + \Delta T/T).$$

For reversible *isobaric* (constant pressure) heating,

$$dH = dU + p\,dV = dQ,$$
$$\Delta S = C_p \ln(1 + \Delta T/T).$$

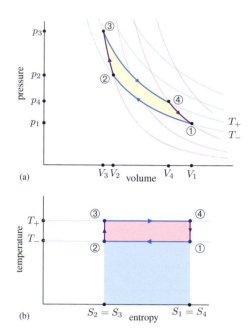

(a)

(b)

Figure 10.8 Graph of a Carnot cycle in the (a) pV- and (b) ST-planes. To help interpret this graph and others of the same type, different colors are used for each fundamental process: blue for isotherms, purple for adiabats, and in subsequent figures, red for isometrics (constant volume) and green for isobars (constant pressure). The conditions are taken from the geothermal engine example in Example 10.3: $T_+ = 100\,°C$, $T_- = 20\,°C$ (so $T_+/T_- = 1.27$), and $V_1/V_3 = 2.5$. An unrealistically large adiabatic index, $\gamma = 2.5$, is chosen in order to make the cycle steps clearer. The shaded areas have the same interpretation as in Figure 10.3. In particular, the yellow (W) and red ($Q_+ - Q_-$) shaded areas are equal and the ratio of the red (upper) shaded area to the sum of blue and red shaded areas (Q_+) is the efficiency of the engine.

An idealized heat engine cycle can be constructed by combining reversible processes to construct a clockwise path in the pV-plane. We describe the Carnot cycle in terms of gas in a cylinder with a piston, though other implementations are possible. The Carnot cycle begins with the working fluid coupled to a heat reservoir at temperature T_-, with the piston out so that the fluid fills the maximum volume V_1 within the cylinder. This is depicted as state ① in Figures 10.8 and 10.9. The working fluid is then subjected to the following sequence of processes, depicted schematically in Figure 10.9:[3]

{12} Isothermal compression at temperature T_-, doing work on the gas, dumping heat to the environment at the lower temperature, and ending in state ② at lower volume and higher pressure.

{23} Adiabatic compression to state ③, doing more work on the gas with no heat transfer, raising the pressure to p_3 and the temperature to T_+ .

{34} Isothermal expansion at temperature T_+ to state ④, with work done by the gas on the piston and heat transfer into the gas from the high-temperature reservoir.

{41} Adiabatic expansion back to state ①, with more work done by the gas on the piston with no heat transfer, ending at temperature T_-.

It is straightforward to verify that the Carnot cycle achieves the maximum possible efficiency – the *Carnot limit* – derived in §8. No heat passes in or out of the system in the adiabatic steps {23} and {41}, and the thermal energy of the working fluid does not change around a full cycle so the 1st Law (energy conservation) gives $Q_{\{34\}} - Q_{\{12\}} = W$, where W is the net work done by the system. The heat input is $Q_{\{34\}} = T_+\Delta S$ where $\Delta S = S_4 - S_3$ is the entropy change in the step {34}. The heat output is correspondingly $Q_{\{12\}} = T_-\Delta S$, with the same ΔS in both cases because the {23} and {41} steps are adiabatic. The efficiency of the engine is then

$$\eta = \frac{W}{Q_{\{34\}}} = \frac{Q_{\{34\}} - Q_{\{12\}}}{Q_{\{34\}}} = \frac{(T_+ - T_-)}{T_+} = \eta_C\,,$$

(10.13)

precisely the Carnot efficiency.

---

[3] We adopt the convention that all thermodynamic cycles begin in the state of lowest temperature and pressure. Most engineering texts use this convention in most cases, though physics texts often use a different convention where the first state has the lowest volume and highest pressure. We denote the heat absorbed or expelled in a process step {ij} by $Q_{\{ij\}}$. Similarly, the work done on or by the device in a given step is denoted $W_{\{ij\}}$. By convention $Q_{\{ij\}}$ and $W_{\{ij\}}$ are always taken to be positive.

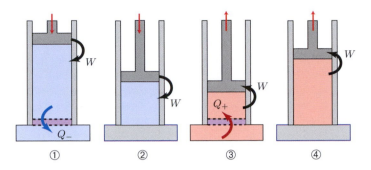

Figure 10.9 The Carnot engine cycle: the condition of the gas is shown at each of the points ①...④. The flow of work and heat and the contact with heat reservoirs that characterize the step immediately following each set point are shown as well. Thus at ①, for example, the gas is shown in contact with the thermal reservoir at T_- and the piston is shown about to do work $W_{\{12\}}$ and transfer heat Q_- to the reservoir.

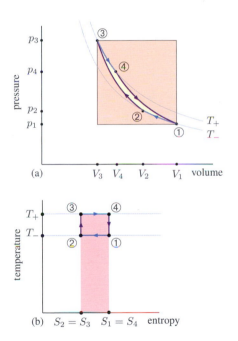

Figure 10.10 A realistic Carnot cycle ($\gamma = 1.4$) graphed in the pV- and ST-planes for the same conditions as Figure 10.8. Note the small amount of work done (yellow area) compared to the range of pressure and volume employed in the cycle ($p_3 - p_1)(V_3 - V_1$) (illustrated by the orange shaded area). The efficiency is the ratio of the yellow shaded area in (a) to the red shaded area in (b) (see §10.1.2)

> ### Carnot Engine
>
> A Carnot engine can, in principle, reach the maximum possible efficiency for a heat engine that extracts thermal energy from a reservoir at temperature T_+ and expels waste heat at temperature T_-.
>
> The Carnot cycle is composed of an isothermal compression at temperature T_-, adiabatic compression to temperature T_+, isothermal expansion at T_+, and finally adiabatic expansion to the starting point of the cycle at temperature T_-.
>
> Carnot engines are impractical, in part because they produce relatively little work per cycle for given extremes of volume and pressure.

and minimum temperature and volume *set points*[4] as Figure 10.8, but with a realistic value of $\gamma = 1.4$. The work per cycle, indicated by the yellow shaded area, is much smaller than that in Figure 10.8. As $\gamma \to 1$ the isothermal portions of the cycle shrink away to nothing, so both Q_+ and Q_- vanish, as does $W = Q_+ - Q_-$. Thus, for γ close to 1, although the efficiency remains high, the amount of work done per cycle becomes very small.

10.4 Stirling Engine

The **Stirling engine** is another gas phase, closed-cycle heat engine that achieves optimal Carnot efficiency in

Despite its high efficiency, the Carnot cycle is not useful in practical applications. Figure 10.8 was produced using an unrealistically large value of $\gamma = 2.5$ in order to make the Carnot cycle larger and more visible. Figure 10.10 shows the Carnot cycle with the same maximum

---

[4] A **set point** is the value of a state function, typically a volume, pressure, or temperature, at a specific point in the thermodynamic cycle, which is chosen by the cycle designer.

Example 10.3   A Geothermal Carnot Engine

Consider a Carnot engine that uses heat from a geothermal water source at $T_+ = 100\,°C$. Assume that the engine uses a cylinder of air at maximum volume $V_1 = 1\,L$ and minimum volume $V_3 = 0.4\,L$, and that waste heat is dumped at an ambient temperature of $T_- = 20\,°C$. The pressure at point ① in the cycle, before the isothermal compression, is taken to be $p_1 = 1$ atm. (These conditions were used in producing the graphs of the Carnot cycle in Figures 10.8 and 10.10.) *What is the engine efficiency and what is the energy extracted per cycle?*

The efficiency is $\eta = (T_+ - T_-)/T_+ = (373 - 293)/373 \cong 0.214$, so the engine is about 21% efficient. The heat energy input Q_+ is equal to the work done on the piston in the isothermal expansion {34},

$$Q_+ = Q_{\{34\}} = W_{\{34\}} = \int_{V_3}^{V_4} p\, dV = p_3 V_3 \ln \frac{V_4}{V_3},$$

as in eq. (10.4). From the ideal gas law,

$$p_3 V_3 = p_1 V_1 \frac{T_+}{T_-} = (1\,\text{atm})(1\,L)(373/293) \cong 129\,J.$$

To compute V_4 we use the adiabatic relation for air $p_4 V_4^{1.4} = p_1 V_1^{1.4}$ and the ideal gas law $p_4 V_4/T_4 = p_1 V_1/T_1$. Dividing the first of these by the second, we find

$$V_4^{0.4} = V_1^{0.4}(T_-/T_+) \quad \rightarrow \quad V_4 \cong V_1 (0.793)^{2.5} \cong 0.547\,L.$$

So the energy input is

$$Q_+ = p_3 V_3 \ln \frac{V_4}{V_3} \cong (129\,J)\left(\ln \frac{0.547}{0.4}\right) \cong 40.2\,J.$$

We can similarly compute $V_2 \cong V_3 (0.793)^{-2.5} \cong 0.731\,L$, so the waste heat output is

$$Q_- = Q_{\{12\}} = p_2 V_2 \ln \frac{V_1}{V_2} \cong (101\,J)\left(\ln \frac{1}{0.731}\right) \cong 31.6\,J.$$

Thus, the total energy output from one cycle of this Carnot engine is

$$W = Q_+ - Q_- \cong 8.6\,J.$$

This matches with our earlier computation of the efficiency as $W/Q_{\{34\}} \cong 0.21$.

---

its idealized thermodynamic cycle, but does significantly more useful work per cycle than the Carnot engine.

Early steam engines were subject to frequent explosions, which caused significant damages, injury, and loss of life. The Stirling engine was developed in 1815, by the Scottish clergyman and inventor Robert Stirling as a less dangerous practical alternative. Stirling engines (then called "air engines") were used fairly broadly over the following century in applications that included pumping water from mines, driving workshop machinery, and powering ships. With the development of the internal combustion engine in the first half of the twentieth century, the use of Stirling engines was reduced to a few niche applications. Because of their high efficiency and other desirable features, however, Stirling engines may play a significant role in energy systems of the future.

Stirling engines are based on the *Stirling cycle*, which we introduce in §10.4.1. In §10.4.2 we describe the practical implementation of the Stirling engine, and in §10.4.3 we summarize some salient features of Stirling engines and provide an example for comparison with the Carnot engine described in Example 10.3. A more detailed introduction to Stirling engines is given in [57].

10.4.1 The Stirling Cycle

Stirling engines are based on a cycle similar to the Carnot cycle, but with the adiabatic expansion and compression steps replaced by isometric heating and cooling. The **Stirling cycle**, depicted in Figure 10.11, consists of the following sequence of processes:

{12} Isothermal compression at temperature T_- with heat output $Q_{\{12\}}$ and work input $W_{\{12\}}$.

{23} Isometric heating from T_- to T_+ with heat input $Q_{\{23\}}$ and no work done.

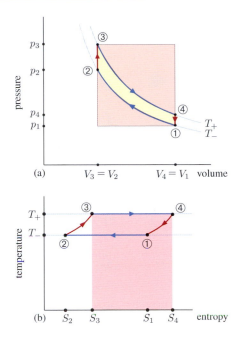

Figure 10.11 Graph of Stirling cycle in the (a) pV- and (b) ST-planes. The temperature and volume set points are the same as Figure 10.10 so that the work done per cycle (area within the pV-curves) can be compared. The significance of the shaded areas is the same as in Figure 10.10.

---

[34] Isothermal expansion at temperature T_+ with heat input $Q_{\{34\}}$ and work output $W_{\{34\}}$.

[41] Isometric cooling from T_+ to T_- with heat output $Q_{\{41\}}$ and no work done.

The net work done in a Stirling cycle is $W = W_{\{34\}} - W_{\{12\}}$. Because the temperature of the fluid changes in isometric heating and cooling processes, heat is added to the working fluid at a continuous range of temperatures. Such a process is challenging to implement in a reversible fashion, since for a process to be reversible the heat must be added at each temperature from a reservoir that is at a temperature only infinitesimally higher than that of the fluid. The key to implementing the Stirling cycle using reversible processes is that the heat $Q_{\{41\}}$ that is output in the isometric cooling process is stored in a device called a **regenerator**, and is returned into the system as $Q_{\{23\}}$ during the isometric heating process. From eq. (10.12), we see that the entropy–temperature relations describing the curves ② → ③ and ④ → ① in the ST-graph in Figure 10.11(b) are identical except for a constant shift in S. The areas under these curves are thus equal, and therefore indeed $Q_{\{41\}} = Q_{\{23\}}$. We describe the physical mechanism of the regenerator in the following section. For

now we simply note that if the heat output at each temperature T between T_- and T_+ is returned to the system from the regenerator as heat input at the same temperature, then there is no net increase in entropy from these processes. All heat input from external sources in the Stirling cycle is added during the isothermal expansion at temperature T_+ and all heat output to an external sink is at temperature T_-. Thus, just as for the Carnot cycle, the Stirling cycle has maximum efficiency

$$\eta = \frac{W}{Q_+} = \frac{Q_{\{34\}} - Q_{\{12\}}}{Q_{\{34\}}} = \frac{(T_+ - T_-)}{T_+} = \eta_C \,. \tag{10.14}$$

10.4.2 Implementation of a Stirling Cycle

A schematic of a simplified Stirling engine is depicted in Figure 10.12(a). The design includes two pistons on opposite sides of a region containing the working fluid. Contained within the space between the pistons is a fixed matrix of material (crosshatched in the figure), something like steel wool, which acts as the regenerator. The regenerator contains a network of voids that are sufficiently connected in space to allow the working fluid to pass through the material, yet the regenerator must also contain sufficient mass to hold the thermal energy output $Q_{\{41\}}$, and must support a temperature gradient without significant heat flow during the time needed for a cycle of the engine. On each side of the regenerator there is a working volume between the regenerator and corresponding piston. These are known as the **expansion space** and **compression space**. The expansion space is kept in contact with a heat reservoir at the higher temperature T_+, while the compression space is kept at the cooler temperature T_-. Generally these temperatures are maintained using efficient *heat exchangers*. For example, the compression space can be cooled by fluid pumped through pipes around the space. This uses some energy and decreases efficiency, but maintains the compression space at the lower temperature required.

With the geometry depicted in Figure 10.12(a), the Stirling cycle is realized by the following sequence of operations, shown in Figure 10.12(b):

[12] Beginning with the working fluid all or almost all in the compression space, so that the expansion volume is very small, the compression piston is moved inward to perform isothermal compression of the working fluid at temperature T_-, while the expansion piston is held fixed.

[23] The compression piston then continues to move inward as the expansion piston moves outward in tandem. This keeps the total volume available to the working fluid fixed, while moving the fluid from the compression region

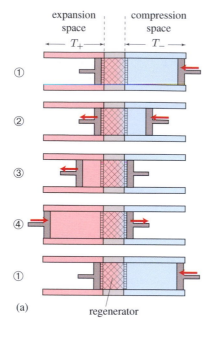

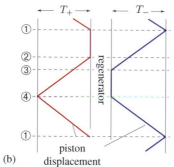

Figure 10.12 A simplified Stirling engine and cycle. (a) A schematic diagram with two pistons, showing the relative positions of the pistons at points ①...④ of the Stirling cycle. (b) A graph of the motion of the pistons as a function of time during one cycle. Note that a more realistic engine could have pistons oscillating sinusoidally with time, which would reduce the thermodynamic efficiency (see text).

to the expansion region. As the fluid passes through the regenerator it is isometrically heated from T_- to T_+ with heat $Q_{\{23\}}$ added from the regenerator matrix.

[34] The compression piston stops moving when the compression volume reaches a minimum at ③, and the expansion piston keeps moving, so that the fluid performs isothermal expansion at temperature T_+.

[41] Finally, after the volume has again reached its original size (at ④), the expansion piston moves back in, and

the compression piston moves out in tandem, again keeping the working volume fixed and performing isometric cooling from T_+ to T_- as the volume passes through the regenerator matrix from the warm region to the cool region. Heat $Q_{\{41\}}$ ($= Q_{\{23\}}$) is passed back into the regenerator matrix.

There are many variants to the specific implementation of the regenerator, associated with many particular designs of Stirling engine, but the general principle is always the same and described by this sequence of processes. Note again that the key is that when the working fluid is heated and cooled isometrically it is done through the regenerator so that the process is reversible.

10.4.3 Stirling Engine: Features and Practical Potential

Unlike most other engines in common use, such as internal combustion engines based on the Diesel and Otto cycles (§11), the Stirling engine can, in principle, reach the Carnot efficiency limit. Stirling engines also perform much more work per cycle than Carnot engines operating with the same set points. This can be seen immediately by comparing the cycles for given limits of p and V, as in Figure 10.13.

In addition to their advantage in efficiency, Stirling engines are potentially cleaner and quieter than many other types of engines in current use, and can operate on a wider range of fuels. *Internal combustion engines* (§11) rely on combustion of fuel added to the working fluid for the heat input. Stirling engines, by contrast, are **external combustion engines**, which operate with an external heat source. As a result, Stirling engines can operate on continuous combustion of fuel, allowing for much quieter operation. Furthermore, internal combustion engines are *open-cycle* engines and must eject the spent fuel on every cycle, often

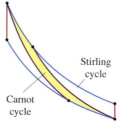

Figure 10.13 The Carnot and Stirling engines depicted in Figures 10.10 and 10.11 have the same minimum and maximum pressure, volume, and temperature. The work per cycle can be compared by superimposing their pV curves as in this figure. As computed in Example 10.4, the Stirling engine produces almost three times as much work per cycle.

Example 10.4   A Geothermal Stirling Cycle

Consider the design of a geothermal heat engine as proposed in Example 10.3. Rather than a Carnot engine, however, consider a Stirling engine operating between the same volume, temperature, and pressure set points.
How much work is performed per cycle?
The work done by the piston in the expansion phase {34} is, analogous to the computation in Example 10.3,

$$W_{\{34\}} = \int_{V_3}^{V_4} p \, dV = p_3 V_3 \ln \frac{V_4}{V_3},$$

where $p_3 V_3$ is again $\cong 129 \, \text{J}$ but now $V_4/V_3 = 2.5$ so

$$W_{\{34\}} \cong (129 \, \text{J})(\ln 2.5) \cong 118 \, \text{J}.$$

Similarly, $W_{\{12\}} \cong (101 \, \text{J})(\ln 2.5) \cong 92.5 \, \text{J}$, so the work done is

$$W = W_{\{34\}} - W_{\{12\}} \cong 25.5 \, \text{J}.$$

This is approximately three times as much work per cycle as is done by the equivalent Carnot engine (though with the same efficiency).

---

including some unburned fraction of the fuel that enters the environment as pollutants. Stirling engines, on the other hand, can burn fuel until combustion is complete. They are thus cleaner than internal combustion engines (though, of course, they still produce the CO_2 resulting from complete combustion of the carbon when powered by fossil fuels). Finally, because the heat source is external, any heat source can be used for power. The Stirling engine can thus be powered by solar thermal energy or other non-fossil heat sources, in addition to arbitrary combustible fuels.

Despite the clear advantages of the Stirling engine in theory, for the most part Stirling engines have not seen widespread use in recent decades. While the theoretical efficiency of the Stirling cycle is very high, realizing this efficiency in practice presents substantial engineering challenges. Existing Stirling engines tend to be larger and heavier than their internal combustion counterparts, though in general they are more robust and require less repair. Some of the particular challenges to constructing an efficient Stirling engine include the following issues:

Materials Because the expansion region of a Stirling engine is constantly exposed to the high temperature, the materials must be more robust than in an internal combustion engine, where the highest temperatures are reached only briefly, and less exotic and expensive alloys can be used in the engine.

Heat exchange For a Stirling engine to cycle quickly and thus achieve maximum power (since the power achieved

by an engine is given by the work done per cycle times the cycle rate), heat must move quickly into the expansion region and out of the compression region. This requires very efficient heat exchangers, which add complexity to the engine. Even with efficient heat exchangers, it is difficult to achieve truly isothermal expansion and contraction. A better model is a **polytropic** process with constant pV^α with α somewhere between 1 (isothermal) and γ (adiabatic).

Regenerator The Stirling engine regenerator must have a fairly high heat capacity to readily absorb heat energy during the isometric cooling process. This requires a reasonably massive material, again raising the cost of the engine.

Currently, Stirling engines are used in several niche markets. In particular they are used in reverse as miniature cryogenic refrigerators (*cryocoolers*) to cool infrared detectors on missile guidance systems and in night-vision equipment. Stirling engines are of particular value in these applications because of their low noise and reliable operation. Stirling engines are also used in some space applications for similar reasons.

As energy efficiency becomes a higher priority and as efforts increase to shift energy sources away from fossil fuels, advances in Stirling engine technology may enable these devices to play an increasingly important role in a wide range of systems from air conditioning to solar thermal power plants to automobile engines.

Stirling Engine

The Stirling engine is another heat engine that can, in principle reach Carnot efficiency. The Stirling engine uses a *regenerator* to implement isometric heating and cooling without additional entropy increase.

The Stirling cycle is composed of isothermal compression at temperature T_-, isometric heating to temperature T_+, isothermal expansion at T_+, and finally isometric cooling to the starting point of the cycle at temperature T_-.

The Stirling cycle produces significantly more work per cycle than a Carnot engine operating between the same extremes of pressure, volume, and temperature.

10.5 Limitations to Efficiency of Real Engines

Due to materials and engineering limitations, no real heat engine can achieve the ideal Carnot efficiency. Issues that decrease the actual efficiency of real engines, including Carnot and Stirling engines, below the thermodynamic ideal include:

Cycles are not ideal In a real engine, the cycle is not composed of distinct idealized processes. For example, in a realistic Stirling cycle, the pistons may undergo sinusoidal motion rather than the motion described in Figure 10.12(b). The resulting thermodynamic cycle encompasses a smaller area in the pV-plane than the idealized cycle, even in the quasi-equilibrium approximation. Since not all heat is taken in at the maximum temperature nor expelled at the minimum temperature in a real cycle, the efficiency is decreased from the ideal Carnot efficiency.

Irreversibility is unavoidable We have assumed reversible quasi-equilibrium processes, where the working fluid is near equilibrium at every point in time. In a real engine, many effects lead to departures from this ideal. *Heat transfer* is never truly reversible, as temperature gradients are necessary for heat to flow. In every realistic working system, *friction* in moving parts converts mechanical energy into heat energy. This heat energy is conducted or radiated into the environment, making the mechanical process irreversible. In some real engines, there are processes that involve *unconstrained expansion*, such as occurs when a valve is opened (§13.2.3), which increases the entropy of the gas as it expands into a larger volume without doing work and without the addition of heat. Finally, in any practical engine, like an automobile engine, where intake, combustion, and exhaust happen thousands of times per minute (e.g. as measured on a tachometer), the gas inside the engine has a spatially inhomogeneous pressure, density, and temperature, and is far from equilibrium at any given instant. To accurately estimate the efficiency of a real engine, engineers often use sophisticated modeling tools that simulate fluid and heat flow within the engine.

Materials limitations In reality, no material is a perfect thermal insulator or conductor. Thus, truly adiabatic expansion and compression, and reversible isothermal heating and cooling are not possible in practice. Deviations from these ideals again reduce the efficiency of real engines. In particular, any part of an engine that becomes extremely hot will transfer thermal energy to the rest of the engine and the environment through conduction, convection, and radiation, leading to energy loss and a net increase in entropy that decreases the overall engine efficiency.

The idealized thermodynamic analysis described here is also inaccurate for a variety of other reasons that are less directly related to efficiency. For example, the heat capacity C_V, adiabatic index γ, etc., are all temperature-dependent. This results in a deviation from the idealized behavior described above. Corrections to the ideal gas law may also affect the behavior of real engines.

Despite these limitations, the thermodynamic analysis of engine efficiency in the ideal limit forms an invaluable starting point for estimating the efficiency of different engine designs. In a real engineering context, however, detailed consideration of many issues including those mentioned above must be taken into account to optimize efficiency.

10.6 Heat Extraction Devices: Refrigerators and Heat Pumps

Up to this point we have only considered half of the story of conversion between thermal and mechanical energy. Heat engines can be considered as devices that move thermal energy from high to low temperature and produce work. **Heat extraction devices** do the opposite: they employ work to move heat from low to high temperature. The same operations that are combined to make a heat engine, if applied in reverse, will act as a heat extraction device. After describing general aspects of such devices, we look at the steps in a cooling cycle based on the Carnot engine run in reverse.

If the primary goal of transferring heat from the colder region to the hotter region is to cool the region at lower temperature, the device is called a **refrigerator** or an **air conditioner**. Although they may be used over different temperature ranges, the basic thermodynamics of air conditioners and refrigerators is the same. Sometimes, for brevity, we refer to both as *refrigerators*. If the primary goal is to increase or maintain the temperature of the hotter region, for example to use thermal energy from the colder outside environment to warm a living space, then the device is called a **heat pump**. Heat extraction devices based on phase-change cycles are described in detail in §13.2 and heat pumps are discussed further in §32 (Geothermal energy).

10.6.1 Coefficients of Performance for Heat Extraction Devices

Before describing heat extraction cycles in more detail, we reconsider the notion of *efficiency* as it applies to this situation. The flow of energy in a heat extraction device is sketched in Figure 10.14. Note the resemblance to Figure 8.4, where the basic structure of an engine was sketched. The heat extraction device uses work W to extract an amount of thermal energy Q_- from a low-temperature reservoir $\mathcal{R}_-$, and delivers heat Q_+ to a high-temperature reservoir $\mathcal{R}_+$. The effectiveness of a heat pump or refrigerator is termed its **coefficient of performance** or **CoP**. In general, the coefficient of performance measures the ratio of the desired effect to the energy input. (For a heat engine, the efficiency is also sometimes referred to as a CoP.) For a heat pump, the CoP is measured by the amount of heat delivered to the warm environment divided by the work needed to deliver it,

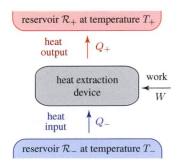

Figure 10.14 A schematic description of a heat extraction device which uses work W to extract thermal energy Q_- from a reservoir $\mathcal{R}_-$ at temperature T_- and expels thermal energy Q_+ to a reservoir $\mathcal{R}_+$ at temperature T_+.

$$\mathrm{CoP_{hp}} = \frac{Q_+}{W}. \qquad (10.15)$$

For a refrigerator, the aim is to remove thermal energy from the low-temperature environment, so its CoP is measured by the amount of heat removed divided by the work needed to remove it,

$$\mathrm{CoP_r} = \frac{Q_-}{W}. \qquad (10.16)$$

The 1st Law relates Q_+ and Q_- to W,

$$Q_+ = Q_- + W, \qquad (10.17)$$

which enables us to relate the coefficients of performance for the same device working as a refrigerator or as a heat pump,

$$\mathrm{CoP_{hp}} = \frac{Q_+}{W} = \frac{Q_- + W}{W} = 1 + \mathrm{CoP_r}. \qquad (10.18)$$

Either way it is used, the 2nd Law limits the magnitude of the CoP of a heat extraction device. Because, as for an engine, the entropy and energy in a heat extraction device do not change over a full cycle, the same argument that led to eq. (8.26) for engines leads to

$$\frac{Q_-}{Q_+} \leq \frac{T_-}{T_+} \qquad (10.19)$$

for heat extraction. Combining this result with the expressions for the CoP of refrigerators and heat pumps, we obtain the *Carnot limit* on the performance of heat extraction devices,

$$\mathrm{CoP}|_{hp} \equiv \frac{Q_+}{W} \leq \frac{T_+}{T_+ - T_-} = \frac{1}{\eta_C},$$
$$\mathrm{CoP}|_{r} \equiv \frac{Q_-}{W} \leq \frac{T_-}{T_+ - T_-} = \frac{1}{\eta_C} - 1, \qquad (10.20)$$

where $\eta_C = (T_+ - T_-)/T_+$ is the Carnot limit on the efficiency of an engine based on the same thermodynamic cycle. Figure 10.15 shows the three (log scaled) coefficients of performance (with $\mathrm{CoP_{engine}} \equiv \eta$) as functions of $\Delta T/T_+ = (T_+ - T_-)/T_+$.

Note that the Carnot limits on $\mathrm{CoP_{hp}}$ and $\mathrm{CoP_r}$ can both be *greater than one*.[5] When the temperature difference $T_+ - T_-$ is very small, engines become very inefficient. It is possible, at least in principle, however, to move large quantities of thermal energy from low to high temperature across a small temperature difference with

[5] Because it can be greater than one, the CoP for a heat extraction device is not referred to as an *efficiency*, a term reserved for a ratio between actual and ideal performance, which must lie between zero and one. In §36 (Systems) we return to the concept of efficiency and extend it so that it can be applied to heat extraction devices as well as engines.

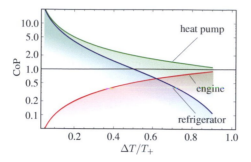

Figure 10.15 The thermodynamic upper limits on the CoPs of an engine, refrigerator, and heat pump as a function of $\Delta T/T$. Note that refrigerators and heat pumps become more efficient as $\Delta T/T \to 0$, while engines become more efficient as $\Delta T/T \to 1$.

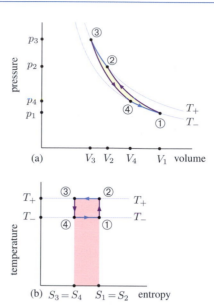

Figure 10.16 A Carnot cycle for a heat pump operating between $T_- = 20\,°\mathrm{C}$ and $T_+ = 100\,°\mathrm{C}$, with a compression ratio of $V_1/V_3 = 2.5$. See Figure 10.10 for comparison.

Heat Extraction Devices

In principle, any reversible engine cycle run backward becomes a heat extraction cycle that uses work W to extract heat Q_- from a domain at low temperature T_- and expel it as heat Q_+ into a domain at high temperature T_+. Such a device can function either as a refrigerator or a heat pump. The 2nd Law limits the coefficient of performance (CoP) of these devices to

$$\mathrm{CoP}|_\mathrm{hp} \equiv \frac{Q_+}{W} \le \frac{T_+}{T_+ - T_-} = \frac{1}{\eta_C},$$

$$\mathrm{CoP}|_\mathrm{r} \equiv \frac{Q_-}{W} \le \frac{T_-}{T_+ - T_-} = \frac{1}{\eta_C} - 1,$$

where $\eta_C = (T_+ - T_-)/T_+$.

These limits can be significantly greater than one for $T_+ \gtrsim T_-$.

very little work. This is the reason why heat pumps are so effective as a source of space heating, particularly in environments where the temperature difference between inside and outside is not large.

10.6.2 A Carnot Heat Extraction Cycle

Figure 10.16 shows the thermodynamic properties of an ideal Carnot device run as a heat pump under the same conditions as the Carnot engine of Figure 10.10. The detailed description of the steps in the cycle and some of their

implications are left to Example 10.5. We note a few crucial features here. First, the cycle is run *in the opposite direction* than an engine cycle, as denoted by the directions of the arrows in Figure 10.16. Thus, the yellow shaded area in the figure signifies the work *done on* the device by an outside agent and the red shaded area signifies the heat *transferred to* the high-temperature reservoir at T_+. The coefficient of performance is the ratio of these areas, which is the inverse of the efficiency of the engine run on the same Carnot cycle. Thus the CoP for an ideal Carnot heat pump or refrigerator becomes infinite in the limit in which $T_+ \to T_-$, as the work needed for a fixed heat transfer goes to 0.

Despite their potentially high coefficients of performance, heat pumps and refrigerators based on Carnot cycles are of little practical interest. Such devices move relatively little heat per cycle compared to other gas cycles. An even more important consideration is that devices that employ fluids that change phase over the cycle have much more desirable heat transfer properties and are much easier to run in near-equilibrium conditions. Thus, even though they usually include an intrinsically irreversible step (free expansion) and therefore cannot reach the Carnot limit, phase-change systems dominate practical applications. We postpone further discussion of heat extraction cycles until

> ### Example 10.5 Using a Carnot Heat Pump to Boil Water
>
> *Describe the steps required for a Carnot engine to be run in reverse as a heat pump to boil water. Use the same minimum and maximum temperatures and volumes as for the Carnot engine example in Example 10.3. Compute the CoP of a heat pump that uses mechanical energy to take in heat at 20°C and output heat at 100°C.*
>
> For the reverse of the Carnot cycle analyzed previously, heat Q_- is taken in at temperature T_-, and combined with mechanical work to produce heat output Q_+ at temperature T_+. The Carnot heat extraction cycle shown in Figure 10.16 begins at ① with the gas at low temperature and low pressure. First it is necessary to heat the gas to T_+. This is accomplished by compressing it adiabatically in step {12}. The work $W_{\{12\}}$ required to compress the gas heats it to T_+. Next, the hot gas is put in contact with the high-temperature thermal reservoir and compressed isothermally from ② to ③. This squeezes heat $Q_{\{23\}}$ out of the gas, which can be used to boil some water. To ready the gas to absorb heat from the low-temperature reservoir, it must be cooled to T_-. This is accomplished by expanding the gas adiabatically in step {34}. Finally, the gas is put in contact with the low-temperature thermal reservoir, and allowed to expand isothermally, sucking thermal energy $Q_{\{41\}}$ out of the ambient environment.
>
> The coefficient of performance of this device acting as a heat pump is
>
> $$\text{CoP} = \frac{Q_+}{W} = \frac{T_+}{T_+ - T_-} = 4.66.$$
>
> With the bounds on temperature and volumes used previously, the work input per cycle is 8.6 J, and heat output is 40.2 J per cycle. Note that heat pumps are generally quite effective, and move much more thermal energy from cold to hot than the work input, as long as the temperature differential is sufficiently small.

§13, after we have explored the thermodynamics of phase change in §12.

Discussion/Investigation Questions

10.1 Why does it take more work to compress a gas adiabatically than isothermally from V_1 to V_2? Why does it take more heat to raise the temperature of a gas isobarically (constant pressure) than isometrically from T_1 to T_2?

10.2 In an engine cycle like the one shown in Figure 10.3, heat is added over a range of temperatures. Explain why even if it were run reversibly, its efficiency would be less than the Carnot limit defined by the high- and low-temperature set points ④ and ② shown in the figure ($\eta_C = (T_4 - T_2)/T_4$). Is this in itself a disadvantage of such a cycle?

10.3 Consider the following ideal gas engine cycle with only three steps: {12} – isothermal compression at T_- from V_1 to V_2; {23} – isobaric (constant pressure) heating from T_- to T_+, where T_+ is chosen so that the volume increases back to V_1; {31} – isometric cooling from T_+ back to T_-, which returns the system to ①. Sketch this system's path in the pV- and ST-planes. Discuss problems with implementing this cycle reversibly.

10.4 An inventor comes to you with a proposal for a new high-tech material containing tiny Stirling engines which exploit the energy difference between the human body and the outside environment. He claims that a complete bodysuit made of his material will generate 15 W of power in steady state when worn by a human

at rest in a space at room temperature of 68 °F. Would you invest in a company based on his invention? Give a scientific argument for your answer.

10.5 The Ericsson engine, invented by Swedish-American inventor John Ericsson in 1833, is based on a cycle that resembles a Stirling cycle except that the isometric steps are replaced by isobaric steps. It uses a regenerator in the same way as a Stirling engine. Sketch the Ericsson cycle in the pV- and ST-planes. What is the efficiency of an ideal, reversible Ericsson engine?

10.6 Since it is not used up in a Carnot cycle, the gas used as a working fluid could be chosen to optimize the properties of the engine. What advantages would helium (a monatomic gas) have over air in a Carnot engine?

Problems

10.1 By how much does the temperature drop in the example of adiabatic expansion in Example 10.2?

10.2 [T] Derive eq. (10.10).

10.3 [T] In isobaric heating (and expansion), heat is added reversibly to a gas kept at constant pressure. Find the change in volume and entropy when a sample of an ideal, monatomic gas, initially at temperature T_1, volume V_1, and pressure p is heated isobarically to T_2. Plot the system's path in the pV- and ST-planes.

10.4 [T] Box 10.1 describes a thermoelectric generator and mentions that the materials used should have large Seebeck coefficient S, small poor thermal conductivity

k, and good electrical conductivity σ. Use dimensional analysis to show that $Z = \sigma S^2 \Delta T / k$ is the unique dimensionless *figure of merit* for the material for thermoelectricity. Show that if the material is a metal that obeys the Wiedemann–Franz–Lorentz law (see Box 6.1) then $Z = (S^2/L)(\Delta T/T)$, so that the Seebeck coefficient is the critical parameter.

10.5 **[T]** Show that the changes in entropy found for isothermal and isometric heating of an ideal gas (eqs. (10.5) and (10.12)) agree with the prediction from the Sackur–Tetrode equation (8.65).

10.6 Consider a Carnot engine cycle operating between a maximum temperature of $T_3 = T_4 = 400\,°F$ and a minimum temperature of $T_1 = T_2 = 60\,°F$. Assume that the lowest pressure attained is $p_1 = 1\,atm$, the highest pressure is $p_3 = 9\,atm$, and the minimum volume of the working gas space is $V_3 = 1\,L$. Take $\gamma = 1.4$. Compute the pressure and volume at each point in the cycle, as well as the heat in, heat out, and work out. Compute the overall efficiency from these numbers and check that this agrees with theoretical Carnot efficiency.

10.7 **[T]** Stirling engines generate more work per cycle than Carnot engines operating under the same conditions. Find the ratio $W_{\text{Carnot}}/W_{\text{Stirling}}$ for engines run with the same compression ratio, $r = V_1/V_3$, and temperature ratio T_+/T_-. Assume the working fluid is an ideal gas.

10.8 Mirrors are used to concentrate sunlight and heat a molten salt mixture to 500 K. A heat engine is then used to convert the thermal energy to useful mechanical form. Compare a Carnot engine to a Stirling engine with the same operating parameters. In each case the engine operates between the maximum temperature of $T_+ = 500\,K$ and a minimum temperature of $T_- = 300\,K$ (ambient temperature). Maximum volume for the engine in each case is 3 L, minimum volume is 0.2 L. Assume that the minimum pressure attained in the cycle is 1 atm, and the working fluid is air with $\gamma = 1.4$. Compute and compare the efficiency and net useful work per cycle for the two engines.

10.9 **[H]** Consider the cycle proposed in Question 10.3 quantitatively. Assume that it is executed reversibly. Show that its efficiency η relative to the efficiency η_C of a Carnot engine operating between the same temperature limits is given by

$$\frac{\eta}{\eta_C} = \frac{1}{\hat{c}_p} \left(\frac{r}{r-1} \right) \left(1 - \frac{\ln r}{r-1} \right),$$

where $\hat{c}_p = C_p/Nk_B$ is the heat capacity per molecule at constant pressure and $r = V_1/V_2$ is the compression ratio. Plot the efficiency ratio versus r. Why is it always less than one?

10.10 **[T]** In section §5.2.2, we gave an example of a situation where an ideal gas is heated to a temperature T_{in} and allowed to expand a small amount against an external ambient pressure, where the external gas has temperature T_{out}. We found that the fraction of thermal energy dU used to do useful work dW precisely realized the Carnot efficiency. Show that the expansion process described there can be completed to a closed loop giving a heat engine with Carnot efficiency by using reversible processes to bring the fluid back to the initial volume so that addition of thermal energy dU restores the initial temperature T_{in}. [Hint: you may find it helpful to make use of a *regenerator* to store thermal energy at a continuous range of temperatures.]

10.11 An air conditioner run on electricity and based on a (ideal) Carnot cycle operates between temperatures $T_- = 25\,°C$ and $T_+ = 40\,°C$. The gas in the air conditioner has $\gamma = 1.75$. The AC is designed to remove 10 000 BTU/h from a living space. How much electric power does it draw? Suppose the gas in the AC has a pressure of 2 atm at its maximum volume, which is 1 L, and the minimum volume is 0.4 L. At what rate must the AC cycle in order to accomplish its mission? [Hint: See Example 10.5.]

10.12 Air conditioners are rated by their **Energy Efficiency Ratio**, or *EER*, defined as the cooling power P_r (in Btu/h), divided by the total electric power input P_e (in watts), measured with $T_+ = 95\,°F$ and $T_- = 80\,°F$: EER $= P_r[\text{BTU/h}]/P_e[\text{W}]$. US energy efficiency standards require central air conditioners to have an EER value greater than $\approx$ 11.7.[6] What is the CoP$_r$ of an air conditioner with EER = 12? How does it compare with the Carnot limit?

---

[6] The US uses a seasonal average EER (*SEER*) which is only approximately related to the EER, and requires SEER $\geq$ 13.

Internal Combustion Engines

The theoretical analysis of heat engines is extended into more applied territory in this chapter. We use the thermodynamic analysis from the previous chapter as a starting point to understand the *internal combustion engines* used to power automobiles, trucks, most ships, and other forms of transport. The study of internal combustion engines takes us close to the world of real systems engineering. Although detailed design and optimization are well beyond the scope of this book, there are several motivations for making such a foray in the case of this particular application. First, it allows us to appreciate how the rather theoretical methods we have developed so far play a role in understanding real-world energy systems. Second, the engines we consider here play a central role in the energy landscape of the early twenty-first century.

As mentioned in §2 (Mechanics), more than 25% of US energy use goes to transportation. This sector is responsible for roughly 33% of US CO_2 emissions [58]. While transportation only accounts for about 20% of global CO_2 emissions, the percentage is expected to grow as personal motor vehicle use increases in countries with expanding economies such as China and India. One of the most important and difficult challenges in converting to renewable energy sources and reducing carbon emissions is to find more efficient and less carbon intensive ways to power transport.

Transportation systems present a unique challenge in a number of ways. First, vehicles such as automobiles and airplanes must carry their own power sources with them (unless a radically new means of distributing power is implemented on a large scale). This means that they must be powered by fuel with a high energy density. Historically, this has led to the widespread use of liquid hydrocarbon fuels for most forms of transport. Second, the engines powering vehicles must be light and compact, with smooth and reliable operation. Third, while there

Reader's Guide

This chapter focuses on the internal combustion engines that dominate road, rail, and most ship transport. The thermodynamic cycles that model the Otto, Atkinson, and Diesel engines are described here, as are some of the properties of the hydrocarbons that fuel them. Gas turbines, which share some common features with internal combustion engines, appear in §13 (Power cycles) along with other cycles used to generate electric power.

Prerequisites: §10 (Heat engines).

Familiarity with internal combustion engines provides context for the discussion of fossil and biofuels in §26 and §33, and frames the discussion of energy storage in §37.

is some hope for capturing carbon emissions from large power plants or industries (§35.4.2), there is no near-term viable approach to sequestration of CO_2 from motor vehicle engines. The particular difficulties of this important part of worldwide energy use justify a more detailed discussion of the existing engines used to power various forms of transport.

Internal combustion engines are open-cycle engines powered by combustion that takes place within the vapor that serves as the working fluid. As mentioned briefly in §10.4.3, internal combustion engines are in general less efficient than closed-cycle external combustion engines, in part because they often release heat and/or unburned fuel in the exhaust step of the cycle. The main types of internal combustion engines in current use can be put into several general classes, each of which is roughly associated with an ideal thermodynamic engine cycle. It should be emphasized that these idealized cycles involve many approximations, particularly of the combustion process, which miss important features of real engine processes. Part of the point of this chapter is to develop a better understanding of the limitations of the idealized thermodynamic analysis by looking at some examples of physical effects occurring in real engines that are missed in the idealized analysis.

Spark ignition (SI) engines Most passenger automobiles and many light trucks run on **spark ignition engines**, in which a fuel–air mixture is compressed and then ignited. The fuel then combusts, releasing chemical energy that is converted into mechanical energy as the combusting mixture expands. As we describe in §11.1, spark ignition engines can be roughly described by a thermodynamic cycle known as the *Otto cycle*, based on an idealization of the combustion as a constant volume process. Much of this chapter is devoted to a basic introduction to spark ignition engines, the Otto cycle, and some discussion of various losses and deviations of real engines from the ideal thermodynamic model.

Compression ignition (CI) engines **Compression ignition engines** (also known as *diesel engines*) are very similar to spark ignition engines, but in a CI engine only air is taken in and compressed. The fuel is injected after the air has been compressed and is at such a high temperature that the fuel ignites immediately upon injection. CI engines are approximately described by the *Diesel cycle*, which we describe in §11.4.2. The Diesel cycle is based on an idealization of the combustion process as a constant pressure process. Some automobiles, most trucks, heavy vehicles, locomotives, and ships use diesel engines.

Gas turbine engines Gas turbine engines are used in some ships and most modern aircraft, and also in modern combined-cycle power plants. Although they are also internal combustion engines, they involve compressors and turbines, two components that are also used in the context of electric power generating cycles. We postpone the discussion of gas turbines to §13 (Power cycles).

The theory and engineering practice of internal combustion engines has progressed to an extremely high level of sophistication. We only touch here on some of the basic themes. Internal combustion engines are described in general terms in many engineering thermodynamics texts such as [19]. A more thorough introduction, which combines the basic theories of thermodynamics, heat transfer, fluid mechanics, and combustion into a systematic introduction to the major issues in internal combustion engines, is given by Milton [59]. The authoritative text by Heywood [60] treats these issues, and others, in much more detail, and will be of great value to anyone who wants to understand the operation of modern internal combustion engines more deeply.

11.1 Spark Ignition Engines and the Otto Cycle

11.1.1 Four-stroke Spark Ignition Engines

The internal combustion engine has an interesting and complicated history [59]. The earliest internal combustion engines were inspired by cannons. In 1857, the Italian engineers Eugenio Barsanti and Felice Matteucci designed a device in which an explosion in a cylinder propelled a heavy piston upward. The cylinder was sufficiently long that the piston stayed within the cylinder throughout its trajectory. After reaching the highest point of its motion, the piston returned downward, engaging a ratchet mechanism that did mechanical work. In 1860, the Belgian engineer Etienne Lenoir patented the first commercial internal combustion gas engine based on a *two-stroke cycle* (see below). Another commercial engine, based on a similar idea to that of Barsanti and Matteucci, was produced by the German inventor Nikolaus Otto and engineer Eugen Langen in 1867. In these early engines, the fuel was ignited at atmospheric pressure, and the resulting efficiency was quite low.

Alphonse Beau de Rochas, a French engineer, emphasized in an 1861 paper the thermodynamic advantages of compressing the working fluid before ignition in an internal combustion engine, which allows for a significantly greater amount of work to be done in the expansion process. The first internal combustion engine incorporating this idea and operating on the now-standard **four-stroke cycle** was the "Silent" engine developed by Otto in 1876. This engine had an efficiency of roughly 14%, and was the forerunner of the engine appearing in modern automobiles. Later improvements by German engineers Gottlieb Daimler and Karl Benz and others led to the first use of the internal combustion engine to power a vehicle in 1886.

The sequence of operations in a four-stroke spark ignition engine is illustrated in Figure 11.1. As in the closed-cycle engines discussed in §10, the basic operating mechanism consists of a cylinder containing a gas. In this case the gas is an air–fuel mixture, with a piston forced to move outward by the expansion of the gas, performing mechanical work. In the internal combustion engine, however, the working fluid does not stay within the cylinder over multiple cycles. Instead, valves allow new air/fuel in at the beginning of each cycle, and allow the products of combustion to be expelled from the cylinder as *exhaust* after the fuel has been burned in each cycle. The linear motion of the piston back and forth as the four-stroke cycle is executed puts engines of this type in the class of **reciprocating engines** – as opposed to *turbines*, which power rotational motion. The linear motion of the piston is transferred via a

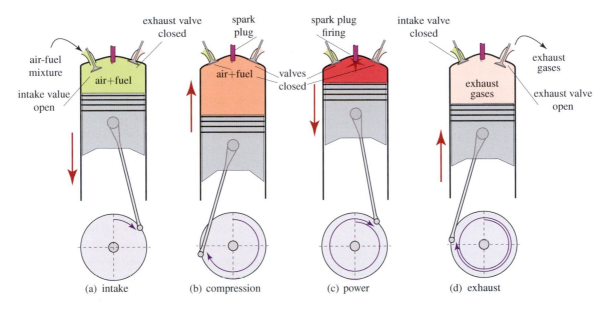

Figure 11.1 The stages of operation of a four-stroke spark ignition engine. (Credit: Adapted from Encyclopaedia Britannica, Inc)

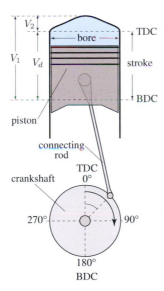

Figure 11.2 Geometry of piston and connection via connecting rod to crankshaft. Figure after [60].

connecting rod to a **crankshaft** that rotates as the piston moves back and forth (see Figure 11.2). The crankshaft in turn drives another rotating shaft (the **camshaft**) that controls the intake and exhaust valves. When the piston is at its highest point (in a vertically oriented cylinder), the connecting rod is at the top of the crankshaft, called **top-dead-center** (TDC). The corresponding lowest point

is called **bottom-dead-center** (BDC). In a single cycle of the four-stroke engine, the piston goes up and down twice, rotating the crankshaft twice. Let us consider in more detail each of the strokes in the four-stroke cycle, as depicted in Figure 11.1:

Stroke one: Intake (Figure 11.1(a))   In the first stroke, the piston moves downward (outward). The intake valve is opened as the volume in the cylinder expands, and new air/fuel is drawn into the cylinder.

Stroke two: Compression (Figure 11.1(b))   In the second stroke, the piston moves upward (inward) again. Both valves are kept closed, and the air/fuel mixture is compressed. An important characteristic of internal combustion engines is the ratio between the maximum volume V_1 of the cylinder (when the piston is at its lowest point so the crank is at BDC) and the minimum volume V_2 of the cylinder (at TDC). This ratio is called the **compression ratio**, $r = V_1/V_2$, and is often written in the form $r : 1$. For example, the original Otto silent engine had a compression ratio of 2.5:1, while for reasons that are discussed below typical modern four-stroke spark ignition engines have a compression ratio around 10:1. The difference between the maximum volume V_1 and the minimum volume V_2 is known as the cylinder's **displacement**. The power that an internal combustion engine can deliver depends critically on its displacement while its efficiency depends critically on the compression ratio.

Stroke three: Power (Figure 11.1(c))   When the compression stroke is almost complete, a spark ignites the compressed fuel–air mixture. This leads to a further increase in pressure, and after the crank passes TDC, the piston is forced downward by the hot, high-pressure gas.

Stroke four: Exhaust (Figure 11.1(d))   Near the end of the power stroke, the exhaust valve opens. The piston then moves upward again as the crankshaft goes through another half revolution. During this stroke, the exhaust gases, consisting of the results of the fuel combustion process, are expelled from the engine.

11.1.2 The Otto Cycle (Constant-Volume Combustion)

The processes occurring in a real four-stroke engine cycle differ in a number of ways from the kind of idealized thermodynamic processes we used in the previous chapter to describe Stirling and Carnot engines. Two differences are particularly important. First, the fuel–air mixture undergoes a combustion process in the cycle, which changes its chemical make-up and thermodynamic properties. Since air is mostly nitrogen, which is not involved in combustion, the thermodynamic changes resulting from combustion are not too extreme. As a first approximation, the working fluid can be treated as an ideal gas with (constant) heat capacity close to that of air at 300 K throughout the cycle, and combustion, which takes place relatively rapidly compared to the motion of the piston, can be modeled as a constant-volume process in which the heat of combustion is added to the working fluid. Second, the still-hot products of combustion are expelled and fresh air is drawn into the cylinder in the exhaust and intake strokes of every cycle. This can be roughly approximated as heat rejection at constant volume, as if the same gas was retained and only its heat content was expelled. This idealization allows us to approximate the actual process by a *closed cycle* in which air is the working fluid throughout, and to perform a standard thermodynamic analysis that forms a good starting point for understanding real engines. Together the assumptions just outlined define the **cold air standard analysis**.[1]

Although the idealized thermodynamic analysis cannot be expected to predict the exact efficiency of a real system, it is useful in providing a first approximation of the actual performance. The idealized thermodynamic analysis is also a good guide as to how systems behave when parameters are changed, and is thus a good differential,

if not absolute, analysis tool. To accurately predict the detailed behavior of a real engine, more sophisticated tools are needed, often involving detailed computer simulations. Some of the additional issues that must be addressed when analyzing engine cycles more precisely are discussed in §11.3.

The thermodynamic cycle that approximates the cycle of a spark ignition engine is called the **Otto cycle**. The hallmark of the Otto cycle is the approximation of combustion as a constant-volume process. The Otto cycle, graphed in the pV- and ST-planes in Figure 11.3, consists of the following sequence of processes, beginning at the point of lowest pressure and temperature just after fresh air has been drawn into the cylinder in stroke one. All processes are taken to be *reversible* in the cold air standard approximation.

[12] Adiabatic compression from volume V_1 to volume V_2. This is a good approximation to the compression process in the second stroke of a four-stroke engine. Since the temperature of the fuel–air mixture is not exceptionally high in this process and the compression occurs quickly, not too much heat is lost to the cylinder walls during compression. Work $W_{\{12\}}$ is done *on* the gas.

[23] Isometric heating of the fuel–air mixture, with thermal energy $Q_{\{23\}}$ added to the gas in the cylinder. This process is used as a thermodynamic model for the actual process of combustion occurring between the second and third strokes, in which chemical energy is liberated from the fuel. In a real engine, the combustion process takes some time, and not all energy is liberated at the instant of minimum volume (TDC). Thus, while this approximation is a good starting point, it does not completely capture the physics of the combustion process. We return to this issue further below.

[34] Adiabatic expansion from volume $V_3 = V_2$ to $V_4 = V_1$. Work $W_{\{34\}}$ is done *by* the gas. This represents the third (power) stroke of the engine. While adiabatic expansion is a reasonable starting point for analysis, this approximation is not at all exact. In particular, since the temperature after combustion is very high, significant heat is lost during this process to the cylinder walls. We discuss this further below.

[41] Isometric cooling back to ambient temperature and pressure, expelling thermal energy $Q_{\{41\}}$ into the environment. As discussed above, this thermodynamic process is used to approximate the effect of expelling the combustion products in the exhaust stroke (stroke four) and taking in new air and fuel in the intake stroke (stroke one). The actual situation is more complicated and is represented by

---

[1] If the thermodynamic properties of the gas, still assumed to be air, are allowed to vary with temperature, the analysis is called *air standard*.

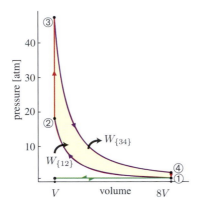
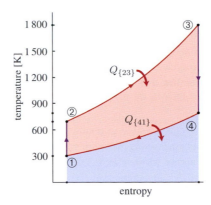

Figure 11.3 The thermodynamic Otto cycle (with cold air standard assumptions) in the pV- and ST-planes. Intake temperature $T_1 = 300\,\text{K}$ and pressure $p_1 = 1\,\text{atm}$ and compression ratio $r = 8:1$ are assumed. See Figure 10.8 for the significance of the shading. The horizontal (isobaric) line attached at ① in the figure is a reminder that the combustion products in the cylinder are actually forced out in the exhaust stroke and replaced with fresh air before the adiabatic compression begins.

additional steps in the pV- and ST-diagrams when going beyond the air standard analysis (§11.3).

While this thermodynamic model is not perfect, the idealized Otto cycle provides a good starting point for spark ignition engine analysis. In the adiabatic compression and expansion processes {12} and {34} there is no heat transfer. Treating the fuel–air mixture as an ideal gas with heat capacity C_V and adiabatic index γ, the heat added in the (isometric) combustion step {23}, is

$$Q_{\{23\}} = C_V(T_3 - T_2),\qquad (11.1)$$

and the heat expelled in the process {41} is

$$Q_{\{41\}} = C_V(T_4 - T_1).\qquad (11.2)$$

Energy conservation gives the total work done by the engine $\Delta W = W_{\{34\}} - W_{\{12\}}$ in terms $Q_{\{23\}}$ and $Q_{\{41\}}$, $\Delta W = Q_{\{23\}} - Q_{\{41\}}$. The efficiency of the engine is then

$$\eta = \frac{\Delta W}{Q_{\{23\}}} = \frac{Q_{\{23\}} - Q_{\{41\}}}{Q_{\{23\}}} = 1 - \frac{T_4 - T_1}{T_3 - T_2}.\quad (11.3)$$

For an isentropic process such as {12}, we have

$$p_1 V_1^\gamma = p_2 V_2^\gamma,$$

or, using the ideal gas law,

$$T_1 V_1^{\gamma-1} = T_2 V_2^{\gamma-1}.\qquad (11.4)$$

Thus, we have

$$T_2 = T_1 r^{\gamma-1},\qquad (11.5)$$

where $r = V_1/V_2$ is the compression ratio of the engine as defined above. Similarly, we have

$$T_3 = T_4 r^{\gamma-1}.\qquad (11.6)$$

Substituting into eq. (11.3), we find

$$\eta = 1 - \frac{1}{r^{\gamma-1}} = \frac{T_2 - T_1}{T_2}.\qquad (11.7)$$

As promised, the crucial feature in maximizing efficiency for SI engines is the compression ratio. The higher the compression ratio, the higher the efficiency. Note that the efficiency (11.7) is the same as the Carnot efficiency for a heat engine operating between the low temperature

Spark Ignition Otto Cycle

Internal combustion engines in most passenger automobiles are spark ignition engines. The thermal physics of an idealized spark ignition engine cycle can be modeled as an *Otto cycle*, in which combustion is modeled as isometric heating of the fuel–air mixture, followed by adiabatic expansion, isometric cooling, and adiabatic compression. The efficiency of the Otto cycle is

$$\eta = 1 - \frac{1}{r^{\gamma-1}},$$

where r is the compression ratio V_1/V_2 of the engine and γ is the adiabatic index of the fuel–air mixture.

T_1 and high temperature T_2, where T_2 is the temperature of the fuel–air mixture after compression but *before combustion*. Since the temperature after combustion is much higher, clearly the Otto cycle is not as efficient as a Carnot cycle operating between low temperature T_1 and the high temperature T_3 achieved by the idealized fuel–air mixture after combustion.

From eq. (11.7), it seems possible to obtain arbitrarily high efficiency by simply increasing the compression ratio. Unfortunately, this is not so. Most spark ignition automobiles currently use engines with compression ratios around 9.5:1 to 10.5:1. To understand why higher compression ratios are not feasible, and to gain further insight into the detailed processes occurring in spark ignition engines, we need to understand the combustion process in somewhat more detail.

11.2 Combustion and Fuels

The **gasoline** (or **petrol**, as it is known in much of the English speaking world) sold for motor vehicle use at service stations around the world is a complex mixture of many molecules. On the order of 500 different *hydrocarbons*, as well as various non-hydrocarbon additives, are combined in this mixture, which has been tuned to allow high engine compression ratios and optimize engine efficiency while minimizing some negative impacts on the environment. In this section we survey some basic properties of hydrocarbons and their combustion. Further discussion of petroleum and fossil fuels in general is given in §33. For a comprehensive description of hydrocarbon combustion in spark ignition engines, see [60]. In §11.2.1 we give a brief review of the molecular structure of hydrocarbons. In §11.2.2 we discuss the combustion process, and in §11.2.3 we discuss the phenomenon of *knock*, or premature combustion, which limits the compression ratio used in real engines.

11.2.1 Hydrocarbons

Hydrocarbons are molecules formed from only carbon and hydrogen atoms. These molecules can have a variety of structures, some of which are depicted in Figure 11.4. Since C has valence 4 and H has valence 1, hydrocarbons are determined by the connected structure of carbon atoms, each with four bonds, forming the "skeleton" of the molecule, with hydrogen atoms attached to each carbon bond not used to form the skeleton. A pair of carbon atoms can be attached by a single, double, or triple bond within a hydrocarbon. The simplest hydrocarbon is methane, CH_4,

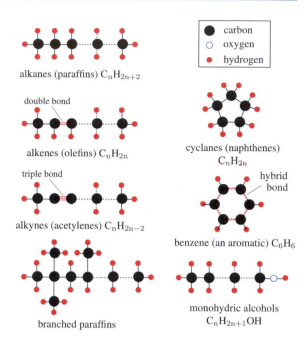

Figure 11.4 Some hydrocarbons and their molecular structures. The *hybrid* CC bonds in benzene share on average 1.5 electrons. Figure after [59].

with a single carbon atom as the skeleton. In general, the following classifications are used for hydrocarbon molecules (see also §33.2.2).

Paraffins/alkanes (C_nH_{2n+2}) **Paraffins**, scientifically known as **alkanes**, are hydrocarbons whose carbon atom skeleton is connected only with single bonds and contains no rings. The skeleton can be straight or branched. Straight paraffins are known as **normal paraffins** or *n*-**paraffins**, and include methane, ethane, propane, *n*-butane, etc., with respectively one, two, three, four, and more carbon atoms. Figure 11.5(a) shows *n-octane* (eight carbons). Branched paraffins are known as **iso-paraffins**. The complete description of a branched paraffin contains information about the branching structure. For example, the molecule commonly referred to as **iso-octane**, and shown in Figure 11.5(b), is described more technically as *2,2,4-trimethylpentane*, as it can be described as a pentane molecule (five linearly connected carbons) with three CH_3 or **methyl groups** attached to positions 2, 2, and 4 in the pentane chain.

Cycloparaffins/cyclanes/naphthenes (C_nH_{2n}) **Cycloparaffins**, like paraffins, have only single bonds, but contain a

---

Example 11.1   An Otto Cycle with Compression Ratio 8:1

Model a spark ignition engine with compression ratio 8:1 by a cold air standard Otto cycle. Assume initial pressure 1 atm and temperature 300 K, and a maximum total volume at BDC of 3.2 L. (These parameters were used to generate Figure 11.3.) Combustion leads to a maximum temperature of 1800 K.

Compute the engine efficiency and the work done per cycle in this idealized model. What is the power at 2500 rpm?

Taking $\gamma = 1.4$ (cold air standard), the efficiency is given by eq. (11.7),

$$\eta = 1 - \frac{1}{r^{\gamma-1}} = 1 - 1/8^{0.4} = 0.56 \,,$$

which also tells us $T_2 = T_1 r^{\gamma-1} \cong 690$ K.

To compute the work done per cycle, we compute $Q_{[23]}$ and use $W = \eta Q_{[23]}$. $Q_{[23]} = C_V(T_3 - T_2) = C_V(1800$ K $-$ 690 K$) = (1110$ K$) C_V$. For air, $C_V = \frac{5}{2} nR$. 3.2 L of an ideal gas at 300 K and 1 atm contains $n = 0.13$ moles, so $C_V = 2.5 \times (8.315$ J/mol K$) \times (0.13$ mol$) = 2.70$ J/K. Thus,

$$Q_{[23]} = (2.7 \text{ J/K}) \times 1110 \text{ K} = 3.00 \text{ kJ} \,,$$

which gives $W = 0.56 \times 3.00 = 1.68$ kJ/cycle. At 2500 rpm this yields power $P = 1.68 \times 2500/60 = 70.0$ kW, or about 94 horsepower.

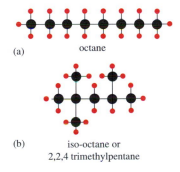

(a)   octane

(b)   iso-octane or
2,2,4 trimethylpentane

Figure 11.5 Molecular structure of octane and iso-octane.

single closed ring in the skeleton of carbon-carbon bonds. *Cyclopentane* is shown in Figure 11.4.

Olefins/alkenes (C_nH_{2n})   **Olefins** (**alkenes**) have a straight or tree-like carbon skeleton like paraffins (no rings), but have one double carbon-carbon bond. The simplest is *ethene*, C_2H_4, more commonly known as *ethylene*.

Acetylenes/alkynes (C_nH_{2n-2})   **Acetylenes** (**alkynes**) also have a straight or tree-like carbon skeleton like paraffins (no rings), but have one triple carbon-carbon bond. The simplest is *ethyne*, C_2H_2, more commonly known simply as *acetylene*.

Aromatics/arenes   **Aromatic hydrocarbons** contain rings of carbon atoms with bonds that can be viewed as hybrids of single and double bonds. The simplest and most important example, **benzene**, is a ring of six carbon atoms (shown in Figure 11.4). The ring structure of aromatics is

very stable, which makes this type of molecule a useful additive to gasoline as an *anti-knock additive* (see §11.2.3).

Depending upon the structure of the molecule, there are many possible uses for hydrocarbons. Crude oil contains a mix of many different hydrocarbons. At refineries, crude oil is separated into parts with different physical properties. The resulting hydrocarbon mixtures are used for a variety of purposes, from motor fuel to plastics (see §33). Commercial gasoline consists of a mixture of these hydrocarbons mostly containing from 5 to 12 carbon atoms.

11.2.2 Combustion

The combustion of hydrocarbons with air is a straightforward application of the chemistry of combustion reactions as discussed in §9. In general, hydrocarbons combine with oxygen to give water (H_2O) and carbon dioxide (CO_2). For example, the complete combustion of two moles of iso-octane with oxygen occurs through the reaction

$$2\,C_8H_{18} + 25\,O_2 \rightarrow 18\,H_2O(l) + 16\,CO_2 + 10.94\,\text{MJ} \,, \tag{11.8}$$

so that the enthalpy of combustion of iso-octane is (see eq. (9.14))

$$\Delta H_c(\text{iso-octane}) \cong -5.47\,\text{MJ/mol} \,. \tag{11.9}$$

Generally the enthalpy of combustion for hydrocarbon fuels is quoted in MJ/kg, which for iso-octane becomes

$$\Delta H_c(\text{iso-octane}) \cong -(5.47\,\text{MJ/mol}) \times (1\,\text{mol}/114\,\text{g})$$
$$\cong -47.9\,\text{MJ/kg} \,. \tag{11.10}$$

This is roughly the same as the enthalpy of combustion for *n*-octane. Note that the quoted enthalpy of combustion is the negative of the *higher heat of combustion* (HHC) (= HHV, see §9.4.3), where the product H_2O is assumed to be in liquid form. As discussed in more detail in §9.4.3, we use the higher heat of combustion for gasoline and other liquid fuels, which represents the theoretical upper bound on energy that could be extracted, even though often the lower heat of combustion is more relevant in realistic situations where the water condenses outside the engine. For gasoline, which contains a mixture of hydrocarbons, the heat of combustion (HHC) is generally taken to be to around 44 MJ/kg.

When the fuel–air mixture is drawn into the engine, the proportion of fuel to air in the cylinder is the most important factor in determining the consequences of the combustion process. The mixture is **stoichiometric** when there is precisely enough oxygen to complete the combustion of all hydrocarbons in the fuel. The mixture is **lean** if there is excess oxygen, and **rich** if there is insufficient oxygen for complete combustion. For gasoline the stoichiometric air:fuel ratio is around 14.7:1 by mass.

The combustion process is modeled in the Otto cycle as if it occurred instantaneously, but of course this is not the case in a real engine. When the spark plug or other ignition mechanism fires, the fuel in the spark region ignites, and a flame front proceeds through the cylinder, igniting the fuel. The timing of the ignition is crucial, and much effort has gone into timing optimization in modern engines. A depiction of the flame front moving through the cylinder is shown in Figure 11.6. The sequence of events in the combustion process is described in detail in [60]. The spark typically discharges towards the end of the compression stroke when the crankshaft is around $-30°$ from TDC. The flame front propagates across the cylinder, reaching the farthest wall at about $15°$ after TDC, with combustion continuing to about $25°$ after TDC, and maximum pressure occurring at about $15°$ after TDC. Thus, the full combustion process takes a time on the order of 1/6th of a complete rotation of the crankshaft. Note that the actual propagation of the flame front is very complex. The flame front has a highly irregular geometry, with total area five to six times the area of a smooth surface. The front propagates at a velocity determined roughly by the characteristic velocity of the turbulent gas in the engine, which is in turn proportional to the rate of crankshaft rotation (and much faster than *laminar* flow would be if the fuel–air mixture in the chamber was not in motion). This dependence explains how it is that the flame front crosses the combustion chamber in the time it takes for the crankshaft

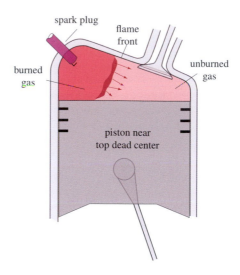

Figure 11.6 Combustion process initiated by spark ignition. After [59].

to rotate through a similar angle, independent of whether the engine is running at 600 or 6000 rpm. During the combustion process the fuel–air mixture reaches temperatures of over 2000 K. The timing of the combustion process is arranged so that the maximum pressure occurs as the volume of the cylinder is expanding. This reduces inefficiency due to heat loss through conduction, since it produces the maximum temperature towards the middle of the power stroke where more pdV work is done. The details of the combustion process have important ramifications for efficiency and engine design. Over the years, engineers have experimented with a wide range of cylinder geometries to optimize various aspects of the engine cycle and combustion processes. Very sophisticated computer models are used to design engines with optimal performance features.

11.2.3 Knock

One of the most important aspects of the combustion process just described is the tendency of hydrocarbon fuels to combust prematurely at high pressures and temperatures. Smooth engine operation depends upon the smooth propagation of the flame front across the combustion chamber. As the flame front progresses across the chamber, however, the fuel–air mixture at the leading edge of the flame front is further compressed and heated. This can initiate a chemical breakdown of some of the fuel molecules in advance of the systematic progress of the flame front, leading to high-frequency pressure oscillations within the cylinder known as "**knock**," for the characteristic sound

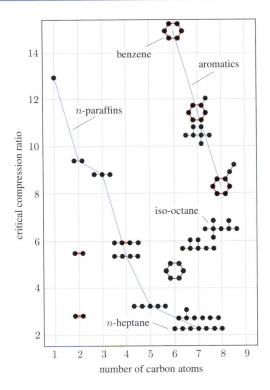

Figure 11.7 Knock tendencies of a selection of hydrocarbons under standard test conditions (600 rpm and 350 °F coolant temperature). The hydrocarbons are represented schematically by their carbon *backbones* with hydrogen atoms not indicated. Note that the critical compression ratio in test conditions may differ substantially from that in real engines. After [61].

produced when this occurs. Knock can also be caused by premature ignition of the high-pressure fuel–air mixture by hot spots in the interior surface of the cylinder combustion chamber. Knock can diminish the efficiency of the engine and can cause damage to the cylinder over time.

Different hydrocarbons have widely different tendencies to knock. The knock tendency of a given substance is measured under fixed conditions (for example, in a given engine at 600 rpm, with coolant temperature 350 °F)[2] by the **critical compression ratio** (CCR) at which knock commences. Critical compression ratios for a variety of

hydrocarbons are shown in Figure 11.7. In general, some different classes of hydrocarbons and other chemical compounds have the following knock tendencies:

n-paraffins The CCR for straight chain paraffins decreases rapidly as *n* increases. Thus, for example, significant *n*-octane content in a fuel leads to knock at a very low compression ratio. This tendency is easy to understand physically – a long chain hydrocarbon with single bonds is relatively unstable and easily split with the addition of thermal energy.

Branching Branching of the hydrocarbon helps to reduce knock. Thus, while *n*-octane leads to knock at a low compression ratio, iso-octane, which is highly branched, is somewhat less susceptible to knock.

Aromatics The benzene ring is a very stable structure, and aromatics in general do not break apart easily. As a result, addition of aromatics to gasoline dramatically reduces knock. For this reason, aromatics have been used widely as fuel additives to increase the viable compression ratio for SI engines.

Anti-knock additives By adding non-hydrocarbon substances to a fuel, knock can be substantially reduced. For example, **tetraethyl lead** ($(CH_3CH_2)_4Pb$ – a tetravalent lead ion bonded to four ethyl groups, each consisting of ethane minus one hydrogen atom) was for many years used as a fuel additive to reduce knock. After the negative health consequences of lead pollution were understood, most countries switched to unleaded gasoline, and instead other additives such as aromatics are included to reduce knock. (Note that aromatics have their own potential environmental issues; for example, evidence suggests that aromatics are relatively carcinogenic.)

The tendency of a fuel to knock is encoded in the single **octane number** (ON) provided at fuel pumps. This number compares the tendency to knock of a given fuel with a mixture of iso-octane (low knock tendency) and *n*-heptane (C_7H_{16}, high knock tendency). An ON less than 100 means that the fuel has the same tendency to knock as a mixture of (ON)% iso-octane and (100–ON)% heptane. Octane numbers higher than 100 are possible, and are defined in terms of iso-octane with a given fraction of lead additive.[3]

Gasoline purchased at pumps in the US generally has octane numbers ranging from 85 to 95. 93 octane

---

[2] These conditions "600–350°" are chosen because they enable many different compounds to be compared. Critical compression ratios observed under normal engine operating conditions are usually higher.

[3] For the precise definition of octane numbers above 100, see [60].

gasoline begins to knock in most SI engines at a compression ratio of roughly 10.5:1 (see footnote 2). Thus, most automobile engines currently in production have maximum compression ratios in the range 9:1–10.5:1. High-performance automobiles push the compression ratio to the maximum possible with existing fuels, and require higher octane fuel for satisfactory performance.

11.3 Real Spark Ignition Engines

While the theoretical thermodynamic analysis we have carried out for the Otto cycle captures various features of the engine accurately, such as the dependence of efficiency on compression ratio, the predicted efficiency (11.7) is much greater than is realized in real SI internal combustion engines. We have explained why current automobiles using SI engines have a maximum compression ratio of around 10:1. Taking $r = 10$ and the cold air standard value of $\gamma = 1.4$, eq. (11.7) would predict an efficiency of slightly greater than 60%. In contrast, a standard 4-cylinder Toyota Camry engine, for example, has a peak efficiency of 35% [62]. In this section we briefly describe some of the issues that reduce the efficiency of real engines below the theoretical ideal.

Before considering inefficiencies we must adopt a more realistic value of the adiabatic index for the fuel–air mixture at the high temperatures attained during the cycle. At 2500 K the adiabatic index of air drops to about 1.29 from the usual value of 1.4 at room temperature. The fuel at the stoichiometric ratio does not significantly affect this number. A standard approximation is to take $\gamma = 1.3$, appropriate to an average cycle temperature of around 1600 K [59]. Substituting a compression ratio of $r =$

Knock

Hydrocarbon fuels can break down or combust at high temperature and pressure before ignition. This phenomenon, known as *knock*, occurs at a compression ratio of around 10.5:1 with conventional gasoline mixtures, and can reduce engine efficiency and cause damage over time. Thus, most spark ignition automobile engines in current production have a compression ratio of between 9:1 and 10.5:1.

10, and $\gamma = 1.3$ into the efficiency formula (11.7), we estimate

$$\eta = 1 - \frac{1}{r^{\gamma-1}} = 1 - \frac{1}{10^{0.3}} \cong 0.499 , \qquad (11.11)$$

a decrease of more than 15% from the cold air standard result.

A semi-realistic four-stroke engine cycle is graphed in the pV-plane in Figure 11.8, and compared to the thermodynamic Otto cycle. This figure serves as a useful guide in reviewing some of the ways in which the real engine cycle deviates from the thermodynamic ideal. Note, however, that the processes in the real engine cycle involve rapid changes that take the gas in the cylinder out of thermodynamic equilibrium, so that the thermodynamic variables are only approximate and this is only a schematic depiction of the real process.

Combustion   As we described in §11.2.2, the combustion process is not instantaneous, and does not occur at constant volume. In a real engine, the combustion occurs over a finite time. Combustion generally begins while the cylinder volume is still decreasing and continues until the volume has increased by a non-negligible amount. Furthermore, even in the absence of knock, the high temperatures to which the fuel–air mixture is exposed cause some molecules to *dissociate*, or recombine chemically, preventing complete and immediate combustion. The combination of these effects keeps the temperature and pressure in the combustion chamber significantly below the extremes suggested by the theoretical Otto cycle. This reduces the area of the actual cycle pV curve in Figure 11.8, reducing the work done and hence the efficiency of the engine.

Heat loss during expansion   The combusted fuel–air mixture is extremely hot during the expansion stroke (as mentioned above, generally over 2000 K after combustion). This leads to rapid heat loss to the cylinder walls and loss of efficiency.

Blowdown   Near the end of expansion, the exhaust valve opens and the high-pressure gas in the cylinder rapidly expands outward and returns to ambient pressure. This is known as **blowdown**. As depicted in Figure 11.8, the blowdown process does not occur at constant volume, and removes some further area from the pV curve.

Exhaust/intake strokes   Finally, the exhaust and intake strokes require work by the engine that is ignored in the ideal Otto cycle. In the ideal cycle the hot exhaust is simply replaced by air at the low temperature and pressure set point, modeled as isometric cooling. In reality the

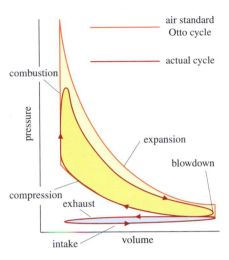

Figure 11.8 Comparison of actual and theoretical Otto cycles in the pV-plane for a four-stroke spark ignition engine. The yellow shaded areas represent work done *by* the engine. The blue shaded part of the cycle, which runs counterclockwise represents work done *on* the engine. After [59].

exhaust gases are forced out of the cylinder and fresh air is drawn in. The exhaust stroke following blowdown therefore occurs at a slightly higher pressure than atmospheric. Similarly, during the intake stroke the pressure must be slightly lower than atmospheric. These two strokes, idealized as a single horizontal line in Figure 11.3, are more accurately described by the blue shaded region of the cycle in Figure 11.8, and lead to an additional nonzero area for the cycle in the pV-plane, which contributes negatively to the total work done by the engine (as this part of the cycle proceeds counterclockwise).

All these deviations from the theoretical ideal combine to significantly lower the maximum efficiency of real spark ignition engines to at most about 80% of the theoretical Otto efficiency [60]. Despite many decades of engineering efforts, this is the state of the art in automobile engine engineering.

Beyond these reductions in efficiency, there is a further issue that compromises SI engine performance. The previous analysis assumed that the engine was operating with a cylinder full of fuel–air mixture on each cycle. In reality, the engine power is adjusted dynamically by the driver using a **throttle** mechanism (actuated by exerting less than maximum pressure on the gas pedal). When the engine runs at less than full throttle, a metal plate (the **throttle plate**) rotates into a position that partially blocks

> **Engine Efficiency**
>
> The ideal Otto cycle efficiency of a spark ignition engine with compression ratio 10:1 is about 50%. Deviations from the idealized cycle, heat losses, and other inefficiencies reduce the peak efficiency of most automobile engines to 35% or below. At less than full throttle, pumping losses cause further reduction in efficiency. Combined with energy lost to the electrical system, typical power delivered to the drive train is 25% of energy from fuel combustion.

the flow of air into the engine and at the same time the flow of fuel into the injection system is decreased, keeping the fuel–air mixture close to the stoichiometric ratio. In this way less energy is released on combustion and the engine generates less power. When the throttle limits the flow of air into the system, the intake stroke brings in less of the fuel–air mixture, resulting in lower pressure than atmospheric at the beginning of the compression stroke. This significantly increases the work done by the engine by increasing the size of the counterclockwise loop formed in the pV-plane by the exhaust-intake strokes. In typical motor vehicle operation, the intake stroke of an SI engine operates at around 0.5 atm. A graph of a throttled Otto cycle is compared with an unthrottled cycle in Figure 11.9. This extra work done by the engine (which can be treated as a negative contribution to the work output) is referred to as **pumping loss**.

Clearly there are substantial issues that reduce the actual efficiency of automobile engines below that of the ideal Otto thermodynamic cycle, which itself is below Carnot efficiency. Finding ways to improve the efficiency of engines would have dramatic commercial and environmental consequences. In the remainder of this chapter we describe some variations on the standard four-stroke SI engine and associated Otto cycle that have been used to increase the efficiency of internal combustion engines.

11.4 Other Internal Combustion Cycles

Up to now we have focused primarily on four-stroke spark ignition engines with an idealized description in terms of the thermodynamic Otto cycle. There are many other designs for internal combustion engines, several of which have found real-world applications. In this section we

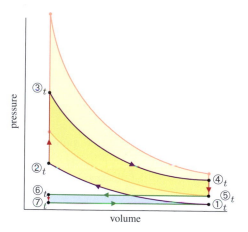

Figure 11.9 Throttled Otto cycle, compared to the unthrottled (orange) cycle in the pV-plane. The exhaust and intake strokes $⑤_t \to ⑥_t$ and $⑦_t \to ①_t$ respectively, are isobaric (constant pressure). The exhaust stroke is above atmospheric pressure and the intake is at one-half atmosphere, requiring the engine to do pumping work on the intake cycle. The scale of the pumping part of the cycle {5671} has been exaggerated for clarity. After [59].

briefly explore two of these alternative cycles: the *Atkinson cycle* used in hybrid automobile engines to increase engine efficiency, and the *Diesel* or *compression ignition cycle*, in which higher compression ratios can be achieved by injecting the fuel after compression.

A more significant departure from the four-stroke paradigm is the **two-stroke SI engine cycle**. Indeed, some of the earliest spark ignition engines used a two-stroke cycle with exhaust and intake incorporated into the end parts of the power and compression strokes. Such engines are simpler than engines based on the four-stroke Otto cycle and are still widely used in lightweight applications such as motorcycle engines, chain saws, and small boat engines. Two-stroke engines tend to produce more pollution than four-stroke engines since the exhaust and intake processes are not as controlled and complete in the two-stroke engine. This has led to the phasing out of two-stroke engines for use in automobiles in many countries including the US. Because two-stroke engines produce power on every stroke, however, they can be more powerful than four-stroke engines of comparable size. There is substantial current effort towards developing a less-polluting two-stroke engine suitable for use in automobiles that would have greater efficiency and power than a similarly sized four-stroke engine.

11.4.1 The Atkinson Cycle

An interesting variation on the standard Otto cycle for spark ignition engines was proposed by English engineer

James Atkinson in 1882. The key feature of the **Atkinson cycle** is that the compression ratio can be different from the expansion ratio. Thus, while the compression ratio is limited by the knock characteristics of the fuel, the expansion ratio can be larger and provide more power. Atkinson's original design, now obsolete, used a novel crankshaft design to implement the cycle. A number of modern automobiles, particularly those that use hybrid engine technology, achieve the same result of different compression and expansion ratios simply by modifying the intake valve timing. The basic idea is that by leaving the intake valve open after the compression stroke begins, some of the air that has just been taken in is expelled again so that the pressure in the cylinder remains constant for some time after the crankshaft passes bottom dead center. The actual compression of the fuel–air mixture begins only after the intake valve closes. The corresponding idealized Atkinson cycle is shown in Figure 11.10. The work performed in the Atkinson cycle is reduced because the amount of air in the cylinder is smaller than in an Otto cycle in an engine with the same displacement. The exhaust and intake strokes are modeled as a constant volume cooling to atmospheric pressure $④ \to ①$, followed by a compression at constant pressure as some air is expelled $① \to ①_b$. Note that these approximations are not very realistic, since in fact the quantity of air in the cylinder increases ($④$ to $①$) and then decreases ($①$ to $①_b$) again during these processes, and in reality there is no cooling in the step $① \to ①_b$, but for the purposes of estimating the work output of the cycle these approximations introduce only relatively small inaccuracies.

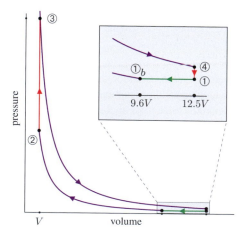

Figure 11.10 Idealized Atkinson cycle with a compression ration of 9.6:1 and an expansion ratio of 12.5:1. The inset shows an enlargement of the part of the cycle where compression begins. The $① \to ①_b$ step is enabled by keeping the intake value open for part of the compression stroke.

Other Engine Cycles

Engine cycles other than the standard spark ignition Otto cycle can improve engine efficiency. Some vehicles use modified valve timing to realize a version of the Atkinson cycle, which gives a higher expansion than compression ratio. This improves efficiency by some 10% at the cost of a reduction in engine power that can be compensated for by the battery in hybrid vehicles.

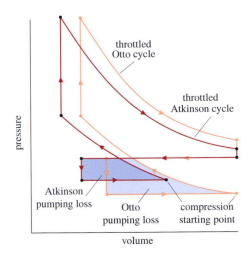

Figure 11.11 Comparison of pumping losses (blue shading) in a throttled Atkinson cycle (red) and a conventional Otto cycle (orange). As in Figure 11.9, the figure is not drawn to scale so that the pumping effects can be made more visible. After [62].

An additional advantage of the intake valve delay in the Atkinson cycle is that it can be used to reduce the pumping losses induced by throttling. By using intake valve delay rather than throttle to reduce the total fuel–air mixture in the cylinder, the reduction in pressure and consequent pumping losses associated with throttling can be mitigated significantly. This is demonstrated in Figure 11.11 for the hybrid Atkinson cycle vehicle described in Example 11.2.

11.4.2 Compression Ignition (Diesel Cycle) Engines

Compression ignition (CI) provides another way of dealing with the problem of knock. The first compression

ignition engine was invented by German engineer Rudolf Diesel in 1893. The four-stroke compression ignition cycle is depicted in Figure 11.12. The primary difference from the spark ignition cycle is that no fuel is taken in with the air in the intake cycle. Thus, compression can proceed

---

Example 11.2   The Atkinson Cycle in the Hybrid Toyota Camry

An example of the use of the Atkinson cycle in a modern vehicle is the hybrid 2007 Toyota Camry. The non-hybrid version of the Camry runs on a four-cylinder 2AZ-FE engine with cylinder displacement $2362 \, \text{cm}^3$ and a compression ratio of 9.6:1. The hybrid version of the vehicle runs on a very similar engine, the 4-cylinder 2AZ-FXE engine with the same cylinder displacement and compression ratio, but an expansion ratio of 12.5:1. The differences between the two vehicles are analyzed in detail in a US DOE report [62]. In the hybrid Camry, the computer-controlled intake valve delay is actuated when conditions indicate it will improve engine efficiency. According to the DOE report, this engine improves peak efficiency from 35% to 38%.

The main disadvantage of the Atkinson cycle is that the peak power of the engine is reduced. The non-hybrid 4-cylinder Camry has peak power of 120 kW (160 hp) at 6000 rpm, while the hybrid only achieves peak power of 110 kW (147 hp). Despite this difference, because the electric motor can provide additional power, the hybrid Camry still has superior acceleration performance to the non-hybrid Camry. The rate of acceleration from 30 to 50 mph is graphed in the figure above for both vehicles. The achievement of higher performance in a vehicle that also has higher fuel efficiency is an important step in moving towards widespread acceptance of hybrid technology, since many consumers are motivated by performance (or perception of performance) as much as or more than by fuel efficiency or environmental concerns.

(Figure: After [62])

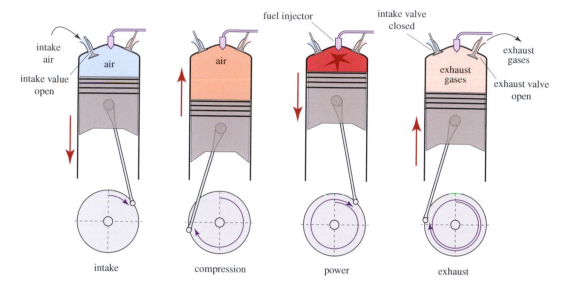

Figure 11.12 Stages in the four-stroke compression ignition (CI) cycle. (Credit: Adapted from Encyclopaedia Britannica. Inc). See Figure 11.1 for comparison and further labeling.

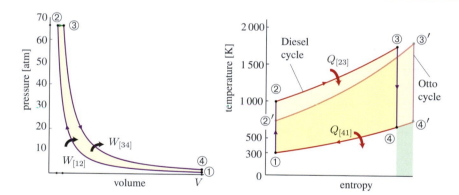

Figure 11.13 An ideal Diesel cycle in the pV- and ST-planes using cold air standard assumptions. The compression ratio is $r = 20$:1 and the cutoff ratio is chosen to be $r_c = 1.8$, so that the heat input, $Q_{\{23\}}$, is identical to the ideal Otto cycle shown in Figure 11.3. Both cycles begin at 300 K and 1 atm at ①. They reach approximately the same maximum temperature. The ideal Diesel and Otto cycles are compared in the ST-plane. Although both cycles have the same heat input $Q_{\{23\}} = Q_{\{2'3'\}}$, the Otto cycle adds more entropy to the engine. The area of the cycles (yellow shaded) shows the work output, which is greater for the Diesel cycle. The additional efficiency of the Diesel cycle is proportional to the area of the green shaded region (Problem 11.6).

to a higher ratio, and to a concomitantly higher temperature and pressure, with no risk of premature combustion. When the crankshaft is at or near TDC, the fuel is injected into the combustion chamber and ignites. In the Otto cycle, combustion occurs roughly symmetrically about TDC and therefore roughly at constant volume. In contrast, in the Diesel cycle the combustion processes occur as the cylinder volume expands from V_2 to V_3 while the pressure is roughly constant. The Diesel cycle is therefore modeled by replacing the isometric heat addition process of the Otto

cycle with an isobaric (constant pressure) heat addition process. A Diesel cycle is depicted in Figure 11.13. While the isobaric approximation is not very accurate, it forms a good basis for beginning to analyze the compression ignition cycle.

The cold air standard thermodynamic analysis of the Diesel cycle is very similar to that of the Otto cycle. The principal difference is that, because the heat addition occurs at constant pressure rather than constant volume, C_V in eq. (11.1) is replaced with C_p, giving

$$Q_{\{23\}} = C_p(T_3 - T_2) = \gamma C_V(T_3 - T_2). \quad (11.12)$$

The efficiency of the engine then becomes

$$\eta_{\text{Diesel}} = \frac{W}{Q_{\{23\}}} = \frac{Q_{\{23\}} - Q_{\{41\}}}{Q_{\{23\}}} = 1 - \frac{T_4 - T_1}{\gamma(T_3 - T_2)}. \quad (11.13)$$

Defining the **cutoff ratio** to be $r_c \equiv V_3/V_2$, we can express η_{Diesel} in terms of r_c and the compression ratio $r = V_2/V_1$. We use again eq. (11.5), $T_2 = T_1 r^{\gamma-1}$, the analogous relation $T_3 = T_4(V_4/V_3)^{\gamma-1} = T_4(r/r_c)^{\gamma-1}$, and the relation $T_3 = r_c T_2$ that follows from the ideal gas law and $p_2 = p_3$, to write the efficiency as

$$\eta_{\text{Diesel}} = 1 - \frac{(r_c/r)^{\gamma-1} r_c r^{\gamma-1} - 1}{\gamma(r^{\gamma-1}r_c - r^{\gamma-1})} = 1 - \frac{1}{r^{\gamma-1}}\left(\frac{r_c^{\gamma} - 1}{\gamma(r_c - 1)}\right). \quad (11.14)$$

As in the Otto cycle, increasing the overall compression ratio r is crucial for maximizing efficiency.

Because there is no risk of knock, diesel engines can operate at much higher compression ratios than spark ignition engines. On the other hand, realizing compression ratios much above 10:1 requires a heavy duty engine with a massive piston and cylinder head in order to withstand the higher pressure. Such engines therefore are most easily accommodated in large vehicles, such as trucks, buses, and heavy equipment, as well as locomotives and large ships. Modern diesel engines generally operate at compression ratios between 15:1 and 22:1. While the theoretical efficiency of these engines can be around 75%, actual efficiency is closer to 45%.

Because knock is not a concern, diesel engines can also operate on a more flexible range of fuels than spark ignition engines. Traditional petroleum-derived **diesel fuel** is a mix of hydrocarbons with 10–15 carbon atoms, with approximately 25% aromatics. Other options for diesel fuel include *biodiesel* and organic oils (§26). In place of octane

---

Diesel Engines

Diesel engines use *compression ignition*, where the air is compressed before fuel injection to avoid knock. Compression ignition engines are modeled by the idealized thermodynamic *Diesel cycle*. Because higher compression ratios are possible, diesel engines are 30–50% more efficient than spark ignition engines. Diesel engines must be more massive, however, and are generally more expensive.

---

number, the combustion quality of diesel fuel is measured by **cetane number**, which is a measure of the time delay between ignition and combustion. A higher cetane number indicates a more easily combustible fuel; typical diesel fuel has a cetane number between 40 and 55, with higher cetane number providing more efficient combustion for high-speed engines. Despite the efficiency advantages of diesel engines, disadvantages include the facts that diesel engines must be more massive and are therefore more expensive than comparable spark ignition engines, and diesel engines can be more difficult to start in cold conditions. In addition, compression ignition can result in less complete combustion, leading to increased emission of pollutants. The high temperatures and pressures reached in CI engines can also result in the formation of *oxides of nitrogen* (typically NO and NO_2, denoted together as NO_x), a further source of pollution. These disadvantages have slowed the development of diesel engines for smaller passenger automobiles, though lighter and more versatile diesel engines are now commercially available.

Discussion/Investigation Questions

11.1 Walk your way around the pV- and ST-diagrams, Figure 11.3, for the Otto cycle, explaining the functional form of the curves, $p(V)$ and $T(S)$. Why is each step vertical in one of the diagrams?

11.2 Research and summarize the advantages and disadvantages of two-stroke engines for personal motor vehicles. What do you see as their future?

11.3 Would you advocate incorporating an Atkinson-like step (isobaric compression after the completion of the power stroke) in a diesel engine? Explain.

11.4 What is the rationale for modeling combustion as a constant pressure process in the Diesel cycle when it was modeled as constant volume in the Otto cycle?

11.5 Why would a diesel engine be hard to start in cold weather?

11.6 A **dual cycle** attempts to model an SI engine more realistically. It models the combustion process in two steps. First some heat is added at constant volume, then the rest is added at constant pressure. Draw the dual cycle in the pV- and ST-planes. How are the Otto and Diesel cycles related to the dual cycle?

Problems

11.1 Assume that an SI engine has the following parameters: displacement ($V_1 - V_2$): 2.4 L; compression ratio 9.5:1; air to fuel mass ratio 15:1; heating value of fuel 44 MJ/kg, pressure at start of compression 90 kPa, intake temperature 300 K. Compute the pressure, volume, and temperature at each of the points ①, ②, ③, and ④ in the idealized cold air standard Otto cycle. Compute the work done per cycle, and estimate the engine efficiency in this idealized model. Compute the power at 3000 rpm.

11.2 The cold air standard value of $\gamma = 1.4$ was based on the heat capacity of a diatomic gas with no vibrational excitation (see §9), $C_V = \frac{5}{2}nR$ and $C_p = C_V + nR$. In reality, the heat capacity increases with increased temperature, and γ decreases accordingly. If we assume, however, that C_V is independent of temperature through the cycle and that $C_p = C_V + nR$ remains valid, what value of C_V is required for $\gamma = 1.3$?

11.3 An SI engine, modeled as an ideal Otto cycle, runs at a compression ratio of 9.6:1 with a maximum cylinder volume (V_1) of 2 L and a corresponding displacement of 1.8 L. Combustion leads to a maximum temperature of $T_3 = 1800$ K. Using the air standard value of $\gamma = 1.3$, find the engine efficiency and the work per cycle. Compute the engine's power at 5000 rpm. Assume that the intake temperature and pressure are 300 K and 100 kPa. Note: you will need the value of C_V from Problem 11.2.

11.4 **[H]** (Requires results from Problem 11.3) Consider the same engine as in Problem 11.3 but now run as an Atkinson cycle. The volume after expansion (V_4) is still 2 L, but the volume before compression (V_{1b}) is 1.54 L. The compression ratio is still 9.6:1, so the minimum volume is $V_2 = 0.16$ L. What is the expansion ratio? Assuming that the amount of air and fuel are both reduced by the ratio of V_{1b}/V_4, find the work done in expansion, $W_{\{34\}}$, the work required by compression, $W_{\{1b\,2\}}$, and the energy input from combustion $Q_{\{23\}}$. Compute the efficiency and the engine's power at 5000 rpm.

11.5 (Requires results from Problem 11.1) Consider a throttled Otto cycle for an engine with the same parameters as Problem 11.1. In the throttled cycle assume that the spent fuel–air mixture is ejected at 1 atm and brought in again at 0.5 atm. Compute the work done per cycle. Compute the associated pumping loss. Subtract from the work per cycle and compute the engine efficiency with throttling.

11.6 Consider the Otto and Diesel cycles shown in Figure 11.13. The parameters have been chosen so both cycles have the same heat input and the same low temperature and pressure set point. Explain why the difference between the Diesel cycle efficiency and the Otto cycle efficiency is proportional to the area of the green shaded region.

11.7 A marine diesel engine has a compression ratio $r = 18$ and a cutoff ratio $r_c = 2.5$. The intake air is at $p_1 = 100$ kPa and $T_1 = 300$ K. Assuming an ideal cold air standard Diesel cycle, what is the engine's efficiency? Use the ideal and adiabatic gas laws to determine the temperature at points ②, ③, and ④. Compute the work per cycle. What is the power output of this engine if its displacement is 10 L and it runs at 2000 rpm? What is its fuel consumption per hour (assuming the energy content of diesel fuel to be 140 MJ/gal)?

11.8 Suppose two engines, one SI, the other CI, have the same temperature range, $T_1 = 300$ K and $T_3 = 1800$ K. Suppose the SI engine, modeled as an ideal Otto cycle, has a compression ratio of 10:1, while the CI engine, modeled as an ideal Diesel cycle, has twice the compression ratio, 20:1. Using cold air standard assumptions, what is the efficiency of each engine? How much more efficient is the diesel engine? [Hint: the fuel cutoff ratio, r_c can be computed as a function of r and T_3/T_1.]

Phase-change Energy Conversion

The Carnot and Stirling engines considered in §10 use a gas as the working fluid, as do the various internal combustion engines described in §11 that dominate transportation. In contrast, for two major other applications of thermal energy conversion, devices are used in which the working fluid *changes phase* from liquid to gas and back again in the course of a single thermodynamic cycle.

Large-scale electric power production is one area in which phase-change systems play a central role. The principal means of electric power generation in stationary power plants, whether nuclear or coal-fired, is the modern steam turbine that uses a phase-change thermodynamic cycle known as the *Rankine power cycle*. As shown in Figure 12.1, almost 80% of the electricity generated in the world is produced from thermal energy (released by fossil fuel combustion or uranium fission). The main exception is hydropower, which provided about 17% of world electricity in 2012 [12]. Wind and solar photovoltaic power generation, which do not involve thermal energy, are still very small components of total electric power generation. Of the 80% of electricity production from thermal energy, nearly all employs the Rankine steam cycle to convert heat to mechanical energy. The rest comes from natural-gas-fueled turbines that make use of a pure gas phase combustion cycle known as the *Brayton cycle*. Increasingly in recent years, the gas turbine is combined with a lower-temperature steam turbine fueled by exhaust gases to make a high efficiency *combined cycle*. This chapter focuses on the basic physics of phase-change systems, and Brayton and Rankine cycles, including the combined cycle, are described in the following chapter. Phase change also plays a role in geothermal power plants (§32), where liquid water from underground reservoirs at high temperature and pressure vaporizes as it reaches the surface where the pressure is lower.

> **Reader's Guide**
>
> This chapter and the next explain the physics of phase-change energy conversion and its application to practical thermal energy conversion devices. In this chapter we explore the thermodynamics of phase change. We first explain why devices where fluids change phase play such a central role in thermal energy conversion. We then describe systems in phase equilibrium using various combinations of thermodynamic state variables, and introduce the state variable *quality*, which describes the mixed liquid–vapor state. Finally, we describe how pressure and temperature are related for water in phase equilibrium.
>
> Prerequisites include: the introduction to phase change in §5 (Thermal energy) and §9 (Energy in matter), the description of forced convection in §6 (Heat transfer), and §8 (Entropy and temperature) and §10 (Heat engines) on thermal energy transfer.
>
> The following chapter, §13 (Power cycles), depends critically on the material presented here. Phase change also figures in thermal energy conversion processes in §19 (Nuclear reactors), §32 (Geothermal energy), and §34 (Energy and climate).

The second major application of phase-change cycles is in heat extraction devices, i.e. air conditioners, refrigerators, and heat pumps, which were introduced in §10.6. Although the scale of refrigeration and air conditioning energy use is smaller than that of electric power generation, it nevertheless accounts for a significant fraction of total electric power consumption. Figure 12.2 shows the energy consumed in US homes in 2009 subdivided by end use. Together, household air conditioning and refrigeration accounted for 13% of all electricity consumption in the US in 2010; energy used for air conditioning and refrigeration in US commercial buildings amounted to another 6% of total electricity consumption

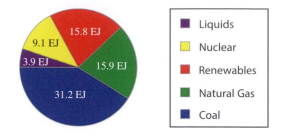

Figure 12.1 2011 world electricity generation by fuel (total 75.9 EJ) [12]. Only hydro, wind, photovoltaic (which are included in "Renewables") and a small fraction of natural gas do not involve phase-change conversion of heat to mechanical and then to electrical energy.

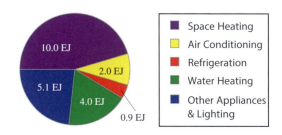

Figure 12.2 US household energy use for 2009. (Adapted using data from US DOE EERE Buildings Energy Data Book)

[12]. Industrial refrigeration and actively cooled transports (i.e. tractor-trailers with refrigerated payloads) also use energy for heat extraction and add to these totals. Heat pumps, which efficiently extract heat from a cooler environment to provide space heating in cold climates, are a small, but rapidly growing energy application that is discussed further in §32.3. Heat extraction devices are described in more detail in the following chapter (§13).

This chapter begins (§12.1) with a discussion of the advantages of phase change in energy conversion. We then review and extend the description of phases begun in §5. In §12.2, we describe the phase diagram of a pure substance from several perspectives – as a surface in the three-dimensional space parameterized by pressure, volume, and temperature, and as seen in the entropy–temperature plane. §12.3 describes how information about phase change is used in concrete applications.

Phase change and the phase structure of pure substances are described in most thermodynamics texts. A more detailed treatment than ours with many examples can be found in [19]. The dynamics of heat transfer and phase change is explored thoroughly in [27].

12.1 Advantages of Phase Change in Energy Conversion Cycles

Whether used in the context of a power plant or a heat extraction device, phase change is not an incidental feature, but is essential to the design of many practical energy conversion systems. Over decades of development, different thermodynamic systems have come to dominate different applications. Internal combustion, for example, dominates transportation, where the versatility and high energy density of liquid hydrocarbon fuels, and the compactness, safety, and reliability of modern internal combustion engines are important. Phase-change cycles have come to dominate large-scale stationary power generation and energy extraction devices, where they have several significant physical advantages. In this section we describe the physics behind four principal practical advantages of phase-change energy conversion.

Energy storage potential in phase change The energy needed to change the phase of a substance is typically much larger than the energy needed to heat it up by a few degrees (§5, §9). Enthalpies of vaporization, in particular, are generally quite large, and particularly so for water (see Table 9.3). For example, it takes 2.26 MJ to vaporize one kilogram of water at 100 °C. A small amount of that energy goes into $p\,dV$-work, and the rest is stored as internal energy in the resulting water vapor. In contrast, one would have to heat about 5.4 kg of liquid water from 0 °C to 100 °C to store the same amount of energy.

The large amount of energy stored when a fluid vaporizes does come with a cost: its volume expands significantly in the transition. In practical devices, the expansion is limited by carrying out the phase transition at high pressure, where the volume difference between liquid and vapor is not so large. The power of an engine or heat extraction system such as a refrigerator is limited by the size of pipes and other mechanisms that transport and store the working fluid and the pressures they can tolerate.

Efficient energy transfer at constant temperature As stressed at the end of §8.5, thermodynamic efficiency of a cycle is highest when heat transfer processes within the cycle occur between parts of the system that are close to the same temperature.

When heat is transferred to a system across a large temperature difference, significant entropy is added, leading to associated losses and inefficiency. In the limit of a very small temperature difference, heat transfer generates no additional entropy and is *reversible*. As we saw in §6, however, the rate of heat flow is proportional to the gradient of the temperature, so heat will not flow

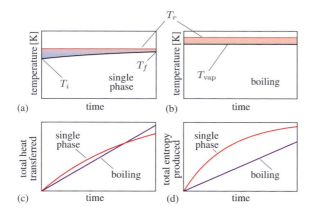

Figure 12.3 A comparison of heat transfer and entropy production in a fluid in a single phase and at its boiling point. The temperatures are those specified in Problem 12.4.

from one system to another unless there is a temperature difference between them. The laws of thermodynamics and heat transfer present us, then, with a conundrum: how can a physical system transfer heat both efficiently and quickly?

Suppose we want to transfer an amount of thermal energy ΔQ from a heat reservoir at a *fixed* temperature T_r to a fluid in a single phase. As shown in Figure 12.3(a), the temperature of the fluid T will rise from its initial value T_i to a final value $T_f = T_i + \Delta Q/C$, where for simplicity we assume that the heat capacity C of the fluid is constant. T_r must be greater than T_f. Early in the process $\Delta T \equiv T_r - T$ is relatively large and the heat transfer is rapid but inefficient. Near the end, ΔT is smaller, which makes the transfer more efficient, but slower. The shaded area in Figure 12.3(a) indicates the temperature difference and is a measure of the thermodynamic inefficiency of the process. To make the transfer both relatively rapid and efficient we would have to constantly adjust the temperature of the heat reservoir upwards as the fluid warms. This kind of heat transfer is accomplished in the Stirling engine (§10), but at the cost of including a massive component – the regenerator – that stores the energy transferred between stages of the cycle. In contrast, if we add thermal energy to a fluid at its boiling point T_{vap} from a heat source at constant temperature, then ΔT remains fixed until all of the liquid turns to vapor (Figure 12.3(b)). We can adjust ΔT to make the process as efficient or as rapid as we require.

Figures 12.3(c) and (d) illustrate the difference in total heat transfer and entropy production between single-phase and boiling heat transfer. To compare the two mechanisms on an equal footing, we have chosen the rate of heat transfer to be the same for a given ΔT, though the rate of

boiling heat transfer is typically much greater (see below). Although single-phase heat transfer is rapid at early times (Figure 12.3(c)), when ΔT is large, so is the rate at which extra entropy is generated (Figure 12.3(d)). Boiling transfers more heat and generates less entropy over the time period shown.

Heat transfer enhancement The dynamics of *boiling* help to make heat transfer both rapid and efficient by combining the efficiency of thermal energy transfer at near-constant temperature with the rapid transfer of heat through forced convection. Forced convection is described empirically by *Newton's law of cooling* (6.15), which gives

$$\frac{dQ}{dt} = \bar{h}A(T - T_0),\qquad(12.1)$$

where $q = \frac{1}{A}\frac{dQ}{dt}$ is the *heat flux*, the heat per unit area per unit time flowing from an object at temperature T into a fluid with bulk temperature T_0. The goal of rapid heat transfer is realized by maximizing the *heat transfer coefficient* $\bar{h}$, which depends on many features of the specific setup including the properties of the fluid and the geometry of the object.

As mentioned briefly in §6.3.1, $\bar{h}$ can be very large in a boiling process. $\bar{h}$ depends sensitively on the dynamics of the fluid in the thin *boundary layer* where it contacts the object. As it boils, the thin liquid layer contacting the hot object's surface vaporizes; the vapor then lifts away by virtue of its buoyancy, and is quickly replenished by the surrounding liquid.[1] Consequently, vapor molecules are produced very quickly at the interface. The heat flux in boiling heat transfer is also enhanced by the magnitude of the liquid's heat of vaporization, as discussed in the first item above; every vapor molecule departing from the boiling surface carries with it the energy needed for vaporization.

Figure 12.4 compares heat transfer in single-phase convection and boiling heat transfer. At the top, a wire maintained at a uniform temperature T is immersed in a stream flowing from left to right that has a temperature T_0 far from the wire. Below, the same heated wire is immersed in a stagnant pool of water that has a uniform temperature of T_{vap}. The wire surface temperature T is maintained above

---

[1]   The phenomenon being described here is **nucleate boiling**. When the temperature difference between object and fluid gets very large, a film of vapor starts to develop between the fluid and the hot object and heat transfer actually decreases with increased temperature difference. At even higher temperature difference, *film boiling* takes over and the heat transfer rate grows once again [27].

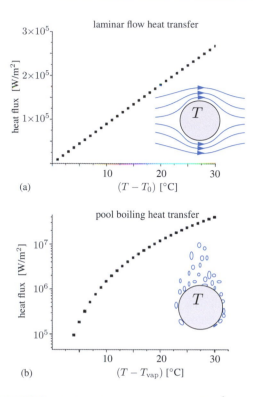

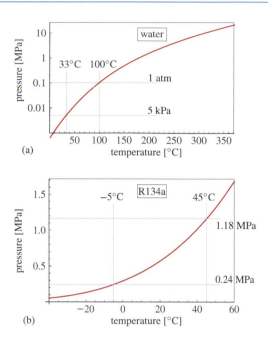

Figure 12.5 The boiling pressure (*vapor pressure*) as a function of temperature for two common fluids used in phase-change energy conversion: (a) water (note that the pressure scale is logarithmic) and (b) the refrigerant R-134a (tetrafluoroethane). Some interesting temperatures and pressures are marked (see discussions in §12.1 and in §13.3.4).

Figure 12.4 Calculated heat transfer rates (in W/m^2) for (a) laminar flow (at a typical velocity) of a single-phase fluid (bulk temperature T_0) over a cylinder held at temperature T; and (b) boiling heat transfer by the same cylinder in a stagnant pool held at the boiling point T_{vap}. Note the vertical scales: the heat transfer rate in case (b) is larger than case (a) by more than an order of magnitude for a temperature difference of 10 °C. Adapted using data from [63].

T_{vap}, so the wire delivers heat to the pool and causes boiling. Vapor bubbles form at the wire surface and then flow upwards. It is possible to compute the rate of heat transfer in these idealized situations [63]; the results are graphed in Figure 12.4, and show that phase-change heat flux is generally orders of magnitude greater than single-phase forced convection heat flux, for similar temperature differentials.

Flexibility in inducing phase transition   The temperature at which a phase transition such as vaporization occurs depends on the pressure. This dependence defines a function $T_{vap}(p)$ (or its inverse $p_{vap}(T)$, known as the **vapor pressure** for a given temperature) that traces out a *curve* in the pT-plane. For example, the pressures at which water and the common refrigerant *R-134a* (tetrafluoroethane) boil are plotted in Figure 12.5 as functions of temperature. The ability to change the temperature at which a liquid boils by changing the pressure (or *vice versa*) is an essential tool for an engineer designing a heat extraction

device or an engine. It is particularly convenient for a heat extraction device such as a refrigerator that needs to operate between temperatures in a specific range. The fluid that runs in the pipes of a refrigerator sucks heat out of the interior by vaporizing and exhausts heat to the room when it condenses. The fluid circulating inside the refrigerator should boil below the desired interior temperature, say at $-5\,°C$, in order to facilitate heat transfer, and the fluid outside the refrigerator should condense somewhat above room temperature, say at 45 °C. Thus, an engineer designing the refrigerator wants the same fluid to boil at two different temperatures, a trick that is accomplished by choosing the appropriate operating pressures. According to Figure 12.5(b), R-134a boils at $-5\,°C$ (45 °C) at a pressure of roughly 0.24 MPa (1.18 MPa). So the desired properties can be obtained by arranging the pressure of the fluid in the interior and exterior of the refrigerator to be approximately 2.4 and 11.6 atm respectively, well within the structural limits of ordinary copper tubing.

It is not an accident that R-134a has a range of boiling temperatures that make it well suited for use in refrigerators. It was developed for that purpose. During

Advantages of Phase-change Energy Conversion

A working fluid that changes phase in the cyclic operation of an engine or heat extraction device enables quick and efficient transfer of relatively large amounts of energy at temperatures and/or pressures that can be adapted to specific applications.

the twentieth century chemists developed fluids with characteristics that are appropriate for different kinds of heat extraction applications. In addition to boiling in the right temperature and pressure range, desirable characteristics include having a large enthalpy of vaporization and being chemically relatively inert. Another feature required in recent years is that the working fluid have a minimal effect on Earth's ozone layer, a topic discussed further in the next chapter.

12.2 Phase Change in Pure Substances

In this section we develop further the thermal physics of phase change in pure substances, a topic first introduced in §5.6. First, in §12.2.1 we clarify what is meant by a "pure substance" and characterize the distinction between evaporation and vaporization. Then, in §12.2.2, we take a more careful look at the phase diagram of water as an example of a pure substance that can exist in solid, liquid, and vapor phases using a variety of different parameterizations in terms of different combinations of two or three state variables. This gives us a picture of the *saturation dome* of water, and provides a framework for describing supercritical steam (§12.2.3) and steam in a liquid–vapor *mixed phase* (§12.2.4).

12.2.1 What is a Pure Substance?

A **pure substance** is a single chemical substance. This seems like a simple enough characterization. It is important, however, to realize that some very familiar and interesting substances are not "pure" by this definition. Water, alcohol, and copper are all pure substances – each molecule of these substances is chemically identical. A solution of salt in water is not a pure substance, nor is an alloy like brass (copper and zinc), nor is air. The materials in the latter set are homogeneous on a macroscopic scale, but they are all mixtures of different types of molecules. Such mixtures have thermodynamic properties that may differ substantially from those of the pure substances of

which they are composed. For example, an alloy of two metals can have a melting point that is lower than that of either pure metal.

It is particularly important to recognize that a substance such as water ceases to be a pure substance when it is placed in contact with air. Suppose a container is initially filled with *dry air*, i.e. air containing no water vapor. If a quantity of pure water is poured into the container, partially filling it, and the container is then sealed, two things immediately happen: some of the air in the container dissolves into the water, and some of the water *evaporates* into the air. **Evaporation** is the phase transition from liquid to vapor that occurs at an interface between a liquid and a surrounding gas when the temperature is below the boiling temperature of the liquid at the ambient pressure. In contrast, *boiling* is a transition that occurs when a liquid is heated to its boiling (or vaporization) temperature at the ambient pressure, $T_{vap}(p)$. When a water–air system comes into equilibrium at, say, room temperature and atmospheric pressure, the water is no longer a pure substance, and the air, which began as a mixture of nitrogen and oxygen, now includes a significant proportion of water molecules as well. If, on the other hand, a vacuum pump were used to remove all of the air after sealing the container described above, then the container would hold both liquid water and water vapor as pure substances. In particular, the pressure would fall to the pressure at which water boils at room temperature (about 0.03 atm), and the container would then include water and water vapor in equilibrium at the boiling pressure $p_{vap}(T)$ of water at temperature T.

12.2.2 Parameterizations of the Phase Diagram for Water

In this section we describe the phase diagram of water in terms of several different combinations of state variables. Different features of the phase behavior of a material are brought out by these different perspectives. We all have everyday experience with the three common phases of H_2O: ice, liquid water, and water vapor. Water is often used in phase-change systems and is typical in many ways[2] of the other fluids we encounter in phase-change energy conversion, so we use it as an example to make the discussion of phase diagrams concrete.

---

[2] One way in which water is atypical is that it expands upon freezing. This plays no role in the use of the liquid and vapor phases in engines and heat extraction devices; however, it plays a major role in Earth's climate because ice floats on water. Also, there are actually several different phases of ice with different crystal structure. That complication is ignored here.

Example 12.1   Partial Pressure and Relative Humidity

When liquid water is allowed to come into thermal equilibrium with the atmosphere at temperature T, it evaporates until the pressure exerted due to molecules of the water vapor above the liquid water is equal to its *vapor pressure* $p_{vap}(T)$, the pressure at which water would boil at that temperature. In general, the contribution to atmospheric pressure that is due to water vapor is known as the **partial pressure** of water vapor. In general, the partial pressure of water in the atmosphere may be less than its vapor pressure. The **relative humidity** ϕ is the ratio (expressed as a percent) of the partial pressure of water in the atmosphere to its vapor pressure at the ambient temperature, and is a measure of the water content of the atmosphere. If the relative humidity reaches 100%, liquid water begins to condense out of the atmosphere forming clouds, dew, frost, and/or rain.

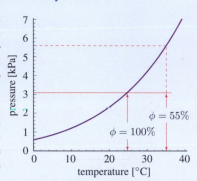

The figure at right shows the vapor pressure of water over a range of temperatures relevant to some weather phenomena. Note that the vapor pressure rises rapidly with increasing temperature – it reaches 1 atm at $T = 100\,°C$ – thus warm air can hold much more water than cold air. The examples below explore some applications of the concepts of relative humidity and partial pressure.

Example 1: At ∼20 °C, people are comfortable when the relative humidity is in the range from ∼25% to ∼60%. In cold climates, when air is brought in from outside, water vapor must be added to reach comfortable levels.
If the outside temperature is $T_0 = 0\,°C$ and the outside relative humidity is $\phi_0 = 50\%$, how much water must be added per cubic meter of air to reach $\phi = 40\%$ at $T = 20\,°C$?
The vapor pressure of water at 0 °C (20 °C) is $p_0 = 0.61\,kPa$ ($p = 2.33\,kPa$). From the ideal gas law, the density of water at temperature T and (partial) pressure p is $\rho = pm/RT$, where $m = 0.018\,kg/mol$ is the molecular weight of water. So the amount of water that must be added to humidify the air is

$$\Delta\rho = \rho - \rho_0 = \frac{m}{R}\left(\frac{p\phi}{T} - \frac{p_0\phi_0}{T_0}\right) = \frac{0.018\,kg/mol}{8.314\,J/mol\,K}\left(\frac{932\,Pa}{293\,K} - \frac{305\,Pa}{273\,K}\right) \cong 0.0045\,kg/m^3\,.$$

Example 2: The **dew point** is the temperature at which water will begin to condense out of air. The dew point depends upon the amount of water vapor in the air. The ambient temperature drops at night due to radiative cooling when skies are clear and winds are calm. When the temperature reaches the dew point, water begins to condense on exposed surfaces, and the temperature falls much more slowly.

If the temperature is 35 °C and the relative humidity is 55% (relatively uncomfortable at such a high temperature), what is the dew point?
From the graph above, $p_{vap} = 5.61\,kPa$ at $T = 35\,°C$. At $\phi = 55\%$, the partial pressure of water is $p_{H_2O} = p_{vap}\phi = 0.55 \times 5.61 = 3.09\,kPa$, which equals the vapor pressure of water ($\phi = 100\%$) at approximately 25 °C, which is the *dew point.*

When we first discussed phase transitions in §5, the phases of water were displayed in Figure 5.11 as a function of pressure and temperature. Utilizing the concept of thermodynamic equilibrium (§8.4) we can say more precisely that this phase diagram – reproduced with somewhat more detail in Figure 12.6 – specifies what phase or phases of water can exist in equilibrium at any given temperature and pressure. At most temperatures and pressures, only one phase is stable. At atmospheric pressure and room temperature (point P in Figure 12.6), for example, ice melts and water vapor condenses; only liquid water

exists in thermodynamic equilibrium at this temperature and pressure. Along the line that separates the liquid and solid phases, however, both can exist in equilibrium. The relative proportion of one phase or the other depends on the amount of energy in the system – the more energy that is added the more ice melts. There is one point in the phase diagram, the **triple point**, where all three phases, solid, liquid, and vapor, can be in thermal equilibrium.

The pT-plane The phase diagram in the pressure–temperature plane should already be familiar from §5.6.

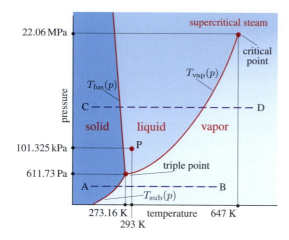

Figure 12.6 A sketch (not to scale) of the important phases of water in the pT plane. The solid (red) lines mark the temperatures and pressures at which two phases coexist, either solid–liquid ($T_{fus}(p)$), solid–vapor ($T_{sub}(p)$) or liquid–vapor ($T_{vap}(p)$). Also shown are the triple point where all three phases exist in equilibrium, and the critical point, beyond which there is no distinction between liquid and vapor phases. The point P identifies liquid as the equilibrium state of water at NTP.

Referring to Figure 12.6, we emphasize a few important points before passing on to other descriptions. First, the temperatures at which phase transitions occur are functions of the pressure. The higher the pressure, the higher the boiling point of water. The reverse is true for melting, but water is unusual in this regard. Second, if the pressure is constant, the temperature remains fixed during a phase transition. As heat is added to ice at a fixed pressure greater than 611.73 Pa, along line CD in the figure for example, the temperature increases until it reaches the melting point. Then the temperature remains fixed until all the ice melts. Similarly when liquid water boils, the temperature remains fixed at $T_{vap}(p)$ until all the water vaporizes. If heat is added to ice at a fixed pressure below 611.73 Pa, say along line AB in the figure, the temperature increases until it reaches the *sublimation temperature* $T_{sub}(p)$, where it remains fixed until all the ice has turned directly to vapor. At the *triple point*, $p = 611.73$ Pa and $T = 273.16$ K, ice, liquid water, and vapor can all exist in equilibrium in any proportions. Third, the *critical point*, which occurs at $p_c = 22.06$ MPa and $T_c = 647$ K, marks the end of the liquid–vapor phase transition line. Beyond the critical pressure or temperature, there is no clear distinction between liquid and vapor phases. At pressures above p_c, for example, water turns continuously from a dense, liquid state to a diffuse vapor as it is heated. The definition

of the *supercritical zone* is discussed further in §12.2.3. Finally, the volume of the water changes dramatically as liquid changes to vapor at fixed pressure. Liquid water can be in equilibrium with vapor at the boiling point with any ratio of liquid to vapor. The volume of a sample of water in liquid–gas phase equilibrium therefore cannot be extracted from the pT-diagram. To expose that information (see also §12.2.4), we must turn to a different representation of the phase diagram, such as the pV-plane.

The pV-plane To understand the way that volume changes in phase transitions, it is useful to plot the phase diagram again, but this time in the pV-plane. Such plots are analogous to the pV-plots for engine cycles introduced in the previous chapter. We plot pressure on the vertical axis and the **specific volume** (volume per kilogram), usually denoted by v, on the horizontal axis. To illuminate the structure of the pV-diagram, it is useful to follow the specific volume as a function of pressure ($v(p)$) along an *isotherm*. For simplicity, we focus on the transition between liquid and vapor, ignoring ice, and consider a sequence of such isotherms in Figure 12.7.

First, consider the 700 K isotherm at the top of the figure. Since 700 K is above the critical temperature $T_c = 647$ K, there is no abrupt transition between vapor and liquid. Along this isotherm, the pressure and volume vary roughly inversely. For an ideal gas, an isotherm would be a hyperbola in the pV-plane, $p = Nk_BT/V$, giving a straight line in logarithmic coordinates, as shown in Figure 12.7(b). Although water vapor is not an ideal gas, the approximate inverse relationship persists: the higher the pressure, the smaller the specific volume at fixed T and N.

Phase Diagram in the pT-Plane

A plot of the phase of a pure substance as a function of pressure and temperature gives a compact display of the phases and phase boundaries between solid, liquid, and gas. Curves specifying the sublimation, melting, and boiling points as functions of temperature separate the phases. At low pressure, ice *sublimes* directly to vapor. All three phases coexist in equilibrium at the *triple point*. Beyond the *critical point*, liquid and gas are no longer distinct. The plot in the pT-plane does not indicate the changes in volume that take place across a phase boundary, when, for example, a liquid changes to a gas.

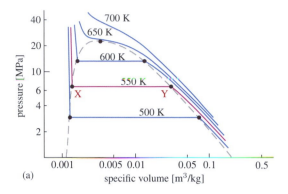

(a)

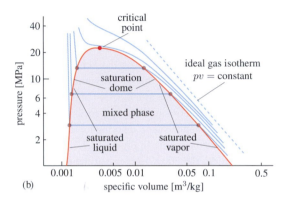

(b)

Figure 12.7 (a) Isotherms in the pV-plane for water. For $T < T_c$ the isotherms have a flat section along which liquid water turns to vapor, significantly increasing its volume. Above T_c the isotherms are smooth. (b) The locus of all the discontinuities in the isotherms defines the *saturation curve* or *saturation dome*. Points on the saturation dome are either saturated liquid or saturated vapor. Inside the saturation dome the liquid and vapor phases are mixed.

Next, consider the isotherm at $T = 550\,\text{K}$, which is below the critical temperature. We start at a very high pressure where the water is liquid (see Figure 12.6). As we decrease the pressure at constant temperature, the volume of the liquid grows very slowly until we reach the boiling pressure $p_{\text{vap}}(T)$, which is around $6\,\text{MPa}$ at $T = 550\,\text{K}$. Then the water begins to boil, and we cannot reduce the pressure any further until all the water has boiled. Thus, *the isotherm remains flat in p while the specific volume v grows dramatically* – from the specific volume of liquid water to that of water vapor. After all the water has vaporized, if we continue to lower the pressure, the specific volume of the water vapor expands as we would expect for a gas. All of this is shown on the $550\,\text{K}$ isotherm (in magenta) in Figure 12.7(a). Note that there is a *kink* (slope discontinuity) in the isotherm at the onset of boiling (point X): the curve that was steep (liquid water is not

very compressible) suddenly becomes flat. At the other end, when all the water has vaporized, there is another kink (point Y) beyond which the isotherm once again falls with specific volume. Between these two points on this isotherm, the liquid and vapor phases of water coexist. The points X and Y on the $550\,\text{K}$ isotherm are special. X corresponds to a liquid just ready to boil. We refer to such a liquid as a **saturated liquid**. Before one can lower the pressure infinitesimally (keeping the temperature fixed), it will all boil. Likewise, the right-hand kink Y represents a vapor about to condense – a **saturated vapor**. To raise the pressure infinitesimally, the vapor must first all liquefy.

If we repeat this exercise at a slightly lower (or higher) temperature, say $500\,\text{K}$ (or $600\,\text{K}$), the scenario plays out as it did at $550\,\text{K}$ except that the onset of boiling is at a lower (or higher) pressure and therefore the volume of the vapor once boiling is complete is larger (smaller). By varying the temperature and tracing many isotherms, as shown in Figure 12.7(a), we can map out the region where the liquid and vapor phases of water can coexist. Just above the critical temperature, the distinction between liquid and vapor disappears and an isotherm such as the one shown at $650\,\text{K}$ no longer has a flat section. The envelope of points like X, which mark the onset of boiling, and those like Y, which mark the onset of condensation, traces out curves that connect smoothly at the critical point, as shown explicitly in Figure 12.7(b). This curve is called the **saturation curve**, or **saturation dome**, since water is either a saturated fluid, ready to boil, on the left boundary, or a saturated vapor, ready to condense, on the right. At points inside the saturation dome, the fluid is a mixture of the two phases.

The phase surface in pVT-space The region of mixed phase in the pV-plane contains the information that is missing when we look at the properties of water in the pT-diagram (Figure 12.6) alone. We can combine

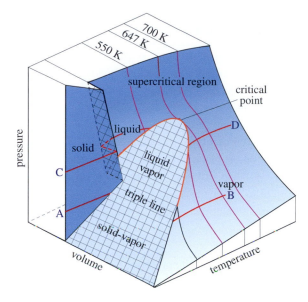

Figure 12.8 A cartoon (not to scale) of the phase diagram of water shown as a surface in the three coordinates, p, V, and T. Every point on the blue surfaces is an allowed equilibrium state of water. Surfaces of phase equilibrium are cross-hatched. Figures 12.6 and 12.7 are projections of this figure onto the pT- and pV-planes. The two isobars labeled AB and CD and highlighted in red here coincide with the paths in Figure 12.6. The three isotherms highlighted in magenta coincide with the isotherms at 700 K, 647 K, and 550 K.

the information in the pT-diagram with the information from the pV-diagram into a *three-dimensional* phase diagram for water, as shown in Figure 12.8. In this three-dimensional picture, the combinations of V, T, and p that are physically possible form a two-dimensional surface within the full three-dimensional coordinate space. Some aspects of the phase structure of water are more transparent in the three-dimensional diagram than in any two-dimensional diagram. Tracing the *isobar* CD from Figure 12.6, for example, in Figure 12.8 one can see that the volume decreases as ice melts (exaggerated in the figure), increases slowly as liquid water is heated, and then increases dramatically as water boils. The approach to ideal gas behavior, $pV \approx Nk_BT$, can be seen at high temperature.

The ST-plane Having followed water through its phase transitions using the state variable combinations pT, pV, and pVT, we now consider the entropy–temperature plane. This perspective, while perhaps less familiar, turns out to be crucial in the practical applications of the following chapter. We plot the temperature on the vertical

axis and the **specific entropy** s (entropy per kilogram) on the horizontal axis. By definition, isotherms are horizontal lines in the ST-plane. The phase transitions of water in the ST-plane can be mapped out using a set of isobars in much the same way that the saturation dome is mapped in the pV-plane using isotherms. At the temperatures and pressures we consider, ice cannot exist in equilibrium. Thus we start with liquid water at temperature T_0 and add thermal energy at a constant pressure p, as in the isobar CD in Figure 12.6. The temperature and entropy grow together (see eq. (5.28) and eq. (8.19)), $ds = dQ|_p/T = c_p dT/T$. Since c_p is approximately constant for water away from the critical point, the entropy grows like the logarithm of T, $s(T) - s(T_0) \cong c_p \ln(T/T_0)$. This continues until the temperature approaches the boiling point $T_{\text{vap}}(p)$, near which c_p increases so that the s–T curve hits the saturation dome at the transition point. Then the temperature remains fixed, and adding heat turns the liquid water to vapor, adding entropy $\Delta s_{\text{vap}} = \Delta h_{\text{vap}}/T_{\text{vap}}(p)$, where Δh_{vap} is the specific enthalpy of vaporization. The isobar at pressure p is thus flat as s increases by $\Delta s_{\text{vap}}(p)$. If we then continue to add heat, the temperature continues to rise and the entropy once again increases by an amount (approximately) proportional to the logarithm of the temperature. If we plot temperature on the vertical axis and entropy on the horizontal axis, an isobar thus consists of a horizontal segment linking two approximately exponential pieces before (at lower temperature) and after (at higher temperature). A set of isobars are plotted in Figure 12.9 corresponding to phase transition temperatures 500 K, 550 K, and 600 K, identical to the values used in Figure 12.7. Note the points X and Y that correspond between the two figures. The kinks in the isobars occur at the onset (saturated liquid) and end (saturated vapor) of the phase transition, and as in the pV-plane, the locus of all these points maps out the *saturation dome*, this time in the ST-plane.[3]

---

[3] Note that the liquid phase region is very narrow in the ST-plane – in fact it has been exaggerated in Figure 12.9(b) in order to make it visible. If the entropy of liquid water were independent of the pressure at fixed temperature then all the isobars in the liquid phase including the critical isobar would lie on top of the left side of the saturation dome. Using the fact that the Gibbs free energy is a function of p and T, one can show that $\partial S/\partial p|_T = -\partial V/\partial T|_p$. (This is an example of a *Maxwell relation*, discussed for example, in [21].) Since the volume of liquid water expands very slowly with increasing temperature at fixed p, the entropy of water decreases very slowly with pressure at fixed T, and the isobars almost coincide in the liquid phase.

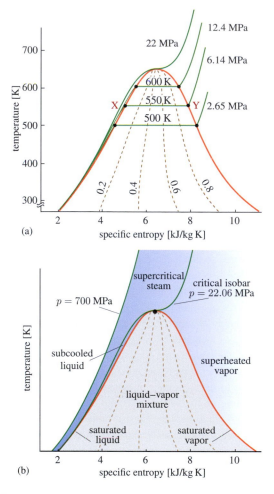

(a)

(b)

Figure 12.9 The phase diagram of water in the ST-plane. (a) A set of isobars (green) are shown, with phase transition temperatures identical to the isotherms shown in Figure 12.7. The points X and Y correspond to the matching points in that figure. The brown dashed lines are contours of constant *quality* (see §12.2.4). (b) Regions and nomenclature are labeled in the ST-plane. Supercritical steam lies above and to the left of the isobar at the critical pressure $p = 22.06$ MPa, shown in green (see §12.2.3 and §12.3).

The ST-plane is the principal playing field for the discussion of phase-change energy conversion cycles in §13, so it is worth spending extra effort getting familiar with it. In addition to the saturation curve, the various phases of water are labeled in Figure 12.9(b), as is the supercritical region where there is no distinction between liquid and gas. The saturation curve separates regions of pure phase (above it) from the region of mixed liquid and vapor phases below it. Below the critical temperature, the liquid phase lies to the left of the saturation curve and the gas lies to the right.

Phase Diagram in the ST-Plane

When the phase diagram is plotted in the ST-plane, the mixed phase lies below a saturation dome similar to that in the pV-diagram. Inside the saturation dome, as heat is added at fixed temperature and pressure, the entropy increases as the saturated liquid turns to saturated vapor. The area under an isobar at pressure p (and fixed $T = T_{\text{vap}}(p)$) that crosses the saturation dome from liquid to vapor measures the specific enthalpy of vaporization at that pressure, $\Delta h_{\text{vap}}(p)$.

12.2.3 The Supercritical Zone

At any pressure below the critical pressure, as a vapor is cooled it will undergo a sharp phase transition and condense to a liquid at $T_{\text{vap}}(p)$. Such a process is denoted by the arrow going directly from point E to point F on Figure 12.10. Similarly, at any temperature below its critical temperature ($T_c = 647$ K for water), a vapor will condense into a liquid as it is pressurized. Beyond the critical pressure and temperature, however, there is no clear distinction between liquid and vapor phases. The physical properties of water vary continuously along any path that does not intersect a phase transition boundary – even one that goes from a vapor phase region to liquid phase by passing above and around the critical point, for example the path P in Figure 12.10. The **supercritical zone** of the phase diagram can be defined as the region where both the pressure and temperature exceed their critical values. This region is denoted with dense cross hatching in the figure. Often,

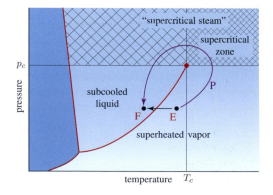

Figure 12.10 Another look at the pT-phase diagram for water showing the supercritical zone and the region of *supercritical steam*. See the text for discussion of the two paths from point E to F.

however, the term **supercritical steam** is applied to denote water above its critical pressure independent of the temperature. The entire hatched region of Figure 12.10, therefore, corresponds to *supercritical steam*.

While the liquid, vapor, and supercritical phases of a fluid are all continuously connected in the region outside the transition curve, the properties of a supercritical substance, and even its chemistry, may be quite different from those of the liquid or vapor phases at sub-critical pressures and temperatures. For instance, unlike its liquid or vapor counterparts, supercritical CO_2 can dissolve caffeine, and is the vehicle for decaffeinating coffee and tea. Moreover, at a density similar to that of liquid water, supercritical CO_2 remains highly compressible, like air. This mixture of liquid-like and vapor-like properties within one substance can have significant advantages as a working fluid for power generation.

12.2.4 Quality and the Mixed Phase

It is apparent that an additional variable beyond temperature and pressure is required to characterize a mixed phase. For example, for water at the boiling point, $100\,°C$ at atmospheric pressure, specifying T and p does not tell us how much of our sample of water is liquid and how much is vapor. Since liquid and vapor have very different enthalpies and entropies, determining these quantities for a given sample depends on knowing the proportion of liquid and vapor in the mixed-phase system.

While knowing the volume provides the additional information needed to characterize a substance in a mixed phase at a given temperature and pressure, it is more convenient to describe the conditions of the substance in terms of the fraction of matter in each of the two phases. This is because, to a good approximation, all the extensive thermodynamic properties of a mixed phase are the mass-weighted sum of the properties of the constituent phases. This simple additive rule seems rather obvious: water in a container in equilibrium with its vapor appears to be two distinct systems with a visible interface between them. It seems natural that its physical properties should be the sum of the properties of each phase weighted by the fraction of the total mass of water in that phase. This is only an approximation, however. At a more precise level of accuracy, the properties of the interface between the phases become important. For example, a mixture of liquid water and water vapor could be in the form of a dense mist of infinitesimal droplets (*fog*). Fog has vastly more surface area at the phase interface than the container of water mentioned above. The internal energy of a system containing many droplets of water suspended in vapor is

Quality

When liquid and vapor are in equilibrium, the quality χ of the mixture is defined to be the mass fraction of vapor. To a good approximation any thermodynamic property of the mixture is the sum of that property of the vapor weighted by χ and that property of the liquid weighted by $1 - \chi$.

different from an equal mass of bulk liquid having a single interface with a corresponding mass of vapor because the molecules at each drop's surface experience different interactions than those in its interior. As long as we do not expect high precision, we can ignore effects like these and consider the thermodynamic properties of a phase mixture to be the weighted sum of the properties of the two phases.

Therefore we generally make the approximation that in a mixed phase each component contributes to the bulk properties according to its mass fraction. We define the **quality** χ of a liquid–vapor mixture in equilibrium at temperature T and pressure $p_{\mathrm{vap}}(T)$ to be the mass fraction of the vapor,

$$\chi = \frac{m_v}{m_v + m_l}. \qquad (12.2)$$

Any extensive thermodynamic property of the mixture, such as internal energy, entropy, enthalpy, or volume, is given by the sum of properties of the vapor, weighted by χ and the liquid, weighted by $(1 - \chi)$. Thus, for example, the specific enthalpy h_m of a mixture is given by

$$h_m = \chi h_v + (1 - \chi)h_l = h_l + \chi \Delta h_{\mathrm{vap}}, \qquad (12.3)$$

where $\Delta h_{\mathrm{vap}} = h_v - h_l$ refers to the specific enthalpy of vaporization. Quality can be used to complete the specification of the state of a two-phase system during its phase transition. In Figure 12.7, for example, points X and Y have the same temperature and pressure, but X is pure liquid and has quality zero, while Y is pure vapor and has quality one. At each point under the saturation dome the mixture has a definite quality. Contours of quality 0.2, 0.4, 0.6, and 0.8, for example, are shown in the ST-plane in Figure 12.9. At any point on the straight line joining X with Y, the properties of the mixture can be computed from known properties of the pure phases, and any single property of the mixture that is known can be used to compute the quality.

Example 12.2 Phase Separation in a Hydrothermal Power Plant

Suppose a container holds a quantity of water as a *saturated liquid* at temperature T_0 and pressure $p_{vap}(T_0)$. If the pressure is suddenly decreased, for example by drawing back a piston, some of the water will vaporize. When the system settles down to a new equilibrium, it will consist of a mixture of saturated liquid water and saturated vapor at a new, lower temperature T and pressure $p_{vap}(T)$. This process is known as *flashing* and plays an essential role in hydrothermal energy plants, where it is used to create steam from the high-temperature, high-pressure liquid water that emerges from a hot water reservoir beneath Earth's surface. As explained in §32.4.2, flashing can be modeled as an expansion that conserves the enthalpy of the water.

At a specific geothermal plant described in Example 32.5, saturated liquid water at $T_0 = 250\,°C$ with specific enthalpy $h = 1086\,kJ/kg$ flashes to a final temperature of $T = 150\,°C$.

What fraction of the final mixture is vapor? What is the specific volume of the final mixture?

To proceed we must know the properties of steam in phase equilibrium at $150\,°C$. In the next section we discuss how to obtain this information. For now, we take $h_l = 632\,kJ/kg$ and $h_v = 2746\,kJ/kg$ as given. First we express the enthalpy of the mixture in terms of the quality and the enthalpy of the two phases, using eq. (12.3),

$$h = \chi h_v + (1 - \chi)h_l \,.$$

We then solve for χ in terms h, h_v, and h_l,

$$\chi = \frac{h - h_l}{h_v - h_l} = \frac{1086 - 632}{2746 - 632} \cong 21.5\% \,.$$

This is the fraction, by mass, of the water that has vaporized in the flashing process.

Knowing χ, we can then compute any other property from eq. (12.3) with h replaced by the property of interest. To find the specific volume, as required, we need the specific volumes of the liquid and vapor in phase equilibrium at $150\,°C$, $v_l = 1.09 \times 10^{-3}\,m^3/kg$ and $v_v = 0.393\,m^3/kg$. Then

$$v = \chi v_v + (1 - \chi)v_l = 0.215 \times 0.393 + 0.785 \times 0.001 \cong 0.085\,m^3/kg \,.$$

12.3 The Real World: Engineering Nomenclature and Practical Calculations

Our principal interest in phase transitions comes from their role in power generation and heat extraction. Over a century of practical experience in developing and optimizing such systems has led to sophisticated methods of calculation and somewhat specialized terminology. To connect the general discussion here to concrete real-world systems it is necessary to clarify some aspects of terminology and to discuss how to find accurate data describing mixed-phase fluids.

12.3.1 Engineering Terminology

In engineering practice, a liquid below its boiling point $T_{vap}(p)$ (for any $p < p_c$) is often referred to as **subcooled liquid** to distinguish it from *saturated liquid*, which describes the liquid at the boiling point. Likewise, **superheated vapor** or **superheated steam** describes the vapor phase above $T_{vap}(p)$ while *saturated vapor* refers to the vapor phase at the boiling point $T_{vap}(p)$. For

water, the regions where these terms apply are shown in Figure 12.9(b).

Steam, in engineering usage, refers to superheated or saturated water vapor *or* the mixed phase of liquid water and water vapor. **Dry steam** is superheated water vapor or saturated water vapor with quality one. The mixed phase consisting of saturated water vapor at temperature $T_{vap}(p)$ together with droplets of liquid water is known as **wet steam**. Superheated steam is a colorless, transparent gas. It is the stuff that emerges from the spout of a tea kettle (see Figure 12.11). In contrast, the clouds of "steam" visible further away from the tea kettle spout, or emerging from the cooling stacks of a generating station, are actually a mixture of air, water vapor, and minute droplets of liquid water, more correctly termed **fog**.

In this chapter and the next we distinguish saturated water vapor versus superheated or supercritical steam, and sub-cooled versus saturated liquid water. In later chapters we use the terms *liquid*, *vapor*, and *steam* more loosely according to conventions of specific areas of application,

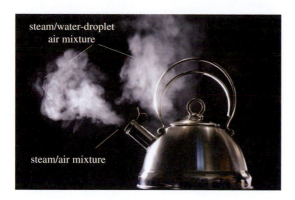
steam/water-droplet
air mixture

steam/air mixture

Figure 12.11 A boiling kettle. A mixture of steam and air coming out of the spout quickly condenses into a mixture of water vapor, a *fog* of minute liquid water droplets, and air.

Engineering Terminology

For this chapter and the next, we adopt standard engineering terminology, where *saturated liquid* and *saturated vapor* denote liquid ready to vaporize and vapor ready to condense respectively. At temperatures below and above the boiling point we use the terms *sub-cooled liquid* and *superheated vapor* respectively. The term *dry steam* applies to superheated or saturated (quality one) water vapor and *wet steam* applies to a liquid–vapor mixture (quality less than one) in equilibrium at $T_{vap}(p)$. Above the critical pressure, water is *supercritical steam*.

and only note the distinctions between, for example, saturated versus superheated when relevant.

12.3.2 Thermodynamic Data on Phase Transitions

When designing real devices, engineers are often interested in domains of pressure and temperature where the behavior of gases is far from ideal. For example, near the critical point of water the density of liquid and vapor are almost the same, and it is a poor approximation to treat the vapor as an ideal gas. For concrete applications it is often necessary to know exact thermodynamic properties (density, internal energy, entropy, enthalpy, latent heats, etc.) as functions of temperature and pressure. These thermodynamic properties have been determined empirically for many fluids across a wide range of temperatures and

Thermodynamic Data

Thermodynamic properties of working fluids such as water, R-134a refrigerant, and ammonia can be found in tables and are conveniently accessed on websites or through software available for computers and smartphones. Typically, specific volume, enthalpy, and entropy for saturated liquid and vapor are available at the phase transition. Data are also available on sub-cooled liquids and superheated vapors.

pressures, and the corresponding numerical data is available in many places. In the pre-computer era, this data was published in voluminous sets of tables. For H_2O, these tables are known as **steam tables**, and this name is often applied to the thermodynamic data for other materials as well. While there are now simple computer programs and smartphone apps that contain this data and can produce any thermodynamic information in the format desired (see e.g. [23]), the computer data is usually used in the same way as data from the traditional steam tables, so for clarity we illustrate the approach with the use of the tables. A sample section of a steam table (for H_2O) is shown in Table 12.1. The main sections of the table are labeled by the pressure, here 1.0×10^5 Pa (very close to 1 atm) and 1.2×10^5 Pa. The boiling point of water is given at each pressure (99.61 °C and 104.78 °C respectively). Below that, labeled by "Sat. Liq." and "Sat. Vap.," the specific volume, internal energy, enthalpy, and entropy of the saturated liquid and vapor are given. On the line between, labeled "Evap.," the differences, $\Delta v = v_v - v_l$, etc., are listed. The remaining rows of the table, labeled by the temperature in °C (in the left-hand column) give the values of thermodynamic properties for the sub-cooled liquid and superheated vapor phases at the relevant temperatures, above and below the horizontal line half-way down the table.

Some entries in the table are redundant. For example the changes in enthalpy, internal energy, and volume recorded in the "Evap." line are related by $\Delta h_v = \Delta u_v + p \Delta v_v$, and the changes in entropy and enthalpy are related by $\Delta s_v = \Delta h_v / T$ (Problem 12.2). It is also interesting to check how far the behavior of the vapor phase differs from what would be expected for an ideal gas (Problem 12.7).

Steam tables can be used in a multitude of ways. Not only can one find thermodynamic properties as a function of temperature or pressure, but also one can work backwards to find the temperature and/or pressure at which

Table 12.1 Section of a *steam table*: thermodynamic properties for water at specified temperatures and pressures. See text for more description. The thermodynamic information summarized in steam tables can be found on-line in convenient calculators and apps. Adapted from [64].

$p(T_{sat})$	1×10^5 Pa (99.61 °C)				1.2×10^5 Pa (104.78 °C)			
Temp [°C]	v [m³/kg]	u [kJ/kg]	h [kJ/kg]	s [kJ/kg K]	v [m³/kg]	u [kJ/kg]	h [kJ/kg]	s [kJ/kg K]
Sat. Liq.	1.0432×10^{-3}	417.40	417.40	1.3028	1.0473×10^{-3}	439.23	439.36	1.3609
Evap.	1.6929	2088.2	2257.4	6.0560	1.4274	2072.5	2243.7	5.9368
Sat. Vap.	1.6939	2505.6	2674.9	7.3588	1.4284	2511.7	2683.1	7.2977
0.00	1.0002×10^{-3}	−0.0404	0.0597	−0.0001	1.0001×10^{-3}	−0.0400	0.0800	−0.0001
10	1.0003×10^{-3}	42.018	42.118	0.1511	1.0003×10^{-3}	42.017	42.137	0.1511
20	1.0018×10^{-3}	83.906	84.006	0.2965	1.0018×10^{-3}	83.905	84.025	0.2965
30	1.0044×10^{-3}	125.72	125.82	0.4367	1.0044×10^{-3}	125.72	125.84	0.4367
40	1.0078×10^{-3}	167.51	167.62	0.5724	1.0078×10^{-3}	167.51	167.63	0.5724
50	1.0121×10^{-3}	209.32	209.42	0.7038	1.0121×10^{-3}	209.31	209.43	0.7038
60	1.0171×10^{-3}	251.15	251.25	0.8313	1.0171×10^{-3}	251.15	251.26	0.8312
70	1.0227×10^{-3}	293.02	293.12	0.9551	1.0227×10^{-3}	293.02	293.14	0.9551
80	1.0290×10^{-3}	334.94	335.05	1.0755	1.0290×10^{-3}	334.95	335.07	1.0755
90	1.0359×10^{-3}	376.96	377.06	1.1928	1.0359×10^{-3}	376.95	377.08	1.1928
100	1.6959	2506.2	2675.8	7.3610	1.0435×10^{-3}	419.05	419.18	1.3072
120	1.7932	2537.3	2716.6	7.4678	1.4906	2535.7	2714.6	7.3794
140	1.8891	2567.8	2756.7	7.5672	1.5712	2566.5	2755.1	7.4800
160	1.9841	2598.0	2796.4	7.6610	1.6508	2597.0	2795.1	7.5745
180	2.0785	2628.1	2836.0	7.7503	1.7299	2627.3	2834.9	7.6643
200	2.1724	2658.2	2875.5	7.8356	1.8085	2657.5	2874.5	7.7499
220	2.2661	2688.4	2915.0	7.9174	1.8867	2687.8	2914.2	7.8320

Example 12.3   Water in a Box

One kilogram of water is introduced into an initially evacuated box of volume 1 m³ and is then heated to 105 °C. *Estimate the resulting pressure, the relative proportion of liquid water and vapor, and the enthalpy.*

Table 12.1 gives data on steam at 104.78 °C, which is close enough to 105 °C for a good estimate. The specific volume of saturated vapor at that temperature is 1.43 m³/kg, and the volume of the liquid is negligible ($\sim$ 1000× smaller) so the fraction of the water that is converted to vapor is $1/1.43 \cong 70\%$. Thus the system is in phase equilibrium with $\chi \cong 0.70$. The pressure is $p_{vap}(105\,°C) \cong 1.2 \times 10^5$ Pa. Finally the enthalpy is obtained by applying eq. (12.3),

$$h(\chi = 0.7) = 0.7 \times 2683 + 0.3 \times 439.4 \cong 2010\,\text{kJ}.$$

Repeat the question supposing that the system is heated to 200 °C.

Table 12.1 does not contain saturation information for $T = 200\,°C$, so we must find data elsewhere. Using an online calculator, we find that the specific volume of saturated vapor at 200 °C is 0.127 m³/kg, which is less than the volume of the container. Thus we conclude that the steam in this case must be superheated in order to fill the entire container. Given the temperature, we seek a pressure at which the specific volume of superheated steam at 200 °C is 1 m³/kg. The online calculator gives $p = 216$ kPa, slightly above 2 atm. As a check on this result, assume that the steam is an ideal gas:

$$p = nRT/V = \left(\frac{1.0\,\text{kg}}{0.018\,\text{kg/mol}}\right) \times (8.314\,\text{J/mol K}) \times (473\,\text{K}) \div (1\,\text{m}^3) \cong 218\,\text{kPa},$$

which is in good agreement with the result obtained from the online data. At this temperature and pressure the enthalpy of superheated steam provided by the online calculator is $h = 2870$ kJ/kg.

Example 12.4   Using Steam Tables to Find the Results of Thermodynamic Processes

Example 1: Suppose that we begin with superheated water vapor at a temperature and pressure of $200\,°C$ and 10^5 Pa. After a device compresses it isentropically we find that its temperature at the outlet of the device has gone up to $220\,°C$. *What is the final pressure?*

According to Table 12.1, the initial specific entropy is ≈ 7.84 kJ/kg K. The outlet temperature is $220\,°C$ and we know that the specific entropy at the outlet remains 7.84 kJ/kg K. Searching the table we see that at $220\,°C$ the specific entropy is 7.83 kJ/kg K at a pressure of 1.2×10^5 Pa, which therefore must be quite close to the outlet pressure. We were lucky that the value of the pressure at which the $220\,°C$ specific entropy was ~ 7.84 kJ kg^{-1} K^{-1} happened to be the one for which data was provided. (Of course the problem was set up this way!) In general one would have to make a rough interpolation, search through the tables, or use a computer application to find the pressure to high accuracy given the final temperature and entropy.

Example 2: Suppose we are given dry steam (i.e. $\chi = 1$) at $p = 10$ atm, and expand it isentropically down to a pressure of 1 atm. *What fraction of the steam condenses to liquid water?*

Table 12.1 does not provide information at 10 atm, so we consult another reference such as [23], where we find that the specific entropy of saturated steam at 10 atm (and $181\,°C$) is 6.59 kJ/kg K. Looking back at Table 12.1, we see this does indeed lie between the specific entropy of water and steam at 1 atm, so the result of the expansion is a mixture of liquid water and steam. Taking the entropy values from the table and setting up an equation like eq. (12.3) for the specific entropy as a function of the quality,

$$6.59 = 7.36\chi + 1.30(1 - \chi),$$

we find $\chi = 0.87$, so 13% of the steam (by mass) has condensed to liquid water.

---

certain values are obtained (software containing thermodynamic information can do this automatically). This is useful in exploring energy conversion cycles. Examples 12.3 and 12.4 illustrate the use of steam tables in pure and mixed phases.

Discussion/Investigation Questions

12.1 What is the role of gravity in boiling heat transfer? Do you think boiling heat transfer would work well on an Earth-orbiting satellite?

12.2 An insulated cylinder closed by a piston is initially filled with one liter of ice at $-20\,°C$. The piston is pulled out, doubling the volume and initially creating a vacuum above the ice. Describe qualitatively what happens as the system comes into equilibrium at $-20\,°C$. Now the system is slowly heated to $110\,°C$. Again describe in words, what is the state of the water in the cylinder?

12.3 Research what a *pressure cooker* is and describe the physical principles that make it work.

12.4 A *heat pipe* is a device that uses heat conductivity and the large latent heat of phase change to transfer heat very efficiently between two solid bodies. Research the subject. What is the role of gravity? Why is the heat pipe evacuated before a carefully measured amount of fluid is added. How can the same heat pipe with the same working fluid be used between quite different temperature objects? Why does a heat pipe fail when it is overheated or sub-cooled?

12.5 On the basis of the pVT-diagram, Figure 12.8, construct a sketch of the phase diagram for water in the VT-plane (specific volume on the horizontal axis, temperature on the vertical) for temperatures at which the liquid and vapor phases are relevant. Label the graph like Figure 12.7. Note that at high temperature and low pressure in the gas phase, water is almost a perfect gas so $T \propto pV$. Use this fact to help you draw some isobars.

12.6 Since the volume parameterizes the points on the phase transition line between liquid and vapor (see Figure 12.7), one could use the density instead of the quality as the thermodynamic variable to characterize the mixed phase. Why do you suppose this is not done?

12.7 Under calm, clear conditions during the summer in a temperate climate similar to Boston's, meteorologists can predict the minimum nighttime temperature quite accurately by making measurements of the relative humidity during the day. Explain why estimates made the same way are much less accurate in the winter, when maximum daytime temperatures are close to 0 °C.

Problems

Note: for a number of the problems in this chapter you will need to use steam table data from a source such as [23] or the equivalent.

12.1 The isothermal compressibility of a fluid is the relative change of volume with pressure at constant temperature $\beta = -(1/V)\partial V/\partial p|_T$. Qualitatively describe and sketch the compressibility of water as a function of temperature from 0 °C to 1000 °C, at a pressure of 1 atm and at the critical pressure $p_c \cong 22.06$ MPa. What is the compressibility at the phase transition from liquid to vapor?

12.2 Check that the Δs_{vap} and Δh_{vap} quoted in Table 12.1 are consistent with $\Delta h = \Delta u + p\Delta v$ and $\Delta s = \Delta h/T$.

12.3 Using the data given in Table 12.1, estimate the enthalpy added to one kilogram of water that undergoes the following transformations: (a) from 0 °C to 90 °C at $p = 10^5$ Pa; (b) from 95 °C to 105 °C at $p = 10^5$ Pa; (c) from 95 °C to 105 °C at $p = 1.2 \times 10^5$ Pa.

12.4 Consider a volume of 100 L of water, initially in liquid form at temperature T_i, to which $\Delta U = 25$ MJ of energy is added from an external reservoir at temperature T_r through a thermal resistance, so that the rate of energy transfer is $\dot{Q} = (1 \text{ kW/K})(T_r - T)$, where T is the instantaneous temperature of the water. The energy is added in two different ways: (a) with the water at initial temperature $T_i = 100$ °C and the external source at temperature $T_r = 150$ °C; (b) with $T_i = 0$ °C, $T_r = 75$ °C. In each case, compute the time needed to transfer ΔU and the extra entropy produced in this process. Do the comparative results match with the expectation from the discussion of energy transfer at constant temperature in §12.1? (You may assume a constant heat capacity for liquid water, independent of temperature.)

12.5 Data on the heat flux for laminar flow of liquid water and for pool boiling of water are shown in Figure 12.4. Take the bulk temperature of the fluid in the case of laminar flow to be $T_0 = 25$ °C and assume that the pool boiling process takes place at 1 atm. (a) Suppose the two methods are used to transfer heat at a rate of $q = 10^5$ W/m². Estimate the rate of extra entropy production (W/m² K) for the two methods. (b) Estimate the heat flux for pool boiling when its rate of entropy production is the same as the rate of entropy production for laminar flow in part (a).

12.6 An industrial freezer is designed to use ammonia (NH₃) as a working fluid. The freezer is designed so that ammonia flowing through tubes inside the freezer at p_- vaporizes at $T = -40$ °C, drawing heat out of the interior. Outside the freezer, the ammonia vapor at $T = +45$ °C liquefies at a pressure p_+ dumping heat into the environment. Find thermodynamic data on ammonia and identify the pressures $p_\pm$ that enable the liquid–vapor phase transition to take place at the required temperatures.

12.7 Using the data given in Table 12.1, find the deviations from the ideal gas law for water vapor just above the boiling point at $p = 10^5$ Pa. For example, does $V/n = RT/p$?

12.8 Revisit Question 12.2 quantitatively. Specifically, what is the pressure in the cylinder when the system comes to equilibrium after the volume has been doubled at -20 °C? What is the pressure after the cylinder has been heated to 110 °C? What fraction of the water in the cylinder is in vapor form at -20 °C and at 110 °C?

12.9 As explained in the following chapter (§13), a steam turbine can be modeled as an isentropic expansion. The incoming steam is superheated at an initial temperature T_+ and pressure p_+. After expansion, the exhausted steam is at temperature T_-. A high-performance turbine cannot tolerate low-quality steam because the water droplets degrade its mechanical components. Suppose that $T_- = 40$ °C and that the turbine requires $\chi \geq 0.9$ throughout the expansion process. If the high-pressure system is rated to 5 MPa, what is the minimum possible initial temperature of the steam?

12.10 Use steam table data to estimate accurately the pressure at which water would boil at room temperature (20 °C). Similarly estimate the temperature at which water boils on top of Mt. Everest, where the air pressure is approximately 1/3 atm?

12.11 A sample of H_2O at a pressure of 1 atm has a specific enthalpy of 700 kJ/kg. What is its temperature? What state is it in, a sub-cooled liquid, superheated vapor, or a mixed phase? If it is in a mixed phase, what fraction is liquid?

12.12 Repeat the preceding question for a sample at 10 atm with the same specific enthalpy.

12.13 On a hot summer day in Houston, Texas, the daytime temperature is 35 °C with relative humidity $\phi = 50\%$. An air conditioning system cools the indoor air to 25 °C. How much water (kg/m³) must be removed from this air in order to maintain a comfortable indoor humidity of 40%? Compare your answer with the result of Example 12.1.

Thermal Power and Heat Extraction Cycles

Steam power and vapor-compression cooling have reshaped human society, by supplying cheap electricity for industry and individuals, and by providing refrigeration to retard food spoilage and efficient cooling for modern buildings. As mentioned in the previous chapter, these technologies rely on the thermodynamics of phase change, specifically between liquid and vapor, to work quickly and efficiently and at high capacity. The thermodynamic cycles that are employed by engines that transform heat to mechanical energy in electric power plants are closely related to the cycles that are used to move heat from low to high temperatures. This chapter describes these cycles and their practical applications. This is also the appropriate place to describe gas turbines. Although they resemble internal combustion engines in their thermodynamics, gas turbines have much in common with steam turbines and are also used for large-scale stationary power generation.

We begin (§13.1) with a brief digression on how to describe thermodynamic processes in which material flows from one subunit to another, a standard issue in phase-change cycles. Then, in §13.2, we turn to the *vapor-compression (VC) cycle* that lies at the heart of air conditioners, refrigerators, and heat pumps. As we did for internal combustion engines (§11), we present ideal cycles that allow us to understand the workings of heat extraction devices at a semi-quantitative level. Since VC cycles differ in some important ways from the gas cycles of earlier chapters, we introduce their complexities in a step-wise manner. We analyze the VC cycle in the ST-plane where it is easiest to understand the thermodynamics. We then look briefly at the physical components of an air conditioner and relate them to the abstract steps in the vapor-compression cycle. To see how heat extraction can work in practice, we design a quasi-realistic air conditioner based on a slightly simplified ideal VC cycle. Finally, we comment on losses

> **Reader's Guide**
> In this last chapter on thermal energy conversion, we describe the thermodynamic cycles that power steam turbines and the related cycles at the heart of modern heat extraction devices. In both cases the key ingredient is a working fluid that changes phase in the course of the cycle. We turn first to vapor-compression cooling cycles and then to the Rankine steam cycle used in electric power plants; we describe the idealized thermodynamic cycles and briefly characterize the physical components and the limitations of real systems. We then consider the Brayton cycle, a gas combustion cycle that also plays a significant role in electric power generation. The chapter closes with a description of combined cycle plants and cogeneration.
>
> Prerequisites: §5 (Thermal energy), §6 (Heat transfer), §8 (Entropy and temperature), §10 (Heat engines), §11 (Internal combustion engines), §12 (Phase-change energy conversion).
>
> Steam power cycles are used in nuclear, geothermal, and solar thermal power systems (§19, §32, §24). Gas and steam turbines are the prime movers that supply power to the electric generators that feed the grid (§38). Heat extraction reappears in §32 (Geothermal energy) (for ground source cooling systems) and §36 (Systems).

and other departures from the ideal situation, and we look briefly at some alternative refrigeration cycles.

In §13.3, we turn to the use of phase-change cycles in heat engines. Specifically, we describe the Rankine cycle that is used to transform the heat generated by nuclear, fossil fuel, geothermal, and solar thermal sources into electrical power. The fluid at the heart of the Rankine cycle is water/steam, and the workhorse of the cycle is the steam turbine. We walk through the steps in an idealized Rankine cycle in the ST-plane, describe the physical components, and specify the parameters of a Rankine cycle for a practical application – in this case a 500 MW coal-fired steam power plant. We briefly mention ways of improving the efficiency and performance of the Rankine power cycle and describe in §13.4 the use of Rankine cycles based on other working fluids to generate electricity from low-temperature heat sources.

We then turn (§13.5) to the high-temperature *gas Brayton cycle*, which is used both in power plants and in jet engines. A Brayton cycle can be combined with a lower-temperature Rankine cycle to make a *combined cycle gas turbine* (CCGT) power plant. This technology is seeing wide adoption because of its high efficiency and relatively low carbon footprint, and because of the relative abundance of natural gas supplies in certain parts of the world. Finally, we describe *cogeneration*, which is the use of the low-temperature steam exiting a power plant to provide heat for industrial processes or space heating and/or cooling for co-located buildings.

In describing various types of electric power plants and related energy resources it is important to distinguish thermal energy production from electrical energy output, since the Carnot bound on efficiency means that in most cases electrical energy output is far lower than the thermal energy produced. We denote by **MWe/GWe** the electric output of a power plant in MW/GW, where by comparison **MWth/GWth** denotes thermal energy output. Thus, for example, the case studied in §13.3.5 is a 500 *MWe* coal-fired power plant.

The subject matter in this chapter is treated in many standard textbooks on thermodynamics for engineers, including [19] and [56]. A presentation specifically targeted at energy applications can be found in [6]. A straightforward and particularly concise presentation can be found in [65].

13.1 Thermodynamics with Flowing Fluids

In devices that employ phase change to transform thermal energy, the working fluid usually moves from one component of the device to another. In an air conditioner, for example, as the coolant cycles around a closed system, it moves from an *evaporator* inside the space to be cooled to a *compressor*, a *condenser*, and a *throttle*, all of which are in the outdoor, high-temperature environment. Before analyzing such devices it is useful to reiterate the way that we idealize the function of cycle steps and also to adapt the 1st Law appropriately.

13.1.1 Idealizing System Components
Although energy and entropy may be passing into or out of the fluid all along its path through a device, it is customary to model the device as a number of discrete components

where well-defined actions occur. At least in the first approximation, small changes unrelated to the principal function of each component are ignored. For example, it is almost impossible to prevent heat from flowing in or out of the fluid as it moves through a compressor that raises the pressure on a fluid. If the compressor works quickly enough, however, the work done will be much greater than the heat transfer, and the action of the compressor on the fluid can be approximated as an adiabatic process. We made a similar approximation in describing the compression stroke of an internal combustion engine. The new twist here is that the fluid actually moves physically from one section of the device to another, each of which is assigned an ideal function. Separating the actions of mechanical devices into isolated components is similar to isolating the properties – resistance, capacitance, inductance – of an electric circuit into individual components.

The assumption of quasi-equilibrium also requires re-examination. When a fluid in a mixed phase passes through a device component, it is rare for the two phases to actually stay in equilibrium through all steps of a realistic energy conversion cycle. The two phases have different heat conductivity and viscosity, so in two-phase flow in a pipe, for example, the different phases may have different temperatures or may flow at different rates. As in simpler cycles, we avoid dealing with the complexities of non-equilibrium processes by examining the system only at specific locations along a cycle, usually at the inlets and outlets of the primary components of the cycle, where we assume that the fluid is close enough to equilibrium that we can compute thermodynamic state properties for the fluid reasonably accurately.[1]

Although practical design must often go beyond the approximations just described, these approximations are good enough for us to explain the way that modern engines and heat extraction devices work.

13.1.2 Following Flowing Fluid
The working fluid flows in and out of device components during the cycles that we consider in this chapter. Each component may add or remove heat, do work on the fluid, or extract work from the fluid. Therefore each device changes the energy and entropy of the fluid. The pressure that keeps the fluid flowing through the system does work on it as well. In this section we examine the implications of

---

[1] We have already made this type of idealization, e.g. for the combustion process, in our study of internal combustion engines in §11.

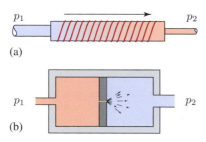

(a)

(b)

Figure 13.1 (a) Fluid flowing through a heating pipe. The temperature of the fluid increases as heat is added. (b) Fluid passing through a throttle. The temperature of the fluid typically changes even though neither heat nor work flows into (or out of) the device.

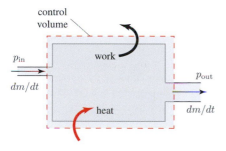

Figure 13.2 A schematic picture of a "black box" device with input and output pipes, which may be of different cross-sectional areas. The dashed line around the device identifies the *control volume* within which we apply conservation of mass and energy.

conservation of energy in this context. Two simple examples to keep in mind are shown in Figure 13.1. The first is a pipe heated by a current-carrying wire that is wound around it. Fluid enters the pipe with density, pressure, and temperature ρ_1, p_1, and T_1, absorbs heat, and leaves with values ρ_2, p_2, and T_2. The second is a **throttle**, a small orifice through which a fluid expands irreversibly from a region of high pressure on the left to low pressure on the right. No external agent does work on the fluid and no heat is added or removed.

In most cases of interest, we can define a **control volume** (Figure 13.2) that contains a given component of a device, and we can balance the energy brought into and taken out of the control volume by the fluid with the heat or work that may be added or removed.

The question we need to answer is: How is energy balanced in a control volume? The simple answer to this question is "enthalpy." More precisely, the *specific enthalpy* (enthalpy per unit mass) of a fluid that comes out of a control volume (h_2) equals the specific enthalpy of the

Energy Conservation in a Control Volume

When a fluid moves through a control volume, the heat q added and the work w done by the fluid change the specific enthalpy of fluid according to

$$h_2 = h_1 + q - w.$$

If the flow rate $\dot{m} = dm/dt$ of the fluid is known, then the rate at which heat is added, $\dot{Q} = dQ/dt = q\dot{m}$, and the power exerted, $P = dW/dt = w\dot{m}$, are related to the enthalpy difference by

$$(\dot{Q} - P) = (h_2 - h_1)\dot{m}.$$

fluid that goes in (h_1) plus the heat per unit mass added to the fluid, $q = dQ/dm$, minus the work per unit mass done by the fluid, $w = dW/dm$, between entering and exiting the device,

$$h_2 = h_1 + q - w. \tag{13.1}$$

The work w is sometimes called **shaft work** to distinguish it from the **flow work** – work done on the fluid by the pressure itself – which is included in the enthalpy balance. If we were discussing an isolated system, conservation of energy would relate the change in *specific internal energy* to the heat and work, $u_2 = u_1 + q - w$. The specific enthalpy, $h = u + p/\rho$, appears for flowing fluids because it takes into account the *flow work* done by the pressure that transports the fluid in and out of the control volume. An explicit derivation of eq. (13.1) appears in Box 13.1.

Here are some elementary, but important, examples that figure in the systems described later in this chapter; the mechanisms that realize these cycle components in real systems are described further in §13.2.4:

Evaporator or condenser: A fluid flows through a long stretch of metal and either absorbs heat from or expels heat to the environment as it changes phase. No work is done on or by external agents, so $w = 0$, but heat is transferred. From the point of view of energy conservation, the change of phase is irrelevant except that it shows up in the enthalpy change. Thus, $h_2 = h_1 + q$, where q is positive/negative if heat is added/removed.

Throttle: A pipe has a restriction in it with an opening that allows the fluid at high pressure to spray into the downstream side of the pipe where the pressure is lower (see Figure 13.1(b)). No work is done, $w = 0$, and if the process occurs quickly, no heat is added, $q = 0$. Thus $h_2 = h_1$, i.e. the process is *isenthalpic*.

Box 13.1   Energy Balance in a Control Volume

Consider a component of a device contained in a control volume, as shown in Figure 13.2, which operates in steady-state, with a continuous flow of fluid through the device. Suppose that in some small time interval a quantity of fluid of mass dm flows into the control volume. Conservation of mass demands that the same amount of mass flows out in the same time interval. The mass that enters has internal energy $dU_1 = u_1 dm$, where u_1 is its specific internal energy upon entering. As the fluid enters the control volume, the pressure p_1 does work $p_1 dV_1 = (p_1/\rho_1)dm$ (using $dm/dV = \rho$), so the *total energy* brought in with the mass dm is $(u_1 + p_1/\rho_1)dm$. Likewise, total energy $(u_2 + p_2/\rho_2)dm$ exits with the fluid that leaves the volume.[a] While the mass dm is in the control volume, heat q is added and work w is done per unit mass. Conservation of energy requires

$$u_2 + p_2/\rho_2 = u_1 + p_1/\rho_1 + q - w.$$

Recognizing that $dH = dU + pdV = (u + p/\rho)dm$ yields the result

$$h_2 = h_1 + q - w.$$

In this derivation we ignored the kinetic and potential energy of the fluid. It is easy to include these effects, leading to the more general result

$$h_2 + gz_2 + \frac{v_2^2}{2} = h_1 + gz_1 + \frac{v_1^2}{2} + q - w,$$

where $z_{1,2}$ and $v_{1,2}$ are the height and speed of the fluid upon entering/leaving the volume. In the devices studied in this chapter the kinetic and potential energy are negligible (Problem 13.7). As discussed in §28 (Wind energy) and §31.1 (Hydropower), these terms are not ignorable in wind and hydropower where the kinetic energy of fluid flow and/or gravitational potential energy are important. For a flowing fluid where no heat is added and no work is done, $q = w = 0$, the equation we have derived here becomes *Bernoulli's equation* (§29.2.2).

---

[a]   Note that if the pressure and density of the incoming and outgoing fluid were the same, the p/ρ terms would cancel.

Pump or compressor: A fluid flows into a component that does (positive) external work $-w$ on it per unit mass, quickly enough so that negligible heat flows; thus $h_2 = h_1 + |w|$.

Turbine: A fluid flows through a component in which it continuously does work w per unit mass. If negligible heat flows, $q = 0$ and $h_2 = h_1 - w$.

13.2 Heat Extraction and the Vapor-compression Cycle

13.2.1 Overview and History

We described the thermodynamics of a generic heat extraction device in §10.6, where we defined appropriate coefficients of performance (CoP) for these systems and pointed out the difference between a refrigerator (or air conditioner) and a heat pump. The only explicit example considered in that discussion was an ideal gas phase Carnot heat extraction device, constructed by running a Carnot engine backwards. Almost all modern heat extraction devices, however, operate on thermodynamic cycles in which a fluid

changes phase. The reasons for this are apparent even in the most primitive cooling devices. After all, the evaporation of sweat – a change of phase – is the fundamental cooling mechanism for the human body. The phenomenon of cooling based on phase change should also be familiar to anyone who has felt the chill of stepping out of a shower or bath into a room with low humidity. This effect was known already in the 1750s to the American statesman, physicist, and inventor Benjamin Franklin. Together with the English chemist John Hadley, Franklin conducted experiments to see how much cooling could be induced by the evaporation of volatile liquids such as alcohol and ether. In the 1820s, Michael Faraday experimented with the cooling potential of the evaporation of ammonia that had been liquefied by compression. The advent of useful heat extraction devices awaited two important developments: first, the incorporation of evaporation into a *cyclic* cooling device, and second, the development of suitable working fluids. Following Faraday's observations, in the 1800s several inventors used a **compressor**, a mechanical device for pressurizing gas, to build cooling loops where the phase-change working fluid could be forced to repeatedly evaporate from a cold region and then condense in a warm region; this was the beginning of vapor-compression cycle cooling technology.

Unfortunately, most of the working fluids available at that time were toxic and/or flammable. These initial working fluids included ammonia, propane, methyl chloride (CH_3Cl), and butane. The modern era of safe and reliable refrigeration and air conditioning began with the development of non-toxic, non-flammable, non-corrosive, stable fluids with a range of boiling points (as functions of pressure) that match the temperature range that applications demand. The first of these fluids was *dichlorodifluoromethane*, CCl_2F_2, developed in the 1920s by American chemist Thomas Midgley, and marketed (along with other similar compounds) under the name *Freon*, or more specifically, *Freon-12* or *R-12*. Freon-12 is a **chlorofluorocarbon**, also known as a **CFC**, i.e. a compound based on light hydrocarbons such as methane, ethane, or propane in which the hydrogen atoms have been replaced with chlorine or fluorine atoms. Freon was followed by a host of other CFCs designed to satisfy the needs (thermodynamic as well as chemical) of air conditioning engineers. Once phase-change cooling technology moved away from noxious working fluids, vapor-compression cooling cycles proliferated in the commercial market.

In the 1970s, chemists F. Sherwood Rowland and Mario Molina discovered that chlorofluorocarbons posed a major threat to the stratospheric *ozone* (see §23, §34) that protects Earth's surface from most of the Sun's ultraviolet radiation. The culprit was the free monatomic Cl-atoms that can be split off from CCl_2F_2 by UV-light, and which catalyze the reactions

$$Cl + O_3 \rightarrow ClO + O_2$$
$$ClO + O_3 \rightarrow Cl + 2O_2 \qquad (13.2)$$

that destroy ozone. International concern over the environmental and human health effects of ozone depletion led to the development of new chlorine-free refrigerants, now mostly fluoro*hydro*carbons (no chlorine) that have negligible impact on atmospheric ozone. Although there are many variants, the most prominent of the modern refrigerants is 1,1,1,2-tetrafluoroethane, known as **R-134a**.[2] This compound, in which four out of six hydrogens in ethane

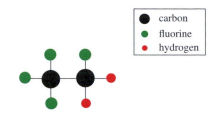

Figure 13.3 A schematic diagram of the molecular structure of the refrigerant 1,1,1,2-tetrafluoroethane, known as R-134a.

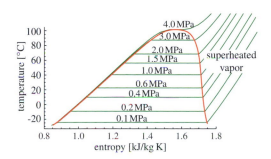

Figure 13.4 The phase diagram of the refrigerant R-134a in the ST-plane. Some isobars are shown in green. Note that the boiling point of R-134a varies over a suitable range for refrigeration or air conditioning ($\approx -25\,°C$ to $+50\,°C$) as the pressure ranges from one to 15 atmospheres (≈ 0.1–1.5 MPa).

have been replaced by fluorine atoms, is illustrated in Figure 13.3. We use R-134a as the working fluid in all of the examples in this section. The phase diagram for R-134a in the ST-plane for the range of temperatures of interest in refrigerators and heat pumps is shown in Figure 13.4.

13.2.2 The Carnot Vapor-compression Cycle

The basic idea behind the **vapor-compression cycle** for heat extraction is extremely simple: a liquid is allowed to vaporize in a cool environment, pulling heat out of the environment. Next, the resulting vapor is compressed, heating it up as its pressure increases. The hot vapor is then allowed to condense at the higher pressure in a warm environment, dumping heat into that environment. Finally, the resulting hot liquid is allowed to expand and cool, whereupon it is put back into the cool environment and the cycle begins once again.

In many ways the VC cycle is for a two-phase system what the Carnot cooling cycle is for a single-phase system. Indeed, the path of the simplest vapor-compression cycle in the ST-plane is identical to that of the Carnot cycle (Figure 13.5(a)). We begin with a look at the highly idealized *Carnot vapor-compression cycle*, and then make a few

---

[2] Although it does not pose a risk to the ozone layer, R-134a is a greenhouse gas that makes a significant contribution to climate change. In October 2016 over 170 countries reached an agreement to phase out the use of R-134a over the next several decades.

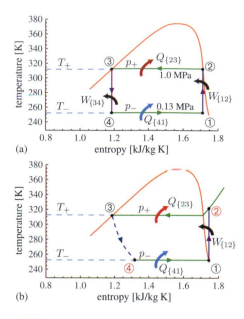

(a)

(b)

Figure 13.5 Vapor-compression cycles in the ST-plane for the coolant R-134a. (a) A Carnot-like VC cycle; the saturation dome is shown in orange and the upper and lower isobars at $p_\pm$ are shown in green. (b) The ideal VC cycle (at the same temperature and pressure set points as (a)) which differs from (a) by replacing the isentropic expansion by throttling and by moving the isentropic compression entirely into the superheated vapor phase. See text for further discussion.

alterations required by the limitations of actual devices. The result is the cycle of Figure 13.5(b), known as the *ideal vapor-compression cycle*. The temperatures chosen in Figure 13.5 are typical for the working fluid in a refrigerator: a low-temperature set point of approximately $-20\,°C$, below the temperature we wish to achieve in the refrigerator interior, and a high-temperature set point of $40\,°C$, high enough to encourage heat transfer out into the surrounding living space. Notice that the choice of temperature set points fixes the pressure set points as well, since liquid and vapor coexist only at one pressure for each choice of temperature.

The **Carnot vapor-compression cycle** shown in Figure 13.5(a) consists of four steps: two isentropic and two isothermal. The steps are assumed to be reversible, making it appear identical to the Carnot cooling cycle of Figure 10.16. The superimposed saturation dome, however, signals an important difference in the behavior of the working fluid around the cycle: the fluid is vaporizing and condensing as the cycle is executed. We trace this vapor-compression cycle, beginning at the low temperature (T_-) and pressure (p_-), high-quality set point ①.

The mechanical devices that implement these steps are described in §13.2.4.

{12} The point ① is in a vapor-rich region of mixed phase (under the saturation dome but close to the saturated vapor side). A compressor isentropically compresses the fluid. The work $W_{\{12\}}$ added to the fluid by the compressor increases the enthalpy and temperature of the fluid as the remaining liquid changes to vapor. This brings us to ② – a hot (T_+), high-pressure (p_+), saturated vapor.

{23} The vapor moves through a **condenser**, where it exhausts heat $Q_{\{23\}}$ into its surroundings as it condenses into liquid. Note that this process takes place at both fixed temperature and fixed pressure. The result, at ③, is a hot, high-pressure, saturated liquid.

{34} The liquid is allowed to expand isentropically, cooling, doing work $W_{\{34\}}$, and partially vaporizing until it reaches ④ – a cold, low-pressure, liquid-rich mixture.

{41} Finally the cold, liquid-rich mixture is circulated into an **evaporator**, where it absorbs heat $Q_{\{41\}}$ from the cold environment, turning to vapor, until the conditions at point ① have been re-established.

Around the cycle, heat $Q_- = Q_{\{41\}}$ has been removed from the low-temperature environment, $Q_+ = Q_{\{23\}}$ has been released into the high-temperature environment, and a net work $\Delta W = W_{\{12\}} - W_{\{34\}}$ is used to make this happen.

Because the path of the cycle in the ST-plane is identical to the Carnot cooling cycle of §10, and every step is in principle reversible (for example, steps {23} and {41} involve heat transfer at a fixed temperature), we conclude that the CoP of this cycle achieves the Carnot limit (10.20) (Problem 13.1). Notice that although the path in the ST-plane is identical to the gas Carnot cooling cycle of §10.6.2, the path in the pV-plane is very different (Problem 13.2). In particular, no work is done on or by the system in the isothermal condensation (②→③) and vaporization (④→①) steps, in contrast to the isothermal compression and expansion of the Carnot cycle.

13.2.3 The Ideal Vapor-compression Cycle

The compression and expansion steps of a mixed phase in the cycle just described present problems for realistic machinery. The liquid and vapor components have quite different mechanical properties, making it difficult to design a mechanical device that can work on both components and keep them in equilibrium throughout. Two straightforward refinements of the VC cycle are designed to avoid having machinery work with a fluid in a mixed phase. The resulting cycle, shown in Figure 13.5(b) and summarized in Table 13.1, is known as the **ideal vapor-compression cycle** – although, as we show below, it cannot reach the Carnot limit on performance.

Compression of Superheated Vapor

In practice, a compressor cannot act efficiently upon a mixed liquid–vapor phase and maintain thermodynamic equilibrium between the phases. Droplets of liquid mixed with vapor also cause excessive wear on the mechanical components of the compressor, reducing its efficiency and lifetime. Consequently the cycle is generally altered so that the fluid is pure vapor at stage ①, and becomes a superheated vapor during the {12} step. Path {23} remains qualitatively unchanged: it takes place at constant pressure, but proceeds in two steps. First the (superheated) vapor cools down to saturation point, and then it condenses fully to a liquid. The cost of this adjustment is that the temperature at ② in Figure 13.5(b) is greater than T_+, indicating that more work was done in the compressor than in the Carnot VC cycle. Note that the asymmetry of the R-134a saturation dome – it is steeper on the right – enhances the quality of the mixture at ① in the Carnot-like cycle, and minimizes the negative effects of this adjustment in the ideal VC cycle (Question 13.3).

Throttling Instead of Isentropic Expansion

Instead of trying to expand a liquid reversibly as it mixes with vapor (the {34} step in Figure 13.5(a)), it is much easier simply to let the liquid expand *irreversibly* through a *throttle*, as shown in Figure 13.1(b). The hot, high-pressure liquid is pushed through a narrow opening into a lower-pressure domain. This step cools the fluid, lowers the pressure, and converts some of the liquid to vapor. Since no heat is added and no external work is done, eq. (13.1) tells us that the enthalpy of the fluid is unchanged in this *throttling process*: throttling is isenthalpic.

The isentropic ③→④ step in Figure 13.5(a) is thus replaced by an isenthalpic ③→④ leg in Figure 13.5(b). As emphasized above, throttling is irreversible. The fluid is not near equilibrium between ③ and ④, and cannot be represented by points along a path in the TS-plane, so we represent the throttling step by a dashed line in Figure 13.5(b). The final point ④ is assumed to be in equilibrium at T_- and p_-; it is a mixture of liquid and vapor with the same enthalpy as the saturated liquid at ③.[3]

---

[3] Although we have not plotted lines of constant enthalpy in the TS-plane, ④ must lie below and to the right of ③ because after throttling it has higher entropy and lower temperature.

The Ideal Vapor-compression Cycle

The *ideal vapor-compression* heat extraction cycle, shown in Figure 13.5(b) and summarized in Table 13.1, begins with adiabatic compression of an initially saturated vapor at temperature T_-, resulting in a superheated vapor at a high temperature T_2. The vapor is condensed at a temperature $T_+ < T_2$, then allowed to expand irreversibly through a throttle. Thermal energy from the colder region is then extracted by allowing the working fluid to vaporize at temperature T_-.

The cost of these practical modifications to the cycle is a reduction in performance. Because throttling is irreversible, it produces an increase in overall entropy and must reduce the coefficient of performance of the ideal VC cycle below the Carnot limit. The change in the compression step, which includes dumping heat at a higher temperature than T_+, also necessarily reduces performance.

Although the expansion through the throttle (step {34}) unavoidably generates some entropy (Problem 13.5), almost all of the heat exchange in the ideal VC cycle occurs at fixed temperature and can be made almost reversible by keeping the temperature difference between the fluid and the thermal reservoirs small. Notice that almost all heat exchanges occur during a change of phase: either vapor condensing to liquid in step {23} or liquid evaporating in step {41}. Therefore all the advantages of phase-change heat transfer described in §12.1 come into play.

13.2.4 Mechanical Components in the Vapor-compression Cycle

In this section we describe the mechanical components of a realistic vapor-compression air conditioner. Figure 13.6 shows the systems representation of the vapor-compression cycle, including the four key components: the compressor, condenser, throttle, and evaporator. All except the compressor are relatively simple devices, which we describe only briefly. We take a slightly more detailed look at the workings of practical compressors because they are more complex machines that play important roles in many everyday devices.

The condenser and evaporator The condenser and evaporator are similar devices designed to change a saturated or superheated vapor to a saturated fluid (condenser) or fluid to saturated vapor (evaporator). These tasks are performed at nearly constant temperature and pressure by rejecting

Table 13.1 Steps in the ideal vapor-compression heat extraction cycle, corresponding to {1234} in Figure 13.5(b). The changes in specific enthalpy are given in terms of the heat transferred and work done per unit mass. The quantities $w_{\{12\}}$, $q_{\{23\}}$, and $q_{\{41\}}$ are all positive.

Step	Initial state	Action	Device	Constant	Enthalpy change
①→②	Saturated vapor	Adiabatic compression	Compressor	Entropy	$h_2 = h_1 + w_{\{12\}}$
②→③	Superheated vapor	Isobaric heat exchange	Condenser	Pressure	$h_3 = h_2 - q_{\{23\}}$
③→④	Saturated liquid	Isenthalpic expansion	Throttle	Enthalpy	$h_4 = h_3$
④→①	Liquid-rich mixture	Isothermal heat exchange	Evaporator	Temperature and pressure	$h_1 = h_4 + q_{\{41\}}$

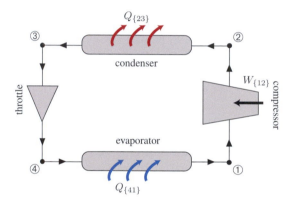

Figure 13.6 The ideal vapor-compression cycle as a system of mechanical components: the *compressor* (stage {12}), *condenser* (stage {23}), *throttle* (stage {34}), and *evaporator* (stage {41}).

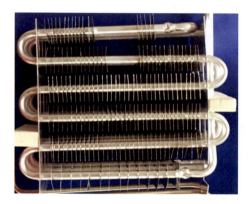

Figure 13.7 An evaporator from an air conditioner. Image: T. Jiangyin, PAWO Electronics, Ltd, China.

heat to or absorbing heat from the environment. A condenser or evaporator is typically a long stretch of metal tubing with smooth interior surfaces and exterior surfaces decorated with metal fins (Figure 13.7). The metal fins on the outside of the condenser or evaporator create additional surface area to increase convective heat transfer. This is necessary in order to keep up with the high heat flux from the two-phase heat transfer in the interior, which operates for condensation in a similar but reversed fashion to the boiling process described in §12.1, with the separation of phases driven by gravity. Often a fan is added to blow air over the heat exchanger and further increase the convective heat transfer rate on the outside. In an AC unit, the condenser is situated in the exterior, warm environment. Sometimes, such as in the interior of a refrigerator, the surface area of the space being cooled is sufficiently large that the walls of that space themselves can be used for the evaporator, eliminating the need for a fan.

The throttle The throttle, or *throttle valve*, is an extremely simple device. It receives a saturated liquid at high temperature and pressure. Denoted in plans as a triangle, the throttle is usually a piece of narrow-diameter tubing or even a valve that is partially opened, forming a restricted channel across which the pressure of the fluid can drop rapidly. The expansion of the hot, high-pressure, saturated liquid across this pressure difference causes some of it to vaporize and the temperature to drop.

The compressor Compressors and pumps both pressurize fluids.[4] Pumps work on liquids, and compressors work on gases. Since gases are much more compressible than liquids, compressors do much more work than pumps (see Problem 5.4), and are sources of much greater inefficiencies in practical machinery. Since a compressor does work on a fluid, it increases the fluid's enthalpy. In stage {12}

---

[4] Pumps are, of course, also used to move liquids up in a gravitational field.

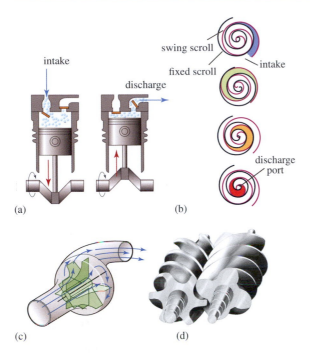

Figure 13.8 Common types of compressors. (a) A reciprocating compressor functions much like a piston in an internal combustion engine. (b) A series of four snapshot diagrams of a scroll compressor. The fluid taken in (top) is compressed as the eccentric scrolls of metal rotate. The stages in fluid compression are denoted in colors ranging from blue to deep red. (c) A centrifugal turbo-compressor uses centrifugal force to compress a gas. (d) In a screw-type compressor the fluid enters axially at the front and is expelled axially at the rear after compression. (Credit: www.grundfos.com)

in automobile engines (§10). Reciprocating compressors are capable of very large compression ratios, but they are noisy and require significant upkeep. **Scroll compressors** are composed of two scrolls of metal that are identically shaped but configured at 180 degrees, with one scroll fixed and the other scroll slightly offset so that it touches the first, with the second scroll orbiting (without rotation) about the central axis of the first scroll so that the point of contact between the two scrolls rotates inwards, driving fluid from the outside of the device towards the center. Since they are easy to manufacture, efficient, relatively quiet, and require little maintenance, scroll compressors are becoming common in air conditioners and refrigerators. **Turbo-machine compressors** include several types of constant throughput devices, which unlike rotating and scroll compressors take in and expel fluid at a constant rate. This class of compressors includes both **centrifugal compressors**, which use rotating blades to accelerate the fluid, producing radial flow and a pressure rise between inlet and outlet, and **axial compressors**, which have both rotating and stationary blades, and use airfoils, with primarily axial flow.

Lastly, the **screw-type compressor** is a hybrid of the turbo-machine and reciprocating machine designs. A screw-type compressor is a positive displacement machine composed of two screws with tightly inter-meshed threads. The axes of rotations of the screws are typically oriented at a slight angle towards each other so that the spacing between the threads decreases as the fluid travels through the threads. Consequently, as the screws turn quickly they drive the fluid through a space that continuously gets smaller, resulting in a higher pressure at the outlet than at the inlet.

of the ideal VC cycle, the compressor takes in fluid at a low pressure and ejects it at a higher pressure and temperature. If the compressor operates quickly enough that there is little heat transfer, then the assumption that the process is adiabatic is a relatively good one.

Compressors are the most complex components of heat extraction devices, and over the years several ingenious designs have been developed. The primary categories are *reciprocating machines*, *turbo-machines*, and *scroll-type compressors* (see Figure 13.8). The reciprocating and scroll-type compressors are the easiest to visualize. They are **positive displacement machines**, meaning that they capture a fixed amount of vapor, compress it mechanically and then discharge it. **Reciprocating compressors** are cylinder-shaped devices containing pistons, with valves that are timed for intake and exhaust of low-pressure and high-pressure fluid, respectively – very similar to those

13.2.5 Example: Specifying the VC Cycle for an Air Conditioner

To get a feel for the scale and performance of a real-world heat extraction system we consider the design of a home air conditioning system that provides 17.5 kW ("5 tons")[5] of cooling power, which would be typical for a $\sim$250 m^2 home in Cambridge, Massachusetts. We are interested in questions such as how much electric power the air conditioner consumes, what its coefficient of performance is, and at what rate coolant must circulate in the device.

First, we choose the refrigerant to be R-134a, for which the phase diagram is shown in Figure 13.4. Next we must

---

[5] A "ton of AC" is an archaic measure used in the US corresponding to the amount of heat required to melt a ton of ice in a day, 1 ton of AC $\cong$ 3.52 kW.

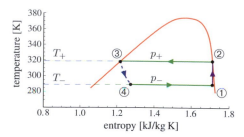

Figure 13.9 The vapor-compression cycle used to model the 5-ton air conditioner in this section. The set points are $p_- = 0.5\,\text{MPa}$, $T_- = 15.7\,°\text{C}$ and $p_+ = 1.2\,\text{MPa}$, $T_+ = 46.3\,°\text{C}$.

Table 13.2 Specific enthalpies at the points ①...④ in the VC cycle of Figure 13.9.

Stage	Specific enthalpy (kJ/kg)	Source (see Example 13.1)
②	422.0	Saturated vapor data
③	266.0	Saturated liquid data
④	266.0	{34} is isenthalpic
①	404.3	{12} is isentropic

$$q_{\{41\}} = h_1 - h_4 = 404.3 - 266.0 = 138.3\,\text{kJ/kg},$$
$$w_{\{12\}} = h_2 - h_1 = 422.0 - 404.3 = 17.7\,\text{kJ/kg}. \quad (13.3)$$

select the high- and low-temperature set points for the refrigerant. In order to obtain rapid heat transfer, the low set point must be colder than the living space it is used to cool. Conversely, the high set point must be hotter than the exterior environment to which the condenser rejects heat. In most buildings we wish to maintain internal room environments at about 20–25 °C, and on a hot day the air outside may be 40 °C or above. It is typically the pressure that can be adjusted externally, so we specify $p_{\pm}$ such that the temperatures suit the conditions. We choose $p_- = 0.5\,\text{MPa}$, corresponding to $T_- = 15.7\,°\text{C}$, and $p_+ = 1.2\,\text{MPa}$, corresponding to $T_+ = 46.3\,°\text{C}$.

Instead of implementing the ideal VC cycle, we simplify our task by choosing a cycle in which expansion is achieved by throttling, but the compression remains in the mixed phase as it was in Figure 13.5(a). The resulting cycle, shown in Figure 13.9, is a good approximation to the ideal VC cycle and slightly easier to analyze since the fluid remains within the saturation dome throughout the cycle. The steps in the cycle differ from the ideal cycle summarized in Table 13.1 in two ways: at ① the fluid is a vapor rich mixture, not a saturated vapor, and at ② the vapor is saturated, not superheated.

First, we determine cooling and work *per unit mass* $q_{\{41\}} \equiv dQ_{\{41\}}/dm$, $w_{\{12\}} \equiv dW_{\{12\}}/dm$, and then choose a *mass flow rate*, $\dot{m} \equiv dm/dt$, in order to produce a desired cooling power $dQ/dt = q_{\{41\}} \times \dot{m}$. The quantities $q_{\{41\}}$ and $w_{\{12\}}$ are expressed in terms of the specific enthalpies at the points ①–④ by the relations in the last column of Table 13.1: $q_{\{41\}} = h_1 - h_4$ and $w_{\{12\}} = h_2 - h_1$. We know the temperature, pressure, and quality at ②, so h_2 can be found in tables of data on R-134a [23]. Example 13.1 explains how h_1 and h_4 are obtained using the methods developed in the previous chapter. The results are summarized in Table 13.2.

With this information we can compute the cooling and work per unit mass,

The coefficient of performance of our air conditioner is $\text{CoP} = q_{\{41\}}/w_{\{12\}} = 138.3/17.7 = 7.8$. Losses and irreversibilities (discussed below) lower this number significantly. What should this CoP be compared to? The appropriate Carnot limit for comparison is not the one computed from the high- and low-temperature set points, $T_+ = 46.3\,°\text{C}$ and $T_- = 15.7\,°\text{C}$, because these temperatures were chosen above and below the ambient temperatures in order to enhance heat transfer. Instead we should use the actual inside and outside temperatures. These might, for example, be $T_{\text{in}} \cong 25\,°\text{C}$ and $T_{\text{out}} \cong 38\,°\text{C}$, in which case the Carnot CoP is $\text{CoP}|_{\text{Carnot}} \cong T_{\text{in}}/(T_{\text{out}} - T_{\text{in}}) = (273 + 25)/(38 - 25) = 298/13 \cong 23$. So this device reaches just over 33% of the thermodynamic limit (Problem 13.4).

Next, we look at the size and power requirements of a practical device based on this system. Since $q_{\{41\}}$ gives the cooling per unit mass, the required cooling power fixes the mass flow rate at $\dot{m} = (17.5\,\text{kW}) \div (138.3\,\text{kJ/kg}) = 0.127\,\text{kg/s}$. Given the density ($\rho \cong 1.21\,\text{kg/L}$ for liquid at 25 °C) of R-134a, this corresponds to about 0.10 L/s, which seems like a reasonable scale for a home-sized device. Finally, assuming that the efficiency of the electrical-to-mechanical conversion in the air conditioner is close to 100% (there is no thermodynamic limit on this process), we find that the power consumption of the 5-ton air conditioner is 17.5/7.8= 2.2 kW – a significant amount of power, but far less than the amount of heat removed from the interior living space – 17.5 kW.

The process we have described is only a cartoon of a realistic design process. A designer might start with an idealized model like this, but would then use the calculated mass flow rate to characterize the compressor and heat exchangers, and make appropriate modifications based on a host of other considerations such as cost, size, and ease of manufacturing. Specific environmental conditions must

---

Example 13.1   A Vapor-compression Cycle for a 5-ton Air Conditioner

To find the heat absorbed and work done in the VC cycle of Figure 13.9, we need to know the specific enthalpy of the refrigerant at points ①, ②, and ④, all of which are either on or within the saturation dome. Typically, tables or software provide the specific enthalpy and specific entropy for both saturated liquid ($\chi = 0$) and saturated vapor ($\chi = 1$) as a function of temperature or pressure. The relevant data for R-134a at $p_- = 0.5$ MPa and $p_+ = 1.2$ MPa are summarized in the table at the right [23].

The specific enthalpy at ② can be read directly from the table because the fluid is a saturated vapor at p_+ at ②, $h_2 = 422.0$ kJ/kg. Throttling is modeled as *isenthalpic*, so the specific enthalpy at ④ is the same as at ③, which can also be read directly from the table, $h_4 = h_3 = 266.0$ kJ/kg.

p [MPa]	T [°C]	χ	Stage	h [kJ/kg]	s [kJ/kg K]
0.5	15.7	0	–	221.5	1.076
0.5	15.7	1	–	407.5	1.720
1.2	46.3	0	③	266.0	–
1.2	46.3	1	②	422.0	1.709

The specific enthalpy at ① is the quality-weighted sum of the specific enthalpies of the saturated liquid and vapor at p_-. To obtain the quality of the fluid at ①, we use the fact that compression is modeled as *isentropic*, thus $s_1 = s_2$. Expressing s_1 in terms of the specific entropies of the saturated liquid and vapor given in the first and second rows of the table,

$$s_1 = \chi_1 \times 1.720 + (1 - \chi_1) \times 1.076 = s_2 = 1.709 \,.$$

This determines the quality at ① to be $\chi_1 = (1.709 - 1.076)/(1.720 - 1.076) = 0.983$. So the mixture at ① is nearly all vapor. h_1 is then the quality-weighted combination of the specific enthalpies of saturated liquid and vapor at p_-,

$$h_1 = 0.983 \times 407.5 + (1 - 0.983) \times 221.5 = 404.3 \,\text{kJ/kg} \,.$$

---

also be considered. For example, if the inside air is humid, then water will condense (and even, perhaps, freeze), on the surface of the inside heat exchanger.

Some final comments on this example. A real air conditioner would have losses and irreversibilities that we have ignored. Lack of reversibility, due for example to local flow patterns in the fluid induced by motion of the compressor, adds entropy to the fluid, while lack of isolation allows the fluid to lose some heat (and therefore entropy) to the environment. In practice, the first effect dominates and the entropy of the fluid increases as it is compressed. Thus, the line from ① to ② slopes to the right as it goes upward in ST-space. Also, friction due to viscous flow in the pipes and heat exchange with the environment causes the pressure to drop during the heat exchange steps ({23} and {41}). The effects of these losses on the cycle of Figure 13.5 are shown schematically in Figure 13.10.

13.2.6 Alternative Refrigeration Cycles

We have explored the vapor-compression cycle in some detail because it is the most widely used refrigeration cycle. There are, however, several other methods of extracting heat from a low-temperature reservoir and expelling it at a higher temperature. Two fairly common alternatives are the *absorption refrigeration cycle* and

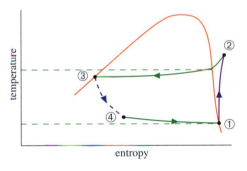

Figure 13.10 A sketch showing some of the most significant losses in a realistic VC cycle. These include irreversibility in compression (raising the entropy of the fluid) and pressure drops during evaporation and condensation.

the *gas refrigeration cycle*. The former uses a liquid to absorb the vapor after evaporation in the cold environment. This liquid is then pumped to higher pressure and heated to release the vapor, which is then condensed and returned to the throttle and evaporator. The absorber is then recycled back to absorb more vapor and so forth. The advantage of the absorption refrigeration cycle is that it eliminates the compressor (the major working component in the vapor-compression cycle). It replaces this component with a relatively low-temperature heat source that evaporates the working fluid out of the absorber and with a pump that expends much less work to pressurize the fluid than the compressor required to pressurize the vapor. This makes devices based on absorption refrigeration generally less expensive and more robust, though generally less powerful, than refrigeration units using a compressor. The absorption refrigeration cycle can be particularly advantageous when a cheap, low-temperature heat source is available, as for example in a steam power plant (as described in the following section), where low-pressure, low-temperature steam is often available in great quantities. In this way the exhaust steam from a power plant can be used to provide refrigeration. This is one aspect of *cogeneration*, which is discussed in more detail in §13.5.3. For gas refrigeration cycles, the simplest, though impractical, example would be a Carnot engine run backwards, as described in §10. Practical gas refrigerators make use of the Brayton power cycle (run in reverse) discussed in §13.5.1. For further description of these and other refrigeration cycles, see [19].

In the next section we reverse the order of operations in the VC cycle and describe heat engines that employ phase change to increase their effectiveness.

13.3 The Rankine Steam Cycle

The same advantages that make phase-change cycles important in heat extraction applications also make them the workhorses for large-scale, stationary thermal power conversion. In this section we introduce the *Rankine steam cycle*, taking a parallel approach to that used for the vapor-compression cycle in the previous section. First we briefly summarize the history of steam energy conversion. Next we describe a highly idealized Carnot-like Rankine cycle in the ST-plane. Then we make some alterations in this cycle required by the limitations of realistic materials, leading to what is known as the *ideal Rankine cycle*. We then give an overview of the mechanical components in a Rankine cycle. We work through the specification of an ideal Rankine power cycle in detail before finally turning to a short discussion of more elaborate and more practical Rankine cycles.

13.3.1 Development of Phase-change Power Generation
The first recorded use of a material undergoing change of phase to convert thermal energy into mechanical energy was the *aeolipile* invented in the first century AD by the Greek engineer Heron of Alexandria (Figure 13.11). In

Figure 13.11 Heron of Alexandria's *aeolipile*, the first device for converting steam to mechanical energy and the first steam turbine. (Credit: Knight's American Mechanical Dictionary, 1876).

Heron's aeolipile, steam created in a *boiler* was channeled into two opposite jets offset from a central axis to produce a torque, causing the device to rotate. Heron's device was also the first *steam turbine*, converting heat energy directly into rotational motion.

The conversion of heat to mechanical energy began in earnest in the earliest days of the industrial revolution, with the steam engines invented by Watt and Newcomen (§10). Although these engines revolutionized industry, they were based on a reciprocating piston paradigm that was quite inefficient and difficult to maintain. Modern steam turbomachinery and power generation technology were born in the 1880s with the development of the first modern *steam turbine* by the British engineer Charles Parsons. This engine was based on a thermodynamic cycle first proposed by the English physicist and engineer William Rankine. Advancement in design of the steam turbine by American engineer and entrepreneur George Westinghouse and by Parsons soon led to a rapid growth in scale to multi-megawatt Rankine power plants.

The *Rankine steam cycle* forms the basis for most of our global power generation infrastructure. Over the past two centuries, the Rankine cycle has been used with a variety of fuel sources, from wood, coal, and sunlight to natural gas, biomass, and nuclear fission.

13.3.2 A Carnot-like Rankine Cycle
Figure 13.12(a) shows a highly idealized steam **Rankine cycle** in the ST-plane. Since the path is followed

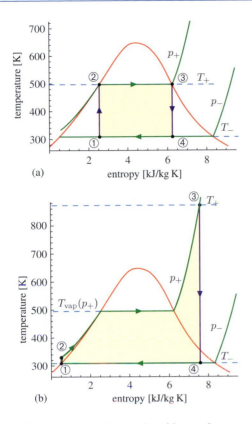

Figure 13.12 Steam Rankine cycles with set points $p_- = 5\,\text{kPa}$ and $p_+ = 2.5\,\text{MPa}$. (a) The simplest Rankine cycle, basically a Carnot cycle for a liquid–vapor system. (b) The *ideal* Rankine cycle, in which the pump {12} works on the liquid phase and the high temperature is raised to 875 K so that the turbine {34} works almost entirely on superheated vapor. Note that the {41} and {23} steps remain isobaric (constant pressure). The shaded area measures the work done in the cycle (see §10.1.2).

isentropic compressions, however, are *totally different*. In this case since point ① is on the left-hand side of the saturation dome, isentropic compression drives it toward quality zero – a pure liquid. In the vapor-compression cycle, where point ① was on the right of the saturation dome, isentropic compression drove it toward quality one – a pure vapor. Physically, only a small amount of enthalpy is added to the fluid in step {12} of the Carnot-like Rankine cycle. The added enthalpy, combined with the latent heat from condensation, produces a saturated liquid at state ② at high temperature and pressure.

Just as in the vapor-compression cycle, process {23} in Figure 13.12(a) is an isothermal heat exchange. Here the heat is transferred *from a high-temperature environment into* the fluid. With the addition of heat, the fluid transitions from a saturated liquid (quality zero) at state ② to a saturated vapor (quality one) at state ③.

In step {34} the high-pressure and high-temperature vapor is allowed to expand against the mechanical components of a turbine, exerting a torque and causing it to rotate. The heat transfer during this phase is small and is neglected in this idealization. Because the pressure of the expanding vapor is working against a force, the expansion can be idealized as reversible, and the entropy can be taken to be constant for an ideal turbine. Thus step {34} is taken to be isentropic. Considerable work is done by the expanding vapor, so the enthalpy of the fluid is considerably reduced. The only way that the enthalpy of a saturated vapor can drop without a change in entropy is if some of the vapor is converted back to fluid. Hence the fluid at ④ is a liquid–vapor mixture with quality less than one.

Finally, in step {41} the high-quality vapor from point ④ undergoes an isothermal condensation, rejecting heat to the (cooler) external environment and returning back to point ①, again as a liquid-rich mixture at the low-temperature operating point of the cycle.

13.3.3 The Ideal Rankine Cycle

The *ST*-diagram for the so-called *ideal Rankine cycle*, shown in Figure 13.12(b), incorporates two essential design changes that respect the limitations of real mechanical systems while improving the overall efficiency of the engine by incorporating some heat transfer at higher temperatures. First, step {34} is shifted to take place almost completely in the superheated vapor phase, while remaining approximately isentropic. Real steam turbines are easily damaged by liquid water; over time even microscopic water droplets cause pitting and erosion of the turbine blades. In applications where it is necessary to operate turbines with lower-quality steam, such as nuclear and geothermal power plants, advanced materials must be

clockwise, it is an engine. Since it is a rectangle in the *ST*-plane, the idealized cycle can in principle attain Carnot efficiency (Problem 13.1). In fact, the cycle looks like the Carnot vapor-compression cycle of Figure 13.5(a) run backwards in both the *ST*- and *pV*-planes. Although this is true at the diagrammatic level, some of the steps are quite different in practice.

It is natural to begin the analysis of the Rankine cycle in Figure 13.12(a) at point ①, with a low-temperature, mixed-phase fluid that is mostly in the liquid phase (low quality). In step {12} the liquid–vapor mixture undergoes an isentropic compression to a saturated liquid (zero quality) at higher temperature and pressure at state ②. The Carnot-like vapor-compression cycle also began with the isentropic compression of a mixed-phase fluid. The results of the two used to strengthen turbine blades. If the heat transfer phase {23} is kept at constant pressure, which is convenient, then the temperature at ③ gets quite high. In the cycle of Figure 13.12(b), for example, the high temperature shifts from 500 K to 875 K to incorporate this modification. Adding heat to an engine at a higher temperature adds less entropy, and therefore tends to increase the efficiency, so a high temperature at ③ is actually desirable. In fact the efficiency of the engine is limited by how high a temperature real machinery can tolerate, typically $T_+ \approx 600\,°C$ in conventional steam turbines.[6] More advanced systems, such as the *supercritical Rankine cycle* (§13.3.6), can improve efficiency beyond that of the ideal Rankine cycle by moving the heat transfer curve upward in the ST-plane and out of the saturation dome. Since, however, it is not possible to raise the temperature of a fluid substantially at realistic pressures without increasing the entropy (see Figure 12.9), much of the heat transfer in any steam cycle engine must occur at temperatures substantially below the maximum temperature T_+ reached in the cycle. Rankine cycles operating at high temperatures (above 500–600 K) thus cannot achieve anything like Carnot efficiency, even though their absolute efficiency can exceed that of a Carnot-like cycle operating within the saturation dome at $T_+ = T_{\mathrm{vap}}(p_+)$.

For turbines that do not work with a mixed phase, the steam at point ④ still contains almost all the enthalpy of condensation, which remains unexploited. If nothing else is done to reclaim this energy, it is all lost to the low-temperature heat reservoir in the condenser. The idea of using the energy contained in the low-pressure, nearly saturated steam at point ④ for further energy generation is the basis for multistage Rankine cycles, which are described below, and for *cogeneration* (discussed in §13.5.3).

The second significant difference between the ideal Rankine cycle of Figure 13.12(b) and the cycle of Figure 13.12(a) occurs in the compression step {12}. As in the vapor-compression cycle, compressors do not work well on a liquid-vapor mixture. The solution in this case is to move ① all the way to a saturated liquid. Because water is

---

[6] For the ideal Rankine cycle, the temperature $T_+ = T_③$ denotes the high temperature reached at point ③; this is above the temperature $T_{\mathrm{vap}}(p_+)$ at which the working fluid changes phase, unlike for the ideal VC cycle, where T_+ denoted the temperature of phase change. This notation matches the physical elements of the system; T_+ for the ideal VC cycle matches the operating temperature of the condenser, while T_+ corresponds to the nominal temperature of the boiler in the Rankine cycle. Note that only a fraction of the heat is added while the fluid is above the vaporization temperature $T_{\mathrm{vap}}(p_+)$.

The Ideal Rankine Power Cycle

The *ideal Rankine power cycle*, shown in Figure 13.12(b) and summarized in Table 13.3, begins with adiabatic pressurization of a sub-cooled liquid by a pump. Fuel burned in a boiler adds heat at constant pressure, vaporizing and superheating the fluid. The fluid then expands adiabatically through a turbine, where it does useful work. Finally the spent, saturated fluid, which is mostly vapor at low pressure and temperature, is condensed back to the liquid state.

nearly incompressible, it requires little work for a pump to compress it to high pressure. Ideally, a pump adds no heat and little work, so it raises the temperature of the water very little. Indeed, the point ② in Figure 13.12(b) has been displaced artificially above ① to make it visible.

A couple of other modifications follow from the first two: First, the condenser must remove almost the entire enthalpy of vaporization, taking the almost-saturated steam at ④ all the way to saturated liquid at ①. Second, in the realistic Rankine cycle the feed to the boiler at ② is a sub-cooled liquid, which must first be heated to boiling, then vaporized, and then superheated to reach point ③, where it is ready for the turbine. The steps in an ideal Rankine cycle are summarized in Table 13.3.

13.3.4 Mechanical Components in the Rankine Cycle

Moving beyond thermodynamic abstraction, Figure 13.13 represents the Rankine cycle in engineering systems form, showing the four key components of the cycle and the four cycle stages that correspond to points ①–④ on the ST-diagram.

The pump While compressors do work to pressurize gases, pumps are designed to pressurize liquids. Just as with compressors, there are several flavors of pump designs, including *centrifugal*, *reciprocating*, and *in-line*. The centrifugal and in-line pumps both use rotating *impellers* to draw and expel liquid. (An impeller is a rotor in an enclosed conduit that acts on a fluid to produce motion or increase/decrease pressure.) In the centrifugal design, liquid is drawn into the pump housing at the center of the impeller and then expelled radially by the moving impeller vanes. Schematically, the centrifugal pump resembles the centrifugal compressor shown in Figure 13.8. An in-line pump is more akin to a paddle water

Table 13.3 Steps in an ideal Rankine cycle, as shown in Figure 13.12(b). The changes in specific enthalpy are given in terms of the heat transferred and work done per unit mass. The quantities $w_{\{12\}}$, $w_{\{34\}}$, $q_{\{23\}}$, and $q_{\{41\}}$ are all positive.

Step	Initial state	Action	Device	Constant	Enthalpy change
①→②	Saturated liquid	Adiabatic compression	Pump	Entropy	$h_2 = h_1 + w_{\{12\}}$
②→③	Sub-cooled liquid	Isobaric heat exchange	Boiler	Pressure	$h_3 = h_2 + q_{\{23\}}$
③→④	Superheated vapor	Adiabatic expansion	Turbine	Entropy	$h_4 = h_3 - w_{\{34\}}$
④→①	Vapor-rich mixture	Isothermal heat exchange	Condenser	Temperature and pressure	$h_1 = h_4 - q_{\{41\}}$

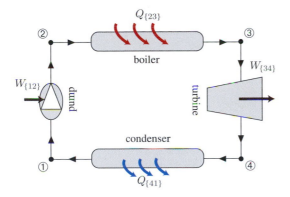

Figure 13.13 The Rankine cycle as a system of mechanical components. The four key components in the cycle are the *pump* (stage {12}), *boiler* (stage {23}), *expander* or *turbine* (stage {34}), and *condenser* (stage {41}).

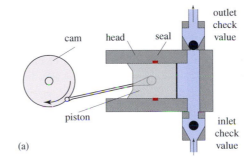

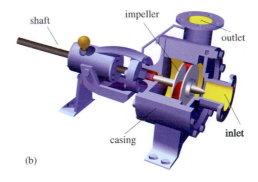

Figure 13.14 (a) A reciprocating pump; (b) a centrifugal pump (Credit: (a) LC Resources, Inc. www.icresources.com (b) Mokery J. Vectorized by Magasjukur2)

wheel: liquid is drawn in from one radial direction and swept through the pump housing by the impeller vanes towards an exhaust port that is 180° from the inlet port. Pressure and flow rate go together in centrifugal and in-line pumps; they cannot sustain high pressure at low flow rates. On the other hand, they are quite efficient at handling large volumes of liquid, so most large pumps are of that design type. Reciprocating pumps, based on the same principle as reciprocating compressors, are not as common as the rotating-type pumps, but are capable of very high pressure ratios at low flow rate. Consequently, reciprocating pumps find much greater use in hydraulic machines where fluids are used to transmit force over long distances; the flow rates can be almost negligibly small but the pressure ratios are enormous. In general, pumps are capable of producing great increases in pressure with relatively little work. In the ideal Rankine cycle shown in Figure 13.12(b), a relatively conservative example, the pump increases the pressure on the liquid from 5 kPa to 2.5 MPa, a factor of 500.

The boiler At point ② the pressurized liquid exits the pump and enters the boiler. In the boiler a heat source is harnessed first to vaporize the fluid and then to superheat the resulting vapor. In a coal-fired steam power plant, for example, the boiler is composed of hundreds of vertical pipes, typically 5 to 8 cm in diameter and 7 to 10 meters

Figure 13.15 A modern steam turbine rotor. (Credit: Siemens Pressebild)

long. The pipes are arranged side by side around the circumference of a cylindrical combustion chamber. Coal in the center of the chamber is burned at a high temperature, generating combustion gases. Convective and radiative heat transfer to the pipe surface boils the water in the pipes. The water enters the bottom of the pipe array as a relatively cool sub-cooled liquid (point ②). Vapor is generated along the length of the pipes and, typically, a liquid–vapor mixture collects in a *steam drum* at the top of the boiler. The saturated vapor is taken off the top of the drum and recirculated through the combustion chamber, where it is superheated and fed to the turbine (point ③). The saturated liquid from the mixture drains down from the steam drum and is used to preheat the water entering the boiler. The back side of the tube arrangement is insulated with materials to form a heat shield that keeps the energy of combustion within the chamber. Boiler designs vary widely depending on the nature of the heat source; nuclear and geothermal boilers, for instance, do not have a combustion chamber.

The turbine The fluid exits the boiler at point ③, as a superheated vapor at high temperature and pressure. In large-scale thermal power conversion systems, a **steam turbine** converts the thermal energy of superheated steam into mechanical energy of rotation. Turbines are complex arrays of blades, some fixed on stationary *stators*, others attached to *rotors*, which rotate about a drive shaft. The steam entering the turbine delivers an impulse to the rotor blades, which, in turn, exert torque on the rotor. A turbine may consist of several different sections, each optimized for the pressure and temperature of the steam that passes through it. The output shaft of the turbine is generally connected directly to the armature of an electric generator.

The pressure at the outlet port of a turbine is an important parameter in system design. Because the steam at point ④ is in phase equilibrium with liquid water, its temperature and pressure are related by $T = T_{vap}(p_-)$. Thus, for example, if the turbine exit is at atmospheric pressure, the steam temperature is $100\,°C$. In electric power applications, where the objective is to obtain maximum efficiency, the turbine exit pressure is maintained below atmospheric pressure, allowing the exit temperature to be lower as well. In the case of Figure 13.12, for example, the exit temperature is $33\,°C$. Looking back to Figure 12.5(a), we see that this requires the pressure to be $5\,kPa$, about 5% of atmospheric pressure. Since the Rankine cycle is closed, keeping the condenser temperature at $33\,°C$ creates a relative vacuum (compared to atmospheric pressure) that maintains low pressure at the turbine exit. In other applications, where, for example, steam is required for industrial processes, the turbine exit pressure might be kept at or above atmospheric pressure, reducing the turbine efficiency but providing steam for other purposes.

The condenser Finally, exiting the turbine at point ④, the fluid enters the condenser. The condenser operates at the low-pressure set point, generally well below atmospheric pressure (as just explained). In a grid-scale electric power plant, the cooling needs of the condenser generally drive the placement of power plants near large bodies of water. In some cases, however, air cooling is used. The condenser is typically a large shell-and-tube heat exchanger, within which the working fluid courses through hundreds of tubes encased in a shell that is flooded with coolant. The coolant is constantly circulated and replenished. The output from the turbine may be condensed in several stages before returning to state ① and the inlet to the pressurization pumps.

13.3.5 Specifying a Rankine Cycle

To get a feeling for the parameters and efficiency of a realistic-scale Rankine power system, we outline the specification of an ideal Rankine cycle for a 500 MWe coal-fired electric power plant. We seek to determine the theoretical efficiency, the rate of fuel consumption, and the amount of water that must circulate through the components of this power plant. The limits on pressures and temperatures are set by the mechanical components of the system. We choose the following set points, relating to the cycle of Figure 13.12(b):

Temperature: System components are typically limited to temperatures below $600\,°C$, so we take $T_+ = T_3 = 600\,°C$.

Table 13.4 Specific enthalpies at the points ①…④ in the ideal Rankine cycle of Figure 13.12.

Stage	h_i (kJ/kg)	Source (see Example 13.2)
①	138	Saturated liquid at T_-
②	140	{12} is isentropic and $p_2 = 2.5\,\text{MPa}$
③	3690	Set point: $T_+ = 600\,°\text{C}$, $p_+ = 2.5\,\text{MPa}$
④	2320	{34} is isentropic and $\chi_4 = 0.90$

Pressure: Although high-performance machinery can tolerate considerably higher pressures, we choose a conservative maximum pressure of ~25 atm, $p_2 = p_3 = 2.5\,\text{MPa}$.

Quality: To protect the turbine from excessive wear, we constrain the quality of the mixture exiting the turbine to be at least $\chi = 0.90$.

The choices of $T_+ = 600\,°\text{C}$ and $p_3 = 2.5\,\text{MPa}$, together with the condition $\chi_4 = 0.90$ determine the required pressure and temperature at the turbine outlet.

As in the case of the vapor-compression cooling cycle, we can compute everything we need if we determine the specific enthalpy at the points ①…④. The relations in the last column of Table 13.3 determine the heat transferred and work performed per unit mass in terms of these specific enthalpies. Then the mass flow rate $\dot{m}$ is chosen to meet the specified power output of the plant. The specific enthalpy at ③ can be looked up in tables because the pressure and temperature have been fixed there. Example 13.2 explains how the specific enthalpies at ①, ②, and ④ are obtained, using similar methods to those in Example 13.1. The resulting specific enthalpies are tabulated in Table 13.4.

With the information in Table 13.4 and the last column of Table 13.3, we can determine the performance of this Rankine cycle. The heat added by the boiler and the work done by the turbine are

$$q_{\{23\}} = h_3 - h_2 \cong 3690 - 140 \cong 3550 \text{ kJ/kg,}$$

$$w_{\{34\}} = h_3 - h_4 \cong 3690 - 2320 \cong 1370 \text{ kJ/kg.}$$

$$\text{(13.4)}$$

Note that in comparison to the work (per unit mass) done by the turbine, the work done by the pump, $w_{\{12\}} = h_2 - h_1 \cong 2.5\,\text{kJ/kg}$, is negligible (as expected). The theoretical efficiency of this system is therefore $\eta = (w_{\{34\}} - w_{\{12\}})/q_{\{23\}} \cong 39\%$. The actual efficiency of a real system would be somewhat lower due to heat losses in the turbine and boiler, resistance in pipes, and

so forth. The Carnot efficiency limit for a thermodynamic system working between 600 °C and 33 °C is $\eta_{\text{Carnot}} = (600 - 33)/(600 + 273) \cong 65\%$. So the ideal Rankine system we have designed reaches roughly $39/65 \cong 60\%$ of its thermodynamic limit. As mentioned earlier, achieving a substantially larger fraction of Carnot efficiency for a phase-change power system is difficult due to the roughly diagonal form of isobars outside the saturation dome in the ST-plane.

To produce 500 MW of power at 39% efficiency the plant must generate $500/0.39 \cong 1280$ MW of thermal energy. Combustion of a ton of coal generates about 25 GJ, so the plant requires about 0.05 tons of coal per second or about 1.6 MT of coal per year.[7] To get a feeling for the size of the components we can estimate the mass of water that must flow around the cycle. To obtain 500 MW when the (specific) work done on the turbine is 1370 kJ/kg requires about 370 kg/s of water flowing through the system. This is quite a large system. Individual turbines for coal-fired power plants range from tens to hundreds of MW, so a 500 MW plant would likely contain several separate turbines each handling a fraction of the total mass flow.

13.3.6 Improvements on the Basic Rankine Cycle

In order to improve the efficiency of steam turbine generators, and to prevent mechanical problems such as condensation of steam in the turbine, several modifications of the Rankine steam cycle have been developed. Real-world implementations usually employ one or more of these modifications. Mechanical engineers have been optimizing steam power cycles for over a hundred years in a variety of ingenious ways; we give here only brief descriptions of some relatively standard variations on the basic Rankine cycle. For further reading see [6, 19, 56, 65].

A common modification of the Rankine cycle that addresses the problem of condensation in the turbine and provides a modest improvement in efficiency by adding additional heat at a relatively high temperature is the **reheat cycle**, shown in Figure 13.16(a). Reheat cycles usually employ two turbines for power extraction. The first turbine, the *high-pressure turbine*, extracts most of the power from the fluid, but leaves it as a saturated vapor. The fluid is then heated back to a high temperature, but at lower

---

[7] Here, and throughout this chapter, we take the energy content of fuels to be the *higher heating value* (HHV) as defined in §9.4.3 and as is customary in the US. Values for the HHV for coal range from 15 GJ/t to 35 GJ/t (§33) depending on the type of coal. For definiteness, we use here a middle value of 25 GJ/t.

Example 13.2   An Ideal Rankine Cycle for a Coal Power Plant

To determine the work done and heat absorbed in an ideal Rankine cycle (Figure 13.12(b)) we need the specific enthalpies of water $h_1 \ldots h_4$. Our cycle is specified by the following information: $T_+ = T_3 = 600\,°C$, $p_+ = p_3 = 2.5\,MPa$, $\chi_1 = 0$, and $\chi_4 = 0.90$.

We use thermodynamic data for water obtained from NIST [23], and recorded in the table at right. First, since we know the temperature and pressure at ③ in the superheated zone, we simply look up the specific enthalpy at p_+ and T_+, which is $h_3 \cong 3690\,kJ/kg$ (keeping three-place accuracy).

To proceed, we next find the temperature T_- and pressure p_- at ④ and ①. Since the turbine {34} is

p (MPa)	T (°C)	χ	Stage	h (kJ/kg)	s (kJ/kg K)
2.5	600	–	③	3687	7.60
0.005	32.87	0.9	④	2318	7.60
0.005	32.87	0	①	137.8	0.476
0.005	32.87	1	–	2561	8.39
2.5	32.93	–	②	140.3	0.476

modeled as isentropic, the specific entropy at ④ must be the same at ③, $s_4 = s_3 = 7.60\,kJ/kg\,K$. Online applications (or tables) allow one to search for p and $T_{vap}(p)$ for given values of s and χ. We find that a mixture with quality 0.90 has specific entropy $7.60\,kJ/kg\,K$ at a pressure of $p_- = 5\,kPa$. At this pressure water boils at $T_- = 32.87\,°C$. The saturation data for p_- and T_- are given in the table, and enable us to compute the specific enthalpy at ④,

$$h_4 = 0.90 \times 2560 + 0.10 \times 138 \cong 2320\,kJ/kg,$$

and at ① where the quality is zero (saturated liquid), $h_1 \cong 138\,kJ/kg$. The next step is to find the fluid properties at ②. In fact, the enthalpy added by the pump is very small, so for a first estimate we can ignore it and set $h_2 = h_1$. To do better, we must keep more than three digits of accuracy (we keep a fourth digit in the table for h and T). Using data from NIST we find that isentropic compression of saturated water from $5\,kPa$ to $2.5\,MPa$ raises its temperature only by $0.06\,°C$ and its specific enthalpy by $2.5\,kJ/kg$, giving $h_2 \cong 140.3\,kJ/kg$.

---

pressure (and therefore higher entropy), and then is circulated through the second, *low-pressure turbine*. In addition to the increase in cycle efficiency and the avoidance of condensation in the turbine, the reheat step increases the power output per cycle substantially.

A second common addition to the basic cycle is **regeneration**, shown in the ST-plane in Figure 13.16(b). In this modification, a fraction f of the steam is intercepted as it expands through the turbine (at point ⑤) and returned to mix with the input water as it exits the first of two pumps at point ②. The enthalpy of condensation of the recycled steam heats the input water significantly, to ⑥, where it enters a second pump that pressurizes the fluid further before passing it to the boiler at ⑦. Regeneration has several advantages: first, the latent heat of condensation of some of the steam is put to use preheating the input water. Second, the fraction of the steam, $1 - f$, that goes through the condenser is reduced, allowing the scale of the condenser to be reduced. Usually these advantages outweigh the costs of extra piping and the second pump. For example, in an ideal Rankine cycle operating between pressures of $p_+ = 2\,MPa$ and $p_- = 10\,kPa$, and with $T_+ = 600\,°C$, if regeneration steam is extracted at $p_5 = 200\,kPa$ and cycled back to ②, the theoretical efficiency increases by about 4.6% (from 35.7% to 37.3%) [65]. There are several variants on this theme, including ones that use only a single pump.

In the real world, several stages of reheating and regeneration may be used in a Rankine steam turbine in order to improve overall efficiency. A combined reheat plus regeneration Rankine cycle is shown in Figure 13.17(a).

The quest for higher-efficiency coal-fired steam generators has led to the development of **supercritical Rankine cycles**. As the name suggests, these cycles work at such high pressure that the heat transfer phase avoids the saturation dome entirely. Since the critical point of water is $T_c = 374\,°C$, $p_c = 22.1\,MPa$, the mechanical components of a supercritical Rankine power plant must be able to withstand fairly extreme conditions. The first supercritical power plants were built in the 1950s, and with increasing concern about efficiency and carbon emissions, supercritical plants are becoming the design of choice for modern large-scale coal-fired power plants. A supercritical Rankine cycle is shown in Figure 13.17(b). In order not to exceed a maximum operating temperature of about $600\,°C$, and avoid condensation in the turbine, reheating is usually necessary in the supercritical cycle. Various stages of regeneration and reheating can be combined with

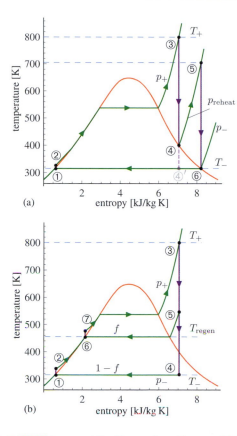

(a)

(b)

Figure 13.16 Improved Rankine cycles with $p_+ = 5$ MPa, $p_- = 8$ kPa, $T_+ = 800$ K, $T_- = 315$ K. (a) A Rankine cycle with reheating: step $\{34\}$ is the high-pressure turbine; $\{45\}$ is the reheating, which takes place at $p_{\text{reheat}} = 250$ kPa, and $\{56\}$ is the low-pressure turbine. The Rankine cycle before adding the reheating step is $\{1234'\}$. (b) A Rankine cycle with regeneration: steps $\{12\}$ and $\{67\}$ are the two pumps (the temperature increase from pumping has been exaggerated). A fraction f of the steam is removed at ⑤ and cycled back to the output of the first pump to preheat the water before it enters the second pump.

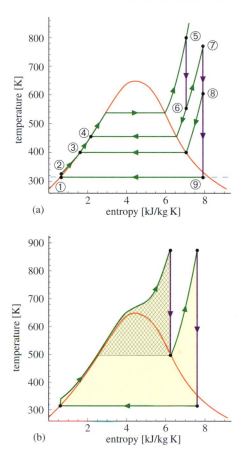

(a)

(b)

Figure 13.17 (a) A Rankine steam cycle with one reheating and two regeneration steps. (b) A supercritical Rankine cycle at 30 MPa with reheating; the area inside the cycle (shaded light yellow) indicates the work performed. The energy output and corresponding efficiency of the cycle can be compared with the ideal Rankine cycle (without cross-hatching).

the supercritical cycle. Comparing the sub-critical Rankine cycle of Figure 13.12(b) with the supercritical cycle of Figure 13.17(b), the increase in efficiency of the supercritical cycle is clear, as indicated by the cross-hatched region. The supercritical cycle adds to the energy in and work done in the ST plane, without modifying the heat extracted in the condenser, hence increasing the efficiency. Modern power plants using supercritical Rankine steam cycles can have efficiencies in the mid-40% range, compared to a Carnot limit of somewhat over 60%. As discussed above, approaching the Carnot limit more closely with a steam cycle is difficult due to the extremely high pressures that

are rapidly encountered when a liquid is heated above the saturation dome without allowing an increase in entropy. Material limitations seem to provide a bound on this type of cycle that is substantially stricter than the Carnot limit.

13.4 Low-temperature Organic Rankine Systems

So far we have discussed Rankine cycles entirely in the context of electric power generation. Over the past century, the cycle has been adapted for numerous applications and fuel sources, ranging from nuclear-fired propulsion systems for submarines to small-scale power generation fueled by combustion of animal waste. In this section we describe a *low-temperature Rankine cycle*, which has several applications, some of which are described in more detail in later chapters.

There are numerous heat sources that reach temperatures above the ambient environment, but below that required by the steam Rankine cycle. For example, there are many easily accessed geothermal resources in the temperature range of 50 °C to 150 °C (§32.2). These lower-grade energy resources can be tapped with a Rankine cycle that utilizes a working fluid other than water.

As the name implies, **organic Rankine cycles (ORC)** utilize organic working fluids; these are often the same refrigerants that are used for vapor-compression cycles. Particularly with energy resources that lie at or below 100 °C, the saturation curve for vapor-compression cycle refrigerants is ideally suited to execute a Rankine cycle. An example using the readily available hydrocarbon isopentane is shown in Figure 13.18. ORC components, because they operate at lower temperature and pressure than conventional steam Rankine cycles, are typically much less expensive than steam Rankine cycle components. Lower-cost components give ORCs an advantage for power generation when financial resources are limited, as in developing countries where low-temperature resources such as biomass and low-temperature geothermal energy may be abundant. As a result, ORCs have been developed for deployment in parts of the world where no prior power infrastructure exists. Though conversion efficiencies of low-temperature ORCs are low due to 2nd law constraints, and range from 10–20%, they have the advantage of utilizing abundant and easily accessed energy resources. While ORC systems are not likely to supply baseload power in the developed world, ORC plants are competitive with solar PV and wind for rural electrification. Significant portions of the globe support rural communities without large-scale electrical distribution systems, so we can expect ORC technology to play a role in a sustainable energy future. The role of the Rankine power cycle in ocean thermal and geothermal energy extraction are discussed in more detail in §27 and §32, respectively.

13.5 Gas Turbine and Combined Cycles

We conclude this chapter with a gas phase internal combustion cycle that has a significant role in electricity generation. The ready availability of natural gas in some countries, and its relatively low carbon emissions compared to coal, have led to an increased interest in turbines fueled directly by natural gas combustion. This interest is enhanced by the attraction of using natural gas in a high

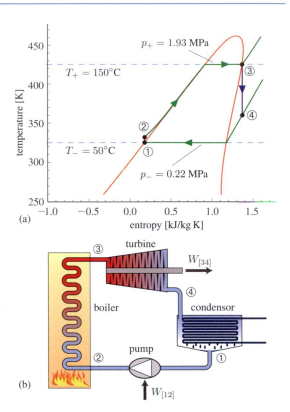

Figure 13.18 (a) An organic Rankine cycle operating between 50 °C and 150 °C, based on the saturation dome for isopentane. The points ①...④ are labeled in the same way as in steam Rankine cycles. Note that the shape of the saturation dome is such that the turbine operates entirely in the superheated vapor phase, without additional heating of the saturated vapor at point ③, which is mechanically advantageous. The shape of the saturation dome also leads to a natural cycle with relatively high efficiency for the given temperature range. (b) The mechanical components, emphasizing the simplicity of the ORC. (Credit: Adapted from Andrew Ainsworth reproduced under CC-BY-SA 3.0 license via Wikimedia Commons)

efficiency *combined cycle gas turbine* (CCGT) and/or in *cogeneration* systems, both of which are described in this section.

Gas turbines operate at very high temperatures, with turbine inlet temperatures that can reach 1600 °C (though typically are closer to 900–1400 °C) and turbine outlet temperatures ranging from 450 °C to 650 °C. Gas turbines alone are not extraordinarily efficient (see below), but the very high temperature of the outlet gases makes them a suitable heat source for a secondary conventional Rankine steam cycle. Together, the gas and steam turbine *combined cycle* typically exceeds 50% efficiency in actual electric power plants. If the steam exhausted at the outlet of the steam turbine is used for industrial processes or for space heating or cooling – a process known as *cogeneration* – then the overall fraction of the thermal energy from combustion that is utilized by the system can be greater than 90%.

In this section we first briefly survey the thermodynamic cycle known as the *Brayton cycle*, which can be used to run a gas turbine. We then describe how the CCGT system combines the Brayton and Rankine cycles. Finally we discuss cogeneration and how it can be integrated into the system.

13.5.1 The Brayton Gas Cycle

Gas turbine power plants burn natural gas to produce a superheated vapor that directly powers a turbine. The process is described by the idealized thermodynamic **Brayton cycle**, which is also used in jet engines for airplanes (with kerosene as the usual fuel). Next to the Rankine cycle, the Brayton cycle is perhaps the most widely implemented thermodynamic cycle in electric power generation.

Natural gas can be combusted at atmospheric pressure for its heat content, but as we saw for internal combustion engines in §11, the thermodynamic efficiency of a combustion engine is enormously enhanced if the fuel–air mixture is compressed before combustion. Thus, the first step in a Brayton cycle is compression of the entering air. Unlike internal combustion engines, which use reciprocating pistons, however, all the components of the Brayton gas cycle use rotary motion.[8] Thus, it is typically a centrifugal compressor or axial-flow compressor of the types described in §13.2.4 that acts to compress the entering airstream. The action of the compressor is modeled thermodynamically as adiabatic compression.

After the air is compressed, natural gas is mixed in and the air–fuel mixture is burned in a combustor at constant pressure. This raises the temperature of the air and the combustion products dramatically. The resulting high-pressure, high-temperature mixture powers a turbine, which is modeled as adiabatic expansion. In a typical gas turbine setup, the gases then exit the turbine back into the environment. Like an automobile engine, this is an *open-cycle* system – the fluid moves through once. True closed-cycle Brayton systems also exist, in which heat is transferred to the gas by some means other than combustion. Like the Otto cycle, the idealized thermodynamic Brayton cycle is modeled as a closed thermodynamic cycle

on a single mass of working fluid. In contrast to automobile cycles, however, where exhaust and intake are constrained by the volume of the cylinder, the expulsion of exhaust and intake of fresh air in the Brayton cycle are modeled as constant pressure heat rejection.

An advantage of the Brayton cycle is that the high temperature of the gases entering the turbine leads to a high Carnot limit on its efficiency even though the temperature of the exit gases is also quite high (450 °C to 650 °C). The Brayton cycle, however, has the disadvantage compared to the Rankine cycle that a great deal of the work produced by the turbine must be dedicated to powering the compressor, since pressurizing gases – unlike pressurizing liquids – requires significant work. This lowers the actual efficiency well below the Carnot limit.

An idealized Brayton cycle operating between 300 K and 1400 K with a compression ratio (see below) of 7:1 is shown in both the ST- and pV-planes in Figure 13.19. Mechanical components in a Brayton engine are shown further in Figure 13.20; the turbine and compressor are shown on the same shaft, illustrating the fact that the turbine supplies the **backwork** required to run the compressor. For the conditions shown in Figure 13.19, the ratio of backwork to useful work, $W_{\{12\}}/W_{\{34\}}$ is 37% (Problem 13.13).

The Brayton cycle and the Otto internal combustion cycle look very similar in the ST-plane even though they differ dramatically in the pV-plane. Like the ideal Otto cycle, the efficiency of the ideal Brayton cycle can be written (see Box 13.2) as a function of the compression ratio $r = p_2/p_1$ alone,

$$\eta_{\text{Brayton}} = 1 - \frac{1}{r^{(\gamma-1)/\gamma}}, \qquad (13.5)$$

where $\gamma = C_p/C_V$ is the ratio of specific heats of the air/gas mixture (assumed constant).

It would seem that increasing the efficiency of a Brayton cycle requires only increasing the compression ratio r. Increasing r, however, increases the temperature T_3 of the gases entering the turbine, which in turn is limited by materials constraints. Also, it is difficult to design axially driven (as opposed to reciprocating piston) compressors with large compression ratios. As a result, the development of relatively high efficiency gas turbines based on the Brayton cycle did not take place until the advent of jet engines for airplanes during and after World War II. Advanced gas turbines now work at temperatures in excess of 1400 °C, and with compression ratios above 15:1. Since the gas in the turbine is mostly air, we can take $\gamma \approx 1.4$ (see §10), and from eq. (13.5) we find a theoretical efficiency

---

[8] Although George Brayton's original device, patented in 1872, employed reciprocating pistons.

Box 13.2 Brayton Cycle Efficiency

The efficiency of the ideal Brayton cycle of Figure 13.19 is computed as follows. By definition, $\eta = (w_{\{34\}} - w_{\{12\}})/Q_{\{23\}}$, but by the 1st Law, $w_{\{34\}} - w_{\{12\}} = Q_{\{23\}} - Q_{\{41\}}$, so

$$\eta = 1 - \frac{Q_{\{41\}}}{Q_{\{23\}}} = 1 - \frac{C_p(T_4 - T_1)}{C_p(T_3 - T_2)} = 1 - \frac{T_1}{T_2}\left(\frac{T_4/T_1 - 1}{T_3/T_2 - 1}\right),$$

where we have used the fact that the {23} and {41} steps are both isobaric, and we have assumed that the heat capacity of the gas is constant throughout. The compression {12} and the expansion {34} are both isentropic, so pV^{γ} is a constant, as shown in §10. If we combine this with the ideal gas law, $pV = Nk_BT$, we find $p^{1-\gamma}T^{\gamma} = $ constant during the adiabatic steps {12} and {34}. Using the fact that $p_2 = p_3$ and $p_4 = p_1$, we find $T_4/T_1 = T_3/T_2$, so that

$$\eta = 1 - \frac{T_1}{T_2} = 1 - \left(\frac{p_1}{p_2}\right)^{(\gamma-1)/\gamma} = 1 - \frac{1}{r^{(\gamma-1)/\gamma}}.$$

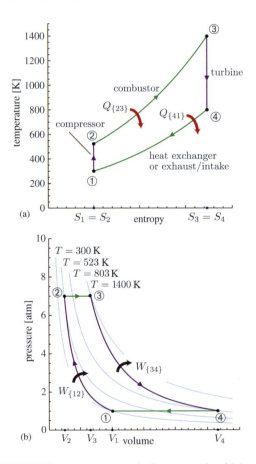

(a)

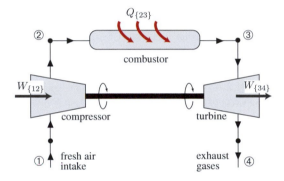

Figure 13.20 Mechanical components of an open Brayton cycle.

Figure 13.19 Representations of a Brayton cycle with intake at 300 K and 1 atm, maximum temperature 1400 K, compression ratio 7:1, and cold air standard assumptions ($\gamma = 1.4$). (a) In the ST-plane, with mechanical components labeled and heat flow shown; (b) in the pV-plane with isotherms and work flow shown.

of $\eta \approx 0.54$ for such an advanced turbine. For a more conservative design, with a maximum operating temperature of $\approx 1200\,°C$, and including the inefficiencies of the compressor and turbine, actual Brayton cycle efficiencies in the range of ≈ 35–$42\,\%$ are reached. The Carnot limit for a cycle operating between $1200\,°C$ and $300\,°C$ is 60%, so these turbines reach 60–70% of the thermodynamic limit.

13.5.2 The Combined Cycle

The combustion products exiting the gas turbine described in the previous section are still very hot. For $T_3 \approx 1200\,°C$, with a typical outlet temperature around $550\,°C$, the gases leaving the Brayton cycle turbine are hot enough to serve as the high-temperature heat source for a separate downstream Rankine cycle, known as a **bottoming cycle**.

The marriage of gas turbines with the Rankine steam cycle has produced the most efficient commercial electric power generation method: the **combined cycle gas turbine**

The Brayton Cycle

The Brayton cycle is a gas phase combustion cycle used in power plants and jet engines. Intake air is first compressed, then heated by combustion of natural gas or other vapor fuel. The hot, high-pressure gas is expanded adiabatically through a turbine. The output gases can be used to heat steam for a co-located Rankine engine (combined cycle). The efficiency of the ideal Brayton cycle,

$$\eta = 1 - \frac{1}{r^{(\gamma-1)/\gamma}},$$

is less than the efficiency of an ideal Otto cycle with the same compression ratio r.

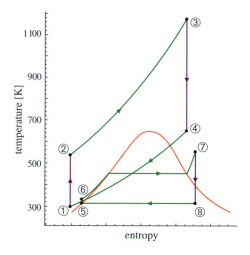

Figure 13.21 The CCGT cycle. An ideal Brayton gas turbine cycle {1234} with a compression ratio of 8:1 and $T_3 = 1200$ K has a theoretical turbine outlet temperature of $T_4 = 662$ K. Although the actual temperature is somewhat lower, it is sufficient to provide heat to an ideal steam Rankine cycle {5678} with minimum and maximum pressures and temperatures of $p_5 = p_8 = 0.008$ MPa, $T_5 = T_8 = 315$ K and $p_6 = p_7 = 1.00$ MPa, $T_7 = 550$ K. The efficiencies of the cycles are $\eta_B \cong 45\%$ and $\eta_R \cong 28\%$, and the efficiency of the (ideal) combined cycle is $\approx 60\%$.

shown schematically in Figure 13.21. The crucial point is that the waste heat from the top (Brayton) cycle serves as the input heat to the bottoming (Rankine) cycle. If the separate efficiencies are η_B and η_R respectively, then the net efficiency (net work divided by the heat provided by the heating value of the fuel), is $\eta_{\text{net}} = \eta_B + \eta_R(1-\eta_B)$ (Problem 13.15). If the cycles are separately, say, 40% and 30% efficient, the net efficiency is 58%, a major improvement! Typical in-service CCGT power plants have efficiencies of over 50% (HHV), and state-of-the-art designs have surpassed 60%. Most modern natural gas-fired power plants are CCGT plants.

The efficiency of CCGT power plants has spurred renewed interest in coal gasification. Coal can be converted to a synthetic gas (*syngas*) that is primarily a mixture of CO and H_2; syngas can be used as fuel for a CCGT plant. The syngas production process and **integrated gasification combined cycle (IGCC)** plants are discussed further in §33. In light of the complexities and energy and carbon cost of the syngas production process (see §33.1.7), there is a vigorous debate as to the relative advantages and disadvantages of IGCC power generation in comparison with direct combustion of coal. There is no question, however, that the efficiency and carbon footprint of the natural-gas-based CCGT are better than either coal based alternative.

13.5.3 Cogeneration

The final step in maximizing the efficiency of a power plant is to make use of the so-called "waste" heat that is rejected from a Rankine steam cycle in the form of relatively low-temperature and low-pressure steam. The concept of using a heat source to provide both mechanical

or electrical energy and useful thermal energy is known as **cogeneration**.

There are two fundamentally different motivations for cogeneration, depending on whether the steam or the mechanical/electrical energy is the byproduct. We have been discussing power plants, where electrical energy is the primary product and the steam would be the by-product. Alternatively, heat is ubiquitous in industrial processes, and mechanical or electrical power can be produced as a by-product. Metals processing, paper manufacturing, oil refining, chemicals manufacturing, are only a few of the industrial processes that make use of **process heat**, which is usually supplied by *steam*, typically at pressures upwards of 5 atm and temperatures of 150 °C or more. Steam for process heat is usually produced in a boiler fueled by fossil fuel combustion, which takes place at temperatures exceeding 1500 K. Even if all the chemical energy in the fuel ends up in the steam, using such high temperatures to produce low-temperature steam is thermodynamically inefficient.[9] Instead, the combustion can be used to produce steam at much higher temperatures

---

[9] We quantify this inefficiency in §36 (Systems).

Combined Cycle and Cogeneration

The combined cycle gas turbine (CCGT) power plant combines the high-temperature, gas phase Brayton cycle with a lower-temperature and lower-pressure Rankine steam cycle to obtain relatively high thermodynamic efficiency. Efficiencies exceeding 60% have been obtained.

Exhaust heat – either gases exiting a Brayton turbine or steam from a Rankine turbine – can be used for industrial processes, space heating, or cooling in *combined heat and power* (CHP) or *cogeneration* facilities. Cogeneration plants that combine a high-temperature Brayton cycle, a low-temperature Rankine cycle, and local space heating and/or cooling can utilize up to 85–90% of the combustion energy of natural gas fuel.

the European Union as a whole produced in cogeneration facilities, and over 40% in Denmark and Latvia. MIT built a cogeneration plant in the 1980s to supply its campus with electricity, heating, and cooling. Its gas turbine utilizes 56 MW of thermal energy from combustion, and generates 21 MW of electricity, for an efficiency of 37.5%. Its **cogeneration efficiency** – the sum of electric power plus energy used for heating or cooling – is stated to be greater than 85%.

In practice, the steam input to most heating and cooling applications must be at a temperature above $100\,°C$ and a pressure of several atmospheres, reducing the efficiency of a Rankine cycle in a cogeneration system. For this reason, cogeneration is particularly attractive when paired with a gas turbine (as MIT's plant is), and the exhaust gases from the Brayton cycle are used directly to produce steam, or even better, when cogeneration is paired with a combined cycle so that the high-quality heat emerging from the Brayton cycle powers a Rankine steam cycle from which emerges the steam for space heating and cooling.

and pressures, which runs a turbine to produce mechanical energy. Then, the steam that emerges from the turbine outlet can be used for the originally intended industrial purposes. In this case the mechanical/electrical energy is the by-product.

The efficiency of a Rankine cycle is a sensitive function of the temperature at the turbine outlet. In §13.3.4 we pointed out that efficient power plants keep the outlet temperature as low as possible. Clearly, designing an industrial heating plant to produce, say, steam at 5 atm and $200\,°C$ must sacrifice some efficiency in the Rankine cycle. Since the mechanical energy is a by-product (i.e. the fuel is being combusted in any event), a lower efficiency is tolerable. Also known as **combined heat and power** (CHP), the practice of generating mechanical power along with process heat has been implemented by efficiency-minded plant designers for years.

Returning to power plants, the idea of cogeneration is to put the exhaust heat from the turbine to useful purposes rather than dumping it into the environment. This heat can provide space heating for homes and businesses, and it can be used as the heat source for an absorption cooling cycle (see §13.2.6). The cost of insulated piping to carry the heated or chilled water to consumers limits this kind of cogeneration to plants located near relatively dense population centers. It is ideal for universities and for compact towns and cities, such as those found in many parts of Europe. Indeed, Europe has the highest incidence of cogeneration, with 11% of all electricity generated in

Discussion/Investigation Questions

13.1 Why is water not a suitable fluid for a vapor-compression kitchen refrigerator? What about for an air conditioner intended to operate between $24\,°C$ and $40\,°C$?

13.2 Discuss the distinction between an adiabatic and an isentropic process. For example, how can it be that a compressor that works very quickly, allowing negligible heat transfer, can nevertheless increase the entropy of a gas?

13.3 The saturation domes for R-134a and isopentane shown in Figures 13.4 and 13.18 are quite asymmetric, with steep (R-134a) or even recurving (isopentane) saturation curves. Water, on the other hand, has a relatively symmetric saturation dome (see Figure 12.9). Explain why this feature makes R-134a and isopentane more suitable fluids for low-temperature heat extraction devices and engines than water. Why is water so widely used for higher-temperature systems such as steam turbines? What kind of shape for the saturation dome would be better for a phase-change fluid in a higher-temperature system?

13.4 As mentioned in §13.2.6, there are several other paradigms for simple refrigerators. Research and describe a gas phase Stirling refrigeration cycle or a gas phase Brayton refrigeration cycle. What does the corresponding thermodynamic cycle look like in the ST-plane? What are the applications of the cycle you study?

13.5 A vapor-compression cycle is specified by a closed, counterclockwise loop in the ST-plane. Show that *if the*

loop is executed reversibly then the area of the loop is equal to the net work performed on the fluid in a cycle. [Hint: This is a corollary to a similar result for engines derived in §10.1.2.] Explain why the area of the ideal VC cycle {1234} in Figure 13.5(b) *is not* equal to the work performed on the fluid.

13.6 Describe in words how you would compute the coefficient of performance for an ideal vapor-compression cycle working between temperatures T_- and T_+. Assume that you have data on the thermodynamic properties of the refrigerant both at saturation and in the superheated vapor phase. Remember that $T_② > T_+$ (see Figure 13.5).

13.7 The efficiency η_a of the idealized Carnot-like Rankine cycle of Figure 13.12(a) is given by the Carnot limit, $\eta_a = \eta_{\text{Carnot}} = (T_+ - T_-)/T_+$. Explain why the efficiency of the *ideal* Rankine cycle of Figure 13.12(b) is higher than η_a for fixed p_+ but lower than the Carnot limit for $T_+ = T_③$.

13.8 **[T]** Consider the modified Rankine cycle of Figure 13.17(a). Describe what happens to a quantity of water as it executes the cycle starting at ①. How many pumps and how many turbines are needed?

13.9 Explain in words what is going on in the Rankine reheat cycle of Figure 13.16(a), and, referring to the figure, explain the statements made at the end of the discussion in the text about the efficiency, power output, and turbine conditions for this cycle.

13.10 Jet engines use a variant of the Brayton cycle in which the gases are only expanded in the turbine to a pressure sufficient to drive the compressor and the plane's mechanical and electrical systems; the resulting gases are expelled from the engine at high velocity. Research jet engines and describe their mechanical components, efficiency, and the *thrust* that they produce.

Problems

13.1 **[T]** Prove that the Carnot-like vapor-compression cooling cycle {1234} in Figure 13.5(a) has a maximum CoP that equals the Carnot limit regardless of the properties of the working fluid. Likewise, prove that the efficiency of the simplest Rankine power cycle {1234} of Figure 13.12(a) equals the Carnot limit.

13.2 **[T]** Sketch the Carnot-like VC cycle of Figure 13.5(a) in the pV-plane (superimposed on the saturation dome), and contrast the resulting shape with that of the Carnot cooling cycle described in §10.

13.3 A heat pump based on the *ideal* VC cycle of Figure 13.5(b) uses the refrigerant R-134a to heat a house. The set points are $T_- = -8\,°\text{C}$, $T_+ = 40\,°\text{C}$. What is the CoP; how does it compare with the Carnot limit? What flow rate ($\dot{m}$) in kg/s is required to provide heat at 50 kW? Relevant thermodynamic data for R-134a

Table 13.5 Data for Problem 13.3.

p (MPa)	T (°C)	χ	h (kJ/kg)	s (kJ/kg K)
0.2169	−8	0	189.3	0.9606
0.2169	−8	1	393.9	1.732
1.0166	40	0	256.4	1.191
1.0166	40	1	419.4	1.711

is provided in Table 13.5. Note that since this is an ideal VC cycle, as opposed to the modified cycle analyzed in §13.2.5, the maximum temperature T_2 must be determined. To find T_2, data in the superheated region are required [23]. Alternatively you can approximate $T_2 \approx T_+$.

13.4 The 5-ton AC unit designed in §13.2.5 reached 33% of the Carnot limit on CoP. Look back at the definition of the "Energy Efficiency Ratio" in Problem 10.12, which employs a different temperature range than the one we specified. Assuming that the unit's fraction of the Carnot limit remains unchanged, compute its EER. How does this compare with the current US standard of EER ≥ 11.7?

13.5 **[T]** Neglecting the work done by the pump (which is a good approximation), show that the efficiency of the ideal Rankine cycle is $\eta = (h_3 - h_4)/(h_3 - h_2)$, where h_j is the specific enthalpy at the point j in the cycle.

13.6 **[T]** Sketch the ideal Rankine cycle on a pV-diagram including the saturation dome. Label the points ①–④. As in Figure 12.7, plot the *logarithm* of the specific volume on the horizontal axis.

13.7 When we designed the Rankine steam cycle in §13.3.5 we ignored the circulating water's kinetic and potential energy, having claimed earlier (see Box 13.1) that they are negligible. Using the work per unit mass done by the turbine to set the energy scale, estimate the importance of the potential energy of the fluid if the height of the boiler is 40 m? Estimate the importance of the kinetic energy if the maximum flow speed in the pipes is a few (say 5) m/s?

13.8 Suppose the condenser in the 500 MWe coal power plant we analyzed in §13.3.5 is cooled by a once-through system, where water taken from the ocean is circulated once through the plant and then returned to the ocean. If regulations permit only a 10 °C rise in the ocean water temperature, how much water (in m³/s) must be used? (Assume that the heat capacity of seawater is the same as that of pure water. Remember that the efficiency of the plant was estimated at 38%.)

13.9 Quantitatively compare the Carnot Rankine cycle of Figure 13.12(a) with the ideal Rankine cycle of Figure 13.12(b). Compare efficiencies and the work done per cycle. All the necessary information about the ideal Rankine cycle can be found in Example 13.2.

Table 13.6 Data for Problem 13.9.

p (MPa)	T (K)	χ	h (kJ/kg)	s (kJ/kg K)
0.005	306.0	0	137.8	0.4762
0.005	306.0	1	2561	8.394
2.5	497.1	0	961.9	2.554
2.5	497.1	1	2800.9	6.254

The required information for the Carnot cycle can be obtained from Table 13.6.

13.10 Modify the coal plant Rankine cycle described in §13.3.5 by including a regeneration cycle similar to the one shown in Figure 13.16(b). A fraction f of the steam is removed from the turbine and returned back to a second pump at a temperature $T_r = 130\,°\text{C}$. Compute the fraction f needed to heat the water from the first pump to $130\,°\text{C}$, and compute the increase in efficiency of the resulting cycle above that computed in the text. [Hint: $s_5 = s_3$ in Figure 13.16(b).]

13.11 The wear on a steam turbine could be decreased by raising the pressure at the turbine outlet so that the quality of the steam at the outlet is one. In the ideal Rankine cycle of Example 13.2 and Figure 13.12, $\chi = 0.9$ at the turbine outlet. Keeping the p_+, T_+, and T_3 unchanged, determine p_- and T_- so that $\chi = 1$ at ④. Compare the efficiency and the power per cycle of the resulting cycle with the cycle of Example 13.2. To simplify the calculation, ignore the work done by the pump. Would you advise taking this step?

13.12 [T] Explain why the Brayton and Otto cycles look so different in the pV-plane even though they are quite similar in the ST-plane.

13.13 [T] Show that the ratio of the backwork (the work necessary to run the compressor) to the total work done by the turbine for the ideal Brayton cycle is $W_{\{12\}}/W_{\{34\}} = T_1/T_4$. [Hint: use the 1st Law and the fact that both the compressor and turbine are adiabatic.] Check your result by comparing to the result stated in the text for the conditions shown in Figure 13.19.

13.14 [T] Calculate and plot the ratio of the efficiency of the Brayton cycle to the efficiency of the Otto cycle with the same compression ratio (with $\gamma = 1.4$).

13.15 [T] Verify the assertion made in §13.5.2 that the net efficiency of a CCGT is $\eta_{\text{net}} = \eta_B + \eta_R(1 - \eta_B)$, where $\eta_{B/R}$ is the efficiency of the Brayton/Rankine cycle.

Part II

Energy Sources

The Forces of Nature

Elementary particles and the forces that act between them are the fundamental ingredients in our most complete theoretical description of microscopic physics. Forces transform energy from one form to another when they act on matter. The study of energy thus inevitably leads to the question: What are the fundamental forces of nature, and what are the elementary particles on which these forces act? Gravity and electromagnetism are the most familiar two of the four forces that store and transform energy in our world. The other two, manifested by the *strong interactions* that hold nuclei together and the more subtle *weak interactions*, are usually the subject of more advanced physics courses.

All four forces contribute to the production and concentration of the forms of energy that we use on Earth. Gravity pulls together hydrogen atoms deep in the Sun. Strong nuclear interactions between these hydrogen atoms, facilitated by crucial weak interactions, result in the release of energy that is radiated from the Sun as electromagnetic waves, some of which hit Earth. Nuclear fission power plants rely on the interplay of strong nuclear forces, electromagnetism, and the weak interactions. Nuclear fuel is found in regions of high concentration in Earth's crust that result from a combination of gravitational pressure and chemical electromagnetic interactions. Geothermal energy, as well, arises from the combined action of the four forces. Indeed, with the exception of tidal energy, which is essentially the result of gravity and electromagnetic forces, all energy sources used by humanity rely on a happy conjunction among all four forces of nature.

The most familiar energy processes that we encounter on a day-to-day basis can be described in terms of electromagnetic and gravitational forces and interactions. The same is true of all the energy systems treated in this book so far. Much of the energy we use, however, originates in the *strong interactions*, which are not encountered in everyday experience (unless, like Homer Simpson, you

Reader's Guide

This chapter forms an "interlude" in which we survey the fundamental particles and forces that determine the structure of matter and its interactions with energy. Because they are less well known than gravity and electromagnetism, special attention is paid to the strong force that holds the nucleus together and to the weak force responsible for many nuclear decays. After discussing the four forces, we describe the properties of all the known elementary particles. We then present a simplified version of particle physics that is adequate for most of our purposes, in which only five particles, the proton, neutron, electron, photon, and antineutrino appear. The chapter closes with an introduction to β-decay and related processes that are important for understanding nuclear energy and related radiation.

Prerequisites: §2 (Mechanics), §3 (Electromagnetism), and §7 (Quantum mechanics), although much of this chapter is accessible with little technical background.

The strong and weak forces are important ingredients in the description of nuclei and nuclear energy in §17–§19, and ionizing radiation in §20. β-decay is key to understanding many nuclear decays. Gravity figures centrally in the large-scale structure of the universe (§21).

work at a nuclear reactor). The drama of the strong interactions involves some relatively familiar players such as the nuclei of atoms, but it also includes a menagerie of exotic particles such as *quarks* and *gluons* that, while they are just as real as electrons and atoms, are never seen outside of the atomic nucleus. In order to explain the origins of solar, nuclear, and geothermal energy, we must first describe some of the basic properties of the strong force. Along the way, we have the opportunity to put gravity and the electromagnetic force into a somewhat broader context.

The fourth force, which is also only relevant in the microworld, is known as the *weak interaction*. Weak

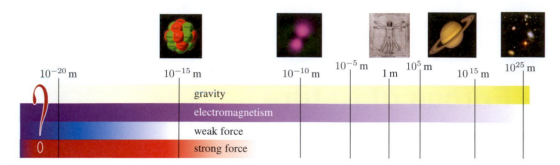

Figure 14.1 An illustration of the relative importance of the four forces over distances ranging from subnuclear (10^{-20} m) to cosmological (10^{25} m).

interactions play only a minor part in the actual generation of energy, but they act as a *facilitator* for many of the most important nuclear energy processes. Weak interactions play a role in fundamental physics a bit like fungi do in the environment. They are often overlooked; they do not contribute to building up great structures (such as nuclei and atoms); and, above all, they are the mediators of decay! Nevertheless we need to understand some of the features of the weak force because so many processes that figure significantly in energy physics would not occur without it.

Fortunately, the list of forces that affect our world stops at four: strong, electromagnetic, weak, and gravitational. There may be other forces locked away at very short distances, or that involve as-yet-undiscovered particles. Even if additional forces exist, however, we already know that they do not influence the dynamics of our familiar world in any significant way at the human scale (see Figure 14.1). We can be sure of this because we can account for essentially *all* the phenomena so far observed in nature in terms of the four known forces.[1] This may sound like an exaggeration: after all there are many common phenomena not yet fully understood by science, from the turbulent flow of liquids to the origins of life. These are generally, however, phenomena characterized by complexity, and there is no evidence that they arise from additional, as yet unknown, fundamental interactions.

If one focuses in on short distances and small systems, processes simplify, and everything that we can observe seems to be explained by a relatively simple set of

---

[1] A few fundamental observed phenomena lie beyond the limits of our understanding. For example, *dark energy* and *dark matter*, which are discussed further in §21 (Energy in the universe), dominate the total energy budget of the universe. Important though they are at the cosmic scale, they play no role at all in the energy transformations of everyday life.

The Standard Model of Particle Physics

The *Standard Model of particle physics* is a *quantum field theory* that, in concert with the classical theory of gravity, describes almost all observed phenomena to a high degree of accuracy in terms of four forces and 17 elementary particles.

fundamental equations. Our understanding is summarized in a meticulously tested and so far perfectly successful set of laws of physics, known as the **Standard Model of particle physics**. The word "*model*" connotes less certainty than "*theory*" or "*law*" – the terms typically used to designate well-established principles of physics. The Standard Model, however, which is formulated in terms of a theoretical framework known as *relativistic quantum field theory*, has described observed physical phenomena to a higher degree of precision than any other physical theory ever developed. Quantum field theory and the Standard Model are on a similar footing to Newton's and Coulomb's laws; they provide a set of simple and basic equations from which the behavior of matter can be deduced by mathematical analysis. The specific structure of the Standard Model dictates that matter is made of essentially pointlike particles with spin 1/2 (quantum spin was discussed in §7). These particles interact with each other by emitting and absorbing photons and other massless and massive quanta that are cousins to the photon. The elementary spin-1/2 particles form families whose membership is limited by symmetry constraints and whose members have common properties. The details are beyond the scope of this book, but some aspects of the Standard Model are summarized in Tables 14.1 and 14.2 and in Figure 14.4.

Despite its successes, the Standard Model is neither a perfect nor a complete theory of nature. One glaring inadequacy is that it has about 20 parameters – numbers like the mass of the electron and the strength of the electric charge – whose origins we do not know. It also has a peculiar redundancy: each type of particle – the electron for example – comes in three versions that differ in mass, but little else. This redundancy, as well as the *raison d'etre* for the particular assortment of particles and fields that appear in nature, and other relationships in the Standard Model, are beyond our present understanding. There are some detailed aspects of the Standard Model, related to *neutrinos* and the *Higgs particle*, which are not fully settled experimentally. And, as mentioned above, it is possible that new forces or other novel phenomena are relevant at very high energies and short distance scales that have not yet been probed experimentally. Finally, and arguably at a more fundamental level, in our current understanding gravity is appended to the Standard Model only as a classical (rather than quantum) field, even though gravity also must exhibit quantum features at very short distance and length scales. Unifying gravity and quantum mechanics has challenged physicists for decades. An approach often called *string theory* provides a framework in which gravity and other forces can be described within a quantum-mechanical theory. Whether and how the Standard Model and the global structure of the universe in which we live is described within the framework of string theory is, however, still an open problem (§21).

All the limitations of the Standard Model that we have just listed are irrelevant, in any case, for the purposes of this book. All known energy sources can be understood with only a subset of the structure in the Standard Model, and it is hard to imagine any way in which physics beyond the Standard Model could have practical consequences for human energy use in the forseeable future. Thus, for all intents and purposes, the fundamental physics underlying any conceivable energy system that may be relevant in this century should be described within the Standard Model.

The aim of this chapter is to take a brief tour of the Standard Model, emphasizing those bits that are most important for the physics of energy and skirting as much as possible the complexities that usually relegate the subject to graduate-level physics courses. Although the concepts discussed here are relatively remote from everyday experience, they are nevertheless relevant to practical energy systems. In fact, the basic ideas presented here are essential for understanding the origins of nuclear and solar energy and the nature of radioactivity. In §14.1 we describe the four fundamental forces in more depth. Starting at the largest distance scales, where gravity is most important, we work our way down to sub-atomic distances, encountering electromagnetic, strong, and weak interactions in succession.

Next (§14.2) we meet the players in the drama of sub-atomic physics, the elementary particles. Before exploring the whole zoo of particles, we introduce a stripped-down version of the Standard Model that includes almost everything needed to understand energy production and flow in the observable universe at the present time. This *Standard Model "lite"* comprises the proton, neutron, electron, photon, and one potentially less familiar addition, the *(anti-)neutrino*. We briefly survey the other particles in the Standard Model and then conclude the chapter (§14.3) with a close look at β-decay, a weak interaction of particular importance in energy physics.

Non-technical introductions to particle physics can be found in many books, for example [66]. Reference [67] provides a more technical introduction, at an advanced undergraduate level.

14.1 Forces, Energies, and Distance Scales

To survey the hierarchy of strengths, distances, and energy scales that characterize the four forces, we start at astronomical distances and work our way into the interior of the atomic nucleus.

Gravitational and electromagnetic potential energies fall off like $1/r$ away from their sources. This is a special case of a more general form

$$V(r) \sim g^2 \frac{e^{-r/b}}{r}, \qquad (14.1)$$

where b is known as the **range** of the force and g is a measure of its *strength*. Equation (14.1) defines what is known as a **Yukawa interaction**, named after the Japanese physicist Hideki Yukawa who proposed it as a model for the strong nuclear force in the 1930s. The $1/r$ potential of gravity and electromagnetism emerges when the parameter b goes to ∞. Hence, these are called *infinite range* forces. In the framework of *quantum field theory*, interactions between massive particles are the result of the exchange of particles called **force carriers** (Figure 14.2). For example, the *photon* is the force carrier for electromagnetism. In general, the exchange of a force carrier of mass m gives rise to a force of the form (14.1), where the parameter b is proportional to $1/m$. Thus, the long-range nature of the electromagnetic force arises from the fact that the photon is massless. We now go through each of the forces in turn and discuss various aspects of each.

14.1.1 Gravitational Interactions
Only gravitational forces are important at astronomical distances. Gravity alone persists at the scale of stars and galaxies not because gravity is strong – in fact gravity

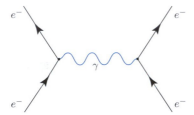

Figure 14.2 In *quantum field theory*, interactions between massive particles are mediated by the exchange of massless or massive *force carriers*. The figure depicts a *Feynman diagram* describing two electrons interacting through exchange of a *photon*, the force carrier for electromagnetic interactions.

$$|\boldsymbol{F}_{\text{gravity}}| = G_N \frac{m_1 m_2}{r^2} = 6.67 \times 10^{-11} \frac{\text{m}^3}{\text{kg s}^2}$$

$$\times \frac{(10^{-3}\,\text{kg})^2}{1\,\text{m}^2} = 6.67 \times 10^{-17}\,\text{N}, \quad (14.2)$$

which is far too weak to observe.

14.1.2 Electromagnetic Interactions

To gauge the strength of electromagnetic forces compared to gravity, we compute the electrostatic (Coulomb's law) force between the same two one-gram lumps of matter described above if the electric charges were not neutralized. That is, we compute the electrostatic force between the positively charged protons in the two lumps. Since about half of the mass of one gram of material comes from protons (the other half comes from neutrons; the electrons are only about 1/2000th of the mass), one gram contains about $N_A/2 \approx 3 \times 10^{23}$ protons, each with charge 1.6×10^{-19} C, for a total charge of $\approx 4.8 \times 10^4$ C. This generates a force of

$$|\boldsymbol{F}_{\text{EM}}| = \frac{1}{4\pi\epsilon_0} \frac{Q_1 Q_2}{r^2} \approx 9.0 \times 10^9 \frac{\text{Nm}^2}{\text{C}^2}$$

$$\times \frac{(4.8 \times 10^4\,\text{C})^2}{1\,\text{m}^2} \approx 2.1 \times 10^{19}\,\text{N}. \quad (14.3)$$

is by far the weakest of the four forces (see Table 14.1). The reason that gravity dominates over electromagnetism on astronomical scales is that *gravity is universally attractive* and its strength is proportional to the mass (or, more generally, the total energy) of an object. In contrast, electromagnetic forces depend on the signs of the electric charges; electromagnetic attraction and repulsion between charged particles are so strong that unless we work very hard to prevent it, matter always relaxes to a state in which it is electrically neutral. So matter at macroscopic scales is essentially electromagnetically uncharged, and the force of gravity can dominate at very large scales. Thus, Newton's law of universal gravitation governs the motions of planets and stars. For very dense and very massive objects, Newton's theory of gravity must be corrected or even replaced by Einstein's General Theory of Relativity, which is described in §21.[2]

As soon as one leaves the astronomical domain, gravity subsides in importance and electromagnetic forces become equally or more important. Of course, gravity holds us here on Earth's surface – again because a very large mass is involved, namely the mass of Earth. To support the claim that gravity is unimportant on human distance scales, consider the strength of the gravitational force between, for example, two one-gram masses, separated by one meter. According to Newton, the gravitational force between them is

Between comparable quantities of stuff, therefore, the electromagnetic force exceeds the gravitational force by ~36 orders of magnitude. Of course, the repulsive Coulomb forces among the protons are so strong that there is no way we could isolate and study a gram of material with charge 4.8×10^4 C. Electromagnetic forces subside to manageable scales because – to very high accuracy – matter is neutral.

Many of the large-scale processes on Earth's surface result from the dynamic balance between electromagnetic and gravitational forces. For example, the electromagnetic repulsion between electrons in different atoms keeps matter from collapsing under huge gravitational pressures in Earth's core, and – at a more human scale – keeps you from falling through your chair. Gravitational and electromagnetic forces maintain the dynamic balance that governs the circulation of the atmosphere and oceans; the Sun's energy is absorbed and transmitted through electromagnetic forces while the overall vertical order is maintained by the pressure sustained by gravity.

There is no specific natural energy scale for the effects of *classical* gravity or electromagnetism. The larger the mass or charge, or the closer the objects, the greater the energy that can be stored in gravitational or electromagnetic potential energy. When quantum mechanics gets into the game, however, energies and charges are quantized,

---

[2] That is not to say, however, that electromagnetism can be ignored in astronomy! For example, we see distant objects only because of the light they give off. Electromagnetism is important for dynamical reasons as well. A cloud of gas cannot collapse to form a star unless it can find a way to give off some potential and kinetic energy, and radiating away energy through electromagnetic waves is the principal way that this happens.

Gravity and Electromagnetism

Gravitational and electromagnetic forces have infinite range and determine the behavior of matter over a huge range of distance scales. With minor exceptions such as natural radioactivity and geothermal energy produced by radioactive decay, natural processes on Earth's surface involve electromagnetic and gravitational interactions alone.

Maxwell's equations and quantum mechanics, along with classical gravity, are sufficient to describe basic dynamical processes from Earth-like distances down to a small fraction of the diameter of an atom.

and the quantum of energy associated with a particular force acting between elementary particles sets the fundamental scale for energies that can be absorbed, liberated, or stored by those forces acting on the quantum level. In §7 we saw that the energy of the electron in a hydrogen atom,

$$E_{\mathrm{Bohr}} = -\frac{1}{2}\left(\frac{1}{4\pi\epsilon_0}\right)^2 \frac{me^4}{\hbar^2} = -13.6\,\mathrm{eV}, \qquad (14.4)$$

sets a scale for the energies of atomic physics. (Remember that the minus sign in eq. (14.4) indicates that the electron in hydrogen is bound – it takes 13.6 eV of energy to remove it far away from the proton.) When a bond is formed or broken, energies of this order of magnitude are absorbed or emitted. The scale of energy available from chemical processes never strays far from the fundamental scale given in eq. (14.4), set by the mass and charge of the electron. When we manipulate materials, whether they are hydrocarbons or more exotic materials, the amount of energy we can get from them per unit mass from chemical processes is roughly the same as that available from burning coal (see Problem 14.1). To find energy sources that have much higher *energy density*, that is, more joules per kilogram, we must consider a stronger force.

14.1.3 Strong Interactions

From the human scale of meters down to the smallest distances that we can see with an optical microscopic, physics is governed by electromagnetism and gravity. Journeying inward to still shorter distances, at somewhat less than 10^{-14} m, things change rather suddenly and dramatically. The nuclei at the centers of atoms are composed of protons and neutrons (collectively referred to as **nucleons**), held together by forces known as **strong nuclear interactions**, which are a hundred times stronger than electromagnetism, and of a different nature.

The strong interactions between nucleons are complex, residual manifestations of simpler and more fundamental forces between elementary particles called **quarks**, of which the nucleons themselves are composed. The interactions among quarks, which are specified by the Standard Model, are known as **chromodynamic** forces in analogy to the *electrodynamic forces*, which they resemble in important ways.[3] Chromodynamic forces between quarks are mediated by **gluons**, massless particles that play a similar role to that of photons in electromagnetism. The forces between nucleons are to quark forces as molecular forces are to the underlying electromagnetic force between a proton and an electron. Just as the primary electric forces cancel out in neutral atoms or molecules, so the primary forces between quarks *cancel out* when the quarks group themselves together into *chromodynamically neutral* protons and neutrons. Like the residual forces between atoms and molecules (see, e.g., §9.3.1), the forces between nucleons left over after this cancellation are complicated and cannot be described accurately by a single simple mathematical formula. The name *strong interactions* was applied to these forces long before quarks and the Standard Model were discovered. Here we reserve the term "strong nuclear interactions" for these residual forces between nucleons and nuclei and distinguish them from the (in some ways even stronger, but conceptually simpler) chromodynamic forces between quarks. The strong nuclear interactions hold nuclei together and are responsible for the dynamics of fission, fusion, and radioactivity, so we explore them further here.

One way in which chromodynamics differs from electrodynamics is that the fundamental forces between quarks are so strong that quarks can never be separated from one another. Only bound states of quarks in which the primary chromodynamic forces are neutralized can exist in nature. These chromodynamically neutral **hadrons**, as they are known, include the proton and neutron and a host of other short-lived particles that can be produced in laboratory experiments. Because the chromodynamic force is so strong, it is very difficult to compute the form of the strong interaction between nucleons analytically from the theoretically simpler interactions between quarks. The best we can do at the present time is to study the nuclear force numerically or with the aid of empirical models.

---

[3] The origin of the "*chromo*" in *chromo*dynamics is explained in §14.2.3.

Chromodynamics and the Strong Nuclear Force

The *chromodynamic force* is a powerful force that binds *quarks* together into nucleons such as protons and neutrons. Residual, short-range *strong nuclear interactions* between nucleons bind nuclei together, just as residual electromagnetic interactions lead to forces between electrically neutral atoms that bind molecules together.

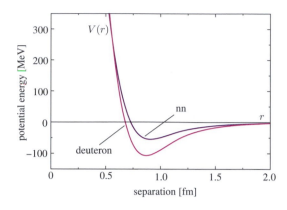

Figure 14.3 The nucleon–nucleon strong interaction potential in two configurations [68]. Roughly speaking, the curve marked *deuteron* is the potential that holds together a proton and a neutron into a bound state called a **deuteron**, while the one marked *nn* is the potential between two neutrons, which is not strong enough to bind a state. Note that while the two potential functions are quite different at ∼ 1 fm, they have the same asymptotic form (14.5) as $r \to \infty$.

The strong nuclear force between nucleons depends upon their separation, their relative velocities, their relative angular momentum, and even on the orientations of their spins (protons and neutrons have spin 1/2 like the electron). This complicated force between two nucleons is often described by empirical formulae with many *fit parameters* (see, for example, [68] and references therein).

At the largest distances where it is detectable, the strong interaction potential between two nucleons at rest takes the simple form of a Yukawa interaction (14.1) with a range $b \cong 1.4 \times 10^{-15}$ m and a strength about 10 times that of electromagnetism,

$$V_{\text{strong}}(r) = -\frac{1}{4\pi\epsilon_0} \frac{g_{\text{strong}}^2}{r} e^{-r/b}, \qquad (14.5)$$

with $g_{\text{strong}}^2 \approx 10\, e^2$, where $e = 1.602 \times 10^{-19}$ C is the proton's electric charge.[4] The unit of distance 1 femtometer $= 10^{-15}$ m, abbreviated 1 fm, is a characteristic unit for strong interaction phenomena. It is also known as *1 Fermi* in honor of the Italian theoretical physicist Enrico Fermi. The strong interaction force between two (chromodynamically neutral) nucleons at large distances is mediated by the exchange of *pi-mesons* or **pions**, which are themselves chromodynamically neutral massive bound states of quarks and antiquarks. This is different from the long-range residual electromagnetic forces between electrically neutral atoms, which arise from exchange of multiple massless photons and have a polynomial form rather than the characteristic exponential suppression of the Yukawa potential.

Note the minus sign in eq. (14.5): the force is attractive. The form given in eq. (14.5) holds (approximately) at separations greater than about 1.5 fm. Below that, the forces

between nucleons become stronger, but they do not have the simple exponential form of eq. (14.5).

Although the strength, and even the sign, of the nucleon–nucleon potential varies depending upon the nucleons' relative angular momentum, spin, and momentum, the nucleon–nucleon interaction is, on average, strong and attractive at distances greater than ∼0.7 fm. At still shorter distances, the nucleon–nucleon force turns strongly repulsive, preventing nucleons from coming too close to one another. Figure 14.3 shows the neutron–neutron potential and the proton–neutron potential; in the latter case, the configuration is the one in which the nucleons can form a bound state, the *deuteron*. Although they look quite different at 1 fm, both potentials reduce to eq. (14.5) with the same strength at large distances. At their minima near 0.8 fm, the potentials shown take values of roughly −50 MeV and −100 MeV. This means that energy in amounts measured in *millions* of electron volts per nucleon is in principle available in nuclear interactions. The depth of the potential is not, however, a good measure of the binding of nuclei. The reason for this is that the potential is very negative only in a relatively thin spherical shell. A quantum bound state in this potential has a wavefunction that is peaked at the spherical shell where $V(r)$ is negative. The second derivative of the wavefunction in the radial direction contributes substantial positive energy in the Schrödinger equation (7.7). This positive *kinetic* contribution cancels much of the 50–100 MeV of negative potential energy, reducing the net binding energy between

---

[4] The factor $1/4\pi\epsilon_0$ is introduced so that the (appropriately scaled) strength g_{strong}^2 of the strong force can be compared with electromagnetism more easily (Problem 14.3).

individual nucleons to a few MeV. The two-nucleon system has a relatively small binding energy per nucleon compared to other nuclei with a few more constituent nucleons, in part since each nucleon is only bound to one other. The deuteron, composed of a proton and a neutron, is bound by only ≈ 2.2 MeV, and the two-neutron and two-proton systems are not bound at all. We investigate the systematics of nuclear binding further in §17.

14.1.4 Weak Interactions

The weak interactions start to make their appearance at roughly the same distance scale as the strong interactions. The weak interactions are characterized by a Yukawa potential energy with a strength not too different from electromagnetism, but with a range that is a factor of about 500 shorter than the range of the strong interactions, so the form of the weak potential is roughly

$$V_{\text{weak}}(r) \approx \frac{g_{\text{weak}}^2}{4\pi\epsilon_0}\frac{e^{-r/b}}{r} \tag{14.6}$$

with $g_{\text{weak}} \sim e$ but $b \approx 2.5 \times 10^{-3}$ fm. Their short range makes the weak interactions far too small to be observed at nuclear distance scales, compared to both strong and electromagnetic interactions. The Yukawa form of the weak nuclear interaction arises from the fact that the fundamental particles ($W^\pm$ and Z vector bosons) mediating this interaction are massive, rather than massless like the photon. The weak interactions would be almost entirely negligible *except for the fact that they facilitate some processes that are forbidden for all the other interactions.* These processes are described in more detail in §14.3.

Are there other forces waiting to be discovered? Perhaps. Answering this question is one of the goals of modern particle physics. The *Large Hadron Collider* (LHC) facility at the European Center for Nuclear Research (CERN) in Geneva, Switzerland is looking for as yet unknown particles and forces that become important at very short distances. Whatever they may find, it will not affect the practical contemporary energy issues we are studying in this book because any new forces that may exist are locked away deep within protons and neutrons, inaccessible to our attempts so far to liberate their energies. We have only sketched the basic properties of the forces that control physics at the subatomic scale. Some of the information about forces accumulated so far is summarized in Table 14.1 along with other facts that appear later in this chapter. Next we turn to the players in the world of subatomic physics – the elementary particles from which the world is built.

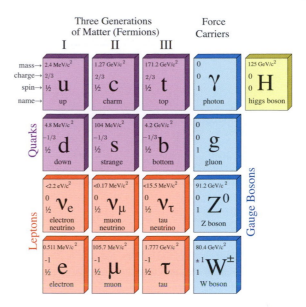

Figure 14.4 A summary of the 17 elementary particles in the Standard Model. Some of the properties of the particles are summarized in Table 14.2. The constituents of matter are shown in the first three columns, the carriers of the forces are shown in the fourth column, and the recently discovered *Higgs boson* appears in the final right-hand column. (Credit: Adapted from HolgerFiedler nach Benutzer: Murphee under CC-BY-SA 3.0 license via Wikimedia Commons)

14.2 Elementary Particles

The elementary particles in the Standard Model come in two types: the constituents of matter and the conveyers of force.[5] The archetype of the first class is the electron, and an example of the second class is the photon. It is a good idea to keep these examples in mind as the names and types of particles proliferate. For reference, Figure 14.4 gives a schematic summary of all the particles in the Standard Model, with more detail in Table 14.2.

The key to understanding the structure of Table 14.2 is the checklist of interactions in the last column. The particles that make up matter can be classified by the forces they feel. Just as electric charge indicates the extent to which a particle interacts electromagnetically, so there are "charges" for each of the other forces. *Quarks* are

---

[5] The recently discovered **Higgs particle** is the lone exception to this rule. It plays a special role in the Standard Model as the breaker of symmetries and the generator of mass. Fascinating though it is to particle physicists, we say little about the Higgs particle here since it plays no direct role in our subject.

Table 14.1 Some properties of the fundamental forces. Although energy is not explicitly listed, all the forces conserve energy. The strengths of the forces are measured relative to the chromodynamic force between quarks.

Force	Mediator	Mediator mass	Relative strength at 1 fm	Conservation laws			
				Baryon number*	Electric charge	Flavor	Parity
Gravity	Graviton	Massless	$\sim 10^{-41}$	✓	✓	✓	✓
Weak	W & Z bosons	$\sim 80\,\mathrm{GeV}/c^2$	$\sim 10^{-8}$	✓	✓	×	×
Electromagnetic	Photon	Massless	$\sim 10^{-2}$	✓	✓	✓	✓
Chromodynamic	Gluons	Massless	1	✓	✓	✓	✓

* Note that while all observed local processes conserve baryon number, subtle *nonperturbative* processes can violate baryon number (§21).

Table 14.2 Properties of the leptons and quarks [69]. Masses are given in units of MeV/c^2 (1 MeV/c$^2 = 1.783 \times 10^{-30}$ kg). Note that each quark can come in three distinct *colors* corresponding to different charges under the chromodynamic force.

Name	Symbol	Mass (MeV/c^2)	Electric charge	Baryon number	Interactions			
					Strong	E&M	Weak	Gravity
Electron	e	0.510 998 92(4)	−1	0	–	✓	✓	✓
Muon	μ	105.658 369(9)	−1	0	–	✓	✓	✓
Tauon	τ	1776.99(29)	−1	0	–	✓	✓	✓
Electron neutrino	ν_e	$0 < m \leq 2 \times 10^{-6}$	0	0	–	–	✓	✓
Mu neutrino	ν_μ	> 0	0	0	–	–	✓	✓
Tau neutrino	ν_τ	> 0	0	0	–	–	✓	✓
Up quark	u	2.3(6)	2/3	1/3	✓	✓	✓	✓
Down quark	d	4.8(4)	−1/3	1/3	✓	✓	✓	✓
Strange quark	s	95(5)	−1/3	1/3	✓	✓	✓	✓
Charm quark	c	$1.275(25) \times 10^3$	2/3	1/3	✓	✓	✓	✓
Bottom quark	b	$4.18(3) \times 10^3$	−1/3	1/3	✓	✓	✓	✓
Top quark	t	$173.2(5) \times 10^5$	2/3	1/3	✓	✓	✓	✓

"charged" with respect to chromodynamic, weak, and electromagnetic interactions. **Leptons** have no chromodynamic charge, so they do not feel the chromodynamic interactions, but all leptons interact by weak forces. Leptons are divided into **charged leptons** and **neutral leptons**, based on whether they are electromagnetically charged or neutral. All particles feel the effects of gravity.

All the particles that constitute observable matter – quarks and leptons – have spin 1/2. In §8, we discussed the division of quantum particles into *fermions* and *bosons*, the distinction being whether only one, or an arbitrary number, of particles can exist in a given quantum state. The fermionic nature of the electron is responsible for the complexity of atomic structure and the richness of

chemistry. All the other leptons and all quarks are also fermions. Indeed, a fundamental theorem of relativistic quantum theory dictates that all particles with half-integral spins are fermions, while all particles with integral spins are bosons. This basic fact is discussed further in §15.5.

The particles that transmit the electromagnetic, chromodynamic, and weak forces are all spin one particles, known as **vector bosons**. The term "vector" refers to the fact that the fields associated with the gluon and the $W^\pm$ and Z are vector fields like the E and B fields associated with the photon. As mentioned earlier, the photon and gluon are massless, while the $W^\pm$ and Z bosons are massive.

The Higgs particle has spin 0, and is known as a **scalar particle** – the Higgs is the first fundamental scalar particle ever observed in nature. (The term *scalar* refers to the fact that the field of which the Higgs particle is an excitation assigns a single number – a scalar – to each point in space-time, unlike a vector field that assigns a vector to each point in space-time.) A consistent quantum theory of gravity would include a further particle associated with a quantum excitation of the gravitational field – a massless **graviton** – which would have to be a spin-2 boson.

An as yet unexplained feature of the spectrum of observed leptons and quarks is that they come in three **generations** that are almost identical except for the particle masses. The *electron* e^-, *electron neutrino* ν_e, and *up* (u) and *down* (d) quarks make up the first generation of fundamental particles. The *muon* is like a heavier version of the electron, and the *charm* (c) and *strange* (s) quarks are heavier versions of the u and d. These, together with the *muon neutrino* (ν_μ), make up the second generation. And the *top* (t) and *bottom* (b) quarks, along with the *tauon* (τ) and τ-*neutrino* (ν_τ), form a similar, but still heavier third generation. The quarks and leptons are grouped into generations in the chart of Figure 14.4.

Clearly, there is an abundance of elementary particles, many of which are difficult to produce, live only briefly before they decay, and have little if any impact on energy production or flow on Earth. Before surveying them all, we strip away the complexity and introduce the **Standard Model "lite"**, which contains the particles responsible for almost all of the phenomena we encounter in this book.

14.2.1 The Standard Model Lite: A $p\,n\,e\,\bar{\nu}_e\,\gamma$ World

For almost all practical purposes, *only five particles* play a role in the world, and one of the five usually hides in the wings, only making an occasional entrance when needed. The five key particles are the proton, neutron, electron, photon, and the (electron anti-) *neutrino*. The electron (e^-) was the first to be identified, having been discovered in the nineteenth century as the **cathode ray** in cathode-ray tubes – the same *CRT*s were later used in television sets and computer monitors. The exact date of discovery of the photon (γ) is debatable. In some sense it was "discovered" by Einstein in 1905, who introduced the concept of a quantum of the electromagnetic field when he used it to explain how light can knock electrons off the surface of metals. The proton (p) and neutron (n) were discovered during the development of nuclear physics in the first third of the twentieth century. The existence of what we now know to

be the electron antineutrino ($\bar{\nu}_e$) was postulated by Austrian physicist Wolfgang Pauli in 1931, though it was not actually discovered until 1956.

In 1930, Dirac pointed out that every particle has an "opposite" or **antiparticle** with the same mass, but opposite charge. So complementing the first three particles in the *standard model lite* are the antiproton ($\bar{p}$), antineutron ($\bar{n}$), and antielectron (e^+, also known as the **positron**). The $\bar{\nu}_e$ is an *antineutrino*, so its opposite is the ν_e, the *neutrino*. The photon is unique in this list in that it is *its own antiparticle*, so there is no separate "antiphoton." Antiprotons and antineutrons play almost no role in our world. They are hard to produce and annihilate quickly into radiation when they encounter any matter. Positrons are emitted in some nuclear decays and have found use in medical diagnostics and cancer treatment. Neutrinos are produced copiously in the Sun (see §22) but pass through our world largely unnoticed. More about positrons and neutrinos can be found in §20 and §22.

The description of the world in terms of the five particles in the *Standard Model lite* is really quite simple:

Protons and neutrons bind together to form nuclei. Nucleon–nucleon forces are complicated, but nuclei are rather simple – at least at the level we encounter them. Nuclei can be described as a quantum version of "liquid drops" in which the protons and neutrons slither about. Note that while protons and neutrons are not truly fundamental particles – they are bound states of quarks and do not appear in Figure 14.4 – only experiments at very high energies can discern the structure within them, so for most practical purposes they can be thought of as primitive constituents of matter.

Electrons are attracted to nuclei by forces that are transmitted by electric and magnetic fields. Together, electrons and nuclei formed from protons and neutrons form atoms; atoms form molecules; and molecules form complex systems like us.

Photons – the quanta of the electromagnetic field – are emitted and absorbed when electrons or protons make transitions from one quantum state to another.

Electromagnetic fields play a dual role: as classical waves that can be broadcast and absorbed, and as quantized photons as already mentioned.

Antineutrinos A neutron can decay into a proton, an electron, and an electron antineutrino. This is a weak interaction known as β-*decay*. Sometimes, when protons and neutrons are bound into nuclei, this reaction is forbidden by energy conservation and instead the opposite reaction

---

The Standard Model "Lite"

Only five particles – the proton, neutron, electron, photon, and the electron antineutrino – are needed to provide a fairly good description of almost all phenomena in our world. Electromagnetism and gravity are the principal macroscopic forces of consequence. Weak and strong interactions operate within nuclei, which are held together by the strong nuclear interaction between nucleons (protons and neutrons).

---

can occur: effectively a proton can decay into a neutron, a positron, and an electron neutrino (ν_e).

Gravity acts on all these particles, even photons which are massless but carry energy.

For a short period of time – roughly the period between 1935 and 1950 – the five particles in the *Standard Model lite* were thought to be the only elementary particles in nature. In fact, this relatively complacent era in fundamental physics was over almost before it began, since the first additional particle, the *muon*, had already been detected in a cloud chamber photograph in 1937. Trying to understand the muon and the plethora of strongly interacting particles related to protons and neutrons that were made in accelerators in the 1950s led eventually to the Standard Model.

The remainder of this section provides a brief introduction to the other particles in the Standard Model.

14.2.2 Leptons

Charged leptons All six known leptons experience the weak, but not the strong interactions. Three of these leptons also carry electromagnetic charge. The electron is the lightest and most familiar of the (electromagnetically) *charged leptons*. We know of two other charged leptons: the **muon** (μ) with a mass of approximately 207 times the mass of the electron, and the **tauon** (τ) with a mass about 3478 times the mass of the electron. As far as we can tell, the μ and the τ are almost identical copies of the electron, except for their masses. They have the same charges, and the same kinds of interactions; they are just heavier. They are short-lived and decay into electrons, neutrinos and, occasionally, other stuff. There is good reason to believe that the electron, muon, and tauon are the only charged leptons.

Neutral leptons Each charged lepton is associated with a *neutral lepton*, called a **neutrino**. For the e there is the

ν_e, for the μ, the ν_μ, and for the τ, the ν_τ. What is meant by "associated with" is explained in more detail in §14.3. Neutrinos feel only the weak force and gravity. They are very light – for many years they were thought to be massless like photons, but late last century it was discovered that they have small but nonzero masses. Their interactions are so weak that they pass through matter almost like it isn't there. To make that statement quantitative: the *mean free path* (the typical distance traveled before an interaction takes place, §20) of a 1 MeV neutrino moving through lead is roughly 1 light-year ($\sim 10^{16}$ m)!

Lepton interactions An important tool in analyzing systems of elementary particles is the existence of additional exactly or approximately *conserved charges* like the total electric charge, which remain unchanged under all, or almost all, physical processes. Each of the three lepton pairs, (e, ν_e), (μ, ν_μ), and (τ, ν_τ) is associated with a discrete *approximately* conserved quantity.[6] This quantity is called **lepton number** (L) generically: **electron number**, **muon number**, and **tauon number** (L_e, L_μ, L_τ) refer to the approximately conserved quantities associated with each of the three lepton types. The charged leptons and their associated neutrinos have lepton number $+1$, their antiparticles have lepton number -1. An example of an interaction that conserves both muon number and electron number is muon decay,

$$\mu \to e\, \bar{\nu}_e\, \nu_\mu. \tag{14.7}$$

The muon ($L_\mu = +1$) decays into a muon neutrino ($L_\mu = +1$), an electron ($L_e = +1$), and an electron antineutrino ($L_e = -1$). Thus the muon number is one on both sides of the equation and the total electron number is zero.

Muons are not particularly exotic. They are made by cosmic rays in the upper atmosphere and travel down well into the ground (on average), before decaying or interacting. The lifetime of the muon is about 2.2 μs, and the flux at sea level is about 200 m^{-2} s^{-1}. So hundreds have passed through you in the time it has taken you to read these words.

---

[6] A subtle phenomenon called *neutrino oscillation* allows one type of neutrino to turn into another. The phenomenon is of great interest to particle physicists, but is extremely difficult to observe and has no impact on the energy processes such as β-decay that we are interested in here. There are also more subtle effects in quantum field theory that can lead to violation of lepton number; these effects are negligible in all circumstances considered in this book.

14.2.3 Quarks and their Interactions

Quarks are massive particles that feel the chromodynamic force. While they are similar in some respects to leptons at a fundamental level, the role they play in nature is quite different. Whether or not a particle feels the chromodynamic force is determined by a "charge" – not electric charge, but similar in that particles with this charge feel the force and particles without it do not. This charge is called *color*, but it has nothing to do with visible color, which is, of course, a characteristic of electromagnetic radiation. Electric charge comes in two types, which we call *positive* and *negative*, with the property that combining positive and negative charges of equal strengths results in a system that is electrically neutral. Similarly, each type of quark comes in three versions which differ by their "color." If three quarks, one of each version, are combined, the result is neutral under chromodynamic forces. Because this resembles the rule for addition of the primary colors of light – when red, green, and blue light are combined, they yield white, or neutral light – physicists chose, somewhat whimsically, to call the chromodynamic charge **color**. The proton and the neutron are both examples of color-neutral three-quark systems. Another successful feature of the analogy is that combining a quark and antiquark also can yield a color-neutral system, in much the same way that combining a visual color with its complement gives color-neutral, i.e. white, light.

The chromodynamic forces are very strong. Furthermore, as discussed further below, gluons themselves carry color charge (unlike photons, which are electrically neutral). These effects combine in such a way that quarks can never escape from the color-neutral bound states that they form. Thus, individual quarks are never seen. Color-neutral states of three quarks are called **baryons** and bound color-neutral states of a quark and an antiquark are called **mesons**. Together, these color-neutral composite particles are known as **hadrons**.[7] Protons and neutrons are the lightest baryons. Almost all the known particles that feel the strong force fit into these two categories: baryons (or antibaryons, like the antiproton) or mesons. The exception being a handful of recently discovered and/or very short-lived states that appear to be essentially composed of two quarks and two antiquarks, or four quarks and one antiquark.

Quark Properties

Baryon number There are literally hundreds of different types of baryons, but all are unstable except for protons (and neutrons when they are bound within certain nuclei).

Baryon Number (Atomic Mass)

An atomic nucleus can be labeled by the number of nucleons it contains. That number is called the *baryon number* or *atomic mass*, and is usually denoted by A. A nucleus of atomic mass A contains $3A$ quarks. Baryon number is conserved in all observed processes.

Eventually, if you follow its life, any baryon decays into a proton or neutron plus a lot of other stuff that may include mesons, leptons, antileptons, and photons. Protons, however, do not seem to decay – ever. Searches for proton decay have come up empty, and the proton lifetime is now known to be greater than 2×10^{29} yr [69]. To account for this, physicists endow baryons with a type of "charge" called **baryon number**, which is +1 for baryons and −1 for antibaryons. So, for example, the proton and neutron have baryon number +1 and the antiproton and antineutron have baryon number −1. The stability of the proton can then be expressed as **conservation of baryon number**.[8] A baryon can decay, but among its decay products one must always find another baryon, and if you wait long enough only protons and/or neutrons will be left. Mesons, leptons, photons, and the rest have baryon number zero. A proton and an antiproton can annihilate into a cloud of mesons and photons because this conserves baryon number: $1 + (-1) = 0$.

The concept of baryon number can be attributed at a more fundamental level to quarks. Since there are three quarks in a baryon, each quark is assigned baryon number 1/3. While quark number conservation implies conservation of baryon number, the name "baryon number" was invented long before quarks were discovered, and the name stuck. Nuclei are bound states of protons and neutrons, so they can be labeled by their baryon number. Carbon, for example, with six protons and six neutrons, has baryon number 12. The baryon number of a nucleus is thus the same thing as its *atomic mass number* in the language of chemistry. We use the terms interchangeably.

---

7 All these names are from the Greek: *hadron* = stout; *baryon* = heavy; *meson* (originally *mesotron*) = intermediate; *lepton* = slight.

8 Some speculative ideas for physics beyond the Standard Model, such as *Grand Unified Theories*, predict a very small probability for proton decay. These ideas have not been confirmed experimentally. Also, as for leptons, subtle processes can give rise to very small violations of baryon number; while negligible for the purposes of energy physics, these processes may have been involved in the early universe in producing the presently observed asymmetry between matter and antimatter (§21).

Quark charges and colors, gluons Aside from the fact that they are never seen in isolation, perhaps at first sight the most bizarre property of quarks is that their electric charges are either $+\frac{2}{3}e$ or $-\frac{1}{3}e$, where $-e$ is the electron's charge. When quarks were originally proposed in 1964 by American physicists Murray Gell-Mann and George Zweig, the idea that a particle could have *fractional electric charge* was considered strange if not silly. Now there is plenty of evidence for these charge assignments, and the only oddity is that fractional charges are not observed directly because the allowed, color neutral, combinations of quarks and antiquarks always have charges that are integer multiples of e.

Given that they are never seen in isolation, how can we be certain that quarks have fractional charge, or indeed, that quarks exist at all? The reason is that when electrons or other particles that do not feel the strong force are scattered from protons or other nuclei, the pattern of scattering depends on the nature and distribution of the constituents of the baryon(s). In the late 1960s American physicists Jerome Friedman, Henry Kendall, and Richard Taylor scattered electrons from protons and observed scattering patterns that could only be accounted for if protons were made of pointlike spin-1/2 particles with electric charges that are third integral fractions of the electron's charge. Physicists now agree that quarks are just as "real" as electrons, even though the nature of chromodynamic force sequesters them permanently inside protons and neutrons.

The gluons that mediate the chromodynamic force couple to color charges in much the same way that a photon couples to electric charge, but in addition, gluons themselves are colored. Two beams of light pass through one another because photons do not scatter significantly from one another.[9] Gluons not only do not pass through one another, they interact with one another strongly by emitting and absorbing more gluons, and those interact by emitting and absorbing more gluons, *ad infinitum*. This is why chromodynamic forces are so complicated.

Flavor On top of all the previous complications, quarks come in six distinct types, more or less in parallel to the six types of leptons (remember (e, ν_e), (μ, ν_μ), and (τ, ν_τ)). The six types are called **up** (u), **down** (d), **charm** (c), **strange** (s), **top** (t), and **bottom** (b). They differ in electric charge: three (u, c, t) have charge $+\frac{2}{3}e$ and three (d, s, b) have charge $-\frac{1}{3}e$. The properties of the quarks, to the

extent they are known, are listed in Table 14.2. Note that they all have electric charge, so they feel electromagnetic forces. They all feel the weak force as well.

These six different types of quarks are known as different **flavors** of quarks. The name "flavor" is even more fanciful than "color." It has no content except to allow physicists to discuss an abstract concept in simple words. Note that the six flavors of quarks can be divided into the three *generations* of matter as discussed above, with up and down quarks in the first generation, charm and strange in the second generation, and top and bottom in the third generation, with each generation progressively more massive than the previous.

The chromodynamic interactions of quarks conserve flavor. The number of u quarks (minus the number of u antiquarks), for example, does not change in a chromodynamic interaction. In contrast, the weak interactions allow quarks to change their flavors. Thus, for example, a weak interaction can enable a d quark to emit an electron and an electron antineutrino and turn into a u quark – a transformation not otherwise possible. These transformations are important in nuclear energy processes, and are described in §14.3.

The quark structure of protons and neutrons

Protons and neutrons are the lightest baryons. The **proton** is a bound state of two u quarks and one d quark. The sum of the electric charges is $2 \times \frac{2}{3} - \frac{1}{3} = 1$. The **neutron** is a bound state of two d quarks and one u quark, with electric charge $\frac{2}{3} - 2 \times \frac{1}{3} = 0$. This structure is illustrated in Figure 14.5. The proton is lighter than the neutron because the u quark is slightly lighter than the d quark. The masses of the u and d quarks are tiny, which at first sight may seem very strange: how can objects of mass $M \approx 940\,\text{MeV}/c^2$ be made of only three particles with masses of $\sim$ 2–5 MeV/c^2 each? Where does all the mass come from? The answer lies in quantum mechanics and the concept of zero-point energy, which was briefly mentioned in §9.2.1. Recall that in each of the potentials that was studied (square-well and harmonic oscillator) in §7, the lowest energy state has energy greater than zero. The smaller the region to which a particle is confined, the greater its energy. For a region as small as 1 fm, the zero-point energy of a massless particle is $\sim$300 MeV (Problem 14.5). The proton contains three such particles, each with a zero-point energy of about 300 MeV, so the energy of a proton at rest is $\sim$900 MeV; by Einstein's $E = mc^2$, that energy contributes 900 MeV/c^2 to its mass.

The visible matter in the universe consists of nuclei and electrons. The nuclei in turn consist of protons and neutrons. The mass of an electron is $\sim$1/2000th that of a

---

[9] There is a tiny quantum effect called *light-by-light scattering*, where a photon briefly materializes as an electron-positron pair, which then interacts with the other photon. This effect is so small that it has not been observed directly.

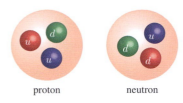

proton   neutron

Figure 14.5 A cartoon of a proton and neutron showing the dominant quarks: *uud* in the proton and *ddu* in the neutron. The quarks are colored to remind us that all three colors must be present to make a color-neutral hadron. In actuality, protons and neutrons are described by quantum states that include fluctuating gluon fields and quark–antiquark pairs, though this complexity does not affect the nuclear reactions relevant to nuclear energy processes.

Mass in the Visible Universe

Almost all of the visible mass in the universe is found in the rest masses of protons (hydrogen nuclei), neutrons, and the nuclei they compose. This mass, in turn, is a manifestation of the quantum zero-point energy of almost-massless quarks confined to the interior of nucleons.

---

proton or neutron. So we conclude that most of the mass of the visible universe is due to the quantum zero-point energy of bound quarks!

Although protons and neutrons have excited states, the energies emitted and absorbed in the nuclear reactions of interest to us here are not great enough to excite the quarks in protons or neutrons. Protons and neutrons can therefore, to a good approximation, be regarded as persistent, unstructured objects whose identity does not change inside a nucleus. The internal quark substructure of protons and neutrons is manifested in the complicated structure of the strong nuclear interactions.

14.2.4 Force Carriers

As summarized earlier, the chromodynamic, weak, and electromagnetic forces each are mediated by spin one *vector bosons*. Each of the vector bosons has its own features and character. The *photon* is most visible because it is massless and carries no charge, so that it propagates freely through space mediating the long-range electromagnetic force. The gluon, while also massless, is itself *charged* under the chromodynamic interactions, so that it cannot propagate freely. Even in the absence of quarks or leptons, gluons can in principle bind together into color-free massive states known as *glueballs*. Up to now, however,

no glueball has been convincingly identified experimentally. The Z^0 and $W^{\pm}$ vector bosons that mediate the weak interactions are massive, and thus weak interactions are intrinsically short-range forces.

14.2.5 The Higgs Boson

The final particle in the standard model is the *Higgs boson*, or *Higgs particle*. Originally predicted in 1964, this particle was found experimentally in 2013. Though this particle does not play a direct role in any energy-related processes, it is a central component of the Standard Model. It is responsible for *electroweak symmetry breaking*, the mechanism that gives rise to masses for the $W^{\pm}$ and Z bosons and differentiates the weak interactions from the closely related electromagnetic interaction. The Higgs particle is also associated with the phenomenon that gives rise to quark and lepton masses.

14.3 The Weak Interactions and β-decay

We have alluded several times to the crucial role that the weak interactions play in the universe. Their fundamental practical importance is that they allow protons to turn into neutrons and *vice versa*. In this section we take a more careful look at this and related processes.

14.3.1 Discovery of β-decay

In the early twentieth century physicists observed that some nuclei emit electrons and change their chemical identity. A simple, illustrative, and important example is the decay of a rare form of potassium into calcium or argon. Potassium is an atom with 19 electrons bound to a nucleus with 19 protons and usually 20 neutrons. This nucleus is stable. About one in ten thousand potassium nuclei contains 21 neutrons rather than 20, giving the nucleus a baryon number (atomic mass number) of 40. An atom with this nucleus at its core has all the chemical properties of ordinary potassium, but it weighs about 2.5% more due to the extra neutron. What is special about potassium-40 (^{40}K in standard notation, §17.1.2) is that it is unstable. It decays with a lifetime of greater than a billion years.

About 88% of the time that the nucleus of a ^{40}K atom decays, it emits an electron, thereby increasing its nuclear charge by one unit. This process turns it into a form of calcium, ^{40}Ca. At first, physicists did not know what particle was emitted in this process. They called the emitted particle a β-ray and they called the decay a **beta decay**. Soon it became clear that the emitted particle was an ordinary electron. But a new puzzle arose: the electron did not always come out with the same energy. According to quantum mechanics, the decay should have taken one quantum state of fixed energy to another, and the electron's energy should have been determined by the difference of the energies of the two states. Even if the transition left the calcium nucleus in one of several excited energy basis states, the energy of the electron would have been limited to one of a few definite values. Instead the electrons' energies measured in many ^{40}K decays were found to vary continuously between zero and a maximum that equaled the difference of nuclear energies.[10] It seemed that energy was leaking away. Further experiments showed that momentum and angular momentum were not being conserved in β-decay either. In 1931, Wolfgang Pauli speculated that an *unobserved* neutral particle was being emitted *with the electron*, and it was carrying off the missing energy, momentum, and angular momentum. To have been overlooked, the neutral particle would have to have had *very weak interactions with ordinary matter*. Fermi named it "a little neutral one" or *neutrino* in Italian. This was the first time (but not the last) that a physicist predicted the existence of a new elementary particle. Further experiments confirmed that Pauli was right and that the reaction that was taking place is

$$^{40}\text{K} \rightarrow {}^{40}\text{Ca} + e^- + \bar{\nu}_e. \tag{14.8}$$

Many nuclei undergo similar decays.

14.3.2 β-decay in the Standard Model

When physicists of the mid twentieth century figured out that the nucleus is a bound state of protons and neutrons, they developed a more universal description of what is happening in ^{40}K and other β-decays. In such decays, a neutron is decaying into a proton,

$$n \rightarrow p + e^- + \bar{\nu}_e. \tag{14.9}$$

The proton remains bound by the strong nuclear force to the nucleus, but the electron and the neutrino are emitted. And when physicists of the late twentieth century understood that the neutron and proton are bound states of quarks, they realized that, at a deeper level still, a d quark is decaying into a u quark by a weak interaction,

$$d \rightarrow u + e^- + \bar{\nu}_e. \tag{14.10}$$

Since quarks are confined to the interiors of the proton and neutron, the observed reaction is

$$n(udd) \rightarrow p(uud) + e^- + \bar{\nu}_e. \tag{14.11}$$

---

[10] Actually the experiments were performed on different, more abundant nuclei with shorter lifetimes. We chose ^{40}K because it both β-decays *and electron captures* (see below) and it plays a role in geothermal energy (see §32).

Weak Interactions Violate Flavor Conservation

The strong interactions appear to conserve flavor absolutely, but weak interactions allow quarks of one flavor to turn into quarks of any other flavor, as long as the quark electric charge changes by ± 1 and the overall reaction conserves baryon number, electric charge, and lepton number. The transformations between u and d quarks lead to β-decay and related processes in nuclei,

$$d \rightarrow u + e^- + \bar{\nu}_e,$$
$$u \rightarrow d + e^+ + \nu_e,$$
$$u + e^- \rightarrow d + \nu_e.$$

The Standard Model gives quite a simple description of what is happening at the quark level. The modifications due to the details of nuclear structure, however, are very important in determining the lifetime of any particular nucleus that can undergo β-decay. Some nuclei that can β-decay have lifetimes that are longer than the age of the universe. The free neutron β-decays to the proton with a lifetime of about 10 minutes, and there are many nuclei with β-decay lifetimes of less than a microsecond. We do not attempt to develop the theory of β-decay here in any detail. Lifetimes for β-decay processes in which we are interested can be looked up online [70].

The fact that quark flavor is not conserved ($d \rightarrow u$) in β-decay confirms that the weak interactions are at work. There are many different types of weak interactions, but the only ones important for our purposes are the one given in eq. (14.10) and a few variations mentioned below. These reactions always conserve electric charge, baryon number, and lepton number as well as the other usual conserved quantities – energy, momentum, and angular momentum – but quark flavor is not conserved. The fact that weak interactions enable reactions that would otherwise be forbidden is the reason that we care about them. When they give small corrections to otherwise allowed processes, as they do under most circumstances, they are usually too small to be noticed.

14.3.3 Processes Related to β-decay

^{40}K β-decays to ^{40}Ca 88% of the time. The other 12% of the time it undergoes a related process that turns it into ^{40}Ar. In this case the nuclear charge must decrease by one unit (argon has 18 protons and 18 electrons). From what

Example 14.1   What Force Was Responsible For That?

Interactions among particles must satisfy the conservation laws that are respected by the type of interaction that is at work. This fact – together with the hierarchy that the strong nuclear interaction is much stronger than electromagnetism, which in turn is much stronger than the weak interactions – makes it possible to identify what force is at work in important processes. In general, the rate at which a quantum process occurs depends on the strength of the coupling, so most strong nuclear interaction processes occur much more quickly than those mediated by weak interactions. (Note that gravity is too weak to drive any measureable reaction among particles or nuclei.)

Some important examples:

- When neutrinos or antineutrinos appear, the weak interactions must be at work, because neutrinos do not couple to the strong or electromagnetic interactions.
- When photons appear, electromagnetism must be at work because photons only interact electromagnetically.
- If protons change into neutrons or *vice versa*, the weak interactions must be at work.

Consider, for example, hydrogen fusion into helium, which is known to be the Sun's primary energy source. Four protons fuse to form one helium nucleus, known also as an α-**particle**, which consists of two protons and two neutrons, as well as two positrons (needed to conserve electric charge). A first guess for the reaction might be

$$4p \rightarrow \alpha + 2e^+ \quad (?)$$

This reaction conserves baryon number and electric charge. Naively it looks like a strong nuclear interaction, which should go very quickly. One wonders why the Sun has lived so long. Closer inspection, however, reveals that two protons have turned into neutrons and that electron number is not conserved. These features indicate both that a weak interaction is involved, and that two neutrinos must be produced. A corrected version of the reaction reads

$$4p \rightarrow \alpha + 2e^+ + 2\nu_e .$$

The reaction can only proceed as rapidly as the weakest force involved permits, so despite the fact that protons and neutrons are involved, the time scale for hydrogen burning in the Sun is set by the weak interactions, which is a good thing for life on Earth.

---

we have learned about the weak interactions of quarks, we might guess that the reaction could be

$$^{40}\text{K} \rightarrow {}^{40}\text{Ar} + e^+ + \nu_e \quad (?) \tag{14.12}$$

where at the microscopic level $u \rightarrow d + e^+ + \nu_e$. This is a good guess, which satisfies all the rules like lepton number and baryon number conservation. This kind of reaction is called $\beta^{\pm}$ *decay* or *positron emission*, but it is *not* in fact how ^{40}K decays to ^{40}Ar. A quick check of the masses of the ^{40}K and ^{40}Ar nuclei reveals that there is not enough energy released in the nuclear reaction to create the rest mass of the positron, 511 keV/c^2. (We explain how to do calculations like this in §17.) Instead, the ^{40}K nucleus *captures* an electron from the surrounding cloud to produce the reaction

$$^{40}\text{K} + e^- \rightarrow {}^{40}\text{Ar} + \nu_e , \tag{14.13}$$

which satisfies all the rules of the weak interactions. This process is called *electron capture* or *K-capture* ("K"

Important β-decay Processes

The summary of interesting β-decay and related processes for nucleons is

β^- decay	$n \rightarrow p + e^- + \bar{\nu}_e,$
β^+ decay	$p \rightarrow n + e^+ + \nu_e,$
electron capture	$p + e^- \rightarrow n + \nu_e.$

The processes are described here in terms of protons and neutrons, but the fundamental processes involve quarks. In applications to nuclear physics, the protons and neutrons are usually bound in nuclei.

because the electron is captured from the innermost atomic shell, known to chemists as the K-shell). This reaction can proceed even if the rest mass of ^{40}Ar is *greater* than that of ^{40}K, as long as the sum of the ^{40}K rest mass plus the electron's rest mass exceeds the rest mass of ^{40}Ar. Many nuclei decay by β^+-decay and/or electron capture, with lifetimes that vary from very short to longer than the age of the universe.

Discussion/Investigation Questions

14.1 Physicists searching to see if the proton decays have looked for the reaction $p \rightarrow e^+\gamma$. Which of the conservation laws explored in this chapter does this decay violate?

14.2 Gravity is a long-range force proportional to the masses of the interacting bodies. Physicists have long searched to see if there might be a long-range force that is proportional to baryon number. Think about the composition of atoms and propose a "thought experiment" that could detect such a force.

14.3 Two possible nuclear fusion reactions are $d + d \rightarrow {}^4\text{He} + \gamma$ and $d + d \rightarrow {}^3\text{He} + p$, where d is deuterium, an atom with a nucleus composed of a proton bound to a neutron. ^{3}He and ^{4}He are helium atoms with nuclei consisting of two protons and one neutron (^{3}He) or two protons and two neutrons (^{4}He). Both of these reactions satisfy energy conservation. Which do you think occurs more quickly and why? (This question and others like it arise when considering fuel for nuclear fusion on Earth.)

14.4 A *baryon* frequently encountered in particle physics experiments is known as the Λ ("*Lambda*"). Its quark content is uds. Sometimes the Λ decays by the process $\Lambda \rightarrow n\gamma$, where n is a neutron (ddu) and γ is a photon. What interactions must be active in this decay?

14.5 The tauon τ decays often to a muon μ, which in turn decays to an electron. How many neutrinos, and of which type are produced in this sequence of weak decays.

Problems

14.1 A rocket fuel should have high energy density (per kilogram) because the weight of the propellant is an important consideration. A favorite fuel is liquid hydrogen plus liquid oxygen. What is the energy density (J/kg) of a stoichiometric mixture of H_2(l) and O_2(l) (assuming they combine to form water in the gaseous state)? The strong nuclear force, in principle, allows the same stoichiometric mixture of two hydrogen atoms and one oxygen atom to fuse to form an *isotope* of

neon, ^{18}Ne, giving off 4.52 MeV in the process. If this reaction were the source of power for the rocket, what would be the energy density of the fuel?

14.2 In principle, and in the absence of electromagnetic or any other non-gravitational interactions, an electron and a proton should form a "gravitational atom," bound by the force of gravity. We know that the energy of the ground state of the ordinary hydrogen atom is given by eq. (14.4). Show that the ground state energy of gravitationally bound "hydrogen" is given by $E_{\text{grav}} = -\frac{1}{2}G_N^2 m_p^2 m_e^3 / \hbar^2$ and evaluate this numerically.

14.3 Show that the combination of constants $g_{\text{strong}}^2 / (4\pi\epsilon_0 \hbar c)$, where g_{strong} is defined in eq. (14.5), is dimensionless. It is taken as a dimensionless measure of the strength of an interaction. Compute this quantity for electromagnetism ($g \rightarrow e$), where it is known as the *fine structure constant* and denoted by α.

14.4 Which of the following interactions are consistent with the conservation laws for baryon number, electric charge, and the three lepton numbers? Don't worry about energy conservation. (A bar denotes an antiparticle. See Question 14.4 for information about the Λ particle.)

$$p \rightarrow n + e + \nu_e \qquad\qquad p \rightarrow n + \bar{e} + \bar{\nu}_e$$
$$\tau \rightarrow \mu + \bar{\nu}_\mu + \nu_\tau \qquad\qquad \bar{\tau} \rightarrow \mu + \bar{\nu}_\mu + \nu_\tau$$
$$\nu_\mu + p \rightarrow \bar{\mu} + n \qquad\qquad \nu_\mu + n \rightarrow \mu + p$$
$$\Lambda \rightarrow p + e + \bar{\nu}_e \qquad\qquad \Lambda \rightarrow n + \nu_e + \bar{\nu}_e$$

14.5 Einstein's energy–momentum relation for a particle of mass m with momentum $\boldsymbol{p}$ is $E = \sqrt{p^2 c^2 + m^2 c^4}$. Estimate the uncertainty in the x, y, and z components of the momentum of a quark confined in a (cubic) box of side length $L = 1$ fm using the Heisenberg uncertainty principle $\Delta p_{x,y,z} \gtrsim \hbar/L$ (Box 7.6). Approximate $\langle p^2 \rangle$ by $(\Delta \boldsymbol{p})^2$, then show that for a quark of mass 5 MeV/c^2, $E \approx |\boldsymbol{p}|c$ and that the zero-point energy of three quarks is of order 900 MeV, the rest energy of a proton.

14.6 To get a sense of the potential practical relevance for energy purposes of one aspect of as-yet-undiscovered physics, consider proton decay. Making the most general assumptions about the nature of its decay, the lifetime of the proton is currently bounded by $\tau_{\text{proton}} > 2.1 \times 10^{29}$ y. Assume that the lifetime is in fact 3×10^{29} y, and that all of the proton's rest energy emitted in the decay can be captured and put to use. Estimate the rate of useful power output from proton decay of a cubic kilometer of hydrogen gas at NTP.

Quantum Phenomena in Energy Systems

Quantum physics predicts many phenomena that violate the laws of classical physics and expectations based on our (classically developed) "intuition." In this chapter we examine more closely three somewhat counter-intuitive consequences of quantum mechanics that are involved in energy processes: first, the fact that events – in particular transitions from one state to another – can only be predicted probabilistically; second, the ability of particles to *tunnel* through barriers in a way that the laws of classical physics would not permit; and finally, the *Pauli exclusion principle*, which forbids multiple electrons or other identical fermions from occupying any given quantum state. We explore these processes because they enter in fundamental ways into the mechanisms involved in nuclear, solar, and geothermal energy. Complex nuclei such as uranium would not exist or decay, the Sun would not shine as it does, and photovoltaic cells would not function without the help of these peculiar aspects of quantum mechanics.

In Box 7.3 we used the radioactive decay of the nucleus plutonium-238 (^{238}Pu) to illustrate the probabilistic nature of events in quantum mechanics. Radioactive decays are only one example of the general phenomenon of *quantum transitions*, where a particle or system emits a packet of energy – a photon, a β-decay electron and antineutrino, or something else – as it changes from one quantum state to another. Atoms of gas in a fluorescent light bulb, for example, are excited by collisions with electrons, and then emit light as they transition to less excited states. The statistical description of quantum transitions is introduced in §15.1 and in Box 15.1.

Tunneling – the ability of quantum mechanical particles to pass through walls that the laws of classical physics would forbid them to penetrate – is among the more exotic consequences of quantum mechanics. The ability of quantum particles to get out of local traps in a potential well is essential to many aspects of energy physics.

Reader's Guide
The goal of this chapter is to describe in some detail three features of quantum physics that arise in energy processes throughout the universe and here on Earth: quantum transitions, quantum tunneling, and the Paul exclusion principle.
    §7 (Quantum mechanics) is an essential prerequisite for this chapter. Readers may find it useful to refer back to the middle sections of that chapter at this point.
    The ideas and results developed in this chapter on both transitions (particularly decays) and tunneling are used throughout our treatment of nuclear energy and solar energy in the next ten chapters.

As described in Box 7.2, nuclear fusion can take place in the Sun's core because two protons can tunnel through the electrostatic potential barrier between them into a region of strong nuclear attraction. The temperature required to make fusion take place in related reactions on Earth is similarly lowered by many orders of magnitude by tunneling. The emission of an α-particle in a nuclear decay process can be understood in terms of tunneling. Electrons propagate through a semiconductor by tunneling from one atom to another. The middle sections of this chapter – §15.2, 15.3, and 15.4 – are devoted to an exploration of the phenomenon of tunneling.

The *Pauli exclusion principle*, the last topic in this chapter, requires that only one electron or proton (or any other spin-1/2 particle) can occupy a given quantum state. This principle does not arise in nonrelativistic quantum mechanics. Quantum mechanics, as described in §7, treats every particle as a separate entity. When quantum mechanics is combined with Einstein's theory of special relativity (§21.1.4) into the framework of *relativistic quantum field theory*, however, a new, powerful concept of "identical particles" emerges. In quantum field theory all elementary particles are quanta of fields in the same way that all photons are quanta of the electromagnetic field. This underlying structure places constraints on the states of many-particle systems. For electrons and other spin-1/2 particles, the wavefunction describing a many-particle state must change sign when any two of the particles are interchanged. This, in turn, leads to a restriction that no two spin-1/2 particles can be in the same energy basis state. The *exclusion principle* forces electrons in atoms, as well as protons and neutrons in nuclei, into ever-higher energy states, giving rise to the richness of atomic and nuclear structure. We briefly summarize the origins and consequences of the Pauli exclusion principle in §15.5.

15.1 Decays and Other Time-dependent Quantum Processes

Quantum mechanics makes probabilistic predictions for measurements. For example, $dP = |\psi(x)|^2 dx$ gives the *probability* dP that a particle will be found in the interval dx in terms of the square of the wavefunction at the point x. Quantum mechanics does not say whether a measurement will or will not reveal a particle at the point x, only how likely it is to be there. The same quantum logic applies to *transitions* such as the α-decay of the radioactive nucleus ^{238}Pu described in §7.5. A system observed to be in state $|a\rangle$ (the ^{238}Pu nucleus) at some initial time may later be found to be in a different state $|b\rangle$ (the daughter nucleus plus the decay α-particle). We refer to events like this, in which a quantum system emits a packet of energy as it changes from one state to another, as *quantum transitions*. In §9 we encountered *molecular transitions* from one vibrational state to another, which are associated with emission or absorption of infrared radiation, and *electronic transitions* in atoms, in which X-rays are emitted or absorbed. The term **decay** is used to describe quantum transitions in which an isolated, unstable system spontaneously transitions to a state of lower energy by emitting particles and/or electromagnetic radiation.

Transitions in a collection of many identical microscopic objects (e.g. atoms) are found to occur randomly but with a predictable probability per unit time, $dP/dt = \lambda$. For a sample of N objects, the number of decays dN in a small time interval dt is proportional to the number $N(t)$ present at that time and to the decay probability per unit time dP/dt. So, on average, the number in the original state decreases according to

$$dN = -\left(\frac{dP}{dt}\right) N dt = -\lambda N \, dt \,. \qquad (15.1)$$

If we integrate eq. (15.1) we get an exponential decay law,

$$N(t) = N(0)e^{-\lambda t} \,. \qquad (15.2)$$

Although the mean number of quantum decays per unit time is given by λN, the actual number of decays varies randomly from time interval to time interval, fluctuating about the mean. The fluctuations themselves can be predicted, but again, only statistically. The pattern that arises from a constant decay probability per unit time is known as a *Poisson distribution*, and is described in Box 15.1. As derived there, the probability of observing exactly k decays in a time interval of length T is given by

$$P_k = \frac{(N\lambda T)^k}{k!} e^{-N\lambda T} \,. \qquad (15.3)$$

It is possible to show that eq. (15.3) predicts that the average number of events in time T is $\lambda N T$. Some data that are distributed according to the Poisson distribution are shown in the figure in the box.

The constant λ has units of $[\text{time}]^{-1}$ and is often replaced by the **lifetime**, $\tau = 1/\lambda$. τ is the time in which the number in a sample decreases to $e^{-1} \cong 0.3679$ of its original value. Often, physicists use a related measure, the **half-life**, denoted $\tau_{1/2}$, which is the time necessary for half of an initial sample to decay:

$$N(\tau_{1/2}) = \frac{1}{2}N(0) = N(0)e^{-\tau/\tau_{1/2}}$$
$$\Rightarrow \tau_{1/2} = (\ln 2)\tau \cong 0.6932\,\tau \,. \qquad (15.4)$$

The study of transitions in quantum mechanics is quite subtle. If $|a\rangle$ were a state of definite energy, it would evolve in time simply by multiplying by a phase (see Axiom 4, eq. (7.23)). So we must conclude that when a decay occurs neither the original state $|a\rangle$ nor the state $|b\rangle$ containing the decay products is an energy basis state. Instead, the energy basis states are linear combinations that include these two states, and the decay probability arises from the different time dependence of the different energy basis states. An estimate of the uncertainty in the energy of state $|a\rangle$ is given by $\Delta E \sim \hbar/\tau$. This **energy/time uncertainty relation** is similar to the position/momentum uncertainty relation described in Box 7.6.

15.2 The Origins of Tunneling

The basic quantum physics responsible for tunneling can be illustrated in the context of the simple one-dimensional harmonic oscillator, which we analyzed classically in Box 2.2 and quantum mechanically in §7.8.1.

Box 15.1   The Poisson Distribution

Although the decay probability per unit time is a constant, quantum systems do not decay at a steady continuous rate. Instead the number of decays in any time interval T is subject to statistical fluctuations, according to a *Poisson distribution*.

Consider a large sample of N systems, each with decay probability per unit time $dP/dt = \lambda$. Divide up the time interval T into many equal subintervals, $\Delta t = T/n$, taking n so large that the likelihood of two decays in the interval Δt (which goes as $1/n^2$) can be neglected. Then the probability of a single decay occurring in the interval Δt is $\Delta P = N\lambda \Delta t = N\lambda T/n \equiv \mu/n$. The probability that *no decays* occur in the interval Δt is $(1 - \Delta P) \cong (1 - \mu/n)$. For no decays to occur in time T, we require that there is no decay in any of the n sub-intervals, $P_0(n) = (1 - \mu/n)^n$. Here we have used the elementary fact from probability theory that the probability of two uncorrelated things happening (or *not* happening, in this case) is the product of the individual probabilities. Taking $n \to \infty$ we obtain the probability for no decays in time T,

$$P_0 = \lim_{n \to \infty} P_0(n) = \lim_{n \to \infty} (1 - \mu/n)^n = e^{-\mu} = e^{-N\lambda T}.$$

What is the probability of exactly one decay? It could occur in any of the n small intervals, so $P_1(n) = n \Delta P (1 - \Delta P)^{n-1} = \mu(1 - \mu/n)^{n-1}$. As $n \to \infty$, the -1 in the exponential can be ignored and we find $P_1 = \mu e^{-\mu}$. For two decays we require events in precisely two Δt-intervals. (Remember that Δt is so small that the possibility of two decays in any one interval is negligible.) There are $\binom{n}{2} = n(n-1)/2$ ways of making two distinct unordered choices from n possible intervals, so $P_2(n) = (n(n-1)/2)\Delta P^2 (1 - \Delta P)^{n-2}$. Taking $n \to \infty$ we find $P_2 = \mu^2 e^{-\mu}/2$. Generalizing this analysis to any fixed larger value of k, the probability of exactly k decays in time T is

$$P_k = \frac{\mu^k}{k!}e^{-\mu} = \frac{(N\lambda T)^k}{k!}e^{-N\lambda T}.$$

This is known as the **Poisson distribution**. Note that the sum over all possibilities in any Poisson distribution is one because $\sum_{k=0}^{\infty} x^k/k! = e^x$.

The figure below illustrates the fluctuations generated by a Poisson distribution. The histograms show how many times k decays were observed in 1000 observations over a time interval $T = \mu/N\lambda$, in which μ decays were expected. Data are shown for $\mu = 5$, 10, and 20.

The mean number of decays $\langle \mathcal{N} \rangle$ in time T, averaged over a large number of experiments on a sample of size N, is

$$\langle \mathcal{N} \rangle = \sum_{k=0}^{\infty} k P_k = \mu = N\lambda T,$$

as expected. A measure of the fluctuations in the number of decays measured in different experiments is given by the *root-mean-squared deviation* from the average,

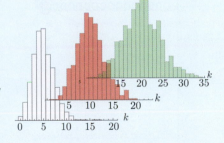

$$\sigma = \left[\langle (\mathcal{N} - \langle \mathcal{N} \rangle)^2 \rangle \right]^{1/2} = \left[\sum_{k=0}^{\infty} k^2 P_k - \mu^2 \right]^{1/2} = \sqrt{\mu} = \sqrt{N\lambda T}.$$

Thus, the fluctuations increase as the rate of decay events grows, but the ratio of the amount of fluctuation to the average decreases, $\sigma/\langle \mathcal{N} \rangle = 1/\sqrt{N\lambda T}$.

Recall that a *classical* harmonic oscillator can be described by a mass m attached to a spring with spring constant $k = m\omega^2$. As explained in Box 2.2 and illustrated in Figure 15.1, the classical particle oscillates sinusoidally about its equilibrium position with displacement $x(t) = \ell \sin(\omega t - \phi)$. Conservation of energy,

$$E = \frac{1}{2}m\dot{x}^2 + \frac{1}{2}m\omega^2 x^2, \tag{15.5}$$

relates the maximum possible displacement ℓ (where $\dot{x} = 0$) to the total energy E,

$$\ell = \sqrt{\frac{2E}{m\omega^2}}. \tag{15.6}$$

We refer to $x = \pm \ell$ as the **turning points** of the classical motion. These are the points where the particle's velocity changes sign and its kinetic energy vanishes.

Example 15.1   Computing a Half-life

A student measures 12 000 alpha-decays per second coming from a gram of *uranium-238* (^{238}U).
What is the half-life of ^{238}U?

One gram of ^{238}U contains $N_A/238 \cong 2.53 \times 10^{21}$ ^{238}U nuclei. According to eq. (15.1), $\tau = |N/(dN/dt)| \cong 2.53 \times 10^{21}/(1.2 \times 10^4 \text{ s}^{-1}) \cong 2.1 \times 10^{17}$ s. The half-life is related to τ by eq. (15.4), so $\tau_{1/2} \cong 0.69 \times (2.1 \times 10^{27} \text{ s}) \cong 1.5 \times 10^{17}$ s $\cong 1.5 \times 10^{17}$ s$/(3.16 \times 10^7$ s/y$) \cong 4.6 \times 10^9$ y. The standard value is 4.5 billion years, so the student's data is pretty good.

Quantum Transitions and Decays

Quantum mechanics predicts the transition probability per unit time $dP/dt = \lambda$ for a system to change states by emitting a packet of energy such as a photon or α-particle. A sample of N_0 objects decays according to

$$N(t) = N_0 e^{-t/\tau},$$

where $\tau = 1/\lambda$ is the *lifetime*. Often the *half-life* $\tau_{1/2} = (\ln 2)\tau = 0.6932\tau$ is used to parameterize the transition rate.

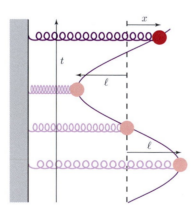

Figure 15.1 A mass on a spring undergoes harmonic oscillation. Classically, the mass is never found outside the region between the classical turning points $x = \pm\ell$.

Recall from §7 that the probability distribution for a quantum system is given by the square of its Schrödinger wavefunction,

$$\frac{dP}{dx} = |\psi(x)|^2 . \qquad (15.7)$$

The energy basis states of the harmonic oscillator have wavefunctions that are solutions to the Schrödinger equation with a potential $V(x) = \frac{1}{2}m\omega^2 x^2$,

$$-\frac{\hbar^2}{2m}\psi''(x) + \frac{1}{2}m\omega^2 x^2 \psi(x) = E\psi(x). \qquad (15.8)$$

In §7 we described these states. Their energies are given by $E_n = (n + 1/2)\hbar\omega$, and the wavefunctions of the ground state ($n = 0$) and a few excited states are given in eqs. (7.44)–(7.47). It is easy to verify by direct substitution that the ground state wavefunction

$$\psi_0(x) = \left(\frac{m\omega}{\pi\hbar}\right)^{1/4} e^{-m\omega x^2/2\hbar} = \frac{1}{(\pi\ell^2)^{1/4}}e^{-x^2/2\ell^2} \qquad (15.9)$$

obeys eq. (15.8) with energy $\frac{1}{2}\hbar\omega$, where $\ell = \sqrt{\hbar/m\omega} = \sqrt{2E_0/m\omega^2}$. Note that the parameter ℓ equals the maximum displacement of a classical oscillator with energy $E_0 = \frac{1}{2}\hbar\omega$.

The probability distribution for the harmonic oscillator ground state,

$$\left.\frac{dP}{dx}\right|_{\text{quantum}} = |\psi_0(x)|^2 = \frac{1}{\ell\sqrt{\pi}}e^{-x^2/\ell^2}, \qquad (15.10)$$

is plotted in Figure 15.2. A striking feature of this plot is that there is a high probability of finding the particle at values of x that are classically forbidden. In fact, the probability of finding the particle with $|x| > \ell$ is ~16%, and there is a non-vanishing probability of observing the particle at arbitrarily large values of $|x|$. All energy basis states of the harmonic oscillator extend beyond the limits of classical motion into the classically forbidden zone (Problem 15.5), though the fraction of the probability distribution with $|x| > \ell$ decreases as the excitation number n goes up. This violation of classical behavior is not special to the harmonic oscillator potential. The wavefunctions of bound states in any finite, piecewise continuous potential extend into the classically forbidden region (where $E < V(x)$). The only exception – as we saw in §7.7.2 – is a quantum particle in a box where infinitely high walls of the potential force the wavefunction to vanish at boundaries and outside the well.

As Figure 15.2 shows, as far as the quantum probability distribution is concerned, there is nothing special about

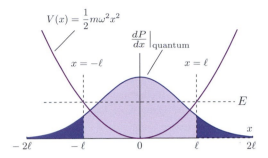

Figure 15.2 The quantum probability distribution (shaded) for a particle in the lowest energy state of a harmonic oscillator. $|x| > \ell$ is the classically forbidden zone, so the area of the dark-shaded regions corresponds to the probability of finding the particle where it is classically forbidden from going.

Quantum Particles in Classically Forbidden Zones

The wavefunction of a bound state with energy E in a potential $V(x)$ always extends smoothly beyond the limits of the classical motion into regions where $V(x) > E$. Thus, quantum particles in energy basis states always have a non-vanishing probability of being found in classically forbidden regions. The only exception is the (idealized) case of an infinitely high potential, which forces the wavefunction to vanish at the hard wall and beyond.

the classical turning point. The smoothness of the quantum probability distribution is a general property of second-order differential equations like the Schrödinger equation: $\psi(x)$ and $d\psi/dx$ are continuous as long as the potential $V(x)$ is piecewise continuous, so quantum wavefunctions generally continue smoothly from classically allowed to classically forbidden regions.

The phenomenon that a quantum particle can be observed in a classically forbidden region is referred to as **quantum tunneling**. Up to this point we have seen that it is possible for a quantum particle to tunnel into a classically forbidden region. When two classically allowed regions are separated by a barrier where $V(x) > E$, the barrier forms an impenetrable wall for classical particles. In contrast, quantum mechanics allows the particle to tunnel into the barrier and then out the other side. Thus tunneling invites us to consider the possibility of **barrier penetration**.

15.3 Barrier Penetration

15.3.1 A Barrier of Constant Height

Barrier penetration occurs when a quantum particle passes through a potential energy barrier from which it would reflect classically (Figure 15.3(a)). As a first example, we consider the case where the barrier has constant height V_0 and thickness d, and the particle is incident from the left, as in Figure 15.3(b).

Since there is no potential on either side of the barrier, the solutions to Schrödinger's equation in the classically allowed regions $x < 0$, $x > d$ are plane waves

$$\psi_{\pm p}(x, t) = e^{\pm ipx/\hbar} e^{-iEt/\hbar} , \qquad (15.11)$$

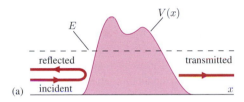

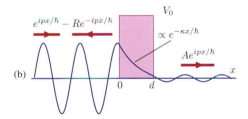

Figure 15.3 (a) A generic potential barrier $V(x)$. A solution to the Schrödinger equation with a wave incident from the left gives rise to both a reflected wave on the left and a transmitted wave on the right of the barrier, even when the energy E is below the height of the barrier. (b) An incident plane wave $e^{ipx/\hbar}$ with energy $E = p^2/2m$ encounters a barrier of constant height $V_0 > E$ and width d.

where $p = \sqrt{2mE}$. Note the $\pm$ sign in $\psi_{\pm p}(x, t)$: the plus sign corresponds to a right-moving particle and the minus sign to a left-mover as described in §4. $\psi_{\pm p}(x) = e^{\pm ipx/\hbar}$ satisfies the free time-independent Schrödinger equation,

$$-\frac{\hbar^2}{2m}\psi''_{\pm p}(x) = E\psi_{\pm p}(x). \qquad (15.12)$$

Suppose, first, that the barrier is infinitely high, $V_0 \to \infty$. As in the case of a particle in a box (§7.7.2), we must find a combination of the plane-wave solutions that vanishes at the barrier, $\psi(0) = 0$. We are considering the case where the particle is incident from the left, so we expect a nonzero wavefunction in the region $x < 0$, but since the wavefunction must vanish everywhere in the classically forbidden region when $V_0 = \infty$, we can take the solution to vanish in the region $x > d$. We have then a solution

$$\psi(x) = 2i \sin(px/\hbar) = \left(e^{ipx/\hbar} - e^{-ipx/\hbar}\right) \text{ for } x < 0,$$
$$\psi(x) = 0 \text{ for } x > 0, \tag{15.13}$$

where we have suppressed the factor of $e^{-iEt/\hbar}$. We can interpret this solution as a superposition of an incident wave ($e^{ipx/\hbar}$), and a wave that is completely reflected ($e^{-ipx/\hbar}$) from the infinitely high barrier. In short, the quantum particle bounces off the barrier.

Next, suppose that the barrier is high, $V_0 \gg E$, but not infinite. Although there is still a reflected wave of large amplitude, the solution to the Schrödinger equation now has a small nonzero amplitude for values of x greater than d, which means that there is a small probability for the quantum particle to tunnel through and end up on the other side of the barrier. To estimate the probability of this happening, we must compute the amplitude of the Schrödinger wavefunction on the far side of the barrier.

We first examine the wavefunction *inside the barrier*, where it obeys the Schrödinger equation

$$-\frac{\hbar^2}{2m}\psi''(x) + V_0\psi(x) = E\psi(x). \tag{15.14}$$

If E were greater than V_0, this differential equation would have oscillatory solutions similar to those of eq. (15.12) but with $p \to \sqrt{2m(E - V_0)}$. Since $V_0 > E$, however, the solutions to this equation are *exponential* rather than *sinusoidal*,

$$\psi_{\pm\kappa}(x) \propto e^{\pm\kappa x/\hbar}, \tag{15.15}$$

where $\kappa = \sqrt{2m(V_0 - E)}$. It is easy to see by direct substitution that this is a solution to the Schrödinger equation inside the barrier (15.14) for either choice of sign.

Because we are assuming that the quantum particle is incident from the left, the solution we seek can only contain a right-moving wave $Ae^{ipx/\hbar}$ in the region $x > d$. This determines both the function ψ and its first derivative ψ' at the point $x = d$, up to an overall phase, in terms of the unknown (real) coefficient A. Both ψ and ψ' must be continuous across the boundary of the barrier at $x = d$. These conditions uniquely determine the linear combination of the two solutions (15.15) that gives the wavefunction in the region inside the barrier in terms of A. The coefficient of each of the components $\psi_{+\kappa}$ and $\psi_{-\kappa}$ is of order A at the boundary $x = d$. $\psi_{-\kappa}$ is exponentially larger at

$x = 0$ than at $x = d$, however, while $\psi_{+\kappa}$ is exponentially smaller. The latter gives a negligible contribution to the total wavefunction at $x \leq 0$, so we need only consider the contribution of the solution $e^{-\kappa x/\hbar}$ in the classically forbidden region. Matching this function to the incoming and reflected waves on the left-hand side of the barrier, and assuming that the incoming wave has magnitude one, we see that the transmitted amplitude A must be suppressed by a factor $e^{-\kappa d/\hbar}$. For more detail on this analysis, see Problem 15.8.

Thus the amplitude of the wavefunction that is incident from the left *decreases exponentially* as it passes through the barrier, reaching a value proportional to $e^{-\kappa d/\hbar}$ by the far end of the barrier.[1] More generally, for a constant barrier of height V_0 in a region $x_1 < x < x_2$, the net decrease in the magnitude of the wavefunction as the particle penetrates from left to right through the barrier is

$$A(E) \sim e^{-\frac{1}{\hbar}\sqrt{2m(V_0-E)}\,(x_2-x_1)}, \tag{15.16}$$

where, as usual, we use "$\sim$" to denote equality at leading order up to a multiplicative coefficient of order one. The squared magnitude of the wavefunction on the right relative to its value on the left gives the probability of finding the particle, which was originally incident from the left, propagating away to the right. We conclude that *the probability $P(E)$ for the particle to tunnel through a constant barrier of height V_0 and width $x_2 - x_1$ is*

$$P(E) = |A(E)|^2 \sim e^{-\frac{2}{\hbar}\sqrt{2m(V_0-E)}\,(x_2-x_1)}. \tag{15.17}$$

Note the factor of two multiplying the exponent – the effect of squaring $A(E)$.

15.3.2 A Variable Barrier: The Semiclassical Approximation

In the more general case where a barrier height varies with x, there is no simple, general formula like eq. (15.17). There is, however, a very useful approximate result that has an appealing physical interpretation. The result is simple: V_0 should be replaced by $V(x)$ in the argument of the exponential in eq. (15.17) and integrated over the barrier rather than multiplied by its width,

$$\sqrt{2m(V_0 - E)}(x_2 - x_1) \Rightarrow \int_{x_1(E)}^{x_2(E)} dx \sqrt{2m(V(x) - E)}. \tag{15.18}$$

---

[1] This exponential suppression factor is multiplied by an energy-dependent coefficient of order one that we do not estimate here. When the exponential suppression is very large, as it is in most applications of interest to us, the precise value of this factor is not important.

Barrier Penetration Factor

When a quantum particle of energy E encounters a *constant* potential barrier of height $V_0 > E$ and thickness d, it has a probability of tunneling through the barrier to the other side

$$P(E) \sim e^{-\frac{2}{\hbar}\sqrt{2m(V_0-E)}d} .$$

In the *semiclassical approximation*, this result can be generalized to a barrier of variable height $V(x)$,

$$P(E) \sim e^{\left(-\frac{2}{\hbar}\int_{x_1(E)}^{x_2(E)} dx\sqrt{2m(V(x)-E)}\right)} .$$

Here, $x_{1,2}(E)$ are the classical turning points, where $V(x_{1,2}) = E$.

Figure 15.4 A fairly complicated potential V(x). For the energy shown by the line labeled E, the regions where the semiclassical approximation holds are shaded in green. In the red shaded regions – either near classical turning points at x_1 and x_2 or in regions where dV/dx is not small – the semiclassical approximation breaks down.

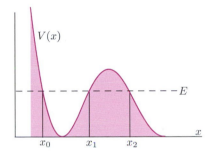

Figure 15.5 A particle with energy E is trapped in a one-dimensional potential well. According to Newton's (classical) laws it can only bounce back and forth between x_0 and x_1. In quantum mechanics, however, it has a small but nonzero probability of tunneling through the barrier and being found to the right of x_2.

Here $x_1(E)$ and $x_2(E)$ are boundaries of the classically forbidden zone – the classical turning points where $V(x_{1,2}) = E$. This result follows from an approximation to quantum physics known as the **semiclassical approximation**, which is derived in Box 15.2.[2] Intuitively, we can think of (15.18) as arising simply from a series of successive constant-height barriers that are taken to approximate the continuous barrier with increasing accuracy.

The semiclassical approximation for tunneling through a classically forbidden region with a varying potential thus gives a **barrier penetration factor**

$$P(E) = |A(E)|^2 \sim \exp\left(-\frac{2}{\hbar}\int_{x_1(E)}^{x_2(E)}\sqrt{2m(V(x)-E)}\right),$$
$$(15.19)$$

in analogy to eq. (15.17).

The semiclassical approximation can fail either when the potential is changing too rapidly or when $p(x) = \sqrt{2m(V(x)-E)} \to 0$ and the (imaginary) De Broglie wavelength (Box 15.2) diverges. $p(x)$ vanishes at the classical turning points, so the approximation is suspect near the limits of the integral in eq. (15.19). Fortunately, the integrand $p(x)$ vanishes at those limits, so the region where the approximation fails makes a negligible contribution to the barrier penetration factor. Figure 15.4 sketches regions

where the semiclassical approximation is or is not valid for a complicated potential. Examples of barriers and barrier penetration factors are given in Box 15.2.

15.4 Tunneling Lifetimes

The barrier penetration factor estimates the probability that a quantum particle encountering a barrier will pass through it. Usually we want to compute something slightly different, namely the *probability per unit time* that a particle initially in a trap will tunnel out of it. Although a full quantum mechanical treatment of this is difficult, it can be estimated in the spirit of the semiclassical approximation simply by multiplying the probability of tunneling by a classical estimate of the frequency of encounters between the particle and the barrier.

Consider a particle trapped in the potential shown in Figure 15.5. Classically, ignoring tunneling, it simply oscillates back and forth in its trap. We can calculate the period

---

[2] More precisely it is the first term in a more complicated approximation. The approximation has many names. In quantum mechanics it is known as the *WKBJ approximation* after Wentzel, Kramers, Brillouin, and Jeffreys.

Box 15.2   The Semiclassical Approximation

In this box we derive the semiclassical approximation, which underlies the estimations of tunneling and barrier penetration found in eqs. (15.18) and (15.19). This analysis is included for the interested reader for completeness; a simpler intuitive explanation is given in the text.

We parameterize the solution to the Schrödinger equation in the form $\psi(x) = e^{i\sigma(x)/\hbar}$. Substituting into the time-independent Schrödinger equation (7.7), we find

$$-\frac{\hbar^2}{2m}\frac{d^2}{dx^2}\exp\left(\frac{i}{\hbar}\sigma(x)\right) + V(x)\exp\left(\frac{i}{\hbar}\sigma(x)\right) = E\exp\left(\frac{i}{\hbar}\sigma(x)\right)$$

$$\Rightarrow \quad -\frac{\hbar^2}{2m}\left(-\frac{1}{\hbar^2}\left(\frac{d\sigma}{dx}\right)^2 + \frac{i}{\hbar}\frac{d^2\sigma}{dx^2}\right) + V(x) = \frac{1}{2m}\sigma'(x)^2 - \frac{i\hbar}{2m}\sigma''(x) + V(x) = E,$$

where, between the first and second line, we cancelled the common factor of $\exp(i\sigma(x)/\hbar)$.

We can ignore higher-order "quantum effects" by dropping the term proportional to $\hbar\sigma''$. Solving for σ' we obtain

$$\sigma'(x) = \pm\sqrt{2m(E - V(x))},$$

$$\sigma(x) = \pm\int^x dx'\sqrt{2m(E - V(x'))},$$

and

$$\psi_\pm(x) \sim \exp\left(\pm\frac{i}{\hbar}\int^x dx'\sqrt{2m(E - V(x'))}\right).$$

This approximation is valid when the magnitude of the dropped $\hbar\sigma''$ term is small compared to the σ'^2 term,

$$\left|\frac{\hbar\sigma''}{\sigma'^2}\right| = \left|\frac{d}{dx}\frac{\hbar}{\sigma'}\right| \ll 1.$$

The semiclassical approximation can be applied not only in a classically forbidden region but also in any region where the potential varies slowly. In a classically allowed region, we can define a *variable DeBroglie wavelength* $\lambda(x) = 2\pi\hbar/p(x)$, where $p(x) = \sqrt{2m(E - V(x))} = \sigma'(x)$. The condition for the validity of the semiclassical approximation can then be expressed as $|d\lambda/dx| \ll 2\pi$. Thus, the approximation is valid if the DeBroglie wavelength is a slowly changing function of position; this condition is intuitively sensible in regions where the particle's energy is substantially larger than the potential, but is also mathematically correct in classically forbidden regions, where the variable DeBroglie wavelength becomes formally imaginary.

---

of this oscillation. Given the particle's energy E, the speed of the particle inside the potential is given by

$$\dot{x} = \frac{dx}{dt} = \sqrt{\frac{2(E - V(x))}{m}}. \tag{15.20}$$

The period of the classical motion is the time it takes the particle to make one full cycle of its motion, or twice the time it takes the particle to go from one turning point to the other. This can be computed by integrating dt/dx between the two turning points and multiplying by 2,

$$T(E) = 2\int_{x_0(E)}^{x_1(E)} dx\,\frac{dt}{dx} = 2\int_{x_0(E)}^{x_1(E)} \frac{dx}{\dot{x}}$$

$$= 2\int_{x_0(E)}^{x_1(E)} dx\sqrt{\frac{m}{2(E - V(x))}}. \tag{15.21}$$

The oscillating particle has a chance to penetrate the barrier once in each time interval T. So we estimate the barrier penetration probability per unit time, dP/dt, to be

$$\frac{dP}{dt} \sim \frac{P(E)}{T(E)} \sim \frac{\exp\left(-\frac{2}{\hbar}\int_{x_1(E)}^{x_2(E)} dx\sqrt{2m(V(x) - E)}\right)}{2\int_{x_0(E)}^{x_1(E)} dx\sqrt{\frac{m}{2(E - V(x))}}}, \tag{15.22}$$

where $P(E)$ comes from eq. (15.19) and $T(E)$ comes from eq. (15.21). Often we are interested in a situation where barrier penetration results in the decay of a system such as a nucleus. In that case, the decay lifetime is given by the inverse of the barrier penetration probability per unit time. Thus, in the semiclassical approximation, the decay

Example 15.2   Barrier Penetration

Tunneling through a barrier of constant height

This is the simplest case. If a particle of energy $E < V_0$ impinges on the rectangular barrier shown in the figure, then the probability that it will pass through is $P(E) \sim \exp\left(-\frac{2}{\hbar}\sqrt{2m(V_0 - E)}\, d\right)$. Note that the tunneling probability decreases dramatically with the mass of the particle as well as the height and thickness of the barrier. Consider, for example, energy and distance scales typical of atomic physics. Imagine an electron of energy 2.5 eV incident on a 5 eV barrier of thickness 5 Å($= 5 \times 10^{-10}$ m).

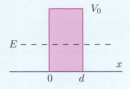

$$P \sim \exp\left(-\frac{2}{1.05 \times 10^{-34}}(5 \times 10^{-10})\sqrt{(5 - 2.5) \times 9.1 \times 10^{-31} \times 1.6 \times 10^{-19}}\right) \approx e^{-6} \approx 3 \times 10^{-3}.$$

Tunneling of electrons through barriers with heights and widths typical of atomic energy scales is thus not so unlikely.

In contrast, under the same conditions, a proton with mass roughly 2000 times the mass of an electron has a probability $P \sim \exp(-6\sqrt{2000}) \sim 10^{-116}$ of tunneling! This has practical consequences: when two pieces of metal are brought very close to one another, electrons can tunnel from one to the other (this is the physical basis of the *scanning tunneling microscope* or STM), but the atoms themselves, which are even heavier than protons, stay put.

Tunneling through a parabolic barrier

The potential $V(x) = V_0 - kx^2/2$ is an approximation to any smooth potential hump near its maximum. Suppose a particle with energy $E < V_0$ comes in from the left. Classical mechanics would require that it bounces back, never getting further than its classical turning point at $x_1 = -a$, where $V(-a) = E$, or $a = \sqrt{2(V_0 - E)/k}$. Quantum mechanics allows the particle to tunnel from x_1 through to the other classical turning point at $x_2 = +a$. From the barrier penetration formula, we estimate a tunneling probability,

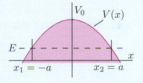

$$P(E) \sim \exp\left(-\frac{2}{\hbar}\int_{-a}^{a} dx\sqrt{2m(V_0 - \frac{k}{2}x^2 - E)}\right) \sim \exp\left(-\frac{2\pi(V_0 - E)}{\hbar\omega}\right),$$

where we have defined $\omega = \sqrt{k/m}$. (Even though the potential well is upside down, $\hbar\omega$ defines a natural energy scale for the quantum mechanics of a particle in the presence of this potential.)

Tunneling Lifetimes

The lifetime of a state that decays by tunneling can be estimated by

$$\tau(E) \sim \frac{T(E)}{P(E)} \sim 2\int_{x_0(E)}^{x_1(E)} dx\sqrt{\frac{m}{2(E - V(x))}}$$

$$\times \exp\left(\frac{2}{\hbar}\int_{x_1(E)}^{x_2(E)} dx\sqrt{2m(V(x) - E)}\right).$$

The *barrier penetration factor* $P(E)$ can be extremely small due to the exponential suppression of tunneling, leading to lifetimes far longer than the period of the classical motion $T(E)$.

lifetime is simply the period of the classical motion divided by the barrier penetration factor,

$$\tau(E) \sim \frac{T(E)}{P(E)} \sim 2\int_{x_0(E)}^{x_1(E)} dx\sqrt{\frac{m}{2(E - V(x))}}$$

$$\times \exp\left(\frac{2}{\hbar}\int_{x_1(E)}^{x_2(E)} dx\sqrt{2m(V(x) - E)}\right).$$

(15.23)

15.5 The Pauli Exclusion Principle

The richness of atomic physics is due in no small measure to the fact that no two electrons can occupy the same state. This *exclusion principle* forces electrons in atoms into ever-higher energy levels with more complex spatial structure. The sequential filling of shells of available states is responsible for the prominent patterns in the periodic table of the elements. It is also responsible for a similar richness in the properties of nuclei and for the electronic properties of metals and semiconductors, which are exploited in photovoltaic devices §25.

Example 15.2   An Approximately Parabolic Trap and Barrier

Consider a particle of mass m trapped in a sinusoidal potential $V(x) = \frac{1}{2} W \sin^2(\beta x)$. We estimate the time it takes for the particle to tunnel from one minimum to the next. We approximate both the trap and the barrier quadratically.[a] Near each minimum the potential looks like a harmonic oscillator potential of the form $V(x) \approx \frac{1}{2} W \beta^2 x^2$, and angular frequency $\omega = \beta \sqrt{W/m}$. We assume that the particle begins in the ground state of this oscillator with $E_0 = \frac{1}{2} \hbar \omega = \frac{1}{2} \hbar \beta \sqrt{W/m}$. The period of its classical motion is $T(E) = 2\pi/\omega$. We estimate the barrier penetration factor by approximating the barrier as an inverted parabola (as in Box 15.2), with $V_0 = W/2$ and $E = E_0 = \frac{1}{2} \hbar \omega$,

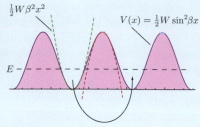

$$P(E) \sim \exp\left(-\frac{\pi}{\hbar \omega}(W - \hbar \omega)\right) = e^{\pi} \exp\left(-\frac{\pi W}{\hbar \omega}\right).$$

So the tunneling lifetime is

$$\tau \sim \frac{T(E)}{P(E)} \sim \frac{2\pi}{\omega} e^{-\pi} \exp\left(\frac{\pi W}{2E_0}\right).$$

The tunneling lifetime grows exponentially with the ratio of the height of the barrier $W/2$ to the ground state energy E_0. Thus, for example, if $E_0 = 1$ eV, the trapped particle's classical period is $T(E) = 2\pi/\omega \sim 2 \times 10^{-15}$ s, but if the trap is 10 eV high, the tunneling lifetime is approximately $e^{9\pi} T(E) \sim 10^{12} T(E) \sim 0.002$ s.

---

[a] Note that this cannot be a good approximation for both the trap and the barrier, since the second derivative of the potential vanishes halfway up. This approximation is good enough, however, for an order-of-magnitude estimate of the tunneling lifetime.

The exclusion principle was discovered empirically by Wolfgang Pauli in 1925 as he attempted to explain the patterns of the periodic table using the (then novel) ideas of quantum mechanics. In fact, however, the exclusion principle itself cannot be derived from nonrelativistic quantum mechanics. The axioms of §7 do not specify the nature of the particles to which they apply. In principle, as far as quantum mechanics is concerned, every electron could be different with, for example, a different mass and/or electric charge.

When physicists in the 1930s began to combine quantum mechanics with Einstein's special theory of relativity, they found that it is necessary to view all particles as quantum excitations of fields that permeate space. We have already introduced the classical electromagnetic field and alluded in §7.3.3 to the fact that *photons* are its quantum excitations. We return to this subject in more depth in §22 when we derive the spectrum of blackbody radiation. Similarly, electrons are excitations of the *electron field*, and their charge and mass are properties of the field. The fact that all electrons must be treated quantum mechanically as identical particles is one of the most powerful implications of *relativistic quantum field theory*, a subject that goes beyond the scope of this book.

The Pauli Exclusion Principle

When relativity and quantum mechanics are combined in *quantum field theory*, it emerges that particles come in two types: *bosons*, with integer spin, whose quantum wavefunctions are symmetric under exchange of particle coordinates; and *fermions*, with half-integer spin, whose wavefunctions are antisymmetric under particle exchange.

Fermions obey the *Pauli exclusion principle*, which requires that only one particle can occupy a given quantum state.

Beyond the identical quantum nature of the excitations of a general quantum field, there is a deep connection between the *intrinsic spin* of a field's quanta and the symmetry of the quantum wavefunctions for these quanta under an exchange of particle labels. Particles with integer spin (the spin one photon, for example) can only exist in states symmetric under particle exchange, and particles with half-integer spin (electrons, protons, and neutrons, for example) can only form states that are antisymmetric under particle exchange. Such particles are known as **bosons** and **fermions** respectively. All the fundamental particles from which matter is built are fermions.

The exclusion principle for fermions follows immediately from the antisymmetric nature of fermion wavefunctions. Consider two electrons labeled by coordinates τ_1 and τ_2. Here τ includes not only the electron's space coordinate x, but also its spin along a fixed axis, which (§7) can take on values $\pm\hbar/2$. Suppose the two electrons occupy two states a and b with normalized wavefunctions $\psi_{a/b}(\tau)$. The two-particle system is described by a two-particle wavefunction, $\Psi_{ab}(\tau_1, \tau_2)$, which must be antisymmetric under exchanging τ_1 with τ_2, namely

$$\Psi_{ab}(\tau_1, \tau_2) = \frac{1}{\sqrt{2}}\left(\psi_a(\tau_1)\psi_b(\tau_2) - \psi_a(\tau_2)\psi_b(\tau_1)\right).$$
(15.24)

It is easy to see that this wavefunction changes sign when τ_1 and τ_2 are interchanged. It is also manifest that *the wavefunction vanishes if both particles are in the same state*, i.e. if $a = b$. Thus, fermions obey the exclusion principle, while any number of bosons can occupy the same state.

Often, to first approximation, the energy of a quantum state for an electron is independent of the particle's spin. Such is the case, for example, for the Coulomb potential summarized in §7.8.2. Each energy basis state can then accommodate *two electrons*, one with spin $+\hbar/2$ and the other with spin $-\hbar/2$. The $1s$ orbital of eq. (7.51) is full when it contains two electrons, explaining the stability of monatomic helium. The next set of states, with $n = 2$, includes the $2s$ orbital and the three $2p$ orbitals. These four orbitals are filled with eight electrons, leading to another highly stable element, neon, with a total of ten electrons. Thus, the sequential filling of atomic orbitals builds up the periodic table of elements. In the following chapter (§17), we apply similar principles to explain some of the stability properties of nuclei.

Discussion/Investigation Questions

15.1 Classical systems also decay. The decay or *failure* rate of many systems follows a *bathtub curve*. Look up and consider this type of decay and contrast it with the decay pattern of a quantum system.

15.2 The *scanning tunneling microscope* (STM) makes use of quantum tunneling to obtain resolutions down to the scale of individual atoms. Investigate the operating principles and applications of STMs.

15.3 It is not hard to balance a (brand new) ordinary pencil vertically upon its eraser. This is a configuration of stable classical equilibrium. Explain why in principle, even under the best of conditions (flat surface, symmetric pencil, no wind currents), the pencil is unstable due to quantum mechanical tunneling. Give a very rough estimate of the tunneling time, and compare to the age of the universe.

Problems

15.1 The first radioactive element discovered by Polish physicist Marie Curie was radium-226. It has a half-life of 1700 years. Suppose you obtained a microgram of pure ^{226}Ra. How long would you have to wait before the number of decays per second from this sample fell to 10^4?

15.2 Uranium-238 and uranium-235 both decay by α-particle emission. Their measured half-lives are 4.5×10^9 years and 7.0×10^8 years respectively. At the present time the ratio of ^{235}U to ^{238}U in naturally occurring uranium is about 0.72%. It is believed that the uranium on Earth was formed in a supernova that took place billions of years ago. It is also believed that the ratio of ^{235}U/^{238}U produced in supernovae is roughly 1.65. Estimate how long ago the supernova that produced the uranium on Earth took place.

15.3 Potassium-40 (^{40}K) is a naturally occurring form of potassium whose β-decays are responsible for some of Earth's geothermal energy (see §14.3 and §32.1.3). ^{40}K can decay in two different ways. The probabilities per unit time for the two decay modes are $\lambda_1 = 4.9 \times 10^{-10}$ y^{-1} and $\lambda_2 = 6.1 \times 10^{-11}$ y^{-1}. What is the lifetime of ^{40}K? What is the ratio of the average number of decays of type one and type two?

15.4 A radioactive source is emitting on average 40 particles per second. In any particular one second interval, what is the probability that it emits 40 particles? 30 particles? None? What is the probability that it emits less than 40 particles?

15.5 Consider the ground state and the first and second excited states of a simple harmonic oscillator as described in §7. Verify the statement made in the text that the probability of finding the particle outside the classically allowed region is 16% when the particle is in the ground state, and compute the corresponding probabilities for the first and second excited states.

15.6 [T] Verify the barrier penetration factor obtained for the parabolic barrier in Box 15.2.

15.7 [T] A particle of mass m and energy E encounters a triangular barrier of width $2w$ and height V_0 (see Figure 15.6). Estimate the probability that it will tunnel through the barrier. How does the probability compare with that for a rectangular barrier of the same height and width?

15.8 [T,H] Compute the exact amplitude for transmission and reflection of a particle from a rectangular potential

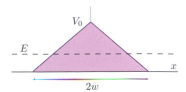

V_0

E

x

$2w$

Figure 15.6 A triangular barrier.

barrier of height V_0 and width d by matching the wave-function and its derivative at the boundary points $x = 0, d$. Note that, for a barrier that is high and wide, the leading term has the form of eq. (15.16) with a multiplicative coefficient as anticipated in Footnote 1. Show that further corrections, which arise from the exponentially increasing solution in the classically forbidden region, are exponentially suppressed.

15.9 An experimenter seeks to use a sinusoidal potential of the form studied in Example 15.2 to make a one-dimensional trap to confine a number of electrons for further study. She places one electron in the ground state of each potential well. Each electron's energy should be 0.1 eV and the electron's tunneling lifetime should be at least five years. Using the approximations developed in Example 15.2, estimate the minimum necessary height of the potential $(W/2)$, the spacing d between minima in the potential, and the root mean square width of each electron's probability distribution.

15.10 **[T]** The energy levels of the one-dimensional harmonic oscillator are given by $E_n = (n + 1/2)\hbar\omega$ with $n = 0, 1, 2, \ldots$. Suppose two spin-1/2 fermions are trapped in a one-dimensional oscillator potential. What are the energies and degeneracies of the first three energy states of this system? (Remember that the degeneracy is the number of independent states with the same energy.)

15.11 **[T,H]** The energy levels of the three-dimensional harmonic oscillator are given by $E_{n_1, n_2, n_3} = (n_1 + n_2 + n_3 + 3/2)\hbar\omega$ with $n_{1,2,3} = 0, 1, \ldots$. Suppose three spin-1/2 fermions are trapped in a three-dimensional oscillator. What is the energy of the ground state? What is its degeneracy? (Remember that the degeneracy is the number of independent states with the same energy.) Extra challenge: What is the energy of the first excited state? What is its degeneracy?

An Overview of Nuclear Power

The nuclei of atoms are bound by the strong interactions. The most tightly bound nuclei are those of intermediate atomic mass. Because of this, energy is given off if very high mass nuclei break apart, or *fission*, into two intermediate mass products or when very low mass nuclei combine, or *fuse*, into heavier nuclei. Although very different in their underlying dynamics and their practical implementation, both nuclear fission and nuclear fusion are potentially sources of great amounts of energy for human use.

Nuclear fission is a high-energy-density, non-carbon energy source that is available with today's technology. In 2014, about 275 GW, or 11% of the world's electricity, was generated by nuclear fission reactors [12]. Nuclear fission energy could in principle supply a significant fraction ($\sim$10–20%) of the world's total energy needs, which are roughly 15 TW, in the near-term future. Nuclear fission is not likely to be a long-term source of energy for human use, however, since the world's economically recoverable reserves of the basic fuel ^{235}U, a rare *isotope*[1] of the heavy metallic element **uranium**, are quite limited.[2] Schemes to *breed* new fuel from the more common form of uranium (^{238}U) or from **thorium**, another heavy metal element, are technically challenging and bedeviled by complex safety

Reader's Guide

In this and the following four chapters we give a comprehensive introduction to nuclear power. This chapter gives an overview of the subject and examines the supply of uranium, the source of the fuel that powers nuclear fission power. §17 introduces the basic ideas of nuclear structure, binding, and decays. In §18 we describe the processes of nuclear fission and fusion that release nuclear energy in reactors and stars. §19 gives an overview of basic aspects of nuclear fission power plants and experiments in controlled nuclear fusion. Finally, §20 describes the nature and effects of ionizing radiation that is produced by nuclear processes and systems, and gives a scientific perspective on the hazards of radiation, nuclear waste, and proliferation.

Prerequisites: §7 (Quantum mechanics), §14 (Forces of nature), §15 (Quantum processes).

and security issues. If, however, it proves possible to convert uranium and/or thorium into fission fuels, and if the technical, social, and political problems that accompany nuclear power can be solved, then nuclear fission energy could in principle supply a large fraction of human energy needs for the foreseeable future.

Although *nuclear fusion* reactions power the Sun and other stars throughout the universe, fusion has proven extremely difficult to control on Earth. Efforts to control fusion processes continue to be the subject of both basic and applied research. If fusion power can be harnessed – and that is a big "if" – it too could supply human energy needs for a very long time.

Nuclear power presents some of the most difficult scientific, political, and ethical issues in the energy field. The environmental and human health hazards from radioactivity and nuclear waste, as well as the risks from nuclear weapons proliferation, are particularly challenging. On the

---

[1] *Isotopes* of an element such as uranium all contain the same number of protons (92 for uranium) but differ in the number of neutrons (143 for ^{235}U). All neutral atoms of uranium have 92 electrons and therefore have the same chemical properties. Isotopes can differ dramatically in their nuclear properties. Isotopes are discussed in more detail in §17 (Nuclear structure).

[2] The term "fuel" is used to refer both to the actual material which undergoes nuclear fission and to the mixture of materials, some of which fission and some of which do not, which is fed into nuclear fission reactors. The meaning should be clear from the context.

other hand, the advantages of a large-scale, carbon-free energy source are also substantial, in view of climate issues discussed in §34. The stakes are high, so it is important that public discussion of nuclear energy (from both fission and fusion) be based on a sound understanding of the scientific facts regarding this energy resource.

The environmental, economic, and political concerns that accompany nuclear power are sufficiently serious that only a few countries have invested significantly in this energy source over the past 30 years. Interest in nuclear energy increased in the early 2000s, as countries have began to confront the environmental consequences of burning carbon-based fuels. The earthquake-triggered tsunami and subsequent nuclear reactor failure at the *Fukushima Daiichi* reactor complex in Japan, however, has in recent years increased the level of scrutiny associated with further development of nuclear power. While development of nuclear power has continued in many countries, several countries have recently decommissioned existing nuclear power plants and curtailed plans for others.

This chapter and the four that follow provide an introduction to the physical principles required to understand nuclear energy and the radiation that is closely associated with it. In this chapter we first describe the basic physical mechanisms behind nuclear fission and fusion and briefly summarize the history of nuclear fission power (§16.1). Next we examine the world's resources of nuclear fission fuel, which may limit the scale to which nuclear fission power can be developed (16.2). Finally, we outline the structure of the four chapters that follow (16.3). Some of the terminology more carefully defined in subsequent chapters is used colloquially in this introduction.

16.1 Overview

16.1.1 Fission: Physics

When a heavy nucleus such as ^{235}U fissions, it gives off about 200 MeV of useful energy, corresponding to about 1 MeV per nucleon. In comparison, when gasoline burns it gives off about 1/2 eV per nucleon (see Problem 16.1). Thus the scale of energy available in nuclear fission is roughly a million times greater than the energy scale of chemistry.

The mechanism that powers fission reactors is a **chain reaction**, in which the fission of one nucleus leads to fission of another. When a nucleus such as ^{235}U fissions, as well as a great deal of energy it also emits a small number of energetic neutrons, which under suitable conditions can go on to induce other uranium nuclei to fission, emitting further neutrons, and so forth. Nuclear weapons exploit the sudden energy release of a run-away chain reaction. Fission power reactors, on the other hand, maintain a tightly controlled reaction at a constant rate, and use the reaction to produce heat and, in turn, electricity.

A fission chain reaction is mediated by neutrons. Normally, electrostatic repulsion keeps the nuclei of atoms too far apart to interact by the strong interactions. Neutrons are the exception: a neutron has no electric charge and can interact strongly when it encounters a nucleus, often with the result that the neutron is absorbed by the nucleus. Since neutrons are bound in nuclei by several MeV, energy is released into a nucleus when a neutron is absorbed, and this energy can cause the nucleus to fission. Some nuclei with large atomic mass, such as uranium and thorium, may fission when they absorb an energetic or *fast neutron*, but the probability of fission is small. Key to fission chain reactions are certain nuclei – ^{235}U is the only one that occurs naturally – that have a very high probability of fissioning when they absorb a very low energy or *slow neutron*, the lower the neutron's energy the better. Such nuclei are termed *fissile*.

Heavy nuclei have many more neutrons than protons. Lighter nuclei have relatively fewer excess neutrons and as a result, when a heavy nucleus fissions several energetic neutrons are typically emitted. Rarely do these fast-moving neutrons directly induce another nucleus to fission. Left on their own, they would either propagate out of the system or be captured by a nucleus, with a subsequent emission of electromagnetic radiation. The fission reaction can be maintained by slowing the neutrons down to low energy where their absorption by a ^{235}U nucleus has a high probability of inducing fission. Thus, to maintain a chain reaction, neutrons must be slowed down quickly, before they can escape or be captured. Neutrons are slowed down by mixing or surrounding the fission fuel with a *moderator*, a material with low atomic mass, usually water or graphite, which slows the neutrons down in multiple two-body collisions and has only a small probability of absorbing the neutrons.

Once slowed down, a neutron must find another fissile nucleus. Since ^{235}U accounts for only 0.72% of naturally occurring uranium, most fission reactors use as fuel uranium that has been *enriched* to a higher percentage of ^{235}U. Enrichment increases the probability that a slow neutron will be absorbed by a ^{235}U nucleus and perpetuate the fission reaction. Once it begins, the rate of a fission reaction is controlled by moving *control rods*, which are made of material with a large appetite for absorbing neutrons, either in or out of the reacting mixture of fuel and moderator. The energy released in the fission reaction is transported out of the reactor by a working fluid, generally referred to as the *coolant*, which is usually water but sometimes carbon dioxide.

Although it may sound complex, a nuclear fission chain reaction is rather easy to establish given a supply of fuel and moderator. Fission reactions even occurred naturally on Earth about two billion years ago when ^{235}U was more abundant and rich deposits of natural uranium were inundated by water. The challenge for nuclear fission energy is to control the fission reaction, safely and reliably convert the heat it generates into useful power, and to safeguard the public and the environment from the radiation that naturally accompanies nuclear power.

16.1.2 Fission: History

Nuclear fission was discovered in the 1930s. The first controlled nuclear fission chain reaction was established in 1942 at the University of Chicago by a group led by Enrico Fermi. Electric power was first generated from nuclear fission at an experimental facility in Idaho in 1951. The first facility to supply grid-scale power was in Obninsky, near Moscow in the USSR, followed soon after by the first commercial nuclear power plant at Sellafield, UK, in 1955. Nuclear fission plant technology developed rapidly in the 1950s, and different types of nuclear reactors proliferated. All nuclear power plants use nuclear fuel to heat the coolant, which then powers a turbine to generate electricity in a fashion similar to a fossil fuel power plant. Early nuclear power plants varied significantly in other aspects of their design. Over time a design known as a *light-water reactor* (LWR) came to dominate the commercial market. LWRs use ordinary ("light") water both as a moderator and as a coolant. In §19 we describe the principles and operation of two varieties of LWRs in some detail, and mention other reactor designs as well.

Nuclear fission was originally thought to be a stable, safe, and virtually limitless source of energy for the world, and nuclear electric power expanded rapidly in the 1960s and 1970s (see Figure 16.1), enjoying substantial government subsidies in the US and other countries. As the technology became more widely deployed, however, concerns regarding safety and security increased. A pair of accidents occurred at nuclear power plants: **Three Mile Island** in the US in 1979, a serious event with no casualties involving a partial reactor meltdown, and **Chernobyl** in the Ukraine in 1986, a reactor explosion that caused significant loss of life and sickness as well as long-term contamination of a sizable then-populated land area. These incidents raised serious questions about the adequacy of nuclear reactor safety systems. The connection between exposure to nuclear radiation and cancer became widely recognized (although it is still not completely understood),

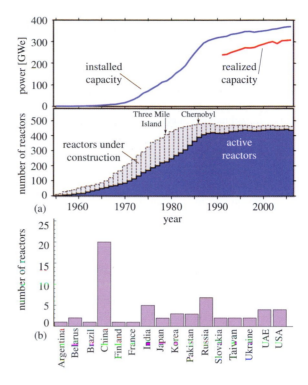

Figure 16.1 (a) Nuclear power generation over time [71]. (b) Nuclear power reactors under construction in 2016 [72].

and contributed to growing concerns over the hazards of nuclear power. By the 1980s the word "nuclear" had been elided from the names of some medical tools such as **[nuclear] magnetic resonance imaging** (now known simply as **MRI**). Lastly, the twin problems of *proliferation* – how to prevent nuclear fuel from being diverted to weapons use by individuals, groups, or countries – and *nuclear waste* – how to safely and securely dispose of the waste products of nuclear energy production – came into focus during this period as well. All of these concerns led to a reconsideration of nuclear power since the 1980s and a significant slowdown in construction of new nuclear power plants. Nuclear power plants have since been held to higher safety and security standards, which have contributed to making them more expensive and less attractive to investors than conventional fossil fuel power plants particularly in areas where fossil fuel resources are bountiful. Nuclear waste continues to be a problem. In the US and elsewhere, growing quantities of radioactive waste are stockpiled in temporary holding areas, with no consensus on long-term treatment or disposal. Finally, the repercussions of the nuclear accident at Fukushima in 2011 have further clouded the future of nuclear energy.

According to the **International Atomic Energy Agency** (IAEA), there are presently 450 nuclear power reactors in operation (as of 2016), with the capacity to produce 392 GW of electricity [71]. As of 2016, there are 60 new reactors under construction [72]. As Figure 16.1(b) shows, the majority of reactors under construction are limited to a small number of countries with rapidly expanding economies, few internal energy resources, or long histories of nuclear power use.

16.1.3 Fusion

Iron, with atomic mass number 56, is the most tightly bound nucleus. If two light nuclei whose mass numbers sum to less than 56 can be made to fuse, then energy is given off. *Coulomb replusion* between the positive electric charges of nuclei, however, keeps them from getting close enough to fuse except at extreme temperatures. The smaller the nuclear charge, the lower the *Coulomb barrier* preventing two nuclei from fusing. Hydrogen – a single proton bound to an electron – is by far the most common element in the universe. It is the dominant constituent of the Sun. The Coulomb barrier to proton–proton fusion is among the lowest for all pairs of nuclei. In the 1930s German-American physicist Hans Bethe explained that the Sun's energy derives from a series of fusion reactions (described in more detail in §22) resulting in the transformation of four hydrogen atoms into a single atom of helium. The nucleus of the helium atom, consisting of two protons and two neutrons, is bound by about 28 MeV. In terms of energy released per nucleon, hydrogen fusion into helium is the most energetic process known for matter occurring in our solar system.

It would require temperatures of order 1.5×10^9 K to raise the kinetic energies of two colliding protons high enough to overcome the Coulomb barrier between them. As explained in Box 7.2, although the temperature in the Sun's core is two orders of magnitude lower than this, fusion takes place there because the protons *tunnel* quantum-mechanically through the *Coulomb barrier*. Fusion reactions later in the life of stars like the Sun produce most of the elements up to and including iron. (Heavier elements are produced in supernovae and other stellar catastrophes.)

Nuclear fusion in the Sun is directly or indirectly responsible for almost all of the energy used on Earth. Only nuclear fission, geothermal, and tidal power do not trace back ultimately to energy produced by nuclear fusion in the Sun.

Despite its prevalence throughout the universe, and despite years of effort, sustained nuclear fusion has never been achieved on Earth. Proton–proton fusion produces a *deuteron* and proceeds by a weak interaction, $p\,p \rightarrow$ $d\,e^+\nu_e$. This reaction is too slow for practical use, and instead, fusion experiments use a mixture of deuterons and *tritons*, the nuclei of the rare, unstable isotope of hydrogen ^{3}H. The resulting fusion reaction $d\,t \rightarrow \alpha\,n$ produces an α-particle, the nucleus of ^{4}He. This is a strong interaction with a low Coulomb barrier and, since it results in the formation of ^{4}He, it produces a great deal of energy. The goal of fusion research is therefore to contain a *plasma* of deuterons, tritons, and electrons at a high enough temperature and density for a long enough period to allow dt-fusion to take place. No material can survive the temperatures necessary to produce fusion. Instead experimenters primarily use specially shaped magnetic fields in devices known as tokamaks to contain the plasma. In §18.4 we discuss the theory of nuclear fusion and in §19.6 we survey experiments in magnetic confinement fusion.

16.2 Nuclear Fission Fuel Resources

Most of the issues that will determine the scale of nuclear fission power deployment in the coming decades are economic and political. The availability of nuclear fuel(s), however, is a potentially limiting factor that is primarily scientific in nature. In this section we explore the world's reserves of uranium (and, more briefly, thorium) and the constraints they may place on the scope of future nuclear power use.

16.2.1 Natural Uranium and Nuclear Fuel Energy

Since the fissile uranium isotope ^{235}U produces on average about 200 MeV for each nucleus that fissions, the fission energy content of a kilogram of this nuclear fuel corresponds to

$$E_f \cong (200 \text{ MeV/nucleus})(2.6 \times 10^{24} \text{ nuclei/kg})$$
$$\times (1.6 \times 10^{-19} \text{ J/eV})/(3.2 \times 10^{13} \text{ J/MWy})$$
$$\cong 2.6 \text{ MWy} . \tag{16.1}$$

The energy generated by the complete fission of 1.15 tonnes of ^{235}U could therefore supply 3 GWth of power continuously for a year. At the 33% thermodynamic efficiency typical of a nuclear power plant, 3 GWth corresponds to 1 GWe.

It takes about 200 t of natural uranium to fuel a 1 GWe nuclear power plant for a year (see e.g. [73]). Since ^{235}U makes up 0.72% of natural uranium, 200 t of natural uranium contains 1.44 t of ^{235}U. Of the 1.44 t, only about half actually fissions in the reactor, accounting for roughly 2/3 of the power generated by the reactor. The other half of the ^{235}U contained in 200 t of natural uranium is lost either in the process of *enriching* the uranium for use in

Uranium Fuel

Approximately 200 tonnes of natural uranium can supply the fuel required by a modern (light-water thermal) nuclear power plant that provides 1 GW of electric power continuously for a year.

the reactor or in the spent fuel removed from the reactor after use (Problem 16.2). The remaining 1/3 of the power is generated by the fission of plutonium created during the operation of the reactor (see §19.3).

Three naturally occurring substances are of interest in the context of nuclear fuels: (1) ^{235}U, which is the primary fuel used in almost all present power reactors; (2) thorium, or more precisely ^{232}Th, a fairly common element that is not itself a fission fuel, but from which the fission fuel ^{233}U can be made; and (3) the common isotope of uranium ^{238}U, which like ^{232}Th can be used to produce a fission fuel, plutonium-239, or ^{239}Pu. If ^{233}U could be *bred* from ^{232}Th with 100% efficiency, 1 GWye could be produced from roughly 1.1 tonnes of ^{232}Th (ignoring losses). Likewise, if ^{239}Pu could be bred at 100% efficiency from ^{238}U then 1 GWye would require roughly 1.2 tonnes of natural uranium. At present only the ^{235}U route is available, so we take 200 tonnes per GWye as a simple conversion factor from natural uranium to energy.[3]

16.2.2 Uranium Resources

At 200 t/GWye, the present complement of operating power reactors (275 GWe) requires approximately 5.5×10^4 tonnes of natural uranium fuel per year. While not a common element, uranium is about as abundant as tin or molybdenum, with an average crustal abundance per unit mass of about 2.7×10^{-6} [74].[4] Both tin and molybdenum, important industrial metals, are produced at a rate of about 2.5×10^5 t/y. The geochemistry of uranium is such that it forms relatively high-quality ores – one large Canadian deposit averages 14% uranium by weight – and is routinely mined from deposits with 0.1% uranium (almost 400 times the average crustal abundance). So it would seem possible in principle to produce of order 10^5 t/y of uranium for many years.

Estimating the world's actual reserves of uranium to any degree of precision, however, is a complex and controversial task, similar to estimating the remaining quantities of fossil fuels. Economic geologists imbue terms like *reserves* and *resources* with relatively precise, technical meanings in their studies, but the terms are often misunderstood or used colloquially, leading to confusion. Box 16.1 introduces some of the technical terms; for a full discussion see [25].

Much information regarding uranium resources is proprietary; assessments are usually based on information voluntarily disclosed by countries, which in turn is assembled from voluntary disclosures by mining companies. See reference [73] for a discussion of the limitations of uranium resource estimates. For many years, the primary reference on uranium resources has been the **Red Book** published jointly by the *Organization for Economic Cooperation and Development (OECD) Nuclear Energy Agency* and the *International Atomic Energy Agency* [76]. The Red Book summarizes resources recoverable at a cost less than some maximum value, in each of three categories: **reasonably assured resources (RAR)**, *inferred resources*, and *undiscovered resources*. RARs correspond roughly to the USGS category of *demonstrated resources* that are recoverable at the stated cost (see Box 16.1).

Table 16.1 shows the 2014 estimated *identified resources* (RAR and inferred) for the countries with the largest reserves as well as the world total and the corresponding number of 1 GWe reactor years that could be supported by these resources. In addition to identified resources, [76] reports both prognosticated and speculative *undiscovered resources* amounting to $\sim 7.7 \times 10^6$ tonnes of uranium, equivalent to 38 500 1 GWe reactor years. This data is based on a recovery cost of $\leq$ \$260/kgU, several times the 2015 market price of \$90/kg.

The total resources estimated by the OECD–IAEA – identified plus undiscovered – would be sufficient to provide 275 GWe of nuclear power (the present rate) for about 150 years. This would not seem to allow for a significant long-term expansion of the world's nuclear electric generating capacity. There is, however, reason to think that this may be a significant underestimate. Like many mineral commodities, estimates of uranium reserves have grown substantially with time (see Box 16.1). The OECD estimate of the uranium resource base (RAR plus inferred, at a fixed price of \$130/kgU) has grown from $\sim$3 Mt to $\sim$6 Mt over the past 40 years – even though over 2 Mt have been extracted during that period – and resource estimates will likely continue to grow.

Many potential sources of uranium, which are considered subeconomic at \$260/kgU, may become available

---

[3] Reference [73] estimates a 25% uncertainty on this number depending on the way the fuel is managed and the thermodynamic efficiency of the plant.

[4] Other estimates range from 9×10^{-7} to 2.8×10^{-6}. See [75].

Box 16.1   Mineral Reserves and Resources

The terms "reserves" and "resources" are used technically in discussions of supplies of minerals and other natural resources. The authoritative definitions are those of the US Geological Survey's yearly *Mineral Commodity Summaries* [25].

Mineral **resources** include both currently and potentially economically recoverable volumes of rock with concentrations of minerals that are higher than typical rocks. Under this heading come both **identified resources**, which may be either **demonstrated** or **inferred** (a distinction based on the certainty of measurements and surveys) and **undiscovered resources** whose existence is postulated on the basis of geological similarity to identified resources and/or geological modeling.

Reserves and resources from USGS	Identified resources			Undiscovered resources	
	Demonstrated		Inferred	Probability range	
	Measured	Indicated		Hypothetical	Speculative
Economic	Reserves		Inferred		
Marginally economic	Reserve base		reserve base		
Sub economic					

Mineral resources are further classified into **economic**, **marginally economic**, or **subeconomic**, depending on whether they can be extracted and processed profitably at the present time and at a certain price.

Mineral **reserves** are a more restrictive category than mineral resources. Reserves are a subset of the **reserve base**, a term that describes those identified resources that satisfy specific quality criteria related to current mining practices. The reserve base is selected from demonstrated resources that are economic, marginally economic, or in some cases, subeconomic. Reserves are that part of the reserve base that can be economically extracted and produced at the present time.[a] Reserves are therefore essentially the same as *demonstrated, economic resources* and are shaded in blue in the figure.

Because it takes significant effort, including drilling and analyzing rock cores, to document reserves, mining companies rarely claim reserves in excess of what they expect to produce in a limited time like 5–10 years. Therefore reserves are not a good measure of Earth's inventory of a mineral commodity. For example, the world's copper reserves were estimated to be 280×10^6 tonnes in 1970 [25]. Between then and 2010 over 400×10^6 tonnes were produced, by which time reserves had grown to about 630×10^6 tonnes. Over the same period the USGS estimate of land-based copper *resources* have grown from 1.6×10^9 to 3×10^9 tonnes.

An extract from a recent summary of the world's uranium resources, both demonstrated and inferred, is given in Table 16.1.

---

[a]   In the fossil fuel industry the term *proven reserves* is equivalent to *reserves* in the nomenclature of [25] (see §33).

---

at higher prices or may become less expensive with economies of scale or with technological advancements. These resources include uranium obtained as a by-product of phosphate mining and direct recovery from seawater, estimated to contain $\sim 4 \times 10^9$ tonnes of uranium (Problem 16.3). See [73] and [76] for discussions of these resources.

This data and probabilistic studies of future resource availability (see e.g. [73]) make it clear that nuclear fission electricity generation using ^{235}U fuel cannot supply a significant fraction of the world's energy needs indefinitely. Models suggest, however, that nuclear fission power could (at $\sim$85% probability) provide ten times the current power output (corresponding to 22% of world energy consumption in 2009) for 100 years at less than three times the current cost of uranium. This could be a significant contribution to a period of transition from the present carbon-based

energy economy to a more sustainable energy source for the next century, if concerns about safety, security, and nuclear waste can be adequately addressed.

The estimates just presented are based on the sole use of ^{235}U as a fission fuel. It is possible that attempts, described in §19.3, to *breed* fuel from the common isotopes of uranium and thorium will eventually succeed. If so, the energy available from natural uranium will increase roughly one hundred fold – recall that ^{235}U is only 0.72% of naturally occurring uranium. Naturally occurring thorium is 100% ^{232}Th, the isotope from which the fission fuel ^{233}U can be bred. Thorium is estimated to be about three times as abundant as uranium in Earth's crust [74, 75]. Since thorium is at present economically less important than uranium, however, thorium reserves and resources have not been studied as carefully as uranium. The OECD–IAEA study quotes world thorium *resources* of 6.2×10^6 t [76].

Table 16.1 Identified uranium resources (both reasonably assured (RAR) and inferred) recoverable at a cost less than $260/kgU as of 2013 [76].

Country	RAR (10^6 t)	Inferred (10^6 t)
Australia	1.208	0.590
United States	0.472	NA
Canada	0.455	0.196
Khazakhstan	0.373	0.503
Niger	0.325	0.080
Namibia	0.297	0.159
Russian Federation	0.262	0.427
South Africa	0.234	0.217
Brazil	0.155	0.121
Greenland	–	0.221
World	4.587	3.048
1 GWe reactor years	22 900	15 200

Uranium Supplies

It is estimated that enough uranium is available from presently recognized resources at an extraction cost below $260/kgU to fuel the world's current suite of fission power reactors for more than 150 years. Models that include uranium from unconventional sources and that allow the cost of uranium extraction to rise above present levels suggest that it may be possible to fuel 5–10 times the present number of fission power reactors for at least 100 years.

If it proves practical to *breed* nuclear fission fuel from natural uranium or thorium, then fuel supplies would expand by more than a factor of 100.

Given the conversion factor of 1.1 t (^{232}Th) $\rightarrow$ 1 GWye, these estimates suggest ample resources for an as yet hypothetical thorium-powered nuclear fuel cycle.

16.3 The Following Chapters

The next four chapters are devoted to a systematic exploration of nuclear physics and nuclear energy. We begin in §17 with an introduction to the basic physics of nuclei.

We focus particularly on the systematics of nuclear binding energies, which determine which nuclei are stable and which decay. In §18 we apply these ideas to study the two types of nuclear reactions that produce energy, nuclear fission and nuclear fusion. Next we turn in §19 to the question of how nuclear fission and fusion reactors work. What are the ingredients in a nuclear fission power plant? How is fission maintained and controlled? What are the safety margins and the operational concerns? We address these questions in §19. Nuclear fusion is still a very speculative possibility – we briefly survey the presently most promising approaches to controlled fusion.

In the last chapter in this sequence, §20, we examine the properties of *ionizing radiation*: how it is measured, how it interacts with matter, and its effects on living matter in particular. We also describe the natural background level of radiation in the environment and the level of manmade radiation exposure in various circumstances. We examine the radioactivity of spent nuclear fuel and how it might be managed. Radioactivity is usually regarded as a harmful agent, but on the other hand thermal energy released in radioactive decays of nuclei deep within the Earth also provides a significant fraction of all geothermal energy. We return to this aspect of radioactivity in §32 (Geothermal energy).

The material in the following chapters is based in part on the following references, in which more detailed treatment of these subjects can be found: [77] for a general introduction to nuclear physics and nuclear energy; [78] and [79] for the more practical aspects of nuclear energy and nuclear reactors; [80] for ionizing radiation and its many effects; and finally [81] for an overview of nuclear power and nuclear weapons.

Discussion/Investigation Questions

16.1 Investigate the proposed mechanisms for recovering uranium from seawater. What do you think are the prospects for this source of uranium?

16.2 Research the present cost of fuel for nuclear fission power plants (in $ per MWhe) and compare with the cost of coal and natural gas for conventional fossil fuel power plants. What are the implications for expanding utilization of nuclear power?

16.3 A possible way to evaluate the extent of uranium resources is to build a probabilistic model that includes both known resources and possible unconventional sources of uranium and their relative cost of extraction, and then to project the cost of uranium as a function of the number of nuclear power plants. As an example, discuss the model presented in §3 of [73].

Problems

16.1 Given the (higher) molar heat of combustion of gasoline (§11.2.2) and knowing its molecular weight, verify that it yields about 1/2 eV per nucleon in complete combustion.

16.2 In one scenario, enrichment of 200 t of natural uranium yields 26 t enriched to 4% ^{235}U that is used to power a reactor for a year. How much ^{235}U is lost in the *depleted uranium* discarded during the enrichment process? The *spent nuclear fuel* removed after a year in the reactor contains about 0.8% ^{235}U. Close to 86% of the ^{235}U consumed actually fissions; the other 14% captures a neutron without fissioning. Verify that approximately half of the ^{235}U contained in the 200 t of natural uranium actually fissions during the year of running the reactor.

16.3 Check the quoted estimate of 4×10^9 tonnes of uranium from seawater quoted in this chapter. For how many years could this amount of uranium supply 100% of the world's energy needs at the level of 2015?

16.4 Assume that thorium can be converted 100% into ^{233}U and used as a fission fuel. According to [76] world thorium reserves are roughly 6×10^6 tonnes. How long could this amount of thorium supply 100% of the world's energy needs at the 2015 level?

Structure, Properties, and Decays of Nuclei

At the core of every atom is a nucleus composed of neutrons and protons (*nucleons*). The number of protons in the nucleus determines the charge – and hence the chemical properties – of the atom. In most familiar circumstances, atomic nuclei contain a fixed set of nucleons. Despite centuries of effort, aspiring alchemists who desired to turn lead into gold were frustrated by the intransigent nature of atomic matter. Nonetheless, in many important energy processes nuclear structure can change, often leading to the release of tremendous quantities of energy. Such processes are central to the energy production mechanisms involved in solar energy, geothermal energy, and nuclear power plants.

While nuclear physics has a reputation as a difficult subject, and nuclei are often regarded as somewhat mysterious, even by practitioners of energy science, in some ways the basic ideas of nuclear physics are similar to the basic ideas of chemistry. Indeed, the classification of nuclei is in a sense much simpler than that of chemical compounds, as the number of fundamental building blocks for nuclei is much smaller.

The binding together of protons and neutrons into nuclei is effected by a balance between two forces. The strong nuclear interactions (§14.1.3), like the residual electromagnetic force between neutral atoms, exert a strong attractive force between nucleons that are in close proximity. Coulomb repulsion, on the other hand, tends to drive protons apart. For large atomic nuclei, the Coulomb repulsion destabilizes the nucleus, so that it can *decay*, breaking apart into smaller, more stable pieces and releasing energy in the process. Depending upon the sizes of the pieces, this process is known as *nuclear fission* or *α-decay*. These processes are responsible for geothermal energy production and can be harnessed in nuclear reactors to provide large amounts of power from small amounts of material. On the other hand, for small nuclei the strong attraction at short distances is great enough that if they come together with

enough energy to overcome the longer-range Coulomb repulsion, the nuclei can *fuse*, again releasing tremendous amounts of energy. This is the energy production mechanism in the Sun; decades of effort have been exerted to produce energy by similar processes here on Earth. In this chapter, we introduce the basic ideas needed to understand nuclear binding and nuclear fission and fusion processes.

Although a detailed exploration of the structure and interactions of nuclei requires considerable quantum mechanics, basic aspects that enter nuclear energy processes can be understood fairly easily. We cannot entirely escape quantum mechanical concepts, but they enter mainly in four relatively straightforward ways that were described in §7 and §15:

Quantized energies Nuclei, like all quantum systems, can exist in various states with specific discrete values of energy.

Pauli exclusion principle Nucleons are fermions. No more than one nucleon of a specific type (proton, neutron) can occupy a single quantum mechanical state.

Random decays Quantum processes such as decays of unstable bound states occur *randomly* with a characteristic natural lifetime τ, or *half-life* $\tau_{1/2} = \ln_e 2 \times \tau = 0.6932\,\tau$.

> **Reader's Guide**
> This chapter introduces the basic structure of atomic nuclei, develops a simple model for nuclear binding energies, and explores the mechanisms by which nuclei decay. This provides essential preparation for the following chapters on nuclear power and radiation (§18–§20). The mechanisms described here also play an important role in solar (§22) and geothermal (§32) energy production.
>
> Prerequisites: §7 (Quantum mechanics), §9 (Energy in matter), §15 (Quantum processes).

Tunneling Some processes that are strictly forbidden classically do occur in the quantum world, usually with a small but not negligible probability. The most famous example is *barrier penetration*.

We begin in §17.1 by introducing the basic terminology of nuclear physics, particularly the standard terms for describing the binding energies of nuclei. Next in §17.2 we use the general properties of the strong nuclear interaction to develop a simple model for nuclear binding energies. Known as the *semi-empirical mass formula* or *SEMF*, we use this fundamental tool in §17.3 and §17.4 to explore the systematics of nuclear binding, stability, and decay mechanisms.

17.1 Basic Nuclear Properties

17.1.1 Nuclear Mass, Size, and Energy Scales

Nuclei are small – no more than 10^{-12} cm in radius – but very dense. The radius of the electron cloud in a typical atom is of order 10^{-8} cm, so the nucleus is only 1/10 000th the radius of an atom. The lightest nucleus, a single proton, has a mass of 1.672×10^{-27} kg, and is almost 2000 times more massive than the electron.

Nuclear reactions often involve emission or absorption of energy in amounts that are a significant fraction of the rest mass energy of the nucleus itself. It is therefore useful to adopt units in which *mass and energy are measured in the same units*. This is done by expressing masses in terms of the equivalent energy through $E = mc^2$ (see §9). A convenient unit for measuring energies and masses on the nuclear scale is the *MeV* (10^6 electron-volts). For a proton, $Mc^2 = 938.27$ MeV. Sometimes the MeV is used as a unit of mass, even though it is actually a unit of energy. Properly it should be divided by c^2, but sometimes the c^2 is not written explicitly. So, for example, the proper statement would be $M_p = 938.27$ MeV/c^2, but this is often simply written as $M_p = 938.27$ MeV. We adopt this convention in the chapters on nuclear energy and measure both masses and energies in MeV. The conversion factor is 1 MeV $= 1.783 \times 10^{-30}$ kg.

Energy is given off when particles form bound systems. For example, the *binding energy $B = 13.6$ eV* is released when an electron combines with a proton to form hydrogen in its ground state. Conservation of energy requires that the mass of the hydrogen atom is less than the mass of the proton plus the mass of the electron by exactly the amount of energy that is given off when they form (divided by c^2)

$$M_H = M_p + m_e - B/c^2 . \qquad (17.1)$$

Note that the binding energy is *positive* for a bound system. Because 13.6 eV is so much smaller than the rest energy of the electron and proton, the mass of the hydrogen atom is only very slightly less than that of its constituents (about 10 parts per billion). The binding energies of nuclei formed from protons and neutrons, however, are about 1% of their rest energy. Therefore the masses of nuclei are typically 1% or so less than the combined masses of the free protons and neutrons from which they are made.

Nuclear energy scales are large compared to the energies available in chemical processes. The rest masses of protons and neutrons are $\sim$1000 MeV, but energies of this order are not available in nuclear energy processes since nuclear reactions do not destroy or create nucleons. Nuclear processes do, however, convert nuclei and nucleons into different forms, thereby routinely emitting energies of order 10 MeV. This is thus the natural scale for nuclear energy. The comparable scale for atomic transformation is a few electron volts, so nuclear processes are inherently about six orders of magnitude more energy intensive than atomic processes. Therein lies both the promise and the threat of nuclear energy.

In §14 we described how protons and neutrons are built out of more fundamental particles called *quarks*. Although we know the fundamental equations that govern the motions of quarks, we cannot solve these equations accurately enough in multi-quark systems to precisely determine the wavefunctions and binding energies of quarks in protons and neutrons. Fortunately though, we do not need to study quarks to learn about most aspects of nuclear physics. The energy required to resolve quarks or to excite quantum quark degrees of freedom within nucleons is much higher even than the already high energies available in nuclear energy processes. So, for our purposes, protons and neutrons can be regarded as structureless building blocks of nuclei.

17.1.2 Isotopes and Isobars

The number of protons in a nucleus is denoted by Z, the number of neutrons by N, and the sum by A. Protons have electric charge $+e$, electrons have charge $-e$, and neutrons are electrically neutral. Z is known as the **atomic number**, and determines the number of electrons in a neutral atom. A is known as the *mass number*, *atomic mass number*, or *baryon number*, as discussed in §14. Since atoms with each value of Z have distinct chemistry, they were distinguished long ago as distinct *elements*.

Two nuclei with the same Z and different N are called different **isotopes** of the same element. Since isotopes have the same nuclear charge, they have the same electronic configuration and behave chemically as the same element. *Isotopes of a single element are therefore almost impossible to separate from one another by chemical means.* Because isotopes do differ in mass, they can be separated by processes that are sensitive to the masses of atoms. A simple example is **distillation**: if a solution containing different isotopes of an element is boiled then the fraction that boils first will be slightly richer in the lighter isotope. Repeated distillation can, with great effort, be used to concentrate one isotope of an element. Other processes that can concentrate one isotope relative to another are (i) centrifugation: a centrifuge mimics an extremely strong gravitational field that separates isotopes by weight; and (ii) gaseous diffusion: the rate at which a gas diffuses through a membrane depends on the mass of the individual molecules, again differentiating between isotopes with the same chemistry but different mass. Chemical separation of elements can be performed in ordinary laboratories and scaled up to industrial plants. Physical separation of isotopes, on the other hand, requires special methods and substantial investment in high-tech equipment. This distinction plays a role in issues related to nuclear reactor fuel cycles and weapons proliferation.

Some elements have only one stable isotope – fluorine ($Z = 9$), for example, comes only with $N = 10$. Others have many – tin ($Z = 50$) has eight stable isotopes, with $N = 64$–70 and 72, and two more with lifetimes far longer than the age of the universe ($N = 62$ and 74). Although they are chemically identical, different isotopes of the same element may have wildly different nuclear properties and must be considered effectively as different species in nuclear energy physics. Occasionally it is useful to refer to nuclei with the same A and different Z and N as **isobars**. A distinct nuclear species, with specific Z and N, is known as a **nuclide**.

Isotopes

Isotopes are nuclei with the same number of protons and different number of neutrons. Two isotopes of the same element have nearly the same chemistry and cannot be separated by chemical means. Separating isotopes requires massive, complex facilities and has only been achieved by a few countries.

To specify a nuclide uniquely we need to know both Z and N. The standard notation ^{A}Z for a nuclide specifies both Z and A while N is obtained by subtraction, $N = A - Z$. Usually the value of Z is denoted by the chemical symbol (H, He, Li, ...) of the element. Thus ^{14}C is the isotope of carbon with 14 nucleons. One must either look up or remember the fact that carbon has six electrons and therefore a nuclear charge of six. Thus it has six protons and, by subtraction, eight neutrons. To save the trouble of looking up or remembering the nuclear charge of the chemical elements, we often specify Z explicitly: $^{14}_{6}$C, where the presubscript gives Z and the presuperscript gives A, and $N = A - Z$ can be extracted when needed.

17.1.3 Using Mass Excesses

To understand energy production in nuclear processes we need to be able to compare the energies (i.e. the masses) of the reactants and products in nuclear reactions. Depending on the situation we may use *binding energies* or *mass excesses*.

The most direct way to quote nuclear masses would be in MKS units (kilograms) or, as we have done for protons and neutrons, in MeV. The binding energy of a nucleus would then be simply the energy that has to be added to decompose the nucleus into its constituent protons and neutrons. Nuclei usually come, however, with atomic electrons that neutralize their electric charge. So it is more convenient to compare the mass of a neutral atom with the mass of Z hydrogen atoms and N neutrons from which it could be formed. We therefore define the **nuclear binding energy** to be

$$B(Z, A) \equiv Z \times M_H + N \times M_n - M_{\text{atom}}(Z, A). \quad (17.2)$$

(Remember that factors of c^2 are understood when converting mass to energy.) There are the same number of electrons in the neutral atom and in Z hydrogen atoms, so the electron rest masses $Z \times m_e$ cancel out on the right-hand side of eq. (17.2). The electron *binding energies*, however, which are different in an atom with charge Z than in Z hydrogen atoms, do not cancel, so $B(Z, A)$

really includes both the nuclear and atomic binding energy. Since the majority of the electrons remain bound during a typical nuclear reaction, the energy defined in eq. (17.2) is the right one to use to compute energy changes in nuclear reactions. If one wanted to study totally ionized nuclei, one would have to remove the atomic binding energies that are included in $B(Z, A)$.

Since the masses of atoms grow roughly linearly in A, it has become customary to subtract a term proportional to A from the mass and call the result the atomic **mass excess**.[1] By convention, the term subtracted is taken to be $A/12$ times the mass of the neutral $^{12}_{6}C$ atom. The mass excess of $^{12}_{6}C$ is thus zero by definition. The quantity subtracted for each nucleon is called the **atomic mass unit** (u), or **dalton**, and abbreviated "u,"

$$1\,u = \frac{1}{12}M(^{12}_{6}C) = 1\,\text{dalton} = 931.494\,\text{MeV}\,. \quad (17.3)$$

In tabulations of nuclear masses (e.g. [70]), data are often presented in terms of atomic mass excesses,

$$\Delta(Z, A) = M_{\text{atom}}(Z, A) - A\,u$$
$$= M_{\text{atom}}(Z, A) - A \times (931.494\,\text{MeV})\,. \quad (17.4)$$

For example, the mass excess of $^{4}_{2}He$ is 2.425 MeV, which is $(2.425 \div 931.494)\,u = 2.603 \times 10^{-3}\,u = 2\,603\,\mu u$. Some fundamental masses and mass excesses useful in nuclear physics are given in Table 17.1.

The energy released (or absorbed) in a nuclear reaction can be computed using mass excesses in much the same way that the reaction enthalpies of chemical reactions are computed from enthalpies of formation using Hess's law. The simplest case is the decay of an initial nucleus into two or more pieces. The initial energy is just the rest energy of the nucleus. The final energy is the sum of the rest energies of the decay products plus the energy released in kinetic energy and photons, known as the **Q-value** of the decay. Energy conservation dictates that

$$Q = M_{\text{initial}} - \sum_{\text{products}} M_i$$

$$= (\Delta_{\text{initial}} + A_{\text{initial}}\,u) - \sum_{\text{products}} (\Delta_i + A_i\,u)\,, \quad (17.5)$$

where M_i, Δ_i, and A_i are the mass, mass-excess, and mass number of the ith decay product. Because the total number of nucleons is conserved in the decay, $A_{\text{initial}} = \sum_{\text{products}} A_i$, and eq. (17.5) simplifies to

---

[1] Often, instead of *mass excess*, one finds **mass deficit** or **mass defect**, which is simply the *negative of the mass excess*.

Table 17.1 Some fundamental mass units for nuclear physics (accurate to 1 KeV).

	Mc^2 (MeV)	Mass excess (MeV)	Mass excess (μu)
Proton	938.272	6.778	7 276
Neutron	939.565	8.071	8 665
Hydrogen	938.783	7.289	7 825
1 dalton	931.494	0	0
Electron	0.511	0.511	549
$^{4}_{2}He$	3728.401	2.425	2 603

Mass Excesses

The mass excess of a neutral atom, $\Delta(Z, A)$, is the atom's mass minus $A/12$ times the mass of the neutral atom ^{12}C. Mass excesses are usually expressed in MeV or *daltons* (u). Energy release in nuclear reactions and decays can be conveniently determined by computing the difference between the mass excesses of reactants and products.

$$Q = \Delta_{\text{initial}} - \sum_{\text{products}} \Delta_i\,. \quad (17.6)$$

The binding energy of a nucleus, defined in eq. (17.2), is simply related to the mass excess (Problem 17.1),

$$B(Z, A) = N\Delta(n) + Z\Delta(^{1}_{1}H) - \Delta(Z, A)\,. \quad (17.7)$$

17.1.4 Conservation Laws

Nuclear energy processes are constrained by fundamental conservation laws that dictate which reactions are allowed and which are forbidden. Of course, conservation of energy (including rest mass energy) is essential. Conservation of electric charge and baryon number (mass number) are also respected by all known interactions. The strong interactions that bind nuclei conserve the number of protons and neutrons separately, as do electromagnetic interactions. As discussed in §14.3, however, the *weak interactions* allow protons and neutrons to transform into one another by β-*decay*, $n \rightarrow p + e^{-} + \bar{\nu}_e$, and related processes. Thus, while the total number of nucleons is conserved in any nuclear reaction, protons and neutrons can be converted into one another by processes mediated through the weak interaction.

A free neutron β-decays with a lifetime of $\tau_n = 885.7 \pm 0.8$ s. Energy equal to the rest mass of the neutron minus

---

Box 17.1   Where to Find Nuclear Data

A convenient place to find data on the mass excesses of nuclei is the online database maintained by the **National Nuclear Data Center** (**NNDC**) at Brookhaven National Lab.

The database contains extensive information on a variety of properties of nuclei and their interactions. A useful summary can be found in what, for historical reasons, are known as the "Nuclear Wallet Cards" [70]. An example, taken from the entry for rubidium, is shown on the right. For each element, the

Nucleus	E(level) (MeV)	Jπ	Δ(MeV)	$T_{1/2}$	Abundance	Decay Modes
$^{85}_{37}$Rb^{FF}	0.0000	5/2−	−82.1673	STABLE	72.17% 2	
$^{86}_{37}$Rb	0.0000	2−	−82.7470	18.642 d 18		β⁻ : 99.99 % ε : 5.2E-3 %
$^{86m}_{37}$Rb	0.5561	6−	−82.1909	1.017 m 3		IT : 100.00 % β⁻ < 0.30 %
$^{87}_{37}$Rb^{FF}	0.0000	3/2−	−84.5977	4.81E+10 y 9	27.83% 2	β⁻ : 100.00 %

Wallet Cards include a list of all known isotopes, their mass excess (Δ), half-life ("$T_{1/2}$"), percentage abundance (if naturally occurring), and decay mechanisms (if unstable). Some other information: a nucleus marked with an "m," as in $^{86m}_{37}$Rb, is a long-lived, or *metastable* excited state. Its excitation energy above the ground state is listed in the column labeled "E(level)." A nucleus labeled with "FF," as in $^{85}_{37}$Rb^{FF}, is a common *fission fragment* in fission of ^{235}U. The column labeled "Jπ" gives the angular momentum J in units of $\hbar$ and *parity* $\pi = \pm$ (see Table 14.1) of the ground state of the nucleus. NNDC data is essential for the problems in this chapter.

(Table credit: National Nuclear Data Center)

Conservation Laws

Nuclear reactions and decays must satisfy conservation of energy, electric charge, and nucleon number. The numbers of protons and neutrons are separately conserved by all processes except those involving the weak interactions, which allow neutrons to turn into protons or *vice versa*.

the rest masses of the proton and electron, appears as kinetic energy on the final particles,

$$Q = M_n - M_p - m_e = 939.56 - 938.27 - 0.51 = 0.78 \text{ MeV} \tag{17.8}$$

(the rest mass of the neutrino is negligibly small, but not zero). A neutron inside a nucleus may undergo this decay, but in many nuclei the decay of the neutron is forbidden by energy conservation. Depending on energetics, the β-decay process can also manifest itself in different ways (§14.3),

$$\text{Proton } \beta^+ \text{ decay: } \quad p \to n + e^+ + \nu_e,$$
$$K\text{-capture: } \quad p + e^- \to n + \nu_e. \tag{17.9}$$

Both of these reactions as written are endothermic, so they do not occur in free space. They can, however, occur when protons are bound inside certain nuclei. β-decay is described in more detail in §17.4.3.

Nuclei can thus transform into one another with the emission of radiation, while leaving the total energy, charge, and mass number unchanged. When we deal with nuclear decays and other forms of radioactivity we have to keep in mind that nuclear reactions must obey these fundamental conservation laws.

17.2 The Semi-empirical Mass Formula

Although the quantum dynamics of nuclei is very complicated and still not entirely understood, the important features of nuclear binding needed to describe nuclear energy processes are summarized in a simple and intuitive model known as the **semi-empirical mass formula** or **SEMF**. In this section we describe different contributions to the SEMF through properties of the forces between protons and neutrons. Although at several points in the motivation of the SEMF it is necessary to take A to be large, the SEMF describes many features of nuclear binding even down to very small A, making it useful for almost the whole range of nuclei that participate in nuclear energy processes.

17.2.1 The Nuclear Force

The force between two nucleons was surveyed briefly in §14. We explore it further here with the aim of extracting those features that dominate the structure of nuclei. As emphasized in §14, the force between nucleons is strong, short-range, and complicated. Despite its complexity, the main contribution to the nuclear force can be summarized *approximately* by a central potential $V(r)$ with the following properties:

- The strong forces between pp, nn, and pn are quite similar. Careful studies indicate that the attractive nuclear force between a proton and neutron (pn) is somewhat stronger than the attraction between pp or nn, which are approximately equal.

- Two protons repel one another electrostatically whereas pn and nn do not.

- The nucleon–nucleon force is attractive at long distances and very repulsive at short distances. Here "long" and "short" are relative terms. Attraction dominates for $r > 0.8 \times 10^{-15}$ m $= 0.8$ fm. The long-range, attractive part is well-approximated by the potential energy function $V_{\text{strong}}(r)$ given in eq. (14.5). At distances less than about 0.8×10^{-15} m, however, the nuclear force turns sharply repulsive. Known as the **hard core**, this repulsive force prevents protons and neutrons from collapsing into one another.

A sketch of the nucleon–nucleon potential energy is given in Figure 14.3, where both the short-range repulsion and the longer range attraction are evident.

17.2.2 The Liquid Drop Model of the Nucleus

The nuclear forces between nn, pn, and pp are attractive in all cases at distances above 1 fm. Thus a collection of protons and neutrons will tend to be bound by the attraction of each nucleon to its neighbors. The nuclear force falls with distance so fast, however, that any one nucleon is not strongly influenced by nucleons more distant than its nearest neighbors. On the other hand, the fact that the nuclear force becomes repulsive at short range keeps nucleons from getting too close to one another. As a result, a collection of protons and neutrons behaves qualitatively like a bunch of impenetrable balls that stick to one another. The protons and neutrons do not occupy fixed positions relative to one another as in a solid; instead they slither about, keeping close, but not too close, to their neighbors. The classical system most resembling a nucleus is a droplet of an incompressible liquid. The hard core repulsion keeps the *density* of the nucleus constant as more and more nucleons are added to it. The longer-range attraction keeps the nucleons close to one another, giving the nucleus a well-defined surface. This structure is illustrated schematically in Figure 17.1. The structure of the nucleus is in striking contrast to atoms, which are organized about the nucleus at their center. Atoms have no recognizable surface, and as

Figure 17.1 A cartoon of a nucleus as a liquid drop. The protons and neutrons are held in close contact by attractive forces, but prevented from overlapping by the hard core repulsion. Like molecules in a liquid, the nucleons are free to move about as long as the density of the nucleus remains approximately constant. This is a cartoon because quantum effects are important in nuclei, so the nucleons must be thought of as de Broglie waves, and cannot be frozen in place for an image like this.

> **A Liquid Drop**
>
> For values of A that are not too small ($A \gtrsim 10$), a nucleus resembles a drop of incompressible liquid with a relatively sharp surface.

Z increases, the increasing nuclear charge pulls the electron cloud inward so that the size of atoms remains roughly constant while the density of the electron cloud increases.

The simple model of a nucleus as a droplet of incompressible fluid is known as the **liquid drop model**, and was first proposed by the Russian physicist George Gamow in the late 1920s. Although it is only a first approximation to the real structure of nuclei, the liquid drop picture is the starting point for deriving the semi-empirical mass formula. First developed by the German physicist Carl von Weizsäcker in 1935, the SEMF gives quite an accurate description of nuclear binding energies.

The word "semi-empirical" needs some explanation. The SEMF is a formula for the binding energy $B(Z, A)$ of a nucleus as a function of Z and A. The *functional form* of the Z and A dependence is based on solid physical principles, but certain numerical parameters in the formula are adjusted to fit experiment. A more fundamental approach would seek to predict both the form and the magnitude of the contributions to $B(Z, A)$ from deeper principles. A totally empirical approach would simply fit $B(Z, A)$ in terms of some convenient set of functions such as a power series. The SEMF sits between these two extremes. It gives an excellent fit to nuclear binding energies, and the terms in the formula all have physical interpretations.

17.2.3 Nuclear Volume and Volume Energy

The interplay between long-range attraction and strong short-range repulsion keeps the density of nuclei approximately constant, independent of the number of nucleons, A. Therefore the nuclear volume grows proportionally to A. Equivalently, the radius grows like $A^{1/3}$. Experiments show that

$$R(A) \approx R_0 A^{1/3}, \qquad (17.10)$$

with estimates of R_0 ranging between 1.1 and 1.5 fm depending on the way it is defined and the process in which it is measured.[2]

A nucleon deep within a nucleus sees a uniform environment and interacts only with its nearest neighbors. Since the interaction is local, each interior nucleon contributes the same amount to the binding of the nucleus. The dominant component of the binding energy of a large nucleus arises from adding these local contributions over all nucleons, and is thus simply proportional to the number of nucleons A, or (equivalently) to the volume of the nucleus.

$$B_V = \epsilon_V A. \qquad (17.11)$$

Fits to nuclear binding energies give $\epsilon_V \cong 15.56$ MeV.[3]

17.2.4 The Nuclear Surface and Surface Energy

Nucleons near the nuclear surface do not have as many attractive "bonds" as nucleons in the interior, so the binding energy of the nucleus is reduced by an amount proportional to the surface area. Nuclei are thus generally spherical, as that shape minimizes the surface area for an object of fixed volume.[4] The area of a sphere grows like the square of its radius, so the surface area of a nucleus is proportional to $A^{2/3}$.

This leads us to conclude that there is a surface energy contribution to nuclear binding that is negative and proportional to $A^{2/3}$,

$$B_S = -\epsilon_S A^{2/3}. \qquad (17.12)$$

Fits give $\epsilon_S \cong 17.23$ MeV.

17.2.5 Coulomb Repulsion

In a typical nucleus, the protons are distributed roughly uniformly. One might think that Coulomb repulsion would push the protons to the outside of the nucleus, but two stronger effects drive the uniform distribution of nucleons. First, as mentioned above, nuclear interactions are more attractive for pn than for nn or pp, which favors a relatively uniform distribution of protons throughout the nucleus. The second effect is the *Pauli exclusion principle* (§15.5), which forces like particles into higher energy states than unlike particles. This effect favors a uniform spatial distribution of proton probabilities in the nuclear wavefunction. Assuming that the nucleus is spherical, its electrostatic Coulomb energy can be approximated as that of a uniformly charged sphere,

$$E_{\text{coulomb}} = \frac{3}{5} \frac{e^2}{4\pi\epsilon_0} \frac{Z^2}{R}, \qquad (17.13)$$

where $R \propto A^{1/3}$. The positive Coulomb energy tends to destabilize the nucleus, so its contribution to the binding energy is negative,

$$B_C = -\epsilon_C \frac{Z^2}{A^{1/3}}. \qquad (17.14)$$

The coefficient $\epsilon_C \cong 0.7$ MeV can be computed from the above parameters and (17.10), and matches well with empirical measurements.[5] Note that ϵ_C is much smaller than ϵ_V and ϵ_S, reflecting the fact that electromagnetic interactions are much weaker than strong interactions. B_C becomes important for large Z, however, because it grows quadratically in Z.

17.2.6 Symmetry Energy

Just as stronger pn binding and the Pauli exclusion principle favor a uniform distribution of protons throughout the nucleus, these effects also tend to favor equal numbers of protons and neutrons within a nucleus, so that nuclear binding energies decrease as $|N - Z|$ increases. This effect is captured by a term in the binding energy known as the **symmetry energy** that is negative and proportional to $(N - Z)^2/A$.

The effect of pp, nn, and pn interactions can be estimated in a simple model in which we simply add up the pairwise interactions. We assume that each proton or neutron interacts with ν nearest neighbors and that the

---

[2] When, in later sections, a definite value is required, we take $R_0 \cong 1.3$ fm.

[3] We take the parameters of the SEMF from [77].

[4] Sometimes other considerations overwhelm the surface energy-driven tendency toward spherical shape, causing nuclei to deform (§17.3.2).

---

[5] The repulsive Coulomb energy of a single proton is already included in its rest mass, so B_C should vanish for $Z = 1$; a more accurate form for eq. (17.14) is $B_C(Z, N) = -\epsilon_C Z(Z - 1)/A^{1/3}$. The difference between this and eq. (17.14) is small and usually neglected when N is reasonably large.

nucleons are uniformly distributed as discussed above. We also make the approximation that the number of nucleons is sufficiently large that we can neglect corrections of order $1/A$. We denote the contribution of a pn pair to the nuclear binding energy as $\epsilon(pn) \equiv \bar{\epsilon} + \delta$ and that of a pp or nn pair as $\epsilon(pp) = \epsilon(nn) \equiv \bar{\epsilon} - \delta$. δ is small and positive, reflecting the fact that pn attraction is slightly greater than pp or nn. It remains only to estimate the number of pp, nn, and pn pairs. The resulting contribution to the nuclear binding energy B can be conveniently expressed in terms of the fraction of protons $f = Z/A$,

$$B = \frac{1}{2}\nu A \left((f^2 + (1-f)^2)(\bar{\epsilon} - \delta) + 2f(1-f)(\bar{\epsilon} + \delta) \right),$$
(17.15)

or in terms of N, Z, and A,

$$B = \frac{1}{2}\nu\bar{\epsilon}A - \frac{1}{2}\nu\delta\frac{(N-Z)^2}{A}.$$
(17.16)

The first term contributes to the volume energy (17.11), while the second is a contribution to the *symmetry energy*, $B_{\text{sym}}^{(1)}(Z, N) = -\frac{1}{2}\nu\delta(N-Z)^2/A$.

The form of $B_{\text{sym}}^{(1)}$ could have been anticipated on general grounds. The interaction energy is most negative (and the binding energy is maximized) when $N = Z$ and the number of pn pairs is maximum. Simply expanding the energy in a power series (similar to the expansion of the potential in Box 2.2) about this extremum leads to the $(N-Z)^2$ dependence. The nearest-neighbor nature of the forces requires that the interaction energy will grow linearly with A for any ratio of Z to N, hence the factor of $1/A$.

The same general principles apply to the second contribution to the symmetry energy, which comes from the Pauli exclusion principle. The essence of this effect can be captured in a simple model where we think of the nucleus as a spherical box in which we place Z protons and N neutrons. The motivation for this model is the observation that nuclei, like atoms, show evidence for *shell structure*. The shell structure of atoms arises from sequential filling of energy levels in the Coulomb field of the atomic nucleus. Nuclear shell structure arises because the nucleons themselves give rise to an average potential or *mean field*, the energy levels of which are effectively filled sequentially by the protons and neutrons in the nucleus. The simplest version of this *shell model* is a spherical container with infinite walls, whose size expands with A so that the nuclear density remains constant. According to the Pauli exclusion principle, each energy state in the mean field can hold two protons (one with spin-up and one with spin-down) and two neutrons. As more protons and neutrons are added, they must be placed in increasingly energetic

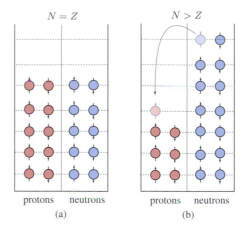

Figure 17.2 Filling energy levels in nuclei. Each single-particle energy level can hold two protons, spin-up or down, and two neutrons. (a) A nucleus with $N = Z$; (b) a nucleus with the same value of A, but $N > Z$. The energy of the nucleus with $N > Z$ is greater than that of the nucleus with $N = Z$. This can be seen by imagining changing one of the neutrons into a proton and placing it into a lower, unoccupied energy level, as shown in the figure.

states. Figure 17.2 illustrates the situation for a nucleus with a fixed value of A for (a) $N = Z$, and (b) $N > Z$. As the figure suggests, the binding energy at fixed A is maximum when $N = Z$. When expanded about the maximum, a negative contribution proportional to $(N - Z)^2$ emerges when $|N - Z| \ll A$. Like $B_{\text{sym}}^{(1)}$, this contribution also decreases proportional to $1/A$ (see Box 17.2). The net effect is a second contribution to the symmetry energy $B_{\text{sym}}^{(2)} \propto -(N - Z)^2/A$.

Combining $B_{\text{sym}}^{(1)}$ and $B_{\text{sym}}^{(2)}$ gives an empirical symmetry energy parameterized as

$$B_{\text{sym}} = -\epsilon_{\text{sym}}\frac{(N-Z)^2}{A}.$$
(17.17)

An important feature of the symmetry energy that one might not have guessed without studying these simple models is its A^{-1} dependence, which diminishes the tendency for N and Z to be equal when A is large. Fits give $\epsilon_{\text{sym}} = 23.28\,\text{MeV}$.

17.2.7 Pairing Energy

One further effect to be included in a model of nuclear binding is the tendency for nuclei with even numbers of protons and neutrons to be more tightly bound than nuclei in which N or Z is odd. When two protons or two neutrons share the same spatial state, they overlap strongly, enhancing their attractive interaction. In a nucleus with odd N and Z, and $N > Z$ as is the case except when A is small, the

---

Box 17.2   The Symmetry Energy of a One-dimensional Nucleus

As a toy model to illustrate the symmetry energy contribution that takes the form $B_{\text{sym}} \propto -(N-Z)^2/A$, we consider a one-dimensional "nucleus" of length L filled with non-interacting protons and neutrons, where L is adjusted to keep the nuclear density constant. The energy levels of a particle of mass M in a one-dimensional box of length L are $E_n = \hbar^2\pi^2 n^2/2ML^2$ (§7.2.3). Because the Pauli principle allows two protons and two neutrons for each value of n, the levels up to $n(\text{proton}) = Z/2$ and $n(\text{neutron}) = N/2$ must be filled to accommodate Z protons and N neutrons. The total energy of this system is

$$E = 2\frac{\hbar^2\pi^2}{2ML^2}\left(\sum_{n=1}^{Z/2} n^2 + \sum_{n=1}^{N/2} n^2\right) \approx \frac{\hbar^2\pi^2}{24ML^2}(Z^3 + N^3),$$

where we have assumed $Z, N \gg 1$ in order to replace the sums by integrals. If the length of our one-dimensional "nucleus" were fixed, the energy would increase cubically with Z and N. The length, however, must grow with A in order that the number density $\rho = A/L$ remains fixed, so replacing L by A/ρ,

$$E \approx \frac{\hbar^2\pi^2\rho^2}{24M}\frac{Z^3 + N^3}{A^2}.$$

As expected this has an extremum at $N = Z$. Rewriting this expression in terms of A and $N - Z$, we find

$$E \approx \frac{\hbar^2\pi^2\rho^2}{96M}\left(A + 3\frac{(N-Z)^2}{A}\right).$$

The first term, proportional to A, gives a contribution to the volume energy $B_V^{(1D)}$. The second term gives a negative contribution to the symmetry energy,

$$B_{\text{sym}}^{(1D)} \propto -\frac{\hbar^2\rho^2}{M}\frac{(N-Z)^2}{A}.$$

Although the counting of states and the numerical coefficients differ in a three-dimensional box (and ρ^2 is replaced by $\rho^{2/3}$), the A and $N - Z$ dependences of the symmetry energy contributions are the same in 1D and in 3D.

---

extra proton and neutron are typically in very different spatial states and their attractive interaction is not enhanced. This effect is parameterized by a term in the binding energy of the form,

$$B_{\text{pairing}} = \eta(Z, N)\frac{\Delta}{A^{1/2}}, \qquad (17.18)$$

where $\eta = 1$ for *even–even nuclei* (Z and N both even), $\eta = 0$ for *even–odd nuclei* (either Z or N odd but not both), and $\eta = -1$ for *odd–odd nuclei*. The $A^{-1/2}$ dependence of this term is empirically determined. Δ is less well determined than the constants ϵ_V, *etc*. We use $\Delta \approx 12$ MeV.

17.2.8 The Semi-empirical Mass Formula

Combining all these contributions to nuclear binding together, we obtain the *semi-empirical nuclear mass formula*,

$$B(Z, A) = \epsilon_V A - \epsilon_S A^{2/3} - \epsilon_C \frac{Z^2}{A^{1/3}}$$

$$- \epsilon_{\text{sym}}\frac{(N-Z)^2}{A} + \eta(Z, N)\frac{\Delta}{A^{1/2}}. \quad (17.19)$$

Table 17.2 Parameters of the semi-empirical mass formula for nuclear binding energies (17.19). Different sources quote different values for these parameters. We follow [77].

Parameter	MeV	Factor
ϵ_V	15.56 MeV	A
ϵ_S	17.23 MeV	$-A^{2/3}$
ϵ_C	0.7 MeV	$-Z^2/A^{1/3}$
ϵ_{sym}	23.28 MeV	$-(N-Z)^2/A$
Δ	12 MeV	$\eta(Z, N)/A^{1/2}$

As will become clear in the following sections, this simple formula provides a remarkably accurate estimate of the binding energy of a nucleus as a function of Z and A, where $N = A - Z$. The parameters in the SEMF, which are determined by an empirical fit to the binding energy of stable and long-lived nuclei, are summarized in Table 17.2.

17.3 Nuclear Binding Systematics

The simple model of nuclear binding energies summarized by the SEMF can explain all the regularities of nuclear

Semi-empirical Mass Formula

The *semi-empirical mass formula (SEMF)* forms the foundation for an understanding of nuclear fission, fusion, and radioactivity. The SEMF expresses the binding energy of a nucleus with Z protons and $N = A - Z$ neutrons in terms of a small number of empirically determined constants,

$$B(Z, A) = \epsilon_V A - \epsilon_S A^{2/3} - \epsilon_C Z^2/A^{1/3}$$
$$- \epsilon_{\text{sym}}(N - Z)^2/A + \eta(Z, N)\Delta/A^{1/2},$$

where the parameters ϵ_V, ϵ_S, ϵ_C, ϵ_{sym}, and Δ are given in Table 17.2, and the function $\eta(Z, N)$ is $\eta = +1(-1)$ when both N and Z are even(odd) and $\eta = 0$ if $N + Z$ is odd.

masses that are of importance in the nuclear energy processes described in the following chapters.

17.3.1 Qualitative Discussion

First we survey some of the qualitative features of nuclear binding predicted by the SEMF. Because many of the terms in the SEMF depend primarily on A, it is natural to consider the mass number A as the primary parameter determining nuclear structure, even though the number of protons Z is the dominant parameter in determining chemical properties of elements. Many important features can be understood by approximating the nuclear binding energy as a smooth function of A and Z, or equivalently of N and Z. The first four terms in the SEMF have just such a smooth form; the pairing term Δ depends on the parity of Z and N, and produces small local fluctuations from the smooth function, which decrease as A increases. In this section we ignore the pairing term and describe qualitatively how the physics of the SEMF affects the shape of the binding energy function and the implications for nuclear stability.

While the volume and surface terms in the SEMF depend only on A, the symmetry and Coulomb energy terms depend on how A is divided up into protons and neutrons. At small A the symmetry energy is more important, because $\epsilon_{\text{sym}} \gg \epsilon_C$. Therefore the Coulomb energy can be ignored for small A and the most stable small nuclides have $Z \approx N$, which minimizes the symmetry energy. If A is even, then $Z = N$, as in $^4_2$He, $^{16}_8$O, and $^{40}_{20}$Ca. If A is odd then $Z = N \pm 1$ as in $^{19}_9$F and $^{39}_{19}$K (Question 17.3).

Light Nuclei

For light nuclei (small A), Coulomb energy can be ignored, the symmetry energy favors $N = Z$, and the SEMF is dominated by the volume and surface terms,

$$B(Z, A) \approx \epsilon_V A - \epsilon_S A^{2/3}.$$

For small A, the SEMF predicts that the binding energy per nucleon grows with A. As a result, energy is given off when two light nuclei fuse.

Neutron-rich Nuclei

As A increases, the Coulomb energy grows in importance relative to the symmetry energy, forcing the most stable nuclei to become neutron rich. Nuclei become less stable for larger A, as it is impossible to keep both the Coulomb and symmetry energy small.

Since the symmetry energy vanishes when $N = Z$, the binding energy of the most stable light nuclei is essentially determined by the volume energy $\epsilon_V A$ and the surface energy $-\epsilon_S A^{2/3}$ alone. As A increases, the effect of the surface diminishes relative to the volume by a factor proportional to $A^{-1/3}$. Thus, the **binding energy per nucleon** $b(Z, A) \equiv B(Z, A)/A$, increases with A for small A. In Example 17.1, for example, we found that ^{16}O is more tightly bound than ^{12}C. The increase of the binding energy per nucleon as the surface to volume ratio decreases explains why it is energetically favorable for light nuclei to fuse.

As Z and N increase it is not possible to ignore the symmetry energy and Coulomb energy because no single choice of Z/N keeps both of these contributions small. The symmetry energy prefers $Z = N$, while the Coulomb energy prefers to keep Z small. The conflict between these two imperatives determines many features of nuclear energy physics. If Z and N remained roughly equal at large A, with $Z \sim A/2$, the symmetry energy would be small, but the Coulomb energy would grow like $A^{5/3}$ and would eventually dominate all other terms. Instead, the most stable nuclide at a given A shifts steadily toward a neutron excess as A grows large, representing a compromise

---

Example 17.1   Mass Versus Mass Excess

The masses of the neutral hydrogen atom and the neutron are 938.783 MeV and 939.565 MeV respectively, so their mass excesses are

$$\Delta(^1_1\text{H}) = M_H - u = M_H - 931.494\,\text{MeV} = 7.289\,\text{MeV}$$

$$\Delta(n) = M_n - u = M_n - 931.494\,\text{MeV} = 8.071\,\text{MeV}.$$

Since carbon is a bound nucleus, its mass is less than the mass of the six protons, six neutrons, and six electrons of which it is built. Therefore the masses of the hydrogen atom (where electron binding energy is negligible at the nuclear scale of MeV) and neutron are both *greater than* one u.

The tabulated atomic mass excess for ^{16}O is $\Delta(^{16}\text{O}) = -4.737$ MeV. Therefore the mass of the neutral ^{16}O atom is

$$M_{\text{atom}}(^{16}\text{O}) = 16(931.494)\,\text{MeV} - 4.737\,\text{MeV} = 14\,899.167\,\text{MeV}.$$

Notice that the *negative* mass excess of ^{16}O indicates that it is more tightly bound than ^{12}C.

Example 17.2   Calculation With Mass Excesses

The α-decay process is a favorite way for heavy nuclei to decay and is discussed in detail in §17.4.2.
How much energy is released in the decay of $^{238}_{92}\text{U}$ into $^{234}_{90}\text{Th}$ and an α-particle, ^4_2He?
The mass excesses of $^{238}_{92}\text{U}$, $^{234}_{90}\text{Th}$, and ^4_2He are 47.304, 40.609, and 2.425 MeV/c^2 respectively, so the energy released in the decay

$$^{238}_{92}\text{U} \rightarrow\, ^{234}_{90}\text{Th} + {}^4_2\text{He}$$

is $Q = 47.304 - 40.609 - 2.425 = 4.270$ MeV. This assumes that all the nuclei remain neutralized by electrons in the atomic ground state. In the real world, the α-particle is emitted fully ionized (no electron) and the uranium nucleus sheds two electrons when it turns into thorium. The differences in atomic binding energy are so small that they can be neglected relative to the 4.27 MeV energy released in the decay. For another example, see Problem 17.2.

The Valley of Stability

The negative of the binding energy has the shape of a valley, known as the *valley of stability*. A slice through the valley at constant A has a parabolic profile (if pairing effects are ignored). The floor of the valley of stability lies along the $Z = N$ line for small A, but bends toward $N > Z$ as A increases. As A increases, the floor of the valley gradually rises.

between the symmetry energy and Coulomb energy. This has two important consequences: heavy nuclei become neutron rich, and the binding energy per nucleon decreases.

To get a feel for the landscape of nuclear binding, consider how the binding energy varies with Z at fixed A. Both the symmetry and Coulomb energy vary quadratically with Z at fixed A. The shape of the binding energy surface is thus *parabolic* in this direction, with maximum binding energy at a value of $N - Z$ that is close to zero for small A and becomes increasingly positive as A increases.

In Figure 17.3, we plot the *negative* of the binding energy per nucleon, with $-b(Z, A)$ on the z-axis and Z and N on the x- and y-axes respectively. The more bound the nuclide, the lower its height in the graph. The resulting shape is picturesquely called the **valley of stability**, with stable nuclei and very long lived nuclei along a line that we refer to as the *valley floor*, and progressively less stable nuclei up the sides. Note that the valley itself, like a river valley, is not exactly flat. As A increases, the floor of the valley, which corresponds to the most tightly bound nucleus for each A, gradually rises.

Another perspective on the valley of stability is given in Figure 17.4, which looks down on the valley of stability from above. Since the depth of the valley measures the *binding energy per nucleon*, the lowest point in the valley

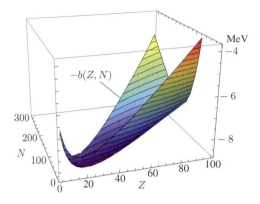

Figure 17.3 A 3D view of the valley of stability. $-b(Z, N)$ is plotted as a function of Z and N for $5 < Z < 100$ and $5 < N < 300$. The pairing energy term, which would modify this smooth shape by small local fluctuations depending upon the parities of N and Z, has been dropped for clarity. Contours of equal binding energy per nucleon are marked. The most tightly bound nuclei correspond to the bottom of the valley. Nuclei at both low A and large A are less bound, as are nuclei that depart from the value of N/Z that yields the tightest binding for fixed A.

(where water would accumulate were this geography rather than nuclear physics) is the most stable nuclide. This turns out to be very close to $Z = 26$, $A = 56$, the most common isotope of iron. Iron (and its neighbors, nickel and cobalt) therefore have a special place in our universe: nuclear processes, both **nuclear fusion** and **nuclear fission**, transform nuclei with small and large A toward iron. Energy can be extracted from nuclei by *fusing* light nuclei or by *fissioning* heavy nuclei. To study these processes more deeply we need to get a quantitative understanding of the valley of stability.

17.3.2 The Valley of Stability and the Systematics of Nuclear Binding

Armed with the SEMF, we take a closer and more quantitative look at the systematics of nuclear stability. To

Iron, Fusion, and Fission

The most stable nucleus is iron, $^{56}_{26}$Fe. Lighter nuclides (smaller Z, A) can *fuse*, combining together with protons or other light nuclei, while heavier nuclides (larger Z, A) can *fission* into smaller nuclides, often releasing additional neutrons. In each case, energy is released as the nuclei approach the stable iron nuclide.

display general trends, we ignore the pairing energy term, thereby essentially averaging over nearby nuclides that differ from one another by only one proton or neutron. We can thus treat Z and A as continuous variables, with respect to which we can differentiate to find extrema of the nuclear binding energy.

The Path Along the Valley Floor   The most stable nuclide for a given A can be found by setting the derivative $\partial b(Z, A)/\partial Z|_A$ to zero and solving for $Z_{min}(A)$. The result is (Problem 17.10)

$$Z_{min}(A) = \frac{A/2}{1 + \frac{\epsilon_C}{4\epsilon_{sym}} A^{2/3}} \cong \frac{A/2}{1 + 7.52 \times 10^{-3} A^{2/3}}.$$

$$(17.20)$$

When A is small, Z_{min} is very close to $A/2$. But as A increases to mass numbers on the order of 100, Z falls significantly below $A/2$. Z_{min} is plotted as a function of A in Figure 17.5. The SEMF estimate of $Z_{min}(A)$ agrees well with data (Problem 17.9).

Nuclear decay processes allow less stable nuclei to transform into more stable ones. Picturesquely, they can roll down the valley of stability, eventually ending up as a stable or very long-lived nucleus near the valley floor. In particular, as described further below, nuclei with fixed A can decay into one another by β-decay and related processes (§17.4.3), moving toward $Z_{min}(A)$ at fixed A unless they first decay by α-particle emission. We examine decays of unstable nuclei in detail in §17.4.

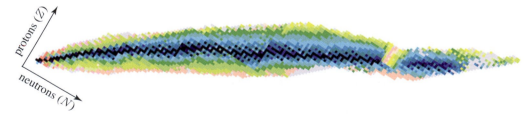

Figure 17.4 A view of the valley of stability from above. The most stable nuclides, the black squares, lie along the floor of the valley. Less stable nuclides, denoted with colors ranging from dark blue (relatively stable) to yellow (very short lived), line the sides of the valley. From [82].

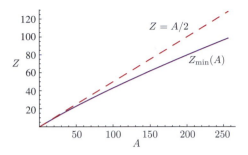

Figure 17.5 The charge Z_{min} of the most stable nuclide (solid) as a function of A, compared to $Z = A/2$ (dashed).

Equation (17.20) is a useful result because it suggests the types of nuclear species – Z as a function of A – that are most likely to be found in our world, and that could be fuels for nuclear energy processes.

Binding Energy Along the Valley Floor If we substitute $Z_{min}(A)$ from eq. (17.20) back into $B(Z, A)$, we obtain the binding energy of the most-bound nuclear species with each A, $B_{max}(A) \equiv B(Z_{min}(A), A)$. It is most convenient to study the binding energy per nucleon, $b_{max}(A) = B_{max}(A)/A$, because this quantity appears in the criteria for nuclear power reactions. After a little simplifying algebra, we obtain

$$b_{max}(A) = \epsilon_V - \epsilon_S A^{-1/3} - \frac{\epsilon_C}{4} \frac{A^{2/3}}{1 + \frac{\epsilon_C}{4\epsilon_{sym}} A^{2/3}}. \quad (17.21)$$

This prediction is compared with data in Figure 17.6 for values of A that occur in nature. The accuracy of the SEMF is confirmed by the fact that $b_{max}(A)$ agrees well with the binding energy of the most tightly bound nuclide as a function of A. Figure 17.6 also illustrates the fact, mentioned at the end of the last section, that the binding energy per nucleon reaches a maximum near ^{56}Fe, with both lighter and heavier nuclei being less bound.

Some Special Light and Heavy Nuclei A closer look at data on $b(Z, A)$, as depicted in Figure 17.6(b), reveals several significant binding energy anomalies for small A. The semi-empirical mass formula does a pretty good job, even for the lightest nuclei, with the very significant exception of ^{4}He. The assumptions that went into the derivation are not valid for nuclei with very small N and Z, however, so we need to look at these cases individually. The only nuclide with $A = 2$, ^{2}H, is the *deuteron* (§14). The deuteron is quite weakly bound, with a binding energy of 2.2 MeV. In contrast, $^4_2$He is quite tightly bound for a light nucleus, with a binding energy of over 28 MeV. Both

---

Light Nuclei

Light nuclei, particularly those with fewer than 10 nucleons, have binding energies that depart significantly from the SEMF. Some of these nuclei, particularly ^{4}He and the isotopes of hydrogen, ^{2}H and ^{3}H, are particularly important in nuclear energy physics.

---

nuclides with $A = 3 - {}^3$H, the **triton** (the nucleus of the **tritium** atom, an unstable, long-lived ($\tau_{1/2} = 12.3$ y), isotope of hydrogen), and ^{3}He, the rare, but stable isotope of helium – are rather weakly bound, as are both isotopes of lithium, ^{6}Li and ^{7}Li. ^{12}C and ^{16}O and a few even heavier nuclei such as ^{208}Pb are anomalously tightly bound. These anomalies are accounted for by the filling of *shells* in the nuclear mean field mentioned in §17.2.6.

The Limit of Nuclear Stability Even though the binding energy per nucleon is still very large around $A = 260$, the plot of measured nuclear binding energies in Figure 17.6 ends abruptly there. There are no stable, or even long-lived nuclides with values of A larger than about 250. In fact, the heaviest nuclide with a lifetime longer than a year is $^{251}_{98}$Cf. (Californium is an artificially created nuclide. This isotope has a lifetime of about 900 years.) The SEMF, however, predicts positive binding energy for nuclei anywhere up to $A \approx 3150$ (see Problem 17.11).

Why are these bound nuclei not found in nature? The answer is that they are classically unstable against deformations that lead to fission. In constructing the SEMF we noted that the nucleus behaves like an incompressible (constant density) fluid and that the surface energy (§17.2.4) is minimized when a nucleus is spherical. A volume-preserving deformation away from the spherical shape can, however, decrease the Coulomb energy of the nucleus. For small A, surface energy dominates because $\epsilon_C \ll \epsilon_S$, and nuclei remain roughly spherical. As A increases the Coulomb repulsion grows like Z^2 and becomes progressively more important. Eventually, for large enough A, Coulomb repulsion wins and an initially spherical nucleus would quickly deform and fission into two smaller, stable nuclei (see Figure 17.7). A nucleus with $A \geq 300$ has no position of stable equilibrium. If it could somehow be created, such a nucleus would almost instantly fall apart. To distinguish such instability from other types of fission, we refer to this as **instantaneous fission**. A simple estimate (Problem 17.12) suggests that the most recently discovered nuclide, with $Z = 118$ and $A = 294$,

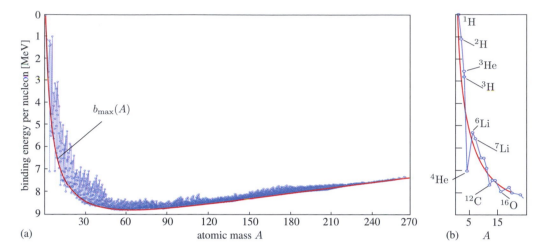

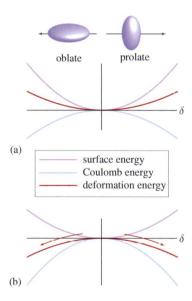

Figure 17.6 (a) The measured binding energy per nucleon of all long-lived nuclides is plotted (increasing downward) versus A. The valley of stability has been flattened into a one-dimensional plot whose lower edge traces the floor of the valley. The SEMF prediction (17.21) for the binding energy $b_{max}(A)$ per nucleon of the most bound nuclide at each value of A is plotted in red. (b) The low-A region of the left-hand plot, with several anomalously bound nuclides indicated.

Figure 17.7 The nuclear surface energy (purple) and Coulomb energy (blue) and their sum (red) as a function of a small deformation away from spherical shape (see Problem 17.12). (a) For nuclei with low to medium A, the increase in surface energy dominates over the decrease in Coulomb energy and the nucleus is stable against deformation. (b) For large values of A, the Coulomb energy dominates over the surface energy and the nucleus is classically unstable, so it instantaneously fissions as indicated by the arrows.

The End of the Table of Nuclides

The table of nuclei that live long enough to be observed ends at about $A = 300$ because larger nuclei would instantaneously fission as the Coulomb repulsion among protons overwhelms the surface energy.

is close to the limit of stability allowed by instantaneous fission.

Instantaneous fission is *not* the kind of fission that generates power in a nuclear reactor. Controlled nuclear fission occurs when certain very long lived nuclides such as ^{235}U are induced to fission by absorbing a neutron (see §18.3).

17.4 Nuclear Decays

Unstable nuclides, whether naturally occurring or created in nuclear fission or fusion reactors, almost always decay in one of three ways: α-particle emission, β-decay (and closely related processes), and (rarely) by *spontaneous fission* (§18.3.1, not to be confused with *instantaneous fission*).

These basic types of decays are often accompanied by emission of electromagnetic radiation as γ-rays. These decays are responsible for the radioactivity that accompanies the production of nuclear power. α- and β-decays are discussed in this section, while γ-radiation is discussed in §17.4.6. Spontaneous fission is addressed in §18.

The naturally occurring or easily produced nuclides of interest to us have A and Z values close to the minimum of the valley of stability. Thus we can use the SEMF, whose parameters were fit to nuclei near the minimum of the valley of stability, to study the decays of the nuclides that are of interest.

The most massive nuclei decay in a sequence of steps known as a *decay chain* (§17.4.5). Along a decay chain, α-decays (§17.4.2) decrease the mass of the nucleus by lowering A by four at each step. At various steps along the way, β-decays (§17.4.3) change the ratio of protons to neutrons to optimize stability at a given value of A. In this section we explore the details of these processes.

17.4.1 Particle Emission in General

The most obvious way that a nucleus can decay is to break into two pieces. We are often interested in a situation where a nucleus with charge Z and baryon number A emits a small nucleus (or even a proton or neutron) with charge z and baryon number a. Fission can be regarded as a limiting class of such particle emission processes in which the "particle" and the residual nucleus are of comparable size, though fission processes are often accompanied by neutron emission to keep the decay products near the valley floor. If an emission reaction is exothermic, the decay can occur. This condition can be expressed as an inequality in terms of binding energies or mass excesses. We adopt a notation $^A[Z]$ for a nucleus with Z protons and $A - Z$ neutrons. Then the decay

$$^A[Z] \rightarrow {}^{A-a}[Z - z] + {}^a[z] \qquad (17.22)$$

is energetically allowed if the products are more bound than the initial nucleus,

$$B(Z - z, A - a) + B(z, a) > B(Z, A), \qquad (17.23)$$

or equivalently, if the mass excess of the products is less than that of the initial nucleus,

$$\Delta(Z - z, A - a) + \Delta(z, a) < \Delta(Z, A). \qquad (17.24)$$

The stability of any given nucleus with respect to such decay processes can be determined by consulting nuclear data tables and checking this inequality (Problem 17.14).

The SEMF enables us to approximate the terms in eq. (17.23) and obtain an estimate of where in the (Z, N) plane a particular kind of particle emission decay is energetically allowed to occur. Since we are interested in naturally occurring or easily created nuclides that cluster around the floor of the valley of stability, we may replace Z by $Z_{min}(A)$ and use $B_{max}(A) = B(Z_{min}(A), A)$. This

approximation ignores the fact that emission of $^a[z]$ typically does not leave the residual nucleus on the floor of the valley of stability. The error is small, however, when a is small, since $\partial B / \partial Z|_{Z_{min}} = 0$ along the floor of the valley. If the mass number a of the emitted nuclide is very small and A is large, then the difference in binding energies in eq. (17.23) can be approximated by the derivative, $B_{max}(A) - B_{max}(A - a) \approx a(dB_{max}/dA)$, leading to

$$\frac{dB_{max}}{dA} < \frac{1}{a}B(z, a) \qquad (17.25)$$

as the condition for an allowed decay. It is convenient to re-write this condition in terms of binding energies *per nucleon*. Substituting $B_{max}(A) = A\,b_{max}(A)$ and $B(z, a) = a\,b(z, a)$, eq. (17.25) becomes

$$b_{max} + A\frac{db_{max}}{dA} < b(z, a). \qquad (17.26)$$

The left-hand side of this inequality can be obtained from eq. (17.21), and is plotted in Figure 17.8. Particle emission is energetically allowed when the curve in Figure 17.8 lies *below* the binding energy per nucleon of the emitted nuclide $^a[z]$.

Nucleon emission Hydrogen ^{1}H and the neutron n have zero binding energy by definition, so a nucleus could emit a single nucleon only if the curve in Figure 17.8 went below zero. This does not occur for any nucleus along the floor of the valley of stability all the way up to $A \approx 300$ where stable nuclei end. So no nuclei of interest to us decay by nucleon emission.[6] The same reasoning applies to emission of ^{3}H or ^{3}He. Their binding energy per nucleon is ~ 3 MeV, below the curve for all relevant A values.

α-emission The nucleus with the smallest a of interest for emission is the *alpha-particle*, $^4_2$He. Its binding energy per nucleon is anomalously large for such a light nucleus (see Figure 17.6(b)), $b(2, 4) = 7.074$ MeV, and this crosses the curve in Figure 17.8 near $A = 155$. So the SEMF predicts that it is energetically possible for a nucleus near the floor of the valley of stability to emit an α-particle if its mass number exceeds about 155.

The lowest-A naturally occurring nuclide that decays by α-emission is an isotope of neodymium, $^{144}_{60}$Nd, which accounts for 24% of naturally occurring neodymium and has a half-life of about 2.29×10^{15} years (a long time!). Below $A = 144$, α-emission is possible only for very unstable isotopes far from the valley of stability. Above $A = 144$ α-emitters become progressively more common.

---

[6] A rare, but important exception occurs when a weak β-decay is accompanied by neutron emission. More about this in §18.

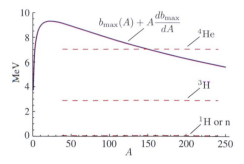

Figure 17.8 The solid curve shows the left-hand side of eq. (17.26) plotted versus A. The horizontal dashed lines give the binding energy per nucleon for a nucleon, tritium (^{3}H), and an α-particle (^{4}He). When the solid curve lies below the dashed ^{4}He line, emission of an α particle is allowed.

Decay by Particle Emission

Many nuclei decay by emitting an α-particle, the nucleus of ^{4}He. The SEMF predicts that nuclides near the floor of the valley of stability become unstable to emission of α-particles for $A \gtrsim 155$. For A just above 155 only sporadic nuclides emit α-particles. As A increases, more and more nuclides are α-emitters, and for values of A in excess of 200 nearly every nuclide is unstable to α-decay unless it decays in some other way first.

Thus the SEMF provides a good guide to the occurrence of α-decay.

Other nuclide emission Figure 17.8 suggests that emission of other relatively light nuclei may also be possible because their binding energy per nucleon is even greater than the α-particle. With a couple of minor exceptions (Problems 17.14 and 17.15), however, virtually no nuclei near the minimum of the valley of stability decay by emitting light nuclei heavier than an α-particle. Note, however, that even larger nuclei can be "emitted" in *spontaneous fission* processes where the nucleus breaks up into two parts of relatively comparable size (§18.3.1).

17.4.2 α-particle Emission

α-particle emission is the principal way that large nuclei reduce their mass number. It is a primary source of both natural and manmade radioactivity. It is responsible for a significant fraction of geothermal energy, and it is the reason why people worry about long-term sequestration of nuclear reactor waste and about radon in their basements.

α-decay Through Quantum Tunneling

α-decay is a classically forbidden process, made possible by quantum mechanical tunneling. The exponential suppression of the barrier penetration factor is responsible for the huge range of α-decay lifetimes.

It is even the source of all the helium that is used to cool superconductors and to fill party balloons.

The lifetimes of nuclei that decay by α-emission vary from tiny fractions of a second up to many orders of magnitude greater than the age of the universe. (Recall ^{144}Nd, with a lifetime greater than 10^{15} years.) Where does this huge variability come from? Due to the relatively strong binding of the α-particle, a nucleus typically contains significant α-particle-like correlations of two protons and two neutrons. Roughly speaking, several α-particles are rattling around inside a heavy nucleus such as ^{238}U at any time. The time scale for their rattling is about 10^{-21} seconds, approximately the time it would take an α-particle moving at about 1/100th the speed of light to cross a nucleus. Yet it takes an α-particle on average 4.5×10^9 years to find its way out of ^{238}U. That is more than 10^{36} trips across the nucleus! This huge suppression for α-decay of ^{238}U arises because α-particle emission is a classically forbidden process, and the decay occurs through the quantum mechanical magic of tunneling.

If α-decay were not suppressed so much by tunneling, all α emitters would have decayed long ago and no fissionable elements would exist on Earth: nuclear weapons and nuclear fission energy would not be possible. This is the first time in this book that tunneling has played such an important role in a physical situation, so it bears careful examination.

The first step in understanding the vast variability of α-decay lifetimes is the observation that both the α-particle and the nucleus are *positively charged*. Therefore they repel one another when separated. The potential energy of an α-particle in the vicinity of a heavy nucleus with charge Z is sketched in Figure 17.9. The two objects repel at large distances according to Coulomb's Law,

$$V_{\text{Coulomb}}(r) = +\frac{e^2}{4\pi\epsilon_0}\frac{2Z}{r}, \qquad (17.27)$$

(where Ze is the charge of the nucleus and $2e$ is the charge of the α-particle) until they get close enough to feel the strong force. In the figure, the strong interaction potential is represented by an attractive spherical potential well of

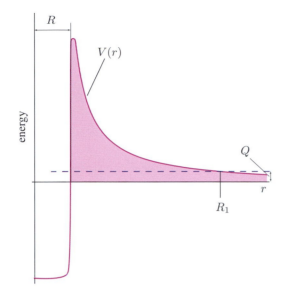

Figure 17.9 A sketch of the potential seen by an α-particle in the vicinity of a nucleus. The Q-value for the decay, which is the energy carried by the α-particle, is denoted by the horizontal dashed line. R, the edge of the attractive nuclear potential, given by eq. (17.28), and R_1, the radius where the Q-value equals the Coulomb potential energy, are the classical turning points for the α-particle's motion.

constant depth. This potential must be attractive – otherwise the α-particle would not be bound to the nucleus in the first place – but its precise form will not matter for the semiquantitative analysis we consider here. We can assume that the rising Coulomb potential cuts off when the surfaces of the nuclei touch, i.e. at a separation of roughly

$$R \approx R_{4\mathrm{He}} + R_{A[Z]} \approx 1.3(4^{1/3} + A^{1/3}) \times 10^{-15}\,\mathrm{m}\,,$$
(17.28)

(see eq. (17.10)), which is about 10^{-14} m for a heavy nucleus like uranium.

From the point of view of an α-particle[7] *on the inside of the nucleus*, the potential shown in Figure 17.9 presents a classical barrier that, absent quantum tunneling, would prevent the α-particle from escaping the nucleus. If the Q-value (remember, Q is the excess energy emitted) for nuclide $^{A+4}[Z+2]$ to decay into $^{A}[Z]$ by α-emission is positive, as denoted by the dashed line in the figure, then the region outside the nucleus is classically allowed. To

---

[7] Or, more correctly, an α-*particle-like* correlation among two protons and two neutrons as described earlier

reach it, however, the α particle must *tunnel through the barrier*.

The lifetime for a decay that occurs by tunneling can be estimated through eq. (15.23) as the ratio of the period of the classical motion $T(Q)$ divided by the *barrier penetration factor* $P(Q)$ (see eq. (15.23)). The lifetime associated with α decay can then be approximated as

$$\tau = T_0 \exp\left(\frac{2}{\hbar}\int_R^{R_1} dr\,\sqrt{2m(V_{\mathrm{coulomb}}(r) - Q)}\right),$$
(17.29)

where R and R_1 are defined in Figure 17.9, and $T_0 \sim 10^{-21}$ s is the time scale mentioned above for an α-particle's motion inside the nucleus.

As emphasized in §15, the tunneling lifetime increases *exponentially* with both the height and width of the barrier. As a result, the probability of α-emission depends dramatically on Q, Z, and R, which affect these parameters. The barrier gets wider if Q *gets smaller*, higher if Z *grows larger*, and wider and higher if R *decreases*. In Box 17.3, we derive an approximate formula for the α-decay lifetime of a nuclide with charge $Z + 2$ and atomic mass $A + 4$. Such results are usually quoted as the base-10 logarithm of the decay lifetime τ (in seconds), i.e.

$$\log_{10}\tau(R, Z, Q) \sim \log_{10}T_0 - 1.3\sqrt{RZ} + 1.7Z/\sqrt{Q}\,.$$
(17.30)

Here R is the separation at which the α particle and nucleus "touch" (17.28), measured in femtometers, and Q is the energy released in the decay, measured in MeV. We can see that eq. (17.30) has the right qualitative behavior: the lifetime increases as R decreases and as Z increases, and becomes infinite as $Q \to 0$, when the barrier becomes infinitely wide. To illustrate the sensitivity to Q, consider nuclides with $Z \approx 90$ and compare the lifetimes for α-particle emissions with $Q = 3\,\mathrm{MeV}$ and $Q = 6\,\mathrm{MeV}$. According to eq. (17.30) the α-decay lifetimes in these two cases would differ by a multiplicative factor of $10^{1.7\times90\times(1/\sqrt{3}-1/\sqrt{6})} \approx 10^{17}$!

A cruder form of eq. (17.30), $\log_{10}\tau = c_1 + c_2 Z/\sqrt{Q}$, where c_1 and c_2 are empirically determined constants, was discovered experimentally by German physicist Johannes Geiger and British physicist John Nuttall in 1910 and is known as the **Geiger–Nuttall law**. Equation (17.30) was derived by George Gamow during the early days of quantum mechanics. It was considered one of the early triumphs of quantum mechanics and in particular to be a validation of the concept of barrier penetration. The accuracy of the Geiger–Nuttall law is displayed in Figure 17.10. The agreement is quite good considering that the lifetimes

---

Box 17.3   Gamow's Calculation of the α-decay Rate

The barrier penetration factor for α-decay is given by (see eq. (15.19))

$$P(Q) = \exp\left(-\frac{2}{\hbar}\int_R^{R_1} dr\,\sqrt{2M(V_{\text{Coulomb}}(r) - E)}\right)$$

$$= \exp\left(-\frac{2\sqrt{2M}}{\hbar}\int_R^{R_1} dr\,\sqrt{\frac{2Ze^2}{4\pi\epsilon_0 r} - Q}\right),$$

where M is the mass of ^{4}He. As defined in Figure 17.9, the upper limit of the r-integration is the radius where the Coulomb potential energy equals the Q-value of the decay, $R_1 = 2Ze^2/4\pi\epsilon_0 Q$. Changing the integration variable to $x = r/R_1$, and defining $z = R/R_1$,

$$P(Q) = \exp\left(-2\sqrt{\frac{2MQ}{\hbar^2}}R_1\int_z^1 dx\,\sqrt{\frac{1}{x} - 1}\right).$$

Usually, when the barrier is wide enough to significantly inhibit the decay, z is much less than unity, so it suffices to approximate the integral in that limit, where one finds $\int_z^1 dx\,\sqrt{(1/x) - 1} = \pi/2 - 2\sqrt{z} + \ldots$, where the omitted terms are proportional to higher powers of z. Introducing this into the expression above for the tunneling probability, and substituting the physical values for e, the mass of the α particle, etc., we obtain

$$P(Q) = e^{\left(3.0\sqrt{RZ} - 4.0Z/\sqrt{Q}\right)} = 10^{\left(1.3\sqrt{RZ} - 1.7Z/\sqrt{Q}\right)},$$

where Q is measured in MeV and R is measured in fm. When combined with the classical period T_0 and substituted into eq. (15.23), we obtain eq. (17.30).

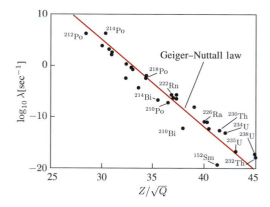

Figure 17.10 The logarithm of the inverse lifetime ($\lambda = 1/\tau$) of 30 heavy nuclei that decay by α-emission are plotted versus $Z/\sqrt{Q}$. According to the Geiger–Nuttall law, they should lie on a straight line with slope -1.7. Although there is some scatter, the fit is pretty good over *thirty orders of magnitude*. From [83].

plotted range from $\sim 10^{-6}$ seconds for $^{214}_{84}$Po to more than 10^{10} *years* for $^{232}_{90}$Th.

According to the semi-empirical mass formula, the energy emitted in an α-decay slowly gets larger as A grows (see Figure 17.8), but the exact value of Q depends on the

α-decay

Many heavy nuclei decay by α-particle emission. α-emission occurs through a quantum tunneling process that depends very sensitively on the Q-value of the decay, and to a lesser extent on the nuclear radius R and charge Z. α-decay lifetimes can differ by tens of orders of magnitude.

fine details of nuclear structure that are beyond the accuracy of the semi-empirical mass formula. Thus, different nuclear species that are important in nuclear fission energy, most of which have $Z \approx 90$, can have wildly different α-decay lifetimes.

17.4.3 Beta Decay and Related Processes

As described in §14, weak interactions allow protons to turn into neutrons and *vice versa*. This enables nuclei to optimize their proton-to-neutron ratio for a fixed mass number A. The fundamental quark processes were described in §14, and the resulting proton–neutron interactions were summarized in §17.1.4. We are interested in cases where these processes occur inside nuclei,

β^--decay $n \to p + e^- + \bar{\nu}_e$ $^A[Z] \to {}^A[Z+1] + e^- + \bar{\nu}_e,$

β^+-decay $p \to n + e^+ + \nu_e$ $^A[Z] \to {}^A[Z-1] + e^+ + \nu_e,$

e^--capture $p + e^- \to n + \nu_e$ $^A[Z] + e^- \to {}^A[Z-1] + \nu_e.$

$$(17.31)$$

On each line the reaction is described both at the nucleon level and in terms of the nuclei. The first of these reactions (β^--decay) may occur when a nucleus is neutron rich compared to the most stable nucleus with a given A. The latter two may occur when a nucleus is proton rich.

Which reaction, if any, may occur is determined by which Q-value is positive. Since the number of electrons changes in these decays, it turns out to be particularly simple to use mass excesses, which include both the mass of the nucleus and the masses of the electrons necessary to make a neutral atom. Thus, for example, in β^--decay, where $^A[Z] \to {}^A[Z+1] + e^- + \bar{\nu}_e$, the Q-value is positive if (remembering that the neutrino mass is small enough to ignore)

$$M_{\text{nucl}}(Z, A) > M_{\text{nucl}}(Z+1, A) + m_e, \qquad (17.32)$$

where M_{nucl} denotes the nuclear mass. To construct neutral atoms we must add the rest masses of Z electrons to both sides of this equation. If we then subtract A daltons (and neglect relatively small changes in the total electron binding energy), the result is simply,

$$\Delta(Z, A) > \Delta(Z+1, A). \qquad (17.33)$$

Note that the electron's rest mass has disappeared from the condition because the electron emitted in the decay is just what is needed to neutralize the product nucleus. Applying the same logic to the cases of β^+-decay and electron capture (Problem 17.19), we find

$$Q(\beta^-) = \Delta(Z, A) - \Delta(Z+1, A),$$
$$Q(\beta^+) = \Delta(Z, A) - \Delta(Z-1, A) - 2\Delta_e,$$
$$Q(EC) = \Delta(Z, A) - \Delta(Z-1, A), \qquad (17.34)$$

where $\Delta_e = m_e c^2$ is the mass excess of the electron.

A couple of inferences from this result are worth noting: (1) β^+ decay should be relatively rare because electron capture connects the same initial and final nuclides, and can occur at a higher Q-value. (2) A nucleus can actually gain mass in electron capture (Problem 17.20). Notice that in $\beta^\pm$-decay, the electron (or positron) energy is not uniquely fixed because the neutrino and electron share the excess energy. In contrast, the energy of the α-particle in α-decay is uniquely determined by the difference of initial and final nuclear masses. This has important practical

consequences. For example, α-decay energies are "fingerprints" that identify the nuclei in a sample of radioactive material. β-decay energies cannot be used in the same way.

17.4.4 Pairing Energy and Stability of Isotopes

According to the SEMF, a cross-sectional cut through the valley of stability at fixed A would be a parabola if the pairing energy term in the semi-empirical mass formula were ignored. This would imply that there would be only one stable nuclide for each value of A, because all the others would either $\beta^\pm$-decay or electron capture to it. There is indeed only one stable nuclide for *odd values of* A, but for even A there are often two and sometimes even more stable nuclides (see Figure 17.11). The origin of this effect

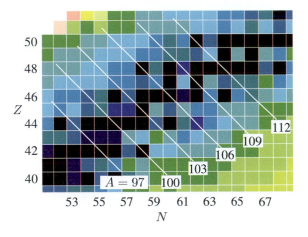

Figure 17.11 A small region of the nuclear mass table around $Z = 45$ and $N = 58$. Stable (lifetimes $> 10^{15}$ seconds) nuclides are denoted in black. The other colors are coded to shorter lifetimes. The values of A are labeled by the diagonal lines. For odd A there is never more than one stable nucleus, while for even A there are generally two. Notice the large number of stable *isotopes* that occur for some values of Z. (Credit: Image adapted from National Nuclear Data Center)

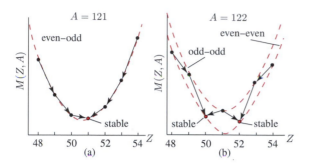

Figure 17.12 Plots of the masses of relatively stable nuclides with $A = 121$ and $A = 122$, along with the predictions of the SEMF (red dashed curves). Note that as a consequence of the pairing energy, there are two different curves for $A = 122$: even–even nuclei (bottom curve) and odd–odd nuclei (top curve). As a result, there are two stable isobars for $A = 122$, as well as for many other even values of A. After [77].

in the pairing energy term is clear in Figure 17.12, where the mass excesses of nuclides with $A = 121$ and $A = 122$ are shown as functions of Z. For $A = 121$, all the nuclear species are Z-odd, N-even or *vice versa*, so the pairing energy plays no role in their relative stability, and as a result, only one nuclide is stable. For $A = 122$, the minimum energy when pairing is ignored occurs at an *odd–odd* nuclide, which is made heavier than its neighbors on either side by the pairing contribution. As a result, the two neighboring even–even nuclides (differing from one another in charge by two) are stable.

As A increases, the curvature of the parabolic cross section through the valley of stability decreases proportionally to $1/A$ (see eq. (17.19)). Since the pairing term decreases more slowly, like $1/A^{1/2}$, stable or long lived isobars (nuclei with the same A) become more common with increasing A (Question 17.4). This effect is most obvious in the range $80 \lesssim A \lesssim 160$. For larger values of A, the overall tendency of nuclei to be unstable counteracts this effect and the number of stable isobars decreases. The proliferation of isobars and the fact that N increases more rapidly than Z give rise to a large number of stable and long lived *isotopes* in the range $80 \lesssim A \lesssim 160$. Thus, for example, tin has ten naturally occurring isotopes with A ranging from 112 to 124, and xenon has nine isotopes with A ranging from 124 to 136.

Pairing energy effects also have the consequence that most stable nuclides are even–even nuclei. Odd–even or even–odd nuclides are the next most numerous types among stable nuclides. Finally, stable or very long-lived odd–odd nuclides are very rare: the only examples are

$^2_1$H, $^6_3$Li, $^{10}_5$B, $^{14}_7$N, $^{40}_{19}$K, and rare isotopes of vanadium, lanthanum, and tantalum.

17.4.5 Decay Chains

The two α-emitters ^{238}U and ^{232}Th are relatively common in our environment because they have lifetimes on the order of ten billion years, which is longer than the time since they were formed in a stellar catastrophe. Together with the somewhat shorter-lived and less common uranium isotope ^{235}U, they are the central players in nuclear fission energy. When an atom of uranium or thorium finally α-decays, it gives birth to a lighter, but still radioactive nucleus. Only after a long sequence of α- and β-decays does the nucleus reach stability. The sequence of α- and β-decays originating from a long-lived heavy α-emitter is known as a **decay chain**.

There are four, and only four, primary decay chains that pass down along the floor of the valley of stability. Any two nuclides that differ by an α-particle are on the same decay chain. The decay chain that includes ^{238}U is shown graphically in Figure 17.13. The existence of decay chains enables us to understand some of the otherwise puzzling features of natural radioactivity: Why are there *very* radioactive substances found in nature? Earth is about 4.5 billion years old, and unstable nuclei originally present in Earth have long ago decayed away, except those with very long lifetimes. When, however, ^{238}U undergoes α-decay, it triggers a sequence of subsequent decays that take place quite rapidly. The ^{238}U decay chain does not end until it reaches $^{206}_{82}$Pb. Since the mass number has changed by 32, eight α-decays have occurred along the chain. Table 17.3 summarizes the lifetimes of those inter-mediate decays (there are several possible branches for the ^{238}U chain, only the dominant one is given in the table). The longest half-life of any descendant in the chain is less than 10^6 years. Many half-lives are much shorter, making those nuclides very radioactive. Radium-226 ($^{226}_{88}$Ra) is the classic example: with a lifetime of 1622 years, it is long-lived enough to be found associated with uranium in ores such as *pitchblende*, and it is short-lived enough to be very radioactive.

One feature of these decay chains, evident in Figure 17.13, is the absence of any long-lived nuclides in the interval between $A = 209$ and $A = 226$: the longest-lived nuclide in this interval is ^{210}Po with a half-life of 140 days. This gap results from the shell structure of nuclei and is not predicted by the SEMF. Among the consequences of this region of instability is the existence of several short-lived isotopes of the noble gas radon, which give rise to potentially hazardous ionizing radiation (see §20.5.1).

Three of the four primary decay chains are initiated by nuclides that occur naturally on Earth, ^{238}U, ^{235}U, and ^{232}Th, leading to an array of radioactive **descendant nuclides** that occur in our environment. The longest-lived element on the fourth chain is ^{237}Np with a lifetime of about 2.14×10^6 years, too short for it to have survived on Earth. All four chains are initiated by nuclei formed in nuclear reactors, however, and are therefore possible constituents of spent nuclear fuel. This makes the composition of spent nuclear fuel quite complex and variable. Since many nuclei on these chains are long-lived – up to hundreds of thousands of years – these components of nuclear waste must either be removed, recycled, or sequestered. These decay chains reappear in §20 (Radiation).

Table 17.3 Dominant decay chain of ^{238}U, including the half-lives of all its descendants.

Nucleus	Decay	Half-life
$^{238}_{92}$U	$^{234}_{90}$Th $+ \alpha$	4.51×10^9 y
$^{234}_{90}$Th	$^{234}_{91}$Pa $+ e^- + \bar{\nu}_e$	24.5 d
$^{234}_{91}$Pa	$^{234}_{92}$U $+ e^- + \bar{\nu}_e$	6.7 h
$^{234}_{92}$U	$^{230}_{90}$Th $+ \alpha$	2.33×10^5 y
$^{230}_{90}$Th	$^{226}_{88}$Ra $+ \alpha$	8.3×10^4 y
$^{226}_{88}$Ra	$^{222}_{86}$Rn $+ \alpha$	1 622 y
$^{222}_{86}$Rn	$^{218}_{84}$Po $+ \alpha$	3.825 d
$^{218}_{84}$Po	$^{214}_{82}$Pb $+ \alpha$	3.05 min
$^{214}_{82}$Pb	$^{214}_{83}$Bi $+ e^- + \bar{\nu}_e$	26.8 min
$^{214}_{83}$Bi	$^{214}_{84}$Po $+ e^- + \bar{\nu}_e$	29.7 min
$^{214}_{84}$Po	$^{210}_{82}$Pb $+ \alpha$	1.5×10^{-4} s
$^{210}_{82}$Pb	$^{210}_{83}$Bi $+ e^- + \bar{\nu}_e$	22 d
$^{210}_{83}$Bi	$^{210}_{84}$Po $+ e^- + \bar{\nu}_e$	5 d
$^{210}_{84}$Po	$^{206}_{82}$Pb $+ \alpha$	140 d

17.4.6 γ Radiation

Electromagnetic radiation often accompanies α- or β-decays. When a nucleus decays, the resulting nucleus is initially frequently in an excited state rather than its ground state. Like an excited atom or molecule, the nucleus can de-excite by emitting a quantum of electromagnetic radiation – a photon. Such a photon typically has an energy in the MeV range. Photons in this energy range emitted

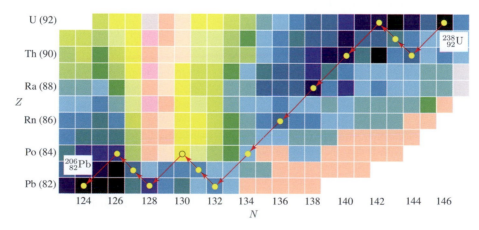

Figure 17.13 The dominant decay chain of ^{238}U superimposed on the chart of the nuclides, on which dark colors correspond to long-lived nuclides. Adapted from [70].

from nuclei are known as γ-*rays* (see §20.1 for a further discussion of the definitions and distinction of γ-rays and X-rays). γ-rays associated with nuclear decays are almost always emitted promptly (within times of order 10^{-19} seconds) after the β- or α-decays that precede them. γ-emission does not change the identity of the emitting nucleus, so γ-emission does not form part of the decay chain of a radioactive nucleus. Nevertheless, γ-ray emissions occur at various stages along the chain of radioactive decays. Since a γ-ray in the MeV energy range is capable of doing quite a bit of damage, they are a major concern in the handling of nuclear materials and waste (§20).

Discussion/Investigation Questions

17.1 How can the neutron be unstable in free space, but stable when inside a nucleus?

17.2 Discuss the ways in which the assumptions behind the SEMF may fail when A is very small.

17.3 For small A, where the symmetry energy overwhelms the Coulomb energy, the most stable nuclei with even A have $Z = N = A/2$. For odd–even nuclei in the same range of A, which is more stable, $Z = N + 1$ or $N = Z + 1$? Why? Check your answer for some nuclei in the range $11 < A < 25$.

17.4 Sketch a curve similar to those of Figure 17.12 for a value of A that results in three isobars that are stable against $\beta^{\pm}$ decay and electron capture. On the basis of the SEMF explain why the number of stable isotopes per element increases with Z.

17.5 Consider the decay chain for ^{238}U shown in Figure 17.13. Why are there β^{-}-decays, but no electron captures, interspersed among the α-decays?

17.6 Radium-226 is a decay product of ^{238}U. ^{226}Ra has a half-life of 1600 years and is found in uranium ores. When it was first discovered and before radioactive decay was well understood, some mineral prospectors claimed to have discovered a rich deposit of radium (a few percent by mass) in a cave in the American midwest [84]. What do you think of this claim? [Hint: you might want to use results of Problem 17.22.]

17.7 The decay chain of ^{238}U is described in some detail in the text. The other naturally occurring heavy nuclide whose decay initiates a decay chain is ^{232}Th. Look up the ^{232}Th decay chain. At what nuclide does it end? How many α-particles and how many electrons are emitted in the entire chain? Roughly how long does it take, after the decay of ^{232}Th, for the decay chain to run to conclusion?

Problems

Note: for a number of the problems in this chapter you will need to use data on nuclides, which is conveniently obtained from the NNDC (Box 17.1) [70].

17.1 [T] Derive the relation quoted in eq. (17.7) between the binding energy of a neutral atom $B(Z, A)$ and its mass excess $\Delta(Z, A)$.

17.2 A typical fission reaction involving ^{235}U is

$$^{235}_{92}\text{U} + n(\text{thermal}) \rightarrow {}^{144}_{56}\text{Ba} + {}^{90}_{36}\text{Kr} + 2n.$$

The kinetic energy of a thermal neutron is $\sim 1/40\,\text{eV}$, small enough to ignore in the energy balance. Show that the energy released in the fission is 179.5 MeV.

17.3 There is no stable nucleus with $A = 5$. This has profound consequences for the way stars burn and form elements. The two candidates for stable nuclei with $A = 5$ are $^5_2$He and $^5_3$Li. Look up or figure out what they decay into and show that their decays are allowed by energy conservation.

17.4 Helium three is an extremely rare but stable isotope of helium made up of two protons and only one neutron. Because ^{4}He is so tightly bound, considerable energy is given off when ^{3}He absorbs a neutron. This makes ^{3}He a very good neutron detector. The nuclear reaction involved is $n + {}^3\text{He} \rightarrow {}^4\text{He} + \gamma$. Neutron detection is a vital aspect of homeland security in the twenty-first century, since nuclear devices inevitably give off very-low-energy neutrons that easily leak out of almost any container. What is the energy of the γ-ray given off when a thermal neutron (with energy $\sim 1/40\,\text{eV}$) is absorbed by a ^{3}He nucleus? (You can ignore the recoil of the nucleus and the kinetic energy of the thermal neutron in this process.)

17.5 ^{3}He is so rare that it must be produced artificially. One way to produce it is through a sequence of reactions that can be carried out in a nuclear reactor, where thermal neutrons are quite common. The reactions are:

$$n_{\text{thermal}} + {}^6\text{Li} \rightarrow {}^4\text{He} + {}^3\text{H}, \text{ followed by}$$
$$^3\text{H} \rightarrow {}^3\text{He} + e^- + \bar{\nu}_e.$$

Verify that both of these nuclear reactions are exothermic and compute the energy released in each. Check that this process is practical by looking up the fractional abundance of ^{6}Li in naturally occurring lithium and the lifetime of ^{3}H. Why should we care about the latter?

17.6 The radius of a nucleon is about $R_N = 0.9$ fm. Compute the volume of a nucleus with a moderately large value of A in the liquid drop model (i.e. using the SEMF) and show that roughly 2/3 of the nucleus is "empty space." In contrast, an atom is almost entirely empty space.

17.7 Estimate the density of a nucleus by assuming it to be a sphere of radius $R_0 = 1.3\,A^{1/3}$ fm. What would be the radius of a nucleus with the mass of Earth?

17.8 Pressure can be defined as the rate of change of energy with volume, $p = \partial E/\partial V$. Similarly, the surface analogue of pressure, **surface tension**, is defined as the rate of change of energy with surface area a, $\sigma \equiv \partial E/\partial a$ at constant volume. Ignoring the effects of symmetry, Coulomb, and pairing energies, compute the surface tension of a nucleus as described by the empirical parameters of the SEMF. The surface tension acts to keep a nucleus spherical.

17.9 Find the SEMF prediction for $Z_{\min}(A = 200)$ from eq. (17.20). Find the actual value of Z that gives the greatest binding energy for $A = 200$. How well does the SEMF do?

17.10 [T] Derive eq. (17.20) by differentiating the SEMF expression for $b(Z, A)$ with respect to Z at fixed A.

17.11 [T] Check the assertion in the text that the SEMF predicts that nuclei have positive binding energy up to $A \sim 3150$.

17.12 [T] The maximum value of A for which nuclei are stable against instantaneous fission can be estimated as follows. When a charged, spherical nucleus, $^A Z$, with surface tension is deformed into an ellipsoid leaving its volume constant, it is possible to compute the change in its binding energy as a function of deformation. The result [77] is

$$\Delta B \approx -\left(\frac{2}{5}\epsilon_S A^{2/3} - \frac{1}{5}\epsilon_C \frac{Z^2}{A^{1/3}}\right)\delta^2, \quad (17.35)$$

where δ measures the deformation (the ratio of major to minor axes is $(1 + \delta)^{3/2}$). Note the signs: the surface contribution tends to stabilize and the Coulomb contribution tends to destabilize the nucleus. Show that eq. (17.35) predicts that instantaneous fission occurs when $Z^2/A \gtrsim 49$. The largest A nucleus observed so far, $^{294}[118]$, has $Z^2/A = 47.4$.

17.13 Explain how to read from Figure 17.8 the Q-value for α-decay predicted by the SEMF.

17.14 Show that the longest-lived isotope of radium, $^{226}_{88}$Ra, is energetically allowed to decay by emission of a $^{14}_{6}$C nucleus. This decay has been observed with a probability of 3.2×10^{-9}.

17.15 ^{235}U can decay by emitting a neon nucleus. How many such decays occur per second per mole of ^{235}U?

17.16 Radon gas $^{222}_{86}$Rn is a serious environmental hazard (see §20). It is a decay product of $^{238}_{92}$U, which is a relatively common constituent of rock. $^{222}_{86}$Rn undergoes α-decay to $^{218}_{84}$Po. When $^{222}_{86}$Rn is created in the foundation of a house, it diffuses out into the basement airspace. If its lifetime were very short, it would decay before getting out of the rock. If it were very

long, it would diffuse away into the atmosphere. Its half-life is 3.8 days, however, just right for causing a great deal of human radiation exposure. Given this half-life, use Gamow's formula (17.30) to estimate the Q-value for this decay and compare with the measured value. Comment on the accuracy of your estimate.

17.17 [T] Examine the derivation of Gamow's formula for α-decay lifetimes, and then derive the equivalent expression for decay by emission of a (small) nucleus of charge z and mass number a. Assume a/A is small enough to ignore recoil. Make sure that the z and a dependences are explicit. Compare your result to an α-decay lifetime with the same Q-value.

17.18 Check that ^{3}He is less tightly bound than ^{3}H. Explain why ^{3}H decays to ^{3}He, and not *vice versa*.

17.19 [T] Verify the expressions for the Q-value in $\beta^{\pm}$-decay and electron capture, eq. (17.34).

17.20 Ignoring small electron binding energies and the very small mass of the neutrino, show that the mass of a nucleus increases when it decays by electron capture if the Q-value of the decay is less than $m_e c^2 \cong 0.511$ MeV. Verify that this is the case for the electron capture decay of the longest-lived isotope of technetium:

$$^{97}_{43}\text{Tc} + e^- \rightarrow ^{97}_{42}\text{Mo} + \nu_e .$$

17.21 97% of naturally occurring calcium is calcium-40, $^{40}_{20}$Ca. This may seem surprising, since if we use the semi-empirical mass formula to estimate the most stable nuclide with $A = 40$ we find $Z \approx 18$. This suggests that $^{40}_{20}$Ca might be unstable to *electron capture*, which would increase its N/Z ratio. Show that $^{40}_{20}$Ca cannot electron capture to $^{40}_{19}$K. Show, however, that it is possible for $^{40}_{20}$Ca to *capture two electrons at the same time*, making $^{40}_{18}$Ar. Look up the information on the lifetime of $^{40}_{20}$Ca and comment on its value. Do we have to worry about exposure to radiation from decay of the $^{40}_{20}$Ca in our bones?

17.22 [H] When one of the naturally occurring radioactive heavy elements like ^{238}U decays, a decay chain follows, leading to the creation of various unstable, radioactive *descendant nuclides*. Consider a sample known to be pure ^{238}U at time zero. Suppose that none of its decay products migrate away from the original sample. Use the fact that the lifetime of the original nuclide, in this case ^{238}U, is much longer than the lifetimes of any of its descendants to show that a quasi-equilibrium develops in which the amounts of ^{238}U and all of its descendant are steadily decreasing at a rate determined by the lifetime of ^{238}U. Show that the ratio of the abundance of each descendant nuclide to the abundance of ^{238}U is very nearly constant in time, and is given by the ratio of the descendant's lifetime to the lifetime of ^{238}U. Show that in this equilibrium every radioactive species contributes equally to the radioactivity (in decays per second) of the sample. [Hint: Write first-order differential equations that govern the rate of change of each nuclear species.]

17.23 In light of the results of Problem 17.22, what is the rate of energy emission (J/kg) of a sample of ^{238}U

that has been around long enough to come into decay equilibrium?

17.24 The uranium isotope ^{234}U accounts for 0.0054% of naturally occurring uranium. Its half-life $\tau_{1/2}(^{234}\text{U}) = 2.455 \times 10^5$ y is much too short for any primordial ^{234}U to still exist on Earth. Explain why ^{234}U occurs naturally and explain its fractional abundance. [Hint: Use the result of Problem 17.22.]

Nuclear Energy Processes: Fission and Fusion

Energy can be derived from nuclei either by *fusing* light nuclei or by *fissioning* heavy ones. The object of this chapter is to explain the basic physical principles involved in these two energy production processes. Mechanisms for capturing this energy in practical nuclear reactors are the subject of the following chapter.

Much of this chapter is devoted to fission, which is already a major contributor to the world's electrical energy generation. Although fusion has not yet been harnessed for energy production on Earth, it is the original source of the energy that we get from the Sun and therefore indirectly responsible for the rest of the energy used on Earth – with the exceptions of geothermal and tidal power.

After an overview (§18.1) of the similarities and differences between fission and fusion energy, and a brief digression (§18.2) to introduce *cross sections*, which encode information about reaction rates, we turn to fission (§18.3) and then to fusion (§18.4) energy processes.

The physics of fission and fusion involves a handful of specific nuclides that play principal roles, ranging from essential fuels to unwanted contaminants. Table 18.1 provides a concise glossary of relevant nuclides, which may serve as a useful reference for this and the following two chapters. Further discussion of fission and fusion energy processes can be found in introductory nuclear physics texts such as [77] and [85], and in [86], which focuses on the physics of nuclear reactors.

18.1 Comparing Fission and Fusion

Fission and fusion processes both produce energy by releasing the binding energy of nuclei. These types of nuclear reactions have many features in common but also differ in profound ways. Common features include, first, the potential to liberate large amounts of energy – tens to hundreds of MeV per reaction. This corresponds to energy densities far larger than any other available energy source.

> **Reader's Guide**
>
> The fundamental physical processes that enable the production of energy through nuclear fission and fusion are described in this chapter. The roles of these processes in fission power reactors and in controlled fusion experiments are explained in the following chapter
>
> After introducing the concept of a cross section, we turn to nuclear fission, how it occurs, what nuclear fragments and other particles it produces, and how slow-neutron-induced fission can give rise to a chain reaction. Phenomena key to controlled nuclear fission such as prompt and delayed neutrons, fission poisons, and neutron moderation, are introduced.
>
> We explain how light nuclei fuse, how quantum tunneling allows fusion to take place at energies much lower than classical mechanics would suggest, and examine possible fuels for controlled nuclear fusion on Earth.
>
> Prerequisites: §7 (Quantum mechanics), §8 (Entropy and temperature), §15 (Quantum processes), §16 (Nuclear), §17 (Nuclear structure).
>
> Aspects of nuclear fission and fusion presented here are essential in §19 (Nuclear reactors). They also inform the discussion of §20 (Radiation). §22 (Solar energy) requires a basic understanding of the nuclear fusion dynamics presented in this chapter.

Second, both fission and fusion use as fuels rare nuclides with unusual properties. Existing commercial fission reactors require the rare isotope ^{235}U, or the manmade isotopes ^{239}Pu or (potentially) ^{233}U. Practical fusion power on Earth is still a dream, but the most plausible candidate reaction uses the artificial, unstable isotope *tritium* (^{3}H) as fuel. Third, both processes lead to the generation of radioactive waste that must be handled judiciously to protect humans and the environment.

The differences between fission and fusion, however, are also striking. Nuclear fission takes place spontaneously on Earth and is self-sustaining under certain circumstances.

Table 18.1 Glossary of nuclei that play important roles in nuclear energy processes and/or environmental radiation. Terms such as *fissile, fertile, fission poison*, etc. are defined later in the chapter.

$^A Z$	Decay	$\tau_{1/2}$	Role(s) in nuclear energy
$^1_1 H$	stable		moderator
$^2_1 H$	stable		excellent moderator
			potential fusion fuel
$^3_1 H$	β^-	12.3 y	potential fusion fuel
			nuclear weapons
$^3_2 He$	stable		potential fusion fuel
			extremely rare on Earth
$^4_2 He$	stable		α-particle
$^6_3 Li, ^7_3 Li$	stable		potential tritium sources
			nuclear weapons
$^{10}_5 B$	stable		neutron absorber
$^{12}_6 C$	stable		moderator
$^{40}_{19} K$	β^-/EC	1.25 Gy	natural radiation source
$^{90}_{38} Sr$	β^-	28.9 y	fission fragment
			radiation hazard
$^{131}_{53} I$	β^-	8.02 d	fission fragment
			radiation hazard
$^{135}_{53} I$	β^-	6.46 h	fission fragment
			$^{135} Xe$ progenitor
$^{135}_{54} Xe$	β^-	9.14 h	fission poison
$^{137}_{55} Cs$	β^-	30.2 y	fission fragment
			radiation hazard
$^{222}_{86} Rn$	α	3.82 d	natural radiation source
$^{232}_{90} Th$	α	14.0 Gy	natural radiation source
			fertile, fissionable
$^{233}_{92} U$	α	159 ky	reactor product; fissile
$^{235}_{92} U$	α	704 My	fissile (naturally occurring)
			nuclear weapons
$^{238}_{92} U$	α	4.47 Gy	natural radiation source
			fertile, fissionable
$^{238}_{94} Pu$	α	87.7 y	reactor product
			energy storage
$^{239}_{94} Pu$	α	24.1 ky	reactor product
			nuclear weapons; fissile
$^{240}_{94} Pu$	α	6.56 ky	reactor product
			spontaneous fission
$^{241}_{94} Pu$	α	14.3 y	reactor product; fissile

Although in the present era it is only possible to produce sustained fission in specially designed reactors, sustained natural fission reactions occurred on Earth roughly two billion years ago and left their traces in the geological record (Box 19.1). In contrast, producing a useful amount of energy from nuclear fusion reactions on Earth requires creating conditions in some ways more extreme than those in the interior of the Sun – conditions that scientists have created so far only for brief moments. Fission is therefore a real candidate for terrestrial power generation, both now and into the future, while fusion power remains a relatively distant possibility. Remarkably, the opposite situation characterizes the rest of the universe: fusion takes place spontaneously and is self-sustaining in the interiors of countless stars, while sustained fission reactions, when they occur at all, take place only rarely and only on rocky planets like Earth.

The radioactive products of fission and fusion differ significantly as well. Spent fission fuel contains highly radioactive *fission fragments*, medium mass nuclei that emit β and γ radiation, as well as highly radioactive *actinides* created in the reactor.[1] Actinides can be sources of new fission fuel or dangerous waste depending on the way they are processed after leaving the fission reactor. In contrast, the principal product of fusion reactions is ^{4}He, which is benign. Fusion also produces vast numbers of neutrons, however, which make the fusion reactor and its surroundings radioactive. Although the problems are less severe than for fission, fusion too generates radioactive waste that must be managed.

We concentrate on fission and fusion where they take place now: fission in controlled nuclear power reactors, and fusion in the interior of stars. In preparation for the next chapter, we also describe here the reactions that might lead to sustained nuclear fusion on Earth.

18.2 Cross Sections

All fusion processes and nearly all fission processes begin with reactions between colliding particles. To describe the rate at which particles collide and react, it is useful to define the notion of a *cross section*. The simplest example arises when a beam of particles B impacts a stationary target particle T, and we look for an outcome C. In this

---

[1] **Actinides** are a series of 15 chemical elements from actinium ($Z = 89$) to lawrencium ($Z = 103$). They are chemically similar because the electrons fill an inner shell throughout the series (much like the **rare earths**). Only thorium ($Z = 90$) and uranium ($Z = 92$) are found in more than trace amounts on Earth. Many actinides are created in nuclear fission reactors and are found in spent nuclear fuel, where they contribute to its radioactivity. The actinides are further classified as **major actinides** – uranium, thorium, and plutonium – and **minor actinides** – the rest. Finally, the term **transuranic elements** applies to all the elements with $Z > 92$. Essentially all the transuranic elements that appear in nuclear waste are actinides.

case the beam can be characterized by a constant *flux*, $\Phi_B = dN_B/dA\,dt$, which is the number of beam particles that pass through a unit area per unit time (see Box 3.1 for an introduction to flux). We are typically interested in the *rate* of some outcome, dN_C/dt. The ratio of the outcome rate to the beam flux is a measure of the likelihood for the reaction to take place,

$$\sigma_{BT \to C} = \frac{1}{\Phi_B}\frac{dN_C}{dt}, \qquad (18.1)$$

where $\sigma_{BT \to C}$ has units of area and is known as the **cross section** for a reaction. σ depends on the nature of the interaction between the projectile and target and on the specific outcome required.

The fact that a reaction is fundamentally characterized by an area may seem unnatural, but common experience suggests that a cross-sectional area is a natural way to compute the probability of a given reaction between particles in random motion. Consider, for example, the game of darts. A dart board (Figure 18.1) assigns different point values to different areas: relatively large and equal areas give 1, 2, 3, ..., 20 points, much smaller areas give double each value, and so on up to the tiny area at the center that is assigned 100 points. If a uniform flux of darts (a fixed number of darts per square centimeter per hour, as might be thrown by a completely unskilled player) is thrown at the dart board, the rate at which each point value occurs is proportional to the corresponding area. The area associated with each outcome is the cross section for that particular outcome. Notice that the cross section is the only fundamental characteristic of the game (scattering process) that needs to be specified in order to compute the rate at which a given score (reaction) occurs when the flux is uniform.

Figure 18.1 A dart board, where different outcomes for a randomly thrown dart are determined by the area dedicated to that outcome.

When an interaction is strong, cross sections are large. In the case of the short-range strong nuclear force, cross sections are typically comparable to the cross-sectional area of a nucleus, which is $(\sim 10^{-14}\,\text{m})^2 \cong 10^{-28}\,\text{m}^2$. Nuclear physicists, who were accustomed to the much smaller cross sections characteristic of electromagnetic interactions, were so impressed by the measured magnitude of these cross sections that they named this unit the **barn** (b), where $1\,\text{b} \equiv 10^{-28}\,\text{m}^2$, recalling the colloquial American expression "*as big as a barn*"!

The wave nature of particles in quantum mechanics can occasionally lead to anomalously large cross sections. For example, when a nucleus $^A[Z]$ has an excited state in which a neutron is just barely bound, the Schrödinger wavefunction associated with that state extends far outside the nucleus. The wavefunction of a very-low-energy neutron scattering from the nucleus $^{A-1}[Z]$ is very similar to the wavefunction of such a bound state. As a result, the nucleus $^{A-1}[Z]$ has an exceptionally large neutron absorption cross section. An example that is important for nuclear fission reactor operation is ^{135}Xe, a short-lived nuclide that is sometimes produced after the fission of ^{235}U. The cross section for ^{135}Xe to absorb low-energy neutrons is greater than 10^6 b, as though the nucleus extended out over 1000 times its actual radius!

Electromagnetic cross sections for reactions involving nuclei are much smaller than strong interaction cross sections. They are typically measured in *millibarns* (mb). Weak interaction cross sections are smaller still, typically *nanobarns* (nb).

Cross sections provide the essential information to describe reactions even in situations more complex than a collimated beam on a stationary target. In a fission reactor, for example, a gas of neutrons interacts with a stationary target and in a fusion reactor, a plasma contains a mixture of two types of nuclei that react with one another. The rates for these reactions can be expressed in terms of the relevant cross sections as described in the following chapter.

18.3 Physics of Nuclear Fission

The aim of this section is to provide an introduction to the basic physics issues associated with nuclear fission. First we study the energetics of fission processes. We estimate the values of mass number A where energy conservation allows fission to occur, and note the characteristic emission of excess neutrons when a heavy nucleus fissions into two fragment nuclei. We then examine three qualitatively different types of fission and their different roles in nuclear energy processes. We focus particularly on *slow-neutron-induced fission*, which powers almost all modern nuclear reactors.

As is often the case in the practical implementation of basic physics principles, for nuclear fission the devil is in the details. The latter part of this section describes some of those details that either enable or complicate the uses of nuclear fission: *delayed neutrons*, without which reactors could not be controlled, which constitute a tiny percentage of all neutrons emitted when a heavy nucleus fissions; and *fission poisons*, which build up in a reactor, particularly just after it is shut down, and prevent the reactor from being restarted until they decay away. Finally we examine how the neutrons emitted in fission reactions can be slowed down by *moderators*, enabling them to initiate further fission reactions effectively. The technical issues associated with fission reactor design and operation are covered in §19.

18.3.1 The Physical Origins of Nuclear Fission

The reason that nuclear fission is a significant energy source can be seen directly from Figure 17.6: when a heavy nucleus splits, energy on the order of hundreds of MeV is given off because the products have greater binding energy per nucleon than the parent. The effect can be seen in eq. (17.21), where the unbinding effects of the Coulomb repulsion between protons grow like $A^{2/3}$, making one heavy nucleus less bound than two lighter ones.[2]

Using the SEMF, we can make a rough estimate of the energy released in fission of a nucleus that lies along the floor of the valley of stability. We assume that the nucleus splits into two identical nuclei, each with half the charge and half the mass number of the original nucleus. With this assumption, the energy released in fission is

---

[2] Note that the effect of the second term in the denominator in eq. (17.21) is small for physically relevant values of A.

$$Q_{\text{fission}} \sim 2B(Z_{\min}(A)/2, A/2) - B(Z_{\min}(A), A),$$

$$(18.2)$$

where $B(Z, A)$ is given by eq. (17.19).

Several significant effects are omitted from this estimate. First, nuclei usually fission *asymmetrically* (see §18.3.5). Assuming symmetric fission *overestimates* Q_{fission}, since the curve in Figure 17.6 is concave upward. Second, if the nucleus were to fission into two identical fragments, the fragments would have more neutrons than the most stable nucleus with mass number $A/2$, because the neutron to proton ratio grows with A. For example, if $^{236}_{92}\text{U}$ split evenly, the result would be two $^{118}_{46}\text{Pd}$ nuclei. The most neutron-rich *stable* isotope of palladium, however, has $A = 110$. In fact, some neutrons are typically given off during the fission process, leaving fission fragments that are still neutron-rich, and which then β-decay until they achieve stability. Ignoring the emission of neutrons also leads to an *overestimate* of Q_{fission}, because the mass excess of a neutron ($\Delta(n) = 8.0713\,\text{MeV}$) is typically larger than the difference in mass defects of two isotopes that differ by $\Delta N = 1$ (Problem 18.6). Finally, (18.2) systematically *underestimates* Q_{fission} because the nucleus $^{A/2}[Z_{\min}(A)/2]$ is generally far from the floor of the valley of stability and the SEMF, whose parameters have been fit to nuclei near the floor of the valley of stability, generally underestimates such binding energies. As a result of these inaccuracies, eq. (18.2) can only be taken as a rough estimate of the A dependence of Q_{fission}.

The estimate (18.2) for Q_{fission} is plotted in Figure 18.2. This estimate suggests that fission becomes exothermic around $A \gtrsim 100$ and that Q_{fisson} reaches $\sim 200\,\text{MeV}$ by $A \sim 250$. Despite the shortcomings of the simple SEMF estimate (18.2), 200 MeV is a good estimate of the energy actually released by the fission of heavy nuclei. Two hundred MeV is large compared to the binding energy of an individual nucleon. In fact, fission of a heavy nucleus

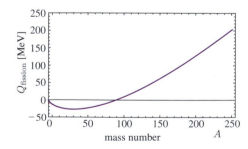

Figure 18.2 A rough estimate (18.2) of the energy released in the fission of a nucleus of mass number A into two nuclei each with $A/2$. The estimate is based on the binding energy per nucleon from the semi-empirical mass formula.

liberates more energy than any other atomic or nuclear event available as a human energy source (see, however, Problem 18.20).

There are at least three qualitatively different mechanisms by which a nucleus can fission, all of which play roles in aspects of nuclear fission energy. Some very heavy nuclei fission *spontaneously* through quantum barrier penetration. Some of these have such long lifetimes that they occur naturally. The other qualitatively different mechanisms for fission are initiated by either low-energy or high-energy neutrons. Quite generally, if a nucleus absorbs a proton, neutron, or another light nucleus (such as an α-particle), binding energy is released, and the added energy can cause a heavy nucleus to fission. Coulomb repulsion keeps charged nuclei far enough away from one another, at least at low energies, to prevent absorption. There is nothing, however, to keep a neutron from being absorbed by a nucleus that it encounters. Neutrons with energies below $\sim$10 eV are known as **slow neutrons**. A small number of nuclei have huge cross sections for absorbing slow neutrons and subsequently undergoing fission, a process known as **slow-neutron-induced fission**.

Neutrons that have come into thermal equilibrium in a fission reactor play a central role in fission energy production. They have energies averaging $\frac{3}{2}k_B T$, where T is the temperature of the reactor core, which may be anywhere from room temperature to upwards of 300 °C. Thus a **thermal neutron**, denoted n_{th}, is often defined as a neutron with energy in the range $\sim$0.02–0.05 eV. When more precision is necessary – some neutron–nucleus interactions depend very sensitively on the neutron energy – the term *thermal neutron* may refer specifically to a neutron with kinetic energy $E_{kin} = 0.0253$ eV, corresponding to $E_{kin} = k_B T$ at $T = 20$ °C. Equivalently, such a neutron may be defined by its speed $v_{th} = 2200$ m/s.

Slow, Fast, and Thermal Neutrons

Although usage varies, a *slow neutron* is typically defined as one with $E_{kin} \lesssim 10$ eV. Similarly a *fast neutron* is one with $E_{kin} \gtrsim 0.5$ MeV. A *thermal neutron* has energies typical of reactor core temperatures, $E_{kin} \sim$ 0.02–0.05 eV. Sometimes the term *thermal neutron* is reserved for a neutron with $E_{kin} = k_B(293\,\text{K}) = 0.0253$ eV, corresponding to $v = 2200$ m/s. Thermal neutrons play a central role in modern fission power reactors.

Finally, many nuclei can undergo **fast-neutron-induced fission** when they are bombarded by energetic **fast neutrons**, which have energies greater than ~ 0.5 MeV. Nuclei that can be induced to fission by neutrons of some energy are termed **fissionable**. Nuclei that often fission after absorbing a slow or thermal neutron are called **fissile**, and are the primary fuels for fission power reactors. Thus fissile nuclei are a subset of fissionable nuclei.

These three different kinds of nuclear fission: *spontaneous fission*, *slow-neutron-induced fission*, and *fast-neutron-induced* fission, are described further in the next three subsections.

18.3.2 Spontaneous Fission

In §17, we argued that when A is very large, the binding energy of a nucleus *increases* as it departs from a spherical shape. Recall that an *increase* in binding energy is equivalent to a *decrease* in total internal energy (or mass excess) of a nucleus. There is, then, no restoring force when such a nucleus deforms, and it would fission instantaneously once formed. "Nuclei" like this cannot be said to exist at all in any meaningful sense and do not figure in fission energy processes.

There is, however, a broad range in Z and A where a small deformation does not increase nuclear binding, but the nucleus can still lower its energy by fission. Figure 18.2 suggests that nuclei with $A \gtrsim 100$ could, in principle, lower their energy by fission. Although all these nuclei are indeed unstable against fission, nevertheless small deformations increase their internal energy, so there is a barrier that must be crossed before fission occurs. The situation for intermediate mass nuclei with $A \gtrsim 100$ is illustrated in Figure 18.3(a). The nucleus is classically stable against ellipsoidal deformations. A large deformation approaching a dumbbell shape must, however, decrease the internal energy of the nucleus, because we know that energy would be given off if the nucleus were to divide in two. The barrier against fission by tunneling is high and broad and the barrier penetration probability is extremely small. Although unstable in theory, these nuclei are never observed to spontaneously fission because their lifetimes are so long.

As A increases, the energy difference between the original nucleus and the fission products grows. By the time A is roughly 230 the situation looks something like Figure 18.3(b), where both the height and the width of the barrier are smaller and tunneling is more likely. Such nuclei are occasionally observed to decay through **spontaneous fission**. ^{235}U, for example, which usually decays by α-emission, fissions spontaneously roughly once in every 10^{10} decays (Problem 18.4). Finally for values of

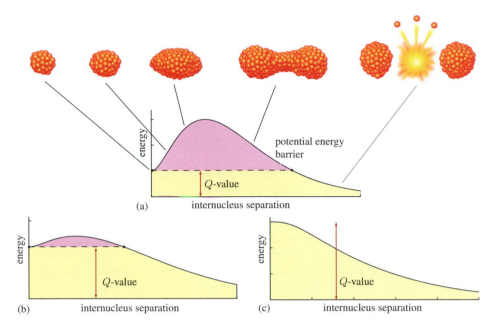

Figure 18.3 Cartoons of instantaneous and spontaneous fission. The progression of nuclear shapes is shown at the top. Below are shown potential energy contours for three nuclei of increasing A: (a) an intermediate mass nucleus that is unstable to fission in principle, but does not fission in practice. The classically forbidden region is shaded above the dashed line, and the red arrows denote the potential fission energy release (Q-value); (b) a heavy nucleus with a smaller barrier and greater energy release. Some nuclei in this category can fission spontaneously with a lifetime short enough to be observed; (c) a very heavy nucleus with no fission barrier. If it could somehow be produced, it would fission instantaneously.

$A \gtrsim 290$ the situation looks like Figure 18.3(c), and fission is instantaneous. Some naturally occurring or long-lived nuclides that have significant probabilities to decay by spontaneous fission are listed in Table 18.2.

18.3.3 Slow-neutron-induced Fission

Neutrons have a large mass excess, so most nuclei can easily absorb excess neutrons, releasing energy in the process even if the resulting nucleus is unstable (Question 18.1). In a heavy nucleus such as uranium, a neutron is bound by ~6 MeV (Problem 18.5). If a neutron is absorbed by such a heavy nucleus, it therefore will liberate considerable energy into the nucleus, even if the neutron has negligible kinetic energy to begin with. What happens to this energy after the neutron has been absorbed? Since strong nuclear interactions are roughly 100 times as strong as electromagnetic interactions, the neutron interacts with the other nucleons and distributes its energy over the bulk of the nucleus, exciting many degrees of freedom before the excess energy can be radiated away as photons. It is as though the nuclear *liquid drop* has been heated. The excitation energy is distributed over many degrees of freedom. Bohr called this *thermally excited nucleus* a

compound nucleus. The thermal energy can feed into various nuclear collective excitations, including vibrations. If the amount of energy added by the neutron exceeds the height of the fission barrier, and if the energy comes to be concentrated in the motion that probes the barrier, the nucleus will fission before the energy can be radiated away.

Consider the specific (and very important) example of ^{235}U. When ^{235}U absorbs a neutron it forms the compound nucleus ^{236}U*. (The asterisk is the standard notation for a compound nucleus.) In this process, enough energy is liberated to allow the ^{236}U* nucleus to fission without tunneling. Thus, slow- or thermal-neutron absorption on ^{235}U can be said to *induce* fission, and ^{235}U is therefore *fissile*. As explained in Box 18.1, ^{235}U *is the unique naturally occurring fissile nuclide*.

Nuclei whose reactions with slow neutrons are important for nuclear power are listed in Table 18.3. Notice the dramatic difference in thermal fission cross section – roughly eight orders of magnitude – when the energy released in neutron absorption (Column 3) exceeds the barrier energy (Column 4) compared to when it does not, since in the latter case quantum tunneling is required.

Table 18.2 The principal examples of nuclei that decay by spontaneous fission are listed in this table. The natural abundance, the half-life, the measured fraction of the decays by spontaneous fission (the dominant decay in each case is α-emission), and the number of fission neutrons emitted per year per mole are tabulated. Data from [70].

Nucleus	Natural abundance	Half-life (years)	Fraction of decays by spontaneous fission	Average number of neutrons per fission	Neutrons per year per mole
$^{235}_{92}$U	0.7204%	7.04×10^8	7.0×10^{-11}	1.86	7.8×10^4
$^{238}_{92}$U	99.2742%	4.46×10^9	5.5×10^{-7}	2.07	1.1×10^8
$^{239}_{94}$Pu	–	2.41×10^4	3×10^{-12}	2.16	1.1×10^9
$^{240}_{94}$Pu	–	6.56×10^3	5.7×10^{-8}	2.21	8.0×10^{12}
$^{252}_{98}$Cf	–	2.65×10^3	3.09×10^{-2}	3.73	1.8×10^{19}

Box 18.1   Fissile and Fertile Nuclei

We can apply our knowledge of nuclear binding as a guide to understanding which naturally occurring nuclei may undergo slow-neutron-induced fission. First, such a nucleus must have very large A to have a relatively low fission barrier. Second, either it must be stable, or if unstable, it must live long enough to still exist on Earth. Third, it helps to have an odd number of neutrons because the added boost of the pairing energy, liberated when a compound nucleus with an even number of neutrons is formed, aids in getting over the fission barrier. And finally, fourth, we can focus on nuclei with an even number of protons because odd–odd nuclei are generally not stable.

Searching through nuclear data, we come to the remarkable conclusion that ^{235}U is the unique candidate. The energy released when a slow neutron is absorbed by ^{235}U is $Q = \Delta(^{235}\text{U}) + \Delta(n) - \Delta(^{236}\text{U}) = 40.922 + 8.071 - 42.448 = 6.545$ MeV. The height of the fission barrier is estimated to be 5.9 MeV (Table 18.3), so slow-neutron-induced fission of ^{235}U is possible. As anticipated above, the energy liberated when a neutron is captured by the naturally occurring *even–even* isotopes of uranium ($^{238}_{92}$U) and thorium ($^{232}_{90}$Th) is not sufficient to overcome their fission barriers. Searching smaller A: radium ($Z = 88$), radon ($Z = 86$), and polonium ($Z = 84$) have no naturally occurring isotopes, and by the time we reach lead ($Z = 82$) the barrier to spontaneous fission is too high. When the naturally occurring even–odd lead isotope ^{207}Pb absorbs a neutron, it liberates 7.368 MeV into the compound nucleus ^{208}Pb*. The fission barrier for ^{208}Pb is, however, estimated to be greater than 20 MeV [85], so the compound nucleus does not fission.

Nuclei heavier than uranium do not live long enough to occur naturally on Earth in useful quantities. A few artificial, manmade nuclides including ^{239}Pu, ^{241}Pu, and ^{233}U do undergo slow-neutron-induced fission. They are made in fission reactors when ^{238}U and ^{232}Th absorb fast neutrons. Since they have relatively long half-lives – 2.41×10^4, 14.3, and 1.59×10^5 years respectively – they can be made in fission reactors, extracted, and processed into new fuel. Because fissile nuclides can be bred from them, nuclides like ^{238}U and ^{232}Th are called **fertile**.

Table 18.3 Slow-neutron capture and fission information for nuclei important in fission reactors. Neutron binding energy is the energy released when the compound nucleus is formed. For fissile nuclei, the average energy released in slow-neutron-induced fission (excluding antineutrinos) and the average number of fast neutrons emitted in fission are also listed. From [77, 87].

Nucleus	Compound nucleus	Neutron binding energy	Fission barrier energy	Thermal neutron fission cross section	Average fission energy release	Average # of fast neutrons
$^{232}_{90}$Th	$^{233}_{90}$Th*	4.8 MeV	6.7 MeV	$< 10^{-6}$ b	–	–
$^{233}_{92}$U	$^{234}_{92}$U*	6.8 MeV	5.85 MeV	531 b	197.9 MeV	2.50
$^{235}_{92}$U	$^{236}_{92}$U*	6.5 MeV	5.9 MeV	585 b	202.5 MeV	2.42
$^{238}_{92}$U	$^{239}_{92}$U*	4.8 MeV	5.8 MeV	2.66×10^{-6} b	–	–
$^{239}_{94}$Pu	$^{240}_{94}$Pu*	6.5 MeV	6.3 MeV	748 b	207.1 MeV	2.88

18.3.4 Other Neutron-induced Reactions

Several different reactions can occur when a neutron collides with a heavy nucleus such as uranium. Furthermore, the cross sections for these reactions have dramatic and complex energy dependence. Three primary processes compete: (a) *scattering* – where the neutron is not absorbed; (b) **radiative capture** – where the neutron is captured and the resulting compound nucleus de-excites by emitting one or more photons; and (c) fission induced by neutrons of any energy. The cross sections for all three are shown (as functions of energy) in Figure 18.4 for ^{238}U.

Scattering, case (a), is not of much interest to us. The neutron just bounces off the heavy nucleus. Radiative capture, case (b), is important in fission power reactors. The radiative capture cross section, shown in Figure 18.4(b) and (b′), is dominated by narrow (in energy) spikes known as **resonances**. These are natural orbits, known as

Fertile, Fissile, and Fissionable Nuclides

A nuclide that undergoes slow-neutron-induced fission is called *fissile*. Fissile materials are those that can be used most easily in power production. A nuclide from which a fissile nuclide can be bred by neutron capture is called *fertile*. A nuclide that can be fissioned by neutrons of some energy is called *fissionable*. ^{235}U is the only naturally occurring fissile nuclide. ^{233}U, ^{239}Pu, and ^{241}Pu are examples of manmade fissile nuclides that can be bred from the fertile (and fissionable) nuclides ^{232}Th and ^{238}U respectively. There are other, heavier, short-lived, fissile *actinides* that are bred in small quantities in fission reactors.

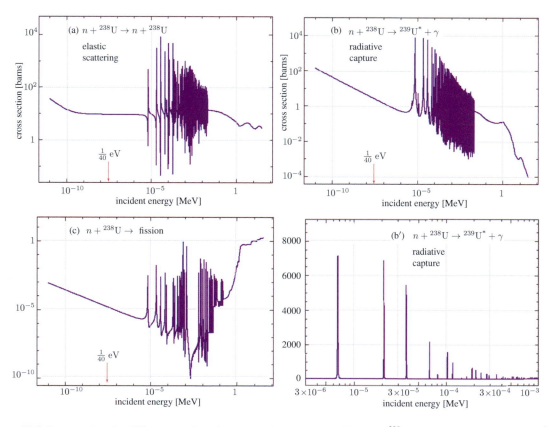

Figure 18.4 Cross sections for different reactions that occur when neutrons collide with ^{238}U with energies between $\sim 10^{-4}$ eV and ~ 10 MeV: (a) elastic scattering, (b) radiative neutron capture, and (c) fission. Note that the vertical scales (cross section in barns) differ for each graph, and they are *logarithmic*. In (b′) the radiative neutron capture cross section of (b) is replotted on a linear rather than logarithmic vertical scale for energies between 3 eV and 1 keV, exhibiting the resonances in the cross section. The red arrows in (a), (b), and (c) mark the energy of a *thermal neutron*, 1/40 eV. From [70].

quasi-bound states, into which a neutron can slip temporarily. Then the neutron either escapes away (in which case the resonances appear in the scattering cross section as in Figure 18.4(a)), or the neutron radiates some energy away as a γ-ray and falls into a truly bound orbit, completing radiative capture. The energies of the quasi-bound states into which neutrons can be absorbed are quantized much like true bound states,[3] hence the spikes.

Radiative capture is important for two reasons. First, this process absorbs a free neutron, eliminating the possibility that the neutron will go on to induce fission in another nucleus. This acts to suppress a chain reaction. Second, neutron absorption changes the identity of the absorbing nucleus. In the case of ^{238}U, it produces ^{239}U. ^{239}U quickly decays to ^{239}Pu, which is long-lived and fissile. In this way, a fission reactor can **breed** new fuel from the dominant uranium isotope ^{238}U. A similar path is possible for ^{232}Th, leading to ^{233}U.

The neutron-induced fission cross section of ^{238}U requires close examination. It is reproduced in Figure 18.5 along with the induced fission cross section for ^{235}U. Note that the ^{235}U-induced fission cross section is huge for very low-energy (e.g. thermal) neutrons (highlighted in red). The thermal-neutron-induced fission cross section for ^{238}U is far smaller. From Table 18.3, we see that the fission barrier in the compound nucleus ^{239}U* (5.8 MeV) is about 1 MeV higher than the energy delivered by a thermal neutron (4.8 MeV), so thermal neutrons are not effective at inducing ^{238}U to fission. On the other hand, neutrons with energy greater than $\sim$1 MeV may induce ^{238}U to fission. From Figure 18.5 we see that this is indeed the case: the fission cross section of ^{238}U rises by over four orders of magnitude as the neutron energy grows past 1 MeV (highlighted in red). Thus, the ^{238}U that accompanies the fissile isotope ^{235}U in a typical nuclear reactor can contribute in two ways to energy production: first, a ^{238}U nucleus can radiatively capture a neutron, producing the fissile nuclide ^{239}Pu; and second, a fast neutron from ^{235}U fission can sometimes induce a ^{238}U nucleus to fission. A pure ^{238}U chain reaction is impossible because the neutrons from fission of ^{238}U fission are not energetic enough to induce further fission. A chain reaction, however, can be sustained by ^{239}Pu created when ^{238}U absorbs a fast neutron. *Fast breeder reactors* exploit the neutron absorption properties

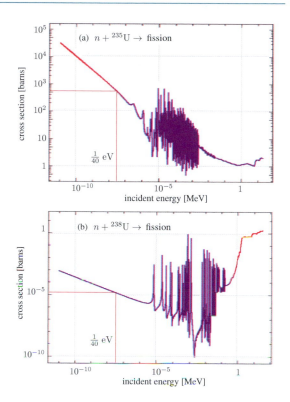

Figure 18.5 Neutron-induced fission cross sections for (a) ^{235}U and (b) ^{238}U for neutron energies from $\sim 10^{-4}$ eV to $\sim$10 MeV. Note: the vertical scales differ by many orders of magnitude. Large cross sections are highlighted in red. From [70].

of ^{238}U to produce more fissile material than they consume (see §19.3).

18.3.5 Fission Products and Energy Release

When a heavy nucleus fissions, many different pairs of neutron-rich nuclei can be produced along with extra neutrons. Here are a few common ^{235}U fission reactions,

$$^{235}_{92}U + n_{th} \rightarrow {}^{144}_{56}Ba + {}^{90}_{36}Kr + 2n,$$
$$^{235}_{92}U + n_{th} \rightarrow {}^{141}_{56}Ba + {}^{92}_{36}Kr + 3n,$$
$$^{235}_{92}U + n_{th} \rightarrow {}^{139}_{52}Te + {}^{94}_{40}Zr + 3n. \tag{18.3}$$

Table 18.4 lists some of the most common **fission fragments**.

Fission is typically asymmetric: one of the fragments from ^{235}U fission generally has atomic mass in the range $A \sim 85$–105 and the other in the range $A \sim 130$–150. The reasons for this asymmetry are complex and do not follow directly from the liquid drop model. The distribution is shown in Figure 18.6. As explained earlier, fission fragments are usually excessively neutron-rich.

---

[3] Unlike true bound states, the energies of these quasi-bound states are not quite discrete. Remember that in quantum mechanics a stationary state of definite energy persists for all time. As mentioned in §15.1 an unstable state with lifetime τ has an uncertainty, or spread, in its energy of $\Delta E \sim \hbar/\tau$.

Table 18.4 Fission products produced with greater than 1% probability in thermal-neutron-induced fission of ^{235}U. The *yield* is the percentage of fission events that lead quickly to this isotope (see Example 18.1). From [88].

Nuclide	Yield	Half-life	Comments
^{133}Cs	6.79%	(2.065 y)	stable*
^{135}I	6.57%	6.46 h	decays to ^{135}Xe, a key *fission poison*
^{93}Zr	6.30%	1.53 My	
^{137}Cs	6.09%	30.2 y	health hazard
^{99}Tc	6.05%	211 ky	
^{90}Sr	5.75%	28.9 y	health hazard
^{131}I	2.83%	8.02 d	health hazard
^{237}Pm	2.27%	2.62 y	
^{149}Sm	1.09%	stable	fission poison

*^{133}Cs is stable, but neutron captures to ^{134}Cs which β-decays with $\tau_{1/2} = 2.065$ y.

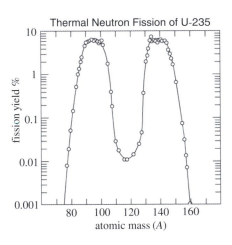

Thermal Neutron Fission of U-235

Figure 18.6 Atomic mass distribution of fragments in ^{235}U fission. After [86].

Typically, the fragments are unstable to β-decay, with lifetimes that range from fractions of a second up to years. Often the resulting decay products are themselves radioactive, leading to a complex decay chain (§17.4.5) with many branches before reaching a stable nuclide (see Example 18.1 and Problem 18.7). Some of the radioactive decay products have serious human health implications, particularly *strontium-90*, *cesium-137*, and *iodine-131*. We discuss these issues further in §20.

Fission Poisons   Among fission fragments are certain nuclides with extremely large neutron absorption cross sections. These nuclides are known as **fission poisons**. In practice, the most important fission poison is ^{135}Xe. Although direct production of ^{135}Xe in ^{235}U fission is not

Neutron Reactions with ^{238}U

Thermal neutrons are more than eight orders of magnitude less likely to induce fusion in ^{238}U than in ^{235}U. Fast neutrons ($E \gtrsim 1/2$ MeV) most often scatter elastically from ^{238}U. The cross section for neutron-induced fission of ^{238}U increases dramatically for neutrons with energies $\gtrsim 1$ MeV, though the cross section is still much smaller than the thermal-neutron fission cross section of ^{235}U. ^{238}U has a large but rapidly varying (in energy) cross section for capturing neutrons with $E \lesssim 10$ keV, leading quickly to the formation of ^{239}Pu.

too common (0.3% of all slow-neutron-induced ^{235}U fissions lead directly to ^{135}Xe), ^{135}Xe is produced in the β-decay of ^{135}I, one of the most common fission products (see Table 18.4). ^{135}I decays to ^{135}Xe with a half-life of 6.46 hours, and ^{135}Xe itself β-decays to ^{135}Cs with a half-life of 9.14 hours (^{135}Cs has a lifetime exceeding a million years). So ^{135}Xe slowly builds up to an equilibrium level in material where a fission chain reaction is occurring. ^{135}Xe has a huge absorption cross section, $\sigma_{\text{absorption}} = 2.75 \times 10^6$ b, for thermal neutrons. Slow neutrons are essential to the perpetuation of a fission chain reaction, so as ^{135}Xe builds up in a reactor and it absorbs an increasing fraction of the available thermal neutrons, it becomes harder and harder to maintain the chain reaction. This affects both the operation and safety of nuclear reactors, as described in §19.1.7.

Prompt and Delayed Neutrons   When ^{235}U undergoes the process of thermal-neutron-induced fission, on average a total of about 2.5 neutrons are given off. To maintain a chain reaction, at least one of those neutrons must be able to induce another fission. Almost all these neutrons are emitted within $\sim 10^{-13}$ seconds of the fission event. These are known as **prompt neutrons**. A tiny fraction are emitted later, over time scales ranging from fractions of a second to minutes. These **delayed neutrons** are emitted with small probabilities during the β-decays of certain fission fragments. For example, the fission fragment $^{87}_{35}$Br β-decays with a half-life of ≈ 55 s. 98% of the time it decays to the ground state of $^{87}_{36}$Kr, but 2% of the time it decays to $^{86}_{36}$Kr plus an additional neutron. The β-decays of ^{235}U fission fragments that can yield delayed neutrons all have half-lives of less than a minute.

In ^{235}U fission about 0.65% of all neutrons are *delayed*, and the average delay time is ~ 12 seconds. As described in §19, these delayed neutrons turn out to be critical to

Example 18.1   What Happens when ^{235}U Fissions?

As an example of the sequence of events following slow-neutron-induced fission, we follow the history of one particular fission reaction,

$$^{235}_{92}\text{U} + n_{\text{th}} \rightarrow {}^{144}_{56}\text{Ba} + {}^{90}_{36}\text{Kr} + 2n \,.$$

The energy released promptly is

$$Q_{\text{fission}} = \Delta(^{235}\text{U}) + \Delta(n) - \left(\Delta(^{144}\text{Ba}) + \Delta(^{90}\text{Kr}) + 2\Delta(n) \right)$$
$$= 40.922 + 8.071 - (-73.937 - 74.959 + 2(8.071)) = 181.747 \,\text{MeV} \,.$$

Most of this energy appears as kinetic energy on the fission fragments; a few MeV is carried by the neutrons and by γ-rays. All of this energy is quickly shared with other atoms in the reactor contributing to the thermal distribution of energies characteristic of the reactor temperature.

Both ^{144}Ba and ^{90}Kr decay by a sequence of β-decays:

$$^{144}\text{Ba} \xrightarrow{11.5\,\text{s}} {}^{144}\text{La} \xrightarrow{40.8\,\text{s}} {}^{144}\text{Ce} \xrightarrow{285\,\text{d}} {}^{144}\text{Pr} \xrightarrow{17.3\,\text{m}} {}^{144}\text{Nd (stable)} \,,$$

$$^{90}\text{Kr} \xrightarrow{32.3\,\text{s}} {}^{90}\text{Rb} \xrightarrow{158\,\text{s}} {}^{90}\text{Sr} \xrightarrow{28.9\,\text{y}} {}^{90}\text{Y} \xrightarrow{64.1\,\text{h}} {}^{90}\text{Zr (stable)} \,,$$

where the β-decay electrons and antineutrinos have been omitted. Notice the long lived isotopes of cerium and strontium (in red). If released into the environment, they represent potential radiation hazards. *Strontium-90* is a particularly insidious hazard since it can find its way into milk and substitute for calcium in bones.

The decays that precede the formation of these long lived nuclides occur over the time scale of minutes. These rapid decays contribute $Q_1 = \Delta(^{144}\text{Ba}) - \Delta(^{144}\text{Ce}) = 6.49\,\text{MeV}$ and $Q_2 = \Delta(^{90}\text{Kr}) - \Delta(^{90}\text{Sr}) = 10.99\,\text{MeV}$ of additional energy (some of which is lost on the antineutrinos) to the energy released in the fission reaction.

The subsequent decays of ^{90}Sr and ^{144}Ce contribute an additional $\Delta(^{90}\text{Sr}) - \Delta(^{90}\text{Zr}) + \Delta(^{144}\text{Ce}) - \Delta(^{144}\text{Nd}) = 6.14\,\text{MeV}$ over their lifetimes. If the original reaction takes place in a fission reactor, this energy continues to heat the reactor fuel long after the fuel is removed from the reactor.

Instability of Fission Fragments

Two consequences of the instability of fission fragments are (1) even after the chain reaction is turned off the material undergoing the chain reaction continues to emit considerable power, and (2) material that has undergone a fission reaction contains fission fragments that are very radioactive and remain so for many years.

the control of a fission power reactor. The fraction of delayed neutrons in ^{239}Pu fission is smaller, only about 0.20%, making control of a plutonium-fueled reactor more difficult.

Energy Release in Fission
The exact amount of energy liberated in slow-neutron-induced fission depends on the specific reaction involved. On average, about 203 MeV is released when ^{235}U fissions. Most of this energy is liberated promptly and appears as kinetic energy

of the fission fragments and neutrons, and as γ-rays. A significant fraction is emitted later as electrons, neutrinos, and more γ-rays when fission fragments β-decay. The average energy budget for ^{235}U fission is summarized in Table 18.5. The emitted (fast) neutrons have energies in the MeV range. The mean neutron energy is just less than 2 MeV. The neutron energy distribution is sketched in Figure 18.7. Table 18.5 shows that about 4.3% of the energy is lost to neutrinos and 6.3% of the energy liberated in fission comes from the radioactive decay of fission fragments. If the fission event takes place in matter, then subsequent radiative capture of some of the prompt neutrons can contribute significantly to the total energy released in fission. In a thermal-neutron reactor this additional energy amounts to about 9 MeV, roughly balancing the energy lost to neutrinos.

18.3.6 Moderators
The energy spectrum of the neutrons emitted in fission (Figure 18.7) is dominated by *fast neutrons* ($E \gtrsim 0.5$ MeV), which have only a small probability of inducing a subsequent fission when absorbed by ^{235}U or ^{238}U.

Table 18.5 Average energy budget for ^{235}U fission. From [87].

Prompt emission	Energy (MeV)
Fission fragments	169.1
Fission neutrons	4.8
γ (and associated electrons)	7.0
Delayed emission (radioactivity)	
β-decay electrons	6.5
β-decay neutrinos (lost)	8.8
γ-emission	6.3
Subtotal	202.5
Neutron capture*	8.8
Total	211.3

* Although not part of the fission reaction itself, energy released when fission neutrons are later radiatively captured contributes to the total budget of energy released when ^{235}U fissions in a reactor.

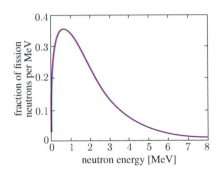

Figure 18.7 The neutron energy spectrum in ^{235}U slow-neutron-induced fission.

Thermal neutrons ($E \sim 0.02$–0.05 eV), on the other hand, have the best chance of inducing another *fissile* nucleus to fission. So it is a priority to slow the neutrons down quickly, before they can be captured.

The mechanism for slowing neutrons down is simple and general. It makes use of the kinetics of two-body collisions between the neutrons and the nuclei of another substance known as a **moderator**. A moderator must be able to slow neutrons down in collisions but must not have a large probability of absorbing them.

First we investigate how a collision of a neutron (mass m) with a stationary nucleus (mass $M \cong Am$) degrades the neutron energy. In the laboratory reference frame, the neutron approaches the target nucleus with energy E, scatters through an angle χ and emerges from the collision with energy $E' < E$ (Figure 18.8(a)). Though E' is a relatively

Summary: Slow-neutron-induced Fission

When a ^{235}U nucleus undergoes slow-neutron-induced fission in a reactor, several things happen: About 180 MeV is promptly liberated in a useful form. On average about 2.5 extra neutrons are liberated. Another ~ 30 MeV is generated by radioactivity of the fission fragments and radiative capture of some neutrons. About 30% of this additional energy is lost on neutrinos, which do not interact. Rarely, a *delayed neutron* is emitted by one of the fission fragments.

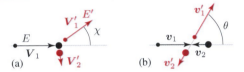

Figure 18.8 Elastic scattering of a neutron from a stationary nucleus, before (black) and after (red) the collision: (a) in the rest frame of the nucleus; (b) in the center-of-mass frame.

complicated function of E and χ, the relation between E' and E is quite simple when expressed in terms of the *center-of-mass scattering angle* θ (see Figure 18.8(b) and Problem 18.10).

$$\frac{E'}{E} = \frac{M^2 + m^2 + 2mM\cos\theta}{(M+m)^2}$$

$$\cong \frac{1 + A^2 + 2A\cos\theta}{(1+A)^2}. \tag{18.4}$$

Nuclear physicists use a measure of the moderating ability of a nucleus with atomic mass A known as the **logarithmic energy decrement** ξ. This quantity is defined as the average value of the decrease in the *logarithm* of the energy in one collision,

$$\xi(A) = \langle \ln(E) - \ln(E') \rangle = \left\langle \ln\frac{E}{E'} \right\rangle. \tag{18.5}$$

Assuming that the scattering is *isotropic* in the center-of-mass frame and averaging over θ, we find (Problem 18.11),

$$\xi(A) = 1 + \frac{(A-1)^2}{2A} \ln\left(\frac{A-1}{A+1}\right) \tag{18.6}$$

$$= \frac{2}{A} - \frac{4}{3A^2} + \mathcal{O}(1/A^3) \text{ as } A \to \infty. \tag{18.7}$$

The logarithmic energy decrement turns out to be the most useful measure of neutron slowing when discussing moderation.

Table 18.6 Neutron energy loss and absorption for selected nuclei, including possible moderators and fission poisons. ξ is the logarithmic energy decrement (18.5). N is an estimate of the number of collisions needed to thermalize a neutron that begins with $E \sim 2\,\mathrm{MeV}$ (Problem 18.12). σ_a is the thermal neutron absorption cross section. Source [70, 77] .

Nucleus	ξ	N^*	σ_a [b]
$^{1}_{1}\mathrm{H}$	1.0	18	0.332
$^{2}_{1}\mathrm{H}$	0.725	25	5.06×10^{-4}
$^{10}_{5}\mathrm{B}$	0.187	–	3.84×10^{3}
$^{11}_{5}\mathrm{B}$	0.171	–	5.50×10^{-3}
$^{12}_{6}\mathrm{C}$	0.158	115	3.37×10^{-3}
$^{16}_{8}\mathrm{O}$	0.120	150	1.90×10^{-4}
$^{135}_{54}\mathrm{Xe}$	–	–	2.66×10^{6}
$^{238}_{92}\mathrm{U}$	0.0084	2200	2.68

* Irrelevant for neutron absorbers

A good moderator should have a large value of ξ, corresponding to large fractional energy loss per collision. ξ has a maximum value of one, realized for $A = 1$, since colliding with a proton (^{1}H) is the most efficient way for a neutron to lose energy. On the other hand, ξ is very small when $A \gg 1$, reflecting the fact that when a neutron collides with a heavy nucleus the neutron simply bounces off, losing little energy. Clearly, light nuclei make better moderators.

A good moderator also must not absorb the neutrons that collide with it. Thus light nuclei with small thermal-neutron absorption cross sections make the best moderators. The thermal-neutron absorption cross sections and other relevant data are given in Table 18.6 for some potential moderators and a few other nuclei of interest. Hydrogen, with the lightest nucleus, is a good, but not ideal moderator, because it has a relatively large neutron absorption cross section. Water moderates like hydrogen – $\xi(\mathrm{H_2O}) = 0.92$ – because the oxygen nucleus is neither a particularly good moderator nor a strong absorber of neutrons. Because water is abundant and can also function as a heat transfer fluid in a reactor, it is the moderator most commonly used in commercial nuclear reactors.

Deuterium ($^2_1$H) is both light and a poor neutron absorber. **Heavy water**, $\mathrm{D_2O}$, is therefore an excellent moderator, with $\xi(\mathrm{D_2O}) = 0.51$. *Deuterated water* HDO occurs naturally in water on Earth at an abundance of approximately 1 molecule in 3200. HDO can be separated from ordinary water, for example by successive distillation, and then further processed either physically or chemically to obtain $\mathrm{D_2O}$. The process is energy intensive and costly,

Moderators

Moderators slow down fast neutrons before they are captured by *fertile* nuclei such as ^{238}U. Once slowed to thermal speeds, neutrons have a high probability of inducing *fissile* nuclei such as ^{235}U to fission. Carbon, hydrogen, water, and deuterium are all good moderators. Deuterium is best since it has both a low mass and a small neutron absorption cross section. Water (effectively hydrogen) and carbon are cheap and common moderators.

but the fact that a reactor fueled with natural uranium and moderated with heavy water can maintain a chain reaction has led to the use of heavy water as a moderator in some power reactors (see §19.4.1). Carbon is less efficient at slowing neutrons down, but has a small neutron absorption cross section and is used extensively as a moderator. Boron, although a light nuclide, is a poor moderator because $^{10}_{5}$B, which accounts for nearly 20% of naturally occurring boron, has a huge neutron absorption cross section, 11 000 times that of hydrogen. Instead, boron is a classic example of a neutron absorber used to control a nuclear reactor (§19).

With this survey of moderators we have finished introducing all the ingredients necessary to understand energy production by controlled nuclear fission reactions. We return to the subject in §19, where we describe how these ingredients, especially the production, moderation, and control of neutrons from fission, can be combined to make a stable nuclear fission power reactor.

18.4 Physics of Nuclear Fusion

From Figure 17.6 it is clear that energy will be given off if light nuclei can be brought to fuse. Light nuclei do not fuse spontaneously, however, because the *Coulomb barrier*

$$V_{\text{coulomb}}(r) = \frac{e^2}{4\pi\epsilon_0} \frac{Z_1 Z_2}{r} \tag{18.8}$$

prevents two nuclei with charges $Z_{1,2}e$ from coming close enough to reach the domain where nuclear forces dominate and lead to fusion. For two protons, the potential energy at the top of the Coulomb barrier is about $V_C \cong 0.2\,\mathrm{MeV}$ (Problem 18.13), which is the average kinetic energy $\langle E_{\text{kin}} \rangle = \frac{3}{2} k_B T$ in a gas at a temperature of $\sim 1.5 \times 10^9\,\mathrm{K}$. The temperature in the core of the Sun is only $1.5 \times 10^7\,\mathrm{K}$, which is far too low to provide the thermal energy necessary to overcome this Coulomb barrier. Yet nuclear fusion takes place spontaneously in stars. The goal of this section is to explain how two key concepts, *quantum tunneling* and the *Boltzmann distribution*, combine to make this possible. The full set of fusion reactions that power the Sun and other stars are described in more detail in the context of solar energy in §22. In closing this section we survey the nuclear reactions that are candidates for controlled nuclear fusion here on Earth. Attempts to harness fusion power are described in the following chapter.

18.4.1 Fusion in a Gas at Temperature T

The fusion rate in a gas at temperature T is given in principle by an integral over energy of the product of three factors: (a) the probability of finding two particles with energy E in their center-of-mass frame; (b) the probability that two particles with center-of-mass energy E get close enough to interact; and (c) a *reaction rate factor* describing the relative likelihood of a fusion event given (a) and (b). The first factor is determined by the Boltzmann distribution, derived in §8; the second factor is given by the barrier penetration factor of §15; and the third factor varies depending on whether the fusion reaction is strong, electromagnetic, or weak. Note that the *cross section* for a particular fusion process is given by the product of (b) the tunneling probability and (c) the reaction rate factor.

The energy distribution in a hot gas The Boltzmann distribution, which determines the probability $P(E_j)$ of finding a particle in a state with energy E_j in a gas at temperature T, was derived in §8

$$P(E_j) = Ze^{-E_j/k_B T} . \tag{18.9}$$

For a hot gas the energy levels are very close together and we can consider the energy as a continuous variable. The sum over states then becomes an integral, giving a probability per unit energy known as the Maxwell–Boltzmann distribution

$$\frac{dP_B}{dE} = \left(\frac{2\pi}{(\pi k_B T)^{3/2}}\right)\sqrt{E}\,e^{-E/k_B T} . \tag{18.10}$$

The first factor in this equation is determined by the fact that the probability of finding the particle with some energy must be unity: $\int_0^\infty dE(dP_B/dE) = 1$. The factor of $\sqrt{E}$ in eq. (18.10) arises from the integral over the different directions $\hat{n}$ in which the particle can be moving (see Problem 22.13). As expected, the mean kinetic energy is $\frac{3}{2}k_B T$: $\langle E \rangle = \int_0^\infty dE\, E(dP_B/dE) = \frac{3}{2}k_B T$ (Problem 18.14).

Although the average energy of a particle in a gas at temperature T is $\frac{3}{2}k_B T$, the Maxwell–Boltzmann distribution predicts that there are many molecules with energies far higher than that. For example, in a mole of gas at temperature T eq. (18.10) predicts that there are $\sim 10^{18}$ particles with ten times the mean energy. There are even thousands

of particles per mole with energies in excess of 30 times the mean energy. This alone, however, cannot explain fusion in the Sun: the probability of finding a particle with energy greater than $V_C = 0.2\,\text{MeV}$ in a gas at 1.5×10^7 K is of order 10^{-69} (Problem 18.15). So effectively, there are no protons in a mole of gas at 1.5×10^7 K with enough energy to overcome the Coulomb barrier.

The Boltzmann Tail

The Boltzmann distribution at temperature T has a long, exponential tail resulting in a significant number of particles per mole with energies as large as 30 times the average energy. This alone, however, is not enough to explain why fusion can occur at temperatures as low as that in the core of the Sun.

The role of tunneling The analysis here is very similar to the one carried out in §17.4.2 to explain α-decay, albeit *in reverse*! There we were interested in the probability of an α-particle tunneling *out* through the very high potential barrier of a heavy nucleus. Here we are interested in the probability of a small nucleus tunneling *in* through the Coulomb barrier. The probability of tunneling P_T is determined by the same *barrier penetration factor* (see Box 17.3). It is determined by the energy of the particle and the form of the barrier,[4]

$$P_T(E) \approx \exp\left(-\frac{2}{\hbar}\int_R^{R_1} dr \sqrt{2\overline{M}\left(\frac{Z_1 Z_2 e^2}{4\pi\epsilon_0 r} - E\right)}\right), \tag{18.11}$$

where $Z_1 e$ and $Z_2 e$ are the charges of the colliding nuclei, $\overline{M} = M_1 M_2/(M_1 + M_2)$ is their reduced mass, and R and R_1 are the inter-nuclear separations that mark the beginning and the end of the forbidden region, as shown in Figure 18.9. As in §17.4.2, we do not worry about the detailed form of the attractive nuclear interaction inside the Coulomb barrier. We can approximate the integral by setting R to zero – the difference is negligible at energies well below V_C. Then changing variables to $x \equiv 4\pi\epsilon_0 r/Z_1 Z_2 e^2$, and using $\int_0^1 dx\sqrt{1/x - 1} = \pi/2$, we find

---

4   The relevant energy is the energy in the center-of-mass frame of the two particles, not the energies of the individual particles. It is not hard to show, however, that the center-of-mass energy follows a Boltzmann distribution with the same temperature as the individual particles.

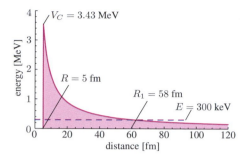

Figure 18.9 The Coulomb potential energy between an α particle and ^{12}C for separations greater than R, where the nuclei touch. The height of the Coulomb barrier is V_C. The dashed line corresponds to a relative energy of 300 keV, which is the energy at which fusion is most probable when the temperature is approximately 2×10^8 K, the point at which α ^{12}C fusion takes place in massive stars. Compare to Figure 17.9.

$$P_T(E) \approx \exp\left(-\pi \frac{Z_1 Z_2 e^2}{4\pi\epsilon_0}\sqrt{\frac{2\overline{M}}{\hbar^2 E}}\right). \qquad (18.12)$$

Note the $1/\sqrt{E}$ in the exponential; it has the same origin as the factor of $1/\sqrt{Q}$ that appeared in the logarithm of the α-decay lifetime (17.30), namely that the width of the barrier becomes infinite as $E \to 0$. Clearly, a small increase in the energy can have a huge effect on the rate. Combining the effects of statistical mechanics and tunneling, the probability of a tunneling event is characterized by the **Gamow distribution**, which is the product of the Boltzmann factor and the barrier penetration factor,

$$dP_{\text{Gamow}} = P_T \times dP_B. \qquad (18.13)$$

In conjunction with the long tail of the Boltzmann distribution, tunneling opens up a window, known as the **Gamow window**, where fusion is possible. The actual probability for a specific fusion process is given by the product of the Gamow distribution (18.13) with the *reaction rate factor*; the reaction rate factor is generally order one for a strong interaction, smaller for an electromagnetic interaction, and much smaller for a weak interaction.

Figure 18.10 illustrates the way that the Boltzmann distribution combines with tunneling to allow fusion to occur at a temperature where the typical particle energy is far below the height of the Coulomb barrier. The figure refers to the pp system ($Z_1 = Z_2 = 1$, $\overline{M} = M_p/2$), with $k_B T = 0.1 V_C$. The Boltzmann distribution falls rapidly with E and is negligible at $E \approx V_C$. The tunneling probability is significant at energies well below V_C,

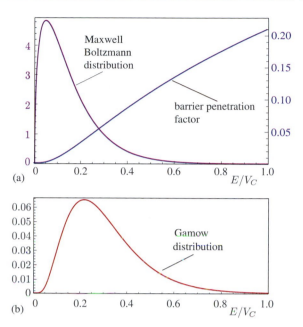

Figure 18.10 The Gamow window for the pp system with $k_B T = 0.1 V_C$. (a) The Maxwell–Boltzmann distribution (purple) and the barrier penetration factor (blue). (b) The Gamow distribution, which is the product of the Maxwell–Boltzmann distribution and the barrier penetration factor, peaks at $E \approx 0.2 V_C$, allowing reactions to take place over a range of energies well above $k_B T$ but well below V_C.

however, producing a *window* where the product of the two probabilities is substantial.

18.4.2 Fusion in the Sun

We now consider again the situation inside the Sun, where the temperature is approximately 1.5×10^7 K, so $k_B T \approx 1.3$ keV. The Sun is fueled primarily by pp fusion, so the Coulomb barrier height, $V_C \cong 0.2$ MeV, is roughly $150 k_B T$. Figure 18.11 shows the Gamow distribution for pp fusion under these conditions. The reaction probability is quite strongly peaked at $E \cong 6$ keV, which is roughly $0.03 V_C$. The area within the Gamow window in Figure 18.11, which is a measure of the fraction of protons that get close enough to fuse, is about 10^{-5}. Considering that there are more than 6×10^{26} protons per kilogram of hydrogen, there are plenty of protons in the Sun with energy sufficient to tunnel through the Coulomb barrier and fuse.

In fact, given this abundance of high-enough-energy protons, it may be surprising that the Sun burns so slowly. After all, it has lasted $\sim 4.5 \times 10^9$ years already and should last roughly as long again. The power output of the Sun is vast, but the power density is quite small, just 0.27 W/m³,

Tunneling and the Gamow Window

Quantum tunneling allows nuclei with energies below the top of the Coulomb barrier to get close enough to interact. The high-energy tail of the Maxwell–Boltzmann distribution combines with the low-energy tail of the tunneling probability distribution to form the *Gamow window* – a range of energies well above $k_B T$, but well below V_C – where there is a significant probability of nuclear fusion reactions.

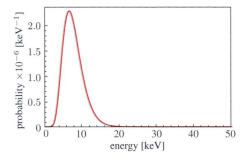

Figure 18.11 The Gamow distribution for circumstances in the interior of the Sun, with $k_B T = 0.0067 V_C$.

---

or about 1.7×10^{12} MeV/m^3 s. A simple estimate based on this power density (Problem 18.16) indicates that a proton in the hot solar core has a probability of engaging in fusion of about 4×10^{-18} s^{-1}, suggesting a lifetime of order 10^{10} years. The fusion probability for an individual proton is so small due to the reaction rate factor, which is very small because *proton–proton fusion* is a weak interaction process (see Table 18.7). We explore solar fusion and solar energy more thoroughly in §22; here we conclude with a look at the roles of reaction rates, Coulomb barrier height, and material abundance in possible fusion reactions here on Earth.

18.4.3 Relevant Fusion Reactions on Earth

To achieve controlled nuclear fusion on Earth it is necessary to raise matter to a high enough temperature and density for a long enough time to make the probability of fusion appreciable. To maximize this probability, the desired fusion reaction should have the lowest possible Coulomb barrier V_C and the highest possible reaction rate factor. Since V_C is proportional to the product of the nuclear charges, and because the tunneling probability decreases exponentially with V_C, only the lightest nuclei,

protons ($p = {}^1_1\underline{H}$), deuterons ($d = {}^2_1\underline{H}$), tritons ($t = {}^3_1\underline{H}$), and ${}^3_2\underline{He}$, are realistic candidates for fusion energy fuels.[5]

Ordinary hydrogen is, of course, abundant on Earth, particularly as a component of water. *Deuterium* accounts for about 0.015% of hydrogen. It cannot be separated from ordinary hydrogen by chemical means, but it is relatively easily separated by physical means because HDO is about 5% heavier than H_2O ("D" is the chemical symbol for deuterium). *Tritium* and ^{3}He are extraordinarily rare on Earth. Tritium does not occur naturally except in minute quantities formed in cosmic ray collisions. It can be manufactured in nuclear reactors by reacting neutrons with ^{6}Li, a rare (7.4%) isotope of *lithium*,

$$n_{\text{th}} + {}^6_3\underline{Li} \rightarrow \alpha + t . \tag{18.14}$$

This is an exothermic reaction with $Q = 4.78$ MeV. Tritium can also be produced in neutron collisions with the common (92.6%) isotope of lithium, ${}^7_3$Li,

$$n + {}^7_3\underline{Li} \rightarrow t + \alpha + n . \tag{18.15}$$

This reaction is endothermic ($Q = -2.47$ MeV). Finally, ^{3}He is common in the universe, but exceedingly rare on Earth. It accounts for about 1.34 parts per million by volume (ppmv) in naturally occurring helium gas. Earth's original endowment of ^{3}He escaped from our atmosphere early in its history; unlike ^{4}He, ^{3}He is not replenished via α-decays. Naturally occurring ^{3}He is so scarce that virtually all ^{3}He used in scientific experiments is obtained from decay of artificially produced tritium.

The laws of nature have not smiled on fusion as a straightforward way to generate power on Earth. Possible fusion reactions among the light elements are listed in Table 18.7. Every fusion reaction listed has at least one serious handicap when considered as an energy source on Earth. Reactions that require weak or electromagnetic interactions have relatively small reaction rate factors and cross sections that are too small. They take place roughly 100 (electromagnetic) or 10^{10} (weak) times slower than strong interactions. This excludes reactions (1), (2), and (3) in the table. Reactions (4) and (5) have low energy yield, requiring relatively high density and temperature and long confinement time (see §19.6). Reactions (6) and (7) require fuels that are extremely rare and/or unstable (^{3}He, ^{3}H). Also, the $d + {}^3_2\underline{He}$ reaction (7) has almost twice the Coulomb barrier of the dt reactions, requiring the reactants to be raised to a temperature about six times

---

5   The notation ${}^A_Z$X introduced in §17.1 referred to a neutral atom. Fusion processes involve fully ionized nuclei, for which we use the notation ${}^A_Z\underline{X}$.

Table 18.7 Characteristics of fusion reactions. The first column gives the reactants and the reaction products. All the reactions involve the strong force, but the rate is limited by the weakest force that must act in order for the reaction to occur. That force is listed in the second column. V_C is an estimate of the height of the Coulomb barrier. The energy Q released per reaction is in the last column.

Reaction	Force	V_C [MeV]	Q [MeV]
(1) $p + p \rightarrow d + e^+ + \nu_e$	weak	0.20	1.43
(2) $p + d \rightarrow {}^3_2\text{He} + \gamma$	EM	0.18	5.49
(3) $d + d \rightarrow \alpha + \gamma$	EM	0.16	23.85
(4) $d + d \rightarrow {}^3_2\text{He} + n$	strong	0.16	3.27
(5) $d + d \rightarrow t + p$	strong	0.16	4.03
(6) $d + t \rightarrow \alpha + n$	strong	0.15	17.59
(7) $d + {}^3_2\text{He} \rightarrow \alpha + p$	strong	0.30	18.35

Candidate Reactions for Controlled Fusion

Only the lightest nuclei are candidate fuels for controlled fusion on Earth. Reaction rate, Coulomb barrier heights, and energy release considerations select *deuterium–tritium* fusion as the most promising candidate. The tritium fuel must be manufactured by irradiating lithium with neutrons, a complication that must be addressed to achieve practical fusion power.

higher than reaction (6). Reaction (6) has the disadvantage that most of the energy is carried off by the neutron (see Problem 18.18), making the energy harder to capture. The highly exothermic reaction $d + d \rightarrow \alpha$ is, unfortunately, forbidden because it cannot conserve both energy and momentum (Problem 18.18).

On nuclear physics grounds the *dt* reaction (6) is the most promising. It has a low Coulomb barrier and one of the highest energy releases. It proceeds by strong interactions. It requires the lowest temperatures, the lowest density, and the shortest confinement time. One drawback, however, is the unfortunate fact that tritium does not occur naturally on Earth. The *dt* reaction is the reaction used in experiments on controlled fusion. Small amounts of tritium have long been made in nuclear fission reactors in connection with nuclear weapons programs. This gives scientists enough tritium to carry out experiments. If *dt*-fueled nuclear fusion power ever becomes practical, large quantities of tritium will be required as fuel. One possibility is to use the neutrons produced in *dt* fusion to manufacture more tritium through the same reactions used to manufacture it in nuclear reactors, (18.14) and/or (18.15). For this to work, neutron economy is essential: each fusion reaction requires one tritium nucleus and produces one neutron. If (18.14) were the only mechanism, every produced neutron would have to react with a ^{6}Li nucleus to make another tritium nucleus in order to sustain the fusion reactor. Neutron reactions with ^{7}Li could, however, in principle yield additional neutrons. Other mechanisms for *neutron multiplication* have been proposed as

well. Alternatively, tritium could be manufactured in a fission reactor using excess neutrons, and harvested for use as a fusion fuel.

The conceptual design for a *dt* fusion reactor is thus to heat and compress a *dt* mixture to sufficient temperature and pressure to initiate fusion, while surrounding the reactor with a *blanket* of ^{6}Li and/or ^{7}Li from which more tritium fuel is subsequently recovered. Progress on meeting the technical and material demands of such a reactor is discussed in the next chapter.

Discussion/Investigation Questions

18.1 When exposed to a flux of thermal neutrons, most common materials absorb one or more neutrons and many become radioactive (β^--emitters). Explain how this is compatible with energy conservation. Can you name a naturally occuring nucleus that cannot absorb a neutron?

18.2 ^{234}U decays primarily by α-emission. It has rarely been observed to decay by spontaneous fission or by emitting a neon or magnesium nucleus. In what sense are all of these decays different examples of the same phenomenon?

18.3 Check that the nuclides ^{239}Pu, ^{233}U, and ^{241}Pu satisfy the four conditions listed in §18.3.3 for being fissile. A few other actinides – nuclei with $Z > 92$ – produced in reactors are also fissile. See if you can find one or two. [Hint: Check out the isotopes of curium ($Z = 96$) and/or californium ($Z = 98$).] Explain what information you would need in order to decide for sure if a given nuclide is fissile.

18.4 Some medical *radioisotopes* – nuclides that produce specific forms of radiation used for medical purposes – are obtained as fission products from thermal-neutron-induced fission of ^{235}U. A particularly important example is technetium-99m, ^{99m}Tc, which is a metastable nuclide with a half-life of only 6 h. Research the production and uses of ^{99m}Tc: how can ^{99m}Tc be produced, separated, transported and used when it has such a short half-life? Why is this particular nuclide well-suited to its application?

Problems

18.1 Suppose a beam of particles with flux Φ is incident on a uniform target of thickness T and density ρ (mass/volume). Assuming that the atoms in the target do not block one another (the *thin-target approximation*), show that the reaction rate dN/dt is given by $dN/dt = I\sigma\rho N_A T/m$, where I is the beam current (i.e., the integral of the flux over the area of the beam), m is the molecular mass of the target, N_A is Avogadro's number, and σ is the reaction cross section.

18.2 Assuming that the particles in Problem 18.1 that react are removed from the beam, use the result of that problem to show that the beam current falls exponentially $I(z) = I_0 e^{-z/\lambda}$ through the target. Find an expression for λ, which is the *mean free path* of the particles passing through the target.

18.3 A beam of thermal neutrons with current $I \equiv \int dA\Phi = 10^8\,\text{s}^{-1}$ is incident on a uniform (thin) target of thickness 1 cm, consisting of 20% ^{235}U and 80% ^{238}U. See Table 18.3 and Problem 18.1. What is the fission rate? What is the thermal power being deposited in the target?

18.4 Given the data in Table 18.2, estimate the number of neutrons emitted by spontaneous fission per second per kilogram of naturally occurring uranium.

18.5 In §18.3.3 it is stated that a neutron in uranium is bound by $\sim$6 MeV. Check this by using data from [70] to compute the binding energy of a neutron averaged over the five isotopes ^{234}U through ^{238}U.

18.6 Verify the statements made after eq. (18.2) about the energy released for the case of thermal-neutron-induced fission of ^{235}U. First, use measured mass excesses to show that the asymmetric fission reaction $n_{\text{th}} + {}^{235}\text{U} \rightarrow {}^{92}\text{Kr} + {}^{144}\text{Ba}$ has a smaller Q-value than the symmetric fission into two ^{118}Pd nuclei. Next show that the reaction $n_{\text{th}} + {}^{235}\text{U} \rightarrow 2({}^{116}\text{Pd}) + 4n$ also has a smaller Q-value than the symmetric fission into two ^{118}Pd nuclei. Finally, show that the SEMF underestimates the binding energy of ^{118}Pd and therefore overestimates the Q-value for the symmetric fission into two ^{118}Pd nuclei.

18.7 An example of a thermal-neutron fission reaction is $n_{\text{th}} + {}^{235}\text{U} \rightarrow {}^{139}\text{Ba} + {}^{94}\text{Kr} + 3n$. Perform the analysis of Example 18.1 for this case.

18.8 When ^{235}U fissions by absorbing a slow neutron, ^{93}Zr is one of the most common fission fragments. Suppose three neutrons are given off in the fission. What is the other nucleus produced in this fission reaction? What is the total energy released in this particular

fission reaction, including subsequent β-decays of the fragments?

18.9 Two sources of delayed neutrons in ^{235}U fission are the fission fragments ^{87}Br and ^{137}I. Find the fraction of their decays that yield a neutron, and the half-life of each. Verify that the Q-value for the β-decay with neutron emission is positive.

18.10 [T] Verify the classical energy loss formula (18.4) for an elastic two-body collision. Use the notation in the Figure 18.8. [Hint: Use the fact that the relationship between a velocity vector V in the target rest frame and the vector v in the center-of-mass frame is $V = v + V_{\text{CM}}$, where V_{CM} is the velocity of the center-of-mass frame.]

18.11 [T] Derive the expression eq. (18.7) for $\xi(A)$ by averaging over angles in the center-of-mass reference frame. [Hint: Take E/E' from eq. (18.4) and remember that the measure for averaging over θ is $\frac{1}{2}\sin\theta\,d\theta$.]

18.12 [T] (a) Derive the large A approximation to the *logarithmic energy decrement* given in eq. (18.5). (b) Show that the number of collisions with a nucleus with atomic mass A required to decrease the energy of a neutron from $E_0 = 2\,\text{MeV}$ to $E_f = 0.025\,\text{eV}$ is $N \cong 18.2/\xi(A)$.

18.13 [T] Estimate the height of the *Coulomb barrier* between two protons as follows: the strong nuclear interaction potential between two protons at large separation is given by eq. (14.5). Add to this the repulsive Coulomb potential given in eq. (18.8). Show that the resulting potential has a maximum at $\approx 5.5\,\text{fm}$, where its value is $V_C \approx 0.2\,\text{MeV}$.

18.14 [T] Use the Maxwell–Boltzmann distribution (18.10) to show that the mean kinetic energy of a particle in a gas at temperature T is $\frac{3}{2}k_B T$.

18.15 Show that the probability of finding a particle with energy greater than $0.2\,\text{MeV}$ in a gas at temperature 15×10^6 K is $\sim 10^{-69}$.

18.16 The net reaction that accounts for the Sun's power takes four neutral hydrogen atoms and fuses them to make a neutral helium atom, $4{}_1^1\underline{\text{H}} \rightarrow {}_2^4\underline{\text{He}}$. Compute the energy released per hydrogen atom. The Sun's total power output is $L_\odot \cong 3.85 \times 10^{26}$ W. How many protons per second are reacting in the Sun? Approximately 10% of the protons in the Sun will fuse during its stable *hydrogen burning* phase. Given the mass of the Sun, $M_\odot \cong 1.99 \times 10^{30}$ kg, and the fact that about 75% by mass of the Sun is hydrogen (almost all the rest is helium), estimate how long the Sun's hydrogen burning phase will last.

18.17 The Coulomb barrier keeping two nuclei apart is given by $V(r) = Z_1 Z_2 e^2/(4\pi\epsilon_0 r)$, where r is the relative separation of the two nuclei. Assume that V reaches its maximum value V_C when $r \sim R_1 + R_2$, where R_j is the radius of the jth nucleus. Using eq. (17.10), estimate the Coulomb barrier between a carbon nucleus and an α particle. Compare your results with Figure 18.9.

18.18 Show that in $d + t \rightarrow n + {}^4\text{He}$ fusion reaction, the ratio of the kinetic energy on the neutron to that on the ${}^4\text{He}$ is 4:1 if you neglect the initial kinetic energies of the reactants. Why is this a good approximation in terrestrial fusion experiments?

18.19 Estimate the quantity of tritium (kg/year) required to fuel a 1 GWe fusion power plant assuming 20% overall efficiency.

18.20 Compare the energy released *per nucleon* for (a) fission of ${}^{235}\text{U}$; and (b) dt fusion.

18.21 Tritium is just barely unstable. How much more tightly would tritium have to be bound in order for it to be stable? Write the nuclear reaction describing tt fusion. How much energy is emitted in tt fusion? Unfortunately the cross section for tt fusion is much smaller than that for dt fusion, so even if tritium were stable, tt fusion would not be an interesting candidate for fusion power.

Nuclear Fission Reactors and Nuclear Fusion Experiments

To provide practical power for humankind, the nuclear energy production processes of fission and fusion must be harnessed in stable and safe reactors. The basic principles of nuclear physics and the mechanisms by which fission (and perhaps fusion) can be harnessed to generate power have been outlined in the previous three chapters. Here we explore in some detail how a fission reactor actually works and then briefly describe experiments in controlled thermonuclear fusion.

In this chapter, we address basic questions about real nuclear reactor systems, such as: What are the constraints on the fuel, the cooling system, the moderator, and the systems that transform the heat from a fission reactor into useful work? How do these constraints influence the safety of the reactor? What are the dominant reactor designs today and how do these reactors transform the energy released in nuclear fission into useful power? How might future reactor designs improve safety and efficiency?

We first describe (§19.1) the neutron budget that governs the fission reaction and energy output of a fission reactor. We emphasize the *neutron multiplication factor*, which measures how many neutrons from each generation survive to the next. This leads to a discussion of possible fuels and moderators, and then to an estimate of the power generated and fuel consumed in a fission reactor. We then turn (§19.2) to the factors that affect nuclear fission reactor control and safety, concentrating on the use of physical principles to guide the design of reactors that are *inherently* safe. In §19.3, we describe *breeder reactors*, which produce more fuel than they use. Next we examine existing and proposed designs for nuclear reactors (§19.4) and the associated thermodynamic cycles and efficiency limits (§19.5). In the final section of the chapter (§19.6) we survey some experiments presently underway to generate controlled nuclear fusion, focusing primarily on *magnetic confinement fusion devices*.

Reader's Guide
In this chapter we present the basics of nuclear fission reactor design and summarize experiments in controlled nuclear fusion. Physical phenomena that affect fission reactor operation, control, and safety are described. Existing electric power reactor designs are described, and advanced designs for potential future reactors are surveyed briefly. We describe the thermodynamics of nuclear reactor power cycles.

We explain the basic concepts of nuclear fusion reactors including the energy balance in steady state, and derive the criteria used to judge the performance of controlled fusion experiments. Finally we present the basic concepts of *magnetic confinement fusion* and mention experiments in *inertial confinement fusion*.

Prerequisites: §7 (Quantum mechanics), §8 (Entropy and temperature), §15 (Quantum processes), §16 (Nuclear), §17 (Nuclear structure), §18 (Fission and fusion).

Concepts developed here figure in the discussion of nuclear waste and proliferation in §20 (Radiation).

In this chapter we provide only a basic introduction to the design and operation of nuclear fission and fusion reactors. We focus on the underlying physical principles and processes that often constrain engineering design parameters.[1] Concise introductions to nuclear reactors from a physicist's point of view can be found in [77] and [85]. Reference [86] provides a modern introduction to nuclear fission reactor physics at the undergraduate

---

[1] As an indication of the relative complexity of practical reactor design and engineering, some years ago Boston Edison operated nuclear-, oil-, and gas-fired plants of roughly similar capacities. The total full-time staffs, excluding contractors, were 600, 97, and 17 people respectively [89].

level. A non-technical, but very readable introduction to nuclear reactors and nuclear power can be found in [81]. Finally, [90] provides an introduction to the physics of magnetic confinement fusion.

19.1 Nuclear Fission Reactor Dynamics

19.1.1 Overview of Fission Reactor Systems

The key to understanding a fission reactor is to follow the neutrons. Consider a block of pure natural uranium, which is a homogeneous mixture of 99.28% ^{238}U and 0.72% ^{235}U. Occasionally a uranium nucleus fissions spontaneously, producing roughly two neutrons on average (see Table 18.2). The neutrons have high energies. As explained in the previous chapter, most of these neutrons get absorbed by the overwhelmingly prevalent ^{238}U before they can come into thermal equilibrium and induce a ^{235}U nucleus to fission. (It takes $\sim$ 2200 collisions with ^{238}U to thermalize a neutron.) At the energies of the neutrons emitted in ^{235}U fission, the induced-fission cross section of ^{238}U is not zero, but it is too low to sustain a chain reaction.

The prospects for fission can be improved by (i) mixing the fuel with a *moderator* to slow the neutrons down more quickly; (ii) *enriching* the fuel by increasing the proportion of ^{235}U; or (iii) *segregating* (*clumping*) the fuel into localized regions within a matrix of moderator. Commercial reactors typically employ both (ii) and (iii).

As described in §18.3.6, a moderator slows the neutrons down quickly, increasing the chance that they thermalize and induce ^{235}U nuclei to fission. If natural uranium and a moderator are mixed homogeneously, only *heavy water*, D_2O, is a good enough moderator to sustain a chain reaction. If the fuel is **enriched**, by increasing the proportion of ^{235}U, the demands on the moderator can be relaxed. At higher levels of ^{235}U enrichment, a chain reaction in a homogeneous mixture with ordinary water or graphite as a moderator becomes possible. The enrichment threshold is 1.6% ^{235}U for a graphite moderator and 2.6% for water.

Another important degree of freedom is **clumping**. If small fuel elements are surrounded by bulk moderator, then a neutron emitted in a ^{235}U fission can escape the fuel element where it was created before being absorbed by ^{238}U, then slow down quickly while propagating through moderator, and finally re-enter a fuel element as a thermal neutron, where it can induce fission of another ^{235}U nucleus. The first sustained nuclear reaction, established in 1942 at the University of Chicago, used lumps of natural uranium embedded in graphite blocks. The importance of

inhomogeneity in reactor design hints at how complicated the subject can be.

The basic figure of merit of a fission chain reaction is the number of thermal neutrons of the $(n+1)$st generation produced by each thermal neutron of the nth generation. This number, known as the **neutron multiplication factor**, or k-**factor**, is the key parameter in designing and controlling a nuclear power reactor.

The behavior of a nuclear reactor over time is determined by how close the neutron multiplication factor k is to one. If k exceeds one, then each generation of neutrons generates more neutrons. The number of neutrons, and therefore the number of fissions $N(t)$, then grows exponentially with time,

$$N(t) = N_0 e^{(k-1)t/t_0} , \qquad (19.1)$$

where t_0 is the time it takes each generation of neutrons to produce the next (§19.1.6). The situation where $k > 1$ is referred to as **supercritical**. If $k < 1$, the reaction dies out according to the same equation (**subcritical**). The art of running a fission reactor is to keep k almost exactly equal to one – to keep the reactor **critical** – and not to let it exceed one other than during the start-up of the reactor. The relative deviation of k from one is known as the **reactivity** ρ,

$$\rho = \frac{k-1}{k} . \qquad (19.2)$$

When k is very close to one, which is often the case, ρ can be approximated by $k - 1$.

From the previous discussion we see that enriching the fuel, improving the moderator, and clumping the fuel all increase the k-factor. The k-factor is decreased by leakage of neutrons out of the core of the reactor, another complex design challenge. All of these are design considerations – they do not allow one to control the k-factor in real time during reactor operation. The principal real-time control mechanism is the ability to insert or remove **control rods** made of material with a very high neutron-absorption cross section. One example is boron-carbide, B_4C, since $^{10}_5$B is an excellent neutron absorber (see Table 18.6). Also, B_4C is an extremely hard and heat-resistant ceramic that can withstand extreme environments. Many other materials, including alloys of silver, indium, cadmium, and hafnium, have been adapted for use in control rods in different reactor environments. Reactors are designed so that complete insertion of the control rods shuts off the fission reaction very quickly. Modern reactors also have inherent responses that are designed to shut the reactor down if the normal control system were to fail, as discussed below.

Finally, the heat generated in the reactor must be conveyed out of it, both to keep the reactor from overheating, and to harvest the useful energy. This is accomplished by circulating a fluid in and around the **reactor core**, where the fuel, moderator, and control rods are located. This fluid is usually referred to as **coolant** even in situations where its primary role is the extraction of useful energy. Usually the fluid is liquid water, which acts as both the moderator and the coolant. In some designs heavy water (D_2O) is used as both moderator and coolant. In other designs CO_2 vapor is used as a coolant. Advanced and experimental reactor designs not yet commercially available employ materials such as helium, liquid sodium, liquid lead, and liquid salt as possible coolants.

19.1.2 The Neutron Cycle in a Thermal Reactor

In this section we briefly summarize the physical processes that determine the *neutron multiplication factor* in a *homogeneous,* uranium-fueled, thermal-neutron reactor. We return to the subject in greater depth in §19.1.4 after learning more about cross sections in the following section.

For simplicity we consider an *infinite* reactor, for which the neutron-multiplicity factor is denoted by k_∞. By limiting ourselves to infinite, homogeneous reactors we are able to obtain a useful schematic understanding of reactor dynamics. All manmade reactors, however, are finite and highly inhomogeneous, so any realistic analysis of reactor design must include neutron leakage and inhomogeneity, which are beyond the scope of this book.

The reactor is assumed to be a homogeneous mixture of *fuel* and *moderator* in a ratio $y = n_M/n_F$, where $n_{M,F}$ is the number density of atoms of moderator/fuel in the mixture. The fuel consists of uranium, with an **enrichment factor** defined by $x = n(^{235}\mathrm{U})/(n(^{238}\mathrm{U}) + n(^{235}\mathrm{U}))$. In a typical electric power reactor $x \ll 1$ and $y \gg 1$.

It is convenient and instructive to express k_∞ as the product of four factors,

$$k_\infty = \eta \epsilon p f \,, \tag{19.3}$$

each of which parameterizes a step in the neutron cycle. The **four-factor formula** was proposed early in the development of nuclear reactors and remains a valuable tool for explaining the neutron cycle in physical terms. In reality the steps are not as clearly separated as the four-factor formula implies, but a more sophisticated treatment requires detailed numerical analysis that goes beyond the scope of this book.

The *reproduction factor* η measures the number of *fast* neutrons that a thermal neutron will generate on average when it encounters a uranium nucleus in the reactor and is either captured or causes fission. Next, the *fast fission factor* ϵ accounts for the fact that occasionally a fast neutron

produced in thermal fission of $^{235}\mathrm{U}$ will initiate fission – principally of a $^{238}\mathrm{U}$ nucleus – before slowing down to thermal velocity. Next comes the *resonance escape probability p*, which is the probability that a fast neutron will be slowed down to thermal velocity by scattering from the moderator without being captured in $^{238}\mathrm{U}$ or some other nucleus with a large radiative capture cross section. Finally, the *thermal neutron utilization factor f* measures the fraction of the now-thermalized neutrons that avoid capture in the moderator or other materials, and interact with the fuel. The original neutron number η is reduced by the factors p and f, which are less than one, and enhanced by ϵ, which is greater than one.

In order to analyze the ingredients in the four-factor formula, we must first go a little more deeply into the subject of cross sections.

19.1.3 More About Cross Sections

When we introduced the notion of a cross section earlier, we imagined a collimated beam of particles with velocity $\boldsymbol{v}$ impacting a target with cross section σ. Inside a reactor, instead, a gas of neutrons moves through a fixed density of scatterers. For this situation the reaction rate is given by

$$\frac{dN_{\mathrm{reaction}}}{dV\,dt} = n_t n_n \langle \sigma_{\mathrm{reaction}} v \rangle, \tag{19.4}$$

where $dN_{\mathrm{reaction}}/dV\,dt$ is the number of reactions per unit time per unit volume, n_n and n_t are the number densities of neutrons and target nuclei, respectively, and $\sigma_{\mathrm{reaction}}$ is the cross section for the reaction of interest. The cross section may vary with the neutron speed v, so the product σv must be averaged over the neutron velocity distribution as denoted by $\langle \sigma v \rangle$. The same formula applies to interactions between two species in a gas or plasma, where in this case v is the relative velocity.

Often the product of the neutron density times the average speed is combined into the **neutron flux**,

$$\Phi_n = n_n v \,. \tag{19.5}$$

Sometimes it is useful to define the cross section *per unit volume*, also known as the **macroscopic cross section**, for a pure material, $\Sigma \equiv \sigma n$. When a variety of different nuclei are present in a material, as is the case inside a fission reactor, we define

$$\Sigma = \sum \sigma(^A[\mathrm{Z}]) \, n(^A[\mathrm{Z}]) \,, \tag{19.6}$$

where the sum is over all nuclear species present. The macroscopic reaction cross section can be defined for any component of the mixture or for any particular reaction $\Sigma_{\mathrm{reaction}} = \sum \sigma_{\mathrm{reaction}}(^A[\mathrm{Z}]) \, n(^A[\mathrm{Z}])$. Combining

Table 19.1 Cross sections (in barns) for thermal-neutron ($v \cong 2200$ m/s) fission, capture, absorption ($\sigma_a = \sigma_f + \sigma_c$), and scattering from for some fertile and fissile isotopes of thorium, uranium, and plutonium, and for natural uranium [70].

Nucleus	Density g/cm^3	σ_f	σ_c	σ_a	σ_s
^{232}Th	11.7	$< 10^{-6}$	7.34	7.34	13.0
^{233}U	18.7	531	45.3	576	12.2
^{235}U	18.7	585	98.7	684	15.1
^{238}U	18.9	1.68×10^{-5}	2.68	2.68	9.30
U$^*_{\text{natural}}$	18.9	4.21	3.37	7.59	9.34
^{239}Pu	19.8	748	271	1019	7.99

* Cross sections for natural uranium are a weighted average of the cross sections of ^{238}U and ^{235}U.

Table 19.2 Properties of moderators relevant for thermal fission reactors, including density (in g/cm^3), atomic mass, cross sections (in barns) for thermal neutrons ($v \cong 2200$ m/s), and logarithmic energy decrement ξ (§18.3.6) [70, 77].

Material	Density	A	σ_a	σ_s	ξ
H_2O	1.0	18	0.664	49.2	0.920
D_2O	1.1	20	1.20×10^{-3}	10.6	0.509
C	1.6	12	3.37×10^{-3}	4.94	0.158

eqs. (19.4)–(19.6) yields a simple expression for the reaction rate per unit volume for a known neutron flux within a target of known composition,

$$\frac{dN_{\text{reaction}}}{dV\,dt} = \langle \Phi_n \Sigma_{\text{reaction}} \rangle . \tag{19.7}$$

In an inhomogeneous reactor, where the fuel and moderator are clumped, the neutron flux, the cross sections, and the rates of various reactions vary as functions of position. It is thus considerably easier to estimate the ingredients in k_∞ in the case of a homogeneous reactor, where all these quantities are independent of location.

The neutron reactions of interest to us include scattering, capture, and fission, with associated cross sections σ_s, σ_c, and σ_f. Often one is interested in the neutron *absorption* cross section, which is the sum of capture and fission cross sections, $\sigma_a = \sigma_c + \sigma_f$. All of these cross sections depend on the neutron energy, often sensitively. For a thermal reactor the cross sections for thermal neutrons are often most useful. Tables 19.1 and 19.2 summarize these cross sections for some components of nuclear fuels and for some moderators respectively.

19.1.4 Components in the Neutron Multiplication Factor

To analyze the four-factor formula $k_\infty = \eta \epsilon p f$, we begin with a thermal neutron of the nth generation and estimate the number of thermal neutrons that it gives rise to in the $(n + 1)$st generation. We enter the cycle when a *thermal* neutron is absorbed by a uranium nucleus.

η – the reproduction factor The **reproduction factor** η measures the number of *fast* neutrons produced by a thermal neutron that is *absorbed* by a uranium nucleus. An absorbed thermal neutron can either initiate a fission or it can be captured. If the neutron is absorbed by a ^{238}U nucleus, it is far more likely to be captured than to cause fission ($\sigma_c/\sigma_f \approx 1.4 \times 10^5$, see Table 19.1). If, on the other hand, the neutron is absorbed by a ^{235}U nucleus, it is quite likely to cause fission ($\sigma_f/\sigma_c \approx 6$), in which case it generates on average 2.42 fast neutrons. So the average number of fast neutrons generated when a thermal neutron is absorbed by a uranium nucleus is given approximately by

$$\eta = 2.42 \left(\frac{x\sigma_f(^{235}\text{U})}{(1-x)\sigma_a(^{238}\text{U}) + x\sigma_a(^{235}\text{U})} \right). \tag{19.8}$$

The factor η is a monotonic function of x, increasing from zero at $x = 0$ to 2.08 for pure ^{235}U. For natural uranium, where $x = 0.0072$, $\eta = 1.33$. Thus for a fission chain reaction to be possible in natural uranium, the product $p\epsilon f$ cannot be smaller than $1/1.33$.

ϵ – the fast fission factor Occasionally a fast neutron produced in thermal fission of a ^{235}U nucleus will initiate fission of a ^{238}U nucleus (recall Figure 18.4(c)) before slowing down to thermal velocity. When a fast neutron induces ^{238}U to fission, it produces on average about 1.7 additional fast neutrons. The resulting enhancement in the number of fast neutrons is measured by the **fast fission factor** $\epsilon \geq 1$. Note that fast neutrons produced in this way are considered as part of the same generation of neutrons as those that initiate the ^{238}U fission. In a heterogeneous reactor, where fuel atoms are packed together in clumps, ϵ may be significantly greater than one. In a homogeneous thermal reactor, the moderator-to-fuel ratio is large, so the chance that a fast neutron will encounter a ^{238}U nucleus and cause it to fission is quite small. We therefore often approximate $\epsilon = 1$ for an infinite homogeneous reactor. A reactor can be *designed* to make use of fast neutrons to initiate fission, in which case ϵ is significantly larger than one. Such *fast reactors* are discussed below (§19.3).

Figure 19.1 provides a visual summary of these first two steps in the thermal-neutron cycle; in the figure $\eta\epsilon = (B + D)/A$.

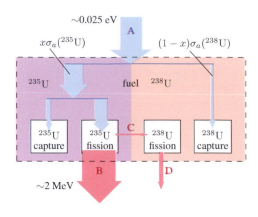

Figure 19.1 A sketch showing the physics behind the neutron *reproduction factor* η and the *fast fission factor* ϵ in a thermal reactor. $\eta = (B + C)/A$ measures the probability that a thermal neutron (blue) produces a fast neutron (red) after encountering a uranium nucleus. $\epsilon = (B + D)/(B + C)$ measures the enhancement in the number of fast neutrons resulting from fast neutrons that cause a ^{238}U nucleus to fission. Note that the likelihood of ^{238}U fission by a thermal neutron is negligible.

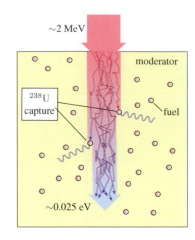

Figure 19.2 A sketch showing the physics behind the *resonance escape probability* p. Fast neutrons propagating through a homogeneous mixture of moderator and fuel slow down to thermal velocities by repeatedly scattering, primarily from moderator nuclei. The neutron flux is reduced by radiative capture at energies that coincide with resonances in the ^{238}U radiative capture cross section.

p – the resonance escape probability To induce fission of another ^{235}U nucleus, a fast neutron must be slowed to thermal velocity by multiple collisions with the moderator. As it slows, the neutron must avoid capture by a ^{238}U nucleus, which has resonances in its radiative capture cross section – especially in the neutron energy range between ~10 eV and ~10 keV (see Figure 18.4). The **resonance escape probability** p measures the probability that the neutron reaches thermal equilibrium without being captured. This factor is not easy to calculate (see Figure 19.2 for a pictorial representation of the process). The analysis simplifies somewhat for a homogeneous reactor, but it is still complicated. An approximate, empirical result for a homogeneous system with $x \ll 1$ is [91]

$$p = \exp\left(-\frac{2.73}{\langle \xi \rangle}\left[\frac{n(^{238}\mathrm{U})}{\Sigma_s}\right]^{0.514}\right),\qquad (19.9)$$

where $n(^{238}\mathrm{U})$ is the number density of ^{238}U in the reactor, Σ_s is the macroscopic scattering cross section for the moderator plus fuel in barns per unit volume, and $\langle \xi \rangle$ is the logarithmic energy decrement for the mixture of moderator and fuel. When $y \gg 1$, the moderator dominates so that $\langle \xi \rangle \approx \xi(\text{moderator})$.

As one would expect for an *escape* probability, p given by eq. (19.9) increases with ξ (better moderation) and with Σ_s (more encounters with the moderator), and decreases with $n(^{238}\mathrm{U})$ (higher density of absorbers).

The resonance escape probability for an infinite, homogeneous reactor increases monotonically as a function of the moderator-to-fuel ratio y but depends only weakly on the enrichment x for the small values of x encountered in thermal reactors.

f – the thermal utilization factor The **thermal utilization factor** f measures the probability that a neutron, once thermalized, is absorbed by a uranium nucleus rather than by the moderator, as illustrated in Figure 19.3,

$$f = \frac{\Sigma_a(\text{fuel})}{\Sigma_a(\text{total})} = \frac{\Sigma_a(\text{fuel})}{\Sigma_a(\text{fuel}) + \Sigma_a(\text{moderator})}. \quad (19.10)$$

For fixed moderator-to-fuel ratio y, f increases with enrichment because ^{235}U is a more efficient absorber than ^{238}U; for fixed enrichment x, on the other hand, f decreases as the moderator-to-fuel ratio y increases, since the higher the fraction of moderator, the greater the chance it will absorb the neutron. With this step, the thermal neutron has entered the fuel and the cycle is complete. Note that (19.10) determines f for an idealized reactor containing only fuel and moderator; if other materials (such as fission poisons, §19.1.7) that absorb neutrons are present, their cross section must be added to the denominator.

Example 19.1 illustrates how the four factors can be used to estimate the k factor and reactivity of a reactor for a realistic choice of moderator-to-fuel ratio and a range of enrichments. Figure 19.4 shows several examples of the

---

Example 19.1   The Reactivity of a Water-moderated, Infinite, Homogeneous Reactor

What is the reactivity of a water-moderated, uranium-fueled reactor with moderator-to-fuel ratio $y = 3$ as a function of enrichment from $x = 0.0072$ to $x = 0.10$?

Recall that the reactivity (19.2) is measure of the neutron multiplication factor and must be zero in order to sustain a chain reaction in steady state. To compute k_∞, we first compute the reproduction factor η, which depends only on the enrichment x. Substituting the necessary cross sections from Table 19.1 into eq. (19.8),

$$\eta(x) = 2.42 \left(\frac{585x}{684x + 2.68(1 - x)} \right).$$

As discussed in the text, for an infinite homogeneous reactor, we approximate the fast fission factor by $\epsilon = 1$. Next, the resonance escape probability p must be estimated using eq. (19.9). Since $\xi_{water} \cong 0.92$ and $y = 3$, we have $\xi = (3/4) \times 0.92 = 0.69$. Thus,

$$p(x, y) = \exp \left(-\frac{2.73}{0.69} \left[\frac{n(^{238}U)}{n(^{238}U)\sigma_s(^{238}U) + n(^{235}U)\sigma_s(^{235}U) + n(H_2O)\sigma_s(H_2O)} \right]^{0.514} \right).$$

Using $n(^{235}U)/n(^{238}U) = x/(1 - x) \approx x$ and $n(H_2O)/n(^{238}U) = y/(1 - x) \approx y$ and substituting cross sections from Tables 19.1 and 19.2, we find

$$p(x, y) = \exp \left(-3.96(9.3 + 15.1x + 49.2y)^{-.514} \right).$$

For $y = 3$, p is only weakly dependent on x and can be approximated by its value at $x = 0$, which is $p \cong 0.73$. Finally, $f(x, y)$ is obtained from eq. (19.10),

$$f(x, y) = \frac{684x + 2.68(1 - x)}{684x + 2.68(1 - x) + .664y}.$$

The figure at right shows the resulting k_∞ and reactivity ρ_∞ obtained from the *four-factor formula*. The reactivity becomes positive around $x \cong 0.013$, which means a fission reaction can, in principle, be sustained in a homogeneous, water-moderated, infinite reactor at enrichment levels above 1.3%.

dependence of k_∞ and its components on enrichment and choice of moderator.

From these analyses of the four-factor formula, we can see some of the tradeoffs involved in the choice of moderator and ^{235}U enrichment level for a reactor. While D_2O is an excellent moderator and works with natural uranium, it is quite expensive. As a result, graphite and ordinary water are the moderators used in most commercial reactors. Figures 19.4(a) and (b) show that it is not possible to have a graphite-moderated, infinite, homogeneous reactor fueled with natural uranium, but that enhancement to 1.6% ^{235}U will work. Note that the ratio of moderator to fuel is very large for a graphite-moderated reactor. This is because moderating a neutron in graphite requires many more collisions than for D_2O or H_2O (Table 18.6), so for the resonance escape probability p to have a high value, the moderator to fuel ratio y must be large. As a result, graphite-moderated reactors are

generally larger than water-moderated reactors, an important design consideration. While water requires fewer collisions to moderate a neutron than graphite, it is also a strong neutron absorber, so f drops rapidly with y. A functioning water-moderated reactor therefore requires a relatively high enrichment level to compensate for the small value of f by increasing η. Figure 19.4(c) shows the components of k_∞ for a water-moderated reactor at 3.7% enrichment, a value typical of modern power reactors. 3.7% was also approximately the fractional abundance of ^{235}U when the Oklo natural reactor went critical some two billion years ago (see Box 19.1).

19.1.5 Power and Fuel Consumption

In this section we give a simplified analysis of the rate of power production and fuel depletion in an infinite homogeneous reactor, to give a sense of the quantities and time

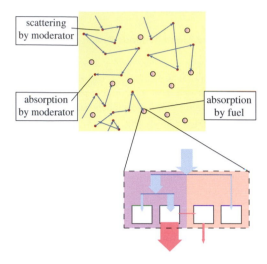

Figure 19.3 Thermal neutrons wander through the moderator–fuel mixture until they encounter a uranium nucleus and begin the neutron reproduction cycle again (Figure 19.1) or they are absorbed by the moderator.

Neutron Multiplication Factor

The neutron multiplication factor k measures the number of neutrons of the $(n + 1)$st generation produced by each neutron of the nth generation. In an infinite reactor,

$$k_\infty = \eta \epsilon p f,$$

where η measures the average number of fast neutrons generated by each thermal neutron that is absorbed by a uranium nucleus; ϵ accounts for extra neutrons produced by fast fission; p measures the probability that a fast neutron escapes capture and thermalizes; and f measures the probability that a thermal neutron is absorbed by the fuel (leading to capture or fission) as opposed to the moderator or other material. Generally, $\eta > 1, \epsilon \geq 1$, while $p, f < 1$.

k must be kept very close to one in order to have a controlled, self-perpetuating chain reaction in a nuclear fission reactor.

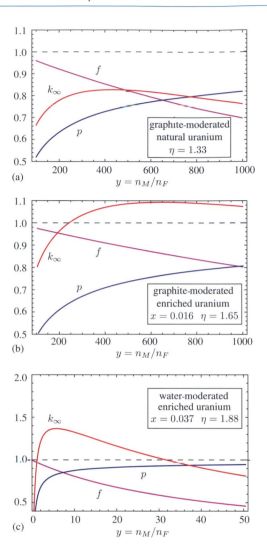

Figure 19.4 k_∞ factors and their components, η, p, and f, for three infinite, homogeneous, uranium reactors ($\epsilon \cong 1$). (a) A graphite-moderated reactor fueled with natural uranium; (b) same as (a), enriched to 1.6% ^{235}U; (c) a water-moderated reactor fueled with uranium enriched to 3.7%, appropriate for the Oklo natural reactor (see Box 19.1). k_∞ must be greater than one for a chain reaction to be possible.

scales involved. When a fission reactor is started up – usually by partially removing control rods – the reactivity ρ is made positive and kept positive long enough for a fission chain reaction to become established. The flux of neutrons in the reactor increases while $\rho > 0$, along with the

fission rate. As the reactor heats up, an increasing amount of thermal power can be extracted from the reactor core. When the reactor reaches the desired power, ρ is decreased back to zero and kept there so that the reactor runs at a steady rate.

At first a uranium-fueled reactor burns only ^{235}U. As time goes on, a significant amount of ^{238}U is converted to fissile ^{239}Pu, which also serves as fuel. A detailed calculation of power output and fuel consumption focuses

Box 19.1   The Oklo Natural Reactor

Uranium occurs naturally in the environment, and ordinary water is a good moderator. The two could combine to produce a **natural reactor** except for the fact that the percentage of ^{235}U in natural uranium is too low (see the figure in Example 19.1) to support a self-sustaining reaction. ^{235}U was more abundant in the past, however; the fraction of ^{235}U in natural uranium has gradually decreased over time because its half-life is a factor of six shorter than that of ^{238}U. In 1956 Japanese nuclear chemist Paul Kuroda [92] suggested that natural fission reactors may have existed in rich uranium deposits in Earth's past. Very early in Earth's history, oxygen was rare. The chemically reduced uranium compounds that existed at that time are not readily soluble in water, limiting the extent to which they might be concentrated by geological processes. As time went on, plants added oxygen to Earth's atmosphere, and by about two billion years ago, soluble uranium salts had become more common, settled out of solution, and accumulated into relatively rich deposits. The stage was set for the ignition of a natural fission reactor.

In 1972 French scientists measuring the uranium enrichment of ore samples from the Oklo uranium deposits in Gabon discovered that the proportion of ^{235}U was 0.7171% rather than the normal concentration of 0.7253%. Even this slight difference is significant because the only known way to reduce the relative abundance of ^{235}U is by induced fission. Further investigation indicated that the abundance of various isotopes of neodymium and ruthenium in the ore was biased toward isotopes that are produced as ^{235}U fission fragments. Subsequent studies have found regions at Oklo where the ^{235}U/^{238}U ratio is a low as 0.292% [93]. There is now little doubt that natural fission reactors evolved in Oklo ores over extended periods beginning about 1.7 billion years ago [94], and thereby depleted the ^{235}U content of the ores.

The Oklo uranium deposits show evidence of fission activity in at least 16 different sites. All the necessary ingredients for a fission reactor were present: the ores were rich in uranium, in places containing over 90% UO$_2$ [93]. The percentage of ^{235}U was 3.67% at the time the reactors began. The ores were embedded in porous sandstones that were periodically submerged or saturated with water in a wet climate. The ores were also relatively low in nuclei with large neutron-absorption cross sections (other than ^{238}U). Furthermore, the ore bodies were large enough that neutron leakage did not prevent the chain reaction. The occasional spontaneous fission of a ^{235}U or ^{238}U nucleus provided the seed neutrons to initiate a fission chain reaction when the water saturation level was high enough to provide the necessary moderation.

The reaction apparently generated temperatures as high as 1000 °C at the core and even 250–450 °C at the boundaries of the reactor zone. These temperatures were high enough to drive away the water moderator, thereby ending the chain reaction. Geologists have evidence that suggest cycles of hydration, reaction, desiccation, and rehydration on time scales of several hours lasting for hundreds of thousands of years. Over the time scale of these cycles, fission poisons presumably were present at relatively constant background levels, limiting the power of the reactor.

The Oklo story represents a remarkable bit of Earth history.

---

on the neutron flux, which determines the rates of depletion of ^{235}U, of conversion of ^{238}U into ^{239}Pu, and of fission of ^{239}Pu. Here for simplicity we consider the situation where only ^{235}U and ^{238}U are present, as is the case just after a uranium-fueled reactor is first powered up. This allows us to circumvent the detailed analysis of the neutron flux, and relate the power generated by the reactor P_f directly to the rate of ^{235}U fission R_f,

$$R_f = \frac{dN_f}{dt} = P_f/E_f. \qquad (19.11)$$

Here $E_f \cong 202.5$ MeV includes all the energy released in the fission of ^{235}U except for the energy that escapes on neutrinos. Note, in particular, that it includes the $\approx$ 8.8 MeV given off (on average) when prompt neutrons are radiatively captured by other nuclei in the reactor. The

rate at which ^{235}U is being depleted in the reactor is proportional to the fission rate

$$\frac{dN(^{235}\mathrm{U})}{dt} = -\frac{\sigma_a}{\sigma_f} R_f, \qquad (19.12)$$

where the factor σ_a/σ_f takes account of the fact that ^{235}U is depleted by radiative capture of thermal neutrons as well as by fission.[2] The cross sections σ_f and σ_a change rapidly with neutron energy (see, e.g., Figure 18.5(a)) and should be averaged over the neutron energy distribution in the reactor. The ratio, σ_a/σ_f, however, varies only slowly over the range of neutron energies found in a thermal-neutron

---

[2] Note that our conventions are such that dN_f/dt is the positive rate of ^{235}U fission while $dN(^{235}\mathrm{U})/dt$ is negative.

reactor, so the cross sections can be taken to be those given in Table 19.1 for purposes of computing this ratio, independent of the precise temperature or neutron flux in the reactor. Combining eqs. (19.11) and (19.12), the reactor power can be related to the rate of depletion of fuel,

$$P_f = E_f \frac{\sigma_f}{\sigma_a} \left| \frac{dN(^{235}\mathrm{U})}{dt} \right| . \qquad (19.13)$$

Since a reactor is typically run with constant power output, the rate of depletion of $^{235}\mathrm{U}$ is constant. Thus, at start-up, when only $^{235}\mathrm{U}$ is being burned, eq. (19.13) allows us to relate the reactor power to the mass ΔM of uranium fuel (enriched to a fraction x of $^{235}\mathrm{U}$) that is used in a time Δt

$$P_f = E_f \frac{\sigma_f}{\sigma_a} \frac{x \, N_A \, \Delta M}{m \, \Delta t} , \qquad (19.14)$$

where m is the molar mass of uranium.

During the course of its operation, $^{239}\mathrm{Pu}$ produced in the reactor increases the supply of fissile material and therefore decreases the rate of $^{235}\mathrm{U}$ consumption. In addition, fast-neutron fission of $^{238}\mathrm{U}$ contributes to the power output without depleting the supply of $^{235}\mathrm{U}$. Together, these effects enhance the power relative to the fuel consumption by a factor of ~ 1.5. Thus the power-to-fuel-consumption ratio over the entire period that the fuel is in the reactor can be approximated by multiplying eq. (19.14) by this factor,

$$P_f \approx 1.5 E_f \frac{\sigma_f}{\sigma_a} \frac{x \, N_A \, \Delta M}{m \, \Delta t} \approx 3.3 \, x \, \frac{\Delta M[\mathrm{t}]}{\Delta t[\mathrm{y}]} \; \mathrm{GW} , \qquad (19.15)$$

where we have substituted values for the cross sections, E_f, and m. For example, running a 3 GWth reactor for one year consumes $\Delta M \approx 0.9\,\mathrm{t}$ of $^{235}\mathrm{U}$ ($x = 1$). This can be shown to be consistent with the statement in §16

Fission Reactor Power

As a rule of thumb, the thermal power per ton of fuel, generated by a fission reactor burning uranium enriched to a fraction x of $^{235}\mathrm{U}$, and designed to have a fuel consumption time constant of Δt years, is roughly

$$P_f \approx 3.3 \, x \, \frac{\Delta M[\mathrm{t}]}{\Delta t[\mathrm{y}]} \; \mathrm{GW} .$$

When losses during enrichment and to incomplete fuel consumption are included, this is consistent with the estimate of 200 t/GWe of natural uranium quoted in §16.

---

that roughly 200 t of natural uranium are required to run a 3 GWth reactor for a year (Problem 19.10).

The fission rate is also directly related to the average flux of thermal neutrons $\langle \Phi_n \rangle$ by eq. (19.7). At startup, when only $^{235}\mathrm{U}$ is present

$$R_f = \int dV \left(\frac{dN_f}{dV dt} \right) = \langle \Phi_n \sigma_f \rangle N(^{235}\mathrm{U}) , \qquad (19.16)$$

where $\langle \ldots \rangle$ denotes an average over the volume of the reactor and the neutron energy distribution. Since R_f is constant when a reactor is run at constant power, eq. (19.16) requires that the thermal-neutron flux must be increased with time as the amount of fissile material in the reactor is depleted.

To estimate the neutron flux in a thermal reactor, consider a 3 GWth reactor fueled with one tonne of $^{235}\mathrm{U}$ at startup, and suppose that the reactor has reached its operating temperature of 320 °C. The fission cross section for $^{235}\mathrm{U}$ drops by roughly 1/3 as the temperature increases from 20 °C to 320 °C (see Figure 18.5), so we take $\sigma_f \sim (2/3)585 = 390\,\mathrm{b}$, and find $\langle \Phi_n \rangle \sim 9 \times 10^{13}\,\mathrm{cm}^{-2}\,\mathrm{s}^{-1}$. Thermal-neutron fluxes of this order can easily be maintained in modern reactors, and there is considerable latitude in adjusting $\langle \Phi_n \rangle$ to keep R_f constant as the amount of fissile material in the reactor decreases.

19.1.6 Reactor Time Scales and Delayed Neutrons

The operator of a nuclear reactor, whether human or machine, must be able to react on the time scale over which conditions in a reactor can change. In particular, since the number of neutrons grows exponentially when the reactivity is greater than one (c.f. eq. (19.1))

$$N(t) = N_0 e^{\rho t / t_0} , \qquad (19.17)$$

the time t_0 that it takes each generation of neutrons to produce the next is the critical time scale for reactor control.[3] We can identify t_0 as the time required for the *longest step* in the chain of events that takes one generation of neutrons to the next. Many of the processes that contribute to t_0 are so rapid that they are negligible by any human standards. For example, the neutrons that accompany fission are emitted within 10^{-15} seconds of the fission event. The other candidates for t_0 are (i) the time t_s that it takes a fast neutron to slow to a thermal speed, or (ii) the time t_p it takes a (prompt) thermal neutron to find another fissionable nucleus as the neutron diffuses through the reactor.

---

[3] In this equation and the following discussion, we have approximated $k - 1 = \rho k \approx \rho$ since k is very close to one throughout.

Delayed Neutrons

A small fraction – less than 1% – of all neutrons emitted in fission are delayed by times ranging from ~1 second to ~1 minute. This small fraction d of *delayed neutrons* allows a reactor to be controlled over those time scales as long as the reactivity ρ does not exceed d.

Typically $t_p \gg t_s$. For example, in a water-moderated thermal reactor, $t_s \approx 5.6 \times 10^{-6}$ s while $t_p \approx 2.1 \times 10^{-4}$ s [79]. We can ignore the shorter time scale and conclude that $t_0 \approx t_p \approx 2 \times 10^{-4}$ s for such a reactor. This sounds like bad news for reactor control: if t_0 were in fact of order 2×10^{-4} s and if ρ were even as small as 0.001, the number of neutrons in the reactor would grow by roughly a factor of $e^5 \approx 150$ in just one second. So an operator would have to be able to adjust control rods on a time scale of milliseconds in order to prevent a runaway chain reaction! Even the most modern feedback and control systems are not up to this challenge.

So far we have examined only the **prompt neutrons** that are emitted in the fission event itself. As described in §18.3.5, however, a small fraction d of the fission neutrons ($d = 0.00641$ for ^{235}U) are emitted by fission fragments long after the fission takes place. In §18.3.5 we mentioned one mode that results in a neutron delayed by ≈ 55 s in 0.14% of ^{235}U fissions. There are in fact six different fission fragments that emit *delayed neutrons* on time scales from 0.23 to 55 seconds. A proper analysis treats each of these delayed neutron groups separately. For a semi-quantitative estimate of the effect of delayed neutrons, it is sufficient to lump them all together with the weighted average lifetime of $t_d \approx 12.5$ s. So there are two distinct time scales that are relevant, one for prompt neutrons (t_p) governing a fraction $(1 - d)$ of the neutrons, and another (t_d) governing the remaining, delayed neutrons. If $\rho_{\text{effective}} = \rho - d < 0$, then the delayed neutrons are not "in play" on the short time scale and the chain reaction does not grow exponentially on the time scale t_p. Only after the time t_d has passed are all the neutrons available (Problem 19.8).

Thus, the existence of delayed neutrons is *absolutely essential to the control of thermal fission power reactors.* As long as a reactor has a thermal-neutron multiplication factor that exceeds unity by less than d, the reactor time scale is set by the delayed neutrons and is long enough for control with modern engineering methods. If, however, the

k-factor in a reactor exceeds $1 + d$, the reactor is **prompt critical**, and is in danger of getting out of control. The Chernobyl disaster was initiated when, through a series of mishaps, a poorly designed reactor was allowed to go prompt critical (see Box 19.2). One aim of novel reactor designs is to include intrinsic safeguards, described below, so that the heat generated by an inadequately controlled reactor would itself automatically reduce k to the point that the fission reaction would cease.

19.1.7 Fission Poisons and Reactor Operation

The notion of a *fission poison* was introduced in §18.3.5. In the context of a fission reactor, a **fission poison** is any nuclide with an anomalously large neutron-absorption cross section. Any material that absorbs neutrons inhibits the fission chain reaction by reducing the thermal neutron utilization factor f. Naturally occurring, stable fission poisons (e.g. ^{10}B) are useful components of control rods, but must be avoided in other reactor materials. Early in World War II, for example, German scientists concluded that graphite would not work as a moderator because their graphite was contaminated with boron that was used in its preparation.

Some fission poisons occur as fission fragments and build up during the course of reactor operation, requiring operators to take compensating measures to keep the chain reaction going. Some decay radioactively or convert to non-fission-poison isotopes when they absorb a neutron, in which case they build up to an equilibrium value during the operation of a reactor. The left-hand side of Figure 19.5, labeled "before shutdown" shows the negative contribution to the k-factor from the build up of a fission poison. Other fission poisons, hafnium isotopes for example, form a sequence of isotopes, all of which have large

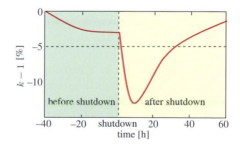

Figure 19.5 Percentage change in k due to xenon poisoning as a function of time before and just after shutdown of a typical reactor. The horizontal dashed line at -5% indicates the maximum decrease in k that typically can be compensated by removal of control rods. After Lilley [77].

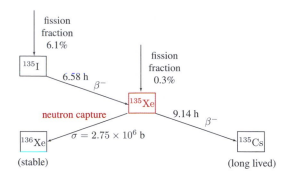

Figure 19.6 Data on the production, decay, and lifetimes of the fission poison ^{135}Xe and its precursor, ^{135}I. Production in ^{235}U fission and half-lives for β-decay are shown. After Lilley [77].

neutron-absorption cross sections, so they effectively build up continuously during reactor burning.

The most important, and perhaps most insidious fission poison is an isotope of xenon, ^{135}Xe, which was mentioned in §18.3.5. The production and decay paths for ^{135}Xe are summarized in Figure 19.6. As a reactor runs, ^{135}Xe begins to build up, mostly via β-decay of ^{135}I, which is a common fission fragment. A ^{135}Xe nucleus either β-decays or absorbs a thermal neutron. While a reactor is operating, neutron absorption converts ^{135}Xe to ^{136}Xe, which is stable and has a relatively small neutron-absorption cross section. β-decay of ^{135}Xe yields ^{135}Cs, which is long-lived and has a small neutron-absorption cross section. In equilibrium, the rate of production of ^{135}Xe is exactly balanced by the rate that ^{135}Xe is eliminated by neutron absorption or β-decay. So a reactor that has been running for a few days has built up an equilibrium concentration of ^{135}Xe. This would cause the neutron multiplication factor k to decrease from its original value at startup. For a reactor burning natural uranium with $\eta = 1.33$, $f = 0.9$, and a neutron flux of $\Phi_n = 10^{14}$ cm^{-2} s^{-1}, for example, k would be reduced by about 3% when ^{135}Xe is in equilibrium [77]. This reduction in k is compensated by partially removing control rods. If the build up of ^{135}Xe becomes too great for the control rods to compensate, or if other considerations require running with the control rods further engaged, then the reactor must be shut down to let the xenon decay away.

It is *after a reactor is shut down* that ^{135}Xe has its greatest impact, because when a reactor is powered down, *the amount of* ^{135}Xe *actually grows dramatically*. Shutting down the reactor eliminates the neutron flux that is deactivating most of the ^{135}Xe, while ^{135}I, the source of most ^{135}Xe, continues to decay with a ~7 h lifetime. The

amount of ^{135}Xe increases and, as shown in Figure 19.5, the reactivity drops dramatically and only recovers over several days after a typical reactor is shut down.

The reactor characterized by Figure 19.5 cannot be restarted for a period of about 30 hours after it is shut down. To be effective at eliminating the ^{135}Xe, the shutdown should last significantly longer than this. Modern *light-water reactors* (see §19.4.1) can be designed to reduce the constraints imposed by the generation of fission poisons following a reduction in power output. Nevertheless, nuclear reactors are most efficient when run at constant power, and have difficulty responding to short-term fluctuations in load demands on a complex power grid.

For another example of the role of fission poisons, see the discussion of the Chernobyl accident in Box 19.2.

19.2 Physics Issues Affecting Fission Reactor Operation and Safety

Before describing specific nuclear reactor designs, we summarize some of the physics principles that underlie questions about the safety of nuclear power. For no other energy source, perhaps, do social, political, and ethical questions more completely dominate public perception and discussion. Nevertheless, the difficulties with nuclear fission power and their possible resolution are highly constrained by the laws of physics.

There are three basic categories of concern regarding the safety and security of nuclear power:

Reactor control and cooling The reactor itself must be controlled, the heat it generates must be carried away, and the structure that contains it must be able to contain radioactive and poisonous materials even under a "worst case" scenario.

Nuclear waste Spent reactor fuel and other reactor components contain many species of radioactive nuclei with a wide variety of lifetimes, decay chains, and types of radiation. The spent fuel, in particular, must be sequestered safely away for a long time, though reprocessing can reduce its activity and volume.

Nuclear proliferation ^{235}U can be fashioned into a nuclear device of great destructive power, as can ^{239}Pu, which is produced copiously in a thermal reactor. These materials, as well as nuclear waste that can be used to produce "dirty bombs," must be kept out of the hands of irresponsible people whatever their political affiliation.

Each is a complex issue with grave consequences if the safety or security goals are not achieved.

The second two issues primarily involve physics that takes place outside of the nuclear reactor itself. Discussion of the *nuclear fuel cycle* and *nuclear proliferation* require some knowledge of the types and properties of radiation, and is postponed until §20 (Radiation). The first of the above categories, reactor control and safety, is discussed here along with some aspects of the second two categories that are internal to the reactor itself. We introduce here only the basic physical concepts underlying the issues of reactor safety and control. For a more extensive non-technical discussion of nuclear fission energy safety, see [81].

19.2.1 Inherent Safety

The nomenclature of nuclear reactor safety analysis requires explanation. A reactor design may have **active safety systems** such as control rods or coolant pumps that must be activated by an operator or automatic control system in order to function. It may also have **passive safety systems** such as pressure release valves or gravity-driven coolant circulation, which function automatically to reduce reactivity in the event of other system failures, independent of human intervention. Finally, features of a reactor may be designed to be *inherently safe*, meaning that the safety features are physical features of the materials themselves. We describe two examples of inherent safety features here, first *negative void coefficient of reactivity* (or *negative coefficient of void* as it is often known), meaning that ρ drops if a void develops in the coolant, and **negative temperature coefficient of reactivity**, i.e. $\partial\rho/\partial T < 0$, meaning that the reactivity drops if the reactor heats up.

Coefficient of Void   The basic ingredients in the *core* of a thermal-neutron reactor are fuel, moderator, and coolant. The moderator slows the neutrons and the coolant transfers the heat generated by nuclear fission to power generating equipment. We distinguish two situations: first, where the same fluid serves as both moderator and coolant, and second, where the moderation and cooling are performed by different materials. Most commercial reactors in the world today are of the first type. They use regular (light) liquid water as both a coolant and a moderator. Water is a good moderator (§18.3.6) and, with high heat capacity and very effective heat transfer in boiling, also makes an excellent coolant. If, due to some system failure, water vaporizes, a *void* is created where the coolant and moderator density is much lower than it had been. When water is the principal moderator and it boils off, the resonance escape probability p drops sharply, sending the reactivity negative and shutting off the fission chain reaction. The reactor remains hot, but the possibility of a runaway chain reaction is averted. Thus a water-cooled, water-moderated

reactor has a **negative coefficient of void**. This is a desirable inherent safety feature of these reactor designs. Note, however, that this is not the end of the story. Although the fission chain reaction may have been shut down, the reactor continues to generate heat from the decay of fission fragments (see §19.2.2). Cooling must be maintained for a long time after shutdown to avert structural damage and release of radioactivity. Thus, for example, the reactors at Fukushima Daiichi were destroyed in the 2011 accident not by the reactors going prompt critical, but rather by heat from residual radioactivity due to failure of the backup core cooling systems after the reactors had been shut down.

The opposite situation can arise in water-cooled reactors that use either graphite or heavy water for moderation. Such reactors can have a **positive coefficient of void**, allowing their reactivity to increase if a void develops in the coolant. How could this happen? Consider a reactor in which the fuel is embedded in a fixed matrix of solid moderator, like graphite, and where the cooling is supplied by pressurized water flowing through channels in the fuel/graphite array. If the cooling water vaporizes, the graphite is unaffected. It continues to moderate the neutrons and the fission chain reaction continues. Since the hydrogen in the cooling water also functions as a neutron absorber (see Table 19.2), vaporizing the cooling water reduces neutron absorption and therefore increases the thermal utilization factor and the reactivity. In short, a void leads to an increase in reactivity and perhaps to a runaway chain reaction. A certain class of electric power reactors of this design (known as *RBMK* reactors), including one in Chernobyl in the Ukraine, were built in the Soviet Union. The positive coefficient of void was a contributing factor to the catastrophic failure of the Chernobyl reactor in 1986 (see Box 19.2). Heavy-water-moderated reactors also have a positive coefficient of void, since the heavy water used as coolant is physically separated from the moderator. Although positive, the coefficient of void is small enough in this case to allow ample margin for control.

Doppler Broadening   Another inherent safety factor, which manifests as a negative contribution to the *temperature coefficient of reactivity*, is known as **Doppler broadening**. This term refers to the fact that as reactor fuel heats up, the width in energy of the peaks in the neutron-absorption cross section in ^{238}U increases. Remember that as a neutron slows down through collisions with the moderator, it must avoid capture into one of the "forest" of narrow absorption resonances in ^{238}U, shown in Figure 18.4(c) and (d). If the neutron's energy coincides with the location of a peak, then it has a high probability of being absorbed in a collision. The peaks are very narrow at

Inherent Safety

Physical features of fission reactor fuel, coolant, and moderator can be exploited to improve the *inherent safety* of a nuclear reactor.

Inherent safety features include the *negative void coefficient of reactivity* of water-cooled and moderated reactors, and the *negative temperature coefficient of reactivity* of uranium fuel caused by *Doppler broadening* of the peaks in the ^{238}U radiative capture cross section.

Decay Heat and Emergency Cooling

When a conventional nuclear power reactor is shut down, radioactive decays of fission products and actinides continue to produce heat in its core. The power level is initially ~6% of pre-shutdown power and decays slowly over a period of months. Emergency core cooling systems are necessary to prevent this residual heat from causing damage to the reactor fuel, the reactor core, and the reactor's surroundings.

normal operating temperatures, so if the neutron's energy differs slightly from the location of the resonance, it evades capture. As the fuel heats up, however, all of the neutron-absorption peaks get wider. This increases the chance that, when a neutron hits a series of ^{238}U nuclei, its energy overlaps an absorption resonance for one of the nuclei, so the radiative capture probability goes up and the resonance escape probability p goes down.

Doppler broadening can be understood quite simply as a consequence of classical kinematics. Thermal energy causes the uranium nuclei to vibrate about their equilibrium positions in the solid fuel: the higher the temperature, the more vigorous the vibrations, and the greater the range of possible velocities of the uranium nuclei. This, in turn, gives rise to a spread in the center of mass energy of the neutron-uranium system, increasing the possibility of overlap with an absorption resonance.

Thus, heating leads to decreased reactivity. Doppler broadening is used as a key inherent safety component in a novel reactor design, the *very high temperature reactor*, one of the *Generation IV* reactors discussed in §19.4.2. Doppler broadening also appears in the context of absorption of electromagnetic radiation in §23.4.

19.2.2 Decay Heat and Emergency Cooling

As described in §18.3.5, a small but significant fraction of the energy produced by a fission reactor comes from the (delayed) β- and γ-decays of fission fragments. For ^{235}U fission, of the total of 202.5 MeV (211.3 MeV minus 8.8 MeV lost in antineutrinos) that is retained as thermal energy in the reactor core, about 12.8 MeV (~6.3%) comes in this form (see Table 18.5). The time dependence of this delayed energy release is not a simple exponential because many different fission products and actinides participate, and the chain of sequential decays is complex. At first the residual power generation decays quickly after the reactor is turned off – it decays by a factor of 10 from its original value of ~6–7% of the full reactor

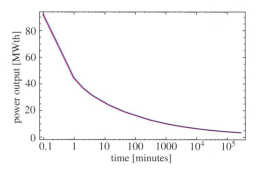

Figure 19.7 Hypothesized dependence of residual power from Fukushima Daiichi 1 Reactor in MWth as a function of time after shutdown in minutes. The reactor was rated at 460 MWe. With an assumed thermal efficiency of 33%, the initial residual power (heat generated) was assumed to have been about $460 \div 0.33 \times 0.063 \sim 90$ MW [96].

power in about 30 hours – and then decays more slowly. As an example, a model calculation of the heat generated by the Fukushima Daiichi 1 Reactor over the months after its abrupt shutdown on March 11, 2011 is shown in Figure 19.7.

The heat generated by a large nuclear power reactor shortly after shutdown can cause major damage to the fuel, the reactor core, and surrounding structures, unless they are designed to withstand very high temperatures and/or adequate cooling is maintained. The important consequence for nuclear safety planning is that a reactor's cooling system must remain operational long after the fission process is shut down. In the event of an accidental loss of primary cooling, emergency cooling systems, either active or passive, are designed to take over. Two of the world's three most serious civilian nuclear accidents, **Three Mile Island** in 1979 and **Fukushima Daiichi** in 2011, were due to loss of cooling and failure of backup systems following an emergency shutdown.

Box 19.2   The Chernobyl Accident

The most serious reactor accident, in terms of loss of life and contamination of property, took place at the **Chernobyl** reactor complex, near Pripyat in the Ukraine in 1986. It occurred when operators were testing an unusual proposal to power the backup cooling system: in the event of loss of electric power, energy stored in the reactor's turbines, functioning like flywheels, was to be used to run the primary coolant pumps until diesel generators could kick in. The idea had been tested several times previously, but it had failed. The new test, which required powering down the 3 GWth reactor to about 700 MWth, was scheduled for the day shift on April 25, 1986, but was postponed until late evening because of power demands on the grid to which the reactor was connected. The night-shift operators were not as experienced with the reactor or the planned test as the day-shift operators.

There is still controversy about the relative importance of design flaws and operator errors in the Chernobyl event [95]. Some of the flaws in the RBMK-1000 reactor are generally agreed upon:

- The graphite-moderated, water-cooled reactor had such a large core – diameter 11 meters and height 7 meters – that an external containment structure was deemed impractical and not constructed. Instead the reactor was capped with a 2000 ton plate to which the entire reactor was connected. The size of the core also meant that local conditions were not well controlled by global responses – it has been said that different parts of the reactor responded essentially independently.
- The reactor was operating in a mode where it had a large, positive coefficient of void (see §19.2.1), so boiling of the water coolant would increase reactivity, potentially leading to a dangerous positive feedback loop.
- The reactor control rods were to be lowered into channels in the graphite core that were normally filled with coolant water. The bottom third of the control rods, known as *displacers*, were made of graphite, which were intended to remain within the reactor core during normal operation. If the rods were totally withdrawn, then reinserting them would displace water (an effective neutron absorber) with graphite, causing an initial increase in reactivity, before the B_4C in the main part of the rods decreased ρ.

Before the test, the reactor's output was reduced below 700 MWth, leading to a build up of ^{135}Xe fission poison, which caused the power to decrease still further to ∼500 MWth. At this point, it appears that control rods were mistakenly inserted, and the reactor reached a near shutdown state of ∼30 MWth, which led to a much larger increase in the ^{135}Xe burden. Attempting to bring the reactor back up to a level where the test could be performed, and apparently not understanding the inhibiting role of the ^{135}Xe, the operators disabled safety constraints, allowing them to withdraw many control rods (including the graphite displacers) completely. This put the reactor in a very unstable mode, and when the test was performed, the sudden drop in cooling initiated a dramatic increase in reactivity. Reinserting the control rods initially displaced neutron-absorbing water with graphite moderator and increased the reactivity still further. It is estimated that the reactor output spiked to 30 GWth seconds after the test began. Such a high reactor output could only have occurred if at least sections of the reactor core went prompt critical. The subsequent explosions and fires released tons of the reactor core material to the atmosphere. Although no more have been built, other RBMK-reactors remain in service in the states of the former Soviet Union. Safety systems and procedures at these reactors have been improved in light of the Chernobyl experience.

19.3 Breeding and Fission Reactors

We have alluded to the fact that ^{238}U and ^{232}Th can be transformed into ^{239}Pu and ^{233}U by neutron absorption. Both ^{239}Pu and ^{233}U are fissile, and therefore either can potentially be used as thermal reactor fuel or to build nuclear weapons. Here we discuss the possibility of using excess neutrons in a reactor to *breed* these isotopes as additional fuel. The possibility that the same material could be weaponized is discussed in §20.6. The reactions leading to ^{239}Pu and ^{233}U are

$$n + {}^{238}_{92}\text{U} \longrightarrow {}^{239}_{92}\text{U} \xrightarrow[23.45\,\text{m}]{} {}^{239}_{93}\text{Np} + e^- + \bar{\nu}_e$$

$${}^{239}_{93}\text{Np} \xrightarrow[2.356\,\text{d}]{} {}^{239}_{94}\text{Pu} + e^- + \bar{\nu}_e$$

$$n + {}^{232}_{90}\text{Th} \longrightarrow {}^{233}_{90}\text{Th} \xrightarrow[21.83\,\text{m}]{} {}^{233}_{91}\text{Pa} + e^- + \bar{\nu}_e$$

$${}^{233}_{91}\text{Pa} \xrightarrow[26.975\,\text{d}]{} {}^{233}_{92}\text{U} + e^- + \bar{\nu}_e .$$

(19.18)

In particular, ^{239}Pu is produced in abundance in existing light-water reactors. ^{239}Pu has a half-life of 24 100 years and decays by α emission. The half-life of ^{239}Pu is sufficiently long that it can be extracted from spent nuclear fuel, processed, and formed into new reactor fuel. If it is not extracted, ^{239}Pu, together with fission fragments and other *minor actinides* produced from the original ^{238}U, remain in the spent fuel and must be sequestered from the environment for a very long time.

The idea of *breeding* more fuel than is consumed in a fission reactor has fascinated nuclear physicists and engineers since the beginning of nuclear power. **Breeder reactors** were much discussed during the 1960s, when the world's uranium reserves were thought to be even more limited than now believed. As described in §16, reserves of natural uranium at an affordable price may be sufficient to fuel existing nuclear reactors for several centuries. Nevertheless, unless extraction of uranium from seawater can be made economical, conventional uranium reserves do not seem to be sufficient to support long-term reliance on an expanded nuclear power sector that would supply a significant fraction of world energy. So the idea of converting ^{238}U or ^{232}Th, both of which are more than 100 times as abundant as ^{235}U, into fissile isotopes, has been considered an attractive possibility by some advocates of nuclear power. Indeed, this would seem to be a necessary component of any effort to significantly increase the role of nuclear power in world energy use over the long term. A fuel cycle incorporating breeder reactors that burn plutonium and other actinides could also in principle substantially reduce nuclear waste. Up to now, however, it has proved impossible to commercialize breeder reactors due to technical difficulties, economic problems, and concerns about their safety.

Technologies that breed ^{233}U from ^{232}Th in a reactor environment are still at the exploratory stage. The ^{239}Pu route has been the primary focus of efforts to construct breeder reactors. In this section we focus primarily on ^{239}Pu breeders, and comment only briefly on the ^{233}U avenue.

Nearly every ^{238}U nucleus that captures a neutron soon ends up as ^{239}Pu. Indeed, about one third of the energy generated in a single fueling cycle of a typical commercial reactor comes from the fission of ^{239}Pu *bred* from the original enriched uranium fuel. A breeder reactor is initially fueled with enriched uranium or with plutonium recovered from spent fuel along with a supply of natural uranium. Neutrons from fission of the fissile isotopes in the fuel convert ^{238}U to ^{239}Pu. As time goes on, more and more of the power is generated by the fission of ^{239}Pu.

All reactors breed new fuel to some extent, so instead of a sharp division into breeder and non-breeder reactors, it is often better to think in terms of the **conversion ratio**

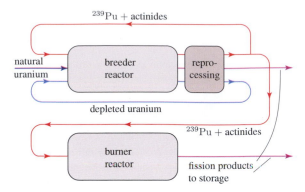

Figure 19.8 In a burner/breeder reactor scheme, some of the ^{239}Pu and other actinides removed from breeder reactors would be recycled and used along with depleted and natural uranium to refuel the breeders, while the rest of the ^{239}Pu plus actinides would fuel burner reactors optimized to minimize the actinide content of their spent fuel and waste could be reduced. In this way, the energy extracted from natural uranium could be multiplied by a factor of over 100.

that measures the amount of new fuel produced by a given reactor in a full fuel cycle. The conversion ratio CR is defined as the number of fissile nuclei produced divided by the number of fissile nuclei destroyed. Note that fissile nuclei produced and then fissioned in a single fuel cycle contribute both to the numerator and to the denominator of the CR. Reactors with CR < 1 are termed **burners** and CR > 1 are **breeders**. By combining these two types of reactors, using breeder reactors to convert natural uranium to plutonium, and burners to use up plutonium and actinides produced by breeders, a fuel cycle could be constructed in which essentially all the energy in the ^{238}U as well as the ^{235}U in uranium would be extracted in useful form, with a minimum of plutonium and actinide waste (Figure 19.8).

Despite the appeal, there are significant technical challenges to realizing a breeder reactor. In a conventional uranium-fueled thermal reactor CR ~ 0.6. Remember that this is averaged over a fuel cycle, during which the rate of ^{235}U fission decreases as the fraction of ^{235}U drops, while the plutonium fission rate increases. To breed fissile isotopes in a reactor with CR > 1, at least two neutrons on average must be used from each fission event. One neutron must go on to trigger the next fission and the other must be absorbed by a fertile nucleus (either ^{232}Th or ^{238}U) to breed a fissile one. Of the two possible breeder fuels, ^{233}U and ^{239}Pu, only ^{233}U has a high enough thermal-neutron *reproduction factor* η to be used practically to breed fuel in a thermal reactor (Problem 19.12). Thermal reactors with high conversion ratios, based on the ^{232}Th$\rightarrow^{233}$U reaction sequence, are being actively researched, particularly in India where supplies of thorium are ample but uranium is not.

While ^{239}Pu cannot be used to breed fuel in a thermal reactor, ^{239}Pu does produce an ample number of neutrons when it fissions by absorbing a *fast neutron*. This has led to the development of **fast breeder reactors**, which dispense with moderators not only to maintain a high level of neutron production in $n + {}^{239}$Pu$\rightarrow$ fission + neutrons (which requires fast neutrons), but also to maximize the probability of neutron absorption in ^{238}U in order to breed new fuel. Most of the proposals for new breeders, as well as the large-scale breeder reactors that are now operating, employ this fast-neutron ^{238}U$\rightarrow$^{239}Pu scheme.

Fast breeders raise some challenging design issues, however. In particular, water cannot be used as a coolant because it is too good a moderator and is also a rather good neutron absorber. Instead, liquid metals such as sodium or lead, with relatively large atomic mass A, have been used in experimental (and in the case of sodium, commercial) reactors. Although liquid metals have an advantage over water, in that they do not have to be kept under high pressure, sodium is (to say the least) highly reactive – it burns in air and explodes in contact with water – and liquid lead is corrosive. The absence of a moderator also makes it impossible for fast breeder reactors to utilize the negative coefficient of void of a water moderator as a safety feature. Many experimental fast breeder reactors have been operated since early in the nuclear power era. The most ambitious was the **Superphénix** reactor built in France and designed to produce 1200 MWe. Superphénix was operated at various power levels between 1983 and 1997, when it was permanently closed. The only fast breeder reactors as of 2017 delivering power to an electricity grid are the sodium cooled *BN-600 reactor* and the *BN-800 reactor* presently being commissioned at the Beloyarsk Nuclear Power Station in Russia, which are rated at 600 and 800 MWe. Fast-neutron breeder reactors represent a significant component in the development of advanced *Generation IV* reactor designs that may play a role in the future of nuclear power. Recently there has also been growing interest in reactors with CR $\approx$ 1 that would extend natural uranium and/or thorium supplies and hopefully avoid some of the technical difficulties of true breeder reactors [73].

19.4 Fission Reactor Design: Past, Present, and Future

Modern nuclear fission power reactors can be characterized by their choice of moderator and coolant. Most

> ### Breeder Reactors
>
> A *breeder* reactor is designed to create more fissile material than it consumes.
>
> Breeder reactors produce ^{239}Pu or ^{233}U from ^{238}U or ^{232}Th. Plutonium-based fast-breeder reactors use fast neutrons and employ no moderator.
>
> In principle, breeder reactors could multiply the energy available in a uranium resource by a factor of 100, and could make thorium useful as a nuclear fuel. The technical difficulties and extra costs of implementing the breeder concept, together with increased estimates of uranium resources, have decreased interest in breeder reactors in recent decades.

Table 19.3 Nuclear power reactors in service as of 2017. From [71].

Reactor type	Number	Capacity (GWe)
Pressurized water	291	274
Boiling water	76	74
Heavy water	49	25
Gas-cooled	14	8
Light-water graphite	15	10
Fast breeder	3	1.4

commercial reactors now in service are moderated and cooled by ordinary water, and use uranium typically enriched to $\sim$3–4% ^{235}U. These are known as **light-water reactors** (LWR) and come in two basic types: **pressurized water reactors** (PWR) and **boiling water reactors** (BWR). These two designs have competed for economic dominance for many years with no clear winner. A third design, pioneered in Canada and known as a **CANDU** reactor, uses natural uranium as a fuel, and heavy water both as a moderator and as a coolant. Finally, a small number of gas- or water-cooled, graphite-moderated reactors are still in operation. Reactors in service in 2017 are summarized in Table 19.3.

Nuclear reactors are sometimes classified into four *generations*. **Generation I** refers to prototypes, developed in the 1940s and 1950s, some of which generated modest amounts of electric power. Most power reactors now operational are **Generation II**, built before the late 1990s. **Generation III** reactors include evolutionary improvements over Gen II designs, such as passive safety systems, and better neutron and heat management. **Generation**

IV reactors represent novel and more advanced reactor designs and are the subject of research efforts. They are loosely grouped into six or seven types. In this section we describe the basic principles behind Gen II and Gen III reactors, and at the end mention some Gen IV ideas.

19.4.1 A Survey of Existing Reactor Types

Light-water Reactors   Light-water reactors use uranium enriched to 3–4% ^{235}U. (This is known as **low-enriched uranium**; much higher levels of enrichment are used in other situations such as for nuclear submarine reactors and nuclear weapons.) This fuel is typically formed into an array of thin tubes clad in a zirconium alloy. Zirconium is chosen because of its low thermal-neutron absorption cross section, its good heat transfer properties, and its corrosion resistance. Hundreds of tubes are grouped into a **fuel bundle** or **assembly**, and interspersed with control rods that can be inserted or removed to control the reactivity (see Figure 19.9).

A reactor might contain hundreds of fuel bundles, amounting to a total mass of approximately 100 tons of

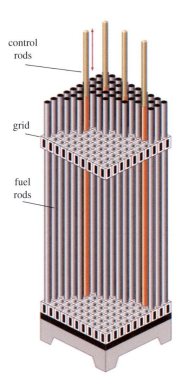

control rods

grid

fuel rods

Figure 19.9 A bundle of tubes containing enriched uranium reactor fuel interspersed with control rods. The drawing is not to scale: the fuel rods typically have a diameter of ∼1 cm and a length of approximately 4 m. (Credit: E. Generalic, http://glossary.periodini.com/glossary.php?en=control+rod)

uranium for a several-year refueling cycle. The fuel and control rods, along with the water that circulates through them, make up the *core* of the reactor. Because water is an excellent moderator, the core of an LWR can be relatively compact, a feature that also serves to reduce neutron absorption (recall that light water is also a significant neutron absorber). The reactor core, which must be kept at high pressure in order to keep at least a fraction of the water in liquid form, is contained in a **reactor vessel**. The reactor vessel and various other parts of the system that become radioactive in the course of operation are situated in a **containment structure** made of steel and/or steel-reinforced concrete. The size and mass of the containment structure is one of the major additional capital costs that differentiate a nuclear power plant from fossil fuel power plants.

Light-water reactors are run at temperatures and pressures somewhat lower than their fossil fuel counterparts because of the physical limitations imposed by the fuel and cladding, and concern about safety at very high temperatures and pressures. The PWR and BWR designs differ not only in the phase structure of the coolant, as the names suggest, but also in the heat-transfer systems required. Diagrams showing the principal systems and structures in a PWR and a BWR are shown in Figure 19.10.

In a pressurized water reactor (PWR), the reactor core is immersed in a water-filled pressure vessel at a temperature of about 320 °C and a pressure of about 150 atm. Since the boiling pressure of water at 320 °C is about 110 atm, the water remains liquid throughout the pressure vessel. This improves moderation since water is denser than steam, but it means that the primary heat source does not directly produce steam for power generation. Pumps circulate the cooling water through the reactor vessel, and a heat exchanger transfers the energy removed from the core to a **secondary cooling loop**. The secondary loop produces saturated steam, which is then used to power one or more turbines (see §19.5).

The water that circulates around the reactor core becomes slightly radioactive, and the secondary cooling loop serves to shield the turbine and other mechanical systems from radioactivity. This is a critical design choice: avoiding contamination of the turbine and associated systems comes at the cost of having two heat transfer stages, which reduces thermodynamic efficiency, and increases the size of the containment vessel.

The pressure vessel in a boiling water reactor (BWR) contains water in a mixed phase. Typically 12–15% of the water in the upper section of the pressure vessel is steam. The core temperature is lower than in a PWR (∼285 °C), as is the pressure (∼68 atm), which is the boiling pressure at the core temperature. Water vapor is a poorer moderator

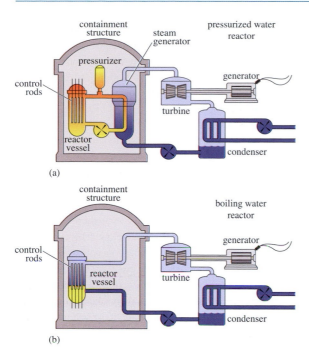

Figure 19.10 Cartoons showing the major systems and structures in (a) a pressurized water reactor (PWR) and (b) a boiling water reactor (BWR). Note the additional coolant loop in the PWR and the radioactive water circulating through the turbine in the BWR. Although shown the same size in the drawing, the containment vessel for the BWR can be significantly smaller than the PWR. (Credit: US NRC)

than liquid water, so less power is produced in the upper core and the fuel and design are adjusted accordingly.

The principal distinction of the BWR design is the direct use of steam generated in the pressure vessel as the working fluid in the turbine. Eliminating the second heat exchanger allows a smaller containment structure and reduces the initial construction cost. The thermodynamic efficiencies of typical PWR and BWR reactors are comparable; the BWR eliminates the second heat exchanger, but PWRs are generally operated at higher core temperatures, resulting in approximately the same steam conditions at the turbine inlet. The fact that the mechanical systems in a BWR become slightly radioactive makes servicing (and decommissioning) them more complicated, however, and diminishes the attractiveness of the system. Another important consideration is that the steam that is removed from the pressure vessel must be "dried" – the liquid water must be removed, raising the *quality* close to one – before it can be used in the turbine. This is accomplished by a **steam separator** (not shown in Figure 19.10) that resides in the reactor vessel.

Both LWR designs share in common the feature that the reactor must be brought off line and cooled before it can be refueled. This is due to the fact that the core resides in the reactor vessel held at high pressure, in contrast to the CANDU design we consider next.

Heavy Water Reactors   During World War II, Canada used its abundant hydroelectric power to produce many tons of heavy water (D_2O) for the Manhattan project. Lacking domestic uranium enrichment facilities, Canadian nuclear engineers developed a reactor designed to run on natural, unenriched uranium, moderated and cooled by heavy water. Because the energy density of natural uranium is much less than enriched uranium (since the ^{235}U content is lower) these *CANDU* reactors required frequent fuel replacement. Thus, the design had to allow for in-service fuel replacement, and the core is held at lower pressure than a typical PWR. The resulting ingenious design, known generically as a **pressurized heavy water reactor** (PHWR), has proved safe and reliable, and even though it relies on large quantities of rare and expensive D_2O, it remains a small but significant piece (see Table 19.3) of the nuclear power inventory, including recently completed plants in China.

When a deuterium nucleus absorbs a neutron it creates tritium, which is radioactive. Because of its value as well as its resulting radioactivity, the heavy water used for cooling is kept within a closed system, and its thermal energy is used to generate steam in a secondary cycle, similar to a PWR. As in LWRs all the radioactive materials are kept within a reinforced containment structure. Because of the relatively low power density of natural uranium, and the structural limits on the overall size of the system, CANDU reactors have typically about $\frac{1}{2}$ to $\frac{2}{3}$ the power rating (600–800 MWe compared to 1000–1300 MWe) of LWRs.

Light-water Reactors

Most commercial reactors employ ordinary water both as a coolant and as a moderator, and use uranium enriched to 3–4% as fuel.

Pressurized water reactors use liquid water. They require a secondary circulation loop, keeping the steam that powers the turbines separate from the radioactive water that cools the core.

Boiling water reactors use water in a mixed phase as both coolant and moderator. Steam from the coolant is dried and used directly to power turbines outside the reactor vessel.

Gas-cooled Reactors   Some of the first power reactors designed in Great Britain and France used natural uranium, were graphite-moderated, and used CO_2 gas for cooling. They were named after the magnesium alloy cladding, **Magnox**, that encased the fuel. The graphite moderator and Magnox cladding, both of which would react with water at high temperature, motivated the use of a non-reactive gas such as CO_2 as a coolant. Since the fuel could be inserted into and removed from channels in the graphite moderator, these reactors could be refueled in service, a requirement for natural uranium fuel (see CANDU reactors above). Since the fuel could be changed as often as desired, Magnox reactors were convenient sources of plutonium for weapons systems. (The reason that frequent fuel changes enhances the ability to weaponize plutonium is discussed in §20.6.3.) Indeed, Magnox reactors were originally designed for this purpose and the use of "waste" heat for electricity generation was considered a by-product.

Several design features, including the thermal instability of metallic uranium fuel, the cumbersome equipment needed for refueling, and the fact that Magnox-clad fuel elements could not be stored under water after use, led to the replacement of Magnox reactors by **advanced gas-cooled reactors** (AGR). AGRs use enriched uranium in the more stable form of uranium dioxide (UO_2), with stainless steel fuel cladding. This allows for higher operating temperatures, leading to better thermodynamic efficiency, less frequent refueling, and easier spent fuel storage. Nevertheless, AGRs continue to suffer from design issues, and have not been widely deployed. **Very high temperature (gas-cooled) reactors** are descendants of this line of development and figure prominently among Gen IV reactor designs discussed below.

Water-cooled, Graphite-moderated Reactors   As mentioned earlier, a graphite-moderated, water-cooled reactor has a potential safety problem: coolant loss can lead to higher reactivity, which could cause the reactor to go prompt critical. Reactors based on this model, known as **RBMK reactors** (a Russian acronym for *high power channel-type reactor*) were, nevertheless, designed in the former Soviet Union in the 1950s and built until 1986, when one of these reactors suffered a catastrophic failure at Chernobyl. The RBMK design was an attempt to capitalize on an existing Soviet military plutonium production reactor design. RBMK reactors typically employ uranium enriched to $\sim$2%, formed into rods that are inserted in channels in the solid graphite moderator. The difficulties with this reactor design are surveyed in the discussion of the Chernobyl disaster (Box 19.2).

19.4.2 Generation IV Nuclear Reactors

Advanced nuclear fission reactor designs attract interest because of the still-unresolved issues that surround nuclear power, including (1) the need for improved safety and proliferation resistance; (2) the low thermodynamic efficiency of existing reactors (see §19.5); (3) limits on ^{235}U resources; (4) the quantity and radioactivity of nuclear waste; and (5) high capital costs for nuclear plant construction. A set of six advanced reactor concepts have been identified and studied by the *Generation IV International Forum* over the past decade [97]. Another concept, known as an *energy amplifier*, is sometimes put into the same category (Question 19.4). A very short description of these reactor concepts is given below. The goal of the *Gen IV* program is to develop these concepts over the next two decades. It is far from clear, however, whether any of these will find widespread deployment, and in any case the development and large-scale commercialization of a radically new reactor design would likely take several decades at least. (It took $\sim$30 years for LWRs to reach widespread deployment in a climate that was quite conducive to the development of nuclear power.) The purpose of this summary is to give some idea of the range of options under consideration.

Generation IV Thermal Reactors   Three of the Gen IV designs are basically thermal reactors and three make use of fast neutrons. The thermal reactors include:

Supercritical water reactor (SCWR)   Similar to a boiling water reactor but operating above the critical point in the water phase diagram. Using a considerably higher temperature and pressure than existing PWRs or BWRs, and a single coolant loop, they aim for thermodynamic efficiencies of $\sim$45% rather than $\sim$30% characteristic of present-day reactors.

Very high temperature reactor (VHTR)   Basically a gas-cooled, graphite-moderated reactor, this design relies on a very sophisticated fuel element design to achieve unparalleled temperatures ($\sim$1000 °C gas temperature at the reactor exit) and important inherent safety features. There are at least two variants: (1) spherical fuel modules, known as *pebbles*, around which the gas coolant circulates; and (2) hexagonal (*prismatic*) fuel columns that include channels through which coolant passes. In the **pebble bed reactor**, the enriched-uranium, plutonium, or thorium fuel (in the form of chemically and thermally stable oxides or carbides) is formed together with graphite moderator into heat-resistant spheres the size of tennis balls. The fuel particles themselves, which are smaller than a millimeter, are coated with several layers, including ceramic silicon

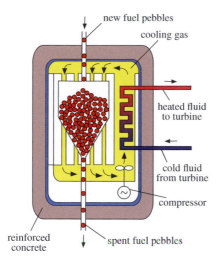

new fuel pebbles

cooling gas

heated fluid
to turbine

cold fluid
from turbine

compressor

reinforced
concrete

spent fuel pebbles

Figure 19.11 A schematic diagram of a pebble bed reactor employing a secondary (steam) loop. Designs in which the coolant gas itself runs the turbine are also being considered.

carbide that is designed to maintain structural integrity up to temperatures above 1600 °C and prevent the release of fission fragments. A hypothetical pebble bed reactor is sketched in Figure 19.11. The individual pebbles are far too small to sustain a chain reaction. When they enter the reactor, the pebbles are assembled into a close-packed array and a chain reaction begins. The fuel quickly heats up, however, and due to its strong negative thermal coefficient of reactivity (§19.2.1), the chain reaction is stabilized at a temperature well below the limit of structural integrity of the pebbles. Only when a gas coolant circulates through the bed, carries away heat, and lowers the temperature of the fuel elements, does the chain reaction increase in power output.

In steady state the reactor produces only as much energy as can be carried away by the coolant. In the event of a catastrophic loss of coolant, the reactor simply heats up and runs at low power, awaiting restoration of cooling or dismantling, without damage to the pebbles or the rest of the system. One coolant under investigation is helium, which has the advantage of not becoming radioactive when exposed to neutrons. Thus, helium could be circulated directly through a gas turbine to produce electricity without making the mechanical systems radioactive.

In a pebble bed reactor, the individual fuel pebbles flow through the reactor bed, are recirculated several times, and eventually are removed for waste management. In addition to inherent safety and convenient fuel management, other advantages of the VHTR design include: (i)

very high thermodynamic efficiency due to the high outlet temperature; (ii) proliferation resistance, since the plutonium isotope mixture in exhausted pebbles is unfavorable (see §20.6.3) and it would be difficult to disassemble the pebbles to obtain enough material for a bomb; and (iii) the possibility of making smaller, modular reactors with ≲100 MWe capacity, which could be produced and assembled with low capital costs. However, neither the pebble bed nor the prismatic alternative has yet been implemented in a commercial reactor.

Molten salt reactor (MSR) Using molten salt as a coolant has the advantage of access to very high temperatures without pressurization because molten salts have low vapor pressures even at temperatures above 1000 °C. One version dissolves the fuel (uranium or thorium) in the circulating coolant. The nuclear reaction occurs when the coolant/fuel circulates through channels in a graphite core, which provides the necessary moderator. Fuel and actinides can then be separated from fission products by chemical processing of the circulating coolant. Other designs resemble pebble bed reactors with the gas coolant replaced by molten salt.

Generation IV Fast-neutron Reactors　The fast-neutron reactor designs included in the Generation IV initiative are all intended to breed more fuel than they consume and thus fall in the domain of *breeder reactors* described in §19.3. The three fast reactor concepts in the Gen IV initiative are labeled by their different cooling systems:

Gas-cooled fast reactor (GFR) The circulating gas acts as a coolant and directly powers a Brayton cycle. Helium or CO_2 have good thermodynamic and neutron-absorption properties. The gas cooling system would allow for high temperatures and good thermodynamic efficiency. A safety concern is that the absence of a moderator and the use of a gas coolant results in low heat capacity and a potential rapid rise in temperature after a loss of forced coolant.

Sodium-cooled fast reactor (SFR) Liquid sodium offers the same advantage as molten salt: low vapor pressure at high temperature, but also has a low neutron-absorption cross section, making it a suitable coolant for a fast reactor. Liquid sodium also has excellent heat transfer characteristics. Designs include many passive and inherent safety mechanisms. However the use of highly reactive liquid metal sodium is a significant concern: liquid sodium burns in dry air and oxidizes explosively in contact with water. The sodium-cooled *BN-600* and *BN-800* reactors at Beloyarsk mentioned above fall into this category and are being studied as prototypes for more advanced sodium-cooled fast reactors currently being designed.

Lead-cooled fast reactor (LFR) Another low-neutron-absorption coolant with low vapor pressure is lead or a lead-bismuth mixture with melting points as low at 123.5 °C. The corrosive nature of liquid lead presents design challenges.

19.5 Nuclear Reactor Power Cycles

Nuclear fission reactors are fundamentally heat sources, like fossil fuel power plants, that produce useable energy through a thermal power cycle. They can therefore be analyzed using the tools of §13. As thermodynamic systems, existing nuclear plants fall into two general classes: single-loop Rankine cycles, such as boiling water reactors (BWR), where the coolant is vaporized and used directly to run a steam turbine, and double-loop Rankine cycles, such as pressurized water reactors (PWR), where the coolant runs in a closed loop and transfers its heat to steam in a second loop that is used to run a steam turbine. Other cycles can be used in specific reactor systems; for example, the very high temperature reactor (VHTR) design described in §19.4.2 could use a single-loop Brayton (or Brayton/Rankine combined) cycle, where the coolant – helium or CO_2 – from the reactor is fed directly into a gas turbine.

The temperature and pressure that can be tolerated by the materials in the reactor core limit the design and the thermodynamic efficiency of nuclear power plants compared to fossil fuel power plants. The structural integrity of the fuel and other components of the reactor core typically limit the water temperature and pressure in LWRs to well below the critical point, $T_C = 374$ °C and $p_C = 22.06$ MPa. For example, the US *Nuclear Regulatory Commission* (NRC) Design Certification [98] for a modern PWR indicates a coolant outlet temperature and pressure of $T_{out} = 321$ °C, $p_{out} = 15.6$ MPa. This in turn heats water in the secondary loop to a saturated vapor (quality one) at $T_+ = 273$ °C, where the pressure is $p_+ = 5.8$ MPa. Boiling water reactors reach somewhat higher temperature and pressure in the turbine loop. The US NRC Design Certification [99] for an advanced BWR quotes $T_+ = 288$ °C, $p_+ = 7.2$ MPa. These values also correspond to saturated steam at quality near to one. A modern version of the CANDU reactor has similar steam characteristics.

In §13 we concentrated on Rankine steam cycles with superheating, as shown in Figure 13.12. In the case of a water-cooled nuclear reactor, superheating is generally impractical because coolant from the reactor core cannot provide a high enough temperature. So the basic Rankine cycle appropriate to an LWR or CANDU reactor is

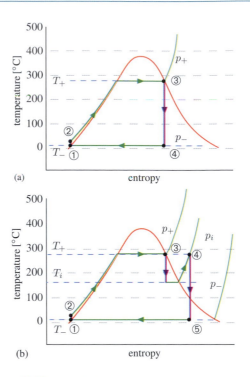

(a)

(b)

Figure 19.12 (a) A Rankine cycle without superheating appropriate to an LWR. Compare with Figure 13.12(b). Note that the steam exiting the turbine is low quality, i.e. it contains considerable liquid water. (b) A Rankine cycle with the same maximum and minimum temperatures as (a), but with reheating. Note that the efficiency of the cycle and the quality of the steam exiting both turbines is higher than (a).

Nuclear Reactor Power Cycles

Commercial nuclear power reactors operate with steam temperatures lower than those achievable in conventional fossil fuel power plants, due to limits imposed by the structural integrity of materials in their cores. Turbines in nuclear plants are therefore designed to tolerate a mixed (water/steam) phase.

Using high-pressure and Rankine cycles that include regeneration and reheating, light-water reactors can reach efficiencies of 33–37%, comparable to conventional coal-fired power plants.

more like the sketch of Figure 19.12(a), with the result that water droplets condense in the steam passing through the turbine as its quality drops below one. A nuclear power plant turbine must be specially designed to handle water droplets that may cause erosion of the metal blades.

Similar considerations apply to geothermal power plants, where the steam drops below the saturation dome in the power cycle (see §32). In practice, LWR Rankine cycles usually employ reheating, as illustrated schematically in Figure 19.12(b). In such a cycle the steam is intercepted at an intermediate temperature and pressure, T_i, p_i, after a high-pressure turbine, and reheated to T_+ before entering a second, low-pressure turbine. As shown in Figure 19.12(b), this not only increases the efficiency of the cycle, but also keeps the steam quality in the turbine relatively high and allows for more flexibility in turbine design.

While nuclear power plants typically operate at lower temperatures and higher pressures than conventional coal-fired power plants, by using regeneration and reheating (see §13), LWRs can reach comparable efficiencies (33–37%). To reach efficiencies comparable to gas-fired combined cycle (CCGT) plants described in §13.5, nuclear plants must run at higher temperatures. This is the aim of the Gen IV VHTR project. With gas-coolant outlet temperatures conceivably in excess of 1000 °C, directly employed in a Brayton cycle, efficiencies approaching those of CCGT power plants may be possible.

19.6 Experiments in Thermonuclear Fusion

The possibility of virtually unlimited energy production from controlled nuclear fusion has raised great hopes, but encountered great difficulties. Notwithstanding the difficulties, the hopes have stimulated a decades-long quest to design and build a practical, economically viable energy source powered by controlled nuclear fusion.

The basic idea of fusion power is to bring matter to a high enough temperature and pressure to achieve a controlled fusion reaction that can be sustained over an extended period of time. As described in §18, deuterium–tritium (dt) fusion is considered most promising on physics grounds, though dd may also be feasible and avoids the use of the artificially created and radioactive isotope tritium. Typically $k_B T$ must be about 15 keV for either dt or dd fusion, but the pressure and thermal energy confinement needed are much greater for dd than for dt fusion. Here we focus on dt fusion, which is much more likely to be achieved in the relatively near term.

At very high temperatures, atoms are completely ionized and the gas of ions and electrons forms a **plasma**. Thus, the study of nuclear fusion takes us into the realm of plasma physics.

The first efforts towards controlled fusion have made use of the fact that a charged particle travels in a helix in a magnetic field. The stronger the field, the smaller the radius of the helical orbit. Based on this effect, a strong magnetic field can confine even a very hot plasma in a relatively small volume. This is the fundamental idea behind *magnetic confinement fusion* (MCF), which is described in §19.6.3 below. Another possible approach, known as *inertial confinement fusion*, is described briefly in §19.6.4. Before describing either specific approach to plasma confinement, we examine the energy balance in a confined plasma undergoing fusion, and develop the criteria used to judge the performance of a fusion reactor.

19.6.1 Power Balance in a Fusion Reactor

In this section we suppose that a plasma consisting of fully-ionized deuterium and tritium with number densities n_d and n_t has been successfully confined and brought to a high temperature T where the fusion reaction

$$d + t \rightarrow \alpha + n + 17.6 \, \text{MeV} \qquad (19.19)$$

takes place in steady state. (The mechanism of confinement does not matter here.) dd and d^3He fusion can be treated similarly. We assume that the deuteron and triton densities are equal, $n_d = n_t = \frac{1}{2}n$, where n is the electron density, and that the plasma can be approximated as an ideal gas with $p = 2nk_B T$. The factor of two comes because the nuclei and the electrons contribute separately to the pressure.

The neutron and α particle are produced with energies in the ratio 4:1, $E_n \cong 14.1 \, \text{MeV}$ and $E_\alpha \cong 3.5 \, \text{MeV}$, as required by conservation of energy and momentum.[4] The charged α particle loses its energy quickly in the plasma, whereas the neutron escapes the plasma and delivers its energy directly to the surrounding material. These energy production processes within the plasma are described in terms of the *power density* P_α produced in α particles and P_n produced in neutrons. Together P_α and P_n sum to the total fusion power production density $P_f = P_\alpha + P_n$. In addition to the energy carried by neutrons, the plasma loses energy to its surroundings by heat conduction and radiation, parameterized by averaged power densities within the plasma given by P_c and P_r respectively. The power deposited in the material surrounding the fusion reactor by neutrons and by radiation and conduction is, in principle, available to heat a working fluid and generate electricity. Finally, some power, parameterized by P_h within the plasma, may be fed into the reacting plasma to maintain its temperature.

---

[4] The kinetic energies of the d and t in the initial state – typically tens of keV – are negligible compared to the energy released in the reaction.

In steady state the power deposited in the plasma by α particles and by external heating must balance the power that the plasma loses to the surroundings,

$$P_c + P_r = P_\alpha + P_h \,. \tag{19.20}$$

The potentially useful thermal power output of the plasma fusion reactor receives contributions from P_n, P_r, and P_c. The ratio of the *net* thermal power output to the power supplied by external heating is known as the **fusion gain factor Q**,

$$Q = \frac{P_n + P_c + P_r - P_h}{P_h} = \frac{P_n + P_\alpha}{P_h} = \frac{P_f}{P_h} \,, \tag{19.21}$$

where we have used the steady-state condition (19.20). The fusion gain factor is thus the total power density produced by fusion P_f compared to the heating that must be externally supplied P_h.

The contributions to the plasma fusion reactor power balance can be described in more detail as follows:

Power added by α particles The power density produced in and delivered to the plasma by α particles is obtained by multiplying the fusion reaction rate by the α particle energy E_α. According to eq. (19.4), the fusion reaction rate per unit volume in a gas consisting of deuterons and tritons with densities n_d and n_t equals $n_d n_t \langle \sigma_f v \rangle$. The fusion process depends upon tunneling and therefore the cross section increases rapidly with energy. The product $\sigma_f v$ is averaged over the Boltzmann distribution at temperature T, as denoted by $\langle \ldots \rangle$ and is dominated by the Gamow window (§18.4.1). The resulting dependence of $\langle \sigma_f v \rangle$ on $k_B T$ is shown in Figure 19.13 for the dt reaction as well as for dd and d^3He. The power density added to the plasma by α particles is then

$$P_\alpha = n_d n_t \langle \sigma_f v \rangle E_\alpha = \frac{n^2}{4} \langle \sigma_f v \rangle E_\alpha \,, \tag{19.22}$$

where we have set $n_d = n_t = n/2$. Note that P_α grows rapidly with temperature between $k_B T = 5$ and $25\,\mathrm{keV}$, which is the temperature range relevant for MCF.

Neutron power density The power density of the neutrons produced by dt fission is four times as large as P_α. This power escapes the plasma and is deposited on the material that surrounds the reactor.

Power lost by radiation The radiative power P_r lost by the plasma is dominated not by thermal radiation, but instead by the radiation emitted when electrons accelerate during collisions with the positive ions in the plasma. This form of radiation, known as *bremsstrahlung* (German for *braking radiation*), is dominantly in the ultraviolet or soft X-ray part of the spectrum, and it escapes from the

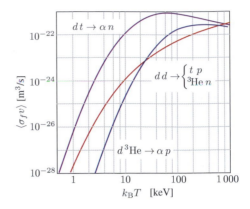

Figure 19.13 $\langle \sigma_{\mathrm{fusion}} v \rangle$ for dt, d^3He, and dd fusion as a function of $k_B T$ (in keV). After [100].

plasma. Bremsstrahlung is described further in §20.2.2. This radiation is absorbed in the material that surrounds the reactor.

Because it originates in two-body collisions, P_r, like P_α and P_n, is proportional to the plasma density squared. Explicit calculation shows that P_r grows with temperature proportional to $(k_B T)^{1/2}$,

$$P_r = C_B n^2 (k_B T)^{1/2} \,, \tag{19.23}$$

where $C_B \cong 5.35 \times 10^{-37}\,\mathrm{Wm^3/(keV)^{1/2}}$ [90]. In the temperature range of interest, $5\,\mathrm{keV} < k_B T < 25\,\mathrm{keV}$, P_r grows much more slowly with temperature than P_α.

Power loss by conduction The energy lost to the surroundings via conduction can be parameterized as a fraction of the energy density in the plasma, which is given by $E = 2(\frac{3}{2} n k_B T) = 3n k_B T$ in the ideal gas approximation, so

$$P_c = E_{\mathrm{plasma}}/\tau_E = 3n k_B T/\tau_E \,. \tag{19.24}$$

The (reciprocal) coefficient of proportionality τ_E is known as the **energy confinement time**, and measures the time over which the plasma would lose an appreciable fraction of its thermal energy by conduction. The parameters P_c and τ_E must be regarded as phenomenological because, among other reasons, conductive losses do not scale with volume.

External heating Finally, in order to maintain its high temperature, some additional power may be supplied to the plasma. In present experiments this power is supplied by external sources, although in principle it would ideally be obtained from the energy delivered to the surroundings by neutrons, conduction, and radiation. The power density supplied by this external heating is parameterized by P_h.

dt Fusion Power Balance

In a steady-state fusion reaction, the *fusion gain factor* is defined as the ratio of the net thermal power output of the system to the power supplied to heat the plasma,

$$Q = P_f/P_h .$$

$Q = 1$ is termed *breakeven* and $Q \to \infty$ defines *ignition*. A fusion reactor need not reach ignition to provide useful thermal power. Q must, however, be much greater than one in order to allow for losses and for reactor power systems.

Equation (19.21) describes the fusion gain factor as the ratio of fusion power output to power input. In a real fusion power system, the net useful thermal output would be further reduced by losses and by the power required to operate the system that confines the plasma. Thus the fusion gain factor Q is an upper limit on the *engineering gain factor* Q_E, which would include these other effects. The performance of experiments in controlled fusion is nevertheless usually gauged by the value of the fusion gain factor. $Q = 1$ is (somewhat arbitrarily) termed **breakeven** and the limit $Q \to \infty$, in which no external power is needed to sustain the fusion reaction, is known as **ignition**. Note that it is not necessary for a controlled fusion device to reach *ignition* in order to provide useful thermal power. In an ideal world without losses, useful power could even be extracted from a system with $Q < 1$. In practice, however, Q would have to be much greater than one in order to allow for losses and for the power necessary to operate the reactor systems.

As of 2017 no controlled fusion device has yet reached breakeven. The most ambitious MCF experiment, the **Joint European Torus**, or **JET**, in Great Britain has reached a fusion gain factor of $Q \approx 0.7$ for a brief period [101]. The new **International Thermonuclear Experimental Reactor**, or *ITER*, project under construction in France aims for $Q \sim 10$ [101].

19.6.2 Fusion Performance Criteria

Plasma density and temperature must be high to produce nuclear fusion. To sustain a fusion reaction, the rate of thermal energy transfer out of the plasma, parameterized by $1/\tau_E$, must also be kept low. The *Lawson criterion*, first defined by British engineer and physicist John D. Lawson in 1955, is a combination of these parameters that is relatively easy to measure, and which serves as a useful

benchmark on the path to a practical fusion reactor. More recently the Lawson criterion has been at least partially superseded by another closely related benchmark known as the *triple product* or *fusion product*. Although they can be adapted to any fusion reaction, we describe these performance criteria for the case of the *dt* fusion reaction.

Both the Lawson and triple product criteria estimate the conditions required for *ignition*, when the plasma increases in temperature or can be maintained in steady state (eq. (19.20)) with no external heating,

$$P_\alpha \geq P_c + P_r . \tag{19.25}$$

Substituting from eqs. (19.22)–(19.24), and using the ideal gas law, $p = 2nk_BT$, to eliminate n, we obtain

$$p\tau_E \geq \frac{24(k_B T)^2}{\langle \sigma_f v \rangle E_\alpha - 4C_B \sqrt{k_B T}} . \tag{19.26}$$

Figure 19.14 shows a plot of the right-hand side of eq. (19.26) as a function of temperature. The red curve includes the contribution from bremsstrahlung losses – the second term in the denominator – the purple curve does not. Ignition is not possible until the temperature exceeds $T \cong 4.4\,\mathrm{keV}/k_B \cong 53 \times 10^6\,\mathrm{K}$. The minimum value of $p\tau_E$ on the curve occurs at approximately $k_B T_{\min} \approx 15\,\mathrm{keV}$, where $(p\tau_E)_{\min} \approx 8.3\,\mathrm{atm\,s}$. Thus a plasma confined at a pressure of 8.3 atm with an energy confinement time of approximately one second can in principle achieve ignition.

From Figure 19.14 it is apparent that bremsstrahlung losses can be ignored for $k_B T \gtrsim 10\,\mathrm{keV}$. In this case eq. (19.26) simplifies to

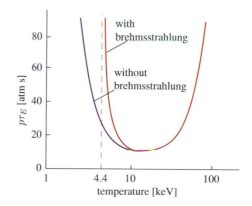

Figure 19.14 The lower limit on the product of pressure × energy confinement time for ignition (19.26), with and without the contribution of bremsstrahlung losses. After Freidberg [90].

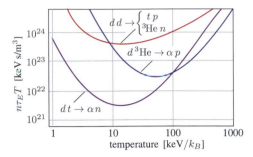

Figure 19.15 The triple product as a function of temperature for dt, dd, and d^3He fusion. (Credit: Dstrozzi reproduced under CC-BY-SA 3.0 license via Wikimedia Commons)

$$nT\tau_E \geq \frac{12k_B T^2}{\langle \sigma_f v \rangle E_\alpha}, \qquad (19.27)$$

where we again have used the ideal gas law to replace p by $2nk_B T$. Equation (19.27) is the **triple product** criterion. The **Lawson criterion** is obtained by canceling the common factor of T on both sides of the equation,

$$n\tau_E \geq L \equiv \frac{12k_B T}{\langle \sigma_f v \rangle E_\alpha}. \qquad (19.28)$$

Clearly the two criteria (19.27) and (19.28) are equivalent – if one is satisfied, so is the other. The triple product owes its ascendence in usage to the convenient fact that $\langle \sigma_f v \rangle$ is approximately proportional to T^2 over the interval $5 \lesssim k_B T \lesssim 25$ keV, so that the right hand side of eq. (19.27) is roughly constant over that interval. Also, the maximum pressure achievable in an MCF reactor, $p \propto nT$, is roughly constant even though n and T can separately vary significantly [90].

Using the data of Figure 19.13, the triple product is plotted in Figure 19.15 for dd and d^3He as well as for dt fusion. The relative difficulty of reaching ignition for these three different fusion fuels can be seen in this figure. Ignition in a dd fusion reactor, for example, would require a product of pressure × energy confinement time roughly two orders of magnitude greater than that required for ignition of a dt plasma.

19.6.3 Magnetic Confinement Fusion

A hot plasma consists of negatively charged electrons and positively charged nuclei. If the plasma is to reach the temperatures needed for dt fusion ($T \gtrsim 10$ keV$/k_B \approx 100 \times 10^6$ K), it must be kept out of contact with any normal matter. The goal of **magnetic confinement fusion** (MCF) devices is to confine the charged particles in a hot plasma with the use of suitably shaped magnetic fields. In

this section we explore how this can be accomplished for individual particles moving in fixed fields. We ignore interactions among particles, as well as radiation and all of the other systems required to confine and stabilize a burning plasma. A thorough overview of MCF at a level not far beyond the level of this book can be found in [90].

The Lorentz force law (3.45),

$$\boldsymbol{F} = q\boldsymbol{E} + q\boldsymbol{v} \times \boldsymbol{B}, \qquad (19.29)$$

dictates that the magnetic force on a particle of charge q is perpendicular both to its velocity and to the magnetic field. In a constant magnetic field, directed in the z-direction the force is in the xy-plane, causing the particle to move in a circular orbit with radius $r = mv/qB$ in the xy-plane (Problem 19.13) as it travels steadily in the $\pm\hat{z}$-direction. The result is a helical trajectory, winding along or opposite to the direction of $\boldsymbol{B}$. In a strong (multi-Tesla) magnetic field, particles with energies of order 10 keV move in orbits with radii r that are small compared to the macroscopic dimensions of a fusion reactor (Problem 19.13). Particles of opposite charge circulate along helices of opposite handedness, as shown in Figure 19.16. Thus, a constant magnetic field does an excellent job of confining both positively charged nuclei and electrons in the directions perpendicular to the field lines. The charged particles are not, however, confined in the direction parallel to $\boldsymbol{B}$.

In the early days of MCF research, many options were considered to confine the plasma motion in the direction along the magnetic field lines. A Russian design in which

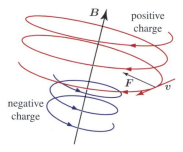

Figure 19.16 Charged particles move in helical orbits in a constant magnetic field. Positive charges (red) orbit opposite to negative charges (blue) and the radii of the orbits are proportional to the square root of the particle's energy. The vectors v and F are shown at one point of the particle's trajectory.

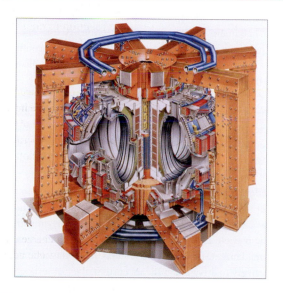

Figure 19.17 A cutaway diagram of the Joint European Tokamak (JET). (Credit: Contains public sector information licensed under the Open Government License v2.0)

the magnetic field lines are wrapped into a *torus*, or donut shape, has emerged as the most promising option. Known as a **tokamak** (a Russian acronym for *toroidal chamber with magnetic coils*), this design is characterized by a plasma and **toroidal magnetic field** that are axisymmetric around the torus (Figure 19.18(a)). A cutaway diagram of the JET, the largest tokamak operated to date, is shown in Figure 19.17. Coils wound around the torus produce a strong magnetic field – 3.45 T at JET, $\gtrsim 10$ T planned at ITER – that forms a closed ring.

Unfortunately, once it has been bent into a toroidal shape, the magnetic field can no longer be uniform.

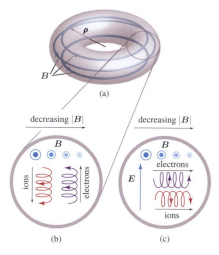

Figure 19.18 Charged particle motion in a toroidal magnetic field. (a) Magnetic field circulates around the toroid and decreases in strength with distance from the axis. (b) With $|B|$ increasing toward the center of the torus, orbits have smaller radius of curvature toward the toroid axis and larger toward the outside, as a result positive (negative) particles drift down (up). (c) In the resulting perpendicular E and B fields, the radius of curvature at the top (bottom) of the orbit is larger for positive (negative) charge, so particles of both charges drift to the right.

Instead, as dictated by Ampere's law (3.48), the strength of the field falls like $1/\rho$, where ρ is the distance from the central axis of the torus (Figure 19.18(a)). In the presence of only the non-uniform toroidal field, the orbit of a particle in the plane perpendicular to the magnetic field would have a smaller radius of curvature ($r = mv/qB$) where $B = |B|$ is stronger, close to the axis of the toroid, and a larger radius of curvature toward the outside of the toroid. As a result, in a toroidal magnetic field that goes clockwise when viewed from above, the positive ions in the plasma would drift down and the electrons would drift up, as illustrated in Figure 19.18(b). The resulting separation of charge would give rise to an electric field in the vertical direction, which opposes the drift. This is not the end of the story, however. A positive charge orbiting in the plane perpendicular to B would then speed up, due to the electric field, near the top of its orbit when it is moving radially outward and slow down near the bottom where it is moving inward. The result, shown in Figure 19.18(c), is that positive ions would drift toward the outer edge of the torus. This phenomenon is known as $E \times B$ drift (see Box 27.2). As the name indicates, it is driven by crossed electromagnetic fields and its direction is perpendicular to

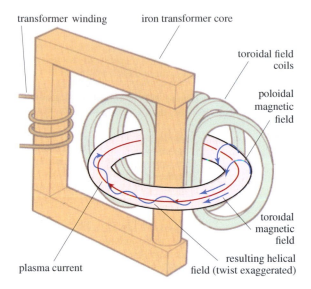

transformer winding   iron transformer core

toroidal field coils

poloidal magnetic field

toroidal magnetic field

resulting helical field (twist exaggerated)

plasma current

Figure 19.19 Schematic diagram showing the toriodal and poloidal magnetic fields in a tokamak, and the resulting helical field in the plasma. Also shown are the coils that produce the toriodal field and the transformer that induces the plasma current, which in turn gives rise to the poloidal field. (Credit: Image adapted from EUROfusion, www.euro-fusion.org)

both E and B.[5] Electrons would drift the same way as the positive ions (see the figure). Thus, both ions and electrons would drift outward, allowing the plasma to leak out of the device.

To prevent $E \times B$ drift and stabilize the plasma, tokamaks must also have an induced current that runs around the torus. This current generates a further contribution to the magnetic field, known as the **poloidal magnetic field**. This field circulates around the smaller cross-sectional circle of the torus, wrapping around the toroidal current, as shown in Figure 19.19. The two components of the magnetic field, *toroidal* and *poloidal*, combine to form a magnetic field whose field lines wind helically around the center of the torus (Figure 19.19). Because of this helical twist, the sidewards drift experienced by the charged particles, as they spiral around the field lines, is alternately directed toward and away from the axis of the torus, and cancels out on average. This can be understood by considering the motion of a charged particle in the poloidal field alone – with an initial outward velocity, the particle will

follow a helical path centered about a circular poloidal field line. The charge separation that would generate the vertical electric field and spell the death of the plasma in the case of a purely toroidal field thus never develops.

Since the plasma has a non-vanishing resistivity, the current that produces the poloidal field also serves to heat the plasma, which is a desirable side-effect. Resistive losses, however, cause the current to die out if it is not constantly replenished. In present plasma fusion experiments, the current is primarily generated inductively by a toroidal *EMF* $\mathcal{E}$ produced by ramping the current I in the windings of a central transformer; by Lenz's law, $\mathcal{E} = -d\Phi/dt \propto dI/dt$. The plasma current and the poloidal magnetic field that it generates can be sustained only as long as the current in the transformer is changing. This forces present tokamaks to operate in a pulsed mode, in which the transformer current is ramped from a maximum in one direction to a maximum in the opposite direction. At the end of the pulse the poloidal field vanishes, and plasma confinement is completely lost. The plasma must be re-established and the transformer current must be reset before the next pulse.

Steady-state operation of a magnetic confinement fusion reactor would require that the plasma current be maintained in other ways, several of which are actively under investigation. Certain antenna structures can launch waves from the plasma edge that propagate toward the core and transfer a net momentum to the electrons, resulting in the generation of current. Alternatively, complex interactions between the ions and electrons, as well as asymmetries in the particle drift directions, can give rise to an inherent plasma current, though these features have yet to be used effectively to drive a significant current.

In a burning plasma, fusion itself supplies the energy to keep the plasma hot. Below ignition, however, some heat must be supplied from other sources. Resistive heating from the plasma current is one source. Others include injecting high-energy beams of neutral particles, and bathing the plasma in intense microwave electromagnetic radiation.

In addition to all the challenges already described, tokamak plasmas are susceptible to internal instabilities, some of which are potentially disastrous for large machines where the engineering tolerances are already extreme. Many of these issues are related to the plasma response in the presence of a very energetic particle population, such as the ^{4}He particles created by the fusion reactions. Finally, particularly at the relatively low pressures used in current practice, the energy density of the *dt* fuel is about 2×10^{-8} times smaller compared, for example, to natural uranium fuel for a fission reactor. Thus very large systems or high

---

[5] A similar effect occurs in the context of ocean currents where the wind and the *Coriolis force* play the role of E and B respectively (see §27.3.1).

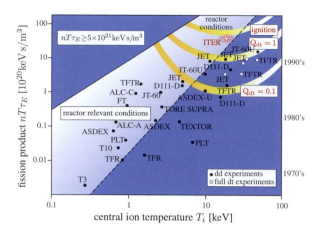

Figure 19.20 Performance of tokamak experiments. The triple product, $n\tau T$, is plotted versus the central temperature in the reactor. dt experiments are open circles, dd are closed. The bands correspond to $Q = 0.1$, 1, and ∞, respectively. The light blue region labelled "reactor relevant conditions" corresponds to ion and electron temperatures more nearly equal, while the dark blue regions have artificially higher ion temperatures [102]. For more information see [103].

throughput of the fuel would be needed to achieve power at the gigawatt scale.

Some very interesting physics will emerge in the coming years with the construction of ITER [101], where burning plasmas will be explored. A measure of the present state of MCF research is summarized in Figure 19.20. ITER is aiming to reach a fusion gain factor of ten in the regime of parameters considered "reactor relevant."

19.6.4 Inertial Confinement Fusion

Another approach to controlled fusion, known as **inertial confinement fusion** (ICF), proposes to use arrays of very-high-power lasers, focused onto a millimeter-size microcapsule or *pellet* containing a few milligrams of a one-to-one deuterium–tritium mixture. When the laser light impacts the pellet, it generates very strong shock waves that compress and ultimately ignite the dt mixture in a blaze of fusion reactions.

The foremost facility for ICF research is the *National Ignition Facility* (NIF) at Lawrence Livermore Labs in Livermore, CA. Starting with approximately 420 MJ of energy stored in a capacitor bank, this device will eventually direct 1.8 MJ of energy in the form of 380 nm wavelength ultraviolet light from 192 separate high-power lasers onto the pellet. So much energy delivered over a few picoseconds corresponds to an instantaneous incident power of about 500 TW, which can produce volume compression ratios

Tokamak

The *tokamak* is a fusion reactor design in which the plasma is confined in a torus (donut shape) by *toroidal* magnetic fields, augmented by *poloidal* fields produced by a toroidal current at the core of the torus.

There are many difficulties in engineering a functioning power plant based on a tokamak with large Q, including maintaining a sufficient plasma current in steady state, avoiding instabilities, refueling on a short time scale, and heating the plasma, along with the additional challenges of producing tritium, maintaining structural integrity in a very high neutron flux, and efficiently extracting thermal energy for power generation. The ITER project aims to initiate dt fusion reactions before 2030, and to achieve a peak of $Q \sim 10$, with a sustained value of $Q > 5$ and 500 MW of thermal energy output over a pulse of roughly 8 minutes.

on the order of 10^4 and ion temperatures on the order of 10 keV.

The concept behind ICF is that very brief and powerful bombardment of the surface of a pellet by radiation causes a sudden vaporization or *ablation* of material from the surface. The escaping material induces an inward-propagating shock wave traveling toward the core of the pellet, which in turn gives rise to compression and heating. With sufficient power to the surface, the core can be heated and compressed by the shock waves to the point at which the dt mixture sublimates into a plasma state and fuses in large quantities.

Major scientific, technical, and economic challenges face ICF. The energy confinement time is so short that even at temperatures of order 10 keV, the target must reach densities more than three orders of magnitude greater than that of normal matter in order to satisfy the triple product criterion. Early in the study of ICF, it was discovered that laser compression leads to various instabilities that can result in asymmetric compression and ignition failure. One of the most common instabilities is the **Rayleigh–Taylor instability**, illustrated in Figure 19.21, which arises when a layer of less dense fluid supports a layer of more dense fluid (as is found in the pellet after it has started its compression). The presence of imperfections in the

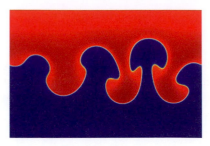

Figure 19.21 A simulation of a Rayleigh–Taylor instability. Here a less dense fluid (blue) supports a denser fluid (red) in a gravitational field. Initially the fluids were separated by a horizontal boundary. An infinitesimal perturbation triggers the instability in which fingers of the less dense fluid erupt as the denser fluid sinks. Such instabilities pose problems for ICF. (Credit: LBM wiki documentation project, wikipalabos.org)

boundary between these surfaces can give rise to an interchange where the less dense fluid erupts to the surface as the more dense fluid rushes in. This instability can grow exponentially and fatally disrupt pellet compression. The surface of the pellet and the energy deposition by the lasers must be exceptionally uniform. The NIF project has taken steps toward achieving this by utilizing a large number of lasers. Despite progress on this problem, Rayleigh–Taylor instabilities remain a serious concern.

Recent work has also focused on the technique of *indirect-drive* that makes use of X-rays generated in a *hohlraum* (German for "hollow space"), which might, for example, be a gold capsule in which the pellet is suspended. Instead of training the lasers on the pellet, indirect-drive instead targets the hohlraum. The intense laser pulse on the gold surfaces creates X-rays that can more uniformly bathe the pellet's surface, resulting in a more even compression.

On top of the scientific and engineering issues mentioned above, there are concerns that the laser-driven system has very low efficiency and there are practical issues associated with the fact that ICF is inherently a pulsed system in which the capacitor banks driving the lasers must recharge between firing.

Discussion/Investigation Questions

19.1 Explain why the resonance escape probability grows and the thermal utilization factor falls with increasing ratio of moderator to fuel at fixed enrichment.

19.2 Explain why the optimal ratio of moderator to fuel is so much larger for a graphite-moderated (infinite, homogeneous) reactor than for a water-moderated reactor at the same enrichment.

19.3 Some designs for breeder reactors use liquid metals such sodium or lead as a coolant. Discuss some of the advantages and disadvantages of such a coolant.

19.4 The *energy amplifier* uses a particle accelerator to feed neutrons into a subcritical reactor. Find suitable sources and give a concise summary of this novel type of reactor, its advantages such as inherent safety and capacity to breed fuel and/or reduce the burden of nuclear waste, and its disadvantages.

19.5 Can you figure out a series of reactions in a neutron-rich environment that would lead from ^{232}Th to ^{232}U? See §20.6.3 for discussion of the role of ^{232}U in possible weaponization of ^{233}U produced in a thorium-fueled reactor. [Hint: there is a $n + {}^{A}Z \rightarrow 2n + {}^{A-1}Z$ reaction involved.]

19.6 Choose one of the Gen IV reactor designs and investigate its advantages and disadvantages. Is there active research and/or development on this design? What is your judgment on its future prospects?

Problems

19.1 Starting from the Maxwell–Boltzmann distribution, eq. (18.10), for the probability of finding a particle with energy E in a gas with temperature T, show that the most probable speed for a particle is $\bar{v} = \sqrt{2k_B T/m}$, but that the average speed of a particle is $\langle v \rangle = \sqrt{8k_B T/\pi m}$. Verify that the most probable speed for a neutron at $T = 20\,°C$ is ~ 2200 m/s.

19.2 Show from eq. (19.9) that in an infinite, homogeneous graphite-moderated reactor, the resonance escape probability is given by $p(y) = \exp\left(-17.28\,(1/(9.3 + 4.9y))^{0.514}\right)$, where y is the ratio of moderator to fuel.

19.3 Consider an infinite, homogeneous reactor fueled with uranium enriched to 3% and moderated by graphite with a ratio of graphite:uranium of 800:1. Find k_∞. To find p, you can ignore the uranium when computing $\langle \xi \rangle$ and in the calculation of the scattering cross section. (You can assume $\epsilon = 1$.)

19.4 Show that an infinite, homogeneous reactor fueled with natural uranium (0.72% ^{235}U) and moderated by heavy water (D_2O) can sustain a fission chain reaction. What is the optimal ratio of moderator to fuel?

19.5 It is estimated that the Oklo reactor occurred when the ratio of ^{235}U to ^{238}U was 3.67%. Given the present ratio of 0.72% and the half-lives of both isotopes, estimate the time when the Oklo reactor was active.

19.6 An approximate description of the Oklo uranium deposit is 90% uraninite (UO_2) by mass, saturated with water that acted as the moderator. Assume that the Oklo deposit was homogeneously saturated with water with a ratio of 3.5 uranium nuclei per water molecule and compute the minimum enrichment (^{235}U/^{238}U) needed to sustain a fission reaction. You can ignore the moderating effect of the extra oxygen in the uraninite (it is negligible). You can also ignore the neutron-absorption cross section of the oxygen in the uraninite and of any other material in the deposit except for the water.

19.7 Suppose a pressurized water reactor is loaded with 200 tons of 3% enriched uranium. The reactor has been designed to run with an average thermal-neutron flux of $\langle \Phi_n \rangle = 1.5 \times 10^{13}$ cm^{-2} s^{-1} for a period of four years before refueling. You can assume that Φ_n is held constant the whole time, ignore neutrons losses due to the finite size of the reactor, and ignore any contribution of ^{239}Pu fission to the reactor power. What is the thermal power output of the reactor immediately after it is started? How does the power output decrease as a function of time over the four years? Use thermal-neutron cross sections from Table 19.1.

19.8 [C,T] The analysis of prompt and delayed neutrons in §19.1.6 leads to the following formula for the time dependence of the neutron density,

$$ n(t) = n_0 \left(\frac{d}{d - \rho} \exp\left(\frac{\rho t}{(d - \rho)t_d} \right) \right. $$
$$ \left. - \frac{\rho}{d - \rho} \exp\left(\frac{-(d - \rho)t}{t_p} \right) \right). $$

Analyze the time dependence of $n(t)$ when $\rho < d$ and when $\rho > d$. For example, take $t_p = 2 \times 10^{-4}$ s, $t_d = 12.5$ s, and $d = 0.0064$, and compute $n(t)/n(0)$ for $\rho = d/2$ and $\rho = 2d$.

19.9 [C,H] When a reactor is turned off, the amount of ^{135}Xe decreases as it decays, but grows as ^{135}I decays to it. Write a pair of differential equations that describe the time rate of change of the numbers N_I and N_X of ^{135}I and ^{135}Xe nuclei respectively, and solve these equations assuming that $N_I = N_I^0$ and $N_X = N_X^0$ at $t = 0$. Show (using the measured half-lives of ^{135}I and ^{135}Xe) that the resulting curve of $N_X(t)$ looks like the $t > 0$ section of Figure 19.5. To simplify your calculations, take $N_X^0 = 0$.

19.10 A uranium enrichment facility produces nuclear reactor fuel enriched to 4% ^{235}U to fuel the reactor whose power to fuel consumption ratio is described by eq. (19.15). For each tonne of natural uranium, the enrichment facility produces 130 kg of fuel. What is the concentration of ^{235}U in the 870 kg of *depleted uranium*? Assume that the percentage of ^{235}U in spent

nuclear fuel removed from the reactor is the same as in natural uranium. How many tonnes of fuel are used in a year? Show that 211 tonnes of natural uranium is required to provide the 0.9 tonnes of ^{235}U consumed by this reactor in one year. Repeat your analysis assuming that the enrichment facility produces 120 kg of fuel enriched to 5% and that the percentage of ^{235}U in the spent fuel remains the same.

19.11 A thermal-neutron reactor is charged every year with enriched uranium containing 1.6 t of ^{235}U. When the spent fuel is removed, it contains 400 kg of ^{235}U and 560 kg of ^{239}Pu. Assuming a conversion ratio CR = 0.6, what fraction of the reactor power was generated by fission of ^{239}Pu? (Ignore radiative capture of thermal neutrons in both ^{235}U and ^{239}Pu, ignore production and fission of any other actinides, and assume that the energy releases in ^{235}U and ^{239}Pu fission are identical.)

19.12 A thermal breeder reactor is fueled with a mixture of a fertile nucleus A (fraction $1 - x$) which breeds a fissile nucleus B (fraction x). To breed more of B it is essential that the *reproduction factor* $\eta(B)$ in thermal-neutron-induced fission of B is greater than two. Express the maximum possible value of $\eta(B)$ in terms of cross sections and the average number of fast neutrons in B-fission. Evaluate $\eta(B)|_{\max}$ for the two possible breeding systems, (^{238}U,^{239}Pu) and (^{232}Th, ^{233}U). Which is the most promising for a thermal breeder reactor?

19.13 Show that a charged particle moving in the plane perpendicular to a constant magnetic field (eq. (19.29)) moves in a circle with $Br = mv/q$. What is the radius of the circle for an electron with energy 15 keV in a 10 Tesla magnetic field (typical of ITER)? What is the radius for a proton?

19.14 A dt plasma fusion reactor is operating in steady state at a temperature $k_B T = 15$ keV and a pressure of 7 atm, where $\langle \sigma_f v \rangle = 3 \times 10^{-22}$ m^3/s. What is the energy confinement time τ_E required to satisfy the triple product criterion? What is the density of the plasma (particles per cubic meter), and how does this compare with the density of air at room temperature and pressure?

19.15 Suppose a fusion power plant operates at the pressure (≈ 7 atm) and temperature ($\approx 150 \times 10^6$ K) planned for ITER, and suppose the plasma volume equals that of ITER (840 m^3) as well. Assuming that the plasma contains equal numbers of d and t nuclei, estimate how long such a fusion reactor could provide fusion power equal to 1 GW before it would have to be refueled.

Ionizing Radiation

Ionizing radiation is inextricably associated with nuclear energy. An understanding of the physical properties, biological effects, and natural and manmade sources of radiation provides a context for the evaluation of the advantages and disadvantages of nuclear power. Such an understanding also informs personal and societal choices associated with nuclear weapons and with the use of radiation in modern medical diagnostics and treatment.

The term **radiation** refers to energy carried by waves or particles propagating through space. The precise meaning of *radiation* has evolved over time, however, as new physical phenomena have been understood and associated technologies have been developed. Originally used in physical science primarily in reference to light emanating from a source, since the early 1800s the meaning of the term radiation has expanded to encompass electromagnetic waves in general, as well as particles such as α-radiation and β-radiation, and even *gravitational radiation* – excitations of the fabric of space-time itself.

Thus, light from the Sun, microwaves used to heat food, and long-wavelength thermal radiation emitted from Earth into space are radiation just as much as the energetic α-particles, electrons, and γ-rays emitted by nuclear waste. In the everyday world, however, the term *radiation* often has negative connotations, suggesting a threat to human health. In common parlance, X-rays, α-, β-, and γ-rays are referred to as *radiation*, but ordinary light is not. *Ultraviolet light* straddles the boundary between innocuous visible light and harmful radiation. *Near-ultraviolet* light (wavelengths between 300 and 400 nm) is known to cause sunburn, eye damage, skin cancer, and other health effects, although it is commonly used in *tanning salons*. Exposure to very low-intensity near-ultraviolet light ("black light") is regarded as relatively safe (Figure 20.1).

An important distinction between dangerous and relatively harmless radiation is whether it can ionize an atom,

Reader's Guide

This chapter describes ionizing radiation, which can damage biological tissues. After surveying the forms of ionizing radiation, the various measures of radiation exposure are explained, as are the mechanisms by which radiation can damage tissues. The focus then shifts to the sources of radiation exposure in the environment, both natural and manmade. The chapter closes with a discussion of radiation and proliferation risks from spent nuclear fuel.

Prerequisites: §7 (Quantum mechanics), §15 (Quantum processes), §16–§19 (Nuclear power).

Figure 20.1 *Black light*, or low-intensity near-ultraviolet light, is absorbed by *optical brighteners* in many fabrics and re-emitted as visible light. Despite evidence that near-UV light causes damage to the skin in large doses, black light is usually not considered "harmful radiation."

leaving behind a charged – and therefore highly reactive – positive ion and a similarly reactive free electron. Not only can the original disruption damage biological systems, but the resulting ions can also transform into *free radicals* (see

§20.4.2) that can do further damage to many molecules. For this reason physicists use the term *ionizing radiation* to describe potentially dangerous radiation. It takes ~13.6 eV to ionize a hydrogen atom, so the realm of ionizing radiation should begin with quanta of about this energy. Though the dividing line is somewhat arbitrary, any emissions of quanta with energy greater than about 20 eV are considered **ionizing radiation**. To get a feel for this energy scale, consider the energies of photons of various types of light. According to the Planck formula, a photon of wavelength λ carries energy $E = h\nu = hc/\lambda$. For green light $\lambda \approx$ 500 nm, so $E(\text{green}) \approx 2.5$ eV. The wavelength of ultraviolet light ranges from 300–400 nm (near-ultraviolet or **black light**) down to about 10 nm (**extreme ultraviolet**) corresponding to 3 eV and 120 eV respectively [104]. So *ionizing radiation* applies to light with wavelengths below about 50 nm, in the extreme ultraviolet. While radiation at lower energies, such as the near-UV radiation in sunlight, can also cause some biological damage by breaking apart chemical bonds such as those in water (which are typically only a few eV), resulting in free radicals, the biological impact of radiation in the 1–20 eV range is much less than that of the higher-energy radiation on which we focus here.[1]

Ionizing radiation is often associated with nuclear energy, where it is a potentially dangerous side effect that must be carefully monitored and managed, and with nuclear weapons, where its effects compound the destruction caused by the blast itself. Ionizing radiation is used in many beneficial contexts as well. In particular, medical scientists and engineers have found many ways to put radiation to good use. Radiation is a primary tool for medical diagnosis and treatment, and its use, especially in the developed world, has expanded rapidly. The present and potential future role of radiation in the human environment is a complex and controversial issue with many social and political ramifications. An understanding of the physics of ionizing radiation helps inform debate on this difficult issue.

In this chapter we first (§20.1, §20.2) summarize the properties of the different forms of radiation, all of which were first encountered in §14 and §17. We pay special attention to the way that radiation interacts with matter.

Different forms of radiation have very different characteristic interactions that dictate how dangerous they are and what sort of shielding is needed to protect people from adverse effects. Next (§20.3) we give an overview of some of the units that are used to measure the effects of radiation. It matters greatly whether radiation is absorbed on the skin or in internal organs, for example, and measurement systems have evolved to take this into account. After these preliminaries, we describe the biological effects of radiation (§20.4) and the sources of radiation in the environment (§20.5). An important lesson of this section is that some radiation is natural and unavoidable. The level of natural radiation provides a base line to which radiation hazards can be compared. Finally (§20.6) we turn to the questions raised by radioactive waste from nuclear reactors and the associated risk of nuclear weapons proliferation. Standard nuclear physics and engineering texts such as those listed in §16 all discuss ionizing radiation. For a more detailed introduction, see [80].

20.1 Forms of Ionizing Radiation: An Overview

We have already met the most important forms of ionizing radiation: the α-particle – the nucleus of the helium atom ${}^4_2\text{He}$; the neutron; the β-particle – otherwise known as the electron; the *muon*, an important component of *cosmic rays* (see §20.5.2); and extreme-UV light, X-rays, and γ-rays, all of which are forms of electromagnetic radiation. X-rays and γ-rays overlap in energy. Usually photons produced by an atomic process are called **X-rays** and photons emitted by a nucleus are called γ-**rays**. Many nuclear processes, however, generate photons with energies much lower than the most energetic atomic X-rays, so the distinction is somewhat artificial. Furthermore, photons of energy greater than 100 keV are often referred to as γ-rays regardless of origin, particularly in astronomy, where the exact source of the radiation may be unknown. Most of these forms of radiation were discovered in the late nineteenth century, including X-rays, β-rays, α-rays, and γ-rays. It took years to correctly identify β-rays as electrons and γ-rays as another form of light somewhat more energetic than X-rays. Neutrons and muons were discovered much later, the neutron in 1932 and the muon in 1936.

There are other forms of radiation that figure in the modern world, but none prevalent enough to warrant detailed treatment here. For example, positrons are used in medical diagnosis. Their interactions with matter are the same as electrons, except that when they finally come to rest, they annihilate with an electron to produce two γ-rays each with energy $m_e c^2 \cong 0.511$ MeV. Protons and pions are

---

[1] There is some limited, and controversial, evidence that under certain circumstances high levels of exposure to certain types of even lower-energy (< 1 eV) non-ionizing electromagnetic radiation can have statistically significant health impacts; we focus here on the much more hazardous and well-established effects of high-energy ionizing radiation, and do not attempt to evaluate or discuss the effects of non-ionizing radiation.

used in cancer therapy. They fall into the general category of *heavy charged particles* like α-particles, which we discuss. No naturally occurring nucleus emits protons or pions, and positron emitters are relatively rare, as discussed in §17.4.3.

The interactions of radiation with matter are complex. Events involving small amounts of energy such as the creation of free radicals can have major consequences in a living organism. Since the typical chemical bond energy amounts to a few electron volts, nuclear radiation with energy in the MeV range has the potential to break apart many thousands of bonds. The amount of ionization depends on the type of radiation, its energy, and its interactions with matter. Depending on the circumstances, physicists use several different quantitative measures to describe how radiation interacts with matter.

The energy loss of heavy charged particles and energetic electrons is usually measured by how much energy they lose per unit distance traveled in matter, that is, by $|dE/dx|$ (where dE/dx is negative). This quantity is known as the **stopping power** and is a property of both the type of radiation and the matter involved. Since $|dE/dx|$ varies almost linearly with the density of the material, the **mass stopping power** $S = (1/\rho)|dE/dx|$ is often used. Energy loss by photons – γ-rays and X-rays – is described by an *attenuation coefficient* introduced below.

The biological effects of radiation are related to the amount of energy per unit distance that ionizing radiation *deposits locally* in material through which it passes. This quantity is known as the **linear energy transfer**, or **LET**. LET and stopping power are often used interchangeably, though the LET is generally less than $|dE/dx|$, since some energy is lost to high-energy particles and photons that transfer energy away from the local region. In the context of materials used for nuclear containment or radiation shielding, the energy loss of charged particles is often described in terms of stopping power. *LET* is used by medical professionals for X-rays and γ-rays as well as charged particles. We use *stopping power* for charged particles traveling through materials in general but *LET* when discussing effects of all kinds of radiation on biological systems.

Different forms of radiation lose energy in quite different ways, which we summarize briefly before going into more detail in §20.2.

Heavy charged particles **Heavy charged particles**, such as α-particles, protons, or fission fragments, knock electrons out of atoms just as a bowling ball would scatter ping pong balls out of its way. We analyzed collisions like these in the context of moderators for nuclear fission reactions

(§18.3.6). The relevant result is eq. (18.4), from which we can obtain the maximum energy loss by a particle of mass M and energy E striking a stationary particle of mass m,

$$\Delta E_{\max} = \frac{4mM}{(m+M)^2}E. \qquad (20.1)$$

When the struck particle is an electron, $M \gg m_e$, so

$$\Delta E_{\max} \approx \frac{4m_e}{M}E. \qquad (20.2)$$

The lightest heavy particle, the muon ($m_\mu = 206.7m_e$), loses no more than 2% of its energy per collision with an electron. Protons, α-particles, and other more massive particles lose much less. A 1 MeV proton must undergo thousands of collisions (Problem 20.2) before it no longer has enough energy to ionize an atomic electron. Stopping a heavy charged particle is therefore a relatively smooth, continuous process.

Electrons Electrons can knock bound electrons out of atoms, but since the masses of the incident and struck particles are the same, a large fraction of the incident electron's energy can be transferred in one collision. Putting $m = M = m_e$ in eq. (20.1) we find $\Delta E_{\max} = E$ – an electron can transfer up to 100% of its energy to another electron. This result holds true even if the initial electron is moving at a significant fraction of the speed of light, as is often the case in β-decay. Because it is so light, and therefore accelerates strongly through interactions with matter, an energetic electron can also radiate energy away as photons, which can in turn ionize more electrons. If a radiated photon has enough energy, in the presence of matter it can create an electron–positron pair. These processes generate a shower of electrons, positrons, and photons that at first grows and then dissipates as it passes through matter.

Energetic photons Photons pass through matter without interacting until they strike an electron or create an initial electron–positron pair, after which the electromagnetic shower of electrons, positrons, and photons progresses in much the same way as a shower initiated by an electron. A photon of a given energy passing through a specific material has a constant probability per unit length for an interaction that initiates an electromagnetic shower. This leads to an exponential decrease of the intensity of light as it penetrates into matter, often referred to as the *Lambert–Beer law* (see §23.4). The absorption of light by matter plays an important role in our treatment of solar energy in §24, ocean energy transport in §27, and climate in §34.

Neutrons Neutrons wander about in matter much as they do in the moderator in a nuclear reactor. They lose energy in collisions with atomic nuclei, usually disrupting the atom in the process. Eventually they get captured by a

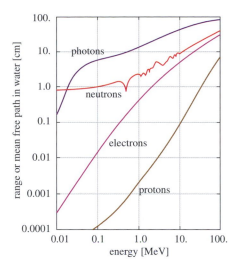

Figure 20.2 The range of electrons and protons in water compared with the mean free path of photons and neutrons. The energies are typical of the radiation emitted in nuclear processes and the most energetic atomic processes. After [112].

nucleus, to which they impart several MeV of excitation energy (see §17.4.1 and Figure 17.8). The excited nucleus may de-excite by emitting a γ-ray, a proton, or an α-particle; it may even fission. All of these lead to secondary radiation.

Figure 20.2 provides a summary of the depths to which different forms of radiation penetrate into matter. In the figure, photons and neutrons are characterized by their **mean free path**, the average distance that they propagate before undergoing an interaction. Electrons and protons are described by their *range*, the characteristic distance over which they deposit energy after entering matter. At the energies typical of nuclear emissions, $\lesssim 5$ MeV, photons and neutrons are the most penetrating radiation and protons are the least. The interactions of different forms of radiation with matter that lead to such great differences in penetrating power are explained in more depth in the following section.

20.2 Interactions of Radiation with Matter

In this section we describe in greater detail the interactions between the most common forms of radiation and matter. The ways in which different types of radiation interact with matter determine to large degree the biological damage they can cause, as well as their medical applications, and their prevalence in the human environment. While the

material in this section is helpful in understanding many aspects of the physics underlying radiation issues, some readers may wish to skim this section on a first reading and return to it as necessary while or after reading the subsequent parts of this chapter.

20.2.1 Heavy Charged Particles in Matter

As *heavy charged particles* pass through matter, their electric fields push electrons out of the way. To compute the stopping power $|dE/dx|$ for a charged particle moving through matter requires the tools of *quantum electrodynamics*, and is a task beyond the scope of this book. There is an approximate expression for $|dE/dx|$ that is relatively simple, however, and its components can be motivated by simple physical arguments. The result was first derived by the German-American physicist Hans Bethe in the early 1930s,

$$-\left.\frac{dE}{dx}\right|_{NR} = \left(\frac{ze^2}{4\pi\epsilon_0}\right)^2 \frac{4\pi nZ}{m_e v^2} \ln\left(\frac{2m_e v^2}{I}\right). \quad (20.3)$$

Here z is the charge of the projectile, and Z is the charge and n is the number density of atoms in the material through which the projectile is propagating, so Zn is the number density of electrons in the material. v is the projectile speed and I is the mean energy required to ionize an electron in the medium. I is hard to determine from first principles and is instead determined empirically. According to [80], a good empirical approximation for $2 \leq Z < 13$ is $I \approx 11.7(Z + 1)$ eV, with $I \cong 19.0$ eV for hydrogen $(Z = 1)$.[2]

The subscript *NR* (nonrelativistic) in Bethe's formula indicates that relativistic corrections that become important for particles moving near the speed of light have been omitted. Nuclear decays produce α-particles with kinetic energies in the 1–10 MeV range and nuclear fission produces fission fragments with tens of MeV of kinetic energy. In all cases the speed of the particle is much less than the speed of light. For example, a 5 MeV α-particle has $(v^2/c^2) \approx 2.7 \times 10^{-3}$. Protons used in medical treatment may have kinetic energies as high as hundreds of MeV, but even then $(v^2/c^2) \approx 0.2$. Certainly for naturally occurring charged particles – except for electrons, which must be studied separately, and muons in *cosmic rays* (see §20.5.2) – the nonrelativistic approximation is sufficiently accurate for our purposes.

---

[2] For compounds, I is estimated by $n \ln I = \sum_j n_j \ln I_j$, where n_j is the electron density of each species and $n = \sum_j n_j$. For water, $\ln I = 4.31$.

The terms in $|dE/dx|_{\text{NR}}$ have simple physical origins. Heavy charged particles exert electromagnetic forces on electrons in the matter through which they propagate. The energy transferred to the electron is proportional to $(\Delta p)^2/2m_e$,[3] where Δp is the momentum delivered to the electron. Δp in turn is proportional to the strength of the electromagnetic force the electron experiences, which is proportional to $ze^2/4\pi\epsilon_0$, times the heavy particle's transit time $\sim 1/v$ through the region of appreciable force. Thus $\Delta E \sim (ze^2/4\pi\epsilon_0)^2/m_e v^2$. The ionization cuts off when the energy transferred to the electron is no longer greater than the mean ionization energy. This feature is accomplished by the logarithm, which vanishes when $2m_e v^2 = I$ (Problem 20.1). Ionization ceases at speeds below this limit. The logarithm arises from an integral over the impact parameter – the distance of closest approach of the heavy particle to the atom. Finally, the number of electrons available to ionize is proportional to the electron density nZ.

The dependence of the stopping power on the material appears in the electron number density nZ and the ionization threshold I. For light elements such as carbon, oxygen, nitrogen, and calcium (but not hydrogen) that dominate the electron density of living tissues, $Z \approx A/2$, so the electron density is proportional to the mass density. Therefore, the *mass stopping power* S, obtained by dividing $|dE/dx|$ by the mass density, is relatively independent of the precise nature of the material, aside from the factor of I in the logarithm. Substituting $A = 2Z$ and introducing the physical values of $m_e, e, 4\pi\epsilon_0$, etc., we have the approximation

$$S(\beta, z, Z) \equiv -\frac{1}{\rho}\frac{dE}{dx}\bigg|_{\text{NR}} \tag{20.4}$$

$$\cong 0.153\frac{z^2}{\beta^2}\left(\ln(1.02\times10^6) - \ln(I[\text{eV}]) + \ln\beta^2\right)\frac{\text{MeV cm}^2}{\text{g}},$$

(where $\beta = v/c$) for light nuclei other than hydrogen. For water, where $Z/A = 5/9$ rather than $1/2$, the coefficient 0.153 is replaced by 0.170. To obtain $|dE/dx|$ (in units of MeV/cm) for a given material, S must be multiplied by the density (in units of g/cm^3). S is plotted as a function of β for α-particles ($z = 2$) for several different materials in Figure 20.3(a), where one can see that the Z-dependence is weak except at small β where I becomes important.

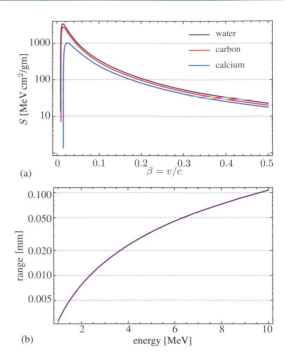

(a)

(b)

Figure 20.3 (a) The mass stopping power of α-particles in water, carbon, and calcium, as a function of the speed of the α-particle. Since the density of water is 1 g/cm^3, the mass stopping power equals the stopping power for water in these units. (b) The range of α-particles in water as a function of energy.

The **range** of a heavy charged particle is the distance it travels before energy loss by ionization brings it to rest. Because a heavy particle ionizes so many atoms as it slows, fluctuations average out, and the range is a quite well-defined quantity that can be computed from $|dE/dx|$ (Problem 20.3). For example, the range of α-particles in water is shown in Figure 20.3(b). As the figure illustrates, the range of low-energy ($v^2 \ll c^2$) α-particles in ordinary matter is very short. Typically a sheet of paper is sufficient to stop the α-particles that are emitted by radioactive nuclei. The US National Institute of Standards and Technology (NIST) maintains a database [105] where the stopping power and range for protons and α-particles in various materials can be found.

Another important feature of heavy charged particles propagating through matter is that they deposit most of their energy near the end of their trajectory. This is due to the factor of $1/v^2$ in eq. (20.3), which grows large as v decreases before it is cut off by the logarithm. Figure 20.4 shows $|dE/dx|$ as a function of distance along the track of a 5.49 MeV α-particle propagating through air.

---

3 According to eq. (20.2), the electron's motion is nonrelativistic, $\Delta E_{\max} \ll m_e c^2$, as long as the heavy charged particle is itself nonrelativistic, $E \ll Mc^2$, as is the case for fission fragments and α-particles emitted in radioactive decays.

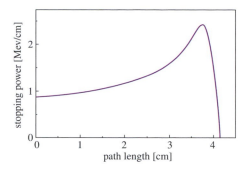

Figure 20.4 $|dE/dx|$ for α-particles in air with an initial energy of 5.49 MeV, as a function of path length as the α-particle slows. The ionization density, which describes the destructive effect of the radiation, follows the same curve.

The pronounced peak at the end of the track is known as the **Bragg peak** after British physicist William Bragg who discovered it in 1903. The number of ionized electrons is proportional to the energy loss, so it too peaks at the end of the charged particle's path. The relative localization of ionization at the end of a heavy charged particle's range has led to use in cancer therapy, where the objective is to deposit as much energy as possible in the tumor to be destroyed (see Box 20.1).

Heavy Charged Particles

Heavy charged particles lose energy rapidly in matter as they ionize large numbers of atomic electrons. They have a well-defined range; for α-particles emitted in radioactive decays, the range is significantly less than a millimeter in water. Their ionization is concentrated in the Bragg peak at the end of their trajectory.

Because α-particles deposit a large amount of energy in a small volume, they can kill cells in living organisms. When they originate outside the organism, α-particles do little damage because they cannot penetrate the dead tissue on the surface of the skin. If α-emitting nuclei are ingested or inhaled, however, they can do terrible damage (see Example 20.2).

20.2.2 Energetic Electrons from β-decay

Energetic electrons are produced in β-decay of unstable nuclei, $^{A}[Z] \rightarrow {}^{A}[Z+1] + e^- + \bar{\nu}_e$. Like α-particles emitted in nuclear decays, β-decay electrons have energies in the few MeV range. Since $m_e c^2$ is only 0.511 MeV, these electrons are often highly relativistic, $v \approx c$ (Problem

20.8). Energy loss via ionization by an electron passing through matter is described by the relativistic generalization of Bethe's formula, which we have not presented here [80]. There are, however, several important differences between electron and heavy particle propagation through matter that complicate matters and make Bethe's formula less useful. First, as described in §14.3.1, the electrons produced in β-decay have a continuous distribution of energies as opposed to a single energy as assumed in Bethe's formula. Second, electrons can lose a much larger fraction of their energy (up to 100%) in any collision, and the secondary electrons produced this way can have energies of the same magnitude as the original electrons. Both primary and secondary electrons can go on to ionize further electrons. Third, for the same force, acceleration is inversely proportional to mass, so colliding electrons accelerate far more than protons or α-particles. Accelerating charged particles radiate *braking radiation*, known by its German name **bremsstrahlung**, proportional to the square of their acceleration. Photons produced in bremsstrahlung, in turn, can go on to ionize other electrons. Radiative energy loss dominates for very-high-energy electrons, but at the energies typical of β-decay, ionizing collisions dominate. In sum, a β-decay electron slowing down in matter gives rise to an **electromagnetic shower** that includes very many electrons and photons. A simulation of an electron shower in a particle detector is shown in Figure 20.5. The rate at which the shower transfers energy to the medium is not described quantitatively by Bethe's formula, eq. (20.3). Although quantitative treatments exist [80], we content ourselves with an empirical description of electron energy loss.

The small mass of the electron makes a huge difference in its rate of energy loss compared to that of a heavy particle with the same energy. Compare, for example, an electron and an α-particle, each with energy 5 MeV.

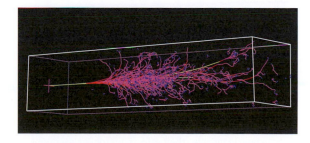

Figure 20.5 A simulation of the electromagnetic shower produced by an 8 GeV electron propagating through a *cesium iodide* (CsI) crystal (outlined in white). (Credit: Sven Menke www.mppmu.mpg.de/~menke/eiss)

Box 20.1   Proton Beam Therapy

The sharpness of the *Bragg peak* and the relatively narrow lateral spread of the ionization path has led physicians to advance the idea of destroying tumors with protons as opposed to the X-rays or γ-rays (photons) that are usually employed in radiation therapy for cancer. The advantage is clear: much of the energy can be deposited directly at the site of the tumor, thus sparing the surrounding tissue. Photons, in contrast, deposit energy in decreasing amounts from one side of the body to the other and the electromagnetic showers they generate spread laterally and damage surrounding tissue. To selectively deposit energy in a tumor using photons, the beam must be rotated with respect to the patient in order that, on average, the tumor receives most of the energy.

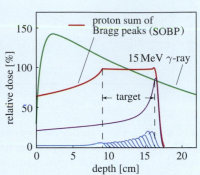

Often in proton beam therapy, to expose the entire tumor to a sufficient dose, protons of several energies are used. As a result the tissue in front of the tumor receives a significant radiation dose, although the tissue beyond the tumor does not. The figure above shows the Bragg peak for a 150 MeV proton beam, which combines with lower-energy doses (blue) to give an integrated dose or *sum of Bragg peaks (SOBP)* approximately 7 cm long (shown in red). For comparison, the relative dose for a 15 MeV photon is shown in green.

A *particle accelerator* known as a *cyclotron* is needed to produce protons of energy sufficient to treat tumors in human subjects (Problem 20.7). For many years the only large-scale proton therapy program was run by physicists at the Harvard Cyclotron Laboratory in Cambridge Massachusetts, in collaboration with the Massachusetts General Hospital in Boston. Advances in magnet technology, however, have allowed the cyclotrons that accelerate the protons to be made smaller and more cheaply, and proton therapy is now available at many locations throughout the world.

As we worked out earlier, $v^2 \approx 2.7 \times 10^{-3} c^2$ for the α-particle. For the electron, however, we find $v^2 \approx 0.99 c^2$. The factor of $1/v^2$ in $|dE/dx|$ (20.3) survives the complications described in the previous paragraph, so a 5 MeV α-particle creates ionization at a rate approximately 400 times greater than an electron of the same kinetic energy. Because of this, electrons lose energy much more slowly than heavy charged particles until they get to energies well below 1 MeV. The total rate of energy loss in water, both collisional and radiative, for electrons with energies between 10 eV and 10 MeV is shown in Figure 20.6(a).

The concept of the *range*, defined for heavy charged particles, must be reconsidered for the case of an electron incident on matter since it is not possible to distinguish the original electron from secondary electrons produced by ionization. Instead the range can be defined as the distance over which the incident energy is deposited. The range for electrons with energies between 10 eV and 10 MeV is shown in Figure 20.6(b). Since individual collisions play such a significant role, their effects no longer average out as they do for a heavy particle, and the point where an individual electron of a given energy *ranges out* is much more variable than the end point of a heavy particle's path, a phenomenon known as *straggling*.

Minimal shielding stops even the most energetic nuclear α-particles. In contrast, it is more difficult and often more important to find ways to provide shielding against energetic electrons. Because $|dE/dx|$ is proportional to Z, elements with large atomic number such as lead are often chosen to shield against β-decay electrons. (They also do a good job on photons – see below.) Lead containers, for

Electrons

Electrons emitted in β-decay penetrate much further into matter than heavy charged particles with the same kinetic energy. A 10 MeV electron, for example, has a range of order 10 cm in water, compared to a range of ~0.1 mm for a 10 MeV α-particle.

Primary and secondary electrons combine into a complex shower. Each electron, near the end of its trajectory, ionizes many atomic electrons. Although each individual electron's ionization exhibits a Bragg peak, the shower does not.

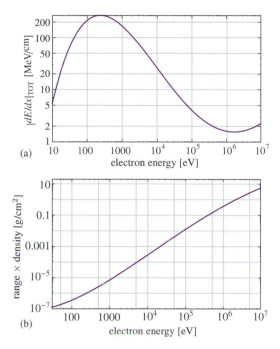

(a)

(b)

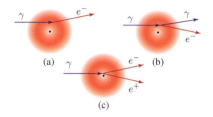

Figure 20.7 Photon interactions in matter: (a) the photoelectric effect – the photon is absorbed by an atomic electron; (b) Compton scattering; (c) pair creation

Figure 20.6 (a) $|dE/dx|$ for electrons in water. Note that high-energy electrons lose energy slowly, but electrons with energies below a few keV lose energy rapidly and create much ionization. The rise of $|dE/dx|$ above 1 MeV is due to relativistic corrections to eq. (20.3) that we have ignored. (b) The range of electrons R in water; since range varies inversely with density, the quantity ρR, which does not depend on density, is plotted. Data from [80].

example, are used to transport both β and γ sources used in medical therapy.

20.2.3 Photons: γ-rays and X-rays

γ-rays are produced in nuclear decays when an α- or β-emission leaves the daughter nucleus in an excited quantum state that subsequently de-excites by emitting a γ-ray. γ-rays are also produced when neutrons are absorbed by nuclei. Also, lower-energy X-rays are produced when energetic electrons pass through matter. This is how X-rays are produced artificially, for example for medical or dental diagnostic use: electrons are accelerated though a $\sim$100 kV potential difference and then directed at a heavy metal target like tungsten. As the electrons scatter off the tungsten atoms, they knock electrons out of inner shells. Outer atomic electrons then emit X-rays as they fall into the now-empty energy levels. The electrons also bremsstrahlung X-rays as they accelerate in the strong atomic electric fields. All these X-rays are then collimated and directed at the object of interest.

Unlike α- and β-particles, photons are electrically neutral so they do not cause ionization directly. Rather,

photons liberate electrons from atoms in one of three possible quantum mechanical processes: *photoelectric effect*, *Compton scattering*, and *pair production*, illustrated schematically in Figure 20.7. Photons with insufficient energy to ionize atoms can also be absorbed into atomic or molecular excitations, as discussed further in §23.4.

In the photoelectric effect, an electron bound in an atom absorbs the photon, ionizing the atom and creating a high-energy electron that produces its own electromagnetic shower. In Compton scattering, a photon scatters from an atomic electron. The electron absorbs some, but not all of the photon's energy. The resulting energetic photon and electron both go on to interact in the material. Finally, in pair production, a photon with enough energy, $E > 2m_ec^2 = 1.022$ MeV, turns into an electron and a positron – one of the mysterious processes that can occur when relativity is combined with quantum mechanics. An isolated photon cannot create an electron–positron pair because it is not possible to conserve both energy and momentum in the process, but pair production can occur in the presence of a nucleus that absorbs the excess energy and momentum. Each of these three mechanisms results in the liberation of an electron, plus in some cases a positron and/or another, lower-energy photon. These particles go on to cause further ionization, generating a shower of electromagnetic energy similar to that generated by an initial electron.

Note that there are significant differences between the ionization pattern due to a photon and an electron. An electron begins to ionize as soon as it enters matter and ionizes according to Figure 20.6(a). If the electron's energy is greater than roughly 10 keV it ionizes little at first, but copiously as it comes to rest. Because photons are neutral, they pass through matter unimpeded until they give rise to energetic charged particles through one of the three processes mentioned above. This occurs with a probability that is independent of the past history of the photon. Thus the photon's survival probability is governed by a Poisson distribution (see Box 15.1) and falls exponentially with distance. Therefore, unlike heavy charged particles and electrons, a beam of photons of a given energy does not have a well-defined range, but is instead characterized by an **attenuation coefficient** μ that sets the scale for the exponential decay of the original beam intensity,

$$I(x) = I_0 e^{-\mu x} . \qquad (20.5)$$

Both absorption and scattering contribute to attenuation of a beam of photons. The attenuation coefficient μ depends on the nature of the material and also on the photon's energy. A beam of photons falls to $1/e$ of its original intensity in a distance $1/\mu$, which is known as the **attenuation length**.[4] The same equation governs the attenuation of longer wavelength (e.g. visible) light passing through an absorbing medium, where it is known as the *Lambert–Beer law* (see eq. (23.11)). In that context, the attenuation coefficient is often referred to as the *absorption coefficient* and is sometimes denoted by κ or α.

Often, instead of μ, photon absorption is characterized by the *half-value thickness* or **half-value layer** (HVL), the depth of material needed for half the photons to be absorbed, $t_{1/2} = (\ln 2)/\mu = 0.693/\mu$. The HVL for a 1 MeV γ-ray in aluminum is about 4.2 cm, which is about 20 times greater than the range of a 1 MeV electron and 10^4 times greater than the range of a 1 MeV α-particle. The attenuation coefficients of photons with energies from 10 KeV to 100 MeV in water are shown in Figure 20.8. While it is possible to completely stop a beam of charged particles by using an attenuator slightly thicker than its range, the same is not true for γ-rays, which attenuate exponentially. An absorber of many half thicknesses is needed to safely shield a strong γ-ray source.

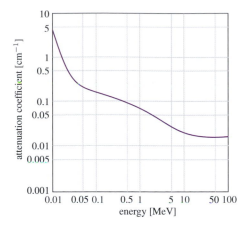

Figure 20.8 The attenuation coefficient μ for photons in water. $1/\mu$ is the distance over which the radiation intensity in water drops by a factor of e. Data from NIST Physical Measurement Laboratory, www.nist.gov/pml.

X-rays and γ-rays

X-rays and γ-rays penetrate farther into matter than electrons and much farther than heavy charged particles. The intensity of a photon beam decreases exponentially through the material, as does the ionization it causes.

20.2.4 Neutrons

Free neutrons are produced in nuclear fission and fusion, but are otherwise rare in nature. They are generated

copiously in nuclear reactors, as discussed in §18. They are emitted with MeV energies. After emission they are either absorbed by nuclei or they may be slowed down by multiple scatterings off other nuclei until they have thermal energies, of order $k_B T_{ambient} \approx 1/40$ eV. The fate of a neutron propagating in tissue is similar to its fate propagating in a carbon or water moderator in a reactor (since living organisms contain primarily carbon, hydrogen, and oxygen nuclei). Thus, neutrons can wander deep into matter, slowing down as they scatter off the nuclei of atoms until eventually most are absorbed by a nucleus. The nucleus then usually de-excites by emitting one or more γ-rays with energies typical of the neutron's binding energy, usually several MeV (Problem 20.11). Neutrons scattering from atoms in materials can damage their atomic structure, introducing defects into otherwise regular crystal lattices or disrupting complex organic molecules. The γ-rays emitted upon capture then do further damage as they ionize electrons and are eventually absorbed. The result is that the effects of neutrons penetrate deep into matter, many centimeters at least.

20.3 Measures of Radiation

The quantitative description of radiation exposure used in various contexts is quite complex. There are many sources of complexity: first, simple quantitative measurements do not capture the variation in effects due to the type of radiation and the different tissues in which it might be absorbed; second, different scientific and technical communities have introduced different measures of exposure appropriate to

---

[4] The simple exponential decay of eq. (20.5) is only an approximation. In practice, Compton scattering of photons and emission of secondary photons by charged-particle brehmsstrahlung modify the beam attenuation. Equation (20.5) is sufficient for estimates of human radiation dose and shielding (see [80] for further discussion).

their own needs; third, the choice of what to measure and how to quantify it is linked to exposure limits mandated by regulatory agencies based in part on economic and political considerations; and finally, traditional units have relatively recently (and somewhat reluctantly) been replaced by SI units that are less well known. For definiteness, we adopt the terminology of [106]. We describe several of the most useful measures of radiation exposure, beginning with the most quantitatively precise units.

20.3.1 Activity

Activity is a measure of the rate of radioactive emission from a source. The SI unit of activity is the **becquerel** (Bq), which is defined as one radioactive decay per second. The concept of *activity* is thus precisely defined, and is independent of the energy or type of emission as well as independent of the material the radiation is impacting. An older – and still often used – measure of activity is the **curie** (Ci), which is roughly the activity of 1 gram of radium-226 (^{226}Ra). The conversion between the two units is now defined as: $1\,\mathrm{Ci} \equiv 3.7 \times 10^{10}\,\mathrm{Bq}$. Therefore the curie is a *much greater* level of activity than the becquerel.

The *activity* of a sample is often used as a measure of the mass of a radionuclide present, since the activity is proportional to the number of nuclei in the sample, $dN/dt = N/\tau$. Thus, for example, a 20 Ci ($= 7.4 \times 10^{11}$ Bq) source made of ^{60}Co ($\tau = 2780$ d), as might be used in cancer treatment, contains $M \cong mN = m\tau\, dN/dt \cong 18$ mg of ^{60}Co, where m is the molar mass of ^{60}Co (Problems 20.10 and 20.13).

20.3.2 Absorbed Dose

Absorbed dose (D) is the measure of radiation energy absorbed per unit mass of absorbing material. Unlike *activity*, which depends only on intrinsic properties of the source of radioactivity, *absorbed dose* and all the other measures of radiation exposure we describe depend on the time-integrated flux of radiation and the interaction cross section between the radiation and the material through which it is traveling.

The SI unit of absorbed dose is the **gray** (Gy), which is defined as the absorption of one joule of radiation energy per kilogram of matter, 1 Gy = 1 J/kg. A joule per kilogram may seem like a small amount of energy. It would raise the temperature of a kilogram of water only 0.000 25 °C. One joule, however, is sufficient energy to ionize $\sim 5 \times 10^{17}$ hydrogen atoms and to do significant damage to living tissue. The gray replaces the **rad** (rad), which is defined in a similar way in the centimeter–gram–second system: 1 rad is defined as the absorption of 100 ergs per gram of matter. Thus 1 Gy = 100 rad.

Absorbed dose, like stopping power, is an imperfect measure of the biological effects of radiation. In general, a given absorbed dose has more severe biological effects when it is delivered in a localized area, though beyond a certain point irreparable damage is inevitable. Thus, high-LET radiation, such as that generated by heavy charged particles, generally has greater biological impact than low-LET radiation. This effect is captured by the *equivalent dose* defined in the following section. Nonetheless, in the most straightforward situation of whole-body exposure to penetrating radiation such as high-energy γ-rays, absorbed dose is often used as a measure of radiation exposure, and is described in units of Gy.

20.3.3 Equivalent Dose

Exposure to ionizing radiation can lead to both immediate and long-term adverse effects on the health of humans and other organisms. The impact of radiation exposure depends upon both the amount of ionization, which varies with the type of radiation, and the particular tissue involved. Thus, the absorbed dose received by a living organism is often labeled with the radiation type R and/or the tissue type T in which the radiation is absorbed, $D_{R,T}$. Although the absorbed dose may be different at each point, we take $D_{R,T}$ to be the average of the absorbed dose over the tissue or organ T.

Experiments show that high-LET radiation produces more irreparable damage to living tissues than the same dose of weakly ionizing radiation. The **relative biological effectiveness** (RBE) of radiation is a quantitative measure of this effect. The RBE of a dose D_R of any type of radiation for a given biological effect, such as cell death, is defined as the ratio RBE $= D_X/D_R$, where D_X is the dose of radiation of a standardized reference type that produces the same biological effect. "X" may be chosen to be a low-LET form of radiation such as 100 keV X-rays, so that the low-LET limit of RBE is one. Although the precise relation between RBE and LET depends on the biological effect of interest and the precise types of radiation involved, RBE typically increases relatively slowly with LET at first, rises to a peak, and then drops again beyond the point where irreparable damage is likely, which typically occurs around 100 keV/μm. A rough sketch of the relation between RBE and LET is given in Figure 20.9.

Since LET (and therefore RBE) is somewhat difficult to measure and quantify precisely outside of a laboratory environment, a much simpler semi-quantitative scheme has been adopted to take into account the biological effects of different types of radiation [106]. The *equivalent dose* to a given organ is defined by weighting absorbed dose $D_{R,T}$

Table 20.1 Radiation weighting factors w_R. We use the system recommended by the International Commission on Radiological Protection [106]. See Figure 20.10 for neutrons.

Type of radiation	Energy range	w_R
γ, $e^{\pm}$, muons	All energies	1
Protons, charged pions	All energies	2
α, fission fragments, heavy nuclei	All energies	20

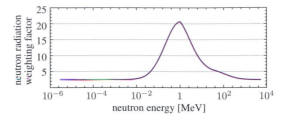

Figure 20.10 Radiation weighting factor for neutrons w_R from 1 MeV to 10 GeV [106].

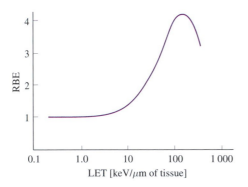

Figure 20.9 Sketch of an example of the relative biological effectiveness of ionizing radiation as a function of LET. The actual RBE varies with the biological effect under investigation. Note the turnover at high LET due to "overkill." After [77].

by a **radiation weighting factor** w_R.[5] The weighting factor reflects the typical *relative biological effectiveness* (RBE) of each specific type of radiation. The weighting factors for common forms of radiation other than neutrons are listed in Table 20.1. The weighting factor for neutrons is more complex and is shown in Figure 20.10. Note that the weighting factors are defined relative to the low-LET standard of e.g. photons and electrons.

To obtain the **equivalent dose** H_T for a given absorbed dose in tissue T, we multiply the absorbed dose $D_{R,T}$ of radiation of each type R by the weighting factor w_R and sum over R,

$$H_T = \sum_R w_R D_{R,T} . \qquad (20.6)$$

In the SI unit system, where the absorbed dose is measured in grays, the equivalent dose obtained from eq. (20.6) is measured in **sieverts** (Sv). An older measure of equivalent dose, still used frequently, is the **rem** (rem) (which originally stood for *Roentgen Equivalent Man*) with a

conversion factor 1 rem = 0.01 Sv. The sievert, like the gray, formally has units of joules per kilogram. The underlying physical connection between a sievert and a joule per kilogram is lost, however, when the energy deposition is weighted by w_R. Thus equivalent dose should be expressed in either sieverts or rems, not in joules per kilogram.

20.3.4 Effective Dose

Studies have shown that the risk of damage leading to cancer or serious genetic disease is greater in tissues with high levels of dissolved oxygen and in tissues where cells divide frequently. Although many radiation sources expose the whole body equally, others affect particular organs preferentially. Radiation therapy for cancer, for example, targets the organ that is the site of the cancer. Also, particular radioisotopes accumulate in specific organs; ingested iodine, for example, is concentrated in the thyroid. The *effective dose* provides a way to convert the equivalent dose to specific organs into a common standard that can be used to assess overall risk for different patterns of exposure.

The **effective dose** is obtained by summing the equivalent doses H_T to various tissues, weighted by a **tissue weighting factor** w_T that expresses the relative sensitivity of different organs. Thus, the effective dose is

$$E = \sum_T w_T H_T = \sum_T w_T \left(\sum_R D_{R,T} w_R \right) . \qquad (20.7)$$

The weighting factors for different organs are summarized in Table 20.2. The factors w_T add to one when summed over all organs, so that the effective dose for the uniform exposure of the whole body is identical to the equivalent dose. Effective dose, like equivalent dose, is measured in sieverts (or traditionally, in rem), again with 1 Sv = 100 rem. The radiation measurement units described here are summarized in Table 20.3. Notice that when an individual is exposed uniformly to an external source of γ- or β-radiation the effective dose and equivalent dose (in sieverts) are equal to one another as well as to the absorbed

---

[5] Formerly known as the *quality factor*.

Example 20.1   Tritium Leak

Some of the hydrogen in cooling water at nuclear reactors absorbs two neutrons and becomes *tritium* (^{3}H), which is radioactive. Tritium decays with a half-life of $t_{1/2} = 12.3$ y into ^{3}He by emitting a β-ray with an average energy of 5.7 keV.

Suppose that a leak at a nuclear power reactor resulted in tritium contamination in well water at a level of 1.2×10^{-10} g/L, and further suppose that a person drinks this water all of their life. *Roughly how much radiation (effective dose) does this person receive in Sv/y? To what fraction of US natural background radiation exposure does this correspond?*

Since the person drinks nothing but tritiated water, a substantial fraction of the water in their body is continuously contaminated. For simplicity we assume that 1/2 of the water in their body is contaminated, with the rest coming from ingested food and respiration. Since roughly 70% of the human body is water and water is relatively evenly distributed throughout the body, we assume that every kilogram of the person's body contains 0.35 L of tritiated water. According to Figure 20.6, the range of a 5.7 keV electron in water is roughly 10^{-4} cm, so it is safe to assume that all of the β-decay energy is absorbed in the person's body.

To compute the effective dose, we first compute the activity (Bq/L) of the contaminated water. With an atomic mass of $A = 3$, 1.2×10^{-10} g of tritium corresponds to

$$N(^3\text{H}) = (1.2 \times 10^{-10}\ \text{g/L})/(3\ \text{g/mol})(6.0 \times 10^{23}\ \text{tritium nuclei/mol}) \approx 24 \times 10^{12}\ \text{tritium nuclei/L}.$$

A half-life of 12.3 y corresponds to a lifetime $\tau = t_{1/2}/\ln 2 \approx 5.6 \times 10^8$ s. Since radioactive decay obeys an exponential law $N = N_0 e^{-t/\tau}$, the activity of the tritium is $A = -dN/dt = N/\tau \cong (24 \times 10^{12}/\text{L})/(5.6 \times 10^8\ \text{s}) \approx 42\,000$ Bq/L. At 42 000 Bq/L and 0.35 L/kg, the tritiated water leads to an absorbed dose

$$P = (0.35\ \text{L/kg})(5.7\ \text{keV/Bq s})(1.6 \times 10^{-19}\ \text{J/eV})(42\,000\ \text{Bq/L}) \approx 1.3 \times 10^{-11}\ \text{J/kg s} = 1.3 \times 10^{-11}\ \text{Gy/s}.$$

The person's annual absorbed dose per is therefore

$$D \cong (1.3 \times 10^{-11}\ \text{Gy/s})(3.2 \times 10^7\ \text{s/y}) \approx 4.2 \times 10^{-4}\ \text{Gy/y}.$$

Finally, since the radiation weighting factor for electrons is one and since all tissues are exposed equally, this translates directly into an *effective dose* of $E \sim 0.42$ mSv/y, which is about 14% of the US yearly natural background radiation dose of $E_{\text{US}} \cong 3.1$ mSv/y (§20.5).

Table 20.2 Tissue weighting factors for individual organs, from [106]. Note that the sum over all organs of w_T is one.

Organ/tissue	Number of tissues	w_T	Total contribution
Lung, stomach, colon, breast, bone marrow, remainder*	6	0.12	0.72
Gonads	1	0.08	0.08
Thyroid, esophagus, bladder, liver	4	0.04	0.16
Bone surface, skin brain, salivary glands	4	0.01	0.04

*Remainder – 13 tissues in each sex – including adrenals, heart, kidneys, pancreas, and spleen. For a complete list, see [106].

dose (in grays). This follows from eq. (20.7), when $w_R = 1$ and $D_{R,T}$ is independent of T.

Prompt (short-term) exposure to large amounts of penetrating radiation such as high-energy γ-rays from external sources is generally expressed in grays or occasionally in sieverts, conveying equivalent information. The long-term risk of cancer or other *stochastic effects* (§20.4.2) of low-level radiation exposure is usually expressed in terms of *effective dose* and measured in sieverts.

An extension of the concept of *effective dose* is useful for ingested radioactive sources. When a quantity of a particular radioactive source is ingested and concentrates in specific organs, its activity decays exponentially with time; it also may be excreted from the body over time. The **committed effective dose** $E(t)$ takes account of these effects. $E(t)$ is the effective dose received by an individual, integrated over a time interval t, from ingesting

Table 20.3 Some radiation measurement units.

Quantity	SI unit	Traditional unit	Conversion
Activity	becquerel (Bq)	curie (Ci)	1 curie = 3.72×10^{10} Bq
Absorbed dose	gray (Gy = J kg^{-1})	rad	1 rad = 0.01 Gy
Equivalent dose	sievert (Sv)	rem	1 rem = 0.01 Sv
Effective dose	sievert (Sv)	rem	1 rem = 0.01 Sv

a specific radioactive source at time zero. When not otherwise specified, the commitment time is taken to be 50 years for adults and 70 years for children. Calculation of $E(t)$ requires information on the way the body metabolizes the radionuclide in question. Values for $E(t)$ are usually given *per unit intake* in units of Sv/Bq and can be found in [107]. An application of *committed effective dose* is given in Example 20.2.

Finally, we mention the **collective effective dose** S, which integrates the effective dose received over a group of individuals exposed during a given event or over a given time period. Collective effective dose is measured in *person-sieverts* and is used to provide estimates of the health impact when a large population is exposed to radiation under well-defined conditions. The use of *collective effective dose* to integrate the effects of very low radiation exposure over a very large population has been subject to some controversy (see Box 20.2).

20.4 Biological Effects of Radiation

Radiation damages living tissues by ionizing atoms and by breaking chemical bonds in complex molecules such as proteins and DNA (deoxyribonucleic acid). The effects of radiation are generally divided into *deterministic* and *stochastic* effects. **Deterministic effects**, such as skin

reddening, tissue damage, and organ failure, arise from prompt (immediate) exposure and increase in *severity* in proportion to the absorbed dose of radiation. **Stochastic effects**, particularly cancer and genetic damage, may arise years after exposure and increase in *likelihood*, but not in severity, with increased exposure. Biological effects are further classified as **somatic** or **hereditary**, depending on whether the effects appear in the individuals exposed (somatic) or in their descendants (hereditary).[6] Broadly speaking, the deterministic effects of moderate to severe radiation exposure are well established and non-controversial. In contrast, stochastic effects – in particular, the extent to which exposure to low radiation doses leads to an increase in the incidence of cancer – remain controversial even after many years of study and debate. We consider these two categories in turn.

20.4.1 Deterministic Effects of Radiation

The deterministic effects of radiation exposure result from the killing of large numbers of cells, leading to tissue damage and in severe cases to organ failure and death. These effects, known as **acute radiation syndrome** (ARS) or **radiation poisoning**, occur when an organism is exposed to a large amount of radiation over a short time – minutes or hours. The first symptoms of ARS are usually nausea and vomiting. Data on deterministic effects comes primarily from the study of ARS after the bombing of Hiroshima and Nagasaki, where approximately 20 000 people died from acute radiation exposure; this data has been augmented by a relatively small number of cases of acute radiation syndrome since then.

The deterministic effects of radiation are generally described as a function of *absorbed dose*, measured in grays (Gy). As discussed above, this is an appropriate standard for whole-body exposure to penetrating radiation such as γ-radiation. Deterministic effects of other types of radiation are more complicated. Various approaches, such

---

Measures of Radiation Exposure

Activity – decays per second – is measured in *bequerels*. *Absorbed dose* – joules per kilogram of absorbed energy – is measured in *grays*. *Equivalent dose* and *effective dose* introduce weighting factors for the type of radiation and the affected tissue respectively. These quantities reflect differences in ionization effects and the potential for an absorbed dose of radiation to lead to subsequent development of cancer or serious genetic damage.

---

[6] We omit effects on embryos or fetuses (*teratogenic*) which fall into a separate category that should be considered in a complete analysis.

as incorporating *relative biological effectiveness* (RBE) weighting factors, are used in such cases, though the exposure is still generally quoted in units of Gy (but *not* J/kg) in evaluating deterministic effects.

The correlation between dose and illness or mortality for acute exposure is accurately known. There is a threshold dose below which no immediate effects of acute exposure have been detected. According to [106], there is no evidence for "clinically relevant functional impairment" to any tissues for acute doses below 100 mGy. Prompt exposure to a higher absorbed dose, however, leads to progressively more serious damage. A **lethal dose**, defined as the whole-body dose from penetrating ionizing radiation that leads to 50% mortality within 30 days (without treatment), is 2.5–4.5 Gy.

The deterministic effects of acute radiation syndrome are well illustrated by the example of exposure to gamma rays, which because of their penetrating power, give an approximately uniform whole-body dose. Exposure to an absorbed dose of less than 250 mGy produces no serious short-term symptoms. Exposures between 250 mGy and 1 Gy may produce symptoms such as nausea and loss of appetite; recovery is typically total. Acute exposure to between 1 and 3 Gy leads to more severe gastrointestinal symptoms as well as a decrease in white blood cells and increased susceptibility to infection. Recovery is probable, but not certain. Between 3 and 6 Gy symptoms rapidly become more severe. Without treatment, fatalities occur at 3.5 Gy, and the death rate reaches 50% (within 30 days) at 4 Gy [80]. With treatment the death rate can be kept below 50% for doses under about 6 Gy, and patients receiving medical treatment have survived exposures up to ~10 Gy [108].

Living organisms have a significant capacity to repair the damage done by acute doses of radiation. As a result, the deterministic effects of radiation exposure are not cumulative. Repair mechanisms operate on a time scale of hours to days, so a dose that would be fatal if delivered abruptly can be tolerated if spread over several days. This effect is illustrated in Figure 20.11, where data on mortality in mice as a function of both dose and dose rate are shown. The reduction of mortality when a dose is spread out over time is an important consideration in medical use of radiation to treat tumors [109, 110]. *Dose fractionation* allows very large doses of radiation to be delivered by spreading the exposure over several days, thereby minimizing the extent of ARS experienced by the patient.

20.4.2 Stochastic Effects of Radiation

Stochastic effects of radiation exposure, particularly cancer and inherited disease, involve complex biochemical

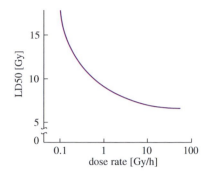

Figure 20.11 Evidence for radiation damage repair in living organisms. The total dose (in Gy) at which mice experience 50% mortality (LD50) as a function of dose rate (Gy/h). Note the steep rise in survival as the dose rate decreases, giving repair mechanisms more time to work. After [111].

processes. In order to survey what is and is not known about the carcinogenic and genetic effects of radiation, it is necessary first to describe some aspects of what is understood regarding the mechanisms by which ionizing radiation disrupts biological systems.

Ionization caused by radiation may directly damage a molecule such as DNA that is critical to cell function. Since mammalian cells consist of 70 to 85% water, however, it is more common that a given radiation-induced interaction causes the ionization or excitation of a water molecule. This does not directly affect cell function, but leads to the production of ions and *free radicals* that can go on to damage proteins and DNA.

When an ionizing particle passes through water, it leaves a trail of ionized water H_2O^+, free electrons e^-, and electronically excited water molecules $H_2O^\star$. Before moving away from the location where they were created, these species very quickly ($\sim 10^{-14}$ s) decay ($H_2O^\star$) or react (H_2O^+ and e^-) with nearby water molecules as follows,

$$H_2O^\star \rightarrow \begin{cases} H_2O^+ + e^- \\ H\bullet + OH\bullet \end{cases}$$

$$H_2O^+ + H_2O \rightarrow H_3O^+ + OH\bullet$$

$$e^- \rightarrow e^-_{aq}, \tag{20.8}$$

where e^-_{aq} refers to an electron in aqueous solution, i.e. surrounded by a cluster of polarized water molecules that are attracted to it. $H\bullet$ and $OH\bullet$ are **free radicals** – atoms ($H\bullet$) or molecules ($OH\bullet$) with an unpaired valence electron. Free radicals can be regarded as having a dangling covalent bond (denoted by the "$\bullet$"), making them chemically very reactive.

As time goes on, these reactive species diffuse away from the original path of ionization. As they diffuse they may react with one another or interact with water molecules, leading in microseconds to an array of active molecules such as hydrogen peroxide (H_2O_2), free radicals, and ions such as OH^-. It has been shown that the subsequent interactions of these molecules, ions, and free radicals with DNA, proteins, and other biological molecules are the primary sources of cell damage. These interactions may lead to cell death or alteration in the cell's genetic code, which may in turn lead to cancer, or if the cell is a germ cell, to a mutation that can be passed on to future generations.

Cancer and heritable abnormalities are thought to be caused primarily by damage to DNA in cell nuclei. The double stranded DNA molecule can be damaged on a single strand or on both strands at a given location encoding a single "letter" of genetic information. As mentioned in connection with the deterministic effects of radiation, the cell clearly has DNA repair mechanisms that operate over time scales of hours. Sometimes damage is irreparable and sometimes the repair mechanisms do not perform correctly. It is believed, however, that these biological systems have sufficient redundancy and fail-safe mechanisms that damage at a single site on one strand of a DNA molecule will virtually never lead to cancer or heritable problems. Even isolated double-strand breaks in DNA are usually repaired by the DNA damage response system. It is believed that multiple independent "hits" producing separate mutations on a cell or its progeny typically occur before a cancer develops; this matches the observed nonlinear increase in cancer rates with age.

There is strong evidence that a cell's repair and redundancy mechanisms can be overwhelmed when the cell suffers acute exposure to high doses of ionizing radiation. This is because such exposure is known to be correlated with development of cancer later in life. The *Radiation Effects Research Foundation's* **Life Span Study** (LSS) has followed approximately 120 000 individuals who suffered *acute* radiation exposure when nuclear devices were detonated in Hiroshima and Nagasaki. More recently about 12 000 of their descendants have been added to the study, enabling the study to track the inherited effects of radiation exposure. One of the primary conclusions drawn from the LSS, and confirmed elsewhere, is that for acute exposure to doses above 100 mSv, there is a linear correlation between the effective dose received and the probability of developing cancer later in life, as illustrated in Figure

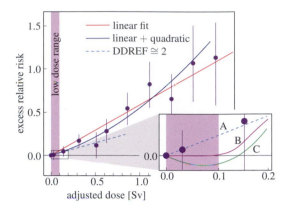

Figure 20.12 Evidence from the LSS (data points) of a linear correlation between cancer risk and effective dose at effective doses above 0.1 Sv. The linear plus quadratic fit and the dashed line illustrate the DDREF. The insert shows possible alternative behaviors at doses below 0.1 Sv discussed in Box 20.2. Adapted from [112].

20.12.[7] It is also believed that there is a similar linear correlation between effective dose and the risk of developing heritable genetic abnormalities, although the risk is now believed to be between one and two orders of magnitude smaller than the risk of developing cancer.

20.4.3 Risk Assessment and Radiation Limits
Measurements and estimates of the biological effects of ionizing radiation form the basis for prescribed limits on radiation exposure both for radiation workers and for the general public. These limits influence the public perception of what constitutes a dangerous level of radiation exposure. They also influence responses to nuclear accidents such as decisions to evacuate populations from areas where radiation exceeds prescribed levels. Several organizations review data on radiation exposure, evaluate risk, and propose limits on radiation exposure. The US *National Council on Radiation Protection and Measurements* (NCRP) Report 116 [113] currently sets the US standard, while the *International Commission on Radiological Protection* (ICRP) Publication 103 [106] sets the international standard. In addition, the US *Committee on the Biological Effects of Ionizing Radiation of the National Research Council* (BIER) and the *United*

---

7 The "excess relative risk" plotted in the figure is the excess above one of the ratio of the lifetime probability of developing cancer after the indicated effective dose to the probability at zero dose (see [112] for further information).

Example 20.2   Radiation Exposure from Ingestion of Radioactive Nuclides

Sometimes individuals or populations are exposed to radiation because they ingest or inhale radioactive materials that are absorbed into their bodies. Usually the amount of ingested material is specified by its *activity* rather than its mass. The two are related by

$$m[\text{kg}] \cong 8.87 \times 10^{-17} A\, t_{1/2}[\text{sec}]\, a[\text{Ci}],$$

where m is the mass of the radionuclide in kilograms, A is its atomic mass number, $t_{1/2}$ is its half-life in seconds, and a is the given activity in curies (Problem 20.10). Thus, for example, ingesting one μCi of ^{90}Sr (see below) corresponds to a mass of

$$m \cong (8.87 \times 10^{-17})(90)(29\,\text{y})(3.2 \times 10^7\,\text{s/y})(10^{-6}\,\text{Ci}) \cong 7.4 \times 10^{-12}\,\text{kg}.$$

The effects of the radiation depend not only on the activity of the radionuclide, the type and energy of the decay, and its radiation weighting factor w_R, but also on the tissue in which it concentrates and on whether the body excretes the radioactive material before it decays. The *committed effective dose $E(t)$*, given in units of Sv/Bq, takes into account all of these factors as well as the time t over which the dose accumulates.

Ingestion of polonium-210: *A man has ingested 10 μg of ^{210}Po. Estimate how long it would take for him to accumulate a lethal dose (see below) of radiation exposure.*
^{210}Po decays with a half-life of 138 days so, using the equation at the top of this example, 10 μg corresponds to $\approx$ 45 mCi $\approx$ 1.7 GBq. ^{210}Po decays by emitting a 5.4 MeV α-particle without any accompanying γ-rays. When they originate outside the body, such α-particles are blocked by air, skin, or clothing, and do essentially no damage to tissues. ^{210}Po taken internally, however, circulates throughout the body and concentrates in the spleen and liver, exposing those organs to highly ionizing α-particles. The absorbed dose rate is the product of the activity and the energy per decay divided by the mass of the exposed tissues. The average masses of the liver and spleen are 1.6 kg and 0.17 kg respectively, so assuming that all the ^{210}Po enters these organs,

$$D \approx (1.7 \times 10^9\,\text{s}^{-1})(5.4 \times 10^6\,\text{eV})(1.6 \times 10^{-19}\,\text{J/eV})/(1.6 + 0.17\,\text{kg}) \approx 8 \times 10^{-4}\,\text{Gy/s}.$$

At this dose rate, the man would have accumulated a dose of 5 Gy to his liver and spleen in just under two hours and would quickly exhibit symptoms of ARS. This scenario is based on a real-life situation. It took the person in question about 21 days to die.

Iodine-131 ingestion and epidemiology: *An adult woman has ingested 1 μCi of ^{131}I from an accident at a nuclear reactor. How does this affect her lifetime risk of developing a fatal cancer? Comment on the number of additional cancer deaths expected if 200 million people suffered the same exposure.*
^{131}I concentrates in the thyroid and β-decays with a lifetime of $\approx$ 8 days. The β-decay electron energy averages 182 keV, and it is usually accompanied by a 364 keV γ-ray [70]. An electron with energy 182 keV has a range of less than a millimeter in water (Figure 20.6), and is therefore absorbed in the thyroid. The γ-ray deposits a fraction of its energy throughout the body. Given this relatively complicated situation, we look in [107] for a calculated value for the (total) *committed effective dose*: $E = 22\,\mu$Sv/kBq. Since 1 μCi = 37 kBq, the person has suffered an effective dose of $(37\,\text{kBq})(22\,\mu\text{Sv/kBq}) \cong 0.81$ mSv, about 1/4 of a year's effective dose from natural background radiation in the US. Given a lifetime excess fatal cancer risk for adults of 4.1×10^{-2}/Sv (see Table 20.4), the exposed person can be estimated to have an additional fatal cancer risk of $(4.1 \times 10^{-2}/\text{Sv})(8.1 \times 10^{-4}\,\text{Sv}) \cong 3 \times 10^{-5}$ or 0.003%. This can be compared to the death rate from cancers of all types in the US, which is $\sim$20%.

If one multiplied this committed effective dose by 200 million individuals of all ages, the result would be a *collective effective dose* of 1.62×10^5 person-Sv, which corresponds to $(1.62 \times 10^5\,\text{person-Sv})(5.5 \times 10^{-2}/\text{Sv}) \sim 9000$ additional cancers. This, however, is the kind of extrapolation that the ICRP and other authorities suggest is questionable due to uncertainty regarding the linear no-threshold model (see Box 20.2).

Nations Scientific Committee on the Effects of Ionizing Radiation (UNSCEAR) regularly review scientific issues affecting radiation standards. A recently updated overview treating both the known biological effects of radiation and the way that radiation standards are set can be found in [80]. A less technical perspective can be found in [81].

The studies mentioned above have established a roughly linear correlation between effective doses of 100 mSv or more delivered in a short time and the lifetime risk of developing cancer or heritable abnormalities, with the probability of cancer increasing by roughly 10% per Sv of radiation exposure. Quantifying the precise implications of this correlation for cancer or heritable abnormalities for individuals exposed to low or moderate doses of radiation over long periods, however, presents a challenging question with no clear, simple, or concise answer: at exposures high enough to measure a statistically significant risk, the risk appears to vary with dose and dose rate and with the type of radiation, the type of cancer, and the organs exposed. Agencies with responsibility for radiation protection standards have attempted, nevertheless, to summarize the risk in relatively simple terms.

The world-average annual natural radiation dose of 2.4 mSv/y [112] is far below the level at which estimates of increased cancer risks have been experimentally established. To assign a level of risk to such low levels of radiation exposure requires a large extrapolation. The latest studies by the US National Academy of Sciences [112], by the ICRP [106], and the NCRP [113] continue to support a long-standing hypothesis known as the **linear no-threshold** (LNT) model. This model assumes that the stochastic risks from radiation exposure – of fatal cancer or severe heritable effects – are linearly proportional to dose even for very small doses. (Some aspects of the LNT model are discussed further in Box 20.2.)

Although the LNT assumes a linear correlation between dose and risk down to zero exposure, the existence of biological repair mechanisms suggests that the risk per unit exposure at low dose or low dose rate is less than that observed at high dose. Thus risk estimates at low dose are typically reduced by a factor known as the **dose and dose rate effectiveness factor** (DDREF) compared to the risk estimated at high doses and dose rates. The latest ICRP study [106] assumes a DDREF of ~2, while the US National Academy [112] study takes a DDREF of 1.5. In Figure 20.12 the curve labeled "linear + quadratic" illustrates a fit to the LSS data that has a slope at the origin

Table 20.4 2007 ICRP assessment [106] of the lifetime probability for stochastic effects ($10^{-2} \times$ probability/Sv) after exposure to radiation at a low effective dose rate. A DDREF of 2 is assumed. 1990 ICRP [114] values are shown for comparison.

Exposed population	ICRP date	Fatal cancer	Heritable effects	Total
All ages	(2007)	5.5	0.2	5.7
	(1990)	6.0	1.3	7.3
Adult	(2007)	4.1	0.1	4.2
	(1990)	4.8	0.8	5.6

Table 20.5 Some standard radiation exposure benchmarks

Effective dose	Consequences
0.29 mSv/y	Annual dose from radionuclides internal to the human body
1.0 mSv/y	Maximum permissible full-body dose to the public from human-made sources, US Nuclear Regulatory Commission (NRC)
2.4/3.0 mSv/y	World/US average annual radiation dose from natural sources
5 mSv/y	Maximum permissible full-body occupational dose for minors (US NRC)
50 mSv/y	Maximum permissible full-body occupational dose for adults (US NRC)
$10 \times Y$ mSv	Maximum permissible *cumulative* full-body dose for radiation workers at age Y (NRC)
100 mSv	Lowest full-body dose at which a statistically significant correlation with cancer has been reported
4 Sv	50% fatalities within 30 days from prompt exposure

(the dashed line) that is 1/2 of the slope measured at higher dose, or roughly 5%/Sv, corresponding to a DDREF of 2. The latest risk coefficients proposed by ICRP [106] are given in Table 20.4 as a lifetime percentage risk per sievert of exposure. The risk coefficients proposed in [112] are similar. Note that the assessment of risk of heritable effects dropped by nearly an order of magnitude between the 1990 and 2007 reports.

The recommended exposure limits put forward by the IRCP and NCRP aim primarily to protect against

---

Box 20.2   The Linear No Threshold (LNT) Model – Consequences and Controversy

The LNT model has stirred controversy for decades. This is not surprising since it is not possible to perform statistically significant, controlled studies of the connection between low doses of ionizing radiation and cancer risks in human populations, and yet such a connection has profound public health consequences. In particular, the LNT model projects many additional deaths when a large population is exposed to a very low dose of radiation. A thorough, non-technical analysis in support of the LNT hypothesis can be found in [81].

To illustrate the implications of applying the LNT model to large populations, consider the case of Finland. The average annual background radiation dose in Finland ($E_{\text{Finland}} \sim 7.5$ mSv/y) is about three times the world average (see Figure 20.13), so the average Finn receives a radiation dose of roughly 200 mSv ($\approx (7.5 - 2.4) \times 40$) beyond the world average over 40 years of adult life. The LNT model predicts an additional lifetime fatal cancer risk of roughly 0.8% (200 mSv $\times$ 4.1%/Sv) per person. In 2011, 23.5% of all deaths in Finland were due to cancer of all forms. So the LNT model attributes $0.8/23.5 = 3.4\%$ or ~ 400 of the $\sim 12\,000$ cancer deaths in Finland in 2011 to the excess background radiation in the country. To verify this prediction it would be necessary to establish a control group of Finns (to control for genetic predisposition) not exposed to background radiation and also to control for other factors such as diet, smoking, and other environmental effects. In fact, Finland has one of the lowest cancer death rates of all European countries.

While it adopts the LNT model, the ICRP [106] urges caution in applying it when a large population is exposed to a low dose of ionizing radiation:

> Collective effective dose is an instrument for optimisation, for comparing radiological technologies and protection procedures. Collective effective dose is not intended as a tool for epidemiological studies, and it is inappropriate to use it in risk projections. This is because the assumptions implicit in the calculation of collective effective dose (e.g., when applying the LNT model) conceal large biological and statistical uncertainties. *Specifically, the computation of cancer deaths based on collective effective doses involving trivial exposures to large populations is not reasonable and should be avoided.* Such computations based on collective effective dose were never intended, are biologically and statistically very uncertain, presuppose a number of caveats that tend not to be repeated when estimates are quoted out of context, and are an incorrect use of this protection quantity. [italics added]

Criticism of the LNT model centers on the question of whether there is a threshold below which repair mechanisms are adequate to remediate all or almost all damage from ionizing radiation. Alternatively, it is sometimes argued that at low dose rate the cancer risk grows quadratically rather than linearly with dose. Some even argue for **radiation hormesis**, the hypothesis that low levels of radiation exposure have a beneficial biological effect. A threshold model (B) and an example showing radiation hormesis (C) are shown along with the DDREF correction to the LNT model (A) in the insert in Figure 20.12. Further discussion of the LNT model can be found in all of the reports referenced in this section [106, 112, 113, 115]. Dissenting opinions are presented in reports from the French Academies [116] and in [117].

---

the stochastic effects of low exposure, though the very highest limits attempt to protect against the deterministic effects of ARS. A selection of radiation exposure benchmarks is given in Table 20.5. A useful number to keep in mind in the following sections is the world average natural dose of radiation exposure: 2.4 mSv/y. In light of the LNT model, regulatory agencies have adopted an attitude that whatever the standard permissible radiation exposure limits may be, the object of radiation safety should be to hold exposure to **as low as reasonably achievable** (the **ALARA** philosophy).

20.5 Radiation in the Human Environment

Ionizing radiation can be found almost anywhere in the natural environment. In fact, humans are subjected to a daily barrage of ionizing radiation not only from the external environment, but also from radioactive isotopes within their own bodies. Understanding the sources and magnitude of natural radiation is helpful in evaluating the dangers of radiation added into the environment as a result of human technology. In this section we describe the many sources of natural and manmade human radiation exposure.

The world average annual effective dose from naturally occurring radiation is approximately 2.4 mSv/y per person. In the US the average dose 3.1 mSv/y is somewhat higher. The dose rate, however, can vary dramatically with location. Cosmic ray exposure increases with altitude, but the largest source of variation stems from the occurrence of natural radionuclides in rock. In 2008 the *UNSCEAR* published an extensive survey of background radiation dose throughout the world [118]. They identify locations with radiation exposure much higher than the world average. For example, people living in areas of India rich in the mineral monazite (which contains thorium) are exposed to a dose rate of ≈ 2800 nSv/h, which translates into about 20 mSv/y – almost ten times the world average background dose. Figure 20.13 shows yearly exposure estimates for a selection of countries (mostly European).

Exposure to manmade radiation arises principally from medical and dental X-rays and from nuclear medicine, and therefore varies considerably from country to country, with greater exposure in developed countries. With the growth of *radiographic* diagnostic tools such as *computerized axial tomography* (CT) scans and the use of radionuclides for diagnosis and treatment, exposure to manmade radiation has grown dramatically in the US in recent years. In 2009 the NCRP estimated that, on average, exposure of patients to radiation associated with medical procedures contributes approximately the same as natural background sources, making the US average annual effective dose $E_{\rm US} \cong 6.2$ mSv/y, nearly twice the world average [119].

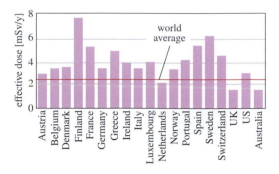

Figure 20.13 Annual dose from naturally occurring radiation in mSv for a set of European countries plus the US and Australia. The dominant contribution to the variability is from radon. (Credit: World Nuclear Association)

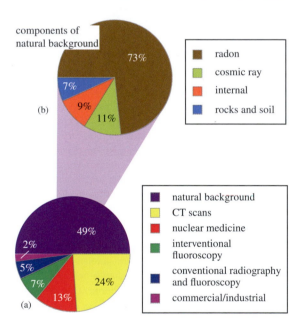

Figure 20.14 Contributions to average annual effective dose per person in the US, $E_{\rm US}$ [119]: (a) all sources totaling 6.2 mSv/y; (b) components of the natural background contribution to $E_{\rm US}$. Note that the contributions from both burning of fossil fuels and from the nuclear fuel cycle are too small to show on this scale.

Biological Effects of Radiation

The biological effects of radiation can be either *deterministic* or *stochastic*. The severity of deterministic effects increases with dose, with a threshold for serious illness and mortality from *acute radiation syndrome* at around 1 Sv and 3.5 Sv respectively. Stochastic effects of radiation exposure include cancer and serious genetic defects. The probability of stochastic effects is known to increase with dose at doses greater than roughly 0.1 Sv. The *linear no threshold (LNT)* model extrapolates the probability of stochastic effects linearly down to zero dose. Although the LNT is the standard model used by agencies charged with monitoring radiation protection, it is not universally accepted.

Figure 20.14 summarizes both the natural and manmade contributions to an individual's average annual effective dose of radiation in the US. We consider the major components of natural background radiation in order of their contribution to the average annual effective dose, after which we turn to manmade radiation sources. In both cases we use data for the US [119]. US and world-average data on radiation exposure from natural sources are quite similar; we focus on the US in order to illustrate the rapidly changing contribution of medical procedures in the developed world. For a recent perspective on worldwide exposure, see [118].

20.5.1 Radon: $E_{US} \cong 2.28 \, mSv/y$

By far the largest single contribution to natural radiation exposure comes from radon, or more precisely, from its decay products (**progeny**). Radon is an inert gas that is produced in the radioactive decay series of ^{238}U, ^{235}U, and ^{232}Th. The isotope ^{222}Rn, which appears in the ^{238}U series, has a long enough half-life – 3.82 days – to accumulate in the environment, and accounts for most of the radon-sourced effective dose. ^{220}Rn, which appears in the ^{232}Th decay chain, has a much shorter half-life of 55.6 s. Thorium, however, is more abundant than uranium in Earth's crust, and despite its short half-life ^{220}Rn contributes a small ($< 10\%$) addition. ^{235}U is much less abundant than either ^{238}U or ^{232}Th and its radon daughter ^{219}Rn has a half-life of only 3.96 s, so its contribution to human exposure can be ignored.

Because it is an inert gas, some radon percolates out of the rock in which it is produced and can migrate long distances. Other radionuclides produced in uranium and thorium decays may be equally or more dangerous in principle, but they are solid at ambient temperatures and remain in the rock where they were created.

Radon enters the atmosphere from the soil; the average emanation rate is about 20 Bq per square meter of soil [120]. Naturally, this rate is highly variable depending on the concentration of uranium in the soil as well as the temperature and humidity. An average value for the activity of radon in the *outside* air, in a layer one to three meters above the ground, is about 40 Bq/m^3 [120].

The danger of radon exposure was first discovered as the cause of an increased incidence of lung cancer among uranium miners. Since then there has been much effort to assess the risks faced by the general population from radon inhalation. Radon levels in the air inside a home can be much higher than the average for outside air if either the building materials or local soil have unusually high uranium content, or if the soil is particularly permeable. The situation is further exacerbated if the air in the house does not circulate very well. According to [121], models suggest that somewhere between 1 in 7 and 1 in 10 of all lung cancer deaths in the US can be attributed to indoor radon exposure. The US *Environmental Protection Agency* (EPA) sets a maximum permissible interior radon level of 4 pCi/L $\cong 148$ Bq/m^3.

The path by which radon damages the lungs is subtle and pernicious. Remember that the range of α-particles is very short. Thus α-particles emitted outside the human body are stopped by a few centimeters of air and cause little harm (though α-decays are often accompanied by γ-rays with much greater range – see below). Because radon is an *inert gas*, nearly all the radon we breathe in is exhaled back into the environment directly after it is inhaled. Thus the annual average effective dose from direct radon decay within the body is estimated to be only 0.05 mSv/y. When ^{222}Rn decays, however, its *progeny* include two short-lived α-emitters, the polonium isotopes $^{218}_{84}$Po and $^{214}_{84}$Po. These are particularly dangerous because (i) they have short half-lives (3.1 m and 164 μs, respectively) and are therefore very radioactive; (ii) they are created as positive ions, which attach to water molecules, dust particles, and aerosols; and (iii) once inhaled, they stick to the linings of the airways of the respiratory tract and the lungs, where they can damage the sensitive cells in those tissues. The contribution of ^{222}Rn *progeny* to the US average annual effective dose is estimated as 2.07 mSv/y. ^{220}Rn progeny are estimated to add another 0.16 mSv/y. Combined with the dose directly from ^{222}Rn decay, the total radon contribution to E_{US} was estimated in 2009 to be $\cong 2.28$ mSv/y [119]. This is an increase from the 1987 estimate of $\cong 1.28$ mSv/y [122].

20.5.2 Cosmic Rays: $E_{US} \approx 0.34 \, mSv/y$

In addition to radiation due to the natural radionuclides on Earth, we are constantly bombarded by radiation from space. Earth's magnetic field is an effective shield against the stream of low-energy charged particles emanating from the Sun. **Cosmic radiation** is the term given to the flux of high-energy particles that originate elsewhere in our galaxy or beyond. The composition of cosmic radiation is about 85% protons and 12.5% α-particles; the remainder includes nuclei of other light elements and electrons. The energy spectrum of cosmic rays, sketched in Figure 20.15, is remarkable: they have been observed with energies in excess of 10^{20} eV, far higher than any energy achieved in man-made accelerators (the Large Hadron Collider in Geneva holds the record with $\approx 7 \times 10^{12}$ eV protons). Physicists study cosmic rays both to understand how the laws of physics might change at such gigantic energies and also to probe the sources of very-high-energy particles in the universe.

The spectrum of cosmic rays shown in Figure 20.15 cuts off below about 10^9 eV. The ionizing radiation associated with cosmic rays with lower energy is essentially completely absorbed in the atmosphere. Furthermore, a significant fraction of cosmic rays with energies below ~ 10 GeV are deflected away from equatorial regions toward Earth's

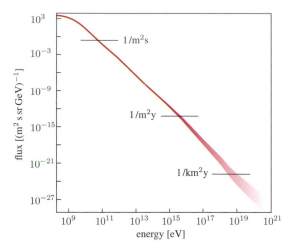

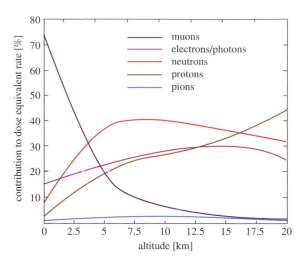

Figure 20.15 The flux of cosmic rays in number per squared-meter per second per unit solid angle per GeV (10^9 eV) energy. The numbers on the graph indicate some benchmark fluxes. Note the magnitude of the flux: for example, one per square meter per second at an incident energy of 10^{11} eV. (Credit: S. Swordy, The Energy Spectra and Anisotropies of Cosmic Rays, 2001, Space Science Reviews 99 with permission of Springer)

Figure 20.17 Variation in radiation dose from cosmic rays with altitude from 1 to 20 km [118].

Figure 20.16 *Aurora borealis*, light emitted when electrons ionized by radiation from space recombine with positive ions in Earth's upper atmosphere; photographed near Fairbanks, Alaska. (Credit: Senior Airman Joshua Strang, US Air Force)

magnetic poles, where they often manifest themselves in the form of *auroras* (Figure 20.16).

A variety of nuclear reactions result when high-energy cosmic rays enter Earth's atmosphere. The primary protons or α-particles undergo strong interactions with nuclei such as ^{16}O and ^{14}N high in the atmosphere. The initial nuclear interactions produce nuclear fragments, protons, neutrons, and unstable strongly interacting particles such as pions. (See §14 for an overview of some of these particles and their interactions.) Many strongly interacting particles

interact further or decay high in the atmosphere, generating a *shower* of particles – photons, electrons, positrons, neutrinos, and muons – that do not interact strongly but which constitute ionizing radiation that affects humans (Figure 20.17). The neutrinos pass right through Earth and out the other side causing no human radiation exposure. Most of the radiation dose at sea level comes from muons, with some additional contribution from electrons, γ-rays, positrons, neutrons, and protons.

The presence of muons in cosmic ray air showers requires some explanation. Muons are produced in the decay of pions ($\pi \rightarrow \mu\bar{\nu}_\mu$), which are themselves produced copiously in the strong interactions of protons and nuclei. With a lifetime of $\tau_\mu = 2.20 \times 10^{-6}$ s, one might think that muons would travel no further than $c\tau \cong 660$ m on average. Relativistic *time dilation* (§21.1.4) increases the lifetime of high-energy muons by orders of magnitude, however, enabling them survive to sea level and beyond (see Problem 20.16). Furthermore, high-energy muons lose energy much more slowly than electrons when propagating through matter. The dominant mechanism for energy loss by very-high-energy particles is bremsstrahlung, but being 207 times more massive than electrons, muons radiate much less. The mass stopping power of high-energy muons is about 2 MeV cm^2/g, and is only weakly dependent on energy or stopping material. At this rate, cosmic ray muons, which average ~6 GeV energy at creation, lose ~ 2 GeV passing through the atmosphere, leaving them with ~ 4 GeV energy at Earth's surface, where they produce significant ionization.

Neutrons become a more important component of cosmic rays at higher altitudes. Because they have a high radiation weighting factor ($w_{neutron} \geq 2.5$) (Figure 20.10), they contribute significantly to an increase in cosmic ray radiation dose with altitude.

The dose rate from cosmic rays approximately doubles with each 1800 m rise in elevation. At sea level the rate is about 0.27 mSv/y; in Denver (elev. 1610 m) it is 0.57 mSv/y; at La Paz, Bolivia (elev. 3900 m), the radiation dose is 2.0 mSv/y, of which 0.9 mSv/y comes from neutrons. Averaged over the US population, the direct effective dose from cosmic rays is 0.34 mSv/y. Flying in an airplane at 11 000 m continuously for a year would result in a dose of 15 mSv.

In addition to the direct effects of cosmic rays, some collisions in the upper atmosphere result in nuclides that are radioactive. Examples include ^{3}H, ^{7}Be, and most importantly ^{14}C, formed by $n + ^{14}_7\text{N} \rightarrow ^{14}_6\text{C} + p$. In fact, most neutrons created in cosmic ray showers are absorbed by a ^{14}N nucleus in the atmosphere. ^{14}C accounts for roughly one out of every 10^{12} atoms of carbon in materials that are in equilibrium with carbon from the atmosphere, including living organisms on Earth's surface. ^{14}C is famous for its application in archaeological dating: once an organism or its tissue (e.g. wood) is removed from contact with the atmosphere, for example by burying, then the abundance of ^{14}C falls exponentially with its 5730 year half-life. Therefore the fractional abundance of ^{14}C compared to the dominant isotope ^{12}C indicates how long a sample has been sequestered from the atmosphere. ^{14}C is also useful in the study of Earth's climate over the past several thousand years (see Box 35.1). ^{14}C decays by β-decay, producing $^{14}_7\text{N}$ and an energetic electron and an antineutrino. Every kilogram of carbon produces about 230 Bq due to the β-decay of ^{14}C. Cosmogenic radionuclei contribute only 0.01 mSv to the global-average human annual radiation exposure.

20.5.3 Radiation from Within the Body: $E_{US} \cong 0.29$ mSv/y

All human beings are unavoidably exposed to radiation from decays taking place within our own bodies. Different isotopes of an element are for the most part chemically equivalent. Therefore, when a biochemical process calls for a certain element, the radioactive isotope will serve just as well as the non-radioactive one. Living organisms contain significant amounts of both carbon and potassium, so the radionuclides ^{14}C and ^{40}K are present in living organisms in the same concentration with which they appear elsewhere on Earth's surface. In addition, people take in

small but not negligible amounts of thorium and uranium in our food and water. ^{40}K is the most important contributor to annual internal radiation exposure, with an activity of ~4000 Bq in a 70 kg person giving rise to an annual effective dose of about 0.16 mSv. ^{14}C contributes roughly the same activity, ~4000 Bq, but the emitted electron has much less energy, resulting in a contribution of only 0.01 mSv/y. Internal uranium and thorium amount to an additional 0.13 mSv/y [119].

20.5.4 Radiation from Rocks: $E_{US} \cong 0.21$ mSv/y

Gamma rays from radionuclide decay are constantly emitted from Earth's surface. Naturally occurring radionuclides (excluding those from cosmic rays, considered above) can be divided into two categories. First, there are those that have half-lives of roughly a billion years or longer and have been present on Earth since its formation. The second category includes short-lived radionuclides that are produced by decays of the long-lived radionuclides.

There are approximately 30 long-lived radionuclides present in Earth's crust, almost all of them in such minuscule amounts that their radiation contributes a negligible fraction to the total background radiation. Four nuclides, ^{40}K, ^{238}U, ^{235}U, and ^{232}Th (all of which we have encountered in earlier chapters), are responsible either directly or indirectly for virtually all of the natural radiation emitted from Earth itself. Important properties of these four nuclei are summarized in Table 20.6. All of them except ^{40}K decay through a long sequence of descendant nuclei, and it is important to remember that these descendants are also radioactive and that the activity due to every descendant species is the same as that due to the parent (Problem 17.22).

α-particles and electrons from decay sequences within rock and soil are largely shielded by the rock in which they are emitted before they impact humans. The γ-rays that often accompany α- and β-decays of ^{40}K, ^{238}U, and ^{232}Th and their progeny are in fact the main source of exposure from radioactive decays in Earth's crust. As emphasized in [118], the abundance of uranium and thorium varies widely over Earth's surface, and there are locations in the US and elsewhere where the radiation exposure from these sources is much greater than the 0.21 mSv average quoted in [119].

20.5.5 Exposure to Manmade Radiation

In addition to naturally occurring radiation, people are also exposed to manmade radiation, particularly in medical diagnostic and treatment procedures. Averaged over the world, the yearly effective dose from manmade sources is approximately 0.60 mSv/y, roughly 20% of the total annual dose of 3.0 mSv/y [118]. Until quite recently, this was

Table 20.6 Long-lived, relatively abundant, radioactive isotopes in Earth's crust: average crustal abundance $\bar{C}$ in parts per million/billion by mass (ppmm/ppbm) [74]. The last column lists the total energy released, including all steps in the decay chain, averaged over branches. The energy listed corresponds to heat deposited (see §32.1.3); the total energy released including neutrinos is given in parentheses.

Nuclide	Crustal abundance ($\bar{C}$)	Half-life (y)	Initial decay	Average energy	Subsequent decay chain
^{238}U	2.7 ppmm	4.46×10^9	$\alpha \to {}^{234}$Th	4.2 MeV	^{238}U $\to {}^{206}$Pb + 8 α + 47.7 (51.6) MeV
^{235}U	19 ppbm	0.704×10^9	$\alpha \to {}^{231}$Th	4.4 MeV	^{235}U $\to {}^{207}$Pb + 7 α + 43.9 (46.2) MeV
^{232}Th	9.6 ppmm	14.0×10^9	$\alpha \to {}^{228}$Ra	4.0 MeV	^{232}Th $\to {}^{208}$Pb + 6 α + 40.5 (42.6) MeV
^{40}K	2.4 ppmm	1.25×10^9	$\beta \to {}^{40}$Ca (89%)	0.56 MeV	
			EC$\to {}^{40}$Ar (11%)	1.46 MeV	

essentially the situation in the US. With the growing use of CT scans and other radiographic diagnostic techniques, as well as the use of radioisotopes in other procedures, the *average* annual effective dose due to medical procedures has grown dramatically in the US. According to [119] it amounts to approximately 3.0 mSv/y, equal to the total effective dose from all natural sources combined, and accounts for almost all manmade radiation exposure. This dramatic growth is highlighted in Figure 20.18. It should be emphasized, however, that this is an average obtained by combining many persons with doses of a fraction of a millisievert from ordinary medical and dental X-rays together with a considerably smaller number of persons who have been exposed to large doses of radiation during specific tests and treatments. With this in mind we summarize US exposure to manmade sources of radiation beginning with exposure for medical procedures.

CT scans: $E_{US} \cong 1.47\,mSv/y$ **CT scans** combine a very large number of X-ray images taken at different angles to produce a three-dimensional image of a specific area within a patient, in essence enabling the diagnostician to "see" inside the patient. CT scans have become an essential diagnostic tool in much of the developed world. Although the dose from each X-ray image is small, the integrated effective dose can range to as much as 32 mSv for CT angiography of the heart [119]. The dominant categories for CT scans are *head* (including brain and neck), *chest*, and *abdomen/pelvis*, with an average of 2 mSv, 7 mSv, and 10 mSv effective dose respectively. In 2006 (the last year included in the analysis of [119]) there were 62 million CT procedures in the US, up from 23 million a decade before.

Conventional X-rays and fluoroscopy: $E_{US} \cong 0.76\,mSv/y$ A conventional *X-ray* administered at a hospital might deposit a total of several mJ into a section of the body that has a mass of several kilograms. Thus the absorbed dose for the given area will be of the order of 1 mGy. X-rays

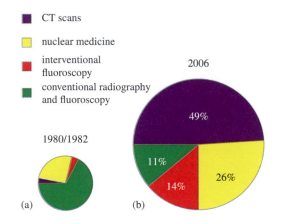

- CT scans
- nuclear medicine
- interventional fluoroscopy
- conventional radiography and fluoroscopy

1980/1982

2006

49%

11%

26%

14%

(a)   (b)

Figure 20.18 Annual average effective dose of radiation from medical procedures in the US (a) in 1980/1982 [122] when the total was 0.53 mSv/y, and (b) in 2006 [119] when the total was 3.00 mSv/y. (Credit: The National Council on Radiation Protection and Measurements, http://NCRPonline.org)

are photons, and the radiation weight factor for all photons is 1, so the equivalent dose from this X-ray is 1 mSv in the exposed region. Tissue weighting factors need to be included in order to convert this equivalent dose into an effective dose. Typically these are of order 0.1, so a typical X-ray results in an effective dose of about 0.1 mSv. Examples from [119] include chest: 0.1 mSv, breast: 0.18 mSv, hip: 0.7 mSv, and dental (full-mouth survey): 0.02 mSv.

Fluoroscopy, which produces moving real-time X-ray images, generally requires greater effective doses, e.g. 2.5 mSv for fluoroscopy of the gastro-intestinal tract. *Interventional fluoroscopy*, in which a medical device is guided through the body with the aid of fluoroscopic images, typically requires even greater radiation exposure than ordinary fluoroscopy. Together, conventional X-rays and fluoroscopy account for an average annual effective dose of 0.33 mSv/y in the US and interventional fluoroscopy contributes 0.43 mSv/y.

Positron emission tomography (PET) scans use a β^+-emitter as a source of positrons that annihilate with an atomic electron to produce two γ-rays of energy 0.51 MeV, which are observed by detectors on opposite sides of the patient. In this way a three-dimensional image, similar to those produced by CT scans, can be generated. PET scans are usually used in conjunction with CT scans, and their small contribution to annual effective dose is included in that category.

Nuclear medicine: $E_{US} \cong 0.77\,mSv/y$ The term **nuclear medicine** applies to the administration of radionuclides in specific chemical compounds to patients for the purposes of diagnosis and treatment. In diagnostic procedures, the emissions from radioactive sources are observed in order to image organs or follow the path of the radionuclide through the body. Examples include the use of γ-emission from ^{123}I to study the uptake of iodine by the thyroid gland and the use of ^{99m}Tc γ-rays to image blood flow in the heart. The 2009 NCRP report estimates an annual average effective dose of 0.77 mSv to the US population from diagnostic nuclear medicine procedures.

In addition, radionuclides may be administered to patients to treat tumors and other diseases. Although these treatments are much rarer than diagnostic nuclear medicine procedures, they usually employ much higher absorbed doses to specific organs (e.g. in treating malignant tumors). They are therefore liable to generate deterministic, and not just stochastic effects, and generally rely on low-LET radiation such as electrons from beta decays, and so are sometimes described in terms of *equivalent* rather than an *effective dose*. Since they involve a relatively small number of treatments and since estimates of the number of treatments and the absorbed doses are not available in the US, this component was omitted in [119]. Radiotherapy with external beams is another source of radiation exposure omitted from [119]. The number of patients receiving these treatments, usually aimed at destroying malignancies, is less than 1% of the number receiving radiation for diagnostic purposes. The absorbed doses are, however, on the order of 5000 to 50 000 times as large. The concept of *effective dose* is difficult to apply to this type of therapy, however, since they aim to produce *deterministic* rather than *stochastic* effects.

Consumer products and activities: $E_{US} \cong 0.1\,mSv/y$ Small amounts of radioactive materials occur naturally in consumer products such as tobacco, water supplies, and building materials. Cigarette smoking, building materials, and commercial air travel account for almost 90% of this source of exposure. In addition, specialty products contain minute amounts of radioactive isotopes; these include, for example, smoke detectors (^{241}Am), self-luminous exit signs (^{3}H), and static-eliminating brushes for film, vinyl records, etc. (^{210}Po). Televisions, computer screens, and other electronic displays that make use of **cathode ray tubes** (CRTs) emit small amounts of X-rays. Altogether these sources add up, on average, to roughly 3% of natural background radiation.

Industrial, medical, and other activities: $E_{US} \cong 0.005$ mSv/y This category is dominated by emissions from fossil fuel consumption, $E_{US} \sim 0.002$ mSv/y, and exposure of the general population to nuclear-medicine patients, also ~ 0.002 mSv/y. Assuming that spent nuclear fuel and other waste is properly contained, nuclear power plants contribute very little radiation to the environment. Uranium mining, however, takes a large quantity of rock that had previously been underground, refines it, and leaves the tailings in piles above ground. Regulatory restrictions on the radioactivity of tailing and other mining waste are stringent. In the years since the 1987 NCRP report, exposure from the entire nuclear fuel cycle has dropped significantly and is now estimated at $\sim 4.5 \times 10^{-4}$ mSv/y.

Fallout from nuclear tests and accidents Now that over 45 years have passed since the last atmospheric nuclear weapons test, radiation from fallout from the world's assorted nuclear bomb tests is very small and not regarded as a source of background radiation. Certain fission products serve as good markers for environmental radiation of reactor and/or bomb origin. A good example is ^{137}Cs, which is a radioactive fission product with a half-life of ~ 30 years. Figure 20.19 shows the measurements of ^{137}Cs β-decay activity in a selection of *Bordeaux wines* over the years from 1950 through 2000. Spikes correspond to nuclear testing during the 1950s followed by exponential decay after the Atmospheric Test Ban Treaty of 1963. ^{137}Cs activity peaked briefly after the *Chernobyl* reactor disaster in 1986. Athough significant in the past, the annual average effective dose to persons in the US both from nuclear tests and nuclear accidents is now regarded as negligible [119].

20.6 Nuclear Waste and Nuclear Proliferation

Having considered the nature and varieties of ionizing radiation, its effects on living systems, and its natural and manmade sources, we turn to one of the most vexing physics issues associated with nuclear energy: nuclear waste – or to be more specific, spent nuclear fuel.

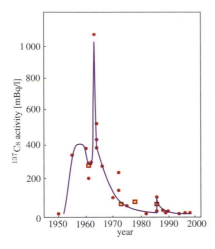

Figure 20.19 Activity of ^{137}Cs in a sampling of Bordeaux wine. Solid circles correspond to reduction of the wine to ash; the open squares were obtained non-destructively [123].

Arguably the single greatest drawback to the utilization of nuclear fission energy is the large quantity of highly radioactive waste that it produces. Spent nuclear fuel remains radioactive essentially forever and remains dangerous for tens of thousands of years. On the other hand, modern civilization generates enormous quantities of other types of toxic waste, some of it extremely dangerous to humans and other organisms. Some of this non-nuclear waste is carefully sequestered, but much of it is released into the environment without adequate precautions. Many types of non-nuclear toxic waste, particularly heavy metals like mercury, cadmium, and arsenic, remain toxic until buried by geological processes. Modern civilizations, however, put radioactive waste in a special category compared to other dangerous waste products. Part of the special concern is related to the possibility that nuclear waste could be used to build destructive devices or even nuclear bombs. Careful consideration of the long and short-term environmental and human health consequences should be a part of any intelligent planning process regarding the use of any technology producing hazardous material as a by-product. Both radioactive and other dangerous toxic wastes should be planned for and managed in an integrated fashion with the industries which produce these wastes.

After this brief introduction we describe the properties of spent nuclear fuel. We then turn to the role of spent fuel in the proliferation of nuclear weapons. Finally we discuss the question of what to do with spent nuclear fuel. Options include reprocessing to reclaim fissile materials and to make the waste less dangerous, permanent sequestration in long-term geological repositories, or managed storage awaiting new technologies and/or political consensus.

Environmental Radiation

On average, human beings living at sea level receive about 3 mSv/y of radiation from natural sources. ^{222}Ra makes the largest contribution to natural radiation – over 70% in the US – followed by cosmic rays, radioactive sources within our own bodies, and radionuclides other than radon in rocks and soil. Radiation exposure from medical (and dental) diagnostics and treatment and from commercial products vary widely. They average globally around 0.6 mSv/y, but are larger in developed countries, and contribute an additional 3 mSv/y on average in the US. Fallout from nuclear tests, radiation from nuclear accidents, and from fossil fuel combustion currently make negligible contributions to yearly exposure.

20.6.1 The Character of Nuclear Waste

We focus on what is known as **high-level nuclear waste**, which is basically the **spent nuclear fuel** (SNF) from nuclear fission reactors. Low- and intermediate-level nuclear waste are much less radioactive. They do not present a severe sequestration problem and do not contain actinides so cannot be used to construct real nuclear weapons. Many schemes and categories exist for classifying nuclear waste, according to its origin, its level of radioactivity, physical state, etc. [124].

Spent nuclear fuel constitutes only ∼1% of radioactive waste by volume, but accounts for over 90% of the radioactivity of the waste. The exact composition of SNF depends on the type of reactor and the fuel it uses. CANDU reactors use natural uranium, which is only 0.7% ^{235}U; most US reactors use *low-enriched uranium*, with about 3–5% ^{235}U at input; and some, like US nuclear submarine reactors, use highly enriched uranium, with ^{235}U concentration as high as 90%. Some reactors, in France for example, use more complex, partially reprocessed fuel that includes fissile nuclides produced in previous reactor burns. To keep things simple, and because it corresponds to the majority of nuclear waste, we consider the spent fuel from a light-water reactor burning low-enriched uranium [81]. We assume that the reactor is on a four year fuel cycle, i.e. 25% of the fuel is removed every year after spending four years in the reactor. Each year a 1 GWe light-water reactor discharges about 21 tons of SNF (Problem 20.21), containing the principal radioactive species summarized in Table 20.7.

Table 20.7 Characteristic composition of $\sim$21 tons of SNF from a light-water reactor fueled with low-enriched uranium. Abundances are approximate; they depend on the specific configuration of the reactor. All the actinides (U, Pu, Np, Am, Cm) are fissionable with fast neutrons. From [81].

Category	Species	Mass	Half-life	Decay
Residue of	^{238}U*	20 t	4.5 Gy	α
original fuel	^{235}U†	162 kg	700 My	α
Plutonium	^{239}Pu†	120 kg	24 ky	α
isotopes	^{240}Pu*§	50 kg	6.6 ky	α
	^{241}Pu†	18 kg	14 y	β^-
	^{242}Pu*	10 kg	375 ky	α
	^{238}Pu*	2 kg	88 y	α
Minor	^{237}Np	10 kg	2.2 My	α
actinides	^{241}Am$^\|$		432 y	α
	^{243}Am	10 kg	7.4 ky	α
	^{244}Cm/^{245}Cm	1 kg	18 y / 8.5 ky	α
Some common	^{99}Tc	18 kg	211 ky	β^-
long-lived	^{93}Zr	16 kg	1.5 My	β^-
fission fragments	^{135}Cs	9 kg	2.3 My	β^-
	^{107}Pd	5 kg	6.5 My	β^-
	^{129}I	3 kg	15 My	β^-
	(total)	760 kg		

* Fertile † Fissile
§ High probability of spontaneous fission
$^\|$ From β-decay of ^{241}Pu

The amount of plutonium in SNF is significant. since it is chemically distinct from the other components of SNF, plutonium can rather easily be separated out of SNF. The odd atomic mass isotopes of plutonium, ^{239}Pu and ^{241}Pu are fissile. Thus SNF is a source of new fuel for power reactors, but also could provide the raw material needed to make nuclear weapons: 120 kg of ^{239}Pu would provide about 10% of the fissile material needed for the yearly operation of a 1 GW$_e$ power reactor. In light of the discussion of uranium reserves in §16, however, a 10% decrease in reactor fuel requirements would not be a decisive factor in the utilization of nuclear power. A reactor designed to breed fuel, like the fast breeders described in §19.4.2, would have an entirely different spent fuel composition. On the other hand, it takes less than 10 kg of ^{239}Pu to make a nuclear weapon, so the SNF sample summarized in Table 20.7 contains the makings of many fission bombs.

Another significant feature of the composition of SNF is the complexity of its physical and chemical makeup. Fission fragments include materials that are gases or liquids at room temperature and above (xenon and iodine for example); SNF also contains highly reactive elements like cesium and rubidium; and it contains toxic metals like cadmium and, of course, plutonium. These features complicate the handling of SNF, but its radioactivity is the foremost concern.

20.6.2 Time Evolution of Spent Nuclear Fuel

Because it consists of so many different radionuclides, and because many of these give rise to further new radionuclides when they decay, the time dependence of the radioactivity of spent reactor fuel is very complicated. As described in §19.2.2 (see Figure 19.7), the radioactivity and heat generated by nuclear fuel are initially very high, decrease quickly over the first few days, and then decrease more slowly over a period of many years. The IAEA quote an estimate that roughly 10 years after discharge from the reactor, the activity of SNF ranges between 5×10^{16} and 5×10^{17} Bq/m^3, generating heat at a rate of 2–20 kW/m^3 [124].

Spent nuclear fuel retains significant activity from both actinides and fission fragments for thousands, even millions of years. Initially those nuclei with the shortest lifetimes dominate. These are mostly fission products. Then over hundreds of years, actinides, particularly ^{241}Am, begin to dominate. The situation switches again around 10 000 years, at which point the long-lived fission fragments such as ^{99}Tc dominate. Eventually only the longest-lived fission fragments and actinides like uranium isotopes remain. A qualitative graph of the activity of spent fuel and its components is given in Figure 20.20. Since ^{239}Pu α-decays to ^{235}U, the anomalous radioactivity of nuclear waste in the very long term resembles that of slightly enriched uranium.

From the point of view of environmental stewardship, the intermediate lifetime components of nuclear waste are the most challenging. One can imagine guarding nuclear waste closely for some number of years (either on the site where it was created or in a repository) while the short-lived, very active and very dangerous components decay away. The components with lifetimes of millions of years, though dangerous, have by virtue of their long lifetimes, lower activity levels. The greatest challenge comes from nuclides with lifetimes so long that their safe sequestration cannot be guaranteed, but so short that they have a high level of radioactivity. Actinides like ^{241}Am, ^{243}Am, and ^{239}Pu fall into this unhappy middle class.

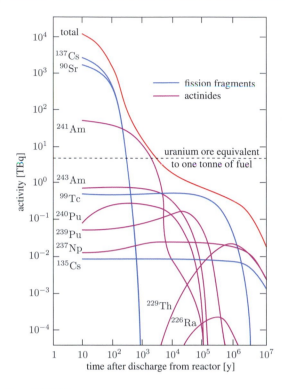

Figure 20.20 Time dependence of the activity of one ton of SNF. The line at $\approx 5 \times 10^{12}$ Bq corresponds to the activity of the amount of uranium ore needed to produce the fuel in the first place. Note that ^{241}Am dominates emissions from SNF during the crucial time scale of hundreds to thousands of years. (Credit: International Atomic Energy Agency (IAEA), Nuclear Power, the Environment and Man, IAEA, Vienna (1982).)

20.6.3 Spent Nuclear Fuel, Nuclear Weapons, and Nuclear Proliferation

The plutonium isotopes ^{239}Pu and ^{241}Pu, which are abundant in SNF, are fissile and could be used in nuclear weapons. Existing nuclear weapons are powered by the fission of ^{235}U or ^{239}Pu. The **critical mass** – the minimum mass of fissionable material required for an explosive chain reaction – of an uncompressed sphere of ^{235}U is about 60 kg and for ^{239}Pu it is only about 10 kg [81]. The power per unit mass of a plutonium bomb is also greater than that of a ^{235}U bomb. Weapons can be made with less fissile material by using various methods to increase the neutron flux within the fissile mass or by dramatically increasing the density by explosive compression. As realistic examples, the first two nuclear bombs used as weapons were made of (1) about 64 kg of ^{235}U enriched to $\gtrsim$ 80% purity, and (2) about 6.2 kg of ^{239}Pu of very high purity. The first of these weapons was dropped on the Japanese city of

Hiroshima, the second on Nagasaki. Both caused massive civilian deaths and destruction.

Note that while there is some concern that low- and intermediate-level nuclear waste (including disused radioactive sources from medical and other uses), could be used to construct **dirty bombs** that disperse radioactive material through conventional explosives, such devices would be significantly less destructive than full-fledged nuclear weapons, and would not represent nuclear proliferation. While such bombs could be extremely disruptive and careful cleanup efforts could be costly and time-consuming, the actual damage caused would likely not be significantly greater than that produced by the underlying conventional explosive.

^{235}**U and Nuclear Weapons**   As discussed in §18, ^{235}U is a small component of naturally occurring uranium and must be separated from the chemically identical dominant isotope ^{238}U by physical means. At present it requires a massive industrial operation using complex and sophisticated physical methods to enrich uranium to the high concentrations necessary for nuclear weapons – typically 85% or greater, though a crude weapon can be made at 20% enrichment, or perhaps less. Although several methods have been developed, the dominant one in use at present is *centrifugation*. The same methods are used to enrich uranium to the $\sim$3–5% concentration suitable for use in a power reactor as are used to enrich uranium to the high concentrations of ^{235}U required for nuclear weapons. It is not easily possible to tell from a distance whether a facility is being used to enrich uranium for peaceful or military use.

Nuclear weapons employing ^{235}U can be conceptually simple and relatively easy to fabricate. In the simplest, **gun-type** weapon, a mass of highly enriched ^{235}U is assembled in such a way that it is slightly subcritical, i.e. it is too small to sustain a fission chain reaction. At the moment of trigger, an equal or smaller mass of ^{235}U is propelled into the larger making the whole mass supercritical and releasing its energy in a massive explosion as illustrated schematically in Figure 20.21(a). This is how the Hiroshima bomb operated. In a simple weapon like this, the fissioning mass expands rapidly, terminating the fission reaction soon after it starts. It is estimated that only 1% of the ^{235}U in the Hiroshima weapon underwent fission. Modern ^{235}U-based nuclear weapons use neutron reflection and implosion techniques (see below) to reduce the required amount of ^{235}U to 20–30 kg and to increase the explosive yield. The only technologically challenging aspect of constructing a gun-type nuclear weapon is the uranium enrichment. No prototype of this bomb was ever

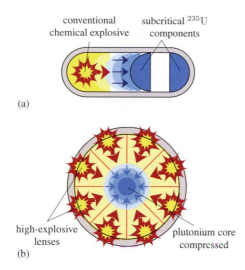

(a)

(b)

Figure 20.21 Schematic sketches of the two dominant designs for fission weapons. (a) A gun-type weapon is technologically unsophisticated and can only be used with ^{235}U. (b) An implosion design can be used either with ^{235}U or plutonium isotopes, and requires high tech shaping and coordination of conventional explosives.

tested during the Manhattan project, since the developers had no doubt that it would work.

Plutonium and Nuclear Weapons

With the exception of minute quantities of ^{244}Pu, no isotope of plutonium has a lifetime long enough to occur naturally on Earth. The large amount of ^{239}Pu, as well as smaller quantities of the other fissile isotope ^{241}Pu, in SNF can, however, be used to fashion nuclear weapons. Unlike ^{235}U, plutonium is both highly radioactive and highly toxic. ^{239}Pu α-decays to ^{235}U with a half-life of 24 000 y. The α-decay is occasionally accompanied by γ-rays as the ^{235}U nucleus settles down to its ground state. The isotopes ^{240}Pu and ^{241}Pu are also α-emitters, and they are more active than ^{239}Pu because they have shorter lifetimes. In fact, most of the radiation coming from the plutonium in spent power reactor fuel comes initially comes from the ^{241}Pu. The radioactivity of plutonium isotopes complicates weaponization of plutonium in several ways that make it difficult to construct a powerful plutonium-based nuclear weapon without substantial resources and technical expertise.

The properties of ^{240}Pu complicate the weaponization of plutonium extracted from SNF. ^{240}Pu is formed when ^{239}Pu absorbs a fast neutron without fissioning. Since ^{240}Pu is not fissile, the ratio of ^{240}Pu to ^{239}Pu grows over the period that a fuel element spends in a reactor. By

the time the fuel is removed in a normal fueling cycle, ^{240}Pu may account for ∼25% of the plutonium present (Table 20.7). ^{240}Pu usually decays to ^{236}U by α-emission. It has a probability 5.7×10^{-4}, however, of undergoing spontaneous fission (see §18.3.2 and Table 18.2). With a half-life of 6561 years, a mole of ^{240}Pu undergoes spontaneous fission with an activity of 4.75×10^{7} Bq (Problem 20.20). When it fissions, ^{240}Pu produces neutrons that can trigger a premature chain reaction in a plutonium-based nuclear weapon. This makes the *gun-type* bomb design ineffective even for plutonium with a relatively small ^{240}Pu content. Unless the critical mass of plutonium is assembled extremely rapidly, premature chain reactions seeded by neutrons from spontaneous fission of ^{240}Pu leads to a **fizzle**, in which a small quantity of the plutonium fissions, causing an explosion that scatters the remaining material. Such a device could still cause immense destruction and radiation effects, but would not be as devastating as a device that achieved more complete fission of the ^{239}Pu. The strategy devised during the Manhattan Project (and basically unchanged to this day) to assemble a critical mass of ^{239}Pu was to use an array of simultaneous conventional explosions to rapidly compress the plutonium. This type of bomb is called an **implosion device** and is sketched in Figure 20.21(b). The famous **Trinity test** at Los Alamos tested this design, which was subsequently used at Nagasaki.

If ^{240}Pu captures another neutron in a power reactor, it forms ^{241}Pu, which accounts for about 9% of the SNF described in Table 20.7. ^{241}Pu is fissile, but its lifetime is short: it β-decays to ^{241}Am with a half-life of $t_{1/2} = 14.3$ years. Fresh SNF, which is both thermally and radioactively hot, contains considerable ^{241}Pu and little ^{241}Am. As time goes on the ^{241}Am content increases. The short lifetime of ^{241}Pu complicates its use in nuclear weapons, and the slow buildup of ^{241}Am presents a radiation hazard because ^{241}Am α-decay is often accompanied by an energetic γ-ray. The activity of ^{239}Pu is low enough that it can be manipulated using only moderate radiation shielding (radiation protected *glove boxes*); however, ^{241}Am produces 10^{3} times the γ-ray dose of ^{239}Pu [125]. It must be manipulated remotely if the workers are to be guarded against significant radiation exposure.

The combination of the high degree of radioactivity of SNF from ^{241}Am and the technical challenges of building an operational implosion device should make it rather difficult for a small group of people or organization with limited resources to use spent nuclear fuel to develop a fully functional plutonium-based nuclear weapon. A nation-state with full control over uranium-fueled nuclear power plant operation and fuel cycles can, however, circumvent these issues. The IAEA classifies plutonium according to the percentage of ^{240}Pu, into **super-grade** (2–3%), **military grade** (< 7%), **fuel grade** (7–18%), and **reactor grade** (> 18%). Plutonium recovered from power reactors using a *normal* fuel cycle, in which the fuel elements are removed after 3–4 years produces reactor grade plutonium, which presents all the challenges just mentioned. Super- or military-grade plutonium, however, can be extracted from a reactor designed for nuclear power production if the fuel elements are removed and chemically reprocessed to extract plutonium after a short time – a few months. Then the isotopes ^{240}Pu and ^{241}Pu are suppressed and weapons-grade plutonium can be obtained. A sophisticated technical community working under the protection of a national government would then have the time and resources to construct a plutonium-fueled implosion device. This is the approach that has been taken in all plutonium-based nuclear devices developed in the past. By monitoring reactor fuel cycles sufficiently closely, it can also thus be determined whether weapons-grade plutonium is being removed from a functioning reactor.

Thorium and Nuclear Weapons

^{232}Th can be used to breed ^{233}U (see §19.3), which can be removed from the reactor and fashioned into a nuclear device. The ^{233}U can be obtained in relatively pure form by chemically separating it from the ^{232}Th fuel. Like plutonium, ^{233}U can be used in an implosion-type bomb. Indeed, the US successfully tested a ^{233}U-based weapon in 1955.

The weaponization of ^{233}U seems to be somewhat more difficult than ^{239}Pu. ^{233}U can easily be *denatured* by mixing it with natural uranium after separation. The resulting mixture would have too little ^{233}U for weapons use, but enough for reactor fuel. It could not be re-separated without complex isotope separation equipment. On the other hand, the potential utility of highly enriched ^{233}U (which can be used in smaller, more efficient, and more easily controlled reactors) compared to a less concentrated form argues against this. Proliferation resistance for ^{233}U comes from another, less obvious, mechanism [126]: when it is produced from ^{232}Th (in one scenario), ^{233}U is accompanied by ^{232}U at an abundance of roughly 0.4%. ^{232}U decays with a 69 year half-life in a decay chain that includes a highly hazardous 2.6 MeV γ-ray that accompanies the β-decay of ^{208}Tl. Even more than the γ-activity of ^{241}Am, this γ-radiation makes the ^{233}U sample too radioactive for easy manipulation. Also, the 2.6 MeV γ-ray, which is hard to shield, makes this material easy to identify. Note, however, that much like the situation with ^{240}Pu, it is possible to design a reactor configuration

---

Nuclear Weapon Materials

^{235}U suitable for a nuclear weapon is difficult to separate from natural uranium, but relatively easy to fabricate into a weapon. ^{239}Pu suitable for a nuclear weapon can be separated by chemical means from spent reactor fuel, but it is more difficult to fabricate into a weapon.

The weaponizability of plutonium from SNF depends on its isotopic composition. The presence of ^{240}Pu, a neutron source, and ^{241}Pu which rapidly decays to ^{241}Am, a γ-emitter, makes weaponization more difficult.

Plutonium extracted from normal power reactor refueling can be used to create crude but effective nuclear weapons, capable of doing considerable damage. Withdrawing reactor fuel on a more frequent basis yields plutonium with relatively less ^{240}Pu and ^{241}Pu and makes weaponization easier.

---

and fueling cycle to minimize the ^{232}U contamination of ^{233}U.

20.6.4 What to Do With Spent Fuel?

There are (at least) three ways to deal with spent nuclear fuel: first, it can be treated as toxic waste and sequestered from the environment with little or no post-processing; second, it can be reprocessed with the twin goals of (i) reducing the radioactivity and/or toxicity of the waste and (ii) extracting fissile actinides that can be used as additional fuel for power reactors; or third – an intermediate option known as **managed storage** (see, e.g., [73]) – it can be stored securely, but retrievably, awaiting development of new reactor designs and a political consensus more favorable to reprocessing or a decision to implement permanent sequestration. Whichever path is followed, when spent fuel is removed from a reactor, it is submerged in pools of water. This is done because the fuel is very hot, both in temperature and radioactivity. The water serves both to cool the spent fuel and to shield the environment from radiation. After 10 to 20 years, the shortest-lived and most active components decay away, leaving the spent fuel cooler and less radioactive.

Most nuclear power plants were designed to have limited pool storage space, with the expectation that after initial cooling, spent fuel would be sent for reprocessing or transferred to long-term storage. At the present time the United States does not reprocess its spent reactor fuel, nor has a long-term geological repository been constructed and commissioned. Instead, after initial cooling, spent nuclear fuel is transferred to dry storage casks, where it is stored on-site indefinitely awaiting the development of secure, long-term storage in geological repositories. This is not a stable situation: dry cask storage above ground requires constant monitoring and high security. Nuclear fission energy cannot be exploited indefinitely without a serious, secure, and reliable long-term repository for spent fuel, or a program for reprocessing the fuel.

Spent reactor fuel can be chemically processed to remove highly radioactive and/or long-lived fission fragments and actinides, particularly plutonium, which can be used as new reactor fuel. The feasibility of this approach has been demonstrated by France, which has been reprocessing spent nuclear fuel for decades. The extracted actinides are combined with uranium into a fuel known as **MOX**, which stands for *mixed-oxides* (principally of uranium and plutonium). Many reactor designs (such as CANDU) can use MOX fuel without any modifications; other designs may require extensive modifications and others still (e.g. fast-neutron reactors) may be designed expressly with MOX fuel in mind. Other highly radioactive components are separated for secure storage. The residual spent fuel after reprocessing is reduced in both volume and radioactivity. Since reprocessing removes the actinides responsible for the intermediate time scale radioactivity of SNF, the problem of geological isolation for the residual spent fuel is somewhat lessened. Earlier in the history of nuclear power the argument for reprocessing was principally an economic one – extending the fissile fuel supply. More recently estimates of uranium reserves and resources have expanded (see §16.2) enough that shortages of natural uranium do not seem imminent. Nevertheless, reprocessing would serve several purposes: it would extend the fissile fuel supply, reduce the danger and quantity of nuclear waste, and reduce the potential for weaponization of spent nuclear fuel.

Whether spent fuel from a reactor is discarded directly after use (as is generally true in the US) or it undergoes one or more rounds of nuclear reprocessing (as is true in much of the rest of the world) the envisaged end of the line is the same: burial. Before burial, the waste is cast with glass (a process called *vitrification*) – though there are alternate forms of encasement – and then sealed within thick, stainless steel containers. At present, in the US, these containers are stored on site at their reactors. The long-term plan is to bury them in a **mined geologic repository**. The requirements for such a repository – that the radiation not leak out into the environment for tens of thousands of years – are daunting. As a comparison, the pyramids at Giza are only

4500 years old and the instructions given regarding their care and maintenance were likely forgotten in short order.

Discussion/Investigation Questions

20.1 Consider the pros and cons of the public health response, including mandatory evacuations, to low levels of environmental radiation following the Fukushima-Daiichi incident. Find estimates of the cost, in terms of both money and human life, of the evacuation. Estimate the expected additional cancer deaths that might have occurred were no action taken.

20.2 Investigate and discuss the physics and public health issues raised by the radium watch-dial painters of the 1920s, the Daghlian criticality accident in 1945, the ^{60}Co contamination accident that occurred in Goiânia, Brazil in 1987, or another incident you choose.

20.3 ^{40}K contributes radiation exposure when it occurs in the human body and in rocks and soil. Why does the internal exposure come dominantly from electrons and the exposure from ^{40}K decays in rocks and soil come dominantly from almost entirely from γ-rays?

20.4 What are *TENORMs*? Discuss some examples such as coal ash, water treatment wastes, or mine tailings. What causes their radioactivity? How significant are the hazards?

20.5 Why do you suppose the author of [123] chose Bordeaux wine for his measurements. Can you suggest other candidates?

20.6 The government of Freedonia claims to be developing nuclear reactors for peaceful purposes using enriched uranium they have produced from their own isotope separation plants. The IAEA (International Atomic Energy Agency), which tracks nuclear proliferation, observes that Freedonia's engineers are refueling their reactors monthly. When protests fail, the IAEA brings this to the attention of the United Nations and ask for sanctions, claiming that Freedonia is aiming to build a nuclear weapon. What is going on?

20.7 An alternative technology for ultimate disposal of spent nuclear fuel, known as *deep borehole disposal*, envisions using well-understood drilling technologies to confine SNF in stable geological formations far below ground water. Investigate and evaluate the pros and cons of this proposal.

20.8 There is some controversy about the effects of moderately increased radiation levels on ecosystem health. Following the Chernobyl disaster in 1986, and the resulting human evacuation of the area, animal populations in the area increased both in size and in diversity. Some evidence has suggested, however, that at least for some species such as barn swallows [127] the area has become a population sink, not only unable to sustain a healthy population but drawing in creatures from adjoining areas and depleting those other populations.

Investigate and evaluate the arguments and evidence on the two sides of this controversy and discuss.

Problems

20.1 Show that the maximum energy transfer by a heavy particle with mass M and speed v to an electron initially at rest is $2m_e v^2$, hence the appearance of this factor in the formula for the stopping power.

20.2 Estimate a lower limit on the number of collisions a 1 MeV proton must make with atomic electrons before its energy is reduced to 10 eV. Why is it reasonable to ignore the atomic binding energy of the electrons in this estimate?

20.3 Show that a particle's range (see §20.2.1) expressed as a function of its initial velocity $R(v)$ is proportional to its mass and inversely proportional to the square of its charge and to the electron density of the material.

20.4 [T] Derive eq. (20.4) from eq. (20.3) and the approximations quoted in the text.

20.5 Show that in any material, the range of an α-particle and a proton with the same speed are the same (in the approximation that $m_\alpha = 4m_p$). Using Figure 20.3, find the range of a 2 MeV proton in water.

20.6 Using the results of Problem 20.3, compare the range of a 2 MeV α-particle in water, air, and bone. (The density of bone is approximately 1100 kg/m^3.)

20.7 To a first approximation for computing stopping power, living tissue can be approximated as water. Using data on the range of protons in water, estimate the energy of a proton required to place its Bragg peak on a tumor at a depth of 10 cm inside a body.

20.8 [T] According to Einstein's theory of relativity, the *kinetic energy* of a particle with mass m and speed $v = \beta c$ is $E_{\mathrm{kin}} = mc^2(\frac{1}{\sqrt{1-\beta^2}} - 1)$. First, verify that when $v \ll c$ this reduces to the familiar result, $E_{\mathrm{kin}} = \frac{1}{2}mv^2$. Then compute β for an electron and an α-particle with 5 MeV kinetic energy.

20.9 The **mass attenuation coefficients** (μ/ρ) for 1.5 MeV photons in concrete ($\rho = 2.35$ g/cm^3) and lead ($\rho=11.4$ g/cm^3) are both about 0.5 cm^2/g. How thick must concrete or lead shielding be in order to absorb 99% of these 1.5 MeV γ-rays?

20.10 In §20.3.1 it is stated that activity a is often used as measure of the mass of radionuclide in a sample m. Show that for a radionuclide with half-life $t_{1/2}$ and atomic mass A, this relationship is $m \cong A\, t_{1/2}[\sec]\, a[\mathrm{Ci}]\, 8.87 \times 10^{-17}$ kg.

20.11 Thermal neutrons wandering through biological tissue are most likely to be absorbed by ^{1}H, ^{12}C, or ^{16}O. If absorption is accompanied by emission of a single γ-ray, what are the γ-ray energies produced in these three cases?

20.12 A nuclear accident has exposed a 70 kg worker to 1 Ci of 2 MeV photons for about one minute. Assuming that all the energy of absorbed photons is deposited in the body, with uniform distribution, what absorbed dose of radiation has the worker suffered? What effective dose? What would have been his effective dose if the exposure had been concentrated entirely in his stomach (mass 0.5 kg)?

20.13 ^{131}I is a dangerous fission product because iodine is preferentially absorbed by the thyroid gland in children. It has a half-life of about 8 days and emits a β-particle with average energy 180 keV. (Ignore the γ-radiation that accompanies this decay (see Example 20.2).) How much mass of ^{131}I (per gram of thyroid tissue) must a child absorb to suffer an equivalent dose of 10 mSv to the thyroid? What is the activity per gram of this radiation dose (in curies) (see Problem 20.10)?

20.14 Rubidium-87, ^{87}Rb is a relatively common, very long-lived nuclide. Look up and compare its abundance and lifetime to ^{40}K. How does ^{87}Rb decay? Why do you think it was omitted from the list of important naturally occurring radionuclei?

20.15 According to the text, roughly 10–15% of lung cancer deaths in the US are due to radon exposure. Use the LNT model and average yearly effective dose from radon to estimate the cancer rate from this source.

20.16 According to relativity, the clocks of an object moving with speed $v = \beta c$ are slowed down by a factor $\gamma \equiv 1/\sqrt{1-\beta^2}$. The total (relativistic) energy of the object is given by $mc^2\gamma$. The mean energy of cosmic ray muons at creation is about 6 GeV. How far can such a muon travel on average before decay?

20.17 The flux of cosmic ray muons at sea level is about one per square centimeter per minute. Given their mass stopping power of 2 MeV cm^2/gm, and their weighting factor $w_\mu = 1$, estimate an average person's (living at sea level) equivalent dose in mSv/y from cosmic ray muons.

20.18 Particle physics experiments are shielded from cosmic ray muons by placement underground, often in old mines. Cosmic ray muons have been detected at 2000 feet below ground at the Homestake Gold Mine in South Dakota. Estimate a lower limit on the energy of these muons when created high in the atmosphere.

20.19 ^{40}K accounts for 0.0117% of naturally occurring potassium and potassium accounts for about 0.2% of the human body mass. Considering only the dominant decay of ^{40}K, which yields a β-decay electron of average energy 0.56 MeV, compute the average annual dose of radiation from potassium inside the body of a 70 kg person. Why can the exposure due to the γ-rays accompanying the electron capture decay be ignored in this estimate?

20.20 Using the data from Table 18.2, compute the spontaneous fission rate (in Bq/kg) for ^{240}Pu. Compare this with the rate in ^{239}Pu and in uranium enriched to 85% ^{235}U. (See Problem 20.10.)

20.21 In §20.6.1 it was stated that approximately 21 t of SNF are removed yearly from a 1 GW$_e$ reactor operating at 33% efficiency. Assuming that the spent fuel contains 0.9% unconsumed ^{235}U and that the reactor derives 1/3 of its power from burning ^{239}Pu created during normal operation, determine the initial enrichment of the fuel (the initial percentage ^{235}U).

20.22 Consider a situation in which, through accident or ill-intention, all the spent nuclear fuel from a year's operation of a 1 GWe light-water reactor were uniformly distributed across one square kilometer of land 1000 years after the waste is sequestered. Using the information in the chapter, estimate the radiation exposure in Sv/year to an unprotected human on this land. For simplicity, assume that the activity is dominated by ^{241}Am. (According to [107] the effective dose (per hour) from a planar ^{241}Am source of γ radiation at a distance of one meter is 1.9×10^{-4} mSv/h for an activity (per unit area) of 1 MBq/m^2.) How broad an area would be needed to reduce exposure below the level of background radiation from within the body (0.29 mSv/y)? Based on this calculation, estimate the area that could be rendered "unsafe/uninhabitable for children" (>10 mSv/year) if all waste from 50 years of nuclear power (from light-water reactors) at 2 TWe were distributed uniformly on the surface 1000 years later.

Energy in the Universe

What precisely *is* energy? We have discussed a wide variety of forms of energy and analyzed many specific energy systems so far in this book without directly addressing this question. For the most part, we have taken a fairly pragmatic view and simply described the different manifestations of energy and their inter-conversion with the understanding that energy exists and is conserved. Colloquial definitions of energy typically state that "energy is the capacity to do work," an assertion that is contradicted by such forms of energy as rest energy ($E = mc^2$) or the vacuum energy that pervades space. Other sources, by way of explanation, enumerate the many forms of energy such as electromagnetic, light, kinetic, thermal, and so forth, and emphasize the capacity of energy to change from one of these forms to another.

In this chapter we take an excursion into some more advanced areas of physics – mechanics beyond §2, and special and general relativity in particular – to shed some light on what precisely energy *is*. In §21.1 we give a general characterization of energy, and identify those systems for which energy can be defined and the circumstances under which it is conserved. We also explain the connection between energy and more general conservation laws. In §21.2 we give a brief history of energy in the universe, from the instant before the big bang up through the present time.

21.1 What is Energy?

At a fundamental level, the concept of energy arises when we consider how a system evolves in time. In particular: *Any isolated physical system with time-invariant quantum or classical physical laws has a conserved quantity, which is called **energy**, associated with the evolution of that system in time.* If a system is not isolated, energy is still a very useful concept even though the system's energy can change

Reader's Guide
This chapter takes a short theoretical diversion to explore the nature of energy and its role in the universe as a whole. We tie up some theoretical loose ends in this chapter that are not resolved elsewhere. Some of the issues addressed here touch on questions at the frontier of current research in fundamental physics. The material in this chapter is not necessary for understanding the more specific energy systems described in the rest of the book, but provides some theoretical background that may help the interested reader to understand energy in a broader context.

Prerequisites: §2 (Mechanics), §7 (Quantum mechanics). Although an effort has been made to make this material self-contained, the perspective of §21.1 is more abstract than that of the rest of the book, and assumes a slightly higher level of mathematical experience; in particular, some familiarity with linear algebra may be helpful in parts of this section. §21.2, however, requires no further background or mathematical sophistication.

as the system exchanges energy with its environment. In some contexts, such as the large-scale structure and evolution of the early universe, we do not have a complete theoretical framework in which the laws of physics can be expressed in a time-independent fashion, so the definition of energy as a conserved quantity becomes problematic. In this section we explain these statements.

The laws of classical mechanics and electromagnetism are approximations to a more fundamental set of physical laws described by quantum theory. In fact, energy appears as a fundamental entity very naturally in quantum theory, and directly governs the time evolution of physical systems. While energy plays a similar role in classical physics, this is less apparent in Newton's and Maxwell's laws of classical physics as we have described them so far.

We therefore turn first, in §21.1.1, to quantum systems, where the role of energy and its relationship to time are

What is Energy?

Energy is a quantity associated with the evolution of a physical system in time. Any closed system described by physical laws that are unchanging in time has a conserved quantity that is called energy.

particularly simple (even if the physics is less familiar). We then describe energy in classical systems in §21.1.2. The association of energy with time-invariant physical laws is a special case of a more general class of relations in physics between conserved quantities and symmetries that we discuss further in §21.1.3. §21.1.4 and §21.1.5 contain brief introductions to some aspects of Einstein's special and general theories of relativity. Throughout these discussions we point out some physical systems that violate the conditions necessary for energy conservation.

21.1.1 Energy in Quantum Physics

The fundamental role played by energy in quantum mechanics is evident in the axioms of §7: for any isolated system governed by physical laws that do not change in time, the system's quantum state can be written as a superposition of its *energy basis states* $|s_i\rangle$, $i = 1, \ldots, N$, and each energy basis state evolves in time through multiplication by a phase proportional to its energy,

$$|s_i(t)\rangle = e^{-i E_i (t - t_0)/\hbar} |s_i(t_0)\rangle. \qquad (21.1)$$

Thus energy is conserved and determines the time evolution of all states of the system.

In §7, we formulated the time evolution of an *arbitrary* quantum state in terms of a matrix differential equation (7.29), the *time-dependent Schrödinger equation*

$$i\hbar \frac{d}{dt} |s(t)\rangle = H |s(t)\rangle. \qquad (21.2)$$

Here the state $|s(t)\rangle$ is represented by a column vector in a (complex) vector space, and H is the *Hamiltonian matrix*, in which the only nonzero entries are the energies along the diagonal.

In a more general basis for the space of states, the matrix H is no longer diagonal. As long as H does not depend explicitly on the time, however, it is always possible to choose a basis of energy states that evolve in time according to eq. (21.1).[1] Wave mechanics, the subject of

§7.2, provides an example of such a situation. The time-dependent Schrödinger wave equation (7.7) has the form of eq. (21.2) with

$|s(t)\rangle \rightarrow \psi(\mathbf{x}, t)$ and

$$H \rightarrow H(\mathbf{x}) = -(\hbar^2/2m)\nabla^2 + V(\mathbf{x}). \qquad (21.3)$$

When we solved for the spectrum of energies for various potentials in §7.8, we were finding the energy basis states in which H assumes a simple diagonal form.

Note that throughout §7 we assumed that we were working with a **closed system**, one that is completely disconnected from other physical processes. Thus we computed the energy levels of a hydrogen atom, for example, isolated from the rest of the world.

Many quantum systems of practical interest – a semiconductor exposed to sunlight, for example – are not isolated. Usually, it is possible to describe such a system's interactions with outside agents in terms of a *time-dependent* Hamiltonian. An electron in a semiconductor exposed to solar radiation, for example, can be described by a Hamiltonian in which external, time-dependent electromagnetic fields appear. A particularly simple example would be replacing $V(\mathbf{x})$ by $V(\mathbf{x}, t)$ in eq. (21.3). The time evolution of such a system is still described by a time-dependent Schrödinger equation like (21.2), but now with an explicitly time-dependent Hamiltonian

$$i\hbar \frac{d}{dt} |s(t)\rangle = H(t) |s(t)\rangle. \qquad (21.4)$$

When the Hamiltonian is time-dependent, it is not in general possible to find a set of states obeying eq. (21.1), and the energy of the quantum system is not conserved. Instead the system can be driven from one energy state to another by the outside agent that is modeled by the time-dependent potential $V(\mathbf{x}, t)$.

If we extend the boundaries of the system to include all of the mutually interacting parts, the Hamiltonian that describes the whole system is once again time-independent and the total energy is conserved. The choice to consider only a part of a system, the energy of which is not conserved, is one of convenience, not a fundamental violation of energy conservation. A different, and more serious, problem with conservation of energy arises in the treatment of gravity in the context of cosmology, as described in §21.1.5.

---

[1] The proof of this assertion depends on the technical fact that in quantum mechanics the Hamiltonian is always a **Hermitian operator**, meaning that it is equal to its transpose conjugate $H_{ij} = H_{ji}^*$.

21.1.2 Energy in Classical Physics

In classical physics, there is also a close connection between energy conservation and time invariance of the physical laws governing the system, though this relationship is not as directly evident in Newton's formulation of classical mechanics as it is in quantum physics. Nonetheless, we can describe energy conservation in classical systems in a simple way that makes this connection apparent.

Consider as an elementary example a system of N particles with masses m_i at positions x_i, $i = 1, \ldots, N$, interacting by a potential $V(x_1, \ldots, x_N, t)$. As in the quantum analysis above, we include the possibility that the potential may have a dependence on time; this may describe, for example, interactions with an outside agent. For simplicity we consider a system in one dimension, though the same analysis works in two or three spatial dimensions. The energy of this system is

$$E(t) = \sum_i \frac{1}{2} m_i \dot{x}_i^2 + V(x_1, \ldots, x_N, t). \quad (21.5)$$

Newton's equations for the system are

$$m_i \ddot{x}_i = F_i = -\frac{\partial V}{\partial x_i}. \quad (21.6)$$

The time derivative of energy is then

$$\frac{dE}{dt} = \sum_i \left(m_i \dot{x}_i \ddot{x}_i + \frac{\partial V}{\partial x_i} \dot{x}_i \right) + \frac{\partial V}{\partial t} = \frac{\partial V}{\partial t}, \quad (21.7)$$

so energy is conserved as long as the form of the potential function V is time-independent.

The laws of classical physics can be formulated in a different way that makes the connection between energy and time development more evident. In general, any classical physical system can be described in terms of a set of $2n$ **dynamical variables**, consisting of n **generalized coordinates** q_i, $i = 1, \ldots, n$, and n **conjugate momenta** p_i associated with variations in the q_i. A particle moving under the influence of a potential $V(x)$ in three dimensions provides a simple example. Here the generalized coordinates can be chosen to be the three Cartesian coordinates $q_i = x_i$ and the conjugate momenta are simply $p_i = m\dot{x}_i$. Newton's laws can be reformulated in terms of the q_i's and p_i's and a **classical Hamiltonian** function $H(q_i, p_i, t)$, which will be interpreted as the energy of the system. In terms of the classical Hamiltonian, the equations of motion are

$$\dot{q}_i = \frac{\partial H}{\partial p_i} \quad \text{and} \quad \dot{p}_i = -\frac{\partial H}{\partial q_i}, i = 1, \ldots, n. \quad (21.8)$$

For example, for a classical harmonic oscillator in one dimension the Hamiltonian is $H = p^2/2m + kq^2/2$, and the equations of motion are $\dot{q} = p/m$, $\dot{p} = -kq$, or $m\ddot{q} = -kq$.

In classical mechanics one is typically interested in the time evolution of the dynamical variables or of functions of the dynamical variables, $\phi(q_i, p_i, t)$, such as the angular momentum or the potential energy. These functions depend implicitly on the time because the generalized coordinates and conjugate momenta vary as the system develops in time; in addition ϕ may also depend explicitly on the time. The time variation of ϕ is therefore given by

$$\dot{\phi} = \frac{d\phi}{dt} = \frac{\partial \phi}{\partial t} + \sum_i \left(\frac{\partial \phi}{\partial q_i} \dot{q}_i + \frac{\partial \phi}{\partial p_i} \dot{p}_i \right)$$

$$= \frac{\partial \phi}{\partial t} + \sum_i \left(\frac{\partial \phi}{\partial q_i} \frac{\partial H}{\partial p_i} - \frac{\partial \phi}{\partial p_i} \frac{\partial H}{\partial q_i} \right). \quad (21.9)$$

This equation can be written succinctly as

$$\dot{\phi} = \frac{\partial \phi}{\partial t} + \{\phi, H\} \quad (21.10)$$

in terms of a quantity $\{\phi, H\}$ called the **Poisson bracket**, which is defined through

$$\{A, B\} \equiv \sum_i \left(\frac{\partial A}{\partial q_i} \frac{\partial B}{\partial p_i} - \frac{\partial A}{\partial p_i} \frac{\partial B}{\partial q_i} \right) \quad (21.11)$$

for any two functions A, B. Thus, the Hamiltonian generates the time evolution of classical systems through a system of first-order differential equations, just as it does for quantum systems. This very general framework for describing classical mechanical systems is known as the **Hamiltonian formalism** and was developed by the Irish mathematician and physicist William Rowan Hamilton in the 1830s. This framework, with some slight generalizations for systems with e.g. periodic coordinates or field degrees of freedom with a continuous index, can be used to describe any classical system, including configurations of particles, waves, electromagnetism, and even systems subject to constraints.

In classical mechanics a quantity ϕ is conserved if $\dot{\phi}$ computed from eq. (21.10) vanishes identically. In particular, if we take $\phi = H$,

$$\frac{dH}{dt} = \frac{\partial H}{\partial t}, \quad (21.12)$$

since $\{H, H\} = 0$ trivially. So H gives a conserved energy whenever H depends on time only through its dependence on the dynamical variables q_i, p_i. This generalizes eq. (21.7), which we derived for a one-dimensional system of particles interacting by a potential.

In classical mechanics, as in quantum mechanics, it is often convenient to analyze systems that interact with an external agent or environment. Since such systems can exchange energy with their environment, their energy is not conserved. The concept of energy nevertheless remains useful in analyzing these systems. A playground swing pushed by an external agent (see Box 38.1) provides a simple example. The swing may be described as an oscillator, but the external agent cannot be represented as a time-independent potential. Thus the energy of the swing as an isolated system is not conserved. Other examples include frictional losses in mechanical systems that can be encoded as external forces, and thermodynamic systems in equilibrium with their environment at fixed temperature, where the internal energy fluctuates, but is conserved on average.

21.1.3 Other Conserved Quantities and Symmetries

Equations (21.10) and (21.12) summarize the connections between energy and time development and between conserved energy and time-independent physical laws in the context of classical physics. Equation (21.10) has even deeper implications, however. It implies a special role for the energy (i.e. the Hamiltonian) in conservation laws in general, since any dynamical variable that does not depend explicitly on time and has zero Poisson bracket with H will be a constant of the classical motion.

When a system is *invariant* under a transformation, it is said to possess a *symmetry*. Symmetries and conservation laws are inextricably related in both classical and quantum mechanics. When a classical system possesses a symmetry, a dynamical quantity associated with the symmetry has vanishing Poisson bracket with the Hamiltonian, and is therefore conserved in time. In the previous section we showed that when a system's Hamiltonian is invariant under time translation $t \to t + \delta$, then the system's energy is conserved. Thus the symmetry of *time translation invariance* implies energy conservation.

Another simple example of the relation between symmetry and conservation is *momentum conservation*. A physical system has a conserved momentum associated with every spatial direction in which the system's Hamiltonian is invariant under translation. For simplicity we consider the classical case; an analogous argument goes through in quantum mechanics. Suppose, for example, that a subset $q_a, a = 1, \ldots, m$, of a system's generalized coordinates are translated

$$q_a \to q_a + \delta, \tag{21.13}$$

and all the other coordinates and all the momenta are left invariant. Such a situation would arise, for example, if the q_a's were the x-coordinates of m particles and the whole system were translated by an amount δ in the x-direction. If the classical Hamiltonian is invariant under eq. (21.13) for all values of δ, then

$$\left(\sum_{a=1}^{m} \frac{\partial}{\partial q_a} \right) H = 0. \tag{21.14}$$

This implies that the Poisson bracket of $p_x \equiv \sum_{a=1}^{m} p_a$ with the Hamiltonian vanishes,

$$\{p_x, H\} = -\sum_{a=1}^{m} \frac{\partial}{\partial q_a} H = 0, \tag{21.15}$$

and therefore, from eq. (21.10), p_x is conserved,

$$\dot{p}_x = \{p_x, H\} = 0. \tag{21.16}$$

Thus, the x component of momentum is conserved if the Hamiltonian has translation invariance in the x-direction.

We used similar logic in a quantum context in §7.7.1 to derive the form of the free particle wavefunction by requiring that the energy of a quantum state not be changed by a translation, so that the free particle wavefunction of definite energy can change only by a phase when $x \to x + \delta$. Translation invariance reappears in a more subtle fashion in the description of semiconductors in §25.2, where we use invariance under certain translations to identify a *lattice momentum* that is conserved in periodic crystals and is useful in understanding quantum physics in photovoltaic cells.

Another case that is widely useful is conservation of angular momentum. Consider for example a particle in three dimensions with Hamiltonian $H(x, p)$, and consider the angular momentum about the z-axis, defined by $L = xp_y - yp_x$. According to eq. (21.10), L is conserved if its Poisson bracket with H vanishes. A direct calculation (Problem 21.4) shows that

$$\dot{L} = \{L, H\} = -\frac{\partial H}{\partial \phi} - \frac{\partial H}{\partial \phi_p}, \tag{21.17}$$

where $\phi = \tan^{-1} y/x$ is the azimuthal angle around the z-axis, and likewise for the azimuthal angle of the momentum, $\phi_p = \tan^{-1} p_y/p_x$. Thus angular momentum is conserved if the Hamiltonian is invariant under a rotation of the system (which rotates both the particle's position and momentum) about the z-axis. This result includes the familiar example that angular momentum is conserved by a central force (where the potential energy depends only on r, the distance from the origin). These ideas are easily generalized to include all three coordinates of angular momentum and arbitrary multi-component systems. Symmetries and conserved quantities play a fundamental role in understanding many physical systems, such as the hydrogen atom discussed in §7.8.2, whose quantum analysis is simplified by the fact that the Coulomb potential, which binds the electron to the proton, is invariant under rotations in three dimensions.

Box 21.1   Newton's Equations From Quantum Mechanics

The Hamiltonian framework for classical mechanics is very naturally related to the Hamiltonian description of quantum mechanics. This connection helps to clarify how classical physics emerges from quantum mechanics. As discussed in Box 7.6, a superposition of quantum states for a particle can describe a wave packet that is relatively well localized in space and has an approximate classical position $x(t)$ and momentum $p(t)$ (but still obeys $\Delta x \Delta p \geq \hbar/2$). For such a wave packet, we can show that the *average* motion of the particle obeys the laws of classical physics, so that Newton's laws of motion follow naturally from quantum mechanics.

As a simple example, consider a one-dimensional quantum particle described by a wavefunction $\psi(x, t)$ moving in a potential $V(x)$. Since $|\psi(x, t)|^2$ is the probability of finding the particle between x and $x + dx$ at time t, the average or *expectation value* of $x(t)$ is given by

$$\langle x \rangle_{\psi(t)} = \int dx \, x \, |\psi(x, t)|^2 = \int dx \, \psi^*(x, t) \, x \, \psi(x, t) \,.$$

If the particle is well localized around $x(t)$, then we can assume that $\langle x \rangle_{\psi(t)} = x(t)$. This quantity then describes the classical position of the particle.

When we associated the Hamiltonian differential operator $H = -\frac{\hbar^2}{2m}\nabla^2 + V(\boldsymbol{x})$ with the energy in §7, we implicitly associated the differential operator $P \equiv -i\hbar\partial/\partial x$ with the momentum. This is evident in the de Broglie relation $p = h/\lambda$, in the one-dimensional Schrödinger wave equation (7.34) where $p^2 \leftrightarrow -\hbar^2\partial^2/\partial x^2$, and in the form of the free particle wavefunction $e^{ipx/\hbar}$, which obeys $(-i\hbar\partial/\partial x)e^{ipx/\hbar} = pe^{ipx/\hbar}$. In fact the expectation value of the momentum in an arbitrary quantum state is given by

$$\langle p \rangle_{\psi(t)} = \int dx \, \psi^*(x, t) \, P\psi(x, t) = \int dx \, \psi^*(x, t) \left(-i\hbar \frac{\partial}{\partial x} \right) \psi(x, t) \,.$$

This statement can be connected to Axiom 4 in a precise fashion, but we focus here on the connection to Newton's laws. It follows directly from the Schrödinger equation that (Problem 21.2)

$$\frac{d}{dt}\langle x \rangle_{\psi(t)} = \frac{1}{m}\langle p \rangle_{\psi(t)} \,,$$

which is equivalent to the classical relation $\dot{x} = p/m$, and

$$m\frac{d^2}{dt^2}\langle x \rangle_{\psi(t)} = \frac{d}{dt}\langle p \rangle_{\psi(t)} = -\langle \frac{\partial}{\partial x}V(x) \rangle_{\psi(t)} \,,$$

reproducing Newton's second law $m\ddot{x} = F = -\partial_x V(x)$. Note that the potential $V(x)$ appears in the expectation value, but as long as the particle is localized we have $\langle V(x) \rangle \sim V(\langle x \rangle)$. Thus, Newton's laws of motion emerge directly from quantum theory when the wavefunction of a particle moving in a potential is localized and describes a classical object.

21.1.4 Special Relativity

Digging more deeply into the fundamental nature of energy and symmetries leads us to the subject of Einstein's special and general theories of relativity. Occasionally throughout this book we quote isolated results from special relativity. Here we briefly describe the essence of this theory, which relates space and time in a fundamental fashion.

The basic idea of Einstein's special theory of relativity is that the laws of physics are the same in coordinate systems, or **reference frames**, that are moving relative to one another at a fixed velocity.[2] For example, the laws of electromagnetism, including Maxwell's equations, should take the same form for a stationary observer and an observer passing by in a rocket ship at 10^8 m/s.

In particular, the velocity of light, which follows from Maxwell's equations (see §4.3), should appear the same to both observers (Problems 21.6 and 21.10). The trajectory of a light ray that passes the point $\boldsymbol{x}$ at time t and passes $\tilde{\boldsymbol{x}}$ at time $\tilde{t}$ obeys $|\tilde{\boldsymbol{x}} - \boldsymbol{x}| = c(\tilde{t} - t)$, or

$$\Delta x^2 + \Delta y^2 + \Delta z^2 - c^2\Delta t^2 = 0 \,, \tag{21.18}$$

---

[2] This applies to *inertial observers*, those for whom Galileo's principle of inertia – that an object not acted upon by a force either remains at rest or persists in rectilinear motion – holds. Inertial reference frames are discussed further in §27.2.

Example 21.1   Two Interacting Particles in Two Dimensions

As a simple example of a system with several conserved quantities, consider a pair of pointlike particles with masses m_1, m_2 at positions $\boldsymbol{x}_1 = (x_1, y_1)$ and $\boldsymbol{x}_2 = (x_2, y_2)$, which interact through a potential $V(r)$, where r is the distance between the particles $r = |\boldsymbol{x}_1 - \boldsymbol{x}_2|$. This system can be described classically by a Hamiltonian

$$H = \frac{\boldsymbol{p}_1^2}{2m_1} + \frac{\boldsymbol{p}_2^2}{2m_2} + V(r).$$

Since H has no explicit dependence on t, $\dot{H} = 0$ and the Hamiltonian defines a conserved energy. A small computation (Problem 21.3) shows that $(\partial_{x_1} + \partial_{x_2})H = 0$, so that the total momentum $p_x = p_{1x} + p_{2x}$ is conserved. Similarly, the total momentum in the y-direction is conserved. The Hamiltonian is also rotationally invariant, so that angular momentum in the plane, $L = x_1 p_{1y} - y_1 p_{1x} + x_2 p_{2y} - y_2 p_{2x}$, is also conserved. This system can be conveniently described in the reference frame in which the center of mass is fixed at the origin, where $m_1 \boldsymbol{x}_1 + m_2 \boldsymbol{x}_2 = 0$; the total momentum vanishes in this rest frame, as can be seen by taking the time derivative of this condition. There is also a simple quantum description of this system, where the wavefunctions for the system are functions $\psi(\boldsymbol{x}_1, \boldsymbol{x}_2)$ and the Hamiltonian is the same as given above, with momenta replaced by the appropriate derivative operators (see Box 21.1).

Conserved Quantities and Symmetries

Conserved quantities are generally associated with symmetries of a (closed) physical system. Energy is conserved in any system governed by physical laws that are invariant in time. Momentum is conserved in any system that is invariant under spatial translation. Angular momentum is conserved in any system that is invariant under rotation.

where $\Delta x = \tilde{x} - x$, etc. for the three Cartesian coordinates and $\Delta t = \tilde{t} - t$. According to special relativity, a light ray must obey the same equation when its trajectory is expressed in terms of the coordinates of any other inertial observer. In other words, eq. (21.18) must be invariant under the transformation that relates one observer's coordinates to another's.

This situation is reminiscent of the invariance of the length of a vector in three-dimensional Euclidean space under rotations. If $\boldsymbol{\Delta x}$ is the vector from $P = (x, y, z)$ to $\tilde{P} = (\tilde{x}, \tilde{y}, \tilde{z})$, then the distance between P and $\tilde{P}$ is given by the Pythagorean relation

$$d^2 = \Delta x^2 + \Delta y^2 + \Delta z^2. \tag{21.19}$$

Under a rotation, the coordinates of P and $\tilde{P}$ change, but the distance d is invariant. It is convenient to view eq. (21.19) as the inner product of a vector $\boldsymbol{\Delta x}$ with itself,

$$d^2 = \boldsymbol{\Delta x} \cdot \boldsymbol{\Delta x}, \tag{21.20}$$

or, in terms of a sum over the Cartesian components of $\boldsymbol{\Delta x}$,

$$d^2 = \sum_{i,j=1}^{3} \Delta x_i \delta_{ij} \Delta x_j, \tag{21.21}$$

where

$$\delta_{ij} = \begin{pmatrix} 1 & 0 & 0 \\ 0 & 1 & 0 \\ 0 & 0 & 1 \end{pmatrix}, \tag{21.22}$$

is the **metric tensor** for three-dimensional Euclidean space. (For a brief introduction to tensors, see Appendix B.3.3.) Two observers who are rotated with respect to one another agree about the length of vectors, and in particular, they agree on the distance between two points given by eq. (21.19).

In special relativity, space and time are combined to form a four-dimensional vector space (**space-time**) with coordinates

$$x^\mu = (ct, x, y, z), \tag{21.23}$$

where the Greek index μ takes values 0, 1, 2, 3 for the four space-time coordinates. Points in this space are known as **events**, with specific coordinates in space and time. The separation Δ in space-time between a pair of events $P = (ct, x, y, z)$ and $\tilde{P} = (c\tilde{t}, \tilde{x}, \tilde{y}, \tilde{z})$ is defined to be

$$\Delta^2 = \Delta x^2 + \Delta y^2 + \Delta z^2 - c^2 \Delta t^2, \tag{21.24}$$

where $\Delta x^\mu = \tilde{x}^\mu - x^\mu = (c\Delta t, \Delta x, \Delta y, \Delta z)$ is known as a **four-vector** in analogy to the 3-space vector $\boldsymbol{\Delta x}$ defined earlier. The speed of light is the same for all observers if the transformations from one inertial coordinate system to another leave Δ^2 invariant. Thus special relativity generalizes the concept of a rotation from space to *space-time* in such a way as to guarantee that all observers agree on the velocity of light.

The coordinate transformations that leave the speed of light invariant mix space and time in much the same way that ordinary Euclidean rotations can mix the x and y coordinates through a rotation around the z-axis. Thus, for example, two events that are simultaneous for one observer appear to occur at different times to an observer moving at a constant velocity relative to the first. Among the consequences of these transformations is the fact that the speed of light is the ultimate limit on the speed of any observer or object. This principle and many other surprising consequences of special relativity such as *time dilation* have been verified by many experiments. The appearance at sea level of muons created by cosmic rays high in the atmosphere is one example already encountered in §20 (see §20.5.2 and Problem 20.18).

Because all observers agree on the **invariant interval** Δ^2 between two events, it can be used to classify the causal relation between events. Δ^2 (21.24) can be positive, negative, or zero. If Δ^2 is zero, the events P and $\tilde{P}$ are on the path of a light ray and light from one event can influence the other. If Δ^2 is positive then there is a reference frame in which the two events occur at the same time and at different places (Problem 21.8). Since no signal can propagate faster than the speed of light, neither of these events can influence the other. This is reassuring since it is also possible to show that the time ordering of such events can be reversed in different reference frames, which would make the notion of causality problematic. Such events are termed *causally disconnected*. If Δ^2 is negative, then there is a reference frame where the two events occur at the same place and at different times (Problem 21.9) and the earlier one can influence the later. In this case the events are termed *causally connected*. Figure 21.1 illustrates these concepts.

A convenient way of characterizing the invariant interval in eq. (21.24) is to generalize the concept of a metric tensor introduced in eq. (21.21). If the **metric tensor of special relativity** is defined by

$$\eta_{\mu\nu} = \begin{pmatrix} -1 & 0 & 0 & 0 \\ 0 & 1 & 0 & 0 \\ 0 & 0 & 1 & 0 \\ 0 & 0 & 0 & 1 \end{pmatrix}, \quad (21.25)$$

then the invariant interval between two points separated by the space-time four-vector Δx^{μ} can be written

$$\Delta^2 = \sum_{\mu,\nu=0}^{3} \Delta x^{\mu} \, \eta_{\mu\nu} \, \Delta x^{\nu} . \quad (21.26)$$

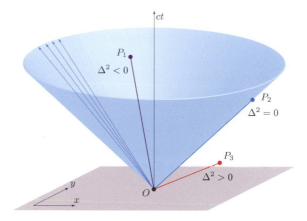

Figure 21.1 Space-time with the z-direction suppressed. The paths of light rays, for example $\overline{OP_2}$, emanating from O are shown in blue. Points O and P_1 are causally connected; the line connecting them represents a signal traveling at $v < c$. O and P_3 are causally disconnected; a signal connecting them would have to travel at $v > c$.

Einstein's Special Theory of Relativity

The *special theory of relativity* combines space and time into a single four-dimensional vector space (*space-time*). The laws of physics, including in particular the speed of light in a vacuum, are the same for observers moving relative to one another at a fixed velocity in space-time. The conserved energy and momentum of a system of mass m satisfy the relativistic relation $E^2 = c^2\mathbf{p}^2 + m^2c^4$.

The metric tensor thus gives a way of characterizing distances in space-time.

The quantities that appear in physical laws can be classified by the way that they transform from one reference frame to another frame traveling at a constant velocity relative to the first. Some quantities, such as mass and electric charge, are *invariant* – they are the same to all observers. Others, such as momentum and energy and electric and magnetic fields, appear differently in different reference frames. Since energy and momentum are related to translations in time and space, it is perhaps not surprising that they together form a four-vector

$$p^{\mu} = (E, cp_x, cp_y, cp_z) . \quad (21.27)$$

This four-vector transforms from one coordinate system to another in the same way as the coordinate four-vector x^{μ}. Therefore the quantity

$$p^2 = \sum_{\mu\nu} \eta_{\mu\nu} p^\mu p^\nu \qquad (21.28)$$

is the same in all reference frames. Substituting from eqs. (21.25) and (21.27) we find

$$-p^2 = E^2 - \boldsymbol{p}^2 c^2 = m^2 c^4. \qquad (21.29)$$

This is the origin of Einstein's famous relation $E = mc^2$ (for an object at rest), which has been referred to in several previous chapters.

Other quantities of interest transform between reference frames in more complex ways. A magnetic field in one reference frame appears as a combination of electric and magnetic fields to an observer moving relative to the first frame. This fact was key to understanding the concept of electromagnetic induction as described in Box 3.6. In special relativity the electric and magnetic fields are encoded in the electromagnetic **field strength tensor** $F_{\mu\nu}$, which is antisymmetric in its indices ($F_{\mu\nu} = -F_{\nu\mu}$). An antisymmetric 4×4 matrix has six independent components, which in this case form the spatial vectors $\boldsymbol{E}$ and $\boldsymbol{B}$. In particular, $c\boldsymbol{E} = (F_{10}, F_{20}, F_{30})$ and $\boldsymbol{B} = (F_{23}, F_{31}, F_{12})$. The transformation rules for four-tensors are straightforward, but algebraically more complicated than the rules for four-vectors. One consequence is that the quantity $E^2 - B^2$, like Δ^2, is the same in all reference frames. Thus although a magnetic field in one reference frame may transform to a combination of electric and magnetic fields in another frame, there is no reference frame in which only an electric field appears.

The metric tensor, $\eta_{\mu\nu}$, which can be regarded as a convenience and a notational simplification in special relativity, plays a central role in Einstein's theory of gravity, the *general theory of relativity*, to which we now turn.

21.1.5 General Relativity

Einstein's **general theory of relativity** provides a geometric description of classical gravity that goes beyond Newton's gravitational force law. While not used elsewhere in this book, it is not possible to discuss the origins of energy in the early universe without referring to some aspects of the general theory of relativity.

Curvature and Einstein's Equations In Newton's theory of gravity, objects at a distance attract one another with a force proportional to $1/r^2$. In electromagnetism, the apparent action at a distance of Coulomb's law is replaced by the local action of the electric and magnetic fields, as discussed in §3. Einstein's theory of general relativity plays a similar role for gravity, but requires a more profound extension of the underlying physics. The basic idea of Einstein's theory of gravity is that space-time itself is

curved. Matter and energy produce local changes in the geometry of space-time, and this geometry in turn affects matter.

It may be difficult to imagine what it means for space-time itself to be curved. The situation is somewhat similar to that of a person walking in a vast desert landscape on our planet, where the ground is flat and appears to form part of an infinite plane. Although Earth's surface is curved, the geometry of the surface looks locally like Euclidean two-dimensional space, where one can impose an infinite Cartesian coordinate system with distances measured by $d = \sqrt{x^2 + y^2}$. In fact, however, the surface of Earth is not flat. A circle of radius r on Earth's surface can be measured to have a circumference that is smaller than $2\pi r$, with the deviation from this Euclidean result increasing as the size of the circle increases.

The *curvature* of space-time is conceptually similar to the curvature of Earth's surface, though more difficult to grasp since our space-time is (as far as we know) not embedded in any higher-dimensional flat space. A massive object such as Earth produces localized curvature in space-time in its vicinity (Figure 21.2). This curvature in turn affects the trajectories of moving objects (such as the Moon). While this may seem like a very complicated story with which to replace the simple Newtonian notion of gravitational attraction, careful calculations show that the motion of the planets and other astronomical objects precisely follow the predictions of general relativity, which increasingly differ from those of Newtonian gravity as space-time curvature becomes more

Figure 21.2 In Einstein's theory of general relativity, the force of gravity arises due to curvature in space-time. A massive object such as Earth generates curvature in the local **metric** describing space-time distances. The consequence of this curvature is that objects moving through space-time in the vicinity of the massive object follow curved trajectories – like the elliptical orbit of the Moon around Earth (Credit: BBC)

pronounced. For example, the orbit of Mercury – the planet closest to the Sun and therefore subject to the greatest space-time curvature – gradually precesses over time in a way that is correctly explained by general relativity and not by Newtonian gravity.

Curvature in space-time is described by local changes in the notion of distance. The constant metric (21.25) that measures space-time distances in special relativity is replaced in general relativity with a metric tensor $g_{\mu\nu}(x)$ that varies depending on the position in space-time. The general metric tensor describes local distances between infinitesimally separated points x^μ and $x^\mu + \delta x^\mu$ through

$$ds^2 = \sum_{\mu,\nu} g_{\mu\nu}(x)\delta x^\mu \delta x^\nu . \tag{21.30}$$

Curvature is then a symmetric two-index tensor that can be defined in terms of (second and products of two first) derivatives of the metric tensor. Einstein's equations of general relativity take the schematic form

$$(\text{curvature})_{\mu\nu} = \frac{8\pi}{c^4} G T_{\mu\nu} , \tag{21.31}$$

where G is *Newton's constant*, and the **energy–momentum tensor** (also known as the **stress–energy tensor**) $T_{\mu\nu}$ encodes the energy and momentum densities, and pressure and stress of matter and radiation, which are sources of curvature in space-time. The component T_{00} describes the energy density, $T_{00} = \rho$; $T_{0i} = T_{i0}$ describes the density of momentum in the ith spatial direction for $i = 1, 2, 3$; and T_{ij} for $i, j = 1, 2, 3$ describes the flux of the ith component of momentum in the jth direction. For an isotropic distribution of matter, the off-diagonal components of $T_{\mu\nu}$ vanish and each of the three diagonal components equals the pressure $T_{ii} = p$ for $i = 1, 2, 3$. Note that in general relativity, unlike in most other physical systems studied in this book, energy enters as an absolute, not a relative quantity. The curvature of space-time determines a definite *zero point* for the measurement of energy.

Classically, energy in general relativity can be understood in terms of curvature. By looking at the metric of space-time on a large sphere, for example, surrounding a region containing matter, it is possible to compute the total energy within the region in terms of the curvature of the surrounding space-time. This gives a sensible way of thinking of energy in general relativity as long as gravitational effects are classical and the system considered is isolated (so that space-time approaches a flat geometry far away from the system). The nature of energy in general relativity becomes more subtle when quantum effects and the global structure of space-time are considered.

The Cosmological Constant

When Einstein first formulated his general theory of relativity in 1916, the prevailing assumption was that the distant galaxies visible from Earth were, on average, moving neither toward nor away from Earth or one another. Einstein's theory, like Newton's, predicts that massive objects such as stars and galaxies attract and should accelerate towards one another. This is not consistent with a static universe. To resolve this apparent inconsistency, Einstein modified his equations by including a **cosmological constant** Λ as an additional source of curvature, replacing eq. (21.31) by

$$(\text{curvature})_{\mu\nu} = \frac{8\pi}{c^4} G T_{\mu\nu} - \Lambda g_{\mu\nu} . \tag{21.32}$$

The cosmological constant Λ has units of m^{-2}. Note that because $g_{00} < 0$, $\Lambda > 0$ acts like a constant, positive energy density throughout space-time that only interacts with gravitational fields. The cosmological constant is thus often referred to as **dark energy**. Unlike matter and radiation, however, which contribute positively to the pressure (T_{ii}), the cosmological constant gives a constant negative pressure throughout space-time, that in the absence of other effects would cause the material in the universe to accelerate apart. Note that here the fact that the *zero point* of energy is not arbitrary in the context of general relativity is essential: the cosmological constant cannot be defined away by shifting all energies by a constant. With this term included, Einstein's equations give a solution in which, by tuning Λ, the galaxies in the universe could maintain fixed distances from one another (though this solution would be unstable).

In the late 1920s, observations by the American astronomer Edwin Hubble showed conclusively that distant galaxies are receding from our own with a velocity proportional to their distance. With the realization that the universe is not static, Einstein's equations were able to describe observed phenomena without a cosmological constant. Late in the twentieth century the cosmological constant re-emerged in two crucial aspects of modern cosmology. Indeed, the cosmological constant, which parameterizes the zero-point energy of space-time, is related to some of the most profound outstanding puzzles in modern physics.

An Expanding Universe

Hubble's observation that the universe is expanding suggested that the observed universe was smaller in the past; this line of reasoning led in time to the **big bang theory**, which proposes that all observed matter has expanded from an original state of very high temperature and density. In such an **expanding universe**, gravitational attraction acts to slow the expansion. More recently, however, evidence has suggested that very early in its history, the universe underwent a period of accelerating expansion, and furthermore that it is entering a similar phase at the present time. In both eras the accelerated expansion can be accounted for by the presence of a cosmological constant, or something functionally equivalent. In particular, over the last decade, careful observations of distant supernovae and other astronomical phenomena have given definitive evidence that the rate at which distant galaxies are retreating from our own is *increasing* over time, not decreasing as would be expected from the gravitational attraction between the material in the universe. This is analogous to observing a ball, thrown upward, which slows down and then speeds up again, moving ever faster away from Earth. The simplest explanation for the increasing expansion rate is a cosmological constant, i.e. *dark energy*.

In the presence of a cosmological constant, the nature of energy in the universe presents some profound puzzles. The simplest solution of Einstein's equations for general relativity (21.32) arises if we assume that space is completely empty except for the *dark energy* associated with a nonzero cosmological constant. In this case, in an appropriate reference frame[3] the metric is given by

$$ds^2 = -c^2 dt^2 + a^2(t)(dx^2 + dy^2 + dz^2), \quad (21.33)$$

i.e. $g_{00} = -1, g_{0i} = 0, g_{ij} = a^2(t)\delta_{ij}$. Here $a(t)$ is a time-dependent *scale parameter*. In this frame the components of Einstein's equation (21.32) become

$$\frac{1}{c^2}\dot{a}^2 = \frac{\Lambda}{3}a^2, \quad \frac{1}{c^2}\ddot{a}(t) = \frac{\Lambda}{3}a. \quad (21.34)$$

The solution to these equations is

$$a(t) = e^{Ht}, \quad (21.35)$$

where $H \equiv \dot{a}/a = c\sqrt{\Lambda/3}$ is known as the **Hubble parameter**. Note that the left-hand sides of the equations in (21.34) describe the curvature, which entails two derivatives of g, so both curvature and Λ have units of $1/[\text{distance}]^2$, while H has units of $1/[\text{time}]$.

The physical interpretation of the solution (21.35) is that the scale of the universe is *inflating* at an exponential rate. In this coordinate frame, the distance between any two

points in the universe increases exponentially with time. To imagine what this means, consider an analogy with an expanding balloon. If the universe is the surface of the balloon, with various galaxies scattered about the surface, then as the balloon expands, each galaxy moves away from each other galaxy. One interesting feature of the solution (21.35) is that it produces an effective "*horizon.*" Points separated by a distance of more than c/H cannot be in causal contact, since the universe expands so quickly that a light ray from one point never reaches the other point (Problem 21.11).

The equations (21.34) can be modified to include matter and radiation. If we assume that the matter and radiation are distributed in a homogeneous (spatially uniform) and isotropic (rotationally invariant) fashion, we can characterize the energy density and pressure by ρ, p. This leads to the **Friedman equations**

$$\frac{\dot{a}^2}{a^2} = \frac{1}{3}\left(8\pi G\rho + \Lambda c^2\right), \quad (21.36)$$
$$\frac{\ddot{a}}{a} = -\frac{4\pi G}{3}\left(\rho + \frac{3p}{c^2}\right) + \frac{1}{3}\Lambda c^2.$$

From the second of these equations, we see that positive energy and pressure density act to slow the rate of expansion, acting in the opposite direction to a positive cosmological constant.

As the universe expands, the relative contributions to the energy density from matter, radiation, and the cosmological constant change over time. For nonrelativistic matter, the total energy remains constant as the volume increases proportionally to a^3, so the energy density of (nonrelativistic) matter scales as

$$\rho_M \sim \rho_M^{(0)}/a^3. \quad (21.37)$$

For radiation (or relativistic matter), the wavelength of each excitation (e.g. photon) increases with a, so that the energy density drops more quickly than for nonrelativistic matter

$$\rho_R \sim \rho_R^{(0)}/a^4. \quad (21.38)$$

For the cosmological constant, on the other hand, the energy density attributed to Λ is independent of a. Here, the difficulty in defining the total energy in an expanding universe begins to become apparent, since both $\rho_R a^3$ and Λa^3 change in time.

Note that for any given initial conditions of matter and energy density, the fraction of the universe's total energy that comes from any nonzero cosmological constant grows as the universe expands, and when Λ is the dominant term on the right-hand side of eq. (21.36) the equations simplify to eq. (21.34). Indeed, expansion of the universe according

---

[3] This form of the metric (and the following Friedman equations (21.36)) assumes that we live in a *spatially flat universe*, which is suggested to a high degree of precision ($< 0.5\%$) by recent experiments. Alternative shapes for the spatial universe could be a three-dimensional sphere of constant positive curvature, or a three-dimensional hyperboloid of negative curvature, which give slightly different forms to the metric and Friedman equations.

to eq. (21.35) is believed to be the mechanism responsible for the increasing rate at which distant galaxies are now moving away from our own.

Quantum Gravity While general relativity is well understood as a classical theory, despite a century of effort there is still no complete quantum theory of gravity that describes our universe. To appreciate the difficulty, note that in a quantum theory, quantities that can be precisely determined classically, such as position and momentum, become uncertain. In quantum gravity, the shape of space-time itself becomes indeterminate. Most of our physical theories use a fixed space-time *background metric* as a scaffolding on which to construct the theory, but this cannot be done for quantum gravity. Furthermore, the realm of quantum gravity is far removed from experimental exploration. In the past, experimenters have guided the development of new domains of physical understanding with critical experimental discoveries such as atomic spectra and the electron spin. The distance scale where the direct effects of quantum gravity are likely to be observed can be estimated by combining Newton's constant with Planck's constant $\hbar$ and the speed of light c to form a quantity with units of length. The resulting unit is known as the **Planck length** and has the value $\ell_{\text{Planck}} \cong \sqrt{\hbar G/c^3} \cong 1.616 \times 10^{-35}$ m, far smaller than the distances, of order 10^{-18} m, that can presently be probed at the highest energy accelerators. Thus quantum gravity must be studied primarily theoretically.

Note that despite the large gulf between the length scales involved in quantum gravity and direct experiment, measurements of some cosmological features may give indirect evidence for aspects of quantum gravity. In particular, upcoming measurements of polarization in the cosmic microwave background (§21.2.1) could suggest the existence of gravitational waves that have their origin in quantum fluctuations in the very early universe. Such indirect evidence, however, does not show any promise of elucidating any detailed aspects of a quantum theory of gravity.

The most successful approach to quantum gravity at this time is **string theory**. String theory has not yet been completely formulated from first principles, and a complete background-independent description of the theory may require the development of new mathematics. Nonetheless, string theory provides a collection of compatible descriptions of quantum gravity in specific limits that appear to tie together into a single unified framework. Furthermore, in more constrained situations with additional space-time dimensions and symmetries (particularly *supersymmetry*, a hypothetical symmetry relating bosons and fermions), any

gravity theory that satisfies known quantum consistency conditions has a description through string theory at high energies. It is not yet known, however, whether and/or how string theory can describe the specific physics observed in our four-dimensional universe including the Standard Model of particle physics and a small positive cosmological constant, or if there are other consistent theories of quantum gravity that may describe our universe.

Energy, Gravity, and Cosmology Defining energy in the context of general relativity and a space-time with cosmological structure and quantum dynamics presents significant challenges. When classical gravity is coupled to classical matter and radiation that inhabits a finite volume, with fixed boundary conditions for the metric at infinity approaching that of a flat space-time, the approach mentioned above of using the curvature of the system at large distances from the source can give a consistent way to define energy for theories with gravity. In a universe where the scale of space-time is inflating according to eq. (21.35), however, describing energy becomes more difficult. The basic problem is that, unlike in flat space, the background space-time metric (21.33) is not invariant under time translation. Since the cosmological constant contributes an energy density Λ, the total amount of dark energy in a given region of space increases as $a^3(t)\Lambda = e^{3Ht}\Lambda$. Thus, even without worrying about issues of quantum gravity, there is no obvious way to define a conserved energy.[4]

If the Standard Model of particle physics, or any other quantum field theory, is described as a quantum theory in a fixed space-time background with a positive cosmological constant Λ, the Hamiltonian of the theory is time-dependent and energy is not conserved. In the spirit of the discussion of closed systems, one should include the contribution to energy from gravity, but the absence of a complete quantum theory of gravity makes it difficult to precisely formulate the concept of energy for quantum theories when gravity is included. In particular, at present we do not have an adequate theoretical framework with which to define energy as a meaningful conserved quantity from the cosmological point of view in the very early universe.

In most practical situations, quantum gravitational effects are irrelevant, and space-time can be described with great accuracy by classical general relativity, with matter

---

[4] Some approaches to defining energy in the context of general relativity lead to a consistent picture where the energy of the universe is identically zero for all time when gravitational energy is included. It is difficult, however, to make sense of this in the framework of quantum gravity.

and radiation treated as quantum systems evolving in a fixed, classical space-time background. Thus this theoretical difficulty is for most practical purposes irrelevant. In our current inflating universe, although dark energy represents some 68% of the total energy in the universe, most of this energy is distributed in the vast empty space between galaxies. The total amount of dark energy in the volume of a sphere with radius equal to Earth's orbital radius, for example, corresponds to a mass of a small fraction of a kilogram. Thus, for any considerations involving terrestrial energy systems, the lack of a precise definition of energy on cosmological scales or the absence of a mathematically consistent theory of quantum gravity is irrelevant. To provide a satisfactory history of the extremely early universe, however, these theoretical problems would have to be addressed.

21.2 A Brief History of Energy in the Universe

Over the centuries since the development of the telescope, humankind has gradually looked deeper into the cosmos and determined the present structure and the history of the visible universe. Over the past 50 years, astrophysicists have been able to piece together a reasonably coherent picture of the history of the universe back to times shortly after the big bang. Modern cosmologists have taken this history back even earlier, perhaps to the era before the big bang, although these ideas remain quite speculative. In this section we follow the evolution of the universe from the perspective of its energy content, beginning with an inventory of the energy distribution in the visible universe as currently understood.

21.2.1 Energy Inventory Today

The precision study of cosmology dates from the 1964 discovery by Americans Arno Penzias and Robert Wilson that the universe is filled with an almost uniform distribution of microwave radiation that has been traveling through space since shortly after the big bang. This **cosmic microwave background radiation** (**CMB**) is now understood to be a remnant of the thermal radiation that accompanied the big bang. The study of the CMB continues with ever increasing precision. The **Wilkinson Microwave Anisotropy Probe**, (WMAP) launched in 2001, provided precise measurements of the large-scale structure of the universe. The **Planck satellite** launched in 2009 refined the WMAP measurements. These measurements, coupled with astrophysical observations of distant galaxies and stars, have led to a consistent picture of the present distribution of mass-energy in the universe.

The measurements by the Planck satellite indicate that roughly 69% of the energy density of the universe is in the

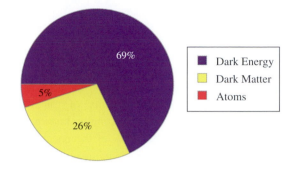

Figure 21.3 Mass-energy distribution in the universe at present

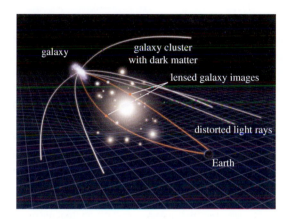

Figure 21.4 A galaxy cluster with both visible and dark matter can act as a gravitational lens, distorting the path of light rays from a distant galaxy and creating multiple images as seen from Earth. Image: NASA/ESA.

form of dark energy. Of the remaining 31% of the mass-energy in the universe, the greatest part – roughly 26% of the total – is composed of another invisible ingredient: **dark matter**. Unlike dark energy, which is uniformly distributed throughout space, dark matter is an invisible form of matter that is clumped in and around large aggregates of visible matter such as galaxies and galactic clusters. The existence of dark matter was first suggested in the 1930s to explain the pattern of galactic motions within clusters, and evidence for the presence of dark matter in galaxies was later given by the observation that the rotation rates of stars around the centers of galaxies are inconsistent with the amount of mass associated with visible stars. Since then, a host of further evidence for dark matter has accumulated from various astrophysical measurements, including measurement of **gravitational lensing** effects, in which the path of light from distant objects is bent by the distribution of mass-energy (see Figure 21.4). Just as the trajectories of objects like the Moon, satellites,

Energy Inventory of the Current Universe

At the present time the energy density of the universe is dominated by *dark energy* that contributes some 68% of the energy in the universe, uniformly distributed at a density of roughly 10^{-9} J/m^3. An unknown form of matter known as *dark matter* gives another 27% of the energy in the universe. Standard *baryonic matter* composed of familiar atoms makes up less than 5% of the energy density in the observed universe.

and meteors are determined by the curvature in space-time due to Earth's mass, the paths of photons passing by regions with extensive dark matter are affected by the curvature of space-time and can be measured by sensitive instruments.

Dark matter is believed to be similar in nature to visible matter, except that it is composed of as yet unknown particles. Dark matter appears to interact minimally with the visible forms of matter such as electrons, neutrons, and protons except through gravity, and it does not emit or absorb electromagnetic radiation (hence the name "dark matter"). Some speculations concerning the nature of dark matter invoke a *weakly interacting massive particle* (WIMP), one source of which might be the symmetry known as *supersymmetry* mentioned in §21.1.5. Other ideas include an elusive, lighter particle known as the *axion*, which is predicted in some extensions of the Standard Model. Experiments are currently underway to directly or indirectly detect dark matter particles and test these various hypotheses.

The remainder of the mass-energy in the universe, only about 4.9%, is composed of matter in the familiar form – atoms composed of quarks (e.g. forming protons and neutrons) and leptons (e.g. electrons) that interact by electromagnetic, strong, and weak forces. This component of the mass-energy of the universe is often referred to as **baryonic matter**. Much of the baryonic matter in the universe today is hydrogen and helium that has been present since the very early universe. In addition, there are substantial amounts of oxygen, carbon, neon, and other light elements up to and including iron, which were formed in stellar fusion processes that we discuss in more detail in the next chapter. There are also small quantities of elements heavier than iron and relatively minute quantities of very heavy elements such as uranium that were produced in stellar death processes such as supernovae.

Only a relatively small fraction, roughly 1/6, of the baryonic matter in the universe is clumped into stars within the more than 100 billion galaxies in the observable universe. Most of the rest is in the form of hot (10^5–10^7 K) intergalactic gas within the huge clusters of galaxies that are the largest known structures in the universe. A small fraction of the baryonic matter remains to be accounted for.

Many galaxies (including our own) are believed to contain extremely massive **black holes** at their cores – regions of space-time into which so much matter has fallen together that even light cannot escape. The total contribution to the energy budget of these massive black holes, along with stellar-mass black holes that are believed to populate typical galaxies, is estimated to be quite small compared to the energy contribution of baryonic matter. A large population of small black holes that may have been produced in the early universe has also been hypothesized, however, as an alternative possibility for dark matter.

21.2.2 Before the Big Bang

The generally accepted history of our universe started with a *big bang*, when a dense, hot gas starting in an extremely compressed state began to expand outward. As it expanded, the material in the universe cooled, and gave rise to the current configuration of galaxies and other astrophysical objects in the universe. There are a number of reasons to believe, however, that the big bang was not the first instant at which the development of our universe proceeded according to understandable physical laws. Evidence from WMAP and the Planck satellite shows a correlation between the microwave background radiation spectra coming from different directions that suggests that a period of *cosmological inflation* preceded the big bang. In this section we summarize briefly the motivation for and implications of inflation, though we emphasize that this idea is still somewhat speculative. An even more speculative, but highly provocative proposal that dark energy may be associated with a structure and history for our universe much richer than that visible since the big bang is described in Box 21.2.

The correlations that are seen in the distribution of microwave background radiation observed arriving from all directions suggest that the traditional big bang scenario is incomplete. The sources of this radiation are sufficiently distant from one another that they have never been in causal contact since the big bang. Nevertheless, the temperature that characterizes the CMB (see §22.2.2) is uniform over the sky to about one part in 100 000, implying that the sources of the radiation had already achieved thermal equilibrium when the radiation was emitted.

Box 21.2 Eternal Inflation in the Multiverse

One of the greatest puzzles in modern physics is why the dark energy density – the cosmological constant – in our universe is so small. We cannot directly compute this number, due to our ignorance of physics at very short distances and high energy scales, but we know no *a priori* reason why this number should be as small as its measured value of $\sim 10^{-122}$ in the natural units for quantum gravity (see Footnote 7). For a typical physical theory, as we understand it, the dark energy arises as a sum of terms, each of which independently is more than 100 orders of magnitude larger than the measured value, so the smallness of this quantity implies an apparently miraculous near-cancellation that is not understood in terms of any known physical principle or mechanism.

Until the turn of the last century, the cosmological constant was either ignored or believed to be constrained by some unknown dynamics to be zero. Years before the cosmological constant was found experimentally to be nonzero, the American physicist Steven Weinberg made an intriguing and profound observation: if the cosmological constant was much bigger than 10^{-118} (in Planck units), the expansion of the universe would have happened so quickly that the galaxies and other structures seen in our universe could not have formed. Weinberg went on to speculate that the smallness of the cosmological constant might be a consequence of the **anthropic principle**, which holds that "...the world is the way it is, at least in part, because otherwise there would be no one to ask why it is the way it is" [128]. To employ anthropic reasoning in a sensible scientific context, there must exist a large ensemble of universes – a multiverse – or an enormous universe containing different vast patches, in each of which the cosmological constant takes a different value. Only in those patches or universes where the cosmological constant is within the limits proposed by Weinberg do galaxies – and observers like us – evolve.

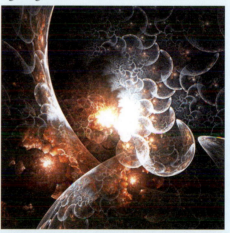

When the American physicist Saul Perlmutter and collaborators determined that the cosmological constant is small, but nonzero, and when no more conventional explanation for its value was forthcoming, cosmologists began to examine Weinberg's anthropic explanation more closely. In the context of cosmological inflation in a multiverse, a region of space-time with a positive cosmological constant will expand exponentially. Within such a region, quantum tunneling can lead to the formation of a patch of another solution with lower energy. If this region also has a positive cosmological constant it also will expand exponentially, and so on. Each of these patches is causally disconnected from the others – a separate universe within the multiverse. (The figure shows an artist's attempt to represent the multiverse. Image: ©Eric Prevost.) Some years ago, theorists working in *string theory* (§21.1.5) recognized that this theory admits many solutions that can give rise to different kinds of physics in widely separated regions of space-time. These solutions are roughly analogous to a permanent magnet that, once heated, can cool with magnetic polarization in a variety of different directions. The number of possible solutions to string theory is so huge that if the value of the cosmological constant is distributed smoothly over the range from zero to one in Planck units, there would be very many in which it takes a value similar to what we observe. Thus inflationary cosmology, which allows for formation of new patches of the universe, and string theory, which can endow each patch with its own value for the cosmological constant, together provide a self-consistent context for an anthropic explanation for the value of the cosmological constant observed in our universe.

On the face of it, this hypothesis seems fantastic, and it remains highly controversial. Resolving the problem of tuning a single number to be very small by invoking an inconceivably vast number of unobservable independent regions of the universe (essentially "parallel universes") seems like an extreme measure. Some scientists reject anthropic reasoning out of hand. Others worry whether it has any predictive value. Nevertheless, cosmological inflation is a relatively well-established fact and string theory is the best candidate for a theory of quantum gravity. Together these ideas lead to the picture of *eternal inflation in the multiverse* – a world vast beyond comprehension, containing local patches with more different kinds of physics than there are atoms in our own observable universe.

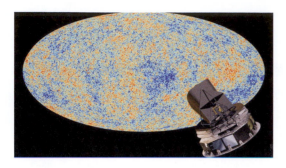

Figure 21.5 A cosmic microwave background radiation map of the whole sky from the Planck satellite. The colors correspond to temperature variations that are only about one part in 100 000 about the average value of 2.726 K. Image: ESA/Planck.

The currently favored and simplest explanation for the correlations is that these regions were once in causal contact, and that the universe underwent a period of exponential expansion known as **cosmological inflation** immediately preceding the big bang. The idea of cosmological inflation was first formulated by the American physicist Alan Guth in 1981. In cosmological inflation models, the universe had a large cosmological constant just before the big bang and expanded by a factor of roughly e^{60} or more through eq. (21.35). This expansion is sufficient to explain why the temperature of the CMB is virtually the same in all directions. Cosmological inflation also resolves a number of other puzzles about the early universe, such as the absence of **relics** – exotic objects such as *monopoles* and *cosmic strings*, which are suggested in many formulations of physics beyond the Standard Model. Though these objects would have been produced in the early universe, cosmological inflation would have diluted their densities exponentially, explaining their apparent absence at the present time. Cosmological inflation also offers an explanation of structure formation in the universe, providing a mechanism by which quantum fluctuations from short distance scales expanded rapidly and became frozen into small density fluctuations that led to the development of structure in the early universe.

The precise mechanism that led our universe to undergo an initial period of inflation is not well understood, nor is the mechanism that ended it. The success of predictions about the present state of our universe that follow from inflation independent of the details of the model, however, provides strong evidence that cosmological inflation did in fact occur in the very early universe. As in any period of inflation governed by eq. (21.35), the concepts of energy and conservation of energy during cosmological inflation

are tied at the quantum level to the mysteries of quantum gravity. Treating gravity classically, there is a sense in which negative gravitational energy balances the positive energy of the cosmological constant during inflation. In most models of cosmological inflation, when the inflationary period ended the energy from the cosmological constant was transformed into hot dense matter that fueled the big bang.

After the end of the period of cosmological inflation, we can consider gravity and the shape of space-time as a classical background. Except for the small cosmological constant that persists today we can treat the universe as a system of quantum matter coupled to classical gravity, with a well-defined energy that is conserved in time.

21.2.3 After the Big Bang

By observing the present configuration of matter in the universe and its velocity distribution, it is possible to reverse the time development equations of classical gravity and build a reasonably accurate picture of the state of the universe at very early times. Starting from the approximate state of the universe at such an early time and moving forward again, we can follow the development of the universe over time using known physical principles. At the beginning of the big bang, presumably just after the universe emerged from a preceding inflationary period, space was filled with a hot gas in approximate thermal equilibrium, with matter and energy packed together at unimaginably high density, pressure, and temperature. The universe contained a thermal distribution of radiation, particles, and antiparticles of all kinds at this very high temperature and density.

The high energy density of matter and radiation in the early universe caused a rapid expansion, which can be understood from the first of the Friedman equations (21.36). From this initial moment, the rate of expansion decelerated due to gravitational forces as the negative contributions from matter and radiation on the right-hand side of the second Friedman equation (21.36) overwhelmed the cosmological constant. As space-time itself expanded, the hot gas cooled as its thermal energy was converted to gravitational potential energy. Very early on, some mechanism that is as yet not understood created an imbalance between the amount of matter and the amount of antimatter in the universe. Though the imbalance may have been very small at first, as the universe cooled particles and antiparticles annihilated, leaving no antimatter and a net abundance of matter that has persisted to the present time. Within seconds, quarks combined into nucleons. After this time, at a temperature around 10^9 K, the time development of the universe followed more or less well understood physics and is on firmer footing than that which preceded. A somewhat delicate balance of nuclear processes protected some of the neutrons from decaying into protons, so that the initial neutron:proton ratio was about 1:6. Within the first few minutes, the universe cooled sufficiently for some neutrons to combine with protons to form helium (^{4}He) (roughly 25% by mass). If the number of neutrons available had been much smaller, much less helium would have been formed. Much smaller amounts of other light nuclei such as ^{2}H, ^{3}He, and ^{7}Li were also formed in **primordial nucleosynthesis**. More massive nuclei were generally not formed due to a lack of excess neutrons and time; as the universe cooled and expanded, the temperature and density of the gas of nuclei dropped below the threshold needed for further fusion processes.

21.2.4 Expansion and Structure Formation

As the universe has expanded and cooled, the relative contributions of radiation (light and relativistic matter), matter (nonrelativistic), and dark energy to the energy inventory of the universe have changed with time, as illustrated in Figure 21.6. In the very early universe, most of the energy density was in the form of radiation. As explained in §21.1.5, however, the energy density of radiation falls like $1/a(t)^4$, whereas the energy density in matter falls less rapidly, like $1/a(t)^3$. Thus the universe became *matter dominated* rather early in its history, about 50 000 years after the big bang (Figure 21.6). At that time, the temperature was still sufficiently high that electrons and nuclei were ionized, and typical photons continued to interact frequently with charged particles. Around 400 000 years after the big bang, this changed. At that time the temperature dropped low enough that electrons began to bind with

nucleons, forming neutral atoms. Most photons present at that time have continued to propagate to the present without interacting directly with matter. This time is called **recombination**,[5] and the radiation observed by WMAP, Planck, and other probes is the residue of the cosmic background radiation that was present at that time. Originally thermal radiation at a high temperature, this radiation has by now been shifted to the microwave part of the spectrum by the expansion of the universe.

After the matter-radiation crossover point, the density of matter provided the largest contribution to the universe's energy budget, and continued to decrease like $1/a(t)^3$. Since the energy density associated with the cosmological constant is independent of the scale factor, eventually dark energy has come to dominate over the energy density in matter. This last crossover occurred roughly 5 billion years before the present time, or about 9 billion years after the big bang. Today, in the dark-energy-dominated era, the universe is once again expanding exponentially as described in eq. (21.35). As far as we know, this will continue into the indefinite future.[6] Of course, since the expansion is exponential, H, and hence Λ, must be extremely small for the universe not to have expanded so quickly long ago that every particle in the universe would be separated from every other particle by trillions of light years. Indeed, the value that has been measured for the cosmological constant corresponds to an energy density of roughly

$$\rho \cong 10^{-9} \text{ J/m}^3 \sim 10^{-122} \frac{c^7}{\hbar G^2}, \qquad (21.39)$$

in natural units for the gravitational physics involved.[7] This corresponds roughly to the mass energy of 6 protons in every cubic meter of space. The small value of this constant explains why it took so long to observe, but also raises a basic challenge for fundamental physics: why is this number so small? The standard model of particle physics (§4)

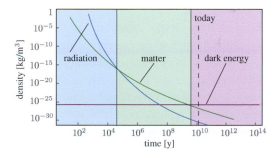

Figure 21.6 As the hot dense gas from the initial big bang has expanded and cooled with the universe, the density of radiation and matter energy have decreased, so that while radiation originally dominated the total energy of the universe, dark energy is now the dominant component. (Credit: Pearson Education 2008)

---

5 The term [re]*combination* is not very accurate, since the electrons and nuclei were never previously bound in a non-ionized state.

6 Though the multiverse picture of Box 21.2 suggests that eventually there may be a tunneling event to another vacuum with an even smaller cosmological constant, which would remove all familiar structure from our local part of the universe.

7 For a quantum theory of gravity the natural units of energy and length are the **Planck mass** $m_{Pl} = \sqrt{\hbar c/G} \cong 2.18 \times 10^{-8}$ kg $\cong 1.22 \times 10^{19}$ GeV/c^2 and *Planck length* $l_{Pl} = \sqrt{\hbar G/c^3} \cong 1.62 \times 10^{-35}$ m; in these units the natural unit of energy density is $m_{Pl}c^2/l_{Pl}^3 = c^7/\hbar G^2 \cong 4.64 \times 10^{113}$ J/m^3.

does not offer any solution to this puzzle. In fact, a naive calculation based on the standard model gives an answer that is too large by over 100 orders of magnitude (a part of this calculation appears in §22.3). One possible resolution of this puzzle is described in Box 21.2.

Current theory and observations suggest that large-scale structures now present in the universe, such as galaxies and galactic clusters, originated in small fluctuations of quantum fields in the very early universe. In the cosmological inflation model, these quantum fluctuations initiated in extremely short-wavelength modes that then expanded until their wavelengths exceeded the horizon size of our locally observable patch of the universe. After this point, these modes became frozen as classical fluctuations in energy density over large scales (scales larger than the horizon scale c/H mentioned above). Given a small fluctuation in energy density, Einstein's equations lead to gravitational attraction, pulling additional energy in the form of matter and radiation towards regions of higher density. Because gravity is universally attractive, the situation is unstable. Small clumps lead to bigger clumps, which eventually lead to large accumulations of matter. With enough hydrogen and helium packed together through gravitational attraction, high temperature and pressure lead to nuclear fusion, releasing radiation. The first stars in our universe were formed in this fashion and began to light the universe through their own radiation around 400 million years after the big bang. From this point on, the universe has continued to expand. Gravitational clumping has continued, leading to the formation of galaxies containing hundreds of millions of stars, and to clusters of galaxies (Figure 21.7). Throughout this process, gravitational potential energy has been converted into kinetic and/or thermal energy of the matter that has been drawn together into stars and galaxies.

As we discuss in more detail in the next chapter on solar energy, as stars burn hydrogen and helium fuel through nuclear fusion, they produce heavier elements such as oxygen, carbon, nitrogen, and others up through the most tightly bound nucleus, iron. Once a star has exhausted its nuclear fuel, its core collapses. In some situations this leads to a sudden massive supernova explosion in which more massive elements such as lead and uranium are formed. The matter expelled into interstellar space by these stellar explosions provides some of the basic material for the next generation of stars. Our own solar system was formed from the residue of material from earlier stars, and contains a liberal abundance of the elements carbon, oxygen, and nitrogen that form the basis of our atmosphere and all terrestrial ecosystems.

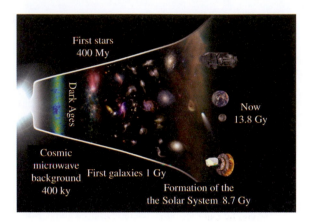

Figure 21.7 Important dates in the history of the universe from the end of inflation to the present time. (Image: NASA/WMAP Science Team)

The History of the Universe

The physical processes that led to the *big bang* are poorly understood, though it is believed that a period of rapid cosmological inflation preceded the formation of a hot dense gas of radiation, matter, and antimatter that expanded outward rapidly after inflation ended. As the universe expanded, it cooled, and roughly 50 000 years after the big bang matter dominated over radiation in overall energy density. At around 400 000 years, electrons bound with nuclei, making the universe transparent to electromagnetic radiation. *Cosmic background radiation* has propagated through the universe since that time, cooling to roughly 2.72 K at the present time and providing an observational window on conditions in the early universe. As the universe expanded further, small fluctuations in density led to accretion of matter through gravitational attraction, leading to formation of the stars and galaxies that now populate our universe.

Discussion/Investigation Questions

21.1 Think of some physical systems in which energy appears not to be conserved. Can you identify a coupling of the system to external processes that add or remove energy from the system? Can you characterize the dynamics of the system quantitatively in a way that makes manifest the flow of energy through the system?

Problems

21.1 **[T]** The equation of motion of a classical harmonic oscillator subject to an external time-dependent force $F(t)$ is $m\ddot{x} = -kx + F(t)$. Show that the oscillator's energy is not conserved, but that its time dependence dE/dt can be related to the power delivered to the oscillator by the agent supplying the force.

21.2 **[T]** Derive the results of Box 21.1 (the last two equations) by using the Schrödinger equation for the time derivative of $\psi(x)$ and $\psi^*(x)$ and integrating by parts.

21.3 **[T]** Confirm by explicit computations the assertions made in Example 21.1 that the Hamiltonian is invariant under translation and rotation so that total momentum and angular momentum are conserved.

21.4 **[T]** Consider motion in the xy-plane governed by a Hamiltonian $H(p_x, p_y, x, y)$. Introduce polar coordinates $x = r\cos\phi$, $y = r\sin\phi$, $p_x = p\cos\phi_p$, $p_y = p\sin\phi_p$, and use the chain rule for partial derivatives, e.g. $\partial H/\partial\phi = (\partial H/\partial x)(\partial x/\partial\phi) + (\partial H/\partial y)(\partial y/\partial\phi)$, etc., to derive eq. (21.17).

21.5 **[T]** Consider a particle of mass m moving in two dimensions, bound in a central potential $V(r)$, so that its Hamiltonian is $H = T(p) + V(r)$, where $T(p) = (p_x^2 + p_y^2)/2m$. Show that the quantity $D = xp_x + yp_y$ is not a constant of the motion by computing $\dot{D} = \{D, H\}$. Compute $\dot{D}$ for the special case that $V(r)$ is a power law, $V(r) = V_0 r^a$. Integrate your result over a time long compared to the period of this bound particle and show that $2\langle T \rangle = a\langle V(r)\rangle$ (where $\langle\dots\rangle$ denotes time averaging). This is the *virial theorem* of classical mechanics. Combine your result with energy conservation to express the total energy of the system in terms of $\langle T \rangle$. What are the implications for the motion of the Moon around Earth (see eqs. (31.29) and (31.30))? [Hint: you will need to use $r\partial/\partial r = x\partial/\partial x + y\partial/\partial y$ and similar relations.]

21.6 **[T]** According to special relativity, the speed of light should be the same as seen by two observers moving at different relative velocities. In Newtonian physics, an observer (2) moving at a velocity v in the x-direction relative to an observer (1) labels the x-coordinate of an event by $x_2 = x_1 - vt$, where x_1 is the coordinate given by the first observer, and both observers use the same time coordinate. Show that with this change of coordinates, the velocity of light traveling in the

x-direction is different for the two observers. If a stationary observer sees light passing by at 3×10^8 m/s, and a second observer is in a rocket ship passing in the same direction at 10^8 m/s, at what speed would the second observer see the light wave moving if the Newtonian transformation were correct?

21.7 **[T]** The relativistic transformation law that relates the coordinates seen by an observer (2) moving at velocity v relative to an observer (1) is $x_2 = (\cosh\phi)x_1 - (\sinh\phi)ct_1$, $ct_2 = (\cosh\phi)ct_1 - (\sinh\phi)x_1$, where $\sinh\phi/\cosh\phi = v/c$. (See Appendix B.4.2 for a summary of properties of the *hyperbolic functions*.) Note the similarity to rotation in Euclidean space by an angle θ. Confirm that under this transformation the space-time distance (21.24) between two events is the same for both observers. Check that in the limit $v \ll c$ the Newtonian transformation rules described in previous problem are reproduced.

21.8 **[T]** Consider two events seen by observer (1), who chooses her coordinates so that the first event occurs at the space-time origin $x_1^\mu = (0, 0, 0, 0)$, while the second event occurs at $\tilde{x}_1^\mu = (c\tilde{t}_1, \tilde{x}_1, 0, 0)$. Suppose Δ^2 for the two events (see eq. (21.24)) is positive. Use the transformation defined in Problem 21.7 to show that there is a reference frame traveling at a speed $v < c$ in which the two events occur at the same time and at different places. Such events are called *causally disconnected*.

21.9 **[T]** Consider the circumstances described in Problem 21.8, but now suppose that $\Delta^2 < 0$. Use the transformation defined in Problem 21.7 to show that there is a reference frame traveling at a speed $v < c$ in which the two events occur at the same place, but different times. Show that the time order of the events is the same in all reference frames related by the transformation of Problem 21.7. In this case the event at the earlier time can influence the later event.

21.10 **[T]** Show that two space-time points $(0, 0, 0, 0)$ and $(t, x = vt, 0, 0)$ that lie on the trajectory of an object traveling at speed v along the x-axis are separated by a distance satisfying $\Delta^2 = 0$ if and only if $|v| = c$. Use the relativistic invariance of Δ^2 to show that the speed of light is the same in all reference frames. Show that if $v > c$ then $\Delta^2 > 0$, so that according to Problem 21.8, the points cannot be in causal contact.

21.11 **[T]** An object moving at the speed of light along the x-axis in the metric (21.35) of an inflating universe satisfies $\dot{x}(t)e^{Ht} = c$. Assuming $x(0) = 0$, compute $x(t)$. Show that $x(t)$ never exceeds the value c/H.

Solar Energy: Solar Production and Radiation

173 000 000 GJ of energy from the Sun enters Earth's atmosphere in the form of electromagnetic radiation every second. This is more than 10 000 times the rate at which the human race uses energy. Almost all the energy we use originally arrived on Earth in the form of solar radiation – the exceptions being nuclear, geothermal, and tidal energy. The energy released when we burn fossil fuels was stored away long ago by plants that used photosynthesis to convert solar energy into chemical energy. The energy in the food we eat has a more obvious solar origin in photosynthesis by the plants that we consume, or that were consumed by animals. Solar energy evaporates water, and heats ocean and land masses, generating wind currents from which power can be harvested. Winds, fueled by solar energy, move evaporated water above mountains and plains where it falls and gathers into rivers and reservoirs from which hydropower is derived.

While most energy we use is solar energy in one form or another, *direct* use of solar energy still represents a very small piece of all human energy use. Solar energy is used in many places for some fraction of water and space heating, but in the critical areas of electrical and transport energy, use of solar energy is still marginal. For example, in 2015 the world derived about 1.3% of its total electrical energy from direct solar sources (over 97% photovoltaic and the remainder solar thermal electric conversion) [129, 130]. Although this is a small percentage, it has increased rapidly from a mere 0.065% in 2008.

One of the greatest challenges of large-scale solar energy use is the diffuse nature of the resource. Unlike fossil fuels or nuclear power, which have very high energy densities, solar energy is broadly distributed across the entire planet. Even in arid deserts near the equator, a substantial land area is needed to gather industrial-scale quantities of energy. Most of the solar-derived energy used by humankind up to the present has been gathered and

Reader's Guide

This chapter and the three that follow describe solar energy from its production in the Sun to its utilization on Earth. This chapter describes the generation of solar energy through nuclear fusion in the Sun, and the transmission of this energy to Earth through electromagnetic radiation. The blackbody spectrum characterizing electromagnetic radiation from a thermal source is described and derived. In §23, we review Earth's orbital mechanics and quantify the solar resource available on Earth. In §24, we discuss solar thermal energy, from small-scale domestic heating applications through large-scale conversion of solar thermal energy to electricity in power plants. We consider photovoltaic conversion of solar energy to electricity in §25.

Prerequisites: §3 (Electromagnetism), §7 (Quantum mechanics), and §18 (Fission and fusion).

In addition to the three subsequent chapters on solar energy, §34 (Energy and climate) and §35 (Climate change) also build on the material presented in this chapter.

concentrated by natural systems into other more readily usable forms such as wood, coal, and petroleum. It may be, however, that in the long term direct use of solar energy will be the simplest and most efficient way to sustainably power our extensive and still-growing energy needs.

In this and the following chapters, we follow solar energy from its production in nuclear processes in the Sun, through radiation to Earth and partial absorption in the atmosphere, to the various technologies that can be used to make direct use of solar energy. In this chapter, we focus on the origins and character of solar energy. First, in §22.1, we examine the origins of solar energy in nuclear fusion processes deep within the Sun. Next, in §22.2, we describe the properties of electromagnetic radiation emitted by a hot object like the Sun. We expand the earlier description of blackbody radiation (§6) by giving a more detailed characterization of the spectrum of thermal radiation (the *Planck radiation law*). This more detailed description of solar radiation plays an important role in understanding how solar energy is gathered on Earth and is also relevant for understanding aspects of Earth's climate. In §22.3, we give a self-contained derivation of the spectrum and intensity of blackbody radiation from basic quantum and statistical physics. The reader primarily interested in applications may wish to skip this derivation on a first reading.

For those interested in more detailed information about the production of energy in the Sun, stellar structure, and solar processes, Carroll and Ostlie [131] present these issues in the context of modern astrophysics at a level that is comprehensive but quite accessible to a non-expert. A clear self-contained introduction to solar physics is given by Stix [132].

In §23, we describe the nature and extent of solar radiation received on Earth. §24 covers the use of solar thermal energy, and §25 treats photovoltaic solar devices. A recent analysis of solar energy from a policy and economic perspective can be found in [133].

22.1 Nuclear Source of Solar Energy

Why does the Sun shine? One early scientific conjecture was that the Sun's energy source might be the potential energy released as the material composing the Sun contracts under its own gravitational attraction. Knowing the Sun's mass, radius, and luminosity,

$$M_\odot = 1.989 \times 10^{30} \, \text{kg} , \qquad (22.1)$$

$$R_\odot = 6.955 \times 10^8 \, \text{m} , \qquad (22.2)$$

$$L_\odot = 3.84 \times 10^{26} \, \text{W} , \qquad (22.3)$$

it is not hard to show that the total energy given off as the Sun contracted to its present size would have been sufficient to power its present luminosity for about 10 million years (Problem 22.6). Geological evidence shows, however, that Earth is about 4.5 billion years old, and that organisms powered by roughly current levels of solar luminosity have been present on the planet for a substantial fraction of that time. Thus, some other energy source must power the Sun. Once quantum mechanics was established and the basics of nuclear physics understood, it became clear, as shown by Hans Bethe in the 1930s, that nuclear fusion of hydrogen to helium is the energy source powering the Sun. Further direct evidence for this conclusion was found when solar neutrinos produced in nuclear fusion reactions were directly detected in the 1960s.

Astrophysicists now have an excellent understanding of solar structure and dynamics. We present here a brief summary of those aspects of solar physics relevant to the basic energy production mechanism.

The Sun is composed of about 74% hydrogen and 24% helium (by mass). The remaining 2% is dominated by the light elements oxygen (0.9%) and carbon (0.3%). Iron, nitrogen, silicon, magnesium, neon, and sulfur are present in concentrations above 0.04%, and none of the other elements contributes more than 0.01%. The Sun formed roughly 4.6 billion years ago from a diffuse cloud of gas that was even more dominated by hydrogen. As the hydrogen cloud collapsed under the effect of gravity, its temperature and pressure steadily increased, until at a temperature of about 1.5×10^7 K nuclear fusion processes turned on, releasing additional energy that raised the temperature and pressure further until the internal pressure was sufficient to halt further gravitational contraction.

22.1.1 Nuclear Reactions in the Sun

The basic physics of nuclear fusion was described in §18. The first step in the solar fusion process involves two protons fusing by means of a weak interaction, resulting in a deuterium nucleus, a positron, an electron neutrino, and a release of 1.43 MeV of energy,[1]

$$^1_1\underline{\text{H}} + {}^1_1\underline{\text{H}} \rightarrow {}^2_1\underline{\text{H}} + e^+ + \nu_e. \qquad (22.4)$$

Since this is a weak reaction, the cross section for it to occur is extremely small. This is the rate-limiting step in solar nuclear fusion.

The temperature at the center of the Sun is approximately $T_c \cong 1.5 \times 10^7$ K, corresponding to an average kinetic energy $E_c = \frac{3}{2}k_B T_c \approx 2$ keV. The height of the Coulomb repulsion barrier between two protons is about 0.2 MeV, corresponding to a temperature of 1.5×10^9 K, some 100 times greater than the Sun's central temperature. As explained in §18.4 and illustrated in Figure 18.11, fusion occurs in the Sun because two effects conspire to allow it to take place even though the energy associated with the temperature is far below the height of the barrier. The exponential tail of the Boltzmann factor $e^{-E/k_B T}$ gives a significant population of protons with energies much greater than $k_B T$, and the tunneling probability $P \propto e^{-b/\sqrt{E}}$ (where b is given by eq. (18.12) with $Z_1 = Z_2 = 1$ and $\overline{M} = M_p/2$) allows fusion between protons with energies well below the height of the barrier. Together, the

---

[1] Recall (from §18) that we use $^A_Z\underline{\text{X}}$ to denote fully ionized nuclei, such as those involved in fusion processes.

effects lead to a reaction probability that peaks at $E_{max} \sim (bk_B T_c/2)^{2/3}$ – about 6 keV (Problem 22.8).

The fusion reaction (22.4) is the first in a sequence of reactions that leads to the production of helium. The reaction sequence that occurs in most cases in the Sun is known as the **PPI chain** (first proton–proton chain)

$$\begin{aligned}
{}^1_1\text{H} + {}^1_1\text{H} &\rightarrow {}^2_1\text{H} + e^+ + \nu_e \\
{}^2_1\text{H} + {}^1_1\text{H} &\rightarrow {}^3_2\text{He} + \gamma \\
{}^3_2\text{He} + {}^3_2\text{He} &\rightarrow {}^4_2\text{He} + 2\,{}^1_1\text{H}.
\end{aligned} \tag{22.5}$$

The positrons rapidly annihilate with electrons into more photons, thus the net reaction (including electrons) is $4\,{}^1_1\text{H} + 2\,e^- \rightarrow {}^4_2\text{He} + 2\,\nu_e + \gamma$'s. The energy given off in the complete conversion of four hydrogen nuclei to two helium nuclei is

$$\begin{aligned}
Q &= 4\Delta({}^1_1\text{H}) - \Delta({}^4_2\text{He}) = 4(7.289\,\text{MeV}) - 2.425\,\text{MeV} \\
&= 26.731\,\text{MeV/reaction} .
\end{aligned} \tag{22.6}$$

The reaction sequence (22.5) accounts for about 69% of the fusion processes in the Sun. Another 31% of the time, the last step of the PPI sequence is replaced by a reaction in which a ${}^3_2\text{He}$ combines with a ${}^4_2\text{He}$, giving ${}^7_4\text{Be}$ (plus a photon). This leads into the PPII sequence

$$\begin{aligned}
{}^3_2\text{He} + {}^4_2\text{He} &\rightarrow {}^7_4\text{Be} + \gamma \\
{}^7_4\text{Be} + e^- &\rightarrow {}^7_3\text{Li} + \nu_e \\
{}^7_3\text{Li} + {}^1_1\text{H} &\rightarrow 2\,{}^4_2\text{He}
\end{aligned} \tag{22.7}$$

in which the ${}^7_4\text{Be}$ captures an electron giving ${}^7_3\text{Li}$ (+ ν_e), which in turn combines with a proton to give two ${}^4_2\text{He}$ nuclei. Although the steps are different, the net reaction and energy released are the same as in the PPI sequence. There is another chain, the PPIII chain, in which the ${}^7_4\text{Be}$ captures a proton, but this occurs in less than 1% of the relevant reactions. The PP reaction sequences dominate fusion energy production in stars of relatively low mass, like our Sun. At slightly higher masses, a different sequence of reactions, the **CNO cycle**, occurs. In this set of reactions, carbon, nitrogen, and oxygen act as catalysts in the fusion sequence producing helium from hydrogen.

22.1.2 Stellar Structure and Solar Luminosity

The details of stellar structure, in particular the interplay between convection, conduction, and radiation mechanisms carrying heat from the star's interior to the outside are rather complex. The fusion reactions generating the Sun's energy occur in the stellar core, within 1/5 of the radius from its center. At every depth within the Sun, density and temperature adjust so that there is a balance

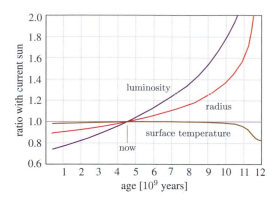

Figure 22.1 Evolution of the effective surface temperature, radius and luminosity of the Sun up to the start of its red giant phase. From [134].

between pressure and the compressive force of gravity. In the interior of the Sun, electromagnetic radiation is the primary transport mechanism carrying energy outward from the core. In the outer regions, where the gas is less dense, convection becomes the principal means of energy transport.

Because hydrogen is gradually being converted into helium, the Sun is not in a static nuclear configuration. As hydrogen is consumed, the solar material near the core compresses inward, leading to an increase in core density and temperature and an overall gradual increase in luminosity. Current stellar models indicate that since its formation 4.6 billion years ago, solar luminosity has increased roughly 40% (see Figure 22.1). The *insolation* – incident solar energy flux – on the early Earth is estimated to have been about 72% of the current value (at the top of the atmosphere). This gradual increase in insolation over billions of years is relevant when interpreting climate information from hundreds of millions of years ago. We return to this issue in §35. Extrapolating into the future, as the Sun burns its hydrogen and heats up, the outer layers will expand dramatically, so that in about 5 billion years it will become a **red giant** star. At this time the core will reach a temperature of roughly 10^8 K, at which point it will begin to burn helium, producing carbon and raising the temperature still further. Even in about 1 billion years, however – less than a quarter of Earth's present age – the Sun's luminosity will increase by about another 10%. Unless other effects intervene, this increase in luminosity will probably raise the temperature on Earth so substantially that life will become untenable in its current form (see Problem 35.4 in §35).

The Sun's total luminosity (22.3) measures the rate at which energy is carried away from the Sun, primarily in the form of electromagnetic radiation.[2] This radiation balances the production of energy in nuclear fusion reactions at the star's core, so that the temperature at the core stays constant over time (except for the gradual rise in temperature mentioned above as the hydrogen fuel is slowly exhausted). Thus, the luminosity depends essentially upon the rate of fusion energy production in the stellar interior.

The Sun radiates electromagnetic energy because it is hot. We first discussed thermal radiation in §6.3.3, where we introduced the concept of a *black body* – a perfect absorber and emitter of radiation. Because of its importance for solar energy physics, the origins and properties of thermal radiation are described in further detail in the remainder of this chapter. As a start we estimate the Sun's surface temperature using the **Stefan–Boltzmann law** (6.17),

$$L = \sigma \varepsilon A T^4, \qquad (22.8)$$

where σ is the Stefan–Boltzmann constant

$$\sigma = 5.670 \times 10^{-8} \, \text{W/m}^2 \, \text{K}^4, \qquad (22.9)$$

A is the emitting object's surface area, and ε is the **emissivity** of the material. We derive this law in §22.3. We can roughly approximate the Sun as a perfect black body with emissivity $\varepsilon = 1$; a more precise description of the solar radiation spectrum is discussed in §23.5 and Box 23.4 (more general discussions of emissivity are given in the following section, §22.2). Given the luminosity and the radius of the Sun, we can estimate the effective temperature of the Sun at its surface to be

$$T_\odot = \left(\frac{L_\odot}{4\pi \sigma R_\odot^2} \right)^{1/4} \cong 5780 \, \text{K}. \qquad (22.10)$$

Interestingly, as can be seen in Figure 22.1, while the radius and luminosity of the Sun have increased over time the surface temperature, which is proportional to $L_\odot / R_\odot^2$, has remained almost constant for the last 4.5 billion years and is expected to remain so for many billions of years to come.

22.2 Blackbody Radiation and Solar Radiation

To understand how solar energy interacts with the atmosphere and terrestrial systems, and how human technologies

---

[2] Neutrinos only carry away about 3% of the energy from the Sun, though there are a lot of them: a flux of roughly 6.4×10^{14} solar neutrinos/m^2 s pass through Earth and everything on it. The luminosity quoted in eq. (22.3) includes only EM radiation and not neutrinos.

Fusion Powers the Sun

Nuclear fusion reactions in the Sun release energy by combining four hydrogen nuclei into a helium nucleus

$$4\,{}^1_1\underline{\text{H}} + 2\,e^- \rightarrow {}^4_2\underline{\text{He}} + 2\,\nu_e + \gamma\text{'s}.$$

Energy emanates from the Sun primarily in the form of electromagnetic radiation. The total power of this radiation is 383.9 YW, corresponding roughly to thermal (blackbody) radiation from the solar surface at a temperature of 5780 K.

---

like photovoltaic cells can be designed to make optimal use of incident solar energy, we need to characterize the structure of solar blackbody radiation in somewhat more detail. In particular, we need to know its *spectrum* – how the power in blackbody radiation is distributed over the range of different frequencies of the electromagnetic waves that carry the energy. We begin by reviewing some of what we learned in previous chapters about the nature of light in §22.2.1. We then give a general characterization of the blackbody spectrum in §22.2.2.

22.2.1 Classical and Quantum Descriptions of Light

As described in §4, light can be viewed classically as electromagnetic waves. There we showed that Maxwell's equations in a vacuum reduce to a simple wave equation for the electric and magnetic fields. This wave equation has plane wave solutions described by sinusoidally oscillating electric and magnetic fields (see Figure 4.5) with wave vector $\boldsymbol{k}$ and (angular) frequency ω (hence wavelength $\lambda = 2\pi/k$ and frequency $\nu = \omega/2\pi$), such as

$$\boldsymbol{E}(\boldsymbol{x}, t) = E_0 \sin(kx - \omega t)\hat{\boldsymbol{y}},$$

$$\boldsymbol{B}(\boldsymbol{x}, t) = \frac{E_0}{c} \sin(kx - \omega t)\hat{\boldsymbol{z}}. \qquad (22.11)$$

These fields describe one of the two solutions (4.21) that correspond to distinct polarizations of a light wave traveling along the $\hat{\boldsymbol{x}}$-axis with speed $c = \omega/k = \lambda\nu$. According to eq. (4.24), either type of wave carries (time-averaged) power per unit area $\langle |\boldsymbol{S}| \rangle = (1/\mu_0)\langle |\boldsymbol{E} \times \boldsymbol{B}| \rangle = \epsilon_0 c E_0^2/2$.

As discussed in §4.2 and §4.3, just as a sound wave can be decomposed into different frequencies, any light wave can be described as a superposition of plane waves of various frequencies propagating in various directions. The different frequency ranges of light waves are summarized in Figure 9.2. Visible light has wavelengths in the range

400–750 nm, while wavelengths shorter than 400 nm correspond to ultraviolet light, X-rays, and gamma rays, and wavelengths longer than 750 nm correspond to infrared light, microwaves, and radio waves.

As for the string described in §4.2, each mode of an electromagnetic wave essentially acts as an independent harmonic oscillator. This is easy to see from eq. (4.20), from which it follows that the electric field for a mode with wave number k satisfies the harmonic oscillator equation

$$\ddot{E} = -k^2 c^2 E = -c^2 \left(\frac{2\pi}{\lambda} \right)^2 E . \qquad (22.12)$$

The angular frequency of the oscillator associated with such a mode is $\omega = kc$. The electromagnetic field can thus be treated both classically and quantum mechanically as a collection of simple harmonic oscillators. This allows us to give a simple description of many aspects of quantum electromagnetic fields. Like any other quantum harmonic oscillator, each mode of the electromagnetic field has quantized energy levels that are separated by $\hbar\omega$. These packets of quantum energy in the electromagnetic field are the *photons* that were first introduced in the study of quantum transitions in §7.3.3 and which figured centrally in our discussion of energy in matter in §9.

Thus, light can be thought of either classically as a superposition of plane waves, or quantum mechanically as photon excitations of the quantum electromagnetic field. In either case, the rate at which energy is transported across a given surface by light can be given a detailed description in terms of the power density $dP(\omega)/d\omega$ as a function of frequency. Here we state the results for the form of this distribution for blackbody radiation, leaving the proof to §22.3.

22.2.2 The Nature of Thermal Radiation

When we first encountered thermal radiation in §6.3.3, we described blackbody radiation as the electromagnetic radiation emitted by an object that is a perfect absorber of incoming radiation. We quoted the Stefan–Boltzmann law, which says that a black body in thermal equilibrium at temperature T radiates energy at a rate proportional to T^4. Using quantum mechanics it is possible to derive this law and to obtain an expression for the *frequency distribution* of thermal radiation, known as the **Planck radiation law**. Planck's law states that the power per unit area emitted in blackbody radiation in a frequency range $d\omega$ around the frequency ω is

$$\frac{dP}{dA\,d\omega} = \frac{\hbar}{4\pi^2 c^2} \frac{\omega^3}{e^{\hbar\omega/k_B T} - 1}. \qquad (22.13)$$

Dimensional analysis shows that the total radiated power, obtained by integrating dP over ω, is proportional to T^4,

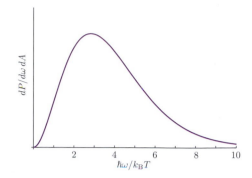

Figure 22.2 Blackbody radiation spectrum as a function of $\hbar\omega/k_B T$ according to Planck's radiation law.

consistent with the Stefan–Boltzmann law. We perform the integration explicitly in the following section, following the derivation of Planck's law.

The power spectrum (22.13) is graphed in Figure 22.2. The peak of this distribution can be found by differentiating (22.13),

$$\omega_{max} \cong 2.82 \frac{k_B T}{\hbar} . \qquad (22.14)$$

No real material is a perfect black body. The *emissivity* of real materials is characterized by a frequency-dependent function $\varepsilon(\omega)$. For a material of emissivity $\varepsilon(\omega)$ the formula (22.13) for the power distribution over frequency is just multiplied by $\varepsilon(\omega)$. Often, when the detailed frequency dependence of radiation is not considered, an average emissivity ε is used so that the total power is related to temperature through eq. (22.8). Emissivity and thermal radiation from non-blackbody sources are discussed further in Box 22.1.

The discovery of Planck's law (22.13) was a key step in the development of quantum mechanics. In the early 1900s, English physicists John Strutt (Lord Rayleigh) and James Jeans used classical reasoning to argue that the rate of energy radiation in the frequency range $d\omega$ should go as $(1/4\pi^2 c^2)k_B T\omega^2$. This **Rayleigh–Jeans law** agrees with eq. (22.13) at small ω (including the overall constant). This hypothesis matched experimental data at the time for low frequencies, but suffered from two problems. First, experimental data showed that the Rayleigh–Jeans law is violated at high frequency. Second, Rayleigh–Jeans predicted an infinite amount of radiated energy when integrated over all frequencies, a problem known as the **ultraviolet catastrophe**. Planck solved these problems by postulating that excitations of the electromagnetic field in a cavity are quantized in units of $\hbar\omega$, leading to eq. (22.13), five years before Einstein postulated the quantization of

---

Box 22.1   More on Black (and Not so Black) Bodies

The interaction of an object with electromagnetic radiation is characterized by its **absorptivity** $\alpha(\omega)$, **transmittance** $\tau(\omega)$, *emissivity* $\varepsilon(\omega)$, and **reflectivity** $\rho(\omega)$, all of which can depend on the frequency ω of the radiation. All are dimensionless numbers describing the fraction of radiation of frequency ω that is absorbed, etc. If an object is opaque (it does not transmit any radiation) then light incident on it must be either reflected or absorbed, so $\alpha(\omega)$ and $\rho(\omega)$ must sum to one when $\tau(\omega) = 0$. German physicist Gustave Kirchhoff, who first introduced the concept of blackbody radiation, used the existence of thermal equilibrium to prove that the absorptivity and emissivity must be equal, since otherwise the object could not reach a thermal equilibrium in the presence of incoming thermal radiation without violating the second law of thermodynamics. Thus, any opaque object has

$$\tau(\omega) = 0, \qquad \rho(\omega) + \alpha(\omega) = 1, \qquad \text{and} \qquad \varepsilon(\omega) = \alpha(\omega). \tag{22.15}$$

A *black body* is defined as a perfect absorber – and therefore perfect emitter – of radiation, with $\alpha = \varepsilon = 1$ for all ω. Blackbody radiation is the radiation emitted by such an object when it is in thermal equilibrium. A close approximation to a black body is given by a small opening in a wall enclosing a cavity held at constant temperature. A glass blower's furnace provides a good example (Figure 6.2). The hole allows all radiation that impinges on it to enter the cavity, where it bounces around until it is absorbed by the walls. Thus, the hole has $\alpha = 1$ to a very good approximation. The radiation that is emitted from the walls and finds its way out of the hole is characteristic of the temperature T of the cavity. Quantum mechanics allows us to view radiation as a *gas* of photons, which is in thermal equilibrium within the cavity, and can be observed when it leaks out of the hole.

If an object has an emissivity less than one but independent of frequency, it is termed a **gray body**. The emissivity of some common materials was mentioned in §6.3.3.

Example 22.1   Blackbody Radiation From Toaster Heating Coils

A household toaster oven is heated by two coils, each composed of a metal cylinder of diameter 0.6 cm and length 1.2 m. If the toaster oven converts 1500 W of electric power into thermal radiation, what is the temperature of the coils and the frequency at which the radiation power spectrum is peaked?

The total surface area of the coils is $A = 2 \times (1.2\,\text{m}) \times \pi \times (0.006\,\text{m}) \cong 0.045\,\text{m}^2$. The radiated power is related to the temperature through

$$P = 1500\,\text{W} = \sigma A T^4.$$

Solving for temperature, we get $T \cong 876\,\text{K} \cong 600\,°\text{C}$. At this temperature, the frequency associated with maximum power is $\omega_{\text{max}} \cong 2.82 k_B T/\hbar \cong 3.2 \times 10^{14}\,\text{s}^{-1}$. This corresponds to infrared radiation of wavelength $\lambda \cong 5.8\,\mu\text{m}$. The tail in the frequency distribution of thermal radiation at this temperature, however, contains almost a milliwatt of radiation in the visible spectrum, so the coil glows red when radiating at this temperature and power (Problem 22.2).

---

electromagnetic energy in photons to explain how light can knock electrons off the surface of metals.

22.2.3 The Solar Blackbody Spectrum

The distribution of solar energy across the electromagnetic spectrum can be estimated from eq. (22.13), where we approximate the Sun as a black body at a temperature of 5780 K. Figure 22.3 shows the distribution of blackbody radiation at 5780 K, both as a function of frequency/photon energy and as a function of wavelength. Solar radiation is strongly peaked in the visible range of 400–750 nm. Because the Sun is not a perfect black body, the spectrum in the figure is only approximate. In fact, though the total luminosity corresponds to a temperature of roughly 5780 K, the shape and location of the peak of solar energy is closer to that of a blackbody distribution at 6000 K. For most purposes, it is sufficient to consider the solar spectrum as a blackbody spectrum at 5780 K. We discuss some further details of the solar radiation spectrum and its absorption in the atmosphere in §23.5.

Thermal Radiation

Any object at temperature T radiates energy in the form of electromagnetic waves. The power per unit area in the radiated energy is distributed among photons of frequency ω according to the Planck radiation law

$$dP = \frac{\hbar}{4\pi^2 c^2} \varepsilon(\omega) \frac{\omega^3}{e^{\hbar\omega/k_B T} - 1} dA d\omega,$$

where $\varepsilon(\omega)$ characterizes the emissivity of the object. If ε is constant, then the total power is given by the Stefan–Boltzmann law

$$P = \sigma \varepsilon A T^4,$$

where σ is the Stefan–Boltzmann constant

$$\sigma = \frac{\pi^2 k_B^4}{60 \hbar^3 c^2} = 5.67 \times 10^{-8} \text{ W/m}^2 \text{ K}^4.$$

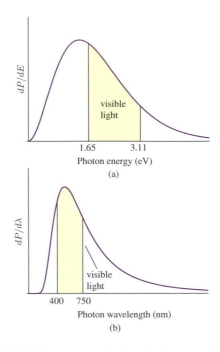

Figure 22.3 The spectrum of solar radiation approximated as a perfect black body at 5780 K, graphed (a) as a function of photon energy, and (b) as a function of wavelength. The band of visible photons is shaded in both graphs.

22.3 Derivation of the Blackbody Radiation Formula

The spectrum of thermal radiation plays a central role in determining Earth's climate and the solar energy available for human use, and warrants a derivation for the interested reader. Furthermore, the Planck radiation law (22.13) can be derived from first principles using the methods of quantum and statistical physics that have been developed in §7 and §8, and provides an excellent example of the power of those methods.

We begin with a brief summary of the argument, which may serve as a road map for the analysis that follows. As described in Box 22.1, we can regard blackbody radiation as the spectrum of radiation emitted from a small opening in a cavity at temperature T. The first step is to enumerate the standing wave solutions of Maxwell's equations in the cavity. These are modes that oscillate sinusoidally in time with certain allowed frequencies. Quantum mechanically, each mode of oscillation with frequency ω behaves like an independent harmonic oscillator with energy levels $E = \hbar\omega(m + 1/2)$, with m a nonnegative integer. Next we turn to statistical mechanics, and in particular to the Boltzmann distribution, which determines the probability of finding each of these field oscillators in the mth energy level when the system is in equilibrium at temperature T. The internal energy of the radiation field is obtained by multiplying the probability for a given mode to be in a given state by the energy of that state and summing over all

states and over all modes of the radiation field. Finally, we imagine that a hole on the side of the box allows electromagnetic radiation to escape, and we compute the power spectrum and total rate of energy loss through the hole. This leads to the Planck radiation law, including an expression for the Stefan–Boltzmann constant in terms of k_B, $\hbar$, and c. The analysis presented here can be found in many texts; see for example [21].

We begin by considering a cubic cavity with sides of length L. We choose this shape to simplify the calculation, though the final result turns out to be an integral over the volume that is independent of the shape of the cavity. We take coordinates x, y, z inside the box to run from 0 to L. We take the box itself to be a conductor, although the final result is independent of the nature of the material as long as it interacts with the electromagnetic field. We wish to consider the quantum electromagnetic field inside the box (see Figure 22.4).

The assumption that the box is made of a conducting material implies that the electric field must be perpendicular to all the boundaries of the box, so no charges are accelerated. For example, at $x = 0, L$, the electric field can only have a nonzero component E_x. The modes of the electromagnetic field are then standing wave solutions of

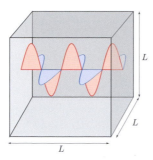

Figure 22.4 To find the blackbody radiation spectrum we quantize the electromagnetic field in a conducting box.

Maxwell's equations in a vacuum that satisfy this boundary condition. Finding these modes is a problem similar to solving the Schrödinger equation for a particle in a 3D box, which we considered in §7. The spatial dependence of each component of the electric field can be written as a product of sine and cosine modes. The boundary conditions restrict the possible combinations for the **E** field, so that the set of mode solutions is given by

$$E_x = E_{x0} \sin \omega t \cos(k_x x) \sin(k_y y) \sin(k_z z) ,$$
$$E_y = E_{y0} \sin \omega t \sin(k_x x) \cos(k_y y) \sin(k_z z) ,$$
$$E_z = E_{z0} \sin \omega t \sin(k_x x) \sin(k_y y) \cos(k_z z) , \quad (22.16)$$

where the modes are parameterized by a 3-tuple (triplet) of positive integers $\mathbf{n} = (n_x, n_y, n_z)$, with $\mathbf{k} = \pi \mathbf{n}/L$. **E** must satisfy Gauss's law $\nabla \cdot \mathbf{E} = 0$, which reduces to a single linear condition $k_x E_{x0} + k_y E_{y0} + k_z E_{z0} = 0$ on the three amplitudes (E_{x0}, E_{y0}, E_{z0}). There are two independent solutions to this equation for each choice of $\mathbf{k}$, corresponding to the two independent *polarizations* of light waves. Each standing wave mode can be written as a linear combination of plane waves. This decomposition uniquely fixes the $\mathbf{B}$ fields of the standing waves, though we will not need the explicit form here. The frequency ω associated with each mode follows from substituting (22.16) into (22.12), giving

$$\omega_{\mathbf{n}} = \pi c n/L = \frac{\pi c}{L}\sqrt{n_x^2 + n_y^2 + n_z^2} . \quad (22.17)$$

Each mode is essentially an independent harmonic oscillator of frequency $\omega_{\mathbf{n}}$. We can therefore think of the quantum electromagnetic field in a conducting box as a system of many independent harmonic oscillators. The oscillators are parameterized by a triplet of positive integers $\mathbf{n}$, and there are two oscillators (polarizations) for each allowed triplet. We know that a quantum harmonic oscillator of frequency ω has a spectrum of states with energy levels $E_m = (m + 1/2)\hbar\omega$. Since the oscillators are independent,

the total energy of the system is simply given by the sum of the energies in all the oscillators.

In §8.6, we determined that in thermal equilibrium the probability of finding a system in a state with energy E is given by the Boltzmann distribution (8.32). As an application, we computed the internal energy of a crystal modeled as 3N independent oscillators (Einstein's model of the specific heat). That result, eq. (8.54), can be adapted to the present situation by removing the factor of 3N. The average energy in the mode with frequency $\omega_{\mathbf{n}}$ is then

$$\langle E_{\mathbf{n}} \rangle = \hbar\omega_{\mathbf{n}} \left(\frac{1}{2} + \frac{1}{e^{\hbar\omega_{\mathbf{n}}/k_B T} - 1} \right) . \quad (22.18)$$

The ground state energy of the oscillators is simply an additive constant, known as the **zero-point energy**, which does not affect the physics, so we drop the factor of $\hbar\omega_{\mathbf{n}}/2$ henceforth.[3]

Given the average energy in each mode of the electromagnetic field, it remains only to sum over all the modes,

$$\langle E \rangle = \sum_{\mathbf{n}} 2\langle E_{\mathbf{n}} \rangle = 2 \sum_{n_1=1}^{\infty} \sum_{n_2=1}^{\infty} \sum_{n_3=1}^{\infty} \frac{\hbar\omega_{\mathbf{n}}}{e^{\hbar\omega_{\mathbf{n}}/k_B T} - 1} , \quad (22.19)$$

where the factor of two comes from the two polarizations of the electromagnetic field. If we assume that the size of the box L is sufficiently large,[4] we can approximate the sums by integrals,

$$\sum_{\mathbf{n}} \Rightarrow \int_0^{\infty} dn_1 \int_0^{\infty} dn_2 \int_0^{\infty} dn_3 = \frac{1}{8} 4\pi \int_0^{\infty} n^2 dn . \quad (22.20)$$

The last step in this equation makes use of a change from cartesian coordinates to spherical polar coordinates (Appendix B.1.2).

We can now use eq. (22.17) to replace n by $L\omega/\pi c$, giving an expression for the average value of the total energy as an integral over the spectrum parameterized by frequency ω,

$$\langle E \rangle \Rightarrow \frac{L^3 \hbar}{c^3 \pi^2} \int_0^{\infty} d\omega \frac{\omega^3}{e^{\hbar\omega/k_B T} - 1} = V \int_0^{\infty} d\omega U(\omega) , \quad (22.21)$$

---

[3] This zero-point energy relates to the discussion in §21.1.5; if we include modes with wavelengths up to the Planck length, this gives a ground state energy that is over 100 orders of magnitude larger than the observed cosmological constant.

[4] The relevant criterion is that the frequencies are closely spaced, which requires $\hbar c/k_B T \ll L$; this condition is satisfied under all circumstances of interest to us.

where we have replaced L^3 by the volume V and where $U(\omega)$ is the **spectral density** of energy,

$$U(\omega) = \frac{\hbar}{c^3 \pi^2} \frac{\omega^3}{e^{\hbar\omega/k_B T} - 1}. \qquad (22.22)$$

$U(\omega)$ describes the density of energy per unit volume contained in modes of the electromagnetic field with frequency between ω and $\omega + d\omega$.

We have thus computed the energy distribution in the electromagnetic field inside a volume V at temperature T. Finally, we must relate this to the energy that is emitted from a hole of area A in the surface of the box surrounding the cavity. Those photons headed towards the hole can continue out of the box. If all photons in the nearby volume were headed directly towards the hole, the rate at which energy associated with photons at frequency ω would leave the box would be $U(\omega)cA$ (density × speed × area). If we put the hole, say, on the face at $z = L$, then a mode $\mathbf{n}$ will be moving towards the hole if $n_z > 0$. The angle of incidence between the mode and the face with the hole is given by $\cos\phi = n_z/n$ (see Figure 22.5). Integrating over all angles of incidence on the hole, the total energy leaving the hole is reduced by a factor of $(\int_0^{\pi/2} d\phi\, 2\pi \sin\phi \cos\phi)/4\pi = 1/4$. Thus, the flux of energy (energy per unit area per unit time) per unit frequency leaving the black body is given by

$$\frac{dP}{dA\,d\omega} = \frac{c}{4} U(\omega) = \frac{\hbar}{4\pi^2 c^2}\left(\frac{\omega^3}{e^{\hbar\omega/k_B T} - 1}\right), \qquad (22.23)$$

which is precisely the Planck radiation law (22.13) for a perfect black body with emissivity $\varepsilon(\omega) = 1$ for all ω.

To compute the total radiation, we simply integrate eq. (22.23); replacing $\omega \to k_B T y/\hbar$, the integral of the term in parentheses in eq. (22.23) is

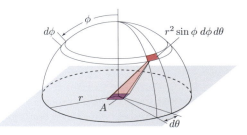

Figure 22.5 A fraction $A\cos\phi/4\pi r^2$ of the photons passing through points in an element of surface area $r^2 \sin\phi\, d\phi\, d\theta$ are heading toward the hole A and will pass out of the cavity.

$$\int_0^\infty d\omega \frac{\omega^3}{e^{\hbar\omega/k_B T} - 1} = \left(\frac{k_B T}{\hbar}\right)^4 \int_0^\infty dy \frac{y^3}{e^y - 1}$$
$$= \left(\frac{k_B T}{\hbar}\right)^4 \left(\frac{\pi^4}{15}\right). \qquad (22.24)$$

This gives us the Stefan–Boltzmann law

$$P = \int \frac{A}{4} cU(\omega)d\omega = \sigma A T^4, \qquad (22.25)$$

where (including the constant factor $\hbar/4\pi^2 c^2$ from eq. (22.23))

$$\sigma = \frac{\pi^2 k_B^4}{60\hbar^3 c^2} = 5.670 \times 10^{-8}\ \mathrm{W/m^2\,K^4}. \qquad (22.26)$$

Thus, we have derived the Planck and Stefan–Boltzmann laws for the electromagnetic field in a conducting cavity. While we have assumed a cubic cavity, note that the energy density scales as the volume of the cavity. Though we have not proven this explicitly, the energy does not in fact depend upon the shape of the cavity walls; the electromagnetic field in any region has an energy density in thermal equilibrium at temperature T that takes the form of eq. (22.22).

Discussion/Investigation Questions

22.1 The interior of a glass blower's furnace may contain shelves and other objects of various shapes and colors. As the furnace is heated, objects in the interior become more and more difficult to discern through a small opening, until finally, when the furnace is hot enough, the interior objects are all but lost in a uniform glow (see Figure 6.2). Explain what is going on here.

22.2 As a source of blackbody radiation becomes hotter, the peak in its radiation spectrum moves from the visible to the ultraviolet and beyond. Does this imply that the object can no longer be seen by the unaided human eye?

Problems

22.1 [C] Compute or estimate the fraction of radiated thermal energy that is in the visible range (wavelength 400–750 nm) for the following two radiation spectra: (a) solar radiation, assumed to be perfect blackbody radiation at 5780 K; (b) EM radiation from a pot of boiling water, again assume a perfect blackbody distribution.

22.2 [C] Compute the power radiated in the visible range from the toaster oven coil radiating with a total power of 1500 W, as described in Example 22.1.

22.3 Estimate the rate of thermal radiation from a household hot-water radiator with a surface area of 1 m² and a temperature of 80 °C.

22.4 The incandescent light bulb is a notoriously inefficient way to convert electric power into visible light. The tungsten filament emits blackbody radiation at a temperature that is limited by its melting point. Define the lighting efficiency as the ratio of the power emitted in the visible range (400–750 nm) to the total power emitted. What is the theoretical maximum efficiency of a tungsten light bulb, given that tungsten melts at 3422 °C?

22.5 Compute the frequency ω_{max} corresponding to the maximum power of radiation for a blackbody at temperature $T = 280$ K. This is roughly the average temperature of Earth's surface. Compute the wavelength corresponding to this frequency.

22.6 **[T]** Estimate the gravitational energy in the Sun, using eqs. (22.1) and (22.2), assuming that the mass is distributed uniformly. Can you confirm the statement in the text that gravitational potential energy could only power the Sun at its present luminosity for $\sim 10^7$ y? At what rate would the radius of the Sun have to be decreasing (in, say, mm/s) if its luminosity came from gravitational potential energy?

22.7 **[T]** Compute the energy released in each step of the solar PPI fusion chain eq. (22.5), and confirm that the total energy released matches eq. (22.6).

22.8 **[T]** Show that the combination of the Boltzmann factor and the tunneling probability give a probability for fusion that is maximized at $E_{max} \approx (bk_B T_c/2)^{2/3}$, as stated in §22.1.1.

22.9 **[T]** Show that the peak of the blackbody spectrum as a function of ω is given by eq. (22.14).

22.10 **[T]** Rewrite the blackbody power spectrum as a function of λ and compute the value λ_{max} where the power density $dP/d\lambda$ is maximized. Find the frequency ω corresponding to λ_{max}. Why is ω not equal to ω_{max}, where ω_{max} is given by eq. (22.14)? Compare λ_{max} to $2\pi c/\omega_{max}$ for a black body at temperature 5780 K.

22.11 **[T]** Use dimensional analysis to show that the wavelength scale of blackbody radiation is given by $\lambda_{th} = hc/k_B T$. The average radiation energy in a cavity depends only on its volume and not on its shape when $\lambda_{th} \ll L$, where L is a typical length scale characterizing the cavity. How low must the temperature be for the shape to matter if $L \approx 1$ cm?

22.12 **[T]** Show that the classical Rayleigh–Jeans law for radiation follows from eq. (22.13) in the limit as $\hbar \to 0$. Show that the total power radiated diverges in this limit (the *ultraviolet catastrophe* that helped lead to the discovery of quantum mechanics).

22.13 **[T]** In §7, we showed that the energy of a particle of mass M in an $L \times L \times L$ box in the state labeled $|n, m, l\rangle$ is $E_{n,m,l} = \pi^2 \hbar^2 (n^2 + m^2 + l^2)/2ML^2$. The probability of finding a particle in the state $|n, m, l\rangle$ at temperature T is given by the Boltzmann distribution $P(n, m, l) \propto \exp(-E_{n,m,l}/k_B T)$. In a hot gas the energy levels are very close together, so E can be regarded as a continuous variable. Show that the probability of finding the particle with energy between E and $E + dE$ at temperature T is given by the *Maxwell–Boltzmann distribution*, $dP/dE = 2\pi \sqrt{E} e^{-E/k_B T}/(\pi k_B T)^{3/2}$. [Hint: Employ the methods used to compute the spectrum of blackbody radiation, including eq. (22.20).]

Solar Energy: Solar Radiation on Earth

Solar radiation propagates outward from the Sun and carries energy to Earth, providing light and warmth to the planet. The amount and character of solar radiation incident at different locations on Earth determine the extent and geographic distribution of the solar energy resource available for human use. In this chapter, we consider the arrival of solar energy on Earth as a two-step process: first, energy radiated from the Sun reaches the top of the atmosphere; second, the radiation makes its way through the atmosphere to ground level. The first half of this chapter (§23.1–§23.3) is devoted to characterizing the solar energy input at the top of Earth's atmosphere as a function of the time of year, time of day, and latitude. This *solar irradiance* varies depending upon Earth's *orbital dynamics* – how the planet rotates about its axis and revolves around the Sun. In the second part of this chapter (§23.4, §23.5), we explore how the Sun's light is absorbed as it passes through Earth's atmosphere. As explained in the previous chapter, the Sun's radiation can be roughly characterized as blackbody radiation at $T = 5780\,\mathrm{K}$. After describing deviations from this theoretical ideal, we consider how light is absorbed and emitted by molecules and then how the predominant absorbing molecular species in Earth's atmosphere modify the solar radiation observed at ground level. Finally, in §23.6 we survey the extent of the solar energy resource on Earth.

23.1 Insolation and the Solar Constant

Given the Sun's average luminosity $L_\odot \cong 384\,\mathrm{YW}$ and the mean radius of Earth's orbit around the Sun, $R_{\mathrm{orbit}} \cong 1.50 \times 10^8\,\mathrm{km}$, it is a straightforward exercise in geometry to compute the power from the Sun that reaches Earth. There is very little matter between the Sun and Earth's orbit to absorb solar radiation, and conservation of energy requires that the rate at which solar radiation carries energy

Reader's Guide

This chapter describes the character and extent of solar radiation incident on Earth. Earth's orbital geometry gives rise to variations of solar energy flux with seasons, with latitude, and with time of day. The interaction of light with matter is reviewed, after which atmospheric absorption of sunlight is characterized and the extent of the terrestrial solar energy resource is considered.

Prerequisites: §7 (Quantum mechanics), particularly absorption and emission of photons; §9 (Energy in matter); §22 (Solar energy).

In addition to providing basic background for the two subsequent chapters on solar power, some of the ideas developed in this chapter play an important role in §34 (Energy and climate) and §35 (Climate change).

through a sphere of any radius about the Sun is $L_\odot$. Assuming that the solar flux is roughly uniform in all directions, the solar **irradiance (insolation)** at the top of Earth's atmosphere is thus given by

$$I_\odot = \frac{L_\odot}{4\pi R_{\mathrm{orbit}}^2} \cong 1366\,\mathrm{W/m^2}, \qquad (23.1)$$

where $4\pi R_{\mathrm{orbit}}^2$ is the surface area of the sphere at a distance R_{orbit} from the Sun. $I_\odot$ gives the solar irradiance at the top of Earth's atmosphere. The solar irradiance $I_\odot$ is often referred to as the **solar constant**, and is measured in units of energy per unit time per unit area.[1] This simple calculation, however, is only the beginning of the story. The

---

[1] We use *irradiance* and *insolation* interchangeably to describe solar energy flux in units of power per unit area. Sometimes, however, insolation is also used to refer to the solar *energy* per unit area incident over a fixed time, as in "daily insolation." The meaning should be clear from the context.

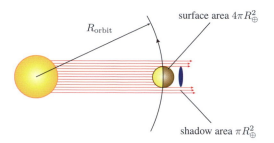

Figure 23.1 The amount of radiation from the Sun (left) incident on Earth (right) is proportional to the cross-sectional area of Earth, seen as the area of Earth's shadow when illuminated by an infinitely distant (plane wave) source.

Solar Constant

The yearly average rate at which solar energy is normally incident at the top of Earth's atmosphere is given by the *solar constant*

$$I_\odot \cong 1366 \, \text{W/m}^2 \, .$$

The total rate at which solar energy impacts Earth is roughly

$$P_{\text{total}} \cong 173\,000 \, \text{TW} \, .$$

Sun is only directly overhead at one point on Earth's surface at any moment, so geometric factors reduce the solar power incident everywhere else on Earth. These geometric factors underlie the principal variations in insolation with latitude, time of day, and time of year that we explore in the next few sections of this chapter. In addition, Earth's distance from the Sun varies over the year, and solar luminosity also fluctuates over time. While these variations are less relevant for human use of solar energy, they play an important role in long-term changes in the planet's climate (§34 and §35).

To compute the total energy incident on Earth's atmosphere, we simply multiply eq. (23.1) by Earth's cross-sectional area $\pi R_\oplus^2$, where $R_\oplus = 6371$ km. Note that we do not multiply by the complete surface area of Earth $(4\pi R_\oplus^2)$, or the complete area of the half that is illuminated at any one time, since in most places the solar energy is not incident perpendicular to Earth's surface. The easiest way to visualize the correct total area to take into account is to think of Earth as casting a shadow of area equal to its cross-sectional area when illuminated by a plane wave of light (see Figure 23.1). The total rate at which solar energy impacts Earth is roughly

$$P_{\text{total}} = \pi R_\oplus^2 I_\odot \cong 173\,000 \, \text{TW} \, . \tag{23.2}$$

As mentioned at the beginning of §22, this is approximately 10 000 times the rate at which the human race uses energy at the present time. Thus, if we can just find a way to use one one-hundredth of one percent (0.01%) of this energy, we could run our entire society with solar energy.

23.2 Earth's Orbit

To understand the sources of variation in insolation we need to use some basic aspects of Earth's **orbital**

dynamics. According to the first of the three **laws of planetary motion** discovered by German astronomer Johannes Kepler early in the seventeenth century, planets like Earth move in elliptical orbits with the Sun at one focus.[2] Some facts about elliptical orbits are summarized in Box 23.1.

Earth's elliptical orbit around the Sun currently has an eccentricity of roughly $e \cong 0.0167$. The resulting variation in the distance between Earth and the Sun over the year leads to an almost 7% difference in incident solar energy between the point of closest approach (*perihelion*, which currently occurs during northern winter, around January 3) and the point of greatest separation (*aphelion*, currently around July 4). As a result, the solar constant ranges from a maximum of around 1412 W/m^2 at perihelion in January to a minimum of around 1321 W/m^2 at aphelion in July. In addition, fluctuations in solar activity can increase or decrease the solar constant by 1 or 2 W/m^2. These variations in incident solar energy are sufficiently small that for the purposes of this chapter we can approximate Earth's orbit as circular. Variations over longer time scales of the eccentricity of Earth's orbit, the relation between perihelion and the seasons, and fluctuations in solar luminosity play a significant role in driving long-term climate change, and are discussed in §34 and §35.

The feature of Earth's orbital mechanics that plays the most significant role in determining the distribution of insolation is the orientation of Earth's axis. The rotational axis of Earth is currently tilted at an angle of $\epsilon = 23.44°$ from the normal to the plane of Earth's orbit (called the **ecliptic plane**). This axis **tilt**, known as the **obliquity**,

---

2   Kepler's other two laws are not important here, but for completeness they are (2) that the line from the Sun to a planet sweeps out equal areas in equal times (illustrated in the figure in Box 23.1), and (3) that the period of a planetary orbit squared is proportional to the length of its semi-major axis cubed.

Box 23.1   Elliptical Orbits

An ellipse is defined to be the set of points in a plane whose distances to two specified points, known as the **foci** of the ellipse, sum to a constant. An ellipse can be described in a Cartesian coordinate system as the set of points (x, y) that satisfy the equation $(x/a)^2 + (y/b)^2 = 1$. Here, a and b are the lengths of the **semi-major** and **semi-minor** axes, respectively. In the figure, the two foci are labelled P and P'; the ellipse is defined to be the set of points Q such that $\overline{PQ} + \overline{P'Q} = 2a$. The distance between the foci is $\overline{PP'} \equiv 2d = 2\sqrt{a^2 - b^2}$.

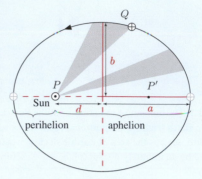

When $a = b$ ($d = 0$), an ellipse reduces to a circle, and when $b = 0$ ($d = a$), it reduces to a straight line segment. The departure of an ellipse from the circular limit is measured by the **eccentricity**,

$$e = \frac{d}{a} = \frac{\sqrt{a^2 - b^2}}{a},$$

which is zero for a circular orbit and goes to one as the ellipse approaches a line segment and becomes degenerate.

For a planet orbiting the Sun, **perihelion** refers to the point of closest approach, at a distance $a - d$ from the Sun, and **aphelion** is the point of greatest distance $(a + d)$ from the Sun.

---

wobbles by several degrees with a period of about 41 000 years, an effect we discuss further in the context of climate change in §35. The tilt of Earth's axis is responsible for the annual cycle of the seasons. The (northern[3]) **winter solstice**, which occurs around December 21, occurs at the point in Earth's orbit when the North Pole[4] is maximally tilted away from the Sun. The **summer solstice**, around June 21, occurs when the North Pole is maximally tilted towards the Sun (see Figure 23.2). The spring (vernal) and fall (autumnal) **equinoxes** occur around March 20 and September 22, when the North Pole lies in the plane perpendicular to the vector connecting the centers of the Sun and Earth. On those dates the length of day and night are equal at all latitudes. (Figure 23.4 gives a reminder of the definitions of latitude and longitude.) As the seasons change, the total daily insolation at a fixed latitude varies in a systematic fashion.

23.3 Variation of Insolation

While the total amount of solar energy incident on Earth is enormous, the quantity of energy available at any particular

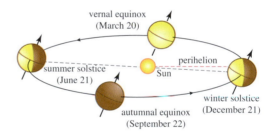

Figure 23.2 Motion of Earth in the ecliptic plane. Winter (summer) solstice occurs when the rotational axis is tilted furthest from (closest to) the direction of the Sun. Perihelion (aphelion), when Earth is closest to (furthest from) the Sun, occurs about two weeks after the winter (summer) solstice.

location is highly dependent upon latitude, time of year, and atmospheric conditions. Ignoring for the moment atmospheric effects (which we discuss in §23.5), insolation depends principally upon the **angle of incidence** of the sunlight. If β is the angle between the normal to the surface and the direction of the incident sunlight, then the insolation is given by

$$I = I_0 \cos \beta, \tag{23.3}$$

where I_0 is the intensity at normal incidence $\beta = 0$ (see Figure 23.3).

At a given point on Earth's surface, the angle of incidence β obviously depends upon the time of day. The angle is 90° at sunrise when the Sun is at the horizon. As Earth rotates, β decreases and then increases again to 90° at

---

[3] We refer to solstices and equinoxes throughout the book using Northern Hemisphere-centric terminology; denizens of the Southern Hemisphere should flip the seasons accordingly.

[4] The North Pole is the point on the surface of Earth in the Northern Hemisphere that lies along Earth's rotational axis. Often referred to as the **geographic North Pole**, this point is distinct from the **magnetic North Pole**, defined in terms of Earth's magnetic dipole field (discussed in §32).

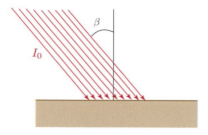

Figure 23.3 Insolation from sunlight incident at angle β to the normal is $I = I_0 \cos \beta$.

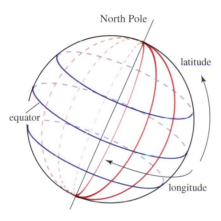

Figure 23.4 **Latitude** (blue) is measured in degrees north or south of the equator, while **longitude** (red) is measured in degrees east (positive) or west (negative) of the **prime meridian**, which for historical reasons passes through the Royal Observatory at Greenwich, England.

sunset. Because of Earth's tilt, the value of β at a particular point on Earth's surface at a given time also depends upon the latitude and the season. Using orbital geometry, we can derive a formula for β as a function of latitude, time of year, and time of day. To do this, it is conventional and convenient to work with the *declination latitude δ*.

23.3.1 Declination

The **declination** δ is defined at any time of year to be the latitude at which the Sun is directly overhead at *solar noon*. **Solar noon** is defined to be the time of day when the Sun is highest in the sky, i.e. when β is minimum for the day. On the vernal (spring) and autumnal (fall) equinoxes, Earth's rotational axis is at $90°$ to the direction of the Sun, and the Sun is directly overhead on the equator at solar noon, so $\delta = 0°$. As Earth moves around the Sun after the vernal equinox, the declination latitude δ gradually increases, reaching a maximum of $\delta = \epsilon = 23.44°$ at the summer solstice. δ then gradually decreases, passing

through $0°$ at the autumnal equinox, reaching a minimum of $\delta = -\epsilon \cong -23.44°$ on the winter solstice, and returning to $\delta = 0°$ again at the following spring equinox. Denoting the time of year by the angular variable

$$\alpha = 2\pi N / 365.24 \,, \tag{23.4}$$

where N is the number of days since the spring equinox, the declination can be related to the time of year α and axial tilt ϵ through

$$\sin \delta = \sin \epsilon \sin \alpha \,. \tag{23.5}$$

It is easy to see that this formula is correct on the equinoxes and solstices where $\sin \alpha = 0$ and ± 1 respectively. More generally, the formula (23.5) follows from some basic spherical trigonometry, as shown in Box 23.2.

23.3.2 Daily Variation of Insolation

We can use the declination, which depends on the time of year, to compute the daily variation of insolation at any latitude. To compute the insolation at a point P on Earth's surface at latitude λ, we need the angle β between the normal $\hat{\boldsymbol{n}}_P$ to the surface and a vector $\hat{\boldsymbol{v}}_{\text{sun}}$ pointing toward the Sun. We first consider the situation at solar noon. At this time, $\hat{\boldsymbol{n}}_P^{(\text{noon})}$, $\hat{\boldsymbol{v}}_{\text{sun}}$, and the unit vector $\hat{z}$ along Earth's rotational axis lie in a common plane; the Sun is at its maximum height and the value of β is therefore smallest. The situation is shown in Figure 23.5.

We define the $\hat{x}$ unit vector to lie along the intersection of the equatorial plane and the plane containing $\hat{\boldsymbol{v}}_{\text{sun}}$ and $\hat{z}$. (Note that this is a different choice of coordinate

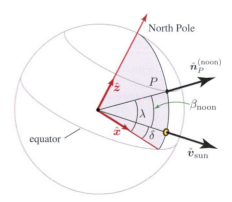

Figure 23.5 The configuration at solar noon in a coordinate system where the vector to the Sun $\hat{\boldsymbol{v}}_{\text{sun}}$ lies in the x–z plane. The normal vector $\hat{\boldsymbol{n}}_P$ at the point P on Earth's surface at latitude λ rotates through the day relative to this coordinate system and the vector $\hat{\boldsymbol{v}}_{\text{sun}}$. The incidence angle for sunlight β takes its minimum value $\beta_{\text{noon}} = |\lambda - \delta|$ at solar noon.

Box 23.2   A Derivation of the Declination

The declination formula (23.5) can be derived by comparing two Cartesian coordinate systems. First, consider a coordinate system based on the configuration of Earth at the vernal equinox, with the x-axis pointing towards the Sun, the z-axis pointing up perpendicular to the plane of the ecliptic, and the y-axis pointing in the plane of the ecliptic perpendicular to the x-axis ($\hat{\boldsymbol{y}}$ would be the direction from Earth to the Sun on the summer solstice for the assumed circular orbit). In this coordinate system, at any given time of year α (eq. (23.4)), the unit vector from Earth to the Sun is in the direction

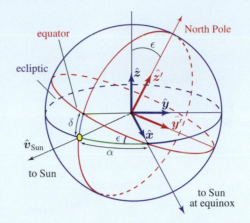

$$\hat{\boldsymbol{v}}_{\text{sun}} = \cos\alpha\,\hat{\boldsymbol{x}} + \sin\alpha\,\hat{\boldsymbol{y}}\,.$$

(In the figure, for clarity, Earth is shown during Southern Hemisphere summer, when $\sin\alpha$ and δ are negative.) To find the declination, we must transform the unit vector $\hat{\boldsymbol{v}}_{\text{sun}}$ into Earth-based coordinates, where we leave the x-axis unchanged, take the z'-axis to point along Earth's rotational axis, and define the y'-axis to lie in the equatorial plane perpendicular to $\hat{\boldsymbol{x}}, \hat{\boldsymbol{z}}'$. The unit vectors $(\hat{\boldsymbol{x}}, \hat{\boldsymbol{y}}, \hat{\boldsymbol{z}})$ are related to the unit vectors $(\hat{\boldsymbol{x}}', \hat{\boldsymbol{y}}', \hat{\boldsymbol{z}}')$ through a rotation by the angle of Earth's obliquity ϵ,

$$\hat{\boldsymbol{x}} = \hat{\boldsymbol{x}}',$$
$$\hat{\boldsymbol{y}} = \cos\epsilon\,\hat{\boldsymbol{y}}' + \sin\epsilon\,\hat{\boldsymbol{z}}',$$
$$\hat{\boldsymbol{z}} = -\sin\epsilon\,\hat{\boldsymbol{y}}' + \cos\epsilon\,\hat{\boldsymbol{z}}'.$$

In Earth-based coordinates (x, y', z'), the vector from Earth to the Sun is then

$$\hat{\boldsymbol{v}}_{\text{sun}} = \cos\alpha\,\hat{\boldsymbol{x}} + \sin\alpha\cos\epsilon\,\hat{\boldsymbol{y}}' + \sin\alpha\sin\epsilon\,\hat{\boldsymbol{z}}'\,.$$

The component of $\hat{\boldsymbol{v}}_{\text{sun}}$ along $\hat{\boldsymbol{z}}'$ is the sine of the declination, from which eq. (23.5) follows.
(Figure Credit: Adapted from [1])

---

frame than was used in Box 23.2.) From the definition of declination we can decompose

$$\hat{\boldsymbol{v}}_{\text{sun}} = \cos\delta\,\hat{\boldsymbol{x}} + \sin\delta\,\hat{\boldsymbol{z}}\,. \tag{23.6}$$

In this coordinate system, the unit normal vector at P at solar noon can be written

$$\hat{\boldsymbol{n}}_P^{(\text{noon})} = \cos\lambda\,\hat{\boldsymbol{x}} + \sin\lambda\,\hat{\boldsymbol{z}}\,. \tag{23.7}$$

Taking the dot product to get $\cos\beta$ we find that

$$\cos\beta_{\text{noon}} = \hat{\boldsymbol{v}}_{\text{sun}} \cdot \hat{\boldsymbol{n}}_P^{(\text{noon})} = \sin\lambda\sin\delta + \cos\lambda\cos\delta$$
$$= \cos(\lambda - \delta)\,. \tag{23.8}$$

Thus $\beta_{\text{noon}} = \lambda - \delta$ when $\lambda > \delta$, while for latitudes below the declination latitude $\beta_{\text{noon}} = \delta - \lambda$. (These results can be seen easily from the geometry depicted in Figure 23.5.) To compute the variation of β through the day, we note that $\hat{\boldsymbol{v}}_{\text{sun}}$ remains fixed while the x-component of $\hat{\boldsymbol{n}}_P$ varies with time as $\cos\omega t$, where $\omega = 2\pi/24\,\text{h}$ and $t = 0$ at solar noon. Since the z-component of $\hat{\boldsymbol{n}}_P$ does not change, this

modifies the inner product (23.8) to give the daily variation of β,

$$\cos\beta(t) = \sin\lambda\sin\delta + \cos\lambda\cos\delta\cos\omega t\,. \tag{23.9}$$

23.3.3 Annual Variation of Insolation

To get the total daily insolation at a given latitude and time of year, we integrate $I = I_0\cos\beta$ from sunrise (where $\cos\beta = 0$) to sunset (where $\cos\beta = 0$ again), being careful not to integrate over the time when $\cos\beta$ is negative, when the Sun is behind Earth. To get the seasonal variation we can compute this daily insolation integral as a function of the time of year α and latitude λ. The result of this integration is graphed for a selection of latitudes in Figure 23.6(a). Note that the daily insolation peaks twice a year at the equator – around the equinoxes – and once a year at higher latitudes. At latitudes with absolute value above $90° - \epsilon \cong 66.6°$ (above the Arctic Circle or below the Antarctic Circle), there is a period of time during the winter when the Sun does not rise above the horizon, and

Example 23.1 Boston at the Winter Solstice (Around December 21)

On the (northern) winter solstice, $\delta = -\epsilon$, since the Sun is directly overhead at noon at latitude $-\epsilon = -23.44\,°$. At solar noon in Boston (latitude $\lambda = 42.3°$), the angle of the Sun from the vertical satisfies

$$\cos \beta = \cos(\lambda - \delta) = \cos(\epsilon + \lambda) \cong \cos 65.75° \cong 0.41\,.$$

We can compute the time of sunrise and sunset (relative to solar noon) by setting $\cos \beta = 0$ (see eq. (23.9)),

$$0 = \cos \beta = \sin \lambda \sin \delta + \cos \lambda \cos \delta \cos \omega t,$$

so in Boston on the winter solstice

$$\cos \omega t = -\tan \lambda \tan \delta \cong -\tan(42.3°)\tan(-23.44°) \cong 0.395.$$

This gives $\omega t \cong 1.16$, and since $\omega = 2\pi/24\,\mathrm{h}$, $t \cong 4.5\,\mathrm{h}$. Thus, sunrise occurs 4.5 hours before solar noon and sunset 4.5 hours after, giving a day that is roughly 9 hours long in Boston on the winter solstice (the shortest day of the year).

the daily insolation is zero. In the summer in these polar regions, there is a period of time during which the Sun does not set. Thus, for example, at the North Pole on the summer solstice, the Sun stays at a constant $\epsilon = 23.44°$ above the horizon, and rotates around the sky during the course of a day.

It is clear from Figure 23.6(a) that the total yearly solar resource in the tropics is considerably larger than that available at larger latitudes. To make this quantitative, Figure 23.6(b) gives the integrated yearly insolation as a function of latitude.

Up to this point we have characterized the incidence of solar energy at the top of Earth's atmosphere. Atmospheric effects reduce incident solar energy significantly and alter its spectrum. In particular, at higher β sunlight must travel a greater distance through the atmosphere before reaching the ground. This increased *atmospheric depth* reduces the insolation below that indicated by eq. (23.3) and Figure 23.6, particularly at higher latitudes. In order to analyze these effects we next explore further the way that light interacts with matter.

23.4 Interaction of Light with Matter

Human use of solar energy depends upon the conversion of energy from electromagnetic waves to thermal or electrical energy. Earth's climate depends crucially on the way in which both incoming solar radiation and outgoing infrared radiation interact with the atmosphere. To understand these processes, it is important to have a basic understanding of the mechanisms by which light interacts with matter. Some aspects of light–matter interactions were introduced in previous chapters – in §7 (Quantum mechanics), we explained

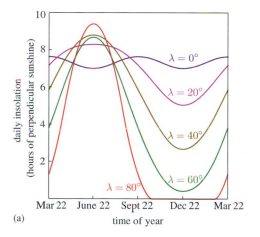

(a)

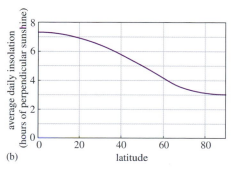

(b)

Figure 23.6 (a) Integrated daily insolation, measured in hours of equivalent perpendicular sunshine, at various Northern Hemisphere latitudes λ as a function of time of year. (b) Daily insolation averaged over the entire year as a function of latitude.

that photons of definite energy and frequency are absorbed or emitted when atomic or molecular systems make transitions from one state to another; in §9 (Energy in matter), we considered molecular vibrations and the wavelengths of light absorbed and emitted in vibrational transitions; and in §20 (Radiation), we showed that the intensity of a beam of high-energy radiation falls off exponentially as it propagates through matter. In this section we apply these results to solar radiation propagating through Earth's atmosphere.

There are three fundamental categories of electromagnetic interaction with matter, **absorption**, **spontaneous emission**, and **stimulated emission**, shown schematically in Figure 23.7. To illustrate these processes, consider a molecule (or atom) with quantum states $|1\rangle$ and $|2\rangle$ having energies E_1 and E_2 respectively, with $E_2 > E_1$. Spontaneous emission occurs when there is a transition from state $|2\rangle$ to the lower-energy state $|1\rangle$, and in the process a photon is emitted with energy $E_\gamma = E_2 - E_1$. Absorption is the *time-reversed* quantum process, in which a photon

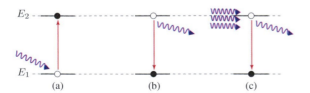

Figure 23.7 (a) Absorption, (b) spontaneous emission, and (c) stimulated emission are quantum processes by which light interacts with matter.

of energy E_γ impinges on the system in state $|1\rangle$ and excites it to $|2\rangle$. The third process, stimulated emission, is less obvious from a quantum point of view: exposing a system in state $|2\rangle$ to an electromagnetic wave with frequency $\omega = (E_2 - E_1)/\hbar$ enhances the probability that the system will de-excite by emitting a photon with the same frequency.

All these processes, like the nuclear decays that we studied in §16, are quantum processes that occur randomly, at a rate determined by characteristic features of the system. For example, photon emission by a molecule is completely analogous to nuclear α-decay or β-decay. A system of N molecules, each in an excited state of energy E_2, will emit photons at a rate $dN/dt = -N/\tau_c$, where τ_c is the characteristic time for the emission decay process. Similarly, light incident on a given material will be absorbed at a rate

$$\frac{dI}{dz} = -\kappa I \,, \tag{23.10}$$

where z is the distance that the light has traveled through the material and κ is the **absorption coefficient** (§20.2.3). κ depends on the nature of the material and the frequency of the light. If, for example, the frequency is close to the frequency associated with a quantum transition in the material, then absorption can be very large (see Example 23.2). Solving eq. (23.10) for a homogeneous material, we obtain an exponential decay in intensity

$$I(z) = e^{-\kappa z} I_0 \tag{23.11}$$

as the light travels into the medium. This is the **Lambert–Beer law**, previously mentioned in connection with the attenuation of high-energy photons traveling through an absorbing medium (see §20.2.3). The (negative of the) exponent in this equation provides a dimensionless measure of the extent of attenuation of light that has penetrated into material. This quantity, $\tau \equiv \kappa z$, is known as the **optical depth**.[5] A material with $\tau \ll 1$ or $\tau \gg 1$ is referred to as **optically thin**, or **optically thick**, respectively.

Atoms and molecules emit or absorb light only at (or near – see Example 23.2) the wavelengths characteristic of transitions between allowed energy states,

$$\lambda_{ij} = \frac{2\pi\hbar c}{|E_i - E_j|} \,, \tag{23.12}$$

where E_i and E_j are the energies of two states that can be connected by photon emission or absorption. When the spectrum of light emitted by an atom or molecule is

---

[5] Although we are primarily concerned with absorption, the *optical depth* is often defined to include the effects of both absorption and *scattering*.

Example 23.2   The $3P$ to $2S$ Transition in Hydrogen

What is the wavelength of the photon emitted when the electron in a hydrogen atom drops from the 3p to the 2s energy level?

According to eq. (7.52), $E_n = -\epsilon_0/n^2$, where $\epsilon_0 \cong 13.6\,\text{eV}$ for hydrogen. The emitted photon has an energy given by the energy difference between the states

$$E_\gamma = \hbar\omega = E_3 - E_2 = (-\epsilon_0/9) - (-\epsilon_0/4) \cong 1.89\,\text{eV}.$$

The corresponding wavelength is

$$\lambda = 2\pi c/\omega \cong 656.281\,\text{nm},$$

which corresponds to light in the red part of the visible spectrum.

   This particular spectral line of hydrogen is used by astronomers to measure the velocity of distant stars relative to our solar system, because the wavelength of the emitted light is more shifted toward the red (or blue) the more rapidly the emitter is moving away from (or toward) Earth. A similar effect, known as the **Doppler shift**, may be familiar from the sound of the whistle of an approaching or receding train. (This is the same effect we encountered in the study of neutron absorption in §19.2.1.)

Absorption of Light by Matter

The intensity of light at frequency ω passing into matter decreases as light is absorbed according to the *Lambert–Beer* law

$$I(z) = e^{-\kappa_\omega z} I_0,$$

where κ_ω is the frequency-dependent *absorption coefficient*. The *optical depth* into the material is $\tau = \kappa z$.

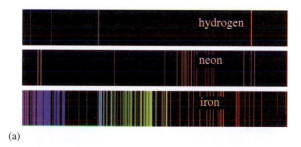

(a)

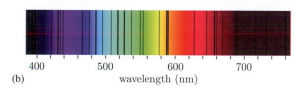

(b)

Figure 23.8 (a) Emission spectra of hydrogen, neon, and iron in the visible range. Note: the more complex the atom, the richer the spectrum. (b) Some of the most prominent absorption lines in the visible part of the Sun's spectrum. Thermal emission from the Sun's surface provides the smooth background, while absorption in the cooler solar atmosphere gives rise to the dark absorption lines.

analyzed, it shows **emission lines** at these characteristic wavelengths. The emission spectra of a few elements are shown in the range of visible light in Figure 23.8. Alternatively, when a light source with a broad spectrum – thermal radiation from a hot object for example – is shrouded by a layer of cool material, absorption by the material gives rise to dark lines in an otherwise smooth emission spectrum. The cooler outer atmosphere of the Sun generates such an absorption spectrum – a section is shown in Figure 23.8. These **spectral lines** act as *fingerprints*, allowing scientists to identify atoms or molecules by the light they emit. The field of **spectroscopy** is concerned with the detailed study of the spectra of different kinds of atoms and molecules. In fact, the element helium was originally discovered in the Sun by analysis of the spectrum of solar radiation (from which helium derives its name).

   Along the lines of the classification of energy scales introduced in §9, there is a hierarchy of wavelength (or frequency) scales in molecular spectroscopy.

Electronic excitations Excitations of electrons within atoms or molecules (see §9.3.3) characteristically have energies of order 1–10 eV, and are associated with photons in the visible and ultraviolet parts of the spectrum. The spectrum of hydrogen has, for example, a gap of 10.20 eV between the ground state and first excited state, corresponding to a photon with $\lambda = 121.6\,\text{nm}$ in the ultraviolet.

Vibrational excitations Excitations associated with vibrational modes (see §9.3.1) typically have energies of order 0.1 eV, associated with photons in the infrared. For example, the vibrational modes of the diatomic molecules O_2 and N_2 have energy gaps of 0.19 eV and 0.29 eV respectively, which would be associated with photons of wavelength 6.5 μm and 4.3 μm respectively (Problem 23.9). Because N_2 and O_2 are symmetric, they do not develop any charge separation (or *electric dipole moment* – see Box 3.2) when their vibrational modes are excited (see §9 Footnote 5). As a consequence they cannot easily absorb single photons of infrared radiation. CO_2, on the other hand, has two vibrational excitation modes that can absorb photons of wavelength $\lambda = 4.26$ μm and 15 μm. These modes play an important role in the greenhouse effect, as discussed further in §34.

Rotational excitations A final class of molecular motions, molecular rotations, are typically associated with energy gaps smaller than 10^{-2} eV, corresponding to photons with wavelengths in the far infrared and microwave region ($\lambda \gtrsim 100$ μm). Transitions from one rotational level to another do not play an important role in atmospheric absorption of sunlight – the subject to which we next turn – because very little of the energy in sunlight is in the microwave region. Highly excited rotational energy levels of water vapor, however, absorb photons with wavelengths as low as 15 μm, but only ~0.02% of the energy in sunlight is found in this part of the spectrum. On the other hand, the outward going thermal radiation from Earth's surface has significant power in the infrared, where absorption by rotational transitions in water vapor is important (see §34). Recall also (§9.3.1) that when a molecule absorbs a photon and changes its vibrational state, the rotational state also changes, giving closely spaced *vibration–rotation* bands.

23.5 Atmospheric Absorption

As solar radiation passes into Earth's atmosphere, molecules in the atmosphere absorb radiation of certain frequencies associated with their excitation spectra. This not only reduces the rate at which direct solar energy impacts Earth, but also modifies the spectrum of the radiation that reaches Earth's surface. The molecules that play the most significant role in atmospheric absorption of solar radiation are **ozone** (O_3), water vapor (H_2O), and carbon dioxide (CO_2).

Ozone has a very strong continuous absorption band, known as the **Hartley band**, at 200–300 nm in the ultraviolet, associated with photodissociation of the ozone molecule, $\gamma + O_3 \rightarrow O_2 + O$. Although the total amount

of atmospheric ozone is rather small (all the ozone in Earth's atmosphere would form a layer less than half a centimeter thick at sea level) and is limited primarily to the upper atmosphere, absorption of UV light by ozone is so strong that almost all ultraviolet light with wavelengths below 295 nm is absorbed by ozone far above sea level. This is fortunate, since – as mentioned in §20 – ultraviolet radiation can be damaging to living organisms. Ozone in the upper atmosphere is destroyed by chlorofluorocarbons used as refrigerants, as discussed in §13.2.1, as well as by natural processes as discussed further in §34.

Water vapor and *carbon dioxide* both strongly absorb radiation in certain frequency bands in the infrared. Many of the relevant infrared absorption bands for these molecules are associated with vibrational modes. Infrared absorption leads to suppression of these frequency bands in the solar radiation that impacts on Earth's surface at sea level. Infrared radiation at wavelengths longer than 14 μm is essentially completely absorbed by carbon dioxide and water vapor in the atmosphere (see Figure 34.4). Just as water vapor and carbon dioxide absorb incoming infrared radiation, they also absorb parts of the outgoing infrared blackbody radiation from Earth, thus trapping heat within Earth's atmosphere that plays an important role in maintaining Earth's climate, as we discuss in §34.

The spectra of solar radiation at the top of the atmosphere and at sea level are shown in Figure 23.9, indicating some of the principal absorption bands of O_3, CO_2, H_2O, and O_2. Under typical conditions, after atmospheric absorption, the total intensity at sea level of normally incident sunlight for an overhead Sun is about 1000 W/m^2.

Note that it is not a coincidence that life on Earth uses light in the visible bands, where solar radiation is peaked, for vision and photosynthesis, and that our organic

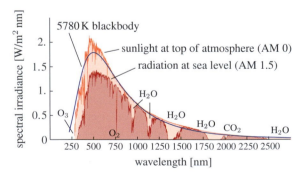

Figure 23.9 Solar radiation spectrum at the top of the atmosphere and at sea level, indicating atmospheric absorption by ozone, carbon dioxide, water vapor, and oxygen [135]. See Box 23.4 for more information.

Box 23.3   Line Width

In absorption and stimulated emission processes, the incoming photon need not have precisely the energy correspond-ing to the transition between the two quantum states, $E_\gamma = E_2 - E_1$. Instead, the *spectral lines* have a *width* in wavelength (or energy). There are several sources of this width. The most fundamental is a **natural line width** associ-ated with quantum mechanics itself. Because the unstable excited state with energy E_2 does not live forever, but instead decays with a lifetime τ, its energy cannot be perfectly known (an example of Heisenberg's time–energy uncertainty principle mentioned in §15.1), and the natural width of the spectral line is proportional to $\Delta E \propto \hbar/\tau$.

In practical situations, the natural line width is usually dwarfed by the effects of temperature and pressure. We encountered *Doppler broadening*, also known as **thermal broadening**, in the study of neutron absorption by heavy nuclei (§19.2.1). The same effect occurs for photon absorption by atoms or molecules. Although a photon might not have exactly the right energy to be absorbed by a molecule at rest, a molecule in motion sees the photon's frequency – and hence energy – as shifted slightly; for the right molecular velocity, absorption can take place. Likewise, a molecule in motion with respect to an observer emits a photon with an energy which is shifted in the reference frame of the observer. This is another example of the *Doppler effect* mentioned in Example 23.2.

Pressure broadening originates in the collisions between molecules, which modify emission and absorption pro-cesses, effectively shortening the lifetime of the excited state and increasing the uncertainty in its energy. Thermal line widths increase with increasing temperature, and pressure broadening increases with both temperature and pressure, since both decrease the mean time between molecular collisions.

Box 23.4   Reference Solar Spectra

To accurately estimate the efficiency and power output of solar energy systems, it is helpful to have a more precise estimate of the solar spectrum than the simple blackbody spectrum. This is particularly necessary for photovoltaic systems, where the response of the system is highly dependent upon the frequency of incident radiation. A widely accepted systematization of solar spectra is given by standard **air mass 0 (AM0)** and **air mass 1.5 (AM1.5)** reference spectra [135]. Air mass 0 refers to the measured solar spectrum at the top of Earth's atmosphere (with solar constant $1366\,\text{W/m}^2$). Air mass 1 designates the spectrum at sea level from normally incident sunlight on a clear day. The air mass A spectrum is the spectrum of solar radiation that has passed through air of optical depth A times greater than that through which normally incident solar radiation passes. For angles substantially less than 90°, Earth's curvature can be neglected, so that the **air mass A spectrum** is given by the spectrum of light with incidence angle β from the normal, where $1/\cos\beta = A$. For $A = 1.5$, $\beta = 48.19°$. The AM1.5 reference spectrum represents terrestrial solar spectral irradiance for the Sun at this angle under very specific atmospheric conditions (for details see [135]).

---

structure is reasonably robust when subjected to visible and infrared light, but easily damaged by UV radiation. Earth organisms have evolved to optimize their survival and reproductive capacity in an environment where visible-wavelength light is dominant and there is little potentially damaging UV light. Perhaps organisms living on a planet orbiting a cooler star might make use of infrared radiation, while life on planets with hotter stars and/or no ozone in the atmosphere would have developed shielding or organic systems suited to UV light that would damage earthly organisms.

Even before encountering Earth's atmosphere, solar radiation differs from a blackbody spectrum. Elements in the Sun's outer atmosphere absorb outgoing radiation, producing dark lines in the spectrum associated with absorption bands of elements such as hydrogen, sodium, and iron (see Figure 23.8). The solar spectrum at the top of Earth's atmosphere varies somewhat depending on solar activity, and the spectrum at the bottom of the atmosphere varies significantly with atmospheric conditions, pollution, and angle of incidence. To normalize analysis of solar power technologies and climate several "standard" spectra of solar radiation have been adopted. They are described in Box 23.4.

We have discussed how the geometry of the Earth–Sun system leads to seasonal variation and latitude dependence of the total daily insolation received at different locations on Earth's surface. Differing weather patterns at different

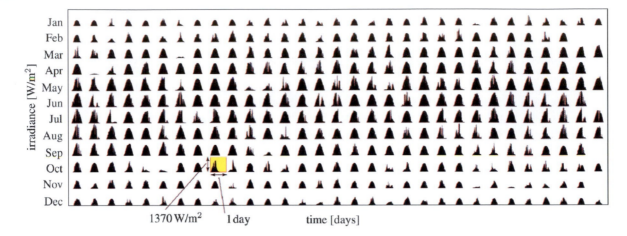

Figure 23.10 Record of insolation over a year in Golden, Colorado, at an altitude of 1700 m in the US Rocky Mountains. Notice the regular seasonal variation as well as more local effects such as summer afternoon thunderstorms (July and August), at this generally sunny location. From [136].

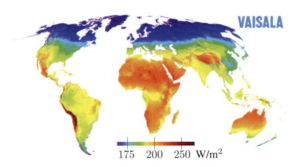

Figure 23.11 World map of yearly average *global horizontal irradiance*, which includes both solar energy directly received on a horizontal surface plus energy from light scattered in the atmosphere that impinges on a horizontal surface (*diffuse horizontal irradiance*). Image: ©Vaisala, Inc.

locations also have a strong effect on the total insolation available in different places. The record of a year's insolation at a particular location, such as Figure 23.10, shows many features, both seasonal and characteristic of the local climate. Maps – like the one shown in Figure 23.11 – of the average annual solar irradiance summarize the effects of both geometry and weather.

The effects of climate variations at locations with the same latitude (and therefore the same average annual insolation above the atmosphere) can be large. For example, Boston, Philadelphia, and San Francisco are all within $\pm 2.5°$ of $40°$ latitude, but a solar panel tilted toward the southern horizon at an angle equal to the latitude receives about 180 W/m², 185 W/m², and 250 W/m² of insolation averaged over the year at the three locations. Seasonal

variations in weather can lead to significant variations in insolation even in locations near the equator where the seasonal effects of Earth's obliquity are minimal. The effects of the annual monsoon in India, for example, with the wet season from June–September, are quite evident in Figure 23.12, which shows the mean insolation on a month-by-month basis.

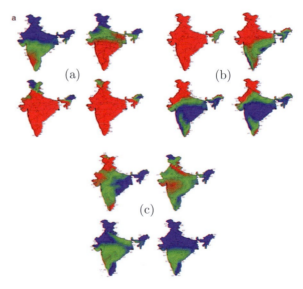

Figure 23.12 Monthly variation in insolation in India. (a) January–April; (b) May–August; (c) September–December. The units are kWh/m² day (> 5.25 red, 4–5.25 green, < 4 blue). From [137].

23.6 Extent of Resource

Before we proceed in the following chapters to analyze various approaches for utilizing solar energy, we explore here briefly the extent of the solar resource and the scale of systems that would be required to harvest it for human use. We can get some perspective on the scale of the solar energy resource by performing a *back-of-the-envelope* estimate of the surface area required to produce a given amount of electrical power using solar energy. Human electric power consumption currently amounts to a steady-state rate of approximately 3 TW, and a large-scale shift towards electric-powered land transport could replace another 20% of current energy use with an additional TW of electric power; what land area would be needed to generate a significant fraction of 4 TW using solar energy?

To make a rough estimate of the area involved, we consider two different scenarios, one corresponding to distributed small-scale power in an urban setting, the other corresponding to large-scale power plant deployment in desert lands with high insolation. In analyzing these scenarios, we consider two different solar energy harvesting technologies that are described in more detail in following chapters. The first approach involves photovoltaic (PV) cells (§25), which convert solar energy directly into electrical energy. Photovoltaics are presently used both in distributed applications such as on residential and commercial rooftops and also in large-scale power plants. We use PVs in both of our scenarios. A second approach employs concentrators to heat water or another medium to sufficiently high temperatures that the energy can be efficiently converted into electricity using a heat engine. Known as solar thermal electricity generation (STE), we describe this technology in §24. At present STE is used primarily in large-scale power plants. We do not consider STE in our urban scenario.

Existing STE plants have **gross solar conversion efficiencies** of about 3–4% (see §24.4). This figure reflects thermodynamic losses in conversion from thermal energy to electricity and also allows for the total land area covered by the system – including support structures, plumbing, generators, and the like, as well as gaps between system components – not just the area of the light-gathering mechanisms. We assume photovoltaic panels produce electricity with an efficiency of 14% – a typical value for panels available to consumers in 2016.[6] We address the gross

Table 23.1 Estimates of certain land types/use categories on Earth.

Land type/use	% land	total area
Cropland + pasture	35%	$53\,\mathrm{M\,km^2}$
Urban	3%	$4\,\mathrm{M\,km^2}$
Desert	15%	$23\,\mathrm{M\,km^2}$
Other	47%	$70\,\mathrm{M\,km^2}$
Total land area	100%	$150\,\mathrm{M\,km^2}$
Total planetary surface		$510\,\mathrm{M\,km^2}$

conversion efficiency for PV installations separately in the two scenarios.

Photovoltaics in populated areas In this scenario we assume that photovoltaic panels with 14% conversion efficiency are installed on urban rooftops. Taking $150\,\mathrm{W/m^2}$ as a rough estimate of average daily insolation in the world's urban areas, a simple calculation (Problem 23.9) gives an approximate surface area of $42\,000\,\mathrm{km^2/TW}$, or about 1% of world urban land area (see Table 23.1) per terawatt.[7]

Large-scale solar power generation in deserts Alternatively, we consider large-scale solar power plants, either PV or STE, located in the world's deserts with high levels of insolation. To be conservative, we take the gross solar conversion efficiency for STE to be 3%, although improved systems may substantially increase this efficiency in the future and some newer plants can reach 4–5%. Gross conversion efficiency for large-scale PV plants must be reduced by a *packing factor* reflecting the space required to optimally orient panels as well as for interconnections, support structures, etc. This packing factor has been variously estimated as a factor of 2 [133] to 2.5 [138]. To be conservative, we take the larger value, and therefore assume a gross solar conversion factor of $14/2.5 \approx 6\%$ for large-scale PV installations.

Estimating an average insolation of $300\,\mathrm{W/m^2}$ in the world's deserts, another simple computation (Problem 23.9) gives $110\,000\,\mathrm{km^2/TW}$ for STE and $55\,000\,\mathrm{km^2/}$TW for PV, corresponding to 0.5% and 0.25% respectively of the world's total desert per terawatt. Thus, using existing technology to produce 4 TW to provide all the world's

---

[6] Module efficiencies of 17% are typical, but converting their low-voltage DC output to higher-voltage AC power along with other factors reduce the overall efficiency by about 20% [133].

[7] Note that here we are considering only the panel surface area, not including the area needed for infrastructure, which is difficult to estimate, but if included would likely reduce the *gross solar conversion efficiency* for urban rooftop installations below that of large-scale solar power plants in rural or desert locations.

current electricity and land transport energy needs using these kinds of systems would require somewhere between 1% and 2% of the world's desert land area.

These estimates of land use provide a measure of the effort required to develop solar power at the terawatt scale. Clearly, to realize either of the scenarios presented above would require an enormous engineering undertaking, together with investment capital, public support, and political leadership. The efficiencies we assumed are appropriate for large-scale implementation with current technology. In fact, improvement of available technology may make these estimates rather conservative in coming years. If either or both of these scenarios could be approximately realized at the level of several terawatts, and if solar energy were used whenever available for space heating and process heat (with much higher efficiency), it is not impossible to imagine that half or more of the world's energy could come from solar sources by the end of the century. In the following chapters we look at each of these approaches to solar energy in turn, discussing the underlying physics and basic processes involved in each.

Discussion/Investigation Questions

23.1 In §23.3 we describe the effect of Earth's axial tilt, or *obliquity* on insolation and, indirectly, on the seasons. Discuss qualitatively how Earth's seasons would change if its obliquity were $\sim 90°$ (as is the case for Uranus) or $\sim 0°$ (as is the case for Mercury or Jupiter).

23.2 A flat panel solar collector is to be tilted at an angle θ to the horizontal to maximize the amount of sunlight it can collect over the whole year. Once tilted, it is oriented toward the south (in the Northern Hemisphere). One might think that θ should be chosen to equal the latitude λ of the installation so that the Sun would be directly above the panel at noon on the equinoxes. Instead, if total insolation were the only concern, θ would be chosen to be less than λ. Explain why. Describe (but do not attempt to compute) how you would compute the optimal angle for a given location to maximize annual collection. Explain why, despite these considerations, in many practical situations θ may be chosen to be equal to or greater than λ.

23.3 In many (Northern Hemisphere) locations, solar PV panels are oriented toward the southwest rather than south. Discuss aspects of electricity demand that might motivate this choice.

23.4 Investigate the distribution of insolation and population in China or another country of your choice, and discuss (in general terms) the prospects for solar power in that country.

Problems

23.1 The average radius of Mars' orbit is 2.28×10^8 km. Compute the solar constant for Mars.

23.2 Show that the insolation averaged over the year and over the entire Earth is $\cong 341$ W/m^2.

23.3 [C] Consider a location for a solar energy installation in Arizona, at latitude 34°. Compute the declination δ on the following dates: (i) March 20 (spring equinox), (ii) April 20. Compute the length of the day (from sunrise to sunset) at this location in Arizona on April 20. Assuming a clear day, compute the insolation on a horizontal surface at 6 a.m., noon, and 3 p.m., and compute the total insolation for April 20.

23.4 For your location, calculate the maximum instantaneous irradiance (at the top of the atmosphere) on the days of the summer and winter solstices. Now suppose Earth's obliquity were 0°. How large would the eccentricity of its orbit have to be to produce the same difference in irradiance between perihelion and aphelion?

23.5 Suppose a gas has an *absorption cross section* per molecule $\sigma(\omega)$ for light of frequency ω (see §18.2 for an introduction to *cross sections*). Show that the gas's absorption coefficient is given by $\kappa(\omega) = \sigma(\omega)\rho$, where ρ is the number density of gas molecules.

23.6 The thickness of Earth's ozone layer at any location is measured in **Dobson Units** (DU), where 1 DU corresponds to a gas layer of thickness $10\,\mu$m if the gas pressure were raised to 1 atm. Show that 1 DU corresponds to 2.69×10^{16} molecules/cm^2. Data on the absorption cross section for ozone are given in Figure 23.13. Normally the thickness of the ozone layer is ≈ 300 DU, but due to ozone destruction by chlorofluorocarbons it had dropped as low as ≈ 90 DU over Antarctica during the 1990s. How low would the ozone level have to drop to allow 10^{-6} of the incident UV flux at $\lambda \approx 250$ nm or 220 nm to reach sea level? (You will need to use the result of Problem 23.5.)

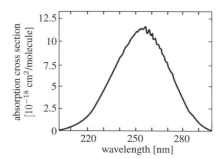

Figure 23.13 Ozone absorption cross section in cm^2/molecule in the Hartley Band [139].

23.7 [**C**] Compute the integrated daily insolation at several different latitudes on July 30 and compare with Figure 23.6.

23.8 The absorption coefficient for crystalline silicon at 300 K for a photon of energy 1.3 eV is roughly 70 cm^{-1}. Compute the thickness of silicon needed to absorb 75% of incoming photons at this energy.

23.9 Verify the estimates of global land use per terawatt in the two scenarios outlined in §23.6.

23.10 The population of the city of Cambridge in the state of Massachusetts is around 100 000. Assume that each resident of Cambridge uses energy at the rate of 1 GJ per day. If the average insolation in the area is 150 W/m^2, compute the area of land needed to supply all the energy for the residents of Cambridge assuming that the conversion efficiency of sunlight is 2%, 4%, 8%, and 16%. Compare to the land area of Cambridge, which is 16.65 km^2.

Solar Thermal Energy

When solar radiation is absorbed by materials on Earth, unless the system absorbing the photons has been carefully tuned by design (photovoltaic cells) or evolution (photosynthesizing plant cells), the absorbed energy is rapidly converted into random thermal energy of the molecules in the absorbing material. The simplest way to use solar energy is to make direct use of this thermal energy. Much of the energy we use goes to heating of buildings and water, and solar thermal energy can be used efficiently for these purposes. Simple solar heating systems do not require particularly high temperatures for effective operation, and can therefore work with fairly simple technologies. Other applications of solar energy, such as industrial heating and cooling, or conversion into mechanical or electrical energy, require higher temperatures, making it necessary to concentrate solar energy using mirrors or other optical systems. In this chapter we discuss some of the principal physical concepts underlying these approaches to the use of solar thermal energy.

Solar thermal energy use can be divided roughly into three categories, depending on the temperature range involved:

Low-temperature passive/active solar heating and lighting
Aside from natural lighting, which is ubiquitous, the simplest use of solar energy is to heat buildings using natural convection. For example, large windows that allow sunlight to shine on a stone floor with large heat capacity can provide substantial thermal energy to a house. This type of direct, passive use of solar energy has been practiced for many centuries around the world. Simple *active* solar collectors, which trap, store, and circulate thermal energy for air and water heating, are now widely used and are an inexpensive and efficient way of using solar energy. To understand the basic principles of solar collectors, we introduce the concepts of radiation balance and heat trapping in §24.1. The same mechanism that traps heat in

Reader's Guide
This chapter covers the use of thermal energy from incident solar radiation. The concept of radiative equilibrium is developed and used to characterize solar thermal energy collection systems. Solar concentrators are studied using geometric optics. A thermodynamic limit on concentration is derived and applied. Finally, systems for conversion of solar thermal to electrical energy are analyzed.

Prerequisites: §6 (Heat transfer), §22 (Solar energy), §23 (Insolation), particularly the topics of radiative and convective heat transfer and insolation; §8 (Entropy and temperature), §10 (Heat engines), particularly regarding the thermodynamic efficiency of heat engines.

Radiative equilibrium plays an important role in Earth's global climate, as developed in §34.

a solar collector is responsible for the *greenhouse effect*, which we discuss further in §34. We describe some basic features of solar collectors in §24.2.

Intermediate-temperature solar heating/cooling Another direct use of solar energy is for industrial process heating. This kind of application requires higher temperatures than domestic hot water applications, generally above 80 °C. To achieve such high temperatures, it is necessary to concentrate the incoming solar radiation and/or to use more sophisticated collector technology to reduce heat loss. Another application of solar energy at these temperatures is for cooling. The intermediate-temperature heat available from concentrated solar collectors is well suited to power the *absorption refrigeration cycle* that was described in §13.2.6.

We do not discuss intermediate-temperature solar thermal applications in detail here; we do, however, briefly describe some intermediate-temperature solar collectors in §24.2, and we give an introduction to the basic physics of solar concentrators in §24.3.

Box 24.1   Solar Thermal Energy

While solar energy still contributes only a tiny fraction of global electrical power, direct use of solar thermal energy provides energy for heating houses, hot water, and swimming pools around the world. The total capacity of water-based solar thermal collectors worldwide was estimated to be $\sim 400\,$GW in 2014, and continues to grow; the power gathered by these systems in 2014 was estimated to be 39 GW on average [140]. Most collectors in place in the first decade of the twenty-first century were low-temperature collectors operating at temperatures below $80\,°$C, providing energy for domestic hot water. Direct use of solar thermal energy represents one of the best opportunities for expanding renewable energy usage. In addition to further opportunities for solar thermal energy use in the domestic sector, a large fraction of energy used by industry is thermal energy for *process heating*. Increased development and deployment of intermediate-temperature solar collectors operating at temperatures from $80\,°$C to $250\,°$C could greatly increase the renewable share of industrial energy use in coming decades.

High-temperature solar thermal electricity conversion

When solar energy is used to heat material to a sufficiently high temperature, the resulting thermal energy can be used to drive a heat engine, generating mechanical or electrical energy. Because this conversion is limited by Carnot efficiency, it is desirable to concentrate the thermal energy to bring a working fluid (or an intermediate substance) to a very high temperature. There are a number of approaches to this kind of solar thermal electricity generation, which may play a significant role in powering the world's electrical grids in future decades. We describe the physics of solar concentrators in §24.3 and give a basic overview of solar thermal electricity technologies in §24.4.

For those interested in more in-depth background on basic solar energy concepts, [141], while older, is a good starting point. For solar thermal electricity conversion, [142] gives some background and [133] provides a recent overview of solar thermal power technologies. Finally, [143] provides a recent and thorough treatment of concentrator technology.

24.1 Solar Absorption and Radiation Balance

24.1.1 Radiative Balance

A key principle governing the temperature that an object reaches when exposed to solar radiation is the concept of *radiative equilibrium*. This concept is helpful for understanding the basic physics of solar collectors, as well as, on a larger scale, the regulation of planetary temperature and Earth's climate. If an object is in thermal equilibrium with its surroundings, the flow of energy into the object exactly balances the flow out. **Radiative equilibrium**, also known as **radiative balance**, refers to a thermal equilibrium in which radiation is the only means

by which the object loses energy. In general, heat can also flow by conduction and convection. Thus, radiative balance applies when the object is doing no work and conduction and convection losses are small or absent. We are often interested in the radiative equilibrium of systems where all or most of the added energy comes from absorbed radiation, in which case radiative equilibrium simply implies that the rate of absorption equals the rate of radiation. The term radiative equilibrium also applies to objects such as stars, where the rate of energy loss to radiation balances the rate at which energy is released through nuclear fusion reactions. In this chapter we often ignore convection and conduction. This is not always a good approximation, and radiative balance is only a rough guide in most actual situations. When a system has radiative and convective losses that together balance the flow of energy in, the system is said to be in *radiative-convective* equilibrium. We discuss such situations further in §34.

The *Stefan–Boltzmann law* (22.8), which relates radiated power to temperature, and the concept of radiative balance, together enable us to find the temperature of an object in radiative equilibrium with a given flux of incoming radiation. To be concrete, consider an idealized flat *black body* with surface area A that radiates at temperature T_b and does not lose energy in any other way. (The back side of the object is assumed either to be perfectly insulated or to be in contact with a substrate at the same temperature as the body, so that the black body only radiates from the side facing the incoming radiation.) According to the Stefan–Boltzmann law, the flux of energy (power per unit area) radiated by the object is $I_b = \sigma T_b^4$, where the emissivity ε has been taken to be one and σ is the Stefan–Boltzmann constant discussed and derived in §22. To remain in equilibrium, the radiated

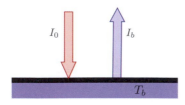

Figure 24.1 Radiative balance between outgoing thermal radiation I_b from a black body at temperature T_b and incoming radiation I_0 fixes the temperature of the black body through $I_0 = I_b = \sigma T_b^4$

power must be matched by an energy input. If the source is an incoming flux of energy I_0, then the principle of radiative equilibrium asserts that

$$I_0 A = I_b A = \sigma T_b^4 A ,$$

or, since A is a common factor,

$$I_0 = \sigma T_b^4 . \tag{24.1}$$

Under these ideal conditions radiative equilibrium determines the temperature of the object,

$$T_b = \sqrt[4]{I_0/\sigma} . \tag{24.2}$$

As a simple example, the average surface temperature of Earth in a simplified homogeneous model without an atmosphere is computed in Example 24.1.

A perfectly absorbing surface in radiative equilibrium with normally incident solar radiation is illustrated in Figure 24.1, and in an idealized approximation would reach a temperature of $T_b \cong 90\,°C$ for $I_0 \approx 1000\,W/m^2$. As explained in Example 24.2, however, convective losses reduce this temperature significantly. In order for solar collectors to reach significantly higher temperatures than the ambient background it is necessary either to trap heat with the help of the greenhouse effect as described in the following section or to concentrate the incident radiation as described in §24.3.

24.1.2 Heat Trapping – The Greenhouse Effect

Since the power of thermal radiation is proportional to T^4, the rate at which an object loses thermal energy back to the environment increases rapidly as its temperature rises. To effectively trap solar radiation, it is useful therefore not only to minimize conductive and convective losses, but also to decrease radiative loss when possible. An obvious way to do this would be to carry the captured heat away from the object quickly enough to keep its temperature low. This, however, conflicts with the goal of capturing

solar energy at a temperature high enough compared to the environment to be useful. So the challenge in solar thermal energy systems is to minimize radiative losses while keeping the object's temperature high.

Incoming solar radiation is primarily concentrated around the wavelength of visible light (see Figure 22.3). In contrast, the power radiated from a terrestrial object at a much lower temperature peaks in the infrared. For example, the power (per unit frequency) radiated by a solar collector at $90\,°C \cong 360\,K$ peaks at a frequency corresponding to a wavelength of around $14\,\mu m$ (see eq. (22.14)), well into the infrared. Thus, one can increase an object's absorption of solar radiation by interposing between the object and the solar source a layer of material that is transparent to visible and near-visible wavelengths, but opaque to the infrared radiation produced by the heated object. A pane of ordinary glass, for example, has the desired properties to trap heat in a simple solar collector. This effect, long used to heat greenhouses above the ambient temperature, is appropriately known as the **greenhouse effect**.

The atmosphere plays the same role relative to the absorption of solar radiation on Earth's surface: **greenhouse gases** such as water vapor and CO_2 admit incoming solar radiation while blocking some of the outgoing infrared radiation from Earth. This is a central topic in §34.

To understand how radiative balance works in such a situation, let us consider again the case of a flat absorbing surface modeled as a perfect black body. The black body radiates, primarily in the infrared, at an intensity

Example 24.1   Earth in Radiative Equilibrium

Consider planet Earth in radiative equilibrium with incoming solar radiation, assuming that Earth is a perfect black body at a uniform temperature and ignoring its atmosphere. As discussed in §23, the effective illuminated area of Earth is its cross-sectional area $A_{\text{eff}} = \pi R_\oplus^2$, so that the total incident radiation is $I_\odot \pi R_\oplus^2$. The surface area of Earth, however, is $A_{\text{total}} = 4\pi R_\oplus^2$. Thus, radiative balance states that

$$I_\odot = 4I_\oplus = 4\sigma T_\oplus^4,$$

where we have cancelled the factor of $\pi R_\oplus^2$ that appears on both sides of the radiative balance equation. Using the solar constant $I_\odot = 1366 \, \text{W/m}^2$, we have

$$1366 \, \text{W/m}^2 = 4\sigma T_\oplus^4 = 4(5.67 \times 10^{-8} \, \text{W/m}^2 \, \text{K}^4)T_\oplus^4$$

$$T_\oplus \cong 279 \, \text{K} \cong 6 \, ^\circ\text{C}.$$

Unlike the parking lot in Example 24.2, Earth does not lose energy to convection or conduction. (There are not enough molecules in space to conduct or convect.) This calculation is nevertheless an oversimplification of the real situation for several reasons. Both the atmosphere and Earth's surface reflect some incoming radiation, so Earth does not absorb or emit as a perfect black body. The atmosphere traps heat through the greenhouse effect, heating the surface. Convection in the lower atmosphere must also be included in a realistic radiative balance computation. Finally, the surface of Earth is not at a uniform temperature; the substantial temperature gradient from equator to poles and smaller local variations in surface temperature modify the flow of energy in radiative balance. We return to these issues in describing Earth's climate in §34.

Example 24.2   Sun on Blacktop

Consider a black body exposed to incident radiation of intensity $I_0 = 1000 \, \text{W/m}^2$. For example, this might describe a parking lot paved with asphalt, subject to direct overhead summer sun on a cloudless day.
Neglecting losses to convection and conduction, at what temperature does the black body reach radiative equilibrium, emitting radiation at a rate $I_b = I_0$?
We compute

$$1000 \, \text{W/m}^2 = \sigma T_b^4 = (5.67 \times 10^{-8} \, \text{W/m}^2 \, \text{K}^4)T_b^4$$

$$T_b \cong 364 \, \text{K} \cong 91 \, ^\circ\text{C}.$$

So the absorbing surface reaches a temperature that is close to the boiling point of water. No wonder it is uncomfortable to walk on an asphalt surface with bare feet in the summer!

Although this example illustrates the concept of radiative equilibrium, it also highlights some of the shortcomings of relying on this principle alone. Actually, the surface of a typical parking lot does not reach temperatures as high as 90 °C, even on the hottest summer days. Furthermore, during cooler times of the year the temperature is much lower, even though I_0 may remain large. This is because heat is also transported away from the surface by conduction and convection. In fact, convection is the primary means of removal of thermal energy from a hot parking lot surface. As we know from §6, convective heat transfer grows with the temperature difference between an object and its environment. This explains both why the asphalt does not reach the temperature predicted by a naive radiative equilibrium calculation, and why it is cooler when the ambient air is cooler. This example also neglects incoming thermal radiation from the atmosphere, discussed further in the text.

$I_b = \sigma T_b^4$, where T_b is the temperature of the black body. We assume that a layer of material (the *covering*) lies above the black body and perfectly transmits the incoming solar radiation I_0 but absorbs all the outgoing radiation I_b. The covering itself has a temperature T_g and emits radiation both upward and downward at an intensity $I_g = \sigma T_g^4$. In the simplest case, we ignore heat transport by conduction and convection both from the covering and within the gap between the absorber and the covering. We also assume that no power is being drawn from the device. This is known as the **stagnant condition**. The resulting contributions to the radiative balance equation are shown in Figure 24.2(a), and give

$$I_b = \sigma T_b^4 = I_0 + I_g \ \text{ and } \ I_g = I_0 , \ \text{ so}$$
$$T_b = \sqrt[4]{2I_0/\sigma} . \tag{24.3}$$

Taking, for example, $I_0 \approx 1000\,\text{W/m}^2$ we find $T_b \approx 160\,°\text{C}$, a significant increase over the temperature achieved without heat trapping.

This example is somewhat unrealistic. The temperature of the covering is $T_g \approx 90\,°\text{C}$, and as in the example of the asphalt parking lot (Example 24.2), such a large temperature difference between the covering and the ambient atmosphere would result in significant convective

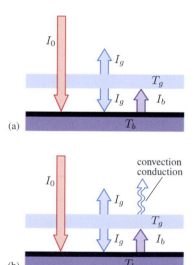

(a)

(b)

Figure 24.2 A covering interposed between incoming solar radiation and an absorber traps heat when the interposed material transmits the visible and near-visible solar spectrum and is opaque to outgoing infrared radiation. (a) An absorber covered by a single pane of glass, in radiative equilibrium with incident solar radiation. (b) A single-glazed collector where the outer pane of glass is held at a constant temperature T_g by conduction and convection to the atmosphere.

and conductive losses. To get a feel for the importance of convective and conductive losses, we assume that they reduce the temperature of the covering to the ambient temperature, $T_g \approx 300\,\text{K}$ (Figure 24.2(b)). Equation (24.3) is then replaced by

$$\sigma T_b^4 = I_0 + \sigma T_g^4 , \tag{24.4}$$

which gives an equilibrium temperature of $T_b \approx 127\,°\text{C}$, still quite a bit higher than the uncovered equilibrium temperature of $90\,°\text{C}$.

So far we have neglected downward infrared radiation from the atmosphere. For an uncovered absorber this radiation mimics the effect of a covering by blanketing the absorber with a downward flux of radiation at a fixed temperature. As discussed in §34, atmospheric temperatures generally decrease with altitude so warming from absorption of atmospheric infrared radiation is less than warming from a glass covering. Including atmospheric radiation, the radiative balance equation for an uncovered absorber (24.1) would include an extra term, $\sigma T_b^4 = I_0 + \sigma T_{\text{atm}}^4$. Assuming an atmospheric temperature of $T_{\text{atm}} \approx 280\,\text{K}$ would give an equilibrium absorber temperature of $T_b \approx 120\,°\text{C}$. Thus, in this simple approximation that ignores convection, the effect of the atmosphere is almost as great $(120\,°\text{C})$ on an uncovered absorber as the effect of a glass covering $(127\,°\text{C})$. In fact, a principal function of the glass plate covering a solar collector is to decrease convective losses by impeding the free flow of air warmed by the absorber. Without the glass plate, convection would substantially reduce the temperature that can be achieved by a collector receiving both solar and atmospheric radiation below the estimate of $120\,°\text{C}$.

Despite the limitations of simple analyses based on radiative equilibrium, we can use this approach to get a qualitative sense of the effects of heat trapping through the greenhouse effect. As another example, consider the effects of a double pane of glass, assuming that there is no convection or conduction between the panes of glass. In this case the outer pane of glass is held at a temperature T_g, while the inner pane of glass is in radiative equilibrium at a temperature T_2. Applying radiative balance and taking $I_0 \approx 1000\,\text{W/m}^2$ and $T_g \approx 300\,\text{K}$, we find $T_b \approx 180\,°\text{C}$, much hotter than with a single pane of glass (Problem 24.5). This temperature is unrealistically high for several reasons. Although we have accounted for convection from the outer covering, convection and conduction within the layers of the system further reduces the absorber temperature. Furthermore, we have neglected reflectivity. Standard glass panes have a reflectivity of roughly 8%. This decreases the total incoming energy more as the number of glass panes is increased, diminishing the heat trapping effect of additional glass panes. In fact [144], using double pane glass in solar thermal collectors does not appreciably increase the efficiency of the collectors, unless non-reflective glass is used (which typically reduces the reflectivity to around 3%). Finally, and most importantly for practical applications, in the scenarios we have described so far, the black body is simply absorbing and reradiating energy, with no energy extracted for human use. In a real solar collector, as we discuss in the following section, energy is constantly extracted, lowering the temperature of the black body and reducing outgoing radiation.

24.2 Low-temperature Solar Collectors

24.2.1 Flat-plate Collectors

The simplest type of low-temperature solar collector in common use is the **flat-plate collector** (see Figure 24.3). In a typical flat-plate collector, pipes are run through a dark absorber plate. The pipes carry a fluid such as water, which is actively pumped through the heating system. Sometimes air – or in cold climates, antifreeze – is used as the heat transport fluid. Transparent glazing over the absorber suppresses convection and traps heat through the greenhouse effect. The enclosure is generally insulated, but standard low-temperature flat-plate collectors still suffer significant losses through conduction and convection. As discussed in §6, heat loss through conduction and convection are roughly proportional to temperature difference, so losses

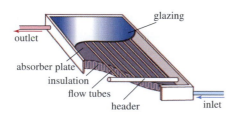

Figure 24.3 A flat-plate collector. (Credit: Southface Energy Institute)

from a collector of this type are roughly linear in $\Delta T = T_{\text{fluid}} - T_{\text{ambient}}$. Thus, as mentioned above in a more general context, for this kind of collector there is a competition between the goals of increasing the temperature – and therefore the usefulness – of the fluid, and decreasing conductive and convective losses. Flat-plate solar collectors generally bring the heating fluid to a temperature well below the boiling point of water, typically in the range 30–70 °C. At a temperature difference of around $\Delta T = 50$ °C, the efficiency of a flat-plate collector can be around 35% (Problem 24.6).

We can analyze a simple model of a solar collector in radiative equilibrium using the methods of the previous section. We assume that the fluid flow is adjusted to keep the collector at a set temperature T_b and that the temperature of the covering T_g is held at the ambient temperature by convection and conduction. As above, we neglect losses to convection and conduction between layers of the collector and assume that no radiation is reflected by the glass. The efficiency of such a flat-plate collector is given by the ratio of the power (per unit area) removed by the coolant, I_{useful}, to the incident solar energy flux,

$$\eta = I_{\text{useful}}/I_0 . \tag{24.5}$$

For a single-pane collector we can modify eq. (24.4) to include the useful power removed from the collector,

$$I_b + I_{\text{useful}} = I_0 + I_g, \tag{24.6}$$

where as usual $I_b = \sigma T_b^4$, $I_g = \sigma T_g^4$. These equations can be solved for η as a function of I_0, T_g, and the absorber temperature T_b, giving $\eta = 1 - (I_b - I_g)/I_0$. For a collector with two layers of glazing, $\eta = 1 - (I_b - I_g)/2I_0$ (Problem 24.7). Taking $I_0 = 1000$ W/m^2 and $T_g = 280$ K and setting, for example, $T_b = 70$ °C, we find a theoretical efficiency with single-pane glass of $\eta = 56\%$, which increases to 78% with double-pane glass.

Since convection and conduction losses are roughly linear in the difference between T_b and the ambient temperature, they decrease the efficiency of solar thermal collectors more at higher operating temperatures. These losses are moderated somewhat compared to the stagnant situation because the temperatures of the absorber and the glass are lower when energy is being extracted. Efficiencies of real double-glazed solar collectors for medium-temperature applications (§24.2.3) that use non-reflective glass and other design optimizations to reduce conduction and convection can approach or exceed 50% at temperature differences $T_b - T_{\text{ambient}} \approx 70$ °C (see Figure 24.4). Note that the efficiency of the collector falls as the incident radiation decreases, for fixed T_b (Problem 24.8).

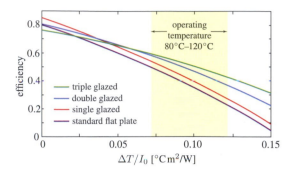

Figure 24.4 Measured collector efficiency for single-, double-, and triple-glazed collectors using "non-reflective" glass compared to a single-glazed flat-plate collector using standard glass [144]. The horizontal axis is the temperature difference $\Delta T = T_b - T_{\rm atm}$ divided by the incoming radiation flux I_0. After [144].

One feature of a solar collector like the flat-plate collector is that it can absorb diffuse radiation, including light scattered from different directions in the atmosphere. This increases the effective insolation, and helps to extend the utility of basic low-temperature solar collectors to a wider range of atmospheric conditions (such as cloudy days).

24.2.2 Fixed-angle Mounting

To maximize energy collected through the day, a flat-plate collector is usually placed at an angle facing south (in the Northern Hemisphere) towards the horizon. The angle chosen depends upon the application. As we learned in §23, the angle of incidence β of solar radiation varies through the day, depending upon the latitude λ and upon the declination δ – which itself varies depending upon the time of year. If a collector is tilted at an angle θ towards the southern horizon, the formula (23.9) giving the angle of incidence at any given time is modified to

$$\cos \beta' = \sin(\lambda-\theta) \sin \delta + \cos(\lambda-\theta) \cos \delta \cos \omega t , \quad (24.7)$$

where β' is the angle between the incoming sunlight and the normal to the collector. The only difference between eq. (23.9) and eq. (24.7) is the replacement of λ with $\lambda - \theta$. Note that not only must $\cos \beta'$ be positive for the Sun to be shining on the collector, but also $\cos \beta$ given by eq. (23.9) must be positive as well, or Earth will block the sunlight from hitting the collector.

A typical choice of θ for a flat-plate collector is **latitude tilt**, $\theta = \lambda$. While the insolation integrated over the year is generally maximized at a tilt lower then λ (see §23, Question 23.2), heating needs are generally greater in the winter months, so flat-plate collectors are often mounted at latitude tilt, or an even greater angle such as $\theta = \lambda + 15°$,

in order to increase energy capture in the winter when the declination is smaller, at the cost of reduced collection in the summer months (Problem 24.9).

24.2.3 Medium-temperature Collectors

For industrial and commercial heating and cooling applications, where higher temperatures are needed, more sophisticated means can be used to improve the efficiency of the basic flat-plate collector. **Intermediate-temperature solar collectors** operate with a working fluid heated to temperatures ranging from $80°C$ to $250°C$, and employ a variety of technical improvements over the basic flat-plate collector. By evacuating a space between the absorber and surrounding glazing, losses to conduction and convection can be substantially reduced. Furthermore, by coating the absorber with a material having selective absorption characteristics, energy lost to radiation in the infrared can be reduced without compromising efficient absorption of incoming solar radiation. Consider, for example, the idealized case of a solar collector without a glass covering that receives incident visible and near-visible radiation from the Sun I_0 and incident infrared radiation from the atmosphere at temperature $T_{\rm atm}$. If the absorber is painted with a material having an emissivity $\varepsilon = 0.9$ for incoming near-visible solar radiation, and an emissivity $\varepsilon = 0.1$ in the infrared, then the radiative balance equation becomes

$$0.1 \times \sigma T_b^4 = 0.1 \times \sigma T_{\rm atm}^4 + 0.9 \times I_0. \quad (24.8)$$

(Note that the emissivity ε controls both absorption and emission rates in a given frequency range, as discussed in §23. The emissivity therefore multiplies both the outgoing radiation rate from the absorber and the incoming radiation rates from the sky and the atmosphere.) Solving eq. (24.8) for T_b, assuming $T_{\rm atm} \cong 280\,{\rm K}$, we get $T_b \cong 640\,{\rm K} \cong 364\,°{\rm C}$ – much higher than the $130\,°{\rm C}$ achieved in the idealized collector without the selective coating. Real collectors do not operate at such high temperatures, not only because of losses, but also because the working fluid is circulated steadily to remove thermal energy for use. Indeed, high-efficiency medium-temperature collectors generally need a load to avoid overheating. A typical example of a collector of this type is an **evacuated-tube collector**, shown in Figure 24.5. Collectors of this type generally operate in a temperature range from $70\,°{\rm C}$ to $180\,°{\rm C}$. Typical stagnation temperatures for such collectors are around $230\,°{\rm C}$. While much more efficient than the basic flat-plate collector described above, they are significantly more expensive, and not generally used in domestic applications. Other issues with collectors of this type include the fact that they are often so well insulated that they do not shed snow well in winter, and the fact

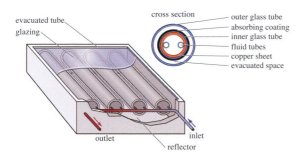

Figure 24.5 In a typical medium-temperature solar collector such as the evacuated-tube collector shown here, a space around the absorber is evacuated, and the absorber is coated with a selectively absorbing material to reduce losses to infrared radiation. Image: US EERE.

that currently produced evacuated-tube collectors such as the one shown in Figure 24.5 are rather fragile and easily damaged. As mentioned above, non-reflective glass is used in medium-temperature collectors to optimize collection efficiency. Efficiencies for medium-temperature collectors with non-reflective glass are compared to a flat-plate collector with normal glass in Figure 24.4. A variety of intermediate-temperature collector designs are described in [144].

24.3 Concentrators

We have described several approaches that can be used to improve the effectiveness of a basic solar collector. These ideas include trapping of heat using the greenhouse effect, treating the absorber with a coating that selectively absorbs incoming solar radiation better than it emits infrared thermal radiation, and eliminating losses due to conduction and convection by evacuating or insulating the region around the absorber. These principles are useful for medium-temperature applications. In order to achieve the much higher temperatures that are desirable for conversion to mechanical or electrical energy, however, it is generally necessary to concentrate the incoming solar energy by means of mirrors or other optical devices, so that the intensity of radiation incident on the absorber is increased.

Concentrators are useful not only in connection with solar thermal energy, but also in combination with photovoltaics. High-efficiency *multi-junction* (see §25.9) PV cells are considerably more expensive than single-junction silicon cells. Since their higher efficiency alone may not justify the additional cost, multi-junction cells are sometimes exposed to concentrated sunlight in order to increase their power output. Also, as we discuss in the

following chapter, the efficiency of photovoltaic cells is generally higher for higher levels of illumination.

In this section we develop some of the basic principles of solar concentrators. We briefly review the optics of mirrors, and give some simple examples of concentrator designs. We discuss the tradeoff between the *concentration factor* and the *acceptance angle* within which incoming radiation is collected; in particular, we use the second law of thermodynamics to derive a simple limit on the concentration possible for a given acceptance angle. We focus on concentrators that use mirrors to reflect incoming light rays to an absorber. Other designs use lenses, particularly for concentration of solar energy on high-efficiency photovoltaic cells. Lens-based concentrators can also be analyzed with the techniques of geometric optics, and obey the same limits on effective concentration.

Whether based on mirrors or lenses, concentrators are designed to focus light incident from a particular direction onto an absorber. Concentrators cannot focus diffuse light such as the infrared radiation emitted by the atmosphere or, more importantly, sunlight diffused through clouds or scattered from blue sky. The fact that concentrators cannot make use of diffuse light is a significant limitation that we return to in the discussion of *solar thermal electricity* in §24.4.

For a recent in-depth text on concentrators, with much more information on both lens- and mirror-based concentrator geometries, see [143].

24.3.1 Geometric Optics of Mirrors

Over distances large compared to the wavelengths involved, light can be treated as a collection of rays that travel along straight lines in vacuum or in a homogeneous (and rotationally invariant) medium. This framework for describing the propagation of light is known as **geometric optics**.

Limiting ourselves to mirror-based concentrators, the only additional principle of geometric optics that we need is the **law of specular reflection**: when a light ray reflects from a surface, its angle of incidence (measured relative to the normal to the surface) is equal to the angle of reflection. This law, and the rest of geometric optics, can be derived from the underlying theory of light as electromagnetic waves, which in turn can be derived from the more fundamental quantum description of light. For our purposes here, however, we do not need these more sophisticated perspectives.

An extremely simple type of solar concentrator is depicted in cross section in Figure 24.6. This concentrator consists of a cylindrical absorber, and two planar mirrors. The mirrors are tilted at 45° from the horizontal, so

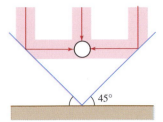

Figure 24.6 A simple solar concentrator uses two mirrors to triple the incident radiation on a cylindrical absorber. The concentrator is shown in cross section; the full 3D geometry is given by translating the figure perpendicular to the page.

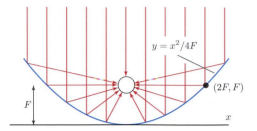

Figure 24.7 A parabolic concentrator with focal length F. All vertically incident light rays are reflected from the parabola to the focal point.

that vertically incident light is reflected horizontally and hits the absorber. Thus, in this configuration the absorber receives three times as much incident radiation as it would without the mirrors.

Although there are a wide variety of concentrator geometries with different characteristics, it is helpful to focus on some general classes of geometries with specific symmetries. While all real concentrators are three-dimensional objects, many concentrators, like the one shown in Figure 24.6, are basically planar shapes that have been extended without change in the third dimension. In other words, they are constructed by linear translation of the cross-sectional two-dimensional (2D) geometry. We refer to concentrators of this type as **2D** or **linear concentrators**. In other cases, the three-dimensional (3D) geometry is produced from a 2D geometry by rotation around an axis of symmetry; we refer to geometries of this type as **3D concentrators**.

24.3.2 Parabolic Concentrator

A simple example of an effective solar concentrator geometry is the **parabolic concentrator**. The key feature of the parabola that makes it extremely useful in concentrator design is that all light incident along a line parallel to its symmetry axis is reflected to a common point, called the *focal point* of the parabola. Demonstrating this fact is a simple exercise in geometry (Box 24.2). A cross section through a parabolic concentrator is shown in Figure 24.7.

The parabolic geometry can be the basis for a 2D concentrator by linear translation into the third dimension. This geometry is used, for example, in **parabolic trough** solar collectors such as the one shown in Figure 24.12(a). The parabola can also be rotated around the symmetry axis to form a **parabolic dish**, another common geometry for solar concentrators, an example of which is shown in Figure 24.12(c).

24.3.3 Effective Concentration and Acceptance Angle

Effective Concentration　To compare the effectiveness of different geometries, it is useful to have a quantitative measure of the concentration capacity of a given concentrator. To this end, we define the **effective concentration** by

$$C = \frac{A_{\text{aperture}}}{A_{\text{abs}}}. \qquad (24.9)$$

In this formula, A_{abs} is the surface area of the absorber, and the area A_{aperture} is defined to be the total area, normal to the direction of the incident radiation, across which all incoming light rays continue to the absorber. The effective concentration C depends upon the angle of incidence of the incoming radiation and the details of the concentrator geometry.

For a specific concentrator geometry, the effective concentration C of the geometry is all that is needed to solve the radiative balance equation and find the temperature of the absorber in radiative equilibrium with a given intensity of incoming light. The total rate of incoming energy $I A_{\text{aperture}}$ must be equal to the total rate of radiation from the absorber $\sigma A_{\text{abs}} T_{\text{abs}}^4$ when the system is in radiative equilibrium,

$$\sigma A_{\text{abs}} T_{\text{abs}}^4 = I \times A_{\text{aperture}} \quad \Rightarrow \quad \sigma T_{\text{abs}}^4 = CI \quad (24.10)$$
$$\Rightarrow \quad T_{\text{abs}} = (CI/\sigma)^{1/4}.$$

Box 24.2 The Focusing Properties of a Parabola

A parabola is defined to be the set of points Q in the plane that are equidistant from a fixed **focal point** f and a line L ($\overline{fQ} = \overline{Q'Q}$, for some point Q' on L). A parabola can be described in a Cartesian coordinate system as the set of points (x, y) that satisfy the equation $y = x^2/4F$. The focal point f of this parabola is at $(x, y) = (0, F)$, and the line L is at $y = -F$. Let us consider the reflection of a ray incident from infinity, traveling down a line parallel to the y-axis, which hits the parabola at the point P with coordinates $(x, y) = (x, x^2/4F)$. The geometry near this point is shown in the figure to the right. The slope of the parabola at the point P is given by the tangent of the angle θ where

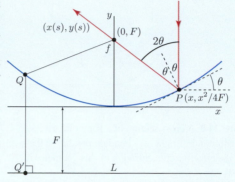

$$\frac{dy}{dx} = \frac{x}{2F} = \tan\theta \, .$$

According to the *law of specular reflection*, the incoming ray is reflected at an angle 2θ from the vertical, so elementary trigonometry gives an equation for the ray reflected from P as a line parameterized by s,

$$(x(s), y(s)) = (x, \frac{x^2}{4F}) + s(-\sin 2\theta, \cos 2\theta) \, .$$

We want to show that this line goes through the focal point at $(0, F)$ regardless of where P lies on the parabola – i.e. regardless of the value of x. Thus, if we choose s such that $x(s) = 0$, then we must show that $y(s) = F$. Setting $x(s_0) = 0$ gives $s_0 = x/\sin 2\theta$, so

$$y(s_0) = x^2/4F + x \cot 2\theta = F \, ,$$

where we have used the trigonometric identity $\tan 2\theta = 2\tan\theta/(1 - \tan^2\theta)$, and $\tan\theta = x/2F$ so $\cot 2\theta = F/x - x/4F$. This proves that any light ray vertically incident on the parabolic concentrator is reflected to the focal point.

Example 24.3 Effective Concentration for a Linear Parabolic Collector

Consider a linear parabolic concentrator in which the absorber has a cylindrical geometry with radius R and length L. The parabolic mirror has focal length F, width W, and length L (cross section shown in Figure 24.8). For this geometry, the total surface area of the absorber is $A_{abs} = 2\pi RL$. The total area over which vertically incident light is reflected to the observer is $A_{aperture} = WL$. The effective concentration is therefore

$$C = (WL)/(2\pi RL) = W/2\pi R \, .$$

Thus, the temperature of the absorber goes as the 1/4 power of the product of the effective concentration and the incoming radiation intensity, with the usual assumptions that the absorber is a perfect absorber, mirrors in the concentrator reflect all light rays perfectly, and there are no conduction or convection losses.

Acceptance Angle

Although a concentrator increases the amount of radiation impinging on the absorber, it does so at a cost in the angular range from which light will be collected. This is clear in the simple case of a parabolic concentrator, where all vertically incident light is reflected to the focal point but the incident light coming from a different angle will not be reflected directly to the focus and may miss the absorber entirely. Thus, concentrators typically accept light from a much more limited range of directions than simple flat absorbers. This effect is measured by the **acceptance angle** of the detector, θ_a, which is defined to be the angle within which all light impinging on a cross-sectional area $A_{aperture}$ hits the absorber.

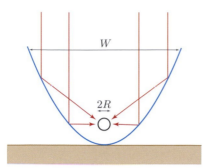

Figure 24.8 A cross section of a parabolic trough concentrator. For light rays incident parallel to the symmetry axis, the effective concentration of this geometry is $C = W/2\pi R$.

Effective Concentration

A solar concentrator with concentration C has the effect of multiplying the incoming radiation by a factor of C in the radiative balance equation, thus raising the radiative equilibrium temperature of the material heated by the concentrator by a factor of $C^{1/4}$.

Maximum Concentration Ideally, one would like to gather as much light as possible by maximizing both the concentration and the acceptance angle of a solar energy collector. The maximum concentration C and the acceptance angle θ_a, however, are not independent. In fact, the second law of thermodynamics gives a simple limit on the maximum concentration that is theoretically possible for any concentrator with acceptance angle θ_a. This result, known as the **Rabl bound**, states that no 3D concentrator geometry with an acceptance angle of θ_a can have an effective concentration that exceeds the bound

$$C_{3D} \leq 1/\sin^2 \theta_a . \qquad (24.11)$$

For a 2D (linear) concentrator geometry, the bound is

$$C_{2D} \leq 1/\sin \theta_a . \qquad (24.12)$$

While we do not give a completely general proof of these limits here, we give a simple argument that if these conditions are not satisfied for a specific source, then it would be possible to violate the second law of thermodynamics. Let us assume that the incoming radiation emanates from a spherically symmetric body of radius r – the Sun for example – at temperature T_s and at a distance

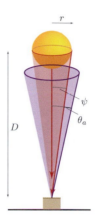

Figure 24.9 A spherical source of radius r at a distance D produces radiation incident upon a concentrator at angles up to $\psi = \sin^{-1} r/D$, where ψ is less than or equal to the acceptance angle θ_a. The second law of thermodynamics shows that the effective concentration of the incoming radiation is bounded by $C \leq 1/\sin^2 \theta_a$.

D from the concentrator. The flux of incoming radiation at the concentrator is then

$$I = \frac{4\pi r^2 \sigma T_s^4}{4\pi D^2} = \frac{r^2}{D^2}\sigma T_s^4 . \qquad (24.13)$$

We assume that light from the center of the source is incident normally on the concentrator. The light rays from the outer edge of the source then hit the concentrator at an angle ψ, where $\sin \psi = r/D$ (see Figure 24.9). Assume further that $\theta_a > \psi$, so that the complete image of the source reaches the absorber and the total incoming radiation flux that impacts the absorber is CI. If no heat is harvested from the absorber, then radiative balance (24.1) relates the temperature T_a to the incoming flux CI,

$$\sigma T_{abs}^4 = CI = C\frac{r^2}{D^2}\sigma T_s^4 = C\sigma \sin^2 \psi T_s^4 , \text{ or}$$
$$T_{abs}^4 = C \sin^2 \psi T_s^4 . \qquad (24.14)$$

This result suggests that for fixed T_s and ψ, by increasing the concentration C, it would be possible to bring the absorber to an arbitrarily high temperature. The 2nd Law, however, says that without doing work, we cannot transfer thermal energy from a body at a lower temperature to a body at a higher temperature. It follows that $T_{abs} \leq T_s$. For a given concentrator geometry, T_{abs} is maximized when $\psi = \theta_a$. From this the upper bound on effective concentration follows,

$$C \leq 1/\sin^2 \psi_{max} = 1/\sin^2 \theta_a . \qquad (24.15)$$

For a linear concentrator, essentially the same argument goes through: by assuming that the source is also linear, for

---

Example 24.4   Acceptance of a Parabolic Concentrator

As a simple example of a concentrator with limited acceptance angle, consider
again a parabolic geometry, where the parabolic reflector terminates at the
point $y_{\text{max}} = F$, so that the height of the reflector is identical to the height
of the center of the absorber. The endpoint of the parabola thus defined is
precisely the point where the slope of the parabola is 1, so that the surface is
inclined at 45°. In this case, it is easy to compute the angle outside which a
light ray will miss the absorber after reflection. The law of specular reflection
states that a light ray hitting the end of the parabolic reflector at an angle of θ
from the vertical will reflect at an angle of θ from the horizontal. This reflected
ray will just touch the absorber of radius R if the angle θ satisfies

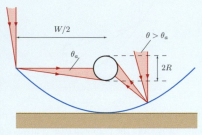

$$\sin\theta = R/(W/2)\,.$$

Thus, for the ray hitting the edge of this parabolic concentrator, the acceptance angle θ_a is given by $\theta_a = \sin^{-1} 2R/W$.
Note that a ray incident on the parabola further in (on right in figure) can hit the absorber even when incident at an
angle greater than θ_a, since the distance to the focus is less than $W/2$.

   As found in Example 24.3, such a concentrator has $C = W/2\pi R$. There is a small subtlety here, since when the
light is coming in at an angle of θ from the normal, the width over which light rays hit the reflecting surface is given
by $W\cos\theta$. Thus, the effective concentration for incoming light at angle θ is modified to $C = W\cos\theta/2\pi R$. If, as is
often the case, the angle θ_a is very small, then $\sin\theta_a \approx \theta_a$ and $\cos\theta_a \approx 1$.

---

example a cylindrical source extended along an axis parallel to that of the linear concentrator, the intensity becomes
$(r/D)\sigma T_s^4$ (in the limit where the linear dimensions of the
source and concentrator are much greater than D and r),
and we find the stronger bound

$$C_{\text{2D}} \leq 1/\sin\theta_a\,. \tag{24.16}$$

   As an example of this limit, consider again the linear parabolic concentrator. We found in Examples 24.3
and 24.4 that a linear parabolic concentrator with height
equal to focal length has $C = W\cos\theta_a/2\pi R$, and $\sin\theta_a =
2R/W$. In this case, $C = \cos\theta_a/\pi \sin\theta_a < 1/\sin\theta_a$,
which clearly satisfies the limit eq. (24.16). In fact, the
linear parabolic concentrator undershoots the maximum
possible effective concentration by a factor greater than
π. Nonetheless, linear and dish parabolic concentrators
are frequently used in practice due to the difficulty of
designing and implementing a practical concentrator with
closer-to-optimal concentration.

   For solar energy applications, the Sun is the source
of the incident radiation on a concentrator. For the Sun,
$r = R_\odot \cong 6.96 \times 10^8$ m and $R = R_{\text{orbit}} \cong 1.50 \times 10^{11}$ m,
so the sun subtends a half-angle in the sky of $\psi_\odot \cong
4.6 \times 10^{-3}$ rad, or about 1/4°. Thus, it is not useful to
have a solar concentrator with acceptance angle below
$\psi_\odot \cong 1/4°$. This limits the effective concentration of a
3D solar concentrator to approximately

$$C_{\text{max}} \cong 1/\sin^2\psi_\odot \cong 40\,000\,. \tag{24.17}$$

24.3.4 Compound Parabolic Concentrators

Several other concentrator designs are of interest either
for their theoretical properties or for their applications to
solar thermal energy. In this section we describe one class
known as **compound parabolic concentrators** (CPCs),
which are of particular theoretical interest because they
can in principle come arbitrarily close to the optimal effective concentration. While based on the parabolic shape,
the CPC combines opposing parts of parabolas at different angles to increase the effective concentration. The key

---

**Acceptance Angle and Maximum
Concentration**

The maximum concentration possible for a concentrator with acceptance angle θ_a is

$$C \leq 1/\sin^2\theta_a\,.$$

For 2D geometries (extended into the third dimension by linear translation of a planar geometry),
the maximum is $C \leq 1/\sin\theta_a$.

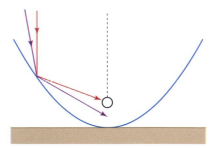

Figure 24.10 Light incident on one side of the parabola at an angle away from the parabola axis is always reflected below the focus.

feature of the parabola that is exploited in the CPC is the fact that light incident on one side of the parabola coming from an angle away from the parabola axis is always reflected below the focus. This feature can be seen in Figure 24.10, where it is clear that it is a consequence of specular reflection.

A (2D) CPC is constructed by mounting identical (congruent) parabolic surfaces on either side of a central symmetry plane, rotated at equal but opposite angles away from the vertical. The basic geometry is depicted in Figure 24.11(a), where the two sides of the concentrator are constructed from two distinct but congruent parabolic segments. The right-hand side of the concentrator follows parabola 1, with focal point A and symmetry axis rotated counterclockwise from the vertical by an angle θ. Similarly, the left-hand side of the concentrator follows parabola 2, with symmetry axis rotated by θ clockwise to the vertical, and focal point B chosen to be the point on parabola 1 at an equal height to point A. By symmetry, focal point A lies on parabola 2. Both parabolic curves continue up to the points where their tangents become vertical, parallel to the collector's axis of symmetry. It follows from simple geometric reasoning (see Problem 24.15) that the line from the top of parabola 1 to its focus (QA) is parallel to the axis of parabola 2 (and the corresponding line on parabola 2 is parallel to the axis of parabola 1). The absorber is placed along the line segment connecting the two foci A and B.

Using the observation that light incident from an angle away from the parabola axis goes below the focus, it is straightforward to show that all incident light within an angle less than θ of normal is reflected by one side or the other of the CPC to the absorber along line $\overline{AB}$. This is illustrated in Figure 24.11(b). Light ray ① grazes the edge of parabola 2 and goes to the focus at B. A light ray such as ②, which is parallel and to the right of ①, hits parabola 1 and goes to its focus at A. The light ray labeled ③ is

incident at an angle less than θ, and is therefore reflected from parabola 1 towards a point to the right of A and hits the absorber along AB. Any such ray incident on parabola 1 at any angle less than θ either hits the absorber along AB after one reflection, or hits parabola 1 again at an angle even further clockwise rotated from the parabola axis, and is reflected again, with the process continuing until the light ray eventually hits the absorber. Finally, a light ray incident at an angle greater than θ – ray ④ for example – misses the absorber and eventually exits the collector. Thus the angle θ by which the parabolas are rotated away from the vertical is also the acceptance angle of the CPC, $\theta_a = \theta$.

The CPC can be adjusted to achieve arbitrarily high concentration C, with correspondingly small acceptance angle. Remarkably, the linear concentrator based on the 2D CPC achieves the ideal concentration limit (24.16), at least in the limit of an infinitely long "trough" in the third dimension. In other words, the ratio $C = \overline{PQ}/\overline{AB}$ (see Figure 24.11(a)) equals $1/\sin\theta$ (Problem 24.17). The 3D CPC is constructed by rotating the 2D CPC cross section depicted in Figure 24.11 around the vertical axis. In the 3D geometry, however, the ray-tracing argument is more subtle, and in fact some incoming rays within the acceptance angle are sent back out of the collector after a number of bounces. Thus, the 3D CPC does not realize the maximal concentration for a given acceptance angle. For a more detailed discussion of CPCs and other novel concentrator designs, see [143].

While CPCs are very intriguing due to their high effective concentration for a given acceptance angle, they have not yet seen widespread use – in part because of the difficulty of inexpensively manufacturing them to the exacting specifications needed for very high concentrations.

24.3.5 Dynamical Adjustment

In §24.2, we discussed the idea that a flat-panel solar collector may be mounted at a fixed angle towards the southern horizon in order to increase effective insolation on the collector. With a solar concentrator, the relative orientation of the concentrator to the Sun becomes even more crucial. For a system with a high effective concentration, the acceptance angle is small, and as the Sun moves through the sky from morning to night over trajectories varying with the seasons, it is necessary to adjust the orientation of the collector in order to keep the Sun within the acceptance angle. The inverse relationship between concentration and acceptance angle leads to a trade-off: for concentrators with large acceptance angle θ_a and small effective concentration C, the concentrator achieves a lower temperature, but does not need much, if any, adjustment. For concentrators with

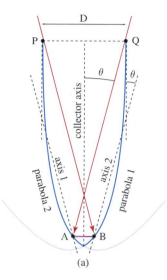

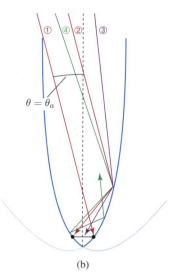

Figure 24.11 A compound parabolic concentrator is constructed from opposing surfaces based on parabolas with different focal points and non-parallel axes. The significance of the geometry depicted in (a) and the light rays traced in (b) is explained in the text.

small acceptance angle θ_a and large effective concentration C, the concentrator can achieve a very high temperature, but needs to be dynamically adjusted to track the Sun through the day. There are a variety of strategies used for adjusting solar collectors to optimize energy collection for minimal adjustment effort. Broadly, these strategies fit into three categories:

Fixed orientation For concentrators with very low effective concentrations ($C \sim 1$–3), and high acceptance angle ($\theta_a \gtrsim 30°$), it is often possible to use a collector with a permanently fixed orientation (*fixed-angle mounting*). Generally such collectors are either flat (as in §24.2.2), or are linear concentrators with the trough axis aligned in the E–W directions and a tilt towards the equator.

Periodic/seasonal adjustment For slightly larger concentrations ($C \sim 3$–6), and smaller acceptance angles ($\theta_a \gtrsim 10°$), a fixed orientation is inadequate to deal with the seasonal variation of the Sun's trajectory across the sky. Often, linear concentrators in this range are periodically or seasonally adjusted, with the trough axis aligned E–W, and the tilt adjusted through the year to match the seasonally changing declination.

Continuous tracking For concentrators with very high effective concentration (for which C could be as high as 200 for a linear concentrator, or 10 000 for a 3D concentrator such as a parabolic dish), and small acceptance angle, it is necessary to continuously track the motion of the Sun through the day. Configurations of this type

typically employ (i) a linear concentrator aligned on a N–S axis with tilt and continuously changing trough angle ranging from east in the morning to west in the evening to follow the Sun (1-axis tracking) or (ii) a 3D concentrator like a parabolic dish or 3D CPC with 2-axis tracking of the Sun's direction from sunrise to sunset. Some energy is needed to run the tracking system for these types of concentrators, but generally the power involved is negligible compared to the total power gathered by the collector.

24.4 Solar Thermal Electricity (STE)

24.4.1 Challenges and Technologies

One of the most promising technologies for large-scale electricity production from solar energy is solar thermal electricity generation, also known as **concentrated solar power** (CSP). This is a simple and relatively low-technology approach to power generation from the Sun. The basic idea is to use concentrators to reflect solar energy at high concentration to a fluid such as water, molten salt, or oil. The fluid is heated to a high temperature, generally between 200 °C and 1000 °C, and can be used to run a relatively high-efficiency heat engine (for example by transferring the heat to water to run a Rankine steam turbine). As mentioned earlier, concentrators require direct exposure to the Sun and cannot make use of light diffused through or reflected from clouds or haze. STE generation is therefore impractical in regions where cloudy days are common and is typically sited in deserts and at relatively low latitudes.

Technology for solar thermal electricity generation was developed in the 1970s and 1980s. A few plants were built at that time, such as the SEGS (*Solar Energy Generating Systems*) facility in California's Mojave Desert. The 9 individual SEGS plants that have been running since that time have a total capacity of 354 MW, with output around 75 MW when averaged over the year. These plants utilize a total land area of about 6.5 km^2, and have a **gross conversion efficiency** (electric power output divided by total insolation on land area used for the plant) of about 3–4%. The actual field area covered by mirrors is somewhat smaller – about 2.25 km^2 – so efficiency relative to insolation on the field area is higher, above 10%. Recently, there has been renewed interest in solar thermal electricity generation. The first commercial solar thermal power plant in Europe was the *Andasol solar power station*, a set of three plants in Andalusia, Spain. Each plant has a 50 MW capacity, with actual output averaging around 20 MW/plant (*capacity factor* $\cong$ 0.4, see Box 24.3). Andasol 1 went online in 2009 and uses a total land area of roughly 2 km^2, with total collector area of roughly 0.5 km^2; efficiency factors are similar to those of the SEGS plants. Construction of many new plants is underway around the world. We briefly summarize here a few of the issues encountered in solar thermal electricity production and some of the main technologies used.

One of the main challenges in solar electricity production is the large area needed. The gross conversion efficiency of ~3–4% realized by existing systems such as SEGS could be pushed higher by newer technologies. Smaller parabolic trough plants have been built with efficiency approaching 20%, and individual parabolic dishes with Stirling engines have surpassed 40% conversion efficiency, but developing large-scale plants with high gross conversion efficiency remains a challenge. At a gross conversion efficiency of 3% and average insolation of 300 W/m^2 in an ideal arid environment, net output of such a plant averages only ~10 W/m^2. So, a plant producing a time-averaged 1 GW of electrical power output would require around 100 km^2. Obviously, constructing a large number of such plants is a substantial engineering undertaking, though not impossible. Related challenges in large-scale solar thermal electricity production include the difficulty of mass-producing highly reflective mirrors and keeping them clean and undamaged over years of exposure to dust, rain, wind, sandstorms, etc. Cooling is also an issue. Standard steam turbine plants require a large amount of water or other resource to act as a heat sink into which to dump entropy, though systems have been developed

Solar Thermal Electricity (STE)

Solar thermal electric power plants use concentrators in various geometries to focus direct sunlight on an absorber. The resulting thermal energy is used to generate electricity. STE has the significant advantage that thermal energy can be stored for later use, thus reducing the variability of solar power. Siting of STE plants is limited to locations with the least cloud cover and haze by the fact that only direct sunlight can be concentrated.

that use less water by employing air cooling to condense steam. Storage and transmission are also substantial challenges, discussed further in §37 and §38. Some recent STE plant designs, such as the Andasol plants (see Figure 37.4), employ a molten nitrate salt mixture that is inexpensive, has a high heat capacity (see Problem 5.8), and can be stored at high temperature for days, allowing nighttime and peak energy distribution independent of the time of energy collection. Other materials with high heat capacities and relatively low cost, such as graphite, have been incorporated in solar thermal storage systems (see §5, Problem 5.9). Research is ongoing for optimal solar thermal storage media (see for example [145]).

Some of the main technologies that have been developed for solar thermal electricity plants involve the following configurations (see Figure 24.12):

Parabolic trough This is the most established technology, used in the SEGS plants in California. A linear parabolic concentrator is used to heat liquid in absorber tubes. The heat transfer fluid that is circulated through the solar field is used to provide thermal energy to a steam turbine. Numerous solar thermal plants using this technology are currently operating around the world.

Power tower In this design, an array of flat mirrors called **heliostats** are arranged on the ground around a central tower. Each mirror is independently oriented with a dual-axis control to reflect the sunlight onto the tower, where a fluid is heated. This design was used in the *Solar 1*, *Solar 2*, and *Solar Tres* (Gemasolar) plants in California and Spain, using high-temperature molten salt as the heating fluid and for storage. As of the time of writing, there were several further plants of this type operating in the US and other countries.

Parabolic dish In this design, 3D parabolic dish concentrators individually track the Sun and reflect to absorbers

Box 24.3 Capacity Factor

The **capacity factor** of a power plant is the ratio of actual output (averaged over time) to its maximum output power capacity. The capacity factors of different power plants vary widely, and depend upon the power source, design features, and usage patterns. Nuclear power plants, for example, which are generally used as 24/7 baseload power, often run at capacity factors of 90% or higher. Due to the daily cycle of insolation, solar thermal power plants without storage tend to run at capacity factors of 30% or lower, while plants with storage can realize significantly higher capacity factors (up to 75% reported for existing plants) by storing thermal energy collected during daylight hours for power generation at night.

(a) (b) (c)

Figure 24.12 Some solar thermal electricity generation technologies: (a) parabolic trough; (b) power tower; (c) parabolic dish. (Credit: (a) DLR, (b) SOLUCAR PS1O reproduced under CC-BY-SA 2.0 license via Wikimedia Commons (c) National Renewable Energy Laboratory, David Hicks)

mounted on each dish. The highest-efficiency solar thermal electricity generators designed so far combine a parabolic dish concentrator with a Stirling engine mounted on the dish. Though several systems using this design have been developed, there has as yet been no large-scale deployment of this technology.

Other technologies Other design approaches for solar thermal electric plants include **compact linear Fresnel reflectors** (CLFRs), in which an array of long rectangular mirrors with single-axis control reflect sunlight onto a linear absorber fixed in place over the reflector field. Though the efficiency of this design is lower than that of the parabolic trough, the implementation is simpler, and it is much easier to mass-produce flat mirrors than parabolic mirrors. Fresnel lenses, which are cheaper than mirrors, can also be used for concentrators. For the most part, solar thermal electricity production is designed as a utility-scale operation, though there is some development of smaller systems, such as individual Stirling engine + parabolic dish configurations.

24.4.2 Efficiency Optimization

The use of solar concentrators makes it possible to heat the working fluid in an STE plant to a high temperature. This increases the maximum (Carnot) efficiency possible for conversion of the solar thermal energy to electricity. On the other hand, at higher temperatures, the absorber reradiates a higher fraction of the incident energy. This

leads to an optimization problem: given a concentrator geometry, the temperature of the absorbing material should be raised high enough for efficient thermal to electric conversion, but not so high that too much incoming radiation is re-emitted.

To analyze this situation quantitatively, we consider an idealized situation where a solar concentrator with concentration C is used to heat the working fluid to a temperature T. We are interested in determining the temperature T at which the theoretical upper bound on overall conversion efficiency is as high as possible, assuming that the absorber radiates as a black body at temperature T. For an ambient temperature of T_-, we assume that the conversion of available thermal energy to electricity can take place at the Carnot efficiency,

$$\eta_{\text{Carnot}} = \frac{T - T_-}{T}. \tag{24.18}$$

If the incident power per unit area is I_0 then the total power hitting the absorber is $P_{\text{incoming}} = CI_0A$, where A is the surface area of the absorber. The power reradiated by the absorber is $P_{\text{radiated}} = \sigma T^4 A$, so the available power is

$$P = P_{\text{incoming}} - P_{\text{radiated}} = \left(CI_0 - \sigma T^4\right) A. \tag{24.19}$$

Assuming Carnot efficiency, the output electric power (per unit absorber surface area) is

$$\frac{1}{A} P_{\text{out}}(T) = (CI_0 - \sigma T^4)(1 - T_-/T) \tag{24.20}$$

$$= CI_0 + \sigma T_- T^3 - \sigma T^4 - CI_0 T_-/T.$$

Example 24.5   2D Concentrator Efficiency Optimization

Consider a solar thermal electric plant built using a system of 2D concentrators that reflect sunlight to a system of pipes carrying molten salt with concentration $C = 200$.
What is the optimal efficiency possible (assuming an ambient temperature of 300 K and incident sunlight at 1000 W/m²)?
Solving eq. (24.23) numerically gives 818 K. At this temperature, Carnot efficiency is approximately 63% and about 13% of incident radiation is reradiated, giving a maximum possible efficiency of $\eta \cong 0.55$.

The overall system conversion efficiency is

$$\eta(T) = P_{\text{out}}/CI_0A. \qquad (24.21)$$

This quantity vanishes at ambient temperature $T = T_-$ (since the efficiency η_{Carnot} goes to zero) and also vanishes at the radiative equilibrium temperature $T = T_{\text{eq}} = (P_{\text{incoming}}/\sigma A)^{1/4}$ (where all incoming radiation is reradiated), and is positive between. For a given concentration and rate of incoming power, then, the theoretical maximum power output available is optimized when

$$\frac{1}{A}\frac{dP_{\text{out}}}{dT} = 3\sigma T_-T^2 - 4\sigma T^3 + CI_0T_-/T^2 = 0, \qquad (24.22)$$

or

$$3\sigma T_-T^4 - 4\sigma T^5 + CI_0T_- = 0. \qquad (24.23)$$

This quintic equation does not have a simple analytic solution, but can be easily solved numerically. The system efficiency $\eta(T)$ is graphed as a function of T in Figure 24.13 for several possible values of concentration C, assuming clear sky overhead sun insolation of $I_0 = 1000\,\text{W/m}^2$. For real systems, the thermal to electric conversion efficiency is substantially less than the Carnot maximum. Knowledge of the actual conversion efficiency η as a function of temperature can be used to optimize temperature and power output for more realistic systems.

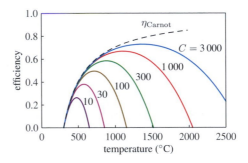

Figure 24.13 The maximum theoretically possible system conversion efficiency $\eta(T)$ for concentrated solar electrical power generation as a function of absorber temperature, assuming ambient temperature $T_- = 300$ K, and incoming power per unit area $I_0 = 1000\,\text{W/m}^2$. The curves are labeled by the concentration C. At low temperatures, efficiency is suppressed by the Carnot limit. (The dashed line gives $\eta_{\text{Carnot}}(T)$.) At high temperatures most energy is re-radiated by the absorbing material.

Discussion/Investigation Questions

24.1 Estimate the energy used for space and water heating in your home. How feasible would it be to get all of this energy from simple low-temperature solar thermal collectors?

24.2 Solar thermal collectors powering small off-grid electric generators have been proposed for the developing world. Research and discuss the advantages and disadvantages of this proposal. Topics might include efficiency, reliability, versatility, and water requirements and availability.

24.3 Discuss some of the possible advantages or disadvantages of the different approaches to solar thermal electricity production: parabolic trough, power tower, parabolic dish. Can you imagine other approaches that might be worthwhile pursuing?

Problems

24.1 The radius of Mars' orbit around the Sun averages roughly $R_{\text{Mars}} \cong 2.28 \times 10^8$ km. Assuming that the surface temperature is roughly uniform over the planet's surface and constant over a Martian day, that the surface is a perfect black body, and that there is no atmosphere, estimate the average surface temperature when the planet is in radiative equilibrium.

24.2 The eccentricity of Mars' elliptic orbit is roughly 0.0935. Estimate the range over which the solar

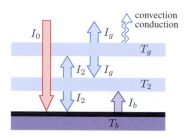

Figure 24.14 A double-pane flat-plate collector.

constant and average surface temperature (in the uniform temperature blackbody approximation) vary during the year.

24.3 Suppose that the surface temperature of the Sun were to increase by 3% from the current value, assuming for simplicity that the radius of the Sun stays fixed. By how much would the solar constant on Earth increase? If, at present, Earth radiates away an amount of energy equal to the current incoming energy flux from the Sun by radiating as a black body at an effective temperature of $T_b \cong 5\,°C$, by what amount would the temperature T_b have to increase to offset the increase in incident solar energy?

24.4 Consider an idealized flat-plate solar collector. Assume that there are no heat losses to conduction or convection. Assume that the insolation is 1000 W/m² incident at an angle of 45° from the vertical. Compute the temperature of the black body in radiative equilibrium assuming that (a) the collector is horizontal, (b) the collector is tilted at 35° towards the Sun.

24.5 Analyze the double pane glass-covered collector described in §24.1.2 and illustrated in Figure 24.14. T_g is assumed to be fixed at $\approx$ 300 K by conduction and convection and $I_0 \approx$ 1000 W/m². Write the equations of radiative equilibrium and show that $T_b \approx 180\,°C$.

24.6 Consider a flat-plate collector covered by a single pane of glass exposed to perpendicular insolation at 1000 W/m². The collector operates at a net efficiency of 35% at a temperature $T_b \cong 65\,°C$. Assuming that the glass covering stays at the ambient temperature of 15 °C, compute the net power radiated and non-radiative losses for the collector. Ignore incident infrared light from the atmosphere.

24.7 Consider the examples of single- and double-pane glass-covered collectors described in §24.2.1. Verify the quoted results for the efficiency as a function of T_b, T_g, and I_0. In the case of double glazing, compute the temperature of the intermediate glass plate. Plot the efficiency as a function of T_b for 25 °C $< T_b <$ 100 °C when $T_g =$ 280 K and $I_0 =$ 1000 W/m².

24.8 Redo the computation of Problem 24.7 assuming incoming radiation at $I =$ 750 W/m², and fixing $T_b =$

70 °C, $T_g =$ 280 K. Show that the efficiency decreases significantly with the reduction in incoming radiation.

24.9 Compare the total (integrated daily) insolation on a flat-plate collector located in Chicago, Illinois (latitude $\lambda =$ 41.98°) on February 1 when the collector is (a) tilted at an angle $\theta = \lambda$ to the south, (b) tilted at an angle $\theta = \lambda + 15°$ to the south.

24.10 Consider an idealized flat-plate solar collector with no conduction or convection losses, and with insolation 1000 W/m² incident at an angle of 45° from the vertical. Assume that the collector is horizontal and is covered with a pane of glass transparent to incoming radiation but opaque to outgoing (IR) radiation. The glass is kept at 300 K by the external environment. Compute the temperature of the absorber in radiative equilibrium. Recompute the temperature if the absorber is coated with a paint that modifies its emissivity in the IR to $\varepsilon \cong 0.2$.

24.11 Consider a parabolic trough where the height of the reflector is identical to the height of the center of the absorber, as depicted in the figure in Example 24.4. The concentrator width is 8 m and the absorbing tube of radius 0.5 m is centered along the focal line at a height of 2 m above the bottom of the trough. Incident sunlight hits the concentrator from a direction parallel to the line of symmetry, with an intensity of 1000 W/m². (a) What is the effective concentration of the concentrator? (b) What is the acceptance angle within which all light hits the absorber? (c) Assume that the tube carries a fluid that is heated and removes some of the incoming energy. The tube then radiates as a black body at a temperature of 150 °C. Compute the net rate at which energy is collected and transferred to the fluid, for each meter of length of the trough.

24.12 Consider a solar thermal "power tower" with a central tower of height 60 m, surrounded by an array of planar mirrors on the ground extending to a radius of 100 m around the tower. At the top of the tower there is a cylindrical absorber with its axis aligned vertically. Each mirror is large enough so that the complete reflection of the Sun can be seen on each mirror from the center of the tower at the height of the absorber. (a) Give an upper limit on the number of mirrors that could be placed around the tower subject to these constraints. (b) How wide does the absorber have to be so that light rays coming from the center of the Sun hit the absorber, no matter which point on which mirror they are reflected from?

24.13 For the power tower geometry from the previous problem, assume that the absorber height is equal to its diameter, and that the diameter is that found in part (b) above. If all the light from an overhead sun that hits the circle of radius 100 m containing the reflecting mirrors were to be reflected to the absorber, what would be the effective concentration of the mirror configuration as a solar concentrator? Assume $1000\,\mathrm{W/m^2}$ insolation from an overhead sun. What would be the temperature of the absorber if it was in radiative equilibrium with the reflected incoming radiation? Do you think that this is a realistic answer? Why or why not?

24.14 [T] Determine the maximum concentration C of a solar concentrator satisfying $\sigma T^4 = C I_0$, where I_0 is the solar constant, directly from the second law of thermodynamics and the surface temperature of the Sun.

24.15 [T] Prove that for the compound parabolic concentrator depicted in Figure 24.11 the line containing points Q and A is parallel to the axis of parabola 2. [Hint: consider the incoming ray that is reflected along $\overline{QA}$.]

24.16 Consider a linear 2D compound parabolic concentrator built from parabolas tilted at $10°$ to the vertical, with a trough of width $3\,\mathrm{m}$, and an absorber width of $0.5\,\mathrm{m}$. Compute the concentration C of the concentrator. If the incoming radiation has intensity $I_0 = 1000\,\mathrm{W/m^2}$ and the (blackbody) absorber is kept at a temperature of $100\,°\mathrm{C}$ by circulation of a thermal fluid, compute the rate of energy transfer to the fluid. Compare this rate of energy transfer to that for a linear parabolic concentrator with the same absorber area and acceptance angle; assume that the parabolic concentrator has height equal to the focal length (as in Example 24.4), and that the

absorber has an area covering the lower half of a cylinder centered on the focal line, and is kept at the same temperature of $100\,°\mathrm{C}$.

24.17 [T] Prove that a linear compound parabolic concentrator realizes the maximum possible concentration for a given acceptance angle θ. Suggestion: following the notation of Figure 24.11, work in a coordinate system where parabola 2 is described by the equation $y = x^2/4F$. Identify B and Q as the points on parabola 2 where the slopes are $\tan\alpha$ and $\tan 2\alpha$, with $2\alpha + \theta = \pi/2$. Then show that $\overline{QA}/\overline{BA} = (1 + \sin\theta)/2\sin^2\theta$, which proves the desired result.

24.18 Consider a parabolic dish concentrator built from a dish with radius $3\,\mathrm{m}$, height equal to the focal length, and an absorber at the focus with spherical shape and radius $0.1\,\mathrm{m}$. Compute the concentration and acceptance angle of this concentrator, and check that the Rabl bound is satisfied.

24.19 For a concentrator with concentration $C = 100$, and incident sunlight at intensity $I_0 = 1000\,\mathrm{W/m^2}$, compute the temperature T of the absorber at which solar to electric conversion efficiency is optimized, assuming Carnot efficiency. How does your answer change if the system has an additional rate of energy loss per unit area to conduction and convection of $(T - T_{\mathrm{ambient}}) \times 20\,\mathrm{W/Km^2}$, with $T_{\mathrm{ambient}} = 300\,\mathrm{K}$?

Photovoltaic Solar Cells

Photovoltaic (PV) solar cells are devices that directly convert solar radiation into electrical power. Over the last 15 years, the worldwide installed capacity of PV power has grown exponentially, doubling roughly every two years. By the end of 2016, installed PV capacity worldwide exceeded 300 GW [146], though delivered power is only about 15% of capacity [133]. Given the enormous solar resource available and the potential for technological improvements and cost reductions in PV devices, solar photovoltaics seem likely to play a significant role in worldwide power generation in coming decades.

In this chapter, we give a basic introduction to the physics of photovoltaic solar cells. PV cells capture energy from electrons that are excited by absorbing individual photons as they propagate through specially designed materials. By structuring a device from these materials in an asymmetric fashion, an electric potential is produced that can drive a current in a preferred direction, allowing the energy of the captured photons to be delivered directly as electrical energy to an external circuit.

The first practical PV cells were developed in the 1950s during the electronic revolution spawned by the invention of transistors. Expensive and inefficient, early silicon solar cells found only niche applications such as powering satellites. Since then, steady increases in efficiency, reductions in material requirements, and improved manufacturing techniques have brought PVs to the point of competitiveness with traditional non-renewable electric power sources in some contexts. Many promising new technologies are currently under development and research on radically new ideas such as organic PVs is quite active. A complete survey of these new technologies is beyond the scope of this book. We focus instead on the basic physical principles of photovoltaic solar cells, particularly the traditional crystalline silicon solar cells that dominate today's PV installations. At the end of the chapter we briefly describe some alternative materials and cell designs. Given the limitations of space, we do not discuss the design and

Reader's Guide

This chapter is devoted to the study of photovoltaic solar cells. Basic ideas from solid-state physics are introduced and the appearance of band gaps in the electronic structure of materials is explained based on quantum mechanics. The principles governing semiconductor photovoltaics are introduced and developed in the context of *p-n* junction-based silicon solar cells. Efficiency limits on solar cells are derived, and some advanced solar cell designs are described.

Prerequisites: §3 (Electromagnetism), §7 (Quantum mechanics), §8 (Entropy and temperature), §9 (Energy in matter), and the three previous chapters on solar energy. Quantum mechanics of a free particle (§7.7) and particles in potentials (§7.8) are particularly important for §25.2.

Material developed in this chapter reappears in §36 (Systems) and §38 (Electricity generation and transmission).

fabrication of specific devices or other engineering issues associated with constructing robust and durable PV cells and systems. We also do not follow the production of electric power outside the boundaries of an individual cell. Converting the intermittent, low-voltage DC power produced by an individual cell into stable high-voltage AC power for practical applications is a subject in its own right, aspects of which are discussed briefly in §38 (Electricity generation and transmission).

We begin with two sections introducing the quantum physics of crystalline materials. A more detailed outline of the rest of the chapter is given at the end of §25.1.

25.1 Introductory Aspects of Solid-state Physics

The basic challenge of the PV device is to design a material that absorbs photons into specific electronic excitations that can drive an electric circuit. To understand how this can be done, it is necessary to extend the discussion of quantum physics beyond individual atoms and molecules to solid materials – a branch of science known as **solid-state physics**.

The quantum states available to an electron in an *atom* have specific, quantized energies. The *Pauli exclusion principle* (see §7.8.2) allows only two electrons (with opposite spins) in each quantum state. When the electrons fill the lowest allowed energy states of the atom, the atom is in its ground state. Incoming photons or other excitations can push one of the electrons into an excited state with higher energy (typically of order ~1 eV), after which the electron can emit this energy as a photon as it drops back to a lower-energy state. We discussed some of the basic mechanisms for such transitions in §23.4. The quantum structure of a *molecule*, where several atoms are held together by electromagnetic forces, is somewhat more complicated, because the quantum state depends not only on the electron configuration but also on the relative positions of the constituent atomic nuclei. Because the nuclei are much more massive than the electrons, their motion occurs on a slower time scale and involves much lower energies. It is usually a reasonable approximation to separate out the quantum excitations of the electrons from excitations of nuclei within the molecule. When we discussed energy in molecules in §9, we focused on the (vibrational) motions of the nuclei, assuming that the electrons remain in the ground state as the nuclei move. In contrast, the spectrum of *electronic excitations* of molecules can be estimated by assuming that the nuclei are in fixed locations corresponding to the ground state of the molecule, and solving the Schrödinger equation for the electrons in the electromagnetic potential produced by the nuclei in this configuration. The result is a spectrum of states with excitation energies at the same scale as atomic excitation energies (i.e. of order ~1 eV), but more complex.

To understand the quantum physics of a *solid*, we can envision it as simply an enormous molecule. For the **crystalline solids** that interest us here, the atoms in the solid are held in place by bonds similar to those that hold atoms together in a molecule. In a solid, however, these bonds hold a large number of atoms together in a macroscopic three-dimensional structure. For crystalline solids, the arrangement of atoms has a regular periodicity, at least at a microscopic scale, which can be described by a space-filling pattern consisting of repeated copies of a basic cell. One of the simplest examples of such a crystalline solid is diamond. Carbon atoms, with valence four, can be configured in a regular pattern in space, known as the **diamond crystal structure**, so that each carbon atom is bonded to

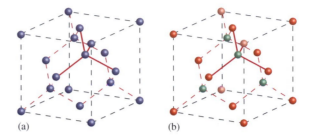

(a)   (b)

Figure 25.1 (a) The *diamond crystal structure*, in which carbon atoms or silicon atoms are configured into a crystalline solid. Each atom has four nearest neighbors to which it is bonded. A single cubic cell of the crystal is depicted, which is repeated periodically in each direction over a macroscopic distance to form a solid material. (b) In the *zincblende* crystal structure, which characterizes several PV materials, two elements combine to produce the same basic pattern as in (a).

four nearest neighbors (see Figure 25.1(a)). As we discuss in more detail in §25.5, the diamond crystal structure can be described in terms of a combination of periodic space-filling *lattices*, which are generated by integer linear combinations of a set of basis vectors in three-dimensional space. The stability of the diamond crystal structure underlies the famous strength and hardness of diamond. Silicon, which lies directly below carbon in group IV[1] of the periodic table (see Figure D.1), also has valence four, and crystallizes in the diamond structure.

In §8.7.3 and §9.2.1, where we introduced Einstein's and Debye's models of the specific heat of solids, we considered the quantum mechanical vibrations of a crystal lattice. Like the vibrations of molecules, these excitations of crystals can be considered as relative motions of entire atoms within the crystal. The electrons adjust their configuration to stay in the electronic ground state as the atoms move. These vibrations can propagate like waves through crystals, and the quanta associated with these waves are known as *phonons* (see §25.5). To understand the *electronic* properties of crystalline solids, on the other hand, we must examine the energy levels available to the electrons in crystals. Just as the electron energy levels of a molecule can be computed with the nuclei held in fixed positions, so the electronic structure of solids can be studied in an approximation where the atoms do not move. After the

---

[1] The **groups** in the periodic table are eight columns of elements with similar chemical properties. Group VIII, for example, are the inert gases; Groups III, IV, V, and VI play important roles in PVs. See [50] for more information about chemical groups.

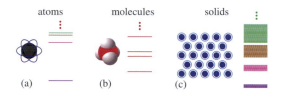

atoms   molecules   solids

(a)   (b)   (c)

Figure 25.2 A schematic representation of the *electron* energy levels in atoms, molecules, and solids. The electron states in a crystalline solid separate into bands of closely spaced states separated by energy gaps with no electronic states.

allowed electron energy levels have been determined with the nuclei in the fixed crystal structure, the exchange of energy between an electron and the crystal vibrations can be considered as a separate problem.

The periodicity of a crystalline solid simplifies the problem of finding the electron energy states in the crystal. In effect, an electron finds itself in a periodic potential generated by the atoms in the crystal lattice. The resulting energy levels are densely packed into intervals or **bands** of energies, where the electron's energy can take essentially any value in a continuous range. These allowed bands are separated by **gaps** in which there are no allowed electron energy states (see Figure 25.2). In §25.2 we give a basic introduction to the quantum mechanics of electrons in a periodic potential and describe the appearance of the band gap structure from two different perspectives.

In silicon, there is a *band gap* of width $E_{\text{gap}} \cong 1.1 \, \text{eV}$ separating the band filled with electrons in the ground state (the *valence band*), from the lowest empty band available for electronic excitations (the *conduction band*).[2] If an electron is excited to the conduction band, it can move freely through the material and conduct electrical current. A photon with energy $E = h\nu$ greater than 1.1 eV can excite an electron into the conduction band.

In §25.3 we describe the basic physics of electrons in semiconductors, giving a general picture of the role of the band gap. In §25.4 we derive a simple limit on the fraction of incoming solar energy that can be collected in PV cells based on a single material (known as *single-junction* PVs).

In §25.5 we give a more detailed description of the band gap in crystalline silicon, and explain the distinction between the *indirect band gap* of silicon and *direct band gaps* that are found in some other semiconductors. In §25.6 we discuss the *doping* of semiconductors with

impurities and the formation of *p-n junctions* that break the symmetry in a crystalline silicon solar cell and form a *photodiode*, which drives a photoelectric current when exposed to light containing photons with energies greater than E_{gap}. We complete the discussion of simple silicon cells in §25.7 with a description of the electronic characteristics of the photodiode and a more complete analysis of efficiency limits on single-junction PV cells. We summarize some features and challenges of silicon photovoltaics in §25.8. In §25.9 we give a brief discussion of some more advanced approaches to solar cell design. Finally, §25.10 contains some general remarks regarding the future role of solar photovoltaics.

For a more in-depth introduction to the physics of solids, see [147]. A comprehensive and comprehensible introduction to solar cells is given in the books by Green [148, 149, 150]; another accessible text that focuses on the physics of solar cells is [151].

25.2 Quantum Mechanics on a Lattice

In this section we describe the basic structure of the energy spectrum for an electron propagating in the periodic potential generated by atoms fixed in a regular crystal lattice. The energy basis states for the electrons can be found by solving the Schrödinger equation in the periodic potential. To keep the math simple we consider electron energy states in a one-dimensional (1D) "crystal," i.e. a particle in a one-dimensional periodic potential. While this analysis is not directly applicable to a three-dimensional (3D) solid such as silicon, which has a more complicated crystal structure, some of the most important qualitative features of the electron energy spectrum of a crystalline solid emerge clearly in the 1D analysis. In particular, the origin of the band and gap structure of the spectrum is clear in the 1D model. Another important feature of the quantum state structure of a solid that is evident in the 1D model is the appearance of a *periodic momentum* on which the electron energy within a band depends. These same concepts appear in the electronic structure of a silicon crystal, though in a somewhat more complicated fashion, as we discuss in §25.5.

We analyze the 1D crystal structure from two complementary perspectives. First we examine the way that the band and gap structure of the energy spectrum emerges in the 1D solid modeled as an array of N adjacent square well potentials. Then we turn to an infinite periodic potential, still in one dimension, and explain how the notion of a periodic momentum emerges. We do not go into the details of exact solutions to the Schrödinger equation; rather, we draw some qualitative insight into how the band structure of the spectrum arises from each of these approaches.

---

[2] More precisely, the band gap is 1.124 eV at a temperature of 300 K; the less precise value 1.1 eV suffices for most of the computations and estimates in this book.

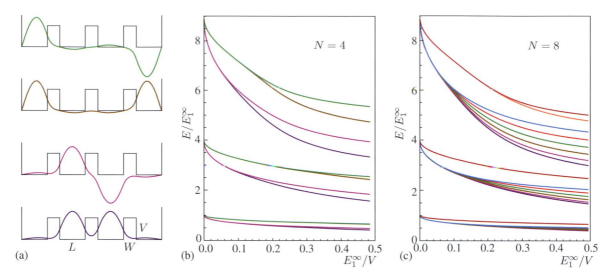

Figure 25.3 (a) A highly simplified model of a crystal: N square wells separated by potential barriers of height V. When $V \to \infty$, there are N independent states at each energy level, corresponding to the usual single-well states in any of the four wells, with tunneling forbidden. For finite but large V, the wavefunctions look like linear combinations of single well states, slightly altered by tunneling. (a) For $N = 4$ the wavefunctions of the four lowest energy states are shown for $V \sim 5E_1^\infty$ where E_1^∞ is the ground state energy of a particle in a single square well. The states are close to linear combinations of square well ground states. The spectra for $N = 4$ (b) and $N = 8$ (c) are plotted in units of E_1^∞ as a function of E_1^∞ / V. As V decreases, the communication between wells increases and the spectra spread. The states at each energy level ($n = 1$, 2, and 3 are shown) which were degenerate when $V \to \infty$, spread out into a band of states separated by gaps. For fixed V, L, and W, the bands of possible energies become essentially continuous at very large values of N.

25.2.1 N Square Wells in 1D: Bands and Gaps

Consider an electron moving in a potential consisting of N separate square wells of width L separated by constant potential barriers of width W and height V, as shown in Figure 25.3(a) for the case $N = 4$. These square wells model the attractive potential wells that atoms provide for an electron in a "crystal" consisting of N atoms. First, imagine that the potential barriers are infinite, $V \to \infty$. In this limit, there is zero probability of finding the electron inside the barriers, so the system behaves like N independent square wells. Remember from §7 that the spectrum of energies for a single 1D potential well with infinite potential barriers at the boundaries is $E_n^\infty = \hbar^2 \pi^2 n^2 / 2mL^2$, for $n = 1, 2, \ldots$, associated with sine wave solutions $\psi_n^\infty(x) = \sqrt{2/L} \, \sin(n\pi x / L)$. These are the possible electron energies for N square wells separated by infinite potential barriers; there are, however, N independent states at each energy level, labeled by the potential well in which the electron is trapped. In this situation, each energy level E_n^∞ has an N-fold degeneracy.

What changes if we lower the height of the potential barriers? As V decreases from infinity, there is a small, but non-vanishing amplitude, even when the barrier is high

and the electron energy is low, for the electron to tunnel between the wells. Thus, even the ground state of the multiple-well system has a nonzero probability for the electron to be in each of the wells. When $V \gg E_1^\infty$, the tunneling rate is very small, but as V decreases there is more and more communication between adjacent wells. With the barriers no longer infinite, the electron energy states spread through the wells, and the degeneracy of the N states at each level n is lifted. When V is large compared to E_n^∞, the solutions to the Schrödinger equation remain close to linear combinations of the separately localized sine wave states, and the energy levels spread only slightly apart (see Figure 25.3(b,c)). As the potential V drops further, the states are less damped in the region between the wells, and the states differ more from the infinite square well states within the wells, with the energy values spread further apart. Notice also that the energy values all decrease as V decreases.

Thus we see that when a sequence of N wells is separated by a high potential barrier, the spectrum looks very much like N copies of a single well spectrum, with the allowed energy values coming in groups of N separated by gaps. As N gets larger and larger (with fixed barrier height

V), the number of almost equal energy states becomes larger, creating an effectively continuous band of allowed energies.

A finely spaced band of allowed energies emerges in essentially the same way in three dimensions when the potential barriers separating attractive wells are decreased. For a real solid, the potential well around each atomic nucleus in the crystal is sufficiently deep to classically trap the electrons. Because the potential barriers separating the attractive potentials around each atom are finite, however, the true quantum states of the system have small amplitudes in the classically forbidden regions between the potential wells, just as in the 1D example considered here. As the number of particles in a solid becomes large (on the order of 5×10^7/cm, for example, in each direction of a diamond crystal), the number of states with energies in each band becomes enormous, so that the band becomes effectively continuous.

25.2.2 A Periodic Potential in 1D: Lattice Momentum

Instead of a series of N potential wells separated by barriers, consider a potential like the one shown in Figure 25.4 that extends over all x and is periodic in x with period Δ,

$$V(x) = V(x + \Delta), \tag{25.1}$$

so that V is *invariant under translation* by a fixed distance Δ. This potential describes a very long 1D crystal composed of locally attractive wells. Note that the potential wells need not be very deep nor the barriers very high.

The periodicity of $V(x)$ gives this problem a symmetry. A similar symmetry helped us sort out the quantum mechanics of a free particle ($V(x) = 0$) in §7.7.1. That system is *invariant under translation by an arbitrary amount* $x \rightarrow x + \delta$, which led us to infer that it must be possible to choose a set of energy basis states whose physical properties are invariant under translation as well. This means that the basis states can change only by a phase under translation. In this way, we discovered that for a free particle in 1D there must exist a set of energy basis states given by the plane wave solutions $e^{ipx/\hbar}$, where p is a conserved momentum that can take any (real) value.

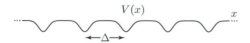

$V(x)$

x

$\leftarrow\!\Delta\!\rightarrow$

Figure 25.4 A potential function $V(x)$ periodic under translation $V(x) = V(x + \Delta)$ forms an idealized 1D model of an electron in a crystalline solid.

A similar argument applies to particle states in a periodic potential. If the potential satisfies eq. (25.1), then we must be able to choose an independent set of energy basis states that change only through multiplication by a phase when $x \rightarrow x + \Delta$,

$$\psi(x + \Delta) = e^{i\alpha}\psi(x). \tag{25.2}$$

Translation by an integer multiple $n\Delta$ gives a phase $n\alpha$ that is linear in n. By analogy with the free particle, it is natural to define $p = \hbar\alpha/\Delta$, so that eq. (25.2) becomes

$$\psi(x + \Delta) = e^{ip\Delta/\hbar}\psi(x). \tag{25.3}$$

This condition is satisfied by a wavefunction that is the product of a *periodic* function and a pure phase,

$$\psi(x) = e^{ipx/\hbar}\phi(x), \text{ where}$$
$$\phi(x + \Delta) = \phi(x). \tag{25.4}$$

Thus, energy basis states spread out over the whole crystal, resembling plane waves modulated by the periodic function $\phi(x)$. Note that this structure does not appear in the wavefunctions shown in Figure 25.3(a) because $N = 4$ is too small to approximate an infinite crystal lattice.

Just as in the case of the free particle, the quantity p defined in this way is interpreted physically as the momentum of the particle, but there are two important differences:

Energy–momentum relation For the free particle, the energy of a state of momentum p is given by the relation $E = p^2/2m$. For a state of momentum p on a periodic lattice, the energy also contains a contribution from the potential V, and thus has a more complicated dependence on p.

Momentum conservation For the free particle, the symmetry of the system under an arbitrary translation implies that p, the momentum, is conserved. In a periodic potential, the fact that the system is only invariant under translations by multiples of the constant Δ corresponds to a weaker version of momentum conservation.

Regarding the second point, note that in eq. (25.3), the phase that multiplies the wavefunction upon translation does not change if we replace p by $p + 2\pi\hbar/\Delta$. Thus, for quantum mechanics of an electron in a periodic potential, any two momenta that differ by a multiple of $2\pi\hbar/\Delta$ are physically equivalent,

$$p \sim p + 2\pi\hbar/\Delta. \tag{25.5}$$

Momentum in this system is only conserved up to this equivalence. Classically, an electron moving through a lattice of atoms would experience forces, given by $-dV/dx$,

Electron Momentum in a Periodic Potential

For an electron moving in a periodic potential, as produced, for example, by atoms arranged in a crystal structure, the momentum p takes values in a finite periodic domain called the *Brillouin zone*. For a particle in a 1D periodic potential that satisfies $V(x) = V(x + \Delta)$, momentum takes values in the range

$$p \in \{-\pi\hbar/\Delta, \pi\hbar/\Delta\}.$$

Momenta that differ by an integer multiple of $2\pi\hbar/\Delta$ are physically identified. This is because these integer quanta of momentum can be transferred to or from the lattice, without affecting the lattice in the idealized limit of an infinite crystal.

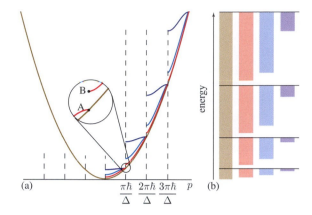

(a) $\dfrac{\pi\hbar}{\Delta}$ $\dfrac{2\pi\hbar}{\Delta}$ $\dfrac{3\pi\hbar}{\Delta}$ p (b)

Figure 25.5 The energy spectrum of a particle in a periodic potential. In (a) the spectrum is shown for $V = 0$ (brown) and a sequence of increasing strengths (red, blue, purple). For clarity only $p > 0$ is shown when $V \neq 0$. Discontinuities appear at $p = \pm n\pi\hbar/\Delta$. The insert shows the behavior of the spectrum at $\pi\hbar/\Delta$. (b) displays the bands of allowed energy levels for the cases shown in (a). As V increases the bands converge to the energies of a single potential well (in this case a square well).

that constantly change its momentum. In effect, the classical electron is constantly exchanging momentum with the rigid block of matter. The quantum electron can also exchange momentum with a rigid crystal lattice, but only in quantized units of $2\pi\hbar/\Delta$. In the infinite limit the crystal lattice has infinite mass and therefore does not recoil when it receives momentum from the propagating electron. In much the same way, when a ball bounces elastically off a brick wall, the brick wall receives momentum from the ball without a measurable recoil. The upshot of this is that for an electron moving in a periodic potential, energy basis states can be characterized by a momentum variable p that takes values in the finite, periodic range

$$-\pi\hbar/\Delta < p \leq \pi\hbar/\Delta. \qquad (25.6)$$

This region of allowed values for p is called the first **Brillouin zone**. We define the **lattice momentum** of a state to be the momentum modulo the equivalence (25.5), that is, a momentum mapped back to the first Brillouin zone. The physical equivalence on a lattice of two momenta that differ by integer multiples of $2\pi\hbar/\Delta$ is important for understanding physical processes like the absorption of a photon by an electron in a silicon solar cell. The restriction to momentum in a finite-size Brillouin zone also occurs in three-dimensional periodic potentials; we discuss this further below in the context of the band structure of silicon.

Keeping these observations about energy and momentum in a periodic potential in mind, it is possible to give a qualitative picture of the spectrum of an electron in a generic periodic potential $V(x)$ satisfying eq. (25.1). First consider the free electron, $V(x) = 0$. As shown by the

(brown) parabola in Figure 25.5, the electron's energy is related to its momentum by $E = p^2/2m$. Even a small periodic potential, however, modifies this relation. The modification, shown by the red curves in Figure 25.5(a), is strongest at the momenta

$$p = \pm n\pi\hbar/\Delta, \quad n = 1, 2, \dots, \qquad (25.7)$$

where the wavefunction has a periodicity compatible with that of the potential. In quantum mechanics, any potential produces a small amplitude for an incident particle/wave to be reflected. When the periodicity of the wavefunction puts these amplitudes for reflection in phase, the effect is amplified, and causes the electron wavefunction to "bounce off" the lattice.[3] In one dimension, the condition for the reflected waves from the crystal to be in phase is that the phase must be the same after the wave travels forward and backward across a distance Δ; this requires periodicity with precisely the momenta in eq. (25.7).

The effect at momenta $p = \pm n\pi\hbar/\Delta$ is evident in the figure: the energy becomes discontinuous and doubly valued at these momenta. Although it is clearer for stronger potentials, this effect occurs for any nonzero

---

[3] This phenomenon is known as Bragg scattering, and can be seen experimentally when a crystalline solid reflects an electromagnetic wave with a wavelength and angle compatible with the periodicity of the solid.

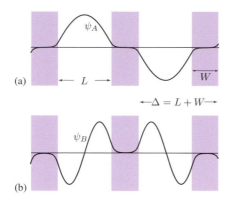

Figure 25.6 The wavefunctions of the energy basis states at (a) the top of the lowest band and (b) the bottom of the second band in a periodic array of square wells. Both wavefunctions are anti-periodic in $x \rightarrow x + \Delta$ because they obey eq. (25.3) with $p = \pi\hbar/\Delta$.

periodic $V(x)$. The effect on the spectrum of allowed states is evident in Figure 25.5(b) – as the magnitude of $V(x)$ increases, the bands of allowed energies narrow and approach the energy levels of a single potential well (in the case of the figure these are square well energies).

A striking feature of Figure 25.5(a) is that the energy is doubly valued at $p = \pm n\pi\hbar/\Delta$. How can it be that there are two states with the same periodicity but much different energies? The answer is that due to the periodic identification of momentum (25.5), there are many states with the same lattice momentum but different energies. This is illustrated in Figure 25.6, where the wavefunctions of two states at $p = \pi\hbar/\Delta$ are shown. These are the points labeled A and B in the spectrum of Figure 25.5(a) (although a somewhat stronger potential was used in Figure 25.6 to accentuate the point). Both wavefunctions shown in Figure 25.6 are anti-periodic under $x \rightarrow x + \Delta$, as required when $p = \pi\hbar/\Delta$. The top wavefunction ψ_A looks approximately like the square well ground state in each separate well (up to a sign), and belongs to the *ground state band*, while the bottom wavefunction ψ_B looks like the square well first excited state in each separate well and belongs in the second band.

The consequence of the momentum equivalence (25.5) is that within each separate band, the energy $E(p)$ is a continuous function of p defined over the first Brillouin zone, considered as a periodic domain. While this is typically the way in which we will describe the relation between energy and momentum for electrons in crystalline solids, sometimes – as in Figure 25.5(a) – it is useful to consider the full range of possible p's, divided up into zones separated by the points (25.7). To make these two different perspectives

clear we have graphed the spectrum for a periodic potential in both ways in Figure 25.7. On the left p is allowed to range over $\{-\infty, \infty\}$, whereas on the right, the entire spectrum is mapped back to the first Brillouin zone. Either way, momenta that differ by eq. (25.5) are equivalent. The periodicity of energy as a function of momentum within each band is clearest in Figure 25.7(b).

The appearance of bands and gaps in a 1D periodic system gives a perspective complementary to the description that was given in §25.2.1 of how bands arise in a system with many wells separated by potential barriers. The edges of bands for a periodic system correspond to the limiting boundary of allowed states for a system of many identical wells. As a simple example of this, one can consider an *infinite* periodic arrangement of wells of the same shape as those considered in Figure 25.3. For this potential, for example, the bottom of the first band of allowed energy states occurs at $E_< \cong 0.584E_1^\infty$ and the top of that band is at $E_> \cong 0.620E_1^\infty$. Thus $E_<$ and $E_>$ define the limits of a band of allowed states of width $\Delta E \cong 0.036E_1^\infty$ in the multi-well system as $N \rightarrow \infty$.[4]

The emergence of band structure in periodic potentials is one of the more subtle consequences of quantum mechanics, arising from the interplay of dynamics (tunneling) and symmetry (under translation by the lattice period). The appearance of bands of allowed energy levels that extend over the whole crystal is responsible for the phenomenon of electric conductivity, since it allows electrons to move fairly freely through the crystal lattice.

Although the analysis in this section was presented in an idealized 1D model, the band and gap structure of the energy spectrum also holds, for essentially the same reasons, for electrons moving in the three-dimensional potential associated with the ions forming a periodic structure in a crystalline solid. Thus, the bands shown in Figure 25.7 can be viewed as a simplified picture of part of the spectrum of a typical crystalline solid.

25.3 Electrons in Solids and Semiconductors

In a neutral solid, just as in an atom, the number of electrons matches the number of protons. Neglecting electromagnetic interactions between the electrons, each electron fits into a specific quantum state in the potential produced by the protons in the solid. As required by the Pauli

---

[4] Note that the top two states in each set shown in Figure 25.3 are almost degenerate and lie slightly above the band boundary. These states are artifacts of the boundary conditions and are not relevant in the infinite periodic limit.

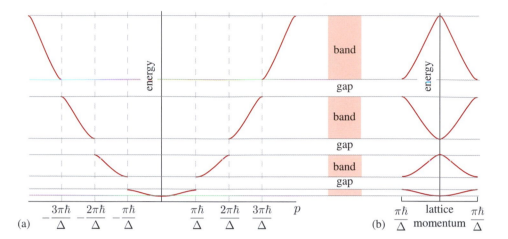

Figure 25.7 Typical band structure of a 1D periodic potential. (a) The energy is shown as a function of p. Bragg scattering at $p = \pm n\pi\hbar/\Delta$ causes discontinuity at those points. (b) All momenta are translated back to the first Brillouin zone $-\pi\hbar/\Delta < p \leq \pi\hbar/\Delta$ using the equivalence under shifts $p \sim p + 2\pi\hbar/\Delta$. This gives a physical picture of the band-gap structure, where within each band, energy is a periodic function on the finite first Brillouin zone.

Spectrum of a Particle in a Periodic Potential

When a quantum particle moves in a periodic potential, the allowed energy levels form continuous bands separated by gaps. The bands can be thought of as periodic functions of the momentum p in the first Brillouin zone defined by eq. (25.6).

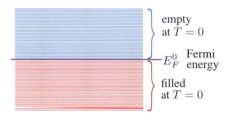

Figure 25.8 At $T = 0$, electrons fill all energy states with energy below the Fermi energy ($E < E_F^0$), and there are no electrons in states with energies above the Fermi energy.

exclusion principle, the electrons fill successively higher energy levels until the electron charge cancels the proton charge and the bulk material is electrically neutral. When the material is at zero temperature, all the electron energy levels are filled up to some maximum energy and all the higher levels are empty. The energy of the highest filled state for a material at zero temperature is called the **Fermi energy**, E_F^0. At temperatures above 0 K, some electrons are thermally excited above E_F^0 and the distinction between filled and empty electron energy levels becomes less sharp. The probability of finding an electron in a particular state is a decreasing function of energy at any given temperature. The **Fermi level** $E_F(T)$ is defined to be the energy where the probability of finding the level occupied by an electron passes 1/2. At $T = 0$ the Fermi level coincides with the Fermi energy ($E_F(0) = E_F^0$), but as we discuss below, these two quantities differ at finite T. The difference is small when $k_B T \ll E_F^0$, which is the case for most materials at room temperature. The Fermi energy of a typical material is depicted in Figure 25.8.

The position of the Fermi level relative to the band structure of a solid plays a central role in determining the physical characteristics of the solid. If the Fermi energy is inside a band, there are empty states with energies just above the ground state, so that only a tiny amount of energy is needed to excite an electron. Such an electron can then move freely through the material. Thus, a material with the Fermi energy in the middle of a band is generally a good electrical (and thermal) conductor, and the band is called a **conduction band**. This situation is depicted in Figure 25.9(a).

On the other hand, if the Fermi energy is between two bands, and there is a large energy separation between these bands, then it is very difficult for an electron to be boosted from the lower, filled, band (known as a **valence band**) to the upper, empty, conduction band. In this case, the material does not easily conduct electrons, and it acts as an insulator. This situation is depicted in Figure 25.9(b).

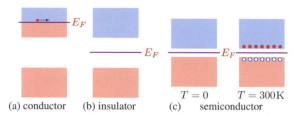

(a) conductor   (b) insulator   (c)   semiconductor

Figure 25.9 Physical characteristics of a material are determined by the location of the Fermi energy (denoted by the solid purple line) relative to the band structure. In (a), the Fermi energy lies inside a band and the material acts as a conductor. In (b) the Fermi energy lies between two bands in a large energy gap and the material acts as an insulator. In (c), the Fermi energy lies between the valence and conduction bands, but the band gap is small and the material acts as a semiconductor. In the figure, empty/filled levels are denoted in blue/red.

In an intermediate situation, it is possible that the Fermi energy may lie between two bands, but that the separation between these bands may not be enormous. In this case, while at zero temperature all electrons are in the (lower) valence band, at room temperature enough electrons may be excited into the (upper) conduction band that the material acts as a reasonable conductor. Such a material is known as a **semiconductor**. The difference in energy between the highest state in the valence band of a semiconductor and the lowest state in the conduction band is known as the **band gap**.

A semiconductor is depicted in Figure 25.9(c), both at $T = 0$ and at room temperature. When an electron moves into the conduction band, it leaves behind a **hole** in the valence band. Thus, we can think of the hole as propagating in the valence band, just as the electron propagates in the conducting band. The phenomenon is analogous to the possible motions of cars in a two-level parking garage, as sketched in Figure 25.10. Since holes move in the direction opposite to the electrons, and represent an absence of negative charge, they behave as *positive* charge carriers. In a semiconductor at room temperature, both electrons and holes conduct electric current.

The sketches in Figure 25.9 are an idealization. Even the most perfect crystal obtainable in practice has **impurities** and **dislocations** – missing or extra atoms or shifts in the crystalline pattern – that give rise to quantum states with energies scattered in the region between bands. Electrons in these states are bound to the impurities and are not free to move through the crystal. When these impurities are randomly distributed in type and position, the extra states in the spectrum are uniformly distributed through the gap. In

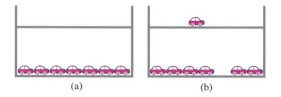

Figure 25.10 Cars in a parking garage mimic electrons and holes in a semiconductor. (a) No cars can move when the ground floor is full and the second floor is empty; (b) When a single car is lifted to the second floor it can move freely and the space it created in the ground floor can move as well, in a direction opposite to the motion of the cars. Adapted from [148].

general, the Fermi level lies roughly midway between the valence and conduction bands. In §25.6 we discuss how the impurities can be manipulated to move the Fermi level of the material in ways that are crucial to the functioning of photovoltaics.

It is possible to compute the probability that an electron is thermally excited to a state with a given energy. This probability is given by the **Fermi–Dirac distribution**, which gives the probability of finding an electron in a state of energy E when the system is at temperature T,

$$p(E) = \frac{e^{-E/k_B T}}{e^{-E/k_B T} + e^{-E_F(T)/k_B T}} = \frac{1}{1 + e^{(E - E_F(T))/k_B T}}. \tag{25.8}$$

This equation is the generalization of the *Boltzmann distribution*, which was derived in §8, to take account of the *Pauli exclusion principle, which requires that only one electron (of each spin) can occupy each quantum state* (Problem 25.19). A simple way to understand the Fermi–Dirac distribution (25.8) is to think of the state of energy E as being in thermal equilibrium with a state located right at the Fermi level $E = E_F$. In this hypothetical two-state system, eq. (25.8) is simply the Boltzmann distribution for the probability that the system is in the state of energy E. At low temperatures, where almost all states of energy $E < E_F$ are occupied, and almost all states of energy $E > E_F$ are empty, the Fermi level is very close to the zero-temperature Fermi energy The temperature dependence of the Fermi–Dirac distribution is illustrated in Figure 25.11.

25.4 The PV Concept and a Limit on Collection Efficiency

The structure of electron energy levels in solids enables us to explain the basic concept underlying a semiconductor photovoltaic cell. When a photon is absorbed in a block

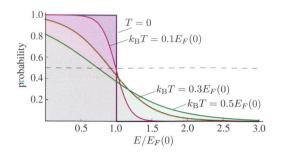

Figure 25.11 The Fermi–Dirac probability distribution for various temperatures. Only at $T = 0$ is there a sharp distinction between filled and empty levels, which occurs at the *Fermi energy* E_F^0. The *Fermi level*, which depends on temperature, is the energy at which the probability is 0.5. Note that the temperature effects shown are much larger than in, say, silicon at room temperature, where $k_B T / E_F \cong 0.023$, and the distribution would be almost indistinguishable from $T = 0$ at the scale of this figure.

Electrons in Semiconductors

A *semiconductor* is a material with a band gap that is small enough that thermal excitations are sufficient to promote some electrons from the valence band into the conduction band.

Crystalline silicon is an excellent example of a semiconductor. The band gap in silicon between the valence and conduction bands is approximately 1.1 eV. At zero temperature the valence band is full and the conduction band is empty.

The Fermi energy E_F^0 divides the filled from empty levels at zero temperature. For nonzero temperature the probability that an energy level is occupied is given by the Fermi–Dirac distribution (25.8), characterized by the Fermi level $E_F(T)$, the energy at which the probability is 1/2.

of a semiconducting material such as silicon, it can excite an electron from the valence band to the conduction band. Once this has occurred, the electron can move freely in the conduction band. Likewise, the hole created in the valence band can propagate like a positively charged electron. The aim of the photovoltaic cell is to use these electron and hole excitations to drive an external circuit.

There are a number of challenges to implementing this approach of conversion of solar energy to electrical energy:

(a) Avoiding recombination In general, an excited electron will rapidly drop back down into the valence band, combining with a hole and converting its energy into some combination of thermal energy and outgoing photons. To use the excited electrons in the semiconductor to drive an external circuit, the excited electrons must not recombine with holes too rapidly. In fact, as we describe in some detail below, a certain amount of recombination is associated with radiative processes, which cannot be avoided in a material that efficiently collects energy from incoming photons. The challenge is to minimize *non-radiative* recombination processes.

(b) Driving the circuit In order to use photo-excited electrons to drive an external circuit, the symmetry of the system must be broken in some way so that the electrons move in a preferred direction. If we simply take a block of silicon, connect it to a circuit, and expose it to sunlight, nothing interesting will happen, since the excited electrons are just as likely to move in one direction as another. In §25.6, we describe how the symmetry can be broken by *doping* different parts of the material with impurities that change the Fermi level. This inhomogeneous modification of the material encourages electrons to move preferentially in one direction (and holes in the other direction) so that the solar cell can drive a circuit.

(c) Minimum photon energy Finally, for a material with a given band gap, only photons with an energy greater than the energy difference across the band gap can excite an electron from the valence band to the conduction band. This limits the *collection efficiency* with which a specific material can convert solar energy to electrical energy. The dependence of this constraint on the band gap guides the choice of physical materials that make efficient photovoltaic cells.

These three issues are not only technical challenges for the construction of functioning photovoltaic cells, but are also responsible for fundamental limits on the efficiency with which photovoltaic cells can convert incoming solar energy into electrical energy. As emphasized in §23, solar radiation is characterized by a source temperature of roughly 6000 K, so the Carnot limit on the efficiency of a solar power device is very high, approximately $\eta_{Carnot} \approx (6000 - 300)/6000 = 0.95$ (see Box 25.1). As in the case of solar thermal power, however, other constraints make this limit inaccessible for existing technologies. In 1961 physicists William Shockley and Hans Queisser made an essential contribution to the study of photovoltaic cells by demonstrating a fundamental limit on photocell efficiency, combining three factors related to the challenges listed above. The third factor (c), the **collection efficiency**, which

---

Box 25.1   Carnot Limit on Solar Conversion Efficiency

The only known absolute physical bound on conversion of energy from solar radiation to electrical energy is the second law of thermodynamics. Since the Sun radiates at $T_\odot \cong 6000\,\mathrm{K}$, incoming solar radiation has an entropy associated with this temperature. The need to dump this entropy to the environment gives an upper limit of

$$\eta_{\mathrm{Carnot}} = 1 - T_{\mathrm{ambient}}/T_\odot \cong 0.95$$

on the conversion efficiency of solar energy to electricity. Current solar thermal and photovoltaic conversion systems are far from this bound. It may be that, in the future, improved technologies may allow us to get closer to this absolute physical limit.

The limit η_{Carnot} may seem confusing. Why is entropy associated with the energy carried by the individual photons in solar radiation? The answer is that the distribution of photons described in §22.3 characterizes *an ensemble in thermal equilibrium at temperature $T_\odot$*. The randomness in this photon distribution is precisely analogous to the randomness in the positions and momenta of molecules of gas in a box. This gives the incoming radiation an entropy of $\Delta S = \Delta E/T_\odot$ that must be dumped to the environment in any solar to electric conversion process in order to satisfy the second law of thermodynamics.

---

limits the fraction of incident sunlight that can be transformed into the energy of excited electrons, is described in this section. The other two factors, one related to unavoidable recombination and the other related to the need to drive electrons around the circuit, are considered in §25.7.2 after introducing the necessary physical concepts, giving the *Shockley–Queisser bound*.

When a photon of energy $E = \hbar\omega$ is incident on a material with a band gap E_{gap} between its (filled) valence band and (empty) conduction band, it can only lead to a conduction electron if $E > E_{\mathrm{gap}}$. Thus, for example, only photons with an energy $\hbar\omega > 1.1\,\mathrm{eV}$ can excite an electron from the valence band to the conduction band of a silicon crystal. Furthermore, any extra energy, $E - E_{\mathrm{gap}}$, above what is needed to excite the electron into the conduction band, is rapidly lost to thermal energy of the material, so only E_{gap} can be extracted from each photon. While some research efforts are directed to finding a way to capture some of this lost energy, circumventing these fundamental physical constraints presents a significant technical challenge that has not yet been surmounted for commercially viable systems.

Knowing the energy spectrum of the photons in solar radiation – which, as discussed in §22, is roughly[5] a blackbody spectrum at 6000 K – we can immediately determine

an upper bound on the collection efficiency of a solar cell with gap E_{gap}. We estimate the limit on collection efficiency using the Planck radiation spectrum (22.13), and then give a more realistic result for the AM1.5 spectrum (see Figure 23.9). First we compute the total incident power

$$P_{\mathrm{total}} = CA \int_0^\infty dE\, \frac{E^3}{e^{E/k_\mathrm{B}T} - 1}, \qquad (25.9)$$

from eq. (22.13), where C is an overall proportionality constant not relevant here, $E = \hbar\omega$ is the energy of a single photon of angular frequency ω, and A is the area exposed to the solar radiation.

The part of this power that is captured by the solar cell, assuming that we get E_{gap} of energy from each photon with $E > E_{\mathrm{gap}}$, is

$$\begin{aligned} P_{\mathrm{max}}(E_{\mathrm{gap}}) &= n(E_{\mathrm{gap}}) E_{\mathrm{gap}} \\ &= CAE_{\mathrm{gap}} \int_{E_{\mathrm{gap}}}^\infty dE\, \frac{E^2}{e^{E/k_\mathrm{B}T} - 1}, \end{aligned} \qquad (25.10)$$

where $n(E_{\mathrm{gap}})$ is the number of individual photons captured per unit time. The fraction of total power included in the integral (25.10) is shown in Figure 25.12 for the case of silicon, where $E_{\mathrm{gap}} \cong 1.1\,\mathrm{eV}$. The usable fraction of energy can then be computed as the ratio of the two integrals. Changing variables to $x = E/k_\mathrm{B}T$ (with, for example, $x \cong 2.17$ for $E_{\mathrm{gap}} = 1.124\,\mathrm{eV}$ and $T = 6000\,\mathrm{K}$), we have for silicon

$$\eta_{\mathrm{collection}}^{\mathrm{max}} = \frac{P_{\mathrm{max}}^{(\mathrm{Si})}}{P_{\mathrm{total}}} = 2.17\, \frac{\int_{2.17}^\infty dx\, x^2/(e^x - 1)}{\int_0^\infty dx\, x^3/(e^x - 1)} \cong 0.439. \qquad (25.11)$$

---

[5] In §22, we found that a temperature of 5780 K matches best to the total solar power output. For photovoltaic analysis, however, a blackbody spectrum at 6000 K is preferable because it matches better the peak and distribution of photon energies. Thus, in this chapter we approximate solar radiation as 6000 K blackbody radiation.

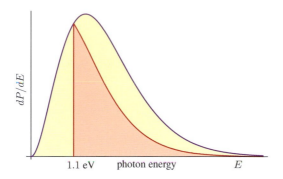

Figure 25.12 The solar power that can be collected by a silicon solar cell with $E_{gap} = 1.1$ V is shown in orange. The total (yellow + orange) area indicates the total power available in a thermal spectrum at 6000 K. The collection efficiency is the ratio of the orange and total areas.

Thus, a silicon photovoltaic solar cell can extract at most 43.9% of the energy in 6000 K blackbody radiation. An upper bound on the collection efficiency of a solar cell can be derived in this way for any material for which E_{gap} is known; the 1.124 eV gap of silicon is very close to optimal for collecting energy from a 6000 K blackbody spectrum (See Problem 25.8).

As described in §22, Earth's atmosphere absorbs much of the incoming ultraviolet and some infrared frequencies out of the solar radiation incident at the top of the atmosphere. As a result, the radiation spectrum seen at ground level – such as the AM1.5 radiation spectrum discussed in §23.5 – contains more energy in the middle of the spectrum, and a slightly higher fraction of incident solar energy (roughly 49%) can be incorporated into excited electrons than is indicated by eq. (25.11) (see Problem 25.9). In §25.9.1, we describe how materials with different band gaps can be used in combination as *multi-junction solar cells* to absorb different pieces of the radiation spectrum, thus significantly increasing the collection efficiency over simple crystalline silicon PV cells.

25.5 Band Structure of Silicon

In this section we describe the band structure of silicon in some detail. In particular, we discuss the *indirect* nature of the band gap between the valence band and the conduction band in silicon. This indirect band gap has the consequence that the photon absorption length in silicon is longer than in materials with direct band gaps. As a result, crystalline silicon solar cells must be more than an order of magnitude thicker than so-called *thin-film* photovoltaics based on materials with direct band gaps (§25.9). The increased

Bound on Collection Efficiency of PV Cells

Photons with energies below the energy of the band gap E_{gap} cannot excite electrons from the valence to the conduction band in a semiconductor. For photons with energies greater than E_{gap}, the excess beyond E_{gap} is lost as electrons thermalize in the conduction band. Together, these give an upper bound on the collection efficiency of simple photovoltaic cells.

For the case of silicon, where $E_{gap} \cong 1.1$ eV the upper bound on the collection efficiency for blackbody radiation at 6000 K is approximately 44%. For the actual spectrum of solar radiation on Earth (e.g. AM1.5), the bound is somewhat higher,

$$\eta_{collection}^{max} \leq 49\% \quad \text{(AM1.5 spectrum)}.$$

absorption length in silicon is at least partially compensated by the correlated suppression of the recombination process.

The band structure of a solid depends upon its crystal structure. Crystal structures can generally be described mathematically in terms of three-dimensional *lattices*. A **lattice** in N dimensions is a set of points that are generated through linear combinations with integer coefficients of a set of N linearly independent *basis vectors*. For example, the simple cubic lattice in three dimensions can be described as the set of points $n\hat{x} + m\hat{y} + p\hat{z}$, where $\hat{x}, \hat{y}, \hat{z}$ are the usual Cartesian unit vectors $\hat{x} = (1, 0, 0)$, $\hat{y} = (0, 1, 0)$, and $\hat{z} = (0, 0, 1)$. The cubic lattice can be envisioned through the infinite replication of the **unit cell** shown in Figure 25.13(a). The **face-centered cubic** (FCC) lattice is the lattice that includes all the points in the simple cubic lattice, but that also includes the center point of each of the (two-dimensional) faces of the unit cube. This lattice is generated by the basis vectors

$$a = (1/2, 1/2, 0), \quad b = (1/2, 0, 1/2), \quad c = (0, 1/2, 1/2). \tag{25.12}$$

Note that the basis vectors of the simple cubic lattice can all be described as linear combinations of $\{a, b, c\}$; for example, $\hat{x} = a + b - c$. A unit cell of the FCC lattice is depicted in Figure 25.13(b). The FCC crystal lattice is described by repeating this cell infinitely in each spatial dimension.

The diamond crystal structure, which describes crystalline silicon and is shown in Figure 25.13(c), consists

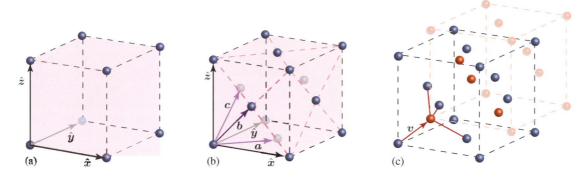

Figure 25.13 Unit cells of (a) a simple cubic lattice with the lattice vectors $\{\hat{x}, \hat{y}, \hat{z}\}$; (b) a face-centered cubic (FCC) lattice with lattice vectors $\{a, b, c\}$; and (c) the diamond crystal structure of the silicon crystal built from two copies of the FCC lattice. Note that the orange FCC lattice is translated by $v = (1/4, 1/4, 1/4)$ relative to the blue lattice and that each point in one lattice has four nearest neighbors in the other. The points in the orange lattice that are not contained in the blue unit cell have been faded. Compare Figure 25.1.

of two copies of the FCC lattice. One copy can be taken to be the lattice generated by the vectors in eq. (25.12), while the second copy is offset from the first by the vector $v = (1/4, 1/4, 1/4)$. Once the diamond crystal structure has been defined in this way, it is straightforward to verify that each point on either of the two intertwined FCC lattices has four nearest neighbors on the other FCC lattice at a distance $\sqrt{3}/4$. For example, as shown in Figure 25.13(c), the point v is separated by a distance of $\sqrt{3}/4$ from the points 0, a, b, and c.

Crystalline silicon is built on the diamond crystal structure. As in one dimension, momentum in a 3D periodic potential is only conserved up to equivalences analogous to eq. (25.5). In a 3D periodic potential the equivalences take the form

$$p \sim p + G, \tag{25.13}$$

where G is a momentum vector with the property that $G \cdot \Delta = 2\pi\hbar n$, where n is an integer for all vectors Δ under translation by which the 3D crystal structure is invariant.[6] Since the periodicity of the diamond crystal structure is the same as the FCC lattice, the vectors Δ can be taken to be the vectors $\{a, b, c\}$ that generate the FCC lattice.

Using the equivalences (25.13), we can define a region of momentum space such that every momentum vector p can be taken to precisely one point in the region by a translation of the form (25.13). This is the generalization of the

first **Brillouin zone** to three dimensions. For the diamond lattice, this region has the shape of a three-dimensional solid bounded by six squares and eight hexagons (a **truncated octahedron**), which is depicted in Figure 25.14(a). Just as the energy bands of the particle in a 1D periodic potential can be thought of as functions in the first Brillouin zone (as shown for example in Figure 25.7), we can describe the energy bands of silicon as functions over the first Brillouin zone. These are therefore functions of three variables $\{p_x, p_y, p_z\}$. A projection of the band structure onto one dimension is shown in Figure 25.14(b).

Precise computation of the band structure of a 3D solid such as silicon is a rather complicated problem in practice, though in principle the problem is essentially that of solving the Schrödinger equation in three dimensions in a specified potential with periodic boundary conditions. The methods needed for such an analysis are beyond the scope of this book. A key feature, however, of the band structure of silicon is clearly evident in Figure 25.14(b): the highest energy state in the valence band (point A) and the lowest energy state in the conduction band (point B) occur at different lattice momenta in the Brillouin zone. This is an example of an **indirect band gap** – as opposed to a **direct band gap** – which occurs when the highest state in the valence band has the same lattice momentum as the lowest state in the conduction band. A direct band gap occurs, for example in the semiconductor *cadmium telluride* (CdTe) (§25.9). Both types of band gaps are illustrated schematically in Figure 25.15.

When a photon excites an electron across a direct band gap, the electron's lattice momentum does not change significantly. The photon does transfer momentum to the

---

[6] Mathematically, this means that G lies in the *dual lattice* to the symmetry lattice of the original crystal containing the vectors Δ.

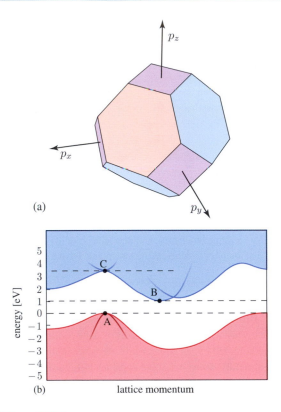

(a)

(b)

Figure 25.14 (a) The shape of the Brillouin zone for silicon; (b) a much simplified graph showing the relevant valence (red) and conduction (blue) energy bands of crystalline silicon in a one-dimensional projection over the Brillouin zone. A few of the band boundaries that overlap in this projection are also shown.

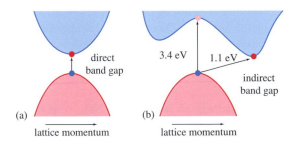

Figure 25.15 (a) A direct band gap (as in CdTe), versus (b) an indirect band gap (specifically in silicon). An indirect band gap is more difficult to excite, but decays more slowly.

electron, but the photon's momentum, $p_\gamma = E_\gamma/c$, is orders of magnitude smaller (see Problem 25.20) than the momentum of an electron with the same energy, and can be neglected. For the same reason, a photon by itself is not sufficient to excite an electron across an indirect band

gap because the electron lattice momentum must change. It is possible, in a material with an indirect band gap, for an incoming photon to excite a direct transition from the top of the valence band to the lowest accessible state in the conduction band at the same lattice momentum. This generally requires more energy, however. In silicon, the direct excitation of an electron requires about 3.4 eV. Only a small percentage of the photons in visible light have energy this large (see Problem 25.4).

In order for an electron to cross an indirect band gap, it must absorb momentum from some other source. In §25.1 we mentioned that vibrations of the ionic cores of the atoms that form the lattice could be described by waves that propagate through the crystal lattice. The quanta associated with these waves are known as **phonons**, in analogy to *photons* which are the quanta of light waves.[7] Phonons also carry lattice momentum, which is conserved along with electron momentum up to equivalences of the form of eq. (25.13). Thus, an electron can be excited across an indirect band gap if a low-energy phonon is emitted or absorbed in the process, thereby conserving lattice momentum. Because it involves an additional process, the quantum amplitude for such an event is smaller than for a comparable excitation of an electron across a direct band gap. As a result, the rate at which electrons are excited across the indirect band gap in silicon is relatively slow. On the other hand, the same physical mechanism also decreases the rate at which an excited electron will drop back from the conduction band into the valence band, giving the charge carriers in indirect band gap materials longer typical lifetimes than in direct band gap materials.

Recall from §23 that light entering a material will be absorbed at a rate $dI/dz = -\kappa I$, where κ is the absorption coefficient of the material. In general, the absorption coefficient is highly dependent upon the wavelength of the incoming light. For a silicon crystal, the dependence of the absorption length $1/\kappa$ on photon energy is shown in Figure 25.16. From the figure, we see that photons at energies above 3.4 eV will generally be absorbed by exciting an electron across the direct gap within 10 nm or so of entering a silicon crystal. On the other hand, to ensure absorption of most photons with energies in the 1.2–1.5 eV range by excitation of electrons across the indirect band gap, the thickness must be on the order of 100 µm ($= 10^5$ nm).

---

[7] The organized vibrations of a solid transmit sound, hence the root *phon-* = sound.

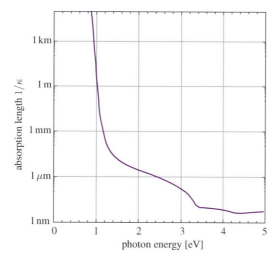

Figure 25.16 Absorption length ($1/\kappa$) in silicon for photons at different energies (at 300 K) [152]. Light with energies above 3.4 eV excites electrons from the valence band to the conduction band without changing lattice momentum (direct gap). The absorption length increases dramatically for photons with energy below $E_{\rm gap} = 1.1$ eV. (Credit: Adapted from data from PVeducation.org)

Direct and Indirect Band Gaps

A *direct band gap* is one in which the lattice momentum of the highest-energy state in the valence band is the same as the lattice momentum of the lowest-energy state in the conducting band. An *indirect band gap* is one where these lattice momenta are different. It is harder to excite electrons through an indirect band gap but also harder for the electrons to return to the valence band. In practical photovoltaic cells, the lower absorption in indirect band gap materials is thus mitigated somewhat by their longer carrier lifetimes and diffusion lengths.

25.6 *p-n* Junctions

Of the initial list of challenges assembled in §25.4, it remains to explain how to break the symmetry of the silicon crystal so that, when exposed to sunlight, a voltage difference spontaneously appears across the semiconductor and electrons are encouraged to move in one particular direction to create a useful electric current. In this section we describe how this can be done by *doping* the silicon using impurities to produce a *p-n junction*.

25.6.1 Doping

Although we have described silicon as if it were a perfect crystal with the diamond crystal structure, as mentioned in §25.3 any real piece of crystalline silicon contains dislocations and small amounts of impurities. These impurities slightly modify the structure of the lattice, and introduce states in the electron spectrum in the gap region between the valence and conducting bands. It is possible to alter the electron spectrum in the gap by intentionally introducing a small proportion of another element into the silicon. This produces *impurities* in the crystal structure either through substitution of another element for one of the silicon atoms (**substitutional impurities**), or by squeezing additional atoms into the crystal structure (**interstitial impurities**). The introduction of impurities in this fashion is known as **doping**.

The most common types of doping used for crystalline silicon are *n*-**type doping** and *p*-**type doping**, implemented by substituting an atom from column V or column III of the periodic table (see Figure D.1) for a silicon atom in the crystal structure. Cartoons of the resulting crystal structure in the vicinity of the impurity are depicted at the top of Figure 25.17. In the case of *n*-type doping, a silicon atom is replaced by an atom from column V, such as phosphorus or arsenic, which has one more electron in its outer orbital than silicon. This extra electron does not participate in any of the bonds to the four neighbors in the crystal, and is only weakly bound to the nucleus of the atom.[8] It therefore takes very little energy to boost this *extra* electron out of its loosely bound state and into the conduction band. In fact, the energy required to lift such an electron to the conduction band is roughly 0.04–0.07 eV for *n*-type impurities from column V atoms (compared to the band gap of $\cong 1.1$ eV). To a good approximation, *n*-type doping can be modeled by adding a large number of electron energy levels just below the bottom of the conduction band. At $T = 0$ all these levels are filled – the extra electrons remain bound to the impurity atoms. At room temperature, however, there is enough thermal energy available to promote most of these electrons into the conduction band, and the Fermi level moves up. The situation is illustrated schematically at the bottom of Figure 25.17(b).

Similarly, substituting into the silicon lattice an atom from column III, such as aluminum or gallium, with one fewer electron than silicon in the outer orbital, has the

---

[8] The binding of this electron to the group V atom to which it belongs is significantly weakened within the silicon crystal, because the positive ionic charge is screened by electrons in the valence band.

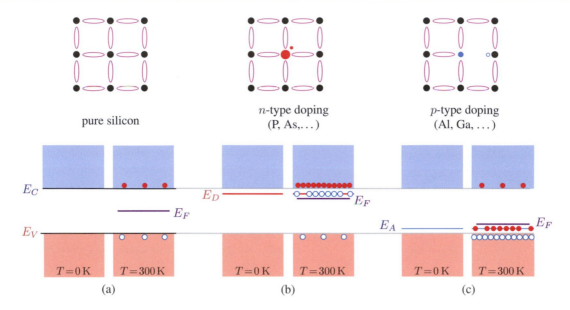

Figure 25.17 Doping a silicon crystal. (a) Undoped: each silicon atom shares four electrons with its nearest neighbors. At $T = 0$ the valence band is filled, and the conduction band is empty. At $T = 300$ K a few electron–hole pairs are thermally excited. (b) n-type doping: donor impurities have five valence electrons, introducing new filled electron states (**donor states**, denoted by the solid red line) just below the conduction band $E_D \lesssim E_C$. At $T = 300$ K most of the donor electrons are thermally excited into the conduction band (leaving behind holes in the donor band) and the Fermi level lies close to E_C (c) p-type doping: acceptor impurities arise from atoms with three valence electrons, introducing new, empty electron states (**acceptor states**, denoted by the solid blue line) just above the top of the valence band, $E_A \gtrsim E_V$. At $T = 300$ K electrons, thermally excited from the valence band, fill almost all of the acceptor states, creating mobile holes in the valence band. The Fermi level lies close to E_V.

opposite effect on the Fermi level. The atom from column III does not have enough electrons to fully bond with all four of the adjacent atoms in the crystal structure. Therefore, one of its neighbors can be thought of as having a hole into which an electron in the valence band can easily be excited. Much like an n-type impurity, the energy difference between the top of the valence band and the new, unfilled state created by the substitution of a group III atom is small. So p-type doping has the effect of moving the Fermi level down towards the top of the valence band (see Figure 25.17(c)).

Thus, by adding n-type and p-type impurities to the silicon crystal the Fermi level can be moved in either direction. Semiconductors that have been doped in this way are known as n-type and p-type materials. At room temperature, electrons in an n-type or p-type material readily move between the Fermi level and the conducting or valence bands. In an n-type material, electrons are easily removed from their loosely bound states into the conducting band, becoming mobile charge carriers, while in a p-type material, electrons easily jump from the valence band into the outer orbitals of the group III atoms that have been substituted in the lattice, producing mobile positive

Doping

The band structure of a crystalline solid can be subtly modified by *doping* with atoms of different elements. Silicon lies in group IV of the periodic table, with four valence electrons per atom. *n-doping* is achieved by inserting atoms of an element from column V that have an *extra* electron. This raises the Fermi level of the semiconductor to near the bottom of the conduction band. *p-doping* is accomplished by inserting atoms of an element from column III, and lowers the Fermi level close to the top of the valence band.

charge carriers (holes) in the valence band. These mobile charge carriers – electrons in an n-type semiconductor and holes in a p-type semiconductor – are known as **majority carriers**, while the opposite charges in each type of semiconductor are **minority carriers**.

These two different ways of modifying the silicon crystal lattice can be combined to generate an asymmetry in the

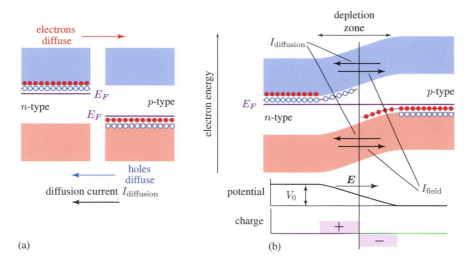

Figure 25.18 Dynamics at a *pn*-junction: (a) Before the *n*-type and *p*-type materials are brought into contact. Note that for simplicity no excitations across the band gap are shown in this figure. When brought into contact, mobile electrons (holes) will diffuse from *n* to *p* (*p* to *n*) regions. This sets up a diffusion current $I_{\text{diffusion}}$ in the direction of the *n*-type material. (b) Electrons diffused from the *n*-type material fill holes in the *p*-type material near the junction, leaving fixed charge densities that create an electric field and a potential difference V_0 across the junction. V_0 drives a field current I_{field} in the direction of the *p*-type material. The system comes into a dynamic equilibrium when the two currents balance and the Fermi levels of the two pieces of matter become equal. In the depletion zone near the interface, there are few mobile charge carriers and an electric field points from the *n*-type material toward the *p*-type material.

spatial configuration of the material, so that electrons travel in a preferred direction, and can drive an external circuit. The basic mechanism for doing this is the *p-n junction*, a fundamental piece of semiconductor technology.

25.6.2 *p-n* Junction Diodes

A ***p-n* junction** is formed when a piece of semiconductor with *n*-type doping is placed adjacent to another piece of semiconductor with *p*-type doping. When these materials are connected (see Figure 25.18(a)), the loose negative charge carriers (electrons) in the *n*-type material diffuse through random thermal motion into the *p*-type material, while the loose positive charge carriers (holes) in the *p*-type material diffuse into the *n*-type material. This diffusion of majority carriers results in a net flow of positive charge from the *p*-type material into the *n*-type material, known as the **diffusion current**, $I_{\text{diffusion}}$.

As charge flows via the diffusion current, **depletion regions** build up on both sides of the interface, in which the density of majority carriers is very small. These regions have opposite net charge density, positive in the *n*-type material and negative in the *p*-type. Like a parallel plate capacitor, these charges generate an electric field pointing from the *n*- to the *p*-type material, resulting in a potential difference V_0 between the two materials. This potential

difference has two effects: first, it suppresses the diffusion current by the *Boltzmann factor* $e^{-eV_0/k_{\text{B}}T}$, when the system is at temperature T. The Boltzmann factor appears because the diffusion current arises from random thermal motion of electrons from the *n*-type region into the *p*-type region (and vice versa for holes), involving an increase in energy of eV_0. And second, the electric field between the regions drives a second current I_{field}, known as the **field current** (also known as the **drift current**), which opposes the diffusion current. The voltage difference between the two regions increases until the diffusion current and field current balance and the system reaches a dynamic equilibrium with

$$|I_{\text{diffusion}}| = |I_{\text{field}}| \equiv I_0. \tag{25.14}$$

At equilibrium the Fermi levels in the two materials are equal, as shown in Figure 25.18(b). This is a general consequence of statistical mechanics: the distribution (25.8) of electron states in equilibrium is characterized by a single value of the Fermi level E_F. If the Fermi level were not constant through the material, charges would rearrange through the material until it was. The matching of Fermi levels in the *p-n* junction is realized, as shown in Figure 25.18, through a shift in electron energy by eV_0, where

V_0 is the potential difference in equilibrium across the junction.

When the *p-n* junction is in equilibrium, away from the interface between the two regions each region acts as a regular semiconductor and conducts current relatively freely. In the depletion region near the interface, however, the material acts more as an insulator than a conductor due to the absence of mobile charge carriers. Without an external field (or incident radiation that could excite electrons), the *p-n* junction is in equilibrium in this configuration. When an external voltage is applied, however, the junction acts as a **diode**, or one-way gate, preferentially allowing current in one direction.

To understand the diode property of a *p-n* junction, consider the effect of applying an external voltage V to the junction, with the *p*-type region on the positive side and the *n*-type region on the negative side of the voltage. For a *p-n* junction this is the **forward bias** direction in which current flows more freely. Note that the voltage V *has the opposite sign* from the equilibrium voltage V_0. The *field current* arises from electrons and holes that enter the interface region and are moved across by the electric field at the interface. The interface region is relatively small, and the field current is relatively independent of the applied voltage since it depends primarily on the rate at which minority carriers migrate into the interface region. We approximate $I_{\text{field}}(V) \approx I_{\text{field}}(0) = I_0$, where I_0 is the equilibrium current introduced above. On the other hand, the Boltzmann factor that suppresses the diffusion current is modified to $e^{-e(V_0-V)/k_{\text{B}}T}$ in the presence of the applied potential V. In equilibrium, with $V = 0$, the field and diffusion currents cancel (eq. (25.14)), so when V is applied the diffusion current increases *exponentially*,

$$I_{\text{diffusion}}(V) = e^{eV/k_{\text{B}}T} I_0 . \tag{25.15}$$

As a result, the total flow of current across the *p-n* junction when a potential V is applied is given by the **Shockley diode equation**,

$$I(V) = I_{\text{diffusion}}(V) - I_{\text{field}}(V) = I_0 \left(e^{eV/k_{\text{B}}T} - 1 \right) , \tag{25.16}$$

where the field current moves against the direction of the applied voltage.

Equation (25.16) describes the behavior of an idealized diode, which acts as a one-way gate for current. In this formula the field current $I_{\text{field}} = I_0$, which contributes negatively to the right-hand side, is known as the **reverse saturation current**; this current is generally quite small and dominates when a negative voltage is applied. When a positive voltage is applied, eq. (25.16) gives a current that grows exponentially. Thus, for a positive applied voltage,

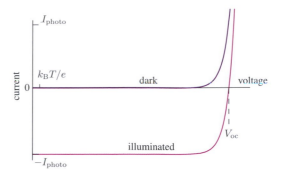

Figure 25.19 The voltage–current characteristics for a typical solar cell photodiode. Both illuminated (magenta) and non-illuminated (purple) diode characteristics are shown. The current grows exponentially with increased external voltage V. Note that when the diode is illuminated and drives a current through an external system with a sufficiently low voltage drop $V < V_{\text{oc}}$, the sign of the current is negative relative to the direction defined by the Shockley diode equation (25.16).

current flows much more readily through the diode than for a negative applied voltage. The voltage–current relation for a typical silicon *p-n* junction diode is shown as the curve labeled "dark" in Figure 25.19.

25.7 The *p-n* Junction as a Photodiode

25.7.1 Illuminated Photodiode

When a *p-n* junction is illuminated with electromagnetic radiation, the incoming photons can lift electrons from the valence band to the conducting band. As long as the recombination rate is not too fast, this increases the number of mobile electrons in the *p*-region that find their way to the interface region and are pushed by the potential gradient into the *n*-region (similarly, holes move from the *n*-region to the *p*-region at an increased rate). This introduces an additional current I_{photo} into eq. (25.16), in the same direction as I_0. The *p-n* junction is now acting as a **photodiode**, and the total current in the presence of an external voltage V is

$$I = I_{\text{photo}} + I_0 - I_{\text{diffusion}} = I_{\text{photo}} - I_0 \left(e^{eV/k_{\text{B}}T} - 1 \right) . \tag{25.17}$$

Note that we have changed the sign of the current relative to the diode equation (25.16) so that the sense of the current is positive when the photodiode drives an external circuit. If the external voltage is not too large, current will flow around the circuit in the direction of the photo current, with positive charge leaving the *p*-type region,

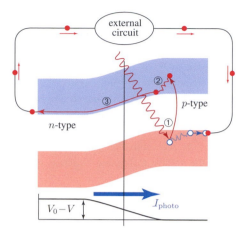

Figure 25.20 In a *p-n* junction photocell, an incoming photon (①) excites an electron into the conduction band leaving behind a hole. When this occurs in the *p*-doped region, after losing excess energy to thermalization (②), the electron is pushed across the junction (③) by the potential difference $V_0 - V$ and leaves the junction on the left. The hole produced by the electron's excitation leaves the junction on the right. These charge motions contribute to the current I_{photo} driving an external connected circuit.

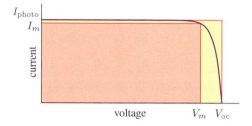

Figure 25.21 The fill factor $IV/I_{photo}V_{oc}$ is the maximum possible value of IV for the photodiode (orange shaded) divided by the product $I_{photo}V_{oc}$. Note that the sign on I is positive in this figure for current flowing in the direction of I_{photo} (the reverse of Figure 25.19).

flowing through the external circuit, and returning to the *n*-type region (shown in Figure 25.20). The voltage–current relation of the diode (25.17) implies a further limit, however, on the fraction of energy that can be extracted from the excited electrons, as described further below.

The idealized model (25.17) of the voltage–current relation for a photodiode is only an approximation to any real system – in particular, the *field current* is not really completely independent of applied voltage. This simplified model, however, captures many of the main physical characteristics of photovoltaic cells.

25.7.2 The Shockley–Queisser Bound on Photovoltaic Efficiency

In the analysis of collection efficiency in §25.4, we assumed that every electron excited across the band gap of a photovoltaic could provide energy E_{gap} to an external circuit, but this is not actually the case. For a device that satisfies the idealized photodiode equation (25.17) and has typical emission and absorption characteristics, it is possible to extend the efficiency analysis of §25.4 to include further constraints on the power output of a photovoltaic cell associated with the voltage–current characteristics of photodiodes and electron–hole recombination, related respectively to issues (b) and (a) discussed in §25.4. When these effects are combined with the limit on collection

---

Current–Voltage Relation for Photodiode

When a photodiode formed by a *p-n* junction is connected to an external voltage V, an external circuit is driven with total current

$$I = I_{photo} + I_0 - I_{diffusion} = I_{photo} - I_0 \left[e^{eV/k_B T} - 1 \right],$$

where I_{photo} is the current contribution from photo-excited electrons, $I_{diffusion}$ is the rate of mobile carrier diffusion across the junction, and I_0 is the reverse saturation current driven by the internal voltage drop across the junction.

---

efficiency (25.11), the resulting limit on the efficiency of an ideal photodiode is known as the *Shockley–Queisser limit* [153].

Current–Voltage Characteristics   Figure 25.21 depicts the voltage–current curve for an *illuminated* ideal photodiode. The maximum current in the power-producing quadrant of the voltage–current graph, I_{photo}, occurs at zero external voltage $V = 0$, and is known therefore as the **short-circuit current**. The maximum voltage occurs when no current is drawn and is known therefore as the **open-circuit voltage** V_{oc}. To induce a current to flow, we must have $V < V_{oc}$. From Figure 25.21 it is clear that the current increases rather quickly as V is reduced below V_{oc}.

At a heuristic level, the open-circuit voltage is essentially the shift in potential needed to match Fermi levels between the *n*-type and *p*-type regions, and eV_{oc} is thus always less than the band gap E_{gap}. Setting $I = 0$ in the ideal photodiode equation (25.17), we obtain a relation between the open-circuit voltage and the short-circuit current,

$$V_{oc} = \frac{k_B T}{e} \ln\left(\frac{I_{photo}}{I_0} + 1\right) \approx \frac{k_B T}{e} \ln\left(\frac{I_{photo}}{I_0}\right),$$
(25.18)

where we have used the fact that I_{photo} is, in general, much greater than the reverse saturation current I_0. For a fixed value of I_{photo}, the open-circuit voltage is thus limited by how small I_0 can be made. We return to this issue below.

Since the total power output of the photodiode is given by the usual relation (3.22),

$$P = IV,$$
(25.19)

the power is maximized when the product IV is maximized. Since to have both $V > 0$ and $I > 0$, we must have $I < I_{photo}$ and $V < V_{oc}$, it is clear that the total power must be less than the product $I_{photo} V_{oc}$. The maximum power $I_m V_m$ occurs at the intermediate point on the I–V curve in Figure 25.21 where the area of the (orange) shaded rectangle is maximized. The ratio of maximum total power to the product $I_{photo} V_{oc}$, is called the **fill factor**, denoted ff,

$$\text{ff} = \frac{I_m V_m}{I_{photo} V_{oc}}.$$
(25.20)

The fill factor approaches one as the I–V curve becomes more rectangular. The fill factor reduces the photovoltaic cell efficiency. For values of V_{oc} relevant for silicon photodiodes, $V_{oc} \cong 0.70$–0.86 V (as discussed below), the fill factor ranges approximately from 0.85 to 0.87 (Problem 25.10).

Electron–Hole Recombination The American physicists William Shockley and Hans Queisser completed their effective bound on photovoltaic efficiency by arguing that simple thermodynamic principles can be used to constrain the open-circuit voltage V_{oc}. The essence of the Shockley–Queisser argument is that the photodiode must radiate energy into the environment as a material at the ambient temperature T and that this puts a lower limit on the reverse saturation current I_0. A lower limit on I_0, in turn, gives an upper limit on V_{oc} from eq. (25.18). This upper limit equals E_{gap} only in the limit that the ambient temperature T goes to zero (Problem 25.21).

The Shockley–Queisser analysis depends on the assumption of **detailed balance**, which states that for a system in thermodynamic equilibrium, microscopic processes occur at equal rates in opposite directions; in the case of the photovoltaic cells of relevance here, this corresponds to the assertion that the rates of absorption and emission of thermal radiation are equal, as in *Kirchhoff's law* (Box 22.1). Thus detailed balance enables the rate at which the photodiode at temperature T radiates into the environment due to *radiative recombination* to be related to the radiation *absorbed* from the environment at that temperature. Assuming that there is no non-radiative recombination gives a bound on the efficiency of a photodiode.

When a photodiode is exposed to incoming thermal radiation at temperature T_i and connected to a load, the current that flows, corresponding to removal of electrons and holes to the external circuit, is proportional to the rate of pair creation from incident photons $F_i(T_i)$ minus the rate of recombination $F_c(V)$, where the latter depends on the voltage difference V across the external circuit,

$$I = e\left(F_i(T_i) - F_c(V)\right).$$
(25.21)

We identify $F_c(V)$ with the radiative recombination rate since forms of recombination that do not result in radiation would only further decrease the photodiode's efficiency. The radiative recombination rate is proportional to the product of hole and electron densities divided by the equilibrium value; the non-equilibrium recombination rate is given by $F_c(V) \approx F_c(0) e^{eV/k_B T}$. Thus eq. (25.21) can be rewritten

$$I = e\left(F_i(T_i) - F_c(0) - F_c(0)(e^{eV/k_B T} - 1)\right).$$
(25.22)

Comparing with eq. (25.17) we identify I_{photo} with $e(F_i(T_i) - F_c(0))$ ($\approx e F_i(T_i)$, since $F_i(T_i) \gg F_c(0)$) and identify the reverse saturation current with the equilibrium recombination rate times the electric charge, $I_0 = e F_c(0)$. These identifications are essentially equivalent to the assumption that all excited minority carriers survive long enough to encounter the junction region and contribute to the relevant current. If the minority carrier lifetime is too short, the photoexcited electrons cannot survive long enough to go across the junction and the photodiode would be less efficient.

Consider first a photodiode in the form of a flat plate of area A_p, exposed to sunlight at normal incidence with a blackbody spectrum characterized by a temperature $T_i = T_\odot$. The photocurrent I_{photo} is essentially proportional to the number of electron–hole pairs created by photons in the incident solar radiation. This, in turn, is proportional to the area A_p of the photodiode, to the number of photons with energy above the band gap, and to a geometrical factor that accounts for the fraction of the Sun's radiation that is incident on the plate,

$$I_{photo} = e F_i(T_\odot) = A_p \frac{\omega_s}{\pi} c \int_{E_{gap}}^{\infty} dE \, \frac{E^2}{e^{E/k_B T_\odot} - 1}.$$
(25.23)

Here ω_s is the solid angle subtended by the Sun,[9] c is a proportionality constant, and the integral over E is proportional to the number of photons with energy above E_{gap} (per unit area, per unit time) emitted by the Sun (see §22.2.2). For maximum efficiency, we assume that the PV is a perfect absorber (and hence emitter) of incoming radiation for photons of all energy above E_g.

Next, to bound the reverse saturation current I_0, consider an unbiased and unilluminated (dark) photodiode in thermal equilibrium with its environment at temperature T. Having identified $I_{diffusion} = I_0$ with the equilibrium radiative recombination rate times the electron charge, we can use detailed balance[10] to bound I_0 by the photocurrent generated by blackbody radiation from the environment at temperature T

$$I_0 \geq A_p c \int_{E_{gap}}^{\infty} dE \, \frac{E^2}{e^{E/k_B T} - 1} . \quad (25.24)$$

Substituting eqs. (25.23) and (25.24) into eq. (25.18) we obtain

$$V_{oc} \leq \frac{k_B T}{e} \ln \left(f \frac{\int_{E_{gap}}^{\infty} dE \, \frac{E^2}{e^{E/k_B T_\odot} - 1}}{\int_{E_{gap}}^{\infty} dE \, \frac{E^2}{e^{E/k_B T} - 1}} \right) , \quad (25.25)$$

where $f = \omega_s/\pi$. Note the appearance of $T_\odot$ in I_{photo} in contrast to T in I_0. The number of photons above the band gap in the incident solar radiation is so much larger than the number of photons above E_{gap} in thermal radiation at $T \approx 300$ K, that despite the small size of $f \cong 2.18 \times 10^{-5}$, the argument of the logarithm is much greater than one. Nevertheless, it is not difficult to show (Problem 25.21) that the right-hand side of eq. (25.25) cannot exceed E_{gap} and that it reaches this limit only as $T \to 0$.

For a band gap of 1.1 eV, the bound from this estimate (25.25) is roughly 0.88 V, using 6000 K incoming thermal radiation (Problem 25.12); the bound for silicon under AM1.5 radiation is often given using $f = \omega_s/2\pi$ as 0.86 V. We use this number as a rough estimate, with

the understanding that slightly different assumptions (e.g. $E_{gap} = 1.1$ vs. $E_{gap} = 1.124$) can affect the precise estimate of the bound to several percent. Currently, optimal solar cells have an open-circuit voltage around $V_{oc} \cong 0.7$ V. Values of V_{oc} significantly above this may be very difficult or impossible to achieve due to material limits associated with charge carrier diffusion and the rate of charge carrier recombination. (For a detailed discussion see [149].) The value $V_{oc} = 0.7$ V corresponds to a factor $eV_{oc}/k_B T$ of around 27 (33 for $V_{oc} = 0.86$ V), confirming the fact that I_{photo} is enormous compared to the saturation current I_0. Note that for a higher band gap E_{gap}, the rate of blackbody radiation at ambient temperatures (appearing in the denominator of eq. (25.25)) decreases rapidly, as the thermal distribution is far out on the exponentially decreasing tail. Thus, the bound on V_{oc} is less constraining for materials with higher band gaps.

Although the electrons excited to the conduction band have extra energy E_{gap}, only eV_{oc} can be extracted from the photodiode. Thus the factor eV_{oc}/E_{gap} decreases the efficiency of the photodiode. This is the final factor in the Shockley–Queisser limit on PV efficiency, to be combined with the *collection efficiency* $\eta_{collection}^{max}$ and the fill factor ff to obtain an overall constraint on photodiode efficiency.

Note that the open-circuit voltage V_{oc} depends logarithmically on I_{photo} (see eq. (25.18)), which in turn is proportional to the power density of incoming sunlight. For this reason, solar concentrators can increase photovoltaic efficiency. In general, the increase in V_{oc} is logarithmic in the concentration C. The best efficiencies for single-junction silicon cells that have been realized to date (see Figure 25.23) use solar concentrators with $C \sim 400$. The effect of concentrators on PV efficiency is explored further in Example 25.1.

The Shockley–Queisser Bound

The **Shockley–Queisser bound** on single-junction photovoltaic efficiency combines the three factors associated with the three issues raised in §25.4: (a) the bound (25.25) on eV_{oc}/E_{gap}; (b) the fill factor from eq. (25.20); and (c) the bound on collection efficiency (25.11), to give an overall efficiency bound

$$\eta = P/P_{solar} \leq \eta_{max} = \text{ff} \times \frac{eV_{oc}}{E_{gap}} \times \eta_{collection}^{max} . \quad (25.26)$$

Multiplying out the factors, the best efficiency possible for a silicon photodiode under AM1.5 conditions is $\eta_{max} \cong 0.87 \times 0.77 \times 0.49 \cong 33\%$. For a silicon photodiode with $V_{oc} \cong 0.7$ V the bound is $\eta_{max} \cong 27\%$. Improvements in materials to increase V_{oc} may push the efficiency somewhat above 27%, though the true bound for

---

[9] The power emitted by the Sun is proportional to its surface area $4\pi R_\odot^2$; the fraction of that power incident on the photodiode is given by the photodiode's surface area divided by the surface area of a sphere of radius equal to Earth's mean orbital distance from the Sun, $4\pi R_\odot^2 A_p/4\pi d^2 = R_\odot^2 A_p/d^2$. The solid angle subtended by the Sun viewed from Earth is $\omega_s = \pi R_\odot^2/d^2$, so the fraction of the Sun's power incident on the photodiode is proportional to $A_p \omega_s/\pi$.

[10] A factor of 2 is sometimes included in the denominator in this relation, assuming that the planar photodiode absorbs and emits radiation on both sides, but the efficiency can be increased by a reflective backing on the photodiode that eliminates this factor [151].

Example 25.1   Concentrated PV and Temperature

A PV system has an open-circuit voltage of 0.7 V when exposed to unconcentrated sunlight. Suppose this system is combined with concentrating optics with $C = 100$. *How much is the efficiency increased?*

The concentrator increases the photocurrent by a factor of 100, so if I_u is the photocurrent without the concentrator, then

$$V'_{oc} = \frac{k_B T}{e} \ln\left(\frac{100 I_u}{I_0} + 1\right) \cong V_{oc} + \frac{k_B T}{e} \ln(100) \cong 0.7\,\text{V} + 0.12\,\text{V} \cong 0.82\,\text{V},$$

where $k_B T/e \cong 1/40\,\text{V}$ and we have ignored the 1 relative to I_u/I_0, which as mentioned in the text is $\sim e^{27}$ for $V_{oc} = 0.7\,\text{V}$. So the efficiency increases by roughly 17% (or slightly more when the fill factor (§25.7.2) is considered) – and the power output of the cell increases proportional to the concentration.

Unfortunately, PV cells exposed to such concentrated sunlight heat up unless they are strongly coupled to heat sinks. *How hot can the PV be allowed to get before the increase in efficiency due to concentration is wiped out?* We assume that the temperature dependence of V_{oc} can be estimated by studying the temperature dependence of the bound on V_{oc} (25.25). We define $x = T/T_0$, where $T_0 = 300\,\text{K}$, and scale the T dependence out of the integrals, leaving

$$V_{oc} \lesssim \frac{x}{40} V\left(-2.9 + \ln\left(\int_{2.2}^{\infty} \frac{dz\, z^2}{e^z - 1}\right) - \ln\left(\int_{44/x}^{\infty} dz\, z^2 e^{-z}\right) - 3\ln(x)\right),$$

where $E_{gap} \cong 1.1\,\text{eV}$ and in the second integral, because the lower limit is so large we have replaced $e^z - 1$ by e^z. The temperature dependence of V_{oc} is plotted in the figure. The efficiency decreases by 17% when the temperature exceeds roughly 150 °C. This rough estimate suggests that cooling is an important consideration for concentrated PV installations.

The Bound on V_{oc} From Electron–Hole Recombination

Thermodynamic considerations lead to a bound on the possible open-circuit voltage of a solar PV,

$$V_{oc} \leq \frac{k_B T}{e} \ln\left(\frac{f \int_{E_{gap}}^{\infty} dE\, \frac{E^2}{e^{E/k_B T_\odot} - 1}}{\int_{E_{gap}}^{\infty} dE\, \frac{E^2}{e^{E/k_B T} - 1}}\right)$$

for a solar blackbody spectrum at $T_\odot$, where f is a geometric factor. For silicon (using the AM 1.5 spectrum for solar radiation) the bound is roughly 0.86 eV. Present silicon photodiodes reach $V_{oc} \cong 0.7\,\text{eV}$.

The power that can be extracted from a solar PV is reduced by the ratio of $e V_{oc}/E_{gap}$; this is one of the three contributions to the Shockley–Queisser efficiency bound. For silicon with $E_{gap} \cong 1.1\,\text{eV}$, this factor is bounded by 0.77, and is roughly 0.64 for practical devices with $V_{oc} \cong 0.7\,\text{PV}$.

a silicon PV may be below 33%. Green [149] uses recombination properties of electrons in silicon to argue that 28.8% is the highest efficiency possible for single-junction silicon PV cells that respond equally to light coming from all directions (**isotropic response**).

The analysis given here is intended to provide a general sense of the physical limits for this technology, and not a precise efficiency bound. As of 2015, the highest efficiency achieved for single-junction silicon photovoltaic solar cells under *one sun* illumination (i.e. without concentrators) was around 25%. Note that while the collection efficiency bound (see §25.4) is maximized for a gap around 1.1 eV, the electron–hole recombination bound (25.25) favors higher band gaps, as mentioned above, so after all factors are included, the optimal band gap is in the range 1.3–1.4 eV, with a Shockley–Queisser bound around 34%. The Shockley–Queisser bound as a function of band gap E_{gap} is graphed in Figure 25.22 (for incoming AM1.5 radiation).

Note that although the bounds on efficiency we have discussed in this and the preceding sections are based on physical principles, each bound is predicated on certain assumptions that may be evaded in various ways. For

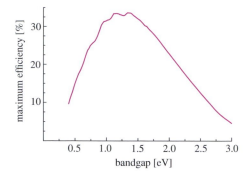

Figure 25.22 The Shockley–Queisser limit on single-junction PV efficiency as a function of band gap (using the AM1.5 spectrum). Image: S. Byrnes, Harvard University.

example, the spectrum bound derived in §25.4 may be evaded if a mechanism is found that can make use of the extra energy in an electron excited by a photon of energy above the band gap. It is possible that open-circuit voltages above 0.7 V may be achieved by a clever combination of device geometry and materials. It is possible that the Shockley–Queisser bound on V_{oc} may be violated, for example if the PV material does not itself radiate as a black body. Nonetheless, exceeding these bounds for a single-junction photovoltaic cell represents a major engineering challenge. In §25.9.1 we discuss the possibility of achieving higher efficiencies in photovoltaic cells using multiple materials (multi-junction PV).

25.8 Silicon Solar Cells

To date, most deployed solar PV systems are based on the crystalline silicon solar cell described in the preceding sections. At present, the most efficient commercially available silicon solar cells have peak efficiencies of roughly 20%. While energy from photovoltaic cells is still (as of 2017) somewhat more expensive for many applications than energy from fossil fuels such as coal, petroleum, or natural gas, PV systems have become competitive with other sources in some markets in recent years. In this section we briefly review some of the practical issues in engineering inexpensive and efficient silicon solar cells.

Material constraints Because of crystalline silicon's indirect band gap, at least $100\,\mu\text{m}$ of material is needed to absorb a significant fraction of incoming solar radiation. Current solar cells have 100–$300\,\mu\text{m}$ thickness of crystalline silicon. While the element silicon is abundant on Earth (crustal abundance 28% by mass), it almost never occurs naturally in pure form. Instead it is usually obtained from quartz (SiO_2), a very common mineral. To form a

The Shockley–Queisser Limit on Silicon Photovoltaic Efficiency

The Shockley–Queisser limit on the efficiency of a photovoltaic cell constructed from a single type of semiconductor with a band gap E_{gap} is

$$\eta_{\max} = \text{ff} \times \frac{e V_{oc}}{E_{gap}} \times \eta_{\text{collection}}^{\max}.$$

Here ff is the *fill factor* determined by the voltage-current relation of the photodiode; V_{oc} is the open-circuit voltage, and $\eta_{\text{collection}}$ is the collection efficiency. This bound limits the efficiency of single-junction silicon solar cells to approximately 33%, assuming AM1.5 radiation and that the cell radiates as a black body. Fundamental materials constraints suggest an upper bound around 29%; current silicon solar cell materials and geometries have achieved 25% conversion efficiency for non-concentrated sunlight.

semiconductor of quality high enough for use in PV solar cells, silicon is refined typically to a purity of 99.9999% ("six nines"), from which **monocrystalline** (single crystals) or **polycrystalline** (many small crystals) stock is grown. Thin wafers are then sawn from the bulk silicon. This process is complex, relatively expensive, and energy intensive. Processing silicon can also release carbon (e.g. in the reduction of silicon dioxide $SiO_2 + 2C \rightarrow Si + 2CO$). Though new methods are being developed that may reduce the cost and energy requirements of producing pure silicon crystals, the complexity of the crystal growing and wafer production process drives up the cost of silicon photovoltaic cells.

Degradation Exposure to UV light and high temperatures gradually damages silicon solar cells, causing defects in the crystal lattice and reducing efficiency. Lifetimes of currently produced solar cells are estimated at about 25 years, so any given solar cell has a finite expected energy production over its lifetime.

Contacts The connection of electrical contacts to the semiconductor junction is a significant engineering challenge in making highly efficient solar cells. Contacts on the front surface can block incoming light, and the interface can reduce cell efficiency. At present, large quantities of silver (Ag) are used for this purpose; on the order of tens of milligrams of Ag are needed per peak watt of solar power. Researchers, however, are actively looking for lower-cost replacements, and the amount of silver used in silicon solar

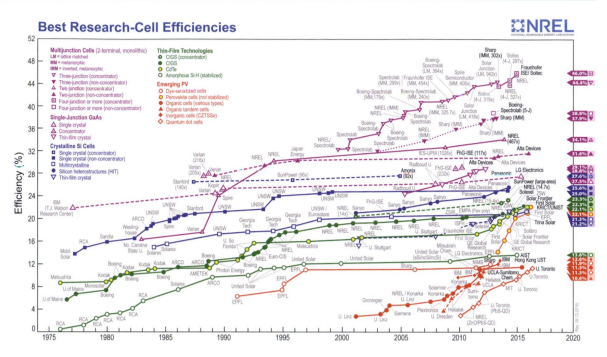

Figure 25.23   Different PV technologies and maximum research-cell efficiencies realized since 1975. From [154].

PV manufacturing is expected to drop significantly over the next decade.

Reflectivity   Bare silicon is highly reflective, so rough surfaces or an antireflective coating must be used to optimize collection efficiency.

Low-voltage DC current   Since the voltage across the cell is typically $V_{oc} \cong 0.7$ V, PV cells are usually connected in series to produce higher-voltage systems. A consequence of this is that shading one cell in a series can reduce the power output of the whole system. Furthermore, all solar cells produce direct current, which must be converted to alternating current to be fed into standard electric power distribution systems (see §38).

Substantial science and engineering efforts have gone into addressing these and other technical issues over the last several decades. Silicon solar cells continue to evolve, with costs steadily decreasing and efficiency gradually increasing. For a substantial jump upward in performance, or downward in cost, however, it is possible that a different technology needs to be developed.

25.9 Advanced Solar Cells

In most of this chapter we have focused on single-junction silicon-based solar cells. A wide variety of alternatives

have been, and continue to be, explored both for improving efficiency and reducing cost of large-scale solar systems. Figure 25.23 gives an overview of some of the technologies that have been explored over the last 50 years, with maximum realized research-cell efficiencies for the different technology types.

While the Shockley–Queisser bound (25.26) represents a limit on a certain class of photovoltaic solar cells, the only known absolute physical bound on conversion from solar radiation to electrical energy is the Carnot limit quoted earlier in this chapter, $\eta_{\text{Carnot}} \cong 0.95$. Present technologies are far from this bound, motivating intense research into novel designs that can operate at higher efficiency (and acceptable cost). Solar PV technologies fall into three categories: first generation, which includes the conventional crystalline silicon cells discussed in depth in this chapter; second generation, including *thin-film* PV cells that employ semiconductors with a *direct band gap*, which account for most of the rest of deployed PV capacity at the present time; and third generation, which includes a variety of approaches that are currently either limited to niche applications (e.g. *multi-junction* PVs, see below) or are still at an early stage in their development (e.g. *dye-sensitized cells*, *organic PVs*, *quantum dots*, and *graphene*). A full discussion of second and third generation PVs goes well beyond the scope of this book. In this section we give a brief description of some features of two of the main threads of development: *multi-junction* cells, which can improve photovoltaic efficiency above the 33% efficiency limit for single-junction PV cells (§25.9.1), and *thin films*, which may significantly reduce the cost of large-scale PV installations (§25.9.2).

25.9.1 Multi-junction Cells

The limit on collection efficiency described in §25.4 assumes that photons can only be excited across one band gap. The idea of a **multi-junction** photovoltaic cell is to evade this limitation by combining several materials with different band gaps. Assume, for example, that two semiconductors with band gaps $E_1 > E_2$ are available. By placing a layer of material 1 above a layer of material 2, a photovoltaic cell can gather more energy than either material alone could collect. Material 1 gathers *more energy* from some photons than could be collected by material 2, and material 2 collects energy from some photons that cannot excite an electron across the band gap of material 1 (see Figure 25.24). For example (see Example 25.2), a two-junction photovoltaic cell composed of materials with band gaps $E_1 = 1.7\,\text{eV}$ and $E_2 = 0.8\,\text{eV}$ can collect over 60% of incoming energy in a 6000 K blackbody spectrum – significantly above the collection efficiency bound for any single-junction cell (based on a material with a single band gap). Even higher collection efficiency can be realized by using three (or more) materials. While the maximum collection efficiency for 6000 K light by a single junction cell is 43.9% (optimized by a band gap $E_{\text{gap}} \cong 1.12\,\text{eV}$), this increases to 60.4% for a double-junction cell and 69.2% for a triple-junction cell (Problem 25.8). In recent

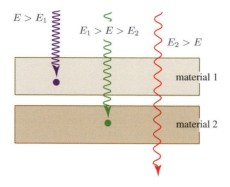

Figure 25.24 A two-junction solar cell containing two materials of energy gaps E_1, E_2. Photons with energy above E_1 excite electrons in the first material. Photons with energy below E_1 but above E_2 excite electrons in the second material. Photons with energy below E_2 do not excite electrons. Collected energy can exceed that of any single-junction PV cell.

years, multi-junction solar cells have been constructed with increasingly high overall conversion efficiencies. Triple-junction cells have been built with efficiencies exceeding 40%.

While multi-junction photovoltaic cells provide the highest conversion efficiencies yet realized, there are a number of challenges to large-scale development of power systems using this technology. Although band gaps can be tuned for certain semiconductors (see Box 25.2) the materials involved are often rare and/or expensive. The world's annual production of **germanium** (Ge), for example, was

---

Example 25.2   Two-junction PV Cell

A solar cell made from two materials with differing band gaps can collect more solar energy than a single-junction cell. Consider, for example, two materials, one with band gap $E_1 = 1.7\,\text{eV}$ and the other with band gap $E_2 = 0.8\,\text{eV}$. By arranging the cell so that incoming solar radiation passes first through material 1 and then through material 2 (see figure at right), collection efficiency can be optimized. Assuming a 6000 K blackbody spectrum, electrons in the first material are excited by photons with $E > E_1$, collecting a fraction of the total incoming energy

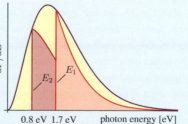

$$P_1 = CA \int_{1.7\,\text{eV}}^{\infty} \frac{E^2 \times 1.7\,\text{eV}}{e^{E/kT} - 1} \cong 0.359\, P_{\text{total}}.$$

The second material then collects energy from the remaining photons that have $E > E_2$,

$$P_2 = CA \int_{0.8\,\text{eV}}^{1.7\,\text{eV}} \frac{E^2 \times 0.8\,\text{eV}}{e^{E/kT} - 1} \cong 0.244\, P_{\text{total}},$$

so that the total energy collected in usable excited electron energy is more than 60% of the incoming solar radiation.

Box 25.2   Materials for Multi-junction PVs

One of the major challenges in constructing multi-junction PVs is to find compatible materials that have band gaps in the specific ranges necessary for optimal collection efficiency. To this end, scientists have been experimenting with a wide range of materials. One broad family of materials that are of interest in this regard are those formed from a combination of elements from columns III and V of the periodic table (see Figure D.1). In general, elements from columns III and V can combine into the **zincblende** crystal structure, which, like the diamond lattice, consists of two interlaced FCC lattices, but with one FCC lattice containing the group III element and the other containing the group V element (see Figure 25.1(b)). Many III–V crystals of this kind form good semiconductors, with band gaps that can be used effectively in multi-junction PVs. A simple example of a III–V semiconductor often used in multi-junction PV cells is **gallium arsenide** (GaAs). This material has a direct band gap of 1.42 eV, which is near the optimum band gap for a single-junction semiconductor, corresponding to a Shockley–Queisser bound around 33.5% (for 6000 K blackbody radiation). The best recorded single-junction cell efficiencies have been realized for GaAs cells – close to 30% efficiency with concentration factors $C > 100$.

An even broader class of materials include alloys that mix elements from one column. For example, **indium gallium phosphide** ($In_x Ga_{1-x}P$), also based on a zincblende crystal, mixes the group III elements **indium** and **gallium** with relative weights x, $1-x$ in one of the FCC lattices. By tuning the relative fraction of the group III elements in the alloy, the band gap can be varied over a range of values, with pure InP and GaP having direct and indirect band gaps of 1.34 and 2.26 eV at the extremes, respectively. One combination of materials that has been used for triple-junction PV cells with realized efficiencies above 40% uses $In_{0.5}Ga_{0.5}P$ for the highest band gap, $In_{0.015}Ga_{0.985}As$ for the middle band gap, and germanium (a diamond crystal structure group IV semiconductor like silicon, but with band gap 0.66 eV) for the lowest band gap. As with silicon, using concentrators increases the efficiency of multi-junction cells by increasing V_{oc} through eq. (25.18).

---

only ~118 tonnes in 2011, which severely limits the production of triple-junction cells using Ge as the lowest band gap element (Problem 25.17). While using concentrators can make multi-junction cells more cost effective, to date most uses of high-efficiency multi-junction PV cells have been in niche applications such as space exploration (Mars rovers) and military systems (for mobile devices) where light weight and high power density are of critical importance. In addition to materials supply limitations, there are also technical issues with multi-junction cells that complicate the engineering of these systems. In particular, combining different semiconductors into a single functioning photovoltaic cell often requires matching the lattice structures of the materials (*heterojunctions*); matching the electrical characteristics of the materials can also be often difficult. In some multi-junction cells, *tunnel junctions* are used in which quantum tunneling plays an important role in connecting the layers of the cell. These technical and engineering issues present a major challenge to large-scale production of cheap and efficient multi-junction photovoltaic cells.

25.9.2 Direct Band Gap Materials and Thin-Film Photovoltaics

Although silicon has a near optimum band gap and is an abundant material, two drawbacks mentioned under the heading of *material constraints* have motivated intense interest in alternative materials. First, the manufacture of silicon PV cells involves a complicated and expensive multistep process. Second, because of its indirect band gap a single-junction cell made of monocrystalline or polycrystalline silicon must be on the order of 100 μm thick to absorb a substantial fraction of incoming solar radiation. In contrast, a material with a *direct band gap* can absorb most solar radiation over a distance of order 1 μm. Though a direct band gap also leads to more rapid recombination, the relative rate of non-radiative recombination can be lower, and in direct band gap materials (such as GaAs) emitted photons can sometimes be reabsorbed in a mechanism known as **photon recycling** that increases the efficiency of the cell. In many cases careful engineering can lead to highly efficient **thin-film** solar cells requiring much less material than traditional silicon solar cells. Furthermore, it has proved possible to deposit thin films using familiar technologies developed in other branches of the semiconductor industry. As a result, some reasonably efficient thin-film PVs are already cost competitive (on a cost per watt basis) with silicon PVs.

The *roll-to-roll* production systems that are used in other parts of the technology industry could, in principle, be used to produce large quantities of inexpensive thin-film PV modules, which could radically alter the large-scale deployment of PV systems. Cheap, lightweight, flexible modules could, for example, be adapted for use in gigawatt-scale desert power systems or building-integrated installations where flexible PV sheets could be wrapped around the exposed surfaces of urban buildings.

Unfortunately, all of the thin-film PV materials that have been developed to a commercial level so far make use of chemical elements that are far less abundant than silicon (see below). The quest for an *earth-abundant*, thin-film PV is one of the most active areas of research in the field of photovoltaic power. This field is developing rapidly, and mostly beyond the scope of this book; we provide a few comments on the main materials that have been used in large-scale commercial photovoltaics of this type.

Cadmium Telluride (CdTe)

Like gallium arsenide (see Box 25.2), **cadmium telluride** forms a zincblende crystalline semiconductor, and has a direct band gap $E_{gap} \cong 1.44$ eV. **Cadmium** is toxic, but relatively plentiful. **Tellurium**, on the other hand, is one of the least abundant solid elements in Earth's crust, rarer than gold. Nevertheless, tellurium is relatively inexpensive because it is produced (in small quantities) as a by-product of copper refining. Research-cell efficiencies for CdTe had reached $21.0 \pm 0.4\%$ by the end of 2016 [156], with commercial efficiencies somewhat lower. Unlike crystalline silicon, CdTe has been deposited and fabricated into an integrated PV cell in a roll-to-roll manufacturing processes. In 2010 CdTe broke the $1/W barrier for PV cell manufacturing cost, and the cost has continued to decline in subsequent years. CdTe photovoltaic cells have entered large-scale production: in 2015 approximately 4% of the global PV market and 60% of thin-film PVs were CdTe [155]. Concerns about the cost and supply limitations on tellurium cloud the future of this otherwise promising technology.

Copper Indium/Gallium (Di)selenide (CIGS)

Another class of thin-film photovoltaics that have been extensively developed in recent years use copper indium/gallium (di)selenide ($CuIn_x Ga_{1-x} Se_2$), also known as **CIGS**. CIGS has a crystal structure known as the **chalcopyrite** structure, closely related to the zincblende crystal, but with copper and indium/gallium appearing in alternate cells so that the overall size of a periodic cell is twice as large. By adjusting the fraction of indium versus gallium, the band gap can be tuned from 1.0 eV ($x = 1$) to 1.7 eV ($x = 0$). As of the end of 2015, the record cell efficiency for CIGS was $21.0 \pm 0.6\%$ [156], almost identical to CdTe. CIGS accounted for about 2% of the global PV production and about 25% of thin-film PVs in 2013 [155]. CIGS presents more manufacturing challenges than CdTe. In particular, it is difficult to control the proportions of the various elements during the deposition process.

Selenium, indium, and gallium, while not as scarce as tellurium, are presently produced in small quantities as by-products of copper, zinc, and aluminum refining respectively. Large-scale deployment of CIGS-based PVs would require significant increases in production of these scarce elements.

Hydrogenated Amorphous Silicon (a-Si:H)

Another type of thin-film photovoltaic is formed from (hydrogenated) **amorphous silicon** (a-Si:H). Unlike the other materials we have discussed, amorphous silicon is non-crystalline, without long-range order. A significant fraction of the silicon atoms are not bonded to four neighbors, and the resulting *dangling bonds* are saturated with hydrogen atoms. The band gap of a-Si:H is not as sharply defined as for crystalline silicon, due to additional states within the gap arising from the irregular structure. Amorphous silicon used for PV cells typically has a band gap of roughly 1.5–1.8 eV, a range that is not well-matched to the solar spectrum (see Figure 25.22). On the other hand, the absorption cross section of amorphous silicon is high, so that a film of a-Si:H only 300 nm thick can absorb ~90% of the photons above its band gap. Nevertheless, record efficiencies reported in 2015 are still low: $10.2 \pm 0.3\%$ for a single junction a-Si:H cell, and $14.0 \pm 0.4\%$ for an a-Si:H/Si/Si triple junction thin film [156]. The low cost and lower amount of

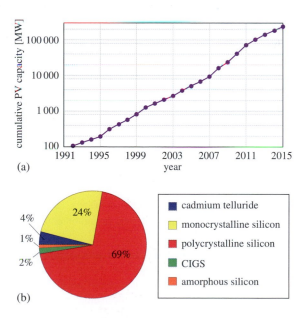

Figure 25.25 (a) Installed capacity of PV systems from 1991–2015. PV capacity grew at a compound annual growth rate of roughly 46% from 1991–2015. (b) PV technology share in 2015 [155].

the raw material required compared with crystalline silicon suggest that a-Si:H, with even modest efficiencies of 6–10%, could eventually become economically competitive with crystalline silicon and the other thin-film technologies. Technological challenges include susceptibility to light-induced degradation and difficulty in engineering the contacts on amorphous silicon sheets [133].

25.10 Global Use of Photovoltaics

Over the years from 2000 to 2015, global production and installed capacity of photovoltaic systems grew exponentially, with a compound annual growth rate (CAGR) of ~46% (see Figure 25.25(a)). Global installed capacity exceeded 1 GWe in 1999, and exceeded 300 GWe by the end of 2016. Actual electric power generated was only a fraction of this. According to [133], "Current annual average PV capacity factors range from 10% in Germany to ~20% in the US." Given the increasing deployment of PV in countries with relatively high insolation, the long-term global average capacity factor for PV is expected to reach 15%. Estimating an average capacity factor of 15%, 300 GWe of installed PV capacity corresponds to 45 GWe, which can be compared with roughly 39 GWth of solar thermal energy *production* in 2014 (Box 24.1).

As described in §22, the available solar energy resource is tremendous, so *energy* resource limitations will likely never constrain the growth of solar power. At the CAGR seen from 2000–2015, by 2030 global installed PV capacity would be enough to supply all of the world's electricity demand. The main obstacles to a continued exponential growth of this type are largely associated with the low power density of the solar energy resource. Covering thousands of square kilometers of land with any kind of solar cell and associated infrastructure is a vast and costly engineering endeavor with substantial environmental and social implications. Further advances in manufacturing technology are needed for solar systems to compete economically with fossil fuel based power systems at current cost levels without subsidies for solar or carbon emissions limitations or taxes.

To summarize, photovoltaic energy has many unique features – PV systems are carbon-free, noise-free systems with no moving parts and low maintenance requirements. Small modular PV units can be deployed almost anywhere, particularly in remote locations, and larger-scale systems can be used, possibly combined with solar concentrators, for utility-scale electricity production. Despite recent steep decreases in cost, the large-scale manufacture of inexpensive, efficient, and robust systems remains a challenge. Storage is also a greater challenge for utility-scale electricity production than for solar thermal electrical plants, where the intermediate thermal energy can be stored more easily than electrical energy from photovoltaics. Storage issues are described further in §37, some aspects of the intermittency of solar power are discussed in §38, and the larger-scale issues involved with TW-scale photovoltaic energy production are discussed further in §36.

Discussion/Investigation Questions

25.1 Discuss qualitatively the origins of the three terms in the Shockley–Queisser bound on PV efficiency and, in particular, explain why they are independent.

25.2 Why does the efficiency of a silicon PV drop as it heats up? [Hint: See eq. (25.25).]

25.3 Suppose a PV with gap E_{gap} were illuminated by a laser that produces monochromatic light with frequency ν_0. How would the collection efficiency of the PV vary with ν_0?

25.4 **Perovskites** are a promising new class of materials for thin-film PV. Perovskites include a wide range of compounds, such as $CH_3NH_3PbI_3$, characterized by a crystal structure first observed in the mineral *perovskite* ($CaTiO_3$). Perovskite solar cells have reached ~20% efficiency under laboratory conditions. They are relatively easy to manufacture and are composed of Earth-abundant elements. Investigate the advantages and the technological challenges of this novel type of photovoltaic material.

Problems

25.1 Show that for each *unit cell* in the crystal shown in Figure 25.1 there are eight atoms, all identical in the case of the diamond and four of each kind in the zincblende case. [Hint: Each atom at a corner is shared among eight cells, etc.]

25.2 Following the logic of Figure 25.6, sketch (no calculation necessary) the wavefunctions of the energy basis states at the *top of the second energy band* and the *bottom of the third energy band* in a periodic square well potential with high barriers. [Hint: First establish that these wavefunctions are both *periodic in $x \to x + \Delta$*. Also remember that inside the potential wells the wavefunctions should resemble the second and third excited states of a square well potential respectively.]

25.3 Estimate the maximum possible collection efficiency $\eta_{collection}^{max}$ for a germanium solar cell ($E_{gap} \cong 0.66\,eV$).

25.4 The direct band gap in silicon is 3.4 eV. What is the maximum possible collection efficiency (for incident thermal radiation at 6000 K) for excitations over this gap?

25.5 Given four materials, with band gaps 0.2 eV, 0.7 eV, 1.5 eV, and 3 eV, which two are likely to make the best double-junction solar cell? Explain your answer.

25.6 [C] Give an estimate of the collection efficiency for the best double-junction cell formed from the materials in the previous problem.

25.7 [C] Estimate the collection efficiency of a triple-junction solar cell where the first layer is an indium gallium phosphide alloy $In_{0.53}Ga_{0.47}P$, with band gap $E_{IGP} = 1.87\,eV$, the second layer is gallium arsenide, and the third layer is germanium.

25.8 [C] Numerically compute the maximum collection efficiency for a single-junction PV cell under 6000 K blackbody conditions as a function of the band gap. Estimate the maximum possible collection efficiency and the associated band gap. Repeat for double-junction and triple-junction cells, confirming the collection efficiencies given in the text and computing the band gaps needed in each case.

25.9 [C] Download the AM1.5 solar spectrum from [135] and numerically compute the maximum collection efficiency expected for a silicon PV cell through a computation analogous to eq. (25.11).

25.10 [C] Use the photodiode equation (25.17) to compute I_{photo}/I_0 for a silicon solar cell with $V_{oc} = 0.7\,V$. Write the power as a function of voltage and compute the *fill factor*, the maximum power IV attainable as a fraction of $I_{photo}V_{oc}$. Repeat the calculation for cells with $V_{oc} = 0.5$ and $0.86\,V$ and compute the maximum efficiency of a single-junction silicon photocell in these cases.

25.11 Consider a single-junction solar cell with $V_{oc} = 0.7\,V$ under *one sun* illumination conditions. Estimate the increase in V_{oc} and the increase in overall efficiency if the incoming light is concentrated by a factor of 400.

25.12 Compute the Shockley–Queisser bound on V_{oc} for a silicon cell with band gap 1.1 eV at temperature $T = 300\,K$. Model the Sun as a black body at 6000 K.

25.13 Given a silicon photovoltaic solar cell with open-circuit voltage $V_{oc} = 0.7\,V$ that operates at an overall 20% efficiency under 1000 W/m² of normal illumination, (a) compute the current density from the collecting surface. (b) How many cells must be connected in series to make a 12 V voltage? (c) What surface area is needed to produce one ampere of current across 12 V?

25.14 Real photodiodes do not precisely obey the ideal photodiode equation (25.17). A more realistic model is the *non-ideal* photodiode equation

$$ I = I_{photo} - I_0 \left[e^{eV/mk_B T} - 1 \right], $$

where $m > 1$ is the *ideality factor*. Consider a non-ideal photodiode with $m = 2$, $V_{oc} = 0.7\,V$. Compute I_{photo}/I_0 and the fill factor. Compare the efficiency with an ideal photodiode with $m = 1$ and the same open-circuit voltage.

25.15 An actual solar cell has internal resistances that affect the behavior of the circuit. These can be modeled as a single *shunt resistance* R_p in parallel with the diode, and another resistance R_s in series with the external load. Derive the modified ideal diode equation in the presence of the resistances.

25.16 Assuming a silicon PV array attains 20% efficiency over a lifetime of 25 years of use, where it is exposed to an average of 250 W/m², what is the total energy output of 1 m² of PV cells (in joules). What quantity of coal would it replace (at 33% net efficiency)?

25.17 The thickness of the germanium (density 5.32 gm/cm³) layer in a triple-junction PV is typically greater than 100 microns. If the overall cell efficiency is 40%, how many watts of PV capacity (nominal insolation of 1000 W/m²) could be manufactured from the world's annual germanium production (165 t in 2014)?

25.18 The absorption coefficient of silicon has a strong dependence on photon energy, as shown in Figure 25.16. For simplicity, consider an idealized material, *material S*, similar to crystalline silicon, with an absorption coefficient of $\kappa = 7 \times 10^3\,m^{-1}$ for light at any wavelength. What fraction of (normally incident) light will be absorbed by a wafer of material S that is 200 μm thick? If we could fabricate a layer of material S just 1 μm thick, what fraction of incident radiation would it absorb? The absorption coefficient of amorphous silicon is significantly higher than that of crystalline silicon. If we had another material, *material A*, with an absorption coefficient $\kappa = 1.5 \times 10^6\,m^{-1}$ (again independent of wavelength), what fraction would be absorbed by a 1 μm layer?

25.19 [T] Derive the Fermi–Dirac distribution (25.8). Start by considering a single electron state of energy E that is either occupied ($n = 1$) or not occupied ($n = 0$) by an electron. Now, consider coupling this two-state system to a thermal reservoir at temperature T so that not only energy but also particles can move between the smaller system and the reservoir. The entropy of the reservoir $S(U, N)$ now depends on the total energy and total number of particles in the reservoir. Define $E_F = -T(\partial S/\partial N)|_U$. Use an argument similar to that used to derive the Boltzmann distribution in §8 to derive the Fermi–Dirac distribution (25.8).

25.20 [T] Compute the momentum of an electron with kinetic energy 2 eV and compare to the momenta of a photon with the same energy. Explain why such a photon cannot excite an electron in silicon without involving a phonon.

25.21 [T] Show that the right-hand side of (25.25) cannot exceed E_{gap} and goes to this value only in the limit $T \to 0$.

Biological Energy

Photosynthesis converts solar energy into chemical energy that is stored in and used by biological systems. This is the principal mechanism that has powered life on Earth for billions of years. It remains the source of almost all energy used by humanity. For thousands of years human society was powered directly by biological energy systems: plants and animals supplied food energy for humans and domesticated animals, which in turn were the primary source of mechanical work, and wood and other *biomass* materials provided fuel for warmth, light, and cooking. The fossil fuels that now provide most of our energy also originated through photosynthesis in the more distant past (§33). Direct use of biological energy sources remains a significant component of human energy use. In addition to continued uses of biological energy for food, heating, and cooking, biomass derived from a variety of sources is used for power generation. Plants such as sugarcane and maize (corn in US terminology) are converted into ethanol and other *biofuels*. And ongoing research seeks to incorporate biological processes into efficient industrial-scale systems for capturing and storing solar energy for human use.

In this chapter we survey some of the primary biological energy systems that currently, or may in the not-too-distant future, provide substantial energy for human use. We also mention some more speculative, but transformative applications of biological energy systems that may be developed in the more distant future. Biological systems are extremely complex, and a detailed analysis of biological energy production mechanisms would go well beyond the scope of this text. Because biologically based energy sources play such an important role in human energy systems, however, we have included in this chapter some basic elements of the subject, focusing on the underlying physical principles and connections with other energy systems described elsewhere in the book, and emphasizing the scope of the resources and processes involved. We focus

Reader's Guide
This chapter surveys the biological conversion of solar energy to chemical energy through photosynthesis and the human use of biological energy as food and through combustion, in the form of biomass, biogas, and biofuels. Food production, which occupys roughly 1/3 of Earth's total landmass, represents the most basic use of biological energy. Biomass – organic material in general – can be burned directly or used to generate biogas. We describe ethanol production from sugarcane and corn (maize), and note the future potential of ethanol derived from cellulose. Vegetable and animal oils can be chemically converted to biodiesel that resembles diesel fuel for which it can be substituted.

Prerequisites: §9 (Energy in matter), §22 (Solar energy), §25 (Photovoltaics).

here primarily on aspects of bioenergy that are purely biological in nature, such as the production of biofuel through the action of enzymes or yeast. Thermochemical methods for conversion of biomass to gaseous and liquid fuels closely parallel analogous methods for coal *gasification* and *liquefaction*, which are described in the fossil fuel context in §33.

Biological energy sources have many advantages that make them an attractive option for renewable energy systems. In terms of human energy use, plants that operate through photosynthesis are basically mechanisms for capturing solar energy. Thus, in some sense biofuels can be thought of as an alternative to solar thermal or photovoltaic energy systems that incorporates both solar energy capture and storage. While, like fossil fuels, biomass and biofuels contain carbon that is released to the atmosphere when burned, this carbon has recently been captured from Earth's atmosphere itself. Thus, the growth and use of biological fuels is essentially carbon-neutral, to the extent that energy and chemical fertilizers derived from fossil fuels are not used as inputs. Biological systems are traditionally self-reproducing and grow autonomously, which reduces the need for external inputs of energy and can dramatically reduce initial capital investment costs compared to comparable-scale solar power systems. Biological systems are also biodegradable, and – in contrast to some other energy systems – generally contain minimal quantities of heavy elements and other toxic materials. Finally, biological systems operate at the molecular level, which in principle gives them an enormous potential for versatility and efficiency.

At the same time, there are drawbacks to biological energy systems that limit their potential contribution to worldwide energy use in the near future. Considered as a type of solar energy, biomass and biofuels are currently much less efficient than solar thermal or PV systems, so that even greater areas of land are needed for substantial power generation. Many biofuel crops require high-quality arable land and therefore compete with crops used for human or animal food. Modern intensive agricultural methods that maximize biofuel production also require substantial energy inputs, which reduces the net efficiency of biofuels. If the energy input comes from fossil fuels, this compromises the effectiveness of biofuels in reducing carbon emissions; intensive agricultural methods also have other deleterious environmental side-effects.

Note that fossil fuels, which provide most of current human energy needs, are also based in biological energy sources. In some sense, there is a continuum of resources that extend from recently harvested biofuels through stored logged wood, peat, etc. to brown coal, depending on the time between the photosynthetic capture of energy and the time the energy is used. We focus in this chapter on biologically sourced energy that comes directly from organic material, leaving sources that involve processing of the organic material through natural systems over relatively long time scales to §33 (Fossil fuels).

At this time, biological energy sources supply more energy to humankind than all other renewable energy resources taken together. Roughly 45 EJ of energy are produced yearly in food crops, and the biomass energy used for heating, cooking, and power worldwide is roughly 56 EJ/y [157], or 10% of worldwide energy use. Even in some industrialized countries, biomass plays a substantial role in the overall energy portfolio; in Sweden, for example, more than 34% of primary energy used in 2014 came from biomass. While at the present time liquid biofuels play a relatively small role in worldwide energy use ($< 1\%$), advances in biotechnology may substantially increase the use of biologically based energy for transport and power production in coming years.

We begin this chapter with a brief introduction to energy capture by photosynthesis in §26.1. §26.2 gives an overview of food energy production, and §26.3 describes the direct use of biomass for heating and power. In §26.4, we describe some of the main approaches currently in use for production and use of biofuels, and in §26.5 we characterize the available biological energy resource and briefly comment on potential future developments in biotechnology related to energy systems.

We only touch superficially here on most aspects of biological energy sources. The biological processes underlying photosynthesis are described in most basic biology texts; a detailed treatment is given in [158]. Perspectives on food production and the history of agriculture are given from a non-scientific point of view in [159, 160]. A good overview and introduction to biofuels, accessible to the non-biologist, is given in the text by Mousdale [161]. Further presentations of biofuels and bioenergy are given in [162, 163].

26.1 Energy and Photosynthesis

Photosynthesis is the biochemical mechanism by which plants capture and store solar energy.[1] The **chlorophyll** molecule plays a key role in the photosynthesis process in most plants (Figure 26.1).[2] This molecule, which comes in several versions, has a structure that suggests a small piece of a regular two-dimensional lattice, with a frame composed of carbon atoms, some nitrogen atoms, and a magnesium atom at the core. Chlorophyll is a *pigment molecule* that has evolved to absorb light optimally in sections of the visible spectrum. One version, known as

---

[1] A class of organisms known as *chemotrophs* do not use photosynthesis but obtain energy from chemical reactions with materials in their environment. For example, in deep oceans near new lava beds or hydrothermal vents, chemoautotrophs derive energy by oxidizing iron. While such organisms might in principle provide an alternative biological pathway for capturing geothermal energy, we do not consider such systems here and focus exclusively on biological systems that get their energy from photosynthesis.

[2] This diagram and other similar diagrams in this chapter are generalizations of the diagrams introduced in §11.2.1 to describe hydrocarbons. Here a carbon atom occupies every vertex. Its four bonds are attached to other carbon atoms (other vertices), to hydrogen atoms (not shown), to other atoms (shown), or to simple radicals such as CH_3. Double lines represent double bonds.

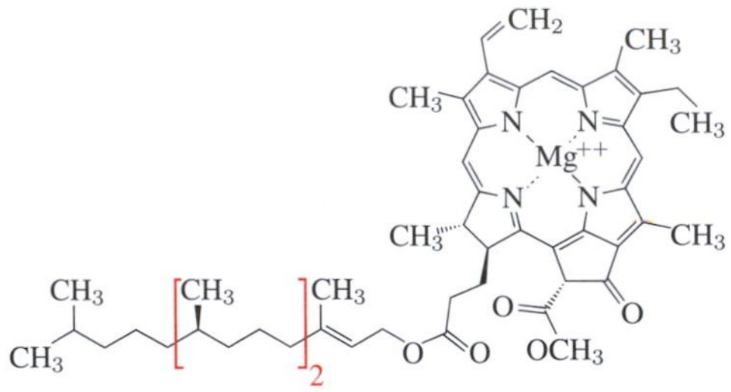

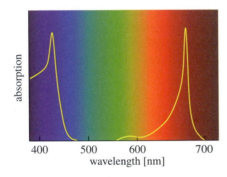

Figure 26.1 The pigment molecule *chlorophyll a*, which captures incoming photon energy in electronic excitations. This molecule operates in photosystem II of green plants; similar molecules play a similar role in other photosystems. The bracketed section is repeated twice. A carbon atom sits at each vertex in this diagram unless otherwise labeled.

Figure 26.2 The absorption spectrum of a *chlorophyll a* molecule is peaked around 430 nm and 680 nm. (Credit: Daniele Pugliesi reproduced under CC-BY-SA 3.0 license via Wikimedia Commons)

chlorophyll a, preferentially absorbs light in the wavelength ranges 650–700 nm and 420–460 nm (Figure 26.2). At a basic physical level, the part of the photosynthesis process in which light energy is captured closely parallels the energy collection process in a photovoltaic cell (§25). Absorbed photons excite electrons in a fashion similar to the excitation of electrons across the band gap of silicon in a photovoltaic solar cell. The excited electrons provide the energy that powers photosynthesis.

Unlike the PV solar cell, however, biological systems store the electrochemical energy gathered from photosynthesis in organic compounds composed primarily of *carbohydrates* such as sugars and starches (Box 26.1). The storage process is implemented through a rather complex series of biochemical reactions. Slightly different photosynthesis mechanisms act in different types of organisms such as algae, bacteria, and green plants. We focus here on the mechanism in green plants. Our description is

only schematic, since even the simplest steps of photosynthesis in a real organism involve a complex variety of molecules and mechanisms, an explanation of which requires familiarity with biochemistry beyond the scope of this book.

26.1.1 Mechanisms of Photosynthesis

The photosynthesis process can be loosely separated into **light reactions** that involve the capture of photon energy and storage in biological intermediates, and **dark reactions** that do not require sunlight and store the energy in more stable molecules for eventual use by the organism.

The light reactions in a green plant occur in two stages, implemented by two separate systems, known as **photosystem II** and **photosystem I**. While photosystem II logically precedes photosystem I, the systems are named in order of their discovery. In these photosystems, electrons are excited through photon capture by chlorophyll and related molecules in a large antenna-like array known as the *light-harvesting complex*. The light-harvesting complex includes a variety of antenna pigments in addition to *chlorophyll a* that have somewhat different absorption spectra. The excited electrons are transferred to a chlorophyll-containing complex known as the *reaction center* of each photosystem. Recent research suggests that this transfer process utilizes *quantum coherence* in the propagation of the electron wavefunction to optimize efficiency [164, 165].

Photosystems II and I are the light-gathering component that result in the production of energy-transporting molecules *adenosine triphosphate* (**ATP**) and *nicotinamide adenine dinucleotide phosphate* (**NADPH**), together with the release of free oxygen O_2. The excited electrons from photosystem II are channelled through an *electron transport chain* and a series of intermediate reactions to photosystem I, where a further set of chlorophyll molecules captures additional photons and once again produces energetic electrons. These electrons energize further reactions that result in the formation of NADPH, a powerful reducing agent.

The electrons removed from chlorophyll molecules in photosystem II are replaced in the *oxygen-evolving complex*, where a reactive structure including several manganese ions catalyzes the reaction

$$2H_2O \rightarrow O_2 + 4H^+ + 4e^- . \qquad (26.1)$$

The electrons neutralize the chlorophyll molecules of photosystem II, and the H^+ ions drive a proton gradient that is used by an enzyme called *ATP synthase* to produce ATP. The reaction eq. (26.1), known as **water splitting**, is responsible for the production of the oxygen in Earth's

Box 26.1   Carbohydrates

A wide range of different types of molecules are involved in biological processes. From an energetic point of view, however, the most important organic molecules are *carbohydrates*. **Carbohydrates** are a relatively simple class of molecules formed from carbon, oxygen, and hydrogen that store and transport energy in biological systems.

Most carbohydrates have the chemical formula $C_n(H_2O)_m$, and can be thought of as bound structures of carbon and water, hence "carbo-hydrate." The simplest carbohydrates, known as **monosaccharides**, or *simple sugars*, consist of four or five carbon atoms, typically linked in a ring containing a single oxygen atom. An additional carbon atom in a CH_2OH unit is linked to one carbon on the ring and hydroxyl (OH) units are attached to the others (see Figure 26.3(a)–(c)). Monosaccharides, as well as more complex *sugars*, are very soluble in water and sweet to the taste. Other arrangements of the atoms (*isomers*), including *open-chain* forms, exist as well. Common six-carbon monosaccharides include *glucose* and *fructose*. Glucose and other sugars are formed by plants from carbon in the atmosphere and water, using energy captured in the process of photosynthesis. There are also simple three-carbon sugars, some of which are mentioned in Box 26.2.

Disaccharides, or *double sugars*, are carbohydrates formed from two monosaccharide units, or **monomers**, joined by a shared oxygen atom through a connection known as a *glycosidic bond*; for example, *sucrose* (Figure 26.3(d)) is a disaccharide produced in many plants for storage and transport of energy.

Complex carbohydrates are polymers formed from larger numbers of monosaccharide monomers joined by glycosidic bonds. Polymers formed from glucose, known as *glucans*, are especially important in biological processes. These include *starches* that are used for long-term energy storage. Starches are further classified as *amylose* if they involve few, if any, side chains, or *amylopectin* if they are highly branched (Figure 26.4). Amyloses may involve hundreds to thousands of glucose monomers, while the number of monomers in amylopectins may be as large as 10^6.

More complex carbohydrates include *cellulose, hemicellulose,* and *lignin*, which together comprise *lignocellulosic* material used by plants for structure (Figure 26.7). While energy can be released directly from all carbohydrates by combustion, a variety of biological processes can also be used to alter carbohydrates to other molecules with useful energy content, for example by converting carbohydrates to liquid *biofuels* such as ethanol that are more convenient for use as motor vehicle fuel.

---

atmosphere. As far as photosynthesis is concerned, however, O_2 is a waste product.

The dark reactions are a set of further biochemical reactions (see Box 26.2) that complete the photosynthesis process by using the energy stored in the light reactions to fix atmospheric carbon from CO_2 into sugars, such as **glucose**, $C_6H_{12}O_6$ (Figure 26.3). Plants store and transport energy using **sucrose** ($C_{12}H_{22}O_{11}$). Sucrose is a *disaccharide* composed of glucose and fructose bound by a shared oxygen atom, a connection known as a **glycosidic bond** (Figure 26.3). In most plants some of the sugars are converted into **starches**, which are polymers built from long chains of glucose units linked by glycosidic bonds. **Amylose** and **amylopectin** (Figure 26.4) are respectively linear and branched glucose polymers that arise as components of generic starches.

Sugars and starches are examples of *carbohydrates*; some properties of carbohydrates are summarized in Box 26.1.

The net result of the photosynthesis process is the conversion of water, atmospheric carbon dioxide, and photons from sunlight into oxygen and stable carbohydrate

Photosynthesis

Photosynthesis is the process by which plants capture and store incoming solar energy in simple organic molecules (sugars). Photosynthesis involves a complicated set of biochemical reactions with the net conversion

$$n(8\gamma + CO_2 + H_2O) \rightarrow C_nH_{2n}O_n + nO_2.$$

The sugars produced from photosynthesis can be combined to form starches and cellulose for energy storage and plant structure formation. The efficiency with which solar energy is converted into biomass for most plants is a fraction of 1%.

molecules that are used to store energy for the photosynthesizing organism

$$n(8\gamma + CO_2 + H_2O) \rightarrow C_nH_{2n}O_n + nO_2. \qquad (26.2)$$

Box 26.2   C$_3$ and C$_4$ Photosynthesis

A special class of plants known as C$_4$ plants are particularly efficient at photosynthesis in Earth's current atmosphere in certain temperature ranges. These plants include sugarcane, maize, and sorghum. Most plants used to produce liquid biofuels such as ethanol are C$_4$ plants. Most plants on Earth, however, use a simpler C$_3$ photosynthesis process. C$_4$ plants have evolved an additional pathway that increases their photosynthesis efficiency at lower levels of atmospheric CO$_2$.

The sequence of reactions in which carbon dioxide is fixed by C$_3$ plants is known as the **Calvin–Benson–Bassham** (CBB) cycle. This cycle, a central component of the *dark reactions* of photosynthesis, includes 13 biochemical reactions catalyzed by 11 distinct molecules. The net result is the conversion of three CO$_2$ molecules into a three-carbon sugar phosphate *triose-P* using energy stored in the ATP and NADPH from the photosynthesis light reactions. Triose-P is then used to produce six-carbon sugars and sugar phosphates that store energy in stable forms such as glucose and sucrose. An enzyme known by the acronym *RuBisCO* catalyzes a critical step, the *carboxylation* of *ribulose-1,5-P$_2$* (*RuBP*), in which CO$_2$ is captured from the atmosphere.

RuBisCo, however, can also catalyze a distinct *oxygenation* reaction in which a molecule of RuBP combines with a molecule of O$_2$. This leads to a reaction sequence that also eventually produces triose-P, but also involves the release of CO$_2$ along the way. The loss of carbon through this process, known as **photorespiration**, decreases the efficiency of the photosynthesis process.

The relative rate at which RuBisCO catalyzes carboxylation of RuBP versus oxygenation depends upon the relative concentrations of oxygen and carbon dioxide present. In C$_4$ plants, carbon is initially fixed in four-carbon acid molecules through a separate mechanism, and shuttled to a region known as the *bundle-sheath*, where the carbon is released forming a region of high CO$_2$ concentration where the RuBisCO-catalyzed step of the CBB cycle occurs. The high CO$_2$ concentration in the bundle-sheath suppresses photorespiration. C$_4$ photosynthesis uses more energy than C$_3$ photosynthesis (20 versus 18 ATP molecules) to synthesize one molecule of glucose, but this is balanced in some circumstances by the lower rate of photorespiration in C$_4$ photosynthesis. The lower rate of photorespiration in C$_4$ plants also substantially reduces the rate at which they lose water in the photosynthesis process, giving them an advantage in arid environments.

The rate of photorespiration in the standard CBB cycle increases when the atmospheric concentration of CO$_2$ decreases and also increases with increased temperature. Thus C$_4$ plants have a greater advantage at lower CO$_2$ levels and at higher temperatures. It is believed (§35) that the level of CO$_2$ in Earth's atmosphere was much higher than at present some 50 million years ago, and that atmospheric CO$_2$ levels have decreased fairly steadily since then (until the recent increase due to human activity). C$_4$ plants are believed to have evolved relatively recently; fossil evidence suggests a dramatic increase in C$_4$ plants between 7 and 5 million years ago. The C$_4$ process may have evolved to compensate for decreasing CO$_2$ levels in the atmosphere. It has been suggested that increasing CO$_2$ levels in the next century may lead to decreased photorespiration and increased photosynthetic efficiency for C$_3$ plants. On the other hand, it is believed that photorespiration may assist in nitrate assimilation, another important energy input to plant life (§26.2.2); it has been suggested that this explains the continued dominance of C$_3$ plants in the current (relatively) low-CO$_2$ environment, as well as a decrease in plant protein concentration as atmospheric CO$_2$ levels rise.

26.1.2 Photosynthesis Efficiency and Extent of the Resource

For most plants, the primary purpose of photosynthesis is to capture energy for the plant's own direct use in growth, survival, and reproduction. In some special cases, such as plants that produce berries, energy is concentrated into parts of the plant that have evolved to be tasty to animals, who assist the plant in reproducing by eating the berries and transporting the seeds to distant locations. For the most part, however, the energy captured by photosynthesis is used by the plant itself. The fraction of energy that may

be useful to humans can be represented by the potential combustion energy that is contained in the added material, or *biomass*, as the plant grows.

For the purpose of understanding human use of bioenergy, it is useful to consider the solar conversion efficiency of photosynthesis to be the fraction of incident light energy that is captured in the energy of combustion of increased biomass. To begin with, only 40% or so of the energy in sunlight lies in the photosynthetically active frequency ranges for chlorophyll. Even when photons are captured, energy is lost in transporting the electrons to the reaction

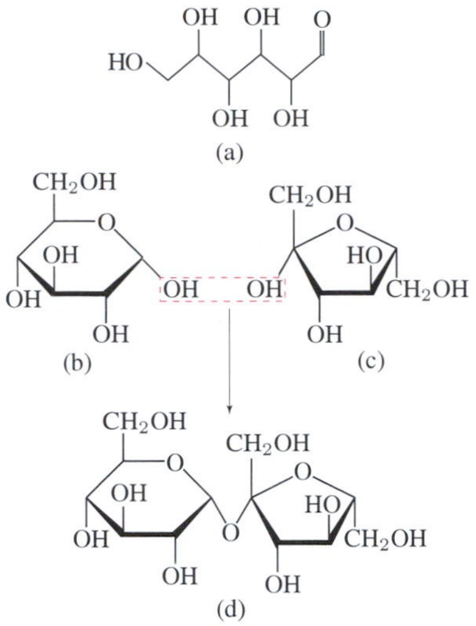

Figure 26.3 Simple sugars used by plants to store energy gathered through photosynthesis: (a) the open-chain isomer of the monosaccharide *glucose*; (b) one of four closed-chain isomers of glucose; (c) *fructose*, another monosaccharide; (d) the disaccharide *sucrose*, formed from glucose and fructose when two OH groups react to form a glycosidic bond and a water molecule.

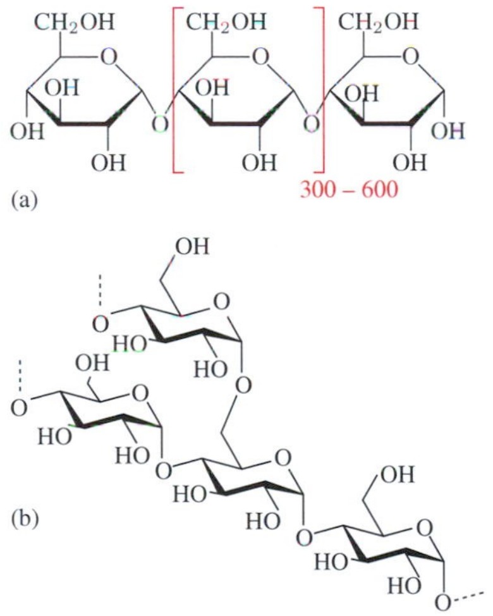

Figure 26.4 The starches amylose (a) and amylopectin (b) are (primarily) straight and branched polymers respectively, composed of multiple glucose units connected by glycosidic bonds.

center and in each stage of chemical reaction, so that the fraction of energy stored in a single CH_2O sugar component from eight incident photons at 680 nm is roughly 30% (Problem 26.1). This gives an upper theoretical limit of 12% on photosynthetic efficiency. Many other factors,

however, drastically reduce this number further in most real plants. Some light is reflected, and not all photons impact on chlorophyll molecules or excite photons. The ratio of reaction centers to antenna chlorophyll molecules is usually relatively small, since reaction centers are energetically costly for the plant to construct. Thus, plants are generally optimized so that at high intensities of sunlight there are not enough reaction centers to process excited electrons from all incoming photons. Under realistic conditions for standard photosynthesis, the maximum efficiency realizable is closer to 5% [166]. Typical efficiencies are much lower. Furthermore, much of the energy transformed to sugars is used quickly for plant metabolism and other purposes. The energy that is stored in structural material within the plant, and which could be extracted by combustion or other means for human utilization, is generally substantially less than 1% of the incident sunlight, and typically closer to 0.25% or less (Example 26.1). This efficiency can be compared, for example, to the efficiency of photovoltaic devices (§25) (see e.g. [167]), or over large areas to the *gross conversion efficiency* of solar thermal electric plants (§24.4). When comparing electric power produced from combustion of biological fuels to solar PV or solar thermal electric power plants, however, further losses from thermal to electric conversion need to be included for a meaningful comparison.

To get a sense of the scope of the energy resource produced by photosynthesis, we can consider the global rate of biomass production through terrestrial photosynthetic activity. This is estimated at 1660 EJ/y [168]. Thus the rate of energy use by humankind, ~550 EJ/y, is a substantial fraction of the total rate of energy captured and stored through photosynthesis by all terrestrial plants.

26.2 Food Energy

26.2.1 Global Food Production

The most basic human use of bioenergy is as food. Roughly 45 EJ of energy per year are produced in the form of food for human consumption (see Problem 26.2), of which roughly one third goes to waste [169]. The production, processing, transport, storage, delivery, and sale of human foods together represent a substantial fraction of most economies; for example, roughly one third of all jobs worldwide are in some form of agriculture.

Roughly one third of the world's total land area, or around 50 M km^2, is used for agriculture. Of this, about 15 M km^2 is used for crops, while 35 M km^2 function as pastures and grassland for grazing. Aside from some use for raw materials and textiles (e.g. 2.4% of cropland is used

Example 26.1   Growing Corn

As a simple example of the rate of photosynthetic energy capture and storage, consider **maize**, one of the most widely farmed crops in America. In the US, this crop is simply referred to as **corn**; we use the two words synonymously in this book. With intensive farming methods it is possible to produce maize with a harvested yield of roughly 10 t/ha y. At an energy density of roughly 15 MJ/kg, this gives an energy capture rate of ∼150 GJ/ha y. Assuming an average insolation of 200 W/m^2 gives a solar input of ∼60 TJ/ha y, for an estimated efficiency of ∼0.25%.

In fact, maize (unlike rice and wheat) is a C$_4$ plant (Box 26.2) and thus is a particularly efficient photosynthesizer. Maize has been bred over the years for strains that grow rapidly and evenly. The high yield of maize and relative ease of harvesting has made this plant one of the most economically effective ways to transform solar energy into carbohydrates. This has led to the dominance of maize in the US food industry. The starches in corn can be processed and broken down easily into simple sugars that are used in industrial food production in a variety of ways.

The economics of the food industry in the US has incorporated corn so completely into the system that a high fraction of calories in most processed foods in that country come from corn: meat is primarily grown on corn, foods are fried in corn oil and sweetened with corn products; most calories in soft drinks come from high-fructose corn syrup. Since C$_4$ photosynthesis has a slightly greater propensity than C$_3$ photosynthesis to incorporate the heavier isotope ^{13}C of carbon, it is possible to analyze organic material to determine the fraction with origins in C$_4$ plants. Carbon isotope analysis of hair samples suggests that more than 50% of the carbon in many North American diets comes originally from corn.

In 2013 roughly 350 Mt of corn were harvested on 35 M ha of land in the US. This represents roughly 5 EJ of energy of which some 70% are in starches, representing several times the number of calories needed to feed the US population. Only a few percent of this harvest, however, was eaten directly by humans in the US: 30% of the corn grown in the US was used for ethanol production (§26.4.1). Roughly another 40% was consumed by beef cattle, poultry, and pigs; factory farming of food animals in the US now relies heavily on corn as a feedstock (Box 26.3). Another 10% of the corn produced was processed into products used in the food industry, such as corn starch, corn oil, and sweeteners such as high-fructose corn syrup.

---

for cotton), most agricultural output goes into the world's food supply, either directly as food crops, or indirectly through food animals. Thus, providing human beings with the food energy necessary to survive and thrive dominates human land use patterns. Given that such a large fraction of available land is already used for food production, the fact that humans use almost 20 times as much energy for other purposes as they consume as food highlights the challenge of using biological energy systems for a significant fraction of all energy uses.

In 2013, roughly 60% of the world's food supply (as measured in kilocalories) came from three crops: maize, rice, and wheat, at 1000, 750, 710 Mt of production respectively [169]. The land used to produce these three crops was over 5 M km^2, roughly one third of global cropland. The global average yields of these three crops were roughly 5.5, 4.5, and 3.2 t/ha y (tonnes per hectare per

Energy in Food

Roughly 45 EJ of food energy are produced globally each year, largely from cultivation of cereal grains such as maize, rice, and wheat. The use of fertilizers containing nitrogen fixed using fossil fuel energy has dramatically increased the yield of major food crops, and is estimated to be responsible for 30–50% of worldwide food production.

year). This represents an efficiency of solar energy capture on the order of 0.1%. A wide range of food crops are raised for direct human consumption; the energy efficiency of these crops varies substantially but rarely exceeds a

---

Box 26.3   Energy and Food Animals

Most of the calories that sustain human life come directly from plant matter, principally grains such as maize, rice, and wheat. Grasses growing on pasture land and some food crops are also used to raise animals for food. Each step in the food chain reduces the fraction of available biomass energy by a factor of 10 to 20. So, for example, grass-fed beef cows provide roughly 20 kg of protein per hectare per year, while soybeans can provide roughly 400 kg of protein/ha y. Milk production from grass-fed cows is intermediate between these extremes and is much more efficient than meat production, as the bacteria in the stomachs of ruminants are quite efficient at converting cellulose to sugars and protein.

In recent years, industrialized agriculture has led to the increased use of **feed lots**, where food animals are confined and fed largely on farmed cereal grains rather than grazing on natural fodder. In the US, corn is a primary feedstock for beef cattle, although it is not a food that they digest naturally. Combined with growth hormones and extensive use of antibiotics, the corn-based feedlot approach leads to cattle that produce biomass very quickly, growing from 200 kg to 500 kg in eight months – one third of the two years needed for similar mass gain by grass-fed cows. This biomass depends, however, in part upon substantial fossil fuel inputs, particularly for nitrogen-based fertilizers generally used for growing the feed corn. The energy in the corn consumed by cows grown in feedlots (even without subtracting fertilizer and other energy inputs) is greater than that contained in the animals' increased biomass over the growth period by a factor of roughly 10 (Problem 26.3). In recent years, laboratory efforts have succeeded in growing artificial animal muscle tissue from stem cells in nutrient pools; while it might mitigate humanitarian concerns regarding animal welfare in concentrated feed lots, this approach is still experimental and extremely energy intensive.

---

small fraction of 1%. Using animals as food is generally far less energy efficient than direct consumption of plants (Box 26.3).

Over the last three centuries, food crop production has shifted from traditional farming methods in which a plot of land supported a variety of crops and livestock in a relatively closed system, to modern **intensive agriculture** methods. In particular, in the last century the extensive use of nitrogen-enhanced fertilizers, along with the development of high-yield varieties of certain crops, irrigation and farm machinery, and widespread deployment of herbicides and pesticides has dramatically increased biomass yield per hectare. A central component of this *green revolution* is the use of fossil fuel energy to *fix nitrogen* for use in fertilizers.

26.2.2 Nitrogen and Fertilizers

While the molecules that store and transport energy in biological systems are primarily carbohydrates and fats that are built from carbon, oxygen, and hydrogen, the functioning of living organisms depends critically upon a more complex set of molecules including DNA and amino acids that incorporate nitrogen in their structure. DNA contains the blueprints from which all functional

components of life are constructed, and the amino acids are the building blocks from which these components, particularly proteins, are built. Each amino acid contains a nitrogen atom in an amine ($-NH_2$) group. Chlorophyll also contains nitrogen (Figure 26.1). Plants thus require nitrogen to grow. While the atmosphere contains abundant nitrogen in the form of N_2, the triple bond in this diatomic molecule is difficult to break. In organic systems, certain microbes (*cyanobacteria*) can **fix nitrogen** from the atmosphere by converting nitrogen to ammonia: $N_2 + 8H^+ + 8e^- \rightarrow 2NH_3 + H_2$. Such microbes are active in soil, coral reefs, and other ecosystems and play a crucial role in the *nitrogen cycle*. In most natural situations, the limited supply of fixed nitrogen is the rate-limiting ingredient for plant growth. Natural methods used in the past to enhance nitrogen availability have included fertilization with bird guano, which is very high in fixed nitrogen, and rotation with crops such as soybeans and other legumes, which have symbiotic bacteria in their root system that significantly boost nitrogen fixation.

In 1909, German chemist Fritz Haber demonstrated that ammonia could be synthesized directly from hydrogen and atmospheric nitrogen at high temperatures in the presence of an appropriate catalyst with a yield above

10%. The *Haber–Bosch process* (§33.3.4) led rapidly to large-scale use of nitrogen-based fertilizers synthesized using hydrogen and energy from fossil fuel inputs. The Haber–Bosch process is energy intensive. It takes 70% more energy per kilogram to synthesize ammonia (roughly 36 MJ/kg) than it does to refine raw steel (~21 MJ/kg). Over 90 Mt of nitrogen is deployed annually in fertilizers, and roughly 1% of global energy use goes to the production of ammonia for use in fertilizer.

The removal of the nitrogen limit on plant growth led to a profound change in agricultural methods. High-yield plant varieties were developed, often with short stalks that can support rapid growth of biomass without energy lost to structural elements. In recent years economic pressures and technological advances have led to dominance of a few food crops that play a central role in modern food production systems (see Example 26.1). Advances in crop yield through cultivation of specific high-yield varieties grown with extensive fertilizer inputs has helped to increase global food supplies greatly. It is estimated that from one third to one half of the food energy consumed by humankind is made possible by the fixed nitrogen in fertilizers. A simple calculation shows that the energy used to produce a given amount of ammonia for fertilizer is much smaller than the estimated increase in total energy captured by plants that are fertilized using that ammonia (Problem 26.4).

Unfortunately, modern agricultural methods also have potentially deleterious long-term consequences. Intensive agriculture and **monoculture** (growing of a single crop over a wide area for several consecutive years) lead to decreased soil quality and increased populations of pests. This, in turn, requires increased use of pesticides and fertilizers, which leads to further environmental problems. Species selected for high yield can be less competitive with other plants, increasing the need for herbicides and genetic modifications to the plants for herbicide and pesticide resistance. The need to feed the planet's still rapidly growing population and immediate economic pressures must be balanced against the long-term benefits of more sustainable agricultural practices.

26.3 Biomass

Biomass refers to any type of organic material derived from living or recently living organisms. Plants of all kinds, including trees, food crops, and other plants that can be used for energy, are forms of biomass. Biomass also includes animals, animal waste, algae, and biological waste from agriculture and the food industry. Biomass

that is not used for food can be used in a wide variety of ways for energy production. Direct burning of biomass and the extraction of *biogas* for use as a fuel are the subjects of this section. The conversion of biomass to liquid *biofuels*, primarily for use in transportation, is described in the following section §26.4.

26.3.1 Direct Burning of Biomass

Humans have burned biomass for heat, light, and cooking since before recorded history. Biomass combustion remains a principal source of energy in many developing countries, and is a significant source of energy in some developed countries, including for example Finland and the US. According to IEA estimates, in 2010 roughly 8% (or ~40 EJ/y) of global energy used came from the combustion of biomass. This includes burning of wood, brush, agricultural by-products, animal dung, and waste materials. Some of the energy produced by biomass combustion is used for electricity generation.

While in many situations biomass is burned in an unsustainable fashion, biomass can in principle be used sustainably. Some typical crops and trees accumulate biomass through photosynthesis at a rate of around 5 t/ha y. At a typical energy density (including moisture content) of 12 MJ/kg, this gives an energy production rate of 60 GJ/ha y. To supply 40 EJ/y sustainably roughly 7×10^8 ha, or 7×10^6 km^2 of land would be needed. This represents roughly one fifth of all forested land area on the planet. At the current time only a fraction of biomass burning is done in a sustainable fashion.

Burning of trash and other waste for energy serves the dual purposes of providing energy and reducing waste mass. In the US, each individual produces roughly 2 kg of *municipal solid waste* per day. This adds up to over 200 million tonnes of waste per year. Over 50% of this waste consists of cardboard, paper, and other cellulosic material that can be burned for energy. At a rough estimate of 10 MJ/kg, this gives on the order of 1 EJ of thermal energy that could be produced from waste-to-energy conversion in the US alone. At present, ~75 PJ of this energy is recovered in electricity production. While the boilers and pollution control equipment needed to cleanly convert waste to electrical energy add somewhat to the cost, much of this waste is burned anyway unproductively in incinerators. Although conservation through recycling, reuse, and reducing packaging represents an even greater potential for energy savings, the inevitable remainder of combustible waste constitutes a substantial available energy resource in many countries. Waste that is simply incinerated adds to atmospheric carbon emissions without any energy benefit. Waste that biodegrades in a landfill produces methane, which in the short term has even greater impact on global climate than carbon dioxide (§34). This methane can, however, be released in a controlled fashion, captured, and used directly for power generation.

26.3.2 Biogas

Aside from direct combustion, other thermochemical and biological processes can be used to extract energy from biomass. Thermal **gasification** is carried out by heating biomass in the presence of oxygen, air, and/or water vapor. In this process, the molecular structure of the biological material is broken down and a gas is produced that contains varying fractions of methane, CO_2, CO, H_2, and water vapor depending upon the precise conditions and materials. If the amount of oxygen present is insufficient for complete combustion, the gas will contain substantial amounts of H_2 and CO. The gas produced in this fashion is known as **biological syngas**. Heating in the absence of oxygen or air, known as **pyrolysis**, similarly releases a gas containing H_2, light hydrocarbons (methane, ethane, etc.), and liquid hydrocarbons depending on conditions. Pyrolysis can also transform solid biomass to *char* by increasing the carbon fraction of the material in a process similar to the production of *coke* from coal (§33.1.6). The gases resulting from biomass gasification or pyrolysis can be combusted for heat or electric power, or can be converted into liquid fuels using liquefaction methods such as the *Fischer–Tropsch mechanism* (§33.4.2). These uses of biologically produced syngas are closely parallel to methods used for gasification and liquefaction of coal and other fossil fuels, and produce similar gaseous and liquid hydrocarbon fuels. These methods are described in the context of fossil fuels in §33. Gases produced thermochemically from biomass differ from their fossil fuel based counterparts in that biogas generally has higher oxygen, hydrogen, and moisture content, and a lower heating value; it also generally has less sulfur than gas and liquid fuels produced from coal.

Biological processing of waste by bacteria also releases methane in the natural process of decomposition that breaks down dead plant matter everywhere. This biological process can be harnessed for power. **Biogas**, as distinct from *biological syngas*, is produced from organic material through **anaerobic digestion** – breakdown without the presence of oxygen – by certain types of bacteria. The main components of biogas are methane and carbon dioxide, with some water vapor and other gaseous components depending on circumstances. The properties and use of methane as an energy source are described in §33.3; we describe in this section only the controlled biological production of biogas. We focus here on methane production, though methods are also being developed for biological production of hydrogen gas from biomass.

The two principal methods for biogas production currently in use are the capture of biogas from landfills and the production of biogas in closed containers known as **anaerobic digesters**. The process of anaerobic digestion occurs in several steps, and depends upon a collection of different organisms with different functionality. In the first stage, enzymes in bacteria break down insoluble organic material into simple sugars and amino acids. In a process known as **hydrolysis**, water is added to the glycosidic bonds (see Figure 26.3) in starches and other complex organic molecules, which are broken down into simple sugars and amino acids. These are broken into simpler molecules by different bacteria in the processes of *acidogenesis* and *acetogenesis*. The resulting *acetic acid*, hydrogen, and carbon dioxide are finally converted into methane and carbon dioxide by primitive organisms known as **methanogens**. The time needed for the complete transformation from solid organic material to biogas depends upon the nature of the organic material and temperature and chemical conditions. There are two standard temperature ranges in which anaerobic digestion is carried out, in which different species of methanogens are operational: 20–45 °C, using **mesophilic bacteria**, and 50–70 °C, using **thermophilic bacteria**. Soluble carbohydrates can be broken down in a few hours, while proteins, fats, and cellulosic material (§26.4.1) such as paper can take several days to break down. Thermophilic digesters act more rapidly, but are less stable and less widely used than mesophilic digesters. Solid woody material is generally not suitable for biogas production as it contains *lignin* (§26.4.1), which cannot be anaerobically digested by standard bacteria. The final ratio of methane to carbon dioxide produced depends on the nature of the original biomass. For carbohydrates there

---

Biomass and Biogas

Biomass is any kind of organic material that can be used for energy production. Biomass has been burned for heat, light, and cooking for thousands of years; such direct use of energy from biomass combustion still represents roughly 8% of global energy use. Many types of biomass can be digested by bacteria, producing *biogas*, primarily methane, which can be used for heat and power production.

is roughly a one-to-one ratio between the number of CH_4 and CO_2 molecules produced,

$$C_nH_{2n}O_n \rightarrow \frac{n}{2}CH_4 + \frac{n}{2}CO_2. \qquad (26.3)$$

For proteins and fats the methane fraction is higher. Typical landfill gas consists of roughly 50–55% methane. After biogas is produced, it can be directly burned, or can be purified to straight methane through removal of CO_2 and hydrogen sulfide. Purified methane can be channeled into the natural gas market for use in homes or in transport.

A wide variety of substances are used to produce biogas, including manure, residential and industrial organic wastes, and dedicated energy crops. These different inputs give rise to substantially different yields. While the yield from manure is relatively low (25–50 m^3 of gas per ton of wet biomass), manure is often co-processed with other material to improve the stability of the process. Fats such as used grease have a very high yield, closer to 1000 m^3/t.

Anaerobic digesters are increasingly used in dedicated biogas production plants in many countries. Germany, in particular, has the largest number of biogas plants in the EU. In 2014, Germany had roughly 8700 biogas-fueled electric power plants with a total electric capacity of 3.9 GW. German plants operate predominantly on a mixture of energy crops – typically maize – with manure. In China, India, and some other countries, small-scale biogas units are used for clean energy production from manure, and from agricultural and household waste (Figure 26.5). Millions of domestic units have been installed across Asia and provide gas for cooking and heat. Biogas thus has potential as a multi-scale technology with a range of applications from domestic to grid-scale.

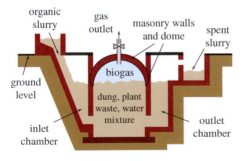

Figure 26.5 A schematic drawing of a simple, small-scale biogas production system. The pressure generated by the biogas pushes the spent slurry into an overflow chamber on the right. Due to the fixed dome, the gas pressure can vary significantly. (Credit: Tutorvista)

26.4 Biofuels

Biofuels are liquid fuels produced from recently living organic material. At present, the primary biofuels in use are *ethanol* and *biodiesel*, though other biofuels are under development and/or produced in more limited quantities.

26.4.1 Ethanol

Ethanol (C_2H_5OH) or **ethyl alcohol**, has been produced and used by humankind for at least 8500 years due to its ease of production and intoxicating properties. Certain strains of yeast can process simple sugars through **fermentation** to produce ethanol. The term *fermentation* in general refers to the biological transformation of sugars to alcohols, acids, and/or gases. Here we use it to refer to the specific case in which the end product is ethanol. For glucose the chemical transformation is

$$C_6H_{12}O_6 \rightarrow 2C_2H_5OH + 2CO_2. \qquad (26.4)$$

A similar process for sucrose relies on the presence of water and gives

$$C_{12}H_{22}O_{11} + H_2O \rightarrow 4C_2H_5OH + 4CO_2. \qquad (26.5)$$

While some energy is extracted from the sugar in the fermentation process, the release of CO_2 removes a larger fraction of the mass, so that the energy density of ethanol is roughly twice that of the simple sugars from which it is produced (Problem 26.6). Because ethanol is a liquid at ambient temperatures and pressures, it is an easy fuel to transport and use. Indeed, ethanol was the primary fuel used for automobiles in the early 1900s. As late as 1925, the American auto manufacturer Henry Ford stated "The fuel of the future is going to come from fruit like that sumac out by the road, or from apples, weeds, sawdust – almost anything. There is fuel in every bit of vegetable matter that can be fermented." Neverthless, as US oil production increased in the 1910s and 1920s, ethanol was gradually replaced by gasoline as the primary fuel for automobiles.

Production and Physical Properties of Pure Ethanol Fermentation of sugars by yeast is the principal mechanism used to produce ethanol from organic material. Because yeast requires the presence of water and cannot tolerate an environment with too high a concentration of ethanol, the fermentation process yields a mixture of water and ethanol with a maximum ethanol fraction of 10–15%. A higher fraction of ethanol can be achieved through the process of *distillation*. The boiling point of ethanol at 1 atm is 78.4 °C, so as the temperature of the mixture is raised, the vapor is initially ethanol-rich, and can be separated from the remainder. Repeated distillation cannot raise the ethanol fraction beyond 96.4% (by volume). Further purifi-cation to 100% ethanol can be carried out in a variety of ways, including the use of *molecular sieves* such as *zeolites* or organic material such as cornmeal or sawdust to separate out the water.

Pure ethanol has a mass density of 0.789 kg/L, and an energy density of 23.4 MJ/L (29.7 MJ/kg, see §9.5.1). While its energy density is thus substantially lower than gasoline (∼32 MJ/l, 44 MJ/kg), ethanol is a more stable molecule with less tendency to knock and has octane number 108.6.

Ethanol as a Motor Fuel

In recent decades, driven by a combination of economic and environmental concerns, ethanol has again come into use as an automobile fuel. In Brazil, ethanol now supplies more than half of all auto fuel following a systematic shift to ethanol produced from sugarcane following the oil crisis of the 1970s. In the US, ethanol production rose from 6 GL (1.6 billion gallons) in 2000 to 54 GL (14.3 billion gallons) in 2014 (see Figure 26.6). While in Brazil the auto market is dominated by *flex fuel* vehicles that can run either on gasoline or straight ethanol, in the US ethanol is primarily used as an additive at the 10% level to standard gasoline supplies.

While reducing the use of fossil fuels and associated atmospheric carbon production may be a primary motivation for expanding the use of ethanol in the future, ethanol has other advantages (and disadvantages) as a motor fuel. The high octane rating of ethanol improves fuel effi-ciency, offsetting the inconvenience of lower energy density. Ethanol combustion also leads to reduced emissions of carbon monoxide and NO_x, though ethanol use generates larger amounts of formaldehyde (CH_2O), which is highly toxic, and other molecules that increase ground level smog compared to gasoline. The expansion of ethanol usage in

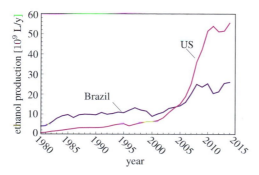

Figure 26.6 Annual ethanol production in Brazil and the US, in billions of liters. (Credit: Data for US – Renewable Fuels Association; data for Brazil – Unica)

Bioethanol

Ethanol produced from fermentation of plant sugars can provide a renewable source of liquid fuel that in principle releases no net carbon to the atmosphere. The *net energy balance* given by the ratio of total output energy to energy inputs is a measure of the effectiveness of a biofuel, and is substantially higher currently for sugar-cane ethanol than for corn ethanol. Developing efficient and economical methods for producing ethanol from *cellulose* could provide a potentially large resource that would significantly lessen con-flict with food supplies.

the US in recent years followed the Clean Air Act of 1970 and 1990 amendments that stated that commercial gaso-line must include oxygen-rich additives, which reduce CO emissions by providing more oxygen to enhance complete combustion. For some years **MTBE** (methyl tertiary-butyl ether, $(CH_3)_3COCH_3$) was used as an additive for its oxy-gen content and anti-knock tendency, though this led to environmental problems related to ground water contam-ination and motivated the switch to ethanol around the turn of the century.

Ethanol from Sugarcane

The first large-scale biofuel industry to develop in recent decades was the production of sugarcane ethanol in Brazil (Box 26.4). Sugarcane is a C_4 plant (Box 26.2) with a high efficiency of photosynthesis. Roughly 30% of the mass of sugarcane consists of sucrose, which can be fermented to produce ethanol. Another 35% of the plant mass consists of leaves and stems, which are removed during harvesting; the remaining 35% of the sug-arcane plant is fibrous material known as **bagasse**. Bagasse contains a substantial amount of cellulose, and is not suit-able for fermentation, but can be used as a biomass fuel after drying. Bagasse has long been used to provide the heat needed for distillation and other aspects of ethanol production, and more recently has been used for electric power production, enabling sugarcane ethanol production to be energetically self-sufficient.

An important metric for biofuels as an energy resource is the ratio of output energy to energy input; this ratio is known as the **net energy balance** for the biofuel. Using modern intensive agricultural methods, sugarcane ethanol can be produced from land at a yearly rate corresponding to roughly 180 GJ/ha y (Box 26.4). This production, however, requires substantial energy inputs. Most of the input energy

comes during cultivation, in particular in the form of fixed nitrogen in the fertilizer (§26.2.2), but also includes the energy cost of labor and fuel for planting and harvesting. The industrial processing in an ethanol plant can give a net energy output when bagasse is burned for power. Finally, some energy is needed for transport and distribution of the resulting ethanol. When these energy inputs and outputs are combined, the net energy balance for sugarcane ethanol in a typical Brazilian sugarcane plant is roughly

$$\mathrm{NEB_{sugarcane\ ethanol}} = \frac{\text{total energy output}}{\text{energy input}} \sim 8.$$

$$(26.6)$$

Thus, the energy output from sugarcane ethanol is substantially greater than the energy inputs, and sugarcane ethanol can be viewed as a means of capturing and storing solar energy in a liquid fuel. Considered as a power density, the 180 GJ/ha y of energy produced in sugarcane ethanol comes to roughly $0.6\,\mathrm{W/m^2}$; assuming a typical insolation in a region suitable for growing sugarcane of $250\,\mathrm{W/m^2}$, this corresponds to a net solar conversion efficiency of roughly 0.25%.

Ethanol from Corn   The largest current contribution to world ethanol production is corn-based ethanol produced in the US, which has increased dramatically since 2000. Almost all of US ethanol production uses corn as a primary feedstock, although sorghum, barley, wheat, and other crops have been used in much more limited quantities. Like sugarcane, corn is a C_4 plant (Box 26.2) with a relatively high photosynthetic efficiency. The industrial production of corn in the US, however, relies on more

energy inputs than Brazilian sugarcane production. Furthermore, while the sucrose in sugarcane can be directly converted to ethanol through fermentation by yeast, the complex carbohydrates in corn require additional steps in processing.

   The energy-containing molecules in corn are primarily starches. The starch in common corn is typically 75% *amylopectin* and 25% *amylose* (Box 26.1). Starches cannot be directly fermented; first the molecules must be broken down into smaller units. The enzyme **alpha-amylase** is used in human digestion to catalyze the breakdown of starches through *hydrolysis* in which water molecules are added as glycosidic bonds between adjacent glucose units are broken (essentially the reverse of the process illustrated in Figure 26.3(b)–(d)). α-amylase is also found in certain seeds that use starches as an energy supply. Hydrolysis has been employed for centuries to produce whiskey and other alcoholic beverages using α-amylase extracted from, for example, germinated barley seeds, to break down larger starch molecules in various grains before fermentation.

   Over time, α-amylase reduces amylose to a mixture of glucose and *maltose* (a disaccharide similar to sucrose, but composed of two glucose units). Glucose and maltose are both fermentable sugars. Because α-amylase does not break the bonds causing the branching in amylopectin, however, when amylopectin is broken down by α-amylase more complicated non-fermentable sugars called *oligosaccharides* (roughly 3–10 glucose sugar units) remain. The partial reduction of starch to sugars using α-amylase is known as **starch liquefaction**. Further enzymes (e.g. *glucoamylase*) must be used to break down the amylopectin completely into fermentable sugars, a process known as **saccharification**.

While a great deal of research continues to be done on the factors required for corn ethanol production, including the required enzymes, yeast, and processing steps, production of corn-based ethanol using current technologies is both less efficient and less productive than sugarcane ethanol. Average corn-based ethanol yield for US production averages roughly 4000 liters/ha y, slightly more than half the yield for sugarcane based ethanol in Brazil. This amounts to roughly 100 GJ/ha y, which is 80% or more of the energy output that could be expected from complete conversion to ethanol of all starch in 10 t of corn crop per hectare (Problem 26.7). Due to the additional processing needed and substantial fertilizer energy inputs, estimates of the net energy balance for corn-based ethanol from US production plants lie in the range

$$\text{NEB}_{\text{corn ethanol}} = \frac{\text{total energy output}}{\text{energy input}} \sim 1.2\text{--}1.6\,.$$

(26.7)

While this is a very rough estimate, and computations of energy inputs for corn ethanol are somewhat controversial, there is no question that this number is substantially below the net energy balance of sugarcane ethanol at the current time.

One of the largest accounting differences in different approaches to computing the energy balance (NEB) for corn ethanol is the treatment of byproducts. Corn is roughly 70% starch, and like sugarcane the remaining material contains useful energy. For dry milled corn, the unfermented residue yields **dried distillers grains with solubles** (DDGS), which is most commonly used as animal feed. The larger estimates for corn ethanol NEB arise when the energy content of DDGS is included.

Substantial research efforts in the biotechnology industry are focused on developing mechanisms for improving the yield, efficiency, and energy balance of starch-based ethanol. In particular, alternative enzymes for starch liquefaction and saccharification have been developed, as well as yeast variants that have higher efficiency and/or can tolerate high levels of ethanol. Much work has been done to optimize the fermentation process to approach the theoretically ideal maximum production rate of two ethanol molecules for each six-carbon sugar unit in an energetically efficient fashion.

Cellulosic Ethanol   Most of the plant material that grows on Earth takes the form of cellulose, which is used for structure and is even more difficult than starch to convert to simple sugars. An efficient mechanism for producing **cellulosic ethanol** from the cellulose materials in non-food plants would make possible the extraction of energy from much of the biomass available on Earth, including grasses, shrubs, and trees that can grow with minimal human intervention in habitats where growing food crops is impractical. While much research is currently ongoing to develop an efficient and economic approach to producing cellulosic ethanol, this technology has yet to achieve broad commercial implementation. Replacing a substantial fraction of motor vehicle fuel used worldwide with ethanol from sugarcane, corn, or other food crops is impractical, however, due to the impact on global food supply, so cellulosic ethanol may represent the most viable pathway to large-scale use of biofuels in transport.

Cellulose (($C_6H_{10}O_5)_m$) is basically a highly connected polymer composed of glucose units. Cellulose has a much more interconnected structure than starches (see Figure 26.7). To break down starches into glucose units, only a single bond must be broken between each unit in the polymer; to similarly decompose cellulose requires breaking many bonds. Closely related to cellulose, **hemicellulose** is primarily built from five-carbon sugars such as *xylose* that cannot be fermented by the same yeast strains that are used for fermentation of glucose. **Lignin** is another, more heterogeneous, class of molecules that along with cellulose and hemicellulose plays a significant role in plant structure. Together, cellulose, hemicellulose, and lignin are referred to as **lignocellulose**. The strength and

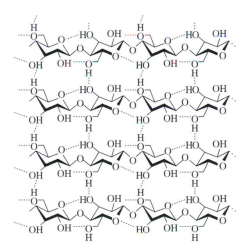

Figure 26.7 *Cellulose* is a straight chain polymer, individual chains of which form hydrogen bonds between OH groups on one chain and hydrogen atoms on another. The resulting strands of polymer are strong and rigid. The figure shows a section of four interconnected (horizontal) chains. (Credit: Luca Laghi)

Box 26.5   Corn-based Ethanol and US Land Use

Large-scale production of biofuels using current technologies can have a significant effect on land available for food crops. The US has roughly 270 million hectares (2.7×10^6 km^2) of arable land. Roughly 15 million hectares, or 5.6%, were used to grow corn for ethanol in 2014. At 100 GJ/ha yr, this represents a yearly energy production of roughly 1.5 EJ, or about 9% of the gasoline energy used in the US in 2014. To replace all gasoline energy used in the US with corn ethanol using current technology, production would need to be scaled up by a factor of 11, which would use over half of all arable land in the country. This would have a significant impact on land used for growing food and plants used for other purposes. The 2007 US Energy Independence and Security Act mandated that by 2022 US ethanol production should increase to 36 billion gallons (production was 13.2 billion gallons in 2013). For this goal to be realized it is expected that almost half of the ethanol produced will need to come from lignocellulosic feedstock (§26.4.1).

Box 26.6   Carbon Reduction from Corn-based Ethanol

The net energy balance (NEB) is not the only consideration in determining the effectiveness of biofuels. A closely related but distinct consideration is the extent to which a given biofuel reduces net carbon dioxide emissions when it is used in place of gasoline. Absent fossil energy inputs, biofuels would be carbon neutral, since the carbon released when they combust was originally captured from the atmosphere. Due to the extensive use of fossil fuel energy in intensive agricultural methods, in fixing nitrogen for fertilizers, transporting, planting, and harvesting the crop, as well as energy inputs in the ethanol production and distillation stages, the carbon cost of producing a liter of corn-based ethanol is fairly close to the carbon savings from the 0.7 liters of gasoline that it replaces. A review of many estimates [170] gives the carbon mass equivalent of greenhouse gas emissions associated with the energy inputs for one liter of corn ethanol at 210 g/L (of which some 60% is associated with nitrogen in the fertilizer), and 370 g/L in the ethanol conversion process. This amounts to roughly 25 g of carbon per MJ of ethanol energy. This is quite close to the rate of carbon emissions of gasoline per MJ of combustion energy. When by-products (e.g. DDGS) are included as an offset, however, the carbon advantage of corn-based ethanol is increased, giving an estimated reduction in greenhouse gas emissions by roughly 18% (with an uncertainty range of -36% to $+29\%$). Note, however, that the numbers entering this estimate are quite controversial and dependent upon specific circumstances. The upshot is, however, that corn-based ethanol is only a marginal improvement at best over direct use of petroleum-based fuel from the point of view of greenhouse gas emissions. If the primary goal is carbon reduction, it has been pointed out that atmospheric carbon would be reduced more over a 30 year time span by allowing a given hectare of land to reforest and simply using gasoline than by using the land to produce corn-based ethanol (Problem 26.8). Of course, when the forest approached its maximum density, the trees would need to be removed and sequestered to maintain the rate of carbon reduction indefinitely.

---

rigidity of lignocellulosic material that enable it to function as an effective structural material in trees and other plants are precisely the features that make its decomposition into simple sugars difficult. Since many bonds must be broken to decompose lignocellulose into molecules that are amenable to standard fermentation processes, more sophisticated biochemical means are needed to produce ethanol from lignocellulose. Mechanisms under consideration are based either on thermochemical methods – for example using acids and high temperatures to initially break down the cellulose – or on biological methods – for example using enzymes derived from tropical fungi or insects such as termites that can break down lignocellulosic material. Ethanol from lignocellulose is often referred to as a **second-generation biofuel**, as opposed to biofuels from oils, sugars, or cereals, which are **first-generation biofuels**.

The enzymes that break down cellulose are known as **cellulases**. The production of ethanol from cellulosic material in principle follows a similar sequence to the production of ethanol from starch. First the plant material is broken down by physical means such as milling to small fragments; the fragments are then treated with cellulase to break down the cellulose and then with enzymes to complete the saccharification process to glucose, followed by a relatively standard fermentation process. Hemicellulose presents additional difficulties due to the other sugar monomers produced, but can be fermented using other yeast variants. One goal of current research related to cellulosic ethanol production is the production of yeast strains that can efficiently ferment both six-carbon and five-carbon sugars. Lignin is particularly difficult to break down, and most approaches to cellulosic ethanol currently envision burning lignin for additional input energy just as bagasse is burned to provide energetic input for sugarcane ethanol production.

Since cellulosic ethanol production is still at an early stage, and has only been carried out in small-scale test projects, the energy balance that will be attained in any large-scale fuel production plant of this type is unclear. Optimistic estimates suggest that the NEB may be even higher than for sugarcane ethanol. Certainly, however, the potential scale of cellulosic ethanol production is quite large.

One crop that is often considered for cellulosic ethanol is **switch grass** (*panicum virgatum*), a tall grass native to North America, which grows naturally from Mexico to central Canada. Switch grass is a C_4 plant that can produce substantial quantities of biomass in marginal conditions and with limited energy inputs. Typical yields of test plots averaged 15 t/ha, with energy inputs less than half of that for corn. Since switch grass can be grown on land that

Figure 26.8 Switch grass, a C_4 plant that can be grown on land otherwise not suitable for agriculture.

is not suitable for food crops, this represents a possible substantial source of biofuel that would not conflict with human food energy needs. The success of this vision, however, depends critically on the development of efficient and economical methods for cellulosic ethanol production.

26.4.2 Biodiesel

Biodiesel is a liquid fuel produced from vegetable oil and animal fat using purely chemical processes. Vegetable seed oils and animal fats include molecules known as **triglycerides** (Figure 26.9). Triglycerides contain three hydrocarbon chains connected by an oxygen-containing structure based on **glycerol**, a simple compound ($C_3H_8O_3$) related to both alcohols and sugars (see Figure 26.9). Triglycerides can be combined with alcohols such as *methanol* (CH_3OH) through a process known as **transesterification**, which breaks the triglycerides into three separate long-chain molecules known as **fatty acid esters**. The resulting fatty acid esters have very similar properties to the hydrocarbons used in diesel fuel (§11.4.2), and can be combusted directly in standard diesel engines. In fact, the original engine constructed by Rudolph Diesel first ran on peanut oil.

> **Biodiesel**
>
> Plant and animal oils can easily be processed chemically to yield *fatty acid esters* with properties very similar to the hydrocarbons in conventional diesel fuel. At present, biodiesel is sourced from oil crops such as rape seed with a net energy balance ranging from below one to three or higher. Some microalgae photosynthesize with high efficiency, and produce oils suitable for conversion to biodiesel. Production of biodiesel by farming microalgae holds promise for future development.

Biodiesel Production Biodiesel can be produced from a wide range of different feedstocks. This leads to substantial variation in the precise distribution of molecules in different batches of biodiesel. Typical biodiesel fuel has an enthalpy of combustion of roughly 37 MJ/kg, 10% lower than standard diesel fuel. The combustion quality of vegetable-oil based biodiesel is relatively good (cetane numbers 46–52), and even higher for biodiesel from animal fat (cetane numbers 56–60). While pure biodiesel (B100) can be used in many diesel engines without problems, biodiesel can dissolve rubber in components such as hoses and gaskets. Blends of 20% biodiesel (B20) or lower

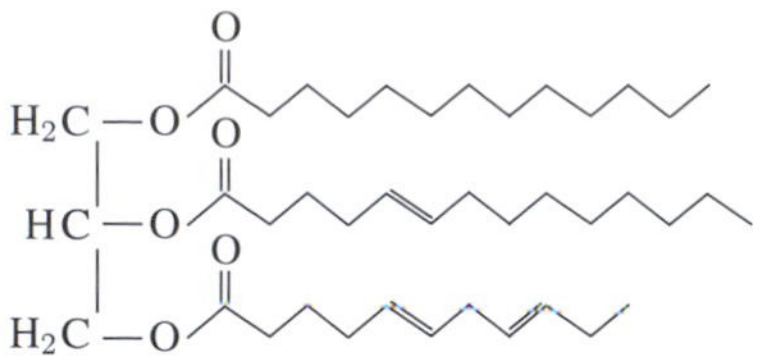

Figure 26.9 Vegetable seed oils and animal fats contain *triglycerides* (a), which contain three *fatty acids* connected to a *glycerol* (b) backbone.

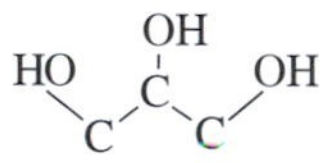

can be used directly in most diesel engines and are more widely distributed.

The purely chemical nature of the transesterification reaction gives biodiesel fuel some advantages over biofuels such as ethanol that are produced through the action of enzymes and yeasts. Unlike the fermentation process used to produce ethanol, with an appropriate catalyst the production of biodiesel is rapid and efficient, and leads to a relatively pure product. The wide range of possible sources of biodiesel also makes this a very flexible fuel source that can rely on different feedstocks depending upon market conditions. At present, biodiesel is primarily produced in Europe using canola oil from rape seed. Biodiesel can be produced from other vegetable seed oils such as soybean or sunflower oil, from animal fats, or even from used cooking oil. In addition to methanol, which is generally produced from fossil fuels, other alcohols such as ethanol can also be used in the transesterification reaction (producing fatty acid *ethyl* esters with slightly different qualities than the fatty acid methyl esters produced by using methanol). A variety of catalysts have been explored for the transesterification reaction, though bases such as sodium or potassium hydroxide (NaOH, KOH) are used in most commercial biodiesel production. The transesterification reaction produces *glycerol as* an additional product, which is used in the pharmaceutical industry and in foods as a sweetener and thickening agent (see also §26.4.3).

Computations of the net energy balance for biodiesel give a range of values ranging from below one to three or higher, depending on the processes involved and how coproducts are treated. Even higher net energy balances have been found in some cases, such as for palm oil biodiesel produced by Brazilian and Columbian plants. In most analyses, for standard biodiesel production from canola, soybean, or sunflower oil, a net energy balance

significantly above one is only achieved by giving full credit for the energy output associated with seed meal and glycerol coproducts. The market for glycerol, however, has already been saturated by output from the biodiesel industry so new uses of glycerol would need to be found to justify including its energy in the energy balance equation. Just as for ethanol (Box 26.6), the carbon reduction realized by biodiesel from farmed plants depends crucially on the level of fossil fuel inputs required for growing and processing the crop.

Green Diesel and Straight Vegetable Oil (SVO)

Vegetable oil can be used as a motor fuel in other ways than through biodiesel production. **Green diesel** (also known as **hydrotreated vegetable oil** (HVO)) is produced from vegetable oil by a refining process in which hydrogen is used to crack the larger molecules in the oil into hydrocarbons much like those produced from petroleum (§33.2.5). The resulting product is sufficiently close to petroleum-sourced diesel that it can be used interchangeably, and differs only in its renewable origin.

In recent years there have also been increasing efforts to directly use straight vegetable oil (SVO) as a liquid fuel for automobiles. The main obstacle to direct use of vegetable oil is its higher viscosity and surface tension compared to standard diesel fuel, particularly at low temperatures. This can be ameliorated by heating the fuel before it enters the engine, though generally it has been found necessary to incorporate a separate tank for standard diesel fuel to be used while the engine is warming up and to ensure that vegetable oil does not remain in the system after the engine is stopped, when it may cool and harden. One source of vegetable oil for SVO automobiles is recycled vegetable oil from the food industry. In the US alone, over 10 billion liters of vegetable oil are recycled yearly. While this represents only a fraction of a percent of the 50 EJ of annual US petroleum energy use, it could contribute to reducing energy use and CO_2 production.

Biodiesel from Microalgae **Microalgae** are single-celled species of algae, which can exist individually or combine into chains or filaments. Microalgae include *diatoms* that are common types of *phytoplankton*[3] and single-celled green algae such as *chlorella*. Microalgae are extremely efficient at converting solar energy to

---

[3] **Phytoplankton** are free-floating, photosynthesizing microorganisms that are common in almost all marine and freshwater ecosystems. Phytoplankton are thought to be responsible for roughly 50% of all photosynthesis on Earth, and are critical to the ocean food chain.

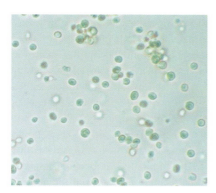

Figure 26.10 *Nannochloropsis*, a microalga under investigation as a biofuel source due to its high content of fatty acids.

biomass. In the 1950s *chlorella* algae was promoted as a "superfood" that could feed the world's population. More recently, microalgae have been considered as a potential source for biofuels. Some species of microalgae build biomass at a rate of 50 t/ha yr, with rates in research tests measured at up to 300 t/ha yr – over 30 times the rate of growth of typical corn plants. This represents a solar energy conversion efficiency that can approach 10%. Microalgae consist of 20–75% lipids containing triglycerides suitable for biodiesel production. Microalgae can be grown in artificial ponds that can be located anywhere sufficient water can be made available; microalgae require relatively little water and can grow in waste water or water with high salinity. Biofuels from microalgae therefore need not conflict with food crops for land use. It has been estimated [171] that enough biodiesel to replace all US transport petroleum use could be grown on roughly 9 million hectares of land, which is approximately 5% of the total cropland in that country, and less than the amount of land currently used for corn ethanol production. The difficulties in mass producing biodiesel from microalgae mainly concern harvesting and extracting the lipids. While a number of mechanisms are under development to do this, large-scale cultivation and biofuel production from microalgae is not yet economical. From an energetic point of view, however, microalgae present a potential for biological capture of solar energy and concomitant biofuel production at an efficiency that could, in principle, compete with solar thermal or PV energy. This technology holds substantial promise for the future.

26.4.3 Other Biofuels

A wide range of other biofuels have been considered, many of which are at an earlier stage of development than the ones mentioned above.

Other alcohols besides ethanol can be produced by fermentation using different organisms. Any low molecular weight alcohol can be used as a combustible fuel. For example, *butanol* (C_4H_9OH) is produced through fermentation by certain species of bacteria in the class *Clostridia*. Butanol has a higher energy content than ethanol (29.2 MJ/L) and has properties closer to standard gasoline; thus, it can be mixed with regular gasoline in larger fractions than ethanol, without requiring engine modification. Butanol can in principle be produced using standard feedstocks (sugarcane, corn, etc.) and potentially also from lignocellulose. Research is ongoing for efficient processes for **biobutanol** production but they have not yet reached commercial scale.

As mentioned in §26.3.2, biomass can be converted to a mixture of gases including H_2 and CO by heating without complete combustion in the presence of oxygen, air, and/or water, or by pyrolysis. The resulting gas can be converted into liquid hydrocarbons through the *Fischer–Tropsch process*, which is described in detail in §33.4.2. This provides a means of producing liquid hydrocarbon fuel directly from a variety of biological sources. Hydrogen production from biological sources may also eventually be a good source of fuel for hydrogen-based power systems (§37.3.3).

Variations on the Fischer–Tropsch process can produce not only pure hydrocarbons but also oxygenated compounds such as alcohols. Some pilot plants have implemented methanol production from biomass. While methanol is not ideal for combustion, with an energy content below that of ethanol, methanol could be used in *fuel cells* that directly convert chemical energy to electrical energy without combustion (§37.3.2).

The reader interested in more detail on these and other novel approaches to biofuel energy is referred to the extensive and rapidly growing literature on biofuels; see [161] for an overview and some further references.

26.5 The Future of Bioenergy

At a fundamental level, biomass and biofuels can be viewed as a way to collect and store solar energy; bioenergy systems can thus naturally be compared with other solar energy technologies. In the very long term, it is possible to imagine successfully integrating the biological features of self-reproduction, self-repair, ecosystem compatibility, and operation at the molecular level into large-scale solar capture and electrical conversion systems, thereby creating a sustainable technology capable of powering human civilization. Current biological systems, however, are far less efficient than the human-engineered macroscopic solar capture systems of photovoltaics and solar thermal electric power plants. At present, sugarcane ethanol represents the most energetically and economically effective approach to liquid biofuel production. The rate of energy production of 180 GJ/ha y for sugarcane ethanol can be compared to the total yearly electricity output of a solar thermal plant such as SEGS (§24.4), which is roughly 3000 GJ/ha y. This corresponds to a yearly average power output of roughly 0.6 W/m^2 for sugarcane compared to 11 W/m^2 for solar thermal energy. When the typical combustion engine efficiency of 30% is factored in, the energy production rate of sugarcane ethanol is smaller than that of the solar thermal plant by a factor of roughly 50. Much of the difference between these numbers arises from the relatively small solar conversion efficiency of the photosynthesis process, compared to solar thermal electricity or photovoltaics. Purely from an energetic point of view, therefore, tremendous improvements in photosynthetic efficiency would have to be realized for biofuels to be competitive with other approaches to solar energy capture and storage.

Existing biofuels have several other disadvantages relative to other solar energy technologies. First, they are restricted to the small fraction of Earth's land surface that is arable land, where they compete with human food production. Second, they require large inputs of fresh water, which is becoming increasingly difficult to provide in many locations. Third, all existing biofuels require significant inputs of energy, either directly or through artificially produced fertilizers, which decreases their net energy balance and their efficacy as a tool for CO_2 reduction. On the other hand, presently existing biofuels have the advantage that they take a form (ethanol) that is compatible with internal combustion engines that dominate automotive transport. For electric vehicles, however (§37.3), solar thermal or photovoltaic systems require significantly less land to gather a comparable amount of energy compared to liquid biofuels. For biofuels based on existing plant species to replace a substantial fraction of transport fuel, an efficient mechanism would need to be developed for making use of lignocellulose as discussed in §26.4.1.

The limitations to producing energy from natural biological processes can also be seen in the fact that human energy use ($\sim$550 EJ/y) represents a substantial fraction of total terrestrial biomass production through photosynthesis ($\sim$1660 EJ/y). For biologically derived energy sources to play a much larger role in future human energy use patterns than the 10% realized by biomass burning at present times, therefore, it seems that a significantly higher photosynthetic conversion efficiency would need to be realized.

In the short term, tremendous efforts are being expended to develop new biotechnological options for energy systems. These range from incremental improvements on existing systems, such as improved enzymes for breaking down starch and cellulose, and more effective yeast strains for fermentation of ethanol, to the development of efficient mechanisms for farming, harvesting, and extracting lipids from microalgae for biodiesel production, to much more radical ideas such as organisms that directly manufacture liquid biofuel without human intervention or organisms that capture carbon from the atmosphere and store it in solid form. The major challenge in making large steps forward in biotechnology is the enormous complexity of biological systems. While evolution can experiment with billions of different approaches to solving a problem, methods for human engineering of biological systems are still at an early stage of development. There has already been substantial progress in recent decades, however, and in the very long term, a hybrid technology combining biological and engineering approaches may enable the development of solar capture and conversion systems with higher efficiency and lower system and maintenance costs than macroscopically engineered solar energy systems like those currently in place. Biotechnology development, perhaps more than any other approach to energy systems mentioned in this book, has a potential for change and evolution in the coming decades and centuries that is difficult to imagine or anticipate at this time.

Discussion/Investigation Questions

26.1 Estimates suggest that 80% of the nitrogen in fertilizer is lost to the environment and does not benefit crops. Investigate the negative side effects of nitrogen fertilizer runoff. If this lost nitrogen were used to produce biomass increase comparable to that attributed to the portion that increases crops, estimate the rate of atmospheric carbon uptake and compare to human emissions.

26.2 Investigate the *raceway pond* and *photobioreactor* approaches to microalgal biomass production. Discuss the relative advantages and challenges of these approaches, and the typical realized rate of biomass production in each approach.

Problems

26.1 The enthalpy of combustion of glucose ($C_6H_{12}O_6$) is roughly 15.6 MJ/kg. Compute the fraction of incident solar energy from 8 photons with wavelength 680 nm stored through the reaction (26.2) in a single CH_2O unit.

26.2 Estimate the total food energy needed for the planet's population assuming a diet of 2400 Calories/day/person. Compare to the global food production rate stated in the text.

26.3 A feedlot cow is fed 12 kg of corn daily for 255 days, and grows from 200 kg to 500 kg in that time. Estimating the energy densities of corn and cow both at roughly 15 MJ/kg (a very rough estimate, actually more of an overestimate for the cow), compute the ratio of corn energy input to cow energy output.

26.4 The fossil fuel energy input currently needed for ammonia production is roughly 36 GJ/t (§33.3.4). Estimate the energy needed to produce ammonia containing 90 Mt of nitrogen. Compare to the biomass energy of 18 EJ in 40% of all food crops that is estimated to be enabled through fertilizer inputs.

26.5 If all land area in the US that is used for growing corn for ethanol were used for solar thermal electric plants operating at a gross conversion efficiency of 3%, estimate the total electrical energy yearly produced assuming an average insolation of 200 W/m^2. Compare to US energy consumption of roughly 100 EJ, assuming 35% efficiency for 65 EJ of petroleum and coal energy used and 100% efficiency of the remainder.

26.6 Look up the standard enthalpies of formation of glucose and ethanol (Table 9.8), and verify that eq. (26.4) is exothermic. Then verify that the gravimetric energy density (relative to combustion) of ethanol is approximately twice that of glucose.

26.7 Compute the maximum energy content of the ethanol produced from corn farmed at 10 t/ha y, assuming that the corn is 70% starch composed of *polysaccharides* (*amylose* and *amylopectin*) built from glucose (*glucan*) units of molar mass 162 g/mol.

26.8 Compute the rate at which carbon is extracted from the atmosphere by a forest growing at the net rate of 4 t/ha y. Assume that the wood is purely composed of cellulose. Compare the carbon savings realized by sequestering this carbon to that realized by using corn ethanol grown on the same land as a motor fuel, according to the estimate in Box 26.6. (See Problem 26.7 for further information.)

26.9 Crop residues (corn stover, straw, or sugarcane bagasse) represent a potentially sustainable source of cellulosic ethanol. Assuming a production of 6 t/ha y and 270 liters of ethanol per tonne of crop residue, estimate the area of farmland necessary to sustainably provide ethanol for one million cars (using 100% ethanol) that travel on average 10 000 km/y at 40 mpg (17 km/L).

Ocean Energy Flow

Energy flows through Earth systems, driving the dynamics of atmosphere, oceans, and climate. In this chapter we begin to study this energy flow and how it can be harnessed for human use. Solar energy, impinging on Earth at a steady rate of 173 000 TW, is the primary source of this energy, but heat escaping from Earth's interior and tidal energy generated primarily by the pull of the Moon also contribute. In this chapter we focus on the solar energy that is absorbed by Earth's oceans.

The world's oceans cover over 70% of Earth's surface (361×10^6 km^2 of a total of 510×10^6 km^2). The immediate effect of much incident solar energy is to warm the waters in the top several hundred meters of the world's oceans, particularly near the equator. The oceans not only store thermal energy from solar radiation; they also play a crucial role in transporting this energy and controlling climate by maintaining the planetary energy balance. The ocean energy balance is described in §27.1.

The world's deep oceans are maintained at a substantially lower temperature than the shallow surface waters. In order to understand why this is true, and more generally how ocean energy transport works, we need to understand something about large-scale ocean currents and atmospheric circulation. Some of the solar energy absorbed by ocean surface waters at low latitudes is transferred back to the atmosphere, producing winds that in turn drive surface currents. Although the complete planetary wind and ocean circulation system is quite complex, the general nature of these currents can be understood in terms of some basic physical principles. One key mechanism that underlies many aspects of global circulation patterns is the *Coriolis force*. The Coriolis force is a *fictitious force*, like the more familiar *centrifugal force*, that is felt by objects in rotating coordinate systems like Earth's surface. In contrast to centrifugal forces, Coriolis forces are felt only by objects that are *moving* in the rotating coordinate system. We introduce

Reader's Guide
In this and the subsequent eight chapters we explore the flow of energy through Earth systems and describe some of the ways that have been devised to harvest this energy for human use. This chapter introduces some basic aspects of ocean physics, focusing on the role of Earth's oceans in gathering and transporting solar thermal energy. We characterize the thermal energy content and transport of the world's oceans, and describe how this energy can be extracted for human use using the technology of *ocean thermal energy conversion (OTEC)*. The Coriolis force is explained and derived; this mechanism affects the behavior of ocean surface currents and helps to explain the large-scale circulation of both the atmosphere and the oceans.

Prerequisites: §2 (Mechanics), particularly rotational motion; §6 (Heat transfer); §10 (Heat engines); and §23 (Insolation).

The concepts developed in this chapter are applied in §28 (Wind energy) and §34 (Energy and climate).

and derive the Coriolis force in §27.2, and apply it to the dynamics of ocean surface currents in §27.3.

Global surface currents transport large amounts of thermal energy from the equator to the poles. As these currents reach colder latitudes, the thermal energy is transferred out of the surface waters, through conduction, convection, evaporation, and radiation. The colder water becomes more saline as some of it evaporates or freezes. The water then sinks, and spreads through deep-water circulation back to the equator. This circulation pattern, known as the **meridional overturning circulation** (MOC) acts as a "global conveyor belt," which transports a significant amount of thermal energy from the equator to the poles[1]

---

[1] Oceans are responsible for transporting roughly half of the energy away from the equator in tropical regions; the

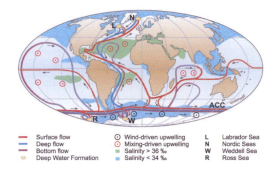

Figure 27.1 Wind-driven surface currents and temperature- and salinity-driven currents transport thermal energy from equatorial region to poles through the *meridional overturning circulation* (MOC). (Credit: Reprinted by permission from Macmillan Publishers Ltd: Nature, S. Rahmstorf, Ocean circulation and climate during the past 120,000 years (Copyright 2002))

and maintains the deep ocean at a relatively low temperature of around 5 °C (see Figure 27.1). The transport of energy across latitudes, known as *meridional heat transport*, plays an important role in moderating the temperature of both equatorial and polar regions, as discussed further in §34. The large-scale structure of ocean and atmospheric circulation is described in §27.4 and §27.5.

Finally, in §27.6 we discuss *ocean thermal energy conversion* (OTEC) – the possibility of tapping the vertical temperature difference between warm tropical surface waters and the colder deep ocean as a source of power. Despite the vast heat content of the ocean, the low Carnot bound on efficiency makes it difficult for OTEC to provide a viable large-scale renewable energy resource. It has also been proposed that the energy in ocean currents may be tapped for human use using underwater turbines. This *marine current* energy source is described in §31.4.

The role of ocean heat transport in regulating climate is explored further in §34. For a more in-depth introduction to physical oceanography and the role of oceans in the global climate system, see the excellent books by Stewart [172] and Hartmann [173]. Much of the presentation here related to ocean surface currents follows that of [172].

27.1 Oceanic Energy Balance and Transport

Approximately 30% of the solar radiation incident on Earth is reflected back into space, by clouds, atmospheric

atmosphere is responsible for the other half of this transport. Outside the tropics, the atmosphere is the dominant medium for energy transport towards the poles.

Ocean Temperature

Surface waters in the tropics are 22–27 °C in the *oceanic mixed layer* (roughly the top 20–200 m), dropping linearly through the *thermocline* to an approximately constant temperature of 5 °C in the *abyss* below 1000 m.

reflection, and surface reflection. Roughly another 20% is absorbed in the atmosphere. A significant fraction of the other half of incoming solar radiation is absorbed into the oceans. Depending upon latitude, season, and local climate, average insolation in the open ocean ranges from 50–260 W/m² at latitudes below 50°. This radiation is absorbed in the upper layers of the ocean. Its intensity falls with depth according to the *Lambert–Beer law* $I(z) = I_0 e^{-\kappa z}$, with an absorption length $l = 1/\kappa$ in seawater that ranges from roughly 1–100 m for light in the visible window (see Problem 27.1). Thus, most incident radiation is absorbed and transformed to thermal energy within the top 200 m of the ocean surface. This region roughly corresponds to the oceanic **mixed layer**, in which water mixes fairly readily through surface convection and has a fairly homogeneous temperature distribution.[2] In the tropics (latitude $|\lambda| \leq \epsilon \cong 23.44°$), the mixed layer stays at a temperature of roughly 22°–27 °C throughout the year.

Below the mixed layer, the temperature steadily decreases to a depth of about 1000 m. This intermediate region in which the temperature has an approximately constant gradient is known as the **thermocline**. Below the thermocline, the ocean at depths greater than 1000 m (also known as the **abyss**) has a fairly constant temperature of roughly 5 °C across the globe.

Given the temperature difference between the tropical mixed layer and the abyss, we can use the methods of §5 to estimate the rate at which heat is conducted across the thermocline. Downward heat transfer in the tropics by convection is negligible since the overlying warm water is less

[2] The oceanic mixed layer is homogenized by the action of wind and waves on the surface, and by convection due to density variations from changes of salinity caused by freezing or evaporation. The precise depth of the mixed layer varies significantly, from tens of meters to hundreds of meters, depending upon the extent of mixing caused by local wind and weather patterns; in some high-latitude locations such as the Labrador and Waddell Seas, the mixed layer can extend to more than a kilometer in depth.

dense than the deeper colder water. The thermal conductivity of water is approximately $k \cong 0.6$ W/mK (Table 6.1). Taking the temperature difference between the mixed layer and the abyss to be 20 °C across a distance of 800 m, the rate of heat conduction is

$$q = k|\nabla T| \cong (0.6 \text{ W/mK}) \frac{20 \text{ K}}{800 \text{ m}} \cong 0.015 \text{ W/m}^2 .$$
(27.1)

This rate of heat transfer is infinitesimal compared to the rate at which thermal energy is added to the mixed layer from solar radiation. At this rate, it would take about 10 000 years to warm the first 1000 meters of the abyss by 1 °C (see Problem 27.3). Two simple conclusions follow from these considerations: first, the solar energy absorbed by the upper ocean in tropical latitudes must go somewhere besides down; and, second, something must keep the abyss cool; otherwise, over geologic time scales it would be warmed to the average surface temperature (or even higher, when the geothermal energy rising up from below (§32) is included).

Where does the thermal energy absorbed by the ocean go? The primary mechanism of heat loss from the ocean is evaporation. As emphasized in §5 and §9.2.2 (see Table 9.2), significant energy is absorbed in the change of phase when water evaporates. A steady rate of evaporation across the ocean surface is responsible for removal of roughly half of the ocean thermal energy absorbed from sunlight. This energy is maintained in the atmosphere when the water vapor subsequently condenses. This **latent heat flux** drives winds and powers storms. Evaporation occurs more readily in areas with low humidity and high winds, and less readily in areas covered with sea ice. Another significant fraction of the thermal energy in the mixed layer is reradiated as infrared radiation. The rate of energy loss to infrared radiation scales as T^4, and is therefore higher in equatorial waters at higher temperatures. The rate of infrared radiation is also affected by cloud cover; more and lower clouds absorb and reradiate the infrared radiation from the ocean surface, trapping the heat by means of the greenhouse effect studied in §23. Note that infrared radiation from the atmosphere is incident everywhere on Earth's surface; this incoming radiation must be included when computing the net thermal radiation lost from the oceans' mixed layers (Problem 27.4). A more detailed discussion of the radiation balance for Earth and its atmosphere is given in §34. In addition to evaporation and radiation, a further contribution to the rate of oceanic heat loss comes from conduction upward into the atmosphere (or overlying ice), which is known to oceanographers and meteorologists as **sensible heat flux**.

The relative strengths of latent heat flux, infrared radiation, and sensible heat flux are highly dependent upon local conditions, including latitude, time of year, and prevailing weather conditions. The global averages of these fluxes are difficult to estimate based on first principles, but can be measured empirically. A detailed discussion of the breakdown of the *oceanic heat budget* is given in [172]. Of the 30–260 W/m² of total insolation, roughly 10–130 W/m² is lost to evaporation through latent heat flux, 30–60 W/m² is lost to infrared radiation, and 2–40 W/m² is lost to sensible heat flux. In the tropics, roughly 15%–20% of the incoming solar radiation is retained as thermal energy in the ocean and is transported through global surface currents away from the equatorial regions.

We can make a rough estimate of the heat carried poleward from the tropics by ocean surface currents. First we estimate the heat absorbed by the 40% of the ocean closest to the equator. With a total surface area of $\sim 140 \times 10^6$ km², and average insolation of ~ 200 W/m², the incoming energy flux amounts to roughly 28 PW. The 15% of the incoming radiation retained and transported by currents amounts to $\cong 4$ PW of power. In Figure 27.2, the estimated heat transport across latitudes is shown, based on satellite data. Around the latitudes of $\pm 20°$, where the rate of energy transport is greatest, roughly 2 PW are transported away from the equator in both the Northern and Southern Hemispheres, in agreement with our rough estimate.

Much of this chapter is devoted to explaining the mechanisms driving the surface currents that carry heat poleward. Wind passing over a rough surface like that of the ocean exerts a force per unit area on the water in the direction in which the wind is blowing. A force per unit area *along* a surface is known as a **stress** – in contrast to a *pressure*, which is a force per unit area perpendicular to a surface. Stress is defined more carefully and described

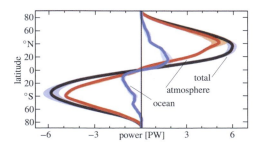

Figure 27.2 Average annual heat transport by ocean and atmospheric circulation, with ± 2 standard deviation range (shaded) [174].

Ocean Thermal Energy Balance

Solar energy is incident on the ocean's surface at an average rate varying from 30–260 W/m². Roughly half of this energy is transferred to the atmosphere through evaporation (*latent heat flux*); additional energy is lost to infrared radiation and conduction (*sensible heat flux*).

In the tropics, roughly 15–20% of the energy is retained in thermal form and transported to higher latitudes by surface currents, resulting in a net poleward flux of roughly 4 PW at latitudes $\pm 20°$.

in §29. One might think that wind stress simply pushes the ocean's surface water along in the direction of wind motion. The actual physics, however, is somewhat more involved and more interesting. In fact, in general surface currents move at an angle to the direction of the wind. This somewhat counterintuitive effect results from the fact that motion on Earth's surface occurs in a *rotating reference frame* due to Earth's spin about its axis. The basic physical mechanism at work is the *Coriolis force*, which also plays an important role in determining global atmospheric circulation patterns.

27.2 Coriolis Force

27.2.1 Inertial Frames and Fictitious Forces

Newton's laws of motion (Box 2.1) state that, in the absence of a force, an object at rest will stay at rest and an object in motion will continue to move in the same direction with constant speed. This famous **principle of inertia**, first formulated by Galileo, does not hold in all reference frames: a ball placed on a car's dashboard will not remain at rest relative to the car as the vehicle turns a corner, although no forces act upon the ball. The reference frames in which Newton's Laws hold are called **inertial frames**. If a reference frame is accelerating when viewed from an inertial frame (like the car turning a corner) it cannot also be an inertial frame. An object viewed in a non-inertial frame may appear to accelerate (like the ball on the dashboard) even though no forces are acting upon it. This acceleration can be described by including *fictitious forces* that act on objects in non-inertial frames. The most familiar example of a fictitious force is the *centrifugal force* experienced by a body at rest with respect to a rotating object or reference frame. The *Coriolis force* is another kind of

fictitious force that acts on bodies that are *in motion* with respect to a rotating reference frame.

A standard class of (approximately[3]) inertial frames are those reference frames in which the nearby stars of our galaxy (known, traditionally, as the **fixed stars**) appear to move at a constant velocity, without acceleration. Earth's surface is not an inertial frame. Because of Earth's rotation about its axis, the fixed stars appear to rotate around the sky every 24 hours.[4] Any object on Earth's surface undergoes rotational motion around Earth's axis, in addition to whatever velocity it may have with respect to Earth. As a result, when Newton's laws are written in terms of Earth-fixed coordinates, they are modified by the addition of fictitious forces.

27.2.2 Centrifugal Force

Before describing the Coriolis force, we first briefly review the more familiar *centrifugal force*. Consider an observer on a rotating object like a merry-go-round, rotating with angular velocity $\boldsymbol{\omega} = \omega \hat{z}$. To an observer at rest on Earth's surface, the person on the merry-go-round is in uniform circular motion and has a constant *centripetal* acceleration $\boldsymbol{a} = -\omega^2 \boldsymbol{r}$ (§2.4). This acceleration is produced by a real force $\boldsymbol{F}$ that is holding her on the merry-go-round – otherwise she would fly off and move in a straight line in the inertial frame. How does the rotating observer describe her situation? She is not accelerating with respect to her frame of reference, yet she must exert a force $\boldsymbol{F} = -m\omega^2 \boldsymbol{r}$ to counteract what she perceives to be a force trying to fling her off the merry-go-round. If she let go, she would appear to accelerate outward from the center of the merry-go-round with acceleration $\omega^2 \boldsymbol{r}$. The rotating observer

---

[3] The fixed stars were used to define an inertial reference frame at the time of Newton, before it was realized that the stars themselves rotate around the center of our galaxy. This subtlety is irrelevant for physics on the scales we consider in this book (Problem 27.5), but suggests that true inertial reference frames are best defined relative to the boundary conditions at infinity in an idealized situation where space-time is infinite. In our universe this is not quite possible due to the curvature of space-time at large distances and the nonzero cosmological constant (see §21). So the notion of an inertial reference frame is always to some extent an approximation.

[4] More precisely, Earth rotates on its axis relative to the fixed stars every 23.9345 hours. This defines the **sidereal day**, the time it takes a given star to repeat a cycle in the sky. This differs from the **solar day**, the time it takes the Sun to repeat its cycle in the sky, because Earth is revolving around the Sun. To a good approximation the number of sidereal days in a year exceeds the number of solar days by one.

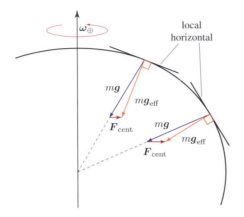

Figure 27.3 A section through Earth showing its oblateness (much exaggerated). In an Earth-fixed reference frame the effective force of gravity $m\boldsymbol{g}_{\mathrm{eff}}$ on an object on Earth's surface is the vector sum of the actual gravitational force $m\boldsymbol{g}$ that points toward Earth's center, and the centrifugal force $\boldsymbol{F}_{\mathrm{cent}}$ that points outward from the axis of rotation.

attributes this urge to accelerate to a *fictitious* **centrifugal force**

$$\boldsymbol{F}_{\mathrm{cent}} = m\omega^2 \boldsymbol{r}\,. \qquad (27.2)$$

The direction of this force is always outward from the axis of rotation.

Since the centrifugal force on Earth's surface due to Earth's rotation is always present and, like gravity, is proportional to mass, it combines with gravity to give a net *effective gravitational force*. Earth's surface formed such that the local horizontal plane is perpendicular to the direction of the effective gravitational force. Since the centrifugal force always points away from the axis, Earth's shape is that of an oblate spheroid, slightly flattened on the poles and fatter around the equator (see Figure 27.3 and Problem 27.6).

27.2.3 The Coriolis Force

The **Coriolis force** is a *velocity-dependent* fictitious force that appears in rotating reference frames and has profound effects on ocean and atmospheric circulations. The Coriolis force, named after nineteenth-century French mathematician and engineer Gustave Coriolis, causes an object moving parallel to Earth's surface in the Northern Hemisphere to accelerate to the right, or clockwise when viewed from above in an Earth-fixed reference frame. In the Southern Hemisphere the effect is reversed; the object is forced to its left, or counterclockwise when viewed from above.

To understand the Coriolis force qualitatively, it is useful to consider several points of view. First, consider an object on which no real forces are acting. The object's trajectory is straight from the point of view of an inertial observer. As shown in Figure 27.4, however, the object appears to curve from the point of view of an observer in a rotating frame as if a force were pushing it to the side.

Alternatively, the presence and nature of the Coriolis force can be deduced from the conservation of angular momentum. Consider an observer at a given latitude $\lambda > 0$, who throws a ball directly to the north in a direction parallel to Earth's surface. The observer is at a distance $r_0 = R_\oplus \cos\lambda$ from Earth's rotational axis. Initially the ball, viewed in an inertial frame, has angular momentum

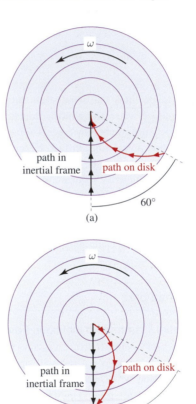

Figure 27.4 The existence of the apparent Coriolis force can be understood from the acceleration that a rotating observer attributes to a trajectory that is perfectly straight from the point of view of an inertial observer. As an example, consider a disk that rotates steadily with angular velocity ω. An object moves on a straight line (black) either inward from the rim to the center (a) or in the opposite direction (b). The disk rotates by 60° during the object's transit. The path marked on the disk (red), corresponding to the trajectory of the object as seen in the coordinate frame of the rotating disk, curves to the right in both cases. In the rotating frame the apparent acceleration would be ascribed to the Coriolis force.

$L = mr_0^2 \omega_\oplus$ around Earth's rotational axis, where $\omega_\oplus$ is the angular velocity of Earth's rotation. As the ball goes north, its distance r from the axis decreases. For the ball's angular momentum $L = mr^2\omega$ to be conserved, as r decreases, the angular velocity ω must increase. Thus the ball moves eastward faster than the ground under it, and it appears to the observer that the object is accelerated to the right. Similarly, if the object is thrown south, r increases, so ω decreases, and the object appears to be accelerated westward, again to the right.

The Coriolis force in a general situation takes the simple form of a vector cross product

$$F_{\text{cor}} = 2m\, \boldsymbol{v} \times \boldsymbol{\omega}, \qquad (27.3)$$

where $\boldsymbol{v}$ is the velocity vector of the object as measured in the rotating frame and $\boldsymbol{\omega}$ is the (vector) angular velocity of rotation. The derivation of this formula is given in Box 27.1. Several features of the Coriolis force are immediately apparent in the expression (27.3). Since $F_{\text{cor}} \cdot \boldsymbol{\omega} = 0$, the Coriolis force only affects motion in the plane perpendicular to $\boldsymbol{\omega}$. And the Coriolis force always acts in a direction perpendicular to the velocity, since $F_{\text{cor}} \cdot \boldsymbol{v} = 0$.

For an object moving on Earth's surface, the Coriolis force has, in general, both horizontal and vertical components. When considering the motions of atmospheric winds or ocean currents, the vertical component of the Coriolis force is usually much smaller than the force of gravity and can be neglected. To separate the horizontal and vertical components we can decompose Earth's angular velocity vector $\boldsymbol{\omega}_\oplus$, which points along its axis of rotation, into components normal and parallel to Earth's surface, $\boldsymbol{\omega}_\oplus = (\omega_\oplus \sin\lambda)\hat{\boldsymbol{n}} + \boldsymbol{\omega}_{\oplus\,\|}$, where $\hat{\boldsymbol{n}}$ is the unit normal. The cross products of these components with $\boldsymbol{v}$ are respectively parallel to and normal to Earth's surface when $\boldsymbol{v} \cdot \hat{\boldsymbol{n}} = 0$, so the horizontal component of the Coriolis force can be written

$$F_{\text{hc}} = 2m\omega_\oplus \sin\lambda\, \boldsymbol{v} \times \hat{\boldsymbol{n}}. \qquad (27.4)$$

When $\hat{\boldsymbol{n}}$ is perpendicular to $\boldsymbol{v}$, horizontal acceleration from the Coriolis force thus always has magnitude $2v\,\omega_\oplus \sin\lambda$. The factor of $\sin\lambda$ makes intuitive sense: at and near the poles where $\sin\lambda = \pm 1$, both horizontal directions are approximately in the plane perpendicular to $\boldsymbol{\omega}_\oplus$ so $|\boldsymbol{v} \times \boldsymbol{\omega}_\oplus| \cong v\omega_\oplus$ and the horizontal Coriolis force is greatest, while at the equator where $\sin\lambda = 0$, $\boldsymbol{v} \times \boldsymbol{\omega}_\oplus$ points vertically, so the horizontal Coriolis force vanishes. Thus, the strength of the horizontal component of the Coriolis force varies significantly with latitude but is independent of the direction of the object's motion.

27.3 Surface Currents

The theory of ocean surface currents was first clearly elucidated by the Swedish oceanographer Vagn Walfrid

Ekman at the beginning of the twentieth century. Ekman's work was motivated by an observation of the Norwegian explorer and scientist Fridtjof Nansen; Nansen observed, while exploring in the Arctic, that icebergs moving under the influence of a prevailing wind do not move directly in the direction of the wind, but rather at an angle of roughly 20° to 40° to the right of the wind direction (in the Northern Hemisphere). This is a consequence of the Coriolis force. The net force on the iceberg that balances the wind force when the iceberg is moving with a constant drift velocity is a combination of the Coriolis force perpendicular to the direction of motion of the iceberg, and the drag force opposite to the direction of motion (see Figure 27.5). A similar principle governs the motion of water on the ocean's surface.

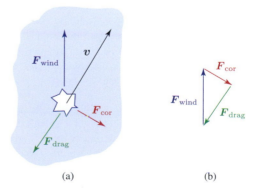

(a)   (b)

Figure 27.5 (a) The Coriolis force causes icebergs to drift in a direction to the right of the direction of the prevailing wind. (b) The force exerted by the wind is cancelled by the combination of Coriolis and drag forces.

Box 27.1　Derivation of the Coriolis Force

To derive the Coriolis force we must relate the rate of change of a vector $\boldsymbol{u}$ in an inertial reference frame Σ to its perceived rate of change in a coordinate system Σ' that is rotating relative to Σ. An observer in the rotating frame Σ' would write $\boldsymbol{u}$ in terms of its components with respect to axes fixed in the rotating frame, $\boldsymbol{u} = u'_x \hat{x}' + u'_y \hat{y}' + u'_z \hat{z}'$, and would measure its rate of change in terms of the time derivatives of those components $d^*\boldsymbol{u}/dt = \dot{u}'_x \hat{x}' + \dot{u}'_y \hat{y}' + \dot{u}'_z \hat{z}'$. We mark this derivative with a * to distinguish it from the time rate of change of $\boldsymbol{u}$ as seen in the inertial frame, which we denote $d\boldsymbol{u}/dt = \dot{u}_x \hat{x} + \dot{u}_y \hat{y} + \dot{u}_z \hat{z}$. The two different time derivatives are related by the effect of rotation

$$\frac{d\boldsymbol{u}}{dt} = \frac{d^*\boldsymbol{u}}{dt} + \boldsymbol{\omega} \times \boldsymbol{u},$$

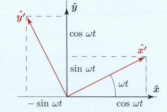

where $\boldsymbol{\omega}$ is the angular velocity of Σ' as observed in Σ. This relation can be derived explicitly by relating the coordinate systems; taking $\boldsymbol{\omega} = \omega \hat{z}$, $\hat{x}' = \hat{x} \cos \omega t + \hat{y} \sin \omega t$, and $\hat{y}' = \hat{y} \cos \omega t - \hat{x} \sin \omega t$ (see the figure at right), so $u_x = u'_x \cos \omega t - u'_y \sin \omega t$ and $u_y = u'_y \cos \omega t + u'_x \sin \omega t$, from which the above relation between the derivatives of $\boldsymbol{u}$ in the different frames follows directly (Problem 27.12).

To learn how an object is perceived to accelerate in the rotating frame Σ' we must apply the equation above twice to the vector $\boldsymbol{r}$,

$$\frac{d^2\boldsymbol{r}}{dt^2} = \left(\frac{d^*}{dt} + \boldsymbol{\omega} \times\right)\left(\frac{d^*\boldsymbol{r}}{dt} + \boldsymbol{\omega} \times \boldsymbol{r}\right) = \frac{d^{*2}\boldsymbol{r}}{dt^2} + 2\boldsymbol{\omega} \times \frac{d^*\boldsymbol{r}}{dt} + \boldsymbol{\omega} \times (\boldsymbol{\omega} \times \boldsymbol{r}),$$

where we have expressed each time derivative in Σ in terms of the perceived rate of change in Σ'. Finally, we use Newton's second law in the inertial frame, $\boldsymbol{F} = m d^2\boldsymbol{r}/dt^2$, and simplify the notation by replacing $d^*\boldsymbol{r}/dt \equiv \boldsymbol{v}$ and $d^{2*}\boldsymbol{r}/dt^2 \equiv \boldsymbol{a}$, where now $\boldsymbol{v}$ and $\boldsymbol{a}$ are understood to refer to the velocity and acceleration measured with respect to axes fixed in Σ',

$$m\boldsymbol{a} = \boldsymbol{F} - 2m\,\boldsymbol{\omega} \times \boldsymbol{v} - m\,\boldsymbol{\omega} \times (\boldsymbol{\omega} \times \boldsymbol{r}).$$

The second term on the right is the Coriolis force and the last term is the centrifugal force (27.2).

Example 27.1　Can Birds Use the Coriolis Force to Navigate?

It has been suggested that migratory birds detect the Coriolis force and use it to navigate. Is this possible?
Since the strength of the Coriolis force does not vary with the direction in which a bird flies, it cannot orient itself using the horizontal component of the Coriolis force alone. It has been suggested that a bird can sense its latitude and the direction of the *meridian* (a line of constant longitude running north–south), by finding a direction of flight in which the Coriolis force vanishes. This occurs when $\boldsymbol{v} \parallel \pm\boldsymbol{\omega}_\oplus$. Choosing $(\hat{x}, \hat{y}, \hat{z})$ to be east, north, and vertical respectively, $\boldsymbol{\omega}_\oplus = \omega_\oplus(0, \cos\lambda, \sin\lambda)$, it is clear that when $\boldsymbol{v} \propto \pm\boldsymbol{\omega}_\oplus$, the bird's pitch (rate of rise or descent) measures the latitude and its direction in the xy-plane determines the meridian. Let us estimate the Coriolis force on a bird in order to see what it is up against if it attempts to navigate in this way.

Consider, for example, a blackpoll warbler, which migrates twice a year between South America and its breeding grounds in Canada. A typical blackpoll warbler mass is 15 g and its ground speed may be as high as 25 m/s. The magnitude of the Coriolis force on the bird is $F_c = 2m\omega_\oplus v = 2(0.015\,\text{kg}) \times (2\pi/(24 \times 3600)\,\text{s}^{-1}) \times (25\,\text{m/s}) \cong 5.5 \times 10^{-5}\,\text{N}$. This is 0.037% of the force of gravity on the warbler and equivalent to a cross-wind of roughly 0.08 m/s (assuming a cross-sectional area of $(0.15 \times 0.05)\,\text{m}^2$ and a drag coefficient of $c_d \sim 0.1$). It seems like a tough task for a flying bird to detect such a small force.

27.3.1 Inertial Currents and the Ekman Spiral

Consider first the horizontal motion of a patch of water on the surface of the ocean set in motion by the wind, but thereafter left to evolve under the (horizontal) Coriolis force (27.4) alone. Unlike the iceberg, drag forces on the patch of water are weak and can be neglected in a first approximation. The horizontal acceleration of an object acted upon by only the Coriolis force (27.4) has a constant magnitude $a_{hc} = 2\omega_\oplus v \sin\lambda$ and points in a direction at right angles to the velocity. Such an acceleration gives rise to uniform circular motion (2.35), with $a = \omega^2 r = \omega v$, so the angular frequency of this motion,

$$\omega = 2\omega_\oplus \sin\lambda, \qquad (27.5)$$

is determined by Earth's rotational frequency $\omega_\oplus$ and the latitude λ. Circular motion of a fluid along such a trajectory is called an **inertial current**. The period of this motion is typically hours or days, since $\omega_\oplus = 2\pi/\text{day}$. At the latitude of Boston, for example, $\sin\lambda \cong 0.67$, so the period of circular inertial current motion is roughly 18 hours. Such inertial currents are a common feature of ocean surface motion, and have been observed at many different locations in Earth's oceans. Without additional forcing, inertial currents generally dissipate over a period of many days due to friction.

Next, we consider the effect of a constant wind on a region of the ocean surface. Naively, one might think that a steady wind blowing from the south would produce a northward current. The example of iceberg drift, however, suggests that the Coriolis force will alter this. Unlike the iceberg, however, where drag leads to a constant drift velocity, for a patch of surface water moving with negligible friction the combination of a steady wind and the Coriolis force leads to a circular motion superimposed over steady drift. The situation is mathematically equivalent to the motion of a charged particle in constant crossed E and B fields in electromagnetism. The equation of motion for a patch of ocean surface is

$$m\boldsymbol{a} = \boldsymbol{F}_{\text{wind}} + \boldsymbol{v} \times (2m\omega_\oplus \sin\lambda\hat{\boldsymbol{n}}). \qquad (27.6)$$

If we make the identifications

$$\boldsymbol{F}_{\text{wind}} \Leftrightarrow q\boldsymbol{E},$$
$$2m\omega_\oplus \sin\lambda\hat{\boldsymbol{n}} \Leftrightarrow q\boldsymbol{B}, \qquad (27.7)$$

then eq. (27.6) becomes the equation for a charge q subject to the Lorentz force (3.45),

$$m\boldsymbol{a} = q\boldsymbol{E} + q\boldsymbol{v} \times \boldsymbol{B}, \qquad (27.8)$$

with constant and perpendicular E and B fields. This field configuration was encountered in connection with

magnetic confinement fusion in §19.6.3, and the resulting motion was discussed qualitatively there. The equation of motion (27.8) is solved in Box 27.2, with the result that the patch circulates at the same frequency derived above ($\omega = 2\omega_\oplus \sin\lambda$) as it drifts on average in a direction at right angles to the wind with speed $v_d = |\boldsymbol{F}_{\text{wind}}|/(2m\omega_\oplus \sin\lambda)$.

Thus, for example, in the Northern Hemisphere a wind blowing to the north gives rise to an average surface velocity $\boldsymbol{v}_d$ in an easterly direction, which in turn gives an average Coriolis force to the south that cancels the force from the wind. The characteristic pattern of periodic circular motion superimposed over uniform drift can be seen in ocean currents as measured by satellite (see Figure 27.6(b)).[5]

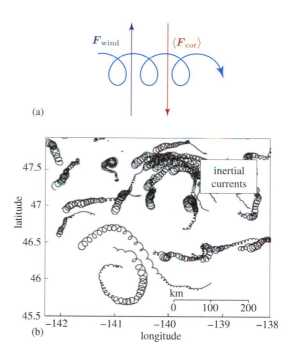

(a)

(b)

Figure 27.6 (a) In a constant wind coming from the south, inertial drift carries water to the east. The average Coriolis force $\langle\boldsymbol{F}_{\text{cor}}\rangle$ cancels the force of the wind. (b) Motion of buoys following surface currents (at 15 m depth) in the North Pacific tracked by NOAA satellites over a one-month period in 1987. Figure from [172].

---

[5] Note that the motion described here refers to homogeneous motion of a large patch of the ocean's surface that is subject to the driving wind, and not to individual eddies within which motion at different points occurs simultaneously in opposite directions, as might be inferred from the figure.

Box 27.2 $E \times B$ and Coriolis Drift

As described in the text, the response of a patch of ocean to a constant forcing wind is mathematically equivalent to the response of a particle of charge q to the Lorentz force produced by constant and perpendicular E and B fields. Here we solve the corresponding electromagnetism problem and then carry the analogy over to the case of ocean currents. To be specific we restrict the motion to the xy-plane, with an electric field $E = E\hat{y}$ in the y-direction and a magnetic field $B = B\hat{z}$ in the z-direction. Working out the cross product, we find $v \times B = B(v_y\hat{x} - v_x\hat{y})$ and the Lorentz force law (27.8) becomes

$$m(\ddot{x}, \ddot{y}) = (Bq\dot{y}, qE - Bq\dot{x}),$$

which has the solution

$$x(t) = r\cos\omega t + v_d t,$$

$$y(t) = -r\sin\omega t,$$

where $\omega = qB/m$ is the **cyclotron frequency** and $v_d = E/B$ is the drift velocity. This describes a clockwise circulation superimposed over a constant drift in the x-direction. Note that, as claimed in the discussion of $E \times B$ drift in §19.6.3, the direction of the drift is independent of the charge q.

Returning to the ocean current problem, and exploiting the analogy of eq. (27.7), we see that ocean currents moving under the influence of wind and the Coriolis force follow the same trajectory as that above, where

$$\omega = 2\omega_\oplus \sin\lambda,$$

$$v_d = |F_{\text{wind}}|/(2m\omega_\oplus \sin\lambda).$$

This description of surface currents is somewhat simplified, though it holds for net motion integrated over depth since there are no other net forces acting on the system; in a more complete analysis, following Ekman's original work, viscosity and the dependence of the surface currents on depth should be considered. A more detailed presentation of Ekman's analysis, including these factors, is given in [172]. The upshot of the analysis is that in a constant wind, the average surface motion over many periods of inertial current rotation is described by linear motion in a direction that varies with depth. At the surface, the average direction of motion (in the Northern Hemisphere) is roughly $\pi/4$ ($45°$) to the right of the wind. Below the surface, the water flows at an angle $\pi/4 + z/z_0$ to the right of the wind, where z is the depth and z_0 is a characteristic depth of the motion. The amplitude of the current at depth z decreases exponentially as e^{-z/z_0}. This structure, in which the angle of the current grows linearly with depth while the magnitude decreases exponentially, is known as an **Ekman spiral**, and is displayed graphically in Figure 27.7.

The **Ekman layer depth** is the depth at which the current velocity is opposite to the surface velocity, $D_E = \pi z_0$. This depth typically ranges from 50 to 300 meters. Empirical measurements indicate that surface currents follow this idealized theoretical model reasonably well in steady-state wind conditions in the deep ocean.

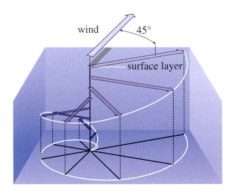

Figure 27.7 An Ekman spiral describes the time-averaged currents driven by the wind. Viscous and Coriolis forces combine to give a progressive rotation of the current direction to the right (in the Northern Hemisphere) with increasing depth. (Adapted from Data Stream Ocean and used with permission of the American Meteorological Society)

27.3.2 Ekman Mass Transport

Ekman's analysis shows that the time-averaged direction and speed of currents driven by a constant wind vary with depth. This does not, however, change the general conclusion that net mass transport is perpendicular to wind velocity, because the integral of the component of velocity parallel to the wind over all depths vanishes,

$$\int_0^\infty e^{-z/z_0} \cos(\pi/4 + z/z_0) = 0. \tag{27.9}$$

Thus, over long periods of time a substantial volume of water will flow on average in a direction perpendicular to the wind direction – to the right in the Northern Hemisphere, and to the left in the Southern Hemisphere (Problem 27.13). Ocean motion of this kind is referred to as **Ekman mass transport**. Oceanic volume transport is typically measured in units of **Sverdrups** (abbreviated Sv), where[6]

$$1 \, \text{Sv} = 10^6 \, \text{m}^3/\text{s} . \tag{27.10}$$

Ekman mass transport of oceanic surface waters plays an important role in regulating climate systems, both locally and globally. Locally, for example, prevailing wind patterns that carry air southward towards the equator along western boundaries of Northern Hemisphere continents (such as prevailing winds from the northwest along the California coast) cause Ekman mass transport to the west, away from the land mass. To replace the surface waters transported away, upwelling of deeper ocean waters brings colder water to the surface. This upwelled colder water generates the foggy and cooler conditions typical of San Francisco and the California coast in general. The upwelled water in such situations is also generally rich in nutrients and supports major fish populations.

27.4 Atmospheric Circulation

To understand the large-scale structure of surface currents in basins such as the Atlantic and Pacific Oceans, it is necessary to have some information about atmospheric circulation and the large-scale wind patterns that drive these ocean currents. This is a large topic, and we only summarize briefly here some of the simplest and most important global features. We delve more deeply into atmospheric circulation and wind patterns in the following chapters on wind energy, and global atmospheric circulation is discussed further in §34 in the context of global climate and radiative equilibrium.

The simplest aspects of global circulation can be seen when atmospheric motion is averaged over time and around the globe along lines of constant latitude. In this **zonal-mean** picture, only the longitudinally averaged motion of air in the vertical and north–south (latitudinal, or **meridional**) directions is considered. The result is known as the **mean meridional circulation**.

The dominant feature in the mean meridional circulation originates when warm moist air produced by evaporation

---

Wind-Generated Ocean Surface Currents

Surface currents generated by a constant wind move on average perpendicular to the direction of wind flow. The *Ekman spiral* describes the average current as a function of depth z, with velocity $v = v_0 e^{-z/z_0}$ and angle $\pi/4 + z/z_0$ from the wind direction. In the Northern (Southern) Hemisphere the current is to the right (left) of the wind direction.

---

of ocean surface waters near the equator rises. This air rises to an altitude of roughly 17 km (the height of the *tropopause* in equatorial latitudes – see §34.2.4), and then moves poleward to roughly latitude ±30°, where it sinks as colder drier air (see Figure 27.8). This circulation pattern is known as the **Hadley cell**, after the English meteorologist George Hadley. The locations of the Northern and Southern Hemisphere Hadley cells oscillate over the year. In the Northern (Southern) Hemisphere winter, the dividing line between the Hadley cells, known as the **intertropical convergence zone**, moves south (north) following the declination latitude. The Hadley cell is larger in the winter hemisphere and smaller in the summer hemisphere. The Hadley cells transport a significant amount of thermal energy from the tropics to the subtropics (latitude $|\lambda|$ between $\epsilon \cong 23.44°$ and $\approx 35°$).

Similar to the Hadley cells, **Polar cells** are formed by rising air at $\sim \pm 60°$ latitude that is drawn towards the poles at the tropopause (which is closer to 9 km in altitude in polar regions). Air from the polar cells drops as cold dry air near the poles. Between the Hadley and Polar cells, **Ferrell cells** operate in the opposite direction, with air at latitudes just below ±60° rising, and returning to ±30° latitude before dropping towards the surface again. Ferrell

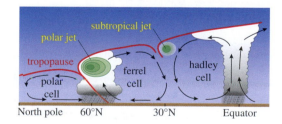

Figure 27.8 Dominant patterns of mean meridional global circulation, showing the Hadley, Ferrel, and Polar cells. (Adapted from US National Oceanic and Atmospheric Administration)

Example 27.2   Ekman Mass Transport

Winds blowing from the south on the open ocean at latitude $+45°$ lead to surface flow along an Ekman spiral with a layer depth $D_E = 100\,m$ and a surface speed of $v_s = 3\,m/s$. What is the rate of mass transport and the wind stress on the ocean's surface?

Let the wind direction (to the north) define the $\hat{y}$-axis, and let $\hat{x}$ point to the east. According to Ekman, the components of the time-averaged water velocity are

$$v_y = \cos(\pi/4 + z/z_0)e^{-z/z_0}v_s,$$
$$v_x = \sin(\pi/4 + z/z_0)e^{-z/z_0}v_s,$$

where $z_0 = D_E/\pi$.

There is no net transport to the north since $\int_0^\infty dz\, v_y = 0$. The net eastward mass transport per unit distance is

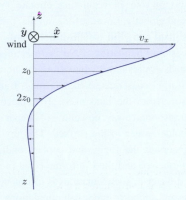

$$\frac{dm}{dy\,dt} = \int_0^\infty dz\,\rho v_x = \frac{\rho v_s z_0}{\sqrt{2}} = \frac{\rho v_s D_E}{\pi\sqrt{2}}$$
$$= (1030\,\text{kg/m}^3) \times (3\,\text{m/s}) \times (100\,\text{m})/(\pi\sqrt{2}) \cong 7.0 \times 10^4\,\text{kg/m\,s},$$

where $\rho \cong 1030\,\text{kg/m}^3$ is the density of seawater. The figure shows the eastward component of the current velocity as a function of depth.

The wind stress (force per unit area) can be obtained from the last equation in Box 27.2. Consider the force dF_wind on a volume $dV = dA\,dz = dx\,dy\,dz$ of the ocean's surface layer. The mass in this layer is $dm = \rho\,dA\,dz$, and has net motion in the x-direction with velocity v_x given above. Thus,

$$dF_\text{wind} = 2\omega_\oplus \sin\lambda\,dA \int_0^\infty dz\,\rho v_x = 2\omega_\oplus \sin\lambda\,\frac{\rho v_s D_E}{\pi\sqrt{2}}\,dA,$$

and the stress $T_\text{wind} = dF_\text{wind}/dA$ is given by

$$T_\text{wind} = \frac{\sqrt{2}}{\pi}\omega_\oplus \rho v_s D_E \sin\lambda = \frac{1}{\pi}\left(\frac{2\pi}{3600 \times 24}\text{s}^{-1}\right) \times (1030\,\text{kg/m}^3) \times (3\,\text{m/s}) \times (100\,\text{m}) \cong 7.2\,\text{N/m}^2.$$

---

cells are a secondary effect, playing a role analogous to that of a ball bearing, and play less of a role in energy transport. In mid-latitudes, circulation of air associated with cyclones and anticyclones plays a greater role in poleward energy transport.

For the purposes of this chapter, the important common feature of Hadley, Ferrell, and Polar cells is the near-surface winds that they produce. To "close the loop" of a Hadley cell, air must travel towards the equator from latitudes around $\pm30°$. The Coriolis force acts on this flow, forcing the winds towards the west as they approach the equator. These prevailing easterly[7] winds, coming from the northeast in the Northern Hemisphere and from the southeast in the Southern Hemisphere, are known as the

trade winds (see Figure 27.9). In a similar fashion, in mid-latitudes from $\pm30°$ to $\pm60°$ the prevailing winds are westerly winds that generally move away from the equator. In high latitudes above $\pm60°$, the prevailing winds are again easterly and carry air away from the poles.

The winds generated by the global atmospheric circulation pattern in turn affect ocean circulation. With westerly winds at mid-latitudes and easterly winds at lower latitudes, the ocean basins in the Northern Hemisphere have dominant wind patterns that contain components of clockwise motion and exert clockwise torque on the ocean surface. Similarly, counterclockwise wind patterns dominate in the Southern Hemisphere.[8] This pattern of wind motion

---

[7] Winds are generally denoted by the direction from which they come, so an "easterly" wind comes from the east.

[8] While air does not travel in closed clockwise or counterclockwise loops near the surface in the Northern and Southern Hemispheres, the average

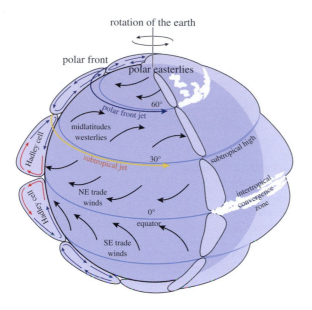

Figure 27.9 Hadley cells drive prevailing winds towards the equator in latitudes below 30°. The Coriolis force pushes these winds to the west, giving the easterly winds known as *trade winds*. Similarly, Polar cells drive easterly winds at high latitudes while at mid-latitudes the winds are westerly and tend away from the equator, following the circulation of the secondary Ferrell cells.

produces corresponding patterns in ocean circulation. We return to other aspects of global wind patterns in §28, when we consider the extent of the global resource of wind energy.

27.5 Ocean Circulation

On a global scale, Ekman mass transport, as described in §27.3, plays a key role in driving the currents that transport thermal energy from the equatorial region to the poles. The detailed structure of wind-driven oceanic surface currents is very complex, and is described by nonlinear equations with complicated boundary conditions. Some core ideas, however, were developed by the oceanographers Harald

clockwise/counterclockwise motion can be described precisely using the notion of *vorticity*. Mathematically, the vorticity of a vector field v such as the wind velocity is defined to be the curl $\nabla \times v$ of the vector field v (see Appendix B.1.3 for further discussion of the curl operation). Vorticity for fluid flow is defined and discussed in more detail in §29.2.3. The average of the local vorticity gives a measure of the net (counter-)clockwise motion of a velocity field.

Global Atmospheric Circulation

The largest-scale features in the circulation of Earth's atmosphere are *Hadley cells* produced by rising warm moist air from equatorial regions that circulates to near ±30° and produces near-surface easterly winds in the lower latitudes known as the *trade winds*. Similar *Ferrell* and *Polar* circulation cells produce westerly winds in mid-latitudes and easterly winds above ±60° latitude. These winds give a pattern of air circulation that is clockwise (counterclockwise) over the large ocean basins in the Northern (Southern) Hemisphere.

Sverdrup (Norwegian), Henry Stommel (American), and Walter Munk (American) in the middle of the twentieth century. These ideas use the basic physics of the Coriolis force and Ekman mass transport, combined with boundary conditions on an ocean basin, to give a qualitative understanding of some principal features of large-scale rotating ocean currents (**gyres**). An excellent and readable account of these ideas is given in [172]. We give a brief summary here of the important points.

We have seen that mass transport is generally at right angles to wind velocity, and is proportional to the strength of the Coriolis force. At latitude λ the horizontal component of the Coriolis force (27.4) is proportional to $\omega_\oplus \sin \lambda$. Thus, the Coriolis force that drives mass transport is greater at higher latitudes. This leads to the conclusion, first reached by Sverdrup, that a wind pattern with clockwise (counter-clockwise) vorticity drives southward (northward) mass transport (see Figure 27.10 and Problem 27.10) As a simple illustration, a wind from the west at latitude $\lambda + \delta\lambda$ drives a southward surface current proportional to $\omega_\oplus \sin(\lambda + \delta\lambda) \approx \omega_\oplus(\sin \lambda + \cos \lambda \, \delta\lambda)$. On the other hand, wind of the same strength from the east at latitude λ will generate a northward surface current proportional to $\approx \omega_\oplus \sin \lambda$. The difference in mass transport will be a southward current proportional to $\omega_\oplus \cos \lambda \delta\lambda$.

It was shown by Stommel that for a fluid in an idealized rectangular basin subject to clockwise wind stress, the Coriolis force (in the Northern Hemisphere) generates an asymmetric steady-state solution with a broad current going south across the basin and a narrow northward boundary current on the west side of the basin, combining to produce a generally clockwise motion of the fluid (see Figure 27.11). Munk added eddy viscosity to the framework developed by Sverdrup and Stommel, showing that

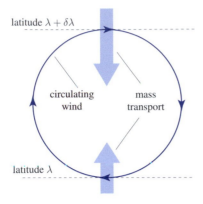

Figure 27.10 A clockwise circulating wind in the Northern Hemisphere drives net southward mass transport. Wind from the west at a higher latitude $\lambda + \delta\lambda$ generates more southward mass transport than the same wind from the east at lower latitude λ, since the Coriolis effect is proportional to $\sin\lambda$.

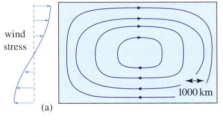

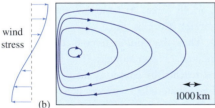

Figure 27.11 Flow in a basin as computed by Stommel, driven by clockwise wind stress: (a) without rotation, (b) with "Coriolis" force varying linearly along the vertical axis. (Credit: Adapted from M. Stommel, the Westward Intensification of Wind-Driven Ocean Currents, Transactions American Geophysical Union **29**, 202 (1948))

in general the flow contains a top layer of wind-driven currents over a relatively static bulk liquid.

Together, these ideas give a fairly complete, if simplified, model of basin-wide wind-driven circulation, and provide a reasonable qualitative picture of real ocean current flow. Although in actual ocean basins, the dynamics is much more complicated, many of the general features are correctly given by this model. In the Atlantic ocean basin,

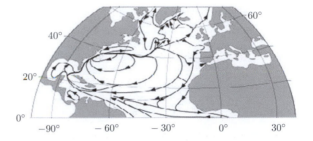

Figure 27.12 The primary ocean circulation pattern in the North Atlantic is a clockwise gyre driven by clockwise wind vorticity. A western boundary current flowing north is balanced by southward flow across the basin on the surface and in the deep ocean. (Credit: Fleming, R.H.; Johnson, M.W.; Hamre, A.; Sverdrup, H.U. *Oceans*; 1st edn, ©1942. Reprinted by Permission of Pearson Education.)

for example, the primary wind pattern has a clockwise vorticity, as discussed in the preceding section. The dominant surface circulation pattern in the Atlantic is a gyre characterized by an asymmetric vortex in which across most of the basin surface currents carry mass south (driven by the clockwise wind vorticity), balanced by a narrow northward current (the **Gulf Stream**) that carries mass back to the north (see Figure 27.12). Another way to think of this balance is in terms of a clockwise torque on the Atlantic by the wind, which is balanced by a counterclockwise torque from friction on the western boundary current. A similar western boundary current known as the **Kuroshio** runs northward along the western edge of the Pacific near the east coast of Asia. Both the Gulf Stream and the Kuroshio carry water volume fluxes on the order of 30 Sv. Analogous southward-running currents arise in the Southern Hemisphere at the east edge of South America, Africa, and Australia. These are not as extensive, however, due in part to the fact that the land masses in the Southern Hemisphere are smaller and do not block the strong eastward **Antarctic Circumpolar Current** (ACC), which is produced by the prevailing westerly winds at mid-latitudes in the Southern Hemisphere. While the ACC has a slow flow rate compared to western boundary currents such as the Gulf Stream, it is very broad and runs deeply, so that its overall flow rate is the strongest current on the planet, with a total flux on the order of 135 Sv.

This gives us a qualitative understanding of the basic physical principles driving many of the major ocean surface currents. Currents such as the Gulf Stream carry large amounts of thermal energy in warm water from the tropics towards the poles, as discussed in §27.1. As this warm water flows to the cooler polar region, it cools through evaporation, radiation, and conduction. Evaporation, and

Ocean Circulation

In Northern Hemisphere ocean basins, clockwise wind vorticity drives a diffuse southward surface current flow over most of the basin, balanced by a strong poleward-moving boundary current along the western edge of the basins. A similar but opposite and less pronounced pattern occurs in the Southern Hemisphere. The boundary currents carry thermal energy towards the poles, where after cooling and evaporation they become more saline and sink, driving the *meridional overturning circulation* that maintains the temperature difference between surface waters and the abyss.

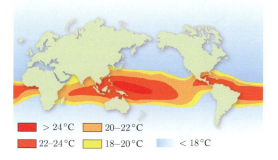

▮ > 24 °C	▮ 20–22 °C
▮ 22–24 °C	▮ 18–20 °C ▯ < 18 °C

Figure 27.13 Temperature difference between surface and depth of 1000 m.

formation of icebergs when the water reaches sufficiently high latitudes increase the salinity of the water. Cooler, more saline water is more dense. This dense water sinks down towards the ocean floor, spreading again towards the equator in the abyssal deep ocean current. The component of global ocean currents that are driven by temperature and salinity in this fashion are known as **thermohaline** currents. Taken together, the poleward-flowing western boundary currents and the thermohaline current drive the *meridional overturning current* (Figure 27.1), which maintains the cold temperatures of around 5 °C in the deep ocean. The poleward mass flow in the western boundary currents returns towards the equator in both the deep ocean current and the diffuse surface current across the ocean basin; in the North Atlantic, for example, the roughly 30 Sv in the Gulf Stream divides roughly evenly into 15 Sv in southward surface mass flow across the Atlantic and 15 Sv in southward mass flow in the deep ocean below 1000 m (see e.g. [175]). We discuss the role of oceanic heat transport in regulating global climate in §34.

27.6 Ocean Thermal Resources and Ocean Thermal Energy Conversion (OTEC)

From the study of heat engines and Carnot efficiency in previous chapters, we know that energy can be extracted whenever there is a temperature difference between two systems. The efficiency with which thermal energy can be extracted from the hotter system depends crucially on the temperature difference. In the case of ocean thermal energy, solar heating and global ocean currents set up a temperature difference between the surface layer and the deep ocean. While the temperature difference is not

large, the vast heat content of the ocean means that the energy potential of this system is immense. **Ocean thermal energy conversion** (OTEC) involves using this ocean heat difference to drive a heat engine.

A global map depicting the temperature difference between the surface and water at a depth of 1000 m is shown in Figure 27.13. Throughout much of the tropics, this temperature difference is in the range 17–22 °C. To obtain a rough order-of-magnitude estimate of the total thermal energy available from this temperature difference, we estimate the area of the tropical waters to be $A \cong 70 \times 10^6$ km^2. We take the average temperature difference to be 20 K, and assume an average depth of the oceanic mixed layer of 100 m, which yields a total energy content in the mixed layer of

$$100 \text{ m} \times A \times 20 \text{ K} \times 4 \text{ MJ/m}^3 \text{ K} \cong 550 \text{ ZJ}. \quad (27.11)$$

This is a tremendous amount of energy. An upper limit on the rate at which energy might be sustainably extracted from the ocean thermal energy in the tropics can be obtained by estimating the rate of solar energy absorption. As discussed in §27.1, the net absorption rate is roughly 30 W/m^2, giving a total of several thousand terawatts over the total area of the tropical oceans. The upper bound could be raised even higher in principle by suppressing energy loss from the ocean surface by evaporation and radiation. Due to the diffuse nature of the resource, however, and the small temperature differences involved, extracting even a small fraction of this potential resource as usable power is quite difficult in practice.

The idea of extracting energy from the ocean was suggested in Jules Verne's fictional book *20 000 Leagues Under the Sea*. A decade or so later, the French physician and physicist Jacques d'Arsonval made a concrete proposal for ocean thermal energy conversion using the heat difference between warm surface waters and deeper colder ocean water. The basic idea is to pump cold water up from the abyss for use as an entropy sink for a heat engine. The first functioning OTEC generator was built in 1930 in Cuba by d'Arsonval's student George Claude, with an output power of 22 kW. Since then, this idea has been pursued in various contexts. The US initiated research on OTEC at the National Energy Laboratory of Hawaii (NELH) in the 1970s, and built a 50 kW **mini-OTEC** device there. A larger 255 kW OTEC system was built at NELH in 1993 and ran successfully for six years. As of 2017, the only OTEC plants in operation worldwide are a 50 kW Japanese plant at Saga University and a 100 kW grid-connected closed-cycle plant in Hawaii that became operational in 2015.

The greatest challenge to OTEC is the small temperature difference. The Carnot limit on the efficiency of a heat engine operating between 27 °C and 5 °C is only

$$\eta_{\text{Carnot}} \cong \frac{300 - 278}{300} \cong 7\% \,. \tag{27.12}$$

Furthermore, the need to pump water from 1000 m, and the difficulty in transferring heat across a small temperature difference, make it hard to achieve anything like even this low efficiency in practice. (Note that since the density of the colder water at 1000 m is only slightly higher than the density of the surface water, and every liter of water pumped up is replaced by a liter of water from above, the minimum energy needed for pumping is negligible; the pumping energy used is primarily due to losses in the system.) Another major challenge to large-scale OTEC implementation is the difficulty of transmitting recovered energy to shore from devices deployed in deep water far from land.

OTEC systems have been developed based on both open- and closed-cycle heat engines. In open-cycle systems (see Figure 27.14), warm surface waters are evaporated in a partial vacuum, and then condensed on a surface chilled by cold water from the depths, driving a small turbine. Closed-cycle systems are based on standard closed heat engine cycles and generally use ammonia or another fluid with desirable phase-change properties. Overall efficiencies realized by existing OTEC systems are in the range of 2–3%.

Despite the low efficiency and technical challenges to widespread deployment, OTEC has some appeal as a renewable resource, particularly in island locations near the equator where deep ocean water is available near shore and other energy options are expensive (as in Hawaii). OTEC also has a number of potential auxiliary benefits. In

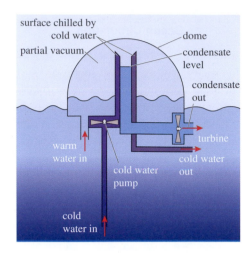

Figure 27.14 Open-cycle OTEC system.

OTEC

Thermal energy is absorbed by equatorial ocean surface waters from solar radiation at a rate of several thousand TW. Extracting this energy is difficult, however, due to the Carnot efficiency limit of roughly 7%, and logistical challenges for off-shore energy harvesting. Functioning OTEC systems developed to date have had efficiencies in the range 2–3% or less.

hot climates, the cold water pumped up from the depths can be used for cooling buildings or machinery. Open-cycle OTEC systems output distilled water, so an OTEC system can double as a desalination plant with negative energy cost. The cold water brought up from the deep ocean is rich in nutrients and can be used for aquaculture. Finally, it has been proposed that extraction of minerals (such as uranium) from the deep sea water brought to the surface by OTEC systems may at some point be economically feasible (see also §16.2).

Discussion/Investigation Questions

27.1 The Coriolis force determines the directions of the trade winds, mid-latitude Westerlies, and Polar Easterlies as described in the text. What about the strength of these winds? How, for example, should the trade winds behave as one approaches the Equator? How should the mid-latitude Westerlies in the Southern Hemisphere vary with latitude? Do your speculations agree with observations?

27.2 Use the Coriolis force to explain the direction of winds circulating around storms (low-pressure areas) and fair weather systems (high-pressure areas) in the Northern and Southern Hemispheres.

27.3 It is often said that the Coriolis force determines the direction that water circulates as it goes down a drain. Do you think this is possible? Perhaps an experiment is in order?

27.4 Discuss and research in more detail the open- and closed-cycle OTEC systems discussed in §27.6. Example questions: Why is such low pressure required for the open-cycle system? Why can the open system double as a desalinator with "negative energy cost"? Can you sketch the cycles in the ST-plane in the fashion of §13?

Problems

27.1 The absorption coefficient of water for light of wavelength 450 nm is roughly $\kappa_{450} \cong 0.0001 \, \mathrm{cm}^{-1}$ while for light of wavelength 750 nm it is roughly $\kappa_{750} \cong 0.03 \, \mathrm{cm}^{-1}$. Estimate the depth by which 90% of light has been absorbed by the ocean in each of these wavelength ranges. Explain the color seen underwater at typical depths encountered by divers.

27.2 Assuming that one half of the energy in solar radiation absorbed at the oceans' surface is lost to latent heat flux (evaporation), give an order-of-magnitude estimate of the rate at which water vapor enters the atmosphere from evaporation on the oceans' surface.

27.3 Consider a location in the tropical ocean where the mixed layer has a uniform temperature of 30 °C to 100 m, and the deep ocean temperature is 5 °C at 1000 m. Estimate the rate of conductive heat flow downward between 100 m and 1000 m, assuming a constant gradient between these depths. At this rate, how long would it take to heat the ocean depths between 1000 m and 2000 m by 5 °C?

27.4 Under the same assumptions as for Problem 27.3, estimate the net rate at which energy is lost to upward thermal radiation. First, compute the rate of upward thermal radiation. Then include a downward IR radiation flux, approximated by thermal radiation from the atmosphere at an altitude of 2000 m at 20 °C, and compute the net upward energy flux.

27.5 Estimate the centrifugal forces acting on a 60 kg person standing at the equator due to (a) Earth's rotation about its axis, (b) Earth's yearly revolution about the Sun, and (c) the Sun's revolution about the center of the Milky Way galaxy.

27.6 Given Earth's radius, mass, and rate of rotation, estimate the difference in radius measured along the equator from that measured to a pole in order that the apparent gravitational potential is the same everywhere on Earth's surface by approximating the gravitational potential as that of a spherical Earth plus a correction term due to centrifugal force. Compare your answer to the actual difference. Can you see how this approximation could be improved?

27.7 If a pitcher throws a baseball at 44 m/s (100 mph), estimate the deviation in position, both in magnitude and direction, due to the Coriolis force when the ball reaches the batter after traveling 18.5 m, when the game is played at a latitude of (a) 30° N, (b) 60° S, or (c) on the equator. (See Problem 2.13 for more about baseball.)

27.8 Compare the gravitational force and horizontal Coriolis force on a person running at 6 m/s at latitude 45°. How fast would an object have to be moving for the two to be equal?

27.9 Assume that a constant wind at $v = 20 \, \mathrm{knots} \, (10.3 \, \mathrm{m/s})$ is blowing in a southward direction on the eastern side of an ocean basin at latitude +35°. Ignore viscous effects and the dependence of surface currents on depth, and instead assume that the top 50 m of ocean surface moves uniformly in response to the stress exerted by the wind. If the wind stress on the ocean's surface is estimated as a horizontal force per unit area $T_{\mathrm{wind}} = \rho c_d v^2$, with ρ atmospheric mass density and $c_d \cong 0.0015$, compute the average drift velocity of the surface water and the rate of mass upwelling along the western side of the continent.

27.10 Estimate the wind in a tropical region at latitude $(30 + y)°$ North as $v = 2y \, \hat{x} \, \mathrm{m/s}$, where $\hat{x}$ is a unit vector pointing eastward. Use the formula from Problem 27.9 to estimate the wind stress on the surface waters. Estimate the rate of southward mass transport assuming $z_0 = 50 \, \mathrm{m}$.

27.11 If an OTEC system could achieve 1/2 of Carnot efficiency, at approximately what rate could energy be extracted sustainably from ocean surface waters in the tropics? What area of ocean would need to be filled with OTEC devices at this rate of energy extraction to produce a sustained gigawatt of power? Compare to the area needed for solar thermal conversion, based on the SEGS power plant (§24).

27.12 [T] Derive the relation $d\boldsymbol{u}/dt = d^*\boldsymbol{u}/dt + \boldsymbol{\omega} \times \boldsymbol{u}$ by an explicit computation in a coordinate system where $\boldsymbol{\omega} = \omega \hat{\boldsymbol{z}}$ as described in Box 27.1.

27.13 [**T H**] When an iceberg is driven across the ocean's surface by a steady wind, it is subject to the combination of forces described in Box 27.2 plus a drag force proportional to $-b\boldsymbol{v}$, where b is a constant. Thus Newton's law reads

$$m\ddot{x} = 2m\omega_{\oplus}\sin\lambda\,\dot{y} - b\dot{x},$$
$$m\ddot{y} = F_{\text{wind}} - 2m\omega_{\oplus}\sin\lambda\,\dot{x} - b\dot{y}\,.$$

Show that the resulting motion consists of transient inertial motion plus uniform drift,

$$\dot{x} = ae^{-\gamma t}\cos\omega t + v_{\infty}\cos\theta,$$
$$\dot{y} = ae^{-\gamma t}\sin\omega t + v_{\infty}\sin\theta\,.$$

Find the constants ω, γ, v_{∞}, and θ in terms of b, m, $\omega_{\oplus}\sin\lambda$, and F_{wind}.

Wind: A Highly Variable Resource

The vast amount of solar radiation that is absorbed by Earth every second is quickly transformed into many different forms of energy. About 30% of the 173 000 TW incident on Earth is reflected back into space by Earth's atmosphere or surface. Almost all of the remaining 120 000 TW is absorbed and transformed into thermal energy in Earth's atmosphere, land, or ocean surface waters. Some of the thermal energy initially absorbed by the land and ocean surface subsequently also enters the atmosphere through conduction, radiation, or evaporation (§27.1). As the air in Earth's atmosphere is warmed and filled with moisture to differing degrees in differing locations, temperature and pressure gradients develop that produce motion of air masses, both on a global and a local scale. The resulting winds redistribute thermal energy and moisture around the planet.

The kinetic energy in the wind can be used to power windmills and other more modern *wind turbines* that transform wind energy into mechanical or electrical power. It has been estimated that between 0.5% and 3% of the 120 000 TW of solar energy entering the Earth system is fed into the wind and is continuously dissipated in the atmosphere (see [176] for a review of these estimates). If all this power were available for human use, it could supply on the order of 1000 TW. Thus, winds dissipate energy at roughly 100 times the rate that humans use it. Much of this energy, however, is inaccessible since it is dissipated in the upper atmosphere, in difficult-to-access locations such as over the deep ocean, in intense localized storms, or in vast areas where the wind speed is too low to provide useful power. Accessible wind energy is nevertheless capable of making a major contribution to human energy needs. In this and the following two chapters, we describe the power in wind and some of the technologies that have been developed to take advantage of this power.

Reader's Guide

This chapter and the two that follow describe wind power, from its origins in the large-scale circulation of the atmosphere to the way it is harvested by modern wind turbines. In this chapter we focus on the wind resource, which is highly variable in space, time, and direction. After surveying winds from a global perspective, we explain how the wind resource may be characterized at a particular site. Finally we examine the geographical distribution of wind power resources and discuss estimates of the total resource.

In §29 we introduce basic concepts from fluid dynamics that govern the flow of energy, pressure, and stress in the wind, and in §30 we apply those concepts to describe the way that wind turbines and wind farms harvest power from the wind.

Prerequisites: §2 (Mechanics), §5 (Thermal energy), §27 (Ocean energy).

Unlike many of the other renewable energy resources described in this book (e.g. solar, hydro, and ocean thermal energy), wind power is characterized by extreme variability both in space and in time. While some aspects of global atmospheric circulation are regular, predictable, and relatively homogeneous, the wind speed and direction at any particular location and time depend sensitively upon rapidly changing weather patterns and upon details of local topography. This is particularly true of land-based wind power locations. This variability presents a number of challenges for wind power exploitation. First, wind power in a given location must be characterized statistically in a more detailed fashion than simply specifying average wind speed. Second, the unpredictable and erratic nature of wind power makes it difficult to integrate substantial quantities of wind-generated electric power into a reliable network without large-scale storage. The latter difficulties are discussed in §38 (Electricity generation and transmission). Finally, like solar power, wind is a diffuse resource and requires extensive land areas for extraction of grid-scale power, although unlike solar power, land used for "wind farms" can also be used simultaneously for agriculture or other purposes.

Despite the challenges of harvesting wind power, it represents one of the most promising renewables for large-scale contribution to the world's energy supply in the coming decades. In favorable locations, the wind has a power density greater than solar energy (§28.3), and can be extracted with relatively low impact on land already used for other purposes. Wind power produces no CO_2 (aside from that generated in the manufacture and installation of equipment) or waste, and is often available in remote locations where other energy resources are scarce. Wind power is a relatively mature technology, and there are no major technological barriers to its efficient exploitation.

Use of wind power has increased dramatically since 2000. After growing at an annual rate of roughly 25% during the decade from 2000 to 2010, the growth rate has slowed to around 17% since 2010 (Figure 28.1). Global wind power capacity at the end of 2015 reached 432 GW. Throughout this period, the *capacity factor* averaged 20%. With recent developments in offshore wind technology, wind power has the potential to grow to a scale of one or more terawatts of capacity within the coming decades. With a capacity factor of roughly 20%, wind power can

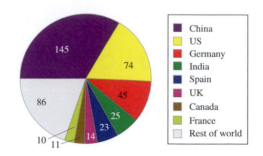

Figure 28.2 Countries with top wind power capacity (in GW) in 2015 [12].

meet ∼10% of global electricity demand per terawatt of installed wind power capacity.

This chapter begins (§28.1) with an extension of the description of atmospheric circulation presented in the previous chapter, emphasizing potential sources of sustained wind power and their variation with latitude, geography, and landforms, all of which affect the accessibility of wind power. Next (§28.2), we describe the variation of the wind with height above Earth's surface and the main features of the wind's variation with time. We then describe methods commonly used to characterize the wind speed, energy, and direction at any particular location. We define *wind frequency distributions*, which characterize the variability of the wind statistically, and introduce *Weibull distributions*, a special class of functions that are often used to characterize wind distributions. We define *wind power classes*, which give a coarse grained estimate of wind power potential, and introduce *wind atlases*, which survey the geographical distribution of wind resources. This chapter closes (§28.3) with an overview of attempts to estimate the magnitude of the world's potentially usable wind power resources.

The following two chapters analyze in more detail the physics relevant to the extraction of energy from wind for human purposes. §29 introduces some basic elements of fluid dynamics, such as mass and energy conservation, laminar flow and turbulence, viscosity, vorticity, and Reynolds number. These ideas enable us to explain the origins of *lift*, a concept essential for understanding the dynamics of wind turbines. In §30 we survey the physics of *horizontal-axis wind turbines (HAWT)*, the dominant technology for harvesting power from the wind. We use conservation of energy and momentum to analyze the flow of air past a wind turbine, leading to the *Betz limit* (which is previewed in this chapter) on the efficiency of a device that harvests wind energy from a freely flowing fluid stream. We explain how the dynamics of lift and drag forces on a

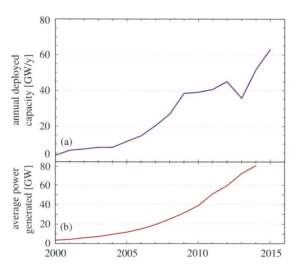

Figure 28.1 (a) Addition to global wind power capacity deployed annually from 2000 through 2015. (b) Annual global electric power generated from the wind from 2000 through 2014, converted to an average continuous power (1 TWh/y = 0.114 GW continuous) [12].

wind turbine blade determine the blade's design. Finally we explore the mechanical design and performance of wind turbines.

More extensive introductions to wind power science and engineering can be found in the monographs by Jain [177] and Mathew [178], and especially in the text by Manwell, McGowan, and Rogers [179], while Emeis [180] provides an introduction to wind power meteorology.

28.1 The Nature of the Wind

We first encountered the structure of Earth's atmosphere in connection with ocean currents and ocean energy transport (§27), and will encounter it again in our study of Earth's climate in §34. Here we focus on those aspects of Earth's atmosphere that affect the geographical variation of wind power potential.

Wind power is distributed unevenly around the globe, and varies over distance scales ranging from millimeters to ten thousand kilometers. The geographical variation of the wind can be divided roughly into three regimes by distance scale. At the largest scale, the **primary atmospheric circulation**, described in §27.4, comprises the Hadley, Ferrell, and Polar cells, including their associated prevailing winds and jet streams. This global circulation operates over distance scales on the order of 10 000 km. Next is the **secondary**, or **synoptic**, regime, where *cyclones* and *anticyclones* (low- and high-pressure areas) control wind patterns that persist for days or weeks across distances of up to 1000 km or more. Also included in the synoptic regime are *monsoons*, seasonal winds driven by land–sea temperature differences that reverse from summer to winter, and *tropical cyclones*. Finally, the *tertiary* scale – of order 10–100 km – includes land and sea breezes, mountain and valley wind systems, and other localized wind regimes.

Variations on each of these scales have significant ramifications for human use of wind energy. In addition to geographical variations, winds vary significantly with height due to interactions with Earth's surface. High in the atmosphere, where the effects of the surface are negligible, winds blow relatively steadily in direction and with intensity determined by a balance between the pressure gradient and the Coriolis force. The layer closer to Earth's surface, where interactions with the ground are important, is known as the *planetary boundary layer* or *atmospheric boundary layer*. The variation of wind speed and direction with height in the planetary boundary layer strongly affects the availability of wind power and the siting of wind turbines. The physics of the planetary boundary layer is described briefly in §28.1.5. This section closes with a survey of the temporal variation of wind power, which reflects the influence of both synoptic and local wind patterns as well as turbulence in the lower part of the planetary boundary layer.

28.1.1 Global Atmospheric Circulation

Earth's **primary atmospheric circulation** was introduced in the previous chapter (see §27.4). This circulation is driven by the excess in solar energy input at the equator compared to the poles. Just as ocean currents carry thermal energy poleward at a rate of several petawatts, atmospheric circulation transports thermal energy away from the equator at an even greater rate through the Hadley, Ferrell, and Polar cells illustrated in Figure 27.8. Together these ocean and atmospheric circulation patterns help to equilibrate temperatures across latitudes so that Earth radiates energy more uniformly back into space than it is received. The consequences of this energy distribution for global climate are discussed further in §34.

Near the equator, where atmospheric circulation is dominated by warm air rising, winds across the surface are weak. Wind power potential is, in general, poor in equatorial regions except where smaller-scale effects such as *monsoon flows* dominate. The Hadley flow returning toward the equator near Earth's surface is shifted westward by the Coriolis force, giving rise to the *trade winds* at latitudes ±5–30°, which are relatively gentle and reliable, and flow from the northeast (southeast in the Southern Hemisphere). Both the Hadley and Ferrell cells contribute to descending, relatively cold air near ±30° latitude (Figure 27.8). These regions of relatively stability, known to sailors as the **horse latitudes**, are areas of weak winds and little precipitation, and are where many of Earth's major deserts are found. Like equatorial regions, in the absence of local effects the horse latitudes do not present good prospects for wind power. In the Ferrell cells – roughly from ±30° to ±60° – poleward heat transport is dominated by synoptic scale cyclones and anticyclones. The resulting surface winds are relatively strong, variable, and on average from the southwest (northwest in the Southern Hemisphere). Wind power potential is often high in these mid-latitudes. The domain of the Polar cells, from ±60° to the poles, recapitulates the Hadley cell dynamics on a smaller scale. The boundary region at +60° and the regions near both poles are relatively calm, although the strong westerly winds just above −60° coupled to synoptic scale disturbances of the mid-latitudes often intrude upon the boundary of the southern Polar cell. From ±60° to the poles are the domain of the **polar easterlies**, relatively gentle and persistent winds flowing outward from the poles and shifted westward by the Coriolis force.

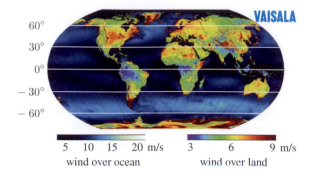

Figure 28.3 Global mean wind speed at a height of 80 m for 2005. Note the relatively weaker wind speeds in the tropics and relatively stronger wind speeds from ±30° to ±60°. Note also the lower wind speeds over continental land masses. The strongest winds are in the 40°–60° band in the Southern Hemisphere (the **roaring '40s** and *furious '50s*), where prevailing westerly winds from the mid-latitude Southern Hemisphere Ferrell cell are relatively unimpeded by large land masses. (Taken from http://www.vaisala.com with permission from Vaisala)

A map of global-average near-surface (80 m above mean ground level) wind speeds is shown in Figure 28.3. The map shows regions of relatively strong and weak winds associated with the global atmospheric circulation as well as other features described below.

The boundaries between the global atmospheric circulation cells are marked by relatively narrow zones where winds blow from the west at very high speeds at high altitudes. These **jet streams**, which persist for thousands of kilometers, can be only a hundred kilometers wide and a few kilometers thick.[1] The **polar** and **subtropical jets** are shown schematically in Figures 27.8 and 27.9. Generally, high atmosphere winds in excess of 50 kts (≈26 m/s) are considered jet stream winds, although speeds more than four times as high have been measured. The polar jet, which lies between the Ferrell and Polar cells, is usually stronger and more pronounced and has a greater effect on the weather below it than does the subtropical jet, which lies between the Hadley and Ferrell cells. The jet streams typically lie at the *tropopause*, which (as mentioned in §27.4 and discussed further in §34.2) delineates the top of the meteorologically active layer of the atmosphere, and which usually ranges from 10 to 17 km for the sub-tropical

jet (around ±30°) and from 7 to 12 km for the polar jet (around ±60°).

The jet streams vary on a time scale of days in location and intensity. Instabilities driven by the variation of the Coriolis force with latitude cause the jet streams to buckle, forming planetary-scale meanders in the jet streams known as **Rossby waves**. The phase velocity (see §4.5.3) of atmospheric Rossby waves always has a westward component relative to the mean westerly (eastward-moving) atmospheric flow. The direction of energy propagation (group velocity) relative to the mean flow is eastward for short wavelengths and westward for long wavelengths; only for very long wavelengths does energy propagate westward relative to Earth's surface.

Rossby waves may cause jet streams to stray far from their usual latitudes and may even cause them to break away as closed centers of circulation that give rise to highly anomalous weather events. An active Rossby wave configuration is shown in Figure 28.4.

Jet stream winds blow at altitudes far too high to be captured by existing land-based wind power systems. Their influence over surface wind patterns is, however, profound, since the interactions of synoptic scale weather systems with jet streams often determine the weather systems' evolution over time and the resulting wind patterns at the

Wind from Global Atmospheric Circulation

Excess solar energy input near the equator drives atmospheric circulation that transports thermal energy toward the poles. The resulting *primary circulation* determines prevailing winds across the globe.

Winds are generally light and variable near the equator and in the *horse latitudes* near ±30°. The Hadley and Polar cells produce relatively gentle and reliable winds known as the *trade winds* (in a band from a few degrees away from the equator to ±30°) and *polar easterlies* (from ±60° to the poles) that blow predominantly from east to west. Winds in the Ferrell cell are stronger, more variable, and predominantly from the southwest (northwest in the Southern Hemisphere).

In the absence of effects due to synoptic and tertiary scale circulations, the global circulation pattern determines which regions of the globe are favorable for wind power.

---

[1] Jet streams originate in the vertical as well as horizontal temperature and pressure differences at the boundaries between primary atmospheric circulation cells, but an explanation of their direction and intensity is beyond the level of this book.

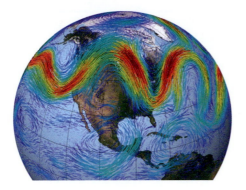

Figure 28.4 Visualization of the polar jet over North America showing a high-amplitude Rossby wave pattern that frequently leads to storms in the US. Regions of high/low windspeed are shown in red/blue. (Credit: NASA/GSFC)

surface. Rossby waves, in particular, play a primary role in driving the synoptic scale weather patterns at middle latitudes.

28.1.2 Synoptic Scale Winds

Above the near-surface planetary boundary layer, air moves in response to pressure gradients in the atmosphere. Pressure gradients, in turn, are driven by temperature differences. When air is warmed from below – by radiation from land warmed by the Sun for example – the air expands upwards, creating low pressure near ground level and raising pressure aloft relative to the atmosphere nearby. Similarly, air cooled from below sinks downward, raising the pressure near the ground and lowering it aloft. Vertical pressure gradients are largely compensated by the forces of gravity, keeping the atmosphere vertically in *hydrostatic equilibrium* (explained in detail in §34.2.1). The pressure differences in the horizontal directions, parallel to Earth's surface, drive winds in the horizontal plane. If the pressure is greater on one side of a parcel of air than on the other, then the air is accelerated in the direction of lower pressure. The net force per unit volume from the changing pressure is equal to the negative gradient of the pressure $f_p = -\nabla p$ (Problem 27.1). Once a pressure gradient sets a parcel of air in motion, the horizontal component of the *Coriolis force* (see eq. (27.4)) deflects it to the side. The strength of the Coriolis force grows with latitude: its effects are absent at the equator and often negligible in the tropics, but they cannot be ignored in the middle and polar latitudes. The dynamics is similar to the wind-driven ocean currents analyzed in the previous chapter. The winds reach a dynamical equilibrium when the pressure gradient and Coriolis forces cancel. The result, as illustrated in Figure 28.5 and explained further in Box 28.1, is that winds above the planetary boundary layer blow approximately

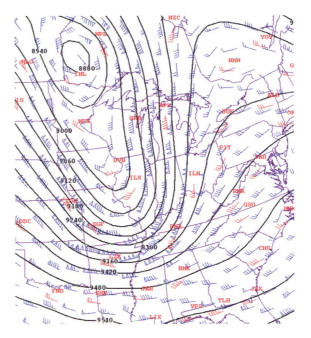

Figure 28.5 A map showing contours (black) of constant altitude (in meters) at which the pressure is 0.3 atm – essentially isobars – over the US on October 16, 1996. Also shown are predicted geostrophic winds (blue barbs), and observed winds (red barbs). *Barbs* show the direction (toward the base of the barb) and the speed (triangle = 25 m/s, full line = 5 m/s). Although geostrophic winds provide a good first approximation, the wind speeds in particular depart significantly from geostrophic where isobars are curved. (Credit: Lynn Mcmurdie)

parallel to the contours of constant pressure (isobars). Such winds are known as **geostrophic winds**.

Note that the air motions due to geostrophic winds preserve rather than reduce the pressure differences that drive them. Thus, in the middle latitudes large (synoptic) scale circulations, driven originally by temperature differences, can persist for long periods. These circulations, counterclockwise around regions of low pressure and clockwise around regions of high pressure (reversed in the Southern Hemisphere), and known as **cyclones** and **anticyclones**, march slowly from West to East in the global atmospheric circulation driven by the Ferrell cell and in particular by the polar and sub-tropical jet streams. They are responsible for the changeable weather and strong and shifting winds of the mid-latitudes.

In addition to the cyclones and anticyclones of the mid-latitudes, two synoptic scale weather features of the tropics, *monsoons* and *tropical cyclones*, impact wind power resources. Tropical cyclones are discussed in the following section. **Monsoons** are atmospheric circulation systems that reverse on a seasonal basis as they are driven

Box 28.1   Geostrophic Winds

Away from the tropics and above the planetary boundary layer where friction with Earth's surface is no longer important, wind directions and speed are determined by the balance of pressure and Coriolis forces in the horizontal plane. The force per unit volume due to a pressure gradient is $f_p = -\nabla p$. Since the gradient points in the direction of increasing pressure, this force alone would act to equalize the pressure by pushing air toward regions of lower pressure. The Coriolis effect completely alters the situation. The horizontal component of the Coriolis force (per unit volume) from eq. (27.4) is $f_{hc} = 2\rho\Omega \sin\lambda\, v \times \hat{n}$, where ρ is the air density, Ω is Earth's angular velocity, λ is the latitude, v is the air velocity, and $\hat{n}$ is the local normal to Earth's surface. The air moves steadily when these two forces cancel, $f_p + f_{hc} = 0$. If we take the vector cross product of the normal $\hat{n}$ with $v \times \hat{n}$ and use the identity (B.5), we find $\hat{n} \times (v \times \hat{n}) = v$ (because $n \cdot v = 0$). Thus the forces balance when

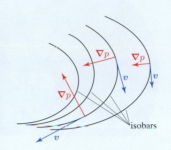

$$2\rho\Omega \sin\lambda\, v = \hat{n} \times \nabla p .$$

As shown in the inset figure, which depicts clockwise winds moving around a local region of high pressure, the gradient of the pressure is perpendicular to isobars, so the cross product $\hat{n} \times \nabla p$ is tangent to the isobars. Thus the wind blows *parallel* to the isobars. Also, the greater the pressure gradient – the more closely spaced the isobars – the greater is the windspeed. Figure 28.5 shows geostrophic winds blowing counterclockwise about a low-pressure trough in southern Canada. The observed winds blow, on average, parallel to isobars and their speeds are greatest where the isobars are most closely spaced.

True geostrophic winds occur only when isobars are straight lines. When isobars are curved, a packet of air following an isobar experiences a *centrifugal force* f_c as well as the Coriolis force and the force due to the pressure gradient. Such winds are known as **gradient winds**; they still blow parallel to isobars but the relation between wind speed and pressure gradient is altered (Problem 28.4).

---

by land and sea surface temperature differences. Water typically has a higher heat capacity than Earth's surface. Furthermore, heat can flow convectively as well as conductively in the ocean's mixed layer, whereas only conduction can transport heat through the soil. As a result, when exposed to summer's high solar irradiance, the land becomes warmer than nearby sea surfaces. Air above the land surface, warmed from below, rises and is replaced by moist air, often saturated with water, drawn in from the nearby sea. If this moist air is forced upward, either by surface heating and subsequent convection, by rising terrain, or by surface convergence (forced, for example, by downslope winds from nearby mountains), it expands and cools, allowing the water vapor to condense as precipitation. The convective loop is closed as the cooled, drier air at high altitude over land is drawn seaward where it sinks. In the winter, the circulation reverses: land cools more quickly than sea surface, so air sinks over land and flows at the surface out to sea, suppressing precipitation in winter. In addition to the often beneficial summer precipitation, monsoon systems bring seasonal winds that are relatively reliable in strength, timing, and direction.

Monsoon circulations are most prominent in the tropics, where solar irradiance is high and land–sea temperature differences are large. The Indian (or Southwest) Monsoon brings moist air from the Arabian Sea and the Bay of Bengal to both sides of the Indian peninsula, leading to heavy late summer rains and dry winters. The East Asian (or Southeast) Monsoon affects areas on the western border of the Pacific Ocean and the South China Sea. It brings warm and wet summers and cool and dry winters to Japan, the Koreas, the Philippines, eastern China, and Indochina. Areas affected by the Asian Monsoons are shown in Figure 28.6. Other monsoonal flows affect Northern Australia, parts of tropical Africa and South America, and southwestern parts of North America. Monsoons provide significant wind power resources to tropical and subtropical areas of the world where the prevailing winds driven by the primary atmospheric circulation are not strong.

28.1.3 Tropical Cyclones

Tropical cyclones, known as **hurricanes** in the Western Hemisphere, as **typhoons** on the Asian Pacific rim, and simply as *cyclones* in the Indian Ocean, transport vast

---

Box 28.2   Power in Tropical Cyclones

Tropical cyclones are extremely powerful, but only a small fraction of their energy is in the form of wind. At any given moment, most of a hurricane's energy is in the form of latent heat of condensation in the warm, moist air that rises in the central regions of the cyclone. Reference [183] computes the rate at which energy is released through condensation by estimating that an average hurricane generates 1.5 cm/day of rainfall within a circle of radius 665 km, corresponding to $\approx 2.1 \times 10^{13}$ kg/day. With a latent heat of condensation of 2.24×10^6 J/kg, this works out to

$$P \approx (2.24 \times 10^6 \text{ J/kg}) \times (2.1 \times 10^{13} \text{ kg/d}) \div (8.64 \times 10^4 \text{ s/d}) \approx 540 \text{ TW} .$$

A large fraction of this energy is transported by the cyclone from the surface layers of the ocean high into the atmosphere, where it eventually finds its way poleward as part of Earth's energy balance.

In steady state, the power in a hurricane's winds is constantly dissipated in friction with Earth's surface. Emanuel [184] estimates the power in a hurricane's winds by integrating this dissipation rate over the wind field of a characteristic hurricane. He finds that a typical Atlantic hurricane with maximum winds of 50 m/s (112 mph) and a radius of maximum winds of 30 km dissipates wind energy at a rate of $\sim$3 TW. A Pacific super-typhoon, with maximum wind speed of 80 m/s (180 mph) and a maximum windspeed radius of 50 km dissipates wind energy at 30 TW, twice the total global average human energy consumption rate of 15 TW.

Geostrophic Winds

Winds high in the atmosphere, blowing predominantly in the horizontal plane, are driven by variations in atmospheric pressure and the horizontal component of the Coriolis force. The resulting *geostrophic winds* blow parallel to isobars. Geostrophic winds blow counterclockwise around regions of low pressure (*cyclones*) and clockwise around areas of high pressure (*anticyclones*) in the Northern Hemisphere. The flows reverse in the Southern Hemisphere. These features dominate the synoptic scale wind regime in middle latitudes.

Geostrophic winds do not equalize pressure differences. As a result, cyclones and anticyclones persist over relatively long time scales and only grow or decay due to additional *ageostrophic* effects arising from frictional (viscosity) effects in the atmosphere.

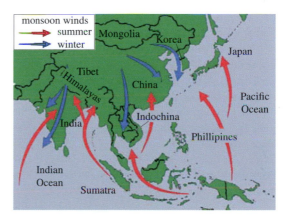

Figure 28.6 Major wind patterns associated with the Indian and East Asian summer and winter monsoons.

amounts of thermal energy from the tropics into middle latitudes. Hurricanes carry large amounts of wind energy (see Box 28.2), but their extreme power as well as their unpredictability makes them virtually useless as sources of wind power for human consumption given current technology. Instead they must be figured among the hazards that wind turbines situated in hurricane-prone locations must be designed to withstand. We mention hurricanes here principally as an example of atmospheric dynamics and to illustrate how a Carnot heat engine can arise in nature.

The conditions that lead to hurricane formation are not completely understood. A deep ($\sim$50 m) layer of very warm ($>26\,^\circ$C) water in the open ocean, some pre-existing convective disturbance, and very cold temperatures at the top of the troposphere seem to be prerequisites. A hurricane begins with the formation of one or more regions of ascending moist air over very warm ocean water. Water condenses out of the rising air, turning its latent heat into sensible heat, and this in turn feeds the ascending column of air. As air at the surface is drawn in to replace

Figure 28.7 Principal air circulations in and around a mature hurricane. (Credit: kelvinsong reproduced under CC-BY-SA 3.0 license via Wikimedia Commons)

the ascending air, the Coriolis force deflects it to the right and causes it to rotate counterclockwise (clockwise in the Southern Hemisphere). Although hurricanes form in the tropics, they are rare within a few degrees of the equator where the Coriolis effect vanishes. As the indrawn air rides over the ocean's warm surface, it becomes saturated with water, ready to fuel the ascending columns near the center of the circulation. Conservation of angular momentum increases the speed of the air as it spirals in. At the top of the hurricane, the air, having cooled by adiabatic expansion in the later stages of its ascent, cools further by thermal radiation into space. This cold, counterclockwise circulating air is forced outward as more air rises. The outflow is once again deflected to the right by the Coriolis force, slowing, and even reversing its counterclockwise circulation. Thus, mature hurricanes show opposite circulations at low and high altitudes as shown in Figure 28.7.

The hurricane absorbs both latent and sensible heat from the underlying ocean at a temperature above 300 K, and radiates and transports heat away as air flows outward at the top of the storm, where temperatures may be below 200 K. The heat absorption and emission processes are roughly isothermal, whereas the ascent in the storm's interior and descent far away can be modeled as adiabatic. Thus the hurricane resembles a *Carnot engine* and because of the large temperature difference, the Carnot efficiency limit for a hurricane may exceed 30% [181]. The Carnot engine powers the hurricane's winds, which are constantly dissipated by friction with Earth's surface; hurricanes also move large quantities of heat from the ocean's surface to the upper reaches of the atmosphere.

28.1.4 Tertiary Circulations

In addition to the effects of global atmospheric circulation and synoptic scale winds, persistent wind patterns on smaller scales can have significant impact on the wind power potential at particular locations. Wind regimes

Synoptic Scale Systems

Winds at the synoptic scale – of order 1000 km – are dominated by regions of high and low pressure that progress from west to east, guided by the polar and tropical jet streams.

Cyclones and *anticyclones* dominate the weather of the mid-latitude Ferrell cells and are responsible for the strong and variable westerly winds of the middle latitudes.

Monsoons are seasonally reversing continental scale flows, driven by land–sea temperature differences. Monsoons can produce relatively strong and reliable winds in the tropics where winds due to primary atmospheric circulation are typically weak.

Tropical cyclones, also known as hurricanes or typhoons, resemble enormous Carnot engines that dissipate large amounts of wind energy as they transport even greater amounts of thermal energy from warm tropical waters to the upper atmosphere.

characterized by distance scales of order 10–100 km are known as **tertiary circulations**. Usually these winds are generated by conditions relatively close to Earth's surface rather than pressure gradients higher in the atmosphere. Because they can be relatively strong and predictable both in direction and speed, these winds often influence plans for wind power development.

Land and sea breezes resemble monsoon flows, but on a diurnal rather than seasonal time scale and on a local rather than a continental distance scale. As shown in Figure 28.8, sea breezes originate in differential heating of the land relative to nearby ocean. During the day, warm air rising over land lowers the pressure near the surface and raises the pressure higher in the atmosphere. Cool air flows inland at the surface and warm air higher in the atmosphere flows seaward in response to this pressure differential. At night radiative cooling differentially lowers the land temperature and reverses the flow. Land–sea breeze circulations are limited both in horizontal and vertical extent. In mid-latitudes they typically penetrate 20–50 km inland, though in the tropics they may reach several hundred kilometers. The vertical extent of the flow may range from several hundred meters to 1–2 km. According to [180], maximum wind speeds in sea breezes reach ≈10 m/s at a height of roughly 100 m. Synoptic scale wind circulations

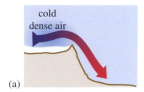
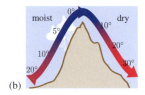

Figure 28.8 Land and sea breezes, driven by diurnal temperature differences, bring relatively cool air inland at low altitude during the day and seaward during the night.

Figure 28.9 (a) Katabatic winds originate when cold, dense air drains downward under the force of gravity. (b) Foehn winds are adiabatically warmed, dry winds that blow down the leeward side of mountains.

can overwhelm land–sea circulations and cloud cover can inhibit the differential heating or cooling that drives the phenomenon. Local circumstances such as seawater temperatures, marine currents, and coastal hills can influence the nature of the land–sea circulation significantly. Nevertheless, land–sea breezes are often reliable and robust enough to act as sources of wind power.

Mountain and valley winds arise from differential heating and cooling of mountains and valleys on a diurnal cycle. During the day, warm winds blow up valleys and mountain sides; at night, the flow reverses as cold air sinks down mountain slopes and down valleys. The mechanisms that drive these winds are somewhat more complex than the simple two-dimensional flows that drive sea breezes. Flows up and down mountain slopes interact with flows up and down the valleys that are perpendicular to the mountain slopes. The diurnal variation of valley winds and the fact that valleys channel the wind along the valley axis are potential advantages for efficiently configuring wind farms based on regular wind patterns with known directionality, but valley winds are often too weak for significant wind power installations.

Mountain ridges and passes Wind flows are accelerated as they pass over ridges or funnel through mountain passes. Ridges and mountain passes provide favorable wind power locations when the prevailing wind direction is perpendicular to the ridge or the pass. One of the earliest extensive wind farms, and still the largest in terms of number of turbines, is located at Altamont Pass in the Diablo Range in California. Temperature differences between the hot California Central Valley and the cool waters of the nearby Pacific drive a land–sea circulation in a direction perpendicular to the Diablo Range, funneling strong and dependable winds through Altamont Pass.

Katabatic winds Several types of mountain–plains or mountain–valley circulation systems generate regular, sometimes intense winds that can be used for wind power. **Katabatic** or **drainage** flows are generated when radiatively cooled – therefore heavier – air formed above mountains, plateaus, or hills drains downward onto plains

or into valleys under the force of gravity. As it sinks, it warms adiabatically. Depending on how warm the air was to start, the result may be hot, like the *Santa Ana winds* of Southern California, cool, like the *Mistral* of Mediterranean France, or frigid, like the katabatic winds that blow outward from the Antarctic ice sheet. Katabatic winds are often too shallow to be used for wind power. **Foehn** winds have features similar to katabatic winds, but different origins. They occur on the down wind or *lee* sides of mountains. Winds originating in synoptic scale circulations that encounter a mountain range are lifted up and cooled, dropping most of their moisture as precipitation on the windward side. On the leeward side of the mountains, the now dry air descends, accelerates, and warms to a higher temperature than at the equivalent altitude on the windward side (see the discussion of dry and moist *adiabatic lapse rates* in §34). The resulting warm dry winds are observed in many locations where they can raise the temperature dramatically in a short period. The *Chinook winds* that are frequent on the east slope of the US Rocky Mountains are famous examples of Foehn winds.

28.1.5 The Planetary Boundary Layer

The wind regimes just described – with the exception of the tertiary circulations – pertain to the **free atmosphere**, far above Earth's surface, where the wind blows primarily in the horizontal plane and primarily in response to gradients in the pressure. Typically the wind in the free atmosphere blows steadily with relatively small fluctuations about its mean speed. Although conditions vary considerably, we assume that the winds in the free atmosphere can be approximated as geostrophic (see Box 28.1). At lower heights, wind interactions with the surface become progressively more important and wind patterns become more complex. This is the region of the **planetary boundary layer** (PBL), also sometimes known as the **atmospheric boundary layer**. The implications of the PBL for wind resources are quite significant. For example, a major advantage of offshore wind turbines comes from the fact

Tertiary Wind Regimes

Local – of order 10–100 km – wind patterns caused by small-scale geographical features frequently can provide wind that is regular and predictable both in strength and direction. These *tertiary scale* winds include two *diurnal circulations*. Land and sea breezes arise from differential heating between land and nearby bodies of water and blow onshore during the day and offshore at night. Mountain and valley winds flow up valleys and mountain slopes during the day and reverse at night.

Other tertiary wind regimes of possible importance for wind power are *katabatic* or *drainage* winds, which result when atmospheric conditions favor the descent of relatively cold air formed above mountains or glaciers, and *Foehn* winds – warm, dry winds which often develop in the rain shadows of mountains.

Tertiary circulations can be overwhelmed by more powerful synoptic scale systems.

that the PBL is much shallower over open ocean than over land, allowing offshore turbines to access the stronger and less variable winds in the free atmosphere.

The planetary boundary layer is generally about 1 km in depth but can range from 30 m to 3 km, depending upon whether local conditions are static or highly convective. The PBL can be subdivided into three layers. The lowest is a thin layer, no more than a few millimeters thick, where intermolecular forces bring the air to rest at Earth's surface. This thin layer is not relevant for wind energy. Directly above this lies the **surface layer** (also known as the **Prandtl layer**) through which the wind increases in speed from zero with little change in direction. The surface layer may be as thin as 10 m in a stable atmosphere above calm open ocean, though a depth of ~100 m is typical for relatively smooth land surfaces. In general the surface layer accounts for roughly 10% of the depth of the PBL. Above the surface layer lies the **Ekman layer**, where wind speed continues to increase, though more slowly, and the effect of the Coriolis force rotates the wind from its surface direction into the direction parallel to isobars. At the top of the Ekman layer the wind attains the geostrophic direction and speed associated with the free atmosphere.

Land-based wind turbines with the tops of their blades below 100 m generally lie entirely within the surface layer

and must be designed to take account of wind speed variation with height. Larger land-based turbines with blade tips above 100 m, and most offshore turbines, penetrate into the Ekman layer and encounter variation with height of both wind speed and direction.

We next describe features of the PBL in more detail, starting at the top and working our way downward. The dynamics of the Ekman layer is dominated by the interplay of the Coriolis force and friction with Earth's surface transmitted by turbulence (see below) through the surface layer; the results are similar to the analysis of *Ekman spirals* in §27.3.1 (though upside-down). The effects of friction are twofold: first, it decreases the wind speed, and second it shifts the wind direction in the direction of the (negative) gradient of the pressure (Problem 28.2). As a result, moving down through the Ekman layer, the wind tends increasingly to flow inward to fill areas of low pressure and outward to diminish the strength of high-pressure areas, as illustrated in Figure 28.10. An analysis similar to that of §27.3.1 gives a wind speed in the Ekman layer that deviates from the geostrophic flow by an amount that increases exponentially with decreasing height, while the angle between the wind direction and the geostrophic flow grows linearly as the height decreases. Thus, the windspeed goes roughly as $v(z) \approx v_{\text{geostrophic}}(1 - e^{-\gamma z})$, where z is the height relative to Earth's surface; this expression is valid for $z > z_p$, where z_p is the height at the top of the surface (*Prandtl*) layer. In practice the decrease in wind

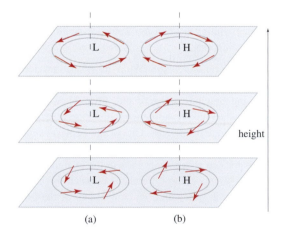

(a)   (b)

Figure 28.10 The effect of friction in the Ekman layer within the planetary boundary layer rotates the geostrophic winds away from isobars toward the (negative) of the pressure gradient and diminishes the windspeed. As a result winds blow (a) inward toward low pressure and (b) outward from high pressure; the deeper into the Ekman layer, the more pronounced is the effect.

Planetary Boundary Layer

The *planetary boundary layer* (*PBL*) is the region of the atmosphere where interactions with Earth's surface are important. The PBL ranges from as thin as 30 m over large expanses of calm water to 3 km when convection is strong.

The PBL is divided into two significant layers, the *surface* or *Prandtl layer* (roughly 10% of the PBL depth) where turbulent mixing is strong and wind speed changes relatively rapidly with height, and the *Ekman layer* where wind speed decreases more slowly with height and Coriolis forces rotate the wind direction. Winds in and below the Ekman layer are diverted away from isobars, and tend to blow inward toward low pressure and outward from high pressure.

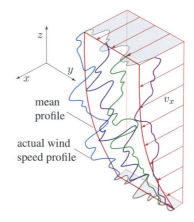

Figure 28.11 A snapshot of the distribution of the *x*-component of the wind velocity along a slice in the *y*–*z* plane in the surface layer. Turbulent fluctuations are superimposed over the mean wind speed profile (red) that increases steadily with height through the surface layer. (Credit: L Jan van der Tempel (2006). Design of Support Structures for Offshore Wind Turbines. PhD Thesis, TU Delft, NL)

speed in the Ekman layer is relatively small compared to the decrease in the surface layer below.

In the surface layer the wind speed decreases from its value at the base of the Ekman layer down to zero at Earth's surface. The windspeed diminishes slowly at first and then more rapidly toward ground level. In contrast to the free atmosphere or the Ekman layer, the surface layer is dominated by *turbulent flow*. **Turbulence** – irregular, quasi-random fluctuations on a broad range of length and time scales – mixes the air in the surface layer both horizontally and vertically. Turbulent eddies transfer energy and momentum from the top of the surface layer downward, until it is dissipated in friction with Earth's surface and in viscous losses. (Viscosity and turbulence are discussed further in the following chapter.) As a result of turbulence, the wind profile in the surface layer is best regarded as a *mean flow* over which turbulent fluctuations are superimposed. Figure 28.11 gives a snapshot of what the wind speed distribution might look like as a function of height along a horizontal slice through the surface layer. Although the component of the wind speed in the *x*-direction fluctuates greatly, the *mean wind speed* increases with height.

Wind turbines respond primarily to the mean wind speed since most of the power in the spectrum of turbulent fluctuations is concentrated at periods of order ~10 s (see Figure 28.14), which is too short to have a significant effect on the power output of a typical wind turbine. The way in which wind speed varies with altitude depends upon the regularity of the surface over which the wind is

passing. Wind flowing at low velocity over a relatively flat surface such as water or tarmac experiences little disruption. In such a situation the wind speed increases rapidly with height. The surface layer is relatively shallow, and relatively little energy is lost to thermal energy through turbulent eddies. Over a more irregular surface, however, there is increased turbulence, the surface layer is thicker, and the wind speed increases more gradually with height, with more energy lost in turbulent dissipation near the surface.

An analysis of mixing in a turbulent boundary layer [179] leads to a semi-empirical *logarithmic profile* for the variation of mean wind speed v with height z above the surface in the surface layer,

$$v(z) = v_0 \ln(z/z_0),\qquad (28.1)$$

where z_0 is known as the **roughness length**, and characterizes the roughness of the surface. The wind speed in eq. (28.1) goes to zero at $z = z_0$ and should be regarded as zero below that point. Very smooth surfaces have small z_0 and the wind speed increases rapidly with height. Rough surfaces have larger z_0 and more slowly increasing wind speed. It is standard to compare the wind speed at a height z with its value v_{ref} at a reference height z_{ref} that is typically taken to be high in the surface layer. Thus

$$v(z) = v_{\mathrm{ref}} \frac{\ln(z/z_0)}{\ln(z_{\mathrm{ref}}/z_0)}.\qquad (28.2)$$

Wind Speed Variation with Height

Wind speed increases with height through the sur-face layer of the atmosphere. The variation is approximately logarithmic,

$$v(z) = v_{\text{ref}} \frac{\ln(z/z_0)}{\ln(z_{\text{ref}}/z_0)},$$

where the *roughness length* z_0 parameterizes the nature of the landscape.

Equation (28.2) allows one to compare the wind power potential of wind turbines of different heights and to inte-grate wind power density over the vertical span of an individual turbine.[2]

The value of the roughness length z_0 depends on the nature of the terrain surrounding the wind turbine. Table 28.1 gives some indication of how roughness lengths are assigned, and Figure 28.12 shows how wind that reaches a value of 10 m/s at a height of 100 m varies at lower heights for several choices of roughness length. Fig-ure 28.12 indicates why the trend in wind turbine design has been toward greater height above the landscape. For $z_0 = 0.0002$ m, appropriate for offshore wind turbines, the power in the wind, which scales proportional to the wind speed *cubed* (see §29), is ~33% greater at a height of 100 m than at 30 m. For $z_0 = 0.03$ m, appropriate for unob-structed agricultural land, it is ~60% greater. Under many circumstances the increase in available power offsets the increased cost of building a taller system.

28.1.6 Temporal Variations in the Wind

Not only does the wind vary from place to place and with altitude, but it also varies significantly with time on many scales. Very short-term fluctuations in wind speed due to turbulence are sources of stress on wind turbines; long-term fluctuations due to daily, seasonal, and annual changes in wind patterns, on the other hand, must be included in analysis of where to site wind turbines. Fig-ure 28.13 shows measurements of wind speed over three different time scales, ranging from seconds up to several years. All three plots show significant fluctuations about the mean wind speed over the time scales covered by the

---

[2] Although it has little theoretical justification, another model for the height dependence of the wind that is often used, especially when surface roughness information is not available, assumes that the wind speed varies as the 1/7th power of the height, $v(z) \propto (z/z_0)^{1/7}$.

Table 28.1 Terrains and roughness lengths. Note that the roughness length increases with the characteristic size of features on the landscape. Note also that this simple classification does not apply to exceptional situations like mountain passes, which require either a more sophisticated model or case-by-case evaluation.

Roughness length (z_0) in meters	Landscape type
10^{-5}	very smooth, ice or sand
0.0002	calm open sea
0.0005	blown sea
0.003	snow surface
0.008	lawn surface
0.01	rough pasture
0.03	fallow field
0.05	crops
0.1	few trees
0.25	many trees, hedges, few buildings
0.5	forest and woodlands
1.5	suburbs
3.0	city centers, tall buildings

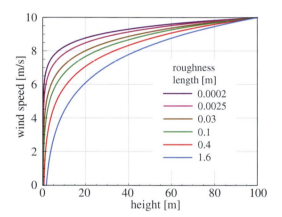

Figure 28.12 Height versus wind speed for various surface roughness lengths, given a wind speed of 10 m/s at 100 m above the surface.

measurements. These fluctuations can be analyzed further using the *Fourier transform* methods introduced in §4.2.2 (see Appendix B.3.5).

Fluctuations about the mean wind speed are given by $\Delta v(t) = v(t) - \langle v \rangle$. The Fourier transform of $\Delta v(t)$ expresses wind speed fluctuations as a function of the char-acteristic frequency of the fluctuation. Thus, for example, if the wind at a particular location rises in the afternoon and evening as a result of sea breezes and falls toward dawn

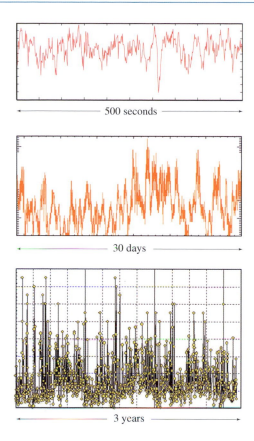

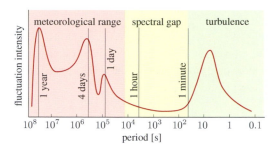

Figure 28.14 A sketch of the wind fluctuation spectrum. Note the importance of annual, synoptic, diurnal, and turbulent effects and also the gap roughly between one minute and one day over which fluctuations are less significant. (Credit: Greenrhinoenergy.com)

Figure 28.13 Time series showing fluctuations in wind speed over three different time scales. Some of the periodicities displayed in Figure 28.14 are evident. (Credit: T. Barszaz, et al., Wind speed modelling using Weierstrass function fitted by genetic algorithm, *Journal of Wind Engineering and Industrial Aerodynamics*, **109**, 68 (2012), with permission from Elsevier.)

and through the morning, then the Fourier transform of $\Delta v(t)$ would show a peak at a frequency of $v = 1 \, \text{day}^{-1}$. A sketch of the *wind fluctuation spectrum* as a function of period (= 1/frequency) at a typical location is shown in Figure 28.14. The wind fluctuation spectrum shows four notable peaks and a notable valley:

Turbulence The peak at times below one minute are due to turbulence. These are the rapid fluctuations that we experience as gusts and lulls in the wind.

Diurnal The peak at one day corresponds to the variations in the wind associated with solar heating and nighttime cooling and the winds that they engender including land and sea breezes and mountain and valley winds.

Fluctuation Time Scales

The wind fluctuates on time scales ranging from fractions of a second to many years. The most important time scales are annual, synoptic (several days), diurnal, and turbulent (less than a minute). There is a gap in the energy spectrum of wind fluctuations between 1 minute and several hours, where driving forces are largely absent. Measurements that average over a period in this gap eliminate turbulent fluctuations, but preserve the variations with longer periods that are significant for wind power.

Synoptic The large peak around 4–5 days is associated with the succession of synoptic systems – cyclones and anticyclones.

Annual Annual cycles, such as the strong winds of winter in middle latitudes and the monsoons of the tropics, give rise to a strong signal in the power spectrum with a period of one year.

Spectral gap There is a noticeable absence of fluctuation energy in the region roughly between one minute and several hours.

The fluctuation power spectrum reflects atmospheric dynamics on the scales of the three regimes introduced earlier in this section. It also suggests that measurements of the wind speed that *average over intervals greater than several minutes* will be insensitive to turbulent fluctuations and therefore provide the information about mean wind speed that is required to evaluate the wind power potential of any site.

Power in the Wind

The power in a fluid flow with velocity v is proportional to the cube of the fluid's speed,

$$\mathcal{P} = \frac{1}{2}\rho v^3.$$

Betz's limit states that, in an unrestricted flow, at most a fraction 16/27 of the power passing through a given cross-sectional area A can be captured by a mechanical device that operates in a plane perpendicular to the fluid flow.

28.2 Characterization of a Wind Resource

The highly irregular nature of wind power presents a challenge for efficient exploitation of this resource for human use. While the general characteristics of global wind patterns and wind variation with altitude described in the previous section can be understood based on meteorological principles, the wind speed and direction at any particular location vary over time in an effectively stochastic way, based on factors at many scales, and are best characterized empirically. In this section we describe some of the tools used for describing wind power at any specific location. These tools can be used to optimize the siting and design of wind turbines or wind farms containing many turbines.

The power in the wind that can be harvested for human use originates in the wind's kinetic energy. The magnitude of the *kinetic energy flux* – the kinetic energy that passes through a unit cross-sectional area per unit of time – in a fluid moving at velocity v is given by the product of the kinetic energy per unit volume and the windspeed (see §29.1.2)

$$\mathcal{P} = \frac{1}{2}\rho v^2 \times v = \frac{1}{2}\rho v^3, \qquad (28.3)$$

where ρ is the fluid's density and $v = |v| \geq 0$ is its speed. For air at sea level, $\rho \cong 1.17\,\text{kg/m}^3$. $\mathcal{P}$, known as the **power density**, is a measure of the mechanical power available in a flowing fluid. Not all of the energy in an unrestricted fluid flow, however, can be transformed directly into mechanical energy by a wind turbine or other device. Under reasonable conditions, at most $16/27 \approx 59\%$ of the power is available (see Box 28.3). This result, known as the *Betz limit*, puts a bound on the efficiency of wind turbines, but it does not change the conclusion that the primary indicator of wind power potential is the *cube of the wind speed*. We analyze the Betz limit and its consequences further in §30.

28.2.1 Wind Speed and Power Distributions

Wind speed has been measured for centuries using a simple device called an **anemometer**, which measures the instantaneous wind speed by allowing the wind to drive rotation of a mechanical system, generally using a propeller or cups. By regularly sampling the wind speed at a given location, the average wind speed can be estimated. For many decades, records have been kept of average wind speeds measured using this method in many locations.

From eq. (28.3) it is clear that the figure of merit for wind power potential of a particular location is not the average wind speed, but instead the average value of the *cube* of the wind speed,

$$\langle v^3 \rangle = \frac{1}{T} \int_0^T dt\, v^3(t). \qquad (28.4)$$

The average power density can be conveniently written in terms of the **root mean cube** of the windspeed,

$$\tilde{v} \equiv \sqrt[3]{\langle v^3 \rangle} = \left(\frac{1}{T} \int_0^T v^3(t)dt \right)^{1/3}, \quad \text{so that}$$

$$\langle \mathcal{P} \rangle = \frac{1}{2}\rho \tilde{v}^3. \qquad (28.5)$$

In general, the cube of an average does not equal the average of a cube, so the average wind speed alone,

$$\langle v \rangle = \frac{1}{T} \int_0^T dt\, v(t), \qquad (28.6)$$

does not give a particularly good estimate of wind power potential. An optimal location would, of course, have consistent high speed wind. When comparing two sites with the *same average wind speed*, however, a site with a broad range of wind speeds is more desirable than one with a narrow range about the average, as illustrated in Example 28.1 (Problem 28.5).

Historically, wind measurements were taken at a given location at regular intervals rather than continuously. To be useful for estimating wind power potential, the measurements should average over a period long enough to eliminate turbulent fluctuations to which a wind turbine cannot respond. On the other end, measurements must be taken often enough to register the variations in wind speed that are significant for power generation. A good choice of averaging period is somewhere within the *spectral gap* shown in Figure 28.14. Modern automated measurements may, for example, average over ten minute intervals and be recorded hourly.

Systematic wind speed measurements form the basis of an evaluation of a site for wind power production. The wind speed measurements can be binned in convenient intervals and compiled into a histogram like the one shown

Box 28.3   The Betz Limit: An Elementary Ballistic Derivation

The power density in a fluid such as the wind, traveling at velocity $\boldsymbol{v}$, is $\mathcal{P}_{\text{wind}} = \rho v^3/2$. The **Betz limit** states that, under fairly general conditions, at most $16/27 \cong 59\%$ of this power passing through a fixed perpendicular cross-sectional area can be extracted by a wind turbine or other mechanical device. A careful derivation of this limit in the context of fluid mechanics is presented in §30. Here, we derive the Betz limit very simply from the basic laws of mechanics in a simplified context [2].

Suppose an object is inserted into a windstream moving with velocity $\boldsymbol{v}$. If the object is held stationary ($\boldsymbol{w} = 0$), it absorbs no energy; if, at the other extreme, the object moves with the same velocity as the wind ($\boldsymbol{w} = \boldsymbol{v}$), the wind exerts no force on the object, and once again the object absorbs no energy. Our goal is to find the intermediate velocity at which the power transferred to the object is maximum.

We consider a simple model, in which all motion is collinear. Air molecules, moving initially with velocity v, collide independently with a massive flat plate moving at velocity w, oriented perpendicular to v (see the figure). The molecules transfer some part of their energy to the plate through elastic collisions. We make the "ballistic" assumption that interactions between molecules in the fluid can be neglected. Below we argue that the conclusion is unchanged if we relax this assumption or consider a more general configuration.

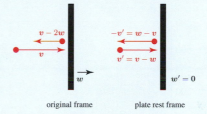

original frame   plate rest frame

Consider the elastic collision of a single molecule with the plate as illustrated in the figure. In the rest frame of the plate, the molecule approaches with velocity $v' = v - w$. Because the plate is far more massive, the molecule bounces back with velocity $-v' = w - v$. In the original reference frame, the final velocity of the molecule is $2w - v$. Therefore the molecule transfers kinetic energy

$$\Delta E_{\text{kin}} = \frac{1}{2}mv^2 - \frac{1}{2}m(2w - v)^2 = 2mw(v - w)$$

to the plate. The number of molecules striking the plate per unit area per unit time is $\Phi = n(v - w)$, where n is the number of molecules per unit volume (see e.g. eq. (19.5)). The power per unit area delivered to the plate is therefore

$$\mathcal{P} = \Phi\,\Delta E_{\text{kin}} = 2(mn)vw(v - w)^2 = 2\rho vw(v - w)^2,$$

where $\rho = mn$ is the mass density of the fluid.

The rate of energy transfer clearly vanishes at $w = 0$ and at $w = v$, as expected. The speed w_{max} at which power transfer is maximized is determined by setting $d\mathcal{P}/dw = 0$,

$$\frac{d\mathcal{P}}{dw} = 2\rho v(v^2 - 4wv + 3w^2) = 2\rho v(v - 3w)(v - w) = 0.$$

The maximum thus occurs at $w_{\text{max}} = v/3$, where

$$\mathcal{P}_{\text{max}} = 2\rho\left(\frac{v}{3}\right)\left(\frac{2v}{3}\right)^2 = \frac{8}{27}\rho v^3 = \frac{16}{27}\mathcal{P}_{\text{wind}}.$$

Thus, the power extracted by a stationary turbine in a constant wind of velocity v is limited to $P \le (8/27)\rho v^3 A$, where A is the perpendicular area of the turbine through which the air moves.

In deriving this limit we made several simplifying assumptions. We assumed that the motions of the molecule and the object are parallel and that both are perpendicular to the surface of impact. If the angle of particle motion is not perpendicular to the impact surface then only one component of the kinetic energy can be transferred to the object, and the Betz limit holds for this component alone. So changing the angle of particle motion cannot increase the rate of power transfer. An elastic collision is not affected by motion of the object in the direction parallel to the impact surface, so including motion of the object in that direction also cannot affect the Betz limit. The "ballistic" assumption that molecules do not interact with one another does not hold for a real fluid such as air or water at terrestrial pressures and temperatures; for example, the fluid velocity must vanish at the interface with a solid object. So the preceding kinematic analysis is not directly valid for a continuous fluid. While the argument given here can be generalized to include collisions at a microscopic level, a more robust derivation of the Betz limit can be developed using the principles of fluid dynamics (§29), as we show in §30.

Example 28.1   Average Wind Speed and Average Wind Power

Compare the wind power available at two locations: at location (A) the wind blows at ~8 m/s roughly 1/3 of the time and is calm 2/3 of the time; at location (B), there is a steady wind at ~5 m/s.

Notice that the average wind speed at location (B), which is ~5 m/s, exceeds that at location (A), where the average is ~2.7 m/s, by almost a factor of two. However, the average of v^3 at location (B) is only ~73% of that at (A). Thus (taking the density of air to be 1.17 kg/m^3 at NTP) the average power density in the wind at location (A) is $\mathcal{P} = \frac{1}{2} \times 1.17\,\text{kg/m}^3 \times \frac{1}{3}(8\,\text{m/s})^3 \cong 100\,\text{W/m}^2$, while that at location (B) is only $\cong 73\,\text{W/m}^2$. So if all other considerations are equivalent, (A) is a better location for a wind farm.

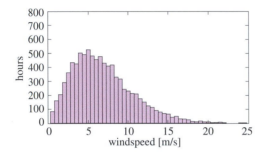

Figure 28.15 Histogram of wind speed measurements for all of 2002 at Lee Ranch, Colorado. The height of each bin records the number of hours during which the wind speed fell in the interval spanned by the bin. (Credit: Lee Ranch Wind Speed Frequency, reproduced under CC-BY-SA license via Wikimedia Commons)

in Figure 28.15. For effective wind power evaluation at a given location, measurements are typically taken over several years. With sufficient data, a histogram such as Figure 28.15 averages out to a smooth curve. If the distribution is rescaled so that the area under it is one, then the result is the wind **frequency distribution** $f(v)$ for the site. For each value of the speed, v, the function $f(v)$ measures the fraction of the time that the wind speed lies between v and $v + dv$.

The wind frequency distribution $f(v)$ can be interpreted as a probability distribution

$$dP = f(v)\,dv.\qquad(28.7)$$

Assuming that other factors – e.g. climate, land use, vegetation – do not change, then $f(v)dv$ gives the probability that at the site in question at a random time, the wind speed will be observed to be between v and $v + dv$. The integral of f from $v = 0$ to v gives the probability that the wind speed is less than v,

$$P(v) = \int_0^v dv'\, f(v'),\qquad(28.8)$$

which is known as the **cumulative function** of the probability distribution.

The average wind speed and mean cube wind speed are given by *moments* of the wind frequency distribution,

$$\langle v \rangle = \int_0^\infty dv\, f(v)v,$$

$$\tilde{v}^3 = \int_0^\infty dv\, f(v)v^3.\qquad(28.9)$$

Examples of wind frequency distributions for two mountain locations in Austria are shown in Figure 28.16.

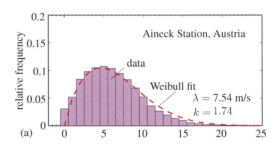

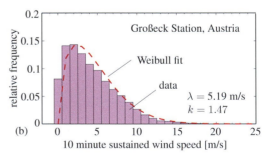

Figure 28.16 Wind frequency distributions from two Austrian Alpine stations. After [182]. Fits to Weibull distributions are shown, as are the fit values of the Weibull parameters λ and k. (a) Aineck station; (b) Großeck Station. The relatively small values of k reflect the relatively high variability of the winds at these mountain locations. (Credit: http://windharvest.meteotest.ch)

28.2.2 Modeling Wind Speed Variations – Weibull Distributions

It is often desirable to find a simpler characterization than the complete set of data in the histogram giving the function $f(v)$. This is generally done by using a small set of numerical parameters that capture the important features of the wind speed distribution at a given location. This makes it easier to compare the wind power potential at different sites and to estimate the power that a specific turbine design can harvest at a particular site. For many types of probability distributions, there are simple functions that capture the important features of the distribution in the limit of a large number of samples. Examples include the *Poisson distribution*, introduced in Box 15.1, which describes radioactive decays, and the *Gaussian distribution* $(1/\sqrt{2\pi\sigma^2})e^{-(v-\langle v\rangle)^2/2\sigma^2}$, which characterizes many probability distributions.

Experience has shown that wind frequency distributions can usually be fit quite well by a distribution function known as the **Weibull distribution**. The Weibull distribution is a simple probability distribution, normalized to one in the range $0 < v < \infty$, which depends on two parameters, a **scale parameter** λ, and a **shape parameter** k

$$f(v, k, \lambda) = \frac{k}{\lambda}\left(\frac{v}{\lambda}\right)^{k-1} e^{-(v/\lambda)^k} = \frac{d}{dv}\left(-e^{(-v/\lambda)^k}\right) = \frac{dP}{dv}. \quad (28.10)$$

The parameter λ sets the scale of the wind speed: a *larger* value of λ indicates a windier location. k, on the other hand, determines the variability of the wind: the *smaller* the value of k, the more variable the wind. The need to describe both the scale and variability of the wind explains why it is necessary to have two parameters; a one-parameter distribution will not suffice. Sometimes, though, when only a crude approximation is needed, or when data are not sufficient to fit both parameters, the parameter k is fixed to $k = 2$. In this case the Weibull distribution reduces to the **Rayleigh distribution**, which is closely related to the Gaussian distribution (Problem 28.8). The choice $k = 2$ implies strong assumptions about the way that wind varies at a given site, and in many cases does not give a good description of the distribution of wind measurements. Weibull functions that fit empirical data on wind frequency distributions are shown in Figure 28.16.

One reason that Weibull distributions are convenient is that $f(v, k, \lambda)$ can be written as a derivative of a simple function, as in eq. (28.10). Therefore the cumulative function $P(v)$ takes a particularly simple form,

$$P(v, k, \lambda) = \int_0^v dv' f(v', k, \lambda) = 1 - e^{-(v/\lambda)^k}. \quad (28.11)$$

Note that $P(v, k, \lambda) \to 1$ as $v \to \infty$, so that the Weibull distribution is properly normalized. Examples of Weibull distributions are shown in Figure 28.17. Figure 28.17(a) shows the variation with k of Weibull distributions all of which share the same value of $\langle v \rangle$. It is clear that small k indeed corresponds to greater variability and implies a longer tail out to large values of v. Figure 28.17(b) shows distributions with the same value of k and different values of λ.

Weibull Distributions

Wind speed distributions at any given location are generally modeled well by the *Weibull distribution*

$$f(v, k, \lambda) = \frac{k}{\lambda}\left(\frac{v}{\lambda}\right)^{k-1} e^{-(v/\lambda)^k}.$$

The parameter λ characterizes the scale of typical wind speeds, while k characterizes the variability at the specific site (smaller k implies greater variability)

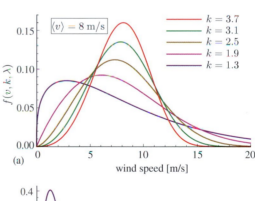
(a)

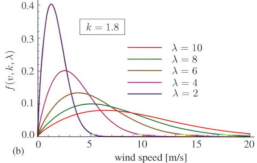
(b)

Figure 28.17 Examples of Weibull distributions. (a) Distributions with the same mean wind speed $\langle v \rangle = 8$ m/s and varying shape parameter k. (b) Distributions with the same shape parameter $k = 1.8$ and various values of the scale parameter λ (in m/s).

Another advantage of the Weibull distribution is that the average and root mean cube wind speed can be expressed simply in terms of the *special function* known as the *Gamma function* $\Gamma(z)$ (Appendix B.4.3). The *m*th **moment** of the Weibull distribution is given by (Problem 28.10),

$$M_m[f(v,k,\lambda)] \equiv \int_0^\infty dv\, v^m f(v,k,\lambda) = \lambda^m \Gamma(1+m/k).$$

$$(28.12)$$

Note that the *m*th moment is proportional to λ^m, in accord with the statement that λ sets the scale of the wind distribution.

Both $\langle v \rangle$ and $\tilde{v}$ can be expressed in terms of Gamma functions:

$$\langle v \rangle = M_1[f(v,k,\lambda)] = \lambda\, \Gamma(1+1/k),$$
$$\tilde{v} = \sqrt[3]{M_3[f(v,k,\lambda)]} = \lambda\, \sqrt[3]{\Gamma(1+3/k)}. \quad (28.13)$$

The ratios $\langle v \rangle/\lambda$, $\overline{v}/\lambda$, and

$$r(k) \equiv \frac{\tilde{v}}{\langle v \rangle} = \frac{\sqrt[3]{\Gamma(1+3/k)}}{\Gamma(1+1/k)}, \quad (28.14)$$

are graphed as functions of k in Figure 28.18. The significance of the shape parameter in determining the power available in the wind is most clearly visible in the ratio $r(k)$ plotted in Figure 28.18(b). $r^3(k)$ measures the power available in the Weibull distribution compared to a steady wind with the same average wind speed. The relative importance of the average wind speed and of its variability are easily estimated with the use of the Weibull distribution parameters. An example is given in Example 28.2.

28.2.3 Wind Direction Distributions

Another feature that is important in siting and designing a wind farm is the direction of the prevailing winds, or, more precisely, the v^3 weighted distribution of wind directions. As described below, wind turbines have significant downstream shadows. An array of wind turbines cannot be arranged so that the turbines each avoid one another's shadows independent of the wind direction, so finding the optimal layout for a given directional distribution of wind is an important consideration. The correlation between the direction and intensity of the wind is important: if the winds blow 70% of the time gently from the southwest, but 30% of the time strongly from the northwest, then despite the prevalence of southwest winds, the wind farm should be aligned to best capture winds from the northwest.

Information on the direction and intensity of the wind is usually summarized in plots known picturesquely as **wind roses**. The compass is divided up into (usually) 8, 12, 16, or as many as 36 wedges, and the probability of observing the wind from each direction is displayed. Usually the data is further subdivided to display the probability as a function of both direction and speed. The example shown in Figure 28.19 is a form that is particularly easy to interpret. For each compass direction, the length of the wedge indicates the probability that the wind will

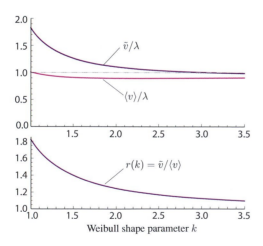

Figure 28.18 Average $\langle v \rangle$ and root mean cube $\overline{v}$ wind speeds, both scaled by the Weibull scale parameter λ, and the ratio $r(k) = \overline{v}/\langle v \rangle$, all as functions of the Weibull shape parameter k.

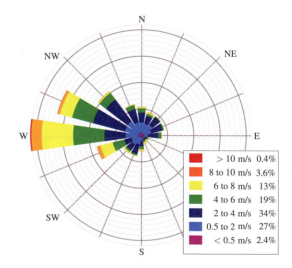

Figure 28.19 Wind rose for Klamath Lake, Oregon from July 2005 through September 2015. The wind speeds are measured at a height of 10 m. The average windspeed over this period was 3.6 m/s. The concentric circles represent increments of 5% probability for the wind coming from the given direction with speed in a specific range. (Credit: US Geological Survey, Oregon Water Science Center)

Example 28.2 Comparing Wind Power at Two Austrian Sites

It is clear from a glance at Figure 28.16 that Aineck Station (A) in Austria is a better location for wind power than Großeck Station (G). How much better? The Weibull distribution parameters – given in the figure – enable us to answer this quantitatively.

The average wind speeds at the two locations are obtained from eq. (28.13),

$$\langle v \rangle_A = 7.54 \, \Gamma(1.57) \cong 7.54 \times 0.89 \cong 6.7 \, \text{m/s},$$
$$\langle v \rangle_G = 5.19 \, \Gamma(1.68) \cong 5.19 \times 0.91 \cong 4.7 \, \text{m/s}.$$

If the power in the wind was proportional to the cube of the average wind speed, Aineck Station would have $(6.7/4.7)^3 \cong 2.9$ times as much power available. The average cubes of the wind speeds, however, are

$$\overline{v_A^3} = (7.54)^3 \, \Gamma(2.72) \cong 428 \times 1.57 \cong 670 \ (\text{m/s})^3,$$
$$\overline{v_G^3} = (5.19)^3 \, \Gamma(3.04) \cong 140 \times 2.08 \cong 290 \ (\text{m/s})^3,$$

and the actual ratio of powers available is $670/290 \cong 2.3$. Thus, as expected, Aineck Station is a better wind power site, but the relatively frequent occurrence of wind speeds greater than the average, indicated by the small value of the Weibull parameter k, makes the Großeck site a somewhat better location than the average wind speed alone would suggest.

---

be found in a given direction with the speed indicated by the color. Thus, for example, the wind at the Klamath Lake site on average blew from the west 21% of the time and it both came from this direction and had a speed between 6 and 8 m/s approximately 6% of the time. Integrating the information in a wind rose over all directions provides a rough estimate of the wind speed frequency distribution, which is given in the key to the figure (Problem 28.14).

28.2.4 Wind Power Classes and Wind Atlases

The wind resource characteristics described in the previous sections must be considered in detail when evaluating any particular site for wind power potential. On a larger scale, however, it is convenient to use a less detailed classification. Governments collect and publish data on wind energy. At the meta-level, the US government groups the information into **wind power classes** ranging from 1 to 7. The classes are defined in Table 28.2. Other governments and international organizations use similar schemes. In some cases, only wind power data is reported, and in some cases only average wind speed at a particular height is reported. The US scheme for conversion between power and average wind speed assumes a Rayleigh ($k = 2$) distribution. Data are typically presented at heights of 10 and 50 meters using the $z^{1/7}$ rule (see Footnote 2) to relate the two. According to the US National Renewable Energy Laboratory (NREL), areas designated as wind power class

3 or above are generally suitable for *utility-scale* wind turbine applications, class 2 is marginal.

Data on wind resources from many sites can be combined to create a **wind atlas**, which makes it possible to visualize wind power potential over large areas. Wind atlases are now available or under development for many regions throughout the world. Figures 28.20 and 28.21 give an overview of the wind power resources available in the US and Europe. Many more detailed atlases are available online. European wind resources (Figure 28.21) are concentrated on its northwest coast and in the British Isles. Wind power potential is relatively low in southern Europe, though local resources in the Pyrenees Mountains and in the South of France (where the *Mistral* blows) are prominent. The US (Figure 28.20) is relatively rich in wind capacity, though some areas such as the Southeast coastal plain and the Southwest deserts have little potential. Much of the resource is in sparsely populated areas such as the upper plains (Oklahoma through the Dakotas), and the Rocky Mountain peak areas. There are, however, some exceptionally felicitous areas where high wind potential coincides with significant population density. Examples include the region just to the East of the San Francisco Bay Area, where the Altamont Pass windfarms were developed many years ago, and (just to the Southeast of MIT) in the shallow waters surrounding Cape Cod, Martha's Vineyard, and Nantucket off the coast of Massachusetts. This local resource can be seen clearly on the more detailed atlas in Figure 28.22.

Table 28.2 Classes of wind power density at 10 m and 50 m [185]. The speed–power relationship assumes a Rayleigh ($k = 2$) distribution. The height variation of the wind between 10 m and 50 m is assumed to follow a $z^{1/7}$ law.

Wind power class	10 m (33 ft)		50 m (164 ft)	
	Wind power density (W/m^2)	Average speed m/s (mph)	Wind power density (W/m^2)	Average speed m/s (mph)
1	< 100	< 4.4 (9.8)	< 200	< 5.6 (12.5)
2	100–150	4.4 (9.8)–5.1 (11.5)	200–300	5.6 (12.5)–6.4 (14.3)
3	150–200	5.1 (11.5)–5.6 (12.5)	300–400	6.4 (14.3)–7.0 (15.7)
4	200–250	5.6 (12.5)–6.0 (13.4)	400–500	7.0 (15.7)–7.5 (16.8)
5	250–300	6.0 (13.4)–6.4 (14.3)	500–600	7.5 (16.8)–8.0 (17.9)
6	300–400	6.4 (14.3)–7.0 (15.7)	600–800	8.0 (17.9)–8.8 (19.7)
7	400–1000	7.0 (15.7)–9.4 (21.1)	800–2000	8.8 (19.7)–11.9 (26.6)

Wind Power Classes

In the US, wind atlases and other large-scale surveys of wind power potential use a simplified system, *wind power classes*, to classify wind power potential. Wind power classes range from 1 to 7, based on the power in the wind referred to a nominal height of 10 or 50 m. Class 3 ($\mathcal{P} = 300$–400 W/m^2 at 50 m) and above are considered suitable for utility-scale wind power development.

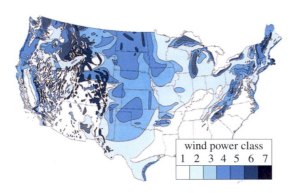

Figure 28.20 An annual average wind atlas of the continental US [185]. Note the absence of wind power in the Southeast.

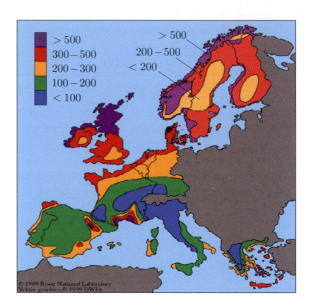

Figure 28.21 An annual average wind power atlas for much of Western Europe. The key gives wind power in W/m^2 at a height of 50 m over an open plain. Note that the data for Finland, Sweden, and Norway are designated differently. The power is greater for coast, open ocean, hills, and ridges, less for sheltered terrain. See [186] for a more detailed key. (Credit: windpower.org)

28.3 The Potential of Wind Energy

We conclude this chapter with a brief survey of local constraints on the exploitation of a wind power resource and an estimate of the global potential for human use of wind power. Where it is abundant, wind energy is a considerably more concentrated resource than solar energy. Unlike solar energy, however, wind energy is not a completely two-dimensional resource distributed across the planet's surface. A given packet of air flowing horizontally over Earth's surface carries energy over many kilometers of distance. Extracting energy from the wind at one location thus depletes the energy supply for a substantial distance

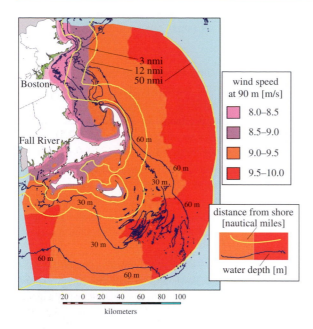

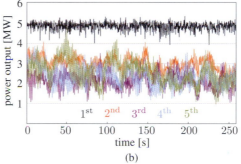

Figure 28.22 An estimated annual average wind power atlas for the east coast of the State of Massachusetts, for points within 50 nautical miles for shore (1 nmi = 1.852 km). (Adapted from map by US NREL)

Figure 28.23 (a) Turbulent wakes at the Horns Rev offshore wind farm in Denmark made visible by unusual weather conditions that created condensation in the turbulence. (b) A simulation of power output from five turbines separated along the flow direction by $3D$ [187].

downstream. This places limits on wind energy extraction both locally and globally. Locally it limits the density of placement of individual turbines in a wind farm. When the required spacing between wind turbines is included, wind and solar power are roughly comparable resources in terms of power produced per unit of land area. On a larger scale, the total energy available in atmospheric wind flow limits the power that can be extracted from winds near Earth's surface. Also, the complexity of modeling energy transport in the atmosphere has made it difficult to estimate the world's total wind power resources accurately.

28.3.1 Local Constraints on Wind Power Density

Table 28.2 indicates that at class 6, the power in the wind at a height of 50 meters above ground level is over $600 \, \text{W/m}^2$. The available power is diminished by efficiency of the wind turbine, which is limited by 16/27 (the Betz limit) and may be as large as 40% under ideal conditions (we discuss wind turbine efficiency in more detail in the next two chapters). Taking 40% efficiency with a wind power of $600 \, \text{W/m}^2$ at a height of 50 m, the recoverable power is $600 \times 0.40 \cong 240 \, \text{W/m}^2$. This power per unit area is almost an order of magnitude greater than that from solar photovoltaics ($\sim 250 \, \text{W/m}^2$ averaged over the year in good locations, gathered with $\approx 15\%$ efficiency), but unlike solar

collectors, wind turbines cannot be densely placed across a two-dimensional plain.

In a wind farm, wind turbines must be sited at some distance from one another in the direction along the flow of the wind to prevent the disturbance in air flow from one turbine from affecting others downstream. Behind each turbine there is a **wind shadow** in which the wind speed is reduced. This effect is illustrated in Figure 28.23(a). The effect is quantified in Figure 28.23(b), which shows the result of a simulation in which five 5 MW rated turbines with 120 m rotor diameters have been separated along the direction of flow by three times their rotor diameter. Only the first turbine performs at its rated power. The power output of the other four is suppressed by a factor of $\sim 1/2$. Thus, wind turbines must be placed at some distance behind one another so that the wind field has space to recover by bringing energy down from higher altitudes so that the wind turbines that are further downstream get the full benefit of the wind power. As a rule of thumb, losses from turbine interactions are believed to be less than 10% when turbines with rotor diameter D are placed at

Constraints on Wind Power Density

Wind turbines disturb the atmosphere in their wake, reducing the power available to downstream turbines in a closely spaced wind farm. Although more research needs to be done, the present state of knowledge indicates that turbines should be spaced by 10 to 15 times their rotor diameter in the direction of prevailing winds and four times their rotor diameter in the crosswind direction.

Because extraction of wind power at one location reduces the energy in the wind for a substantial distance downstream, wind power does not scale with area as solar power does.

a distance of $4D$–$5D$ apart in the direction perpendicular to wind flow, and on the order of $10D$ apart along the direction of flow [178, 179], though recent work that includes land and turbine cost considerations suggests that greater distances ($\sim 15D$ in the flow direction) may avoid inefficiency due to excess shadowing effects [188]. If, for example, turbines are separated by $4D$ in the crosswind direction and $10D$ in the downwind direction, then each turbine with blade covering an area $\pi D^2/4$ is allocated a land area of $40D^2$. Thus the overall power produced per unit of land area is reduced by a factor of $\pi/160 \cong 0.02$, so that in the example of the previous paragraph the power output would be roughly 5 W/m^2, somewhat smaller than could be realized for a solar field of similar area in a desert location. Thus, the land area needed for large-scale wind farms even in a high-quality (wind power class 6, for example) location is larger than that needed for solar power. In a wind power class 3 or 4 location, the area needed is several times that needed for a comparable solar power plant. On the other hand, solar arrays require covering the allocated land area densely with solar collectors, whereas most of the wind farm area is left undeveloped. Thus land used for wind turbines can simultaneously be used for other purposes such as agriculture, while total ground coverage by solar thermal mirrors or photovoltaic panels makes other uses more difficult or impossible.

On larger scales, wind can in principle only recover from the shadowing effect of turbines a finite number of times, since the height of the mass of moving air is finite. Thus, in a sense wind may best be thought of as a high-density *linear* resource, which cannot scale with area over arbitrary regions. This consideration is relevant when estimating the

limit to global wind power potential, which we consider next.

28.3.2 Global Wind Power Potential

As mentioned at the beginning of this chapter, roughly 1 PW of wind power is continuously dissipated in the atmosphere. The uncertainty in this number – a factor of two or three – is small compared with the range of current estimates of the wind power available for human use, which vary over roughly two orders of magnitude, from ~ 1 TW [176] to ~ 100 TW [189]. Since this subject remains in flux and is controversial, it is not appropriate to review it in depth here. Instead we outline and comment upon the methods of analysis that lead to such divergent estimates.

Estimates of the global wind power potential fall into two categories: *top-down* approaches that study energy flow in the atmosphere and try to estimate the kinetic energy dissipated in regions of the atmosphere available to human exploitation; and *bottom-up* approaches that survey wind resources at a local and regional level, make assumptions about accessibility, and then aggregate the harvestable power to obtain a global estimate. Different groups using these two approaches also include different constraints in their estimations. For example, some include the Betz limit, and others factor in the efficiency of existing wind turbines; some exclude the large geographical areas where the wind power resource is below some particular threshold, others consider the entire non-glaciated portion of Earth's land surface; some include offshore resources, others include only land-based wind power, etc.

A recent top-down analysis by Miller *et al.* gives a total available resource of 18–68 TW [190]. The higher value can be obtained from what they call a "back of the envelope" calculation: they estimate ≈ 900 TW wind power dissipation in the atmosphere, of which about 1/2 is dissipated in the free atmosphere where it is not available. Of the remaining ≈ 450 TW, roughly 3/4 is dissipated over ice or deep ocean, leaving ≈ 112 TW. Miller *et al.* then use the Betz limit to further reduce this estimate by a factor of 16/27 to ≈ 68 TW; since, however, the Betz limit applies only to single turbines over a given cross-sectional area, and does not imply that the remaining 11/27 of the wind energy in that area is irrevocably lost to downstream turbines, such a reduction may not be warranted. A more detailed analysis of the balance between momentum extraction and dissipation in the lower atmosphere reduces the estimate of Miller *et al.* – also including the reduction due to the Betz limit – to ≈ 21 TW; integration of their methods into a general climate circulation model gives ≈ 18–34 TW. Note that this estimate includes all regions where wind energy is dissipated over or near land outside the polar regions. Other workers have combined estimates of the wind power dissipated in the lower atmosphere with more detailed assumptions about viable regions for realistic wind power usage, organization of wind farms, and efficiency of wind turbines, and have obtained even lower estimates of the total available wind resource. de Castro *et al.*, for example, obtain an estimate of global wind-generated electrical power of ≈ 1 TW [176].

An example of a bottom-up analysis was performed by Archer and Jacobson in 2005 [189]. They estimated the global onshore wind power resource by combining observations from all seven continents. They took data from a wealth of measurements at 10 m above ground level and extrapolated to 80 m, which is the hub height for a typical modern 1.5 MW wind turbine. One of their wind atlases is shown in Figure 28.24. They selected those sites with mean annual wind speeds at 80 m exceeding 6.9 m/s (corresponding to wind power class 3 and above) and integrated over the regions where these favorable winds were observed. They assumed that favorable areas could be covered with turbines at a density of $4D \times 7D$ and estimated the actual power output from the turbines under local conditions. Archer and Jacobson did not exclude inaccessible areas such as Antarctica, but such locations do not dominate their survey. They estimated that if fully exploited, wind power at the onshore locations they identify could provide approximately 72 TW of mechanical power.

One criticism of the bottom-up approach is that it neglects the constraints associated with the limited total energy in any wind system. As we have emphasized, wind is not a two-dimensional resource like solar energy that

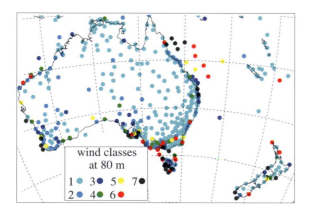

Figure 28.24 Measurements of mean annual wind power classes at 80 m for sites in Australia and New Zealand and Tasmania [189].

Global Wind Power Potential

There is no scientific consensus yet on a precise estimate for global wind power potential. Assessments range from 1 TW up to nearly 100 TW. Top-down approaches, which follow the energy budget of the atmosphere and attempt to compute the wind power dissipated in parts of the atmosphere accessible to human technology, are on the low side, and including all relevant constraints seems to push the bound down to a small number of terawatts. Bottom-up approaches, which identify areas with high-quality resources and cover those areas with a high density of wind turbines without considering global constraints give higher estimates. Future research should clarify the matter; at the rate that deployment is currently increasing, wind power production may approach or pass the 1 TW benchmark by mid-century.

can be exploited locally in any region on Earth's surface without affecting nearby regions. Consider for example a prevailing westerly wind sweeping across an area of plains measuring 300 km by 300 km at 8 m/s. The power density in this wind is $\mathcal{P} = \frac{1}{2}\rho v^3 \cong 326$ W/m^2. The total power available, however, is not given by multiplying this number by the area of the region ((300 km)$^2 \times$ 326 W/m$^2 \cong$ 30 TW). Rather, since the wind is flowing from west to east, the total power available may be bounded by multiplying the north–south extent of the wind field (300 km) by the height of the wind field, which may be generously taken to be on the order of the height of the troposphere, say 10 km. This reduces the total power available by a factor of 30, to about 1 TW. Furthermore, removal of wind energy from the flow on this scale could affect the wind energy available for many hundreds of kilometers downstream of this area. Note, for example, as shown in Figure 28.3, that the wind speed in the Southern Hemisphere 40–50° latitude belt gradually increases over thousands of kilometers east of the southern tip of South America. Similar effects are seen in the recovery of the wind fields on the lee sides of other continents.

A more definitive estimate of world wind power potential awaits further research. Ongoing work aimed at understanding the wake of a wind turbine (which can have measurable impact extending over many kilometers) and the total energy lost to the wind field through the combined effect of the wind turbine and turbulence in the wake, combined with a more detailed understanding of the total energy flow and availability in specific wind patterns and wind locations, may make a more precise estimate of wind power limitations possible in the coming years. In the meantime we quote Emeis's conclusion [180] that "Probably a single-digit number given in terawatts is a realistic estimate for the wind energy available from ... Earth's atmosphere." Recent projections of world wind power capacity in 2050 by the IEA [191] range from 2300–2700 GW. At an anticipated 31% capacity factor, this corresponds to an average of 0.70–0.83 TW of electric power generated from the wind. Thus the lowest estimates of Earth's wind power resource may be tested in practice before the end of this century.

Discussion/Investigation Questions

28.1 What is the most promising site for wind power within a 100 kilometer radius of your location? Can you identify the meteorological origin of its favorable position? If you can find data, make an estimate of its potential.

28.2 The Bahrain World Trade Center is a prominent example of *building-integrated* wind power. What are the parameters of this installation? What are its pros and cons? What is your opinion of building-integrated wind power in general?

28.3 Investigate the history of the Cape Wind project in the state of Massachusetts and discuss the arguments made in favor of and against this project.

Problems

28.1 [T] Assume that the pressure in a volume of air can be expressed to first order in a local coordinate system (x, y, z) as $p = p_0 + fx + \mathcal{O}(x^2, y^2, xy, \ldots)$. Show from the balance of forces on a small volume element dV that the force per unit volume is f in the $-x$-direction. Using the fact that the coordinate system can be chosen arbitrarily, show that more generally $\boldsymbol{f}_p = -\nabla P$.

28.2 [T,H] The forces on a parcel of moving air in the Ekman layer include the pressure force $\boldsymbol{f}_p = -\nabla p$, the Coriolis force $\boldsymbol{f}_{hc} = f\boldsymbol{v} \times \hat{\boldsymbol{n}}$ (where $f = 2\Omega\rho \sin\lambda$), and a frictional force directed opposite to the wind velocity $\boldsymbol{f}_f = -\alpha(z)\boldsymbol{v}$, where $\alpha(z)$ arises from the velocity gradient and turbulent viscosity. The coefficient function $\alpha(z)$ decreases with increasing height z through the Ekman layer and vanishes at the top of the layer. Find the wind velocity by demanding that the sum of the forces on the parcel vanishes. Express $\boldsymbol{v}$ in terms of components along ∇p and $\hat{\boldsymbol{n}} \times \nabla p$. Show that your result reproduces qualitatively the

wind pattern (both in direction and magnitude) shown in Figure 28.10.

28.3 Evaluate the Coriolis factor $f = 2\Omega\rho \sin\lambda$ at $\pm 45°$ latitude. How large a pressure gradient (in millibars/km) is necessary to sustain a geostrophic wind at 30 m/s at this latitude?

28.4 Suppose a gradient wind (see Box 28.1) is flowing along a circular isobar with radius R at a latitude of $45°$. Compare the strength of the Coriolis and centrifugal forces on this flow as a function of wind speed. At what speed (as a function of R) are the two inertial forces equal?

28.5 [T] The wind speed at any site can be written as the sum of its average $\langle v \rangle$ plus a fluctuating term that averages to zero, $v(t) = \langle v \rangle + \delta v(t)$. Show that the effect of the fluctuations is always to make $\langle v^3 \rangle$ greater than $\langle v \rangle^3$.

28.6 Compute the energy density and power density in a steady wind at 8 m/s and at 16 m/s.

28.7 What is the maximum power that could be extracted by a mechanical device of cross-sectional area $10\,\text{m}^2$ in a steady wind at 10 m/s?

28.8 [T] Suppose the wind speed obeys a Gaussian frequency distribution in the two components of the horizontal velocity $\boldsymbol{v} = (v_1, v_2)$,

$$f(v_1, v_2) = \frac{d^2 P}{dv_1 dv_2} = \frac{1}{2\pi\sigma^2} e^{-(v_1^2 + v_2^2)/(2\sigma^2)}.$$
(28.15)

Show that the *wind speed distribution*, $f(v) = dP/dv$, where $v = \sqrt{v_1^2 + v_2^2}$, is a Rayleigh distribution (Weibull parameter $k = 2$). What is the scale λ?

28.9 [T] Show that $\Gamma(n) = (n-1)!$ for integer n.

28.10 [T] Derive the formula for M_n in eq. (28.12).

28.11 A steady wind is blowing at 8 m/s at 200 m above ground level. Estimate the wind power at a height of 10 m above a large open cropped field. Approximately what height is needed for a turbine in a small town to reach wind flowing with the same power?

28.12 Consider building a wind turbine at a wind power class 5 site. Assume that the hub height is 50 m and the blade length is 15 m, and that the turbine operates at 50% of the maximum theoretical efficiency. Estimate the average power output of the turbine.

28.13 Reference [192] studies wind power prospects for several locations in Saudi Arabia, among them Yenbu (Y) and Al Qaysumah (A). They report Weibull parameters measured at a height of 10 m: $\lambda_Y \cong 5.9$ m/s, $k_Y \cong 2.25$; $\lambda_A \cong 5.1$ m/s, $k_A \cong 4.38$. Estimate the maximum available wind power density and the wind power class for each site at a height of 80 m assuming a roughness length $z_0 \approx 2 \times 10^{-5}$ (sand). Can you explain (on the basis of geography) why the winds at Yenbu are so much more variable than at Al Qaysumah?

28.14 Use the data given in the wind rose Figure 28.19 to make a crude estimate of the average $\langle v \rangle$ and root mean cube $\bar{v}$ wind speed at Klamath Lake, Oregon. From these, estimate the Weibull parameters for this distribution. Is your result close to a Rayleigh distribution ($k = 2$)? Is this a promising wind power location?

28.15 The following wind frequency distribution data were obtained in the month of January at a height of 10 m above an airport runway in Buffalo, New York. Calm: 2.0%; 0–5 kt: 5.3%; 5–12 kt: 40.1%; 12–20 kt: 39.3%; >20 kt: 13.3% (1 knot $= 0.514$ m/s). Assume that the wind above 20 kt has an average value of 25 kt. Make a crude estimate of the power in the wind at 10 m. Is this a promising location for wind power based on the January data alone?

Fluids: The Basics

Wind and flowing water are moving fluids that carry kinetic energy in the motion of their constituent molecules. Wind turbines capture this energy for human use. Other systems attempt to capture the kinetic energy in ocean waves or marine currents. Up to this point, it has not been necessary to go beyond a colloquial description of fluids, but in order to explain how some of these systems work, we must describe some basic properties of fluids in more detail. Although the behavior of fluids can become very complicated – indeed, the analysis of turbulent flow in many situations requires tremendous computational resources – the basics are relatively simple and give considerable insight into how it is possible to harvest energy efficiently from wind and moving water.

Two key attributes of fluids have been introduced in earlier chapters. In the previous chapter we found that the *power density* in a fluid flow is $\mathcal{P} = \rho v^3/2$, and is a good measure of the quality of a wind power resource. The notion that fluids exert a *drag force* on objects that move relative to the fluid was introduced in §2, where we studied *air resistance*, which consumes much of the energy expended by automobiles. One principal aim of this chapter is to introduce and explain the physical origins of *lift*, another force that a flowing fluid can exert on an object. Lift plays an essential role in the physical mechanism that powers wind turbines and therefore it merits careful consideration. One other force that even a static fluid can exert on an object is the *buoyant force*, mentioned in §5 and §6.

We begin (§29.1) by defining a fluid and describing how a flowing fluid can be characterized in terms of local variables such as velocity, energy density, mass and energy flux, in addition to the familiar thermodynamic variables of pressure, density, and temperature. In the following section (§29.2) we explore the implications of conservation of mass and energy in fluid flow. We specialize to *steady*

Reader's Guide

This chapter develops some of the basic principles of fluid mechanics. Conservation laws, Bernoulli's principle, vorticity, circulation, viscosity, and Reynolds number are introduced. The phenomenon of *lift* is described in some detail. Lift plays a central role in understanding the operation of wind turbines.

Prerequisites: §2 (Mechanics), §5 (Thermal energy), §28 (Wind energy).

This material is used in the following chapter (§30) to describe aspects of wind turbines and their design.

flows of fluids, which are simpler than general fluid flow, yet general enough to describe most situations of interest to us. Putting aside frictional losses for the moment, we use conservation of energy to derive *Bernoulli's equation*, an important basic result of fluid dynamics that relates the pressure and flow velocity in a moving fluid. This section ends with a discussion of fluid *circulation* and *vorticity*. The next section (§29.3) is devoted to friction in fluids, which is parameterized by *viscosity*. After defining viscosity and explaining how it is measured, we introduce *Reynolds number*, a dimensionless parameter that determines the relative importance of viscous and inertial forces on objects immersed in fluid flows. With these ingredients in hand, we turn to the discussion of lift in §29.4. We explain the connection between lift and circulation embodied in the *Kutta–Zhukovskiĭ theorem*, and apply it to airfoils, where viscous forces establish a steady fluid flow in which viscous losses are small. Finally, we discuss the importance of vorticity in the performance of wings and wind turbine blades.

As mentioned at the outset, fluid dynamics is a large and complex subject. A basic introduction from a physics perspective can be found in [193]. Issues in fluid dynamics are treated from the perspective of energy science in [7].

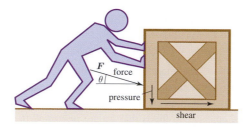

Figure 29.1 When a person pushes with a force F on a crate, which is held motionless on the floor by static friction, they are exerting both pressure and shear stress.

29.1 Defining Characteristics of a Fluid

29.1.1 What is a Fluid?

Fundamentally and intuitively, a fluid is a large collection of molecules that are free to move. More precisely, a **fluid** can be defined as a substance that cannot support a **shear stress** in mechanical equilibrium. Shear stress is a concept similar to pressure. Both are forces per unit area. Pressure is a force per unit area that acts perpendicular to a surface; shear is a force per unit area that acts in a direction *parallel* to the surface on which it is applied. When the person in Figure 29.1 pushes on the crate held stationary by friction, the force applied at an angle θ to the horizontal has a downward component $F \sin\theta$ that generates a pressure on the floor and a horizontal component $F \cos\theta$ that exerts a *shear stress* on the floor. Like pressure, shear stress is measured in *pascals* (1 Pa = 1 N/m^2) in SI units.

While a solid object like the floor in Figure 29.1 stays fixed when acted upon by the shear stress exerted by the crate, the same is not true of water or honey. These materials respond to a shear stress by flowing. Fluids can withstand pressure in equilibrium without motion, but not shear stress.[1] Water and gases such as air and steam, which are of primary interest to us, are classic examples of fluids.

29.1.2 Local Characterization of a Fluid

The molecules in a fluid are so small that the smallest volume of interest to us still contains a vast number of molecules. A cubic millimeter of liquid water, for example, contains more than 3×10^{19} water molecules. This

---

[1] There are many interesting materials, known collectively as *non-Newtonian fluids*, that share properties of both fluids and solids. A mixture of corn starch and water, known as *oobleck*, for example resists shear like a solid when considerable pressure is exerted on it. When the external pressure is removed, oobleck flows like a fluid under the force of gravity.

number is so huge that we can idealize a fluid as a *continuous medium* even at scales much smaller than a millimeter. We make this **continuum assumption** throughout our discussion of wind and water power. A property such as the mass density of the fluid is therefore naturally described as a continuous function $\rho(x, t)$ taking values at every point x within the fluid at every time t. This continuous function should be understood as the limit of the mass Δm in a finite volume of size ΔV centered at x, as the volume ΔV becomes small compared to any macroscopic scale of interest, but still large enough to contain an immense number of molecules,

$$\rho(x, t) = \lim_{\Delta V \to 0} \frac{\Delta m}{\Delta V}. \tag{29.1}$$

Although all materials are to some degree compressible, it is often a good approximation to neglect small changes in density and replace $\rho(x, t)$ by a constant ρ_0 and assume that a fluid is **incompressible**. We indicate when our results apply only to incompressible fluids and when they are more general.

The *local properties* ascribed to fluids should be understood in the same fashion as the density: their values at any point and time are defined as averages over a macroscopically small volume that nevertheless contains very many molecules. In addition to the mass density, there are several other properties of fluids that enter the description of wind and water power.

Velocity field The average velocity of the molecules in a fluid at a point x at time t is described by a *velocity field* $v(x, t)$. Like the electric and magnetic fields encountered in §3, $v(x, t)$ assigns a vector to each point in space and time.

Mass flux The flux of mass is the amount of mass crossing a unit area per unit time. Like the current density (3.27) in electromagnetism, which is a flux of charge, the mass flux is defined to be the product of the local density and velocity fields

$$\phi(x, t) = \rho(x, t) v(x, t). \tag{29.2}$$

In analogy to Figure 3.12, the mass flowing through a small surface dS in a time dt is given by $dm = \phi \cdot dS\, dt$.

Kinetic energy density Moving mass carries kinetic energy. In the case of a flowing fluid, the mass Δm in a volume ΔV about a point x carries kinetic energy $\Delta E_{kin} = \frac{1}{2} \Delta m\, v(x, t)^2$, leading, in the limit $\Delta V \to 0$, to a **kinetic energy density** $\varepsilon_{kin}(x, t)$,

$$\varepsilon_{kin}(x, t) = \lim_{\Delta V \to 0} \frac{\Delta E_{kin}}{\Delta V} = \frac{1}{2} \rho(x, t) v(x, t)^2. \tag{29.3}$$

Note that the velocity and kinetic energy density of a fluid depend on one's frame of reference. A fluid flowing rapidly past a stationary observer may possess considerable kinetic energy, while to an observer moving along with the fluid, the fluid is at rest and possesses no kinetic energy. Generally we are interested in harvesting power from a fluid moving with respect to an observer stationary on Earth's surface, so this is the frame in which kinetic energy density is relevant for wind power.

Kinetic energy flux and power density The flux of kinetic energy $J(x, t)$ is defined in analogy to the mass flux, as the kinetic energy per unit time crossing a unit area normal to the fluid's direction of motion,

$$J(x, t) = v(x, t)\varepsilon_{\text{kin}}(x, t) = \frac{1}{2}\rho v(x, t)^3 \hat{v}(x, t).$$
$$(29.4)$$

The *power density*, introduced in the previous chapter, is the magnitude of J,

$$\mathcal{P}(x, t) = |J(x, t)| = \frac{1}{2}\rho(x, t)v(x, t)^3.$$
$$(29.5)$$

For a comparison of the power density in moving air and water, see Figure 29.2.

Pressure Within a moving fluid, the pressure $p(x, t)$ at a point x is the force per unit area on a hypothetical surface in a reference frame that is *at rest with respect to the fluid*. This pressure is often referred to as **static pressure** in the context of fluid mechanics. Air and water, like most fluids, are *isotropic*: at any point in the fluid the static pressure is the same in all directions. It is the static pressure that is related to the fluid's temperature and density by an equation of state. The static pressure in a moving fluid decreases and increases as the flow velocity increases and decreases according to *Bernoulli's equation* (§29.2.2). For the fluids

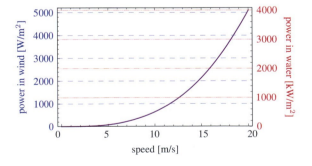

Figure 29.2 The power density in a fluid grows as the cube of the velocity. Wind speed $\sim$12 m/s gives 1000 W/m^2, comparable to insolation from overhead sun on a clear day. Water moving at a speed of 1.25 m/s has the same power density.

of interest here, the range of variation in the static pressure in the flowing fluid is generally much smaller than the absolute value of the pressure (see Example 29.3).

Temperature The intermolecular interactions that bring a fluid into local thermal equilibrium are typically rapid compared to the time scales over which the fluid moves macroscopically, so fluids that we are interested in maintain local thermal equilibrium as they flow and can be characterized by a temperature $T(x, t)$ that is a local function of x. Note that a fluid contains thermal energy – a result of the random thermal motion of the molecules – that is distinct from the fluid's bulk motion. We assume that the temperature of the fluid when at rest is fixed and equal to the temperature of the environment. The temperature of a moving fluid varies slightly, however, along with the pressure, as the velocity of flow changes (Example 29.3). Since moving wind and water are typically at or very close to the same temperature as their environment, these small fluctuations in thermal energy do not represent a resource that can be extracted for human use. Although frictional forces can dissipate the kinetic energy of fluid motion, the resulting change in temperature is often negligible as shown in Example 29.1, so we ignore the thermal energy content of flowing fluids here and in the study of wind turbines (§30) and water power (§31).

The *equation of state* (§5) provides a relationship among the static pressure, density, and temperature of a fluid. For air we assume an ideal gas law, $p = \rho k_{\text{B}} T/m$, where m is the average mass of a molecule of air. The thermal conductivity of air is small and the temperature fluctuations involved are also small in the regimes of interest to us. Therefore, as small fluctuations occur in the density during the motion of the fluid, heat flow can be neglected and the local change in pressure and temperature can be modeled *adiabatically*. In general, when a fluid is compressed adiabatically both the temperature and pressure increase. In particular, for adiabatic expansion or compression of air (see §10.2.2), the adiabatic gas law, $pV^\gamma = $ constant ($\gamma > 1$), implies $p \propto \rho^\gamma$; this combined with the ideal gas law gives $T \propto \rho^{\gamma - 1}$.

The dynamics of a flowing fluid can be viewed either from a fixed coordinate system (the *Eulerian approach*) or by using variables that move along with the fluid (the *Lagrangian approach*). These two approaches are essentially different methods of *bookkeeping* to keep track of the motion of the fluid. The dynamics of the fluid is, of course, independent of which method is used to describe it, although the dynamics may look quite different in these two formalisms. In this book we are primarily interested in how Earth-based devices harvest energy from wind or

---

Example 29.1   Thermal Effects in the Flow of Wind and Water

Suppose the kinetic energy in a flowing fluid was dissipated as thermal energy. *How much would the temperature of the fluid rise?*

The kinetic energy per unit volume is $\frac{1}{2}\rho v^2$. If this energy is converted to thermal energy, it warms the same volume of water by $c\rho\Delta T$, where c is the specific heat (per unit mass). The resulting increase in temperature is

$$\Delta T = \frac{v^2}{2c} \,.$$

For wind, $c \cong 1.0035\,\text{kJ/kg K}$, and

$$\Delta T_{\text{air}}\,[\text{K}] \approx \frac{1}{2\times 10^3}\, v^2[\text{m}^2/\text{s}^2]\,.$$

Thus, converting the kinetic energy in hurricane winds blowing at $\sim$100 knots ($\sim$50 m/s) to thermal energy would raise the air's temperature by only $\sim$1 K. For water, where $c \cong 4.186\,\text{kJ/kg K}$, the temperature rise is roughly 1/4 as large. Note that the kinetic energy in fluid motion is much more useful than the thermal energy into which it could be converted (Problem 29.2).

---

water as it flows by. Therefore, we generally use the **Eulerian approach** to the description of fluid motion, in which all variables, such as the density and velocity, are considered as functions in a fixed coordinate system. Thus $v(x, t)$, for example, measures the velocity at time t of the wind at a point x fixed on Earth's surface. In §31.2.1, where we describe energy flow in ocean waves, however, it is convenient to switch to Lagrangian coordinates.

29.2 Simplifying Assumptions and Conservation Laws

The dynamics of fluid flow can be quite complicated, as evinced for example by the complicated flow of water as it cascades down a stream bed or over a waterfall. Fortunately, this full complexity need not be confronted in order to give a satisfactory introductory description of wind (or water) power. For most of our purposes, it suffices to study *ideal* (i.e. frictionless) fluid flows that are *steady*, i.e. unchanging in time. We explain here first how the assumption of steady flow combines with conservation of mass to simplify the geometry of a fluid flow. We next make the further assumption that a fluid flow is ideal or *inviscid*, and derive *Bernoulli's equation*, which expresses conservation of energy in the absence of friction. Finally we introduce *vorticity*, a measure of the angular momentum in fluid flow, which is conserved in the steady flow of an incompressible, ideal fluid.

29.2.1 Steady Flow of a Fluid

We first consider the implications of conservation of mass for fluid flow. The situation is analogous to the discussion of electric charge conservation and electric current in

§3.2.2. If the mass of fluid in a small region R is changing, then conservation of mass requires that a net mass must be flowing through the surface S that bounds R. The local form of this conservation law relates the time derivative of the mass density to the divergence of the mass flux in exact analogy to eq. (3.33),

$$\frac{\partial \rho(x, t)}{\partial t} = -\nabla \cdot \phi(x, t) = -\nabla \cdot (\rho(x, t) v(x, t))\,. \quad (29.6)$$

In fluid dynamics, eq. (29.6) is known as the **continuity equation**; it applies to wind and water flow independent of the simplifying assumptions we introduce below.

The usefulness of the continuity equation is limited by the irregularities of fluid flow. At a given point x fixed in space, the velocity of the flowing fluid can vary with time in a highly irregular fashion. The situation is much simpler in the case of **steady flow**. In a steady flow, in the *Eulerian* description, all the characteristics of the fluid, $\rho(x, t)$, $v(x, t)$, $\mathcal{P}(x, t)$, etc. are independent of time. Each individual element of fluid changes its position and velocity with time as it moves, but it simply replaces the element of fluid that formerly occupied the space that it moves into, leaving the pattern unchanged. Steady flow contrasts with **turbulent flow**, where the flow pattern changes over time and may appear chaotic. The flow of water from a tap (Figure 29.3(a)) is a familiar example of steady flow, which contrasts with the turbulent flow shown in Figure 29.3(b) The conditions under which steady or turbulent flow of a fluid is favored are explored in §29.3.

The continuity condition simplifies in the case of steady flow because

$$\partial \rho(x, t)/\partial t = 0 \quad (29.7)$$

(a)   (b)

Figure 29.3 Steady (a) and turbulent (b) flow. (a) Water flowing steadily from a faucet. (b) Hot air rising from a cigarette, made visible by the smoke, is initially a laminar flow, but becomes turbulent. (Credit: (b) Jessie Murphy – http://www.uglyhedgehog.com)

at every point in space. When this is combined with the continuity equation, we obtain

$$\nabla \cdot (\rho(\boldsymbol{x}, t)\boldsymbol{v}(\boldsymbol{x}, t)) = 0, \text{ or} \qquad (29.8)$$

$$\boldsymbol{v} \cdot \nabla\rho + \rho\nabla \cdot \boldsymbol{v} = 0, \qquad (29.9)$$

where in the last line we used eq. (B.19).

Note that the continuity equation simplifies further when we specialize to *incompressible fluids*, for which $d\rho/dp = 0$ is a good approximation over the relevant range of pressures. For water flowing at the range of pressures experienced on Earth's surface, this is an excellent approximation. During steady flow of an incompressible fluid, the density does not change from point to point, so $\nabla\rho = 0$ and eq. (29.9) reduces to

$$\nabla \cdot \boldsymbol{v} = 0. \qquad (29.10)$$

This is the same condition we obtained for the electric field in charge-free space (see §3). In electromagnetism, $\nabla \cdot \boldsymbol{E} = 0$ implies that electric field lines can never begin or end in a charge-free region of space. By analogy, the *velocity field* $\boldsymbol{v}(\boldsymbol{x})$ that describes the steady flow of an incompressible fluid can be envisioned in terms of **streamlines** that neither begin nor end. The streamlines are the paths followed by individual fluid elements. For a *compressible fluid* in steady flow, the mass flux vector $\boldsymbol{\phi} = \rho\boldsymbol{v}$ is divergenceless; the streamlines of this flow are the same as those of the field $\boldsymbol{v}$.

By definition, fluid flows along the streamlines. Since $\boldsymbol{v}(\boldsymbol{x}, t)$ has a unique value at each point, streamlines cannot cross. This enables us to define the useful concept of a **flow tube**, which is the domain bounded by the streamlines that cross a closed curve $\mathcal{C}$ fixed in space (Figure 29.4). The

Conservation of Mass in Fluid Flow

Conservation of mass in fluid flow leads to the *continuity equation*, which relates the density and velocity fields,

$$\frac{\partial\rho(\boldsymbol{x}, t)}{\partial t} + \nabla \cdot (\rho(\boldsymbol{x}, t)\boldsymbol{v}(\boldsymbol{x}, t)) = 0.$$

If the flow is *steady* then the mass flux vector $\boldsymbol{\phi} = \rho\boldsymbol{v}$ is divergenceless,

$$\nabla \cdot (\rho\boldsymbol{v}) = 0.$$

This means that the *streamlines* – the paths followed by individual fluid elements – neither begin nor end.

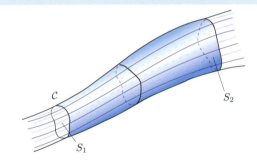

Figure 29.4 A flow tube showing some of the streamlines that form its boundary.

material inside a flow tube, cannot – by definition – ever get out.

29.2.2 Energy Conservation in an Ideal Fluid and Bernoulli's Equation

In general, kinetic energy is not conserved in fluid flow, even when no external forces act on the fluid. Frictional forces between adjacent regions of fluid moving with slightly different velocities transform kinetic energy into thermal energy, in somewhat the same way that friction dissipates kinetic energy when solid surfaces slide by one another. This is how an agitated fluid comes to rest. Common experience suggests that different fluids dissipate energy at very different rates. *Viscosity* is the measure of the strength of dissipative forces in a liquid. We postpone further study of viscosity until §29.3.

When viscosity can be ignored, a flow is termed **inviscid** or **ideal**.[2] In many aspects of the study of wind and water

---

[2] Note that an *ideal fluid*, which has no viscosity, should not be confused with an *ideal gas* (§5.1.3).

power, air and water may be regarded as ideal. Under these circumstances, conservation of energy provides a powerful relation, known as *Bernoulli's principle* or *Bernoulli's equation*, that relates the static pressure and the kinetic and potential energy along any streamline in the fluid.

The basic framework of Bernoulli's equation was described already in Box 13.1 where we implemented energy conservation within a *control volume* as shown in Figure 13.2. In that context, we consider a flow tube as the control volume, as illustrated in Figure 29.4, and choose the flow tube narrow enough that the fluid's static pressure and speed can be considered constant over the cross section. We can analyze the flow of the fluid, as done in Box 13.1, by following a small volume, or *slug*, of fluid passing along the flow lines within the tube. Assuming that no heat is added (adiabatic, $q = 0$) and no work is done on any external system ($w = 0$), conservation of energy gives the result that

$$u_1 + p_1/\rho_1 + gz_1 + \frac{1}{2}v_1^2 = u_2 + p_2/\rho_2 + gz_2 + \frac{1}{2}v_2^2$$
(29.11)

at any two points along the flow tube, where u represents the internal energy per unit mass. Thus

$$u + p/\rho + gz + \frac{1}{2}v^2 = \text{constant} \qquad (29.12)$$

along a flux tube. This is **Bernoulli's equation** for a compressible fluid. If the fluid is incompressible, ρ and u are constant, and we have *Bernoulli's equation* for an incompressible fluid,

$$p/\rho_0 + gz + \frac{1}{2}v^2 = \text{constant}. \qquad (29.13)$$

A principal consequence of Bernoulli's equation is that in regions where the fluid velocity is higher, the (static) pressure is lower. The quantity $\rho v^2/2$ is sometimes referred to as **dynamic pressure**, although it does not express a physical pressure exerted as a force per unit area.

Essentially the same relation as (29.13) holds when the flow is compressible but adiabatic, as long as the fluctuations δp, $\delta \rho$, δu in the pressure, density, and specific internal energy are small compared to their ambient values. To see this, expand p, ρ, and u in eq. (29.12) about their ambient values p_0, ρ_0, and u_0,

$$u_0 + \frac{p_0}{\rho_0} + \frac{\delta p}{\rho_0} + gz + \frac{1}{2}v^2 + \left(\delta u - p_0\frac{\delta \rho}{\rho_0^2}\right) = \text{constant}.$$
(29.14)

For adiabatic changes, the first law of thermodynamics for a slug of fluid of mass m reduces to

$$\delta U = m\delta u = -p\delta V = -p\delta\left(\frac{m}{\rho}\right) = mp\frac{\delta \rho}{\rho^2}, \quad (29.15)$$

Bernoulli's Principle

Applying conservation of energy to the steady flow of an *ideal* fluid leads to Bernoulli's equation, which relates the pressure in the fluid to the fluid velocity and the potential energy in a (constant) gravitational field. For an incompressible fluid,

$$\frac{1}{2}v^2 + \frac{p}{\rho} + gz = \text{constant}$$

along any streamline in the fluid.

For a compressible adiabatic ideal fluid where the fluctuations in pressure are small compared to the ambient pressure p_0, we have the analogous relation

$$\frac{1}{2}v^2 + \frac{\delta p}{\rho_0} + gz \approx \text{constant}.$$

so the term in parentheses in eq. (29.14) vanishes. So for small adiabatic fluctuations in a compressible fluid, Bernoulli's equation takes essentially the same form as for an incompressible fluid

$$\frac{p}{\rho_0} + gz + \frac{1}{2}v^2 \approx \text{constant}. \qquad (29.16)$$

This simplified form of Bernoulli's equation is useful in describing air flow in situations such as around the blade of a wind turbine (see §30.1.1).

The analysis here shows that the quantity in eq. (29.16) is approximately constant along streamlines. If all streamlines for a given flow originate from points with a common value of v, p, ρ, and z, then the constant is uniform across the flow. Note that the analysis of flow around an object such as a turbine blade or airplane wing must be performed in a frame where the object is stationary; otherwise the flow cannot be described as steady (Question 29.5).

29.2.3 Circulation and Vorticity

Two closely related but distinct features of fluid flow that arise in several aspects of the study of wind power are *vorticity* and *circulation*. **Vortices** and eddies are phenomena familiar from bathtubs and rivers. Whirlpools are large-scale vortices. Vortices can appear in steady flow, as when water drains from a bathtub, or in unsteady flow, where vortices break loose from a steady flow and wander downstream in quasi-regular way (Figure 29.5). A vortex signals a net circulation of fluid. Circulation can be quantified by constructing the *line integral* of the velocity around a closed curve $\mathcal{C}$

Example 29.2 A Venturi Flow Meter

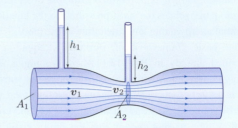

Bernoulli's equation (29.13) predicts that the faster a fluid flows, the lower its pressure. This, together with conservation of mass, makes it possible to construct a simple and accurate meter to measure the flow rate of an incompressible fluid, like water flowing through a pipe. The result, known as a **Venturi flow meter**, is shown in the figure at right. The object is to measure the mass per unit time $Q = \rho v_1 A_1 = \rho v_2 A_2$ of a fluid flowing in the pipe by measuring the difference in the fluid's pressure in the sections of pipe with cross-sectional areas A_1 and A_2. The pressure and velocity are assumed to be constant across the pipe. The pressures are measured by the height of a column of liquid $p = \rho g h$ (a simple application of Bernoulli's equation) in the two sections of pipe. Bernoulli's equation for the flow of an incompressible fluid states that

$$p_1 + \frac{1}{2}\rho v_1^2 = p_2 + \frac{1}{2}\rho v_2^2 \quad \text{or} \quad \rho g h_1 + \frac{1}{2}\rho v_1^2 = \rho g h_2 + \frac{1}{2}\rho v_2^2 .$$

Combining this with mass conservation leads directly to

$$Q = A_1 A_2 \sqrt{\frac{2g(h_1 - h_2)}{A_1^2 - A_2^2}} .$$

Example 29.3 Variations in Pressure, Temperature, and Density of Air in Motion

Consider airflow around the tip of a rotating wind turbine blade. Relative to the blade, the ambient air in the local wind is moving at a speed of $v \cong 50$ m/s. *How do the pressure, temperature, and density of the air vary between different locations in the flow?*

According to Bernoulli's equation (29.16), the (static) pressure of a fluid increases as the velocity of the fluid decreases. The change in static pressure is given by the (negative of the) *dynamic pressure* $\rho_0 v^2/2$. To get an upper bound on the magnitude of the fluctuations, we consider the pressure difference between the flow at $v \sim 50$ m/s and $v = 0$ m/s.

The fractional variation in its pressure is roughly

$$\frac{\delta p}{p} \sim \frac{\rho_0 v^2}{2 p_0} \sim \frac{(1.17 \, \text{kg/m}^3) \times (50 \, \text{m/s})^2}{2 \times 10^5 \, \text{Pa}} \sim 0.015 .$$

Assuming that the air is a compressible ideal gas with temperature and pressure that vary adiabatically with fluctuations in volume, pV^γ is a constant for a given mass of air, with $\gamma \cong 1.4$. It follows that

$$\gamma \frac{\delta V}{V} \sim -\frac{\delta p}{p} \quad \Rightarrow \quad \frac{\delta \rho}{\rho} = -\frac{\delta V}{V} \sim 0.010 .$$

From $T \propto pV$, we have

$$\frac{\delta T}{T} \sim \frac{\delta p}{p} + \frac{\delta V}{V} \sim 0.005 .$$

The proportional changes in all these quantities are therefore relatively small, with the smallest fractional change in the temperature.

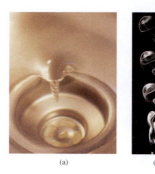

(a)   (b)

Figure 29.5 (a) A vortex formed in the steady flow of water down a drain. (Credit: Jessica Rose) (b) A *vortex street* moving downstream from an obstruction. (Credit: Jürgen Wagner reproduced under CC-BY-SA 4.0 license via Wikimedia Commons)

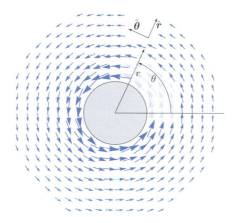

Figure 29.6 (a) A vortex circulating around a cylindrical obstruction in a fluid. $\oint d\boldsymbol{x} \cdot \boldsymbol{v} = \Gamma$ is nonzero but the vorticity $\boldsymbol{\omega} = \nabla \times \boldsymbol{v}$ vanishes everywhere in the fluid. The polar coordinates (r, θ) are defined in the figure, as are the unit vectors $\boldsymbol{\theta}$ and $\boldsymbol{r}$.

$$\Gamma(\mathcal{C}) = \oint_{\mathcal{C}} d\boldsymbol{x} \cdot \boldsymbol{v} . \qquad (29.17)$$

$\Gamma(\mathcal{C})$ is known as the **circulation** around the curve $\mathcal{C}$. The sign of $\Gamma(\mathcal{C})$ depends on the direction chosen for the integration around the curve $\mathcal{C}$.

A particularly simple example of circulation is a symmetric flow of an incompressible, ideal fluid around a circular cylinder (see Figure 29.6),

$$\boldsymbol{v} = \frac{\Gamma \hat{\boldsymbol{\theta}}}{2\pi r} , \qquad (29.18)$$

where we have chosen the radial dependence so that the circulation is constant – and equal to Γ – on all curves that surround the cylinder (Problem 29.10). The figure shows

a cross section through the cylinder and the fluid circulating around it. It is not hard to show that this velocity field satisfies eq. (29.10) as an incompressible fluid must (Problem 29.10).

Suppose that, in contrast to the previous example, a surface S can be found that spans the curve $\mathcal{C}$ in eq. (29.17) and lies *entirely within the fluid*. In that case, the integral that defines $\Gamma(\mathcal{C})$ can be converted to an integral over the surface S by use of *Stokes theorem* (B.27),

$$\Gamma(\mathcal{C}) = \oint_{\mathcal{C}} d\boldsymbol{l} \cdot \boldsymbol{v} = \int_{S} d\boldsymbol{S} \cdot \nabla \times \boldsymbol{v} . \qquad (29.19)$$

Under these circumstances the circulation around a curve in a fluid can be regarded as a surface integral of a *local* quantity

$$\boldsymbol{\omega} = \nabla \times \boldsymbol{v} , \qquad (29.20)$$

which is known as the **vorticity**. $|\boldsymbol{\omega}|$ measures the magnitude of local circulation of the fluid and $\hat{\boldsymbol{\omega}}$ points along the axis around which the fluid is circulating. Two examples of flows with vorticity are shown in Figure 29.7. The flow on the left is obviously a vortex; the flow on the right does not appear to obviously circulate, but increases in velocity from left to right. When $\boldsymbol{v}$ is integrated around the circle $\mathcal{C}$, the result is nonzero because the (positive) contributions on the right are larger than the (negative) contributions on the left.

If a fluid flow has nonzero vorticity at a given point $\boldsymbol{x}$, then the circulation around any sufficiently small closed curve encircling that point and the axis of vorticity will be nonzero. Thus, vorticity implies circulation. The converse, however, is not true. The symmetric flow around a cylinder shown in Figure 29.6 provides a counter-example: the velocity defined in eq. (29.18) satisfies $\nabla \times \boldsymbol{v} = 0$ everywhere outside the cylinder (Problem 29.10), so it has zero vorticity even though its circulation Γ_0 is not zero. This is possible because the fluid velocity flow vector $\boldsymbol{v}$ is not

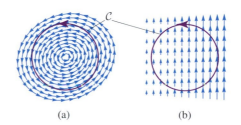

(a)   (b)

Figure 29.7 Two flows with vorticity, illustrated by velocity vectors of the flow. In (a) the fluid is clearly circulating. In (b) the flow is linear, but the velocity increases to the right, making the integral around a closed curve nonzero.

defined as a continuous field extending into the interior of the cylinder. A fluid flow in which the vorticity is zero is called **irrotational**. The fact that an irrotational flow can nevertheless have nonzero *circulation* around an obstructing object figures centrally in the explanation of lift in §29.4.

For a steady flow, vorticity, like all other properties of the fluid, is time-independent. The time dependence of vorticity in a non-steady flow can be more complex, as illustrated in Figure 29.5(b) for example. The British physicist William Thomson (Lord Kelvin, for whom the absolute temperature scale is named) proved that any flow of an ideal (inviscid) *barotropic*[3] fluid that is irrotational will remain irrotational. Furthermore, **Kelvin's circulation theorem** states that the vorticity of a packet of fluid is constant as the packet moves through the flow. Hence vorticity, once created, is persistent. Viscous effects eventually dissipate vorticity in real fluids, as when, for example, a vortex stirred up in a glass of water eventually comes to rest.

Circulation and Vorticity

The *circulation*

$$\Gamma(\mathcal{C}) = \oint_{\mathcal{C}} d\boldsymbol{x} \cdot \boldsymbol{v}$$

measures the extent to which a fluid circulates around a closed curve $\mathcal{C}$. If the *vorticity* $\boldsymbol{\omega} = \nabla \times \boldsymbol{v}$ is non-vanishing at a point within the fluid, then there are curves surrounding that point about which the circulation is nonzero.

A fluid in which $\boldsymbol{\omega} = 0$ is termed *irrotational*. When a curve $\mathcal{C}$ surrounds an obstruction in the fluid such as a cylinder, the circulation can be nonzero even if $\boldsymbol{\omega} = 0$ everywhere in the fluid. Irrotational flows can have non-vanishing circulation in the presence of obstructions.

29.3 Viscosity

Internal friction plays an essential role in the motion of fluids. The term *viscosity* refers both to the phenomenon of friction in fluid flow and to the parameter that measures

---

[3] A fluid is **barotropic** when the density can be written purely as a function of the pressure. This condition is satisfied both for incompressible fluids and for a compressible fluid where pressure and density changes are adiabatic.

Figure 29.8 Dust accumulates on a ceiling fan despite its rapid rotation because the air comes to rest in the boundary layer at the blade's surface. (Credit: Noelle Talley, Homemakerchic.com)

the tendency of a fluid to experience internal frictional forces. A fluid's viscosity is an intrinsic property like its density or heat capacity. In everyday experience we associate viscosity with the *thickness* of a fluid as well as with its resistance to flow. A relatively low-viscosity liquid such as water splashes when it is poured and stays in motion for a relatively long time when it is stirred. In contrast, a high-viscosity liquid such as honey pours smoothly and comes quickly to rest when stirred.

Viscosity can never be entirely ignored when considering the interaction of objects with a flowing fluid. Molecules of the fluid interact strongly with the molecules of the object at the interface, leading to the **no-slip condition**: the fluid must come to rest relative to the object at the object's surface. A commonly encountered, though counter-intuitive, example of this phenomenon is the build-up of dust on the surface of a fan blade (Figure 29.8). Even though the blade may move quickly through the air when turning, dust that has accumulated on the blade is not swept away because the air is at rest at the blade's surface. The layer directly above an object's surface through which a fluid recovers its free-stream velocity is known as the **boundary layer**; one example being the *planetary boundary layer* that was analyzed in the previous chapter. Viscosity cannot be ignored in the boundary layer because it is responsible for bringing the fluid to rest. If a fluid has low viscosity (or more precisely, a high *Reynolds number*, §29.3.2), however, the boundary layer is thin and viscosity is not important in the bulk flow.

29.3.1 Defining Viscosity

Viscosity is associated with *shear stress*. We have already encountered this concept under static conditions, but *viscous shear* characterizes a fluid in motion rather than at rest. Suppose a fluid is flowing steadily in the x-direction above the plane $z = 0$, where there is a solid surface, as shown in Figure 29.9. This could illustrate water flowing in a pipe, a wind stream blowing over Earth's surface as

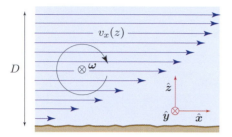

Figure 29.9 A viscous flow past a plate in the xy-plane. High above the plate the fluid flows with a bulk velocity $v\hat{x}$, while at the plate's surface the fluid is at rest. The fluid recovers its bulk velocity over a distance D. Note that this *laminar flow* has nonzero vorticity, as indicated in the figure.

discussed in §28.1.5, or the air passing over the blade of a wind turbine. Far above the plane the fluid is flowing at a speed v, but at the surface the fluid is at rest as required by the no-slip condition. The *boundary layer* is conventionally defined as the domain in z over which the fluid recovers 99% of its bulk velocity $v = v\hat{x}$. A flow such as the one shown in Figure 29.9, where layers of fluid slip past one another, is known as **laminar flow**, because the fluid moves in quasi-parallel layers with (ideally) no disruption. At a small enough velocity v_x, all fluids flow in this way. Without viscosity, however, laminar flow is unstable since the slightest perturbation would never damp out, but would instead propagate indefinitely downstream of its origin.

In a boundary layer, such as the one shown in Figure 29.9, the fluid at each value of z exerts a drag force on the layers of fluid directly above and below, which are moving at different speeds. The force on the layer above acts in the $-\hat{x}$-direction, opposing the motion of the fluid, much like the frictional force familiar from particle mechanics. The frictional force per unit area dF_x/dA acts in the xy-plane, so it is a *shear stress*. The **viscosity** η of the fluid is defined as the coefficient of proportionality between the frictional force and the velocity gradient $\partial v_x/\partial z$ that gives rise to the force.

$$\frac{dF_x}{dA} = \eta \frac{\partial v_x}{\partial z} . \tag{29.21}$$

(A partial derivative is necessary because v_x can, in principle, depend on the other coordinates x and y.)

The units of viscosity are momentum per unit area, or kg/m s, and the SI unit is the pascal-second (1 Pa s = 1 kg/m s). A cgs unit, the **poise** (abbreviated P) is commonly used (1 P = 1 gm/cm s = 0.1 Pa s). Common liquids have viscosities on the scale of 10^{-2} poise = 1 *centipoise*. For example, at 20 °C, the viscosities of water, di-ethyl ether, and olive oil are 1.002, 0.233, and 84 centipoise,

respectively. Gases typically have viscosities on the scale of *millipoise*. For example, air at STP has $\eta_{air}(STP) = 0.184 \times 10^{-3}$ poise. The viscosities of gases depend only weakly on density and pressure.

Viscosities can be determined by studying laminar flow in a straight pipe of circular cross section. In this configuration the total mass flowing through the pipe per unit time $\dot{m} = dm/dt$ is related to the pressure drop Δp from one end of the pipe to the other by **Poiseuille's law**,

$$\dot{m} = \int_0^R dr\, 2\pi r \rho v(r) = \frac{\pi \rho \Delta p R^4}{8\eta l} , \tag{29.22}$$

where R and l are the radius and length of the pipe and ρ and η are the density and viscosity of the fluid. Poiseuille's law, which is derived in Example 29.4, summarizes many commonly observed aspects of viscous flow. For example, the fact that $\Delta p \propto l$ with all other variables fixed expresses the phenomenon that pressure drops along the length of a pipe through which water is flowing. The fact that $\Delta p \propto 1/R^4$ agrees with the observation that fire hoses, which must maintain high pressure, are very thick. Poiseuille's law provides a simple method to estimate the viscosity of different fluids, by measuring the flow rate as a function of pressure in a pipe of fixed length and radius.

29.3.2 Reynold's Number and the Variation of Viscous Effects with Length Scale

When an object is placed in a fluid flow with a homogeneous velocity v far from the object, viscous shear stress gives rise to a drag force on the object. The viscous drag on the object is proportional to both the viscosity and the gradient of the velocity in the vicinity of the object (eq. (29.21)), $f_{viscous} \propto \eta v/K$, where K is some relevant characteristic dimension of the object. The object also experiences a drag force proportional to the *dynamic pressure* of the fluid $f_{dynamic} \propto \frac{1}{2}\rho v^2$. (See the discussion of drag forces in §2.3.) The dimensionless ratio of these two sources of drag,

$$\text{Re} = \frac{\rho v^2}{\eta v/K} = \frac{\rho v K}{\eta} , \tag{29.23}$$

defines the **Reynolds number** (Re) named after the Anglo-Irish engineer Osborn Reynolds.

The character of fluid flow in the vicinity of an object is determined by the Reynolds number. At large Reynolds number, viscous drag is negligible. Minute fluctuations in fluid motion do not damp out, and lead to turbulence. At small Reynolds number, viscous damping dominates and fluid flow is laminar. For a fluid flowing in a long cylindrical pipe, for example, where K is taken to be the diameter and v is the average speed of flow, the flow is always laminar if Re $\lesssim$ 2300 and always turbulent if Re $\gtrsim$ 4000. In

Example 29.4   Poiseuille's Law

Poiseuille's law relates pressure, viscosity, and flow rate in a cylindrical pipe. It illustrates how viscosity affects flow in a practical situation and also shows how it can be measured. Consider the uniform flow of an incompressible fluid through a straight pipe with circular cross section as shown in the figure at right. We assume that the flow is steady and laminar, that the pipe is very long, and that we can ignore

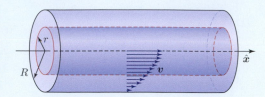

boundary effects at the ends of the cylinder. From the translational and rotational symmetries along and around the pipe's central axis, it follows that the fluid velocity only has an $\hat{x}$ component, and the speed depends only on the distance r from the center of the pipe. Thus, $v(x, t) \rightarrow v(x) = v(r)\hat{x}$. The no-slip condition requires the fluid to be at rest at the walls, $v(R) = 0$. Consider a (mathematical) cylinder of radius r and length l in the fluid. From the definition of viscosity, the shear stress on this cylinder is $\eta\, dv/dr$. The total force on the cylinder's horizontal surface (with area $2\pi rl$) is then $F(r) = -2\pi rl\eta\,(dv/dr)\,\hat{x}$. The minus sign indicates that the viscosity opposes the motion. Since the flow is steady, this force must be cancelled by the force generated by the pressure difference between the ends of the cylinder, which is $\pi r^2 \Delta p$,

$$\pi r^2 \Delta p = -2\pi rl\eta\frac{dv}{dr}.$$

This simple differential equation can be integrated, remembering that $v(R) = 0$, to get an equation for $v(r)$,

$$v(r) = \frac{\Delta p}{4\eta l}(R^2 - r^2).$$

v vanishes at $r = R$ and increases proportionally to the pressure that drives the flow; it decreases with the viscosity (increased friction) and with the length of the cylinder (also increased friction).

The total mass of fluid flowing through the pipe per unit time, $\dot{m} = dm/dt$, can be obtained by integrating the mass flux in the $\hat{x}$-direction, $\phi = \boldsymbol{\phi} \cdot \hat{x} = \rho\boldsymbol{v} \cdot \hat{x}$, over the cross section of the pipe,

$$\dot{m} = \int_0^R dr 2\pi r\rho v(r) = \frac{\pi\rho\Delta pR^4}{8\eta l},$$

which is Poiseuille's law.

---

between is a transition region, where both laminar and turbulent flow are possible depending on details such as the roughness of the pipe's inner surface. Examples of flows at low and high Reynolds numbers are shown in Figure 29.10. The same fluid behaves very differently depending on the speed v and the scale K of the flow. A person ($K \sim 1$ m) moving through water at $v \sim 1$ m/s, for example, experiences a turbulent environment since Re $\sim 10^6$ under these conditions. In contrast, the flow of water at $v \sim 1$ cm/s past a small creature ($K \sim 1$ cm) is predominantly laminar since Re ~ 100 under these conditions. At human scales ($v \sim 1$ m/s, $K \sim 1$ m) water would have to be replaced by honey (η(honey) $\sim 10^4\eta$(water)) to have Re ~ 100! Note that at low enough velocities and/or small enough scales any fluid flow is laminar and dominated by viscous forces.

The length scale K that enters the definition of Reynolds number is somewhat arbitrary. For a pipe of circular cross section, for example, either the radius or the diameter could

be used. (It is conventional to use the diameter.) For a pipe with some other cross section, K is chosen so that the transition from laminar to turbulent flow occurs close to the same values of Re as for a circular pipe. This leads to some ambiguity in the definition of Reynolds number in any specific isolated situation. This ambiguity cancels out, however, when Reynold's number is used to compare the behaviors of two geometrically similar objects of different overall size. For example, the flow of air with speed v around a sphere of radius R will have the same character as the flow of air at speed $2v$ around a sphere of radius $R/2$ since the Reynolds numbers of the two flows are the same.

The similarity of flows at the same Reynolds number makes it possible to use the results of tests at a small scale to predict the behavior of full-scale systems. Consider, for example, the aerodynamics of a large wind turbine blade. Early in the design process, it is impractical to test a variety of potential designs at full scale. Reynolds number scaling

Reynolds Number and Viscosity

The *Reynolds number* of a particular fluid flow is given by the ratio of inertial forces to viscous forces within the fluid, and is a characteristic of a flow pattern. For fluid flow relative to a surface,

$$\mathrm{Re} = \frac{\rho v K}{\eta},$$

where v is the maximum velocity of the fluid relative to the surface, K is a length scale characteristic of the surface, and η is the *viscosity* of the fluid, given by the ratio of shear stress to velocity gradient. Flow at low Reynolds number is typically smooth and steady laminar flow, while flow at high Reynolds number is turbulent and chaotic. For air at STP,

$$\eta_{\mathrm{air}}(\mathrm{STP}) \cong 0.184 \times 10^{-3}\,\mathrm{P} \cong 0.018 \times 10^{-3}\,\mathrm{Pa\,s}.$$

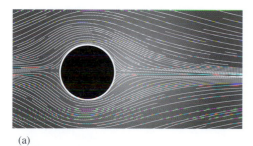

(a)

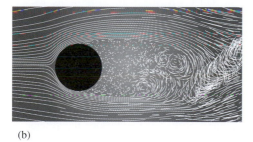

(b)

Figure 29.10 Flows past a cylinder (in cross section) at (a) low and (b) high Reynolds number. At low Re the flow is everywhere laminar. At high Re there is a turbulent wake behind the cylinder, but the flow is laminar outside the wake.

implies that useful information can be obtained by testing a scaled down version of the blade, provided that the density and velocity of the airstream are adjusted to give a Reynolds number corresponding to actual operating conditions. Thus, for example, the aerodynamic characteristics

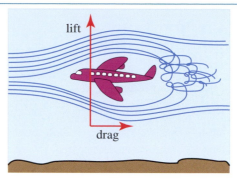

Figure 29.11 A cartoon showing both lift and drag forces experienced by a plane as it flies through the air.

of a wind turbine blade of length 50 meters, can be studied on a 1/10th scale model if the product of the density and velocity of the airstream is increased by a factor of 10, so Re is unchanged.

29.4 Lift

In §2 we introduced the notion of *drag*, the force opposite to the direction of motion experienced by an object moving through a fluid. **Lift**, on the other hand, is a force on a moving object that acts in a direction *perpendicular* to its direction of motion. Thus, a plane moving horizontally through the air experiences an upward force that can counteract the downward force of gravity (see Figure 29.11). Lift is an essential ingredient in many energy-related technologies, from high-performance gas turbines and jet engines whose blades are driven by lift, to airplane and ship propellors.

Wind turbines are powered by *lift*. Naively one might think that the wind bounces off the blades of a wind turbine, forcing them to rotate in the direction that the wind is moving. Indeed, the earliest wind turbines (see §30) were driven this way, but by the middle ages, the wind turbines of Northern Europe were using lift to power the blades of a wind turbine, greatly improving their efficiency. Just as a sailboat can move diagonally into the wind, a moving turbine blade is pushed along in its direction of motion by lift forces that are perpendicular to the wind direction relative to the moving blade.

Lift can be understood as arising from a pressure difference in the fluid on opposite sides of an object around which a fluid is flowing. This pressure difference arises from Bernoulli's principle, and occurs when the flow of air on the two sides has different velocities. For example, the air flowing over an airplane wing moves faster than the air below the wing, leading to a lower pressure above than below, which produces a lift force on the wing.

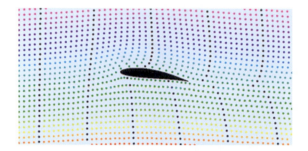

Figure 29.12 Air flows faster over the top of an airplane wing than across the bottom, resulting in a higher pressure below than above and a lift force that keeps the plane up. Note that nearby molecules of air that diverge at the wing's leading edge reach the trailing edge at different times. (Credit: Kraalennest reproduced under CC-BY-SA 3.0 license via Wikimedia Commons)

This mechanism of lift production is sometimes described in an oversimplified fashion, where it is assumed that the air moving over the wing reaches the tail of the wing at the same time as the air below, but this is in general not the case (Figure 29.12, Question 29.4). A precise computation of the force on an airplane wing or a wind turbine requires a detailed solution of the fluid dynamics equations in the presence of the wing or turbine blade; in practice, such computations are done numerically using methods of *computational fluid dynamics* (CFD) (Box 29.1). A useful qualitative understanding of lift forces can, however, be given by relating lift to the circulation of the fluid around the object. In this section we describe this connection; the application to turbine blades is given in the following chapter.

29.4.1 Lift, Circulation, and the Kutta–Zhukovskiĭ Theorem

Lift is intimately related to circulation. If there is a net circulation of fluid about an object placed in an asymptotically uniformly flowing fluid, the object experiences lift. That is not to say that circulation causes lift. Rather the two come together, so if one can identify circulation about an object in a flowing fluid, the object is sure to experience lift. This connection is summarized by the *Kutta–Zhukovskiĭ theorem*, discovered independently by the German mathematician Martin Kutta and the Russian aerodynamicist Nikolai Zhukovskiĭ; this theorem gives a precise relation between circulation and lift for the steady flow of an inviscid, incompressible fluid around an object.

The connection between circulation and lift can be seen immediately in an idealized example illustrated in Figure 29.13, where we consider a flat "wing" that extends a finite length L in the x-direction and is effectively infinite in the

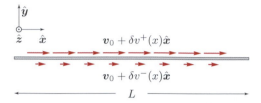

Figure 29.13 A simple case of the Kutta–Zhukovskiĭ theorem, with fluid flowing just above and just below a wing modeled as an infinitesimally thin flat segment.

z-direction out of the page. We assume that the wing moves with a steady velocity $-v_\infty \hat{x}$ through a volume of air with ambient pressure p_0. The air is treated as *incompressible* with density ρ_0. Far from the wing the air is stationary. In the rest frame of the wing, the wing extends from $x = 0$ to $x = L$, and the air far away is moving with steady velocity $v_0 = v_\infty \hat{x}$. Thus we can treat the flow as two-dimensional flow across a finite flat segment. If the wing did not disturb the flow stream, then the velocity of the fluid everywhere would be $v(x, y) = v_0$.

We assume that the flow just above and below the wing deviates slightly from the homogeneous flow, so that $v(x, \pm\epsilon) = v_0 + \delta v^\pm(x)\hat{x}$ (Figure 29.13), and that this flow pattern near the wing smoothly approaches v_0 at increasing distances from the wing. The detailed behavior of the flow away from the wing need not concern us here. At a point x between 0 and L just above/below the wing, the pressure is given using Bernoulli's equation (29.13) by

$$p_0 + \delta p^\pm(x) + \frac{1}{2}\rho_0(v_\infty + \delta v^\pm(x))^2 = p_0 + \frac{1}{2}\rho_0 v_\infty^2 , \tag{29.24}$$

so to leading order in $\delta v^\pm$,

$$\delta p^\pm(x) = -\rho_0 v_\infty \delta v^\pm(x). \tag{29.25}$$

The lift force on the wing in the vertical (y) direction is obtained by integrating the pressure over the upper and lower surfaces of the wing and taking the difference. The lift force per unit length along the z-direction is therefore given by

$$\frac{d\mathbf{F}_l}{dz} = \mathbf{f}_l = \hat{y}\int_0^L dx\,(\delta p^-(x) - \delta p^+(x))$$
$$= \rho_0 v_\infty \hat{y}\int_0^L dx\,(\delta v^+(x) - \delta v^-(x))$$
$$= -\rho_0 v_\infty \Gamma \hat{y}, \tag{29.26}$$

where Γ is the (counterclockwise) circulation around the wing.

Equation (29.26) is a simple example of the **Kutta–Zhukovskiĭ theorem**, which relates the lift per unit length

---

Box 29.1   Navier–Stokes and Computational Fluid Dynamics

A system of fundamental equations known as the **Navier–Stokes equations** govern the behavior of a general class of fluids. The Navier–Stokes equations arise in the continuum limit for fluid systems that satisfy mass, energy, and momentum conservation, and are independent of the microscopic details governing fluid behavior at the molecular level.

The simplest of these equations is simply the conservation of mass equation (eq. (29.6))

$$\frac{\partial \rho}{\partial t} + \nabla \cdot (\rho v) = 0 \,.$$

As discussed in the main text, for an incompressible fluid in steady flow, ρ is constant and this equation simplifies to $\nabla \cdot v = 0$. For an incompressible fluid, conservation of momentum gives the Navier–Stokes equation

$$\frac{\partial v}{\partial t} + (v \cdot \nabla)v = -\frac{\nabla p}{\rho} + \frac{\mu}{\rho}\nabla^2 v + f(x, t) \,.$$

Here the second term on the left-hand side describes the convection of momentum, the first term on the right-hand side describes forces from pressure gradients, the second term on the right-hand side describes the effects of viscosity as diffusion of momentum, and the last term describes external forces on the system. For a compressible fluid the system of Navier–Stokes equations becomes correspondingly more complex.

The Navier–Stokes equations are a complex system of partial differential equations that are challenging to understand analytically. Proving that smooth, physically relevant solutions to these equations exist and are well-behaved is an open problem in mathematics – indeed this is one of the Clay Mathematics *Millennium Prize Problems*, for which a one million dollar prize is offered for the solution. Nonetheless, these equations can be solved numerically. **Computational fluid dynamics** (CFD) codes that numerically integrate these equations can effectively solve the Navier–Stokes equations in a variety of circumstances. For turbulent systems, numerical solution is difficult due to the mixing of length scales, so that a very fine grid would be needed for a standard numerical solution. Advanced techniques including *Reynolds-averaged* Navier–Stokes flow and models of turbulence have made the effective numerical solution of complex fluid flow around systems such as airplanes and turbine blades practical and efficient. Numerical work now contributes significantly to turbine blade modeling for the design of efficient wind turbines and wind farms.

---

on a fixed object immersed in the (asymptotically) uniform flow of an incompressible fluid to the circulation about the object,

$$f_l = -\rho_0 v_\infty \Gamma \hat{y} \,, \tag{29.27}$$

where $v_\infty = v_\infty \hat{x}$ is the fluid velocity far from the object and Γ is the counterclockwise circulation around the object. In Box 29.2 we give a derivation of this theorem for arbitrary shaped objects immersed in the steady flow of an incompressible ideal fluid. As explained in Example 29.3, although air is in principle compressible, the large magnitude of atmospheric pressure at sea level makes the variations in density quite small ($\sim 1\%$) for the conditions experienced by wind turbines.

Another simple and instructive example of the Kutta–Zhukovskiĭ theorem is given by the flow of an incompressible fluid around a circular cylinder. A simplifying feature of steady flows of ideal, incompressible fluids is that the equations of fluid mechanics become linear, so two flows can be superposed. This allows us to construct a steady flow around a cylinder with arbitrary circulation.

Figure 29.14(a) shows a cylinder immersed in a fluid flow that goes to $v = v_\infty \hat{x}$ far away from the cylinder and flows symmetrically on both sides of the cylinder. The fluid velocity is given by

$$\begin{aligned}
v_a &= v_\infty \hat{x} + v_\infty \frac{R^2}{r^2}\left(\hat{x} - 2(\hat{x}\cdot\hat{r})\hat{r}\right) \\
&= v_\infty \left(\hat{r}\cos\theta\left(1 - \frac{R^2}{r^2}\right) - \hat{\theta}\sin\theta\left(1 + \frac{R^2}{r^2}\right)\right),
\end{aligned} \tag{29.28}$$

where, in the second line, we have introduced polar coordinates as defined in the figure. This velocity field is divergenceless, has zero vorticity, and goes asymptotically to $v_\infty \hat{x}$ far from the cylinder (Problem 29.12). It also has zero circulation about the cylinder since it is symmetric about the x-axis (Problem 29.12).

Figure 29.14(b), on the other hand, shows a fluid flow circulating around the cylinder. It is, in fact, the same velocity field as Figure 29.6 except with the sense of the circulation reversed,

Box 29.2   A Proof of the Kutta–Zhukovskiĭ Theorem

The Kutta–Zhukovskiĭ theorem is a key ingredient in understanding lift. To better understand the origins of the theorem we derive it for irrotational steady flow of an incompressible ideal fluid. Consider an object with boundary S shown in two-dimensional cross section at right. The object is immersed in a steady flow in the xy-plane, $\boldsymbol{v}(x, y) = v_\infty \hat{\boldsymbol{x}} + \boldsymbol{w}(x, y)$, where $\boldsymbol{w}(x, y)$, the distortion of the flow caused by the object, goes to zero at large distances. The force on the object is obtained by integrating the pressure over its surface,

$$\boldsymbol{F} = -\oint_S dS\, p\, \hat{\boldsymbol{n}} = \frac{1}{2}\rho_0 \oint_S dS\, v^2 \hat{\boldsymbol{n}}\,,$$

where $\hat{\boldsymbol{n}}$ is the *outward* normal. We have used Bernoulli's equation (29.16) to replace the pressure by $-\rho_0 v^2/2 + \rho_0 v_\infty^2/2 + p_0$ and dropped the latter two constant terms, which integrate to zero.

Now consider a large volume V of fluid in the region between S and a cylinder S' (also shown in two-dimensional cross section) at a large distance from S. The *divergence theorem* (B.28) allows us to write for any vector field $\boldsymbol{U}(x, y)$,

$$\oint_S dS\,\hat{\boldsymbol{n}} \cdot \boldsymbol{U} = \oint_{S'} dS\,\hat{\boldsymbol{r}} \cdot \boldsymbol{U} - \int_V d^3x\, \boldsymbol{\nabla} \cdot \boldsymbol{U}$$

where $\hat{\boldsymbol{r}}$ is the outward normal on the cylinder S'. To select the component of the force in the ith coordinate direction ($i = x$ or y), we choose $\boldsymbol{U} = (1/2)\rho_0 v^2 \hat{\boldsymbol{i}}$, where $\hat{\boldsymbol{i}}$ is the unit vector in the i direction ($\hat{\boldsymbol{i}} = \hat{\boldsymbol{x}}$ or $\hat{\boldsymbol{y}}$),

$$F_i = \frac{1}{2}\rho_0 \oint_S dS\, v^2 \hat{n}_i = \frac{1}{2}\rho_0 \oint_{S'} dS\, v^2 \hat{r}_i - \rho_0 \sum_{j=x,y} \int_V d^3x\, v_j \partial_i v_j\,.$$

Since the fluid flow is irrotational $\boldsymbol{\nabla} \times \boldsymbol{v} = 0$, $\partial_i v_j = \partial_j v_i$, and since the fluid is incompressible $\boldsymbol{\nabla} \cdot \boldsymbol{v} = 0$, $\sum_{j=x,y} \partial_j v_j = 0$. Substituting these relations into the above equation we find

$$F_i = \frac{1}{2}\rho_0 \oint_{S'} dS\, v^2 \hat{r}_i - \rho_0 \sum_{j=x,y} \int_V d^3x\, \partial_j (v_i v_j) = \frac{1}{2}\rho_0 \sum_{j=x,y} \oint_{S'} dS\, \left(\hat{r}_i v_j v_j - 2\hat{r}_j v_j v_i\right),$$

where we have used the divergence theorem once again to replace the volume integral by integrals over S and S'. The integral over the object's surface S vanishes since $\hat{\boldsymbol{n}} \cdot \boldsymbol{v} = 0$ on S. The integral over the cylinder can be decomposed into the integral over its length in the z-direction and an integral over the circle C' in the xy-plane. Dividing out the length in the z-direction gives the force per unit length f_i on the object. When we substitute for $\boldsymbol{v} = v_\infty \hat{\boldsymbol{x}} + \boldsymbol{w}$, the terms depending only on v_∞ cancel and give no force. Since the cylinder can be taken arbitrarily far from the object, $\boldsymbol{w}$ is small and we need keep only terms linear in $\boldsymbol{w}$. For the vertical (y) component, we have

$$f_y = \rho_0 v_\infty \oint_{C'} dl\, \left(w_x \hat{r}_y - w_y \hat{r}_x\right) = -\rho_0 v_\infty \oint_{C'} dl\, \hat{\boldsymbol{r}} \times \boldsymbol{w} = -\rho_0 v_\infty \Gamma\,,$$

since $\hat{\boldsymbol{r}} \times \boldsymbol{w}$ projects out the component of $\boldsymbol{w}$ tangent to the cylinder. This is the Kutta–Zhukovskiĭ theorem. For the horizontal (x) component of the force, we have

$$f_x = -\rho_0 v_\infty \oint_{C'} dl\, \left(w_x \hat{r}_x + w_y \hat{r}_y\right) = -\rho_0 v_\infty \oint_{C'} dl\, \hat{\boldsymbol{r}} \cdot \boldsymbol{w} = 0\,,$$

where the final expression vanishes due to conservation of mass for steady flow. This shows that for the steady, irrotational flow of an incompressible ideal fluid around an object, there is in general no drag force. The drag force arises solely due to viscosity and the associated dissipative effects.

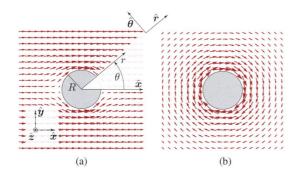

(a)   (b)

Figure 29.14 (a) A circular cylinder immersed in a free flowing fluid with velocity $v = v_0\hat{x}$ far from the cylinder. (b) The same cylinder surrounded by a circulating fluid with circulation Γ.

$$v_b = \frac{\Gamma \hat{\theta}}{2\pi r}, \qquad (29.29)$$

with $\Gamma < 0$.

When the two flows shown in Figure 29.14 are combined, $v = v_a + v_b$, the result is shown in Figure 29.15(a). Note that adding the circulation to the symmetric flow of Figure 29.14(a) does not change the fact that the fluid velocity far away from the cylinder is $v_\infty \hat{x}$, since the circulating flow falls like $1/r$ at large distance r from the cylinder. When the two flows are combined, the fluid flows much faster just above the cylinder, where velocities of the two flows add, than below where they partially cancel. The force per unit length on the cylinder is determined by the integral of the pressure over the cylinder's surface,

$$f = -R \int_0^{2\pi} d\theta \, p\,\hat{n} = \frac{R\rho_0}{2} \int_0^{2\pi} d\theta \, v^2 \hat{r}, \qquad (29.30)$$

where we have used Bernoulli's equation (29.16) and dropped the constant terms $p_0 + \rho_0 v_\infty^2/2$, the integrals of which vanish. Note that the sign in eq. (29.30) arises from the choice of $\hat{n} = \hat{r}$ as the *outward* normal on the cylinder. When $v^2 = v_a^2 + v_b^2 + 2v_a \cdot v_b$ is substituted into the integral, the first term does not contribute because v_a is symmetric about the x-axis and the second term does not contribute because v_b^2 is constant on the cylinder's surface. Thus,

$$f = R\rho_0 \int_0^{2\pi} d\theta \, v_a \cdot v_b \,\hat{r} = -\frac{\rho_0 \Gamma v_\infty}{\pi} \int_0^{2\pi} d\theta \, \sin\theta \,\hat{r}$$

$$= -\frac{\rho_0 \Gamma v_\infty}{\pi} \int_0^{2\pi} d\theta \, \sin\theta \left(\hat{x}\cos\theta + \hat{y}\sin\theta\right)$$

$$= -\rho_0 \Gamma v_\infty \,\hat{y}, \qquad (29.31)$$

which verifies the Kutta–Zhukovskiǐ theorem for this case. Note also that there is no drag force. As shown in Box 29.2, this is a general result for the steady flow of an ideal fluid.

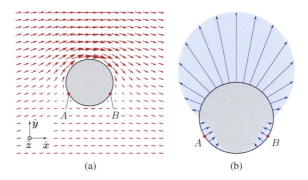

(a)   (b)

Figure 29.15 A circular cylinder immersed in a flow which is the superposition of the two flows shown in Figure 29.14. (a) The velocity field around the cylinder. A and B label the stagnation points. (b) The pressure on the cylinder surface relative to the ambient pressure p_0.

Notice that there are two points on the surface of the cylinder in Figure 29.15 (actually lines in three dimensions), where the fluid is at rest. Marked A and B in the figure, these are **stagnation points**. According to Bernoulli's principle, the pressure on the cylinder is maximum at the stagnation points. In this example the location of the stagnation points depends on the ratio $v_c/v_\infty = |\Gamma|/2\pi R v_\infty$ (Problem 29.13) and can be altered by changing Γ. The existence and location of stagnation points plays a central role in the explanation of the origin of lift on real airfoils.

29.4.2 Lift on Airfoils

The Kutta–Zhukovskiǐ theorem correlates lift with circulation. It does not explain, however, how circulation develops in a flow that is initially circulation free. In fact, *Kelvin's*

Lift and Circulation

If an object (of any cross-sectional shape) is positioned perpendicular to a uniformly flowing incompressible, ideal fluid with density ρ and free-stream velocity $v = v_\infty \hat{x}$ far from the object, then the lift per unit length $f_l = f_l\,\hat{y}$ experienced by the object is given by the *Kutta–Zhukovskiǐ theorem*

$$f_l = -\rho v_\infty \Gamma,$$

where Γ is the fluid circulation about the cylinder. In the truly inviscid limit the drag force vanishes under these circumstances.

circulation theorem forbids the development of vorticity in an initially irrotational flow of an incompressible fluid in the absence of viscosity. Nevertheless, for a suitably shaped *airfoil*, an initially irrotational flow necessarily gives rise to viscous effects that generate vorticity, leading to circulation and lift, and to a subsequent steady, irrotational, and nearly inviscid flow. How this happens is the subject of this section.

An **airfoil** is a geometric shape that can be idealized as extending infinitely in a direction perpendicular to an ambient fluid flow, with an invariant two-dimensional cross-section. For a typical airfoil, there is a blunt leading edge facing the fluid flow, and a sharp trailing edge. A cross section through an airfoil is shown in Figure 29.16, along with associated nomenclature. The asymmetry of the airfoil shown in the figure about its *chord line* – the line joining its leading and trailing edges – is a useful design feature, but is not necessary to obtain lift.

Suppose an airfoil is placed in a fluid that is initially at rest. Then suppose that airfoil begins to move – or, preferring to work in the airfoil rest frame, that the fluid begins to flow. The airfoil is oriented as shown in Figure 29.17(a), inclined at an angle α (the **angle of attack**) to the direction of fluid flow. Initially the flow is irrotational. There is no circulation (Kelvin's theorem) and no lift. As in the case of a circular cylinder, two stagnation points appear on the airfoil, as sketched in Figure 29.17(b). Due to the shape of the airfoil and its orientation relative to the direction of fluid flow, one stagnation point (A_0) lies near the leading edge (A_0) and the other (B_0) appears on the upper surface a short distance in front of the trailing edge (B_0). This situation is not stable, however.

If the fluid were completely inviscid, the initial situation (Figure 29.17(b)) would persist without circulation

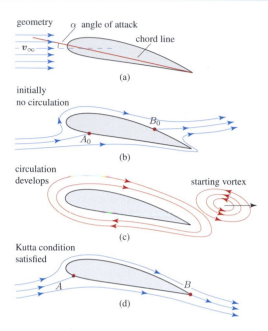

Figure 29.17 Sketches of the development of lift. (a) The airfoil is inclined at an angle α to the direction of the free airflow. (b) When fluid flow begins there is no circulation about an airfoil. Stagnation points at A_0 and B_0 are shown. (c) Viscous forces lead to the development of (clockwise) circulation compensated by shedding the starting vortex. (d) With the addition of circulation, the stagnation point moves to the trailing edge B and steady flow develops.

and without lift. The fluid velocity *around* the trailing edge in Figure 29.17(b), however, is extremely large. (Formally, it is infinite for a sharp trailing edge with a discontinuous tangent!) In reality, the fluid must come to rest on a material surface (the no-slip condition), so viscosity cannot be ignored in the boundary layer just outside the surface of the airfoil between B_0 and the trailing edge. The large velocity gradient causes large viscous forces, which generate (counterclockwise) vorticity in the fluid above the trailing edge. The appearance of vorticity in a viscous flow in the boundary layer above a surface was illustrated in Figure 29.9. As this vorticity develops it is compensated by a developing clockwise circulation around the airfoil (Kelvin's theorem again). The counterclockwise **starting vortex** is shed in the airfoil's wake, leaving the steady, clockwise irrotational circulation around the airfoil (Figure 29.17(c)). When combined with the original irrotational flow of Figure 29.17(b), the rear stagnation point shifts to B at the trailing edge of the airfoil. The result is a steady, laminar, inviscid flow except within a thin boundary layer around the airfoil and a very narrow region of turbulent wake, directly behind its trailing edge.

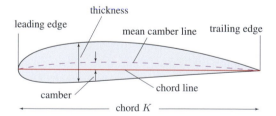

Figure 29.16 A section through an airfoil. The **chord line** is a straight line that connects the **leading** and **trailing edges**. Its length, the **chord length** or simply **chord** K, sets the scale of the airfoil. The **mean camber line** passes halfway between the upper and lower surfaces. The **camber** is the largest gap between the chord line and the mean camber line. Camber results from the asymmetric curvature of the airfoil.

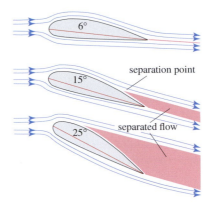

When an airfoil with a sharp trailing edge is immersed in a uniform fluid flow, viscous forces in the boundary layer establish a circulation of exactly the magnitude necessary to keep the rear stagnation point at the trailing edge of the airfoil.

The result of this complex process, driven by viscous forces, is steady, nearly inviscid flow with circulation around the airfoil and therefore with lift. It can be viewed as a "fortunate circumstance," to quote Batchelor, "that the effect of viscosity acting in the boundary layer initially is to cause the establishment of precisely the value of the circulation that enables the effects of viscosity to be ignored in the subsequent motion" [193].

The observation that a body with a sharp trailing edge immersed in a steady flow of a fluid will typically create around itself a circulation that will *move and hold the rear stagnation point to the trailing edge* was elevated to the level of a principle by Kutta (and independently Zhukovskiĭ and Russian physicist S. A. Chaplygin) and is known as the **Kutta condition**. While no actual physical airfoil has a true mathematically sharp edge, and the Kutta condition is violated increasingly for flow at low Reynolds number, it is a good general rule of thumb for realistic airflows.

The magnitude of the lift experienced by an airfoil depends on the velocity of the airflow and on the shape and orientation of the airfoil. For a given airfoil and flow velocity, lift can be increased – up to a point – by increasing its angle of attack. This is intuitively sensible from the microscopic ballistic picture of fluid flow. For small α, the airflow is laminar except for the narrow turbulent wake behind the trailing edge of the airfoil. As α increases, however, the point at which the steady flow separates from the top of the airfoil moves forward along the top of the airfoil, leaving behind an enlarged region of turbulence. The lift continues to increase as the separation point moves forward. Eventually, the lift reaches a maximum at the **critical angle of attack**, beyond which the lift decreases rapidly and the airfoil has **stalled**. Figure 29.18 illustrates schematically the progression from normal lift to the critical angle to a stall condition. The physical processes that lead airfoils to stall are quite complicated and need not be considered here. In the discussion of wind turbine dynamics it is sufficient to parameterize lift and drag forces in terms of empirically measured *lift* and *drag coefficients*, which the subject of the next section.

Figure 29.18 Progression from lift to stall. At an angle of attack of 6° the airflow around this particular airfoil is almost entirely laminar, with only a small turbulent wake behind a separation point near the trailing edge. At the critical angle of attack (15° for this illustration) lift is maximum despite a significant separate, turbulent flow. At 25° the airfoil has stalled, with very little lift and a great deal of drag due to the turbulent wake.

29.4.3 Parameterizing Lift and Drag

In a completely inviscid fluid flow, an airfoil will experience only lift, and no drag, as shown in §29.4.1. For realistic flow, however, viscous forces come into play, and there will be significant drag as well as lift. The lift and drag forces experienced by an airfoil depend in detail on its shape, on the Reynolds number of the flow, and on the angle of attack α. Given these complications, these parameters are usually not computed from first principles, but are instead parameterized in terms of a **lift coefficient** c_l and the *drag coefficient* c_d as defined in §2.3.1. c_d and c_l are defined by decomposing the force per unit length (along the axis) on an airfoil into components parallel and perpendicular to the airstream as illustrated in Figure 29.19,

$$f_d = \frac{d\boldsymbol{F}_d}{dz} = \frac{1}{2}c_d\rho v^2 K \hat{\boldsymbol{v}},$$

$$f_l = \frac{d\boldsymbol{F}_l}{dz} = \frac{1}{2}c_l\rho v^2 K \hat{\boldsymbol{a}} \times \hat{\boldsymbol{v}}, \qquad (29.32)$$

where the dynamical pressure in the airstream $\frac{1}{2}\rho v^2$ sets the scale for these forces and K is a standard length scale that characterizes the airfoil, universally chosen to be the chord length. Here $\hat{\boldsymbol{a}}$ denotes the unit vector along the (idealized) infinite axis of extension of the airfoil. Data on the lift and drag coefficients of specific airfoils at relevant values of Re are generally determined by experimental studies of models in wind tunnels, where the density and speed of the fluid have been adjusted to

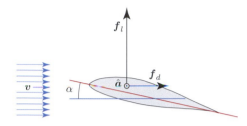

Figure 29.19 Lift and drag forces (per unit length) on an airfoil mounted at an angle of attack α with respect to an airflow v.

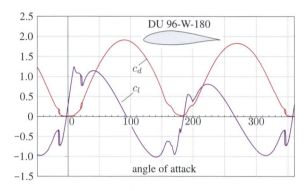

Figure 29.20 Lift and drag coefficients c_l and c_d for Delft University airfoil DU 96-W-180, with cross section as shown in the figure, as a function of the angle of attack α for a Reynolds number of 700 000. This relatively narrow airfoil is designed to be used in the outer portion of modern wind turbine blades. From [194].

obtain the same Reynolds number that will apply to the actual object in its application. In many cases these numbers can be predicted accurately by numerical simulations using Reynolds-adjusted CFD codes with turbulence models (Box 29.1). Figures 29.20 and 29.21 give data on two airfoils developed specially for horizontal-axis wind turbines at the University of Delft [194]. In Figure 29.20, c_l and c_d are plotted as functions of the angle of attack α. In Figure 29.21, c_l is plotted versus α and also versus the ratio c_l/c_d. The red arrows in the figure illustrate how to read c_l and c_l/c_d for a given angle of attack from the figure. Both airfoils shown in the figures have very large lift to drag ratios – greater than 100:1 in the case of Figure 29.21 – at angles of attack of order 5–10°. Both airfoils also show a precipitous drop in the lift to drag ratio above the angle of maximum lift, where the airfoils *stall*. Although these airfoils were specifically designed for use in wind turbine blades, these properties are common to airfoils used in airplane wings, propellors, and elsewhere.

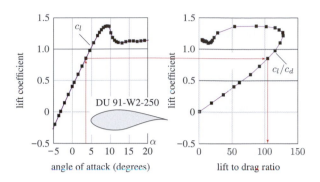

Figure 29.21 Performance information on the Delft University DU 91-W2-250 airfoil at Reynolds number 3.0×10^6. The cross section of the airfoil is shown in the insert. On the left c_l is plotted versus α, the angle of attack. On the right c_l is plotted versus c_l/c_d parametrically in α. The red arrows indicate how to read lift and drag information off the graph. From [194].

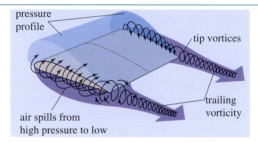

Figure 29.22 Wing tip vortices form when air at relatively high pressure below an airfoil is able to flow to the low-pressure region above. The vorticity is shed as the wing or turbine blade moves forward. (Credit: pilotfriend.com)

29.4.4 Lift and Vorticity

The close association of circulation and lift has some unfortunate consequences. The starting vortices that are shed from the trailing edge of an airfoil are familiar to pilots and air traffic controllers. Their persistence dictates the minimum spacing between planes that take off from a busy airport. Starting vortices do not produce drag in steady flight, but **wing tip vortices** do. The pressure difference between the upper and lower surfaces of a wing or wind turbine blade that is responsible for lift induces an unwanted circulation at the end of the wing or blade. As illustrated in Figure 29.22, the pressure difference causes air to circulate up from below around the wing tip; this circulation would not be possible if the wing continued indefinitely. The circulation spills off the wing tips, forming a trailing vortex that carries away energy and generates

Figure 29.23 Wing tip vortices made visible by their effect on clouds. (Credit: Steve Morris)

a turbulent wake that persists many kilometers behind the trajectory of a jet liner, as illustrated in Figure 29.23. The vorticity generated at the tips of wind turbine blades reduces their efficiency and contributes to the downstream wake that requires turbines to be separated by relatively large distances.

Discussion/Investigation Questions

29.1 Why does the stream of water in Figure 29.3(a) get thinner as it falls? Why, if left to fall long enough, would it break up into droplets?

29.2 *Aspirator pumps* are used in many applications, ranging from emptying fish tanks to providing vacuum for industrial processes. How do they work?

29.3 What is the *Magnus effect* and what does it have to do with the phenomenon of lift?

29.4 There are many flawed "elementary explanations" of lift. One (mentioned in §29.4) is based on the idea that air takes an equal time to pass above and below an airfoil. What is the argument? What is wrong with it?

29.5 Consider the flow of air across an airplane wing considered in the ground-based reference frame where the air is stationary far from the airplane (no wind). Is Bernoulli's equation relevant here? Why?

29.6 Since the pressure is symmetric with respect to reflection across the yz-plane, the cylinder shown in Figure 29.15(b), which is immersed in the uniform flow of an ideal, incompressible fluid, experiences no drag force. Nevertheless, a real cylinder immersed in such a flow does experience a drag force. The contradiction is known as *d'Alembert's Paradox*. Investigate this phenomenon and describe qualitatively how the paradox is resolved.

Problems

29.1 Angel Falls in Venezuela, at a height of 979 m, is the highest waterfall in the world. Ignoring air resistance, how much warmer is the water at the bottom of the falls, where the kinetic energy it acquired by falling has been dissipated as heat, than it was at the top?

29.2 An entrepreneur claims to have devised a system for extracting thermal energy from the wind by cooling it. His device lowers the temperature of the air that passes through it by 1 °C. He claims to generate 800 W/m² from air flowing at 1 m/s. Does this device conserve energy? Would you invest money in the entrepreneur's scheme?

29.3 A stream of water emitted by a tap at velocity v_0 (as in Figure 29.3) has a circular cross section with radius r_0. Use conservation of mass to predict how its radius changes as it falls.

29.4 Supply the missing steps in the derivation of the expression for the flow rate Q in a Venturi flow meter in Example 29.2.

29.5 Estimate the speed of water emerging from an opening at the bottom of a dam when the water in the reservoir behind the dam is held at a height h above the opening. Compare the results of using conservation of energy directly to the application of Bernoulli's principle.

29.6 The maximum takeoff mass for a Boeing 777-300ER passenger jet is 3.5×10^5 kg. The plane's wing area is about 440 m². Compute the average air pressure on the plane's wings required to overcome the plane's weight and lift the plane off the ground. How far does this pressure deviate from average atmospheric pressure at sea level? Repeat the calculation for a Lear jet with mass 3300 kg and wing area 21.5 m². Comment on the implications for the assumption that air behaves as an incompressible fluid in these situations.

29.7 Show that for the flow of a viscous fluid through a pipe, as analyzed in Box 29.1, the energy per unit mass dissipated as thermal energy by viscous forces is $dQ/dm = \Delta p/\rho$, where Δp is the pressure difference between the ends of the pipe. Note that the answer is independent of the viscosity and the cross-sectional shape of the pipe. Suppose water is forced through a pipe by a pressure difference of 10 atm. How much does it heat up?

29.8 [T] The dependence of Reynold's number on the parameters of fluid flow can be inferred from dimensional analysis. First show that it is not possible to construct a dimensionless number out of the viscosity η, density ρ, and speed v of a fluid flow. Next show that if a length scale K characterizing the object is introduced, the dimensionless combination is that of eq. (29.23).

29.9 A large wind turbine blade is 50 m long and has a maximum chord length of 4 m at a point one-third of the way out from the wind turbine hub. When rotating at its design rate, the blade sees an airstream moving at 100 m/s at the blade tip. What is the Reynold's number of the air flow at the point of maximum chord length? Suppose the aerodynamic properties of the blade are to be studied in a wind tunnel that can accommodate at most a blade scaled down to a length of 7.5 m. The wind tunnel operates at atmospheric pressure. To study the properties of the blade near its point of maximum chord length, what wind speed should be used in the wind tunnel?

29.10 [**T**] Show that the flow $\boldsymbol{v}$ around a cylinder given in eq. (29.18) has the same functional form as the magnetic field $\boldsymbol{B}$ around a long, straight wire (3.50) and therefore Maxwell's equations imply that $\boldsymbol{\nabla} \cdot \boldsymbol{v} = \boldsymbol{\nabla} \times \boldsymbol{v} = 0$. (Note that the radial coordinate in the plane is denoted ρ in §3.) Then use Stoke's theorem (B.1.5) and/or direct integration to show that the circulation around any curve that surrounds the cylinder is Γ_0.

29.11 [**T,H**] Suppose that in a region of steady flow the velocity of a fluid is given by $\boldsymbol{v} = \boldsymbol{\Omega} \times \boldsymbol{r}$, where $\boldsymbol{r}$ is the vector from a fixed point in space. Describe the streamlines of this flow. Show that $\boldsymbol{\nabla} \cdot \boldsymbol{v} = 0$, so this flow is consistent with the fluid being incompressible. Compute the circulation Γ on a circle of radius R about the center of the flow. Show that the fluid is rotating with angular velocity $\boldsymbol{\Omega}$ and that the vorticity is $\boldsymbol{\omega} = 2\boldsymbol{\Omega}$.

29.12 [**T,H**] In §29.4.1 it is asserted that eq. (29.28) describes the steady, circulation-free flow of an incompressible ideal fluid around a cylinder. To confirm this, verify (a) that $\boldsymbol{v}_a \rightarrow v_\infty \hat{\boldsymbol{x}}$ as $r \rightarrow \infty$; (b) that no fluid flows into or out of the cylinder;

(c) that the circulation is zero around any circle of radius $r > R$; and (d) that $\boldsymbol{\nabla} \times \boldsymbol{v}_a = 0$ (zero vorticity) and $\boldsymbol{\nabla} \cdot \boldsymbol{v}_\infty = 0$ (incompressibility) everywhere outside the cylinder. [Hint: For part (d), note that $\boldsymbol{v}_a$ can be written as the gradient of a scalar function, $\boldsymbol{v}_a = \boldsymbol{\nabla}\Phi$, where

$$\Phi = v_\infty \hat{\boldsymbol{x}} \cdot \boldsymbol{r} + v_\infty R^2 \frac{\hat{\boldsymbol{x}} \cdot \boldsymbol{r}}{r^2} .$$

Also make use of the identities eq. (B.18) and eq. (B.20), and the facts that $\boldsymbol{\nabla}(1/r^p) = -p\hat{\boldsymbol{r}}/r^{p+1}$ and $\nabla^2(1/r^p) = p^2/r^{p+2}$. The identity $\nabla^2(fg) = f\nabla^2 g + g\nabla^2 f + 2\boldsymbol{\nabla} f \cdot \boldsymbol{\nabla} g$ may be useful as well.]

29.13 [**T**]Investigate the location of the stagnation points for fluid flow around a cylinder as described in §29.4.1. Show that for small Γ there are two stagnation points on the cylinder and find their location. What is the critical value of Γ/Rv_∞ for which there is only one stagnation point? For larger values of Γ are there stagnation points in the flow? If so, where are they?

29.14 A modern medium size commercial jet plane has a wing span of $L = 30$ m and a mean chord length of $K = 4$ m. Its lift coefficient is $c_l = 0.83$ at its cruising angle of attack $\alpha \sim 2°$. What is the lift force generated by the airplane wings when cruising at $v \cong 800$ km/h? While cruising, the engines exert a thrust of ~ 80 kN. Estimate the wings' drag coefficient c_d and the lift to drag ratio.

Wind Turbines

Humankind has captured power from the wind since the beginning of history. Sailboats were the principal means of long-distance transportation across the seas from ancient times until they were supplanted by steamboats in the mid-nineteenth century. Windmills also have ancient origins. Perhaps they were known to the Babylonians in the seventeenth century BC or in India by 400 BC [178, 179]; possibly Heron of Alexandria used wind power to run a machine in the first century AD [195]. Certainly by the middle of the first millennium AD, Persians were using vertical-axis wind-driven machines to grind grain. Known as **panemone**, these machines consisted of an array of fixed vertical wooden (or cloth) sails connected to a vertical central axis, which is free to rotate. A photograph of such a machine, still in use in Iran, is shown in Figure 30.1. Notice that in order to drive rotation, the sails must be sheltered from the wind through half of each cycle. Panemone derive their power chiefly from drag forces – the sails are pushed by the wind. As we explain later in this chapter, most modern wind turbines are driven by lift forces rather than drag, and can reach greater efficiency than drag-powered machines.

Horizontal-axis windmills seem to have first appeared in medieval Europe, making use of gearing to transmit power from the horizontal shaft of the rotor to a vertical post that turned a grindstone or pumped water at or below ground level. The aerodynamic shape of the blades (Figure 30.2(a)) used on these windmills indicates that they make use of lift as a driving force, increasing their efficiency. Vast wetlands of northern Europe were drained for farming and pasture with the help of pumps powered by such windmills. Many of the features of modern airfoils – camber, blunt leading and sharp trailing edges, etc., were developed as these windmill designs were improved through the eighteenth and nineteenth centuries. Simple, relatively inexpensive horizontal-axis windmills were once ubiquitous throughout the American West, where they

Reader's Guide

In this chapter we explore the dynamics of wind turbines, which use *lift* forces to extract power from the wind. We focus on horizontal-axis wind turbines (HAWT), which are by far the most commonly deployed designs. After introducing axial-momentum theory, we use it to derive the Betz limit, which limits the efficiency of an idealized planar device that extracts power from a free flowing fluid. We examine the dynamics of wind turbine blades, show how they are powered by lift, and explain how their shape must change along their length in order to function optimally. We explore some aspects of how HAWT are operated in order to maximize their efficiency.

Prerequisites: §2 (Mechanics), §5 (Thermal energy), §28 (Wind energy), §29 (Fluids).

Figure 30.1 An Iranian *panemone*. Note the adobe wall sheltering the sails on half of their circuit. (Source: www.flickr.com/photos/rahnama-ir/sets/72157622791457394, © Reza-ir)

were used to pump water for livestock and for household use (Figure 30.2(b)).

Figure 30.2 (a) A traditional horizontal-axis windmill of Northern European design. (b) A traditional US designed wind turbine used to pump water for livestock.

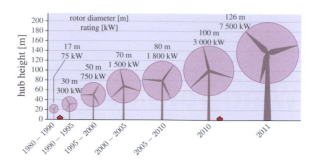

Figure 30.3 Growth of HAWT size and height between 1980 and 2011. (Credit: ROPATEC, France)

The first use of wind power to generate electricity dates to the late nineteenth century; by 1910, hundreds of wind turbines were generating electricity for villages in Denmark. Interest in wind power grew significantly in the decade after World War II, when electricity prices were relatively high. Many technical advances were made during this period, including the development of low *solidity*, high *tip-speed ratio* rotors (see below). Falling electricity prices in the 1960s and 1970s interrupted the progress of wind power development and deployment, but interest in renewable energy sources has led to rapid development since 2000, as can be seen in Figure 28.1. As interest in wind power has increased, so has the size of wind turbines. Figure 30.3 shows the progression toward greater size since the 1980s, driven largely by economies of scale and the desire to exploit the stronger winds available at greater hub heights and in offshore locations.

In this chapter we describe the physical principles that govern the performance and limit the efficiency of **horizontal-axis wind turbines** (HAWT), which are the dominant technology today. Toward the end of the chapter we mention other designs, though none are commercially competitive with HAWT at present. In §30.1 we develop a basic description of how power is harvested from the wind by an idealized version of a wind turbine. This analysis uses a technique known as *axial-momentum theory*, which ignores many possible inefficiencies of wind turbines, but gives a systematic approach to deriving the *Betz limit* on wind turbine efficiency. Improvements on the Betz limit from a more sophisticated analysis that includes *wake rotation* are mentioned and the efficiency of a panemone is analyzed as an example of a *drag-powered* wind machine. In §30.2 we analyze the dynamics of lift and drag on a wind turbine blade, describe some of the constraints on blade design required by the circular motion of HAWT blades, and explain how the performance of turbine blades dictates the way power can be extracted from the wind by HAWTs. Finally in §30.3 we briefly survey some of the many design considerations that affect the operation of real wind turbines. More extensive introductions to the dynamics of wind turbines can be found in [177, 178, 179]. These references also include material on wind turbine engineering issues, as well as on wind power economics, and on the environmental impact of wind power.

30.1 Axial-momentum Theory and Betz's Limit

30.1.1 Axial-momentum Analysis

At the most basic level, a horizontal-axis wind turbine can be modeled as an idealized planar machine, called an **actuator disk**, that slows down and removes energy from an airstream flowing perpendicular to the plane of the disk. By making simplifying assumptions about the properties of the airstream and without specifying the details of what goes on in the plane of the disk, it is possible to obtain an upper limit on the fraction of the power in the airstream that such an actuator disk can extract.

The airstream is assumed to be an ideal fluid flowing steadily in the $\hat{x}$-direction. Standard axial-momentum analysis uses Bernoulli's equation and therefore applies to incompressible fluids or to adiabatic flows of compressible fluids with small variations in pressure and density. These conditions are well satisfied by the winds that interact with wind turbines at atmospheric pressure p_0 and density ρ_0. Both vorticity and turbulence are ignored. The air that will pass through the disk begins at the point labeled ① in Figure 30.4, within a *flow tube* of circular cross section A_1. It moves uniformly with speed v_1 at ambient atmospheric pressure $p_1 = p_0$. Three other points are labeled in the figure: ② is situated directly in front of the disk where the cross-sectional area of the flow tube $A_2 = A_T$ is the area of the actuator disk itself; ③ is directly behind the disk, and ④ is far downstream where the pressure has once again reached p_0. It is assumed that the actuator disk interacts with the airstream in a uniform way, altering its velocity by the same amount throughout the cross section of the flow tube. If the turbine interacts with the airstream in non-uniform way, for example if the blades turn so slowly that some of the air passes through the disk without being affected, then the assumption of uniform interaction would be invalid and the turbine efficiency would decrease. We return to this issue toward the end of the chapter.

Conservation of mass and energy in a fluid, as developed in the previous chapter, together with Newton's second law, enable us to find the energy per unit time per unit area extracted by the actuator disk in terms of a single parameter a known as the *axial induction factor*, which measures how much the actuator disk slows down the airstream. Conservation of mass sets $v_2 = v_3$, since no matter is accumulating at the location of the disk, and also requires

$$v_1 A_1 = v_2 A_2 = v_4 A_4 \equiv (Av). \tag{30.1}$$

Allowing for the possibility that the pressure is discontinuous at the actuator disk, conservation of energy in the form of *Bernoulli's equation* (29.16), requires

$$p_0 + \frac{1}{2}\rho_0 v_1^2 = p_2 + \frac{1}{2}\rho_0 v_2^2,$$

$$p_0 + \frac{1}{2}\rho_0 v_4^2 = p_3 + \frac{1}{2}\rho_0 v_3^2, \quad \text{whence}$$

$$p_2 - p_3 = \frac{1}{2}\rho_0(v_1^2 - v_4^2). \tag{30.2}$$

The pressure must therefore drop across the actuator disk as it removes kinetic energy from the airstream.

The rate at which energy is removed from the airflow, corresponding to the maximum power of a turbine in the actuator disk model, can be computed by taking the difference between the rates of energy flow (energy density times area times fluid velocity) between points ① and ④,

$$P = \frac{1}{2}\rho_0(v_1^2 - v_4^2)(Av). \tag{30.3}$$

The power extracted by the actuator disk can also be computed in an independent but similar fashion from the momentum taken from the airstream. In a time dt momentum $\rho_0 v_1(A_1 v_1 dt) = \rho_0 v_1(Av\,dt)$ enters at ①, and momentum $\rho_0 v_4(A_4 v_4 dt) = \rho_0 v_4(Av\,dt)$ exits at ④. Using eq. (2.5), the power harvested by the turbine can be expressed as the difference in momentum per unit time between points ① and ④ multiplied by the velocity v_2 at the turbine,

$$P = \rho_0(v_1 - v_4)(Av)v_2. \tag{30.4}$$

Comparing these two expressions for the power, (30.3) and (30.4), we find

$$v_2 = \frac{1}{2}(v_1 + v_4). \tag{30.5}$$

The airspeed at the disk is thus the average of the upstream and downstream speeds.

The **axial induction factor** a is defined as the fractional drop in airspeed at the face of the actuator disk,

$$a \equiv \frac{v_1 - v_2}{v_1}. \tag{30.6}$$

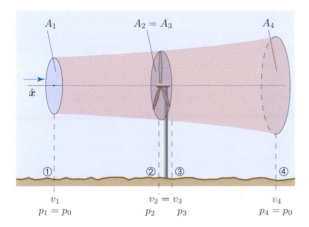

A_1 $A_2 = A_3$ A_4

①   ② ③   ④

v_1 $v_2 = v_3$ v_4
$p_1 = p_0$ p_2 p_3 $p_4 = p_0$

Figure 30.4 The ingredients in axial-momentum analysis. A flow tube of air is followed from far upstream ①, to the front of the actuator disk ②, and then from the back of the disk ③ to a place far downstream ④. As the actuator disk harvests power, the airflow slows down and the flow tube expands.

It is now an easy algebraic exercise (Problem 30.4) to solve for all the other quantities of interest in terms of a, v_1, and $A_T = A_2$, the area of the actuator disk,

$$v_2 = (1 - a)v_1, \qquad\qquad v_4 = (1 - 2a)v_1,$$

$$A_1 = (1 - a)A_T, \qquad\qquad A_4 = \frac{1 - a}{1 - 2a}A_T,$$

$$P = \frac{1}{2}\rho_0 v_1^3 A_T 4a(1 - a)^2. \qquad (30.7)$$

Axial-momentum analysis provides insight into several aspects of wind turbine behavior: the airstream expands and slows down as the wind approaches the turbine; it expands even more as the stream slows down further in the wake. In this idealized model, as $a \rightarrow 1/2$ the downstream airspeed goes to zero and the wake area diverges. This signals a breakdown of the axial-momentum analysis, indicating that effects such as turbulence and vorticity cannot be ignored. More sophisticated methods are required to analyze wind turbine performance as a approaches and exceeds 1/2 [179]. Note that the power extracted from the wind by the turbine vanishes for $a = 0$, when the airstream is not slowed by the turbine, and at $a = 1$, when no air passes through the turbine. It must, therefore, have a maximum in between – this is the *Betz limit*.

Note that in the actual flow far behind the wind turbine, nonzero viscosity transfers momentum into the downstream flow, and eventually the wind field recovers to approximately its asymptotic flow rate; the energy extracted by the turbine is lost, but the remaining $\geq 11/27$ of the energy that passed through the actuator disk merges again with the energy in the rest of the fluid, and can still in principle be extracted (at least in part, subject to the Betz limit again) by further downstream turbines.

30.1.2 Betz's Limit on Wind Turbine Efficiency

The **power coefficient** C_P of a wind turbine is defined to be the ratio of the power it harvests from an airstream to the power that would have passed through the area swept out by the turbine if the turbine were not present, $P_0 = \frac{1}{2}\rho_0 v_1^3 A_T$. Axial-momentum analysis (30.7) gives

$$C_P = P/P_0 = 4a(1 - a)^2, \qquad (30.8)$$

which is plotted in Figure 30.5. C_P has a maximum at $a = 1/3$ where it takes on the value $C_P(\text{max}) = 16/27 \cong 0.59$. This result,

$$C_P \leq \frac{16}{27}, \qquad (30.9)$$

known as the **Betz limit**, was illustrated in an elementary, ballistic model in Box 28.3. This limit was derived first by the British engineer Frederick Lanchester in 1915, but named after German aerodynamicist Alfred Betz, who

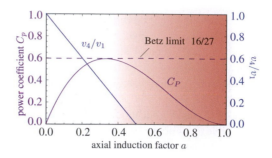

Figure 30.5 Results from axial-momentum analysis: the power coefficient C_P and the ratio of exiting and entering windspeeds v_4/v_1, both as functions of the axial induction factor a. The shading indicates that the axial-momentum analysis breaks down as $a \rightarrow 1/2$.

rediscovered it in 1920. It provides an upper bound on the power that a single planar device with fixed area can extract from a free flowing fluid. While often referred to as an *efficiency limit*, note that the actual power contained in the area A_1 of the airflow entering the turbine is $P_1 = (1 - a)P$, which is $2P/3$ when $a = 1/3$, so a turbine that realizes the Betz limit is really extracting energy with an efficiency of $(16/27)/(2/3) = 8/9 \cong 89\%$ from the affected airstream. This efficiency could be increased further by increasing a towards 1/2, but this would decrease the power produced by the turbine. Going forward we sometimes follow common usage and refer loosely to C_P as the efficiency of a turbine, but this caveat should be kept in mind. Equation (30.9) is an upper limit because the processes that were ignored in deriving it can only subtract from the turbine's power output. These include turbulence generated by the flow over the wind turbine blades, vorticity produced at the blade tips, viscous losses, and losses from air that passes through the turbine disk without encountering a blade.

The Betz limit was here derived under the assumption that the airflow in which the turbine is immersed is uniform and infinite in all directions. It is possible to obtain a higher power coefficient by changing these conditions, for example by surrounding the turbine by a **shroud** (Figure 30.6), although this can also simply be interpreted as increasing the effective area A over which the wind intercepts the turbine. While shrouded turbines have aerodynamic advantages, they present structural challenges, especially for grid-scale wind turbines with megawatt power ratings and actuator disk diameters as large as 125 m. Research on shrouded turbines and similar devices continues although at present no such machines have been widely deployed.

It is important to note that the Betz limit is a qualitatively different type of limit from other bounds on energy

Figure 30.6 Artist's conception of a Japanese design known as a *wind lens* includes both an inlet *shroud* on the upstream side and a *diffuser* on the downstream side of the turbine. (Credit: Research Institute for Applied Mechanics, Kyushu University)

efficiency that we have considered, such as the Carnot limit (§8.5) or the Shockley–Queisser limit (§25.7.2). While the Carnot and Shockley–Queisser limits give an absolute upper bound on the fraction of energy in a given resource that can usefully be extracted with a certain type of technology, with the residual energy always being lost to the environment, for the Betz limit the energy not extracted from the flow returns to the wind field and can be extracted by further turbines in a wind farm or elsewhere. Thus, the Betz limit is primarily of concern with respect to understanding the rate at which energy can be extracted by a device with certain dimensions, and is relevant for engineering and practical aspects of wind farm development, but is not particularly relevant in evaluating the total energy resource of a wind system. In particular, as discussed in §28, the Betz limit should *not* be used in estimating the total wind resource availability on a planet-wide scale.

30.1.3 Wake Rotation and an Improved Betz Limit

When air passes through a HAWT, the air exerts a torque on the blades, the rotation of which drives a generator that produces electrical energy. In order to transfer angular momentum to the rotating blades, the airstream must be deflected from the axial direction. To take account of rotation of the airstream, one can analyze the trajectory of air entering the wind turbine in an annular region of radius r and width dr. If the angular velocity of the blades is Ω, then the incoming air has a tangential velocity $v_t = \Omega r$ when viewed from the blade rest frame. After passing through the turbine, the tangential velocity of the air in this frame must have increased to $v_t' = (\Omega + \omega)r$ in order to transfer angular momentum to the

blades. ω is parametrized by the **angular induction factor** a', defined by $\omega \equiv 2a'\Omega$. Taking into account this **wake rotation** effect leads to a reduction in the maximum possible extracted power below the Betz limit. We give only a brief overview of this analysis here; a more detailed description is given in [177, 179].

Conservation of energy, momentum, and angular momentum can be implemented in each annular region at radius r, with the pressure p, wake rotation ω, and induction factors a and a' all taken to depend on r. The torque on the blades and the power captured by the turbine in each annular segment can then be computed in terms of a, a', and the ratio of the tangential velocity of the blade Ωr to the incoming axial wind speed v. The ratio $\lambda = \Omega R/v$, known as the **tip-speed ratio**, where R is the length of the blade, is an important parameter in the analysis of wind turbine design and performance (see below). Integrating the power transferred from the wind and optimizing with respect to a and a', one obtains a refinement of the Betz limit in which the maximum power coefficient C_P depends on λ. The resulting limit is plotted in Figure 30.7. Also shown in Figure 30.7 are estimates of the maximum power coefficient for a number of alternative wind turbine designs as functions of the tip-speed ratio. Note that C_P vanishes as $\lambda \to 0$. This reflects the fact that the turbine is not turning in this limit and cannot harvest any power from the airstream.

Example 30.1   Checking a Wind Turbine's Power Rating

A wind turbine with a 3.3 m rotor radius is rated at 10 kW in a 10 m/s wind. Is this consistent with the Betz limit?
From eq. (30.9), the maximum power allowed by the Betz limit is $P(\text{max}) = \frac{8}{27}\rho_0 v^3 A$, which evaluates to

$$P(\text{max}) = \frac{8}{27} \times (1.275\,\text{kg/m}^3) \times (10\,\text{m/s})^3 \times (\pi(3.3\,\text{m})^2) \cong 12.9\,\text{kW},$$

under the specified conditions. The turbine is thus rated as capable of reaching $10/12.9 \cong 77\%$ of the Betz limit, consistent with the limit.

Suppose this is a fixed-speed *turbine (see below) with a frequency of 60 rpm. Is the rating then consistent with the Betz limit as refined in §30.1.3?*
The frequency 60 rpm corresponds to a tip-speed ratio of $\lambda = 2\pi \times (1\,\text{Hz}) \times (3.3\,\text{m})/(10\,\text{m/s}) \cong 2.0$ at the specified wind speed. From Figure 30.7, the Betz limit should be corrected to ≈ 0.51 at this value of λ, corresponding to a maximum power of $12.9\,\text{kW} \times (0.51/0.59) \cong 10.5\,\text{kW}$. Thus the turbine is rated to reach 96% of the theoretical limit – an ambitious goal.

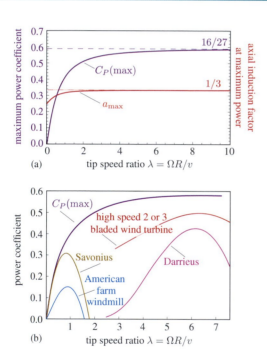

(a) tip speed ratio $\lambda = \Omega R/v$

(b) tip speed ratio $\lambda = \Omega R/v$

Figure 30.7 (a) Refinement of the Betz limit taking account of the angular momentum delivered to the wake of a rotating wind turbine. (b) Estimates of maximum power coefficients for some alternative wind turbine designs. After [196].

30.2 Turbine Blades and Power

Horizontal-axis wind turbines (HAWT) are the workhorses of modern wind power production. They harvest power from the wind by exploiting the lift force exerted on their airfoil-shaped blades as they circulate in the plane perpendicular to the wind direction. Vertical-axis wind turbines (VAWT) have also been developed. Some relative advantages and disadvantages of HAWT and VAWT are summarized in Example 30.3; most deployed modern wind turbines are of the horizontal-axis type, and we focus on them here.

Wind turbine blades are critical dynamical components in wind power systems, and are as central to wind power as photodiodes are to solar PV power. The way that the blades are oriented relative to the direction of the airstream, together with a blade design optimized for a very large lift-to-drag ratio, enable the blades to capture power efficiently. In this section we study how lift and drag forces act on a rotating blade to harvest power from the wind. After establishing the basic mechanism that drives the blades forward, we introduce a simple model related to *axial-momentum theory* that gives insight into some of the main features of blade design, in particular the way that blades must change their shape and orientation as a function of distance from the hub about which they rotate.

30.2.1 Geometry

We assume that the turbine is oriented so that it faces into the wind, which blows in the $\hat{x}$-direction. As shown in Figure 30.8, the vertical defines the y-axis, and the z-axis is horizontal and perpendicular to the wind direction. The wind velocity at the turbine is $v = (1 - a)v\hat{x}$, where v is the windspeed seen by a stationary observer away from the turbine and a is the axial induction factor (30.6). With this orientation, the airstream seen by an observer at rest on the blade is independent of the blade's angle in the yz-plane, so for convenience we analyze a blade at the instant that it is vertical, pointing in the $\hat{y}$-direction. The blade

---

Example 30.2   Power Coefficient of a Drag Machine

The *panemone* mentioned at the outset of this chapter and shown in Figure 30.1 is one example of a wind power device powered by drag forces. Such machines rely on the drag force of wind bouncing off their surfaces to power their rotors. The analysis of Box 28.3 suggests that such devices will still be subject to a Betz-type limit, although they are not as closely related to the *actuator disk* model as a horizontal-axis wind turbine.

To illustrate the analysis of a drag-powered machine, we consider the simple case of a flat rectangular plate of height h and width $2R$ (area $A = 2hR$) rotating at angular velocity Ω about an axis perpendicular to a wind stream with windspeed v – a model for the panemone with two vanes shown in the figure. Note that although the machine harvests wind power from an area $A/2 = hR$, the power in an equal area is unavailable due to the shielding required for the wind to exert a net torque on the machine. We estimate the power at the instant that the plate is perpendicular to the wind stream, when the power is a maximum. Consider an element of area $dA = hdr$ at a distance r from the axis. As we learned in §2.3.1, drag pressure can be parameterized by a drag coefficient c_d, with $p_d = \frac{1}{2}c_d\rho_0 v_{\rm rel}^2$, where $v_{\rm rel}$ is the speed of the fluid relative to the surface on which it acts. The relative velocity is $v_{\rm rel} = v - \Omega r$, which we assume is positive for all $r < R$, and the torque due to the drag is $d\tau = r p_d dA = r\frac{c_d}{2}\rho_0(v - \Omega r)^2 h\, dr$. The total power is obtained by multiplying the torque by Ω (see eq. (2.44)) and integrating over r,

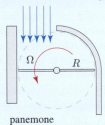

panemone

$$P = \frac{c_d h\rho_0}{2}\int_0^R dr\,(v - \Omega r)^2\Omega r = \frac{1}{2}P_0 c_d\left(\frac{\lambda^3}{4} - \frac{2\lambda^2}{3} + \frac{\lambda}{2}\right),$$

where $P_0 = \frac{1}{2}\rho_0 v^3(A)$ is the power in the airstream, and λ is the tip-speed ratio $\lambda = \Omega R/v$. (Note that the factor of one half to account for the shielded area is included in P_0.)

The power coefficient for this drag machine $C_P = P/P_0$ is maximum at $\lambda \cong 0.54$ and takes the value $C_P = 0.058c_d$. To give an upper bound on C_P, we take a drag coefficient of $c_d = 2$ corresponding to air reflecting ballistically from a flat plate oriented perpendicular to an airstream. Thus the maximum power coefficient in the most favorable configuration of this machine is $2 \times 0.058 \cong 0.12$, which is 20% of the Betz limit. Furthermore, the power drops off as the plate's orientation changes. A panemone can be made more efficient by removing the material close to the axis of rotation, since the drag pressure produces little torque in this region. Indeed, Figure 30.1 indicates that this effect was known to the designers of that machine.

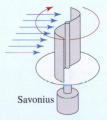

Savonius

Other designs of drag-based machines have higher efficiencies. The **Savonius rotor** shown in the figure is one example. Savonius machines are simple, inexpensive, and robust. The "trough-shaped" rotor not only has a high drag coefficient, but also generates lift when the rotor turns out of the wind, with the result that the power coefficient increases [179]. Figure 30.7 shows $C_P \sim 0.3$ for a tip-speed ratio of order one for a Savonius rotor. (Figure credit: Adapted from The Renewable Energy Website (REUK))

---

is assumed to be rotating uniformly with angular velocity Ω. Thus, a point on the blade at radius r is moving in the $-\hat{z}$-direction with velocity $-\Omega r\hat{z}$.

A cross section through the blade at this instant is shown in Figure 30.9, along with other aspects of the geometry needed in the analysis below. An observer at rest on the blade sees an airstream with velocity $\boldsymbol{w}$ given by the vector sum of the wind $(1 - a)v\hat{x}$ and the headwind $\Omega r\hat{z}$ due to rotation,

$$\boldsymbol{w} = \Omega r\hat{z} + (1 - a)v\hat{x}. \qquad (30.10)$$

Since the resulting airstream varies in strength and direction with r, it is necessary to consider the forces

acting on each section of the blade as a function of r. It is conventional to characterize each section of the blade by the lift and drag coefficients of an infinite cylinder with the same cross section as the blade at that value of r. This set of approximations defines **blade-element theory** [177, 179]. To simplify the analysis still further, we ignore the fact that the turbine imparts angular momentum to the airstream, as in the *axial-momentum analysis* of §30.1.1. Without *wake rotation*, blade-element theory is consistent with axial-momentum analysis and yields the same total power (eqs. (30.7)). A more thorough treatment including wake rotation refines but does not qualitatively change the results of this analysis [179].

Example 30.3   Horizontal- versus Vertical-axis Wind Turbines

Although we focus on horizontal-axis wind turbines in this chapter, **vertical-axis wind turbines** (VAWT) continue to interest researchers and entrepreneurs. Two VAWT designs are shown at right. The **Darrieus turbine** was patented by French engineer George Darrieus in the 1931. The unusual design results in purely tensional stress on the blade when rotating at a high angular velocity. The **giromill** replaces the complex blades of Darrieus's design with straight vertical airfoils. The tangential and normal forces on the blades of a VAWT due to lift and drag change periodically through a rotation as the angle of attack moves through all 360°. Although the actuator disk

Darrieus   giromill

model cannot be used directly to demonstrate that VAWTs have the same Betz limit as HAWTs, a related analysis can be carried out [179] and in practice VAWTs can achieve comparable efficiencies. Because the geometry of the wind configuration changes as a VAWT rotates, the analysis of these turbines must be carried out as a function of angle and then integrated.

Advantages of VAWTs compared to HAWTs include (see later sections in this chapter for some terms used here):

- with no need for *yaw control* to orient them toward the wind, they respond smoothly to rapid changes in wind direction and perform well in turbulent winds
- Darrieus mechanical systems are at ground level, convenient for maintenance
- constant *chord* and no *twist* in some VAWTs such as giromills make blade fabrication easier
- rapid rotation rate allows high efficiency.

Disadvantages of VAWTs compared to HAWTs include:

- lower rotor height (Darrieus) prevents VAWTs from reaching stronger winds at greater height
- greater blade mass per unit swept area increases costs
- cables are required to stabilize the rotor shaft against torques that would cause wobble
- torque on blades varies periodically with rotation, causing periodic stresses that reduce efficiency and fatigue components
- without *pitch control* on blades, *stall control* must be used to limit power in high winds
- no torque at zero wind speed, so they are not self-starting
- greater mass at largest distance from shaft requires more robust mechanical components
- VAWTs experience more drag than HAWTs since the drag must be averaged over all angles of attack as the blade rotates.

At present HAWTs are favored especially at large scale, but research on VAWTs remains active and future developments may change the situation.

The blade is free to move only in the $-\hat{z}$-direction, but its chord line can be *twisted* with respect to this direction in order to better orient the airfoil with respect to the airstream. As shown in Figure 30.9, the *angle of attack* α is determined by the difference between the **twist angle** θ and the angle of the apparent wind direction ϕ, so $\alpha = \phi - \theta$. Trigonometry relates ϕ to the windspeed $(1-a)v$, angular velocity Ω, and radius r,

$$\tan\phi = \tan(\theta + \alpha) = \frac{(1-a)v}{\Omega r} = (1-a)\frac{R}{\lambda r}, \quad (30.11)$$

where $\lambda = \Omega R/v$ is the *tip-speed ratio* introduced earlier in this chapter.

To maximize the lift, which powers the turbine, and to reduce drag, it is desirable to keep the angle of attack small, positive, and approximately constant along the length of the blade. Under fixed conditions (constant v and Ω), eq. (30.11) then requires that for α to stay constant, θ must increase as r decreases. In other words, *the blade must twist into the wind* as r decreases. The twist angle (as a function of r) is determined by the tip-speed ratio λ for which the blade design has been optimized. The optimal tip-speed ratio λ determines a relationship between the angular frequency of rotation and the wind speed. To maintain the ideal tip-speed ratio, as the wind velocity

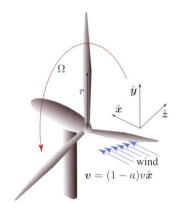

Figure 30.8 Coordinate system for analyzing a HAWT. The wind defines the x-axis, the vertical defines the y-axis. The turbine is shown at a point where one blade is vertical, the configuration depicted in Figure 30.9.

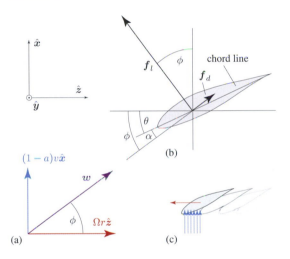

Figure 30.9 Lift and drag forces on an element of a rotor blade at a distance r out from the hub. (a) Components of the airstream seen by an observer at rest on the blade. (b) Angles defining orientation of the blade with respect to the airstream and the blade's direction of rotation. (c) Motion of the blade viewed from above in an Earth-fixed reference frame. The lift $\boldsymbol{f}_l$ and drag $\boldsymbol{f}_d$ force vectors are not drawn to scale.

increases the rate of rotation of the blade must increase correspondingly. If this is not possible, then the angle of attack all along the blade grows, reducing the efficiency of the blade. We discuss this issue in more detail below.

30.2.2 Forces on a Blade Element

From Figure 30.9 it is clear that the lift force drives the blade forward, powering the turbine, while the drag force opposes the turbine's motion. The key to an efficient blade

design is therefore to maximize the lift and minimize the drag on the blade. Figures 29.20 and 29.21 show that these wind turbine airfoils have been designed to have large lift and a large lift-to-drag ratio at angles of attack that are small and positive. In practice, ratios of order 100:1 can be obtained in the vicinity of 6–8°. This enables us to make the additional approximation of ignoring c_d compared to c_l.

The lift force per unit length $\boldsymbol{f}_l$ has components tangential to the direction of rotation, $\boldsymbol{f}_T = -|\boldsymbol{f}_l|\sin\phi\,\hat{\boldsymbol{z}}$, and normal to the direction of rotation, $\boldsymbol{f}_N = |\boldsymbol{f}_l|\cos\phi\,\hat{\boldsymbol{x}}$. The normal force or **thrust** is opposed by the forces exerted by the structure that supports the turbine and is responsible for removing momentum from the axial flow of the wind. The tangential force creates the torque that keeps the turbine blades rotating, and the corresponding reaction on the wind produces wake rotation. We can compute the power harvested by the wind turbine, which is equal to the power delivered to the blades through torque, by using the thrust to determine the momentum removed from the wind.

The tangential and normal components of the force per unit length on a blade element (ignoring drag, as discussed above) can be obtained from the components of $\boldsymbol{f}_l$ defined in eq. (29.34),

$$\boldsymbol{f}_T = -\frac{1}{2}\rho_0 K B w^2 c_l \sin\phi\,\hat{\boldsymbol{z}} = -\frac{\rho_0 K B v^2 (1-a)^2 c_l}{2\sin\phi}\,\hat{\boldsymbol{z}},$$

$$\boldsymbol{f}_N = \frac{1}{2}\rho_0 K B w^2 c_l \cos\phi\,\hat{\boldsymbol{x}} = \frac{\rho_0 K B v^2 (1-a)^2 c_l \cos\phi}{2\sin^2\phi}\,\hat{\boldsymbol{x}},$$

$$(30.12)$$

where we have used $w = v(1-a)/\sin\phi$, K denotes the chord length of the turbine blade, and we have multiplied by the number of blades B to obtain the net force per unit radial length on the blade assembly. From axial-momentum analysis of the pressure difference across the turbine, we can compute the thrust per unit area, T/A, on the turbine blades,

$$T/A = p_2 - p_3 = \frac{1}{2}\rho_0(v_1^2 - v_4^2) = \frac{1}{2}\rho_0 v^2\,(4a(1-a))\,.$$

$$(30.13)$$

Applying this force on a radial section of area $2\pi r\,dr$ gives the axial-momentum approximation to the normal force per unit length along the blade,

$$\boldsymbol{f}_N\,(\text{axial momentum}) = \frac{1}{2}\rho_0 v^2\,8\pi r a(1-a)\hat{\boldsymbol{x}}\,. \quad (30.14)$$

Equating this result to $\boldsymbol{f}_N$ from eq. (30.12), we obtain a constraint between the axial induction factor a and the design parameters K, B, λ, and ϕ,

Blade Design

Wind turbine blades must twist and thicken toward the rotor hub in order to maintain a fixed, small angle of attack and to generate sufficient thrust as the airstream direction seen by the blade changes with radius r. In a simple model which assumes axial-momentum flow and ignores drag forces, at each value of radius the *twist* θ and *chord* K are related to the axial induction factor a, the lift coefficient c_l, and the angle of attack α by

$$\theta = \tan^{-1}\left(\frac{(1-a)R}{r\lambda}\right) - \alpha,$$

$$\frac{K}{R} = \frac{8\pi a(1-a)}{\lambda c_l B}\frac{1}{\sqrt{(1-a)^2 + \lambda^2 r^2/R^2}}.$$

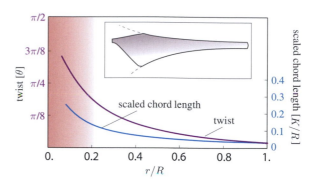

Figure 30.10 Variation of a wind turbine blade's ideal chord length and twist with distance from the hub for the simple model described in the text. The analysis breaks down in the red shaded region. See the text for parameter values. The inset shows a blade profile based on this optimal design (after [197]). After [179].

$$\frac{K}{R} = \frac{8\pi a}{\lambda c_l B}\sin\phi, \tag{30.15}$$

where we have used eq. (30.11) to introduce the *tip-speed ratio*.

Consistency of axial-momentum theory requires that the tangential component $\boldsymbol{f}_T$ of the force per unit length $\boldsymbol{f}_T$ must generate the torque that powers the wind turbine. Indeed, combining the expression for $\boldsymbol{f}_T$ in eq. (30.12) with the constraint (30.15), it can be shown that the power generated by the torque is the same as the power (30.7) harvested from the wind stream (Problem 30.7). Thus the energy harvested by the wind turbine can be viewed either as the kinetic energy removed from the axial flow of the wind or as the work done turning the rotor. Note that in this analysis we have essentially assumed that the turbine blades are dense enough in the area of the actuator disk that all air passing through the disk encounters a blade. This is discussed in further detail below.

30.2.3 Blade Design
Under the simplifying assumptions of axial-momentum flow and no drag forces, eqs. (30.11) and (30.15) relate the chord length K and twist angle θ as functions of r to the tip-speed ratio λ, the axial induction parameter a, and the angle of attack α,

$$\theta = \tan^{-1}\left(\frac{(1-a)R}{r\lambda}\right) - \alpha, \tag{30.16}$$

$$\frac{K}{R} = \frac{8\pi a(1-a)}{\lambda c_l B}\frac{1}{\sqrt{(1-a)^2 + \lambda^2 r^2/R^2}}. \tag{30.17}$$

The power coefficient C_P (30.8) depends only on a in the axial-momentum analysis, and reaches the Betz limit

at $a = 1/3$. An ideal design, therefore, would adjust θ and K as functions of r subject to the constraint that $a = 1/3$, while keeping α in the range where c_l is large and c_d/c_l is small. In practice, eqs. (30.16) and (30.17) must be solved iteratively since an optimal value of α cannot be chosen without knowing the lift and drag coefficients, which in turn depend on the chord length. A designer typically works within a given family of blade profiles (see, for example, Figure 30.11) with the aim of keeping a as close to 1/3 as possible.[1] To illustrate the general result that both the twist and the chord length grow with decreasing radius, θ and K/R are graphed in Figure 30.10 for $c_l = 1$, $B = 3$, $\alpha = 6°$, and $\lambda = 8$, and $a = 1/3$. A sketch of a blade profile based on this model is shown in the insert in Figure 30.10. Fabricating light and durable blades with such complex profiles is one of the significant challenges of wind turbine design.

When the effects of wake rotation and drag are included, the relations among K, θ, λ, r, α, and a become more complex but their qualitative features are preserved. An example of the variation of chord length and twist with distance from the hub for a realistic wind turbine blade is shown in Figure 30.11.

The choice of K and ϕ depends on the tip-speed ratio. Therefore if a turbine has been designed for optimum power coefficient at one tip-speed ratio, it will be suboptimal at both larger and smaller values. An example is shown in Figure 30.12. This turbine has been designed to function

---

[1] Note that the approximations leading to these results break down when $KB \sim 2\pi r$ since the blades would overlap. For $a = 1/3$ this limits the validity of the analysis to $\phi \lesssim 40°$ (Problem 30.8).

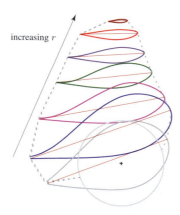

Figure 30.11 Variation of a wind turbine blade profile as a function of distance from the hub, showing increasing chord length and twist closer to the hub. From [194].

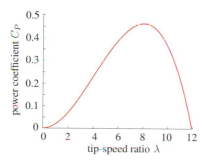

Figure 30.12 Variation in the power coefficient with tip-speed ratio for a blade optimized for $\lambda = 8$. (Credit: Prof. dr. Gerald J.W. van Bussel, TU Delft, NL)

optimally at $\lambda = 8$, where it reaches approximately 78% of the Betz limit. In addition to vanishing as $\Omega \to 0$, C_P vanishes as λ becomes large (v small) because the wind is not blowing strongly enough to activate the turbine (see Question 30.1).

30.2.4 Wind Turbine Power

A HAWT blade can maintain an optimal power coefficient if the tip-speed ratio λ can be held fixed at the design value. Since $\lambda = \Omega R/v$ changes with the wind speed, λ can be held fixed only if Ω is adjusted as v changes. Wind turbines with variable Ω are known as **variable-speed turbines**. Historically, however, most wind turbines have operated at a fixed value of Ω, dictated by the frequency of the electric grid to which the turbine is connected (see §38). Such **fixed-speed turbines** only reach their design power coefficient at one value of wind speed. Modern developments in power electronics have made it possible to convert the

> ### Wind Turbine Operation and Power
>
> Wind turbines can be operated at fixed or variable speed and fixed or variable pitch. Fixed-speed, fixed-pitch turbines reach their design efficiency only at a single wind speed. Variable-speed turbines adjust their angular velocity to keep the tip-speed ratio near optimal as the wind speed changes. They shed power at high wind speeds by pitching their blades away from the wind.
>
> The wind power potential of a given turbine at a given site is obtained by integrating the product of the turbine's power coefficient $C_P(v)$ and the site's wind frequency distribution $f(v)$,
>
> $$P = \int_{v_{in}}^{v_{out}} dv \, \frac{1}{2} \rho_0 v^3 \, f(v) C_P(v) A_T.$$

power generated by variable-speed turbines to the desired line frequency, and deployment of this type of turbine has increased in recent years.

The power output of a specific wind turbine design at a given location can be predicted from measurements of the power density in the wind and a measured power coefficient similar to that shown in Figure 30.12,

$$P = \frac{1}{2}\rho_0 v^3 C_P(\lambda) A_T , \qquad (30.18)$$

where the dependence of C_P on the tip-speed ratio λ has been displayed explicitly. The power output of an ideal *variable-speed turbine* in a steady wind is shown in Figure 30.13(a). At speeds below the **cut-in speed** v_{in}, typically 3–5 m/s, the torque is assumed to be insufficient to turn the blades and they are held fixed. Above v_{in}, Ω is adjusted to give the optimal tip-speed ratio, and the power output grows proportional to v^3. At the *rated* wind speed v_r, typically 12–15 m/s, the power output reaches the power rating of the generator. At higher wind speeds, adjustments are made to prevent the power growing further (see below) and the power remains fixed at the rated value until the wind speed reaches the **cut-out speed** v_{out}, typically around 25 m/s, above which braking mechanisms stop the turbine rather than risk damage to the rotor. Thus, in the interval $v_r \leq v \leq v_{out}$, we have, effectively, $C_P(v) = (v_r/v)^3 C_P(v_r)$. The actual power curve of the Vestas V90 3 MW turbine shown in Figure 30.13(b) is reasonably well approximated by the ideal case.

The power output of a fixed-speed turbine differs significantly from that of a variable-speed turbine. In particular, the rated power is achieved only at the design wind speed.

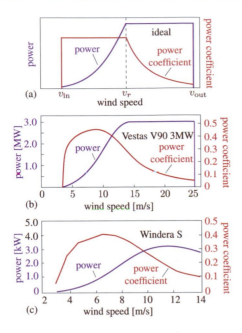

Figure 30.13 The power output and power coefficient for HAWTs: (a) ideal variable-speed HAWT, with constant power coefficient between v_{in} and rated speed; (b) actual Vestas V90 3 MW variable-speed, variable-pitch wind turbine; (Credit: (b) Wind Power Program, www.wind-power-program.com) (c) actual Windera S fixed-speed, fixed-pitch wind turbine rated for 3.2 kW at 11 m/s. (Credit: (c) Ennera)

At higher or lower speeds the falloff of C_P shown in Figure 30.12 reduces the power significantly. The smallest HAWTs, which employ both fixed speed and fixed pitch to minimize cost and maximize simplicity of operation must sacrifice efficiency at wind speeds both below and above their rated speed.

To determine the wind power potential of a given turbine at a given location, the power curve of the turbine must be combined with the *wind frequency distribution* $f(v)$ that characterizes the location (see §28.2.1),

$$P = \int_0^\infty dv \frac{1}{2} \rho_0 v^3 f(v) C_P(v) A_T . \tag{30.19}$$

In general such calculations are carried out numerically, using actual data for both $f(v)$ and $C_P(v)$. Estimates can be made by combining a Weibull parameterization (§28.2.2) with the power coefficient of an ideal variable-speed turbine (Problem 30.4).

30.3 Some Design Considerations

The design and manufacture of wind turbines is a sophisticated branch of mechanical engineering. References [177]

and [179] provide introductions to the subject. We close our examination of wind power by discussing briefly a few of the design issues that involve relatively simple physical principles.

Control at high wind speeds Since high winds are relatively rare, most wind turbines are designed for optimum efficiency at a relatively low speed, typically 12–15 m/s. At higher wind speeds, power must not be allowed to grow beyond the rated power of the generator, and turbines must be protected from damage in very high winds. There are two approaches to solving these problems: **pitch-regulated** and **stall-regulated** control. Pitch-regulated turbines have motors that rotate the blades along their radial axes to change the angle of attack. As α is increased, lift decreases, drag increases, and the power output of the turbine decreases. In very high winds the blades can be *furled*, i.e. brought edge on to the wind, stopped, and a brake applied to the rotor. Pitch regulation is also used to keep the output of a variable-speed turbine at the rated level when the wind is in the range between the design speed and v_{out}.

Stall regulation is accomplished by designing the wind turbine blades to *stall* at a predetermined wind speed that increases with the distance from the hub. As the wind speed increases, the blade stalls at progressively larger r, spilling more and more power into the rotor's turbulent wake. Stall regulation, which requires no active control, is most effective for fixed-speed turbines where, as eq. (30.16) shows, the angle of attack grows with v at fixed r.

Currently, the trend is toward pitch control, especially for large, megawatt-scale, variable-speed wind turbines. Smaller wind turbines in the tens of kilowatt range typically employ stall control for power regulation.

Yaw control As the wind direction changes, a HAWT must maintain its axis along the wind direction by rotating about the vertical direction. Rotation about the vertical axis is known as **yaw**, originally a nautical term. A turbine designed for downwind operation follows the wind direction like a weather vane. This orientation had the disadvantage that the supporting tower *shadows* the blades when they pass behind it, reducing the power and causing periodic stress on the mechanical systems. Downwind systems must also include a damping mechanism to prevent the turbine from oscillating about the downwind position. Most modern turbines and all grid-scale turbines are oriented upwind, which requires a **yaw motor** to maintain the orientation. The improved power and stability outweigh the additional cost.

Blade number Most modern HAWTs that produce electric power have three blades. *Windmills* designed to pump water, known as **wind pumps**, on the other hand, have many broad blades (see Figure 30.2(b)). The distinction originates in the different mechanical constraints on the two applications. A wind turbine that produces electric power typically has a sophisticated gear box that allows the rotor to be decoupled from the load at start-up. The turbine is connected to the generator only after it has "spun up" to a high angular velocity. Thus, with little initial resistance, it requires relatively little torque to start a wind turbine. *Wind pumps*, on the other hand, are simpler and less sophisticated devices. Wind pumps have found wide application providing water for livestock on the great plains of the US and Australia, and in other isolated locations, where they must be designed to function unattended for long periods of time. The majority of wind pumps are therefore purely mechanical devices in which the rotational motion of the windmill powers a mechanical pump. Wind pumps typically require high torque at startup because the pump must work against a standing head of water.

Thus the distinction between wind turbines and wind pumps lies in the torque they must produce. Torque and power are related by eq. (2.44), $P = \tau\Omega$, where Ω is the angular velocity of the windmill. As mentioned above, windmills are usually optimized for a particular *tip-speed ratio* $\lambda = \Omega R / v$. For a given power and wind speed, then, the torque produced by a windmill varies inversely with tip-speed ratio,

$$\tau = \frac{P}{\Omega} = \frac{PR}{v\lambda}. \tag{30.20}$$

Wind pumps are typically optimized for $1 < \lambda < 3$ to produce high torque, in contrast to electric power producing HAWTs, which are designed for $4 < \lambda < 10$. Referring to Figure 30.7(a), we see that wind pumps must sacrifice efficiency for higher torque.

When Ω is small, the windmill blades rotate slowly and intuitively it seems obvious that more power will be generated if the number of blades is large. This relationship is encoded in eq. (30.15), which we rewrite as an equation for the number of blades as a function of the blade design parameters K, ϕ, and c_l, assuming that the optimal value of $a = 1/3$ has been obtained,

$$B = \frac{1}{\lambda}\left(\frac{8\pi R \sin\phi}{3c_l K}\right). \tag{30.21}$$

For a given blade design, eq. (30.21) indicates that the number of blades is inversely proportional to the tip-speed ratio and therefore directly proportional to the torque. Table 30.1 gives suggested number of blades for various tip-speed ratios (from [179]).

Horizontal-axis wind turbines that produce electric power require lower torque and can be powered by fewer

Table 30.1 Suggested number of blades B for various values of the tip-speed ratio λ [179].

Tip-speed ratio λ	Number of blades B
1	8–24
2	6–12
3	3–6
4	3–4
>4	1–3

Figure 30.14 Smoke emitted from the tips of helicopter blades trace the *wingtip vortices* which contribute to aerodynamic losses of wind turbines. (Credit: Henn Werlé, and ONERA, the French Aerospace Lab.)

blades of narrower chord. Since the large, complex blades of massive wind turbines are expensive, a small number of blades helps keep costs down. Why not one or two blades for such turbines? A single blade would require a counterweight and would have to rotate very rapidly, adding to stress and increasing noise from the turbine. One- and two-bladed turbines have the additional problem that their moment of inertia about the vertical axis changes as the blades rotate. This produces a rapidly varying load and additional stress on the *yaw motor* as it works to keep the turbine headed into the wind. Three is the smallest number of blades that, when arranged symmetrically, give a moment of inertia that remains constant as the blades rotate (Problem 30.9).

Tip vortices Wind turbine blades experience significant aerodynamic losses due to the shedding of vortices at the tips of the blades (see Figure 30.14). Vortices form as air migrates from the lower, high-pressure side of the airfoil to the upper, low-pressure side around the wingtip. Efforts are made to mitigate these losses, but they remain one of the principal limitations on wind turbine efficiency and contribute to the persistent downstream wake that makes it necessary to space wind turbines by many diameters in a wind farm.

Discussion/Investigation Questions

30.1 In the text we explained the shape of Figure 30.12 by taking $\Omega \to 0$ and $v \to 0$. What about the other limits: $\Omega \to \infty$ at fixed v and $v \to \infty$ at fixed Ω? Why does C_P vanish in those limits?

30.2 Investigate the *Darrieus* and closely related *giromill* VAWT designs described in Example 30.3. What are their histories and their advantages and disadvantages relative to the conventional HAWT design?

30.3 The *Savonius* VAWT shown in Example 30.2 is mainly driven by drag. As shown in Figure 30.7(b), the Savonius design is capable of reaching roughly half of the Betz limit, and for tip-speed ratio close to one, it comes close to saturating the limit on C_P including the effects of wake rotation. Research on Savonius turbines continues and they are used in niche applications. Investigate and discuss the advantages and disadvantages of Savonius turbines.

Problems

30.1 A variable-speed HAWT has a power coefficient of 0.45 at its rated wind speed v_r and maintains its rated power for wind speeds between this value and its cut-out speed of 25 m/s. If the rotor diameter is 100 m, and the rated power is 3 MW, what is v_r?

30.2 A wind pump based on the "American farm windmill" design (see Figures 30.2(b) and 30.7) with rotor diameter 4.5 m and tip-speed ratio $\lambda \sim 1$ claims to be able to pump 2400 L/h of water from a depth of 30 m when the wind speed is 4 m/s. Do you think this is possible?

30.3 The maximum power of a Savonius wind turbine (see Example 30.2) is claimed to be $P = 0.36\, hrv^3$, where P is in watts, and the height h and radius r of the Savonius rotor, and the wind velocity v, are all in SI units. What is its maximum power coefficient as a fraction of the Betz limit?

30.4 [C] Wind frequency data obtained at a height of 10 meters at a particular location is summarized by a Weibull distribution with $\lambda = 9$ m/s and $k = 1.3$. The land use is best characterized as "rough pasture." Your wind power company is considering placing a variable-speed wind turbine with a hub height of 105 m, a rotor diameter of 90 m, and a rated power of 3 MW

at this site. The cut-in, rated, and cut-out speeds are $v_{in} = 3$ m/s, $v_r = 12$ m/s, and $v_{out} = 25$ m/s respectively. Calculate the wind turbine's power coefficient at its rated wind speed. Assume a constant power coefficient between v_{in} and the design speed, and assume that the turbine produces its rated power between the rated speed and v_{out}. How much total energy (in MWh) would you estimate this turbine can produce per year?

30.5 [T] Verify the results of eq. (30.7), and evaluate them at the Betz limit.

30.6 [T] Derive eq. (30.15) by equating the normal force per unit length on a blade element given in eq. (30.12) with the thrust per unit length obtained from the axial-momentum analysis (30.14).

30.7 [T] Axial-momentum analysis led to a power coefficient $C_P = 4a(1 - a)^2$ for a HAWT (eq. (30.7)). Compute the power coefficient another way: start from the tangential force per unit length f_T in the simple model that led to eq. (30.12); implement the axial-momentum theory constraint (30.15); find the torque per unit length $d\tau/dr$ and compute the power $P = \int_0^R dr\, \Omega(d\tau/dr)$. Show that the result is the same as the result of the axial-momentum analysis (30.7).

30.8 [T] The axial-momentum theory approximation to blade design breaks down when r/R becomes too small. Use eqs. (30.11) and (30.15) to show that the condition $KB < 2\pi r$ limits the angle ϕ to $\sin\phi\tan\phi < c_l(1 - a)/(4a)$. For the choice of parameters in Figure 30.10 what is the smallest value of r/R allowed by this condition?

30.9 [T] Model the blades of a HAWT rotor as rods of equal length and mass, equally spaced around a circle, rotating with angular frequency Ω. Compute the moment of inertia of the rotor about the *vertical axis*. Show that the moment of inertia is independent of time as long as there are three or more blades.

30.10 [H] Use blade-element theory to compute the correction to the axial-momentum theory expression for the power (30.7) to first order in c_d/c_l for an optimized blade. First, recompute f_T and f_N including the drag force and show that the right-hand side of eq. (30.15) is modified by a factor $(1 + (c_d/c_l)\tan\phi)^{-1}$. Then compute the power as in Problem 30.7, keeping only terms to first order in c_d/c_l. How big is this correction for a turbine with $c_d/c_l = 0.01$, $\lambda = 10$, and $a = 1/3$?

Energy from Moving Water: Hydro, Wave, Tidal, and Marine Current Power

The motion of water has played a crucial role in shaping Earth's surface and climate over the 4.5 billion years since the planet's formation. Because Earth's surface temperature is generally in the narrow range between the freezing and boiling points of water, most water on Earth is in liquid form and much of it is in constant motion. Water has a much higher mass density than air – by a factor of nearly 800 – so that moving water has a much higher power density than moving air for comparable flow volumes and velocities. Ocean currents, surface waves, tidal flow, and flow of water in rivers all contain substantial amounts of energy, which can in many cases be extracted at locations where the flow is concentrated.

Ocean currents, surface waves, and hydropower are all essentially third-order effects of solar energy. Because all these forms of energy are at some level driven by wind and/or latent heat flux, which are in turn generated by solar energy, the total energy available in these water resources is substantially less than the energy available in Earth's atmosphere. The maximum power potential of each of these water-based resources amounts to no more than a few terawatts. All these forms of water energy are, however, highly concentrated, so that to the extent that they are available they can be tapped more easily and economically for human use than the more diffuse forms of wind and solar energy. Tidal power, unlike essentially all the other sources of energy that we consider in this book, which ultimately arise from nuclear energy processes either in the Sun or on Earth, is produced purely mechanically from gravitational energy.[1] Although most tidal energy is dissipated in the deep ocean, in particular locations this energy source also becomes highly concentrated and can be efficiently harnessed for human use.

Reader's Guide

This chapter describes a set of water-related energy sources: hydropower, ocean surface wave energy, tidal energy, and marine currents. Though all relate to the motion of water, from which mechanical energy can be harvested at relatively high efficiency, each of these sources involves a different set of physical principles. The origins and propagation of ocean surface waves are explained, as is the origin of the tides. Some technologies that have been developed to extract energy from these resources for human use are described and the total power available from them is estimated.

Prerequisites for this chapter include §2 (Mechanics), §4 (Waves and light), §27 (Ocean energy), §28 (Wind energy), and §29 (Fluids).

At the present time, among these resources only hydroelectric power is harvested in significant quantities, with nearly 15 EJ (4000 TWh) produced yearly from over 1 TW of installed capacity. In contrast, all ocean-based power sources including tidal, wave, and marine current produced just over 1 TWh (4 PJ) of electric power from 0.5 GW of installed capacity in 2014 [198]. Although it is unlikely that they will ever be exploited to the same extent, tidal, wave, and marine current power represent resources of magnitude similar to hydropower, and they will likely be harnessed by human beings more extensively in the future. Also, understanding these energy sources leads us to consider a variety of physical phenomena and aspects of energy flow on Earth's surface that are interesting in their own right. We describe hydropower, wave power, tidal power, and marine current energy in the four sections of this chapter.

31.1 Hydropower

Anyone who has seen a substantial waterfall, from the rushing cascades of a swollen mountain stream during

---

[1] Note that a fraction of the geothermal energy flowing outward at Earth's surface comes from gravitational contraction (see §32.1.2).

Figure 31.1 A twelfth-century watermill in the Belgian village of Braine-le-Château.

spring snowmelt to the thundering majesty of Niagara or Victoria Falls, can appreciate the tremendous power carried by moving water. The physics of this resource is quite elementary compared to other renewable energy technologies. Moreover, hydropower is the only renewable energy source that currently contributes substantially to global electricity production, and is the only current means of large-scale storage of energy from the electric grid.

Hydropower has been used by humans for more than two thousand years. Water wheels, consisting of a wheel rotating on a horizontal axis with many blades or paddles that are pushed by moving water, have been used since antiquity to power mills for grinding grain to produce flour. There is evidence that water wheels were in use in Greece in the third century BC, and by the first century AD they were found throughout the Mediterranean world and in ancient China.

Hydropower is essentially solar energy that has been transformed by natural processes into potential energy of water. As discussed in §27.1, roughly half of the solar energy incident on the ocean surface is lost through latent heat flux to evaporation, producing tremendous quantities of water vapor. Winds and buoyant forces, driven by solar energy, carry the water vapor aloft, high into Earth's atmosphere and above the world's mountains, where some of the water condenses and falls to Earth as rain and snow, retaining some of the original solar energy in its gravitational potential energy.

While the total energy resource available in hydropower is substantially less than the total available wind or solar resources, hydropower is highly concentrated spatially and can be very cost-effective to exploit. The hydrological and meteorological processes that transform heat originating in the upper layers of the ocean into the more concentrated

(and zero entropy) mechanical potential energy of water at high altitudes can be viewed as a vast, planetary heat engine (see §10). Although its efficiency is undoubtedly low, this engine operates continually without human intervention, transforming solar energy into a form that is easily harvested for human use. Starting at high altitude, the natural gravitationally induced flow of precipitated water down the local topographical gradient feeds streams and rivers that carry water from large geographical regions through narrow channels. Obstructing such a channel by a dam (Question 31.2) with surface area measured in hundreds or thousands of square meters makes it possible to extract energy from water that has fallen over regions measured in hundreds of thousands of square kilometers. For example, the *Hoover Dam* in the arid southwest US has a height of 221 m and a width of 379 m, but a catchment basin of area 435 000 km^2 (see Figure 31.2). This geographical localization through natural processes is what makes hydropower such an effective resource.

31.1.1 Physics of Hydropower

The basic physics of hydropower is very simple. A mass m of water at a height h has gravitational potential energy

$$V = mgh,　　(31.1)$$

from eq. (2.9). **Hydropower** is produced by the conversion of this gravitational potential energy into useful mechanical energy. In a hydroelectric plant, the force of the water is used to turn a turbine. The turbine in turn drives a generator that produces electromagnetic power, as described in §3 and §38.

In a typical hydropower setup, a dam is built to hold water in a reservoir (Figure 31.3). An enclosed channel carries water from the bottom of the reservoir through a turbine (Figure 31.4), and releases it into a river or other outflow channel. The difference in height Z between the top of the reservoir and the position of the turbine is called the **hydraulic head**. When a small volume $\Delta \mathcal{V}$ of water having mass density $\rho \cong 1000 \, \text{kg/m}^3$ passes through the turbine, the potential energy lost to the reservoir is $\Delta V = gZ\Delta m$, from eq. (31.1), where $\Delta m = \rho \Delta \mathcal{V}$. Thus, the power output of the turbine can be described by the equation

$$P = \epsilon \, d\mathcal{V}/dt = \rho g Z \epsilon Q,　　(31.2)$$

where $Q = d\mathcal{V}/dt$ is the **flow rate** of the water and ϵ is the efficiency of the turbine.

Because the power is proportional to the hydraulic head Z, the power output of the dam is essentially proportional to its height. Note that the power of the dam depends upon the potential energy of the water at the *top* of the reservoir,

Figure 31.2 Left: Hoover Dam on the Colorado River in the Mojave Desert of the US has a hydraulic head of 180 m and has 2.08 GWe of generating capacity. Right: Hoover Dam (marked with the star) has a catchment area that includes most of the Colorado River Basin (shaded) except for the drainage of the Gila River in the south. (Credit: Left, Shannon reproduced under CC-BY-SA 4.0 license via Wikimedia Commons)

Hydropower

The power that can be extracted from a hydroelectric dam is given by

$$P = \rho g Z \epsilon Q,$$

where ρ is the mass density of water, Q is the water flow rate (volume/time), g is the gravitational constant, Z is the **hydraulic head** (the height differential between the top of the reservoir and the turbine), and ϵ is the turbine efficiency. Turbine efficiencies for hydropower are not bounded by the Betz limit and can reach around 95% in practice.

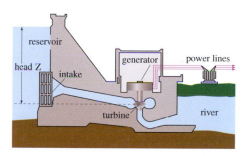

Figure 31.3 Schematic of a hydroelectric dam showing the head Z. When a small volume of water $\Delta \mathcal{V}$ flows through the turbine, the reduction in total potential energy of the water in the reservoir is given by $\Delta V = \rho g Z \Delta \mathcal{V}$. (Credit: Tomia reproduced under CC-BY-SA 3.0 license via Wikimedia Commons)

even though the water passing through the turbine is flowing from the *bottom* of the reservoir. This follows directly from conservation of energy; when a volume of water $\Delta \mathcal{V}$ is removed from the reservoir the drop in water depth shows that the loss in potential energy is $\rho g Z \Delta \mathcal{V}$, assuming that the volume of water removed is small enough that the change in depth is much smaller than the depth ($\Delta Z \ll Z$). This result can be understood physically from the increased pressure at the bottom of the reservoir that drives the turbine (Problem 31.1).

Another feature of hydropower dams is that they can be used for energy storage; by pumping water into the reservoir, energy can be efficiently converted from electromagnetic to potential form. We discuss the storage aspect of hydropower further in §37.

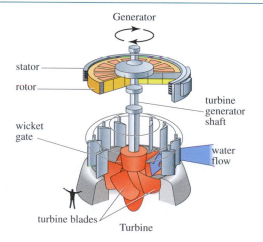

Figure 31.4 Schematic of a modern hydro turbine connected to an electric generator (see §38 (Electricity generation and transmission)). (Credit: top-alternative-energy-sources.com)

31.1.2 Hydropower Technology

Unlike a wind or marine current turbine that extracts kinetic energy from a moving fluid, a hydropower turbine extracts gravitational potential energy from a volume of water by releasing deep water that initially has no velocity from a region of high pressure to a region of much lower (atmospheric) pressure. The Betz limit is not relevant under these circumstances, since the force from the water's pressure is associated with mechanical potential energy and not with the bulk kinetic energy of a moving fluid, so there is no theoretical bound on the efficiency of a hydropower turbine.

Large turbines used in modern hydroelectric facilities such as the Three Gorges Dam on the Yangtze (Chang Jiang) River in Hubei, China can achieve efficiencies of 96% or greater. As of 2012, the Three Gorges power plant (Figure 31.5) was the largest hydroelectric facility in the world. With a dam height of 101 m, a hydraulic head of $Z = 80.6$ m, a flow rate of up to $Q = 900$ m^3/s for each generator, and an average turbine efficiency near 95%, each turbine has a maximum power of around $\rho Q g Z \epsilon \cong$ 700 MW. With 32 turbines, the facility has a capacity of 22.5 GW.

Large-scale hydropower facilities raise a number of practical and environmental concerns. Dams disrupt local ecosystems; for example, the presence of a dam reduces the downstream flow of sediment carrying valuable nutrients, can lead to excess methane production by rotting vegetation in the reservoir area, and can interfere with fish spawning. In recent years many dams in the American West have been removed in order to restore natural riverine ecosystems. Also, large-scale hydropower installations flood large land areas, leading to land use efficiency that is comparable to large-scale solar power installations (see Problem 31.2); the riverine ecosystems disrupted are generally both ecologically richer and rarer than desert locations where solar power can be harvested. Accumulation of blocked sediment behind the dam can interfere with water flow through the turbines and cause practical difficulties with plant operation. The water weight in large reservoirs may lead to a shift in land mass. In some situations there is concern that such an effect may increase the chances of an earthquake in seismically active areas and/or increase the chances of dam failure. Catastrophic dam failures can result in significant human casualties and loss of property. On the positive side, the role of dams in flood control and irrigation can have significant social and economic value. Furthermore, in some situations such as the Three Gorges dam, dams can play an important role in expanding transport channels through water-borne navigation.

Smaller-scale hydropower systems based on water wheels or underwater turbines like those discussed in §31.4 have fewer environmental impacts. **Microhydro** facilities with generating capacity from a kilowatt to several megawatts can be built on dams with a low hydraulic head and can produce renewable energy with reduced environmental impact. While not practical for large-scale power generation, such implementations provide a carbon-free renewable energy resource that is particularly useful in remote locations with regular water flow.

31.1.3 Extent of Resource

As shown in Figure 31.6, global hydroelectricity generation has been growing steadily for many years. By 2013, world hydropower *capacity* passed 1 TW and hydropower actually provided about 3750 TWh/y (13.5 EJ/y), or about 430 GW of electric power averaged over the year, close to 1/6th of world electricity. This corresponds to a capacity

Figure 31.5 Left: *Three Gorges Dam* on the Yangtze (Chang Jiang) river in Hubei, China; in 2012 this was the largest hydroelectric facility in the world with capacity of 22.5 GW. Right: One of 32 Francis turbines used in the Three Gorges Dam, with an average efficiency of nearly 95% and capacity 700 MW. (Credit: Voith Gmbh)

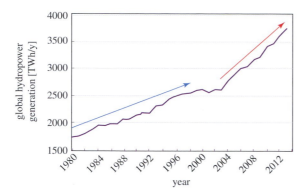

Figure 31.6 After a brief hiatus at the beginning of the century, hydroelectricity generation has been growing more rapidly than its earlier historical average. After [168].

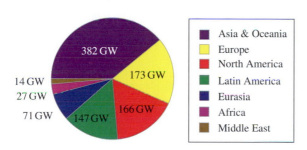

Figure 31.7 Hydroelectric generating capacity in 2012 by region [12].

factor of ~43% [168]. The recent relatively rapid growth of hydropower has been dominated by new installations in Asia and, to a lesser extent, Europe and Latin America. Hydropower deployment in North America has been relatively static for many years. The present distribution of hydroelectric power capacity over the world's major geographical regions is summarized in Figure 31.7.

The total potential for hydropower is limited. The World Energy Council and the IPCC estimate that ~10 000 TWh/y (36 EJ/y) of hydropower remains that could in principle be exploited, potentially increasing world hydropower by a factor of 3.7 to an annual average of about 1.6 TW [168, 199]. The greatest potential for new hydropower deployment is in Asia, Africa, and Latin America. In Asia alone, the theoretical undeveloped hydropower potential has been estimated at 7000 TWh/y (25 EJ/y), and 92% of potential hydropower resources in Africa are undeveloped. Whether these resources will be developed depends upon many economic, environmental, and political considerations, particularly those related to competition for land use and water resources.

Because hydropower arises as a third-order effect from solar power, it is natural that the total power available from gravitational potential energy of water is substantially less than the total amount of power available in wind, which is itself a second-order effect of solar power. Indeed, 1.6 TW is substantially less than the ~500 TW of wind energy dissipated in the lower atmosphere, but may not be too different from the amount of wind energy that can be successfully harvested (see §28.3.2). The difference in the fraction of recoverable energy between these resources lies in the fact that, as discussed above, hydropower is concentrated by natural forces from a diffuse two-dimensional resource to an essentially pointlike resource.

31.2 Wave Power

Wind blowing across the surface of the ocean magnifies small disturbances on the sea surface, giving rise to waves that increase in size with the strength and duration of the wind. Waves produced at one point in the ocean propagate outward, carrying energy often as far as thousands of kilometers until the waves break on the shore of a continent, island, or shallow undersea feature such as a reef. Along many ocean coastlines, incident wave energy provides a potential renewable energy source with relatively high power density. In this section we describe the propagation of ocean waves, the life cycle of a wave from production to breaking, and some technologies for extracting ocean wave energy.

The basic equations that govern surface waves and simple surface-wave solutions are described in §31.2.1. The reader more interested in energy applications may skip this section on a first reading; the only result needed from this section is the final form of the propagating wave (31.9). In §31.2.2 we use conservation of energy to characterize the *dispersion* relation between frequency and wavelength for propagating surface waves, and to describe the transport of energy in surface waves. The life cycle of a wave from its initial production by wind to breaking on the shore is described in §31.2.3, and the extent of wave energy resources and practical mechanisms for harvesting wave energy are described in §31.2.4 and §31.2.5.

For a more detailed introduction to surface wave physics see [200]. A nice description of the current understanding of wave processes is given in [172]. The physics of some of the basic principles for wave energy capture are described in [201], and a more recent overview of wave energy conversion technologies is given in [202]. Finally, for a concise introduction to water waves in an introductory physics text, see [203].

31.2.1 Propagating Surface Waves

A complete analysis of **surface waves** at the ocean–air boundary is very complicated. The fluid equations are non-linear, and the boundary conditions themselves depend upon the solution, since the position of the air–water interface depends upon the motion of the wave. Fortunately, many of the key physical properties of propagating ocean waves can be understood in a simplified analysis in which water is assumed to be an *incompressible, ideal* fluid. Recall that an ideal fluid is one in which viscous losses can be ignored and vorticity is conserved (see §29.2). We further make the approximation that the displacement of the water from its equilibrium position is small. We describe the consequences of the first two assumptions below. The last approximation allows us to ignore terms that are quadratic (or of higher order) in the water's displacement. The result is a *linearized* description of propagating surface waves that captures many of their most important properties from the point of view of waves as an energy resource. Note that during the processes of wave production and breaking, nonlinear effects become relevant.

Imagine starting with a body of water at rest, where the surface is smooth and horizontal. We describe the fluid by following the motion of each packet of water relative to its position x in the equilibrium rest configuration. When a disturbance arises, a small packet of water initially at the point x is displaced to a new position $R(x, t)$, where

$$R(x, t) = x + s(x, t). \qquad (31.3)$$

The variation of s over time describes the dynamical motion of the water (see Figure 31.8), where $s(x, t)$ tracks the displacement of the packet of water that would be at position x in the original equilibrium configuration. This is the *Lagrangian approach* to fluid dynamics, in which the relevant variables move along with the fluid, as discussed in §29.1.2. The velocity of the packet of water with respect to fixed (Eulerian) coordinates is $v(R(x, t), t) = \partial s(x, t)/\partial t$. The essence of the linearized description is to retain only terms linear in s or v in the equations of motion. This approximation is valid as long as any given packet of water does not migrate too far from its equilibrium position. This would not work, of course, if we were attempting to describe the bulk motion of the fluid, as in a flowing river or blowing wind. Appearances notwithstanding, water waves do not result in cumulative bulk motion over time, and the linearized description is a good starting point.

In our analysis we consider idealized waves moving in a direction $\hat{x}$, with $\hat{z}$ the vertical direction and the undisturbed water's surface at $z = 0$ (see Figure 31.8). For

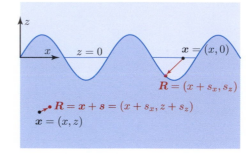

Figure 31.8 A slice in the xz-plane through a water surface wave moving in the x-direction with height above the resting surface given by the coordinate z. The idealized waveform is independent of the third coordinate y in the direction out of the page. The configuration of the water is described by the displacement vector $s(x, z, t)$ of an infinitesimal packet of water from its rest position $x = (x, z)$ at time t. The magnitude of the z-displacement is exaggerated.

such waves, there is no displacement of the fluid in the y-direction, and the motion is invariant under translation in this direction, so that all physical phenomena of interest can be described in terms of the two coordinates x and z. In the linear approximation, the assumptions that water is incompressible, inviscid, and irrotational lead to two constraints on the displacement s

$$\nabla \cdot s = \partial_x s_x + \partial_z s_z = 0, \qquad (31.4)$$
$$\nabla \times s = (\partial_z s_x - \partial_x s_z)\hat{y} = 0. \qquad (31.5)$$

These constraints are derived in Box 31.1, and simplify the description of water waves enormously. In particular, eq. (31.5) allows us to write the displacement s as the *gradient* of a scalar function, $\psi(x, t)$,

$$s(x, t) = \nabla \psi(x, t), \qquad (31.6)$$

in exactly the same way that the vanishing of $\nabla \times E$ in electrostatics (see §3.1) allowed us to write E as the gradient of the electrostatic potential. Furthermore upon substituting eq. (31.6) into eq. (31.4), we obtain a single, linear, second-order differential equation for ψ,

$$\nabla \cdot \nabla \psi(x, t) = \nabla^2 \psi(x, t) \qquad (31.7)$$
$$= \left(\frac{\partial^2}{\partial x^2} + \frac{\partial^2}{\partial z^2} \right) \psi(x, z, t) = 0,$$

where we have simplified the *Laplacian* ∇^2 defined in eq. (B.17), using the fact that ψ is independent of y.

An immediate consequence of these constraints is that there are no waves in an infinite volume of water without an interface. This is the analog of the statement that in

---

Box 31.1   Constraints on Water Waves in the Linear Approximation

We assume that water is an incompressible, ideal fluid. In §29 we showed that incompressibility implies $\nabla \cdot \boldsymbol{v} = 0$, where $\boldsymbol{v}$ is the velocity of the fluid at any point. As discussed in §29.2.3, vorticity $\boldsymbol{\omega} = \nabla \times \boldsymbol{v}$ is conserved in the flow of an ideal fluid – a vortex once formed would never dissipate – so if an ideal fluid begins at rest, then it will not develop vorticity when set in motion by wind blowing in a straight line. Thus, for the water waves we study, it is a good approximation to take the flow to be *irrotational* with zero vorticity, $\nabla \times \boldsymbol{v} = 0$. Roughly speaking, these two conditions, $\nabla \cdot \boldsymbol{v} = 0$ and $\nabla \times \boldsymbol{v} = 0$, together with the linear approximation, are the origins of the constraints (31.4) and (31.5).

Another approach exposes the importance of the linear approximation. Since the fluid is incompressible a small packet of fluid should have the same volume in the equilibrium coordinates (x, z) and in the physical coordinates defined by eq. (31.3) $(R_x, R_z) = (x + s_x, z + s_z)$. (For simplicity of notation we have suppressed the third coordinate y.) In other words, $dx\, dz = dR_x\, dR_z$. But we can regard the transformation from equilibrium to physical coordinates as an ordinary change of variables. According to the rules of calculus,

$$
dR_x\, dR_z = \begin{vmatrix} \partial R_x/\partial x & \partial R_x/\partial z \\ \partial R_z/\partial x & \partial R_z/\partial z \end{vmatrix} dx\, dz
$$

$$
= \begin{vmatrix} 1 + \partial s_x/\partial x & \partial s_x/\partial z \\ \partial s_z/\partial x & 1 + \partial s_z/\partial z \end{vmatrix} dx\, dz
$$

$$
= \left(1 + \nabla \cdot \boldsymbol{s} + \mathcal{O}(s^2) \right) dx\, dz \,.
$$

So the volume is preserved if $\nabla \cdot \boldsymbol{s} = 0$ and we ignore terms quadratic in $\boldsymbol{s}$. A similar analysis yields $\nabla \times \boldsymbol{s} = 0$.

A final comment: incompressibility precludes the possibility of waves where the *density* of the bulk fluid oscillates sinusoidally in time. These are, of course, **sound waves**, and while sound certainly propagates in water, these are not the waves that we see traveling along the ocean's surface.

---

electrostatics, the potential is constant if there is no charge anywhere in space. Electrostatics is governed by eq. (3.12), $\nabla \cdot \boldsymbol{E} = -\nabla \cdot \nabla V = -\nabla^2 V = \rho/\epsilon_0$, so eq. (31.7) corresponds to electrostatics with $\rho = 0$. Thus, the water–air interface is crucial to the dynamics of water waves, and we should expect that the effect of a wave propagating on the surface will die out with depth.

We thus seek oscillatory solutions of eq. (31.7) for $z \le 0$ that vanish as $z \to -\infty$. Since we expect a wave-type solution with some wave number k and (angular) frequency ω, we anticipate that a propagating wave will be described by a function of the general form

$$
\psi(x, z) = f(z) \sin(kx - \omega t) \,, \tag{31.8}
$$

where $f(z)$ is a function of depth that vanishes as $z \to -\infty$. Substituting eq. (31.8) into eq. (31.7), we find $d^2 f(z)/dz^2 - k^2 f(z) = 0$, so the solution that decreases as $z \to -\infty$ is $f(z) = \tilde{a} e^{kz}$, where $\tilde{a} = a/k$ is a constant. From eq. (31.6) we find the displacement vector $\boldsymbol{s}(\boldsymbol{x}, t)$,

$$
s_x = \frac{\partial}{\partial x} \psi = a e^{kz} \cos(kx - \omega t),
$$

$$
s_z = \frac{\partial}{\partial z} \psi = a e^{kz} \sin(kx - \omega t). \tag{31.9}
$$

The characterization of the wave in terms of the single function ψ gives a convenient description of the problem and solution, though it is also possible to check directly that the displacement $\boldsymbol{s}(x, z)$ given in (31.9) solves the constraint equations (31.4) and (31.5) (Problem 31.4).

The solution described by eqs. (31.9) has a simple geometric structure (see Figure 31.9). The wave form has wavelength $\lambda = 2\pi/k$, and travels to the right at velocity $v = \omega/k$. This velocity is the *phase velocity* (see §4.1.2), the velocity at which the phase of the oscillation advances, as distinguished from the *group velocity*, which is discussed in the next section. Note that each small packet of water moves in a circle, confirming the assertion that there is no bulk motion of the fluid and supporting the use of the linear approximation. The radius of the circle for each particle is $a e^{kz}$, decreasing exponentially with depth.

31.2.2 Energy and Energy Transport in Deep-water Waves

While we have found the general form for a surface wave, we have not yet determined the *dispersion relation* between the frequency and wave number for ocean surface waves. This relation is needed to complete our description of surface wave propagation.

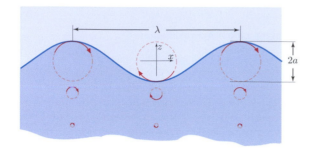

Figure 31.9 Linearized solution for surface waves in an incompressible fluid. Individual particles move in circles with radii exponentially decreasing with depth.

Figure 31.10 Short-wavelength surface tension, or *capillary* waves spread over the surface of much longer wavelength gravity waves. (Credit: ©2014 by Jacky Kwok)

Before getting into specifics, we investigate what dynamical mechanisms might be responsible for water waves. The oscillations in the waves that dominate energy flow on Earth's oceans are driven by gravity, so these are known as **gravity waves**.[2] When a packet of water rises above its equilibrium location, gravitational forces proportional to ρg (where ρ is the density of water) act to pull it back down. The incompressibility of water and the absence of dissipation (viscosity), however, require that if one packet moves down, water elsewhere must move up, and wave motion results. Surface wave oscillations can also be driven by the phenomenon of **surface tension** (denoted by σ). We encountered surface tension indirectly in the discussion of the surface energy of nuclei (see Problem 17.8) as a potential energy per unit area stored in the surface of a medium. Just as the potential energy per unit length of a spring causes the spring to oscillate once it is set in motion, so surface tension can cause an air–water interface set in motion to oscillate. The resulting waves are known as **capillary waves**. The importance of surface tension compared to gravity is measured by the ratio $\sigma/\rho g$, which has units of [length]^2. For water $\sqrt{\sigma/\rho g} \sim 3$ mm; this is roughly the scale below which surface tension dominates over gravity as the dynamical driver of waves (Problem 31.5).

While the general form (31.9) of a propagating surface wave is fixed by the constraints (31.4) and (31.5) alone, the relationship between the wave number k and frequency ω, i.e. the *dispersion relation* (see §4.5.3), is fixed by the physical mechanism that drives the oscillations. The appropriate dispersion relation for gravity waves can be

derived using Newton's law and the gravitational force law; in the spirit of this book, however, we derive the relation here using conservation of energy. To simplify the argument, we begin by considering a *standing wave* solution composed of superimposed left-moving and right-moving propagating solutions

$$\psi(\boldsymbol{x}) = \frac{a}{k} e^{kz} \cos kx \cos \omega t . \qquad (31.10)$$

It is easy to check that this standing wave solution satisfies eq. (31.7), and is a linear combination of solutions of the form (31.9).

To most clearly keep track of energy in the standing wave, we consider wave motion in a finite horizontal domain so that we can be certain that energy is not flowing into or out of our system. The standing wave solution (31.10) can describe motion in a domain bounded by walls at $x = 0$ and $x = L$ if $k = N\pi/L$ (for integer N), since then the horizontal displacement s_x vanishes at the walls, as required. We assume that the width W in the y-direction of the region of water considered is arbitrary but finite.

The strategy we follow is parallel to the treatment of the harmonic oscillator in Box 2.2. We compute the kinetic and potential energy of the solution (31.10) and demand that the sum remain constant in time.

The kinetic energy in the standing wave solution is obtained by integrating $\frac{1}{2}\rho \dot{s}^2(x, y, z, t)$ over all $z < 0$,

$$E_{\text{kin}} = \int_0^L dx \int dy \int_{-\infty}^0 dz \frac{1}{2}\rho a^2 \omega^2 e^{2kz} \sin^2 \omega t$$

$$= \frac{\rho a^2 \omega^2 A}{4k} \sin^2 \omega t, \qquad (31.11)$$

where $A = LW$ is the area in the xy-plane.

---

[2] Not to be confused with *gravitational waves*, which are waves of the gravitational field predicted by Einstein's general theory of relativity, and recently observed by the *Laser Interferometer Gravitational-Wave Observatory (LIGO)*.

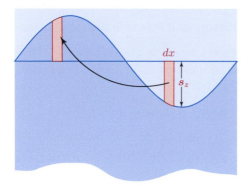

Figure 31.11 The potential energy of a surface wave can be computed by imagining lifting water in an interval dx from the trough to the corresponding point in the peak.

The potential energy of the wave arises because water at the surface has been lifted from the troughs to the peaks. Although water moves around below the surface, it generates no net change in potential energy (the density remains constant due to the assumption of incompressibility). As shown in Figure 31.11, we can keep track of the potential energy by imagining that the water in each interval dx of the trough has been lifted to the corresponding interval of the peak. Thus a quantity of water $dx\,\rho|s_z(x,0,t)|$ has been moved upward a distance $|s_z(x,0,t)|$, for values of x in the trough. This gives rise to a total change in potential energy relative to the equilibrium configuration

$$E_{\text{pot}} = \int_0^L dx \int dy \frac{1}{2}g\rho|s_z(x,0,t)|^2 = \frac{\rho g a^2 A}{4}\cos^2\omega t,$$
(31.12)

where the first factor of 1/2 comes from the fact that troughs occupy only half the interval from $x = 0$ to $x = L$, and the second factor comes from $\langle\cos^2 kx\rangle = 1/2$. Combining E_{kin} and E_{pot} we obtain the total energy,

$$E = \frac{\rho a^2 A}{4}\left(\frac{\omega^2}{k}\sin^2\omega t + g\cos^2\omega t\right).$$
(31.13)

As in the case of the simple harmonic oscillator, E can only be time independent if the coefficients of $\sin^2\omega t$ and $\cos^2\omega t$ are identical, which leads to the *dispersion relation*

$$\omega = \sqrt{gk}.$$
(31.14)

Notice that the dispersion relation for gravity waves (31.14) has a functional form different from the dispersion relation for light ($\omega = ck$). Finally, note that the energy per unit surface area (integrated over z) in the standing wave is

Surface (Gravity) Waves

In the linearized approximation for an irrotational flow of an incompressible, ideal fluid, there is a simple surface wave solution in terms of the displacement vector $s(x,t)$,

$$s_x = ae^{kz}\cos k(x-vt); \qquad s_z = ae^{kz}\sin k(x-vt),$$

for a plane wave traveling in the $\hat{x}$-direction.

The solution is parameterized by amplitude a and wave number k. The phase velocity of the wave is

$$v = \omega/k = \sqrt{g/k} = \sqrt{g\lambda/2\pi},$$

where g is the gravitational constant and ω is the angular frequency of the wave.

$$\mathcal{E} = \frac{E}{A} = \frac{1}{4}\rho g a^2.$$
(31.15)

We now return to the traveling surface wave solution (31.9). Traveling ocean waves transport energy as the wave propagates. To compute the power in a surface wave we first determine the energy density, and then the rate at which the energy is transported in the direction of wave motion. This is roughly analogous to the computation of energy density and power density in the wind in §28. Note, however, that unlike for wind the individual water molecules do not experience any net motion in the direction of the wave (at least in the linear approximation), though there is still a net energy flux for surface waves.

The calculation of the kinetic energy density proceeds along the same lines as for standing waves. We integrate $\frac{1}{2}\rho\dot{s}^2$ over z. The result is a constant kinetic energy per unit area, $\mathcal{E}_{\text{kin}}$ (with units J/m^2),

$$\mathcal{E}_{\text{kin}} = \frac{1}{4}\rho a^2\omega^2/k = \frac{1}{4}\rho g a^2.$$
(31.16)

Note that the kinetic energy per unit area is constant, independent of position and time, at least in the linear approximation. The calculation of the potential energy (per unit area) also proceeds similarly to the case of the standing wave (eq. (31.12)). Averaging over a wavelength in x we have

$$\mathcal{E}_{\text{pot}} = \frac{\rho g}{2L}\int_0^L dx|s_z(x,0,t)|^2$$
(31.17)

$$= \frac{\rho g a^2}{2L}\int_0^L dx\sin^2(kx - \omega t) = \frac{1}{4}\rho g a^2.$$

So $\langle \mathcal{E}_{\text{pot}} \rangle = \mathcal{E}_{\text{kin}}$ and

$$\langle \mathcal{E} \rangle = \langle \mathcal{E}_{\text{pot}} \rangle + \mathcal{E}_{\text{kin}} = \frac{1}{2} \rho g a^2 , \qquad (31.18)$$

which is twice the energy density in a standing wave *of the same amplitude a*. Notice that the superposition of propagating waves in the $+\hat{x}$- and $-\hat{x}$-directions yields a standing wave of the form (31.10) with amplitude $2a$, so that the energy in the standing wave is equal to the sum of the energies in the two propagating waves – as it must be.

While the traveling wave transports energy in the x-direction, the rate at which the energy in the wave moves in the x-direction is not simply the energy density times v. A simple way to see this heuristically is that the kinetic energy associated with a given column of water is unchanging in time, while the location where the potential energy is stored can be thought of as moving along with the wave peaks at a velocity v. Since the potential energy represents 1/2 of the total, this suggests that the rate at which energy density is transferred in the x-direction is given by $\mathcal{E}v/2$. We can think of this as the full energy density of the wave moving at a velocity $v/2$ in the x-direction. Indeed, for a localized group of waves given by a superposition of sine wave solutions, the group propagates with velocity $v/2$ even though the peaks and valleys in the detailed waveform propagate at velocity v. The velocity with which a wave group and its energy density travel is known as its **group velocity**. The group velocity of deep-water waves is given by

$$v_g = \frac{d\omega}{dk} = \frac{1}{2}\sqrt{g/k} = v/2 = g/2\omega . \qquad (31.19)$$

This group velocity describes the rate at which energy is carried forward by the wave, matching the preceding heuristic argument. The group velocity is derived and described further in Box 31.2.

From eq. (31.19), the power density of a deep-water surface wave is given by

$$\mathcal{P} = \mathcal{E}v_g = \mathcal{E}v/2 = \frac{\rho a^2 g^2}{4\omega} = \frac{1}{8\pi} \rho g^2 a^2 T , \quad (31.20)$$

where T is the period of the wave. This power density has units of J/m s, or power per unit length. We have integrated over the vertical height of the water column, so this is the power per unit of length parallel to the wave crest.

31.2.3 The Beginning and End of Waves

To understand how waves can be used for power generation, it is helpful to know something about how waves are produced and how waves behave in shallow waters. In this section we explore these aspects of waves and describe some important ways in which the approximations made in

---

Energy and Power in Surface Waves

The energy density in a traveling surface (gravity) wave, per unit area on the surface (integrated over all depths) is

$$\mathcal{E} = \frac{1}{2} \rho g a^2 ,$$

where ρ is the mass density of water and a is the wave amplitude. This energy is effectively transported with the *group velocity* $v_g = v/2 = g/2\omega$ so that the power density is

$$\mathcal{P} = \mathcal{E}v_g = \frac{\rho a^2 g^2}{4\omega}.$$

---

the idealized deep ocean, linearized analysis break down in these processes for real waves. A more detailed description of the origin and structure of ocean surface waves, with further references, is given in [172].

Wave Production   Waves are produced as wind blows across the surface of the open ocean. Even when the sea is extremely calm, tiny fluctuations on the surface, beginning as capillary waves, are amplified as wind stress transfers energy from the air to the water. Over time, in a sustained wind the fluctuations grow into larger and larger waves. Like any function, even in a regime where the dynamics are nonlinear, the motion of the surface of the sea can be described by a linear superposition of sine waves with different wavelengths, frequencies, and phases. The linearized deep-water wave solution (31.9) gives a reasonable first approximation to the motion of each of the component waves generated by wind stresses. Nonlinear aspects of the fluid motion are important, however, during wave production.

Over the years, much effort has been put into observing and predicting the behavior of ocean waves. Because ocean surface waves are not, in fact, regular, but are a superposition of fluctuations at many different wavelengths propagating in various directions, a measure of typical wave amplitude is desirable to compare observations of different irregular sequences of waves. The most commonly used metric for wave size is **significant wave height**, H_s. For many years, significant wave height was defined by convention to be the average $H_{1/3}$ of the wave heights of the highest third of the waves measured over a period of minutes, as it was found that this gave the best match to reported observations. The wave height is defined to be the vertical distance between the bottom of a trough

Box 31.2   Dispersion and Group Velocity in Waves

As discussed in §4, in realistic situations waves are not described by sinusoidal fluctuations of infinite extent. Rather, physical wave excitations are generally finite in extent, like the one shown in Figure 4.4, and described by linear superpositions of sinusoidal modes, as in eq. (4.15). Such solutions are known as wave **packets** or **groups**. They may look like a sine wave over a number of wavelengths but decrease gradually in amplitude at the ends of the wave packet. The propagation of such a wave group is depicted in Figure 31.12. As the figure shows, the rate at which the individual wave peaks move forward – the *phase velocity* – is different from the rate at which the group as a whole moves ahead – the *group velocity*. The relationship between group and phase velocities depends upon the relationship between wave number and frequency for the waves in question. If the frequency is proportional to the wave number, $\omega = ck$, as it is for electromagnetic waves, then all frequency components of a wave propagate with the same velocity, $v = \omega/k = c$. In general, however, ω depends upon k in a nonlinear fashion, and different components of a wave travel with different velocities, causing the shape of the wave to change as it propagates. This phenomenon is known as *dispersion* (§4.5.3) and the functional relation $\omega(k)$ is known as a *dispersion relation*. For deep-water surface waves, as we see from eq. (31.14), the dispersion relation is $\omega(k) = \sqrt{gk}$.

To understand the relationship between the dispersion relation and the group velocity, consider a simple linear combination of two plane wave solutions with nearly identical wave numbers. To simplify the analysis we use complex solutions of the form $e^{i(kx-\omega(k)t)}$. (This is just a linear combination of (real) cosine and sine solutions. A real solution can always be extracted by taking the real part of the complex solution (Re $e^{i(kx-\omega(k)t)} = \cos(kx - \omega(k)t)$).) A linear combination of two such waves with equal amplitude but differing wave numbers k and $k + \delta k$ is given by

$$A(x, t) = ae^{i(kx-\omega(k)t)} + ae^{i((k+\delta k)x-\omega(k+\delta k)t)} \, .$$

We have chosen coordinates so that at $t = 0$, $x = 0$, these waves are in phase, since both complex exponentials are 1. When the waves are in phase, the total wave has a maximum amplitude, associated with a peak in the wave "packet." More generally, the two components of eq. (31.2) are in phase when

$$e^{i(kx-\omega(k)t)} = e^{i((k+\delta k)x-\omega(k+\delta k)t)} = e^{i(kx-\omega(k)t)}e^{i(\delta k)(x-(\partial\omega(k)/\partial k)t)} + \cdots, \text{ or}$$

$$e^{i(\delta k)(x-(\partial\omega(k)/\partial k)t)} \cong 1 \, ,$$

where we drop terms of order $(\delta k)^2$. Thus, the peak of the wave packet moves with a velocity

$$v_g = \frac{\partial\omega(k)}{\partial k} \, .$$

This determines the group velocity in terms of the dispersion relation connecting wave number and frequency.

A linear combination of only two modes is not a truly localized wave packet; the wave form fits within a larger envelope that repeats periodically as shown in the figure below, and each of the periodic packets within the overall wave form moves with the group velocity. By including more modes, as in eq. (4.15), the same general analysis holds, forming a truly localized wave packet that moves with the group velocity v_g, such as that shown in Figure 31.12.

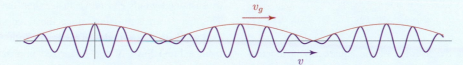

For the deep-water surface wave, we have $\omega(k) = \sqrt{gk}$, so the group velocity is

$$v_g = \frac{\partial\sqrt{gk}}{\partial k} = \frac{1}{2}\sqrt{g/k} = v/2 \, ,$$

as we found from the more heuristic argument in the main text.

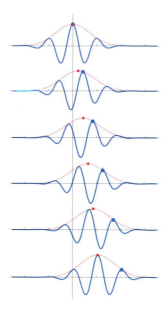

Figure 31.12 A group of surface waves moving to the right depicted at several successive moments in time. The actual wave is shown in blue; the envelope that defines the wave-packet is shown in red. The motion of the wave packet (*group velocity*) is denoted with the red dots, while an individual crest (which moves at the *phase velocity*) is followed with the blue dots. The individual wave crests move to the right twice as fast as the group as a whole. Thus, new waves appear to form at the back of the group and pass forward through the group. This phenomenon can be seen as ocean waves approach the shore from deep water, or in the pattern of waves produced when a rock is thrown into a lake.

and the top of the following peak ($H = 2a$ for the idealized wave solution (31.9)). A more modern (and useful) definition of significant wave height is

$$H_s \equiv 4\sqrt{\langle h^2 \rangle}, \qquad (31.21)$$

where $\langle h^2 \rangle$ is the average of the square of the fluctuation of sea level (for the idealized surface wave solution, $\langle h^2 \rangle = a^2/2$). This definition agrees reasonably well with the older $H_{1/3}$ definition in typical circumstances. The definition (31.21) is useful since wave energy density is given by $\mathcal{E} = \rho g \langle h^2 \rangle$; hence, H_s^2 is a proportional measure of energy contained in a general wave spectrum.

As waves produced by a sustained wind become higher, the cross-sectional area of the peaks perpendicular to the wind velocity grows, and the wind stress increases the rate at which energy is transferred into the motion of the water near the surface. This would appear to lead to a runaway growth in wave energy. As this process continues, however,

several of the approximations we have made break down. For one thing, we have neglected friction and dissipation. For smooth, long-wavelength waves the rate of dissipation is quite small, but in a turbulent sea the rate of energy dissipation is greater. As waves grow, eventually they become sufficiently steep that they begin to break, forming **white caps**. At this point energy is lost from the coherent wave motion to random turbulence and eventually converted to thermal energy.

An early model for wave production was developed by Pierson and Moskovitz in 1964. Their model was based on the assumption that after a long period of time in a constant wind, the rate of energy dissipation would match the rate of energy input through wind stress, and the spectrum of waves would approach a fixed distribution that they termed a **fully developed sea**. They used empirical data to construct a model for this state. According to their model, significant wave height in a fully developed sea is given roughly by $H_s \cong 0.22 w^2/g$, where w is the wind speed measured at 10 m above the sea surface.[3] Note that unlike surface currents or objects on the water's surface, which move in a direction perpendicular to the wind as discussed in §27, wind-driven waves generally propagate and transport energy in the same direction as the wind, with some variation in angle. The difference is that a small packet of water in the wave experiences no net motion (at least in the linear approximation) over long times and therefore feels no Coriolis force. The peak of the spectrum in the *Pierson–Moskovitz model* is around $\omega \cong 0.85\, g/w$. Thus, according to this model, a wind blowing at 10 m/s over a period of several days produces a fully developed sea with significant wave heights of roughly 2 m. The phase velocity of these waves, describing the rate at which the peaks move, is $v \cong g/\omega \cong 12$ m/s. This corresponds to empirical measurements, which show that waves produced by sustained winds can move faster than the wind itself. This may seem surprising, since the wind stress pushes backward on these modes. This result follows, however, from a breakdown of the linearized approximation. As described in more detail in Box 31.3, nonlinear interactions in the wave-development process transfer energy from higher-frequency, shorter wavelength modes into lower-frequency, longer wavelength modes, so that over time the wave energy is concentrated in long-wavelength modes

---

[3] Although Pierson and Moskovitz obtain this equation from a more complicated formula for the spectrum, up to the empirically determined dimensionless constant 0.22, this formula can be determined by dimensional analysis, since the only combination of w and g with units of height is w^2/g.

Box 31.3 The Effect of Nonlinearities on Ocean Wave Development

In the linearized approximation wave solutions can be superposed: two independent solutions ψ_1 and ψ_2 to the wave equation (31.7) add to give another solution. The exact equations describing fluids with a surface boundary, however, contain nonlinearities that modify this conclusion. In particular, a combination of waves at different wavelengths and frequencies can interact, producing new contributions to the total solution. Interactions of this type are particularly relevant during wave production. The consequences of such interactions were worked out in the 1960s by the German physicist and oceanographer Klaus Hasselmann using the same methods used to analyze interactions of elementary particles in high-energy accelerators [204]. Hasselmann showed in particular that interactions of three or more waves can transfer energy into lower-frequency excitations associated with waves of longer wavelength. Since longer wavelength modes have faster phase velocities, these interactions explain the phenomena seen in the Pierson–Moskovitz spectrum, that long-wavelength modes with phase velocity greater than the wind velocity in fact carry much of the wave energy. Hasselmann also found, through analysis of further empirical data, that the sea continues to develop over very long periods of time while the energy in the waves increases approximately linearly with the *fetch F*. As the fetch increases, the dominant frequency decreases correspondingly. This analysis has led to improved empirical models for the wave spectrum. One example is the *JONSWAP (Joint North Sea Wave Project)* spectrum [205] (see Figure 31.13). According to this model, the surface energy density in the waves grows linearly with the fetch and is given by

$$\mathcal{E} \sim \rho (1.67 \times 10^{-7}) w^2 F,$$

(where ρ is the density of seawater) and the significant wave height is given by

$$H_s \sim \left(1.6 \times 10^{-3}\right) w \sqrt{\frac{F}{g}}.$$

Neither the JONSWAP model nor the Pierson–Moskovitz model give completely accurate descriptions of the distribution of waves produced by a given wind, but both are nonetheless useful tools to get an order-of-magnitude estimate of wave production from sustained winds.

Example 31.1 Energy in a Typical Surface Wave

Consider a moderate ocean swell that contains regularly spaced waves of amplitude $a = 1$ m with a 10 second period. *What is the energy density in these waves, and what is the power density along the wave front?*
To compute the energy per unit surface area on the ocean we use eq. (31.18),

$$\mathcal{E} = \frac{1}{2}\rho g a^2 \cong \frac{1}{2}(1025\,\text{kg/m}^3)(9.8\,\text{m/s}^2)(1\,\text{m})^2 \cong 5.0\,\text{kJ/m}^2$$

Given the period, we can compute the angular frequency, wave number, and the wavelength between peaks, as well as the phase velocity v,

$$\omega = 2\pi/T \cong 0.63\,\text{s}^{-1},$$
$$k = \omega^2/g \cong 0.04\,\text{m}^{-1},$$
$$\lambda = 2\pi/k \cong 156\,\text{m},$$
$$v = \sqrt{g/k} \cong 15.6\,\text{m/s}.$$

The group velocity $v_g = v/2$ characterizes the rate of energy transport by the waves, so the power density of these waves in deep water is

$$\mathcal{P} = \mathcal{E}v_g = \mathcal{E}v/2 \cong (5.0\,\text{kJ/m}^2)(7.8\,\text{m/s}) \cong 39\,\text{kW/m}.$$

Compared, for example, to the solar constant $I_0 \cong 1366\,\text{W/m}^2$, this is a substantial power density.

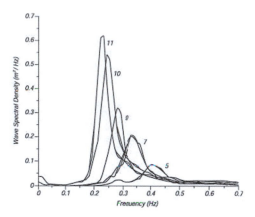

Figure 31.13 Wave spectrum measured by Joint North Sea Wave Observation Project (JONSWAP). Measured data and best-fit analytic spectra are graphed for a variety of fetches. Labels 5–11 correspond to different measuring stations with increasing fetches. From [205].

with phase velocity faster than w. This nonlinear analysis also shows that the spectrum continues to develop over time in a constant wind, and that the total energy in the waves continues to increase over long time periods roughly linearly with the **fetch** F, which is the distance over which the constant wind acts. The spectrum of waves produced by a constant wind for different fetches is shown in Figure 31.13.

A consequence of the frequency dependence of the wave velocity in eq. (31.19) is that waves at lower frequencies (and longer wavelengths) move more quickly. Thus, when an initial disturbance like a storm at sea produces a distribution of waves of different frequencies, the disturbance of the sea surface gradually spreads, or *disperses*, as the energy in the waves at each frequency moves according to the corresponding group velocity $v_g = g/2\omega$. Thus, the energy from any disturbance of the sea surface will over time be distributed over a large area. Much of the energy from a substantial storm or sustained wind is transferred to low-frequency, long-wavelength wave modes that travel more quickly away from the source of the disturbance than higher-frequency, short-wavelength waves. Propagating surface waves that have dispersed far from their source tend to be close to monochromatic, and are described quite accurately by the linearized theory. Such smoothly propagating surface waves are known as **swell**. For example, many of the most significant large swells that hit the western coast of the US (and that generate excellent surfing waves there) originate in large storms many thousands of kilometers away in the Southern Pacific ocean.

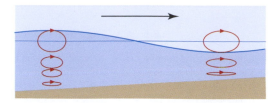

Figure 31.14 Motion of a surface wave in shallow water. The water moves in ellipses with almost constant horizontal length at each x and height that decreases with depth. The cases shown correspond to $h = 1/k$ and $h = 0.7/k$.

Modern wave-forecasting models combine empirical observations with numerical models of the propagation of hundreds of different wave modes, including components with different direction and wavelength. Using models of energy transfer from local winds and Hasselmann's nonlinear formulation of mode interactions (Box 31.3), current wave-forecasting software can make fairly accurate predictions days in advance.

Waves in Shallow Water

The deep-water wave solution (31.9) is only valid in the approximation that the water is infinitely deep. This approximation is valid when e^{-kh} is small, where h is the depth of the water and $k = 2\pi/\lambda$ is the wave number. Thus, the deep-water approximation is good roughly when $\lambda < h$. As a wave moves into shallower water, the presence of the sea floor modifies the boundary conditions for the wave. Specifically, since the water cannot move into the seabed, we require $s_z(x, -h, t) = 0$, which is not satisfied by eq. (31.9). Incorporating this boundary condition, the solution for a surface wave in shallow water (where $h \ll \lambda$) becomes[4]

$$s_x \approx \frac{a}{kh}\cos(kx - \omega t) + a\,\mathcal{O}(k(z + h)),$$

$$s_z \approx a\frac{z + h}{h}\sin(kx - \omega t) + a\,\mathcal{O}(k^2(z + h)^2),$$

$$(31.22)$$

and the shallow-water dispersion relation is

$$\omega(k) = k\sqrt{gh}, \qquad (31.23)$$

so the phase and group velocity of shallow-water waves are equal

$$v = \omega/k = \sqrt{gh} = \partial\omega/\partial k = v_g. \qquad (31.24)$$

Equation (31.22) and the associated dispersion relation eq. (31.23) can be obtained as approximations to a more general analytic solution valid at any depth (Problems 31.8

---

[4] Recall that z is negative below the equilibrium ocean surface.

Example 31.2   Waves from Steady Wind

Consider a constant wind at 8 m/s that blows over the ocean's surface for a distance of 100 km. Use the JONSWAP model to estimate the energy density and significant wave height of the resulting waves. Compare to the Pierson–Moskovitz model for a fully developed sea.

The energy density is

$$\mathcal{E} \cong \rho (1.67 \times 10^{-7})(8\,\text{m/s})^2 (10^5\,\text{m}) \cong 1\,\text{kJ/m}^2\,.$$

The significant wave height is

$$H_s^{(\text{JONSWAP})} \cong \left(1.6 \times 10^{-3}\right) w \sqrt{\frac{F}{g}} \cong 1.3\,\text{m}\,.$$

According to the Pierson–Moskovitz model, significant wave height would be

$$H_s^{(\text{PM})} \cong 0.22\,w^2/g \cong 1.4\,\text{m}\,.$$

So in this case the two models are in quite close agreement.

---

and 31.9). The shallow-water solution is a leading order approximation in the parameter h/λ, and is valid to ~5–10% for water depth $h < \lambda/20$.

From eq. (31.24), we see that as a wave enters shallower water, its group (and phase) velocity decreases proportionally to $\sqrt{h}$. The frequency must stay the same, since the number of peaks that passes any point in a given time must be the same. Thus, the wavelength must decrease. To the extent that dissipation can be neglected, energy conservation requires that the power density of the wave does not change as it enters shallow water. According to eq. (31.20), $\mathcal{P} = \mathcal{E} v_g$, and therefore the energy density in the wave increases proportionally to $1/\sqrt{h}$. The increase in energy density is realized by an increase in wave height ($a \sim 1/h^{1/4}$). Thus, while the power density stays the same (or decreases slightly due to dissipation on the ocean floor), the wave height and energy density increase significantly as the water depth decreases (Figure 31.15).

Underwater topography also affects the direction of wave propagation. Just as light *refracts* (§4.5.2) when it passes into a medium such as glass in which the speed of light is different from that in air, a wave passing over a shallow sea floor with variable depth will refract, with wave energy being directed towards regions with slower wave velocity (i.e. lesser depth). An underwater ridge, for example, often associated with a coastal outcropping, slows the waves, focusing the wave energy along the ridge, as shown in Figure 31.16. Underwater channels allow faster wave motion so that wave energy is dispersed to the sides of the

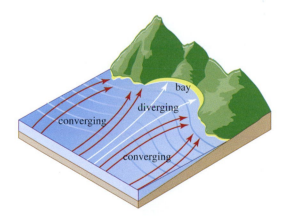

Figure 31.16 Waves focus in shallow water and defocus in deep water with the familiar result that they attack headlands and leave broad embayments relatively calm. Refraction of water waves has profound consequences for coastline geography and ecology. (Credit: Image by Byron Inouye, used with permission of the Curriculum Research & Development Group, University of Hawaii)

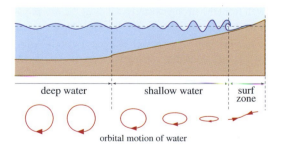

Figure 31.15 As a train of waves enters shallow water, the distance between peaks decreases and the wave height and energy density increase until the wave breaks.

channel. Waves can also reflect when incident on almost-vertical boundaries, and can diffract out into a large bay or estuary after passing through a narrow channel. Thus, water surface waves can be described by many of the same techniques that are used to describe the propagation of light. Such analysis is useful in developing and describing mechanisms for wave energy capture. For example, some devices use reflection of incident waves or refraction to focus increased amounts of wave energy on the wave energy collection device.

In shallow water, nonlinear effects become increasingly important. Wave peaks become more narrow and troughs more broad. Such waves are no longer well described by the linear theory. Several more complete theories that incorporate some nonlinearity are described in [201]. As the water becomes progressively more shallow, eventually the phase velocity decreases to the point that the velocity of the particles at the wave peak exceeds the phase velocity, and the wave begins to break. How the wave breaks depends critically upon the gradient of the sea floor/beach where the wave is breaking, and also on the steepness of the wave (see Figure 31.17) and on the direction and speed of the wind. When the gradient of the sea floor is very shallow, the wave begins to gently crumble as the top of the wave just exceeds the phase velocity. When the gradient is steeper, the wave becomes steeper more rapidly, and the top of the wave shoots past the forward surface of the wave before plunging downwards (Figure 31.18). When the gradient is extremely steep, the wave does not

Figure 31.18 Surfing a breaking wave.

really break at all, but surges up the beach in a mass of whitewater. While crumbling breakers are excellent surfing waves for beginners and longboarders, plunging breakers that produce a long tube form an ideal playground for expert surfers. Since much of the energy is dissipated in the breaking of the wave, however, with some exceptions most technologies for extraction of wave energy are not designed for the surf zone.

31.2.4 Extent of Wave Resource

The annual average wave power in oceans and near oceanic coastlines, shown in Figure 31.19, ranges from around 10 kW/m to over 60 kW/m. Regions with substantial wave power are clearly correlated with high average wind velocities as seen in Figure 28.3. In regions with prevailing westerly winds, substantial fetch leads to a great deal of wave power at the western boundaries of continents. As seen in the figure, on the western coasts of North America and Europe, and on the southern coasts of South America and Africa outside the tropics, wave power is in the 50 kW/m range. The southern coast of Australia, exposed to swell from the strong westerly winds of the *roaring '40s* with a fetch extending to South America, have extremely high annual mean wave power in excess of 60 kW/m. (It is not a coincidence that surfing is a very popular activity among Australians!) In the tropics, on the other hand, wave power is generally below 20 kW/m.

Like wind power, wave power at any location is highly variable. While 1 m swells with a 10 second period carry a power density of 40 kW/m (see eq. (31.20)), a 10 m swell with a 12 second period carries 120 times this much power, or almost 5 MW/m (Problem 31.11). Devices for wave energy extraction must be extremely robust to handle such a range of conditions. In practice, like wind power, the

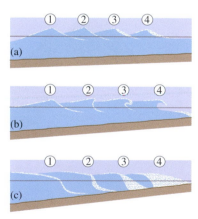

Figure 31.17 Four stages in the breaking of a wave on beaches of increasing slope. (a) A shallow gradient leads to gently spilling breakers; (b) an intermediate gradient gives steeper plunging breakers; and (c) a very steep gradient gives surging waves. (Credit: Horikawa, K. (ed.) (1988): Nearshore Dynamics and Coastal Processes, Univ. of Tokyo Press)

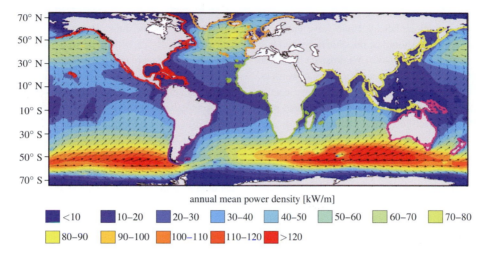

annual mean power density [kW/m]

■ <10   ■ 10–20   ■ 20–30   ■ 30–40   ■ 40–50   ■ 50–60   ■ 60–70   ■ 70–80

■ 80–90   ■ 90–100   ■ 100–110   ■ 110–120   ■ >120

Figure 31.19 Global annual mean wave power density in kW/m (colors) and annual mean direction (arrows). Note that the prevalence of westerly winds favors the western edges of continents for wave power capture. After [206].

distribution of wave power observed at a typical location is reasonably accurately modeled by a Weibull distribution. The ratio of the variance to the mean value of the wave power distribution tends to be large when the mean is small and is small in the regions of the largest wave power potential.

As for wind power, good global wave power atlases are available from which assessments of the wave power resource can be extracted. Total wave power worldwide has been estimated at around 2–3 TW [206, 207]. The order of magnitude of this number can be confirmed by estimating the length of coastline in Figure 31.19 with annual average wave power over 30 kW/m to be roughly 50 000 km. Multiplying this figure by an average over this coastline of perhaps 50 kW/m, gives 2.5 TW. As we discuss in the next section, wave energy extraction devices can have high conversion efficiency. It may in principle be possible to capture a substantial fraction (meaning 10% or more) of the world's available wave energy [208], but this would, of course, require engineering, manufacturing, and deploying large arrays of wave capture devices. These might be an obstacle to navigation, and furthermore, removal of a substantial fraction of wave energy from the global ocean systems may pose environmental concerns for local ecosystems. In the near term, however, wave energy provides a promising source of alternative energy at a local scale, due to the high linear power density of waves at many locations.

31.2.5 Wave Power Technologies

Designing practical systems to harvest wave energy presents several substantial challenges. First of all, the

Wave Energy

Wave energy is produced when winds blow across the surface of the open ocean. Nonlinear wave dynamics shifts energy into long-wavelength, faster-moving waves (*swell*) that can transport the energy across thousands of kilometers until the waves break. Average wave power density along the shore can range from a few kilowatts per meter to over 60 kW per meter. Wave energy conversion devices can capture a substantial fraction of incoming wave energy in favorable locations. Total wave power incident on shores worldwide is on the order of magnitude of 2–3 TW, only a small fraction of which could ever be practically utilized.

wide range of power densities that can arise at a single location requires wave energy systems to be robust and able to withstand powerful stresses. Because of the energy lost when waves break and the more turbulent environment in the surf zone, most wave energy systems are designed for offshore use, where they can derive most of their energy from low-frequency, high-energy swells. Anchoring such devices in place and transferring energy from them to the shore pose substantial engineering challenges. Most wave energy capture devices contain mechanical structures or have resonances that work best with swell in a particular frequency range. It is a challenge to build a device that can capture a large part of the wave energy across a range of swell frequencies. Nonetheless, since wave energy is essentially mechanical, it should be possible to convert it to useable form with high efficiency. This prospect has led to many promising approaches for wave energy converters. Although, at the time of writing, deployed wave energy conversion systems are principally experimental and pilot devices, we briefly describe some of the main ideas and mention a few devices presently under development.

As we showed in §31.2.2, energy in deep-water surface waves is distributed half in kinetic and half in potential form. Some mechanical wave energy converters are designed to extract the kinetic energy, by using it to move a rotating wheel, for example. Other designs extract the potential energy, for example by using the variation in pressure at a depth below the surface to force air through a turbine. Another simple design that uses potential energy is a buoy on the surface connected to the sea bottom by a cable that drives a generator as the buoy is lifted on wave crests. The transfer of energy back and forth from kinetic to potential form as the wave oscillates allows for the capture of significantly more than half of the wave's total energy, even for a device that relies only on the potential or kinetic energy content of the waves.

One early technology for mechanical wave energy conversion is the **Salter's duck,** or **nodding duck,** developed by British engineer Stephen Salter in 1974. This wave energy converter has a vaguely duck-shaped cross section and rotates under the influence of passing waves (Figure 31.20). The "beak" of the duck is both pushed forward by the kinetic motion of the water as the wave passes, and lifted by the rising swell, thus extracting both kinetic and potential energy from passing waves. Experiments with the duck in a two-dimensional flow pattern show that this system can capture on the order of 90% of the energy of passing waves. In a three-dimensional wave environment, a Salter's duck of finite width can capture about 1.5 times as

much wave energy as is contained in the waves that impact over the width of the duck. This result demonstrates that a wave capture device with small dimensions in each direction can capture more energy than is directly incident upon it. This at first unintuitive phenomenon can be explained by recognizing that a wave capture device acting in reverse is a wave generation device that produces waves that move outward in a circular fashion. When acting as a wave capture device, the device in effect produces waves that precisely cancel the waves that would have passed through it. Since the produced waves move out in more than one linear direction, the device can cancel, and thus capture, more energy than would appear in its geometrical "shadow." Thus the duck in some sense acts as an antenna, capturing not only the incident plane wave energy but also wave energy from nearby directions. Because of this feature of wave capture devices, it is possible to imagine deploying an array of wave capture devices that can capture a large fraction of the wave energy from an area without completely blocking the area to navigation. For a mathematical analysis of the Salter's duck, see [201]. A more thorough discussion of the duck in the larger context of wave technology is given in the chapter by Salter in [202].

Another system for wave capture is based on a configuration of multiple linear segments whose angle relative to one another changes as the swell passes through the system. An analysis of one of the first such systems, a system of attached rafts of differing lengths, can be found in [201]. To optimize the capture energy from a variety of wavelengths, it is desirable to give the attached rafts differing lengths. A modern device, the **Pelamis,** based on similar principles, is depicted in Figure 31.21. In this device the motion of the segments drives hydraulic rams that pump oil through motors, which in turn convert the mechanical energy into electrical energy. A key feature of the Pelamis device is that its resonant frequency can be tuned by adjusting the resistance in the attachments. The design capacity of a single Pelamis device is 750 kW.

Another class of wave capture devices relies upon the potential energy in the tops of waves. When waves are incident on such an **overtopping device** (Figure 31.22), the water at the top of the wave passes up a curved ramp and into a reservoir. The water then drains into the sea beneath through hydro turbines that convert the potential energy into electrical energy. Water may be channeled into the device by extended "wings" that reflect incident waves into the collection area. Overtopping devices must be held relatively stationary with respect to wave motion. This can be accomplished either by tethering the device to the sea floor or by making it very massive. The 4 MW rated unit shown in the figure is specified to have a mass of 22 000 t.

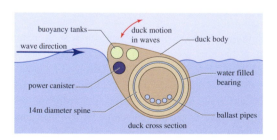

Figure 31.20 Cross section through a Salter's duck wave energy conversion device. (Adapted from P. Lynn, *Electricity from Wave and Tide*, John Wiley and Sons Ltd (2014))

Figure 31.21 The Pelamis wave capture device's multiple segments convert their relative motion in response to wave action into electrical energy. (Credit: Rob Ionides)

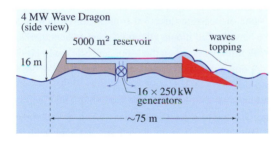

Figure 31.22 Design for a 4 MW, 22 000 t Wave Dragon wave capture device, which operates by capturing water at the top of waves with large potential energy. The "wings," shown in red, are highly foreshortened in this side view. (Credit: Wave Dragon)

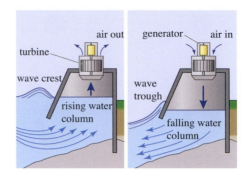

Figure 31.23 Schematic of an oscillating water column (OWC) wave power plant, showing periodic reversal of flow.

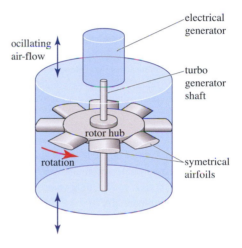

Figure 31.24 Wells turbine blade used in oscillating water column (OWC) wave power plant.

The systems described above are primarily designed for use in deep water. One technology that has been developed for use along the shore line is the **oscillating water column** (OWC). The idea of the OWC is that a chamber containing air is separated from the open ocean by a wall descending below the low-water line (Figure 31.23). The water at the bottom of the chamber moves up and down with incoming waves, causing a periodic compression of the air inside the chamber. The pressurized air drives a turbine to extract energy from the wave motion. In some installations, such as the LIMPET plant in the UK, a **Wells turbine** is used. By using a symmetric rotor blade, the Wells turbine can turn in a single direction despite the periodic reversal of the airflow. Although it sacrifices some efficiency, the Wells

turbine avoids an elaborate and expensive valve system to rectify the air flow (Figure 31.24).

A more detailed description of current technologies for wave energy capture, including Pelamis, Wave Dragon, and the oscillating water column with Wells turbine system, can be found in [202].

31.3 Tidal Power

Aside from nuclear and geothermal power, tidal power is the only terrestrial source of energy that does not ultimately come from solar radiation. Tidal power instead taps into Earth's store of rotational energy via the gravitational interaction between Earth's oceans and the Moon (and to a lesser extent, the Sun). Tidal forces cause motion of the oceans, which dissipates some of Earth's rotational energy through friction and also transfers some of that energy to the Moon.

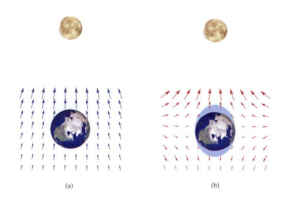

Figure 31.25 (a) The Moon's gravitational force at points near Earth; (b) the same as (a) except that the gravitational force on Earth's center of mass has been subtracted away. What remains is the tidal force that pulls material on the sides of Earth facing and opposite to the Moon upward, causing bulges on both sides of Earth. The sizes of the Moon and Earth have been exaggerated and the distance between them has been decreased to make the tidal effect, which otherwise would be too small to see, visible. (Credit: NASA/JPL/ASF and artdesigner.lv)

While the total amount of tidal energy available is small compared to other renewable resources such as wind or solar, in certain locations it can be highly concentrated and may be effectively commercially exploited. In this section we describe the physical origins of tidal power and some of the technologies for harvesting this resource.

31.3.1 Physical Origin of Tidal Power

Tidal fluctuations at any given location on Earth's surface depend upon a complex interplay between gravitational forces from the Moon and Sun and the motion of oceans constrained by Earth's geography. The details of tidal motion are complicated, but with a few simplifying assumptions it is possible to understand the most important general features of tidal dynamics.

At the simplest level, tides arise because the gravitational force from the Moon (and Sun) acting on material at different points on or within Earth varies according to the Newtonian force law (2.21) as the inverse square of the distance from the source of the gravitational force. Thus, water on the surface of the ocean at the point closest to the Moon is more strongly attracted to the Moon than the mass at the center of Earth, while water on the surface of the ocean furthest from the Moon is attracted less. The net tidal force on any point on or within Earth can be computed by subtracting the average gravitational force on Earth from the force at that point (see Figure 31.25). In a simplified static picture, where the Moon and Earth are held fixed and

Earth does not rotate, this would lead to two *bulges* on the ocean's surface on opposite points of the globe. This picture was the basis for the first quantitative analysis of tides by Newton and others. In this static approximation of the lunar tidal force, the ocean's surface would take the form of an ellipsoid (an ellipse in cross section), with a difference in height between *high tide* and *low tide* of about 0.5 m (Problem 31.13). A similar calculation for the solar tidal force gives an additional tidal perturbation of about half this amplitude.

This simplified picture provides a qualitative understanding of some aspects of the nature of tides. As Earth rotates, and the positions of the Moon and Sun change relative to Earth, the tidal bulges move around Earth's surface. Since the Moon is the dominant influence in tides, high and low tides are correlated with the position of the Moon. Combining the 27.3 day orbital period of the Moon with the 24 hour rotation period of Earth, the time at which the Moon reaches the highest point in the sky shifts by approximately 24 h/27.3 $\cong$ 50 minutes each day, which is roughly the daily change in time of high tide. This picture also correctly predicts that when the Moon and Sun are aligned in the same or opposite direction from Earth (which occurs at the new and full moons) the tidal amplitude is increased (**spring tide**), while when the Moon and Sun are in perpendicular directions from Earth the tidal amplitude decreases (**neap tide**).

While this simple static Newtonian picture captures some features of tides (Problem 31.13) it is incomplete. The simple static picture would predict that high tide would occur when the Moon is either directly overhead or on the opposite side of Earth. This is not generally observed; in most locations high tide occurs hours after the Moon is at its zenith. Also, the *phase* of the tide is highly dependent upon location. Finally, the simple Newtonian picture does not answer the basic question: where precisely does tidal energy come from and how much is there?

For a more detailed characterization of tides, further physical effects must be taken into account. In the eighteenth century, the French mathematician Pierre-Simon Laplace formulated a set of partial differential equations for fluid flow on the surface of a rotating sphere, incorporating Coriolis forces as well as tidal effects to provide a dynamical theory of tides. These equations were further developed by others, who analyzed the oscillation modes of water on the sphere. Current tidal prediction software incorporates hundreds of modes in such analysis, combined with geography, observations of ocean depth (**bathymetry**) and tidal observations to provide detailed predictions of tides across the globe.

A key feature of the dynamical theory of tides is that it correctly models the phase lag between the lunar forcing and observed tides. Improving on the static picture, one can picture the tides as a surface wave moving around the planet. In an idealized model with no land, uniform ocean depth, and constant tidal forcing from a lunar source in the equatorial plane, such a wave would have a single mode of wavelength $\pi R_\oplus \cong 20\,000$ km around the equatorial great circle. We can estimate the natural period of such a surface wave using eq. (31.24); though this calculation is not completely accurate since the wave is traveling on a sphere, this gives a good qualitative estimate. If we assume an average ocean depth of $h = 3.5$ km, the surface wave velocity is $v = \sqrt{gh} \cong 185$ m/s, so the period T_0 is on the order of 30 hours. This mode of oscillation is driven by the Moon with a period $T = 12$h 25 m. Thus, in this simple, idealized model, the tidal oscillations represent the response of a damped oscillator with natural frequency $\omega_0 = 2\pi/T_0$ to a periodic driving force with a higher frequency $\omega = 2\pi/T$. In steady state a damped, driven oscillator oscillates at the driving frequency, but it can be shown that because $\omega > \omega_0$, the response (i.e. the tidal bulge) *lags* the driving force by an angle that lies between $45°$ and $90°$, depending on the amount of dissipation [209]. This corresponds to a three to six hour interval between the time the Moon is at its zenith and the time of high tide.

The phase lag of the surface bulges causes a net torque on the oceans from the tidal forces, and a corresponding opposite torque from the gravitational field produced by the deformed oceans on the Moon (see Figure 31.26). The effect of this torque is to gradually slow the rotation of Earth while transferring angular momentum and energy to the Moon. We now turn to a quantitative analysis of this effect, which gives an estimate of the total tidal energy dissipated on Earth.

31.3.2 Quantitative Estimate of Tidal Energy Dissipation

A first-principles calculation of the total energy transfer due to tidal effects is impractical. While in principle it may be possible to imagine computing the instantaneous or time-averaged torque on Earth from knowledge of the topography of Earth's oceans, a simpler approach is to measure one of the consequences of this torque and use this information to deduce the rate of tidal energy transfer. Lunar ranging experiments performed over the last 35 years by bouncing lasers off reflectors placed by the Apollo missions show that the average Earth–Moon distance is increasing by roughly $\delta r \cong 38$ mm each year. By combining this with basic data about the Earth–Moon system, we can compute the increase in energy and angular

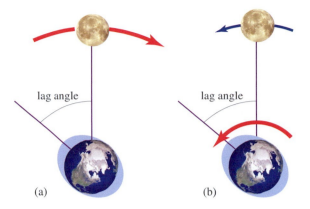

Figure 31.26 An idealized model of the Earth–Moon system showing tidal bulges, including the phase lag. (a) From the point of view of an Earth-fixed, non-inertial observer, the Moon circles overhead and the tidal bulge *lags* behind. (b) From the point of view of an observer at rest in an inertial frame centered on Earth, the tidal bulge on the rotating Earth *leads* the Moon, which in turn revolves around Earth (blue arrow). The bulge exerts a torque on the Moon, which increases its angular momentum and energy. The Moon's torque on the bulge decreases Earth's angular momentum and rotational energy.

momentum of the Moon and the corresponding decrease in angular momentum and rotational kinetic energy of Earth. This allows us to compute the total energy dissipated in tidal friction. This is a straightforward exercise using the equations of rotational mechanics described in §2.4. We need to know the masses m, M of the Moon and Earth, the radius r of the Moon's orbit, and the radius R of Earth,

$$m = M_{\text{moon}} = 7.348 \times 10^{22} \text{ kg},$$

$$M = M_\oplus = 5.972 \times 10^{24} \text{ kg},$$

$$R = R_\oplus = 6.378 \times 10^6 \text{ m},$$

$$r = r_{\text{lunar orbit}} = 3.834 \times 10^8 \text{ m}. \tag{31.25}$$

We neglect for simplicity the (5%) ellipticity of the Moon's orbit and assume that the Moon orbits in the equatorial plane (though actually the Moon's orbit is inclined at roughly $5°$ to the ecliptic and therefore between $18°$ and $28°$ from Earth's equatorial plane). At the end we quote corrections for these effects.

The first step is to compute the change in the moon's orbital angular momentum in terms of δr. First we equate the Moon's centripetal acceleration (see eq. (2.35)) to the acceleration due to Earth's gravitational attraction and solve for the Moon's angular velocity,

$$r\omega^2 = \frac{GM}{r^2} \Rightarrow \omega = \sqrt{\frac{GM}{r^3}} \cong 2.66 \times 10^{-6}\,\text{s}^{-1}\,,$$
$$\tag{31.26}$$

(which agrees with the observed lunar orbit period of 27.3 days) and use this to write the Moon's orbital angular momentum L in terms of r,

$$L = I\omega = mr^2\omega = m\sqrt{GMr}\,. \tag{31.27}$$

The change of angular momentum of the Moon in a year is then given by

$$\delta L = \frac{dL}{dr}\delta r = \frac{1}{2}m\sqrt{\frac{GM}{r}}\,\delta r \cong 1.42 \times 10^{24}\,\text{kg m}^2/\text{s}\,. \tag{31.28}$$

Next we compute the yearly change in the Moon's orbital energy. Its orbital kinetic and potential energies are

$$E_{\text{kin}} = \frac{1}{2}mr^2\omega^2 = \frac{1}{2}\frac{GMm}{r}\,,$$
$$E_{\text{pot}} = -\frac{GMm}{r}\,, \tag{31.29}$$

so the total orbital energy of the Moon is

$$E_{\text{moon}} = E_{\text{kin}} + E_{\text{pot}} = -\frac{1}{2}\frac{GMm}{r}\,. \tag{31.30}$$

Thus, the increase in the Moon's total orbital energy over a year is

$$\delta E_{\text{moon}} = \frac{1}{2}GMm\frac{\delta r}{r^2} \cong 3.8 \times 10^{18}\,\text{J}\,. \tag{31.31}$$

Since the total angular momentum of the Earth–Moon system is conserved, Earth must lose angular momentum each year equal to δL. Expressing Earth's rotational angular momentum in terms of its moment of inertia $I_\oplus$ and its angular velocity $\omega_\oplus \cong 7.29 \times 10^{-5}\,\text{s}^{-1}$, $L_\oplus = I_\oplus\omega_\oplus$, we have

$$\delta L_\oplus = I_\oplus\delta\omega_\oplus = -\delta L\,. \tag{31.32}$$

Finally, we compute the reduction over a year in Earth's rotational kinetic energy due to tidal forces

---

Tidal Power

Tidal power comes from dissipation of Earth's angular momentum through motion of ocean water pulled by the Moon (and to a lesser extent the Sun). Roughly 3.75 TW is dissipated, with a small fraction going to an increase in the Moon's orbital energy. Most of this is lost in the deep ocean but in places with large tidal estuaries fed by narrow channels, power is highly concentrated and can be harvested effectively.

$$\delta E_\oplus = \delta(\tfrac{1}{2}I_\oplus\omega_\oplus^2) = I_\oplus\omega_\oplus\delta\omega_\oplus = -\omega_\oplus\delta L$$
$$\cong -1.04 \times 10^{20}\,\text{J}\,. \tag{31.33}$$

Comparing with eq. (31.31), it appears that only about 4% of Earth's energy loss is transferred to the Moon; the remainder is dissipated through tidal friction. Thus, the total rate of tidal energy dissipation is on the order of

$$P_{\text{tidal}} \cong 10^{20}\,\text{J/y} \cong 3\,\text{TW}\,. \tag{31.34}$$

A more careful calculation, including the relative angle between the Moon's orbit and the equatorial plane (which contributes a factor of $1/\cos\epsilon \sim 1.1$) and solar tidal torque, refines this estimate to around 3.75 TW.

One can also directly measure the rate at which Earth's rotation is decreasing. Atomic clocks show that the length of the day is currently increasing by roughly 1.7 ms/century. The increase in length of day due to the tidal effect alone, however, would be about 2.3 ms/century (see Problem 31.14). The discrepancy between these numbers arises from the fact that Earth's shape is still rebounding from polar compression in the last ice age, so that its moment of inertia is decreasing slightly with time, leading to an increase in angular velocity that partly offsets the slowing due to tidal effects. (The effect is like that of a figure skater who can spin more quickly by pulling in her arms, decreasing the moment of inertia.)

Most of the 3.75 TW of dissipated tidal energy is lost to friction on the ocean floor. Along most of the world's coastline, tidal energy is too diffuse to harvest efficiently (see Problem 31.16). At locations where tidal flow fills a large basin or estuary through a narrow channel, however, the power density is higher and extraction of the tidal energy can become practical. The existing systems for taking advantage of tidal energy in such situations are the *tidal barrage* and *tidal stream turbines*. We briefly describe each of these technologies in turn.

31.3.3 Tidal Barrage

A **tidal barrage** is essentially a low-overhead dam that can operate in both directions, capturing energy both during the incoming **flood tide** and during the outgoing **ebb tide**. In flood tide operation, the dam is opened when the tide is high and water drives turbines as it fills the basin. After high tide, as the water recedes, the dam is kept closed. In ebb tide operation, the water that was retained is allowed to flow out through the dam at low tide, again driving turbines. As the water flows in and out, the hydraulic head diminishes during each process, until the water level in the basin matches that of the outside ocean. Then the dam is closed, and the basin is held at the high/low tide level as the outside tide drops/rises.

Consider an idealized tidal basin of area A with a **tidal range** (height differential between high and low tide) of h. If the basin is filled at the moment of high tide, and emptied at the moment of low tide, the potential energy that can be captured during the outflow is

$$E = \int_0^h dz\, A\rho g z = \frac{1}{2}\rho g A h^2 , \qquad (31.35)$$

where for simplicity we have assumed that the walls of the basin are vertical, so that the surface area A is unchanged between high and low tide. As with hydropower installations discussed above, this potential energy must be multiplied by a turbine efficiency ϵ, which can be quite close to one.

Real tidal barrages differ from this idealization in several ways. First, in realistic tidal basins, the surface area of the basin is larger at high tide than at low, so more energy can be extracted from the ebb tide than the flood tide. Thus, some tidal barrage installations forego flood tide generation and generate power on the ebb tide only. Second, extra energy output can be gained by pumping additional water in at high tide, increasing the height of the reservoir at low energy cost (Problem 31.18). And third, the tide may flow in or out substantially during the period of time over which a tidal basin can be filled or emptied. In this case the *hydraulic head* is less than the tidal range h during much of the in- or out-flow, and the potential energy available for capture is reduced. Such a situation is illustrated for the case of *ebb tide* generation in Figure 31.27. In this example, generation takes place over half the tidal period T, beginning at $t_1 = T/8$ and ending at $t_2 = 5T/8$, and the energy available is approximately 50% of the energy estimated in eq. (31.35) (Problem 31.19).

There are a number of practical considerations that are relevant for tidal barrage installations. Since the potential energy captured by the barrage scales as h^2, it is clearly

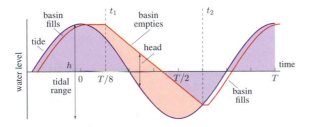

Figure 31.27 The water level of the tide (purple) and the water level in a tidal barrage (red) showing power generation over half of the tidal period. (Credit: Alternative Energy tutorials www.alternative-energy-tutorials.com)

desirable to site such installations in locations with large tidal fluctuations. Typical tidal fluctuations are of order 2–3 m, but in some particular locations, like the Bay of Fundy on the US/Canada border, tidal fluctuations can reach 16 m or more. Tidal barrages share some of the ecological and practical difficulties of dams on freshwater rivers, but also present unique challenges. Salt water is more corrosive than fresh water, and can damage machinery over time. Channels presenting good opportunities for capturing tidal energy are also often heavily used for navigation, for which barrages present an obstacle. Tidal barrages also affect the salinity of the tidal basin, and have issues with sediment, turbidity in the tidal basin, and impact on fish and other sea organisms.

31.3.4 Tidal Stream Energy

Because water is significantly more dense than air, a mass of flowing water carries much more kinetic energy than a comparable volume of air moving at the same velocity. In locations where large tidal basins are regularly filled and emptied by flow through narrow channels, it is possible to capture significant amounts of tidal energy by placing underwater turbines along the channel. The physics underlying such underwater turbines is essentially the same as the physics of a wind turbine at a theoretical level, but there are number of practical differences.

Since the density of ocean water of typical salinity is roughly 800 times that of air, the power density (see eq. (28.3)) $\mathcal{P} = \rho v^3/2$ of water moving at velocity v is roughly 800 times that of air moving at the same velocity. Another way to characterize the difference is that water moving at a velocity v has the same power density as wind moving at a velocity

$$v_{\text{air}} \cong \sqrt[3]{800}\, v \cong 9.3 v . \qquad (31.36)$$

So, for example, a tidal flow at 3 m/s carries a power per unit area of

$$\mathcal{P} = \rho v^3/2 \cong 14\,\text{kW/m}^2 , \qquad (31.37)$$

comparable to a steady wind at 28 m/s (55 knots). This power density is more than an order of magnitude greater than the highest wind power class (7) (see Table 28.2).

In §30.1, we derived the *Betz limit*, which determines the maximum fraction of the energy in a fluid that can be extracted by an idealized mechanical device of fixed area. The same constraint applies to underwater turbines extracting energy from moving liquid,

$$\mathcal{P}_{\text{max}} = \frac{16}{27}\left(\frac{1}{2}\rho v^3\right) = \frac{16}{27}\mathcal{P}_{\text{fluid}} . \qquad (31.38)$$

The high power density of tidal flow can be exploited by placing tidal stream turbines in locations with strong,

Example 31.3   Rance Tidal Barrage

The world's first large-scale tidal power station was the Rance experimental tidal plant in Brittany, France. This tidal barrage consists of a dam of length 750 m with twenty-four 10 MW turbines powered by both flood and ebb tides. The tidal basin has an area of approximately $A = 22.5 \text{ km}^2$, and the average tidal range is 8 m.

The maximum energy that could be extracted by either emptying or filling the lagoon starting from an 8 m height differential would be

$$E = \frac{1}{2}\rho g A h^2 \cong \frac{1}{2}(1025 \text{ kg/m}^3)(9.8 \text{ m/s}^2)(22.5 \text{ km}^2)(8 \text{ m})^2 \cong 7.2 \times 10^{12} \text{ J},$$

where we have used $\rho \cong 1025 \text{ kg/m}^3$ as the average mass density of seawater. Extracting this much energy on each ebb and flood tide would give an average power of

$$P = 2E/12.42 \text{ h} \cong 320 \text{ MW}.$$

Average power output of the Rance plant is 68 MW, or about 20% of this theoretical maximum. Tidal barrages cannot achieve the same overall efficiencies as hydroelectric facilities on dammed rivers, since, for example, in the time it takes for water retained in the basin to flow out through the barrage, the sea level outside the barrage will rise, reducing the hydraulic head.

---

steady tidal stream flow. While the number of favorable locations and total potential for this resource is much more limited than for wind power, in the right location individual tidal stream turbines can produce more power than comparably sized wind turbines. Compared to tidal barrages, underwater turbines may be easier to put in place, and are less disruptive to navigation and local ecosystems, as they do not block the entire channel. There are, however, numerous practical challenges to implementing tidal stream turbines. Installing underwater turbines and power cables is a nontrivial operation. Salt water and organic materials can rapidly degrade the operation of marine turbines, and servicing underwater turbines is more complex logistically than servicing wind turbines in a field.

Beyond logistical considerations, at the level of basic physics, one of the most significant differences between tidal stream turbines and wind turbines is the much greater force of the water on a tidal stream turbine. According to eq. (30.13), for a blade of fixed geometry with given drag and lift coefficients, the pressure (force per unit area) p on the blade is proportional to ρv^2 whereas the power per unit area $\mathcal{P}$ is proportional to ρv^3, thus

$$\frac{p_{\text{water}}}{p_{\text{air}}} = \frac{\mathcal{P}_{\text{water}}}{\mathcal{P}_{\text{air}}} \frac{v_{\text{air}}}{v_{\text{water}}}. \tag{31.39}$$

This makes it more difficult to take full advantage of the increased power density in the water. If, for example,

we compare water and air flows with velocities such that $v_{\text{water}}/v_{\text{air}} \approx 1/6$, then the power density in the water is roughly 3.7 $(800/6^3)$ times as great, but the force per unit area on the turbine blades would be roughly $3.7 \times 6 \approx 22$ times as great. Thus, underwater turbines must be built with extremely strong support towers in order to take advantage of the high power density. (On the other hand, water flow is not subject to the same extreme variations as wind, so there is greater certainty in the range of conditions to which an underwater turbine will be subjected.)

A related difference is that because water typically flows more slowly than wind, tidal stream turbines turn more slowly than wind turbines. The efficiency of underwater turbines therefore is lower than wind turbines (since the rate at which the blades pass through a given point in space is lower, more of the water flows freely between the blades). On the other hand, the slower motion of the blades makes tidal stream turbines less of a hazard for fish than wind turbines are for birds. The fact that some of the water passes through the turbine unaffected reduces the net force on the structure, reducing the engineering demands on the system. Furthermore, the higher viscosity of water and decreased disruption of flow makes it possible to place underwater turbines in closer proximity to one another than wind turbines.

While the horizontal-axis design is conventional, as it is for wind turbines, other designs may eventually prove advantageous. Other designs that have been considered and

Example 31.4   A Tidal Stream Turbine

The first commercial tidal stream turbine was a SeaGen turbine installed in 2008 in Strangford Lough, Ireland. This is an excellent location for tidal stream energy, with tidal flow velocity ranging up to 3.5–4 m/s. The tidal basin has a surface area of roughly 150 km^2. One of the earliest known tidal mills, the Nendrum Monastery mill, which dates to 787 AD, has been excavated on an island in this area.

The **SeaGen turbine** has two rotors, each with a diameter of 16 m, for a total swept area of approximately 400 m^2. The rotors are pitch-controlled, so that the lift/drag ratio can be maximized in a variety of flow conditions, and the blades can be stopped on short notice. The rotors can operate on both ebb and flood tides. The rotors are mounted at a fixed angle on a vertical pile firmly embedded in the sea floor, and the rotors and power units can be lifted out of the water for easy maintenance.

The SeaGen turbine has a capacity of 1.2 MW. According to Marine Current Turbines, the company that produces the SeaGen turbine, the turbine can operate at 48% efficiency (close to the Betz limit of 59%), and can operate at rated capacity with a flow of at least 2.4 m/s. Because the flow rate goes below 2.4 m/s for some time between tides, the average output of the turbine is slightly below 1 MW.

At a flow velocity of 2.4 m/s, the power density of seawater is $\mathcal{P} = \rho v^3/2 \cong 7\,\text{kW/m}^2$. Multiplying by the swept area of 400 m^2 gives a total power of 2.8 MW in the water flowing past the rotors at this velocity. The Betz limit reduces the maximum possible power that can be extracted to $P_\text{max} \cong (16/27) \times 2.8\,\text{MW} \cong 1.7\,\text{MW}$. For the SeaGen to operate at 1.2 MW, the required efficiency is $1.2/2.8 \cong 42\%$, which compares well with the stated efficiency.

(Image credits: Top, Map data ©2017 Google; Bottom, Altantis Resources corporation)

---

built in prototype are shown in Figure 31.28. These include (a) vertical-axis rotors, such as the *Gorlov vertical-axis turbine*, (b) oscillating devices, and (c) shrouded turbines.

The Gorlov helical turbine is a variant of the Darrieus wind turbine design mentioned in §30, and depicted in Example 30.3. An advantage of the vertical-axis turbine is that the turbine can be driven equally well by flow in any horizontal direction. The Gorlov design incorporates a helical shape for the blades, so that the forces on the blades are essentially unchanged through a rotation of 360°. (In the Darrieus design, the force varies periodically through a rotation, which can cause vibrations that are difficult to damp effectively.)

By placing a shroud around a turbine, water flowing across an increased surface area can be channeled through the turbine. Because this increases the amount of water flowing through a fixed area, the flow velocity is increased. This has the potential to substantially increase the power output of a turbine system, and may allow for the use of tidal stream turbines in locations with otherwise marginal

flow rates. The increased power density brings increased force on the turbine, however, so that turbines with shrouds must be emplaced more firmly in the sea floor beneath the tidal channel. Shrouded turbines are currently in an experimental stage of development. Note that to apply the Betz limit when a shroud is present, the full area including the shroud must be considered.

31.3.5 Extent of the Tidal Resource

As described earlier, Earth's tides behave like a driven oscillator in dynamical equilibrium. Energy fed in by torque from the Moon (and Sun) on Earth's tidal bulge is dissipated by the ebb and flow of the tides. The total rate of tidal energy dissipation, which is less than 4 TW, therefore sets a theoretical upper limit on the power that can be harvested for human use. Since only a small fraction of this amount can be practically recovered, clearly tidal power cannot ever provide more than a small fraction of the 15 TW currently used globally by humanity. Even compared to wave energy, tidal energy is much more

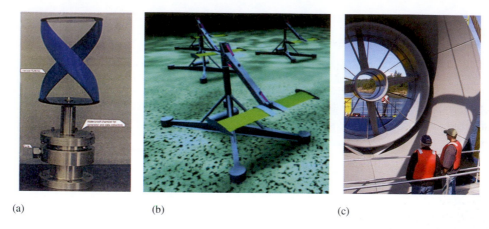

(a)   (b)   (c)

Figure 31.28 Alternatives to standard horizontal-axis rotors for underwater turbines include (a) Gorlov vertical-axis turbine; (b) Stingray oscillating tidal generator; and (c) shrouded turbine. (Credit: (a) https://grabcad.com; (b) Royal IMC; (c) Gary Fletcher reproduced under CC-BY-SA 3.0 license via Wikimedia Commons)

limited practically since much of the energy is lost far from shore. As discussed above, tidal power is a very location-specific resource. In specific spots where a combination of a large tidal range and a restricted channel force large volumes of water into substantial tidal basins, however, the high power density of tidal energy makes it a very attractive resource. In particular, for island nations such as Great Britain or New Zealand, where there is a relatively high ratio of coastline to area, tidal power may eventually play a substantial part in the energy portfolio. Total global potential for tidal power utilization has been estimated to be at most 3% of the total rate of dissipation, i.e., on the order of 500–1000 TWh/y (1.8–3.6 EJ/y), corresponding to an average power of 60–120 GW [210]. In fact, only a fraction of this power would be likely to be developed due to economic, environmental, and other constraints. Since only a tiny fraction of the energy dissipated through tidal friction can be captured, the potential impact of human activity on natural global tidal systems or Earth's rate of rotation would seem to be negligible.

31.4 Marine Current Energy

Just as underwater turbines can be used to capture tidal energy, similar systems could, in principle, capture some of the enormous amount of energy in large-scale oceanic currents. While yet to be demonstrated in practice, in recent years there has been increased interest in this renewable energy resource. Although the total energy available in ocean currents is large, extracting this energy poses additional significant challenges beyond tidal stream energy.

For one, oceanic currents flow in places where the water is much deeper than a typical tidal channel; and for another, the mean flow rate of the large global currents is significantly lower than at optimal tidal stream locations. As discussed in §27, the world's largest current, the *Antarctic Circumpolar Current*, which circulates around the South Pole at a latitude of approximately 60°S, carries a water volume flux of about 135 Sv $= 1.35 \times 10^8$ m^3/s. While this represents a tremendous flow of energy, the average speed of the current is around 1 m/s, and the ocean is quite deep at that latitude, so it is probably impractical to imagine tapping into this energy source.

The *Gulf Stream*, however, may present a more practical opportunity for extracting marine current energy. While the volume flux of this current is only around 30 Sv, in the *Florida Straits* the flow of the current is channeled through a relatively narrow passage between Florida and Cuba (Figure 31.29), where the speed of the flow reaches around 4 m/s. This represents a total power of roughly 250 GW (Problem 31.21). It has been suggested that emplacement of marine current turbines in the Florida Straits could extract some fraction of this energy for human utilization. While the Straits are quite long ($\sim$500 km), note that 250 GW represents an absolute maximum to the power that could be extracted, since even if marine current turbines were placed at many locations along the channel, they would be extracting energy from the same flow. Realistically, it is probably impractical to extract more than a few percent of the total 250 GW, perhaps up to on the order of 10–20 GW [208]. This might represent a substantial contribution to the electric power needs of the state of Florida, for example, but the total global potential for

Tidal Stream and Marine Current Energy

Underwater turbines can extract energy from currents arising from tides or oceanic circulation. Because of the higher density of water, the power density of ocean currents is comparable to that of wind at approximately 10 times the speed of the ocean current.

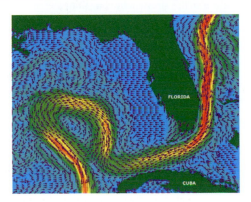

Figure 31.29 The Florida straits, at the beginning of the Gulf Stream, compresses the flow of 30 Sv = 3×10^7 m^3/s through a narrow channel, at a speed of roughly 4 m/s, for a total power of about 250 GW. (Credit: Arthur J. Mariano, University of Miami Rosentiel School of Marine and Atmospheric Science)

practical marine current energy extraction in the near term is probably at most a few times that available in the Gulf Stream. In addition, removing more than a few percent of the total energy from a major global current such as the Gulf Stream would have consequences for global ocean systems, as well as global energy balance and climate.

Discussion/Investigation Questions

31.1 Consider how the circumstances of a hydro turbine differ from those of a wind turbine, and explain why the Betz limit does not apply for hydropower.

31.2 In contrast to traditional hydropower plants where water is stored behind large dams, **run of the river** (ROR) hydro-plants use flowing water with little or no water storage (*pondage*). Investigate and discuss the relative advantages and disadvantages of ROR versus traditional hydropower. Include environmental, social, as well as technical issues.

31.3 Using the results of §31.3.2 discuss the eventual fate of the Earth–Moon system: Is the length of the month increasing or decreasing as a result of tidal forces? How

about the length of the day? In the absence of other effects, what would be the ratio of the length of the day to the length of the month? Find out what experts believe will actually happen.

31.4 Suppose a spherically symmetric body with some angular momentum L begins to contract under its internal gravitational attraction. Assume that the object's angular momentum is conserved during its contraction. Show that the body's rotational kinetic energy must increase. Where does this energy come from? Show that the body's angular velocity of rotation must also increase.

31.5 Explain the following statement made in the discussion of tidal barrage power: "In realistic tidal basins, the surface area of the basin is larger at high tide than at low, so more energy can be extracted from the ebb tide than the flood tide."

Problems

31.1 Compute the pressure at a depth Z below the surface in a reservoir behind a hydroelectric dam. Compute the work done by a volume of water ΔV as it passes from this pressure on one side of a turbine to essentially zero pressure on the other side. Show that this analysis yields the same formula (31.2) for the power output as the energy analysis presented in §31.1.1.

31.2 Compare land use of hydropower to solar power by computing the ratio of power output to reservoir surface area for the Three Gorges Dam (maximum capacity 22 500 MW, reservoir area 1084 km^2) and the Hoover Dam (maximum capacity 2080 MW, reservoir area 680 km^2), and comparing to the power density of a typical large-scale solar thermal or solar photovoltaic power plant in the desert.

31.3 Estimate the power output possible for a hydroelectric dam with hydraulic head $Z = 30$ m, flow rate $Q = 300$ m^3/s, and turbine efficiency 90%.

31.4 [T] Show that the displacement s given by (31.9) satisfies the constraints eqs. (31.4) and (31.5).

31.5 Show that the quantity $l = \sqrt{\sigma/\rho g}$ has the dimensions of length, where the *surface tension* σ has units of energy per unit area. Look up the surface tension of water and evaluate l for water. When a small drop of water is placed on a porcelain surface it forms a round bead, but as the drop gets bigger, it flattens out. Explain what l has to do with this. Does the value of l agree with your experience?

31.6 [T] Use dimensional analysis to determine the dispersion relation $\omega(k)$ for *surface tension waves* (up to a multiplicative constant), which can only depend on the surface tension σ, the density ρ, and the wave number k. Find the phase velocity and the group velocity, and show that $v_g = 3v_p/2$. Describe how you could distinguish visually between gravity waves and surface waves by observing the way that individual wave crests move through a wave packet.

31.7 Consider a deep-water wave with amplitude $a = 1$ m and period 12 s. Compute the energy density and power. When the wave moves into shallow water, what is the height and energy density of the wave at 2 m and at 1 m depth?

31.8 [T] Show that the function

$$\psi(\boldsymbol{x}, t) = \frac{a}{k \sinh kh} \cosh(k(z+h)) \sin(kx - \omega t)$$

describes a (plane) gravity wave propagating in the $\hat{\boldsymbol{x}}$-direction in a uniform ocean of depth h. In particular, show that $\psi(\boldsymbol{x}, t)$ satisfies eq. (31.7) and the boundary condition that $s_z = 0$ at $z = -h$. (See Appendix B.4.2 for properties of the hyperbolic functions $\cosh x$ and $\sinh x$, and in particular, for their expansions near $x = 0$.) Show that for the limiting cases $kh \ll 1$ and $kh \gg 1$ this correctly reproduces eq. (31.22) and eq. (31.9). For waves of fixed period, graph H/H_0, λ/λ_0 as functions of depth, where H_0, λ_0 are the wave height and wavelength in the limit of infinite depth.

31.9 [T] The function

$$\psi(\boldsymbol{x}, t) = \frac{a}{k \sinh kh} \cosh(k(z+h)) \cos kx \cos \omega t$$

is a standing wave solution analogous to eq. (31.10) for a surface gravity in an ocean of depth h. Following the derivation in the text, construct the kinetic and potential energy per unit area and find the dispersion relation for arbitrary h. Show that your solution has the correct limit for deep and shallow water. (Answer: $\omega = \sqrt{gk \tanh(hk)}$, where $\tanh x = \sinh x / \cosh x$.) Show that the power density in this wave is the same as in a deep-water wave, eq. (31.20).

31.10 A *tsunami* is a very long-wavelength wave produced by a sudden movement in the sea floor over a distance of hundreds of kilometers. The wavelengths in a tsunami are generally much greater than ocean depth so that a tsunami can be described accurately in the shallow wave approximation, and is non-dispersive. Consider a tsunami described by a wave with approximate wavelength 100 km and amplitude 1 m. Compute the velocity and period of this wave and the power density per unit length of shoreline perpendicular to the incident wave. (Note, you may approximate ocean depth as $h \sim 3.5$ km. Also you will need the power density in the shallow wave approximation, see Problem 31.9.)

31.11 Derive the statements made about the power density of 1 m and 10 m deep ocean swells in §31.2.4.

31.12 In the JONSWAP model, the energy transferred to the moving ocean surface waters is proportional to the distance over which wind travels (fetch, F). For a steady wind at 10 m/s compute the rate of energy transfer per unit area to the ocean's surface. What fraction of the energy of the wind in the 100 meters over the ocean's surface is transferred to the ocean with a fetch of 200 km?

31.13 [H] Imagine that Earth rotated about its axis once per month and that the Moon moved in a circular orbit above Earth's equator. Under these *tidally locked* conditions, the Moon and the resulting tidal bulge in Earth's oceans would remain fixed in an Earth-based reference system. Give an estimate of the tidal displacement Δ in this situation by computing the gravitational potential along a line passing through the centers of the Moon and Earth and comparing to the potential at a second point also on Earth's equator but perpendicular to this line. [Hint: Expand the potentials in R/r and Δ/R (in the notation of eq. (31.25)) and discard the term linear in R that is associated with the Moon's gravitational force acting on the full Earth.]

31.14 Show that energy loss from Earth's rotational kinetic energy at a rate of 3.9 TW (including dissipation and energy transferred to lunar orbit) corresponds to an increase in length of day of 2.3 ms/century. You may take Earth's moment of inertia about its rotational axis to be $I = 8.0 \times 10^{37}$ kg m^2.

31.15 Is tidal power a renewable resource? Assuming the rate of energy loss due to the tides remains constant at 3.9 TW, estimate the time it will take for Earth to lose 10% of its rotational energy to tidal losses. If humans were to extract additional tidal energy at the rate of 100 MW in such a way that friction increases in the system and this additional energy was also lost by Earth, how much sooner would Earth lose 10% of its rotational energy? (See Problem 31.14 for Earth's moment of inertia.)

31.16 Consider building a *tidal barrage* along a straight section of beach in California. Assume that the average tidal range is 3 m, with two high tides a day. Assume that the tidal basin is 20 m wide and extends for 1 km along the coast. Assume that all available energy from both incoming and outgoing tides is extracted. Compute the average power output of this installation. Do you think this installation would be worth building?

31.17 A basin under consideration for a tidal barrage power installation has an area that grows quadratically with height z, $A(z) = A_0 z^2 / h^2$, where h is the difference between high and low tide. Calculate the maximum power that could be extracted (a) if the full basin is emptied at low tide, and (b) if the empty basin is filled at high tide.

31.18 Consider a tidal barrage constructed to enclose a lagoon of surface area 5 km^2, with typical tidal range of 4 m. What is the maximum possible average power output of such a barrage? How much would the net output be increased if an additional 1 m of water were pumped in at every high tide and pumped out at every low tide? (Remember to subtract energy used for pumping.)

31.19 When a tidal barrage follows the water release profile shown in Figure 31.27, it only captures a fraction of the energy (31.35) available in the water stored at high tide. Note that for simplicity the discharge is indicated continuing until the outside and inside water levels are equal. Show that this fraction is given by

$$f(t_1, t_2) = \frac{1}{4}\left(\sin^2 \omega t_2 - 2(1 - \cos \omega t_2)\frac{(\sin \omega t_2 - \sin \omega t_1)}{\omega(t_2 - t_1)}\right),$$

where $\omega = 2\pi/T$ and $T \cong 12\,\text{h}$. For the parameters shown in the figure, where $\omega t_1 = \pi/4$ and $\omega t_2 = 5\pi/4$, show that $f \cong 0.5$.

31.20 Consider an idealized tidal reservoir of surface area $A = 30\,\text{km}^2$ that fills and drains through a narrow channel of width $w = 60\,\text{m}$ and depth $d = 20\,\text{m}$. Assume that the tidal range is $h = 2.5\,\text{m}$. For simplicity, assume that the water flows uniformly throughout the channel and varies sinusoidally in time, with a 12 hour period, at a velocity $v(t) = v_0 \sin \omega t$, where $\omega = 4\pi/24$ hours. Compute the maximum speed v_0 needed for the reservoir to drain and fill completely on each tidal cycle and the power density of the water passing through the channel. Compute the total tidal stream energy passing through the channel daily. Consider installing four tidal stream turbines in the channel. Assume that they operate at 20% of the Betz limit, and that each has two rotors of radius 8 m. Estimate the average power output of the turbines. Compare this result with the energy extracted by a tidal barrage that utilizes both the ebb and flood tide with 50% efficiency.

31.21 Estimate the total power of the Gulf Stream in the Florida Straits, given that the mass flux is roughly 30 Sv. Take the speed of the flow to be 4 m/s.

Geothermal Energy

The energy in sunlight, wind, and water all originates from the Sun. Solar energy does not, however, penetrate far into Earth's interior. From §5, we know that surface temperature variations over the seasons only affect ground temperatures to a depth of a few meters. As we discuss further in §34, incident solar energy is essentially all reradiated to space from Earth's surface and atmosphere, maintaining a net energy balance between incoming and outgoing radiation.

Beneath Earth's surface, however, an entirely different energy system is at work. Descending beneath the surface, the temperature rises steadily, until at around 100 km depth it is sufficient to melt rock. Within the planet's interior, large-scale dynamics of molten rock and metal transport enormous quantities of thermal energy and produce motion of the continental plates on Earth's surface. Some fraction of this thermal energy is continuously conducted to the surface, while some drives earthquakes and volcanic activity. At the same time, the thermal energy within the planet is continuously replenished by the release of energy through decay of radioactive isotopes.

Thermal energy released to Earth's surface through hot springs and other geothermal features has been used for warmth in many cultures for hundreds or thousands of years. In Iceland, which has a rich endowment of surface geothermal features, almost 90% of home heating is provided by geothermal energy. In the last century, geothermal power plants have produced a steadily growing supply of energy to electrical grids around the world. In principle, the potential of geothermal power generation is tremendous – millions of exajoules of thermal energy are stored in the crystalline basement rock beneath the US alone, enough to supply world energy needs for millennia. Extracting this energy, however, poses difficult technical challenges. In this chapter we describe the basic geophysics underlying geothermal energy, and some of the ways in which humans

can currently, and may in the future, harvest this energy resource.

The basic geophysics relevant for understanding thermal energy in the Earth is described in §32.1. The various kinds of geothermal energy systems currently in use or envisioned for the future are described in §32.2–32.5, including ground source heat pumps, which are actually a form of solar energy. An overview of available geothermal resources is given in §32.6.

Further background on Earth physics can be found in modern geophysics texts such as those by Lowrie [211] and Stacey and Davis [212]. The latter text, in particular, presents a detailed description of the current understanding of the structure and dynamics of Earth's interior and plate tectonics. A general treatment of geothermal energy is given in the book by Gupta and Roy [213], and an insightful and extensive analysis of geothermal power plants, with description of many particular installations, is given in the book by DiPippo [214]. A 2006 report by a group at MIT [215] gives a detailed assessment of the future of geothermal energy, with particular focus on *enhanced geothermal systems*.

Reader's Guide

This chapter describes the origin, distribution, and utilization of geothermal energy. Geothermal resources described include conventional hydrothermal resources, ground source heat pumps (really a solar energy resource), low-enthalpy geothermal resources, and potential future resources such as hot dry rock (enhanced geothermal systems).

Prerequisites: §5 (Thermal energy), §6 (Heat transfer), §10 (Heat engines), §12 (Phase-change energy conversion), §13 (Power cycles), §17 (Nuclear structure).

The material developed in this chapter forms a basis for the further study of Earth systems relevant to fossil fuels (§33) and Earth's climate (§34, §35).

Figure 32.1 Left: Iceland's *Blue Lagoon*, where tourists and geothermal energy mix; Right: A *fumarole* on the side of an Alaskan volcano, where superheated steam and other gases vent to the atmosphere. (Credit (Right Image): Fumaroles Fourpeaked Volcano, by Cynus Read, Courtesy of the US Geological Survey)

32.1 Thermal Energy in Earth's Interior

To understand how to tap into the tremendous store of thermal energy in Earth's interior, it is helpful to know something about the internal structure of the planet. In this section we give a brief introduction to some basic aspects of geophysics. §32.1.1 presents a simplified picture of Earth's internal structure. The sources of thermal energy in Earth's interior are described in §32.1.2 and §32.1.3. Some aspects of the dynamics of Earth's interior, plate tectonics, and the processes that bring energy from Earth's interior to the surface are described in §32.1.4.

32.1.1 Earth's Internal Structure

Seen from its surface or from space, Earth might appear to be an enormous mass of solid rock, covered in places with a thin veneer of liquid water, ice, or organic material. In fact, however, Earth's crust is only a thin layer of solid material surrounding a dynamic and complex hot interior. Despite its proximity, the structure and dynamics of Earth's interior remain only partially understood. The material composition of the planet, as well as temperature and pressure, vary with depth beneath the surface. Direct probes have only been able to measure to a depth of a dozen or so kilometers beneath the surface – a small fraction of a percent of the 6370 km to the center of the planet. The **seismic waves** of propagating elastic deformation created by earthquakes provide a powerful tool for studying the structure of the planet in regions too deep for direct observation. The velocity of seismic waves depends on the density, temperature, and rigidity of the material through which they propagate. Like light and water waves (§4, §31.2), seismic waves that pass through media with varying propagation velocities refract, and can also reflect

from sharp boundaries between media. By measuring the arrival times of seismic waves that pass through the planet, the speed of wave propagation through different parts of the interior can be inferred. Combined with empirical and theoretical understanding of how the behavior of different materials changes as a function of pressure and temperature, seismological studies have led to an approximate picture of the structure of Earth's interior. Artificially generated seismic waves play an important role in exploration for fossil fuels, and they are described in more detail in the following chapter (see Box 33.1).

Earth's inner structure, as it is currently understood, is depicted in Figure 32.2. The main layers of structure are commonly divided into the **crust**, **mantle**, and **core**, with

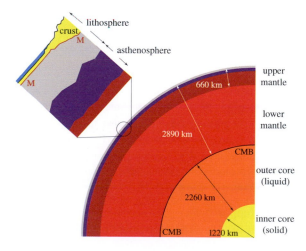

Figure 32.2 Earth's interior, with the outer ≈ 200 km enlarged in the inset. M marks the *Mohorovičić Discontinuity* and CMB marks the *core–mantle boundary*.

finer divisions into layers, primarily defined by seismic discontinuities, as follows:

Crust The solid outer surface of the planet is known as the *crust*. *Continental crust* can vary in thickness up to 60 km or more beneath massive mountain ranges, with an average thickness of 35 km. Very roughly speaking, continental crust near the surface consists largely of **granite**, a relatively light, crystalline rock with a large proportion of silicon and oxygen. Granite is an **igneous rock**, formed from the gradual cooling of underground magma. Surface rocks also include sediments, volcanics, and **metamorphic rocks**, which have been altered by heat and/or pressure. *Oceanic crust* is much thinner, generally 6–8 km thick. Oceanic crust is dominated by **basalt**, a heavier silicate with more iron and magnesium than granite.[1] Continental crust also contains basalt at greater depths; the lower part of most continental crust is believed to be primarily composed of basalt. The bottom of the crust is marked by the **Moho**, or **Mohorovičić discontinuity** (after the Croatian seismologist Andrija Mohorovičić), below which the velocity of seismic *P-waves* (see Box 33.1) suddenly increases from roughly 5.5 km/s to roughly 8 km/s. The Moho discontinuity is caused by a combination of a change of composition and change of crystalline phase of the material involved. The relative thickness of Earth's solid crust is sometimes compared to the shell of an egg, though the crust is several times thinner in proportion to the diameter of Earth than an eggshell is to the diameter of the egg.

Lithosphere The uppermost portion of the mantle (sometimes called the **lid**) is rigid, associated with high seismic wave velocities. The term **lithosphere** refers to the region containing both the crust and this rigid outer mantle layer that supports the crust. The lithosphere extends to a depth of between 50 and 200 km.

Asthenosphere Below the lithosphere, the mantle becomes more plastic, and seismic wave velocities decrease. The portion of the mantle down to approximately 220 km is called the **asthenosphere**. The asthenosphere resembles a viscous fluid that allows slow convection and decouples the motion of the lithosphere from the dynamics of Earth's interior regions.

Mantle (upper/lower) The mantle continues to a depth of roughly 2890 km. The **upper mantle**, including the lid,

asthenosphere, and a transition zone (which is sometimes included as part of the asthenosphere) extends down to roughly 660 km. The upper mantle is more rocky than metallic, and is primarily composed of iron and magnesium silicates. As depth and pressure increase, minerals in the mantle undergo phase transitions into higher-density crystalline forms. These transitions are marked by boundaries where seismic wave velocities change, which makes them visible to seismic imaging. At the bottom of the mantle, just above the **core–mantle boundary** (CMB), is a boundary layer (referred to as the D″ layer) that is believed to play an important role in transporting heat outward from the core and determining mantle dynamics. The strongest seismic discontinuity inside Earth occurs at the CMB. This is called the **Gutenberg seismic discontinuity** after the German physicist Beno Gutenberg who in 1914 used seismic data to locate the depth of the core–mantle boundary to within 10 km of the currently estimated value.

Core (outer/inner) Earth's core has a radius of roughly 3480 km. The core is believed to be composed primarily of iron, with some 10% nickel and small amounts of other trace elements. The inner core, with a radius of roughly 1220 km, is solid, and is surrounded by the liquid outer core. The density of the core is much higher than that of the mantle. Density is estimated at $5500 \, \mathrm{kg/m^3}$ just above the core–mantle boundary, around $9900 \, \mathrm{kg/m^3}$ just below the boundary, and around $13\,000 \, \mathrm{kg/m^3}$ near Earth's center. It is believed that this density is somewhat lower than would be achieved by a pure iron core, so that some additional element such as silicon or sulfur should also be present in the core in alloy form. The early conditions during which Earth was formed are not sufficiently well enough understood to determine the composition with certainty.

As depth beneath the surface increases, both density and pressure increase. Earth's average density is given by

$$\rho_\oplus \cong M_\oplus/(4/3 \, \pi \, R_\oplus^3) \cong 5.975 \times 10^{24} \, \mathrm{kg}/1.08 \times 10^{12} \, \mathrm{m^3}$$
$$\cong 5500 \, \mathrm{kg/m^3} \, . \tag{32.1}$$

Estimates for density and pressure as functions of depth based on a simple Earth model (PREM, or *Preliminary Reference Earth Model* [216]) are shown in Figure 32.3. The density distribution of material inside the planet has a significant effect on the variation of gravitational force with depth. For a solid sphere of uniform mass density and radius R, the gravitational acceleration at a point inside the sphere at radius r from the center is given by $g_{\mathrm{uniform}}(r) = g_0 r/R$, where g_0 is the gravitational acceleration on the surface. This follows simply from the $1/r^2$ variation of Newton's gravitational force law, combined with the fraction of total mass $(r/R)^3$ inside a sphere of radius r.

---

[1] **Silicates** are chemical compounds that include silicon as a negative ion, usually in the form $(SiO_4)^{4-}$. A silicate common in basalt is *olivine*: $(Mg,Fe)_2SiO_4$.

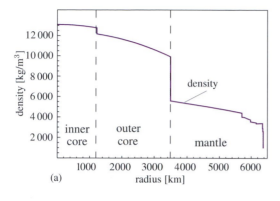

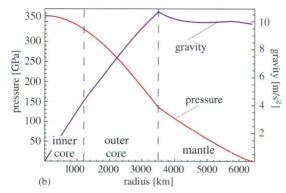

Figure 32.3 (a) Density and (b) pressure and the acceleration due to gravity as functions of distance from Earth's center in a simplified model of Earth's interior. After [216].

Thus, in a sphere of uniform mass, the gravitational force decreases linearly with decreasing r. Within Earth's interior, in the upper mantle the density is so much smaller than the core density that the $1/r^2$ term dominates, however, and the gravitational force actually *increases* with increasing depth, decreasing only within the radius of the core (Problem 32.1). The radial dependence of gravitational acceleration is depicted in Figure 32.3(b).

32.1.2 Sources of Thermal Energy

Along with density and pressure, temperature also increases with depth beneath Earth's surface, rising rapidly through the lithosphere and more slowly through the mantle and towards the center of the core. This thermal gradient is associated with a constant flow of heat out to Earth's surface. The relatively steep temperature gradient in the lithosphere arises from the low thermal conductivity of solid rock. Within the mantle and fluid outer core, convection plays an important role in moving thermal energy outward, so that a similar rate of energy flow outward can

be sustained by a weaker thermal gradient than that of the lithosphere.

Averaged over Earth's surface, geothermal energy flows outward at a rate of

$$\bar{q}_{\text{geo}} \cong 87 \, \text{mW/m}^2 . \tag{32.2}$$

While this is many orders of magnitude smaller than the average solar energy input, integrating over Earth's surface area of roughly $500 \times 10^6 \, \text{km}^2$ gives a total power of roughly 44 TW.

Where does this thermal energy come from? One substantial source for Earth's internal energy is radioactivity. Measurements of radioactive isotope densities in near-surface rocks, combined with other considerations, suggest that in continental crust, roughly 60% of the thermal energy flowing outward has its source in radioactive decays in the crust. The density of radioactive elements is much lower, however, deeper in the continental crust and in oceanic crust.

The source of the fraction of geothermal energy flux that does not originate in crustal radioactive decay is not entirely understood. A complete picture would require a better understanding of the conditions present at the initial formation of the planet and more information about the present distribution and properties of material in Earth's interior. Laboratory experiments on iron and minerals at high temperatures and pressures, coupled with seismic and other empirical studies have, however, progressed to the point that a reasonably coherent picture of the approximate interior configuration and temperature profile of the planet can be drawn.

It is generally agreed that most of the geothermal energy flux through the surface that does not come from radioactive elements in the crust and mantle is associated with the gradual cooling of the planet's interior. At the time of Earth's formation, as a mass of dust was pulled together through gravitational attraction, the gravitational potential energy released as the material contracted heated the growing planet, presumably raising its temperature until the material became a molten mass. Heavier elements such as iron gravitated to the core, releasing more energy. After some hundreds of millions of years, the surface of the newly formed planet cooled sufficiently to form a crust, which substantially slowed the rate of release of internal energy. This storehouse of initial energy is still largely present within the planet, maintaining the high temperatures that fuel the internal dynamics of the core and mantle.

It may seem surprising that thermal energy could be stored within the planet for many billions of years, but the relatively small thermal gradients and associated heat flow in the crust give a very long time scale for cooling of the interior. This is similar to the slow rate of downward heat flux between the ocean surface layer and the deep ocean estimated in eq. (27.1). The total thermal energy content of Earth is currently estimated at 12.6×10^{30} J (Problem 32.2). At a rate of outward energy flow of 44 TW, it would take on the order of ten billion years for this thermal energy to dissipate. Given that a substantial fraction of the outward energy flow is replaced by radioactive decay, the time scale for cooling of the planet is clearly longer than the time since Earth's formation.

In addition to the energy coming from radioactive decay and residual thermal energy from the initial formation of the planet, it is likely that further thermal energy is continuously added in the interior as further stratification, thermal contractions, and phase changes occur in the core and mantle. The precise details of these processes are the subject of current research.

Direct borehole measurements of temperatures at depths of up to several kilometers beneath the surface, as well as empirical data on thermal conductivity of rock, show that typical temperature gradients in continental crust range between roughly 10 °C/km and 50 °C/km. *Thermal conductivity* in the crust is typically in the range 2–4.5 W/mK. For example, granite at surface temperature and pressure has a thermal conductivity of

$$k_{\text{granite}} \cong 2.2 \, \text{W/mK}. \qquad (32.3)$$

From this we see that a temperature difference of 30 °C over a 1 km thickness of granite would give a heat flow of $q = k\Delta T/\Delta z \cong 66 \, \text{mW/m}^2$. Since granite also contains radioactive elements, the added heat from nuclear decays within the granite must be added in a more realistic calculation (see Example 32.1).

32.1.3 Radioactive Decay in the Crust

The four isotopes that contribute most substantially to radioactive heat production in the crust are long-lived isotopes of uranium (^{238}U, ^{235}U), thorium (^{232}Th), and potassium (^{40}K). The half-lives, primary decay channels, and average (near-surface) crustal abundances $\bar{C}$ of these isotopes are given in Table 20.6. Note that potassium is quite abundant, composing some 2% of the mass of typical crustal rock, but the radioactive isotope ^{40}K is rare. The abundance of the radioactive elements is in general highly variable, depending not only on the type of rock but also upon specific location. Granite, for example, has particularly high concentrations of all these radioactive elements. The concentration of radioactive isotopes diminishes sharply below the upper crust. It is often convenient to express the net radioactive heat production in

rock $\dot{Q}$ (in units of power/mass) as a function of the mass concentration ratios C_{element} through (Problem 32.3)

$$\dot{Q} \cong 98 \, C_{\text{U}} + 26.5 \, C_{\text{Th}} + 0.0035 \, C_{\text{K}} \quad [\mu\text{W/kg}]. \quad (32.4)$$

The flow of thermal energy through rock containing radioactive isotopes is described by an extension of the usual heat equation in which a source is explicitly included. In an idealized situation where the conducting material is homogeneous in the horizontal directions and has a thermal conductivity $k(z)$ depending only on the coordinate z in the direction of heat flow, a slight modification of the argument in §6.2.4 gives the heat flow equation

$$\frac{d}{dz}\left(k(z)\frac{dT(z)}{dz}\right) = -\dot{Q}(z)\rho(z), \qquad (32.5)$$

where $\dot{Q}(z)\rho(z)$ is the rate of thermal energy added through radioactive decay, expressed in units of power/volume.

As mentioned above, in continental crust, roughly 60% of the thermal flux reaching the surface comes from radioactive decay in the lithosphere, with most of the radioactive material residing in the upper crust. The more recently formed oceanic crust contains much less radioactive material; most of the thermal flux from beneath the ocean comes from cooling of the lithosphere, with only some 10% of the heat flow coming from beneath the lithosphere and 5% from radioactivity.

Of the total of 44 TW of geothermal energy passing out to the surface, current estimations indicate that roughly 12 TW comes from the decay of ^{238}U and another 13 TW from the decay of ^{232}Th. (The contribution from decay of ^{235}U is negligible.) The contribution from ^{40}K in the crust is substantially smaller, on the order of 3 TW. While the contribution from potassium in the lithosphere and mantle is relatively small, it has been hypothesized that the core may contain a substantial quantity of potassium in the form of a potassium-iron alloy, which would contribute to geothermal flux and could impact core dynamo dynamics, though recent work [217] suggests that this quantity is actually fairly small. The estimates for radioactive fractions in standard Earth models are corroborated through neutrino measurements by particle physics experiments on Earth's surface, and by comparison to fractions of radioactive material in **chondrite meteorites** that are believed to come from asteroids in the early solar system and to have element fractions representative of the material forming the solar system as a whole.

32.1.4 Dynamical Processes in Earth's Interior

Extracting energy in usable quantities from the average rate of thermal energy flow at the surface, or from the

---

Example 32.1   Flow of Heat Through Granite

Consider as an example a 1 km thick slab of granite, assuming for simplicity that the thermal conductivity $k(z) = 2.2\,\mathrm{W/mK}$ and mass density $\rho = 2.7 \times 10^3\,\mathrm{kg/m^3}$ are uniform throughout the slab. Assume that the temperature on the bottom of the slab is 330 K and the temperature at the top is 300 K. If the concentrations of uranium, thorium, and potassium are given by typical values for granite (ppmm = parts per million by mass)

$$C_U = 4.6\,\mathrm{ppmm}, \qquad C_{Th} = 14\,\mathrm{ppmm}, \qquad C_K = 33\,000\,\mathrm{ppmm},$$

then *what is the rate of heat flow at the top and bottom of the slab?*
The rate of heat production from radioactivity is, from eq. (32.4),

$$\dot{Q} = 0.94 \times 10^{-9}\,\mathrm{W/kg}.$$

The heat flow equation (32.5) then becomes

$$k\,d^2T(z)/dz^2 = -\rho\dot{Q},$$

so

$$d^2T(z)/dz^2 \cong -\rho\dot{Q}/k \cong -1.15 \times 10^{-6}\,\mathrm{K/m^2}.$$

The solution to this equation is (taking $z = 0$ at the surface, z negative below the surface, and measuring z in units of kilometers)

$$T(z) = T_0 + T_1 z - (\rho\dot{Q}/2k)z^2 \cong 300 - 30.6z - 0.58z^2,$$

where we have solved for T_0, T_1 using the boundary conditions at $z = 0$ and $z = -1$. The rate of heat flow on the top and bottom surfaces is

$$-k\,dT/dz|_{z=0} = 67.3\,\mathrm{mW/m^2}, \qquad -k\,dT/dz|_{z=-1\,\mathrm{km}} = 64.8\,\mathrm{mW/m^2}.$$

The difference between these rates of heat flow is $2.5\,\mathrm{mW/m^2}$, equal to the heat added from radioactive decay

$$\dot{Q}\rho\Delta z = (0.94 \times 10^{-9}\,\mathrm{W/kg})(2.7 \times 10^3\,\mathrm{kg/m^3})(10^3\,\mathrm{m}) \cong 2.5\,\mathrm{mW/m^2}.$$

Geothermal Heat Flux

The average global rate of geothermal heat flux is

$$\bar{q}_{\mathrm{geo}} \cong 87\,\mathrm{mW/m^2}.$$

Some of this is energy from radioactive decay of elements in Earth's crust, while some is residual thermal energy in the planet's interior from gravitational energy released during Earth's formation. The rate of heat flow is higher through oceanic crust than through continental crust and highest through freshly formed crust, and near plate margins and *hotspots*.

distributed release of energy through radioactive decay is clearly impractical. Exploiting geothermal energy for human use requires identifying locations where high-temperature reservoirs of thermal energy are accessible to existing technologies. The distribution of thermal energy beneath Earth's surface is far from uniform, so to understand where and how geothermal energy can be extracted we need to know something more about how processes within Earth affect the local distribution of geothermal energy.

Both conduction and convection play important roles in the transport of energy within the planet. In the solid lithosphere, conduction is the dominant means of outward heat transport. In the mantle, though motion is slow on human time scales, convective processes are nevertheless most effective in transferring heat outward. In the fluid outer core, convection dominates, though conduction is also important since the metallic outer core is highly electrically and thermally conducting. We give here a brief summary of the role of dynamics within the core and the mantle, and how these dynamics affect the lithosphere, temperature distribution, and magnetic field of the planet.

Core Dynamics and Earth's Magnetic Dipole Field

Understanding the dynamics of Earth's fluid outer core represents a significant challenge for geophysicists. In particular, while it has been believed for decades that the dynamics of the core is responsible for generating Earth's magnetic field, the details of this process are still poorly understood. We outline some of the basic features of Earth's magnetic field here; a much more detailed description, along with a summary of the current understanding of the source of the geomagnetic field, can be found in [218].

The precise distribution of the magnetic field is rather complicated and time-dependent, but the dominant contribution is from a magnetic dipole field (Box 3.2). The strength and orientation of this magnetic dipole change with time. The dipole moment at present is approximately 8×10^{22} A m^2, with a tilt of about $11°$ relative to Earth's axis of rotation. Geological evidence, which we discuss further below, shows that Earth's magnetic field occasionally flips its direction of polarity. (A flip of polarity from the current configuration would mean, for example, that all compasses would point south instead of north.) Over the last 80 million years such polarity reversals have occurred on average several times every million years, with a frequency that has increased since a preceding period of almost 40 million years in which polarity flips apparently did not occur (the **Cretaceous superchron**). Polarity reversals occur relatively quickly, and with no obvious regularity.

The magnetic dipole of Earth arises primarily from electric currents circulating through its liquid outer core. The material in the outer core is mostly iron, which conducts well at the high temperatures and pressures achieved thousands of kilometers beneath the surface. From the estimated resistivity of this material, the currents that produce Earth's magnetic field would dissipate over less than 20 000 years, so they must be continually generated by a self-exciting **dynamo** mechanism. The basic idea of the dynamo is similar to the electromagnetic generator (§3), though in the dynamo the current produced by motion of charges through a magnetic field follows a trajectory that feeds back into the magnetic field. A simple model of a rotating disk dynamo is shown and described in Figure 32.4. Earth's dynamo is, however, significantly more complex than this simple model. In particular, the dynamics of the core (like the observed magnetic field) does not have an exact axial symmetry, and must operate in a regime where the observed polarity flips are possible. The basic idea is that internal torques within the core combined with the Coriolis effect drive the dynamo. Even the source of the energy driving the dynamo is not generally agreed upon, however. The power driving the dynamo is estimated at between 0.1 and 1 TW. Energy of this order of magnitude is in principle available from several sources: tidal friction in the

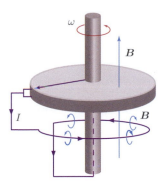

Figure 32.4 A very simple dynamo based on a rotating, conducting disk. An external magnetic field initially runs parallel to the axis of the disk. A torque acting on the disk causes it to rotate. The rotation of the disk in the magnetic field produces a Lorentz force that generates an outward-directed current in the disk. The current is conducted to wires that carry it back to the central axis after following a circular path. This circulating current enhances the magnetic field. Even after the original magnetic field is removed, this results in a self-sustaining system in which energy input through the external torque is converted to resistive losses as the current flows through the dynamo.

core, thermally driven convection, and convection driven by compositional changes. While for some time it was believed that the dynamo is driven by thermal convection, current understanding suggests that the primary driver is energy released through compositional convection – such as energy released when iron from the outer core solidifies and increases the size of the inner core. Irrespective of the energy source, the detailed dynamics of the dynamo process in the core are rather poorly understood. The *magnetohydrodynamic equations* describing electromagnetic fields in a moving fluid become quite complicated when applied to a rotating system in a gravitational field with an inhomogeneous temperature distribution, and have solutions of a somewhat chaotic nature. Analytic models and numerical simulations are not yet able to reproduce the observed time-dependence of Earth's magnetic field even in a probabilistic sense, although simulations can produce regimes in which polarity reversal occurs with varying frequencies.

Over the last 500 years Earth's *dipole moment* has decreased by some 15%, with the rate of decrease accelerating over the last century. Our understanding of core dynamics at this time is insufficient to know whether this indicates an impending polarity flip over the next few thousand years, or just a random fluctuation in the magnitude of the dipole.

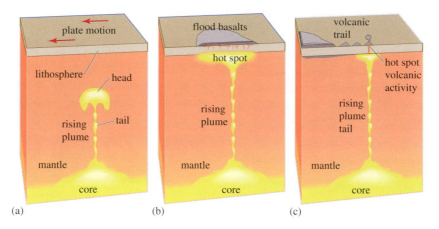

Figure 32.5 Stages of mantle plume development. (a) A plume rises from the D'' layer just above the core–mantle boundary; (b) the plume generates volcanism at Earth's surface; (c) the plume forms a *hot spot* that can generate a chain of islands in oceanic crust. (Credit: Dennis Tasa, Tasa Graphic Arts, Inc.)

Earth's Magnetic Field

Earth's magnetic field is dominated by a dipole field, currently tilted at 11° to Earth's rotation axis. The magnetic field is believed to be produced by dynamic currents in the outer core through a complex dynamo action. The magnitude and direction of the field change dramatically over geological time scales.

Heat Transport Within the Mantle Although the viscosity of the mantle is much greater than that of the outer core, its thermal conductivity is also much lower. Convection within the mantle occurs over long time periods, and is believed to dominate the transfer of energy from the core–mantle boundary to the lithosphere. From the material properties and approximate temperature gradient it is believed that convective flow in the mantle gives rise to motion of the mantle material at velocities on the order of 5–10 cm/y. While slow from the point of view of the human time scale, mantle convection plays an important role in driving the motion of the crustal plates. The details of mantle convection are only partially understood. Heat added from radioactive decay is distributed throughout the mantle. Thus, unlike a boiling pot where convection is driven by heating from below, mantle convection is largely driven by surface cooling. This is discussed further below in the context of *plate subduction*. In addition, there are phase boundaries associated with changes in crystal structure in the mantle at depths of around 410 km and 670 km, which also affect the structure of mantle convection.

Mantle plumes are thought to play an important role in transporting thermal energy from the core to the lithosphere (see Figure 32.5). Mantle plumes are highly localized streams of mantle material that are believed to flow from the core–mantle boundary to the bottom of the lithosphere and to be driven by buoyancy effects, as is the Rayleigh–Taylor instability described in §19.6.4. As mentioned above, the D'' layer at the bottom of the mantle plays an important role in determining mantle dynamics. Reconstructions of three-dimensional images of Earth's interior from information obtained by the scattering of seismic waves, a process known as **seismic tomography**, indicate that this layer is highly irregular, and may play a boundary role similar to the crust. Thick regions within the D'' layer are referred to as **crypto-continents** and thin layers as **crypto-oceans**. Mantle plumes are believed to originate from areas within crypto-oceans, and transport heat relatively quickly via convection to the bottom of the lithosphere, where they form **hotspots** often associated with volcanic activity on the surface.

Plate Tectonics Earth's lithosphere is broken up into large connected regions called **tectonic plates** (see Figure 32.6). There are major plates associated with many continent-sized land masses. Plate sizes range from ~1000 to 10 000 km. The plates move somewhat independently over the asthenosphere. The rate of relative motion of adjacent plates varies from 2 to 16 cm/y. This motion has in recent years been confirmed and extensively measured using satellite global positioning systems (GPS). Most seismic activity at Earth's surface occurs at plate boundaries. There are several different types of plate

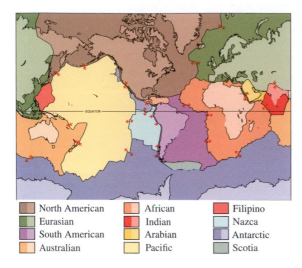

North American	African	Filipino
Eurasian	Indian	Nazca
South American	Arabian	Antarctic
Australian	Pacific	Scotia

Figure 32.6 Lithospheric plates move over the asthenosphere, with relative velocities at plate boundaries ranging from 2 to 16 cm/yr. (Credit: http://pubs.usgs.gov/gip/dynamics/slabs.html, Courtesy of the US Geological Survey)

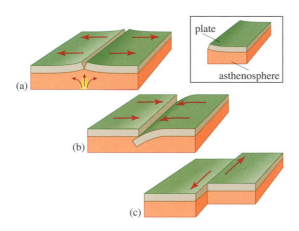

Figure 32.7 Different kinds of plate boundaries: (a) divergent, showing new crust created from rising magma; (b) convergent, showing the oceanic plate subducted beneath the continental plate; and (c) transform.

boundaries, distinguished by the relative direction of motion of the plates (see Figure 32.7).

Constructive (divergent) plate margins At divergent margins, two plates are pushed apart from one another, and new crust is formed between the plates. Divergent margins within continental land masses give rise to rift valleys, like the East Africa Rift, that can over time grow and deepen to become long narrow lakes or shallow seas. At undersea plate boundaries where two plates are moving apart

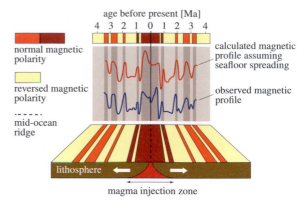

Figure 32.8 Reversals of Earth's magnetic field are captured in the magnetization of new sea floor created as magma is injected along ocean ridges. (Credit: US Geological Survey)

from one another, new lithosphere is formed as magma rises from the mantle to fill the opening between the plates. These regions are marked by significant *seismic activity*, i.e. earthquakes. The newly formed rock is relatively light and hot, and produces ocean ridges. As the rock moves away from the **spreading center**, it gradually cools and becomes more dense; further from the plate margin the lithosphere sinks deeper into the asthenosphere and the ocean becomes deeper. Near ocean ridges, the oceanic lithosphere is fairly thin, but older oceanic lithosphere approaches 100 km in thickness far from ridges. Some minerals in the basaltic lava forming the new crust have magnetic properties. When the material is molten and hot, the magnetic fields in the mineral are randomized and point in all directions. As the mineral cools, the magnetic dipoles in the rock can achieve a lower energy configuration by aligning with Earth's magnetic field. This leaves a permanent imprint in the rock recording the direction and strength of the magnetic field at the time of solidification of the rock (see Figure 32.8). As discussed above, Earth's magnetic field has experienced repeated reversals over many millions of years. By matching time-dated geomagnetic records in continental lava flows with sea floor records, the rate of spreading of the sea floor at a constructive plate margin can be estimated. The half-rate of spreading (describing the rate of motion of one plate) in the North Atlantic ocean is roughly 1 cm/yr, while the half-rate of spreading in the Pacific ocean is roughly 4–6 cm/yr. We discuss sea floor spreading further in §34, in the context of the climate history of the planet and the long-term carbon cycle.

Destructive (convergent) plate margins When two plates are in relative motion towards one another, dramatic things occur along the plate margin. Continental plates are thicker and less dense than oceanic plates, so when a continental plate collides with an oceanic plate the oceanic plate is **subducted** – the associated lithosphere sinks into the mantle beneath the oncoming continental plate. Similar subduction occurs when two oceanic plates collide; one plate is subducted and the other rides up above the subducted plate. The subduction of one oceanic plate beneath another leads to a deep trench at the boundary between the plates. An example is the *Mariana trench*, formed as the Pacific plate is subducted under the small western-lying Mariana plate in the Western Pacific. At $10\,994 \pm 40$ m deep, the Mariana trench is the deepest point in the world's oceans. When one plate is subducted under another, the solid crust and lithosphere from the subducted plate melts as it sinks into the asthenosphere. This produces hot magma, which rises beneath the overriding plate, producing volcanic activity and island arcs just beyond the convergent plate margin. Seismic activity is greatest near plate margins at which subduction occurs. The time scale for recirculation of material in oceanic plates between formation at a divergent margin and subduction at a convergent margin is on the order of 100 My. When two continental plates collide along a convergent plate margin, neither plate can easily subduct, so the plates tend to stop and/or crumple, giving rise to mountain ranges – such as the Himalayas, formed as the Indian continent collided with Eurasia over the last 50 million years.

Conservative (transform) plate margins A conservative plate margin is one in which the plates are in relative motion in a direction parallel to the plate boundary. The *San Andreas fault* along the western edge of California is an example of a conservative plate margin. Conservative plate margins, like convergent plate margins, are often associated with seismic activity.

Evidence from a variety of sources verifies the basic picture of plate tectonics. Geomagnetic records make it possible to reconstruct the motion of the continents back about 200 million years. We discuss some of this history further in §34. The precise mechanisms responsible for generating the motion of the plates, however, are understood only in a qualitative fashion. Because the details of convective flow within the mantle are poorly understood, it is not possible to describe the specific forces currently at work in driving plate motion, or to extrapolate plate motions far into the future. Some basic physical principles, however, are sufficient to explain the general nature of the relevant forces. Convection of the mantle leads to transverse motion (parallel to the surface) just beneath the lithospheric plates. This causes a drag force on the plates that drives them in a direction parallel to the motion of the mantle material. Further driving forces behind plate motions arise from plate interactions. In particular, the **slab pull force** acts on the solid lithosphere of a subducted oceanic plate drawn into the less dense mantle. This force pulls the plate further towards the subduction zone. The descent of the cooler, dense subducted plate material into the mantle also plays a role in driving mantle convection, as mentioned above. Other minor forces acting on plates include the resistance along transform boundaries, the forces away from spreading ridges, and localized forces from *hotspots*.

Seismic Events As plates move, stress builds up in the crust along the plate boundaries, where motion is impeded by friction between the plates. When the stress becomes sufficiently high, the plates slip, releasing large amounts of energy in a **seismic event**. Most of the world's seismic activity occurs along plate boundaries. While the energy released in seismic events represents only a small fraction of the total energy flow from inside the planet, individual events can involve significant amounts of energy. Only a small part of the energy released is transformed into seismic waves associated with an earthquake; the remaining energy goes into breaking, deforming, and heating the material in the crust. The magnitude of seismic events is now primarily measured according to the *moment magnitude* scale, which replaces the *Richter scale* used in the past. The Richter scale was based on seismometer measurements at a fixed distance from the epicenter, and did not provide an accurate measure for extremely strong earthquakes. The moment magnitude is based on the *seismic moment* M_0 for a given event, which is the product of the area A of the rupture between plates, the distance d over which the plates move, and the shear modulus μ, a parameter that describes the strength of the faulted rock (see Box 33.1),

$$M_0 = \mu A d \,. \tag{32.6}$$

M_0, which has units of energy, is a measure of the mechanical energy released when the plates rupture. The **moment magnitude** is then defined by

$$M_W = \frac{2}{3} \log_{10}(M_0) - 6.0 \,, \tag{32.7}$$

where M_0 is measured in Newton-meters and the subscript W denotes *mechanical work*.[2] The moment magnitude scale was developed to approximately match the Richter scale for earthquakes of intermediate scale (magnitude 4–7), but gives a better metric for comparing extremely

---

[2] Geophysicists measure M_0 in dyne-cm, in which case
$M_W = (2/3) \log_{10}(M_0[\text{dyne-cm}]) - 10.7$.

powerful seismic events of magnitude 7 or greater. A complementary magnitude scale based on energy released in seismic waves is the **energy magnitude**

$$M_E = \frac{2}{3} \log_{10}(E) - 2.9, \qquad (32.8)$$

where E is the energy released, measured in joules. The energy magnitude is estimated from measurements of seismic waves across the full spectrum, while the moment magnitude is essentially determined by the low-frequency part of the spectrum. These two magnitudes are not generally the same, and provide complementary perspectives on the effects of a given earthquake or other seismic event. From eq. (32.7), it is clear that increasing either magnitude by one represents an increase in seismic moment/released energy by a factor of $10^{1.5} \cong 31.6$.

The number of earthquakes throughout the world over a fixed time period increases by a factor of roughly 10 for each unit decrease in magnitude, an empirical observation known as the **Gutenberg–Richter law** and illustrated in Figure 32.9. Typically there is one earthquake of magnitude 8–9 each year worldwide, and on the order of millions of seismic events of magnitude 2–3. Since the energy grows more rapidly than the frequency decreases as magnitude rises, most energy dissipated in earthquakes is associated with the largest seismic events. Because of the basic geographical limitations on the size of a fault fracture and the bounded magnitude of the shear modulus of rock, earthquakes have a maximum moment magnitude in the

vicinity of 9.5. Figure 32.10 compares the moment magnitudes of the three largest earthquakes between 1906 and 2005 with the total seismic moment release of all quakes with $M_W \geq 6$ during that period. Larger seismic events have occurred; for example, the meteorite impact 65 million years ago in the Yucatán Peninsula (discussed further in §34) is estimated to have had a moment magnitude of 12.25.

Thermal Energy Flow at Earth's Surface   The rate at which thermal energy flows to Earth's surface varies widely depending upon local conditions. For oceanic crust, most of the thermal energy conducted to the surface comes from the cooling of the lithosphere. The highest rate of heat flow occurs near spreading ridges, where the lithosphere and crustal material is recently formed and very hot. As the crust cools and thickens, moving away from the spreading ridge, less heat is conducted to the surface. While lithospheric cooling plays less of a role in continental crust, this cooling still contributes some 20% on average of the thermal energy flow through continental crust. For much older Precambrian crust, formed over 800 million years ago, the total rate of energy flow can be as low as 40–50 mW/m^2, while in young continental crust the flow rate can be 70–80 mW/m^2. Average heat flow rates are around 100 mW/m^2 for oceanic crust and 65 mW/m^2 for continental crust. In those regions with higher heat flow, the temperature gradient is higher. A map of estimated underground temperatures across the US at several depths is shown in Figure 32.11.

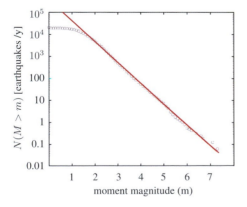

Figure 32.9 The number $N(M > m)$ of earthquakes in the Southern California region with moment magnitude M greater than m during the period 1984–2000. The data follow the *Gutenberg–Richter* law, $N(M > m) = a - b \log_{10} m$, with $b = 1$ (red line). Data from [219]. (Credit: Copyright (2002) National Academy of Sciences)

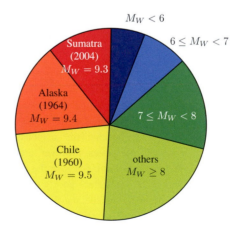

Figure 32.10 The moment magnitudes of the three largest earthquakes during the period 1906–2005 compared with the total seismic moment release of all quakes above $M_W = 6$. The total seismic moment of all earthquakes during this period was approximately 10^{24} J. (Credit: Richard Aster)

Example 32.2   2011 Tōhoku Earthquake

The powerful *Tōhoku earthquake* occurred in Japan on March 11, 2011. The epicenter of the event was 70 miles off the eastern coast of Japan. This seismic event was associated with the ongoing subduction of the Pacific plate. The earthquake involved a plate motion of roughly 30 m across a fault length of roughly 300 km, with a seismic moment of $M_0 \sim 4 \times 10^{22}$ Nm. The map at right shows the slip in meters as well as the location of significant aftershocks (gray circles).

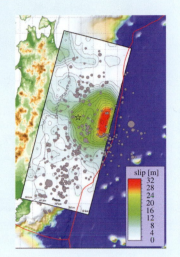

slip [m]
32
28
24
20
16
12
8
4
0

The moment magnitude of this event was 9.0, while the energy magnitude was 8.6. From eq. (32.8), the energy in seismic waves was

$$E = 10^{\frac{3}{2}(8.6+2.9)} \text{ J} \cong 1.8 \times 10^{17} \text{ J} \cong 180 \text{ PJ}.$$

The energy released in seismic waves was thus roughly equivalent to the thermal energy flowing outward to the surface from the center of the planet over about one hour, or the energy used worldwide by humans over more than three hours (assuming a rate of 15 TW). Note that only a fraction of the total energy released in the earthquake was transformed into seismic waves; the total energy released in this event was much larger.

(Credit: US Geological Survey)

Near oceanic ridge axes, the heat flow can be extremely high. As magma comes close to the surface along the ridge, channels and fractures in the rock can bring magma or very hot rock into contact with seawater. This produces a strong hydrothermal circulation, which transfers substantial thermal energy to the surface. This hydrothermal circulation is greatest right along the ridge axis, but also is present in irregular locations further from the spreading ridge. This kind of hydrothermal heat transfer is responsible for ∼1/3 of oceanic heat flow and ∼1/4 of global heat flow.

The most promising locations for extraction of geothermal energy are places where hot material is close to Earth's surface. These are generally the locations where thermal energy flow to the surface is highest. In addition to plate boundaries, hotspots – which, as discussed above, are believed to originate from magma plumes beneath the lithosphere – can be locations where substantial thermal energy resources are close to the surface. Hotspots are characterized by anomalous volcanic activity in a region away from plate boundaries, and in the case of oceanic hotspots, often by an upward deformation of the lithosphere. In such regions the ocean is more shallow than expected from ocean plate spreading models. There are roughly 40 recognized hotspots around the globe. Most of these are in oceans, but there are also a number of continental hotspots, including in Iceland, Yellowstone in the US, and a number of locations in Africa (see Figure 32.12). Hotspots persist over geologically long periods of time,

and move more slowly than continents. Thus, as the continents move over Earth's surface, the relative location of the hotspots gradually changes. The Hawaiian islands are an archetypical example of an oceanic hotspot. As the oceanic plate has moved over the Hawaiian hotspot, a chain of islands has been formed. By using radiometric dating methods (see §34), the age of formation of the rock composing the islands can be determined, indicating that the Pacific plate has been moving over the Hawaiian hotspot at roughly 10 cm/y for the last 40 million years.

32.2 Geothermal Energy Resources

Geothermally heated water has been used by humans since long before the Romans, who made direct use of geothermal energy for heating homes as well as public baths. Currently, low-temperature geothermal resources are used for space, agricultural, and process heating in locations around the world. Over the last century, geothermal energy has also been used to produce electrical power.

One challenge in utilizing geothermal power is the wide variation in available geothermal resources. Geothermal energy is very different in this regard from, for example, wind energy, where the variability of the resource from place to place can be captured in a few simple parameters (as in the Weibull distribution), and essentially identical turbines can be used in a variety of locations. Potential geothermal resources are embedded in a wide range

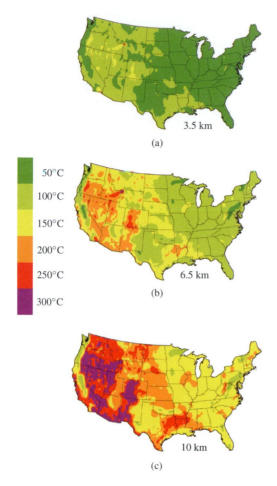

Figure 32.11 Temperatures across the US at depths of (a) 3.5 km, (b) 6.5 km, (c) 10 km. (Credit: US Department of Energy)

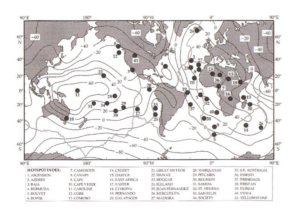

Figure 32.12 41 hotspots where mantle plumes bring hot magma close to the surface. As continental plates move, the relative location of the hotspots changes over time. After [211]. (Credit: Elsevier (1980))

of different geological situations, and can vary in many ways. The approach taken to exploit geothermal energy in a particular location may depend upon the depth of the high-temperature resource, the types of rock above and surrounding the resource, the permeability of the rock, the presence or absence of water and/or steam and the associated flow pattern, the chemical composition of impurities in the water or steam, and many other factors. Despite the range of possibilities, some general classes of geothermal resources can be distinguished. Generally, the temperature of a resource is a key indicator, but pressure is also important, so geothermal resources are often characterized by their *enthalpy*. The major types of geothermal energy resources listed here are described in more detail in the following sections.

Ground source heat pumps/cooling systems Ground source heat pumps and heat extraction devices exploit the approximately constant ambient temperature several meters below Earth's surface to drive efficient heat transfer into or out of buildings or other systems. As explained in §6.5.3, it is actually solar energy that maintains the roughly constant near-surface ground temperature. Thus, while often described as a type of geothermal energy, ground source heating and cooling systems are actually a form of solar energy.

Low-enthalpy wet geothermal resources The vast majority of easily accessible geothermal resources contain fluids heated to temperatures below 150 °C. Such reservoirs occur in a wide range of geological circumstances. Near plate boundaries or hot spots, magma intrusions that reach within several kilometers of the surface can generate low-enthalpy as well as higher-temperature geothermal fluids. Low-enthalpy geothermal resources can also result from heating by rock formations in continental crust with high radionuclide densities, or in regions where porosity and fractures in the rock bring **meteoric water** – water originating as precipitation – far beneath the surface into contact with hot basement rock, with a circulation pattern that brings the heated water back close to the surface.

Low-enthalpy geothermal resources can be used directly for space and industrial heating. More recently, *binary cycle* technology has evolved, where heat exchangers are used to run a phase-change cycle based on a working fluid with a relatively low boiling point, as discussed at the end of §32.4.3.

Medium to high-enthalpy hydrothermal resources Regions with geothermal fluids heated to temperatures above 150 °C within a kilometer or so of the surface present the most desirable geothermal resources for power production. Such geothermal resources generally only occur in regions with ongoing or recent tectonic or volcanic activity, associated with plate boundaries or hotspots. In optimal high-temperature geothermal locations, temperatures up to 300–350 °C can be reached within 2–3 km of the surface. Such resources can be fluid or steam dominated depending upon the pressure and configuration of the location.

Hot dry rock In many locations the thermal gradient is sufficiently steep that temperatures of 200–300 °C are attained at depths of 5–10 km in the hot basement rock below the surface of continental crust. This *hot dry rock* represents a tremendous potential energy resource that may be within reach of the next generation of geothermal technologies, known as *enhanced geothermal systems* (EGS).

Magma Molten rock at temperatures of 600–1400 °C provides the original thermal energy source for most near-surface geothermal energy resources. In some places on Earth's surface, particularly near active volcanic areas, magma comes within several kilometers of the surface. It has been proposed that directly tapping the thermal energy within such magma intrusions could provide a tremendous source of geothermal energy. The high temperatures involved, however, place such an endeavor beyond the reach of existing technology. If materials are eventually developed that make drilling and extracting thermal energy directly from magma feasible, this could represent a substantial energy resource in the long-range future. Magma comes particularly close to the surface in the vicinity of ocean spreading ridges, but the logistical difficulties of accessing such thermal energy resources are difficult to imagine surmounting at the present time.

Geopressured resources In addition to the types of geothermal resource just described, there are also **geopressured resources** in locations where heavy sedimentation has compressed a volume of fluid or gas contained within a region bounded by impermeable rock. Such systems were first discovered beneath the Gulf coast of the US near the Gulf of Mexico, where a mixture of hydrocarbons and water are buried beneath clay sediments, with a high thermal and pressure gradient. Such locations represent another potential future energy resource.

It can be rather complicated to develop any geothermal energy system other than the simplest near-surface low-temperature resources. First, exploration must identify feasible locations for geothermal energy exploitation. Second, a systematic assessment of a potential resource must be carried out. The process of assessing potential resources is quite challenging due to the substantial distance beneath the surface at which measurements must be made. Third,

a location-specific system must be engineered to exploit the resource. Finally, there are a variety of technical challenges to building and maintaining geothermal systems, including the fact that as substantial amounts of energy are extracted from most underground geothermal systems, pressure, temperature, and fluid concentrations can change, necessitating a dynamical approach to utilizing the resource.

We describe the main types of geothermal energy that can be used for large-scale power systems in somewhat more detail below after a brief discussion of ground source heat pumps in the following section.

32.3 Ground Source Heat Pumps

As mentioned above, **ground source heat pumps** are often referred to as "geothermal energy," even though they are primarily powered by solar energy. Just as solar energy is absorbed in ocean surface waters providing a reservoir of thermal energy (§27.6), solar energy is absorbed and stored by the ground, and can be put to use by ground source heat pumps and cooling systems. In §6.5.3, we used the heat equation (6.22) to give a simple estimate of the ground temperature at a depth of a few meters beneath the surface as a function of time. The basic result is that diurnal (daily) fluctuations are damped in 10 or 20 cm of soil, and seasonal fluctuations are damped in several meters of soil. The solar energy absorbed during the summer months at mid-latitudes propagates slowly downward, with an effect that decreases exponentially with depth so that at a depth of 3–6 m beneath the surface the temperature is quite constant year round in most locations and is approximately equal to the yearly average surface temperature (see Figure 6.16). Due to the small flux involved, the flow of geothermal heat upward has very little effect on the temperature of the surface.

Since the ground temperature at several meters depth is roughly constant through the year, it can be used effectively as a source or sink for thermal energy to drive a heat pump. Recall from §10.6 that a heat engine run in reverse can be used to transfer thermal energy from a cooler region to a warmer region. This makes it possible, for example, to warm a house in the winter using thermal energy from the ground with a much smaller input of electrical energy than would be required by electric resistive heating. The coefficient of performance (CoP) of a heat pump (10.15) is the ratio Q_+/W of the heat output over the work done. This is bounded above by the Carnot limit $T_+/(T_+ - T_-)$. For example, to keep a house at 20 °C when the ground is at a temperature of 10 °C, the Carnot bound on the CoP is around 30. While realistic heat pumps cannot come close to this performance, it is possible to achieve CoPs of 2–5 for commercial systems; this is significantly better than the ratio of one achieved by an electric heater. Ground source cooling systems work in a similar fashion, by pumping heat from the warmer above ground environment down into the cooler ground in the summer.

Several approaches have been developed for ground source heat pumps. A standard configuration currently in use employs a *closed-loop system* in which water (mixed with an antifreeze agent such as monopropylene glycol) is circulated through a horizontal system of pipes (usually polybutylene, chosen because of its high thermal conductivity) under the ground at a depth of 2–4 meters. Thermal energy flows from the ground into the fluid in the pipes. The energy is then transferred to a phase-change working fluid in a second loop and is extracted by a heat pump that deposits the thermal energy in the space to be heated. The principal limitations to this form of heating system are the thermal conductivity and heat capacity of the soil (or rock) through which the pipes run. The lower the heat capacity of the soil, the larger the field from which energy must be drawn. If the soil's thermal conductivity is low, heat cannot be captured from far away and the temperature of the soil in the immediate vicinity of the pipes falls, decreasing the CoP of the heat pump.

If the ground has substantial water content, its thermal conductivity and heat capacity are higher, and the system is more effective. In some cases, a body of water can be used as the primary thermal energy source. Variations on the approach just described include **open systems** – where water is extracted from and returned to an underground channel – and deeper vertical systems extending to a depth of 30–200 m. Vertical systems are sometimes used when insufficient horizontal area is available. For both open- and closed-loop systems based on water as a heat transfer fluid, a separate loop is needed for the refrigerant used in the heat pump itself. Simpler **direct exchange systems** circulate the refrigerant itself through the ground in copper pipes.

The key to a successful ground source heat pump installation is ensuring that the volume of ground from which energy is extracted is sufficient to cover the needs of the system. To illustrate this we perform a simple estimate in Example 32.3 of the area that must be covered by the pipe loop to heat air and water in a typical single-family house in a location with relatively cold winters. In general, the area becomes prohibitively large unless the ground has sufficient moisture content and/or groundwater flow. These issues are explored further in the problems.

Ground Source Heat Pumps

Ground source heat pumps use the ground several meters below Earth's surface as a thermal energy source/sink for a heat pump that heats or cools buildings or water. Substantial soil volume is needed, and water content in the ground improves heat transport to replace energy in the region utilized. The coefficient of performance (CoP) for a ground source heat pump-based heating system can be in the range 2–5.

32.4 Hydrothermal Energy

The first geothermal electric power plant went into operation in Larderello, Italy in 1904, where by 1913 a 13 MW power plant was converting geothermal steam into electricity. Geothermal electric power production developed slowly in the first half of the twentieth century, but over the last several decades has expanded to a capacity of over 10 GWe. Most of the present electricity production from geothermal power taps fluid-dominated hydrothermal resources at medium to high temperatures, but the use of binary cycle systems to produce electricity from low-enthalpy geothermal systems has been rising.

32.4.1 Hydrothermal Resources
Optimal hydrothermal resources occur in locations where the following features are present:

Steep thermal gradient The most easily exploited hydrothermal resources have temperatures around 300 °C occurring within a kilometer or so of the surface. It is easy to drill to this depth using well-established technology. Such a steep thermal gradient, compared to the typical value of 30 °C/km, occurs only in locations such as faults or hotspots where magma has come close to the surface.

Water resource To bring the thermal energy to the surface, water must be present in the geothermal field. In most situations the water is under sufficient pressure that it exists in liquid form even at high temperature. The fluid in the geothermal field often contains significant chemical impurities that can complicate the process of extracting energy.

Permeable rock The rock in the geothermal field must be sufficiently permeable to allow water to flow through the reservoir. Water must be able to flow from throughout the reservoir to the **production well** where the

Example 32.3 Ground Source Heat Pump in Boston

To illustrate the need for a large ground area and adequate moisture content we perform a back-of-the-envelope calculation to spec a ground source heat pump with CoP = 4 in a mid-latitude location such as Boston, where ground temperatures at 6 m are around 10 °C. Consider trying to use a ground source heat pump for air and water heating in a single-family house that requires a total of 24 GJ (6600 kWh) of thermal energy in the month of February when outside temperatures average around 0 °C. With a CoP of 4, 18 GJ must be taken from the environment. If the ground is dry and sandy, the volumetric heat capacity is roughly $c \approx 1.2 \, \mathrm{MJ/m^3 \, K}$. So, providing the required energy for the month of February would require lowering the temperature of 2000 m^3 of ground by around 7.5 K. The heat capacity of the ground increases with moisture content, which can reduce the volume needed by a factor of 2–3. Since even with moist ground this is a very large volume, thermal conductivity is needed to draw thermal energy from the ground away from the immediate vicinity of the pipes and reduce the quantity of piping needed. For dry sandy earth, the thermal conductivity is around 0.3 W/m K. Covering a horizontal area of 200 m^2 with a piping network, a temperature gradient of 10 K/m below the pipes would give a thermal energy flow rate of 600 W towards the pipes. This is an order of magnitude less than the 10 kW average needed for the house in February. (Assuming the horizontal piping is installed at a depth of 3–6 m, thermal energy would soon cease to flow from above; if the piping network lay in a vertical plane, the flow rate estimated here would be doubled as thermal energy would come in from both sides.) Soil with increased moisture content can have a thermal conductivity of 1–3 W/m K, making the deployment of an adequate piping system for a ground source heat pump a more tractable proposition. The situation is much easier in locations with some flow of groundwater, where thermal energy transport in the ground is greatly enhanced.

This illustrates some issues involved in designing and implementing ground source heat pumps. Clearly, in locations with cold winters and dry soil, good home insulation plays an important role in reducing the total energy needed to make a ground source heat pump system tractable.

(Image credit: Mark Johnson)

---

water is extracted. The reservoir must also be sufficiently permeable to allow either recharge of water from natural sources or re-injection from the geothermal plant.

Cap For the water to flow easily to the surface through the production well, the reservoir must be maintained at a high pressure. This requires an impermeable rock cap over the permeable rock that contains the fluid in the reservoir.

An idealized hydrothermal resource is depicted in Figure 32.13.

Geothermal regions where some of these features are absent, or only present in limited form, may still be productively exploited in some cases. For example, if there is not a natural system for replenishing the geothermal fluid, fluid can be artificially pumped down into the reservoir through an **injection well**. In many geothermal systems, after some time the fluid in the original resource becomes depleted, and the pressure drops, complicating extraction. In some such locations, such as the Geysers geothermal

field in California, waste water is pumped back into the system to replenish the fluid.

Identifying a geothermal region suitable for economically viable energy extraction involves substantial research and development before wells are drilled. A variety of approaches, many based on geophysical methods, are used to assess a potential geothermal resource. A simple survey of the surface can identify obvious features such as *steam vents*, *fumaroles*, or *geysers*, which indicate a highly active geothermal region. But more detailed assessment of the configuration below ground requires a number of indirect methods. The volume of fluid in the reservoir, temperature, pressure, permeability of rock, and chemical composition of the fluid must all be evaluated. A hydrologic survey, taking into account inputs to the region from rainfall and surface water flow, and analyzing chemical impurities in surface fluids, gives some information about the properties of the geothermal fluid. Measurements of near-surface heat flow and rock conductivity give some estimate of the thermal profile of the region. The electrical resistivity

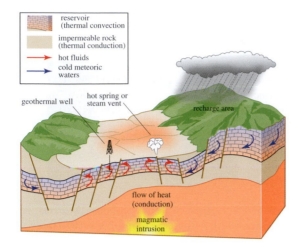

Figure 32.13 An idealized hydrothermal resource, containing a geothermal field at moderate depth composed of porous rock containing fluid water at high temperature and pressure. Water is kept at high temperature by a magma intrusion into the crust relatively near the surface, and at high pressure by an impermeable rock cap above the reservoir. Meteoric (rain) water recharges the reservoir. (Credit: *Geothermal Energy: Utilization and Technology*, edited by Mary H. Dickson and Mario Fanelli (United Nations Educational, 2003))

of the ground, measured by passing current through a high voltage between electrodes spaced several kilometers apart, can indicate the water content of the rock down to a depth of over 1 km, with lower resistivity indicating higher water content. Salinity and higher temperature of the geothermal fluid decrease the resistivity. Imaging methods using electromagnetic and seismic waves can also give useful information about the structure of the underlying formation, as can local variations in the gravity field associated with the density of subsurface material. Finally, if the initial assessment is positive, drilling exploratory wells gives more direct information about the potential resource. Many of these methods, particularly seismic surveys and *well logging*, have been extensively developed in the oil and gas industry and are discussed in more detail in the following chapter.

Once a promising geothermal resource has been identified, a great deal of physics and engineering is involved in designing a system to optimize the utilization of the resource. Experience in the oil and gas industry has led to a sophisticated set of theoretical and practical tools for modeling fluid flow through complex underground rock formations, drilling production and injection wells, and for constructing above-ground systems that connect the wells. In a typical fluid-dominated reservoir, permeability of the

rock plays a crucial role in the reservoir dynamics. For a fluid of viscosity η flowing in a porous medium and subject to a pressure gradient ∇p, the **permeability** K is defined through **Darcy's law**

$$\boldsymbol{u} = -\frac{K}{\eta}\nabla p \,, \qquad (32.9)$$

where $\boldsymbol{u}$ is the flow velocity of the fluid. This relation can be thought of as parallel to the local form of Ohm's law (3.29) or Fourier's law of heat conduction (6.1). The unit of permeability is the **darcy** (D)

$$1\,\text{darcy} = 1\frac{\text{cP cm}^2}{\text{atm s}} = 9.87 \times 10^{-13}\,\text{m}^2 \,. \qquad (32.10)$$

A typical geothermal reservoir has a permeability of 10–70 mD. Of course, real geological configurations do not have constant permeability; fracture patterns in rock are irregular and may change over the time scale of utilization of a given resource. Nonetheless, eq. (32.9) gives a simple description of local fluid flow in a permeable medium that can be used as part of a more detailed analysis of a complete system.

In a fluid-dominated geothermal system, even though the temperature of the reservoir may be well over 100 °C, water will stay in liquid form as long as the pressure is sufficiently high. While the detailed flow of a fluid through a permeable medium is rather complicated to analyze analytically, a simple model of a reservoir using eq. (32.9) illustrates the issues involved in extraction of the geothermally heated fluid from the reservoir. If a well is drilled that opens into the top of the reservoir, the high pressure within the reservoir will force the fluid out through the well. As the fluid flows, the pressure will begin to decrease in the part of the reservoir near the well. A pressure gradient will develop consistent with conservation of the fluid and eq. (32.9). From eq. (32.9), it follows that in steady state the flow rate will be proportional to the difference in pressure between a point in the reservoir at the bottom of the well and other points in the reservoir far away from the well. This difference is characterized by the **drawdown coefficient** C_D, which is defined through

$$p_r - p_w = C_D \dot{m}, \qquad (32.11)$$

where $\dot{m}$ is the mass flow rate, p_r is the ambient reservoir pressure far away from the well, and p_w is the pressure at the point where the well reaches the reservoir. A simple model for C_D is given in Problem 32.6. As more fluid is extracted from the well, the pressure near the well opening within the reservoir decreases, as does the pressure through the **wellbore** (the channel between the **wellhead** at the surface and the bottom of the well at

---

Example 32.4 Pressure in a Hydrothermal Reservoir

In many circumstances, at a reasonable flow rate through a hydrothermal production well, the flash horizon moves into the wellbore. Consider for example a hydrothermal reservoir at a depth of 1.7 km containing a geothermal fluid that is essentially water at a temperature 220 °C. The pressure in the reservoir before it is tapped is about 17 MPa. Note that this is very close to the pressure produced by a column of cold water 1.7 km deep; we can think of the reservoir pressure as being maintained by cold water flowing into the reservoir from the periphery outside the geothermal region and capstone.

Now assume that a well is drilled. When the well reaches the reservoir, the hot water shoots upward through the wellbore. The density of water at 220 °C is considerably less than the density of cold water (roughly 850 kg/m^3 versus 1000 kg/m^3). Thus, even when the well is full of water the weight of the water in the wellbore is generally not sufficient to stop the flow. A pressure gradient develops as water flows out of the reservoir. In steady-state flow, the rate of water flow is related through eq. (32.11) to the difference in pressure between the bottom of the wellbore and the ambient pressure $p_r \cong 17$ MPa far from the wellbore.

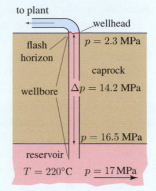

Making the simplifying assumption that the pressure in the wellbore decreases vertically according to the weight of overlying water ($dp/dz \cong \rho g$), we can compute the difference in pressure between the top of the well and the bottom when the well is full of liquid at 220 °C

$$p_w - p_{top} \cong (850 \text{ kg/m}^3 \times (9.8 \text{ kg m/s}^2)) \times 1.7 \text{ km} \cong 14.2 \text{ MPa}.$$

Using steam tables, we determine that the vapor pressure at $T = 220$ °C is $p_v \cong 2.32$ MPa. Thus, if the pressure at the bottom of the well drops below 16.5 MPa (= 14.2 MPa+ 2.3 MPa), then the flash horizon will move into the wellhead.

If the drawdown coefficient takes the value $C_D \cong 0.02$ MPa/(kg/s), then the maximum flow at which the flash stays outside the wellbore will be

$$\dot{m} = \frac{p_r - p_w}{C_D} \cong \frac{0.5 \text{ MPa}}{0.02 \text{ MPa/(kg/s)}} \cong 25 \text{ kg/s}.$$

The figure depicts the configuration at this flow rate, where the flash horizon is right at the top of the wellbore.

Typically, a larger flow rate is desired, which moves the flash horizon into the wellbore. We have also neglected friction here in the motion of the fluid through the well; explicit inclusion of a friction term (see, for example, [214]) increases further the difference $p_w - p_{top}$, and moves the flash horizon further into the wellbore.

---

the opening to the reservoir). When the pressure in the wellbore decreases to the saturation vapor pressure for water (see §12.1, Figure 12.5 and Example 12.1) at the temperature of the fluid, the water begins to vaporize, or **flash**. The point where this occurs is called the **flash horizon**. As the resource becomes depleted and the pressure decreases, the flash horizon moves down through the wellbore and can enter the formation (Problem 32.8). This complicates the flow of the water/vapor mixture through the porous rock in the reservoir. The analysis of **two-phase flow** through a porous medium involves more sophisticated simulation techniques. Managing this process is part of the general problem of geothermal system engineering. This illustrates the dynamical nature of geothermal systems; any power system designed to optimize extraction of geothermal energy must evolve over time to match the changes in the resource as it is depleted.

As mentioned above, often injection wells are used to replace the fluids within the reservoir. Since the fluid entering the injection well is cold, it impacts the hydrodynamics within the formation. Placement of injection wells can play an important role in optimizing geothermal systems, and improper placement can lead to inefficiency; for example, at the Hatchobaru geothermal field in Japan, an injection well was directionally drilled and terminated close to a production well, leading to a cooling of the fluid beneath the production well that made that well unusable [214].

Another example of an issue that must be dealt with in geothermal systems engineering is the tendency of calcium carbonate ($CaCO_3$) to be deposited in the well casing just above the flash horizon. The fluids in many hydrothermal reservoirs are saturated with calcium carbonate, which precipitates out as soon as the fluid vaporizes. This substance can rapidly build up within the wellbore, restricting flow. This issue must be balanced with the desire for greater flow rates, which tends to move the flash horizon into the wellbore (see Example 32.4).

32.4.2 Single-flash Power Plants

The simplest kind of power plant used to convert energy from a hydrothermal resource into electricity is a **single-flash steam power plant**. The idea of the single-flash plant is to use steam derived from the hydrothermal system to power a turbine. The water may flash into steam in the reservoir, in the wellbore, or in the above-ground piping as the pressure decreases. In any case, a *separator* is used to separate out the steam from liquid water, usually by spinning the fluid so that the two phases separate due to their different densities. The steam is then fed into a turbine. Unlike a typical Rankine power plant, the steam entering the turbine is a saturated vapor. Thus, as in nuclear power plants, the steam begins to condense and the quality decreases as it powers the turbine. Small droplets of water hit the turbine blades; engineering a turbine that is not damaged by constant exposure to condensed liquid requires using more elaborate materials for the turbine plates than is needed for turbines driven by superheated steam. After passing through the turbine, the steam is condensed to a low temperature and pressure (see the discussion in §13.3.4). This reduces back-pressure on the turbine and maximizes the fraction of the enthalpy contained in the steam that is turned to work. The fluid is generally then pumped back underground through an injection well, along with the remaining liquid coming from the separator.

An idealization of the S–T state path of the working fluid in a single-flash plant is depicted in Figure 32.14. Although the working fluid itself does not generally go through a complete closed cycle, the cycle can be treated as closed for the purpose of computing the energy output and

thermodynamic efficiency of the system. In principle, if the underground hydrothermal reservoir is considered part of the system and the injection rate matches the extraction rate, then the system is actually closed, with the underground porous rock heating the injected fluid before it returns to the surface.

The flow of energy and the efficiency of the idealized single-flash plant cycle can be analyzed in a similar fashion to the other thermodynamic cycles we have considered in previous chapters. The most salient new features of the single-flash cycle are the separation of a fraction of the working fluid as steam and the powering of the turbine completely below the saturation dome. Following through the cycle, we begin at point ①, with a high-temperature saturated liquid at the flash horizon. As the pressure drops further the fluid begins to vaporize and enter a mixed phase. The specific enthalpy (enthalpy per unit mass) h_1 of fluid at point ① depends only on the initial temperature T_+, and can be determined from steam tables. Although the flash horizon may be in the wellbore and below the level of the power plant, we can neglect gravitational potential energy, which cancels out over a full cycle. The point ② is the mixed phase of the working fluid after it is partially vaporized. Vaporization, like throttling (see §13.2.2), can be modeled as an isenthalpic process, so

$$h_1 = h_2 . \tag{32.12}$$

Knowing h_2, at a given temperature T_s for the separator, we can determine the quality χ describing the mass fraction of the fluid in vapor form. This is the steam that drives the turbine. The specific enthalpy of the steam going into the turbine is h_4, associated with saturated vapor at temperature T_s. In an ideal situation where the turbine is run isentropically, the curve from ④ → ⑤′ is vertical, so the specific entropies at these points are equal, $s_4 = s_{5'}$. The energy output per unit mass entering at ① is then

$$e_{\text{out}} = \chi (h_4 - h_5') . \tag{32.13}$$

After the turbine, the working fluid is then condensed at the low temperature T_-. As mentioned above, this condensation maintains low pressure on the downstream side of the turbine, needed to maximize the energy extracted in the step ④ → ⑤′. The fluid is often re-injected into the reservoir not only to maintain fluid and pressure in the reservoir, but also because geothermal fluids contain many impurities that are environmentally damaging if released.

The total thermodynamic efficiency of the idealized single-flash plant is then given by the energy output divided by the total available enthalpy $h_1 - h_6$

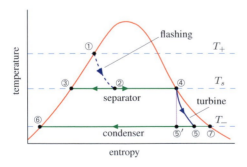

Figure 32.14 A single-flash geothermal power plant "cycle." After [214].

$$\eta \cong \chi \frac{h_4 - h_5'}{h_1 - h_6} . \qquad (32.14)$$

For the temperatures reached by typical geothermal fluids this efficiency is generally below 15% (see Example 32.5). Furthermore, the efficiency is reduced below this ideal value by many deviations from the ideal thermodynamic cycle just outlined. In particular, the turbine is not truly isentropic, so the true curve ④ → ⑤ goes to a higher entropy state (labeled ⑤ in the figure) than the point ⑤'. In addition, the condensation of the fluid below the saturation dome while driving the turbine decreases the turbine efficiency further. An engineering rule of thumb known as the **Baumann rule** states that for every 1% of liquid in the steam turbine, the efficiency is decreased by roughly 1%. Thus, for example, if the average quality of the steam in the turbine during the power cycle is 90%, the overall thermodynamic efficiency is reduced by an extra multiplicative factor of 0.9.

An important design consideration for a single-flash plant is the choice of temperature T_s for the separator. If T_s is very close to T_+, then the flash provides very little steam, so the quality factor χ entering eq. (32.14) is small. On the other hand, if T_s is very close to T_-, then $h_4 - h_5$ becomes very small, again giving a small overall efficiency. In general, the optimum separator temperature is fairly close to the mean

$$T_s^{\text{optimal}} \cong \frac{T_+ + T_-}{2} . \qquad (32.15)$$

The optimal separator temperature can be proven to be precisely the mean (32.15) in an ideal approximation where the enthalpy is linear in s and T, and the relevant points (③, ④, ⑦, ⑥) on the saturation dome lie on a trapezoid (Problem 32.10).

For thermal resources such as geothermal systems, it is often useful to compare the realized conversion efficiency to the Carnot limit. In general, the ratio of the useful work performed by a system to the maximum work allowed by the laws of thermodynamics is known as the *2nd law efficiency* (denoted ϵ). This concept and the closely related concept of *exergy* or *available work* are described further in §36. In the case of the geothermal system described in Example 32.5, for example, the realized efficiency is 13.6%, while the Carnot efficiency limit for the given temperatures T_+, T_- is $\eta_{\text{Carnot}} = 38\%$, so the 2nd law efficiency is

$$\epsilon = \eta/\eta_{\text{Carnot}} \cong 36\% . \qquad (32.16)$$

The rate at which energy can be extracted from a geothermal resource depends upon the product of the efficiency of the power plant and the mass flow rate through the production well(s). The pressure at the wellhead can be

decreased so that the mass flow rate increases, but only to a point of saturation where the fluid flow rate achieves a limiting value (known as *choked flow*) and the well is said to be **choked**. Reservoir engineering involves balancing these various factors, including the rate at which the geothermal resource is replenished or depleted.

Single-flash hydrothermal plants produced over 40% of geothermal power in 2015, with individual plant outputs in the range 3–100 MW. One significant feature of the single-flash plant design is that there is significant enthalpy in the liquid leaving the separator. This liquid can be fed into a secondary system for additional power generation (see below). In many installations, particularly in Japan and Iceland, the enthalpy from this fluid is used directly for heating.

32.4.3 Other Hydrothermal Power Plant Systems

Double-flash power plants

The double-flash power plant concept is closely related to the single-flash plant. In a double-flash plant the high-temperature fluid coming out of the separator is allowed to flash at a lower pressure, providing a flow of steam to a second generator. The idealized thermodynamic paths of the two streams of geothermal fluid in a double-flash plant are shown in Figure 32.15. Including the second flash increases the power output of a plant with given production flow of geothermal fluid by on the order of 20% (Problem 32.12). In some favorable cases, triple-flash plants have been constructed, though these are often first constructed as single-flash plants with other flash/turbine systems added later.

Dry-steam power plants

Though most hydrothermal resources are liquid-dominated, in some locations the geological factors combine to provide vapor-dominated geothermal resources. In this case, saturated vapor can be directly extracted from the ground

> **Single-flash Hydrothermal Power Plants**
>
> The most common type of hydrothermal power plant is based on a *single-flash* design. As hot water is forced out of the ground by high pressure below, it begins to vaporize, or *flash* at lower pressure. A *separator* removes the steam, which is fed into a Rankine cycle plant. Efficiencies are in the range 10–15%, limited by the moderate temperature of the hydrothermal fluid and condensation in the turbine.

Example 32.5   Single-flash Hydrothermal Plant

Consider a single-flash hydrothermal plant where the fluid begins as a saturated liquid at the temperature $T_+ = 250\,°C$, and the low temperature at which the turbine output is condensed is $T_- = 50\,°C$.

What is the plant's idealized thermodynamic efficiency?

T	$h_{\text{liquid}}(T)$	$h_{\text{vapor}}(T)$	$s_{\text{liquid}}(T)$	$s_{\text{vapor}}(T)$
250	1 086	2 801	2.794	6.072
150	632.2	2 746	1.842	6.837
50	209.3	2 591	0.7038	8.075

We assume that the separator is at $T_s = 150\,°C$. The thermodynamic data needed is given in the above table of specific enthalpy (kJ/kg) and entropy (kJ/kg K) for saturated liquid and vapor at the relevant temperatures.

At the points in the ST-plane as labeled in Figure 32.14, we have then

$$h_1 = h_2 = 1086\,\text{kJ/kg}, \qquad h_4 = 2746\,\text{kJ/kg}, \qquad h_6 = 209.3\,\text{kJ/kg}.$$

Interpolating linearly between liquid and vapor values of h at T_s (as described in §12.2.4), we find $\chi \cong 0.215$. Similarly, we can use $s_4 = 6.837\,\text{kJ/kg K}$ and approximate the turbine as isentropic to obtain an approximate value of $h'_5 \cong 2191\,\text{kJ/kg}$. The thermodynamic efficiency of the ideal cycle is then

$$\eta \cong \chi\,\frac{h_4 - h'_5}{h_1 - h_6} \cong 13.6\%\,.$$

If the step ④ → ⑤′ were perfectly isentropic, then quality at point ⑤′ would be $\chi_{5'} \cong 0.83$, so the average quality going through the turbine would be $\cong 0.92$, so by the Baumann rule of thumb, the condensed fluid in the turbine would decrease the efficiency to $< (0.92)(13.6) \cong 12.5\%$.

A typical production well can provide a rate of mass flow of roughly 50–100 kg/s. At a flow rate of 100 kg/s and an efficiency of 11%, taking into account pressure losses in piping and various other thermodynamic inefficiencies, such a single-flash power plant would generate power at a rate of about 10 MW.

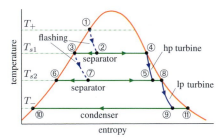

Figure 32.15 A double-flash plant allows the fluid coming out of the separator to flash, providing steam at a lower pressure for use in a secondary low-pressure (lp) turbine. After [214].

and fed through a turbine after impurities are removed. This kind of **dry-steam geothermal plant** is simpler than a flash plant and generally more efficient, since no flash and associated drop in temperature is required. Two major dry steam fields are known, one in Larderello, Italy where the first geothermal electric plant was installed, and the

other in Geysers, California. Along with a few other smaller locations around the world, dry-steam resources represented roughly 25% of worldwide geothermal energy production in 2007.

Binary cycle geothermal plants

Rather than using the geothermal fluid directly to drive a turbine, a heat exchanger can be used to transfer thermal energy to a working fluid in a standard closed-cycle Rankine system. Geothermal power plants of this kind are very similar to other kinds of power plants that use a secondary loop for the working fluid, such as nuclear *pressurized water reactors* (§19.4.1). Binary cycle systems have recently been developed that can generate electrical power from geothermal fluids as cool as 80–90 °C, using organic Rankine cycles or the **Kalina cycle** – a phase-change cycle with a working fluid that is a solution of two distinct fluids with different boiling points, so that the liquid–vapor transition occurs over a range of temperatures. Such binary cycle plants have great potential for wide use around the world for relatively modest (less than a few MWe [220]) power production from low-enthalpy geothermal resources.

Binary cycle systems can also be used to extract additional energy from the fluid output from the separator in a traditional single- or double-flash geothermal plant.

32.5 Enhanced Geothermal Systems (EGS)

In hydrothermal systems, natural features combine to provide a reservoir of water held underground in proximity to a high-temperature source of thermal energy. While such locations are optimal for geothermal power generation, this confluence of factors is relatively uncommon. A much more widespread potential source of geothermal power is the crystalline basement rock lying many kilometers beneath the surface, with no associated hydrothermal resource, known as **hot dry rock**. In many locations, a geothermal gradient above 50 °C/km gives temperatures of 250 °C or greater within 5 km or so of the surface. Tapping into this resource could provide a substantial fraction of the world's needed energy. There are technical challenges to extracting this energy that push the limits of existing engineering technology, however, as well as some possible associated risks that must be carefully evaluated before large-scale exploitation of this resource might become possible.

Extraction of energy from hot dry rock deep beneath Earth's surface requires more human intervention than most near-surface hydrothermal resources. First, wells must be drilled to depths of 5 km or more. The densities and temperatures encountered at such depths place extreme stress on drilling equipment. Second, the rock must be fractured to allow water to pass through it to absorb the thermal energy. This is done by pushing water into injection wells at extremely high pressures. This method, known as **hydrofracturing**, is commonly used in the oil and gas industry to increase the flow of oil or gas in underground reservoirs to facilitate recovery and is discussed further in the following chapter. Geothermal systems using artificial injection of water for a geothermal fluid and/or fracturing of the rock to improve permeability are known as **enhanced geothermal systems** (EGS) (see Figure 32.16).

A pilot project to explore the feasibility of extracting energy from hot dry rock was carried out at Fenton Hill, New Mexico from 1973–2000, near the Valles Caldera, a dormant volcano. Wells were drilled to over 4 km depth, reaching rock with temperatures above 300 °C. After hydraulic fracturing, connectivity was established between an injection well and an extraction well. This verified the

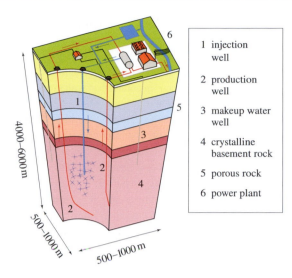

Figure 32.16 Enhanced geothermal systems to extract energy from *hot dry rock* involve drilling to several kilometers depth and hydrofracturing rock to provide permeability connecting injection and extraction wells. (Credit: Siemens AG)

feasibility of this approach in principle. Tests of enhanced geothermal systems and demonstration plants have since been carried out in several other locations in the US, Europe, and Japan. While these preliminary efforts demonstrate that it is feasible to extract energy from hot dry rock using existing technology, a potentially significant challenge in developing large-scale systems for commercial power production is developing a better understanding and control of the fracturing pattern and flow patterns in the fractured volume. In particular, at high injected flow volumes a large fraction of injected water may flow out through the fractured rock and not return to the surface. This has been a problem with some of the test sites. More recently, a demonstration EGS plant in Cooper Basin, Australia successfully produced 1 MWe of power over a sustained six-month period in 2013 [221]. The Cooper Basin plant used fractured rock at a depth of 4.2 km, and had a reported wellhead temperature of 215 °C, and a record closed-loop flow rate of 19 kg/s. The success of this and other early test systems suggest that large-scale power production from EGS may be feasible in the coming decades.

Another concern – observation of seismic events associated with the hydraulic fracturing of underground rock – led to the early termination of an EGS test in Basel, Switzerland. Even away from tectonic margins, there are substantial stresses on and strains in all underground rock structures. The forces involved become greater at greater depths. Fracturing of rock or other manipulation of geological formations and reservoir pressure leads to small seismic events, or **micro-earthquakes**. Such seismic events are not unique to geothermal energy facilities. Similar micro-earthquakes occur in connection with many other energy-related alterations of underground structure. In particular, similar seismic events occur around many major oil fields and mining operations in connection with hydrofracturing and other modifications of geological structure. Most micro-seismic events associated with EGS tests and other similar operations have relatively low magnitudes in the range of 2–3, with rare events at magnitude 4 or above; the events that led to the cancellation of the Basel project were of magnitude 2.5–3.4. In that case the area where the tests were conducted supports a substantial population, and Basel lies over a fault in a seismically active area (the city was destroyed in 1356 by a magnitude 6.5 earthquake), so local concern about such seismic activity contributed to a prudent cessation of the tests. Seismic activity produced by deep geothermal drilling is certainly an issue that must be further evaluated. This risk must be weighed, however, against the risks and costs of any other approach to large-scale energy production. While an earthquake of magnitude 4 can be felt on the surface, there are hundreds of thousands of such quakes around the world every year. Some geological analyses of probable seismic activity associated with deep geothermal energy extraction suggest that it is unlikely that earthquakes of magnitude much greater than 4 would be caused directly by enhanced geothermal systems. It is also believed (at least by some experts) that local seismic activity would likely have little impact on distant faults. As long as enhanced geothermal systems are constructed in regions away from major tectonic faults and away from population centers, the risks of this technology may be much lower than the risks associated with other energy resources, and the risk of seismic activity associated with large-scale geothermal energy production may be no worse than the risk of seismic activity associated with current oil extraction activities. Further research is needed, however, to increase the certainty level associated with these risks.

The energy available in hot dry rock is substantial. The thermal energy difference in one cubic kilometer of granite between the temperatures of 300 °C and 200 °C is on the order of 200 PJ, enough energy to run a 50 MW power plant for 20 years even assuming less than half of the Carnot efficiency is realized (Problem 32.13).

32.6 Magnitude of Geothermal Resources

In 2015, worldwide installed capacity for geothermal electric power generation was 12.6 GWe, with 73.6 TWh

Enhanced Geothermal Systems

There is enough thermal energy in the hot dry rock in the US within 5–10 km of the surface to supply all of world energy needs for hundreds or thousands of years. Extracting this energy would require deep drilling and hydrofracturing, pushing the limits of current technology.

(0.26 EJ) of energy produced, giving a capacity factor of 67% [222]. Forty percent of this capacity was in single-flash plants, 23% in dry steam, 20% in double flash, and 14% in binary plants. In 2014 worldwide total installed capacity for direct thermal use of geothermal energy was 70.3 GWth (of which 55% was from ground source heat pumps), growing at an annual rate of 7.7%, and performing at a capacity factor of 27% [223]. Each of these numbers has grown steadily over the last century, and has potential to grow significantly further. We briefly summarize the situation with regards to different aspects of geothermal power.

Direct Use   Direct use of geothermal energy for heating is the most efficient way to use low-enthalpy geothermal resources. In Iceland, roughly 70% of the nation's energy needs are met from geothermal energy, primarily through the direct use of geothermal heating. Many low-temperature geothermal resources around the world have not yet been tapped. The current level of direct usage could increase many times over with broader use of these low-temperature resources. A 2006 NREL study estimated that direct use of geothermal energy in the US could expand to 60 GWth, not including ground source heat pumps that could provide up to 1 TWth of additional heating energy.

Hydrothermal   Optimal hydrothermal resources are distributed most heavily in favorable locations near plate boundaries. Nonetheless, many of these resources are as yet untapped. The US has the largest geothermal electric power production capacity, at 3.45 GWe [223], of which a substantial fraction is the 1.58 GWe dry-steam Geysers installation. Another 0.88 GWe is provided by double-flash plants, and 0.87 GWe comes from binary plants. The second largest producer of geothermal electric power is the Philippines, with capacity 1.87 GWe, roughly 2/3 from single-flash plants. Many countries have hydrothermal resources that are not tapped, and global production could be increased several-fold. NREL estimates that the US could expand to greater than 100 GWe if all hydrothermal resources, including those as-yet undiscovered, were tapped [224].

Hot Dry Rock   Much of the Western US has a geothermal heat flux of 60–100 mW/m² (Figure 32.11), with a geothermal gradient above 40 °C/km and has hot dry rock at temperatures above 250 °C within 5 km of the surface. An MIT report [215] on enhanced geothermal systems estimates that there are some 100 000 EJ of energy in sedimentary rock under the US, and 13 000 000 EJ in crystalline basement rock. If the technology can be developed and deployed to extract this energy at large scale, the US resources alone could supply all the world's energy needs for thousands of years at the current rate of energy use. Such an enterprise would, however, require systematically accessing hot dry rock over a large area, with new drilling and/or plants likely required on at least a decadal time scale (Problem 32.13).

Discussion/Investigation Questions

32.1   Find out how the *Kalina cycle* works and discuss its application to geothermal power. To what kind of resources is it best suited?

32.2   Both nuclear and geothermal power plants use saturated steam in Rankine cycles. What are the fundamental differences between the thermodynamics of these two energy sources?

Problems

32.1   Use Newton's law to compute the gravitational acceleration $g(r)$ as a function of radius r within a sphere of total mass M that is uniformly distributed through the volume. Compute $g(r)$ in a simple model of Earth with a core of radius 3500 km with density 1.2×10^4 kg/m³ surrounded by a mantle of depth 3000 km with density 5×10^3 kg/m³. Graph this function and compare with Figure 32.3(b).

32.2   Make a very rough estimate of the thermal energy content of Earth, assuming that the core has radius 3480 km, temperature 4000 K, density 11 000 kg/m³ and heat capacity 800 J/K kg, and that the rest of the planet, dominated by the mantle, has radius 6370 km, temperature 2000 K, density 4500 kg/m³ and heat capacity 1200 J/K kg. Numbers for a more detailed model of the planetary interior can be found in the appendices to [212].

32.3   Check eq. (32.4) by estimating the radioactive heat production rate using data from Table 20.6. Take the fractional abundances of ^{235}U, ^{40}K to be 0.72%, 0.012%. Take the average energy release in ^{40}K decay to be 0.71 MeV.

32.4   Assume that a region of continental crust has a typical surface heat flux of 65 mW/m², crustal density of 2750 kg/m³, and typical crustal abundances of radioactive nuclides (as given in Table 20.6). If 60% of the

surface heat flux comes from radioactivity in the upper crust, to what depth does this typical crustal composition extend?

32.5   Consider a house in a northern location that needs a total of 100 GJ of thermal energy over a winter season from November through March. Assume that the house has a ground source heat pump that operates with a CoP of 3. The ground around the house has reasonable moisture content, a heat capacity of 3 MJ/m³ K, and a thermal conductivity of 1.5 W/m K. Ground temperatures average 10 °C at 6 m. Compute (a) the volume of ground whose temperature must be reduced by 8 K in order to provide thermal energy to supply the season's heating energy needs for the house, and (b) the area of a horizontal piping field needed for the rate of thermal energy replacement through ground conduction to match the average rate of power needed over the winter, assuming a gradient near the pipes of 8 K/m.

32.6   Compute the drawdown coefficient (32.11) in a simple model where the geothermal reservoir is a large cylindrical volume with height h much smaller than the radius r (see Figure 32.17). Assume that the pressure is constant at p_w in a small cylinder of the radius of the wellbore r_w, and that flow is radial and vertically uniform outside that volume. Show that

$$C_D = \frac{\eta}{2\pi K h \rho} \ln\left(r/r_w\right).$$

[Hint: Use conservation of mass to show that $|u| \propto 1/r$, and note that $\nabla f(r) = \hat{r} f'(r)$.]

32.7   Consider a geothermal reservoir (see Figure 32.17) at a depth of 1.5 km, modeled as a cylindrical volume of radius $r = 5$ km and height $h = 500$ m, with pressure at the periphery of $p_r \cong 15$ MPa, pressure at the bottom of the well of $p_w \cong 12$ MPa, and permeability of $K \cong 50$ mD. Use the answer to Problem 32.6 to estimate the mass flow rate through a wellbore of radius $r_w \cong 0.125$ m. Assume that the fluid is water at roughly 180 °C, where the viscosity is $\eta \cong 0.15$ cP.

32.8   Given a geothermal reservoir containing water at temperature $T = 200$ °C at a depth of 1.5 km, use steam tables to determine the density and vapor pressure of the water. Assuming that pressure in the wellbore

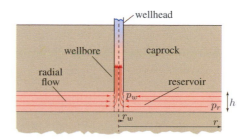

Figure 32.17 Thermal reservoir for Problems 32.6 and 32.7.

decreases according to the overlying weight of water, at what well pressure P_w will the flash occur in the wellbore?

32.9 For a geothermal well that has been running for some time, pressure at the bottom of the well is $p_w \cong 15$ MPa at a depth of 2 km. Assume the fluid is water at 280 °C. Estimate the location of the flash horizon. (You may neglect the contribution to the pressure from the weight of the vapor above the flash horizon.)

32.10 [T] Assume that enthalpy is a linear function $h(s, T) = as + bT$. Consider a single-flash cycle, assuming that the saturated liquid and vapor curves between T_- and T_+ form an isosceles trapezoid, i.e.

$$s_{\text{liquid}}(T_+) - s_{\text{liquid}}(T_-) = s_{\text{vapor}}(T_-) - s_{\text{vapor}}(T_+).$$

Prove that the optimal separator temperature is

$$T_s = (T_+ + T_-)/2.$$

32.11 Consider the efficiency computed in the example single-flash cycle in Example 32.5. Use thermodynamic data on water to show that running the separator at $T = 140$ °C or $T = 160$ °C will decrease the efficiency of the ideal cycle.

32.12 Compute the (ideal) efficiency of a double-flash plant with the same conditions as the example single-flash plant described in Example 32.5. The cycle is labeled in Figure 32.15. You will need to choose temperatures for the two separators. Try using $T_{s1} = 183.3$ °C, $T_{s2} = 116.7$ °C. Why do you think this would be better than, for example, $T_{s1} = 150$ °C, $T_{s2} = 100$ °C?

32.13 Estimate the energy extracted from 1 km^3 of granite if cooled from 300 °C to 200 °C. Assume that a power plant runs between the (time-varying) temperature T_g of the granite and an ambient temperature of 20 °C and reaches 1/2 of the Carnot efficiency at each T_g. For how long could a 100 MW power plant run off this energy?

Fossil Fuels

Fossil fuels have driven the development of the modern world, and they will continue to play a major role in human energy use for many decades to come. Coal-powered steam engines enabled the industrialization of Western Europe and the US in the nineteenth century. With the development of coal-powered ships and trains, and the birth of coal-based electricity production, by 1900 fossil fuels had become the dominant global source of energy. Use of coal for electricity continued to grow through the twentieth century, while petroleum-based products powered further advances in transportation and agriculture. The internal combustion engine significantly increased the efficiency, safety, and flexibility of land and sea transport in the early part of the century, while the development mid-century of the gas turbine made possible rapid air transport. Fossil fuel derived energy and fossil fuel based agricultural methods have made possible the development of a society in which individuals can move quickly and easily both locally and internationally, consuming food and other products produced cheaply and efficiently around the globe. In 2014, 87% of the roughly 550 EJ of energy used by humanity was derived from fossil fuels (see Figure 33.1). A graph of world energy use since 1820 (see Figure 33.2) shows the dominance of fossil fuels in the industrial age.

A principal feature of fossil fuels that has made these developments possible is their high energy density. Additionally, coal is very stable and liquid fuels, such as gasoline, diesel fuel, and kerosene, are highly versatile and easily transferred from refinery to end-use locations such as the gas tank of an automobile. The energy density (HHV)[1] of coal ranges from 15 to 35 MJ/kg, and that

Reader's Guide

Fossil fuels – coal, petroleum, and natural gas – have dominated human energy use for the last century. This chapter surveys the origins, nature, extraction, uses, carbon footprint, and availability of these resources from a scientific perspective. This topic leads naturally into the later chapters on energy and climate change.

Prerequisites: §9 (Energy in matter), §11 (Internal combustion engines), §12 (Phase-change energy conversion), §13 (Power cycles), §26 (Biofuels), §32 (Geothermal energy).

of gasoline is around 47 MJ/kg. By comparison, a typical lead-acid battery has an energy density around 0.1 MJ/kg, and even modern lithium-ion batteries still have energy densities below 1 MJ/kg (see §37). The high energy density of fossil fuels makes them efficient not only to transport and to use, but also to extract from the rich deposits in which they occur. The coal that may be extracted in a matter of a week or two from a coal seam of area 1 km^2 and thickness 1 m yields of order 10 PJ of electricity at the 30% conversion efficiency of a coal power plant; a solar thermal plant covering the same area of desert land and with a typical conversion efficiency and insolation would need to operate for several decades to produce a comparable amount of electrical energy (Problem 33.1).

The high energy density of fossil fuels derives from the enthalpy of combustion of *hydrocarbon* molecules. As

---

[1] Recall, as discussed in §9.4.3, that the lower (upper) *enthalpy of combustion*, also known as the lower (higher) *heating value*, abbreviated LHV (HHV), or net (gross) *calorific value* of a

hydrocarbon fuel is computed assuming that the heat of condensation of the produced water is not (is) included in the enthalpy balance (Problem 33.3). Unless explicitly noted, throughout this chapter we use the *higher heating value*, as this represents the full energy content that can in principle be extracted from a given fuel.

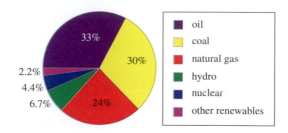

Figure 33.1 2014 world primary energy consumption by fuel [9]. Petroleum, coal, and natural gas together account for 87% of all primary energy.

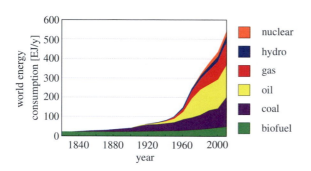

Figure 33.2 Global energy consumption by source since 1820, in EJ. After [225].

Table 33.1 Enthalpies of combustion: lower and higher heating values for a variety of hydrocarbon molecules. (HHV from [22], LHV computed from HHV as in Example 9.2.)

Hydrocarbon	LHV (MJ/kg)	HHV (MJ/kg)	Molar mass (g/mol)
Methane (CH_4)	50.0	55.5	16.0
Ethane (C_2H_6)	47.5	51.9	30.1
Propane (C_3H_8)	46.3	50.3	44.1
Butane (C_4H_{10})	45.7	49.5	58.1
Pentane (C_5H_{12})	45.3	49.0	72.1
Hexane (C_6H_{14})	44.7	48.3	86.2
Octane (C_8H_{18})	44.1	47.5	114.2
Dodecane ($C_{12}H_{26}$)	43.0	46.4	170.3
Benzene (C_6H_{12})	38.5	41.8	78.1
Graphite (C)	32.8	32.8	12.0

discussed in §11, energy is released from hydrocarbons through the combustion reaction

$$C_nH_m + (n + \frac{m}{4})O_2 \rightarrow n\,CO_2 + \frac{m}{2}\,H_2O\,. \quad (33.1)$$

A table of lower and higher heating values for a variety of simple hydrocarbons is given in Table 33.1. These combustion energies can be roughly understood in terms of the constituent bond energies within the molecule (Problem 33.2).

Although fossil fuels are a finite resource, they will continue to dominate human energy use for years to come. Existing petroleum reserves combined with anticipated future discoveries seem adequate to supply at least a substantial fraction of world oil demand for many decades into the future. Alternative sources of oil including tar sands and oil shale are also abundant, though their use produces substantially greater carbon emissions per joule of usable energy than use of oil from conventional petroleum resources. There are tremendous reserves of coal, which could continue to power global electrical systems at the current rate for over a century. Natural gas, which emits the least carbon dioxide per unit of energy released of all the fossil fuels, is playing an increasingly significant

role in power production, and is likely to increase further in importance in the coming decades. Technologies that enable the conversion of coal to liquid or gaseous fuels, could in principle eventually replace oil and natural gas, which are less abundant than coal. Natural gas can also be converted to liquid fuel. Thus, were it not for the impact on climate of carbon emissions from extensive fossil fuel use (§34, §35), and other environmental effects and economic considerations, it would be possible for fossil fuels to continue to dominate human primary energy consumption for at least many decades into the future.

Given the role that fossil fuels have played in past and present energy systems, and their likely continued importance in the future, these energy sources must play a central part in any discussion of energy policy or economics. There is a tremendous body of scientific and engineering knowledge relating to the discovery, extraction, and utilization of fossil fuels. Over the last two centuries a sophisticated set of technologies have been developed by fossil-fuel-related industries. Many physicists and other scientists devote their careers to identifying new deposits of fossil fuel resources using a variety of probes of underground structure such as *seismic surveys* and *well logs*, modeling the details of reservoir dynamics and the extraction process, and optimizing the extraction of energy from fossil fuels in systems from power plants to automobile engines to fertilizers. We touched on some of the issues involved in exploration and reservoir modeling in the related context of geothermal energy in §32. Modeling of oil fields is now an extremely sophisticated and computation-intensive enterprise. We do not go deeply into the physics of fossil fuel discovery or extraction in this book. We give a scientific overview of these topics, as well as of the origins, energy content, uses, and transformations of fossil fuels. In §11 and §13 we analyzed the physics behind the transformation of thermal energy from fossil fuel combustion into kinetic energy (in automobile engines) and electromagnetic energy (in power plants). This chapter puts those analyses into the context of available fossil fuel resources.

We describe in this chapter each of the main categories of fossil fuels: coal (§33.1), petroleum (§33.2), and natural gas (§33.3). Emerging fossil fuel resources such as *light tight oil* and *shale gas* are included in the discussion, as are unconventional fossil fuels such as *tar sands* and *oil shale*. In §33.4 we describe conversion from solid and gas fossil fuels to liquid hydrocarbons. We conclude in §33.5 with a brief summary of the essential energy, carbon, and resource information for fossil fuels.

For further reading and a more in-depth technical study of the science of fossil fuels, there are number of good texts covering each of the major fossil fuel sources. Thomas [226] contains a clear description of the geology of coal, including geophysical methods used in exploration and coal mining. Miller [227] gives a comprehensive treatment of coal energy systems, including emission issues. A concise introduction to modern developments in coal-fired power plants, along with a brief history of power plant designs is given by Termuehlen and Emsperger [228]. Gluyas and Swarbrick [229] gives an introduction to petroleum geoscience, with emphasis on exploration, while Bjorlykke [230] takes a more physics-based approach. Smil [231] gives a good non-technical overview of the oil industry, including the history and geology of oil, as well as political and economic commentary. An overview of the current status and future prospects for natural gas is given in an MIT report [232]. A particularly useful source of statistics on fossil fuel use can be found in [9].

33.1 Coal

Coal (Figure 33.3) is an abundant and easily accessible fossil fuel that is widely distributed around the world. Coal currently plays a central role in electric power generation worldwide. In 2014, coal provided 30% of the energy consumed worldwide, roughly 160 EJ. The global rate of coal production[2] since 1980 is shown in Figure 33.4. **Proven**

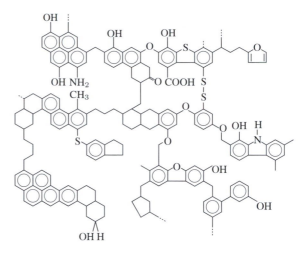

Figure 33.3 A sample of the chemical structure of coal. From [50].

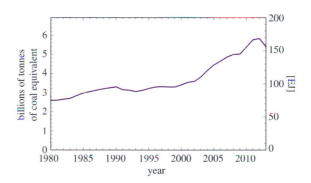

Figure 33.4 Global coal production since 1980 [12]. 1 tonne of coal equivalent $\cong$ 29.31 GJ.

reserves of coal would be sufficient to sustain the current level of use for over 100 years.[3] Thus, without a substantial change to the status quo, coal will continue to play a central role in power generation for the foreseeable future. Current

---

[2] Note that in the context of coal, oil, and other natural resources, the word "production" refers to the extraction of the resource from its natural location for human use.

[3] The term *proven reserves*, used in the fossil fuel industry, corresponds to *reserves* in the nomenclature summarized in Box 16.1. Thus *proven reserves* are those resources that can be economically extracted and produced at the present price. Other less certain resources, which may be classified as *demonstrated*, *inferred*, or *undiscovered*, are usually considerably greater than proven reserves. Quoted proven reserves therefore considerably underestimate ultimately recoverable resources as prices rise in response to increasing scarcity.

statistics regarding coal can be found, for example, on the website of the World Coal Association (WCA) [233].

33.1.1 Origins of Coal

Fossil fuels are formed when organic material, deeply buried under sediments and deprived of oxygen, is subjected to high temperature and pressure at increasing depth. Over tens or hundreds of millions of years, the complex organic molecules that compose living organisms break down, release moisture, oxygen, and other volatile substances, and recombine into simpler hydrocarbons.

The exact process by which organic material is transformed into coal depends both upon the nature of the original material and the sequence of conditions to which the material is subjected. Coal is classified in several ways. Because it is closely linked with the process by which coal is formed, we focus first on the classification by **rank**. The rank of a particular coal specimen refers to the extent to which the original organic material has undergone **coalification**, and ranges through peat to lignite, sub-bituminous, bituminous, and anthracite. Note that the precise definition of the different ranks is not universally standardized or agreed upon; thus the numbers we give here for carbon and energy content for the different ranks are intended as a rough characterization of the different ranks, and may differ from values used by specific organizations.

Peat The first stage in this series of transformations is the formation of **peat**. Peat is organic matter with elevated carbon content, produced from mosses or other plants. While plant matter is typically 50% carbon, peat typically contains 60% carbon by mass. Peat forms most readily in wet areas such as bogs or marshes, where dead plant material is often submerged in an environment without adequate oxygen for complete decomposition by aerobic bacteria, which would release the carbon again to the atmosphere. When dried, peat burns readily, and is used for heat and cooking in many places around the world. The energy content of completely dry peat composed purely of matter of organic origin with minimal mineral content is somewhat below 20 MJ/kg. While this energy/mass ratio is comparable to some coals, peat is much less dense, so that the energy content per unit volume is lower than that of any coals by a factor of four or five. Most peat is not dried completely, and the residual moisture content lowers the calorific value by a factor as large as two, giving a typical energy content of 10–15 MJ/kg. Furthermore, in most locations with large quantities of peat, the organic content is not 100%, further lowering the energy density. There are substantial deposits of peat throughout the world (estimated at $\sim 4.5 \times 10^6$ km^2 [234]) ranging up to 20 m in thickness. Finland,

for example, which has one of the highest proportions of wetlands of any country in the world, has substantial peat resources; there cogeneration is used to provide both electricity and heat from burned peat, which provides 6% of the country's total energy supply.

Lignite When peat is buried more deeply, pressure and temperature increase. This leads to dehydration and compaction of the material, with associated loss of oxygen. As the O/C ratio and moisture content decrease, the material begins to form the lowest grade of coal, known as **lignite**. Lignite ranges from 60–80% carbon, and has an energy content below roughly 18 MJ/kg.

Bituminous coal At greater depths and pressures, the coalification process progresses further. As the O/C ratio decreases further and the structure of the organic molecules continues to break down, **bituminous coal** is formed, with 80–92% carbon and an energy content ranging up to 35 MJ/kg; coal with energy content in the range 18–28 MJ/kg is generally described as **sub-bituminous coal**. The process of **bituminization** is believed to occur at temperatures around 100–150 °C. In most locations, the degree of coalification increases progressively with depth beneath the surface, with the rate of increase correlated with the geothermal gradient. For a geothermal gradient of 30 °C/km, lignite is formed in the two kilometers directly beneath the surface, and increasing bituminization occurs from 2 to 6 km beneath the surface.

Anthracite Bituminous coal that is subjected to substantially higher temperature and pressure undergoes further metamorphic change in which the molecular structure breaks down further, and remaining oxygen, sulfur, and other **volatiles**, as well as some hydrogen, are released. The remaining carbon-rich material forms the most dense type of coal, known as **anthracite**. Anthracite is the purest type of coal, with a carbon fraction ranging from 92.1% to 98%. The energy content of anthracite can be in the range 28–33 MJ/kg, though sometimes the term *anthracite* is reserved for coal with energy content >32.5 MJ/kg. Anthracite and bituminous coal are often referred to as **black coal**, while **brown coal** refers to sub-bituminous coal as well as lignite.

Graphite If coal is subjected to even higher temperatures and pressures, the remaining hydrogen atoms are released, mostly in the form of methane. Under sufficiently high pressure and temperature and subject to other geological stresses, the carbon atoms begin to align into a regular lattice, through the process of **graphitization**. **Graphite**, a crystalline form of pure carbon, can be considered the end point of the evolution of coal – although graphite can be formed in other ways and is even found in meteorites. As

Figure 33.5 A rendering of a *coal forest* from the Carboniferous by the German illustrator Heinrich Harder.

Figure 33.6 Primary world coal deposits. From [235].

the number of hydrogen atoms decreases, coal becomes less easy to ignite and the energy content asymptotes to 33 MJ/kg for pure graphite.

Coal is generally found in **seams**, typically ranging from less than a meter to several meters in thickness. Particularly thick seams can be 10–20 m or more thick. Irregularities in seams arise from many geological processes. In particular, nearby intrusions of magma can cause more rapid coalification in localized regions, leading, for example, to isolated deposits of anthracite in regions generally dominated by brown coal.

The oldest substantial coal deposits date back to the **Carboniferous period** around 300 Ma (Ma = million years ago, see Appendix D.2). Before this time terrestrial plant cover was not sufficient for large-scale coal formation. During the Carboniferous period, the eastern US and much of Europe were located near the equator, and great swamp forests known as *coal forests* gave rise to large coal deposits.

Most of the known anthracite deposits are in coal dating to the Carboniferous and subsequent **Permian** periods. Coal deposits dating to the **Jurassic** and **Cretaceous** (206 Ma–66 Ma) are less extensive, and contain substantial amounts of bituminous coal. More recent coals from the **Tertiary period** (66 Ma–1.6 Ma) are dominated by lignites, or brown coal. These coals are more varied than the older coals, due in part to the increased diversity of plants and in part because newer coals have been subject to less metamorphosis, so that distinguishing characteristics associated with their different sources are still present.

The seams of Tertiary coals are generally thicker than those of the older coals.

Coal is broadly distributed geographically, with the most significant deposits in North America, Asia, and Australia. As shown in Figure 33.6, fewer coal deposits are found in South America and north and west Africa.

33.1.2 Coal Properties and Classification

In addition to *rank*, coal is classified according to a number of different properties. The **type** of coal refers to the materials of which the coal is composed. This is determined in part by the kind of organic material from which the coal was formed. The different types of basic organic units, called **macerals**, are somewhat analogous to the minerals that make up inorganic rock. Macerals are generally divided into three categories: **huminite/vitrinite**, woody materials such as tree trunks, roots, or bark; **liptinite**, algae, spores, resins and cuticles; and **inertinite**, oxidized plant material. Coal is often referred to through one of these types or a combination of types; vitrinite has the highest tendency to form coal, while liptinite has greater hydrogen content and favors petroleum production, and inertinite has little potential for fossil fuel production. Another important feature characterizing the type of coal is the fraction of inorganic minerals, such as clay minerals, that have mixed into the coal. The reflectance of coal, particularly vitrinite, generally increases with rank, and is often used as a rank indicator in coals with high carbon content such as anthracite.

The **quality** of coal refers to the overall utility of the coal, which depends upon rank, type, and other characteristics. The **ash content** of a coal refers to the fraction of the coal that remains as inorganic residue after the coal is combusted. The **volatile content** of coal represents the fraction that is liberated at high temperature, including oxygen and sulfur in addition to various volatile hydrocarbons. In general, the quality of the coal increases with its rank and with lower ash and volatile content. Anthracite contains only a few percent volatile matter and ash, while typical brown coal can have 50% or more volatile matter.

Coal is generally measured in tonnes (1000 kg), or in **short tons** (907.2 kg). The energy content of coal can vary widely, ranging from 15 MJ/kg to 35 MJ/kg depending upon the rank of the coal, ash content, and other features. As a unit of energy, a **tonne of coal equivalent** (TCE) is defined as

$$1 \text{ tonne of coal equivalent} = 29.3076 \text{ GJ}. \quad (33.2)$$

Note that a typical tonne of coal contains somewhat less than this amount of energy. The 924 Mt of coal produced in the US in 2013, for example, corresponded to 21.1 EJ of primary energy, averaging roughly 25.2 GJ/tonne.

33.1.3 Coal Exploration

Geophysical methods play an important role in exploration for and identification of new fossil fuel resources. Some of these methods were described in the context of exploration for geothermal energy resources in the previous chapter. Seismic surveys are of particular value in coalfield exploration, though other surface-based methods such as gravitational surveys, resistivity mapping, and magnetic surveys are also useful. Geophysical surveying or *logging* of boreholes also provides detailed information about the structure of rock and coal seams in a given area. There is a great deal of physics involved in all these techniques; we mention briefly just a few of the relevant issues involved.

Seismic Surveys   A primary method used for coalfield exploration is **seismic surveying**. Coal is significantly less dense and has a lower seismic wave velocity than most other rock structures; seismic velocity in hard coal is generally between 1.8 and 2.8 km/s compared to ∼3.6 km/s in sandstone and ∼5.5 km/s in limestone. Thus, seismic waves reflect strongly when passing between coal seams and more dense rock (see Box 33.1). By producing seismic waves at the surface using either explosions or an earth compacter, and then measuring the reflected waves, coal seams can be accurately identified to a depth of 1– 1.5 km. Additional features such as faulting, washed out regions of the seam, or changes in the thickness or structure of the seam can also be identified using this approach. As technology for seismic surveys and computer analysis software has progressed over recent decades, this has become a highly accurate method for determining coalfield structure from the surface. Seismic methods using horizontally propagating waves are also used to determine the

structure of specific coal seams. Since the density and wave velocity within a coal seam are substantially lower than in the rock above and below the seam, the coal seam acts as a **wave guide**.[4] By generating a seismic wave that propagates along the seam starting from an underground point in the seam, faults or other obstructions in the seam can be accurately identified. This technique is heavily used in *longwall mining* (see below).

Geophysical Borehole Logging   A more precise picture of the strata in a given location is given by **borehole logging**. After a borehole is drilled, a measuring device called a **sonde** is lowered into the hole. As the sonde is gradually lifted through the borehole, it can make a variety of measurements including determination of geophysical parameters using sonic, resistivity, and caliper (borehole diameter) probes. Among the most useful measurements in coalfield exploration are radiation-based measurements. A **gamma ray log** records the level of γ-rays coming from radioactive isotopes, particularly ^{40}K, in the surrounding rock. Since coal is generally low in radioactive isotopes, a low γ-ray count can be indicative of the presence of a coal seam. Within the coal seam, an increase in γ-ray count can correlate with increased ash content, though this is highly dependent on the type of inorganic material mixed into the coal; the radioactive isotope content of different impurities can vary significantly.

A **density log** is produced using a sonde with an energetic γ-ray emitter on one end and two scintillation detectors at different distances from the emitter. The detectors measure γ-rays that are scattered back to the sonde by electrons in the rock around the borehole. The two detectors together make it possible to calibrate for the borehole configuration and to measure the density of the surrounding rock. The number of γ-rays that reach the detectors varies *inversely* with the density of the surrounding rock since higher density gives rise to more scatterings, which reduce the γ-ray energy below the detectors' threshold. Since coal has a much lower density than most other rock types, coal seams are easily identified as local regions in the log with high counts of returning gamma rays.

33.1.4 Coal Extraction Methods

One reason why coal is inexpensive relative to other energy sources is that highly efficient extraction technologies have been developed over the last century. In general, coal

---

[4] A wave guide is designed to constrain the propagation of a wave to a confined channel by imposing appropriate boundary conditions on the degrees of freedom supporting the wave.

Box 33.1   Geophysical Methods: Seismic Waves

Just as sound passes through the air as a wave of oscil-lating pressure, energy can be transmitted through liquid and solid material by the propagation of waves. Waves propagating within or on the surface of Earth are called seismic waves (see §32.1.1). As illustrated in the figure at the right, waves propagating in the bulk material beneath the surface are divided into primary **P-waves**, associ-ated with variations in pressure, where the motion of the material is in the direction parallel to wave propagation, and secondary **S-waves**, associated with shear stress and motion transverse to the direction of wave propagation. The designation primary/secondary comes from the early

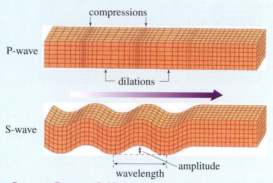

observation that P-waves travel faster and arrive sooner than S-waves. Because fluids cannot support shear stress, only P-waves propagate in fluids. Seismic waves have played an important role in understanding the internal structure of Earth, as discussed in the previous chapter, and play an important role in coal and oilfield exploration and analysis.

A full treatment of seismic waves involves some details from the theory of elasticity, but the basic structure is quite simple and intuitively clear. A solid is characterized by two **elastic moduli**, which are conceptually similar to the spring constant k of an oscillator, in that they measure the restoring force that develops in response to a deformation of the material. The two elastic moduli give the ratio of *stress* to *strain* for compressions and shear deformations. As discussed in §29, pressure and shear stress have units of force/area. **Strain** is a fractional deformation in a dimension of a solid; for example, a rod that is uniformly compressed to 99% of its original length has a strain of $u = 0.01$ along the axis of the rod. The **bulk modulus** K and the **shear modulus** μ give the ratios of pressure and shear stress over the corresponding strains. The elastic modulus governing the motion of P-waves is a combination of the two moduli K and μ, $\chi = K + 4\mu/3$ (Problem 33.5). The wave equation for P-waves analogous to eq. (4.2) essentially follows directly from Newton's law $F = ma$,

$$\frac{\partial^2 u}{\partial t^2} = \frac{\chi}{\rho}\frac{\partial^2 u}{\partial x^2} = v^2 \frac{\partial^2 u}{\partial x^2} \, ,$$

and describes waves with velocity $v = \sqrt{\chi/\rho}$. A seismic P-wave in limestone travels at roughly 5.5 km/s, in sandstone at roughly 3.6 km/s, and in hard coal at between 1.8 and 2.8 km/s.

Like the light waves described in §4.5.2, when seismic waves pass from one medium to another medium with dif-ferent properties, part of the wave energy is reflected and part continues to propagate into the second medium. When a seismic wave propagates from sandstone or limestone into coal, the amount reflected is determined by the differ-ence in **acoustic impedance** $v\rho$ between the two media, which is quite substantial. This property is used in coal exploration; a seismic wave with a gradually changing fre-

quency may be produced on the surface, for example by a Vibroseis™ truck, as shown at right. The truck lifts into the air and vibrates, producing powerful sound waves that are reflected by underground structures to receivers arrayed across the surface. The vibration begins at a low frequency and proceeds continuously to higher frequencies, so that the part of the waveform measured by the receivers can easily be identified and matched to the source to determine the time of propagation of each component of the reflected wave. This information is used to locate underground coal deposits (Figure 33.7).

(Image credits: (Top) US Geological Survey; (Bottom) NEES hub)

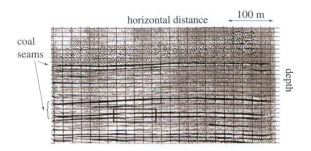

Figure 33.7 Seismic profile showing two coal seams, Wyodak, Wyoming, USA. Depth beneath the surface can be estimated from round-trip travel time. From [226]. (Credit: John Wiley and Sons (2002))

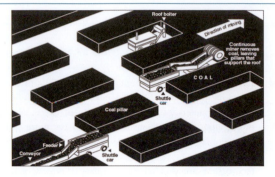

Figure 33.9 In a room and pillar coal mine, pillars to support the mine roof are left behind as the coal is removed. Often, the pillars are later worked as the roof collapses behind. (Credit: Peabody Energy Corporation)

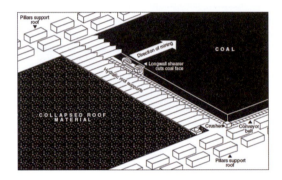

Figure 33.8 A longwall mining operation is highly automated. A longwall shearer moves back and forth across the face being mined, cutting coal that falls onto a conveyer system for removal. As the shearer advances, supports are moved in tandem, and the mine is allowed to collapse in the region behind the shearer. (Credit: Peabody Energy Corporation)

Figure 33.10 An opencast lignite coal mine at Garzweiler in Germany. The mine covers 48 km². (Credit: ©Raimond Spekking ICC-BY-SA 4.0 license via Wikimedia Commons)

extraction methods can be divided into underground and surface methods.

Underground methods are generally used for extracting coal more than 100 m beneath the surface. The main underground methods are **longwall mines** and **room and pillar mines**; these approaches can also be combined in a single mine. Underground mines are often the only way to access seams of the highest quality black coal.

Longwall mining In a longwall mine, all coal is removed from a roughly rectangular "panel" within a seam of coal (Figure 33.8). This method is usually used for seams 1.5–4 m thick, though variations are possible for thicker seams. The panel typically has a width of 100–300 m, and may extend for as much as several kilometers. Longwall mining is highly automated; a single *longwall shearer* can mine

at a rate on the order of 50 t/h, and production rates in longwall mines are on the order of 10 kt/employee-year.

Room and pillar mining In this approach, a pattern of tunnels is dug leaving substantial intact pillars between the tunnels to support the roof (Figure 33.9). This method is cheaper and easier to implement than longwall mining, but less productive. Room and pillar mining has also been largely automated in some places.

Opencast and strip mining Surface extraction methods known as **opencast** and **strip mining** proceed by excavating all material from the surface downward, including material above and between coal seams (Figure 33.10). Surface mining is generally only viable to depths of less than 100 m. Surface mining has fewer technical complications than underground methods, though extra costs are involved in even partial restoration of areas impacted by surface mining. Surface mining is almost exclusively used for extraction of brown coals. As some easily accessed

---

Box 33.2 Extraction of Coal Energy

One way to appreciate the highly compact and effective nature of fossil fuels is to consider the rate of extraction of energy in the form of coal from a standard coal mining operation. In a typical US longwall mining operation, the rate of coal extracted per employee is on the order of 10 kt/employee-year. Assuming that the coal being mined is anthracite with an energy density of 30 MJ/kg, this represents some 300 000 GJ of energy extracted in one year from the efforts of each person involved. It is illuminating to consider the efforts needed to extract this quantity of energy from other systems (see Problem 33.4).

---

black coal reserves have been exhausted and technology for surface mining and using brown coal effectively has advanced, new mining has largely shifted to surface methods in recent years. Large brown coal surface mines can produce up to 40 Mt/yr.

Both underground and surface mining methods can have numerous environmental impacts. Surface mining can affect large areas of land, though in many developed countries reclamation efforts are mandatory that can at least in part restore affected landscapes. Underground mining can also impact substantial land areas through subsidence. Surface and underground coal mining produces a substantial amount of liquid effluents (**acid mine drainage**) that can affect surface and ground water. Coal mining releases trapped methane, which acts as a strong greenhouse gas (see §34). There are also many health and safety issues associated with coal mining. For a much more detailed discussion of health and environmental issues related to coal extraction, see [227].

33.1.5 Uses of Coal

Coal has been burned as a source of heat for warmth and cooking for thousands of years, and has been used in forges and for metalworking since about 1200 AD. The primary contemporary uses of coal in industrialized nations are for electrical power generation, steel and iron production using *coke*, and in cement production. In 2015, 92% of US coal use was in the production of electrical power. Globally, however, other uses of coal (including ongoing direct use in many countries for domestic heating and cooking) represent substantial energy use and contribute significant amounts to global CO_2 emissions. As mentioned in §9, some 5% of global CO_2 emission is from cement production, of which 40% comes from coal

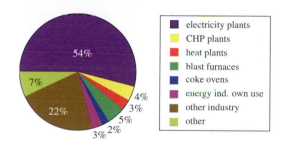

Figure 33.11 World coal use in 2014 (includes peat and oil shale) [236].

combustion. Coal can also be converted to a gas (§33.1.7) or liquid hydrocarbon fuel, as discussed in §33.1.7 and §33.4, though currently coal-to-liquid fuel conversion is not economically favorable or widely practiced. In the past, coal was used as a feedstock for production of various chemicals. Petroleum now has replaced coal in this role in most situations, though a resurgence of coal as a chemical feedstock may occur as oil supplies become less plentiful.

33.1.6 Coal and Coke in Steel and Iron Production

Coke is produced by heating coal without oxygen to a temperature in the vicinity of 1000 °C for an extended period of time, generally from one to several days. In this process, volatiles are released and the coal increases in strength and carbon density, producing the material known as coke.

Coke is primarily used in steel and iron production. In the smelting of iron ore, coke acts not only as a heating fuel but also as a reducing agent (see §9.5.3) for the iron. Because most of the volatiles are removed in coke production, burning coke produces very little smoke; for this reason, coke is often used in furnaces and other combustion environments where smoke needs to be minimized.

There are a number of constraints on the types of coal that can be used in coke production. Coal used for coke must be low in ash and sulfur content, and should have a relatively high (> 40%) vitrinite content; this type of maceral fuses well into a stronger coke material. The rank of coal used is also fairly constrained, coke is generally made from bituminous coal but not from lignite or anthracite.

33.1.7 Coal and Electrical Power Generation

In 2014, coal produced nearly 40% of the world's electricity. Typical coal power plants currently use pulverized coal as a fuel. Reducing the coal to a fine powder increases the surface area to volume ratio of individual particles, allowing the coal to burn more quickly. Before the widespread adoption of pulverized coal, *stoker*, or *fixed-bed*, designs were used, with much larger fuel particles having a size of a centimeter or more. Coal of all ranks, from lignite to black coal, is used in power plants, though most coal used for electricity production is in the middle of the range, with energy content typically from 20–25 MJ/kg. The thermal energy from the combusted coal is transferred to steam that drives a turbine in a Rankine cycle as described in §13. As of 2014, average coal plant efficiency worldwide was roughly 33%, though ultra-supercritical (USC) plants (see below) have efficiencies of up to 45%. A number of features of the coal supply must be considered for use in any given plant. Ash and volatile content are important, as is moisture content; greater ash and volatile content reduces the energy density of the fuel, and moisture increases the difference between higher and lower heating values of the fuel. Levels of trace elements such as sulfur affect emissions, as discussed below.

Over the last 100 years, the efficiency of conversion of thermal energy from the combustion of coal into electrical energy has increased steadily, from steam turbine generators with thermal-electric conversion efficiencies of 15% in the early 1900s to modern USC plants operating at three times that efficiency. The principal advances that have made this improvement possible are a significant increase in steam turbine efficiency (from 60% or so 100 years ago to around 92% currently), and an increase in Rankine cycle efficiency due in large part to an increase in the temperature and pressure at which the steam cycle operates. The improvement in turbine efficiency, like the improvement in wind turbine efficiency discussed in §28, has come in part from an increased ability to understand, model, and design complex 3D turbine geometries. Rankine cycle efficiency was improved in the early part of the twentieth century by implementing systems to preheat the air used for combustion and the water feeding the boiler using waste heat from the system, as well as by the shift to pulverized coal. We review briefly some of the different coal plant technologies that have enabled a further increase in overall efficiency through improved operating conditions.

Supercritical and Ultra-supercritical (USC) Plants

As discussed in §13.3.6, Rankine cycle efficiency is improved when the pressure is sufficiently high that the heat transfer phase of the cycle is supercritical, bypassing the saturation dome. To handle pressures above 22.1 MPa and temperatures of 550 °C and above, improved steel alloys were developed in the 1950s and 1960s. Modern **ultra-supercritical** (USC) power plants operate at pressures of 25 MPa or higher and at temperatures of 600 °C or above, with a conversion efficiency of up to 45%. Further developments in material technology, for example using nickel alloys, could make it possible to operate plants at pressures of >30 MPa and temperatures of 700 °C or higher, which could raise coal plant efficiencies to 47–48%. Efficiency as high as 50% may be attainable with further advances in turbine technology [228]. Note that the steam temperatures relevant for efficiency are generally much lower than the temperature at which the coal itself is combusted, which can exceed 1200 °C in pulverized coal combustion systems.

Fluidized Bed Combustion (FBC)   A new technology for coal plants known as **fluidized bed combustion** has been under development since the 1980s, and began to see widespread commercial deployment in the 1990s. In a fluidized bed combustion system, strong jets of air are passed upward through the coal or other fuel, keeping the fuel particles in a turbulent "fluid-like" motion during the combustion process. This technology was originally developed to reduce power plant emissions. Because heat transfer is enhanced in the fluidized bed, combustion can occur at a lower temperature than in a standard pulverized coal operation, generally 800–900 °C. This suppresses the formation of nitrogen oxides, which form at substantially higher temperatures. Limestone placed in the fluidized bed can absorb sulfur, reducing emissions of sulfur dioxide as well. A further benefit of the lower combustion temperatures is that a variety of fuels can be used. Fluidized bed plants can operate on a mixture of coal with wood, biofuels, coke, petroleum, or other combustible fuels. Current efficiencies of FBC plants are over 40%. This technology is not extremely expensive, and is a promising direction for future development, though the low combustion temperature in current FBC systems limits the efficiency that can be attained without further development of the technology.

Integrated Gasification Combined Cycle (IGCC)

In §13, the combined cycle plant was described, in which a high-temperature gas turbine is combined with a lower-temperature standard Rankine turbine in a high-efficiency power plant. Coal cannot be used directly in a combined cycle plant, since even if it is pulverized, the combustion products contain *fly ash* (see below) that would destroy the turbine. As mentioned in §13, coal can be combined with superheated steam to produce a combustible synthetic gas known as **syngas**, which contains principally a mixture of hydrogen and carbon monoxide and may also include carbon dioxide. Syngas can be used as a fuel in a combined cycle plant. Beginning in the 1970s, coal gasification and combined cycle technologies have been combined into a single power plant design. In an **integrated gasification combined cycle** (IGCC) plant, a coal gasification system is integrated into a combined cycle plant, and optimized to fit the parameters required by the gas turbine. IGCC systems have been built that achieve efficiencies as high as the best standard steam cycle plants. This technology is quite expensive, but has the potential for reaching higher efficiencies than standard steam cycle plants.

Coal gasification begins with **steam reforming**, in which high-temperature steam is passed over a bed of coke forming a mixture of hydrogen and carbon monoxide known as **water gas**. The chemical reaction

$$C + H_2O(g) \rightarrow H_2(g) + CO(g) \qquad (33.3)$$

is endothermic $\Delta H_r = 131 \, \text{kJ/mol}$ and has positive reaction free energy $\Delta G_r = 92 \, \text{kJ/mol}$, so it does not occur spontaneously at room temperature. ΔG_r goes negative in the neighborhood of 1000 K, so the reaction requires energy both because it is endothermic and because the reactants must be raised to a high temperature for it to proceed. This energy can be provided by allowing into the gasifying reactor a limited amount of oxygen (or, at the cost of dilution with nitrogen, air), which reacts exothermically to produce more carbon monoxide. If additional steam is present, the water gas will react *exothermically* to produce hydrogen and carbon dioxide, in the **water gas shift reaction**

$$CO(g) + H_2O(g) \rightarrow H_2(g) + CO_2(g). \qquad (33.4)$$

Since this reaction is both exothermic $\Delta H_r = -41 \, \text{kJ/mol}$ and has negative reaction free energy $\Delta G_r = -28 \, \text{kJ/mol}$, it quickly comes into equilibrium in the reactor. The gas that exits the reactor is therefore a mixture of H_2, CO, and CO_2.[5] In practice coal gasification is complicated by

the need to drive off volatiles from the coal to form coke and by the need to integrate reactions (33.3) and (33.4) with introduction of oxygen to provide process heat for the reactor. As discussed below in §33.3.4, a process similar to coal gasification is used to produce hydrogen from methane for use in fertilizer production. More generally, steam reforming of fossil fuels is used to produce hydrogen for a variety of purposes, including hydrogen inputs used for the refining of crude oil.

Depending on the way that the syngas is produced, IGCC plants may offer advantages for carbon capture and sequestration. If the syngas is made by steam reforming coal without mixing with air, then the gas contains little nitrogen, so the final combustion products have a greater CO_2 concentration than the products of burning coal directly in air, making CO_2 capture and sequestration easier. Carbon capture and sequestration is discussed further in §35.4.2.

One drawback of the IGCC type of plant is that existing coal gasification processes have relatively low conversion efficiency. The fluidized bed and IGCC concepts have been combined in some plants. A pressurized fluid bed (PFBC) in combination with a partial gasifier can be used to balance the strengths and weaknesses of these two technologies; high-temperature gas turbines can be used with the output of the gasifier, with the residual fuel from the gasifier being used in the fluid bed. While this type of plant in principle can achieve very high efficiency, it also presents technical challenges, requiring a variety of extra plant components; the *topping* combustor for the syngas is more complicated than in a standard FBC system, and various cleanup systems add to the costs of this technology.

33.1.8 Coal Power Plant Emissions

Environmental impact from emissions represents a substantial issue for coal plants. While a portion of the ash remains in the combustion chamber (*bottom ash*), the remainder of the ash, known as **fly ash** leaves the combustion chamber with the flue gases. The fly ash can contain toxic trace elements such as lead, cadmium, and arsenic. The flue gases contain not only CO_2, but other gases such as sulfur dioxide (SO_2), and nitrogen oxides (NO_x) that can have negative environmental impacts. Flue gases also contain substantial amounts of mercury. Emissions control systems that capture fly ash and sulfur and nitrogen oxide emissions have been developed over the last 50 years. These systems can dramatically reduce the environmental

---

[5] If the water gas shift reaction is driven to completion, the resulting mixture of H_2 and CO_2 is a source of H_2, which has many industrial applications (see below).

impact of coal plants, although they add to the cost of coal power and somewhat reduce the efficiency of coal plants in which they are used. We very briefly review some of the main types of coal plant emissions and the basic types of primary emission control systems here, leaving discussion of possible carbon capture mechanisms to §35.4.2. A much more detailed discussion of coal plant emissions and related issues can be found in [227].

Fly ash and trace particles Particles can be removed from flue gas in a variety of ways. The most common mechanism used is an **electrostatic precipitator**. This is a device in which the flue gases are passed through a substantial electric field (voltage differences used can range up to $100\,\mathrm{kV}$). Ionized dust particles are then collected on an electrode and removed. Over 99.5% of the particles in flue gas can be removed with a sufficiently large electrostatic precipitator system when the flue gas is moving sufficiently slowly.

Sulfur Sulfur dioxide causes respiratory problems, and leads to acid rain and smog. A number of mechanisms can be used to reduce sulfur emissions from coal plants. Using coal with lower sulfur content reduces the level of emissions directly. In the US, for example, coal used in power plants has a maximum allowed sulfur content dictated by local regulations, generally in the range 0.8–1%. In recent decades, mining of high-sulfur coals in the eastern US has decreased in favor of lower-sulfur coals from western states. SO_2 can also be removed from flue gases. There are a variety of ways of doing this; one widely used approach is to use an alkali material such as limestone to absorb the SO_2. *Scrubbing* of flue gases with limestone produces calcium sulfite ($CaSO_3$), which can be removed and further oxidized to form **gypsum** ($CaSO_4 \cdot 2(H_2O)$), a commodity valuable in the building industry. Seawater is also sometimes used to scrub flue gases of SO_2 at coal plants in coastal locations.

Nitrogen oxides Nitrogen oxides (predominantly NO and NO_2 from coal combustion, but also including N_2O, N_2O_2, etc., collectively denoted NO_x) form by oxidation of nitrogen by combustion air at temperatures of $1300\,°C$ and above. Nitrogen oxides lead to ozone production in the lower atmosphere, a major component of smog. Approaches such as fluidized bed combustion (FBC) described above, which reduce the temperature in the combustion environment, can reduce the formation of nitrogen oxides in coal plants. NO_x can also be removed from flue gases by **selective catalytic reduction** (SCR). In this approach, ammonia is mixed with the flue gas, which is then passed through a system in which it comes in contact with a catalyst on a set of plates or a honeycomb-type lattice. The catalyst promotes the reaction

$$6\,NO_x + 4x\,NH_3 \rightarrow (3 + 2x)\,N_2 + 6x\,H_2O, \quad (33.5)$$

removing the ammonia along with the nitrogen oxides in favor of nitrogen gas and water.

Mercury Mercury-containing compounds are contained in flue gases from coal combustion. Mercury is a hazardous pollutant that can cause neurological problems in people and animals even at low doses. Mercury can be removed from power plant emissions, at least in part, by mixing an absorbent powder into the flue gases; the powder reacts with the mercury to form inert particles that can be removed by the electrostatic precipitator system.

Other environmental control systems have been developed for coal plants that combine several of these systems in various ways. Although many coal plants still operate without emission controls, as the technology improves it is in principle possible to remove most pollutants from coal plant emissions (at a significant cost) without substantially impacting plant efficiency. As we discuss in §35.4.2, however, the situation is somewhat different for carbon emissions. Because carbon dioxide represents the primary combustion product from burning coal, and contains all the carbon content of the original fossil fuel, capturing and sequestering all carbon emissions from a coal plant is technically challenging and requires using a sizable fraction of the energy produced by the plant.

Coal

Coal is classified by rank, with lower rank lignite and sub-bituminous coal (*brown coal*) having lower energy/carbon content and higher rank bituminous coal and anthracite (*black coal*) having higher energy/carbon content. Energy content of coal is in the range

$$E_{\mathrm{coal}} = 15 - 35\,\mathrm{MJ/kg}\,.$$

Standard Rankine steam plants transform combustion energy from coal to electrical energy with roughly 30–35% efficiency. Supercritical and ultra-supercritical plants (USC) operate at higher pressures and temperatures and can achieve efficiencies up to 40–45%. Alternative designs including fluidized bed combustion (FBC) and integrated gasification combined cycle (IGCC) have positive features including the possibility of integrated pollution/carbon control systems.

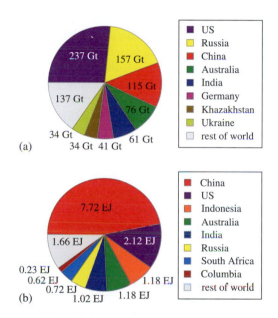

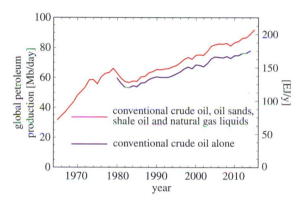

Figure 33.13 Global petroleum production, conventional crude [12] compared to all petroleum including unconventional sources such as tar sands, oil shale, and natural gas liquids [9].

Figure 33.12 (a) Proven coal reserves of the top eight countries in 2014 in billions of tonnes; (b) 2014 coal production by the top eight producing countries in exajoules [9].

33.1.9 Coal Resources

Coal is distributed across all continents of the globe. Identified coal resources (in the nomenclature of Box 16.1) exceed several trillion (10^{12}) tonnes. Proven reserves in 2014 were 891 billion tonnes, slightly over half of which were sub-bituminous and lignite, with over 400 billion tonnes of anthracite and bituminous coal [9]. The proven reserves of the top eight countries are shown in Figure 33.12. These reserves represent enough for over 100 years at the current rate of production (8140 Mt/y in 2013 [12]). More detailed descriptions of the distribution of coal resources can be found in [9, 226, 227].

33.2 Petroleum

It is difficult to overemphasize the role that petroleum plays in the modern world. The energy density of most liquid crude oil is in the range 42–44 MJ/kg, higher than that of coal by a substantial fraction. Furthermore, fossil fuel in liquid form is much easier to transport (e.g. through pipelines), store (e.g. in tanks), and extract from the ground (it is generally under pressure). Petroleum products not only underlie contemporary vehicular and air transport, they are also used in asphalt for roads, as lubricants, in the production of fertilizers and pesticides for agriculture, and are ubiquitous in the form of plastic, polymers used for clothing, and a wide range of other products. While petroleum resources are limited, and production may peak in the near future, sufficient resources remain to dominate many sectors of the energy economy for decades to come.

33.2.1 Origins of Petroleum

The geophysical origin of petroleum is less transparent than that of coal. The hydrocarbon molecules in crude oil[6] are simpler and bear fewer traces of their organic origins than those of coal; furthermore, liquid hydrocarbons, being fluids, do not generally remain fixed where they are formed. Like coal, most oil is believed to have been produced from organic material after exposure to high temperatures and pressures over long time periods, though liquid hydrocarbons can also have inorganic origins (see Box 33.4). The process of producing oil is somewhat more delicate than for coal.

It is believed that most petroleum originates from phyto- or zooplankton (such as *foraminifera*, which we encounter again in §34), and some plants and sea animals. As in the generation of peat and coal, the organic material must be deprived of oxygen and buried before it is decomposed by aerobic bacteria. The precise combination of anoxic environment and rapid sedimentation necessary for deposition of a sufficient quantity of organic material to form the basis for oil **source rock** is relatively rare. This can be seen from the fact that even in areas with rich oil fields, the sediment layers containing the source rock are generally quite thin, perhaps 3–10 m out of 1000 m of sediment. The precise range of environments that can produce oil source rock is not completely understood, but is believed to include broad shallow and stagnant seas or lakes that maximize biological productivity and limit oxygen levels at the sea/lake bottom. River deltas may also be a viable environment for source rock production.

The formation of petroleum begins with **kerogen**. Kerogen is solid organic material, containing a variety of organic compounds. Some oxygen is removed when kerogen is formed from organic material, so that kerogen has a lower O/C ratio than the original organic material. The precise composition of kerogen varies widely, depending as for coal on the dominant type of macerals. *Type I kerogen*, composed primarily of algal liptinites, is relatively rare but has the highest hydrogen/carbon ratio (H/C > 1.25) and the greatest potential for oil production. *Type II kerogen* contains non-algal liptinites (sometimes referred to as **exinite**) and an H/C ratio that is high but generally less than about 1.25; type II kerogen also has great potential

for petroleum production. *Type III kerogen*, which is dominated by vitrinite, may produce coal and natural gas, but rarely forms petroleum. *Type IV kerogen* is formed largely from inertinite and does not form petroleum.

As the source rock is buried more deeply, at higher temperatures the large hydrocarbon molecules in the kerogen begin to break down and form shorter hydrocarbons that become liquid and then gas with extended exposure to higher temperatures. This breakdown of hydrocarbons from molecules containing more to fewer carbon atoms is known as **cracking**. The extent to which the kerogen in source rock has been cracked is referred to as the **maturity** of the source rock. As the source rock matures and the kerogen it contains releases hydrocarbon liquids and gases, the H/C ratio of the kerogen decreases. When thermally mature, the source rock kerogen has essentially exhausted all of the hydrogen available for further petroleum or natural gas production. The *oil window* in which petroleum is produced lies between temperatures of roughly 60 °C and 150 °C. In a typical region with geothermal gradient around 30 °C/km, the oil window is between approximately 2 and 5 km in depth. As source rock containing kerogen enters the oil window, over millions of years the kerogen solids release (primarily) liquid hydrocarbons. Below the oil window, the kerogen releases gaseous hydrocarbons and the liquid hydrocarbons are cracked further, generating *wet gas* composed of methane and heavier alkanes such as ethane and propane. Finally, at the highest temperatures remaining liquids and heavier alkanes are cracked into methane and much of the remaining hydrogen in the kerogen is emitted as methane, resulting in *dry gas*. Thus, for production of liquid petroleum, source rock with appropriate kerogen content must be produced by deposition of organic rich material in an anoxic environment, and the source rock must then be buried for millions of years within, but not below, the oil window (Figure 33.15(a)).

In the first stages of transformation (**diagenesis**), water, CO_2, and other oxygen-bearing compounds are released from the kerogen. In the next stage (**catagenesis**), as oil and gas are released, the H/C ratio of the remaining kerogen decreases (see Figure 33.14). As the H/C ratio of the kerogen decreases, the extractable petroleum tends to contain a higher fraction of aromatics such as benzene C_6H_6, and even multiple-ring aromatics, which have a lower H/C ratio than paraffins and cycloparaffins that are found when the kerogen is at lower maturity. At this later stage, some of the material is converted into *bitumen* (often referred to as *asphalt*), a highly viscous mixture of heavy hydrocarbons, which is soluble in petroleum. During catagenesis, Types I and II kerogen typically produce both oil and gas while Type III kerogen produces mostly gas.

---

[6] Following common practice, we use the terms *petroleum*, *crude oil*, and *oil* interchangeably to refer to liquid hydrocarbons from natural sources. We use the term *conventional crude oil* specifically to describe mixtures of hydrocarbons that exist in liquid phase in underground reservoirs.

Box 33.4 Evidence for Biogenic vs. Abiogenic Origin of Petroleum

While the majority of geoscientists believe that all or almost all substantial liquid fossil fuel deposits originated in organic material, some scientists contend that all or some of the currently existing large petroleum reservoirs around the world were generated by inorganic processes at high temperatures and pressures deep within Earth. There are a number of independent pieces of evidence for the organic origins of fossil fuels, including residual molecules known as **biomarkers** found in petroleum deposits that match structures found in organic material, trace amounts of nitrogen in many crude oil deposits, and a carbon isotope ratio $^{12}C/^{13}C$ characteristic of organic material (carbon isotope ratios are discussed further in §34). Biomarkers include organic molecules such as porphyrins and steroids that are commonly found in petroleum. The hydrocarbons in crude oil have the property that they rotate polarized light in a specific fashion, which is a characteristic of organic material. Furthermore, the biogenic theory of the origin of petroleum has successfully predicted locations of new oil fields, unlike the abiogenic theory, which has failed to make such predictions. We follow the orthodox biogenic theory of petroleum generation in this book, but leave open the possibility that some deep hydrocarbon deposits, particularly of methane, may have an abiogenic origin.

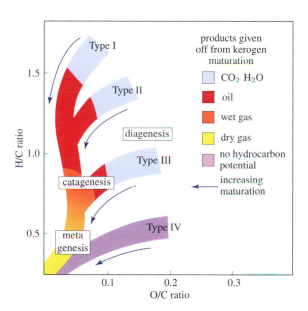

Figure 33.14 A schematic *Van Krevelen diagram*, showing the maturation process and decrease in H/C ratio for different kerogen types. After [237]. (Credit: © Schlumberger 2011)

In the final state (**metagenesis**) prolonged heat and high pressure convert more complex molecules into methane, leaving behind a carbon residue. The precise rate at which kerogen is converted into liquid and gas hydrocarbons depends upon the type of the kerogen, and the time over which the source rock is exposed to high temperatures. The basic physics of this process depends upon the Boltzmann equation in the form of the **Arrhenius clock** (see Box 33.5). The reflectivity of small quantities of vitrinite contained within the source rock increases with exposure to high temperatures, as in coal beds, and is often used as a measure of maturity.

Typical oil source rock contains less than 10% organic material. A kerogen content of > 0.5% is needed for production of economic quantities of petroleum. Once liquid hydrocarbons are formed, given sufficient permeability the liquid can be squeezed out of the rock. Since oil is lighter than rock or water, petroleum generally flows upward. In many cases the oil reaches the surface and decomposes there. For a petroleum reservoir to form, the oil must be trapped under some impermeable barrier before reaching the surface, and the underlying rock must be sufficiently porous to contain a substantial density of the liquid oil. The combination of *source rock*, **trap**, and **reservoir** is key to the formation of any substantial petroleum deposit; identification of these features plays a central role in petroleum exploration (Figure 33.15(b)).

33.2.2 Physical Properties of Petroleum

Crude oil contains a wide mix of different hydrocarbon molecules. As introduced in §11.2.1, common types of hydrocarbons encountered are the paraffins, cycloparaffins, olefins, and aromatic hydrocarbons. These terms are generally used in the petroleum industry; the corresponding scientific terms for these molecules are alkanes, cycloalkanes, alkenes, and arenes. Crude oil deposits exhibit wide variation in their precise hydrocarbon content. Naturally occurring petroleum reservoirs are dominated by paraffins, cycloparaffins, and aromatics, with relatively few olefins. In general, crude oil contains roughly 83–87% carbon by weight, 11–15% hydrogen, and a few percent of other elements. Sulfur is a common contaminant in crude oil; oil

---

Box 33.5 Arrhenius Clock

The rate r at which a given chemical reaction occurs depends upon the **Arrhenius equation**

$$r = ae^{-E/k_B T},$$

where E is the **activation energy** of the reaction and a (which depends weakly on the temperature T) is an overall rate constant. The Arrhenius equation is essentially an empirical relation, though – as its form suggests – it is closely related to the Boltzmann distribution that governs the probability of a state having energy E at temperature T. For most processes of interest, $E \gg k_B T$, so the reaction rate roughly doubles when the temperature increases by $\Delta T \cong (k_B T^2 \ln 2)/E$ (Problem 33.8). Typical activation energies for chemical processes involved in petroleum generation are on the order of 200–300 kJ/mol ($E \cong 3$–5×10^{-19} J). For such a process, for temperatures around 130 °C, roughly a 3–5 °C increase in temperature gives a doubling of the reaction rate. (A 10 °C increment for reaction rate doubling is often used as a rule of thumb, but this is more appropriate for reactions with somewhat lower activation energies.) It is easy to compute from the Arrhenius equation that a process with an activation energy of 200–300 kJ/mol that takes 10 million years at 130 °C can run in a matter of hours at a temperature of 430 °C (Problem 33.9). Petroleum production processes have been reproduced in laboratories at such temperatures and time scales. The Arrhenius equation is used by geologists to understand the thermal maturation history of oil source rocks.

---

with > 2% sulfur is denoted *sour*, and oil with < 0.5% sulfur is denoted *sweet*.

The lightest paraffins, methane (CH_4) and ethane (C_2H_6), are gases even within underground reservoirs, while propane (C_3H_8) and butane (C_4H_{10}) become liquid at moderate pressure, and are known as **liquid petroleum gases** (LPG). Paraffins with more carbon atoms are liquid, transitioning toward solid for molecules with 17 or more carbon atoms.

The density of crude oil can range from 0.75 kg/L to over 1 kg/L, with typical densities in the range 0.8–0.9 kg/L. In the petroleum industry, density is measured in units of °API (**API gravity**), with

$$°\text{API} = (141.5 \text{ kg/L})/\rho - 131.5, \tag{33.6}$$

where ρ is the density in kg/L, measured at 60 °F. Thus, a very heavy oil with the density of water has °API 10, while a light oil with density 0.8 kg/L has °API 45. *Heavy crude* denotes an °API below 22.3, and *light crude* denotes °API above 31.1. Lighter crudes have a larger fraction of lighter hydrocarbons. In general, lighter crude is more expensive as less refinement is needed for production of gasoline and kerosene from light crude.

The viscosity of oil can also range widely. Though in general heavier oil is more viscous, the viscosity depends on the exact hydrocarbon decomposition and is generally strongly dependent upon temperature. Viscosity at 20 °C can range from 10 cP for light crude to above 1000 cP for heavy crude. Oil with a viscosity above 10 000 cP is classified as **bitumen** or **asphalt**. Motor oils are made

from hydrocarbons with 18 or more carbon atoms, and are classified by the Society of Automotive Engineers (SAE) according to viscosity, with, for example, SAE 10 having a viscosity of 65 cP and SAE 40 having a viscosity of 319 cP at 20 °C.

A standard measure of quantity for oil is the **barrel**, denoted by bbl (for *blue barrel*) or b, with

$$1 \text{ b} = 42 \text{ US gallons} \cong 159 \text{ L}. \tag{33.7}$$

Oil production is often measured in b/d (barrels per day). Since different deposits of crude oil have different hydrocarbon composition and different energy content, the energy content of a barrel of oil can vary significantly. The **barrel of oil equivalent** (BOE) is, however, often used as a unit of energy and is defined in the US as

$$1 \text{ BOE} = 5.8 \times 10^6 \text{ BTU} \cong 6.118 \text{ GJ}. \tag{33.8}$$

The **tonne of oil equivalent** (toe) is also used frequently and is defined in various ways by different organizations; the IEA uses the definition 1 **toe** = 41.868 GJ.

33.2.3 Petroleum: Discovery and Extraction

The development of new and more effective methods for identifying and exploiting petroleum resources was a driving force behind some of the largest economic and industrial shifts of the twentieth century. The extent of this scientific and engineering effort, and the dominant role of petroleum in the modern energy economy, justify delving into some details of the scientific aspects of these endeavors.

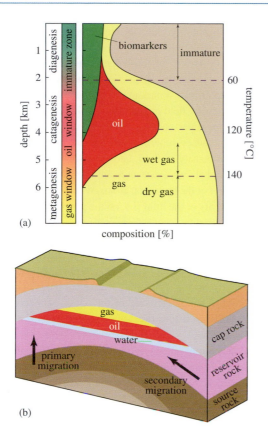

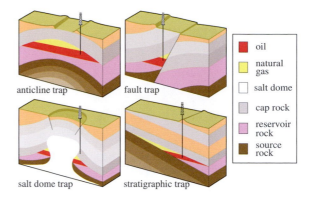

Figure 33.16 Geological formations that trap oil: anticlines, faults, salt, and stratigraphic traps. (Credit: Magenta Green, reproduced under CC-BY-SA 3.0 license via Wikimedia Commons)

Figure 33.15 (a) Oil maturation: a kerogen source generates oil in the *oil window*; at higher temperatures the cracking process is more complete, yielding natural gas. (b) Oil formed in source rock migrates upwards to reservoir rock, and is eventually trapped, along with natural gas and water, by cap rock. (Credit: Earth Science Australia, reproduced under CC-BY-SA 3.0 license via Wikimedia Commons)

Many of the earliest oil field discoveries followed either from direct observation of oil seeping to the surface or a combination of good fortune and intuition. Over the years, the search for new oil fields has developed into a sophisticated scientific and engineering enterprise. The primary goal in discovering new oil fields is identifying the combination of source rock, reservoir rock, and geological trap described above and sketched in Figure 33.15. Furthermore, these geological features must have come into existence with the correct timing so that the trap was in place at the time that the source rock came to maturity and the petroleum became mobile.

The first key to identifying regions with potentially viable petroleum resources is the identification of source rock strata. While aerial surveys can be used to identify possible candidates for geological features (Figure 33.16) such as **anticlines** (concave-downward rock folds), fault

structures, or salt domes that may act as traps, a more detailed description of the underlying geology involves a combination of surface-based geophysical measurements and measurements through test wells (*well logging*).

Seismic and Gravitational Surveys The *reflection seismic survey*, introduced here in the context of coal exploration, was primarily developed for the petroleum industry. In current oilfield exploration, **three-dimensional (3D) seismic surveys** are a routine part of assessing any new region for potentially viable reservoirs. In a 3D seismic survey, receivers are deployed across the surface of a land or sea region of interest, so that the entire 3D geometry of the underlying geology is probed by the reflected seismic waves. The physics and mathematics by which the reflected signal is produced from a given geometry (known as the *forward problem*) is straightforward. Inverting this process, however, and reconstructing the underground geometry from the returned signal, is a much more difficult problem in signal processing (the *backward problem*). Substantial computational resources are devoted to analyzing 3D seismic data in the petroleum industry.

In addition to seismic surveys, other surface geophysical methods such as gravitational and magnetic surveys are used to determine aspects of subsurface geology. For example, salt is less dense than most other rock types, and thus tends to deform upwards and form dome shapes that form excellent traps for petroleum reservoirs (see Figure 33.16). Because of the lower density of the salt, the gravitational field above a massive salt dome is slightly lower than would otherwise be expected (see Problem 33.7).

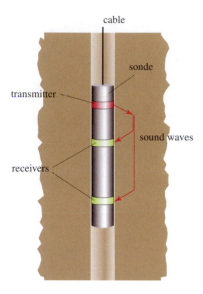

cable

sonde

transmitter

sound waves

receivers

Figure 33.17 A sonde (probe) used to measure geological features at varying depths along a borehole. In this figure, the sonde is equipped with a sonic probe that measures the porosity of surrounding rock.

Well Logging Early petroleum exploration relied heavily on **mud logging**, in which the mud and rock chips brought up during well drilling are analyzed for clues about the underlying geology. Now, a wide variety of geophysical measurements are made by *borehole logging* (§33.1.3), which, like seismic surveying, was pioneered in the petroleum industry. The first borehole measurements of this type were done by the French physicist and engineer brothers Conrad and Marcel Schlumberger in the early 1900s. The Schlumbergers used a probe to measure the electrical conductivity of the rock surrounding a drilled well. Because oil and gas are poor conductors, while water is a good conductor, this leads to ready identification of many oil reservoirs. The Schlumberger corporation and others have developed a wide variety of useful measurements and analysis tools, and now *well logging* is a core element of the petroleum industry.

Commonly used well logging tools include conduction/resistivity, temperature, and caliper probes to directly measure physical parameters of the borehole and surrounding rock. As discussed in §33.1.3, many of the most useful well logging tools involve measurements of radiation, such as the gamma ray and density logs, which measure radiation produced either by the surrounding rock or by the logging tool itself. Along with the density probe, sonic and neutron tools are used to measure the **porosity** of rock layers (i.e. the fraction of empty space within the rock).

Because permeability plays an important role in the formation of petroleum reservoirs, substantial effort has been put into development of tools that measure this feature. In recent years, **nuclear magnetic resonance** (NMR) has been used to provide a measure of permeability. In NMR, a strong magnetic field is generated, which causes hydrogen atoms in water and hydrocarbons to align with the field. Radio pulses superimposed on the magnetic field measure the density of fluid, and gradual variation of the magnetic field allows a measurement of relaxation time of the alignment of the hydrogen atoms, which depends upon the connectivity of the fluid (i.e. on the permeability of the surrounding rock).

Reconstruction of Reservoir Geology and Geological History In modern oilfield exploration, the various available exploration techniques are used to build a coherent picture of the 3D geology of the region. This picture is in turn used to reconstruct the geological history of the underlying strata.

Identification of source rock determines the possibility of substantial petroleum production in a given region. Identification of cap rock and trap geometries suggests possible locations for reservoirs. Some reservoirs are trapped under hard impermeable rock, such as salt domes or *anhydrite* ($CaSO_4$), which is gypsum out of which all H_2O has been squeezed. In other cases, mudstone or shale can act as a cap rock; even if the rock is slightly permeable, if the pore size is very small the surface tension at the oil–water interface may be sufficient to retain substantial quantities of petroleum under the rock.

Measurement of porosity and permeability determines whether rock is suitable as a reservoir; viable reservoir rock generally has a porosity in the range 15–30%, and a permeability of 100–500 mD (millidarcys), with a minimum permeability of 10 mD necessary. Note that higher permeability is needed for extraction of petroleum than for geothermal fluids, due to the higher viscosity of oil. For natural gas reservoirs, a permeability of 1 mD is sufficient. The reservoir rocks containing most oil deposits are sandstone and carbonate rocks.

Detailed computer simulations of the geological history of a region based on seismic and well log data are used to reconstruct the sequence of burial, uplift, and erosion processes that determine the structure of the underlying rocks. As rock is buried more deeply, its porosity decreases; rock with porosity of 60–80% near the surface will generally have a porosity of 5–15% at a depth of roughly 4 km. Understanding the geological history of the region determines whether the timing of source rock maturation and trap formation occurred in the proper sequence to enable substantial reservoirs to form. Combining all these pieces of information suggests locations for further test wells, which refine the understanding of the local geology.

Reservoir Management Once a substantial reservoir is identified, the process of optimizing extraction of petroleum from the reservoir involves a great deal of sophisticated science and engineering. Determining the optimal locations for extraction wells depends on a detailed understanding of the underlying geology. As oil is extracted and the oil/water configuration in the reservoir changes, often time-dependent ("4D") seismic data is used to model the detailed evolution of the reservoir and optimize the extraction process. Much physics and computation is involved in this analysis. In particular, modeling the hydrodynamics of oil and water (two-phase) flow in a porous medium with complicated boundary conditions is an extremely difficult computational problem, and continues to be the subject of extensive research both in industry and academia.

The basic tool for extracting petroleum from an underground reservoir is the rotary drilling rig. Early drills used *fishtail bits*, which resemble smaller wood or metal drill bits and which penetrate slowly through rock by scraping. In 1908 American Howard R. Hughes invented the **roller cone bit** (Figure 33.18), which employed two and later three conical cutters that chip, crush, and pulverize even hard-rock formations. The development of the roller cone bit spurred the rapid growth of oil production

Figure 33.18 A modern *tricone* roller cone bit. Roller cone bits have evolved substantially since the original design by Howard Hughes, and are crucial in drilling oil wells through rock to a depth of several kilometers.

in the first half of the twentieth century. Once the well is drilled, logged, and lined with steel casing, explosive charges puncture the casing at the location of the petroleum reservoir(s), and production begins. Originally, all wells were drilled as close to vertical as possible. Over time, however, directional drilling methods have been improved, and now horizontal and extended directional drilling are used widely to expand the range of reservoir accessible from a single wellhead, to reach otherwise inaccessible reservoirs, and to more efficiently extract petroleum from existing reservoirs. Drilling depths for oil wells reached 6 km in 1950. Though much deeper wells have since been drilled, most production wells are at depths within or above the oil window, and rarely below 6 km. The average depth for production wells is around 1.6 km.

In a freshly drilled well, the reservoir pressure is often sufficient to force oil and gas out through the well. Typical fluid pressure at the level of the reservoir is equivalent to that at the bottom of a column of water of the corresponding depth. For oil with density substantially less than water, this pressure is enough to force the oil out through the well without additional intervention. As the reservoir pressure decreases, pumps can raise additional oil from the reservoir. Extraction using natural pressure and/or pumping is known as **primary oil recovery**. At one time, oil wells were capped and abandoned once primary recovery was exhausted, often after only 30–40% of the petroleum in the reservoir had been recovered. Now, **secondary recovery methods** are widely used. Water or gas injection wells are used to restore reservoir pressure. In some cases the water mixes with the petroleum, requiring removal after extraction. Other **tertiary recovery** or **advanced recovery methods** have been developed, including steam flooding, chemical flooding, and injection of CO_2; in the future the latter approach may be combined with carbon capture at point sources such as coal plants (§35.4.2).

Petroleum Sources

The key features needed for a petroleum deposit are:

Source rock: rock formed from sedimentation millions of years ago that captured organic material in an oxygen-free environment.

Trap: a non-permeable cap keeping the petroleum from leaking to the surface.

Reservoir rock: rock that is sufficiently porous (typically 15–30%) to contain substantial amounts of petroleum, and sufficiently permeable (typically 100–500 mD) to allow the petroleum to flow out of the reservoir.

Much of the effort in oilfield exploration is devoted to identifying regions containing these three features.

33.2.4 Tight Oil, Tar Sands, and Oil Shale

In addition to standard petroleum resources, there are several related **unconventional oil resources** that are beginning to play a significant part in the energy landscape and are likely to become increasingly important in coming decades. *Tight oil*, also known as *light tight oil*, has played an important role in the dramatic increase in US oil production in recent years. *Tar sands* and *oil shale* can also be used to produce petroleum products, though at a higher energy and environmental cost compared to standard crude oil reservoirs.

Light Tight Oil

Light tight oil (LTO), also known as **tight oil** or (sometimes) **shale oil**, is light crude oil that is encased in low-permeability geological formations, so that it cannot be extracted using standard primary or secondary oil recovery methods. Light tight oil can be contained in a variety of formations, including shale, sandstone, and carbonates. Advanced recovery methods have been used in recent years along with advanced horizontal drilling and/or hydraulic fracturing to increase permeability for extraction of light tight oil. In particular, in the early 2010s, following the commercial development of shale gas production in the 2000s (see below), extensive development of new light tight oil producing wells in the US played an important role in dramatically increasing the country's rate of production and extent of proven petroleum reserves (Figure 33.19). In 2012, tight oil contributed almost 30% of US crude production. The EIA estimated in 2013 that technically recoverable light tight oil in shale alone worldwide could contribute in excess of 345 billion barrels to the global petroleum reserves [238]. This does not include possible tight oil production from sandstone or carbonate formations that often lie adjacent to shale oil deposits. One limitation on shale oil production is that at least roughly

15–25% of the fluids within the shale must be gaseous in nature to allow for expansion that will drive the oil to the wellbore once the permeability is increased through hydrofracturing and the well is bored. Uncertainty regarding this and other factors make it difficult to precisely estimate the potential productivity of a given area without somewhat extensive site testing.

Tar Sands

Tar sands, also known as **oil sands** or **bituminous sands**, appear in regions where an oil reservoir has come close to the surface and biodegraded, leaving only a tar residue (*bitumen*) containing heavy hydrocarbon molecules with many carbon atoms. Because the oil does not flow freely, extracting the extra-heavy oil requires injecting steam to reduce the oil's viscosity or (where surface mining is viable) heating and processing tremendous quantities of sand to separate out the hydrocarbons, which are then subjected to catalytic cracking and other processing. The energy used in heating and processing tar sands can be up to 30% of the energy eventually recovered from the resulting oil. Tar sands are generally rich in sulfur, which must be removed before the resulting products are used in most applications. Extensive tar sands exist in several locations, particularly in Athabasca (Alberta, Canada) and in Venezuela's Orinoco Belt. The Athabasca tar sands lie close to the surface, making it possible to access the sands using surface methods; most other major tar sand repositories lie deeper and have not been accessed by surface mining. The resource in Athabasca alone is estimated at roughly 1.7 Tb, though only 10–20% is likely to be economically recoverable. (The 2014 BP *proven* reserves estimate for Athabasca was 167 Gb, and for the Orinoco belt was 220 Gb [9].) Fossil fuels from tar sands have begun to play a significant role in Canada's energy economy; the recent growth of production from tar sands plays a role in generating the difference between the two curves (total oil versus conventional crude oil) in Figure 33.13.

Oil Shale

Oil shale, not to be confused with shale oil (above), is essentially oil source rock that has not spent sufficient time in the oil window for the hydrocarbons to break down sufficiently to form a fluid and leave the source rock. Typical oil shale contains on the order of 10% hydrocarbons, primarily composed of heavy molecules with many carbon atoms. To recover petroleum products from oil shale, the rock must be crushed and then heated to separate the hydrocarbons. This process is very energy intensive since the kerogen in the oil shale is only partially broken down. While several pilot plants have been constructed, and shale oil is used in some places (e.g. Estonia) for electricity production, large-scale commercial production

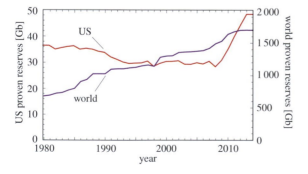

Figure 33.19 US and world proven oil reserves from 1980 to 2014 in billions of barrels [9]. Note the 70% increase in US reserves between 2008 and 2014.

of petroleum from oil shale has not yet become economically favorable. The oil shale resources known to exist in the western US contain a quantity of energy comparable to the complete world supply of petroleum ($\sim$1.7 Tb). Oil shale often contains only 5–10% oil, so that the volume of waste produced can be roughly 20 times the volume of oil extracted. Significant quantities of shale are burned to heat the remaining shale from which oil is extracted, giving liquid fuel from oil shale a very high CO_2 footprint compared to conventional crude oil.

33.2.5 Petroleum Refining

As discussed above, different crude oils contain different combinations of hydrocarbons, ranging from methane and liquid petroleum gases to extremely large molecules with many carbon atoms. Commercial applications require a distribution of hydrocarbon inputs that differs from the makeup of crude oil. In particular, the demand for light (low molecular mass) hydrocarbons exceeds the supply in crude oil. Refineries therefore not only separate out light hydrocarbons, but also use several methods to crack heavier hydrocarbons to produce more useful petroleum products.

The first step in refining crude oil is the removal of salts, water, and other non-hydrocarbon trace components. The crude is then distilled. By raising the temperature gradually, hydrocarbons with relatively few atoms per molecule are separated out, followed by progressively heavier molecules. First the lightest petroleum gas molecules (methane, ethane) are distilled out, then the liquid petroleum gases (propane and butane). Some of the other useful components that are separated out in refineries include:

Naphtha Light naphtha consists of hydrocarbon molecules with between 5 and 7 carbon atoms, which boil off at temperatures between 27 °C and 93 °C. Heavy naphtha has 6–10 carbon atoms per molecule and boils off at temperatures from 93–177 °C. Heavy naphtha contains molecules such as octane, and is used in production of gasoline.

Kerosene Kerosene consists largely of paraffins and aromatic hydrocarbons containing 10–14 carbon atoms, and boils off at temperatures above 175 °C. In the early days of the oil industry, kerosene was the principal desired product, used for lighting in the 1860s and 1870s. Due to its relatively low cost, relatively high ignition temperature, and dual use in jet turbines both as fuel and as a lubricant, kerosene is now used primarily for jet fuel.

Fuel oil The heavier fractions of crude oil are known as fuel oils. Fuel oils are generally graded from 1 through 6, with the lightest, number 1 fuel oil, referring to kerosene.

The precise range associated with different numbers of fuel oil varies according to the standards of different organizations and institutions. Number 2 fuel oil is generally used for diesel fuel and home heating. Number 4 and number 5 fuel oils are commercial and industrial heating oils. Numbers 1–3 and some number 4 fuel oils, known as **distillate fuel oils**, are obtained from the distillation process while some number 4 and numbers 5 and 6, known as **residual fuel oils**, are obtained from the residue after distillation.

After the lighter fractions have been removed from crude oil through distillation, the remaining heavier components are subjected to **catalytic cracking**, in which temperature and chemical catalysts are used to rapidly break down the large hydrocarbon molecules into smaller more useful units. Either hydrogen must be added or carbon must be removed in this process, as the desirable shorter-length hydrocarbon chains have much higher H/C ratios than the residue after distillation. Both methods are used: *hydrocracking* reacts residual oils with hydrogen at very high pressure and temperature in the presence of a catalyst; *fluid catalytic cracking* contacts the residual oil with a catalyst at high temperature and moderate pressure, producing lighter hydrocarbons and coke, which is deposited on the catalyst and must be removed in a *regenerator*. Thermal cracking in vacuum or with steam is also used in *cokers*, which produce both lighter hydrocarbons and specialized forms of coke used for refining metals. *Catalytic reforming* and *isomerization* are also used to rearrange molecules into more desirable forms; for example, *n*-paraffins can be rearranged into branched hydrocarbons with higher octane numbers. Catalytic cracking, reforming, isomerization, and other methods for modifying the molecular composition of petroleum fractions play an important role in matching the crude oil supply to the demand for different petroleum products; in particular, reforming and isomerization are used extensively in producing gasoline mixtures from heavy naphtha. Current refining techniques can turn 80% or more of light crude oil into immediately useful light petroleum products. The remainder, containing heavier, more complicated hydrocarbon molecules, is used to produce lubricants, waxes, and asphalt among other things.

Oil refining is a critical part of the petroleum supply chain. It is also a relatively energy-intensive process; over 10% of the energy content of crude is usually used in refinement.

33.2.6 Uses of Petroleum

Petroleum has a tremendous range of uses. In addition to the principal uses as transport fuel and heating oil, petroleum products can be used directly in various ways, including as lubricants and asphalt. Furthermore, the chemical composition and energy of the hydrocarbons in petroleum makes them an excellent **chemical feedstock** for production of many chemicals and materials. It has been argued, in fact, that the hydrocarbon molecules contained in crude oil are so useful that they should not be wasted as transport fuel, but saved for future generations to use in ways some of which have not yet been invented.

Transportation The principal use of petroleum products is as fuels for transportation. In the US in 2014, roughly 47% of the 19 million barrels (116 PJ) of petroleum products consumed per day was in the form of gasoline, with another 21% used for aviation and diesel fuel. In other parts of the world, a smaller fraction of refinery output is used for gasoline, but in most countries transport applications use more than half of all petroleum products produced. Energy use in internal combustion engines powered by gasoline or diesel fuel and in the kerosene-powered Brayton cycles that power jet airplane engines is described in earlier chapters of this text (§11, §13). In addition, liquid petroleum gases see more limited use as a transport fuel.

Heating/Cooling The second largest use of petroleum products is space and process heating, including home heating as well as commercial and industrial heating. As mentioned above, fuel oils represent the main fractions of refinery output that are used for heating purposes. Liquid petroleum gases, particularly propane, are also used for heating, cooking, and cooling in Europe and many rural locations.

Direct Use As mentioned above, the components of the crude oil containing larger hydrocarbons than the fuel oils are used for lubricants and asphalt. **Paraffin wax**, containing hydrocarbons roughly ranging from 20 to 40 carbon atoms, has a variety of uses, from candles and crayons to surf wax.

Petrochemicals Another important application of petroleum products is in the industrial production of chemicals and materials including fertilizer, pesticides, pharmaceuticals, solvents, and plastic. The simplest olefins (alkenes), ethylene (C_2H_4) and propylene (C_3H_6), are the progenitors of the common plastics, polyethylene and polypropylene. Another class of hydrocarbons that constitute a few percent of typical crude oil, and that are produced in greater quantities in refineries through catalytic reforming are the aromatic molecules *benzene*, **toluene**, and **xylene**, often referred to together as **BTX**. Toluene and xylene are also known as methylbenzene and

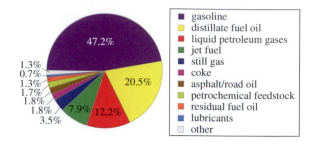

Figure 33.20 Petroleum uses in the US in 2015 [12]. Liquid petroleum gases are primarily ethane/ethylene through isobutane/isobutylene; jet fuel is primarily kerosene; *still gas* is primarily methane and ethane used as refinery fuel or as petrochemical feedstock.

dimethylbenzene, having the structure of a benzene ring (see Figure 11.4) with one or two hydrogen atoms replaced with a methyl group CH_3. The BTX hydrocarbons are the principal feedstock for producing a vast range of useful chemicals including polystyrene, nylon, polyurethane, and polyesters. Other intermediate chemical feedstocks are produced by steam cracking of naphtha and fuel oil components of refinery output. The many uses of petroleum are summarized in Figure 33.20.

33.2.7 Petroleum Resources

Like coal, petroleum resources are widely distributed across the globe. There are significant petroleum reservoirs in all continents except Australia and Pacific Asia outside the Middle East, which have relatively small reserves. Global proven reserves (including unconventional oil) are estimated by [9] to be approximately 1.70 Tb (see Footnote 3). The proven reserves of the top ten countries are shown in Figure 33.21.

Estimating the world's ultimately recoverable petroleum resources has been a challenging and controversial issue for many decades. Numerous authors have predicted that world oil production will peak at one or another particular time. Many of these predictions have already proved false, as production has continued to increase past the predicted date of "**peak oil**." Nonetheless, petroleum is clearly a limited resource. Given the millions of years required for generation of new oil fields, the current rate of petroleum production, and the extent of the efforts that have already gone into identifying viable petroleum reservoirs, it is inevitable that the rate of petroleum production will flatten and begin to decrease within decades. But precisely determining how much oil can be made available for human use is a rather tricky subject. New techniques and technologies continue to expand the range of petroleum resources

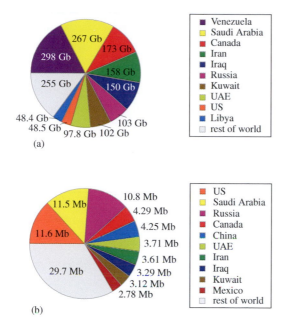

(a)

Venezuela
Saudi Arabia
Canada
Iran
Iraq
Russia
Kuwait
UAE
US
Libya
rest of world

(b)

US
Saudi Arabia
Russia
Canada
China
UAE
Iran
Iraq
Kuwait
Mexico
rest of world

Figure 33.21 (a) Proven oil reserves, including unconventional sources, in billions of barrels (Gb) of the top ten countries in 2014; (b) 2014 oil production by the top ten producing countries in millions of barrels per day (Mb) [9]. See Footnote 3 for more information.

that can be accessed economically. Furthermore, estimation of existing oil resources is made more complicated by the fact that reporting of existing reserves by nation-states and corporations is subject to intentional manipulation for economic and political reasons, and such estimates are not easily verified.

In 1956, American geophysicist M. King Hubbert used historical data on oil production, combined with estimates of total available petroleum resources, to predict the date of peak oil production for the US and the world. Hubbert's basic technical assumption was that the rate of oil production would follow a derivative logistic probability distribution (see Box 33.6). The shape of this distribution reflects the emergence, maximum exploitation, and eventual exhaustion of a finite resource extracted with a fixed technology. Fitting this distribution to the data, Hubbert predicted that US oil production would peak in the early 1970s, and that world oil production would peak before the end of the twentieth century. Subsequent work by others using Hubbert's method revised his estimate of global peak oil to between 2004 and 2008. Readable more detailed summaries and critiques of the predictions of Hubbert and others can be found in [239] and [231].

Hubbert's prediction that US oil production would peak in the early 1970s turned out to be roughly correct, in the context of petroleum resources and extraction methods as they were understood in his time. Historical data on the rate of US oil production through 2010 is depicted in Figure 33.22. While the production curve indeed reached a peak in the 1970s, the dramatic increase in US production in the 2010s has wildly violated Hubbert's projection. Hubbert based his prediction on an estimate of a total of 200 Gb of recoverable oil in the US. Discovery of additional large oil fields in Alaska and the Gulf of Mexico has substantially increased estimates for ultimately recoverable oil for the US. In fact cumulative US oil production already exceeded 200 Gb in 2010. The spike in US oil production that began around 2010 was due in part to the development of a technology (§33.2.4), unavailable in Hubbert's time, for extracting *light tight oil (LTO)*. The widespread deployment of this extraction technology was hastened by economic factors and political imperatives.

The prediction that world oil production would peak between the years 2004 and 2008 has also proved to be incorrect. With the economic downturn at the beginning of the twenty-first century, global crude oil production dropped from 2006 to 2007 and from 2008 to 2009, but has reached a new peak every year from then through 2016.

Current estimates of total recoverable oil worldwide range are significantly above Hubbert's 1981 estimate of 2 Tb, and generally exceed 3 Tb, including more than 1 Tb already produced. When unconventional fossil fuels are added, this number becomes substantially greater, and may exceed 4 Tb – twice Hubbert's estimate, and much greater than the *proven* reserves of 1.70 Tb quoted at the outset of this section.

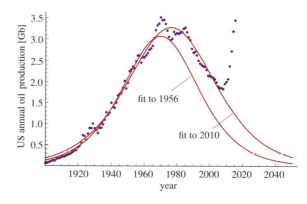

Figure 33.22 Yearly US oil production from 1900 to 2015, in billions of barrels [12]. Compared to logistic distributions fitting data to 1956, following Hubbert, and fitting data to 2010.

Box 33.6   Logistic Functions and Distributions

A variety of simple mathematical functions are used to model natural and complex phenomena statistically. The familiar normal (Gaussian) distribution of the form $ae^{-(x-\mu)^2/s^2}$ characterizes many probability distributions of independent random events. We have also encountered the Weibull distribution that is used to model wind speed and wave power at specific locations (§28). For modeling processes and events that represent cumulative progress towards a fixed and finite end point, such as the depletion of a finite resource, a different class of distributions are needed. To model, for example, the fraction of a finite

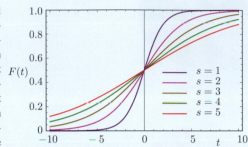

resource that is consumed at time t, such as petroleum used by humans, or a supply of sucrose consumed by yeast cells in a petri dish, the cumulative consumption will be described by a function $F(t)$ that goes to 0 as $t \to -\infty$ and goes to a fixed value A as $t \to \infty$. A simple function with these properties is the **logistic function**

$$F(t) = A \frac{1}{1 + e^{-(t-\mu)/s}},$$

where μ and s are parameters that dictate the center and width of the region of variation of the function. $F(t)$ is plotted above for $A = 1$, $\mu = 0$, and various values of s. (Note that this is the same form as the Fermi–Dirac distribution (25.8), where $t, \mu, s \to E, E_F, T$.)

The instantaneous rate of consumption is given by the derivative of $F(t)$, which in the case of the logistic function gives the *derivative logistic function* or *logistic distribution*

$$f(t) = F'(t) = \frac{Ae^{-(t-\mu)/s}}{s(1 + e^{-(t-\mu)/s})^2}.$$

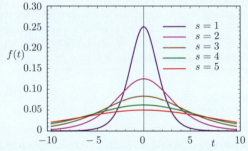

$f(t)$ is plotted at right for $A = 1$ and $\mu = 0$. This parameterized distribution serves as a good statistical model for many resource depletion scenarios, where the coefficients A, μ, and s are matched to empirical data. In 1956, Hubbert used this model to predict the peak of US oil production. Other parameterized functions with similar properties are sometimes used for statistical models, including for example the **lognormal distribution**, $f(t) = ae^{-(\ln t - \mu)^2/s^2}$. There are heuristic arguments in favor of different choices of modeling function, but in any given situation the best modeling function is generally determined empirically. As discussed in the main text, any resource depletion estimation using this kind of parameterized distribution relies heavily on assumptions regarding a given set of technologies and sociopolitical constraints that define the economically accessible resource.

As more and more sophisticated methods are used to discover and develop new oil fields, in particular those in deeper waters, as secondary and tertiary methods are used to extract more petroleum from existing reservoirs, and as unconventional oil sources are increasingly developed, estimates of current reserves have continued to grow, matching and exceeding the rate of production. A useful measure is the ratio of reserves to production (R/P), which is measured in years and gives a minimum number of years at which current production could be maintained using existing reserves. As it has for many mineral commodities,

this number for oil has grown steadily for many years (see Figure 33.23).

In summary, attempts to predict the peak production of petroleum are complicated by the difficulty of estimating the extent and quality of undiscovered resources, which may be very large, and by the impact of unanticipated technological advances, which may be disruptive. Certainly, existing conventional crude oil reserves are sufficient for many decades of production at current levels. Unconventional fossil fuels including shale oil and tar sands could extend the dominance of petroleum as an energy source

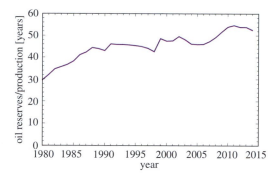

Figure 33.23 Ratio of world oil reserves to production from 1980 through 2014. The ratio R/P indicates years remaining of proven reserves at the current production rate. (Reserve and production estimates from [9] include crude oil, shale oil, oil sands, and natural gas liquids.)

for further decades if their exploitation becomes economically viable and proves environmentally acceptable. Thus, while it may be approaching its peak, global petroleum production is likely to decline more slowly than it has increased unless other energy resources begin to be more cost effective than petroleum in critical areas such as transportation.

33.3 Natural Gas

Natural gas refers to any naturally occurring combination of gaseous hydrocarbons. Most natural gas resources are primarily composed of methane but also contain smaller

fractions of ethane, propane, and butane, and trace amounts of higher hydrocarbons, as well as in some cases varying amounts of nitrogen, CO_2, water vapor, oxygen, hydrogen sulfide, and even helium.[7] For many years, natural gas occurring in oil fields was not considered a commercially useful product and was burned off during oil production. More recently, natural gas has been used widely both for heating and power generation. Because there are few gaseous impurities, natural gas generally needs little processing and burns very cleanly. Natural gas emits less carbon dioxide per unit of energy released than other fossil fuels, due to its higher hydrogen to carbon ratio relative to heavier hydrocarbons. Also, natural gas powered electric plants that use CCGT technology operate at higher efficiency than coal plants. These features make natural gas much more environmentally friendly than other fossil fuels. In the US, constrained natural gas supplies in the 1970s limited the range of use of this fuel. With the discovery and development of new natural gas resources and extraction techniques, along with the deregulation of natural gas prices, the use of natural gas has risen steadily in the US in recent decades. Global natural gas production since 1970 is graphed in Figure 33.24.

33.3.1 Natural Gas Sources

Natural gas is generated in rock containing organic material when continued exposure to pressure and temperature crack hydrocarbons to smaller and smaller units. Such **thermogenic natural gas** is present in oil fields and coal beds and also in oil shale and other geological formations; in general, natural gas is produced when hydrocarbon-bearing material is buried below the *oil window* for substantial time. In the absence of a cap rock

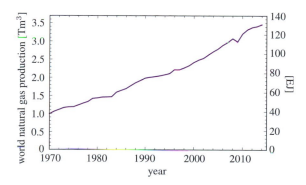

Figure 33.24 Global natural gas production since 1970 [9], measured in trillion cubic meters (Tm^3) and exajoules.

---

[7] In fact all the world's supply of *helium* is obtained as a by-product of natural gas production (Question 33.3).

Petroleum

Petroleum is one of the most useful and versatile substances known. Its high energy density 42–44 MJ/kg and the wide range of hydrocarbon molecules it contains make it useful in a wide range of applications as an energy source and for products including lubricants, asphalt, fertilizers, and clothing. While conventional crude oil resources are finite and production is likely approaching a peak, sufficient reserves exist to maintain the current level of production for many decades. This time can be extended further by exploiting unconventional oil resources such as tar sands that carry a higher environmental cost.

configuration, natural gas easily passes through permeable rock and can escape to the atmosphere. In places where there is a geological trap above an oil reservoir, natural gas accumulates. Natural gas can also migrate from its point of origin, for example by following a slanted cap rock uphill, and accumulate in a geological trap elsewhere, giving an isolated gas resource. Various traps for conventional natural gas are shown in Figure 33.25 (see also Figure 33.16). Natural gas can also remain trapped in impermeable rock, such as in coal beds and oil shale. As discussed in §26, natural gas can also be produced biogenically, e.g. through organic decomposition in marshes and landfills.

Conventional natural gas resources are large discrete accumulations of natural gas in permeable rock. Gas in conventional resources can generally be extracted (*produced*) through one or more extraction wells. Many conventional deposits are associated with petroleum reservoirs and are extracted in the process of petroleum recovery. Other **non-associated gas resources** are found in locations without oil, and are extracted directly.

Unconventional natural gas resources include gas deposits in coal beds (**coal bed methane**), oil shale (*shale gas*), and gas trapped in sandstone formations (*tight gas sands*). There are also substantial accumulations of methane near the ocean floor and frozen under continental tundra (*methane hydrates/clathrates*). Each different type of unconventional natural gas resource presents different challenges for extraction.

Associated natural gas   Most petroleum reservoirs contain some amount of associated natural gas. Some of this natural gas may be separated from the oil, typically forming a **gas cap** at the highest point in the reservoir. There is also generally natural gas in solution within the petroleum. The gas cap can be extracted separately, while the dissolved gas is separated from the crude oil and water extracted from the reservoir at the surface in a **separation plant**.

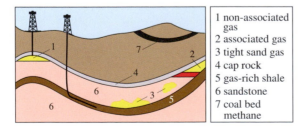

1	non-associated gas
2	associated gas
3	tight sand gas
4	cap rock
5	gas-rich shale
6	sandstone
7	coal bed methane

Figure 33.25 A schematic depiction of the geology of different types of natural gas resource. After [12].

Non-associated natural gas   Non-associated natural gas fields can be generated when natural gas escapes during the formation of an oil field, or when a petroleum reservoir goes too far below the oil window for too long and the hydrocarbons are all cracked, producing only methane and some liquid petroleum gases. Non-associated natural gas is easier to extract than gas that is dissolved in a petroleum reservoir. Because gas flows freely through rock with even a relatively small permeability (a few millidarcys), a substantial fraction of the gas present can be extracted from most conventional resources without undue difficulty.

Coal bed methane   Coal bed methane refers to natural gas that is produced either thermogenically during the process of coalification, or through bacterial (biogenic) processes when low rank coal comes close to the surface. Methane is easily adsorbed onto the surfaces of coal grains, and fills micropores within the coal. The methane is held in place by water pressure. Coal bed natural gas tends to have a purer methane content and a corresponding higher energy density than other natural gas resources. Extraction of coal bed methane is carried out by removing water from the coal bed, reducing the pressure so that the methane flows out; in general this requires multiple wells for each production stream. Extraction of coal bed methane is significantly less efficient than extraction from conventional resources; on a per-well basis, roughly one fifth as much natural gas is removed from a coal bed methane well as from a conventional natural gas well.

Shale gas   Shale gas is natural gas embedded in shale rock with very low permeability. This gas cannot be extracted in the same way as a conventional natural gas resource, since the gas does not flow freely through the rock. Hydrofracturing is used to increase the permeability of the rock. In this process, high-pressure water mixed with chemicals is pumped into rock and creates fractures, allowing the gas to move more freely through the rock so that it can be extracted using one or more extraction wells. In recent years, hydrofracturing in combination with horizontal drilling methods has greatly increased the exploitation of shale gas resources, particularly in the US.

While natural gas is less carbon intensive than petroleum or coal, the hydrofracturing process releases some methane to the atmosphere, which acts in the short term as a potent greenhouse gas (§34). Other concerns about hydrofracturing in the extraction of shale gas include the possibility that methane and/or chemicals used in the process may contaminate ground water or that some of the large quantity of hydrofracturing fluids that return to the surface as wastewater may not be properly treated, leading to environmental contamination. Such concerns have been particularly acute in recent years in the US, where shale gas extraction has proceeded more quickly than the appropriate health and safety regulations have been developed. There has also been some concern about low-magnitude earthquakes associated with natural gas hydrofracturing, similar to the analogous concerns in the context of *enhanced geothermal systems* (§32.5). As mentioned in §32, hydrofracturing is also used in petroleum extraction, where it has generated fewer concerns regarding seismic safety, though the process is essentially the same.

Tight gas sands Natural gas accumulations in tight gas sands, typically siltstone, limestone, dolomite, or chalk, are similar to shale gas described above.

Methane hydrates/clathrates At relatively low temperature and high pressure, water and natural gas can combine to form a **gas hydrate**, in which the water molecules form a cage-like structure wherein the hydrocarbons are embedded. Such a compound, where one molecule traps and contains another, is known as a **clathrate**. Methane hydrates/clathrates form most easily with pure methane – higher-weight hydrocarbons tend to destabilize the hydrate structure. The phase diagram for the water-methane system in the clathrate region of temperature and pressure is sketched in Figure 33.26.

Methane hydrates are known to occur naturally in large quantities in shallow marine environments and at relatively shallow depths (less than 1000 m depth) in the lithosphere. Methane hydrates in the ocean generally occur below 300 m depth at temperatures around 2 °C. Continental methane hydrates occur at higher latitudes in beds of sandstone or siltstone. Substantial quantities of methane

hydrate are trapped beneath the permafrost in the Northern Hemisphere. While methane hydrates could potentially represent an enormous energy resource, large-scale extraction is not yet commercially viable. Possible release of methane from clathrates as permafrost melts in response to climate change is discussed in §35.

Methane hydrates can also form in wells and pipelines in the process of natural gas extraction and transport, when the gas cools sufficiently at high pressure in the presence of water. Such hydrates can block the flow of gas and must be removed carefully. Antifreeze or specially designed hydrate inhibitors are sometimes used to prevent the formation of hydrates within equipment.

33.3.2 Natural Gas Properties

The precise composition of different natural gas resources varies substantially. Natural gas containing only methane, with minimal liquefiable hydrocarbon content, is referred to as **dry gas**, while **wet gas** contains a substantial amount of higher-weight hydrocarbons. Natural gas with substantial sulfur content is referred to as **sour gas**. In general, all impurities and higher-weight hydrocarbons are removed from natural gas before it is sold as fuel. Ethane, propane, butane, and higher-weight hydrocarbons constitute **natural gas liquids** and are separated out for other uses. Commercially sold natural gas is *dry gas* that is at least 85% methane by weight.

The energy content (HHV) of pure methane is 55.5 MJ/kg, while that of commercial natural gas is roughly 50 MJ/kg, and that of typical wet gas is generally somewhat lower. Natural gas is often measured in units of cubic meters (m^3 at 0 °C, 1 atm), or cubic feet (1 cf $\cong$ 0.0283 m^3); the energy content of commercial natural gas in these units is generally in the range 37–39 MJ/m^3 (1.1 MJ/cf $\cong$ 1000–1050 BTU/cf).

Since dry gas is primarily methane, the mass fraction that is carbon is slightly below 12/16, or 75%. The mass of carbon released with combustion of methane per joule of energy released (carbon intensity) is lowest of all fossil fuels, at around 14 kg/GJ. This is roughly 40% less carbon for a given amount of energy than coal, which has a ratio of roughly 25 kg/GJ.

Natural gas becomes a liquid at approximately −162 °C at one atmosphere of pressure. **Liquefied natural gas** (LNG) has a volume that is reduced by a factor of roughly 600, so that the energy density (HHV) of LNG is about 24 MJ/L.

Pure methane is odorless, colorless, and tasteless. Small trace amounts of chemicals known as *thiols* with a strong odor are often added to consumer natural gas to aid in leak detection. Methane is only flammable in air at

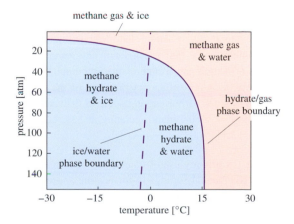

Figure 33.26 Sketch of diagram for methane hydrate (the dependence on the ratio of methane to water is suppressed). After [240]. Note that the pressure decreases downward. (Credit: Sarah E. Harrison, Stanford University)

concentrations between 5% and 15%. Liquid methane is not flammable except at high pressures.

33.3.3 Natural Gas Transport and Storage

There are a variety of ways in which natural gas can be transported from source to destination. For transport between fixed locations, pipelines are often used. In typical pipelines, natural gas is compressed to 50–100 atm. This pressure both reduces the volume of the gas substantially and serves to force the gas through the pipeline. Generally compressors are added at intermediate points along long pipelines.

Natural gas is also often transported as a liquid. The substantial reduction in volume makes it possible for a refrigerated LNG tanker to carry a large quantity of gas (some carry over $200\,000\,m^3$ of liquid, equivalent to over $10^8\,m^3$ of natural gas), but it requires substantial investment in facilities for liquefying and refrigerating the gas in transport.

Gas can also be transported as **compressed natural gas** (CNG) at pressures of 200 atm or more. Large-scale transport of CNG is currently under development, and requires less sophisticated technology than LNG; CNG holds particular promise for smaller-scale transport and distribution.

In many locations where natural gas is used for heating, demand has significant seasonal variation, so substantial facilities are needed to store off-season production. Large-scale long-term storage of LNG is expensive due to the

need for refrigeration. Instead, most natural gas storage is achieved by pumping the gas into underground reservoirs – usually depleted gas reservoirs, but sometimes aquifers or salt caverns are used. Such large gas storage facilities must contain unrecoverable gas in order to maintain a base level of pressure; depleted gas reservoirs are particularly suitable for storage because they already contain this base level of gas. There is concern, however, that leakage of methane from underground storage sites contributes significantly to anthropogenic greenhouse gas emissions.

It has also been proposed that gas could easily be transported and/or stored in solid (hydrate) form, though this is not yet a commercially viable approach. Gas can also be converted directly to electric power, or other materials (ammonia as mentioned below, or other chemicals) before transport or storage.

33.3.4 Uses of Natural Gas

Like petroleum, natural gas has a wide variety of end uses. It serves as a fuel for electricity generation and for production of thermal energy for heating and other uses in residential and commercial settings. A significant fraction of natural gas is also used in the industrial sector, for fueling boilers, and for providing process heat and combined heat and power. Natural gas also serves as a chemical feedstock, especially in the production of ammonia (§26.2.2) and hydrogen. Natural gas plays only a very small role in the transportation sector in the US, although in some countries it is an essential fuel for small-scale urban transportation.

Electricity Generation As described in §13, natural gas-powered combined cycle plants are currently the most efficient large-scale fossil fuel power plants in general use. Burning natural gas generates far fewer pollutants than coal. Combining the higher efficiency and lower carbon intensity of natural gas, a state-of-the-art combined cycle plant operating at 60% efficiency produces on the order of one third as much carbon per gigajoule of delivered electrical energy as a coal plant operating at 35% efficiency (Problem 33.11).

Heating A primary use of natural gas both in residential and commercial settings is for heating space, water, and food. Natural gas lines running directly to houses and businesses provide a relatively clean and safe source of thermal energy. Gas-powered home heating systems convert the enthalpy of combustion of natural gas to useable heat with losses of less than 5%, and gas-powered boilers, stoves, clothes dryers, and other appliances are used in many places.

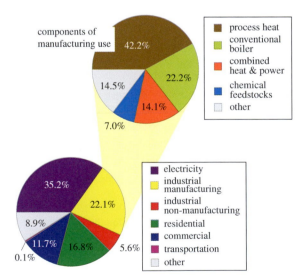

Figure 33.27 Uses of natural gas in the US in 2015 [12]. The more detailed breakdown of manufacturing uses data from 2011 [232].

Hydrogen and Fertilizer Similar to the steam reforming of coal described in §33.1.7, with an appropriate catalyst (such as nickel oxide) and heat, methane and steam combine to form syngas through the reaction

$$CH_4 + H_2O \rightarrow CO + 3H_2. \tag{33.9}$$

Natural gas is a principal source of *hydrogen* used for a variety of purposes. In particular, large amounts of hydrogen are required in petroleum refineries when long chain hydrocarbons are cracked to form more desirable short chain molecules with higher H/C ratios.

Another major use of hydrogen derived from the reaction (33.9) is as a feedstock for fertilizer production. As discussed in §26.2.2, living organisms require nitrogen to synthesize amino acids, proteins, and other essential biological molecules, and intensive agriculture uses a large supply of nitrogen. The synthesis of *ammonia* (NH_3) is an essential step in fixing nitrogen in a form that can be used by growing plants. In the **Haber–Bosch process**, ammonia is created by reacting nitrogen with hydrogen from natural gas through the reaction

$$N_2\,(g) + 3H_2\,(g) \rightarrow 2\,NH_3\,(g) \tag{33.10}$$

with enthalpy of reaction $\Delta H_r = -92.22$ kJ/mol. Ammonia can be used to produce a variety of nitrogen compounds used in *fertilizer*. The use of natural gas to fix nitrogen for agriculture can be thought of as a way of using fossil fuel energy to boost biological energy production. In 2002, about 5% of global natural gas use went to producing ammonia for fertilizer. This use of fossil fuel energy in fertilizer production is discussed further in §26.

33.3.5 Natural Gas Resources

As of 2014, proven natural gas reserves worldwide were around 190 Tm^3 [9]. This represents a reserves to production ratio (R/P) of about 55 years. This ratio has been relatively constant over the last three decades (Figure 33.28). Figure 33.29(a) shows proven natural gas reserves held by the top eight countries. The ranking of countries by *production* of natural gas is quite different, and is shown in Figure 33.29(b). Note, as in the similar figures for coal and oil reserves, estimates of *total recoverable natural gas* are in most cases much larger than the proven reserves shown in the figure. Also, in the case of natural gas, frequent discoveries and announcements of new resources have and most likely will continue to move the proven reserve estimates upward.

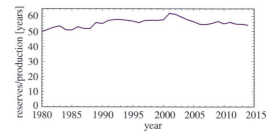

Figure 33.28 Reserves to production ratio variation over the last 30 years for natural gas [12]. As production has increased, additional proven reserves have kept the ratio between 50 and 60 over this time period.

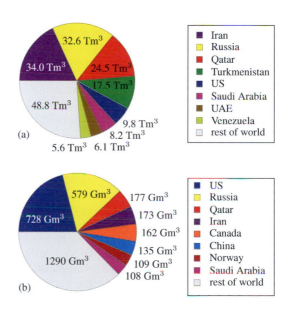

Figure 33.29 (a) Proven natural gas reserves, including unconventional resources, of the top eight countries in 2014 in trillion cubic meters; (b) 2014 natural gas production by the top eight producing countries in billion cubic meters [9]. See Footnote 3 for more information.

Natural Gas

Natural gas has the highest energy density (∼50 MJ/kg) and lowest carbon intensity of all the fossil fuels. Natural gas is widely used for heating, and is used increasingly for electric power production.

Box 33.7   Offshore Oil and Gas at Northern Latitudes

While in recent years new oil and gas field discoveries on land have decreased substantially, offshore exploration and production are continuing to expand. One major region that was developed over the last four decades is the North Sea area off Norway. In October, 1969, discovery of the *Ekofisk oil field* 200 miles southwest of Stavanger, *Norway* initiated an era of tremendous activity in offshore oil production in the North Sea. At its peak in 2002, the Ekofisk field was producing 300 000 b/d. Many other large oil fields were discovered in the North Sea from 1969 to 1984. With the discovery of large gas fields as well, Norway rapidly became one of the world's largest exporters of oil and gas.

The challenge of recovering large quantities of oil and gas from deep beneath the North Sea has helped to drive the development of increasingly sophisticated offshore technologies. From enormous stationary rigs attached to the sea bottom, to floating platforms anchored in hundreds of meters of water and designed to withstand harsh North Sea winter storm conditions, to complex undersea processing stations and pipelines, offshore oil recovery in this area has involved some of the most extreme engineering efforts ever undertaken. The Troll A platform (pictured left) in the Troll gas field 100 km northwest of Bergen, Norway, rests on massive concrete pillars in 303 m of water, weighs roughly 600 Mt, and is the largest structure ever moved by humankind.

While oil production has peaked in the North Sea (production from oil fields on the Norwegian continental shelf is shown at right), a steady decrease in Arctic sea ice in recent years (§35.3) has begun to open up major new areas for exploration, in the Barents Sea and the Arctic Sea. In the Barents Sea, the Norwegian Snohvit ("Snow White") gas field began production in 2006 and has estimated recoverable reserves of roughly 200 Gm^3 of natural gas. The much larger Russian Shtokman field, also in the Barents Sea, north of the Kola peninsula has estimated reserves of almost 4 Tm^3, and is currently under development. And these fields may only be the tip of the iceberg. The USGS has estimated that one quarter or more of all undiscovered oil and gas

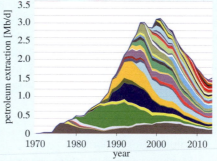

resources worldwide may be in the Arctic region. As summer Arctic sea ice diminishes year after year, many companies and countries are considering the exploitation of these resources. The sensitivity of relatively pristine Arctic ecosystems has raised controversies, however, regarding the extent to which oil and gas production in this region should be promoted or tolerated. Thus far, a combination of high costs, regulations, and fear of the consequences (both for the environment and for the company and/or nation involved) of a catastrophic spill in the Arctic have kept development in this area to a minimum.

(Map credit: Imagery © 2017 Landsat/copernicus, IBCAO, Data SIO, NOAA US Navy, NGA GEBCO, US Geological Survey; Image: Harald Petterson/Statoil; Graph: Rune Likvern Fractional Flow)

33.4 Hydrocarbon Conversion

A variety of chemical processes can be used to convert solid fossil fuels (e.g. coal) to gas, and solid or gas fuels to liquid form. Though conversion costs energy, there are a variety of reasons that such conversion may be useful. Gas and liquid fuels are easier to transport and handle than coal. And as new petroleum resources become more scarce, the economic incentive for producing liquid fuels from solid or gas becomes stronger.

33.4.1 Gasification

Heating coal in the absence of air produces a gas known as **coal gas**, composed primarily of hydrogen gas, methane, and carbon monoxide, as well as small amounts of other hydrocarbons, nitrogen, and carbon dioxide. Coal gas is one of a variety of manufactured gases known as **town gases** that have been used as fuel over the years. The simple carbonization process that produces coal gas converts roughly 20% of the mass of typical bituminous coal to a combustible gas. In the early eighteenth century, town gas produced in this fashion – originally a by-product of the production of coke for industrial applications – was used broadly for gas lighting systems in Europe and the US.

The carbon in coal can also be converted to gas form by passing air over a very hot carbon fuel (either the coke remaining after production of town gas or anthracite, for example). The carbon is partially oxidized to carbon monoxide, $2C + O_2 \rightarrow 2\,CO$. The resulting mixture of CO and residual N_2, known as **producer gas**, is quite toxic and has very low energy density, but was nevertheless used for many years in industry.

The most effective approach currently used to convert coal to combustible gas form is the method of steam reforming described in §33.1.7, which produces the variety of syngas known as **water gas** – largely a mixture of carbon monoxide and hydrogen. Water gas has a higher energy density than producer gas in part because it is undiluted by nitrogen. It is used as an input in various industrial processes. When the carbon monoxide is converted to further hydrogen and carbon dioxide by the water–gas shift reaction (33.4), the carbon dioxide can be removed, leaving pure hydrogen gas that has a high energy content and a variety of uses (see for example §37.3.2). A variety of systems are currently used for coal gasification, including fixed-bed, fluidized-bed, and *entrained-flow* systems (see [227]).

In addition to the fuel source flexibility provided by gasification techniques, the combustible gas fuels produced from coal are generally much more environmentally friendly, since most impurities are left behind in the gasification process, giving a clean and clean-burning output fuel. This advantage must be balanced against the energy cost of the conversion process.

33.4.2 Liquefaction

Just as coal can be converted to a combustible gas, it is possible to convert solid and gaseous hydrocarbons to liquid fuel. The **Fischer–Tropsch process** converts syngas into liquid hydrocarbons by using an iron or cobalt catalyst and increased pressure and temperature in the range 150–300 °C leading to the reaction

$$(2n + 1)H_2 + nCO \rightarrow C_nH_{(2n+2)} + nH_2O\,. \quad (33.11)$$

The distribution of the hydrocarbons produced varies depending upon the precise conditions. In general, higher pressure increases the likelihood of forming longer hydrocarbon chains and increases the reaction rate. The distribution of hydrocarbon lengths is given roughly by the **Anderson–Flory–Schulz distribution**, which states that the fraction of mass in hydrocarbons of length n is roughly

$$w_n = n(1 - \alpha)^2 \alpha^{n-1}\,. \quad (33.12)$$

The distribution peaks at $n \sim -1/\ln\alpha$ and the average value of n is $\bar{n} = (1 + \alpha)/(1 - \alpha)$, so as α approaches one the hydrocarbon chains become longer. The parameter α can be tuned to be near one by adjusting the pressure, catalyst, and other conditions of the reaction. The hydrocarbons produced by Fischer–Tropsch are generally straight-chain paraffins, though a small fraction of alkenes and other hydrocarbons, as well as alcohols and other oxygen-containing molecules, are formed.

The Fischer–Tropsch process can be used to produce liquid fuel from coal, or from natural gas (**gas to liquid**). Historically, Fischer–Tropsch synthesis was used in Germany during the Second World War for fuel production from coal. Currently Sasol in South Africa runs a number of plants that produce liquid fuel from coal and natural gas using this method.

The overall conversion efficiency of the Fischer–Tropsch process is generally below 50%, including both the hydrocarbon and thermal energy inputs to the process. Thus, for example, running automobiles on fuel produced through Fischer–Tropsch synthesis from coal is much more carbon intensive than using petroleum-derived gasoline. The possibility of large-scale coal-to-liquid or gas-to-liquid conversion, however, makes fossil fuels essentially interconvertible, at the price of some loss in efficiency. If market conditions raise the price of petroleum-derived liquid fuel sufficiently, it may become economically advantageous to produce liquid fuel from coal or natural gas using this mechanism. This is particularly likely to occur with natural gas, where an oversupply may lead to decreased prices, increasing the economic advantage of gas-to-liquid technology.

The Fischer–Tropsch process is an **indirect liquefaction** method for converting coal to liquid fuel, as it uses syngas as an intermediate stage. It is also possible to convert coal to liquid fuel using **direct liquefaction** methods. The basic idea is that coal has a lower H/C ratio than liquid hydrocarbon fuels, so coal can be liquefied by adding hydrogen with appropriate chemical reactions (**hydrogenation**). Combining gaseous hydrogen with coal at high temperatures and pressures leads to liquefaction, but at a relatively low efficiency. More sophisticated techniques have been developed in which the coal is mixed with a hydrogen-rich solvent and/or catalysts are used to activate the hydrogenation process. Direct liquefaction was used by Germany in the 1930s and 1940s, and is currently a subject of active research and development.

33.5 Fossil Fuel Summary

Known reserves of fossil fuels are quite extensive. For coal and natural gas, the proven reserves are sufficient to maintain close to the current level of use for over 50 years. For petroleum, estimates are slightly less certain and reserves are more limited, but it is likely that conventional crude oil reservoirs will be sufficient to maintain close to the current level of use for at least several decades, with light tight oil, further offshore discoveries, and unconventional oil likely to significantly extend availability.

Table 33.2 summarizes the energy content, carbon intensity, and resource availability of the main fossil fuels. For each fossil fuel, an estimate is given of the number of years for which the proven reserves could maintain the current rate of consumption. Since reserves of petroleum and natural gas have been growing at least as fast as consumption, this provides a reasonably conservative estimate for the time over which current fossil fuel consumption rates could be maintained. We also tabulate for each resource the number of years for which the established reserves could provide 500 EJ/year, the current rate of global energy use, to give a sense of how these resources compare to the totality of human energy use.

Although some of this data is quite rough, the clear conclusion is that there are enough fossil fuel reserves available to supply all human energy needs for many decades. While conventional crude oil reserves may run short within a few decades, even if extended somewhat by increased shale oil recovery technologies, the coal-to-liquid and gas-to-liquid conversions described in §33.4 are in principle capable of replacing all liquid fuel needs as long as coal and natural gas reserves are available. Tar sands and oil shale can also produce liquid hydrocarbon fuel. Neither contributes significantly to the current estimated oil reserves in Table 33.2 but they represent a much larger potential resource that could be tapped as conventional crude becomes scarce, so that there is likely no obstacle in principle to continuing use of liquid fuel for air and land vehicles through the twenty-first century. Furthermore, the relation between *proven reserves* and *eventually recoverable resources* must be kept in mind. Not only are new reserves continually being proven, but as real prices of fossil fuels rise, presently subeconomic resources become economic reserves, leading to ever increasing estimates of proven reserves.

One of the principal drawbacks of fossil fuels is that when combusted they release carbon dioxide into the atmosphere. Unlike the carbon dioxide released from biofuel combustion, the carbon from fossil fuels is added into the global carbon cycle from underground repositories where it has been sequestered for millions of years. Net carbon emissions from various fossil fuel sources are shown in Figure 33.30. In the next few chapters (§34, §35) we explore the consequences of these carbon emissions for Earth's climate.

Fossil Fuels

Over 80% of energy presently used by humans comes from fossil fuels. In the absence of policies or other circumstances that would lead to an abrupt redirection of resources, this level of fossil fuel usage is likely to continue for the immediate future. Fossil fuels are extremely energy-dense and relatively plentiful, though they are a finite resource. Proven reserves of coal and natural gas are sufficient for over 50 years at current usage rates. While conventional crude oil is not quite as plentiful, new offshore discoveries, enhanced recovery methods, and unconventional resources provide an abundant resource that could supply liquid fossil fuels for many decades to come. Coal and natural gas can also be converted to liquid fossil fuel with some expenditure of energy and at some cost to the environment. For all fossil fuels, eventually recoverable resources are likely significantly greater than present proven reserves.

Table 33.2 Comparing energy density, carbon intensity, and available reserves for the primary fossil fuels, in round numbers. World average coal carbon intensity from [12]. Estimates of production and proven reserves (in common units of EJ) as of 2015 from [9]. Oil energy and carbon density is for typical conventional crude, though production and reserves numbers include a fraction from unconventional sources with greater carbon footprint. Natural gas numbers include proven economically recoverable shale gas reserves but do not include methane hydrates/clathrates.

Resource	Energy density (LHV) (MJ/kg)	Carbon intensity (kg C/GJ)	Production (EJ/y)	Reserves (EJ)	R/P (y)	Years at 500 EJ/y
Oil	42–44	20	205	10 390	51	21
Coal	15–35	25	157	17 830	113	36
Natural gas	50	15	134	7 100	53	14

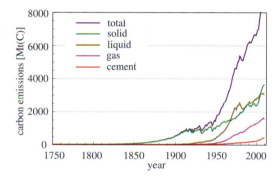

Figure 33.30 Global carbon emissions from fossil fuel use [241].

Discussion/Investigation Questions

33.1 Determine the current relative costs of crude oil, gasoline, coal, and natural gas on a per unit of energy basis.

33.2 Explain why petroleum is currently the primary source of fuel for land and air transport, while coal is the primary fuel for electric power plants.

33.3 Investigate the production of helium. How was it formed? Why is it found in natural gas? How much of it is recovered? Discuss conservation of the world's helium resources.

33.4 Estimate the relative carbon output of heating a home using (a) natural gas, and (b) electrical heating powered by electricity from coal-fired power plants. How much do your assumptions about what technology is used affect your answer?

Problems

33.1 Estimate the energy content of a coal seam measuring 1 km² by 1 m deep. Estimate the electrical power output

of a coal plant burning this fuel over one year, and the area required for a solar thermal plant in a high insolation location to produce a comparable power output. Take the density of coal to be $\rho = 1300\,\text{kg/m}^3$.

33.2 A rough measure of the enthalpy of combustion of a hydrocarbon can be obtained by assuming that all C-C bonds contribute the same energy to the combustion process when broken, and that the same is true for all C-H bonds. This is particularly appropriate for alkanes and cycloalkanes, where all C-C bonds are roughly equivalent *single bonds*. Using this simple model, write a general formula for the combustion energy of an alkane with n carbon atoms and m hydrogen atoms in terms of two variables characterizing the energies of the two bond types. Fit for the two variables by matching with the energy of methane and dodecane, and compare the intermediate energies to those of Table 33.1.

33.3 When computing efficiencies, European power plants sometimes use the LHV of a fuel, while for US power plants the HHV is standard. Consider a power plant burning bituminous coal with HHV/LHV of 31.1/30 MJ/kg. If the power plant's efficiency is 35% when computed from the HHV, what is its efficiency when computed from the LHV?

33.4 Estimate how many solar panels or wind turbines are needed to produce the same amount of electrical energy as the 10 kt of coal that can be extracted in a longwall mining operation in one year by one person. (Make reasonable assumptions and approximations.)

33.5 [T] The deformation in a solid can be described by a vector field $\boldsymbol{u}(\boldsymbol{x}) = (u_1, u_2, u_3)$ describing the displacement of the solid away from the equilibrium point $\boldsymbol{x}$. This is similar to the displacement vector $\boldsymbol{s}$ used to analyze waves in §31.2. For small deformations, the **strain tensor** σ is a 3×3 *matrix* with elements $\sigma_{ij} = (\partial u_i/\partial x_j + \partial u_j/\partial x_i)/2$. For example, a rod compressed uniformly to 99% of its original length in the x-direction would have $\sigma_{11} = -0.01$. The energy of compression is given by

$$U = \mu \sum_{i,j=1}^{3} \left(u_{ij} - \frac{1}{3}\delta_{ij} \sum_{k=1}^{3} u_{kk} \right)^2 + \frac{1}{2} K \sum_{i=1}^{3} u_{ii}^2 ,$$

(33.13)

where μ is the shear modulus, K is the bulk modulus, and $\delta_{ij} = 1$ if $i = j$ and 0 otherwise. (a) Show that for uniform hydrostatic compression, where $u_{11} = u_{22} = u_{33} = u$, and all other components vanish, the energy of compression depends only on K. (b) Show that for P-wave compression, where $u_{11} = u \sin(kx)$ and all other components vanish, the energy of compression is proportional to the elastic modulus $\chi = K + 4\mu/3$. For a further introduction to the *theory of elasticity*, see [242].

33.6 Compute the fraction of Carnot efficiency realized by a coal plant operating at 600 °C and 45% efficiency (assume ambient temperature of 300 K). If another plant operates at the same fraction of Carnot efficiency at a temperature of 700 °C, what is the thermal efficiency of the second plant?

33.7 An underground mass of salt may form a dome trapping a petroleum reservoir. Compute the gravitational anomaly (deviation of gravitational acceleration) produced by an underground salt mass, where for simplicity we assume that the salt (density 2220 kg/m³) forms a sphere of radius 0.5 km centered 0.5 km beneath the surface, and compute the gravitational anomaly by determining the difference from a corresponding mass of granite (density 2640 kg/m³). Express your answer in mGal = 10^{-5} m/s². (Gravity data can be measured to a precision of $\sim$ mGal.)

33.8 [T] Use the Arrhenius equation in Box 33.5 to show that a reaction rate doubles under a temperature shift of $\Delta T \cong (k_B T^2 \ln 2)/E$, assuming $E \gg k_B T$.

33.9 Using an activation energy $E = 250$ kJ/mol, compute the change in rate of petroleum generation when temperature increases from 130 °C to 430 °C. If petroleum generation through such a process takes 10 million years at 130 °C, estimate the time it would take in the laboratory at 430 °C.

33.10 Bitumen extracted from tar sands in Athabasca, Canada, has an API gravity of °API 8. Compute the mass of one liter of such bitumen and compare to that of a liter of light crude at °API 35.

33.11 Compute and compare the amount of carbon released per GJ of delivered energy by (a) a coal plant operating at 35% efficiency, (b) a natural gas combined cycle plant operating at 60% efficiency, and (c) an automobile engine running on gasoline at 25% efficiency.

33.12 Compare the energy density of oil and geothermal fluid at 250 °C pumped from oil/geothermal wells at a rate of 30 L/s. To make a fair comparison, assume that the oil is transformed into useful mechanical energy at 20% net efficiency, and that the geothermal fluid emerges from the well as saturated liquid and is converted into electricity in a single-flash geothermal power plant as described in Example 32.5.

Part III

Energy System Issues and Externalities

Energy and Climate

Isolated energy systems have been the focus of most of the earlier chapters of this book. We have followed energy that originates in solar, nuclear, and gravitational sources through to its end uses by humans, and have analyzed the physics of the processes involved in this flow. Almost all energy that reaches Earth from space or that is released in terrestrial processes is eventually reradiated out into space. The flow of this energy through the atmosphere and oceans determines Earth's climate. **Climate** refers to the statistical distribution of meteorological phenomena such as temperature, humidity, wind, and precipitation in a given region over many years (while *weather* refers to the configuration of these conditions on a time scale of hours or days). Earth's climate depends upon a complex interplay between Earth's atmosphere and its surface, including the **hydrosphere** (water), **cryosphere** (ice), and **biosphere** (living systems). Over the 4.5 billion years of Earth's lifetime, changes in the composition and mass of the atmosphere, orbital geometry, solar radiation levels, and the configuration of Earth's land masses have led to a wide variation of climate. In this chapter we introduce some basic notions of climate science and relate the planet's climate to the flow of energy in Earth systems.

We begin in §34.1 with a highly simplified global model of Earth's energy budget, and review how the greenhouse effect traps outgoing infrared radiation, helping to provide a climate conducive to life on Earth. We then describe some of the basics of atmospheric physics in §34.2, and global energy flow in §34.3. This material gives us a foundation for understanding some central mechanisms underlying Earth's climate systems.

In §34.4 we describe in more detail the carbon cycle and the role that CO_2 plays as an atmospheric greenhouse gas. Observational data on recent changes in CO_2 levels give unequivocal evidence that human activity is significantly increasing the quantity of this gas in Earth's atmosphere.

Reader's Guide
This chapter focuses on the physics of climate. Beginning with the basics of albedo and atmospheric physics, the primary mechanisms underlying energy flow in Earth's atmosphere are developed. The carbon cycle is described and the greenhouse effect is analyzed. The following chapter (§35) presents a systematic discussion of past, present, and future climate.

Prerequisites: §6 (Heat transfer), §24 (Solar thermal energy), §27 (Ocean energy), §28 (Wind energy), §33 (Fossil fuels).

Given the known parameters of Earth's present climate, atmosphere, and ocean currents, the direct effect of this increase is a computable increase in downward radiation flux (*radiative forcing*) from atmospheric trapping of infrared radiation. Human activities also directly affect climate in a variety of other ways. For example, fossil fuel combustion leads to the formation of aerosols that affect the rates of reflection and absorption of incoming solar radiation, and it also produces particulate matter that has accelerated the melting of glaciers and sea ice at high altitudes and northern latitudes, as described in §34.4.4.

In §34.5 we describe positive and negative feedbacks that increase or decrease the effects of a change in radiative forcing such as that produced by an increase in atmospheric CO_2. Accurately determining the sign and magnitude of feedbacks proves to be one of the most difficult aspects of predicting the response of Earth's climate to changing conditions. We describe the principal sources of climate feedback and computerized *general circulation models* that are used to estimate the effects of these feedbacks and the resulting consequences for Earth's climate.

Evaluating the long-term impact on climate of changes in radiative forcing requires extrapolating away from the current relatively well-understood climate configuration. Because of the complexity of Earth's global energy balance and climate system, such an extrapolation is difficult. In addition to computer climate models, study of past changes in climate over a range of time scales is helpful in understanding how Earth's climate systems evolve in varying circumstances. In the following chapter, §35 (Climate change), we review some basic aspects of Earth's climate history, and describe some of the physics that underlies both previous changes in Earth's climate and the tools used in paleoclimatology to determine past climate. This provides us with background for a discussion of current efforts to predict future climate in §35.2. We conclude §35 with some discussion in §35.3 of the consequences of global warming and possible mechanisms for mitigating large-scale climate change.

For further reading, good references on basic aspects of atmospheric physics are [173, 243, 244]. *Climate Change 2013: The Physical Science Basis*, the Working Group I Contribution to the Fifth Assessment Report of the Intergovernmental Panel on Climate Change (**IPCC5-WG1**) [245] contains an extensive summary of the state of knowledge (circa 2013) regarding Earth's climate and climate change due to anthropogenic production of greenhouse gases.

While the basic science described in most of this book has been understood for decades or centuries, and is well established and verified by many independent research programs, the study of many aspects of Earth's climate is still at a relatively early stage. Scientific understanding of the complex interplay between the atmosphere, oceans, cryosphere, and biosphere, and the effects of human activities on these systems, is evolving rapidly at present. The material in this chapter is, by and large, almost universally accepted in the scientific community and is not controversial. The primary exception is the magnitude and sign of certain feedbacks, which remain highly uncertain, as noted in §34.5. Both the accuracy of the present understanding of past climate and the ability to make predictions of Earth's climate future are important topics in the following chapter.

34.1 Albedo and the Greenhouse Effect

The notions of *thermal radiation* and *radiative equilibrium* were first introduced in §6.3.3 (*c.f.* §6, Problem 6.16) and discussed at some length in §24.1. Radiative equilibrium expresses conservation of energy in the context of systems that gain and lose energy through radiation. Radiative equilibrium is particularly important when the other methods of heat transfer (conduction and convection) are negligible

---

Climate

Climate refers to the statistical distribution of meteorological phenomena such as temperature, wind, and precipitation over many years. At any given time, Earth's climate is determined by interactions between Earth's atmosphere, hydrosphere (water), cryosphere (ice), land masses, and biosphere. Over longer times, climate is affected by many factors, including Earth's orbital dynamics, solar output, geological processes, and human activities.

– as they are at a global level for planet Earth, which is surrounded by the near-perfect vacuum of interplanetary space. Under such circumstances, the thermal energy that a body radiates must equal the amount of incoming radiation that it absorbs. If not, the temperature of the body would rise or fall until equilibrium was established. Note that while Earth considered as a whole is in radiative equilibrium with incoming solar radiation, Earth's surface and atmosphere are not locally in radiative equilibrium – convection and conduction in the atmosphere and ocean play an important role in distributing thermal energy around the planet.

A very simple model of Earth was described in Example 24.1. Treating the planet as a uniform spherical black body that perfectly absorbs incoming solar radiation at $I_\odot = 1366\,\mathrm{W/m^2}$ and radiates it back into space at a temperature $T_\oplus$, *radiative equilibrium* requires that $\pi R_\oplus^2 I_\odot = 4\pi \sigma R_\oplus^2 T_\oplus^4$, or $T_\oplus = \sqrt[4]{I_\odot/4\sigma}$. This model gives an equilibrium temperature of Earth of $6\,°\mathrm{C}$. To construct a slightly more realistic model, and to develop a qualitative understanding of the role of the atmosphere, we need to include two more ingredients: reflectivity, or *albedo*, and the greenhouse effect.

34.1.1 Albedo

The **albedo** of an object is defined to be the fraction of incident electromagnetic energy that is reflected by the object, given a specific frequency or spectrum of incoming radiation. For our purposes, we define albedo with respect to the spectrum of solar radiation, though in other contexts other choices of spectrum are used. The terms *albedo* and *reflectivity* (see Box 22.1) are essentially equivalent, though used in somewhat different contexts.[1]

---

[1]   Often the same symbol α is used for *albedo* as well as for *absorptivity*. In some climate literature, a is used for albedo. We use the latter notation to distinguish it from absorptivity.

Albedo

The *albedo a* of an object or material is defined to be the fraction of incident radiative energy that is reflected.

The albedo a of any particular surface is therefore a number between 0 (completely absorbant) and 1 (completely reflective). Objects or materials that are highly reflective in the visible frequency range, such as fresh snow and ice, have relatively high albedos, while highly absorbant objects or surfaces, such as water or trees, have low albedos (Figure 34.1). The albedo of a specific surface may depend upon the angle of incidence of incoming radiation.

In Earth's current state, the average albedo is about

$$a_\oplus \cong 0.30 . \qquad (34.1)$$

This albedo arises from a combination of effects. About 6% of incoming radiation is reflected by Earth's atmosphere, so the albedo of the atmosphere is around 0.06. The surface albedo is around 0.1, and clouds contribute another 0.14. Although there are seasonal fluctuations in albedo and short-term fluctuations from weather events, this yearly average albedo is stable and well established to within a few percent. The reflection of incoming radiation, described by Earth's albedo, lowers the average

Figure 34.1 Objects such as trees that absorb most incoming radiation have low albedo. Objects such as sheep and dry grass that reflect a significant fraction of incident radiation have higher albedos.

temperature needed for the planet to maintain radiative equilibrium (see Example 34.1).

34.1.2 The Greenhouse Effect

Although Earth's atmosphere is fairly transparent to incoming solar radiation, which is peaked in the visible part of the electromagnetic spectrum, the atmosphere contains significant amounts of water vapor and other gases such as CO_2 that have absorption spectra in the infrared.

Example 34.1 Earth with a Transparent Atmosphere

Consider a simplified model of Earth in which the atmosphere contains only O_2 and N_2. In this simplified model we assume that the atmosphere is transparent to outgoing infrared radiation, and that the albedo is $a = 0.16$ (no clouds). What would the equilibrium temperature of the planet be?

In this situation the average over the surface of the rate at which Earth absorbs incoming radiation is (incorporating the ratio of 4 between Earth's surface area and cross-sectional area as in Example 24.1)

$$\langle I_{in} \rangle = (1 - a) I_\odot / 4 = 0.84 \, I_\odot / 4 \cong 287 \ \text{W/m}^2 .$$

In radiative equilibrium, the surface temperature T_s satisfies

$$I_{out} = \sigma T_s^4 = \langle I_{in} \rangle ,$$

which yields $T_s \cong 267 \, \text{K} \cong -6\,°\text{C}$. This is colder than the earlier estimate found in Example 24.1, which ignored albedo. Both estimates are, however, much colder than the current average surface temperature of $14\,°\text{C}$, which is higher due to the heat-trapping effects of Earth's atmosphere.

Note that if all infrared-absorbing gases in the atmosphere were really removed, as the temperature decreased formation of additional ice would increase the albedo further, leading to a further temperature drop. It has been suggested that at times in Earth's history something like this has indeed occurred; we discuss this *snowball earth* hypothesis again briefly in §35.1.4.

In §24, we described the phenomenon of heat trapping, also known as the *greenhouse effect*, which occurs when a material that absorbs incoming solar radiation is covered by a layer of another material that is transparent to the incoming radiation but that absorbs infrared radiation. In such a situation, the covering layer – in this case Earth's atmosphere – collects some of the outgoing infrared radiation and reradiates it back downwards, increasing the temperature of the absorbing material – in this case Earth's surface.

To see how this effect works in the context of the global Earth system, we modify the toy model of Earth's climate to include an idealized atmospheric layer that absorbs and reradiates all the radiation it receives from Earth's surface, which is in the infrared part of the spectrum. We assume that Earth's total albedo is given by the current value of 0.3, and that the atmospheric layer is otherwise transparent to incoming solar radiation. The flow of energy in this version of the toy model is depicted in Figure 34.2. Two coupled radiative balance equations determine the two temperatures T_a, T_s of the atmospheric layer and the surface. Radiative equilibrium for the entire Earth–atmosphere system requires

$$\sigma T_a^4 = \langle I_{\text{in}} \rangle = 0.7\, I_\odot/4 \cong 239\,\text{W/m}^2\,, \qquad (34.2)$$

where the average incoming radiation is given (as in Example 34.1) by $(1-a)I_\odot/4$. Radiative equilibrium at Earth's surface requires

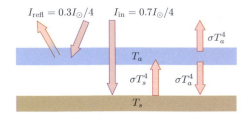

Figure 34.2 Energy flow in a simplified model of Earth with albedo 0.3 and a single-layer atmosphere that absorbs (and reradiates) all infrared radiation, but transmits all visible light.

$$\sigma T_s^4 = \langle I_{\text{in}} \rangle + \sigma T_a^4 = 2\sigma T_a^4 = 2\langle I_{\text{in}} \rangle\,. \qquad (34.3)$$

The solution of these equations is

$$T_a \cong 255\,\text{K}, \qquad T_s \cong \sqrt[4]{2}\, T_a \cong 303\,\text{K}\,. \qquad (34.4)$$

Thus, although we have an increased albedo compared to the simplest model (Example 34.1) in which Earth's atmosphere is completely transparent, the surface temperature in this model is substantially higher, due to the greenhouse effect.

To get a qualitative sense of the effect of increasing the quantity of greenhouse gases in the atmosphere, consider a model with an n-layer atmosphere, as described in Example 34.2. In this model the surface temperature grows like $\sqrt[4]{n+1}$ (Problem 34.2). While such simplified radiative equilibrium models give some qualitative sense of the

---

Example 34.2   A Multi-layer Model of Earth's Atmosphere

A slightly more realistic model of Earth's atmosphere than the one described in eqs. (34.2)–(34.4) contains several layers. In a model with a two-layer atmosphere, the upper layer of atmosphere acts as a blanket over the lower layer, which in turn blankets the surface. We then get a series of three equations, which we can solve for the surface temperature,

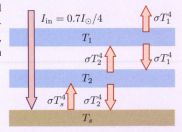

$$\sigma T_1^4 = \langle I_{\text{in}} \rangle = (1-a)I_\odot/4,$$
$$\sigma T_2^4 = \langle I_{\text{in}} \rangle + \sigma T_1^4 = 2\sigma T_1^4,$$
$$\sigma T_s^4 = \langle I_{\text{in}} \rangle + \sigma T_2^4 = 3\sigma T_1^4,$$
$$T_s \cong \left(\sqrt[4]{3}\right) 255\ \text{K} \cong 336\,\text{K}\,.$$

The three equations here are derived by using radiative equilibrium for volumes with three different boundaries: outside each of the two atmospheric layers and just above the surface. One can repeat the process for an arbitrary number of layers (Problem 34.2), which leads to the result that a model with n atmospheric layers that are perfectly absorbing in the infrared gives a surface temperature of

$$T_s = \left(\sqrt[4]{n+1}\right) \times 255\,\text{K}\,.$$

increase in surface temperature as atmospheric absorption of infrared radiation is increased, a more detailed model is needed to obtain quantitatively meaningful results.

In Earth's actual atmosphere, greenhouse gases such as water vapor and CO_2 are distributed through the atmosphere. These greenhouse gases are not perfect absorbers of all infrared radiation. Rather, they absorb certain frequencies very effectively and allow other frequencies of infrared radiation to pass through. The dependence on the number of levels in the simplified n-level model given in Example 34.2 is unrealistically strong because the gaps in the real absorption spectrum allow certain frequency ranges of infrared radiation to escape freely through the atmosphere (Problem 34.3). To obtain a more realistic picture of the radiative equilibrium of the atmosphere, we should include information about the details of the atmospheric constituents and their absorption spectra as well as their distribution in the atmosphere. This distribution, in turn – as well as the flow of radiation from the surface – depends upon the temperature and pressure profiles of the atmosphere, which vary from place to place on the globe. In particular, the distribution of the most significant greenhouse gas, *water vapor*, also depends crucially on non-radiative heat transfer through convection, giving a complex coupling between the radiative trapping of heat through greenhouse gases and other heat transport mechanisms.

We therefore would like to have a more detailed model of Earth's atmosphere and the large-scale distribution and transport of thermal energy in Earth's climate system. We proceed in §34.2 to develop some of the basic principles of atmospheric physics. This will enable us to describe in more detail the flow of energy in the vertical direction, governing Earth's radiative equilibrium. In §34.3 we describe some basic aspects of large-scale global energy flow in the horizontal directions across the surface of the planet. Any detailed understanding of Earth's climate and temperature must incorporate all of these aspects of global energy flow.

34.2 Atmospheric Physics

In this section we analyze the vertical structure of the atmosphere, treating the atmosphere as a one-dimensional system. We describe the equilibrium pressure and density profiles of the atmosphere and the role of various atmospheric constituents in the vertical structure of the atmosphere in radiative equilibrium. The vertical temperature gradient, called the *lapse rate*, and its connection to convective instability are discussed. This leads to a one-dimensional model of the *radiative–convective*

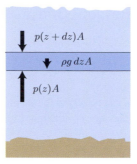

Figure 34.3 In hydrostatic equilibrium, the change in pressure over a height interval dz must compensate the force of gravity on the mass of air in that interval.

equilibrium of the atmosphere. The notion of *radiative forcing* is developed as a key concept for comparing quantitatively the effects of various phenomena on Earth's net radiative equilibrium.

34.2.1 Hydrostatic Equilibrium

The pressure of the planetary atmosphere arises from the weight of the column of gas above the point where the pressure is measured.[2] In equilibrium, the vertical change in pressure over a small height differential dz (Figure 34.3) is given by the gravitational force on the volume of gas contained within that height differential

$$dp = -g\rho\, dz, \qquad (34.5)$$

where ρ is the atmospheric mass density. To proceed we make the simplifying assumption that the atmosphere is an ideal gas composed of identical molecules of mass $m \cong 29u$. This is a relatively good approximation, because the atmosphere is mostly oxygen ($m \cong 32u$) and nitrogen ($m \cong 28u$) with similar masses and the relative concentration of N_2 and O_2 is close to 4:1 throughout the atmosphere, so that most properties – such as density – can be accurately estimated by taking a weighted average over these primary atmospheric constituents. Assuming identical molecules of mass m, the ideal gas relation can be used to relate pressure to density through

$$pV = Nk_BT = V\frac{\rho}{m}k_BT, \qquad (34.6)$$

where $\rho = Nm/V$. It follows that

$$\rho = mp/k_BT. \qquad (34.7)$$

---

[2] For the purpose of describing hydrostatic equilibrium, we approximate the atmosphere as a volume over a planar surface (Problem 34.4).

Combining eq. (34.5) and eq. (34.7) gives

$$dp/p = -dz/H, \qquad (34.8)$$

where $H = k_B T/mg$ is the (temperature-dependent) **scale height** of the atmosphere.

Equation (34.8) can be solved for the pressure at altitude z in terms of the pressure p_0 at height $z = 0$,

$$p(z) = p_0 e^{-\int_0^z dz/H} . \qquad (34.9)$$

In a region of constant temperature, H is constant and we have

$$p(z) = p_0 e^{-z/H} . \qquad (34.10)$$

In this case, the scale height is the height over which the atmospheric pressure drops by a factor of $1/e$. Near Earth's surface, estimating for example $T = 290\,$K, we have $H \cong 8.5\,$km. The density can similarly be obtained from eq. (34.7),

$$\rho(z) = \rho_0 e^{-z/H} = \rho_0 e^{-mgz/k_B T} , \qquad (34.11)$$

in a region of constant temperature. This is precisely the density profile that would be obtained from the *Boltzmann distribution* for the molecules in a gas at constant temperature T in a gravitational potential (see §8.6.1). In general, the temperature T – and therefore the scale height H – depends on z, and we must use the more general formula (34.9).

Hydrostatic Equilibrium

In *hydrostatic equilibrium* the pressure at any point in the atmosphere is determined by the weight of the column of air above a unit area parallel to the surface. In the approximation that the atmosphere consists of a single molecular species with mass m in hydrostatic equilibrium, the pressure varies with the height z as

$$dp/p = -dz/H ,$$

where $H = k_B T/mg$ is the *scale height*. In a region of constant temperature, pressure varies as

$$p(z) = p_0 e^{-mgz/k_B T} .$$

From eq. (34.8), it follows that pressure is a decreasing function of height. Since many physical properties of the atmosphere depend primarily upon pressure, p is often used as a convenient proxy for height.

34.2.2 Atmospheric Temperature and the Lapse Rate

The temperature of the atmosphere is not constant but, at least in the lower atmosphere, generally decreases with increasing altitude. The rate at which T decreases is known as the **lapse rate** Γ, defined as

$$\Gamma = -dT/dz . \qquad (34.12)$$

Depending upon the lapse rate, the vertical temperature profile of the atmosphere can lead to convective instability. In general, if the temperature drops too quickly with altitude, the warmer air from below tends to rise and reduce the lapse rate. To understand this in a quantitative fashion, consider a parcel of air that moves upward adiabatically, i.e. without heat exchange with the surrounding air. Then the first law of thermodynamics states that

$$dQ = dU + dW = C_V dT + p dV = 0 . \qquad (34.13)$$

Since p decreases with height, if a packet of air moves upward, it expands and dV is positive, as is $p\,dV$. Thus from eq. (34.13), the change in temperature dT is negative. Suppose that the decrease in temperature of the packet is less than the actual temperature change of the atmosphere. From the ideal gas law, a warmer packet of air at the same pressure as its surroundings will have lower density, and therefore will continue to rise, so convection will continue. If, on the other hand, the atmosphere cools more slowly with height than an adiabatically rising air packet, then the atmosphere will be stable against convection. The **adiabatic lapse rate** is defined to be the lapse rate such that the temperature change in a parcel of air moving upward adiabatically matches the change in temperature of the surrounding air. This determines the maximum rate at which the temperature can decrease vertically without convective instability.

To compute the adiabatic lapse rate, first take the derivative of the ideal gas law

$$pdV + Vdp = Nk_B dT$$
$$= (C_p - C_V)dT , \qquad (34.14)$$

then combine this with eq. (34.13) and eq. (34.5) to obtain

$$C_p dT = Vdp = -Vg\rho dz . \qquad (34.15)$$

The **dry adiabatic lapse rate** Γ_d (we have ignored water vapor) is then

$$\Gamma_d = \frac{dT}{dz} = -\frac{gV\rho}{C_p} = -\frac{g}{c_p} . \qquad (34.16)$$

Near Earth's surface at $T = 290\,$K the specific heat capacity of dry air is

$$c_p \cong 1.0035 \, \text{kJ/kg K}, \qquad (34.17)$$

so the dry adiabatic lapse rate is

$$\Gamma_d \cong 9.8 \,^\circ\text{C/km}. \qquad (34.18)$$

When the surface of Earth is heated by solar radiation, thermal energy is transferred to the air just above the surface. This produces a temperature gradient in the vertical direction. When this temperature gradient is larger than the adiabatic lapse rate Γ_d, convection occurs and thermal energy is transferred upward until the temperature gradient is reduced to (or below) the adiabatic value. Thus, in the lower part of Earth's atmosphere called the **troposphere** – generally up to a height of 9 to 17 km – the temperature generally falls with altitude,[3] and convection makes an important contribution (roughly 60% [173]) to the upward transport of energy. The energy transported in this fashion includes not only the thermal energy of dry air but also latent heat in water vapor. Thus, a quantitative understanding of vertical energy flow in Earth's lower atmosphere involves not only radiative heat transfer but also convection. Convection of energy and water vapor give rise to the rich variety of weather phenomena that characterize the troposphere. The troposphere ends rather abruptly at the *tropopause*, where, for reasons explained below, the temperature begins to increase with altitude, so that convection and weather diminish dramatically in the region above the tropopause known as the **stratosphere**.

The adiabatic lapse rate given in eq. (34.18) is the value for dry air. When the air includes water vapor, the situation is slightly more complicated, and convective effects generally determine the vertical distribution. If the air is saturated, then as it rises and the temperature decreases, some of the water vapor will condense, releasing latent heat. This raises the temperature of the parcel of air, and can lead to convective instability at a smaller value of the lapse rate. Thus, for air saturated with water vapor the **moist adiabatic lapse rate** is smaller. Since convection occurs on shorter time scales than the convergence to radiative equilibrium, the lapse rate in radiative–convective equilibrium is close to the moist adiabatic lapse rate. In the lower atmosphere, the observed lapse rate typically varies between 3 °C/km and 10 °C/km, with a mean value of around 6.5 °C/km.

---

[3] Under unusual conditions pools of warm, less dense air can move or develop over colder, denser air low in the atmosphere causing an **inversion** in which temperature actually rises with height. Such inversions suppress convection and can trap pollution near the surface.

Lapse Rate

The *lapse rate*

$$\Gamma = -dT/dz$$

describes the rate of decrease of temperature with altitude in the atmosphere.

For dry air, the maximum lapse rate at which the atmosphere is stable to convection is the *dry adiabatic lapse rate*

$$\Gamma_d \cong 9.8 \,^\circ\text{C/km}.$$

The observed lapse rate in the lower atmosphere ranges from 3 °C/km to 10 °C/km, with a mean value around 6.5 °C/km.

Table 34.1 Some relevant atmospheric constituents as of 2015 (ppm/ppb = parts per million/billion by volume, see Box 34.3).

Molecule	Fraction by volume
Nitrogen (N_2)	78.08%
Oxygen (O_2)	20.95%
Argon (Ar)	0.934%
Water vapor (H_2O)	variable
Carbon dioxide (CO_2)	400 ppm
Methane (CH_4)	1.83 ppm
Nitrous oxide (N_2O)	328 ppb
Ozone (O_3)	variable

34.2.3 Atmospheric Composition and Radiative Absorption

While Earth's atmosphere is primarily composed of nitrogen and oxygen, many other types of molecules appear in trace amounts. The absorption properties of many of these other molecules play an important role in the flow of energy through absorption and radiation within the atmosphere. Some of the most significant atmospheric constituents are listed in Table 34.1.

In the preceding subsection, we made the approximation that the atmosphere consists of a single type of molecule with a mass equal to the weighted average of the masses of nitrogen and oxygen. Since they make up 99% of the atmosphere this is a reasonable first approximation. In principle, the relative density of different molecules in the atmosphere could depend upon altitude. For an atmosphere in equilibrium *without convection* and at constant temperature T, the density of each type of molecule would vary as the Boltzmann factor $e^{-mgz/k_B T}$ (eq. (34.11)), so heavier molecules generally would be more prevalent in the lower atmosphere. The time period over which convection mixes Earth's atmosphere is so short, however, even above the tropopause, that the atmospheric composition is actually quite homogeneous for most molecules up to an altitude of about 100 km. Thus, the approximation of treating the atmosphere as a single (ideal) gas with effective mass $m \cong 29u$ is an excellent one. Certain constituents such as ozone and water vapor that are added to and removed from the atmosphere by local processes on time scales short compared to the atmospheric mixing time are exceptions to this result. They are generally out of equilibrium over large ($\sim$100 km) distance scales and their density distribution is generally not described by eq. (34.11). The density of water vapor in the atmosphere, in particular, is affected by many factors, and depends upon rates of evaporation, convection, and condensation as well as atmospheric circulation.

The absorption spectra of the minority constituents in the atmosphere affect both the downward flow of solar radiation and the upward propagation of infrared radiation. As discussed in §23.4, the diatomic molecules N_2 and O_2 cannot easily absorb single photons and are effectively transparent to most incoming solar radiation and to outgoing infrared radiation. Oxygen can, however, be dissociated by UV photons of wavelength below 246 nm,

$$O_2 + h\nu \rightarrow O + O. \qquad (34.19)$$

This occurs in the *stratosphere* at altitudes of $\sim$15 km or more; below this altitude the level of UV radiation is significantly decreased. *Ozone* is produced when a single oxygen atom resulting from this dissociation combines with an oxygen molecule through a three-body collision with a third molecule M that conserves momentum and energy,

$$O + O_2 + M \rightarrow O_3 + M. \qquad (34.20)$$

Little ozone is produced above about 60 km since the density of material is so low that three-body collisions of this type are rare.

Ozone in turn is readily dissociated by UV photons in the continuous *Hartley band* between approximately 200 nm and 300 nm,

$$O_3 + \gamma \rightarrow O_2 + O, \qquad (34.21)$$

as discussed in §23.5. Because ozone is created and then destroyed within the stratosphere, the density of ozone has a local maximum at high altitude, around 25 km, and is far from hydrostatic equilibrium. The absorption of high-energy UV photons in the stratosphere by ozone and

oxygen dissociation deposits significant solar energy in the middle atmosphere, heating the stratosphere, so that the temperature actually rises through the stratosphere. This absorption of UV radiation also helps to protect life on Earth, which is easily damaged by UV radiation.

In the troposphere, below roughly 15 km altitude, water vapor, CO_2, and other greenhouse gases absorb both incoming and outgoing infrared radiation and can influence the temperature profile of the lower atmosphere, though generally in radiative–convective equilibrium the lapse rate is determined by convection. Water vapor is the most significant atmospheric constituent for radiative absorption in the IR; it has a number of absorption bands that capture incoming solar radiation as well as outgoing thermal radiation from Earth. Because the distribution of water vapor in the atmosphere is not uniform, and is affected by many aspects of climate and weather over short time scales, incorporating water vapor accurately into atmospheric models is complicated. Water vapor enters the atmosphere when it evaporates from the ocean or land surface, or is transpired by plants. Roughly 87% of water vapor comes from the ocean surface; because the rates of evaporation and plant transpiration are difficult to evaluate separately, generally the combined rate of **evapotranspiration** is measured. Most water vapor stays within the bottom 3 km of the atmosphere, and individual water molecules are resident in the atmosphere for on the order of 10 days before precipitating back to the surface. Some of the large absorption bands of water vapor in the solar wavelength spectrum can be seen in Figure 23.9. The infrared absorption spectra of water and a selection of other atmospheric gases, as well as the aggregate absorption, are depicted in more detail in Figure 34.4. This graph shows the percentage of solar radiation that has been absorbed from

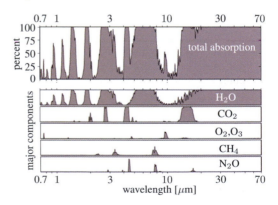

Figure 34.4 Infrared absorption for various atmospheric constituents. (Credit: Robert A. Rohde, Global Warming Art Project, reproduced under CA-BY-SA 3.0 license via Wikimedia Commons)

radiation that reaches sea level for wavelengths between 700 nm and 70 μm. These absorption bands are associated with vibrational excitations combined with rotational excitations at closely spaced frequencies, which with line broadening (see Example 23.4) produce effectively continuous absorption bands. Water vapor has a number of (primarily vibrational, see §9.3.1) absorption bands in the near-infrared part of the spectrum with wavelengths between 1 μm and 4 μm that absorb incoming solar radiation, as well as a large vibration–rotation band around 6.3 μm in the infrared (associated with the symmetric bend mode in Table 9.4). There is a window from about 8 μm to 12 μm in which solar radiation passes relatively intact through the atmosphere (except for an ozone absorption band around 9.6 μm). Beyond 12 μm, there is a densely spaced set of pure rotation bands for water. Around 15 μm there is a significant vibration–rotation absorption band for CO_2 (associated with the vibrational bending mode), which we discuss further in §34.4.

As discussed in §23.4, radiation of a specific frequency ν passing through a gas or other medium is absorbed at a rate

$$\frac{dI}{dz} = -\kappa_\nu I \,. \tag{34.22}$$

The *absorption coefficient* κ_ν for radiation of a particular wavelength by a particular material or atmospheric constituent can be expressed as the product

$$\kappa_\nu = k_\nu \rho \,, \tag{34.23}$$

where ρ is the density of the absorber and k is the **absorption cross section** (which has the same units – area per unit mass – and a similar physical interpretation as the cross sections encountered in §18.2). To describe absorption of radiation in the atmosphere, it is therefore sufficient to know the absorption cross section of each atmospheric constituent (as a function of frequency), and the density profile $\rho(z)$ of each as a function of altitude, which can be determined as described in §34.2.1. The *optical depth* (see §23.4) of the atmosphere in a particular frequency range is then given by

$$\tau_\nu(z) = \int_z^\infty dz' \kappa_\nu(z') = \int_z^\infty dz' k_\nu \, \rho(z') \,. \tag{34.24}$$

Thus, for example, when the Sun is directly overhead, the flux of solar radiation in a frequency range around ν that penetrates to a height z is

$$I_\nu(z) = (1-a) I_\odot^{(\nu)} e^{-\tau_\nu(z)} \,, \tag{34.25}$$

where $I_\odot^{(\nu)}$ is the intensity of solar radiation at the frequency ν.

34.2.4 1D Atmospheric Model

The ingredients just described can be combined into a simple one-dimensional (1D) model of Earth's atmosphere that gives the global average temperature profile as a function $T(z)$ of the height above Earth's surface. Such a model makes many simplifications: It ignores variations depending on latitude and longitude, assumes that all radiation propagates only in a vertical direction, ignores scattering, and neglects the curvature of Earth. Nonetheless, such a 1D model gives a good general picture of the global structure of Earth's atmosphere. For simplicity, we assume that reflection of incoming radiation associated with albedo occurs at the top of the atmosphere, so that within the atmosphere only absorption and radiation need be considered.

In the 1D approximation, the density variation of the atmosphere with height can be determined from the condition of hydrostatic equilibrium (§34.2.1). All atmospheric constituents are assumed to be uniformly distributed with a constant mass fraction, except for stratospheric ozone and water vapor.

The transport of radiant energy can be studied separately for each frequency ν. First consider the downward-heading flux of radiation with frequency ν, $I_\nu^-(z)$, which at height z impinges on a sheet of atmosphere of thickness dz. According to eq. (34.22), the radiant power per unit area dI_ν^- absorbed by this infinitesimally thick sheet is

$$dI_{\nu\,\text{abs}}^- = -I_\nu^- \kappa_\nu dz = -k_\nu \rho I_\nu^- dz \,, \tag{34.26}$$

where the minus sign indicates that absorption decreases the downward flux. In addition, the sheet emits blackbody radiation (both upwards and downwards) at a rate given by eq. (22.23), modified by the absorptivity $\kappa_\nu dz = k_\nu \rho dz$ (see Box 22.1),

$$dI_{\nu\,\text{emit}}^- = B_\nu(T) k_\nu \, \rho dz \,, \tag{34.27}$$

where

$$B_\nu(T) = \frac{2\pi h \nu^3}{c^2} \frac{1}{e^{h\nu/k_B T} - 1} \,. \tag{34.28}$$

The net change in downward radiation flux with frequency ν is given by the sum of these two effects,

$$dI_\nu^- = (-I_\nu^- + B_\nu(T)) k_\nu \rho dz \,. \tag{34.29}$$

This differential equation, which governs the flow of radiative energy, is known as the **Schwarzschild equation.**[4] An identical equation holds for the upward-moving radiation flux I_ν^+.

---

[4] The appearance of $k_\nu \rho$ in both the absorption and emission terms corresponds to *Kirchhoff's law* that absorptivity and emissivity are equal (see Box 22.1).

When the global mean atmosphere is in radiative equilibrium at a height z at which the atmosphere is stable to convection ($\Gamma \leq \Gamma_d$), then at that height the total rates of absorption and emission cancel,

$$\int dv \, \left(dI_v^+(z) + dI_v^-(z)\right) = 0. \tag{34.30}$$

The integrand here depends on $T(z)$ and on the density ρ, which is implicitly a function of T (eq. (34.11)). The condition that eq. (34.30) vanishes for all z establishes radiative equilibrium throughout the 1D atmosphere and can be used in principle to determine the temperature profile $T(z)$. Simultaneously solving the Schwarzschild equation (34.29) for incoming and outgoing radiation and the equilibrium equation (34.30) is a nontrivial problem analytically due to the coupling between the equations and due to the integral form of the equilibrium equation. A numerical procedure that can be used in practice to find the equilibrium temperature profile is to begin with an approximate temperature profile close to the expected solution, impose a fixed boundary condition on $I^-(z)$ at a large value of $z = z_+$ outside Earth's atmosphere such as $I_v^-(z_+) \cong (1-a)I_\odot^{(v)}$ and then solve eq. (34.29) and the analogous equation for I_v^+ for the net inflow/outflow of radiative energy at each altitude. By then increasing/decreasing the temperature at each altitude z (including the surface, which radiates as a black body at temperature $T(0)$) by a small constant multiple of the net rate of energy inflow/outflow at height z, and iterating, it is possible in typical situations to converge to a global mean solution of eq. (34.30).

If only the radiative equilibrium conditions (34.30) are solved, given current conditions for $I^-(z_+)$ and atmospheric composition, then the resulting lapse rate at low altitudes exceeds by a significant margin the observed global average lapse rate in the troposphere of 6.5 °C/km. This corresponds to the fact that in the real atmosphere, as mentioned in §34.2.2, convection is responsible for more than half of upward energy transport in the troposphere. This can be incorporated into a simple 1D **radiative–convective equilibrium** model by simply fixing the lapse rate artificially to the global mean value at altitudes where the lapse rate exceeds the global mean. This

convective adjustment is a simple way of effectively accounting for convection without explicitly modeling the complexities of convective flow. The full radiative–convective equilibrium problem is complicated by the fact that radiation drives convective instability, which changes the water vapor distribution, which in turn modifies radiative properties, giving a complex nonlinear dynamical system. Temperature profiles for a 1D radiative equilibrium computation, including convective adjustment to both dry adiabatic and global mean lapse rates, are shown

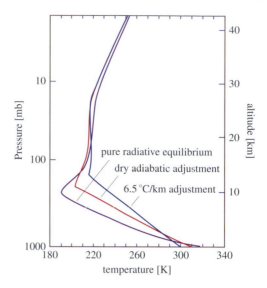

Figure 34.5 Temperature profiles for numerical solution of 1D radiative equilibrium (purple) and radiative–convective equilibrium with convective adjustment to dry adiabatic (red) and global mean (blue) lapse rates. (Credit: S. Manabe and R.F. Strickler, Thermal equilibrium of the atmosphere with a convective adjustment, *Journal of the Atmospheric Sciences*, volume 21, page 361 (1964). Figure 4, page 370. © American Meteorological Society. Used with permission)

in Figure 34.5. While not completely accurate, models of this type give a good general picture of the structure of the atmosphere and can be used to give simple estimates for the effect of various changes in the atmosphere or modifications to the climate or other Earth systems that modify the radiative structure of the atmosphere. For example, including different combinations of atmospheric constituents with UV and IR absorption spectra modifies the solution of the 1D global mean model significantly. Figure 34.6 shows the radiative–convective equilibrium when only water vapor is present, and the equilibria when CO_2 and CO_2 and ozone are included. Without ozone or CO_2, the temperature continues to decrease indefinitely as altitude increases. When CO_2 is included, the atmospheric temperature has a similar profile but with a constant temperature shift of roughly 10 °C. When photochemically produced ozone in the upper atmosphere is also included, the temperature rises in the stratosphere due to absorption of solar UV radiation by ozone. The region in which the direction of the temperature gradient reverses between the stratosphere and the troposphere is the **tropopause** (see also §27.4). The solution with CO_2 and ozone included qualitatively reproduces the general features of the observed global mean temperature profile, shown as the dashed black curve in Figure 34.6, though in Earth's actual atmosphere there

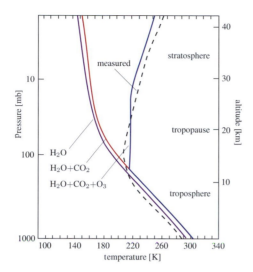

Figure 34.6 Temperature profile in a 1D atmospheric model with different atmospheric constituents included, but no clouds. The measured temperature profile from Figure 34.16 (dashed black) is shown for comparison. (Credit: S. Manabe and R.F. Strickler, Thermal equilibrium of the atmosphere with a convective adjustment, *Journal of the Atmospheric Sciences,* volume 21, page 361 (1964). Figure 4, page 370. (c) American Meteorological Society. Used with permission)

is a small increasing temperature gradient that begins just above the top of the troposphere, and the tropopause is more localized. This temperature *inversion* (see Footnote 3) at the tropopause suppresses convective activity above this point, so the tropopause serves as an effective cap on the region of dynamic atmospheric convective flow.

To go beyond the simplified 1D atmospheric model described in this section it is necessary to include a more detailed description of radiative–convective equilibrium and more information about the variation of insolation and other relevant parameters, as well as horizontal energy transport, across latitude and longitude. Of particular importance is the extensive heat transport from the tropics to the polar regions that is mediated by large-scale ocean and air currents. We described some aspects of this global energy flow in §27.4 and return to the subject in §34.3. *General circulation models*, which attempt to numerically model the complete Earth climate system by including all relevant factors, are discussed in §34.5. In any case, empirical measurements can be used to reduce the complex global structure of energy flow in Earth's atmosphere to global mean values. A summary of vertical global mean energy flow is shown in Figure 34.7.

34.2.5 Radiative Forcing

With some understanding of the temperature and pressure profile of Earth's atmosphere, and of the

radiative–convective equilibrium of the atmosphere as a whole, we can introduce the concept of *radiative forcing*, which is used in climate science as a quantitative measure of the effect of various perturbations to the global climate system. **Radiative forcing** is defined to be: "The change in net (down minus up) irradiance (solar + longwave; in W/m^2) at the tropopause after allowing for stratospheric temperatures to readjust to radiative equilibrium, but with surface and tropospheric temperatures and state held fixed at the unperturbed values" [246].[5]

The essence of this definition is that for a given phenomenon it is often feasible to do a fairly accurate computation of the solution to the radiative balance equations when all conditions other than the single perturbation of interest are kept intact from the surface up to the top of the troposphere. The conditions above the tropopause can generally be solved to determine radiative equilibrium through the stratosphere; in many situations the stratospheric adjustment is negligible. This gives a

> **Radiative Forcing**
>
> *Radiative forcing* is defined as a net increase in downward radiation at the tropopause, with the state of the surface and atmosphere held fixed below the tropopause but allowing for the stratosphere to adjust to equilibrium. Radiative forcing is a widely used quantitative measure of the impact of different perturbations on climate, as it can be accurately estimated using knowledge of the existing climate and does not depend upon complicated and poorly understood feedback mechanisms.

---

5 This definition is used by the Intergovernmental Panel on Climate Change in its Fourth Assessment in 2007 (**IPCC4**). In its Fifth Assessment in 2013 (*IPCC5*) [245], the IPCC introduced a modification of this concept, which they refer to as **effective radiative forcing** (ERF). In much the same way that RF allows for short-term adjustments in the stratosphere, ERF is defined to incorporate short-term troposphere adjustments in temperature, water vapor, and clouds, while holding fixed factors that respond over longer time scales, such as global mean surface temperature, ocean temperature, and Arctic sea ice. ERF is more difficult to compute but is argued to be a better indicator of global mean temperature response. The numerical differences between *radiative forcing* and ERF values are relatively small (except in the case of anthropogenic aerosols), and since radiative forcing is conceptually simpler, we primarily focus on radiative forcing here and not ERF.

Box 34.1   Clouds

There are many approximations made in the 1D global mean model described in §34.2.4 that miss important features of Earth's full 3D atmosphere. One of the most significant shortcomings of the simplified 1D model we have described here is the absence of clouds. Clouds are formed from suspended droplets of water or particles of ice that have condensed as air rises and the temperature and pressure drop. Clouds play an important role in the radiative flow of energy in the atmosphere. While some models of radiative–convective equilibrium have been developed that incorporate clouds in approximate ways, the role of clouds in the atmosphere in general is quite complex. Depending upon their altitude and composition, clouds can produce either a net upward or net downward change in radiation flux. The size of individual cloud particles can range from a few microns to a millimeter, above which point the particles have a high enough mass to surface area ratio to begin to precipitate. Clouds composed of small water droplets have a higher albedo since the surface area to volume ratio is higher for a fixed water content. On average, clouds contribute about 1/2 of Earth's total albedo. The water in clouds also, however, strongly absorbs terrestrial thermal radiation, and contributes to the greenhouse effect, partly offsetting the cooling effect from the increased albedo. Because the effects of clouds are quite significant for global energy balance, and yet highly sensitive to detailed atmospheric conditions, accurate incorporation of clouds into atmospheric models is a difficult and challenging (yet important) problem. For a more detailed introduction to the role of clouds in the atmosphere see [173] or [243]. We return to the question of clouds later in this chapter.

(Image credit: Jeffery Perloff)

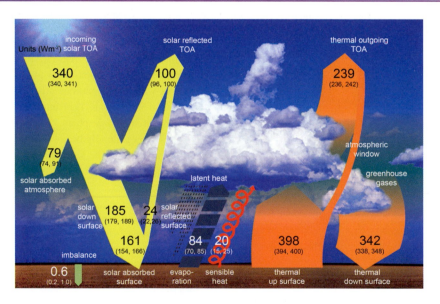

Figure 34.7 Actual global mean vertical energy flow, indicating magnitude of both downward and upward energy transport in W/m^2 from solar radiation, terrestrial thermal radiation, latent and sensible heat transport by convection, representing climate conditions at the beginning of the twenty-first century. Numbers in parentheses give the range of values consistent with observational constraints. (Credit: [247])

Example 34.3 Radiative Forcing from Albedo Change

Consider a modification of Earth's average surface albedo from 0.1 to 0.11, as might occur for example with substantial ice sheet accumulation. *What would be the approximate radiative forcing from this albedo change?*

As explained earlier in this chapter, the average insolation at the surface is roughly $250\,\mathrm{W/m^2}$ and the combined albedo of atmosphere and clouds is roughly $a_{ac} \cong 0.21$. Increasing surface albedo by 0.01 would thus increase the intensity of reflected radiation by $\cong 2.5\,\mathrm{W/m^2}$. This radiation would be distributed according to the (ground level) solar spectrum. Apart from a fraction $a_{ac} \cong 0.21$ reflected by atmosphere and clouds, and some losses to infrared absorption in the atmosphere, the effect of this change would be seen in a net upward radiative forcing at the tropopause. The net radiative forcing would thus be on the order of $-(1 - 0.21) \times 2.5 \cong -2\,\mathrm{W/m^2}$, where the negative sign indicates upward radiation flux. Note that if the albedo change were due primarily to an increase in ice sheets near the poles, the net magnitude of radiative forcing would be somewhat smaller due to latitudinal variation in insolation.

Box 34.2 Radiative Forcing by Clouds

Clouds in Earth's atmosphere contribute significantly to both albedo and heat trapping. While the details of cloud effects depend on local cloud properties, the net contribution to the albedo and greenhouse heat trapping has been measured experimentally, for example by using satellites to measure local cloud reflectivity contributions to Earth's albedo and averaging over space and time. One such analysis [248] shows that in Earth's current climate configuration, clouds contribute a radiative forcing of roughly $-44\,\mathrm{W/m^2}$ (upward), corresponding to an albedo contribution of roughly $a_c \cong 0.13$. At the same time, clouds contribute some $31\,\mathrm{W/m^2}$ of downward radiative forcing due to absorption and downward reradiation of outgoing terrestrial thermal radiation. Thus, in Earth's current climate the net radiative forcing contribution from clouds is roughly $-13\,\mathrm{W/m^2}$. If clouds ceased to form in Earth's atmosphere, more solar radiation would hit the surface, and the average surface temperature would increase. As mentioned previously, clouds play an important role in Earth's radiative equilibrium, and are one of the most difficult components of the climate system to model accurately.

---

first-order description of the effect of any modification to the climate system while removing from the question the complicated issue of feedbacks (discussed in §34.5) that make the ultimate effect of any change very complicated to determine. Radiative forcing is used by climate scientists because it is a fairly clean and simple way to measure the overall first-order impact of various modifications of Earth's radiative equilibrium, allowing different climate effects to be compared in a quantitative fashion.

Any change in Earth's climate system can lead to a net radiative forcing F. Positive radiative forcing leads to increased downward radiation at Earth's surface, which generally increases the surface temperature, while negative radiative forcing leads to cooling. Examples of effects that lead to radiative forcing are given in Example 34.3 and Box 34.2.

A particularly important ingredient in the analysis of climate systems is the radiative forcing that would be generated simply by a change in Earth's temperature,

$$F_0 = \lambda_0 \Delta T \,. \tag{34.31}$$

This forcing is known as the **uniform temperature response**, and λ_0 is sometimes referred to as the **Planck feedback parameter** (§34.5). If the planet's temperature increases, then with everything else fixed, the upward flux of energy increases and the radiative forcing as defined above decreases. Thus λ_0 is negative. For given global atmospheric conditions, λ_0 can be computed, and has been estimated under current conditions to be [245, 253][6]

$$\lambda_0 \cong -3.2 \pm 0.1\,\mathrm{W/m^2\,^\circ C} \,. \tag{34.32}$$

In the absence of climate feedback, an externally imposed radiative forcing F would give rise to a change

---

[6] When we quote data from [245, 246, 249], we follow their convention that error estimates represent 90% uncertainty intervals. Thus λ_0 has 90% probability to lie between -3.1 and $-3.3\,\mathrm{W/m^2\,^\circ C}$. Sometimes this interval is quoted in parentheses as a range of values.

in temperature ΔT_0. The change ΔT_0, in turn, would give a uniform temperature response that cancels the original radiative forcing,

$$F + \lambda_0 \Delta T_0 = 0, \text{ or}$$
$$\Delta T_0 = -F/\lambda_0. \tag{34.33}$$

Thus, radiative forcing by $1\,\text{W/m}^2$ would lead to a uniform temperature increase of roughly $0.3\,°\text{C}$ in the absence of feedbacks. The value in eq. (34.32) is determined quite accurately by reasonably simple climate models, and is a relatively well-established and non-controversial result. Indeed, we can derive a simple estimate of eq. (34.32) directly from the Stefan–Boltzmann law. From the observed albedo of Earth, $a \cong 0.3$, it follows that the effective temperature at which Earth's atmosphere radiates thermal energy to space is approximately given by the solution to $\sigma T^4 \cong 0.7 \times 343\,\text{W/m}^2$, yielding $T \cong 255\,\text{K}$. If the temperature at which Earth radiates is roughly $255\,\text{K}$, then a change of temperature by ΔT gives rise to a direct change in outgoing infrared radiation energy by

$$- F_{0\,\text{approx}} \cong \Delta(\sigma T^4) \cong 4\sigma T^3 \Delta T \cong 3.76\,\Delta T\,\text{W/m}^2\,\text{K}^{-1}. \tag{34.34}$$

This gives $\lambda_0 \approx -3.76\,\text{W/m}^2\,\text{K}$. The more accurate value (34.32) is slightly smaller in magnitude since not all the outgoing infrared radiation is emitted at the same temperature. At wavelengths with stronger atmospheric IR absorption, the outgoing thermal radiation from Earth is emitted from higher altitudes and therefore lower temperature, as depicted in Figure 34.8. Distributing the total outgoing energy over a range of effective temperatures for different wavelength bands always has the effect of lowering the magnitude of λ_0, making $|\lambda_0| = 3.76\,\text{W/m}^2\,\text{K}$ an upper bound (Problem 34.8). Thus, eq. (34.34) gives a simple estimate of the uniform temperature response that is an upper bound on $|\lambda_0|$, and is correct to within about 20%. Note that the definition of the Planck feedback parameter can vary, depending upon, for example, whether tropospheric response is considered as a feedback or a forcing. The value used here includes tropospheric response, in accord with the analysis of [245, 253] and references cited there.

If the uniform temperature response were the only consequence of a given radiative forcing, then understanding climate change would be relatively straightforward. In fact, however, the indirect consequences of radiative forcing associated with feedbacks in the climate system are much more complicated and difficult to understand. We discuss feedbacks and climate modeling in §34.5.

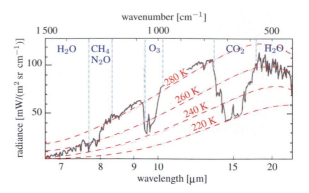

Figure 34.8 Power spectrum of outgoing infrared radiation from Earth as measured by the Nimbus 4 IRIS satellite. Note that in frequency bands with substantial atmospheric IR absorption the outgoing radiation is emitted from higher in the atmosphere, corresponding to a lower effective blackbody temperature. (Credit: J.T. Houghton, *Global Warming: The Complete Briefing*, Cambridge University Press (2004))

Uniform Temperature Response

The *uniform temperature response* is the radiative forcing F_0 that would be generated by a change in Earth's temperature by ΔT. λ_0 is negative because an upward shift in Earth's temperature produces a net upward change in radiative flux at the tropopause. λ_0 has been estimated to be

$$\lambda_0 \cong -3.2 \pm 0.1\,\text{W/m}^2\,°\text{C}.$$

34.3 Global Energy Flow

The description of Earth's atmosphere and radiative balance in the preceding section was carried out in a highly simplified *1D model* in which all dependence on latitude and longitude was averaged.[7] In fact, however, the distribution of incoming radiation is highly latitude-dependent, and horizontal transport of energy by Earth's atmosphere and oceans plays an important role in determining the planet's climate.

From the analysis of insolation in §23.3, we can determine average insolation as a function of latitude. As shown in Figure 23.6(b), average daily insolation on the equator (latitude 0°) amounts to 7.3 hours of perpendicular sunlight, compared to about 3 hours at the poles

[7] Note, however, that the estimate in eq. (34.32) was made using global models.

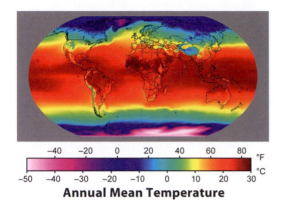

Annual Mean Temperature

Figure 34.9 Global annual average surface temperatures. (Credit: Robert A. Rohde. Global Warming Art Project, reproduced under CA-BY-SA 3.0 license via Wikimedia Commons)

(latitude $\pm 90°$). A simple local radiative balance computation shows that in the absence of local differences in atmosphere or albedo, and in the absence of transport, this would give a temperature difference of about 55 °C between the equatorial and polar regions. (For example, assuming a simplified model of the atmosphere and homogeneous albedo of 0.16 as in Example 34.1, the equatorial and polar temperatures in equilibrium would be 280 K and 225 K respectively – see Problem 34.11.) Including the difference in albedos between the equator and icy polar regions would greatly accentuate this difference. But in fact, the temperature contrast between latitudes is significantly weaker than would be suggested by local radiative balance and albedo alone. Annual average surface temperatures are shown in Figure 34.9, and demonstrate an equator–polar temperature difference closer to the naive estimate above with homogeneous albedo than a calculation including albedo variation. A detailed understanding of latitudinal temperature differences requires incorporating thermal energy transport from the equator to the poles.

The large average temperature gradient across latitudes helps drive global circulation patterns that transport large quantities of energy from the equator towards the poles. As discussed in §27 and §28, warm air at low latitudes rises, driving atmospheric circulation that carries energy towards the poles, while wind-driven ocean circulation carries substantial quantities of energy poleward on western boundary currents such as the Gulf Stream and Kuroshio currents. By analyzing the excess of outward minus inward radiation at various latitudes, the net energy transport can be computed, and approaches a rate of 6 PW towards

each of the poles near $\pm 40°$ latitude. Estimates of the fraction of energy transport from atmosphere and ocean were discussed in §27 and are shown in Figure 27.2. There is also a difference in the flow of sensible and latent heat (see §27.1). Latent heat, contained in water vapor carried by near-surface (below 3 km) air flow, flows towards the equator at low latitudes and away at higher latitudes, while sensible heat is carried away from the equator by warm air masses at all latitudes. During winter months in each hemisphere, the latitudinal temperature gradient is larger, driving a stronger atmospheric circulation pattern.

While the latitudinal dependence of the radiation budget and atmospheric transport can be computed based on observations, ocean transport is not as well understood. One of the best estimation approaches currently available is to simply approximate the transport as the difference between the known net radiation flux and the best estimation of the better understood atmospheric transport. As we discuss in the following chapter, a more detailed understanding of ocean energy transport is needed for models to accurately reproduce past climates and to make accurate estimations of climate under different radiative forcing conditions. Understanding and accurately simulating the mechanisms behind energy transport over Earth's surface is a principal outstanding challenge for climate scientists (see Box 35.3).

34.4 CO$_2$ and the Carbon Cycle

As discussed in §34.2.3, several atmospheric gases have absorption bands in the infrared, and play the role of greenhouse gases. Significant greenhouse gases include water vapor, CO$_2$, methane, tropospheric ozone, N$_2$O, and chlorofluorocarbons (CFCs). The relative abundances of these constituents (excepting CFCs) are listed in Table 34.1. Water vapor is the most important greenhouse gas, contributing a large fraction of the radiative forcing that currently keeps Earth's temperature some 20 °C warmer than the −6 °C it would be if the atmosphere were composed purely of N$_2$ and O$_2$ (Example 34.1). Water evaporates from the oceans and land, and precipitates from the atmosphere back to the surface at a rapid rate, so that the time scale within which the average water vapor content of the atmosphere comes into equilibrium with the climate is very short – about two weeks. For other greenhouse gases, however, the time scale over which the atmospheric concentration reaches equilibrium with the climate is longer. Methane, for example, is a potent greenhouse gas that lasts in the atmosphere for a time scale on the order of a decade (lifetime $\cong 12$ y) before breaking down.

After water, carbon dioxide is the next most important greenhouse gas and contributes significantly to radiative forcing. While excess atmospheric methane generates greater radiative forcing per unit mass than CO_2, it is less abundant and therefore generates less total radiative forcing than does CO_2. Modifications in atmospheric CO_2 levels also take much longer than methane to reach equilibrium with the rest of the climate system. For this reason, CO_2 is likely to be the primary driver of climate change over the time scale of the next one or two centuries. In this section we describe the role of CO_2 in Earth's climate system in some detail, and summarize recent data on the increase of CO_2 levels in Earth's atmosphere. The role of CO_2 and other greenhouse gases in Earth's past and future climate is discussed further in §35.1 and §35.2.

34.4.1 The Natural Carbon Cycle

Carbon forms the basis for all life on Earth, and the emission and absorption of CO_2 is a natural and essential ingredient in biological life cycles. There is a vast quantity of carbon on Earth that circulates over long periods of time between terrestrial biomass, Earth's atmosphere, and the oceans, much of it transported in the form of carbon dioxide. Over geological time scales, carbon circulates between these systems and Earth's crust through sediment formation and volcanism. Currently, there are roughly 450–650 Gt of carbon tied up in terrestrial biomass, 800 Gt of carbon in atmospheric CO_2, and approximately 38 000 Gt of carbon in CO_2 dissolved in Earth's oceans. Soil, ocean sediments, permafrost, and underground repositories of

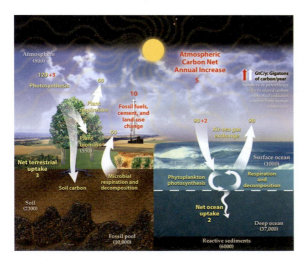

Figure 34.10 Primary carbon stores and transfers in Earth's carbon cycle. Transfers in Gt/y; stores in parentheses; red indicates anthropogenic carbon. (Credit: © American Meterological Society)

fossilized carbon are estimated to contain another 1500–2400, 1500–6000, ~1700, and ~10 000 Gt of carbon respectively. These existing *carbon stores* are shown in Figure 34.10. Note that the estimates on some of these carbon repositories are quite rough, though the atmospheric carbon content is known quite accurately.

In the *natural* **carbon cycle** large quantities of carbon are transferred between different components of the Earth system each year. Roughly 90 Gt/y of carbon in the form of CO_2 is transferred from the atmosphere into the ocean. This transfer is effected by the mixing of air with water when waves break on the ocean's surface. At higher latitudes, where the surface water is colder and more saline, and sinks downward, the meridional overturning current (§27.5) mixes the dissolved CO_2 into the deep ocean. As CO_2 from the atmosphere mixes into the ocean, a roughly equivalent quantity of CO_2 is released out of solution from the ocean into the atmosphere. The dissolved CO_2 in the ocean provides a carbon source for marine life. Plankton and other marine organisms near the surface carry out photosynthesis, binding solar energy into organic molecules using carbon derived from oceanic CO_2. This carbon then passes through the marine food chain. Some carbon is bound into creatures that perish and fall to the bottom of the ocean before being consumed. This transports some carbon to the sea floor. Most of this carbon is used again in the deep-ocean organic food chain, but a small fraction is permanently buried beneath the sea floor in sediment, and removed from the carbon cycle for tens or hundreds of millions of years until it is returned to the atmosphere over very long time scales by volcanic emissions (§35.1.4). In natural equilibrium, the average transfer of CO_2 in each direction between ocean and atmosphere is equal, so that the stored quantities remain roughly constant in time, with minor variations from year to year.

Large quantities of carbon are also exchanged between the atmosphere and terrestrial biomass. Each year a net of roughly 60 Gt of carbon is extracted from the atmosphere by terrestrial plants through photosynthesis, and a corresponding quantity of carbon is released through decay of plant and other organic matter on land. Because there is more vegetated land in the Northern Hemisphere, in northern spring and summer CO_2 is extracted from the atmosphere by terrestrial biomass more rapidly than it is returned through decay processes, while in northern fall and winter the balance is reversed. Thus, there is a yearly cycle to the level of CO_2 in the atmosphere, with a minimum in October and a maximum in May, differing by roughly 10 Gt. Within the Northern and Southern Hemispheres, the mixing time for the lower atmosphere is on the order of several months, and the inter-hemispheric mixing time for the troposphere is on the order of 1–2 years,

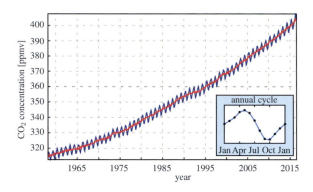

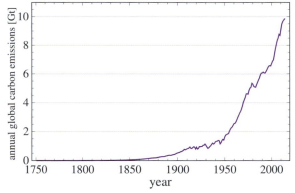

Figure 34.11 Atmospheric CO_2 concentration over the last 50 years, measured at the Mauna Loa Observatory in Hawaii. The increase from pre-industrial levels of 280 ppm (by volume) is attributable to human causes, including fossil fuel usage, cement production, and land usage. Inset shows the annual CO_2 cycle described in the text. (Credit: Sémhur/Wikimedia Commons/CC-BY-SA-3.0, or Free Art License)

Figure 34.12 Annual anthropogenic emissions of carbon in CO_2 from fossil fuel combustion and cement production. (Data from Carbon Dioxide Information Analysis Center (CDIAC), Oak Ridge National Lab.)

so changes in atmospheric composition from localized sources generally equilibrate fairly rapidly.

34.4.2 Anthropogenic CO_2

Highly accurate measurements of atmospheric CO_2 levels taken over the last 50 years indicate a steady increase in CO_2 concentration in the atmosphere. Atmospheric CO_2 levels measured since 1958 by the Mauna Loa Observatory in Hawaii are graphed in Figure 34.11. A variety of other measures (discussed in more detail in §35.1) show that pre-industrial levels of CO_2 were close to 280 ppm (by volume, see Box 34.3) before about 1900, while in

2015 the average CO_2 level exceeded 400 ppm. Paleoclimate data indicates that CO_2 levels have been below a maximum of around 300 ppm over at least the last 600 000 years (Figure 35.7).

Human activity modifies the natural flow of CO_2 between atmosphere, ocean, and biomass described in the previous section. In 2010, as indicated in Figure 34.12, roughly 9 Gt of carbon in the form of CO_2 was released into the atmosphere from the use of fossil fuels and cement production. Furthermore, human land use patterns such as deforestation contribute to increasing atmospheric CO_2 levels. This impact is harder to determine reliably, but estimates suggest that human land use patterns contributed approximately an additional 1 Gt of carbon into the

---

Box 34.3   Units: CO_2 Concentration by Mass or Volume

Note that, in some contexts, the CO_2 concentration in the atmosphere is quoted on a per mass basis, in parts per million by mass (ppmm), while in most contexts it is given on a per volume basis. The molecular mass of CO_2 is close to 44 while the average atmospheric molecular mass is roughly 29. Thus, the units of ppm per volume and mass for CO_2 in the atmosphere are related through

$$1\,\text{ppmv} \cong 44/29\,\text{ppmm} \cong 1.52\,\text{ppmm}.$$

For example, as of 2015, atmospheric CO_2 levels were 400 ppmv, or equivalently $400 \times 1.52 = 608$ ppmm.

Given the total atmospheric mass of roughly 5.14×10^{18} kg, we can estimate the total mass of atmospheric carbon, noting that carbon represents 12/44 of the mass of carbon dioxide

$$(5.14 \times 10^{18}\,\text{kg}) \times (608\,\text{ppmm}) \times 12/44 \cong 850\,\text{Gt}.$$

We generally denote gas fractions of parts per million by volume by simply ppm, with ppmm indicating parts per million by mass.

atmosphere in 2010, for a net anthropogenic CO_2 increase of roughly $10\,\mathrm{Gt\,C/y}$. The net yearly increase in atmospheric CO_2 is only about half of this – amounting to around $5\,\mathrm{Gt\,C/y}$. This is because natural sinks have been absorbing about half of the extra human CO_2 output over recent years. There is much controversy about the exact rate at which natural sinks operate, and the capacity of these sinks to continue operating over long time scales, but current best estimates suggest that the net uptake of $5\,\mathrm{Gt/y}$ of carbon is roughly evenly divided between land and ocean sinks.

As described above, CO_2 is dissolved into the ocean and sinks at high latitudes into the deep ocean, where it is circulated through the meridional overturning current. While atmospheric CO_2 levels come into equilibrium with the oceanic mixed layer fairly quickly, the time scale for water to circulate through the deep oceans is on the order of a thousand years, so this is the approximate time scale over which we may expect that changes in CO_2 levels are equilibrated with CO_2 levels in the deep ocean. It is estimated that currently a net of about $2.3\,\mathrm{Gt/y}$ of carbon is absorbed into the world's oceans. Whether this rate of uptake can be sustained over decades or hundreds of years is a question currently subject to debate. In particular, as discussed further in §35.1.3, paleoclimate data suggests that over longer periods of time, warming of the climate in the past has led to a net *release* of CO_2 into the atmosphere from the oceans. Regarding the land uptake of CO_2, estimates suggest that roughly $3\,\mathrm{Gt/y}$ of carbon are currently absorbed into terrestrial biomass. There is substantial uncertainty about the precise mechanism of this process, but some studies suggest that, at least in North America, some of this carbon uptake corresponds to reforestation of northern forest previously cleared for farming. Another component of the land uptake of CO_2 arises from general increased plant growth due to higher atmospheric CO_2 levels (augmented by increased use of nitrogen fertilizers, §26.2.2). Note that while CO_2 levels are a limiting factor for terrestrial biomass growth, oceanic biomass is primarily limited by light and nutrient levels. Without a better understanding of the precise nature of the processes involved, it is difficult to estimate how the land CO_2 sink will operate over the next several decades as atmospheric CO_2 continues to increase. In any case, certainly the potential for CO_2 absorption through reforestation and general plant growth are subject to finite limits. Some discussion of these issues and further references can be found in [250].

34.4.3 Radiative Forcing from CO_2 and Other Anthropogenic Sources

The role of CO_2 as a greenhouse gas was first discovered in 1861 by Irish scientist John Tyndall. Over 100 years

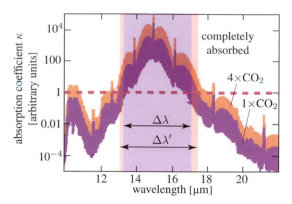

Figure 34.13 The absorption factor κ for CO_2 in the region around $15\,\mu\mathrm{m}$. κ is normalized such that regions of the spectrum where $\kappa > 1$ are essentially completely absorbed by the atmosphere. κ for present CO_2 levels (purple) and $4\times$ present levels are compared. The range of wavelengths that are completely absorbed grows logarithmically because κ falls exponentially in the wings of the absorption band. (Adapted from 'Part II: What Ångström didn't know', raypierre and Spencer Weart, 26th June 2007)

ago, Swedish scientist Svante Arrhenius pointed out that CO_2 from fossil fuel combustion could play an important role in Earth's climate [251]. The primary impact of CO_2 as a greenhouse gas in Earth's atmosphere comes from a particular vibration–rotation absorption band in the CO_2 spectrum around 15 microns. The absorption coefficient for CO_2 in the part of the spectrum near $15\,\mu\mathrm{m}$ is shown in Figure 34.13. As discussed in Example 23.4, the individual lines in the quantum spectrum of any molecule are broadened by several effects, including thermal and pressure broadening, into bands within which photons with a continuous range of frequencies can be absorbed. The absorption cross section of CO_2 peaks near $15\,\mu\mathrm{m}$ and decays approximately exponentially in *wings* on either side of the peak. Such behavior is typical of absorption bands produced by line broadening. Near the peak the absorption factor is so large that at the present level of atmospheric CO_2 essentially 100% of the radiation in this wavelength region (denoted $\Delta\lambda$ in Figure 34.13) is absorbed by the atmosphere. Increasing atmospheric CO_2 does not substantially change the radiative balance for wavelengths in this region, but widens the range of wavelengths over which absorption is essentially saturated (to $\Delta\lambda'$ in the figure). Because the absorption factor decays exponentially in the wings, the width of the spectral window over which outgoing radiation is absorbed grows logarithmically as a function of the quantity of CO_2 in the atmosphere (Problem 34.7). Consequently, the radiative forcing from atmospheric CO_2 grows logarithmically with the quantity of CO_2.

The radiative forcing from an increase in CO$_2$ can thus be described using a simple mathematical approximation. If an initial quantity M^* of CO$_2$ produces radiative forcing F_*, then increasing the quantity of CO$_2$ to M_{CO_2} will increase the radiative forcing to

$$F \approx F_* + F_{2\times} \log_2 \left(M_{CO_2}/M^* \right) , \qquad (34.35)$$

where the coefficient $F_{2\times}$ is the radiative forcing arising from a doubling of CO$_2$ levels. The radiative forcing from an increase in CO$_2$ levels can be estimated by leaving the existing temperature profile and other features of the atmosphere unchanged except for the increase in CO$_2$ levels and solving the Schwarzschild equation (34.29) approximately over the full 3D Earth atmosphere to compute the change in net downward radiation flux at the tropopause. This computation leads to the estimate [246, 249]

$$F_{2\times}^{(RF)} \approx 3.7 \, \text{W/m}^2 . \qquad (34.36)$$

In its 2013 assessment, the IPCC considered two somewhat more sophisticated ways to extract the value of $F_{2\times}^{(ERF)}$ associated with *effective radiative forcing* (see Footnote 5) from climate models, which give values of $F_{2\times}^{(ERF)}$ of $3.4 \pm 0.8 \, \text{W/m}^2$ and $3.7 \pm 0.8 \, \text{W/m}^2$. Since the uncertainties are large compared to the difference between these estimates and the simple radiative forcing value of eq. (34.36), we generally use the RF value here. Given $F_{2\times}$ and eq. (34.35), we can estimate the radiative forcing due to the 2015 CO$_2$ levels (Table 34.1) compared to the pre-industrial level of 280 ppm,

$$F_{2015} \cong F_{2\times} \log_2(400/280) \cong 1.9 \, \text{W/m}^2 . \qquad (34.37)$$

The IPCC estimated the radiative forcing from CO$_2$ in 2011 (391 ppm) at $1.8 \pm 0.2 \, \text{W/m}^2$.

Carbon dioxide is not the only anthropogenic source of radiative forcing. Other changes that can be attributed to human activities give a variety of sources of radiative forcing, both positive and negative. A chart of some of the main anthropogenic sources of radiative forcing is shown in Figure 34.14. These sources include:

Methane The atmospheric methane level in 2011 was 1800 ppb, up from a pre-industrial value estimated to have been between 300 and 700 ppb over the last half million years (§35.1.3). Methane levels are measured from ice core data to have been 700 ppb in 1750, and increased rapidly in the twentieth century, though the rate of increase slowed after about 1980. Atmospheric methane stabilized near 1800 ppb around the turn of the century, and grew only slightly through the first decade of the twenty-first century. As mentioned above, methane is a very strong infrared absorber through several vibrational modes. Thus

even at a much smaller concentration than that of CO$_2$, methane gives rise to significant radiative forcing, estimated for 1800 ppb to be about 0.5 W/m^2 (relative to forcing at 700 ppb).

Other well-mixed greenhouse gases (WMGHG) N$_2$O and CFCs, as mentioned above, are relatively long-lived greenhouse gases that also contribute to positive radiative forcing. Ozone in the troposphere also absorbs in the IR, so that anthropogenic ozone low in Earth's atmosphere contributes to radiative forcing. The contribution to radiative forcing from these other WMGHGs including tropospheric ozone is estimated at around 0.5 W/m^2.

Aerosols Small particles in Earth's atmosphere contribute radiative forcing in several ways. They can reflect and scatter sunlight, directly increasing Earth's albedo and producing negative radiative forcing. Aerosol particles can also nucleate clouds that change the albedo and contribute to radiative forcing. The impact of aerosol-induced clouds is very difficult to determine with accuracy, so the error bars on this effect are quite large. Other particles, such as black carbon aerosols, can absorb incoming radiation and give a positive contribution to radiative forcing.

One source of aerosols is associated with burning of fossil fuels, in particular coal power plants. As noted in §33.1, coal contains a substantial amount of sulfur (*medium-sulfur coal* is defined as coal containing between 1% and 3% by mass). When coal is burned, the sulfur is released as sulfur dioxide (SO$_2$). Sulfur dioxide reacts with water molecules in the atmosphere to form sulfuric acid (H$_2$SO$_4$) vapor, which condenses on dust particles. This leads to the formation of sulfate aerosols, which both scatter incoming solar radiation directly and nucleate reflective clouds. The atmospheric lifetime of the sulfate particles is rather short, on the order of weeks to months for particles in the troposphere or lower stratosphere. Thus, this negative contribution to radiative forcing from fossil fuel usage is a relatively short-term effect that masks in part the increased radiative forcing from anthropogenic greenhouse gases (though this effect is present only as long as coal is burned without limiting sulfur emissions).

The RF and ERF methods differ most significantly in their estimation of radiative forcing from aerosols. The total RF value for aerosols was estimated in the IPCC 2013 report at about -0.35 (-0.85 to $+0.15$) W/m^2, including both direct effects and aerosol–cloud interactions. The ERF estimate of aerosol radiative forcing, which includes short-term effects on clouds and carbon particulates, gives -0.9 (-1.9 to -0.1) W/m^2. Following [245] we use the ERF value for aerosol forcing (see Figure 34.14).

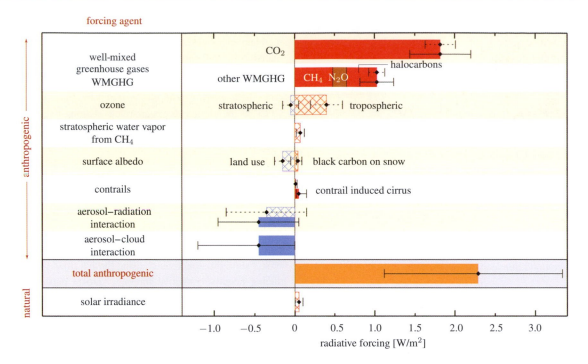

Figure 34.14 Changes in radiative forcing since 1750, as compiled in IPCC5 [245]. Solid bars are *ERFs* (see Footnote 5) while hatched bars are *radiative forcings* (both are shown for several sources, indicating the difference between the two).

In summary, CO_2 emissions and a number of other anthropogenic sources lead to substantial radiative forcing. IPCC5 estimate the radiative forcing due to CO_2 in 2011 to be $\cong 1.8 \pm 0.2\,\mathrm{W/m^2}$, consistent with eq. (34.37). The total effect of all WMGHGs including CO_2 in 2011 was estimated to be $\cong 2.8 \pm 0.3\,\mathrm{W/m^2}$ [245]. Aerosols originating in particulate emissions give negative contributions to radiative forcing of order $-0.9\,\mathrm{W/m^2}$. Tropospheric ozone contributes a positive forcing of order $0.5\,\mathrm{W/m^2}$. Combining all these effects and other small forcings such as changes in albedo, the best current estimate of the change in total (effective) radiative forcing between 1750 and 2011 is [245]

$$F_{2011}^{(\mathrm{ERF})} \approx 2.3 \ (1.1 \text{ to } 3.3)\,\mathrm{W/m^2}\,. \qquad (34.38)$$

34.4.4 Increased Ice Melt from Carbon Particulates

While, as discussed above, aerosols formed from coal plant emissions cause a net negative forcing, the products of biological and fossil fuel combustion also contain particulate matter known as **black carbon** that absorbs visible light very effectively. Black carbon is produced particularly heavily in the burning of biomass fuels, diesel fuels,

Radiative Forcing from Anthropogenic CO_2

The principal driver of climate change over the next 200 years is expected to be anthropogenic CO_2. Radiative forcing from CO_2 grows approximately logarithmically with CO_2 added to the atmosphere

$$F \approx F_* + F_{2\times} \log_2\left(M_{CO_2}/M^*\right),$$

where the increase in radiative forcing from a doubling of atmospheric CO_2 can be estimated as $F_{2\times} \cong 3.7 \pm 0.8\,\mathrm{W/m^2}$. Considering only the immediate (*uniform temperature*) response, Earth's surface temperature would need to rise by roughly $\Delta T_{2\times} \approx 1.2\,°\mathrm{C}$ to offset this forcing. This aspect of climate science is clearly understood and relatively non-controversial among those with scientific understanding of the issues. Feedbacks, which are less well understood, are believed to amplify this temperature change to $\Delta T_{2\times} \approx 3.2 \pm 1.3\,°\mathrm{C}$.

Figure 34.15 Greenland ice sheet: black soot containing black carbon from biofuel and fossil fuel combustion blown onto a glacier accelerates melting and resulting sea level rise. (Credit: James Balog, Extreme Ice Survey)

and in (non-pulverized) coal-fired boilers. Black carbon in the atmosphere leads to positive radiative forcing. Furthermore, black carbon is a primary constituent of **black soot** that accumulates on ice surfaces at high altitudes and higher latitudes. Its (small) effect on Earth's albedo is included in the radiative forcing summarized in eq. (34.38) and Figure 34.14. In addition, surface accumulations of black soot absorb solar radiation and increase the melt rate of glaciers and northern sea ice, an effect not included in its contribution to radiative forcing. In recent years, black soot has been identified as an important factor in the rapid melt rate of the world's glaciers and Arctic sea ice, contributing to global sea level rise, as we discuss in the subsequent chapter.

34.5 Feedbacks and Climate Modeling

34.5.1 Climate Feedbacks

As discussed in §34.2, the *uniform temperature* direct response of Earth's temperature to a given external radiative forcing F can be estimated to a fair degree of accuracy using the current atmospheric configuration to be $\Delta T_0 = -F/\lambda_0$, with $\lambda_0 \cong -3.2\,\mathrm{W/m^2\,^\circ C}$, as in eqs. (34.32) and (34.33). Thus, disregarding any other changes in the Earth system that might result from the warming, the direct effect of anthropogenic radiative forcing circa 2011 would be to increase Earth's temperature compared to 1750 by roughly

$$\Delta T_0 = -F_{2011}/\lambda_0 \approx 0.7 \ (0.3 \text{ to } 1.0)\,^\circ C. \quad (34.39)$$

This is the change in temperature that would be needed to increase the rate of outgoing thermal radiation to compensate for the estimated total 2011 anthropogenic forcing of ≈ 2.3 (1.1 to 3.3) $\mathrm{W/m^2}$. Note that this response corresponds to a hypothetical new equilibrium configuration in which all indirect effects are ignored. Earth's current

climate is not in equilibrium, as many parts of the system have not had time to adjust to the changes in CO_2 level that have already occurred.

To fully describe the reaction of Earth's climate to a given radiative forcing is very difficult. The climate is a very complicated and nonlinear dynamical system, with many components operating on different time scales. Because the heat capacity of many components of the Earth system, particularly the oceans, is enormous, it would take many years for the temperature of the planet to change even to match the direct uniform temperature response. But what makes the system particularly complex is the existence of many *feedbacks* that themselves contribute to radiative forcing.

If, for example, Earth's surface begins to warm in response to radiative forcing, then ice may melt at high latitudes, decreasing Earth's surface albedo. The decreased albedo leads to further radiative forcing, which in turn leads to further warming, further ice melt, etc. This is an example of a **positive feedback**. There are many feedbacks, some positive and some negative, that may affect the climate system. To get a realistic sense of how the climate responds to radiative forcing it is essential to incorporate these feedbacks.

Once they have been identified and quantified, feedbacks can be incorporated in a simple mathematical fashion, if their effects are assumed to be linear at each step in the process. Any radiative forcing F can be cancelled by a *uniform temperature response* $\Delta T_0 = -F/\lambda_0$, defined to be the change in temperature that directly produces a radiative forcing that would exactly cancel F. In the absence of other effects it is natural to expect that the planet would warm by roughly this amount until radiative equilibrium is reestablished. Suppose, however, that the temperature change ΔT_0 has effects that give rise to additional radiative forcing – as in the example of melting ice – and further *suppose that the additional forcing is linearly proportional to the temperature change*, $F_1 = \lambda \Delta T_0$. Then the system will come into radiative equilibrium with a total surface temperature change ΔT such that the uniform temperature response $\lambda_0 \Delta T$ exactly cancels the original radiative forcing F plus the additional forcing generated by the feedback, $\Delta F = \lambda \Delta T$,

$$(\lambda_0 + \lambda)\Delta T + F = 0. \quad (34.40)$$

This result can also be understood as a sum of terms in a geometric series describing the iterated feedback process (see Box 34.4).

Equation (34.40) provides a method for treating climate systems with feedbacks mathematically, assuming that all feedbacks give rise to radiative forcing in a linear fashion.

Box 34.4   Iterated Feedbacks as a Geometric Series

Suppose that some process gives rise to an initial radiative forcing F. The surface temperature changes according to eq. (34.33),

$$\Delta T_0 = -F/\lambda_0 \,.$$

This temperature change generates radiative forcing that cancels the initial forcing F. Because of feedback, however, ΔT_0 leads to an additional forcing, $F_1 = \lambda \Delta T_0$, where λ is a parameter that characterizes the feedback. The surface temperature responds to this additional forcing as it did to the original F,

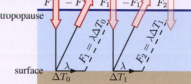

$$\Delta T_1 = -F_1/\lambda_0 = -\frac{\lambda}{\lambda_0}\Delta T_0 = \frac{\lambda}{\lambda_0^2}F \,.$$

Feedback driven by this temperature change engenders further radiative forcing $F_2 = \lambda \Delta T_1 = (\lambda^2/\lambda_0^2)F$, and temperature response

$$\Delta T_2 = -\frac{\lambda^2}{\lambda_0^3}F \,,$$

and so on *ad infinitum*. The total temperature change is the sum of all these effects,

$$\Delta T = \Delta T_0 + \Delta T_1 + \Delta T_2 + \cdots = -\frac{1}{\lambda_0}\left(1 - \frac{\lambda}{\lambda_0} + \frac{\lambda^2}{\lambda_0^2} + \ldots\right)F = -\frac{1}{\lambda_0}\frac{1}{(1+\lambda/\lambda_0)}F = -\frac{1}{\lambda_0+\lambda}F \,,$$

where we have used the fact that $\sum_{k=0}^{\infty} x^k = 1/(1-x)$ (see eq. (B.63)).

This may be true for small perturbations around a stable climate configuration, but is not guaranteed to be true generally. Indeed, the climate system in general is highly nonlinear and may have multiple distinct local equilibria associated with the same external conditions. In some situations, it may happen that small changes in the climate system lead to a sudden shift from one local dynamical equilibrium to another. We discuss this issue further in §35.3.7.

The analysis that led to eq. (34.40) generalizes to the situation where multiple feedbacks are relevant, provided all feedbacks can be assumed to be linear. For N distinct feedbacks with parameters $\lambda_1, \lambda_2, \ldots \lambda_N$, and λ_0 denoting the *Planck feedback parameter* (34.32), the general solution is

$$(\lambda_0 + \lambda_1 + \cdots + \lambda_N)\Delta T + F = 0 \,. \tag{34.41}$$

The net effect of *positive* feedbacks ($\lambda > 0$) is to amplify the warming that would occur from the uniform temperature response alone by a factor of

$$f = \Delta T/\Delta T_0 = \lambda_0/(\lambda_0+\lambda) = 1/(1-\lambda/|\lambda_0|) \,, \tag{34.42}$$

where $\lambda = \lambda_1 + \cdots + \lambda_N$ is the total feedback parameter.

The temperature shift ΔT and the radiative forcing F can be combined to define the **climate sensitivity parameter** σ_c,

$$\Delta T = \sigma_c F \,. \tag{34.43}$$

In the linear-feedback approximation, σ_c is related to the sum of feedbacks,

$$\sigma_c = -f/\lambda_0 = -1/(\lambda_0 + \lambda) \,. \tag{34.44}$$

Whether or not the linear-feedback approximation is adopted, however, the climate sensitivity parameter, defined through eq. (34.43), remains a useful parameter that links changes in radiative forcing to temperature shifts.

Note from eq. (34.42) that the climate sensitivity parameter σ_c is *nonlinear* in λ, even in the linear feedback approximation. In fact, if the total feedback parameter λ were greater than $|\lambda_0|$, then each term in the geometric series in Box 34.4 would be greater than the previous term, and the series would diverge. This corresponds to a runaway feedback loop, which continues until the change in climate runs beyond the regime where it can be treated linearly. Such a runaway feedback may have occurred early in the life of Venus, where an atmosphere composed 96.5% of CO_2 now maintains the surface temperature at approximately 460 °C. As we discuss in more detail in §35.1, it seems unlikely that Earth's atmosphere could produce such a runaway climate condition in the next few hundred million years without extreme forcing, since over Earth's history there have been periods with significantly higher temperatures and greater concentrations of greenhouse gases than are currently present or anticipated. Over the next several billion years, however, as the Sun converts more of its hydrogen fuel to helium and solar output steadily increases, rising temperatures will move Earth's climate into warmer and warmer regimes until liquid water becomes impossible on the surface, well before there is any chance that the expanding red giant Sun could engulf the planet in five billion years or so.

34.5.2 Global Climate Models

One way to estimate the effects of feedbacks in the climate system is to attempt to model all relevant parts of the entire climate system numerically on a large computer. Programs that are designed to compute the dynamics of the atmosphere and oceans over the surface of the entire planet are known as **general circulation models** (GCMs). We summarize some of the key elements of GCMs here; more detailed descriptions of computer climate models can be found in [244, 245, 252]. While some components of GCMs – principally atmospheric circulation – are simulated in local detail using approximations to the underlying equations of physics, other components (such as clouds and aspects of ocean circulation) are not described by simple, well-understood physical equations and are instead *parameterized*, meaning that the effects of these components are characterized by a small set of variables that couple to the other parts of the system and that are estimated by matching to empirical data for some range of the variables. When the models are used to study climate response to changing conditions, these parameterizations are often extrapolated into regimes beyond where they can be independently verified. Although no better approach is currently available to construct climate models, parameterization of incompletely understood climate system components may introduce significant error or uncertainty in the outcome of a simulation.

Atmosphere model A numerical model of 3D atmospheric dynamics lies at the core of a general circulation model. In the most basic approach, the atmosphere is divided into cells by latitude, longitude, and height, with typical GCMs in the early 2000s having a resolution of 1–2° in latitude and longitude and 20–50 vertical points in the atmosphere. The discrete set of points in the horizontal directions is often chosen so that the distribution of points is relatively uniform across Earth's surface. (Whereas simply using longitude and latitude as a coordinate system would give many more points near the polar regions.) In more sophisticated models, alternative coordinatizations are used, such as a 2D analog of sine waves over Earth's spherical surface (known as *spherical harmonics*). Surface features such as coastlines, mountain ranges, surface roughness, etc., are incorporated into the model, though the limited precision means that local detail is generally lost. The basic differential equations for the atmosphere, including energy and momentum conservation, surface friction, and radiative balance, are solved numerically using discrete methods analogous to those described in §6 for the heat equation. Models generally include water vapor content as a dynamical variable, as well as the densities of various trace constituents such as carbon dioxide, sulfate aerosols, etc. The atmospheric model is coupled to components of the program that describe various other aspects of the climate system including oceans, cryosphere, biosphere, etc., as described below. A typical simulation might evolve Earth's climate through 50 years using discrete time steps of several minutes or tens of minutes.

Oceans Ocean models in early GCMs were much simpler than atmospheric models, taking account only of surface effects, and often without dynamical variables characterizing the ocean configuration. More recent models include global ocean dynamics, in which the full ocean system is discretized in a way similar to the atmosphere, and time-dependent ocean circulation is modeled, including freezing and melting, salinity, and energy transport. Models that include global ocean dynamics are generally used to simulate longer time scales covering one or more centuries, in which deep ocean circulation becomes relevant.

Cryosphere Coupling the cryosphere (sea ice, ice packs, and glaciers) to general circulation models is a major challenge for GCMs. As discussed elsewhere in this and the following chapter, the dynamics of ice, such as the rate of melting of large glaciers and Arctic sea ice, is not completely understood. Older GCMs included only very limited treatment of the cryosphere. In some more recent models, formation and melting of Arctic sea ice, the dynamics of glaciers in Greenland, Antarctica, and major mountain ranges, and the effects of black soot on ice are all treated, generally in a parameterized fashion. This component of climate models is still a source of significant uncertainty in both short-term and long-term predictions.

Biosphere GCMs generally couple the biosphere to the atmosphere through effects such as albedo, surface roughness, and the rate of uptake and release of CO_2 and water vapor through photosynthesis, transpiration, and decay processes. These couplings are generally averaged over the large grid cells of the atmospheric circulation model, and parameterized. Since the dynamics of the biosphere depends in a very complex way on climate variables and on future human activity, this is a very difficult part of the GCM to control precisely.

Clouds Clouds are difficult to model accurately in a GCM. Formation of clouds depends on local meteorological conditions that generally vary over spatial and temporal scales too short to be described in any detail by the model. Furthermore, the impact of clouds on incoming radiation through scattering, absorption, and emission, as well as the rate of precipitation, depends on the details of cloud structure, such as the size of water droplets within the cloud. Thus, clouds are generally parameterized in a fairly simple way in GCMs, and are a source of significant uncertainty.

Other climate components There are many other climate components that are included to a limited extent or not at all in many climate models. For example, soil moisture plays an important role through evaporation and plant growth rates, and marine biota can play an important role in the CO_2 cycle. Many such effects are as yet only partially understood and are incompletely integrated into climate models. As scientific understanding of the complex interplay between component systems improves, and computer systems increase in scale, the potential for accurate climate modeling should continue to improve in the coming decades.

GCMs are tested by comparing their predictions with observed present and past climate data, and the various parameterizations within the model are tuned until acceptable agreement is obtained. For example, as shown in Figure 34.16, the global average temperature profile of the atmosphere predicted by models can be compared with present measurements. Current measurements are used to define initial conditions for simulations of future climate. Finally assumptions regarding future human activity are included and the output of the model is interpreted as a prediction of the response of the climate to those activities.

Note that to predict climate over periods of many years, GCMs must generally be much more sophisticated than weather models, which address shorter time scales and therefore are focused generally on the short-term atmospheric dynamics. Weather models also do not need to incorporate many features such as ice and ocean dynamics that are crucial for climate models. Weather models, however, attempt to predict much more precise local effects such as the extent of precipitation in a given region over a period of several hours, while climate models focus on more general questions about average behavior over a period of years. For certain classes of initial conditions, atmospheric dynamics is *chaotic* and has a very short *mixing time* (see §8.4.1), making accurate weather prediction

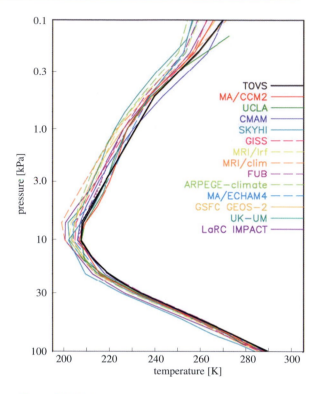

Figure 34.16 The global mean temperature of the atmosphere as a function of altitude. Actual value (black) is compared to output of several climate models. (Credit: Pawson, S., et al. The GCM-reality intercomparison project for SPARC: Scientific issues and initial results, *Bull. Am. Meteorol. Soc.,* 81, 781–796, 2000. Figure 1. © American Meteorological Society. Used with permission)

over periods greater than about five days extremely difficult. For this reason, the output of weather models is often presented as a statistical distribution of possible outcomes, and the spread of the distribution gives important information regarding the range of potential variability and uncertainty in the predicted conditions. While climate models deal in general with meteorological phenomena averaged over longer times, they are subject to similar kinds of uncertainty. The predictions from a climate model for a particular scenario should always be accompanied by an estimate of uncertainty that includes both the predictability of the dynamics and the accuracy and methodology of the model components.

Because of the extensive computational complexity of a general climate model, the most accurate models require the largest and fastest available computer systems, and are generally run on large *clusters* containing many coupled processing units. At present, climate model accuracy is limited both by available computer power and by the limited level of understanding and implementation of many of the components such as ocean, clouds, biosphere, etc. mentioned above. While certain features are incorporated by tuning parameters in the code to observed data sets, this does not necessarily mean that the models are accurate outside the range of the "training" data set. It is also worth noting that some modules of code, particularly for parameterized subsystems, may be shared between different general circulation models, so there may be systematic errors shared by otherwise independent GCMs.

34.5.3 Feedback Estimates from GCMs

In recent years, a number of general circulation models have been developed and used to estimate the full impact of various changes driving Earth's climate system. The International Panel on Climate Change, in their 2013 report, summarized the results of these GCMs for climate change engendered by human activities and their estimates for the feedback parameters for some of the most important climate feedbacks (see Table 9.5 in [245]). The results of this analysis are shown in Figure 34.17. We expect that in coming decades improved modeling and matching with empirical data will substantially improve the estimates for these feedbacks and the climatic consequences of anthropogenic additions to radiative forcing.

We briefly comment on some of the feedbacks:

Water vapor As the climate warms, water evaporates at a greater rate from the ocean surface, and the capacity of

air to hold water increases as the air temperature rises. Increased water vapor in the atmosphere increases the greenhouse effect and leads to additional radiative forcing. The IPCC5 2013 summary of climate models suggests that the resulting feedback parameter is in the range $\lambda_{wv} \approx 1.6 \pm 0.3 \, \text{W/m}^2 \, °\text{C}$.

Lapse rate As the temperature rises and the water vapor content of the atmosphere increases, the adiabatic lapse rate decreases, as discussed in §34.2. This decreases the actual lapse rate at low altitudes, leading to higher temperatures in the troposphere and increased upward thermal radiation, giving a negative radiative forcing. Thus, the lapse rate feedback is negative, $\lambda_{lr} \approx -0.6 \pm 0.4 \, \text{W/m}^2 \, °\text{C}$ [245]. The lapse rate feedback is difficult to predict precisely, and there are some discrepancies between climate models and observed data, so there is substantial uncertainty in this feedback.

Clouds As has been discussed above, clouds are one of the most difficult parts of the climate to simulate, and lead to substantial uncertainty in GCMs. Clouds can lead to negative feedback through increased albedo, or positive feedback through increased greenhouse effect. According to [245], most GCMs indicate a positive feedback, $\lambda_c \approx 0.3 \pm 0.7 \, \text{W/m}^2 \, °\text{C}$, though there is a wide range of results.

Albedo Albedo generally decreases with increased temperature due to ice melt. The consensus of climate models is a small positive albedo feedback with relatively small uncertainty $\lambda_a \approx 0.3 \pm 0.1 \, \text{W/m}^2 \, °\text{C}$ [245]. Note that ice melt due to increased absorption by black soot contamination (§34.4.4) counts as a forcing, not a feedback.

Combining all the feedbacks listed in the 2013 IPCC report, the total feedback parameter λ based on the average of general circulation models is [254]

$$\lambda = \lambda_{wv} + \lambda_{lr} + \lambda_c + \lambda_a$$
$$\approx +1.6 - 0.6 + 0.3 + 0.3$$
$$\approx 1.6 \pm 0.3 \, \text{W/m}^2 \, °\text{C}. \tag{34.45}$$

With $\lambda_0 \approx -3.2 \, \text{W/m}^2 \, °\text{C}$, the corresponding climate sensitivity parameter (34.44) in the linearized feedback model would be

$$\sigma_c = -\frac{1}{\lambda_0 + \lambda} \approx -\frac{1}{-3.2 + 1.6} \approx 0.6 \, °\text{C/(W/m}^2). \tag{34.46}$$

The **equilibrium climate sensitivity** is defined to be the temperature change if equilibrium were established after a doubling of CO_2 levels. Since the temperature shift is roughly logarithmic in the atmospheric CO_2 level, as discussed above, this is a useful benchmark in gauging the impact of anthropogenic greenhouse gases on climate. The

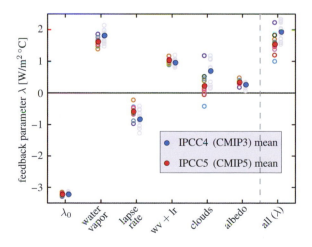

Figure 34.17 Estimates of climate feedbacks from 2007 and 2013 IPCC reports. For each feedback type, an ensemble of 2013/2007 models (CMIP5/CMIP3) is shown on the left/right. The total feedback parameter (λ) is estimated as $\lambda \sim 1.6$ for the 2013 models, compared with $\lambda \sim 2.0$ in 2007. From [245].

equilibrium climate sensitivity can be roughly estimated in the linearized model from eq. (34.43) and the estimated value of $F_{2\times}$ (34.36) and σ_c

$$\Delta T_{2\times} = \sigma_c F_{2\times}^{(RF)} \approx 0.6 \times 3.7 \approx 2.2\,^\circ\text{C}. \qquad (34.47)$$

In [245] the IPCC directly estimated the equilibrium climate sensitivity from general circulation models [255]. CO_2 levels in the model simulations were instantaneously increased by a factor of four and kept constant for 150 years of simulation time. The equilibrium climate sensitivity was then extracted by dividing the resulting temperature shift by two. This approach gave an estimated equilibrium climate sensitivity substantially larger than the linearized estimate would suggest (Problem 34.12),[8]

$$\Delta T_{2\times}(\text{IPCC5}) \cong 3.2 \pm 1.3\,^\circ\text{C}. \qquad (34.48)$$

Comparing $\Delta T_{2\times}(\text{IPCC5})$ with $F_{2\times}$, the IPCC [245] obtained an estimated value of the climate sensitivity parameter

$$\sigma_c(\text{IPCC5}) = \Delta T_{2\times}(\text{IPCC5})/F_{2\times}^{(ERF)}$$
$$\approx (3.2 \pm 1.3)/(3.4 \pm 0.8)$$
$$\approx 1.0 \pm 0.5\,^\circ\text{C}/(\text{W/m}^2). \qquad (34.49)$$

Note that using the other value of $F_{2\times}^{(ERF)}$ or the value of $F_{2\times}^{(RF)}$, both of which are 3.7 W/m², would give

$$\sigma_c \approx 0.9\,^\circ\text{C}/(\text{W/m}^2). \qquad (34.50)$$

Although there are clearly many uncertainties, for definiteness, going forward we will use this latter value.

Between the linearized estimate (34.46) and the estimate (34.49) for climate sensitivity, the eventual shift in temperature that might be expected from the central value of the current level of anthropogenic radiative forcing (34.38) would be in the range $\Delta T \approx 1.4$–$2.2\,^\circ\text{C}$. As we discuss in the next chapter, however, reaching such a new equilibrium would take on the order of 100 years, and it is essentially inevitable that the level of anthropogenic forcing will change significantly over that time frame.

While there has been much controversy over the predictions of climate change due to atmospheric CO_2, the IPCC report represents the best scientific understanding (circa 2013) of the impact of atmospheric greenhouse gases on climate. Following the publication of the previous IPCC report in 2007, an independent task force of scientists assessed the methodology used by the IPCC in compiling their report and found it sound [256]. The 2013 report had very similar quantitative conclusions regarding feedbacks and estimated temperature change to the 2007 report, but with reduced error bars. Thus, while the results are subject to a considerable amount of uncertainty, these conclusions do represent the best understanding of the scientific community.

The global circulation models described in this section can be checked against current conditions and basic physics understanding. But their predictions can also be compared with climate data from past times. Climate models, which are most reliable in the context of small perturbations about the current climate configuration, have little to say about nonlinearities or qualitative changes in climate. In fact, climate data from earlier periods of Earth's history has the potential to shed light on how Earth's climate has responded to conditions that are significantly different from the present, and to give some insight into effects that cannot be studied with current climate data and our limited understanding of climate processes. Thus, we turn next to the subject of climate in Earth's history and then to predictions for the future.

Climate Feedbacks

Climate *feedback* occurs when a surface temperature change ΔT triggers a process that gives rise to additional radiative forcing, $F = \lambda \Delta T$, where λ is the *feedback parameter*. Examples include changes in lapse rate, atmospheric water vapor, albedo, and clouds. Feedback can either amplify or suppress the temperature response to an initial radiative forcing,

$$\Delta T = \sigma_c F,$$

where σ_c is the *climate sensitivity parameter*. In the linear-feedback approximation $\sigma_c = -1/(\lambda_0 + \lambda)$ includes both the uniform temperature response λ_0 and the sum of all feedbacks $\lambda = \lambda_1 + \lambda_2 + \cdots$.

The feedback parameters are difficult to estimate accurately and introduce significant uncertainty into climate change predictions. Best estimates from current models [245] give $\sigma_c \cong 1.0 \pm 0.5\,^\circ\text{C}/(\text{W/m}^2)$.

---

[8] For further discussion of these issues see §9 of [245], and in particular §9.7.2.4. Note also that a somewhat more conservative result, that $T_{2\times}$ "is *likely* between 1.5 °C and 4.5 °C" is quoted in the *Technical Summary* in TFE.6 in [245], where a more extensive discussion of the uncertainties in this estimate can be found. The relevance of nonlinear effects is also discussed in [254].

Discussion/Investigation Questions

34.1 The term *inversion* describes a weather event in which the lapse rate near the ground is *negative* – the temperature actually increases with height. Explain why inversions are often associated with episodes of severe air pollution and are also responsible for the phenomenon of *freezing rain* in the winter.

34.2 Hydrogen and helium are the two most common elements in the universe. What are their concentrations in Earth's atmosphere? Why are they so rare?

34.3 The particles of black soot in Figure 34.15 seem to *drill down* into the surrounding ice. Can you think of a geometrical condition that might give a good first-order estimate of how far down they go? Can you estimate the depth as function of cross-sectional area of the particle for summer conditions in Greenland, at latitude 75°N?

Problems

34.1 Given the following approximations for planetary albedos and given approximate radii of their orbits, estimate the average effective temperature of the planets Mars and Neptune. (The effective temperature is the temperature at which the planet radiates thermal radiation into space, which is not necessarily equal to the surface temperature). You may assume that the planets have circular orbits. Mars: albedo $\cong 0.16$, $r_{orbit} \cong 2.28 \times 10^8$ km; Neptune: albedo $\cong 0.29$, $r_{orbit} \cong 4.5 \times 10^9$ km.

34.2 Show that an n-layer atmosphere of the type described in Example 34.2 gives a surface temperature $T_s = \left(\sqrt[4]{n+1}\right) \times 255$ K.

34.3 Consider a slight improvement on the two-layer toy model of Earth's atmosphere shown in the figure in Example 34.2. Assume that there are two atmospheric layers, each transparent to incoming solar radiation with incoming solar radiation intensity $I = \langle I_{in}\rangle \cong 0.84 I_\odot/4$. Assume that all outgoing infrared radiation (from the surface and from each atmospheric layer) consists of two equal parts. One part is in a frequency range that is completely absorbed by each atmospheric layer, and the second part is in a frequency range that passes through all atmospheric layers. Thus, for example, 1/2 of the IR radiation emitted by the surface is absorbed completely by the lower atmosphere layer (layer 2) and the other 1/2 passes through both atmospheric layers and escapes to space. Set up the three linear equations for the temperatures T_s, T_1, T_2 of the surface and the two atmospheric layers. What is the temperature at the surface? [Hint: to simplify the algebra you may want to first solve for the values $\tau_i = \sigma T_i^4$, for $i = s, 1, 2$.]

34.4 We have described hydrostatic equilibrium in §34.2.1 under the approximation that the atmosphere lies above a planar surface. Compute the pressure in hydrostatic equilibrium as a function of height in an atmosphere around a spherical planet, and compare to the Boltzmann distribution.

34.5 The bond energy of the double bond in the O_2 molecule is 5.06 eV. Verify that light must have wavelength less than $\cong 246$ nm in order to break this bond. Assuming sunlight has a blackbody spectrum at $T = 6000$ K, what fraction of the photons in the sunlight incident on the upper atmosphere can dissociate O_2? [Hint: see eqs. (22.23) and (22.24).] Repeat this calculation for the triple bond in N_2, which has a bond energy of 9.61 eV. Compare the relative importance of O_2 and N_2 as absorbers of ultraviolet light in Earth's upper atmosphere.

34.6 Use the value of atmospheric pressure at sea level to estimate the total mass of the atmosphere. Estimate the number of molecules in the atmosphere and check for agreement with recent numbers for the total mass of CO_2 in the atmosphere and parts per million by volume.

34.7 Prove that radiative forcing grows logarithmically with CO_2 concentration under the circumstances illustrated in Figure 34.13. Specifically, assume that for wavelengths in the vicinity of the 15 μm absorption peak, the CO_2 absorption coefficient $\kappa(\lambda)$ exceeds the value κ^* at which absorption saturates. Further, assume that $\kappa(\lambda)$ falls exponentially $\kappa(\lambda) = \kappa_0 C_{CO_2} e^{-c(\lambda-\lambda_0)}$, beyond some λ_0 in the *right wing* of the 15 μm absorption peak. Here κ_0 is a constant and C_{CO_2} is the concentration of atmospheric CO_2. Show that the range of wavelengths over which absorption is saturated grows proportional to the logarithm of C_{CO_2}.

34.8 Show that radiation at a distribution of temperatures decreases the estimate (34.34) of the uniform temperature radiative response λ_0 by proving the result in the case where the radiation comes from two different temperatures. Assume $I = a\sigma(T-y)^4 + b\sigma(T+y)^4 = 240$ W/m², for fixed values of $a+b = 1$ and y, and repeat the computation leading to the estimate (34.34). (a) Show that the distributed estimate decreases for the specific values ($a = b = 1/2$, $y = 1$ K); (b) prove analytically that the distributed estimate decreases for any values of a, b, y [Hint: start with $a = b = 1/2$]; (c) argue that if the result for λ_0 decreases in magnitude when the radiation is distributed between two temperatures, the same is true when the radiation is distributed between any number of different temperatures for fixed outgoing total radiation flux.

34.9 Earth's albedo is currently estimated to be $a_\oplus \approx 0.30$. By what percentage would $a_\oplus$ have to increase to offset the estimated increase in forcing eq. (34.36) due to doubling of CO_2 since pre-industrial levels?

34.10 Assume that ice albedo feedback gives a feedback parameter $\lambda = 0.5\,\mathrm{W/m^2\,°C}$. Estimate the corresponding addition to the change in temperature under a doubling of atmospheric CO_2 in the absence of other feedbacks. Assume that water vapor and the lapse rate feedback together contribute a feedback parameter $\lambda = 1\,\mathrm{W/m^2\,°C}$. Estimate the temperature change with this feedback alone, and compare to the combined temperature change when both feedbacks are included.

34.11 Using the simplified model of Earth's atmosphere as in Example 34.1, with no absorption, no meridional energy transport, and an albedo of 0.16, use the yearly average integrated insolation computed in §23.3 to compute the temperature in radiative equilibrium on the equator and at the poles.

34.12 Assume that the linearized analysis of climate leading to (34.46) is correct, and $F_{2\times} \cong 3.7\,\mathrm{W/m^2}$. Assume also that climate models correctly predict that quadrupling of CO_2 levels would give a temperature shift of $\Delta T_{4\times} \cong 6.4\,°\mathrm{C}$. Use a quadratic function to fit the change in temperature as a function of the logarithm of the multiplicative increase μ in atmospheric CO_2 levels to estimate the change in temperature with a doubling of pre-industrial carbon dioxide levels.

Earth's Climate: Past, Present, and Future

The foundations of climate science introduced in the previous chapter are, by and large, well established. Predictions of future climate and of the impact of human activity on global climate, in contrast, necessarily involve many uncertainties and unknowns. Indeed, due to the complexity of the climate system and the dependence on uncertain future events and processes, the determination of Earth's future climate cannot be a completely predictive science. The goal rather is to use existing data and the current understanding of climate science to project possible future climate configurations, with as good an understanding as possible of the range of possible outcomes and of the dependence of future climate on specific human actions.

Analysis of Earth's past climate plays an important role in predictions for the future. As emphasized in the previous chapter, climate models become more uncertain as the predicted excursion from current conditions grows larger. Study of Earth's past climate can help us to understand the effects of changes in conditions outside the range of validity of the models described in §34. In addition to providing insight into the range of climate variation and possible extremes, Earth's past climate provides a testing ground for many ingredients in global climate models and gives a context for projections of future climate change.

We begin this chapter with a survey of Earth's past climate and the conditions that may have been responsible for significant periods of warmth or cold. We start with historical data, where there are indications of links between global mean temperatures and fluctuations in solar activity. Next we look back over the past 3 million years or so, when Earth has on average been roughly 5 °C colder than it is now. Finally we look back into deep time, across which variations in solar output, Earth's atmosphere, and the arrangement of continents have caused significant changes in climate. We then turn to the future in §35.2 and examine predictions for future climate change in light of projected patterns of human activity. This leads us

Reader's Guide
This chapter presents an overview of what is known about Earth's climate history and then considers the problem of predicting future climate. Time scales ranging from historical to tectonic provide complementary perspectives on the forces that have driven climate change in the past. Models of future climate that include greenhouse gas emissions and other anthropogenic effects are explored. The potential effects of significant warming over the coming centuries, such as sea level rise, ice melt, and ocean acidification, are described. Finally some proposals to mitigate climate change are presented.

Prerequisites: §34 (Energy and climate) is essential; see also §27 (Ocean energy), §28 (Wind energy), §33 (Fossil fuels).

to consider implications of climate change in §35.3, such as the global distribution of temperature change, the effects of rising seas and melting ice, and impacts on weather, ecosystems, and ocean acidification. Finally, in §35.4, we briefly describe some ways that have been proposed to mitigate or adapt to climate change.

It is not in the spirit of this book to attempt to make definitive statements regarding Earth's future climate. Rather, the scenarios and models described in the later parts of this chapter are intended to give a sense of the scope of the effects of anthropogenic climate change and the time scale and range of issues involved, not to make specific quantitative predictions. We expect that analysis of climate systems will improve rapidly in the coming decades so that in future years the impact of human activity on climate will be understood more clearly and precisely. While the basic principles described in this chapter should continue to be valid, new perspectives will inform this subject as the effects of human activity on Earth's climate unfold through the twenty-first century.

35.1 Past Climate

Records of Earth's past climate can provide insight into the effects of changes in conditions that are too large to be accurately predicted by a simple linear extrapolation from the present circumstances. There are clear indications that over the 4.5 billion years of Earth's history the planet has experienced wide climatic fluctuations. These changes in climate have been driven by changing conditions that range from fluctuations in Earth's orbit and in the greenhouse gas content of the atmosphere to wholesale rearrangement of the continents and oceans due to continental drift. For example, over most of the last million years, the climate has been dominated by long glacial periods (**ice ages**) with average temperatures some 5 °C below current values, while 50 million years ago, when modern mammals first came to prominence in the Eocene epoch,[1] average temperatures may have been 7 °C or more above current values. Although these temperature differences may not seem enormous, the climate during those periods differed significantly from Earth's present condition.

The scientific study of past climate on Earth involves a wide range of measurements and analyses. In this section we explore some of what is known about Earth's climate in the past, focusing on a number of different time scales. Over each of these time scales, different approaches are required to obtain information about past climatic conditions, and different mechanisms are responsible for driving climate changes. The **historical time scale**, covering the last few hundred years, is the period over which we have the best understanding of climate history, although even over this period the precise mechanisms driving the observed climate changes are not completely understood. The study of Earth's climate before the historical period is known as **paleoclimatology**, and involves using indirect, or **proxy measurements** of climatic variables such as temperature over Earth's history. The **orbital time scale**[2] covers the period of time over the last 600 000 years or so, during which it is believed that regular oscillations of the parameters of Earth's orbit about the Sun have played a primary role in driving the climate. Over much longer periods of time, on the order of hundreds of millions of years, motion of Earth's tectonic plates has affected the global circulation pattern and the carbon cycle, and has dominated long-term climate effects. This longest time scale

is referred to as the **tectonic time scale**. We are primarily interested here in the more recent historical and orbital time scales, about which the most is known, though we also describe some aspects of tectonic scale climate change.

The subject of paleoclimatology is complex and fascinating, and provides many mysteries waiting to be solved by aspiring young scientists. We only touch the surface of the subject here, and refer the interested reader to several excellent recent texts for a deeper introduction to the subject. The textbook by Ruddiman [257] provides a clear and pedagogical introduction to the ideas of paleoclimatology. Cronin's slightly more advanced book [258] is also excellent and provides extensive references to the recent literature. Another good, slightly more technical, reference is [259]. The 2013 IPCC report on climate change [245] also has a detailed section on paleoclimate data and its relation to the question of modern anthropogenic climate change.

35.1.1 Historical Climate Data

While the term *climate* encompasses a wide range of phenomena, the temperatures of Earth's surface, atmosphere, and oceans play a central role in defining climatic conditions. At a basic level, temperature is a measure of (thermal) energy, and energy flow controls climate. Direct measurements of terrestrial temperature have been recorded in some locations for hundreds of years in a reasonably systematic fashion. The German physicist Daniel Fahrenheit proposed the temperature scale named after him in 1724, a few years after developing the relatively high precision mercury thermometer that made accurate measurements of temperature possible. Systematic temperature records have been reconstructed (for locations in central England) as far back as the mid-seventeenth century, and temperatures have been measured at a sampling of points around the world regularly since about 1850.

Over the last 150 years, many different methods have been used for measuring temperature. For example, in the late 1800s, sea temperatures were measured by dropping a wooden bucket over the side of a sailing ship, hauling the bucket up, and measuring the temperature of the water. In later years, metal buckets were used, and now sea temperature is measured from the water passing through the engine intake of ocean vessels. Obviously, consistently integrating data from such different measurements is a significant challenge. Different approaches to measurement suffer from different systematic errors and uncertainties; for example, the thermal energy transferred between a sample of seawater and a bucket dipped into the ocean depends upon the heat capacity and thermal conductivity of the material forming the bucket, as well as the ambient

---

[1] See Appendix D.2 for a timeline that identifies the geological eras mentioned throughout this chapter.

[2] More generally, the *orbital time scale* refers to time periods on the order of tens or hundreds of thousands of years that are relevant for periodic changes in orbital geometry.

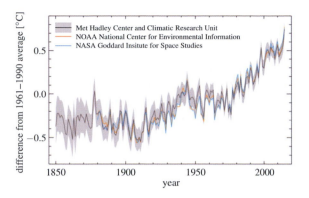

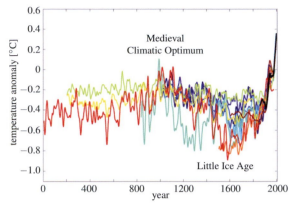

Figure 35.1 Estimated average global surface temperature from 1850–2015 from three sources [260]. The 95% confidence range for the Hadley data is shown in gray. Black line: HadCRUT dataset produced by the Met Office Hadley Center in collaboration with the Climatic Research Unit at the University of East Anglia (gray area shows the 95% confidence range). Blue line from the NASA Goddard Institute of Space Studies' GISTEMP data set. Orange line from the global average temperature anomalies from the National Climatic Data Center's MLOST data set. Anomalies are defined relative to the 1961–1990 average.

Figure 35.2 A variety of reconstructions of global surface temperature in the Northern Hemisphere over the last 2000 years, using a variety of proxies for temperature. Consistent features include the warmer Medieval Climatic Optimum around 1000–1300 AD and the cooler Little Ice Age around 1300–1900 AD, as well as the substantial rise in temperature over the last century. (Credit: Robert A. Rohde, Global Warming Art Project, reproduced under CA-BY-SA 3.0 license via Wikimedia Commons)

temperature and the time elapsed between the extraction of the bucket and the measurement of the water temperature. Nonetheless, with some attention to how the data was taken, and cross-correlation of observations using different methods, it is possible to obtain a reasonably accurate estimate of atmospheric and ocean temperatures since 1850. A plot of the estimated **global mean surface temperature** since this time is shown in Figure 35.1. It is evident that over this time, global mean temperatures have increased by close to 1 °C.

Before about 1850, data from direct measurements becomes less reliable and is available only from a limited set of locations and times. To reconstruct information about temperatures or other aspects of climate before 200 years ago, it is necessary to make indirect measurements by drawing inferences from information stored in physical form such as layers of rock or ice. For times in the not-too-distant past, data can also be extracted from biological systems such as tree rings. By correlating the behavior of these proxies with recent instrumental temperature records and with other known effects, the statistical significance of proxy data for determining the climate record can be estimated.

The use of *tree rings* as a proxy in estimating past climate conditions is known as **dendroclimatology**. Using data from long-lived trees, this method can be extended back several thousand years. Generally, higher temperatures are correlated with a longer growing season and greater tree ring growth. There are, however,

other influences, such as drought, disease, other weather conditions, and many other local factors, which can impact the size of tree rings. The correlation with temperature is therefore imperfect, and large numbers of samples must be combined statistically to estimate past surface temperatures. With a limited data set, such estimations are necessarily subject to a substantial degree of uncertainty.

Focusing on the temperature record over the last two thousand years, a variety of estimates of average global temperature produced using different combinations of proxies are shown in Figure 35.2. Several features evident on this graph are associated with climatic conditions that are also noted in the historical record. The first 300 years or so of the second millennium AD is known as the **Medieval Climatic Optimum**. During this time, at least in Europe, prevailing climatic conditions were relatively warm. This is the time period during which Greenland was named and colonized by the Vikings. (Although, according to Norse legend, Eric the Red gave Greenland its name more for the purpose of encouraging potential settlers than because it was particularly green, even at that time.) Following the Medieval Climatic Optimum, temperatures dropped substantially in Northern Europe during the **Little Ice Age**, which lasted from about 1300 to 1900 AD. During this time the Norse died out or left Greenland, and Europe suffered from very harsh winters as well as shorter growing seasons that led to substantial suffering and famine. There is also evidence of a cooler climate during this period in other regions, particularly in the Northern Hemisphere.

Because the reconstruction of past temperatures from proxy data is subject to many uncertainties, there is a fairly broad range of variation and statistical uncertainty in the estimates shown in Figure 35.2. There has been some controversy about the accuracy of these estimates, specifically regarding how proxy data, particularly from tree rings, are correlated with the recent direct temperature record and how data from different sources are spliced together. The different reconstructions of past climate shown in Figure 35.2 are based on many different proxies, including isotope levels in ice cores and ocean sediment cores, and sea coral growth patterns. These other proxies form independent checks on the estimations of past surface temperatures. We discuss ice core data in more detail in §35.1.3, as well as some other methods for measuring climate in the distant past.

35.1.2 Drivers of Climate Change Over the Last Millennium

To understand the possible impact of human activity and anthropogenic CO_2 on Earth's climate, it is important to distinguish the relative influence of various natural sources of climate variability. From analyses such as those depicted in Figure 35.2, and from historical records of the Little Ice Age in Europe, we believe that, at least in the Northern Hemisphere, climate fluctuations have occurred over the last thousand years that have included temperature variations on the order of several tenths of a degree Celsius. What are the possible sources of climate variation on a time scale that is so short compared to geological processes? While disentangling the precise effects of various factors is difficult, there are a number of known phenomena that may contribute to short-term climate variation.

Volcanic activity releases sulfur dioxide into the stratosphere, generating sulfate aerosols that produce negative radiative forcing, as described in §34.4.3. Individual large volcanic eruptions can lead to global cooling of roughly 0.1–0.3 °C for a period of several years. The eruption of Mt. Pinatubo in the Philippines in 1991, for example, created a global cooling effect that is believed to have lowered the global average temperature by several tenths of a degree Celsius over the following year and a half. Volcanic activity, as far as we understand it, is essentially a random stochastic process. While individual volcanic events can have significant impact on climate over some years, such events are not correlated through any known mechanism and thus believed not to lead in general to large systematic long-term climate effects when averaged over a period of many decades. There is some evidence, however, that a sequence of massive volcanic eruptions in the thirteenth century, and further eruptions up to the nineteenth

century, may have contributed to initiating and maintaining the lower temperatures of the *Little Ice Age*.

Significant **El Niño** events, which are the warm phase of the **El Niño Southern Oscillation**, a cycle of warm and cold sea surface temperatures in the tropical Pacific ocean, can also lead to warming episodes lasting one or several years. Arguments have been put forth that El Niño events have had significant impact on human society; for example, a strong El Niño in 1876–77 may have contributed to droughts and famines that occurred in China, India, and other countries at that time.

Available information is not sufficient to determine definitively the precise extent to which closely spaced volcanic eruptions or extensive ocean/atmospheric activity may have influenced climatic changes over the last millennium. While there were few major volcanic eruptions during the period from 1912–1963, measured volcanic and ocean activity (or lack thereof) during the last century do not seem to have a strong enough influence on climate to have played a dominant role in the overall warming trend since 1900 evident in Figure 35.1, particularly that seen over the last half century.

Natural fluctuation in the level of solar radiation is another factor that may play a significant role in affecting climate on various time scales. As discussed in §22.1, the intensity of solar radiation has increased by roughly 40% since the Sun began burning its nuclear fuel almost 4.6 billion years ago. This gradual steady increase in luminosity is relevant over time scales of hundreds of millions of years. On shorter time scales, however, there are also changes in the level of solar activity. Increases in solar luminosity are associated with the appearance of greater numbers of dark **sunspots**, regions of the Sun's surface where intense magnetic fields slow the convective flow carrying thermal energy to the surface. Higher sunspot activity is associated with an increase in the appearance of bright regions called **faculae**, which radiate at a higher temperature ($\sim$7000 K) than most of the solar surface. Sunspots are also correlated with **solar flares**, in which explosions within the Sun produce large bursts of energy in a short period of time.

Solar **sunspot activity** and the associated increase in total solar luminosity vary with a fairly regular period of about 11 years. Over many cycle times, the amplitude of sunspot activity varies in a pattern with less obvious regularity. Sunspots were noted by the Chinese as early as the fourth century, and have been observed with telescopes since Galileo (1609). An estimate of the number of sunspots as a function of time over the last 400 years is graphed in Figure 35.3. From this graph it is clear that there is at least some correlation between sunspot activity and the temperature history estimated in Figure 35.2.

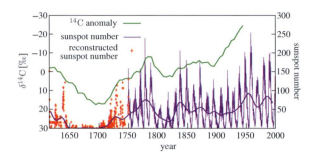

Figure 35.3 Sunspot records (red, reconstructed; purple, recorded) and atmospheric ^{14}C variation (green) from 1610 to 2000 (see Box 35.1). (Credit: Leland Mcinnes, reproduced under CA-BY-SA 3.0 license via Wikimedia Commons)

In particular, there was an almost complete cessation of sunspot activity during the **Maunder Minimum** from 1645 to 1715, in the middle of the Little Ice Age. An earlier such period, the **Sporer sunspot minimum**, occurred from about 1460 to 1550, also during the Little Ice Age. Although direct observations of sunspots provide less complete data for earlier times, particularly before the use of the telescope, solar activity can also be correlated with carbon-14 (^{14}C) levels in long-lived trees (see Box 35.1), corroborating the low level of sunspot activity during these periods. Sunspot activity, as well as temperature, apparently increased from 1700 to 1800, decreased slightly just after 1800, and then both rose significantly in the twentieth century. This suggests a measureable link between solar activity and terrestrial temperature. Given the notable increase in sunspot activity through the first half of the twentieth century, this raises the question of what part of the increase in temperatures over the last century can be attributed to solar activity and what part can be attributed to anthropogenic carbon dioxide increase. This topic has been a subject of some debate in recent years.

The difference in solar luminosity between sunspot minima and maxima has been measured very precisely by satellites, and in recent years has been less than 0.1%, or $< 1.4\,\text{W/m}^2$ difference in the solar constant at the top of Earth's atmosphere.[3] Since climatic response time is much longer than the sunspot period of 11 years, one would expect the climate to respond most clearly only to solar activity averaged over the sunspot cycle. From tree ring studies of atmospheric ^{14}C and beryllium-10 (^{10}Be) (see Box 35.1 and Question 35.1) levels over the last

millennium, it has been determined that at the time of the Maunder Minimum, total solar luminosity was below the current value by on the order of 0.1%, corresponding to a virtual cessation of sunspot activity.

We can get an order-of-magnitude estimate of the impact of changes in sunspot activity on terrestrial temperature by using the estimate for climate sensitivity from the previous chapter and using $1.4\,\text{W/m}^2$ as an upper bound on radiative forcing from the change in solar luminosity. This should give an upper bound on the direct impact of solar luminosity changes on temperature. A $1.4\,\text{W/m}^2$ difference in solar constant corresponds to a difference in radiative forcing of

$$F_{\text{sun}} = \frac{1}{4}(0.7)(-1.4\,\text{W/m}^2) \cong -0.25\,\text{W/m}^2 , \quad (35.1)$$

where the factor of 1/4 comes from the usual ratio of Earth's cross section to surface area, and $0.7 = 1 - a$ comes from the albedo. Note that eq. (35.1) is almost a factor of ten smaller than the radiative forcing (34.37) attributed to the anthropogenic CO_2 increase above pre-industrial levels circa 2010. Given an estimated climate sensitivity of $\sigma_c \cong 0.9$, the radiative forcing (35.1) gives an estimated effect on surface temperature of

$$\Delta T = \sigma_c F_{\text{sun}} \cong 0.9\,°\text{C}/(\text{W/m}^2)(0.25\ \text{W/m}^2) \cong 0.2\,°\text{C} .$$
$$(35.2)$$

This is the correct order of magnitude to have contributed to the temperature difference between the Maunder Minimum around 1700 and temperatures in the mid-twentieth century, as seen in Figure 35.2, but is much less than the total measured temperature increase of over $0.8\,°\text{C}$ during the last 100 years. This can be seen as corroboration of the interpretation given by many climate scientists that the temperature rise in the first half of the last century was in some part a recovery from the Maunder Minimum/Little Ice Age, while the continued temperature rise in the second half of the twentieth century was due primarily to increased atmospheric CO_2; indeed, most anthropogenic CO_2 has been added to the atmosphere since 1950 (see Figure 34.12).

Some scientists have argued that sunspot activity has played a greater role in twentieth-century warming. For this to be true, however, it would seem that the climate sensitivity parameter σ_c would have to be much greater than the value used in eq. (35.2) (or that earlier in the century the magnitude of fluctuations in solar luminosity was significantly greater). This would suggest that doubling CO_2 levels, associated with a radiative forcing of $\sim 3.7\,\text{W/m}^2$, would produce a terrestrial temperature shift well above the $3.2\,°\text{C}$ mean estimate arrived at by the IPCC. Thus, since the radiative forcing from both solar activity changes

---

[3] It is generally assumed that this fluctuation range in luminosity over the solar cycle, which has been empirically measured only in the last few decades, has held over longer periods of time.

Box 35.1　Carbon-14 and Paleoclimatology

Carbon has two stable isotopes, ^{12}C and ^{13}C, with relative natural abundances of roughly 99% and 1%. A third (unstable) isotope of carbon, ^{14}C, plays an important role in understanding many aspects of climate over the last 50 000 years or so.

As described in §20.5.2, ^{14}C is produced when neutrons generated by cosmic rays react with nitrogen in Earth's atmosphere. ^{14}C decays with a 5730 year half-life. In a living organism, or in biologically active soil, carbon is regularly exchanged with the atmosphere, so that the abundance of ^{14}C in living matter is the same as in the atmosphere – about one part per trillion. Once an organism dies, however, the ^{14}C is no longer replenished and begins to decay. By measuring the abundance of ^{14}C in once-living matter or soil, the time since the organism died can be determined. ^{14}C is used in paleoclimatology in many ways, often as a way of assigning a fixed date to an event in the geological record. Because eventually the fraction of ^{14}C becomes too small to measure, this method is only useful for material up to roughly 50 000 years old.

The variation of ^{14}C in past times can also be correlated with the level of solar activity. The *solar wind*, a stream of charged particles – primarily electrons and protons – flowing outward from the Sun, ebbs and flows with solar activity. These charged particles deflect many of the incoming cosmic rays, leading to an anti-correlation between ^{14}C production and solar activity. Figure 35.3 shows both (reconstructed) atmospheric ^{14}C variation and sunspot activity that exhibit this correlation. Variations in solar luminosity over the last thousand years due to changes in sunspot activity levels can be deduced by comparing ^{14}C levels from known times to tree ring data as shown in the figure below.

Carbon-14 data confirms that solar activity was weaker during the Sporer and Maunder sunspot minima in the latter parts of the fifteenth and seventeenth centuries, some of the coldest periods during the Little Ice Age.

(Credit: Leland Mcinnes, reproduced under CA-BY-SA 3.0 license via Wikimedia Commons)

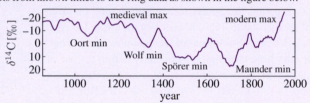

and CO_2 increases is quite well quantified, the only scenario in which solar activity could be responsible for all of the warming over the last century is one in which anthropogenic CO_2 changes will have even more dramatic impact on climate than is currently anticipated. The majority scientific consensus at this point, however, as summarized by the IPCC, is that this is not the case and that solar luminosity is unlikely to have affected terrestrial temperatures over the last century (or to affect terrestrial temperatures in the coming century) by more than a few tenths of a degree Celsius.

To summarize our understanding of temperature changes over the last millennium, while natural[4] effects including volcanic activity and changes in solar activity levels were likely primarily responsible for the Little Ice Age, the rapid temperature change over the last 50 years can only reasonably be attributed to the effects of

anthropogenic changes. In particular, the well-understood radiative forcing from anthropogenic CO_2 is greater by a substantial factor than the forcing from any known natural source, and the observed temperature change is completely consistent with past climate response to other less significant forcings. Figure 35.4 gives a perspective on the estimated contributions of different drivers to global mean surface temperature variations over the period 1870–2010. The role of natural drivers and the increasing dominance of the anthropogenic component are evident. The relative roles of solar activity and other natural forcings and atmospheric greenhouse gases in determining global warming/cooling may be much more clearly understood over the next decade or two with further data and an improved understanding of climate dynamics.

Before turning to evidence of climate changes older than the past millenium, we take a brief look at the past 15 years of climate records. As can be seen in Figure 35.1, global temperatures rose from about 1900–1940, were relatively flat from 1940–1970, and have risen again fairly steadily from 1970–2015. One estimate of the global mean

---

4 Some arguments have also been put forth that human activity may have played some part in cooling during the Little Ice Age through changes in forestation.

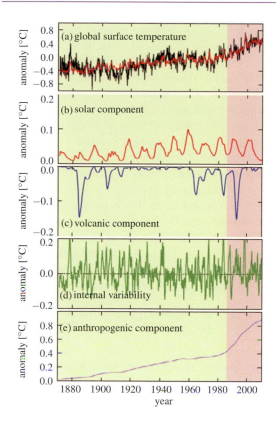

Figure 35.4 (a) Global surface temperature anomalies from 1870–2000 and the (b) solar, (c) volcanic, (d) internal, and (e) anthropogenic factors (including greenhouse gases and aerosols) that influence them. In each case, the temperature response is calculated from estimated forcing. The sum of these factors is given by the red curve in (a). Adapted from [245].

surface temperature during the latter period is shown in Figure 35.5. Although the decade from 2000–2010 was the warmest in the instrumental record, in the early 2010s there was some discussion of an apparent slowdown, or *hiatus* in warming over the roughly 15 year period between the very warm year 1998 and 2013.[5] After 2014, 2015, and 2016 each broke the previous record for the warmest year in recorded history, however, the "hiatus" has appeared simply to be a small fluctuation in an overall monotonic warming trend.

---

[5] Some explanations for such a hiatus even in the presence of forcing from anthropogenic CO_2 include decreased solar activity, negative radiative forcing from sulfates in increased coal emissions, and cooling from ocean oscillations. Whether there was in fact a true hiatus has been questioned by a careful reanalysis of sea and land surface temperature records [262].

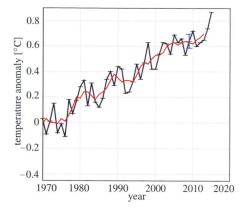

Figure 35.5 Global mean surface temperature from 1970 to 2015 [261]. The red line is smoothed over five years. The blue bar is an estimate of the uncertainty of recent data.

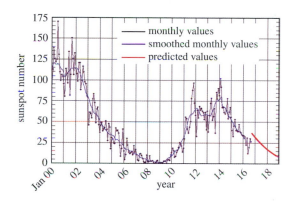

Figure 35.6 Detailed sunspot data from 2000–2016, with predicted values through 2019. (Public domain from Space Weather Prediction Center of the National Oceanic and Atmospheric Administration)

During the period from 2000 to 2015, the warmest 15 year period in the instrumental record, solar activity has been relatively low (Figure 35.6). After a peak around 2001, sunspot numbers decreased steadily to a minimum around the end of 2008. The most recent sunspot cycle (known as *Solar Cycle 24*) has been quite weak, with a (smoothed) peak of roughly 75 sunspots in early 2014 (compared with a smoothed peak of roughly 120 at the 2001 maximum). Since 2014, sunspot activity has decreased, and is expected to hit the next minimum around 2019, while 2014 and 2015 both set records as the warmest years in recorded history. Thus the most recent increase in global mean temperature seems to have occurred at a time of low forcing from solar luminosity.

Recent Climate Change

In the last century, global temperatures have risen by roughly 0.8 °C. Roughly half of this rise occurred in the first part of the century and may have been due in part to a return to higher solar activity levels and other natural forcings. The increase of roughly 0.4 °C over the period from 1970–2010 is primarily attributed to positive radiative forcing from anthropogenic atmospheric CO_2.

It should be emphasized that attempting to analyze or predict climate on time scales shorter than one or two decades is much more difficult than assessing the long-term impact of major changes in forcing such as anthropogenic carbon emissions. In general, both the mechanisms driving the climate and the climate's response over such short time periods are so poorly understood that simulations and predictions should be interpreted with limited confidence and large error bars. Natural *internal variability* (Figure 35.4(d)), associated with fluctuations within the system not associated with specific identified global drivers, also plays a significant role on shorter time scales, and makes precise predictions on the scale of decades essentially impossible; the rule of natural variability in climate prediction is discussed further in §35.2.

35.1.3 Climate Over the Last 3 Million Years and Orbital Forcing

To put recent climate changes into context, it is helpful to consider how climate has varied over periods longer than the 2000 year span described in the previous section. Over periods of tens and hundreds of thousands of years, larger-scale climatic fluctuations have occurred with a fairly regular pattern. For the last several million years, the dominant climate on Earth has been substantially colder than recent times, with typical surface temperatures up to 5 °C colder than the present. Over the last half million years, long periods of extensive glaciation have been periodically interrupted by warmer **interglacial periods**, of which the last 10 000 years represents one of longest and most stable.

Reconstructing Climate Over the Last 600 000 Years To learn about global climate conditions more than several thousand years ago, it is necessary to go beyond the data found in living biological records such as tree rings, and to consider data from persistent sources

such as deep sea sediments and ice cores. One key indicator found in deep-sea sediment cores is the oxygen isotope ratio in fossilized *foraminifera* shells. Foraminifera are small (generally ∼1 mm) protozoa, of which a significant population live in the **benthic** (sea-bottom) ecosystem. Their shells form a substantial component of many marine sediments – the limestone used to build the Egyptian pyramids, for example, is largely composed of fossilized foraminifera shells. While most oxygen atoms in Earth systems are the common stable oxygen-16 (^{16}O) isotope, some 0.2% of naturally occuring oxygen is oxygen-18 (^{18}O), which is also stable. When shells containing calcium carbonate ($CaCO_3$) are formed, the heavier ^{18}O is slightly more likely to be incorporated when ambient temperatures are cooler. Thus, the $^{18}O/^{16}O$ **ratio** serves as a proxy for temperature. Variations in ^{18}O are generally measured in parts per thousand (**per mil** or ‰), so

$$\delta^{18}O \ (‰) = \left(\frac{(^{18}O/^{16}O)_{\text{sample}}}{(^{18}O/^{16}O)_{\text{reference}}} - 1 \right) \times 1000 \ . \quad (35.3)$$

An increase of $\delta^{18}O = 1‰$ corresponds to roughly a 4 °C decrease in water temperature.

The $^{18}O/^{16}O$ ratio is also affected by the total amount of water locked up in ice around the planet and by shifts in global temperature. Because ^{18}O is heavier than ^{16}O, ^{18}O evaporates less readily and precipitates out of the atmosphere preferentially before reaching the Arctic, so ^{18}O appears in polar ice at a lesser fraction than in seawater. This effect is exacerbated at lower temperatures since there is less energy in the system to evaporate the ^{18}O. As the temperature drops and global ice volume on land increases, therefore, the $^{18}O/^{16}O$ ratio in the oceans increases while the ratio in polar ice decreases. For climate conditions over the last several million years, a change in ice volume corresponding to a 10 m drop in sea level corresponds to a change in ^{18}O of roughly +1‰. Thus, ^{18}O levels in sea sediments form a combined measure of temperature and ice volume, with increased ^{18}O indicating cooler temperatures and more ice. Similarly, ^{18}O levels in ice cores are lower from periods of lower temperature.

While sea sediment cores have been used for many years in study of prehistoric climate, in recent decades the use of ice core data has been particularly valuable in attaining a clearer picture of Earth's climate over the last half million years. In Greenland and Antarctica, yearly accumulation of snow and ice has produced ice packs many kilometers thick that preserve abundant information about past climate conditions. Bubbles of air trapped in the ice core contain information about the history of the atmospheric composition. Dust, ash, and sulfates are indicators of past volcanic activity. Furthermore, ratios of hydrogen isotopes D/H (deuterium to hydrogen, or $^2H/^1H$) as well as $^{18}O/^{16}O$ in the ice are indicators of past temperature. Even wind speeds can be estimated from the size of ice particles. As mentioned above, ^{14}C (and ^{10}Be, see Question 35.1) can be used as measures of past solar activity.

Proxy information contained in more recent ice cores can be dated by identifying individual layers associated with each year's snow accumulation. In Greenland, where the accumulation is fairly rapid, ice core data reaches back around 100 000 years (100 ka), and layering can be used to date samples back as far as 10 ka. In Antarctica, where the accumulation is slower, the data reaches back much further – close to 800 ka – but layering becomes indistinct much more quickly. Since ^{14}C dating only works for 40 000 years or so, dating earlier ice core samples becomes more difficult, and is estimated using models for ice flow behavior over millennia. Dating atmospheric information from ice core samples is even more tricky, since air flows fairly freely through the top several tens of meters of ice and snow cover. In Greenland, air captured in ice cores may date from 100 years or more after the time that the surrounding ice was laid down as snow; in Antarctica, it can take 2000 years before air flow is shut off at a depth of 50 m or so and the atmospheric conditions are permanently encapsulated in the ice. This can lead to significant uncertainty in the temporal relationship between changes in temperature and atmospheric composition.

Although there are some discrepancies in the details between different proxies, there is enough agreement and correlation between the data contained in the different long-term climate records to assemble a reasonably consistent picture of the basic aspects of Earth's climate and atmosphere over the last 650 000 years. A graph of several different kinds of proxy data over this time period is shown in Figure 35.7. This data indicates a glacial cycle at roughly 100 000 years, with longer glacial periods (represented in the temperature proxy data by δD and $\delta^{18}O$) interrupted at fairly regular periods by interglacial periods lasting anywhere from 5000 to 30 000 years. Independent data from ancient coral reefs confirms the last 100 000 years of this record, and shows that at the last glacial maximum 20 000 years ago sea level was roughly 110 m below current values. Combining this information about sea level with the $\delta^{18}O$ data, it is estimated that the global surface temperature at this time was roughly 5 °C below current values. Compared to the prevailing conditions over the last 600 000 years, the period from around 9 ka–1200 AD was part of a particularly warm and stable interglacial period.

The data from ice cores and other sources show that CO_2 levels have fluctuated between roughly 180 ppm during

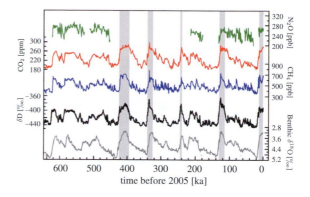

Figure 35.7 Data on levels of atmospheric constituents (CO_2, CH_4, N_2O), and temperature proxies (δD, $\delta^{18}O$) over the last 650 000 years. Scale of ^{18}O is inverted, so all proxies have higher values on the vertical axis correlated with higher global temperature. Gray areas indicate warmer interglacial periods, with temperature fluctuations of roughly 5 °C between glacial maxima and interglacial periods. From [246].

glacial maxima and 280 ppm during interglacial periods. Similarly, methane levels have fluctuated between 300 and 700 ppb. Present atmospheric concentrations of these greenhouse gases are significantly higher than at any time in the 650 000 year data record. It is of course striking that atmospheric greenhouse gas concentrations over this period have tracked global temperatures so closely. This gives rise to the natural question of whether the change in climate was caused by changes in greenhouse gas concentrations, or whether the change in greenhouse gas content in the atmosphere was itself caused by climate changes, or indeed whether both were driven by another, as yet undetermined agent. In fact – though there is some uncertainty regarding the precise relative timing of the changes in atmosphere and temperature because of the lag in sealing off air in ice sheets – careful analysis of ice core data indicates that the atmospheric changes in carbon dioxide and methane shown in Figure 35.7 in fact *lagged behind* changes in the climate by some hundreds or thousands of years. We discuss this issue further later in this section.

Orbital Forcing of Climate There is good evidence that a primary driver of climate change over the last several million years has been the variation in the seasonal and latitudinal distribution of insolation, arising from regular oscillations in the shape of Earth's orbit and the tilt of Earth's axis relative to the ecliptic. The role of orbital changes in influencing climate has been appreciated by climate scientists since shortly after the Swiss geologist Louis Agassiz found evidence for extensive glaciation at

---

Box 35.2   The Vostok and EPICA Ice Cores

A significant step forward in our understanding of Earth's climate conditions over the last half million years was made when a team of scientists from Russia, the US, and France extracted an ice core 3623 m long at Vostok Station in East Antarctica in the 1990s. This ice core contains information from the last 420 000 years, covering four major glaciation cycles [264]. The comprehensive data from this ice core has yielded substantial new insights into climate history over this period.

More recently, the European Project for Ice Coring in Antarctica (EPICA) extracted a core at Concordia Station, some 560 km from Vostok, which reaches back more than 700 000 years [265]. Annual snowfall at this location is currently roughly 25 mm/yr, and the core taken is 3270 m long. The data from this ice core complement and extend the Vostok data.

The core shown above is a section of an ice core from Greenland, showing distinct layering.

(Image credit: National Ice Core Laboratory)

---

mid-latitudes of the Northern Hemisphere and proposed in 1837 the theory of ice ages. A clear and succinct account of the historical development of the **orbital theory of climate change** is given in [258]. The Serbian engineer and mathematician Milutin Milanković put these ideas on a concrete footing in the first half of the twentieth century by systematically analyzing the effects of changes in orbital parameters on insolation. The orbital theory of climate change, also sometimes referred to as the **Milankovitch theory**, is now widely accepted as the primary principle underlying the pattern of Earth's climate over the last 3 million years or so.

There are three key parameters describing Earth's orbit and rotational axis that play a role in modifying insolation:

Tilt As discussed in §23.2, the current tilt of Earth's axis (*obliquity*) is $\epsilon \cong 23.44°$. Over thousands of years, however, the obliquity gradually changes. The pull of large astronomical bodies – primarily Jupiter – on Earth's equatorial bulge gives rise to torque that causes a fairly regular oscillation of the axis tilt between a minimum of 22.05° and a maximum of 24.50°, with a period of about 41 000 years. At the present time the tilt is decreasing at a rate of roughly 46 arc-seconds/century, and will reach a minimum in roughly 10 000 years.

Eccentricity As described in §23.2, Earth's orbit at present is elliptical with an eccentricity of $e \cong 0.0167$. Due to a combination of forces within the Solar system, the eccentricity of Earth's orbit varies over time between roughly 0.005 and 0.061. This variation is not as regular as the oscillations in obliquity; a significant component of the variation arises from four distinct cycles with

periods between 95 000 and 131 000 years; these oscillations combine into a pattern with a dominant periodicity around 100 000 years. There is a further component of the oscillation with period around 400 000 years. The eccentricity of Earth's orbit is decreasing at this time.

Precession of the equinoxes The Sun and Moon exert a torque on Earth because of its equatorial bulge, which causes the direction of the rotational axis to gradually **precess**, or rotate about the normal to the plane of the ecliptic, with a period of about 26 000 years. At the same time, Earth's elliptical orbit itself precesses due to effects from other planets. The combination of these effects is a regular change in the relationship between seasons on Earth and the times when Earth is closest to (perihelion) and furthest from (aphelion) the Sun, with a period of around 22 000 years. The precession of the equinoxes around the orbit is parameterized by the angular variable $\tilde{\omega}$, where $\tilde{\omega} = 0$ when the (Northern) vernal equinox coincides with perihelion. As discussed in §23.2, currently perihelion occurs during Northern winter, around January 3, corresponding to $\tilde{\omega} \approx -\pi/2$. (Note that a positive value of $\hat{\omega}$ is associated with closer proximity to the Sun on the Northern summer solstice.) The variation in distance between the Sun and Earth on the Northern (summer) solstice is then given by the precession index $e \sin \tilde{\omega}$, which currently has a value of roughly -0.016.

The three effects just described – tilt, eccentricity, and precession of the equinoxes – do not significantly change total insolation on Earth averaged over the yearly cycle (at least at the top of the atmosphere; the latitudinal differences in albedo affect the rate of absorption, however). These effects do, however, have significant impact on the seasonal and latitudinal distribution of insolation. When the tilt ϵ is near its maximum value, latitudes further from the equator receive significantly more summer insolation. And when the precession index $e \sin \tilde{\omega}$ is at a maximum, so that Northern summer coincides with perihelion while eccentricity is at a maximum, the Northern Hemisphere also receives more summer insolation.

Permanent ice sheets, or **glaciers**, are formed on land when the rate of snow/ice accumulation over a year in a given region is faster than the rate of melting, or **ablation**. While the rate of accumulation in any region is rarely more than 0.5 m/year, independent of insolation, the rate of ablation rises sharply as summer temperatures rise above $0\,°C$. Thus, as Milankovitch realized, summer insolation at high northern latitudes is a crucial factor linking orbital forcing to ice sheet formation. As the orbital parameters change, total summer insolation at any given latitude varies, leading to a predictable variation of the latitude at which ice sheet formation can occur and a corresponding motion of the ice sheet boundary. Since temperature decreases with altitude, ice sheet formation can occur at lower latitudes at higher altitude.

At high latitudes, the greatest effect on summer insolation comes from changes in axis tilt. Using the methods discussed in §23.3, it is possible to compute, for example (see Problem 35.1), that at 65° latitude, the change in total summer insolation between a tilt of 22.05° and 24.5° is about 6%.

The 41 000 year cycle in the tilt parameter leads to a change in summer insolation in both hemispheres. The Northern Hemisphere has a greater preponderance of land masses at relevant latitudes, and has experienced major ice sheet fluctuations over the last several million years. The strongest evidence for regular ice sheet fluctuations with a 41 000 year period comes from sea-core data on ^{18}O levels between 2.75 Ma and 1 Ma. The data indicate that during this time, regular glaciations swept southward across the Northern Hemisphere and retreated with a periodicity of about 40 000 years.

Production and ablation of substantial ice sheets takes many thousands of years. For example, if net ice accumulation occurs at a rate of 0.3 m/year, it will take on the order of 10 000 years to build up an ice sheet of 3000 m thickness. Similarly, a simple calculation (see Problem 35.5) shows that a comparable time is needed to ablate such an ice sheet given a modest excess of summer insolation. Thus, Northern Hemisphere ice sheets are expected to lag insolation by roughly a quarter of the 41 000 year period for tilt oscillations. Though astronomical calculations can precisely describe Earth's axis tilt as a function of time

over the last several million years, fixing equally precise dates for sea-core samples is more difficult. Nonetheless, this lag is roughly reproduced by the data.

The orbital forcing due to precession of the equinoxes does not have as strong a systematic effect on Earth's climate as tilt. Increased tilt produces greater summer insolation in both hemispheres. In contrast, the effect of precession is opposite in the two hemispheres: when perihelion coincides with the Northern summer solstice, summer insolation in the Northern Hemisphere is greater but summer insolation in the Southern Hemisphere is maximally reduced. Thus, while perihelion precession also correlates with changes in Northern Hemisphere glaciation, the net effect of precession tends to cancel and does not contribute as strongly to net ice buildup or ablation. Nonetheless, a distinct signal with period 22 000 years can be seen in the ^{18}O records of the time period 2.75 Ma–1 Ma, superimposed over the 41 000 year cycle; this signal is attributed to precessional orbital forcing.

In fact, the fluctuations of ^{18}O levels over the last several million years are so well correlated with orbital forcing (Figure 35.8) that the best method for dating sea core data currently is to match it to astronomical computation of orbital parameters. The combination of the effects of regular tilt fluctuation and precession modulated in strength by changes in eccentricity provides distinctive variations that enables identification of each distinct tilt cycle in the sea core data.

A graph of sea-core ^{18}O levels matched to variations in orbital forcing over the last 5 million years is shown in Figure 35.9. This figure shows a gradual cooling over the last 3 million years, during the first part of which ice sheets and global temperature exhibit periodic oscillations that have been matched to the 41 000 year oscillations of Earth's axial tilt. Close to 0.9 Ma, however, this regular behavior changed and climate response became more complex. Over the last half-million years, glacial periods have been substantially longer, punctuated only every 100 000

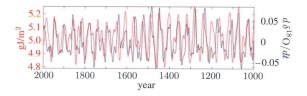

Figure 35.8 Levels of ^{18}O measured in deep-sea cores (black) compared to summer insolation at high latitudes (red) in the early Pleistocene (2 Ma to 1 Ma). ^{18}O data is dated independently of orbital fluctuations. From [266].

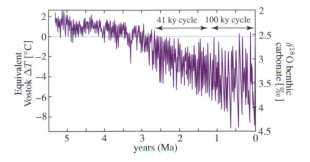

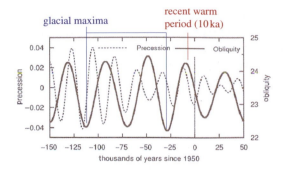

Figure 35.9 Levels of ^{18}O measured in deep-sea cores show the combined effects of changes in temperature and ice sheets. In the figure, ^{18}O fluctuations have been dated by matching to a model of orbital forcing based on astrophysical computation of orbital parameters. Between 2.75 Ma and 1 Ma, the 41 000 year oscillation period of Earth's axial tilt dominates. In the last million years, reaction to orbital forcing is more complicated and exhibits a 100 000 year approximate periodicity. After [173].

Figure 35.10 Orbital parameters obliquity (tilt) and precessional index $e \sin \tilde{\omega}$ in the past and to 50 ka in the future. The last two glacial maxima coincide with simultaneous minima in 41 000 year tilt cycle and 22 000 year precessional cycle; the recent long warm interglacial period coincides with simultaneous maxima in both cycles. After [173].

years or so by warmer interglacial periods. The precise mechanism for the appearance of this approximate 100 000 year cycle (and whether the last four interglacial periods really suggest a regular cycle) is a subject of current debate among climate scientists. Nonetheless, even over this recent period of time there is still a clear signal of the basic mechanisms of orbital forcing. A graph of obliquity ϵ and the procession index $e \sin \tilde{\omega}$ computed from 150 000 years before the present to 50 000 years in the future appears in Figure 35.10. From the figure, it is clear that both glacial maxima and the recent interglacial are correlated with reduced and increased Northern Hemisphere insolation respectively.

Thus, while the precise magnitude and timing of changes in climate over the last half-million years are not completely understood, it seems that a primary cause of these changes is the variation in seasonal and latitudinal insolation distribution associated with fluctuations in Earth's axis tilt and orbit.

Paleoclimate and the Carbon Cycle

We return to a puzzle mentioned in §35.1.3: Why do changes in the atmospheric greenhouse gases CO_2 and CH_4 lag behind corresponding changes in temperature and ice sheet extent, and what mechanism effects these changes? The answers to these questions are not completely understood. It is, however, well established that during glacial maxima both atmospheric CO_2 levels and carbon in terrestrial biomass were significantly reduced. Atmospheric CO_2 levels were lower by roughly 100 ppm than during interglacials. Estimates of the decrease in the carbon reservoir in surface

vegetation at times of glacial maxima are on the order of 25% of the carbon content during interglacials. This means that during periods of glaciation tremendous amounts of carbon were transferred out of the atmosphere and terrestrial biomass. The only plausible place for this carbon to have been stored is in the deep ocean; thus, ocean dynamics, which may include biological and chemical effects in addition to circulation, must have played a role in this carbon transfer process. Using sea-core analysis of the carbon isotope ^{13}C, which is less easily incorporated by organic systems than the prevalent isotope ^{12}C, climate scientists have indeed confirmed that during periods of extensive glaciation, substantial amounts of carbon (on the order of 500 Gt) have moved into the deep ocean. Mechanisms that may have effected this transfer include **ocean carbon pumping** by high rates of photosynthesis in surface waters [267], and changes in deep-water circulation patterns. CO_2 is also more soluble in colder water, but this effect in isolation could only have contributed to a relatively small fraction of the change in oceanic carbon content during glacial periods. Regarding the relative timing of temperature and CO_2 changes, one current line of reasoning is that temperature changes in middle and high latitudes drove the carbon dioxide change, which in turn drove the temperature change in the tropics, where otherwise the orbital forcing has the opposite sign from the poles, since as discussed above, the net insolation on Earth is not affected as significantly by orbital effects as the latitudinal distribution.

Regardless of precisely how climatic shift gives rise to increased atmospheric CO_2 levels during interglacials, the associated increase in radiative forcing leads to a substantial positive feedback. Estimating atmospheric CO_2 levels to be 180 ppm during glacial maxima and 280 ppm during (pre-industrial) interglacial periods, the resulting radiative forcing is roughly

$$F \cong F_{2\times} \log_2(280/180) \cong 2.4 \, \text{W/m}^2 \, . \qquad (35.4)$$

Using a climate feedback sensitivity estimate of $\sigma_c \approx 0.9$ (§34.5.3) suggests that this change in CO_2 levels may have been responsible for approximately $\Delta T \cong 2\,°C$ of the roughly $5\,°C$ of temperature difference between glacial maxima and interglacial periods. This would mean that the increase in atmospheric CO_2 during the warming following glacial periods contributed an additional 50% or more to the temperature change that would have occurred in the absence of atmospheric changes. As mentioned above, the change in CO_2 may have actually been the primary driver of temperature changes in the tropics. Because these processes are poorly understood, however, it is hard to disentangle the precise interplay among the sources of climatic change during substantial shifts of this kind. It has been suggested that the fairly rapid shift between glacial maxima and interglacial warm periods represents a shift between two relatively stable local equilibria (see §35.3.7) in the climate system, catalyzed by relatively minor changes in seasonal and latitudinal insolation from shifts in orbital parameters.

To summarize, it is clear that orbital forcing plays an important role in determining climate change over periods of hundreds of thousands of years. Changes in seasonal and latitudinal insolation due to orbital forcing are believed to be the primary influence driving the pattern of glacial and interglacial periods over the last several million years. Orbital forcing, however, does not lead to a significant change in net radiative forcing. Rather, it alters the temporal and geographic distribution of radiation on Earth's surface. Orbital forcing alone is not enough to explain the large changes in temperature and ice extent experienced over the last several million years. Feedback mechanisms within the climate system, including ice albedo feedback and a poorly understood CO_2 feedback, appear to have amplified the changes caused by orbital forcing and produced substantial climatic change with temperature changes on the order of $\Delta T \cong 5\,°C$, as has been experienced in the last 20 000 years since the last glacial maximum.

35.1.4 Tectonic Scale Climate Change

To put our understanding of Earth's climate over the historical and orbital time scales into context, it is helpful to expand our focus one step further and to consider the history of Earth over hundreds of millions of years. In this section we give a brief summary of what is known of the history of Earth's climate over this much longer time scale. We emphasize in particular the last 100 million years, during the last half of which Earth has experienced a gradual cooling since the greenhouse climates of the late Mesozoic era and the Eocene epoch.

Planet Earth formed roughly 4.5 billion years ago (4.5 Ga) as dust and gas containing debris of ancient stars clumped together to form the Solar System. The billions of years of Earth's early history are often referred to as **deep time**, emphasizing the difficulty of comprehending such a time scale from the human perspective. While the geological record provides only limited information about climate conditions on Earth over the first few billion years of the planet's existence, measurements of oxygen isotope levels of ancient carbonates, as well as other clues, suggest that early Earth was quite warm. Some estimates suggest surface temperatures averaging over $50\,°C$. While at that time incoming solar radiation provided almost 30% less energy than current insolation levels, it is believed that high levels of methane and/or CO_2 (20–1000 times current CO_2 levels) maintained the high surface temperature. Life is believed to have first evolved in liquid water around 3.5 Ga, with multi-cellular organisms evolving around 1 Ga. Photosynthesis began around 3 Ga, and began slowly to replace atmospheric CO_2 with oxygen. The **snowball earth hypothesis** suggests that between 1 Ga and 0.5 Ga Earth experienced one or more periods of extreme glaciation, with most or all of Earth's surface covered with ice. Such episodes, if they occurred, would have been terminated by a gradual increase in atmospheric CO_2 from volcanic activity, increasing the greenhouse effect and warming the planet with the assistance of ice albedo feedback. Around 0.5 Ga, the oxygen content of Earth's atmosphere reached sufficiently high levels that an ozone layer formed, blocking ultraviolet radiation and enabling life to colonize land.

It is believed that a primary driving force behind climate change over time scales of hundreds of millions of years is the fluctuation of atmospheric CO_2 levels driven by effects associated with tectonic motion. Two important effects of this type are **plate spreading and volcanism** and **uplift and weathering**. These two processes cycle CO_2 between Earth's crust and atmosphere over tens and hundreds of millions of years. Uplift occurs when continental plates collide and buckle. This produces mountain ranges with steep slopes and broken terrain. The exposed rocks are largely composed of silicate minerals such as *quartz* (SiO_2) and **feldspar** ((K, Na, Ca)$Al \, Si_3 O_8$). These minerals react with groundwater containing carbonic acid (H_2CO_3) formed when atmospheric CO_2 dissolves in rainwater. The affected rock weathers into insoluble **clay minerals** (consisting largely of Si, Al, H, and O) that form soils. This process also produces carbon-containing ions such as HCO_3^- and CO_3^{--} that are carried by runoff to the ocean. There, the carbon is incorporated as calcium carbonate ($CaCO_3$) into the shells of marine organisms, ranging from microscopic plankton to molluscs. Some of this carbon is eventually buried at the ocean bottom, where it may stay buried as ocean sediment for millions of years. This process of **chemical weathering** gradually removes CO_2 from the atmosphere and transfers the carbon into the oceanic crust. The rate of weathering and associated carbon transfer has been found to decrease roughly exponentially with the age of the mountain range. Young mountains expose more crust and form steeper slopes that increase precipitation patterns and erode more rapidly. It has been suggested that weathering provides a natural thermostat for Earth's climate on the million-year time scale: if Earth's temperature increases, then precipitation, wind, and weathering will also increase, lowering atmospheric CO_2 levels and producing radiative forcing that would act to lower the temperature. This process occurs over a very long time scale, however; it would likely take millions of years for this process alone to remove the excess CO_2 that humans have added to the atmosphere in the last century.

The carbon transferred into oceanic crust from uplift and weathering is returned to the atmosphere when oceanic plates impact other plates along convergent margins and are subducted downward. The carbon in the crust is released back into the atmosphere through the volcanic activity associated with melting of the subducted crust. CO_2 is also released into the atmosphere along divergent plate boundaries where new sea floor is produced as carbon-carrying magma rises to the surface. The **spreading rate hypothesis**,[6] suggests that through this mechanism the rate of sea floor spreading has played an important role in controlling climate over the last several hundred million years. This hypothesis is difficult to test, however, since a substantial fraction of ocean crust is subducted and destroyed within less than 100 million years of creation.

As Earth's tectonic plates have moved over the surface of the planet, the configuration of the continents has been in constant flux. The past positions of plates can be estimated by **paleomagnetism** – that is, by measuring the

---

6 Also known as the *BLAG* hypothesis after its originators, American geologists R. A. Brenner, A. C. Lasaga, and R. M. Garrels.

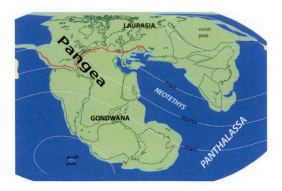

Figure 35.11 Over 200 million years ago, the land masses of Gondwana and Laurasia collided, producing the supercontinent Pangaea. It is believed that the polar position of Gondwana and reduction in CO_2 from extensive weathering on mountains formed by the uplift from the collision contributed to an extended *icehouse* period with very low global temperatures. Later, the two supercontinents separated once again along the fracture line denoted in red. (Credit: Modified from 'The formation of Pangea', G.M. Stampfli, C. Hochard, C. Vérard, C. Wilhem and J. vonRaumer, *Tectonophysics* 593, 1–19, (2013))

effect of Earth's magnetic field on ancient rock at the time of formation from a molten state – and by the mapping of geological formations across present continental boundaries. Between 325 Ma and 240 Ma, a massive supercontinent called **Gondwana**, containing the land masses now forming South America, Africa, Arabia, India, Antarctica, and Australia, moved across the South pole and collided with the land mass of **Laurasia** that now forms North America, Europe, and Asia (Figure 35.11). The resulting supercontinent **Pangaea** contained almost all of the continental land mass on the planet. During the period from 325 Ma to 240 Ma, there was an extended **icehouse** period with extensive glaciation on Gondwana. This glaciation period is generally attributed to a combination of the polar position of Gondwana and extensive weathering associated with uplift from the collisions forming Pangaea resulting in lower atmospheric CO_2 levels.

Around 180 Ma, Pangaea began to break up and many continents began to move towards their present positions, though Antarctica and Australia remained connected. There is strong evidence that 100 million years ago, in the middle Cretaceous period, global temperatures were substantially warmer than today. Average temperatures at the North and South poles are estimated to have been 20 °C and 40 °C warmer than current values. There were no ice caps or ice sheets, and sea levels were perhaps 100 m above current sea level. There is fossil evidence of dinosaurs and plants adapted to warm climates living at polar latitudes. There is some uncertainty about temperatures at mid-latitudes. The geological record suggests equatorial temperatures only a few degrees warmer than current values. Global circulation models given boundary conditions appropriate for this time period produce temperature profiles with a larger difference between low and high latitudes than is suggested by the geological record. This may indicate that temperatures at low latitudes were higher than some scientists have previously believed; but this may also indicate that global circulation models are missing some key aspects of ocean dynamics underlying the mechanism for poleward heat transport. Life was abundant in the mid-Cretaceous; over half of all current oil reserves come from this period. CO_2 levels from this time are estimated at having been from 3 to 12 times current levels.

Around 65 Ma, a cataclysmic event occurred that led to the extinction of 70% of all species living at that time. It is widely believed that at this time a large asteroid impacted Earth, probably at the site of a large crater in the area of Mexico's Yucatán Peninsula. This event produced dramatic short-term effects, likely including widespread fires that burned most terrestrial vegetation, and dust and particulate matter that blocked sunlight for a number of years. While the effects of this impact contributed to the extinction of the dinosaurs, and emptied many ecological niches for population by new species, it does not seem to have had a long-term effect on climate. Climate changes since the beginning of the Cambrian era are summarized in Figure 35.12, where *icehouse* and exceptionally warm eras are highlighted.

Past Climates

Past climates are studied using a variety of *proxy* measures of temperature and atmospheric conditions. Over the last 50 million years Earth's climate has gradually cooled as CO_2 has been removed from the atmosphere by *chemical weathering* and reduced rates of volcanism and sea floor spreading. In the last 3 million years, the climate has oscillated between long ice ages and shorter interglacial periods, driven primarily by changes in Earth's orbital parameters. Over the last thousand years, a mild cooling (the *Little Ice Age*) was caused in part by volcanic activity and a decrease in solar activity. In the last century, solar activity has returned to higher levels and atmospheric CO_2 has increased dramatically, driving rapid warming.

A graph of estimated global surface temperature averages over the last 65 million years, based on benthic ^{18}O and other geological records, is shown in Figure 35.13. A particularly warm period occurred around 55 Ma, at the beginning of the Eocene epoch. This is often referred to as the **Early Eocene climatic optimum**. CO_2 levels at that time are estimated to have been at least five times current values. Temperatures are estimated to have been

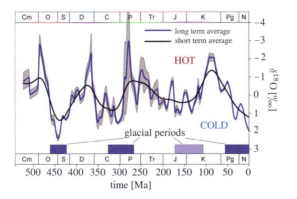

Figure 35.12 Oxygen isotope ratios in fossils, a proxy for Earth average surface temperature, since the beginning of the phanerozoic era ≈ 550 Ma. Periods of extensive glaciation including the present are highlighted. (Credit: Robert A. Rohde, Global Warming Art Project, reproduced under CA-BY-SA 3.0 license via Wikimedia Commons)

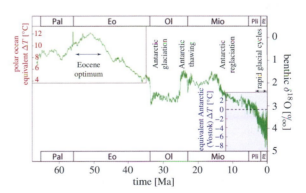

Figure 35.13 Earth average surface temperature since 65 million years ago (65 Ma), estimated by proxy data. Following the Early Eocene climatic optimum at 55 Ma, temperatures and CO_2 levels have decreased steadily, with periodic upward fluctuations. Note that the zero of $\delta^{18}O$ differs between this figure and Figure 35.12. (Credit: Robert A. Rohde, Global Warming Art Project, reproduced under CC-BY-SA 3.0 license via Wikimedia Commons)

higher than those in the mid-Cretaceous by perhaps $5\,^\circ C$ at low latitudes and up to $10\,^\circ C$ at higher latitudes. Several other extremely warm periods occurred during the Eocene. The Eocene was the time when mammals began to proliferate and occupy many ecological niches. As in the Cretaceous period, during the Eocene the temperature gradient between low and high latitudes was small, and polar regions were quite warm.

Since the early Eocene, the planet has gradually cooled. Deep sea temperatures have dropped by an estimated $12\,^\circ C$. Around 35 Ma, following the separation of Antarctica and Australia, Antarctica developed permanent glaciation. Antarctic ice sheets experienced major growth around 13 Ma, and permanent ice sheets developed in Greenland between 7 Ma and 3 Ma. As described above, around 2.75 Ma substantial glaciation in the Northern Hemisphere began, with ice ages punctuated by periodic interglacial periods. While the precise timing and mechanisms responsible for the cooling that has occurred over the last 50 million years are not fully understood, several factors are believed to have played an important role in driving this gradual climatic change. The collision of India with Asia, producing the enormous Tibetan Plateau, is probably the largest occurrence of uplift from a collision of continental plates since the formation of Pangaea. This collision began about 50 million years ago. The corresponding chemical weathering process has led to a depletion of atmospheric CO_2 that may have exceeded the replacement rate from volcanism since the Eocene. The rate of sea floor spreading also decreased markedly from about 50 Ma. According to the *spreading rate hypothesis*, this would have led to a decreased rate of CO_2 addition to the atmosphere. The spreading rate has, however, begun to increase again over the last 15 million years, so this cannot be the sole driver of the temperature decrease since 50 Ma. It has also been suggested that changes in ocean circulation arising from the opening and closing of seaways has played a role in cooling the planet. For example, the separation of Australia and South America from Antarctica enabled the formation of the Antarctic circumpolar current, discussed in §27. This strong current may have deflected warmer gyre currents, reducing heat transport to Antarctica, and exacerbating the tendency of the southern polar continent to develop permanent glaciation. There is some disagreement among climate scientists as to the impact of changes in seaways on long-term climate; the current consensus is that such changes may have contributed, but have not played a dominant role in controlling climate change. In any case, CO_2 levels have fairly steadily decreased since the Eocene, due presumably to an excess of weathering-related removal of CO_2 from the atmosphere over the replacement

from volcanism and sea floor spreading. Combined with other tectonic-scale effects and ice albedo feedback, this has driven a gradual decrease in temperatures leading to the recent glacial/interglacial cycle. This transition from Eocene to recent (Pleistocene) conditions is often referred to as a **greenhouse–icehouse transition**.

35.2 Predicting Future Climate

Having described some aspects of past and present climates, we turn to the question of what the future may hold for Earth's climate.

On the very long time scale – of billions of years – the Sun will continue to increase in luminosity. Well before the Sun reaches the red giant stage, perhaps in one or two billion years, the temperature on Earth will increase sufficiently to boil off all water, and life in its current form on the planet will become unsustainable.

On the tectonic time scale of hundreds of millions of years, we do not understand Earth's systems well enough to make an accurate prediction. While the overall trend for the last 50 million years has been towards lower temperatures and lower CO_2 levels, we do not understand the precise way in which the relevant mechanisms are operating well enough to extrapolate sensibly into the future. It may be that the cooling trend will continue for several more tens of millions of years, or it may be that an increased rate of release of CO_2 from sea floor spreading may overcome the rate of atmospheric depletion from chemical weathering and lead to a long-term CO_2 increase and a return toward conditions more like those of the Eocene. A much better understanding of the precise details of tectonic-scale climate change and carbon cycling would be needed to make reliable predictions for the next 50 million years of Earth's history.

On the orbital scale of tens of thousands of years, we are now 10 000 years past the recent coincidence of maximal positive contributions to northern summer insolation from axial tilt and equinox precession. From the dominant pattern over the last million years, the default expectation should be that over the next several thousand to ten thousand years another ice age will ensue. Human influences aside, there is no known mechanism that would give reason to doubt this prediction.

On the shorter time scale of decades to hundreds of years, considering only external influences, the planet's climate may be influenced by an expected near-term decrease in solar activity. This may produce a small cooling effect. Given the scale of human impact on atmospheric greenhouse gases including CO_2 and methane, as well as other climate forcing mechanisms such as particulates, it seems

---

Box 35.3   Climate Puzzles and Challenges

While our understanding of Earth's climate has progressed rapidly in recent decades, there are still some significant unresolved questions that limit our understanding of Earth's past and future climate and the consequences of human activity on Earth's climate. Three puzzles that represent challenges for climate scientists in the coming decades are:

Ocean energy transport Transport of energy from equatorial to polar regions helps balance temperature extremes on the planet. Paleoclimate data suggests that 50 million years ago when the planet was much warmer, the temperature difference between poles and the equator was relatively small. Current climate models do not correctly reproduce this difference, suggesting that current models do not accurately describe ocean energy transport. Even the meridional ocean energy transport in Earth's current configuration is not well understood either observationally or theoretically. It is important to understand this transport in order properly to model changes in temperature distribution in global warming scenarios.

Arctic sea ice loss Over the 15 years from 2000–2015, Arctic sea ice cover diminished substantially. 2012 saw the lowest level of summer Arctic sea ice on record (through 2016): $3.6\,\mathrm{M\,km^2}$ compared to an average minimum of $7\,\mathrm{M\,km^2}$ over the 20 years from 1979–2000. While some part of the ice melt is understood in terms of effects of temperature change, black soot, and changes in wind and currents, climate models have consistently underestimated the rate of Arctic ice loss, and the precise mechanisms driving the melting are not well understood quantitatively. Since the Arctic is the region in which the greatest anthropogenic climate change is expected, it is important to develop a better understanding of the processes involved in the melting of Arctic sea ice.

Ocean and land biomass CO_2 uptake At the present time, Earth's oceans are absorbing roughly one quarter of excess human CO_2 emissions, slowing the rate of growth of atmospheric CO_2. Paleoclimate data suggest, however, that in earlier interglacial warming periods, an increase in radiative forcing from changes in orbital parameters and associated warming of the planet led to a net release of CO_2 from the oceans. The mechanisms that maintain the ocean/atmosphere CO_2 balance are not well understood, but they are key to understanding the longevity of CO_2 in Earth's atmosphere and the extent and duration of climate change caused by anthropogenic CO_2. Terrestrial biosystems currently absorb approximately another quarter of anthropogenic CO_2 emissions. Some studies suggest that both ocean and land uptake will slow or even reverse direction as CO_2 levels increase further; in particular, the fractional rate of ocean uptake of excess CO_2 above pre-industrial levels will likely decrease as CO_2 saturation and temperatures increase [268, 269], and may already be slowing [270], but there is as yet no scientific consensus on the precise future trajectory of either the ocean or land carbon sinks over the coming decades or centuries.

---

that the future of Earth's climate in the coming centuries depends primarily upon human influences. The time scale for climate change driven by a change in radiative forcing is relatively long compared to the rate at which atmospheric CO_2 levels have recently changed. Thus the impact of anthropogenic greenhouse gases already emitted still lies largely in the future, and the effects of greenhouse gases over the coming centuries will depend strongly on human activities in the near-term future. To make any kind of prediction, it is necessary to make some assumptions regarding future human activity – for example, by estimating future levels of greenhouse gas emissions.

Any estimation of future human activity related to climate effects involves assumptions about social behavior, economics, and politics, which are outside the scope of this book. While the per capita rate of energy use from fossil fuel sources is clearly an important component in

determining total human greenhouse gas emissions, as discussed in §1 (Introduction), the rate of population increase is also an important factor. All these ingredients are subject to substantial uncertainty and are difficult to estimate. Nevertheless, it is worth attempting to estimate future climate change based on a range of possible scenarios in order to estimate how human behavior over the next few decades will affect Earth's climate for several centuries into the future. Since methane and aerosols are removed from the atmosphere on a much shorter time scale than CO_2, anthropogenic CO_2 is a primary driver of century- to millennium-scale climate effects, and provides a single simple parameter on which one can base a discussion of some future effects on the time scale of the next century or two.

In the remainder of this section we focus on the time scales over which CO_2 emissions, atmospheric CO_2, and global temperature change may reach their respective peaks, and put this in the context of scenarios for future climate considered by the IPCC [245]. We must emphasize the heuristic nature of this discussion given the many uncertainties involved; the goal of this analysis is to give a sense of the relevant time scales involved, not to make specific predictions.

CO_2 emissions and atmospheric concentration From Figure 34.12, it is clear that anthropogenic CO_2 emissions have increased rapidly in the first decade of the twenty-first century. Whether government regulation, economic imperatives, and other human choices or actions are leading to a leveling out or will result in a decrease in emissions in the coming decades is impossible to know now. In the most optimistic scenario, emissions might peak in the coming decade and drop precipitously well before the end of the century. In a pessimistic scenario, emissions could rise for on the order of one hundred years, and tail off slowly as all available fossil fuel resources are used. Obviously, the difference in resulting radiative forcing between these scenarios is enormous. The IPCC in 2013 considered four overall scenarios for radiative forcing [245]. These were based not on specific socioeconomic scenarios but rather on internally consistent scenarios of radiative forcing and atmospheric concentrations of greenhouse gases and aerosols. These **Representative Concentration Pathway** scenarios, RCP 2.6, 4.5, 6.0, and 8.5, are labeled by the peak or stabilization level of radiative forcing reached in the twenty-first century. The RF levels in these models are graphed in Figure 35.14, and the CO_2 emissions in each model are shown in Figure 35.15.

In addition to the uncertainty in human activity and the rate of greenhouse gas emissions, the poorly understood nature of the land and ocean carbon sinks further complicates predictions. As mentioned in §34.4, part of the carbon uptake from terrestrial biomass is associated with reforestation of previously cleared regions in the Northern Hemisphere, and part is from a general increase in plant growth due to increased accessibility of CO_2. It is unclear for how long either of these processes will continue to remove CO_2 from the atmosphere at current rates. Similarly, our understanding of oceanic absorption of CO_2 is insufficient to make reliable predictions for the future. We emphasize that as the climate has been warming over the last 50 years oceanic uptake of CO_2 has increased, whereas the geological data on the last several hundred thousand years shows that warming in the past appears to have led to a *net increase* of atmospheric CO_2 associated with decreased oceanic levels of carbon. Since neither process is yet well understood, predictions of future oceanic CO_2 uptake are subject to substantial uncertainty. As a result, even were the rate of CO_2 and other greenhouse gas emissions in the future known with complete certainty, there would still be considerable uncertainty as to the effects on radiative forcing over the next several centuries. The IPCC5 RCP scenarios focus on the future trajectory of radiative forcing, and do not make specific assumptions regarding anthropogenic emissions and the future behavior of ocean and land sinks.

Time scale of atmospheric CO_2 Even after net CO_2 emissions peak, atmospheric *levels* will continue to rise for quite some time. Atmospheric CO_2 concentrations will only begin to drop when the rate at which CO_2 is removed by natural systems such as oceanic absorption *exceeds* the anthropogenic excess in CO_2 emissions. For example, since currently (circa 2015) roughly one half of the excess emissions are removed by natural systems, emissions would need to drop by a factor of two for atmospheric levels to begin to decrease. Given the uncertainties mentioned above, the most optimistic assumption is that uptake by both terrestrial biosystems and the oceans will continue to remove a similar fraction of the CO_2 excess above the pre-industrial equilibrium concentration. A projection of future atmospheric CO_2 based on this assumption is probably an underestimate, since recent studies suggest that both land and ocean sinks will slow or even reverse with increased CO_2 levels [268, 269]. Slowing of terrestrial biomass or ocean uptake would lead to a longer persistence of atmospheric CO_2 levels and higher levels in the future.

The time scale for excess CO_2 to be removed from the atmosphere is thus not well understood. Estimates range from several decades to on the order of centuries or possibly thousands of years. The terrestrial biomass carbon sink

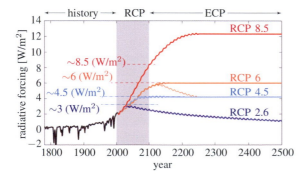

Figure 35.14 Total radiative forcing for Representative Concentration Pathway (RCP) scenarios used in IPCC5 report [245].

has a finite capacity that may be saturated well before the pulse of excess CO_2 from human fossil fuel usage has been removed from the atmosphere. And even as CO_2 levels begin to drop again biological systems such as terrestrial plants will presumably begin to release net carbon to the atmosphere again to restore equilibrium. While chemical weathering would presumably eventually restore atmospheric CO_2 levels to their current natural equilibrium even in the absence of other mechanisms, this process occurs over very long time scales.

In the short term the primary mechanism for removing most anthropogenic CO_2 from the atmosphere is assumed to be ocean absorption, leading to deep ocean mixing. Assuming that the mixed layer is eventually saturated at a higher CO_2 level in equilibrium with the atmosphere, the rate for further CO_2 removal from the atmosphere depends in part upon the rate of deep water formation in areas such as the North Atlantic. It is possible that this process may slow as, for example, melt water from the Greenland ice pack is released into the North Atlantic, reducing the salinity of that water. Notwithstanding these uncertainties, a rough estimate based on the (optimistic) assumption that current rates of ocean and land biomass absorption of excess CO_2 will continue, suggests that once emissions have ceased, the atmospheric lifetime of excess atmospheric CO_2 will be on the order of 50 years or more. Thus, even in an optimistic scenario such as RCP 2.6, the effects of human fossil fuel emissions on atmospheric CO_2 levels will almost certainly last at least into the next century. In a less optimistic scenario, if the ocean and land sink absorption does not continue indefinitely, the lifetime of CO_2 in the atmosphere may be closer to the deep ocean mixing time of on the order of one thousand years.

Time scale for temperature change Different parts of the climate system have very different time scales for reacting to externally forced climate change. As discussed above in the context of orbital-scale climate change, large ice sheets react to changes in temperature or radiative forcing on a time scale of hundreds to thousands of years. Thus, for example, it is expected that the large Antarctic ice sheets will remain largely intact over the several hundred or thousand years during which the anthropogenic pulse of added atmospheric CO_2 impacts the climate. The climate system, therefore, will not have time to come into a new global equilibrium. Instead, relatively faster-reacting parts of the system will shift dramatically in response to the rapid changes in radiative forcing, while other parts of the climate system will be less affected. Since there is no clear geological record of such rapid changes in external climate forcing at any point in the past, paleoclimate

Time Scales of Anthropogenic Climate Effects

Simple considerations suggest that atmospheric CO_2 levels will continue to rise for many decades after anthropogenic CO_2 emissions peak, and that global mean temperatures will continue to rise over even longer time scales. This, together with the logarithmic response of radiative forcing to CO_2 levels, suggest that past and near-term future emissions will lock in climate changes that will persist over at least the next century.

records provide only a partial analogy. Thus, climate modeling, while imperfect, may be our best guide for what to expect in the next several hundred years for a given anthropogenic climate forcing. Since climate models are generally used to simulate geologically short periods of time evolution, slower-reacting components such as the Antarctic ice sheets are often incorporated as fixed boundary conditions. As discussed in §34.5, climate models suggest that the change in global surface temperature will be related to the change in radiative forcing through the climate sensitivity parameter $\sigma_c \cong 0.9 \, °C/(W/m^2)$, so that a doubling of CO_2 would lead in steady-state to an increase in temperature of something like $\Delta T_{2x} \cong \sigma_c F_{2\times} \cong 3.2 \pm 0.7 \, °C$. This change will lag by a time on the order of many decades following a given increase in radiative forcing. For example, a simple computation (see Problem 35.6) shows that it would take several decades for a radiative forcing of $3.7 \, W/m^2$ to raise the temperature of the oceanic mixed layer by $3.2 \, °C$, even if all extra energy input from radiative forcing contributed to this warming.

Putting all these pieces together, the picture we arrive at is that for a given profile of anthropogenic CO_2 emissions, atmospheric CO_2 levels will continue to rise for some time after the emissions peak, and temperatures will in turn continue to rise for some time after the atmospheric CO_2 level itself peaks. Thus, there will be a time lag of perhaps 50–100 years or more between the time when human emissions begin to decrease and the time of maximum deviation in global surface temperature. A very simple model of these effects and time scales is developed in Example 35.1. The same effects can also be seen clearly in the more detailed projections used in the IPCC5 report, shown in Figure 35.15.

Note that the logarithmic dependence of radiative forcing and temperature change on CO_2 levels (as in

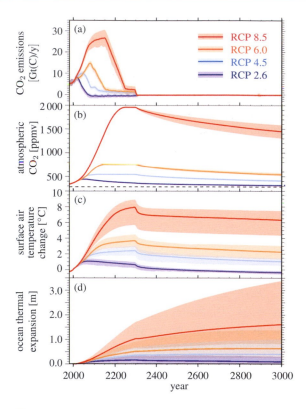

Figure 35.15 Scenarios of (a) CO_2 emissions, (b) atmospheric CO_2, (c) projected surface temperature change, and (d) ocean thermal expansion for Representative Concentration Pathway (RCP) scenarios used in IPCC5 report. Adapted from [245].

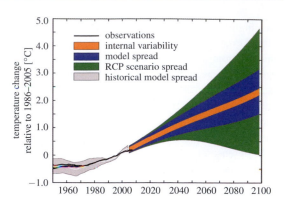

Figure 35.16 IPCC estimate of uncertainties in temperature projections through the end of this century [245].

eq. (34.35)) means that the effect of increasing CO_2 levels on global temperature is greatest for the initial deviation from pre-industrial levels. While doubling atmospheric CO_2 content by adding 280 ppm above pre-industrial levels may lead to 3.2 °C of temperature change in a new steady-state configuration, to generate a mean 6.4 °C temperature increase would require going to four times pre-industrial levels, or 1120 ppm. This sublinear growth can be seen clearly in the different RCP scenarios shown in Figure 35.15. This logarithmic dependence of climate on greenhouse gas content can be regarded either as a reassuring sign that the anthropogenic increase in CO_2 emissions is unlikely to lead to a catastrophic runaway climate situation, or as a call to immediate action because the emissions with the greatest relative climatic impact are the ones occurring now.

It is well established that global mean surface temperatures will rise as a consequence of past and future anthropogenic greenhouse gas emissions. Nevertheless, many uncertainties remain in current analyses of the expected

future consequences of these emissions. Climate models, which provide the best available estimates of climate sensitivity, are imperfect in a number of ways. The models are unable to fully incorporate important feedback such as from clouds with certainty, may be missing some important features such as increased ice melt from particulate emissions, and disagree in some ways (such as latitudinal temperature distribution) with paleoclimate data about some particular past climatic conditions. As a result, as emphasized in the reports of the IPCC, the assessment of the uncertainty in projections of future climate is at least as important as the specific predictions made. An illustration of the relative uncertainties in the RCP models and simulations is shown in Figure 35.16. The figure shows several sources of uncertainty in projected temperature for RCP models [245]: *internal variability* refers to uncertainty within individual models, arising from natural variability in the climate system, and represents a fundamental limit on the precision of predictions; *model spread* indicates differences among different simulations, and gives some indication of uncertainty due to imperfect modeling of natural systems; and *RCP scenario* spread refers to the variation in input parameters, including uncertainty about future human actions, and is responsible for the greatest uncertainty in temperature predictions.

It is also possible that abrupt changes in radiative forcing may generate climate change that cannot be anticipated by near-equilibrium modeling. At times in the past, such as the glacial–interglacial transition points, fairly small changes in external forcing seem to have switched the global climate from one class of local equilibria to another qualitatively different class of local equilibria. Such changes in Earth's climate cannot be ruled out over the next several hundred years in which anthropogenic

---

Example 35.1   A Simple Model of Time Scales for CO_2 Concentrations and Climate Response

A highly simplified model of the relationship between CO_2 emissions, atmospheric CO_2 levels, and global temperature change illustrates time lags between these processes and provides a qualitative understanding of some of the processes underlying the simulation results illustrated in Figure 35.15. CO_2 emissions are modeled as a logistic distribution (see Box 33.6) with peak fixed in 2050 and width and height chosen to fit emissions data from the last 50 years. CO_2 emissions are measured in units of 2010 emissions; 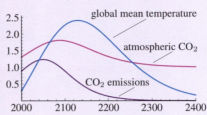 CO_2 levels are graphed as a multiple of pre-industrial levels (280 ppm); and temperature is plotted in °C above the 1961–1990 mean. The graphs in the figure are based on the assumption that land and ocean absorption remove 2% of excess atmospheric CO_2 yearly (roughly the current rate), and that radiative forcing is controlled by the temperature of the oceanic mixed layer, which in turn absorbs 50% of excess incoming radiation (as in Problem 35.6), assuming that the radiative forcing associated with doubling of CO_2 is $F_{2\times} \cong 3.7\ \mathrm{W/m^2}$, as in eq. (34.36).

This model is intended to give a heuristic sense of the relative time scales involved for peaking of CO_2 emissions, atmospheric concentration, and climate effects, and should not be taken as quantitatively accurate. The object is to formulate a simple calculable model that can illustrate the scale of the time lags involved and that anyone can easily reproduce. The reader is invited to improve the assumptions and to compute and analyze the results him/herself (Problem 35.7). Note that the assumption made for the land and ocean sinks is almost certainly overly optimistic, so the true lag of atmospheric concentration and climate response is likely longer than this model suggests.

---

forcing may play a dominant role in determining short-term climatic change.

35.3 Effects of Climate Change

We summarize some of the main consequences that may be associated with anthropogenic climate change over the next 100–200 years, keeping in mind the caveats in the final paragraphs of the preceding section.

35.3.1 Global Distribution of Temperature Change

While we have focused so far primarily on global average temperature differences, climatic change will not be uniformly distributed around the planet. Greater warming will occur over large continental land masses than over oceans. This effect arises only in small part from the different heat capacities of land and ocean surface waters, and has more to do with atmospheric and surface effects, such as clouds and soil moisture changes [245]. Paleoclimate data from warmer periods such as the Cretaceous and Eocene, as well as global climate models, indicate that as global temperature rises the temperature gradient between equatorial and polar regions decreases. Thus, high latitudes are expected to be more strongly affected by global warming than lower latitudes. Ice albedo feedback makes this effect particularly

pronounced in the Northern Hemisphere. While, as mentioned above, climate models and paleoclimate data are not in complete agreement regarding the extent of heat transport to high latitudes in warmer climates, combining these perspectives makes some predictions possible.

The predicted distribution of temperatures across the planet in the RPC scenarios is shown in Figure 35.17. In each of the four scenarios, warming is greater in the interior of large continental land masses than over the ocean, and the maximum warming occurs at high latitudes. The bulk of the Antarctic ice sheet remains intact, which along with the strong Antarctic circumpolar current keeps the southern high latitudes from warming as much as the northern high latitudes. In all scenarios the Arctic region experiences the greatest local warming, with temperature increases of more than twice those of equatorial regions.

One way to characterize the temperature increase due to global warming is to relate the resulting local temporary climate situation to earlier times with similar global surface temperatures. As described above and depicted in Figure 35.13, Earth's climate has been gradually cooling for the last 50 million years. As the temperature rises due to CO_2-induced radiative forcing, for parts of the planet the climate effect may be similar to moving backward in time toward the Eocene epoch, with a greater increase in temperature corresponding to an earlier time period in Earth's history. If the average global temperature

Annual mean surface air temperature change

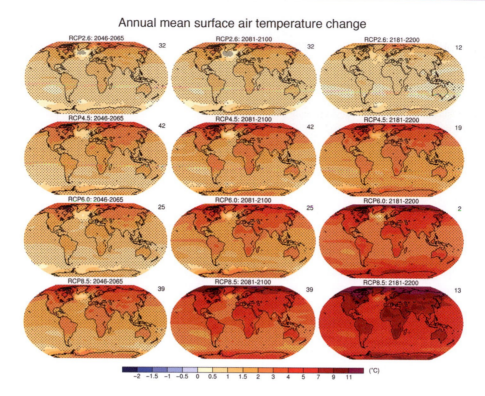

Figure 35.17 Distribution of local temperature changes in the four RCP scenarios [245].

were to rise by 2100 by on the order of 3 °C, the most recent corresponding time in Earth's history would be over 3 million years ago, in the mid-Pliocene. At that time, atmospheric CO_2 levels are believed to have been around 360–400 ppm, mean global surface temperatures were 2–3 °C above current values, and sea levels are believed to have been roughly 25 m higher than the present day. As in the predictive climate model results, temperatures at high latitudes were significantly higher, though tropical temperatures may have been only slightly warmer than current values. The analogy to earlier times is, however, imperfect, since many parts of the climate system – most notably the cryosphere – would not have time to reach a new equilibrium, so sea levels presumably would not rise nearly as high as in the Pliocene before atmospheric carbon levels would have dropped again. Of course, however, organisms would have no time to adapt through evolution to the changed climate so there would be very strong impacts on many ecosystems. Raising temperatures beyond 3 °C would run the clock further backward in time towards the Eocene, when global temperatures are believed to have been more than 7 °C above current values.

At mid-latitudes, the change in temperature will be most notable at the seasonal extremes. Winters will be warmer

and shorter, summers will be longer and hotter. A simple way of roughly characterizing the effect, stated clearly by Ruddiman [257], is that a doubling of CO_2 would likely lead to warming at mid-latitudes that might cause a shift in usual seasonal conditions by roughly one month. So with a doubling of atmospheric CO_2 levels by 2100, at mid-latitudes in the Northern Hemisphere, spring would come one month earlier, May would be like June of 100 years earlier, etc. Only in the middle month of summer would temperatures tend to reach unfamiliar regimes. Of course, however, the effects of warming will impact different local regions in very different ways, and will change the distribution of precipitation and extreme weather events, as we discuss in further detail below in §35.3.4.

35.3.2 Arctic Climate

The Arctic region is likely to be the area of the planet most affected by climate change. While global average temperatures did not increase significantly in the decade from 2000–2010, the extent of sea ice in the Arctic ocean diminished substantially during this period and has continued to decrease on average as the temperature has risen since 2010. The extent of sea ice reached the lowest level ever recorded in 2012 (see Figure 35.20(a)), going below

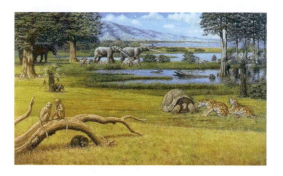

Figure 35.18 With a doubling of atmospheric CO_2 and a temperature rise of 2–3 °C by 2100, terrestrial conditions could roughly correspond to those of the mid-Pliocene over 3 million years ago, though the analogy is imperfect as ecosystems and many components of the climate will not have time to adapt to a new equilibrium. (Credit: Mauricio Anton)

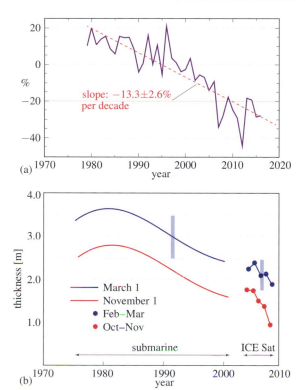

Figure 35.20 Trends in extent and thickness of Arctic sea ice. (a) Arctic sea ice extent in the month of September from 1978 through 2016. The red dashed line shows the trend. (b) Spring and autumn ice average thickness measured by submarine and satellite. Shaded bars indicate quality of data. (Credit: (a) NSIDC; (b) R. Kwok and D. A. Rothrock, Decline in Arctic sea ice thickness from submarine and ICEsat records: 1958–2008, *Geophysical Research Letters*, **36**, L15501 (2009))

Figure 35.19 A map of the extent of Arctic sea ice from September 2012 compared with the long-term average (red). (Credit: The National Snow and Ice Data Center (NSIDC), University of Colorado)

$4\,M\,km^2$ compared to an average minimum of $7\,M\,km^2$ over the 20 years from 1979–2000. The rate of summer Arctic ice melt is increased not only by radiative forcing from atmospheric CO_2, but also through ice albedo feedback, black soot, increased current and wind patterns, and weaker ice as the ice sheets diminish in size and smaller regions are covered with multi-year ice that has survived the summer thaw from previous years. Submarine and satellite measurements of ice thickness over the last 25 years have shown a steady decrease in ice thickness (see Figure 35.20(b)).

In addition to the positive ice albedo feedback, decreased Arctic sea ice exposes more ocean surface directly to the atmosphere. Increased melting and evaporation increases the water vapor content in the air, increasing the net (positive) water vapor plus lapse rate feedback. As the ice cap shrinks, heat is also released from the ocean to the atmosphere more rapidly, leading to more melting and warmer temperatures in the late summer and fall, and providing another positive feedback. In recent years the rate of Arctic ice melt has been significantly faster than had been predicted by most models and by the IPCC 2007 report. The models used in the 2013 report come closer to matching observed data but still underestimate the rate of Arctic ice melt. Roughly half of the RCP 4.5 models project that the Arctic will be essentially free of summer sea ice by 2100, and some recent estimates predict that this may occur substantially sooner, with predicted dates of ice extent below 1 million km^2 ranging from 2020 to 2050.

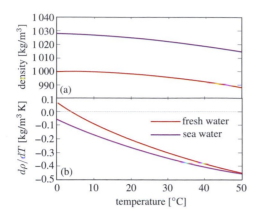

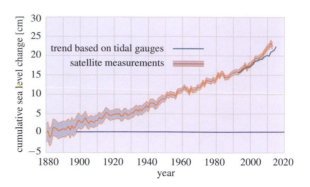

Figure 35.22 Global mean sea level rise since 1880. (Credit: US Environmental Protection Agency)

Figure 35.21 The (a) density (b) derivative of density with respect to temperature for fresh and seawater. Note that seawater is more dense than fresh water below 50 °C but that its density decreases more rapidly with temperature. Note also that the density of fresh water actually increases with temperature between 0 °C and 4 °C, which is why ice floats on fresh water.

35.3.3 Rising Seas and Melting Ice

Rising temperatures lead to rising sea levels for two main reasons: **thermal expansion** and *ice melt*. The density of salt water decreases with temperature, as does the density of fresh water above about 4 °C (see Figure 35.21). This behavior is typical of most fluids, since increased molecular activity increases the mean distance between molecules. Therefore the volume of seawater expands as it warms. (When salt water freezes, the salt content of the ice is much reduced; ice floats on both salt and fresh water, a fact that has profound consequences for climate and life on Earth.) Global sea levels rose by about 20 cm between 1900 and 2000 (see Figure 35.22). Over half of this sea level rise has been due to thermal expansion. A simple estimate (see Box 35.4) shows that warming a 200 m vertical column of water by 1 °C leads to a rise in sea level of roughly 4 cm, which suggests that thermal expansion is unlikely to lead to more than 10–20 cm of sea level rise in the next century. This conclusion is compatible with the RPC simulation results shown in Figure 35.15. The contribution of thermal expansion and other factors to sea level rise over the next century in the different RPC scenarios is broken down in Figure 35.23; in all these scenarios, thermal expansion contributes the largest component, between roughly one third and one half, of projected sea level rise.

The other primary source of rising sea levels is the melting of ice over land. Melting of sea ice, such as much of the Northern polar ice cap, does not affect sea levels since the ice is floating on the ocean surface, displacing almost

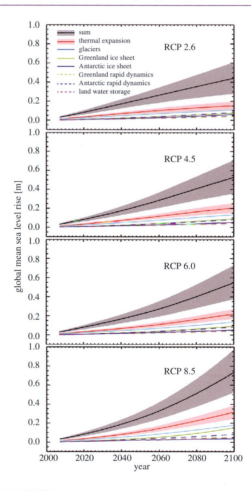

Figure 35.23 Projected contributions of different factors to sea level rise over the coming century in four RCP scenarios [245].

exactly the volume that it will fill after melting. Melting of ice over land, however, has led to substantial changes in sea level over geological time scales. As mentioned

---

Box 35.4   Sea Level Rise

We can get a rough estimate of the impact of thermal expansion on sea levels by considering the expansion of the oceanic mixed layer in the tropics. The coefficient of thermal expansion α_V, defined by

$$\frac{\Delta V}{V} = \alpha_V \Delta T,$$

determines the amount by which water expands with temperature, and a simple calculation relates α_V to the density ρ,

$$\alpha_V = -\frac{1}{\rho}\frac{d\rho}{dT}.$$

For seawater near 15 °C, $\alpha_V \cong 213 \times 10^{-6}/\mathrm{K}$ (see Figure 35.21). For simplicity we treat this coefficient as constant. If the temperature increases by 1 °C, then a mixed layer of depth 200 m will expand vertically by $200\,\mathrm{m} \times (213 \times 10^{-6}) \cong$ 4.25 cm.

If the oceanic mixed layer warms by 3.2 °C over a period of 30 years (as in Problem 35.6), then a total rise of ~14 cm is expected. Although the temperature increase in the twentieth century has been less than 1 °C, a rise in deep ocean temperatures as well as the surface mixed layer has increased the effects of thermal expansion beyond our simple estimate.

---

above, it is estimated that sea levels have risen about 120 m since the last glacial maximum 20 000 years ago. And in the much warmer climates of the Cretaceous and Eocene, when there were no continental ice sheets, sea levels are believed to have been 100–200 m above current levels. Currently the Antarctic ice sheets contain the greatest mass of ice on land, and there are substantial quantities of land ice over Greenland and in mountain glaciers at high altitudes around the world. As temperatures warm, each of these masses of ice will react to the changing climate. Much of the rise in sea levels over the last century that was not due to thermal expansion came from melting glaciers, not including the primary Greenland and Antarctica ice sheets (an **ice sheet** is generally defined as a glacial mass of ice over land having area above 50 000 km^2). In the RCP projections (Figure 35.23), glacial melting is expected to constitute the next-largest contribution to sea level rise over the next century, following thermal expansion. In these projections, 15% to 85% of the volume of glaciers outside Antarctica are expected to be lost by 2100.

The energy needed to melt continental ice sheets is tremendous, and to complete such a transformation takes many thousands of years. A simple back-of-the-envelope calculation gives a sense of the rate at which energy used to melt ice can lead to sea level rise. If we assume that 1% of the excess energy from a radiative forcing of 3.7 W/m^2 went to melt ice, then a short calculation (see Problem 35.8) shows that sea levels would rise roughly 5 cm/decade. It is useful to consider separately the various possible contributors to substantial sea level rise through ice melt and related changes in the cryosphere.

Antarctica Antarctica is covered by ice sheets several kilometers thick over an area of roughly 14 M km^2. The total volume of ice is roughly 30 M km^3, enough to raise global sea levels by well over 50 m. In East Antarctica, the ice sheets lie on ground above sea level. The West Antarctic ice sheet rests largely on the sea bottom, with edges that extend into large floating ice shelves. Most of the ice mass added in precipitation in Antarctica is offset by loss of ice mass to the ocean, with only a small fraction directly melting. As temperatures in the Southern Hemisphere rise, Antarctica receives increased precipitation. In East Antarctica, this may be enough to keep total ice volume roughly constant, or even to increase ice volume in the short term. In West Antarctica, melting increases the rate of flow of ice sheets into the ocean, and currently the rate of ice loss exceeds the accumulation rate. The Scientific Committee on Antarctic Research (SCAR) combines ongoing scientific research results into regular publications that document the current scientific understanding of climate change to Antarctica. In their 2016 update [271, 272], they summarize several studies that suggest that with an increase in temperature of roughly 2 °C in the Southern Ocean, there is a substantial risk of a West Antarctic ice sheet collapse over the next thousand years, and that high emissions scenarios could lead to ice loss giving a sea level rise of 1–3 m by 2300. A large-scale collapse of the West Antarctic ice sheet could lead to a rapid rise in sea levels by 5–6 m. Current models do not suggest that the conditions leading to such an event will be reached in the next century, but these models may not cover the range of deformation in the ice sheet that could be reached with rapid ice shelf collapse, and observations made to date do not rule out a significant tail in the probability distribution that would include several meters of sea level rise on a century time scale. A better understanding is needed for a high-confidence projection.

Greenland Greenland has a total ice volume of roughly $2.85\,\mathrm{M\,km^3}$, less than 10% of the ice volume of Antarctica. Because the Arctic region is expected to experience greater warming than the southern polar region, however, it is expected that a larger fraction of Greenland's ice sheet will melt in the coming centuries. The rate of ice melt in Greenland is accelerated by additional anthropogenic effects, particularly *black soot* (§34.4.4) from fossil fuel emissions that absorbs incoming sunlight at a high rate. Melted water on the surface passes rapidly down through channels in the glacier to the ground beneath, where it has the effect of lubricating the system and increasing the rate of flow of the glacier, further increasing the melt rate. Recent estimates indicate that the rate of ice melt in Greenland rose from roughly $55\,\mathrm{Gt/y}$ in the 1990s to roughly $290\,\mathrm{Gt/y}$ in 2011. In the IPCC RCP scenarios, the contribution of Greenland ice melt to sea level rise grows faster than linearly, and by 2500 contributes at a comparable rate to thermal expansion. If Greenland's ice sheet were to melt completely, it would raise global sea levels by roughly $7\,\mathrm{m}$. In any scenario currently contemplated, however, this would take many centuries, and is deemed unlikely to occur unless temperature increase goes beyond a threshold of around $4\,^\circ\mathrm{C}$.

Mountain glaciers Most of the world's mountain glaciers are small enough to be significantly affected by global warming over a period of decades. Glaciers form a relatively easy to measure signal of climate change; during warmer periods, glaciers retreat, during cooler periods they advance. Following the Little Ice Age, glaciers retreated from about 1850–1950. They advanced again during the slight cooling from 1950 to 1980, and almost all of the world's glaciers have been retreating steadily during the warming trend since 1980. Many glaciers, such as those in the Himalayas and elsewhere in the Northern Hemisphere, are experiencing accelerated melting due to black soot particulates. If temperatures rise by $3\,^\circ\mathrm{C}$ or more, as is predicted with a doubling of pre-industrial CO_2 levels, it is likely that many of the world's glaciers will be gone by the end of the century. This contributes to a significant fraction of the sea level rise predicted by the IPCC. It has been estimated that a complete melting of all alpine glaciers would lead to a sea level rise of roughly $0.5\,\mathrm{m}$. Since glaciers are a primary source for fresh water in many parts of the world, the melting of glaciers will also compromise water supplies in many countries.

Incorporating all factors, the IPCC 2013 report projected that total sea level rise in the twenty-first century would likely be between $40\,\mathrm{cm}$ and $100\,\mathrm{cm}$, depending upon the scenario. The largest uncertainty in these projections is the possibility of an unexpectedly large contribution from rapid dynamics of the West Antarctic ice sheet. A rise in sea level of $1\,\mathrm{M\,km^2}$ would cover roughly $1\,\mathrm{M\,km^2}$ of coastal land area [273]. While this is less than 1% of the $150\,\mathrm{M\,km^2}$ of total terrestrial land area, it would include many sensitive ecosystems and population centers. A rise in sea level of $6\,\mathrm{m}$ would cover a total of roughly $2.2\,\mathrm{M\,km^2}$ of land; note that this land area is comparable to the desert area computed in §22 that would be needed to supply all current and future world energy needs at 3–6% solar conversion efficiency.

35.3.4 Changes in Weather

As the climate warms, many changes may occur in local weather patterns. Such local effects are difficult to predict with accuracy. A number of global trends, however, are easier to understand and predict. As mentioned above, there will be more warming over land than over oceans, and the Arctic region will experience the greatest warming due to a combination of increased Northern heat transport, ice albedo feedback, and increased ice melt from particulates and changing weather and current patterns.

The increased greenhouse effect leads to an increase in downward IR radiation flux from the atmosphere as a fraction of total incoming radiation. Since the time scale for the atmosphere to reach thermal equilibrium is measured in hours or days, this will even out diurnal fluctuations and increase daily minimum temperatures more than daily maximum temperatures. Due to the seasonal shift described above, there will be more very hot days and heat waves and fewer very cold episodes during the seasonal extremes.

Increased flow of energy through the ocean surface layer and increasing temperatures over land will lead to an overall increase in evaporation, leading in turn to a net increase in precipitation. Many desert and arid land areas exposed to greater warming will become drier as the increase in evaporation exceeds the increase in precipitation. Warming will cause precipitation intensity to increase in other areas but occur with reduced frequency, with greater energy in the system producing more extreme weather events.

35.3.5 Ecosystem Impact

Impact on ecosystems goes somewhat outside the purview of this text. A few observations, however, are relevant and help to complete the picture of the possible effects of future warming.

At mid-latitudes the increased availability of CO_2 may lead to increased plant growth and food production. As mentioned above, as of 2015 some 30% of human carbon emissions are absorbed in increasing terrestrial biomass. It is not well understood how far this trend will continue or whether it may reverse beyond a certain point.

In some dry and tropical regions increased desertification from increased evaporation will place additional pressure on water supplies, with impact both on human and natural ecosystems.

The Arctic ecosystem will likely experience the greatest impact; substantial habitat loss due to changing conditions may destroy parts of this ecosystem completely.

Because the pulse of warming from anthropogenic carbon emissions is projected to persist over a period of at most several thousand years, species and ecosystems will not have sufficient time to adapt substantially to changes. Many species whose range and habitat is already limited may be pushed to extinction by the short-term change in their environment. According to IPCC projections [246], if global temperature rise is between $1.5\,°C$ and $2.5\,°C$, 20% to 30% of species worldwide will be at an increased risk of extinction. With temperature rise of $3.5\,°C$ or greater, over 40% of species may go extinct. Note that projections of this kind are highly uncertain, and this risk should be interpreted in a context in which many species are already threatened with extinction by habitat loss and other human impact. The populations of a large fraction of the non-domestic animals weighing more than a few kilograms are already severely impacted by human activities such as land use and fishing. Even in the absence of global warming, it is likely that many large land and marine animal species will be driven to extinction by human activities in the twenty-first century. Global warming will hasten the extinction of those species already living at the limit of their resiliency.

35.3.6 Ocean Acidification

As described above, some of the excess CO_2 produced by human activity is absorbed by the world's ocean waters, at a level corresponding to roughly 2.3 Gt of carbon per year. As the CO_2 comes into equilibrium in the ocean, reactions with water produce carbonic acid (H_2CO_3), which in turn can release hydrogen ions (H^+) and bicarbonate ions (HCO_3^-). The **acidity** of a solution is measured by **pH**, which is roughly proportional (precisely in the original definition) to the (negative of the base-10) logarithm of the density of H^+ ions. Over the last several hundred years, the pH of ocean surface waters has decreased from roughly

8.25 to 8.14, corresponding to an increase by roughly 30% in hydrogen ions. As the ocean becomes more acidic, carbonate ions (CO_3^{--}) combine with free hydrogen ions to form bicarbonate ions. As discussed in §35.1.4, carbonate ions are used in production of shells and skeletal materials by many marine organisms, including in particular corals as well as mussels, clams, oysters, snails, sea urchins, and calcereous phytoplankton. Most of these organisms live at levels in the ocean that are oversaturated with carbonate ions. As the ocean pH drops, these areas become undersaturated and it is more difficult for the organisms to build shells. In recent years, many coral and shellfish populations have suffered extreme degradation from the decrease in pH as well as other human impacts. With CO_2 levels approaching or exceeding 560 ppm, an even more dramatic decrease in pH, by 0.2 or 0.3, may be expected over the next century. While, as discussed previously, atmospheric CO_2 levels were higher many millions of years ago, the rate of change of ocean pH from the current anthropogenic CO_2 increase is several orders of magnitude faster. Although the populations of marine organisms affected could presumably migrate over periods of many thousands of years or more to depths with an appropriate carbonate ion level, such migration is much more difficult over a period of a few decades. Thus, many marine scientists have great concerns regarding the impact of rising atmospheric CO_2 levels on marine ecosystems.

35.3.7 Nonlinear Climate Effects

Most modeling and climate analysis is based on the assumption that for given external conditions there is a unique equilibrium climate configuration, and that small fluctuations in the external conditions will lead to a linear climate response. The analysis of climate feedbacks in §34.5, for example, is based on this assumption of linearity. The complete set of equations governing atmosphere and ocean circulation, ice melt, deep water formation, interaction with biological systems, and all the other complexities of climate systems, however, are highly nonlinear. In general, nonlinear dynamical systems like this can have many different local equilibria, even for fixed external conditions. As external parameters shift, such a system can undergo a rapid shift between regimes where solutions have very different qualitative features. Over the last several million years there have been periods of rapid warming at the beginning of interglacials following long glaciated periods. Some climatologists believe that this rapid shift is evidence for a *bistable* system in which small perturbations can cause the climate system to rapidly move between one locally stable equilibrium and another. It is possible that anthropogenic CO_2 emissions and the associated warming may move the climate system far enough from its current equilibrium for some nonlinear effects to kick in that could take the system into a new equilibrium. Such events are difficult to describe accurately in computer models due to a limited understanding of the precise mechanisms for such processes. There is currently no consensus among climate scientists regarding the likelihood of events of this type.

One scenario of this type that has been discussed extensively is the possibility that fresh water melting into the North Atlantic from Greenland and other sources may reduce the salinity of the water sufficiently to reduce the rate of deep water formation and slow the rate of northward heat transport in the Atlantic Ocean, affecting global circulation patterns. Such a slowdown of the meridional overturning current may have been responsible for North Atlantic cooling during the **Younger Dryas** 12 000 years ago, and oscillations in temperature during earlier glacial periods measured from Greenland and Antarctic ice cores may also have been caused by this mechanism. While recent analysis suggests that this possibility is not likely to play a large role in climate changes over the next few hundred years, research in this area is ongoing.

Another nonlinear effect that could lead to a rapid climate shift is the sudden collapse of the West Antarctic ice sheet, as discussed above, which would lead to a significant contribution to sea level rise and increased albedo feedback in the southern polar regions.

One other contribution to climate change with some nonlinear features may come from methane clathrates (§33.3) under frozen permafrost and in ocean sediments that could be released extensively after a certain temperature threshold is passed. It is believed that this mechanism may have contributed to the increase in methane levels following warming at the end of glacial periods identified in the geological record. Because methane is a highly potent greenhouse gas (see §34.4), the release of significant amounts of methane could provide a substantial positive feedback for warming. On the other hand, because of the long time scale of warming surface ice and ocean sediments, this process may not occur particularly suddenly or quickly; furthermore, methane's short lifetime in the atmosphere would mitigate this effect.

While the particular effects just mentioned are not believed likely to lead to dramatic climatic shifts, these and many aspects of climate systems are not well understood. From a statistical point of view this indicates that the distribution of possible outcomes of the climate "experiment" may be broader than a typical normal distribution. Practically speaking, this means that significant deviations in either direction outside the estimated error bars around the average predicted values cannot be ruled out with high

certainty. From a policy point of view, it is important to incorporate the uncertainty in climate predictions as well as the mean of the distribution. In particular, it may be prudent for society to take significant precautions to preclude the occurrence of events at the tail of the probability distribution if those events would have catastrophic consequences, even if the estimated probability of such events is relatively small.

35.4 Mitigation and Adaptation

Given the fact of ever-increasing anthropogenic CO_2 emissions and our scientific understanding of the role of CO_2 in increasing radiative forcing, and notwithstanding our less-than-complete understanding of how the carbon cycle and global climate will react in the coming decades, it seems inevitable that Earth will experience a warming climate over the next century, though the precise extent of this warming is still unclear. In this section we discuss some of the scientific aspects of options that humans may adopt in responding to these circumstances.

35.4.1 Options for the Future

Faced with rising atmospheric CO_2 levels from fossil fuel usage, there are four logical possibilities for how humans may react:

1. Move to non-carbon based power sources and/or reduce energy use substantially.
2. Capture carbon, from the atmosphere and/or at the point of release, and sequester it away somewhere.
3. Reduce radiative forcing directly by geoengineering.
4. Do nothing and deal with whatever consequences arise.

Of course, these options are not mutually exclusive; more than one will almost certainly come into play in the next century. Much of this book has been dedicated to describing the physical principles of non-carbon based power sources. In this section we briefly discuss the other three options.

35.4.2 Carbon Capture and Sequestration

One way to mitigate the excess carbon dioxide emissions caused by fossil fuel usage is to capture the CO_2 and remove it from the global carbon cycle by placing it in some kind of long-term storage, a process known as **carbon capture and sequestration** (CCS). The easiest way to capture the CO_2 is at the point of emission; large point sources such as coal plants currently account for roughly 40% of human CO_2 emissions. It has also been proposed that carbon can be removed directly from the atmosphere, a process known as **direct air capture** (DAC).

Figure 35.24 The first large-scale CO_2 sequestration facility in operation: Sleipner Vest natural gas field, Norway. Beginning in 1996, this Statoil facility sequestered roughly 1 Mt of CO_2 per year. (Credit: Bloomberg/Bloomberg/Getty Images)

Several options have been suggested for sequestration. One possibility is to compress the CO_2 and pump it into geological repositories such as oil or gas fields from which original fossil fuels have already been extracted. It has also been suggested that CO_2 can be stored in the deep ocean, either by dissolving it into the ocean at a depth of several thousand meters, or injecting it below 3000 m depth, where carbon dioxide becomes more dense than water and will pool at the bottom for a long time before re-entering the carbon cycle. Methods have also been suggested for fixing the carbon into carbonates that can be stored in compact solid form.

There are many technical challenges to the capture and sequestration of carbon that go beyond the purview of this book. Here we consider some aspects of CCS from point sources where the CO_2 is relatively concentrated. DAC is explored in Problems 35.11 and 35.12. A useful lower bound on the effort entailed in CCS can be obtained by considering the entropy change that must occur when CO_2 is separated from other gases with which it had been mixed. If molecules of CO_2 are originally mixed with other gases at a fractional concentration of c, then when the CO_2 is separated out, the entropy of the system is reduced by the entropy of mixing as described in Example 8.1,

$$\Delta S = -\Delta S_{\text{mixing}} = Nk_B \left(c \ln c + (1-c) \ln(1-c)\right).$$
$$(35.5)$$

But the second law of thermodynamics requires that the entropy of the universe must always increase; thus at least this entropy must be released to the environment. If the ambient temperature is T then the separation of the gases requires that energy

$$\Delta E_{\text{separation}} \geq T|\Delta S| \qquad (35.6)$$

be expelled into the environment. An equivalent way to think about this calculation is through Gibbs free energy

(8.68); eq. (35.6) is just the statement that the Gibbs free energy increases when the separation is performed.

One currently favored approach to long-term sequestration of CO_2 is storage in compressed form in geological repositories. If captured CO_2 is stored at high pressure, additional work must be done to compress the gas. A small calculation (Example 35.2) shows that capturing the CO_2 from the burning of coal in air and compressing it to 100 atm reduces the power available from the coal-fired power plant by at least 17%. This calculation only gives the theoretical minimum energy needed; actual energy costs would be significantly greater.

Some pilot plants have been developed with post-combustion CO_2 capture. Estimates of actual energy cost for post-combustion CO_2 capture are in the range of 25%–40% of plant output. The corresponding increase in fuel required for a given usable energy output in turn increases the energy used for mining and transporting the coal fuel to the plant, which in turn produces additional CO_2, reducing the effective offset from the carbon capture in the plant. By incorporating pre-combustion carbon capture into an IGCC plant (see §33), carbon emissions can in principle be reduced at lower energy cost, because the CO_2 is present at higher concentrations and higher pressure in pre-combustion syngas than in post-combustion emissions. While the chemical engineering needed to capture carbon at high pressure and temperature is challenging, it has been estimated that pre-combustion carbon capture may be possible with energy requirements in the range 11–22% of plant output.

In addition to capture from point sources and direct air capture by conventional chemical processes (see Problems 35.11 and 35.12), a number of alternative approaches to direct air capture of CO_2 have been considered. These include using biological systems – terrestrial biomass, for example – to concentrate carbon, and sequestering the mass thus produced. It has been suggested that seeding the ocean with iron or other nutrients could enhance the rate of growth of phytoplankton over large areas, increasing the rate at which the biological pump transfers the carbon from atmospheric CO_2 to the deep ocean; while experiments have confirmed that added iron can stimulate plankton blooms, the efficacy of this approach for carbon sequestration is still under debate. Concerns about the widespread disruption of marine ecosystems, among other things, have diminished interest in this approach [274]. More speculative approaches include ideas like bioengineering organisms to optimize conversion of atmospheric CO_2 to carbonates that can be compactly stored over long periods with small degradation rate.

Example 35.2   Energy Cost of CO_2 Capture and Sequestration

A coal power plant with carbon capture and sequestration captures CO_2 from coal that is burned in air, compresses it to 100 atm, and stores it in an underground repository. We estimate a lower limit percentage of the power plant's output required to separate the CO_2 from flue gases and to compress it (isothermally) to 100 atm at an ambient temperature of 300 K. We assume that the coal is burned with the minimal amount of air required for complete combustion and that the efficiency of the power plant (without considering the energy required for CCS) is 30%.

First consider the energy required to satisfy the second law of thermodynamics. Every molecule of O_2 in the combustion air is converted to a molecule of CO_2. Thus the concentration of CO_2 in the flue gases is $c = 0.21$, the same as the concentration of O_2 in air. Equation (35.5) enables us to compute the entropy reduction by separation (per mole of flue gas), and eq. (35.6) relates this to the minimum energy that must be expelled to the environment by the separation process. Burning 1 kg(C) produces $1/0.012 = 83.3$ mol(CO_2), which in turn, generates $83.3/0.21 = 397$ moles of flue gas. The minimum separation energy per mole of flue gas at 300 K is

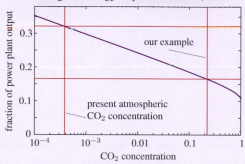

$$\Delta E_{separation}/\operatorname{mol}(\text{flue gas}) \geq RT\,(0.21\ln(0.21) + 0.79\ln(0.79)) = 1.28\,\text{kJ/mol(flue gas)},$$

or per kilogram of carbon, $\Delta E_{separation} = 397 \times 1.28 \cong 508\,\text{kJ/kg(C)}$. Compressing the resulting 83.3 moles of CO_2 to 100 atm isothermally requires $\Delta E_{compression} = nRT \ln V_2/V_1 = 958\,\text{kJ/kg(C)}$. Thus the total energy required for CCS is $958 + 508 \cong 1470\,\text{kJ/kg(C)}$. Burning a kilogram of coal produces $\cong 29.3$ MJ of thermal energy, of which 30% or 8.8 MJ is available for useful work. Thus CCS requires at least $1.47/8.8 \cong 17\%$ of the power plant's useful output.

Since the concentration of CO_2 in the effluent is the principal variable in CCS, we show the minimum percentage of power plant output required for sequestration as a function of CO_2 concentration in the figure above. The case considered above and the case of direct air capture are marked on the figure. Note that for direct air capture of a comparable quantity of CO_2 after emission from the plant, over 30% of the plant's output would be used at least. Other difficulties with DAC are explored in Problem 35.12.

35.4.3 Counteracting Radiative Forcing through Geoengineering

If fossil fuels continue to be used in large quantities, and rising atmospheric CO_2 levels lead to warming with undesirable consequences, and if removal of CO_2 directly from the atmosphere is too difficult, another option that has been raised is the direct manipulation of radiative forcing using large-scale geoengineering – option 3 on the list above. This approach may be less expensive in the short term than large-scale carbon capture or investment in low-carbon energy sources. It has several disadvantages, however. Any geoengineering effort that only affects the radiative forcing will not change the forcing due to CO_2 directly, only offset it using another mechanism. Thus, the geoengineering technique must be maintained as long as atmospheric CO_2 levels remain elevated. If CO_2 levels continue to rise, then the geoengineering project must be increased commensurate to the increase in radiative forcing (which is, however, logarithmic in the CO_2 level, as discussed previously). Finally, geoengineering to offset radiative forcing would not address other aspects of elevated CO_2 levels, including in particular ocean acidification.

We briefly summarize a few of the methods that have been suggested for geoengineering to modify radiative forcing.

Reflection of incoming radiation One possibility that has been suggested is the emplacement of large systems of mirrors or reflecting material to effectively increase Earth's albedo. It has been suggested that these mirrors be placed in space, though this would be prohibitively costly. A more economical approach would be to cover large tracts of land, perhaps in desert areas, with inexpensive reflecting material. For example, to offset a radiative forcing of $3.7\,\text{W/m}^2$ as expected for a doubling of CO_2, a simple estimate of the land coverage needed assuming additional reflection of 30% of an incoming daily average of $300\,\text{W/m}^2$ would be $3.7/90 \cong 4\%$ of the planet's surface area. This is roughly the area ($\cong 20\,\text{M km}^2$) of all the world's deserts. For comparison, the rough estimate in

§23.6 of the fraction of desert that would need to be covered by solar systems to generate all world energy needs is roughly 10% of desert area or $2\,M\,km^2$.

Sulfate aerosols Another approach that has been advocated is the large-scale release of sulfate aerosols into the atmosphere, leading to both a direct albedo effect and to nucleation of clouds giving a secondary albedo effect, as discussed in §34.4.3. At some level, this mechanism is already in use, because coal plant emissions of sulfur dioxide are currently producing a net negative radiative forcing, offsetting some of the positive radiative forcing arising from increasing CO_2 levels. Since the atmospheric lifetime of sulfates is much shorter than CO_2, sulfates would have to be injected into the atmosphere continuously in order to prevent return to greenhouse conditions.

Cloud seeding Another simple type of geoengineering that has been suggested would be to seed low-lying clouds by injecting other kinds of particulate matter into the low atmosphere. For example, injecting ocean salt into the air above the ocean could increase cloud cover and increase albedo. If done at a sufficiently large scale to counteract substantial radiative forcing on the order of several W/m^2, this could lead to unexpected changes in precipitation and circulation patterns.

35.4.4 Adaptation

In the absence of finding technically, economically, and politically acceptable mechanisms for implementing options 1, 2, or 3, humankind will have to adapt to changing climate and weather patterns beginning in the near future and continuing over the next several hundred to thousand year time frame. While very fast on the geological time scale, significant changes in seasonal weather and sea levels will only occur over decades, relatively long on the time scale of the human attention span. These changes will necessitate substantial migration of human populations from areas affected by changes such as desertification or sea level rise, and/or projects to provide irrigation, dikes, and the like. Natural ecosystems will not have time to adapt through evolution, and in many places already are constrained by human land use. Climate change will add to the strong external stresses already imposed on many terrestrial and ocean ecosystems by human activity.

Discussion/Investigation Questions

35.1 The radio-isotope beryllium-10 (^{10}Be) is often employed in studies of paleoclimates. How is ^{10}Be formed? With which aspects of paleoclimate is ^{10}Be correlated? How is ^{10}Be used, for example, to extract climate information from ice cores?

35.2 In §35.3.3 we remarked that the fact that ice floats in seawater has far-reaching consequences for climate and life on Earth. Discuss what Earth's surface might look like if ice did not float.

35.3 Data are available on the location of coastlines on several continents during the Pliocene era when temperatures were at levels projected to be reached over the next century. Locate data on a region of interest to you and consider some of the implications.

35.4 Fertilizing phytoplankton in the deep ocean was mentioned briefly as a proposed way to sequester large amounts of CO_2. Explore this proposal. What do its advocates claim? What criticisms have been voiced? What is your opinion of the present state of affairs?

Problems

35.1 Compute the insolation at 65° latitude on the summer solstice for tilt of 22.05° and compare to the value at 24.5°.

35.2 Variation in sunspot activity leads to variation in the solar constant by roughly 1–2 W/m^2 with a roughly 11 year cycle. Assuming an albedo of $a = 0.3$, estimate the radiative forcing arising from an increase in the solar constant by 2 W/m^2. Using the IPCC mean result for feedback, what increase in surface temperature would this cause?

35.3 Since the formation of the solar system 4.6 billion years ago, the net solar luminosity has increased roughly 40% from its initial value. Assuming that luminosity has increased at a constant rate, estimate the solar constant 100 million years ago, during the Cretaceous period. Assuming current climate conditions and an albedo of $a = 0.3$, what radiative forcing would be needed to keep Earth in its current state (as of say 1950) if the Sun had the luminosity of the Cretaceous period? What level of atmospheric CO_2 would achieve this radiative forcing?

35.4 In another billion years solar luminosity will increase by roughly another 10%. Estimate the resulting radiative forcing and change in terrestrial surface temperature.

35.5 Compute the time to melt 3000 m thickness of ice if there is an extra 5% of 200 W/m^2 average daily insolation for four months out of the year beyond the insolation that would give an ablation rate matching the accumulation rate from fresh snowfall.

35.6 Compute the time for the oceanic mixed layer (350 M $km^2 \times 200$ m) to warm by 3.2 K, using 50% of the energy coming from a change in radiative forcing of 3.7 W/m^2. Advanced version: assume that radiative forcing is proportional to $T - T_{eq}$ as the surface warms and describe the time-history of the surface water by writing and solving a differential equation.

35.7 Reproduce the graphs in Example 35.1 by performing the numerical calculations on a computer. Find a way of improving the assumptions and redo the calculation.

35.8 Assume that 1% of a net radiative forcing of 3.7 W/m^2 worldwide goes to melting ice over land. Estimate the rate of sea level rise from this melting.

35.9 (a) Assume that the rate of carbon emission at 10 Gt/year increases at a constant rate to a maximum of 20 Gt/year in 2050, and then decreases at the same rate. Assume that ocean and land biomass together absorb excess CO_2 at a constant rate of -5 Gt/year. In what year will atmospheric CO_2 levels peak? What will be the level of atmospheric CO_2 in that year? (b) Repeat the analysis under the more optimistic assumption that ocean and land biomass together absorb 2.5% of the excess in atmospheric carbon above 600 Gt each year, assuming for simplicity that atmospheric carbon is 800 Gt in 2010.

35.10 Estimate the contribution to global sea level rise over a century if Greenland's glaciers continue to melt at a rate of 290 Gt/y.

35.11 Using the methods developed in Box 35.4.2, find a lower limit on the energy per kilogram of carbon for removing carbon directly from ambient air. Compute the energy cost of removing all 10 Gt/year of current carbon emissions and compare to total electric production worldwide.

35.12 A concern with plans for capture of CO_2 directly from ambient air (DAC), is the sheer volume of material that must be processed. One design requires fans to blow air at 2 m/s through 2.8 m thickness of absorber and other materials and aims to absorb 50% of the CO_2 from the airstream. Assume a rectangular installation of length L, height 100 m (and thickness 2.8 m). How large must L be in order for this DAC plant to remove the CO_2 produced by a 1 GW$_e$ coal-fired power plant operating at 30% efficiency? How large a population (in the US or in your country) could have its carbon footprint eliminated by this DAC plant assuming the plant to be powered by a non-carbon source? Use the results of Problem 35.11 to redo this estimate if the coal power plant itself powers the DAC facility.

Energy Efficiency, Conservation, and Changing Energy Sources

Energy conservation and energy efficiency can decrease the total quantity of energy used by humanity, reducing pressure on finite resources and minimizing undesirable side effects of energy use such as climate change, pollution, and habitat destruction. The finite nature of fossil fuels and other non-renewable resources motivates conservation and efficiency and will eventually necessitate a shift to renewable energy resources. The impact of carbon emissions on global climate also motivates conservation and efficiency as well as a more rapid change to renewable resources. In this chapter we consider these issues, integrating material from the preceding parts of the book.

Although they are conceptually quite different, energy efficiency and conservation are often discussed together and sometimes conflated. Here we use **energy efficiency** to refer to the relative amount of useful output, such as mechanical work, that can be obtained by a device or system from a given input, such as thermal energy. The quest for greater energy efficiency is primarily a scientific and engineering challenge. **Energy conservation**, on the other hand, involves choosing among systems or altering behavior with the goal of using less energy. Energy conservation involves economic, social, and policy choices that can be informed by, but not solely determined by technical considerations. A simple physics analysis can determine how much energy can be saved by turning down a home thermostat (§36.5.3), but it cannot predict the willingness of a family to set the thermostat to 18 °C and don sweaters rather than T-shirts indoors in midwinter.

The first parts of this chapter are devoted to a quantitative reconsideration of the notion of *efficiency*. So far in this book we have introduced several measures of effectiveness for devices that transform or transfer energy, such as the efficiency of a heat engine or a photodiode. These are known as *1st law efficiencies*: the amount of a desired

Reader's Guide

This chapter explores the related concepts of efficiency and conservation and the long-term prospects for changing energy sources. The concepts of 1st and 2nd law efficiencies are introduced and applied to transformations of energy from one form to another. *Exergy* or *available work* is introduced as a measure of the useful work that can be extracted from a given energy resource, and exergy is related to 2nd law efficiency. A few case studies are presented to illustrate the use of physics to provide input to energy conservation analyses. Finally, an overview is given of large-scale and high-density energy resources that may in the future replace fossil fuels as primary energy sources.

Prerequisites: §5 (Thermal energy), §8 (Entropy and temperature), §10 (Heat engines), §9 (Energy in matter), §13 (Power cycles).

The concept of exergy arises in §37 (Energy storage).

output such as work or thermal energy divided by the energy input. In §36.1 we briefly summarize the 1st law efficiencies and other related concepts introduced throughout the book along with the fundamental physical limits that constrain them.

When the 1st law efficiency of a device is quoted, it gives no indication of how well the device performs compared to the best performance allowed by the laws of thermodynamics. This information is provided by the *2nd law efficiency*, which is introduced in §36.2. There we define and tabulate 2nd law efficiencies for many of the most commonly encountered processes. To illustrate 2nd law efficiencies we consider the practical example of the most efficient way to provide space heating in §36.3.

The notion of 2nd law efficiency leads naturally to the question of how to quantify the maximum amount of *useful energy* that can be provided by a given device or system located in a particular environment. This maximum energy is known as the *available work* or *exergy* of a device or system relative to the environment. Exergy is a very useful concept, which enjoys widespread use in engineering fields and in *industrial ecology*, the study of the flow of materials and energy through industrial systems. Both exergy and 2nd law efficiency were mentioned briefly in §32.4.2 in connection with the performance of geothermal power plants. In §36.4 we examine exergy more closely, show how it can be used to evaluate the utility of various energy sources, and return to the question first raised in §1 (Introduction) about the meaning of the phrase "energy consumption" in light of the fact that energy is conserved. We point out that exergy is a quantitative measure of the value of energy, and that the transition from "useful" to "useless" energy can be understood as the destruction of exergy.

No special physics concepts are required to evaluate proposals for energy conservation, which we consider in §36.5. Instead, the role of physics in energy conservation is usually to provide technical input to decisions where economic and policy considerations are of fundamental importance. In keeping with the scope of this book, we do not attempt to treat the economics of energy conservation here. Instead we present several examples to illustrate how the principles developed in this book provide the scientific input necessary to evaluate the potential for energy conservation and to enable informed economic choices among competing technologies. The American Physical Society has produced two excellent reports on energy efficiency where further information on efficiency and conservation can be found [275, 276].

We conclude this chapter with some global perspective on energy resources that is relevant when considering the prospects for changing energy sources over the long term, or in the shorter term motivated by decarbonization of energy systems. In §36.6, we summarize the extent of available resources that have potential for large-scale power production or that may be particularly economically favorable due to high energy density. Finally, we examine the potential of these resources for replacing fossil fuels in the long run.

36.1 First Law Efficiency

Many measures of performance have been introduced throughout this book to indicate how effectively a device or system transforms energy from one form to another. The ratio of the desired energy output or transfer, which is often work or thermal energy, divided by the energy input, which

may come in any form, defines the **1st law efficiency** of the device or system,

$$\eta = \frac{\text{energy transfer (of a desired form)}}{\text{energy input}} . \qquad (36.1)$$

Some transformations are by their nature quite efficient. For example, electrical energy can be transformed into mechanical energy by a motor with losses of only a few percent. Others are necessarily very wasteful, notably the transformation of heat into work, which becomes very inefficient when the heat source is only slightly warmer than the ambient environment.

There are obvious issues with eq. (36.1) as a definition of efficiency. In particular, although it is natural to think of "efficiency" as a quantity that is less than one, η can be much greater than one in the case of heat pumps and other heat extraction devices, where η coincides with what we earlier called the *coefficient of performance* (CoP, §10.6.1). This and other issues can be resolved by introducing the concept of *2nd law efficiency*, which we do in the following section.

In this section we briefly review 1st law efficiencies and some related ideas for various energy transformation processes and collect the limits on these efficiencies. Although many of these ideal limits are rarely approached (and even then only in laboratory devices), they are the benchmarks to which practical devices should be compared. The 1st law efficiencies reviewed in this chapter are summarized in Table 36.1. We also discuss several other types of energy limits in this context that are not precisely 1st law limits, namely the Betz limit on energy extraction from wind by a system modeled as an actuator disk, and the limit on how much computation can be performed with a given amount of energy.

36.1.1 Transformations Involving Mechanical and Electrical Energy

Transformation from one form of mechanical energy to another, for example from the gravitational potential energy of water stored behind a dam to the rotational motion of a turbine, or from the reciprocating motion of pistons in an automobile to the rotation of the wheels, can in principle be carried out with very little waste. Losses are mainly due to friction between moving parts or to resistive drag by a fluid, typically air or water. Friction and air/water resistance convert entropy-free mechanical energy into the random vibrations of solids or the random motion of a fluid and therefore increase the entropy of the system.

For convenience, we refer here to all macroscopic forms of mechanical energy as "work," and define the 1st law efficiency for transformation between one form of mechanical energy and another as

$$\eta_{w \to w} = \frac{\text{work out}}{\text{work in}} . \quad (36.2)$$

To the extent that frictional losses can be eliminated, mechanical energy can be transformed from one form to another with nearly perfect efficiency,

$$\eta_{w \to w} \leq 1 . \quad (36.3)$$

The transformation of the kinetic energy in a flowing fluid into useful mechanical energy requires further consideration. As discussed in §28 and §30, the *Betz limit* requires that a device such as a *horizontal-axis wind turbine* (HAWT) that can be modeled as an *actuator disk* of area A, operating in a plane perpendicular to a steady, unidirectional fluid flow, can harvest no more than 16/27 or 59% of the power that passes across an area A in the flowing fluid far from the device. The rest of the energy remains in the flowing fluid, however, and can, in principle, be captured by devices placed further downstream. Although it is therefore not an absolute efficiency limit like Carnot's limit on heat engines, Betz's limit plays an important role in judging the effectiveness of wind and tidal stream turbines.

Transformations of mechanical to electrical energy, and from one form of electrical energy to another, have characteristics similar to transformations from one form of mechanical energy to another. Resistive losses in imperfect conductors are analogous to friction in mechanical systems, converting ordered electrical energy into heat. The 1st law efficiencies for processes that convert mechanical energy into electrical energy or *vice versa* can approach one to the extent that resistive losses and friction can be eliminated.

36.1.2 Transformations Involving Heat and Work

Heat engines transform thermal energy into mechanical energy ("work") and heat extraction devices such as air conditioners, refrigerators, and heat pumps use work to transfer thermal energy. The efficiencies and coefficients of performance of these devices defined in §8–§10 were, in fact, 1st law efficiencies. The first and second laws of thermodynamics lead to universal limits, such as the Carnot limit, on the 1st law efficiency of any cyclic device that converts thermal energy to work or uses work to extract heat from a low temperature and move it to a high temperature. The definitions and thermodynamic limits on 1st law efficiencies of heat engines and heat extraction devices are summarized in Table 36.1.

36.1.3 Other 1st Law Efficiencies

In the course of studying the various primary sources of energy, we defined other measures of performance for devices or systems that inter-convert different forms of energy. These take the form of 1st law efficiencies (36.1). Often we have been able to derive limits on these efficiencies from the first and second laws of thermodynamics.

Transformation Between Light and Other Forms of Energy Light comes in many forms; light emitted by a hot object is qualitatively different from laser light, and has different limits on its utility. First, we consider thermodynamic limits on conversion of light to work; we then review the limits on solar thermal collectors and solar photovoltaic devices from §24 and §25.

Thermal radiation The thermal radiation source of greatest interest is sunlight, which is approximately described by a blackbody spectrum at a temperature of about $T_\odot \approx$ 6000 K. In a solar power system, electromagnetic radiation effectively transfers thermal energy from a reservoir (the Sun) at $T_\odot$ to a device that does work or directly generates electrical energy while expelling some thermal energy to the environment at temperature T_0. Solar radiation carries entropy in the distribution of photons (22.13). When a device absorbs an amount of energy Q from sunlight, it also absorbs entropy $S \geq Q/T_\odot$. To convert this energy to useful work without building up entropy in the device itself, it must dump this entropy into the environment at the temperature T_0. As explained in Box 25.1, this means that any device that converts sunlight into mechanical (or electrical) energy has a Carnot limit on its efficiency of

$$\eta_{\text{sunlight}} = \frac{W}{Q_+} \leq \eta_C = 1 - \frac{T_0}{T_\odot} . \quad (36.4)$$

If $T_0 = 300$ K, then the maximum achievable efficiency is 95%. This limit applies to solar thermal and solar photovoltaic power systems. Note that the efficiency of a device that uses thermal radiation given off by an object at lower temperature is considerably more constrained.

Laser light All the photons in laser light have (almost) the same frequency, and their oscillations are in phase. Very little is therefore unknown about the quantum state of the photons, and as a result the entropy of laser light is very small. The conversion of laser light to electrical or mechanical energy can thus in principle be extremely efficient, though if it is done using photovoltaic cells there are still device limitations related to, for example, the fill factor and open-circuit voltage parts of the Shockley–Queisser efficiency bound (25.26).

Table 36.1 First law efficiencies for some familiar processes and devices, adapted from [275]. Note: $T_+(T_-)$ is the temperature of a high- (low-)temperature heat source, and T_0 is the ambient temperature in which the device functions.

Type of device or system	Numerator in defining η	Denominator in defining η	η_{max}	Standard nomenclature
Electric motor	Mechanical energy (work) out	Electrical energy in	1	Efficiency
Heat engine	Mechanical or electrical energy out	Heat at T_+ in	$1 - T_0/T_+ < 1$	Efficiency
Electric heat pump	Heat added at T_+	Electrical energy in	$\dfrac{1}{1 - T_-/T_+} > 1$	CoP
Air conditioner or refrigerator	Heat removed at T_-	Electrical energy in	$\dfrac{1}{T_+/T_- - 1}$	CoP
Solar collector	Electrical or mechanical energy	Solar radiation in	$1 - T_0/T_\odot < 1$	Efficiency
Solar PV	Electrical energy out	Solar radiation in	$34\%^a$	Efficiency
Electrochemical cell	Electrical energy out	Chemical energy in	$|\Delta G_r/\Delta H_r|^b$	Efficiency

^{a}For optimal choice of band gap of a single-junction device.
^{b}See §37.

Example 36.1   1st Law Efficiency of an Absorption Refrigerator

An *absorption refrigerator*, described in §13.2.6, uses thermal energy from a high-temperature source T_+ to extract heat from the interior of the refrigerator (temperature T_-), and expel the heat into the ambient environment at temperature T_0.

Define a 1st law efficiency for an absorption refrigerator and find the appropriate Carnot bound.

The natural measure of the performance of an absorption refrigerator is the ratio of heat removed from the cold environment Q_- to the heat that must be supplied from the high temperature source Q_+,

$$\eta = Q_-/Q_+ \, .$$

To find the Carnot bound on η, we follow the argument applied to engines and refrigerators in §8–§10. Energy conservation, the 1st Law, requires

$$Q_+ + Q_- = Q \, ,$$

where Q is the heat expelled to the ambient environment. To obtain the best possible performance, we assume that the device operates *reversibly*, absorbing entropy $S_+ = Q_+/T_+$ from the heat source and $S_- = Q_-/T_-$ from the interior of the refrigerator, and expelling entropy $S = Q/T_0$ to the ambient environment. Since this cyclical device cannot accumulate any entropy, the entropy expelled must equal the entropy absorbed,

$$\frac{Q}{T_0} = \frac{Q_+}{T_+} + \frac{Q_-}{T_-} \, .$$

Combining these two equations, solving for Q_-/Q_+, and identifying this with the upper (Carnot) bound on η, we find

$$\eta \le \eta|_{\text{Carnot}} = \frac{1 - T_0/T_+}{T_0/T_- - 1} \, .$$

Notice that this is the same bound that would have been obtained if we regarded the absorption refrigerator as a two-step process, first running a reversible heat engine between T_+ and T_0 to produce mechanical energy and then using this energy to run a reversible mechanical heat pump between T_- and T_0.

Solar concentrators In §24 we used the thermodynamic limit (36.4) to derive a limit on the effective concentration of two- or three-dimensional solar concentrators, eqs. (24.15) and (24.16),

$$C_{2D} \le 1/\sin\theta \qquad \text{2D concentrator,}$$

$$C_{3D} \le 1/\sin^2\theta \qquad \text{3D concentrator.} \qquad (36.5)$$

Solar photovoltaic devices The 1st law efficiency of a photovoltaic device may be defined as the ratio of the electric power generated to solar power incident on the device,

$$\eta_{PV} = \frac{P_{out}}{P_{incident}(AM1.5)} , \qquad (36.6)$$

where *AM1.5* labels a reference solar spectrum (see Box 23.4). Beyond eq. (36.4), more restrictive limits can be framed for certain specified solar conversion technologies. For single-junction photovoltaic cells, for example, with some assumptions about the standard structure of the device, there is an extensive discussion in §25. This discussion culminates in the Shockley–Queisser bound on the efficiency of a single-junction PV as a function of the band-gap of the semiconductor, which for an optimal choice of band-gap gives a limit of approximately 34%.

Transformation Between Chemical and Other Forms of Energy

The 1st law efficiency of a device or system that derives its energy from a chemical reaction is defined to be the ratio of the work or thermal energy transferred divided by the negative of the enthalpy of reaction $|\Delta H_r|$. If a chemical reaction is used merely to supply space heating, then the 1st law efficiency can approach one. Chemical energy can be transformed into electrical energy at high efficiency by a *battery* or a *fuel cell*. As shown in §37, the most electrical energy that can be harvested from a chemical reaction in a battery or fuel cell is given by the negative of the *Gibbs free energy* of reaction $|\Delta G_r|$. Thus the upper limit on the 1st law efficiency of a battery or fuel cell is given by $\eta_{battery} = |\Delta G_r/\Delta H_r|$. If, on the other hand, the enthalpy of reaction is used to supply heat at temperature T_+ as an intermediary towards electric power production, as for example when methane is combusted in a gas-fired electrical power plant, then the laws of thermodynamics apply and the Carnot limit $(1 - T_0/T_+)$ determines the maximum possible efficiency. Although it might at first seem that the efficiency could approach one as T_+ increases to infinity, as shown in §36.4 the 1st law efficiency cannot ever exceed $|\Delta G_r/\Delta H_r|$. In practical situations, the temperature T_+ is sufficiently low that the Carnot limit for energy obtained through combustion processes is much stronger than the Gibbs free energy limit. Further discussion of the efficiency limits on batteries and fuel cells can be found in §37.

Unavoidable Energy Loss in Computation

Computing devices provide another example of an efficiency of energy use that is bounded by the laws of thermodynamics. Devices that compute have become ubiquitous; not only computers *per se*, but cell phones, DVD players, and even coffee makers contain chips that process information. Computers perform operations on the *bits* (0 or 1) of information (see §8.2) that they store. A computer's energy efficiency can be defined in terms of the number of operations that it can perform with a given energy input,

$$\eta_{computer} = \frac{operations}{energy\ in} . \qquad (36.7)$$

Conventional computers discard bits when they compute. A simple example of an irreversible loss of information is a replacement $C \leftarrow A$, in which the value of the bit C is replaced by that of A. One bit of information has been lost, so the 2nd Law requires that the entropy of the computer and its environment increase by at least $S = k_B \ln 2$ (eq. (8.10)). Adding this entropy to an environment at temperature T_0 adds heat $Q = T_0 S = k_B T_0 \ln 2$. This is a lower limit on the energy that must be supplied by the device that powers the computer. At $T_0 = 300$ K, $Q \approx 4 \times 10^{-21}$ J, so that the thermodynamic limit is $\eta_{computer} \lesssim 2.5 \times 10^{20}$ single bit replacements per joule. A typical desktop computer with a clock speed of 3 GHz can act on roughly 10^{13} bits per second, and the corresponding thermodynamic lower limit on its power consumption is $(10^{13}) \times (4 \times 10^{-21}) \approx 4 \times 10^{-8}$ W. Present-day computers generate heat at a rate far greater than this 2nd law limit. The most efficient supercomputer operating in 2015 was the Japanese Shoubu supercomputer, which can operate on roughly 2×10^{11} bits per joule (~7G single-precision FLOPs); the PEZY-SC graphics processing unit processors used by this machine have a claimed energy efficiency of 3 TFLOPS/70 W, or roughly 10^{12} bits/J.

The thermodynamic efficiency of computers has become an important practical matter in recent years as the problem of transporting heat away from ever more compact computational devices has become a practical limit on their size and power. As we have just seen, the efficiency of computation can be vastly improved before the theoretical limit on computational efficiency becomes relevant. The efficiency of computation has doubled every 1.6 years since the 1940s [277], and if this keeps up, the 2nd law limit is not that far off. As computation becomes more pervasive, the energy wasted in computation has become a more important target for increased efficiency.

In principle it is possible to do **reversible computing**, in which no information is lost, so no entropy is ejected to the environment. For example, if no bit is ever overwritten unless its value is 0, and all intermediate results are stored, any computer program can be made reversible. After any computation, the reverse sequence of computations can be performed, putting the computer back in its original state. Existing logic gates are not reversible, as some spurious outputs are dropped, but in principle it should be possible

1st Law Efficiency

The *1st law efficiency* of a device that transfers or transforms energy is defined to be the amount of energy of a desired form that is transferred by the device divided by the energy input to the device.

When the input or output of the system is in the form of thermal energy, the 2nd Law places limits on the 1st law efficiency.

to build reversible computing devices and use reversible software algorithms so that the only bits that need to be ejected are the bits of user input, which must be replaced to make way for the next computation. While practically it will always be necessary to dump some information, this means that there is no in-principle physical limitation on the efficiency of large-scale information processing beyond the need to jettison obsolete user data.

36.2 Second Law Efficiency

As a measure of performance of a device or system, the 1st law efficiency has several shortcomings. First, the 1st law efficiency does not indicate how well the device or system performs the task that it was intended to accomplish relative to the best that is allowed by the laws of physics. Second, the 1st law efficiency need not be less than one. It is also potentially confusing or misleading to work with 1st law efficiencies for processes with multiple steps where intermediate energies are in both mechanical/electrical and thermal form (Problem 36.4).

These shortcomings can be remedied by incorporating information from the 2nd Law into the concept of efficiency. The **2nd law efficiency** of a device or system, denoted by ϵ, is defined as the ratio of the performance of a device or system to the best performance that can be achieved by any system performing the same function consistent with the laws of thermodynamics, or more specifically [275],

$$\epsilon = \frac{\left\{\begin{array}{l}\text{useful energy transferred by}\\ \text{a given device or system}\end{array}\right\}}{\left\{\begin{array}{l}\text{maximum possible useful energy transferable for}\\ \text{the same function by any device or system using}\\ \text{the same energy inputs as the given system}\end{array}\right\}}. \quad (36.8)$$

We were employing 2nd law efficiency when we characterized a thermodynamic cycle by saying that it "reaches such-and-such percent of the Carnot limit." Thus, for example, the 1st law efficiency of a particular heat engine

W/Q_+ is bounded by the 1st law efficiency of a Carnot engine, $(1 - T_-/T_+)$, operating between the same temperature extremes. So the 2nd law efficiency of a heat engine is given by

$$\epsilon_{\text{engine}} = \frac{W}{Q_+(1 - T_-/T_+)} = \frac{\eta_{\text{engine}}}{1 - T_-/T_+}. \quad (36.9)$$

Note that the 2nd law efficiency is defined for an energy input that is either in the form of thermal energy at a fixed temperature or an entropy-free energy type such as mechanical or electrical energy. In the former case, the ambient temperature T_-, at which waste heat is dumped must also be specified for an unambiguous definition of 2nd law efficiency. This condition is discussed further in §36.4.

For a heat pump or refrigerator the "best that can be achieved" is again given by the Carnot limit (Table 36.1) and consequently we define the 2nd law efficiency of a mechanically driven heat pump or refrigerator by

$$\epsilon_{\text{hp}} = \frac{Q_+}{W/(1 - T_-/T_+)} = \frac{\eta_{\text{hp}}}{1/(1 - T_-/T_+)}, \quad (36.10)$$

$$\epsilon_{\text{r}} = \frac{Q_-}{W/(T_+/T_- - 1)} = \frac{\eta_{\text{r}}}{1/(T_+/T_- - 1)}. \quad (36.11)$$

Note that in all these examples the 2nd law efficiency is the ratio of the 1st law efficiency of the device or system to the Carnot limit on the 1st law efficiency.

The 2nd law efficiency of an absorption refrigerator, derived in Example 36.1, is also given by the ratio of its 1st law efficiency to the Carnot limit,

$$\epsilon_{\substack{\text{absorption}\\ \text{refrigerator}}} = \frac{Q_-(T_0/T_- - 1)}{Q_+(1 - T_0/T_+)}. \quad (36.12)$$

This result illustrates some important properties of 2nd law efficiencies. Instead of computing the 2nd law efficiency of the absorption refrigerator directly, one could instead choose any reversible (and therefore maximally efficient) set of devices that accomplishes the same energy transfer and use them to compute the denominator in eq. (36.8). For the absorption refrigerator one could first use a heat engine running reversibly between T_+ and T_0 to transform thermal energy Q_+ to work W, and then use W to run a mechanical refrigerator running reversibly between T_- and T_0. Because this two-step system has the net effect of using thermal energy Q_+ to remove Q_- from the interior of the refrigerator, and because the system is run reversibly at each stage, it gives the same 2nd law efficiency as the direct calculation in Example 36.1.

Note that care must be taken when defining 2nd law efficiency to be clear about the form of the energy input. For example, eq. (36.12) is defined in terms of eq. (36.8) with respect to *thermal energy* as the energy input. This

Table 36.2 2nd law efficiencies for simple thermo-mechanical processes. T_0 is the ambient temperature; $T_\pm$ are the hot and cold reservoir temperatures; T is the temperature at which heat is provided for space heating. An example is given for each choice of energy source and energy use. Adapted from [275].

Energy use	Energy source		
	Mechanical or electrical energy W_{in}	Fuel with combustion free energy $\|\Delta G_c\|$	Heat Q_+ from a reservoir at T_+
Mechanical or electrical energy W_{out}	$\epsilon = \dfrac{W_{out}}{W_{in}}$ Electric motor	$\epsilon = \dfrac{W_{out}}{\|\Delta G_c\|}$ Fossil fuel power plant	$\epsilon = \dfrac{W_{out}}{Q_+(1 - T_0/T_+)}$ Geothermal power plant
Add heat Q to a warm reservoir at T	$\epsilon = \dfrac{Q}{W_{in}}(1 - T_0/T)$ Electrically driven heat pump	$\epsilon = \dfrac{Q}{\|\Delta G_c\|}(1 - T_0/T)$ Gas-powered heat pump	$\epsilon = \dfrac{Q(1 - T_0/T)}{Q_+(1 - T_0/T_+)}$ Solar hot water heater[a]
Extract heat Q_- from a cool reservoir at T_-	$\epsilon = \dfrac{Q_-}{W_{in}}(T_0/T_- - 1)$ Electric refrigerator	$\epsilon = \dfrac{Q_-}{\|\Delta G_c\|}(T_0/T_- - 1)$ Gas-powered air conditioner	$\epsilon = \dfrac{Q_-(T_0/T_- - 1)}{Q_+(1 - T_0/T_+)}$ Absorption refrigerator

[a] Note that for the solar hot water heater, the reservoir is the volume of heated water in a solar collector, not the original solar energy source.

The 2nd law efficiency of a device such as a motor, engine, refrigerator, or heat pump that outputs either heat or work (but not both) is defined as the useful heat or work produced by the device divided by the maximum possible useful heat or work for the same function using the same energy input in the same environment. The maximum possible value of ϵ is unity in all cases.

limit does not apply to a gas (e.g. propane) powered refrigerator when the gas is considered as the input. Although one may choose to burn the propane to produce heat to run an engine, it is also possible in principle to turn the propane's chemical energy directly into electricity – using a fuel cell for example – making use of the full Gibbs free energy of combustion, $|\Delta G_c|$. So the 2nd law efficiency of a gas-powered refrigerator that operates at temperature T_0 is given by eq. (36.11) with W replaced by $|\Delta G_c|$,

$$\epsilon = \frac{Q_-}{|\Delta G_c|/(T_0/T_- - 1)}. \tag{36.13}$$

The 2nd law efficiencies of simple thermodynamic processes are summarized in Table 36.2. To illustrate the use of 2nd law efficiencies and in particular to examine the

meaning of "the best that can be achieved," we take an extended look at the most efficient way to provide space heating.

36.3 Example: The Efficiency of Space Heating

The single greatest energy use in residences is to provide space heating, most often by burning fuel in a furnace. Space heating also figures significantly in commercial and industrial energy consumption. Furnaces are rated by the percentage of the heat generated by the fuel that is delivered to the space to be heated. The best systems reach 95% or more. The quoted 95% is a 1st law efficiency: the heat delivered compared to the total heat of combustion, $\eta_{furnace} = Q/|\Delta H_c|$.[1] When this is compared to "the best that can be achieved," however, we find a 2nd law efficiency $\epsilon_{furnace} \approx 13\%$ for a furnace with "95% efficiency"! At least the furnace is better, however, than what can be achieved with resistive electric heating ($\epsilon_{resistive} \lesssim 7\%$) if the electricity is generated by a combustion-fueled power plant.

We suppose that the fundamental energy source is a quantity of chemical fuel that can be burned in air with an enthalpy of combustion ΔH_c. The 2nd law efficiency of a

---

[1] In this discussion we do not differentiate between the higher and lower heating values of the fuel.

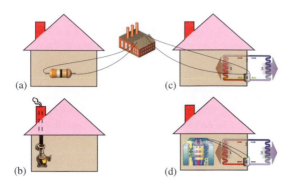

Figure 36.1 Four ways to heat your house, with increasing 2nd law efficiency: (a) resistive heating; (b) a furnace; (c) an electrically powered heat pump; (d) a heat pump powered by a fuel cell.

heating system is the ratio of the space heating Q delivered for a given ΔH_c to the largest value of Q that the laws of thermodynamics allow.

Rather than simply derive the 2nd law efficiency, we work through as examples a series of improving methods of converting the chemical energy in the fuel into heat. In each case we assume that the steps are carried out reversibly so that they are as efficient as possible. In constructing these estimates we encounter several different temperatures. We consider providing heat at a temperature T when the temperature of the environment is T_0. For numerical computation, we take $T_0 = 0\,^\circ\text{C}$ and $T = 40\,^\circ\text{C}$. When a heat engine is part of the system, we take the low-temperature set point T_- of the heat engine also to be $40\,^\circ\text{C}$, while the high-temperature set point T_+ of the heat engine is somewhat arbitrarily set at $400\,^\circ\text{C}$.

Option 1: Resistive electrical heat A particularly inefficient approach would be to burn the fuel in a power plant to supply thermal energy at temperature T_+ to a heat engine that in turn produces electricity, and then use the electricity to power a resistive heater. Although the resistive heater is nearly 100% efficient at converting electricity into heat, the power plant can at most reach the Carnot efficiency, so in this case the space heat provided is bounded by[2]

$$Q_{\text{resistive}} \leq (1 - T_-/T_+)\,|\Delta H_c| \cong 0.53|\Delta H_c|\,. \quad (36.14)$$

Option 2: A good furnace Clearly it is better simply to burn the fuel directly in a furnace at the location where the heat is required, with a bound

$$Q_{\text{furnace}} \leq |\Delta H_c|\,. \quad (36.15)$$

As already mentioned, it is not unusual to find modern furnaces reach 95% of this limit.

Option 3: An electrically powered heat pump It is possible to do better by using electricity from the power plant to power a heat pump that brings in heat from the environment, since the CoP of the heat pump can be larger than one. In this case the limit is

$$Q_{\substack{\text{electric} \\ \text{heat pump}}} \leq \frac{1 - T_-/T_+}{1 - T_0/T}|\Delta H_c| \cong 4.1|\Delta H_c|\,. \quad (36.16)$$

The numerator factor, $1 - T_-/T_+$, is the penalty for using a heat engine to generate electricity, while the denominator factor, $1 - T_0/T$, is the benefit of using the electricity to pump heat in from the environment rather than converting it directly to heat in a resistor.

Option 4: A chemically powered heat pump Finally, we can use a fuel cell to convert to convert the fuel's (Gibbs) free energy of combustion $|\Delta G_c|$ directly to electricity, bypassing the need for a heat engine.[3] Then the amount of heat supplied to the living space would be at most

$$Q_{\substack{\text{fuel cell} \,\& \\ \text{heat pump}}} \leq \frac{1}{1 - T_0/T}|\Delta G_c| \approx 7.8|\Delta G_c|\,. \quad (36.17)$$

It is not possible to do better than this with any system. Note that $\Delta G_c(T_0)$ is very close to ΔH_c for most common fuels, but generally slightly smaller. Although it appears naively that Option 3 could lead to an output arbitrarily close to $Q = |\Delta H_c|/(1 - T_0/T) \approx 7.8|\Delta H_c|$ by working with an arbitrarily high temperature T_+, the *Gibbs condition* (§9.4) for the combustion reaction to go forward generally imposes a maximum possible combustion temperature, which along with other thermodynamic considerations (see §36.4.1) leads to the same limit as eq. (36.17) for an arbitrary combustion process. Furthermore, there are practical limitations from materials that make it difficult to approach the maximum allowed efficiency of a chemically powered heat pump by using combustion. By avoiding thermal energy entirely as an intermediate stage, the energy $|\Delta G_c|$ can be converted into work with minimal waste, and a reversible heat pump can use that work to transfer thermal energy from T_0 to T in the thermodynamically most efficient way. So eq. (36.17) is the "best that can be achieved" in principle, and the 2nd law efficiency of any device that converts fuel into space heating is given by

$$\epsilon_{\text{fuel}} = \frac{Q}{Q_{\substack{\text{fuel cell} \,\& \\ \text{heat pump}}}} = \frac{Q}{|\Delta G_c|}(1 - T_0/T)\,. \quad (36.18)$$

---

[2] See Example 36.3 for some subtleties associated with thermodynamic bounds on the useful energy that can be obtained by combustion of a fuel.

[3] See §36.4.1 and §37.3.1 for an explanation of why ΔG_c rather than ΔH_c appears here.

The limits on the 2nd law efficiencies of the systems considered are then

$$\epsilon_{\text{resistive}} \le (1 - T_-/T_+)(1 - T_0/T)\,|\Delta H_c/\Delta G_c| \approx 0.067,$$

$$\epsilon_{\text{furnace}} \le (1 - T_0/T)\,|\Delta H_c/\Delta G_c| \approx 0.13,$$

$$\epsilon_{\substack{\text{electric} \\ \text{heat pump}}} \le (1 - T_-/T_+)\,|\Delta H_c/\Delta G_c| \approx 0.53, \quad (36.19)$$

$$\epsilon_{\substack{\text{fuel cell \&} \\ \text{heat pump}}} \le 1,$$

for the temperatures specified in this example and setting $|\Delta H_c/\Delta G_c| \approx 1$.

On the face of it, the furnace looks like a rather poor way to provide space heating. In practice, however, the situation is more complicated. Stable, reliable, and affordable fuel cells for home use have only recently entered the market and are still a developing technology. The next best choice seems to be an electrically powered heat pump. Commercially available heat pumps typically reach only 30–40% of their 2nd law limit [278], so in practice

$$\epsilon_{\substack{\text{electric} \\ \text{heat pump}}} \approx (0.3 - 0.4) \times 0.53 = (0.16 - 0.21). \quad (36.20)$$

In contrast, an off-the-shelf home furnace can come close to its 2nd law efficiency limit of $\epsilon_{\text{furnace}} \approx 0.13$. Although a heat pump wins, it does not win by much. The choice would likely hinge on other considerations such as the relative costs of the systems, the cost of the heating fuel versus electricity, and the desire to decrease CO_2 emissions from the primary energy source for the electrical energy. In new construction the heat pump has an additional advantage of providing air conditioning in the cooling season.

36.4 Exergy

The denominator in the definition of 2nd law efficiency requires calculation of the maximum amount of useful work that can be extracted from a given system under specific circumstances. This quantity defines a new thermodynamic function that is a measure of the quality of an energy source. The original name for this function was *available work* or *availability*, but the name *exergy* now has gained wide acceptance.

Exergy, or **available work**, often denoted by B, is defined as *the maximum amount of usable work that can be provided by a system or a material as it is brought into thermodynamic equilibrium with its environment.* Exergy is not a *state function*, because unlike a state function it refers both to the system and to its environment. When a system is in equilibrium with its environment, at the same temperature (T_0) and pressure (p_0), etc., it is said to be in the **dead state** because no further work can be extracted from it. Thus, for example, our atmosphere

Exergy

Exergy (B) is the maximum amount of usable work that can be provided by a system or a material as it is brought into thermodynamic equilibrium with its environment. The 2nd law efficiency of a device that performs work as it proceeds from state 1 to state 2 is

$$\epsilon_{1 \to 2} = \frac{W_{\text{actual}}(1 \to 2)}{B_1 - B_2}.$$

Conversely the 2nd law efficiency of a process that consumes work as it proceeds from state 1 to state 2 is

$$\epsilon_{1 \to 2} = \frac{B_2 - B_1}{W_{\text{actual}}(1 \to 2)}.$$

(at rest) contains a great deal of thermal and mechanical energy by virtue of its temperature and pressure, but we cannot extract energy from the quiescent atmosphere, and its exergy is zero. Exergy proves to be a very useful way to compare the *quality* of different energy resources.

36.4.1 The Exergy of Various Forms of Energy

Systems that are out of equilibrium with their environment have exergy to the extent that they can be harnessed to do work.

Exergy of mechanical or electrical energy Both electrical and mechanical energy can in principle be converted completely into useful work. Thus the exergy of a mechanical or electrical device is equal to its stored energy. The exergy of a mechanical system, however, depends on its location and velocity relative to the reference frame in which the *dead state* of minimal energy is defined. Thus a rock of mass m at a height z falling with velocity $v = -v\hat{z}$ has exergy $B = mgz + \frac{1}{2}mv^2$ relative to an observer at rest at $z = 0$. Similarly, the exergy of stored electromagnetic energy depends upon the relevant value of the zero of electrostatic potential.

Exergy of a heat reservoir A geothermal energy source, for example, can be modeled as a heat reservoir at a fixed temperature T. Assuming that the heat reservoir is large enough that its temperature does not change appreciably when a quantity Q of thermal energy is removed, the maximum amount of useful work that can be extracted from Q is given by the Carnot limit,

---

Example 36.2   2nd Law Efficiency of a Solar Hot Water Heater

A typical intermediate-temperature solar thermal energy collector of the type described in §24.2.3 is used to heat hot water for a home hot water supply system. The solar collector uses a double layer of non-reflective glass. The temperature of the water exiting the collector is $T_+ = 125\,°C$ when the incoming solar radiation flux is $1000\,W/m^2$. Approximately 10% of the thermal energy captured is lost by heat conduction as the hot water circulates to the storage tank, where it transfers its thermal energy to the stored water, keeping it at a temperature of $T = 60\,°C$. The ambient temperature is $T_0 = 25\,°C$.

What is the approximate 2nd law efficiency of this hot water heater?

According to Figure 24.4, since $\Delta T/I_0 = (T_+ - T_0)/I_0 = 100/1000 = 0.1\,°C\,m^2/W$, the collection efficiency of the collector is roughly ≈ 0.45. Including 10% conductive losses, the 1st law efficiency of this hot water heater is

$$\eta_{\text{solar hw heater}} = Q/Q_{\text{in}} \approx 0.40\,,$$

where Q_{in} is the incident solar thermal energy and Q is the heat delivered to the stored hot water.

Given a supply of thermal energy at 125 °C in an environment at 25 °C, a thermodynamically reversible way to generate thermal energy at 60 °C is first to run a reversible heat engine between high and low temperatures of 125 °C and 25 °C, and then use the work from this engine to run a reversible mechanical heat pump to extract heat from the environment at 25 °C and transfer it to the stored water at 60 °C. A reversible heat engine takes solar thermal energy Q_{in} at $T_+ = 125\,°C$ and produces mechanical energy $W = (1 - T_0/T_+)Q_{\text{in}}$. This work, in turn, can be used to extract $Q = W/(1 - T_0/T)$ from the environment at T_0 and add it to the stored water at T, so the maximum possible Q consistent with the laws of thermodynamics is

$$Q_{\text{max}} = \frac{1 - T_0/T_+}{1 - T_0/T}\,Q_{\text{in}},$$

and the 2nd law efficiency of the solar hot water heater is

$$\epsilon_{\text{solar hw heater}} = \eta_{\text{solar hw heater}}\frac{1 - T_0/T}{1 - T_0/T_+} \approx 0.40\left(\frac{1 - 298/333}{1 - 298/398}\right) \approx 17\%\,.$$

$$B_{\text{hot reservoir}} = (1 - T_0/T)Q \quad \text{for } T > T_0\,. \quad (36.21)$$

If, instead, an effectively infinite "cold" reservoir is available at a temperature T below the ambient temperature T_0, then work can be extracted by running a reversible heat engine between T_0 and T. The maximum useful work that can be extracted from a quantity of thermal energy *delivered to the cold reservoir* is

$$B_{\text{cold reservoir}}(T) = (T_0/T - 1)Q \quad \text{for } T < T_0. \quad (36.22)$$

Ocean thermal energy conversion (§27.6) provides an example of such a situation, where cold water at depth plays the role of the cold reservoir. If a reservoir is not large enough to be assumed to maintain constant temperature, then as the relevant quantity Q of energy is moved in or out, the reservoir must be treated as a finite quantity of hot or cold matter.

Exergy of a finite mass of hot or cold matter A finite quantity of matter at a temperature T above or below the ambient temperature T_0 can supply useful work as it relaxes to the ambient temperature. The maximum useful work is obtained by using the thermal energy to run a reversible heat engine. For definiteness, we suppose that $T > T_0$ and that the process takes place at constant pressure. If an amount of thermal energy dQ at temperature T' is transferred via a heat engine running reversibly, an amount of work $dW = (1 - T_0/T')\,dQ$ can be extracted and the material cools by $dT' = -dQ/C(T')$, where $C(T')$ is the material's heat capacity at constant pressure. The exergy of the hot matter is obtained by combining these two relations and integrating from T to T_0,

$$B_{\text{hot matter}} = \text{Max}\{W_{\text{useful}}\} = -\int_T^{T_0} dT'\,C(T')(1 - T_0/T')\,. \quad (36.23)$$

If the heat capacity can be assumed to be a constant C over the range of temperatures of interest, then

$$B_{\text{hot matter}} = C\left(T - T_0 - T_0\ln(T/T_0)\right)$$
$$= \Delta H\left(1 - \frac{1}{(T/T_0 - 1)}\ln(T/T_0)\right)\,, \quad (36.24)$$

Box 36.1   Exergy – Examples

Exergy of compressed air What is the exergy of n moles of compressed air at temperature T_0 and pressure p? Treat air as an ideal gas.

To reach the dead state, let the gas expand isothermally and reversibly. The work done by the gas is

$$B = \int_V^{V_0} dV\,(p - p_0) = nRT_0\left(\frac{V}{V_0} - 1 + \int_V^{V_0}\frac{dV'}{V'}\right) = nRT_0\left(\ln\frac{p}{p_0} - 1 + \frac{p_0}{p}\right),$$

where we have used $p_0V_0 = pV = nRT_0$.

Exergy of a vacuum A cylinder of volume V is evacuated by pulling out a piston. What is the exergy of the empty cylinder relative to an environment at temperature T_0 and pressure p_0?

To reach the dead state, let the atmosphere push the cylinder back in. Thus $B = p_0V$.

Exergy of cold water What is the exergy per kilogram of water at temperature $T_c < T_0$? Take the specific heat capacity of water c to be constant.

Although the water absorbs energy as it comes into thermal equilibrium, its exergy is positive. To extract work from the cold water, run a reversible heat engine between the environment and the water. The maximum work dW that can be extracted by taking an amount of thermal energy dQ from the environment at T_0 and expelling heat to the cold water at temperature T' is $dW = (1 - T'/T_0)dQ$; the rest of the thermal energy, $dQ_c = dQ - dW = (T'/T_0)dQ$, goes into heating the water, $dQ_c = cdT' = (T'/T_0)dQ$. Combining these equations, $dW = (1 - T'/T_0)(cT_0/T')dT'$, which we integrate from T_c to T to obtain the exergy of the cold water at temperature T,

$$B = c\int_{T_c}^{T_0} dT'\left(\frac{T_0}{T'} - 1\right) = c\left(T_0\ln\frac{T_0}{T_c} - T_0 + T_c\right).$$

For a generic quantity of cold matter at temperature $T < T_0$, constant heat capacity C, and $\Delta H = C(T_0 - T)$,

$$B_{\text{cold matter}} = \Delta H\left(\frac{1}{1 - T/T_0}\ln(T_0/T) - 1\right),$$

in analogy to eq. (36.24).

---

where $\Delta H = C(T - T_0)$ is the enthalpy lost by the material as it cooled. A similar result holds for $T < T_0$ (see Box 36.1). In Figure 36.2, $B_{\text{hot/cold matter}}(T)$ is compared with the exergy of the same amount of enthalpy transferred as heat at constant temperature T (see eqs. (36.21) and (36.22)). As the figure shows, energy stored in a finite hot or cold object is less useful than the same amount of energy extracted from a hot or cold reservoir at constant temperature.

Exergy of compressible systems The exergy of a gas or other compressible system is equal to the useful work that can be extracted by bringing the system to its dead state by thermodynamically reversible processes. B is a function of the state variables $(U, S, V, ...)$ that characterize the system in its given state and those (U_0, S_0, V_0) that characterize its dead state. The exergy of a compressible system is[4]

$$B \equiv \text{Max}\{W_{\text{useful}}\} = (U - U_0) + p_0(V - V_0) - T_0(S - S_0). \tag{36.25}$$

To derive eq. (36.25), consider a closed system initially at (T, p) in an environment with (T_0, p_0). Suppose the system emits heat Q into the environment[5] and does work W as it relaxes to the conditions of the environment. According to the 1st Law, $\Delta U = -Q - W$. Only some of the mechanical work is useful, however: if the system's

diffusive equilibrium after it has come into dynamical and thermal equilibrium at p_0 and T_0. If, for example, the gas is dry air and the environment has a nonzero relative humidity, then work can be extracted as the dry air mixes with the humid air in the environment. In the case of combustion of a fuel at high temperature, the useful work that could be extracted from mixing of exhaust gases into the atmosphere is about 5% of the total useful energy released. In any event, it is difficult to see how useful work could be extracted from the mixing of gases. See [275] for further discussion (see also Question 36.2).

---

[4] This result and several that follow ignore the relatively small amount of work that can be provided by a gas as it comes into

[5] $Q < 0$ if the system absorbs heat from the environment.

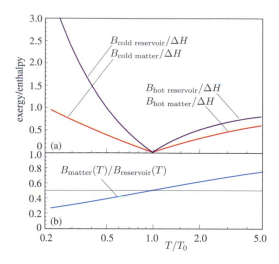

Figure 36.2 (a) The ratio of the exergy to enthalpy for a finite quantity of matter at temperature T and the same ratio for thermal energy extracted from a thermal (infinite) reservoir at the temperature T, as functions of T/T_0. (b) The ratio of the exergy of hot matter at temperature T to the exergy of energy delivered at temperature T as a function of T/T_0. Note the higher exergy of energy delivered at temperature T compared to the exergy of hot matter at the same temperature.

volume changes by ΔV then only the work $W_{\text{useful}} = W - p_0\Delta V$ can be used. The rest, $p_0\Delta V$, is employed to move the atmosphere out of the way. Substituting into the 1st Law, and solving for the useful work,

$$W_{\text{useful}} = -\Delta U - p_0\Delta V - Q. \qquad (36.26)$$

The total entropy change, ΔS_{total}, can be decomposed into the entropy change of the system, ΔS, and the entropy change of the environment, $\Delta S_{\text{environment}} = Q/T_0$,

$$\Delta S_{\text{total}} = \Delta S + Q/T_0, \qquad (36.27)$$

which can be solved for Q and substituted into eq. (36.26),

$$W_{\text{useful}} = -\Delta U - p_0\Delta V + T_0(\Delta S - \Delta S_{\text{total}}). \quad (36.28)$$

The useful work is maximized if the process is reversible, in which case $\Delta S_{\text{total}} = 0$. To obtain the exergy we maximize W_{useful} and proceed to the dead state, so $\Delta U = U_0 - U$, $\Delta V = V_0 - V$, and $\Delta S = S_0 - S$, leading to eq. (36.25).

Exergy of chemical reaction A chemical reaction taking place at the ambient temperature and pressure corresponds to a special case of eq. (36.25), in which U, V, and S refer to properties of the reactants and U_0, V_0, and S_0 correspond to properties of the reaction products. Since the temperature and pressure are held fixed at T_0 and p_0

throughout the reaction, the terms in eq. (36.25) combine to give the change in the Gibbs free energy (see §9.4),

$$\begin{aligned} B_{\text{reaction}} &= (U + p_0 V - T_0 S) - (U_0 + p_0 V_0 - T_0 S_0) \\ &= G_{\text{reactants}} - G_{\text{products}} \\ &= -\Delta G_{\text{reaction}}(T_0, p_0). \end{aligned} \qquad (36.29)$$

Exergy of a fuel In combustion, a fuel reacts chemically with air that can be supplied from the ambient atmosphere.[6] Since combustion is a chemical reaction, the exergy of the fuel is simply the negative Gibbs free energy of combustion $-\Delta G_c$ at the ambient temperature and pressure,

$$B_{\text{fuel}} = -\Delta G_c(T_0, p_0). \qquad (36.30)$$

As mentioned earlier, a fuel cell (§37) operating at ambient temperature and pressure is not subject to the Carnot limit, and an ideal fuel cell can in principle convert the full exergy content of a fuel into electric energy.

It is also possible in principle to extract the full exergy of a fuel through combustion, though in practice existing materials limit the temperature of combustion so that the Carnot limit provides a more stringent bound on the energy that can be extracted. Even for combustion at very high temperatures, however, where the Carnot efficiency approaches one, the Gibbs free energy of combustion at T_0 provides the true theoretical bound on energy that can be extracted.

Consider in particular the situation when $|\Delta H_c(T_0)| > |\Delta G_c(T_0)|$ (so $\Delta S_c(T_0) < 0$); this is true for typical fuels such as methane. One might attempt to take advantage of the Carnot bound on a heat source at high temperature T, $W_{\text{isothermal}}(T) \leq |\Delta H_c(T)|(1 - T_0/T)$ by combusting the fuel *isothermally* at a temperature so high that $W_{\text{isothermal}}(T) > |\Delta G_c(T_0)|$ in apparent violation of the relation $B_{\text{fuel}} = |\Delta G_c(T_0)|$. This is not possible, however, because T is bounded by the condition that $\Delta G_c(T) < 0$, which is required for combustion to proceed. In a simplified situation where ΔH_c and ΔS_c are independent of temperature, for example, the maximum combustion temperature T_M is determined by $\Delta G_c(T_M) = \Delta H_c(T_0) - T_M\Delta S_c(T_0) = 0$, where for simplicity we have ignored the temperature dependence of ΔH_c and ΔS_c. Substituted into the definition of $W_{\text{isothermal}}(T)$, this condition gives $W_{\text{isothermal}}(T_M) = |\Delta G_c(T_0)|$, saturating the exergy bound. To extract all the exergy in this idealized situation, it would also be necessary to transfer the thermal

---

[6] If combustion takes place in pure oxygen or a different mixture of gases, then the work required to prepare this mixture must be subtracted from the work obtained from combustion.

energy from the cooling products of the reaction to the fuel, heating it to the combustion temperature T_M; otherwise, entropy is produced in heating the fuel (Example 36.3). A complete analysis in which the heat capacities of products and reactants are allowed to vary with temperature leads to the same result (Problem 36.13), that in general the exergy (36.30) provides the proper bound on energy that can be extracted from any combustion process.

Nuclear exergy The slow-neutron-induced fission of a fissile nucleus such as ^{235}U is analogous to a chemical reaction with an enthalpy of reaction equal to the fission energy release $\Delta H_r = Q_{\text{fission}}$, with a few important differences. First, fission leads to a variety of different decay products, many of which are themselves unstable, over which we must average. Second, the neutrinos emitted during the reaction cannot contribute in any meaningful way to useful work and should be excluded in practice if not in principle from useful energy. Both of these issues can be resolved by taking the reaction enthalpy to be the average fission energy release excluding neutrinos, which may be found in Table 18.5: $\Delta H_r \approx 200\,\text{MeV}$. The reaction free energy differs from ΔH_r by $T_0 \Delta S$. Since ΔS is of order k_B, $T_0 \Delta S_r$ is much smaller than ΔH_r, and this difference can be ignored. We can therefore take the exergy to be $B_{\text{fission}}(^{235}\text{U}) \approx \Delta H_r \approx 200\,\text{MeV}$. Since there is no analog of a nuclear fuel cell available, the only way to extract useful work from nuclear fission is by using the kinetic energy of the decay products to provide thermal energy to a run a heat engine. Thus the Carnot limit and material limits on the temperature in the reactor core provide a practical limit on the ability to extract the full exergy of nuclear fission.

An unstable nucleus can also be viewed as a source of exergy. For example, certain spacecraft are powered by *thermoelectric generators* (see Box 10.1) where the heat is provided by the α-decay of ^{238}Pu.

Exergy of phase change When a material undergoes a phase transition at temperature T and pressure $p(T)$, it acts as a source of thermal energy at constant temperature. Thus, the maximum useful work that can be obtained from a change of phase is given by eqs. (36.21) and (36.22) with Q replaced by $\Delta H(T)$, the enthalpy change of the phase transition at temperature T and pressure $p(T)$,

$$B_{\text{phase change}}(T) = (1 - T_0/T)\,\Delta H(T), \quad \text{for} \quad T > T_0,$$
$$= (T_0/T - 1)\,\Delta H(T), \quad \text{for} \quad T < T_0. \tag{36.31}$$

In all the cases just considered, which can be thought of as essentially special cases of the general formula eq. (36.28), the actual useful work that can be performed by

a device is reduced from the exergy by a term proportional to the amount of entropy generated by the process,

$$W_{\text{useful}} = B - T_0 \Delta S_{\text{total}}. \tag{36.32}$$

This makes it clear that the production of entropy is what reduces the work that can be obtained from a system. Anything that generates excess entropy – friction, heat transfer across a nonzero temperature difference, free expansion – reduces the useful work that a system can perform. Thus the quantity $T_0 \Delta S_{\text{total}}$ is often called **irreversibility** or **lost work**, and is a measure of the inefficiency of a system. Note that the notion of entropy also clarifies the definition of the term useful work. **Useful work** can be thought of as work that increases the exergy of some other system.

36.4.2 Exergy and 2nd Law Efficiency

When a system or device makes a transition from a state 1 to a state 2 the useful work done is related to the exergy change by

$$W_{\text{useful}}(1 \to 2) = B_1 - B_2 - T_0 \Delta S_{\text{total}}(1 \to 2). \tag{36.33}$$

This result can be derived by viewing the transition as a transition from 1 to the dead state followed by a transition from the dead state back to state 2. While we have focused on a simple class of thermodynamic systems that can be described as in (36.25), it is possible to define exergy and to derive the relation (36.33) under much more general conditions. There are some additional subtleties, however, for more complicated systems such as those where the entropy of mixing becomes relevant (Question 36.2).

We have defined exergy as the maximum useful work that can be provided by a system as it relaxes from some given condition to the dead state. It should be clear that the exergy could equally well have been defined as the *minimum* amount of work that is required to raise a system to a given condition when it begins in the dead state, since this minimum is achieved when the process occurs reversibly, i.e. with no excess entropy generation. So the actual work consumed to bring a device or system from the dead state to a given configuration through a given process with total entropy production S_{total} is

$$W_{\text{consumed}} = B + T_0 \Delta S_{\text{total}}, \tag{36.34}$$

and the work consumed by a system when it makes a transition from state 1 to state 2 is

$$W_{\text{consumed}}(1 \to 2) = B_2 - B_1 + T_0 \Delta S_{\text{total}}(1 \to 2). \tag{36.35}$$

Whether work is produced or consumed, the performance of an actual device or system departs from the ideal because of irreversibility.

Example 36.3   Exergy in Adiabatic Combustion

The exergy content of a fuel is the negative of the Gibbs free energy of combustion at the ambient temperature (and pressure) $-\Delta G_c(T_0)$. In realistic combustion processes, however, entropy is created as materials are heated to the combustion temperature, and furthermore the Carnot efficiency associated with the temperature of the combustion chamber further reduces the 2nd law efficiency of combustion systems. In particular, when a fuel is burned adiabatically at high temperature T additional entropy is created, reducing the exergy of the combustion products significantly below the Carnot limit $|\Delta H_c|(1 - T_0/T)$.

As an initial state, we take a quantity of fuel together with the amount of air necessary for complete combustion, at the ambient temperature T_0 and pressure p_0. The final state after combustion consists of the reaction products at temperature T_f. (T_f is determined below.) The hot gases that result from combustion provide power for a heat engine. To maximize the thermal energy of the combustion products, we assume that the combustion takes place adiabatically, so that no heat leaks out into the environment, and at constant pressure, so that the combustion products do no useful work. Thus from eq. (36.33) with $W_{useful} = 0$, the change in exergy during the combustion reaction is given by $B_{products} = B_{fuel} - T_0 \Delta S_{total}$. Since entropy is created when the system is heated (irreversibly) to T_f, the exergy of the reaction products is less than the exergy contained in the original fuel at the ambient temperature.

As an example [275], we consider combustion of methane in air (21% O_2 and 78% N_2),

$$CH_4 + 2O_2 + 7.5N_2 \rightarrow CO_2 + 2H_2O(g) + 7.5N_2 , \quad \text{with } \Delta H_c^0 = -800\,\text{kJ/mol} .$$

Note that the nitrogen, although inert, is the major component of the hot gases that exit the combustion chamber. Since the combustion process is adiabatic and no useful work is done, enthalpy is conserved (see eq. (13.1)),

$$H_{products}^0 - H_{reactants}^0 = -\Delta H_c^0 = \int_{T_0}^{T_f} dT' \sum_j n_j \hat{C}_{p,j}(T') = ncR(T_f - T_0) ,$$

where the integral gives the enthalpy required to heat the combustion products to temperature T_f. For simplicity we have assumed that the heat capacities of the combustion products are constant over the temperature range of interest and defined the parameter c by $\sum_j n_j \hat{C}_{p,j} \equiv ncR$, where $n = 10.5$ is the total number of moles of gas produced, R is the gas constant, and a careful calculation gives $c \cong 4.75$. The temperature defined by this equation is known as the *adiabatic flame temperature*, $T_f = T_0 - \Delta H_c^0/ncR$. Taking $T_0 = 300$ K, we find $T_f \cong 2240$ K.

Knowing the temperature T_f, the total change in the entropy is easily computed (see Example 8.2),

$$\Delta S_{total} = S_{products}^0 - S_{reactants}^0 + ncR \int_{T_0}^{T_f} \frac{dT'}{T'} = \Delta S_c^0 + ncR \ln(T_f/T_0) .$$

Substituting this into the equation defining $B_{products}$ ($B_{products} = B_{fuel} - T_0 \Delta S_{total}$),

$$B_{products} = -\Delta G_c^0 - T_0 \Delta S_c^0 - ncRT_0 \ln(T_f/T_0) = |\Delta H_c^0| \left(1 - \frac{T_0}{T_f - T_0} \ln(T_f/T_0)\right) \cong 800\,\text{kJ/mol} - 250\,\text{kJ/mol} ,$$

where the last term is the exergy destroyed by the combustion process, in this case equal to $250/800 \cong 31\%$ of the enthalpy of combustion.

Note that the same result can be obtained by regarding the products of combustion as hot matter at temperature T_f with enthalpy $|\Delta H_c^0|$, from which work is recovered as it cools (see eq. (36.24)). Thus, the reduction in $B_{products}$ compared to $|\Delta H_c|(1 - T_0/T_f)$ comes from the unavoidable drop in Carnot efficiency as the combustion products cool from T_f to T_0.

Realistic combustion cycles such as internal combustion or Brayton cycles use some of the output work of the engine or turbine to compress and heat the air/fuel mixture before combustion. What they thereby avoid in exergy destruction during combustion they lose by diverting engine or turbine work to preheating and precompression of the reactants. Thus the fact that the 1st law efficiency of an internal combustion engine or a gas Brayton cycle falls short of the Carnot limit operating between T_+ and T_0 can be regarded in part as a consequence of exergy destruction in combustion. Some or all of this exergy destruction could in principle be avoided by (reversibly) transferring thermal energy from the exhaust products to the fuel before combustion (Problem 36.13).

We can now give a more quantitative definition of the 2nd law efficiency. For a system or device that does work, it is the ratio of the actual useful work performed to the maximum possible, which is ΔB, and for a system that requires work, it is the ratio of the minimum possible, which is ΔB, to the actual work required,

$$\epsilon = \frac{W_{\text{useful}}}{\Delta B} = 1 - \frac{T_0 \Delta S_{\text{total}}}{\Delta B} \quad \text{(work producing)}, \quad (36.36)$$

$$\epsilon = \frac{\Delta B}{W_{\text{consumed}}} = \left(1 + \frac{T_0 \Delta S_{\text{total}}}{\Delta B}\right)^{-1} \quad \text{(work consuming)}. \quad (36.37)$$

The departure of ϵ from one is governed by $T_0 \Delta S_{\text{total}} / \Delta B$, which is known as the **inefficiency** of the device or system.

36.4.3 Exergy and the "Quality" of Energy

The concept of exergy provides a quantitative measure of the quality of an energy resource. Given two energy resources with the same internal energy, the one with the higher exergy is capable of producing more useful work and can be considered to be of higher quality.

Indeed, the ratio of the exergy to the internal energy of a resource, both referred to the same dead state, is sometimes referred to as the *quality* of an energy resource. Thus, for example, Figure 36.2(a) can be interpreted as showing the quality of energy provided as hot matter or as a heat reservoir as a function of the ratio of T/T_0.

Exergy considerations shed some light on the question raised in §1 regarding the notion of "energy consumption": *The colloquial concept of energy consumption corresponds to destruction of exergy.* Useful work is itself a form of mechanical exergy – it could, for example, be converted without loss to gravitational potential energy – so eq. (36.33) can be regarded as a statement of exergy conservation:

$$B_{\text{final}} + W_{\text{useful}} = B_{\text{initial}} - T_0 \Delta S_{\text{total}}. \quad (36.38)$$

Exergy is conserved in the absence of entropy generation. Since the 2nd Law requires that $\Delta S_{\text{total}} \geq 0$, total exergy cannot increase. When friction brings a moving object to rest, when a hot stone is thrown into a bucket of cold water, or when a balloon pops, energy is conserved, but exergy – the usefulness of the kinetic, thermal, or compressional energy – is destroyed. These ideas are explored quantitatively in Problems 36.9 and 36.10. Exergy – in the form of incident solar radiation, in the collective motion of air and water in wind, waves, tides and currents, in Earth's internal heat, and in chemical and nuclear energy stored in fossil fuels, biofuels, and nuclear fuels – rather than energy

---

Exergy and the Quality of Energy

Exergy is a measure of the *quality* of an energy resource. Of two resources with the same energy, the one with the greater exergy can provide more useful work, and is in this sense, of higher quality. Energy is conserved; but the 2nd Law requires that exergy can only decrease. "Energy consumption" should, more correctly, be referred to as "exergy destruction."

---

per se, is the resource that humankind relies on to power modern societies.

36.5 Efficiency and Conservation Case Studies

There are a variety of ways in which humans can modify their behavior and energy choices to reduce energy use or the potentially harmful impact of their energy use. As we have defined it, *energy efficiency* relates to changes of technology or usage patterns that decrease the amount of a specific energy resource needed for a given desired energy output. In contrast, *energy conservation* refers to situations in which one forgoes or modifies some desired output utility in order to reduce energy consumption. For example, adding regenerative braking to an automobile model that is otherwise unchanged results in a gain in energy efficiency, while two neighbors carpooling to work at the same office is an example of energy conservation.

Many other situations are commonly termed as energy efficiency or conservation, although they do not fit these more precise definitions. For example, purchasing a car with a smaller coefficient of drag c_d (§2.3.1) results in reduced energy use; since the net mechanical energy delivered to the drive train is also decreased this is not precisely *energy efficiency* as we have defined it, but the net utility to the user is also not compromised, so it does not match the narrow definition of *energy conservation* given above. This type of situation may be regarded as a way of reducing the destruction of exergy. The utility of the automobile has stayed the same, while the energy lost to extra entropy production through air resistance has been decreased. In the terminology of the previous section, the *irreversibility* has been reduced. Avoided destruction of exergy turns out to be a useful way of characterizing some reductions in energy consumption that do not fall neatly into the categories of increased efficiency or energy conservation as we have defined them. We refer to changes that result in the same utility with less destruction of exergy as *energy savings*. In this section we elaborate on a few case studies of energy conservation, efficiency, and energy savings, including scenarios that lie somewhere between these definitions. Note that for most of the fuels we consider – fossil fuels, biofuels, nuclear fuel, etc. – the exergy of the fuel is generally fairly close to the enthalpy of combustion or energy content, so the extent to which destruction of exergy is avoided is quite closely correlated with the 1st law efficiency of energy utilization. In this section we thus use these notions somewhat interchangeably.

It can be tricky to evaluate energy savings that arise from substituting one primary energy source for another. In general, fossil fuels, nuclear power, and other energy resources that are typically utilized through a thermal energy intermediate lead to greater destruction of theoretically available exergy, and thus have lower 1st law efficiency, in performing a given function than resources such as hydropower that are already available in mechanical or electrical form. One might, therefore, claim an energy savings by switching from a coal power plant to a hydropower plant. This is somewhat misleading, however, since we have no technology that can efficiently convert the exergy in coal to an entropy-free form such as mechanical or electrical energy without first converting it to thermal energy and thereby destroying a large fraction of the exergy originally contained in the coal. Thus, it is not necessarily useful to refer to such resource choices as providing specific energy efficiency or savings, particularly when the energy from each resource is being utilized to the maximum extent possible by current systems and there is no significantly better technology on the horizon.

In evaluating energy options, there are also important considerations that go beyond energy efficiency, conservation, and savings. The substitution of one primary energy source for another is one example of many energy choices that can modify externalities such as carbon emissions or habitat destruction associated with energy use, even when the net amount of energy used is unchanged. Precisely determining the balance between efficiency, conservation, and other impacts such as *decarbonization* of energy use can be rather complex and can depend upon systems issues. For example, when switching from a gas vehicle to an electric vehicle, the carbon output as well as the energy savings depend upon the source of energy used by the electric power plant.

It is easier to maximize energy efficiency for a given resource choice when the power is produced centrally at a large power plant rather than in small engines in individual vehicles, so generally the switch from gas to electric vehicles can effect some energy savings and concomitant reduction in carbon emissions, even when the power plant uses fossil fuels. If the power plant is a coal plant using outdated technology, however, the energy savings may be minimal or even negative and the carbon output may be greater. If, on the other hand, the power plant uses hydro or solar PV power, there is a dramatic decrease in the carbon emissions per unit of useful work done by the car, while as noted above the value of the reduction in exergy destruction taken in isolation is debatable.

Because the relative merits of using different energy sources with different 1st law efficiencies is difficult to evaluate and depends on many technological and economic factors, we do not focus on that question here. Rather, we consider the relative carbon output of these choices, which is clearer and easier to evaluate quantitatively, and which may provide the principal motivation for changing energy sources in the near term. Continuing with the example of powering an automobile, the carbon output from a natural gas plant producing electricity that is transmitted through the grid to an electric car can be compared quantitatively to the carbon output from the combustion engine, giving a precise measure of effectiveness in reducing the externality of carbon emissions with a given choice of energy sources. There are some subtle issues in precisely computing these numbers; factors including methane leakage in natural gas extraction, transmission losses, etc., must all be taken into account. Nonetheless, it is easy to arrive at a rough estimate of the reduction in carbon emissions by switching to e.g. an electric car powered through the grid by solar energy, since to first approximation the carbon emissions of the solar resource are negligible compared to those from fossil fuel combustion as an energy source. We return to the question of changing energy sources in §36.5.3.

Many studies have pointed out that energy saved by conservation or efficiency functions effectively like a new energy source with very little environmental impact [276]. Saving energy that was otherwise wasted or used to minimal effect allows us to employ that energy to our benefit without mining fossil fuels or uranium, and often with relatively small capital investment – much less, for example, than required to harvest an equivalent amount of renewable energy from the Sun or the wind.

In fact, energy obtained from conservation often has a **negative marginal cost**. For example, turning off lights in unused rooms saves both energy and the cost of electricity. The capital cost of the sensors and electronics necessary to do this automatically and reliably is far less than the recurring cost of purchasing the energy needed to keep the lights on. Since energy conservation reduces the demand for primary energy, most of which comes from burning fossil fuels, energy conservation can also provide CO_2 emissions *abatement* at a negative marginal cost. McKinsey [279], in particular, has studied the CO_2 abatement opportunities as a function of cost per tonne of avoided CO_2 emissions and identified many opportunities for CO_2 abatement at negative marginal cost. For readers interested in a systematic approach to the economic, technical, and political aspects of energy conservation and greenhouse gas abatement, the McKinsey study makes a good starting point.

Rather than attempt to cover the vast spectrum of issues, most of which are fundamentally economic and political, associated with energy conservation, we present a few case studies designed to highlight the use of physics principles – and simple reasoning – to analyze interesting problems in energy efficiency and conservation. In the following sections we compare different energy resources and discuss issues related to decarbonization through resource changes.

36.5.1 The US Electric Power Sector

One of the more striking features of the 2016 US energy flow diagram in Figure 1.2 is how much of the primary energy input to electricity generation goes to waste. Of the 39.6 EJ[7] of primary energy that went to electricity generation, only 13.3 EJ or 33.6% was provided as electric power. What became of the other 26.3 EJ of primary energy input labelled "rejected energy" in the figure? Some losses are essentially unavoidable, the consequence of the Carnot limit on the 1st law efficiency of the conversion of thermal to electrical energy given materials and technology constraints on operating temperatures T_+. Some losses are due to the less-than-optimal performance of existing systems (2nd law inefficiencies) that use a given technology operating at a given T_+. Finally, the 1st law efficiency of power production can in many cases be increased by changing the resource used. In these latter two categories, economic or political priorities or historical traditions have in many cases led to the use of less than optimal technologies or energy sources. Here we analyze US electric power production to determine opportunities for energy efficiency and savings and for CO_2 abatement.

More detailed data from the US EIA helps to clarify the origin of losses in electric power generation [12]. One-third of US net electricity production in 2015 came from coal, with an average 1st law efficiency (electrical energy produced divided by thermal energy input) of 32.7%; one-third came from natural gas with an average

1st law efficiency of 43.2%; 20% came from nuclear plants with an average 1st law efficiency of 32.6%; 6% came from conventional hydroelectric sources with a 1st law efficiency that the EIA estimates at about 90%; and most of the remaining 8% came from wind, biomass, solar, and geothermal sources with a 1st law efficiency that is difficult to estimate without information on individual resources. The EIA further breaks down the natural gas contribution, with about 90% coming from combined cycle (CCGT) plants with a 1st law efficiency averaging 44.6% and the rest coming from gas-fired steam cycle or gas turbine systems with an efficiency of about 30–33%.

As discussed in §33.1.7, modern *ultra-supercritical* coal-fired power plants achieve 1st law efficiencies in the mid-40% range. Thus there is considerable room for improvement in *efficiency* in US coal plants. For natural gas, as also noted in §13, state-of-the-art CCGT 1st law efficiencies in excess of 60% have been achieved. So even though natural gas-fired power plants have higher efficiency than coal plants on average, there is still room for improvements in CCGT plant efficiency. Finally, as explained in §19.5, Generation III light-water reactor power plants that include reheating and regeneration are able to reach 33–37% efficiency. So although there is still room for improvement, nuclear electric power plants are operating closer to their current technical limits than either coal or combined cycle gas plants.

The total opportunities for improving energy efficiency (without changing energy sources) across the US electricity generating sector are significant. Together, coal, natural gas, and nuclear power plants account for 22.3 EJ/y of the 26.8 EJ/y of "rejected energy" in US electricity generation.[8] If these power plants were improved to state-of-the-art systems with the best realizable 1st law efficiency for each primary energy source, the rejected energy would decrease from 22.3 EJ/y to 14.1 EJ/y, saving 8.2 EJ/y of primary energy per year. The associated abatement of CO_2 emissions would be about 0.5 Gt(CO_2)/y, which is almost 10% of total US CO_2 emissions (Problem 36.15).

The opportunity for further reductions in "rejected energy" and for CO_2 abatement through changing energy sources are also significant. As discussed above, in the absence of consideration of externalities such as carbon emissions, there is no clear motivation to reduce rejected energy *per se* if all systems being used deliver the largest fraction of available exergy possible using existing

---

[7] Figure 1.2 measures energy in *quads*, which convert to exajoules by 1 quad = 1.055 EJ.

[8] The remaining 4.5 EJ/y of "rejected energy" reflects transmission losses and the way that the EIA assigns losses to renewable energy source such as hydropower and wind energy.

technology. The motivation for CO_2 abatement, however, provides an incentive to reduce rejected energy from fossil fuel sources, however, since this has the side effect of also reducing carbon emissions. The average 1st law efficiency of coal-fired power plants is less than 3/4 of the average 1st law efficiency of US CCGT plants. Even more striking, the CO_2 intensity (mass of CO_2 emitted per unit of electricity produced) of coal power plants is over 2.5 times the CO_2 intensity of natural gas power plants. Replacing all US coal power plants with natural gas power plants at the current average efficiency would reduce rejected energy by 9.4 EJ/y and reduce CO_2 emissions by over 0.8 Gt(CO_2)/y, abating nearly 15% of US yearly CO_2 emissions. Upgrading natural gas power plants to CCGT plants at 60% efficiency would further reduce rejected energy by 6.4 EJ/y and further abate CO_2 emissions by another 0.3 Gt/y. Of course, shifting away from fossil fuels entirely could in principle reduce electricity production related CO_2 emissions nearly to zero. As emphasized in the introduction to this chapter, decisions such as these among competing technologies require economic and policy analyses that go beyond the technical aspects we have considered.

36.5.2 Automobile Efficiency and Mileage

The transportation sector was the second largest source of rejected energy in the US in 2016 (Figure 1.2). This was also the sector with the lowest 1st law efficiency (21%). This is particularly remarkable because, as discussed in §2, people and goods can in principle be transported across Earth's surface with no losses at all; when transporting mass between two points at sea level, there is no change in potential energy and any kinetic energy transferred to the mass in transit can in principle be recuperated through regenerative braking at the end of the transport process. Most energy used in transport is lost to drag through air resistance, representing an *inefficiency* in the language of the preceding section.

The idea of **vacuum tube trains** or *vactrains* that eliminate air resistance by moving vehicles through evacuated tunnels or pipes has been contemplated for many years. Such transport devices could eliminate mechanical friction in motion by using magnetic levitation or other comparable mechanisms, and could use regenerative braking to recapture energy used in acceleration. Taken together, these measures could reduce the energy cost of transport tremendously. There are numerous practical obstacles to implementing such systems at large scale, particularly for proposals that involve tunneling channels for the vacuum train at great depth over large distances, so that gravitational potential energy can be used as the energy source. At this point there is no realistic plan to implement vacuum

trains in any commercially viable context, but these ideas illustrate the possibility of essentially eliminating energy costs in transport and highlight the absence of any real physical limit that mandates a minimum energy cost for transport processes.

At a more immediately practical level, tremendous energy savings can be realized simply by continuing to decrease the mass and drag coefficient of standard passenger cars and light trucks. The idea of the *hypercar* has been championed by the American physicist Amory Lovins and others for many years, with idealized designs that could reach 130 km/L (300 mpg) by using highly streamlined aerodynamic designs, lightweight but strong composite materials, and a hybrid engine with regenerative braking. In recent years some concept cars entering production have begun to approach some of these ideals. The *Volkswagen XL-1* is a two-person vehicle with a diesel-electric hybrid engine that is designed to optimize fuel economy. The XL-1 entered production in 2013, weighs roughly 800 kg, and has a drag coefficient of 0.189, roughly half that of typical commercial cars. Volkswagen quotes a fuel economy of 110 km/L (260 mpg), though that includes charging the battery roughly every 75 km; the pure diesel fuel economy is closer to 50 km/L (120 mpg).

The average fuel economy for new passenger cars in the US was 17 mpg in 1978, and reached 25 mpg ($\sim$10 km/L) in 2015. This amounts to more than a 30% reduction in fuel use and carbon emissions per mile traveled. Increasing fuel economy to 50 or 75 mpg, which seems in principle feasible within decades, would further reduce carbon emissions per mile traveled by a factor of 2 or 3, with 75 mpg giving a reduction to less than 25% of 1978 CO_2 emissions per mile traveled.

While kilometers per liter or miles per gallon are often used as an approximate measure of efficiency, they are not a good measure of energy consumption or a good comparative measure across vehicles. Because a given driver tends to drive a given distance in a week or year, rather than use a given quantity of fuel, what matters on average is the amount of fuel used (L/km) or CO_2 generated (kg/km) to go some fixed distance, which is *inversely proportional* to the mileage. In fact, outside the US automobile energy efficiency is usually quoted in units of liters per 100 km. Consider, for example, two drivers with cars that respectively get 5 km/L and 15 km/L, and suppose each drives 15 000 km in a year. The first driver uses 3000 L of gasoline and the second uses 1000 L. On average, the two drivers use 2000 L to drive 15 000 km per year. Thus $(2000/15 000) \times 100 = 13.3$ L/km is an accurate measure of the yearly gasoline use of the average driver, corresponding to an average of $100/13.3 = 7.5$ km/L, much less than the average of $(5 + 15)/2 = 10$ km/L that would apply when each uses the same quantity of fuel.

One particular utility of using L/km rather than km/L, is to compare the relative increase in fuel economy between different vehicles. Suppose, for example, each car in the previous example made improvements leading to a savings of 500 L in its annual fuel consumption. For the inefficient car, this corresponds to an increase in mileage from 5 km/L (12 mpg) to 6 km/L (14 mpg), a relatively modest 17% increase in efficiency. For the second car, however, a similar fuel savings would require doubling its fuel economy from 15 km/L (35 mpg) to 30 km/L (70 mpg), a much more difficult challenge using current technology. The comparison is illustrated graphically in Figure 36.3. This example makes it clear that for global reduction in carbon emissions, the easiest route to immediate progress is to increase the fuel economy of those vehicles that are least fuel efficient, rather than aiming for increasingly high fuel economies for the vehicles in the fleet that are already most fuel efficient.

We do not attempt to describe or analyze in detail the opportunities for improved energy savings in the full transportation sector; such a task is made quite difficult by the complex mix of modes – land, sea, and air – and scales – local, national, and intercontinental – that characterize transportation. For each transportation mode, however, there is a great potential for energy savings and carbon emission reduction. Beyond improving fuel economy, a switch from fossil fuel sources to electric cars powered by solar or wind energy would clearly provide ground transformation at minimal cost in carbon emissions. While it is unlikely that large passenger aircraft could be run from electric power alone, due to constraints on the mass

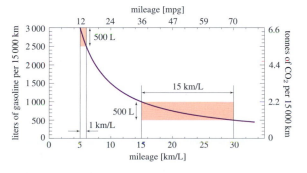

Figure 36.3 Amount of gasoline required to drive 15 000 km as a function of mileage. 500 L are saved by increasing mileage from 5 to 6 km/L. An equivalent savings for an automobile that gets 15 km/L would require doubling its mileage to 30 km/L.

of batteries, one option for a carbon-free approach to air travel would be to use biofuels. Conservation measures for transport such as pooling vehicle use through public transit, trains, carpools, etc., offer the prospect of further reductions. And at a broader level, changes in socioeconomic configurations that reduce commute distance provide opportunities for further savings and carbon emission reduction.

36.5.3 Extended Example: Home Heating

As mentioned in §36.3, home heating is the single largest contributor to home energy consumption, about 4.5 EJ of primary energy in the US [12]. Most of this energy comes in the form of natural gas and is employed with high 1st law efficiency through direct use of the thermal energy from combustion. Nonetheless, homeowners, renters, or landlords concerned about energy consumption or CO_2 emissions related to residential heating can take many steps to reduce both. Some steps, such as turning down the thermostat, turning off the heat at night or lights in unoccupied rooms, represent *energy conservation* methods that come at zero capital cost. Other energy saving measures, such as adding insulation, upgrading windows, or reducing infiltration, provide the same level of heating performance with lower energy consumption, come at relatively low cost, and pay back the investment quickly. Some of these approaches to energy savings were described in §6.4. Still other steps, such as installing solar thermal collectors or a ground source heat pump, require significant investments. In many countries policy makers have advocated, and governments have adopted, incentives to promote one approach or another. How should these measures be compared? Ultimately choices depend on the cost of investments and the relative availability and cost of different forms of energy. The foundation of any comparison, however, lies in the basic physics that determines the amount of energy used, saved, or generated by a given behavior. In this section we compare the results of various different approaches to energy conservation and savings in the context of home heating.

Heat Losses from a Residence

To illustrate the possibilities we analyze as a simple model a "typical" family dwelling with a floor area of 100 m², and an effective exterior wall area (including ceilings, floors, and 15 m² of windows) A_H of 100 m².[9] To make quantitative estimates we also need an estimate of the heat capacity C_H of the dwelling including walls, furniture, air, etc.,

---

[9] This dwelling corresponds roughly to the average in the UK. Dwellings in the US are 50% larger [280, 281].

which is known in this context as the **thermal mass**. We take $C_H = 10$ MJ/K, corresponding to about 2.5 t of water.

Heat must be supplied to interior spaces to make up for the heat that leaks out by conduction, convection, and radiation though walls, ceilings, floors, and windows. Air infiltration through cracks and other openings is also significant. Since all of these losses are proportional to the difference between the interior temperature T and the temperature of the environment T_0, heat leakage can be summarized by a single parameter, the **effective thermal resistance** R_H of the dwelling (see §6.2.5), which enters the equation

$$\dot{Q} = \frac{dQ}{dt} = \frac{A_H}{R_H}(T - T_0) . \qquad (36.39)$$

In principle, R_H could be calculated from an assessment of the materials and construction of the dwelling; in practice this is quite complicated and R_H, or more precisely A_H/R_H, is instead determined empirically. In steady state, some source must provide heat at the rate (36.39) in order to keep the interior temperature constant. Some of this heat may come from normal household activities like cooking and thermal radiation from the living inhabitants, some may come from solar gain through windows, and some may be generated by a furnace or other system installed for the explicit purpose of heating the dwelling. We define $\dot{Q}_F = dQ_F/dt$ to be the heating rate provided by all sources. By observing the heating rate required to maintain a temperature difference $T - T_0$, the quantity A_H/R_H can be determined. We take the effective thermal resistance for the entire exterior shell (including windows) of our dwelling to be $R_H = 0.7$ K m^2/W ($R_H = 4$ ft^2 °F h/BTU in US units), so $A_H/R_H \cong 140$ W/K.

If $\dot{Q}$ and $\dot{Q}_F$ are not equal, the interior temperature will change at a rate proportional to the difference and inversely proportional to the thermal mass of the dwelling,

$$\frac{dT}{dt} = \frac{1}{C_H}(\dot{Q}_F - \dot{Q}). \qquad (36.40)$$

This relation provides a way to empirically determine C_H, which, like R_H, is difficult to compute from first principles.

The yearly total amount of thermal energy required to keep the living space at a constant temperature T can be computed by setting $\dot{Q} = \dot{Q}_F$ and integrating eq. (36.39) over the period when the exterior temperature $T_0(t)$ is less than T,

$$Q_F(T) = \frac{A_H}{R_H} \int dt\, (T - T_0(t))$$

$$= \frac{A_H}{R_H} \tau (T - \overline{T}_0) \equiv \frac{A_H}{R_H} D(T) , \qquad (36.41)$$

where the integral is over the period of the year (the **heating season**) when the integrand is positive, which is taken to have duration τ, $\overline{T}_0$ is the average of the exterior temperature $T_0(t)$ over the heating season, and $D(T)$ has units of °C-days and is known as the number of **heating degree days**. In many locations l the average outside temperature $T_l(j)$ is recorded on each day j, allowing $\overline{T}_0$ to be estimated as the average of the values $T_l(j)$ over the days j in the heating season.

At a given location l, the number of heating degree days $D_l(T)$ thus depends on the assumed interior temperature T. A standard working assumption, which we follow here, is that 20 °C (68 °F) is comfortable and that appliances and occupants generate enough heat to raise the interior temperature by about 1–2 °C (3 °F), making $T_s = 18.3$ °C (65 °F) a standard value.

When averaged over years, the daily outside average temperature $T_l(j)$ in a given location l can be taken to be a smooth function, $T_l(j) \rightarrow T_l(t)$, and $\overline{T}_{0,l}$ can be taken to be the average of this function over the heating season,

$$\overline{T}_{0,l} = \frac{1}{\tau_l} \int_{-\tau_l/2}^{\tau_l/2} dt\, T_l(t) , \qquad (36.42)$$

where τ_l is the length of the heating season in days and $\tau_l = 0$ corresponds to the midpoint of the heating season.

As explained in Box 36.2, in temperate latitudes the function $T_l(t)$ is well approximated by

$$T_l(t) = \overline{T}_l - \Delta T_l \cos(\omega t) , \qquad (36.43)$$

where $\omega = 2\pi/365$ days^{-1}, $\overline{T}_l$ is the average annual temperature, and ΔT_l is the maximum excursion of the average daily temperature from $\overline{T}_l$.

Substituting eq. (36.43) into eq. (36.42) gives an analytic estimate for $\overline{T}_0$

$$\overline{T}_{0,l} \cong \overline{T}_l - \frac{2\Delta T_l}{\tau_l} \int_0^{\tau_l/2} dt\, \cos(\omega t)$$

$$= \overline{T}_l - \frac{2\Delta T_l}{\omega \tau_l} \sin(\omega \tau_l/2), \qquad (36.44)$$

where $\cos(\omega \tau_l/2) = (\overline{T}_l - T_s)/\Delta T_l$ relates the length of the heating season to the parameters that describe the yearly temperature variation at the location in question.

For example, at Boston, $\overline{T}_{\text{Boston}} = 10.7$ °C and $\Delta T_{\text{Boston}} = 12.4$ °C, so that $\tau_{\text{Boston}} = 260$ days, and the sinusoidal approximation of eq. (36.44) yields $D_{\text{Boston}}(T_s) \cong 3110$ °C-days, compared to the actual (1996–2010 average) result of 3060 °C-days (Problem 36.16). With these basic ingredients in hand we can evaluate different strategies to conserve or produce energy.

Box 36.2   Sinusoidal Yearly Temperature Variation

To study the effects of changing the interior temperature, it is useful to have an analytic form for $T_l(j)$. As Figure 23.6(a) indicates, the insolation at temperate latitudes (e.g. $\pm 40°$) varies roughly sinusoidally with a period of 12 months, while at the equator the variation is sinusoidal with a period of 6 months (and smaller amplitude). One expects, therefore, that the temperature at a location l in the temperate latitudes, averaged over many years, would be approximately described by a sinusoidal function

$$T_l(j) = \overline{T}_l - \Delta T_l \cos(\omega j) ,$$

where $\omega = 2\pi/365$ so that the cosine function goes through one oscillation in a year. Here $\overline{T}_l$ is the average annual temperature, ΔT_l is the maximum (average) excursion from $\overline{T}_l$, and days of the year are counted starting at midwinter where T_l is lowest. At locations closer to the equator a component with frequency 2ω should become significant. Monthly average temperature data from two US cities, Boston (latitude 42.3°) and Atlanta (latitude 33.7°), are shown in Figure 36.4, along with a sinusoidal fit. For Boston the $\cos(2\omega j)$ term is negligible (< 0.002 compared to the $\cos(\omega j)$ term). For Atlanta the $\cos(2\omega j)$ term is about 6% of the dominant $\cos(\omega j)$ term. The fit with both $\cos(\omega j)$ and $\cos(2\omega j)$ is shown in green on the Atlanta figure.

Energy Conservation and Savings Strategies

Turning down the thermostat Lowering the setting on the thermostat is a simple and often proposed conservation measure to save energy. How effective is it? If T_s is lowered by one degree, from 18.3 °C to 17.3 °C, D_l is reduced by one degree day for every day of the heating season, so (ignoring the small resulting change in the length of the heating season)

$$\Delta D_l(T_s) = D_l(T_s - 1) - D_l(T_s) = -\tau_l . \quad (36.45)$$

In Boston one saves 260 °C-days for a 1 °C thermostat set back – a very significant savings of 260/3110 = 8%. In warmer climates such as Atlanta, where $\Delta D_l/D_l = 12\%$, the energy savings is less, but the percentage effect is even greater. To get a feel for the numbers, for our standard dwelling as described above we find $Q_F \approx 40$ GJ for Boston (Problem 36.17). An 8% saving amounts to 3 GJ, the energy equivalent of nearly 80 L of home heating oil. We conclude that turning down the thermostat is an effective, cost free, and relatively painless way to conserve energy.

Turning off the heat If turning down the thermostat is a good idea, then turning off the heat entirely might be even better. Not permanently, of course, but turning down the thermostat during the night when residents are asleep is a widely recommended way to conserve energy. Naively, by turning off the heat entirely for eight hours at night one saves ~33% of a day's heating energy, or even more when one considers that exterior temperatures are generally lowest overnight. A considerable amount of energy must, however, be used to reheat the dwelling back to its

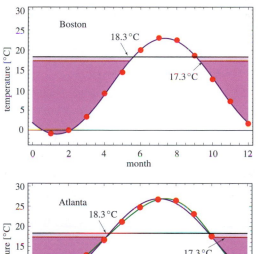

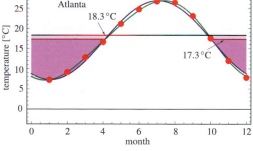

Figure 36.4 Monthly average temperature in Boston and Atlanta. The total shaded area represents the heating degree days with $T_1 = 18.3$ °C; the dark shaded area corresponds to $T_1 = 17.3$ °C. The significance of the green line is explained in Box 36.2. Lowering the thermostat set point by one degree centigrade decreases the seasonal heating requirement by about 8% in Boston and 12% in Atlanta.

normal set point T_s in the morning. It is clear that there must be some net savings, since the energy lost is linear

Example 36.4   Reheating a Residence After an Overnight Setback

Suppose a homeowner's strategy to conserve energy is to turn the heat off for eight hours at night and then reheat her residence to T_s quickly in the morning. For simplicity, assume that the exterior temperature remains fixed at T_l throughout the day and night. *What is the net energy savings as a fraction of the energy required to maintain T_s throughout the whole day? Express your answer in terms of the effective area A_H, thermal mass C_H, and thermal resistance R_H of the dwelling.*

First, the energy saved by turning off the heat for a time Δt can be computed by integrating eq. (36.39),

$$\Delta Q_0(\Delta t) = \frac{A_H}{R_H}(T_s - T_l)\Delta t .$$

To compute the energy required to reheat the residence, we first determine the temperature that the residence reaches after the heat has been off for time Δt. With $\dot{Q}_F = 0$, eqs. (36.39) and (36.40) combine to give

$$\frac{dT}{dt} = -\frac{A_H}{R_H C_H}(T - T_l),$$

which predicts an exponentially decreasing interior temperature. After time Δt, the interior temperature has subsided to

$$T_f \equiv T(\Delta t) = T_l + (T_s - T_l)e^{-A_H \Delta t / R_H C_H} .$$

The energy necessary to reheat the dwelling ΔQ_1 can be computed by integrating eq. (36.40) with $\dot{Q} = 0$ over the short period of reheating, $\Delta Q_1 = C_H(T_s - T_f)$. Substituting for T_f, subtracting the reheat energy from the saved energy, and dividing by the heat required to maintain a temperature T_s for the full 24 hours, we find the fraction f of saved energy,

$$f = \frac{\Delta Q_{\text{TOT}}}{3\Delta Q_0} = \frac{\Delta Q_0 - \Delta Q_1}{3\Delta Q_0} = \frac{1}{3}\left(1 - \frac{R_H C_H}{A_H \Delta t}\left(1 - e^{-A_H \Delta t / R_H C_H}\right)\right),$$

where $\Delta T = T_s - T_l$. To interpret this result, it is helpful to express f as a function of the overnight temperature fall $T_s - T_f$ divided by the inside–outside temperature difference $T_s - T_l$,

$$x \equiv \frac{T_s - T_f}{T_s - T_l} = 1 - e^{-A_H \Delta t / R_H C_H} ,$$

$$f = \frac{1}{3}\left(1 + \frac{x}{\ln(1-x)}\right).$$

This result is plotted in Figure 36.5 and discussed in the text.

in the temperature difference between the interior and the exterior, which is decreased during the eight nighttime hours in this strategy. But how large exactly is the energy savings in this scenario?

To analyze the essential features of this strategy, we assume that the heat is off for $\Delta t = 8$ hours (e.g. from one hour before bedtime to one hour before waking) and that the outside temperature T_l remains fixed over the period (a more accurate estimate would include the variation of T_l overnight). The energy required to reheat the residence back to T_s in the morning is computed in Example 36.4. The result is shown in Figure 36.5 where the percentage of the day's heating energy saved is plotted against the variable $x = (T_f - T_l)/(T_s - T_l)$, which measures the fractional falloff of the interior temperature overnight compared to the difference of interior setpoint

and exterior temperature. If, for example, the residence is poorly insulated and the interior temperature falls by $10\,°C$ when the set interior–exterior temperature difference is $20\,°C$, then $x = 0.5$. From Figure 36.5 it is clear that the poorer the insulation, the greater are the savings ($x \to 1$ as $R_H \to 0$) and *vice versa*. For our dwelling ($A_H = 100\,\text{m}^2$, $C_H = 10\,\text{MJ/K}$, and $R_H = 0.7\,\text{m}^2\,\text{K/W}$) we find $x = 0.34$ and the savings amount to about 6% of the day's heating energy. Note that the true savings when the variation in daily temperature is included may be substantially larger since the day's greatest energy losses occur at night when the outside temperature is minimal (Problem 36.18). Methods similar to the ones developed in this example can be used to explore other scenarios such as lowering the thermostat by a fixed amount during the night (Problem 36.19).

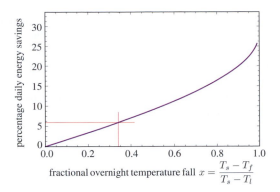

Figure 36.5 Percentage of daily heating energy saved as a function of fractional temperature decrease for a conservation scenario where the heat is turned off overnight and the temperature falls to T_f, and the dwelling is reheated in the morning to the setpoint temperature T_s, in a simplified model with constant exterior temperature T_l. The red lines indicate the fractional temperature loss and energy savings for our typical dwelling.

Insulation The use of insulation to reduce heat flow was the subject of §6.4.1. Since the energy required for heating, eq. (36.41), varies inversely with the effective thermal resistance R_H, any improvement in insulation directly impacts heating energy requirements. As in the case of automobile mileage, greater energy savings with less effort are obtained by improving the insulation of homes with the worst insulation. The cost of insulation in new construction is a minor component of total building cost, and its long-term effect on energy consumption is substantial. When improving R_H it is important to remember that distinct pathways for heat flow contribute to thermal resistance like resistors in parallel. Thus improving one aspect of home insulation can have limited effect if other problems remain unaddressed (Problem 36.21). Some explicit examples of the effects of different insulating materials on the rate of heat loss in a simple building are described in Example 6.2.

Double glazing Windows in general have much lower thermal resistance than walls, particularly walls that have been well insulated. Eliminating windows is not an attractive option: not only do they contribute to occupants' quality of life, but they also add important solar gain on sunny winter days. The thermal resistance of windows, however, can be increased relatively easily.

Suppose that the single pane windows of our standard dwelling were replaced with double glazing. Single glazed windows have a U-factor of $6\,\mathrm{W/Km^2}$. Double glazing improves this to $3\,\mathrm{W/Km^2}$ (with 7 mm air

space).[10] "State of the art" window systems with many layers, special gas filling, and infrared reflective films, may have U-factors close to or even less than one (in SI units), but they may sacrifice solar heat gain on sunny days. The heating energy saved by the replacement can easily be calculated for any location where the average heating degree days are known. For Boston for our standard dwelling with an assumed window area of $15\,\mathrm{m^2}$, replacing single by double glazing saves $\approx 12\,\mathrm{GJ}$ of the average of $40\,\mathrm{GJ}$ of thermal energy required for heating in an average year (Problem 36.22).

The preceding examples describe ways in which energy can be saved and carbon emissions reduced by using simple technologies to reduce the energy needed for home heating. For a homeowner interested, for example, in reducing their energy usage and carbon footprint, the cost and benefit of these types of systems should be compared with other ways of achieving those goals through other avenues. For comparison, therefore, we turn to a much more expensive and active system: electricity generation by solar photovoltaic panels.

A solar photovoltaic system Our standard dwelling could accommodate about $30\,\mathrm{m^2}$ of solar panels on its roof. Assuming that the panels face south at latitude tilt, and have 16.7% net efficiency, this system would be rated at 5 kW, where the "rating" refers to a reference value of $1\,\mathrm{kW/m^2}$ insolation. A check of insolation data indicates that a solar installation in Boston at latitude tilt averages $189\,\mathrm{W/m^2}$, or $4.53\,\mathrm{kWh/m^2\,d}$, averaged over the year. So the system actually produces $\approx 0.189 \times 5000 = 945\,\mathrm{W}$ average power throughout the year. This integrates to $\approx 30\,\mathrm{GJ/y}$ of electrical energy. In comparison, turning down the thermostat $1\,°\mathrm{C}$ would save 3 GJ/yr, replacing single by double glazing would save $\approx 12\,\mathrm{GJ/y}$, and doubling the effective thermal resistance from 0.7 to $1.4\,\mathrm{m^2\,K/W}$ would save $\approx 20\,\mathrm{GJ/y}$. The same PV system in Atlanta would produce $\approx 34\,\mathrm{GJ/y}$, where turning down the thermostat would save about 2.5 GJ/y, turning the heat off overnight would save $\approx 1.2\,\mathrm{GJ/y}$, and doubling the effective insulation would save 10.5 GJ/y.

This comparison between solar PV electricity and measures to conserve home heating energy has ignored the difference in utility between these two forms of energy. Which is more desireable depends both on the particular circumstances and on the homeowner's objective. Suppose, for example, that the homeowner's objective is to

---

[10] These U-factors include the sum of conductive, convective, and radiative heat flow and are determined empirically (see Table 6.3 and the associated discussion).

reduce their carbon footprint. If the homeowner heats with electric resistive heating, then the avoided emissions are the same whether energy is saved through conservation or additional energy is generated by a PV system. If, on the other hand, the dwelling is heated by a gas furnace operating at 80% (1st law) efficiency and the dwelling receives electric power from a gas-fired power plant operating at 40% efficiency, then per joule, the PV-electricity (applied to the home's electric power needs) is twice as effective at reducing the homeowner's carbon footprint as energy saved through conservation measures that reduce the home's heating requirements. The situation regarding carbon emissions is reversed, however, if the home's electric power is generated primarily by a hydroelectric or nuclear power plant.

To directly compare the PV system to the other energy-saving measures for the home heating system, we could imagine using electrical energy from the PV system to run a heat pump. Similar to the cases described in §36.3, this is a very energy-efficient way to heat the house. The detailed analysis of the energy savings is straightforward but slightly more complicated than the case in §36.3 due to the temporal dependence of the temperature difference. Note, however, that the question of whether to use an electric heat pump to heat the house is logically independent of whether installing PV to provide electricity has greater benefit in terms of cost savings or carbon reduction than the other measures discussed here.

One might also consider reducing heating costs by installing a solar thermal heating system. Naively, one could compute based on the PV analysis above that a $20\,\text{m}^2$ area latitude-tilt flat-plate collector working at a typical 33.3% efficiency could provide on average the necessary $40\,\text{GJ/y}$ of thermal energy needed for heating the home. This neglects, however, the seasonal dependence of both insolation and temperature. Most of the energy from such a system would be provided in the summer, when little space heating is needed. While the exact tradeoffs depend on the details of the systems involved, in a location such as Boston with cold winters and reduced winter insolation, a PV system with heat pump is likely a better choice; in summer months the excess electrical power could be used to run a heat pump in reverse or run the other electrical systems in the household.

36.6 Energy Systems: Scales and Transformations

The conservation and energy saving approaches described in the previous section focus primarily on systems at the individual scale. If human society continues to persist on Earth over many more thousands of years in anything like its current form and extent, wholesale changes must take place in the sources of energy used to power human energy systems.

Currently, fossil fuels are the source of roughly 85% of the energy used by humankind. As discussed in §33, this finite resource base may be able to power most human energy needs for a few centuries more at most, even incorporating the use of unconventional fossil fuels such as tar sands and oil shale, and assuming that fossil fuel substitutions can be used for whatever energy systems require, such as through coal–liquid conversion. Changes already underway may, however, transform the energy supply on a shorter time scale than centuries. The pressing issue of climate change, due in large part to greenhouse gas emissions associated with fossil fuel combustion (§34), may catalyze a more rapid change to renewable energy resources. Improvements in technology for extracting solar, wind, geothermal, or other smaller-scale renewable resources combined with increasing scarcity and difficulty of extraction/conversion for fossil fuels may also simply make fossil fuels less economically viable well before the fossil fuel resource base is exhausted. Whatever precise sequence of events unfolds, it seems inevitable that somewhere between 30 and 300 years from now human energy systems will be primarily powered by non-fossil fuel resources.

In this section we give a brief overview of the resources that may play a role in this inevitable change. After a brief review of the origins of renewable and non-renewable energy resources in §36.6.1, we summarize in §36.6.2 the main large-scale resource bases that are available to replace fossil fuels and point out that there is no real risk of humankind "running out of energy." In §36.6.3 we briefly describe a few high-energy-density resource bases – including some that are quite speculative – that may play a role in a low-carbon future. The content of this section is essentially a top-level summary of some of the information contained in the previous chapters on individual resources, emphasizing their scale and availability as future energy sources.

36.6.1 Basic Energy Sources

All of the energy involved in systems on Earth originates either in solar energy (§22), as heat released from radioactive decay (§18, §32), or in Earth's rotational energy dissipated in tidal friction (§31). Solar energy, in particular, cascades into other often more concentrated energy resources through transformations described in earlier chapters. A substantial fraction of the $173\,000\,\text{TJ}$ of solar energy incident on the planet every second eventually ends up as thermal energy stored in the ocean mixed layer, as kinetic energy in winds, ocean waves, and currents, as potential energy of water that could be used for hydropower, or as chemical energy in biological systems (see Box 36.3). In earlier chapters we have analyzed the processes that lead to these transformations and the technologies that have been developed or proposed to harvest energy from such secondary solar resources. Solar power and all of its derivatives, as well as tidal power, are essentially *renewable* resources, providing power at a more-or-less constant rate for the indefinite future.

In addition to the renewable resources, geological processes acting over Earth's long history have resulted in the concentration of otherwise diffuse resources in particular locations where it is economical to extract them. These *non-renewable* resources include fossil fuels, the uranium and thorium fuels for nuclear fission power, and lithium, which is potentially useful in producing fuel for nuclear fusion reactors. High-quality hydrothermal energy sources associated with plate boundaries and/or hotspots, and regions of hot dry rock suitable for enhanced geothermal systems have also formed as a result of geological processes acting over long times and should be considered non-renewable energy resources. Some of the features of renewable and non-renewable resources are summarized in Table 36.3, which serves as a guide to this section.

36.6.2 Large-scale Non-fossil Energy Resources

Focusing on the largest-scale non-fossil energy resources, we summarize the situation for solar, wind, ocean thermal energy, geothermal energy, and biofuels. As can be seen from the figure in Box 36.3, where these resources are found at the top of the graph, and from Table 36.3, these are the only sources among those studied in this book that have sufficient total power available to in principle power human civilization in perpetuity (or for thousands of years in the case of geothermal). Only solar and wind power could be harvested using existing technologies at the scale required to replace a significant fraction of the roughly 15 TW of energy used by humanity in 2016 or the 30 TW that may be needed within a few decades if current trends continue. The technology for exploiting enhanced geothermal systems has not yet reached the level of maturity of solar and wind energy systems, and the exploration of energy production by biological systems in ways that approach the efficiency of existing solar systems is in its infancy; these energy sources could in principle provide terawatts of power, however, in the intermediate-term or long-term future.

Solar power (§22, §25) Earth receives solar energy at a rate of 173 000 TW at the top of the atmosphere. This averages to 343 W/m^2 over Earth's surface. After reductions due to atmospheric absorption and scattering, including weather induced intermittency, about 190 W/m^2 (or ∼180 W/m^2 of exergy) is available on average for capture. Solar power is a *two-dimensional resource* (Box 36.3); the recoverable energy scales with the area of the devices that are deployed to harvest the power. Existing solar power extraction systems, however, can capture only a small fraction of the exergy incident on a given area of solar collectors. Current solar thermal energy systems can convert only a net 3–5% of incident solar energy to useful electrical power over large areas of land. Using currently economical technology, large-scale power stations using photovoltaics can achieve gross efficiencies of around 6%. Even so, as discussed in the solar energy chapters, much of human energy needs could be supplied by covering a small fraction (1% or less per TW) of the world's desert areas and/or urban areas with solar power systems. While the most immediate potential use of solar electric power is to supply energy to the electrical grid, or to supply power for individual residences or commercial buildings, more generally solar electric power could be used to power land transport and heat pumps for residential and commercial heating, so that solar electric power could be used to cover a broad spectrum of human energy uses; this and related issues are discussed further in §36.6.4.

Wind (§28) Winds are driven by temperature and pressure gradients that originate in the uneven pattern of solar energy absorption over Earth's surface. Of the 120 000 TW of solar power absorbed by Earth (assuming an albedo of 0.3), between 1/2% and 3% is believed to be fed into winds. Of this ∼1000 TW, estimates of the power that can be extracted from the wind for human use range between ∼1 and ∼70 TW. (The uncertainty in these estimates is discussed at length in §28.3.2.)

Although wind energy is distributed over Earth's entire surface, a wind harvesting system cannot be extended over an arbitrarily large area. On the one hand, wind energy can be extracted from more than one single line of turbines placed along a direction perpendicular to the wind, as more energy mixes down from above after a distance of several times the turbine height. This general idea governs the optimal placement of wind turbines in the downstream direction as discussed in §28. On the other hand, placing wind turbines over an area many hundreds of kilometers in each direction will have the potential to deplete the wind up to near the top of the troposphere, so that downstream turbines will not have power. Uncertainty in the way that

Box 36.3   Renewable Resources, Dimensionality, and Scaling

At each step in the cascade of energy away from its original source, the energy in any given subsystem is reduced, often by a factor of 50 to 1000, from the initial energy source. While this picture is very rough, it gives some perspective on the relative scale of different energy resources. For example, a substantial fraction of the 173 000 TW of solar energy incident on the planet is absorbed as thermal energy in the mixed layer of the ocean, and the rate of transport of ocean thermal energy from the tropics is on the order of 4000 TW (§27), down by a factor of 50. In the figure below, the various renewable energy resources are graphed using the two axes of power density (W/m^2) and total power (W). Resources that derive their energy from another more fundamental resource are connected by an arrow; for example: solar energy heats the upper ocean, giving rise to ocean thermal energy; ocean thermal energy heats the air giving rise to wind, and wind produces waves.

For each of these resources the power density is measured in terms of a unit area. For solar, ocean thermal, biomass, geothermal, and tidal energy this represents horizontal area over Earth's surface; for wind, waves, hydro, marine and tidal current, this represents a vertical area perpendicular to the flow of the energy in the resource. For all but hydropower, the power density given is that which occurs naturally in the resource, while for hydropower the power density assumes a dam concentrating the resource so that in all cases the power density matches the extraction technology.

A possible way of organizing the various energy resources plotted in this figure is in terms of their **dimensionality**, referring to the number of terrestrial-scale physical dimensions, measured in hundreds or thousands of km, across which the resource is extended. In particular, solar energy and several of the resources that cascade from it, namely ocean thermal energy and biomass, are all *two-dimensional resources*, in the sense that they are distributed over a substantial fraction of the surface of the planet, with areas measured in hundreds of millions of square kilometers. These resources, along with tidal energy and geothermal flow (highlighted in the figure), lie roughly along a diagonal where the ratio of total power to power density is on the order of 10^{14} m^2, corresponding to the square of the natural terrestrial distance scale 10^7 m.

On the other hand, wave energy is essentially a one-dimensional resource, since waves traveling from thousands of kilometers away all carry their energy to a one-dimensional shoreline (with a vertical dimension measured in meters); thus for wave energy the power to power density ratio is roughly the terrestrial distance scale 10^7 m $\times$ (1 m). From this perspective, hydropower and tidal current power are essentially zero-dimensional resources as the energy is concentrated across areas measured in tens of meters in the horizontal and vertical dimensions; for such resources the power to power density ratio corresponds to the product of the typical power-harvesting area and a multiplicity factor (Problem 36.3). Wind lies somewhere between one and two full terrestrial dimensions in extent; like wave energy,

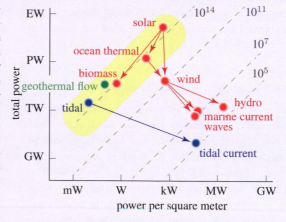

wind energy from thousands of kilometers away all passes through a vertical plane passing through a one-dimensional line on Earth's surface. The height of this plane across which wind energy passes, however, is on the order of the tropopause, roughly 10^4 m. So it is natural that wind energy lies on a diagonal (with power/power density ratio $\sim$10^{11} m^2) slightly more than halfway between the diagonals that pass through the 2D resources and 1D wave energy.

We can think of non-renewable resources such as fossil fuels, uranium, and geothermal energy in the energy flow/cascade and dimensionality framework by considering non-renewable resources as renewable resources that are concentrated in time rather than in space.

Table 36.3 Characteristics of existing and potential low-carbon energy resources. Data on fossil fuels are included for comparison. Resource base estimates are rough order-of-magnitude estimates only. Where question marks appear, they indicate quantities that are impossible to estimate at this time. High-quality resources produce mechanical energy directly or thermal energy at very high temperature; medium-quality resources produce thermal energy at temperatures in the range 300–1000 °C; low-quality resources exploit temperature differences of 30 °C or less.

Renewable resource	Solar cascade	Maturity[a]	Resource base[b]	Power density[b]	Scale dimension d^c	Quality (exergy/energy)
Solar	Primary	1	10^5 TW^d	10^2 W/m^2	2	High
Wind	Tertiary	1	10^2 TW	10^2–10^3 W/m^2	1–2	High
Hydropower	Fourth order	1	1–2 TW	$\lesssim 10^{10}$ W^e	0	High
OTEC	Secondary	2	10^2 TW	2 W/m^2	2	Low
Conventional biofuels	Secondary	1	1 TW	1 W/m^2	2	Medium
Advanced biological	Secondary	3	10^5 TW^f	10^2 W/m^{2f}	2	??
Ocean wave	Fourth order	1–2	2–3 TW	10–100 kW/m	1	High
Tidal stream	Fourth order	1–2	0.1 TW	10 kW/m^2	0	High
Marine current	Fourth order	2	< 1 TW	10 kW/m^2	1	High

Non-renewable resource		Maturity[a]	Total resource base[b]	Energy density[b]	Scale dimension d^c	Quality (exergy/energy)
Fossil fuels		1	135 ZJ^g	50 MJ/kg	–	Medium
Hydrothermal		1	1 TW	$\lesssim 10^9$ W^e	0	Medium
Enhanced geothermal		2	??	~200 PJ/km^2	2	Medium
Nuclear fission		1	> 1 ZJ	0.5 TJ/kg(U)	0	Medium
Nuclear (dt) fusion		3	>10 YJ	220 TJ/kg(Li)	0	Medium

[a] 1→ Deployable with existing technology

   2→ Demonstration technology exists, but technical problems, perhaps major, make deployment impractical at present

   3→ Facing technical challenges of unknown difficulty that may never be overcome

[b] Order-of-magnitude accuracy unless more precision is shown

[c] System for harvesting energy scales with size as [length]^d

[d] Over land between latitudes 60° and −60°

[e] Output of largest existing facility

[f] Effectively a form of solar power

[g] From [282].

the finite vertical nature of the resource impacts power captured by large-scale wind harvesting systems is one reason for the wide range in the estimates of total power that can be extracted from the wind. A consequence of the spacing needs between wind turbines is that land use (measured in km^2/MW) is higher for wind turbines than for solar plants. Wind turbines, however, can more easily be combined with other uses of land than can solar installations.

Wind turbine technology has reached a state where wind farms placed in favorable locations could likely supply a substantial fraction of current electric power needs. A primary technological challenge facing wind (and solar) electric power is the need for large-scale storage (§37) to better match the supply of renewable electric power with demand (§38).

Geothermal (§32) While standard hydrothermal systems probably cannot ever replace a substantial fraction of human energy use, geothermal energy available in deep hot dry rock could in principle power humanity for thousands or tens of thousands of years. The technology necessary to effectively extract this enhanced geothermal energy is not yet in place, though it is not far beyond the current range of closely related drilling and extraction technologies developed in the oil industry. If concerns about seismic impact of deep drilling and fracking can be ameliorated, if underground fluid losses can be controlled, and if efficient and cost-effective systems can be developed, in principle enhanced geothermal systems could provide a non-renewable replacement for fossil fuels that would last for many hundreds or thousands of years.

Biofuels (§26) In discussing potential large-scale energy resources that may eventually replace fossil fuels, energy from biological systems cannot be neglected. A present limitation is that the total rate of conversion of solar energy to terrestrial biomass is only a factor of three or so greater than total current human energy use. A large fraction of biomass is not easily accessible to human use, and much land is already used for growing crops needed for food. Thus, existing biofuels cannot be scaled up to provide a substantial fraction of human energy use. On the other hand, the study of biological energy conversion, and biosystems in general, is at a very early stage of development; in principle biological systems working at the molecular level have much finer local control over solar energy conversion processes than simpler homogeneous materials like a silicon cell. One can imagine hybrid semi-biological systems that combine relatives of photosynthesizing organic material with an engineered physical substrate to provide cheap but highly efficient self-reproducing systems that capture solar energy more effectively than current photovoltaic or solar thermal systems. In a general sense, biofuels can be thought of as solar energy capture devices, a category that also contains photovoltaics and solar thermal power plants. At present, the best biofuels are about a factor of 100 less efficient than a photovoltaic array as a source of electricity (§26.5). Thus several orders of magnitude increase in efficiency would be necessary for advanced biological systems to compete with solar energy as a source of electricity. Nevertheless the ultimate potential of biofuels is unknown and their potential to provide liquid hydrocarbons directly from a renewable source should not be ignored.

Ocean thermal (OTEC) (§27) Ocean thermal energy conversion seeks to use the temperature difference between the surface waters and water at depth in tropical oceans to drive a heat engine. As such, OTEC is clearly a two-dimensional resource, spread around Earth's equatorial zone. A reasonable upper bound (§27.6) on the energy density that could be available for exploitation by OTEC is given by the rate at which solar energy is absorbed at the ocean's surface, roughly $30\,\mathrm{W/m^2}$. Estimating the area of near-equatorial waters to be $140 \times 10^6\,\mathrm{km^2}$, we find a rough estimate of the total OTEC resource to be $4000\,\mathrm{TW}$, corresponding to the rate of ocean thermal energy transport away from the equatorial region; this total renewable resource power is exceeded only by solar power itself.

Unfortunately, ocean thermal energy is a low-quality resource (in the technical sense, i.e. (exergy/energy)) because the temperature differences involved are no more than 17–$22\,^\circ\mathrm{C}$. The Carnot limit on the 1st law efficiency

of an OTEC system is therefore in the range of 5–7%, and the exergy available is on the order of $200\,\mathrm{TW}$, or roughly $2\,\mathrm{W/m^2}$. Given irreversibilities and other losses, 2nd law efficiencies are also relatively low. While the vast extent of the resource has sustained interest in further developing OTEC systems, the low exergy density, which is roughly two orders of magnitude smaller than the original solar resource, makes it clear that terawatt-scale utilization of OTEC would be impossible without covering a sizable fraction of the ocean's surface with low-efficiency energy capture devices. So far OTEC installations have been limited to research and demonstration facilities.

In summary, even if fossil fuels were to completely vanish from the planet over the next 20 years, humanity would have the capacity to replace fossil power sources through solar, wind, and other resources. Technical issues such as storage and transmission remain, but the primary obstacle to such a transition currently is economic. Because of their high energy density and easy extraction, fossil fuels are still the least expensive energy resource for most purposes, though solar and wind energy systems are rapidly becoming competitive in many markets. It seems inevitable that over time fossil fuel energy costs will increase as supply and accessibility decrease, while solar energy, in particular, is likely to continue to decrease in cost over the coming decades. Once solar or wind (or geothermal) energy becomes cheaper on a per watt basis than fossil fuels in a sufficiently wide range of contexts, society may begin a rapid transformation away from fossil fuels to non-carbon based energy resources.

36.6.3 Resources with High Energy Density

For replacement of a large fraction of current fossil fuel based energy sources on the scale of many terawatts in the near term the only viable options are large-scale resources such as solar and wind, and potentially geothermal. These large-scale resources are, however, generally quite diffuse. From the economic point of view, it is desirable to identify carbon-free energy resources with high power density. A glance at Table 36.3 and the right-hand side of the figure in Box 36.3 identifies hydropower, wave power, and marine and tidal current power in addition to nuclear energy as such resources. Even though some of these resources are not sufficiently extensive to produce many terawatts of power, it may be economically advantageous to exploit them in favorable locations. Even producing a fraction of a terawatt with each of these resources could help provide a substantial quantity of high-quality carbon-free electrical power in an economically optimal fashion. We briefly comment on each of these resource options.

Hydropower (§31.1) Even though the total resource base is at most a few terawatts, three characteristics make hydropower a very attractive high-power-density resource. First, hydropower can be concentrated in small areas that are essentially pointlike on the terrestrial scale. Hydroelectric generating facilities that are capable of producing energy at rates exceeding 10 GW can therefore be relatively compact, although the area needed for the associated reservoirs makes hydropower less efficient in overall land use than solar power. Second, the resource has high *quality* (exergy/energy) so conversion to electricity can be very efficient. And third, the resource takes a simple mechanical form – gravitational potential energy of stored water – which can be converted to electricity with well-understood technologies. Hydropower is currently the only renewable energy source that contributes substantially (~3%) to world energy use, providing roughly 0.43 TW of electrical power in 2013. It has been estimated that this contribution could be increased to 1 TW or more over the next century.

Wave energy (§31.2) Although ocean wave energy is distributed over the surface of the world's oceans, as mentioned in Box 36.3, wave power should be considered a *one-dimensional* resource. For example, placing a line of wave energy extracting devices running north–south in an area where waves move in an eastward direction could extract all wave energy over a substantial surface area of ocean.

Annual average wave power near oceanic coastlines ranges from roughly 10 kW/m to over 60 kW/m. Assessments of the total wave power resource are around 2–3 TW, though for reasons discussed in §31.2.4, only a small fraction of this would likely ever be accessible to human use. Nevertheless, the high energy density may make this resource economically favorable in some locations as technologies for wave energy capture develop.

Tidal and marine current energy (§31.3, §31.4) The tides dissipate about 4 TW continuously, mostly through friction with the ocean floor. Only a small fraction of tidal power, perhaps 60–120 GW, is available for human utilization through tidal barrages (§31.3.3) or turbines placed in tidal streams (§31.3.4). The energy density and scaling properties of tidal barrages are similar to hydropower; since there are fewer good locations and since tidal barrages have lower hydraulic heads than dams used for hydropower, their potential as energy sources is a small fraction of that of hydropower. In favorable locations, however, existing tidal barrages are capable of generating as much as 250 MW at peak power.

Because water is almost 1000 times as dense as air, moving water carries much more energy than air moving at a similar speed. Where tidal flows are constrained to narrow streams, underwater turbines can harvest significant amounts of power. Power densities in favorable tidal locations are greater than 10 kW/m^2. In some ocean locations, large-scale marine currents can give similar power density. Despite the logistical challenges, the high power density may make devices such as the tidal stream underwater turbines described in §31.3 economically favorable in locations with substantial tidal flow and appropriate coastline geography. Marine currents such as the Gulf Stream may also be a source of relatively dense power. Like tidal power, total marine current potential is also substantially less than one terawatt.

Nuclear fission power (§16) Nuclear power, while not renewable, is also carbon-free. The attraction of nuclear power is the extremely high energy density of this resource. In §16 we concluded that roughly 200 tonnes of natural uranium could fuel a 1 GWe (assumed ~3 GWth) once-through fission reactor for a year. This corresponds to an energy density of about 500 GJ/kg. So the energy content of a single kilogram of natural uranium is equal to the energy content of about 15 000 L of gasoline.

Uranium is widely distributed over Earth's crust at an abundance of about 2.7 ppmm, so Earth's crust can be viewed as an energy resource with about 1.3 MJ/kg due to its uranium content, although the cost to extract uranium at this concentration is prohibitive. Like fossil fuels, uranium for nuclear fission power is a resource that has been "integrated over time," forming areas of high energy density that are pointlike (zero dimensional) on a terrestrial scale. At present, economically viable uranium deposits have about 0.1% uranium on average, about 400 times the average crustal abundance.

As discussed in detail in §16, world uranium resources recoverable at a cost of $\leq$ \$260/kg U (see [76] for details) are currently estimated at 7.7×10^6 t. With a total energy content of about 3.6 ZJ, this uranium could provide about 270 GWe of nuclear electrical power (the present value) from once-through reactors for about 150 years (§16.2.2). While the cost of uranium is not a major component of the cost of a nuclear fission reactor, and economically recoverable reserves could grow significantly with increased demand, expanding the contribution of nuclear fission power to electric power production beyond that of hydropower would be difficult to achieve with once-through fission reactors, and impossible to sustain over the long term. Much more energy could become available, however, if *breeder reactors* that produce fission fuel from the common isotope of uranium (^{238}U) or from thorium can be made safe and economically viable.

The high energy density of nuclear fission power is offset by the technical complexity of nuclear power plants and by externalities including the environmental impact of uranium mining and enrichment, the hazard presented by radioactive waste, and the threat of nuclear proliferation. The extent to which nuclear energy will be used as part of the world's energy portfolio in the twenty-first century will depend to a large extent upon how the risk of nuclear power is weighed by society relative to the risks and associated costs of fossil fuels and other energy options.

Nuclear fusion power (§19) Nuclear fusion power based on deuterium–tritium (dt) fusion resembles fossil fuels and nuclear fission power in that it is a high-energy-density resource that relies on a naturally occurring substance (in this case lithium) that has been concentrated into enriched ores over geological time. Nuclear fusion, however, differs in one critical way: the technology to exploit nuclear fusion does not yet exist. As explained in §18.4.3, only dt fusion is under consideration at the present time. Tritium, however, does not occur naturally on Earth; it must be created by bombarding lithium with neutrons. Each atom of lithium yields one atom of tritium, which fused with deuterium gives off 17.6 MeV of energy. This corresponds to a potential energy density of about 220 TJ/kg of lithium (Problem 36.25), higher even than that of uranium. Lithium, like uranium, is in limited supply. According to [25] a rough estimate of the world's lithium resources in 2016 was 40 Mt, corresponding to a total fusion energy potential of about ∼10 YJ or approximately 3×10^5 TWy. If even a small fraction of this energy could be captured by a practical fusion reactor, it is enough to provide 30 TW of power to humanity for hundreds if not thousands of years. If, eventually, dd fusion is developed, then the abundance of deuterium (1‰ of the hydrogen in Earth's oceans) would assure an adequate energy supply for an even longer time.

Despite this potential, experiments to demonstrate the practical use of nuclear fusion power face daunting technical challenges, some of which are described in §19.6.3. So for now, the prospect of essentially unlimited (carbon-free) fusion power remains elusive, and is unlikely to play a major role in energy systems in the coming decades.

36.6.4 Transforming Energy Sectors

As has been emphasized throughout the book, there is a major distinction between entropy-free (high-quality) energy sources such as hydro or PV solar power, and lower-quality thermal energy sources that suffer a Carnot limit on the (1st law) efficiency of conversion to electrical or mechanical form. This basic principle plays an

important role in guiding energy choices for different use sectors. Other constraints limit the role that certain energy sources can play in powering systems such as air transport. In the context of the energy resources described above, we briefly summarize how human energy needs may eventually be matched with sustainable energy sources. The numbers used here are very rough orders of magnitude that suggest the scale and type of transformations that may be possible.

Currently, roughly one third (∼5 TW) of global energy use goes to producing roughly 2 TW of electric power. Two thirds of the electric power produced comes from fossil fuel combustion. Roughly an additional 20% (∼3 TW) of total global primary energy consumption goes to powering cars and light trucks, primarily through combustion of petroleum products. Together, the electric power sector and small vehicle land transport sectors account for roughly half of all world energy consumption, and are responsible for the majority of energy losses through 1st law inefficiency (as illustrated in the US economy in Figure 1.2). With conversion to electricity-based vehicles, in principle all of this energy could be supplied by electricity from solar or wind sources, though this would require the development of new storage methods for electric energy (§37) in the cases of solar PV and wind power production. With development of more fuel-efficient vehicles as discussed in §36.5.2, global electric and light vehicle needs could potentially be kept below 4 TW, even with substantial growth in vehicle use. As discussed in §22, this level of electric power production could be achieved with solar installations on a few percent of world desert area.

Much of the remaining half of world energy use involves direct use of thermal energy, for residential, commercial, and industrial space, process and material heating. In favorable locations, some fraction of these systems already directly use solar thermal energy (§24), which is the renewable energy source most naturally suited to provide thermal energy requirements at a large scale, and which has the capacity to expand substantially to cover many moderate-temperature heating needs. In some locations, geothermal energy can also play a large role. In locations and for heating applications where neither solar thermal energy nor geothermal energy are practical, solar- or wind-generated electric power can be used to run heat pumps, so that in principle renewable energy could provide the power for almost all human energy uses, although again storage methods must be improved for renewable electric power to play a dominant role in powering human society. In the short term as the gradual transition towards renewables proceeds, natural gas represents the least carbon-intensive fossil fuel source, and may continue as a thermal energy source for some time. In the long run, biofuels represent an alternative net carbon-free source that may play an important role in providing high-temperature thermal energy sources for industrial processes, airplane engines, and other applications where a combustible fuel is difficult to replace.

A principal technical obstacle to replacing most fossil fuel use with renewable energy from solar and wind power is therefore the need for large-scale energy storage. Long-distance transmission is also an issue, but as discussed in §3, transmission losses simply reduce available energy at distant locations by on the order of 10% and do not represent a conceptual obstacle to replacing most energy sources with renewables. On the other hand, storage is a serious issue. Pumped hydro is the only mechanism currently used for grid-scale energy storage. Natural hydro facilities cannot provide the level of storage needed to run global energy systems from renewable sources like solar and wind. New large-scale energy storage systems will be needed. Some of the alternatives are described in §37: compressed air energy storage is difficult to achieve efficiently at large scale, and the energy density of batteries is limited by basic chemistry. Barring a dramatic development in battery technology, reversible fuel cells, or other novel storage devices, some of the most promising directions for large-scale energy storage in the future may involve simple mechanisms like pumped hydro in artificial caverns or above-ground installations. Solar thermal electric power, as discussed in §24, has the advantage that storage can be integrated in the system, which may become more significant as renewables begin to penetrate more deeply into the energy market. Although there is no obstacle in principle to storing grid-scale quantities of energy in a future driven by renewable energy, unless better systems can be developed, large-scale storage will represent a substantial additional cost that can delay or obstruct the transformation to renewable energy. Thus it may be necessary for natural gas and nuclear power plants to continue to provide baseload power for some time until adequate storage is developed.

With the transformations described above in energy sources, it would seem there is no obstacle in principle to replacing most human energy use with energy from renewable carbon-free energy sources within the coming 30–50 years, though the need for expanded capacity for grid storage may slow this transition and make it more expensive. The principal obstacles to such a transformation are economic, political, and social considerations that lie beyond the scope of this book.

Discussion/Investigation Questions

36.1 Since the conversion of laser light to electricity can be so efficient, why not use low-temperature heat to power a laser and then convert the laser light to electricity at high efficiency?

36.2 Consider a system 1 where a container is divided into two parts by an impermeable membrane. The two equal-size parts contain nitrogen and oxygen gases at the same pressure and temperature. The membrane is removed and the gases are allowed to mix, giving a system 2 in thermodynamic equilibrium. Describe the change in exergy between system 1 and 2. What does this mean about how much useful work could be extracted as the gases are allowed to mix? Can you devise a device that extracts this energy?

36.3 Consider the power/power density ratio $10^5 \, \mathrm{m}^2$ for the "zero-dimensional" hydro resource in the figure in Box 36.3, compared to the "1D" waves resource ($10^7 \, \mathrm{m}^2$) and the "2D" solar resource ($10^7 \, \mathrm{m}$)^2. Try to explain the relative ratio in the hydro case using the scale of a typical hydro power installation (which is very small compared to the terrestrial length scale) and a rough estimate of the number of independent hydropower installations that could be constructed.

36.4 Investigate *vacuum tube trains* as an efficient, low-carbon transportation option. To what extent could they use gravity to accelerate/decelerate? Where would such train lines make sense? Would they be practical for transport of materials or only for people? What are some of the pros and cons of such a transportation system?

36.5 Assume that in a given scenario with no significant increase in nuclear power usage and gradually increasing reliance on renewables over the next century, atmospheric CO_2 levels reach a maximum of 700 ppmv and then stabilize. Now assume that this scenario is varied by building a thousand 1 GW nuclear power plants at a steady rate over the next century. Estimate the decrease in radiative forcing and average surface temperatures assuming that these nuclear plants replace coal power plants. How would you compare the risks and environmental hazards associated with the nuclear power plants against the risks posed by the marginal warming offset by the nuclear plants?

36.6 Consider and discuss the relative benefits in the contexts of efficiency, energy savings, and decarbonization, as well as the practical and economic challenges, for switching from coal power to natural gas or solar power for 50% of your country's total electric power needs.

36.7 What is the exergy of sunlight (in units of W/m^2 at normal incidence)? Discuss how exergy is destroyed when sunlight is converted to electricity by photovoltaic cells, by solar thermal electric power plants, and by direct burning of biomass.

Problems

36.1 Estimate the thermodynamic limit on the 1st law efficiency of a device that turns heat radiation from a wood fire into electricity.

36.2 In §13.3.5 we specified a realistic Rankine cycle working between $T_+ = 600\,°C$ and $T_- = 36\,°C$. We found its efficiency to be 38%. What is its 2nd law efficiency?

36.3 Verify eqs. (36.14)–(36.17).

36.4 A heating system uses thermal energy from a reservoir at temperature T_+ to run a heat engine that in turn powers a mechanical heat pump that takes thermal energy from an environment at temperature T_- and delivers it for space heating at temperature T. Define a suitable 1st law efficiency for this system. First assume that the heat engine dumps its waste heat into the environment at T_-. Show that the limit on the 1st law efficiency of this system is the product of the Carnot limits on the efficiency of the heat engine and the CoP of the mechanical heat pump. Next, assume that the heat engine dumps its waste heat into the interior space at temperature T. Compute the limit on the 1st law efficiency of the system and show that it is *not* the product of the Carnot limits on the efficiency of the heat engine and the CoP of the mechanical heat pump.

36.5 A geothermal resource outputs saturated (liquid) water at $250\,°C$ (as in Example 32.5). Consider two alternative ways to use this resource to deliver space heating at $T = 20\,°C$. First, the hot water is circulated through a heat exchanger that transfers its thermal energy with 90% (1st law) efficiency to the space to be heated. Second, the geothermal resource is used to generate electric power at 13% (1st law) efficiency. The electric power in turn runs a heat pump with a 2nd law efficiency of 33% that operates between T and an outside temperature T_0. Further, the waste heat from the geothermal power plant in the form of saturated water at $50\,°C$ is circulated through a heat exchanger at 90% (1st law) efficiency to provide additional space heating. What is the minimum ambient temperature T_0 for which the second system outperforms the first? [Hint: You will need to look up the specific enthalpy of saturated (liquid) water at various temperatures.]

36.6 A homeowner is trying to decide whether to heat with a furnace rated at 95% efficiency or by an electrically powered heat pump. She lives in a town where electricity is produced by a coal-fired power plant that claims to operate with 1st law efficiency that is 55% of the Carnot limit. The heat pump's CoP is advertised to be 40% of the Carnot limit. Take the values of T_+, T_-, and T from §36.3. For what range of outside temperatures T_0 would the 2nd law efficiency of the furnace be greater than that of the heat pump?

36.7 Reproduce the results in Example 36.4 using eq. (36.25) rather than by the direct calculation of available work that was used in the example. [Hint: you can use the

Sackur–Tetrode formula (8.65) for the entropy of an ideal gas, expressed as a function of temperature and pressure.]

36.8 Systems that are colder than their environment have positive exergy. Suppose a quantity of monatomic gas originally at ambient temperature and pressure (T_0, p_0) is adiabatically and reversibly expanded until its temperature has dropped to T. Compute its exergy from eq. (36.25) and show that it is positive. How can this gas do useful work?

36.9 What is the internal energy and the exergy of a one ton block of stone ($c = 0.8\,kJ/kg\,K$) at $100\,°C$ relative to an environment at $20\,°C$? What would be the mass and internal energy of a block of stone with the same exergy at $60\,°C$? This problem illustrates exergy as a measure of the quality of stored energy.

36.10 Suppose a quantity of gas, held in a fixed volume V and initially at the ambient temperature T_0, is heated by a resistor to temperature T. Let Q be the heat delivered by the resistor and let C_V be the (assumed constant) heat capacity of the gas at constant volume. How much exergy is destroyed in this process? Show that when $T - T_0 \ll T_0$ essentially all of the initial exergy is destroyed.

36.11 In §5.2.2 we described a simple system in which a piston is used to extract energy from a hot gas with efficiency (5.20). Show that the energy extracted precisely corresponds to the *exergy* of the heated gas.

36.12 [**H**] A colloquial measure of power used in the air conditioning industry is the *ton of air conditioning* = $3.517\,kW$ (see §9, Problem 9.5). It is the average power required to melt one ton of ice in one day. Check this number. If melting ice was actually used to provide cooling, it would not only be allowed to melt but would also be allowed to warm to room temperature. Compute the 2nd law efficiency of this method of air conditioning a home with indoor temperature $27\,°C$ and outdoor temperature $37\,°C$. To do this you must find the maximum reversible work that can be done by a ton of ice under these conditions.

36.13 [**H**] Consider an *isothermal* and *isobaric* combustion system that operates as follows: a stream of fuel and air originally at T_0 and p_0 absorbs thermal energy reversibly from a *regenerator* reaching temperature T. It then enters a combustion chamber at temperature T and pressure p_0, where the fuel is combusted yielding enthalpy of combustion $|\Delta H_c(T)|$ per mole. This energy drives a reversible heat engine with $T_+ = T$ and $T_- = T_0$. The combustion products leave the combustion chamber at temperature T and cool to temperature T_0 transferring some of their energy to the regenerator used to heat the incoming fuel–air stream. Assume that the (molar) heat capacity of the products exceeds the heat capacity of the reactants ($f(t) = \hat{C}_{products}(t) - \hat{C}_{reactants}(t) > 0$ for all t between T_0 and T), so some residual energy in the products can also be used to run a reversible heat engine as the products cool. Assume also that $|\Delta H_c(T_0)| > |\Delta G_c(T_0)|$, where $\Delta G_c(T)$ is the Gibbs free energy of combustion. Show that the exergy (maximum useful work) of the fuel combusted in this system is given by

$$B(T) = -\Delta G_c(T_0) + \Delta G_c(T).$$

Since $\Delta G_c(T)$ must be negative for the combustion reaction to occur, $B(T)$ reaches its maximum at T_M where $\Delta G_c(T_M) = 0$.

36.14 Look up the molar enthalpy of combustion $\Delta H_c(T_0)$ and Gibbs free energy of combustion $\Delta G_c(T_0)$ for hydrogen under standard conditions. Notice that $|\Delta H_c(T_0)| > |\Delta G_c(T_0)|$, suggesting that one might be able to extract more than $|\Delta G_c|$ from combustion of hydrogen at high temperature, contradicting the claim that the exergy of a fuel is $|\Delta G_c(T_0)|$. Suppose that the hydrogen is combusted isothermally at temperature T in an engine that runs at the Carnot limit in an environment at T_0. Assume for simplicity that ΔH_c and ΔS_c are independent of temperature. Show that the maximum amount of work that can be extracted from such an engine is in fact $W_{\text{isothermal}}(T_M) = |\Delta G_c(T_0)|$ and find the temperature T_M at which the engine must run to reach this limit.

36.15 According to the US EIA, 1350 Mt(CO_2) were emitted from combustion of coal and 530 Mt(CO_2) were emitted from combustion of natural gas for electricity production in 2015. Using the data in §36.5.1 show that improving the efficiency of coal and natural gas power plants to 45% and 60% respectively would abate roughly 500 Mt/y of CO_2 emissions.

36.16 Given $\overline{T}$ and ΔT for Boston, show that the heating season is on average ≈ 260 days long and evaluate eq. (36.44) to obtain the average number of heating degree days in Boston. Then choose a city for which monthly average temperature data are available, compute $\overline{T}$, ΔT, and D_{city}. Compare your answer with the (commonly available) actual value for D_{city}.

36.17 Show that ≈ 1000 L of home heating oil with energy density of about 40 MJ/L are required to provide heat for our typical dwelling ($A_H = 100\,\text{m}^2$, $C_H = 10\,\text{MJ/K}$, $R_H = 0.7\,\text{m}^2\,\text{K/W}$) in Boston. Assume that the (1st law) efficiency of the oil furnace is 95%.

36.18 [C] Perform a numerical estimate of the savings in heating energy for the residence modeled in §36.5.3 in Boston when the thermostat is turned off for eight hours, assuming that the outdoor temperature over the day varies sinusoidally with a mean temperature of $0\,^\circ\text{C}$ and a maximum variation of $\pm 5\,^\circ\text{C}$. You may assume the interior temperature setpoint is $T_s = 18.3\,^\circ\text{C}$ and that the thermostat is turned off over the coldest eight hours of the day.

36.19 A homeowner turns down the thermostat from $T_s = 18.3\,^\circ\text{C}$ to $T'_s = 16.3\,^\circ\text{C}$ for $t_{\text{total}} = 8\,\text{h}$ every night. Including the effects of reheating, compute the fraction of heating energy saved compared with leaving the thermostat set at $18.3\,^\circ\text{C}$ throughout. Assume that the outside temperature remains fixed at $0\,^\circ\text{C}$. First calculate the time Δt that the heat is off while the dwelling cools by $2\,^\circ\text{C}$. Then compute the energy saved at the low thermostat setting, and finally compute the energy required for reheating. Evaluate your result for the reference dwelling with $A_H = 100\,\text{m}^2$, $R_H = 0.7\,\text{m}^2\,\text{K/W}$, and $C_H = 10\,\text{MJ/K}$. How significant are the effects of reheating?

36.20 A frequent traveler, who lives alone in Washington DC ($\overline{T} \cong 13.2\,^\circ\text{C}$, $\Delta T \cong 12\,^\circ\text{C}$), is away from home every other week throughout the year. Normally she keeps her thermostat at $20\,^\circ\text{C}$, and turns it down to $10\,^\circ\text{C}$ when she is away. Estimate her fractional savings in heating energy first ignoring the need to reheat her home upon return. Then estimate the correction for reheating. Take $A_H = 100\,\text{m}^2$ and $C_H = 10\,\text{MJ/K}$, and assume that she has insulated her home so that $R_H = 2.1\,\text{m}^2\,\text{K/W}$.

36.21 A rural cottage sits on pilings with a breezeway underneath. It has insulated wooden walls ($A_{\text{walls}} = 50\,\text{m}^2$, $R_{\text{walls}} = 3\,\text{m}^2\,\text{K/W}$), a plywood floor ($A_{\text{floor}} = 35\,\text{m}^2$) 2.5 cm thick, and a simple peaked slate roof (slope 30°) covering the floor area. The slates are roughly 2 cm thick, laid over plywood sheathing 2.5 cm thick. The cottage has four double pane glazed windows ($\frac{1}{4}$" air gap), each $1\,\text{m}^2$. Estimate the effective R value for the dwelling as a whole. Now imagine doubling the thermal resistance of any component, walls, floor, or roof. Recompute the effective R-value and decide which would be most effective for energy conservation.

36.22 Verify the claim that the yearly heating requirement of the dwelling described in §36.5.3 drops from $\approx 40\,\text{GJ/y}$ to $\approx 28\,\text{GJ/y}$ when the $15\,\text{m}^2$ of windows are changed from single to double glazing.

36.23 The cottage described in the previous problem sits in a location with average mean temperature $\overline{T} = 8\,^\circ\text{C}$, and maximum average annual variation, $\Delta T = 16\,^\circ\text{C}$. Estimate the number of heating degree days at this location. If you solved the previous problem, calculate the annual heating energy requirement of the cottage before any improvements.

36.24 Check the claimed rating and actual energy production of the solar photovoltaic system described in §36.5.3.

36.25 Lithium is the source of tritium, which is the primary "fuel" for a dt fusion reactor. (The deuterium, which is abundant and relatively cheap, can be considered as analogous to the air required for combustion of a fossil fuel.) Compute the energy density (J/kg) of lithium with respect to dt fusion. First, ignore the difference between isotopes ^{6}Li and ^{7}Li, assume that one lithium atom produces one tritium atom at no energy cost, and use the fact that 17.6 MeV is emitted in *dt* fusion. Next improve your estimate by noting that 7.6% of naturally occuring Li is ^{6}Li and 92.4% is ^{7}Li. Note that the reactions eq. (18.14) and eq. (18.15) are exothermic and endothermic respectively, thereby allowing you to estimate the energy required (or emitted) to convert lithium to tritium. Show that this correction to the energy density of naturally occurring lithium is $\approx$ 10% and that the result agrees with the number quoted in Table 36.3.

Energy Storage

Energy does not always flow directly from its source to its end use in an uninterrupted chain of dynamical physical processes. Instead it is often necessary or useful to store the energy for some time at intermediate points between its natural source and final human use. Energy storage is particularly crucial in several areas:

Grid-scale storage Energy storage is critical for variable energy resources such as solar and wind energy. For these renewable resources to produce a significant fraction of global electricity generation, large-scale facilities capable of storing gigawatt hours (terajoules) of energy over periods from hours to days are required. In §37.2 we give an overview of some of the main candidates for grid-scale energy storage. In the following chapter (§38) we explore other aspects of electric power grids.

Transport Land, water, and air transport represent one of the most difficult energy storage challenges. Gasoline and other fossil fuels provide a convenient, compact, and relatively safe way to deliver energy for transport. Concern regarding the consequences of CO_2 emissions motivates an ongoing search for low-carbon alternatives to fossil fuels in transportation. Powering vehicles with energy from renewable sources requires novel mechanisms for energy storage. Energy storage systems for vehicles must be compact and light, and yet must be extremely safe; when a large amount of energy is stored in a small volume, a disruption to the system (such as an automobile accident) can lead to a catastrophic release of all the stored energy. In §37.3 we describe a variety of possible systems for energy storage in mobile environments.

Load balancing/uninterrupted power supply Another function of energy storage is to provide reliable power when an energy supply may be uneven or undependable. This applies to load balancing on the grid as a whole, and on a smaller scale to guaranteeing a steady power supply for

Reader's Guide

This chapter covers a variety of methods for energy storage, focusing on high-capacity systems appropriate for electric grid-scale storage and on mobile storage systems appropriate for land and air transport. At the grid scale, pumped hydro, compressed air, and thermal energy storage are described. For mobile energy storage, electrochemical batteries, fuel cells, and flywheels are discussed. Energy storage in the electric fields of capacitors and the magnetic fields of superconductors is also described. Energy and power density, total and instantaneous power, round-trip storage efficiency, degradation, and longevity are used to benchmark the performance of different energy storage systems.

Prerequisites: §2 (Mechanics), §3 (Electromagnetism), §5 (Thermal energy), §8 (Entropy and temperature), §9 (Energy in matter).

critical applications (such as hospitals, transport hubs, delicate experiments, etc.). We do not focus particularly on this aspect of storage, though we do describe briefly in §37.4 a few storage mechanisms that are useful in this context.

In the first part of this book, we described the many different physical forms of energy: kinetic and potential mechanical energy, thermal, electromagnetic, chemical, nuclear, and even mass energy. Each of these modalities can in principle be used for energy storage. Flywheels (§37.4.1) store kinetic energy; *pumped hydro* (§37.2.1) stores mechanical potential energy; capacitors and *supercapacitors* (§37.4.2) store energy in electric fields while *superconducting magnetic energy storage* (*SMES*, §37.4.3) stores energy in magnetic fields; many systems, such as tanks of molten salt, store energy in thermal form (§37.2.3), and energy can be stored in compressed air (§37.2.2); batteries store energy in chemical form (§37.3.1), as do fossil fuels, biofuels, and hydrogen that are used in combustion engines and fuel cells (§37.3.2). At this time there are no plausible proposals for storing energy from other sources in nuclear form or in mass energy. While using energy to produce antimatter, which can be converted back into energy by combining with matter in drives like those that power the Star Ship *Enterprise,* would represent a theoretical extreme in compact energy storage systems, for the present such technology remains the stuff of science fiction.

For a more extensive introduction to the science of energy storage systems, the recent text by Huggins [283] is a good starting point.

37.1 Performance Criteria for Energy Storage

Energy storage systems can be characterized by many different performance criteria. Which parameters are important can vary widely depending upon the application. Some significant parameters, and their associated units, include:

Energy density [J/kg or J/m^3]   The quantity of energy that can be stored for a given mass or volume (gravimetric or volumetric energy density), ε, is crucial for mobile energy storage applications.[1]

For most systems – for example, gasoline in the fuel tank of an automobile – energy density (HHV) is computed based on the total stored energy, regardless of the efficiency with which the energy can be practically extracted. In some cases, however, such as batteries, it is conventional to use energy density to refer to the quantity of energy that can actually be extracted. For clarity we use the term **effective energy density**, or **effective specific energy** – denoted $\varepsilon_{\text{effective}}$ – in the latter case.

Total energy capacity [J]   Total energy storage capacity is important for many applications, particularly for grid-scale energy storage.

Round-trip storage efficiency [%]   The fraction of energy that is recovered after storage is important in all applications, particularly in situations where energy is moved in and out of storage frequently on a short time scale.

Instantaneous power [J/s]   The rate at which energy can be removed from storage is crucial in many applications such as car batteries, where rapid acceleration requires large instantaneous power.

---

[1] As mentioned in §3 (Footnote 1), throughout this text energy density generally, but not always, refers to gravimetric energy density (specific energy); the two types of energy density can be distinguished by their units.

Energy Storage

Energy storage is particularly crucial for integrating variable renewable sources into the grid, and for transport applications.

The parameters that are important for energy storage systems depend greatly upon the application. Energy and power density, storage efficiency, safety, cost as well as instantaneous power, degradation rates, and cycle life can all be important parameters in different contexts.

Power density [W/kg]   Power density, or **specific power**, refers to the maximum power that can be stored for a given mass, and is important in many specific applications where short-term bursts of power are needed; power density is less important for grid-scale storage.

Degradation rate [J/s]   The rate at which energy is lost from the storage system should be small relative to the time scale of energy storage.

Longevity [# cycles]   Longevity, or **cycle life** of a storage system indicates the number of times that energy can be stored and recovered from the system before the system becomes less efficient or ceases to function according to original specifications.

Safety issues   Safety issues depend upon application. Systems that can potentially cause a great deal of ancillary damage either by combustion or disintegration must be avoided, for example, in automobiles and airplanes.

Cost   Overall cost, or cost per joule of storage, is important for all applications, more important for some than for others.

The effectiveness of different energy storage devices is often compared by plotting the range of values for a set of storage technologies on a pair of axes characterized by different parameters. For example, the **Ragone plot** (Figure 37.1) graphs storage using the axes of (gravimetric) energy density and power density, and is useful for characterizing different options for mobile storage.

37.2 Grid-scale Storage

Variations in electricity demand have long been met by bringing marginal, usually expensive, generating capacity on- or offline on time scales ranging from minutes to days. Known as *peaking generators*, these systems must

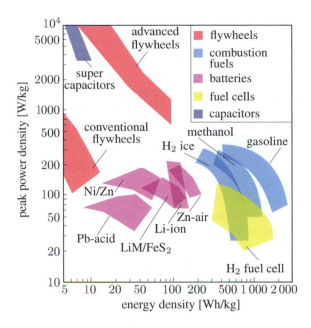

Figure 37.1 A Ragone plot compares energy density to power density for a range of mobile energy storage technologies. Ordinary capacitors (not shown) have power densities as large as 5–10 kW/kg, comparable to supercapacitors, but their energy densities, $\lesssim 0.1$ Wh/kg, are far below those of the other storage technologies shown. After [284]. (Credit: Elsevier (2011))

be kept in standby mode, ready to be dispatched as needed to match changes in load (see §38). As variable renewable energy resources such as solar and wind power come to provide a significant fraction of electric power, the level of variability will increase and the problem of *load matching* will be exacerbated.

Large-scale energy storage becomes increasingly necessary as variable renewable energy sources play a greater role in grid-scale electricity production. To appreciate the scale of the storage requirements, imagine that solar power were to provide 30% of the US electricity supply. In this case, the total solar energy generated each day would be roughly 12 PJ, a substantial fraction of which might need to be stored to guarantee adequate power availability on a 24-hour basis. To store this order of magnitude of energy on the type of lithium-ion batteries used for typical laptops, hundreds of batteries would be required for every adult in the United States (Problem 37.1). While more general aspects of electric grids are the subject of the following chapter (§38), in this section we focus on the principal candidates for grid-scale energy storage.

For grid energy storage, crucial factors are net capacity, round-trip storage efficiency, degradation rate, and cost.

Access rate (instantaneous power) also must be considered, to the extent that storage is used to displace backup generating capacity that responds on short time scales. Energy density, however, is a less important consideration for grid storage than for mobile applications. In particular, storage of gigawatt-hours of energy requires large quantities of material, and is generally only economically viable if the materials are relatively simple and inexpensive. This leads naturally to storage of energy in systems such as pumped water (§37.2.1), compressed air (§37.2.2), or thermal storage in simple bulk materials (§37.2.3). Another strategy for storing excess grid energy would involve **distributed storage**, such as in car batteries or other storage media distributed throughout commercial and residential areas. *Smart grid* technologies that could coordinate such distributed storage use may form a part of grid-scale storage in the future, but we focus here separately on large-scale storage systems appropriate for storing grid-scale energy quantities.

37.2.1 Hydro Storage

The only mechanism currently used for storing substantial quantities of excess electrical energy from the grid is **pumped hydro storage**. The physics of pumped hydro storage is very simple. Electrical energy is used to pump water uphill into a reservoir. Water can be released from the reservoir and used to run a turbine when needed. As discussed in §31.1, hydropower turbines are very efficient. Pumping can also be done quite efficiently, so that the overall round-trip storage efficiency for hydro storage can be above 80%; typical efficiencies of 70–85% are generally quoted. Pumped hydro storage suffers minimal degradation. Energy losses over time are only to evaporation, and for reservoirs that already are used for hydropower, the increase in evaporative loss with an increased water level is relatively small. In 2014, the power capacity of worldwide hydro storage systems was approximately 143 GW, with over 23 GW in the US, and 50 GW in Europe.

While locations for pumped hydro storage are limited, proposals have been made for using existing or artificially excavated underground reservoirs for **underground pumped hydro** (UPH) **storage**. Water is held in the underground reservoir, and pumped upward to a separate reservoir on the surface to store energy. When the energy is needed, the water is allowed to flow downward through vertical pipes to turbines, efficiently transforming the mechanical potential energy of the water back into electrical power. While no large-scale storage facility of this type has yet been built, systems have been considered for installation in old mines, aquifers, and artificially excavated reservoirs. While UPH systems would be more expensive

Pumped Hydro Storage

Energy can be stored by pumping water uphill into a reservoir. The water is released and passed through a turbine when the energy is needed. At present, pumped hydro is the only kind of system used for large-scale storage of energy from electrical grids. Worldwide pumped hydropower capacity in 2014 was roughly 143 GW. The specific energy of pumped hydro is of order 1 kJ/kg per 100 m of hydraulic head.

Figure 37.2 The Hohenwarte II pumped hydro storage facility on the Saale River in Germany has a capacity of 318 MWe and an upper storage basin capable of storing 3×10^6 m^3 of water. (Credit: Vattenfall)

to construct than an above-ground dam in a favorable location on a large river, UPH systems would have less environmental impact, and could dramatically increase the potential range of locations for pumped storage facilities. For a reservoir at a depth of 100 m, the energy density of the stored water would be $\rho g h \cong 1$ MJ/m^3 (volumetric), or $\varepsilon \cong 1$ kJ/kg (gravimetric). While significantly less than the energy density of a typical battery (§37.3.1), the

simplicity of UPH systems may make them economically favorable in some circumstances. Smaller-scale surface pumped storage systems have also been developed using seawater rather than fresh water.

For renewable energy storage, particularly for wind energy, it is possible to power the hydro pumps directly from the renewable source. This helps to even out irregularities in the power source, and eliminates the inefficiency of sequential conversions from mechanical wind energy to electrical energy to mechanical pumping energy.

37.2.2 Compressed Air Energy Storage (CAES)

Another approach to grid-scale energy storage is **compressed air energy storage** (CAES). The concept behind CAES is to use pumps to compress air into large storage facilities such as underground caverns. In principle, the compressed air can then be used directly to drive turbines to return energy to the grid when needed. In existing installations, however, the compressed air is mixed with natural gas and combusted. The resulting high-pressure, high-temperature gas powers a turbine that provides electricity to the grid. In effect the stored energy is used to increase the efficiency of a gas turbine generator. While CAES is often discussed as an option for grid-scale storage, there are substantial technical challenges to achieving high round-trip efficiency for this approach to energy storage. A recent review by Elmgaard and Markussen [285] provides an overview of several approaches to CAES.

CAES schemes differ with respect to the way that they handle the heat that is generated when air is compressed. Two idealized limiting cases are isothermal and adiabatic compression. In either case the efficiency could, in principle, reach 100%. According to the 1st Law (10.1) $-p\,dV = dU - dQ$. When the gas is compressed, $p\,dV$ work is done on it ($p\,dV < 0$). If the process is isothermal, heat is expelled to the environment and the temperature of the gas does not change. Ideally, the system moves along an isotherm as in Figure 10.5, discharging energy into the environment in the compression phase. In principle, the energy can be reclaimed without loss through isothermal expansion. If instead the compression process is adiabatic ($dQ = 0$), the internal energy of the gas increases, and it heats up. Ideally, as it is compressed or expanded, the system can move back and forth along an adiabat as in Figure 10.6 with no loss of energy. Thus, the energy used to do the work can either be transferred to the environment (isothermal) or stored as internal energy of the compressed gas (adiabatic) and fully retrieved when required in an ideal world. Neither isothermal nor adiabatic compression/expansion can be achieved with high efficiency in practice, however. Isothermal compression is

Example 37.1 Dinorwig Pumped Hydro Facility in Wales, UK

When Great Britain anticipated deploying large amounts of nuclear electric power generation in the 1970s, several pumped hydro facilities were constructed to provide rapidly dispatchable peak power, which nuclear plants are unable to supply. The *Dinorwig Power Station* in North Wales, completed in 1983, combined *Llyn Peris*, a natural lake with its surface at approximately 100 m, with a reservoir created by damming the outlet of an abandoned quarry above the lake. The surface of the resulting *Marchlyn Mawr Reservoir* is at roughly 635 m. Its capacity is 9.2×10^6 m^3. Each of the six turbines at Dinorwig is capable of generating up to 288 MWe or pumping at 275 MW. The turbines can go from standstill to full load in less than two minutes. The round-trip efficiency of the pumped hydro storage is 75%.

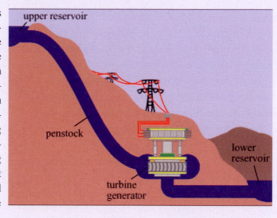

What is the flow rate from the reservoir at maximum power? The total storage capacity is stated to be 9 GWh (32 TJ). To what fraction of the reservoir volume does this capacity correspond?

Assuming that the efficiency of generation and production are equal, electricity is generated at $\sqrt{0.75} = 87\%$ efficiency. The required flow rate of water would be

$$f = \frac{P}{\eta \rho g h} = \frac{6 \times 288\,\text{MW}}{(0.87)(1000\,\text{kg/m}^3)(9.8\,\text{m/s}^2)(535\ \text{m})} \cong 380 \ \text{m}^3/\text{s}.$$

Ignoring the drop in the level of the reservoir, we estimate the volume of water required to produce a total energy of 9 GWhe to be

$$V = \frac{E}{\eta \rho g h} = \frac{(9 \times 10^9\,\text{GWh})(3600\,\text{s/h})}{(0.87)(1000\,\text{kg/m}^3)(9.8\,\text{m/s}^2)(535\ \text{m})} \cong 7.1 \times 10^6 \ \text{m}^3,$$

which is 77% of the reservoir's capacity.

The discovery of the North Sea gas fields and concerns about nuclear safety derailed the UK's anticipated switch to nuclear power. Nevertheless, the Dinorwig plant is still used to meet sudden surges in electricity demand. According to Wikipedia, one such scenario occurs during advertising breaks in popular television shows when British consumers make tea and demand surges by as much as 2.8 GW. Assuming 2–3 kW per kettle, one can estimate the number of viewing households to be of order 10^6.

(Image credit: RWE Power AG)

---

difficult to accomplish rapidly without extremely efficient heat exchangers. Adiabatic compression requires preventing heat exchange with the environment, and even if that were possible, at high compression ratios adiabatic compression would raise the gas to temperatures too high for most storage systems.

Consider first the isothermal idealization, in the approximation that air is an ideal gas, so $dU = 0$. The work done to isothermally compress air initially at pressure p_1 and volume V_1 to a pressure p_2 and a volume $V_2 = p_1 V_1 / p_2$ is, from eq. (10.4),

$$W = N k_B T \ln \frac{V_1}{V_2} = p_2 V_2 \ln \frac{p_2}{p_1}. \qquad (37.1)$$

In a CAES storage facility, typically the air is kept at a minimum pressure p_L after discharge. Additional air that is originally at pressure p_0 is pumped in from the outside until finally the storage facility volume V is filled with air at a maximum pressure p_H. The total work done is the difference between the work required to pressurize *all* the gas from p_0 to p_H and the work that would have already been done pressurizing the initial volume of gas to p_L,

$$W^{\text{isothermal}} = V \left(p_H \ln \frac{p_H}{p_0} - p_L \ln \frac{p_L}{p_0} - (p_H - p_L) \right). \qquad (37.2)$$

The third term in this expression comes from the work supplied by the outside pressure p_0 in the compression process, $p_0(V - V p_H / p_0) - p_0(V - V p_L / p_0)$.

Thermal energy equal to $W^{\text{isothermal}}$ is expelled into the surrounding environment during compression, and this is the maximum work available via subsequent isothermal expansion of the stored air. A particular CAES facility is analyzed in Example 37.2. For the parameters of that facility ($p_H = 75$ atm, $p_L = 45$ atm), the isothermal work is ~ 130 kJ/kg, but due to inefficiencies only a fraction of this energy can be extracted in the re-expansion process.

For an idealized adiabatic system, a similar analysis (Problem 37.5) can be used to compute the total energy stored. The energy stored as internal energy of the gas in the adiabatic case is less than the energy stored in the idealized isothermal case for an equal storage volume and maximum pressure p_H. This can be seen from the fact that at each stage in the compression, as the pressure increases by a given amount, the change in the volume dV is smaller in the adiabatic case (since the temperature increases, increasing the pressure more rapidly with the change in volume). Thus, compression to a final pressure p_H gives a smaller integral $\int p \, dV$ in the adiabatic case. Furthermore, adiabatic compression of air to high pressure gives rise to high temperatures. Compressing a volume of air from pressure p_0 to $p_H = rp_0$ raises the temperature by a factor of $T_H/T_0 = r^{(\gamma-1)/\gamma}$, where $\gamma \cong 1.4$ is the adiabatic index of air (eq. (10.9)). For $r = 75$ and $T_0 = 300$ K, this gives $T_H > 1000$ K; this rise in temperature makes pure adiabatic CAES impractical in most storage environments (Problem 37.6).

Existing compressed air energy storage facilities use mechanical devices known as *intercoolers* to expel thermal energy from the gas as it is compressed. The net result is that the compressed air is stored close to the ambient temperature. For a given V, p_H, and p_L, the stored energy is the same as the isothermal case (37.2). The energy used by the intercoolers, however, must be taken into account when determining the complete efficiency of round-trip storage. Sometimes this form of *irreversible isothermal* compression is referred to as **diabatic compression**, to contrast with *adiabatic* compression in which no heat transfer takes place.

Because it was computed assuming *reversible* isothermal compression, $W^{\text{isothermal}}$ (37.2) expresses the exergy (§36.4) of a volume V of gas at temperature T_0 and pressure p_H *minus* the exergy of the same volume of gas at the same temperature and pressure p_L (Problem 37.7). Note that there is no extra internal energy stored in the compressed air in the chamber. By virtue of its exergy, however, the compressed air can still be used to produce useful work. If the gas were expanded *reversibly* and isothermally back to the initial conditions, energy equal to $W^{\text{isothermal}}$ would be available for conversion to mechanical or electrical

---

Compressed Air Energy Storage (CAES)

Compressed air energy storage is often discussed as a potential mechanism for grid-scale energy storage. Although energy densities achieved in principle can be greater than 100 kJ/kg, this energy is difficult to extract with high efficiency due to the need for rapid and efficient heat exchange. Existing systems use the exergy of compressed air to augment the output of gas turbines, providing only a modest fraction of the stored energy as net output energy.

---

energy. Unfortunately, it is impractical to extract all the available work by allowing the gas to expand isothermally and reversibly, while absorbing heat from the surrounding rock. If, instead, the gas were allowed to expand *adiabatically* from p_H back to p_L a certain amount of work could be extracted (see Example 37.2 and Problem 37.5), but the gas would cool far below T_0. Further work could, in principle, be extracted by running a heat engine between the ambient temperature and the temperature of the adiabatically expanded gas. Indeed, if both the adiabatic expansion and the heat engine were run reversibly, the full amount of available work could be extracted from the compressed gas. Such a complicated reversible process is also impractical. In practice, the compressed gas is used as input to a natural gas powered turbine, thereby increasing the efficiency of the turbine (see Example 37.2), but capturing only a fraction of the exergy stored in the compressed gas.

As of 2016, two utility-scale diabatic CAES plants were in operation: a 290 MW plant in Huntorf, Germany built in 1978, and a 110 MW plant in McIntosh, Alabama built in 1991 (see Example 37.2). Some smaller (1–2 MW) systems are also operating using dedicated manufactured storage vessels. A number of other plants are currently in various stages of design and development.

In order to increase the efficiency of CAES beyond what is possible in existing diabatic plants, in recent years there has been increased interest in **advanced-adiabatic CAES** (AA-CAES) systems, which are designed to remove and store the thermal energy produced when air is compressed adiabatically and then release this energy when the air is expanded. Several systems of this type are currently in design/development phases. In Germany, a project known as ADELE plans to store the thermal energy of the compressed gas in a large heat storage facility containing stone or ceramic bricks with a high heat capacity. Another

---

Example 37.2   McIntosh Compressed Air Energy Storage Plant

In McIntosh, Alabama, a solution-mined salt cavern with volume $270\,000\ \mathrm{m^3}$ is used for compressed air energy storage. Initially the cavern is filled with air at 45 atm. Additional air is pumped in until the pressure reaches a maximum of 75 atm.

In the McIntosh plant, the heat from compression is jettisoned to the external environment. After compression, eq. (37.2) tells us that approximately 3.4 TJ (130 kJ/kg) is stored in the compressed gas. Because the compression is diabatic rather than truly isothermal, however, a substantially larger amount of energy is actually needed to compress the gas into the cavern and expel excess heat to the environment.

The total exergy of 3.4 TJ could only be extracted if the gas were expanded isothermally, which is not practical. Simply allowing the air to expand adiabatically, leaving the original volume of air in the cavern at 45 atm, and allowing the remaining air to expand to 1 atm would do 1.4 TJ of work (Problem 37.8), recapturing about 41% of the energy available in the stored gas. This would not use all the exergy originally stored in the reservoir, since the expanded air would be very cold. (The air expelled to the environment would be at 87 K and the air left in the mine would be at 260 K.) One could, for example, operate a heat engine between the ambient temperature and the cold air and thereby extract more work. In fact if the adiabatic expansion were performed reversibly and a reversible Carnot engine were used to extract further work, all of the 3.4 TJ could be extracted from the stored gas. This is not feasible in practice, however.

To extract useful energy from the compressed air at the McIntosh facility, rather than directly using the work done by expansion, the compressed air is used as input to a natural gas powered turbine. As explained in §13.5.1, in a standard natural gas power plant, much of the combustion energy is used to compress the gas–air mixture. Thus, using the compressed air from CAES increases the efficiency of a natural gas plant. According to [287], at the McIntosh plant, 0.82 MJ of energy used for compression combine with 1.2 MJ of energy from natural gas to provide 1 MJ of output energy. The same amount of natural gas used to fuel a state-of-the-art CCGT power plant at 60% efficiency would produce 0.72 MJ of electric energy. The McIntosh plant can thus use 0.82 MJ of excess wind or solar energy to enhance the energy available from natural gas by 0.28 MJ. This represents a $0.28/0.82 \cong 34\%$ round-trip storage efficiency. A more detailed exergy-based efficiency analysis quotes 36% cycle efficiency for the McIntosh plant, 29% cycle efficiency for the Hundorf plant, and estimates over 40% for a more modern diabatic design proposal [285].

---

possibility is to use water – actually a fine, dense mist – to absorb the heat of compression, forming steam which is then stored and later extracted when it is needed to reheat the air during expansion. An exergy-based analysis of AA-CAES [286] suggests that a cycle efficiency of 50% or more may be possible with AA-CAES, though such systems will be somewhat complicated and may be more expensive than underground pumped hydro facilities of comparable size, which are capable of reaching cycle efficiencies of 75% or more.

37.2.3 Thermal Storage

Thermal energy is much easier to store than mechanical or electromagnetic energy. It may be impractical, however, to store electrical energy by converting it to thermal energy, since the 2nd law penalty that must be paid when converting back to electrical energy would make the storage quite inefficient. Thus thermal energy storage is not a good option for storing the output of solar PV arrays or wind farms. On the other hand, as discussed in §24, *solar thermal energy (STE)* collected for conversion to electrical

Thermal Energy Storage

Storing energy in thermal form makes most sense when the energy originates as thermal energy. For solar thermal power plants, thermal storage is an effective and efficient way of storing energy for days or weeks to match the intermittent supply to variations in demand. Molten salt mixtures have been developed with high heat capacity, low cost, and minimal toxicity, that can store solar energy for days with small losses before it is used to power generators.

power can be stored in thermal form for hours or days before conversion.

A variety of materials have been proposed for storage of solar thermal energy at STE plants. Some existing plants use a *molten salt* composed of a binary mixture

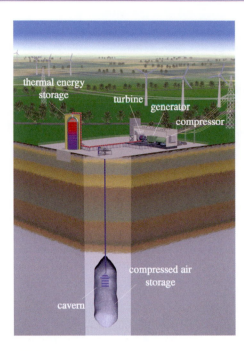

Figure 37.3 Sketch of the proposed ADELE AA-CAES facility. Thermal energy generated when air is compressed is held in a heat storage facility and used to reheat the air when it is expanded and used to drive the turbine. (Credit: RWE Power AG)

Figure 37.4 Molten salt storage tanks at the Andasol 1 power plant in Spain store thermal energy acquired during the day for power generation at night. (Credit: Solar Millenium)

of sodium nitrate ($NaNO_3$) and potassium nitrate (KNO_3) that can be used effectively both for absorption of solar thermal energy and storage (Figure 37.4). This material is relatively inexpensive and non-toxic. Its heat capacity is roughly 1500 J/kg K. The binary nitrate salt has a melting point in the range 130–230 °C, depending upon the precise ratio of nitrate salts used. It can be heated by solar energy to a temperature in the range of 400–550 °C, and can be stored in insulated tanks at this temperature for several

days with limited degradation. Thermal energy from the salt can then be transferred to a working fluid that drives a turbine, leaving the salt molten at a temperature around 250–300 °C. The energy density of molten salt storage between these temperatures is thus on the order of $\varepsilon \cong$ 300–400 kJ/kg. Research at Sandia National Laboratory [288] and elsewhere on a variety of materials suggests that a ternary mixture of nitrate salts may provide an even better solar thermal medium, with a lower melting point closer to 100 °C, and potentially accessible higher temperatures approaching 600 °C. This would improve the efficiency and storage potential of concentrated solar power systems.

Other options that have been suggested for solar thermal energy storage include solid materials such as *graphite*. Graphite has a specific heat capacity that is about 700 J/kg K at room temperature, and increases at higher temperatures (see Problem 5.9). Graphite can be heated safely to temperatures of 1800 °C or higher, providing an energy storage capacity per unit mass of 1–2 MJ/kg. At least one prototype solar thermal plant that uses graphite blocks as a storage medium is under construction, and other storage media are also being investigated.

A variant of thermal energy storage known as **pumped heat electricity storage** (PHES) uses a heat pump to convert electricity into thermal energy that is stored at high temperature and then used to generate electricity when required [289]. If both the heat pump and the generator ran at their Carnot efficiency, the round-trip efficiency of PHES would be 100%. Temperature differences within the system, along with other irreversibilities, reduce the efficiency, but recent analyses suggest round-trip efficiencies as high as 65–70% could be obtained, making PHES systems potentially competitive with pumped hydro and exceeding existing CAES systems. A further variant known as *pumped cryogenic electricity storage* (PCES) uses a refrigeration cycle to create a store of exergy in a cold reservoir.

37.3 Mobile Energy Storage

For mobile systems that use substantial amounts of energy, such as automobiles, airplanes, and laptop computers, energy density is a principal consideration. In particular, the energy stored per unit mass combined with functional mass limits provides a basic bound on the range of an electric automobile or use time of a laptop computer and limits air transport to the use of fossil fuels for the forseeable future.

Petroleum-derived fuels including gasoline, kerosene, and diesel fuel dominate vehicle applications because of their extremely high specific energy of 40–44 MJ/kg. Even

---

Example 37.3   Molten Salt Energy Storage

A nitrate salt mixture with a density of $\rho = 2200\,\text{kg/m}^3$ and heat capacity of $1500\,\text{J/kg K}$, is raised to a storage temperature of $500\,°\text{C}$, and used in a power cycle through which it is cooled to a temperature of $250\,°\text{C}$. *Assuming an average conversion efficiency of 33%, what volume of salt must be stored to produce 100 MW of power for 12 hours?* The volumetric energy density is

$$\varepsilon = (1500\ \text{J/kg K})(250\,\text{K})(2200\,\text{kg/m}^3) \cong 820\,\text{MJ/m}^3.$$

The total thermal energy required is

$$E \approx (12\,\text{h})(3600\,\text{s/h})(10^8\,\text{J/s})/(0.33) \approx 13\,\text{TJ},$$

so the total volume required is

$$V \approx (13 \times 10^{12}\,\text{J})/(8.2 \times 10^8\,\text{J/m}^3) \approx 16\,000\,\text{m}^3.$$

Using storage cylinders of diameter 10 m and height 10 m, roughly 20 storage containers would be needed.

---

at 25% engine efficiency, this gives an effective storage density of over 10 MJ/kg. For comparison, bulk energy storage, described in the previous section, based on physical properties such as air pressure or thermal energy storage provides energy storage densities on the order of 1 MJ/kg or less. Furthermore, energy stored in such systems is not easily converted to electrical or mechanical energy in vehicles or small mobile devices. Most portable energy storage systems use chemical energy sources, which have energy densities ranging from below 1 MJ/kg for batteries to above 100 MJ/kg for hydrogen. Nuclear and mass energy, of course, have higher energy density than chemical storage systems, and nuclear-powered submarines and ships are in regular use. Nuclear-powered automobiles and spaceships have also been designed but never put into use. For common applications, the complexity of nuclear power systems is too great for small-scale implementation. Thus, we focus in this chapter on chemical means of energy storage – batteries (§37.3.1), fuel cells, and hydrogen energy storage (§37.3.2, 37.3.3), and combustible materials (§37.3.4). One other possible approach to energy storage that may yield high enough energy densities for vehicular use is flywheel technology; this is discussed in the following section (§37.4.1).

For chemical storage of energy, the specific energy of the substances involved can be expressed as

$$\varepsilon = \frac{-\Delta H}{\text{molecular mass}}, \qquad (37.3)$$

where ΔH is the molar enthalpy of reaction for the overall chemical reaction that releases energy, and the molecular mass is the sum of the masses of the reactants, *not including* gaseous inputs from the external environment.

The 2nd Law limits the fraction of the energy in eq. (37.3) that can be extracted. For combustion processes, this is the Carnot limit (8.28). For chemical reactions this limit comes from the fact that the entropy ΔS produced in the chemical reaction must be dumped into the environment. Thus only the *Gibbs free energy of reaction*, $\Delta G = \Delta H - T\Delta S$, can be put to useful work, and the ratio

$$\eta = \Delta G / \Delta H \qquad (37.4)$$

represents the fraction of the stored chemical energy that can be extracted as electrical or mechanical energy. Note that both ΔH and ΔG are negative for such reactions. If, in a chemical reaction, the entropy of the products is greater than the entropy of the reactants so that ΔS is positive, the reaction absorbs heat $T\Delta S$ from the environment, and it is possible for η to be greater than one. For fuel cells in particular, where $\Delta G / \Delta H < 1$, this ratio gives a bound on thermodynamic efficiency that can be compared to the analogous 2nd law bound for combustion processes. For batteries, the Gibbs free energy gives an upper bound on the *effective* specific energy possible for given chemical constituents

$$\varepsilon_{\text{effective}} \leq \frac{-\Delta G}{\text{molecular mass}}. \qquad (37.5)$$

The ratio $\varepsilon_{\text{effective}}/\varepsilon = \Delta G/\Delta H$ also thus provides a limit on the storage efficiency of batteries and materials used in fuel cells.

While the energy/mass ratio is a primary consideration in many energy storage applications, for automobiles the power density and maximum instantaneous power are also important, since high power is needed for the starter motor that must be run before combustion is initiated.

37.3.1 Batteries

A **battery** is a device that directly converts stored chemical energy into electrical energy. Batteries are composed of **electrochemical cells**. The principal components of an electrochemical cell are the electrolyte, anode, and cathode:

Electrolyte The electrolyte is a material, often an aqueous solution but sometimes a solid, that allows ions to pass freely through but does not conduct free electrons.

Anode The anode contains molecules that can readily release one or more electrons through *oxidation* and simultaneously release the residual ion into the electrolyte.

Cathode The cathode contains molecules that are formed by *reduction* of ions from the electrolyte, by combining each ion with one or more electrons.

A particularly simple type of electrochemical cell is the copper-zinc cell, known as the **Daniell cell**, shown schematically in Figure 37.5. The zinc anode is in a zinc sulfate solution, and the copper cathode is in a copper sulfate solution. Oxidation of a zinc atom in the anode produces two electrons that flow from the anode to the cathode along an external electrical circuit, while the zinc ion is released to the zinc sulfate solution. This **half-reaction** is denoted $Zn(s) \rightarrow Zn^{++}(aq) + 2e^-$, where "(aq)" indicates that the zinc ions are in an *aqueous solution*. Reduction of a copper ion at the cathode takes place when the ion combines with a pair of electrons, $Cu^{++}(aq) + 2e^- \rightarrow Cu(s)$,

depositing solid copper on the cathode. These reactions proceed spontaneously because the net Gibbs free energy released to the environment in the sum of the two half reactions is positive. The passage of a sulfate ion (SO_4^{--}) from right to left through the electrolyte balances the net flow of charge. If the two electrodes were placed in a copper sulfate solution, the copper sulfate would react directly with the zinc anode without producing any current in a wire connecting the two electrodes. In practice, the zinc sulfate and copper sulfate solutions are separated by a porous barrier that allows allows only the sulfate ions to pass and forces the electrons to flow through the wire.

The net chemical reaction is the sum of the two *half-reactions*

$$Cu^{+2}(aq) + Zn \rightarrow Cu + Zn^{+2}(aq). \qquad (37.6)$$

The energy that can be extracted from a Daniell cell is the negative of the Gibbs free energy $-\Delta G^0 = 212\,kJ/mol$ of the reaction (37.6); the enthalpy of reaction $-\Delta H^0 = 219\,kJ/mol$ is slightly larger and includes energy that must be dissipated as heat due to 2nd law considerations. The superscript "0" on ΔG and ΔH indicates *standard conditions*, $T = 298\,K$, $p = 1\,atm$ (which differ slightly from *NTP*), and the concentration of all aqueous solutions equal to one mole per liter.

Each reaction drives two electrons through the external electrical circuit. Thus the free energy of reaction $-\Delta G^0 = 212\,kJ/mol$ is carried by $2N_A$ electrons, and the potential difference across the circuit can be computed from eq. (3.7) $\Delta V = \Delta E_{EM}/q$, or in this case $\Delta V = -\Delta G^0/2eN_A = 1.10\,V$. This is the theoretically ideal **cell voltage** $\mathcal{E}^0_{cell}$. In general if an electrochemical reaction involving the transport of n electrons releases $-\Delta G^0$, then the *cell voltage* is given by

$$\mathcal{E}^0_{cell} = \frac{-\Delta G^0}{neN_A} \equiv -\frac{\Delta G^0}{nF}, \qquad (37.7)$$

where $F = eN_A \cong 9.648 \times 10^4\,C/mol$ is the **Faraday constant**, equal to the charge of a mole of electrons. Standard cell voltages for electrochemical cells are computed by combining tabulated standard **reduction potentials**. In the zinc-copper cell the reduction potentials are $-0.76\,V$ for zinc and $0.34\,V$ copper,

$$Zn^{++}(aq) + 2e^- \rightarrow Zn(s) \quad \mathcal{E}^0 = -0.76,$$

$$Cu^{++}(aq) + 2e^- \rightarrow Cu(s) \quad \mathcal{E}^0 = 0.34. \qquad (37.8)$$

Since zinc is *oxidized* at the anode it contributes $+0.76\,V$ to the cell voltage. These voltages are measured under *open-circuit* conditions; when current flows, the voltage falls due to irreversibilities that are usually parameterized as an internal resistance of the battery cell.

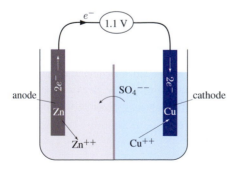

Figure 37.5 A *Daniell cell*. A zinc anode is immersed in a zinc-sulfate solution; a copper cathode is immersed in a copper-sulfate solution. Oxidation of zinc at the anode liberates electrons that flow through the external circuit to the cathode where copper is reduced. A permeable membrane allows SO_4^{--} ions to flow from right to left.

Electrochemical Thermodynamics

Batteries and fuel cells transform chemical energy into electrical energy without the intermediate step of conversion to thermal energy. Therefore they are not subject to the Carnot limit. If ΔH and ΔG are the enthalpy and free energy of the net chemical reaction, then $\mathcal{E}^0 = -\Delta G/nF$ is the ideal, open-circuit voltage of the electrochemical cell and $\eta = \Delta G/\Delta H$ is the fraction of the chemical energy in the cell that can be converted into useful work.

Table 37.1 Typical performance characteristics of rechargeable batteries. (Data from [290]).

Type	Effective energy density (kJ/kg)	Cycle life
NiCd	150–250	1500
NiMH	200–400	300–500
Lead-acid	100–180	200–300
Li-ion	400–600	500–1000
Rechargeable alkaline	250	50

From the free energy and corresponding cell voltage we can determine a simple upper bound on the effective energy density of a copper-zinc battery using eq. (37.5). The molecules involved (one each of Zn and Cu) have total mass 129 u, so the maximum possible effective specific energy is approximately 212 kJ/mol ÷ 0.129 kg/mol $\cong$ 1.65 MJ/kg, though this does not include the mass of the aqueous sulfate solution that serves as the electrolyte and the non-reacting parts of the anode, the cathode, and the containing structure. Thus this is a weak upper bound on the effective energy density of this type of battery. When the remaining parts of the battery are included, the effective energy density decreases significantly. A variety of more complicated chemical reactions are used in real batteries, with energy densities that depend similarly upon the (Gibbs) free energy of reaction and the masses of the constituents.

A single *cell* refers to a system containing a single anode and cathode connected to an electrolyte. In most situations, the voltage across a single cell is too low for desired applications, so a battery combines a number of cells in series. An exception to this occurs in some cellular phones, which use a single lithium-ion cell with voltage 3.7 V.

Batteries that can be used only once are called **primary cells**, while batteries that can be recharged by reversing the chemical reactions are called **secondary cells**. Note that in primary cells electric current always flows from the cathode to the anode (opposite to the flow of electrons); for secondary cells the current reverses when the cell is being recharged. In general, effective energy densities of rechargeable batteries are lower than for primary cells. A few performance characteristics for some common types of secondary cells are given in Table 37.1.

Different battery types have features that make them desirable for different applications. As battery technology has improved over time, and performance criteria have evolved, the materials used in common batteries have changed, with a gradual change towards chemical systems that support higher energy densities. We describe briefly some of the principal battery storage technologies:

Zinc-carbon dry cell A simple and inexpensive type of non-rechargeable battery employs a zinc anode and a graphite cathode surrounded by a layer of manganese oxide (MnO_2). The electrolyte is ammonium chloride (NH_4Cl). The net reaction is

$$Zn(s) + 2\,MnO_2(s) + 2\,NH_4Cl(aq) \tag{37.9}$$
$$\rightarrow Mn_2O_3(s) + Zn\,(NH_3)_2Cl_2(aq) + H_2O(l)\,.$$

The **zinc-carbon dry cell** has an open-circuit voltage of ≈ 1.5 V and $n = 2$ electrons per reaction. A calculation similar to the one we performed for the Daniell cell shows that the reaction free energy is $\Delta G^0 = -2F\mathcal{E}^0 \approx -290$ kJ/mol, and the upper bound on the effective energy density is ≈ 0.84 MJ/kg. Real zinc-carbon batteries have a typical effective energy density of around 0.13 MJ/kg when the electrolyte, casing, carbon, and other extra materials are included.

Alkaline batteries **Alkaline batteries** replace the electrolyte of the zinc-carbon dry cell with a *base*, potassium hydroxide (KOH), which supplies *hydroxide ions* (OH^-). In the presence of KOH the following half-reactions occur at the anode and cathode respectively,

$$Zn(s) + 2OH^-(aq) \rightarrow ZnO(s) + H_2O + 2e^-,$$
$$2MnO_2(s) + H_2O + 2e^- \rightarrow Mn_2O_3(s) + 2OH^-(aq)\,, \tag{37.10}$$

with reduction potentials of -1.28 V and $+0.15$ V. The net reaction

$$Zn(s) + 2MnO_2(s) \rightarrow ZnO(s) + Mn_2O_3(s) \tag{37.11}$$

therefore has an ideal cell voltage of $\mathcal{E}^0 = 1.43$ V (Problem 37.11). Although the cell voltages are nearly the same, alkaline batteries have higher effective energy densities than zinc-carbon dry cells, typically 0.4–0.6 MJ/kg, because the components can be more compact.

Nickel-cadmium batteries Sealed **nickel-cadmium** rechargeable batteries (**NiCd**, or *NiCad*) have seen widespread use since the 1950s. NiCd batteries are particularly valued for their long lifetime – typically these batteries have a cycle life on the order of 1500 charging cycles before their energy storage capacity drops below 80% of its original value. One drawback of NiCd batteries is that to function well they must occasionally be completely discharged. Otherwise "memory" crystals form within the cell that degrade the performance of the battery. The toxic nature of the cadmium in NiCd batteries means that disposal and environmental impact is also an issue for these batteries.

Nickel-metal hydride batteries **Nickel-metal hydride (NiMH)** batteries began to replace NiCd batteries for many applications in the 1990s. NiMH batteries have a higher energy density and a better environmental footprint than NiCd batteries, although the cycle life is generally shorter and the charging time is generally longer for NiMH batteries.

Lead-acid batteries Standard automobile batteries are rechargeable **lead-acid batteries**. This battery technology has been used for well over 100 years. The anode of a lead-acid battery is lead (Pb) while the cathode is lead oxide (PbO$_2$), and the electrolyte is a mixture of roughly 1/3 sulfuric acid (H$_2$SO$_4$) and 2/3 water. The effective energy density of typical lead-acid batteries ranges from 100 to 180 kJ/kg (Problem 37.12). Round-trip storage efficiency is in the range 50%-90%, and the battery can be recharged 500–1000 times. Although the energy density is low (due in large part to the water needed for the electrolyte), high discharge power densities – around 180 W/kg – can be sustained for relatively brief periods.[2] The high power density is what made this battery a mainstay of vehicles through the twentieth century.

Lithium-ion batteries Lithium is an attractive component for batteries because the standard potential, $\mathcal{E}^0 = -3.05$ V,

---

[2] The range of "maximum power densities" shown for lead-acid batteries in Figure 37.1, which are lower, were obtained from plots of power versus voltage similar to Figure 25.21 (for a photovoltaic cell) under standard laboratory conditions. Higher power densities, such as the value quoted in the text, can be obtained in brief and relatively inefficient discharges [291].

Batteries

Batteries store energy in chemical form and use electrochemical reactions to convert energy to and from electrical form. The lead-acid battery was developed in 1859, and is still widely used in automobiles and other devices. Many other batteries have been developed that use other electrochemical processes, with increasing energy density, efficiency, and reliability. Energy storage density in existing batteries is generally limited to below 1 MJ/kg due to limits on the binding energy to mass ratio of constituents involved in electrochemical processes. Lithium-ion and other lithium-based batteries achieve relatively high energy density due to lithium's large standard reduction potential and low density.

---

for reduction of atomic lithium, Li$^+$(aq) + e$^-$ → Li(s), is the most negative of all the elements. Furthermore lithium, with a density of 0.534 gm/cm^3, is the least dense of all elements that are solid at room temperature. Thus batteries based on lithium electrochemical cells offer the possibility of both high voltage and high gravimetric energy density. Rechargeable batteries with pure metallic lithium as the anode were developed in the 1980s, but had safety issues. After many cycles, the lithium electrode could become thermally unstable, with explosive results. Since then, materials have been developed such as graphite intercalated with lithium that can be used more safely (though even more modern lithium-based batteries have been known to develop thermal instabilities in certain circumstances). **Lithium-ion batteries** are a family of battery types in which both the anode and cathode contain lithium ions. In 2016, lithium-ion batteries dominated the market for mobile phone batteries, laptop computer batteries, and hybrid and electric automobile batteries. The energy density of modern lithium-ion batteries typically is around 500 kJ/kg, with a power density of 250–350 W/kg. Round-trip storage efficiency is very high, 80–90%, and lithium-ion batteries can have a cycle life of 1000 or more cycles. Although lithium-ion batteries do not need to be completely power cycled, and their performance is relatively independent of the use pattern, a drawback of these batteries is that they suffer a significant degradation in capacity on a time scale of one to three years whether or not the battery is used.

Box 37.1   Voltage and Efficiency in Real Batteries

In real batteries, the voltage is generally less than the ideal EMF, $\mathcal{E}^0_{cell}$, for the relevant chemical reaction, due to internal resistance and other deviations from ideality. Furthermore, the voltage varies depending upon the state of charge of the battery. One important aspect of this dependence is captured in the entropy contribution to the Gibbs free energy. Consider a battery based on a (discharge) reaction of the form $A + B \rightarrow C + D$. When the battery is almost fully discharged, the concentrations a, b of reactants A, B become very small. From the information-theoretic view of entropy (§8) we know that the contribution to the entropy from a particle mixed at concentration c is $-k_B \ln c$. The Gibbs free energy of the reaction therefore includes contributions of the form $k_B \ln Q$, where Q is the **reaction quotient** given by the product of the concentrations of the reaction products divided by the product of the concentrations of the reactants, in this case $Q = cd/ab$. This contribution to the Gibbs free energy modifies the cell potential to

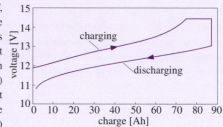

$$\mathcal{E}_{cell} = \mathcal{E}^0_{cell} - \frac{k_B T}{ne} \ln Q = \mathcal{E}^0_{cell} - \frac{RT}{nF} \ln Q \,.$$

This relation is known as the **Nernst equation**. The effect of this modification is that the voltage of the battery suffers a marked decrease as the cell approaches a discharged state, and an increase as the cell approaches a fully charged state. This charge-dependence of the voltage is depicted in the accompanying figure (after [292] for a typical 12 V lead-acid battery). The voltage drop of the battery is also different when the battery is being charged than when it is discharged. This is related to the necessary thermodynamic losses associated with the difference between ΔH and ΔG. More energy must be put into the battery than can be extracted. While the difference, which is given by the area of the *hysteresis* curve shown in the figure, can be reduced by an intelligent charging schedule, there are always some losses, so that no battery storage system is 100% efficient.

While energy density has gradually improved over the years of battery development, this progress has been incremental and may be difficult to extend very far beyond the energy density of the lithium-ion battery, due to the fundamental limit on energy density coming from the energy/mass ratio of the chemical constituents involved. One approach that has been explored is to use multivalent ions such as magnesium, calcium, or aluminum,

$$Mg^{++}(aq) + 2e^- \rightarrow Mg(s) \quad \mathcal{E}^0 = -2.36 \text{ V},$$
$$Ca^{++}(aq) + 2e^- \rightarrow Ca(s) \quad \mathcal{E}^0 = -2.87 \text{ V},$$
$$Al^{+++}(aq) + 3e^- \rightarrow Al(s) \quad \mathcal{E}^0 = -1.66 \text{ V}, \quad (37.12)$$

in place of lithium. Since the cell free energy is proportional to the number of electrons released (eq. (37.7)), the effective energy density of a battery could be increased modestly by increasing the number of charges involved per reaction at the cost of an increase in atomic mass. Magnesium is also more abundant and more stable than lithium, which could make magnesium-based batteries both cheaper and safer than lithium-ion batteries. For mobile applications, however, *fuel cells* may present the

greatest promise for a significant jump in portable energy storage densities.

37.3.2 Fuel Cells

Fuel cells are conceptually very similar to batteries, with an anode, a cathode, and an electrolyte that conducts ions. The big difference is that in fuel cells there is a continuous flow of consumables entering the device. A simple hydrogen fuel cell is shown schematically in Figure 37.6.

The chemical reaction in a typical fuel cell is equivalent to a combustion reaction. For example, for a *(PEM)* hydrogen fuel cell, a continuous flow of hydrogen enters the cell. The anode contains a catalyst that increases the rate of hydrogen dissociation. Hydrogen ions then pass through an electrolyte medium, and combine at the cathode with oxygen that also flows in continuously. The net reaction for a hydrogen fuel cell is that of hydrogen combustion,

$$H_2 + \tfrac{1}{2}O_2 \rightarrow H_2O \text{ (l)} \,. \quad (37.13)$$

The most important difference between a fuel cell and combustion-based power systems is that in a fuel cell the reaction occurs chemically, not through combustion,

---

Box 37.2   Carbon-free Cars

Automobiles present one of the greatest challenges to reducing fossil fuel emissions. It is hard to develop an energy storage device that matches the energy density and convenience of gasoline fuel. Several approaches, however, may lead to widespread deployment of carbon emission-free vehicles in coming decades.

Many of the first automobiles in use around 1900 were electric, powered completely by lead-acid batteries (right). Electric-powered cars have several advantages. No gears are needed for an electrically powered engine, since the power profile is much broader than that of a gasoline engine. Furthermore, braking energy can be easily stored for later use (see §2). The greatest challenge for battery-powered vehicles is the limited energy density of existing batteries. With improved battery technology, and more streamlined vehicles, however, the range of modern commercial electric vehicles has increased to the point that they have a range on the order of 200–400 km and are a realistic replacement for the transport needs of most individuals, particularly for commuting and local transport. By the middle of 2016 over 1 500 000 all-electric cars and light trucks had been sold worldwide. Fast three-phase or direct current charging stations can provide a full charge of roughly 70 MJ in under 30 minutes to electric vehicles such as the Nissan Leaf (below).

Alternatives to electric vehicles include hydrogen fuel cells and flywheel storage systems. Hydrogen fuel cells can in principle be very efficient, and hydrogen has a high energy to mass ratio, but a lower energy to volume ratio than gasoline even when compressed or liquefied. The challenges of hydrogen production and storage make hydrogen fuel cells a complex but potentially attractive route to carbon-free vehicles. A cheap and efficient, reversible hydrogen fuel cell would represent a major step forward in this direction. Flywheel storage systems have been used in prototype automobiles but have potential safety issues that have not been satisfactorily resolved.

---

and the chemical energy is directly converted to electric potential energy. Thus, there is no thermal energy intermediate, and the Carnot limit on efficiency does not apply. There is still a bound on the efficiency coming from the 2nd Law, however, $\eta \le \Delta G^0/\Delta H^0$, as discussed above. The enthalpy of reaction for combustion of hydrogen is $-\Delta H^0 \cong 285.8$ kJ/mol, while the Gibbs free energy is $-\Delta G^0 \cong 237.1$ kJ/mol, so the 2nd law efficiency bound is about 83% at 298 K and 1 atmosphere.

Because fuel cells have no moving parts and minimal startup complications, they are extremely reliable. This makes fuel cell technology attractive for power systems in remote locations, and for applications such as military, medical, and space operations where reliability is of paramount importance. The weaker 2nd law constraint on the efficiency of fuel cells compared to combustion makes fuel cells a promising approach for power plants that can use hydrogen from steam-reforming of fossil fuels. In power plants, the waste heat from fuel cells could be used for heating, potentially increasing the fraction of useful energy above the 2nd law bound of 83%. At the present time, most uses of fuel cells are in stationary systems. In particular, for example, methane fuel cells are now used in homes in Germany and elsewhere for *combined heat and power*, with electrical energy output used for home power and the remaining released thermal energy used for heating.

Fuel cells can also be considered for many mobile applications that have traditionally used batteries. If the fuel cell uses oxygen from the atmosphere, this component does not enter the energy/mass calculation. Furthermore, for a fixed fuel cell mass, as the size of the storage unit for the consumable fuel increases, the energy/mass ratio approaches that of the fuel alone. As a result, the energy/mass ratio of fuel used for a fuel cell is computed in a similar fashion to fuel for combustion engines, while the efficiency can be significantly higher due to the lack of the Carnot constraint.

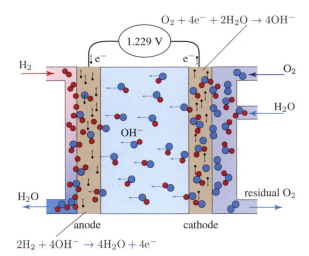

$$O_2 + 4e^- + 2H_2O \rightarrow 4OH^-$$

1.229 V

H_2

O_2

H_2O

OH^-

H_2O

residual O_2

anode   cathode

$$2H_2 + 4OH^- \rightarrow 4H_2O + 4e^-$$

Figure 37.6 An alkaline (Bacon) hydrogen fuel cell. Hydrogen molecules enter on the left, dissociate on the anode and combine with OH^- ions to form water. This reaction releases electrons, which flow through the external circuit to the cathode, where they combine with oxygen, which enters from the right, and water to form OH^- ions, which migrate to the anode. The reactions are summarized in eq. (37.14).

For hydrogen gas the energy density is roughly $\varepsilon \cong 140$ MJ/kg. The 2nd law limit of 83% suggests that fuel cells may be able to realize effective energy densities that approach $\varepsilon_{\text{effective}} \cong 120$ MJ/kg. Noting that two electrons are liberated at the anode in the reaction (37.13) and that the reaction free energy is 237.1 kJ/mol, eq. (37.7) predicts an ideal standard cell voltage of $\mathcal{E}^0_{\text{cell}} = 1.229$ V. In real hydrogen fuel cells the open-circuit voltage is typically in the range 0.85–1.05 V. With current flowing, the voltage of a typical fuel cell is closer to 0.7 V, and therefore existing fuel cells realize efficiencies in the range 40–60%. This still represents a very high effective energy density. Hydrogen gas as a fuel, however, presents some additional challenges discussed in the following section (§37.3.3).

While the first working fuel cells were built in the nineteenth century, the concept was not considered for practical use until 1959, when the English engineer Francis Thomas Bacon improved the technology and patented the *Bacon cell*, which uses an alkaline electrolyte. The first generation of commercial fuel cells used in the 1960s were based on alkaline electrolytes, following the example of the Bacon cell. As in an alkaline battery, potassium hydroxide provides OH^- ions that facilitate the following half reactions at the anode and cathode respectively,

$$2H_2(g) + 4OH^-(aq) \rightarrow 4H_2O(l) + 4e^-,$$
$$O_2(g) + 4e^- + 2H_2O(l) \rightarrow 4OH^-(aq). \quad (37.14)$$

Fuel Cells

Fuel cells, like batteries, convert chemical energy directly to electrical energy. Fuel cells, however, use a continuous flow of consumables. Hydrogen fuel cells combine hydrogen with oxygen from the air to form water, with no other significant emissions. The Carnot limit does not apply, but 2nd law considerations limit hydrogen fuel cell efficiency to roughly 83%. Significant challenges to large-scale use of hydrogen fuel cells for automobiles include the problem of hydrogen storage and the challenge of efficiently producing hydrogen through electrolysis.

In the net reaction, $2H_2 + O_2 \rightarrow 2H_2O$, the KOH is not consumed.

One significant challenge for fuel cells with alkaline electrolytes is the tendency of KOH to react with CO_2 in air. To avoid this in a hydrogen-based fuel cell, CO_2 must be separated from the oxygen that participates in the fuel cell reaction. Many more recent hydrogen fuel cells use a solid membrane known as a **PEM** (*proton exchange membrane*, or *polymer-electrolyte membrane*) as an electrolyte. The PEM allows protons (H^+) to pass through, but not electrons.

Other kinds of fuel cells can be designed, in close parallel with combustion reactions. One type under development is a methanol fuel cell. Since methanol is a liquid, it is easier to store and use as a vehicle fuel than hydrogen gas. In a direct methanol fuel cell with an acidic electrolyte, methanol combines with water in the anode,

$$CH_3OH(l) + H_2O(l) \rightarrow CO_2(g) + 6H^+ + 6e^-. \quad (37.15)$$

The protons pass through the electrolyte membrane and combine with oxygen in the cathode

$$\frac{3}{2}O_2(g) + 6H^+ + 6e^- \rightarrow 3H_2O(l). \quad (37.16)$$

Other cells have been designed, including methane fuel cells that "burn" natural gas with high efficiency. A disadvantage of fuel cells using methanol or other carbon-based fuels is the release of CO_2 in the combustion process. In principle the carbon dioxide can be captured and not released to the air. Research is currently underway to develop carbon capturing fuel cells for power plant applications, but this approach is not practical in the near future in mobile fuel cells.

One of the main challenges in fuel cells is realizing a high rate of reaction. The catalyst used at the anode must produce a high rate of dissociation. Most current fuel cells use platinum as a catalyst at both the hydrogen and oxygen electrodes, which significantly increases the cost of the cells. The power output of a fuel cell is often measured in terms of power per unit area of electrode (in W/cm^2). For example, **PEMFCs** (PEM fuel cells) based on technology developed in the 1990s operate at temperatures around $80\,°C$, and have a power density of 0.6–$0.75\,W/cm^2$. Earlier alkaline fuel cells reached comparable power densities only at much higher temperatures ($\sim 250\,°C$).

One promising variation on a fuel cell, the **solid oxide fuel cell** (SOFC), uses a solid ceramic material as the electrolyte. In SOFCs, unlike standard fuel cells, a negative ion travels through the electrolyte from the cathode to the anode. These fuel cells are run at high temperatures, up to $1000\,°C$, which allows for high efficiency but requires additional energy input and specialized materials.

In recent years there has been a great deal of effort focused on research and development on various fuel cell technologies, with potential uses ranging from power plants to electric cars to laptop computers and digital watches. While fuel cells are commercially available that are reasonably robust and have high energy densities, for fuel cells to achieve broad commercial success, further advances must be made in reducing cost, improving longevity, and increasing power density. In particular, fuel cell production, platinum-based catalysts, and PEMs are all currently quite expensive. Furthermore existing fuel cells have limited lifetimes. Improving on each of these aspects represents a significant research challenge. The field of fuel cell technology is rapidly developing; a good starting point for the reader further interested in this subject is [293].

37.3.3 Hydrogen Energy Systems and the "Hydrogen Economy"

Due to its high energy density, hydrogen is a very attractive option for mobile energy storage. This has led to proposals for a **hydrogen economy** in which large amounts of energy would be stored in hydrogen, which could be used as a clean and energy-dense fuel for automobiles and other devices. There are a number of challenges, however, that must be overcome before stored hydrogen could provide a safe and efficient mechanism to power transportation. In addition to the need for powerful and efficient hydrogen fuel cells, there are many steps and associated inefficiencies in existing processes for transforming other forms of energy into stored hydrogen energy, and there are safety issues involved in hydrogen storage. These problems have slowed commercial development of hydrogen-based

energy systems, though in recent years (as of 2016) several hydrogen-powered cars have entered limited commercial production.

Hydrogen Production While hydrogen is sometimes referred to as a "fuel," and certainly acts as a nuclear fuel to power the Sun, molecular hydrogen (H_2) is not found naturally on Earth, and is not a primary energy source for human use. Hydrogen should rather be thought of as an energy storage medium, or **energy carrier**. To use hydrogen as a power source for a vehicle or other machine, energy must first be gathered from another source and then converted into chemical energy of molecular hydrogen.

Currently, the primary method used to produce pure molecular hydrogen is steam reforming of natural gas, and to a lesser extent coal and other natural hydrocarbons, as described in §33. This approach has several drawbacks. First, unless the carbon dioxide produced in steam reforming is captured and sequestered, using hydrogen energy transformed from fossil fuels does not reduce anthropogenic carbon emissions. Second, this process has limited efficiency. The overall energy conversion efficiency in producing hydrogen from methane is roughly 60% or below, with even lower conversion efficiencies for other hydrocarbon sources.

Another approach to producing hydrogen is through **electrolysis**, where electrical energy is used to *split* water into hydrogen and oxygen. This essentially amounts to running a fuel cell in reverse. Indeed, electrolyzers are constructed in a very similar way to fuel cells, though some of the materials used are generally somewhat different. For example, PEMs are used in electrolyzers just as in fuel cells. In a fuel cell, however, the **gas diffusion layer** (GDL) that allows the reactants to access the catalyst is generally based on *hydrophobic* materials (to keep out water), while for an electrolyzer water is the reactant and the GDL must be more *hydrophilic*. In general, different catalysts can be used in electrolyzers than in fuel cells. While platinum alloys are often used in electrolyzers, there are promising results using other catalysts such as cobalt and phosphate [294]. Typical electrolysis systems require an input voltage around 2 V, significantly above the idealized 1.23 V voltage needed to reverse the reaction (37.13), so that the efficiency of energy conversion is roughly 60%. Though conversion efficiencies of up to 80–90% have been realized, it is currently more costly to produce hydrogen by electrolysis than by steam reforming of fossil fuels. One advantage of hydrogen produced by electrolysis is that it is much purer than hydrogen produced by steam reforming; high purity is desirable for low-temperature fuel cells where contaminants like CO can be absorbed and interfere with catalysts on the electrodes.

In principle, an ideal setup for using hydrogen fuel cells in automobiles and other mobile systems would involve integrating an electrolyzer and a fuel cell so that the fuel cell could be recharged by providing electricity from the grid to perform electrolysis by running the cell backwards. There are several challenges to building efficient systems of this type. One problem arises from the different materials constraints, such as the hydrophilic versus hydrophobic nature of the gas diffusion layer in proton-exchange membrane-based electrolyzers versus fuel cells. There are also problems with the catalysts – while platinum is widely used as a catalyst for hydrogen fuel cells, when run backwards as an electrolyzer the platinum tends to become permanently oxidized, creating a layer of platinum oxide that degrades efficiency as a fuel cell. The development of a reversible fuel cell system is currently a very active area of research.

Hydrogen can also be produced in a sequence of chemical reactions known as the **sulfur–iodine cycle** if a source of high-temperature thermal energy is available. The necessary chemical reactions,

$$I_2 + SO_2 + 2H_2O \rightarrow 2HI + H_2SO_4,$$
$$2H_2SO_4 \rightarrow 2SO_2 + 2H_2O + O_2,$$
$$2HI \rightarrow H_2 + I_2, \qquad (37.17)$$

do not consume either sulfur or iodine. In fact the net reaction is the same as electrolysis of water. The advantages of the sulfur–iodine cycle are a predicted efficiency as high as 50% and the absence of carbon emissions. The drawbacks include the high temperature required to drive the reactions – the second of the three reactions requires temperature in excess of 850 °C – and the many corrosive reagents. If these materials problems can be overcome, the sulfur–iodine cycle is considered as a candidate for hydrogen production at high-temperature nuclear reactors and solar thermal power plants. One class of nuclear reactors that may reach sufficiently high temperatures for hydrogen production to be efficient are the *very high temperature reactors (VHTR)* described in §19.4.2.

Hydrogen Storage

Although the gravimetric energy density of hydrogen is very high, at standard temperature and pressure hydrogen is a gas, with very low volumetric energy density. This is one of the major drawbacks of hydrogen energy storage: hydrogen either takes up a very large volume or it must be compressed under very high pressure. For mobile use (other than for enormous air vehicles such as dirigibles), volume constraints require compressed hydrogen.

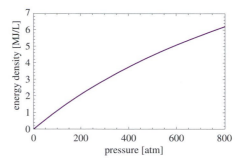

Figure 37.7 Energy density of compressed hydrogen (using HHV) as a function of pressure. Note that hydrogen deviates from the ideal gas law at high pressures, so that the pressure increases faster than linearly with the density. Data from NIST [22].

Compressed hydrogen At 1 atm pressure and 300 K, hydrogen has a density of roughly 8.2×10^{-5} kg/L, for an energy density of roughly 12 kJ/m^3. Hydrogen gas is often compressed and stored at pressures of up to 700 atm, where the energy density is ~5.6 MJ/L (see Figure 37.7). While this is significantly lower than the volumetric energy density of gasoline (~30 MJ/L), the higher efficiency of fuel cells compared to combustion engines puts this in the range where portable use in automobiles may be practical.

Liquefied hydrogen Liquefied hydrogen has been used as a fuel for rockets and in the power plants of the US space shuttle. Hydrogen liquefies at 20.28 K, below which point it has a density of ≈ 71 kg/m^3 with a corresponding volumetric energy density of roughly 10 MJ/L, substantially higher than that of compressed hydrogen at 700 atm. While liquid hydrogen stored in cryogenic tanks has been successfully used in space, the possible use of this technology in automobiles is still under debate. A significant amount of energy is needed to liquefy hydrogen, and to keep the tanks cool, particularly in a small system. There are many inefficiencies involved in refueling small passenger automobile tanks with liquid hydrogen, and safety issues have not yet been resolved.

Metal hydrides Another approach to hydrogen energy storage makes use of *metal hydrides*. Certain metal alloys such as LaNi$_5$ bind strongly with hydrogen and are liquids at normal operating temperatures. While the energy density of hydrogen bound in metal hydrides is generally substantially below that of fossil fuel liquids, ongoing research is seeking metal hydrides that might be used for safe and efficient hydrogen energy storage.

Advantages of Hydrogen Fuel Cell Power Systems

Despite the substantial obstacles to developing efficient

The "Hydrogen Economy"

Since molecular hydrogen is not found on Earth, hydrogen is not a primary energy source, and is properly regarded as an energy storage system. The gravimetric energy density of hydrogen, approximately 140 MJ/kg, is extremely high, but as a gas at STP, its volumetric energy density, 12 kJ/m^3, is low. To be useful in transportation systems, hydrogen must be compressed to hundreds of atmospheres or liquefied.

Hydrogen can be produced by electrolysis or by chemical means, but most hydrogen today is produced by steam reforming of hydrocarbon fuels. Widespread use of hydrogen as a transportation energy storage system awaits more efficient production mechanisms and safe and efficient high-density storage. A reversible hydrogen fuel cell would be a significant advance.

hydrogen production and storage mechanisms, the absence of pollution or carbon dioxide emissions from hydrogen fuel cells, and the non-toxic nature of hydrogen, make the development of vehicles based on hydrogen energy systems an attractive possibility that has driven continued interest in these systems.

Furthermore, although hydrogen releases substantial energy when it reacts with oxygen, it is in many ways safer than hydrocarbon fuels. Hydrogen is lighter than air and therefore quickly disperses if inadvertently spilled or vented. The auto-ignition temperature of hydrogen (580 °C) is much higher than that of gasoline (280 °C), and hydrogen will not burn in concentrations less than ~4% (the limit for gasoline is ~1%).

Efficiency of Hydrogen Energy Systems

To determine the efficiency of a hydrogen-based energy system, all the steps in the hydrogen production and storage processes must be included. If the hydrogen used is produced from electrical energy through electrolysis, then the overall efficiency depends upon:

- $\eta_{conversion}$: the efficiency for conversion of the original energy source to electricity, including transmission losses from the source to the electrolysis plant;
- $\eta_{electrolysis}$: the efficiency of the electrolysis process;
- $\eta_{storage}$: efficiencies associated with infrastructure (recovery and compression of the H_2, distribution, transport, and storage);

- η_{engine}: the efficiency of the fuel cell or engine that converts the hydrogen into useful energy.

The product of all these efficiencies,

$$\eta = \eta_{conversion} \times \eta_{electrolysis} \times \eta_{storage} \times \eta_{engine}, \quad (37.18)$$

must be compared with the efficiency of alternatives. For example, one could compare the overall efficiency of an automobile run on a hydrogen fuel cell using compressed hydrogen produced by electrolysis to a plug-in electric vehicle with a chemical battery storage system. The large number of steps in the hydrogen process make it difficult for hydrogen-based systems such as this to compete in efficiency. The principal advantage of the hydrogen fuel cell vehicle compared to a battery-powered plug-in is the substantial difference in energy density, allowing for a much greater range for a vehicle with limited mass devoted to the energy system. If an efficient and inexpensive reversible hydrogen fuel cell is developed in the future, then many of the additional efficiency constraints of the hydrogen-based system would be mitigated.

37.3.4 Combustible Storage

While most combustible fuels, such as fossil fuels, are derived from natural sources, these fuels are also in a sense energy storage devices. Indeed, as discussed in §26, the energy in hydrocarbons and related biological materials is essentially stored solar energy, captured by the process of photosynthesis. By manipulating biological and chemical systems it is possible to transform energy from other forms into the chemical energy of combustible molecules, and to use this as a form of energy storage. As a simple example, consider using solar or wind energy to produce methanol through electrolysis – essentially by running a methanol fuel cell backwards. This converts a natural source of energy to a chemically stored form that can be released through combustion as well as through a fuel cell. Much of the material in previous chapters on combustion-based power plants and fossil fuels applies to artificially produced combustible materials as well as to natural energy sources. As discussed in the previous section, the tradeoff between extracting energy from a given fuel through combustion or through a fuel cell is that the energy extracted from combustion is limited by the 2nd Law, while fuel cells require substantial area and material for high power production.

37.4 Other Energy Storage Systems

37.4.1 Flywheels

Pumped hydro stores energy in the form of mechanical potential energy. Energy can also be stored in mechanical

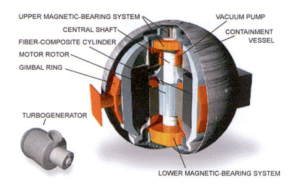

Figure 37.8 A modern flywheel, with magnetic bearings and vacuum pump to minimize frictional losses (Credit: Slim Films)

kinetic energy. One simple way of using kinetic energy for storage employs a *flywheel*. Flywheels are often used to provide energy at a continuous rate when the source is intermittent. Flywheels have also been proposed for energy storage in hybrid automobiles.

Flywheels: Basic Principles

Flywheels store energy in the rotational kinetic energy of a massive spinning object. The physics of a flywheel is quite simple (see Example 2.4). The rotational kinetic energy of an object is given by

$$E_{kin} = \int d^3x \, \frac{1}{2}\rho r^2 \omega^2 = \frac{1}{2}I\omega^2 \,, \qquad (37.19)$$

where ω is angular velocity and the moment of inertia of the object is

$$I = \int d^3x \, \rho r^2 \,. \qquad (37.20)$$

For example, for a solid disk of mass M and radius R, the moment of inertia is $I = \frac{1}{2}MR^2$.

Frictional losses are an important concern for flywheel energy storage. A poorly engineered flywheel will rapidly lose energy to friction at the point of contact between the rotating hub and its support structure. Energy is also lost to air resistance if the flywheel is rotating in air at atmospheric pressure. Modern flywheels use magnetic bearings to minimize friction of the axle, and incorporate a vacuum pump to minimize air resistance.

Flywheels are simple mechanical devices that can be extremely robust, and relatively efficient. Current designs are made of strong carbon-composite material, can be used for 10^5–10^7 cycles without breakdown, and can have a round-trip storage efficiency of up to 85–90% The energy density of advanced flywheel designs can be comparable to the best battery technologies (up to 500 kJ/kg), although the energy densities of commercially available flywheels are typically an order of magnitude lower. The instantaneous power can be very high for flywheels. Thus, flywheels are of particular use in applications where short power bursts are needed, such as for testing circuit breakers. Flywheels are often used for short-term load leveling and uninterruptible power supplies. Due to their high reliability, long lifetime, and minimal maintenance requirements, flywheels are used for space applications.

Flywheels in Automobiles

Flywheels have been explored for possible use in automobile energy storage. The principle of a hybrid automobile engine is to divide the power source into two components. The first component is an efficient but relatively small fossil fuel powered engine, such as an engine based on the Atkinson cycle as discussed in §11. The small internal combustion engine is complemented by an additional power source used for rapid acceleration; generally the additional power source gets its energy from regenerative braking. A flywheel has several advantages as a temporary storage system for kinetic energy from regenerative braking. The transformation of motional kinetic energy into a flywheel's rotational kinetic energy and back requires only a mechanical coupling. With one efficient transformation needed for storage and another for recovery, flywheel round-trip storage efficiency can easily be engineered in the range $\eta \sim (0.8\text{–}0.9)^2 \sim 0.65\text{–}0.8$ and can be above 0.8 for precision-engineered systems. Battery storage, in contrast, requires more steps – the vehicle's kinetic energy must be transformed into electrical energy using a generator, then stored in the battery, recovered, and transformed again into kinetic energy using a motor. Even if each stage in this process were 80% efficient, the overall round-trip storage efficiency can be below 50% ($0.8^4 \sim 0.4$). Compared to batteries, flywheels are not only more efficient, but can have higher power density and survive many more cycles than any existing battery technology.

So why are flywheels not already the technology of choice for short-term energy storage in hybrid automobiles? While prototypes have been built (Example 37.4), and some engineers argue that flywheel technology offers substantial advantages and has only been held back by industrial inertia, flywheels raise some safety issues if they are used in automobiles. One important issue is that when a flywheel fails – due either to material failure or damage to the system – it can rapidly disintegrate, sending fragments of metal in all directions at velocities above that of a bullet (Problem 37.17). Use of flywheels in passenger automobiles would require safety systems to protect the driver

Flywheels store rotational kinetic energy. The effective energy densities of flywheels are of the same order as batteries. Flywheel power densities, however, are higher than any other conventional energy storage system except for supercapacitors, which have much lower effective energy density. Flywheels also have very high instantaneous power, high round-trip storage efficiency, and great longevity – of order 10^5–10^7 cycles. Safety issues and gyroscopic effects have so far prevented widespread adoption of flywheels for mobile energy storage.

and passengers from flywheel disintegration in case of an accident.

Another issue for flywheels concerns gyroscopic effects. A rapidly rotating flywheel has a significant amount of angular momentum about a specific axis. Any rotation of a vehicle containing a flywheel that changes the orientation of the flywheel's angular momentum axis requires an additional torque, according to eq. (2.43). For turns in the plane of the car's motion, the required torque can be minimized by a vertical flywheel axis, but other common maneuvers can lead to unexpected effects on the car's behavior – if, for example, the car tilts forward, the flywheel's angular momentum must tilt with it, requiring a torque that is produced when the car's center of gravity shifts to one side, adversely affecting its stability (see Example 37.4). Although the torque required to change the flywheel angular momentum is generally small compared to the torque generated by shifting a substantial part of the weight of the car from the wheels on one side to the other, this effect can be large enough to influence the car's handling and safety. This effect has been dealt with in some existing designs by including two counter-rotating flywheels and ensuring that their angular velocities are equal in magnitude and opposite in direction, thereby canceling all gyroscopic effects.

37.4.2 Capacitors

Storing electromagnetic energy effectively is particularly challenging. The simplest direct approach to storing electromagnetic energy is to store it as electrostatic potential energy; most other approaches to electromagnetic energy storage involve conversion to chemical or mechanical form. In §3.1.3 we reviewed the basic physics of energy storage in capacitors. While capacitors cannot store a lot of

energy, they can release their stored energy very quickly. In addition, they are relatively light, very durable, do not need to contain exotic or toxic components, and do not heat up during use. Disadvantages of capacitors as energy storage devices, beyond their relatively low energy storage capacity, are that they are intrinsically low-voltage devices, that their voltage drops linearly as they discharge (remember $V = Q/C$), and that the stored charge leaks away over relatively short time scales (on the order of hours or days for typical standard capacitors).

Traditional Capacitors Traditional capacitors are based on the plate–dielectric–plate design described in §3.1.3. Energy in a capacitor can be thought of as being stored in the electric field as expressed in eq. (3.20)

$$E_{\text{EM}} = \int d^3x \, \frac{\epsilon}{2} \, |\boldsymbol{E}(\boldsymbol{x})|^2 \ . \qquad (37.21)$$

While the volumetric energy density of a capacitor that uses air as a dielectric is limited to roughly $40\,\text{J/m}^3$, this can be increased significantly by using more sophisticated materials as the dielectric. The storage capacity $E = CV^2/2$ of a capacitor is increased both by increasing the dielectric constant $\kappa = \epsilon/\epsilon_0$, since C is proportional to κ, and by using a dielectric material with a higher breakdown voltage than air, since the maximum possible V is proportional to the breakdown voltage. Even with optimal materials, however, the energy density of conventional capacitors is on the order of $400\,\text{J/kg}$, smaller than that of available batteries by a factor of 1000.

While the energy density of capacitors is generally low, the power density can be very high. The internal resistance of a capacitor can be quite small. This allows a large fraction of the charge stored in the capacitor to be released very quickly. For most batteries, on the other hand, the internal resistance is relatively high due to the kinetics of the electrochemical reactions involved. Power densities of electrochemical batteries generally range from $\sim$1 W/kg (rechargeable AA batteries) to $\sim$500 W/kg (peak power of some lead-acid batteries powering automobile starter motors), while conventional capacitors can have power densities of $\sim$5–10 kW/kg or higher. Capacitors are also very robust and reliable due to their simple construction. Thus, capacitors are used in many situations where short high-power energy release is needed. Typical applications include boosters for starter motors and powering electronic devices during short downtimes while batteries are changed.

Super- and Ultracapacitors In recent years, more sophisticated materials and technologies have been used to create **supercapacitors** and **ultracapacitors**. These

Example 37.4   Flywheels in Automobiles and Gyroscopic Effects

Prototype automobiles have been constructed that use flywheels for temporary energy storage. In 1997, for example, Rosen Motors successfully road tested a gas turbine powered car that used a 55 000 rpm flywheel for regenerative energy storage and to supplement acceleration. The car, however, was never brought to market.

Gyroscopic effects can affect the handling of a car with flywheel energy storage. Consider a typical passenger car with mass 1800 kg, height 1.4 m, and width 1.6 m. Suppose that this car is traveling at 100 km/h and that its (single) flywheel – a solid disk of mass 5 kg and radius 0.25 m – is spinning at its maximum angular velocity $\omega = 2\pi \times 55\,000/60 \cong 5800\,\text{s}^{-1}$. The flywheel has angular momentum $L \cong 900\,\text{kg m}^2/\text{s}$ and energy 2.64 MJ. (For this and other calculations in this box, see Problem 37.18.) The energy stored in the flywheel is several times the vehicle's kinetic energy, and would be enough to keep the car running at 100 km/h for about 4 minutes.

A torque must act on the automobile if it turns in any way that causes the flywheel angular momentum to change. To avoid changing the flywheel angular momentum every time the car turns left or right in the horizontal plane, the flywheel axis can be aligned vertically. If the car tilts forward to descend a hill, however, the direction of the flywheel's angular momentum is forced to change. The initial torque from the gravitational force as the front wheels enter the hill is along an axis pointing out the left-hand side of the car. In response to this torque, the angular momentum axis of the flywheel shifts in the direction of the torque axis through eq. (2.43), giving the car a tendency to roll onto its left side (looking along the direction of motion) as it tips down the hill. This effect, in which a rapidly rotating object responds counterintuitively to a torque by rotating to align the axis of rotation towards the torque axis, is known as *gyroscopic* behavior, and can also be seen, for example, when a rapidly rotating bicycle wheel is tilted and appears to rotate around an axis perpendicular to the torque axis.

Suppose, for example, that the car tilts forward by 10° over a distance of 10 meters to descend a hill. For the car to tilt with the hill, the angular momentum must acquire a component $\Delta L = \sin(10°)L \cong 160\,\text{kg m}^2/\text{s}$ in a time Δt of order 0.36 s, requiring a *further* torque $\tau = \Delta L/\Delta t \cong 450\,\text{Nm}$ along the car's direction of motion, in addition to the original torque mentioned above produced from the difference in pressure between the front and back wheels as

 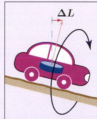 

the car enters the hill. As shown in the figure, the required torque to shift the flywheel angular momentum vector forward is provided by the weight of the car shifting onto its left side. A simple estimate indicates that a roughly 6% increase/decrease in the weight on the left/right side of the car (looking along the direction of motion) would supply the necessary torque. This is a significant effect, comparable to the imbalance caused by a sidewind blowing at a speed of tens of kilometers per hour, and increasing the likelihood of the car's rolling during a turn if it is simultaneously increasing or decreasing its forward angle of inclination.

The gyroscopic problem can be solved in various ways, either by including counterrotating flywheels or using gimbals, to decouple the direction of the angular momentum from the direction of motion of the car. Such systems, however, increase the complexity and cost of flywheel energy storage. Failure of these systems could subject the vehicle to strong torques. The gyroscopic problem, potential problems with catastrophic flywheel destruction in automobile accidents, and the inertia of the auto industry, have kept flywheels from playing a major role so far in commercial hybrid automobile technologies.

devices combine aspects of electrostatic and electrochemical storage to achieve both relatively high energy density and high power density.

Unlike conventional capacitors, which generally use a solid dielectric, ultracapacitors store electrostatic energy at an interface region between a conducting electrode and an electrolyte. The charge separation distances involved are on the order of $d \sim 1\,\text{nm}$ or less, which enables the capacitance $C = \epsilon A/d$ to be quite large. This mechanism is known as **double-layer capacitance**.

Supercapacitors can also employ electrochemical storage mechanisms involving the electrodes, known as

Relatively small amounts of energy can be stored in the static electric fields of capacitors and magnetic fields generated by currents in superconductors. An ordinary capacitor with an air dielectric stores approximately $40 \, \text{J/m}^3$. Supercapacitors can store roughly $100 \, \text{MJ/m}^3$, and magnetic energy storage gives $400 B^2 \, \text{kJ/m}^3$ (B in tesla).

Despite their small energy density, capacitors and supercapacitors find many applications where their very high power density, reliability, and cycle life are essential. Superconducting magnetic energy storage is used in a few installations to control electrical power quality.

pseudocapacitance. This substantially enhances capacitance without generating excess resistance that would compromise power density.

Different super- and ultracapacitor designs combine these electrostatic and electrochemical mechanisms in different ways to achieve capacitors that can have power densities comparable to conventional capacitors but much higher energy densities (see Figure 37.1). While the terms *supercapacitor* and *ultracapacitor* are often used interchangeably, the term *ultracapacitor* can be used specifically for capacitors where the double-layer electrostatic energy is dominant and *supercapacitor* can be used to describe capacitors dominated by pseudocapacitance. For supercapacitors with large pseudocapacitance, the voltage–charge relation deviates from the simple linear form of a capacitor and begins to look more like that of a battery. The specifications of an example ultracapacitor are described in Example 37.5. While ultra- and supercapacitors have energy/mass densities that are much higher than conventional capacitors, they are still below batteries by roughly one order of magnitude, so supercapacitors will not be used as primary power sources for laptops or automobiles in the immediate future. In particular, the constraints on electrostatic energy density per unit volume from breakdown field and electrochemical energy density per unit mass from fundamental chemical constraints suggest that supercapacitors will never surpass electrochemical batteries in energy density.

37.4.3 Superconducting Magnetic Storage

Just as energy can be stored in electric fields through eq. (37.21), energy can be stored in magnetic fields through eq. (3.71)

$$E_{\text{EM}} = \frac{1}{2\mu_0} \int d^3 x \, |\boldsymbol{B}(\boldsymbol{x})|^2 \,, \qquad (37.22)$$

where μ_0 is the magnetic permeability of the vacuum. Equation (37.22) gives an energy density of approximately $400 B^2 \, \text{kJ/m}^3$, with B the magnetic field strength in tesla.

Unlike electric fields produced by static charge configurations, magnetic fields are only produced by moving charges. In most situations, therefore, the current producing a magnetic field will lose energy rapidly by Joule heating at the rate $I^2 R$ so that, as in a typical solenoid, energy must constantly be provided to maintain the field.

For certain materials known as **superconductors**, however, below a **critical temperature** quantum effects eliminate the resistance entirely, so that currents can flow with no resistive losses at all. The first superconductor was discovered in 1911, when it was found that below $4.2 \, \text{K}$ mercury conducts electric current without resistance. Further developments over the years have led to the discovery of **high-temperature superconductors**, such as $\text{YBa}_2\text{Cu}_3\text{O}_7$, which has a critical temperature of $92 \, \text{K}$, above the temperature of liquid nitrogen ($77 \, \text{K}$). Until quite recently all high-temperature superconductors involved copper oxide or **cuprate compounds**. In 2006 iron-based high-temperature superconductors were discovered, but none of these have critical temperatures above the temperature of liquid nitrogen. At this time the highest confirmed critical temperatures realized for superconductors at atmospheric pressure are around $138 \, \text{K}$ for some mercury-based cuprates. But there is no known reason that superconductors could not be found with significantly higher critical temperatures.

By forming a superconducting material in the shape of a solenoid or toroidal loop and passing a current through the loop, a steady magnetic field can be produced that stores energy. Due to the absence of resistivity and Joule heating, this current will continue as long as the material is kept below the critical temperature, with decay time measured in years. This is the mechanism of **superconducting magnetic energy storage** (SMES). Energy storage via SMES is limited by the fact that superconductivity is only possible for currents below a limiting critical value that depends upon the material. Since no chemical reactions are involved, SMES has a rapid response time, high power density, and can have high round-trip storage efficiency.

Due to the expense of superconducting materials and the energy required to maintain them below the critical temperature, SMES has very few applications at present. SMES is generally used in applications such as for controlling power quality where ultra clean power is required, for example in chip fabrication facilities. A safety issue with SMES is that if the superconducting material exceeds the critical temperature or the current gets too high, then the

Example 37.5   An Ultracapacitor

Recent advances in the understanding of materials at the nanometer scale have made possible the construction of *ultracapacitors* that have very large capacitance while maintaining small size and weight. Ultracapacitors and supercapacitors combine electrostatic and electrochemical energy storage mechanisms to combine the high power density typical for capacitors with relatively high energy density per unit mass.

A Maxwell Technologies BCAP0310 ultracapacitor, for example, is roughly the same size as an ordinary D-cell battery, has a capacitance of 310 F, and works with voltages up to 2.85 V [295]. This ultracapacitor has a cylindrical shape, roughly $H = 60$ mm high and $R = 16.5$ mm in diameter. The internal resistance of the BCAP0310 is $r \cong 2.2$ mΩ.

Fully charged, the BCAP0310 stores $\frac{1}{2}CV^2 = 1130$ J and weighs 0.06 kg, so its energy density is ~ 19 kJ/kg. This can be compared to a modern rechargeable Li-ion battery, which stores $\sim$500 kJ/kg. Similarly, the volumetric energy density of the BCAP0310 is easily computed to be roughly 90 MJ/m^3, much lower than other energy storage technologies. The high power density of ultracapacitors, however, makes them very useful in applications needing a rapid release of energy, including smoothing irregularities in power supplies, amplifiers in automobile stereo systems, and pulsed power releases used in systems including lasers, particle accelerators, fusion research, and weapons systems. Like conventional capacitors, ultracapacitors are very robust and can be cycled hundreds of thousands of times, far more than electrochemical batteries. Maxwell Technologies specifies an "absolute maximum current" of 500 A (compared to the short-circuit current of $V/r \sim 1200$ A), so the complete charge of $Q = CV \cong 900$ Coulombs can be discharged in a matter of seconds, with a power output of up to 0.9 kW and a corresponding power density of $\sim$15 kW/kg (Problem 37.19).
(Image credit: Maxwell)

---

material develops a resistance, and the stored energy can rapidly be dissipated as Joule heating. This could be an issue for large-scale storage of energy via SMES if high-temperature superconductors are ever developed and used for this purpose. The energy density of SMES can be relatively high in principle, but is limited by the critical current of the superconducting material.

Discussion/Investigation Questions

37.1 For each of the energy storage parameters described in §37.1, identify an application for which this parameter is critical. For each such application, discuss some good options for energy storage or limitations of current technologies.

37.2 Discuss the merits of different storage types for (a) storage of off-peak electrical power on large-scale energy grids, (b) automobiles without combustion engines, and (c) onboard power systems for airplanes.

37.3 Investigate and discuss *pumped heat electricity storage* and/or *pumped cryogenic electricity storage* [289]. Consider the effects that reduce the efficiency of these systems below the Carnot limit. Discuss potential advantages of PHES and PCES over pumped hydro and

CAES including, for example, freedom from geographical constraints and material flexibility.

37.4 A *silver cell* is an alkaline dry cell in which the manganese/carbon cathode is replaced by silver-oxide (Ag$_2$O), so the net reaction is Zn + Ag$_2$O → ZnO + 2Ag. Investigate the properties of silver cells and explain why they are well suited to use in hearing aids, implanted pacemakers, and similar applications.

37.5 Consider the advantages and potential limitations of a purely electric vehicle with a 200 MJ battery energy storage system. For what fraction of your vehicle use would this be practical?

Problems

37.1 A typical lithium-ion battery for a laptop has a storage capacity of 200 kJ and a mass of 0.5 kg. Compute the number of such batteries required and total mass in order to store 30% of US daily electric energy consumption of 12 PJ.

37.2 Consider the scope of pumped hydro storage needed to store and release 12 PJ of energy per day. Compare to the total US hydropower output. If the full 12 PJ were stored using local modest-sized UPH reservoirs of volume 20 000 m^3 at a depth of 100 m, how many such reservoirs would be needed?

37.3 The world's largest pumped hydro storage facility, located in Bath County, Virginia, has a generating capacity of about 3 GW. When generating power, the turbine flow rate is roughly 850 m³/s. Estimate the difference in elevation between the upper reservoir and the turbines. The area of the upper reservoir is 110 ha and its surface level fluctuates by approximately 30 m during operation. Estimate the total energy storage capacity of the facility.

37.4 A company is proposing to build an *advanced rail energy storage* system: excess electric energy will power engines that will transport a heavy mass up a slope. During periods of peak demand, the process will be reversed, converting the gravitational potential energy of the mass back into electric power. The company proposes to build a first commercial-scale facility that will store 12.5 MWh using a height difference of 500 m. The railroad cars have a loaded mass of 130 tonnes. How many railroad cars are required for the proposed installation? The company suggests a grid-scale storage facility might have a capacity of 16–24 GWh. For a height difference of 1000 m, how many of these railroad cars would be needed?

37.5 [T] Show that the energy stored when air is compressed adiabatically from pressure p_0 to pressure p_H in a final volume $V_H = V$ is given by

$$E_{stored}^{(adiabatic)} = p_H V \left(\frac{\gamma}{\gamma-1} \left(1-(1/r)^{(\gamma-1)/\gamma}\right) - (1-1/r) \right)$$

after the work done by/on the atmosphere is subtracted. Here $r = p_H/p_0$ and $\gamma = 1.4$ is the adiabatic index of air.

37.6 Consider using the McIntosh salt mine described in Example 37.2 for adiabatic storage. Compute the energy stored if a volume of air initially at $p_0 = 1$ atm is compressed to $p_H = 75$ atm, and compare to the energy stored if the air were compressed isothermally. Take $\gamma = 1.4$ for air. Compute the temperature of the air at pressure p_H assuming that the initial temperature was 300 K.

37.7 Confirm explicitly that the work (37.2) done in isothermal compression of a CAES system with volume V, and low, high, and ambient atmospheric pressures p_L, p_H, p_0 is equivalent to the exergy (36.25) of the system.

37.8 [H] Verify the computations in Example 37.2 describing the McIntosh CAES plant. In particular, use eq. (37.2) to compute the energy stored through isothermal compression; verify that 1.4 TJ of work would be done if the stored air were allowed to expand adiabatically as described in Example 37.2 and verify the temperatures of the air left in the mine and the air released into the environment after adiabatic expansion.

Table 37.2 Reaction data for Problem 37.11.

Compound	$-\Delta H_f^0$ (kJ/mol)	$-\Delta G_f^0$ (kJ/mol)
ZnO	348	318
MnO_2	521	466
Mn_2O_3	972	890

37.9 Estimate the specifications for a solar thermal energy plant and storage system. The plant should output a steady 100 MW, 24/7. Assume that the plant is built in a desert area with an average 250 W/m² insolation. Assume a gross conversion efficiency of 3%. Tanks of molten salt at 450 °C (specific heat capacity 1.5 kJ/kg K) should be able to store 24 hours of power. You can assume that thermal energy from the salt is converted to electrical power with 33% efficiency and that the salt ends up at a temperature of 250 °C after transferring its energy to the working fluid that drives the turbine. How big does the plant need to be, and how many kilotonnes of molten salt are needed per square kilometer of the power plant?

37.10 In a lithium-iodine (LiI) battery, a solid lithium anode is surrounded by a polymer impregnated with iodine. The net reaction, $Li + \frac{1}{2}I_2 \rightarrow LiI$, has a standard reaction free energy of $\Delta G^0 = -266.9$ kJ/mol and a standard reaction enthalpy of $\Delta H^0 = -270.08$ kJ/mol. Estimate the cell voltage $\mathcal{E}^0$, the upper limits on the thermodynamic efficiency η, and the upper limit on the specific energy $\varepsilon_{effective}$ of a LiI battery.

37.11 Compute the free energy of reaction $-\Delta G^0$, ideal cell voltage $\mathcal{E}^0$, maximum possible effective energy density $\varepsilon_{effective}$, and thermodynamic efficiency limit $\eta = \Delta G^0/\Delta H^0$ for an alkaline dry-cell battery (37.11) given the enthalpies and free energies of formation in Table 37.2.

37.12 Compute an upper bound on the energy density of a lead-acid battery in which the overall chemical process is

$$Pb + PbO_2 + 2H_2SO_4 \qquad (37.23)$$
$$\rightarrow 2PbSO_4 + 2H_2O$$

and for each such reaction two electrons pass around the external circuit, assuming that the only material needed is that involved in the reaction. The standard cell potential for this battery is 2.04 V. The actual energy density is significantly lower than the upper bound due to the need for additional water, as well as other battery components, for operation of the battery.

37.13 Confirm the 2nd law bound $\eta = 83\%$ quoted in the text on the efficiency of a hydrogen fuel cell.

37.14 Compute the 2nd law bound on the efficiency of a methanol fuel cell.

37.15 Estimate the relative range in highway miles of a passenger vehicle with comparable size and drag coefficient to the Camry described in §2, assuming it is operating using (a) a 40 L gasoline tank and 25% efficient engine, (b) a 150 kg lithium-ion battery, and (c) a compressed hydrogen fuel tank at 700 atm with volume 60 L and a hydrogen fuel cell with twice the energy loss required by the 2nd Law.

37.16 According to Figure 37.7, the energy density of hydrogen compressed to 700 atm is approximately 5.6 MJ/L. Assuming hydrogen behaves as an ideal gas throughout, estimate the energy density of hydrogen compressed to 700 atm by adding the enthalpy of combustion and the energy required to compress the hydrogen isothermally starting at 1 atm. How does your answer compare with the quoted figure of 5.6 MJ/L? Compare the total energy stored in hydrogen compressed to 700 atm with the energy density of gasoline.

37.17 Compute the speed of a fragment of metal released from the rim of the example flywheel described in the box in Example 37.4. Compare to the speed of a bullet fired by a handgun.

37.18 For the conditions described in Example 37.4 compute the moment of inertia of the flywheel, its angular momentum, and kinetic energy of rotation. Assuming that the car loses kinetic energy only to wind resistance (with a drag coefficient of $c_d = 1/3$) and that the

flywheel's energy can be converted to kinetic energy of forward motion at 90% efficiency, find how long the flywheel could keep the car moving at 100 km/h. Finally, show that left/right imbalance of approximately $\pm 6\%$ in the weight on the tires is sufficient to supply the torque needed when the car tilts forward $10°$ in 0.36 s.

37.19 For the Maxwell BCAP0310 ultracapacitor described in Example 37.5, the capacitance is 310 F, mass is 0.06 kg, maximum voltage is 2.85 V, and internal resistance is $r = 2.2\,\text{m}\Omega$. Consider connecting the ultracapacitor to an external load with resistance R_L. Compute the maximum possible instantaneous power delivered to the load when the capacitor is fully charged by optimizing over R_L.

37.20 Suppose that the Maxwell BCAP0310 ultracapacitor described in Example 37.5 used "conventional" capacitor technology. Specifically, assume that it can be described essentially as a parallel plate capacitor of area A, plate separation d, and dielectric constant $\epsilon = k\epsilon_0$ rolled into a spiral to fit into a cylinder of radius $R = 16.5\,\text{mm}$ and height $H = 60\,\text{mm}$. First compute the plate separation d in terms of k. Then estimate the insulator/dielectric's breakdown voltage in terms of k, assuming that the breakdown voltage sets the limit $V \leq 2.85\,\text{V}$ quoted in the box. Ordinary dielectrics have k in the range 1–100. Are the distance scales and dielectric properties of the BCAP0310 consistent with your expectations for a conventional capacitor?

Electricity Generation and Transmission

Roughly one third of the world's primary energy consumption goes to produce electric power. Almost all of this power reaches consumers over networks of high-voltage *transmission lines* and lower-voltage, shorter-range *distribution systems*. An integrated system of generators, transmission and distribution lines, transformers, and other related electrical devices, as well as the customers (*loads*) they serve forms an **electric grid**. Complex systems of measurement and control, some automated and some requiring active operator intervention, are needed to keep electric grids stable and safe. Electric grids must respect not only physical and technical limitations but also a complex system of economic, regulatory, and political constraints that have evolved in different ways in different localities. This chapter focuses on the basic physical principles that dictate how electric power is produced, transformed, transmitted, and distributed. A clear understanding of these principles is needed for sensible development of electric grids within the larger socioeconomic framework.

Before embarking on a detailed study of electric grids and their components, we begin with a brief overview in §38.1. In §38.2, we introduce the tools used to analyze alternating current (AC) circuits with the three primary circuit elements of resistance, capacitance, and inductance that were introduced in §3. We focus on the relative phase between current and voltage, which is essential for understanding the flow of power in electric grids operating on AC voltages. With these tools in hand, in §38.3 we revisit the subject of AC power generation. We give a brief description of the most common types of generators: *synchronous generators* and *asynchronous (induction) generators*. In this section we introduce the basics of *three-phase power*, which has significant advantages over the *single-phase* AC power described in §3.4.2. We next explore how generators interact with loads; in particular, we show how a generator's output can be stabilized as a load fluctuates.

> **Reader's Guide**
> This chapter describes some of the principles underlying the generation, transmission, and distribution of electric power over large-scale electric grids. Synchronous grids are introduced. Basic concepts of alternating current (AC) circuit dynamics are developed, and the dynamical principles of both synchronous and asynchronous electric generators are explained. Three-phase electric power is introduced and some of its advantages are explained. Systems issues including the operation, stability, design, and control of transmission and distribution networks are described, as are challenges raised by the need to incorporate variable energy resources such as wind and solar power into electric grids.
>
> Prerequisites: §3 (Electromagnetism), §4 (Waves and light), §37 (Energy storage).

In §38.4 we characterize other key physical components of the transmission/distribution system. These include transmission lines, transformers, and voltage support systems. We also describe some basic aspects of the layout of transmission and distribution networks, and advantages that three-phase power brings to power distribution and use by consumers. We then turn to systems issues that arise when many generators serve many loads, including measurement, control, and stability.

Finally, in §38.5 we examine how electric grids may be affected by the widespread deployment of intermittent and/or distributed renewable energy sources.

Electric power systems are a complex subject. von Meier [296] provides a particularly clear and insightful analysis of electric power systems at a technical level appropriate for readers of this book. Another accessible introduction can be found in [297], while [298] provides a policy-oriented study of US electric grids that integrates technical with economic, social, and regulatory issues.

38.1 Overview of Electric Grids

38.1.1 Grid Connectivity and Energy Flow

The physical structure of an electric grid is summarized schematically in Figure 38.1. At the core of the system are high-voltage, long-distance transmission lines that connect generating units and carry power to distant locations for commercial and residential use. The *transmission network* often contains loops to increase connectivity and provide redundancy. Transformers step up the voltage from generators to the high voltage carried by the transmission lines, and then step down the transmission voltage to an intermediate level used in the *distribution network*. Finally, other local transformers lower the voltage to various levels provided to industry, businesses, and homes.

Electric grids have two features that distinguish them from more conventional energy delivery systems like natural gas pipelines. First, they have very little storage capacity. While a limited amount of pumped hydro storage (§37.2.1) is available in some locations, for the most part if electrical energy is not used when it is produced, it circulates through the grid, potentially causing instabilities, until it is lost in resistive heating. Keeping the production and use of electrical energy closely balanced is therefore one of the principal objectives of grid management. Electricity demand changes and requires response on many time scales, from fractions of a second as loads fluctuate, to minute-by-minute as loads come on or offline, to hours-long peak demand periods that may occur only a few times during the year, to seasonal swings that are large and usually predictable. A second unusual feature of electric grids is that the route that energy follows from production to consumption is not as clearly defined as it is for a material commodity such as oil. The flow of current and energy through a complicated and multiply connected

transmission network is determined by the laws that govern electric circuitry. In some situations, for example, electric power flows between generators rather than from generators to loads, causing undesirable losses. The complexity of energy flow in the grid complicates the electricity market because consumers cannot purchase power directly from suppliers. In light of these complexities it is perhaps surprising that electric grids are as efficient as they are. Figure 38.2 shows the history of electricity transmission and distribution losses in the US through 2010 and the losses for selected countries between 2011 and 2015.

The scope of electric grids has evolved over time. Electric power transmission systems were initially limited to single producers serving a network of consumers, much like the sketch in Figure 3.14(a). Soon, however, electrical engineers learned that they could connect many different generators, even when widely separated, and that natural features of electric circuits could be used to keep the system stable. There are many economies of scale that drive greater integration. By integrating more generators into a single system, greater reliability and control are possible. Other generators can take over when one fails or must be removed from service for maintenance. By integrating more customers, average fluctuations in demand can be

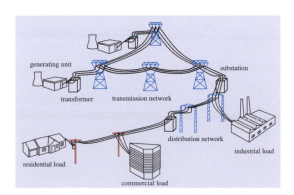

Figure 38.1 A schematic depiction of an electricity generation, transmission, and distribution grid. From [298].

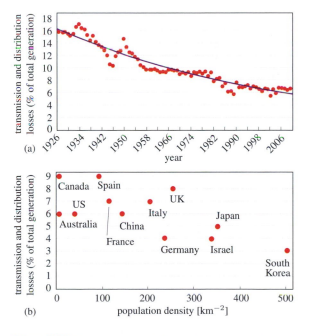

Figure 38.2 Transmission and distribution losses on electric grids. (a) Losses in the US from 1926 to 2010. After [298]. (b) Losses (including theft) in a selection of countries in 2011–2015 [299]. Note the inverse correlation with population density, which is a proxy for the line length.

reduced, allowing managers to program the use of the available generators more economically. Having a variety of generators available on a single grid enables some to be allocated for *baseload* use and others to serve periods of peak demand, thereby increasing the overall efficiency of the system. The existence of multiple transmission pathways on a multiply connected grid makes the network more robust against unplanned outages. Finally, limited resources like transmission rights-of-way can be most fully utilized in a large-scale grid. In light of these advantages of scale, it is not surprising that electric distribution networks have expanded to the limit that historical accidents or national boundaries have allowed. Future developments, such as the incorporation of substantial generating power from fluctuating renewables like wind and solar, or using electric vehicles for grid storage, may be facilitated by the large scale of modern electric grids.

38.1.2 AC Power and Synchronous Grids

With few exceptions, electric power is transmitted as AC current. As a result, the voltage and current at each point in an interconnected electric grid oscillate sinusoidally in proportion to $\cos(\omega t - \phi)$, where ω is a fixed (angular) grid frequency and ϕ is a phase shift that takes a definite value at each point on the grid. We explore the origins of ϕ in §38.2. Since the current seen at a point on a grid is the superposition of currents driven by many different generators, the sum would be incoherent unless the generators provide power at the same frequency and at fixed relative phase. Thus all points on an interconnected grid must be synchronized both in frequency and phase, leading to the term **synchronous grid** for such systems. As we saw in §3.4.2, the frequency of the EMF created by a simple AC generator is determined by the rate at which a current loop rotates in a magnetic field. For each generator, the phase and precise frequency must be constantly adjusted by feedback systems to maintain synchrony with the entire interconnected grid.

In practice, two synchronous grids using different frequencies cannot be connected directly. Once connected, two grids with the same frequency must be kept synchronized with one another [300]. If power must be transferred from one synchronous grid to another, it is generally first converted to direct current (DC) and then converted back to AC at the frequency and phase of the second grid. As shown in Figure 38.3, the US has evolved three independent synchronous grids, or *interconnects*: the Eastern, Western, and Texas Interconnects, with *high-voltage DC* (HVDC) lines between them. The map of high-voltage AC transmission lines in Figure 38.3 gives some idea of the complexity of the grid. Europe is served by a number of

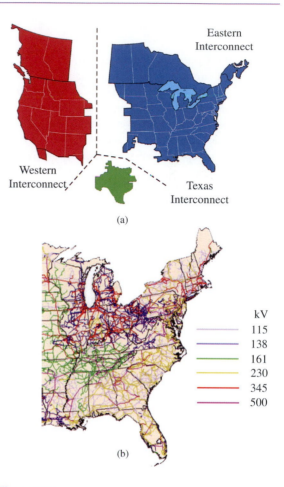

(a)

(b)

	kV
——	115
——	138
——	161
——	230
——	345
——	500

Figure 38.3 (a) In the US and southern Canada electricity is transmitted over three principal synchronous grids. In addition to Eastern and Western regions, the *Electric Reliability Council of Texas* (ERCOT) manages the grid for most of Texas. From [301]. (b) High-voltage transmission lines in part of the Eastern Interconnect.

independent grids, as shown in Figure 38.4, also connected by HVDC lines.

Grid Voltage The motivation for using high-voltage transmission over long distances is to minimize resistive losses in the transmission lines, which, as we saw in §3.2.3, scale as $1/V^2$. Typical transmission lines operate at voltages above $100\,\mathrm{kV}$ and up to $\sim 1\,\mathrm{MV}$.[1] Safety requires that generators and distribution lines operate at much lower voltages of order 10–20 kV, so it is necessary

---

[1] The highest voltage transmission lines in the US and Europe are 765 kV and 380 kV respectively, while AC lines crossing Siberia have reached ∼1.25 MV [296].

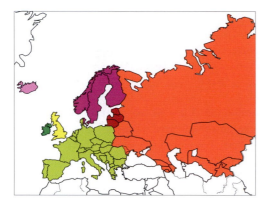

Figure 38.4 The principal synchronous grids of Europe. (Credit: 'Goldztajn' reproduced under CC-BY-SA 3.0 license via Wikimedia Commons)

Electric Grids

Almost all electric power is carried from producers to consumers over vast, interconnected synchronous AC grids. An electric grid includes generators, transformers, high-voltage transmission lines, distribution networks, and other components, all carrying AC currents at the same frequency and at fixed relative phase. Frequencies of 50 Hz and 60 Hz dominate, and are usually associated with consumer voltages of 220–240 V and 110–120 V respectively. Transmission line voltages may vary within a single synchronous grid. Power can be transferred from one grid to another over high-voltage DC lines.

to transform electric power efficiently between very different voltages. Transformers perform this function reliably and efficiently and at relatively low cost. Solid-state voltage converters are now common in low-power applications – cell-phone chargers are a familiar example – but have not yet achieved the reliability, efficiency, or low cost necessary to replace transformers in high-voltage, high-power applications. Because transformers work on the principle of induction, they require AC currents. When electricity transmission was first commercialized, there was a famous battle between backers of DC power transmission (principally the American inventor, Thomas Edison) and champions of AC power generation and transmission (including American industrialist George Westinghouse, Serbian-American engineer Nikola Tesla, and others). DC has several advantages; in particular, there is no frequency or phase to match when connecting multiple DC power sources. Absent a system to transform DC voltage, however, Edison's system required transmission at the same low voltage provided to consumers, resulting in unacceptable transmission losses, and the AC paradigm eventually won out (Problem 38.1).

The frequency of AC electricity in standardized systems is either 50 or 60 Hz, chosen to be high enough that no disturbing flicker is perceived in electric lighting, and low enough to minimize stresses on mechanical components that must rotate at a rate proportional to the AC frequency. The US and parts of South America favor 60 Hz, while most of the rest of the world uses 50 Hz. Japan is divided into 50 and 60 Hz regions. Countries using 60 Hz typically have chosen consumer level voltage (V_{RMS}) of 110–120 V, while 50 Hz power is usually associated with 220–240 V. Appropriately designed transformers make it possible to

supply other voltages to industrial customers as needed. Modern solid-state electronics make conversion between AC and DC easier and more efficient than in the past, simplifying the trading of power between different synchronous grids. As a result, there is limited incentive for synchronization to spread beyond its current limits.

38.2 LRC Circuits

Large-scale electric power transmission networks are driven by generators that produce sinusoidally varying EMFs. In order to understand these systems, we present the basic features of electric circuits driven by sinusoidal EMFs.

38.2.1 The Series LRC Circuit

Considered abstractly, the cables that transmit the power and the loads that they drive include the three primary circuit elements that we introduced in §3: capacitors, resistors, and inductors. Present-day transmission networks inevitably involve resistive loss in transmission lines.[2] Devices that use electric power often involve motors and electromagnets that introduce inductance as well as resistance. Significant capacitance is less often encountered in typical loads, but capacitance is often used to modify the flow of power through an electrical network (see §38.4). Other circuit elements such as transistors and diodes are

---

[2] The use of high-temperature superconductors with zero resistance (§37.4.3) is a possibility for future power transmission.

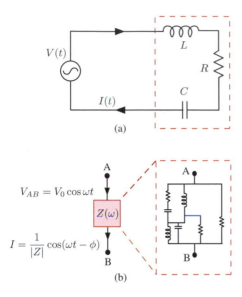

Figure 38.5 (a) A series LRC circuit. Its response to a sinusoidally oscillating EMF $V_0 \cos \omega t$ can be described by a complex impedance $Z(\omega)$ (see eq. (38.4)). (b) Any collection of resistors, capacitors, and inductors can be described by an *effective impedance $Z(\omega)$*.

important in specific devices, but typically do not affect the large-scale structure of transmission systems.

One of the simplest examples of a circuit with the features of interest is a **series LRC circuit**, which consists of an inductance L, a resistance R, and a capacitance C, connected in series with an EMF $V(t)$, as shown in Figure 38.5(a). The equation that determines the current flowing through such a circuit can be obtained by summing the voltage drops around the circuit. Instead, in the spirit of this book, we analyze the circuit by following the storage and dissipation of energy. Energy is stored in the capacitor ($Q^2/2C$) and in the inductor ($L\dot{Q}^2/2$); power $R\dot{Q}^2$ is dissipated in the resistor; and power $V(t)\dot{Q}$ is added by the external agent $V(t)$. (Note that the current I has been denoted by $\dot{Q}$.) Energy conservation requires that the rate of change of the stored energy must equate to the rate that energy is being added by the external agent minus the rate that it is being burned up in the resistor. From this follows an equation for the time dependence of the circuit,

$$\frac{dU}{dt} = \frac{d}{dt}\left(\frac{L}{2}\dot{Q}^2 + \frac{1}{2C}Q^2\right) = V(t)\dot{Q} - R\dot{Q}^2$$

$$L\dot{Q}\ddot{Q} + \frac{1}{C}Q\dot{Q} = V(t)\dot{Q} - R\dot{Q}^2$$

$$L\ddot{Q} + R\dot{Q} + \frac{1}{C}Q = V(t), \qquad (38.1)$$

where in the last line the common factor of $\dot{Q}$ has been cancelled, since the solution where $\dot{Q} = 0$ (no current) is not interesting. Note that the final form of eq. (38.1) simply expresses the total voltage drop across the circuit using $V_{\text{resistance}} = IR$, $V_{\text{capacitance}} = Q/C$, and $V_{\text{inductance}} = L\dot{I}$.

Equation (38.1) is a second-order differential equation that can be solved given values for Q and $\dot{Q}$ at some initial time. Equation (38.1) can be understood through a mechanical analog (Problem 38.2), where the inductor and capacitor store kinetic and potential energy respectively, and the resistor dissipates energy like a frictional force. Problems 38.3 and 38.4 explore some of the simple features of the LRC circuit described by eq. (38.1).

38.2.2 Sinusoidally Driven LRC Circuits

When an LRC circuit is driven by an AC generator that produces an EMF $V(t) = V_0 \cos \omega t$, the circuit responds by oscillating at the same frequency as the generator,[3] but with an amplitude and phase that are determined by L, R, and C.

A simple way to understand this behavior is to solve eq. (38.1) for a *complex* oscillating voltage $\tilde{V} = V_0 e^{i\omega t}$, where we have chosen the time coordinate and phase so that V_0 is real. In the steady-state solution, the current, as well as Q and $\ddot{Q}$, oscillate with the same frequency as the voltage,

$$\tilde{I} = \tilde{I}_0 e^{i\omega t},$$

$$\tilde{Q} = \tilde{Q}_0 e^{i\omega t} = -(i/\omega)\tilde{I}_0 e^{i\omega t},$$

$$\dot{\tilde{I}} = \ddot{\tilde{Q}} = i\omega\tilde{I}_0 e^{i\omega t}. \qquad (38.2)$$

Since eq. (38.1) is linear in V and Q with real coefficients L, R, and $1/C$, a solution for real $V(t)$ can be found by taking the real part of the terms corresponding to the left-hand side of (38.1). In particular, the real part of $\tilde{I}$ gives the current in the circuit driven by the real part of the voltage $\tilde{V}$, $V(t) = \text{Re}\tilde{V} = V_0 \cos \omega t$. Substituting from eqs. (38.2) into eq. (38.1), we find

$$Z\tilde{I}_0 = V_0 \quad \text{or} \quad \tilde{I}_0 = V_0/Z, \qquad (38.3)$$

where the **impedance** Z is given by

$$Z(\omega) = |Z|e^{i\phi} = R + i\left(\omega L - \frac{1}{\omega C}\right) \equiv R + iX. \quad (38.4)$$

The real part of the impedance is the resistance, while the imaginary part X is known as the **reactance**. The magnitude $|Z|$ and phase ϕ of the impedance are given by

---

[3] There is also a transient response, but it typically dies off quickly compared to the time scales of interest.

$$|Z| = \sqrt{R^2 + \left(\omega L - \frac{1}{\omega C}\right)^2}, \qquad (38.5)$$

$$\phi = \tan^{-1}(\omega L - 1/\omega C)/R. \qquad (38.6)$$

The physical current driven by the real voltage $V(t) = V_0 \cos \omega t$ is then the real part of $\tilde{I}$,

$$I(t) = \text{Re}\,(\tilde{V}/Z) = \frac{V_0}{|Z|} \cos(\omega t - \phi). \qquad (38.7)$$

Thus, when driven by an AC voltage, the current oscillates at frequency ω but is shifted from the voltage by a relative phase ϕ given by eq. (38.6). The phase angle ϕ can range from $-\pi/2$, for a purely capacitive load, to $+\pi/2$, for a purely inductive load. The amplitude of the current is $V_0/|Z|$, where $|Z|$ is given by eq. (38.5).

The salient feature of resistance, capacitance, and inductance for the preceding analysis is that they are all **linear circuit elements**, in the sense that the contribution to the voltage drop in eq. (38.1) is linear in the charge Q and/or current I for each element. In general, any combination of linear circuit elements can be replaced by a single frequency-dependent complex impedance $Z(\omega)$ that relates the current to the voltage across the system through eq. (38.7). This relationship is exhibited schematically in Figure 38.5(b).

The impedances of an isolated resistor, inductor, or capacitor are R, $i\omega L$, and $-i/(\omega C)$ respectively. Resistance, impedance, and reactance are all measured in ohms. One convenient feature of complex impedances is that the rules for combining them are identical to the rules for resistances (Problem 38.10): two impedances Z_1 and Z_2 in series add, $Z_{\text{eff}} = Z_1 + Z_2$, and two impedances in parallel add reciprocals, $Z_{\text{eff}}^{-1} = Z_1^{-1} + Z_2^{-1}$. These simple rules suffice to compute the impedance of many (but not all) complicated circuits formed from the basic LRC circuit elements.

38.2.3 Phases and Energy Transfer in AC Circuits

The fact that in AC circuits the voltage and current need not be in phase has significant effects on the way that power flows through the circuit. A national, or even local electric grid requires integrating many generators together with many loads, each with its own characteristics. Keeping the currents and voltages in the right phase relation to maximize efficiency is a task that demands sophisticated analysis and control.

In an AC circuit with only resistance, the current is **in phase** with the voltage, meaning that the phase difference vanishes, $\phi = 0$. Averaged over a period, the power consumed in the resistance is

---

LRC Circuits

When a network of resistors, capacitors, and inductors is driven by a sinusoidal EMF, $V(t) = V_0 \cos \omega t$, the response of the circuit can be described by a *complex impedance* $Z = Z(\omega) = |Z(\omega)|e^{i\phi(\omega)}$. In particular, the current is given by

$$I(t) = \frac{V_0}{|Z|} \cos(\omega t - \phi).$$

Impedances can be added in series or in parallel according to the same rules as resistances. The impedance of an isolated resistor, inductor, and capacitor are R, $i\omega L$, and $-i/(\omega C)$ respectively.

$$\langle P \rangle = \langle I(t)V(t)\rangle = I_{\text{RMS}} V_{\text{RMS}} = I_{\text{RMS}}^2 R, \qquad (38.8)$$

where, as in §3.2.3, the RMS voltage and current are defined by $V_{\text{RMS}} = V_0/\sqrt{2}$, $I_{\text{RMS}} = I_0/\sqrt{2}$, and $I_0 = V_0/R$ is the amplitude of the current. In a purely inductive or purely capacitive circuit, the current and voltage are $\pi/2$ out of phase, and although energy flows in and out of the inductor or capacitor, no power is used. Power that sloshes back and forth in an AC circuit without being consumed is characterized by *reactive power* (Problem 38.5).

If $\omega L > 1/\omega C$ then inductance dominates over capacitance, as is usually the case in electrical power networks. In this situation, $\phi > 0$, and in each cycle the current reaches its maximum somewhat after the voltage (but while the voltage is still positive). The current is said to **lag the voltage**. Conversely, if capacitance dominates (as it would at very low frequency in any circuit with both capacitance and inductance), $1/\omega C > \omega L$, so $\phi < 0$ and the current **leads the voltage**. These three cases are illustrated in Figure 38.6(a)–(c). Note that the effect of a capacitance on ϕ in a series LRC circuit increases as C decreases (an open-circuit corresponds to a limit of zero capacitance). In contrast, when capacitance is added in parallel to a load with impedance Z, the effect on the phase increases with increasing capacitance (Problem 38.8).

When an electric circuit is viewed from the perspective of a generator producing an RMS voltage V_{RMS} and a load drawing an RMS current I_{RMS}, the **real power** transferred to the load is given by

$$P_{\text{real}} = \langle V(t)I(t)\rangle = V_{\text{RMS}} I_{\text{RMS}} \cos \phi = I_{\text{RMS}}^2 R, \qquad (38.9)$$

in close analogy to the last equation in Box 38.1. This real power should be compared to the **reactive power** given by

Box 38.1   Phases

Phase differences between action and response occur in many places in physics, and are most simply understood though a simple mechanical analogy. In mechanics, the instantaneous power delivered by an agent to an object moving in one dimension is given by eq. (2.12), $P(t) = F(t)v(t)$. For simplicity, we suppose that the force and the velocity are parallel. To make the analogy to AC voltage and current, suppose that the force and velocity oscillate sinusoidally and that there is a *phase difference* ϕ between them, $F(t) = F_0 \cos \omega t$ and $v(t) = v_0 \cos(\omega t - \phi)$. The time-averaged power supplied by the force to the object is given by

$$\langle P(t) \rangle = \langle F(t)v(t) \rangle = \langle F_0 v_0 \cos(\omega t) \cos(\omega t - \phi) \rangle$$
$$= F_0 v_0 \langle \cos \omega t (\cos \omega t \cos \phi + \sin \omega t \sin \phi) \rangle$$
$$= \tfrac{1}{2} F_0 v_0 \cos \phi,$$

because the average of $\cos^2 \omega t$ is 1/2 (see Figure 3.13) and that of $\cos \omega t \sin \omega t$ is zero. The power delivered from the agent to the object is maximum when $\phi = 0$, i.e. when the force and velocity are *in phase*; it is zero when $\phi = \pm \pi/2$; and the object actually transmits power to the agent when $\cos \phi$ is negative. All this is illustrated graphically for a selection of choices of phase in Figure 38.6 for the electrical case where $F(t) \to V(t)$ and $v(t) \to I(t)$.

If we define the RMS force and velocity by $F_{\text{RMS}} = F_0/\sqrt{2}$ and $v_{\text{RMS}} = v_0/\sqrt{2}$, then the time-averaged power, known in power electronics as the *real power* P_{real}, is given by

$$P_{\text{real}} = F_{\text{RMS}} v_{\text{RMS}} \cos \phi,$$

where $\cos \phi$ is known as the *power factor*.

An example familiar from childhood is shown in the figure: the woman supplies the oscillating force and the swing responds. If the woman pushes most strongly when the child is at the bottom of her trajectory moving forward, even a small effort can transfer considerable power to the child on the swing; if the woman pushes only at the extremes of the swinger's motion, when the swing is at rest ($\pi/2$ out of phase), no power is transferred and pushing is ineffective; and if the woman pushes forwards as the child is moving backwards, then the energy in the swing is transferred back to the woman.

$$P_{\text{reactive}} = V_{\text{RMS}} I_{\text{RMS}} \sin \phi = I_{\text{RMS}}^2 X, \qquad (38.10)$$

and the **apparent power**,

$$P_{\text{apparent}} \equiv V_{\text{RMS}} I_{\text{RMS}}. \qquad (38.11)$$

The apparent power is useful because, as the product of I_{RMS} and V_{RMS}, it is easily measured, and because it is a better indicator of line losses than the real power (see Example 38.1). Note that the real and reactive power add in quadrature like perpendicular components of a vector, $P_{\text{apparent}}^2 = P_{\text{real}}^2 + P_{\text{reactive}}^2$, rather than linearly. The factor of $\cos \phi$ by which real power is smaller than apparent power is known as the **power factor**. Although real power is measured in watts, it is customary to quote apparent power in *volt-amperes* (VA) and reactive power in *volt-amperes-reactive*, denoted VAR, to emphasize the fact that these quantities are not available to do work.

Earlier, in §3.2.3, we investigated resistive losses in transmission lines assuming the load to be purely resistive.

In practice, loads usually include motors and electromagnets, which use induction to generate currents, so loads often have significant reactance. Therefore, the circuit consisting of the generator, transmission lines, and load may have a power factor $\cos \phi$ considerably less than one if nothing is done to correct it. In the presence of reactance, the analysis that led to eq. (3.40) under the assumption that the transmission line resistance R is much smaller than the load resistance, yields instead (see Problem 38.6),

$$\frac{P_{\text{lost}}}{P_L} \cong \frac{1}{\cos^2 \phi} \frac{R P_L}{V_{\text{RMS}}^2}. \qquad (38.12)$$

Here R is the resistance of the *transmission line*, which usually consists of several separate *conductors*, whose losses must be added to obtain the total line loss. Thus, for example, in a circuit with $\cos \phi = 0.8$, the resistive losses in the conductors are 56% greater than if $\cos \phi$ were one. Since in a system with substantial reactance the currents

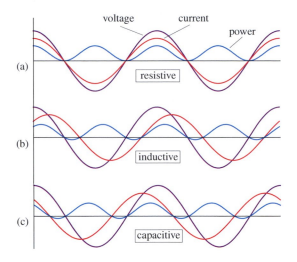

Figure 38.6 Phase relations between applied AC voltage and current. The voltage, current, and power are shown for three cases. (a) Purely resistive circuit, where current and voltage are in phase $\phi = 0$. The time-averaged power transferred to the load is maximum. (b) A circuit with significant inductance, where $\phi = \pi/4$ and current *lags* voltage. (c) A circuit with significant capacitance, where $\phi = -\pi/4$ and current *leads* voltage.

Power Factor and Transmission Losses

A nonzero phase angle ϕ between AC voltage and current, caused by reactance in the system, increases resistive losses and adversely impacts other aspects of electric power transmission. The real power consumed by a load is

$$P_{\text{real}} = V_{\text{RMS}} I_{\text{RMS}} \cos\phi,$$

where $\cos\phi$ is the power factor, while $P_{\text{apparent}} = V_{\text{RMS}} I_{\text{RMS}}$ is a measure of the total power flowing in the circuit. The power lost in transmission is

$$P_{\text{lost}} \cong \frac{P_L^2}{\cos^2\phi\, V_{\text{RMS}}^2} R = \frac{P_{\text{apparent}}^2}{V_{\text{RMS}}^2} R,$$

where R is the total resistance of the conductors that make up the transmission line. It is also useful to define the reactive power $P_{\text{reactive}} = V_{\text{RMS}} I_{\text{RMS}} \sin\phi$, which is a measure of power flowing in the circuit that is not delivered to the load.

throughout the circuit are larger than required to deliver the real power used in the load, wires and other components must be scaled up, adding to the cost of the system (see Example 38.1). In practice, banks of capacitors or other devices are introduced in electricity distribution networks in order to compensate for inductive loads and bring the power factor back close to one (Problem 38.8).

38.3 Grid-scale Electricity Generation

Generators convert mechanical energy into electromotive force. The basic principle was explained in §3.4.2: a source of mechanical power – a steam turbine, for example – referred to as the **prime mover**, moves a wire loop in an external field or causes the magnetic flux through a loop to change with time. In either case, an EMF is produced in the loop, which causes a current to flow. The current and/or the magnetic field it generates act to oppose the motion of the prime mover. Thus, mechanical power is translated into electric power flowing in the loop.

Generators are described using a terminology common with motors. The *rotor* is the rotating mechanical component, while the *stator* is the stationary component. The **armature**, on the other hand, refers to the component of the generator that develops the EMF and is responsible for producing electric power, while the *field component*

produces the magnetic field. Thus, rotor and stator refer to mechanical parts, while armature and field component refer to the parts with specific electric functionality. Depending on the generator geometry, the armature may be either the rotor or the stator. For most real generators, the armature is stationary and a time-varying magnetic field is produced by the rotor, inducing a current in the stator. The reason for this is that the armature wires carry higher voltages than those of the field component and require more insulation and more robust structure. In general, as for motors, the armature is wound many times to increase the EMF. The field component is also often an electromagnet, in the form of a solenoid wound around an iron core, though sometimes it is a permanent magnet. In a typical generator with stationary armature, the stator coils are embedded in a magnetic material like iron, which serves to amplify the field and also to guide the magnetic field lines. The way that the magnetic field lines sweep over the coils of the stator in such a generator is illustrated in Figure 38.7.

Electric generators come in many varieties suited to different conditions and applications. Two basic designs for AC generators are known as *synchronous* and *asynchronous* generators. Each of these types can be elaborated in various ways. We describe each of these designs assuming that the armature is stationary, as is the case in

Example 38.1 Reactive Power

In a circuit with substantial reactance, extra current flows back and forth through the circuit without delivering power. This extra current gives rise to additional transmission losses.

As a simple example, consider a circuit containing a resistance $R = 2\Omega$ associated with transmission wires, and a load that has both a resistance R_L and an inductance L. Assume that the power is provided by a generator that operates at a fixed voltage $V_{\text{RMS}} = 120\,\text{V}$, and that the total power consumed by the load is $P_L = 100\,\text{W}$.

First assume that there is no inductance. Then the current is $I_{\text{RMS}} = V_{\text{RMS}}/(R + R_L)$, there is no phase difference ($\phi = 0$) between the current and the voltage, and the power delivered to the load is $P_L = I_{\text{RMS}}^2 R_L$. Combining the two equations for R_L and I_{RMS} in terms of the data given gives a quadratic equation for R_L; solving gives

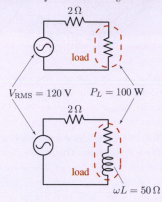

$$R_L \cong 140\,\Omega, \quad I_{\text{RMS}} \cong 0.845\,\text{A}.$$

In this case the power lost in transmission is $I_{\text{RMS}}^2 R \cong 1.43\,\text{W}$.

Now, assume that the load still consumes $P_L = 100\,\text{W}$ of real power, but that it has substantial inductance, with $\omega L \cong 50\,\Omega$. We now have the equations

$$I_{\text{RMS}} = \frac{V_{\text{RMS}}}{\sqrt{(R + R_L)^2 + \omega^2 L^2}},$$

$$P_L = I_{\text{RMS}}^2 R_L.$$

Solving, we find

$$R_L \cong 119\,\Omega, \quad I_{\text{RMS}} \cong 0.917\,\text{A}.$$

In this case the power factor is

$$\cos\phi = \frac{R + R_L}{\sqrt{(R + R_L)^2 + \omega^2 L^2}} \cong 0.92.$$

The power lost in transmission is $P_{\text{lost}} = I_{\text{RMS}}^2 R \cong 1.68\,\text{W}$. Thus, adding a moderate inductance in this case increases the current by roughly 8% and increases transmission losses by roughly 18% (in agreement with the approximation that power losses go as $1/\cos^2\phi$ when $R \ll R_L$). Note that the lost power can be written simply in terms of the *apparent power and the RMS voltage (which is typically fixed)* $P_{\text{lost}} = (P_{\text{apparent}}/V_{\text{RMS}})^2 R$. Thus the apparent power is the critical factor in determining resistive losses and the associated thermal loads on system components.

Because of the increase in current and consequent additional transmission losses associated with reactance, power suppliers try hard to reduce reactance in electric grids. Some power companies now impose additional charges to customers whose loads have a power factor that is significantly different from unity.

most real generators. In a **synchronous generator**, a DC current *excites* the magnetic field of the rotor. Known as the **rotor** or **excitation current**, this DC current is independent of the currents induced in the stator. As the rotor magnetic field turns, it induces a current in the stator that oscillates at the rotor's frequency of rotation – hence the term *synchronous*. In an *asynchronous* or *induction generator*, described in detail below, the current in the rotor (which again acts as the field component) is itself *induced* by the AC currents in the stator, so the magnetic field of the rotor does not necessarily oscillate at the frequency at which the rotor spins – i.e. it is asynchronous. Typically, generators that use steady power sources such as fossil fuels, nuclear, and hydropower are of the synchronous type. Induction generators are often used with variable sources of mechanical power such as wind turbines.

38.3.1 Synchronous Generators and Three-phase Power

The synchronous generator shown in Figure 38.7, with a single set of windings on the stator, is known as a **single-phase generator**. As the magnetic field rotates, both the

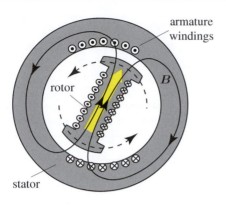

Figure 38.7 Cutaway view of a single-phase generator. Cylindrical windings on the rotor, shown in cross section, produce a magnetic field B (denoted by a yellow arrow) along its axis. As the rotor spins, its magnetic field lines (only two of which are shown in the figure) cut through the wire coils mounted in the stator. The windings are suitably shaped so that the EMF in the stator coils oscillates sinusoidally with time. After [296].

EMF and the current induced in the stator coils go through one cycle for each complete rotation of the rotor. Note that unlike the generator design described in §3.4.2, where the armature is in the rotor, here the current in the stator coils is induced through Faraday's law of induction (3.64) and not the Lorentz force law (3.43). The stator current in turn gives rise to a magnetic field pointing along the axis perpendicular to the windings. The stator field interacts with the magnetic field of the rotor to produce a torque that opposes the motion of the rotor (as in Box 3.4). This is the torque that the prime mover must overcome to keep the rotor spinning and to generate electric power. The time dependence of the flux through the stator, the stator current and magnetic field, and the torque are plotted in Figure 38.8.

Single-phase generators are rarely used in grid-scale electric power generation. Their principal shortcoming is that the torque on the drive mechanism is not constant, but instead oscillates between zero and its maximum twice on each cycle. This leads to excessive wear on mechanical systems. A relatively simple extension known as **three-phase power** resolves this problem and has several other advantages in transmission and distribution. We describe the implementation of three-phase power in the context of a synchronous generator although similar considerations apply to asynchronous generators as well.

The simplest implementation of three-phase power, illustrated in Figure 38.9, has three coils wound around the stator at angles of 120°. This results in three separate sinusoidally oscillating EMFs in separate wires, each 120° out of phase with one another. The magnetic field produced by the current induced in each of the three coils oscillates along the axis of the coil just like the field of a single-phase generator. While the stator remains fixed, as the three magnetic fields oscillate, their vector sum simply rotates at the same frequency as the rotor (Problem 38.9). This is displayed graphically in Figure 38.10. Thus, the three-phase generator operates under a constant torque from the mechanical drive mechanism.

As depicted in Figure 38.10, the stator field lags 90° behind the field of the rotor. This phase relationship holds in the case of a purely resistive load.[4] If the load has an impedance characterized by a phase angle ϕ, then the current in the windings and the resulting stator magnetic field either lags ($\phi > 0$ – a dominantly **inductive load**) or leads ($\phi < 0$ – a dominantly **capacitive load**) the EMF. The angle between the rotor and stator fields is $90° + \phi$. These situations are illustrated in Figure 38.11. The torque exerted by the stator field on the rotor is of the form (eq. (3.53)) $\boldsymbol{\tau} = \boldsymbol{m} \times \boldsymbol{B} = |\boldsymbol{m}||\boldsymbol{B}|\sin(\pi/2 + \phi)$, where $\boldsymbol{m}$ is the magnetic moment of the rotor and $\boldsymbol{B}$ is the stator field. Thus, for fixed $|\boldsymbol{B}|$ and $|\boldsymbol{m}|$, corresponding to fixed excitation and line currents, when $\phi = 0$ the torque is maximized and the generator is maximally effective in driving power through the system. In the extreme case when inductance dominates, $\phi \to 90°$, and the stator and rotor fields are nearly antiparallel. Likewise, for a dominantly capacitive load $\phi \to -90°$, and the stator and rotor fields are nearly parallel. In these cases, to maintain a high level of torque and power delivery, the stator field, and therefore the current, must be very large. A large current flowing in the windings and the attached transmission lines increases losses, and requires more robust components throughout the electric network. To the extent that ϕ differs from zero, *reactive electric power* circulates in the system without any benefit.

Note that the stator field itself induces an EMF in the stator armature. While we do not treat this effect here, a full description of a generator involves both the self- and mutual inductances of the armature coils. These effects modify the relation between the rotor field and the total induced EMF in the armature. This is particularly relevant when an inductive or capacitance load is placed on the system, as discussed in Box 38.2.

---

[4] By "load" here we mean the entire system of transmission and distribution lines, transformers, other system components, and the consumer devices seen by the generator.

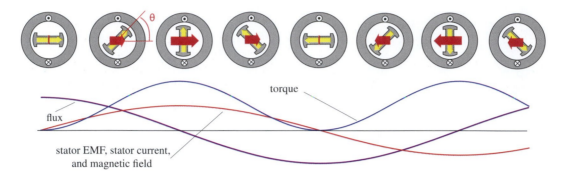

Figure 38.8 Views of a single-phase generator at regular intervals in the period of the rotor. Only one loop of the stator coil is shown. As the rotor field (yellow) rotates, the flux through the stator coil varies like $\cos\theta$. In the case shown here, a resistive load connected to the stator windings leads to a current and reaction magnetic field in phase with the stator EMF, both varying like $\sin\theta$. The torque on the rotor is proportional to the product of the stator current and EMF and is proportional to $\sin^2\theta$. Note that the stator magnetic field (red) is always in the direction perpendicular to the stator coils.

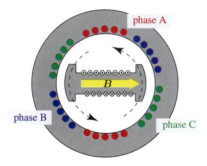

Figure 38.9 A generator with an armature wound for three-phase power. The three independent windings – phases A, B, and C – are shown in red, blue, and green, respectively. As the rotor turns, its magnetic field generates a sinusoidally oscillating EMF in each winding, with relative phases offset by 120°. After [296].

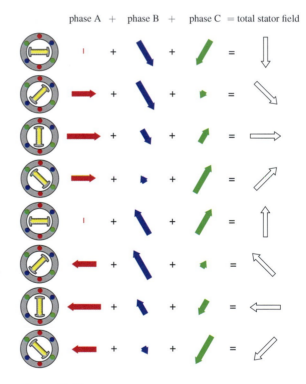

Figure 38.10 Armature magnetic fields in an ideal three-phase generator. As the rotor turns, it induces an oscillating magnetic field parallel to the axis of each of the three windings. The orientation of the rotor is shown along with the direction and magnitude of the field through each winding and the vector sum of the three magnetic fields. Although each magnetic field simply oscillates back and forth without changing direction, the total stator field, which is the vector sum of the three fields, rotates at the frequency of the rotor. After [296].

There are several different ways to connect three-phase generators to the power grid. One conceptually simple and widely used approach is to connect one end of each of the three coils to a common, *neutral* wire. In this case, as sketched in Figure 38.12, the output of the generator is carried by four conductors, one registering the EMF in each coil and one at a common voltage that we can take to be ground, $V = 0$. This is known as the **wye configuration** because of its resemblance to the letter "Y." The resulting EMFs in each conductor are

$$V_A = V_0 \sin\omega t,$$
$$V_B = V_0 \sin(\omega t - 2\pi/3),$$
$$V_C = V_0 \sin(\omega t - 4\pi/3), \qquad (38.13)$$

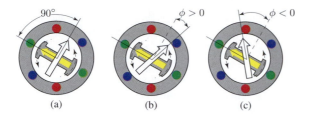

Figure 38.11 The angle between the rotor (yellow) and stator (white) magnetic fields for (a) purely resistive, (b) predominantly inductive, and (c) predominantly capacitive loads. ϕ is the phase angle defined in eq. (38.6). After [296].

Three-phase Power

In three-phase power, three separate conductors carry voltages that are 120° out of phase from one another. If the load on each conductor is the same, the total power in the three conductors is time independent.

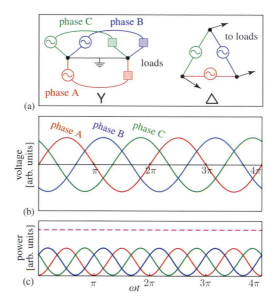

Figure 38.12 Three-phase power as a function of time: (a) the wye and delta circuits; (b) voltage in the three phases for the wye circuit; (c) power delivered by the three phases with a purely resistive load; the sum of all three, shown by the dashed purple line, is constant.

---

as shown in Figure 38.12(a). Note that if the three conductors with the three phases are attached through equal loads to ground, the total current vanishes

$$
I_A + I_B + I_C = V_0 \, \mathrm{Re} \left(\frac{1}{Z} \left(e^{i\omega t} \right. \right.
$$
$$
\left. \left. + e^{i(\omega t - 2\pi/3)} + e^{i(\omega t - 4\pi/3)} \right) \right) = 0 .
$$
$$
(38.14)
$$

Thus, if the loads are well balanced, no return is needed and the fourth wire can literally be connected to ground. An alternative approach to connecting a three-phase generator to the grid is the "Δ" configuration, also sketched in Figure 38.12, where the coil outputs are connected pairwise in a triangular configuration.

In addition to the advantages in generation mentioned above, three-phase power has several other advantages where power is delivered to the load, as described in §38.4.2.

38.3.2 Asynchronous (Induction) Generators

An alternative to the synchronous generator design is the **asynchronous**, or **induction generator**. Unlike synchronous generators, an induction generator does not have a separate source of rotor current. Instead, the rotor current and associated magnetic field are induced by the changing flux through the rotor winding and the rotor's motion

through the stator magnetic field via a combination of Faraday's law of induction and the Lorentz force law. Although this may seem counterintuitive (if not impossible) at first sight, since the field from the rotor is itself what drives the stator current, the basic principle can be understood from the device geometry and Lenz's Law. The principal advantage of the asynchronous generator is that the mechanical frequency of rotation of the rotor is independent of the line frequency. Thus, asynchronous generators function well with variable power sources. While synchronous generators are standard for most types of grid-scale power production, asynchronous generators are useful for small wind turbines and other variable prime movers such as small-scale hydropower.

The rotor of a simple induction generator contains no permanent magnets and has no external electrical connections. Instead, it consists of a set of current loops arranged around the surface of a cylindrical frame. The simplest configuration, shown in Figure 38.13, is known as a **squirrel cage**, for obvious reasons. The bars parallel to the axis and the connecting sections of the end discs form the loops through which current can circulate.

An induction generator can also function as an **induction motor**. We begin by describing the mechanism of the induction motor, which is conceptually somewhat simpler. Suppose that a stator is wound for three phases and

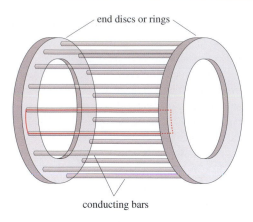

Figure 38.13 A sketch of the squirrel cage rotor of an induction generator. The bars, together with the end discs, form loops (one is shown in red) through which current can circulate.

is connected to three-phase AC power. Stator currents then give rise to a magnetic field that rotates at the line frequency $\nu = \nu_0$.[5]

If the rotor is stationary, the changing stator field induces an EMF in the rotor, which in turn induces currents in the rotor loops. The bars carrying these currents experience a Lorentz force in the stator magnetic field and the resulting torque drives the rotor to follow the direction of rotation of the stator field. In this way the effective rate of change of magnetic flux through the rotor loops decreases as required by Lenz's law. If the rotor is coupled to a load, then the machine functions as a motor.

As long as the frequency of rotor rotation is less than ν, the torque continues to push the rotor in the direction that the stator field is rotating and the device acts as a motor. Suppose, however, that the rotor is turning at exactly the frequency ν. In this case the flux through the moving rotor loops is constant. Thus, no current flows in the rotor, there is no rotor magnetic field or torque, and no power is transferred from the stator to the rotor or *vice versa*.[6] This is known as the **no-load condition**, and occurs when the rotor spins at the synchronous speed. Except for mechanical friction and resistive losses in the stator, this state could be maintained indefinitely.

If a prime mover exerts torque on the rotor in the direction of rotation, then the rotor will rotate at a higher frequency than ν. In this case, the interaction between the rotor field and the stator field will provide a backward torque against which the prime mover does work. In this way, the prime mover transfers energy from the rotor's rotation to the currents flowing in the stator coils and the device functions as an *induction generator*.

In either motor or generator mode, the mechanical rate of rotation of the rotor differs from that of the stator magnetic field. Thus, the induction generator is *asynchronous* – the rotational frequency of the rotor is decoupled from the line frequency. If the rotor is spinning at an angular frequency ν_r, then the frequency at which the magnetic flux through the rotor coils changes is known as the **slip speed** $\Delta\nu \equiv \nu - \nu_r$. Often the ratio $S \equiv \Delta\nu/\nu$, known as the **slip**, is used in place of the slip speed to describe the state of an induction generator/motor. $\Delta\nu$ is positive in motor operation and negative as a generator. Note that the rotor currents and the magnetic field they create rotate at $\Delta\nu$ relative to the rotor, so with respect to the stator, the rotor field rotates at $\nu_r + \Delta\nu = \nu$, the same frequency as the stator field. The EMF the rotor field induces in the stator coils therefore oscillates at the same frequency as the line voltage, so an induction generator always produces power at the line frequency.

The torque exerted by the prime mover on the rotor and the power supplied by the generator are positive in generator mode, zero at the synchronous speed, and negative in the motor mode. For small slip speeds, either positive or negative, the torque and power depend linearly on the slip speed, as shown in Figure 38.14. This allows for stable operation over a range of $\Delta\nu$ (see Problem 38.3). When the slip speed becomes large, the rotor and stator begin to decouple and the torque falls. This is known as an **overspeed condition** that leads to unstable operation and eventually damage to the machine. An extreme example is when a child's toy, run by an induction motor, stalls. The slip is 100%, the torque is nonzero, but the rotor does not turn.

As a consequence of their drive mechanism, induction generators have several special limitations. First, they must be connected to an external source of AC voltage since some stator current is needed initially to induce the rotor current. Second, they typically have lower and upper cut-offs on the rotor speed. At too low a speed the machine would be operating in the motor mode and absorbing line power, while at too high a speed the generator can reach the overspeed condition that would damage its components. Third, an induction generator must see a dominantly capacitive load. Since the rotor and stator magnetic fields

---

5 Usually, the stator is wound with multiple poles per phase; with n poles, the stator field rotates at a frequency $\nu = \nu_0/n$.

6 While Faraday's law of induction cannot really be applied in a rotating reference frame, as noted in §3, Problem 3.20, the EMF on a rotating loop from the combined effects of the Lorentz force and induction is equivalent to that on a stationary loop with the same rate of change of magnetic flux.

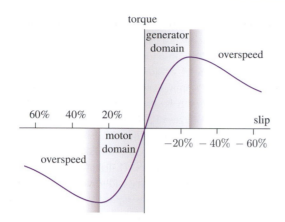

Figure 38.14 The torque exerted by the rotor in an induction machine. When the slip speed is negative, the rotor is turning faster than the stator field, and electric power is generated. When the slip speed is positive, the stator field spins faster, and electric power is consumed.

spin at the same speed in an induction generator, the angle between them is fixed as it is for a synchronous generator. At the no load condition the two fields are parallel, corresponding to the absence of torque. Acting as a motor, the stator field is ahead of the rotor field – it drags the rotor along – and acting as a generator, the rotor field is behind. The angle between the two fields is proportional to the slip speed, which is typically quite small. Thus for an induction generator the phase angle ϕ shown in Figure 38.11 is negative and only slightly less than $-90°$, corresponding to a dominantly capacitive load. Most loads, however, are resistive and inductive corresponding to $\phi \gtrsim 0$. Therefore an induction generator must in general be accompanied by a bank of capacitors before it can be coupled to the grid. These restrictions are manifest, for example, in the operation of wind turbines that employ asynchronous generators.

38.3.3 Characteristics and Control of Synchronous Generators

Synchronous generators dominate grid-scale electric power generation. A generator is typically powered by a turbine, which in turn is powered by steam, combustion gases, or water power. The power output of utility-scale generators is generally on the order of 100–1000 MW. The output (RMS) voltages of utility generators are designed to be on the order of 10–25 kV. This choice represents a compromise between the desire for higher voltages to minimize resistive losses, just as in transmission lines, and the increasing cost of insulation and safety systems that comes

with very high voltages. Even so, a 10 kV utility-scale generator producing 100 MW has instantaneous currents close to 4700 A in each of its three phases. Resistive losses of as little as one percent within the generator would produce over a megawatt of heat in its internal workings, requiring a high-performance cooling system.

To produce power at a frequency of 60 Hz, a generator using the **two-pole** (one north, one south) rotor described in §38.3.1 would have to turn at 3600 rpm. While this is achievable for steam and combustion gas powered turbines, it is not practical for hydropower. In that case and in other situations where the rotational speed of the turbine is constrained, more complex rotor magnetic fields are employed to increase the frequency at which the field changes compared to the frequency of rotation. A simple example is the **four-pole rotor** shown in Figure 38.15. A rotor of this type spinning at 1800 rpm would generate 60 Hz power. Some of the generators at the Grand Coulee Dam, the largest hydroelectric power plant in the US, use 60-pole rotors like the one shown in Figure 38.16 and therefore rotate at 120 rpm.

Figure 38.15 A sketch of a three-phase generator with a four-pole rotor.

Figure 38.16 A rotor from one of the hydroelectric generators at Grand Coulee Dam in the US. It is wound with 60 poles and rotates at 120 rpm. (Credit: US Bureau of Reclamation reproduced by CC-BY-SA 2.0 license via Wikimedia Commons)

Control of a Synchronous Generator

The state of a synchronous generator is specified by the output frequency, voltage, and current at the busbar. The current is determined by the voltage and the attached load.

As demands on the generator change, the control system maintains constant line frequency and voltage by modifying the input power to match real power demand, and by modifying the excitation current through the rotor to compensate for changes in reactive power requirements (inductance and capacitance).

Aside from the line frequency, the principal operational characteristics of a generator are the output voltage and current in each phase. These are measured at a single location, known as the **busbar**, or **bus**. This is a metal bar where all (hence "omnibus" or "bus" for short) outgoing and incoming conductors are gathered together. The response of a generator to fluctuations in load or to the behavior of other generators is measured in terms of the voltage and current measurements "at the busbar."

The many quantities that characterize a generator – output voltage, currents in the rotor and stator, magnetic field strengths, and frequency of rotation, to name only a few – are highly interdependent. For a consistent and versatile AC power supply, a generator must operate at a constant output voltage and frequency. As the load fluctuates, however, this requires that the operator of the generator dynamically adjust the generator inputs to compensate. The variables that are available for adjustment are the power input from the turbine and the excitation current. By modifying these parameters, the operator can maintain an AC output at a constant frequency and voltage as the load changes. In particular, as the real power required by the load goes up or down, the operator (which is generally automated) adjusts the power input from the turbine. If the impedance of the load changes, the power factor $\cos\phi$ changes and the excitation current and power output must be modified to maintain constant voltage and offset increased line losses. Some details of this control system are described in Box 38.2.

38.4 Transmission and Distribution of Electric Power

In broad terms, electric power networks are divided into *transmission* and *distribution* systems. The **transmission** **network** carries electric power from a set of interconnected generators over long distances to the locales where the power is used. Transmission networks usually employ voltages on the order of hundreds of kilovolts and span distances ranging from tens to hundreds, sometimes thousands of kilometers. It is often useful to distinguish **subtransmission** at voltages of 30–130 kV from long-distance transmission at hundreds of kilovolts. Local **distribution** **systems** convert and deliver power to industrial, commercial, and residential customers. Distribution systems can be further divided into **primary** and **secondary distribution systems**. The former are typically at 2–30 kV and carry power over urban distance scales, serving industrial and some commercial customers. The latter run at 100–240 V and serve residential and small commercial customers.

In this section we look at some of the components of electricity transmission and distribution systems, the three-phase power they deliver, and the way the components are organized into networks.

38.4.1 Physical Components of the Grid

Classification of Generating Units   The generators described in §38.3 provide power to electric transmission and distribution networks. Variations in power demands over short time scales are met by the response of individual generators, but variations on the time scales of hours or longer are met by bringing generating capacity on or offline. Some types of generating units are better suited for rapid response, while others run most efficiently when run continuously. Generating units are classified in three categories, **baseload**, **intermediate**, and **peaking**.

Baseload generators are designed to run year-round, producing a relatively constant output and serving the base needs of the network. They go offline only for scheduled maintenance or for repair. Systems with the lowest fuel costs are preferred. Nuclear power is the classic example since, as we saw in §18, nuclear fuel is relatively cheap and nuclear reactors take considerable time to power up or down. Other steam-driven units, particularly coal-fired, fall in the baseload category. Hydropower plants can also provide baseload power, particularly those with large reservoirs that are not subject to seasonal or short-term fluctuations in flow.

Intermediate units are used to respond to regular fluctuations in demand over the day, week, and year. These generating units should be able to run for long periods and vary their output. Combined-cycle gas turbine plants are typical examples.

Peaking units handle the extremes of demand. They must be able to turn on or off quickly. Because they are used less frequently, less efficient plants or those with

---

Box 38.2   Dynamics of Synchronous Generator Control

As the load on a synchronous generator fluctuates, the generator's input parameters must be changed dynamically to keep the supplied voltage and frequency close to constant. This is generally done on short time scales by automated systems that can alter the input power from the prime mover and the excitation current through the rotor, with human control involved on longer time scales. Although response to arbitrary changes in the load requires adjustments in both these input parameters, changes in the real power demand primarily require changes in input power, and changes in reactive power primarily require adjustment of the excitation current. To see this, it is instructive to examine simple cases in which either P_{real} or $P_{reactive}$ changes, but not both.

For simplicity, we consider a system with a single synchronous generator initially supplying power P_{real} to a *purely resistive load* R_0 and we neglect line losses. The operator fixes the input power P that matches the demand from the load, and an excitation current I_E that generates the rotor field and produces the desired RMS voltage V_{RMS} when the rotor is turning at the proper frequency ν. The stator current is V_{RMS}/R_0 and $P_{real} = V_{RMS}^2/R_0$. Note that the voltage that arises from the EMF in the stator due to the changing flux of the rotor field is proportional to νI_E, and the voltage is in phase with the stator current.

Suppose, first, that an additional resistive load with resistance R_1 is placed in parallel with the original load, for example by adding incandescent lighting or resistive heating. The effective resistance is $R' = R_0 R_1/(R_0 + R_1) < R_0$ and the real power demanded by the load increases from V_{RMS}^2/R to $P'_{real} = V_{RMS}^2/R'$. Since the resistance drops, the stator current rises accordingly, increasing the backward torque on the rotor. This causes the rotor to slow down, reducing ν and therefore also V_{RMS} proportionately. The system operator must add power, to counterbalance the torque on the rotor so the frequency returns to the line frequency ν and V_{RMS} returns to the desired line voltage. The net result is that the prime mover now supplies real power P'_{real} to the load.

Next suppose that an inductive load with reactance $i\omega L$ and resistance R_0, such as a motor or a fluorescent lighting system, is connected in place of the original resistive load R_0. It is not hard to show (Problem 38.12) that the real power demand from the load is unchanged at $P_{real} = V_{RMS}^2/R_0$, but reactive power now flows back and forth between the generator and the load.

The current in the circuit now lags the voltage by an angle $\phi = \tan^{-1}(\omega L/R_0)$ and the stator magnetic field lags the rotor field by $90° + \phi$. The stator field component perpendicular to the rotor field is proportional to $I_{RMS}\cos\phi = P_{real}/V_{RMS}$ and does not change; the torque remains constant and there is no direct effect on rotational frequency. The stator field component parallel to the rotor field (shown in orange in the figure to the right (after [296])), however, induces a back-EMF on the stator that counteracts part of the effect of the rotor field. This acts to reduce the voltage output of the generator. To restore the desired voltage, the operator must increase the excitation current I_E. This mechanism works as long as the excitation current stays within the design limits set by the rate at which resistive heating in the rotor can be dissipated. Capacitive loads are less common, but their

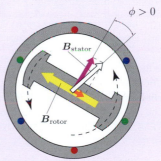

$\phi > 0$

B_{stator}

B_{rotor}

effect can be compensated by decreasing the excitation current. We have neglected line losses in this analysis; when these are included, even adding a pure reactance to the system requires additional power from the prime mover, since resistive line losses require increased real power.

---

higher fuel costs can be used. Examples include gas turbine and diesel generators.

Intermediate and peaking plants are sometimes referred to as **dispatchable reserves** because they are activated on demand by operators to match changes in the load. Among dispatchable reserves the terms **spinning** (also known as **cycling**) and **non-spinning** are encountered. Spinning generators are already running in synchrony with the grid and ready to supply power on time scales of a few minutes. Non-spinning reserves are offline, but can be synchronized quickly. In general, the order of dispatch of different generating systems as the load increases is in order of increasing fuel costs. Incorporating **variable energy resources** (VERs) such as wind, solar power, and hydropower plants with limited or no reservoir (*run-of-the-river* plants) presents additional challenges due to the intermittent nature of these resources. We return to these issues at the end of this chapter.

Transmission Lines High-voltage transmission lines, carried on *lattice steel towers* (LST) or *tubular steel poles* (TSP), are the most visible manifestations of the electric power grid. Some examples are shown in Figure 38.17. Three-phase high-voltage AC power is carried on three separate conductors with typically one or two circuits per tower. The cables (Figure 38.18) and the towers that carry them are designed for safety and efficiency. The height of the lowest cable is dictated by minimum ground clearance (allowing for sag between towers). The cables are not insulated due to the added expense and weight of insulation. Instead the separation between cables, of order 10–15 m in the examples shown, serves to prevent arcing.

A number of constraints must be considered in configuring transmission lines. As described in §3, transmission lines must be designed to minimize resistive losses. It

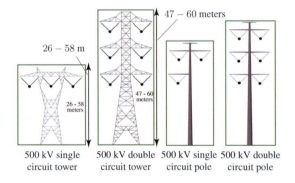

26 – 58 m

47 – 60 meters

47 - 60 meters

26 - 58 meters

| 500 kV single | 500 kV double | 500 kV single | 500 kV double |
| circuit tower | circuit tower | circuit pole | circuit pole |

Figure 38.17 500 kV electric power transmission TSPs and LSTs. Both single circuit (three cables) and double circuit (six cables) structures are shown. The heights shown for the LSTs are typical. Cable locations are marked. In addition, each tower carries a ground wire (not shown) [302].

Figure 38.18 Aluminum conductor steel reinforced (ACSR) high-voltage transmission cables. Note that the cables are braided from many strands making them more flexible and easier to manipulate. (Credit: Lisa Cable)

is also important to minimize the reactance of the lines, which reduces the power factor and results in wasteful reactive power. Excess reactance in transmission lines connecting generators can also destabilize the system.

Estimating resistive losses is quite straightforward. According to eq. (38.12), the resistive heating per unit length dP_{lost}/dl in a transmission cable is approximately given by

$$\frac{dP_{\text{lost}}}{dl} = \frac{\rho}{A} \left(\frac{P_L}{V_{\text{RMS}} \cos\phi} \right)^2, \qquad (38.15)$$

where ρ and A are the resistivity and cross-sectional area of the conductor. Limiting resistive losses places stronger constraints on longer transmission lines, since total resistive loss grows linearly with length. There is also a constraint associated with eq. (38.15) that is independent of the length of the line. If dP_{lost}/dl gets too big, the cable may heat enough to stretch and sag below its allowed ground clearance or past its limit of elastic deformation. Whatever the length, the cross-sectional area of a cable and its response to resistive heating can be designed for the anticipated load (and power factor).

The constraints on transmission lines due to complex impedance effects are less transparent. The magnetic fields created by the currents in the conductors give rise to self-inductance in each conductor and mutual inductance between conductors. The mutual inductance dominates because it is proportional to $\ln D/r$, where D is the distance between the conductors and r is their radius. As mentioned above, D/r is typically kept large to insulate the conductors from one another. This **line inductance** adds an imaginary component to the real impedance from the line resistance, effectively acting in series. Both terms grow linearly with the line length, but typically the line inductance dominates. For example, a typical arrangement of aluminum conductors used in a 345 kV transmission line may have a resistance per unit length of $R' = R/l = 0.035\ \Omega/\text{km}$, and an inductive reactance per unit length of $\omega L' = \omega L/l = 0.37\ \Omega/\text{km}$, more than ten times as large (Problem 38.15). In fact, for many purposes – in particular in the analysis of grid stability where phases are of central importance – it is a good approximation to ignore line resistance compared to inductive reactance and consider **lossless lines**.

Similarly, electric fields generated by charges on the wires give rise to **line capacitance** between the conductors. The capacitance is proportional to $1/(\ln D/r)$, so the large separation between conductors tends to diminish the importance of *line capacitance*. In high-voltage transmission, line capacitance can be ignored entirely for lines with length less than ~80 km. At greater lengths, capacitance

Constraints on High-voltage Transmission Lines

Both resistive and reactive effects limit the power-carrying capacity of high-voltage AC transmission lines. Resistance causes power losses and also heating in the conductors. Reactance, primarily due to induction between conductors, reduces the power factor and can lead to instabilities. Line inductance, like resistive losses, grows linearly with the length of the transmission line; inductance is often the limiting factor for long-distance power transmission.

must be included along with line resistance and inductance in order to obtain an accurate computation of line impedance.

Line inductance becomes problematic when it makes a significant contribution to the total impedance. For example, for a purely resistive load with $P_L = V^2/R$, line inductance gives a substantial contribution to the impedance – lowering the power factor – when $\omega L \sim V^2/P_L$. When the inductance is too large on a line connecting two generators, it not only affects reactive power but can actually destabilize the network. This is described further in §38.4.4. We have considered here limits on steady-state performance. In some cases the ability of the system to respond to transients places more severe limits, particularly on the longest transmission lines [298].

For very long lines, one cost-effective way to avoid inductance problems is to convert from AC to DC at the generator, transmit the power as direct current, for which inductance is meaningless, and then convert back to AC before the step-down transformer at the head of the distribution system.

Transformers *Transformers* are required in order to raise or lower AC voltages, and are generally used in at least three stages along an electric power network. First, **step-up transformers** convert electricity generated in the 10–20 kV range to voltages suitable for long distance transmission. These transformers are usually located at the power plant. **Step-down transformers** are needed to convert from transmission to distribution voltage levels. These transformers are typically found at **substations** (see Figure 38.19), along with switches, circuit breakers, capacitor banks (see below), and other electronic devices to monitor and control electric power. Finally, smaller transformers are located at the interface between primary and

secondary distribution networks. Transmission at low voltage is wasteful, so these final transformers are located as close to the retail customer as possible. Different patterns of settlement have led to different secondary distribution systems in the US and Europe (see below) that utilize transformers differently.

For electricity transmitted as three-phase power, each phase requires a separate transformer. Transformers therefore generally occur in banks of three, except at the final stage of distribution, where a single phase may be brought to a residential customer (see Figure 38.20). Transformers used in the electric transmission system are typically more than 99% efficient. Nevertheless, because they handle hundreds of megawatts of power, transformers must be designed to dissipate heat efficiently. This requirement raises its own set of environmental problems (Question 38.4).

Voltage Support As electric power flows through transmission and distribution networks, its voltage degrades. Line resistance decreases the magnitude of the voltage. Reactive effects reduce the power factor, which increases transmission current and leads to further line voltage reduction. Electrical devices are designed to accept a certain range of supplied voltage. In the US, for example, the standard is ±5%, so a device rated for 120 V is designed to function with voltages between 114 and 126 V. To meet these standards, voltage support devices are added to the grid, usually located at the same substations where transformers are needed. **Variable transformers** have an output voltage that can be adjusted by tapping the output coil at different locations, effectively changing the number of windings. These are the most straightforward and inexpensive voltage support devices, and are used to adjust the magnitude (but not the phase) of the voltage as needed. Traditional transformer taps must be physically moved and typically have some limited number of settings, making this a cumbersome way to support voltage; settings are typically either set to fixed positions or pre-programmed to change in response to anticipated daily fluctuations in load. **Capacitor banks** allow operators to correct the phase of the line voltage. Adding a capacitance in parallel with inductive loads corrects the power factor, and provides a local source of reactive power that reduces the transmission current needed, consequently decreasing the resistive voltage drop in the network (Problem 38.8). Advances in technology in recent years continue to improve the level of detail and automation that is achievable in grid control in areas such as voltage support; in particular, more modern devices known as **static VAR compensators** or SVCs use semiconductor electronics to provide voltage

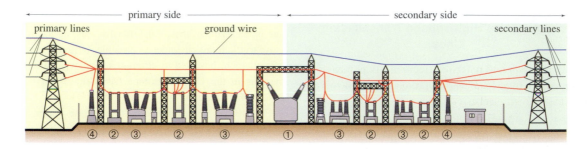

Figure 38.19 Diagram showing components at an electric power distribution substation. The *step-down transformer* ① converts power from the primary to the secondary voltage. Some of the other components shown are ② disconnect switches, ③ circuit breakers, and ④ lightning arrestors. (Credit: Shigeni23 reproduced under CC-BY-SA 3.0 via license Wikimedia Commons)

Figure 38.20 Left: A large-capacity three-phase generator-step-up (GSU) transformer for nuclear power plants manufactured by Mitsubishi Electric. It is rated at 760 MVA (see §38.2.3) and transforms electric power generated at 25 kV to 345 kV transmission voltage. Note the three phases entering and the forced air cooling fans. Right: A pole-mounted single-phase distribution step-down transformer for supplying power at 120 V to a few residences in the US. Note the single-phase input on top. This canister transformer is oil cooled. (Credit: (Left) Mitsubishi Electric Power Products, Inc; (Right) Steve Mann)

support over short response times. These devices can rapidly compensate for changes in reactive load without operator intervention, though they are more expensive than mechanically switched capacitors.

Other Components: Safety and Control   A variety of other components, including switches, circuit breakers, lightning arresters, and many types of measurement and monitoring equipment are employed throughout an electric grid. Circuit breakers are placed so as to isolate other components such as generators, transformers, or loads when tripped due to overloads in the system. Switches allow operators some control over how power flows through a network, although as already mentioned it is not possible to dictate the flow of power in complete detail.

38.4.2 Distribution of Three-phase Power

Three-phase electricity can be connected to power loads in a number of different ways. This versatility, along with reduced resistive losses, provide additional benefits of three-phase power beyond the virtues in power generation mentioned earlier.

Direct three-phase use   The simplest way to connect and use three-phase power is to directly connect all three phases to compatible devices. Three-phase power is perfectly matched to motors with the same phase structure, allowing them to provide constant torque.

Wye connection – phase to ground   With three phases running along distribution lines, three customers or devices can each be supplied with the output of a single phase, with voltage $V_{\mathrm{RMS}} = V_0/\sqrt{2}$ (120 V in the US or 230 V in Europe) relative to ground. This is the *wye configuration* shown in Figure 38.12(a). Naively, it might seem that the Wye configuration would require six wires: three feeds and three returns. If the loads on each phase are equal, however, then the phases cancel and the sum of the three currents is zero as in eq. (38.14). In this situation, no additional return wires are needed. Even if the loads are not perfectly balanced, the sum of the return currents is generally small enough to be carried through the ground (literally). Thus, when three-phase power is distributed to customers, utilities strive to balance the loads on the three phases. A glance at overhead wires carried by many utility poles in residential neighborhoods will confirm the presence of three conductors. In some cases, a fourth common neutral wire is used to return excess current when loads are out of balance; this can be necessary particularly in local distribution networks where a small number of varying loads have larger fluctuations as a fraction of total power.

Delta configuration – phase to phase An alternative way to connect to three-phase power is to tap one phase with respect to a second. As illustrated in Figure 38.21(a), if all three combinations are used, the result is a Δ *connection* analogous to the Δ generator configuration of Figure 38.12. The interphase voltage, plotted in Figure 38.21(b), has $V_{\mathrm{RMS}} = \sqrt{3/2}V_0$ (208 V in the US or 398 V in Europe). Higher voltage is often useful for customers with heavier machinery. The customer provided with two phases also has the option of connecting either of the supplied phases to ground to obtain single-phase power at $V_{\mathrm{RMS}} = V_0/\sqrt{2}$.

In addition to voltage versatility, three-phase power reduces resistive losses in transmission lines. Suppose, for example, that a single-phase generator supplies power P_L to a purely resistive load R_L over two conductors, one carrying current $I_{\mathrm{RMS}} = P_L/V_{\mathrm{RMS}}$ in each direction. Total resistive losses are $P_{\mathrm{lost}} = 2R(P_L/V_{\mathrm{RMS}})^2$, where R is the resistance of a single conductor. In contrast, if a three-phase generator provides $P_L/3$ to each of three resistive loads through three conductors, the current in each conductor is one third as large as in the single-phase case, and the resistive losses in the three conductors are $P_{\mathrm{lost}} = 3R(P_L/3V_{\mathrm{RMS}})^2$. Thus using three-phase power it is possible to transmit the same power with only 1/6 of the resistive losses (Problem 38.17). This advantage of three-phase power could alternatively be exploited by using smaller-diameter conductors, while tolerating the same resistive losses as would occur for two-phase power. Although this analysis suggests that even

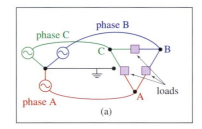

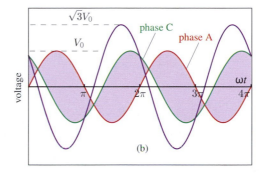

Figure 38.21 (a) Delta connection for three-phase power. (b) Voltage in phases A (red) and C (green) and the difference $V_C - V_A$ (purple), which would be accessed if one were tapped relative to the other.

more phases would lower resistive losses even further, fixed costs and additional complexity associated with the additional phases have limited practical implementation to three phases.

38.4.3 The Topology of Transmission and Distribution Systems

Transmission and distribution systems take quite different forms. Transmission systems are typically highly interconnected **mesh networks** in which power can flow along many alternative routes. The transmission network for a section of a regional power system is shown in Figure 38.22. Since many generating plants, scattered over a large region, must be able to serve a complex and distributed set of loads, interconnectivity is essential. It must be possible to take power lines and generators out of service without disconnecting any of the distribution systems or other generators from the transmission network, even if in some cases the direction of current flow in transmission lines reverses when sources and loads change.

The interconnected structure of transmission networks leads to a number of complications in managing the flow of energy through the network. Energy from a given generator can flow along a variety of paths, often passing through one or more other generators, before reaching the load(s).

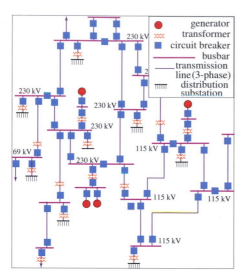

Figure 38.22 Diagram of a section of a transmission system. Voltages at busbars are labeled. After [296].

The details of how power flows in a given network with constantly varying sources and loads can be analyzed by solving a system of linear equations analogous to eq. (38.1) for each segment, with the additional constraints that at each junction the incoming current is equal to the outgoing current.[7] The resulting energy flow can have unexpected features.

Decreasing the output of a generator may actually increase the current being carried on a line connecting that generator to another. There can also be **loop flow**, where there is a net flow of energy around a closed loop within the circuit. Loop flows contribute to resistive losses. In addition, loop flows can result in power inadvertently crossing jurisdictions, confounding regulatory and business practices.

Electricity **distribution networks** are somewhat simpler in principle than transmission networks. They are typically **radial** or **star networks** where primary feeders emerging from a distribution substation are in turn stepped-down to secondary distribution voltages that feed retail customers. Distribution networks may also contain loops in order to provide redundancy, particularly in densely populated areas. Only one of the loop paths is likely to be active at any given time, however, with the other inactivated by an open switch. The reason for limiting service to one line is

to avoid a phenomenon known as **islanding** in which current continues to flow around a distribution loop after the upstream connection has been broken either intentionally or in a power outage. Islanding poses a safety hazard to utility workers who expect to find no activity in lines that have been isolated from the grid.

Distribution networks have distinctly different characters in North America compared to Europe and many other countries, as illustrated in Figure 38.23. North American distribution networks were established when population densities were low, and feeder lines had to be long. To minimize resistive losses, higher voltage primary lines were extended far down the distribution network, with transformers – typically of low capacity and mounted on poles – providing the final retail voltage at locations very near to the customer. Thus, secondary distribution lines in the US tend to be short. When new services are needed, an additional transformer mounted on a nearby pole initiates a new secondary distribution network. Electrification in Europe occurred when population densities were higher, so primary lines are shorter, and transformers, that are typically in vaults, have higher capacity and serve more customers. Secondary lines are longer, but secondary line losses are less because the voltage is higher than in the US. In the European model, adding more service requires running more secondary lines from the transformer and is easy until the capacity of the transformer is saturated, at which point a new primary line must be run. The European system has the advantage that the distribution lines are more easily put underground, because low-voltage secondary lines can more easily be run underground than the high-voltage primary lines. Thus the overhead electric power lines that

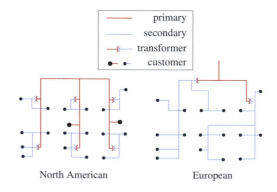

Figure 38.23 Comparison of North American and European distribution systems. The North American system propagates primary distribution further down the line and also provides power at the primary line voltage to some (commercial) customers. After [296].

---

[7] These relations together are known as **Kirchhoff's laws**, which govern the behavior of any linear circuit.

are ubiquitous in North America are rare in Europe. As noted, however, the North American model accommodates growth more easily than the European model.

38.4.4 Measurement, Stability, and Control of Electric Power Networks

As electric grids have grown more complex, so too has the problem of keeping them up and running. Potential problems exist on multiple time scales. On the shortest scale, fractions of a second, a grid must be stable both against small fluctuations about its equilibrium operating conditions and against large transients caused by sudden changes in loads or by unanticipated outages. As power needs change throughout the day, reserve generators must respond accordingly. Over the course of the year, the highest power demands occur for only a relatively small number of hours. These are the needs met by *peaking generators* or by negotiating contracts with consumers to change their habits, so-called **demand management**.

We consider only the issue of stability and control over short time scales, since the longer time-scale issues are questions more of policy and economics than of physics. The goal of stable operation is to keep all the generators connected to the grid spinning at the same frequency[8] and to keep them "in phase." Due to impedances in transmission lines and loads, the phase of the voltage is not the same throughout the system. The relative phases between different generators play an important role in the stability of the system.

Multi-generator Stability   In §38.3.3 we described the feedback system needed to adjust the power input and excitation current in a single generator to maintain constant frequency and voltage. Networks with many generators and many loads are, of course, more complicated, so management of large electrical networks requires a more complex system of measurement and control. Some aspects of network stability, however, are accomplished naturally when multiple generators are connected and maintained at relatively close phases.

In a network, it is useful to characterize each generator by its phase relative to a standard that is advancing at the nominal frequency of the whole network. These phase angles are known as **power angles** and are denoted δ_j, where j labels the generator or other measurement

location. The power angle difference between two locations is then $\delta_{ij} = \delta_i - \delta_j$. If the network is in a steady (equilibrium) state, all the generators are spinning at the same frequency and all the power angles are constant. When something changes, for example when a new load appears or a generator or transmission line joins or leaves the network, the generators may desynchronize, and begin to produce power at slightly different frequencies, with changing power angles. For a *stable equilibrium*, the dynamics of the network will automatically exert restoring forces that push the generators back to an equilibrium configuration. This kind of stable equilibrium is precisely analogous to the equilibrium configuration of the simple harmonic oscillator (Box 2.2), where a small fluctuation from the equilibrium at the bottom of the potential well gives rise to a restoring force. A system of many connected AC generators is stable as long as the inductances of the lines connecting the generators are not too great.

As a simple example, consider two AC generators that are connected in series by a line of negligible inductance. This would be the case, for example, for two generators at the same power plant. Suppose for some reason the power angle of generator (1) gets ahead of the power angle of generator (2) (Figure 38.24). This causes a slight oscillating voltage difference between the generators, which drives an oscillating current circulating between the generators. Since the stator coils are principally inductors, the current lags 90° behind the voltage difference. As a result, the current reaches its maximum magnitude at approximately

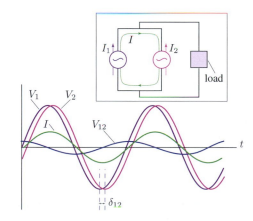

Figure 38.24 A phase difference δ_{12} between two voltages of equal magnitude and frequency leads to an oscillating voltage $V_{12} = V_1 - V_2$ across a line. The current I is approximately in phase with V_1 and V_2 when δ_{12} is small. The inset shows the configuration of the generators and the direction of the loop current I.

---

[8] Actually this refers to the frequency of the stator voltage, which may be a multiple of the frequency at which the rotor is spinning, depending on how the stator and rotor are wound.

the same time (to the extent $\delta_{12} \ll 1$) that the voltage in each generator is maximum. This circulating current adds to the current flowing through the stator coils of (1) and subtracts from the current in the stator coils of (2) (see inset in Figure 38.24). As a result generator (1) is producing more real power than generator (2), so generator (1) naturally slows while generator (2) speeds up and the system returns to equilibrium. A similar mechanism acts to stabilize the system should the voltage at one generator fluctuate, leading to a fluctuation in reactive power.

These stabilization mechanisms also work for a system of many generators connected in a network where generators are connected by lines with non-negligible inductance, as long as the power angles of generators that can send power to one another are not too different. Ignoring line resistance, the power transferred across a line with inductive reactance $X_{12} = \omega L_{12}$ between voltages V_1 and V_2 at relative phase δ_{12} is (see Problem 38.13)

$$P_{12} = \frac{V_1 V_2 \sin \delta_{12}}{X_{12}}. \tag{38.16}$$

Note that no power flows when $\delta_{12} = 0$, so some phase difference must be maintained between connected generators for power to flow on a line between them. Two generators are tightly coupled when a change in relative power angle, δ_{12}, leads to a proportional change in power flow P_{12}. The coupling breaks down, and the system becomes unstable, if $|\delta_{12}|$ gets too large; at $90°$ where δ_{12} is maximum, a change in δ_{12} has no effect on the power flow and the stabilizing mechanism breaks down. In practice, power engineers regard $40°$ to $50°$ as reasonable limits on δ_{12}.

Grid Measurement and Power Flow Analysis

Operators, whether human or automated, are able to control the frequency of generators by calling for more or less power from the prime mover. They are also able to control the voltage, both in magnitude and phase, at the generator busbars and at substations where voltage support is available. Given frequencies and voltages, the loads on the network determine the currents, both in magnitude and in phase. To control an electric grid, it is desirable to know the magnitude and phase of the voltage and current at critical locations throughout the network. Computers can use this information to reconstruct the state of the system and predict its future development. This is the subject of **power flow analysis**. The measurement challenge is *how*, *when*, and *where* to measure the current and voltage throughout the network. In the past this was a difficult art as limited measurement tools were available, but new technologies have made it possible to obtain

A "Smart Grid"

The term *smart grid* refers to a number of ideas "only loosely connected to the grid's intelligence" [298]. The ideas cluster around the expanded use of new sensing, communication, and control systems across all levels of the grid, with the goal of making the grid more robust, fault-resistant, and secure, as well as more efficient both economically and energetically.

far more information on operationally useful time scales. Relatively new devices known as **phasor measurement units**,[9] or PMUs, invented in the 1980s, provide the "how." Industry standards require that PMUs report at a rate of at least 30 times per second, and GPS coordination allows the information from PMUs to be synchronized to give a snapshot of the status of the entire network at each time. As for "where," it is sufficient to measure voltage and current (both magnitude and phase) at the busbars where the system components – generators, transformers, and primary distribution lines – interface with the network. With this information and powerful computers it is possible to reconstruct, monitor, and predict the evolution of the state of an electric grid in real time. PMUs are, however, currently somewhat expensive systems and their deployment is often resource limited. Ongoing research towards developing smaller and cheaper **micro-PMUs** provides one possible route to efficient future grid monitoring.

Fully deployed, such a system would be the backbone of a **smart grid**. Essential features of a smart grid could include automatic matching of electric production with demand, optimal network routing and local voltage and phase adjustments, rapid rerouting in case of network outages or congestion. More advanced aspects of smart grids that have been considered would include monitoring usage at the level of individual consumers in real time and adjusting both pricing and usage patterns to optimize network performance, as well as potentially incorporating distributed power generation through localized renewables such as solar or wind (§38.5), and/or distributed storage such as through grid-connected electric vehicle batteries.

---

[9] **Phasors** are vector quantities describing voltages and currents that contain both magnitude and phase information, essentially equivalent to the complex numbers used in §38.2.2.

38.4.5 High-voltage DC

High-voltage direct current (HVDC) does not suffer from the synchronization requirements that afflict AC power, nor does inductance limit the length of DC power lines. The equipment necessary, however, to transform between AC and DC power, and to transform DC power from one voltage to another, is expensive and relatively inefficient. As a rule of thumb, AC transmission substation losses are of order 0.5%, while the combined losses at the two ends of a AC→DC→AC conversion are of order 1.5% [298]. HVDC has other advantages, such as the lower capital cost of two conductors compared to the three required for three-phase AC, and the absence of the **skin effect**, which is the propensity for AC current to flow dominantly on the surface of conductors. Another disadvantage of HVDC, in addition to conversion cost, is that it cannot be tapped at intermediate locations along a transmission line.

The advantages of HVDC outweigh its disadvantages in (at least) three relatively common situations. First, for exchanging power between different synchronous grids – for example between the three large interconnects in the US. HVDC is essential in this case. Second, for undersea power cables where *capacitive impedance* is a major problem for AC cables. In this case DC wins the cost battle for distances greater than about 50 km. As a result, many undersea HVDC cables connect European countries under the Mediterranean, Baltic, and North Seas. Third, for very long lines, where the lower per-kilometer cost of HVDC wins out over the conversion costs at the cable ends. At present, the longest DC connections are the two 2500 km long 600 kV lines carrying 3150 MW each from hydropower generators near Porto Velho in northwest Brazil to São Paulo.

38.5 Renewables: Variable and Distributed Energy Resources

Since the beginning of electric power generation, almost all electricity has been generated either from fossil fuel, nuclear, or hydroelectric power plants. All these sources are well matched to modern grid demands: they can power synchronous generators matched to grid frequencies, they can be concentrated at relatively few locations where they can be actively managed, and their availability is generally high and predictable. Some alternative energy sources like geothermal, solar thermal, and tidal provide power with characteristics similar to those classic sources. Geothermal energy is continuously available. Concentrated solar thermal energy plants can use thermal storage to buffer fluctuations in insolation; although their output varies diurnally and with the seasons, these variations are quite predictable.

Tidal power, while variable, is even more reliable and predictable than solar thermal.

Other renewable power sources, however, such as wind, photovoltaic, and wave energy are quite different in character. They are variable on many time scales; the power they produce is not in a form that can be directly integrated with the grid; they often are located far from the loads they serve; and they may occur as relatively small, broadly distributed installations, where direct, active control is difficult and where they are more likely to connect to the distribution system than directly to a transmission network. In short, they are *variable, unpredictable, and distributed* – all features that can, in principle, pose problems for the structure and stability of electric power networks. Here, we concentrate on wind and PV, the most extensively deployed non-hydro renewables. In the previous chapter, §37, we described grid-scale storage mechanisms that could be used to ameliorate the problems of these variable energy sources. In this chapter we focus on how these resources can be integrated with the grid in the absence of large-scale grid storage.

First, in §38.5.1 we consider the general problems of variability and (un)predictability as they apply to wind and PV. Next, in §38.5.2 and §38.5.3 we examine the steps that must be taken to convert the power output from these energy sources to a form that can be integrated into a large-scale synchronous grid. Finally, in §38.5.4 we look briefly at the issues particular to small-scale, distributed power generation.

38.5.1 Variable Energy Resources

Wind and PV power are the prime examples of *variable energy resources*. They have common features as well as significant differences. Before getting into details, it is useful to note that electric grids have much experience dealing with *variable loads*, which are in many ways the mirror image of *variable sources*. A few relatively small, rapidly varying loads have little impact on a large network. The automatic control systems on baseload generators can react on time scales of a fraction of a second to adjust to these transients. At the other extreme, load variations over days, weeks, or months, that fall within the limits of available dispatchable resources, can be folded into the planning of system operators. These statements are as true of variable sources as they are of variable loads.

We therefore focus on ways in which the variability introduced into the grid by VERs differs from existing variability. First and foremost is the question of scale. If VERs represent but a small fraction of the grid capacity then their variability is not a significant problem. The relevant measure is *total demand minus VER* for a given network; fluctuations in this quantity must be compensated by adjustments in conventional power sources. In most current circumstances, fluctuations in demand minus VER are small enough to fall well within system capabilities (with some exceptions, see Box 38.3). Matching generating capacity to demand is expected to become more difficult in the future, however, as variable renewable energy sources expand to supply a greater fraction of grid power.

The size of the domain over which supply and demand must be balanced, known as the **balancing region**, can mitigate the impact of VERs. First, consider short-term variability. Both wind and PV plant outputs can vary dramatically over quite short time scales. Photovoltaics show the greatest short-term variability. When puffy cumulus clouds are blown by strong winds aloft, casting intense shadows that stream across the array, PV installations can change their power output by ±70% over time scales of a few minutes (see Figure 38.25). Wind farms vary somewhat less dramatically, but it is not unusual to find excursions of greater than 20% at a large (∼15 MW) wind plant over time scales of an hour [304].

Some of the effects of short-term variability can be ameliorated by increasing the balancing region. The more individual PV arrays or wind turbines are integrated, the more the fluctuations average out. The *law of large numbers* (§8.4.3) dictates that the effect of fluctuations due to turbulence over distance scales smaller than the spacing between wind turbines decreases with the size of a wind farm. The same is true for fluctuations due to variations in cloud cover over distances that are small compared to the area of a solar PV plant. Also important, as a larger physical area is covered, it becomes more likely that different weather conditions will be sampled at different locations, which further smooths the response. This effect is illustrated in Figure 38.26, where the output from wind turbines averaged over five progressively larger areas, starting with a single 2.8 kilometer square, are compared. The larger the balancing region, the less significant are the power fluctuations on short time scales. As described in §38.5.3, very short-term power fluctuations by VERs have further effects on the quality of the power injected into the grid that are not reduced by expanding balancing regions, but instead require electronic voltage regulation (see §38.5.3).

Increasing the size of the balancing region is less effective at smoothing out VER variations over time scales of hours or days. Over those time scales, VERs typically have variations that are correlated across large areas. Weather systems spread clouds and wind across large areas, winds are systematically stronger at night, and of course PVs shut off when the Sun sets. These effects require significant

---

Variability

As more VERs are added to electric grids the problem of variability will grow more serious. Short-term fluctuations average out over large balancing regions and primarily present power quality issues. Longer time scale variations will require either additional ramping capacity or enhanced grid-level storage.

---

changes in conventional power supplies, known as **ramps**, that can strain the resources of even large-scale networks. Operators have always had to deal with daily and seasonal ramps, but the scale of variations in demand minus VER are more severe than for demand alone (see Box 38.3). Thus, the expanded deployment of VERs will require either additions to conventional generating resources or enhanced grid-level storage (§37.2), or both.

38.5.2 Shaping the Electric Output from Wind Power

The generators used in conjunction with wind turbines face unique design constraints. Wind turbine blades turn slowly. Although small turbines rated at tens of kilowatts may run at 300–500 rpm, a 300 kW turbine operates at 50–60 rpm, and large offshore turbines rated at ∼10 MW run only at 5–10 rpm.[10] Wind turbines can be designed to run at fixed or variable speed. Choosing fixed speed sacrifices efficiency (§30.2.4), while choosing variable speed requires a more complicated power conversion system. Either way, the rotational frequency of the blades is far lower than the desired line frequency (either 3600 or 3000 rpm for 60 Hz or 50 Hz power), so raw wind power must be converted to higher frequency.

Small, grid-connected wind turbines can be used with simple induction generators. These turbines rotate relatively rapidly, making it relatively easy to use gearboxes to raise the rotor frequency high enough so the machine operates as a generator. (Recall that the rotor frequency must exceed ν_0/n, where ν_0 is the line frequency and $2n$ is the number of magnetic poles on the stator. On a 60 Hz line with a four-pole stator, for example, the required frequency is 1800 rpm, a number that is accessible with a simple gearbox. Below this frequency, the induction machine must *cut*

---

[10] In §28 we followed convention and called the blade system the wind turbine *rotor*. Here, however, we reserve the term *rotor* for the generator component, which spins at a different frequency from the blades.

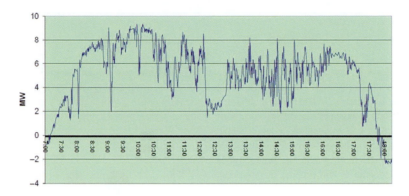

Figure 38.25 Variation in the output of a relatively large-scale PV installation in Nevada, US on a partly cloudy day [305].

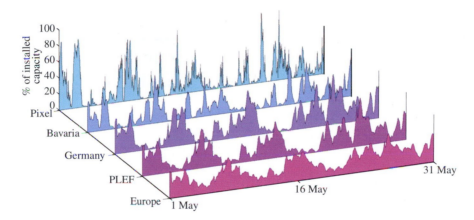

Figure 38.26 Variations in power output for five levels of integration of wind turbines in Europe over a period of one month. One "pixel" is equivalent to an area of 2.8 × 2.8 km. *PLEF* denotes Pentalateral Energy Forum (Benelux, Germany, and France). (Credit: Fraunhofer IWES (2015): The European Power System in 2030: Flexibility Challenges and Integration Benefits. Analysis on behalf of Agora Energiewende)

out to avoid drawing power as a motor.) Since small induction generators are relatively cheap and reliable, this is a good solution for a small, grid-connected turbine. If, however, the turbine is isolated from grid power, an induction generator cannot be excited without an independent power supply.

For larger wind turbines (and small turbines not connected to the grid) simple induction generators are not the best solution. When the blade rotation frequency is low, then either a more complex gearbox or a rotor with many magnetic poles would be required. Complex gearboxes are a major source of reliability problems in wind turbines, and rotors with a large number of magnetic poles are challenging to design and have significant resistive losses. More advanced induction generators have been developed [305], but synchronous generators are preferred. In particular, **permanent magnet synchronous generators** (PMGs) are

now being used in many medium to large wind turbine designs. PMGs use an array of high-strength permanent magnets on the rotor, and stators wound with many poles, to reduce demands on the gearbox.

Synchronous generators are very efficient for wind power when operating with a variable blade rotation rate, but their output frequency varies with the blade rotation rate. The wind turbine output must therefore be fed into a **power conversion system** that shapes the frequency and the voltage, generally by converting AC → DC → AC before the electricity is raised to the transmission line voltage by a step-up transformer. The combination of reliability, efficiency, and electrical output shaped by modern power electronics makes PMGs the preferred option for wind turbines at this time. Note, however, that PMGs require many massive high-field permanent magnets. For example, a 3.5 MW rated wind turbine uses in excess of

Box 38.3   Large Ramps from Wind Power

When variable renewable resources generate a significant fraction of the power in an electric network, there can be large *ramps* in which the difference between demand and the supply from the variable source change dramatically over a period of hours or days. To cope with such ramps, a grid must either have a large pool of dispatchable reserves (§38.4.1) or grid-level storage (§37.2).

As an example of such a situation, a large balancing region serving an average load of about 12 GW is described in [305]. As shown in the figure, in the absence of wind generation, the demand in this region varied over the course of an average day from 9600 MW to 14 100 MW, requiring a ramping capability of about 4500 MW. With the addition of wind generation, the demand for conventional power was reduced to about 7000 MW at night and about 13 600 MW during the day, requiring a ramping capacity of 6600 MW. Thus, in this case incorporating wind power decreased the average conventional generating power needed substantially, but increased the need for dispatchable reserves by roughly 45%. This type of situation will become more common as renewables play a larger part in global electricity production.

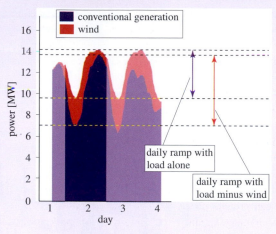

2 tonnes of neodymium-boron-iron magnets. The availability of sufficient supplies of neodymium, a rare earth element, has become a serious issue for wind turbine development.

38.5.3 Shaping the Electric Output from Photovoltaic Power

Whether on the roof of a home and rated at a few kilowatts, or covering many square kilometers in the desert and rated at hundreds of megawatts, photovoltaic plants produce DC power at a variable voltage. As described in §25, the maximum voltage from a single solar cell is generally quite small, around $V_{oc} \cong 0.7\,\mathrm{V}$ for a single-junction silicon solar cell. When current is drawn, as shown in Figure 25.21, the voltage drops, and for fixed temperature and illumination conditions there is a combination of current and voltage that realize the maximum power point. As conditions vary, so do the voltage and current outputs at the maximum power point.

Individual cells are combined in series to produce higher voltages, in order to minimize resistive losses. Even in small (few kilowatt) installations, DC voltages as high as a kilovolt are preferred. Electronic devices known as **inverters** convert the DC output of solar panels to AC power. Inverters can be stand-alone or grid-connected.

Grid-connected inverters sense the AC waveform to which they are connected and adjust the frequency and phase of their output to match. They are also designed to constantly adjust the power that they draw so as to track the maximum power point as conditions at the solar array change.

The extreme variability of the power output of photovoltaic installations causes some additional problems for interfacing with a network. Although the inverter can maintain a roughly constant voltage, very short-term power variations, like those shown in Figure 38.25, cause small but rapid voltage fluctuations that can have an adverse affect on the grid to which the installation is connected [306]. Similar effects come from highly variable loads – arc furnaces are an example – which are known to adversely impact networks. As PV power is more broadly deployed, these problems become more severe. It has been suggested [306] that voltage regulators for PV installations could be installed to provide reactance that can cancel rapid fluctuations and provide more stable power to the network.

38.5.4 Distributed Power Generation

The term **distributed generation** (DG) refers to relatively small power sources that produce power close to intended consumers and are connected to the grid on the distribution rather than transmission side. DG may be as small as a few kilowatts of PV power generated on a residential rooftop, or as large as a combined heat and power cogeneration plant generating tens of megawatts. In many situations, DG sources are designed to provide local power as needed, and feed excess supply back into the grid. In 2009 the EIA reported approximately 13 000 industrial and commercial DG units in the US, with a combined capacity of about 16 GW [307]. Direct generation by PV, wind, and other renewables at the commercial and residential scale has grown exponentially in recent years. A measure of this activity in 2014 is the number, almost 700 000, and capacity, over 6.5 GW, of **net metering**[11] customers on the grid [308]. DG using renewable energy resources is expected to continue to grow rapidly, particularly as small-scale PV installations proliferate in response to government subsidies and falling prices.

In principle, distributed generation offers some advantages, particularly to consumers. These include reliability (DG can provide power when the main network experiences an outage), reduction of congestion and peak power demand, reduced emissions (if the DG comes from renewables), and reduced line losses, to name a few. There are also substantial challenges in integrating extensive DG sources into the grid. Some of these challenges are economic and sociopolitical in nature. Traditional power companies are based on a paradigm of central distribution, and new models must be developed for DG to become widespread.

Beyond the regulatory and economic concerns that usually dominate the discussion of DG, three practical issues can be mentioned. First, distribution networks based on a radial paradigm were typically designed to carry power outward from substation to primary and on to secondary lines. DG often results in power flowing the other way, which requires redesigning switches and circuit breakers. Second, unless corrected by electronic voltage regulation, the kind of voltage fluctuations described in the previous section can degrade the power quality in the network to which the DG is connected. It is generally recognized that advanced voltage regulation electronics should be integrated into DG systems to resolve this problem.

Finally, one of the principal arguments for DG also presents a significant problem. If DG continues to function in the event of a system outage it creates a *power island*, energizing lines that would otherwise be "dead." While this

may serve the needs of local customers, it can be deadly for utility workers who unexpectedly encounter a live line. At the present time, US regulations require that DG systems disconnect from the grid when faults occur or when the voltage or frequency at their interconnect falls outside expected ranges. Considerable discussion and study continues on this subject, since islanded operation is clearly desirable for reliability purposes.

Discussion/Investigation Questions

38.1 When you turn on a powerful motor in your home, such as an air conditioner compressor, you will notice the incandescent lights dim briefly. Explain why.

38.2 Some utilities charge customers for the apparent power delivered (i.e. including reactive power) rather than real power consumed. Discuss the possible justifications for this.

38.3 Discuss the stability of an induction machine operating in the generator domain (see Figure 38.14). For example, suppose more power from the prime mover causes the slip speed to increase. How does the torque on the rotor due to its interaction with the stator magnetic field change? Contrast this with the situation when the generator is in the overspeed condition.

38.4 Make a list of the properties that you would expect to characterize a liquid used as a transformer coolant. Investigate: what was the dominant coolant used prior to the 1970s? Why was its use discontinued, and what coolants are used now?

38.5 Figure 30.12 shows the power coefficient for a wind turbine blade as a function of *tip-speed ratio*. In light of this, explain why a wind turbine constrained to operate at fixed rotor speed is less efficient than one with variable speed.

38.6 If they are visible, find and follow the electric wires that connect your residence to the grid. Can you identify where three-phase power is split into separate phases to serve residential customers? Can you identify the transformers that step down the voltage on the way to your residence?

Problems

38.1 Thomas Edison's first DC power plant on Pearl St. in New York City produced roughly 100 kW at 110 V (DC), which was distributed over two copper wires spanning distances of order 1 km. Suppose resistive transmission losses were limited to 5%. Estimate the diameter of the copper wires necessary to distribute this electric power. Compare to the wire diameter if the same conditions were served by a single-phase 20 kV AC distribution line.

---

[11] *Net metering* refers to a pricing system that credits DG producers at the retail rate for electricity that they supply to the grid.

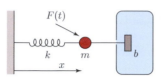

Figure 38.27 For Problem 38.2.

38.2 **[T]** Suppose a mass m is connected to a rigid support by a spring with spring constant k. A frictional force $-b\dot{x}$ is applied by a damper that provides viscous friction, and an external force $F(t)$ may also be applied (see Figure 38.27). Show that Newton's laws give an equation of motion for the mass that is precisely analogous to the equation for an LRC circuit, eq. (38.1). Identify the electromagnetic analogs of m, k, b, and $F(t)$.

38.3 **[T]** Suppose a capacitor C, initially charged to a voltage V, is connected to an inductor L. Show that the current in the resulting LC circuit oscillates, $I(t) = I_0 \sin \omega t$. Find I_0 and show $\omega = 1/\sqrt{LC}$. Sketch the current in the circuit and the voltage across the capacitor as functions of time. An (idealized) circuit like this without resistance would slowly radiate energy away as electromagnetic radiation with frequency $\nu = \omega/2\pi$. Suppose you wanted to construct an oscillator to produce radiation with a frequency of $\nu = 880$ kHz. You have an inductor with inductance 1.5μH. What size capacitor should you choose for your oscillator?

38.4 **[T]** Suppose a small resistance R is added in series to the LC circuit of Problem 38.3. Assume that $Q(t)$ has the form $Q(t) = \overline{Q}(t) \cos \omega t$, where $\overline{Q}(t)$ is a slowly changing function that would be constant if there were no resistance and $\omega = 1/\sqrt{LC}$ was determined in Problem 38.3. Compute the energy stored in this circuit $U(t)$ as a function of $\overline{Q}(t)$ in the approximation that $\dot{\overline{Q}} \ll \omega \overline{Q}$. Then compute the average (over a cycle of oscillation) rate of energy loss in the resistor $\dot{U}$, also in terms of $\overline{Q}$. Combine these results to show that $\dot{U} = -R\langle U \rangle / L$ and therefore $U = U_0 e^{-Rt/L}$, when R is small.

38.5 **[T]** Consider an AC circuit with only a capacitor of capacitance C, driven by an AC voltage $V(t) = V_0 \cos \omega t$. Compute the average $\langle |P| \rangle$ of the absolute value of the rate of energy transfer back and forth from the power source to the capacitor. Compare to the apparent (= reactive) power for this system.

38.6 Repeat the analysis that led to eq. (3.40) for a load with impedance described by a power factor $\cos \phi$. Show that the fraction of power lost compared to real power consumed in the load is approximately given by eq. (38.12), when resistive losses are much smaller than power consumed in the load.

38.7 Fluorescent light bulbs have significant impedance. Measurements on a compact fluorescent bulb with luminosity equivalent to a 40 W incandescent indicate that it consumes 9 W real power. Measurements of

RMS current and voltage, however, indicated that it drew an apparent power of 14 VA. What is the bulb's power factor? If the RMS voltage is 110 V, what is the bulb's resistance? What is its inductance?

38.8 **[H]** A 150 kV (RMS) power supply is providing 300 MW of (real) power at 60 Hz to a load with power factor $\cos \phi = 0.8$. What capacitance would you have to install in parallel with the load to restore the power factor to one? Given a large number of individual 310 μF capacitors rated to a maximum of, say, 1100 V_{RMS}, how would you construct the needed capacitance by grouping them in parallel and in series?

38.9 **[T]** Show that the stator magnetic field of a three-phase generator has a constant magnitude and rotates with constant angular frequency. You can assume that the magnetic fields produced by the stator windings A, B, and C, are $\boldsymbol{B}_A = \hat{\boldsymbol{e}}_A \sin \omega t$, $\boldsymbol{B}_B = \hat{\boldsymbol{e}}_B \sin(\omega t - 2\pi/3)$, and $\boldsymbol{B}_C = \hat{\boldsymbol{e}}_C \sin(\omega t - 4\pi/3)$, where $\hat{\boldsymbol{e}}_{A,B,C}$ are unit vectors rotated by 120° with respect to one another. Note: if you are comfortable with manipulating complex numbers, this may be the simplest approach to take, though the problem can be solved without complex numbers.

38.10 **[T]** A three-phase generator delivers voltages $V_A(t) = V_0 \cos \omega t$, $V_B(t) = V_0 \cos(\omega t + 2\pi/3)$, and $V_C(t) = V_0 \cos(\omega t - 2\pi/3)$. First, assume that all three phases are connected to identical, purely resistive loads. Show that the total power delivered is $3V_0^2/2R$, independent of time. Next, assume that the loads have impedance $|Z|e^{i\phi}$, and show that the total power remains time independent. What is the role of the power factor, $\cos \phi$?

38.11 The rating of each generator at the Three Gorges Dam (TGD) in China is quoted as 778 MVA of *apparent* power (see §38.2.3) produced at 20 kV and 50 Hz. Given that the quoted real power is 700 MW, what power factor has been assumed? The TGD generator rotors turn at 75 rpm. Assuming a three-phase stator winding, as in Figure 38.9, what is the pole structure of the rotors? What is the current in the stator windings?

38.12 **[TH]** Suppose a synchronous generator is supplying power at (RMS) voltage V to a load characterized by impedance $Z_0 = R_0 + iX_0 = z_0 e^{i\phi_0}$, when an inductive load with negligible resistance $Z_1 = i\omega L$ is added in parallel with the original load. Show that the real power delivered by the generator does not change. What is the change in reactive power? Similarly show that if a purely resistive load $Z_1 = R_1$ is added in parallel to the original load, the reactive power in the circuit does not change and compute the change in the real power.

38.13 **[T]** Suppose, as in Figure 38.28, a generator is supplying (time-averaged) power P_L to a purely resistive load R. The voltage at the generator is $V_1(t) = \sqrt{2}V_1 \cos \omega t$. The power is carried to the load along a single conductor with inductance L and negligible resistance. $V_2(t)$ is the voltage at the end of the conductor, where it connects to the load. This is an

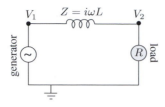

$$V_1 \qquad Z = i\omega L \qquad V_2$$

generator

R load

Figure 38.28 For Problem 38.13.

example of an LR circuit and can be analyzed using complex impedance as described in §38.2.2. Show that the power delivered to the load is $P_L = V_1^2 \cos^2 \phi / R$ and that $V_2(t) = \sqrt{2} V_2 \cos(\omega t - \phi)$ with $V_2 = V_1 \cos \phi$, where the phase difference ϕ between the voltages satisfies $\tan \phi = \omega L / R$. Combine these results to show that $P_L = V_1 V_2 \sin \phi / \omega L$, thereby deriving eq. (38.16).

38.14 A 200 kV transmission line is being designed to carry power 400 km from one power plant to another as part of a transmission network. The line has inductive reactance per unit length $\omega L' = 0.44 \, \Omega/\text{km}$. Assuming that the maximum phase difference that can be tolerated between the sending and receiving ends of the line is 25°, what is the maximum power that can be sent on this line?

38.15 A 345 kV transmission line consisting of three conductors has a resistance per unit length of $R' = R/l = 0.0351 \, \Omega/\text{km}$, and an inductive reactance per unit length of $\omega L' = \omega L / l = 0.371 \, \Omega/\text{km}$, as

described in §38.4.1. Assess the utility of this line for transmission of three-phase power over a distance of 200 km. Assume that the line delivers (real) power $P_L = 350 \, \text{MW}$ to a purely resistive load. Estimate the fraction of power lost in transmission and the effect on the power factor of line impedance.

38.16 **[H]** Suppose a power plant is supplying 300 MW of real power over a 150 kV (RMS) transmission line to a load that has an impedance characterized by a phase $\phi = 26°$ (corresponding to a power factor of $\cos \phi = 0.9$ lagging). The (three-phase) power is supplied over lines with inductive reactance and resistance per unit length $\omega L' = \omega L / l = 0.49 \, \Omega/\text{km}$ and $R' = R/l = 0.11 \, \Omega/\text{km}$ respectively. Assume that an equal fraction of the load is supplied by each phase. Compute the power factor seen at the busbar of the generator as a function of the length l of the line. How long a transmission line would reduce the power factor to 0.75? [Hint: First compute the impedance of the load.]

38.17 A load consists of a large number N of identical resistors R (e.g. incandescent lights, toasters,...) in parallel. Suppose they are powered by an AC generator with RMS voltage V. Compare the line losses if (a) the load is connected to the generator by single-phase power, or (b) the load is divided in three and each group is connected (in a wye configuration) to a single phase of three-phase power. Assume that all the conductors have resistance R_0 and ignore line inductance. Finally, suppose the N resistors are connected in parallel to a DC generator. Compare the line losses if they are powered at the same level as the previous two cases.

Notation

The following is a list of most of the notations used in this book. The numbers in parentheses refer to the chapter where the symbol is first defined. Constants are indicated by "[constant]"; the specific values of these constants can be found in the following appendices. Standard abbreviations for SI units can be found in Appendix D.

a	acceleration (§2), thermal diffusivity (§6), scale parameter (§21), axial induction factor (§30), albedo (§34)	Ci	curie [constant] (§20)
		CoP	coefficient of performance (§10)
		d	distance (§3), down quark (§14)
$\boldsymbol{a}$	vector acceleration (§2)	$\boldsymbol{d}$	electric dipole moment (§3)
a_0	Bohr radius (§7)	D	distance (§3), absorbed dose (§20), (heating) degree days (§36)
A	area/cross-sectional area (§2), amplitude (§4), atomic mass (baryon) number (§14)	δ	declination (§22), power angle (§38)
α	absorption coefficient/absorptivity (§20), angle of attack (§29)	δ_{ij}	metric tensor (3-space) (§21)
		Δ	atomic mass excess (§17)
α_V	coefficient of thermal expansion (§35)	e	electron (§14), eccentricity (§23), electron charge [constant] (§25)
b	bottom quark (§14), binding energy per nucleon (§17)	E	energy (§1)
B	binding energy (§9), number of blades (§30), exergy (§36)	E_F^0	Fermi level (§25)
		E_F	Fermi energy (§25)
$\boldsymbol{B}$	magnetic field (§3)	E_{kin}	kinetic energy (§2)
β	beta decay (§14), angle of incidence (§22)	E_{EM}	electromagnetic energy (§3)
c	speed of light [constant] (§1), concentration (§8), charm quark (§14)	$\boldsymbol{E}$	electric field (§3)
		$\mathcal{E}$	electromotive force (§3), energy per unit area (§31)
c_d	drag coefficient (§2)		
c_l	lift coefficient (§29)	ϵ	permittivity (§3), fast fission factor (§19), obliquity (§23), turbine efficiency (§31), 2nd law efficiency (§36)
c_n	Fourier coefficient (§4, Appendix B.3)		
c_V, c_p	specific heat capacity at fixed volume, pressure (§5)		
		ϵ_0	vacuum permittivity [constant] (§3)
$\hat{c}_V$	degrees of freedom per molecule in ideal gas (§5)	ε	specific energy density (§3), emissivity (§6)
		ε_0	hydrogen binding energy [constant] (§7)
C	capacitance (§3), effective concentration (§24)	$\varepsilon_{\mathrm{kin}}$	kinetic energy density (§29)
C_D	drawdown coefficient (§32)	η	(1st law) efficiency (§5), (neutron) reproduction factor (§19), viscosity (§29), 1st law efficiency (§36)
C_V, C_p	heat capacity at fixed volume, pressure (§5)		
$\hat{C}_V, \hat{C}_p$	molar heat capacity at fixed volume, pressure (§5)		
		$\eta_{\mu\nu}$	metric tensor (special relativity) (§21)
C_P	power coefficient (§30)	f	thermal neutron utilization factor (§19)
$\mathcal{C}$	closed curve (§3)	$\boldsymbol{f}$	force per unit volume (§28), force per unit length (§29)
χ	quality (§12)		

ff	fill factor (§25)
F	force (§2), Helmholtz free energy (§8), fetch (§31), radiative forcing (§34), Faraday [constant] (§37)
$\boldsymbol{F}$	vector force (§2)
g	gravitational acceleration [constant] (§2)
$g_{\mu\nu}$	metric tensor (general relativity) (§21)
G	Newton constant [constant] (§2), Gibbs free energy (§8)
γ	adiabatic/isentropic index (§5), photon (§9)
Γ	Gamma function (§28, Appendix B), circulation (§29), lapse rate (§34)
h	height (§2), specific enthalpy (§5), local heat transfer coefficient (§6), Planck constant [constant] (§7)
$\bar{h}$	heat transfer coefficient (§6)
$\hbar$	Planck constant (reduced) [constant] (§7)
H	enthalpy (§5), Hamiltonian (§7), effective dose (§20), Hubble parameter (§21), scale height (§34)
H_s	significant wave height (§31)
I	moment of inertia (§2), electric current (§3), irradiance/insolation (§23)
I_{RMS}	root mean square current (§38)
$\boldsymbol{j}$	current density (§3)
$\boldsymbol{J}$	kinetic energy flux (§29)
k	spring constant for harmonic oscillator (§2), wave number (§4), thermal conductivity (§6), neutron multiplication factor (§19), absorption cross section (§34)
$\boldsymbol{k}$	wave vector (§4)
k_B	Boltzmann constant [constant] (§5)
K	chord (§29), permeability (§32), bulk modulus (§33)
κ	dielectric constant (§3), absorption coefficient/absorptivity (§20)
l	length (§3)
ℓ	quantum oscillator length scale (§7)
ℓ_{Planck}	Planck length [constant] (§21)
L	angular momentum (§2), inductance (§3), length (§4), luminosity (§22)
λ	wavelength (§4), probability per unit time (§15), latitude (§23), tip-speed ratio (§30), (climate) feedback parameter (§34)
λ_0	Planck feedback parameter (§34)
Λ	cosmological constant (§21)
m	mass (§1)
$\dot{m}$	mass flow rate (§13)
$\boldsymbol{m}$	magnetic dipole moment (§3)
M	mass (§2), mutual inductance (§3)
M_p, M_n	proton, neutron mass [constant] (§1, §17)

μ	(magnetic) permeability (§3), muon (§14), attenuation coefficient (§20), shear modulus (§33)
μ_0	vacuum permeability [constant] (§3)
n	number per unit length (§3), number of moles (§5), energy level (§7), neutron (§14)
$\hat{\boldsymbol{n}}$	unit normal vector (§3)
N	number of windings (§3), particles (§5), ...
$\mathcal{N}$	number of microstates (§8)
N_A	Avogadro's number [constant] (§5)
$\boldsymbol{\nabla}$	gradient, divergence, curl operator (Appendix B)
ν	frequency (§4), neutrino (§14)
Nu_L	Nusselt number (§6)
ω	angular velocity/frequency (§2)
$\tilde{\omega}$	orbital precession parameter (§34)
$\boldsymbol{\omega}$	angular velocity vector (§2), vorticity (§29)
Ω	angular velocity (§28)
p	pressure (§5), proton (§14), resonance escape probability (§19)
$\boldsymbol{p}$	vector momentum (§2)
P	power (§2), probability (§7), power density (§19), path (Appendix B)
$\mathcal{P}$	power per unit area (§28), closed path (Appendix B)
ϕ	angular variable (§2), phase (§2)
$\boldsymbol{\phi}$	mass flux (§29)
Φ	flux (§3)
ψ	quantum wavefunction (§7)
q	electric charge (§3), heat per unit mass (§13)
$\boldsymbol{q}$	heat flux (§6)
Q	electric charge (§3), heat transfer (§5), fusion gain factor (§19), flow rate (§31)
r	radial distance (§2), compression ratio (§11)
$\boldsymbol{r}$	radial separation vector (§2)
$\boldsymbol{r}_{12}$	vector from object 1 to object 2 (§3)
R	radius (§2), resistance (§3), gas constant [constant] (§5), thermal resistance (§6), rate (§19), region of space (Appendix B)
$\mathcal{R}$	heat reservoir (§8)
Re	Reynolds number (§29)
ρ	mass density (§2), charge density (§3), radial (cylindrical) coordinate (§3), reactivity (§19), reflectivity (§22)
ϱ	resistivity (§3)
s	quantum state (§7), strange quark (§14)
$\boldsymbol{s}$	displacement (of fluid) (§31)
S	surface (Appendix B), entropy (§8), Seebeck coefficient (§10), mass stopping power (§20)
$\boldsymbol{S}$	Poynting vector (§4), spin angular momentum (§7)
$\mathcal{S}$	set of symbols (§8), closed surface (Appendix B)

σ	conductivity (§3), Stefan–Boltzmann constant [constant] (§6), information entropy (§8), root-mean-squared deviation (§15), cross section (§18), surface tension (§31)	$\mathcal{V}$	volume (§3)
		w	width (§3), work per unit mass (§13), wind speed (§28)
σ_c	climate sensitivity parameter (§34)	$w_{R/T}$	radiation/tissue weighting factor (§20)
Σ	macroscopic cross section (§19)	W	work (§2)
t	time (§2), top quark (§14)	x	spatial coordinate (§2), enrichment factor (§19)
T	period (§4), temperature (§5)	$\boldsymbol{x}$	vector coordinate (§2)
$T_{\mu\nu}$	energy–momentum tensor (§21)	$\hat{\boldsymbol{x}}$	unit vector in x-direction (§2)
τ	torque (§2), tension (§4), tauon (§14), lifetime (§15), transmittance (§22), optical depth (§22)	X	reactance (§38)
		ξ	logarithmic energy decrement (§18)
		y	spatial coordinate (§2)
τ_E	energy confinement time (§19)	$\hat{\boldsymbol{y}}$	unit vector in y-direction (§2)
$\tau_{1/2}$	half-life (§15)	z	spatial coordinate (generally vertical axis) (§2), single-particle partition function (§8), complex number (Appendix B)
$\boldsymbol{\tau}$	torque vector (§2)		
θ	angular variable (§2)		
u	energy density (§4), specific internal/thermal energy (§5), up quark (§14), atomic mass unit (Dalton) [constant] (§17)	z_0	roughness length (§28)
		$\hat{z}$	unit vector in z-direction (§2)
U	thermal energy, internal energy (§5), U-factor (thermal conductance) (§6)	Z	height (§2), partition function (§8), atomic number (§17), impedance (§38), hydraulic head (§31)
v	speed (§2), specific volume (§5)	‰	parts per thousand (§35)
$\tilde{v}$	root mean cube wind speed (§28)	$\odot$	Sun/solar (§18)
$\boldsymbol{v}$	vector velocity (§2)	$\oplus$	Earth (§2)
V	potential energy (§2), electrostatic potential (§3), volume (§3)	$X\bullet$	free radical (§20)
		$\lvert\cdot\rangle$	quantum state (§7)
V_{RMS}	root mean square voltage (§3)	$[\ldots,\ldots]$	commutator (§7)
		$\{\ldots,\ldots\}$	Poisson bracket (§21)

APPENDIX

B

Some Basic Mathematics

This appendix contains a brief summary of aspects of vector calculus, complex numbers, vector spaces, tensors, Fourier transforms, and other basic mathematical facts that are used in the text. While the review here is intended to be somewhat self-contained, there are no detailed explanations or rigorous proofs. For further background the reader should consult any of the many introductory mathematics or mathematical physics texts available; specific suggestions are made for some topics.

B.1 Vector Calculus

This section summarizes the fundamentals of vector calculus. We assume familiarity with the basics of vectors in two- and three-dimensional space and with the fundamentals of many-variable calculus. A more complete introduction including proofs and exercises can be found in a standard text such as [16].

B.1.1 Vectors

We denote **vectors** in n-dimensional space by bold font x or coordinates $(x_1, x_2, \ldots, x_n)$; when $n = 2$ or 3 we often denote the coordinates by (x, y) or (x, y, z). The addition of two vectors $v + w$ gives another vector

$$(v_1, v_2, \ldots, v_n) + (w_1, w_2, \ldots, w_n)$$
$$= (v_1 + w_1, v_2 + w_2, \ldots, v_n + w_n). \quad (B.1)$$

The linear space spanned by such vectors is known as a **vector space**. The **dot product** between two vectors v and w is

$$v \cdot w = v_1 w_1 + \cdots + v_n w_n = vw \cos\theta, \quad (B.2)$$

where θ is the angle between the two vectors. The **norm** of a vector v is denoted $|v|$ or simply v and is given by

$$v = |v| = \sqrt{v \cdot v}. \quad (B.3)$$

A **unit vector**, denoted $\hat{v}$, is a vector of norm 1, $|\hat{v}| = 1$. To every nonzero vector v there is an associated unit vector $\hat{v}$ pointing in the same direction as v, $\hat{v} = v/v$.

In three dimensions, the **vector cross product** ($\times$) associates a new vector to every pair of vectors v, w

$$v \times w = (v_2 w_3 - w_2 v_3, v_3 w_1 - w_3 v_1, v_1 w_2 - w_1 v_2)$$
$$= vw \sin\theta\, \hat{n}, \quad (B.4)$$

where, when $vw \sin\theta$ is nonzero, $\hat{n}$ is a unit vector perpendicular to both v and w, with direction defined by the **right-hand rule** for vector cross products (v, w, $\hat{n}$ have the same relative orientation as the thumb, index finger, and middle finger of the right hand when placed perpendicular to each other). Two simple identities involving vector dot and cross products

$$a \times (b \times c) = (a \cdot c)b - (a \cdot b)c, \quad (B.5)$$
$$a \cdot (b \times c) = b \cdot (c \times a) = c \cdot (a \times b), \quad (B.6)$$

can easily be verified by explicit computation in coordinates.

B.1.2 Partial Derivatives and 3D Integrals

We often denote partial derivatives of a function $f(x_1, \ldots, x_n)$ by

$$\partial_i f \equiv \frac{\partial}{\partial x_i} f(x_1, \ldots, x_n). \quad (B.7)$$

In three dimensions we sometimes abbreviate $\partial_x \equiv \partial/\partial x$, etc. When the variables x_i depend in turn upon a parameter τ, the **total derivative** of f with respect to τ is

$$\frac{df}{d\tau} = \partial_1 f(x)\left(\frac{dx_1}{d\tau}\right) + \cdots \partial_n f(x)\left(\frac{dx_n}{d\tau}\right). \quad (B.8)$$

Integration of a function f over multiple variables $x_1, \ldots, x_n$ may be denoted by

$$\int d^n x\, f(x) = \int dx_1 dx_2 \cdots dx_n f(x_1, \ldots, x_n). \quad (B.9)$$

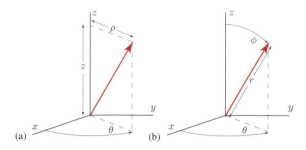

Figure B.1 (a) cylindrical coordinates (ρ, θ, z); (b) spherical polar coordinates (r, ϕ, θ).

In three dimensions we sometimes work in **spherical coordinates** r, θ, ϕ with[1]

$$\boldsymbol{x}(r, \theta, \phi) = (r \sin\phi \cos\theta, r \sin\phi \sin\theta, r \cos\phi) \quad (B.10)$$

for $0 \leq \theta < 2\pi$, $0 \leq \phi < \pi$ (see Figure B.1).

The rotationally invariant measure for integration of a function in spherical coordinates is

$$\int d^3\boldsymbol{x} f(\boldsymbol{x}) = \int_0^{2\pi} d\theta \int_0^{\pi} d\phi \, \sin\phi \int_0^{\infty} dr \, r^2 \, f(r, \theta, \phi) \,. \quad (B.11)$$

For a function invariant under 3D rotations (i.e. independent of θ, ϕ), the integration becomes

$$4\pi \int_0^{\infty} dr \, r^2 f(r) \,. \quad (B.12)$$

Here, 4π is the total solid angle of the sphere,

$$\Omega = \int d\Omega = \int_0^{2\pi} d\theta \int_0^{\pi} d\phi \, \sin\phi = 4\pi \,; \quad (B.13)$$

solid angles are measured in **steradians**, which are dimensionless units.

B.1.3 Vector Fields and Operations

A **vector field** associates a vector $\boldsymbol{v}(\boldsymbol{x})$ with every point $\boldsymbol{x}$ in some domain in an n-dimensional space.

Gradient The **gradient** of a function $f(\boldsymbol{x})$ is a vector field defined by

$$\boldsymbol{\nabla} f = (\partial_1 f, \ldots, \partial_n f) \,. \quad (B.14)$$

---

[1] Note that in some physics texts the role of θ and ϕ is reversed from the notation we use here, which is conventional in mathematics and has a choice of azimuthal angle θ that matches the **cylindrical coordinate system** $\boldsymbol{x}(\rho, \theta, z) = (\rho \cos\theta, \rho \sin\theta, z)$.

Divergence The **divergence** of a vector field $\boldsymbol{v}(\boldsymbol{x})$ is a function

$$\boldsymbol{\nabla} \cdot \boldsymbol{v} = \partial_1 v_1 + \partial_2 v_2 + \cdots + \partial_n v_n \,. \quad (B.15)$$

Curl In three dimensions, the **curl** of a vector field $\boldsymbol{v}(\boldsymbol{x})$ is another vector field

$$\boldsymbol{\nabla} \times \boldsymbol{v} = (\partial_2 v_3 - \partial_3 v_2, \partial_3 v_1 - \partial_1 v_3, \partial_1 v_2 - \partial_2 v_1) \,. \quad (B.16)$$

Laplacian The **Laplacian** of a function $f(\boldsymbol{x})$ in n dimensions is the function

$$\nabla^2 f = \partial_1^2 f + \cdots + \partial_n^2 f \,. \quad (B.17)$$

The following identities can be verified easily by computation in coordinates:

$$\boldsymbol{\nabla} \cdot (\boldsymbol{\nabla} f) = \nabla^2 f, \quad (B.18)$$

$$\boldsymbol{\nabla} \cdot (f\boldsymbol{v}) = \boldsymbol{v} \cdot \boldsymbol{\nabla} f + f \, \boldsymbol{\nabla} \cdot \boldsymbol{v}, \quad (B.19)$$

$$\boldsymbol{\nabla} \times (\boldsymbol{\nabla} f) = 0, \quad (B.20)$$

$$\boldsymbol{\nabla} \cdot (\boldsymbol{\nabla} \times \boldsymbol{v}) = 0, \quad (B.21)$$

$$\boldsymbol{\nabla} \times (\boldsymbol{\nabla} \times \boldsymbol{v}) = \boldsymbol{\nabla}(\boldsymbol{\nabla} \cdot \boldsymbol{v}) - \nabla^2 \boldsymbol{v}, \quad (B.22)$$

$$\boldsymbol{\nabla} \cdot (\boldsymbol{v} \times \boldsymbol{w}) = (\boldsymbol{\nabla} \times \boldsymbol{v}) \cdot \boldsymbol{w} - (\boldsymbol{\nabla} \times \boldsymbol{w}) \cdot \boldsymbol{v} \,. \quad (B.23)$$

B.1.4 Line and Surface Integrals

A **line integral** of a vector field $\boldsymbol{v}$ is an integral along a path P from a point $\boldsymbol{a}$ to a point $\boldsymbol{b}$ (or along a closed path $\mathcal{P}$ describing a loop) of the component of $\boldsymbol{v}$ parallel to the path (see, for example, Figure 2.5). Parameterizing the path as $\boldsymbol{x}(\tau)$ in terms of a parameter τ with $\boldsymbol{x}(0) = \boldsymbol{a}$ and $\boldsymbol{x}(T) = \boldsymbol{b}$, the line integral is given by

$$\int_P d\boldsymbol{x} \cdot \boldsymbol{v} = \int_0^T d\tau \, \frac{d\boldsymbol{x}(\tau)}{d\tau} \cdot \boldsymbol{v}(\boldsymbol{x}(\tau)) \,. \quad (B.24)$$

The vector $d\boldsymbol{x}(\tau)/d\tau$ is tangent to the path at the point $\boldsymbol{x}(\tau)$. For a line integral around a closed path, the symbol $\oint_{\mathcal{P}}$ is used.

A surface integral of the normal component of a vector field $\boldsymbol{v}$ in three dimensions over a two-dimensional surface S is similarly denoted by (see Figure 3.6)

$$\int_S d\boldsymbol{S} \cdot \boldsymbol{v} \,, \quad (B.25)$$

where $d\boldsymbol{S} = dS\hat{\boldsymbol{n}}$ denotes an infinitesimal region of surface with area dS and unit normal $\hat{\boldsymbol{n}}$. (Note that this definition only makes sense when the surface is *orientable* so that there exists a global choice of direction for $\hat{\boldsymbol{n}}$.) If the surface is parameterized by $\boldsymbol{x}(\xi, \eta)$, then the normal vector $\boldsymbol{n}$ is parallel to $\partial_\xi \boldsymbol{x} \times \partial_\eta \boldsymbol{x}$, and the surface integral is computed as

$$\int_S d\boldsymbol{S} \cdot \boldsymbol{v} = \int d\xi d\eta \, (\partial_\xi \boldsymbol{x} \times \partial_\eta \boldsymbol{x}) \cdot \boldsymbol{v} \,. \quad (B.26)$$

Example B.1   A Line Integral

Consider the vector field

$$\boldsymbol{v}(x, y, z) = \frac{a}{x^2 + y^2}(-y, x, 0).$$

What is the line integral around the closed path described by a circle of radius r around the origin in the xy-plane?
We use the parameterization $\boldsymbol{x}(\theta) = (r \cos \theta, r \sin \theta, 0)$ for $0 \leq \theta < 2\pi$. Then

$$\boldsymbol{v} = \frac{a}{r^2}(-r \sin \theta, r \cos \theta, 0) \quad \text{and} \quad \frac{d\boldsymbol{x}}{d\theta} = (-r \sin \theta, r \cos \theta, 0).$$

Thus,

$$\oint_{\mathcal{P}} d\boldsymbol{x} \cdot \boldsymbol{v} = \int_0^{2\pi} d\theta \, \frac{d\boldsymbol{x}}{d\theta} \cdot \boldsymbol{v} = \int_0^{2\pi} d\theta \, (-r \sin \theta, r \cos \theta, 0) \cdot \frac{a}{r^2}(-r \sin \theta, r \cos \theta, 0) = 2\pi a.$$

Such a vector field describes the magnetic field around a long, straight wire carrying a current I (see eq. (3.50)) and also describes the velocity field of an incompressible fluid circulating around a cylinder (see § 29.2.3 and eq. (29.17)).

Example B.2   A Surface Integral

Consider the vector field

$$\boldsymbol{v}(x, y, z) = \frac{a}{(x^2 + y^2 + z^2)^{3/2}}(x, y, z).$$

What is the surface integral $\oint_{\mathcal{S}} d\boldsymbol{S} \cdot \boldsymbol{v}$ over the closed surface $\mathcal{S}$ described by a sphere of radius r?
We choose ϕ and θ as coordinates over the sphere and use the parameterization (B.10),

$$\oint_{\mathcal{S}} d\boldsymbol{S} \cdot \boldsymbol{v} = \int_0^{2\pi} d\theta \int_0^{\pi} d\phi \left(\frac{d\boldsymbol{x}}{d\phi} \times \frac{d\boldsymbol{x}}{d\theta} \right) \cdot \boldsymbol{v},$$

where

$$\boldsymbol{v} = \frac{a}{r^2}(\sin \phi \, \cos \theta, \sin \phi \, \sin \theta, \cos \phi),$$

$$\frac{d\boldsymbol{x}}{d\phi} = (r \cos \phi \, \cos \theta, r \cos \phi \, \sin \theta, -r \sin \phi),$$

$$\frac{d\boldsymbol{x}}{d\theta} = (-r \sin \phi \, \sin \theta, r \sin \phi \, \cos \theta, 0),$$

so that

$$\oint_{\mathcal{S}} d\boldsymbol{S} \cdot \boldsymbol{v} = \frac{a}{r^2} \int_0^{2\pi} d\theta \int_0^{\pi} d\phi \, ((r \cos \phi \, \cos \theta, r \cos \phi \, \sin \theta, -r \sin \phi) \times (-r \sin \phi \, \sin \theta, r \sin \phi \, \cos \theta, 0))$$

$$\cdot (\sin \phi \, \cos \theta, \sin \phi \, \sin \theta, \cos \phi)$$

$$= a \int_0^{2\pi} d\theta \int_0^{\pi} d\phi \, \sin \phi = 4\pi a.$$

Such a vector field describes the electric field around a point charge q as described in eq. (3.3), with $a = q/4\pi\epsilon_0$. Note that this integral could also be computed by dividing the sphere into two patches with $z \geq 0$, $z \leq 0$, using parameterizations $x = \xi$, $y = \pm \eta$, $z = \pm \sqrt{r^2 - \xi^2 - \eta^2}$, with (ξ, η) in the unit disk $\xi^2 + \eta^2 \leq 1$. The details of this computation are left as an exercise to the reader.

If the surface is **closed** (i.e. has no boundary) then we denote it by $\mathcal{S}$, and the symbol $\oint_{\mathcal{S}}$ is used. Note that sometimes it is not easy to find a single parameterization of a surface and it is easier to divide the surface into several patches with different parameterizations. The total surface integral is the sum of the surface integrals over the patches, and is independent of the division of surface and the choice of parameterization.

B.1.5 Stokes' Theorem and the Divergence Theorem

Stokes' theorem relates the integral of a vector field around a closed curve $\mathcal{P}$ to the integral of the curl of v over any surface S with boundary $\mathcal{P} = \partial S$ (∂ is the standard mathematical notation for the boundary of a surface or region)

$$\oint_{\mathcal{P}=\partial S} v \cdot dx = \int_S dS \cdot (\nabla \times v).$$ (B.27)

By convention, the orientation of the curve $\mathcal{P}$ and the surface element dS are related by another use of the *right-hand rule*: if the fingers of the right hand are wrapped in the direction associated with the orientation of $\mathcal{P}$ then the surface element dS points in the direction of the thumb (as shown, for example in Figure 3.17(a)). The relation (B.27) holds independent of the choice of surface S.

Closely related to Stokes' theorem is the **divergence theorem** that relates a surface integral of the normal component of a vector field v over a closed surface $\mathcal{S}$ to the divergence of the vector field inside the region R with boundary $\mathcal{S}$

$$\int_{\mathcal{S}=\partial R} dS \cdot v = \int_R d^3x\, \nabla \cdot v(x).$$ (B.28)

Derivations of Stokes' theorem and the divergence theorem can be found in any standard vector calculus text such as [309]. In the modern approach, these theorems are easily proven and unified through the language of *differential forms*, see [310] for an introductory treatment.

B.2 Complex Numbers

Complex numbers play a central role in the theory of quantum mechanics (§7) and enable a *spectral decomposition* of waves (§31.2) that is useful for understanding many physical systems. We use only the most basic properties of complex numbers in this book.

Complex numbers are an extension of the more familiar concept of *real* numbers. Any real number squares to a positive number (for example, $(-1)^2 = +1$), so among

real numbers there is no value that squares to -1. If, however, we simply define a new "number" i with the property that

$$i^2 = -1$$ (B.29)

then we can use standard algebraic operations to extend our usual set of real numbers to a larger set of **complex numbers**, which are defined to be all numbers of the form

$$x + iy$$ (B.30)

where x, y are real and i satisfies eq. (B.29). Often x is referred to as the **real part** of the complex number eq. (B.30) while iy is the **imaginary part**. For example, $1 + 3i$ is a complex number, with real part 1 and imaginary part $3i$.

Complex numbers can be added together algebraically

$$(x + iy) + (a + ib) = (x + a) + i(y + b).$$ (B.31)

They can be multiplied, using eq. (B.29) to get another complex number of the form (B.30)

$$(x + iy) \cdot (a + ib) = xa + i(ya + xb) + yb(i^2)$$
$$= (xa - yb) + i(ya + xb).$$ (B.32)

The letters z, w are often used to describe complex quantities, such as $z = 3 + 2i$.

Complex numbers can be visualized as vectors in the two-dimensional xy-plane (Figure B.2). Often it is convenient to describe a complex number $z = x + iy$ in terms of the magnitude or **norm**

$$r = |z| = \sqrt{x^2 + y^2}$$ (B.33)

of the complex number and the angle $\theta = \arctan(y/x)$, which is called the **phase** of the complex number z. In terms of these quantities,

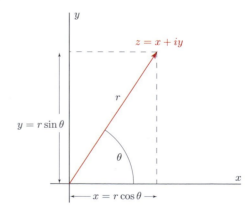

Figure B.2 The complex number $x + iy = re^{i\theta}$.

$$z = x + iy = r(\cos\theta + i\sin\theta) = re^{i\theta}, \quad \text{(B.34)}$$

where we can identify

$$e^{i\theta} = \cos\theta + i\sin\theta \quad \text{(B.35)}$$

by using a Taylor series expansion of both sides. So, for example, the complex function of time $f(t) = e^{i\omega t}$ for real ω can be thought of as a point $z = \cos\omega t + i\sin\omega t$ in the complex plane that moves around the unit circle in the complex plane with constant angular frequency ω.

In terms of r, θ, multiplication of complex numbers is straightforward

$$(re^{i\theta})(se^{i\psi}) = rse^{i(\theta+\psi)}. \quad \text{(B.36)}$$

One final very useful property of complex numbers is the notion of *complex conjugation*. Given a complex number $z = x + iy$, the **complex conjugate** of z, usually noted $\bar{z}$ and sometimes z^*, is given by negating the imaginary part of z

$$\bar{z} = x - iy. \quad \text{(B.37)}$$

Geometrically, as can be seen in Figure B.2, complex conjugation amounts to reflecting the vector associated with z across the *real axis* where $y = 0$. Note that with this definition

$$|z|^2 = z\bar{z} = (x + iy)(x - iy) = x^2 + y^2. \quad \text{(B.38)}$$

A useful property of complex numbers is that $\overline{(zw)} = \bar{z}\bar{w}$, which follows from just flipping the signs on all imaginary parts.

B.3 Vector Spaces and Fourier Transforms

The concepts of *linear spaces*, including function spaces and Fourier decomposition, appear in the analysis of the frequency structure of waves and the energy states of quantum systems. In particular, an understanding of some introductory aspects of linear algebra helps to clarify the phenomenon of superposition in quantum mechanics.

B.3.1 Vector Spaces

In Appendix B.1.1, we introduced n-dimensional vector spaces. These can be incorporated in a more general framework that can also be used to describe the infinite-dimensional linear spaces of functions that are relevant for waves and quantum physics.

Real Vector Spaces   In general, a **real vector space** of n dimensions is defined to be the set of all possible linear combinations of a set of n independent **basis vectors** $e_1, e_2, \ldots, e_n$ with real coefficients $v_1, \ldots v_n$,

$$v = v_1 e_1 + \cdots + v_n e_n. \quad \text{(B.39)}$$

For example, a 2D vector $x = (x, y)$ can be written as $x = x\hat{x} + y\hat{y}$, where $\hat{x} = e_1$, $\hat{y} = e_2$ are the two independent basis vectors. Vectors are often written in column form

$$v = \begin{pmatrix} v_1 \\ v_2 \\ \vdots \\ v_n \end{pmatrix}. \quad \text{(B.40)}$$

Vectors add in the obvious way:

$$v + w = (v_1 + w_1)e_1 + \cdots + (v_n + w_n)e_n. \quad \text{(B.41)}$$

The vector spaces we are interested in for physics applications generally have an **inner product** $v \cdot w$, also known as the *dot product*. The inner product defines a *norm* through eq. (B.3). The inner product is *symmetric*, $v \cdot w = w \cdot v$, and the norm is generally **positive definite**, $v \cdot v > 0$ for $v \neq 0$, though in some cases such as special relativity (§21) it is useful to introduce a norm of indefinite sign.

When the inner product is positive definite we can choose the basis vectors e_i to have unit norm and to be orthogonal to one another.

$$e_i \cdot e_j = \delta_{ij} = \begin{cases} 1, & i = j, \\ 0, & i \neq j, \end{cases} \quad \text{(B.42)}$$

where δ_{ij} is known as the **Kronecker delta**. When (B.42) is satisfied, the e_j are said to form an **orthonormal basis**, and in this case the inner product can be written in the form of eq. (B.2). Note that the existence of an inner product allows one to determine the coefficients in the expansion of a vector v in an orthonormal basis,

$$v_i = e_i \cdot v. \quad \text{(B.43)}$$

While we have focused on n-dimensional vector spaces where n is a finite integer, we can generalize the notion of vector spaces to infinite-dimensional spaces, where for example the index i can be any nonnegative integer. The notions of inner product and orthonormal basis can be generalized to the infinite-dimensional case in a relatively transparent fashion. The main new feature is that the coefficients v_i in the expansion (B.39) of a vector v must give a finite norm, so $\sum_i v_i^2$ must converge to a finite value for any vector in the space. There are some further technical mathematical issues that we do not bother with here; the interested reader should consult any beginning book on real analysis, such as [311].

Complex Vector Spaces   For many purposes, in particular for quantum mechanics and for some aspects of wave analysis, it is helpful to work with vector spaces where the coefficients v_i in eq. (B.39) take values in the complex numbers (Appendix B.2) instead of being real numbers. Such a vector space is called a **complex vector space**.

Just as for a real vector space, an orthonormal basis of a complex vector space e_i can be chosen that satisfy eq. (B.42). For a complex vector space with an orthonormal basis, the inner product is often defined as

$$\langle \mathbf{w}, \mathbf{v} \rangle = w_1^* v_1 + w_2^* v_2 + \cdots + w_n^* v_n \,. \qquad \text{(B.44)}$$

This inner product is linear in the second argument $\mathbf{v}$, but **antilinear** in the first argument $\mathbf{w}$, meaning that it is linear in the complex conjugates w_i^* of the components of $\mathbf{w}$.

B.3.2 Linear Transformations and Matrices

An important class of operations that can be carried out on vectors are **linear transformations**, which associate to every vector $\mathbf{v}$ a new vector $\mathbf{v}'$ with coefficients given by linear combinations of the coefficients of $\mathbf{v}$

$$v_i' = \sum_{j=1}^{n} M_{ij} v_j \,. \qquad \text{(B.45)}$$

The coefficients in a linear transformation can be arranged into a square **matrix**

$$M = (M_{ij}) = \begin{pmatrix} M_{11} & M_{12} & \cdots & M_{1n} \\ M_{21} & M_{22} & \cdots & M_{2n} \\ \vdots & \vdots & \ddots & \vdots \\ M_{n1} & M_{n2} & \cdots & M_{nn} \end{pmatrix} \,. \qquad \text{(B.46)}$$

Matrices can be multiplied together, corresponding to a sequential application of the linear transformations (from right to left)

$$(MN)_{ik} = \sum_{j=1}^{n} M_{ij} N_{jk} \,. \qquad \text{(B.47)}$$

A simple way of describing matrix multiplication is to note that the ik entry in the product matrix corresponds to the inner (dot) product of the ith row of the first matrix with the kth column of the second matrix. Matrices are used in various places in the book, for example to describe the time evolution of quantum systems.

Another use of linear transformations arises in considering a change of basis for a vector space. The components v_i of a vector expressed in terms of one orthonormal basis through eq. (B.39) are related to the components v_j' in a different basis through a linear transformation (B.45) where the orthonormality of the basis implies that M is a **special orthogonal matrix** satisfying $(M^T M)_{ik} = \delta_{ik}$ if the vector space is real, or a **unitary matrix** satisfying $(M^\dagger M)_{ik} = \delta_{ik}$ if the vector space is complex, where M^T is the *transpose* matrix with $M_{ij}^T = M_{ji}$, and $M^\dagger$ is the *adjoint* or transpose conjugate matrix $(M^\dagger)_{ij} = M_{ji}^*$.

B.3.3 Tensors

A **tensor** is a generalization of a vector to an object that carries multiple vector-type indices; the **rank** of the tensor is the number of indices. A *scalar* (simple number) is a tensor of rank zero, and a *vector* is a tensor of rank one. A tensor of rank two is written T_{ij}, and transforms under a linear transformation such as (B.45) through

$$T_{ij}' = \sum_{k,l=1}^{n} M_{ik} M_{jl} T_{kl} \,. \qquad \text{(B.48)}$$

Formally, a tensor can be written as a linear combination of basis elements that are *tensor products* of basis vectors for the linear space

$$\mathbf{T} = T_{11}(\mathbf{e}_1 \otimes \mathbf{e}_1) + T_{12}(\mathbf{e}_1 \otimes \mathbf{e}_2) + \cdots T_{nn}(\mathbf{e}_n \otimes \mathbf{e}_n) \,. \qquad \text{(B.49)}$$

Tensors of rank two are often written in matrix form, though tensors of higher rank, e.g. S_{ijk}, cannot be described in this way. Tensors appear in §21 and §33.

B.3.4 Vector Spaces of Functions: Fourier Series

A key application of the theory of vector spaces is to spaces of functions. In general, the set of functions $f(x)$ form a linear space, where the sum of two functions is given by $(f + g)(x) = f(x) + g(x)$. The **domain** of a function is the set of possible values for x. When the functions are limited to those with a proper set of boundary conditions,[2] then an inner product between two functions can be defined as an integral over the domain on which the functions are defined,

$$f \cdot g = \int dx \, f(x) g(x) \,, \qquad \text{(B.50)}$$

if f and g are real, or

$$\langle f, g \rangle = \overline{\langle g, f \rangle} = \int \bar{f}(x) g(x) \,, \qquad \text{(B.51)}$$

if f and g can take on complex values. In such cases, the general technology of vector spaces can be used to describe the space of functions.

As an example, consider the set of real-valued square-integrable functions $f(x)$ on the interval $0 \leq x \leq L$, with

---

[2] Strictly speaking, the functions must be reasonably well-behaved mathematical functions for the integral to be well defined; the requisite technical condition is that the functions be *square integrable* on the domain in question, so that $\int dx \, |f(x)|^2$ exists and is finite. Most of the functions that we work with in this book, which are at least piecewise continuous, with piecewise continuous first derivatives, etc., and which take bounded values on bounded domains, are also well behaved in the sense that they are square integrable.

Example B.3   A Fourier Transform

Consider the function $f(x) = \cos(3\pi x)e^{-x^2/4}$, which is continuous and differentiable for all x and is square integrable, $\int_{-\infty}^{\infty} dx\, f(x)^2 < \infty$. Waves of this form can be encountered in the propagation of light (§4), quantum particles (§7), and waves on the ocean surface (§31). $f(x)$ is plotted at right. Note that $f(x)$ is a cosine wave with wavelength $\lambda = 2/3$ modulated by a gaussian envelope.

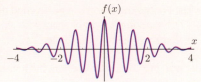

What is the Fourier transform $F(k)$ of this function?
The Fourier transform of $f(x)$ is defined by eq. (B.60),

$$F(k) = \frac{1}{2\pi} \int_{-\infty}^{\infty} dx\, \cos(3\pi x)\, e^{-x^2/4}\, e^{ikx} = \frac{1}{\pi} \int_{0}^{\infty} \cos(3\pi x)\, e^{-x^2/4} \cos(kx)$$

$$= \frac{1}{2\sqrt{\pi}} \left(e^{-(k-3\pi)^2} + e^{-(k+3\pi)^2} \right).$$

$F(k)$ peaks at $k = \pm 3\pi$ corresponding to $\lambda = 2\pi/k = \pm 2/3$ indicating the dominance in the function $f(x)$ of components with this wavelength.

---

boundary conditions $f(0) = f(L) = 0$. The inner product between two such functions $f(x), g(x)$ can be defined to be

$$f \cdot g = \int_0^L dx\, f(x)g(x). \tag{B.52}$$

We can define an *orthonormal basis* for this set of functions

$$e_n(x) = \sqrt{\frac{2}{L}} \sin\left(\frac{\pi n x}{L}\right). \tag{B.53}$$

It is easy to check that these functions satisfy eq. (B.42). Thus, any (square integrable) function $f(x)$ on the interval obeying the desired boundary conditions can be written as a **Fourier series expansion**

$$f(x) = \sum_{n=1}^{\infty} c_n e_n(x) = \sum_{n=1}^{\infty} c_n \sqrt{\frac{2}{L}} \sin\left(\frac{\pi n x}{L}\right). \tag{B.54}$$

This result is known as **Fourier's theorem**. Just as for finite-dimensional vector spaces, the coefficients in the Fourier series expansion can be determined by

$$c_n = e_n \cdot f = \int_0^L dx\, \sqrt{\frac{2}{L}} \sin\left(\frac{\pi n x}{L}\right) f(x), \tag{B.55}$$

in analogy to eq. (B.43) The norm of f can be written

$$f \cdot f = \int_0^L dx\, f(x)^2 = \sum_{n=1}^{\infty} c_n^2. \tag{B.56}$$

The Fourier series expansion can be generalized in various ways. In particular, if the boundary conditions are

periodic, so that $f(L) = f(0)$, then the Fourier series expansion contains both sines and cosines

$$f(x) = \sum_{n=1}^{\infty} a_n \sqrt{\frac{2}{L}} \sin\left(\frac{2\pi n x}{L}\right)$$

$$+ \sum_{m=0}^{\infty} b_m \sqrt{\frac{2}{L}} \cos\left(\frac{2\pi m x}{L}\right). \tag{B.57}$$

This expansion can be generalized to *complex* periodic functions f, for which the Fourier series expansion is

$$f(x) = \sum_{n=-\infty}^{\infty} z_n \sqrt{\frac{1}{L}} e^{2\pi i n x/L}. \tag{B.58}$$

The complex modes in this expansion are just combinations of sines and cosines, using $e^{i\theta} = \cos\theta + i\sin\theta$ (eq. (B.35)). For real functions, eq. (B.57) and eq. (B.58) are equivalent representations, with $z_0 = b_0$, $z_n = (b_n - ia_n)/\sqrt{2}$, and $z_{-n} = (b_n + ia_n)/\sqrt{2}$ for $n > 0$. The analog of eq. (B.55) is then

$$z_n = \int_x^{x+L} dy\, \sqrt{\frac{1}{L}} e^{-2\pi i n y/L} f(y).$$

B.3.5 Fourier Transforms

An important generalization of the Fourier series expansion is the *Fourier transform*, which applies to general continuous functions $f(x)$ defined on the infinite real line $-\infty < x < \infty$. Any function $f(x)$ (subject to some technical mathematical conditions irrelevant for physics) can be written as an integral over sine and cosine waves, or more generally in terms of complex periodic functions e^{ikx}

$$f(x) = \int_{-\infty}^{\infty} dk \, F(k) e^{ikx} . \tag{B.59}$$

The function $F(k)$ is called the **Fourier transform**[3] of $f(x)$. The Fourier transform can be inverted through

$$F(k) = \frac{1}{2\pi} \int_{-\infty}^{\infty} dx \, f(x) e^{-ikx} . \tag{B.60}$$

In order that $f(x)$ be a real function, the Fourier transform must obey $\bar{F}(k) = F(-k)$.

Fourier analysis, which enables us to decompose a wave into a set of periodic modes with characteristic frequencies, plays a central role in analyzing many energy-related phenomena, from solar and wave energy to electricity transmission, geophysics, and Earth's climate.

B.3.6 Differential Operators

On infinite-dimensional spaces of functions, finite-sized *matrices* are replaced by **differential operators**. For example, $-i \, d/dx$ is a differential operator that acts on the space of functions on the real line. This operator acts on the Fourier transform through multiplication by k. Such differential operators play an important role in quantum physics (§7,§21).

B.4　Miscellaneous Mathematical Tools

B.4.1　Power Series Expansion

A sufficiently well-behaved function $f(x)$ can be expanded in a *Taylor series*, or **power series expansion**, around any point x_0

$$f(x) = c_0 + c_1(x - x_0) + c_2(x - x_0)^2 + \cdots , \tag{B.61}$$

where the coefficients in the expansion are related to the derivatives of $f(x)$ at the point $x = x_0$,

$$c_n = \frac{1}{n!} \frac{d^n f}{dx^n} \Big|_{x=x_0} . \tag{B.62}$$

Thus,

$$f(x) = f(x_0) + (x - x_0) \frac{df}{dx}\Big|_{x_0} + \frac{1}{2}(x - x_0)^2 \frac{d^2 f}{dx^2}\Big|_{x_0} + \cdots .$$

The **radius of convergence** of a series describes the range of values of x around x_0 within which the series

is convergent. As a simple example, the function $f(x) = 1/(1-x)$ is easily seen to have the series expansion around $x = 0$

$$f(x) = \frac{1}{1 - x} = 1 + x + x^2 + x^3 + \cdots \tag{B.63}$$

by multiplying both sides by $1 - x$. This series converges for all x such that $|x| < 1$. In contrast, the exponential function e^z has a power series expansion,

$$e^z = \sum_{n=0}^{\infty} \frac{z^n}{n!} = 1 + z + \frac{1}{2}z^2 + \frac{1}{6}z^3 + \cdots , \tag{B.64}$$

which converges for any z in the complex plane.

A function of n variables, $f(\boldsymbol{x}) = f(x_1, \ldots, x_n)$ can be expanded as a power series around a point $\boldsymbol{x} = \boldsymbol{x}^{(0)} = (x_1^{(0)}, \ldots, x_n^{(0)})$ as

$$f(\boldsymbol{x}) = f(\boldsymbol{x}^{(0)}) + \sum_i (x_i - x_i^{(0)}) \frac{\partial f}{\partial x_i}\Big|_{\boldsymbol{x}=\boldsymbol{x}^{(0)}}$$
$$+ \sum_{i,j} \frac{1}{2}(x_i - x_i^{(0)})(x_j - x_j^{(0)}) \frac{\partial^2 f}{\partial x_i \partial x_j}\Big|_{\boldsymbol{x}=\boldsymbol{x}^{(0)}} + \cdots . \tag{B.65}$$

A particularly useful example of a power series expansion is the *binomial expansion*

$$(1+x)^a = 1 + ax + \frac{a(a-1)}{2!}x^2 + \frac{a(a-1)(a-2)}{3!}x^3 + \cdots . \tag{B.66}$$

B.4.2　Hyperbolic Functions

The **hyperbolic functions** sinh, cosh, and tanh, known as the **hyperbolic sine**, **hyperbolic cosine**, and **hyperbolic tangent** respectively, are defined by

$$\sinh x = \frac{e^x - e^{-x}}{2},$$
$$\cosh x = \frac{e^x + e^{-x}}{2},$$
$$\tanh x = \frac{\sinh x}{\cosh x} = \frac{e^x - e^{-x}}{e^x + e^{-x}} . \tag{B.67}$$

Like trigonometric functions, they obey various useful identities,

$$\cosh^2 x - \sinh^2 x = 1,$$
$$\cosh(2x) = \cosh^2 x + \sinh^2 x,$$
$$\sinh(2x) = 2\cosh x \sinh x, \tag{B.68}$$

and relations among their derivatives,

---

[3] Note that unlike Fourier series, Fourier transforms do not have a simple description in the standard language of vector spaces with norm. In particular, the "basis" functions e^{ikx} on which the Fourier transform is based are not normalizable functions, and have inner products leading to divergent integrals. A formal mathematical treatment of Fourier transforms involves the theory of *distributions*.

$$\frac{d}{dx}\cosh x = \sinh x,$$

$$\frac{d}{dx}\sinh x = \cosh x,$$

$$\frac{d}{dx}\tanh x = 1/\cosh^2 x \equiv \operatorname{sech}^2 x. \qquad \text{(B.69)}$$

For small x, hyperbolic functions can be expanded in Taylor series in x,

$$\sinh x = x + x^3/6 + \mathcal{O}(x^5),$$

$$\cosh x = 1 + x^2/2 + \mathcal{O}(x^4),$$

$$\tanh x = x - x^3/3 + \mathcal{O}(x^5).$$

B.4.3 Factorials, Gamma Functions, and Stirling's Approximation

The factorial function $n!$ assigns the value

$$n! = n(n-1)(n-2)\cdots 1 \qquad \text{(B.70)}$$

to any positive integer n. The **Gamma function** $\Gamma(x)$ is defined by the integral

$$\Gamma(z) = \int_0^\infty dt\, t^{z-1} e^{-t}, \qquad \text{(B.71)}$$

which converges for any complex number z with real part greater than zero. It is easy to show using inte-

gration by parts that when z equals a positive integer, then

$$\Gamma(n) = (n-1)!. \qquad \text{(B.72)}$$

Thus $\Gamma(z)$ can be regarded as an extension of the factorial function from the integers to the complex numbers.

Gamma functions are a convenient tool in the analysis of many physical systems, including wind power distributions (§28.2.2). The factorial appears often in statistical mechanics (see for example §8.7.4) where its argument is often much larger than one. In that case a useful estimate is given by Stirling's approximation,

$$n! \sim \sqrt{2\pi n}\left(\frac{n}{e}\right)^n \quad \text{or}$$

$$\ln(n!) \sim n\ln n - n + \frac{1}{2}\ln(2\pi n), \qquad \text{(B.73)}$$

where e is the base of the natural logarithm and the $\sim$ denotes that the difference between the left- and right-hand sides vanishes as $n \to \infty$. Stirling's approximation can be derived by writing the Gamma function as

$$\Gamma(n+1) = \int_0^\infty dt\, e^{n\ln t - t} \qquad \text{(B.74)}$$

and expanding the argument of the exponential to second order about its maximum at $t = n$.

Units and Fundamental Constants

Table C.1 Units of Energy and Power.

Colloquial unit	SI equivalent
1 electron volt [eV]	$\cong 1.602 \times 10^{-19}$ J
1 eV per molecule	$\cong 96.49$ kJ/mol
1 erg	$\equiv 10^{-7}$ J
1 foot pound [ft lb]	$\cong 1.356$ J
1 calorie$_{IT*}$ [cal$_{IT}$]	$\equiv 4.1868$ J
1 calorie$_{th*}$ [cal$_{th}$]	$\equiv 4.184$ J
1 BTU$_{IT*}$	$\cong 1.055$ kJ
1 kilocalorie$_{IT*}$ [kcal] or Calorie$_{IT*}$ [Cal]	$\equiv 4.1868$ kJ
1 kilowatt-hour [kWh]	$\equiv 3.6$ MJ
1 cubic meter natural gas	~ 36 MJ
1 therm (US)	$\cong 105.5$ MJ
1 tonne TNT [tTNT]	$\equiv 4.184$ GJ
1 barrel of oil equivalent [BOE]	$\cong 6.118$ GJ
1 ton of coal equivalent [TCE]	$\equiv 29.3076$ GJ
1 tonne of oil equivalent [toe]	$\equiv 41.868$ GJ
1 quadrillion BTU [quad]	$\cong 1.055$ EJ
1 terawatt-year [TWy]	$\equiv 31.56$ EJ
1 watt [W]	$\equiv 1$ J/s
1 foot pound per second [ft lb/s]	$\cong 1.356$ W
1 horsepower [hp] (electric)	$\equiv 746$ W
1 ton of air conditioning	$\cong 3.517$ kW

$\equiv \leftrightarrow$ definition
$\cong \leftrightarrow$ four significant figures
$\sim \leftrightarrow$ actual value varies

*th $\equiv$ thermochemical
*IT $\equiv$ International Table

Table C.2 Useful conversion factors (to four significant figures).

Unit	Colloquial	SI unit
Mass	1 ounce$_{avoirdupois}$ [oz]	0.028 35 kg
	1 pound$_{avoirdupois}$ [lb]	0.4536 kg
	1 ton (US)	907.2 kg
	1 metric tonne [t]	1000 kg*
Length	1 foot [ft]	0.3048 m*
	1 mile [mi]	1609 m
Time	1 year [y]	3.156×10^7 s
Force	1 pound [lb]	4.448 N
Area	1 acre	4047 m^2
	1 hectare [ha]	10 000 m^2*
Volume	1 liter [L]	0.001 m^3*
	1 fluid ounce [fl oz] (US)	0.029 57 L
	1 gallon [gal] (US/Imperial)	3.785/4.546 L
	1 oil barrel [bbl]	159.0 L
Speed	1 mile per hour [mph]	0.4470 m/s
	1 knot [kn]	0.5144 m/s
Pressure	1 atmosphere [atm]	101 325 Pa*
	1 psi	6896 Pa
Temperature	°Celsius [°C]	K $-$ 273.15*
	°Fahrenheit [°F]	32 + 1.8 °C*
Thermal resistance	1 ft^2 °F hr/BTU	0.1761 K m^2/W
Magnetic field	1 gauss [G]	0.0001 T*
Radiation	1 rad	0.01 Gy*
	1 rem	0.01 Sv*
	1 curie [Ci]	3.7×10^{10} Bq
Viscosity	1 poise [P] (cgs)	0.1 kg/m s*
Permeability	1 darcy [D]	9.869×10^{-13} m^2

* $\equiv$ exact

Table C.3 SI units for various physical quantities.

Mass	kilogram	kg	
Length	meter	m	
Time	second	s	
Force	newton	N	$kg\,m/s^2$
Energy	joule	J	$kg\,m^2/s^2$
Power	watt	W	J/s
Pressure	pascal	Pa	N/m^2
Charge	coulomb	C	A s
Current	ampere	A	
EM potential	volt	V	J/C
Resistance	ohm	Ω	V/A
Capacitance	farad	F	C/V
Inductance	henry	H	V s/A
Magnetic field	tesla	T	$V\,s/m^2$
Amount	gram-mole	mol	
Temperature	kelvin	K	
Activity	becquerel	Bq	s^{-1}
Absorbed dose	gray	Gy	J/kg
Dose equivalent	sievert	Sv	J/kg

Table C.4 Prefixes.

yocto (y)	10^{-24}	kilo (k)	10^3
zepto (z)	10^{-21}	mega (M)	10^6
atto (a)	10^{-18}	giga (G)	10^9
femto (f)	10^{-15}	tera (T)	10^{12}
pico (p)	10^{-12}	peta (P)	10^{15}
nano (n)	10^{-9}	exa (E)	10^{18}
micro (μ)	10^{-6}	zetta (Z)	10^{21}
milli (m)	10^{-3}	yotta (Y)	10^{24}

Table C.5 Fundamental constants and useful physical quantities.

π	$3.141\,592\,653\dots$
e	$2.718\,281\,828\dots$
Planck's constant (reduced) ($\hbar = h/2\pi$)	$1.054\,571\,628(53) \times 10^{-34}$ J s
Speed of light (c)	$2.997\,924\,58 \times 10^8$ m/s (exact)
$\hbar\,c$	$1.973\,269\,631(49) \times 10^{-7}$ eV m
Newton's constant (G_N)	$6.674\,28(67) \times 10^{-11}$ $m^3/kg\,s^2$
Vacuum permeability (μ_0)	$4\pi \times 10^{-7}$ $N\,A^2$ (exact)
Vacuum permittivity (ε_0)	$(\mu_0 c^2)^{-1} = 8.854\,187\,817\dots \times 10^{-12}$ F/m (exact)
Avogadro constant (N_A)	$6.022\,141\,79(30) \times 10^{23}$ mol^{-1}
Boltzmann constant (k_B)	$1.380\,650\,4(24) \times 10^{-23}$ J/K
Stefan–Boltzmann constant (σ)	$5.670\,400\,(40) \times 10^{-8}$ $W/m^2\,K^4$
Electron charge (e)	$1.602\,176\,487(40) \times 10^{-19}$ C
Electron mass (m_e)	$9.109\,382\,15(45) \times 10^{-31}$ kg $\cong 0.511$ MeV/c^2
Proton mass (m_p)	$1.672\,621\,637(83) \times 10^{-27}$ kg $\cong 938.272$ MeV/c^2
Atomic mass unit, or dalton (u)	$1.660\,538\,782(83) \times 10^{-27}$ kg $\cong 931.494$ MeV/c^2
Rydberg energy	$13.605\,691\,93(34)$ eV
Mean radius of Earth's orbit (1 AU)	$1.495\,978\,706\,60(20) \times 10^{11}$ m
Earth mean equatorial radius	$6.378\,1366(1) \times 10^6$ m
Mass of Earth	$5.972\,3(9) \times 10^{24}$ kg
Average solar constant above atmosphere	1366 W/m^2
Global geothermal flux / average gradient	44.2 TW / 25 K/km
Standard gravitational acceleration	$9.806\,65$ m/s^2 (exact)
Standard temperature and pressure (STP)	273.15 K, 100 kPa
Molar volume at STP	22.71×10^{-3} m^3/mol
Gas constant ($R \equiv N_A k_B$)	$8.314\,472(15)$ J/ mol K
Water – latent heat of melting/vaporization	334 kJ/kg/2.26 MJ/kg
Specific heat capacity of water (15 °C) / air (STP)	4.186 kJ/kg K / 1.0035 kJ/kg K
Mass density of water (15 °C) / air (STP)	999.1 kg/m^3 / 1.275 kg/m^3
Half-life of ^{235}U / ^{238}U / ^{239}Pu	7.0×10^8 / 4.5×10^9 / 2.4×10^4 y
Half-life of ^{232}Th / ^{40}K / ^{222}Rn	1.4×10^{10} y / 1.25×10^9 y / 3.8 d
Average annual environmental radiation exposure	3×10^{-3} Sv /y

Table D.1 Energy, power, and CO_2 quantities.

Solar power output / incident on Earth	384 YW / 174 PW
Rest mass Energy of 1 kilogram	90 PJ
Energy to refine 1 barrel of oil	1.2 GJ
Estimated energy to produce 1 tonne of	
raw steel / aluminum / cement / glass	21.3 / 64.9 / 5.1 / 5.3 GJ
synthetic nitrogen / phosphate / potash fertilizer	78.2 / 17.5 / 13.8 GJ
Approximate energy content of one gallon of	
diesel / gasoline / ethanol / LNG	140 / 120 / 84 / 78 MJ
energy from complete fission of 1 kg ^{235}U	77 TJ
Hydrogen/methane/iso-octane/graphite/ethanol	
lower heating value (net calorific value)	242 / 800 / 5050 / 394 / 1330 kJ/mol
	121 / 50 / 44 / 33 / 29 MJ/kg
higher heating value (gross calorific value)	284 / 888 / 5445 / 394 / 1366 kJ/mol
	142 / 56 / 47 / 33 / 30 MJ/kg
CO_2 concentration in atmosphere (annual mean 2016)	404.21 ± 0.12 ppm
Atmospheric CO_2 annual growth rate (2010-2016 average)	2.44 ± 0.11 ppm

Table D.2 Global and national energy, power, and CO_2. Data for 2014 from US EIA [12].

	World	China	US	Europe	Russia	Japan	India	Africa
Primary energy consumption (EJ)	**576**	130	103	82	33	20	26	20
Total petroleum consumption (EJ)	**196**	24	37	31	7.8	9.1	7.9	8.5
Total coal consumption (EJ)	**156**	81	19	12	4.3	4.4	13	4.3
Dry natural gas consumption (EJ)	**134**	7.1	29	18	16	5.2	2.1	5.2
Net nuclear electricity generation (EJ)	**8.7**	0.45	2.9	3.1	0.61	0	0.12	0.053
Total net electricity consumption (EJ)	**75**	18	14	12	3.2	3.5	3.3	2.2
Energy related CO_2 (10^9 t)	**33.7**	9.6	5.5	4.0	1.8	1.2	1.9	1.2
Per capita primary energy consumption (GJ)	**80**	95	326	132	227	158	20	17
Per capita energy related CO_2 (t)	**4.7**	7.0	17.3	6.4	12.2	9.3	1.5	1.1

Annual world renewable electricity production in EJ. Data for 2014 from US EIA [12].				
Hydro	Wind	Biomass and Waste	Geothermal	Solar
13.89	2.57	1.75	0.27	0.72

D.1 Periodic Table

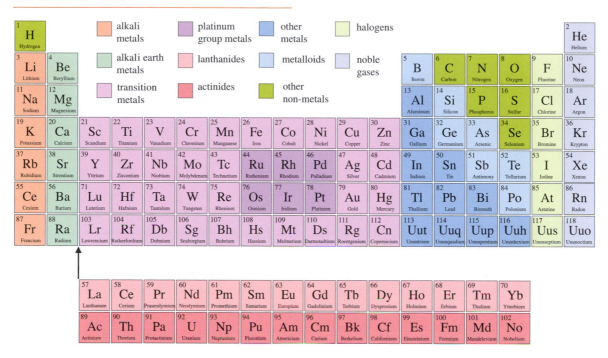

Figure D.1 The Periodic Table of the elements.

D.2 Geological Timeline

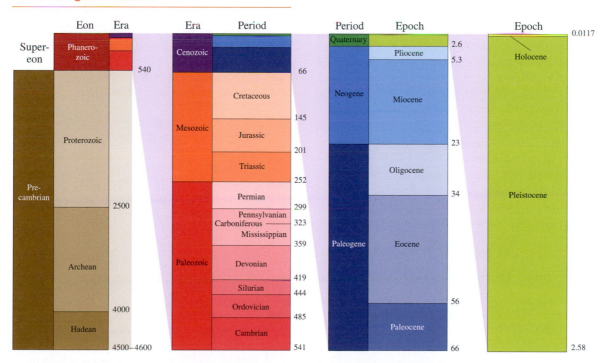

Figure D.2 Geological time scale. Each column expands upon the latest time period in the preceding column, allowing finer time resolution for more recent time periods than in the distant past. (Eras are not shown in the Precambrian Supereon.) Times are given in Ma or *millions of years ago*.

References

[1] E. Boeker and R. van Grondelle, *Environmental Physics*, 2nd edn (Hoboken, NJ: Wiley & Sons, 1999).

[2] A. V. Da Rosa, *Fundamentals of Renewable Energy Processes* (Burlington, MA: Elsevier Academic Press, 2005).

[3] D. J. C. MacKay, *Sustainable Energy – Without the Hot Air* (Cambridge, UK: UIT, 2009).

[4] J. W. Tester *et al.*, *Sustainable Energy: Choosing Among Options* (Cambridge, MA: MIT Press, 2012).

[5] V. Smil, *Energy Transitions: History, Requirements, and Prospects* (Santa Barbara, CA: Praeger, 2010).

[6] F. M. Vanek, L. D. Albright, and L. T. Angenent, *Energy Systems Engineering: Evaluation and Implementation*, 2nd edn (New York, NY: McGraw Hill, 2012).

[7] J. Andrews and N. Jelley, *Energy Science*, 2nd edn (Oxford, UK: Oxford University Press, 2013).

[8] US Census Bureau International Programs, website www.census.gov/population/international/.

[9] BP, *BP Statistical Review of World Energy June 2016*, available online at http://bp.com/statisticalreview.

[10] Population Division, Department of Economic and Social Affairs, UN Secretariat, *The World at Six Billion*, ESA/P/WP.154, 1999, available online at www.un.org/esa/population/publications/sixbillion/sixbillion.htm.

[11] Lawrence Livermore National Laboratory (LLNL), *2016 Energy Flow Chart*, LLNL-MI-410527, available online at https://flowcharts.llnl.gov.

[12] US Energy Information Administration (EIA), Independent Statistics & Analysis, website www.eia.gov.

[13] US Environmental Protection Agency (EPA) Transportation and Air Quality, website www3.epa.gov/otaq/.

[14] D. Kleppner and R. Kolenkow, *An Introduction to Mechanics* (Cambridge, UK: Cambridge University Press, 2013).

[15] E. Mazur, *Principles and Practice of Physics* (New York, NY: Pearson Education, 2014).

[16] G. B. Thomas, M. D. Weir, and J. R. Hass, *Thomas' Calculus*, 13th edn (New York, NY: Pearson Education, 2014).

[17] National Electrical Manufacturers Association Standards, website www.nema.org/StandardsPages/Motors-and-Generators.aspx.

[18] E. M. Purcell and D. J. Morin, *Electricity and Magnetism*, 3rd edn (Cambridge, UK: Cambridge University Press, 2013).

[19] Y. A. Çengel and M. A. Boles, *Thermodynamics: An Engineering Approach*, 7th edn (New York, NY: McGraw Hill, 2011).

[20] S. J. Blundell and K. M. Blundell, *Concepts in Thermal Physics*, 2nd edn (Oxford, UK: Oxford University Press, 2009).

[21] C. Kittel and H. Kroemer, *Thermal Physics*, 2nd edn (New York, NY: W. H. Freeman, 1980).

[22] National Institute of Standards and Technology (NIST), *Chemistry Webbook*, website http://webbook.nist.gov.

[23] NIST, *Thermophysical Properties of Fluid Systems*, website http://webbook.nist.gov/chemistry/fluid/.

[24] *The Engineering Toolbox*, website, www.engineeringtoolbox.com.

[25] US Geological Survey (USGS), *Mineral Commodity Summaries 2016* (Washington, DC: US Geological Survey, 2016).

[26] R. A. McDonald, Heat content and heat capacity of an extruded graphite from 341 to 1723 K. *Journal of Chemical Engineering Data*, **10**, 243 (1965).

[27] J. H. Lienhard IV and J. H. Lienhard V, *A Heat Transfer Textbook*, 4th edn (Boston, MA: Phlogiston Press, 2015), available online at `http://ahtt.mit.edu`.

[28] American Society of Heating, Refrigerating, and Air-Conditioning Engineers (ASHRAE) *2009 ASHRAE Handbook – Fundamentals* (ASHRAE, 2009), available online at `www.ashrae.org/resources-publications/bookstore/handbook-online`.

[29] W. M. Haynes, ed., Section 6, Fluid properties: thermal conductivity of gases. In *CRC Handbook of Chemistry and Physics*, 96th edn (Boca Raton, FL: CRC Press, 2015), available online at `www.hbcponline.com`.

[30] S. P. Arya, *Introduction to Micrometeorology*, 2nd edn (Cambridge, MA: Academic Press, 2001).

[31] R. E. Peierls, quoted in F. Bonetto, J. L. Lebowitz, and L. Rey-Bellet, Fourier's Law: a challenge for theorists. In *Mathematical Physics 2000* (London: Imperial College Press, 2000), arXiv: math-ph/0002052.

[32] K. Brenden, Why *R* is not simply the inverse of *U*. *Door & Window Manufacturer Magazine*, **11**, 10 (2010), available online at `industry.glass.com/Door_and_Window_Maker/Backissues/2010/September2010/AAMAAnalysis.htm`.

[33] G. C. Vliet and G. Leppert, Forced convection heat transfer from an isothermal sphere to water. *Journal of Heat Transfer*, **83**, 163 (1961).

[34] M. D. Mikhailov, Natural convection between two vertical plates, Wolfram Demonstrations Project, available online at `demonstrations.wolfram.com/NaturalConvectionBetweenTwoVerticalPlates/`.

[35] M. Isaac and D. P. Van Vuuren, Modeling global residential sector energy demand for heating and air conditioning in the context of climate change. *Energy Policy*, **37**, 507 (2009).

[36] US EIA Consumption and efficiency, website `www.eia.gov/consumption/residential/reports/electronics.cfm`.

[37] US Department of Energy Office of Energy Efficiency and Renewable Energy (EERE), *Energy Saver: Tips on Saving Money & Energy at Home*, Report DOE/GO-102014-4405, available online at `http://energy.gov/energysaver/downloads/energy-saver-guide`.

[38] A. P. French and E. F. Taylor, *Introduction to Quantum Physics* (Boca Raton, FL: Chapman & Hall, 1978).

[39] C. Cohen-Tannoudji, B. Diu, and F. Laloë, *Quantum Mechanics* (Hoboken, NJ: Wiley, 1977).

[40] R. P. Feynman, R. Leighton, and M. Sands, *The Feynman Lectures on Physics* (Reading, MA: Addison-Wesley, 1964).

[41] C. Jönsson, Electron diffraction at multiple slits. *American Journal of Physics*, **42**, 4 (1974).

[42] P. W. Atkins, *The Second Law: Energy, Chaos, and Form* (New York, NY: W. H. Freeman, 1984).

[43] D. V. Schroeder, *An Introduction to Thermal Physics* (San Francisco, CA: Addison Wesley Longman, 2000).

[44] C. E. Shannon, A mathematical theory of communication. *Bell System Technical Journal*, **27**, 379; **28**, 623 (1948).

[45] R. G. Gallager, *Information Theory and Reliable Communication* (Hoboken, NJ: Wiley, 1968).

[46] D. McKay, *Information Theory, Inference, and Learning Algorithms* (Cambridge, UK: Cambridge University Press, 2004).

[47] T. M. Cover and J. A. Thomas, *Elements of Information Theory* (Hoboken, NJ: Wiley-Interscience, 2006).

[48] R. C. Nave, Einstein model of a solid. In *HyperPhysics*, available online at `http://hyperphysics.phy-astr.gsu.edu/hbase/hph.html`.

[49] D. Larsen, *LibreTexts Chemistry*, available online at `http://chem.libretexts.org/Core/Physical_Chemistry/Statistical_Mechanics/Heat_Capacity_of_Solids`.

[50] P. Atkins, L. Jones, and L. Laverman, *Chemical Principles*, 6th edn (New York, NY: Macmillan, 2013).

[51] L. M. Shulman, The heat capacity of water ice in interstellar or interplanatary conditions. *Astronomy & Astrophysics*, **416**, 187 (2004).

[52] H. C. Ko, N. Ahmed, and Y. A. Chang, *Thermodynamics of the Calcination of Calcite*, Bureau of Mines Report of Investigations RI-8647 (Avondale, MD: US Bureau of Mines, 1982).

[53] US DOE EERE Hydrogen analysis resource center: lower and higher heating values of fuels, website `http://hydrogen.pnl.gov/tools/lower-and-higher-heating-valuesfuels`.

[54] BCS Inc., *Mining Industry Energy Bandwidth Study*, US DOE Industrial Technologies Program, available at `www1.eere.energy.gov/manufacturing/resources/mining/pdfs/mining_bandwidth.pdf`.

[55] D. D. Konowalow, *Morse Potential Molecular Interactions* (Madison, WI: University of Wisconsin Press, 1961).

[56] K. Wark and D. E. Richards, *Thermodynamics*, 6th edn (New York, NY: McGraw-Hill, 1984).

[57] G. Walker, *et al., The Stirling Alternative* (Yverdon, Switzerland: Gordon and Breach Science, 1994).

[58] US EPA climate change website, www3.epa.gov/climatechange/index.html.

[59] B. E. Milton, *Thermodynamics, Combustion, and Engines* (Boca Raton, FL: Chapman & Hall, 1995).

[60] J. B. Heywood, *Internal Combustion Engine Fundamentals* (New York, NY: McGraw-Hill, 1988).

[61] W. G. Lovell, Knocking characteristics of hydrocarbons, *Industrial and Engineering Chemistry*, **40**, 2388 (1948).

[62] Oak Ridge National Laboratory (DOE), Engineering Science and Technology Division, *Subcontract Report: Technology and Cost of the MY2007 Toyota Camry HEV*, ORNL/TM-2007/132, available online at www.osti.gov/scitech/biblio/921781.

[63] S. Samouhos, private communication.

[64] J. H. Keenan, *et al., Steam Tables: Thermodynamics Properties of Water Including Vapor, Liquid, and Solid Phases, SI Units* (Hoboken, NJ: Wiley, 1978).

[65] M. C. Potter and C. W. Somerton, *Schaum's Outline of Thermodynamics for Engineers* (New York, NY: McGraw-Hill, 2006).

[66] F. E. Close, *Particle Physics: A Very Short Introduction* (Oxford, UK: Oxford University Press, 2004).

[67] D. J. Griffiths, *Introduction to Elementary Particles*, 2nd edn (Hoboken, NJ: Wiley, 2008).

[68] R. B. Wiringa, A. Arriaga, and V. R. Pandharipande, Quadratic momentum dependence in the nucleon-nucleon interaction. *Physical Review C* **68**, 054006 (2003).

[69] K. A. Olive, *et al.* (Particle Data Group), The review of particle physics. *Chinese Physics C*, **38**, 090001 (2014).

[70] J. K. Tuli, *et al. National Nuclear Data Center Nuclear Wallet Cards* (Upton, NY: Brookhaven National Laboratory, 2011) available online at www.nndc.bnl.gov/wallet/wccurrent.html.

[71] International Atomic Energy Agency (IAEA) Power Reactor Information System Database on Nuclear Power Reactors, website www.iaea.org/pris/.

[72] European Nuclear Society, website www.euronuclear.org/info/encyclopedia/n/nuclear-power-plant-world-wide.htm.

[73] M. Kazimi, E. J. Moniz, *et al., The Future of the Nuclear Fuel Cycle* (Cambridge, MA: MIT Energy Initiative, 2011).

[74] W. M. Haynes, ed., Section 14, Geophysics, astronomy, and acoustics: Abundance of elements in the Earth's crust and in the sea, in *CRC Handbook of Chemistry and Physics*, 96th edn (Boca Raton, FL: CRC Press, 2015), available online at www.hbcponline.com.

[75] For references, see Wikipedia Abundance of the elements data page, website https://en.wikipedia.org/wiki/Abundances_of_the_elements_(data_page).

[76] Organization for Economic Cooperation and Development (OECD) and IAEA, *Uranium 2014: Resources, Production and Demand (Red Book)*, OECD NEA No. 7209, 2014, available online at www.oecd-nea.org/ndd/pubs/2014/7209-uranium-2014.pdf.

[77] J. Lilley, *Nuclear Physics: Principles and Application* (Hoboken, NJ: Wiley, 2001).

[78] J. R. Lamarsh and A. J. Baratta, *Introduction to Nuclear Engineering*, 3rd edn (New York, NY: Pearson Education, 2001).

[79] T. Jevremovic, *Nuclear Principles in Engineering* (New York, NY: Springer, 2005).

[80] J. E. Turner, *Atoms, Radiation, and Radiation Protection*, 3rd edn (Hoboken, NJ: Wiley, 2007).

[81] R. L. Garwin and G. Charpak, *Megatons and Megawatts: The Future of Nuclear Power and Nuclear Weapons* (Chicago, IL: University of Chicago Press, 2002).

[82] National Nuclear Data Center, *Chart of the Nuclides*, available online at www.nndc.bnl.gov/chart.

[83] R. B. Leighton, *Principles of Modern Physics* (New York, NY: McGraw-Hill, 1959).

[84] V. H. Schoffelmayer, Radium mine in the Ozarks? *Technical World Magazine*, **18**, 523 (1912).

[85] K. S. Krane, *Introductory Nuclear Physics* (Hoboken, NJ: Wiley, 1988).

[86] E. E. Lewis, *Fundamentals of Nuclear Reactor Physics* (Cambridge, MA: Academic Press, 2008).

[87] G. Kaye and T. Laby, Nuclear fission and fusion, and neutron interactions, §4.7.1 Nuclear fission, in *Table of Physical and Chemical Constants*, 16th edn, available online at www.kayelaby.npl.co.uk/atomic_and_nuclear_physics/4_7/4_7_1.html.

[88] A. L. Nichols, D. L. Aldama, and M. Verpelli, U-235 chain fission yields, Table C-1.3 in *Handbook of Nuclear Data for Safeguards: Database Extensions, August 2008* (Vienna: IAEA, 2008), available online at www-nds.iaea.org/sgnucdat/c1.htm.

[89] A. D. Hodgdon, private communication.

[90] J. Freidberg, *Plasma Physics and Fusion Energy* (Cambridge, UK: Cambridge University Press, 2007).

[91] D. J. Bennet and J. R. Thomson, *The Elements of Nuclear Power*, 3rd edn (Hoboken, NJ: Wiley, 1989).

[92] P. K. Kuroda, On the nuclear physical stability of uranium minerals. *Journal of Chemical Physics*, **25**, 781 (1956).

[93] K. A. Jensen and R. C. Ewing, The Okélobondo natural fission reactor, southeast Gabon: Geology, mineralogy, and retardation of nuclear-reaction products. *Geological Society of America Bulletin*, **113**, 32 (2001).

[94] A. P. Meshik, The workings of an ancient nuclear reactor. *Scientific American*, **293**, 82 (2005).

[95] International Nuclear Safety Advisory Group of the IAEA, *INSAG-7 The Chernobyl Accident: Updating of INSAG-1*, (Vienna: IAEA, 1992).

[96] MIT Nuclear Science and Engineering Nuclear Information Hub, What is decay heat?, website `https://mitnse.com/2011/03/16/what-is-decay-heat/`.

[97] OECD Nuclear Energy Agency and the Generation IV International Forum, *Technology Roadmap for Generation IV Nuclear Energy Systems* (Paris, France: OECD, 2014), available online at `www.gen-4.org/gif/upload/docs/appli cation/pdf/2014-03/gif-tru2014.pdf`.

[98] US Nuclear Regulatory Commission (NRC), *Issued Design Certification – Advanced Passive 1000*, available online at `www.nrc.gov/reactors/ new-reactors/design-cert/ap1000 .html`.

[99] US NRC, *Issued Design Certification – Advanced Boiling-Water Reactor*, available online at `www .nrc.gov/reactors/new-reactors/ design-cert/abwr.html`.

[100] D. Keefe, Inertial confinement fusion. *Annual Reviews of Nuclear and Particle Physics*, **32**, 391 (1982).

[101] ITER Science, 60 years of progress, website `www.iter.org/sci/beyonditer`.

[102] S. Cowley, private communication.

[103] Eurofusion, 50 years of the Lawson criterion, website `www.euro-fusion.org/2005/12/50- years-of-lawson-criteria/`.

[104] W. K. Tobiska and A. A. Nusinov, *Summary of ISO 21348 Wavelengths and Compliance Criteria*, available online at `www.spacewx .com/pdf/SET_21348_2004.pdf`.

[105] NIST Physical Measurements Laboratory, *estar, pstar, and astar*, website `http://physics .nist.gov/PhysRefData/Star/Text/ intro.html`.

[106] International Commission on Radiological Protection (ICRP), The 2007 recommendations of the international commission on radiological protection; ICRP Publication 103. *Annals of the ICRP*, **37** (2007).

[107] D. Delacroix *et al.*, Radionuclide and radiation protection data handbook 2002. *Radiation Protection Dosimetry*, **98**, 1 (2002).

[108] Merck, Radiation exposure and contamination. In *The Merck Manual for Health Care Professionals*, available online at `www.merckmanu als.com/professional/injuries-pois oning/radiation-exposure-and-conta mination`.

[109] D. J. Strom, *Health Impacts from Acute Radiation Exposure*, Pacific Northwest National Lab Report PNNL-14424 UC-610 (2003), available online at `www.pnl.gov/main/publications/exter nal/technical_reports/PNNL-14424 .pdf`.

[110] National Council on Radiation Protection and Measurements (NCRP), *Influence of Dose and Its Distribution in Time on Dose-Response Relationships for Low-LET Radiations*, NCRP Report 64, 1980.

[111] J. F. Thomson and W. W. Tourtellotte, The effect of dose rate on the LD50 of mice exposed to gamma radiation from cobalt 60 sources. *American Journal of Roentgenology Radium Therapy and Nuclear Medicine*, **69**, 826 (1953).

[112] Committee on the Biological Effects of Ionizing Radiation of the National Research Council, *Health Risks from Exposure to Low Levels of Ionizing Radiation: BEIR VII – Phase 2* (Washington, DC: National Academy Press, 2006).

[113] NCRP, *Limitation of Exposure to Ionizing Radiation*, NCRP Report 116, 1993.

[114] ICRP, 1990 Recommendations of the international commission on radiological protection, ICRP Publication 60. *Annals of the ICRP*, **21** (1991).

[115] United Nations Scientific Committee on the Effects of Atomic Radiation (UNSCEAR), *Sources and Effects of Ionizing Radiation* (New York, NY: United Nations, 1993).

[116] M. Tubiana and A. Aurengo, Dose effect relationship and estimation of the carcinogenic effects of low doses of ionising radiation: The joint report of the Academie des Sciences (Paris) and of the Academie Nationale de Medecine. *International Journal of Low Radiation*, **2** (2006).

[117] W. Allison, *Radiation and Reason: The Impact of Science on a Culture of Fear* (Oxford, UK: Wade Allison, 2009).

[118] UNSCEAR *Sources and Effects of Ionizing Radiation, Annex B: Exposures of the Public and Workers from Various Sources of Radiation* (New York, NY: United Nations, 2008).

[119] NCRP, *Ionizing Radiation Exposure of the Population of the United States*, NCRP Report 160, 2009.

[120] J. Hála and J. Navratil, *Radioactivity, Ionizing Radiation, and Nuclear Energy* (Brno, Czech Republic: Konvoj, 2003).

[121] Committee on the Biological Effects of Ionizing Radiation of the National Research Council, *Health Effects of Exposure to Radon: BEIR VI* (Washington, DC: National Academy Press, 1999).

[122] NCRP, *Ionizing Radiation Exposure of the Population of the United States*, NCRP Report 93, 1987.

[123] Ph. Hubert, From the mass of the neutrino to the dating of wine. *Nuclear Instruments and Methods in Physics Research A*, **580**, 751 (2007).

[124] IAEA, *Classification of Radioactive Waste, General Safety Guide* IAEA Publication 1419 (Vienna: IAEA, 2009).

[125] D. Albright and K. Kramer, Neptunium 237 and Americium: World Inventories and Proliferation Concerns, *Institute for Science and International Security (ISIS) Reports* (2005), available online at http://isis-online.org/uploads/isis-reports/documents/np_237_and_americium.pdf.

[126] J. Kang and F. von Hippel, U-232 and the proliferation-resistance of U-233 in spent fuel. *Science & Global Security*, **9**, 1 (2001).

[127] A. P. Moller, *et al.*, Condition, reproduction, and survival of barn swallows from Chernobyl. *Journal of Animal Ecology*, **74**, 1102 (2005).

[128] S. Weinberg, The cosmological constant problem. *Reviews of Modern Physics*, **61**, 1 (1989).

[129] IEA, *2015 Snapshot of Global Photovoltaic Markets*, IEA Report PVPS T1-29:2016 (Paris: IEA, 2016).

[130] IEA *Technology Roadmap: Solar Thermal Electricity* (Paris: IEA, 2014).

[131] B. W. Carroll and D. A. Ostlie, *An Introduction to Modern Astrophysics*, 2nd edn (Boston, MA: Pearson/Addison-Wesley, 2007).

[132] M. Stix, *The Sun: An Introduction*, 2nd edn (New York, NY: Springer, 2002).

[133] R. Schmalensee *et al.*, *The Future of Solar Energy* (Cambridge, MA: MIT Energy Initiative, 2015).

[134] I. Ribas, The sun and stars as the primary energy input in planetary atmospheres. *Proceedings of the International Astronomical Union, IAU Symposium*, **264**, 3 (2009).

[135] American Society for Testing and Materials (ASTM) and National Renewable Energy Laboratory (NREL), *Reference solar spectral irradiance: air mass 0 and air mass 1.5 standards*, available online at http://rredc.nrel.gov/solar/spectra/.

[136] J. Jean, *et al.*, Pathways for solar photovoltaics. *Energy and Environmental Science*, **8**, 1200 (2015).

[137] T. V. Ramachandra, R. Jain, and G. Krishnadas, Hotspots of solar potential in India. *Renewable and Sustainable Energy Reviews*, **15**, 3178 (2011).

[138] V. Fthenakis and H. C. Kim, Land use and electricity generation: a life-cycle analysis. *Renewable and Sustainable Energy Reviews*, **13**, 1465 (2009).

[139] M. Griggs, Absorption coefficients of ozone in the ultraviolet and visible regions. *Journal of Chemical Physics*, **49**, 857 (1968).

[140] F. Mauthner, W. Weiss, and M. Spörk-Dür, *Solar Heat Worldwide* (Vienna: IEA, 2015), available online at www.iea-shc.org/data/sites/1/publications/Solar-Heat-Worldwide-2015.pdf.

[141] D. Rapp, *Solar Energy* (Englewood Cliffs, NJ: Prentice-Hall, 1981).

[142] P. De Laquil III, *et al.*, Solar-thermal electric technology. In T. B. Johansson, *et al.*, eds., *Renewable Energy: Sources for Fuels and Electricity* (Washington, DC: Island Press, 1993).

[143] R. Winston, J. C. Miñano, and P. Benítez, *Nonimaging Optics* (Amsterdam: Elsevier, 2005).

[144] W. Weiss and M. Rommel, *Process Heat Collectors: State of the Art within Task 33/IV* (Feldgasse, Austria: AEE INETEC, 2008), available online at http://archive.iea-shc.org/data/sites/1/publications/task33-Process_Heat_Collectors.pdf.

[145] D. A. Brosseau, P. F. Hlava, and M. J. Kelly, *Testing Thermocline Filler Materials and Molten-Salt Heat Transfer Fluids for Thermal Energy Storage Systems Used in Parabolic Trough Solar Power Plants*, Sandia National Laboratories report SAND2004-3207 (2004), available online at http://prod.sandia.gov/techlib/access-control.cgi/2004/043207.pdf.

[146] International Energy Agency (IEA), *Photovoltaic Power Systems (PVPS) Report: Snapshot of Global PV 2017* (Paris: IEA, 2017), Report IEA-PVPS T1-31:2017.

[147] C. Kittel, *Introduction to Solid State Physics*, 8th edn (Hoboken, NJ: Wiley, 2005).

[148] M. A. Green, *Solar Cells: Operating Principles, Technology and System Applications* (Sydney, NSW: University of New South Wales Press, 1998).

[149] M. A. Green, *Silicon Solar Cells: Advanced Principles & Practice* (Sydney, NSW: University of New South Wales Press, 1995).

[150] M. A. Green, *Third Generation Photovoltaics: Advanced Solar Energy Conversion* (New York, NY: Springer, 2006).

[151] J. Nelson, *The Physics of Solar Cells* (London: Imperial College Press, 2004).

[152] M. A. Green and M. J. Keevers, Optical properties of intrinsic silicon at 300 K. *Progress in Photovoltaics: Research and Applications*, **3**, 189 (1995).

[153] W. Shockley and H. J. Queisser, Detailed balance limit of efficiency of *p-n* junction solar cells. *Journal of Applied Physics*, **32**, 510 (1961).

[154] NREL National Center for Photovoltaics, *Research Cell Efficiency Records*, available online at `www.nrel.gov/pv/assets/images/efficiency_chart.png`.

[155] Fraunhofer Institute for Solar Energy Systems, ISI *Photovoltaics Report* (Freiburg, Germany: Fraunhofer ISE, 2016), available online at `www.ise.fraunhofer.de/en/downloads-englisch/pdf-files-englisch/photovoltaics-report-slides.pdf`.

[156] M. A. Green, *et al.*, Solar sell efficiency tables. *Progress in Photovoltaics*, **25**, 3 (2017)

[157] IEA, *World Energy Outlook 2014* (Paris: IEA, 2014).

[158] P. M. Dey and J. B. Harborne, eds., *Plant Biochemistry* (Cambridge, MA: Academic Press, 1997).

[159] M. Pollen, *The Omnivore's Dilemma* (London: Penguin Press, 2006).

[160] T. Standage, *An Edible History of Humanity* (New York, NY: Walker, 2009).

[161] D. M. Mousdale, *Introduction to Biofuels* (Boca Raton, Florida: CRC Press, 2010).

[162] S. Lee and Y. T. Shah, *Biofuels and Bioenergy: Processes and Technology* (Boca Raton, Florida: CRC Press, 2012).

[163] W. Soetaert and E. J. Vandamme, eds., *Biofuels* (Hoboken, NJ: Wiley, 2009).

[164] M. Mohseni *et al.*, Environment-assisted quantum walks in photosynthetic energy transfer. *Journal of Chemical Physics*, **129**, 17 (2008).

[165] E. J. O'Reilly and A. Olaya-Castro, Non-classicality of the molecular vibrations assisting exciton energy transfer at room temperature. *Nature Communications*, **5**, 3012 (2014).

[166] D. A. Walker, Biofuels, facts, fantasy, and feasibility. *Journal of Applied Phycology*, **21**: 509–517 (2009).

[167] R. E. Blankenship *et al.*, Comparing the efficiency of photosynthesis with photovoltaic devices and recognizing opportunities for improvement. *Science*, **332**, 805 (2011).

[168] World Energy Council, *World Energy Resources: Charting the Upsurge in Hydropower Development* (London: WEC, 2015).

[169] FAOSTAT, Food and Agriculture Organization of the United Nations, statistics division, website `http://faostat3.fao.org/home/E`.

[170] A. E. Farrell *et al.*, Ethanol can contribute to energy and environmental goals. *Science*, **311**, 506 (2006); data available at `http://rael.berkeley.edu/ebamm/`.

[171] Y. Chisti, Biodiesel from microalgae. *Biotech. Advances* **25**, 294 (2007).

[172] R. H. Stewart, *Introduction to Physical Oceanography*, available online at `www.colorado.edu/oclab/sites/default/files/attached-files/stewart_textbook.pdf`.

[173] D. L. Hartmann, *Global Physical Climatology* (Cambridge, MA: Academic Press, 1994).

[174] J. T. Fasullo and K. Trenberth, The annual cycle of the energy budget. Part II: Meridional structures and poleward transports. *Journal of Climate*, **21**, 2313 (2008).

[175] C. Wunsch and P. Heimbach, Two decades of the Atlantic meridional overturning circulation: anatomy, variations, extremes, prediction and overcoming its limitations, *Journal of Climate*, **26**, 7167 (2013).

[176] C. de Castro, *et al.*, *Energy Policy*, **39**, 6677 (2011).

[177] P. Jain, *Wind Energy Engineering* (New York, NY: McGraw Hill, 2011).

[178] S. Mathew, *Wind Energy: Fundamentals, Resource Analysis and Economics* (New York, NY: Springer, 2006).

[179] J. F. Manwell, J. G. McGowan, and A. L. Rogers, *Wind Power Explained: Theory, Design, and Application*, 2nd edn (Hoboken, NJ: Wiley, 2009).

[180] S. Emeis, *Wind Energy Meteorology* (New York, NY: Springer, 2013).

[181] K. Emanuel, The theory of hurricanes. *Annual Review of Fluid Mechanics*, **23**, 179 (1991).

[182] G. A. Mayr, ed., *Alpine Windharvest, Work Package 6*, available online at `www.alpine-space.org/2000-2006/uploads/media/AWH_WP6_Wind_measurements_and_wind_potential_Final_report_26P_EN.pdf`.

[183] National Oceanographic and Atmospheric Administration (NOAA) Hurricane research Division, Frequently asked questions, website, `www.aoml.noaa.gov/hrd/tcfaq/D7.html`.

[184] K. A. Emanuel, The power of a hurricane: An example of reckless driving on the information superhighway. *Weather*, **54**, 107 (1999).

[185] D. L. Elliott, *et al.*, National Renewable Energy Laboratory (NREL), *Wind Resource Atlas of the United States*, available online at `rredc.nrel.gov/wind/pubs/atlas`.

[186] I. Troen and E. L. Petersen, *European Wind Atlas* (Roskilde, Denmark: Risø National Laboratory, 1989), available online at `http://orbit.dtu.dk/files/112135732/European_Wind_Atlas.pdf`.

[187] R. Linn, Los Alamos National Laboratory, Earth & Environmental Sciences, Computational Earth Science, Wind Turbine and Turbulent

Flow Modeling, website www.lanl.gov/orgs/ees/ees16/windblade.shtml.

[188] J. Meyers and C. Meneveau, Optimal turbine spacing in fully developed wind farm boundary layers. *Wind Energy*, **15**, 305 (2012).

[189] C. L. Archer and M. Z. Jacobson, Evaluation of global wind power. *Journal of Geophysical Research: Atmospheres*, **110**, D12110 (2005).

[190] L. M. Miller, F. Gans, and A. Kleidon, Estimating maximum global land surface wind power extractability and associated climatic consequences. *Earth System Dynamics*, **2**, 1 (2011).

[191] IEA, *Technology Roadmap for Wind Energy 2013* (Paris: IEA, 2013), available online at www.iea.org/publications/freepublications/publication/Wind_2013_Roadmap.pdf.

[192] H. M. Farh, A. M. Eltamaly, and M. A. A. Mohamed, Wind energy assessment for five locations in Saudi Arabia. *International Proceedings of Chemical, Biological & Environmental Engineering*, **28**, 48 (2012).

[193] G. K. Batchelor, *An Introduction to Fluid Dynamics* (Cambridge, UK: Cambridge University Press, 1967).

[194] R. van Rooij, *Airfoil Design and the Consequences for the Rotor Blade*, presented at the GCEP Energy Workshop, Stanford University, April, 2004, available at http://gcep.stanford.edu/pdfs/energy_workshops_04_04/wind_van_rooij.pdf.

[195] A. G. Drachmann, Heron's Windmill. *Centaurus*, **7**, 145 (1961).

[196] R. E. Wilson and P. B. S. Lissaman, *Applied Aerodynamics of Wind Power Machines*, Oregon State University Report (1974), unpublished, available online at http://ir.library.oregonstate.edu/xmlui/bitstream/handle/1957/8140/WilsonLissaman_AppAeroOfWindPwrMach_1974.pdf. Note that the *Savonius* and *American farm windmill* curves were inadvertently interchanged in this reference.

[197] R. Gasch, ed., *Windkraftanlagen* (Stuttgart, Germany: Teubner, 1996).

[198] IEA, *Renewable Energy: Medium-Term Market Report 2015* (Paris: IEA, 2015).

[199] A. Kumar, T. Schei *et al.* Hydropower. In O. Edenhofer, *et al.* eds., *Renewable Energy Sources and Climate Change Mitigation: Special Report of the Intergovernmental Panel on Climate Change* (Cambridge: Cambridge University Press, 2011).

[200] M. E. McCormick, *Ocean Engineering Wave Mechanics* (Hoboken, NJ: Wiley-Interscience, 1973).

[201] M. E. McCormick, *Ocean Wave Energy Conversion* (New York, NY: Dover, 2007).

[202] J. Cruz, ed., *Ocean Wave Energy: Current Status and Future Perspectives* (New York, NY: Springer, 2008).

[203] H. Georgi, *The Physics of Waves* (Englewood Cliffs, NJ: Prentice-Hall, 1993).

[204] K. Hasselmann, Feynman diagrams and interaction rules of wave-wave scattering processes. *Reviews of Geophysics*, **4**, 1 (1966).

[205] K. Hasselmann, *et al.*, Measurements of wind-wave growth and swell decay during the Joint North Sea Wave Project (JONSWAP). *Ergnzungsheft zur Deutschen Hydrographischen Zeitschrift Reihe*, **A(8)**, 95 (1973).

[206] K. Gunn and C. Stock-Williams, Quantifying the global wave power resource. *Renewable Energy*, **44**, 296 (2012).

[207] G. Mørk, *et al.*, Assessing global wave energy potential. In *Proceedings of the 29th International Conference on Ocean, Offshore Mechanics and Arctic Engineering, 2010* OMAE2010-20473, available online at www.oceanor.no/related/59149/paper_OMAW_2010_20473_final.pdf.

[208] A. Louis, *et al.*, Ocean energy. In O. Edenhofer, *et al.* eds., *Renewable Energy Sources and Climate Change Mitigation: Special Report of the Intergovernmental Panel on Climate Change* (Cambridge, UK: Cambridge University Press, 2011).

[209] E. I. Butikov, A dynamical picture of the ocean tides. *American Journal of Physics*, **70**, 1001 (2002).

[210] A. C. Baker, *Tidal Power* (London: Peter Peregrinus, 1991).

[211] W. Lowrie, *Fundamentals of Geophysics*, 2nd edn (Cambridge: Cambridge University Press, 2007).

[212] F. D. Stacey and P. M. Davis, *Physics of the Earth*, 4th edn (Cambridge: Cambridge University Press, 2009).

[213] H. Gupta and S. Roy, *Geothermal Energy: An Alternative Resource for the 21st Century* (Amsterdam: Elsevier, 2007).

[214] R. DiPippo, *Geothermal Power Plants: Principles, Applications, Case Studies and Environmental Impact*, 3rd edn (Amsterdam: Elsevier, 2012).

[215] J. W. Tester, *et al.*, *The Future of Geothermal Energy* (Cambridge, MA: MIT Energy Initiative, 2006).

[216] A. M. Dziewonski and D. L. Anderson, Preliminary reference Earth model. *Physics of the Earth and Planetary Interiors*, **25**, 297 (1981).

[217] W. Watanabe, *et al.*, The abundance of potassium in the Earth's core. *Physics of the Earth and Planetary Interiors*, **237**, 65 (2014).

[218] R. T. Merrill, M. W. McElhinny, and P. L. McFadden, *The Magnetic Field of the Earth: Paleomagnetism, the Core, and the Deep Mantle*, International Geophysics Series, Volume 63 (Cambridge, MA: Academic Press, 1996).

[219] K. Christensen, *et al.*, Unified scaling law for earthquakes, *PNAS*, **99** (supp 1), 2509 (2002).

[220] J. W. Lund and T. Boyd, Small geothermal power project examples. *Geo-Heat Centre Quarterly Bulletin*, **20**, 9 (1999).

[221] R. A. Hogarth and D. Bour, Flow performance of the Habanero EGS closed loop. *Proceedings, World Geothermal Congress, Melbourne Australia, 2015*, available online at `https://pangea.stanford.edu/ERE/db/WGC/papers/WGC/2015/31006.pdf`.

[222] J. W. Lund and T. L. Boyd, Direct utilization of geothermal energy 2015 worldwide review. *Proceedings, World Geothermal Congress, Melbourne Australia, 2015*, available online at `https://pangea.stanford.edu/ERE/db/WGC/papers/WGC/2015/01000.pdf`.

[223] R. Bertani, Geothermal power generation in the world 2010–2104 update report. *Proceedings, World Geothermal Congress, Melbourne Australia, 2015*, available online at `https://pangea.stanford.edu/ERE/db/WGC/papers/WGC/2015/01001.pdf`.

[224] B. D. Green and R. G. Nix, *Geothermal – The Energy Under Our Feet: Geothermal Resource Estimates for the United States*, NREL Technical Report NREL/BR-840-40665 (2006).

[225] G. Tverberg, Our Finite World, website `http://ourfiniteworld.com`.

[226] L. Thomas, *Coal Geology*, 2nd edn (West Sussex, UK: Wiley-Blackwell, 2012).

[227] B. G. Miller, *Coal Energy Systems* (Cambridge, MA: Academic Press, 2004).

[228] H. Termuehlen and W. Emsperger, *Clean and Efficient Coal-Fired Power Plants* (New York, NY: ASME Press, 2003).

[229] J. Gluyas and R. Swarbrick, *Petroleum Geoscience* (Malden: Blackwell, 2003).

[230] K. Bjorlykke, *Petroleum Geoscience: From Sedimentary Environments to Rock Physics* (New York, NY: Springer, 2010).

[231] V. Smil, *Oil, a Beginner's Guide* (Oxford: Oneworld Publications, 2008).

[232] E. J. Moniz, A. J. Meggs, *et al.*, *The Future of Natural Gas* (Cambridge, MA: MIT Energy Initiative, 2011).

[233] World Coal Association, website `www.worldcoal.org`

[234] World Energy Council, *World Energy Resources: 2013 Survey* (London: WEC, 2013).

[235] Maps of the World, available at `www.mapsofworld.com/business/industries/coal-energy/world-coal-deposits.html`.

[236] IEA, *Key World Energy Statistics* (Paris: IEA, 2009).

[237] K. McCarthy *et al.*, Basic petroleum geochemistry for source rock evaluation. *Oilfield Review*, **23**, 32 (2011).

[238] US EIA, *Technically Recoverable Shale Oil and Shale Gas Resources: An Assessment of 137 Shale Formations in 41 Countries Outside the United States* (Washington, DC: EIA, 2013).

[239] K. S. Deffeyes, *Hubbert's Peak, The Impending World Oil Shortage* (Princeton, NJ: Princeton University Press, 2001).

[240] K. A. Kvenvolden, Gas hydrates – geological perspective and global change. *Review of Geophysics*, **31**, 173 (1993).

[241] T. A. Boden, G. Marland, and R. J. Andres, *Global CO2 Emissions from Fossil-Fuel Burning, Cement Manufacture, and Gas Flaring: 1751-2009*, Carbon Dioxide Information Analysis Center, Oak Ridge National Laboratory Report 10.3334/CDIAC/00001_V2010 (2010), available online at `www.cdiac.ornl.gov/trends/emis/tre_glob.html`.

[242] L. D. Landau and E. M. Lifshitz, *Theory of Elasticity* (New York, NY: Pergamon, 1975).

[243] J. T. Houghton, *The Physics of Atmospheres*, 3rd edn (Cambridge: Cambridge University Press, 2002).

[244] R. G. Barry and R. J. Chorley, *Atmosphere, Weather and Climate*, 9th edn (New York, NY: Routledge, 2010).

[245] T. F. Stocker, *et al.*, eds., *IPCC, 2013: Climate Change 2013: The Physical Science Basis. Contribution of Working Group I to the Fifth Assessment Report of the Intergovernmental Panel on Climate Change* (Cambridge: Cambridge University Press, 2013).

[246] S. Solomon, *et al.*, eds., *IPCC, 2007: Climate Change 2007: The Physical Science Basis. Contribution of Working Group I to the Fourth Assessment Report of the Intergovernmental Panel on Climate Change* (Cambridge: Cambridge University Press, 2007).

[247] M. Wild, *et al.*, The global energy balance from a surface perspective. *Climate Dynamics*, **40**, 3107 (2013).

[248] V. Ramanathan, *et al.*, Cloud-radiative forcing and climate: results from the Earth Radiation Budget Experiment. *Science*, **243**, 57 (1989).

[249] V. Ramaswamy, *et al.*, Radiative forcing of climate change. In J. T. Houghton, *et al.*, eds., *IPCC,*

2001: *Climate Change 2001: The Physical Science Basis. Contribution of Working Group I to the Third Assessment Report of the Intergovernmental Panel on Climate Change* (Cambridge: Cambridge University Press, 2001).

[250] J. L. Sarmiento and N. Gruber, Sinks for anthropogenic carbon. *Physics Today*, **55**, 30 (2002).

[251] S. Arrhenius, On the influence of carbonic acid in the air upon the temperature of the ground. *Philosophical Magazine and Journal of Science, Series 5*, **41**, 237 (1896).

[252] L. Donner, W. Schubert, and R. Somerville, eds., *The Development of Atmospheric General Circulation Models* (Cambridge: Cambridge University Press, 2011).

[253] B. J. Soden and I. M. Held, An assessment of climate feedbacks in coupled ocean-atmosphere models. *Journal of Climate*, **19**, 3354 (2006).

[254] J. Vial, J.-L. Dufresne, and S. Bony, On the interpretation of inter-model spread in CMIP5 climate sensitivity estimates. *Climate Dynamics*, **41**, 1725 (2013).

[255] T. Andrews, *et al.*, Forcing, feedbacks and climate sensitivity in CMIP5 coupled atmosphere-ocean climate models. *Geophysics Research Letters*, **39**, L09712 (2012).

[256] InterAcademy Council, *Climate Change Assessments: Review of the Processes and Procedures of the IPCC* (Amsterdam: InterAcademy Council, 2010).

[257] W. F. Ruddiman, *Earth's Climate: Past and Future*, 3rd edn (New York, NY: Macmillan, 2014).

[258] T. M. Cronin, *Paleoclimates: Understanding Climate Change Past and Present* (New York, NY: Columbia University Press, 2010).

[259] T. J. Crowley and G. R. North, *Paleoclimatology* (Oxford, UK: Oxford University Press, 1991).

[260] UK Met Office Hadley Center, Global average temperature series, website `www.metoffice.gov.uk/hadobs/hadcrut3/diagnostics/comparison.html`.

[261] NASA Goddard Institute of Space Studies, *Surface Temperature Analysis*, available online at `http://data.giss.nasa.gov/gistemp/graphs/`.

[262] T. R. Karl, *et al.*, Possible artifacts of data biases in the recent global warming hiatus. *Science*, **348**, 1469 (2015).

[263] R. K. Kaufmann, *et al.*, Reconciling anthropogenic climate change with observed temperature 1998-2008. *PNAS*, **108**, 11790 (2011).

[264] J. R. Petit, *et al.*, Climate and atmospheric history of the past 420 000 years from the Vostok ice core, Antarctica. *Nature*, **399**, 429 (1999).

[265] L. Augustin, *et al.*, Eight glacial cycles from an Antarctic ice core. *Nature*, **429**, 623 (2004).

[266] P. J. Huybers, Early Pleistocene glacial cycles and the integrated summer insolation forcing. *Science*, **313**, 508 (2006).

[267] W. S. Broecker, Ocean chemistry during glacial time. *Geochimica et Cosmochimica Acta*, **46**, 1689 (1992).

[268] M. R. Raupach, *et al.*, The declining uptake rate of atmospheric CO_2 by land and ocean sinks. *Biogeosciences*, **11**, 3453 (2014).

[269] W. R. Wieder, *et al.*, Future productivity and carbon storage limited by terrestrial nutrient availability. *Nature Geoscience*, **8**, 441 (2015).

[270] S. Khatiwala, F. Primeau, and T. Hall, Reconstruction of the history of anthropogenic CO_2 concentrations in the ocean. *Nature*, **462**, 346 (2009).

[271] J. Turner *et al.*, *Antarctic Climate Change and the Environment.* (Cambridge: Scientific Committe on Antarctic Research (SCAR) Cambridge University Press, 2009).

[272] Scientific Committee on Antarctic Research (SCAR) *Antarctic Climate Change and the Environment: 2016 update*, available online at `www.scar.org/scar_media/documents/policyadvice/treatypapers/ATCM39_ip035_e.pdf`.

[273] R. J. Rowley, *et al.*, Risk of rising sea level to population and land area. *EOS Transactions, American Geophysical Union*, **88**, 105 (2007).

[274] A. Strong, *et al.*, Ocean fertilization: Time to move on. *Nature*, **461**, 347 (2009).

[275] K. W. Ford, *et al.*, eds., *Efficient Use of Energy: The American Physical Society Studies on the Technical Aspects of the More Efficient Utilization of Energy*, American Institute of Physics, Conference Series, **25** (New York: AIP, 1975), available online at `http://scitation.aip.org/content/aip/proceeding/aipcp/25`.

[276] B. Richter, *et al.*, *Energy Future: Think Efficiently*, American Physical Society Report (Washington: APS, 2008), available online at `www.aps.org/energyefficiencyreport/`.

[277] J. Koomey, *et al.*, Implications of historical trends in the electrical efficiency of computing. *IEEE Annals of the History of Computing*, **33** (3), 46 (2011).

[278] COP variation with output temperature table on website `http://en.wikipedia.org/wiki/Heat_pump`.

[279] P-A. Enkvist, J. Dinkel, and C. Lin, *Impact of the Financial Crisis on Carbon Economics: Version 2.1 of the Global Greenhouse Gas Abatement Cost Curve* (McKinsey & Co, 2010).

[280] D. Hafemeister, *Physics of Societal Issues: Calculations on National Security, Environment, and Energy* (New York, NY: Springer, 2007).

[281] B. Boardman, *et al.*, *40% House*, Environmental Change Institute, University of Oxford Report (2005), available online at `www.eci.ox.ac.uk/research/energy/downloads/40house/40house.pdf`.

[282] C. McGlade and P. Ekins, The geographical distribution of fossil fuels unused when limiting global warming to 2 °C. *Nature*, **517**, 187 (2015).

[283] R. A. Huggins, *Energy Storage* (New York, NY: Springer, 2010).

[284] A. Ghoniem, Needs, resources, and climate change: clean and efficient conversion technologies. *Progress in Energy and Combustion Science*, **37**, 15 (2011).

[285] B. Elmegaard and W. B. Markussen, Efficiency of compressed air energy storage. In *ECOS2011: The 24th International Conference on Efficiency, Cost, Optimization, Simulation and Environmental Impact of Energy Systems*, available online at, `http://orbit.dtu.dk/en/publications/efficiency-of-compressed-air-energy-storage(9ce24503-affd-42c1-9ac5-879f73019c1a).html`.

[286] W. F. Pickard, N. J. Hansing, and A. Q. Shen, Can large-scale advanced-adiabatic compressed air energy storage be justified economically in an age of sustainable energy? *Journal of Renewable and Sustainable Energy*, **1**, 033102 (2009).

[287] H. Pollak, *History of First US Compressed Air Energy Storage Plant Volume 2: Construction*, Electric Power Research Institute (EPRI) Report TR-101751-V2 (1994), available online at `http://publicdownload.epri.com/PublicDownload.svc/product=TR-101751-V2/type=Product`.

[288] R. W. Bradshaw and N. P. Siegel, Molten nitrate salt development for thermal energy storage in parabolic trough solar power systems. In *ASME 2009 3rd International Conference on Energy Sustainability*, **2**, 615 (2009).

[289] A. Thess, Thermodynamic efficiency of pumped heat electricity storage. *Physics Review Letters*, **111**, 110602 (2013).

[290] I. Buchmann, *Batteries in a Portable World*, 3rd edn (Richmond, BC, Canada: Cadex Electronics, 2011).

[291] A. Ghoniem, private communication.

[292] C. Glaize and S. Genies, *Lead and Nickel Electrochemical Batteries* (Hoboken, NJ: Wiley, 2012).

[293] V. S. Bagotsky, *Fuel Cells: Problems and Solutions*, 2nd edn (Hoboken, NJ: Wiley, 2009).

[294] M. W. Kanan and D. G. Nocera, In situ formation of an oxygen-evolving catalyst in neutral water containing phosphate and Co^{2+}. *Science*, **321**, 1072 (2008).

[295] Maxwell Technologies, website `www.maxwell.com/products/ultracapacitors/docs/bcseries_ds_1017105-4.pdf`.

[296] A. von Meier, *Electric Power Systems: A Conceptual Introduction* (Hoboken, NJ: Wiley/IEEE Press, 2006).

[297] S. W. Blume, *Electric Power System Basics: For the Non-electrical Professional* (Hoboken, NJ: Wiley/IEEE Press, 2007).

[298] J. G. Kassakian, R. Schmalensee, *et al.*, *The Future of the Electric Grid* (Cambridge, MA: MIT Energy Initiative, 2011).

[299] World Bank, *Indicators: Electric Power Transmission and Distribution Losses*, available online at `www.data.worldbank.org/indicator/EG.ELC.LOSS.ZS?view=chart`.

[300] United Nations Department of Economic and Social Affairs, *Multi Dimensional Issues in International Electric Power Grid Interconnections* (New York, NY: United Nations, 2006).

[301] US DOE, available online at `www.energy.gov/sites/prod/files/oeprod/DocumentsandMedia/NERC_Interconnection_1A.pdf`.

[302] California Public Utilities Commission, *Factsheet: Types of Transmission Structures*, available online at `ftp://ftp.cpuc.ca.gov/gopher-data/environ/tehachapi_renewables/FS8.pdf`.

[303] US Department of Labor Occupational Health and Safety Administration, *Electric Power Generation, Transmission, Distribution eTool*, available online at `www.osha.gov/SLTC/etools/electric_power/illustrated_glossary/substation.html`.

[304] H. Holttinen, *et al.* *Design and Operation of Power Systems with Large Amounts of Wind Power, IEA Wind Task 25* (VTT Research Center of Finland, 2009); available at `http://www.vtt.fi/inf/pdf/tiedotteet/2009/T2493.pdf`.

[305] North American Electric Reliability Corporation (NERC), *Accommodating High Levels of Variable Generation* (Princeton, NJ: NERC, 2009), available at `www.nerc.com/files/ivgtf_report_041609.pdf`.

[306] R. A. Walling and K. Clark, Grid support functions implemented in utility-scale PV systems. *2010*

IEEE Power & Energy Society Transmission and Distribution Conference and Exposition.

[307] US EIA *Electric Power Annual 2011* (Washington, DC: EIA, 2013).

[308] US EIA *Electric Power Annual 2014* (Washington, DC: EIA, 2016).

[309] M. Spivak, *Calculus*, 4th edn (Houston, TX: Publish or Perish, 2008).

[310] M. Spivak, *Calculus on Manifolds: A Modern Approach to Classical Theorems of Advanced Calculus* (Boulder, CO: Westview Press, 1971).

[311] H. L. Royden and P. M. Fitzpatrick, *Real Analysis*, 4th edn (New York, NY: Pearson Education, 2010).

Index